TECHNOLOGY TRANSFER IN A GLOBAL COMMUNITY

SOCIETY FOR THE ADVANCEMENT OF MATERIAL AND PROCESS ENGINEERING

SAMPE®

28th International SAMPE Technical Conference

VOLUME 28

TECHNOLOGY TRANSFER IN A GLOBAL COMMUNITY

Westin Hotel
Seattle, Washington

November 4-7, 1996

Edited by
John T. Hoggatt
Anthony Falcone
Dr. Karl M. Nelson
Sylvester G. Hill

Additional copies of this publication may be obtained from
Society for the Advancement of Material and Process Engineering

SAMPE International Business Office
P O Box 2459
Covina, California 91722

ISBN 0-938994-76-X

INTERNATIONAL OFFICERS

President	Ronald B. Keyson
Executive Vice President	Dr. A. Brent Strong
Senior Vice President	John Willis
Vice President	Allan B. Goldberg
Vice President-Europe	Mark A. Erath
Vice President-Asia	Nobuo Ohashi
Secretary	Dr. Scott W. Beckwith
Treasurer	Dr. Paul Cheng
Immediate Past President	Ray A. Cull

OFFICERS OF SPONSORING CHAPTER

(Seattle Chapter)

Chairman	William B. Avery
First Vice Chairman	Karl Nelson
Second Vice Chairman	Christopher C.M. Eastland
Secretary	Collette M. Fuller
Treasurer	Bill Skelly
Senior International Director	Monte Martinez
Junior International Director	Don W. Chamberlain

28TH INTERNATIONAL SAMPE TECHNICAL CONFERENCE COMMITTEE

(top row l-r) John Hoggatt, Mort Kushner, Noel Gerken, Anthony Falcone, Lou Bergan; (front row l-r) Karl Nelson, Sylvester Hill, Caryn Banasky, Don Chamberlain, Monte Martinez, Stacy Biel, (not pictured) Pete George.

Co-General Chairman	John T. Hoggatt	The Boeing Company
Co-General Chairman	Anthony Falcone	The Boeing Company
Co-Technical Program Chairman	Sylvester G. Hill	The Boeing Company
Co-Technical Program Chairman	Karl M. Nelson	The Boeing Company
Finance Chairman	Stacy L. Biel	Hexcel Corporation
Assistant Finance Chairman	Noel T. Gerken	The Boeing Company
Audio Visual Chairman	Monte Martinez	Courtaulds Aerospace
Arrangements/Security Chairman	Monte Martinez	Courtaulds Aerospace
Registration Chairman	Don W. Chamberlain	Chamberlain Industries
Membership Chairman	Don W. Chamberlain	Chamberlain Industries
Publicity Chairman	Lou A. Bergan	ATACS Products Inc
Paper Evaluation Chairman	Morton Kushner	The Boeing Company (retired)
Special Events Chairman	Morton Kushner	The Boeing Company (retired)
Secretary	Caryn K. Banasky	Heatcon Composite Systems
Transportation Chairman	Pete E. George	The Boeing Company

PREFACE

Unlike any period before, the Twentieth Century has been an unprecedented era for science and technology. Our libraries and technical journals are brimming over with inspired work from literally millions of scientists and researchers around the world. Dissemination of information is exceedingly fast, especially via electronic data access through the World Wide Web. The problem now has become one of where the information exists rather than whether it exists. Regardless of former politics, corporate global alliances are forming having the capitalistic goal of producing a product better and less-expensive than anywhere else in the world. Furthermore, government agencies of the United States, unlike never before, are seeking corporate partnerships to transfer technology to the private sector. Truly for all disciplines of science and engineering, now is a time of *"Technology Transfer in a Global Community."* Of all the ways information is passed from one person to another, speech, despite being the oldest, will always be the best. Verbal communication is an interactive process where ideas flow both ways. New concepts and scientific discoveries are held up to the scrutiny of an audience and ultimately either validated or refuted. On the other hand, printed text is an indelible record of our work and an avenue of technology transfer into the future. Indeed, there is no better way of transferring ideas than in a technical conference having associated with it a published proceeding.

With this said, the Seattle Chapter of SAMPE takes exceptional pride in hosting the 28th Technical Conference in Seattle Washington, and would like to express sincere thanks to all authors, speakers and session chairman for their success in bringing together the best people in material and process engineering from around the world. Special thanks are due to the people of DARPA, NASA, Wright Patterson AFB, and the Navy for assembling special sessions associated with their organizations.

John T. Hoggatt
General Co-chairman

Sylvester G. Hill
Program Co-chairman

Anthony Falcone
General Co-chairman

Karl M. Nelson
Program Co-chairman

CONTENTS

TECHNICAL CONFERENCE
PAPERS, AUTHORS, SESSIONS AND SESSION CHAIRMEN

TECHNICAL CONFERENCE
PAPERS, AUTHORS, SESSIONS AND SESSION CHAIRMEN

TECHNICAL CONFERENCE
PAPERS, AUTHORS, SESSIONS AND SESSION CHAIRMEN

TECHNICAL CONFERENCE
PAPERS, AUTHORS, SESSIONS AND SESSION CHAIRMEN

TECHNICAL CONFERENCE
PAPERS, AUTHORS, SESSIONS AND SESSION CHAIRMEN

TECHNICAL CONFERENCE
PAPERS, AUTHORS, SESSIONS AND SESSION CHAIRMEN

TECHNICAL CONFERENCE
PAPERS, AUTHORS, SESSIONS AND SESSION CHAIRMEN

TECHNICAL CONFERENCE
PAPERS, AUTHORS, SESSIONS AND SESSION CHAIRMEN

TECHNICAL CONFERENCE
PAPERS, AUTHORS, SESSIONS AND SESSION CHAIRMEN

TECHNICAL CONFERENCE
PAPERS, AUTHORS, SESSIONS AND SESSION CHAIRMEN

28th International SAMPE Technical Conference
November 4-7, 1996

PROCESS DEVELOPMENT AND MECHANICAL PROPERTIES OF IM7/LaRC PETI-5 COMPOSITES

M. L. Rommel and L. Konopka
Northrop Grumman Corporation
Hawthorne CA. 90250-3277

P.M. Hergenrother
NASA Langley Research Center
Hampton, VA. 23681-0001

ABSTRACT

A solution prepregging process was used to produce high quality unidirectional tape with unsized Hercules IM7 fibers and a phenylethynyl terminated imide oligomer (LaRC PETI-5) in the form of an amide acid. The optimized prepreg was then used to conduct a study to examine the effects of cure parameters, devolatization conditions and robustness of the consolidation process. The mechanical and physical properties of laminates produced from the optimized composite with IM7 fiber are also presented and compared to properties obtained with other intermediate modulus graphite fibers. The results of a toxicity study conducted on the PETI-5 is presented which demonstrates that the material is safe to use.

KEY WORDS: Graphite, Manufacturing/Fabrication/Processing, Polyimides

1. INTRODUCTION

Polyimides are attractive for aerospace applications because of their excellent thermooxidative stability and mechanical performance at elevated temperatures. However, polyimides exhibit difficult processability due to the high temperature and pressure generally required during fabrication of parts. This has limited their structural use in the aerospace industry to specialized applications that demand high service temperatures. Several new polyimides with improved processability have been developed at NASA Langley Research Center as part of a program directed towards high performance aerospace applications. Recently, considerable work has concentrated on a new phenylethynyl-terminated oligomer designated LaRC PETI-5 [1,2]. This polymer cures upon elevated temperature exposure without volatile evolution at temperatures greater than 300°C to yield a polymer with an excellent balance of mechanical and physical properties [3,4].

This paper describes the mechanical property screening of IM7/LaRC PETI-5 and the development of processing parameters and mass transport limitations. The data forms a preliminary basis for the development of this system for aerospace applications. Conclusions regarding the balance of properties afforded by the IM7/LaRC PETI-5 composites are discussed. The results of a study to examine the toxicity of the oligomeric system and the 4-phenylethynylphthalic anhydride (4-PEPA) are presented and demonstrate that the resin is safe to use.

2. EXPERIMENTAL

Unidirectional tape using unsized IM7 fiber from Hercules was prepared by Fiberite using their solution tape machine with the oligomeric amide acid solution (35 percent solids in NMP). The solution prepregging required modification in order to ensure that adequate wet-out of the tows was achieved. The prepreg was produced by Fiberite per their product specification FMS 30010-01. The resin was synthesized by Imitec, Inc., of Schenectady, New York.

SCREENING OF MECHANICAL PROPERTIES. The unidirectional tape was cut and laid-up into laminates and cured per the schematic shown in Figure 1. The laminates were ultrasonically inspected using Northrop Grumman specification IT-89. Only laminates that passed ultrasonic inspection were machined and used for mechanical property characterization. Composite mechanical properties were determined using either standard ASTM procedures, Boeing Document D6-5587 or Boeing Specifications. The test matrix and laminate stacking sequences for this effort are shown in Figure 2. Values for resin content and fiber volumes were determined via acid digestion in accordance with ASTM D2734 and D3171. The calculations were based on a 1.77 g/cm^3 fiber density and a 1.32 g/cm^3 resin density. All mechanical properties are normalized to 62 percent fiber volume except those where the fiber volume was below 56 percent.

PROCESSING STUDIES. The approach consisted of definition of the minimum pressure required to consolidate laminates ranging in thickness from 10 to 40 plies, and the size limitations when processing large panels. The successful manufacture of large areal size parts from polyimide composites is dependent on the ability to remove most of the volatiles prior to consolidation. The removal of volatiles during autoclave processing follows mass transport principles. The system consists of the "wet" part, porous breather/bleeder plies, and a vacuum system. Volatile transport can either occur through the thickness of the laminate or along the fibers. In large parts, the areal size exceeds the critical distance available for mass transport, therefore devolatization must occur through the thickness. As the part size increases either in thickness or area, the breather/bleeder system can become overtaxed and ineffective. A method of assessing the effectiveness of producing large areal sized parts using small scale models was developed by DuPont for Boeing as part of the Avimid K3B wing skin program [5]. This approach was adopted for use by this program and is described in the following paragraph.

Mass transport in large parts is shown schematically in Figure 3. The transport of volatiles occurs through the part thickness and then through the porous bleeder/breather. Movement within the bleeder/breather system is governed by differences between the center and edge of the system that is closest to the vacuum port. This can be simulated with a small section of the part as shown in Figure 4. A narrow, half-width portion of the part/bleeder/breather system is taken and the edges are sealed to prevent transport of volatiles through the edges. This allows transport through the thickness only and out the vacuum port along the edges. Therefore the edge farther away from the port simulates the center of a very large part. The part selected to simulate for was 549-cm 549-cm and 0.56-cm thick. This was judged a size representative of large composite parts that are used in the aerospace industry at this time. This led to a model panel that was 15.2-cm wide by 274-cm in length and 0.56-cm (40 plies) thick. The bleeder/breather and devolatization hold were variables that were investigated during these studies.

The study to define the minimum consolidation pressures required was conducted using 30.4-cm by 30.4-cm laminates with a 0/90 lay-up. The laminate thicknesses evaluated were 10, 20, and 40 plies, which corresponded to nominal panel thicknesses of 0.14-cm, 0.28-cm, and 0.56-cm. The 0/90 configuration was selected because it represented the most difficult to consolidate flat panel configuration. All three laminate thicknesses were cured in the same bag using either 45, 75 and 0.69 MPa. The cure cycle used for processing each of the laminates is the same cycle that was used to cure the panels for mechanical property characterization. The quality of the panels were determined via ultrasonic inspection techniques and photomicrographic examination at 15 and 50X.

3. RESULTS AND DISCUSSION

SCREENING OF MECHANICAL PROPERTIES. The IM7/LaRC PETI-5 was selected for scale-up and further mechanical property evaluations based on the results of the tests reported in reference 3. The selected approach was based on the production of amide acid oligomers of the appropriate molecular weight. The chemistry of the LaRC PETI-5 is shown in Figure 5. The amide acid route was selected as an approach because NASA LaRC and some portions of the aerospace community felt that having all of the diamine reacted into an oligomer presented the least threat to worker safety.

The mechanical performance of the IM7/PETI-5 is summarized in Figure 6. All laminates were ultrasonically inspected prior to testing and were found to be of excellent quality.

Compression Properties. The notched compression properties of a quasi-isotropic laminate showed a reduction in strength from RTD to 177°C/wet of 30 percent. The reduction in notched compression performance due to moisture pickup at 350°F is only 7 percent. The notched quasi-isotropic compression strength at room temperature of T650-42/LaRC PETI-5 and IM9/LaRC PETI-5 was 348 MPa and 333 MPa, respectively. This data is similar to that obtained with the IM7 fiber. The 350°F/wet open hole compression strength of T650-42/LaRC PETI-5 and IM9/LaRC PETI-5 was 233 MPa and 232 MPa, respectively. The IM7/LaRC PETI-5 gave a 177°C/wet strength of 238 MPa which is similar to the other intermediate fibers. The

0° compression strength of the IM7/LaRC PETI-5 was 1655 MPa. The 0° compression strength obtained with T650-42/LaRC PETI-5 was slightly lower at 1427 MPa. The same trend was evident with the IM9 fiber which gave a 0° compression strength of 1427 MPa.

Tension Properties. The notched tension strength of a quasi-isotopic laminate was 437 MPa at room temperature and 455 MPa at 177°C/wet. The IM9/LaRC PETI-5 notched tension strength was 547 MPa at room temperature which demonstrates that the higher strength fiber did yield an improvement in tension properties. The T650-42/LaRC PETI-5 had a room temperature notched tension strength of 490 MPa, slightly higher than the IM7 graphite fiber. The 0° tension strengths showed the same trends in performance that the notched tension strengths revealed.

Compression Strength After Impact. The IM7/LaRC PETI-5 showed excellent compression after impact properties. The room temperature dry strength after impacting with a 270 in-lb impact energy level was 316 MPa. The IM9 and T650-42 yielded higher strengths with the IM9 giving a strength of 355 MPa and the T650-42 of 393 MPa. The effect of impact energy on the compression strength and damage area is also reported in Figure 6. As expected, the strength decreases with increasing damage area and impact energy level.

Bearing Properties. The single shear pin bearing yield strength of a quasi-isotropic laminate was 641 MPa and of a (35/50/12) laminate was 551 MPa.

Physical Characteristics. The fiber volume of the laminates ranged from 57.6 percent to 59.2 percent by volume. The resin content ranged from 33.9 percent to 35.9 percent by weight . The glass transition temperatures of the laminates measured by DMA were found to be about 277°C (taken as the drop in storage modulus).

HEALTH AND SAFETY. Toxicity tests were performed at Cytec by the Toxicology and Product Stewardship Department to ascertain the safety and toxicity of the LaRC PETI-5 and 4-PEPA. The LaRC PETI-5 polymer and the 4-phenylethynyl phthalic anhydride passed all tests. A summary of the data is provided in Figure 7. The data presented shows that the dermal limits established for both the PETI-5 and 4-PEPA are greater than 2000 mg/gram of body weight. The acute oral limit was greater than 5000 mg/kg of body weight. The Ames Samonella Assay tests showed that neither the LaRC PETI-5 and the 4-PEPA were mutagenic.

PROCESSABILITY OF IM7/PETI-5. The processability of the scaled-up IM7/PETI-5 prepreg was evaluated. The processability centered on defining a process for large areal size parts and the minimum pressure required for fabricating laminates up to 40 plies thick. Through the use of relatively small mass transport model parts (e.g., 15.2-cm x 274-cm x 0.56-cm thick), it was demonstrated that this prepreg has the potential for fabricating excellent quality large area (549-cm x 549-cm x 0.56-cm thick) parts at pressures as low as 0.35 MPa. Longer devolatilization holds (4 hours maximum) in the autoclave cycle, and increased porosity (up to 8 plies of 120 glass) in the breather stack were both found to be effective. A total of four trials were performed which represented different cure pressures and after the first panel longer devolatilization holds. The panels fabricated represented panels cured at 1.4, 0.69 and 0.35 MPa. A summary of the results of the four mass transport panels is shown in Figure 8. The panel quality is ascertained via the c-scan and photomicroscopic examination. Photomicrographs from the representative panels are shown in Figures 9 and 10.

The photomicrographs shown in Figure 9 indicate that the primary devolatization mechanism is through the thickness of the panel based on the panel photomicrographs, which showed cracks in the plies that were not healed during consolidation because of the lack of pressure on the part. High quality, well-consolidated, thick (40 plies) laminates were fabricated simulating 549-cm x 549-cm x 0.56-cm thick in the next three trials at pressures of at 1.4, 0.69 and 0.35 MPa. Typical photomicrographs which represent the quality of these panels were shown in Figure 101. Very good consolidation was noted in all cases. The dB loss values were all acceptable per Northrop Grumman standards. The difference in sound attenuation between the panels could not be attributed to differences in panel quality. Part quality was consistent regardless of the relative position of the vacuum port. The cure cycle used to produce these laminates included a four hour devolatization hold at 288°C and a one hour consolidation hold at 371°C. Pressure was applied at 316°C during the ramp from the devolatization hold to the final dwell of 371°C. Eight plies of 120 glass was used as breather pack. No further optimization work was performed to determine if shorter devolatization holds or less breather plies could be used to effect good parts.

The results of the study to examine the minimum pressure required to consolidate laminates is summarized in Figure 11. Photomicrogarphs of the panels showed that the 0.14-cm thick (10 plies) and 0.28-cm thick (20 plies) panels processed at 0.31 MPa exhibited good consolidation and low void contents. The 40 ply panel appeared to show some evidence of lack of consolidation and was not acceptable. Based on these results, it appears that the IM7/LaRC PETI-5 can be consolidated at 0.31 MPa for thicknesses up to 0.28-cm thick. These results indicate that the minimum consolidation pressure for PETI-5 is approximately 0.35 MPa for flat laminates as thick as 0.56-cm (40 plies).

Work was also conducted to determine the robustness of the cure process. These are important data for manufacturing since anomalies in the cure process can lead to scrapped parts if information is not available on when a cure can be aborted or when differences in cure parameters can be expected to affect the quality of the part. This effort was performed using the baseline PETI-5 cure cycle shown in Figure 1. Parameters such as heat-up rates, temperature and time were evaluated to determine the effect on laminate quality and mechanical performance. A summary of the results of this investigation is presented in Figure 12. The effect of heat-up rate showed that varying the rate from 0.6° to 1.7°C had no effect on laminate quality in either 0.22-cm (16 plies) or 0.34-cm (24 plies) thick laminates. We also showed that the cure could be aborted after imidization at 232 or 288°C and still achieved excellent quality laminates. To simulate a blown bag during the climb to the final consolidation temperature, the effect of stopping the cure at 343 and 377°C was evaluated and it was found that good quality laminates could be produced after cooling down and restarting the cure cycle using a similar cure cycle with no devolatization hold. The final hold temperature was also found not to effect laminate quality when varied from 354 to 377°F but lower mechanical properties were noted for the laminates cured at 354°C. This seems to correlate with a lower glass transition temperature, which showed that the 354°C dwell produced laminates with a Tg of 263°C while the glass transition temperatures of the laminates with the higher temperature consolidation holds were all greater than 270°C. The need for a postcure was also evaluated by examining laminates postcured at 354°C for 4 hours and at 374°C for 2 hours. The postcure resulted in a slight increase in the glass transition and no effect on open-hole compression strength. A porous panel produced by processing a laminate with a poor quality vacuum bag was used to assess the reprocessability of the IM7/LaRC PETI-5. The original panel had an overall attenuation loss of 27 dB and after reprocessing at 1.4 MPa and 371°C the overall attenuation loss was 6 dB indicating nearly acceptable laminate quality.

4. CONCLUSIONS

This paper describes the screening of the composite properties of the IM7/LaRC PETI-5 polyimide composite. The mechanical properties obtained showed that this composite system affords an excellent balance of mechanical properties. Processability of the IM7/LaRC PETI-5 was excellent as evidence by a cure cycle study which showed the versatility of the cure parameters and processes. Mass transport studies indicated that very large areal sized parts could be produced using a cure cycle with a long devolatized hold which did not compromise the final consolidation of the laminates using pressures as low as 0.35 MPa. The effect of laminate thickness on the pressure required for consolidation showed that panels as thick as 0.28-cm could be consolidated at pressures as low as 0.31 MPa.

5. ACKNOWLEDGEMENTS

The authors are grateful to Mr. Robert Skidmore, Northrop Grumman, for the fabrication and cure of laminates, Mr. Al Galius, Northrop Grumman, for machining of specimens, Mr. Joe Dominick, Northrop Grumman for mechanical testing and Mr. Oscar DeCastro, Northrop Grumman for photomicroscopy.

6. REFERENCES

1. P.M. Hergenrother, R.G. Bryant, B.J. Jensen, and S.J. Havens, J. Polym. Sci.: Polym. Chem. Ed., 32, 3061 (1994).
2. R.G. Bryant, B.J. Jensen, J. G. Smith, Jr. and P.M. Hergenrother, IBID., (Closed Paper), 39, 273, SCI. ADV. MATL. PROC. Eng. Series (1994).
3. P.M. Hergenrother and M. L. Rommel, IBID., SAMPE Proceedings, Vol. 41, 1061 (1996).
4. P.M. Hergenrother, and J.G. Smith, Jr, Polymer, 35, 4857 (1994).
5. R.J. Boyce, Dupont, Private Communication, September 1995.

7. BIOGRAPHY

Monica Rommel is an integrated technology team leader in Structures Research at Northrop Grumman's Military Aircraft Systems Division. She received a B.S. degree in Chemistry-Material Science from University of California, Los Angeles in 1985. Her work at Northrop Grumman has focused on development and characterization of advanced materials and processes for aircraft applications.

Lou Konopka is a senior research analyst in the Composites Research and Development Department, Military Aircraft Systems Division, Northrop Grumman Corporation. He received an AS degree in Metallurgy from El Camino College in Torrance, California. Mr. Konopka has been involved in the characterization of composite materials and materials analysis for use in aircraft and spacecraft applications.

Paul M. Hergenrother is a senior polymer scientist in the Composites and Polymers Branch of the Materials Division at NASA Langley Research Center (LaRC). He received a B.S. degree from Geneva College and took graduate work at the University of Pittsburgh and Carnegie-Mellon University. Prior to coming to NASA, he held various research and management positions at the Koppers Co., The Boeing Co., Whittaker Corp. and Virginia Polytechnic Institute and State University. His work at NASA is focused on the development of high performance polymers for aerospace applications.

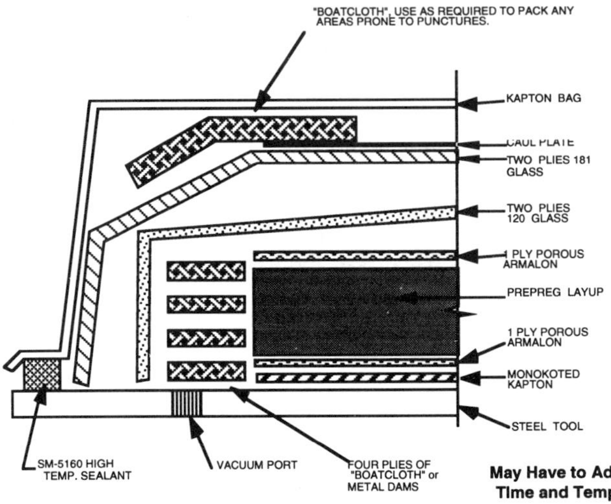

1. ROOM TEMPERATURE, 0 MPa, APPLY 68 to 76 CM "VACUUM.
2. HEAT AT 1.7°C/MINUTE TO 288 ± 4°C.
3. HOLD AT 288 ± 4°C FOR 120 MIN.
4. HEAT AT 1.7°C/MIN. TO 377 ± 4°C
5. APPLY 0.56 MPa AT 300 ± 4°C.
6. CONTINUE HEATING AT 1.7°C/MIN. TO 377 ± 4°C
7. HOLD AT 377 ± 4°C FOR 60±10 MIN.
8. COOL AT 1.7°C/MINUTE (±1°C) TO BELOW 66°C UNDER PRESSURE.
9. RELEASE VACUUM AND PRESSURE AND DEBAG LAMINATE.

May Have to Adjust Intermediate Temperature Hold Time and Temperature for Large Thick Laminates

FIGURE 1. CURE CYCLE FOR IM7/LaRC PETI-5

PROPERTY	TEST METHOD	SPECIMEN LAYUP
Open Hole Compression Strength	BSS 7260, Type I	$[+45/0/-45/90]_{2S}$ & $[\pm45/0_2/\pm45/0/90]_S$
Compression Modulus	BSS 7260, Type I [no hole)	$[+45/0/-45/90]_{2S}$ & $[\pm45/0_2/\pm45/0/90]_S$
Open Hole Tension Strength	BSS 7260, Type I	$[+45/0/-45/90]_{2S}$ & $[\pm45/0_2/\pm45/0/90]_S$
Compression Interlaminar Shear Strength	D6-55587	0_{30}
Unnotched Tension Strength & Modulus	BSS 7320, Class I	0_8 & 90_8
Unnotched Compression Strength	BSS 7260, Type III	0_8
Compression After Impact Strength	BSS 7260, Type II	$[+45/0/-45/90]_{4S}$ & $[\pm45/0_2/\pm45/0/90]_{2S}$
In-Plane Shear Modulus	BSS 7320, Class I	$[+45/-45]_{2S}$
Pin Bearing Strength (single pin, double shear)	2"x8" specimen with 0.25" hole	$[+45/0/-45/90]_{4S}$ & $[[+45/0_2/-45/0_2]_2+45/90_2/-45]]_S$
Thermal Cycling	D6-55587	$[\pm45/0/90_4/\pm45/90_3]_S$

FIGURE 2. TEST METHODS AND SPECIMEN LAYUPS

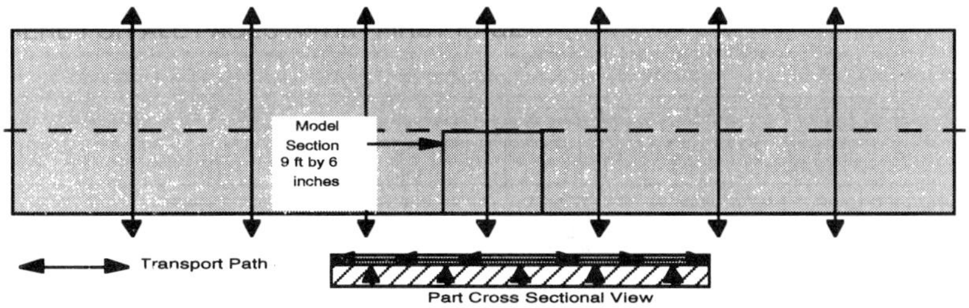

FIGURE 3. MASS TRANSPORT PATHS IN LARGE PARTS

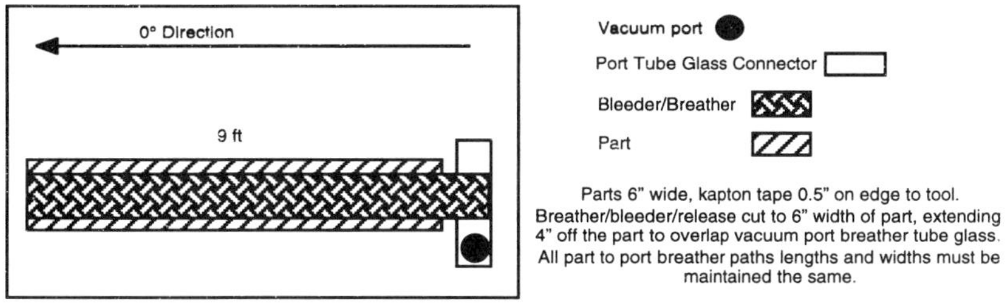

FIGURE 4. MODEL PANEL CONFIGURATION

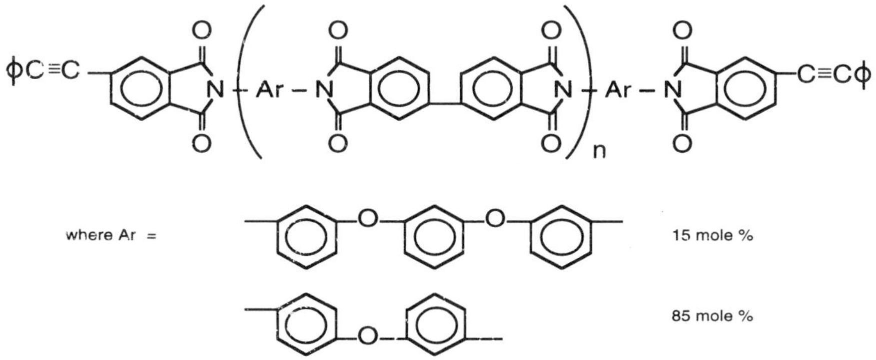

Calculated Molecular Weight 5000 g/mole
Made Via Amic Acid Which is Converted to the Imide
Prepreg is Made from Amic Acid Solution

FIGURE 5. CHEMISTRY OF IM7/LARC PETI-5

PROPERTY (LAYUP)	TEST CONDITIONS [1]	TEST RESULTS [2] ACTUAL [3]	NORMALIZED [4]
Open Hole Compression Strength, MPa (25/50/25)	RT 177°C 177°C/wet	311 237 221	335 255 238
Compression Modulus, MPa, (25/50/25)	RT	53,090	49,950
Open Hole Compression Strength, MPa (38/50/12)	RT 177°C/wet	348 279	369 296
Compression Modulus, MPa, (38/50/12)	RT	64,800	64,123
Open Hole Tension Strength, MPa, (25/50/25))	RT 177°C/wet	437 434	461 452
Open Hole Tension Strength, MPa (38/50/12)	RT 177°C	521 528	557 564
Compression Strength After Impact (25/50/25), Impact Energy:		Strength (MPa)	Damage Area (cm)
135 in-lb	RT	444	0.8
270 in-lb	RT	316	2.7
360 in-lb	RT	296	4.1
450 in-lb	RT	257	6.4
545 in-lb	RT	234	7.0
Compression Strength After Impact (38/50/12), Impact Energy		Strength (MPa)	Damage Area (cm)
135 in-lb	RT	469	0.8
270 in-lb	RT	393	3.2
360 in-lb	RT	351	6.6
450 in-lb	RT	325	8.8
545 in-lb	RT	272	9.1
Compression Interlaminar Shear Strength, MPa (100/0/0)	RT RTW 177°C/wet	86.1 85.5 46.9	-- --
In-Plane Shear Modulus, Msi (0/100/0)	RT 149°C/wet	4.2 3.2	-- --
Compression Strength and Modulus, MPa (100/0/0)	RT	1650 137000	1655
Pin Bearing Yield Strength, MPa,(25/50/25)	RT	641	--
Pin Bearing Yield Strength, MPa,(38/50/12)	RT	552	--
Tension Strength, Mpa, (100/0/0)	RT	2165	2294
Tension Modulus, MPa (100/0/0)	RT 177°C	157,000 159,000	167,000 169,000
90-Degree Tension Strength, MPa (0/0/100)	RT 177°C	72 49	-- --
90-Degree Tension Modulus, MPa (0/0/100)	RT 177°C	9700 7600	-- --
T_g, °Cvia DMA @ 10 °C/Min	N/A	277	--

1. Wet condition: 2 week soak in 71°C water.
2. Batch No. 53151C.
3. Actual data.
4. Data normalized to 62 percent fiber volume.

FIGURE 6. FIBERITE IM7/PETI-5 MECHANICAL PROPERTIES

FIGURE 10. PHOTOMICROGRAPHS AT 15 X OF MASS TRANSPORT PANEL CURED AT 0.69 MPa, REPRESENTATIVE OF 1.28 AND 0.35 MPa.

Pressure	*Laminate Thicknesses*		*Results*
1.28 MPa	10 plys 20 plys 40 plys	0°/90° Layup	Excellent Quality Laminates
0.69 MPa	10 plys 20 plys 40 plys	0°/90° Layup	Excellent Quality Laminates
0.35 MPa	10 plys 20 plys 40 plys	0°/90° Layup	10 to 20 Ply Laminates are Well Consolidated But 40 Ply Laminate Showed Poor Consolidation

FIGURE 11. EFFECT OF PRESSURE ON LAMINATE QUALITY FOR IM7/LaRC PETI-5

COMPONENT TESTED	TEST	RESULT
LaRC PETI-5	Acute Oral LD50 Acute Dermal LD50 Primary Skin Irritation Primary Eye Irritation Mouse Ear Swell Test Ames Salmonella Assay	Greater than 5000 mg/kg Greater than 2000 mg/kg Non-Irritating Minimal Non-Sensitized Not Mutagenic
4-Phenylethynyl Phthalic Anhydride	Acute Oral LD50 Acute Dermal LD50 Primary Skin Irritation Primary Eye Irritation Mouse Ear Swell Test Ames Salmonella Assay	Greater than 5000 mg/kg Greater than 2000 mg/kg Non-Irritating Minimal Non-Sensitizer Not Mutagenic

FIGURE 7. TOXICOLOGY TEST RESULTS SUMMARY FOR LARC PETI-5 AND 4-PEPA

PRESSURE	C-SCAN RESULTS	CONCLUSIONS
0.35 MPa	Db loss level ranged from 2 to 4 along length of panel	Panel quality excellent, no voids along the entire length
0.69 MPa	Db loss level was 3 along the length of panel	Panel quality excellent, no voids along the entire length
1.28 MPa	Db loss level ranged was 2 along length of panel	Panel quality excellent, no voids along the entire length

FIGURE 8. RESULTS OF MASS TRANSPORT STUDY

FIGURE 9. PHOTOMICROGRAPHS AT 15X OF MASS TRANSPORT PANEL WITH BLOWN BAG

Processing Variables, Product Form -IM7/LaRC PETI-5 Unidirectional Tape		
Condition	***Parameters***	***Results***
Heat-Up Rates	0.6°C/Minute, 1.1°C/Minute, 1.7°C/Minute } 1.4 MPa	Excellent Quality 24 ply and 16 ply Laminates
Intermediate Dwell/Pressure Application	• 249°C,1Hr/288°C Pressure • No Dwell, 371°C Pressure • 249°C, 2 Hrs/371°C Pressure • 288°C, 2 Hrs/ 371°C Pressure	Excellent Quality 24 ply and 16 ply Laminates
Final Hold Temperature	354°C, 360°C, 366°C, 371°C } 1.4 MPa	Excellent Quality 24 ply and 16 ply Laminates, Equivalent Mechanicals for All Cures Except 354°C Final Dwell
Post Cure	349°C/4 Hours, 371°C/2 Hours } Free-Standing	Increase in Tg and Slight Increase in OHC Strength, Can Be Compensated by Increased Final Dwell Time
Cure Aborts	• After Imidization - 232°C - 288°C • After 343°C • After 371°C } 0.69 MPa	Excellent Quality 24 ply Laminates Cure with No IntermediateTemperature Hold

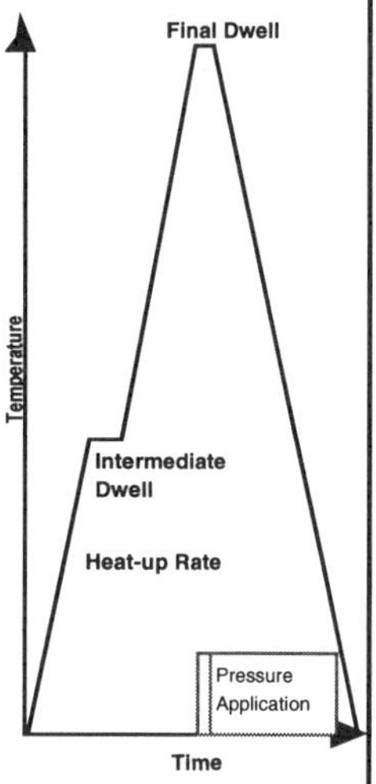

FIGURE 12. CURE CYCLE ENVELOPE FOR IM7/LARC PETI-5

28th International SAMPE Technical Conference
November 4-7, 1996

COMPOSITE PROPERTIES OF CURED PHENYLETHYNYL CONTAINING IMIDE OLIGOMERS

J. W. Connell, J. G. Smith, Jr. and P. M. Hergenrother
National Aeronautics and Space Administration
Langley Research Center
Hampton, VA 23681-0001
and
Monica Rommel
Northrop Grumman Corporation
Hawthorne, CA 90250-3277

As part of a program to develop high performance/high temperature structural resins for aeronautical applications, imide oligomers with calculated number average molecular weights of ~5000 g/mole containing terminal phenylethynyl groups, pendent phenylethynyl groups and pendent and terminal phenylethynyl groups were prepared, characterized, thermally cured and the cured resins evaluated as adhesives and composite matrix resins. The objective of this work was to develop a resin system that is processable using conventional equipment into void free laminates, exhibits high mechanical properties with long term durability at ambient and elevated temperatures, and is unaffected by exposure to moisture and common aircraft fluids. The imide oligomers containing phenylethynyl groups were fabricated into titanium adhesive specimens, and composite specimens using IM-7 fiber under 1.4 MPa for 1 hr at 350-371°C. The mechanical properties measured at room and elevated temperature varied depending on the phenylethynyl content. The neat resin and composite properties of these oligomers and their cured polymers will be presented.

KEY WORDS: High Temperature Polymers, Polyimides, Phenylethynyl Containing Imides, Adhesives, Carbon Fiber Composites, Oligomeric Thermosetting Imides

1. INTRODUCTION

High temperature/high performance structural resins with unique property combinations are needed for potential use on advanced aircraft. The materials must be readily processable under pressures and temperatures that are compatible with the capabilities of manufacturing equipment presently in use or in development. In addition, they must meet minimum requirements for mechanical performance during operating conditions which include high temperature, high air flow, high stress, impact damage, thermal cycling, exposure to moisture and aircraft fluids such as hydraulic fluid, deicing fluid, paint stripping solvents, high pH cleaning solutions and aircraft fuels.

Oligomers and polymers containing phenylethynyl groups have received considerable attention over the last twenty years. For example, 2,2'-bis(phenylethynyl)biphenyl moieties were placed within the backbone of phenylquinoxaline[1] and imide[2] polymers in an attempt to achieve intramolecular cycloaddition to rigidize the polymers. Subsequently, phenylethynyl groups were incorporated pendent along the chain of polyphenylquinoxalines to provide sites for thermally induced crosslinking[3]. These groups have been placed on the ends of imide oligomers as a means of obtaining processable (compression moldable) materials[4-19]. Polyimides containing pendent phenylethynyl groups[20-23] and both pendent and terminal phenylethynyl groups have also been reported[24]. This versatile group has also been used to terminate other oligomers such as Schiff-base type compounds[25], arylene ethers [26-29], benzoxazoles[30], and benzimidazoles[31-32]. Arylene ether polymers containing pendent phenylethynyl groups have also been reported[19,33]. The use of pendent phenylethynyl groups on polyphenylenes to obtain fully conjugated ladder polymers has been attempted[34].

Recent developments in the area of phenylethynyl terminated imide oligomer technology have produced promising results. These materials, presumably due to their oligomeric form, exhibit good melt flow and are readily processable as adhesives and composite matrix resins under pressures of 1.4 MPa or less. They are thermally cured for ~1 hr at 350-371°C causing the latent phenylethynyl groups to react and provide a pseudo three-dimensional network. The chemical structure of the cured material is not well known; however, the materials exhibit an excellent combination of properties including high glass transition temperature (T_g), good thermal stability, good moisture and solvent resistance, good toughness and high mechanical properties.

The focus of this work was to maximize certain mechanical properties and the retention of those properties at elevated temperature under wet conditions. Specifically, improvement of compressive properties without drastically affecting other mechanical properties was the goal. It was postulated that this may be achievable by increasing the crosslink density (i.e. distance between crosslinking sites) and accordingly modulus. Previous work in our laboratories has shown that simply limiting the molecular weight of the phenylethynyl terminated oligomer did not produce the desired results.

The approach taken was to prepare imide oligomers containing both pendent and terminal phenylethynyl groups of comparable molecular weight to those previously reported which contained only terminal or only pendent phenylethynyl groups. The crosslink density could be adjusted by the amount of pendent phenylethynyl groups incorporated randomly along the oligomer backbone. This approach required the preparation of a diamine (or dianhydride) which contained pendent phenylethynyl groups. An aromatic diamine containing pendent phenylethynyl groups was previously prepared in a three step procedure in relatively good yield[23]. Imide oligomers with a calculated number average molecular weight ($\overline{M}_n$) of 5000 g/mole containing both pendent and terminal phenylethynyl groups were subsequently prepared. The oligomers were characterized and mechanical properties of unoriented thin films, neat resin moldings, titanium (Ti)-to-Ti lap shear adhesive specimens and IM-7 carbon fiber reinforced composite laminates were determined. These properties were compared with those of oligomers of comparable chemical composition and $\overline{M}_n$ with only terminal and only pendent phenylethynyl groups. The changes in properties, as a result of increased crosslink density, are discussed.

2. EXPERIMENTAL

2.1 Starting Materials The following chemicals were obtained from the indicated sources and used without further purification: 3,4'-Oxydianiline (3,4'-ODA, m. p. 84°C, Mitsui Petrochemical Ind., Ltd.), 3,3',4,4'-biphenyltetracarboxylic dianhydride (BPDA, m. p. 227°C, Allco Chemical Co.), 1,3-bis(3-aminophenoxy)benzene (APB, m. p. 107-109°C, National Starch and Chemical Co.), 4-phenylethynylphthalic anhydride (PEPA, m. p. 152°C, Imitec, Inc.), 3,5-diamino-4'-phenylethynylbenzophenone (DPEB, m. p. 154-156°C Daychem Laboratories, Inc.) and N-methyl-2-pyrrolidinone (NMP, Fluka Chemical Co.).

2.1.1 Synthesis of Imide Oligomers Containing Pendent and Terminal Phenylethynyl Groups Imide oligomers containing pendent and terminal phenylethynyl groups (PTPEIs) were prepared using a variety of different diamines and dianhydrides and endcapped with PEPA. In order to narrow the focus of this paper, only oligomers comprised of 3,4'-ODA, APB, DPEB, BPDA and PEPA will be discussed. The oligomers were prepared by first dissolving the appropriate quantities of diamines (3,4'-ODA, APB and DPEB) in NMP at room temperature under nitrogen. The diamine solution was subsequently cooled to ~10°C and the dianhydride (BPDA) and endcapper (PEPA) added in one portion as a slurry in NMP and the solids concentration adjusted to either 30 or 35% (w/w) with additional NMP. The reactions were allowed to stir for ~24 hours at room temperature under nitrogen and an aliquot was subsequently removed to determine inherent viscosity. This solution was used to prepare thin films, adhesive tape and carbon fiber prepreg. In order to prepare imide powder, the flask was subsequently fitted with a moisture trap and condenser, toluene was added, and the mixture heated to reflux overnight to affect imidization. The imide oligomers generally precipitated from solution during the imidization process. The powders were isolated by adding the reaction mixture to water, washing in warm water and methanol. The yellow powders were dried to constant weight with yields typically >95%.

2.1.2 Films NMP solutions (15-20% solids) of the amide acid oligomers containing pendent and terminal phenylethynyl groups were centrifuged, the decantate doctored onto clean, dry plate glass and dried to a tack-free form in a low humidity chamber. The films on glass were imidized and cured by heating at 100, 225 and 350°C for 1 hr each in flowing air. Thin film tensile properties were determined according to ASTM D882 using four specimens per test condition.

2.1.3 Molded Specimens Powdered imide oligomer was compression molded in a 3.2 cm^2 stainless steel mold under 1.4 MPa by heating initially to 300°C for 0.5 hr and subsequently to 350°C for 1 hr. Miniature compact tension specimens (1.6 cm x 1.6 cm x 0.95 cm thick) were machined from the moldings and subsequently tested to determine fracture toughness (K_{Ic}, critical stress intensity factor) according to ASTM E399 using four specimens per test condition. G_{Ic}, (critical strain energy release rate) was calculated using the mathematical relationship $G_{Ic} = (K_{Ic})^2/E$, where E is the modulus of the material.

2.1.4 Adhesive Specimens Adhesive film was prepared by multiple coating of 112-E glass (A-1100 finish) with a NMP solution of the amide acid oligomer (20% solids) and stage dried in a forced air oven to 225°C after each coat. After sufficient thickness had been achieved, the film was dried at 250°C for 0.5 hr. The final volatile content of the tape was ~2%. Standard lap shear adhesive specimens (bond area 2.54 cm wide x 1.27 cm overlap) using titanium (Ti, 6Al-4V) adherends with a PASA Jell 107 (Products Research and Chemical Corp., Semco Div.) surface treatment were fabricated under 1.4 MPa by heating initially to 300°C for 0.5 hr and subsequently to 350°C for 1 hr. Tensile shear adhesive strengths were determined according to ASTM D1002 using four specimens per test condition. The adhesive specimens were exposed to solvent soaks (immersion) for 72 hours in boiling water, hydraulic fluid and jet fuel.

2.1.5 Composite Specimens A 35% solids solution of the amide acid oligomer (PTPEAA-1 or PTPEAA-2) in NMP ($\overline{M}_n$ = 5000 g/mole, inherent viscosity ~0.25-0.30 dL/g) with a Brookfield viscosity of ~15,000-20,000 cps at 25°C was used to coat unsized IM-7 carbon/graphite fiber on a multi-purpose prepregging machine. The unidirectional tape (21.6 cm wide) exhibited a volatile content of ~19-23% and a fiber aerial weight of ~145g/m^2. Laminates were fabricated in a vacuum press under 76.2 cm mercury during the entire process cycle. The laminates were fabricated by heating to 250°C over ~1 hr period and holding at 250°C for 1 hr. Pressure (1.4 MPa) was subsequently applied and the temperature ramped up over a 0.5 hr period to 371°C and held for 1 hr. The laminates were cooled under pressure to about 100°C and the pressure subsequently released. The composite panels were ultrasonically scanned (C-scanned), cut into specimens and tested for mechanical properties according to ASTM procedures.

2.1.6 Gel Permeation Chromatography Gel permeation chromatographic (GPC) analyses were conducted on amide acid solutions in NMP. The analyses were performed using distilled NMP containing 0.02M LiBr. Chromatography was performed on a Waters 150°C GPC, which was equipped with a differential refractive index detector in parallel configuration with a Viscotek model 150R differential viscometer. A Waters Styragel HT 6E linear column covering a molecular weight range from 10^3 to 10^7 g/mole was used in series with a Styragel HT 3 column, which covers the range from 10^3 to 10^4 g/mole. A universal calibration curve was generated with narrow molecular weight distribution polystyrene standards with molecular weights ranging from 5 x 10^2 to 2.75 x 10^6 g/mole.

2.1.7 Other Characterization Inherent viscosities (η_{inh}) were obtained on 0.5% (w/v) solutions in NMP at 25°C. Differential scanning calorimetry (DSC) was conducted on a Shimadzu DSC-50 thermal analyzer at a heating rate of 20°C/min with the glass transition temperature (T_g) taken at the inflection point of the ΔT versus temperature curve. Dynamic thermogravimetric analyses (TGA) were performed on a Seiko 200/220 instrument on cured polymer powder samples at a heating rate of 2.5°C/min in air and nitrogen at a flow rate of 15 cm^3/min. Brookfield viscosity was obtained on a Brookfield LVT Synchro-Lectric viscometer at 25°C.

3. RESULTS AND DISCUSSION

3.1 Synthesis of Amide Acid and Imide Oligomers Containing Pendent and Terminal Phenylethynyl Groups Although many different PTPEI compositions were prepared and characterized only two will be addressed herein. Amide acid oligomer designated as PTPEAA-1, was prepared using 85 mole percent of 3,4'-ODA, 15 mole percent DPEB, 91 mole percent BPDA and 18 mole percent PEPA at a calculated $\overline{M}_n$ of 5000 g/mole (Figure 1). The reactions were carried out in NMP at either 30 or 35% solids. Initially the diamines were placed in NMP, allowed to dissolve and the solution cooled to ~10°C. The anhydrides were subsequently added in one portion as a slurry in NMP and the solution allowed to gradually warm to room temperature. Amide acid oligomer, designated as PTPEAA-2 was prepared in a similar manner as PTPEAA-1 using 70 mole percent of 3,4'-ODA, 15 mole percent APB, 15 mole percent DPEB, 91 mole percent BPDA and 18 mole percent PEPA at a calculated $\overline{M}_n$ of 5000 g/mole. The resulting amide acid oligomer solutions were used to prepare thin films, adhesive film and unidirectional carbon fiber tape. Imidized powder of PTPEI-1 and PTPEI-2 was prepared by cyclodehydration of the amide acid oligomer by heating to reflux with toluene under a moisture trap for ~24 hours. During the imidization process, the imide oligomers precipitated from solution. Oligomer characterization is provided in Table 1. For comparative purposes, data for two other imide oligomers that have either terminal (PETI-5) or pendent (PPEI-1) phenylethynyl groups are included. The composition of PETI-5 is 85 mole percent 3,4'-ODA, 15 mole percent APB, 91 mole percent BPDA and 18 mole percent PEPA with a calculated $\overline{M}_n$ of 5000 g/mole[14].

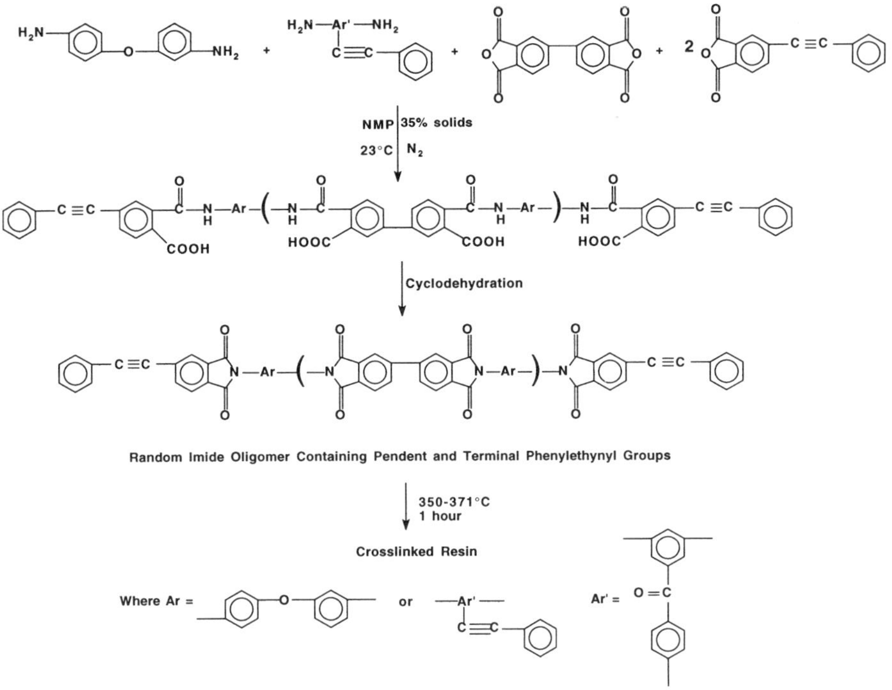

Figure 1. Preparation of PTPEI-1.

Table 1. Oligomer Characterization

Oligomer[1]	Cal'cd $\overline{M}_n$, g/mole	η_{inh}, dL/g[2]	Tg, °C Initial[3] (Tm)	Tg, °C Cured[4] (Tm)	Temp of 5% Wt Loss, °C[5]
PTPEI-1	5000	0.32	231(282)	313	532
PTPEI-2	5000	0.29	224 (284)	289	511
PPEI-1[23]	5000	0.31	209	279	505
PETI-5[14]	5000	0.27	210 (286)	270	503

1. PTPEI-1: 0.85 3,4'-ODA, 0.15 DPEB, 0.91 BPDA, 0.18 PEPA
   PTPEI-2: 0.70 3,4'-ODA, 0.15 DPEB, 0.15 APB, 0.91 BPDA, 0.18 PEPA
   PPEI-1: 0.85 3,4'-ODA, 0.15 DPEB, 0.91 BPDA, 0.18 PA
   PETI-5: 0.85 3,4'-ODA, 0.15 APB, 0.91 BPDA, 0.18 PEPA
2. Inherent viscosities determined on 0.5% (w/v) amide acid solution in NMP at 25°C.
3. Determined by DSC during initial heat-up on oligomeric imide powder.
4. Determined by DSC after 1 hr cure at 371°C in static air in sealed aluminum pan.
5. Determined by TGA in air on imide powder after curing 1 hr at 371°C in air.

The composition of PPEI-1 is 85 mole percent 3,4'-ODA, 15 mole percent DPEB, 91 mole percent BPDA and 18 mole percent phthalic anhydride (PA) with a calculated $\overline{M}_n$ of 5000 g/mole[23]. The PTPEIs exhibited initial T_gs of 224 and 231°C with melting transitions (T_m)

at ~283°C. The T_ms were not recoverable by annealing. After curing for 1 hr in a sealed DSC pan at 371°C, the T_gs increased ~65-82°C. This increase in T_g is comparable to that exhibited by PETI-5 (60°C) and PPEI-1 (70°C). The exothermic onset and peak due to the reaction of the phenylethynyl groups occurred at ~390 and 445°C, respectively. However, the cure reaction appeared to be complete after a 1 hr hold at 371°C as evidenced by no further increase in T_g upon curing an additional hr at 375°C. The temperatures of 5% weight loss as measured by TGA in air on imide powder cured for 1 hr at 371°C ranged from 503-532°C and compares well with that of imide oligomers containing only terminal phenylethynyl groups[14] or pendent phenylethynyl groups[23].

3.1.1 Films Unoriented thin films were cast from NMP solutions of the amide acid oligomers and cured in flowing air to 350°C. In general, the cured films exhibited slightly higher T_gs than imidized powders that were cured in sealed DSC pans. This phenomenon, where air cured phenylethynyl containing imides exhibited higher T_gs than those cured under minimized air exposure, has been previously described[14]. Thin film tensile properties are presented in Table 2. For comparative purposes, the properties of PETI-5[14] and PPEI-1[23] are included. In comparing the data in Table 2, PPEI-1 exhibits the highest tensile modulus at 23°C. Whereas the modulus measured at 177°C is comparable to that of the other oligomers. Both PTPEI-1 and PTPEI-2 unoriented films exhibited comparable tensile properties at 23 and 177°C to PETI-5 with the exception of elongation to break. The high elongation to break exhibited by PETI-5 is more typical of a linear thermoplastic than a thermoset and indicates that in this case the thermal curing of the terminal phenylethynyl groups gives rise to a high degree of chain extension. PETI-5 contains 18 mole percent of terminal phenylethynyl groups while PPEI-1 contains 15 mole percent of pendent phenylethynyl groups distributed randomly along the oligomer backbone, yet the room temperature moduli and elongations to break are significantly different. Both PTPEI-1 and

Table 2. Tensile Properties of Unoriented Thin Films Cured 1 hr at 350°C in Air

Oligomer	Test Temp., °C	Strength, MPa	Modulus, GPa	Elong. @ Break, %
PTPEI-1	23	139.3	3.4	10
	177	78.6	2.2	9
PTPEI-2	23	140.7	3.4	15
	177	77.2	2.1	24
PPEI-1[23]	23	117.2	3.9	7
	177	64.1	2.6	9
PETI-5[14]	23	129.6	3.1	32
	177	84.1	2.3	83

PTPEI-2 have similar tensile strengths and moduli to PETI-5 yet they contain significantly higher concentrations of phenylethynyl groups (33 mole percent). The fact that PPEI-1, PTPEI-1 and PTPEI-2 exhibit relatively low elongations suggest that they have lower molecular weight between crosslinks (i.e. higher degree of crosslinking) than PETI-5. The nature of the cured polymers from PETI-5, PPEI-1 and the PTPEIs are thought to be significantly different presumably due to the concentration and location of the phenylethynyl groups prior to cure.

3.1.2 Moldings Fracture toughness was determined on cured moldings of PTPEI-1 and PTPEI-2 at room temperature. The molding was prepared from solution imidized powder in a stainless steel mold under 1.4 MPa by heating to 300°C for 0.5 hr and 350°C for 1 hr. The moldings were well consolidated and exhibited some flash indicating good flow. PTPEI-1 exhibited a K_{Ic} of 1318 $Pa{\cdot}m^{1/2}$ and a G_{Ic} of 511 J/m^2. PTPEI-2 exhibited a K_{Ic} of 1920 $Pa{\cdot}m^{1/2}$ and a G_{Ic} of 1085 J/m^2. PPEI-1 exhibited a K_{Ic} of 2066 $Pa{\cdot}m^{1/2}$ and a G_{Ic} of 1090 J/m^2. PETI-5 cured under the same conditions exhibited a K_{Ic} of 3388 $Pa{\cdot}m^{1/2}$ and a G_{Ic} of 4025 J/m^2. The difference in toughness exhibited by PTPEI-1 and PTPEI-2 is notable since they each contain the same concentration of phenylethynyl groups and is likely attributable to the 15 mole percent APB present in PTPEI-2. This monomer imparts chain flexibility and thereby a mechanism to dissipate impact energy via molecular motion. PTPEI-2 exhibits toughness comparable to PPEI-1 which contains less than half the concentration of phenylethynyl groups. The high toughness exhibited by PETI-5 lends further support that the thermal cure reaction of the terminal phenylethynyl groups gives rise to significant chain extension and perhaps branching whereas the curing of pendent and pendent and terminal phenylethynyl groups appears to generate a greater degree of crosslinking.

3.1.3 Adhesives Titanium tensile shear specimens of PTPEI-1 and PTPEI-2 were fabricated under 1.4 MPa by heating to 300°C for 0.5 hr and 350°C for 1 hr. The adhesive films were prepared by multiple coating of 112 E-glass and stage-dried up to 225°C in air after each coat. After the final coat was applied the film was stage-dried up to 250°C to reduce the volatile content to ~2%. This multiple coating/drying process appeared to cause some reduction in melt flow compared to that exhibited during neat resin molding. Multiple stage-drying up to 300°C caused significant reduction in melt flow behavior of the adhesive film of PPEI-1 and consequently poor tensile shear properties[23]. This suggests that repeated exposure to temperatures of 275-300°C causes some reaction of the phenylethynyl groups and consequently a reduction in melt flow. Specimens were fabricated using Ti adherends with a PASA Jell 107 surface treatment. The tensile shear strengths are presented in Table 3. PTPEI-1 exhibited lower tensile shear strengths at 23°C and adhesive failures. The strengths exhibited by PTPEI-2 at 23°C were good and remained unchanged at 177°C. The failure modes were mixed although predominantly cohesive. PETI-5 processed under 0.7 MPa and a final cure of 1 hr at 350°C exhibited higher strength at room temperature, and a lower strength at 177°C as compared to PTPEI-2. PTPEI-2 adhesive specimens were subjected to

Table 3. Preliminary Ti-to-Ti Tensile Shear Properties of PTPEI-1 and PTPEI-2

Surface Treatment	Test Temperature, °C	Strength, MPa	Cohesive Failure, %
PTPEI-1*	23	27.6	0
PTPEI-2*	23	34.5	50
	177	34.5	75
	200	32.4	90
	232	27.6	100
	260	17.9	50
PETI-5**[14]	23	49.0	100
	177	29.7	100

* Processed at 350°C for 1 hr under 1.4 MPa

** Processed at 350°C for 1 hr under 0.7 MPa

immersion in boiling water, hydraulic fluid and jet fuel for 72 hrs to determine the effect on tensile shear strengths (Table 4). The specimens exposed to the water boil exhibited a drop in shear strength and those tested at 23°C exhibited predominately adhesive failures indicating a degradation in the PASA jell 107 surface treatment caused by the exposure to moisture. Exposure to hydraulic fluid and jet fuel had negligible effects on the shear strengths, however the failure modes on the specimens tested at 23°C changed to predominately adhesive.

Table 4. Ti-to-Ti Tensile Shear Properties of Solvent Exposed PTPEI-2

Solvent Soak*	Test Temp., °C	Tensile Shear Strength, MPa	Cohesive Failure, %
Water Boil	23	26.9	5
	177	26.2	70
	200	24.1	80
Hydraulic Fluid Hyjet IV	23	32.4	15
	177	29.7	75
	200	29.6	85
Jet Fuel JP V	23	33.1	20
	177	31.7	80
	200	30.3	85

*Immersion for 72 hrs

3.1.4 Composites IM-7 carbon fiber composites of PTPEI-1 and PTPEI-2 were fabricated from solution coated unidirectional tape in a vacuum press under 1.4 MPa with a final cure at 371°C for 1 hr. The processing cycle for these laminates is presented in Figure 2. This cycle was originally developed for PETI-5 and since it provided good quality laminates from prepreg of PPEI-1, PTPEI-1 and PTPEI-2 no optimization work was performed on processing conditions. The laminates generally exhibited excellent C-scans indicating good consolidation and little or no void content. The composite properties are presented in Table 5. For comparative purposes, IM-7 composite properties of PETI-5 and

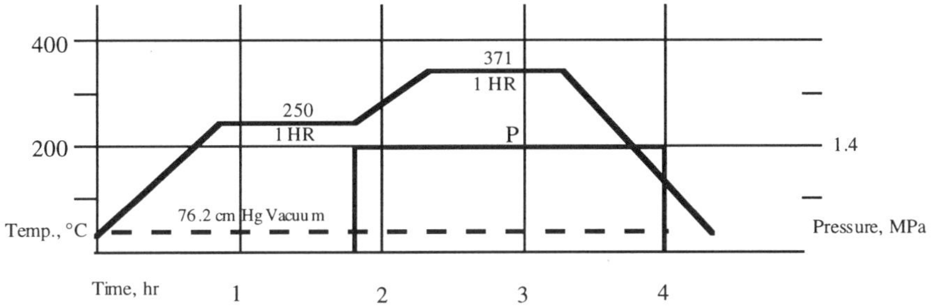

Figure 2. Laminate Processing Cycle for PETI-5, PPEI-1, PTPEI-1 and PTPEI-2.

PPEI-1 are also included. In most cases, PTPEI-1 exhibited equal or higher properties than PETI-5 at both room temperature and at 177°C. The open hole compressive strengths (OHC), moduli and retention of these properties at 177°C (wet) for PTPEI-1 were excellent. They were approximately 15% higher than those of PETI-5 and PPEI-1. However, a trade off was evident in the compression after impact strength (CAI) for PTPEI-1 which is 20-25% less than that exhibited by PPEI-1 and PETI-5, respectively. PTPEI-2 exhibited high room temperature OHC strength, but the strength dropped significantly when tested at 177°C wet. Unfortunately, the data for PTPEI-2 could not be normalized as the fiber volume was too low (~55-56%), therefore this data is not directly comparable to the data for the other matrix resins listed in Table 5. However, based on comparisons of unnormalized data, the CAIs exhibited the same trend as the neat resin fracture toughness. PTPEI-1 had the lowest CAI and G_{Ic}, PPEI-1 and PTPEI-2 had identical CAIs and G_{Ic}s even though they contain significantly different concentrations of phenylethynyl groups. It is clear that the laminate properties exhibited by these resin systems can not be easily explained using a simple crosslink density/toughness relationship. The chemical structures of the product(s) formed from the thermal reaction of phenylethynyl groups are complex and depend on concentration, location, conformational restrictions and molecular mobility and therefore are not easily predicted.

Table 5. Preliminary IM-7 Carbon Fiber Composite Properties

Property (Lay-up)	Test Temp., °C	PETI-5*	PPEI-1*	PTPEI-1*	PTPEI-2#
Flexural Strength, MPa (Unidirectional)	23 177	1787.5 1441.6	1795.2 1513.1	1786.2 1379.3	TBD
Flexural Modulus, GPa (Unidirectional)	23 177	144.7 133.7	148.3 142.0	142.0 137.9	TBD
Open Hole Compressive Strength, MPa (42/50/8)	23 177 (wet)	393.1 317.2	396.6 307.6	462.0 365.5	441.4 303.5
Compressive Strength, MPa (Unidirectional)	23 177	1371.9 -----	1600.0 1544.8	1401.7 -----	TBD
Compressive Modulus, GPa (Unidirectional)	23 177	138.5 -----	159.3 151.7	144.2 -----	TBD
Short Beam Shear Strength, MPa (Unidirectional)	23 177	106.5 62.8	93.1 73.1	TBD TBD	TBD
Compression After Impact Strength, MPa (Quasi-isotropic)	23	324	296	241	297

*Normalized to 62% fiber volume.

#Not normalized due to low fiber volume.

TBD=To be determined.

3.1.5 Gel Permeation Chromatography GPC analyses were performed on as-prepared amide acid solution of PTPEI-1 and PTPEI-2 to assess molecular weight and distribution. In general, the GPC data agreed well with the calculated $\overline{M}_n$ and the amide acid oligomers exhibited Gaussian distributions with polydispersities of ~2.

To assess the effects of aging on changes in molecular weight distribution, samples of PTPEAA-1 (35% solids in NMP) were stored at 5°C and at 23°C and analyses performed periodically over a 220 day period (Figure 3). The trend exhibited by the samples stored at 23°C over this exposure period was a reduction in the $\overline{M}_n$, weight average ($\overline{M}_w$) and z average ($\overline{M}_z$) molecular weights. The samples stored at 5°C exhibited the same trend, but to a lesser degree. After seven days at 23°C, the molecular weight distribution was unchanged (Figure 3). The effect of 23°C aging on intrinsic viscosity was evident as well, it dropped from 0.255 dL/g to 0.222 dL/g after 220 days. The sample stored at 5°C exhibited a slight change from 0.255 to 0.252 after 220 days. The effects of these changes in molecular weight distribution on laminate properties is not known.

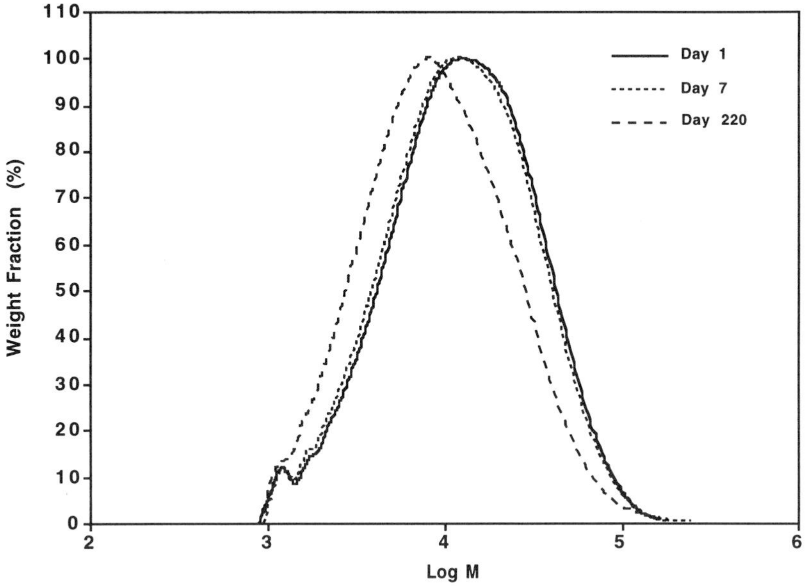

Figure 3. Effect of 23°C exposure on molecular weight distribution of PTPEAA-1.

4. SUMMARY

Amide acid and imide oligomers containing pendent and terminal phenylethynyl groups were prepared, characterized and preliminary neat resin, adhesive and carbon fiber reinforced laminate properties were determined. The cured polymers exhibited high T_gs and high unoriented thin film tensile properties. Adhesive panels and carbon fiber reinforced composites were readily fabricated under 1.4 MPa. The adhesive properties of PTPEI-2 were good with excellent retention of tensile shear strength at elevated temperatures. PTPEI-1 exhibited excellent open hole compression strengths at 23 and 177°C (wet). The properties of the cured resins were controlled to a large extent by the phenylethynyl group content and configuration.

5. ACKNOWLEDGMENT

The authors are grateful to Dr. Emilie J. Siochi, Lockheed-Martin Engineering and Sciences Company, for the gel permeation chromatographic work.

6. REFERENCES

1. F. L. Hedberg and F. W. Arnold, J. Polym. Sci. Polym. Chem. Ed., 14, 2607(1976).
2. F. L. Hedberg, R. F. Kovar, and F. E. Arnold, Contemporary Topics in Polymer Science (ed., E. M. Pearce), vol. 2, Plenum Pub. Co., NY, 1977, p. 235.
3. P. M. Hergenrother, Macromolecules, 14, 898(1981).
4. F. W. Harris, A. Pamidimukkala, R. Gupta, S. Das, T. Wu, and G. Mack, Polym. Prepr., 24(2), 324(1983). J. Macromol. Sci.-Chem., A21(8 and 9), 1117(1984).
5. S. Hino, S. Sato, K. Kora and O. Suzuki, Chem. Abstr., 110, 115573w (1989). Japanese Patent 63,196,564 (1989).
6. C. W. Paul, R. A. Schultz, and S. P. Fenelli, U. S. Patent 5,138,028(1992), To National Starch and Chemical Corp. In "Advances in Polymer Science and Technology" (Eds., C. Feger, M. M. Khoyasteh, and M. S. Htoo), Technomic, Lancaster, PA, 1993, p. 220.
7. S. Jayaraman, G. Meyer, T. M. Moy, R. Srinivasan, and J. E. McGrath, Polym. Prepr., 34(1), 513(1993).
8. R. G. Bryant, B. J. Jensen, and P. M. Hergenrother, Ibid., 34(1), 566(1993).
9. G. W. Meyer, S. Jayaraman, and J. E. McGrath, Ibid., 34(2), 540(1993).
10. B. J. Jensen, R. G. Bryant, and S. P. Wilkinson, Ibid., 35(1), 539(1994).
11. S. J. Havens, R. G. Bryant, B. J. Jensen, and P. M. Hergenrother, Ibid., 35(1), 553(1994).
12. R. G. Bryant, B. J. Jensen, J. G. Smith, Jr., and P. M. Hergenrother, Sci. Adv. Matl. Proc. Eng. Ser., 39 273(1994). (Closed Paper Proceedings).
13. P. M. Hergenrother, R. G. Bryant, B. J. Jensen, J. G. Smith, Jr., and S. P. Wilkinson, Ibid., 39, 961(1994).
14. J. G. Smith, Jr. and P. M. Hergenrother, Polym. Prepr., 35(1) 353(1994). Polymer, 35(22), 4857(1994). U. S. Patent Application Serial # 08/330,773 (1994) to NASA.
15. J. A. Johnston, F. M. Li, F. W. Harris, and T. Takekoshi, Ibid., 35(22), 4865(1994).
16. T. Takekoshi and J. M. Terry, Ibid., 35(22), 4874(1994).
17. P. M. Hergenrother, R. G. Bryant, B. J. Jensen, and S. J. Havens, U.S. Patent 5,412,066(1995), to NASA.
18. G. W. Meyer, T. E. Glass, H. J. Grubbs, and J. E. McGrath, Polym. Prepr., 35(1), 549(1994). J. Polym. Sci.: Pt. A: Polym. Chem., 33, 2141(1995). U. S. Patent 4,493,002 (1996), to Virginia Polytechnic Institute and State University.
19. G. W. Meyer, Y. J. Lee, S. J. Pak, and J. E. McGrath, Polym. Mat'l. Sci. Eng. Proc., 70, 496(1994).
20. F. W. Harris, S. M. Padaki and S. Varaprath, Polym. Prepr. 21(1), 3 1980.
21. B. J. Jensen, P. M. Hergenrother, and G. Nwokogu, Polym. Prepr., 33(1), 914(1992).

22. P. M. Hergenrother and B. J. Jensen, U.S. Patent 5,344,982(1994), to NASA.
23. J. W. Connell, J. G. Smith, Jr. and P. M. Hergenrother, Sci. Adv. Matl. Proc. Eng. Ser., 41 1102 (1996).
24. J. G. Smith, Jr., Adhesion Society Proceedings, 19, 29 (1996).
25. R. Rossi, U.S. Patent 4,730,032(1988), to National Starch and Chemical Corporation.
26. R. G. Bryant, B. J. Jensen, and P. M. Hergenrother, Polym. Prepr., 33(1), 910(1992).
27. B. J. Jensen, R. G. Bryant, and P. M. Hergenrother, U.S. Patent 5,268,444(1993), to NASA.
28. S. Jayaraman, G. Meyer, T. M. Moy, R. Srinivasan, and J. E. McGrath, Polym. Prepr., 34(1), 513(1993).
29. B. J. Jensen and R. G. Bryant, Ibid., 35(1), 513(1994).
30. W. D. Joseph, J. C. Abed, T. H. Yoon, and J. E. McGrath, Ibid., 35(1), 551(1994).
31. J. G. Smith, Jr., J. W. Connell, E. J. Siochi, and P. M. Hergenrother, High Perf. Polym., 7, 41(1995).
32. J. G. Smith, Jr., J. W. Connell, and P. M. Hergenrother, Polym. Prepr., 34(1), 431(1993).
33. B. J. Jensen and P. M. Hergenrother, J. Macromol. Sci.-Pure Appl. Chem., A30(6 and 7), 449(1992).
34. M. B. Goldfinger and T. M. Swager, Polym. Matl. Sci. Eng. Prepr., 73, 256(1995).

6. Biographies

John W. Connell is a senior polymer scientist in the Composites and Polymers Branch of the Materials Division at NASA Langley Research Center (LaRC). He received B.S. and Ph. D. degrees from Virginia Commonwealth University in 1982 and 1986, respectively. Prior to joining NASA LaRC in January 1988, he was a research associate at Virginia Commonwealth University. Since coming to NASA, his work has focused on the development of high performance polymers for aerospace applications.

Joseph G. Smith, Jr. is a polymer scientist in the Composites and Polymers Branch of the Materials Division at NASA Langley Research Center (LaRC). He received a B.S. degree from High Point College in 1985 and a Ph.D. from Virginia Commonwealth University in 1990. Prior to joining NASA LaRC in September 1994, he held postdoctoral research positions with the University of Akron and Virginia Commonwealth University. His work at NASA has focused on the development of high performance polymers for aerospace applications.

Paul M. Hergenrother is a senior polymer scientist in the Composites and Polymers Branch of the Materials Division at NASA Langley Research Center (LaRC). He received a B.S.

degree from Geneva College and took graduate work at the University of Pittsburgh and Carnegie-Mellon University. Prior to coming to NASA he held various research and management positions at the Koppers Co., The Boeing Co., Whittaker Corp. and Virginia Polytechnic Institute and State University. His work at NASA has focused on the development of high performance polymers for aerospace applications.

Monica Rommel is an integrated technology team leader in Structures Research at Northrop Grumman's Military Aircraft Systems Division. She received a B.S. degree in Chemistry-Material Science from University of California, Los Angeles in 1985. Her work at Northrop Grumman has focused on development and characterization of advanced materials and processes for aircraft applications.

28th International SAMPE Technical Conference
November 4-7, 1996

SYNTHESIS AND CHARACTERIZATION OF HIGH PERFORMANCE THERMOSETTING POLYIMIDES FOR STRUCTURAL ADHESIVES AND COMPOSITE MATRIX SYSTEMS

J. E. McGrath, B. Tan, V. Vasudevan, G. W. Meyer, A. C. Loos and T. Bullions
Virginia Polytechnic Institute and State University
Departments of Chemistry and Engineering Science & Mechanics
NSF Science and Technology Center:
High Performance Polymeric Adhesives and Composites
Blacksburg, VA 24061-0344

ABSTRACT

Synthesis and characterization of polyether imides fitted with functional endgroups derived from (mostly) phenylethynyl phthalic anhydride or phenyl maleic anhydride were prepared. The resulting materials show a wide processing window that is defined by the difference between the glass transition temperature of the oligomer and the exotherm from either of the reactive endgroups, which occurs in the range of 350–400°C. Poly(ether imide)s (PEI), based on commercially available monomers were prepared in controlled molecular weight from about 2,000 to 20,000 g/mole. Emphasis was placed principally on oligomers which had M_n values about 2,000-5,000 g/mole, which produced oligomers with Tg's from about 175 to over 200°C. The cured materials were prepared at 350°C or 380°C for 30-90 minutes and their network formation and structure were characterized by rheology and gel fraction measurements. In favorable cases, very high, e.g., 98-99% gel fractions were generated that allowed for good solvent resistance. Tg values for the cured networks were as high as about 265°C, even if the starting oligomers were of 2,000-3,000 g/mole molecular weight. The resulting materials and their compatible physical blends with thermoplastic PEI were characterized as structural adhesives for titanium and very good values were obtained. The oligomers were also utilized as the matrix for continuous carbon fiber composites via resin infusion molding. Indications are that low void materials with good flexural modulus, strength, and high fracture toughness are possible. The synthesis and characterization of these materials will be reviewed and the mechanical behavior of both the structural adhesives and composite matrix systems is discussed.

KEY WORDS: Polyetherimide thermosets, structural adhesives, composite matrix

1. INTRODUCTION

High performance thermosetting polyimides for structural adhesives and composite matrix systems, as well as other applications, such as electronic packaging materials are of great interest and have been reviewed in a number of books and recent literature citations (1-5).

Polyimides are leading candidates in view of their established high performance mechanical and thermal characteristics. Nevertheless, processability is a major consideration, along with the resulting durability of the structural adhesives and polymer matrix systems. There is considerable interest in techniques that will minimize evolution of volatiles during the processing and fabrication steps and this suggests utilization of powder processing or related techniques, such as resin infusion molding. Generation of soluble polyimide precursors is thus desirable to employ techniques that do not require evolution of large volatile fractions, such as alcohols, NMP, water, etc. One system that has been of considerable interest to this research group is variations on the well-known poly(ether imide) available from the General Electric Company, known as Ultem™.

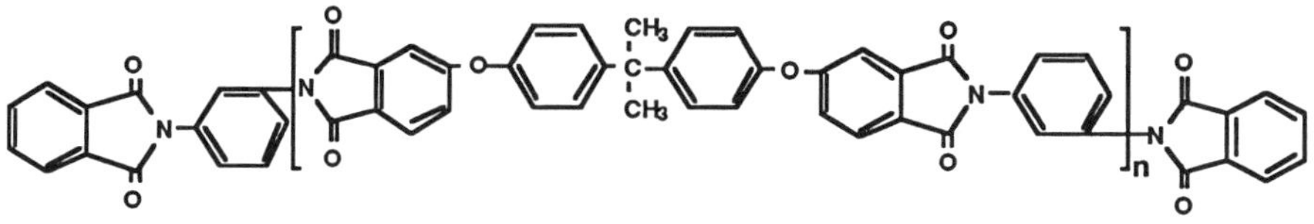

Ultem™-1000

The thermoplastic material shown above has a glass temperature of about 215°C, good mechanical properties, but suffers from relatively poor solvent resistance to at least some of the aggressive environments that could be expected in the aerospace arena, including hydraulic fluid, MEK, toluene, and possible de-icers, such as ethylene glycol. In order to circumvent this solvent resistance problem, our own group and others (3,6-10) have reported that utilizing phenylethynyl endcapped polyimide can afford curable, solvent resistance materials with good mechanical and adhesive properties.

The poly(ether imide) provides one of the critical needs, namely the ability to fully cyclize the material prior to subsequent processing. This has been demonstrated in earlier publications and the second issue is how to provide a reactive endblocker, which would allow for network thermoset formation and the accompanying improvement in solvent resistance. A number of groups have studied the phenyl ethynyl material (3,6-10), which offers major advances in processability since the exothermic heat curing temperature is quite high (350-400°C). This paper focuses on materials that were prepared in conjunction with General Electric scientists as a model high performance thermosetting imide that might have attractive properties suitable for chemically resistant structural adhesives and also composite matrix materials. The approach investigated thus far has been to examine the 3,000 g/mole molecular weight oligomers in structural adhesive applications and the shorter, more highly crosslinked, networks derived from the 2,000 g/mole molecular weight oligomers as matrix resins for carbon fiber composites by resin infusion.

2. EXPERIMENTAL

2.1 Polymer Synthesis and Characterization Synthesis and characterization of the phenyl ethynyl poly(ether imide) oligomers have already been well described in several publications (3,6,7) and will only be briefly reviewed here. Essentially, the meta-linked oligomers can be prepared by reacting the endcapper and the bisphenol-A based dianhydride in o-dicholorobenzene(o-DCB) with m-phenylene diamine at about 180°C (Scheme 1). The resulting polymerizations were allowed to react for 20 hours at 170°C and then were recovered by methanol coagulation and drying. This process produces a powdery precursor that is suitable for resin infusion or dry powder prepregging processing, as described later. Phenyl maleic-anhydride capped systems were prepared as described earlier (6).

The resulting oligomers were characterized by size exclusion chromatography and quantitative carbon NMR spectroscopy to confirm that the target molecular weights of about 2,000-3,000 g/mole were, in fact, achieved. Intrinsic viscosity measurements were also conducted to further verify the nature of the molecular weights of the oligomer. Curing studies were conducted using differential scanning calorimetry that also established the glass temperature of the oligomer, as well as the anticipated curing temperature interval. Rheological studies using a parallel plate, Rheometrics RMS-800 viscometer, demonstrated the suitability of the resulting oligomers for rapid processing, including thus far, resin infusion molding and potentially resin transfer molding. Polymer blends with linear thermoplastics were made via solution blending and drying, although it is anticipated that melt blends could also be satisfactorily prepared. This latter work enabled somewhat better handleability to be achieved, which is important for adhesives. It may also offer improved economics, since a portion of the materials would involve already commercially produced PEI's. In this regard, the material of choice for the blends was commercially available Ultem™ 1000.

2.2 Adhesive Tests The Ti-Ti single lap shear specimens were prepared and tested according to ASTM-D1002. A solution of the oligomer [15%(w/v)] in chloroform was coated onto glass cloth (112E) and dried at 180°C for 6-12 hrs under reduced pressure. This procedure was repeated (~4 times) until the scrim contained 85 wt.% oligomer. The scrim was cut into rectangles slightly larger than 1.27 cm × 2.54 cm.

Titanium coupons were sandblasted, washed with water, dipped in Pasa-Jell for 20 minutes, and, then washed with water and dried at 100°C. Next, the coupons were primed with the same polymer solution and dried at 180°C under reduced pressure. A piece of scrim was placed between two titanium plates placed end-to-end with an horizontal overlap of 1.27 cm × 2.54 cm. The specimens were placed in a melt press, and the temperature was raised from 25°C to curing temperature (350-380°C) at 10°C/minute. After reaching the curing temperature, the specimens were cured for 45-90 minutes at 25-75 psi.

The Ti-Ti lap shear specimens were mounted between two grips of an Instron model 1123 and tested in tension with a crosshead speed of 0.05 in/minute.

2.3 Composite Preparation Composite panels based upon Hercules AS-4, 3K, plain-weave, G-sized fabric were prepared with a matrix of either 2,000 or 3,000 g/mole molecular weight reactive endcapped oligomer using resin infusion techniques (Figure 1). Four-point flexural tests (ASTM D790-92) have been conducted on a panel with a 2,000 g/mole matrix; the panel had dimensions of 15.24 cm x 15.24 cm x 0.300 cm. The early experimental consolidation conditions resulted in a fiber volume fraction estimated at about 66%. Five test specimens with dimensions of 12.07 cm x 2.54 cm x 0.300 inches were cut from the panel with the 12.07 cm dimension of the specimen running parallel to the warp direction of the fabric in the panel. The axes of the loading noses ran perpendicular to the 12.07 cm length of the specimen. Each specimen had a support span of 9.53 cm to afford about a 32 to 1 span to depth ratio. A 1.27 cm overhang on each end of the specimen was allowed. Control panels utilized Shell Chemical Company's RSL-1895 epoxy, with a Tg of about 204°C, and the material known as PR-500, available from 3M, which has a Tg of about 200°C.

3. RESULTS AND DISCUSSION

3.1 Polymer properties The polyetherimide oligomer was synthesized at a target M_n of 2,000 and 3,000 g/mole. It was synthesized under homogeneous conditions by using only *o*-DCB as the solvent. Molecular weight control was achieved by utilizing 4-phenyl-ethynylphthalic anhydride as an endcapping agent as shown in Table 1. The intrinsic viscosity measurements showed the correct trend; number average molecular weight values were also estimated from end group analysis via quantitative ^{13}C n.m.r. spectroscopy by ratioing the integrals of the carbons in the backbone (-CH_3, δ=30 ppm) and in the endgroup -C=C- (δ=88 and 94 ppm). The experimental molecular weights obtained from ^{13}C n.m.r. analysis matched well with the theoretical molecular weights. The soluble all *m*-PDA based imide oligomer was analyzed by GPC, which had a viscosity detector, using a universal calibration (11). The results also satisfactorily matched the theoretical values.

A typical DSC thermogram is shown in Figure 2. For this M_n=3,000 g/mole BPADA/*m*-PDA 4-PEPA capped oligomer, a glass transition (T_g) of 183°C was observed at 10°C/min heating rate. No melting endothermic peaks were found indicating it is an amorphous system and a curing exothermic peak was observed that had an onset around 350°C and peak at 396°C. This curing exothermic peak was easily assigned to the reactive phenylethynyl endgroups. Precise definition of the resulting molecular network junction structure has been challenging, but efforts are continuing. The amorphous nature of this type of material ensures a wide processing temperature window between T_g and the cure exotherm; thermal results are summarized in Table 2.

Studies were conducted to determine the gel fraction of the cured samples. Films cured at various times and temperatures were extracted with chloroform for five days and then dried for five days at 120°C under vacuum. The solvent extraction data for the 3,000 g/mole oligomer are listed in Table 3. It was concluded that curing this oligomer at 350°C for 90 minutes or 380°C for 30 minutes achieved high gel fractions (>95%). Longer curing time at 380°C (>60 minutes) can achieve almost 100% gel fraction.

3.2 Adhesive Properties and Blend Systems Experiments have been initiated to evaluate the adhesive properties on titanium substrates. All our tests were done based on titanium (6V-4Al) alloy as matrix, Pasa-Jell surface treatment and ASTM-D1002 method; the initial results are shown in Table 4. For reactive oligomers, the bonding time-temperature-pressure played an important role for adhesive test results. Even without optimizing the bonding condition, the lap shear strengths were high and most (M_n=3,000 g/mole) were higher than the thermoplastic control. These values, coupled with solvent resistance, suggest that this type of material may be a candidate structural adhesive for many high performance applications.

Independently, a Mn=3,000 g/mole, BPADA/m-PDA 4-PEPA capped polyamic acid solution in NMP was prepared and sent to CYTEC Inc. for adhesive evaluation (Figure 3). The results are quite encouraging. The M_n=2,000 and 3,000 g/mole oligomers as fully imidized powders are not easily melt pressed to form self supported films for adhesive bonding, as expected for low molecular weight oligomers. Thus, the adhesive "green strength" was weak and scrim cloth was utilized in our adhesive tests. This scrim functioned as a support for the oligomers and also served to control thickness. However, the glass scrim may complicate adhesive failure mode and sample preparation since there is another glass-polymer interface generated. We developed a method to solve this problem by simply blending the thermosetting oligomers with a thermoplastic to improve the adhesive "green strength". Such blend systems also showed some interesting results as discussed below.

The blend system is based on M_n=3,000 BPADA/*m*-PDA 4-PEPA oligomers which have a high gel fraction (>99%) after a 370°C, 45-minute cure and the polyetherimide (PEI) thermoplastics which have M_n=20,000 g/mole with good adhesive strength, but modest solvent resistance. Interestingly, the gel fraction of the cured blend was higher than our expectation and even higher the 25 wt.% PEI thermoplastic blend had a gel fraction as high as 94%. Samples with longer extraction time (up to 10 days) were 94% insoluble, indicating good retention of solvent resistance (Figure 4). This good retention may be due to tight semi-IPN formation which would make it difficult for the solvent to extract out the linear, but miscible thermoplastic. The glass transition temperatures of the cured samples are a function of the PEI wt.% as shown in Figure 5. All the samples showed single T_gs indicating they are miscible systems. We are particular interested in PEI wt.% in the 10-25 range since the T_gs and gel fractions are reasonably high and the material can be melt-pressed to form self-supporting films for adhesive tests. Adhesive lap shear tests were conducted and the results are also listed in Table 4. The good adhesive strength of the blends suggests an improved way to prepare adhesives and lower the cost.

3.3 Rheological and Composite Properties The flow behavior was examined by rheological parallel plate studies. The results of a 3°C/minute temperature ramp of these oligomers is shown in Figure 6. The oligomers flow well at certain temperature-time ranges and a resin infusion method was applied successfully to make the M_n=2,000 g/mole matrix composites. Warp flexural test results based on AS4-PW-3K fabric showed flexural strength of 790±66 MPa and flexural modulus of 73.8±1.7 GPa. Compared to other systems (Figure 7), such results show promise in applying these types of materials as carbon fiber matrix systems.

4. CONCLUSIONS

Imide oligomers based on modified Ultem™ with 4-phenylethynylphthalic anhydride endcapper were synthesized which yielded highly imidized, controlled molecular weight phenylethynyl functionalized materials. Upon thermal curing, insoluble, ductile films with the anticipated T_g values were obtained that exhibit excellent thermooxidative stability. The successful behavior is related to the wide processing window afforded by the phenylethynyl endcap. Excellent solvent resistance and adhesive behavior to titanium were obtained from this polyimide matrix system. Carbon fiber composites have been successfully produced by resin infusion and optimization is continuing.

5. ACKNOWLEDGMENTS

The authors would like to gratefully acknowledge the National Science Foundation and the Center for High Performance Polymeric Adhesives and Composites for partial support under contract DMR-9120004. In addition, support from the Boeing Commercial Aircraft Company is also acknowledged. The authors thank Dr. R. Mayhew of the CYTEC Co. for some of the adhesive results. They also thank Dr. J. Rich, et. al, of the General Electric Co. for use of the scale-up facilities and partial support.

6. REFERENCES

1. C. Feger, M. M. Khojastech and J. E. McGrath, editors, Polyimides: Materials, Chemistry, and Characterization, Elsevier (1989).

2. D. Wilson, P. M. Hergenrother and H. Stenzenberger, Editors, *Polyimides*, Chapman and Hall, (1990)

3. G. W. Meyer, B. Tan and J. E. McGrath, "Solvent-resistant Polyetherimide Network Systems Via Phenylethynylphthalic Anhydride Endcapping", *High Performance Polymers*, **6**, 423-435, (1994).

4. P. M. Hergenrother, Editor, *Advances in Polymer Science*, **117**, (1994).

5. P. M. Hergenrother, *Trends in Polymer Science*, **4**, No. 4, 104-105, (1996).

6. G. W. Meyer, et al., *Polymer* (London), accepted, (1996).

7. G. W. Meyer, T. E. Glass, H. J. Grubbs and J. E. McGrath, *J. Polym. Sci., Part A: Polym.Chem.*, **33**, 2141, (1995).

8. P. M. Hergenrother and J. G. Smith Jr., "Chemistry and Properties of Imide Oligomers End-capped with Phenylethynylphthalic Anhydrides", *Polymer*, **35**, 4857-4864, (1994).

9. J. A. Johnston, F. Li, F. W. Harris and T. Takekoshi, "Synthesis and Characterization of Imide Oligomers End-capped with 4-(phenylethynyl)phthalic Anhydrides", *Polymer*, **35**, 4865-4873, (1994).

10. T. Takekoshi and J. M. Terry, "High-Temperature Thermoset Polyimides Containing Disubstituted Acetylene End Groups". *Polymer*, **35**, 4874-4880, (1994).

11. M. Konas, T. M. Moy, M. E. Rogers, A. R. Schultz, T. C. Ward and J. E. McGrath, *J. Polym. Sci. (Physics)*, **33**, 1441-1448, (1995).

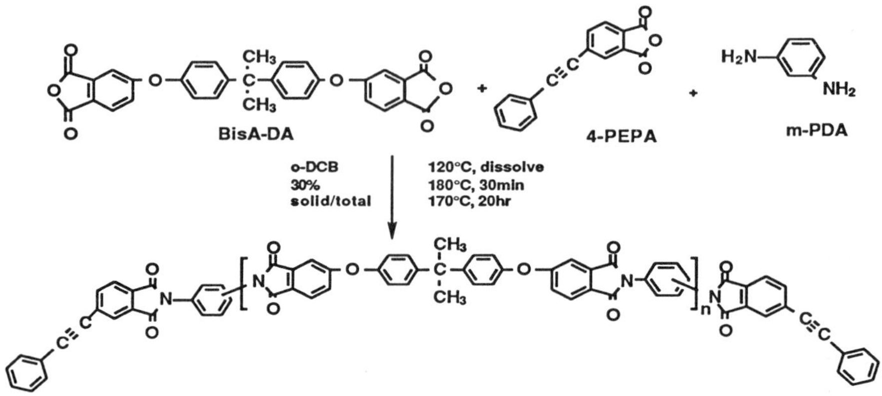

Scheme 1

Synthesis of 4-phenylethynyl phthalic anhydride (4-PEPA terminated Ultem™-like imide oliogmers

Table 1

Molecular Weight Characterization of Phenylethynylphthalic Anhydride-Terminated Polyimides

System	Target Mn (g/mole)	[h] (dlg^{-1}, NMP 25°C)	Mn by ^{13}C NMR (g/mole)	Mn by GPC (g/mole)
BPADA/*m*-PDA	2,000	0.15	2,000	1,500; Mw/Mn=1.5
	3,000	0.19	2,900	3,700; Mw/Mn=1.7

Table 2

Thermal Characterization of Phenylethynylphthalic Anhydride-Terminated Polyimides

System	Target Mn (g/mole)	Tg before cure (°C)[a]	Tg after cure (°C)[a]	Cure exotherm (°C)[a]	TGA 5% wt. loss (°C)[b]
BPADA/	2,000	172	265	392	525
m-PDA	3,000	184	252	395	520

[a]Heating rate of 10°C/minute in nitrogen

[b]5% weight loss after cure, measured in air at 10°C/minute

Table 3

Solvent Extraction Data in Chloroform of BPADA/m-PDA 4-PEPA Capped Polyimide (Mn=3,000 g/mole)

Curing Temperature (°C)	Curing Time (minutes)	Gel Fraction[a]
350	60	88
350	90	97
350	30	96
380	60	99
380	90	99

[a]Samples were extracted for 5 days and then dried for 5 days at 120°C under vacuum

Table 4

Ti-Ti (6Al-4V) Single Lap Shear Test

System	Adhesive Strength (Psi)	Curing Conditions
BPADA/*p*-:*m*-(7:3)PDA 4-PEPA capped, 3,000 g/mole	4,200±400	380°C, 90 min, 75 Psi
BPADA/m-PDA 4-PEPA capped, 2,000 g/mole	3,700±300	370°C, 45 min, 25 Psi
BPADA/m-PDA 4-PEPA capped, 3,000 g/mole (3K)	5,300±210	370°C, 45 min, 25 Psi
3K (90%)-PEI (10%)*	4,200±670	370°C, 45 min, 35 Psi
3K (75%)-PEI (25%)*	4,700±520	370°C, 45 min, 50 Psi
PEI	4,100±320	270°C, 15 min, 75 Psi

* Samples prepared from self-supported films.

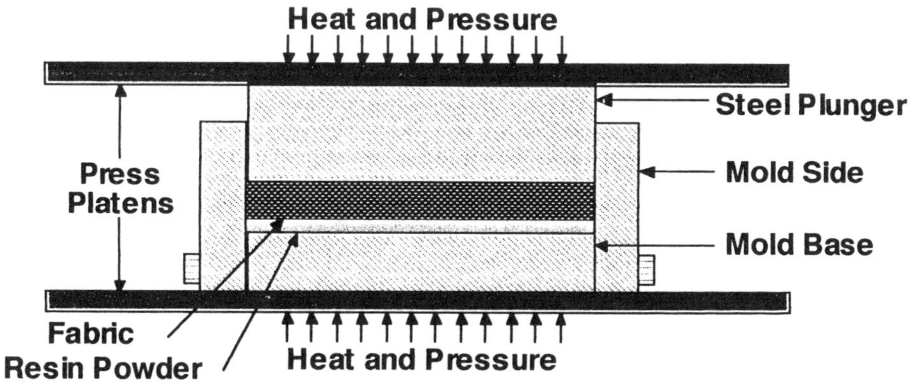

Figure 1. Resin Powder Infusion Technique

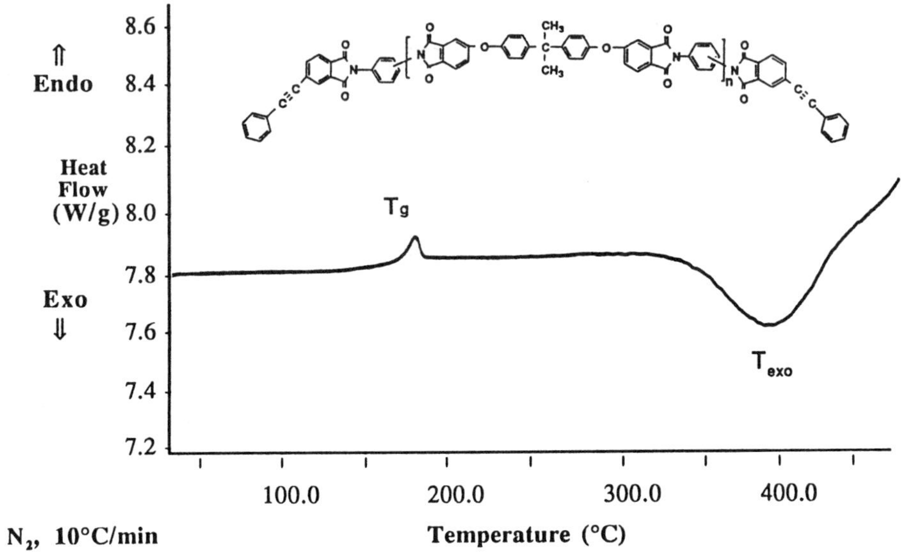

Figure 2. DSC Thermogram of Mn=3,000 g/mole, BPADA/m-PDA 4-PEPA Capped Polyetherimide Oligomer

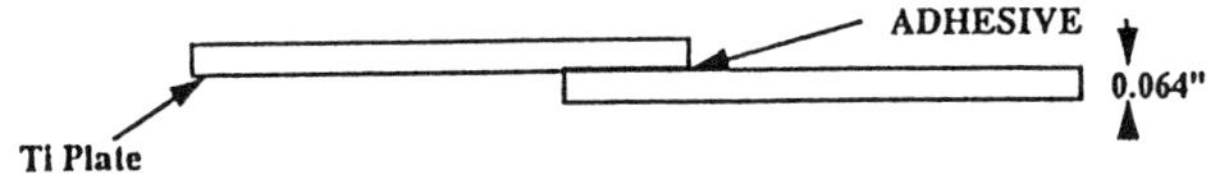

Substrate:	0.050" 6A1-4V Titanium
Prep:	Chromic acid anodized, Boeing method
Primer:	BR® x5 (based on PETI-5 resin) primer cured 60 min at 660°F
Bond Cycle:	Autoclave cure, 60 minutes at 660°F, 50 psi

Lap Shear Strength		
	RT	6300 psi (80% cohesive)
	350°F	5250 psi (80% coh)
	400°F	4400 psi (80% coh)

Courtesy of Dr. R. T. Mayhew at CYTEC ENGINEERED MATERIALS INC.

Figure 3. Single Lap Shear Test of Mn=3,000g/mole, BPADA/m-PDA 4-PEPA Capped Polyetherimide Oligomer From Polyamic Acid Solution

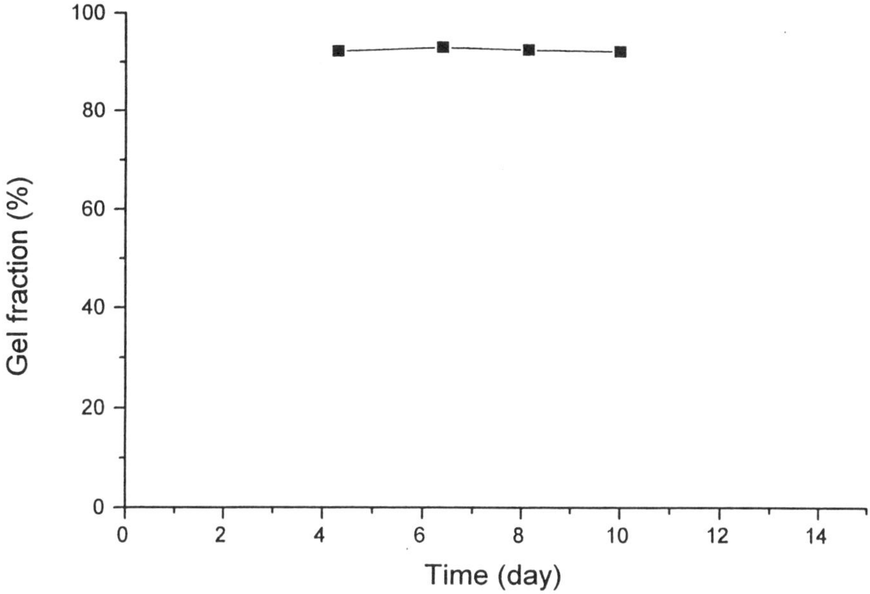

Figure 4. Long Term Solvent Extraction of 3,000 g/mole Oligomer-PEI (25 wt.%) Blend (cured at 370°C/45 min. and Extracted From $CHCl_3$)

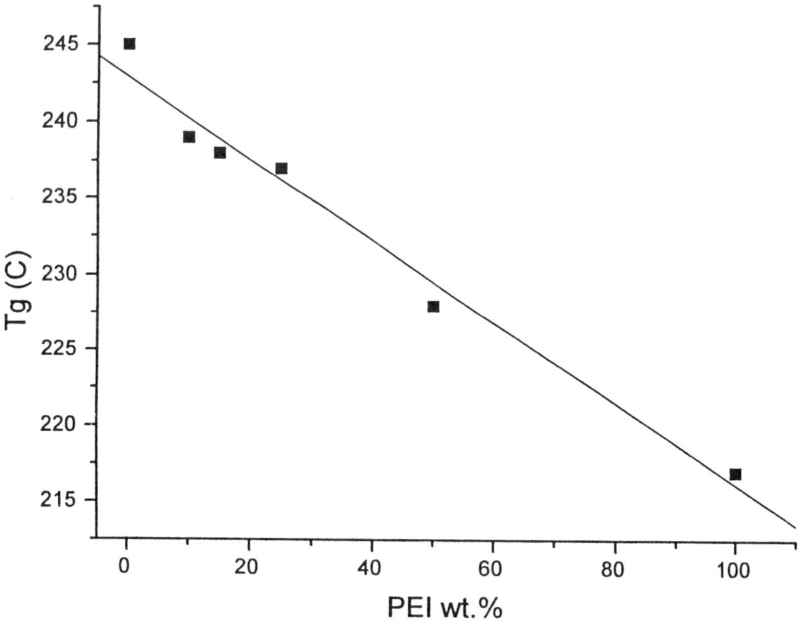

Figure 5. Influence of PEI Weight Fraction on the Tg of Cured 3,000 g/mole Oligomer-PEI Blends

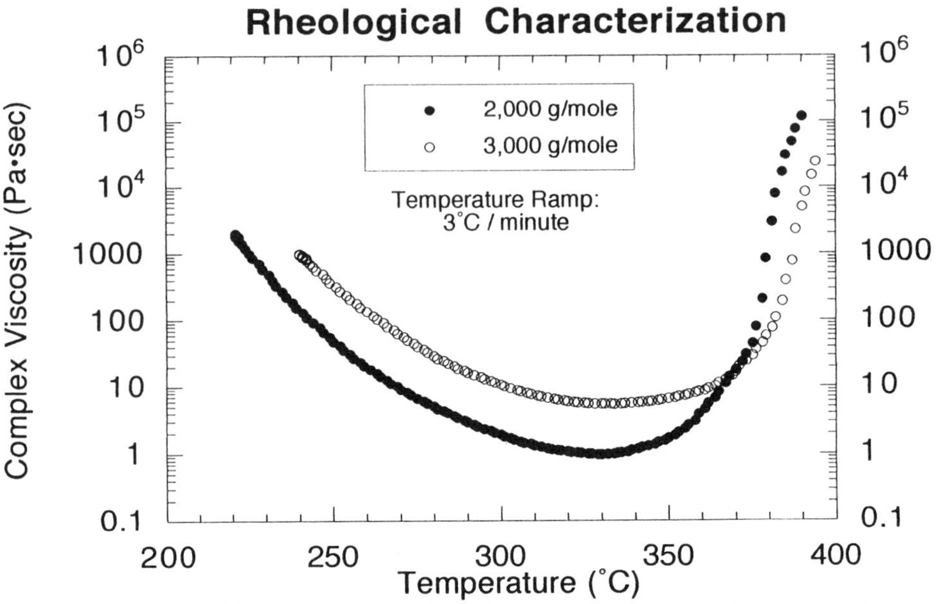

Figure 6. Rheological Characterization of BPADA/m-PDA 4-PEPA Capped Polyetherimide Oligomers

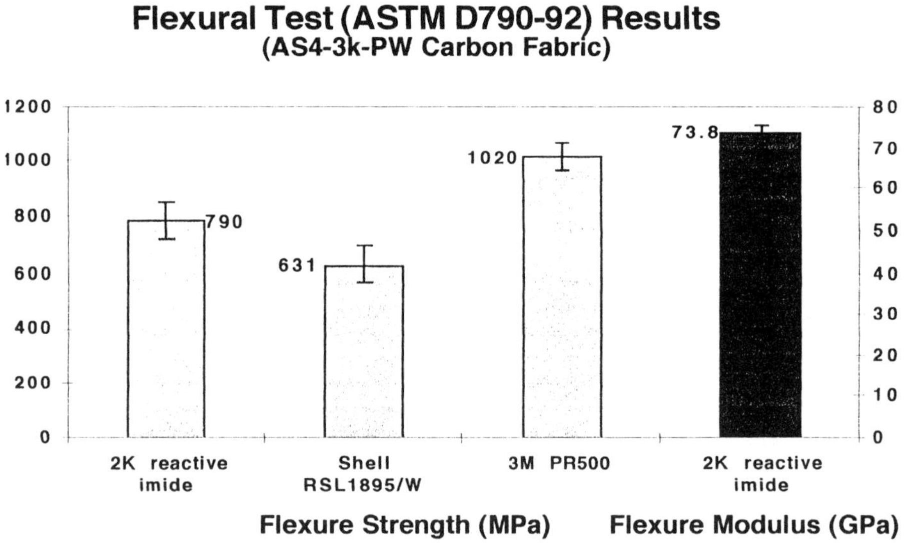

Figure 7. Flexural Test (ASTM D790-02) Results Based on AS4-PW-3K Carbon Fabric

28th International SAMPE Technical Conference
November 4-7, 1996

IMIDE OLIGOMER MODIFIED POLYIMIDES FOR IMPROVED COMPOSITE PROCESSING

Eleonora Vaccaro and Daniel A. Scola
University of Connecticut
Insitute of Materials Science, U-136
Polymer Program
Storrs, CT 06269-3136

ABSTRACT

The high melt viscosities of high temperature polyimide resins require high pressures (1000 psi) for consolidation to produce low void composites. It would be desirable to have a low melt, low viscosity polyimide system which can be processed at lower pressure (100 to 300 psi) to produce low void composites in high yield. A series of low molecular weight, low viscosity imide oligomers have been synthesized and blended with polyimides. Graphite fiber composites were fabricated from the oligomer modified polyimide resins/graphite fiber prepreg. The quality of the composites will be discussed in terms of the resin flow and void contents.

KEY WORDS: Imide Oligomers, Polyimide, Composites, Viscosity Lowering, Thermoplastic

1.INTRODUCTION

The need for polymeric materials for application in hot air environments, such as in gas turbine engines over the temperature range from 316°C to 400°C is increasing, particularly for military engines and more recently, for commercial gas turbine engines. The requirement for such materials is the ability to resist oxidation or degradation in a flowing air environment at pressures from 15 to 90 psi and at temperatures which can exceed 370°C. Present polyimide systems such as 3F-PDA, 36F-PDA, Avimid-N and AFR-700B exhibit high melt viscosities at process temperatures, thereby requiring high pressures (1000 psi) for consolidation to produce low void composites. It would be desirable to have a low melt, low viscosity polyimide system which can be processed at lower pressures (100 to 300 psi), to produce low void composites in high yield. This is the object of the present work.

A review of the use of low molecular weight additives in high molecular weight thermoplastic polyimides has been published by T.L. St.Clair et.al., (1). Low molecular weight bis-amide acids or bis-imides (2-8) have been shown to be effective in lowering the viscosities of thermoplastic polyimides formed from 3,3'- diaminobenzophenone (DABP) and 3,3'4,4'-benzophonone tetracarboxylic dianhydride (BTDA), or 4,4'-bis(3,4-dicarboxyphenoxy)diphenyl sulfide (BDSDA) and 2,2-bis[4-(4-aminophenoxy)phenyl] hexafluoropropane (BDAF). St.

Clair et al (1) also discuss how the additives affect other polyimide properties such as moisture absorption, free volume and dielectric constant. Two mechanisms (1) of how these additives function to reduce the viscosity were proposed: (1) the additives destroy polymer chain-chain interactions because additive-polymer interactions are stronger than polymer chain-chain interactions, (2) above the glass transition temperature (Tg), the polymer and additive interact either through transamidation or transimidization reactions to decrease the molecular weight of the polymer. Neither one of these mechanisms has been substantiated.

In the present work, two series of low molecular weight imide oligomers with meta and para orientation were synthesized. The viscosity of several of these was determined, and one oligomer was selected to evaluate its ability to improve the composite processing characteristics of a thermoplastic polyimide (3F-PDA), derived from 4,4'-(2,2,2'-trifluoro-1-phenylethylidene) biphthalic anhydride and p-phenylene diamine.

2.EXPERIMENTAL

2.1 Materials The synthesis of oligomer 6F-m-PDA-29 where n=2 will be described. All other oligomers were synthesized in the same manner.

2.1.1 Phenyl End-Capped 6F-PDA Imide Oligomer Synthesis In a 3-necked flask attached to a condenser, stirrer motor and nitrogen gas inlet a solution of 6FDA (30g, 0.0675 mole) in 100 ml of dimethylacetamide (DMAc) was added over a half-hour period to a solution of m-phenylene diamine (5.488g, 50.75 in DMAc (150 ml) while stirring in a nitrogen atmosphere. The solution was allowed to stand at room temperature for three hours while stirring under a nitrogen atmosphere. A sample was removed for characterization by infrared spectroscopy. The IR spectrum (Fig. 1) showed absorptions in the ranges 3300-2700 cm^{-1}, and 1700-1660cm^{-1}, typical of polyamic acids. A Dean - Stark tube was attached to the flask, and 50 ml of liquid was removed by distillation, o-xylene (30 ml) was added, and an additional 50 ml of liquid was removed by distillation . Aniline (8.687g, 8.5 ml, 0.0933 mole) and 30 ml of o-xylene was added. The solution was brought to reflux and additional liquid was removed by distillation via the Dean - Stark tube. This liquid contained approximately 0.7 cc of water as a separate layer. The final volume of material in the flask was about 50 cc. The flask was cooled to room temperature. The infrared spectrum (Fig. 2) showed strong absorptions in the regions 3400-2700 cm^{-1} and at 1676 cm^{-1}, typical of polyamic acid, and strong absorptions at 1782 and 1722 cm^{-1}, typical of an imide. The oligomer was therefore a mixture of amic acid and imide. No absorption due to anhydride was present. This oligomer solution was poured into 2 liters of water, filtered, air-dried and then vacuum dried at 150°C overnight to yield 37.0g of yellow powder, mp 196-200°C (foaming). Foaming indicates that amic acid and solvent are still present. The solid was further treated at 225°/2hrs/vacuum to yield yellow solid (33.58g), mp 190-195°C, (red glassy, viscous melt). The melt flows at 220-240°C, with slight gas evolution at 260°C. The infrared spectrum (Fig.3) showed strong absorptions at 1784 and 1718 cm^{-1} due to imide. No absorption due to amic acid or anhydride impurity were present.

2.1.2 3F-PDA Resin Synthesis for Prepregs 4,4'(2,2,2-trifluoro-1-phenylethylidene) biphthalic anhydride (3FDA) (32.5392 g, 0.07916 mole) was added to 75 ml of absolute ethanol. The mixture was refluxed until the dianhydride was completely dissolved. Para - phenylenediamine (p - PDA) (6.9887 g, 0.066313 mole was added, and the mixture was warmed until the diamine dissolved. The solution was transferred to a bottle, and the flask was washed in absolute ethanol to affect complete transfer. The solution contained 28.76 wt% of 3F - PDA resin in ethanol. 3F - PDA resin solutions prepared in this manner contained variable resin solids (25-30 wt%).

2.1.3 Preparation of 3F-PDA/6F-PDA Oligomer Blends To a solution of 3F-PDA resin in ethanol prepared as described above was added a solution of 6F-PDA-29 oligomer (4.02g) in 49 ml of dimethylacetamide (DMAc). The total solids content after blending in the oligomer solution was 23.8 wt%. The resin solution contained 10 wt% oligomer. 3F-PDA/Oligomer blends containing 5 and 7.5 wt% were prepared in a similar manner.

2.1.4 Fabrication of G40-600(u)/6F-PDA-Oligomer Prepregs and Composites Unsized G40-600(u) 12k graphite fiber tows were passed through an acetone bath and then wound onto a 45 cm (17 in.) diameter drum to form a 11.43 cm (4.5 in.) wide tape. The tape was impregnated with a 3F-PDA-Oligomer solution by brush application as the drum rotated slowly to yield a prepreg containing 44 to 46 wt% resin. The solvent was allowed to evaporate at room temperature, then the tape was cut into 18.5 cm (6.5 in.) lengths. The tape was then laid-up as follows: to a teflon mold treated with mold release was placed 2 layers of 0.12 mm Kapton, followed by 2 layers of Armalon. Five graphite prepreg plies were stacked over this followed by a mirror image arrangement of the release plies used on the bottom of the mold. The mold lay-up was treated in an oven at 125°C/2hrs + 200°C/1hr and at 250°C/1hr in vacuum. Then the mold was placed in the press or autoclave and processed as described below.

2.1.5 Fabrication of Composites by the Compression Molding Process

a. temperature raised to 250°C (482°F) at 2.7°C (5°F)/min with contact pressure and at temperature, 1000 psi was applied and held for 1 hour.

b. temperature raised to 275°C (527°C) at 2.7°C (5°F)/min, 1000 psi, held 1 hour, the pressure was release and then re-applied.

c. temperature raised to 316°C (600F) at 2.7°C (5°F)/min, 1000 psi, pressure released then re-applied and held 12 hours.

d. temperature raised to 343°C (649°F) at 2.7°C (5°F)/min, 1000 psi, pressure released and held 2 hours.

e. temperature raised to 371°C (700°F) at 2.7°C (5°F)/min, 1000 psi, pressure released and then re-applied and held 24 hours.

f. temperature raised to 416°C (781°F) at .6°C (1°F)/min, 1000 psi, pressure released and then re-applied and held 8 hours.

g. cooled to 280°C (536°F) at 2.7°C (5°F)/min, 1000 psi, then pressure was released and mold was allowed to cool to room temperature.

2.1.6 Autoclave Processing of G40-600(u)/3F-PDA/Oligomer Composites The 11.45 cm (4.5 in.) in. x 16.5 cm (6.5 in) G40-600(u) plies were laid up as follows: Kapton film (5 mils) was placed on the bottom of the mold followed by 2 layers of Armalon release cloth. Five graphite prepreg plies were stacked over this followed by one layer of Armalon release cloth and 3 layers of 7781 fiberglass cloth. Kapton film 0.005 cm (0.002 in.) thickness was placed over the entire lay-up and sealed at the edges with DC 18C adhesive primer and A 800 tape. A tube leading from this lay-up to a vacuum pump was used to evacuate the system. The composite was placed in the autoclave and processed as follows:

1. RT to 100°C partial vacuum, 1°C/min.

2. 100°C to 150°C full vacuum, 2°C/min.
3. 150°C to 250°C/2°C/min full vacuum, 400 psi applied at 250°C, held 2 hours.
4. 250°C to 316/2°C/min, full vacuum, 400 psi held 2 hours.
5. 316°C to 371°C, full vacuum, 400 psi, held 2 hours.
6. It was post cured in the free state at 371°C/24 hours in a nitrogen atmosphere.

2.1.7 Rheology Measurements Rheology measurement were made on a Rheometrics II Rheometer. Sample discs 2.54 cm diameter x 0.16 cm (thickness) were placed in the Rheometer in the parallel plate mode. The temperature was increased to close to the softening temperature of the sample as determined by capillary tube melt and DSC experiments. The force on the sample were adjusted and the changes in the complex viscosity of the sample was measured as a function of temperature at 3.3°C/min temperature rise.

2.1.8 Thermal Analysis Differential scanning calorimetry (DSC Module 912) was performed on TA 9900 Thermal Analyzer.

3. RESULTS AND DISCUSSION

The oligomers synthesized and characterized are shown in Figure 4. Some properties of the imide oligomers are listed in table 1. Inspection of table 1 reveals that the para isomer series exhibit considerably higher melting or softening range than the meta isomer series. For the isomers where n=1, the complex viscosity as determined by parallel plate rheometry, the meta isomer exhibits a much lower viscosity than the para isomer. This is an interesting observation and it suggests that the meta isomers may have a greater influence in viscosity reduction of the thermoplastic polyimide than the para isomers. Therefore for this series of experiments, one imide oligomer, 6F-PDA-29 was selected for evaluation in the fabrication of 3F-PDA/Oligomer/G-40-600(u) graphite composites.

The 3F-PDA/Oligomer/G40-600(u) composites fabricated by the compression molding and autoclave methods are listed in Table 2. The oligomer containing composites fabricated by the compression molding process were essentially void-free as determined by optical cross sections and Non-Destructive Evaluation (NDE) using the ultrasound method. The non-oligomer containing composite, however, contained a higher void level than the oligomer blended composite systems. The higher density for the oligomer blended composite over the non-blended composites is additional evidence for composite quality. This suggests that the presence of low viscosity, low molecular weight imide oligomer offers some advantage in the fabrication of composites by the compression molding cycle. The experiments further suggest that the use of lower pressure is possible with the oligomer blended polyimide.

Composites fabricated by the autoclave method were processed simultaneously in one vacuum bag to avoid variables associated with composites processed separately. This first experiment to demonstrate that oligomer blended 3F-PDA polyimide composites could be processed more easily than non-oligomer blended 3F-PDA polyimide, required modification during the run due to failure of the Kapton bag and subsequent vacuum loss. To prevent bag failure, many layers of glass bleeder cloth were added to the top of the ply lay-ups. This altered the dynamics of the resin flow particularly in the low viscosity blended composites, and caused these composites to exhibit excessive bleeding, creating a condition for void formation. Therefore the oligomer free composites, contained a higher void content than the oligomer free composite, despite the much greater flow, as determined by weight loss. This is the opposite result expected. The greater degree of resin flow for the oligomer blended composites demonstrates that oligomer additives can effect the fabrication process, and further demonstrates that fabrication of large area composites is feasible by the autoclave process. However, much greater control of the process parameters are required to optimize low viscosity imide applications in the composite fabrication process. This is currently under investigation.

4. CONCLUSIONS

Low molecular weight, low viscosity, thermally stable 6F-PDA imide oligomers were blended at 5, 7.5 and 10 wt% levels in 3F-PDA thermoplastic polyimide to determine effects on process ability. Results show that excellent quality, essentially void free, composites were fabricated by the compression molding process from 3F-PDA compositions containing 5, 7.5 or 10 wt% oligomers. The composite fabricated using the oligomer-free 3F-PDA polyimide composition contained more void than the oligomer containing polyimide compositions, as determined by optical cross sections and NDE. However, the oligomer 3F-PDA compositions generated high flow in composites fabricated by the autoclave process producing composites with a low concentration of voids between ply layers. The use of several layers of glass bleeder cloth served to magnify the flow behavior, by providing a wicking action to remove resin at a fast rate, causing voids to form. The improved flow characteristics however demonstrates that fabrication of large components is feasible by the autoclave process using oligomer modified 3F-PI polyimide.

5. REFERENCES

1. J.L. St. Clair, J.R. Pratt, D.M. Stoakley and H.D. Burks: Polyimides: Materials, Chemistry and Characterization, edited by C. Fezer, M.M. Khojasteh and J.E. McGrath, Elsevier Science Publishers B.V. Amsterdam, 243-250 (1989)

2. H.D. Burks, and T.L. St. Clair: The Effect of Molecular Weight on the Melt Viscosity and Fracture Energy of BDSDA/APB, J. of Appl. Polym. Sci., 29, 1027-1030 (1984)

3. D.M. Stoakley, T.L. St. Clair, A.K. St. Clair, J.R. Pratt, and H.D. Burks: Effect of Diamic Acid Additives on the Processability of Polyimides, ACS Polymer Preprints, 27 (2), 406-407 (1986)

4. J.R. Pratt, T.L. St. Clair, H.D. Burks and D.M. Stoakley: Polyimide Processing Additives, SAMPE Symposium Preprints, 32, 1036-1050 (1987): US Patent Application No. 084, 064

5. N.J. Johnston,. and T.L. St. Clair: Thermoplastic Matrix Composites, SAMPE Technical Conference Preprints, 18, 53-67 (1986)

6. J.J. Singh, T.L. St. Clair and J.R. Pratt: Investigation of the Effect of Short Chain Processing Additives on Polymers, NASA TM 89023, November 1986

7. D.M. Stoakley, and A.K. St. Clair: The Effect of Diamic Acid Additives on the Dielectric Constant of Polyimides, ACS PMSE Preprints, 29 (2), 33-36 (1998)

8. A.K. St. Clair, and T.L. St. Clair: A Multipurpose Thermoplastic Polyimide, SMAPE Quarterly, 13 (1), 20-25 (1981)

Table 1
Some Properties of 6F-PDA Imide Oligomers

6F-PDA Oligomers	n	Theoretical MW,g/mole	Melting or Softening Range, °C	*Initial Flow Temp, °C	Complex Viscosity Pa.s @ 315°C	Complex Viscosity Pa.s @ 340°C
Para-Isomers						
119-1	1	1627	~ 300	>300	3.15×10^3	18
118-2	2	2143	~ 300	>300		
129-2	4	3175	> 300	>300	1.7×10^2	17
130-2	5	3692	> 300	>300		
Meta-Isomers						
30.2	0	1110	165-170	190		
151-2	1	1627	225-230	220	1.5	0.65
29-2	2	2143	196-198	230		
152-2	2	2143	236-240	265	26	9.6
189	4	3175	240-260	280		
192	5	3692	240-260	280		

*Determined by melt capillary observation

Table 2
G40-600(u)/3F-PDA/Oligomer Composites Fabricated by Compression and Autoclave Methods

No/Composite System	Oligomer Content %	Quality as Determined by Optical Cross Sections/NDE[1]	Fabrication Process / Pressure	Density g/cc	% resin flow[2]
F-208/G40-600(u)/ 3F-PDA-Oligomer	0% 6F-PDA-29	Excellent, low void	Compression 1000 psi	1.49	
F-190/G40-600(u)/ 3F-PDA-Oligomer	5% 6F-PDA-29	Excellent, very low void	Compression 1000 psi	1.52	
F-205/G40-600(u)/ 3F-PDA-Oligomer	7.5% 6F-PDA-29	Excellent, very low void	Compression 1000 psi	1.52	
F-182/G40-600(u)/ 3F-PDA-Oligomer	10% 6F-PDA-29	Excellent, very low void	Compression 1000 psi	1.52	
F-263/G40-600(u)/ 3F-PDA-Oligomer	0% 6F-PDA-29	very low void on two surface layers	Autoclave 400 psi	1.55	22.5
F-261/G40-600(u) 3F-PDA-Oligomer	5% 6F-PDA-29	Low void between ply layers	Autoclave 400 psi	1.49	34.1
F-262/G40-600(u)/ 3F-PDA-Oligomer	10% 6F-PDA-29	Low void between ply layers	Autoclave 400 psi	1.52	34.8

1. NDE (Non Destructive Evaluation), ultrasonic attenuation determined by through transmission measurements ("C" scan image).

2. Defined as: $\frac{\text{weight of resin due to flow}}{\text{weight resin in composite before processing}} \times 100$

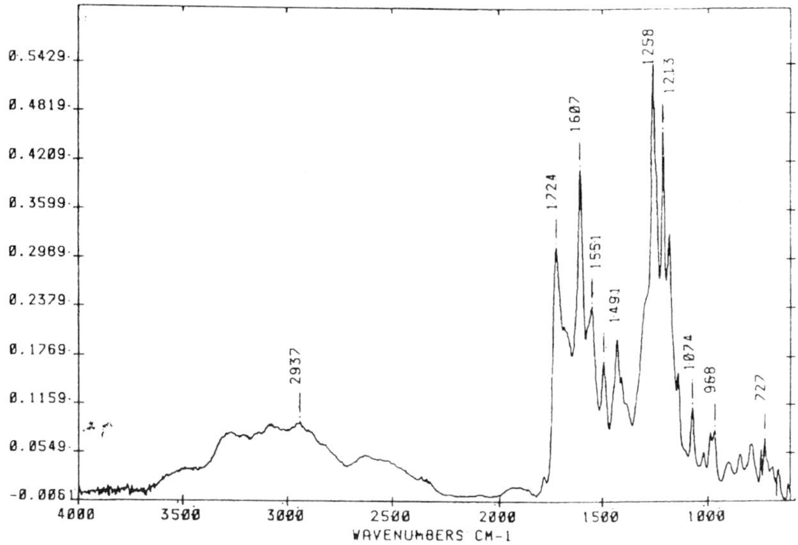

Fig. 1 6F–PDA–29 Polyamic Acid

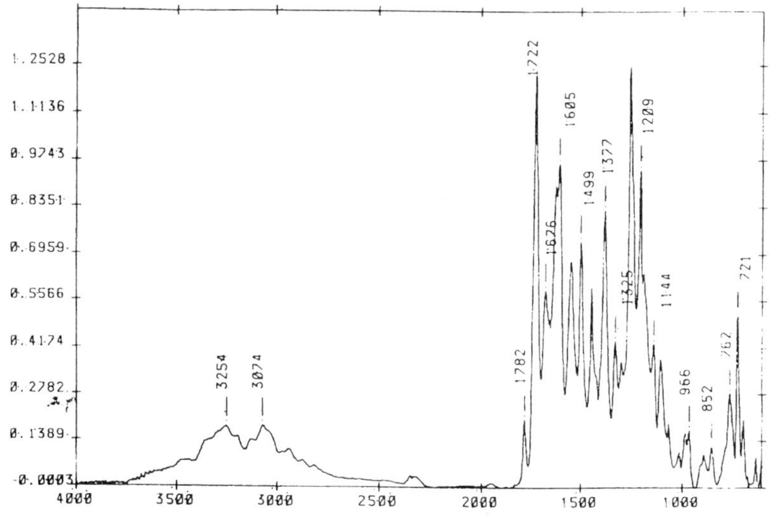

Fig. 2 6F–PDA–29 After Aniline End–Capping

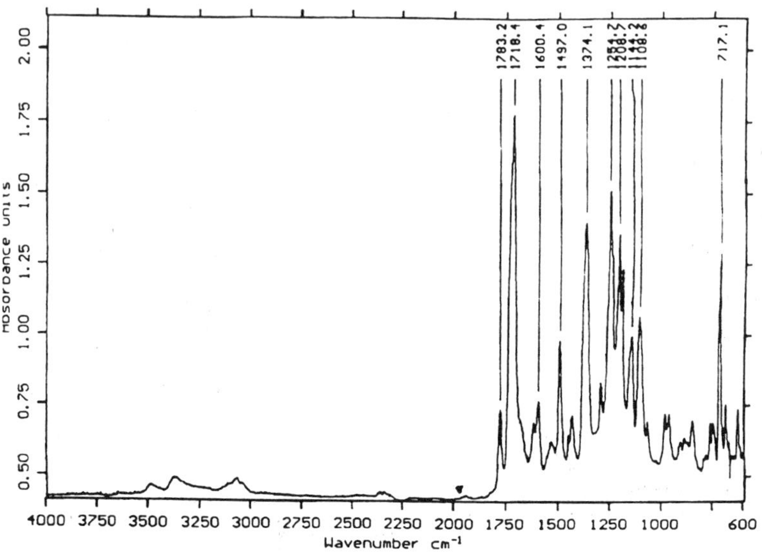

Fig. 3 Imidized 6F-PDA-29 (225°C/2 hrs/vacuum)

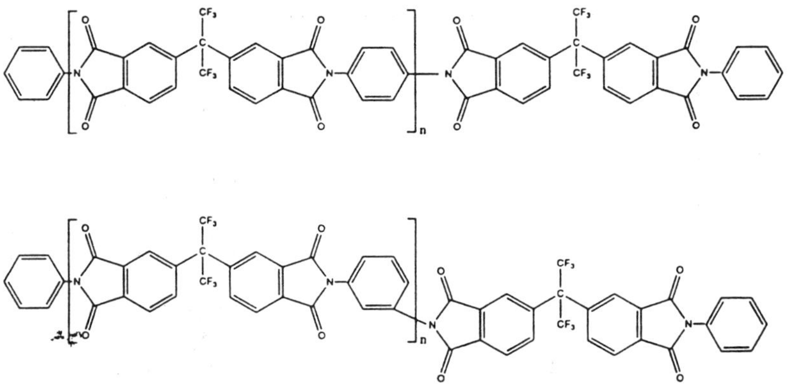

Fig. 4 6F-PDA Imide Oligomers
para series: n=1,2,4,5
meta series: n=0,1-5

BIOGRAPHIES

Dr. D.A. Scola received his BA degree from Clark University in 1952 and his M.S. from Williams College in 1954, and his Ph.D. in chemistry from the University of Connecticut, Storrs, CT in 1959. Dr. Scola joined Monsanto Research Corp. In Everett, MA in 1958, and as a group leader, directed the synthesis of organic compounds from various applications. He joined United Technologies Research Center (UTRC) in 1967, and after a brief tenure at Norton Co. Between 1968 to 1969, rejoined UTRC as a manager of Organic Materials Research and later as a Senior Materials Scientist. At UTRC Dr. Scola conducted research in advanced composite materials for aerospace applications. He is the inventor of a family of high temperature polyimides and has recently been elected as a fellow in the Society of Plastics Engineers. Presently Dr. Scola is a Professor at the University of Connecticut, Polymer Science Program, Storrs, Connecticut.

Eleonora Vaccaro received her B.S. in Industrial Chemistry form University of Genova (Italy) in 1993. She worked as a Research Fellow for the University of Sassari and University of Napoli (both in Italy), dealing with synthesis of aromatic polyamides and gas permeation through ionomeric membranes. In September 1994 she joined the University of Connecticut as an one-year exchange visitor research scholar, working on immiscible polymer blends. In September 1995 she joined the Ph.D. program in Polymer Science at the University of Connecticut.

28th International SAMPE Technical Conference
November 4-7, 1996

THE CHARACTERIZATION OF A THERMOPLASTIC POLYIMIDE COMPOSITE

J. Timothy Hartness
CYTEC Engineered Materials Inc.
1440 N. Kraemer Blvd. Anaheim, CA 92806

ABSTRACT

A program initiated to develop a new generation of commercial aircraft capable of supersonic speeds will require composite materials with exceptional performance and durability for the life of the aircraft. In addition to the material requirements there exists a need to reduce manufacturing costs over current practices.

In response to this challenge a considerable number of matrix resins have been screened. The initial results of this effort indicates that a new generation of matrix resins will be needed to fulfill this requirement. The matrix needs to have excellent toughness, thermal and chemical resistance, good processability, as well as, be cost effective.

One of the matrix resins that has shown the promise to meet this challenge is a thermoplastic polyimide, namely PIXA.

In the paper, the resin and prepreg development is characterized. Mechanical properties have been generated along with processing. Several resin versions will be discussed, such as a polyamic acid, as well as, a fully imidized powder. A discussion of composite manufacturing concepts will also be explored.

KEY WORDS: Composites, thermoplastic, polyimide

I. INTRODUCTION

The next generation of commercial aircraft are now in the initial design phase. The objective of this effort will be an airplane that will provide transportation times one half that of current commercial jets. A design that would travel at supersonic speeds would allow two round trips per day from the west coast of the United States to the Pacific rim. The weight and speed of the aircraft will put severe demands on materials. It has been determined that a considerable portion of the aircraft must be constructed out of composite materials in order to satisfy the weight and speed requirements. It is anticipated that the aircraft structure may have high sustained temperatures. It is also anticipated that composites may be exposed to these temperatures for many thousands of hours.

Efforts are on-going to identify and evaluate a matrix material that would satisfy these requirements. The current thinking within the community suggests that a tough lightly cross-linked polyimide(1) or thermoplastic polyimide will be required to satisfy this objective.

One such matrix resin under evaluation is a thermoplastic polyimide called PIXA. This resin is supplied from Mitsui Toatsu Chemicals and is currently being evaluated in its polyamic acid form and as a fully imidized powder for dry prepreg forms.

Shown in Table 1 are the neat resin properties of PIXA as supplied by Mitsui Toatsu. The resin has excellent toughness and strength, as well as, excellent chemical resistance and thermal performance. PIXA also has good resistance to moisture with equilibrium gains under 0.5%.

II. MATRIX FORM CONSIDERATIONS

Consideration of a matrix resin in regards to PIXA has evolved as several composite processes have emerged.

Essentially two material forms are currently being explored. One uses a more traditional wet solvent prepreg that is autoclaved while the other is a dry, stiff and boardy material that is currently being explored as a material that would be insitu consolidated in a robotic process called Advanced Tape or Two Placement (ATP). This process would eliminate the labor intensive hand lay-up and may also eliminate the entire expensive autoclave consolidation that is currently being used for composites.

In order to generate mechanical properties on PIXA it was determined that an amic acid solvent system would be the best choice as a dry material database would take considerably more time to develop.

III. EXPERIMENTAL

3.1 Prepregging Procedure - The prepregging of PIXA amic acid using unsized Hercules IM-7 (12K) carbon fiber was accomplished at NASA Langley using their multipurpose tape machine. An amic acid solution in NMP with a 31% by weight of solid was received from Mitsui Toatsu. This solution was further diluted to 29% solid resulting in a Brookfield viscosity of 15400 centipoise at RT. The solution was prepregged at 30°C (86°F) at a viscosity of 7300 centipoise. A total of 48 pounds of 10-inch wide prepreg were produced. The final prepreg characteristics are shown in Table 2. As shown, a target fiber aerial weight (FAW) of 140 gm/m^2 was obtained. The prepreg was delivered to Cytec for processing and cure cycle development. As shown in Table 2, the volatile content is very high. This is required to provide adequate tack and drape. The solid content of the prepreg was determined by taking an average of three samples across the width of the tape from 3 rolls. The prepreg was dried for 20 minutes at 350°C (662°F) to remove solvent and convert the amic acid to imide. During this process approximately 6% water loss occurs. By this determination the actual NMP level ranged from 19% to 24% by weight.

The as received prepreg handle well with adequate tack and drape. The choice of using a coated release paper allowed some additional drying upon standing at room temperature.

3.2 Process Cycle Characterization - A process cycle was chosen based on the necessity to remove all NMP prior to the application of autoclave pressure, as well as, the conversion of amic acid to imide. It was determined that a hold period of 2 hours at 288°C (550°F) would be very adequate. A full vacuum was applied initially in the cycle and remained throughout.

Results from test panels indicated a resin loss of approximately 3% by weight. The fully imidized PIXA powder has a transient crystalline melt point between 361°-368°C (680°-698°F). This is verified by DSC, as shown in Figures 1-2. It was determined that by taking the tape to 382°C (720°F), 20°C above Tm, the full melting of any crystalline material would be complete.

Based on the above discussion a process cycle was chosen and is shown in Table 3. The hold period at 382°C (720°F) can be extended or shortened depending on the part thickness.

3.3 Mechanical Property Development - Shown in Table 4 is a test matrix. In addition to this matrix, flexural and short beam shear properties were determined to compare to other materials being evaluated.

All fiber dominated properties were normalized to 62% fiber volume. All panels were C-scanned prior to testing. In addition, samples were taken for photomicrographs, as well as, acid digestion to determine fiber volume and void content. All panels resulted in void contents of less than 1%. The photomicrographs indicated good resin distribution with no fiber bundle wetout problems.

Shown in Table 5 are mechanical properties generated to date.

3.3.1 Flexural and Short Beam Shear Properties - Flexural strength was determined by using a Mod ASTM D790 test in which a small rubber pad is placed under the load nose in the three point configuration. This helps prevent local compression failure and distributes the load better through the specimen resulting in more complete involvement of all the fiber and thus resulting in better failure modes and more representative strength numbers. This local crushing problem has been observed with other thermoplastic matrix composites.

The short beam shear specimens also showed some local plastic deformation with no large visible shear cracks as is often observed with highly cross-linked resin no large visible shear cracks as is often observed with highly cross-linked resin systems. This somewhat explains the lower values obtained with this resin system, as well as, the fact that properties are not normalized for this resin dominated test.

3.3.2 Open Hole Compression (OHC) - Panels were fabricated with three different lay-ups in terms of 0° plies. The 58% - 0° lay-up is a panel designed for maximum stress between plies and is also used to determine if the resin is prone to microcracking after being exposed to thermal cycling from -54°C (-65°F) to 177°C (350°F).

The specimens used for OHC testing measure 3" x 1" and come from a Northrop requirement. Mode of failure was, as normal, through the hole. No delamination was noticed extending beyond a very local compression area.

3.3.3 Compression After Impact (CAI) - Panels fabricated for CAI were quisi-isotropic 32 ply lay-ups. Specimen size was 4" x 6". Specimens were impacted with 1500 in.lbs/in. All specimens were C-scanned after impact with the damage measured. A significantly small delamination area averaging 0.48 sq.in. was measured. This small damage area is indicative of the very high CAI values obtained with this material.

3.3.4 Solvent Sensitivity Testing - Solvent testing was carried out on OHC specimens as shown in the test matrix. The specimens were not under load during exposure. Figure 3 shows the results of the exposure on OHC properties. No drop in performance was noted as compared to 177°C (350°F) wet tested specimens.

3.3.5 Composite Physical Properties - Fracture surface analysis was completed by SEM on tested flexural specimens at NASA Langley. Shown in Figure 4 the SEM reveals that all of the unsized IM-7 fiber seems to be completely covered with resin.

The glass transition of the composite was determined by DMA. A DuPont Model 983 was used in the torsion mode. Samples were taken from a 16 ply unidirectional panel and heated at 10°C/minute. Moisture aged samples exposed at 71°C (160°F) at 98% RH for 14 days were also run. The samples gained an average of 0.48%. The Tg was determined by extrapolating from the linear portion of the G' curve. The dry and wet Tg were determined to be 274°C and 247°C, respectively. The DMA plots are shown in Figures 5 and 6.

IV. DRY PRODUCT FORMS

In addition to the wet PIXA amic acid solvent prepreg a "dry" product is also under development and evaluation. The objective behind this development is to have a material that can be consolidated on the fly with automated robotics. This will eliminate the labor intensive hand lay-up often needed. There is also a major drive to eliminate the costly autoclave process especially for very large parts. It is anticipated that with optimized time, temperature and heat, this could be achieved with a thermoplastic matrix such as PIXA.

Currently a fully imidized PIXA powder has been developed by Mitsui Toatsu Chemical. This material is impregnated into unsized IM-7 fiber using a Cytec proprietary process. This towpreg is then used to make a dry tape or ribbon for what is known as Advanced Tape or Tow Placement (ATP).

This process is currently under development with limited amounts of 0.25 inch and 3.0 inch material being made. Material specifications for a material of this type are quite demanding as time at temperature and pressure at the robotic head are less than one second. The material must have very good resin to fiber distribution, uniform thickness and width, as well as, a void content expected to be less than 3%. Essentially a single ply composite.

A high quality ribbon has been achieved using PIXA impregnated IM-7 fiber at Cytec and then ribbonized[2] at NASA Langley. Shown in Figure 7 is a cross-sectional photomicrograph of a 0.25 ribbon showing excellent resin to fiber distribution, good thickness and width control, as well as, low voids. This type of material has been successfully ATP up to speeds of 180" per minute. The resulting composite properties are approaching 90% of an autoclaved PIXA composite.

Efforts are on-going at Cytec to develop a 3" tape with similar attributes.

V. SUMMARY

Initial composite properties have been generated on a thermoplastic polyimide that shows considerable promise with an overall good balance in performance.

The matrix has shown considerable versatility as being able to be processed from both the amic acid, as well as, from a fully imidized powder.

VI. REFERENCES

1. M. L. Rommel, P. M. Hergenrother, (proceedings of 29. Int. SAMPE Symp., 1996).

2. D. A. Sandusky, J. M. Marchello, N. J. Johnston, R. M. Baucom, (proceedings of the Int. SAMPE Tech. Conf., 1994).

BIOGRAPHY

J. Timothy Hartness is a Research and Development Contract Specialist at Cytec Engineered Materials Inc.. He joined BASF in 1987 after working at the University of Dayton Research Institute on-site at the Air Force Materials Lab at WPAFB, OH., since 1968. His primary work involves composite development in support of a NASA Contract.

TABLE 1

NEAT RESIN PROPERTIES

	PIXA	
Tg (°C)	252	
Tensile Strength		
RT	910	Kg/cm^2
	13	(Ksi)
Tensile Modulus		
RT	290	Kg/cm^2
	412	(Ksi)
Elongation		
RT	66	%
Flexural Strength		
RT	1360	Kg/cm^2
	19.4	(Ksi)
150°C	890	Kg/cm^2
	12.7	(Ksi)
Flexural Modulus		
RT	300	Kg/cm^2
	427	(Ksi)
150°C	280	Kg/cm^2
	399	(Ksi)
177°C	270	Kg/cm^2
	384	(Ksi)

TABLE 2

CHARACTERISTICS OF IM-7 / PIXA PREPREG

PREPREG CHARACTERISTIC	NOMINAL VALUE
FAW, g/m^2	140
VOLATILES, % Wet Wt.	25-30
SOLIDS, % Dry Wt.	38-43

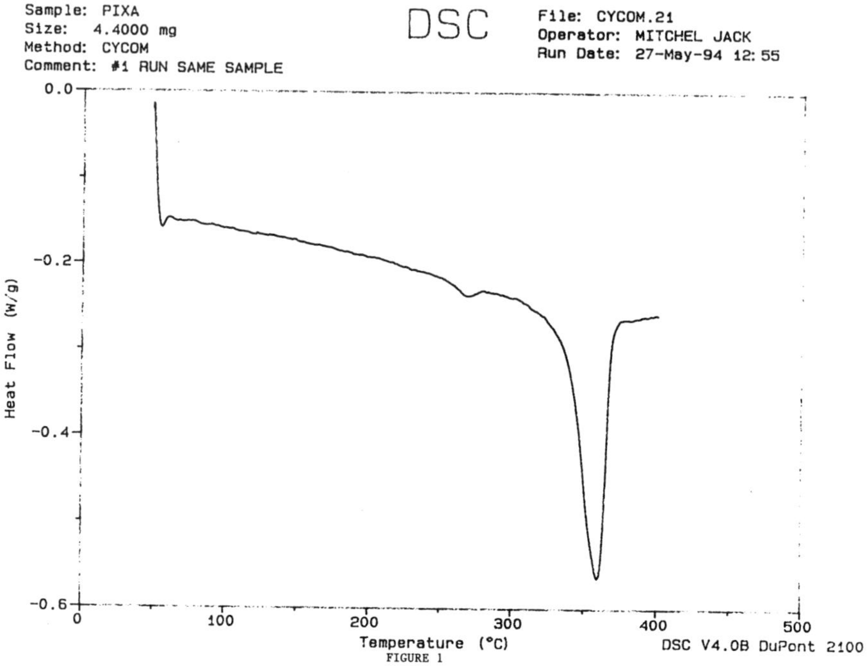

FIGURE 1

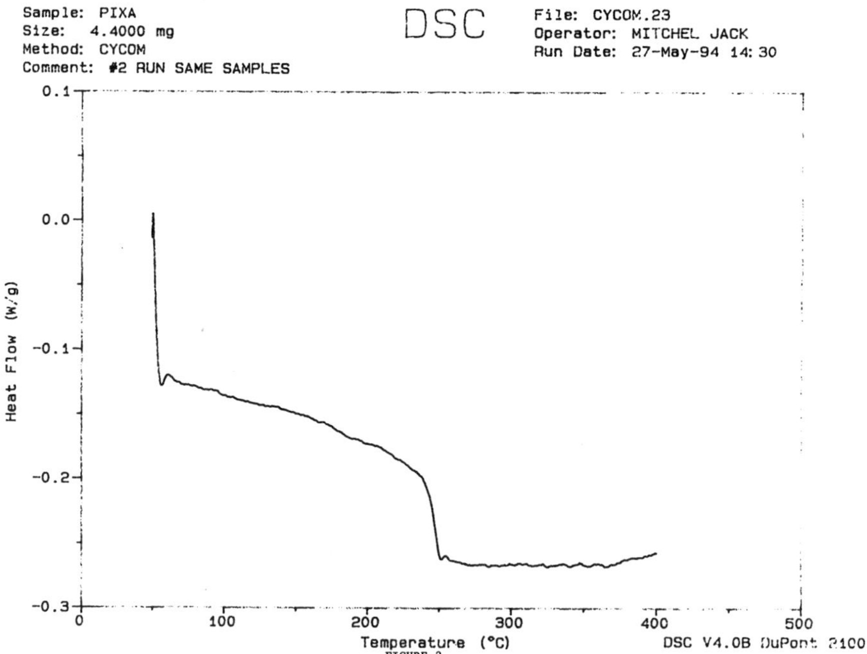

FIGURE 2

TABLE 4

PROPOSED COMPOSITES TEST MATRIX

PHASE	TEST TYPE	NO. AT EACH CONDITION	TEST TEMPERATURE (°C)	TEST METHOD	TEST SPECIMEN
I. Composite Quality Optimization	Open Hole Compression (58/34/8) Lay-up[1]	10	RT & 177°C/Dry	Northrop 3" x 1"	Northrop 3" x 1"
	Interface Test • RT 90° Flexure	5	RT	ASTM D790	4" x 0.5", 16 ply
	Laminate Physicals • Resin Content	3	RT	ASTM D2734	1" x 1"
	• Fiber Volume	3	RT	ASTM D3171	1" x 1"
	• Void Content	3	RT	ASTM D792	1" x 1"
	Dry and Wet Tg via DMA	3	RT to 288°C	Company Spec.	2.5" x 0.5"
II. Screening	Open Hole Compression [25/50/25] Lay-up	10	RT & 177°C/Wet	Northrop 3" x 1"	$[+45/0/-45/90]3_S$
	[42/50/8] Lay-up	10	RT & 177°C/Wet	Northrop 3" x 1"	[+45/0/-45/90]
	Quasi Compression After Impact	5	RT	BSS 7260	4" x 6",$[+45/0/-45/90]4_S$
	Thermal Cycling	3	-54°C to 177°C	McAir	4" x 4", see 1
	Solvent Sensitivity Testing[2]	6 Each Fluid	RT for G12 177°C for OHC 90° Flexure	BSS7320 Northrop 3" x 1" In Fluid	0.5" x 9",$(+45/-45]2_S$ $[+45/0/-45/90]3_S$ [90]16
	Compression Modulus [25/50/25] Lay-up	10	RT & 177°C	BSS 7260	$[+45/0/-45/90]3_S$
	[42/50/8] Lay-up	10	RT & 177°C	BSS 7260	$[\pm45/90/04/\pm45/03]_S$
	Compression Interlaminar Shear	10	RT & 177°C	D6-55587	[0]30
III. Final Screening	Open Hole Compression [25/50/25] Lay-up[3]	15	RT,177°C,177°C/Wet	BSS 7260	12" x 1.5",$[+45/0/-45/90]2^S$
	[42/50/8] Lay-up	10	RT & 177°C	BSS 7260	12" x 1.5", LSS TBD
	Open Hole Tension [25/50/25] Lay-up	15	RT & 177°C	BSS 7320	12" x 1.5",$[+45/0/-45/90]2^S$
	42/50/8] Lay-up	10	RT & 177°C	BSS 7320	12" x 1.5", LSS TBD
	Compression After Impact[4] [25/50/25] Lay-up & 5 Impact Energy Level Energies	2 Each Energy Level	RT	BSS 7260	4" x 6",$[+45/0/-45/90]4^S$
	[42/40/8] Lay-Up at 270 in-lb	2	RT	BSS 7260	4" x 6",$[+45/0/-45/90]4^S$
	0°/90° Tension	10	RT & 177°C	BSS 7320	9.0" x 0.5",[0]8
	0° Compression	10	RT & 177°C	BSS 7260	3.2" x 0.5",[0]8
	In-Plane Shear	10	RT & 149°C/Wet	BSS 7320,Class I	0.5" x 9",$[+45/-45]2^S$
	3000 Hours of Isothermal Aging	1 Set	204°C/101 Pa	OHC, CAI and In-plane Shear Modulus, & Weight Loss	Stress If Possible

1 (58/34/8) lay-up is $[\pm45/90/0/0/0/0/\pm45/0/0/0]_S$ to test for process induced microcracking

2. Solvent Sensitivity Testing is as follows:
- 14 days: 71°C water
- 14 days: 60°C hydraulic fluid
- 14 days: 60°C Jet A
- 7 days: RT MEK
- 7 days: RT toluene
- 7 days: RT deicer

3. Modulus Measured Prior to OHC Test

4. Five Impact Energies = 135 in-lb, 270 in-lb, 360 in-lb, 450 in-lb, and 545 in-lb

TABLE 3

PIXA / IM-7 AMIC ACID TAPE

AUTOCLAVE PROCESS

1. APPLY FULL VACUUM
2. HEAT RT ——> 288°C @ ?°C/ MINUTE
3. HOLD @ 288°C FOR 120 MINUTES
4. APPLY 200 PSI (1.38 mPa)
5. HEAT 288°C ——> 382°C @ ?°C/ MINUTE
6. HOLD @ 382°C FOR 120 MINUTES
7. COOL TO BELOW 66°C @ ?°C/ MINUTE
8. REMOVE VACUUM AND PRESSURE
9. DEBAG LAMINATE

TABLE 5

PIXA / IM-7 MECHANICAL PROPERTIES

PROPERTY	CONDITION	VALUE*		METHOD
		GPa	(KSI)	
SBS STRENGTH	RT 177°C DRY	.1 .05	(14) (7.5)	ASTM D-2344-84
0° FLEXURAL STRENGTH	RT 177°C DRY	2.14 1.4	(311) (203)	MOD ASTM D-790 (RUBBER NOSE)
90° FLEXURAL STRENGTH	RT	.10	(14.3)	STD ASTM D-790
QUASI - OHC	RT 177°C WET	.32 .21	(46.3) (29.7)	NORTHROP
OHC 42% - 0°	RT 177°C WET	.34 .25	(49.9) (36.7)	NORTHROP
OHC 58% - 0°	RT 177°C DRY	.40 .29	(58.4) (41.5)	NORTHROP
CAI 1500 IN LBS/IN	RT	.36	(52)	BOEING BSS 7260
THERMAL CYCLING	-54°C TO 177°C 200 CYCLES	NO MICROCRACKING		McAIR
350°C OHC AFTER				
SOLVENT EXPOSURE		COMMENT		
14 DAYS 60°C HYDRAULIC FLUID		NO EFFECT		
14 DAYS 60°C JET A FUEL		NO EFFECT		
7 DAYS RT MEK		NO EFFECT		
7 DAYS RT TOLUENE		NO EFFECT		
7 DAYS RT DEICER		NO EFFECT		
GLASS TRANSITION HEAT RATE 5°C / MIN 10°C / MIN 10°C / MIN	 261°C DRY 274°C DRY 247°C WET			DMA-G'

* NORMALIZED TO 62% FV

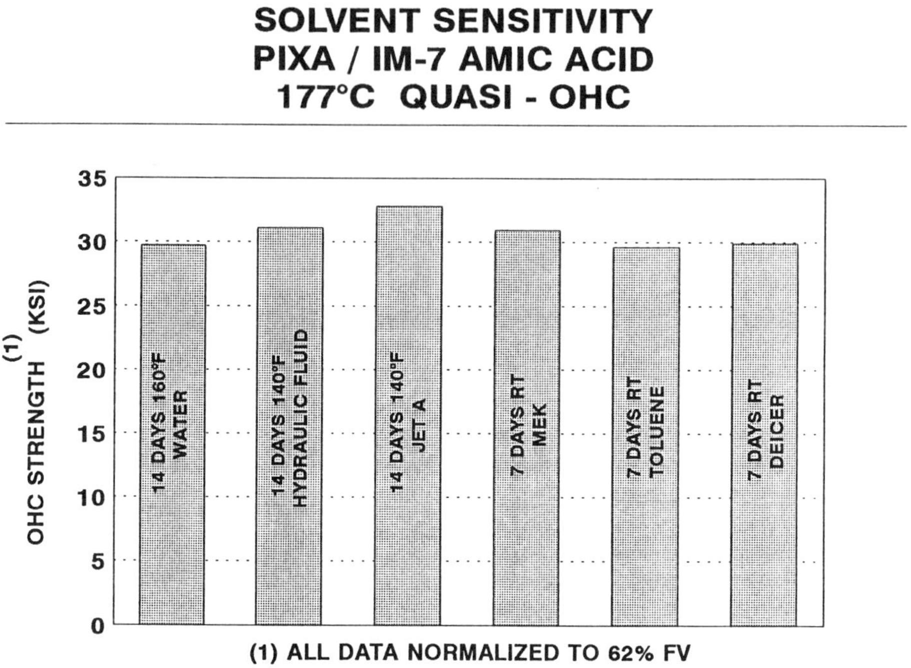

FIGURE 3

CYTEC

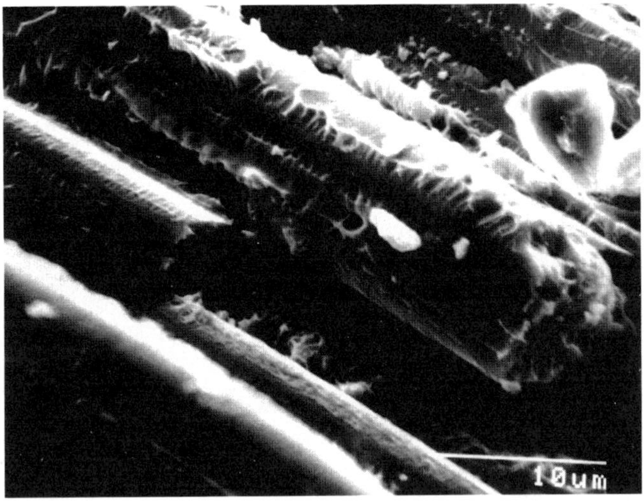

FIGURE 4 SEM of PIXA/IM-7 Fracture Interface

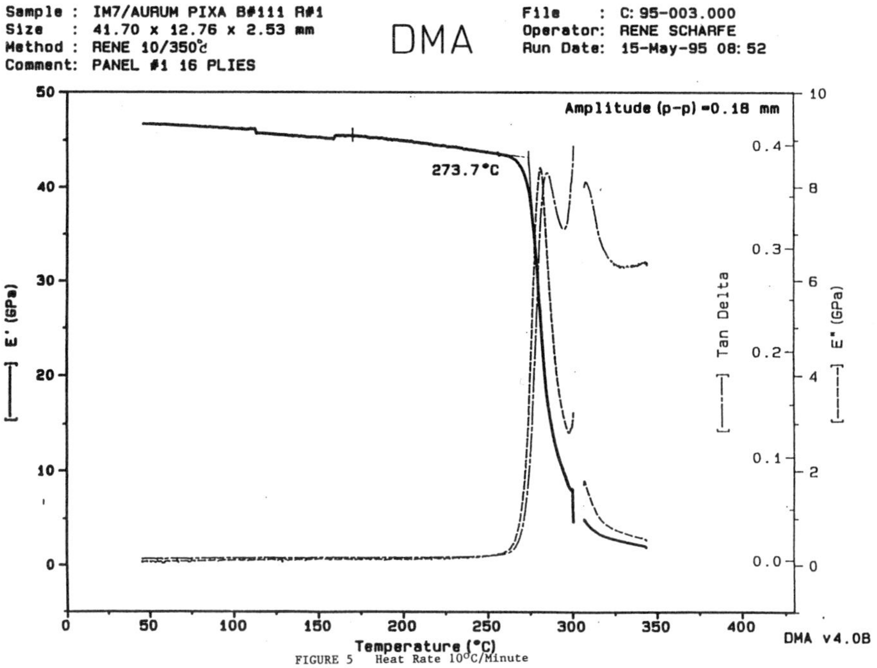

FIGURE 5 Heat Rate 10°C/Minute

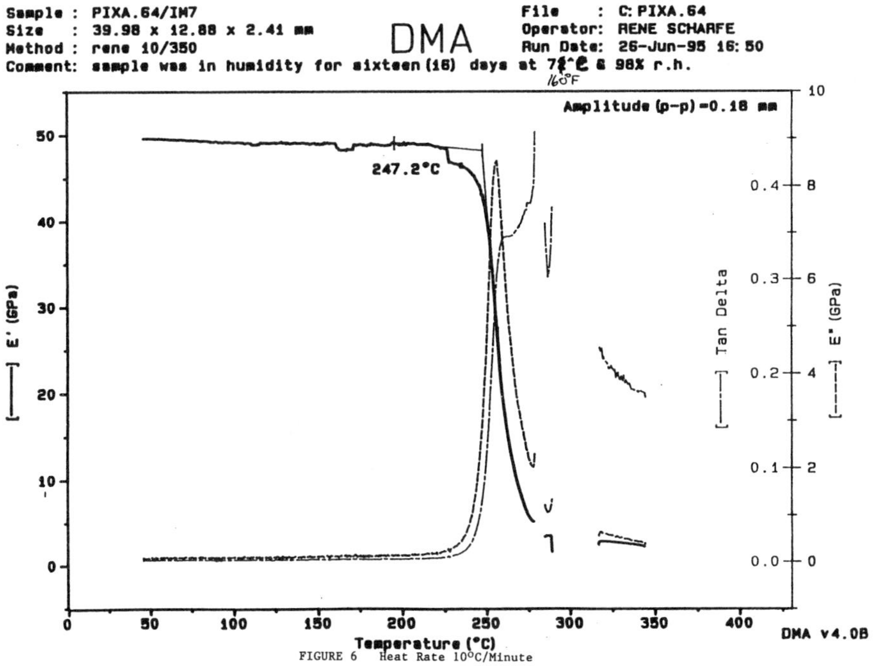

FIGURE 6 Heat Rate 10°C/Minute

FIGURE 7 Photomicrograph of PIXA/IM-7 Single Layer 0.25" Ribbon (400X)

28th International SAMPE Technical Conference
November 4-7, 1996

POLYIMIDE FILMS FOR LIQUID CRYSTAL DISPLAYS USED IN AEROSPACE APPLICATIONS

Fuming Li, Edward P. Savitski, Jyh-Chien Chen, Yeocheol Yoon,
Stephen Z.D. Cheng and Frank W. Harris
Maurice Morton Institute and Department of Polymer Science
The University of Akron, Akron, Ohio 44325-3909, USA

ABSTRACT

Organo-soluble aromatic polyimides with varying chemical structures and molecular weights were synthesized in refluxing *m*-cresol. Modifications to the dianhydride and diamine monomers were carried out in order to obtain aromatic polyimides that meet the requirements of liquid crystal display (*LCD*) applications. Solution-cast films exhibited linear optical anisotropy (*LOA*) that is commonly referred to as *uniaxial negative birefringence* (*UNB*), which is characterized by a larger refractive index along the in-plane direction than in the out-of-plane direction. The *UNB* was critically associated with the backbone linearity and rigidity as well as the intrinsic polarizability of the polyimides. A polyimide synthesized from 2,2'-bis(3,4-dicarboxyphenyl)hexafluoropropane dianhydride and 2,2'-bis(trifluoromethyl)-4,4'-diaminobiphenyl was used to study the effect of molecular weight on the *LOA*. For films prepared from fixed molecular weight samples, the refractive indices were constant for film thicknesses below 15 μm. The refractive index along the out-of-plane direction decreased while the in-plane refractive index increased as the polyimide molecular weight was increased to a critical level. Films with *UNB* have been used as negative birefringent compensators in avionic *LCDs* to improve the display off-angle viewability.

KEY WORDS: Polyimides, Films, Optical Testing

1. INTRODUCTION

It has been recognized since the 1960s that aromatic polyimide films exhibit structural anisotropy in the directions parallel (in-plane) and perpendicular (out-of-plane) to the film surface. This phenomenon has been defined as "in-plane orientation" *(1-5)*. Recently, it has been recognized that such anisotropic structure leads to anisotropic thermal, mechanical, dielectric and optical properties along the in-plane and out-of-plane directions *(5)*.

Our interests are particularly focused on the linear optical anisotropy (*LOA*) in solution-cast polyimide films and their applications. Since aromatic polyimide molecules commonly tend to align parallel to the film surface during the film forming process *(1-5),* the in-plane refractive index is larger than the out-of-plane refractive index. The degree of in-plane orientation and the resultant extent of *LOA* in the films can be estimated readily using refractive index measurements. Therefore, the *LOA* can be expressed by the birefringence which is the difference in the refractive indices along the in-plane and out-of-plane directions. In the field of optics, this phenomenon is defined as *uniaxial negative birefringence* (*UNB*). One of the applications for the *LOA* in polyimide films is that they may be utilized in negative birefringent compensators for twisted and super-twisted nematic liquid crystal displays (*TN*- or *STN-LCDs*) used in aerospace applications. Such films can be used in both active and passive forms to improve *LCD* viewing angles *(6-8)*.

It is well known that most aromatic polyimides are difficult to process, since they do not flow before decomposition and are insoluble in conventional solvents. The traditional approach to improving processability is to use a two-step polymerization route in which processing is made possible through the generation of soluble polyimide precursors, i.e. poly(amic acid)s. For example, polyimide films are generally produced by solution-casting or spin-coating poly(amic acid) solutions that are either thermally or chemically imidized. The imidization process affects the ultimate structure, morphology and properties of the films *(9,10)*. Conventional aromatic polyimides also often possess strong optical absorption in the low *UV*-visible wavelength region (<450 nm). This is caused by electronic conjugation involving the phenylene and imide groups, which results in dark yellow or brown films. Crystalline and ordered structures in the films can also introduce density fluctuations, which result in light scattering and, consequently, increase the optical loss. Different degrees of imidization in such films may also give rise to varying refractive indices along both the in-plane and out-of-plane directions. All of these complications with aromatic polyimide films have precluded them from applications in optical devices.

Over the past eight years, a family of organo-soluble aromatic polyimides has been designed and synthesized in our laboratory *(4-8,11-13)*. These polyimides were synthesized in refluxing *m*-cresol, such that the intermediate poly(amic acid)s were not isolated *(11-13)*. Many of the polyimides can be easily dissolved in common organic solvents such as acetone, chloroform, 2-pentanone, cyclopentanone, methyl ethyl ketone, tetrahydrofuran, dimethylformamide, N-methylpyrrolidinone and dimethyl sulfoxide. Solutions of the imidized polymers can be directly processed to form films, which are used as negative birefringent retardation compensators in *LCDs (6-8)*. The objective of this research was to determine the effects of film thickness, the molecular backbone structure and molecular weight on the *UNB* behavior of these thin films.

2. RESULTS AND DISCUSSION

2.1 Effects of Polyimide Structure on Linear Optical Anisotropy

Our work on organo-soluble aromatic polyimides for optical applications has focused on the interruption of the electronic conjugation along the polymer backbone. This has been shown to increase light transmittance in the low *UV*-visible light wavelength region, decrease the amount of ordered structures and loosen the chain packing. Our approach has been to introduce substituents at the 2- and 2'-positions in 4,4'-diaminobiphenyls. Substituents that have been used include methyl, trifluoromethyl, phenyl, cyano, and halogens. Steric hindrance between these groups causes the two phenylene groups in the diamine to twist to angles close to 90° from each other. This twist between the phenylenes leads to nonplanar conformations. For example, various aromatic polyimides have been prepared from 2,2'-bis(trifluoromethyl)-4,4'-diaminobiphenyl (PFMB) and different dianhydrides. The dianhydrides include 3,3',4,4'-tetracarboxylicbiphenyl dianhydride (BPDA), 2,2'-bis(3,4-dicarboxyphenyl)hexafluoropropane dianhydride (6FDA), 3,3',4,4'-benzophenonetetracarboxylic dianhydride (BTDA) and 4,4'-oxydiphthalic anhydride (ODPA). The in-plane (n_e) and out-of-plane (n_o) refractive indices and *UNB* values ($\Delta n = n_o - n_e$) of thin films of these polyimides were determined at a wavelength of 632.8 nm (Table 1). Their *UV*-visible transmittance (80%) was also determined. It can be seen that the refractive indices (1.52-1.64) as well as the *UNB* values (0.046-0.094) were critically affected by the structure of the dianhydrides. The more rigid and linear the dianhydride the higher the *UNB*. All of the *UNB* values (Δn) are approximately one order of magnitude higher than that of films solution-cast or spin-coated from flexible macromolecules.

Table 1. Linear Optical Properties of Aromatic Polyimides Based on PFMB

Polyimides	n_o	n_e	Δn	*80% trans. UV-Vis. (nm)*
BPDA-PFMB	1.540	1.634	0.094	445
BTDA-PFMB	1.576	1.639	0.063	426
ODPA-PFMB	1.587	1.639	0.052	408
6FDA-PFMB	1.516	1.558	0.046	388

Optical birefringence dispersion is also an important factor in the design of compensation films. The optical birefringence dispersion of a thin film of 6FDA-PFMB, which has been found to dissolve in a variety of common organic solvents, was determined (Figure 1). The open circles in Figure 1 are the experimental data and the solid line represents a Cauchy function, which theoretically represents optical dispersion *(14)*. The dependence of the *UNB* values on wavelength for the 6FDA-PFMB film was also compared to that of a liquid crystal mixture developed for *LCD* applications (Merck ZLI-1840) (Figure 1). Although the magnitude of the Δn of 6FDA-PFMB films is about three times smaller, the Δn dispersion of the film is similar to that of the *LC* mixture. A good match of birefringence dispersion between the *OC* and the *LC* mixture over the entire visible range leads to a null phase retardation for all three primary colors (red, green and blue) employed in *LCDs*. As a result, a high contrast ratio can be achieved significantly improving the off-angle viewability *(15,16)*.

2.2 Effect of Film Thickness on Linear Optical Anisotropy

6FDA-PFMB polyimide films of varying thickness were prepared in order to study the effect of the thickness on the refractive indices. Figure 2 shows that the refractive indices were fairly independent of the film thickness in the range of 1 μm to 15 μm. For thicknesses above 15 μm, the difference between the in-plane and out-of-plane refractive indices (n_e, n_o) gradually decreased as the film thickness increased. This indicates that the thin film linear optical properties gradually approach the bulk properties, which are three-dimensionally isotropic.

2.3 Effect of Molecular Weight on Linear Optical Anisotropy

The *UNB* values of the 6FDA-PFMB polyimide films exhibited an intrinsic viscosity (molecular weight) dependence as shown in Figure 3. The thickness of the films could be controlled to about 4 μm. The *UNB* values of the films initially increased with increasing intrinsic viscosity and then leveled off when a critical intrinsic viscosity of about 1.1 dl/g was reached. It is speculated that the molecular weight differences affected molecular packing due to changes in the chain end-group density, which affects free volume. When the molecular weight reached a critical value, the effects of end-groups on the packing became less important compared to other factors.

2.4 Application to *LCDs*

The thin films were incorporated in *UNB* compensators on avionic *LCDs* by a major American manufacturer. The films dramatically improved the off-angle viewability of the displays. This is a major advantage for *LCD* avionics applications where cross-cockpit viewing is required. The compensators will be commercially available in the near future.

3. CONCLUSIONS

Polyimides can be prepared from 2,2'-disubstituted-4,4'-diaminobiphenyls that are soluble in common organic solvents. Thin films of these polyimides can be solution cast that display *UNB*. The *UNB* values of the films can be varied by changing the dianhydride monomer used. The more rigid and linear the dianhydride the higher the value of the *UNB*. The polymer molecular weight and the film thickness can also affect the *UNB*. However, the value of the *UNB* becomes constant when the polymer viscosity is above 1.1 dl/g and the film thickness is less than 15 μm. The films can be used in *UNB* compensators on *LCDs* to dramatically improve the display off-angle viewability.

4. ACKNOWLEDGEMENT

The initial support of this research by the NASA-Langley Research Center under Grant NAG. 1-448 is gratefully acknowledged. We would like to express our gratitude to the Material Research Division of the National Science Foundation (DMR-8920147) who supported this research through the Science and Technology Center on Advanced Liquid Crystalline Optical Materials (ALCOM) at Kent State University, The University of Akron, and Case Western Reserve University.

5. REFERENCES

1. I.M. Ikeda, J. Polym. Sci., Polym. Lett. Ed., 4, 353 (1966).
2. T.P. Russell, H. Gugger and J.D. Swallen, J. Polym. Sci., Polym. Phys. Ed. , 21, 1745 (1983).
3. N. Takahashi, D.Y. Yoon and W. Parrish, Macromolecules, 17, 2583 (1984).
4. S.Z.D. Cheng, F.E. Arnold, Jr., A. Zhang, S.L.-C. Hsu and F.W. Harris, Macromolecules, 24, 5856 (1991).
5. F.E. Arnold, Jr, D. Shen, C.J. Lee, F.W. Harris, S.Z.D. Cheng and S.F. Lau, J. Mater. Chem., 3, 353 (1993).
6. U.S. Pat. 5,344,916 (1995) F.W. Harris and S.Z.D. Cheng.
7. U.S. Pat. 5,480,964 (1996) F.W. Harris and S.Z.D.Cheng.
8. F. Li, F.W. Harris and S.Z.D. Cheng, Polymer, 37, xxx (1996).
9. D.A. Hoffman, H. Ansari and C.W. Frank, Material Science of High Temperature Polymers for Microelectronics, D.T. Grubb, I. Mita and D.Y. Yoon, Eds.; MRS Symp. Ser., Materials Research Society, Pittsburgh, Pennsylvania, 227, 125 (1991).
10. L.A. Hornak, Polymers for Lightwave and Integrated Optics: Technology and Applications, L.A. Hornak, Ed., Marcel Dekker, New York, New York, pp 207 (1992).
11. F.W. Harris and S.L.-C. Hsu, High Perform. Polym., 1, 1 (1989).
12. U.S. Pat. 5,071,997 (1991) F.W. Harris.
13. F.W. Harris, S.L.-C. Hsu and C.C. Tso, ACS Polym. Preprints, 31(1), 342 (1990).
14. L. Pohl and U. Finkenzeller, Liquid Crystals. Applications and Uses, B. Bahadur, Ed, World Scientific, Singapore, 1, 139 (1990).
15. S.-T. Wu and A.M. Lackner, Appl. Phys.Lett., 64, 2047 (1994).
16. S.-T. Wu, Phys. Rev. A, 33, 1270 (1986), see also, J. Appl. Phys., 69, 2080 (1991).

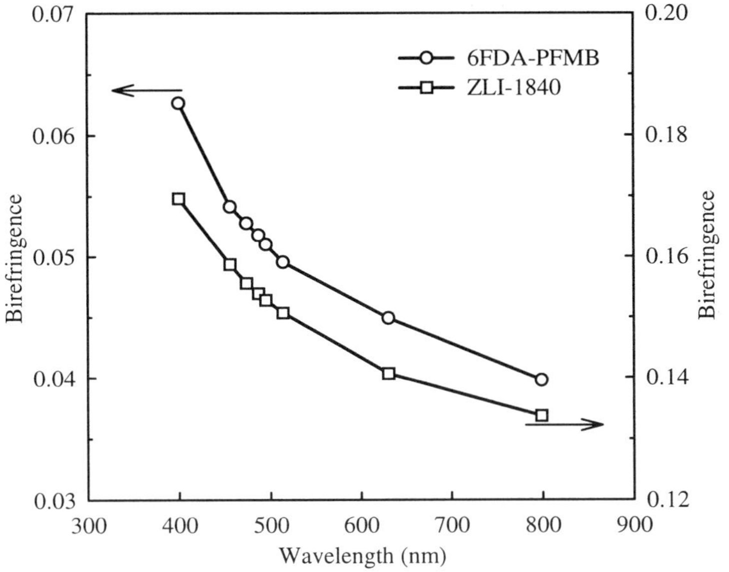

Figure 1 Comparison of birefringence dispersion of 6FDA-PFMB film with that of a liquid crystal mixture (Merck ZLI-1840) in the visible wavelength region

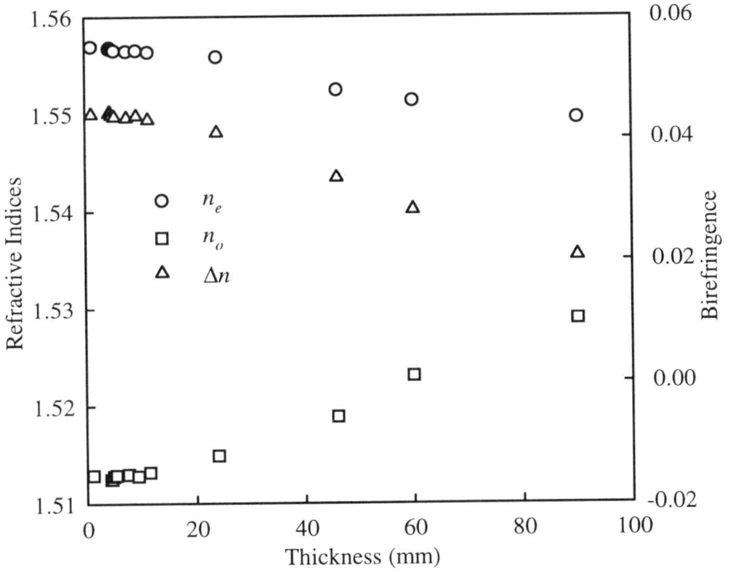

Figure 2 Effect of film thickness on the refractive indices and *UNB* values for 6FDA-PFMB films

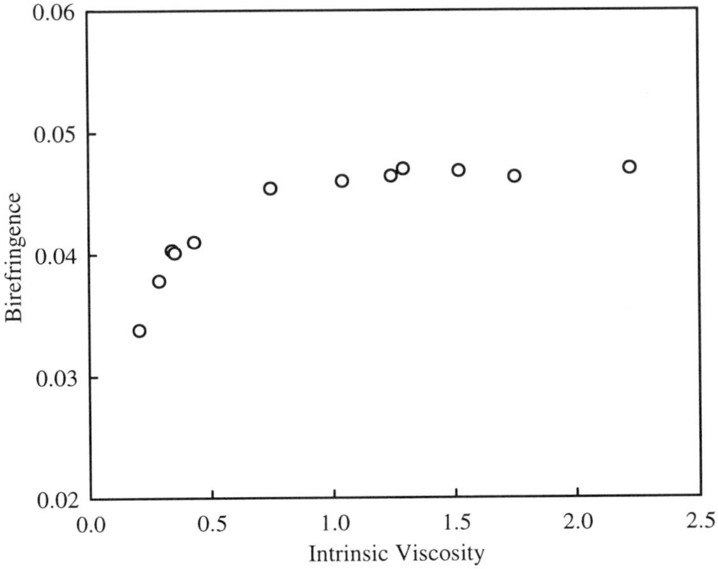

Figure 3 Effect of intrinsic viscosity on the *UNB* values for 6FDA-PFMB films

28th International SAMPE Technical Conference
November 4-7, 1996

NATURAL CONVECTION SUPERCRITICAL FLUID CLEANING APPLICATIONS

Mary Marshall, Paul Hauwiller*, David Morales, Tom Roberds, Gordon Pollard
Southwest Research Institute, 6220 Culebra Road, San Antonio, TX 78238
*Anteon Corporation, 5100 Springfield Pike, Suite 509, Dayton, OH 45431-1231

ABSTRACT

This patented technology has been scaled to a pre-production system under USAF funding. Successful cleaning has been accomplished for removing low molecular weight oils (hydrocarbon-based) independent of geometry (crevices and non-line-of-sight surfaces). Inorganic fillers used in greases cannot be solubilized by supercritical carbon dioxide, but polishing aids (mechanical impact systems) can knock-off insoluble surface particulates. Technology transfer activities have included participation from two Air Logistics Centers (Warner Robins and Ogden). Testing with filters, wheel bearings, gyroscopes, and contaminated gloves has defined the operating conditions for particular contaminants and part geometries.

KEY WORDS: Cleaning, Ozone Depletion, Supercritical Fluids

1. INTRODUCTION

Many chlorofluorocarbons (CFCs) destroy the ozone layer of the earth's stratosphere. The Montreal Protocol calls for a worldwide effort to phase out the use of all fully-halogenated CFCs in an effort to preserve our environment. The original Montreal Protocol was signed in 1987 and ratified by 64 countries. The Clean Air Act Amendments of 1990 mandated a production ban on Class 1 ozone-depleting substance (ODS) materials effective 31 December 1995 in the United States.

ODS materials being banned from production have numerous applications in the USAF including refrigerants, fire extinguishers, solvent cleaners, paint solvents, and adhesives. The USAF, EPA, DOE, and production shops in both small and large businesses have been substituting non-ODS materials for these applications. Substitution successes have been numerous as evidenced by the reduction in 1,1,1-trichloroethane purchases in 1995; the USAF purchased only 14% of the 1,1,1-trichloroethane purchased in 1992. The USAF estimates that 19% of the ODS usage for which adequate substitutes are unavailable is for cleaning. Of the cleaning applications, 35% is for metals, 40% for electronics, 12% for oxygen systems, and 13% for standards. In addition, five of the top fifteen solvent-related USAF needs (see Table 1) directly deal with cleaning.[1]

TABLE 1. TOP ODS SUBSTITUTION NEEDS FOR USAF CLEANING APPLICATIONS

Need No.	Description	Rank in Top 15 Solvent-Related Needs
129	Heavy degreaser on aluminum surfaces, carbon removal, and a pre-paint and pre-adhesive cleaner	2
991	Non-ODS/Non-EPA 17 drop-in replacement for 1,1,1-trichloroethane used in oxygen-tube cleaning	5
103	Replacement for 1,1,1-trichloroethane and MEK as wipe solvents and inclusion into Tech Data and MIL-SPECs	6
429	Replacement cleaner/drying agent for applications where parts could be exposed to nitrogen tetroxide (oxidizer)	10
203	Eliminate or reduce VOC emissions and/or hazardous waste generation from solvents used to clean metal parts	11

The DoD recognized several years ago that specialized requirements for weapon system readiness might not have acceptable technology to eliminate the need for ODS materials, so two approaches were taken: (1) invest funds into research for new technologies, and (2) stockpile ODSs to establish a reserve for "mission-critical" requirements through and beyond the year 2000, allowing time for research and development advances. To maintain mission readiness, the DoD Reserve will contain 350,000 pounds of 1,1,1-trichloroethane for USAF cleaning requirements and 500,000 pounds of CFC-113 for USAF and Navy cleaning needs.[2] Consequently, the need for new technology is crucial to the USAF. The subject of this paper is one of the new technologies investigated for a non-ODS cleaning system.

2. SUPERCRITICAL FLUID CLEANING TECHNOLOGY

Supercritical carbon dioxide (SC-CO_2) as a cleaning solvent has both environmental and performance improvements over CFCs. SC-CO_2 is a nontoxic and nonflammable solvent, produces no solvent waste, is not a reportable emission, is not a CFC, is not an ODS, and is not on EPA's list of 17 most toxic chemicals. SC-CO_2 is advantageous over other solvents including aqueous cleaning systems since there is no solvent residue and no parts drying. It can also perform better than CFCs for ultra-small pores and interstices since the supercritical fluid has very low solvent surface tensions. Operating temperatures for SC-CO_2 are moderate (30-80 °C).

Much has been published on the cleaning efficiency of SC-CO_2 and the components which have been examined. Pacific Northwest Laboratory (operated by Battelle Memorial Institute for the U.S. DOE) has cleaned metal parts, bearing assemblies, optical and laser components, precision mechanical parts, electromechanical systems, and medical devices[3]. Phasex Corporation has removed hydrocarbons, machining and lubricating oils, esters, silicones, Krytox, fluorocarbons, perfluoropolyethers, organosilicones, polychloro- and bromotrifluoro-ethylenes, and halocarbon-substituted-triazines from parts including gyroscopes, accelerometers, porous ceramics and metals, thermal switches, electrical components, nuclear valve seals, laser optics parts, polymeric containers, electromechanical assemblers, and camera lenses. Phasex has also identified the highly compatible materials for SC-CO_2 to include stainless steel, beryllium, gold, silver, copper, ceramics, Teflon®, silicone, and epoxy potting compounds.[4]

Many researchers have been studying the use of SC-CO_2 for a cleaning solvent to replace CFCs without relying on EPA 17 or ODS materials. Researchers include Rockwell International, Hughes Aircraft Company[5] and the "superScrub" device, the Joint Association for Supercritical Technologies (JAST) which includes Autoclave Engineers, CF Technologies, National Forge, Liquid Carbonic, and Los Alamos,[6] GCG Technologies Inc.,[7] and a cooperative European group consisting of Lund University and the Institute for Engineering Research in Goteborg.[8] All of these researchers have worked with a flow-through system for cleaning the parts.

SC-CO_2 cleaning technology is not eureka since it does have disadvantages which include high capital costs associated with the stainless steel pressure vessel (operating pressures from 1,000 to 6,000 psi), corrosion in the presence of water (requiring stainless steel construction), and parts damage from the pressure cycles (either crushing sealed chambers or de-laminating coatings). Consequently, the cleaning industry typically recommends supercritical fluids as a last resort after proving that aqueous cleaners and no-clean options are not feasible.

3. PROCESS DESCRIPTION

Southwest Research Institute (SwRI) invented a process[9,10] based upon natural convection for using SC-CO_2 in a cleaner without continuously compressing carbon dioxide and without pumping it past the dirty part until all of the contaminant is solubilized. The SwRI alternative technology is schematically presented in Figure 1 and is called the SC-CO_2 natural convection cleaner.

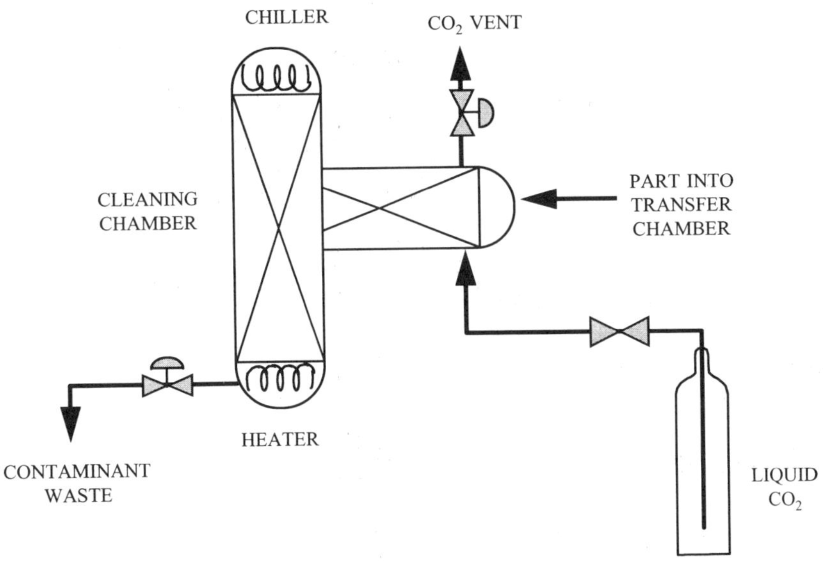

Figure 1. Process Schematic of Natural Convection SC-CO2 Cleaner

At pressures near the critical pressure, the solubility of a compound in SC-CO_2 decreases with increasing temperature. Also, the density of the SC-CO_2 decreases with increasing temperature at a given pressure. Thus, at isobaric conditions, the contaminant is solubilized in the cold region. The cold fluid with the contaminant then falls by gravity to a bottom heat exchanger at which point phase separation occurs (i.e., the contaminant separates from the hot, lower saturation concentration carbon dioxide). The hot SC-CO_2 rises since it is less dense, and the contaminant pools at the vessel floor since the contaminant (oils and greases) is more dense than SC-CO_2. Baffles have been added in the vertical column to effect greater temperature differentials between the hot and cold zones.

The solubility temperature dependence at constant pressure and the resultant natural convection currents are thus exploited to clean the part and regenerate the carbon dioxide within a single vessel. In this fashion, the CO_2 usage is independent of cleaning time (it is based solely on the amount of CO_2 used during parts introduction and removal), and energy usage is less than other SC-CO_2 devices since only heat exchangers are employed without a compressor.

The SC-CO_2 cleaner is shown in Figure 2. It consists of both a horizontally and vertically oriented 11-inch inside diameter cylindrical vessel connected by a cross. The vertical portion is where the cleaning process occurs, and the horizontal portion contains the transfer chamber and parts basket transportation mechanism. The cleaning and transfer chambers are separated from one another by a ball valve.

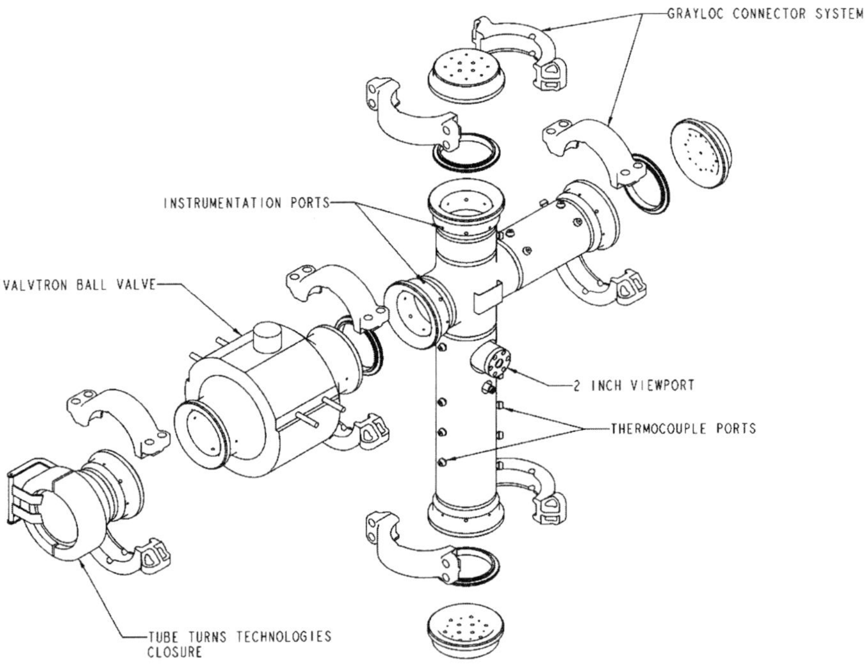

Figure 2. Exploded View of Pre-production Unit

A parts basket and a chain-driven rail transport system are used to transport the parts from the transfer chamber into the cleaning chamber. The horizontal extension of the cleaning chamber, opposite the transfer chamber, houses the mechanisms used to move and rotate the parts basket. Clamped connections are used to connect the end closures and ball valve to the chamber.

Liquid carbon dioxide is syphoned into the cleaner by creating a slight pressure drop via the CO_2 vent valve. Once sufficient liquid CO_2 is added to the vessel, electrical resistance heaters raise the temperature and, consequently, the pressure into the supercritical fluid zone. Thus, the expensive compressors and support equipment necessary in flow-through SC-CO_2 cleaners are eliminated. In addition, no hearing protection is needed since noise levels during normal operation are low.

The first prototype was designed and fabricated under the SwRI internal research and development program. The prototype vessel began operation in 1993 for parts less than 1.5 inches in any dimension. Under USAF funding from Wright Laboratory, Materials Directorate, a preproduction device has been constructed for parts as large as 8.25 inches in diameter and 12 inches long. Operation has successfully demonstrated cleaning without a compressor or supercritical fluid pump.

4. MATERIALS SELECTION

Materials selection for SC-CO_2 applications is a difficult task since very little published data exists (i.e., materials compatibility charts do not include SC-CO_2). The *Encyclopedia of Polymer Science and Engineering*[11] does give two lists of soluble and insoluble polymers in SC-CO_2, but several common materials used for seals, coatings, and adhesives are not included. Material selection is further complicated since the co-existence of liquid and gas phases of CO_2 can cause cavitation problems and extreme temperature excursions. Many equipment suppliers have experience with SC-CO_2 and can advise in seal selection, but other suppliers don't understand the difficulties and often supply unacceptable materials.

The first internal baffle was machined from polycarbonate. The heat from the external band heaters and exposure to gas, liquid, and SC-CO_2 damaged this baffle. The polycarbonate was "foamed", badly warped, and discolored (transformed from clear to yellowish-white). Nylon was also used for the baffle, but the localized heat zones upon start-up caused the nylon to melt. The baffle is now constructed from Teflon.®

Polyvinyl chloride (PVC) wire insulation becomes very brittle when used inside the SC-CO_2 system. Failures with the electrical control systems were attributed to the PVC flaking off and causing electrical shorts. The interior wiring has been replaced with Teflon®-insulated wire and has been working well. The electrical penetrators have polyimide insulation which has withstood the SC-CO_2 environment.

Vendors could not supply pressure-rated switches for SC-CO_2 operation; commercially available switches are either hermetically-sealed or lubricated with hydrocarbons. SwRI fabricated leaf-spring switches made of beryllium-copper attached to a printed circuit board base (reinforced epoxy). The actuator was initially made of Teflon® which warped from mechanical strain. The preferred and currently used material for the actuators is Delrin® which is more mechanically stable than Teflon®. The switches and actuators are used in the control system for identifying the parts basket location and the transporting mechanism.

Adhesives for mounting switches, mechanical stops, and wiring have been a constant source of operating failures. The adhesives must be chemically compatible and resistant to thermal cycling from the cooling of depressurization (temperatures of -20 °C have been achieved). The pressure vessel expansion from atmospheric pressure to supercritical conditions creates a mechanical stress on the adhesive which can be compensated by creating "expansion joints" in the adhesive, i.e., instead of a line of adhesive, use a broken line. Over an eight-month period, the two epoxy adhesives in the system became brittle; these were EPOXI-PATCH Kit #615 from The Dexter Corp. and Devcon 5-minute epoxy by ITW Devcon. (The Devcon epoxy becomes milky in appearance upon initial exposure to SC-CO_2 whereas the EPOXI-PATCH #615 has no color change. SwRI has successfully used JB Weld from J-B Weld Company in Sulphur Springs TX, a 2-part epoxy found in auto supply stores. Dexter Corp. recommended their high performance epoxy, EPOXI-PATCH #1C which is designed to withstand more temperature and pressure swings. In side-by-side comparisons, the JB Weld epoxy is very difficult to remove after exposure to SC-CO_2 while the EPOXI-PATCH #1C can be plied off of the stainless steel, so JB Weld is the preferred adhesive.

Buna-n rubber eventually fails under SC-CO_2 exposure. Mechanical deterioration is visually identifiable and "popping" sounds can be heard from explosive decompression of the rubber. In the flange seal of the TubeTurns® closure, there is an "O" ring about one foot in diameter ("O" ring size is as568a-453) . The original material supplied was 70-durometer buna-n costing \$1.75 each; this material failed from explosive decompression after a single use. They have been successfully replaced with a 90-durometer buna-n (Parker compound n552-90) for \$4.73 each and are replaced after severe stretching (typically after a dozen cleaning cycles). The expensive option of Teflon® "O" rings (approximately \$150 each) is not necessary since seal failure has multiple redundancies in the closure design and replacement is simple ("O" ring fitment is an essential operating step every time a part is added to the transfer chamber).

Similar issues apply to all wetted parts of the SC-CO_2 cleaner. For example, the diaphragm of pressure transducers should be specified as stainless steel or Teflon®, not the standard buna-n rubber diaphragm. For critical devices such as pressure transducers, the more expensive diaphragm material is necessary for safety.

5. EXPERIMENTAL

Parts (hydraulic filters, wheel bearings, gyroscopes, and contaminated gloves) received from Warner Robins and Ogden Air Logistics Centers were weighed, attached to the parts basket with plastic tie wraps, and placed into the transfer chamber. The automated cleaning sequence consisted of purging the transfer chamber of oxygen and moisture, filling it with liquid CO_2, pressurizing, moving the part into the supercritical CO_2 cold zone, cleaning with continuous part rotation, moving the part out of the SC-CO_2, and depressurizing the part in the transfer chamber. The part was then weighed and visually inspected for cleanliness determinations.

The thermocouples and pressure transducers were all calibrated prior to the experiments; thermocouple measurements are good to ±0.5 °C and pressure measurements to ±10 psig. The carbon dioxide used was technical grade supplied by Liquid Carbonics in standard laboratory liquid-syphon pressurized cylinders. Cosolvents were not employed during these experiments.

6. RESULTS AND DISCUSSION

Two parts of a gyroscope were placed into the SCF preproduction device and cleaned for 60 minutes at 1010 psig, 43 °C for the hot zone, and 34 °C for the cold zone conditions. The pieces had inks, dust, and oil visually present before cleaning. After cleaning, only the ink remained. The amount of oil and contaminant on the parts was unknown, but about 1 gram and 2 grams, respectively, were removed from the two parts. There was oil inside one of the parts after the cleaning cycle which recontaminated the exterior surface during shipment.

OO-ALC/LIT operations include cleaning parts for hydraulic systems, gear boxes (including the B-2 box whose paint cannot be damaged), and servo valves. Most of their parts can fit within the preproduction unit. Several parts were tested in the natural convection supercritical fluid cleaner using a cleaning time of 10 minutes at about 1500 psig with a hot zone temperature of 55 °C, cold zone around 48 °C, and continuous rotation. The overall cleaning cycle was 45 minutes including purging out air, filling the transfer chamber with liquid CO_2, moving the parts in and out of the cleaning chamber, and depressurizing.

The servovalve filters were sintered rods about one inch long and one-quarter inch diameter plugged with aircraft hydraulic residue and covered with dust. After immersion in SC-CO_2, the residue was loose and could be brushed off. This is acceptable cleanliness since compressed air could be used as a secondary treatment to blow off the particles. A valve body about two inches in length and in width and one inch tall had numerous cavities which were plugged with dry hydraulic oil; after SC-CO_2 cleaning, the dry oil was dissolved and seeped out of the small openings. There was also some baked on oil which appeared to be partially cleaned by SC-CO_2.

Numerous experiments have been conducted with flat aluminum panels using continuous basket rotation. A trace contaminant was added to each panel to determine whether high precision cleanliness was possible. The contaminants were typical materials found on aircraft parts and included a machine oil, a hydraulic oil, a water-soluble lubricant, and masking tape adhesive. The panels were numbered using a paint marker so that as many as four panels could be tested in a single batch. In all instances, the paint marker dried during exposure to the SCF and the paint partially chipped off the panel, but sufficient paint remained to identify the panel number.

The aluminum panels were cleaned under a wide range of operating conditions. Pressures ranged from 980 psig to 1720 psig. Cold zone temperatures ranged from 31 °C to 59 °C; hot zone temperatures ranged from 40 °C to 78 °C; and the temperature differential was varied from 6 °C to 31 °C. Cleaning times studied were 2, 40, 60, 120, and 140 minutes while the amount of liquid CO_2 added to the transfer chamber ranged from 9 to 32 kg. Some contaminant removal was observed in all cases; however, residue could be seen for all but two panels. No residue was detectable for the machine oil and hydraulic oil cleaned in SC-CO_2 for 60 minutes at 1450 psig, a cold zone temperature of 46 °C, hot zone of 54 °C, and 23 kg liquid CO_2 in the transfer chamber.

A wheel bearing contaminated with grease from a C-141 was tested. The grease organic components solubilized in the supercritical carbon dioxide and were removed; however, the inorganic fillers were left behind. The non-line-of-sight segments of the bearing system were stripped of the lubricating organic fluids. Marbles were added to the basket and the wheel bearing was added back to the cleaning chamber. The basket was rotated and much of the inorganic fillers were removed by the mechanical action. The cleanliness is insufficient since the marbles did nothing to remove the inorganic material from the non-line-of-sight sections.

Hydraulic filters (one-inch diameter disks with metal mesh sandwiched between brass rings) from the C-130 which were contaminated with oil and particulates were tested under a couple of conditions, but the results do not meet the cleanliness level required. The carbon particulates are not soluble in supercritical carbon dioxide and remain imbedded in the filter mesh.

A contaminated glove was placed into the SCF preproduction cleaner with operating conditions of 1050 psig, 43 °C in the cold zone, and 52 °C in the hot zone. Cleaning time was 15 minutes, and no basket rotation was used. The glove weighed less after exposure to the SC-CO_2 but was visually still dirty (black coloring was present), so it was not successful.

7. LESSONS LEARNED

The pressure vessel is stainless steel which was selected for corrosion-resistance since carbonic acid is formed when water contaminates the carbon dioxide (carbonic acid is highly corrosive as a dilute solution in CO_2). It is virtually impossible to prevent water from entering the cleaner, especially since many water-soluble lubricants are being introduced in the production shops under solvent substitution efforts.

The Grayloc® connector system for the pressure vessel made assembly easier and less costly. The device weight was dramatically reduced, and its size less forbidding for commercial applications than had flanges been used. The TubeTurns® closure has proven highly reliable with mechanical safety features which will minimize catastrophic failures such as the recently-reported fatality at a Massachusetts biotechnology firm employing supercritical fluids.[12]

To replace all needs for pumps, the CO_2 is provided via syphon fill from compressed gas cylinders. This system has proven reliable, but the cold temperatures from the initial rapid expansion of the liquid created start-up problems including water condensation in the transfer chamber during the purging sequence. The automated cleaning system was modified to compensate for thermal expansion cooling.

8. TECHNOLOGY TRANSFER

Review of USAF depot cleaning operations which demand a high level of cleanliness or use undesirable cleaners (e.g., TCA, perc, MeCl, and CFC 113) indicate that these operations could be conducted in a cleaning chamber with dimensions of 24-in by 18-in by 10-in. The materials needing cleaning are aluminum, steel, stainless steel, magnesium, and titanium; some parts have chrome coatings. Inspection methods are primarily visual.[13]

Based upon the initial cleaning experiences with the preproduction device, the operating conditions need further refinement before application to a depot operation. The current research has resolved many scale-up issues and identified many reliability and operability improvements. These will be included in the economics and employment decisions before proceeding with a larger pressure vessel design for accomodating the remaining cleaning operations requiring environmentally friendly solvent substitutions.

9. CONCLUSION

Successful cleaning has been accomplished for removing low molecular weight oils (hydrocarbon-based) independent of geometry (crevices and non-line-of-sight surfaces). Inorganic fillers used in greases cannot be solubilized by supercritical carbon dioxide, but polishing aids (mechanical impact systems) can knock-off insoluble surface particulates. The natural convection system has provided much insight into the material compatibility issues and pressure vessel design unique to supercritical fluids.

10. ACKNOWLEDGEMENTS

This research has been sponsored by the USAF Wright Laboratory Materials Directorate. The authors thank the following USAF personnel for their assistance in this program: Ted Reinhart, Phil Mykytiuk, Harvey Paige, Osbin Clark, Bill Best, and Frank Collins.

11. REFERENCES

1. "Managing the Implementation of the Transition: An Air Force Perspective," Eric L. Stephens, Director, Environmental Management, Human Systems Center, Brooks AFB, Sixth Annual International Workshop on Solvent Substitution, December 6, 1995.

2. "The DoD Ozone-Depleting Substances Reserve," Ron Sibley, Defense General Supply Center, Sixth Annual International Workshop on Solvent Substitution, December 7, 1995.

3. "Precision Cleaning with Supercritical Carbon Dioxide," Max Phelps, August 1995.

4. "Precision Parts Cleaning: CO_2 Replaces CFCs for Many Applications," Phasex Corporation Newsletter: Notes on Supercritical Fluids, Number 1.

5. "Hughes, Los Alamos to Work on Alternative to CFC Cleansers," New Technology Week, June 8, 1992.

6. "Newly Commercial Systems Clean Parts with Supercritical CO_2," Chemical Engineering, p. 17, September 1993.

7. "Precision Cleaning," Aviation Week & Space Technology, p. 13, September 14, 1992.

8. "Development of an Industrial De-greasing Process based on Supercritical Carbon Dioxide," Press Release from Swedish Work Environment Fund, Stockholm, 1992.

9. U.S. Pat. 5,401,322, Mary Marshall, "Apparatus and Method for Cleaning Articles Utilizing Supercritical and Near Supercritical Fluids," March 28, 1995.

10. U.S. Pat. 4,433,538, Mary Marshall, "Apparatus for Cleaning Articles Utilizing Supercritical and Near Supercritical Fluids," July 9, 1996.

11. McHugh, Mark A., and Val J. Krukonis, "Supercritical Fluids", Volume 16, pp. 390-391, Encyclopedia of Polymer Science and Engineering, John Wiley & Sons, New York, 1989.

12. "Pilot Plant Operator Killed in Pressure Vessel Release at a Massachusetts Biotechnology Company", Massachusetts Fatality Assessment Control and Evaluation (FACE) Report, Investigation #94-MA-019, June 10, 1995.

13. *Evaluation of Metal Cleaning Processes at Air Logistics Centers*, Technical Report under Contract Number F41624-94-D-8135, Delivery Order 0009, Prepared for HQ HSC/YAQ, March 1996.

28th International SAMPE Technical Conference
November 4-7, 1996

SOLID STATE, VAPOR PHASE, AND AQUEOUS CLEANERS FOR AEROSPACE MANUFACTURING OPERATIONS

Rodney F. Patterson, Components & Materials Center, Hughes Aircraft Company
Philip D. Mykytiuk, Air Force Pollution Prevention R&D Office, Wright Laboratory

ABSTRACT

This paper focuses on six examples of what the Government and industry are doing to replace environmentally unfriendly cleaning agents used in various manufacturing cleaning operations:

(1) Wright Laboratory at Wright-Patterson Air Force Base is funding industry efforts to develop long-term solid state cleaning technologies to leapfrog liquid cleaning technology.

(2) Sandia National Laboratory has gone through a step-by-step program to replace trichloroethylene (TCE) degreasing and isopropyl alcohol (IPA) rinsing previously used (2).

(3) Rocketdyne Division of Rockwell developed an excellent in-depth plan to identify effective all-purpose aqueous cleaners for replacing their ozone-depleting cleaners (3).

(4) Northrop conducted a screening test program that narrowed 50 candidate aqueous degreasing cleaners to one that successfully replaced 1,1,1-trichloroethane (TCA) degreasing (4).

(5) McDonnell Douglas Aerospace methodically removed and replaced all Class I ozone-depleting chemicals (ODCs) from the F-15 support documents and put all their findings in an electronic bulletin board database (5).

(6) Hughes Aircraft has developed four innovative environmental technologies that are being used to preclude the need for ODCs in a number of cleaning applications (6-9).

Solid state, vapor phase and aqueous cleaners were found in all cases to be successful in a variety of applications for aerospace manufacturing operations.

KEY WORDS: Cleaning, Environmental/Pollution Prevention, Ozone Depletion

1. INTRODUCTION

The aerospace industry has done an excellent job of complying with the government phase-out schedule for ozone-depleting chemicals (ODCs) and the tightening regulations on volatile organic compounds (VOCs). These schedules and regulations have driven the aerospace industry in particular to invest a great deal of effort toward the replacement of the restricted chemicals.

2. U.S. MILITARY HELP

Wright Laboratory at Wright-Patterson Air Force Base is funding industry to develop long-term, environmentally acceptable solid state metal cleaning processes with minimal waste. Wright Laboratory is keenly aware that cleaning is a major operation at military depots. All military vehicle (planes, trucks, helicopters, tanks, etc.) components that require maintenance must be cleaned prior to inspection. Components that need painting, bonding or plating must first be cleaned. The parts range in size from nuts and bolts to wing skins. The materials are primarily 2024-T6 aluminum, 7075-T3 aluminum and titanium 6Al-4V. A few specialty applications, such as gyros and oxygen tubing, also require a very high cleanliness level. The traditional cleaners have been trichlorotrifluoroethane (CFC-113), TCA, and methyl ethyl ketone (MEK).

While industry has developed several alternatives involving nonhazardous solvents and aqueous cleaners that reduce toxic wastes, these cleaners increase liquid waste. Regulations concerning waste water are becoming more restrictive, however. Therefore, Wright Laboratory began a solid-state metal cleaning initiative to develop technologies that leave only solid waste, inclusive of the soils being removed. The successful development of such technologies will reduce the sources of waste and allow depots to implement a long-term cleaning solution. The five solutions being developed are ultraviolet/activated oxygen cleaning, biodegradable solid media blasting, supercritical CO_2 cleaning, oxygen plasma cleaning of oxygen tubing, and laser cleaning and coating removal.

2.1 Ultraviolet/Activated Oxygen Cleaning The ultraviolet/activated oxygen cleaning process, being developed at the McDonnell Douglas Corporation, combines CO_2 pellet blasting, ultraviolet light, and activated oxygen to clean small-to-medium size metallic surfaces prior to bonding, plating, and painting. The blasting process reduces contamination to a thin film. Parts are then placed inside a 0.68 cubic meter (24 cubic foot) chamber containing a wide-area ultraviolet light. This 185-and 254-nm wavelength light generates 10 ppm of ozone which, together with the light, reacts with the remaining contamination. A positive air flow carries the released contamination from the chamber.

The project engineer for the process, Joan Deffeyes, points out that with either step in the process, there is no requirement for further processing for separation—the UV light transforms the contamination on the surface of the part into harmless gas, and the dry ice sublimes away. The final UV light processing removes the last trace of contamination and leaves a very clean, water-break-free surface.

2.2 Biodegradable Solid Media Blasting The biodegradable solid media blasting process, under development at Boeing, cleans small to large metallic surfaces prior to painting, plating, and bonding. The parts are placed in a cabinet similar to that used for abrasive grit blasting. The biodegradable media, composed of wheat starch, corn cob, and/or sodium bicarbonate, is loaded into the blasting system where pressure ranges from 0.21 to 0.41 Mpa (30 to 60 psi). These materials are blasted across the surfaces, with the residual media removed by blowing nitrogen gas through the cabinet.

2.3 Supercritical CO_2 Cleaning The supercritical CO_2 process precision cleans a variety of items, including tubing, gyroscope components, bearings, optics, and rags. The removed contaminants are the only pollutants. Supercritical CO_2 has very good solvating properties. Because its surface tension is nearly zero, it can clean small and hidden orifices. It also produces no solvent waste, is not an ODC or VOC, and is not flammable.

The preproduction prototype cleaning unit, developed at the Southwest Research Institute, is being used to test industry- and depot- donated components and can accommodate items that measure 22.9 cm in diameter by 30.5 cm long, and weigh less than 32 kg. The items to be cleaned are placed in a basket in the unit transfer chamber. The chamber is pressurized to approximately 10 MPa (1400 psi), equal to the cleaning chamber pressure. Once the pressure is equalized, a ball valve is opened, and the parts basket is moved into the cleaning chamber.

The bottom of the cleaning chamber is heated to 60°C, and the top is cooled to 30°C, creating a 30°C gradient that produces convection flow within the chamber. The colder supercritical CO_2 fluid has greater solvating properties than hot fluid; therefore the contaminants are removed in the cold zone and circulate down to the hot zone, where they drop out of solution. The contaminants are removed from the bottom of the chamber through a valve. The equipment eliminates the need for a high pressure CO_2 pumping system.

2.4 Oxygen Plasma Cleaning of Oxygen Tubing The manufacture of oxygen tubing leaves a residual tube-bending oil contaminant which must be removed. Laboratory testing of an oxygen plasma cleaning process at Rockwell International has resulted in the cleaning of this tubing to a high precision level (less than 10 mg/sq m). This technology can be applied to such tubing on military and commercial aircraft, as well as medical oxygen systems.

In the full scale cleaning process, the tubing is initially heated to reduce the viscosity of the oil. The oil is drained from the tube to minimize the amount of contaminant that must be removed. A flexible device is either pushed or pulled through the tube to reduce the oil to a thin film. A wire is threaded through the tube, connecting the tube to the plasma system via quick connect couplings. A plasma is ignited and sustained within and along the entire tube length. Charged particles, ultraviolet radiation, and atomic oxygen break down the tube bending oil, and the by-products are carried away by a vacuum pump.

2.5 Laser Cleaning and Coating Removal A laser cleaning and coating process, being developed by Craig Walters Associates will deliver a reliable system to handle the most challenging cleaning and coating removal problems for which there are no environmentally acceptable alternative solutions. The process removes soils and hard coatings (metals, ceramics) on all types of substrates, including metals, nonmetals on avionics, propulsion, and armament components.

The laser beam sweeps parts as large as 122 x 305 x 122 cm (4 x 10 x 4 ft), removing organic and inorganic contaminants. Both an infrared and UV laser will most likely be required, owing to the wide range of coatings and contaminants identified by DoD depots personnel. To prevent the laser from damaging the substrate, a feedback control system will determine when the unwanted material has been removed.

The laser system will be developed and rigorously tested under harsh conditions to verify its robust performance. All effluents from the process will be collected, treated, and captured. The final system will be delivered to a depot.

3. FINDING ALTERNATIVES

A reader of Precision Cleaning magazine from McClellan Air Force Base in California commented recently on the plethora of cleaning agents (1). He wrote in the April 1995 issue that he had built up his own database of 690 cleaning product manufacturers. He said, however, that his list did not include 58 of the 100 presented by Precision Cleaning in its January 1995 Cleaning Agent Directory issue. He reasoned that, if Precision Cleaning's list is a random sample of all manufacturers, simple statistics hold that, for a 95 percent probability, there must be at least 2,050 cleaner manufacturers. Assuming at least 10 cleaning products per manufacturer, he reasoned that a minimum of 20,000 cleaners are available.

Finding alternatives for hazardous and ODC solvents can present a formidable challenge to industrial activities. Fortunately, most companies are more than willing to share what they have found. It is quite easy to learn who is replacing what solvent with what environmentally acceptable product in what particular application. Because this type of information is valuable, the authors' findings are given in Table 1; the information source for each entry is footnoted. As valuable as this information may be, however, it might be still more valuable to learn exactly how a company evaluates and determines which cleaning agent or product is best (or at least acceptable) for its own purposes. Along this line, the experiences of a government agency and three companies are presented next. Finally, work that Hughes Aircraft Company has done to develop alternative cleaning technologies is presented.

Table 1. Sampling of Solvent Substitutions at Companies and Government Agencies

Company or Agency (Sources Follow)	Cleaning Application	ODC Cleaner Used & Soils Removed	Alternative Solvent
Beech Aircraft Co., Wichita, KS **(A)**	Aircraft components; multimetals, plated, conversion coated	TCA used to remove oily-type soils and shop soils	6000-gal aqueous cleaning tank at 66°C-71°C, air agitated, plus spray wand
Comment: Aqueous cleaning provided comparable or better cleaning at less cost.			
General Electric, Evandale, OH **(A)**	High-nickel-alloy turbine vanes and blades for aircraft engines	CFCs in five vapor degreasers used to remove oil and graphite from electrodes	400- to 2600-gal aqueous cleaning tanks maintained as 60°C baths
Comment: Tanks were converted vapor degreasers; avg. life of baths exceeded 6 months			
Martin-Marietta, Denver, CO **(A)**	High-technology aerospace components	TCA in two vapor degreasers used to remove gross soils from metal parts prior to various chemical processing	18,000-gal aqueous alkaline immersion bath, steam heated to maintain 54°C, with agitation by cleaning fluid recirculation
Comment: As of May, 1995, the bath life to date was more than 3 years.			

Company or Agency (Sources Follow)	Cleaning Application	ODC Cleaner Used & Soils Removed	Alternative Solvent
Hughes Aircraft Co., LaGrange, GA **(A)**	Missile components	TCA vapor degreaser used to remove oily contamination and shop soils	85-gal aqueous unit in 3 separate tanks: 66°C wash tank, spray rinse over-tank at 2% conc. for rust control, and oil separation tank
Comment: Oil separation tank removes floating tramp oils overflowed from the wash stage, using bag and carbon filters. Better cleaning at less cost obtained.			
Schott Fiber Optics, Southbridge, MA **(B)**	Multiple fiber-optics rods in 25 x 61 cm carriers	TCA used to remove dust and dirt	Aqueous cleaning followed by high-velocity air knife/ sonic in-line filter
Corpus Christi Army Depot, Corpus Christi, TX **(A)**	Variety of missile components involved in military refurbishment and repair	CFC vapor degreaser used to remove various field and shop soils	Aqueous technology using existing 864-gal vapor degreaser tank with agitating high-pressure turntable and ultrasonic washers
Comment: Cost of converting existing vapor degreaser totaled just over $5000 ($2000 labor and $3054 materials).			
Manth-Brownell Inc., Kirksville, NY (automotive industry) **(B)**	Screw machine parts and various metals	Mineral spirits pre-cleaning followed by Freon vapor degreasing	Aqueous cleaner formulations and console supplied by Aqueous Cleaning Technologies
Black & Decker, Towson, MD **(B)**	Parts used in the production of power tools	TCA vapor degreaser used to remove shop soils	Multi-stage aqueous spray wash cabinet supplied by Better Engineering Mfg., Inc.
Hamilton Support Systems unit, Hamilton Standard, Div. of United Technologies, East Windsor, CT **(C)**	Variety of aerospace components and control systems	TCA and perchloroethylene vapor degreasers used to remove soils ranging from baked-on carbon to oils, coolants, and preservatives	Aqueous cleaners in 6 cabinet-type spray washers and 4 ultrasonic cleaning systems
Comment: To extend life of cleaning baths and rinses and to minimize waste generation all equipment had oil skimmers, coalescers, and water conditioning systems.			
Sandia Nat'l Lab, Albuquerque, NM **(D)**	Metal piece part degreasing applications	TCE used to remove machining lubricants	Aqueous alkaline cleaner; Brulin 815 GD was favored

Company or Agency (Sources Follow)	Cleaning Application	ODC Cleaner Used & Soils Removed	Alternative Solvent
TRW Power Systems Dept., Redondo Beach, CA **(E)**	Soldering of interconnects to bare silicon cells	Freon TES (CFC-113/ethanol) used to remove RMA flux residue	Saponifier-based aqueous/semiaqueous solvent, including Alpha 2110, Ionox FC, and Aquanox 101
Gillette Co., Boston, MA **(F)**	Razor & blade preparation for Pt/Cr film sputtering &PTFE coating	TCE solvent used to remove oil, dirt, and debris	Aqueous cleaners
Comment: Gillette preferred aqueous cleaners in spite of their drawbacks of slow drying and possible metal staining and/or corrosion.			
Los Alamos Nat'l Lab, Los Alamos, NM **(G)**	Various, from paint stripping to optics cleaning	TCA, among others, used to remove various shop soils	Hydrocarbon hybrid cleaners; X-Caliber favored
Comment: Los Alamos National Laboratory rated Inland Technology's X-Caliber best, with P-T Technologies' PF Solvent and 3D Inc.'s "Siege" next best.			
Los Alamos Nat'l Lab, Los Alamos, NM **(G)**	Cleaning for electroplating	TCA used to remove shop soils	Bluegold aqueous detergent by Carroll Co.
Abex Corp., Aerospace Division, Oxnard, CA **(H)**	Aircraft hydraulic systems	TCA and CFC-113 used to remove shop soils	Chem Tech CT-1 aqueous biodegradable cleaner, and hydrocarbon cleaner CT-2
GenCorp Aerojet, Azusa Operations, Azusa, CA **(I)**	Paint and epoxy stripping	Methylene chloride used to remove coatings	Alcohol or acetone along with alcohol- or acetone- soluble epoxy
GenCorp Aerojet, Azusa Operations Azusa, CA **(I)**	General cleaning	TCA used to remove various soils	Closed-loop aqueous cleaning, or use of alcohol or acetone
San Antonio Air-Logistics Center, Kelly AFB, TX **(J)**	Aircraft engines, turbine parts	TCA and CFC-113 vapor degreasing use to remove various soils	ARMEX blast media, and aqueous cleaners in part washers
McDonnell Douglas Corp., St. Louis, MO **(K)**	Cleaning prior to metal finishing operations	Various solvent cleaners used to remove various soils	UV light process after pre-cleaning with dry ice pellet blasting
Howmet Corp., La Porte, IN **(L)**	Removal of fluorescent penetrant from aerospace castings	TCA used to remove dye penetrant	Steam at 127°C in superheated water line of wet steam

Company or Agency (Sources Follow)	Cleaning Application	ODC Cleaner Used & Soils Removed	Alternative Solvent
Delta Airlines, Atlanta, GA **(M)**	Depainting entire airframes	Methylene chloride used to remove paint	Acidic benzyl alcohol strippers, including Turco 6776 and EZE 540
Naval Aviation Depot, Jacksonville, FL **(M)**	Depainting airframes	Methylene chloride used to remove paint	Alkaline benzyl alcohol strippers, including Turco 6813 and EZE 541
Five aircraft depots: 1. Lockheed, Fort Worth, TX and Marietta, GA. 2. Grumman, Bethpage NY 3. Warner-Robins Air Logistics Center at Robins AFB, Warner-Robins GA 4. Northwest Airlines, Minneapolis, MN 5. Naval Aviation Depot, San Diego, CA **(M)**	Depainting airframes	Methylene chloride used to remove paint	Plastic media blasting (polycarbonate)
Beech Aircraft Co., Wichita, KS **(M)**	Depainting airframes	Methylene chloride used to remove paint	Wheat starch blasting
Draper Laboratories, Cambridge, MA **(N)**	Parts cleaning	CFC-113 used to remove shop soils	System using supercritical (SC) carbon dioxide
Thiokol Corp., NASA Space Flight Center Huntsville, AL **(O)**	Removing preservative grease in Refurbishment Center	TCA vapor degreaser used to remove grease	High-pressure water wash system
Aerospace Guidance and Metrology Ctr., Newark AFB, OH (a major repair facility) **(P)**	Aircraft and missile systems, some with complex components extremely susceptible to contamination	CFC-113 used up to 1250 metric tons/yr, plus smaller amounts of TCA to remove various soils	90% replaced with aqueous technologies and 10% nonaqueous, including SC fluids of CO_2 and ethane

Company or Agency (Sources Follow)	Cleaning Application	ODC Cleaner Used & Soils Removed	Alternative Solvent
Briggs & Stratton Corp. Wauwatosa, WI **(Q)**	Various metal cleaning and lubricating operations for engines	TCA (one of top 5 TCA users) used to remove shop soils	Aqueous washing methods, plus blowing off excess water
Raytheon Company, Lexington, MA **(R)**	Printed circuit board cleaning	TCA, used mainly to remove flux	Semiaqueous solvent cleaning: Petroferm's Bioact EC-7R favored
Rocketdyne Division, Rockwell International, Canoga Park, CA **(S)**	Defluxing of PWAs for Space Station program	CFC-113 used to remove flux	Semiaqueous solvent cleaning: favored Petroferm Bioact EC-7R and Axerel 32
Rocketdyne Division, Rockwell International, Canoga Park, CA **(T)**	Rocket engine hardware	CFC-113 and TCA used to remove shop soils	Aqueous cleaning agents: 3D Supreme XP and Turco 3878 LF-NC favored
Alcatel Network Systems, Assembly Operations Div., Richardson, TX **(U)**	Printed circuit board assembly processes; removing flux residues	CFC-113/methanol (This use ended in 1992.)	Low-residue no-clean flux with aqueous cleaner installed off-line as a contingency
Motorola Govt. and Systems Technology Group, Scottsdale, AZ **(V)**	Printed circuit board defluxing and cleaning	CFC solvents to remove RMA flux	Aqueous cleaning to remove water-soluble flux
General Dynamics Land Systems Div., Tallahassee, FL **(W)**	Printed circuit board cleaning	CFC-113 to remove flux	Water/saponifier process
Comment: Their $173K/year cost to support freon over three years now reduced to $33K per year for the water process.			
Digital Equipment Corp., Springfield, MA and other plants **(X)**	Disk drive manufacturing	CFC solvents	Aqueous cleaning systems with a vacuum final dry process
GTE Government Systems Corporation, Taunton, MA **(Y)**	Printed circuit board assembly cleaning and pre-conformal coat cleaning	Aqueous-cleaning solvent with amine glycol saponifier	Aqueous cleaning with a zero-VOC saponifier that is non-chelating
Allied Signal Commercial Avionics System Division, Ft. Lauderdale, FL **(Z)**	Defluxing printed wiring assemblies	CFC-113 + DI water and IPA cleaning process (baseline)	Aqueous solvent cleaning with inorganic based aqueous saponifier @ 100% conc.

Sources of Information referred to in the above table by letter:

A - Joann A. Quitmeyer, W.R. Grace & Co., Lexington, MA, “Converting to an Aqueous Cleaning System,” *Products Finishing Magazine*, May, 1995.

B - Special Advertising Section, “Company Compliance Profiles,” *Precision Cleaning*, 3 (11), 57 (1995).

C - Robert W. Jenkins, “Closing the Loop on Aqueous Cleaning at Hamilton Standard,” *Precision Cleaning*, 3 (10), 18 (1995).

D - E. P. Lopez, Sandia National Laboratories, “Identifying a Suitable Alternative Cleaner for Mechanical Piece Part Applications,” *Precision Cleaning*, 3 (10), 23 (1995).

E - L. E. Ryan, “Environmental Technology Projects for the Reduction of Air Emissions - Subtask 1: CFC-113 Based Solvent From Solar Cell Defluxing Operations,” *California Manufacturers Association Clean Air Conference*, Long Beach, CA, May 5, 1994.

F - Dr. Manohar S. Grewal, “Switch to Aqueous Technology gives Gillette Edge in Blade Manufacture,” *Precision Cleaning*, 3 (4), 15 (1995).

G - Newsletter, “Pollution Prevention Advisor,” Defense Programs Publication: *Quarterly Newsletter for the Nuclear Weapons Complex*, Spring, 1993, p.3.

H - Telecon with representatives at Chem-Tech International, Inc. phone 1-800-268-6189 on November 6, 1995.

I - Gary Robertson, Aerojet Azusa, “Targeted Chemicals Minimization in Design and Manufacturing,” *AIA Conference Proceedings*, Cincinnati, OH, September 5-8, 1995.

J - Jamie Tiley, WPAFB, “The Elimination of ODS’s Associated with the F-100 and F-110 Engine Families,” *AIA Conference Proceedings*, Cincinnati, OH, September 5-8, 1995.

K - J.E. Deffeyes, McDonnell Douglas Corporation, “UV Light Process for Metal Cleaning,” *AIA Conference Proceedings*, Cincinnati, OH, September 5-8, 1995.

L - Robert Evans, R. G. Hanson Co., Inc., “Vapor Degreaser Replacements for Removing Fluorescent Penetrants from Aerospace Castings,” *AIA Conference Proceedings*, Cincinnati, OH, September 5-8, 1995.

M - Jerry Bauer, Burns & McDonnell Engineering Co., “Aircraft Depainting: Alternatives for Environmental Compliance,” *AIA Conference Proceedings*, Cincinnati, OH, September 5-8, 1995.

N - Dr. John W. (Bill) Agopovich, Draper Laboratory, “Fluorocarbons and Super Critical Carbon Dioxide Serve Niche Needs,” *Precision Cleaning*, 3 (2), 15 (1995).

O - Larry A. Caddy, NASA, Ross Bowman, Thiokol Corp., and Dr. Rex A. Richards, Interdivisional Coordinator for Thiokol, “New Partnerships, Technologies, Approaches,” *Precision Cleaning*, 3 (2), 40 (1995).

P - Dr. Stephen O. Anderson, Deputy Director of the Stratospheric Division of the U.S. Environmental Protection Agency, “Flowing Forward with Aqueous Technologies,” *Precision Cleaning*, 3 (2), 34 (1995).

Q - Glen Singer and George Bryant, Briggs & Stratton, Corp., "Shifting Gears for Solvent Replacement,” Stratton Corp., *Precision Cleaning* , 3 (2), 36 (1995).

R - Industry contact with Mr. Robert Whitcomb, Raytheon Co., Lexington, MA.

S - S. R. Khoshbin, "Freon Replacement Alternative for Defluxing of Space Station PWA's," *California Manufacturers Association Clean Air Conference*, Long Beach, CA, May 5, 1994.

T - J. F. Becker, "Evaluation of Aqueous Cleaning Agents as Alternative to Chlorinated Solvent Degreasing," *California Manufacturers Association Clean Air Conference*, Long Beach, CA, May 5, 1994.

U - Chris Castenada, R. Glenn Robertson and Jane White, "Eliminating Ozone-Depleting Chemicals," *Circuits Assembly*, 4 (6), 34 (1993).

V - David A. Dietz, "Aqueous Cleaning and Water-Soluble Flux Qualification," *Circuits Assembly*, 4 (6), 62 (1993).

W - Edward Sanchez, "A Viable Alternative for the Defense Electronics Industry: Aqueous Cleaning," *Circuits Assembly* 4 (6), 66 (1993).

X - Russell C. Manchester, "Precision Aqueous Cleaning System and Process Design," *Precision Cleaning* , 2 (6), 11 (1994).

Y - Barrie G. Candeas and Daniel Z. Gould, "Implementing a New PCA Cleaning Solvent," *Circuits Assembly* , 5 (6), 38 (1993).

Z - Michael Weidman and Brad Nunn, "A New Cleaning Evaluation Benchmark, *" Circuits Assembly* 5 (6), 30 (1994).

4. EXPERIENCE AT SANDIA NATIONAL LABORATORY (2)

Sandia fabricates many different assemblies using metals that, after cleaning, often are subsequently bonded, painted, or sealed. Examples of the metals they clean are copper, aluminum, molybdenum, Fe-Ni-Co alloy, gold, and Au-Cu-No alloy.

The main criteria used by Sandia to screen and select cleaning alternatives included cleaning efficacy, materials compatibility, and etch rate. Other crucial factors, such as worker safety, recycling, waste disposal issues, and cost, were also addressed.

Sandia's first step was to establish the criteria for cleaner selection. Certainly, the alternate cleaner had to clean as well or better than the existing baseline process which used trichloroethylene (TCE) and isopropyl alcohol (IPA). (That process involved ultrasonic and vapor degreasing in TCE followed by a rinse in IPA, with a final blow dry using filtered nitrogen.) Next, it had to meet various material compatibility requirements, including corrosion, dissolution, and absorption with respect to different mechanical assemblies. This criterion required measuring material attack in terms of weight loss or gain, and change in appearance. It also required measuring the etch rate of each candidate cleaner using an inductively coupled, plasma-atomic emission spectroscope (ICP/AES), an instrument sensitive to 1 ppm.

Sandia selected aqueous alkaline cleaners for this study. Eight of these cleaners were chosen as potential candidates based on previous experience with them or on manufacturer's recommendations. An IPA ultrasonic/vapor degreasing process and an IPA/cyclohexane immersion/vapor degreasing process were also chosen as potential candidates.

Eight machining lubricants were identified as potential contaminants. Cleaning efficacy was determined by visual examination, surface wettability analysis with a contact angle goniometer, analysis of the residual surface composition by Auger electron microscopy and x ray, photoelectron spectroscopy, and measurement of organic contamination using a MESERAN surface analysis technique.

Two cleaners, Brulin 815 GD and Oakite NST, were identified as the best choices for further materials compatibility evaluations, based primarily on the cleaning efficacy results. Both cleaners are of an aqueous alkaline composition with a pH in the 10 to 11 range. They are designed for tank immersion cleaning of aluminum and its alloys. Brulin 815 GD was selected as the best choice for replacing TCE and IPA. Important to Sandia is the fact that Brulin 815 GD is biodegradable and contains no hazardous chemical substances.

5. EXPERIENCE AT ROCKETDYNE (3)

The course chosen by the Rocketdyne Division of Rockwell International in 1993 for solvent substitution in their manufacturing operation is worth reporting. Rocketdyne developed an excellent in-depth plan for identifying cleaning agents to meet admittedly narrow requirements. The data were gathered in a form that facilitates selection of appropriate cleaners for a broad range of future applications. Exact details of Rocketdyne's tests and cleaner selections are not given here because the information would not have wide interest, but their methodology is presented as an excellent example of how one may establish a cleaner evaluation program. (See Figure 1.)

The purpose of Rocketdyne's 1993 test program was to evaluate and identify effective all-purpose aqueous cleaners as replacements for the solvent degreasing agents CFC-113 and TCA in cleaning rocket engine hardware.

Rocket engine components require all levels of cleanliness from visually clean to meeting liquid oxygen system requirements. These components are fabricated from a variety of alloys, such as nickel, cobalt, stainless steel, titanium, aluminum, and low alloy steels. Some of the soils that must be removed are machining and tapping lubricants, hydraulic fluids, drawing oils, adhesive tape residues, and dye penetrant fluids.

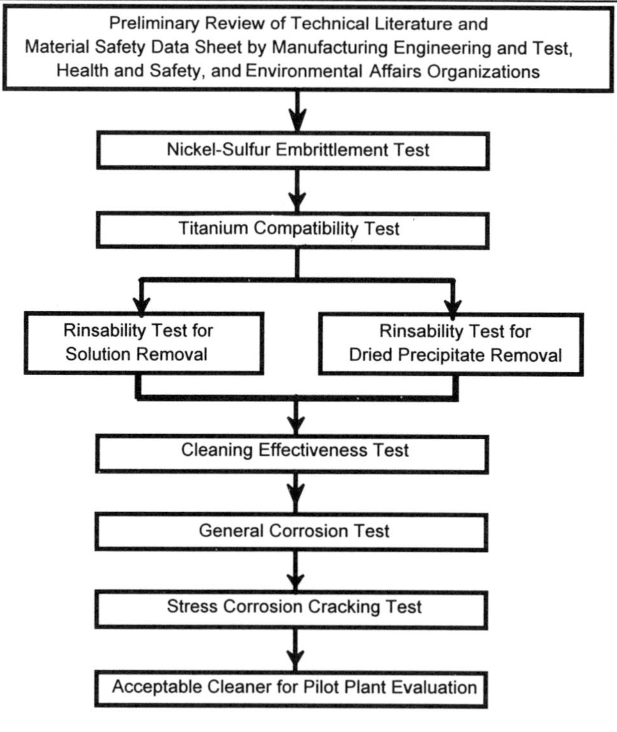

Figure 1. Rocketdyne cleaner evaluation process flow diagram.

The cleaning agents first had to meet a number of "given" requirements from outside agencies. To satisfy Government agency requirements, the cleaners could not contain chromated compounds or sulfides, and had to have a low VOC content. ("No chromated compounds" is a requirement from California State Proposition 65 which restricts certain chemicals known by the State to cause cancer or reproductive toxicity. "No sulfides" is a Los Angeles City limit on the sulfide content in effluent water discharged into the city sewer system; and "low VOC" is a local EPA requirement.)

Next, the cleaner had to be compatible with the hardware alloys in use at Rocketdyne. This required the cleaner to pass corrosion test requirements, including a test for sulfur embrittlement of nickel, a stress corrosion cracking (SCC) test for titanium, and general corrosion and SCC tests for each alloy. Also, because of very low tolerance requirements for dye penetrant inspection, the cleaners could not contain silicates. An exception to the no-silicates requirement was made when a cleaner had sufficient "rinsability" to remove all trace of silicates. This exception points up how variations in process can greatly affect results. Rocketdyne found that rinse formulations containing sufficiently high concentrations of surfactants allowed the silicate residue to be rinsed off easily with ambient or hot deionized water.

Last, the cleaning agent had to be an effective cleaner of Rocketdyne's manufacturing soils; therefore, a cleaning effectiveness test had to be established.

The test requirements were ordered by putting the most important first. This method provided a systematic approach and eliminated unnecessary testing by excluding a cleaner from further testing if it failed early in the sequence.

With all of the above in mind, Rocketdyne solicited and received 200 recommendations from 112 cleaning agent manufacturers. After a preliminary review of the technical literature and Material Safety Data Sheet for each recommendation, 95 were screen tested. Of those tested, 18 cleaners met the evaluation criteria presented in Figure 1.

6. EXPERIENCE AT NORTHROP (4)

A study similar to Rocketdyne's was conducted by Northrop Aircraft Division. Northrop also desired to replace TCA vapor degreasing with aqueous degreasing. Their applications included vapor degreasing of metal fabricated parts prior to specified metal finishing operations such as plating, conversion coating, and painting. After identifying approximately 50 potential aqueous degreasing cleaners, they selected five.

Their screening test program included soil removal, salt spray, etch rate, climbing drum peel, wedge crack extension, and wet tape adhesion. For the soil removal test, test panels were contaminated with shop soils, immersed in the test cleaner, and mechanically agitated for 10 to 15 minutes at the recommended temperature. Panels were then rinsed and checked for cleanliness using the water-break-free technique. Only the Brulin 815 GD (10 percent solution) provided water-break-free results equal to their control methodology [TCA vapor degrease plus a 10-minute immersion in Isoprep 44 (MacDermid Inc.) at 60°C] for the full range of contaminants.

Mr. Calvin Fong, the author of the Northrop study, noted that the Brulin 815 GD succeeded in not only replacing the TCA vapor degreasing step prior to initial FPL etch, but also replaced the initial alkaline cleaning step in the FPL etch process. He said the Brulin 815 GD tank immersion process was eventually used in Northrop production for all-purpose cleaning of metal parts; it consisted of a 15 to 30 minute immersion at 71°C plus two water rinses at 43°C.

A recent comment by Hughes' Mr. Wen Tay Cheng on the Brulin 815 GD cleaning process is that he found it required a better rinsing process. He changed to Turco 4215 NC-LT aqueous cleaner and has continued to use it.

7. EXPERIENCE AT MCDONNELL DOUGLAS AEROSPACE (5)

Mr. Glenn DeBrucq at McDonnell Douglas Aerospace describes a very straightforward, six-step approach for identifying and removing Class I ODCs from all their F-15 support documents. This work was recently completed under a USAF F-15 System Program Office contract. The contract's six steps or tasks consisted of the following:

1. Review support documentation for all consumable materials.
2. Identify Class I ODCs, Class II ODCs, and EPA-17 materials.
3. Identify the applications for the above identified materials (where used.)
4. Identify suitable replacements for the Class I ODCs.
5. Revise support documentation to remove the Class I
6. Put results in database to share with the entire F-15 support community.

The database they developed contains the location (document number and paragraph number) of each consumable material in all F-15 support documents. The application of each hazardous material is listed as well as the targeted ingredients in the product. For Class I ODCs, the substitute is listed along with scheduled and actual document revision data. The customer, Warner Robins Air Logistics Center, refined an electronic Bulletin Board System (BBS) to communicate this information worldwide. This BBS allows maintenance personnel to determine if a suitable substitute has been identified for their specific ODC application. (The BBS originally went on-line in August, 1995.)

The database is not only helping the F-15 support community with ODC elimination, it is positioning the F-15 to eliminate Class II ODCs and EPA-17 materials as the need arises. For example, if MEK is prohibited (as anticipated) for hand cleaning in aerospace manufacturing and maintenance operations, the database can be queried to list all MEK applications and locations in the support documents. As substitutes are identified, knowing which documents need revising eases the task. Also, new chemicals will likely be identified as environmentally unfriendly in the future. The F-15 database will ease identifying those new chemicals because the database lists all consumable materials, not just materials deemed hazardous today.

When he presented his paper, Glenn DuBrucq said that the BBS electronic bulletin board is available to all interested parties, and that you may call Todd Lavender at (912) 926-6634 for information to flow it down to your PC.

8. TECHNOLOGY DEVELOPMENTS AT HUGHES AIRCRAFT

Within the past several years, Hughes has developed and patented three new solvent cleaning technologies and a water soluble soldering flux material. Two of the cleaning technologies take advantage of the unique solvent properties of carbon dioxide under supercritical (SC) conditions, while the other, known as Reactive Aqueous Defluxing System (RADS™), involves aqueous cleaning. The Hughes flux material, known as HF 1189™, is water rinsable, and eliminates the need for CFCs in post-solder cleaning operations.

8.1 Supercritical CO_2 Technology (6) CO_2 is said to be in the supercritical state when its temperature and pressure are close to or above the critical points of 31.3°C and 7.4 Mpa, respectively. Under these conditions, CO_2 possesses solvent characteristics equal to or better than CFCs. It is also inert, ozone friendly, non-combustible, naturally occurring, inexpensive, and it does not contribute to smog.

Hughes Aircraft initiated research into SC CO_2 cleaning processes about 13 years ago to meet the company's internal production needs. In 1990, a project was undertaken to implement a commercial-scale SC CO_2 cleaning system in a manufacturing environment. The system, designed jointly by Hughes and Autoclave Engineers Division of Snap Tite, Inc., was given the name SUPERSCRUB™ Precision Cleaning Equipment. The first unit, installed in November 1991, has successfully cleaned flex cables, cryogenic detector components, precision bearings, hybrid micro-circuits, and navigational gyros. For more information, contact Larry Rosio at Autoclave Engineers, (814) 838-5796.

Many companies and Government agencies have followed the Hughes SC fluid project with great interest. Several of these organizations have joined with Hughes in forming the Joint Association for the Advancement of Supercritical Fluid Technology (JAAST). Other charter members include Boeing, IBM, Los Alamos National Laboratory, and Sandia National Laboratory.

8.2 Hughes Eco-Snow™ CO_2 Jet Spray (7) The jet spray technique called Eco-Snow™ depends on storing liquid CO_2 at ambient temperature and 5.76 Mpa (835 psi) at which point CO_2 becomes a liquid. Upon leaving a jet spray nozzle, the liquid expands very rapidly, causing cooling and the lowering of pressure. At what is called a triple point in the CO_2 phase diagram, the solid, liquid, and gas states exist simultaneously, producing CO_2 snow.

The CO_2 jet spray effectively removes particulates and thin molecular films. Dusty silicon wafers are returned to their original pristine optical performance. Contaminant films of Krytox and pump oil are successfully removed from various substrates. Dust, debris, and carbon soot are completely removed from metal and glass optical components.

Eco-Snow™ was the only method able to completely clean an aircraft optical alignment flight window; and the cleaning time was reduced from 3 days to 1 hour at a cost savings of $1.3 M a year. Eco-Snow™ has also been effective in applications ranging from the delicate cleaning of the UV mirror coating on space telescope optics to the commercial removal of sanding residue on automobile body panels. For more information, contact Marjorie Burtch, Hughes Aircraft Company, (310) 616-7323.

8.3 Hughes Reactive Aqueous Defluxing System (RADS™) (8) (RADS™) is a fully aqueous cleaning system for removing various types of flux from electronics hardware. The feature that distinguishes RADS™ from other aqueous solvents is the unique Ecosolve™ solutions that decompose flux rosin through an oxidation chemical reaction. The system has successfully demonstrated superior performance, as compared to use of ODCs, in removing flux from printed wiring board assemblies at stand-offs as small as 2 mils.

RADS™ consists of four stages: flux removal, neutralization rinse, final rinse, and drying. The system is operated in a fashion similar to existing aqueous cleaning solutions (e.g., saponifiers) currently being marketed. Ecosolve™ is commercially available from Alpha Metals Inc.; contact Chris Bafbastecki at (201) 324-3609.

8.4 Hughes HF 1189 Flux (9) HF 1189 is a halide-free soldering flux comprised of citric acid, deionized water, and a polyglycol-free foaming agent. It has been used on military programs since September 1990 with excellent results. It completely eliminates CFCs, VOCs, vapor degreasers, rosin flux, and their associated waste products. Additionally, users report dramatic improvements in solder joint quality and a major reduction in touch labor from the post-solder cleaning operations.

HF 1189 is qualified to Joint Industry Standard ANSI/J-STD-004, "Requirements for Soldering Fluxes." It is commercially available from Kester Solder Company; contact Kester Technical Services at (800) 2 KESTER.

9. REFERENCES

1. Editor's Readers' Forum, "A Call to Clean Up Inconsistent Cleaning Agent Identifiers," Precision Cleaning, 3 (4), 10 (1995).
2. E. P. Lopez, "Identifying a Suitable Alternative Cleaner for Mechanical Piece Part Applications," Precision Cleaning, 3 (10), 23 (1995).
3. J. F. Becker, "Project XVIII Evaluation of Aqueous Cleaning Agents as Alternative to Chlorinated Solvent Degreasing," Final Report to the South Coast Air Quality Management District, El Monte, CA, October, 1993.
4. Calvin Fong, "Aqueous Degreasing Replacement," Final Report to the South Coast Air Quality Management District, El Monte, CA, March, 1992.
5. Glenn DuBruq, "Identification and Removal of Hazardous Materials from F-15 Support Documents," AIA Conference Proceedings , September 5-8, 1995.
6. John McHardy, et.al., "Progress in Supercritical CO_2 Cleaning," SAMPE Journal, 29 (5), 20 (1993).
7. Anon. "Eliminate CFC Cleaning with Eco-Snow™," Hughes Aircraft publication, copyright 1991, 1993.
8. Richard C. George and Ronald E. Robbins, "A Reactive Defluxing System for Replacement of ODC Processes," Proceedings, Electronic Interconnect Technology, A Technical Symposium, El Segundo, CA, November 11, 1993.
9. Ray L. Turner, "HF 1189, The Environmentally Green Flux," Hughes Aircraft publication, (not dated).

10. BIOGRAPHY

Rodney Patterson is currently head of the Materials Development Team in the Components and Materials Center of Hughes Aircraft Company, El Segundo, California. In his previous position at Hughes as head of Nonmetallics Engineering in Fullerton, California, he led the effort to phase out ozone depleting chemicals (ODCs) at the Fullerton site, meeting a self-imposed deadline of December 31, 1994.

Mr. Patterson has a BA in Chemistry from Pacific Lutheran University and a BS in Chemical Engineering from the University of Washington. He has been teaching various courses in materials engineering at California State University, Fullerton since 1982.

Philip Mykytiuk has been employed by the United States Air Force for 19 years and has held several engineering positions at the Aeronautical Systems Center, Wright-Patterson Air Force Base, Ohio. He is a member of the Environmental Technology Group within the Materials Directorate, USAF Wright Laboratory where he has been a primary architect of the Pollution Prevention Program. He is also a program manager in the Materials Directorate, involved in planning and guidance of in-house work preliminary to the establishment of contractual and R&D efforts.

Mr. Mykytiuk has a BS in chemistry and a BS in engineering management from the University of Massachusetts, an MS in polymer chemistry from Wright State University, Dayton, and an MS in materials engineering from the University of Dayton.

28th International SAMPE Technical Conference
November 4-7, 1996

DRY ICE - UV LIGHT PROCESS FOR METAL CLEANING

J.E. Deffeyes, H.V. Lilenfeld, and J.J. Reilly, McDonnell Douglas Aerospace
P.D. Mykytiuk, Air Force Pollution Prevention Office, Wright Laboratory

ABSTRACT

In order to reduce pollution generated by metal cleaning operations, a water-free, solvent-free cleaning process is being developed. The approach being investigated utilizes pellets of dry ice and ultraviolet light. The cleaning process is intended for use prior to metal finishing operations such as ion-vapor-deposited (IVD) aluminum, conversion coating or anodize. Lab testing is on-going, and a limited shop trial may occur as soon as 1997.

SOURCE REDUCTION IN CLEANING OPERATIONS

Pollution prevention is most effective when source reduction can be employed to eliminate a waste stream, rather than using control technology after the waste stream has been created. Incorporating this philosophy into cleaning technology leads one to the conclusion that a process which does not involve water or other solvents is highly desirable. A cleaning process which does not use organic solvents has the advantage that it does not produce harmful air emissions. The aerospace industry, along with many other industries, has been forced to change it's cleaning processes due to tighter environmental regulations, especially those regarding stratospheric ozone depleting substances. However, the switch to new solvents which have less potential to deplete the ozone layer brings with it concerns about the green house warming potential of the new solvents. Thus a cleaning process which does not use organic solvents at all is desirable.

For these reasons, aqueous cleaners have become increasingly popular in applications where corrosion is not a significant concern. However, the aqueous cleaner solution eventually gets loaded with contamination and must be rejuvenated or replaced. Depending on the detergent package used, the types of contamination being cleaned, and local environmental regulations, the solution must either be treated before discharge to the sewer or else hauled out as hazardous waste. Thus a non-aqueous cleaning process is also desirable from an environmental standpoint.

Water is a precious natural resource, and concern about its short supply is increasing. Proposed regulations, such as EPA's proposed Metal's Processing and Machining rule, would severely limit the total hydrocarbon dischargable to a public water treatment plant, making treatment of water discharge more expensive. "Polluter Pays" type legislation imposes a tax or fee on discharges of aqueous oils and greases, such as a clean water act reauthorization bill which was introduced in 1993 (HR 2199)[1]. Thus we anticipate even more stringent requirements in the future because of strong public support for environmental regulations, although how soon that future arrives depends on the whims of the political system.

The solid state cleaning process is conceptually quite different from other cleaning technologies. Here "solid state cleaning" refers to a cleaning process which does not use organic solvents or aqueous solutions to dilute surface contamination. Traditionally, cleaning technologies work on the principle of dilution of the contamination from the part to be cleaned, which is accomplished at the expense of contamination of a solvent or other dilutent. The dilutent may be an organic solvent, or it may be an aqueous detergent solution. Once it becomes excessively contaminated, the dilution media must then either be disposed of or treated to separate out the contamination. In contrast, solid state cleaning processes do not generate a contaminated liquid cleaner. Thus no further processing of a contaminated cleaning solution is required.

CLEAN CLEANING TECHNOLOGIES AND THE BOTTOM LINE

As discussed in a recent review of cleaners for aerospace manufacturing operations[2], there are a wide range of technologies available for cleaning operations. This wide range of technologies exist because there is no one drop-in replacement for all the ozone depleting substances, but rather a number of different niches to be filled. Several of the technologies listed produce no waste except for the actual contamination removed. For example, for critical cleaning of the interior surface of tubing, atomic oxygen cleaning is a promising technology. For light duty cleaning of delicate surfaces, carbon dioxide snow can be used. Supercritical carbon dioxide cleaning will work well for cleaning organics out of very intricate shapes or assemblies, provided that the structure can survive the very high pressure required. For less complex shapes, such as most structural aerospace parts (which comprise a large percent of the surface area we need to clean in our shop), we are developing a two-step cleaning process which we believe can be scaled up to clean large parts in a cost efficient manner. While the new process is estimated to be about 2.5 times more expensive per-square-foot than our current aqueous cleaning process (with sewer discharge of the spent cleaner after minimal treatment), a lowering of the limit on dischargable oils and greases and/or a tax on discharges of oils and greases to public water treatment facilities has the potential to remove that cost differential.

---

[1] R. Dhonau, "Clean Water Act Reauthorization: Round Two," *Metal Finishing*, Vol. 91, #11, P.35-39, November 1993.

[2] R.F. Patterson and P.D. Mykytiuk, "Solid State, Vapor Phase, and Aqueous Cleaners for Aerospace Manufacturing Operations," SAMPE Journal, May-June 1996.

OVERVIEW OF THE DRY ICE/UV LIGHT PROCESS

The process we are developing removes contamination on the surface in two steps. Most of the dirt, oils and greases on a surface are first physically removed with a dry ice pellet blast pre-clean step. Then the small amount of organic contamination which can not be physically removed is oxidized at low temperature by the ultraviolet light process to produce small amounts of gaseous carbon dioxide (CO_2) and water, which are normal constituents of air. An artist's conception of how the process might be implemented in a production facility is shown in Figure 1. A higher degree of automation would be necessary to reduce the cost of the process because labor costs to manually move the parts around as shown would be high. An artist's conception of this type of system is shown in Figure 2.

An pre-cleaning step is necessary because the UV light process can only remove thin layers of contamination. Thick layers of organic contamination tend to crosslink under ultraviolet irradiation rather than oxidize. A dry ice pellet pre-cleaning process is used to reduce the thickness of the contamination layer prior to UV-activated oxygen exposure. After cleaning, the dry ice sublimes. Thus no further processing is required for separation, an advantage over plastic media or wheat starch type blasting technologies. The dry ice blasting removes the bulk of the contamination. However, it leaves enough organic contamination on the surface to fail a water-break-free test. The UV light processing removes the last trace of contamination and leaves a very clean, water-break-free surface. (The water-break-free test involves holding the sample under running de-ionized water to wet it. The surface is then examined to see how the water flows off of the surface. A water "break" at some point - i.e., where the surface becomes de-wetted - is taken as evidence of contamination which is hydrophobic in nature. Thus a "water-break-free" surface is reasonable clean from hydrophobic contamination.)

The ultimate test of any practical metal part cleaning technology is whether the metal part is suitably cleaned for the next step in the process. Much of the cleaning of parts in the aerospace industry is in preparation for the application of a corrosion protective coating. Cleanliness is critical because residual contamination on the part will interfere with the application of the corrosion resistant coating, leaving the part vulnerable to corrosion and thus potentially reducing the service life of the part. A number of corrosion protection coatings are used in the aerospace industry, but typically they consist of an inorganic barrier layer/adhesion promoter covered by a primer/topcoat paint system. The critical cleaning step typically occurs immediately before the application of the inorganic layer because it is at this step that much of the machining oils, forming lubricants, etc. deposited during manufacture of the part must be removed. We have used the application of inorganic coatings to determine whether or not the solid state cleaning process is effective. Three standard inorganic coatings which have been used in this study include anodize, conversion coating, and IVD aluminum.

Anodizing is the electrochemical oxidation of the outermost surface layer of an aluminum part. The part is immersed in an acid solution and a voltage applied with the part serving as the anode. The oxidation which occurs at the anode surface deposits a layer of aluminum oxide on the surface of the part. This oxide layer is then "sealed" by immersion of the part into a hot water bath, possibly containing chromium compounds to enhance the corrosion resistance of the coating. The sealing process chemically hydrolyzes some of the oxide and thereby closes up much of the porosity which exists on the anodize surface. The application of the anodize coating requires that the surface be reasonably clean, although the acid in the anodize bath can clean small amounts of contamination.

Conversion coating is applied by immersion of the part in an aqueous solution containing, among other things, hexavalent chromium ions. The chromium deposits on the surface of the part as a very thin, partially hydrated oxide layer. Again, contamination on the part can interfere with the surface chemistry necessary to deposit the coating.

An inorganic coating which can be used as an environmentally friendly replacement for cadmium plating is IVD aluminum, commercialized for use in the aerospace industry by McDonnell Douglas. Aerospace aluminum alloys contain high amounts of copper, magnesium and other alloying elements to improve the mechanical properties of the material. However, the alloying elements also degrade these materials' corrosion resistance. The IVD process coats the parts with a layer of pure aluminum which is much more corrosion resistant than the high-strength alloys. IVD aluminum is also a good choice for a corrosion protective coating on alloy steel. The aluminum coating is applied in a large vacuum system and thus there is no aqueous waste stream generated by the process. The over-spray, high purity aluminum, is contained by the vacuum system and thus has minimal environmental impact. Thus the IVD aluminum process is a good match with the solid state cleaning for maximum reduction in water pollution. However, adhesion of IVD aluminum is quite sensitive to surface contamination.

DETAILS OF THE CLEANING PROCESS

Dry Ice Pellet Pre-Cleaning System The equipment used for the dry ice pre-clean was manufactured by Cold Jet, Inc. in Cincinnati, and has been modified to operate robotically. The process uses dry ice pellets, and thus is much better able to remove tough soils than carbon dioxide "snow" processes. Dry ice pellets are driven through a nozzle at the surface to be cleaned by compressed air. Dry ice pre-clean operating parameters have been initially established as follows:

stand off distance (nozzle tip to surface):	4 inches
dry ice use rate:	400 lbs/hr
nozzle inlet pressure:	175-180 psi
travel rate of nozzle:	0.8 inches/minute
nozzle angle (w.r.t. panel surface):	about 45°
number of passes:	4 (with high-density pellets)
	6 (with low-density pellets)

These parameters have been found to clean well, but we have not yet worked on optimizing this part of the process. Other parameters or equipment may work just as well. The dry ice pre-clean may also be done manually, as shown in Figure 4. However, this is a difficult process to do by hand, and shop implementation would be greatly benefited by using a robotics system for the dry ice pre-clean.

Ultraviolet Light Cleaning Equipment The ultraviolet cleaning system in which we performed our initial testing to-date is manufactured by Ultraviolet Ozone Cleaning Systems (UVOCS). It is a commercial device based upon the Vig. et. al.[2] cleaning methodology. The device consists of a 10" by 10" serpentine, low pressure quartz-mercury vapor lamp mounted on the ceiling of an interlocked enclosure with a 0 to 99.0 minute timer. Samples are set on a tray which is mounted about 2" below the lamp.

More recently, we have modified a commercial oven structure by mounting UVOCS manufactured UV lights on the top and sides of the interior oven structure. The equipment is shown in Figure 3. Even with all these extra lights, we find that the cleaning time necessary to obtain a water-break-free surface is the same as in the smaller cleaning system. However, larger parts can be cleaned in the larger cleaning cell. Additionally, since lights are mounted on all sides, parts with complex geometries can be cleaned. Parts have to be turned once in order to clean those areas shaded by the shelf on which the part rests during cleaning. Very intricate geometries and assemblies will likely not be able to be cleaned by this technique, but we anticipate that most of the large structural parts manufactured in the production of aircraft will be able to be cleaned.

Ultraviolet Light Cleaning - How it Works We should start this section by stating that the cleaning mechanism is not well understood at this point in time. However, by reading the literature in the area of atmospheric photochemical reactions (e.g. Seinfeld[3]), certain reaction mechanisms can be postulated. The low pressure mercury lamp discharge lamp in use in the cleaning process emits UV light strongly at wavelengths of about 254 nm and 185 nm. The energy and intensity of the light emitted by the UV lamp is a function of the light generated by the mercury discharge, and is also a function of the optical transmission properties material used to make the lamp.

Ozone, O_3, is generated by the photodissociation of O_2 near 185 nm by the reaction

$$O_2 + h\upsilon(185\text{ nm}) \rightarrow 2\text{ O (atomic)}$$

followed by the three-body reaction,

$$\text{O (atomic)} + O_2 + M \rightarrow O_3 + M$$

(where M can be any molecule present in the air). At normal atmospheric pressure, the steady-state concentration of O_3 is much, much larger than the concentration of atomic oxygen. Hydroxyl radical may also form under these conditions by reaction of ozone and/or atomic oxygen with water vapor. Future plans include testing water as a chemical accelerator for the process. The water may be introduced either as humidity in the air or as a mist applied to the surface.

Thus one role of the 185 nm light in the cleaning process is to create ozone molecules from diatomic oxygen. However the 254 nm UV light would be expected to interact more efficiently with a wide variety of organic molecules. The following list shows absorption data in the UV range for several types of organics[4].

---

[3] J. H. Seinfeld, Atmospheric Chemistry and Physics of Air Pollution, John Wiley & Sons, Inc.

[4] F.A. Carey and R.J. Sundberg, Advanced Organic Chemistry, Part A Structure and Mechanisms, Plenum Press, New York, 1977, p.466.

Table 1. Absorption Maxima of Various Organics

Type of organic	Absorption Maxima (nm)
simple alkanes	190-200
alicyclic dienes	220-250
cyclic dienes	250-270
styrenes	270-300
saturated ketones	270-280
α,β - unsaturated ketones	310-330
aromatic ketones and aldehydes	280-300
aromatic compounds	250-280

The hydrocarbon molecule on the surface absorbs a photon to create a short-lived electronically excited state. The excited state might decompose, it might polymerize with other surface organics, or it might demonstrate enhanced oxidation reactivity if oxidizing species are present. Additionally the excited state may simple relax back to the ground state if no reactants are present and if no energetically favorable dissociation mechanism exists. The 254 nm UV light has been shown[5] to exhibit some cleaning action itself, but the combination of UV light with ozone present greatly enhances the cleaning effectiveness of the process. Vig, et. al. observed an enhancement in cleaning activity when UV light and ozone are used together which would not have been predicted from the cleaning rates of either UV light or ozone alone.

While the gas phase concentration of atomic oxygen is negligible, most (if not all) of the oxidation processes occur while the organic is attached to the surface. Dissociation of ozone on the surface could lead to chemically significant concentrations of adsorbed atomic oxygen on the surface. Reaction of this oxygen with surface hydrocarbon may be an important mechanistic pathway in the cleaning process. This the surface itself might be acting as a catalyst for the cleaning reaction, as it allows adsorbed oxygen and hydrocarbon to come in contact with each other. This mechanism is one explanation for why thick layers of hydrocarbon can not be cleaned by ultraviolet light. Exposed metal surface sites may be necessary to dissociatively adsorb the ozone and generate atomic oxygen. Additionally, the 254 nm light may be enhancing the surface dissociation of ozone, in addition to (or instead of) enhancing the reactivity of the hydrocarbon.

We have found that introducing excess ozone from a corona discharge system during the cleaning process actually reduces the cleaning effectiveness of the process. Presumably the drop in effectiveness was due to absorption of the 254 nm light by the excess ozone, as this wavelength happens to be very near the absorption maximum for ozone. Thus there appears to be a delicate balance in the process optimization between the amount of ozone and the intensity of the 254 nm light. Some ozone enhances the process but too much is detrimental. However, because air was used as the feed gas for the corona discharge rather than an oxygen/inert gas mixture, we cannot rule out that the NOx generated in addition to the ozone may have been contributing to the lower cleaning efficiency.

---

[5] J.R. Vig, "UV/Ozone Cleaning of Surfaces," Research and Development Technical Report SLCET-TR-86-8, AD-A169.024. U.S. Army Laboratory Command, May 1986

OZONE POLLUTION

One issue which should be discussed is the environmental impact of production of ground-level ozone by this process. Ozone is present naturally at ground level at about 0.02 to 0.08 ppm due to mixing of ground-level air with air from the stratosphere. However, one important component of ground level air pollution is increased levels of ozone. The reason ground-level ozone is a problem is because it is created as a component of photochemical smog. Photochemical smog is a mixture of nitrogen oxides, hydrocarbons, and their photochemical reaction products including ozone. Ozone is so reactive that it is removed from the air naturally by reaction with other gas-phase species and by reactions at surfaces. The ground-level ozone problem is a result of NOx and hydrocarbon emissions which react photochemically to create ozone even faster than the ozone decomposes. Due to the chain reactions involved, one organic molecule can be responsible for the creation of many ozone molecules[4].

Thus from an environmental perspective, the release of hydrocarbons from a cleaning operation is more detrimental to the ground-level ozone problem than the release of ozone itself. However, it is still desirable to reduce ozone emissions from the process because ozone is quite toxic, even if its lifetime in the air is relatively short. Because ozone is so reactive on surfaces, it can be easily removed from the effluent of the process by a high-surface-area device such as an activated carbon filter.

The main concern with the ozone process is worker exposure, especially where the ozone is most concentrated, i.e., near the box. Thus proper air-handling equipment must be in place to use this process. The equipment enclosure serves the dual purpose of containing the ozone as well as preventing worker exposure to the UV light.

RESULTS

We have successfully cleaned a wide variety of contaminants from surfaces using the UV process. The contaminants were chosen because of their extensive sue in our shop (e.g., coolant for machining processes) or because other cleaning studies have found them particularly difficult to remove (e.g., Castrol 140 stick wax). The following have all been tested and successfully cleaned. "Successful cleaning" means that we were able to achieve a water-break-free surface, excellent adhesion of IVD aluminum coating, and demonstrate good corrosion resistance (per ASTM B117 neutral salt spray test) of conversion coat, sulfuric acid anodize and chromic acid anodize coatings, applied after cleaning with the UV process. We have cleaned them both in the "as applied" state, and also in a simulated aging state. The contamination was applied to a test panel, and then baked in an oven at 120°F for two weeks to simulate a very long aging time. The aged specimens were only tested for a water-break-free surface.

Table 2 - Contaminants Successfully Cleaned with the Dry Ice - UV Light Process

Deburring Wax (Castrol stick wax #140)
Al Tube Forming Wax (Aldraw J-2)
Machining Coolant (MMS 611 Coolant)
Preservation Oil (VV-L-800)
Liquid Penetrant (Zyglo Penetrant/Developer)
Simulated Dirt (ASTM standard)
Grease (Mobile Grease #28, MIL-G-81322)
Lubricant/Preservative Oil (Cosmoline 1116)
Corrosion Preventative Compound (MIL-C-16173, Grade 2)
Forming Oil (Mobile DTE 25)

One contaminant which we have tested does not clean entirely, but leaves a visible residue. The contaminant is Dykem Blue toolmakers ink, a cellulose base with methyl violet, and is a strong blue color. Dykem Blue is not as strongly hydrophobic as the other greasy contaminants in our test matrix, but it will still cause a failure in the water-break-free test. The residue remaining after exposure to UV light is a chemically converted material. It is no longer blue, but rather black after a 15 minute exposure, and turns to a grayish color after longer exposure. The residue is hydrophilic, as indicated by a pass in the water-break-free test. It does not allow good paint adhesion on conversion coat, but does not seem to interfere with the corrosion protection of either conversion coat or anodize, as indicated by one week standard neutral salt spray testing (ASTM B117). A standard grit blast used as a pre-treatment in the IVD processing removes this residue, so the residue does not interfere with IVD aluminum adhesion.

SUMMARY

An environmentally friendly process for metal cleaning has been successfully demonstrated. The process involves a pre-clean step using dry ice pellet blasting, followed by a final cleaning by exposure to ultraviolet light. We were able to successfully clean a number of different contaminants which are typically found in our shop. "Successful cleaning" means that we were able to achieve a water-break-free surface, excellent adhesion of IVD aluminum coating, and demonstrate good corrosion resistance of conversion coat, sulfuric acid anodize and chromic acid anodize. We are currently working to test the robustness of the process, characterize the degree of geometric complexity of a part which can be cleaned with the process, and to improve the affordability of the process.

ACKNOWLEDGMENTS

This work is funded by the Strategic Environmental Research and Development Program (SERDP) throughout the Wright Laboratories Materials Directorate, prime contract #F33615-92-D-5812, administered by Lawrence Associates, Inc., subcontract #LAI-94-5812-66-10.

Figure 1 - Artist's conception of shop implementation

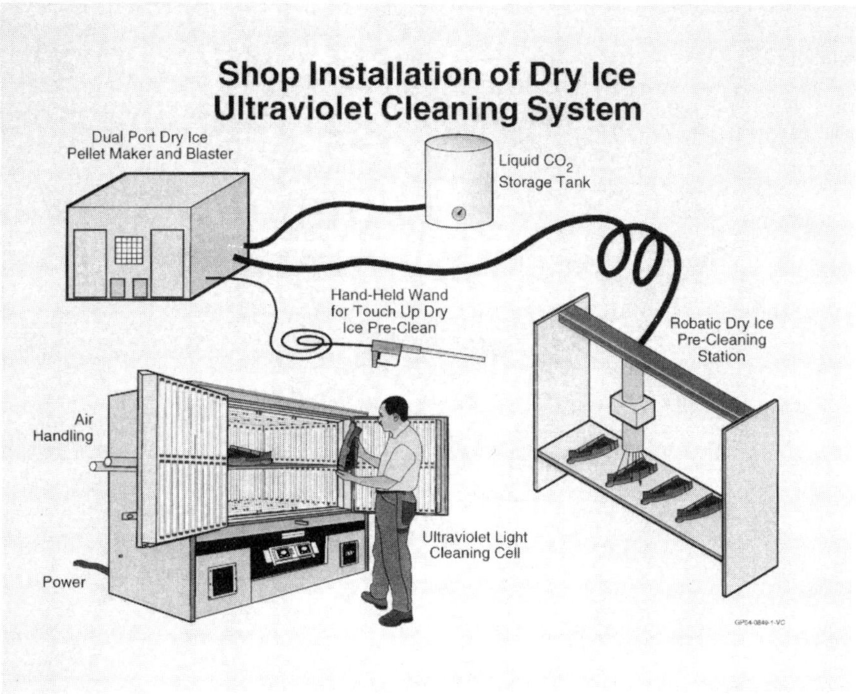

Figure 2 - Artist's conception of a highly automated process

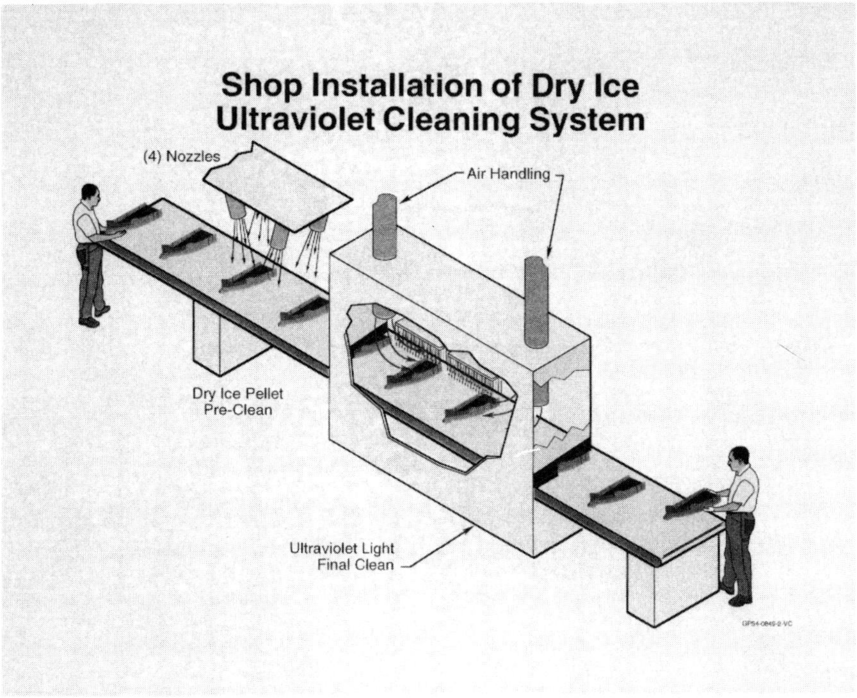

Figure 3 - McDonnell Douglas Technician works on installation of the electronics in the ultraviolet light cleaning cell

Figure 4 - The dry ice pre-clean can be done robotically, or manually as shown.

28th International SAMPE Technical Conference
November 4-7, 1996

CLEANING PERFORMANCE AND NEW TECHNOLOGIES BASED ON VOLATILE METHYL SILOXANES.

Stephen Swanson, Ray A. Cull, Donovan Bryant, John Moore
Dow Corning Corporation
Midland, MI 48686

ABSTRACT

Linear volatile Methyl siloxanes (VMS) are the #1 preferred cleaning materials for substrate cleaning, precision cleaning, and electronics cleaning by the California South Coast Air Quality Management District. This new class of solvents are VOC exempt, non ozone depleting, SNAP approved, are not hazardous air pollutants, and have negligible global warming potential due to their less than 30 day atmospheric lifetimes. They have negligible odor, very low viscosity, low surface tension, good toxicology properties, and 100 % evaporation for effective cleaning, drying, and rinsing. Ultra high purity versions of these products have been commercialized that have very low nonvolatile residue, making them suitable for operations involving secondary bonding and painting. Straight VMS will clean non polar contaminates including silicone residue, greases, and oils. These materials will also soften and lift cured silicones, allowing easy removal, as well as removing any remaining residue. This allows bonding of organic adhesives and paints to surfaces previously coated with silicone. New azeotrope formulations have been developed which enable the cleaning of rosin flux from circuit boards, and liquid crystal display residue, and optics cleaning.

KEY WORDS: Solvents, precision cleaning, volatile methyl siloxane

1. INTRODUCTION

Many materials and process engineers are scurrying to find alternatives to replace the solvents and cleaners being banned by changing government regulations. No direct drop in alternatives have been found for previous industry-standard chlorinated and fluorinated solvents. CFCs, which are being phased out in accordance with the Montreal Protocol, were as close to ideal solvents as possible. They were inexpensive, non flammable, non toxic, and effective cleaners. There are no replacements that offer all these advantages.

Since all of the solvent alternatives have drawbacks or limitations, it is important for those looking for new cleaning procedures and materials to be familiar with many of the alternatives, so they can pick the best for their application. This paper will highlight the pros and cons of a new class of solvents that can be used as cleaners and carriers based on ultra pure low molecular weight linear volatile methyl siloxanes (VMS). Table 1 shows some of the pros and cons of linear VMS.

Table 1
PROS AND CONS OF LINEAR VMS

Pros	Cons
non ozone depleting	Low-moderate flashpoint
non VOC, not a HAP	moderate cost
negligible global warmer	mild solvency
good toxicological profile	
very low odor	
readily available at moderate cost	
fast evaporation without residue	
tailored evaporation rate	
compatible with most materials	
mild solvency which can be enhanced with azeotropes	
#1 preferred cleaner by the California South Coast Air Quality Management District	

Key definitions for terms used in this paper:

Silicone polymers: high molecular weight polydimethyl siloxane (PDMS) polymers used as mold release agents, lubricants, greases, and sealants. These materials can leave a slippery residue that is difficult to remove, affecting painting and adhesion.

Volatile methyl siloxane (VMS) - a class of materials based on very volatile, low molecular weight polydimethyl siloxane. Approved by the United States of America Environmental Protection Agency (EPA) as substitute for ozone depleting substances under a program known as SNAP (Significant New Alternative Policy), also granted VOC exception status from the EPA.

OS Fluids: Product name for a highly distilled, ultra pure form of linear VMS and azeotropes manufactured by Dow Corning. These represent the simplest linear permethylated siloxane molecules, including the disiloxane, trisiloxane, tetrasiloxane, and an azeotrope of disiloxane and glycol ether. Dow Corning is marketing these materials as OS-10, OS-20, OS-30, and OS-120 respectively. Each of these products is distilled to 99.9% minimum of a single component. Products are tested and certified to <1 ppm silicone residue for every lot. OS-10, OS-120, and OS-20 are flammable. OS-30 has a flash point in the combustible range. Table 2 shows the structure and nomenclature for commercial linear VMS products.

Table 2
LINEAR VMS STRUCTURES AND NOMENCLATURE

Structure	Chemical Name	Dow Corning Trade Name
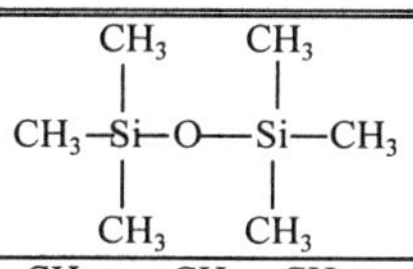	disiloxane, hexamethyl	OS-10
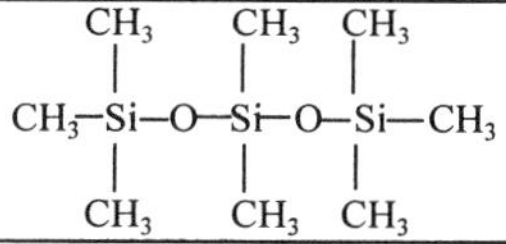	trisiloxane, octamethyl	OS-20
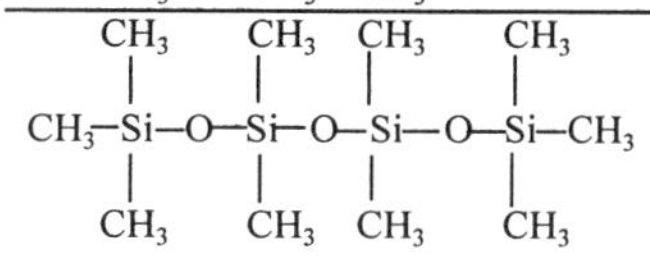	tetrasiloxane, decamethyl	OS-30

2. WHAT ABOUT SILICONE CONTAMINATION?

Most people are familiar with high molecular weight silicone polymers, including greases, lubricants, mold release agents, waxes, sealants, adhesives, rubbers and coatings. Since VMS materials dissolve high molecular weight silicones, they can dissolve these contaminants during distillation, storage, and packaging. When the VMS evaporates, the dissolved oils can be left behind as a contaminant residue. It is imperative that special test procedures, and handling precautions be established to assure the silicones remain pure during distillation, storage, and packaging. If these precautions are not taken, then a residue ranging from several hundreds to thousands of parts per million can be detected. These levels will cause the fish eyes in paints and bond line failures in adhesives that many are familiar with.

To ensure that the OS Fluids are free of PDMS or other residues, each lot, after distillation, is tested for NVR, and must meet a specification of <1 ppm before being approved for shipment. Various surface analysis and performance tests have been conducted to assure the OS fluids do not affect adhesion or leave a residue. Sensitive surface analysis techniques, including ESCA and contact angle, were used to examine metal surfaces after the OS Fluids had been allowed to evaporate from them. No evidence of silicone residue was found in either experiment, as compared to untreated controls. In a second study, paintability was checked after steel and aluminum panels had been cleaned with OS fluids. Standard coating rating tests, including appearance and cross-hatch adhesion (ASTM D 3359 - 78 B), were made and found to be at least as good as panels cleaned with toluene.

A quantifiable test was done to assure that the Dow Corning OS fluids do not leave any residue that may affect adhesive bonding operations. Clean metal panels were dipped in OS fluids and allowed to air dry. Lap shear specimens were then made from the OS exposed panels with 3-M 2216 gray epoxy, 3 M 8101 Urethane, and Bondini "Super Glue" adhesives. As demonstrated in Table 3, in all cases the panels exposed to OS fluids had as good or better adhesion than the clean control panels, which were wiped with isopropyl alcohol. This shows that not only do ultra pure versions of linear VMS not hurt adhesion, but they actual prepare the surface for improved adhesion.

Table 3

LAP SHEAR ADHESION TESTING WITH OS FLUIDS

Sample	3-M 2216 Epoxy	3-M 8101 Urethane	Bondini2 Super Glue
Control	1213 PSI	344 PSI	475 PSI ± 50
OS-10	1233 PSI ± 20	494 PSI ± 20	581 PSI
OS-20	1440 PSI ± 50	383 PSI	640 PSI
OS-30	1478 PSI ± 100		507 PSI

Lap shears are based upon soaking the substrate in the OS Fluids, then allowing them to air dry

The control samples were cleaned with IPA and wiped

3. CLEANING CAPABILITIES

With the exception of their flash points, the physical properties of VMS are very favorable for precision cleaning. Like CFCs, surface tensions and viscosity's are very low, providing rapid wetting, even into tiny crevices and complex geometry. These properties also lead to rapid draining and drying, especially for disiloxane.

VMS solvents are not aggressive, with low Kauri-Butanol and solubility parameter values. Their solubility characteristics match non-polar materials best, such as mineral and silicone oils. Their relative mildness results in very few materials compatibility issues, as is discussed in more detail elsewhere in this paper. Figure 1 shows a comparison of linear VMS and a VMS azeotrope with CFC 113 on key properties required in precision cleaning.

Figure 1

PHYSICAL PROPERTIES (25C) FAVORABLE FOR PRECISION CLEANING

	Disiloxane (OS-10)	(OS-120)	Trisiloxane (OS-20)	Tetrasiloxane (OS-30)	CFC-113
Viscosity (cp)	0.49	0.49	0.82	1.3	0.68
Surface Tension (dyne/cm)	15.2	16.3	16.5	17.3	17.3
Density (g/ml)	0.76	0.77	0.82	0.85	1.57
Solubility Parameter $(MPa)^{1/2}$	12.7	13.2	12.1	12.1	14.7
Kauri-Butanol Value	16.6	--	15.1	13.4	31

Given the mild nature of VMS in terms of compatibility, we find that they are not aggressive solvents. However, for non-polar soils such as light oils, and of course silicones, VMS are very effective. Simple coupon studies, analyzed gravimetrically, show complete removal of light oils. These trials were done by gently shaking at room temperature for three minutes. For more viscous soils, such as thick fluids and greases, linear VMS are still a good match in terms of solubility, but either more vigorous agitation, increased temperature, or longer cleaning times would be needed over the conditions used here to completely remove the soils. Figure 2 shows the cleaning performance of linear VMS for various contaminants.

Figure 2

CLEANING PERFORMANCE			
Percent Soil Removed from Steel Coupons	OS-10	OS-20	OS-30
Motor Oil	99	100	99
Quenching Oil	100	100	100
Cutting Oil	100	100	100
Gyroscope Fill Fluids*	100	100	100
200® Fluid, 350 cst	100	100	100
200® Fluid, 100,000 cst	94	88	89
Grease HD #2	84	75	76
DC 550 Fluid	99	99	99
DC 710 Fluid	97	96	100
Uncured Silicone Conformal Coating	100	99	99

*Results from 4 fluids used at Newark AFB

Under funding from the US Air Force Defense Construction Supply Center, Bob Kaiser from Entropic Systems has evaluated the performance of VMS as particle removers, in his Entroclean™ ultrasonic cleaning system in which parts are cleaned and the fluid is monitored continuously for particle concentration. These studies showed that trisiloxane and tetrasiloxane were more effective than disiloxane for displacing particles, and were better than all other candidates he tried (including CFC-113, HFC-43-10, PF5070 and IPA), with the exception of a mixture of fluorocarbon surfactant in PFC, which removed the most particles (Reference 1). See Table 4 for the summary of the particle removal data.

Table 4
DOW CORNING® OS FLUIDS
PARTICLE REMOVAL DATA

Summary of Data for Stainless Steel Nuts & Bolts Cleaning Tests
(Particles greater than 2um removed in first 15 minutes ultrasonic cleaning)

Cleaning Solution	Ultrasonic Power (watts)	Liquid Flow Rate (ml/min)	Particles Removed
CFC-113	15	300	160,000
IPA	15	300	140,000
HPC-43-10 mee	15	300	180,000
PF 5070	240	300	300,000
PF 5070 + surf.	60	300	335,000
		900	435,000
OS-10	60	30	250,000
	240	30	350,000
OS-20	60	30	360,000
OS-30	60	30	400,000

Test pieces: 2 each 1/4 x 20 x 1/2 inch long nut & bolt

An azeotrope of OS-10 and glycol ether (OS-120) has been commercialized because polar contaminates, such as rosin flux are not effectively cleaned with pure VMS. Work is underway to develop additional azeotropic blends of OS fluids with organic cosolvents for use in cleaning polar contaminates.

OS-120 has been shown to effectively remove rosin flux and some cured silicone conformal coatings from circuit boards without harming the board or its components. By immersing a circuit board coated with Dow Corning(R) 3-1753 Conformal Coating into OS-120, the coating will begin to lift from the board in less than a minute. After 10 minutes to one hour, the coating will be loose enough over the entire board to be blown clean with an air hose.

3.1 Materials Compatibility

As stated previously, VMS solvents are relatively mild, and thus are safe for use with most common materials of construction. We have conducted compatibility testing using an approach similar to the protocol described by the National Center of Manufacturing Sciences, who have created a computerized data base of compatibility data after testing of seven ozone-depleting chemical alternatives. Incidentally, this group plans testing of six additional solvents, including Dow Corning's OS-30.

The Dow Corning program looked at 16 plastics and 10 elastomers. The procedure is to suspend 1" x 2" specimens in the solvent for three hours at 50°C. For disiloxane, exposures were also done at reflux (100°C). Weight, volume, and appearance changes were negligible for each of the plastics listed in Figure 3. In the case of acrylic and polycarbonate, which are susceptible to crazing, no change in optical properties was observed after immersion in linear VMS, using a colorimeter. The compatibility with acrylic and polycarbonate, even at 100°C, is especially noteworthy, since these materials are quite susceptible to this problem with many solvents.

Figure 3

MATERIALS COMPATIBILITY

- Plastics linear VMS are compatible with:

ABS	Nylon	Polyethylene
Acetal	Polytetrafluoroethylene	Polyethylene Terephthalate
Acrylic	Polyvinylidenefluoride	Polypropylene
Chlorinated PVC	Polycarbonate	Polystyrene
Epoxy	Polyetherimide	Polysulphone
Ionomer	Polyvinylchloride	

Figure 4 shows the elastomers tested where durometer hardness plus weight and volume changes were measured. The Vitons and Hypalon are essentially unaffected, even in refluxing disiloxane. Those listed in the second category are not swelled by VMS solvents, but we find these solvents are effective in removing extractables, such as plasticizers and other additives. Hardness values are typically changed a few percent in these cases.

VMS solvents will swell neoprene rubber and natural rubber. Once the solvent has been dried from the elastomer, its properties return to normal except that some non-crosslinked species are apparently extracted. Due to the large degree of swell, these two materials should not be exposed to linear VMS unless such swelling is shown to be of no consequence in the application.

Figure 4

MATERIALS COMPATIBILITY (continued)

- Elastomers - Immersion on linear VMS at 50°C produced following:

Little or No (<2%)	Fluoroelastomers (Viton A & Chlorosulfonated Polyethylene)
Slight Change: (2-5%)	Polydimethylsiloxane Fluorosilicone Ethylene-Propylene Terpolymer Isobutylene-Isoprene Acrylonitrile-Butadiene (Buna)
Not (>5%)	Natural rubber Polychloroprene

A compatibility study funded by the US Air Force, conducted at Battelle Institute, was recently completed. It was designed to study the effects of linear VMS on metals and adhesive bond integrity. 1,1,1 trichloroethane was used as the control, and linear VMS were concluded to result in no effects greater than those of the control for eight different adhesives and five metals commonly used in aerospace inertial guidance assemblies. (Reference 2)

4. CLEANING PROCESSES

Processes using volatile methyl siloxanes require designs which are safe for flammable solvents. One approach is pneumatically controlled machines, which require no electrical components. Air driven pumps are used to circulate the cleaning fluid, and the compressed gas supply also can be directed to the final stage to speed the drying step. Depending on the parts and soils involved, separate cleaning and rinsing chambers, plus a drying section can be used. When using OS-10 or OS-20, no cosolvents or additional rinsing or drying agents is necessary, since these dry completely, spot free at room temperature. One example of this equipment is the Jet-Clean machines from Martin Marietta, which are being marketed and are available for demonstration through Micro Care. Micro Care also has OS-120 under the name Vericlean in pressurized cans equipped with trigger grips and brushes.

A second process which has been demonstrated for OS-30 is to convert a vapor degreaser to a vat cleaner. By allowing heating to no more than 45°C, the vapor pressure is maintained below 25% of the LEL for this product. There are vapor phase degreasers which are engineered for safe operation with flammable chemistries, most notably the

S&K IG-200. S&K has successfully demonstrated the use of OS-10 in this machine in the vapor phase degreasing mode.

Similarly, a multi-tank machine designed for flammables is a good choice for the OS Fluids. One demonstration in the Forward Technologies A series machine, using three tanks plus a drying region, resulted in cleaning silicone oils from metal parts to a degree of cleanliness ten times better than the customer's spec. The process consisted of a 70°C wash in a tank with ultrasonics, a rinse by immersion in 40°C fluid, and a final rinse in the condensing vapor region of a boil sump. FTI designs the machine so that the dirtiest fluid moves to the boil sump, which produces pure OS-10 vapor that cascades into the rinse tank after condensing. The rinse in turn overflows to the wash tank, thus maintaining a process in which the parts see increasingly clean solvent as they are moved from tank to tank.

The initial cost of VMS Fluids is higher than for some alternative chemistries. However, the total operating cost for our customers is very competitive, since they are able to reuse these fluids due to their ease of distillation, whether operated as part of the cleaning process, or off-line in a distillation central unit. The use of adsorbent beds, desiccants, and filters also are effective in returning spent OS fluids to fresh cleaning quality.

5. THERMAL AND FIRE PROPERTIES

The vapor pressure of disiloxane at 20°C is 33 mm, and at 25°C is 42 mm. It would therefore not be restricted by the proposed 45 mm (20°C) ceiling in the new NESHAP standard for volatile solvents used in the aerospace industry, but it is sufficiently high to provide rapid drying. Figure 5 shows the thermal properties of linear VMS compared to CFC-113.

Figure 5

THERMAL PROPERTIES

	Disiloxane (OS-10)	(OS-120)	Trisiloxane (OS-20)	Tetrasiloxane (OS-30)	CFC-113
Vapor Pressure @ 25°C (mm Hg)	42.2	44.6	3.9	0.43	331
Boiling Point (°C)	100	98	152	194	47.5
Heat of Vaporization at the b.p. (BPU/gal)	528	705	450	398	819
Drying Rate (n-butyl acetate =1)	3.8	3.5	0.7	0.15	14.7

A characteristic of volatile methyl siloxanes that has made them a common choice as carriers for personal care products such as antiperspirants and deodorants is their comparatively low heats of vaporization. In personal care products the benefit is that they evaporate from the skin without a cooling sensation. In the case of precision cleaning, once at the boiling point, the heats of vaporization of the VMS solvents are among the lowest of any CFC replacement candidates. The energy cost to maintain reflux of disiloxane for a given tank size can be expected to be 35% lower than for CFC-113. As another comparison, isopropyl alcohol exhibits a significantly higher energy cost per volume in the vapor degreaser mode, with a delta H_{vap} near 2000 BTU/gallon.

Drying rates, based on ASTM 1901, have been measured in which solvent is allowed to drain and evaporate from an aluminum panel. With the rate for butyl acetate defined as 1, we find OS-10 to be 3.8 times faster, and OS-20 to be slightly slower. This test provides more realistic drying data than a simple evaporation from the bulk, and would be expected to be dependent on not only the vapor pressure, but the viscosity, surface tension, and relative humidity. For example, we have observed a tendency for hydrophilic solvents such as acetone and alcohol to attract condensed moisture as they evaporate, thus slowing the drying process compared to what would be expected strictly based on their respective vapor pressures.

Volatile methyl siloxanes are able to readily support combustion, and can burn more intensely than hydrocarbon fires. The disiloxane and trisiloxane flash points put them in the DOT flammable classification, while with a closed cup flash of 57°C, the tetramer is rated combustible. Data for two other commonly used flammable solvents is included in Figure 6 for comparison. With lower flammable limits of about 1%, it only takes a vapor pressure of around 7 mm for VMS to reach the explosive limit. Thus, disiloxane reaches the LEL without heating, and trisiloxane only needs to reach 7°C above room temperature. The autoignition temperatures of the linear VMS are well over 300°C.

Figure 6

FLAMMABILITY PROPERTIES

	Disiloxane (OS-10)	(OS-120)	Trisiloxane (OS-20)	Tetrasiloxane (OS-30)	IPA	Cyclohexane
Flash Point °C (°F)	-3 (27)	-4 (24)	34 (94)	57 (135)	12 (53)	-20 (-4)
Lower Flammable Limit (%)	1.25	1.25	0.9	0.9	2	1.3
Upper Flammable Limit (%)	18.6	--	13.8	>3.6	12	8
Autoignition Temperature (°C)	341	365	350	350	399	260

6. ENVIRONMENTAL AND TOXICITY PROPERTIES

In early 1994 the EPA granted the volatile methyl siloxanes approval under the SNAP program as alternatives to ozone depleting solvents for precision cleaning (Reference 4). Additionally, VMS are not regulated as Hazardous Air Pollutants (HAPs) nor controlled under NESHAP requirements. In addition, they are not controlled by the Clean Air Act National Ambient Air Quality Standards.

Studies carried out for Dow Corning by the Statewide Air Pollution Research Center at the State University of California at Riverside were cited in a petition to EPA to exempt VMS materials from being regulated as VOCs. This exemption was granted in late 1994. Since VOCs are regulated at the state and regional levels, we are now working with these agencies to implement the federal exemption. As of June 1, 1996, 40 states have implemented the VOC exemption, including California, and none have turned it down.

VMS chemicals exhibit unique behavior in the atmosphere. It is rare that a chemical breaks down rapidly in the atmosphere, but does not participate in the free radical chemistry leading to ozone formation, which is the main cause of lower atmosphere smog. The siloxane molecule has a lifetime of from 10 to 30 days in the atmosphere. A series of chemical reactions occur in which methyl groups are replaced by hydroxyls, with the ultimate degradation products being CO_2, water, and water dissolved silica, which of course are abundant, naturally occurring compounds.

To summarize the environmental aspects, VMS are not ozone depleters, are not VOCs, and by virtue of their conversion to CO_2, water and silica, have negligible contribution to global warming. They are truly environmentally friendly, and can be selected for cleaning processes with the assurance that they will be environmentally and regulatory acceptable for years to come. In fact, the California South Coast Air Quality Management District has selected linear VMS as their #1 preferred cleaner for precision cleaning and substrate cleaning, second only to no-clean solder flux for electronics cleaning in their January 11, 1996 report for Ozone Depleting Compounds Replacement Guidelines (Reference 5).

VMS materials are low in toxicity as evidenced by results of acute oral, inhalation and dermal exposures; subchronic oral and inhalation studies including a recently completed 90 day study on OS-10; and tissue culture biocompatibility and genotoxicity evaluations. In the acute studies, no adverse effects were observed even at the maximum achievable doses in laboratory animals. Contact of the fluid with eyes may result in temporary eye discomfort. Subchronic oral and inhalation studies did not reveal any toxicological responses significant to human health.

Based on the recently completed 90 day study, Dow Corning's toxicologists and industrial hygienists have set an internal Industrial Hygiene Guideline (IHG) at 200 ppm for OS-10, OS-20, and OS-30. This is defined as the maximum average exposure level to which workers may be exposed over an eight hour work day. Additional testing is being done on OS-120, the azeotrope of 91% OS-10 and 9% glycol ether. The glycol ether has an IHG of 100 ppm. Field testing with 14 users at two different sites using pressurized dispensing cans of OS-120 as a bench top cleaner showed 8 hour time weighted average exposures to range from less than one ppm to 16 ppm.

When the vapor pressure and exposure limits of volatile solvents are combined, a measure of the actual inhalation hazard for workers is obtained, the Inhalation Hazard Ratio (IHR)*. The higher the number, the higher the risk of getting close to allowable limits. Listed in Figure 7 are some of the well-known cleaning solvents and their IHR, and we find that disiloxane falls near the lower risk region of the list.

Figure 7

INHALATION HAZARD RATIOS OF SELECTED VOLATILE SOLVENTS	
Solvent	IHR
hexane	3816
methanol	796
toluene	703
methyl ethyl ketone	630
1,1,1 Cl_3 ethane	451
CFC 113	434
cyclohexane	396
acetone	386
ethyl acetate	288
OS-10	278
IPA	140

* IHR is defined as the saturated vapor concentration of a chemical at 25°C divided by the permissible exposure concentration. The IHR should not be used as an index of toxicity.

7. SUMMARY

Linear volatile methyl siloxanes (VMS) are a new class of solvents comprised of permethylated linear siloxanes. VMS's low surface tension and viscosity make them ideal for precision cleaning operations especially for non-polar soils such as light oils and silicone residue. Azeotropes are now available which enable cleaning of more polar soils such as rosin flux. VMS are mild cleaners and thus compatible with most materials including plastics and elastomers. Cleaning processes can range from hand wiping and cold bath soaking to ultrasonics and vapor degreasers specially rated for flammable chemistries. VMS are not regulated as Hazardous Air Pollutants (HAP) nor do they fall under NESHAP requirements. They are not ozone depleting compounds and are rated VOC exempt by the EPA and most states. VMS have low toxicity and negligible odor. These materials are available at very low non-volatile residue levels as Dow Corning ® OS-Fluids which will not adversely affect painting and bonding operations. In fact, the OS fluids are especially effective at cleaning silicone contaminated surfaces.

8. REFERENCES

1. Ambati, R.R. and Kaiser, R, Silicone Particle Removal Study for Gyro Components, ESI Report 135-1, June 1994.

2. Prause, R. H., Experimental Evaluation of the Adhesion Degradation and Corrosion Potential of Silicone Fluids, Jan 1995, Contract F09603-90-D-2217-Q805.

3. Cull, R.A., Moore, J.A., Williams D.E., Volatile Methyl Siloxanes: A New Class of VOC Exempt Solvents, SAMPE International Technical Conference Proceedings, October 1995.

4. Federal Register; EPA:59, Fed Reg. No. 192; October 5, 1994, pp 50693-50696

5. Iwata, Todd, Motavassel, Frank, Perryman, Pamela, Ozone Depleting Compounds Replacement Guidelines, California South Coast Air Quality Management District, Office of Stationary Source Compliance, January 1996

28th International SAMPE Technical Conference
November 4-7, 1996

PROTECTIVE COATINGS ON ALUMINUM BY SPONTANEOUS POLYMERIZATION

Rajat Agarwal and James P. Bell
Polymer Program
Institute of Materials Science, U-136
University of Connecticut, Storrs, CT 06269-3136

ABSTRACT

A novel process to form protective polymer coatings on aluminum will be described along with its advantages in comparison to conventional surface treatment processes, such as anodization and conversion coatings. The coatings formed were of uniform thickness and conformed to the shape of the substrate. Up to 50 micron thick coatings could be formed and were found to have excellent thermal stability and protective properties. Effects of important process variables such as polymerization time, temperature, and monomer feed concentrations were examined. Characterization results of the polymer coatings are discussed.

KEY WORDS: Corrosion Resistant Coatings, Coating Metals, Polymer Coatings, Coatings on Aluminum, Spontaneous Polymerization

1. INTRODUCTION

Aluminum, by volume, is the most widely used metal today. Applications range from aerospace and marine to architectural areas. In all cases a surface treatment and finishing process is applied. Anodization and conversion coatings are the two most common treatments for aluminum[1]. Anodization is an energy intensive process and uses phosphates. Conversion coatings use chromates and other heavy metal ions. Both of these treatments are environmentally hazardous. Thus, there is a need to develop alternate surface treatment processes which are environmentally green. This paper describes a novel process for forming polymer coatings on aluminum and its alloys. The same process can also be used for other technologically important metals such as iron[2] and copper.

Spontaneous polymerization on metals is a new process by which polymer coatings can be synthesized directly on the metal surface. The process is conducted at room temperature and no external driving force is required. The key difference of this process from other conventional coating processes such as dip coating or electrophoresis is that here, the polymer chains "grow" at the metal surface. Some significant advantages of the process are that the coatings formed are conformal and pinhole free. Thus uniform coatings can be formed on objects with complex topographies. Such a process also results in better adhesion of the metal substrate to the polymer coating, because wetting of the metal surface is easier by monomers, compared with polymers. It is not necessary to evaporate large quantities of solvent during the drying of the polymer coating.

Figure 1 shows the structures of the monomers used in this research, along with their roles when present in the coating. Copolymer coatings of Styrene (St) and N-phenyl maleimide (NPMI) were spontaneously polymerized on aluminum. Bis-Maleimide (BMI) was incorporated in the polymer coating for cross-linking, and 2-(methacryloyloxy) ethyl acetoacetate (MEA) was incorporated as adhesion promoter. Copolymers of St with NPMI and its derivatives have been synthesized in solution and electropolymerized in the past.[3,4] Such systems form alternating copolymers when polymerized with free radical initiators like azo-bis-isobutyronitrile (AIBN). Many different initiation and polymerization mechanisms for such donor-acceptor (D-A) monomer systems have been proposed and are reviewed.[5,6,7] The aspect of the process is that under the proper conditions, a St-NPMI copolymer will spontaneously form on the aluminum surface in absence of any initiator.

Polymer coatings upto 50 microns in thickness were formed. The coatings were uniform and conformal in nature and showed excellent adhesion to the substrate. Such coatings can be used in a variety of applications ranging from coating of circuit board components to automotive components. This room temperature method of polymerization requires minimum surface pretreatment of the substrate, and has the advantage of being economical and environmentally friendly.

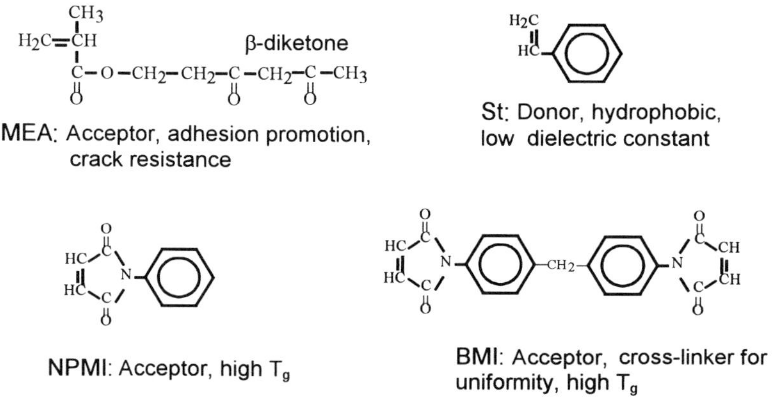

Figure 1 Monomer Structures and Their Roles

The focus of the current research has been twofold. , (i) To study the effects of various process parameters such as aluminum surface pretreatment, monomer feed composition, polymerization time and drying temperature, on the properties of the resultant polymer coating and (ii) To develop a process to synthesize useful protective coatings on aluminum for electronic applications. The desired properties for such applications are high temperature resistance and stability, good electrical breakdown resistance, low dielectric constant, corrosion protection, uniform thickness and conformity to the substrate. The measurements of the above properties are reported. The spontaneous initiation of polymerization on the aluminum surface was also investigated and a reaction mechanism is proposed herein.

2. EXPERIMENTAL

Styrene from Aldrich Chemical Co. was vacuum distilled at 40 °C. N-Phenyl Maleimide (NPMI) and bis-maleimide (BMI) were purchased from Mitsui Toatsuo Chemical Co., Japan. NPMI was recrystallized from cyclohexane and BMI was used as obtained. 2(methacryloyloxy)ethyl acetoacetate (MEA) from Aldrich Chemical Co. was purified by passing through DHR-4 inhibitor removal columns from Scientific Polymer Products. When the monomer is passed through such columns, the inhibitor is trapped by the micro porous beads which fill the column while the monomer flows through. N-methyl pyrrolidinone was purchased from Fisher Scientific Co. Aluminum used was 1100, 2024, 6061, or 7071 alloy.

2.1 Aluminum Pretreatment Aluminum was given to two types of pretreatments: (i) Degreased with 5% aqueous Micro® (an alkaline soap) solution in an ultrasonic bath for 5 min., rinsed with distilled water and treated with 5% hydrofluoric acid for 15 sec., followed by a distilled water rinse. (ii) Grit blasted using 170 mesh silica and rinsed with distilled water.

2.2 Monomer feed solution NPMI, BMI and St were dissolved in (NMP). Dilute aqueous sulfuric acid (0.025M) was then added slowly to the solution while stirring, until a 57/43 volume ratio of NMP/Water was obtained. MEA was then added dropwise. A clear, yellow colored solution of pH 3.3 was obtained by this method.

2.3 Polymerization The monomer solution was first purged with nitrogen. Dissolved oxygen content was kept to less than 2 ppm. When the pretreated aluminum was immersed in the monomer bath, a white, swollen polymer coating began to form. Polymerization time for samples in the bath varied from 10 min. to 120 min. The coated sample was then immersed in a gently stirred 10% NMP solution for 1 hour, to remove excess NMP, and was oven dried at 150 °C for 1 hour and then at 225 to 250 °C for 4-6 hours.

2.4 Characterization Fourier transform infrared (FTIR) spectra for the polymer coatings were recorded using a Nicolet 60X FTIR spectrometer, by pressing KBr pellets from mixtures of about 150 mg KBr powder and 10 mg polymer. Spectra of resultant polymer coatings were recorded for varying monomer feed compositions. A Perkin Elmer DSC-7 was used to evaluate the glass transition temperature (T_g) of the polymer coatings made from varying monomer compositions in the feed. Thermal stability was measured using a Perkin Elmer TGA-7. A heating rate of 20 °C/min. was used in the DSC, while the TGA was run at 2 °C/min in nitrogen. Molecular weights relative to polystyrene standards were measured using a Waters 550C-150 Gel Permeation Chromatograph, using NMP as solvent and mobile phase.

2.5 Corrosion Studies Corrosion protection properties of the polymer coatings were studied by exposing samples to a 5% NaCl salt fog, following the ASTM B-117 test method.

3. RESULTS AND DISCUSSION

3.1 Polymerization Process The solvent quality of the 57/43 NMP/water solution is close to the solubility limit of the monomers. The polymer formed on the aluminum surface is insoluble in the solution but is in a swollen state. This condition is essential, as it allows the initiation and formation of the polymer at the aluminum surface without dissolution of the formed polymer. The swollen nature of the polymer coating permits the diffusion of monomers through the polymer to reach the aluminum surface and also to react with the propagating chain ends. Thus, coatings up to 50 μm thick can be obtained. Coating thickness can be controlled by varying the polymerization time, or the monomer concentration.

Reaction Mechanism: Requirements for spontaneous polymerization to work are that at least one electron donor monomer (styrene) and one electron acceptor monomer (NPMI, MEA) be present together in a solution of pH less than 4. The electron donating character of styrene comes from the presence of the phenyl ring next to the C=C bond. The electron withdrawing character of the NPMI comes from the presence of the two carbonyl groups attached to the C=C bond. Spontaneous copolymerization between the two monomers can only take place if there is a substantial polarity difference between the two reaction partners. In the Styrene-NPMI system under investigation, such spontaneous copolymerization was not seen in the free standing monomer solution from which the polymer coatings on aluminum were formed. The fact that the polymerization was initiated only on the aluminum surface indicates that the metal surface plays an essential role in the initiation process.

Also, polymerization is initiated only in the presence of acid. Solutions at neutral pH showed no spontaneous polymerization in case of most substrates tested. This indicates that the polarity difference between St-NPMI or St-MEA monomer pair is not strong enough to cause spontaneous polymerization, and a further driving force is needed.[8] This driving force comes from the aluminum surface where Lewis acids such as Al^{3+} are generated when it is immersed in an acidic solution. The Al^{3+} ion is classified as a hard Lewis acid and therefore has a strong affinity for an electron pair.[9] In other studies[10] Lewis acids like $ZnCl_2$ have been shown to cause or promote spontaneous reactions of Acrylonitrile with substituted 1,3-Dienes. In those cases, the Lewis acids increase the electrophilicity of the electron-poor monomer by complexing to its lone pair electrons. This in turn increases the electron disparity between the reaction partners. It is proposed here that these aluminum ion Lewis acids interact with NPMI and increase the electrophilicity of the monomer, consequently resulting in the spontaneous formation of an NPMI-St tetramethylene diradical.[5] Formation of tetramethylene diradicals due to the presence of Lewis acids has been observed in similar systems.[6] The polymerization then proceeds by alternate addition of the donor (Styrene) and acceptor (NPMI) monomers to the growing radical. This kind of alternating addition of monomers is also called cross-propagation. Figure 2 shows the proposed mechanism for spontaneous initiation and polymerization. More experiments are currently underway to verify the proposed mechanism.

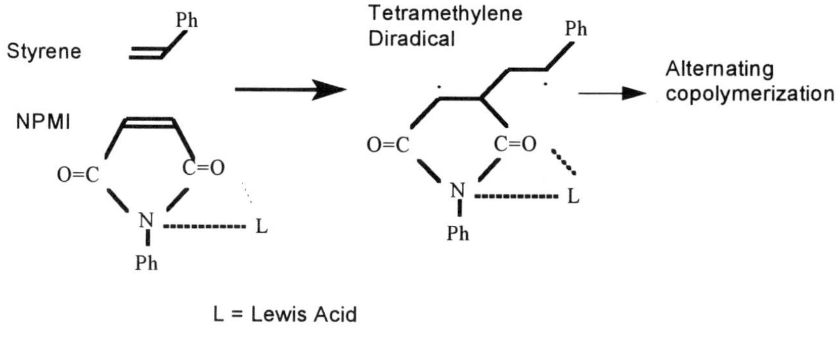

Figure 2 Proposed spontaneous initiation and propagation reaction.

3.2 Polymerization Kinetics Figure 3 shows a plot of coating thickness obtained as a function of polymerization time for various monomer compositions in feed. The ordinate axis is plotted as the square root of time, which results in linear fits of the measured coatings thickness. This indicates that such a polymerization process is limited by rate of diffusion of monomers to the propagating radicals. These observations are consistent with the equations obtained using a flat plate surface reaction model[11], where the rate limiting step is the process of diffusion of monomers in the swollen polymer coating, being formed on the aluminum surface. In such a model the thickness of the polymer coating is proportional to the square root of reaction time. As the ratio of MEA to NPMI in the feed was increased, the rate of polymerization decreased. This suggests a lower activity of the MEA monomer compared to the NPMI monomer. Sand-blasting and HF etching pretreatments gave similar kinetic results.

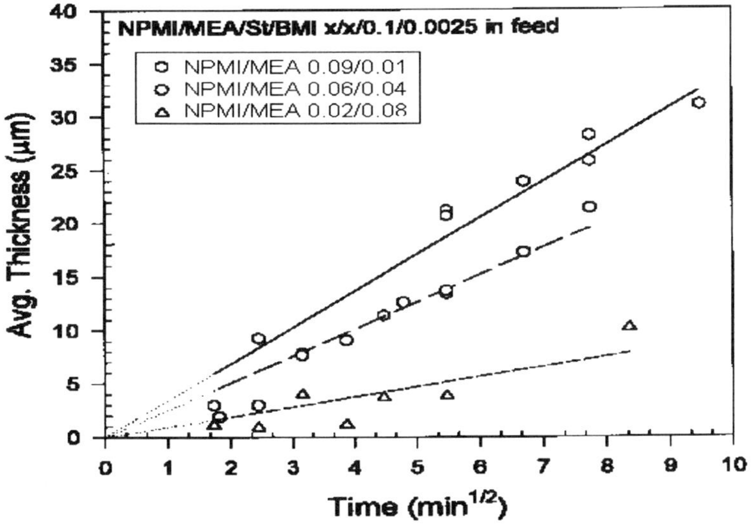

Figure 3 Effect of polymerization time on the resulting coating thickness for various monomer feed compositions.

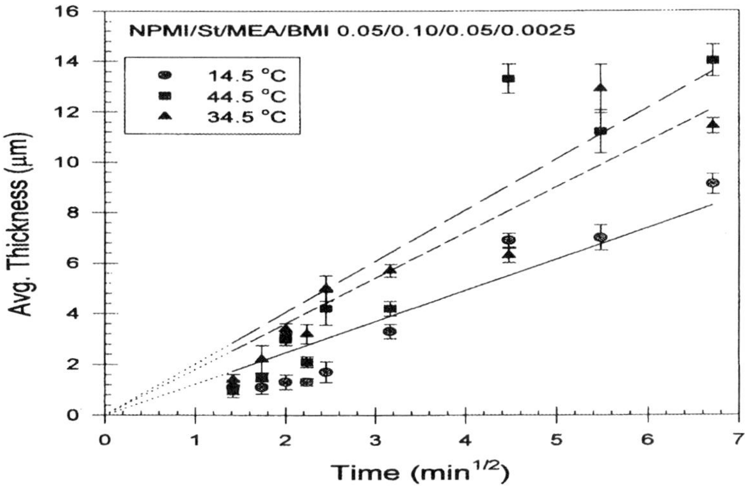

Figure 4 Effect of polymerization temperature on the measured coating thickness at various polymerization times.

Figure 4 shows the coating thickness obtained at different polymerization temperatures and times. The increase in the polymerization rate is due to the increase in diffusion coefficients of the monomers in the solution, consistent with the diffusion limited model. Regression fits show that the rate of polymerization increases with temperature. The effect, however, is not very dramatic. This also indicates that the polymerization process is limited by the monomer diffusion and not reaction rate limited

3.3 Molecular Weight Determination GPC measurements were performed on NPMI-St polymers spontaneously polymerized at various solution pH's. At a low pH, one would expect greater number of Lewis acid sites generated at the surface, resulting in a high rate of initiation. This would give a relatively low molecular weight polymer in keeping with classical free radical polymerization equations. The polymerization reaction could be quenched by the addition of 2,2-Diphenyl-1-picrylhydrazyl hydrate (DPPH) which is a free radical scavenger, confirming its free radical nature. As the pH is increased though, the number of Lewis acids formed should be less, consequently the molecular weight of the polymer obtained should be higher.

Fig. (5) shows the weight average molecular weight (M_w) of the spontaneously polymerized coating plotted as a function of total monomer concentration, for solutions of different pH. As expected, higher monomer concentration gave higher molecular weight. At any given concentration, molecular weight increased with solution pH. Polydispersity index (PDI) ranged from 2.5 to 3.5 in all cases. These results are in agreement with typical free radical polymerization reactions.

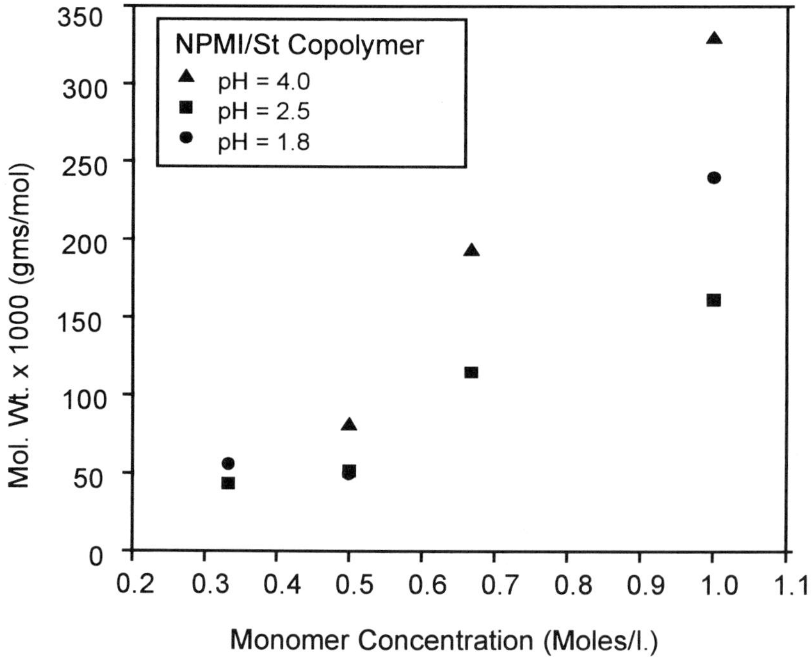

Figure 5 Effect of total monomer feed concentration on the weight average molecular weight on the resulting coating at various solution pH.

3.4 Incorporation of MEA The copolymer coatings of NPMI/St formed on aluminum exhibited a tendency to crack while drying. This problem occurred when the coating thickness was greater than 10 microns. To solve this problem, another monomer had to be incorporated. Organic compounds with β-diketone functionalities have been used in adhesives and anti-corrosion primer compositions.[12,13] The objective here was to eliminate the cracks in the coatings and improve the adhesion of the polymer coatings formed on the aluminum. The MEA monomer is an electron acceptor due to presence of electron withdrawing carbonyl groups next to the C=C bond, similar to NPMI (fig 1). It is expected that it would be incorporated in the polymer alternating with styrene. MEA was introduced in the feed solution, and NPMI/MEA ratio was varied while keeping styrene concentration constant. This way an equimolar ratio of acceptor and donor monomers was maintained in the solution. Inclusion of MEA in the polymer resulted in crack free coatings and a significant improvement in the adhesion of the swollen coating to the aluminum substrate. Corrosion resistance of the coated samples was also improved by addition of MEA.

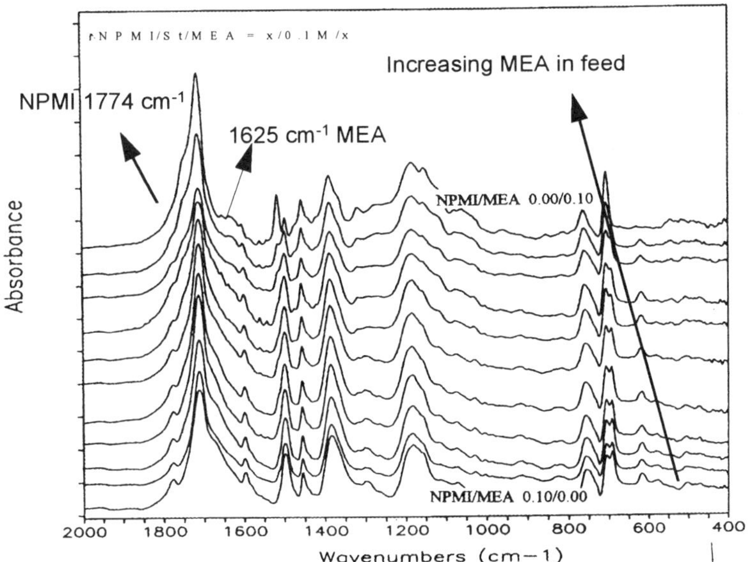

Figure 6 FTIR overlaid spectra showing formation of NPMI/St/MEA polymer.

3.4.1 Detection of MEA in polymer To detect the inclusion of MEA in the polymer coating, FTIR spectra were collected on a series of polymer coatings made by varying the NPMI/MEA ratio from 0.10M/0.00M to 0.00M/0.10M in increments of 0.01M. The total concentration of NPMI + MEA was kept at 0.10M. St and BMI concentrations were 0.10M and 0.0025M respectively. Spectra for the series are shown in fig. 6. As the concentration of MEA in feed is increased, a carbonyl peak in the 1625-1635 cm^{-1} region appears, and increases in intensity, relative to the 1774 cm^{-1} imide peak from NPMI. The low wave number from the MEA peak is due to the hydrogen bonded enol form of the beta-diketone linkage. Also a general broadening of the carbonyl bands in the 1720 cm^{-1} region occurs, due to the overlapping from MEA and NPMI. The 1598 cm^{-1} peak is due to the phenyl ring from NPMI and BMI. It also decreases in intensity as MEA content relative to NPMI increases in the resulting polymer.

3.5 Incorporation of BMI After the sample is coated with the polymer in the bath, it is withdrawn and then needs to be dried. During the drying process, water evaporates faster due to its lower boiling point, (B.P. of NMP = 210 °C) and the relative content of NMP in the polymer coating increases. Consequently, the swollen polymer coating is plasticized and some flow of the coating occurs before all of the solvent evaporates. This phenomenon occurs frequently when poly(NPMI-St-MEA) coatings are dried. To arrest this flow, BMI was incorporated to cross-link the coating and increase the molecular weight. Addition of 0.0025M BMI in NMP/water-monomer feed solution arrested the flow of the coatings during drying and gave coatings with uniform thickness. This point also illustrates the versatility of the spontaneous polymerization process. A four component coating can be formed with ease on the substrate. This has significant implications, that the properties of the coatings can be modified simply by altering the structures of the monomers or the ratios in which they are added.

3.6 Thermal Properties Glass transition temperatures were measured using differential scanning calorimetery (DSC). A single T_g was observed for all NPMI/St copolymers and all other compositions of NPMI/St/MEA and NPMI/St/MEA/BMI. A separate T_g of poly(MEA) (25 °C) was not observed. This indicates that MEA-St acceptor-donor monomers were included in the NPMI/St copolymer in random units. The T_g decreased only slightly with increasing MEA content in the polymer coatings, both in presence and absence of BMI. (Table 1). All transition temperatures were greater than 200 °C. Glass transition temperatures of the coating also increased with addition of BMI in feed. This is a very significant outcome. It demonstrates that the auto polymerization process is suitable for making high temperature resistant coatings by polymerization at room temperature. Thermal stability of the coatings was found to be very good. Onset of degradation took place at more than 350 °C under nitrogen atmosphere and was not affected by incorporation of MEA. (fig 7).

Monomer feed composition (moles/lit) NPMI/MEA/St	T_g (°C)	Monomer feed composition (moles/lit) NPMI/MEA/St/BMI	T_g (°C)
0.10/0.00/0.10	216.2	0.10/0.00/0.10/0.0025	219.8
0.08/0.02/0.10	-	0.08/0.02/0.10/0.0025	218.4
0.07/0.03/0.10	207.7	0.07/0.03/0.10/0.0025	215.8
0.06/0.04/0.10	-	0.06/0.04/0.10/0.0025	215.5
0.05/0.05/0.10	205.5	0.05/0.05/0.10/0.0025	-
0.04/0.06/0.10	197.2	0.04/0.06/0.10/0.0025	215.1
0.03/0.07/0.10	200.4	0.03/0.07/0.10/0.0025	210.3
0.02/0.08/0.10	200.6	0.02/0.08/0.10/0.0025	-
0.01/0.09/0.10	198.6	0.01/0.09/0.10/0.0025	201.8

Table 1 Glass transition temperatures measured for polymer coatings formed from various monomer feed compositions.

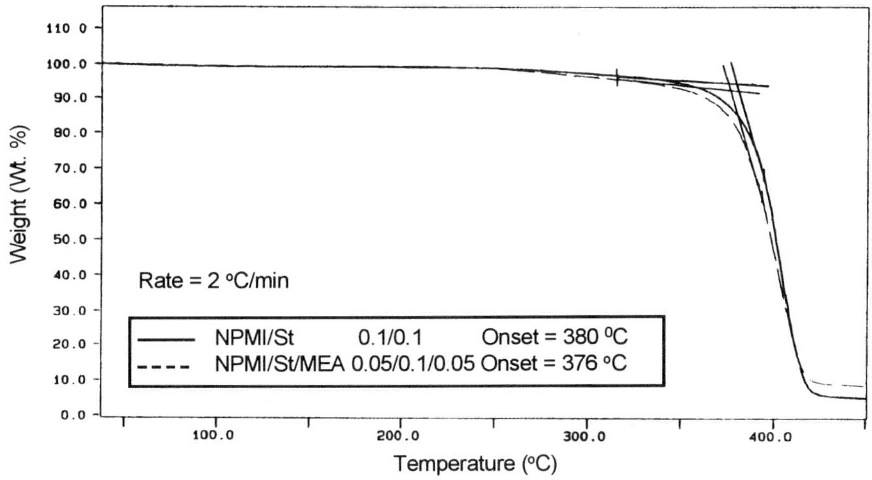

Figure 7 Thermogravimetric analysis of polymer carried out at 2 °C/min in nitrogen atmosphere.

3.7 Electrical Resistance The coatings exhibited good resistance to DC potential. Samples coated with 20 μm thick polymer, spontaneously polymerized from a feed solution of NPMI/St/MEA/BMI ratio of 0.05/0.10/0.05/0.0025M, did not show any significant current leakage up to an applied potential of 1800 V.

3.8 Corrosion Studies Fig. 8a shows an unexposed 6061 aluminum sample coated with 20 μm thick NPMI/St/MEA polymer coating on the right half. Fig. 8b shows the sample after 1500 hours of exposure to salt fog (ASTM B-117). The coated side is unaffected, while the uncoated side is corroded. No propagation of corrosion is seen occurring at the polymer-metal interface. No corrosion is seen around any of the edges of the sample, illustrating the uniformity of the polymer coating. Similar results are seen for sample exposed for 3000 hours (fig. 8c). This is an inherent property and advantage of the dip auto polymerization process. Since polymerization is initiated only at the surface, metal sample with complex surface topographies can be uniformly coated. Coating also forms inside holes and on edges.

3. CONCLUSIONS

1. The spontaneous polymerization process can be used successfully to form uniform conformal coatings on aluminum.

2. Spontaneous initiation of polymerization takes place, possibly due to the formation of NPMI-St tetramethylene diradicals at the metal surface. The propagation takes place by free radical mechanism.

3. Such coatings have good adhesion, corrosion resistance and electrical insulating properties. Very high molecular weights in the range of M_w = 250,000 were obtained.

4. This process is versatile and can be adapted to form polymer coatings with other desired properties by modifying the structures of the donor and acceptor monomers.

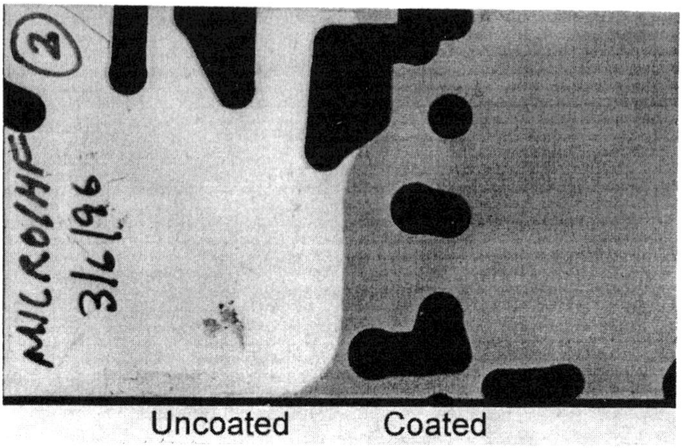

Figure 8a Unexposed sample

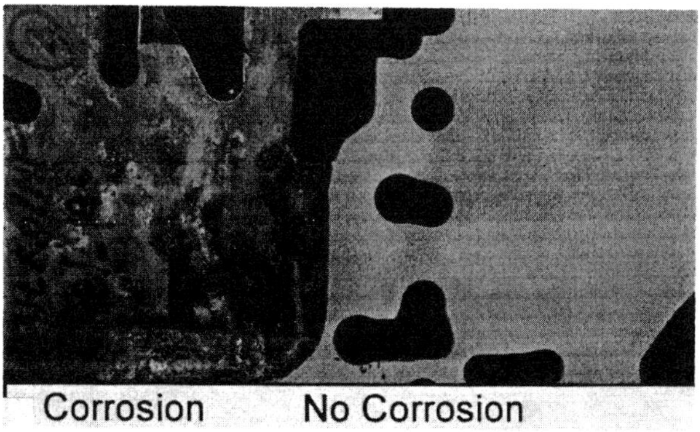

Figure 8b Sample exposed for 1500 hours

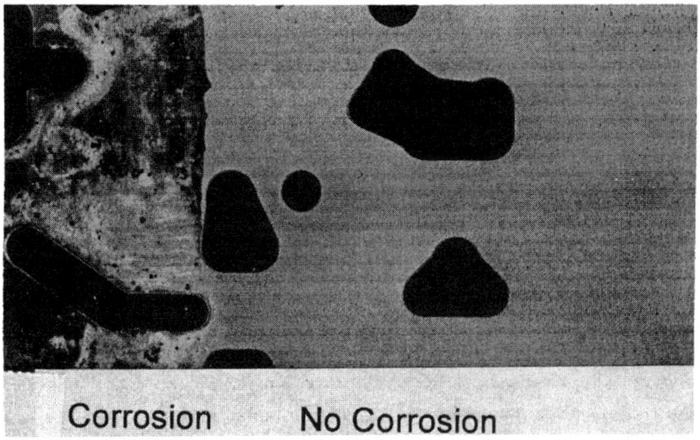

Figure 8c Sample exposed for 3000 hours

4. REFERENCES

(1) S. Wernick, R. Pinner and P. G. Sheasby, *The Surface Treatment and Finishing of Aluminum and its Alloys*, ASM International, Metals Park, Ohio, U.S.A., Finishing Publications Ltd., Teddington, Middlesex, England, 1987

(2) X. Zhang and J. P. Bell, *Manuscript in preparation*

(3) S.Iwatsuki , M.Kubo, M. Wakita, Y. Matsui, and H. Kanoh *Macromolecules* , **24**, 5009 (1991)

(4) J. O. Iroh, J. P. Bell, and D. A. Scola *J. Appl. Polym. Sci.*, **47**, 93 (1993)

(5) H. K. Hall Jr. *Angew. Chem. Int. Ed. Engl.*, **22**, 440 (1983)

(6) J. M. G. Cowie, ed. *Alternating Copolymers*, 1985, Plenum Press, New York

(7) A. Matsumoto, T. Kubota, and T. Otsu *Macromolecules*, **23**, 4508 (1990)

(8) M. G. Michael, A. B. Padais , and H. K. Hall Jr. *Macromolecules*, **27**, 7499 (1994)

(9) W. B. Jenson, *The Lewis Acid Base Concepts, An Overview*, Wiley-Interscience, 1979

(10) E. Tsuchida, T. Tomono, and H. Sano *Die Makromolekulare Chemie*, **151**, 245 (1972)

(11) O. Levenspiel, *Chemical Reaction Engineering, 2nd ed.*, Chapter 12, pp. 367-373 Wiley, 1972,

(12) V. P. Wystrach, F. C. Rauch *U.S. Pat* 3,615,888 (1971)

(13) H. S. Gilbert, *U.S. Pat* 4,324,676 (1982)

28th International SAMPE Technical Conference
November 4-7, 1996

INTEGRATING PROCESS ANALYTICAL CHEMISTRY INTO POLLUTION PREVENTION SYSTEMS

Ralph E. Taylor-Smith
Bell Laboratories
Lucent Technologies
Murray Hill, New Jersey 07974

ABSTRACT

Process analytical tools are becoming essential elements in manufacturing operations. The need for rapid, precise non-invasive sensors and monitors is being recognized, particularly in the area of pollution prevention and control. Waste reduction systems are being developed to effectively incorporate analytical diagnostics for effluent detection, identification, and quantification in a variety of industrial environments. In particular, infrared spectroscopy holds great potential as an accurate on-line analytical monitor of vapor-phase waste emissions. Our current focus of implementation in this arena involves ceramics processing; one critical step in production of high-performance optical and opto-electronic components and devices involves purification via pyrolytic treatment. Organic additives/processing aids are removed via volatilization and oxidative decomposition, thus generating a complex process stream of waste organic effluents. As part of a case-study, fourier-transform infrared spectroscopy was utilized as a quantitative tool to characterize the effluent evolution process at the laboratory scale. Infrared spectroscopy was also incorporated into a modular catalysis-based environmental control scheme designed to limit hazardous emissions from the process equipment; spectra acquired in-situ were utilized to evaluate control efficacy. Results from this work will be used to illustrate the manner in which process analytical monitors are being integrated into environmentally-conscious manufacturing practice.

KEYWORDS: Environmental; Process Monitor; Infrared Spectroscopy

1. INTRODUCTION

Process analytical chemistry is being recognized as an integral part of environmentally-responsive manufacturing[1]. Product and process sensors are now utilized for various applications, including - process monitoring and control, leak detection, environmental health and safety, pollution prevention and waste minimization. Such sensor systems must simultaneously satisfy multiple criteria: reliability; speed of action; robustness; low-cost; sensitivity; multi-component species differentiation; non-invasive/non-destructive capability; and permit in-situ, on-line implementation. Whilst the present generation of sensors and monitors have typically been applied as process diagnostics, to provide input data for control or optimization strategies, or to analyze and quantify fugitive emissions for regulatory compliance purposes, it is expected from current trends[2] that environmentally-benign manufacturing processes of the future will necessitate a several-fold increase in sensor quality, quantity and diversity. Instead of "single-point" sensors, which enable characterization of process parameters or contaminants at one single location, "multi-point" sensing systems which scan over several locations will be required, to essentially map out parametric distribution functions over wide regions in real-time. Various process analytical tools are currently under development for such applications[3-6]. In particular, infrared spectroscopy (IR) holds great promise as an accurate on-line analytical monitor in manufacturing. While the technique has traditionally been applied for elucidation of molecular structure and bonding information, recent advances in data analysis procedures such as chemometrics[7] enable extraction of quantitative information from complex multicomponent data sets in a variety of sampling environments.

We are working towards integrating infrared spectroscopy and similar tools into manufacturing for pollution prevention and waste minimization. As process technology is scaled up and transferred to manufacturing, it is necessary to define environmental control schemes appropriate for handling effluent streams at the factory level. For optimal design and configuration of such controls, detailed quantitative characterization of the effluent stream is required, specifically waste component identification, determination of relative amounts, evolution kinetics etc. One current focus of implementation in this arena includes ceramics processing; a critical step in production of high-performance optical and opto-electronic components and devices involves purification via pyrolytic treatment. Organic additives/processing aids are removed via volatilization and oxidative decomposition, thus generating a complex process stream of waste organic effluents (VOCs). Detailed quantitative characterization of the effluent stream is required in design and implementation of an effective environmental control for this process, specifically organic component identification, determination of relative amounts, evolution kinetics etc. Here, Fourier-transform infrared (FTIR) spectroscopy is utilized as a quantitative tool, to characterize the organic effluent stream at the laboratory scale via application of a simple three-component model. It is demonstrated that the complex effluent evolution profile can be well-described by an elementary combination of three organic compounds - trimethylamine, methanol and methyl formate assumed to be generated from volatilization and/or decomposition of the organic additives/processing aids. Thus, with implementation of this simple approximation and detailed unequivocal spectral assignments, analysis of FTIR spectra permits quantitative characterization of effluent evolution kinetics over the complete range of process temperatures. Infrared spectroscopy was also incorporated into a modular catalysis-based environmental control scheme designed to control hazardous emissions from the process equipment directly at the site of generation; spectra acquired in-situ were utilized to evaluate control efficacy. Results from this work will be used to illustrate the manner in which process analytical monitors are being integrated into environmentally-conscious manufacturing practice.

2. EXPERIMENTAL

Details of the laboratory-scale experimental setup utilized in this investigation have been fully discussed in an earlier publication[8]. Data acquisition was performed utilizing a BOMEN Michelson 100 FTIR spectrometer, via flowing process effluent through a 10 cm path-length gas cell. Time-resolved infrared spectra were acquired at 10 minute intervals over a 12 hour period corresponding to a temperature range spanning ~ 75 - 600 °C. The effluent carrier gas was purified nitrogen, introduced at sufficient flowrate so as to eliminate spectral artifacts from pressure broadening[9]. Teflon lines bearing the effluent stream to the FTIR cell were heat-traced with electrically-controlled heating cable to minimize pre-analysis condensation, and for similar reasons, the interconnection distance was limited to the minimum practical length. Methodology utilized to generate quantitative calibration standards for the simple three-component model applied here to approximate the effluent stream composition is briefly outlined below. Essentially, infrared spectra for each individual component were obtained under carefully controlled standard conditions at which gaseous concentration (partial pressure) could be accurately determined or estimated. Thus, absolute absorbance intensities of characteristic absorption bands of a particular chemical compound could be correlated with specific gas-phase concentrations. Experimentally derived vapor pressure - temperature correlations for each of the three components were found in the literature[10] and utilized to determine specific vapor pressures at certain experimentally-accessible, easily-controllable temperatures. Thus, the vapor-phase FTIR spectrum of methanol was acquired at 0 °C (vapor pressure = 3.82 kPa) via equilibration of a nitrogen-methanol bubbler delivery system in an ice-water bath. Similarly, the vapor-phase spectrum of methyl formate was obtained at -78 °C (vapor pressure = 0.13 kPa) via equilibration of a nitrogen-methyl formate bubbler delivery system in an acetone/dry ice slurry. The trimethylamine component vapor-phase spectra acquired at a temperature of -70 °C (vapor pressure = 2.00 kPa) was taken from the literature. The calibration data is tabulated in Table I.

3. RESULTS AND DISCUSSION

Figure 1 shows time-resolved FTIR spectra acquired from the laboratory-scale ceramics pyrolytic-purification setup over the approximately 12 hour processing period. It is essentially a complete effluent evolution profile of the process and such data has the potential for yielding a detailed description of the amounts, relative rates and evolution kinetics of effluent components. However, this would be a non-trivial task due to the large number and complex mix of components present believed present[11]. Nevertheless, a simplified picture of the waste stream can be obtained via application of an elementary three-component approximation of the effluent composition. The validity of this approximation can be gauged from a comparison of the spectral data in Figure 2. The top plot, denoted A, is a single spectrum simply extracted from the center (approximately 22,000 sec.) of the multiple spectra data set shown in Figure 1; this single "snap-shot" from the cross-section of the evolution profile exhibits all representative features of the effluent data. Conversely, the bottom plot, denoted B, is a composite plot of infrared spectra of the three selected model components. This was generated artificially by extracting respective spectra from the National Institutes of Standards and Technology (NIST) computer database[12], digitizing each spectrum and overlaying the separate curves on common axes. It is clear that, at least in qualitative terms, spectral features of the composite model data, plot B, are in good agreement with those of the experimentally-obtained effluent cross-section shown in plot A. The relative absorbance intensities are obviously not in quantitative agreement since, for ease of presentation, a simple 1:1:1 component ratio was assumed in the generation of plot B. However, locations of characteristic absorption bands are in excellent congruence. In fact, the degree of fit is so significant that application of this simple three-component model is deemed sufficiently valid for this preliminary study.

One characteristic absorption band specific to each component was selected such that there was minimal overlap from the remaining two (see Figure 2) and the absolute intensity of the deconvoluted band taken as a measure of that component concentration. For trimethylamine, the intense in-phase -N-CH_3 stretch (2774 cm^{-1}) is shifted somewhat lower in frequency and thus easily distinguishable from other aliphatic C-H bands (due to the proximity of the nitrogen non-bonding electron pair[9]); thus, this band was utilized as the reference. For methanol, the strong intense alcohol C-O stretch (1032 cm^{-1}) was utilized as the reference, and similarly for methyl formate, the strong intense carbonyl C=O stretch (1765 cm^{-1}). Absorbance absolute intensity as a function of data acquisition time was obtained via interpolation from Figure 1 for the respective model component reference bands. It is clear that the trimethylamine peak is slightly distorted or "clipped" at the top due to signal saturation, indicating that results obtained for that component will not be absolutely quantitative over the complete data acquisition range. Table I shows the one-point calibration data obtained for the model components. Using this information, Beer's Law[9] was applied to compute concentrations (kPa), process temperatures were calculated from data acquisition times through the heating schedule and normalization was performed with regard to sample weight.

Results of these manipulations are presented in Figure 3. The plot is essentially a quantitative map of effluent evolution over the programmed heating schedule. From inspection of the data, the methanol:trimethylamine ratio is observed to approximate unity; thus, it seems straightforward to conclude that these two components result from direct decomposition of tetramethylammonium formate (a intermediate assumed generated from inter-reaction of two process reagents). This confirms some initial speculations regarding a reaction pathway for effluent decomposition chemistry and the premise is reinforced by the fact that evolution kinetics of these two components are almost identical and peak at exactly the same process temperature (~ 238 °C), instead of the more volatile species (the amine) evolving at a relatively more rapid rate. In contrast, the methyl formate has significantly different evolution kinetics and peaks at a much lower process temperature (~ 215 °C). The simple three-component model utilized to describe the effluent composition neglects the presence of many compounds earlier shown to be present at presumably low concentrations via several qualitative and semi-quantitative analytical techniques. A large fraction of these are various aldehydes, acetonitriles, amides and carboxylic acids. However from cross-referencing, the results obtained here via infrared spectroscopy were found to be in good agreement with those from an earlier effluent analysis by gas chromatography. This preliminary investigation on application of infrared spectroscopy as a vapor-phase effluent monitor utilized a simple one-point, three-component calibration with off-line quantification of waste composition and evolution kinetics. We are developing a simultaneous, multi-point, multi-component statistical calibration based on a chemometrics partial least-squares multivariate algorithm to permit real-time characterization of the effluent stream, with simultaneous quantification of multiple waste components. This will enable a quantitative description of the identity, relative composition and evolution kinetics of the waste as a function of process conditions.

The effluent control under development for this application involves "point-of-use" catalytic technology[8]. Wastes are converted to relatively benign compounds directly at the point of generation, eliminating the need to pipe process exhaust streams from diverse locations within the factory to a central point for consolidation prior to treatment. Figure 4 schematizes the modular two-stage catalysis-based abatement system currently being investigated. A laboratory-scale setup is being utilized as a benchmark for a feasibility study, to provide preliminary data for characterizing process efficacy, evaluation of scale-up criteria, guidelines for optimization and eventual implementation. An infrared spectrometer was integrated into the catalytic reactor exhaust line; spectra acquired in-situ were utilized to evaluate control efficacy.

4. CONCLUSIONS

In this investigation, Fourier-transform infrared spectroscopy was utilized as a quantitative tool to monitor effluent evolution from a manufacturing process. In this preliminary study, a simple three-component model was initially applied to describe the effluent system. Quantitative information on effluent composition and evolution kinetics was extracted from complex spectra of the process exhaust. This work illustrates the potential of infrared spectroscopy and similar process analytical tools for rapid on-line real-time characterization of process streams in manufacturing operations, particularly in the arena of pollution prevention and control.

5. ACKNOWLEDGMENTS

The author would like to acknowledge J.B. MacChesney for his assistance and input throughout all phases of this project; his efforts in many ways are much appreciated. Informative discussions with T. Stockert, L. Seibles, D. Qin, G. Cadet, and J. Valdes proved invaluable.

6. References

1. H. Baltes, W. Gopel, J. Hesse ed.s, Sensors, VCH Publishing Group, NY (1996).
2. J. Osteryoung and T.S. King in Basic Research Needs for Environmentally-Responsive Technologies of the Future: An Integrated Perspective of Academic, Industrial and Govermental Researchers, an NSF-DOE co-sponsored report, P.M. Eisenberger ed., published by Princeton Materials Institute/Princeton University (1996).
3. J.R. White, *Pollution Eng.*, 27(6), 46 (1995).
4. "Emissions Monitoring",*Chem. Eng. Prog*, Special Issue, 92(2) (1996).
5. J.A. O'Neill, M.L. Passow, T.J. Cotler, *J. Vac. Sci. & Tech.*, 12(3), 839 (1994).
6. L.L. Tedder, G.B. Lu, B.F. Conaghan, G.W. Rubloff, *Proc. Rapid Thermal Processing 95*, Amsterdam, Nederland, Aug 30 - Sep 1 (1995).
7. T.M. Niemczyk, J.E. Franke, B. Wangmaneerat, D.M. Haaland in Computer-Enhanced Analytical Spectroscopy Vol 4, C.L. Wilkins ed., Plenum Press NY (1993).
8. R.E. Taylor-Smith and J.B. MacChesney, *SAMPE J.*, 32(2), 55 (1996).
9. Colthup N.B., Daly L.H., Wiberley S.E., Introduction to Infrared and Raman Spectroscopy, 3rd Ed., Academic Press, New York (1990).
10. Perry R.H., Green D.W., Maloney J.O., Perry's Chemical Engineering Handbook, 6th Ed., McGraw Hill, New York (1984).
11. J. Bihnlein-Mass, W. Sigmund et al., *Adv. Materials*, 4(2), 73 (1992); and references cited therein.
12. National Institutes of Standards & Technology (NIST) Standard Reference Database Series 35: Gas-Phase Infrared Spectra, NIST, Gaithersburg MD 20899.

Table I: Model Component Calibration Data

Compound	Characteristic Absorption Band	Spectral Location (cm^{-1})	Absolute Absorbance Intensity	Conditions Pressure (kPa)	Conditions Temperature (°C)
Trimethylamine	$-N-CH_3$	2774	0.830	2.00	-70
Methanol	C-O	1032	0.947	3.82	0
Methyl formate	C=O	1765	0.074	0.13	-78

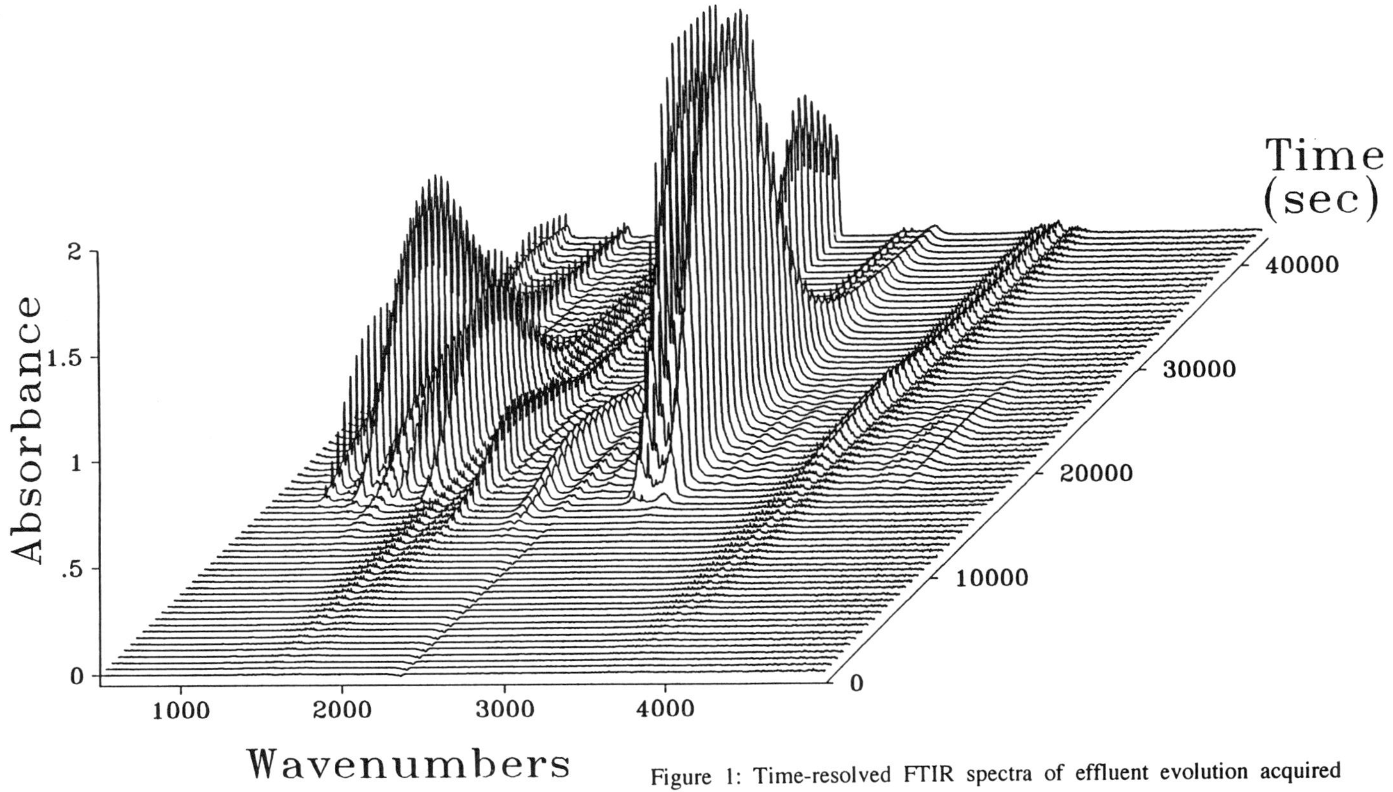

Figure 1: Time-resolved FTIR spectra of effluent evolution acquired during monitor of process exhaust over 12 hour period

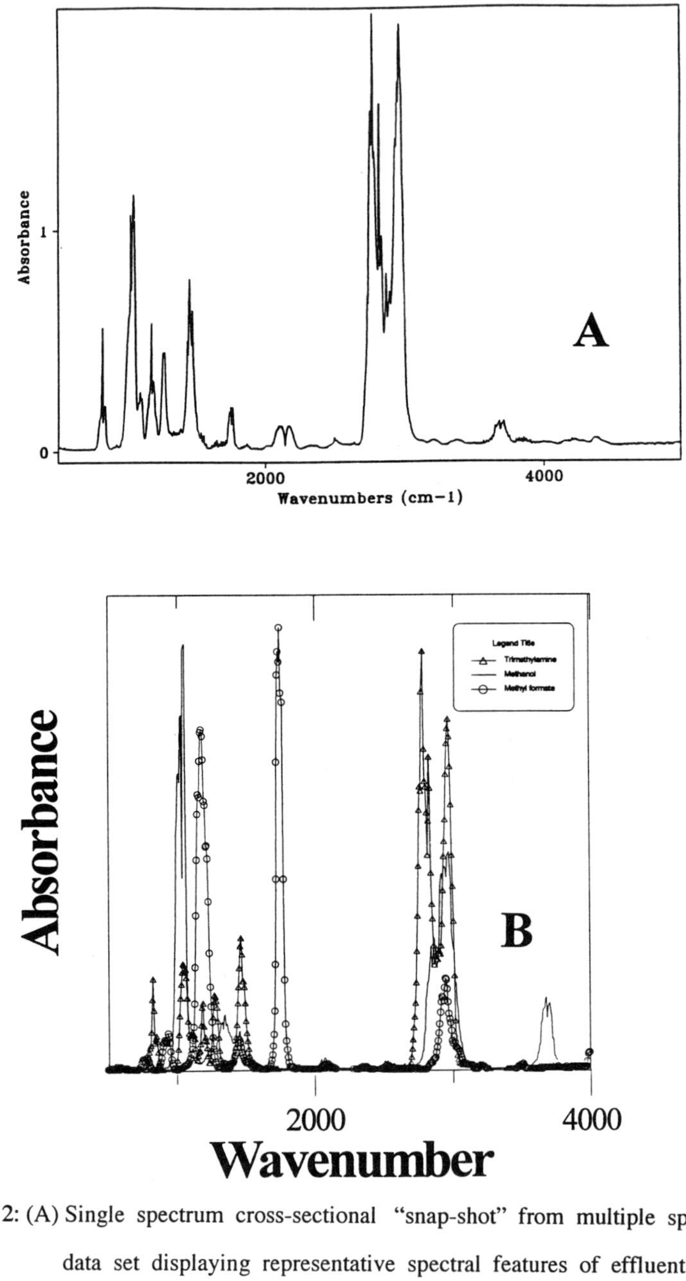

Figure 2: (A) Single spectrum cross-sectional "snap-shot" from multiple spectra data set displaying representative spectral features of effluent; (B) Artificially-generated composite plot displaying infrared spectra of selected model components

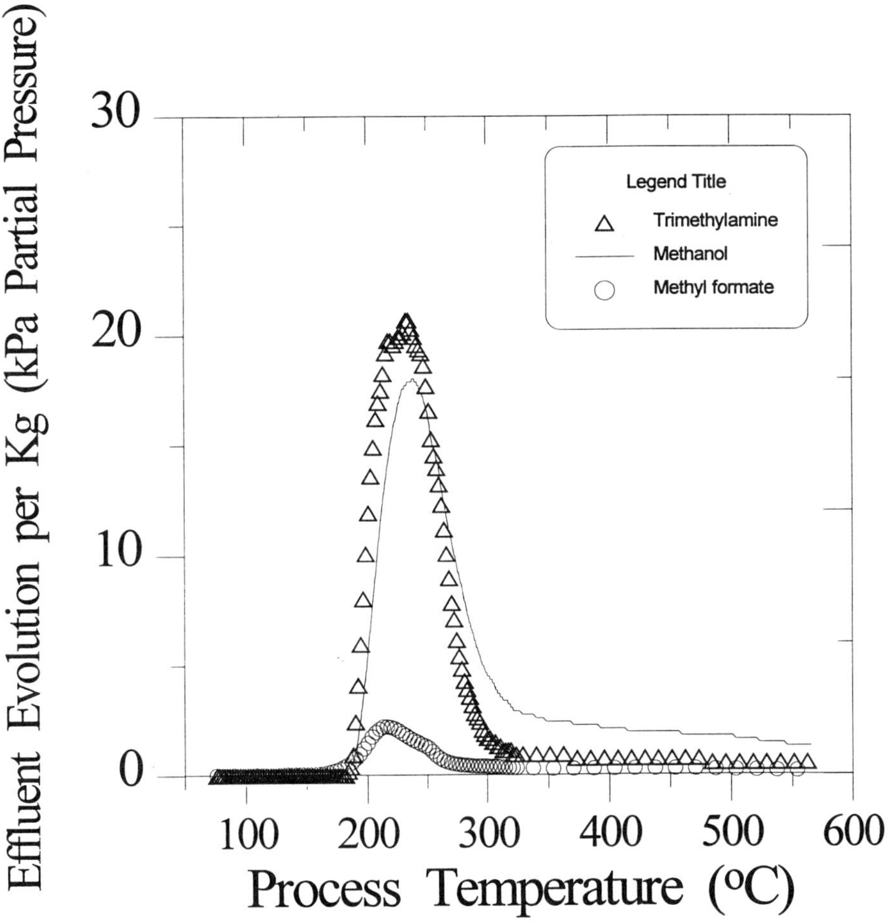

Figure 3: Quantitative measure of organic effluent evolution over range of process temperatures (in terms of model composition)

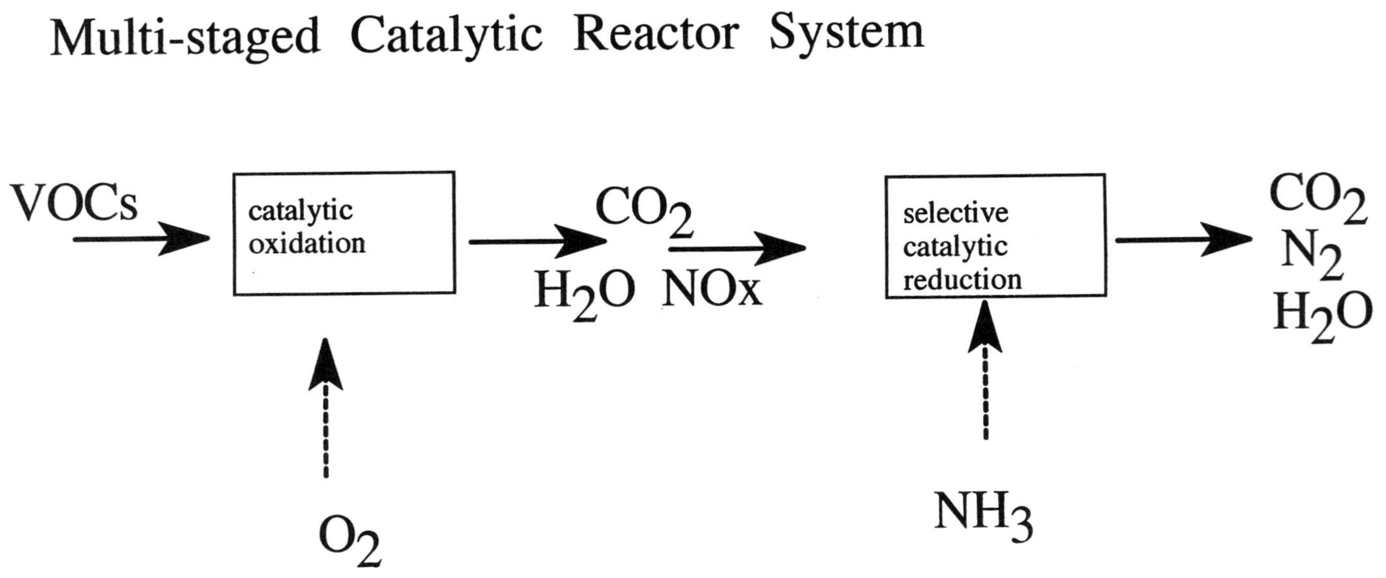

Figure 4: Schematic of "point-of-use" catalysis-based environmental control

28th International SAMPE Technical Conference
November 4-7, 1996

PROPERTIES OF CARBON FIBERS RECLAIMED FROM COMPOSITE MANUFACTURING SCRAP BY TERTIARY RECYCLING

Ronald E. Allred
Andrew B. Coons
Adherent Technologies, Inc.
9621 Camino Del Sol NE
Albuquerque, NM 87111*

and

Robert J. Simonson
Sandia National Laboratories
Albuquerque, NM 87185-5800**

ABSTRACT

Tertiary recycling is rapidly evolving as the process of choice for reclaiming otherwise difficult to recycle materials such as composites, electronics, and plastic mixtures. In the Adherent Technologies, Inc. tertiary recycling system, polymer fractions are converted to a gaseous mixture of low molecular weight hydrocarbons with a low-temperature (≈200°C) catalytic process termed catalytic conversion. Reinforcing fibers are recovered from the process for reuse in molding compounds or as milled fibers. X-ray photoelectron spectroscopy spectra of carbon fibers recovered from unidirectional prepreg show nearly identical surface chemical composition as virgin fibers. Carbon fibers recovered from cured fabric/epoxy composites show only a 9% loss in tensile strength compared to virgin unwoven fibers. Some of that loss can be attributed to the weaving process, which indicates that the tertiary recycling process causes little degradation of carbon fibers. These data indicate that the reclaimed carbon fibers from tertiary recycling should have substantial value that will lead to positive economics for the overall recycling process for composite manufacturing scrap.

* Supported under DARPA Contract DAAH01-95-C-R004.
** The Sandia National Laboratories are operated for the U. S. Department of Energy under Contract Number DE-AC04-94AL85000.

KEY WORDS: Recycling, Scrap Composites, Prepregs, Carbon Fibers, XPS, Surface Chemistry

INTRODUCTION

Tertiary recycling is rapidly emerging as an economical method for reclaiming composite manufacturing scrap [1-3]. In this process, polymeric waste is converted into reusable hydrocarbon fractions for reincarnation as polymers, monomers, fuels, or chemicals, leaving behind the valuable fibers for reuse. Initial results show that most types of matrices, thermosets as well as thermoplastics, can be converted in high yields with this novel catalytic process. Conversion times are rapid and the process is closed and, thus, nonpolluting. The resultant hydrocarbon products vary with the plastic feedstock, but a surprisingly small number of compounds form the majority of products from a given plastic. A schematic showing this recycling approach is given in Figure 1.

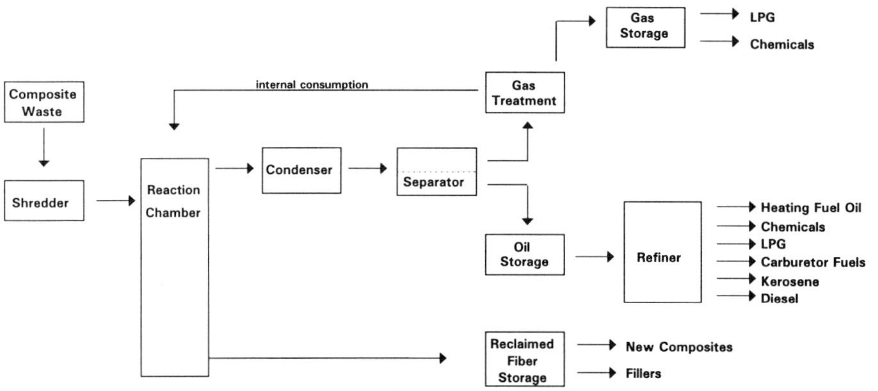

Figure 1. Schematic Flow Diagram for Composite Tertiary Recycling Process

Analysis of the flow diagram given in Figure 1 shows that the following steps comprise the tertiary recycling process:

- collection of composite manufacturing scrap
- size reduction to fit reactor transport system and specify reclaimed fiber length distribution
- processing through the reactor system
- collection, separation, characterization, and resale of the converted hydrocarbons
- characterization and resale of the reinforcing fibers.

Each of these steps must be thoroughly investigated to determine the overall economics of tertiary recycling composite manufacturing scrap. Previous papers have described size reduction, reactor process parameters, hydrocarbon conversion products, and fiber surface appearance from this recycling process [1-3].

The main value of tertiary recycling of carbon-reinforced polymer-matrix composite materials comes from reuse of the carbon fibers. Fiber form varies considerably with end use. Molding compounds require loosely held fiber assemblies. This implies that the epoxy or other matrix should not be completely removed during the catalytic conversion process for some applications. Other applications use completely clean fibers for milling as fillers in carbon/carbon. In that case, even tangled fibers would work well. The results reported in this work are part of a set of experiments to verify that the catalytic conversion recycling process under development can be tailored to produce both types of recycled fiber assemblies. Once carbon-reinforcing fibers are recovered from composite scrap prepregs and trim, it is essential to characterize their mechanical properties and surface chemistry to determine their potential value for use in new applications. This paper presents initial characterization results of tensile strength and surface chemistry on reclaimed carbon fibers from a low-temperature, catalytic tertiary recycling process.

EXPERIMENTAL PROCEDURES

Materials

Two types of composite scrap feedstock were examined: (1) out-of-date unidirectional carbon fiber prepreg with silicone-treated kraft paper release ply supplied by Ciba-Geigy Composites (Anaheim, CA) designated as RAC 6350 and identified as resin batch 305618825/Mfg. 11-23-87, and (2) cured carbon fabric/epoxy trim from Heath Techna Corp. (Seattle, WA).

Tertiary Recycling System

The tertiary recycling studies were conducted in a custom-made continuous feed reactor. The process flow diagram is shown schematically in Figure 2.

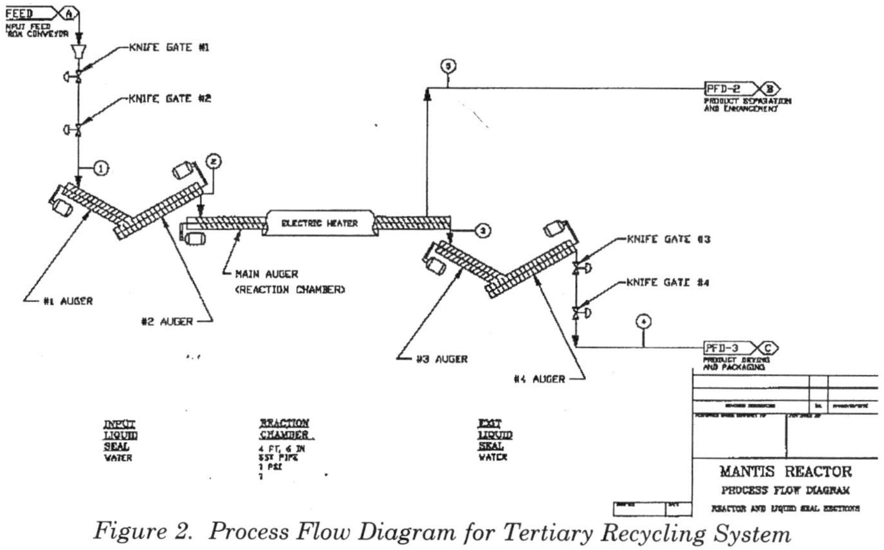

Figure 2. Process Flow Diagram for Tertiary Recycling System

A photograph of the reactor is shown in Figure 3.

Figure 3. Appearance of the Tertiary Recycling Reactor System

The reactor, named the Mantis from the appearance of the liquid seals, uses a 15-cm-diameter auger drive for feedstock material transport and can process 5-10 kg of scrap composite per hour. Conversion times varied between 5 min for unidirectional prepreg to 15 min for cured fabric trim at temperatures near 200°C. As previously reported [3], the silicone-treated release ply in the RAC 6350 was almost completely consumed in the recycling process with only a very light ash observable in addition to the fibers. The reclaimed carbon fibers from the RAC 6350 prepreg were completely separated and showed only isolated small particles ($< 0.1\mu$) of residual epoxy. Once process parameters were optimized for the Heath Techna fabric composites, the fabric began to lose its identity and fell into individual fiber bundles. The appearance of the reclaimed carbon fibers from the Heath Techna trim is shown in Figure 4.

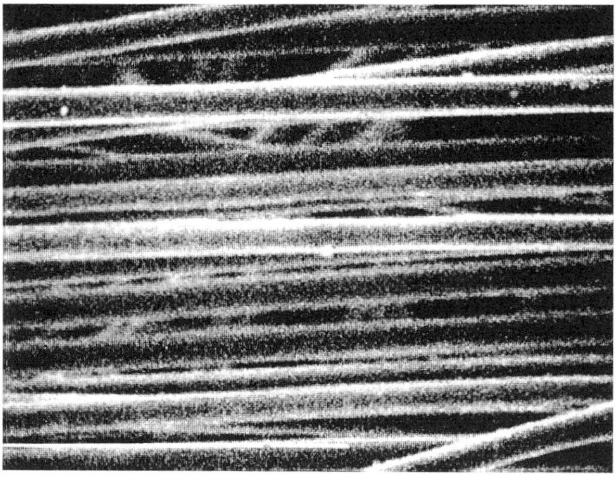

Figure 4. Appearance of Carbon Fibers Reclaimed from Cured Fabric/Epoxy (1200X)

Single Fiber Tensile Tests

Single fiber tensile tests were performed on recycled fibers recovered from the Heath Techna composite trim material. Virgin fibers obtained from Ciba Composites (now Hexcel) that were identical to those used to make the Heath Techna prepreg (Tenax-HTA 5131 Lot 50211) were tested as controls. Fifty of each of the recycled and control fibers were tested. The test procedure and results are discussed below.

Fiber Separation

Single fibers were separated from the recycled material by first identifying fiber bundles or tows that were of sufficient length to be mounted in the test tabs that were to be used (see Fiber Mounting, below). The fibers were then separated by hand by first separating a few fibers from the tow, and then separating a single fiber from the smaller group. An effective method of separating the fibers was to hold the fiber bundle between the fingers of one hand and lightly pass the index finger of the free hand over the ends of the fibers until a single fiber or a few fibers could be separated. This technique was repeated on successively smaller bundles until a single fiber was obtained.

The control fibers were separated with the same technique after cutting a bundle of the appropriate length from the end of a fresh roll of fiber. Separation of the control fibers was much easier than the recycled fibers, which contained a small amount of residual resin and some kinking from being woven.

Fiber Mounting

Each of the single fibers was mounted longitudinally in a paper tab as described in ASTM D3379-75 [4]. The tabs used for these tests were 5.7 cm long by 1.26 cm wide, with a gauge length cutout of 2.54 cm. The fibers were cut to a length of about 4 cm, such that each end of the fiber would lie approximately at the midpoint between the end of the tab and the end of the gauge length cutout. The fibers were then approximately centered on the tab, and fixed in place by applying a small amount of household cement at either end, mostly covering each free end of the fiber. Since the fiber's strength is extremely sensitive to surface flaws, fiber mounting was carried out under a low-power (10X) microscope to reduce the chance of accidentally nicking the fiber with the edge of the applicator and breaking it. Care was taken to ensure that the adhesive was applied up to the edge of the gauge length cutout at either end, so that the gauge length would be constant throughout the fiber tests.

Test Procedure

After the fibers were attached to the test tabs, each fiber was given an identification number that was written on both ends of the tab. In addition, a coded label was written on both ends of each tab to identify the test run. (The information must be put on both ends because the test specimens will be in two pieces after the testing is done.)

The single fiber tensile tests were performed in accordance with ASTM D3379-75 using an Instron Model 1101 mechanical testing machine. The fixed crosshead of the mechanical tester was fitted with a Sensotec Model 31 load cell with a 1000-g capacity. A Precision Digital Model PD691 digital strain gauge meter was connected to the load cell to record maximum load applied up to the point of fiber failure.

Tensile testing of each fiber was carried out as follows: The moving crosshead was fitted with a mechanical grip that was suited to the tabs that were used. A mechanical grip was custom fabricated from aluminum to fit both the test tabs and the Sensotec load cell that was attached to the fixed crosshead. The distance between the grips was set such that each grip terminated at the approximate midpoint between the end of the tab and the end of the gauge length cutout. The tester was set up so that the moving crosshead would return to this position after each test. Each tab with the mounted fiber was carefully placed into the upper and lower grips so that the fiber was aligned as closely as possible with the direction of pull of the tester. The grips were then tightened down on either end of the tab, and the side sections of the tab were carefully cut with scissors, leaving only the fiber connected between the grips. The strain gauge meter was set to the mode that would display peak load in grams and zeroed. The moving crosshead of the tester was set to travel at 0.5 cm per minute. Once the meter was zeroed, the moving crosshead was activated until the readout on the meter stopped changing, indicating fiber failure. The peak load for each fiber was recorded in a lab notebook before proceeding to the next fiber test.

XPS Analysis

The surface chemical composition of reclaimed carbon fibers from the RAC 6350 prepreg was analyzed using X-ray photoelectron spectroscopy (XPS). XPS spectra were obtained using a Physical Electronics 5400 system with a chamber base pressure of $<1 \times 10^{-9}$ Torr. XPS data were acquired using Mg K_α incident radiation with emitted photoelectrons collected at 45° from the surface normal. Total power applied to the Mg K_α X-ray source was 350 W. Under these conditions, no evidence of irradiation-induced sample degradation was detected in the spectra. Survey spectra were acquired using a pass energy of 89.45 eV, yielding a constant analyzer energy resolution $\Delta E = 1.34$ eV. For the higher resolution spectra (presented in Figures 8-10), the analyzer pass energy was 35.75 eV, yielding an analyzer resolution $\Delta E = 0.54$ eV. Binding energies were referenced to the C_{1s} peak at 284.6 eV.

RESULTS AND DISCUSSION

The results from the tensile tests of recycled (Figure 4) and control fibers are shown in Table I. Comparison of the mean values from each data set indicates a reduction in mean tensile strength of 8.8% of the recycled fibers from that of the virgin fibers. A small number of both the control and recycled fibers failed at what seemed to be an abnormally high load. Under the suspicion that two fibers may have been mounted accidentally, those samples were examined with an optical microscope at magnifications up to 625 X to determine if this was the case. All of the samples in question were verified to be single fibers. The peak load values were not considered high enough to be statistical outliers and were left in the data set.

Table I. Tensile Strength Results for Tenax Single Fibers

	Control	Reclaimed
Tensile Failure Load (g)	7.41	6.76
Standard Deviation	± 3.29	± 2.59

Figure 5 shows the Weibull analysis of the raw tensile test data. In this plot of Peak Load *versus* Cumulative Failure, the recycled fiber plot follows the control fiber plot very closely, indicating the slight loss in tensile strength discussed earlier. Table II gives the Weibull parameters from a regression fit of the Weibull plot. The scale parameter corresponds to mean of the cumulative failure distribution and the shape parameter corresponds to the shape of the distribution.

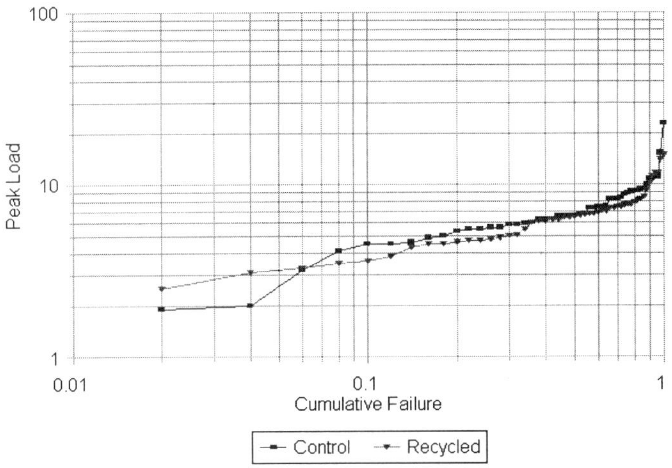

Figure 5. Weibull Analysis of Single Fiber Tensile Test Data from Heath Techna Recycled and Control Fibers

Table II. Weibull Parameters for Tenax Single Fiber Tensile Tests

Property	Control Fiber	Reclaimed Fiber
Scale Parameter	8.173	7.439
Shape Parameter	2.855	3.255

The Weibull parameters given in Table II show that the failure distribution for both fiber types is extremely flat meaning that there is a large sample-to-sample variation in strength. Both fiber types display anomalous data points on the high and low ends of the spectrum. As seen in the average failure loads, the scale parameter for the reclaimed fibers is 9% lower than the virgin fibers.

Considering the rigorous life cycle that the recycled fibers have been through -- from weaving, molding, and part fabrication in the original Heath Techna composite material and subsequent size reduction before reactor processing -- the fibers show a remarkably high degree of retention of their original tensile strength. Weaving alone often causes a 5-10% drop in tensile strength. This data gives strong support to the validity of the tertiary recycling process for reclaiming composite manufacturing scrap.

XPS Analysis

Because the performance of reinforcing fibers in composite materials is very sensitive to their surface chemistry, it is imperative that the surface composition of the reclaimed carbon fibers be characterized. An additional potential concern when prepreg is recycled with the silicone-treated paper release ply is that the silicone could be deposited in some form on the fiber surfaces. Such a layer could potentially reduce the value of the recycled fibers if it adversely affected composite properties. To check that concern, recycled AS4 and virgin, unsized AS4 fibers were analyzed using XPS. Survey spectra of unsized and recycled AS4 carbon fibers are shown in Figures 6 and 7. No substantial differences are seen in the two survey spectra.

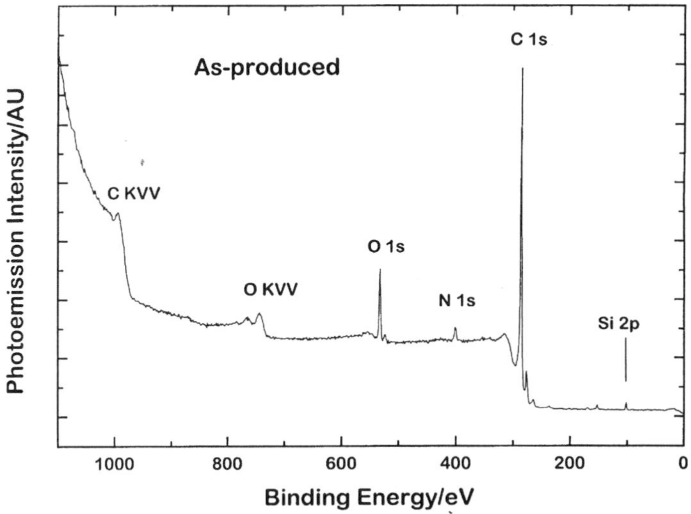

Figure 6. Survey Spectrum of Virgin, Unsized AS4 Carbon Fiber

Integration of the survey spectra peaks in Figures 6 and 7 yields the atomic concentration results given in Table III. These values were calculated using relative sensitivity factors provided by the instrument manufacturer and are therefore subject to possible large errors in absolute accuracy due to chemical matrix effects and sample topography. However, the relative changes in surface concentration measured between these similar substrates given in Table III are useful for comparison.

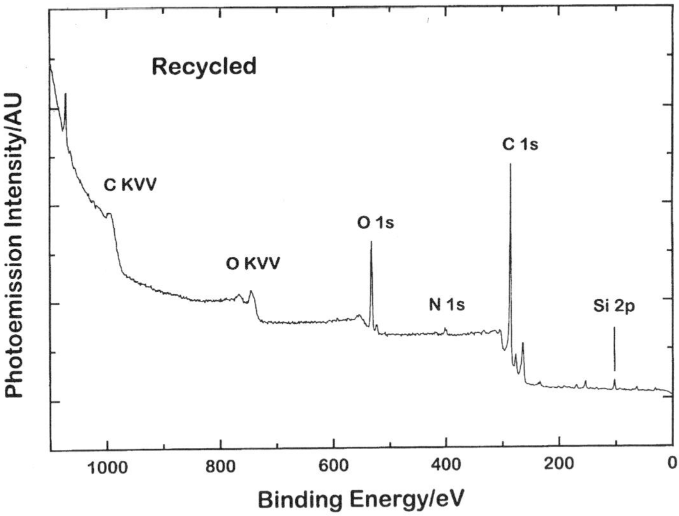

Figure 7. Survey Spectrum of Recycled AS4 Carbon Fiber from Ciba RAC 6350 Prepreg

Table III
Atomic Composition of Virgin and Recycled AS4 Carbon Fibers

	Virgin AS4	Recycled AS4
Element	**Concentration (at. %)**	**Concentration (at. %)**
C_{1s}	85.77	75.49
O_{1s}	8.47	15.53
S_{2p}	0.14	0.60
Si_{2p}	1.54	2.81
N_{1s}	3.89	2.71
Na_{1s}	0.19	2.85

The data given in Table III show that there is remarkably little difference between the two fibers, especially given the complex and violent history seen by the recycled fiber. The recycled fiber has been sized, prepregged with epoxy, molded, size reduced, and recycled using the low-temperature catalytic conversion process.

Additional evidence for the similarity between the two fibers is found in the high resolution photopeaks for carbon, oxygen, and silicon given in Figures 8-10. The high resolution C_{1s} photopeak (Figure 8) shows no difference between the two fiber types. This means that the increase in oxygen seen in Table III on the recycled fibers is either not bound to carbon or is in an equivalent chemical form to the original oxidized surface from the shear treatment applied by Hercules.

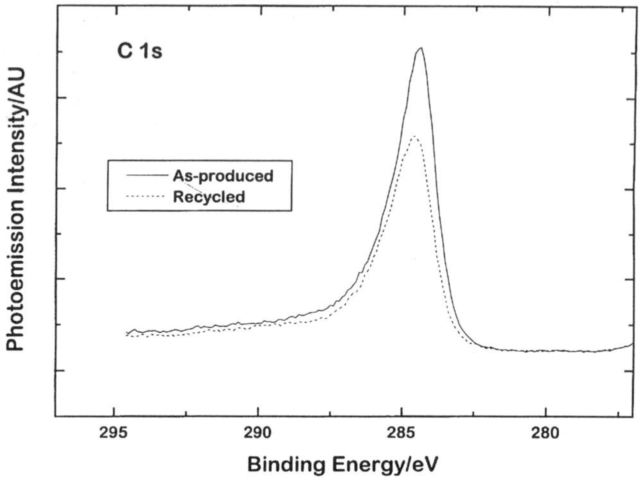

Figure 8. High Resolution C_{1s} Photopeaks for Virgin and Recycled AS4 Fibers

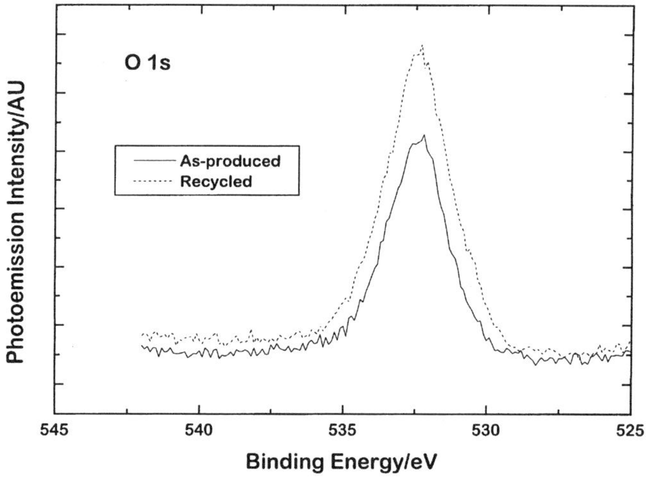

Figure 9. High Resolution O_{1s} Photopeaks for Virgin and Recycled AS4 Fibers

Similarly, the O_{1s} photopeaks are nearly identical with only a small low binding energy tail shift seen in the recycled material (Figure 9). A small deviation on the high binding energy tail of the Si_{2p} is also seen in Figure 10. These are surprisingly small changes. Apparently, the silicon that is present on the recycled fibers is well oxidized and appears similar to that seen on the original fibers from dust (silicate) contamination. However, the resolution of these spectra is insufficient to distinguish between silicate contamination and silicone compounds [5]. Thus, the possibility of a small surface concentration of silicone species cannot be completely excluded. The additional oxygen seen on the recycled fibers can be attributed to a mixture of hydroxyls, carbonyls, and carboxyls with a distribution similar to the oxygen moieties on the virgin fibers. These XPS data indicate that overaged prepreg can be recycled with the release plies without sacrificing fiber quality, which should be a cost savings in operating a recycling plant.

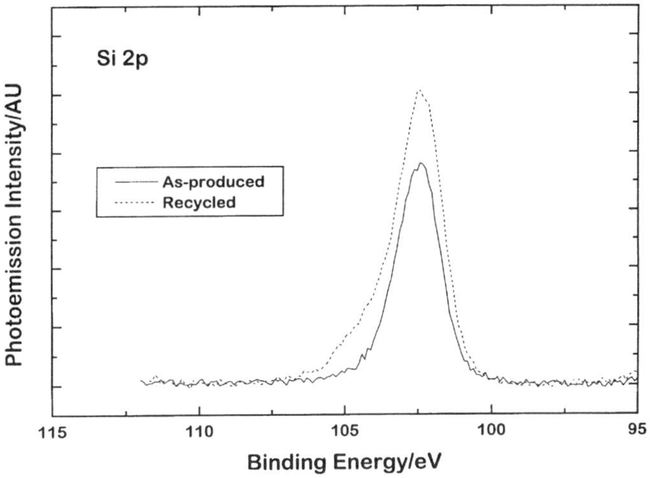

Figure 10. High Resolution Si_{2p} Photopeaks for Virgin and Recycled AS4 Fibers

SUMMARY AND CONCLUSIONS

Results show that the catalytic conversion process under development reclaims carbon fibers from scrap prepregs and cured composites in near virgin condition. Single fiber tensile strengths from recycled fabric/epoxy composites display a 9% loss compared to virgin, unwoven fibers and show similar Weibull parameters. Much of the observed strength loss may be due to the weaving process rather than the recycling process. Carbon fiber surface chemistry is strikingly similar when comparing reclaimed and virgin fibers. An increase in surface oxygen is seen on the recycled fibers that has a nearly identical distribution of XPS binding energies for the C_{1s} and O_{1s} high resolution peaks. Thus, the oxygen on the carbon fiber surfaces corresponds to the same types and ratios of moieties (hydroxyl, carbonyl, carboxyl) as produced by the shear treatment on the virgin fibers. Carbon fibers with near virgin properties that are reclaimed from scrap composite materials should have substantial value for use in molding compounds or as feedstock for milling.

ACKNOWLEDGMENTS

Mr. Tom Doak and Mr. Lincoln Busselle conducted the recycling trials and Mr. Brent Gordon conducted the Weibull analysis at Adherent Technologies, Inc. Mr. Roy Chandler at Heath Techna and Mr. Ian Stoddart at Ciba Composites (Hexcel) generously provided the scrap composite feedstocks.

REFERENCES

1. R. E. Allred and R. M. Salas, "Chemical Recycling of Scrap Composites," *Proc. AIA, SACMA, and NASA Joint Conference on Environmental, Safety, and Health Considerations--Composite Materials in the Aerospace Industry*, Phoenix, AZ, October 20-21, 1994.

2. R.E. Allred, R. M. Salas, and B.W. Gordon, "Tertiary Recycling Process for Plastics, Composites, and Electronic Materials," *Proc. 40th Intl. SAMPE Symp. and Exhib.*, Anaheim, CA, May 8-11, 1995.

3. R. E. Allred, "Recycling Process for Scrap Composites and Prepregs," *Proc. 41st Intl. SAMPE Symp. and Exhib.*, Anaheim, CA, March 24-28, 1996.

4. ASTM D3379-75, "Test Method for Tensile Strength and Young's Modulus for High-Modulus Single-Filament Materials," *Annual Book of ASTM Standards*, 1982.

5. J. F. Moulder, W. F. Stickle, P. E. Sobol, and K. D. Bomben, *Handbook of X-ray Photoelectron Spectra*, Physical Electronics, Inc., Eden Prairie, MN, 1995.

BIOGRAPHIES

Dr. Allred earned his ScD in the Polymerics Panel of the Department of Materials Science and Engineering at MIT. He has a BS in Chemistry and an MS in Nuclear Engineering, both from the University of New Mexico. Before establishing Adherent Technologies in 1990, Dr. Allred was Director of the Materials Development Department at PDA Engineering and a Member of the Technical Staff at Sandia National Laboratories. Dr. Allred's research in addition to polymer recycling has included interfacial adhesion and environmental effects on composite materials, heat damage in aircraft composite systems, materials for electronic packaging, chemical microsensors, and the development of specialty polymers, adhesives, coatings, and foams. Through his 27 years in the materials industry, Dr. Allred has over 60 publications in technical journals and conference proceedings and has 20 patent disclosures and 50 company reports in polymer and composite technology.

Mr. Coons earned his B.S. in Materials Engineering from the New Mexico Institute of Mining and Technology. Mr. Coons' experience is primarily in the areas of materials fabrication and testing, process design and engineering, mechanical design, and mechanical testing. Before joining Adherent Technologies in 1996, Mr. Coons was a process engineer with Merrick and Company under subcontract to Los Alamos National Laboratory.

Dr. Simonson is a Senior Member of the Technical Staff in the Surface and Sensor-Controlled Processes Department at Sandia National Laboratories. He obtained a B.S. degree in chemistry from Bradley University and a PhD. in Physical Chemistry from the Massachusetts Institute of Technology. Dr. Simonson specializes in the spectroscopy and analysis of solid surfaces and interfaces. His research interests include contamination control in semiconductor processing, chemical modification of surface properties, and adsorbate effects on optical scattering. He previously held research positions at the Dow Chemical Company, Los Alamos National Laboratory, and the U.S. Air Force Phillips Laboratory before joining Sandia National Laboratories in 1994.

28th International SAMPE Technical Conference
November 4-7, 1996

SELECTION AND DEMONSTRATION OF A LOW-EMISSION SOLVENT FOR CLEANING PAINT APPLICATION EQUIPMENT

Jenni M. Elion, Research Triangle Institute, Research Triangle Park, NC
Dan Gillum, Marine Corps Logistics Base, Albany, GA

ABSTRACT

The Marine Corps Logistics Base (MCLB) in Albany, GA, carries out maintenance activities on a wide variety of equipment from small arms to tanks, trucks, and other vehicles. The processes for paint stripping, repainting, and cleaning of paint equipment release significant amounts of hazardous air pollutants (HAPs). By Executive Order 12856, the MCLB is required to reduce these air emissions by 50% from 1992 levels. The MCLB desires to accomplish this goal by implementing pollution prevention (P2) technologies. The purpose of this project was to demonstrate on a full production scale a replacement for methyl ethyl ketone (MEK) for cleaning the paint application pumps, hoses, and guns at the MCLB. Based on the results of materials compatibility and paint removal efficiency tests, the cleaner selected for demonstration at the base was a blend of 40% propylene carbonate and 60% benzyl alcohol (PC/BA), by weight. MCLB used this solvent as a direct replacement for MEK. No capital investment was required. Based on the demonstration, PC/BA cleans green Chemical Agent Resistant Coating from the pumps as well as MEK, and cleans epoxy primers better than MEK. This substitution will lower emissions from HAPs 21% from 1992 levels.

KEY WORDS: Cleaning, Environmental/Recycling/Pollution Prevention, Solvent Alternatives

1. INTRODUCTION

1.1 Background The demonstrations in this project were performed primarily at the Marine Corps Logistics Base (MCLB) in Albany, GA. This base carries out maintenance activities on a wide variety of equipment from small arms to tanks, trucks, and other vehicles. Much of the maintenance on the vehicles requires removing the existing paint prior to the repair procedures, and application of new paint once the maintenance has been performed. The processes for paint stripping, repainting, and cleaning of paint equipment release significant amounts of hazardous air pollutants (HAPs). By Executive Order 12856, the MCLB is required to reduce these air emissions by 50% from 1992 levels. The MCLB desires to accomplish this goal by implementing pollution prevention (P2) technologies.

The MCLB primarily uses air-assisted airless spray systems to apply paint; cup sprayers are used for touch-up paint. The coatings used by the MCLB can be classified as polyurethane-based plural-component Chemical Agent Resistant Coatings (CARCs) and single-component CARCs, epoxy primers, and undercoatings.

MEK needs to be replaced because it is a hazardous air pollutant (HAP) and is one of seventeen toxic chemicals identified in the Environmental Protection Agency (EPA) 33/50 Program. The 33/50 Program is a voluntary pollution prevention initiative to reduce national pollution releases and off-site transfers. The seventeen toxic chemicals chosen were selected because they pose environmental and health concerns, are high-volume industrial chemicals, and may be reduced through pollution prevention technologies.

The EPA Air Pollution Prevention and Control Division initiated this project with funding from the Strategic Environmental Research and Development Program and worked with Research Triangle Institute and the MCLB to identify and demonstrate a suitable alternative.

1.2 Objectives The objective was to demonstrate a process that would reduce the emissions of HAPs resulting from cleaning the paint application equipment. This objective was achieved by replacing the MEK with a less hazardous, less volatile solvent and a combination of work practices. These technologies and work practices were tested at RTI and then demonstrated on-site at the MCLB over a four-week period. The MCLB continued to use the new process after completion of the demonstration.

2. TECHNICAL EVALUATION OF ALTERNATIVES TO MEK

2.1 Preliminary Screening The first step in finding a replacement solvent was to test the efficacy of a group of solvents that was most likely to be capable of removing the CARCs and primer. The list of capable solvents and solvent mixtures was narrowed to those that best met environmental, cost, and safety criteria. These criteria included high flash point, low vapor pressure, and removal of coatings at ambient use temperature.

RTI sent four aluminum 15.2 cm x 15.2 cm x 0.15 cm (6 in. x 6 in. x 0.060 in.) panels to the MCLB, where they were each coated with one of four coatings: one-component CARC, two-component CARC, epoxy primer, or undercoating. The panels were returned to RTI. The panel coated with the undercoating arrived at RTI in poor condition and was not included in the preliminary screening. Each of the other panels was sheared into 1.3 cm x 1.3 cm (½ in. x ½ in.) squares. The square coupons were sent to Huntsman Specialty Chemical Corporation in Austin, TX, for preliminary screening tests.

Preliminary screening consisted of visual examination of the effect of different solvent formulations on the three coatings. Although the coatings cleaned at MCLB are usually uncured or only partially cured, the preliminary screening was conducted on fully cured coatings to ensure that a robust cleaner would be selected. Sixty-five different cleaners were tested. The cleaners were various proportions of propylene carbonate and other compounds, including methyl isoamyl ketone, dibasic esters, n-methyl pyrrolidone (NMP), and benzyl alcohol.

For each test, one test square coated with one-component CARC was placed in a test vial, one test square with two-component CARC was placed in another test vial, and one test square with epoxy primer was placed in a third test vial. Approximately 2 ml of the test blend was placed in each of the three vials. This was repeated for all sixty-five cleaners. MEK was also included to provide a baseline for comparison. Each coupon was visually inspected after twenty-four hours and again after forty-eight hours. Samples were rated "L" for little or slight attack as evidenced by small pock marks or blisters in the coatings. Samples were rated "M" for moderate attack as evidenced by wrinkling or blistering of the coating. Samples were rated "S" for severe attack, indicated by heavy blistering, wrinkling, or removal of the coating. No rating indicated no observed effect on the coating. MEK, the current cleaner, had no observed effect on the three coatings.

Only one screening formulation, #48, severely attacked all three coatings after forty-eight hours. Three other formulations severely attacked two of the three coatings and moderately attacked the third. Based on the results of the screening study, five formulations were selected for further tests. These five formulations are listed in Table 1. Formulations with high percentages of NMP or benzyl alcohol appeared to be the most effective. For this reason, formulations #5 and #48 were chosen as likely candidates to replace MEK. Noting that the screening study with cured panels was a "worst-case" condition, formulation #64 was chosen to determine whether a formulation with a lower percentage of benzyl alcohol would prove effective. Formulation #65 was chosen to represent the propylene carbonate/NMP blends. Propylene carbonate (PC), the common constituent, was also included for further testing even though screening tests were not done with 100% PC.

TABLE 1. CLEANERS SELECTED FOR TECHNICAL EVALUATION

Product	Formulation
Formulation #5	Propylene carbonate (40%) Benzyl alcohol (60%)
Formulation #48	Propylene carbonate (10%) Dibasic esters (19.2%) γ-Butyrolactone (10.8%) Benzyl alcohol (60%)
Formulation #64	Propylene carbonate (33-1/3%) Dibasic esters (33-1/3%) Benzyl alcohol (33-1/3%)
Formulation #65	Propylene carbonate (50%) n-Methyl pyrrolidone (50%)
Propylene carbonate	Propylene carbonate (100%)

2.2 Material Compatibility The choices of materials for compatibility testing were based on wetted materials in the paint application pump systems then in use at the MCLB. The MCLB provided RTI with the instructions and parts list for the Graco plural component mixer manifold then in use to spray coatings. The manifold parts that would come in contact with the paint and paint cleaner included chrome alloy, 303 stainless steel, Delrin™, nylon, and Teflon™. The test materials chosen were aluminum, brass, nickel, 303 stainless steel, Delrin™, nylon, and Teflon™.

Test coupons measuring 1.9 cm x 5.1 cm x 0.3 cm (3/4 in. by 2 in. by 1/8 in.) thick were made for each material. Each coupon was weighed to the nearest 0.01 mg and the weight recorded. Replicate coupons of each material were placed in three separate test vials. Approximately 4 ml of cleaner were added to each vial, which covered about half the coupon. This allowed exposure of the coupon to the liquid phase, the liquid/vapor interface, and the vapor phase of each cleaner. Each vial was then capped and stored at room temperature. The test coupons were removed and weighed to the nearest 0.01 mg after one day immersion. They were returned to the vials and reweighed after three days immersion and again after seven days immersion. A fourth coupon, the control, was also reweighed after one, three, and seven days exposure to ambient conditions.

When the coupons were removed from a cleaner to be weighed, they were patted dry with absorbent towels for 60 seconds and then immediately weighed. Excessive contact with air and delays in weighing were avoided to minimize the effects of ambient exposure. Coupons were returned to their individual vials as soon as possible after weighing. Weight changes indicate loss of material or absorption of cleaner.

Visual examination of the test coupons during weighing revealed no pitting, flaking, or discoloration on any coupons. No signs of corrosion were visible on the metal coupons. No shape distortion was visible on the plastic coupons.

A comparison of the gravimetric analyses of the compatibility tests for the five cleaners and MEK is shown graphically in Figure 1. This figure shows only the three plastics because there was no significant change in the weights of any of the metals with any of the cleaners. The data in Figure 1 indicate that Teflon™ is the most stable of the plastics over the seven-day exposure. It absorbed a small percentage of MEK, but not the other cleaners. Percent change in coupon weight for the other cleaners is very close to the limits of detection for the method. The Delrin™ absorbed the most of each of the five cleaners and absorbed more MEK than any of the other cleaners. The data indicate that nylon lost a small weight percentage upon seven-day immersion to the five cleaners. Cleaner formulations #5 and #48 were chosen for further evaluation because they showed the least weight gain or loss for the plastics.

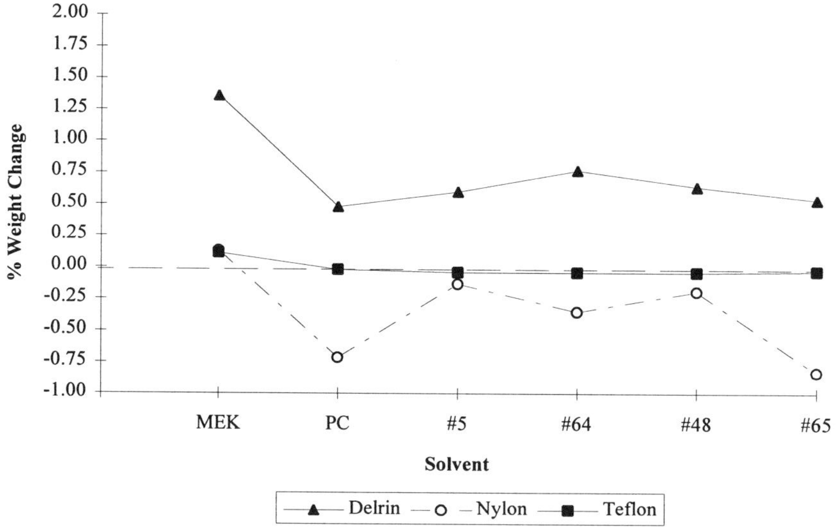

FIGURE 1. GRAVIMETRIC RESULTS OF MATERIALS COMPATIBILITY TESTS

2.3 Paint Removal Efficiency Cleaners #5 (40% PC/60% benzyl alcohol) and #48 (10% PC/19.2% dibasic esters/10.8% γ-butyrolactone/60% benzyl alcohol) were selected from the material compatibility study to be evaluated further for paint removal efficiency. MEK was used as a baseline for comparison.

In this test, cleaner was pumped through paint coated tubes to simulate the way the hoses for the paint spray equipment are flushed at the MCLB. A small section of glass tubing was connected between two sections of hose to allow visual examination and gravimetric analysis of the paint removal by the cleaners. In this set of tests, paint was not allowed to cure before removal to more closely reproduce actual cleaning conditions at the MCLB.

The 7.6 cm (3 in.) long glass tubes had an inside diameter of 6 mm (0.220 in.) and an outside diameter of 9.5 mm (0.375 in.). All the tubes were precleaned and weighed to the nearest 0.01 mg to obtain $W_{initial}$. The paints were stirred before use. For the dual-component CARC and epoxy primer, the two components were mixed according to the manufacturer's instructions. The tubes were contaminated, reweighed, and cleaned one at a time to minimize paint curing. Paint was applied to the interior of each glass tube using a cotton swab. The tubes were reweighed to the nearest 0.01 mg to obtain the coated weight, W_{coated}. To clean, approximately 200 ml of the appropriate cleaner was flushed through each glass tube from a low-pressure pot. The tubes were dried overnight and reweighed to the nearest 0.01 mg to obtain the cleaned weight, $W_{cleaned}$. Each paint was tested with three replicates and a control for each cleaner.

The means and standard deviations are shown graphically in Figure 2. Cleaner performance was determined by averaging all samples with a given coating for a given cleaner. Neither of the two cleaners was as good at removing the undercoating (black) as MEK. Further discussions with the MCLB revealed that the black undercoating is thinned before applying and sometimes heated to reduce its viscosity. These factors would significantly effect its ability to be cleaned. Formulation #5 was nearly as effective at removing the dual-component CARC (green) as MEK. Neither of the two cleaners was as good at removing the single-component CARC (tan) as MEK. Although MEK was the current standard at the MCLB, it did not successfully remove all four coatings as expected. Both of the cleaners were significantly better than MEK at removing the epoxy primer (white).

Based on this data, formulation #5, the blend of propylene carbonate and benzyl alcohol, was chosen for demonstration at the MCLB. The formulation is a custom-made blend with no commercial name. It will be referred to through the rest of this paper as PC/BA (propylene carbonate/benzyl alcohol) rather than #5.

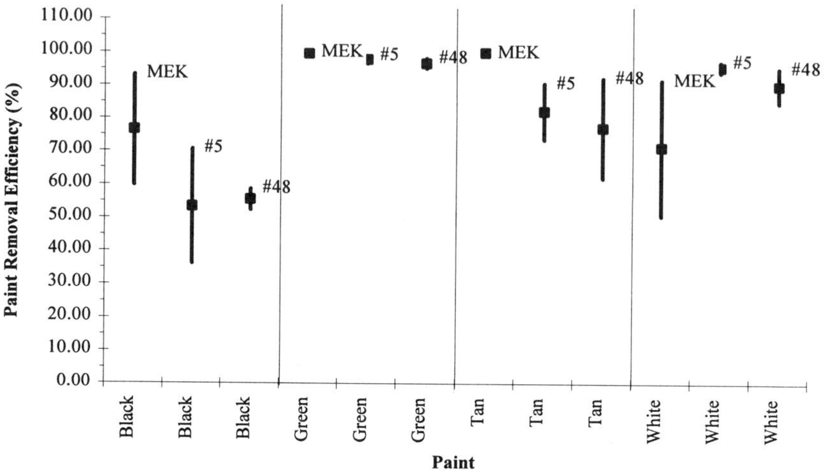

FIGURE 2. PAINT REMOVAL EFFICIENCY RESULTS OF MEK ALTERNATIVES

3. ON-SITE DEMONSTRATION OF ALTERNATIVE PAINT EQUIPMENT CLEANER PROPYLENE CARBONATE/BENZYL ALCOHOL

3.1 Current Process More than 27,220 kg (60,000 lbs) of MEK was purchased by the MCLB in 1992 for cleaning paint application equipment. Equipment is cleaned at the end of each shift and between color changes. Previously used solvent is flushed through the hoses and guns as a prewash to remove remaining paint. The equipment is flushed with prewash solvent until no paint is visible in the wash effluent. The prewash is collected in a hazardous waste container which is later emptied into a hazardous waste barrel. The paint pickup tube then is placed in a container of fresh solvent for a final wash. Fresh solvent is flushed through the hoses and guns and collected in the prewash bucket. The pump filter is then backflushed with fresh solvent and the waste disposed of in the hazardous waste container. Finally, fresh solvent is recirculated through the system for several minutes.

3.2 Setup of Demonstration Before starting the official on-site demonstration phase, RTI met with the MCLB painters to describe the test results and new cleaner, and answer any questions. The painters' main concerns were that the new cleaner would not remove the paint as well as MEK or that it would work so well that it would cause large chunks of paint to be released from the hoses, causing the pumps to plug and then seize. To address these concerns, RTI conducted preliminary tests at the MCLB with the cooperation of the painters. The painters first sprayed a sample of the cleaner through the pumps and confirmed that the cleaner could withstand the high pressures with no adverse reactions.

Next, three pumps were cleaned prior to the quantified portion of the demonstration. The first pump, a Graco Senator, had been flushed with MEK and was considered clean by the operators. The area leaderman placed a filter in line in one of the hoses to see if the new cleaner would cause sloughing of chunks of paint. About 19 liters (5 gallons) of PC/BA were continuously recirculated through the pump for about two hours. The PC/BA removed additional paint from the equipment, but there were no large chunks caught in the filter that might have caused the pump to seize. Some paint residue was left on the main filter, and the in-line filters were clean. A catch sample taken from the bucket showed small green particles in suspension, which later settled. The cleaner warmed slightly when recirculated for over an hour and heated the pump slightly as well. The area leaderman did not consider this a problem as the cleaning process generally takes less than half an hour. Air pressure of 310 kPa (45 psi) was required to circulate the cleaner at the beginning of the test. Required air pressure increased to 414 kPa (60 psi) after half an hour. At the end of the test, only 276 kPa (40 psi) was required. The painters were pleased that the new cleaner was able to remove paint left behind by the MEK and that it broke down the old paint residue into small pieces that did not seize the pumps.

The second pump cleaned, also a Graco Senator, had recently been torn down, thoroughly cleaned, and rebuilt. The spray system was used for a single shift and then cleaned with PC/BA. The cleaning procedure was the same as that for cleaning with MEK. The dirty PC/BA from the first pump served as the prewash for the second pump. A final wash of fresh PC/BA was recirculated through the lines. Again, the cleaner removed the paint from the pump satisfactorily and did not cause any large clumps of paint to be released that might cause the pump to seize.

The third pump cleaned was a Graco 5-gallon model used only for spraying white primer. The cleaner dissolved the paint and flushed it completely. A catch sample taken from the waste container showed no settling or paint flakes; instead, the white primer appeared to dissolve completely.

The painters and RTI were pleased with the results of these tests and proceeded to the full-scale demonstration of replacing MEK with PC/BA for cleaning the paint spray equipment.

3.3 Demonstration of Alternative Cleaner A scale and flow monitoring system were used to monitor the amount of cleaner and paint flushed through the system. Accessories to the system included a tape printer, a remote Tare and Print push-button station, and a multiport controller. The system was used for analysis purposes only and is not required to implement the new cleaner.

Collecting data did not interfere with the painters' ability to clean the equipment. For the prewash, the painters placed the hazardous waste container on the scale and pressed <*Tare*> to zero the display. The painters continued the prewash as normal, with the paint pickup tube in the prewash container. The painters flushed the cleaner through the hoses and guns into the hazardous waste container on the scale until the cleaner was visibly free of paint. For the final wash, the painters switched the paint pickup tube from the prewash bucket to the final wash bucket. Then they replaced the hazardous waste container on the scale with the prewash bucket, and pressed <*Tare*>. The filter wash used fresh cleaner from the final wash bucket, but the waste was dumped into the hazardous waste container, not the prewash bucket. Some painters included the filter wash in the prewash by simply switching the paint pickup tube from the prewash bucket to the final wash and leaving the hazardous waste container on the scale. Other painters took the extra step of switching the prewash bucket with the hazardous waste container again and pressing <*Tare*>. Either was acceptable since it was the total amount used that was recorded. Painters printed the data block before and after the prewash, the final wash, and the filter wash, totaling the amount of cleaner flushed through the system.

All data were collected from one spray booth. Although both tan and green CARC are sprayed in this booth, during the course of data collection, only green CARC was used. The painters also continued to use PC/BA to clean the white epoxy primer from the Graco 5-gallon pumps.

The amount of PC/BA used to flush the Graco Senator clean of green CARC is shown in Table 2. The operators considered the Senator to be clean when no paint was visible in the wash solvent as it came out of the spray gun. Because the printer included the date and time with each data block, it was also possible to determine how long the new cleaner took to flush the equipment. Some painters included the filter wash as part of the prewash, as described earlier. This is indicated by an asterisk in the filter wash column in Table 2. Cleaning data for the training sessions is included in averaging the amount of cleaner used to flush the equipment clean. However, the time for the training sessions is not included in the average for process time. The training sessions took longer than the normal cleaning procedure because time was taken to answer questions as they arose.

Although data were not collected for the Graco 5-gallon pump used only for spraying primer, the operators have successfully used the blend to clean it since November 14, 1995.

The amount of MEK used to flush the Graco Senator clean of green CARC is shown in Table 3. As can be seen from the table, these data were collected after the completion of the demonstration. Although area leadermen reported that the cleaning process takes two operators using 19 liters (5 gallons) of MEK about fifteen to twenty minutes per paint spray system to clean the equipment, data collected for verification indicate less MEK is actually used and the cleanup process takes less time. Because data for MEK usage were collected after the PC/BA demonstration, it is possible that the operators had gotten used to using less solvent and spending less time to clean the pumps. For this reason, the MEK figures in Table 3 may be artificially low.

No direct correlation can be made between volume of cleaner and time required to clean the equipment. Various factors affect both these measures. Some of these factors are the time between completing a paint job and cleaning the equipment, ambient conditions, and human variability. These factors were not controlled during the demonstration.

The values in Tables 2 and 3 form the basis of assumptions regarding usage in the economic analysis.

TABLE 2. VOLUME AND TIME TO CLEAN EQUIPMENT WITH PC/BA

DATE	*START*	*STOP*	*PREWASH (liters)*	*FINAL (liters)*	*FILTER (liters)*	*TOTAL (liters)*	*ELAPSED TIME (h:mm:ss)*
11/28/95	3:21:03 PM	3:30:13 PM	3.41	2.46	*	5.87	0:09:10**
11/29/95	3:17:17 PM	3:26:38 PM	4.62	1.29	1.14	7.05	0:09:21**
11/30/95	3:26:08 PM	3:30:24 PM	5.49	1.02	*	6.51	0:04:16
12/05/95	3:33:31 PM	3:44:35 PM	3.26	4.39	*	7.65	0:11:04
12/13/95	3:23:51 PM	3:28:54 PM	3.79	2.35	1.10	7.24	0:05:03
12/14/95	10:26:35 AM	10:30:33 AM	4.58	3.67	2.16	10.41	0:03:58
12/14/95	3:25:43 PM	3:34:23 PM	3.14	2.23	2.12	7.49	0:08:40
12/15/95	11:32:28 AM	11:35:22 AM	4.69	2.61	2.73	10.03	0:02:54
					AVERAGE:	7.78	0:05:59
					STANDARD DEVIATION:	1.61	0:03:11

* Indicates filter wash was included with prewash

** Not included in average and standard deviation because time was taken to answer questions

TABLE 3. VOLUME AND TIME TO CLEAN EQUIPMENT WITH MEK

DATE	*START*	*STOP*	*PREWASH (liters)*	*FINAL (liters)*	*FILTER (liters)*	*TOTAL (liters)*	*ELAPSED TIME (h:mm:ss)*
12/19/95	3:23:09 PM	3:36:53 PM	4.50	1.29	3.60	9.39	0:13:44
12/20/95	3:18:59 PM	3:24:11 PM	4.35	2.01	4.43	10.79	0:05:12
12/21/95	3:17:51 PM	3:23:49 PM	4.81	2.16	1.74	8.71	0:05:58
03/06/96	4:34:35 PM	4:38:56 PM	6.62	3.71	1.82	12.15	0:04:21
03/06/96	9:22:07 AM	9:26:07 AM	5.07	2.27	1.89	9.23	0:04:00
03/06/96	1:06:22 PM	1:11:51 PM	5.34	2.73	1.97	10.04	0:05:29
03/07/96	4:27:56 PM	4:34:24 PM	6.06	5.34	2.16	13.56	0:06:28
					AVERAGE:	10.55	0:06:27
					STANDARD DEVIATION:	1.75	0:03:19

Feedback from the painters was generally positive. The painters reported that the new cleaner did not have an objectionable odor like MEK. They believed it removed the white primer from the Graco 5-gallon pumps much more effectively than MEK. They also believed that the "oily" nature of the PC/BA blend would reduce the binding that occurs when pumps are cleaned with MEK. The Senator pumps still must be torn down approximately every two weeks because paint builds up between the seals and packing, causing the pump to seize. However, the new cleaner was just as effective at cleaning the components of the torn-down pump as MEK. As of mid-March 1996, the MCLB reported that the Graco pump for primer has not required maintenance. This is a significant improvement over the maintenance required when cleaning the pump with MEK. Like the Senator pumps, it had to be torn down and rebuilt approximately every two weeks.

On the negative side, the operators were concerned about unknown health risks. They also did not like the fact that they cannot thin the CARCs with the new blend. Material Safety Data Sheets (MSDSs) for the individual components of the blend were provided to the MCLB Safety and Health. These MSDSs were more detailed than the MSDS developed for the blend. The blend is not intended to be used as a thinning agent; it is to be used for cleaning only. The MCLB uses only approved thinning agents with CARC paints.

4. IMPLEMENTATION OF NEW CLEANER

4.1 Regulations Affecting the Alternative Cleaner Replacing MEK with PC/BA decreases the burden of tracking and reporting for the MCLB. MEK is listed as a Hazardous Air Pollutant (HAP) under the Clean Air Act Amendments of 1990 (CAAA) Title III, section 112. Neither benzyl alcohol or propylene carbonate are classified as HAPs. Other regulations impacting the use of MEK are summarized in Table 4. Benzyl alcohol and propylene carbonate are not subject to these regulations.

TABLE 4. REGULATIONS IMPACTING THE USE OF MEK AND PC/BA

CAS	Chemical Name	Hazardous Substances CERCLA	Toxic Chemical SARA	RCRA Code	CAA, Title III Section 112
78-93-3	Methyl ethyl ketone	2268 kg†	313	U159	yes
100-51-6	Benzyl alcohol	Not listed	Not listed	Not listed	no
108-32-7	Propylene carbonate	Not listed	Not listed	Not listed	no

†The regulation cites reportable quantities in English units; for MEK, the reportable quantity is 5,000 lbs.

MEK is subject to reporting requirements under Title III of the Superfund Amendments and Reauthorization Act (SARA). Releases of CERCLA hazardous substances, in quantities equal or greater than their reportable quantity, are subject to reporting to the National Response Center under CERCLA. Such releases are also subject to state and local reporting under section 304 of SARA Title III. MEK is classified as a CERCLA hazardous substance; its reportable quantity is 2,268 kg (5,000 lbs). Neither benzyl alcohol nor propylene carbonate are classified as CERCLA hazardous substances.

MEK is also classified as a toxic chemical under SARA Title III, section 313. Emissions, transfers, and waste management data of MEK must be reported annually as part of the community right-to-know provisions of SARA Title III. Neither benzyl alcohol nor propylene carbonate are classified reportable under SARA Title III, section 313.

Because MEK is flammable, it is classified as a RCRA hazardous waste. Neither benzyl alcohol nor propylene carbonate are subject to RCRA. Because the paint used at the MCLB is not RCRA-regulated either, the PC/BA-paint waste stream is not RCRA-regulated.

4.2 Safety From a safety standpoint, PC/BA has several advantages over MEK. First, PC/BA is nonflammable. It is classified as combustible because its flashpoint is 101°C (213°F). Because the flash point of MEK, -6°C (20°F), is well below ambient temperatures, the use of PC/BA poses significantly less risk of fire than the use of MEK. Second, its vapor pressure is only 9 Pa (0.07 mmHg) at 20°C. This is less than MEK's vapor pressure of 10.4 kPa (78 mmHg) under equivalent conditions by three orders of magnitude.

The Threshold Limit Value for MEK is 200 ppm. Overexposure to MEK vapors may result in nausea, dizziness, and headache, among other symptoms. MEK is also a skin and eye irritant. Although no TLVs have been established for either propylene carbonate or benzyl alcohol, proper ventilation is still necessary to prevent buildup of vapors. Also, the very low vapor pressure means PC/BA will not evaporate quickly if accidentally splashed on skin. Gloves should be worn when handling the cleaner to prevent skin contact. Benzyl alcohol is a severe eye irritant. Safety glasses are necessary to prevent eye contact with the cleaner. Operators should also wash their hands before smoking or eating.

4.3 Economic Analysis One aspect of changing process operations for pollution prevention purposes is the effect on capital and annual cost. Pollution prevention changes are more likely to be embraced if these costs stay the same or decrease.

The methodology for estimating costs is taken from the EPA methods described in the OAQPS Control Cost Manual[1], which allows a convenient means of comparing different processes based on their annualized costs. Where actual costs are not known, factors applied to base equipment costs are used to estimate the remaining costs. These factors have been developed from a wide variety of sources associated with pollution control systems. The method provides a study estimate, which is intended to give a cost estimate within an accuracy range of ±30% of the actual cost when all the information affecting the costs is reasonable well known. This accuracy range is typical for EPA's estimates of pollution control systems when assessing cost impacts on existing or model facilities. Greater accuracy can be obtained with budget authorization estimates (±20%), definitive estimates (±10%), or contractor's estimates (±5%). Improved accuracy of the estimate is obtained only by improving the detailed knowledge of items that make up the

---

[1]OAQPS Control Cost Manual, EPA-450/3-90-006 (NTIS PB90-169954), January 1990.

estimate. In the present case, unit costs, rates, and consumptions are not known with sufficient accuracy to go beyond a study estimate.

The estimated annualized cost for using PC/BA is about $97,094/year. The top three costs are for the solvent, disposal costs, and labor hours (including overheads). The estimated annualized cost for using MEK as a cleaning solvent is about $93,947/year or about 97 percent of the cost for PC/BA. The major cost elements are for waste disposal, solvent and labor hours.

PC/BA has a lower usage rate and lower disposal costs (both lower quantities and lower unit cost). However, because it costs over 3 times that of MEK, the annualized cost for PC/BA is higher than that of MEK. Returning the waste cleaner to the manufacturer for reclaiming, even at zero credit, would reduce the annualized cost of using PC/BA below that of MEK. Buying the components and blending the cleaner on-site rather than buying the blended cleaner from a chemical supplier will also reduce the raw material cost. More cost savings could be realized by filtering and reusing the cleaner. Reduced pump maintenance may also be indicative of long-term savings in equipment costs due to longer equipment life.

5. DISCUSSION OF OBJECTIVE

As discussed previously, the MCLB is required to reduce emissions from hazardous air pollutants by 50% from 1992 levels.

The MCLB has already replaced 1,1,1-trichloroethane vapor degreasers with aqueous parts washers. This change alone will reduce emissions from HAPs by 16%. By eliminating the MEK used for cleaning purposes, the MCLB can further reduce emissions from HAPs another 21%. The MCLB has also replaced methylene chloride with n-methyl pyrrolidone for immersion paint stripping, reducing emissions from HAPs an additional 11%. These three changes combined result in a reduction of emissions of 48%.

The MCLB plans to replace solvent-borne CARCs with water-borne CARCs in 1996 to achieve over 50% reduction in emissions from HAPs.

6. CONCLUSIONS

Based on this demonstration, the following conclusions may be drawn:

- PC/BA cleans green CARC from the pumps as well as MEK.
- PC/BA cleans epoxy primers from the pumps better than MEK.
- PC/BA removes additional green CARC from pumps previously cleaned with MEK.
- PC/BA will not remove black undercoating.
- The use of PC/BA neither increases or decreases downtime of the pumps used for applying CARC.
- The use of PC/BA decreases downtime of the pumps used for applying primer.
- PC/BA does not dry out the seals and packings of the pumps between maintenance as did MEK.
- The use of PC/BA does not cause clumps of paint to form, which could cause the pumps to seize.
- The raw material cost for the PC/BA blend is higher than for MEK. The higher cost may be offset by recovery and reclamation of the PC/BA.

Other advantages of PC/BA for the MCLB are

- replacing MEK with PC/BA will reduce the HAP emissions at MCLB by 21%.
- PC/BA-paint waste generated at the MCLB is not RCRA regulated. (The waste would be subject to RCRA regulations if the paint contained high concentrations of heavy metal pigments.)
- PC/BA is not subject to SARA Title III reporting, and is not a HAP under the CAA, Title III, section 112.

DISCLAIMERS

Although the research described in this article has been funded wholly by the Environmental Protection Agency under contract 68-D4-0120 to Research Triangle Institute, it has not been subject to the Agency's review and therefore does not necessarily reflect the views of the Agency, and no official endorsement should be inferred.

The use of trade names and company names in this paper does not signify recommendation for use or endorsement by either the EPA or Research Triangle Institute.

ABOUT THE AUTHORS

Jenni Elion is a Research Environmental Engineer at Research Triangle Institute working in the Center for Engineering and Environmental Technology's Surface Cleaning Program. She is responsible for engineering research and development pertaining to pollution prevention in coating processes and precision cleaning of surfaces. Other activities include leading projects aimed at solvent elimination through process development and support of environmental information systems. Before joining RTI in 1993, Ms. Elion was associated with General Dynamics Space Systems in San Diego, CA. As Coordinator of the Composites Processing Laboratory, Ms. Elion was responsible for supervising tasks required to support the Atlas and Titan/Centaur programs and research and development work. Ms. Elion has a B.S. in materials engineering from Virginia Polytechnic Institute.

Dan Gillum is an Environmental and Safety Engineer in the Maintenance Directorate at the Marine Corps Logistics Base in Albany, GA. The Directorate is responsible for environmental and safety programs at MCLB Albany and MCLB Barstow, CA, and leads pollution prevention efforts at both facilities. Before joining the Maintenance Directorate, Mr. Gillum directed the environmental program at Reese Air Force Base in Lubbock, TX, and at MCLB Albany. Mr. Gillum holds a B.S. in mechanical and aerospace engineering from the University of Missouri–Columbia and has 25 years' experience in environmental engineering. He is a registered Professional Engineer in Iowa and Indiana and is the recipient of the Thomas D. White Environmental Award (1988), Professional Engineer Environmental Award (1989), Marine Corps Environmental Award - Industrial Division (1993), and the Marine Corps Pollution Prevention Team Award (1995).

28th International SAMPE Technical Conference
November 4-7, 1996

A Simple Model for Predicting the Bending Response of Partially Delaminated Composite Plates

X. Quan, R. Vaziri
Composites Group, Depts. of Civil Engineering and Metals and Materials Engineering
The University of British Columbia, Vancouver, Canada
D. Ursenbach
CHT Aerospace Inc., Langley, British Columbia, Canada
and
D. Delfosse
EMPA Thun, Feuerwerkerstrasse 39, CH-3602 Thun, Switzerland

ABSTRACT

Delamination is a common failure mechanism in composite laminates when subjected to out-of-plane loading such as occurs during impact. Since delamination results in significant reduction of structural stiffness it is vital to understand and predict the response of composite laminates consisting of internal delaminations. This is particularly important in assessing the ultimate load carrying capability of a laminate.

In the present investigation, a simple analytical model is developed for the response of a delaminated plate to static punch loading. In this approach the plate is assumed to consist of a central delaminated region embedded in an otherwise undamaged plate. The delaminated region is modelled as a split plate with free parameters built in to account for the number and location of delamination interfaces through the thickness. Experimental measurements and damage detections using C-scans and sectional micrographs are used to provide an estimate of these parameters. The predicted load-displacement curve and the total energy absorption during punching are in reasonable agreement with the experimental results.

KEY WORDS: Carbon Fibre-Reinforced Composite; Delamination Modelling; Punch Test.

1. INTRODUCTION

Laminated composite structures are known to have a low resistance to out-of-plane loading such as impact. Internal delamination is a major damage mode that occurs during impact on these structures and although it contributes to absorption of impact energy it results in a substantial reduction of structural stiffness which is undesirable if these structures are to remain in service. Therefore, understanding and predicting the highly non-linear behaviour of laminates due to delamination is of paramount concern.

Modelling impact induced delamination has been tackled by several authors. A selective number of the previous work are reviewed here. Assuming a rectangular shape of impact induced delaminations, Wu and Springer (1) proposed a failure criterion based on the concept of dimensional analysis to predict the location and size of delaminations. The same procedure was also used in (2, 3) to extend the delamination model to a broader range of impact conditions, where a failure criterion based on strain energy release rate was proposed. However, these criteria usually require a detailed knowledge of stress distribution from extensive two- or three-dimensional finite element analysis of the problem which can often be a daunting task.

Experimental study conducted by (4-6) indicated that in static punch tests on carbon/epoxy laminates the relationship between the contact force and projectile displacement showed approximately the same trends for various laminate spans. It was inferred from the punch load-displacement curves that the load at onset of damage and the ultimate load were about the same. Impact test results also showed that the dynamic failure modes were very similar to those obtained under static loading (4,5). This close agreement in damage modes indicated the possibility of using a static punch curve to characterise the dynamic penetration process.

Davies *et al.* (7) constructed a simple circular plate model where all delaminations were collapsed onto a singular circular patch as inner portion of the plate. A classical plate analysis based on the concept of strain energy release rate was carried out to determine the critical contact force at which delamination crack starts to propagate. Sun and Potti (6) proposed a simple ring model to predict damage propagation and energy absorption during the static penetration process. The transverse shear modulus was degraded to model the progressive reduction of plate stiffness appearing in static punch curves.

Although all these models are simple and can essentially be used to predict energy absorption due to delamination, no information about the location and size of delaminations were provided.

In the present investigation, a simple analytical model is developed for the non-linear response of a partially delaminated plate under static punch loading. In this approach the plate is assumed to consist of a central delaminated region embedded in an otherwise undamaged plate. The delaminated region is modelled as a split plate with free parameters built in to account for the number and location of delamination interfaces through the thickness. Experimental measurements and damage detections using C-scans and sectional micrographs are used to provide an estimate of these parameters. A methodology is devised for predicting the load-displacement curve during punching of laminated plates. This theoretical strategy is then validated by comparing the predicted shape of the load-displacement curves and energy absorption with corresponding test results for quasi-isotropic carbon fibre-reinforced plastic (CFRP) laminates of various thicknesses.

2. EXPERIMENTAL OBSERVATIONS

The present experimental study is an extension of the previous work by Ursenbach *et al.*. (8). The tests consisted of the static penetration of CFRP laminates (T300H/F593) with a lay-up of $[-45/90/45/0]_{ns}$, where n is equal to 2, 3, 4, and 5 corresponding to plate thicknesses of 3.35, 5.00, 6.70, and 8.30 mm, respectively. The specimens were loaded with a cylindrical punch of 7.6 mm diameter and simply-supported over rectangular (76.2 x 127.0 mm) and circular (114.3 mm diameter) openings.

Figure 1 shows the measured load-displacement curves for the CFRP laminates under displacement-controlled punch tests. An idealised schematic of a typical load-displacement curve is shown in Figure 2. Three distinct stages of response can be clearly identified in this figure. The first stage from O to A represents the elastic deflection of the plate. The drop in load at point A represents the sudden onset of delamination around the indenter. The curve from point B to C and D continues to act linearly, but with a reduced stiffness. Intermediate drops at point C are associated with further delamination and occur only in thicker plates, such as the 6.70 and 8.30 mm thick plates. The plate finally fails at point D with the formation and ejection (point D to E) of a plug.

The present experimental investigation focused on the growth behaviour of delamination. Specifically of interest were the through-thickness location and size of the delamination. Intensive C-scan and pulse-echo ultrasound (PEUS) studies revealed that delaminations form in pie-shaped sectors in a staircase helical pattern through the whole sublaminate thickness, as shown in Figure 3. It was also found that for quasi-isotropic lay-ups, one full revolution (circle) of the helical delamination pattern will occur in each sublaminate. Cross-sectional views, such as that shown in Figure 4, confirmed that the transverse distance from one delamination site to the next is constant for a given plane and equal to one sublaminate thickness. The delamination pattern is further characterised by the overall conical shape of the delaminations through the thickness, as represented in Figure 5.

Further information was gathered through interrupted load-unload tests of the thicker specimens. Figure 1 shows the load-displacement curve for the 8.30 mm thick plate with successive load drops occurring before final failure (point D in Figure 2). The specimen was unloaded (not shown) after each of these drops and C-scanned to determine the extent and location of delamination growth. Figure 6 shows the cross-section of the 8.30 mm thick plate for each successive load drop.

It is important to note in Figure 6 that the most significant delamination occurs near the bottom surface of the laminate. The succession of helical delaminations form, for the most part, during subsequent load drops. It is this initial delamination which seems to have the most significant overall effect on the reduced bending stiffness of the laminate. Figure 7 shows a plot of through-thickness location of initial delamination as a function of laminate thickness. Both the outermost top and bottom delamination locations as measured from C-scans are indicated.

3. ANALYTICAL MODEL

Clearly, the complex and stochastic nature of the delamination formation does not lend itself to simple analysis. In order to model the behaviour of a partially delaminated plate, Ursenbach *et al.* (8) assumed an embedded isotropic plate of reduced stiffness to represent the region of delamination. The present study seeks to build on this assumption and represent the actual delaminated region as a series of split plates. The aim of the analysis is to simulate the reduction of plate stiffness that stems from progressive delamination and hence predict the essential features of the load-deflection response. However, in order to bridge the gap between experimental observation and analytical solution, some simplifying assumptions has to be made: (*i*) one full circular delamination is assumed to occur at each delamination site, rather than the complex helical pattern, and (*ii*) the delaminated interfaces (splits) have the same projected dimension, i.e. cylindrical rather than the observed conical through-thickness shape shown in Figure 5.

For simplicity, the governing equations for circular isotropic plates are used to model the bending response of the test specimens. For rectangular coupons, the central plate stiffness is calculated using the Whitney-Pagano first order shear deformation theory and subsequently the classical isotropic circular plate theory (9) is adopted to obtain the diameter of an equivalent isotropic circular plate which yields the same bending stiffness at the plate centre. Thus the whole circular plate shown in Figure 8 is divided into two regions; one without delamination (called the unsplit region) and the other with delamination (called the split region). The diameters of the entire circular plate and the split region are denoted by $2a$ and $2b$, respectively, and the plate thickness is h_0. At present, in the absence of a model for predicting the size of delamination, the maximum delamination size obtained from C-scan measurements is used to provide an estimate of the delamination diameter, $2b$, which is then used as an input to the analysis.

Owing to the quasi-isotropic nature of the laminates under consideration, the in-plane elastic modulus, E and Poisson's ratio, ν for all four test coupons are identical and can be calculated from the classical laminate plate theory applied to the repeating unit or sublaminate $[-45/90/0/45]_s$.

It is assumed that each part (substructure) of the split plate can be treated as an individual disk and that the deflection curves of all substructures are compatible. Analysis of split beams using similar assumptions were previously carried out by other authors (10, 11).

In reality, the split plate is a complicated three-dimensional problem. The static elasticity analysis conducted in (10) suggested that the large stresses at the split tip are highly local and do not substantially affect the global plate response. Therefore, the detailed behaviour at the split tip is not considered here.

In the present model, the classical bending theory of plates is used for unsplit as well as split regions. Wang *et al.* (11) pointed out that for large split sizes ($b > 0.3a$) and quasi-static bending deformations, the in-plane (longitudinal) motions have negligible effect on natural frequencies of the split beam. Therefore, in the present analysis it is reasonable to ignore the effect of membrane actions.

The continuity conditions used in the analysis are: transverse deflection w, deflection slope dw/dr, bending moment M_r at a diametrical section, and transverse shear force resultant Q through the whole thickness. These are the same as those used in (10, 11).

Referring to the Appendix, the central plate deflection can be expressed as

$$w_d = \frac{P}{16\pi D_d}(A_2 - A_1 A_3 b^2) \tag{1}$$

Details of the derivation and expressions for A_1, A_2, and A_3 are given in the Appendix. In Eqn (1) D_d is the bending rigidity of the split region. If there are k split interfaces, the plate is divided into $k+1$ split parts through the thickness and the bending rigidity of the split region is then the sum of the bending rigidities for each individual substructure, i.e.,

$$D_d = \sum_{m=1}^{k+1} I_m \tag{2}$$

where

$$I_m = \frac{Eh_m^3}{12(1-\nu^2)} \text{ and } \sum_{m=1}^{k+1} h_m = h_0 \tag{3}$$

Each substructure thickness, h_m, is necessarily bounded as follows

$$h_{\min} \le h_m \le h_0$$

where $h_{\min}$ is the sublaminate thickness.

Guided by experimental observations, the location of the first delamination through the thickness, z_0, is assumed to be related to the plate thickness as

$$z_0 = (n-1)\frac{h_0}{2n}, \quad n \ge 2 \tag{4}$$

where n is the total number of symmetric sublaminates. A plot of non-dimensional location of first delamination, z_0/h_0, versus plate thickness and its comparison with experimental C-scan results is shown in Figure 7. It can be seen that for the range of plate thicknesses considered here, Eqn (4) provides a reasonable estimate of the through-thickness location of the first delamination interface in the lower half of the laminate.

In the present analytical approach it is assumed that subsequent delaminations occur at the sublaminate interfaces below the first delamination. When all such interfaces have delaminated, the laminate is considered to have reached its maximum energy absorption capability in bending. At this stage, the analysis dealing with the progressive stiffness reduction terminates and the shear plugging phase commences. Figure 9 shows schematically the assumed pattern of delamination progression for the 5 and 8.3 mm thick laminates.

To complete the analysis required for predicting the load-displacement profile, the critical load levels at which new delaminations form have to be determined. Based on the present experimental observations and previous work (4), it is reasonable to assume that the initial delamination and all the subsequent delaminations occur at the same maximum load level. Therefore, for the purposes of this work it is sufficient to establish only the peak force at onset of delamination. Jackson and Poe (12) showed experimentally that the contact force can be used as a parameter to predict maximum delamination size in case of quasi-static loading or low-velocity impact on CFRP laminates. They also showed that there is some relationship between the peak contact force and the plate thickness h_0 and that a linear relationship exists between the peak contact force and maximum delamination diameter. On the basis of these findings it is conceivable to construct an empirical relation between the critical delamination force and the plate thickness for a given CFRP system. Since data over a wide range of thicknesses is not currently available for a reliable empirical equation, the measured ultimate load (at the onset of plugging) is used here as the peak force.

4. Model Predictions

Four rectangular and two circular laminated specimens are considered. The material properties and geometrical dimensions of the equivalent circular plates and the delamination sizes are listed in Table 1.

Using the measured values of the ultimate loads and the initial delamination radii as input, the analysis outlined in the preceding section was carried out for all the specimens. Figure 10 shows a comparison between the measured and predicted load versus central plate deflection for two laminate thicknesses $h_0 = 5$ and 8.3 mm. It should be noted that the indentation deformation due to contact between the punch and the plate has been subtracted from the punch displacement to yield the measured load-deflection curves shown in Figure 10. It can be seen that the present model captures the overall features of the load-deflection curves including the saw-tooth characteristics associated with the successive load drops.

A comparison between the predicted and measured energy absorption (area under the load-deflection curve) for all the plates tested is shown in Figure 11. Note that the frictional energy dissipated in plug shear-out is ignored in the energy calculations. The generally close agreement between the experimental and analytical results is encouraging particularly if the model is to be used to assess the laminate perforation resistance under punch or impact loading.

5. CONCLUSIONS

A simple semi-analytical model for predicting the progressive delamination and resulting stiffness reduction of CFRP laminates subjected to quasi-static punch loads has been presented. The model, which treats the delaminated region as a stack of split circular plates, captures the essence of the measured load-displacement curves for various laminate thicknesses considered in this study. The reasonable agreement between the predicted and measured energy absorption demonstrate the potential usefulness of the approach in assessing the resistance of CFRP laminates to out-of-plane loading such as impact. Further work is required to minimise the level of empiricism in the model.

6. ACKNOWLEDGEMENTS

This work has been supported by the Canadian Department of National Defence and the Natural Science and Engineering Research Council of Canada. The authors wish to thank The Boeing Airplane Company, Seattle, for providing the CFRP materials. Thanks are also extended to Dr. A. Poursartip, Dr. M. D. Olson and Dr. R. O. Foschi for valuable discussions. The assistance of Mr. R. Bennett and Mr. S. Milaire with the experiments is greatly appreciated.

7. REFERENCES

1. Wu, H.-Y. T., and Springer, G. S. Journal of Composite Materials, 22 (6), 533 (1988).
2. Finn, S. R., and Springer, G. S. Journal of Composite Structures, 23, 191 (1993).
3. Finn, S. R., and Springer, G. S. Journal of Composite Structures, 23, 177 (1993).
4. Lee, S.-W. R., and Sun, C. T. Journal of Composite Materials, 27 (3), 251 (1993).
5. Razi, H., and Kobayashi, A. S. AIAA Journal, 31 (8), 1498 (1993).
6. Sun, C. T., and Potti, S. V. International Journal of Impact Engineering, 18 (3), 339 (1996).
7. Davies, G. A. O., Zhang, X., Zhou, G., and Watson, S. Composites , 25 (5), 342 (1994).
8. Ursenbach, D.O., Vaziri, R., and Delfosse, D. Journal of Composite Structures, 32, 197 (1995).

9. Timoshenko, and Woinowsky-Krieger, S. Theory of Plates and Shells, 2nd ed. McGraw-Hill Book Company (1959).

10. Farris, T. N., and Doyle, J. F. Journal of Sound and Vibration, 130 (1), 137 (1989).

11. Wang, J. T. S., Liu, Y. Y., and Gibby, J. A. Journal of Sound and Vibration, 84 (4), 491 (1982).

12. Jackson, W. C., and Poe, C. C. Jr. Journal of Composite Technology & Research, 15 (4), 282 (1994).

8. APPENDIX: DEFLECTION FUNCTIONS

Let w denote the downward plate deflection; M_r the bending moment per unit length acting along a diametrical plate section; and Q the transverse shear force resultant through the entire plate thickness.

The differential equation for a circular plate under a central concentrated force is

$$D(\frac{d^2}{dr^2}+\frac{1}{r}\frac{d}{dr})(\frac{d^2w}{dr^2}+\frac{1}{r}\frac{dw}{dr})=0 \qquad \text{(A-1)}$$

where D is bending rigidity of the plate and r is the radial distance from the plate centre. Its general solution can be expressed by a combination of the functions of $\ln r$, $r^2 \ln r$, r^2, and a constant.

In unsplit region, the plate deflection is assumed to be in the form

$$w_1(r)=s_1 \ln\frac{r}{a}+s_2 r^2 \ln\frac{r}{a}+s_3 r^2+s_4 \qquad \text{(A-2)}$$

where s_1, s_2, s_3, and s_4 are unknown constants. The boundary conditions for the unsplit region are that w_l and M_{rl} vanish at the outer edge of the plate.

In the split region, the plate deflection is assumed to be given by

$$w_2(r)=t_1 r^2 \ln\frac{r}{b}+t_2 r^2+t_3 \qquad \text{(A-3)}$$

where t_1, t_2, and t_3 are unknown constants. The logarithmic term $\ln\frac{r}{b}$ is not used in equation (A-3) because it can cause a singularity in the solution at the plate centre.

Across the boundary between the split and the unsplit regions, the following continuity conditions should apply:

$$w_l=w_2,\quad \frac{dw_l}{dr}=\frac{dw_2}{dr},\quad M_{rl}=M_{r2},\quad Q_l=Q_2 \qquad \text{(A-4)}$$

For a given concentrated load P, the equilibrium condition on the diametrical plate section at $r \neq 0$ yields

$$Q_1=\frac{P}{2\pi r},\quad b\le r\le a\ ; \qquad Q_2=\frac{P}{2\pi r},\quad 0<r\le b \qquad \text{(A-5)}$$

Therefore, the deflection functions for a circular plate with an embedded circular split patch can be written as

$$w_1 = \frac{P}{16\pi D_0}\{b^2(1+\nu)[(\frac{D_0}{D_d}-1)(1+2A_1)]\ln\frac{r}{a}+2r^2\ln\frac{r}{a}-(a^2-r^2)[\frac{b^2}{2a^2}(\frac{D_0}{D_d}-1)$$
$$(1-\nu)(1+2A_1)-\frac{3+\nu}{1+\nu}]\} \tag{A-6}$$

$$w_2 = \frac{P}{16\pi D_d}[2r^2\ln\frac{r}{b}+2A_1r^2+(A_2-b^2A_3A_1)] \tag{A-7}$$

where

$$A_1 = \frac{-\frac{1}{2}(1-\frac{D_d}{D_0})(1-\nu)(1-\frac{b^2}{a^2})+2\frac{D_d}{D_0}\ln\frac{b}{a}-\frac{3+\nu}{1+\nu}\frac{D_d}{D_0}}{2-(1-\frac{D_d}{D_0})[1+\nu+(1-\nu)\frac{b^2}{a^2}]}$$

$$A_2 = b^2(1-\frac{D_d}{D_0})(1+\nu)\ln\frac{b}{a}+2b^2\frac{D_d}{D_0}\ln\frac{b}{a}-[\frac{b^2}{2a^2}(1-\frac{D_d}{D_0})(1-\nu)-\frac{3+\nu}{1+\nu}\frac{D_d}{D_0}](a^2-b^2)$$

$$A_3 = 2+(1-\frac{D_d}{D_0})[(1-\frac{b^2}{a^2})(1-\nu)-2(1+\nu)\ln\frac{b}{a}]$$

At the plate centre, the deflection can be expressed as

$$w_d = \frac{P}{16\pi D_d}(A_2 - A_1A_3b^2) \tag{A-8}$$

When the delamination radius b approaches zero, or the bending rigidity of the split region, D_d, approaches the bending rigidity of the unsplit region, D_0, Equations (A-6), (A-7), and (A-8) revert to the same form as that in (9), i.e.

$$w(r) = \frac{P}{16\pi D_0}[2r^2\ln\frac{r}{a}+\frac{3+\nu}{1+\nu}(a^2-r^2)] \text{ and } w_0 = \frac{3+\nu}{1+\nu}\frac{Pa^2}{16\pi D_0}$$

Table 1. Quasi-isotropic material properties and sizes of equivalent circular plate and central delamination

Geometry size (mm)				Material properties
Designation	h_O	a	b	$E = 45.51$ GPa
A	3.35	43.56	17.50	$\nu = 0.31$
B	5.00	46.70	27.25	
C	6.70	50.74	33.75	
D	8.30	54.54	45.00	
E*	5.00	57.15	28.00	
F*	8.30	57.15	46.50	

* Rectangular test coupons supported over circular openings

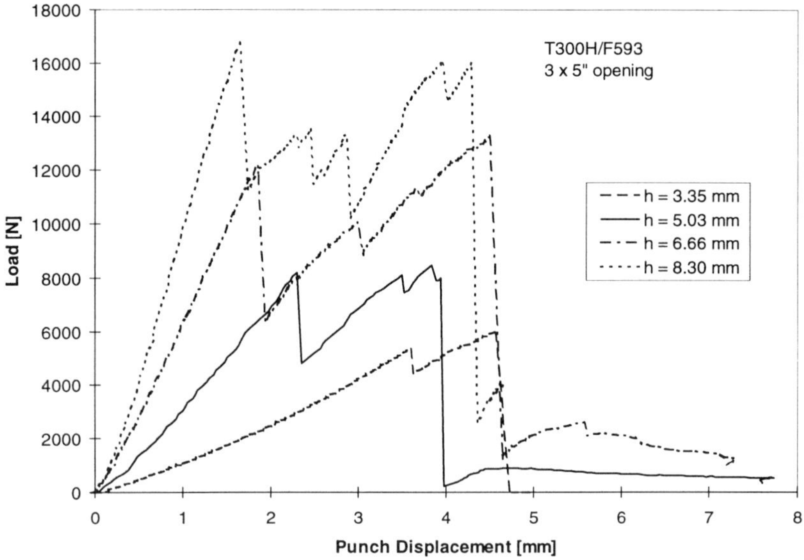

Figure 1 - Static punch load-displacement curves for various CFRP laminate thicknesses.

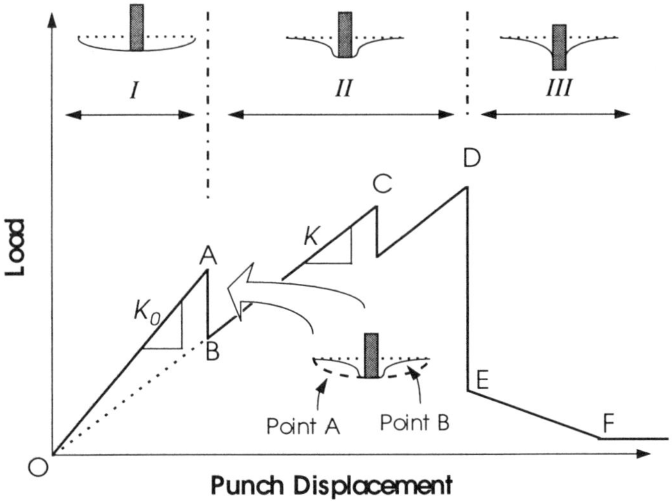

Figure 2 - Idealised load-deflection diagram showing the various stages of CFRP plate response during a quasi-static punch test.

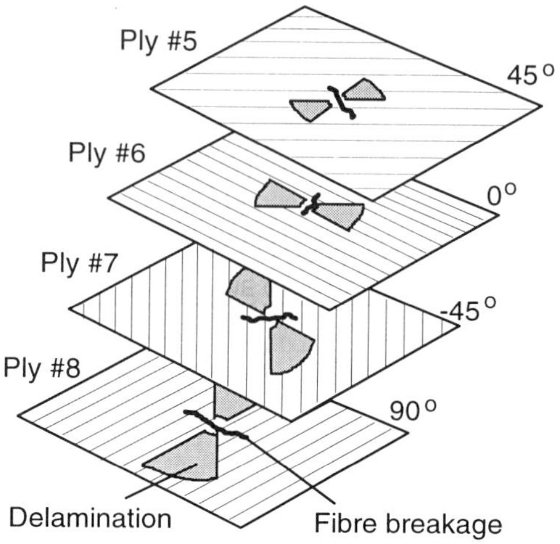

Figure 3 - Schematic of delamination patterns in a CFRP laminate with a stacking sequence of $[45/0/-45/90]_{ns}$ (different from the present system) caused by static or dynamic out-of-plane loading.

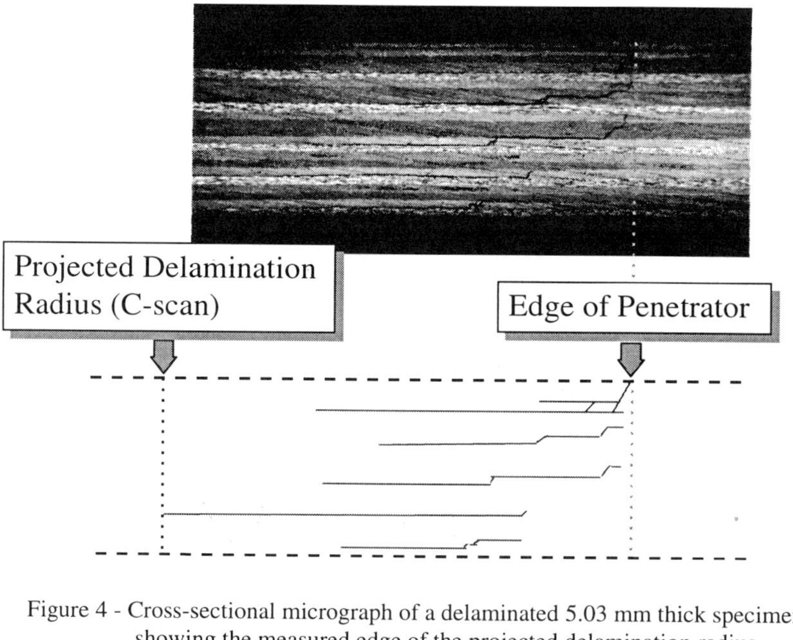

Figure 4 - Cross-sectional micrograph of a delaminated 5.03 mm thick specimen showing the measured edge of the projected delamination radius.

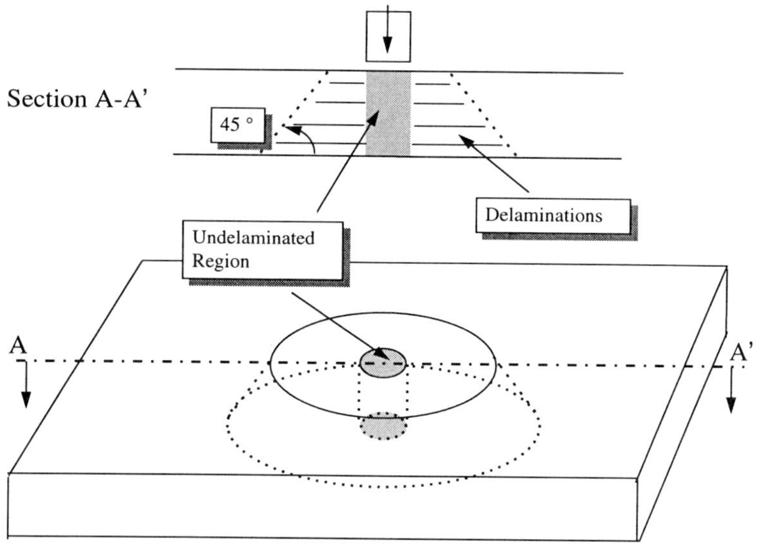

Figure 5 - Observed delamination pattern through the thickness.

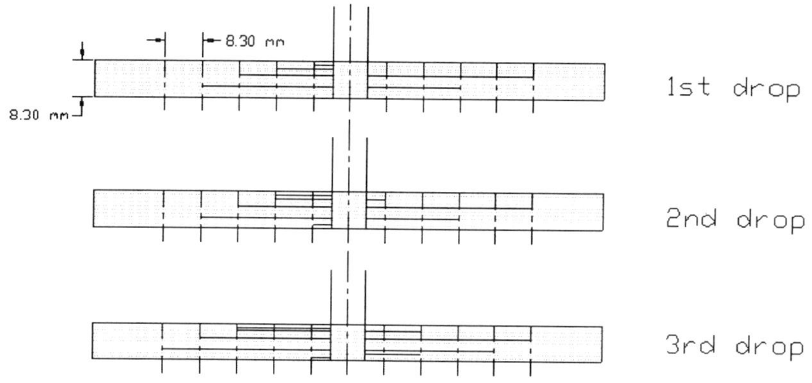

Figure 6 - Cross-sectional schematics of the 8.30 mm thick plate showing delamination progression over a series of load drops.

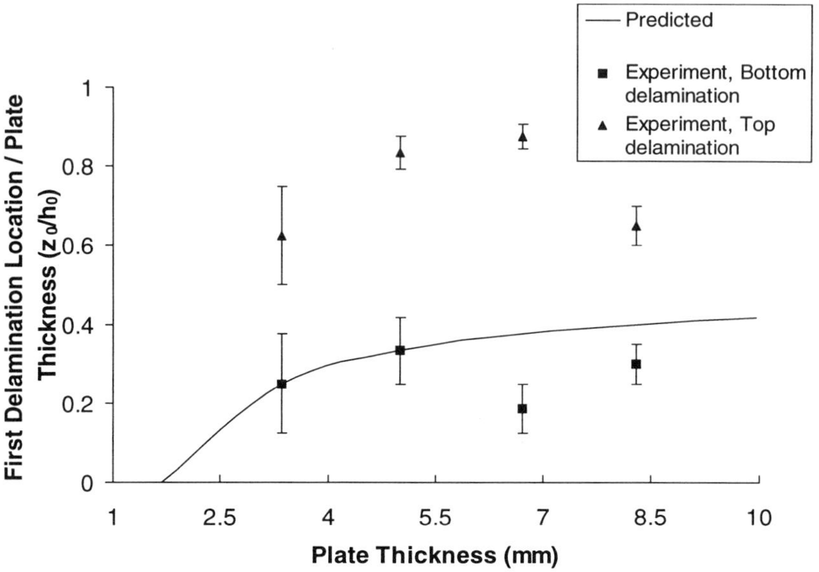

Figure 7 - Initial delamination location as a function of laminate plate thickness.

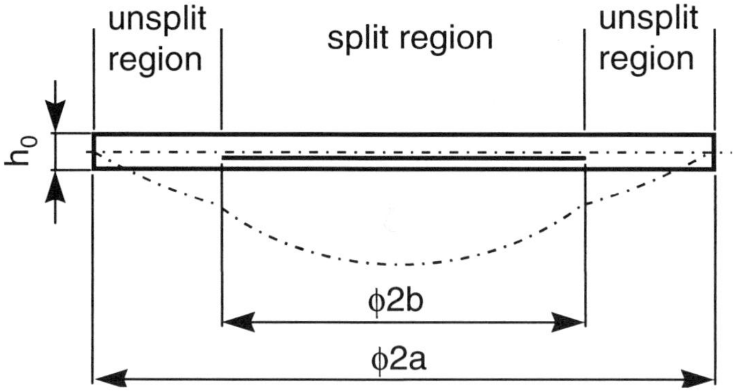

Figure 8 - Schematic illustration of the circular plate model with an embedded split plate region.

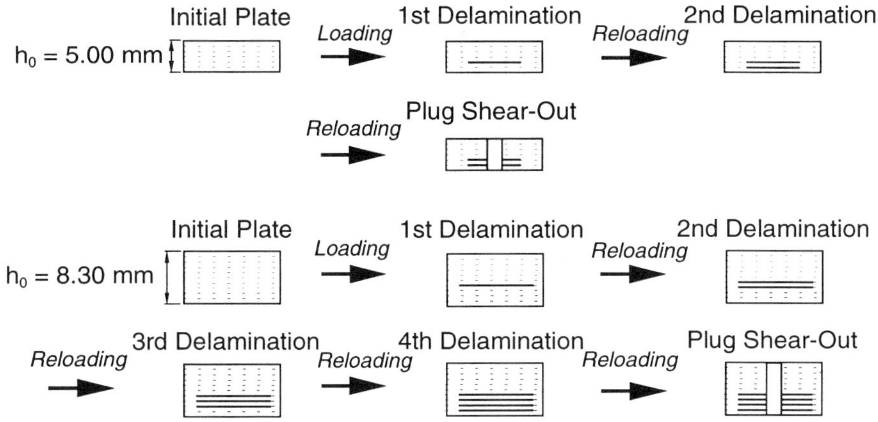

Figure 9 - Illustration of the assumed delamination progression in the equivalent circular plate model of 5.0 and 8.3 mm thick laminates.

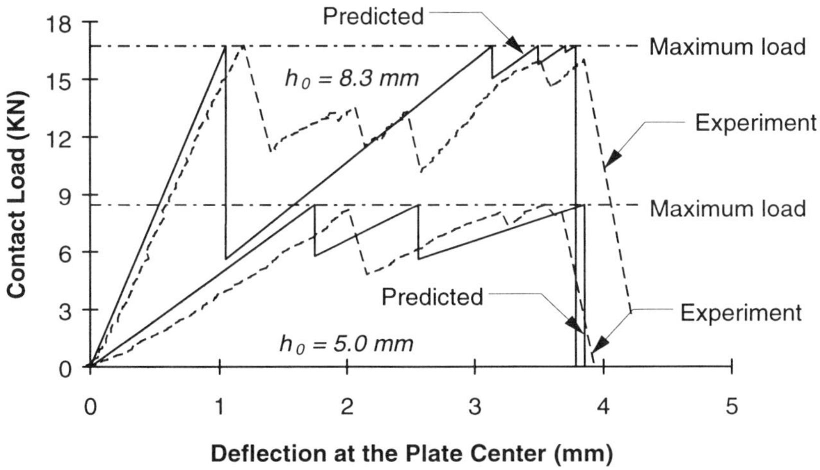

Figure 10 - Comparison between the measured and predicted punch load-plate deflection curves for two laminates of thicknesses 5.0 and 8.3 mm.

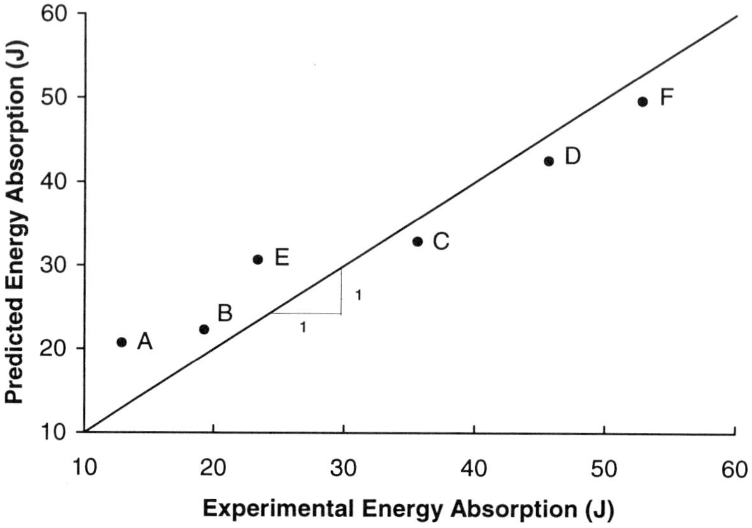

Figure 11 - Comparison between the measured and predicted energy absorption of the laminates subjected to static punch loading. The letters correspond to the specimen designations listed in Table 1.

28th International SAMPE Technical Conference
November 4-7, 1996

EFFECT OF THICKNESS ON THE COMPRESSION RESIDUAL STRENGTH OF NOTCHED CARBON FIBER/EPOXY COMPOSITES

Stephen H. Ward and Hamid Razi
Boeing Commercial Airplane Group
Seattle WA 98124

ABSTRACT

This paper discusses the results of a test program conducted using a toughened epoxy CFRP material. Notched compression tests were conducted on solid laminates of various thicknesses ranging from 16 to 80 plies. Several different notch sizes were investigated to establish strength versus notch size curves. Test results demonstrate clear trends of increased strengths and flatter strength versus damage size curves for thicker laminates. Data from local strain gages at the notch tip are presented to demonstrate the progressive damage propagation away from the notch. This data suggests a material strain softening behavior may be used to model the failure mechanics. These results are consistent with those previously reported for similar material systems. The increase in notched strength with thickness is compared to the observed effect of lamina stacking sequence on unnotched compression strength. Analytical predictions using a Damage Zone Model incorporating a linear strain softening law are compared with the test results.

Key Words: Carbon Fiber Composites, Notched Compression Strength, Fracture Analysis

1. INTRODUCTION

Carbon fiber/epoxy composites have been used in several major primary structure applications on commercial aircraft. With increases in material toughnesses exhibited by current generation carbon fiber reinforced plastics (CFRP), "barely visible" impact damage has, for thicker gage structure, become less of a design driver than large notch residual strength. Consequently, test data and analysis methods are needed to support residual strength analyses of notched CFRP laminate structure. For wing and empennage type structure, notched compression strength can be the critical parameter affecting the sizing of skin panels.

Recent studies (1-5) have investigated the residual strength of notched laminates and sandwich panels. These studies were undertaken as part of the NASA/Boeing composite fuselage development program (6). Extensive databases of notched laminate strength for several material systems were developed for tension and compression loadings. The results of these data were used to develop analytical failure predictions using a strain-softening progressive damage approach (7). Notched compression panel results reported in (1), along with unnotched and notched compression data for similar materials tested at Boeing, indicated that notched compression strength tended to increase with increasing laminate thickness.

Numerous residual strength models have been proposed for notched composites (8). In general, these models do not include failure mechanisms such as damage growth and stress redistribution. The Damage Zone Model (DZM) has been shown to give good predictions of tensile fracture loads for various notch geometries in composite laminates (9-12). These predictions have used a linear strain-softening material behavior in the damage zone.

An experimental program was conducted to investigate the effect of laminate thickness on notched compression strength of a toughened carbon fiber/epoxy material. Preliminary correlations with predictions using the FRACOM1 implementation of the DZM (12, 13) are presented.

2. EXPERIMENTAL RESULTS

2.1 Test Specimens Compression tests were conducted on specimens from six different laminates containing a range of notch sizes. The laminates were fabricated with a toughened carbon fiber/epoxy prepreg material, T800H/3900-2, manufactured by Toray Industries. The material form was unidirectional tape with a fiber areal weight of 190 gm/m^2 and a nominal resin content of 35% by weight. The resulting nominal lamina thickness is 0.188 mm (0.0074 in).

The test matrix is shown in Table 1. Lamina stacking sequences for each laminate are listed in Table 2. The 0° fiber direction was oriented parallel to the length of the specimen. All of the laminates have similar lamina percentages (25%-30% 0° plies, 50% +/-45° plies). Notch dimensions, and specimen lengths and widths are indicated in Table 1. Holes and slits were mechanically routed into the specimens.

Table 1: Notched Laminate Test Matrix

	Notch Type	Hole (mm)		Slit (mm)				
	Notch Size	6.4	12.7	25.4	38.1	50.8	63.5	76.2
	Width (mm):	38	254	254	254	254	254	254
Number of Plies	Layup	Number of Tests:						
20	30/50/20	3		1 ▷				
40	30/50/20	3	2 ▷	1		1	1	2 ▷
60	30/50/20	3	1 ▷	1		1		1 ▷
80	30/50/20	3	1			1 ▷		
16	25/50/25	30			1	1 ▷	1	
72	25/50/25				1	1 ▷	1	

Layups: (X/Y/Z) represents X% 0° plies, Y% +45° & -45° plies (total), and Z% 90° plies
Lengths: 6.4 mm hole specimens: 305 mm
Other notch sizes: 762 mm for 30/50/20 specimens,
1016 mm for 25/50/25 specimens
Slit Width: 6.4 mm
▷ One specimen had strip of strain gages at edges of notch.

Table 2: Lamina Stacking Sequences

Number of Plies	Layup	Stacking Sequence:
20	30/50/20	[(45/90/-45/0)2 /0/ (45/-45)]sym
40	30/50/20	[(45/90/-45/0)3 /0/45/90/-45/0/45/-45/0]sym
60	30/50/20	[(45/90/-45/0)5 /0/45/90/-45/0/0/45/-45/0/ (45/-45)]sym
80	30/50/20	[(45/90/-45/0)6 /0/45/90/-45/0/0/45/90/-45/0/0/45/-45/0/45/-45]sym
16	25/50/25	[(45/90/-45/0)2]sym
72	25/50/25	[(45/90/-45/0)9]sym

2.2 Test Procedures All tests were conducted using restraining fixtures to prevent global buckling of the specimens. The 6.35 mm (0.25 in) hole specimens were tested in a typical open hole compression fixture (14). Larger hole and slit tests were conducted in the fixture shown in Figure 1. A gap was provided between the restraining plates so that local failure modes would not be suppressed, while still being short enough that column buckling would not occur. The gap ranged from 3.18 cm (1.25 in) to 6.35 cm (2.50 in) depending on the thickness of the test article. The ends of these specimens were restrained between wedges and end loaded with flat plates.

Strain gages were used on all specimens with notches larger than 6.35 mm (0.25 in) to verify the uniform application of compression loading. In addition, eight specimens had a series of 10 closely spaced strain gages ("strip gage") located adjacent to the edge of the notch. These gages were used to detect damage propagation at the notch tip (see Figures 2 and 3).

Figure 1: Compression Test Fixture

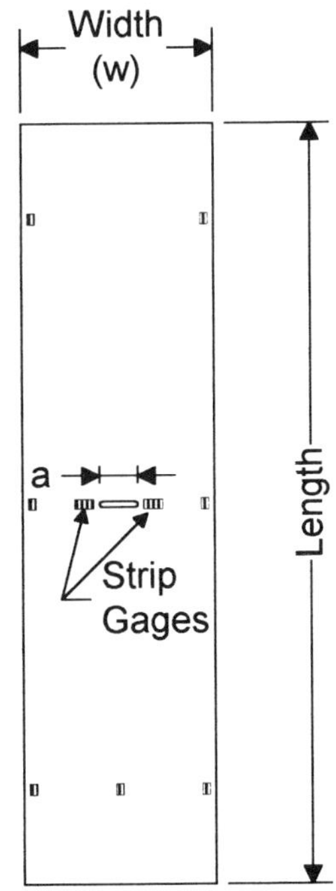

Figure 2: Large Specimen Configuration

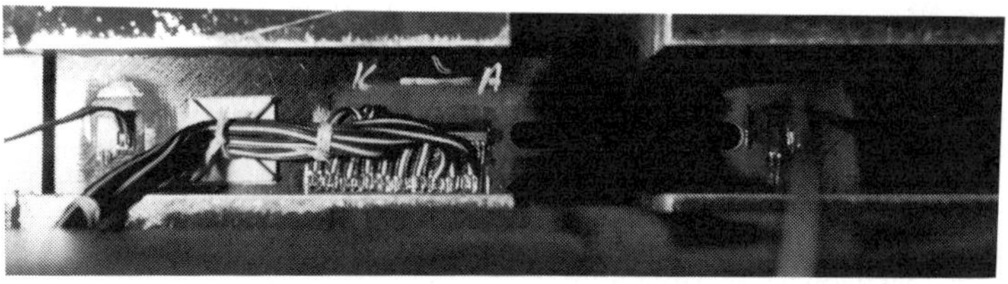

Figure 3: Strain Gages at Notch Tip

2.3 Test Results Test results in terms of gross area failure stress are listed in Table 3. The failure mode of a typical specimen is shown in Figure 4. Failure stresses for selected notch sizes are plotted in Figure 5 versus number of plies in the laminate. A clear trend of increasing strength with increasing thickness is evident for notch sizes at and above 25.4 mm (1.0 in). The failure stress data is shown in Figure 6 plotted against notch size.

Stable damage propagation was observed in most specimens with notch widths greater than 12.5 mm (0.50 in). Figures 7-9 show the notch tip strain data at progressively increasing load levels for three representative specimens. The extent of the damage zones prior to specimen fracture, Figure 10, were determined from the strain data near the notch tips. Six of the seven strain gaged specimens showed stable damage propagation, the exception being the 40-ply laminate with a 12.7 mm (0.50 in) hole. Peak strains prior to damage propagation, at a distance of 2 mm (0.08 in) from the notch, are compared in Figure 11. Peak strain values for specimens with slits were in the range of 0.007 to 0.008 mm/mm, while values for specimens with holes were in the range of 0.0095 to 0.011 mm/mm. This limited data suggests that the damage zone may vary in shape depending on the radius of the notch.

Table 3: Notched Laminate Test Results

	Notch Type	Hole (mm)		Slit (mm)				
	Notch Size	6.4	12.7	25.4	38.1	50.8	63.5	76.2
Number of Plies	Layup	Failure Stress (MPa):						
20	30/50/20	407		208				
		371						
40	30/50/20	332	311	260		193	181	159
		335	327					160
		327						
60	30/50/20	369	301	293		216		183
		380						
		380						
80	30/50/20	376	341			234		
		371						
		372						
16	25/50/25	(305)			164	135	134	
72	25/50/25				249	217	198	
20	30/60/10	351				191		

(xx) = average value

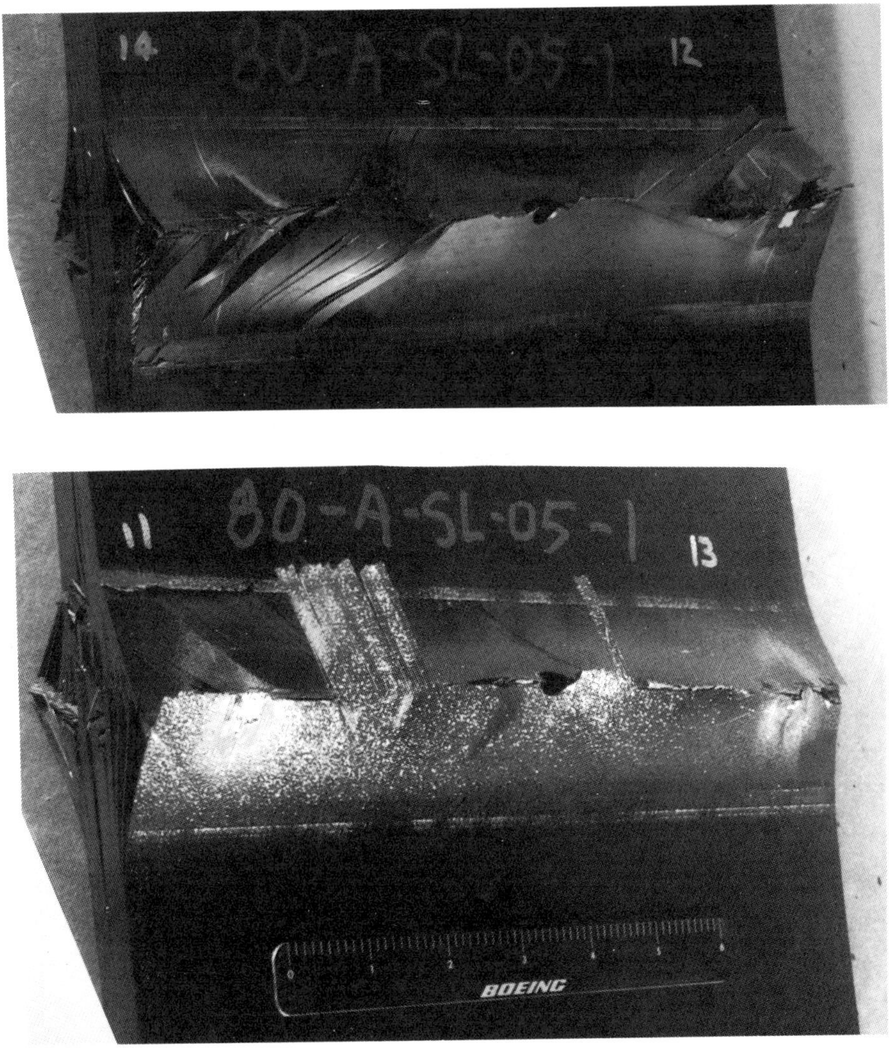

Figure 4: Front and Back of Failed 80-Ply, 12.7 mm Hole Specimen

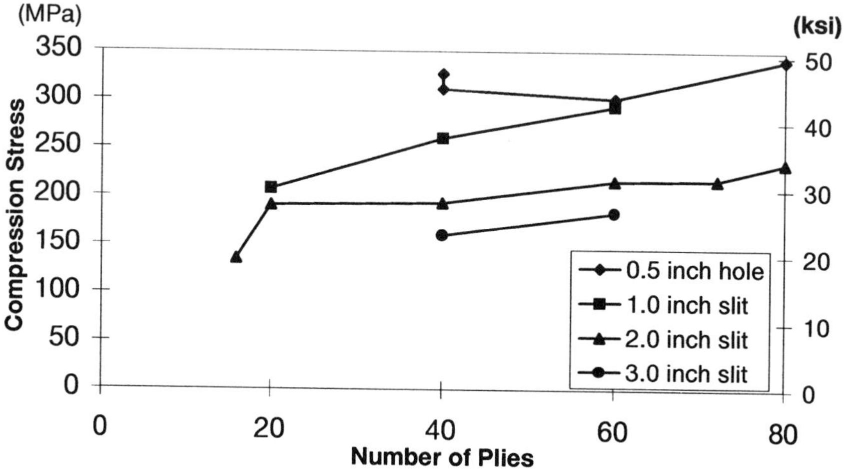

Figure 5: Failure Stresses Showing Thickness Effect

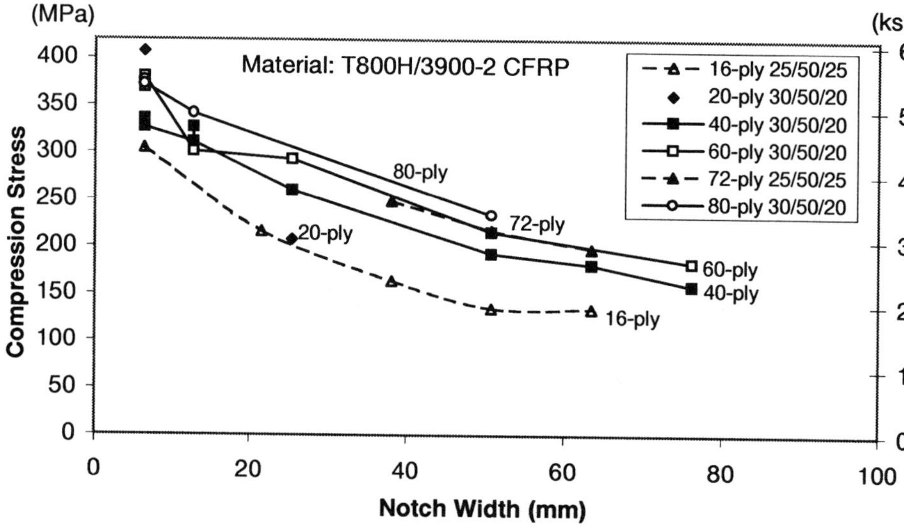

Figure 6: Failure Stresses Showing Notch Size Effect

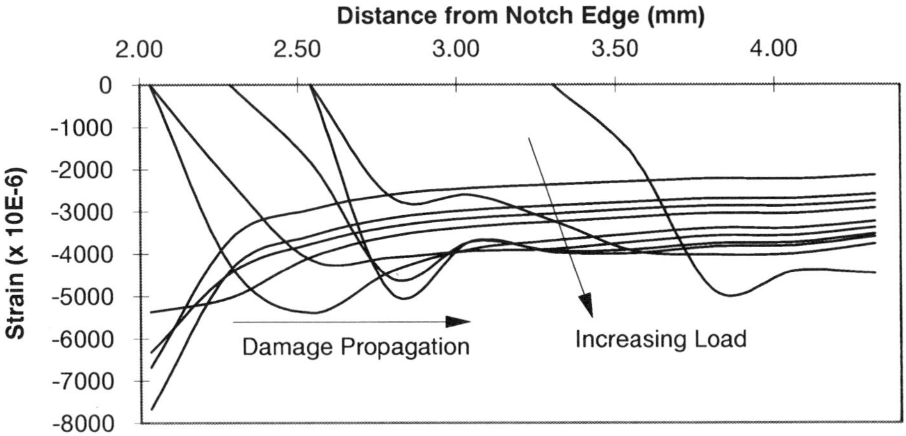

Figure 7: Damage Propagation in 40 Ply Laminate With 76 mm (3.0 in) Slit

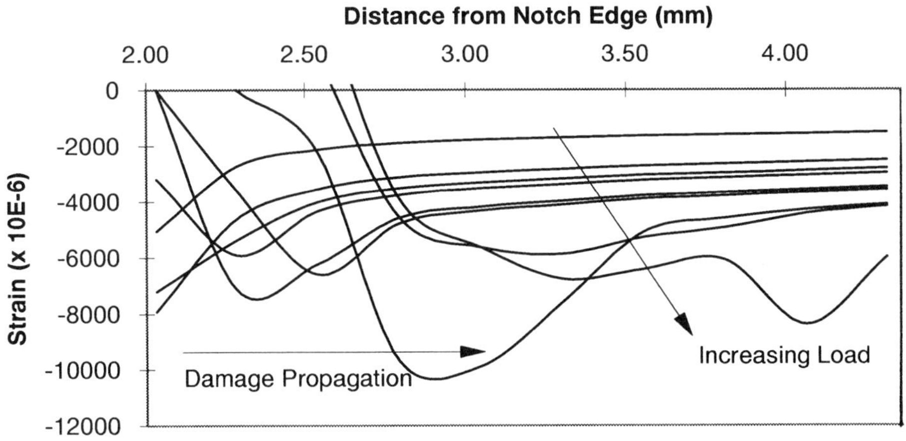

Figure 8: Damage Propagation in 60 Ply Laminate With 76mm (3.0 in) Slit

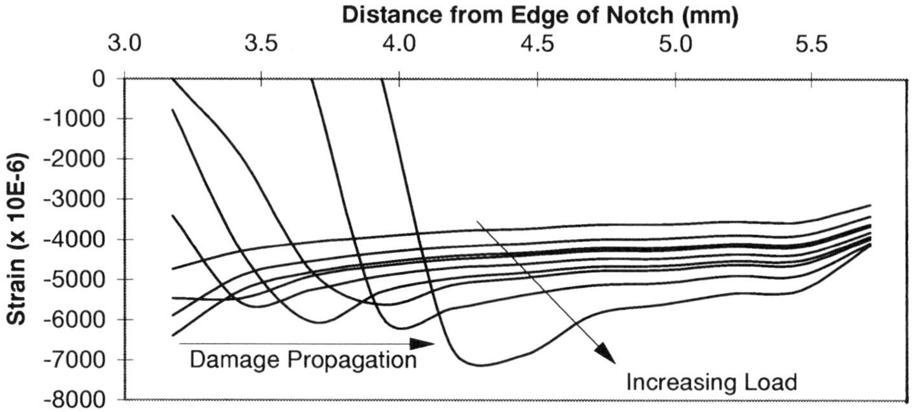

Figure 9: Damage Propagation in 72 Ply Laminate With 51mm (2.0 in) Slit

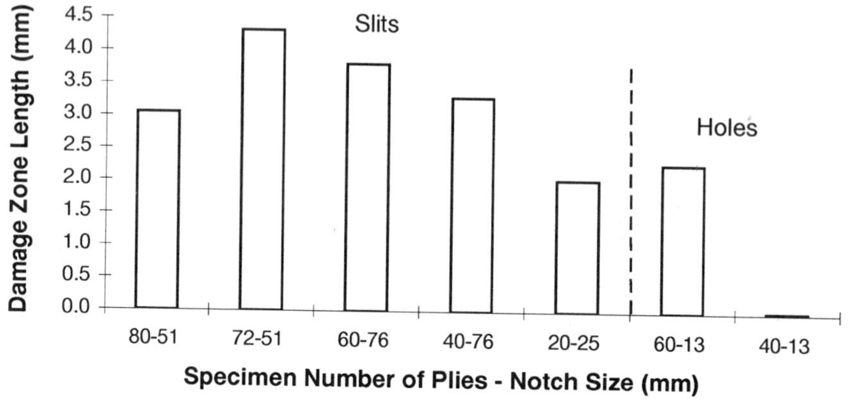

Figure 10: Damage Zone Lengths at Notch Tips as Indicated by Strain Gages

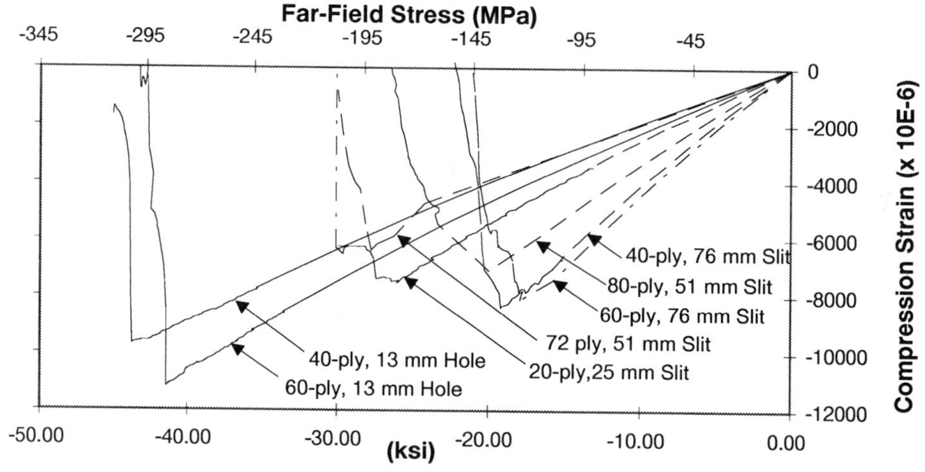

Figure 11: Peak Strains at Notch Tip - Holes versus Slits

3. DISCUSSION

Notched compression data generated for different NASA ACT composite fuselage program material systems are shown in Figure 12. Trends seen in the ACT data have been confirmed by the data reported in this paper.

It is postulated that the increase in notched compression strength with increasing laminate thickness is due to increased global stability combined with three-dimensional damage propagation. Increased thicknesses will result in a reduced tendency for the laminate to bend near the notch tip, resulting in lower stresses in the load carrying fibers.

Localized failure of the load carrying 0° fibers is theorized to be a function of the number of supporting plies outside of the 0° plies. These supporting plies influence the micro-level stability of the loaded fibers. For a thin laminate the number of discrete 0° plies is small;

therefore failure of one ply will lead to rapid damage propagation. However, as the laminate thickness increases there is a greater number of discrete 0° plies, each potentially with different degrees of ply waviness. This allows for more stable local damage growth, whereby the damage propagates through the thickness, progressively failing the 0° plies from the laminate surfaces inwards, as well as in the plane of the laminate.

Corroborating evidence for the supporting ply theory is shown in Figure 13. Plotted are unnotched compression failure strains for five different stacking sequences of a quasi-isotropic (25/50/25) layup. These data are from the T800H/3900-2 material and tested using a straight-sided compression coupon. As the load carrying 0° plies are moved towards the center of the laminate, resulting in more plies located on either side of the 0° plies to delay fiber buckling, the failure strains increase. The drop in the strain value for the fifth layup is attributed to having all four 0° plies adjacent at the laminate midplane. This layup has a lower bending stiffness, possibly resulting in global buckling of the test specimen.

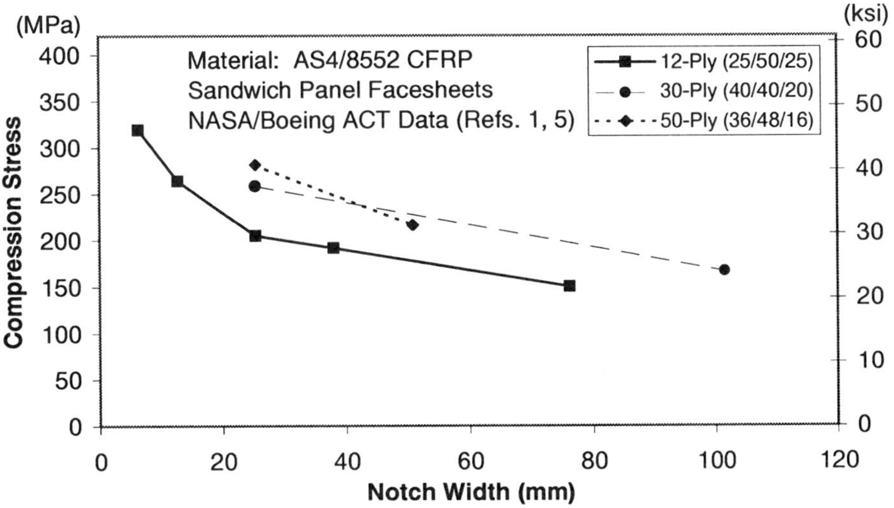

Figure 12: Failure Stresses - NASA/Boeing ACT Data (1, 5)

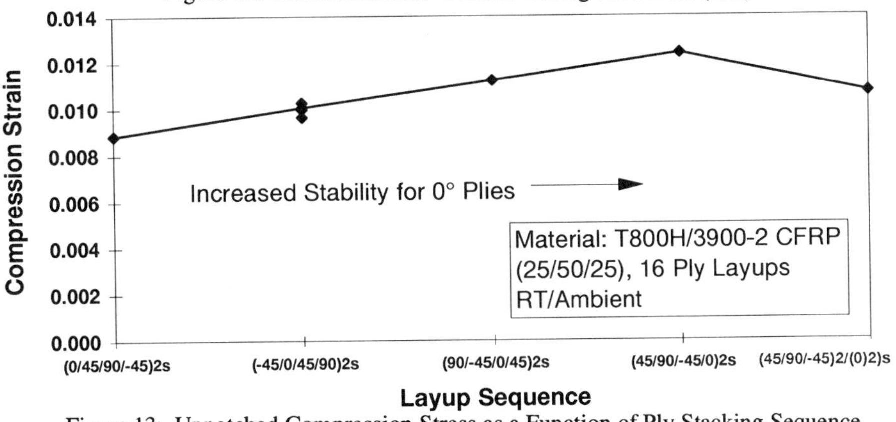

Figure 13: Unnotched Compression Stress as a Function of Ply Stacking Sequence

4. APPLICATION OF DAMAGE ZONE MODEL

The Damage Zone Model (9-11) was used to make analytical predictions for several of the laminate / notch combinations through the use of the FRACOM1 code (12, 13). A linear strain softening material law is incorporated into the code. The first step in the analytical process was to model the test specimens with a linear elastic finite element model. A reduced stiffness matrix from the model with degrees of freedom along the damage zone was then input into FRACOM1. Curves of notch sensitivity (σ_N/σ_O) versus normalized characteristic damage size (l_{ch}/w) were generated for the specific laminates and notch geometries. The l_{ch} parameter is a characteristic damage zone size, and is a function of laminate layup and thickness.

Test results from specimens with the smallest slit width were used to determine the l_{ch} value for each laminate configuration. This l_{ch} value was then used to generate predictions for the larger notch sizes. The results of this analysis are summarized in Table 4. The test data points and FRACOM1 analytical predictions are compared in Figure 14. The correlation between experimental and analytical results indicates the analysis tends to underpredict the test results for larger notch sizes. While these results are encouraging, additional work is required to improve the analytical method.

Previous studies (1) concluded that notched compression failure is controlled by local instabilities, and that strain softening and non-local material responses are active in the compression damage propagation. The non-local response at the notch tip is postulated to be due in part to a characteristic length over which ply buckling occurs. Non-linear strain softening capability will be incorporated into the FRACOM1 code with the goal of improving the analytical predictions.

Table 4: Results of Damage Zone Model Predictions

Number of Plies	Layup	Notch Width (mm)	Lch/W	Fn/Fo	Pred Fn/Fo	%error
16	25/50/25	38.1	0.016	0.263	0.263	
		50.8	0.016	0.216	0.214	1%
		63.5	0.016	0.215	0.182	18%
72	25/50/25	38.1	0.045	0.398	0.398	
		50.8	0.045	0.348	0.338	3%
		63.5	0.045	0.317	0.294	8%
40	30/50/20	25.4	0.024	0.382	0.382	
		50.8	0.024	0.283	0.259	9%
		63.5	0.024	0.263	0.222	18%
		76.2	0.024	0.233	0.198	18%
60	30/50/20	25.4	0.034	0.431	0.431	
		50.8	0.034	0.316	0.300	5%
		76.2	0.034	0.269	0.231	16%

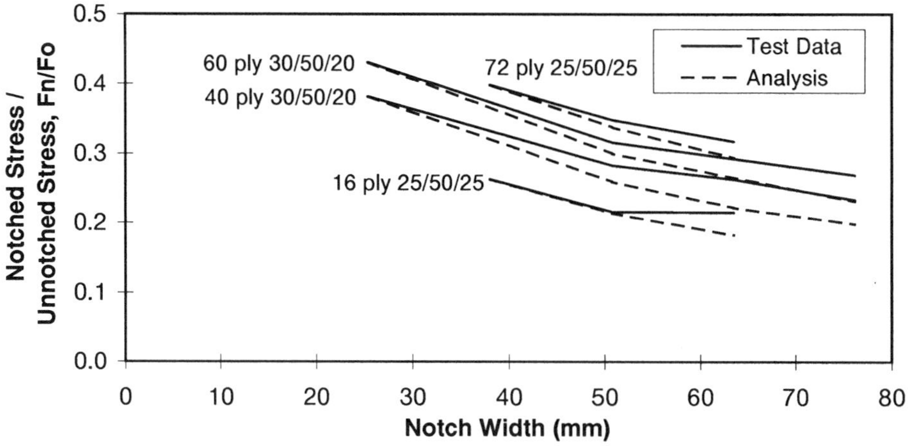

Figure 14: Test Results versus Analytical Predictions using FRACOM1

5. CONCLUSIONS

A thickness effect for compression loaded notched CFRP laminates has been verified over a range of notch sizes. Stable damage propagation was observed, and gives further evidence to the theory of local fiber instabilities controlling compression strength.

For wing and empennage type aircraft structure, notched compression strength is often the critical sizing parameter. Therefore large notch tests covering the expected range of laminate thicknesses anticipated for the design should be included in material screening programs.

The effects of thickness on compression notched strength should be investigated for additional laminate layups. Additionally, the effects of lamina stacking sequence on strength, damage zone size and damage propagation require further experimental and analytical development.

Analytical predictions using the Damage Zone Model incorporating a linear strain softening law have been compared with notched compression strengths. Enhancements to the damage zone model incorporating non-linear strain softening and non-local material responses should be developed to improve correlations with test data.

6. ACKNOWLEDGMENTS

The authors gratefully acknowledge the assistance and contributions of D. Bickford, L. Ilcewicz, A. Tanoura, T. Walker and D. Wilson of BCAG Structures Engineering and E. Campbell of the BCAG Materials Test Lab.

7. REFERENCES

1. T. H. Walker, et. al., "Damage Tolerance of Composite Fuselage Structure", *6th NASA/DoD/ARPA Advanced Composites Technology Conference*, Anaheim CA, August 1995.

2. L. B. Ilcewicz, et. al., "Tension Fracture of Laminates for Transport Fuselage Part 4: Damage Tolerance Analysis", *4th NASA/DoD/ARPA Advanced Composites Technology Conference*, NASA CP 3229, Salt Lake City UT, June 1993.

3. T. H. Walker, et. al., "Tension Fracture of Laminates for Transport Fuselage Part 3: Structural Configurations", *4th NASA/DoD/ARPA Advanced Composites Technology Conference*, NASA CP 3229, Salt Lake City UT, June 1993.

4. T. H. Walker, et. al., "Tension Fracture of Laminates for Transport Fuselage Part 2: Large Notches", *3th NASA/DoD/ARPA Advanced Composites Technology Conference*, NASA CP 3178, Long Beach CA, June 1992.

5. D. Scholz, et. al., Advanced Technology Composite Fuselage--Materials and Processes, NASA CR-4731 (1996).

6. L. Ilcewicz, et. al., "Advanced Technology Composite Fuselage",*6th NASA/DoD/ARPA Advanced Composites Technology Conference*, Anaheim CA, August 1995.

7. B. Dopker, et. al., "Composite Structural Analysis Supporting Affordable Manufacturing and Maintenance", *6th NASA/DoD/ARPA Advanced Composites Technology Conference*, Anaheim CA, August 1995.

8. Awerbach and M. S. Madhukar, J. Reinf. Plastics and Composites, 4 (3) 159 (1985)

9. Aronsson and J. Bäcklund, J. Compos. Mater., 20, 259 (1986).

10. Bäcklund and G-G. Aronsson, J. Compos. Mater., 20, 287 (1986).

11. Aronsson and J. Bäcklund, Composite Materials: Fatigue and Fracture, ASTM STP 907, 134 (1986).

12. Hollmann and J. Bäcklund, Eng. Fracture Mechanics, 31 (4), 577 (1988).

13. Hollmann, et. al., FRACOM1 User Manual, Report No. 85-10, Dept. of Aeronautical Structures and Materials, The Royal Institute of Technology, Stockholm, Sweden (1985).

14 SACMA Recommended Test Method for Open Hole Compression Properties of Oriented Fiber-Resin Composites, SRM 3-88, Suppliers of Advanced Composite Materials Association

28th International SAMPE Technical Conference
November 4-7, 1996

COMPONENTS OF MANUFACTURING DISTORTION IN CARBON FIBER/EPOXY ANGLE BRACKETS

T.S. Rennick and D.W. Radford
Composite Materials, Manufacture and Structures Laboratory
Mechanical Engineering
Colorado State University
Fort Collins, Colorado 80523-1374

ABSTRACT

Manufacturing distortion, or warpage, has been a problem for autoclave-cured composites for many years. Often tooling geometry is modified in an iterative fashion to compensate for these fabrication-induced distortions. Previous research efforts have shown these distortions to be related to differences between in-plane and through-thickness properties of laminated composites. While it is generally agreed that cooldown from process temperature is a major contributor to warpage, isothermal cure shrinkage of the epoxy matrix may also play a role. However, the exact sources of the problem have not been completely nor clearly shown. Additionally, no single work has directly measured the relative contributions of these effects. In this research, distortion in carbon fiber/epoxy angle bracket specimens has been experimentally measured to determine the thermoelastic (reversible) and non-thermoelastic (irreversible) contributions to warpage. A new technique, incorporating a laser source, was developed to obtain precise measurements of the included angle of the specimens at temperatures ranging from room temperature to above 170°C. Differences between the measured thermoelastic contribution and the tooling geometry allow the remaining non-thermoelastic contribution to be determined. Calculations indicate that both components contribute significantly to manufacturing distortion; although, the relative amounts show sensitivity to variations in specimen geometry.

KEY WORDS: Composites, Manufacturing / Fabrication / Processing, Design

1. INTRODUCTION

1.1. Advanced Composite Materials Manufacturing Advanced composite materials are well known for their superior properties, and their tailorability generally allows them to be incorporated into most designs. These attributes have led to a significant increase in the use of composite materials in all areas, ranging from sporting goods to aerospace components. However, difficulties in manufacturing and associated manufacturing costs are still often strong deterrents. Autoclave processing is a proven source of high performance, reproducible composite parts. Difficulties arise when trying to create, without extensive tooling iteration, a unique or prototypical composite part that is of precise, specified shape. This iteration of the tooling can be costly, both in materials and time. If composites are to be successfully implemented, it is vital that cost-effective manufacturing be a reality. A solid understanding of distortion development during composite fabrication is needed for this to occur. For the present project, research is specifically aimed at quantifying the factors that affect warpage to aid in the overall understanding and development of cost-effective composite manufacturing, avoiding unnecessary tool iteration.

1.2. The Existing Problem of Manufacturing Distortion Manufacturing distortion results from the internal balancing of process-induced residual stresses (1) and can be severe in many cases, leading to out-of-tolerance dimensions and rejected parts or even difficulty extracting the part from the tooling (2). Spring-in is a term referring to the type of manufacturing distortion experienced in curved composite parts. Spring-in results in closure of a curved composite included angle. Examples of spring-in are seen in wing leading edges (3), U-channel cross-sections (4), and waveguides (5). Spring-in of simple angle bracket specimens will be the focus of this study.

1.3. Current Solutions to the Problem Manufacturing distortion is usually dealt with by force-fitting the part into the system and then securing it into position, or an iterative approach to the design is taken that results in non-net shape tooling. Force-fitting leads to increased installation forces (2,3,6) and possibly damaging stresses. Only flexible parts can be forced into a system. A distorted, stiff part is unusable (3). Iterative modification of the tooling geometry is a solution to the problem (6-9). However, if material or laminate design is subsequently changed, the modified tool may then be ineffective. The modified tooling becomes a case-specific mold that is not the exact shape of the desired end-product. Some manufacturers consider spring-in during their initial tool design, but only on an experience-based level.

Additionally, increasing the bending stiffness of the part, through the use of attached stiffeners, to reduce warpage is often practiced. However, when increasing the number of stiffeners to reduce warpage, residual stresses and assembly forces are increased. Support structures, such as wing spars, may place limits on these stresses and forces (2).

Many of the currently employed solutions to overcoming manufacturing distortion do not directly address the causes of distortion. This could be attributed to not understanding the problem fully or not recognizing the problem. A general misconception, which sometimes leads to these indirect solutions, is the belief that quasi-isotropic, symmetric laminates are distortion-free. In a laminate such as this, stresses are balanced across the thickness and in-plane properties are essentially isotropic. However, the laminate still has anisotropic qualities when comparing through-thickness properties to in-plane properties. This anisotropy is occasionally neglected when transitioning from two-dimensional flat plate designs to more complex three-dimensional designs and leads to unexpected warpage. When more complex part geometries are involved, out-of-plane distortions should be expected during processing .

2. BACKGROUND

2.1. Contributors to Warpage Manufacturing distortion can be seen in many types of fiber reinforced composites, with the properties of the constituent materials generally governing the severity. Warpage has been measured to characterize residual stresses in metal matrix composites (MMC), polymer matrix composites (PMC), and even superhybrid composites consisting of alternating metallic and polymeric composite plies (10). One of the worst cases is the carbon fiber/epoxy composite. This is due to a low coefficient of thermal expansion (CTE) in the longitudinal direction of the reinforcing fiber and a much larger CTE in the transverse direction.

Many factors must be considered in solving the problem of manufacturing distortion. Anisotropic thermal contraction and isothermal cure shrinkage of the epoxy matrix are two commonly cited factors affecting warpage. While anisotropic CTE is generally accepted as a major contributor to warpage, acceptance of isothermal cure shrinkage as a significant contributor is mixed. In some studies, cure shrinkage has not been considered in warpage predictions (2,10,12). In other studies, it is believed to dissipate through viscoelastic relaxation during the curing process (1,13,14). There are many researchers who differ in opinion and consider isothermal cure shrinkage significant (7-9,15-17). Other contributing factors that have been briefly studied are tooling material properties (6), fiber misalignment (10,18), and fiber volume fraction gradients (18,19).

Several models exist that take into account many of the above-mentioned factors (1,12,13,20-23), attempting to predict the development of process-induced stresses and, thus, warpage. Most place an emphasis on the analysis of thick-section composites. Severe through-thickness property gradients, such as variations in degree-of-cure, are predicted to develop in these thick composites, resulting in a stress imbalance and warpage. Thinner, two-dimensional composite laminates may not see these extensive property gradients.

To satisfy an apparent need to resolve the factors affecting manufacturing distortion, an experimental approach has been taken in this work to quantify the various contributions. These contributions have been categorized into two principle components:

(1) thermoelastic (reversible) CTE contributions, and

(2) non-thermoelastic (irreversible) cure and manufacturing contributions.

2.1.1. Thermoelastic Component The thermoelastic component represents the reversible thermal expansion and contraction of the angle bracket specimen, which leads to its changing included angle. Previous research efforts have shown manufacturing distortions to be related to differences between in-plane and through-thickness properties of laminated composites, and simple closed-form equations have been developed to predict the thermoelastic change in shape for simply curved composite laminates (2,5,7-9). The initial, uncured and final, cured geometry for an angle bracket corner radius are depicted in Figure 1.

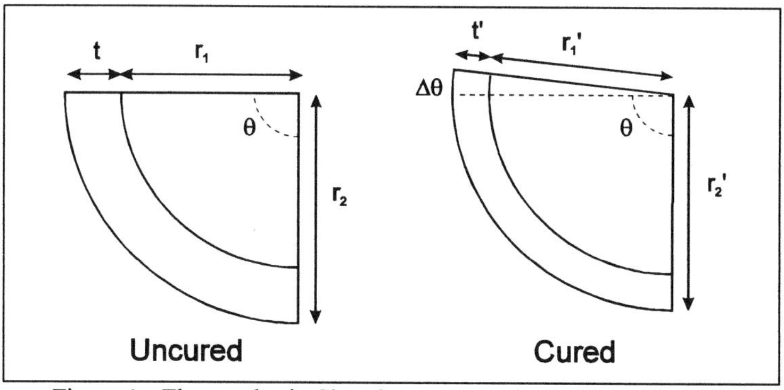

Figure 1 - Thermoelastic Closed-Form Equation Geometries (7,8)

The change in section angle or included angle, $\Delta\theta$, has been shown to be predicted by (7,8):

$$\Delta\theta = \theta\left[\left(\frac{t}{t'}\right) - 1\right], \text{ or} \qquad [1]$$

$$\Delta\theta = \theta\left[\frac{(\alpha_I - \alpha_T)\Delta T}{(1+\alpha_T \Delta T)}\right], \qquad [2]$$

where α_I = in-plane coefficient of thermal expansion,
α_T = through-thickness coefficient of thermal expansion, and
ΔT = change in temperature.

Equation [2] is derived from equation [1] and shows the predicted dependence of the thermoelastic component on the initial section angle, the difference between in-plane and through-thickness CTE values, and the temperature change. It also shows the predicted thermoelastic independence of corner radius. By bringing the temperature change term, ΔT, to the left side of equation [2], equation [3] is developed:

$$\Delta\theta / \Delta T = \theta\left[\frac{(\alpha_I - \alpha_T)}{(1+\alpha_T \Delta T)}\right]. \qquad [3]$$

2.1.2. Non-Thermoelastic Component The non-thermoelastic component of manufacturing distortion is a "one time only" warpage contribution. It occurs during fabrication and is irreversible. Many factors are assumed to affect the non-thermoelastic component, such as isothermal cure shrinkage and factors associated with the manufacturing technique. In previous work (7,8), it was shown that fabrication shrinkage could be accounted for in an extension of equation [2]:

$$\Delta\theta = \theta\left\{\left[\frac{(\alpha_I - \alpha_T)\Delta T}{(1+\alpha_T \Delta T)}\right] + \left[\frac{(\phi_I - \phi_T)}{(1+\phi_T)}\right]\right\}, \qquad [4]$$

where ϕ_I = in-plane fabrication shrinkage, and
ϕ_T = through-thickness fabrication shrinkage.

In equation [4], fabrication shrinkage is treated as a strain. If fabrication shrinkage values for the constituent materials are available, anisotropic composite fabrication shrinkage values could be obtained from a micromechanics-based approach. This would allow the addition of a predicted cure shrinkage induced non-thermoelastic contribution to the already developed thermoelastic prediction.

3. EXPERIMENTAL

3.1. Angle Bracket Specimens Carbon fiber/epoxy angle brackets were autoclave-processed at 178°C on aluminum right-angle convex tooling. They were divided into groups of three specimens each, investigating stacking sequence and corner radius. Prior to testing, all specimens were thermally cycled once to dehydrate and then placed in a dessicator for storage to eliminate any conflicting instability caused by moisture content variation. The stacking sequences are referenced with 0° parallel to the angle bracket axial direction and 90° following the curvature around the angle bracket corner.

3.2. Measuring Equipment Prior research in the area of manufacturing distortion in autoclave-cured, carbon fiber/epoxy composites developed an approach to accurately measure the thermal response of angle brackets and determine the thermoelastic contribution to manufacturing warpage (7,8). A similar technique has been utilized by others (11). The technique requires a laser beam be directed at an angle bracket, with a mirror affixed to one arm, and the resulting reflection is displayed on a calibrated screen. The changing position of the reflected beam on the screen determines the variation in the included angle. With the specimen mounted inside an environmental chamber, this procedure enables the measurement of the change in included angle over a range of temperatures. This reflection technique was developed to verify equation [3], relating angle bracket thermal properties to their resulting thermoelastic response, $\Delta\theta/\Delta T$. Unfortunately, the technique, as originally implemented, was not capable of directly measuring the absolute included angle at room temperature and at specified elevated temperatures and, thus, was not able to account for non-thermoelastic contributions to the warpage.

For the present research project, an experimental setup was developed, which directly measures the included angle through the use of mirrors mounted on each angle bracket specimen arm, a rotational fixture for the specimen, and a laser source. The approach uses the laser source to accurately monitor the specimen's rotation and a shaft encoder to measure the amount of rotation. This precludes the need to measure the displacement of a reflected laser ray at a specific distance and then convert it to an angular change, which was required with the reflection technique. Further, this technique measures the actual included angle rather than just the change in angle. This experimental setup incorporates a heated chamber, enabling measurements to be obtained over a wide temperature range. The rotation technique simplifies the measurement of thermoelastic response, without sacrificing precision or accuracy, and allows the included angle to be measured directly at any desired temperature. A conceptual view of a specimen prepared for measuring, comparing the reflection and rotation techniques, is shown in Figure 2.

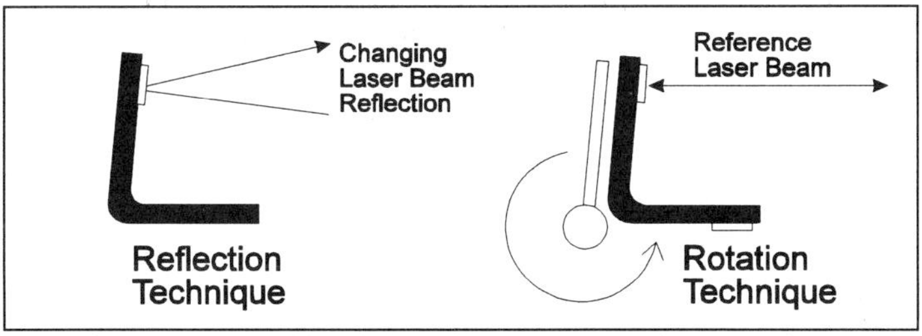

Figure 2 - Comparison of Reflection and Rotation Techniques

3.3. Experimental Procedure The general approach used to measure the included angle as a function of temperature involves the laser rotation technique described above. In short, the technique involves mounting the angle bracket specimen by one arm in the heated chamber with small mirrors affixed to both arms. The laser source is aimed through the chamber glass door at the mirror on one arm and the resulting reflection point is recorded on the vertical reflection board. The specimen is then rotated until the reflection from the second arm aligns with the previously recorded position. The shaft rotation is measured and from this value the included angle is determined. This measurement is repeated in the reverse direction to bring the specimen back into its original position and allow a cross-check of the measured included angle. The chamber temperature is then adjusted for the next data point and allowed to stabilize prior to repeating the measurement process.

Angle bracket specimen measurements were made at room temperature (considered to be between 24-28°C due to day-to-day variations) and then at elevated temperatures on all angle bracket specimens evaluated. Each specimen was then measured with the rotation technique in temperature increments to 122°C. Following this, one specimen from each group was remeasured to 178°C in a similar fashion. Once the maximum temperature was reached, measurements were repeated back down to room temperature to observe the repeatable, thermoelastic response of the angle bracket.

4. RESULTS AND DISCUSSION

4.1. Separating Thermoelastic and Non-Thermoelastic Components Thermoelastic response, $\Delta\theta/\Delta T$, measured with the rotation technique, was similar to previously obtained values (7,8,11), which were collected using reflection techniques on comparable curved composite specimens. Additionally, thermoelastic response closely followed the response predicted by equation [3], and values obtained when measuring to 122°C matched values obtained when measuring to 178°C. The data collected indicate that all of the angle bracket specimens respond linearly and in a repeatable fashion to temperature changes up to their cure temperature. This linear behavior allows calculations to be made, determining the relative contribution of the thermoelastic component of distortion:

$$(\Delta\theta/\Delta T) \times (-150^\circ C) = \Delta\theta_{\text{thermoelastic}} \qquad [5]$$

Equation [5] is obviously dependent on a reference temperature, which for simplicity is assumed to be the cure temperature (178°C), where the matrix is essentially fully cured and cooldown commences. The non-thermoelastic component is then equal to the total spring-in with thermoelastic contributions subtracted out:

$$\Delta\theta_{\text{total}} - \Delta\theta_{\text{thermoelastic}} = \Delta\theta_{\text{non-thermoelastic}} \qquad [6]$$

The determination of the non-thermoelastic component, equation [6], is dependent on an accurate room temperature measurement of the included angle, which is now possible with the laser rotation technique.

Using equations [5] and [6], relative non-thermoelastic contributions were computed for the specimens and found to range from approximately 30% to 95% of the total room temperature spring-in, depending on stacking sequence and corner radius. The highest fractional non-thermoelastic contributions were for specimens with stacking sequences that resulted in 90° in-plane CTE's very similar to the through-thickness values, yielding negligible thermoelastic distortion.

Further data manipulation and comparison showed the two components of manufacturing distortion could also be separated graphically in charted form. Figures 3 and 4 indicate total spring-in as a function of temperature. On the left ordinate is the value of total room temperature spring-in. On the right ordinate at 178°F, the chosen reference temperature, thermoelastic contributions are removed and any remaining spring-in is due solely to non-thermoelastic contributions.

4.2. Stacking Sequence and Radius Effects As can be seen in figures 3 and 4, both thermoelastic and non-thermoelastic components contribute to the overall warpage of angle brackets. However, their absolute and relative contributions vary, depending on stacking sequence and corner radius.

Figure 3 displays the experimental thermoelastic responses obtained for six different stacking sequences. Most of the laminates show similar thermoelastic response values, which are viewed on the charts of figures 3 and 4 as the slope, $\Delta\theta/\Delta T$. The $[(0)_8]_T$ and $(0/+30/0/-30)_S$ laminates show very little thermoelastic response due to their large in-plane CTE in the 90° direction, which is similar to the through-thickness CTE value as suggested in the previous section. All of the laminates, though, show some non-thermoelastic contribution (on the right ordinate), which ranges from about 0.7 - 1.2 degrees for the 8 ply, 6.4 mm radius specimens.

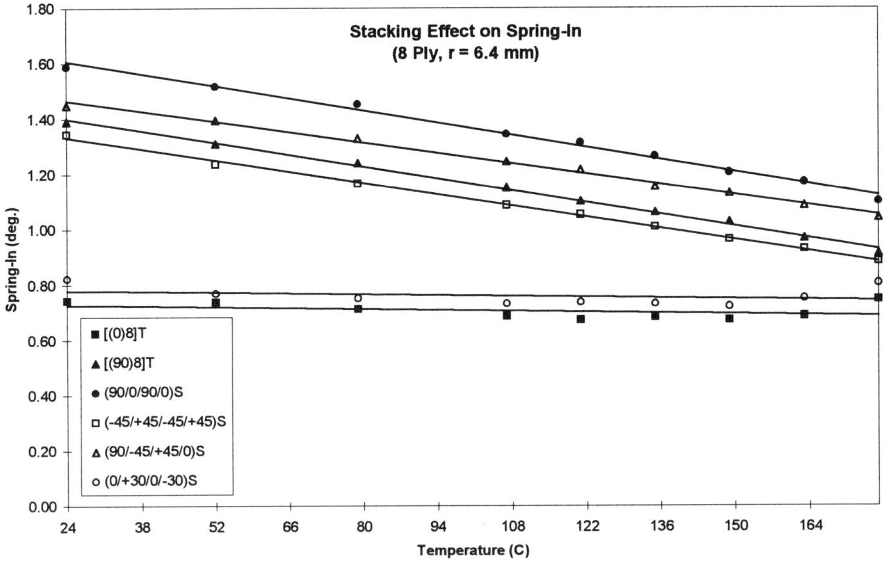

Figure 3 - Experimental Stacking Sequence Effects

Experimental values of thermoelastic response closely follow predictions, using equation [3]. The necessary values of the in-plane and through-thickness CTE were determined using a micromechanics-based model, which considers constituent materials and ply orientations.

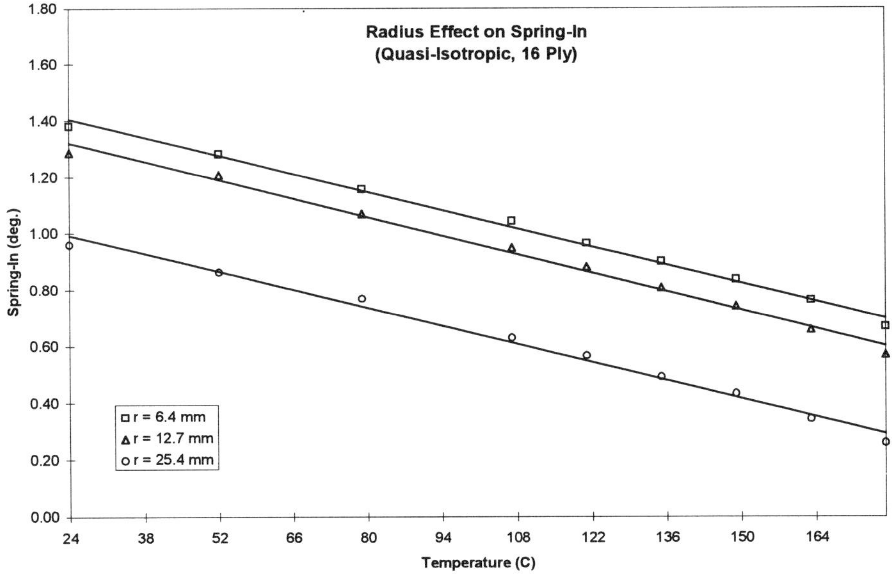

Figure 4 - Specimen Corner Radius Effects on Spring-In

Varying non-thermoelastic contributions can also be seen when comparing specimen corner radii. Figure 4 displays corner radius effects on spring-in. The corner radius directly determines the non-thermoelastic contribution to warpage, without affecting thermoelastic response. As the radius increases, non-thermoelastic contributions decrease. There is essentially no change in the thermoelastic response for the three different radii as indicated by the similar slopes in figure 4. This independence of thermoelastic response on radii was predicted by equations [1-3]. Possible causes of the variation in the non-thermoelastic effect may be attributed to sources of manufacturing distortion other than a predictable cure shrinkage induced non-thermoelastic contribution.

5. CONCLUSIONS

In this research, spring-in of carbon fiber/epoxy angle bracket specimens has been experimentally measured to determine the thermoelastic (reversible) and non-thermoelastic (irreversible) components of warpage. A new technique, incorporating a laser source, was developed to obtain precise measurements of the included angle of angle bracket specimens at temperatures ranging from room temperature to above 170°C. The data collected indicate that all of the specimens respond linearly and in a repeatable fashion to temperature change. Differences between the measured thermoelastic contribution and the tooling geometry allow the remaining non-thermoelastic contribution to be determined.

Relative non-thermoelastic contributions were computed for the specimens and found to range from approximately 30% to 95% of the total room temperature spring-in, depending on stacking sequence and corner radius. This is a significant amount of the total specimen distortion. Further data manipulation and comparison showed the two components of manufacturing distortion, thermoelastic and non-thermoelastic, could also be separated graphically in charted form.

Trends in thermoelastic response, $\Delta\theta/\Delta T$, with stacking sequence variation were predictable, based on a simple closed-form solution. The measured independence of thermoelastic response on changing corner radii also matched predictions. However, the varying non-thermoelastic contributions for different radii were not predicted. As specimen geometry changed with increasing corner radius, non-thermoelastic contributions decreased, indicating sources of manufacturing distortion other than a predictable cure shrinkage induced non-thermoelastic contribution.

Future efforts are planned to involve more experimental work with this proven technique to investigate other variables, such as laminate thickness, included angle, tooling material, ply misorientation, volume fraction gradients, and manufacturing and processing techniques. Additionally, more in-depth data analysis is suggested to attempt to separate the non-thermoelastic component into sub-components.

6. REFERENCES

1. S.R. White and H.T. Hahn, Journal of Composite Materials, 26 (16), 2402 (1992).
2. G.E. Mabson, et al., Third NASA Advanced Composites Technology Conference, Long Beach, California, June 1992.
3. S.C. Nolet and P.M. Sandusky, SAMPE Quarterly, 17 (4), 46 (1986).
4. M. Konig, Proceedings of International Conference on Composites Engineering, 797 (1994).
5. M.G. Hammond and K. Farrell, Proceedings of ICCM II, 1392 (1978).
6. S. Pagliuso, Proceedings of ICCM IV, 2, 1617 (1982).
7. D.W. Radford, Shape Stability in Composites, Ph.D. Thesis, Rensselaer Polytechnic Institute, Troy, NY, 1987.

8. D.W. Radford and R.J. Diefendorf, Proceedings of the Fourth Japan - United States Conference on Composite Materials, 439 (1988).
9. R.H. Nelson and D.S. Cairns, SAMPE International Symposium, 2397 (1989).
10. C.C. Chamis, Proceedings of ICCM II, 221 (1978).
11. L.E. Doxee and G.S. Springer, Journal of Composite Materials, 25, 1340 (1991).
12. A.C. Loos and G.S. Springer, Journal of Composite Materials, 17, 135 (1983).
13. S.R. White and H.T. Hahn, Journal of Composite Materials, 26 (16), 2423 (1992).
14. H.T. Hahn. and N.J. Pagano, Journal of Composite Materials, 9, 91 (1975).
15. L.P. Martin, T.A. Bogetti, and J.W. Gillespie, Jr., Proceedings of the American Society for Composites, Fifth Technical Conference, 415 (1990).
16. J.D. Russell, SAMPE Quarterly, 24 (2), 28 (1993).
17. M.S. Madhukar, R.P. Kosuri, and K.J. Bowles, Proceedings of ICCM X, 3, 157 (1995).
18. P. Kim, et al., International Workshop on Advanced Materials for High Precision Detectors, Archamps, France, September 1994.
19. D.W. Radford, Composites Engineering, 5 (7), 923 (1995).
20. P. Hubert, et al., Proceedings of ICCM X, 3, 149 (1995).
21. T.A. Bogetti and J.W. Gillespie, Jr., Journal of Composite Materials, 25, 239 (1991).
22. T.A. Bogetti and J.W. Gillespie, Jr., Journal of Composite Materials, 26 (5), 626 (1992).
23. S.C. Tseng, ASME International Mechanical Engineering Congress and Exposition, ASME Paper 95-WA/HT-26, 1995.

7. ACKNOWLEDGEMENT

The authors would like to acknowledge Lockheed-Martin Tactical Aircraft Systems for support in this work.

8. BIOGRAPHIES

Timothy S. Rennick received his Bachelor of Science in Ocean Engineering from the United States Naval Academy in 1989. Concurrent with graduation, he received a commission as an officer in the US Navy's Nuclear Submarine Force. During his service, he completed assignments at the US Naval Coastal Systems Center, at Los Alamos National Laboratory, and upon two submarines. In 1994, he resigned his commission as an officer in the US Navy to pursue graduate studies at Colorado State University. He will be receiving his Master of Science in Mechanical Engineering and starting employment at Southwest Research Institute in San Antonio, Texas, during the Fall of 1996. He is currently a student member of SAMPE.

Donald W. Radford is an Associate Professor of Mechanical Engineering at Colorado State University and is the Director of the Composite Materials, Manufacture, and Structures Laboratory. He received his Ph.D. in Materials Engineering at Rensselaer Polytechnic Institute, his MASc. in Metallurgical Engineering and BASc. in Mechanical Engineering at the University of British Columbia. He has a long history of research in the area of manufacturing induced distortion in composites. Other recent research programs have included approaches to dual resin bonded carbon fiber reinforced PEEK/PEI composites, impact fatigue of metal-composite joints, ultra-lightweight EMI shielding composites and morphological studies of composites.

28th International SAMPE Technical Conference
November 4-7, 1996

Thermal Characterization of New High Performance Semi-Interpenetrating Polymer Networks

R. H. Pater*
NASA Langley Research Center
Hampton, VA 23681-0001

Asit K. Ray**
Department of Chemical Engineering
Christian Brothers University
Memphis, TN 38104-5581

ABSTRACT

Thermosetting and thermoplastic polyimides have complementary physical/ mechanical properties. While thermosetting polyimides are generally easier to process but more brittle, thermoplastic polyimides are harder to process but tougher. A combination of these two types of polyimides may help produce polymers more suitable for aerospace and other non-aerospace applications that require both easy processing and good toughness. Semi-interpenetrating polymer networks (semi-IPN's) of thermoset LARC™-RP 46 and thermoplastic LARC™-IA polyimides were prepared in weight percent ratios ranging from 100:0 to 0:100. Thermal analysis and dynamic mechanical analysis were made on these samples to correlate their chemical composition with physical/ mechanical/morphological properties. As expected, the glass transition temperature increases with an increase in thermoset concentrations. The opposite is true for the thermo-oxidative stability. All the semi-IPN's showed unusually good phase-mixing. The resin density goes through a maximum at about 50:50 weight ratio, suggesting that the semi-IPN samples are not merely a solid solution of the two polymeric components. This non-linear behavior may be interpreted in terms of the existence of an electrostatic interaction between the two constituent polymers.

Key Words: High Performance semi-interpenetrating polymer networks, LARC™-RP 46, LARC™-IA

*To whom all correspondence should be addressed.
**NASA-ASEE Summer Faculty Fellow, 1995.

1. INTRODUCTION

Due to an increasing emphasis on easy processing and good fracture toughness of composite matrix resins for aerospace structural applications, we continue to search for new, inexpensive, environmentally friendly high performance semi-interpenetrating polymer networks (semi-IPN's). Thermosetting polyimides are generally easy to process, but are brittle. In contrast, thermoplastic polyimides are tough, but difficult to process. Combining them would lead to tough and easy to process high performance semi-IPN's (1). Based on this concept, we have synthesized a series of semi-IPN's by combining various weight ratios of LARC™-RP 46 (2), LARC™-IA (3) polyimides. Both of these polyimides were recently developed at NASA Langley Research Center. Their outstanding physical and mechanical properties show promise for a wide variety of aerospace and non-aerospace applications. Both polymers are now commercially available.

LARC™-RP 46 is a thermosetting polyimide prepared from the monomethyl ester of 5-norbornene-2,3-dicarboxylic acid (NE), dimethyl ester of 3,3',4,4'-benzophenone tetracarboxylic acid (BTDE) and 3,4'-oxydianiline (3,4'-ODA). NE is an end-capping agent through which crosslinking reactions occur. The formulated molecular weight (FMW) of the uncrosslinked oligomer was stoichiometrically controlled to be 1,500 g/mole. Figure 1 shows the synthesis of LARC™-RP 46 pre-polymer. LARC™-IA is a thermoplastic polyimide prepared from 4,4'-oxydiphthalic anhydride (ODPA), 3,4'-ODA and 8 mole percent of phthalic anhydride. Phthalic anhydride acts as a non-reactive end group to control the molecular weight of the polymer to be 9,000 g/mole. Figure 2 shows the synthesis of LARC™-IA polyamic acid, which is a precursor of the polyimide. The semi-IPN's were then prepared by combining LARC™-RP 46 imidized oligomer with LARC™-IA polyamic acid. Upon heating, LARC™-RP 46 undergoes crosslinking reactions in the immediate presence of LARC™-IA, which is undergoing simultaneous cyclo-imidization and linear chain extension. Figure 3 shows the synthesis of the semi-IPN.

It is well known that morphology plays an important role in controlling physical and mechanical properties of a semi-IPN. The degree of phase separation depends strongly on the chemical compatibility of the component polymers from a thermodynamic standpoint. In the earlier works, much of the efforts have been devoted to developing chemically compatible semi-IPN's using thermosetting and thermoplastic polyimide components which have an identical repeating unit, in order to achieve a good phase mixing (a favorable entropy term due to a small difference in solubility). However, for semi-IPN's that contain chemically dissimilar polymer components, a gross phase separation was almost invariably observed because of a very small gain in entropy upon mixing polymers of large chain lengths (an unfavorable positive value for the Gibb's free energy of mixing). The extent of the phase separation can sometimes be kinetically controlled, if mixing occurs on a low molecular weight level and a simultaneous synthetic method is adapted, as exemplified by the present semi-IPN synthesis.

Microstructural characterization of the present semi-IPN's by positron lifetime spectroscopy has been reported (4). The preliminary results suggest that a strong inter-molecular electrostatic interaction exists between the LARC™-RP 46 and LARC™-IA. Accordingly, the semi-IPN's should have a good phase mixing, despite the absence of an identical repeating unit in the constituent polymer components. To further clarify this unusual morphological behavior, thermal characterization was made on seven resin systems that contained 100:0, 75:25, 65:35, 50:50, 35:65, 25:75, and 0:100 percent weight ratios of LARC™-RP 46 and LARC™-IA. The effects of the chemical composition on resin density, glass transition temperature, thermo-oxidative stability and dynamic mechanical properties are discussed below.

2. EXPERIMENTAL

2.1 Synthesis

LARC™-RP 46 molding powder (5) and LARC™-IA polyamic acid solution having 30% solids in N-methyl pyrrolidinone (NMP) (6) were prepared following the procedures reported earlier.

A typical synthesis of the semi-IPN is described below. The LARC™-RP 46 oligomers were dissolved in a minimum amount of NMP at 70°C and then added to the LARC™-IA polyamic acid solution in NMP. The mixture was stirred for 1 hr. at room temperature followed by heating for 2 hrs. at 160°C in vacuum and then 2 hrs. each at 200°C and 250°C without vacuum. This heating sequence produced a B-staged molding powder. Approximately 5.0 grams of the B-staged molding powder was placed in a 3.175 cm x 3.175 cm stainless steel mold, which was preheated with a Frekote™ release agent. The mold was heated to 290°C at 13.8 MPa pressure and then held for 1 hr. at 325°C. Because the 100% LARC™-IA molding powder had greater resin flow than the semi-IPN's and the 100% LARC™-RP 46, it was cured at 355°C for 1 hr. under 2.07 MPa pressure. All the resin systems prepared showed no detectable voids and had exceptionally high densities.

2.2. Characterization

Density was measured by a standard water displacement technique. Each density reported in Table 1 is an average of three determinations. Glass transition temperatures (Tg's) were determined using a DuPont 990 Thermal analyzer in connection with a DuPont Model 943 Thermal Mechanical Analyzer (TMA) at a heating rate of 5°C/min. and with a load of 5.0 g. Tg's were also recorded on a DuPont 1090 Thermal Analyzer in connection with a Model 910 Differential Scanning Calorimeter (DSC) run at 20°C/min. Thermo-oxidative stability of the resins was evaluated based on the weight loss determined by a Seiko 220 DTA Thermogravimetric Analyzer (TGA) at a heating rate of 2.5°·C/min. in nitrogen. Dynamic mechanical analysis (DMA) was made using DuPont DMA 982 at a heating rate of 5°C/min.

3. RESULTS AND DISCUSSION

Table 1 summarizes the physical properties of the semi-IPN's. The densities range from 1.34 to 1.38 g/cc. These values are considerably higher than 1.32 g/cc, which is normally obtained for a high performance polymer. Of particular interest is the correlation between density and chemical composition as shown in Figure 4. With the exception of the sample that contains a 25:75 percent weight ratio of LARC™-RP 46 and LARC™-IA, the resin density goes through a maximum at about the 50:50 weight ratio. It is unclear at the present time why the sample with the 25:75 composition exhibits an exceptionally high density, and this is currently being investigated. A somewhat similar correlation is also observed for the Tg with composition (Figure 4). The Tg also goes through a maximum at about the 50:50 weight ratio for the portion of the compositions that contain up to 65:35 LARC™-RP 46: LARC™-IA. The Tg ranges from 291°C for the 100% LARC™-RP 46 to 210°C for the 100% LARC™-IA material (Table 1). Despite a significant difference (80°C) in Tg between LARC™-RP 46 and LARC™-IA, all the semi-IPN samples studied showed a single Tg in each composition, which was determined by TMA and confirmed by DSC (Table 1). As shown in Figure 5, the 100% LARC™-RP 46 sample exhibits a rather pronounced thermal transition at around 370°C (Tm). This behavior was also observed in our earlier work on the film properties (7). The x-ray diffraction data supports the presence of a thermoplastic phase in LARC™-RP 46. This most likely arises from the inability of BTDE to compete with NE for 3,4'-ODA, due to the significantly higher reactivity of NE than that of BTDE toward a diamine. Thus, the thermoplastic polyimide from BTDA and 3,4'-ODA is responsible for the thermal transition at 370°C. It is also one of the reasons why there is a broad Tg transition peak seen at a temperature of 298°C for the 100% LARC™-RP 46 sample. There is no detectable Tm at a temperature above 350°C for the 100% LARC™-IA.

Each of the semi-IPN's shows a Tm, which varies from 350°-377°C. The area under the Tm peak decreases, when the LARC™-RP 46 concentration is decreased, suggesting that the Tm transitions in the semi-IPN's are due to the semi-crystalline thermoplastic phase in the LARC™-RP 46.

Figure 6 shows percent weight retention as a function of temperature determined by TGA. The temperature at which 5% weight loss occurs (T5%) is used to compare thermo-oxidative stability between materials (Table 1). The T5% decreases when there is an increase in the LARC™-RP 46 concentration. The weight loss observed occurs primarily because of crosslinking reactions in the nadic end group in the LARC™-RP 46.

Figure 7 compares the loss modulus (E") peak in the glass transition region (α-transition) between samples. The LARC™-RP 46 has the broadest α-transition peak. As explained earlier, the LARC™-RP 46 material contains a broad distribution of crosslinked and linear structural components. The heterogeneity is undoubtedly responsible for the broadening of the α-transition peak. As the LARC™-RP 46 concentration decreases, the α-transition peak progressively increases its sharpness. This unexpected result suggests that the molecular mixing increases as the LARC™-IA is added to the LARC™-RP 46. As shown in Figure 8, the half-width of the α-transition peak appears to correlate to the LARC™-RP 46 concentration.

Attempts were made to use scanning electron microscopy to show more clearly the phase domain sizes and shapes. However, no phase separation can be detected by this technique.

In this study, the cured thermosetting and thermoplastic polymers have almost the same molecular weights in their repeating units (486 vs. 458). Thus, their weight ratio is essentially equal to their molar ratio. The conventional theory as to how a semi-IPN behaves is that there is a mutual interpenetration between the constituent polymers, which is responsible for semi-IPN characteristics. Based on this theory, the physical interlocking progressively increases until the constituent polymers reach a 50:50 composition at which a maximum overlapping of the molecular chains occurs. Therefore, it would be expected that the semi-IPN's will gradually increase their physical/mechanical properties until they reach a 50:50 composition. However, such a linear concept does not take into account the interatomic (electrostatic) forces that also maximize under the same conditions. The electronic charge distribution in the LARC™-RP 46 molecules induces a corresponding charge distribution in the LARC™-IA molecules, when they overlap each other, through a donor-acceptor type of mechanism. This produces strong electrostatic bonding between these molecules. The combination of the physical and electrostatic interactions results in a non-linear behavior with a maximum density and a maximum Tg at the 50:50 composition as discussed above.

4. CONCLUDING REMARKS

We have investigated the thermal and dynamic mechanical behavior of semi-IPN's prepared from LARC™-RP 46 and LARC™-IA polyimides. The results cannot be explained solely by the mutual physical interpenetration of the constituent polymers. Their non-linear characteristics require the presence of both inter-molecular mechanical interlocking and electrostatic interaction to explain the observed links between the polymers' density, glass transition temperature and dynamic mechanical properties with their chemical composition.

5. REFERENCES

1. R. H. Pater in S. M. Lee, ed., International Encyclopedia of Composites, VCH Publishers, Inc., New York, Vol. 2, 1990, pp. 377-401.

2. R. H. Pater: U.S. Patent 5,171,822, 1992.

3. T. L. St. Clair: U.S. Patent 5,147,966, 1992.

4. J. J. Singh, R. H. Pater, and A. Eftekhari, presented at the ACS Meeting, March 24-29, 1996, New Orleans, LA.

5. R. H. Pater, SAMPE International Symposium, 36, Closed Papers, 78 (1991).

6. T. H. Hou, N. J. Johnston, and T. L. St. Clair, High Performance Polymers, 7, 105-124 (1995).

7. M. D. Soucek and R. H. Pater, J. of Plastic Films and Sheetings, 9, 67 (1993).

Table 1. Physical Properties

Resin System LARC™-RP 46: LARC™-IA	Density, g/cc	Tg, °C By TMA	Tg, °C By DSC	T5%[a] Weight Loss, °C
100:0	1.3572 ± 0.0020	291	298	458
75:25	1.3546 ± 0.0037	238	250	474
65:35	1.3629 ± 0.0033	231	244	480
50:50	1.3713 ± 0.0041	234	245	479
35:65	1.3668 ± 0.0020	233	243	517
25:75	1.3815 ± 0.0029	215	212	504
0:100	1.3402 ± 0.0030	210	209	516

[a]Temperature at 5% weight loss determined by TGA in nitrogen.

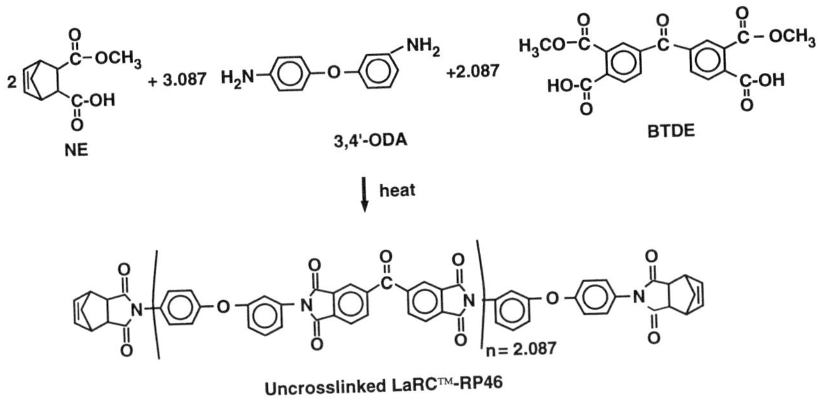

Figure 1. Synthesis of LARC™-RP 46 Prepolymer.

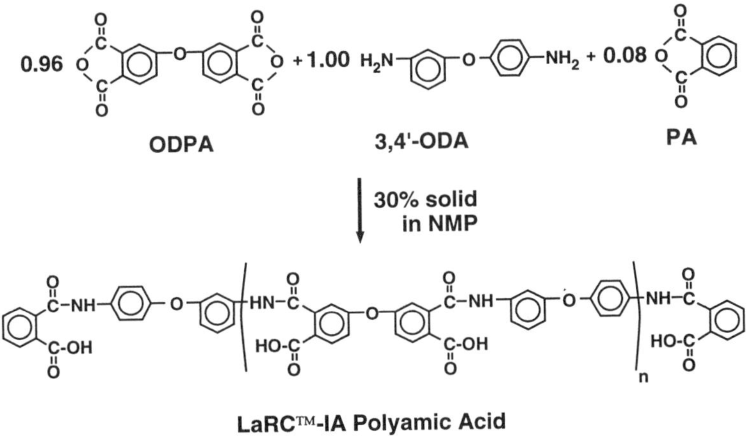

Figure 2. Synthesis of LARC™-IA polyamic acid.

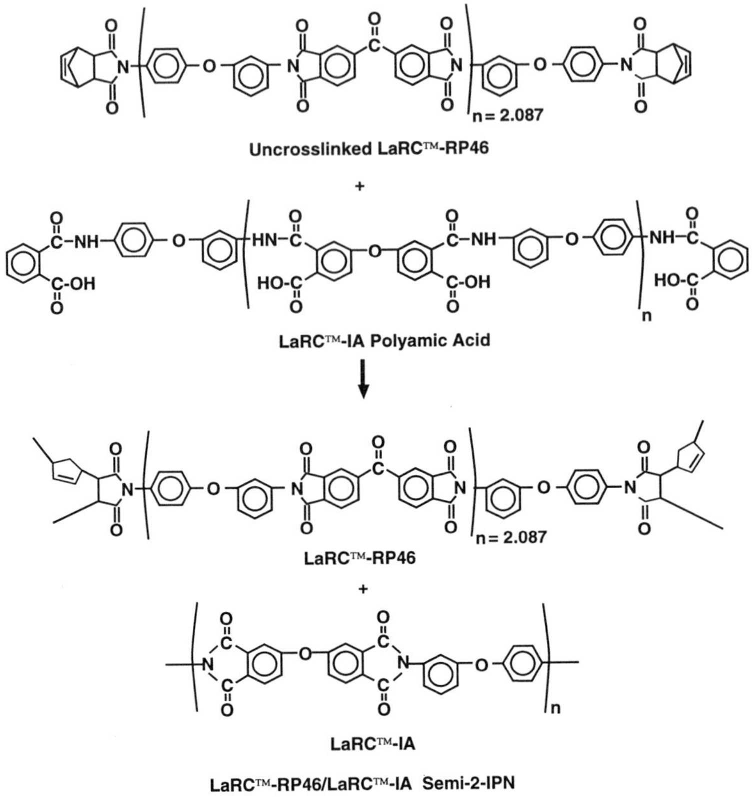

Figure 3. Synthesis of semi-IPN.

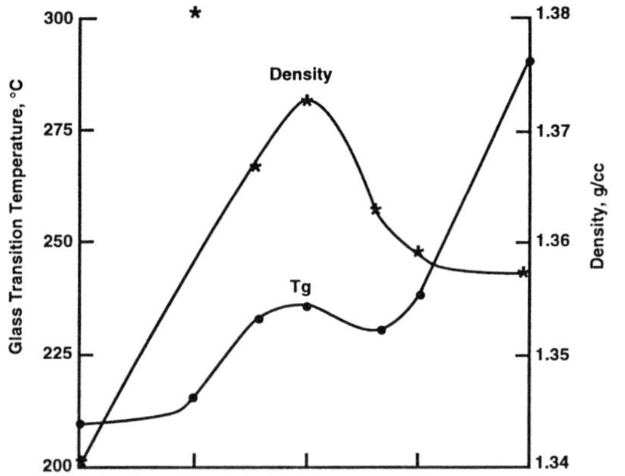

Figure 4. Density and glass transition temperature vs. concentration of thermoset component.

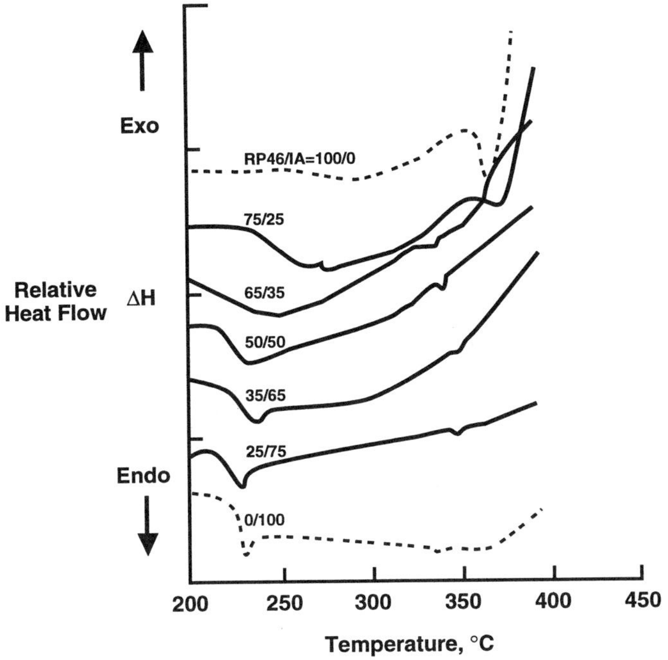

Figure 5. Differential scanning calorimetry curves.

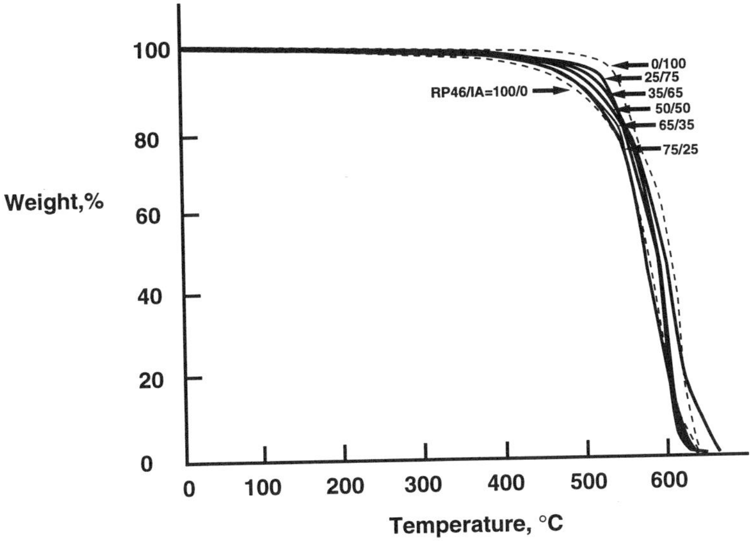

Figure 6. Weight loss vs. temperature.

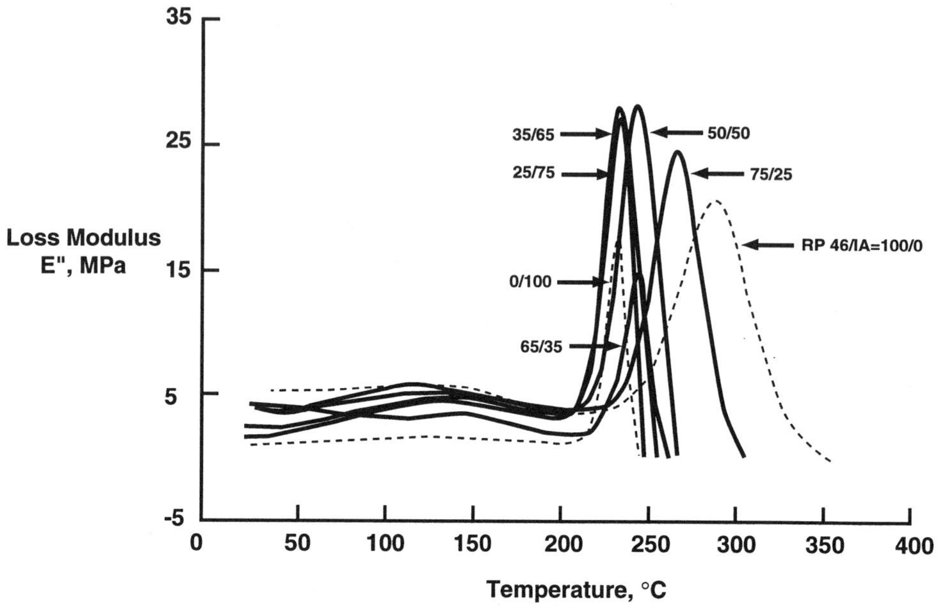

Figure 7. Dynamic mechanical properties.

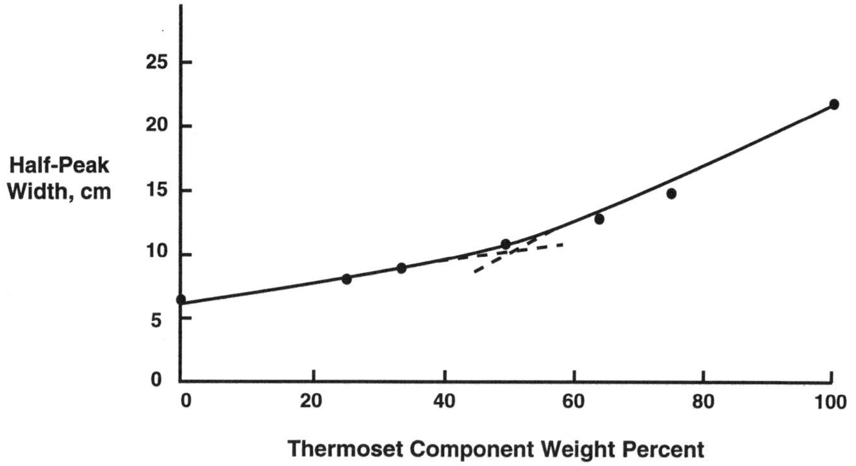

Figure 8. Half width of the α-transition peak vs. concentration of thermoset component.

28th International SAMPE Technical Conference
November 4-7, 1996

STUDY ON TOUGHENING BISMALEIMIDE RESIN WITH THERMOPLASTICS I. BMI/DP RESIN TOUGHENED WITH PEK-C

Wang Rumin, Zheng Shuirong, Shi Yutan, Yang Li, and Lan Liwen
Department of Chemical Engineering, Northwestern
Polytechnical University, Xi' an, Shaanxi 710072, P. R. China

ABSTRACT

To toughen p,p' -bismaleimidodiphenylmethane (BMI) copolymerized with o,o' -diallylbisphenol A (DP), it was blended and modified with rigid thermoplastics polyaryletherketone with phenolphthalein (PEK — C) as the toughening agent. The results show that addition of small amount of PEK—C can further increase the toughness of BMI/DP resin, e. g. , fracture surface energy (G_{IC}) of modified resin rises from 214 J/m^2 to 404J/m^2 after addition of 7.9% of PEK—C. Higher molecular weight of PEK—C is better in toughening effects, and hydroxyl-terminated PEK—C is benificial to the strength and heat resistance of modified resin compared with common PEK—C. With the increase of proportions of PEK—C, modified resin presents two kinds of different morphological structures, thermosetting matrix with dispersed small size thermoplastic particles and thermoplastic matrix with collected great size thermosetting spheres.

KEY WORDS: Bismaleimides, Polyaryletherketone with Phenolphthalein, Blending

1. INTRODUCTION

Bismaleimide (BMI) resin is a new generation of important thermosetting matrices for advanced composite materials applications after epoxy resin because of its excellent heat-resistant properties (1). But most BMI resins are brittle and need to be toughened for extending their uses. The most effective way to toughen thermosetting resin now is to be blended with heat-resistant rigid thermoplastic resin (2). Research works on toughening BMI resin with thermoplastics began in the late eighties (3-4). Papers on toughening BMI s with modified polyaryletherketone have not come forth in the literatures. In this paper polyaryletherketone with phenolphthalein (PEK—C) was blended to increase the toughness of p,p' -bismaleimidodiphenylmethane/o,o' -diallylbisphenol A (BMI/DP) copolymerized resin.

2. EXPERIMENTAL

2. 1 **Materials** BMI resin system in this study included two components, p,p' -bismaleimidodiphenylmethane (BMI) and o,o' -diallylbisphenol A (DP). Polyaryletherketone with phenolphthalein, named as PEK — C, was used as toughening agent. PEK — C is an amorphous

thermoplastics powder with rigid backbone and 231℃ of glass transition temperature. The molecular structure of BMI, DP and PEK—C is as follows.

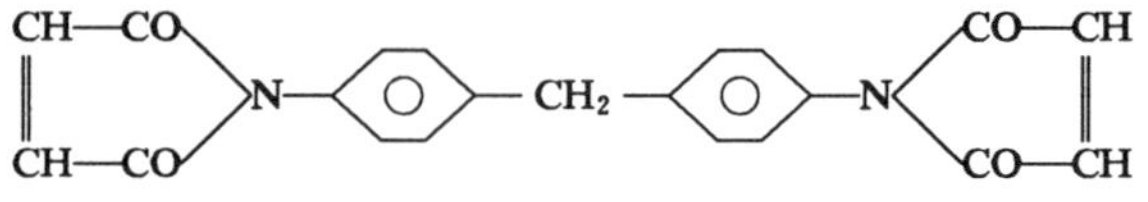

(BMI)

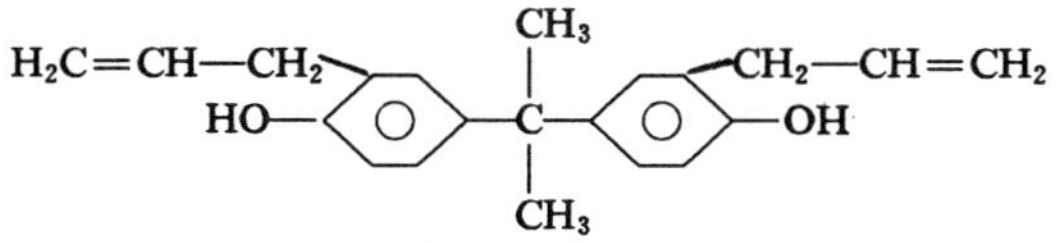

(DP)

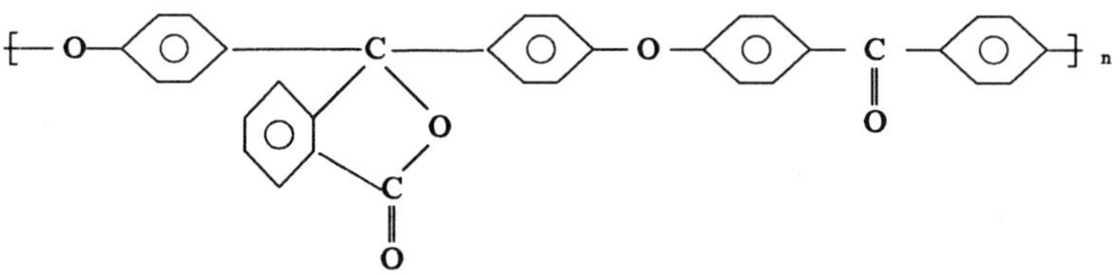

(PEK—C)

2.2 **Preparation of Specimens** After PEK—C was dissolved in liquid DP at about 170℃ and cooled to 150℃, BMI crystal powder was added to and prepolymerized with DP for 25 to 30min at 130 to 140℃. Then mixed BMI resins were poured into a pre-heated mould, and degassed in a vacuum oven at 120℃. Resin cast plate was prepared ofter cure cycle of 160℃ /2h+180℃/2h+200℃/2h plus postcure cycle of 230℃/4h in an air circulating oven. Specimens with desired size were obtained by sand cutter machining on cast plate.

2.3 **Test Methods** Heat distortion temperature (HDT) was measured in accordance with ASTM 648—56 using resin cast specimen. Flexural strength (σ_F) and impact strength (σ_I) of resin casts were tested by non-standard method of cantilever beam load. Fracture surface energy (G_{IC}) of casts was measured using double cantilever beam (DCB) split method (5) on Instron 1195 tester. After having sprayed with gold on specimen morphological observation of fracture surface was carried out on a AMARY 1000 scanning electron microscope.

3. RESULTS AND DISCUSSIONS

3.1 **Contents of PEK—C on Properties of Blended BMI/DP Resin** BMI/DP copolymerized resin is a typical BMI one which has excellent thermal, mechanical properties and good toughness. As a result Ciba-Geigy XU 292 is one of two winners of new materials appraised by American magazine "Research & Development" in 1984. This paper chose BMI/DP resin system as an object of study on toughening. For phenolphthalein to be introduced to molecular structure of polyaryletherketone, toughening agent, PEK—C loses crystallinity and becomes amorphous plastics with good processabilities. Table 1 lists the experimental results of contents of

PEK—C on properties of blended BMI/DP resin casts. With BMI fixed in 100 parts in weight and DP fixed in 75 parts in weight, PEK—C changes in 0,5,10 and 15 parts in weight respectively. The examined properties includes flexural strength (σ_F), impact strength (σ_I), fracture surface energy (G_{IC}) and heat distortion temperature (HDT) of resin casts. From table it is knownthat after addition of 5 parts (2.8%) of PEK—C, flexural strength and impact strength increase by a factor of 14.4% and 66.3%, but HDT decreases by a factor of 8.1%, which show that toughness of blended resin has been improved. With the increase of addition amount of PEK—C, 10 parts (5.4%) and 15 parts (7.9%), properties of blended resin have not changed obviously, but as compared with 0 parts of PEK—C flexural strength and impact strength of the resin are improved greatly and HDT falls a little. Addition of 7.9% of PEK—C makes fracture surface energy of the resin increase by a factor of 88.8%, which show that ability to resist crack propagation, i. e., toughness of the resin is enhanced greatly. Improvement of toughness is related to morphological structure of second phase particulates. Why addition amount of PEK—C hasn' t led to continuus increase in toughness is related to more viscous of resin system which will make degassing and well-dispersing more difficulty. Glass transition temperature of PEK—C is 231℃ which is much lower than HDT of BMI/DP resin, so heat resistance of BMI/DP resin blended with PEK—C decreases.

Table 1 Addition amount of PEK—C on properties of blended resin casts

PEK—C		Flexural strength		Impact strength		Heat distortion temp		Fracture surface enrngy	
parts	%	σ_F, MPa	↑ %	σ_I, KJ/m²	↑ %	HDT, ℃	↓ %	G_{IC}, /m²	↑ %
0	0	108		8.9		297		214	
5	2.8	123.5	14.4	14.8	66.3	273	8.1		
10	5.4	130.6	20.9	11.1	23.6	262	11.8		
15	7.9	122.7	13.6	15.6	75.3	271	8.8	404	88.8

3.2 **Influence of Functionalized PEK—C on Properties of Modified Resin** In order to examine the role of end group of thermoplastics, non-terminated PEK—C, i. e., hydroxyl-terminated PEK—C was used as toughening agent of BMI/DP resin. Experimental results of influence of addition of different parts of hydroxyl-terminated PEK—C on flexural strength, impact strength and HDT of blended BMI/DP resin are listed in table 2. Data in table 2 show that at the same addition amount BMI/DP resin blended with functionalized PEK—C has higher flexural

Table 2 Influence of hydroxyl-terminated PEK—C on properties of blended resin

HO—PEK—C		Flexural strength		Impact strength		Heat distortion temp	
parts	%	σ_F, MPa	↑ %	σ_I, KJ/m²	↑ %	HDT,	↓ %
0		108		8.9		297	
5	2.8	128.5	19.0	12.0	34.8	300	−1.0
10	5.4	133.5	23.6	9.4	5.6	290	2.4
15	7.9	131.9	22.1	8.6	−3.4	275	7.4

strength, lower impact strength and higher HDT than that with terminated PEK—C (table 1). Active hydroxyl is possible to increase adhesion of interface between thermosets and thermoplastics, and strong interface leads to high strength, low deformation and small fracture surface area.

3. 3 Molecular Weight of PEK—C on Properties of Blended Resin Three kinds of different molecular weights of PEK—C, with intrinsic viscosity number 0. 32, 0. 57 and 0. 72 (marked as J, K and L respectively), were chosen to study the influence of molecular weight of thermoplastics on toughening effect. Table 3 lists experimental results of flexural strength, impact strength and HDT of BMI/DP resin blended with different parts and different molecular weights of PEK—C. From table 3 it is generally seen that at the same addition amount of PEK—C, with the increase of molecular weight of PEK—C, flexural strength increases, impact strength increases a little and decreasing rate of HDT goes down. The reason for this may be related to the increase of molecular entanglement between thermosetting and thermoplastic network with the increase of molecular weight of PEK—C.

Table 3 Molecular weight of PEK—C on properties of blended resin

PEK—C		Flexural strength		Impact strength		Heat dsitortion temp	
parts	types	σ_F, MPa	↑ %	σ_I, KJ/m^2	↑ %	HDT, ℃	↓ %
0		108		8. 9		297	
5	J	109. 6	1. 5	12. 2	37. 1	276	7. 1
	K	128. 5	19. 0	12. 0	34. 8	300	−1. 0
	L	128. 7	19. 2	11. 1	24. 7	292	1. 7
10	J	113. 7	5. 28	9. 2	3. 4	281	5. 4
	K	133. 5	23. 6	9. 4	5. 6	290	2. 4
	L	137. 5	27. 3	10. 5	18. 0	292	1. 7
15	J	108. 3	0. 3	7. 6	−14. 6	270	9. 1
	K	131. 9	22. 1	8. 6	−3. 4	275	7. 4
	L	134. 5	24. 5	9. 3	4. 5	270	9. 1

3. 4 Morphology and Toughening Mechanisms of BMI/DP Blended with PEK—C In experiments it has been discovered that PEK—C can be thermally dissolved in DP and BMI/DP prepolymerized resin and produced homogeneous transparent melt, which show that PEK—C has good compatibility with BMI/DP copolymerized resin, which is beneficial to molecular diffusion and adhesion between PEK—C and BMI/DP. As cure reaction of BMI and DP goes on, molecular weight of thermosetting parts increases, interaction between thermoplastics and thermosets changes, and compatibility between them becomes little. Phase separation comes forth when cure reaction reaches to some extent. The better compatibility between them, the later phase separation, and the smaller size with greater numbers of second phase particulates, the thicker and the more indistinct phase boundary. As content of PEK—C is 5 parts (2. 8%), morphological structure of fracture surface of blended resin is evenly distributed at low magnification (Fig 1a), and at great magnification there present many dispersed second phase nodules (about 0.

8μm in diameter) on crack propagation surface (Fig 1b). Most of nodules initiate new microcracks, and lead to the increase of resin toughness. According to the proportion of thermosets and thermoplastics in blended resin, nodules here should be thermoplastics, and toughening mechanism is similar to that of traditional epoxy resin toughened with rubbery particulates. Nodular particles have roles of resisting crack propagation, changing crack propagation direction, initiating new cracks and branched cracks. With the increase of content of PEK—C viscosity of blended resin increases rapidly, so it is difficult to make blended resin homogeneous by mechanical stir in spite of transparent by the eye. Fig. 2 is SEM photographs of fracture surface of blended resin with 15 parts (7.9%) of PEK—C. Different from content of 5 parts (2.8%) of PEK—C, there can be seen two types of morphological area on fracture surface at low magnification (Fig 2a), ralative smooth area (named as S) and collected nodules area (named as N). In S area

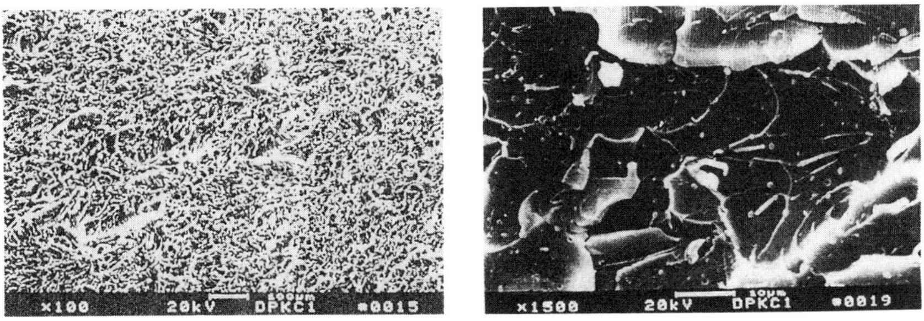

(1a) ×100 (1b) ×1500

Fig. 1 SEM photographs of fracture surface of BMI/DP blended with 5 parts of PEK—C

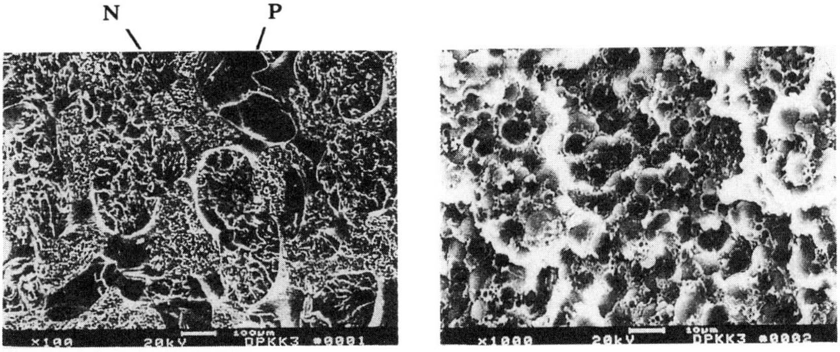

(2a) ×100 (2b) N area ×1000

Fig. 2 SEM photographs of fracture surface of BMI/DP blended with 15 parts of PEK—C

relative content of thermoplastics is lower, so its morphological structure is that dispersed thermoplastic nodules with size smaller than 1μm present in thermosetting matrix with many river markings, which is similar to morphological structure of 5 parts (2.8%) of PEK—C (Fig 1b). Relative content of thermoplastics is higher in N area than in S area, so there presents unlike morphological structure, many nodules with great size (about 10μm) collected in low proportion of continuous phase (Fig 2b). According to the proportions in blended resin, collected nodules in N area should be thermosetting resin and the matrix should be thermoplatics. Formation of N area owes to the increase of content of PEK—C in bulk or local area, and there is only a S area morphological structure in blended resin as proportion of PEK—C is low. Flexural strength and impact strength of resin with S, N area couple structure are similar to those with only S area structure, but ability to resist main crack propagation is greater because slitting from crack surface of S,N area couple structure consumes more energies.

4. CONCLUSIONS

BMI/DP resin toughened with PEK—C were successfully prepared by method of hot melt blending, and its morphological structure, toughening mechanism and macroscopic properties were also studied. The results show that addition of small amount of PEK—C can further increase the toughness of BMI/DP resin, but its heat resistance decreases a little. Higher molecular weight of PEK—C is better in toughening effects because of enhanced moleculer entanglement, and hydroxyl-terminated PEK—C is beneficial to the strength and heat resistance of modified resin because of increased adhesion between two phases compared with common PEK—C. With the increase of proportions of PEK—C blended resin presents two kinds of different rnorphological structures, thermosetting matrix with dispersed small size thermoplastic particulates and thermoplastic matrix with collected great size thermosetting nodules.

5. REFERENCES

1. D. A. Scola, in W. C. Harrigan, ed. , Proceedings of 5th International Conference on Composite Materials, California, 1985, pp. 1601—1621.
2. Lixin Yin, Acta Materiae Compositae Sinica, 8, (1), 37—41(1991).
3. H. D. Stenzenberger, SAMPE International Symposium, 33, 1549—1560(1988).
4. R. J. Morgant, etal, Polymer, 34, (4), 835—842(1993).
5. S. M. Wiederhorn, J. Applied Physics, 39, (3), 1569—1572(1968).

28th International SAMPE Technical Conference
November 4-7, 1996

THE STRUCTURE-PROPERTY RELATIONS OF BISMALEIMIDE COMPOSITE MATRICES AND DURABILITY OF HIGH TEMPERATURE POLYMER MATRIX COMPOSITES

Roger J. Morgan, Eugene E. Shin, and Jason E. Lincoln
Advanced Materials Engineering Experiment Station
Composite Materials and Structures Center
2203 Eastman Avenue, Midland, MI 48640
Jin Choi and Andre Lee
Department of Materials Science and Mechanics
Michigan State University

ABSTRACT

Present and future polymer-matrix fibrous composite aerospace applications, for bismaleimide-(BMI) and polyimide (PI)-carbon fiber composites, involve combined complex synergistic stress, thermal, moisture, time, radiation and oxygen service environment-induced degradation mechanisms of composite performance. In this paper we report and characterize the critical aging mechanisms that control damage initiation in BMI and PI carbon fiber composites in aerospace environments.

For BMI-carbon fiber composites systematic Fourier Transform infrared spectroscopy, FTIR, and differential scanning calorimetry, DSC studies reveal that these resins are not fully cured and can continue to react in service environment as a result of dehydration induced ether crosslinks. This further crosslinking causes T_g increases, mechanical property decreases and further resin shrinkage with associated microcrack development. The relations between network structure and BMI thermal and mechanical properties are discussed.

In addition the two most critical hygrothermal-induced degradation mechanisms of BMI- and PI-carbon fiber composites, namely (i) matrix blistering and composite delamination and (ii) permanent "dry" T_g decreases are critically reviewed.

KEYWORDS: High Temperature Composites, Durability, Bismaleimides and Polyimides.

1. INTRODUCTION

Utilization of high temperature polymer matrix-carbon, C, fiber composites for structural applications in future commercial, such as High Speed Civil Transport (HSCT), and military aircraft will be exposed to prolonged and often extreme service environment conditions (1,2). These complex service environment conditions of combined stress, time, temperature, moisture, chemical and atmospheric environments require a thorough understanding of potential physical, chemical and mechanical-induced composite degradation mechanisms, and their synergistic effects that could lead to the most probable critical failure path. An understanding of the critical fundamental aging mechanisms is necessary for credible long-term composite performance predictions from experimentally observed shorter time service environment induced composite performance deterioration mechanisms.

Two of the leading high temperature polymer matrix-carbon fiber composites consist of either (i) crosslinked bismaleimide (BMI) thermoset and (ii) thermoplastic polyimide (PI) matrices. Such composites are utilized in present aerospace structures and are being evaluated for structural applications in future commercial and military aircraft. The thermoplastic polyimides have exhibited both processing and toughness limitations whereas the thermoset bismaleimides are more processible but have exhibited a more brittle mechanical response.

Based on our previous studies (2-10) we have identified the most likely potential, critical aging mechanisms that control damage initiation in BMI and PI carbon fiber composites based on a systematic durability evaluation methodology. The critical aging mechanisms involve (i) further cure of BMI-carbon fiber composites with associated T_g increases, mechanical property decreases and enhanced microcrack development in service environment exposure conditions, and (ii) hygrothermal induced thermal and mechanical property deterioration as a result of physical and chemical structural changes in BMI- and PI-carbon fiber composites.

In the case of BMI-carbon fiber composites, the BMI cure reactions are incomplete after standard composite fabrication and post-cure procedures and further cure can occur in service environment conditions resulting in T_g increases and associated composite mechanical property deteriorations. Loseth and Rothschild (11) have observed real time T_g increases for commercial BMI-carbon fiber composites (BASF 5260) of up to 70°C after isothermal exposure in the 150-200°C temperature for up to ~ 10^4 hours. Li (12) reported a 50% decrease in impact energy after thermal aging of BMI-carbon fiber composites for 1,000 hours at 190°C. Pederson et al.. (13) report an increase surface ply crack density after isothermal 177°C air exposure for 750 hours for [0/90] laminates. Also Burchan et al. (14) report transverse microcrack density increases for 16 ply quasi-isotropic BMI-carbon fiber composites aged isothermally in the 150-204°C range for 16,000 hours as a function of time, temperature and ply depth.

Commercial BMI resins, such as Matrimid 5292, Ciba Geigy and BASF 5260 are based on the 4, 4'-bismaleimidodiphenyl methane (BMPM)/0, 0'-diallyl alcohol of bisphenol A (DABPA) system (Figure 1). In our previous studies (4) we conducted systematic Fourier transform infrared spectroscopy, FTIR, and differential scanning calorimetry, DSC, studies of the cure reactions of the BMPM/DABPA BMI resin system as a function of chemical composition and temperature-time cure conditions and together with literature data (15-17) it was concluded:

(i) In the 100-200°C range the BMPM and the DABPA monomers react via the "ene" reaction to form the "ene" adduct, illustrated in Figure (2). The "ene" adduct is pentafunctional as a result of three double -C=C- bonds, capable of chain extension and crosslinking, designated (A), allyl, (B), propenyl and (C) maleimide in Figure (2) and two hydroxyl groups capable of etherification by hydroxyl dehydration.
(ii) The principal cure reactions occur in the 200-300°C range via the -C=C- double bonds with etherification occurring above 240°C at DSC heating rates of 10°C/min.
(iii) Cure was incomplete due to glassy-state diffusion restrictions at 250°C and further cure at 300°C for 1 or 2 hours produced a constant T_g near 350°C. Residual DSC exotherm experiments indicate <10% unreacted groups remain after 2 hour at 200°C, 6 hour at 250°C and 2 hour at 300°C cure schedule.

Recent isothermal BMI cure studies by Mijovic and Andjelic (18) by remote fiber optic near infrared spectroscopy in the 140-250°C range have been conducted. They report the relative rates of allyl and maleimide consumption and that essentially complete conversion of these groups occurs at 200°C. Also, they observe the onset of dehydration at low temperatures, 140° C, over periods of hours.

Hygrothermal exposure of BMI- and PI-carbon fiber composites has resulted in reported (i) matrix blistering and composite delamination (8) and (ii) permanent "dry" T_g decreases as a result of possible matrix physical and chemical modifications (19-22).

In this paper we have extended our systematic FTIR and DSC studies to 1 to 14 hour isothermal exposures in the 130° to 300°C range. These studies were carried out in order to ascertain (i) the sequence and relative rates of the cure reactions, and (ii) the nature of chemical reactions that are capable of further reactions, with associated thermal and mechanical property modifications, in future aerospace service environment conditions. In addition, we critically address hygrothermal induced composite degradation mechanisms in terms of potential fundamental physical and chemical matrix and fiber-matrix interface modifications that could cause composite property deterioration and describe a series of controlled experiments that could identify the fundamental degradation mechanisms.

2. MATERIALS

The BMI resin system utilized was the BMPM/DABPA (Ciba-Geigy) system (Figure 1). The PI system considered was Avimid K3B, Dupont, formed from an aromatic diethyl ester diacid and an aromatic diamine in N-methyl-2-pyrrolidone.

3. RESULTS AND DISCUSSION

3.1 Structure-Property Relations of Bismaleimide Composite Matrices

3.1.1. *F.T.I.R. Characterization of the Cure Reactions.* In Figure 3, we show typical FTIR spectra of the BMPM-DABPA BMI system as a function of cure conditions. The IR bands at 820cm^{-1}, 915cm^{-1}, 1183cm^{-1} and 3473cm^{-1} are assigned to the maleimide, allyl, ether and hydroxyl groups respectively, that are associated with the cure reactions (10). The band intensities were normalized against the CH_3 group in DABPA. This CH_3 group has two characteristic regions of absorption at 2970cm^{-1} and a doublet for the dimethyl group at 1365 and 1385cm^{-1} (10).

From the normalized IR bands we plot, in Figure 4, reaction percentages of the maleimide and allyl double C=C bonds and the ether appearance caused by hydroxyl dehydration as a function of sequential, accumulative cure conditions from 130° to 250°C. The maleimide and allyl double C=C bonds associated with the "ene" adduct are completely reacted after 3 hours at 250°C. The commercially recommended cure of the BMPM-DABPA BMI system is 6 hours at 250°C. The maleimide and allyl double C=C bonds react at very nearly the same rates but for certain cure sequences the maleimide consumption can be up to 10% greater than the allyl consumption.

The relative rates of the allyl, propenyl and maleimide double C=C bond cure polymerization reactions can be complex and depend on the cure history sequence. Previous DSC studies of the reaction exotherms of the BMPM and DABPA monomers and their mixtures showed that the BMPM exotherm was centered at 210°C, whereas the BMPM-DABPA mixture exotherm was 45°C higher at 255°C. The lower BMPM exotherm is equivalent to a 25x faster reaction rate for the maleimide-maleimide double C=C bond polymerization compared to other double C=C bond polymerization combinations that involve allyl, propenyl and maleimide combinations. However, as the viscosity increases during cure intermolecular reactions will become restricted and intramolecular reactions will be favored. Previous molecular modeling characterization of the BMI cure reactions showed the intramolecular "ene" allyl-propenyl cyclization is possible (4). The (A)-(B) double C=C bond radical cyclization structure, Figure 2, is considerably more reactive than the maleimide double C=C bond. Hence, cure path sequences and associated viscosity-time-temperature histories may affect observed relative reaction rates of the three "ene" double C=C bonds and the resultant BMI network structure.

The ether crosslink reaction via hydroxyl dehydration is only 50% complete after the sequential isothermal cure and 9 hours at 250°C as illustrated in Figure 6. This crosslink reaction is inhibited by glassy-state diffusion and molecular network topography restrictions. Further post-curing at 300°C for 9 hours results in a further ~10% increase in ether crosslinks, as monitored by the intensity decrease of the IR hydroxyl band at 3473cm^{-1}, and an associated T_g increase from 280 to 350°C. However, at the 300°C cure conditions the ether crosslinks become thermally unstable and partially dissociate into other species, as indicated by the lack of further ether IR band intensity increase that correlated directly with the hydroxyl IR band decrease at lower cure temperatures. Previous systematic FTIR studies of diaminodiphenyl sulfone cure tetraglycidyl 4,4' diaminodiphenyl methane epoxies as a function of cure conditions and initial monomer ratio show that dehydration formed ether crosslinks start to thermally dissociate in the 300°C temperature range (23).

The FTIR studies indicate the BMI network is not fully cured under standard recommended composite cure conditions. The cure reaction rates and the resultant network structure could be dependent upon cure path history sequence, with intramolecular reactions favored at higher viscosity conditions.

3.1.2 ***DSC Characterization of the Cure Reactions.*** In Figure (5) the increase in the T_g as monitored by thermal mechanical analyses (TMA), as a function of isothermal exposure temperatures from 130 to 300°C for up to 30 hours is illustrated for the BMPM-DABPA (1:1 molar) BMI system. For all isothermal cure temperatures the T_g increases rapidly at short times in the liquid state and only increases slowly at longer times in the glassy state as a result of diffusion restrictions. There is an abrupt increase in T_g between isothermal 200°C and 250°C cure temperatures over similar time periods, which is consistent with the maximum rates of the double C=C polymerization cure reactions occurring in this range.

The degree of cure, α, determined from the DSC exotherm $\Delta H_t/\Delta H_\infty$ ratios versus T_g as a function of isothermal cure temperature is illustrated in Figure (6) for a constant cure time of 10 hours. This plot clearly indicates the T_g for a specific degree of cure, α, is cure history path dependent, as the isothermal T_g-α plots do not fall on one common plot. For example, for a 90% degree of cure at isothermal cure temperature 250°C the T_g is 280°C; however, for the same degree of cure at isothermal cure temperatures of 280°C and 300°C, the T_g is 70°C higher at 350°C. The lack of sensitivity of the degree of cure, α, to the T_g, particularly at the higher cure temperatures indicates that the crosslink reactions in this range do not exhibit as strong ΔH_t exotherms as a lower temperatures. We associate this lack of sensitivity of the $\Delta H_t/\Delta H_\infty$ ratio to the T_g increase to the lower heat of reaction of the dehydration reaction that exclusively occurs in the latter stages of cure in the 280-300°C range. A DSC scan of the DABPA monomer exhibits an endotherm followed by an exotherm in the 250-300°C range with resultant little overall heat evolution.

In the 130-250°C temperature region, we observed the T_g at final cure temperature to be essentially independent of cure history path. The T_g is, therefore, not sensitive to any network structural differences that may result from a range viscosity-time cure history profile differences that would affect intermolecular versus intramolecular cure reactions.

3.1.3 ***BMI Network Structure Interrelations with Mechanical and Thermal Properties***
We are conducting a systematic study of the effects of BMPM:DABPA stoichiometric ratio and cure conditions on the mechanical and thermal properties of BMI resins and composites. The objective is to correlate BMI network structure with these properties.

In Table (1), we illustrate the test matrix for this study. In Table (2), we illustrate early data of the room-temperature tensile mechanical properties for low temperature cured BMI neat resin as a function of stoichiometry. Excess BMPM resins exhibit superior mechanical properties as a result of lower crosslinked density networks. In Figure (7) we illustrate the T_g of BMI systems for isothermal cure temperatures in the 170-300°C range as a function of BMPM-DABPA ratios. At lower cure temperatures in the 170-220°C range the T_g increases with BMPM concentration as a result of the more aromatic nature of the BMPM structure. However, at cure temperatures above 300°C at the higher BMPM concentrations the T_g decreases with increasing BMPM concentration because there are less hydroxyl groups available for formation of ether crosslinks whole formation dominate T_g increases at these high cure temperatures.

3.2 Hygrothermal Induced Degradation Mechanisms. In this section we discuss the two most critical hygrothermal induced degradation mechanisms of BMI-and PI-carbon fiber composites as related to future aerospace service environments, namely (i) matrix blistering and composite delamination and (ii) permanent "dry" T_g decreases.

3.2.1 ***Matrix Blistering and Composite Delamination*** When the trapped moisture vapor pressure exceeds the local polymer matrix yield stress during dynamic hygrothermal exposure matrix cavitation will occur, Zheng and Morgan (24). Such matrix cavitation and associated delamination depends on a series of variables such as (i) previous humidity-time-temperature exposure and associated moisture concentration profiles; (ii) component thickness; and (iii) rapid heat-time service environment exposures, such as thermal spikes. We will report the effect of thermal spikes on the onset of blistering of BMI-and PI-carbon fiber composites. Moisture saturated K3B-carbon fiber composites blister at 230°C for a 250°C thermal spike. However, BMI-carbon fiber composites exhibit no evidence of blistering for a similar thermal spike. A thermoset is more resistant to cavitation and associated macroscopic blistering than a thermoplastic because thermoset cavitation requires rupture of co-valent crosslinked molecular segments.

3.2.2 ***Permanent "Dry" T_g Decreases*** Cornelia (19) has reported significant decreases in PI-carbon fiber composite dry T_g's >50°C after exposure to high temperature hygrothermal environments. Thorp and Crasto (20) also report smaller permanent dry T_g decreases. Curliss (21) reports significant modification of the nuclear magnetic resonance spectra of hygrothermal exposed P.I.'s that are subsequently dried. These disturbing reported permanent T_g decreases have been attributed to possible imide ring hydrolysis at high temperatures with reversion to a polyamic acid linkage, which in itself is susceptible to chain scission via hydrolyses (9). However, the higher T_g decreases reported would require 30-40% formation of polyamic acid linkages.

An alternative mechanism of the reported “dry” T_g decreases is a result of moisture molecular “lock-in.” At higher temperatures new matrix H-bonding sites become available to moisture as the molecular structure of the matrix expands with temperature. Upon relatively rapid cool-down water molecules can become locked into molecular cages and can only diffuse out upon subsequent exposure to high temperatures. Such water molecules can not be released at low temperatures but upon release at higher temperatures act as plasticizers and lower the measured matrix T_g.

We will report controlled isothermal temperature-time weight loss measurements presently underway of both initially wet and dry BMI and PI matrices and their composites in order to ascertain if the water molecular lock-in mechanism is occurring in these matrices.

ACKNOWLEDGMENT

The authors thank Air Force Office of Scientific Research (AFOSR), sponsored by Dr. Charles Lee, and State of Michigan Research Excellence Funds for financial support and encouragement of this work.

REFERENCES

1. P.M. Hergenrother and M.E. Rogalski, Polymer Preprints (A.C.S.), 33, 354 (1992).
2. R.J. Morgan, E.E. Shin, C. Dunn, R.J. Jurek, and A. Jurek, Proc. of 39th SAMPE Conf., 39, 1564, (1994).
3. H.H. Man, and R.J. Morgan, Proc. of 7th ASM/ESD Advanced Composites Conf., Detroit, MI, 555, (1991).
4. R.J. Morgan, R.J. Jurek, A. Yen, and T. Donnellan, Polymer, 34, 835, (1993).
5. E.E. Shin, R.J. Morgan, Proc. of ANTEC 93 on Plastic Engineering, SPE, II, 1357, (1993).
6. E.E. Shin, Q.Zheng, and R.J. Morgan, Proc. of SEM 50th Anniversary Spring Conf. on Experimental Mechanics, SEM Inc., 366, (1993).
7. E.E. Shin, R.J. Jurek, L.T. Drzal, R.J. Morgan, J.K. Choi and A. Lee, Proc. of ASME Meeting, San Francisco, 183, (1995).
8. E.E. Shin, C. Dunn, E. Fouch, R.J. Morgan, M. Wilenski and L.T. Drzal, Proc. of ASME Meeting, San Francisco, 191, (1995).
9. R.J. Morgan, Thermal Characterization of Composites in Thermal Analysis of Polymers, 2nd Edition, Ed. E. Turi, Academic Press, Chap. 9, (1996).
10. R.J. Morgan, E.E. Shin, B. Rozenberg and A. Jurek, Polymer, (in press).
11. C. Loechett and R. Rothschild, Boeing Commercial Airplane Co., Seattle, WA, (1992) (Private Communication).
12. L. Li, Polymer Plast. Technol. Eng., 29, 549, (1990).
13. C.L. Pederson, J.W. Gillespie, R.L. McCullough, R.J. Rothschilds and S.L. Stanek, Polymer Composites, 16, 154, (1995).
14. L.J. Burchan, R.F. Eduljee and J.W. Gillespie, Polymer Composites, 16, 507, (1995).
15. M. Chaudhari, T. Galvin, and J. King, SAMPE Journal, July/August, 17, (1985).
16. B.J. Lee, M.A. Chaudhari, and V. Blyakham, Polymer News, 13, 297, (1988).
17. S. Zahir, M.A. Chaudhari, and J. King, Makromol. Chem. Macromol. Symp., 25, 141, (1989).
18. J. Mijovic and S. Andjelic, Macromolecules, 29, 239, (1996).
19. D. Cornelia, Proc. of 39th SAMPE Conf., 39, 917, (1994).
20. K.E.G.Thorp and A.S. Crasto, Proc. of Am. Soc. of Composites 10th Technical Conf., 601, (1995).
21. D.B. Curliss, Proc. of 28th Central Region ACS Meeting, Dayton OH, Paper III-C-6, (1996).
22. A. Lee, J. High Performance Polymer, (in press).
23. R.J. Morgan and E.T. Mones, J. Appl. Polym. Sci., 33, 999, (1987).
24. Zheng and R.J. Morgan, J. Composite Materials, 27, 1465, (1993).

Table (1)

TEST MATRIX TO STUDY BMI NETWORK STRUCTURE INTERRELATIONS WITH MECHANICAL AND THERMAL PROPERTIES

<table>
<tr><th colspan="2">COMPOSITIONS
A to B RATIO in pbw
(MOLAR RATIO)</th><th>SYSTEM 1
100:85
(1:1)</th><th>SYSTEM 2
113:85
(1:0.87)</th><th>SYSTEM 3
120:85
(1: 0.82)</th><th>SYSTEM 4
87:85
(1:1.13)</th></tr>
<tr><td rowspan="8">CURE CYCLE</td><td rowspan="2">STANDARD TEMP. CURE</td><td colspan="4">1 hr. @ 180°C + 2 hrs. @ 200°C + 1 hr. @ 250°C</td></tr>
<tr><td colspan="4">1 hr. @ 180°C + 2 hrs. @ 200°C + 6 hrs. @ 250°C</td></tr>
<tr><td rowspan="3">LOWER TEMP. CURE</td><td colspan="4">1 hr. @ 177°C + 1 hr. @ 200°C</td></tr>
<tr><td colspan="4">1 hr. @ 177°C + 3 hrs. @ 200°C</td></tr>
<tr><td colspan="4">1 hr. @ 177°C + 10 hrs. @ 200°C</td></tr>
<tr><td rowspan="3">HIGHER TEMP. CURE</td><td colspan="4">1 hr. @ 180°C + 2 hrs. @ 200°C + 6 hrs. @ 250°C + 1 hr. @ 300°C</td></tr>
<tr><td colspan="4">1 hr. @ 180°C + 2 hrs. @ 200°C + 6 hrs. @ 250°C + 2 hrs. @ 300°C</td></tr>
<tr><td colspan="4">1 hr. @ 180°C + 2 hrs. @ 200°C + 6 hrs. @ 250°C + 6 hrs. @ 300°C</td></tr>
</table>

*** Postcure under N_2 atmosphere.**

Table (2)

Room Temperature Tensile Properties vs. Composition of Low Temperature Cured BMI Neat Resin (Cure cycle: 1 hr @177°C + 1 hr @200°C)

Properties / Composition	YOUNG'S MODULUS MPa	YIELD		FAILURE	
		σ_y, MPa	ϵ_y, %	σ_f, MPa	ϵ_f, %
A:B = 1.00:1	2293±202	N/A	N/A	69±2	3.4±0.9
A:B = 1.15:1	1678±21	12.22±2	0.92±0.2	96±13	6.2±0.7
A:B = 1.22:1	2056±47	N/A	N/A	94±5	5.2±0.3
A:B = 0.88:1	1809±42	N/A	N/A	72±14	4.0±0.7

Figure 1. The chemical structures of BMPM and DABPA monomers.

4,4' Bismaleimidodiphenylmethane (BMI)   *O,O'* -Diallyl Bisphenol A (DABPA)

Figure 2. The chemical structure of the BMPM/DABPA "ene" adduct prepolymer. (A), (B) and (C) are double bonds capable of polymerization.

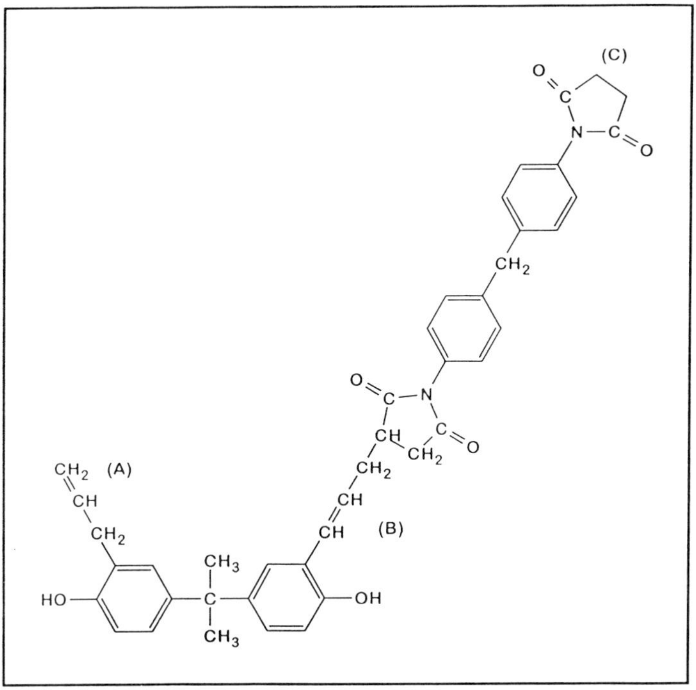

Figure 3. FTIR spectra of BMDPM-DABPA (1:1 molar) system as a function of sequential cure conditions. 1:-prepolymer 0.5 hour, 130°C mixing; 2,3,4,5,6 :-1 hour, at sequential temperatures of 130°C, 150°C, 175°C, 200°C, 225°C respectively; 7,8 :-3 hour and 9 hour at 250°C respectively.

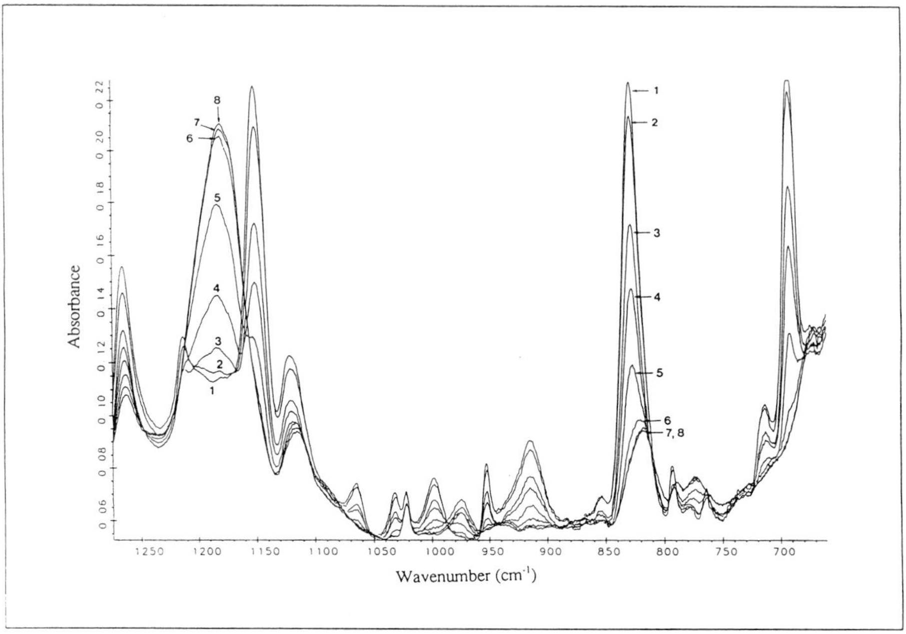

Figure 4. The consumption of the maleimide (✻) and allyl (□) groups and appearance of the ether group for the BMDPM-DABPA (1:1 molar) system as a function of sequential, isothermal cure conditions.

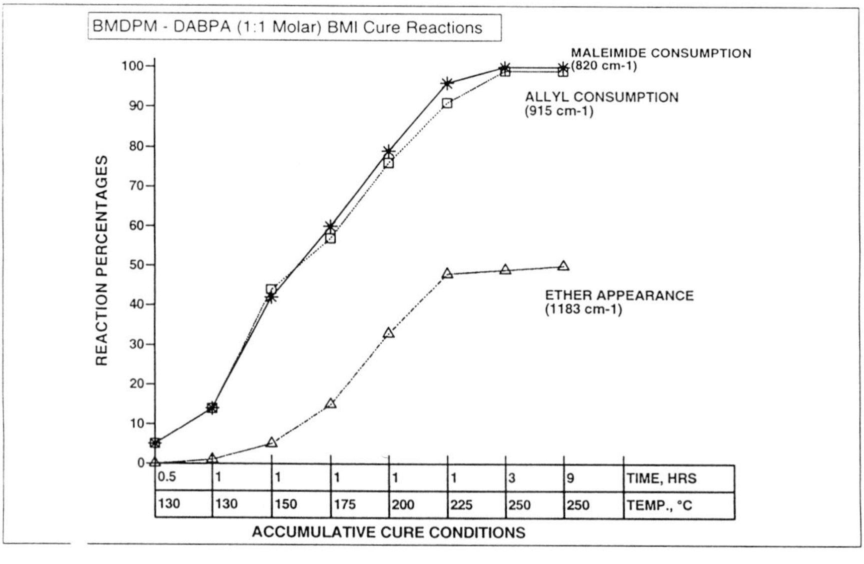

Figure 5. T_g versus isothermal cure time for temperatures ranging from 130°C to 300°C for the BMPM-DABPA (1:1 molar) BMI system.

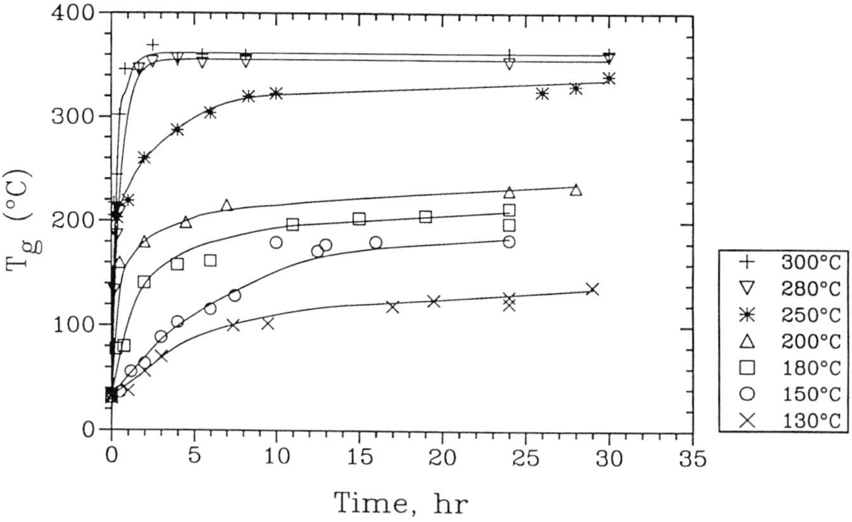

Figure 6. T_g versus degree of cure, α, ($\Delta H_t/\Delta H_\infty$) for isothermal cure temperatures in the 130°C to 300°C range for a constant cure time of 10 hours for the BMPM-DABPA (1:1 molar) BMI system.

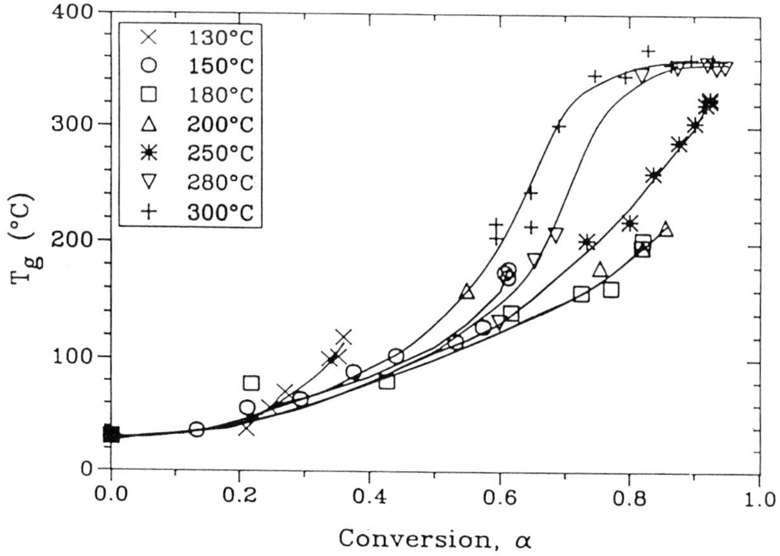

Figure 7. T_g of BMI formulations cured for 12 hours isothermally in the 170-300°C range as a function of BMPM-DABPA ratios.

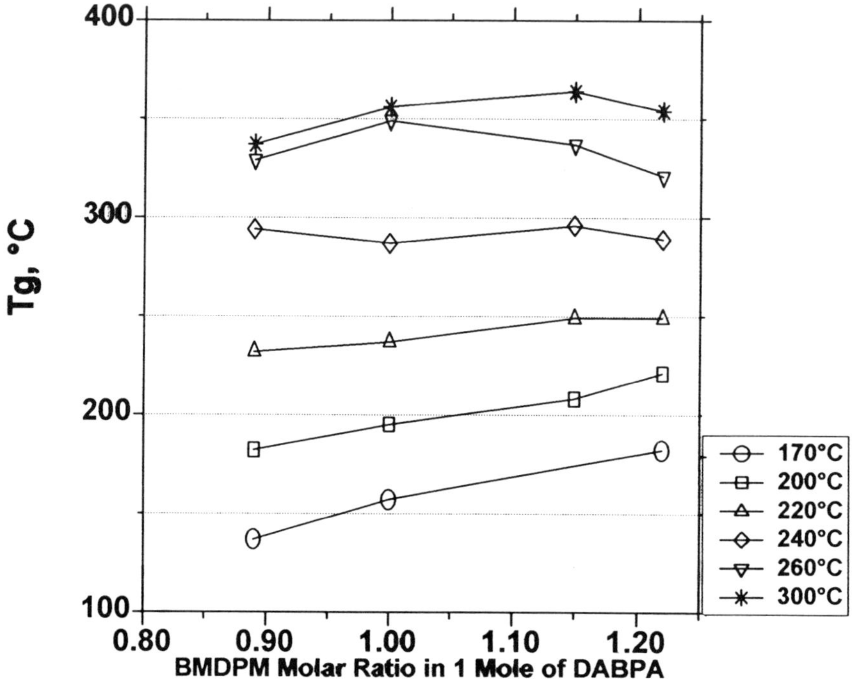

28th International SAMPE Technical Conference
November 4-7, 1996

THE EFFECTS OF PROCESSING CONDITIONS AND SERVICE ENVIRONMENTAL EXPOSURES ON HIGH TEMPERATURE POLYMER MATRIX CARBON FIBER COMPOSITES

E.Eugene Shin[1], Roger J. Morgan[1], Mark Wilenski[2],
Jiming Zhou[1], Jason Lincoln[1], and Lawrence T. Drzal[2]
Advanced Materials Engineering Experiment Station (AMEES)[1],
Composite Materials and Structures Center[2],
Michigan State University,
2203 Eastman Avenue, Midland, MI 48640.

ABSTRACT

Studies of the characterization of (i) the critical fundamental damage initiation mechanisms on the molecular and microscopic level and (ii) the macroscopic failure modes of bismaleimide (BMI) and polyimide (PI) - carbon fiber composites in future aerospace stress-time-temperature-moisture -chemical service environments are reported. Matrix microcracking behavior and the fiber matrix interfacial integrity of BMI as a function of cure conditions and subsequent service environment exposure conditions were studied with various fiber volume fraction composites ranging from 35 to 65 v%. The synergistic effects of moisture and thermal exposures, such as a simulated thermal cycling from -54 to 177°C and a thermal spiking up to 250°C, upon microscopic and macroscopic composite damage and residual mechanical properties of BMI and PI-C fiber composites are presented. The onset of blistering in moist PI-C fiber composites occurred at 230 °C during the thermal spiking. The damage is discussed in terms of moisture vapor-induced cavitation, plasticization of molecularly locked-in water and hygrothermal chemical degradation. Attempt is also made to identify possible mechanisms of the reported "dry" T_g decreases in PI-C fiber composites after exposure to high temperature hygrothermal environments by a systematic and controlled isothermal temperature-time weight loss measurement of initially wet PI matrix under a saturated steam environment; 150°C and 140psi.

KEY WORDS: BMI and PI-Carbon Fiber Composites, Aerospace Service Environments, BMI Microcracking, Hygrothermal Spiking, Moisture-Induced Damage

1. INTRODUCTION

Present and future high temperature polymer matrix/carbon fiber composite aerospace applications involve combined stress, time, temperature, moisture, chemical and atmospheric service environment exposure which can lead to complex synergistic environmental-induced degradation of composite performance as described earlier in this publication (1). These complex service conditions require a thorough understanding of potential physical, chemical and mechanical-induced composite degradation mechanisms, and their synergistic effects that can lead to the most probable critical failure path. Future BMI and PI-carbon fiber composites that contain absorbed moisture can be exposed to service environment that involve rapid temperature changes. If the moisture cannot diffuse-out of the composite in sufficient time prior to moisture vaporization then the trapped moisture vapor could cause damage to the composite in the form of matrix microcracking and/or blistering and/or fiber-matrix interfacial failure. In addition, the moisture vapor pressure may cause the moisture plasticized matrix to swell which may then be permanently frozen-into the glassy-state upon cool-down. The swollen matrix will have a higher glassy-state free volume and exhibit higher associated oxygen and moisture diffusion.

In this paper, our progresses on durability and critical fundamental aging mechanisms of BMI and PI-Carbon fiber composites is reviewed and some current results of the extended studies on microcracking behavior, effects of hygrothermal spiking, and effects of "lock-in" moisture is presented.

2. MATERIALS

The composite materials studied include BMI 5292 (Ciba Geigy)-IM7 carbon fiber, BMI 5260 (BASF)- G40-800 carbon fiber, and PI (Avimid® K3B, DuPont)-IM7 carbon fiber composites. BMI 5292-IM7 composite panels were fabricated at MSU following a standard procedure, a 177°C cure for 1hr followed by a post cure for 2 hrs at 200°C and for up to 6 hrs at 250°C (2,3) and/or other predetermined cure cycles. BMI 5260 and PI-carbon fiber composite panels had been fabricated at BOEING using an autoclave vacuum bagging technique. The cure cycle used for BMI 5260-carbon fiber composites was 4 hrs at 190°C and post cure at 215°C for 6 hrs. K3B-carbon fiber composite was consolidated at 360°C.

The BMI neat resin panels were casted in clamped two tempered glass plates separated by a high temperature gasket with the BMPM/DABPA (Ciba-Geigy) system using a suggested standard cure cycle. In the case of the PI neat resin plaques, fully imidized K3B powder was compression-molded at 350°C(670°F) in a deep metal mold.

3. RESULTS AND DISCUSSION

3.1 Overview Our durability program plan initially involves a series of core test programs that separates distinct service environment conditions and their effect on composite performance. The most likely synergistic critical failure path is identified from these core test programs. This durability program is illustrate in Figure 1. The overall program goal is to: (i) identify the synergistic critical failure path and its associated performance deterioration mechanisms; (ii) conduct synergistic service environment accelerated performance testing based on the physics and /or chemistry and/or mechanics of the aging mechanisms of high temperature polymeric composites of potential utilization in the high speed aircraft; and (iii) develop fabrication and material optimization procedures to alleviate any performance deficiencies.

Based on our previous studies (1-10) we have identified the most likely potential, critical aging mechanisms that control damage initiation in BMI and PI carbon fiber composites based on the systematic durability evaluation methodology. The critical aging mechanisms involve (i) further cure of BMI-carbon fiber composites with associated T_g increases, mechanical property decreases and enhanced microcrack development in service environment exposure conditions, and (ii) hygrothermal induced thermal and mechanical property deterioration as a result of physical and chemical structural changes in BMI- and PI-carbon fiber composites. The mechanism (ii) was identified from residual property determination of the composites after various thermal cycling simulation programs as a function of temperature, humidity, moisture content, heating/cooling rate, and applied stress (3,6,8). Models on the moisture absorption behavior and hygrothermal induced damage mechanisms, such as matrix blistering and delamination, of the composites and their interrelationship had been suggested (8,11).

In addition, the extensive creep rupture experiments have been conducted at AMEES 600 specimen test facility to ascertain the effects of long time-stress/thermal environments, in the absence of chemical degradation (i.e., oxidation) on the service environment-induced composite dimensional and damage growth mechanisms and, subsequently, to predict service lifetimes from shorter time data using stress and temperature as acceleration vehicles (8,12). Detailed comparison between BMI and PI-Carbon fiber composites on short-term strength, creep rupture and failure mode has been made as a function of time under stress, test temperature, geometry factor, stress level, prior aging at high temperature, prior mechanical damage. Another core test simulation program was conducted to determine principal composite degradation mechanisms and to predict composite residual strength deterioration from exposure to combined stress, air flow, and composite temperature service environments.

Based on our understanding in durability performance of the BMI and PI-carbon fiber composites in various core service environment exposure conditions, more fundamental and systematic studies on thermal-moisture induced "damage" mechanisms have been developed, Figure 2. Our latest progress in this program will be presented.

3.2 Microcracking Behavior The objectives of this task are to ascertain the fiber/matrix adhesion level and the microcracking threshold conditions as a function of cure and service exposure. Their ultimate effect on the mechanical response of these composites will be determined and a strategy for eliminating fiber/matrix interphase cracks will be developed.

Microcracking in cross-ply laminates has been studied extensively, and a variety of analyses have been used including finite element (13), analytical (14), experimental (15), and Monte-Carlo simulation (16). It has been suggested that the stress state in these composites results from the thermal expansion/resin shrinkage mismatch between the 0° and 90° plies. A shear stress transfers the load from one layer into the next, causing a combination of shear stresses at the ply interface and tensile stresses at the ply mid-plane (17) as shown schematically in Figure 3(a) and as evidenced by the ESEM (Environmental Scanning Electron Microscope) image of a cure induced microcrack in Figure 3(b). In addition to this macroscopic stress, a microscopic stress is present due to the resin shrinkage and mismatch between the radial thermal expansion of fibers and resin.

We have used both finite element analysis and a closed form analytical solution to model the combination of macroscopic and microscopic stresses which result from thermal loading. In addition to the modeling, $[0°/90°]_S$ BMI/IM7 composites have been analyzed using MSU's ESEM. It was found that all laminates were microcracked regardless of cure/post-cure cycle. As expected, crack widths grow with decreased fiber volume fraction, and increase with cure advancement. Observation in the ESEM revealed poor adhesion between the IM7 fiber and the BMI resin for both cure induced (inset of Figure 3(b)) and mechanically induced cracks Figure 4(a). No cure induced microcracks were found in the K3B/IM7 system, but poor adhesion was observed after mechanical loading as seen in Figure 4(b).

Recently, an effort has been made to determine cure-induced composite matrix and fiber matrix interfacial microcracking threshold conditions for BMI-carbon fiber composites, using 4 layer foil, 0°/90°, as a function of cure and subsequent service environment exposure conditions with a newly developed in-situ ESEM monitoring technique. The threshold resin microcrack resistance and fiber-matrix interface integrity will be correlated with models and reported literature data that correlates composite residual stresses with thermal expansion/resin shrinkage mismatch. These studies will reveal the optimal composite conditions that maximizes composite toughness and minimizes composite fabrication induced damage for expected specific composite service environment conditions as a function of cure conditions or matrix/interface toughening modification.

3.3 Effects of Hygrothermal Spiking The effects of thermal spikes on the physical and mechanical integrity of both dry and moisture saturated BMI and K3B resins and their composites (IM7 fiber) have been investigated. Critical thermal spike issues are summarized in Figure 5. Both resin specimens and composites of BMI and K3B with IM7 fibers have been manufactured and machined for Four Point Flex, DMA, Open Hole Compression, Iosipescu Shear testing, and density measurement. A schematic block diagram in Figure 6 illustrates a detailed thermal spiking procedure including preconditioning and conditioning conditions and schedules for weight and optical microscopic damage evolution monitoring. Both a resin specimen with a thermocouple embedded in its center and a 16-ply BMI composite with thermocouples embedded between each ply were manufactured to determine the amount of time necessary to achieve thermal equilibrium during a thermal spike. It was found specimens are within 95% of final temperature after about five minutes, Figure 7, so that the duration of 15 min in hot stage will be used for spiking to minimize moisture loss during the cycle.

Figure 8 shows moisture absorption during wet preconditioning and desorption-re-absorption during thermal spiking in terms of weight gain vs. square root of time normalized by the thickness of the specimen for both composites and neat resin specimens. It is of interest to note that the BMI composite and PI resin reached the equilibrium moisture saturation within a relatively short exposure time per thickness while the PI composite and BMI resin were still absorbing moisture after the extended exposure time. It is clear that this is not just thickness dependent but possibly due to the existing cure-induced microcracks in BMI composites. This is also indicated from the desorption behavior of composites and resins during thermal spiking. An attempt has been made to ascertain whether the microcracks affect the diffusion rate of water molecule in composites using an unidirectional BMI-carbon fiber composite which did not contain any microcracks.

Perhaps, the most important results observed from the hygrothermal spiking experiment is that the severe blistering occurs on moisture saturated K3B/IM7 composites, Figures 9 and 10, after the first spike. The blistering always started at the center in the middle layers and grew to edges. However, no damage or structural morphological changes observed in other specimens including both wet and dry K3B neat resin and all of the BMI neat resin and composites, wet or dry. We have proposed a possible mechanism of the blistering and delamination and also suggested the differences between thermoset and thermoplastic systems (1,8,11). The onset of blistering was determined by monitoring a dimensional change e.g., thickness direction, of composite test specimen using a high temperature precision dial indicator (0.01mm resolution). Surprisingly enough, the wet K3B-carbon fiber composites blister at 229°C for a 250°C thermal spike, Figure 11. It should be also noted that from this experimental set-up the T_g of wet K3B composite was detected to be 195°C which is consistent to the literature value.

Upon completion of the thermal spiking experiment up to 12 cycles, the hygrothermal induced property degradation and the related mechanisms will be determined and presented in detail.

3.4 Effects of "Lock-In" Moisture As described earlier in this publication (1), the real mechanism of the permanent "Dry" T_g decrease in the hygrothermal exposed PI's needs to be identified. A simple pressure vessel has been built to perform an accelerated hygrothermal aging experiment for more than 30 days. A systematic and controlled isothermal temperature-time weight loss measurement of initially wet PI matrix under a saturated steam environment; 150°C and 140psi, has been carefully designed.

5. ACKNOWLEDGMENT

The authors thank Air Force Office of Scientific Research (AFOSR), sponsored by Dr. Charles Lee, and State of Michigan Research Excellence Funds for financial support and encouragement of this work.

6. REFERENCES

1. Roger J. Morgan, E.Eugene Shin, Jason E. Lincoln, Jin Choi, and Andre Lee, Proceedings of the 28th International SAMPE Technical Conference, Seattle, Washington, November 4 -7, (1996).
2. R.J. Morgan, R.J. Jurek, A. Yen, and T. Donnellan, Polymer, 34, 835, (1993).
3. E.E. Shin, Q.Zheng, and R.J. Morgan, Proc. of SEM 50th Anniversary Spring Conf. on Experimental Mechanics, SEM Inc., 366, (1993).
4. R.J. Morgan, E.E. Shin, C. Dunn, R.J. Jurek, and A. Jurek, Proc. of 39th SAMPE Conf., 39, 1564, (1994).
5. H.H. Man, and R.J. Morgan, Proc. of 7th ASM/ESD Advanced Composites Conf., Detroit, MI, 555, (1991).
6. E.E. Shin, R.J. Morgan, Proc. of ANTEC 93 on Plastic Engineering, SPE, II, 1357, (1993).
7. E.E.Shin, R.Jurck, L.T.Drzal, R.J.Morgan, J.K.Choi, and A.Lee, Proceedings of the ASME Materials Division, MD-Vol. 69-1, 1995 IMECE, ASME P183-189, (1995).
8. E.E.Shin, C.Dunn, E.Fouch, R.J.Morgan, M.Wilenski, and L.T.Drzal, Proceedings of the ASME Materials Division, MD-Vol. 69-1, 1995 IMECE, ASME P191-200, (1995).
9. R.J. Morgan, Thermal Characterization of Composites in Thermal Analysis of Polymers, 2nd Edition, Ed. E. Turi, Academic Press, Chap. 9, (1996).
10. R.J. Morgan, E.E. Shin, B. Rozenberg and A. Jurek, Polymer, (in press).
11. Zheng and R.J. Morgan, J. Composite Materials, 27, 1465, (1993).
12. C.M.R. Colin and R.J. Morgan, Proceedings of SEM Spring Conference, June, Grand Rapids, MI, (1995).
13. Donald F. Adams, Journal of Reinforced Plastics and Composites, **6**, pp. 66-88, (1987).
14. Y. M. Han, H. T. Hahn, and R. B. Croman, Composites Science and Technology, **31**, pp. 165-177 (1988).
15. Catherine Wood and Walter Bradley, "A New Technique to Study the Interfacial Strength and Transverse Cracking Scenario in Composite Materials," in STP 1290, "Fiber, Matrix, and Interface Properties", August (1996).
16. A. S. D. Wang, P. C. Chou, and S. C. Lei, Journal of Composite Materials, **18**, pp. 239-254, (1984).
17. Norman Laws and George J. Dvorak, Journal of Composite Materials, **22**, pp. 900-916, (1988).

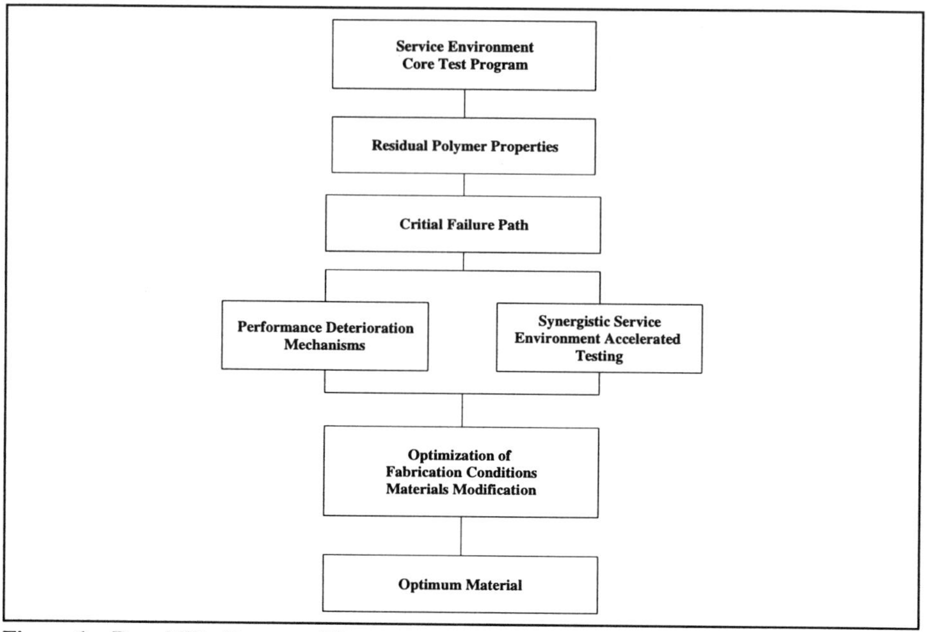

Figure 1. Durability Program Plan

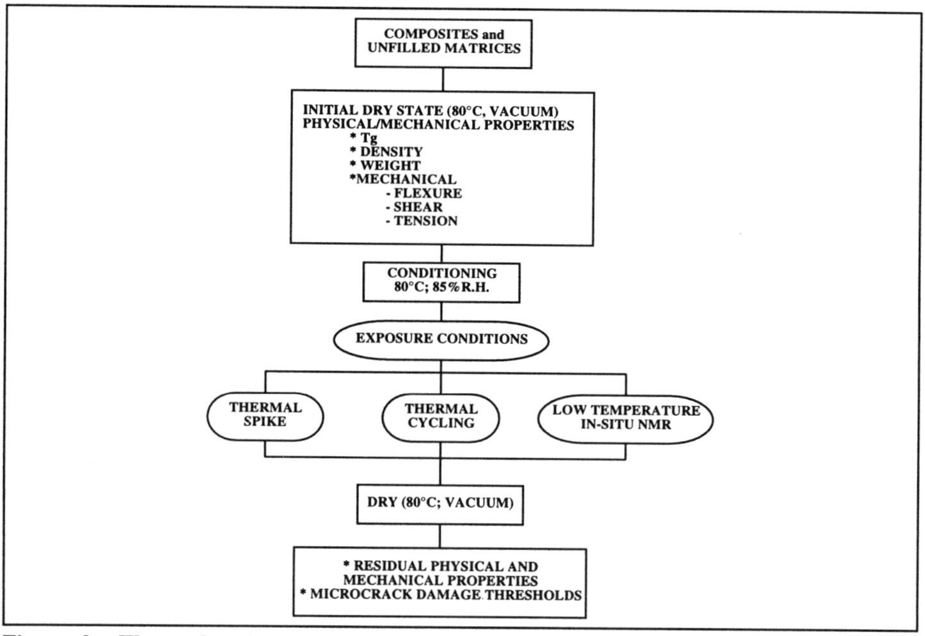

Figure 2. Thermal-moisture Induced "Damage" Mechanisms of BMI and PI Matrix Composites.

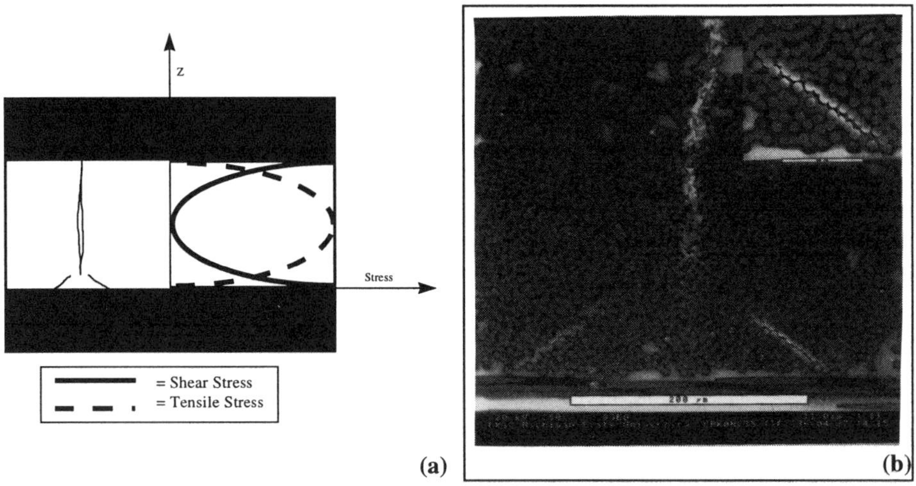

Figure 3. The Stress State in Cross-Ply Composite Laminate, (a) as Shown by a Schematic Diagram, and (b) as Evidenced by a Cure Induced Microcrack in $[0/90]_{4S}$ BMI/IM7 Composite.

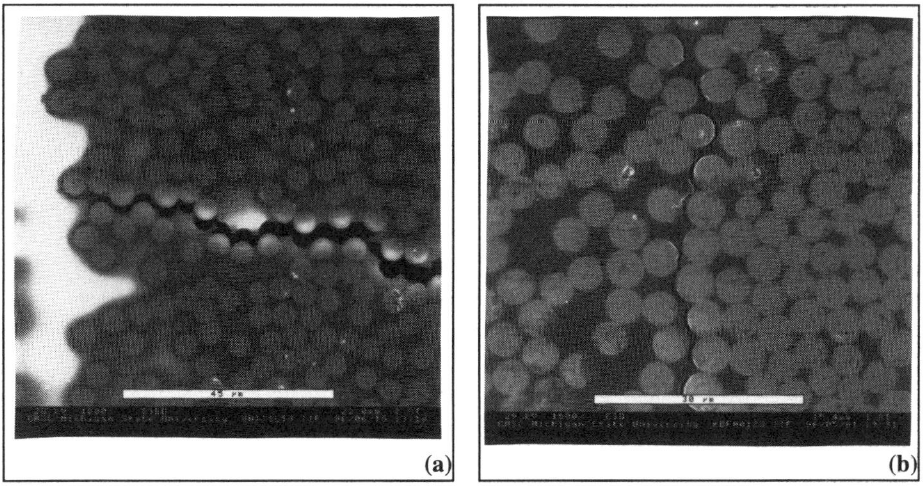

Figure 4. ESEM Micrographs Showing Mechanically Induced Microcracks in (a) BMI/IM7 and (b) K3B/IM7 Composites.

THERMAL SPIKE ISSUES

INITIAL MOISTURE PROFILE

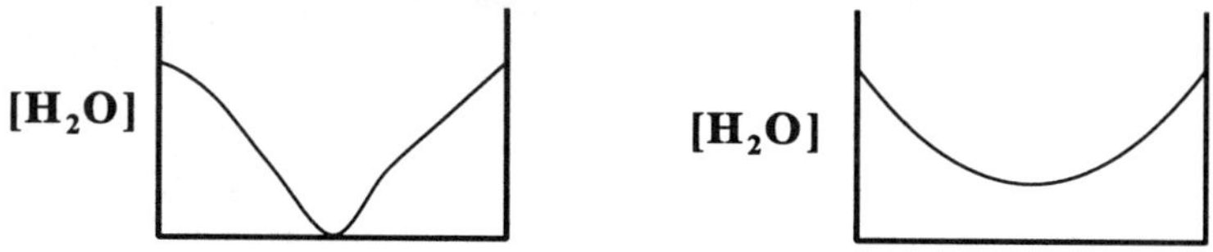

THERMAL SPIKE

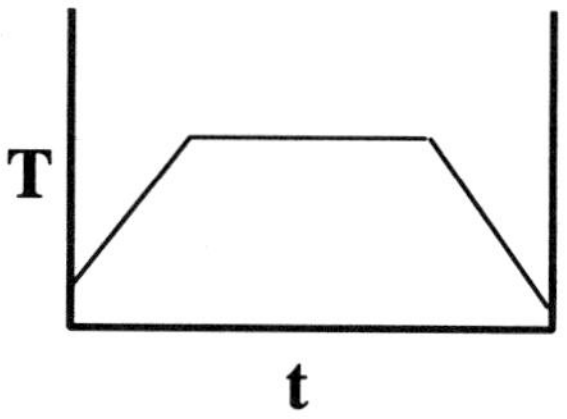

FINAL MOISTURE PROFILE

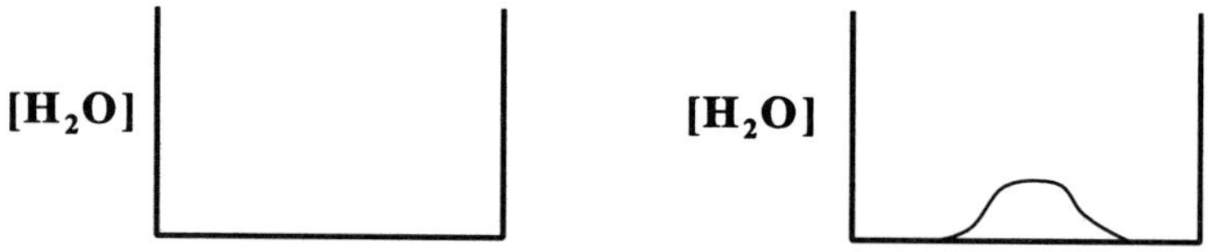

VARIABLES

- **Exposure History**
- **Specimen Thickness**
- **Thermal Spike Characteristics**

PERFORMANCE DETERIORATION MECHANISMS

- **Dependent on High Temperature $[H_2O]$-T-t Profile**
  - **PHYSICAL MOLECULAR**
    - **Packing Modifications**
    - **Cavitation**
    - **Covalent Bond Fracture**
    - **H_2O Lock-In**
  - **CHEMICAL**
    - **Dehydration**
    - **Depolymerization**

Figure 5. Schematic Description of Thermal Spike Issues.

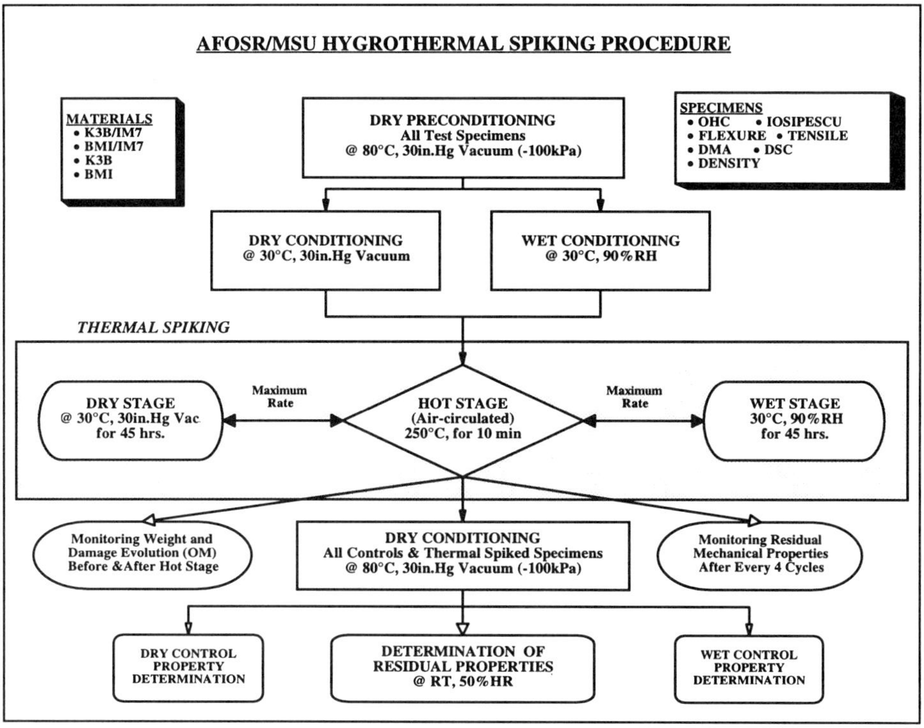

Figure 6. Hygrothermal spiking procedure

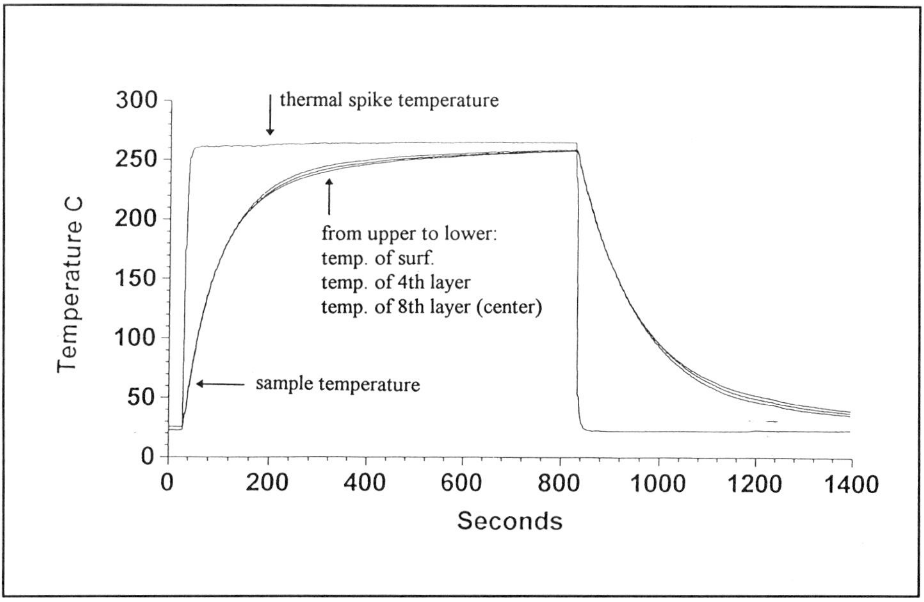

Figure 7. Thermal spiking temperature profile and the actual composite temperature during spiking

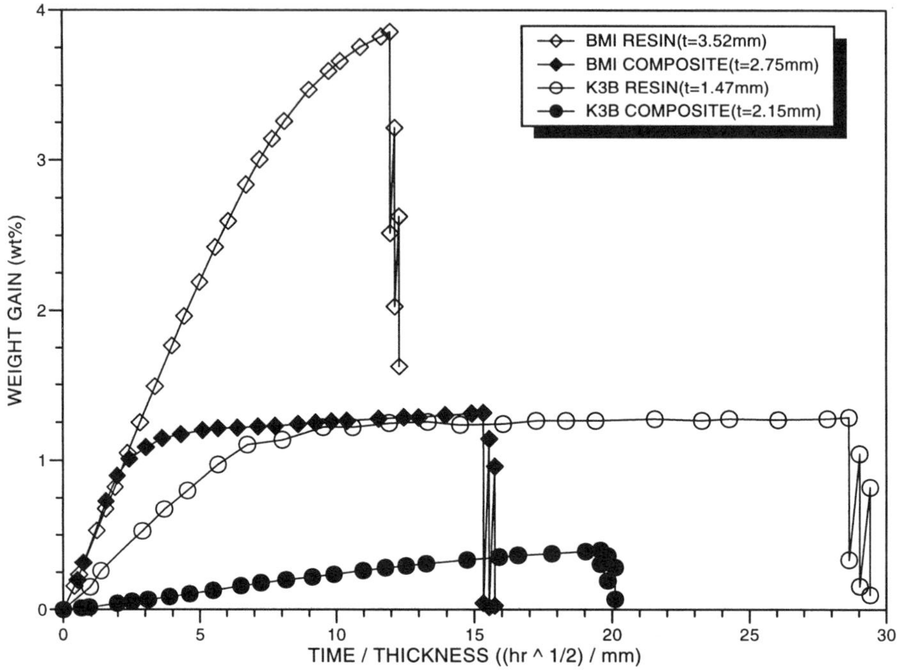

Figure 8. Moisture Absorption Behavior During Wet Pre-conditioning and Desorption-Reabsorption Behavior During Thermal Spiking in Various Specimens.

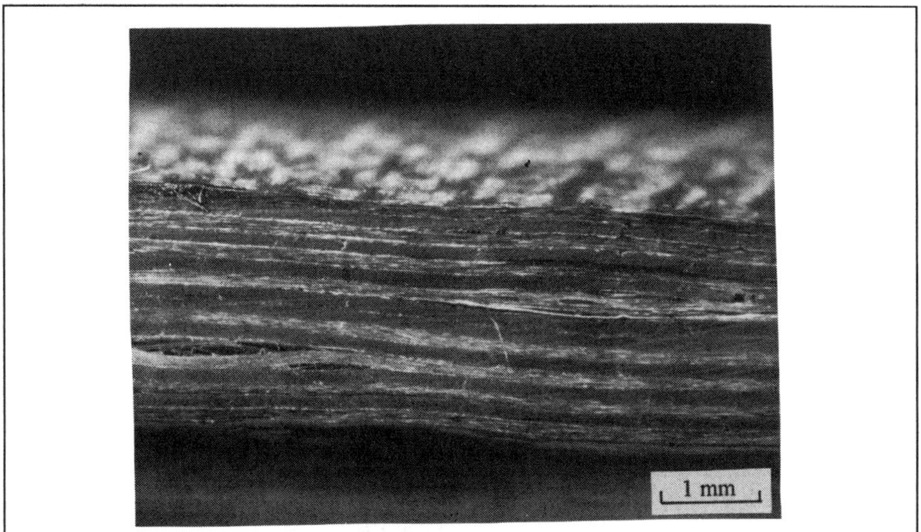

Figure 9. Optical Micrograph Showing Edge Delamination Due to Blistering After the First Thermal Spiking in Wet K3B/IM7 Composite.

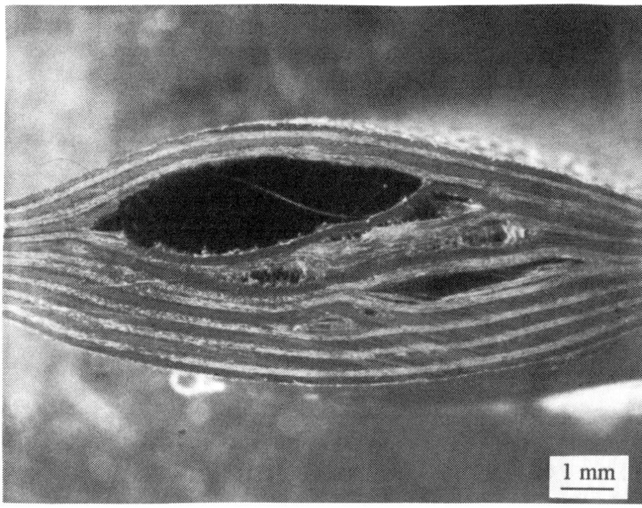

Figure 10. Optical Micrograph of the Middle Cut Cross-section Showing Large Blisters in Wet K3B/IM7 Composites after the First Thermal Spiking.

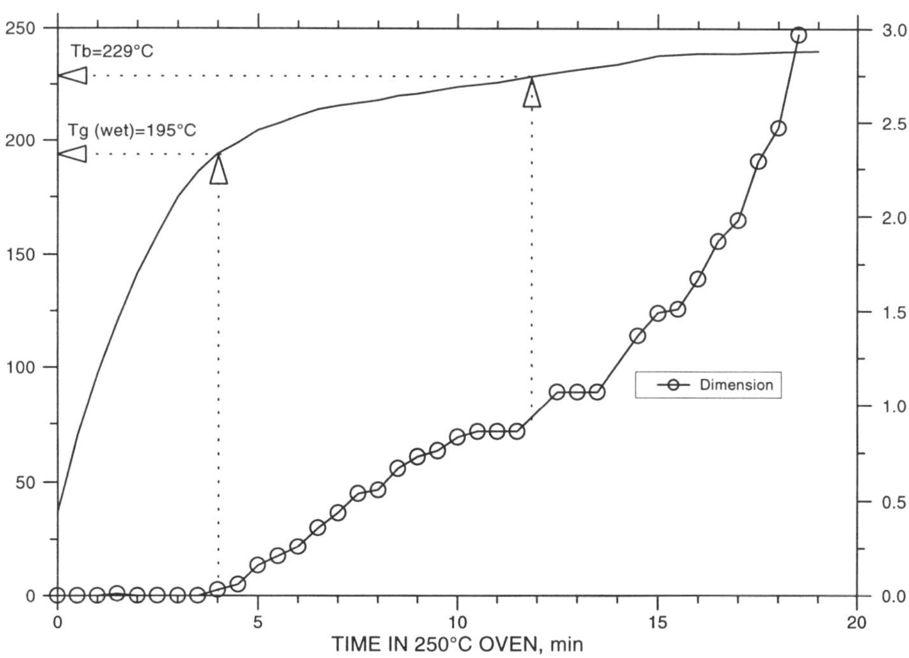

Figure 11. Dimensional Change of the Wet K3B/IM7 Composite as a Function of Composite Temperature During Thermal Spiking in a Pre-heated Oven at 250°C

28th International SAMPE Technical Conference
November 4-7, 1996

Novel Tougheners for Epoxy-based Composites

Louis Boogh, Bo Pettersson†, Peter Kaiser and Jan-Anders Månson*
Laboratoire de Technologie des Composites et Polymères
Ecole Polytechnique Fédérale de Lausanne, CH-1015 Lausanne, Switzerland
† Perstorp Polyols, Application Technology, S-28480 Perstorp, Sweden

ABSTRACT

Increased matrix toughness is a main objective of matrix system development in order to improve composite properties such as delamination resistance, impact response, fatigue behavior and damage tolerance. A novel toughening system based on a hyperbranched polymer which shows no effect on the thermomechanical properties of the matrix is presented. The viscosity of the matrix is unaffected which allows to fully maintain the good processability of epoxy resins. Furthermore, the size and the spherical configuration of the hyperbranched molecule eliminates detrimental particle filtering effect observed with other conventional toughening systems. Thus, the system is well suited for both conventional prepregging techniques and for RTM, where high shear rates and long flow distances occur. The toughness properties of both glass and carbon fiber reinforced composites have been increased. The critical energy release rate, G_{Ic}, of carbon fiber reinforced epoxy has been increased from 1400 J/m^2 to 2500 J/m^2 with the addition of only 5% of a hyperbranched polymer toughener. For glass fiber reinforced composites the addition of 5% hyperbranched polymer toughener increased the G_{Ic} from 500 J/m^2 to 1150 J/m^2.

KEY WORDS: Epoxy Resins, Composites, Mechanical Properties.

1. INTRODUCTION

The matrix has been shown to play an important role in the performance of fibrous composite materials [1]. The matrix will not just control the maximum service temperature, due to its glass transition, but also the moisture sensitivity and the ageing properties of the composite. Furthermore, defects induced during processing can often be avoided by adjusting and controlling the resin viscosity and its fiber wetting characteristics. The anisotropic nature of composites further emphasizes the importance of the mechanical properties of the matrix. Interlaminar properties are mainly affected, as are intralaminar properties, where low level damage propagation occurs through matrix cracking. Therefore, matrix properties are of primary interest in the development of new composite materials.

The matrix toughness is important when considering damage initiation and fracture of composites. It will affect properties such as the interlaminar crack resistance, impact energy absorption, damage propagation and fatigue resistance. These are often limiting characteristics for the materials in high performance applications. Reinforced epoxy composite materials are widely used as structural load bearing elements where low toughness properties will affect the durability of the components and impose severe limits on design parameters.

(*) to whom correspondence should be addressed

Toughness implies energy absorption before failure occurs. The energy absorption at a crack front can be achieved through various deformation mechanisms [2] which are most effectively induced by the presence of a second phase in the form of particles. The secondary phase is often introduced in the form of rubber particles which normally leads to a considerable increase in the resin viscosity.

The development of new matrix materials requires that processing considerations be taken into account. During the initial impregnation stage of a fiber bed, the resin penetration into the bed is controlled by the initial viscosity of the thermosetting resin. This determines, in an RTM process for example, the processing temperature, pressure, and times required. Furthermore, filtration of the toughening component by the composite plies, or fiber strands, must be avoided during the impregnation, so that toughening can be ensured at all structural levels, even within the fiber bundles. If toughening is induced by particles larger than the average fiber spacing (usually 5-10 μm), these will be filtered out by the proximity of the fiber strands in the reinforcement plies. Thus, within the plies no toughening will occur, and in the interply region weak zones might appear due to the agglomeration of toughening particles. This filtering effect is avoided when using an initially homogeneous material and where the particles are formed by a phase separating process occurring during the resin cure.

The proposed toughening system includes a controlled phase separation process of a low viscosity modifier, a hyperbranched polymer (HBP), which allows the toughness of high performance epoxy matrices to be increased without affecting the thermomechanical properties and the processability.

Both the fiber-matrix and the interlaminar interfaces play an important role in the fracture properties of the laminate. Thus, it is important to control the morphology of the matrix in the interfacial region. The fibers can influence the phase separation process by acting as nucleation sites. They also can induce migration of the toughener onto the fiber surfaces which may lead to poor interfacial properties. These effects have been studied using various fiber surface treatments.

2. HYPERBRANCHED POLYMERS

A liquid toughening agent, compatible with the uncured thermosetting resin, must be used in order to have a homogeneous system during impregnation. Subsequently, the phase separation should only be induced upon curing. A relatively high molecular weight toughening agent will facilitate the control of the phase separation process and increase its efficiency as a toughener. However, high molecular weights in general imply a high viscosity which is to be avoided. In order to satisfy both requirements, low viscosity and high molecular weight, a spherical molecular structure is preferable to a linear one. Dendritic polymers possess this spherical structure (see Figure 1) and have shown lower viscosities than linear polymers of comparable molecular weight [3]. Similar properties are also achieved for some hyperbranched polymers with core/shell structures [4] resembling the architecture of dendrimers. The mechanical properties of these molecules will largely be determined by the interior architecture of the hyperbranched polymer. On the other hand, the chemical structure of the shell will define the phase separation process. The kinetics of both the curing and the phase separation influence the processing conditions of the composites, as well as the structure of the secondary phase in the matrix. Consequently, the obtained structure will determine the mechanical performance of the cured resins. The chemical architecture of the shell can be adapted to various thermosetting resins without affecting the mechanical performances of the bulk resin. Reactive groups tailored for the thermosetting matrix are grafted onto the HBP shell in order to ensure the desired stress transfer between the phase separated domains and the matrix.

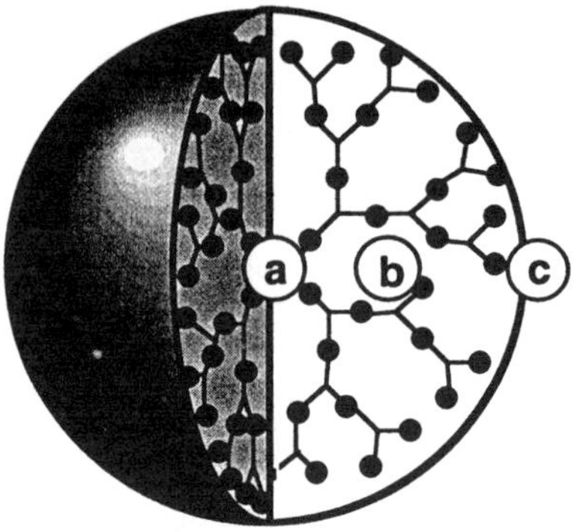

Figure 1. A dendritic molecular structure characterized by the core (a), the bulk (b), and the shell (c).

3. HBP MODIFIED RESIN PROPERTIES

Previously [5] it has been shown that the K_{Ic} values of epoxy resin modified with 5% of HBP presented a 2.5 fold increase from 0.63 MPa$\sqrt{m}$ for the unmodified resin to 1.54 MPa$\sqrt{m}$ for the modified resin, and a 6 fold increase in the critical energy release rate G_{Ic} from 0.12 kJ/m^2 to 0.72 kJ/m^2 respectively. Increasing the volume content of the modifier to 10% and to 15%, the K_{Ic} values were further increased to 1.75 and 2.03 MPa$\sqrt{m}$ respectively. The initial viscosity of the 5% modified epoxy is very low and remains under 100 mPas at room temperature.

The recommended processing temperature range for the resin is 20°C - 100°C. Figure 2 shows the time-temperature-transformation diagram of the modified epoxy resin. The successive events occurring during isothermal processing are determined following horizontal lines in the TTT diagram. The events represented are the phase separation behavior as measured by turbidimetry, followed by the gelation and the vitrification boundaries. Superimposed are isoconversion curves as measured by DSC under isothermal conditions. At the different processing temperatures, the times available for particle nucleation and growth become considerably shorter as the temperature is increased. Furthermore, by raising the temperature the compatibility is increased and at 80°C phase separation is only initiated at a conversion of 60%, as opposed to a conversion of 30% at 40°C.

RTM processing requires low viscosities which are obtained by using higher temperatures. At 80°C the time available for processing is less than 15 minutes. Although this should be sufficient, it is possible to increase this time by multistage processing. Mixing and resin preparation can be done at lower temperatures and the resin can be heated just prior to injection. In order to determine the time available at 80°C for injection the isoconversion lines can be followed. If the resin preparation is done at 40°C during 70 minutes, the resin has reached a conversion of 30%. Heating up to 80°C for the injection, an equivalent time is obtained of 4.5 minutes following the isoconversion line. This leaves about 10 minutes for the injection before phase separation occurs. The processing of the composite laminates used in this investigation has been done using a multistage process and the times to phase separation have been visually verified.

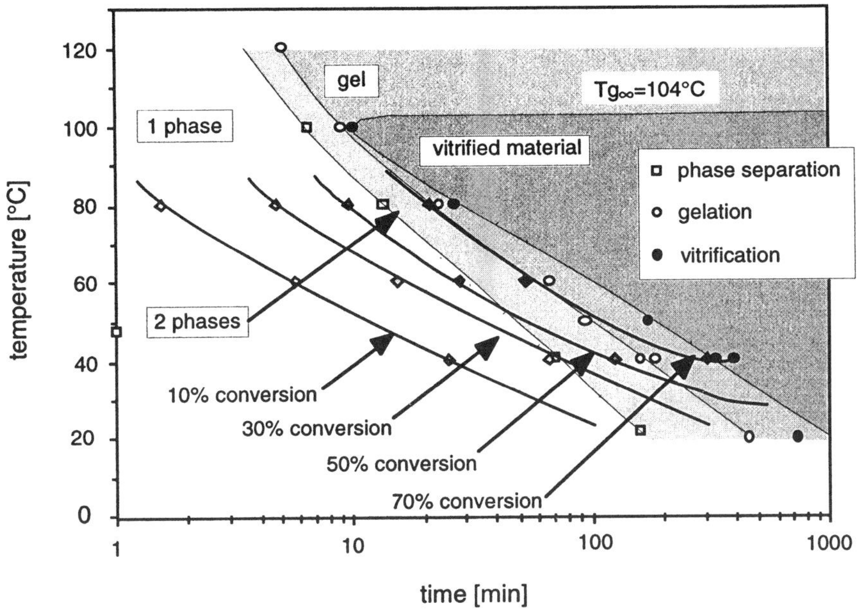

Figure 2. Time-Temperature-Transformation (TTT) diagram of the epoxy resin with 5% of the reference HBP modifier. Phase separation, conversion, gelation and vitrification curves are defined under isothermal curing conditions.

4. EXPERIMENTAL

Various epoxy functional hyperbranched molecules were synthesized following a procedure described elsewhere [5,6,7]. A three generation hyperbranched molecule with a tetra-functional core was used as modifier, and its shell was functionalised with hydroxyl groups and secondary epoxy groups for which the epoxy ring was shielded. In this study, two different HBP's were used. These materials were blended with a commercially-available DGEBF epoxy resin (LY5082, Ciba-Geigy). The modified resin was cured isothermally with a stoichiometric ratio of an isophorondiamine hardener.

Two different plain weave glass fabrics were used as reinforcement in a $[0/90]_{10}$ configuration (glass1 and glass2). Various silane based surface treatments from OSI specialties have been tested in order to optimize the interfacial properties. The results presented are mainly based on an amine silane A-1100. Furthermore, unidirectional carbon fibers have been used in a $[+45/-45]_6$ lay-up. Both reinforcements yielded a fiber weight fraction of about 55%.

Composites are processed using a Plastech Hypaject II RTM equipment. The resin is degassed under vacuum for 5 minutes, injected at a pressure of 2.5 bar and cured at 80°C for 12 hours. A pre-notch for interlaminar fracture testing is introduced at the mid plane using a Kapton tape.

Flexural tests were performed at room temperature according to ASTM standard D790M. Double cantilever beam (DCB) tests were performed according to the ESIS protocol [8]. All tests were performed on a conventional tensile testing equipment (UTS-100).

In order to evaluate the adhesive properties, fragmentation tests were carried out on 150μm thick samples in which one monofilament was embedded. Tests were performed on a tensile testing equipment at 50°C in order to reach the saturated fragmentation which was measured to be at 15% elongation.

5. RESULTS AND DISCUSSION

Composite interlaminar properties have shown an increase of over 100% when adding only 5% of HBP modifiers to the matrix. This increase is however strongly related to the interfacial strength which must be sufficient to induce cohesive failure and promote matrix rupture mechanisms. The presence of the HBP modifier can affect the interfacial strength during the phase separation mechanisms. If the phase separation is initiated on the fiber surfaces, a migration of the HBP towards the fiber-matrix interface is induced. This will decrease the interfacial strength and thus the extent of matrix rupture. Toughening of the matrix will in this case not be effective and the interactions during curing between the fiber and the HBP phase have to be minimized.

The importance of the fiber surface characteristics on the phase separation process are illustrated in Figure 3. Figure 3(a) shows the migration of the HBP onto the fiber surface. A concentration of particles is observed in the fiber bed (right) compared to the resin fracture surface (middle). This can be attributed to the higher compatibility of the HBP with untreated fiber surface than with the epoxy. This leads to poor interfacial properties and adhesive failures in the composite. In this case the G_{Ic} values for the modified resin composite were lower than for the epoxy resin composite.

Figure 3(b) shows another type of glass fiber (glass fiber 2) which is uncoated. Only a limited migration of the HBP towards the fiber interface can be observed. However, a reduction in the interfacial strength is still measured by the fragmentation tests. This will enhance adhesive failure in the composites. In this case, the HBP modified composite G_{Ic} values increased compared to the unmodified epoxy resin, but the enhancement of adhesive failure does not allow to fully exploit the advantage of the HBP modified resin toughness properties.

Figure 3(c) shows the absence of migration of the HBP towards silane treated fibers. The presence of cohesive failure at the fiber-matrix interface is clearly observed. In this case, the full advantage of the HBP modified resin toughness properties can be exploited and the composite G_{Ic} values are increased by over 100%.

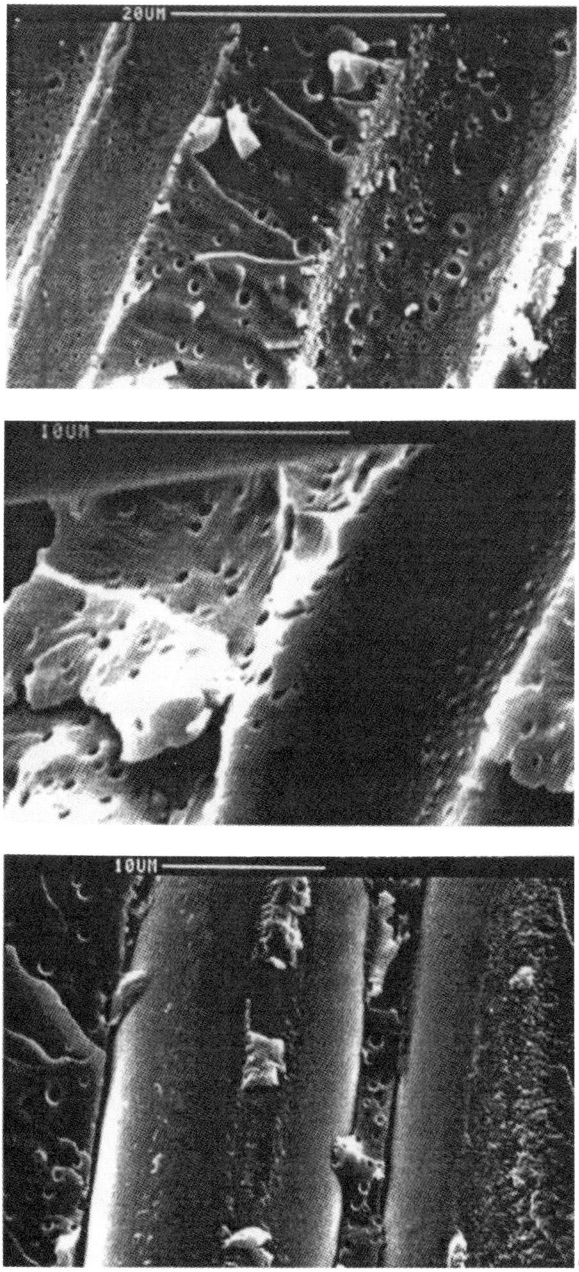

(a)

(b)

(c)

Figure 3. SEM micrographs of laminate fracture surfaces showing various degrees of HBP migration towards the fiber surface. (a): untreated glass fiber 1, (b): untreated glass fiber 2, (c): surface treated glass fiber 2.

Since the resin mechanical properties are unaffected by the presence of the HBP modifier [5], the composite modulus and strength are also expected to be unaffected. Flexural tests have indeed shown that the mechanical properties of the different laminates are unaffected by the presence of the HBP modifier. The strength and strain at break show that the type of fiber has an important effect which can be related to the fiber strength. The surface treatment has a very limited effect which is due to a better wettability of the fibers and an increased stress transfer capacity between the fibers to the matrix. The presence of the HBP modifier has no influence, nor has the chemical nature of the HBP on the flexural properties.

Figure 4 shows the critical energy release rates G_{Ic} obtained from the DCB tests. The type of glass fiber and surface treatments have no effect on the G_{Ic} values for unmodified epoxy composites. With the presence of the HBP modifier on the other hand, the effect of the surface treatments is clearly observed. Almost no HBP particles are present at the interface on the surface treated fibers, and interfacial properties are maintained. This allows full advantage of the modified matrix toughness properties to be achieved since adhesive failure is inhibited.

It should be noted that the nature of the HBP plays an important role as well. These two modifiers induce the same resin toughness properties but differ in their physical nature. The interaction parameters between the surface treated fibers and HBP-2 compared to HBP-1. This results in an increase of the compatibility between the resin and the modifier. Therefore, less of the HBP will tend to migrate to the fiber surface and the interfacial properties are further enhanced.

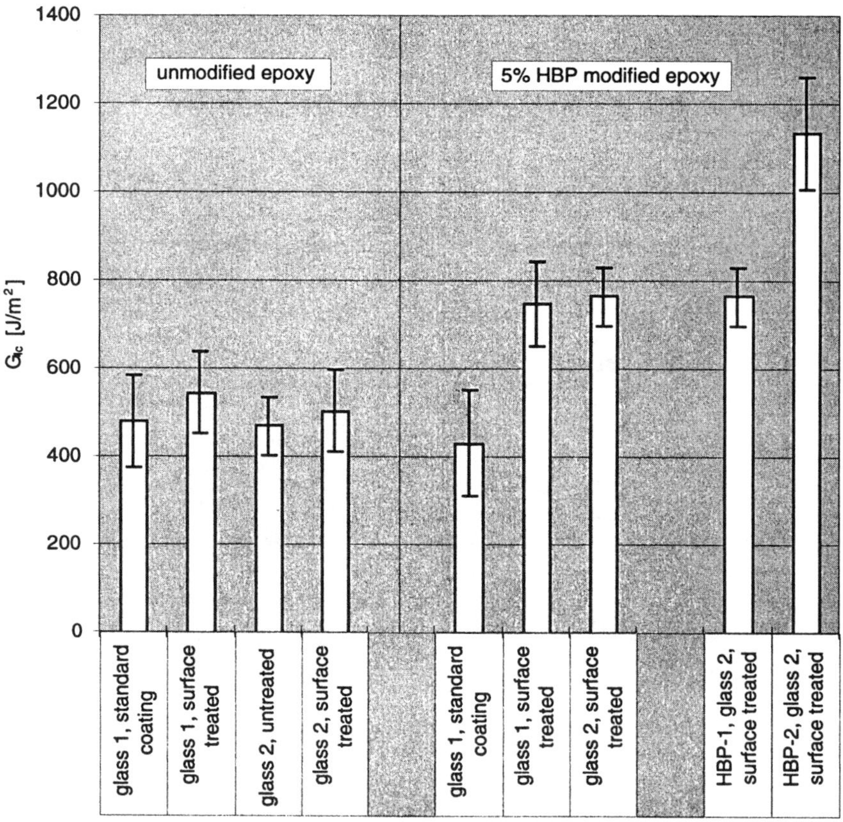

Figure 4. Critical energy release rates for unmodified and 5% HBP modified epoxy in glass fiber reinforced laminates. The error bars represent ± one standard deviation.

The optimum combination of silane treated fibers and the HBP-2 results in an increase of G_{Ic} of over 100% from 500 J/m^2 to 1150 J/m^2 for the unmodified and the 5% HBP modified epoxy resin composites respectively, as can be seen in Figure 5.

Carbon fibers are less sensitive to surface treatments. Fragmentation tests have shown that the interfacial properties are unaffected by the presence of HBP modifiers. With carbon fibers the composite G_{Ic} values are increased from 1400 J/m^2 to 2500 J/m^2 with the addition of only 5% HBP modifier.

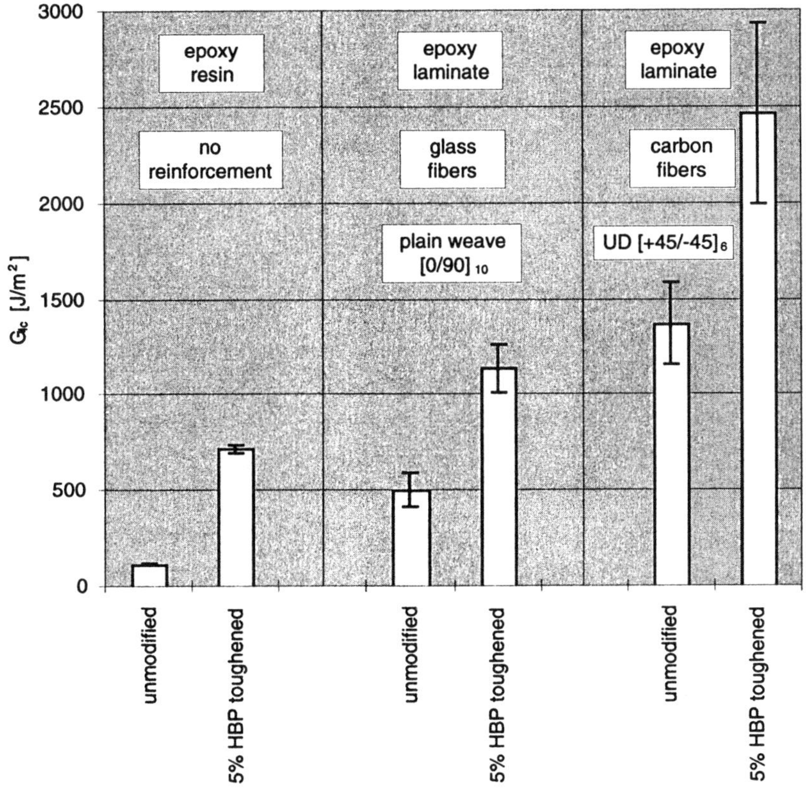

Figure 5. Critical energy release rates of the unmodified and 5% HBP modified epoxy resin, and their composites. The error bars represent ± one standard deviation.

6. CONCLUSION

An increase of over 100% of the G_{Ic} values of both glass fiber and carbon fiber reinforced epoxy composites is obtained with the addition of only 5% of an HBP modifier. This results in an increase of the toughness of the epoxy resin from a K_{Ic} value of 0.63 $MPa\sqrt{m}$ for the neat epoxy resin to 1.54 $MPa\sqrt{m}$ for the modified epoxy resin.

The interfacial properties of the glass fiber reinforcements have been shown to play an important role in the migration of the HBP towards the fiber-matrix interface and in the fiber-matrix interfacial properties. The influence of the surface on the composite laminate critical energy release rates as measured by DCB is due to the enhancement of cohesive failure allowing full advantage to be taken of the HBP modified resin toughness. The HBP chemical and physical nature has an identical effect on its migration towards the fiber-matrix interface and also influences interfacial and energy release rate properties.

The low viscosity of the HBP modifier, and its phase separation characteristics which avoid a detrimental filtration effect as seen with other conventional tougheners, lead to excellent processability characteristics of the modified epoxy resins.
The chemical architecture of the shell, namely its polarity and reactivity, controls the kinetics of the phase separation process, and can be tailored for different epoxy resins and to maintain a wide processing window for all conventional composite processing techniques.

7. REFERENCES

1. J. C. Seferis and L. Nicolais, The Role of the Polymeric Matrix in the Processing and Structural Properties of Composite Materials, Plenum Press, New York, 1983.
2. R. A. Pearson, Sources of Toughness in Modified Epoxies, Ph.D. thesis, University of Michigan, 1990.
3. B. Pettersson and K. Sörensen, High-solid alkyds based on hyperbranched polymers, Proceedings of the 21st Waterborne Higher-Solids and Powder Coatings Symposium, vol. 2, pp. 753-764 , 1994.
4. D. A. Tomalia, A. M. Naylor and W. A. Goddard III, Angew. Chem. Ed. Eng. 29, pp. 138-175, 1990.
5. L. Boogh, B. Pettersson, S. Japon and J.-A. E. Månson, Proceedings of the 10th International Conference on Composite Materials, august 14-18, 1995, Whistler, Canada, vol. 4, pp. 389-96, 1995.
6. PCT Pat./SE93/00148 (1993), A. Hult, M. Johansson, E. Malmström, K. Sörensen
7. PCT Pat./SE94/04440 (1994), B. Pettersson, K. Sörensen, L. Boogh, J.-A. E. Månson
8. Interlaminar Fracture Testing of Composites. Protocol for Joint Round Robin. Mode I: DCB. Mode II: ENF, European Group of Fracture, 1990

ACKNOWLEDGMENTS

Perstorp Polyols is acknowledged for the financial support of this work.

28th International SAMPE Technical Conference
November 4-7, 1996

ON THE SPRINGBACK EFFECT OF CFRP STRINGERS: AN EXPERIMENTAL, ANALYTICAL AND NUMERICAL ANALYSIS

A. Stephan*, E. Schwinge
Fraunhofer-Institute for Applied Materials Research, Lesumer Heerstr. 36,
28717 Bremen, Germany

J. Müller, Daimler-Benz Aerospace Airbus GmbH, Hünefeldstr. 1-5,
28199 Bremen, Germany

H. Öry, University of Technology Aachen, Institute for Aerospace Structures,
Wüllnerstr. 7, 52060 Aachen, Germany

ABSTRACT

The cure process of carbon fiber reinforced plastics (cfrp) induces residual stresses which lead to an angular distortion of curved laminates (springback effect). An initial angle of 90° between the web and flange of a T-stringer is reduced primarily by anisotropic thermal expansion and matrix shrinkage. To compensate for this phenomenon the tool angle is adjusted accordingly (trial-and-error approach).

Besides the primary parameters *anisotropic thermal expansion* and *shrinkage*, other variables influence the angle change. These secondary parameters, e.g. fiber orientation, cooling rate etc., are mainly a function of the chosen manufacturing technique. In this paper the experimental investigation as well as the analytical and numerical simulation of the springback effect for a typical CFRP stringer (Fibredux 6376-HTA) used in aircraft structures are discussed. The T-shaped stringers were manufactured in an autoclave. A complete and intensive variation of all previously determined variables was conducted in order to verify the simulation methods. The analytical simulation is based on classical lamination theory and a compatibility analysis. The finite-element calculation was carried out with the software package PATRAN/ABAQUS. The results indicate that the primary parameters can be well implemented into the simulation. The secondary parameters are included by the introduction of correcting factors.

KEY WORDS: Dimensional Stabilty, Residual Stresses, Thermal Analysis

---

*currently visiting research scholar at Virginia Tech, Dept. of Engineering Science and Mechanics

1. INTRODUCTION

The difference between curing-temperature and room-temperature in the manufacturing process of composite materials lead to residual stresses within the laminate. These induced stresses result in a thermal distortion in the case of an unsymmetric lay-up or curved laminate. In this paper the latter effect is investigated.

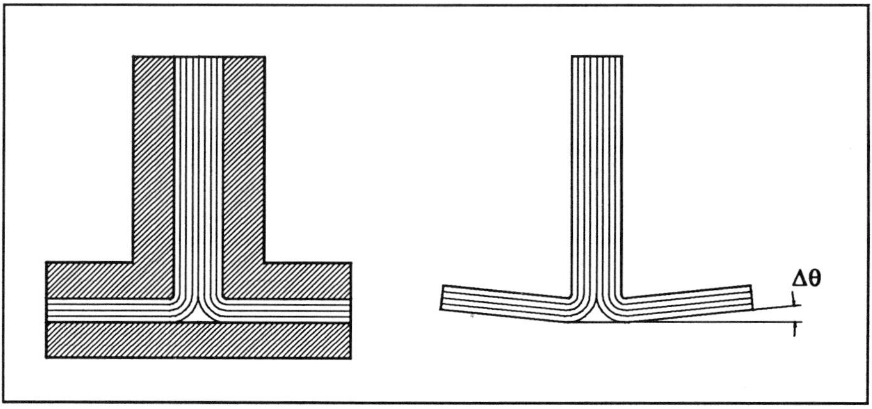

Figure 1: Angle Change

In industrial applications the magnitude of the angle change is assessed by using an trial-and-error approach. Additional costs and a delayed serial production are associated with this procedure. A quantitative simulation of the distortion would eliminate these drawbacks (Rapid Tooling) and would possibly increase the shape accuracy of the stringers.

For the new CFRP wing box, currently under devolopment at Daimler-Benz Aerospace Airbus GmbH, a new manufacturing process is envisaged. The proven but expensive module design should be replaced by using pre-cured components (differential design). This concept requires the exact determination of the angle change.

2. FUNDAMENTAL EFFECTS

The follwing mechanisms contribute differently to the springback effect.

- anisotropic contraction
- unsymmetry of residual stresses
- partial inhomogenities and local geometry changes

2.1 Anisotropic contraction Anisotropic contraction is the dominant factor affecting thermal distortion. The polymer matrix has a significant positive coefficient of thermal expansion (CTE), whereas the fibers are characterized by very small or even negative CTE. The isotropic contraction of the matrix is surpressed in the fiber direction by the stiff fibers; hence the thermal expansion behaviour of a unidirectional layer is dominated by the CTE of the fiber in the fiber direction. In the transverse direction, the in-plane and out-plane behaviour of the laminae are governed by the thermoelasticity of the matrix material. The thermal anisotropy leads to a reduction of the angle of curved laminates. In order to fully understand this phenomenon an isotropic curved material under a thermal load is considered. Due to the isotropy the thermal expansion is constant in each direction. The component angle remains constant while the sector length changes.

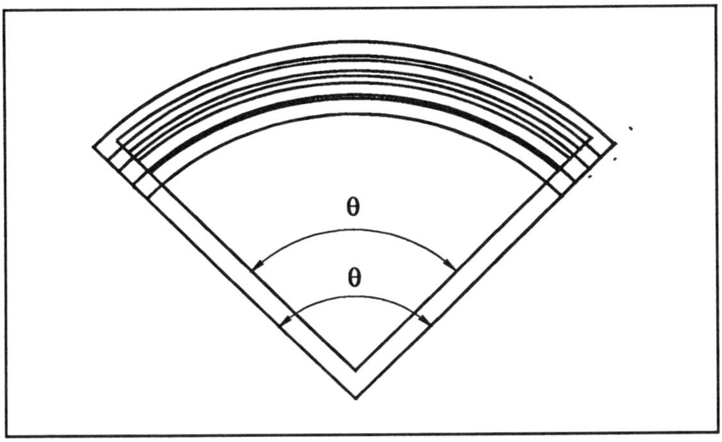

Figure 2: Isotropic expansion

The anisotropic multidirectional curved laminate shows a different thermal behaviour. Cooling results in a thermal expansion in the thickness direction which is significantly bigger than the fiber dominated in-plane expansion. The sector length of a multidirectional laminate is almost constant while the thickness experiences a significant change.

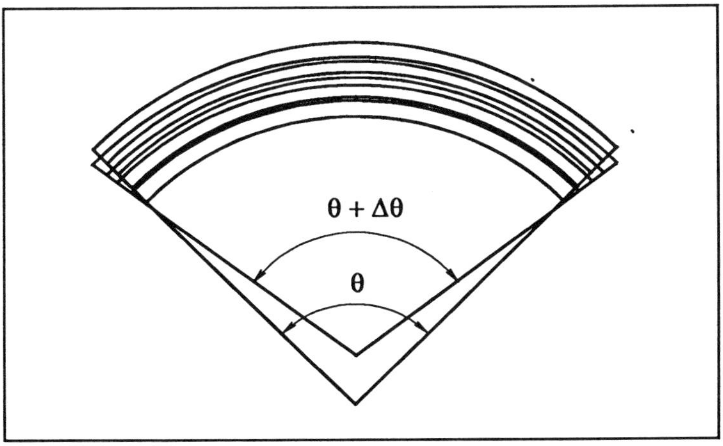

Figure 3: Anisotropic expansion

The influence of the material parameters and the manufacturing process can already be derived from these simple thoughts. The relevant CTE are a primary function of the chosen constituents and the lay-up. A lower fiber volume content would increase the effect of the matrix material properties.

In addition to the above mentioned effect the irreversible matrix shrinkage cannot be neglected for thermoset matrices. This additional anisotropic contraction can be assessed by a similar geometric relationship. The transformation of the volume shrinkage into a virtual CTE for a given temperature difference was applied here.

2.2 Unsymmetric Stresses The material chosen for the tooling has an additional effect on the thermal distortion. Unsymmetric residual stresses can develop due to the interaction of the tooling and the laminate. The tooling fixture normally applies a tensile force on the laminate during heating of the autoclave. The higher CTE of the tooling material which is often aluminum is the reason for this effect. The resultant stress distribution in the thickness is nonhomogeneous because the residual normal stresses are counteracted by interlaminar shearing deformations (1).

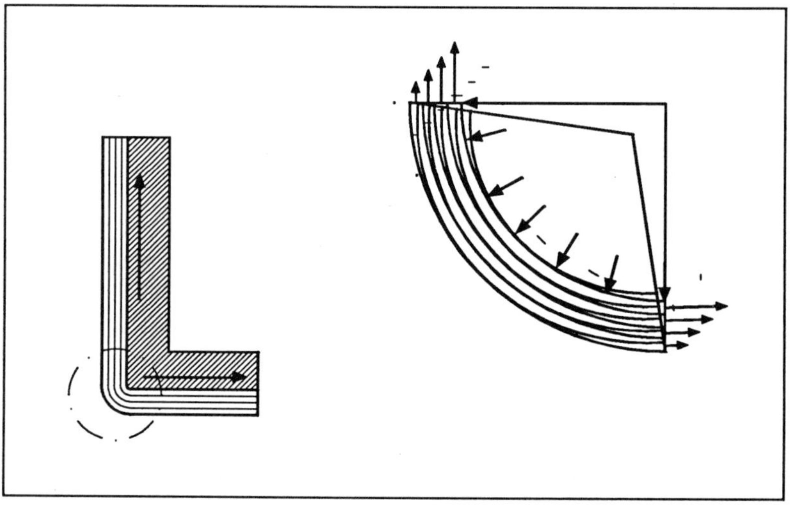

Figure 4: Tooling

Furthermore unsymmetric stress fields can be initiated by a nonhomogeneous temperature field in the autoclave (e.g. unmatched heat capacity).

2.3 Partial inhomogenties The influence of changes in the geometry of the laminate or of local nonhomogenities on the springback effect can be severe. Particularly, the local variation of relevant material properties is noteworthy. Investigations on curved laminates by Kim et. al. (2) have shown a varying fiber volume fraction or a resin rich zone through the thickness of the laminates. This leads to a through-the-thickness gradient of the thermal and mechanical material properties (material unsymmetry). The so-called ply-thinning (transverse fiber flow) initiated through the static pressure may be another reason for a local geometry change (3).

3. EXPERIMENTAL ANALYSIS

The determination of the experimental parameters was based on experience from industrial composite design and manufacturing. Two groups were identified:

1. Design parameter
- fiber orientation and stacking sequence
- woven fabric vs. tape
- web and flange thickness
- usage of "noodle"

2. Process parameter
- heating and cooling rate
- fiber-resin system

A classical stringer with a fiber orientation $[\pm 45°]_{12}$ for torsion loads as well as a $[45°,0°,-45°,0°,90°,0°]_s$ stringer were manufactured. In order to supplement the analysis the two unrealistic stringer lay-ups $[0°]_{12}$ and $[90°]_{12}$ were cured. Furthermore, a comparative study between a woven fabric and a tape material was conducted. The objective was to study the influence of the fiber waviness on the thermal distortion. The laminate thickness of the web and the flange of the T-shaped stringer were other parameters of the experimental analysis. Prepreg materials with different resins (6376-HTA(12K)-5-35% versus 913C-HTA(12K)-5-35%), and hence different curing temperatures, were used in order to determine their effect on the springback effect. The fiber volume fraction was another process parameter not to be neglected. A low fiber volume fraction results in a bigger influence of the resin material on the thermal expansion behaviour. Prepregs with 35% and 40% nominal resin mass fraction were used to study this effect.

The investigation of the process parameter focused on the variables relevant for industrial manufacturing. The process parameter range as given by the material supplier was never exceeded. Herein the main variables were the heating and cooling rate. Due to the size of the autoclave and the molds (air flow shade through big components) used in serial production, a nonhomogeneous temperature rate distribution can be expected. This analysis is especially valuable for the transformation of the laboratory results into industrial manufacturing. The assessment of the reproductability is also related to these tests.

The experiments were carried out on T-shaped stringers with a web height of 40mm and flange width of 35mm. The stringer length was approximately 500mm. The autoclave was operated with the following parameters:

- temperature: 125 or 180°C
- pressure: 7 bar
- heating rate: 2.5°C/min

All stringers were cured with a "noodle" which is a string of wrapped tape material in the fillet area. This additional material is mandatory in order to achieve a homogeneous fiber orientation in the fillet. Each stringer was manufactured twice in order to assess the influence of different cure cycles. For each variation of the parameter, 2 x 4 angle measurements were taken. The displayed values are the average of these measured angle changes. A projector with a magnification factor of 20 was used for the measurements. The accuracy was 1 arc minute. The following table gives an overview of the experimental results (reference values are in brackets).

Variation of parameter	Springback in arc minutes
±45°-lay-up	30
multidirectional lay-up (50/40/10)	28
90°-lay-up	44
0°-lay-up	4
double web and flange thickness	41 (32)
woven fabric vs. tape	34 (35)
±45°-lay-up 6376-HTA(12K)-35%	30
±45°-lay-up 913C-HTA(12K)-35%	32
±45°-lay-up 913C-HTA(12K)-40%	35
low heating rate (0.3°C/min)	10(25)

Table 1: Experimental results

As expected the biggest angle change was recorded for the 90° stringer. The difference between the in-plane CTE in the relevant direction and the out-of-plane CTE is a maximum for this fiber orientation. The angle change of 4 mintes for the 0° stringer is due to the different transverse and thickness CTE. All other lay-ups are in between this range. Obviously the heating rate has a big influence on the springback. A low heating rate leads to more than a 50% decrease in the thermal distortion (stress relaxation).

In addition to the above mentioned parameter variations two other tests were conducted. To fully describe the thermo-elastic behaviour of composite materials it is necessary to determine the CTE in the thickness direction. Due to the suppression of the Poisson effect the CTE of multidirectional laminates in the out-of-plane direction is bigger than the CTE in the transverse direction (approximately 40%). In addition, interferometric studies of the thickness CTE at the Fraunhofer-Institute for Machine Tools and Forming Technology also recorded a higher thickness CTE for the unidirectional 6376-HTA (12)-35%. This CTE (α_{33}) was found to be 18% higher than in the transverse direction (4).

The curing process of laminates based on thermoset resins is accompanied by the typical matrix shrinkage. This irreversible volume change is not included in the CTE and therefore has to be added to the thermal expansion process. The influence on the springback is similar to the reversible thermal expansion. The matrix shrinkage was determined using the bimetal theory by Timoshenko (5). A unidirectional 4 layer strip was cured in two steps so that the resulting curvature was only due to the matrix shrinkage; then a virtual "CTE" was calculated.

$$\alpha_{22shrink} = 19\% \cdot \alpha_{22therm}$$

$$\alpha_{11shrink} = 6\% \cdot \alpha_{11therm}$$

These results were used for the analytical and numerical modelling.

4. Analytical Modelling

The analytical modelling was performed using the classical lamination theory (CLT) and a geometric compatibility condition (6). The primary parameters *anisotropic expansion* and *matrix shrinkage* are directly used for the calculation whereas the secondary parameter can be considered as correction factors.

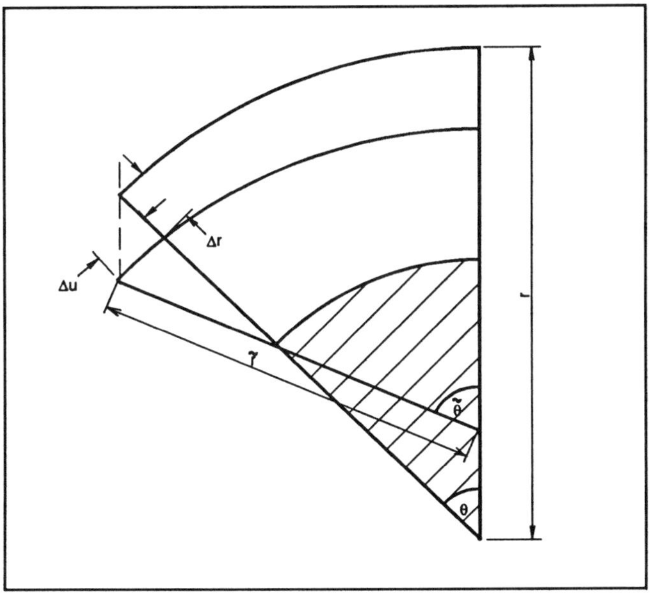

Figure 5: Geometry

With the compatibility condition the equation for the angle change becomes

$$\Delta\theta = \theta \cdot \frac{|\varepsilon_r - \varepsilon_t|}{1-|\varepsilon_r|}.$$

ε_t describes the thermal strain in the tangential direction. The radial strain is represented by ε_r. Using the known symbols α and ΔT for the thermal expansion, ξ for the matrix shrinkage and Ψ as the correcting factor, the equation for ε_r is

$$\varepsilon_r = \alpha_{33} \cdot \Delta T + \xi_r + \Psi_r = \Delta T \cdot (\alpha_{33} + \alpha_{22shrink}) + \Psi_r$$

The strain ε_t in the tangential direction was calculated with the CLT (7).

$$\begin{Bmatrix} \varepsilon_t \\ \kappa \end{Bmatrix} = \begin{bmatrix} A & B \\ B & D \end{bmatrix}^{-1} \begin{Bmatrix} N_t \\ M_t \end{Bmatrix}$$

with
$$\begin{Bmatrix} N_{xt} \\ N_{yt} \\ N_{xyt} \end{Bmatrix} = \sum_{k=1}^{n} \begin{bmatrix} \bar{Q}_{11} & \bar{Q}_{12} & \bar{Q}_{16} \\ \bar{Q}_{12} & \bar{Q}_{22} & \bar{Q}_{26} \\ \bar{Q}_{16} & \bar{Q}_{26} & \bar{Q}_{66} \end{bmatrix}_k \begin{Bmatrix} \alpha_x \\ \alpha_y \\ \alpha_{xy} \end{Bmatrix}_k \cdot \Delta T$$

and
$$\begin{Bmatrix} M_{xt} \\ M_{yt} \\ M_{xyt} \end{Bmatrix} = \sum_{k=1}^{n} \begin{bmatrix} \bar{Q}_{11} & \bar{Q}_{12} & \bar{Q}_{16} \\ \bar{Q}_{12} & \bar{Q}_{22} & \bar{Q}_{26} \\ \bar{Q}_{16} & \bar{Q}_{26} & \bar{Q}_{66} \end{bmatrix}_k \begin{Bmatrix} \alpha_x \\ \alpha_y \\ \alpha_{xy} \end{Bmatrix}_k \cdot \Delta T \cdot (z_k - z_{k-1})$$

The influence of moisture on the springback effect was neglected due to practical reasons. Nevertheless it can be included into the Mathematica® program which was set up for the analytical modeling. The results are shown in the following table.

Variation of parameter	Springback in arc minutes
±45°-lay-up	27
multidirectional lay-up (50/40/10)	29
90°-lay-up	36
0°-lay-up	5
double web and flange thickness	32
woven fabric vs. tape	34
low fiber volume fraction	34

Table 2: Analytical results

In general the analytical simulation compares well with the experimental results. Only the results for the doubled web and flange thickness and for the 90° fiber orientation show slight deviations. It is assumed that unsymmetric stresses or fiber misalignment are the cause of these distortion.

5. Finite Element Analysis

Due to the midplane symmetry only one side of the stringer was modeled with PATRAN. Three brick elements (type C3D20R) each containing four layers were used for the thickness direction. In the axial direction five elements were used. Nodes in the symmetry plane were fixed in order to satisfy the boundary conditions. Then a temperature difference was applied in order to simulate the thermal load.

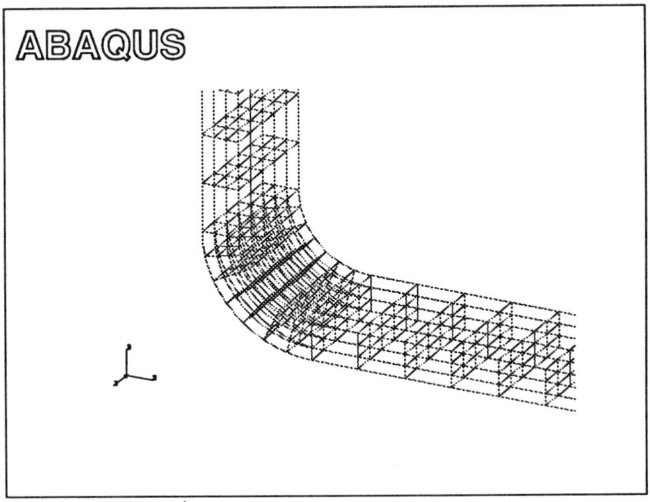

Figure 6: FE-Mesh

Table 3 shows the results of the FEA.

Variation of parameter	Springback in arc minutes
±45°-lay-up	36
multidirectional lay-up (50/40/10)	35
90°-lay-up	41
0°-lay-up	5
woven fabric vs. tape	34
low fiber volume fraction	36

Table 3: Numerical results

The comparison of the experimental and numerical results do not show significant deviations except for the 45° fiber orientation and the multidirectional laminate. This might be due to the difficult and cumbersome determination of the spatial material data.

6. Conclusion

The primary cause of the sprinback effect, namely the anisotropic expansion and the matrix shrinkage, were determined experimantally. Both theoretical approaches are capable of calculating the thermal distortion due to these parameters. However, the cooling and heating rate impact could not be determined theoretically. The stress relaxation at elevated temperatures is probably the most important factor in this case. The implementation of curing models in the simulation might overcome this problem. This is especially true for thick laminates. The residual stresses induced by the mold could not be obtained. Numerically this can be done by modelling the mold and the laminate for the FE analysis.

In contrast to the analytical simulation, the FEA can also be applied to complex shaped components. In addition, an unsymmetric laminate can be modeled easily.

7. References

1. R.H. Nelson, Prediction of Dimensional Changes in Composites during Curing, 34th International SAMPE Symposium, May 8-11,1989

2. C.G. Kim et. al., Spring-In of Composite Laminated Bends, Proc. ICCM-VII, 1989

3. J.A. Barnes et. al., Dimensional Stability Effects in Thermoplastic Composite-towards a Predictive Capability, J. of Composite Manufacturing, Vol. 2, No. 3/4, 1991

4. P. Aswendt, Personal Communication, March 1996, FhG-IWU, Chemnitz

5. S. Timoshenko, Analysis of Bi-Metal Thermostats, J. Opt. Soc. Amer., II 1925

6. D.W. Radford et.al., Shape Instabilities in Composites resulting from Laminate Anisotropy, Proc. 4th Japan-US Conference on Composite Materials, 1988

7. R.M. Jones, Mechanics of Composite Materials,Taylor & Francis, 1975

28th International SAMPE Technical Conference
November 4-7, 1996

THE NUMERICAL SIMULATION OF PRESS FORMING OF CONTINUOUS FIBER REINFORCED THERMOPLASTICS

P. de Luca[1], P. Lefébure[2], A.K. Pickett[3], A. M. Vodermayer[4], W. Werner[5]

[1]Engineering Systems International, 20 Rue Saarinen, F-94578 Rungis Cedex, France.
[2]Dassault Aviation, B.P. 12, F-78141 Vélizy-Villacoublay Cedex, France.
[3]Engineering Systems International, 13-15 Frankfurter Strasse, D-67560 Eschborn, Germany.
[4]Sulzer Innotec, Advanced Composites 1524, CH-8401 Winterthur, Switzerland.
[5]Dornier Luftfahrtgesellschaft, EK15, D-88039 Friedrichshafen, Germany.

ABSTRACT

Numerical simulation of the press forming of continuous fiber reinforced thermoplastics (CFRTP) has so far been limited. Most analytical investigations to date have either concentrated on very specific aspects of the process or have used geometric mapping codes to predict the feasibility of forming a fabric to a given shape. This paper outlines a European research project which has developed computer aided design tools to simulate the pre-heating, forming and cooling phases of the thermoforming process.

The pre-heating analysis undertakes a three dimensional orthotropic heat analysis of the laminate for a variety of heat sources and provides an initial temperature distribution for the subsequent forming analysis. The forming simulation predicts temperature distribution during forming and has advanced material laws to represent the complex deformations of the laminate. Phenomena such as ply buckling, fiber strains and void content distributions can be predicted. During cooling dimensional stability is the main issue and a model to predict residual deformations from the thermal stresses has been developed. Example results for industrially relevant parts are presented.

KEY WORDS: Thermoforming, Numerical Simulation, Composites Manufacturing.

1. INTRODUCTION

The press forming of continuous fiber reinforced thermoplastic sheets, which also have the possibility to be recycled, can provide an economical and fast forming process for the manufacture of lightweight high strength structural components. For these reasons CFRTP materials have received considerable interest from industry. However, successful thermoforming requires a detailed understanding of the complex deformation mechanisms that occur during forming and the influence that the different processing conditions have on the final part performance and quality. These factors, and the relatively high material costs, have so far hindered their industrialization.

Material costs have shown a steady and continuous decline, however further reductions will still be needed before these materials can penetrate markets such as the automotive industry. This paper is concerned with the material processing aspects and shows how computer simulation can provide an effective tool to avoid costly and inefficient 'trial and error' practices. The work has been undertaken through a European research project (1) to develop and industrially validate simulation software to numerically predict the full thermoforming process.

Briefly, three key phases in the press forming process may be identified. Firstly, a laminate of pre-consolidated plies is heated by some form of external heating to a temperature above the matrix melt temperature which is then quickly transferred to the colder press. Secondly, the hot sheet is press formed to the required shape usually using either matched metal tooling, ‘hydroforming’, or the ‘rubber pad’ forming technique. Finally, the part is cooled in the tools to a temperature where it has sufficient internal stiffness to be dimensionally stable during extraction from the mold and further cooling in air. The forming pressure, application time and sheet and tool temperatures must all be carefully selected to obtain a satisfactory compromise between fast cycle times and quality consolidated parts.

For the CAE design of the thermoforming process various numerical analysis tools are needed:

- Firstly, a pre-heating software is required to predict the three dimensional temperature distribution in the sheet and provide the initial temperatures for the forming analysis. The tool should be suitable to ‘optimize’ the pre-heating setup.
- Secondly, a forming simulation is required to predict the deformations in the CFRTP sheet and predict such phenomena as temperature evolution, delamination, ply buckling and void content distribution.
- Finally, analysis methods are needed to predict the post-forming cooling and build-up of internal stresses which largely determine final part quality and shape.

Work has so far concentrated on commercially available pre-consolidated continuous fiber reinforced thermoplastics appropriate to the aerospace and automotive industries. These materials include UD carbon fiber reinforced APC2-AS4, carbon fabric reinforced PEI-CETEX, and glass fabric reinforced PA-6 and PA-12 (Vestopreg).

2. PRE-HEATING

A pre-heating analysis is the important first step in the CAE design of thermoforming. The three dimensional temperature evolution in the CFRTP sheet must be predicted considering the laminate thermal properties, the laminate size and lay-up, and the heating source. Various types of heating are commonly used in the industry including Infra-red, convection oven and contact platens; each of which has very different heating characteristics. Any general purpose analysis tool must allow the different laminate and equipment parameters to be investigated and be suitable for optimization studies.

As a part of a the research project (1) a PC based analysis software has been developed to investigate the preheating (2). The analysis uses a conventional finite difference solution to obtain either a coupled 1D though-the-thickness and 2D inplane solutions, or a full 3D solution. The software has been industrially validated and has proven particularly useful to avoid 'trial and error' setup methods. The following example considers a pre-consolidated 150mm*130mm*2mm thick APC2-AS4 sheet heated by three banks of infra-red heaters. Three thermocouples were imbedded between the lower two plies, the middle two plies and the top two plies. Figure 1 shows the good agreement between experiment and analysis for each position through the thickness.

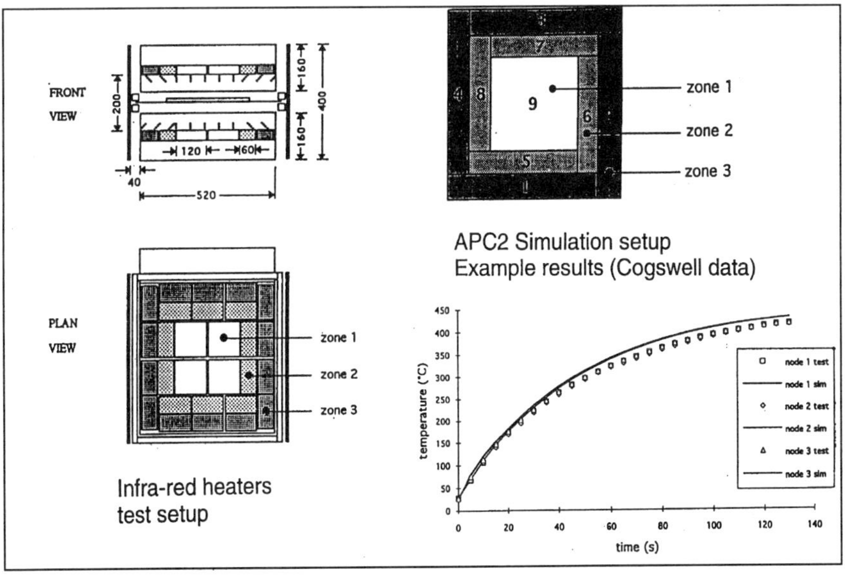

Figure 1: Comparison of model prediction and test for an APC2-AS4 panel with thermocouples distributed through the thickness.

3. MODELING OF THE FORMING PROCESS

3.1 Finite Element Modeling A finite element software, originally dedicated to the simulation of sheet metal stamping (PAM-STAMP™ (3)), has been further developed to simulate the hot press forming of CFRTP sheets. The PAM-STAMP™ code uses conventional finite elements to discretize the problem which is then solved using 'explicit' integration (4). This approach has proven particularly suitable for highly non-linear transient dynamic problems involving contact. Typical fields of application have included impact, crash analysis, and metal forming.

There are obvious analogies between the forming of metal and CFRTP sheets. In both cases an initially flat sheet is press formed using devices such as matched metal tooling, hydroforming, or rubber pad forming, to the required shape. Thermoforming simulation however requires additional analysis capabilities including heat transfer modeling, new material laws and viscous-friction laws. These features are briefly described in the next sections.

3.2 Laminate Modeling In metal stamping simulation a single layer of elasto-plastic shell elements is sufficient to describe the sheet inplane, bending and shear deformations. During the arbitrary shaping of CFRTP laminates the predominant deformation modes are single and double curvature forming. For single curvature forming interply shearing is required, whereas double curvature forming will require both inter- and intra-ply shearing (Figure 2). These deformations cannot be modeled using a single layer of shell elements and, consequently, new techniques have been developed in which each ply is modeled as one layer of discrete shell elements with viscous friction laws to control the sliding between the stacked plies. This approach also enables potential delamination to be identified and modeled.

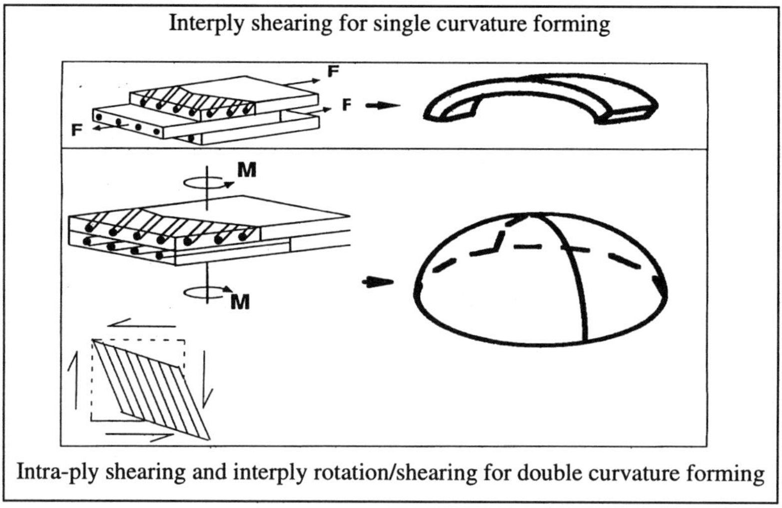

Figure 2: The predominant deformation modes in laminated CFRTP composites.

The material law for a single ply is based on a general theory for idealized fiber reinforced fluids first proposed by Rogers (5). This has recently been extended by Johnson (6) for woven fabrics. In its simplest form, for the modeling of a UD prepreg ply, the plane stress form of the constitutive law is given by,

$$\begin{pmatrix} \sigma_{11} \\ \sigma_{22} \\ \sigma_{12} \end{pmatrix} = \begin{bmatrix} 4\eta_L & 2\eta_T & 0 \\ 2\eta_T & 4\eta_T & 0 \\ 0 & 0 & 2\eta_L \end{bmatrix} \begin{pmatrix} \dot{\varepsilon}_{11} \\ \dot{\varepsilon}_{22} \\ \dot{\varepsilon}_{12} \end{pmatrix} + \begin{pmatrix} E\varepsilon_{11} \\ 0 \\ 0 \end{pmatrix}$$

where conventional notation is used and η_L and η_T are the longitudinal (fiber direction) and transverse sheet viscosities respectively. Temperature dependency is introduced using the Cross equations to describe the viscosities.

The viscous friction between plies has been investigated by Murtagh (7). Experimental studies have investigated the influence of normal pressure, temperature, and the relative fiber orientation and velocity between adjacent plies for APC2-AS4 and PEI-CETEX materials. From this work governing equations have been established and used in the forming simulation to determine the interply viscous friction. Similar studies have also been undertaken to investigate the friction between the CFRTP materials and tooling steels.

3.3 Heat transfer modeling For fast production cycle times the tool temperature is usually held well below the laminate forming temperature. Due to the rapid heat flow, and the temperature dependent nature of the material and inter-ply sliding, it is essential to predict temperature evolution as an integral part of the solution.

The dominant heat transfer in thermoforming is through mechanical contact between the sheet and tools, and between the different plies. Less important, but significant, is the anisotropic in plane heat conduction (8). These two heat transfer processes are modeled using a 1D solution for though-the-thickness ply and tool contact, coupled with a 2D anisotropic in-plane heat conduction analysis. The heat transfer between the tools and the plies requires surface heat transfer coefficient values (9), whereas the inplane treatment requires temperature depend specific heat, and longitudinal (along the fibers) and transverse conductivity values.

This work has been validated against analytical solutions and experimental tests. An example of experimental validation is shown on Figures 3 and 4. The set-up (Figure 3) shows 8 PEI-CETEX plies to be formed into a 'U' shape. The metal tooling (blankholder, die and punch) have an initial temperature of 250°C and the composite initial temperature is 360°C. Figure 4 shows an exploded view of the laminate at an intermediate state of forming. The cooling in the plies is clearly limited to regions in contact with the colder punch, blankholder and die. Experimental measurements using thermocouples and simulation results have shown good agreement.

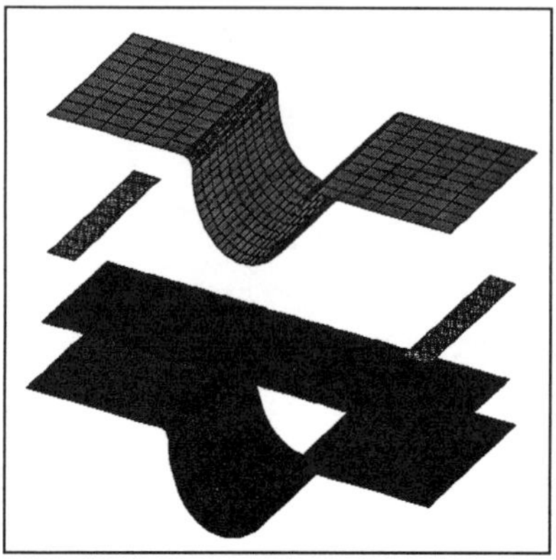

Figure 3: Numerical 'setup' showing the plies, punch, die and blankholders.

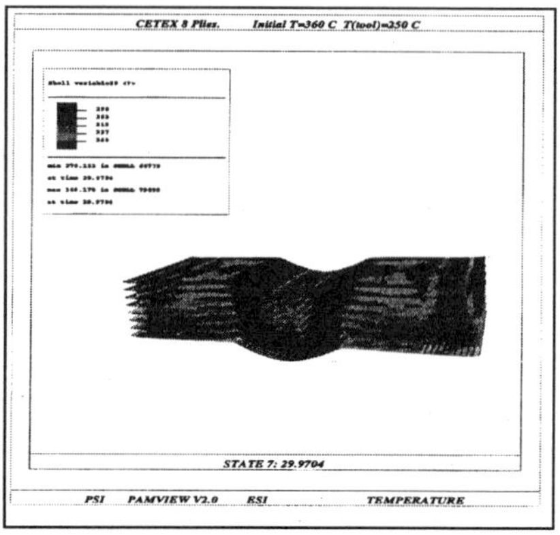

Figure 4: Temperature contour. 8 plies of PEI-CETEX. (exploded view).

4. EXAMPLE ANALYSES

4.1Forming of a Sikkens As an example application of the use of the simulation, the forming of a part using different processing conditions (punch velocity) is studied. The part studied is a Sikkens (a half cylinder with two quarters of a sphere at both ends) made of a quasi-isotropic layup of 8 plies woven PEI-CETEX. Figure 5 shows the finite element numerical setup. Figures 6 shows the final part formed with a punch velocity of 40 mm/s to be compared with Figure 7 where a punch velocity of 5 mm/s is used. The part formed using the higher velocity exhibits a lot of unacceptable wrinkles whereas the low velocity formed part gives a good quality part. This numerical result has also been observed experimentally (10). This last reference addresses the influence of various other forming parameters from both a numerical and an experimental point of view.

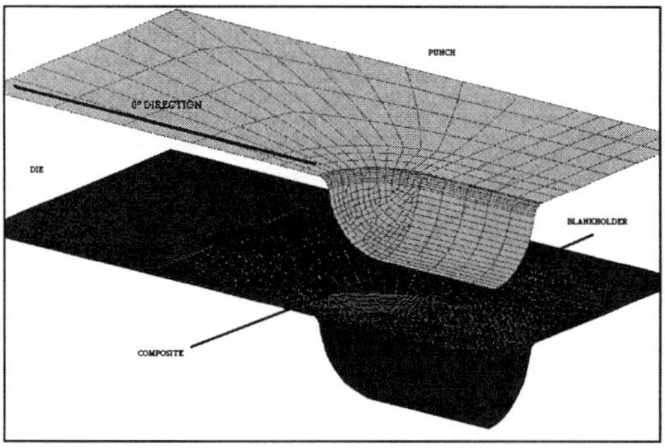

Figure 5: Numerical 'setup' of the sikkens forming simulations.

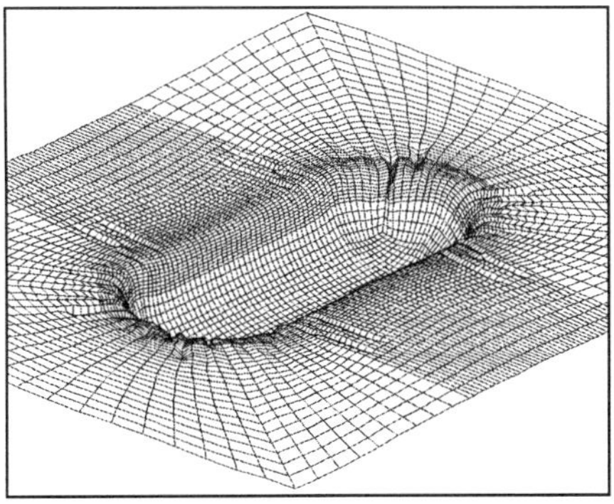

Figure 6: The quasi-isotropic PEI-CETEX part. Forming velocity = 40 mm/s.

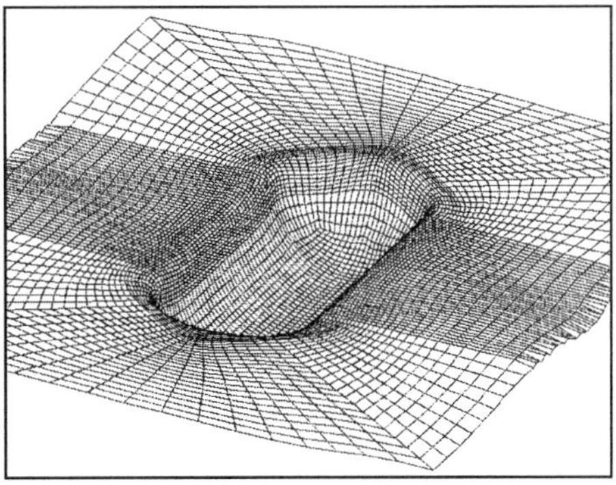

Figure 7: The quasi-isotropic PEI-CETEX part. Forming velocity = 5 mm/s.

4.2 Post-forming Analysis: Part Quality, Cooling and Dimensional Analysis

During the preheating, and possibly forming, the laminates will deconsolidate and void. These voids must be removed by the application of presssure at sufficiently high temperatures while the material is still 'soft'. The pressure levels and rates of cooling in the mold will therefore determine the final void content and part quality. A model to predict final void content has been developed and implemented in the forming software (11).

Prediction of cooling rates, particularly while the part is in the mold, provides essential information to indicate final part quality. A design engineer also needs to predict the thermal stresses generated in the part due to the cooling and the residual deformations (dimensional stability) that result as these stresses relax when the part is removed from the mold. Prediction of cooling and dimensional stability is demonstrated here for the case of a Rapier wheel (Figure 8, 9); used in weaving machines. The part is essentially a low curvature sheet having a radial tape wound construction. The central part has about 40 overlapping tapes with a total thickness of 4.0 mm, this decreases to 4 overlapping tape at the outer edge and a total thickness of 0.5mm.

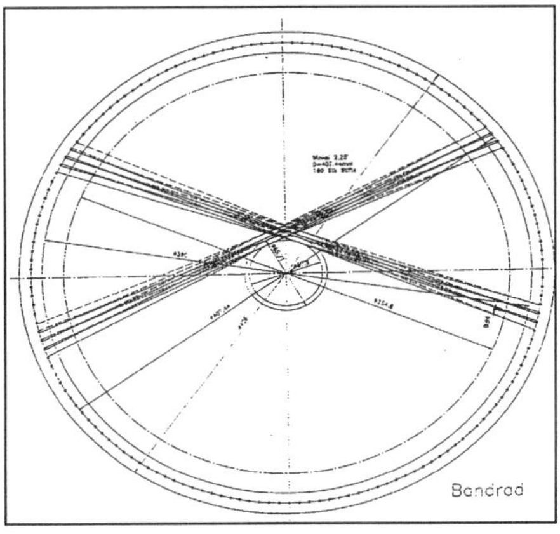

Figure 8: Construction pattern of the Rapier wheel.

Figure 9: The Rapier wheel.

4.2.1 Cooling Analysis Generally long duration events such as 10 to 60 seconds cooling process are uneconomical for the explicit integration method due to the physical restriction of the constant timestep. However, during cooling the part is constrained between the tools and does not undergo further deformations. The expensive element computations may therefore be ignored and a pure temperature analysis may be made. This gives an economical analysis that can predict the full cooling process within moderate computing times.

The following example shows a cooling analysis for the Rapier wheel.Thermocouples were embedded in the test to monitor temperature evolution. Three of these measuring points are compared with simulation results which, in each case, gives a good agreement.

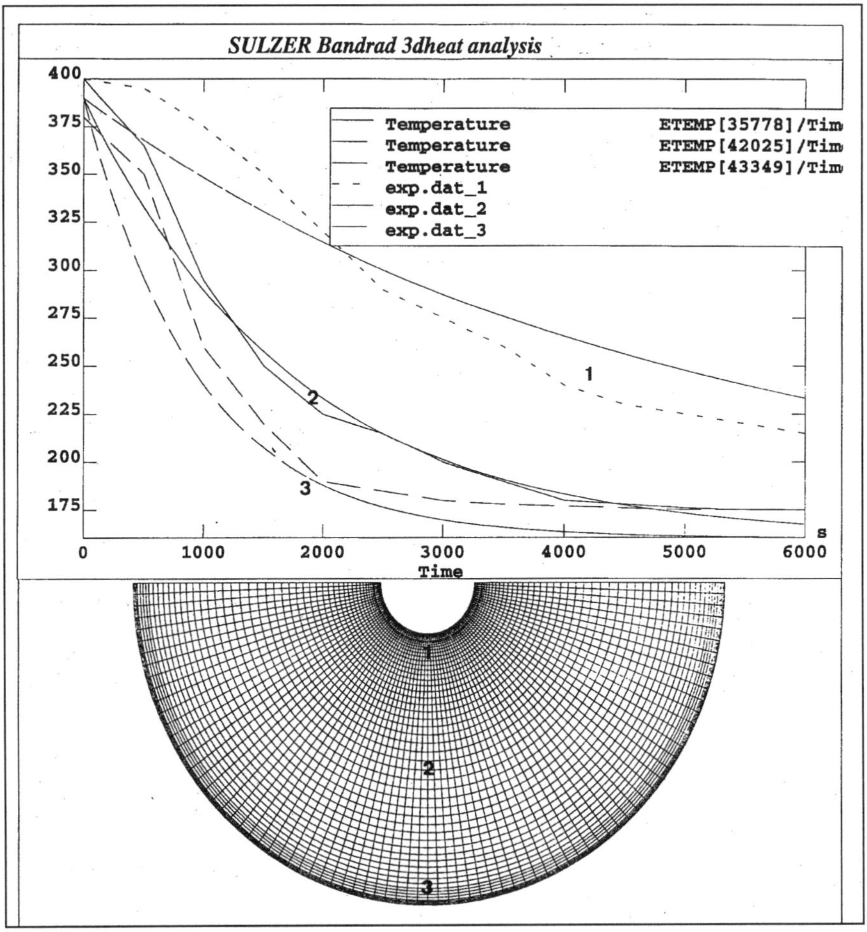

Figure 10: Comparison (simulation/measurements) of temperature evolution for the Rapier wheel cooling analysis.

4.2.2 Dimensional Stability During cooling thermal stresses develop in the part which may lead to matrix cracking and damage. As the part is extracted from the mold these stresses try to relax by deforming the part. Analytical solutions have been proposed for a laminated channel section (12). For more general shapes a finite element solution is necessary.

In general the internal stresses are a combination of the thermal (cooling) stresses and 'locked-in' elastic fiber stresses created by the forming. Some stress relaxation will occur due to the viscous nature of the materials, however further investigations (11) have shown this contribution to be small and thus may be ignored. Furthermore, including a full visco-elastic relaxation analysis considerably complicates the analysis. Therefore the chosen analysis uses a conventional thermo-elastic solution where an elastic anisotropic material model represents the laminated composite to which the anisotropic thermo-elastic stresses are imposed.

The following model is used to compute the thermal stresses. For APC2-AS4, the material is assumed to have two distinct phases separated by the glass transition temperature Tg (143°C). Above 285°C the material has insufficient stiffness to sustain stresses. It is assumed that the lower temperature limit is the room temperature (T_{Room}). The expressions to determine thermal stresses are given for the fiber and transverse to the fiber directions as,

$$\sigma_F = E_F^1 \alpha_F^1 (285-143) + E_F^2 \alpha_F^2 (143 - T_{Room})$$

$$\sigma_T = E_T^1 \alpha_T^1 (285-143) + E_T^2 \alpha_T^2 (143 - T_{Room})$$

where the subscripts F and T denote the fiber and transverse directions respectively. Superscripts 1 and 2 denote the phases above and below Tg respectively. The following material data has been used,

	Fiber Direction (F)	Transverse Direction (T)
E^1	130 GPa	2.9GPa
E^2	130 GPa	8.8GPa
α^1	$1.0\ 10^{-06}$	$80.0\ 10^{-06}$
α^2	$0.2\ 10^{-06}$	$28.8\ 10^{-06}$

Table 1: Numerical values used in the dimensional stability analysis.

The thermo-elastic analysis predicts a maximun deformation at the outer perimeter of 14mm which agrees to within 5% of test results. Studies have been conducted to determine the required initial forming shape so that a desired end shape, after cooling, may be proposed.

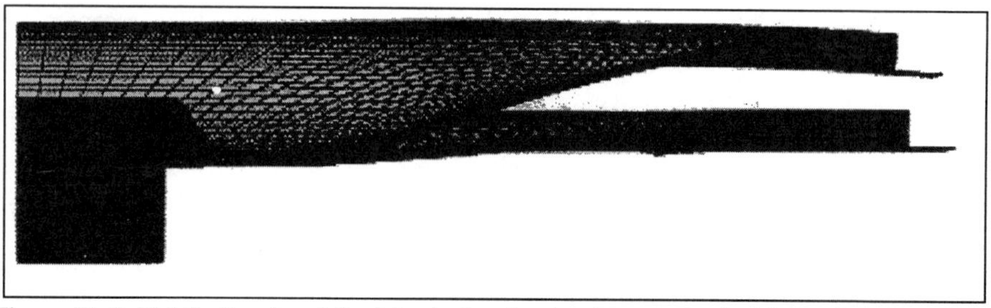

Figure 11: Initial (bottom) and final (top) plots for the thermo-elastic (dimensional stability) analysis of the Rapier wheel problem.

5. CONCLUSIONS

New simulation tools for the pre-heating, press forming and post-cooling analysis of pre-consolidated CFRTP sheets have been presented. The capabilities of these softwares include:

- Pre-heating analysis: These covers the heating prediction for numerous commonly used CFRTP materials using either infra-red, conduction or convection heating. Numerous validation studies have shown the software to be accurate and reliable.
- Forming analysis: Numerous industrial studies have been undertaken to date. Phenomena such as fabric wrinkling, temperature and pressure evolution, and the prediction of void content distribution have agreed well with test results.
- Post-forming analysis: A technique to predict dimensional stability has been established. Further work will be needed to validate this for more complex parts in which the 'locked-in' fiber stresses, created during the forming, play an important role in the post-forming residual deformations and stresses.

6. REFERENCES

(1) "BRITE-EURAM BE-5092, Industrial Press Forming of continuous fiber reinforced thermoplastics sheets and the development of numerical simulation tools", July 1996.

(2) J.E. Cunningham and P.F. Monaghan, Proceedings of the 4th International Conference on Automated Composites, Notthingham, England, 1995, pp. 237.

(3) The PAM-STAMP™ code, Engineering Systems International, 20 Rue Saarinen, F-94578 Rungis Cedex, France.

(4) R.D. Cook, D.S. Malkus and M.E. Plesha, Concepts and Applications of Finite Element Analysis, Wyley Publish., 1989.

(5) T.G. Rogers, Composites, 20 (1), 1989, pp. 21.

(6) A.F. Johnson and E. Costalas, in Proc. 4th International Conference on Automated Composites, Notthingham, England, 1995, pp. 341.

(7) A. Murtagh, J.J. Lennon, Proceedings ICCM-10, Whistler, Canada, 1995.

(8) S.D. Gilmore and S.I. Güceri, Polymer Composites , 11 (6), 1990, pp. 406.

(9) S.F. Cassidy and P.F. Monaghan, Composites Manufacturing, 5 (4), 1994, pp. 225.

(10) P. de Luca, P. Lefébure and A.K. Pickett, Proceedings of the 4th International Conference on Flow Processes in Composites Materials FPCM'96, 7-9 September 96, Aberyswyth, Wales.

(11) J.M. O'Neill, T. Rogers and A.J.M. Spencer, Math. Engineering., 2(1), 1988, pp. 65.

(12) R. Phillips, Ph.D thesis, Ecole Polytechnique Fédérale de Lausanne, Switzerland, 1996.

28th International SAMPE Technical Conference
November 4-7, 1996

PICTURE-FRAME RHEOLOGICAL SHEAR TESTING FOR COMPOSITES SHEET-FORMING PROCESSES

C.M. Ó Brádaigh, G.B. McGuinness, R.A. Canavan and E.R. Simon,
Composites Research Unit,
Department of Mechanical Engineering,
University College Galway, Ireland

ABSTRACT

Composites sheet forming processes have been developed in both thermosetting and thermoplastic matrix materials in order to automate manufacturing methods. Forming of double-curvature shapes requires that fabrics or unidirectional laminates shear in the plane of the sheet, a deformation mode known as intraply shear. Finite element process simulations of composite sheet-forming have been developed for which the major viscosities of the prepregs at forming temperature are required as inputs. The objective of this work is to directly measure the intraply shear viscosities of high volume-fraction impregnated polymer melts by using a specially-developed picture-frame apparatus which is loaded in a test machine at the composite process temperature. The material is pinned to the picture frame in a manner which leaves it free to rotate, yielding a homogenous large-deformation. Measurement of the load history at various test displacement rates allows the material to be characterised using various model responses. Materials tested in this work include two types of glass-fabric reinforced Nylon-12 (VESTOPREG), a carbon-fabric reinforced PEI (CETEX) and a unidirectional carbon-fibre PEEK (APC-2). These materials are characterised as either anisotropic viscous or viscoelastic fluids, with intraply shear viscosities typically of the order of MPa.sec, many orders of magnitude greater than the equivalent matrix viscosities. Conclusions as to the critical buckling angles and loads of the materials in shear are also made from the apparatus.

KEYWORDS: Hot Drape-Forming, Shear Viscosity, Fabrics/Tapes

INTRODUCTION

Sheet-forming processes [1] are seen as potentially cost-efficient methods of automating hand lay-up procedures for reasonably complex curvature structures. For thermoplastic composites, pre-consolidated flat sheets are heated in an infra-red oven and then rapidly press-formed and cooled between matched-metal, or rubber-metal tooling, as shown in Figure 1. The advantages are the low cycle-times and high performances available, with the process being particularly suitable for smaller components in the automotive and general industrial sectors.

In the aerospace industry, hot-drape forming of thermoset prepreg involves flat sheets of tape-laid materials being vacuum or positive-pressure formed over tooling. The prepreg may be heated to an intermediate temperature in order to decrease material viscosity, but generally, no curing is effected during the drape-forming process. The prepreg may be formed using only a top diaphragm, or between two diaphragms, or using a top and removable bottom diaphragm. In general, the formed laminate is subsequently cured in the same manner as hand laid-up structures. The advantages of the process are in the economic savings made by automation of the lay-up procedure.

In either the thermoplastic or thermoset case, the deformation required of the composites to allow them to conform to a particular tool surface is identical. Consider the classes of tools show in Figure 2 and the fundamental forming mechanisms of Figure 3. The simplest deformation is that of *resin percolation* through the fibre bed and this is all that is required when a laminate is processed by application of pressure across a compliant diaphragm as shown. The next mechanism required is that of *transverse flow* of fibres and resin, and this is required when matched-metal tooling is used. Neither of the above two mechanisms can be said to be sheet-forming mechanisms as they are encountered in simpler processes also.

For forming onto a single-curvature tool, however, individual plies are required to slip over each other as shown in Figure 3. This is a mechanism known as *interply slip*, or *interply shear*. The radii of curvature of different plies through the thickness are different after forming and, as the fibres cannot stretch to accommodate this difference, the plies slip over each other like a pack of cards when bent. The interply shearing behaviour of thermoplastic composites has been studied elsewhere [2] and is reported to be a lubricated-frictional type behaviour.

Sheet-forming of double-curvature shapes is where the real benefits of the process are to be seen, however, though the more complex the shape is, the more likely it is that process failures such as ply splitting, buckling and wrinkling may be encountered. The fact that the continuous fibre reinforcement cannot stretch (they are effectively inextensible) means that adjacent fibres must shear past each other in order to accommodate the double-curvature surface of the tool, as shown in Figures 2 and 3. This shearing mechanism is known as *intraply shearing* in unidirectional materials, and as *trellis shearing* in fabric materials.

Previous characterisation of the intraply shearing mechanism in unidirectional fibre melts has used surface shear loading between plates, particularly the torsional rheometer [3], but this method is thought to be unsatisfactory and appears to be completely impractical for fabric

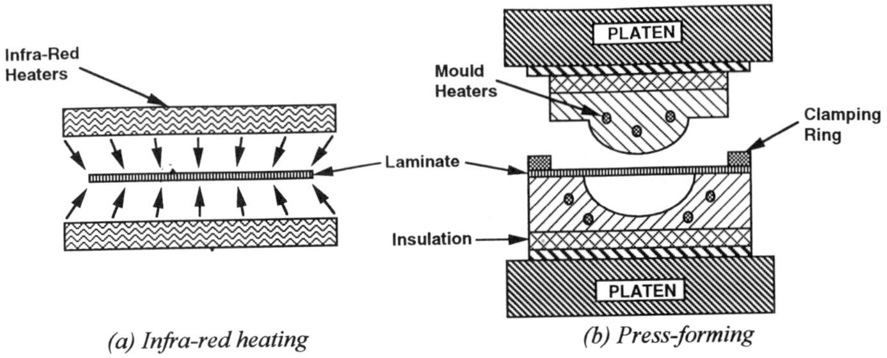

Figure 1. Pre-heating and press-forming of thermoplastic composites

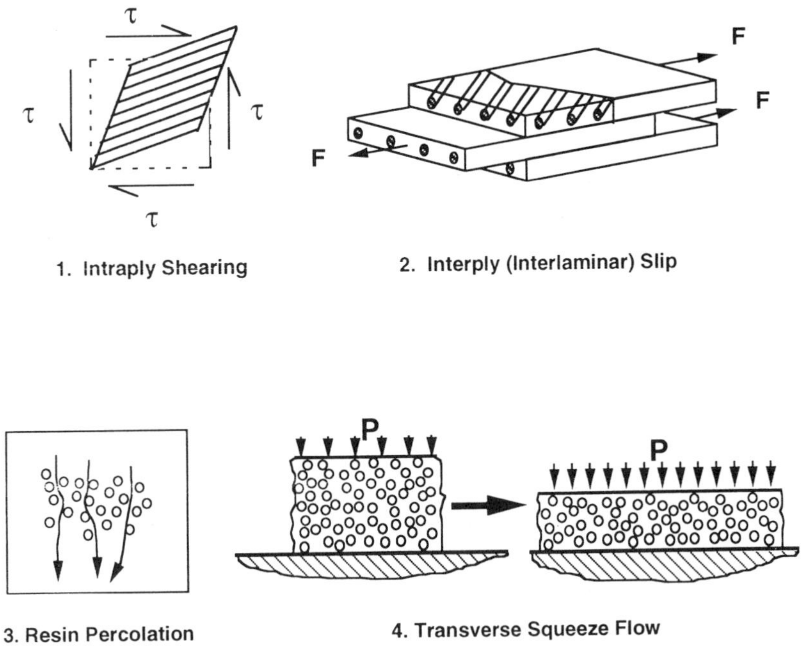

Figure 2. Forming mechanisms for continuous-fibre reinforced materials

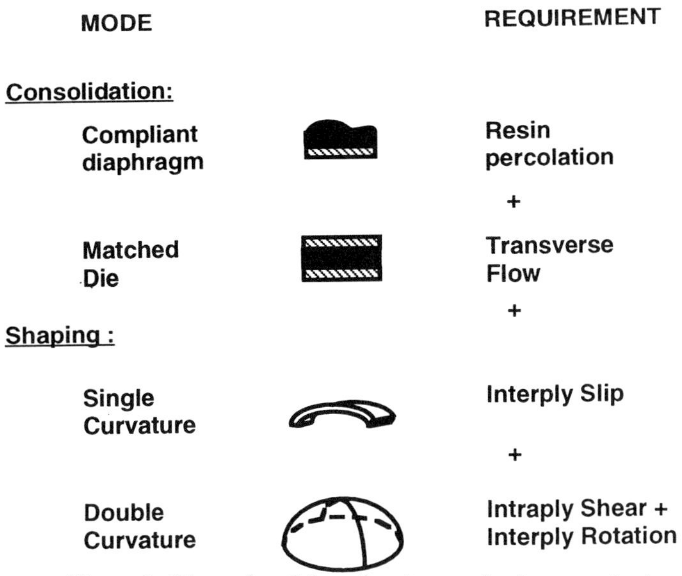

Figure 3. Hierarchy of sheet-forming mechanisms required

reinforced melts. Prior to the work reported in this paper, no method had been found to characterise the intraply shear viscosities of fabric reinforced materials.

The first step in modelling the sheet-forming process is to assemble the various deformation mechanisms into a properly invariant continuum model of the composite fluid/solid at forming temperatures. The measured viscosities for each distinct deformation mode are then input as material variables into the continuum model, which can be used to predict the deformation of the composite sheet under any loading history.

The most commonly used model for unidirectional or fabric reinforced melts is known as the Ideal Fibre Reinforced Material (IFRM), which assumes the melt to be a fluid or solid which is inextensible in one or two fibre directions, and is also incompressible [4]. The material behaviour itself may be that of a linear elastic solid, a Newtonian or generalised Newtonian fluid, or a viscoelastic fluid or solid [5].

The numerical simulation codes which have been developed to solve the general problem of the deformation over time of a composite IFRM sheet onto a tool surface have taken two finite element approaches. The first is the implicit formulation [6,7] which was shown to be useful in simple predictions of deformation and buckling sites in the problem of forming of a small female indentation in the centre of a large composite sheet. The second approach is the explicit finite element fomulation, which was employed by a consortium of industry and universities under a European Union Brite-EuRam research contract [8]. The simulation code, known as PAM-STAMP is capable of modelling the pre-heating, forming and cooling of unidirectional fibre and fabric-reinforced thermoplastic aerospace and industrial components [9,10].

Figure 4 illustrates two PAM-STAMP simulation outputs for the simple case of forming of an initially square fabric preform over a male hemisphere, showing the deformed sheet shape at two steps in the process, with superimposed intraply shear-strain contours. The necessary inputs to this code involve the tool and sheet geometry, as well as process parameters, and the interply and intraply shear viscosities.

This paper describes a completely new test which has been developed to directly measure the intraply shear viscosity of both unidirectional and fabric reinforced composite melts. The test involves the large-deformation shearing at high-temperature of the composite in a picture-frame apparatus. The experimental apparatus will be introduced and the response of various anisotropic fluid models to the deformation predicted. Finally test results will be presented for four thermoplastic composite materials and the shear behaviour interpreted accordingly.

EXPERIMENTAL

The picture-frame experiment involves subjecting a flat pre-consolidated composite specimen to a uniform in-plane deformation using a specially designed four bar linkage and a hydraulic INSTRON Universal Testing Machine. A schematic representation of its operation is shown in Figure 5. The apparatus is enclosed within an environmental chamber which can be heated to a maximum of 450° C to allow testing at the composite processing temperatures. The chamber contains three windows so that material buckling may be visually observed during the test. The joints of the apparatus are treated with a high-temperature lubricant in order to minimise its frictional response. Experiments are carried out at pre-set constant actuator displacement rates, and force versus displacement data is obtained for each test using a personal computer based data acquisition system. The full details of the apparatus design are contained in [11].

All thermoplastic composite materials tested are preconsolidated at a pressure of 0.4 MPa and at a recommended process temperature. Each specimen is cut to the shape shown in Figure 6 and has holes drilled in it at the locations shown. The length L of each side of the square piece is 200 mm, while the thickness t will be different for different test materials and lay-ups. The holes are then fitted with small metal bushings and the specimen attached to the apparatus by means of appropriately located pins. The bushings prevent any tear-out of the pins during loading and the material must be completely free to rotate around the pins in order that the required homogenous deformation is produced. Care is taken with fabric materials to note the direction of the warp and weft fibres with respect to the test axes, which can be an important test parameter.

MODELLING AND TEST INTERPRETATION

The models employed in this work follow the definition of the Ideal Fibre Reinforced Material [4] in which the unidirectional melt is assumed to be incompressible and reinforced

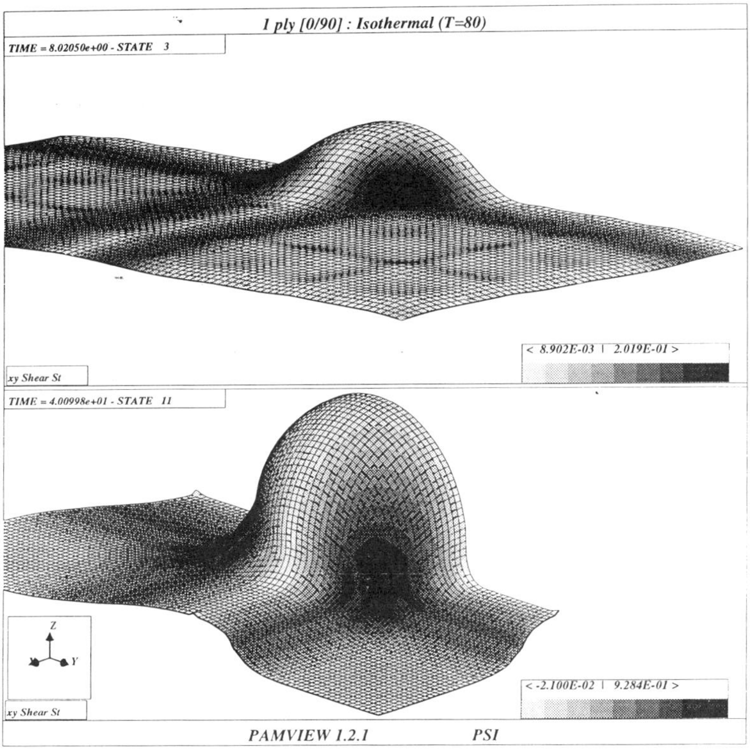

Figure 4. PAM-STAMP simulation results showing a [0/90] deformed fabric sheet with shear-strain contours after 8 seconds (top) and after 40 seconds (bottom) of forming over a male hemisphere.

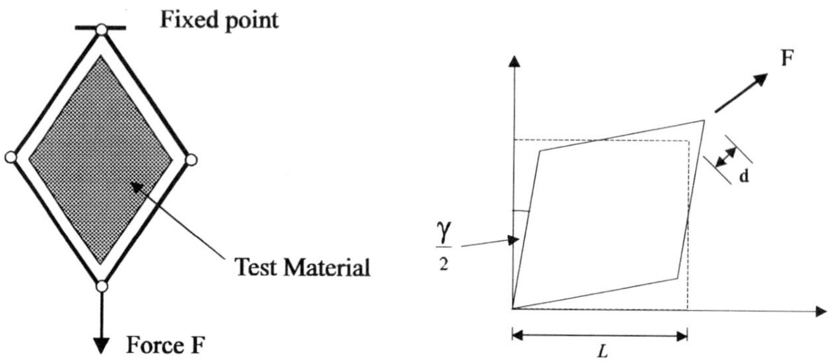

Figure 5. Schematic of picture-frame shear apparatus and nomenclature used in tets interpretation

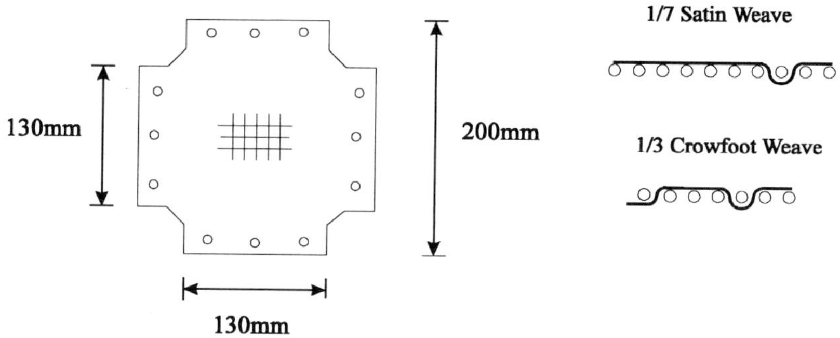

Figure 6. Picture-frame specimen shape, dimensions and hole locations Also, two types of weave architectures used in VESTOPREG laminates

with one family of fibres, represented by an inextensible direction. In a similar fashion, the fabric materials are considered to be incompressible and reinforced by two initially orthogonal fibre directions, each of which is represented by an inextensible direction.

The formulation of properly invariant constitutive relations for these types of materials was originally presented for unidirectional and fabric reinforced elastic materials [4] and polymers [5]. Recent work by the authors in the context of the picture-frame test [12,13,14] and by researchers who have investigated +/-45° fabric tensile tests [15,16] have suggested some simplifications. These theories are beyond the scope of the present paper, the purpose of which is to discuss the design and results gained from the new test. The important material models and relevant constants may, however, be summarised as follows:

Unidirectional Polymer Melts

- Anisotropic Newtonian Fluid: Longitudinal shear viscosity, η_L (indep. of shear rate)
  Transverse shear viscosity, η_T (indep. of shear rate)

- Anisotropic Power-Law Fluid: Longitudinal shear viscosity, η_L (function of shear rate)
  Transverse shear viscosity, η_T (function of shear rate)

$$\eta_L = m_L\left[\sqrt{2d_{ij}d^{ij}}\right]^{n-1} \qquad \eta_T = m_T\left[\sqrt{2d_{ij}d^{ij}}\right]^{n-1} \qquad (1)$$

where m_L and m_T are the zero shear-rate viscosities and n is the power-law index.

Fabric Reinforced Polymer Melts

- Anisotropic Newtonian Fluid: Trellising shear viscosity, η_{12} (indep. of shear rate)
  Through-thickness viscosity, η_{33} (indep. of shear rate)

- Anisotropic Power Law Fluid: Trellising shear viscosity, η_{12} (function of shear rate)
  Through-thickness viscosity, η_{33} (function of shear rate)
  similar to relationships above.

- Isotropic Viscoelastic Solid: Trellising shear viscosity, η (function of shear rate)
  Trellising shear modulus, β (indep. of shear strain and rate)

The relationship between applied force *(F)* and measured displacement *(d)* for the simplest model above, the unidirectional anisotropic Newtonian fluid, for the picture-frame test (see Figure 5 for nomenclature) has been calculated [13,21] as follows:

$$F\dot{d} = L^2 t\left(\dot{\gamma}^2\left(\eta_L\left(\frac{1+\sin^2\gamma}{\cos^2\gamma}\right)+2\eta_T\left(\frac{\sin^2\gamma}{\cos^2\gamma}\right)\right)\right) \qquad (2)$$

where γ is the change in angle of the specimen, and a superimposed dot indicates the time rate of change of a quantity. Equation (2) can be reorganised to give

$$\left(\frac{F\dot{d}}{\dot{\gamma}^2 L^2 t}\frac{\cos^2\gamma}{\left(1+\sin^2\gamma\right)}\right)=\left(\eta_L+\eta_T\left(\frac{2\sin^2\gamma}{1+\sin^2\gamma}\right)\right) \quad (3)$$

which is of the form of a line of slope η_L and intercept η_T, when plotted on X and Y axes:

$$Y=\eta_L+X\eta_T$$

This permits a graphical interpretation of the picture frame experiment results since all force and kinematic quantities can be determined from measured data. If the material being tested in the picture-frame apparatus does indeed behave as an anisotropic Newtonian fluid, then data at all displacement rates will collapse onto the single line represented by Equation (3). This ideal line is depicted in Figure 7 below.

Similar graphical interpretations of all the other models above can be found, as will be shown in the results section which follows. More details of these interpretations and their derivations can be found in refs [12,13,14 & 21].

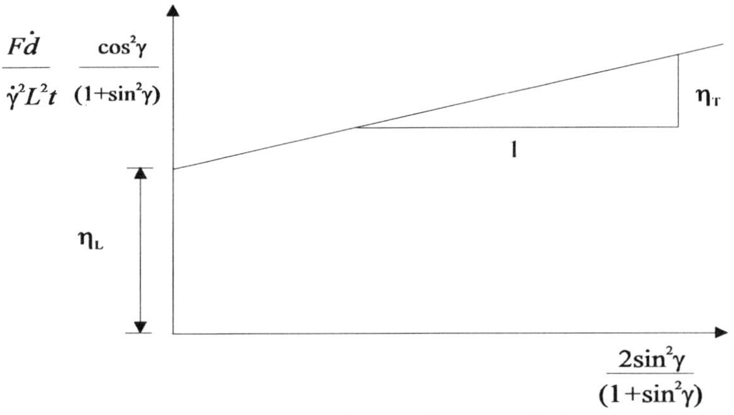

Figure 7. Interpretation of picture-frame results in terms of an anisotropic Newtonian fluid model.

RESULTS AND DISCUSSION

Four thermoplastic composite materials were tested in the picture-frame apparatus at various displacement rates and temperatures. The first is APC-2 [17] from Fiberite, a unidirectional AS-4 carbon-fibre reinforced PEEK material, usually processed at temperatures between

360° and 390°C. The second and third materials are Vestopreg [18] materials from Hüls, both glass-fabric reinforced PA-12s. Vestopreg G101 has a 1/7 satin weave, as shown in Figure 6, with 50% of the reinforcing fibres at each orientation. Vestopreg G201 has a 1/3 Crowfoot weave and a 90% to 10% bias in the number of fibres for each direction, and may thus be said to be a "unidirectional" type of fabric. The final material is CETEX [19] from Ten Cate, a carbon-fabric reinforced PEI with a 5H satin weave.

APC-2 Results

Cross-ply laminates of APC-2 were shear tested at actuator displacement rates between 10 mm/min and 500 mm/min at temperatures between 350° and 380°C. Tests were conducted when the material had been at temperature for fifteen minutes. The material temperature was measured using thermocouples attached to the bars of the picture-frame apparatus. The force versus displacement results for 360°C, which are representative of typical results are shown in Figure 8 below.

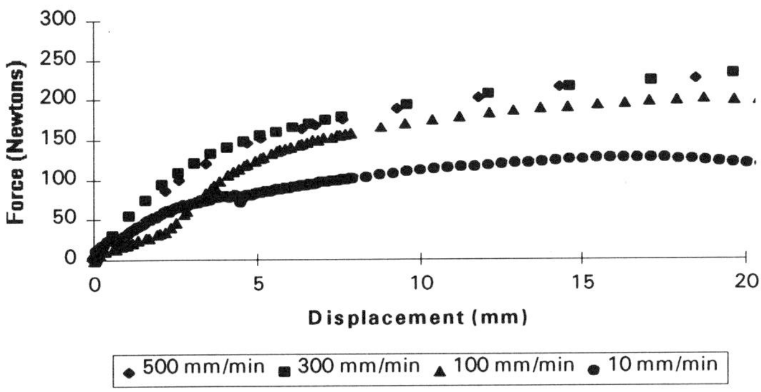

Figure 8. Force response of APC-2 at 360°C.

The graphical interpretation of the APC-2 experimental data at 360°C for the anisotropic viscous unidirectional material is shown in Figure 9. It can be seen that, apart from an initial start-up region, the power-law model gives a good fit for the response of the material. Beacuse of practical problems in loading the specimen, it was necessary to test cross-ply laminates instead of unidirectional plies of APC-2. Experiments have also been carried out on APC-2 laminates tested between diaphragm sheets of Upilex film, which lead to the conclusion that the response of APC-2 cross-plies is somewhat higher than unidirectional plies [13,21]. The difficulty in subtracting the composite response from the response of the films, however, renders this method of testing impractical. The direct approach with cross-plies is therefore the only practical way to test unidirectional materials in this apparatus.

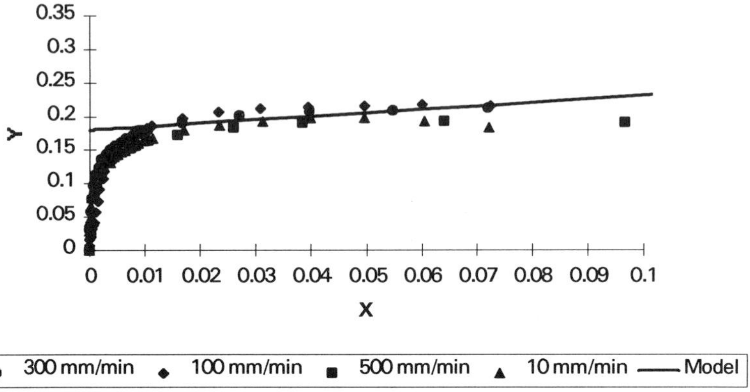

$$X=\left(\frac{2\sin^2\gamma}{1+\sin^2\gamma}\right) \quad ; \quad Y=\left(\frac{F\dot{d}}{\dot{\gamma}^2L^2t}\frac{\cos^2\gamma}{\left(1+\sin^2\gamma\right)}\right)\left(\dot{\gamma}\sqrt{\frac{1+3\sin^2\gamma}{\cos^2\gamma}}\right)^{1-n}$$

Figure 9. Interpretation of APC-2 response at 360°C.
Anisotropic Power-Law UD Model : $n = 0.15$, $m_L = 0.18$ MPa.s^{n-1}, $m_T = 0.5$ MPa.s^{n-1}

Vestopreg G-101 Results

Hüls Vestopreg G101 has been tested at four different displacement rates and at several temperatures between 180° and 240°C. It has been discovered [20] that the direction of shearing of this material has an important effect on its response in the test. Two specimen orientations are possible in the picture-frame apparatus. The warp fibre directions may make an angle of either +45° or -45° with the axis of loading, as viewed from the front of the picture-frame apparatus. If, for example, a -45° test is taken to be shearing of a specimen in the positive sense, then a +45° test will be a negative shearing deformation.

Tests at both possible specimen orientations at 240°C are shown in Figure 10 below. A minimum of three tests were conducted for each set of conditions and averaged force versus displacement results for each displacement rate at each specimen orientation presented. The materials response is observed to be rate dependent as expected. Even more pronounced, however, is the effect of specimen orientation. For a positive shear deformation the loads recorded are roughly twice those for a specimen which experiences a negative shear deformation under the same conditions. This dependence on specimen orientation has practical significance, since most complex curvature moulds will require a fabric to shear in both directions, at different locations, as shown in the PAM-STAMP simulations of Figure 4.

An isotropic viscoelastic model is found to characterise both the positive and negative shear of Vestopreg G-101 at 240°C using a graphical method similar to that illustrated above. The interpretation for negative shear, together with appropriate axis quantities, is shown in Figure 11. The performance of this model is also found to describe the behaviour of the material in stress relaxation experiments [13,14] The best models for each material and parameters used are summarised in Table 1(c).

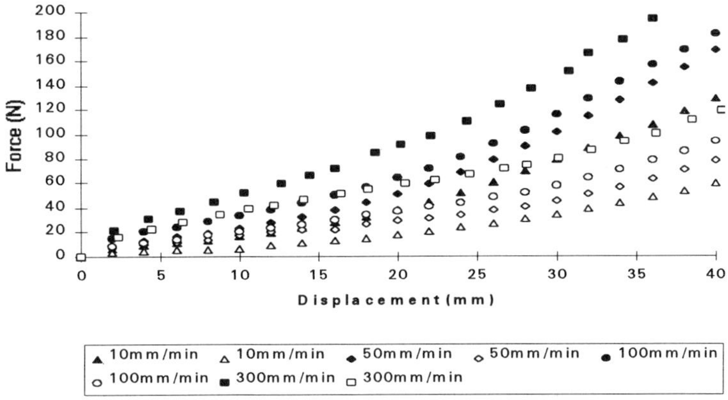

Figure 10. Force versus displacement response for G101 in positive and negative shear at 240 C over a range of displacement rates. Black symbols represent positive shear, white symbols represent negative shear.

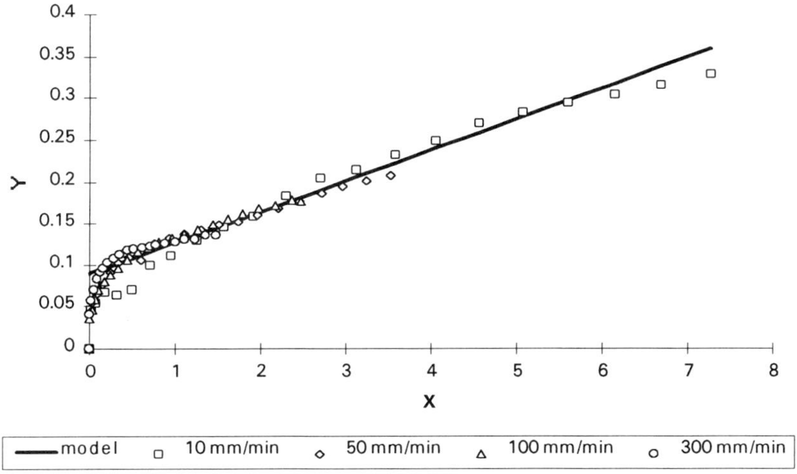

$$X = \dot{\gamma}\left(\frac{\sin^2\gamma}{\cos\gamma} + \frac{2\sin^2\gamma}{\cos^3\gamma}\right)\left(\dot{\gamma}\sqrt{\frac{1+3\sin^2\gamma}{\cos^2\gamma}}\right)^{-(n+1)} \qquad Y = \left(\frac{F\dot{d}}{L^2 t}\right)\left(\dot{\gamma}\sqrt{\frac{1+3\sin^2\gamma}{\cos^2\gamma}}\right)^{-(n+1)}$$

Figure 11. Interpretation of negative shear of G101 using an isotropic viscoelastic fabric model. Model parameters : $\eta = 0.45$, $m = 0.09$ MPa.s^n, $\beta = 0.037$ MPa.

Vestopreg G-201 Results

Experiments have been conducted with the G201 material at both possible specimen orientations and at various displacement rates, but no directional effect is observed. The response is well described by a power-law viscous fabric model and no significant elastic effects have been seen. This is believed to be a consequence of the symmetric nature of the reinforcement architecture and due to the practically unidirectional nature of the material. The viscosity constants for the material are listed in Table 1(b).

CETEX Results

Picture-frame shearing experiments for CETEX have been conducted at 100 mm/min for the two possible material orientations in the apparatus at 320°C with the conclusion that the effect of the direction of shear is small [22]. Tests conducted at three different displacement rates, 10 mm/min, 100 mm/min, 300 mm/min, are shown in Figure 12. Once again, the materials response is seen to be rate dependent, but the shape of the response curve is very different from those observed with the glass fabric materials, with the force response reaching a peak at approximately 10 mm of displacement and dropping off for the remainder of the test.

During the course of the tests conducted using CETEX, each specimen deconsolidated significantly. Since there is no consolidation pressure during press forming there is no reason to provide it in the experiment. It is felt that the deconsolidated state of the CETEX specimens could be a contributing factor to the unusual response curves.

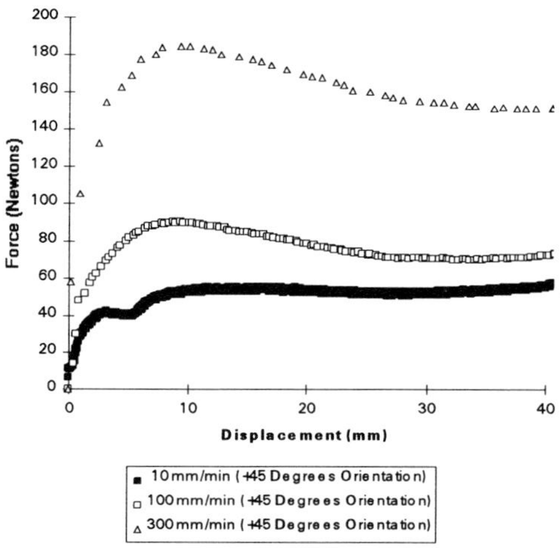

Figure 12. Force versus displacement response for CETEX over a range of displacement rates at 320°C.

The CETEX material is found to be modelled best using the anisotropic power-law fabric model. Though shear-thinning viscous behaviour seems evident, a single power-law rule does not correctly model the shear rate dependency. Two approximate power-law models are presented for CETEX in Table 1(b) below, giving reasonable representation of the viscosity levels of the molten sheet.

Summary of Results

The following tables list the parameters of the models that best describe the behaviour of the four thermoplastic composite materials investigated using the picture-frame test.

	n	m_L (MPa.s^n)	m_T (MPa.s^n)
APC-2	0.15	0.18	0.5

Table 1(a). Anisotropic UD viscous model parameters for APC-2 at 360°C

	n	m_{12}(MPa.s^n)	m_{33}(MPa.s^n)
G201	0.08	0.0019	0.115
CETEX (i)	0.7	1.8	0.45
CETEX (ii)	0.23	0.28	0.01

Table 1(b)Anisotropic viscous fabric model parameters for G-201 and CETEX.
CETEX (i) Interpretation of results for 100 mm/min and 300 mm/min.
CETEX (ii) Interpretation of results for 100 mm/min and 10 mm/min.

	n	m (MPa.s^n)	β (MPa)
G101 (+ve shear)	0.4	0.1	0.11
G101 (-ve shear)	0.45	0.09	0.037

Table 1(c). Isotropic viscoelastic fabric model parameters for Hüls Vestopreg G101 in positive and negative shear.

CONCLUSIONS AND ACKNOWLEDGEMENTS

A new rheological test has been presented for measurement of the intraply shear viscosities of molten composite materials. The test is carried out within an environmental chamber and comprises a picture-frame apparatus to which the material is pinned. Testing is carried out at controlled displacement rates using a test machine which allows direct measurement of the load applied over the time of the test. The test results are interpreted using various models of unidirectional and fabric reinforced polymer behaviour, which assume the fibre directions in the materials to be inextensible.

The intraply shearing behaviour of four thermoplastic composite materials are characterised using the new picture-frame test. APC-2 and Vestopreg G-201 are shown to exhibit viscous shear-thinning behaviour, whereas Vestopreg G-101 is best characterised using a viscoelastic model but its behaviour is very different in different directions of shear. CETEX is shown to have a non-linear shear-thinning behaviour. The models developed can all be used as input to an explicit finite element simulation code which has been developed to model the sheet-forming of composite materials.

The authors would like to acknowledge the support of the European Union, who sponsored this work through Brite-EuRam contract No. BE-5092. The cooperation and collaboration of our industrial and academic partners in this research project is also much appreciated.

REFERENCES

1. C.M. Ó Brádaigh in S.G. Advani, ed., Flow and Rheology in Polymer Composites Manufacturing, Elsevier Science Publishers, Amsterdam, pp. 517. (1994).
2. A.J. Murtagh and P.J. Mallon, Composites Manufacturing, 6(3-4), 169 (1995).
3. D.J. Groves and D.M. Stocks, Composites Manufacturing, 2(3/4), 179 (1991).
4. A.J.M. Spencer, Deformation of Fibre-Reinforced Materials, Clarendon Press, Oxford, 1972.
5. T.G. Rogers, Composites, 20(1), 21 (1989).
6. G.B. McGuinness and C.M. Ó Brádaigh, Composites Manufacturing, 6(3-4), 269 (1995).
7. C.M. Ó Brádaigh, G.B. McGuinness and Pipes, R.B., Composites Manufacturing, 4,(2), 67 (1993)
8. Brite-EuRam Contract No. BE-5092, European Commission, 1993-1996.
9. P. De Luca et al., Proceedings of ICAC '95, Nottingham, pp. 183 (1995).
10. A.K. Pickett et al., Composites Manufacturing, 6(3-4), 237 (1995).
11. T.A. Nestor, "Experimental Investigation of the Intraply Shearing Mechanism in Thermoplastic Composites Sheet-Forming", M.Eng.Sc. Thesis, Department of Mechanical Engineering, University College Galway, Ireland (1995).
12. G.B. McGuinness et al., Proceedings of the ASME Materials Division, MD-Vol. 69-2, IMECE, pp. 1107 (1995).
13. G.B. McGuinness, "Characterisation and Modelling of Plane Flows in Composite Materials Sheet-Forming", PhD Dissertation, National University of Ireland, 1996.
14. G.B. McGuinness and C. M. Ó Brádaigh, J. Non-Newt. Fluid Mechs, submitted, 1996.
15. A.F. Johnson, Composites Manufacturing, 6(3-4), 153 (1995).
16. A.F. Johnson and E. Costalas, Proceedings of ICAC '95, Nottingham, pp. 341 (1995).
17. ICI Fiberite Data Sheet 5, Fabricating with APC-2 Composites, Fiberite Corporation (1986).
18. Vestopreg Data Sheet, Hüls Aktiengesellschaft AG, Germany, (1994).
19. CETEX Design Guide CD0282/8463 Carbon Fabric Reinforced Polyetherimide, Ten Cate Advanced Composites, The Netherlands.
20. R.A. Canavan, G.B. McGuinness and C.M. Ó Brádaigh, ICAC '95, Nottingham, (1995).
21. G.B. McGuinness and C. M. Ó Brádaigh, Flow Processes in Composite Materials '96, Aberystwyth, (1996).
22. G.B. McGuinness and C. M. Ó Brádaigh, Polymer Composites, submitted, (1996).

28th International SAMPE Technical Conference
November 4-7, 1996

MECHANICAL PROPERTIES OF PVC STRUCTURAL FOAM

G.-Q. Li
Institute of Lightweight Structures & Ropeways
Swiss Federal Institute of Technology Zürich (ETH)
CH-8092 Zürich, Switzerland

ABSTRACT

A simple analytic approach in microscopic scale for determining mechanical properties of PVC structural foam materials used in modern sandwich structures is presented. Varying from conventional institute in which Poisson's ratio is described as a function of apparent Young's modulus and shear modulus for solid isotropic materials, a relation among moduli and Poisson's ratio of structural foam materials, which is in proportion to the volume content of the cell wall inclusions in the total volume of the foam, is derived by treating the foam as a dispersed isotropic composite body. The results reveal that the properties of PVC structural foams strongly depend on their initial cell geometry. To get valid data of the Poisson's ratio which may significantly change the local behavior of the sandwich structures in the case of local load problems, an uniaxial compression test is conducted for certain type of Airex's PVC foams. It shows that the Poisson's ratio of the selected foam varies from 0.3 to zero during the deformation except the region of plasticity where the Poisson's ratio becomes 0.45 approximately.

KEY WORDS: PVC forms, Closed cell, Mechanical properties

1. INTRODUCTION

Polyvinyl chloride (PVC) structural foams have received increasingly applications in today's engineering practices. Their dominant use is in sandwich constructions where the PVC foams are selected as core materials to save structural weight. Differing from other cellular materials, PVC structural foams possess some unique features of their mechanical properties. In classic contitunity mechanics, based on the theory of symmetric stress tensor deduced from some primary hypotheses (theory of elasticity), the isotropic constituent is described phsically only with two independent elastic constants. With this theory a negative ratio of the lateral to the axial strain, known as Poisson's ratio μ, is derived from the relation

$$G = \frac{E}{2(1+\mu)} \tag{1}$$

where E and G represents the Young's modulus and shear modulus, respectively. In many cases, however, the structural foams do not meet this condition. Table 1 shows some data of typically elastic linear PVC foams obtained through laboratory tests [1,2] with exception that of data in last row where the values of Poisson's ratio derived directly from eqn (1) do not agree with reality.

Table 1. Mechanical properties of Airex PVC foams

	R62.80	R63.50	R63.80	R90.300	R90.400
E	55	29	64	280	350
G	18	12	21	170	240
μ	0.37-0.44	0.1	0.1-0.3		
μ †	0.527	0.208	0.523	-0.1765	-0.271

† data are issued from the eqn (1)

There are several explanations for this confused feature: effect of cell shape contained in the foam, distinct properties of the cell wall material from that of the foam, and inadequate hypothesis in elasticity theory to Poisson's ratio which seems to be an independent property for the structural foams even they behave, in macroscopically, as isotropic materials. In the literatures, there are several theories which use more than three elastic constants to run into special problems, for example, the couple-stress theory [3-5]. However, it is seen that the theories are very complicated for those multiphasical materials [6].

As one of the salient mechanical properties, Poisson's ratio may be important with remarkable effect on certain performance of the sandwich structures, especially in the case of a slab of sandwich rested on a rigid base subject to locally lateral loads, for instance [7]. For this reason, there are several analytical approachs [6,8,9], based on the elasticity theory, on the mechanics of foams. The recent development is reviewed in the book [6] by Gibson and Ashby who used a simple procedure with scaling ideals to elucidate the mechanisms of polymeric foams analytically and results are tested by experimentation. The present work extents this approach [6] to investigate Airex PVC structural foams. To establish a relation among Young's modulus, shear modulus and Poisson's ratio, which dose not exist yet in the litertures, a simplified closed-cell model is used. Unlike the eqn (1) for solid isotropic materials, the new relation includes the effect of volume fraction of cell wall to total volume of the foam while Young's modulus and shear modulus are also derived in terms of the volume fraction and Poisson's ratio. Because the fact that test results of Poisson's ratio from early works [1,2,6] are very scatter, in the speical cases it covers a range of 0.1-0.45 for a selected PVC foam, an uniaxial compression test is carried out to help to understand this feature.

2. BASIC FORMULATION

Figure 1 shows scanning photos of two PVC structural foams which are identified as isotropic materials constructed through uniform and regular arrays of cells in all directions.

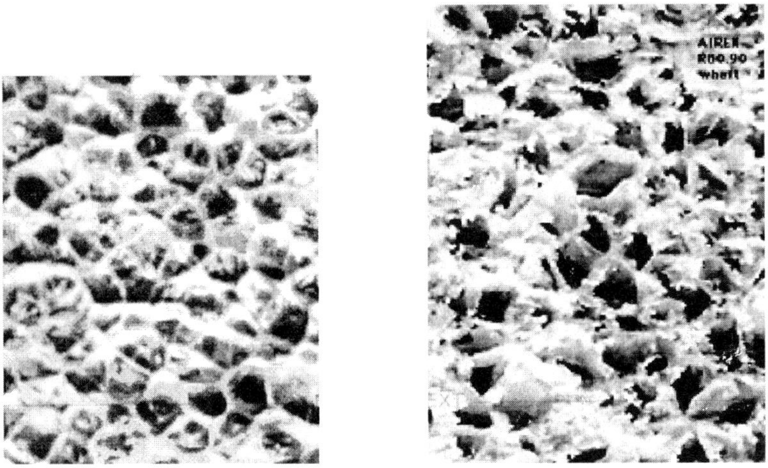

(a) R 60.50 (b) R 80.90

Figure 1. Airex PVC structural foams with closed cell

Based on this observation, a simplified closed-cell model which consists of cell surfaces and cell edges in a regular hexhedron form (Fig. 2) is employed to study PVC foams in micromechanical level, according to [6]. It is known that the mechanics of the foams are highly related to the cell configuration and material properties of the cell wall that gives two important contributions to the moduli: bending stiffness of the cell edge and membrane stiffness of the cell face. In the present work, the effect of gas pressure (p_o) which is embraced in the cells during foaming is ignored, the foams are supposed having linear elastic behaviour when stress level is low and their shear non-linearity is also neglected.

2.1 Young's modulus of the closed cell foam

Consider a closed cell of cubic form in Fig. 2 where t_f and t_e are thickness of the thin faces and the cell edges, respectively,

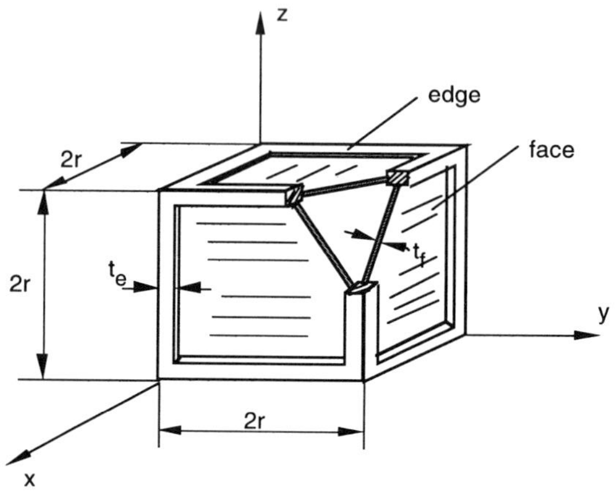

Figure 2. Schematic showing the geometry of a closed cell

the moduli of the foam are calculated from the approach [6] modified in a little different way, as follows. A force P causes the cell to deflect by w (Fig. 3), potential energy U_c of the cell is the sum of the contributions of the cell edges and the cell faces without regard to influence of shear.

$$U_c = U_e + U_f \tag{2}$$

and

$$U_c = U_{ce} + U_{cf} \tag{3}$$

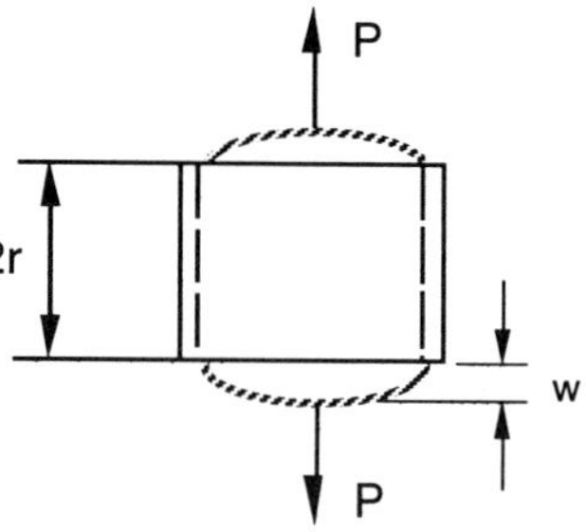

Figure 3. Schematic showing the deformation of a closed cell due to the load P

In the eqn (3) U_{ce} equals to work $\frac{1}{2}Pw$, while corresponding strain energy U_e due to the edge bending is proportional to $\frac{E_e I_e w^2}{(2r)^3}$ where E_e is Young's modulus of cell edge and I_e is inertia moment of cell edge with a value of t_e^4, approximately. Equating U_{ce} with U_e and using $E_c = \frac{P}{2rw}$, we obtain a proportional relation for Young's modulus due to the cell edge bending , thus

$$\frac{E_c}{E_e} = \phi^2 f_w^2 \tag{4}$$

where ϕ is a fraction of solid contained in the cell edges (for the cell faces it is $1-\phi$), f_w is a volume fraction of the cell wall to the cell. Note that in the present work the subscript "c" refers to the closed cell, "w" refers to the cell wall, "e" refers to the cell edge and "f" refers to the cell face as well.

To derive U_f arised by membrane stretch in the cell faces we consider a face sheet under in-plane uniaxial loading condition (Fig. 4) with regard to the effect of Poisson's ratio of the cell face μ_f and the cell μ_c.

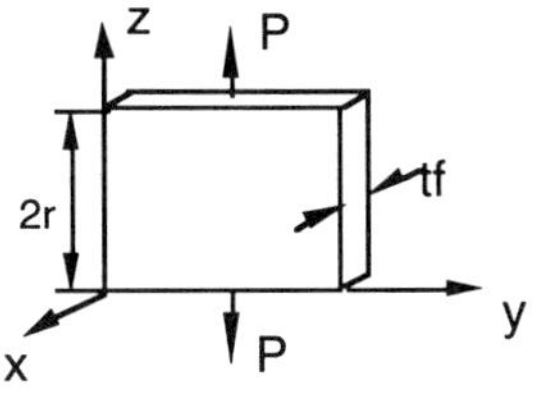

Figure 4. Cell face under in-plane uniaxial loading condition

Note that P_e in Fig. 2.4 is the stretching force in the cell face caused by the cell edge. From Hook's law, normal stress components of the selected face can be written as

$$\begin{aligned} \sigma_{fz} &= \frac{E_f}{1-\mu_f^2}(\varepsilon_z + \mu_f \varepsilon_y) = \frac{1-\mu_c\mu_f}{1-\mu_f^2} E_f \varepsilon_z \\ \sigma_{fy} &= \frac{E_f}{1-\mu_f^2}(\varepsilon_y + \mu_f \varepsilon_z) = \frac{\mu_f - \mu_c}{1-\mu_f^2} E_f \varepsilon_z \end{aligned} \tag{5}$$

where ε_z and ε_y are in-plane strains of the closed cell in z and y direction, respectively. Provide that $\varepsilon_y = -\mu_c \varepsilon_z$. Thus, U_f takes the form as

$$U_f = \frac{1}{2}\int_f 4\left(\frac{\mu_c - \mu_f}{1-\mu_f^2}\mu_c + \frac{1-\mu_c\mu_f}{1-\mu_f^2}\right)E_f\varepsilon_z^2 dxdydz \qquad (6)$$

Substituting $\frac{1}{2}\int_f 4E_c\varepsilon_z^2 dxdydz$ for left-side of the eqn (6) which is correspondent to U_{cf} in eqn (3), gives a proportional relation for Young's modulus due to the cell face strecthing by

$$\frac{E_c}{E_f} = 1 + \frac{(\mu_c - \mu_f)^2}{1-\mu_f^2} \qquad (7)$$

Provide that $E_w = E_e = E_f$, $G_w = G_e = G_f$ and $\mu_w = \mu_e = \mu_f$, an approximately expression for Young's modulus of the closed cell can be found in terms of the volume fraction and Poisson's ratio by combining eqns (4) and (7), it gives

$$\frac{E_c}{E_w} = \phi^2 f_w^2 + (1-\phi)\left(1 + \frac{(\mu_c - \mu_w)^2}{1-\mu_w^2}\right) \qquad (8)$$

where a proportional factor (1-ϕ) is imposed to consider the condition that the closed cell becomes an isotropic solid material when ϕ goes to its limits of 0 or 1.

2.2 Shear modulus of the closed cell foam

Shear modulus G_c, of the closed cell, like its Young's modulus, is contributed by the cell faces and the cell edges (Fig. 5).

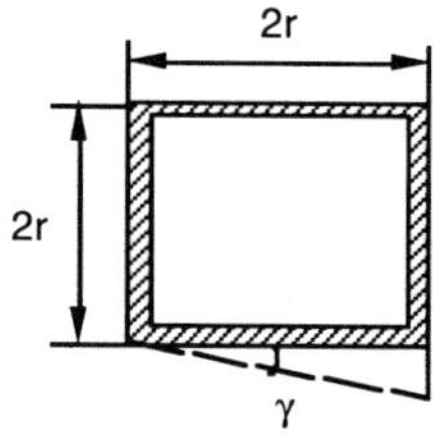

Figure 5 Shear deformation of a closed cell

The shear strain can be described by

$$\frac{\tau_c}{G_c} = \frac{c_w}{G_w}(\tau_e + \tau_f) \qquad (9)$$

Using $\tau_e = \frac{P}{t_e^2}\phi,\ \tau_f = \frac{P}{2rt_f}(1-\phi),\ and\ \tau_c = \frac{P}{(2r)^2}$, by following the procedure indicated in the reference [6], the relation (9) becomes

$$G_c = 2\pi\, f_w \frac{G_w}{c_w} \tag{10}$$

where G_w is shear modulus of the cell wall and c_w is an unknown which reflecting the geometry constants of proportionalities. To solve c_w, consider a speicial case by setting $f_w = 1$ (isotropic solid material), we obtain the unknown by $c_w = 2\pi$. For the cell wall the conventional relation (eqn 1) remains avaliable, giving

$$G_w = \frac{E_w}{2(1+\mu_w)} \tag{11}$$

Joining eqns (8), (10) and (11) together, a new expression which links Young's modulus, shear modulus, Poisson's ratio and the volume fraction of cell wall for a closed cell of cubic form is derived by

$$\frac{E_c}{G_c} = \frac{2(1+\mu_w)}{f_w}\left[\phi^2 f_w^2 + (1-\phi)(1+\frac{(\mu_c-\mu_w)^2}{1-\mu_w^2})\right] \tag{12}$$

Thus, the properties for PVC structural foams can be expressed by substituting E, G and μ for E_c, G_c and μ_c in eqns (8), (10) and (12), to a certain degree approximation, respectively (eqns 13-15). Provide that by setting $f_w = 1$ with the limits of $\phi = 0$ or 1, the relation (15) degenerates to the eqn (1) while $E = E_w$ (eqn 13) and $G = G_w$ (eqn 14) for solid isotropic materials, identically.

$$E = \left\{\phi^2 f_w^2 + (1-\phi)\left[1+\frac{(\mu-\mu_w)^2}{1-\mu_w^2}\right]\right\}E_w \tag{13}$$

$$G = f_w G_w \tag{14}$$

$$\frac{E}{G} = \frac{2(1+\mu_w)}{f_w}\left[\phi^2 f_w^2 + (1-\phi)(1+\frac{(\mu-\mu_w)^2}{1-\mu_w^2})\right] \tag{15}$$

From the above expressions it is observed that the mechanical behaviour of a foam material is deeply associated with its cell geometry and material properties of the cell wall. There might be a limitation for the potential application of such formulations (eqns 13-15) because the properties of the cell wall is rarely known in a precise level.

3. EXPERIMENTAL DETERMINATION OF POISSON'S RATIO OF AIREX PVC STRUCTURAL FOAM

To determine Poisson's ratio of the structural foams, a simple test is conducted by using an uniaxial compression model shown in Fig. 6 where a cylinderical specimen loaded by an uniaxial compression in z direction. The foam-made specimen is allowed to deform only in z direction so that during the test the variable to be measured is deformation w only.

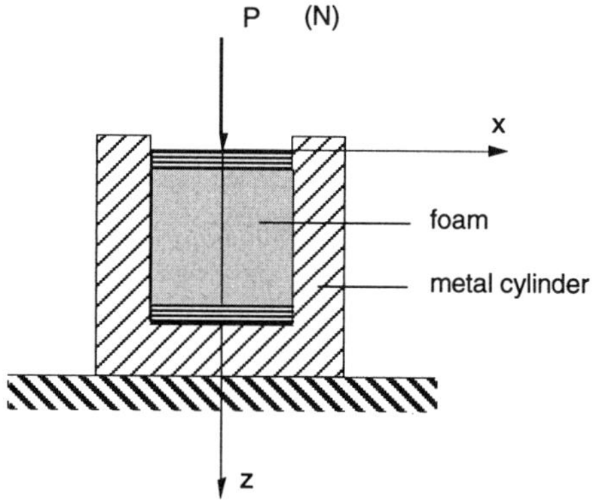

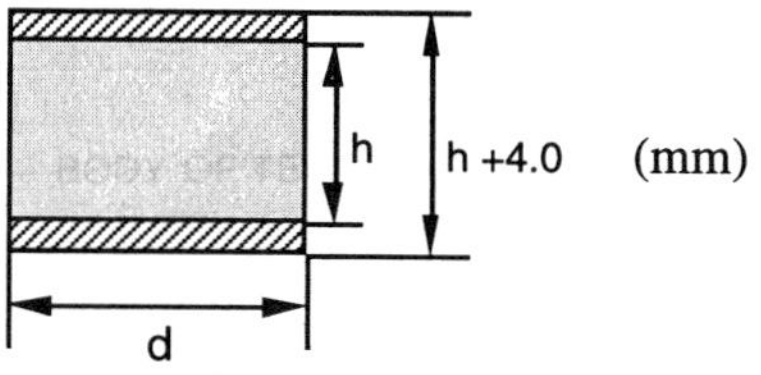

Figure 6. Test layout and the specimen

In this case the stress-strain relation is defined by

$$\sigma_z = \frac{E}{(1+\mu)(1-2\mu)}[(1-\mu)\varepsilon_z] \tag{16}$$

where $\varepsilon_y = \varepsilon_x = 0$. Based on this assumption Poisson's ratio can be written in terms of applied load P and the deformation w by

$$\mu = \frac{1-c \pm \left[(1-c)^2 - 8c(1-c)\right]^{\frac{1}{2}}}{4c} \tag{17}$$

where c is a parameter with $c = \frac{h}{4\pi d^2 E}\frac{P}{w}$, in which P and w are measured directely from the experiment. Thus, Poisson's ratio can be calculated out by introducing c into eqn (17). It is notice that during the test a friction force P_f between the specimen and the metal cylinder is also measured so that the final result is corrected by eliminating the effect of the friction force. To assess the validation of the data an rubber-like material is also tested under the same condition.

4. RESULTS AND DISCUSSION

Typical results for closed cell foams are shown in Fig. 7-12. In Fig. 7 a modulus ratio of G/E for a foam containing a volume fraction of 10% for the solid is curved via Poisson's ratio of the cell wall (solid cell edges and faces) by using eqn (15). The plot adds up to that there is almost no effect of Poisson's ratio of the cell wall on the properties of the foam when most of its volume is occupied by gas primarily. In this case, the modulus ratio is about 0.04 while the difference caused by changing Poisson's ration of the foam is very small. Apparently, the mechanics of the foam is mainly contributed by its shape of the cell rather than by its material. Figure 8 presents the data of a foam with its 50% volume occupied by the solid. The same trend mentioned above is observed.

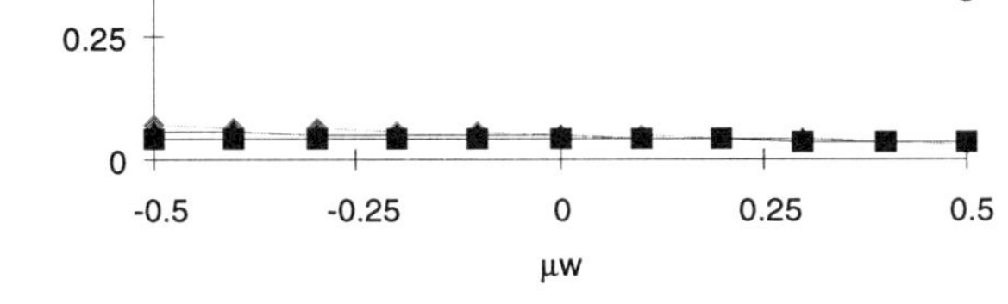

Figure 7. Modulus ratio of an undeformed foam with 10% of the volume occupied by the solid (eqn 15), plotted against Poisson's ratio of the cell wall μ_w

The results (Fig. 7-8) clearly indicate that since Poisson's ratio, μ_w, of the cell wall, which is difficulty to be measured, has no significant influence on the foam's properties when their volume fraction is small ($f_w \leq 0.5$), the eqn (15) could help to determine the mechanics of the structural foams analytically by estimating μ_w approximately. With increase of the volume fraction, f_w, the curves of the modulus ratio move up. They are perfectly incorporated in the line for the isotropic solid material when the foams become as homegeneous solid materials too (Fig. 9). Consider that the structural foam which contains a high modulus ratio through foaming process has an advantage than others in the applications of sandwich constructions, this parameter (G/E) can be used to control the final volume of the polymeric structural foams during the manufacture.

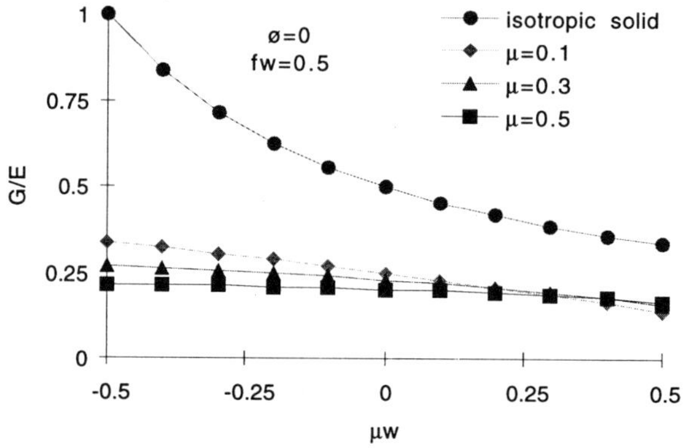

Figure 8. Modulus ratio of an undeformed foam with 50% of the volume occupied by the solid (eqn 15), plotted against Poisson's ratio of the cell wall μ_w

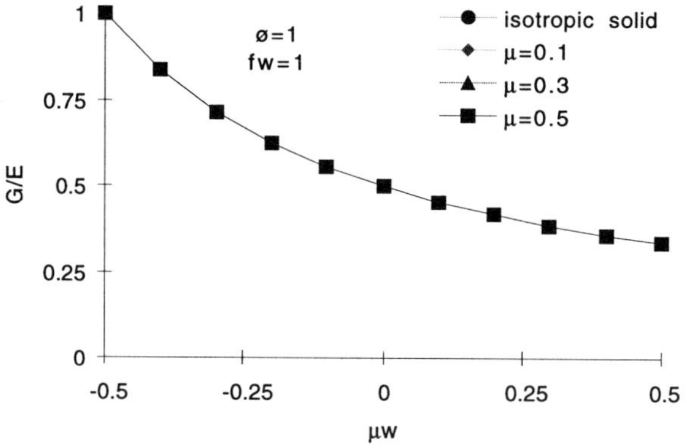

Figure 9. Modulus ratio of an undeformed isotropic solid material (eqn 15), plotted against Poisson's ratio of the cell wall μ_w

The result of mechanical properties of a selected Airex foam obtained via the experiment described in section 3 is given in Fig. 10-12. The nature of such materials may be better understood by considering a relation between poisson's ratio of the foams and their volume change over entier deformation region. To this purpose, the test result of Poisson's ratio of the foam Airex 62.80 is plotted vers strain in Fig. 10 where the date of a rubber is also presented. In the elastic region, Poisson's ratio radically varies from its initial value of about 0.3 down to a value of 0.1 while the foam is deformed from zero strain to its elastic limit. Beyond the elastic limit the material is yielded with collapse of the cell, in this region, Poisson's ratio of the foam becomes about 0.45, approximately. Corresponding stress-strain curve for the foam with fixed/free boundary conditions is displaced in Fig. 11-12 where a linear elastic deformation of the foam is observed when the stress level is low.

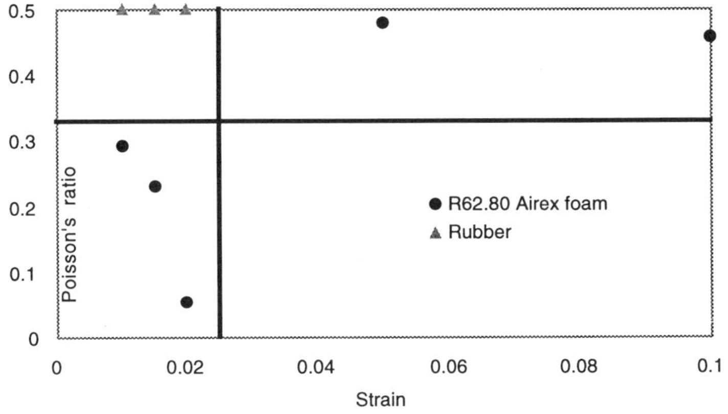

Figure 10. Poisson's ratio, plotted against the uniaxial strain in z direction over entire elasto-plastic region

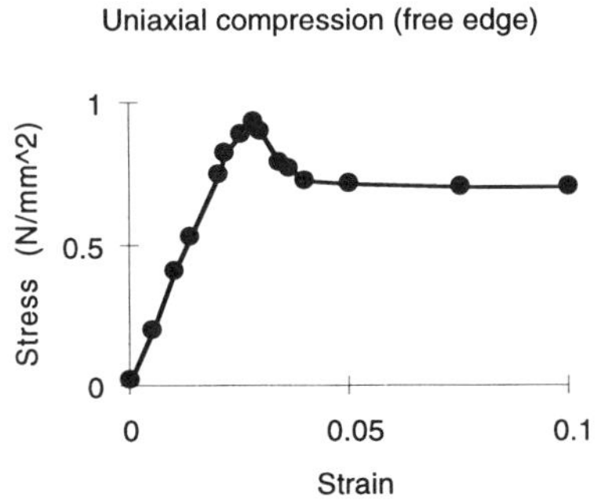

Figure 11. Stress-strain curve in z direction for the specimen with free edge

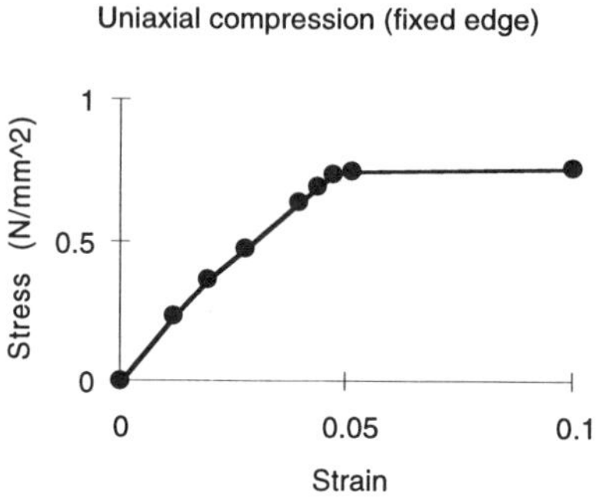

Figure 12. Stress-strain curve in z direction for the specimen with fixed edge

5. CONCLUSION

PVC structural foams were studied, a simple analytic approach was applied to investigate mechanical properties of them. A closed-cell model was used to derive a relation among Young's modulus, shear modulus and Poisson's ratic of structural foam materials, which dose not exist yet in the litertures, in terms of the volume fraction and Poisson's ratio of the cell wall included in the foams. An uniaxial compression test is carried out to determine Poisson's ratio of Airex's structural foams. The results show that the moduli of PVC structural foams strongly depend on their initial cell geometry while Poisson's ratio varies as a result of the deformation history.

6. REFERENCES

1. Manz, H., Versuche mit Sandwichproben zur Ermittlung der Materialdaten von Airex-Schaumstoff, Report No.TM208, ILS ETHZ, June 1989.
2. Tatasciore, P. and H.-R. Meyer-Piening, Measurement of Poisson's Ratio for Structural Foam Material by Speckle Interferometry, Report No.TM228E, ILS ETHZ, Sept. 1991
3. Tan, Z.-T., ACTA Mechanica Solida Sinica, 4, (1982).
4. Grioli, G., Ann. Mat. Pura. Appl., Ser IV 50, 389 (1960).
5. Mindkin, R., Arch. Rational. Mech. Anal.,16, 1 (1964).
6. Gibson, L.J. and M.F. Ashby, Cellular solids - Structure & Properties, Pergamon Press, Oxford, 1988, pp.120-200.
7. Li, G.-Q., "Concentrated Forces on Composite Laminates/Sandwich Plates with Large deformation", Ph.D. Thesis, ETH-Zuerich, (1996) in press.
8. Tsai, S.W., Composites Design - fourth edition, Think Composites, Dayton, 1987.
9. Nordstand, T., L.A. Carlsson and H.G. Allan, Composite Structure, 27, 317 (1994).

28th International SAMPE Technical Conference
November 4-7, 1996

APPLICATIONS FOR PMI FOAM-CORED SANDWICH MATERIAL - WHY PMI?

Wolfgang Pip
Röhm GmbH
64293 Darmstadt
Germany

ABSTRACT

Closed-cell polymethacrylimide rigid foam with high strength, high moduli of elasticity and shear is presently not exceed by any other foam plastic of the same gross density. Cost-effective production of sandwich parts by co-curing in autoclaves at temperatures of 175°C and pressures up to 0.7 MPa, or in closed molds where the cure pressure is generated by the PMI foam itself, are the main advantages of these products. Very high fatigue strength, good water absorption properties and excellent electrical values are other determining factors for the use of PMI foam plastics.

KEY WORDS: Composites, Foams, Manufacturing / Fabrication / Processing

1. INTRODUCTION

Highly load-bearing sandwich structures for lightweight parts made from fiber composites are nowadays state-of-the-art. This engineering technique has enabled the aeronautic and aerospace industry not only to develop weight-optimized structures, but simultaneously to obtain surfaces with excellent aerodynamic properties such as are virtually impossible to achieve using conventional metal constructions. I. Terceno from Sikorsky Aircraft (1) thus established the following design criteria for surfaces as early as 1985 (Fig. 1):

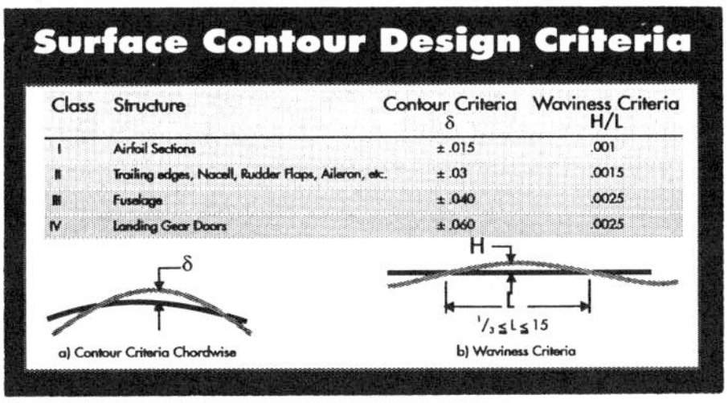

Fig. 1

With thin cover skins and conventional honeycomb core material, considerable losses in strength must sometimes be expected when co-curing in the autoclave, since the skin not in contact with the mold is pressed into the cavities of the honeycomb by the cure pressure. Under lateral pressure, the honeycombs will yield unless edge reinforcement is provided. These problems can be solved using isotropic rigid foams made from polymethacrylimide (PMI). Rigid foams of this type support the cover skins over the whole surface, thus uniformly transmitting the cure pressure (Fig. 2).

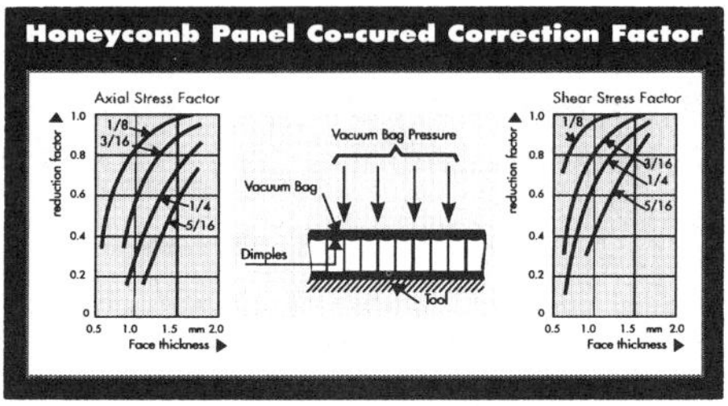

Fig. 2

These foams are thermoformable, which means that complex structures can be manufactured at reasonable cost.

In closed molds, the pressure built up by PMI rigid foam cores during curing of the cover skin material provides impeccable sandwich layers. This possibility therefore opens up completely new and inexpensive structures or curing processes.

PMI rigid foams are produced from cast copolymer sheets of methacrylic acid and methacrylonitrile at temperatures around 200°C (Fig. 3).

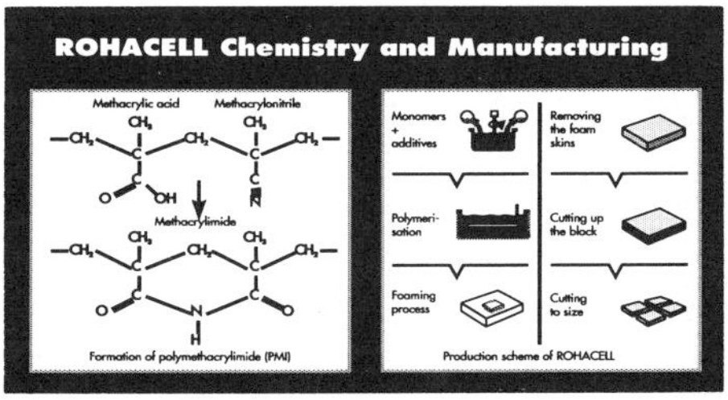

Fig. 3

The copolymer is thereby converted to PMI. The density is between 30 and 300 kg/m³, which corresponds to a modulus of elasticity ranging from 36 to 360 MPa. Different grades are available for specific applications.

2. APPLICATIONS AND MANUFACTURE OF COMPOSITES WITH A PMI FOAM CORE

The earliest application in aircraft dates back to 1973. The French helicopter Ecureuil (Squirrel) AS 350, whose maiden flight was on June 27, 1974, has cabin fairing and engine cowling (except for the nose) manufactured from thermoformed PMI foam cores (Fig. 4). Both cover skins were co-cured. These parts are still being produced today and give impeccable service.

For the Cessna Citation, Sikorsky Aircraft (1) developed landing flaps which are also co-cured in a closed aluminum mold (Fig. 5). The core consists of PMI foam of various densities or strength levels. The cure pressure is generated by inlaid silicone rubber tubes, at 125°C/0.35 MPa.

For the Saab SF 340, Fairchild developed landing gear doors which are nowadays produced by Saab itself (Fig. 6). The PMI foam core is formed in a one-part mold under vacuum film and pressure at approx. 170°C. After machining, both cover layers are cured in one step in the same mold.

The stringers for the rear engine cowling of the MD 11 are cut out of the foam panels as straight parts by means of simple woodworking machines, then thermoformed or curved at 210°C (Fig. 7). The cowling panels are cured in one-part molds in the autoclave under vacuum film at 175°C and 0.35 MPa. The engine cowling for the Airbus is manufactured in a similar way but at a cure pressure of 0.7 MPa.

The ease of fabrication combined with high machining speeds and the simultaneous application of the cover skins were contributing factors for the choice of PMI foam as the core material for the flap vanes of the MD 11 and the C 17 (Fig. 8).

For manufacturing helicopter rotor blades, the PMI foam core is manufactured with a slight excess tolerance. Prepreg layers and core, for example, are placed inside aluminum molds, which are then closed (Fig. 9). The elastic deformation of the PMI core inside the closed mold generates the pressure required to obtain parts free from entrapped air. A cure cycle is shown in Fig. 10.

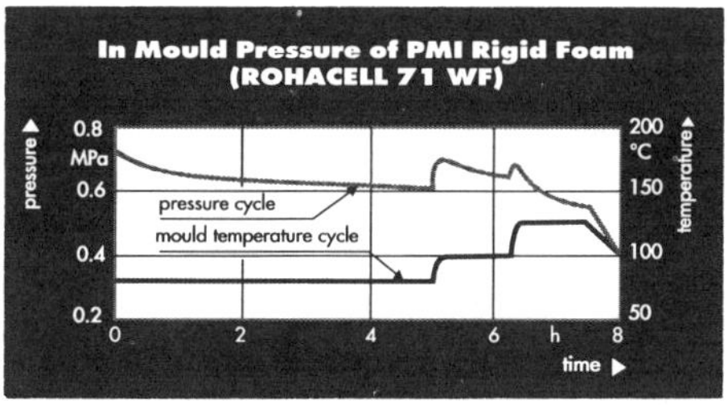

Fig. 10

Flying times of 20 thousand hours are nowadays realistic values which correspond to the service life of the helicopter. If a bullet penetrates the blade, it only leaves a hole but does not destroy the whole blade. A look at the difference in cost using conventional honeycomb structures or PMI cores shows a cost saving of more than 80 % for the blade with a PMI core (Fig. 11).

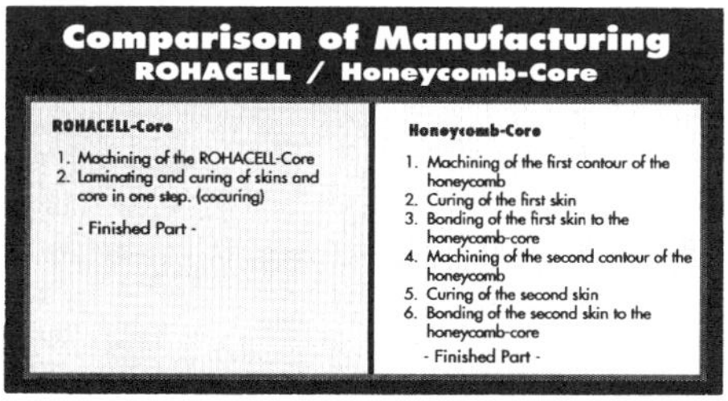

Fig. 11

The wings of the Penguin Missile from Kongsberg Aerospace are manufactured by the same method (Fig. 12).

Sikorsky Aircraft (1), again, developed the canard flap for Piaggio, with the PMI foam blank being shaped to exactly the required contour by compression of the outer layer in the heated mold (Fig. 13). After placing prepregs in the same mold, the part is cured in one step (175°C, 0.35 MPa). The cure pressure in the closed mold is generated by the existing oversize of the core, the thickness of the cover skin and the elastic memory during heating to cure temperature.

The very low absorption and diffusion of radiation by the PMI foams are decisive for their use as core materials for x-ray tables and mammography cassettes (Fig. 14). For use in radomes and antennas (Fig. 16, 17), the excellent dielectric properties are of interest (Fig. 15).

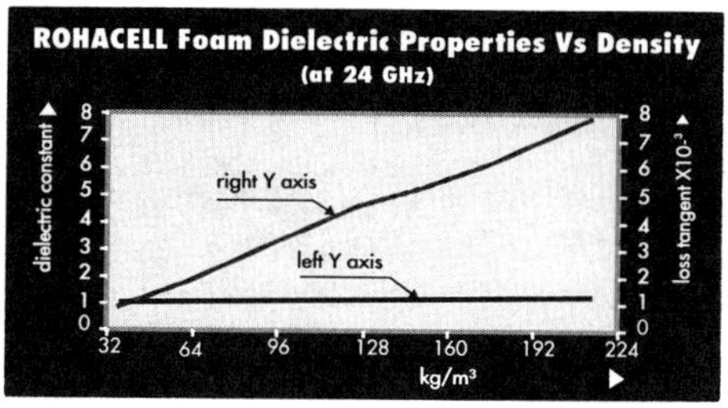

Fig. 15

In particular, it should be mentioned that there is virtually no change in these properties upon absorption of moisture (Fig. 18)!

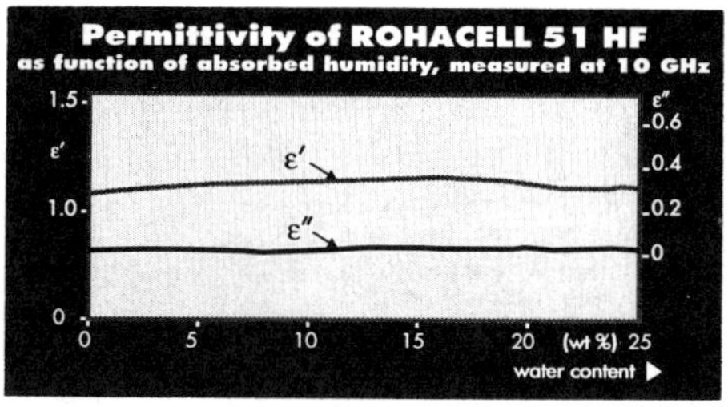

Fig. 18

Radomes with a high degree of stretching present different wall thicknesses after forming. The formed blank is therefore heated in the same closed mold beyond the foaming temperature, which provokes post-foaming and thus enables exact control of the wall thickness. For thinner cover skins, the same mold can also be used for co-curing. The pressure build-up results from the excess tolerance of the PMI core according to the cover skin thickness.

There is a wide range of applications for antennas and radomes, for example thermoformed large-scale radomes on ships (Fig. 19).

A few words should at this point be said about the behavior of closed-cell PMI rigid foams in general.

In none of the tests on composite samples or original parts was the composite construction method found to have any disadvantages as compared with the conventional technique (Fig. 20).

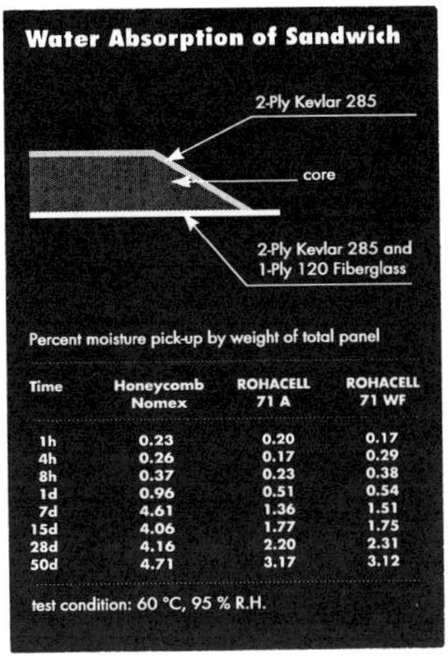

Time	Honeycomb Nomex	ROHACELL 71 A	ROHACELL 71 WF
1h	0.23	0.20	0.17
4h	0.26	0.17	0.29
8h	0.37	0.23	0.38
1d	0.96	0.51	0.54
7d	4.61	1.36	1.51
15d	4.06	1.77	1.75
28d	4.16	2.20	2.31
50d	4.71	3.17	3.12

Fig. 20

Sikorsky Aircraft (1) tested sandwich parts made from the PMI product ROHACELL 51 WF and carbon-fiber epoxy cover skins according to MIL-STD-810C:

75 cycles: (a) sea level: 95 % RH, 63°C, 1 hr
(b) altitude of 15,545 m, -60°C, 1 hr

Result: no deterioration of ROHACELL, no signs of delamination of cover skin/ ROHACELL.

Northrop Corporation, Division B2 (2) compared structural elements of 1 m in length made with a honeycomb core and a PMI rigid foam core in thermal/humidity cycling tests and at the differences in altitude and temperatures which occur in ultrasonic aircraft. The honeycomb structure absorbed more moisture than the part with the PMI rigid foam (Fig. 21).

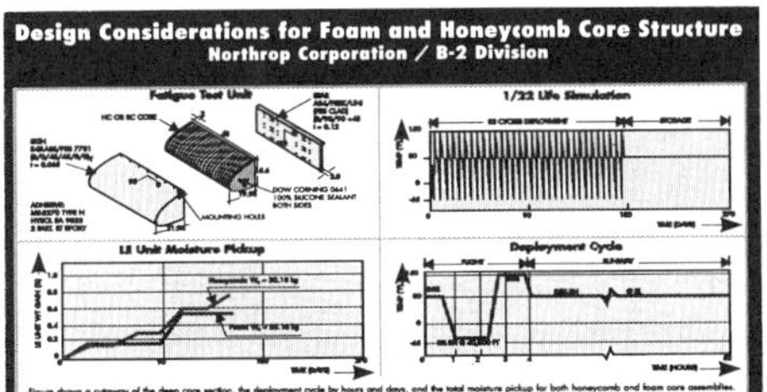

Fig. 21

Even in large-scale shipbuilding, particularly of high-speed ships of the so-called SES type (surface effect ships), composite structures have become indispensable. As such ships now reach speeds of more than 50 kn, the actual floating bodies are exposed to extreme stress levels (Fig. 22). With regard to possible repairs, the question of the water penetration behavior under damaged cover skins was to be clarified. Since PMI foam is a 100 % closed-cell material, water penetrates through diffusion only.

A corresponding test showed that after 40 months' exposure to water, the PMI foam was only damp up to a diameter of 300 mm around the centre of the artificial fracture in the cover skin (Fig. 23).

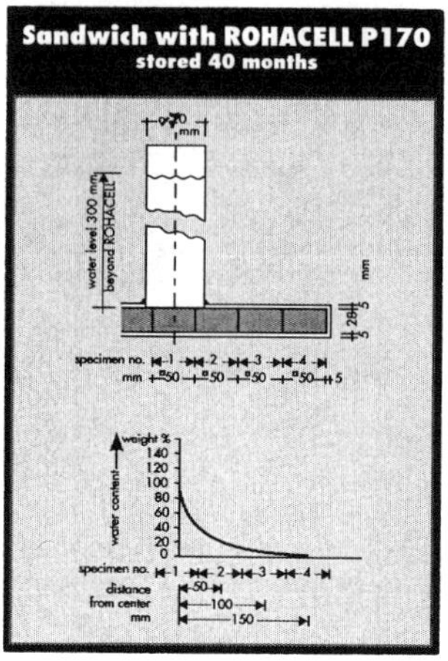

Fig. 23

Two further criteria were crucial for the use of PMI rigid foam as a core material for this application. In the event of fire, the core material was not to release any toxic smoke fumes which might prove harmful to the crew and, also essential, the electronic system was on no account to be damaged by corrosive vapors (Fig. 24).

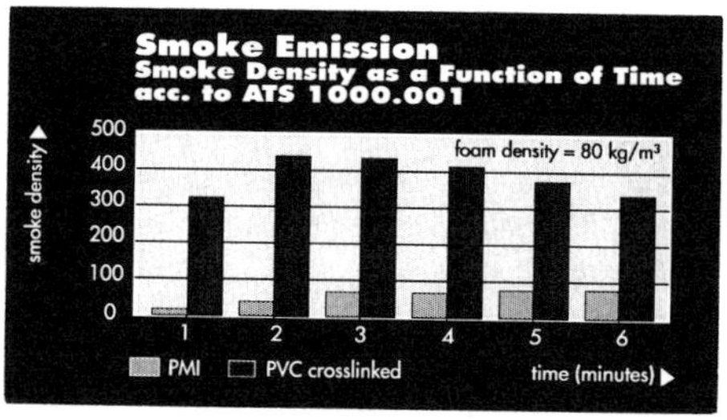

Fig. 24

Both conditions are met by the PMI rigid foams, which do not release any HCl and are free from halogens.

These points also helped to tip the scales in favor of the use of this material in railway vehicles. Schindler Waggon AG in Switzerland manufactures complete railway coaches as self-supporting composite structures with PMI rigid foam cores by the filament winding method (Fig. 25).

In France, the prototype of the "TGV New Generation" is built using C/G fiber hybrid prepreg with a PMI rigid foam core. Here too, the question of the toxicity of smoke fumes was decisive. Naturally, another prerequisite was that the co-curing process should be easy to perform.

PMI rigid foam is also an excellent core material for parts manufactured by the RTM method. Aérospatiale in France thus manufactures the fairings of the ATR 42 by RTM after thermoforming the PMI core (Fig. 26).

Eurocopter manufactured the first rear rotor blade by this method (Fig. 27).

3. SUMMARY

PMI rigid foams have become well-established in aeronautic and aerospace engineering, the construction of antennas and radomes, x-ray technology, shipbuilding and train construction. The reservations initially voiced concerning water absorption have been refuted in every single case. The ease of thermoforming, the isotropic structure of the core material and the high heat distortion resistance combined with excellent mechanical properties open up completely new possibilities for designers of composite parts and offer process engineers reasonably priced manufacturing methods.

4. REFERENCES

1. I. Terceno, Sikorsky Aircraft, Stratford, Connecticut 06602, Cocured, Foam filled Structural Component Design Considerations, Presented at the American Helicopter Society Northeast Region National Specialists Meeting on Composites, June 1985

2. Ron Banuk, Northrop Corporation B-2 Division, 8900 East Washington Boulevard, Pico Rivera, Ca 90660-3783, U.S.A.

Figure 4

Figure 5

Figure 6

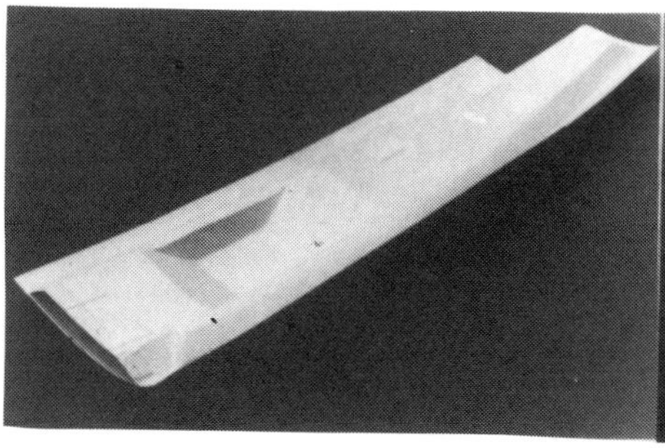

Figure 7

Figure 8

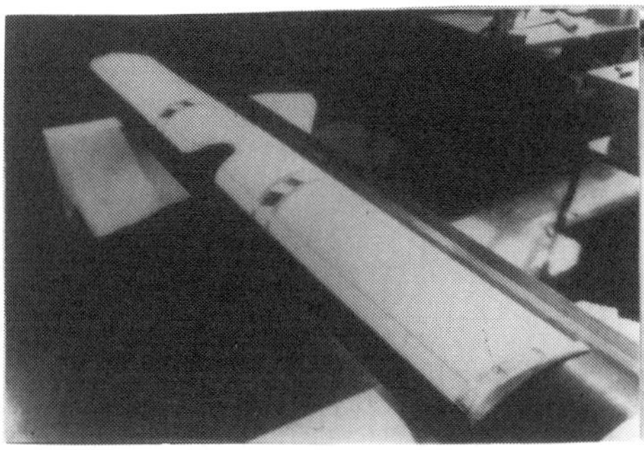

Figure 9

Figure 12

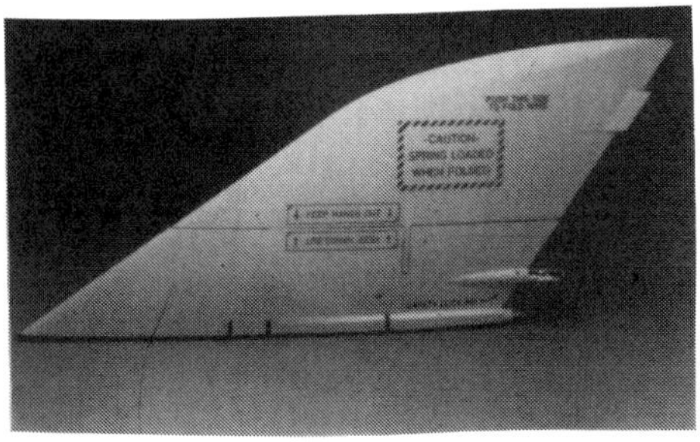

Figure 13

Figure 14

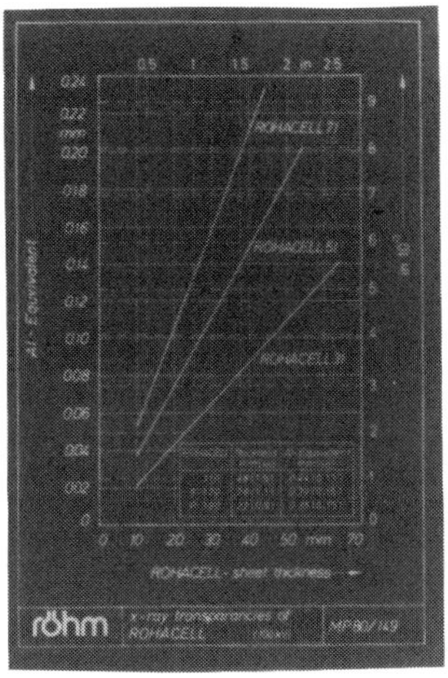

Figure 16

Figure 17

Figure 19

Figure 22

Figure 25

Figure 26

Figure 27

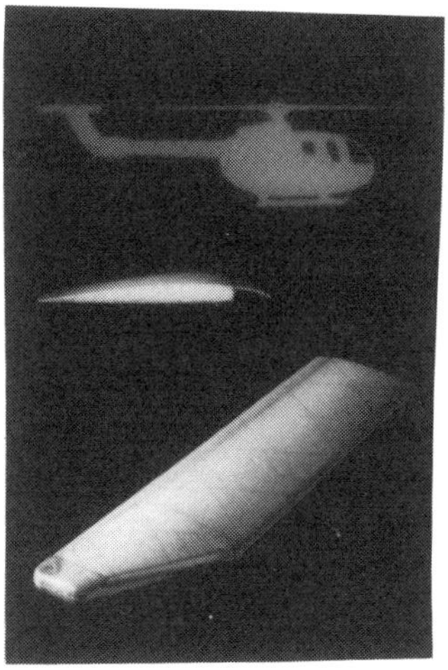

28th International SAMPE Technical Conference
November 4-7, 1996

INFLUENCE OF FRICTION ON THE BRANCH CRACK

Chiheb CHAKER and **M**ichel **B**ARQUINS

Laboratoire de **P**hysique et **M**écanique des **M**ilieux **H**étérogènes
Ecole **S**upérieure de **P**hysique et de **C**himie **I**ndustrielles de la Ville de Paris
URA au CNRS n° 857, 10 Rue Vauquelin - 75231 Paris Cedex 05 - FRANCE

ABSTRACT

The physical conditions of such propagations are not fully understood, controlling factors would include loading conditions, stress orientations, rock types, or the angle between the defect front and the direction of relative movement. In order to understand the branching process from defects in geological conditions, a set of branch fractures, formed under uniaxial loading at the tip of a pre-existing oblique slot inserted with thin metal-plates, was studied in PMMA (polymethyl methacrylate) plates. An analysis of the stress distribution around a crack indicates the points of fracture initiation as well as the initial direction of crack propagation, as a result of the change in stress distribution associated with fracture propagation in order to predict the final path of the propagating crack. An elliptical slit is cut in a plate of photoelastic material, to compare actual dynamic behavior in the same material, plates were loaded in compression until the slits actually grew unaided. Here, the slits were made sharper than in the previous method in order to decrease the applied stress necessary to cause crack growth. During crack growth, the applied load apparently remained constant, within the sensitivity of measurement of load, which was a few per cent. The cracks always grew by the development of branch fractures which curved into the direction of applied compression. The growth of branch cracks was slow and could be stopped or started at will. Most of the cracks could be stopped before they reached the loaded ends of the PMMA plates. By assuming that a natural flaw or crack has the shape of an ellipse, stress on the surface of a crack of any orientation can be calculated.

KEY WORDS: Failure Analysis, Fracture.

1. INTRODUCTION

Various brittle geological structures tend to form at the tip of pre-existing joints submitted to shearing elementary faults movements. In actual experimental fracture of brittle plate, it is hard to observe details of formation of a fault because of the explosive nature of fracture, even under moderate confining pressure. In geological conditions, an increasing number of observations have shown that some kinds of fault on different scales originate from shearing movements along pre-existing joints. We studied the evolution of planar defects with sliding in a set of plates of a brittle analogous material PMMA subjected to oblique uniaxial compressive loading.

We observed the mode II crack propagation of two classical branch cracks. At constant strain rate, the increasing load first gave rise to the microscopic growth of some many microcracks induced by sawing at the slot tip. As soon as the pressure reached the required level, a pair several seconds as already observed, corresponding to a velocity jump with acoustic emission, this was marked by a slight stress drop and followed by a gradually slowing propagation.

2. MODEL

For understanding the physical origin of these structures and to investigate the problem of the initiation location and branching path and to study the growth of inclined cracks in uniaxial compression, we have tested the PMMA (Young's modulus E = 2.9 GPa, Poisson's ratio $\nu = 0.4$) plates (50 x 32 x 5 mm), each with a 0.3 mm wide and 10 mm long central slot was sawn and inserted with very thin steel-plates. This slot is oriented in five directions ranging from 15 to 75° (15, 30, 45, 60 and 75°) with respect to the loading axis (Fig.1).

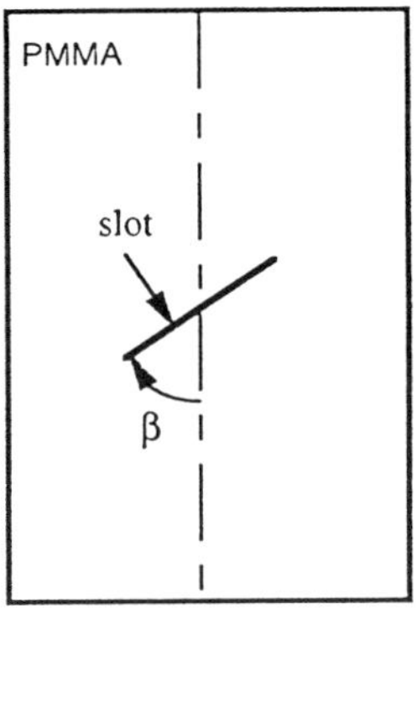

Fig.1 : PMMA plate with an oblique slot inserted with thin steel plates and making an angle ß with the loading axis.

3. EXPERIMENTAL SET-UP

In a first set of experiments, the specimen was held in place at the sides by two guiding devices in order to prevent buckling and the pressure is transmitted by two jaws (Fig.2). The application of compression effort is given by Compression-Traction machine with servo-command hydraulic-system.

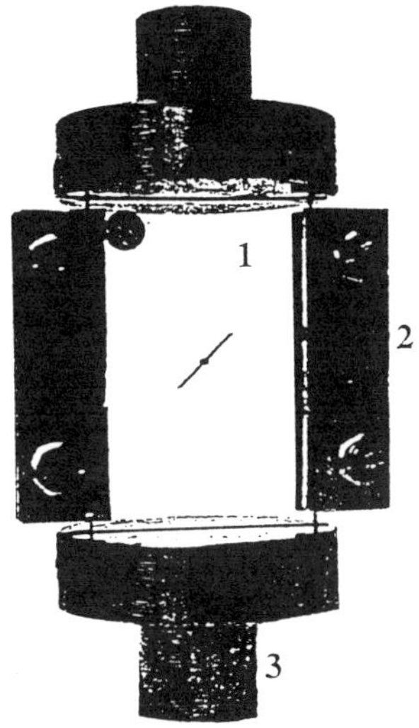

Fig.2 : Experimental device
(1) test-plate
(2) guiding devices
(3) jaws

4. EXPERIMENTAL PRINCIPLE

Under moderate load in uniaxial constant loading conditions, the propagation of branch crack well known in various brittle or semi-brittle materials (1-8) was observed in the material, followed at higher stress levels by the formation of a shear band more or less in the extension of the defect (9).

In order to analyse the fracture initiation points and the initial crack propagation direction, we have determined the branching angle θ (angle between the direction of the slot and the branch crack root) and the distance δ (from the branch crack root to the end of slot) which is mesured parallel to the slot axis (Fig.3).

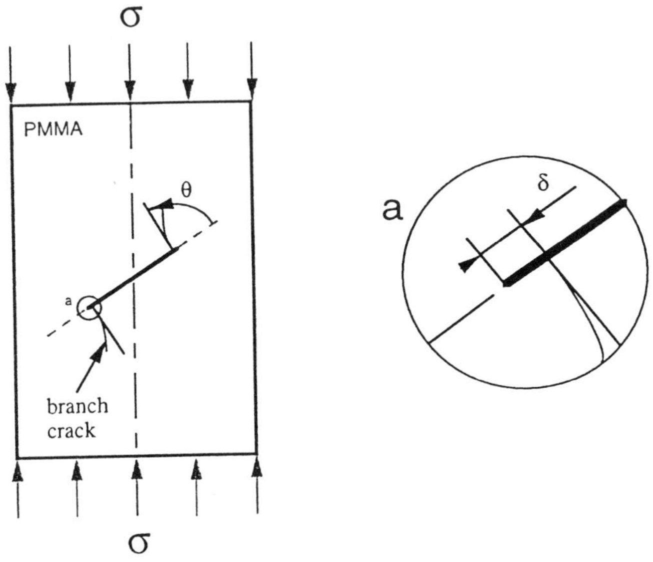

Fig.3 : Sketch showing the branching angles θ and the distance δ when branch cracks are formed.

5. RESULTS AND DISCUSSION

5.1 Branching angle θ   Here, we are interesting on the branch crack problem where the path (initiation and propagation) is not understood in spite of extensive experimental data. For ß = 15 to 45°, the branching angle θ appeared less (Fig.4), there is the tendency of the branch crack to initiate at or near the irregularity shaped slot tip.

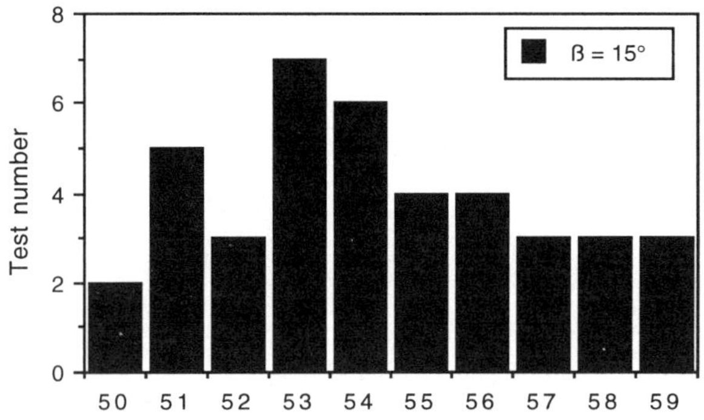

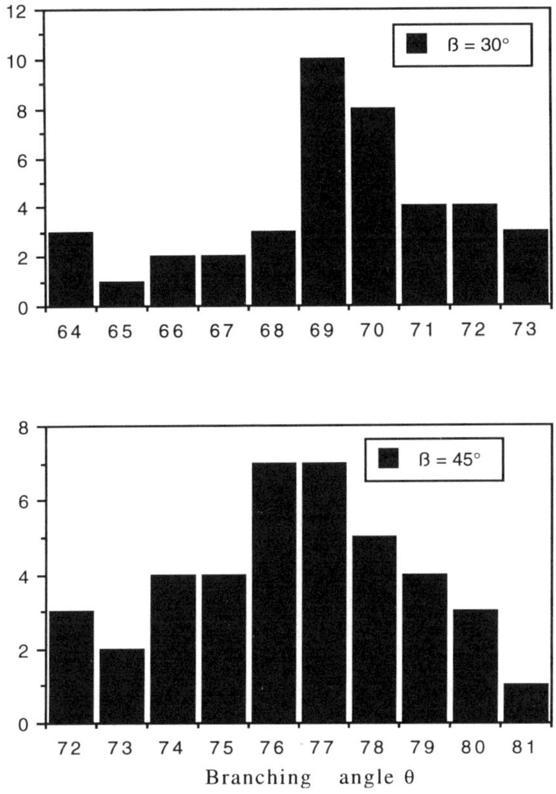

Fig.4 : Branching angles θ measured for ß = 15, 30 and 45°

When ß = 60 and 75°, the branching angle θ was at 80 ± 3° to the slot axis (Fig.5).

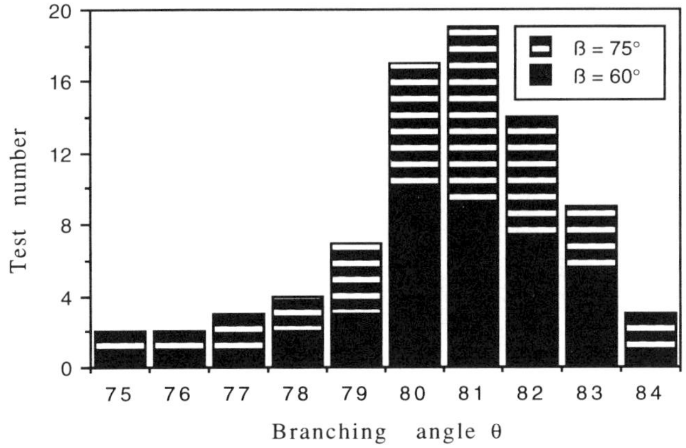

Fig.5 : Branching angles θ measured for ß = 60 and 75°

The mean value of the branching angle θ increase when the angle ß was grew (Fig.6).

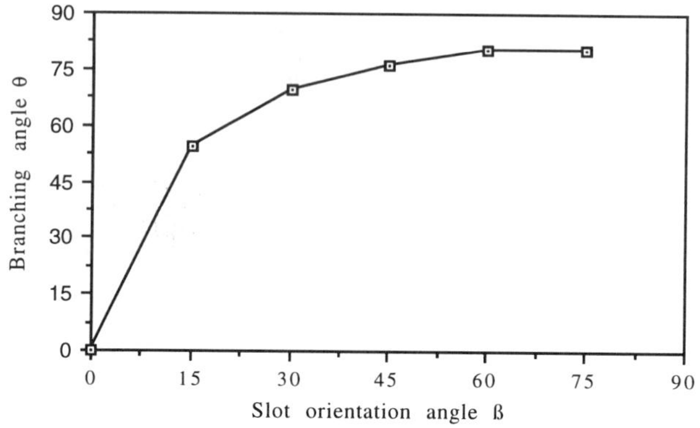

Fig.6 : Mean branching angle θ for different orientations ß of the slot

5.2 Distance δ Statistical observations on the root of the branch crack showed that for ß = 15 or 30°, initiation occured at proximity of the crack front $\delta \leq 0.3$, while for ß = 45, 60 or 75°, the distance δ tended to be greater than the slot width, i. e. $\delta \geq 0.3$ mm (Fig.7).

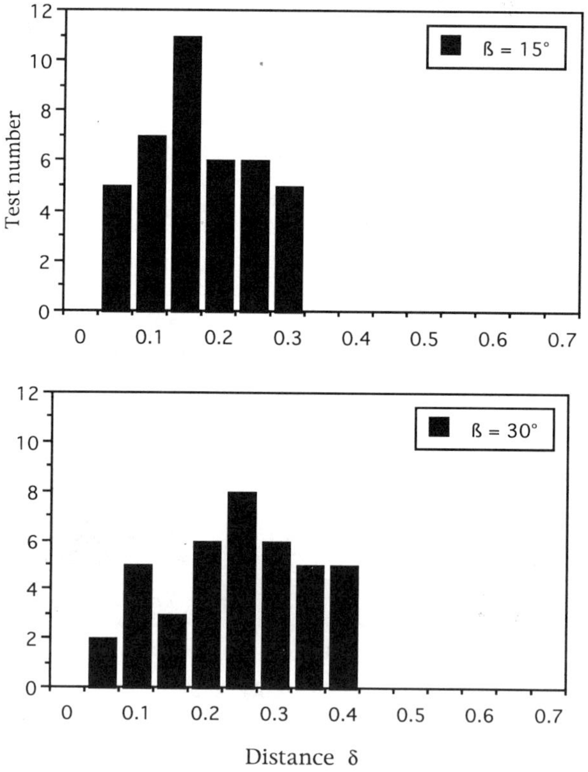

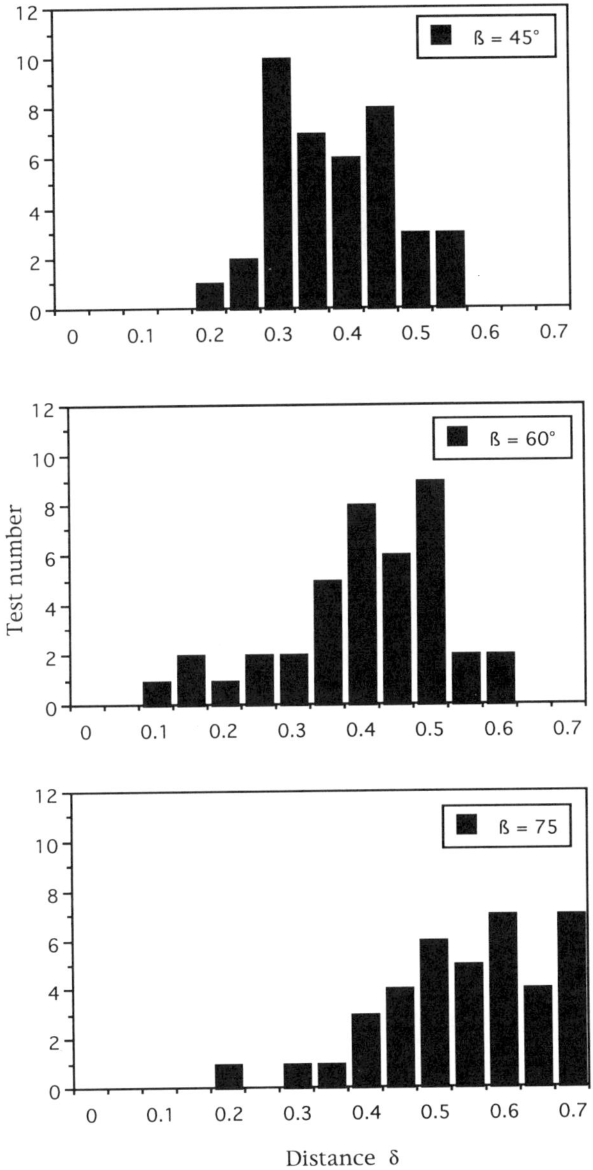

Fig.7 : Distance δ measured from the slot tip to the branch crack root for the various valus of the slot angle ß.

5.3 Strength required Now, we are looking the evolution of the strength, with the distance compressed (displacement) of our specimen (Fig.8), and with the time during experiment (Fig.9) for different orientations ß of the slot.

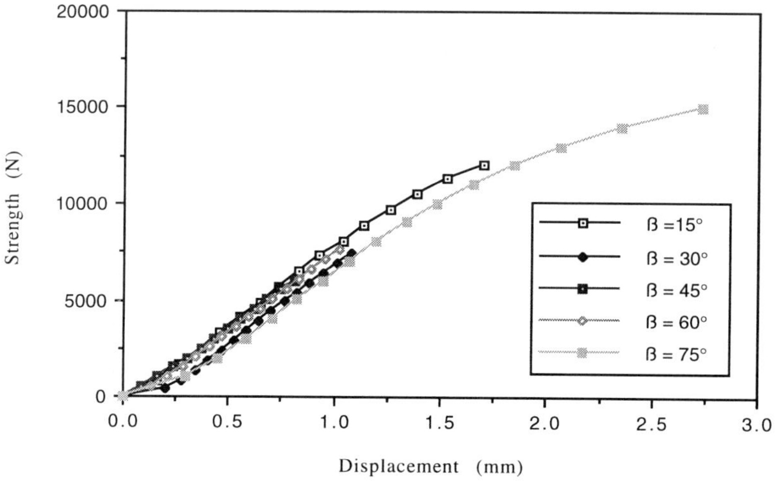

Fig.8 : Evolution of the strength with the compressed distance of the specimen for ß = 15 to 75°.

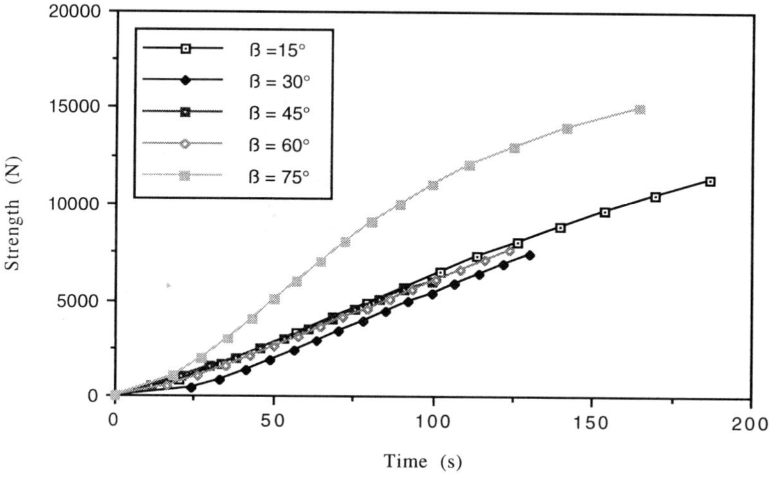

Fig.9 : Evolution of the strength with the time during experiment for ß = 15 to 75°.

5.4 Evolution of cracks The branch crack propagation occurs rapidly and the evolution of cracks at slot tip during uniaxial loading tests was showed on Fig.10.

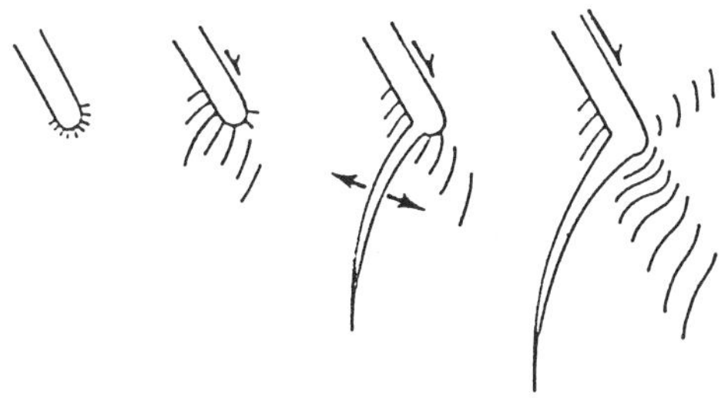

Fig.10 : Evolution of cracks at slot tip during uniaxial tests

5.5 Analysis of the stress Under uniaxial compressive loading, the relative movement of the slot edges corresponds to mode II plane shearing. The applied stresses on a small part situated at a distance r from the slot front are showed on Fig.11.

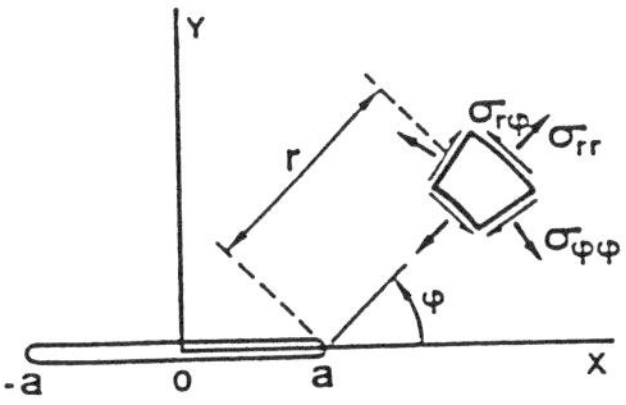

Fig.11 : State of stress on a small part situated at a distance r from the slot tip.

Our specimen with a slot whose axis makes an angle ß with the direction of inposed uniaxial compressive stress may be compared to an infinite plate with an elliptical slot where the long axes $2a$ is equal to the slot length and the short axis $2b$ equal to the width. The stress tensor for an elliptical oblique slot was calculated by Wu and Chang (10) using the curvilinear coordinates system, defined by two families of ellipses and homofocal hyperboles, with the same focuses as the ellipse representative of the slot (Fig.12).

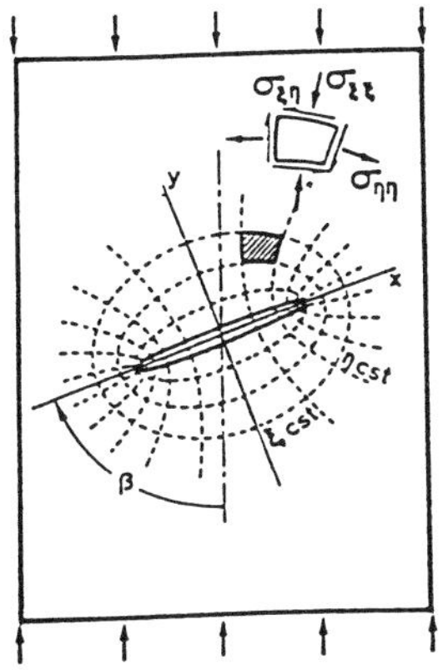

Fig. 12 : State of stress near an elliptical crack tip in curvilinear coordinates.

6. CONCLUSIONS

The experimental tests has been shown that two classical branch cracks, formed from the tip of a central pre-existing oblique slot in PMMA plate submitted to an uniaxial loading, are symmetrical with the defect center and rejoined the direction of the applied load. We found that the branch crack root-crack tip distance and the branching angle increases with the slot-loading axis angle.

The stress field around an elliptical slot confirms that the branch crack is initiated at the point where the tensile principal stress is maximum and that the branch crack propagates along the trajectory of the other principal stress passing at the same point. The study of the stress field linked to an elliptical slot under pure shear conditions is in good agreement with our experimental observations.

7. REFERENCES

1. W. F. Brace and E. G. Bombolakis, Journal of Geophysical Research, 68, 1963, pp. 3709,3713.
2. E. Hoek and Z. T. Bieniawski, International Journal of Fracture Mechanics, 1, 1965, pp. 137,155.
3. B. Cotterell, International Journal of Fracture Mechanics, 8, 1972, pp. 195,208.
4. D. K. Sketty, A. R. Rosenfeld and W. H. Duckworth, Engineering Fracture Mechanics, 26, 1987, pp. 825,840.
5. S. Nemat-Nasser and H. Horii, Journal of Geophysical Research, 87, 1982, pp. 6805,6821.
6. M. F. Ashby and S. D. Hallam, Acta Metallurgica, 34, 1986, pp. 497,510.
7. E. Z. Lajtai, Tectonophysics, 11, 1971, pp. 129,156.
8. P. Tapponnier and W. F. Brace, International Journal of Rock Mechanics, Mining Science and Geomechanics Abstracts, 13, 1976, pp. 103,112.
9. J.-P. Petit and M. Barquins, Comptes rendus Academy of Sciences Paris, 305, série II, 1987, pp. 55,60.
10. H.-C. Wu and K.-J. Chang, Journal of Applied Mechanics, 45, 1978, pp. 258,262.

28th International SAMPE Technical Conference
November 4-7, 1996

INFORMATION CONTENT OF THE SINGLE FIBER COMPOSITE TEST

Korabelnikov Y.G., Rashkovan I.A.
UVICOM Co.Ltd.,Kolontsov 5, Mytischi, Moscow reg. 141009,Russia

ABSTRACT

Single fiber composite tests (SFC-test) as a rule have been used to obtain information about the critical length of the reinforcing fiber. The study of the kinetics of the build-up of the single fiber breaks in such experiments made it possible to get also the information concerning : the fiber strength-length dependence, the strength distribution of the fiber with different length up to the critical length, the critical length distribution of reinforcing fiber. High tenacity PAN-based carbon fibers (untreated and electro-chemically treated) have been studied. The features of the investigation of boron fibers with 90 mcm diameter by SFC-test are discussed.

KEY WORDS: Reinforcing Fibers, Adhesion, Strength Scale Dependence, Critical Length, Distribution .

1. INTRODUCTION

SFC-test first proposed by Fraser and co-authors [1] in the middle of seventies for the determination of rigid reinforcing fibers critical length and their adhesion to polymer matrices has gained currency among researchers. The fast investigation of effects of fiber surface treatments, hardening degree of the resin and other factors on interphase interaction in composites has been potentiated by its simplicity and good replication of data under the tension of the SFC. The consecutive bursts of filament occur in polymer block under the SFC-test. The fragmentation degree is mainly determined by adhesion of the fiber to matrix. The lengths of fiber fragments formed as a result appear limiting for a given system polymer-fiber and their average value is used for the determination so-called critical length l_c.Then shear adhesion strength τ at the polymer-fiber interface is calculated by the famous expression [2]

$$\tau=\frac{\sigma_c d}{2l_c} \qquad (1)$$

σ_c - the strength of the fiber with a critical length,
d - the fiber diameter.
However the widest information is laid in the fragmentation process and the possibility of the direct watch on it by the optical and acoustic emission equipments. So the strength scale effect of fibers , fragments lengths distribution, thermal residual stresses were studied under the SFC-test [3-7]. The interfiber interaction was investigated under the test

of the model specimen contained some filaments [8]. This paper is concerned with pecularities of the determination of the fiber strength scale dependence, interpretation of fragments length distributions of untreated and surface treated carbon fibers at the final stage of tests, getting of strength distributions of reinforcing fibers with length less 1 mm.

2. EXPERIMENTAL

Initial and surface treated high-tenacity carbon fibers UKN-01 based on polyacrilic fibers and boron fiber with 90 mcm diameter were selected as the subject of inquiry. Thermoreactive epoxy resin EDT-10 and thermoplastic polyamide PA-12 and polysulfone PS-TP were used as binders. Specimens were dogbone shaped with measures calculated individually for every polymer-fiber system under study. The cross sections of specimens have been chosen such that the proportion of the load which is necessary for the filament deformation and its breakage could be ignored. The single carbon filament composites with EDT-10 and PA-12 were respectively: dogbone with 33 mm gauge length and 3x3 mm cross-section and a strip with 40 mm gauge length and 5x0,4 mm cross-section. The specimens contained the boron filament were dogbone shaped with 50 mm gauge length and cross-sections of 4x5,6 mm and 6x7 mm for epoxy binder and polysulfone matrix respectively. The preparation technology of specimens with epoxy and polyamide matrices is described in [4]. The preparation of polysulfone specimens with a boron fiber was carried out in two stages. The plates with 1 mm thickness were pressed from polysulfone granules in advance. The boron individual fibers with 80 mm length were arranged between plates and this blank was positioned between bulk beds of granules. The pressing was conducted under 280 ^{0} C during 1 hour. The compression molding pressure was 1 MPa. The dogbone shaped specimens were stamped from produced plate. The testing carried out by Instron tester at a strain rate of 0,1 mm/min . The fiber breaks monitoring was carried out by acoustic emission installation "PAC 3100/300" The elongation was measured by the extensometer fixed on a specimen. The measurement of fragments lengths of single carbon fiber in PA-12 and study of fracture surface specimens were performed by the optical microscope.

3. RESULTS AND DISCUSSION

The approach to the prediction of the unidirectional fiber composite strength taking account of effects of adhesion and statistical characters of reinforcing fibers strength addition to another factors was suggested in [9] . As it was mentioned above SFC-test is the most attractive for an estimation of adhesion of fine and rigidity fibers to polymer matrices. It is necessary to know the strength of the fiber with critical length σ_c for calculation of adhesion strength by the expression (1). Since the critical length of a majority fine reinforcing fibers in polymer matrices are less 1 mm the direct determination of the strength σ_c is technically impractible. As a rule the value of the strength of the fiber with critical length is obtained by the results of tests of fibers with large gauge lengths. The consideration of the kinetics of the build-up of single fiber breaks by the increase of a model specimen deformation allowed to get the information about the critical length, strength-length dependence and hence the strength of the fiber with critical length [4,5]. In contradistinction to [5] the SFC-tests in [4] were performed by the Instron tester with the constant strain rate and continuous monitoring of acoustic emission signals accompanying the fragmentation. Consider some features of the determination of strength-length dependence by this method.

3.1. Specimens Dimensions. For the reliable maintanance of the fragmentation up to critical lengths it is necessary maximal share of the load applied to specimen accounts for the specimen deformation. The force for a filament deformation must be negligible. Calculating the rupture load of the fiber under study with the safety factor 1,5-2 assume it is equal 1% of the total load applied to a specimen. Defining the total load and knowing the strength of matrix we can easily calculate the specimen cross-section. The matter of the right selection of the specimen cross-section is specially acute at the investigation of high-tenacity fiber with diameter more than 50 mcm. The rupture of such fiber is accompanied by the release of energy producing the impact wave. Besides the formed microvoid becomes the stress concentrator and the site of the trunk crack formation and growth. The similar mechanism was observed under the testing of a single boron fiber in epoxy resin EDT-10. The first breakage of the boron filament caused the premature fracture of the specimen. The fractography analysis (See Fig.1) confirmed foregoing proposals about the causes of such character of the fracture. The Figure 1 illustrates that trunk crack originated in the site of the fiber rupture and propagated quickly into epoxy matrix. Evidently the safety factor used under the calculation of the specimen cross-section ignoring the stress concentration and impact wave was not enough . Moreover it is necessary to take account of relaxation-dissipative properties of matrix for reception of information about the strength scale dependence and another features of polymer system contained the rigidity high-tenacity thick fiber. The surface rupture of single boron fiber composite with polysulfone is shown in Figure 2. In this case the fragmentation proceeds completely. These specimens was characterized by the better relaxation-dissipative properties and a larger cross-sections shown in Figure 2 the specimen rupture occured after completion of fragmentation as a result of the growth of the crack originated at the specimen edge. There was obtained the critical length, strength scale dependence , strength of fiber at critical length and adhesion strength. For boron fiber under study the critical length and adhesion to polysulfone were 4,8 mm and 50 MPa respectively . The strength-length dependence in log-log co-ordinates appeared linear, and values of its parameters A and B were equalised 3,845 and 0,19 respectively.

3.2. Correction for "Looseness" Of Fiber. The fragmentation of the filament up to the critical length takes place under the SFC-test. In response to increase of stresses the filament bursts occur in the sites of the most dangerous flaws. Hence the fiber fragmentation is controlled by the strength scale dependence. The mean length of fiber fragments at any stage of testing may be determined by the amount of acoustic emission signals from filament breaks. High modulus brittle fibers hold a linear relation between stress and elongation up to the moment of the failure. The equality or closeness of the fiber and resin deformations are due to their adhesion interaction. Thus it is enough to fix the specimen deformation at the rupture moment to determine breaking stresses in the filament. However in a number of cases the equality of the fiber and resin deformations is disturbed. This fact may be an outcome of the looseness or pretension of the filament under the specimen preparation. Stresses also can originate in fiber during the process of hardening or pressing of specimens due to the distinction of fiber and matrix coefficients of thermal expansion. To avoid the errors in the determination of fiber stresses a correction is introduced. The magnitude of the correction $\Delta\varepsilon$ can be found by comparing the deformation of the model composite specimen at which the filament is fragmentated into parts with length l_i, on the one hand, with the elongation at the break for a fiber with same length in a traditional tensile test. In such calculations it is convenient to use certificate data of fiber under study.

3.3. The Dependence Of The Average Length Of Fiber Fragments On The Specimen Deformation. The dependence of the average length of fragments $\bar{l}_f(\varepsilon)$ on the specimen elongation lets find out a number of mechanisms for fiber-polymer systems. The typical dependences $l_f(\varepsilon)$ in log-log co-ordinates are represented in Figure 3. A characteristic feature of this these relationships is a bend at a certain value of the mean length of fragments l_k. As shown in [4] the slope of the flattened area (See Figure 3) is determined by the strength scale dependence of the fiber. The bend at the plot can be a consequence of the change of the fiber strength scale dependence or of the fracture mechanism in the SFC. The strength scale effect of fibers is related to variations of surface defects on dangerous degree and irregularity of their spread on the length. In the case of the change of the fiber strength scale effect in the interval of short lengths that is possible under the occurrence of different types of flaws in fibers [10] the value l_k must not depend on the kind of the polymer matrix. Actually l_k (See Figure 3) is various for specimens with different binders. Thus the hypothesis about the bend as a result of the change of the fiber strength scale effect does not find justification. On the other hand the adhesion interaction breaks when the fragments lengths reach the critical value and the plastic shear happens. So far as fragments with same length can be of different strength, also the fiber diameter is characterized by some spread, the critical length must have a field of spread. Thus while some fragments keep bursting according to the continued strain, other with length, which is not enough to transmit the fracture force begin to "slide" The number of bursting fragments progressively decreases that causes the bend of the relationship at some mean length of fragments l_k. The break-up of single filament discontinues when the mean length of fragments approaches to δ. The difference in value of l_k and δ indicates the wide distribution of the critical length. This is verified by results of optical measures of fragments formed after the completion of the filament break-up in SFC-test.

3.4. The Distribution Of The Filament Fragments Lengths In Block Of Polymer After Completion Of SFC-test. While studying the effect of surface treatment on adhesion of UKN-01 to polyamide-12 we have established not only decreasing of average fragments length, that is quit natural, but the change of the trend of the length distribution. As following from distribution curves (Figure 4) bimodality is a characteristic of untreated carbon fibers. One of distribution modes (Fig 4, curve 1) corresponds to 1,3 mm average length and the other one to 0,25 mm. The fragments with 1,3 mm average length ($l/d>200$) are typical of the systems with low adhesion, in which interface interaction is caused principally by mechanical cohesion of the reinforcing fiber with matrix. The fragments 0,25 ($l/d \leq 50$) average length generate chemical bonds with polymer due to presence of chemical active centres on the fiber surface. thus, the distribution character represents the contribution of chemical and mechanical types of interaction into the formation of adhesion of fiber to matrix. The bimodality of the distribution is explained by the carbon fiber heterogenity, poor amount of chemical active atoms and groups, and their irregular allocation on the fiber. The increasing of chemical active groups concentration and surface defects levelling result from electrochemical treatment of carbon fiber. It fits more narrow unimodal distribution of treated carbon fiber fragments (See Figure 4, curve 2). So the adhesion heterogenity of carbon fiber surface along with scatter of strength and diameter is a cause of wide critical length distribution and existence of different levels of adhesion suitable to physicochemical and mechanical interactions. In view of all these facts it is important to choose rightly the magnitude of the critical length for τ calculation by expression (1). So long as l_k corresponds to the beginning of the adhesion fault the application of just this value as a critical length l_c is represented rational. Experiments have shown that fiber surface treatment icreasing adhesion caused the relationship bend (See Figure 3) and consequently decrease of l_k. The values of l_k for the same fibers in different matrices are discerned under different adhesion. Evidently l_k is a complex characteristic of system and sensitive to the change of properties of its components (reinforcement and binder). The replication of l_k is significantly more than one of δ under SFC-test that is explained by the dependence of δ on the specimen breaking strain. Moreover the determination of fiber strength at length l_k used for calculation does not demand the extrapolation of the strength scale dependence. The comparison of calculated values τ with shear strength of real composites (See Table 1) is a cause for l_k application as a critical length.

Following the theoretical analysis of possible approaches to the calculation of the adhesion strength authors in the [11] give preference to value l_k too.

3.5. The Strength Distribution Of The Carbon Fiber With Short Length. Parameters of Weibull distribution obtained from tests of fibers at large gauge lengths are usually used for the determination of the fiber strength at the critical length σ_c. The errors in the determination of σ_c and as result the in the estimation of the adhesion by the expression (1) are linked with the possible polymodality of real fiber strength distributions [10] and distinction in fracture mechanisms under fiber tensile tests and SFC-tests [6]. The feature of the fracture under the SFC-test is a multiple consecutive breakups of the initial filament. The formed fragments are stronger than the initial filament because they have not contained rough flaws already. The further break-up of the filament proceeds by the rejection of fragments with strength which corresponds the reached stress level. If the single filament in the block of polymer is considered as a chain consisting of individual links, the strength of every link is determined by the most dangerous flaw. The strength of different links of the chain is unequal. To plot the strength distribution of fibers with selected length it is enough to test 3-4 single fiber composites. To obtain the strength distribution of fragments-links with length $l \geq l_k$ it is necessary to divide up all range of strengths realized to the moment of the formation of fragments of the given length into equal intervals and to calculate the number of breaks in each interval. The strength distributions of initial and surface treated fibers UKN-01 with 0,75 mm length in EDT-10 are presented in Figure 5. These distributions are close each to other that is explained by the slight distinction in strength scale dependences. Their feature is enough full information about the low strength tail and the level of mean strength. To get the strength distributions with length $l_k > l \geq \delta$ it is necessary to extrapolate the strength scale dependence to δ. Such extrapolation is legitimate if the scale dependence does not change in the interval from l_k to δ (See Figure 6.1). The increase of adhesion as a result of the electrochemical surface treatment of fiber caused the decrease of the value δ and respectively extended the range of realizable strengths in SFC-test. The splash of the fragmentation of the surface treated fiber at final stage of SFC-specimen deformation is noteworthy. This fact seems particular unexpected if we consider the logical reduction of the number of breaks at the stage of adhesion fault. Almost all the breaks of treated fiber occurred in the interval of stresses 8500-9500 MPa (See Figure 6.2) It is possible to suppose that the weakening of the strength-length dependence takes place in this range. Hence in the last case the extrapolation of the strength scale dependence to length δ cannot be admitted correct. However the obtained qualitative information is very useful.

4. CONCLUSION

In the present paper the possibility of an application of the SFC-test for study of the strength scale dependence and adhesion of high-tenacity fibers with a large diameter to polymer matrix is shown. Analysis of experimental data permitted to suggest the adaptation of the value l_k in which the bend of the dependence $l(\varepsilon)$ occurs as a critical length. The method of a construction of the strength distribution of fibers with short length is developed. It gives possibility to determine the fiber strength at critical length very trust worthy. Analysis of the fragments length distribution at the final stage allows to form an opinion about the nature of adhesion interaction between fiber and matrix.

5. REFERENCES

1. W.A. Fraser, F.H. Ancker and A.T. Dibenedetto Proceedings of 30th Aniversary Technical Conference 1975 (Reinforced plastics Institute The Society of Plastic Industry Inc.) Section 20 -C .
2. A.Kelly and W.R.Tyson J.Mech. Phys. Solids, 13, 329 (1965).
3. USSR A.C. 1704015 Yu.G. Korabelnikov, I.A. Rashkovan et.al. Appl. For July 6 1989, Otkrytiya Izobreteniya, 1, pp. 168-169.
4. V. P. Tamuzh, Yu.G. Korabelnikov et.al. Mech. Compozitn. Mater. 4 (1991), pp. 641-647.
5. H.D. Wagner and A. Eitan, Appl. Phys. Lett. 56, 1965 (1990).
6. Yu.G.Korabelnikov and I.A. Rashkovan Chem. Fibers, 2 (1996) pp. 39-43.
7. H.D. Wagner J. Adhesion,V.52(1995), pp.131-148.

8. A. Eitan and H.D. Wagner, Appl. Phys. Lett.,56, 1965 (1990)
9. Yu.G. Korabelnikov, O.F. Siluyanov et.al. Mech. Compozitn. Mater. 3 (1987) pp. 270-274.
10.I. Kalnin, Mech. Compozitn. Mater., 3 (1979), pp. 397-406.
S.F. Zhandarov and E.V. Pysaniva Mech. Compozitn. Mater. V.31, 4, (1995), pp. 444-461.

Table 1

Surface treatment	Matrix	δ, mm	l_k, mm	Calculated τ, MPa			Short beam shear strength, MPa
				$l_c=\delta$	$lc=4/3\delta$	$l_c=l_k$	
Untreated	PA-12	0,6	0,933	37,8	26,5	20,0	15-20
	EDT-10	0,387	0,613	58,6	41,3	33,4	30-35
Electrochemical treated	PA-12	0,53	0,88	42,3	29,7	23,0	20-25
	EDT-10	0,186	0,422	135,0	96,0	51,0	50-60

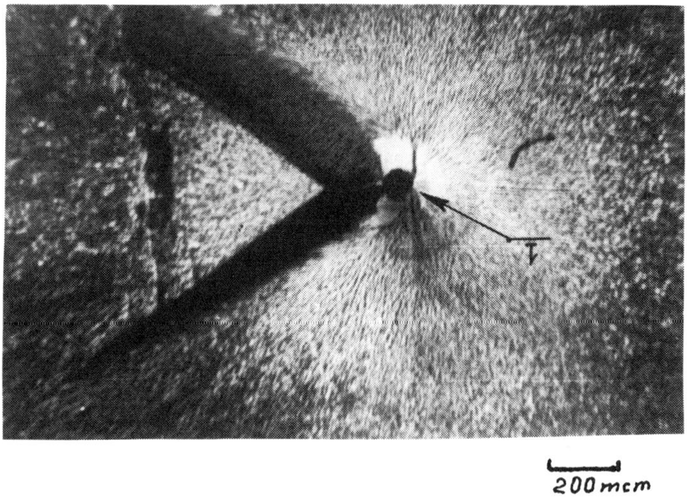

Figure 1. The fracture surface of the single boron fiber composite with EDT-10 , 1-boron fiber

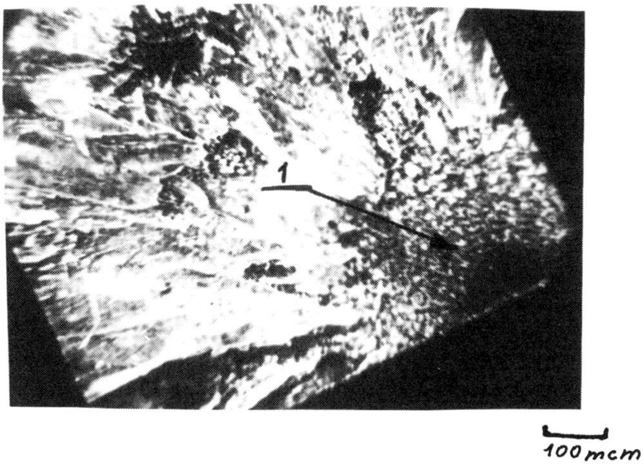

Figure 2. The fracture surface of the single boron fiber composite with polysulfone PS-TP, 1- the edge of the spesimen.

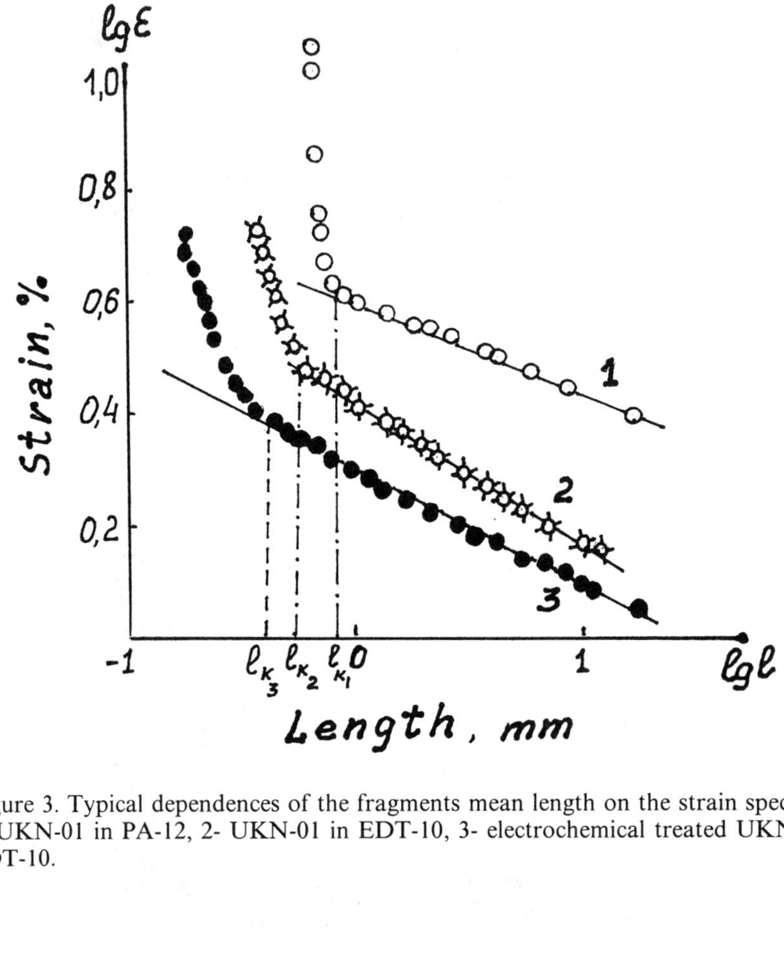

Figure 3. Typical dependences of the fragments mean length on the strain specimens: 1- UKN-01 in PA-12, 2- UKN-01 in EDT-10, 3- electrochemical treated UKN-01 in EDT-10.

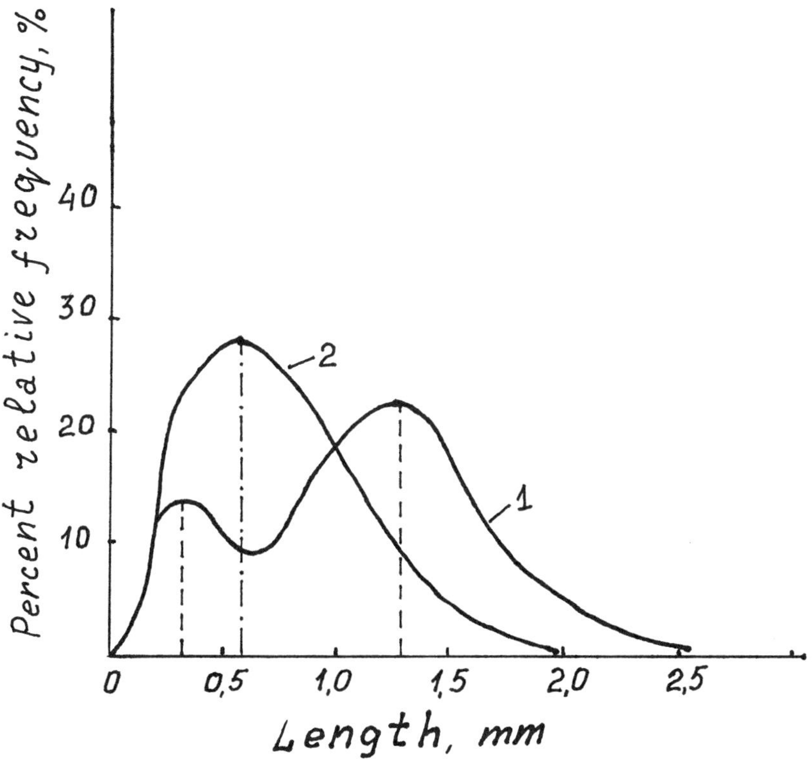

Figure 4. Fragments length distributin from5 specimens at the final stage of SFC-test. 1-UKN-01 initial in PA-12, 2- electrochemical treated UKN-01 in PA-12.

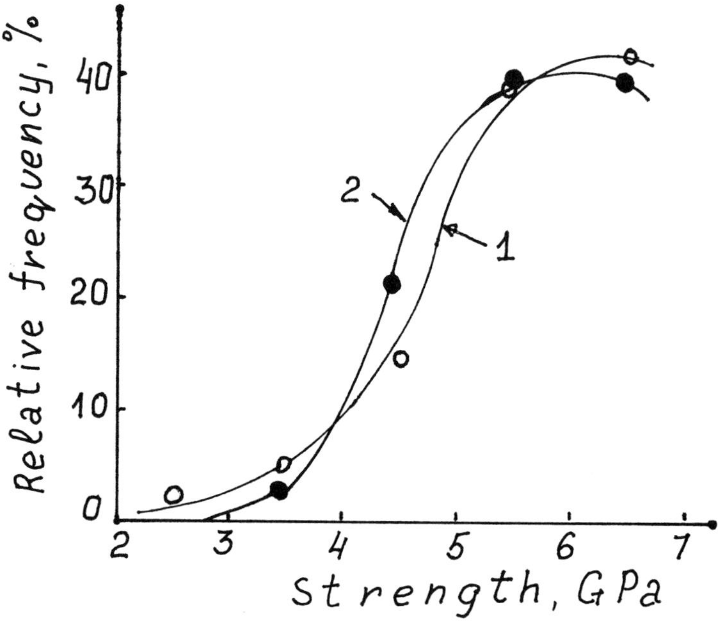

Figure 5. Fragments strength distributions of initial (1) and treated (2) fibers UKN-01 in EDT-10 at the stage preceding the adhesion fault l=0,75 mm.

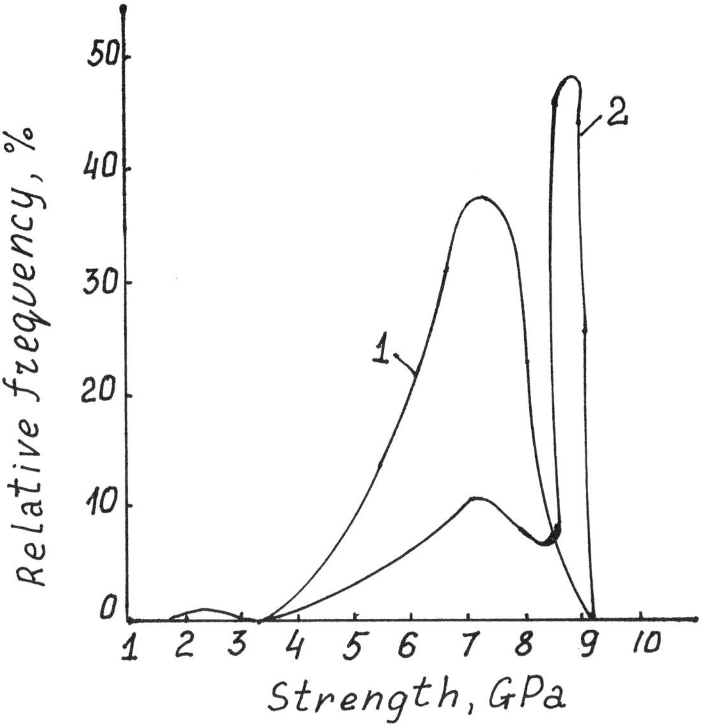

Figure 6. Fragments strength distribution of carbon fibers UKN-01 at the stage of the break-up completion: 1-initial fiber l= δ_1 =0,387 mm, 2 - treated fiber l= δ_2 = 0,186 mm.

28th International SAMPE Technical Conference
November 4-7, 1996

MECHANICS METHODOLOGY FOR TEXTILE PREFORM COMPOSITE MATERIALS

Clarence C. Poe, Jr.
M.S. 188E
NASA Langley Research Center
Hampton, VA 23681-0001

ABSTRACT

NASA and its contractors have completed a program to develop a basic mechanics underpinning for textile composites. Three major deliverables were produced by the program:

1. a set of test methods for measuring material properties and design allowables
2. mechanics models to predict the effects of the fiber preform architecture and constituent properties on engineering moduli, strength, damage resistance, and fatigue life
3. an electronic data base of coupon type test data

This report describes these three deliverables.

KEY WORDS: Textile Composites, Mechanical Properties, Test Methods, Analysis, Data Base

1. INTRODUCTION

NASA and its contractors have completed Phases A and B of the Advanced Composites Technology (ACT) program to develop composite wings and fuselages for a commercial transport airplane with costs that are competitive with those of current metal airplanes. Textile composites were considered for many components to improve structural performance and to reduce costs. Boeing developed advanced composites for making large crown, side, and keel panels for a fuselage, Lockheed-Martin teamed with Boeing to evaluate textile composites for fuselage frames, window-belt reinforcements, and various components in the keel of the fuselage. Northrup Grumman evaluated textiles concepts for making stiffened skins using 3-D textile composites, and McDonnell Douglas evaluated knitted/stitched textile fabrics for a wing box. To support these activities, a NASA Langley in-house team planned and implemented a program to develop a basic mechanics underpinning for the textile composites. Summaries of that work were published in (1).

The geometry of periodic textiles can be described by spatially translating copies of unit cells, without rotations or reflections. The unit cells of braids and weaves can be 2 to 20 mm in length and width and up to the thickness in depth. On the other hand, the length and width of a unit cell of a laminate made from prepreg tape is on the order of a fiber diameter, 5 to 8 microns. Thus, the inplane displacement fields of a textile composite will be macroscopically inhomogeneous compared to a tape laminate. The field of inplane normal displacements in the loading direction of a tension coupon is shown in Figure 1 for a triaxial braid (2). In a triaxial braid, the braider yarns arebraided about the fixed yarns, which are relatively straight. The fixed yarns in Figure 1 are perpendicular to the loading direction. For uniform displacements (constant strain), the wavy lines (fringes) would be straight and equally spaced. The locations of the yarns are evident in the waviness of the fringe patterns. The braider yarns are associated with the joggles in the fringes, which are caused by large, intense shearing between the yarns in regions of high resin content. The rotations (shearing) reverse where the braider yarns cross one another on the surface. Normal strain varies inversely with line spacing. Thus, the normal strains are highest over the fixed yarns and lowest where the braider yarns cross. The average strains in these two regions differ by a factor of three. The inhomogeneity in Figure 1 raised concerns about strain measurements on test specimens and about adequate specimen size to measure bulk or average mechanical properties and to understand notched strength and damage tolerance behavior of textile composites. This large level of inhomogeneity guided planning of the program to develop a basic mechanics underpinning for the textile composites.

The participants in the Mechanics of Textile Composites Program included aerospace companies, universities, and NASA. A review of this program with technical highlights is given in (3). The program had three primary objectives. First, test methods were developed or modified to establish a set of test methods for measuring material properties and design allowables. Second, mechanics models were developed to predict the effects of the fiber preform architecture and constituent properties on engineering moduli, strength, damage resistance, and fatigue life. Third, coupon-type test data that were generated on textile composites in the ACT program was compiled into an electronic data base.

The program objectives have been largely accomplished, and the results integrated together into engineering guidelines for mechanics of textile composites, which contain the essential engineering methodology required to design airframe-type structural components using textile preform composites. These guidelines are contained in three NASA reports: Standard Test Methods For Textile Composites (4), Handbook of Analytical Methods for Textile Composites (5), and Database of Mechanical Properties for Textile Composites (6). These reports are also available in electronic format from the Langley Technical Report Server (LTRS) using the Uniform Resource Locator (URL): http://techreports.larc.nasa.gov/ltrs. This paper summarizes the contents of these three reports.

Several out-of-plane and impact test methods were not sufficiently mature to be included in (4). Thus, a review of that work is included in this paper for completeness. Another important investigation that should not be overlooked is an experimental comparison of mechanical propertes of 2-D triaxial braids and equivalent prepreg tape laminates (7).

2. STANDARD TEST METHODS FOR TEXTILE COMPOSITES

Strain gages are commonly applied to composite coupons to obtain a stress-strain curve, to measure elastic constants, and to check alignment of test coupons. Strain gages are also applied to structural components to make strain surveys. The inhomogenious field of displacements in Figure 1 for a triaxial braid suggests that strains should be averaged over several unit cells. Thus, a study was made to determine appropriate strain gage size for textile composites (2). When strain gages were longer than the unit cell, the coefficient of variation in modulus was less than five percent and strain gages and extensometers were in good agreement.

2.1 Test Methods Evaluated by Lockheed Engineering & Sciences Co. Aerospace companies have developed test methods to derive design allowables for laminated prepreg tape composites. This work is ongoing in organizations such as American Society for Testing Materials (ASTM), MIL-HDBK-17, and Suppliers of Advanced Composites Materials Association (SACMA). Because many test methods used by industry are not formal standards, a survey was made of aerospace companies to determine the test methods used most commonly to derive design allowables for laminated prepreg tape composites. These companies have significant experience designing and manufacturing structures made of high-performance composite materials. The objective of the survey was to determine the minimum test methods that should be evaluated for textile composites. Table 1 lists the different types of test methods that were compiled from the survey.

The Boeing Defense & Space Group made textile composites and conducted experiments to evaluate the test methods in Table 1. The results were reported in (8). The textile composites were made using 2-D triaxially braided fabric, stitched uniweave fabric, and 3-D interlock woven fabric. The fabrics were made of AS4 fiber, and the composites were resin-transfer molded using Shell 1895 epoxy. In addition, Lockheed-Martin conducted similar coupon tests to develop allowables for use in ACT fuselage trade studies, the University of West Virginia conducted open-hole tests on some of the triaxial braids to develop failure models (9), and NASA LaRC conducted four-point bend tests on the braids and weaves to develop an improved method to measure out-of-plane tension strengths (10). Lockheed Engineering & Sciences Co. evaluated all these results to determine the best of the test methods in Table 1 (11-15). Criteria such as low coefficient of variability, appropriate failure mode, and simplicity were used. Those selections, which are listed in the right-hand column of Table 1, are reported in (4).

None of the inplane shear test methods were totally acceptable. The methods gave reasonably similar values of shear modulus for a given material, but the coefficients of variation were high for some materials. The coefficients of variation were even larger for strength. Surprisingly, the coefficients of variation for tubes typically were larger than those for the rail shear and compact specimens. Of course the manufacturing methods for the tubes were also significantly different from those of the flat plates. Thus, some of the variability in the tube data may have resulted from the manufacturing method. In some of the rail shear specimens, bearing failures occurred where the rails are bolted to the specimen. The compact specimen, which was clearly the cheapest and most simple to test and gave relatively low coefficients of variation, was generally the best of the group. Because of variations between manufacturing methods for plate and tubes, the tube should probably be tested for tubular applications. Additional work is necessary to develop a shear test method.

Several bend specimens and test methods were evaluated for determining interlaminar tension strength. Boeing evaluated C- and L-Shaped Bend specimens (8) and NASA LaRC evaluated the Four-Point-Bend Specimen (10). The four-point-bend test method was recommended because of the simplicity associated with a section of constant bending moment. Values of interlaminar tension strength were calculated using the same expressions as those used for tape laminates. The four-point-bend test method produced interlaminar failures in the 2-D braids and in plain and 5-harness-satin 2-D weaves. However, the four-point-bend test method did not produce interlaminar failures in the 3-D weaves. Instead, premature failures resulted from radial matrix cracks associated with the bending.

The relationship between unnotched tension strength and specimen size was investigated for the unnotched tension specimens made of the triaxial braid. For specimen widths between 25 and 63 mm (1.2 and 6.0 times the unit cell widths), mean strengths and coefficients of variation exhibited little correlation with specimen width. (Specimen lengths varied from 16.9 to 54.6 times the unit cell lengths.) Also, strengths of 3.2- and 6.3-mm thick specimens were within 10% of one another.

2.2 Other Interlaminar Test Methods Interlaminar test methods are desirable for optimizing the type and amount of through-the-thickness reinforcement in textile composites. For a constant fiber volume fracture, an increase in through-the-thickness reinforcement results in a decrease in inplane reinforcement and consequently a decrease in inplane stiffness and strength. The interlaminar test methods in Table 2 were evaluated. Although these test methods were not sufficiently mature to be included in the report on Standard Test Methods For Textile Composites, a review of that work is included in this paper for completeness.

2.2.1 Interlaminar Fracture Toughness Boeing conducted Double Cantilever Beam (DCB) and End Notch Flexure (ENF) tests on four of the 2-D triaxial braids to measure Mode I (G_I) and Mode II (G_{II}) interlaminar fracture toughness, respectively (8). 2-D braids are quasi-laminar in nature. The values of G_{II} where two to three times those of G_I, and both G_I and G_{II} were considerably greater than those for prepreg tape laminates. Coefficients of variation, which were 13 to 40%, were also greater than those for prepreg tape laminates.

University of Florida conducted DCB and ENF tests on stitched uniweave composites to measure G_I and G_{II} (16). The specimens were unidirectional except for a layer of fiberglass cloth on each surface to prevent the stitches from pulling through. Values of G_I and G_{II} were much greater than those of unstitched specimens. Because of crack bridging, G_I and G_{II} increased with crack length, especially for G_{II} where the crack front propagated without breaking any stitches. Thus, the toughness values do not represent a material property.

2.2.2 Interlaminar Tension Two types of flat-wise tension test methods were evaluated by University of Delaware for 3-D weaves - a bi-material specimen and a specimen with surface pile (17). Special 25-mm-thick versions of the 3-D weaves were used. Tension load was applied to the bi-material specimen by gripping long metal beams that were bonded to each face of the composite and were bonded to the surface pile of the other specimen by gripping a composite made of the pile yarn. Because failures occurred at the interface, neither specimen was suitable for measuring tension strength. However, both methods were suitable for obtaining interlaminar elastic constants.

2.2.3 Interlaminar Compression An IITRI specimen configuration was evaluated by University of Delaware (17). The same 25-mm-thick 3-D weaves were used as those used for the bi-material tension specimen. The specimen also resembled the bi-material tension specimen except that the metal beams were replaced by Glass/epoxy beams to produce a more uniform stress state. The specimen dimensions were chosen to minimize the stress singularity at the corners of the bond interface. Elastic constants and strengths were reasonable. Failure appeared to be associated with shear damage in the warp and weft tows rather than shear kinking of the weaver tows. Again, the lower limit of thickness for this specimen was not evaluated.

2.2.4 Interlaminar Shear Compact and double-notch (ASTM D-3846) shear specimens were evaluated by University of Delaware (17). Dimensions of both were modified to test the same 25-mm-thick 3-D weaves as those tested with the bi-material tension specimen. Also, the ASTM D-3846 double-notch shear specimen was used to test the 6-mm-thick 3-D weaves. The weaves were tested with two orthogonal directions to obtain τ_{23} and τ_{32} properties, where directions 1, 2, and 3 are associated with warp, weft, and thickness directions. Shear moduli were obtained satisfactorily in all cases, but not strengths. Strengths τ_{32} were obtained satisfactorily for the 25-mm-thick weaves using the modified compact specimen and for the 6-mm-thick weaves using the modified double-notch shear specimen. These strengths were also in good agreement for the two thicknesses. However, strengths τ_{23} were not obtained satisfactorily; the modified compact and double-notch specimens failed outside the test section. Although the modified compact specimen is less complex than the double-notch specimen, it cannot be used for thin composites.

Boeing also evaluated the Short Beam Shear (SBS) and Compression Interlaminar Shear (CIS) Specimens (8). The CIS specimen is similar to the Double-Notch Shear Specimen evaluated by University of Delaware. (For the SBS specimens, the load is applied transverse to the plane of the plate; but for the CIS specimens, the load is applied parallel to the plane of the plate.) The 2-D braids and 3-D weaves were tested. Interlaminar shear strengths by SBS were 20% greater than those by CIS on average. Also, coefficients of variation were lower for SBS than for CIS. Strengths were similar for the 2-D braids and 3-D weaves. (The median value was 43 MPa.)

2.3 Impact The work on impact test methods is not sufficiently mature to be included in the report on Standard Test Methods For Textile Composites. Although additional work is needed, a review of the work is included here for completeness.

2.3.1 Quasi-Static Impact Behavior First, a discussion of quasi-static impact behavior will be useful. Assuming that the force-displacement relationship for a falling weight impact test is the same as that of a quasi-statically loaded plate (quasi-static behavior), the following "energy balance" equation can be derived (18).

$$\frac{1}{2} m_i v_i^2 = \frac{1}{2}\frac{F_{max}^2}{k} + \frac{2}{5}\frac{F_{max}^{5/3}}{n_o^{2/3} R_i^{1/3}} \qquad [1]$$

F_{max} is the impact force, m_i is the mass of the impacter, v_i is the velocity of the impacter, R_i is the radius of the spherical impacter, and k is the flexural stiffness of the plate for a quasi-statically applied impact force. The term on the left hand side of equation 1 is the kinetic energy of the impacter, and the first and second terms on the right hand side of equation 1 account for flexural displacements and for Hertzian indentation, respectively. For thin plates, the Hertzian term is negligible, and impact force increases in proportion to the square root of the product of kinetic energy and flexural stiffness.

Dynamic finite element analyses were made of 7.0-mm-thick AS4/3501-6 $[45/0/-45/90]_{6s}$ plates impacted with various masses having a 12.7-mm-diameter tup (18). Viscoelastic effects and damage were not taken into account. In Figure 2, the maximum calculated values of impact force are plotted against the frequency ratio $\omega^2/(k/m_i)$ for a constant value of impacter kinetic energy, where ω is the natural frequency of the plate and $(k/m_i)^{1/2}$ is a pseudo natural frequency of the impacter mass and plate. For homogeneous uniformly thick plates, $\omega^2/(k/m_i)$

$= (m_i/m_p)/\alpha^2$, where $\alpha = 0.471$ and 0.371 for simply supported and clamped boundaries, respectively. Results are plotted for three square plates with various sizes and boundary conditions. In general, impact force decreases with increasing impacter mass and plate size and is less for simply supported than clamped boundary conditions.

The time of contact from the dynamic finite element analyses decreased with decreasing impacter mass. Thus, the number of reflections from the boundaries before the impact force reached a peak decreased with decreasing impacter mass. When impacter mass was to the left of the vertical dashed lines in Figure 2, no reflections returned before the impact force reached a peak, and impact force was not affected by plate size nor boundary conditions. The horizontal dashed lines in Figure 2 represent calculations with equation 1 and $n_o = 4.52$ GPa. For large values of impacter mass, the dynamic analysis curves asymptotically approach equation 1 as the number of reflections become large, indicating that the force-displacement relationships are quasi-static.

2.3.2 Impact Test Methods In Current Use The NASA and SACMA impact test methods in Table 3 are the most commonly used for falling weight impact tests. For specimens 6-mm-thick or less, which is usually the case, $(m_i/m_p)/\alpha^2 > 70$ for the specimen sizes and impacter masses in Table 3. Thus, the analytical results in Figure 2 indicate that impacts are essentially quasi-static for both methods.

It is useful to think of impact response as having two aspects: damage resistance (size of damage for a given impact force) and damage tolerance (strength for a given size of damage). Neither of the test methods in Table 3 specifies the measurement of impact force, but impact force can be measured in a falling weight test with an instrumented tup. With measurements of impact force, both damage resistance and damage tolerance can be determined; without impact force, only damage tolerance can be determined.

2.3.3 Recommended Damage Resistance Test Method Metrics for damage resistance were developed in (18). Damage diameters from C-scans are plotted against impact force in Figure 3 for laminates made of AS4/3501-6 and IM7/8551-7 prepreg tape. The 3501-6 is a brittle epoxy, and the 8551-7 is a toughened epoxy. Data are plotted for both quasi-static indentation (QSI) and falling weight impact tests. The QSI tests were conducted by mounting an instrumented tup with 12.7-mm diameter in a universal testing machine and forcing the indenter against a specimen clamped over a 10.2-cm-diameter circular opening. For the falling weight tests, the same tup was used and the specimens were clamped over a 12.7-cm-square opening. The damage initiated when a critical value of impact force F_I was exceeded; then the damage diameter increased in proportion to impact force. (The diameter was calculated for a circle equal in area to the C-scan image.) The change in damage diameter with impact force was inversely proportional to a transverse shear strength per unit length Q^*. Damage from the QSI tests and falling weight tests were approximately the same size for a given impact or contact force. (Other investigations have shown similar results.) The values of Q^* and F_I for IM7/8551-7 are significantly greater than those for AS4/3501-6. Values of Q^* and F_I were also found to increase significantly with increasing thickness (10). Both Q^* and F_I can be used as metrics for damage resistance.

Values of Q^* and force F_I for the 2-D braids and 3-D weaves and for stitched uniweave composites supplied by McDonnell Douglas are given in (10). The stitched uniweave materials are described in (19).

Both falling weight and QSI tests can be used to measure damage resistance. However, QSI tests have the following advantages:

1. The test can be conducted using a universal testing machine, which is available in most laboratories and can readily measure contact force.
2. Contact force can be readily controlled, giving more direct control over damage.
3. Damage tolerance criteria tend to be based on a visibility metric such as dent depth rather than impacter kinetic energy. (The force-displacement relationship can be integrated to give a reasonably accurate estimate of impacter energy under quasi-static conditions. Thus, impact energy levels to produce specific levels of damage can be determined efficiently.)

A subcommittee of ASTM committee D30 on composites is in the final stages of developing a standard QSI test method.

The ASTM QSI method and a modified NASA method are recommended in Table 4 for measuring damage resistance. (The weak clamping apparatus and the 32-mm-thick plywood baseplate of the SACMA method may result in impact forces and consequently damage sizes that are less than those for the NASA method.) A circular opening is recommended to give polar symmetry for the apparatus. The specimens should be securely clamped over the opening, and the diameter of the opening should be greater than the diameter of the damage. (No significant difference was found in (20) among values of Q^* in QSI tests with fixture opening diameters varying between 5.1 cm and 10 cm and with damage diameters ranging between 30 and 60% of fixture opening diameters.) The MIL-STD-1530A specifies a 25.4-mm-diameter impacter. (Larger diameter tups have been shown to give larger nonvisible damage (21).) Thus, the standard tup diameter should be increased to 25.4 mm. Of course, other tup diameters could be used as necessary. For the modified NASA method, the tup should be instrumented to measure impact force. (An instrumented tup can also be used for the QSI test if the load cell of the testing machine is too large to give accurate readings.) To assure quasi-static behavior for the modified NASA method, the impacter mass should be at least 10 times that of the specimen (based on the unsupported size of the specimen). The reader should be aware that the impact forces in Figure 2 are sometimes smaller for the quasi-static situation than for small masses. Thus, damage may be smaller for the quasi-static situation than for small masses.

2.3.4 Recommended Test Method For Measuring Damage Tolerance A modified NASA method is recommended in Table 5 for measuring post-impact tension and compression strength (damage tolerance). The recommended specimen width is at least 3 times the damage diameter, and the length is two times the width plus additional length for introducing load. The compression method uses the NASA type of edge and end supports to prevent Euler buckling. The tension method does not require edge or end supports; the length λ will be determined by the load introduction method.

The specimen width in Table 5 was chosen using the following analysis. The material damaged by impact carries much less load than the undamaged material, and can be conservatively represented as an open hole of equal diameter. The stress at the edge of an open hole in an isotropic sheet of finite width was normalized by three times the far-field stress and plotted against ratio of hole diameter to sheet width in Figure 4 (22). A ratio of hole diameter to specimen width of 0.3 would result in a stress ratio of only 1.12. Thus, one would expect the far-field stress at failure for an infinite specimen to be no more than 112% of a specimen with a ratio of hole diameter to specimen width of 0.3. For a 2.5-mm dent (MIL-STD-1530A), the damage diameter for a 6-mm-thick specimen was 7 cm (23), which is 54% and 70% of the specimen widths in Table 3. For a 7-cm-diameter hole and the specimen widths in Table 3, the stress ratios in Figure 4 are 1.55 and 2.30. Thus, post-impact strengths would likely be extremely conservative.

3. HANDBOOK OF ANALYTICAL METHODS FOR TEXTILE COMPOSITES

The report on Handbook of Analytical Methods for Textile Composites includes sections on

- Overview of textiles
- Choice between textiles and tape laminates
- Failure mechanisms
- Predicting elastic constants and thermal expansion
- Nonlinear stress-strain behavior and strength
- Fatigue life
- Summary of available computer codes

A bibliography is also included. A brief overview of these sections follows.

3.1 Overview of Textiles Textiles were categorized by their two and three dimensional fiber architecture and by their quasi-laminar and nonlaminar characteristics. Processes were described for weaving, braiding, knitting, and stitching fiber preforms and for infiltrating the preforms with resin and for molding to net shape. Powder coating and commingling methods were also described for combining resin and fiber prior to and during the textile process.

Ideal unit cell geometries were described for various textile processes. Equations were given that relate fiber volume fraction and textile geometry in terms of process parameters. Irregularities in ideal geometry were described. Examples were given of integral structure made by textile processes.

3.2 Choice between Textiles and Tape Laminates Textiles and tape laminates were contrasted with regard to handling, manufacturability, and mechanical properties. Dry textile preforms can be handled in net form, stored indefinitely, and draped over and into forms of complex curvature. However, the final fiber architecture can be disturbed significantly by handling and draping. Stiffness and unnotched strength of textiles tend to be less than those of tape laminates because of the yarn crimp introduced by interlacing in the textile process - strength more than stiffness. On the other hand, notched strengths of textiles and tape laminates can be nearly equal for equivalent areal weights of fiber. For 3-D (nonlaminar) type textiles, delaminations and failures associated with secondary bonds can be reduced or eliminated. The size of impact damage can be greatly reduced by stitching.

3.3 Failure Mechanisms Failure mechanisms for nominal shear, compression, and tension were discussed. In shear and tension, failure typically begins in the resin where shear and tension stresses are large, much as in tape laminates. But failures tend to occur at lower nominal strains in textiles than tapes because yarns of textiles are much larger than plies of tape. Delaminations and kink bands can form in textiles much as in tape laminates. However, delaminations are not likely to form in the nonlaminar textiles. Even in quasi-laminar textiles, crimp in yarns almost assures kink band formation.

In both tension and compression, nonlinear processes at the notch ameliorate the local stresses and make textiles remarkably notch insensitive. The large yarns enhance the nonlinear processes. Preliminary experiments did not reveal any alarming degradation of strength due to fatigue. Damage developed more quickly in compression fatigue than in tension fatigue, much as in tape laminates. The failure mechanisms in monotonic and fatigue loading were similar.

3.4 Predicting Elastic Constants and Thermal Expansion Mathematical modeling concepts were reviewed. Isostrain and isostress assumptions were contrasted. Tows are typically assumed to be homogeneous and transversely isotropic. Because tows are more tightly compacted than tape laminates, elastic constants of tapes cannot be used. Thus, micromechanics theories are used to calculate elastic constants for the tows. (Fiber volume fractions for tape laminates are typically 0.6. Even though fiber volume fractions for textiles composites are typically 0.5 to 0.6, the fiber volume fraction in the individual tows may exceed 0.7.)

Analyses are usually made of representative volumes of textiles called unit cells. Unit cells, appropriate boundary conditions, and length scales for macroscopic behavior were discussed. Equations for average strain, stress, and virtual work theorems were derived. Also, equivalent stiffness and compliance matrices were derived by orientation averaging.

The capabilities of computer codes listed in Table 6 were reviewed. The mechanics assumptions and methods used to represent textile geometry were described for each code. Applications of the codes to quasi-laminar textiles was discussed.

Plate stiffnesses (A, B, and D matrices of Classical Lamination Theory (CLT)) and elastic constants were calculated for a carbon/epoxy plain weave. Experimental values and calculations by CLT and a finite element method (FEM) were compared. The CLT calculations were made for a laminate with equivalent areal weights of fiber and equal overall fiber volume fraction. Some of the values of plate stiffness did not agree well. The differences were attributed to the manner in which bending-stretching coupling was taken into account. These differences should diminish with increasing number of layers. Except for ν_{12}, the elastic constants were in general agreement for the various codes and experiments. The value of ν_{12} calculated using μTex-10 was in agreement with the test value, but the other calculated values were about one third of the test value.

3.5 Nonlinear Stress-Strain Behavior and Strength Matrix plasticity, matrix cracking, and rotations of crimped tows (straightening) typically preceded catastrophic failure of the tows. This progression of damage results in nonlinear stress-strain behavior. The capabilities of the computer codes to predict local stresses, nonlinear stress-strain behavior, and strength are given in Table 6. None of the models can analyze notched textile composites.

3.6 Fatigue Life A kink band formation model for compression fatigue was reviewed. The stress to cause kink band formation is given by

$$\sigma_k = \tau_c / \phi \qquad [2]$$

where τ_c is the flow stress and ϕ is the misalignment angle. Values of ϕ were determined by examining photomicrographs of the textile composite. The flow stress was assumed to deteriorate with cycles of load according to

$$\frac{d\tau_c}{dN} = -A(\Delta\sigma_s\phi)^m \qquad [3]$$

where $\Delta\sigma_s$ is the cyclic stress amplitude and A and m are material constants. Integrating equation 4,

$$N_k = \frac{\tau_c - \Delta\sigma_s\phi}{A(\Delta\sigma_s\phi)^m} + 1 \qquad [4]$$

For load controlled fatigue tests, fatigue failures followed formation of just a few kink bands. Values of m were 30 for carbon/epoxy 3-D interlock weaves and 15 for glass/urethane 2-D triaxial braids. For strain controlled fatigue tests of carbon/epoxy 3-D interlock weaves, fatigue failure did not follow formation of a few kink bands. Instead, the number of kink bands increased gradually.

3.7 Summary of Available Computer Codes A summary and user's guides of the computer codes in Table 6 was provided. The authors names, references, modeling assumptions, source code language, and description of input and output are given.

4. A DATABASE OF MECHANICAL PROPERTIES FOR TEXTILE COMPOSITES

The test data was assembled from the NASA Advanced Composite Technology (ACT) program and from the public domain. The data documentation requirements of MIL-HDBK-17, Section 8.1.2 were satisfied. Properties were included for

- 2-D triaxial braids
- 3-D multiaxial braids
- stitched uniweaves
- 8 harness weaves
- 3-D interlock weaves
- stitched multiaxial warp knits
- uniweaves

Materials and product forms are given in Table 7. Each material and product form is characterized by a set of textile parameters. For example, braids are characterized by axial and braider tow sizes and spacings, by braid angle, etc.

The database include the following elastic properties and strengths:

- Tension and compression moduli
- Shear moduli
- Poisson's ratios
- Unnotched tension and compression strengths
- Open-hole tension and compression strengths
- Filled-hole tension and compression strengths
- Bearing
- Through-the-thickness tension
- Tension and compression after impact
- Inplane shear
- Fatigue
- Biaxial tension and compression

Not all elastic properties and strengths were available for all material subclasses and product forms.

MSC/MVISION, which requires a work station with a UNIX operating system, was chosen for the principal database manager. The schema hierarchy structure for the MSC/MVISION database is as follows:

Class (material class, schema level...)
  Material (material name, material form...)
    Constituent (fiber/matrix class, fiber size...)
      Preform (textiles preform architecture, preform descriptions...)
        Process (process method, tackifier information...)
          Molding (panel layup, NDE method...)
            Test (test type, test facility, test fixture...)
              Coupon (nominal coupon geometry)
                Conditioning (test environment, coupon conditioning...)
                  Actuals (failure loads/stresses, measured values...)

The MSC/MVISION database, which is of moderate size (≈17 megabytes), has two useful features. First, using the ***Query panel***, information to be displayed can be limited to specific materials or properties, for example only stitched AS4/RSL-1895 material. Second, values of specific attributes can be selected using the ***Select*** command. Data for multiple materials and attributes can be selected using operators like ***and*** and ***or***. When multiple tests were conducted, the data for each test is included. Thus, statistical analyses can be performed by the user. Load-deformation curves were included for some tests.

Also, Microsoft Excel was chosen for a summary or executive database manager for PC and Apple Macintosh desktop computers. Much less information about the material, test variables, and test results is given in the Microsoft Excel databases than in the MSC/MVISION database. The Microsoft Excel database has tabs to select specific materials with average properties distributed among the columns.

5. CONCLUDING REMARKS

NASA and its contractors have completed a program to develop a basic mechanics underpinning for textile composites. Three major deliverables were produced by the program:

1. a set of test methods for measuring material properties and design allowables
2. mechanics models to predict the effects of the fiber preform architecture and constituent properties on engineering moduli, strength, damage resistance, and fatigue life
3. an electronic data base of coupon type test data

6. REFERENCES

1. Mechanics of Textile Composites Conference. NASA CP 3311, Parts 1 & 2, Oct. 1995.
2. John E Masters, NASA CR-198286, February 1996.
3. C. C. Poe, Jr. and Charles E. Harris NASA CP-3326, Vol. 1, 1996.
4. John E Masters and Marc A. Portanova, NASA CR-4751, 1996.
5. Brian N. Cox and Gerry Flanagan, NASA CR-4750, 1996.
6. Jerry Delbrey, NASA CR-4747, 1996.

7. Pierre J. Minguet and Christian K. Gunther, NASA CR-4610, July 1994.
8. Pierre J. Minguet , Mark J. Fedro, and Christian K. Gunther, NASA CR-4609, July 1994.
9. Timothy L. Norman, Colin Anglin, David Gaskin, and Mike Patrick, NASA CR-4676, June 1995.
10. Wade C. Jackson and Marc A. Portanova, NASA CP 3311, Part 2, October 1995, pp. 315-348.
11. M. A. Portanova and J. E. Masters, NASA CR-198262, December 1995.
12. M. A. Portanova and J. E. Masters, NASA CR-198263, December 1995.
13. M. A. Portanova, NASA CP-198264, December 1995.
14. M. A. Portanova and J. E. Masters, NASA CP-198266, December 1995.
15. John E. Masters, NASA CP-198285, February 1996.
16. Suresh K. Sharma and Bhavani V. Sankar, NASA CR-195042, February 1995.
17. Dru Hartranft, Azar Pravizi-Majidi, and Tsu-Wei Chou, NASA CP 3311, Part 2, October 1995, pp. 251-314.
18. W. C. Jackson and C. C. Poe, Jr., Journal of Composites Technology & Research, Vol. 15, No. 4, Winter 1993, pp. 282-289.
19. Marc Portanova, NASA CP 3311, Part 2, October 1995, pp. 391-424.
20. Young S. Kwon and Bhavani V. Sankar, NASA CR-187624, March 1992.
21. C. C. Poe, Jr., Proceedings of the American Society for Composites, Third Technical Conference, 1988, pp. 334-343.
22. Raymond J. Rourk, and Warren C. Young, Formulas for Stress and Strain. Fifth Edition. McGraw-Hill Book Co., 1975.
23. M. A. Portanova, NASA CR-198265, December 1995.
24. Peter G. Ifju, NASA CP-3311, Part 1, October 1995, pp. 141-176.
25. NASA Reference Publication 1142, June 1985.
26. SACMA Recommended Test Method for Compression after Impact Properties of Oriented Fiber-Resin Composites, SRM 2-88.

TABLE 1. TEST METHODS EVALUATED AND SELECTED BY LOCKHEED ENGINEERING & SCIENCES CO.

Test method type	Test methods evaluated	Best test methods
	Inplane	
Unnotched tension	Straight sided (ASTM D3039) and dog-bone specimens	Straight sided (ASTM D3039) specimens
Unnotched compression	NASA Short Block, Modified IITRI, Boeing OHC, Zabora, Boeing CAI, NASA ST-4, and sandwich column	Zabora
Open-hole tension	Straight sided specimen (ASTM D5766)	Straight sided specimen (ASTM D5766)
Open-hole compression	NASA Short Block, Modified IITRI, Boeing OHC, Zabora, Boeing CAI, NASA ST-4, and sandwich column	Zabora
Filled-hole tension	Straight sided specimen (ASTM D5766)	Straight sided specimen (ASTM D5766)
Bolt bearing	Boeing Unstabilized Single Shear, Boeing Stabilized Single Shear, and Boeing Double Shear	Boeing Stabilized Single Shear
Shear	Tube Torsion, Rail Shear, and Compact Shear (24)	None totally satisfactory
	Out-of-plane	
Interlaminar tension	C- and L-Shape Bend and Four Point Bend	Four Point Bend (strength only)

TABLE 2. OTHER OUT-OF-PLANE TEST METHODS EVALUATED

Test method type	Test methods evaluated	Comments
Interlaminar fracture toughness	DCB (Mode I)	Reasonable for 2-D braids and stitched uniweaves
Interlaminar fracture toughness	ENF (Mode II)	Reasonable for 2-D braids
Interlaminar tension	Flat-Wise Tension	Reasonable for elastic constants
Interlaminar compression	Modified IITRI	Reasonable for elastic constants and strength
Interlaminar shear	Compact, Double Notch Shear, and Compression Interlamnar Shear	Compact for thick & Double Notch for thin composites
Interlaminar shear - transverse	Short Beam Shear	Reasonable for 2-D braids and 3-D weaves

TABLE 3. IMPACT TEST METHODS IN CURRENT USE

Test method	Unsupported specimen size, cm	Post-impact specimen size, cm	Edge support during impact	Edge support during compression test	[a]Impacter tup diameter, mm	Impacter mass, kg
NASA (25)	13x25	13x25	Clamped	Simply supported	12.7	4.5-5.4
SACMA (26)	7.6x13	10x15	[b] Clamped	Simply supported	15.9	5.0

[a]Hemispherical shaped tups.

[b]The prescribed clamping apparatus more nearly represents simply supported conditions than clamped conditions.

TABLE 4. RECOMMENDED DAMAGE RESISTANCE TEST METHODS

Name	[a]Minimum circular opening dia., cm	Edge support during impact	[b]Impacter tup diameter, mm	[c]Minimum impacter mass, kg
QSI	D	Clamped	25	-
Modified NASA	D	Clamped	25	$2.5\pi\ (\rho h)_p x D^2$

[a]D should be larger than the diameter of the impact damage to avoid boundary effects and should be much larger than the specimen thickness to minimize shear deformation effects.

[b]Hemispherical shaped tups.

[c]$(\rho h)_p$ is the product of density and thickness of the specimen.

TABLE 5. RECOMMENDED POST-IMPACT STRENGTH TEST METHOD (DAMAGE TOLERANCE)

Name	[a]Specimen width	[b]Specimen length	Edge support	End support
		Compression test		
Modified NASA	$\geq W$	$\geq 2W + \lambda$	[c]Simply supported	Clamped
		Tension test		
Modified NASA	$\geq W$	$\geq 2W + \lambda$	None	Clamped in grips or pin loaded

[a]$W = 3d_o$, where d_o is the diameter of the impact damage.

[b] λ represents the additional specimen length required for load introduction.

[c]When appropriate, Euler buckling of large, thin specimens can be prevented by supporting the faces of the specimen.

TABLE 6. SUMMARY OF COMPUTER CODE CAPABILITIES

Code	Textile forms	3-D stiff-ness	Ther-mal expan-sion	Plate stiff-ness	Local stress	Non-linear-ity	Strength
PW	Plain weaves						
SAT5	5 harness satin						
SAT8	8 harness satin		Yes	Yes			
CCM-TEX	3-D weave and 2- and 4-step braids	Yes					Yes
μTex-10	General	Yes	Yes	Yes	Yes		
μTex-20	User defined	Yes	Yes	Yes	Yes		
SAWC	plain weave (FE code)	Yes			Yes	[a]Yes	Yes
TEXCAD	2-D weaves and 2-D braids	Yes	Yes	Yes	Yes	[b]Yes	Yes
WEAVE	3-D weaves	Yes					
BINMOD	3-D weaves	Yes				Yes	Not directly

[a]Stiffness discounting.

[b]Shear hardening and stiffness discounting.

TABLE 7. MATERIAL SUBCLASSES AND PRODUCT FORMS IN DATABASE

Material subclass	Available product forms
AS4/3501-6	Uniweaves, knits, fabric, & tape
AS4/8551-7	Tape
AS4/CET-3	8H satin fabric
AS4/E905-L	3-D braid & 8H satin fabric
AS4/IM7/3501-6	Stitched knit
AS4/PR-500	Braids, weaves, & fabrics
AS4/PT-500	3-D weaves
AS4/RSL-1895	Braids, weaves, uniweaves, & fabrics
AS4/RSS-1952	8H satin fabric
IM7/8551-7	Tape

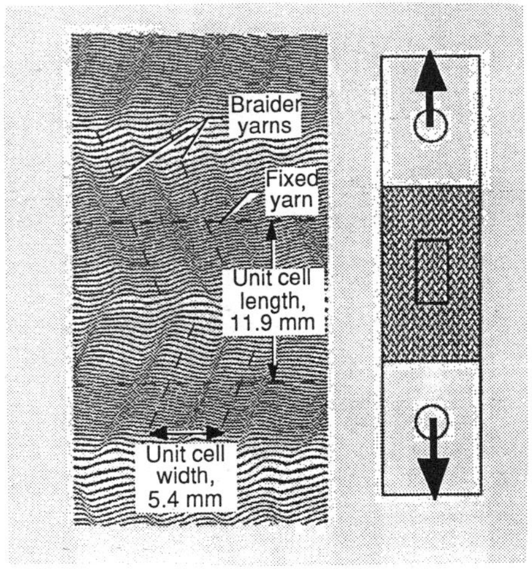

Figure 1. Vertical displacement field in triaxial braid by Moiré.

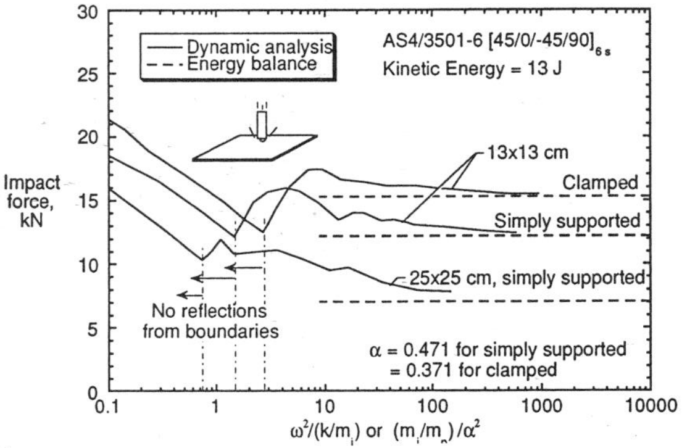

Figure 2. Comparison of dynamic finite element analysis and energy balance equation.

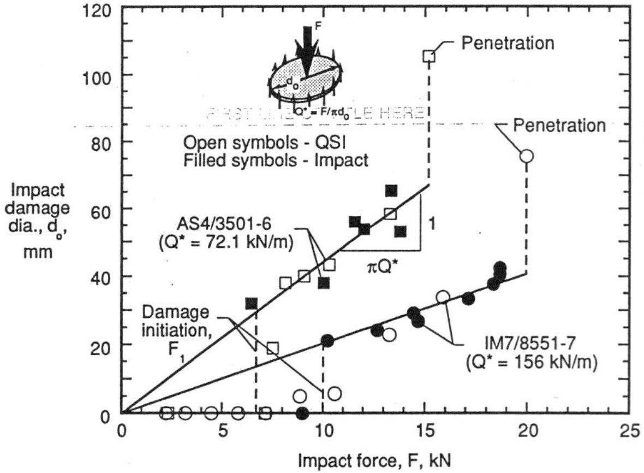

Figure 3. Damage for AS4/3501-6 and IM7/8551-7 $[45/0/-45/90]_{6s}$ prepreg tape laminates (7.1 mm thick) caused by quasi-static indentation and falling weight impact tests.

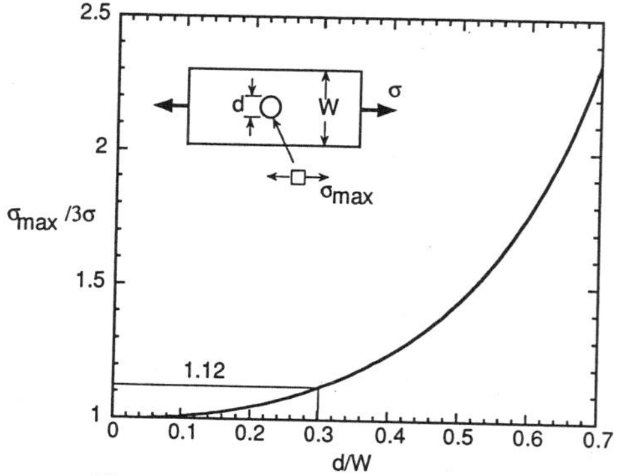

Figure 4. Finite width effect for open hole.

28th International SAMPE Technical Conference
November 4-7, 1996

IMPACT DAMAGE RESISTANCE OF TEXTILE COMPOSITES

Wade C. Jackson
Vehicle Structures Directorate, Army Research Laboratory, Hampton, VA

Marc A. Portanova
Naval Air Warfare Center, Aircraft Division, Patuxent, MD

ABSTRACT

Composites made from textile preforms are currently being evaluated for use in aircraft structures. Textile composites have the potential for significant cost savings through automated preform fabrication and low-cost resin transfer molding (RTM). Another benefit of textile composites may be improved impact properties over composites made from unidirectional tape. Laminated composites made from unidirectional tape are very susceptible to delamination from out-of-plane loads due to their relatively low interlaminar properties. Textile composites may improve the through-the-thickness properties through nesting of preform layers or direct reinforcement in the thickness direction. This report summarizes a series of tests that was run to characterize the impact damage resistance of a variety of textile composites that included 2D braids, 3D weaves, and stitched and non-stitched uniweaves. All materials had a nominal thickness of 6.35 mm (0.25 in.). The specimens were evaluated using a quasi-static indentation (QSI) test method. The impact damage resistance was summarized in terms of the contact force required to initiate large delaminations and a delamination resistance parameter. The delamination resistance parameter, Q*, related delamination diameter to impact force over a range of delamination sizes.

KEY WORDS: Composites, Impact Resistance, Textile Reinforcement

1. INTRODUCTION

To improve the weight saving benefits of carbon/epoxy materials, there is a need to improve the damage tolerance and delamination resistance of these materials. Textile composite materials have the potential for better through-the-thickness properties than traditional laminated composites. Traditional laminated composites are very susceptible to delamination from out-of-plane loads that may occur from impact loading or around structural details such as curved geometry, ply drops, or fasteners. With improved through-the-thickness properties, the susceptibility to damage from out-of-plane loads should be greatly reduced. Textile composites may have a 2D construction where discrete layers are stacked to produce a desired thickness or a 3D construction where a single layer is manufactured to a desired thickness. In the 2D materials, the preform layers are expected to nest snugly together to improve the through-the-thickness properties. In the 3D materials, reinforcement is placed in the thickness direction to directly improve these properties.

This investigation focused on measuring impact damage resistance. Existing impact test methods were not suitable to measure damage resistance. Compression-after-impact (CAI) tests incorporate elements of both damage resistance and damage tolerance. The only results typically reported from CAI tests are compression strengths. Since the compression strength is a function of both damage resistance and damage tolerance, the user is unable to determine which properties were improved or not improved. An improvement in CAI strength could even be caused by an improvement in damage tolerance but a decrease in damage resistance. All CAI test results are based on impacter kinetic energy. Consequently, the amount of damage is a function of target and impacter configuration (i.e., plate size, BC's, indenter diameter, etc.) [1]. In addition, since the results are configuration dependent, the results from coupon tests cannot be easily scaled to plates of different sizes and boundary conditions.

Impact tests can be performed with a large range of impacter mass and velocity combinations (air gun, falling weight, pendulum, etc.). A falling-weight impact test typically uses a large mass dropped from a relatively low height. The air-gun impact test uses a small mass projectile fired down a barrel. Depending on the mass and velocity combinations, the impact specimen can respond in various manners. It has been shown for many cases that a relatively small plate impacted by a large mass at a low velocity will deform in a quasi-static manner [1-3]. This response is typical for falling-weight impacts on CAI test specimens. A quasi-static indentation (QSI) test uses a universal test stand to slowly apply a transverse load to simulate this response.

The quasi-static indentation (QSI) test was selected to measure damage resistance for this test program. This test method is the easiest of the impacting methods to use since it requires no specialized testing equipment. In addition, the force is applied very slowly under displacement control which allows for additional measurements during the test. Since a quasi-static deformation occurs, the test closely simulates the falling-weight impact test in many cases. By using impact force rather than impact energy, the damage is much less dependent on the impact configuration. If desired, the impact energy can be obtained by integration. The force where large delaminations initiate was used as one measure of damage resistance. Since the delamination size is a function of the applied force, a damage resistance parameter, Q*, which relates delamination diameter to impact force over a range of delamination sizes was used.

The objective of this study was to characterize the impact damage resistance of a variety of textile composites that included 2D braids, 3D weaves, and stitched and non-stitched uniweaves. Laminates made from unidirectional tape were also included for comparison. All materials had a nominal thickness of 6.35 mm (0.25 in.) and were manufactured with AS4 fibers and similar epoxies. The specimens were evaluated using a quasi-static indentation (QSI) test method. The damage resistance was summarized and compared in terms of the contact force required to initiate large delaminations and the Q* delamination resistance parameter.

2. MATERIALS

A variety of textile and prepreg tape materials were used in this investigation. These included 2D triaxial braids (four different braid geometries), 3D interlock weaves (three architectures), stitched and nonstitched uniweaves, and toughened and untoughened prepreg tape materials (AS4/8551-7 and AS4/3501-6). All specimens were manufactured to a nominal thickness of 6.35 mm (0.25 in.) with Hercules AS4 carbon fibers and Hercules 3501-6 epoxy or similar epoxies. The toughened epoxy (8551-7) prepreg was included for comparison purposes. All materials, except for the prepreg tapes, were impregnated with epoxy resin using the resin transfer molding (RTM) process [4]. After manufacture, all laminates were ultrasonically c-scanned to ensure that the panels were of high quality and free from manufacturing defects.

The 2D braids were formed on cylindrical mandrels and incorporated longitudinal yarns to create a triaxial construction with a $0°/\pm\theta°$ orientation (Fig. 1). Four different braid geometries were manufactured [3]. The braid angle, the yarn sizes, and the longitudinal yarn content were varied to determine sensitivity to these parameters. The desired thickness was obtained by overbraiding layers. Since all specimens had the same nominal thickness of 6.35 mm, the number of layers decreased with increasing yarn bundle size. The preforms were impregnated with Shell RSL-1895/W epoxy resin using the RTM process [4]. The 1895/W system was developed for RTM and has similar properties to 3501-6 epoxy.

Type	Braid angle	Braider size	% 0° yarns
$[0_{30k}/\pm70_{6k}]_8$	±70	6k	46
$[0_{75k}/\pm70_{15k}]_6$	±70	15k	46
$[0_{36k}/\pm45_{15k}]_6$	±45	15k	46
$[0_{6k}/\pm45_{15k}]_{10}$	±45	15k	12

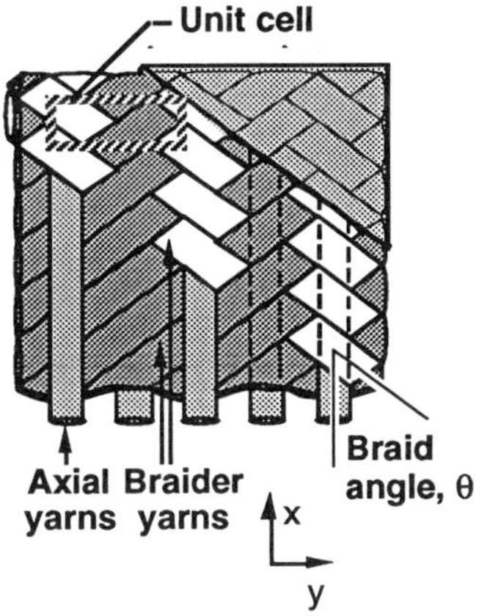

Figure 1. 2D triaxial braid architecture

The 3D weaves were all interlock woven fabrics where yarns are woven through the thickness to provide direct resistance to delamination. The interlock tows ran parallel to the warp (0°) yarns and wrapped around the weft (90°) yarns. Three different weave architectures were investigated: through-the-thickness orthogonal interlock (OS), through-the-thickness angle interlock (TS), and layer-to-layer angle interlock (LS). The weave architectures are shown schematically in Fig. 2. For each architecture, one panel was woven using small fiber bundles (-2), and one panel was woven using large fiber bundles (-1). These preforms were also impregnated with Shell RSL-1895/W epoxy resin.

Stitched and nonstitched uniwoven composite materials were manufactured with quasi-isotropic lay-ups of $[45°/0°/-45/90°]_{6S}$. These laminates were made from layers of dry uniweave carbon fabric that consisted of 97% Hercules AS4 carbon fibers and 3% fiberglass fill yarn by weight. The fill yarns were 225-denier fiberglass and were woven normal to the carbon tows to hold the fibers together while stitching. The preforms for the stitched laminates were lock stitched at eight penetrations per 25.4 mm with a 1250 yd/lb fiberglass yarn. The stitch rows were in the 0° direction with 3.18-mm (1/8 inch) row spacing. The preforms were impregnated with Hercules 3501-6 epoxy by Douglas Aircraft Company using the RTM process.

The quasi-isotropic tape specimens were 48 plies thick with a layup of $[0/45/90/-45]_{6S}$ and manufactured using prepregs of AS4/3501-6 or AS4/8551-7. To make direct comparisons with 2D braids and 3D weaves, tape equivalent specimens were made from AS4/3501-6. Tape equivalents were produced of the $[0_{6k}/\pm45_{15k}]$ triaxial braid with a layup of $[(45/-45)_2/0/(45/-45)_3/0/(-45/45)_3/0/(-45/45)_2]$ and of the TS-2 through-the-thickness angle interlock weave with a layup of $[(90_3/0_4)_3/90_3/(0_4/90_3)_3]$.

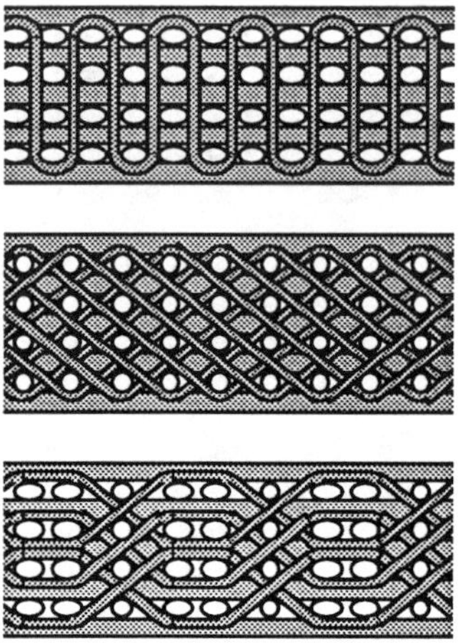

Code	Warp #	Warp size	Weft #	Weft size	Weaver size
-1	4	24k	5	12k	6k
-2	6	12k	7	6k	3k

Figure 2. 3D weave architecture

3. TEST PROCEDURE AND RESISTANCE PARAMETERS

The QSI tests were performed in a servo-hydraulic load frame under stroke control at a ramp rate of 0.51 mm/min. All specimens were clamped firmly in an aluminum test fixture during testing. Several specimen and test fixture sizes were used in this study. For indenter displacements less than the plate thickness, the damage resistance parameters that were measured in this study were not affected by changes in specimen and fixture dimensions [1, 5]. An instrumented tup attached to a 12.7-mm-diameter hemispherical indenter was used to measure load. The tup was mounted in the grips of the load frame such that the indenter traveled normal relative to the face of the specimen. The indentation load and stroke output were recorded at a rate of one data point per second throughout the loading history using a digital data storage oscilloscope. The relevant damage and test parameters were measured and recorded after each test. Many of the test coupons were indented more than one time. Initially, each specimen was loaded to a pre-determined force and then unloaded. The damage and test parameters were then measured and recorded. After this initial series of measurements was taken, the specimen was then reloaded to a higher force and the series of measurements was repeated. This process was repeated at approximately 2.22-kN (500-lbf) increments until the delaminations had grown close to the edge of the test fixture.

To gain a better understanding of the impact damage mechanics, particular damage events were related to the impact force where the event occurred. The force where large damage initiates, F_1, was used as one measure of impact damage resistance. This force was defined as the force at which the contact force versus indenter displacement curve changes because of damage formation or growth. At this force, a sharp load drop often occurs which is associated with unstable delamination growth. This unstable delamination growth typically causes a sudden loss of transverse stiffness and an audible "pop." This force was used because of its ease of identification. The F_1 force may not represent the initiation of damage since matrix cracks and small delaminations may form prior to reaching F_1. A comparison of F_1 forces for all of the materials was made.

One objective of this study was to measure the damage resistance of different material forms over a range of damage sizes. CAI tests are generally performed at a single impact energy representing a single damage size. To quantify damage resistance over a range of damage sizes, a new method was used that relates damage size to impact force [1]. In this method, a single damage resistance parameter called "Q*" provides a means to quantitatively rank the damage resistance. The analysis used to develop Q* was based on experimental observations involving impacts of quasi-isotropic plates. The Q* parameter represents the average transverse shear force per unit length associated with the edge of the delamination (Fig. 3). This parameter was observed to be constant between initial delamination growth and indenter penetration for quasi-isotropic laminated composites [1]. Since Q* is calculated from equilibrium, it is not affected by the damage state. By making Q* calculations, each materials ability to resist delamination growth can be assessed.

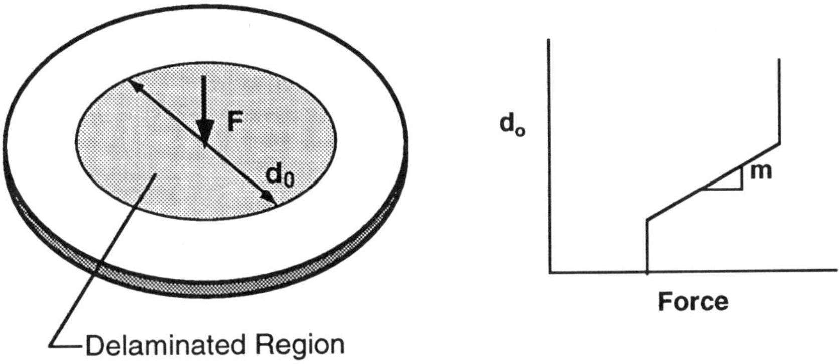

From equilibrium:

$$Q_r^* = \frac{F}{\pi d_0}$$

$$Q_r^* = \frac{1}{\pi m}$$

Figure 3. Calculation of the Q* damage resistance parameter.

To calculate Q*, the relationship between impact force and maximum delamination diameter, d_0, was determined. Q* is then calculated from the slope of the force - delamination diameter line between delamination initiation and indenter penetration (Fig. 3). The extent of the delamination was determined ultrasonically using c-scans which provides a projection of all the various regions of delamination. The delamination area was calculated using image analysis and converted to diameter assuming a circular delamination area.

An example of an impact force - delamination diameter plot is shown in Figure 4 for a 48-ply AS4/3501-6 quasi-isotropic laminate. A Q* of 75.0 kN/m was calculated from the slope of the line. There were no significant effects on the delamination diameter between running single tests per specimen versus multiple tests per specimen.

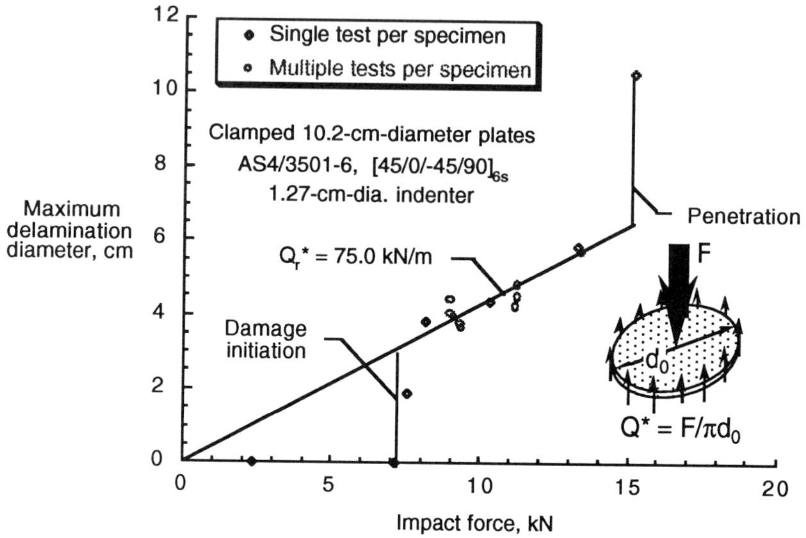

Figure 4. Delamination diameter as a function of impact force for AS4/3501-6.

4. RESULTS

4.1 Force-Displacement Response and the F_1 Damage Resistance Parameter
The indenter displacement and contact force were digitally recorded during all tests for subsequent analysis. Sample force-displacement responses are presented in this section along with a summary of F_1 values.

4.1.1 Stitched and Nonstitched Uniweaves The force - indenter displacement is shown for a 48-ply stitched and nonstitched uniweave in Figure 5. Each of these coupons was 102-mm square and contained in a fixture with a 76.3-mm square opening. The initial slopes of both uniweaves are nearly identical which indicates that the stitching did not affect the transverse stiffness. The average values of F_1 along with the high and low values are also shown in Figure 5. The average value of F_1 for the stitched laminates was approximately 17% lower than for the nonstitched laminates. The stitches may have been the source of a stress concentration that allowed damage to develop at a lower contact force. Typical of quasi-isotropic tape laminates, the nonstitched laminate has a sharp load drop at F_1 accompanying unstable delamination growth. Stitching, however, appears to have prevented this initial unstable delamination growth since a sharp load drop did not occur. As the force was increased, the stitched laminate maintains a higher transverse stiffness. The higher transverse stiffness was attributed to smaller delaminations that were held tightly together by the stitches. The sharp load drop that occurred at the higher loads in the stitched laminate may have been caused by stitch failures.

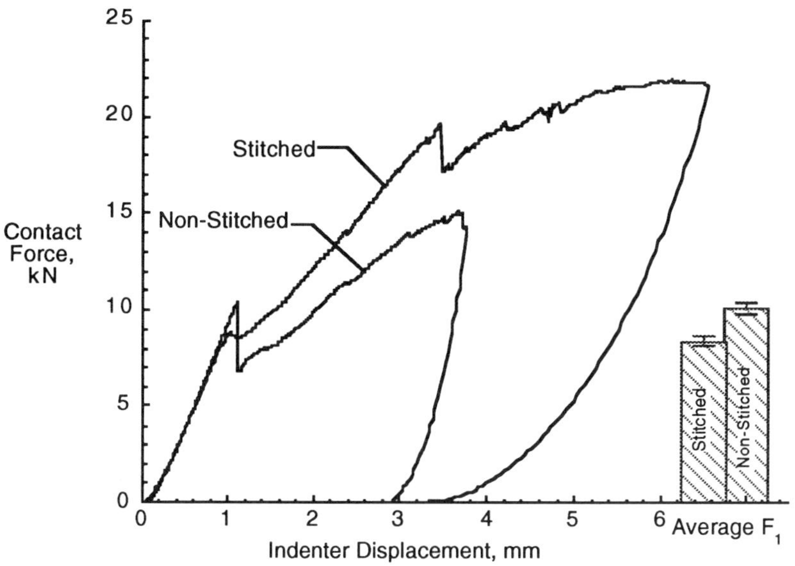

Figure 5. Quasi-static indentation of 48-ply stitched and non-stitched uniweaves

4.1.2 2D Triaxial Braids Each of these coupons was 76.2-mm square and contained in a fixture with a 50.8-mm square opening. A smaller specimen was used due to a shortage of material. Previous studies have shown that the value of F_1 was independent of plate size. The force - indenter displacement is shown for the braids and a tape equivalent braid in Figure 6. The differences in transverse stiffness between the different braids reflect the different fiber architectures which result in different flexural properties. The values of F_1 were easily identifiable since the layers of braided preform delaminated causing a loss of the local transverse stiffness. There was no sharp drop in force when F_1 was reached as occurs with composites made from unidirectional tape. Consequently, the nesting of the layers may have prevented unstable delamination growth. The tape equivalent braid, however, did exhibit unstable delamination growth. A summary of the average values and high/low values of F_1 is also shown in the figure. The $[0_{75K}/\pm70_{15K}]$ braid had a lower value of F_1 (lower by a minimum 17% in all cases) than the other three braids. This lower value may be caused by low transverse strengths in the large 75k fixed yarns. The tape equivalent laminates had an average F_1 that was more than 25% greater than the actual braid. The other three braids had very similar values of F_1. After delaminations formed, the braids had a slightly higher transverse stiffness relative to the tape equivalent laminate. Again, the higher stiffness may be attributed to smaller delaminations.

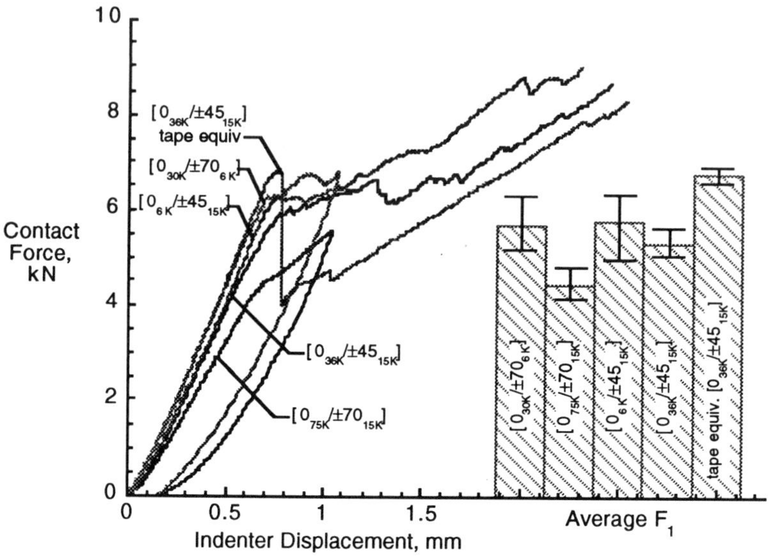

Figure 6. Quasi-static indentation of 2D triaxial braids.

4.1.3 3D Interlock Weaves Again, small 76.2-mm square coupons contained in a fixture with a 50.8-mm square opening were used due to a shortage of material. The loading history is shown for three of the weaves and a tape equivalent weave in Figure 7. All woven panels were cured to the same thickness and, consequently, the differences in stiffness reflect the different textile architectures. The value of F_1 was not easily identifiable in several of the weaves. In the OS-2 weave, a distinct point could not be identified at all since the material gradually lost stiffness. A summary of the average and high/low values of F_1 for all the 3D weaves and the tape equivalent is also shown in Figure 7. The LS and TS weaves have very similar values of F_1. The OS-1 weave has a slightly lower value of F_1 than the other weaves. Again, a distinct value of F_1 could not be identified for the OS-2 weaves. The TS-2 tape equivalent laminate had a 35% larger value of F_1 than the actual weave. The tow size did not affect the F_1 values for the LS and TS weaves. However, the smaller tows and the corresponding higher density of through-the-thickness reinforcement in the OS-2 weave may have significantly reduced or eliminated delaminations. Only the tape equivalent laminates showed a sharp drop in force when F_1 was reached. Therefore, the through-the-thickness reinforcement in the weaves was able to prevent unstable delamination growth. The slope of the force-displacement curve was considerably less in all the weaves except the OS-2 after force F_1. This loss in transverse stiffness is typically the result of delamination formation. Consequently, the 3D reinforcement appears to be unable to prevent delamination in these weaves. In comparison to the tape equivalent laminate, all the weaves have a significantly higher transverse stiffness after F_1. Consequently, the 3D reinforcement appears to have limited the size of the delaminations and/or held the delaminations tightly together.

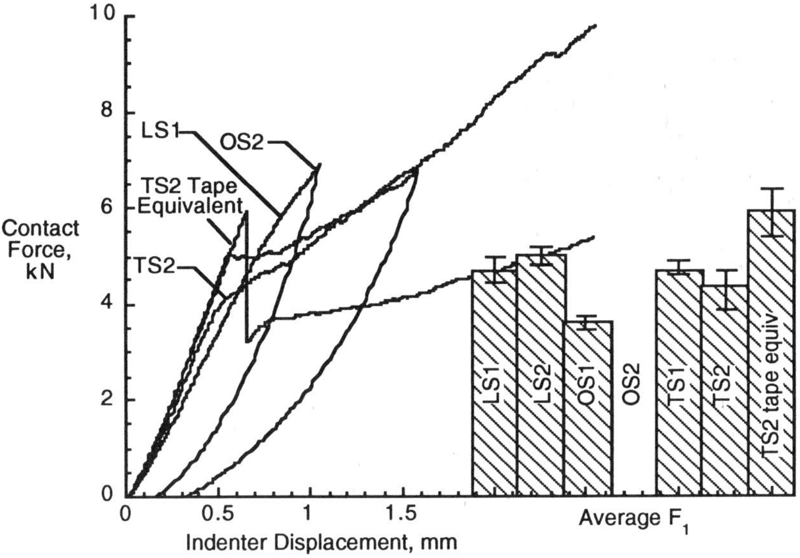

Figure 7. Quasi-static indentation of 3D weaves.

4.1.4 Summary of F1 Values An overall comparison of average F_1 values is shown in Figure 8 for all the different material systems except for the OS-2 weave. As previously stated, the value of F_1 provides a measure of initial large delamination formation. All of the comparison materials have a nominal thickness of 6.35 mm. In general, the materials made from unidirectional layers performed significantly better than the weaves and the braids. The toughened tape (8551-7) had much higher values of F_1 than the non-toughened tape (3501-6). Stitching lowered the F_1 value in the uniweave material. The tape equivalent laminates had lower values of F_1 than the quasi-isotropic tape laminates. The low F_1 values of the weave equivalent laminates may have been caused by the presence of thick layers that were created by stacking multiple layers of tape on top of one another in the same orientation to simulate the thick layers in the weaves. In summary, the out-of-plane performance of the textile materials did not improve as measured by F_1. The value of F_1 for the OS-2 weave could not be determined since the through-the-thickness reinforcement appear to have suppressed the formation of large delaminations. Consequently, the F_1 force may not be a suitable parameter to evaluate this material.

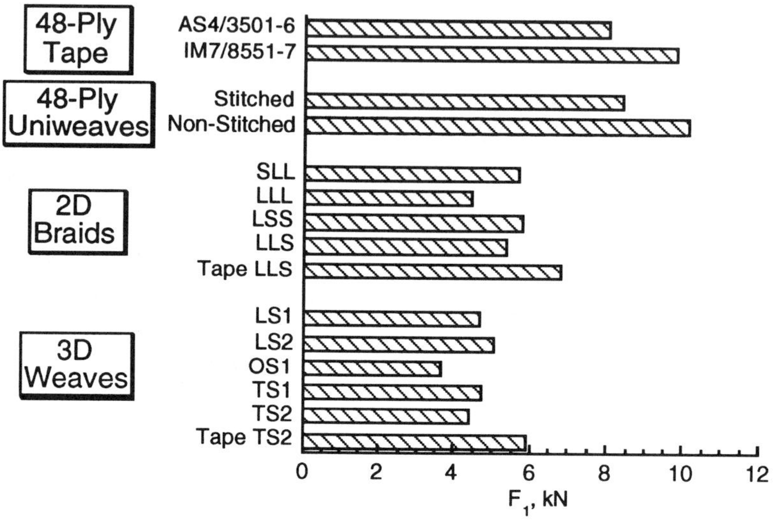

Figure 8. Overall comparison of average F_1 values for all material forms.

4.2 The Q* Damage Resistance Parameter A comparison of Q* values is made in Figure 9 for all the materials. Each value of Q* was obtained from a linear curve fit through all the data from each material form. For the 48-ply quasi-isotropic laminates, the toughened epoxy (8551-7) laminate had a significant increase in delamination resistance over the untoughened epoxy (3501-6) laminate. For a given force, the overall delamination diameter in the untoughened laminate would be more than twice the diameter of the toughened laminate. A similar improvement in Q* was observed by uniweave laminates which contained stitching. Stitching increased the damage resistance as measured by Q* by 90%.

All the 2D triaxial braids, except for the $[0_{75k}/\pm 70_{15k}]$ braid, have Q* values within 1% of one another. The Q* value of the $[0_{75k}/\pm 70_{15k}]$ braid was approximately 25% lower than the other braids. The $[0_{36k}/\pm 45_{15k}]$ tape equivalent laminate had a 22% lower value of Q* than the actual $[0_{36k}/\pm 45_{15k}]$ braid. Consequently, the nesting of the braided layers may have increased the damage resistance over a similar tape laminate.

In Figure 9, the results of the Q* calculations for each of the 3D weaves and for a tape equivalent weave are also shown. In the LS and TS weaves, there was no significant effect on Q* as a result of using the smaller tows. However, in the OS weave, Q* increased by more than a factor of two as a result of using the smaller tows in the OS-2 weave. The OS-2 weave was difficult to c-scan which may be caused by the absence of distinct delaminations between layers. The OS-1 weave (larger tows) had Q* values comparable to the other weaves. The LS weaves had approximately 25% lower values of Q* than the TS weave. The TS-2 tape equivalent laminate performed worse than the TS-2 weave by about 50% indicating that the through-the-thickness reinforcement was effective relative to a non-reinforced laminate with a similar construction.

Overall, the best performing materials were the toughened tape (IM7/8551-7), the stitched uniweave, and the OS-2 through-the-thickness orthogonal interlock weave. The delamination resistance of the braids (excluding the $[0_{75k}/\pm 70_{15k}]$ braid) and the weaves (excluding the OS-2 weave) was very similar to the quasi-isotropic tape (AS4/3501-6) and nonstitched uniweave laminates. The tape equivalent laminates and the $[0_{75k}/\pm70_{15k}]$ braid were the least resistant to delamination. The tape equivalent laminates performed worse than the same material stacked in a quasi-isotropic layup. The thick layers of the TS-2 tape equivalent laminate, which were formed by stacking multiple layers in the same orientation, may have reduced the delamination resistance.

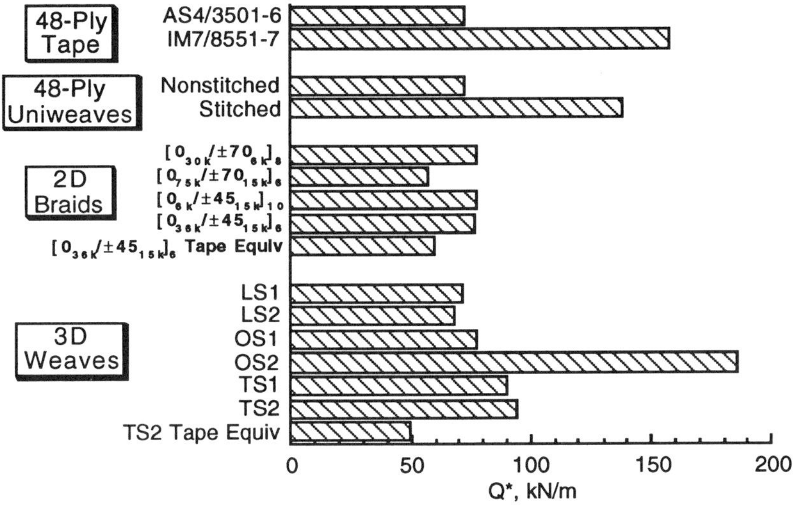

Figure 9. Overall comparison of Q* values for all materials.

5. SUMMARY

In 2D braided and 3D woven textile composites, the force to initiate large delaminations (F_1) was significantly lower than in the tape or uniweave materials. The lower initiation forces may be caused by the thicker layers used in these materials. The $[0_{75k}/\pm70_{15k}]$ braid which had large axial yarns (75k tows) had the lowest values of F_1. The through-the-thickness orthogonal interlock weave (OS-2) that was woven using small fiber bundles was the only material where a distinct value of F_1 could not be determined. The architecture of this material may have prevented large delaminations from forming. The equivalent laminates had higher values of F_1 than the textiles which may indicate that fiber waviness also plays a role in the low values of F_1. The average value of F_1 for the stitched laminates was slightly lower than for the nonstitched laminates. The stitches may have been the source of a stress concentration that allowed delaminations to develop at lower contact forces. The delamination resistance of these materials was measured by Q* which relates contact force to maximum delamination diameter over a range of delamination diameters. Using this parameter, the toughened epoxy laminates, stitched uniweaves, and OS-2 weaves had the best performance in terms of Q* (smaller damage size for a given force). Through-the-thickness tensile strength testing in the OS-2 orthogonal interlock weaves using curved beam specimens failed to produce a delamination in this material [6]. Consequently, large distinct delaminations may not have formed in this material. The delamination resistance of the braids and the weaves (excluding the OS-2 weave) was very similar to the quasi-isotropic tape (AS4/3501-6) and nonstitched uniweave laminates. However, compared to tape laminates formed to simulate the textile architecture, the textile materials did show a greater resistance to delamination growth.

6. REFERENCES

1. Jackson, W. C. and Poe, C. C. Jr., Journal of Composites Technology & Research, 15 (4), pp. 282-289, (1993).
2. Sjöblom, Peter, 32nd International SAMPE Symposium, pp. 529-539, (1987).
3. Elber, W., "Failure Mechanics in Low-Velocity Impacts on Thin Composite Plates," NASA TP 2152, May 1983.
4. Falcone, A., Dursch, H, Nelson, and Avery, W., "Resin Transfer Molding of Textile Composites," NASA CR 191505, July 1993.
5. Kwon, Y. S. and Sankar, B. V., "Indentation-Flexure Damage in Graphite/Epoxy Laminates," NASA CR-187624, March 1992.
6. Jackson, W. C. and Ifju, P. G., "Through-the-Thickness Tensile Strength of Textile Composites," NASA TM-109115, May 1994.

28th International SAMPE Technical Conference
November 4-7, 1996

IMPACT DAMAGE TOLERANCE OF TEXTILE COMPOSITES

M.A. Portanova
Naval Air Warfare Center/Aircraft, Patuxent River, MD

ABSTRACT

To fully exploit the weight saving benefits associated with graphite/epoxy composite materials there is a need to improve the damage tolerance of these materials. Composites made from textile preforms have the potential of offering the increased damage tolerance needed for application to today's advanced aero-structures. Their are various forms of textile composites, each differing not only in architecture but in the benefits associated from their design. These textile composites vary from simple stitched uni-woven laminates to complex 3-dimensionally braided or woven architectures.

This report summarized the results from a series of experiments intended to characterize the response of these materials to impact damage. Stitched and non-stitched uniweaves, 2-D braids, and 3-D weaves were evaluated. All of the specimens were subjected to either falling weight or quasi-static impacts and then loaded in tension or compression to failure to measure residual strength. The ratio of damaged to undamaged strength was used to evaluate their damage tolerance.

KEY WORDS: Damage Tolerance, Textile Composites, Impact Damage

1. INTRODUCTION

Holes, notches, or impact induced damage can severely degrade the structural integrity of conventional laminated graphite fiber reinforced epoxy composites. Because of this, relatively low design strains are imposed. To obtain the weight saving benefits associated with graphite/epoxy materials there is a need to improve the damage tolerance of these materials. Attempts to accomplish this through the introduction of toughened epoxy resins has been very expensive. Another approach has been to incorporate through-the-thickness reinforcement such as in textile composites.

This paper will address the impact damage tolerance of three different textile composite material forms. The material forms tested were stitched and non-stitched uniweaves, 2-Dimensional braids, and 3-Dimensional weaves. The uniweave was evaluated in both stitched and non-stitched configurations. Several variations of braiding and weaving parameters were assessed. All the materials were subjected to either quasi-static indentation or low velocity (large mass) impacts and then loaded in tension or compression to measure residual strength.

All specimens were impacted or indented using an instrumented tup containing a 0.5 inch diameter hemispherical tip. The low velocity impacts were done with a instrumented falling weight, dropped from a predetermined height. The static indentations were performed in a servo hydraulic load frame under stroke control. All specimens were clamped firmly in an aluminum picture frame test fixture for impacting. Damage extent was determined through the application of ultrasonic through transmission C-scans. Damage tolerance was determined by measuring the materials ability to support load for a given damage state.

2. DESCRIPTION OF MATERIALS

2.1 Dimensional Triaxial Braids

All of the 2-D fabric preforms were braided by Fiber Innovations Inc., Norwood, MA. The test panels were resin transfer molded (RTM) using Shell RSL-1895 epoxy resin and cured at Boeing Defense and Space Group in Seattle, WA. Details of Boeing's manufacturing process can be obtained from Ref. [1], "Resin Transfer Molding of Textile Composites". Four different braided architectures were evaluated but at only one laminate thickness. The desired plate thickness was achieved by stacking layers of braided fabric to obtain a nominal 0.25 inch thickness after curing. Axial tow size and content, expressed as the percentage of total yarn content, as well as braided tow size were varied in an attempt to evaluate the effect of yarn size. The intent was to allow comparisons between two or more braids based on a single variation in their construction.

2.2 3-Dimensional Interlocking Weaves

Three different 3-D woven composites were evaluated in this investigation. These were an Orthogonal Interlock (OS), an Angle Interlock (TS), and an Layer to Layer Interlock (LS). Each 3-D architecture was woven to its desired thickness in a single pass, thus the weaves did not contain interfaces between plies. All three of these architectures provide true through-the-thickness reinforcement by interlacing yarns in the z direction. The preforms were produced by Textiles Technologies Inc. and were resin transfer molded at Boeing using Shell RSL-1895 epoxy and cured.

2.3 Stitched Uniweaves

The uniweave fabric was produced by Textile Technologies Inc. and then resin transfer molded at Boeing. The materials tested were constructed from AS4/3501-6 graphite epoxy. Stitching of the uniweaves was performed by Cooper Composites in Seattle, WA. with S2 Glass at a stitch density of 8 stitches per inch, spaced 0.125 inch apart. The stitch tow size used was 3K. A 48 ply quasi-isotropic $[+45/0/-45/90]_{6S}$ lay-up was tested in both stitched and unstitched forms. The stitched uniweaves had an average thickness of 0.295 inch while the unstitched averaged 0.267 inch. The 48 ply Lay-up was chosen because it approximately matched the thickness of the 2-D and 3-D materials.

3.0 TEST SPECIMEN CONFIGURATION & TESTING METHODOLOGY

Both static indentation and falling weight impact tests were performed for this investigation. All specimens were 4.0 inch wide. The static indentation specimens were square while the CAI specimens were 6 inches long. The TAI specimens were 9.0 inches long to allow for griping in the load frame. Both testing methods used the same instrumented tup and 1/2 inch diameter indenter. In general, the method employed in this program was to perform repeated static indentation tests to obtain various amounts of damage. This damage was then documented and impact energies were determined for falling weight impacting. After falling weight impacting, each specimen was loaded to failure in either tension or compression to determine the damage resistance of the material.

The data taken from the static indentation tests were used to determine the falling weight energies required to produce pre-determined dent depths. In most cases multiple loading cycles were performed on a specimen until the desired, pre-determined amount of damage was attained. C-scans and dent surface dent measurements were taken between each loading excursion to ascertain the damage extent.

Two impact energies were determine from the static indentation tests. The first produced on average a 0.10 inch deep dent in the specimen. The second produced a barely measurable surface dent. These shallow dents were only a few thousandths of an inch deep. For the purpose of this paper, the term "Visible Impact Damage" or "VID" refers to an impact energy sufficient to produce an average dent depth of 0.10 inch. The term "Barely Visible Impact Damage" or "BVID" refers to an impact energy sufficient to produce damage barely measurable by C-Scan. All specimens were impact at 15.39 ft•lbs to produce the BVID and 62.86 ft•lbs to produce VID.

Each panel was ultrasonically C-scanned before impact to ensure that the panels were of high quality, free from manufacturing defects that would lead to premature failure. C-scans after impact were used to determine the extent of the impact damage. Area calculations of the damage were made from the through-the-thickness ultrasonic projections.

After impacting, each of the falling weight impact test specimens were loaded to failure in a closed-loop servo-hydraulic testing machine run in stroke-control. All testing was conducted at room temperature, dry conditions. Load cell output and machine stroke were recorded using a digital data storage oscilloscope that allowed dynamic measurements during the test. The compression specimens were mounted in a special test fixture to prevent global buckling of the test coupon during loading. This standard NASA compression after impact test is described in Ref. [2]. A through description of each material architecture and testing methodology is given in Ref. [3].

4. DISCUSSION OF RESULTS

The effectiveness of textile architectures in improving damage tolerance is discussed in the following sections. Damage tolerance, or the materials ability to support load once damaged, will be the discriminator. Each of the materials is evaluated with various extents of damage and compared

4.1 Damage Tolerance of Stitched and Unstitched Uniweaves

Damage Tolerance is defined as a materials ability to support load after being damaged. Compression after impact and tension after impact strengths were used as a discriminator of the damage tolerance in this investigation. Evaluations will be made comparing both strength and residual strength where residual strength is the post-impact strength normalized by its unnotched strength.

Figures 1 and 2 show an evaluation of the compression response of the stitched & unstitched uniweaves. Results are given for experiments at both the lower "Barely Visible Impact Damage" (BVID) energy and the more sever “Visible Impact Damage” level (VID). Figure 1 is a plot of the compression after impact (CAI) strength for each of the material forms. The CAI data, normalized by the respective unnotched strength, is reported in Figure 2. Each data point is an average of two experiments.

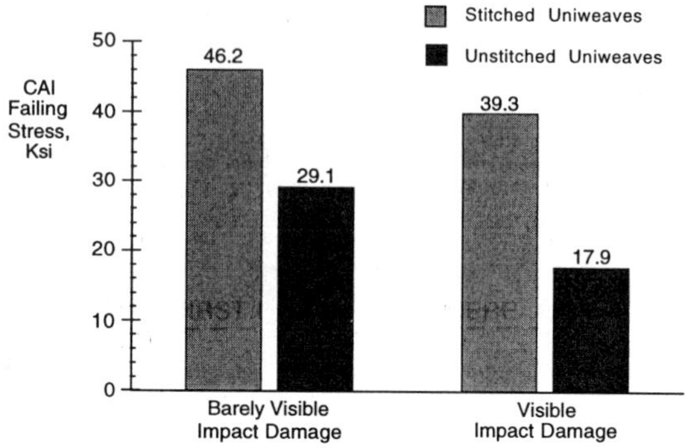

Figure 1. Compression After Impact Strength of Stitched and Unstitched Uniweaves

A review of the data indicates that the compression strengths were improved significantly with stitching. Note that the stitched laminates compression strength approached or exceed 40 ksi, a current industry target for CAI strength. Only at the severe impact level of 62.86 ft•lbs was the result marginal. Even at this severe impact energy level, compression strengths averaged better than 39 ksi. Even though these specimens were severely impacted, this average is just slightly below the acceptance level for cc design consideration.

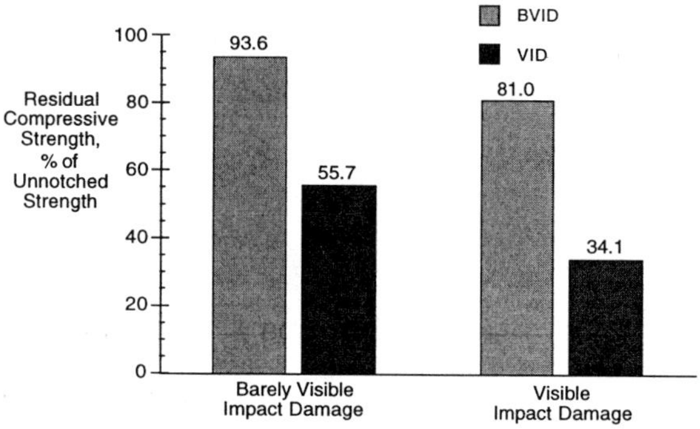

Figure 2. Damage Tolerance of Stitched and Unstitched Uniweaves

In general, stitching appears to enhance this material's damage tolerance. For example, a review of Figure 2 shows the stitched laminates impact at BVID averaged more than 93% strength retention compared to 56% for the unstitched material. The difference in CAI strength with and without stitching appears to be more significant at this higher impact energy. A review of Figure 1 data shows that the CAI strength of the 48 ply material was improved 59% with stitching. At the higher VID impact energy, the CAI strength of the 48 ply stitched uniweave is 119% greater than the unstitched uniweave. Thus, damage tolerance may improve with increasing impact energy.

Figure 3 and 4 show an evaluation of the tension response of the stitched and unstitched uniweaves impacted at the lower "Barely Visible Impact Damage" energy level (BVID) and the more severe "Visible Impact Damage" level (VID). Figure 3 is a plot of the tension after impact (TAI) strength. Figure 4 is a plot of the residual strength where the TAI data for each thickness has been normalized by its unnotched tensile strength. Two experiments were averaged for each data point shown.

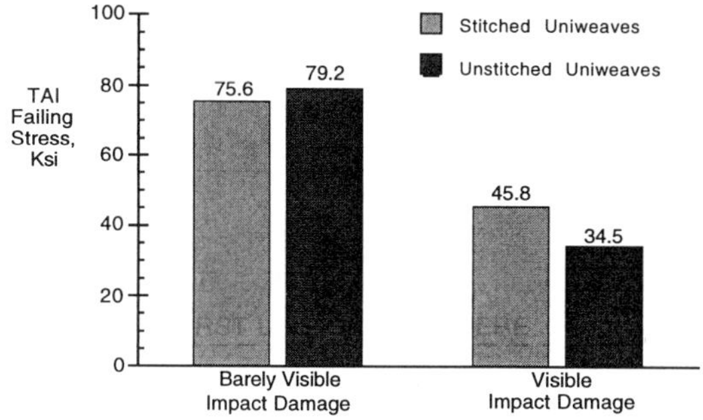

Figure 3. Tension Strength of Stitched and Unstitched Uniweaves

An examination of Figure 3 and 4 shows that stitching doesn't appear to significantly enhance the tension after impact strength. Some improvement to the residual tensile strength is apparent but may be due to low unnotched strength. It's important to realize that because of the stitch yarns, the stitched uniweaves were slightly thicker than the unstitched laminates. Because stress is calculated based on specimen thickness, and tension strength in composites is 0° fiber dominated, the stitched uniweaves with BVID damage tended to have slightly lower load carrying capability than the unstitched

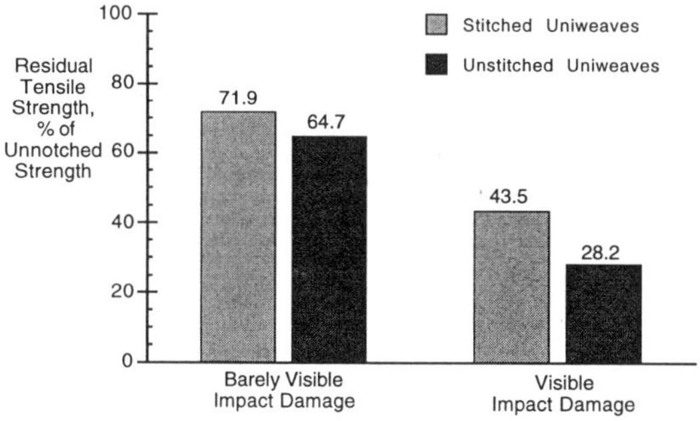

Figure 4. Damage Tolerance of Stitched and Unstitched Uniweaves

At the lower "Barely Visible Impact Damage" level, the stitched uniweaves had an average failure strength of 76 ksi and an average of 72% strength retention while the unstitched averaged 79 ksi failure strength and only a 65% retention of unnotched strength. Thus, even though the failure strengths were similar, the percent retention of residual strengths was slightly less.

Damage tolerance of the stitched and unstitched uniweaves to impact at the higher "Visible Impact Damage" energy level is also shown in Figures 3 and 4. Stitching continues to improve the damage tolerance of the uniweaves although strengths have been significantly reduced by this severe impact. Failing stresses were again greater for the stitched uniweaves than for the unstitched materials. Failure stresses were all above 40 ksi for the stitched specimens but averaged fell well below this threshold for the unstitched with VID.

Although stitching has improved the uniweaves TAI strength, the stitched laminate strengths have been reduced to around 43% percent of their unnotched value by the VID impact. Recall the data in Figure 2, a plot of the residual strength of the compression after impact response of the uniweaves. Strength retention averaged approximately 94 percent of its unnotched value with BVID and 81 percent with VID. Thus, improvements in damage tolerance are not as significant in tension as they are in compression.

4.2 Damage Tolerance of 2-D Braids.

Damage tolerance of the 2-D braids are shown in the following figures. Again the compression after impact (CAI), tension after impact (TAI), and residual strength of each will be used to make the comparisons. Each figure shows results from both the "Barely Visible Impact Damage" impact energy of 15.39 ft•lbs and the "Visible Impact Damage" impact energy of 62.86 ft•lbs. Because the braids differed only in architecture, not thickness, impact response should be an artifact of the differing architecture.

Compression after impact strength is plotted in Figure 5 while residual strength, expressed as a percentage of the unnotched strength, is shown in Figure 6 for each of the four 2-D Braided architectures. Data for both the lower "Barely Visible Impact Damage" level and the upper "Severe Impact Damage" level are shown in each figure. Each data point is an average of two experiments.

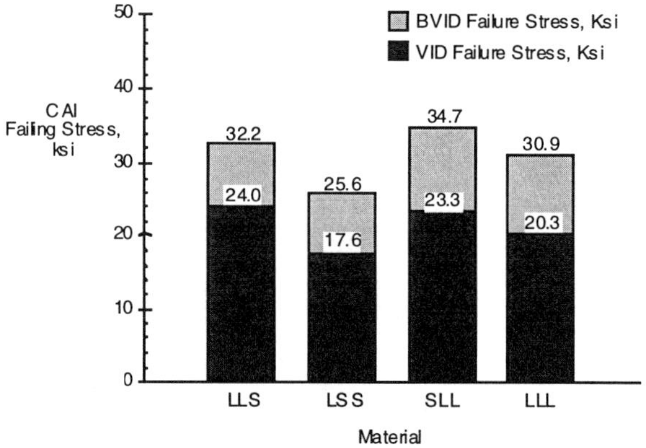

Figure 5. Compression Strength of 2-D Braids Impact

An examination of Figures 5 and 6 shows that the strength of the 2-D braids are significantly reduced by impact damage. None of the 2-D braids retained more than 63% of their unnotched strength. Failing strengths, even at the lower impact energy, were all below the acceptance guideline value of 40 ksi.

Recall that some of the braids were designed to allow an evaluation of tow size effects. Specifically, the SLL and LLL architectures were constructed exactly the same with the exception of tow size. The LLL contains tow bundles 2.5 times the size of the SLL. An examination of the impact response of these two material architectures shows some interesting effects. Strengths were greater for the SLL material, regardless of impact energy level. Thus it would seem that the smaller tow size in the SLL material must contribute to a strength improvement. Notice, however, that in Figure 6, the residual strength ratio of the LLL braid is better than that of the SLL material. Therefore, although tow size may lend itself to improving damage tolerance, it may contribute to a decrease in unnotched strength of the material.

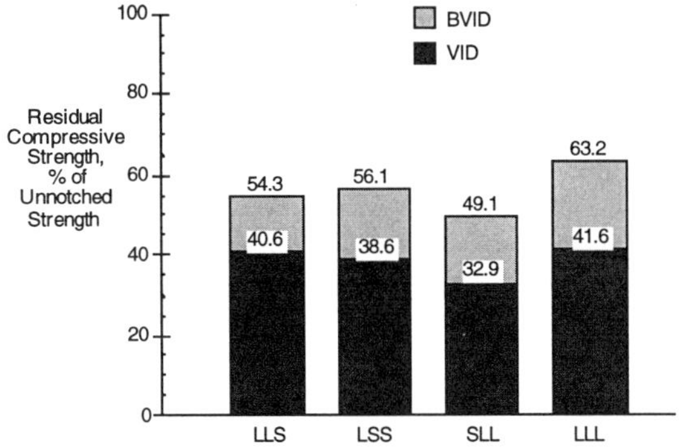

Figure 6. Damage Tolerance of 2-D Braids Impact

Figure 7 is a plot of the tension after impact (TAI) response of each of the 2-D braids. Data for both the "BVID" and "VID" impact energy levels are given. Data shown are averages for two experiments.

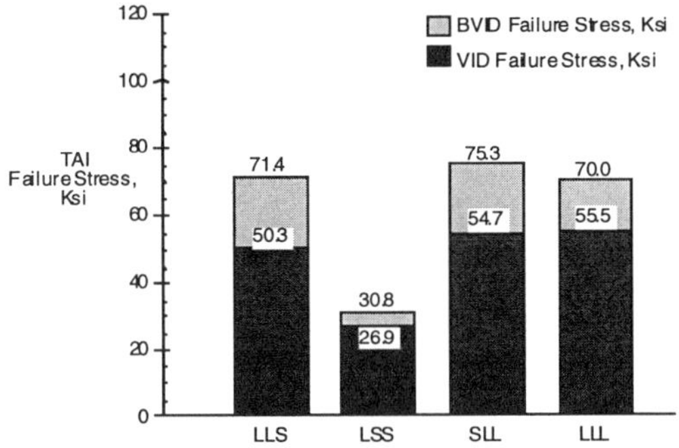

Figure 7. Tension Strength of 2-D Braids Impact

Examination of Figure 7 shows the 2-D braids response to impact damage is better in tension than in compression. With the exception of the LSS material, strengths averaged above 70 ksi with Barely Visible Impact Damage and above 50 ksi with Visible Impact Damage. Note that the LSS has only 12% axial yarns. This accounts for its significantly lower failure stress.

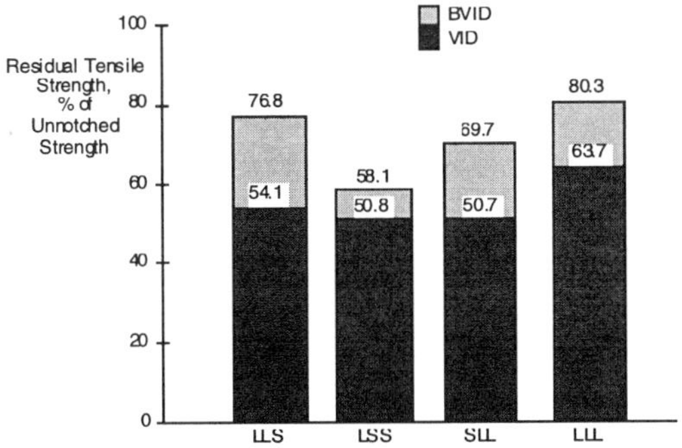

Figure 8. Damage Tolerance of 2-D Braids Impact

Residual Strength as a percentage of unnotched strength is plotted in Figure 8 for each of the four 2-D Braided architectures. Data for both the lower Barely Visible Impact Damage level and the upper Visible Impact Damage level are shown in this figure. Data shown are averages for two experiments.

The LLL again has the greatest percentage retention of unnotched strength. At the Visible Impact Damage impact energy, the average percent residual strength is only 16% less than at the lower BVID impact energy level. Although the 2-D Braids performed significantly better in tension that in compression, their overall strength is still rather low. An examination of tow size effects in tension reveals a response similar to that found in compression with these materials. The SLL has higher strength but a lower percentage residual strength than the LLL material. Thus, again it is demonstrated that the smaller tows may improve strength while the larger tows tend to improve damage tolerance, as measured by the percent of the undamaged strength retained by the impacted coupon.

4.3 Damage Tolerance of 3-D Weaves

Compression after impact (CAI) strength is plotted in Figure 9 and residual strength as a percentage of unnotched strength is shown in Figure 10 for each of the six 3-D weaves. Data from both the lower "Barely Visible Impact Damage" level and the upper "Visible Impact Damage" level are shown in each figure. Data shown are averages for two experiments.

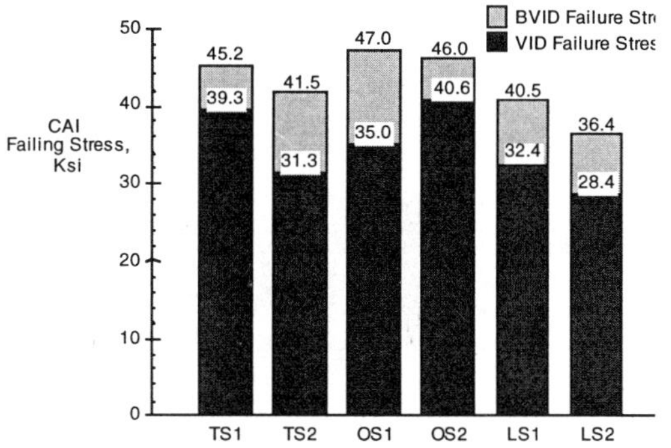

Figure 9. Compression Strength of 3-D Weaves Impact

The data indicate that the weaves outperformed the 2-D braids and the unstitched uniweaves at both impact energy levels in compression. The woven laminates compaired about equally with the stitched uniweaves, having a failing stress of greater that 40 ksi with the OS2 architecture at the highest impact energy

The data also indicate that the woven materials' residual strengths (expressed as a percent of their unnotched strength) were better that those of the 2-D Braids, regardless of the impact energy level. The unstitched uniweaves retained a higher percentage of their unnotched strength than the 3-D weaves at the lower impact energy level but not at the more severe VID level. However, the Stitched Uniweaves outperformed the 3-D Weaves in compression, regardless of the impact energy level.

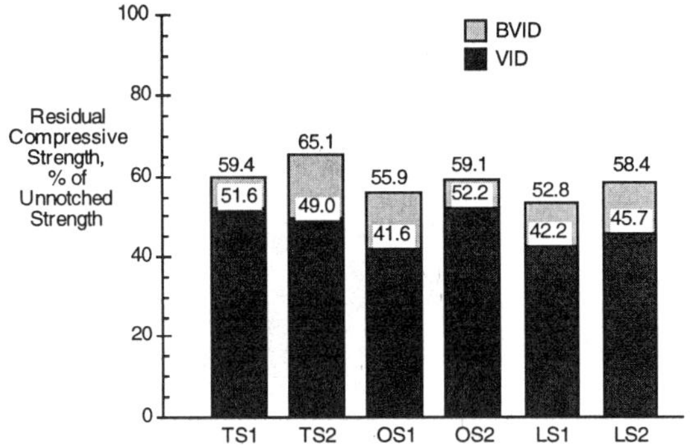

Figure 10. Damage Tolerance of 3-D Weaves Impact

Recall that the 3-D weaves were constructed in pairs. The "1" materials had yarn bundles twice as large as the "2" materials. Although the specimens made of the larger yarns had slightly greater compression strength after being impacted at the BVID level, no consistent pattern was evident in the materials impacted at the more critical VID level.

The tension after impact response of the 3-D weaves is shown in Figure 11. Data from both the lower Barely Visible Impact Damage level and the upper Visible Impact Damage level are shown in each figure. Averages from two experiments are represented by each data point.

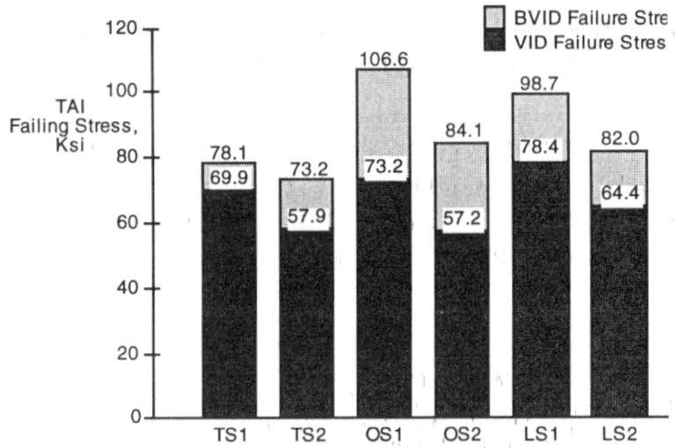

Figure 11. Tension Strength of 3-D Weaves Impact

The 3-D Weaves tended to exhibit their best performance in tension. Although the extent of damage from impact was generally higher than that of many of the other material architectures, the failing stresses obtained from these materials were typically higher. Failing stresses for the OS1 specimens averaged 106.6 ksi, better that any other material form examined in this study. Only the 32 ply Stitched Uniweave came within 10% of this value. The best performing 2-D Braid offered 30% less load carrying capability at the "Barely Visible Damage" impact level.

Residual Strength is shown in Figure 12 for each of the six 3-D Weaves. Data from both the lower "Barely Visible Impact Damage" level and the upper " Visible Impact Damage" level are shown in each figure.

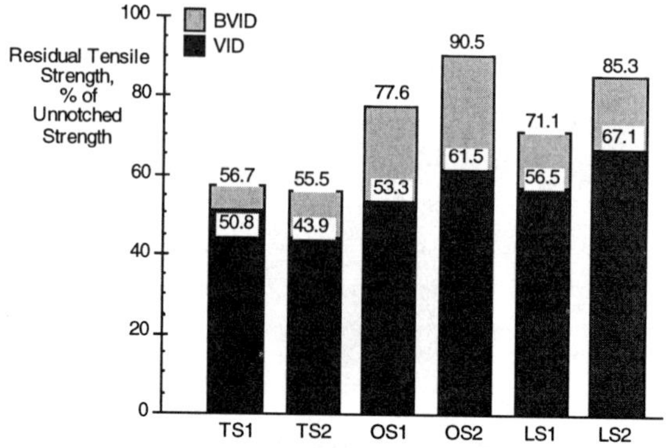

Figure 12. Damage Tolerance of 3-D Weaves Impact

The residual strength comparison revealed that the OS2 material retained better than 90% of its unnotched strength. Again, outperforming any other architecture evaluated in tension.

A comparison of the effect of fiber bundle size in tension produces somewhat different observations than those obtained in compression. The failing stress was always greater for the materials with the larger fiber bundles but the residual strength retention trends were generally reversed. This response is opposite that observed in the 2-D braids. The OS and LS materials continue to show significant improvements in damage tolerance with the smaller tows while the TS material did not. Residual strength was not only about equal between this pair of materials, it was also significant lower than the other architectures.

Figures 13 and 14 allow a comparison of the best and worst response to damage tolerance for all the material architectures in both tension and compression. Figure 13 shows the results of the compression testing while Figure 14 illustrates the tension data. Both figures display results obtained for the "Visible Impact Damage" and "Barely Visible Impact Damage" impact levels.

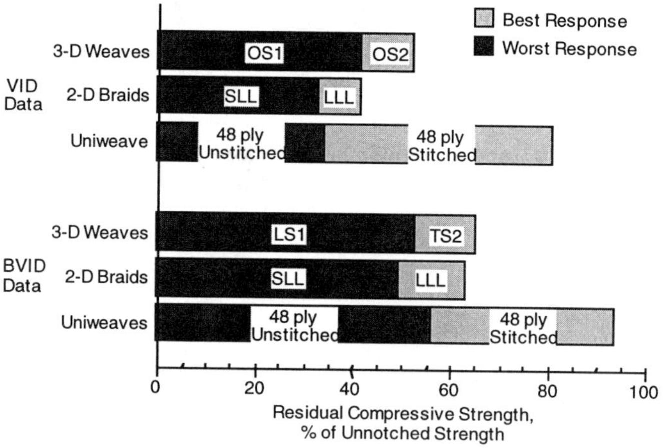

Figure 13. Summary of the Compression After Impact Response of 3-D Weaves, 2-D Braids, and Uniweaves

An examination of Figure 13 shows that stitching excels at reducing the effect of damage at both the Visible Impact Damage and Barely Visible Impact Damage impact levels. The 2-D braids and 3-D weaves show little improvement in damage tolerance over the unstitched uniweaves

Examination of Figure 14 shows that stitching, which excels at reducing damage growth in compression, does not appear to enhance damage tolerance as significantly in tension. Overall, the 2-D braids offer little tolerance to impact damage in compression but have moderately good response in tension. The 3-D weaves, which offered reasonably good damage tolerance in compression, outperform all the other textile architectures compared in tension. This result is surprising, given the poor damage resistance of the 3-D weaves.

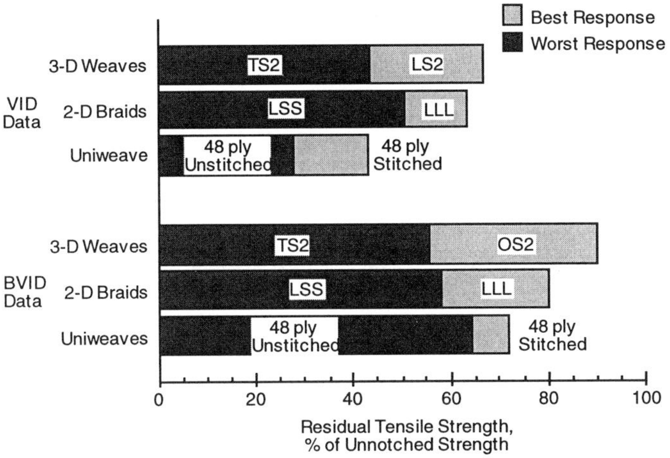

Figure 14. Summary of the Tension After Impact Response of 3-D Weaves, 2-D Braids, and Uniweaves

5. Findings and Conclusions

An evaluation of the impact damage tolerance of stitched and unstitched uniweaves, 2-D braids, and 3-D weaves was conducted. The uniweave materials were tested both with and without stitching. Several variations of the braiding and weaving perameters were tested. All of the materials were subjected to either quasi-static indentation or low velocity (large mass) impacts and then loaded in tension or compression to measure residual strength.

5.1 Damage Tolerance of Stitched and Unstitched Uniweaves

Stitching improved the compression after impact (CAI) strength at both impact energy levels. Stitched 48 ply CAI specimens retained 94% of their unnotched strength. Stitching improved strength in tension (TAI) at the VID impact level but not significantly at the lower BVID impact level. Stitched 48 ply TAI specimens retained 72% of their unnotched strength. In all CAI and TAI experiments, strength was generally above 40 ksi, the current rule-of-thumb acceptance level for CAI strength of tape laminates. Damage tolerance of the uniweaves improved in all cases, regardless of the impact energy with stitching. It was also noted that damage tolerance tended to increase with increasing plate thickness.

5.2 Damage Tolerance of 2-D Braids

The residual strength of the 2-D braids was better in tension than in compression. At the BVID impact damage energy level, the LLL retained 80% of its unnotched strength in tension, but only 63% of its unnotched strength in compression. The material had the least damage area but there was no significant improvement in strength retention over the other braids.

5.3 Damage Tolerance of 3-D Weaves

The 3-D weaves also had better damage tolerance in tension than in compression. At the BVID impact level, the OS2 retained 90% of its unnotched strength in tension but only 65% in compression. None of the weaves had outstanding performance in compression. In tension the TS2 and OS2 materials were noticeably better that any of the other weaves. The OS2 specimen retained 90% of its unnotched strength. Only the 48 ply stitched uniweave came within 10% of this performance.

5.4 Effect of Tow Size

The effect of fiber bundle size was compared wherever possible. Two of the 2-D braids were constructed with the same percentage of axial yarns and the same braid angle but using braided tows 2.5 times different in size. The 3-D weaves were constructed in pairs with the "1" materials having yarn bundles twice as large as the "2" materials.

It was found that tow size may affect the damage tolerance of the 2-D braids and 3-D weaves. With the 2-D braids, the larger tow size resulted in an improvement in residual strength but a decrease in post-impact strength. The 3-D weaves' response was opposite that of the 2-D braids. In this case the large fiber bundles resulted in an increase in post-impact strength but a reduction in the percent of strength retention.

The behavior of the 2-D braids with varying tow size was always consistent. Regardless of the impact energy or whether the specimen was tested in compression or tension, residual strength always improved and post-impact strength always decreased in the specimens with the larger fiber bundles. The behavior of the 3-D weaves was also very consistent in both tension and compression, although some aberrations to these trends were noticed. Specifically, the OS2 material had higher strength in compression and the TS2 had lower residual strength. These responses were the opposite of the other weaves. The cause of this response is unclear but may be architecturally dependent.

6. REFERENCES

1 Falcone, A., Dursch, H., Nelson, K., Avery, W., "Resin Transfer Molding of Textile Composites," NASA CR. 191505, March 1993.

2. "NASA/Aircraft Industry Standard Specification for Graphite Fiber/Toughened Thermoset Resin Composite Material", NASA Reference Publication 1142, June 1985

3. Portanova, M.A., "Evaluation of the Impact Responce of Textile Composites", NASA Contractor Report 198265, December 1995

28th International SAMPE Technical Conference
November 4-7, 1996

RESIN TRANSFER MOLDING OF CONTINUOUS FIBER-REINFORCED, PARTICLE-FILLED CERAMIC-CERAMIC COMPOSITES: PARTICLE FILTRATION

James Bozza, Merve Erdal and Selcuk Guceri
Mechanical Engineering, University of Illinois at Chicago
Chicago, Illinois 60607

ABSTRACT

The development of preceramic polymers (PCP's) has enabled many liquid processing techniques, such as resin transfer molding (RTM) to be applicable to the processing of ceramics, specifically ceramic-ceramic composites. A major issue in RTM of ceramics is the void formation due to densification/mass loss of the PCP during pyrolysis. This necessitates multiple reimpregnation-pyrolysis cycles that increase the processing cost. An effective way of reducing the number of cycles is to introduce filler particles to the resin during injection.
The introduction of filler particles leads to particle filtration during injection. Since the degree of cure and properties of the end product are highly affected by the filler distribution at the end of fill, the filtration mechanics needs to be understood and controlled. A combined flow-filtration model is developed as a first approximation to the problem. Experiments are performed for 1-D infiltration to determine various model parameters. The developed model is then tested on a 2-D irregular mold geometry. The current model is capable of generating results for highly complex geometries with anisotropic domain characteristics.

KEY WORDS: Resin Transfer Molding (RTM), Ceramic Composites, Particle Filtration

1. INTRODUCTION

The development of ceramic precursors or preceramic polymers (PCP) has enabled conventional polymer processing methods such as injection molding and resin transfer molding to be easily adaptable to the processing of ceramic matrix composites (CMC). Although injection molding has been applied to ceramic processing before, the method was accompanied by the tedious binder burn-out phase. Continuous fiber reinforced ceramics are novel and rare and are mostly processed through chemical means (sol-gel processing, chemical vapor deposition) requiring high temperatures and long processing times. Resin transfer molding (RTM) is one of the most promising and cost-effective methods for manufacturing continuous fiber reinforced, high temperature composite structures with reliable properties. It has the advantage of processing higher strength parts compared to its whisker/particle reinforced injection molded counterparts (1), without the complexities arising from high solid loadings and whisker orientations in injection molding (2). A main advantage of PCP's in resin transfer molding is that the polymer is converted to ceramic during the process without the necessity to burn out a binder suspending the particulate ceramic as in more conventional ceramics processing.

One drawback of resin transfer molding using preceramic polymers is that during the pyrolysis stage when the polymer is converted to ceramic, considerable mass loss is observed accompanied by shrinkage due to the densification of the part (3). The resulting porous structure can be densified through a series of reimpregnation-pyrolysis cycles which raise the processing cost.

Inclusion of particle fillers into the process along with the infiltrating polymer reduces the effective polymer volume leading to reduction in void formation and increase in the packing density of the pyrolysed part. In the case of active particles, the filler particles experience a volumetric expansion after reaction with the products of pyrolysis and become converted to ceramic (4).

A two-dimensional process model is presented to simulate the injection of particle-filled preceramic polymers into fiber preforms during resin transfer molding of continuous fiber reinforced ceramic-ceramic composites. The inclusion of particle fillers introduce additional complexities to the process which requires the modification of various system parameters. The present model can easily be extended to three-dimensions and is capable of taking into account the anisotropy of the flow domain.

2. PROCESS MODEL

The process consists of three stages: In the first stage, the particle-filled preceramic polymer (resin) is injected into the fiber preform in the mold cavity (Figure 1). The preceramic polymer is mixed with filler particles to form a suspension before injection. The injection is performed at room temperature. One particular advantage of most preceramic polymers is that the injected resin can be specifically prepared to exhibit viscosities in a wide range of values. After injection, the mold is placed in a low-temperature oven until the resin is fully polymerized to form a rigid part. The pyrolysis stage is the final step in processing the composite where the fully polymerized part is placed in a pyrolysis furnace and burnt at temperatures around 1000 °C, where the polymer is converted to ceramic. During this stage, the part density increases (the part shrinks) due to the conversion of the polymer to ceramic. In addition, mass loss is observed as a result of series of chemical reactions leading to additional voids in the final part. Filler particles are introduced to increase solids loading in the part in order to reduce the number of reimpregnation-pyrolysis cycles, and therefore, the cost.

The process is highly complicated due to difficulties associated with infiltrating the preform with uniform particle concentrations, i.e. avoiding or minimizing particle filtration well enough to achieve homogenous mechanical properties. In the case of functionally gradient materials, an uneven particle distribution could be desirable, however, the process would still have to be highly controlled, as carefully designed as the case for homogeneous distribution.

The focus of the current model is the impregnation stage where the filtration of particle fillers forms the utmost complexity. In order to control (and minimize) the rate of filtration, the relations between various processing parameters and filtration need to be understood and modeled. The proposed model consists of two simultaneously occurring physical mechanisms; Flow through porous medium and particle filtration.

2.1 Flow Through Porous Medium The basis of this model is taken from Coulter (5) and Coulter and Guceri (6). The slurry flow in the fibrous preform is essentially flow through porous media, stated by Darcy's Law as

$$\underset{\sim}{V} = -\frac{1}{\mu}(\underset{\approx}{K} \cdot \nabla \underset{\sim}{P}) \qquad [1]$$

where $\underset{\sim}{V}$ is the bulk flow velocity, μ the viscosity and $\underset{\approx}{K}$ the anisotropic permeability tensor of the preform. In the current study, the permeability term is further complicated with the inclusion of particle fillers.

The conservation equation for the bulk flow is given in terms the streamfunction formulation of continuity equation,

$$\nabla^2 \Psi = 0 \qquad [2]$$

with

$$\frac{\partial \Psi}{\partial y} = u(1 - v_f) \qquad [3.a]$$

$$-\frac{\partial \Psi}{\partial x} = v(1 - v_f) \qquad [3.b]$$

where Ψ is the streamfunction, u and v the velocity components in x and y directions, and v_f the total solid volume fraction (preform fiber + filtered particles on the preform) of the flow domain. For the case of a diagonal permeability tensor where the domain coordinate axes coincide with the principle orthogonal axes of the medium, equations (1-3) can be combined to yield the governing equation for numerical solution,

$$\frac{\partial}{\partial x}\left(\frac{\mu}{K_y}\frac{\partial \Psi}{\partial x}\right) + \frac{\partial}{\partial y}\left(\frac{\mu}{K_x}\frac{\partial \Psi}{\partial y}\right) = 0 \qquad [4]$$

where K_x and K_y are the principal permeabilities in x and y directions. The above equation can similarly be extended to three dimensions.

The process is modeled isothermally since the injected resin does not require preheating and no reaction takes place during the injection stage. The main flow solution is performed by the previously developed numerical code TGMOLD (7), a comprehensive finite-difference polymer processing simulation package.

2.2 Filler Particle Filtration The success of the described slurry injection process depends on whether the end product will result in a homogenous particle distribution with sufficiently high particle packing or not. Before actual operation, the relations between various process variables need to be investigated which necessitates development of filtration models for the current process.

Unlike cake filtering which is most frequently modeled for filter-pressing or slip-casting techniques in ceramic processing (8-10), particle filtration in the current process is similar to *deep-bed filtration* where a contaminated fluid travels through a filter medium, such as a fiber preform. The filler particles considered in the current process are in the order of 1 μm for

these particles are filtered the least efficiently (11,12). For ease of formulation, in the current study, particle-to-particle interactions in the slurry and effect of particle size are neglected although these factors are most probably important elements in the actual process.
The unsteady species equation for filler particles is given as,

$$\frac{\partial [(1 \quad v_f) C]}{\partial t} + \underset{\sim}{V} \cdot \nabla [(1 - v_f) C] + \frac{\partial \sigma}{\partial t} = 0 \qquad [5]$$

where C is the particle concentration in the resin (volume by volume) and σ is the specific filtered particle volume deposited to the fiber preform. Note that

$$v_f = v_{fo} + \sigma \qquad [6]$$

where v_{fo} is the fiber volume fraction. Equation (5) is a multi-dimensional form of the material balance equations for filtration reported by Herzig et al. (13).
The rate of particle deposition onto the fiber preform is assumed to be first order with respect to the particle concentration and is presented by the filtration kinetic equation (14),

$$\frac{\partial \sigma}{\partial t} = \alpha V C - \beta \sigma \qquad [7]$$

where V is the local scalar velocity, α and β the accumulation and detachment coefficients, respectively. The first term on the right-hand side of equation (7) reflects the filtration rate, i.e. the rate of deposit of particles onto the fiber preform, and the second term reflects the portion of the filtered particles reintroduced to the flow due to increased local velocities in clogged pores.
A closer inspection of filtration physics suggests that eventually the filter medium (fiber preform) will be saturated and no further filtration will take place. At this point, the specific deposit will have reached its ultimate value, σ_u, regardless of the initial particle concentration value, and the rate of filtration will be zero. Therefore, α and β can not be independent of each other,

$$\beta = \frac{\alpha V C}{\sigma_u} \qquad [8]$$

The ultimate specific deposit value, σ_u can not exceed the domain porosity $1\text{-}v_{fo}$. As a result, the filtration kinetic equation is modified as,

$$\frac{\partial \sigma}{\partial t} = \alpha V C \left[1 - \frac{\sigma}{\sigma_u} \right] \qquad [9]$$

The model is kept at its simplest form, without relating filtration coefficient to other process variables such as specific surface of the prorous domain (14), permeability or viscosity of the particle-filled resin (15). The preliminary experiment results that will be discussed in the next section are used to determine the current model parameters as a first approximation to the nature of filtration., incapable of generating detailed information that will lead to a thorough understanding of the complex interactions between various process parameters.

The preform permeability is clearly affected from particle concentration and needs to be characterized accordingly. The fundamental effect of filtration on permeability arises from the narrowing of passages due to clogging. From the Kozeny-Carman model, it is seen that the permeability is inversely proportional to flow surface area and directly proportional to the porosity of the filter (16). With this assumption, the permeability is expressed as,

$$K = K_o\, R_h \left[\left(\frac{1 - v_f}{1 - v_{fo}} \right)^2 \left(\frac{v_f}{v_{fo}} \right)^{-1} \right] \qquad [10]$$

where K_o the clean (unfilled) preform permeability and R_h the ratio of filled to unfilled domain Kozeny constants. The Kozeny constant depends on the length of fluid path and the shape of the flow passages and clearly, during filtration, both of these factors are prone to vary. The scalar permeability in eqn.(10) is the composite value determined by,

$$K_o = \frac{V^2}{\dfrac{u^2}{K_{xo}} + \dfrac{v^2}{K_{yo}}} \qquad [11]$$

3. EXPERIMENTAL PROCEDURE

A rectangular geometry mold with a line gate is tested to simulate one-dimensional injection. Warner-Electrak 2000 linear actuator system coupled with a injection piston-cylinder apparatus is used to induce motion capable of withholding up to 80 lb. of force at 2 in/sec of velocity. The mold cavity has a flow length of 14 cm with 5 cm of width and a thickness of 0.16 cm. Any flow in thickness direction can safely be ignored. Furthermore, the line gate extending along the width of the flow area induces a uniform flow front across the mold, resulting in one-dimensionality of the flow.

The injection fluid used for the current experiments is Dow Corning 510 Silicon oil with a nominal viscosity of 50 cP (at 25 °C). Two layers of Owens-Corning M8610 continuous filament E-glass mat constitute the fibrous domain for injection. With the current mold thickness, the domain fiber volume fraction is 22 %. For injection with particle fillers, H.C.Starck's alpha phase SiC particles (UF-15) with a nominal diameter of 1.3 μm (narrow size distribution with 90 % below 1.3 μm) are used.

The resin viscosity change with varying filler content is presented in Figure 2. The profile follows the equation,

$$\mu = \mu_o \left[1 - \frac{C}{A} \right]^{-2} \qquad [11]$$

as suggested by Kitano et al.(17). The constant A was found to be 0.13.

The first part of the experimentation involves the injection of neat resin (no fillers) into the fibrous preform. The constant injection velocity is 1.29 m/s and the measured neat resin viscosity is 58.6 cP (at 20 °C, the injection environment). From the filler-free injection results, the permeability of the preform is determined through matching the experimental results with the numerical results.

Next, silicon oil is injected with 2 % (by volume) SiC filler particles into the preform. The volume fraction of particles are determined through taking samples at the inlet gate. Although the initial mixture contained 4 % particles by volume, some portion of the particles was observed to settle on the bottom of the injection cylinder while experiencing a drastic flow area change. The particle volume fraction at the outlet vent was measured to be 0.13 %.

4. RESULTS AND DISCUSSION

The inlet gate pressure profile for one-dimensional injection without fillers is presented in Figure 3. The profile follows a linear trend as expected, which suggests a constant permeability value for the preform. The numerical model was run for no filler case to match the experimental pressure profile and the resulting permeability value was determined to be $0.00017*10^{-6}$ m^2. The initial pressure values carry some degree of uncertainty due to difficulty encountered in pinpointing the exact starting time. However as the flow progressed, the numerical pressure slope matched considerably well with the experimental profile.

The clean preform permeability was next used in the permeability model for filled systems along with the viscosity model to obtain numerical results for particle filtering. Specifically, various model parameters were sought for the particular process. The inlet gate pressure profile and outlet filler concentration level are taken as the convergence criterion for the numerical model. That is, when these two conditions are reached, the parameters that were used along with the given input conditions such as injection velocity and viscosity model are the ones matching this particular process window. The resulting pressure and flow front profiles are presented in Figures 4 and 5. The experimental pressure profile is not exactly linear especially towards the end of injection. The resin flow near the endwalls were observed to move slightly faster due to lower flow resistance associated with the misalignment of the preform with the mold wall at this region. As a result, the increase rate in pressure dropped. The flow front progression results from the numerical model matched satisfactorily. These results were generated through a series of runs in which the model variables α, R_h and σ_u were varied for the pressure and front profiles to match the experimental profiles as well as the resin outlet particle concentration (0.13 %). The determined values were 0.17 for α, 7.8 for R_h and 0.5 for σ_u. As discussed earlier, it should be noted that neither these values nor the current model are unique but are utilized to constitute an initial approach to the problem.

The presented model is next tested on a two-dimensional irregular mold geometry (Figure 6), with the same fiber preform architecture and particle-resin characteristics. The depth of the mold is 0.2 cm therefore the flow can be treated as two-dimensional. In order to observe the effectiveness of the numerical stability of the solution, the mold geometry is chosen to induce both converging and diverging flow patterns. The model variables determined through the one-dimensional mold experiments are used in the numerical analysis. The injection velocity is chosen as 2 cm/s with the inlet particle concentration level again at 2 %.

The instantaneous flow front locations are presented in Figure 6. The flow front near the mold walls after the flow starts converging around the corners is observed to lag behind the majority of the front. The wall flow accelerates and eventually catches up with the rest of the front. The inlet pressure rise during injection is presented in Figure 7. The pressure initially rises quite steeply due to the narrow inlet passage. Due to the divergence of the mold, the flow relaxes soon after and the pressure increase rate is observed to fall down. Towards the end of injection, as the flow converges, the pressure is again observed to increase rapidly.

Figure 8 depicts the filler particle concentration levels at the instant of fill. For the given process parameters, there is considerable filtration with the maximum filtration efficiency observed to be around 90 %. The effect of converging wall flow around the corners is apparent. Ives (14) claims that increasing local velocities adversely affects the filtration efficiency. In this case, the wall front moves at a faster rate than the flow next to it, beginning at the converging corners of the mold. As a result, particles at these regions are filtered at a slower rate than the mid mold region.

Lastly, the results of the dynamic permeability model are presented in Figure 9. There is extensive particle deposit initially. Consequently, a drastic increase in permeability is observed near the inlet with the minimum permeability level resulting at the inlet gate. As expected, flow front is the highest permeability region due to lack of deposited particles.

5. CONCLUSIONS

An important result obtained through the experiments was that there is considerable cake filtration of particles at the inlet gate as well as in-preform filtration. This will necessitate the development of cake-filtering models to be incorporated into the current formulation. Compression molding in which the excess resin is bled out after injection is complete increasing the solids packing in the part, is a promising technique to overcome this problem. The proposed model would again be an effective tool in detecting the final particle distribution and designing the process variables

Another issue to be resolved is the instability of the particle-polymer suspension. Due to the density differences between the filler particles and the polymer, particle sedimentation and agglomeration were observed. For a viable processing method in which the proposed model can be utilized properly, a homogenous suspension formation is a key factor. Use of surfactants for the particles, higher viscosity polymer and mechanical excitation of the flow field during injection are some of the options that will be implemented to the system.

The model as it is, is not complete. The dependence of filtration rate on system variables such as permeability, viscosity, particle size distribution will be implemented utilizing information gained through a series of detailed experiments. Furthermore, the permeability model needs to be improved to include effects of different preform architecture and particle size. The determination of the permeability distribution along the flow path as well as particle distribution characterization (visualization of sections of the part after cure) are therefore, of paramount importance in order to develop sound models. With the stated problems resolved, a low-cost processing method development is a likely outcome, enabling continuous fiber reinforced ceramic-ceramic composites to become affordable materials for structural, high temperature applications

6. REFERENCES

1. C.L.Tucker in T.G.Gutowski, ed., The Manufacturing Science of Composites-Volume IV, ASME Publications, New York, NY, 1988, pp.95-104.
2. C.M.Altan, et al., Polymer Engineering and Science, 30, (14), 848, (1990).
3. P.Greil and M.Seibold, Journal of Material Science, 27, 1053, (1992).
4. P.Greil and M.Seibold, Adv. Comp. Mater. Ceram. Trans., 19, 43, (1991).
5. J.P.Coulter., 1988, Resin Impregnation During the Manufacturing of Composite Materials, Ph.D. Thesis, University of Delaware, Newark, DE, 1988.
6. J.P.Coulter and S.Güçeri, Journal of Reinforced Plastics and Composites, 7, 200, (1988).
7. S.Güçeri, TGMOLD-A Computational Code for the Analysis of Injection Molding, RIM, RTM, and SRIM, University of Illinois at Chicago, Chicago, IL, (1995).
8. J.H.D.Hampton, S.B.Savage and R.A.Drew, Br. Ceramics Transactions Journal, 91, (4), 103, (1992).
9. J.H.D.Hampton, S.B.Savage and R.A.Drew, J. Amer. Cer. Soc., 75, (10), 2726, (1992).

10. A.P.Philipse, B.C.Bonekamp and H.J.Veringa, J. Amer. Cer. Soc., 73,(9), 2720, (1990).
11. J.E.Tobiason, et al., Journal of Environmental Engineering, 119, (3), 520, (1993).
12. C.R.O'Melia and W.Ali, Prog. in Water Technology, 10, (5/6), 167, (1978).
13. J.P.Herzig, D.M.Leclerc and P.Le Goff, Ind. Eng. Chem., 62, (5), 8, (1970).
14. K.J.Ives in K.J.Ives, ed., Proceedings of NATO Advanced Study Institute on The Scientific Basis of Filtration, Noordhoff Int. Pub. Co,, Leyden, 1975, pp. 183-224.
15. A.Adin and M.Rebhun, American Water Works Assoc., 69, 444, (1977).
16. M.F.Tiller and N.B.Hsyung, J. Amer. Cer. Soc., 74, (1), 210, (1991).
17. T.Kitano, et al., Rheo.Acta., 17, 149, (1978).

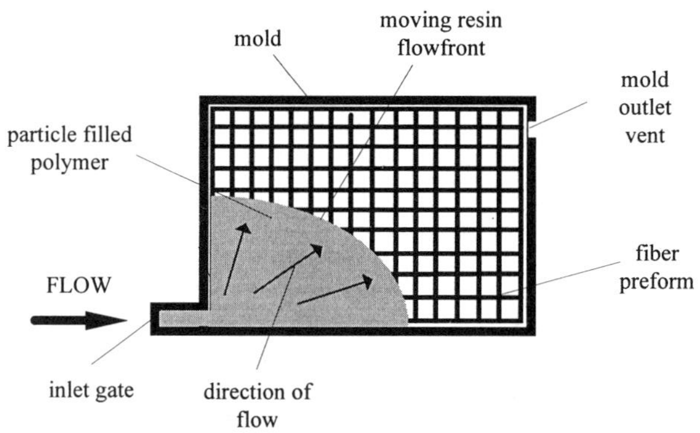

Figure 1. Injection of Particle Filled Resin into the Fiber Preform

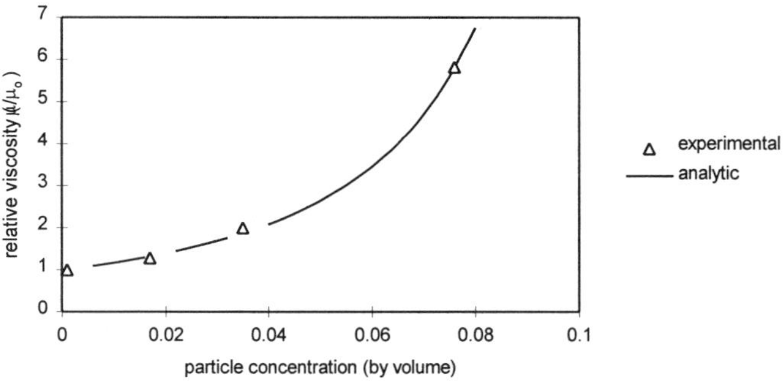

Figure 2. Silicon Oil- SiC Particle Mixture Viscosity Variation (Eqn.11)

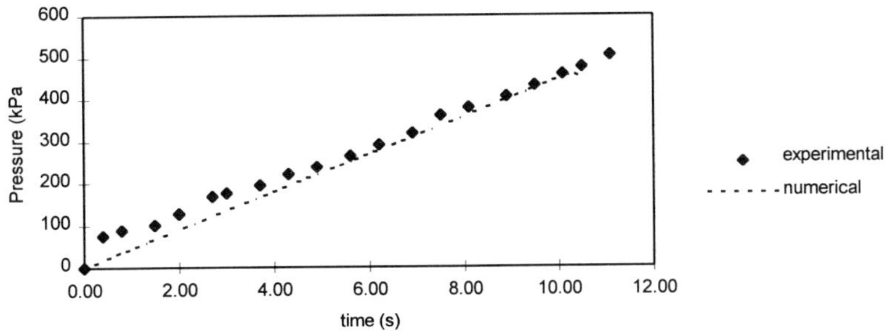

Figure 3. Pressure Variation with Injection Time at the Inlet Gate - No Filler Particles ($(K_{numerical} = 0.00017*10^{-6}\ m^2)$)

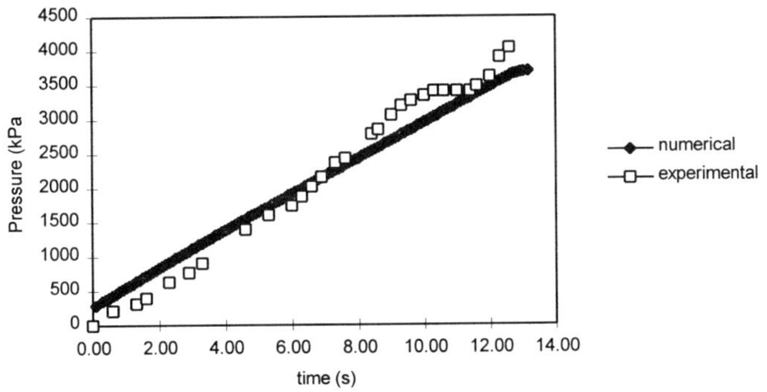

Figure 4. Pressure Variation with Injection Time at the Inlet Gate (Filled Injection)

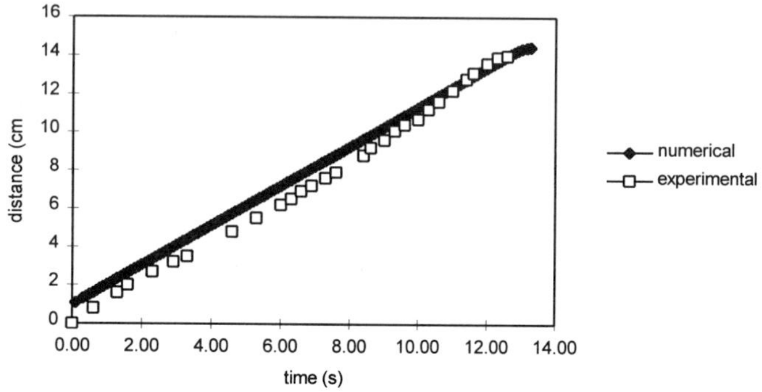

Figure 5. Comparison of Flow Front Progression with Injection Time (Filled Injection)

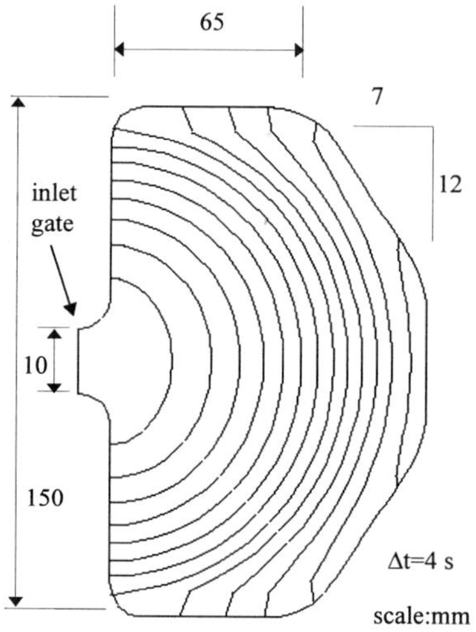

Figure 6. Instantaneous Flow Front Locations During Filling (total fill time=53.6 s)

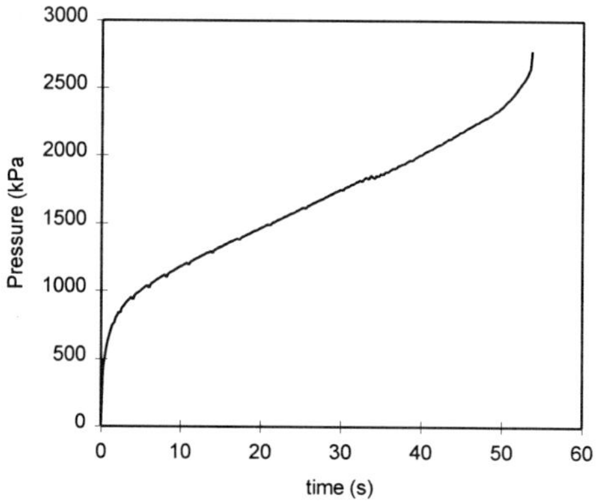

Figure 7. Inlet Pressure Variation with Injection Time for the Irregular Shaped Mold

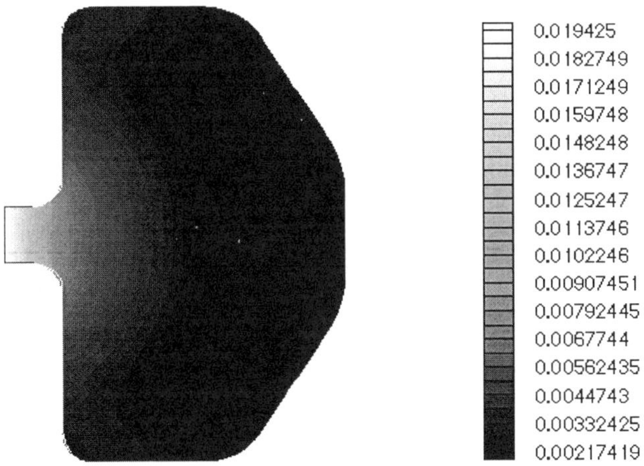

Figure 8. Filler Particle Concentration (by volume) Profile at the Instant of Fill (C_{inlet} = 2 %)

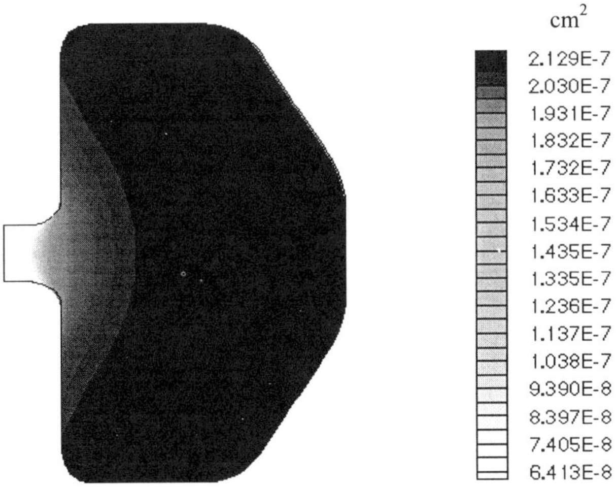

Figure 9. Domain Permeability at the Instant of Fill

28th International SAMPE Technical Conference
November 4-7, 1996

VACUUM ASSISTED RESIN TRANSFER MOLDING OF A LAYERED STRUCTURAL LAMINATE FOR APPLICATION ON GROUND COMBAT VEHICLES

T. Pike, M. McArthur, and D. Schade
United Defense, L.P.
Ground Systems Division
San Jose, California 95103

ABSTRACT

A Vacuum Assisted Resin Transfer Molding (VARTM) technique was developed as a potential low cost method for producing structural laminates with integrated armor for ground combat vehicle hulls. An integrated armor and structure design offers the best combination of ballistic protection and structural performance at minimum weight, but requires the layering of several material forms, including fiberglass, ceramics, and energy absorbing materials. Existing procedures for integrating multiple layers of materials require several elevated temperature cure cycles and adhesive layers. A VARTM technique has been demonstrated that cocures epoxy prepregs, ceramic armor tiles, and liquid resins in a single elevated temperature cure cycle. Structural and ballistic performance of a low cost VARTM epoxy laminate with integrated armor was compared to a similarly configured toughened epoxy laminate fabricated using an automated fiber placement process. This VARTM technique offers an affordable approach for manufacturing layered laminates tailored for structural and ballistic performance.

KEY WORDS: Vacuum Assisted Resin Transfer Molding (VARTM), Armor, Glass Laminates

1. INTRODUCTION

The use of composite technology on ground combat vehicles is not new. Ceramic armor appliqués are currently being used on a variety of military vehicles, as well as proof-of-principle technology demonstrators. The Composite Armored Vehicle Advanced Technology Demonstrator (CAV-ATD) is an armored ground combat vehicle that demonstrates the first use of a fully integrated armor and structure composite upper hull as shown in Figure 1. A key exit criteria for the CAV-ATD program is to meet an affordability goal of no more than 1.4 times the cost of a metallic vehicle with equivalent armor protection. The affordability goal is a significant challenge which requires the pursuit of low cost alternatives to the current baseline manufacturing process.

VARTM is a vacuum-assisted version of resin transfer molding, which uses a vacuum bag for one tool surface. Liquid resins are infused into dry layers of fibrous reinforcements using only vacuum pressure. VARTM processing is currently being used commercially to fabricate large composite structures such as boat hulls, wind turbine blades and transit buses.

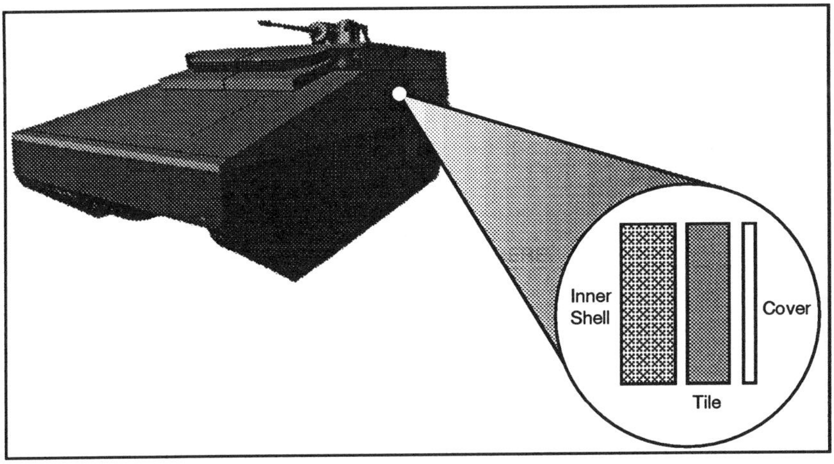

Figure 1. Composite Armored Vehicle Advanced Technology Demonstrator features a fully integrated armor and structure upper hull.

A Vacuum Assisted Resin Transfer Molding (VARTM) technique has been developed as an alternative low cost method for manufacturing integrated armor and structural laminates. Integrated armor and structure laminates fabricated using the VARTM process offer several advantages over conventional composite fabrication techniques; including the use of low cost liquid molding resins, low temperature and pressure processing, low capital equipment and tooling costs and flexibility in the selection of fibrous reinforcement materials (woven fabrics, preforms, chopped fiber mats, etc.)

2. BASELINE MANUFACTURING PROCESS

Composite armor systems are based on combinations of various layers of fiber reinforcements, energy absorbing media and ceramic materials. Materials of high hardness and stiffness are layered near the exterior surface for ballistic performance, while materials with high toughness and strength are layered on the interior surface for structural performance and damage tolerance.

The baseline manufacturing process for the CAV-ATD upper hull is comprised of three distinct steps, each requiring an elevated temperature vacuum bag oven cure and several ambient vacuum bag debulks. The first step builds up the inner shell of the upper hull laminate, including a thin layer of S-2 Glass®/epoxy prepreg, specifically formulated for low smoke and flammability properties and Automated Fiber Placement (AFP) layers of a toughened S-2 Glass®/epoxy (Figure 2a). After cure, the inner shell is removed from the AFP mandrel and placed on a support structure for the remaining steps of the manufacturing process.

The tile application is the second step of the baseline upper hull manufacturing process as shown in Figure 2b. A thin layer of a compliant material is adhesively bonded to the inner shell, followed by hand placing ceramic hexagonal tiles. A specially formulated epoxy resin system is used to bond both the compliant material and ceramic tiles to the inner shell. The tile application step is completed with a vacuum bag and oven cure.

The final step in the baseline upper hull manufacturing process is to lay-up and cure the outer cover (Figure 2c). The outer cover includes a ground plane layer and several plies of specially treated S-2 Glass®/epoxy prepreg for reduced vehicle radar signature. An adhesive is also added between the tiles and the outer cover to improve the bond strength. The entire laminate is then vacuum bagged and oven cured.

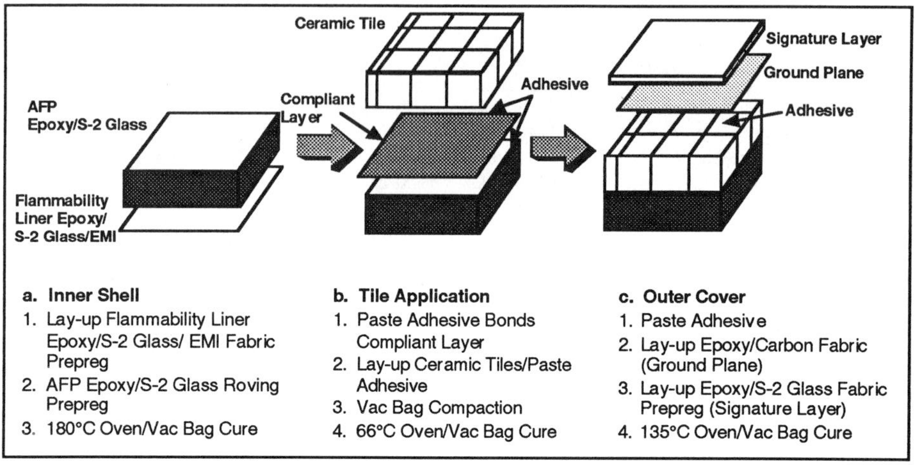

Figure 2. Fabrication of an integrated armor and structure laminate using the AFP baseline manufacturing process.

Cost modeling was performed by Draper Laboratories to estimate unit costs of the composite components of the CAV-ATD based on production rates of 25, 100, and 300 units per year for five years[1]. Cost drivers for each of the components (labor, material, and amortized capital/tooling) were also identified by the cost model. Figure 3 shows the unit costs for fabricating a single CAV aft upper hull assuming an annual production rate of 300 units per year and the baseline manufacturing process as described above. For all production rates, material is the most significant cost driver (Figure 3a). Further break-down of the material costs indicates that the toughened S-2 Glass®/epoxy roving used in the AFP processing of the structural layers account for 57% of the total material costs and raw material costs of $23/lb (Figure 3b).

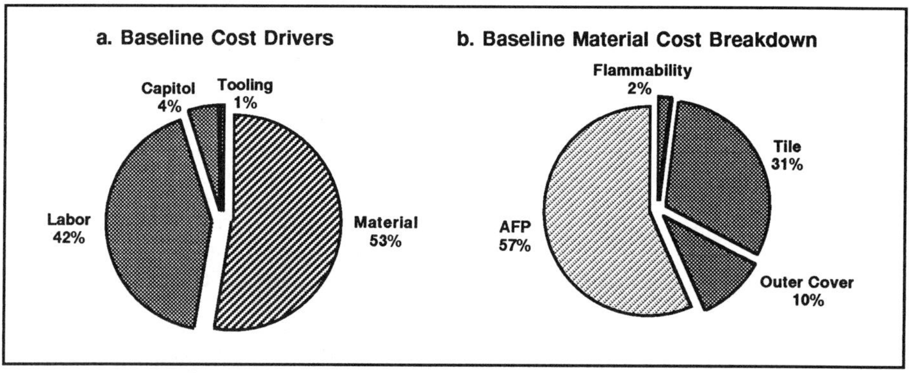

Figure 3. Baseline AFP processing unit cost for fabricating the aft upper hull for the CAV-ATD assuming an annual production rate of 300 units per year for 5 years.

3. LOW COST ALTERNATIVE PROCESS

The VARTM process for the CAV-ATD upper hull is comprised of four distinct lay-up steps and only one final elevated temperature cure cycle. The first step includes hand laying the outer cover prepreg material into a female tool (Figure 4a). Although the VARTM process could be performed using male or female tooling, a female tool was selected to enhance the exterior surface quality of the upper hull laminate, which in turn helps to reduce the signature profile of the vehicle. The use of female tooling is an added benefit of using the VARTM process over the baseline AFP process which required a male tool.

The ceramic armor tiles are then placed directly on top of the uncured outer cover and a vacuum bag is placed over the entire lay-up (Figure 4b). The same resin system used to bond the ceramic armor tiles in the baseline process is infused into the tile layer using the VARTM process. After the resin has staged, typically 6-12 hours at ambient conditions, the vacuum bag is removed and dry S-2 Glass® woven roving fabric is laid-up on top of the tile layer (Figure 4c). A vacuum bag is placed over the dry fabric and a structural epoxy resin system is infused into the laminate using the VARTM process. An ambient staging of the structural resin is performed prior to the removal of the vacuum bag. Finally, the flammability liner prepreg material is laid on top of the staged structural layers (Figure 4d). The entire assembly of layers is vacuum bagged and oven cured in a single elevated temperature cure cycle.

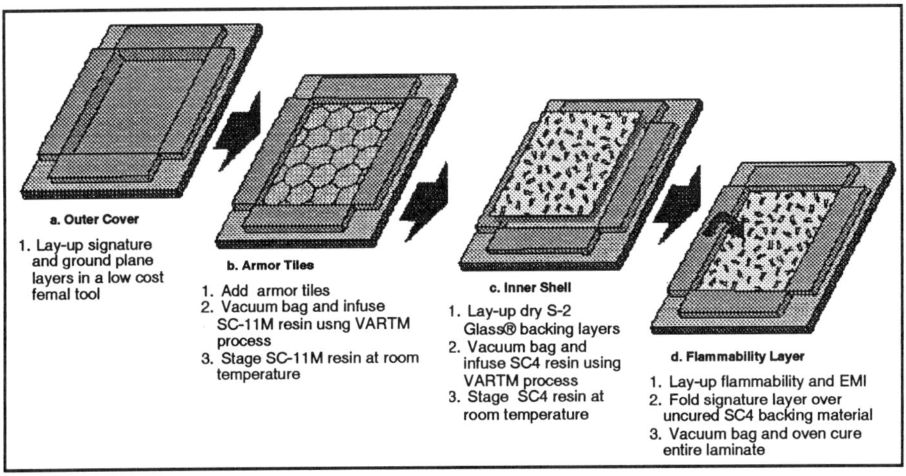

Figure 4. Fabrication of an integrated armor and structure laminate using the VARTM manufacturing process.

The VARTM process uses a low cost liquid epoxy resin manufactured by Applied Poleramic Inc. (API), Benicia, CA and commercially available S-2 Glass® woven roving fabric. The combined estimated cost of the VARTM resin and glass fabric is less than $10/lb for production quantities of 300 vehicles per year for five years. The VARTM process provides a reduction in material costs for the structural layer of approximately 56% over the baseline process. A comparison of the costs for fabricating a CAV aft upper hull using the VARTM process to the current baseline AFP process is shown in Figure 5. The overall cost savings using the VARTM process is approximately 21%, which reflects the reduction in material costs as well as significantly reduced in capital equipment costs.

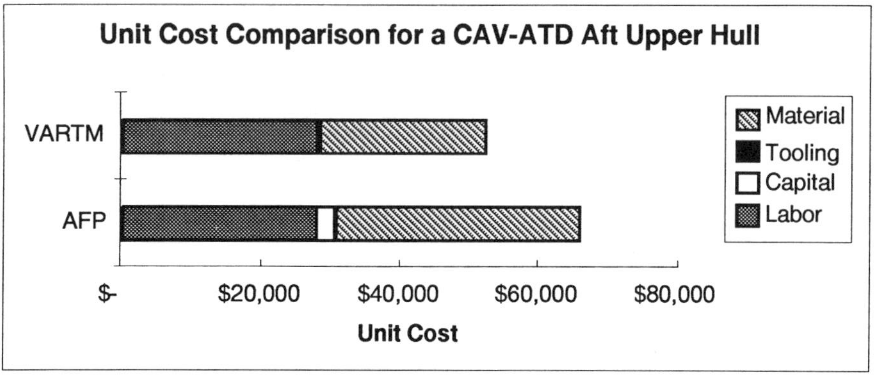

Figure 5. Comparison of costs associated with a VARTM manufactured aft upper hull and the baseline AFP manufactured hull shows a 21% improvement in overall costs.

Although the VARTM process requires additional labor to hand lay the dry glass, this increase is offset by the reduced number of handling steps required in the VARTM process over the baseline AFP process (i.e. vacuum debulking the AFP plies, additional elevated temperature cures, transferring inner shell from mold to a supporting structure, etc.). Further cost reductions are anticipated (but are not included in this cost model) using the VARTM process and replacing the labor intensive hand lay-up of dry fabric with a fibrous preform.

4. PERFORMANCE

Structural Test Results

To compare the structural performance of the current baseline AFP process to a VARTM fabricated laminate, flat 16-ply laminates were manufactured. Unidirectional and quasi-isotropic AFP laminates were fabricated using Hercules' 8553-40 toughened epoxy/ S-2 Glass® roving and cured to a maximum temperature of 180°C per manufacturer's recommendations. A quasi-isotropic VARTM laminate was fabricated using API's SC-4 epoxy VARTM resin and 6781 style S-2 Glass® fabric and cured to a maximum temperature of 105°C. Open hole compression (OHC) and short beam shear (SBS) tests were performed in accordance to SACMA recommended test methods SRM3-88 and SRM8-88 respectively. A summary of the structural test results is presented in Table 1.

Table 1. Structural test results comparing AFP processed laminates to a low cost VARTM processed laminate.

Test Condition/Property	AFP Process	VARTM Process	Design Requirement
RT/Dry OHC Strength, kPa (C.V.%)	253 (7.1)	223 (4.8)	217
RT/Dry SBS Strength, kPa (C.V.%)	66 (7.1)	36 (2.6)	34
82°C/Wet OHC Strength, kPa (C.V.%)	134 (5.0)	104 (2.8)	103
82°C/Wet SBS Strength, kPa (C.V.%)	31 (4.6)	19 (6.7)	17

The structural performance of the VARTM laminates met all of the design requirements for the CAV-ATD upper hull. VARTM open hole compression strengths were slightly lower than the baseline AFP laminates. The interlaminar shear strengths (SBS) of the VARTM laminates were significantly lower than the AFP laminates at both test conditions. The VARTM SBS results were obtained on 16-ply quasi-isotropic fabric laminates, which may contribute to the low values reported for this material. Also, the VARTM laminates were cured at a significantly lower temperature than the AFP laminates and a slight improvement in hot/wet performance of the VARTM laminates is anticipated with the higher cure temperature which is required to cure the outer cover.

Ballistic Test Results

Ballistic test results of a VARTM processed integrated armor and structure laminate were compared to a similarly configured toughened epoxy laminates fabricated using the baseline AFP process. All panels were ballistically tested using an artillery fragment simulating projectile (FSP). Six ballistic rounds were fired at each panel and the resulting damage/penetration was recorded. Figure 6 shows the resulting damage to the VARTM processed panel. Four partial penetrations and two complete penetrations were recorded for the six shots fired at the VARTM processed panel, which is also typical of the results obtained on AFP processed panels. Although the penetration resistance of the VARTM processed panel is equivalent to the AFP processed panel, the damage mechanisms were different in the two panels. The AFP panel contained a high degree of fiber breakage and localized delamination, while the VARTM panel had larger areas of delaminations, but very little fiber breakage.

Figure 6. Ballistic performance of a VARTM processed panel was equivalent to an AFP processed panel.

5. CONCLUSIONS

The Vacuum Assisted Resin Transfer Molding (VARTM) technique developed is a viable low cost method for producing structural laminates with integrated armor for ground combat vehicle hulls. The VARTM process offers potential overall unit cost savings of approximately 21% as compared to a Automated Fiber Placement baseline manufacturing process. Commercially available low cost epoxy VARTM resins have been shown to meet the structural design requirements for the CAV-ATD upper hull laminate and the ballistic penetration resistance of a VARTM processed laminate is equivalent to the typical performance of an AFP processed panel.

VARTM processing also offers several other advantages over the current baseline AFP process:

- Requires only a single elevated temperature vacuum bag oven cure, unlike the three cure cycles required for the baseline method.
- Permits the use of a female mold for enhanced exterior surface finish.
- Permits the use of a variety of fiber forms.
- Requires a minimal capital equipment investment.

The existing performance of laminates fabricated using the VARTM process is acceptable for manufacturing the CAV-ATD upper hull, however additional testing is recommended to fully validate the structural and ballistic performance. Performance may also be enhanced by additional resin development and/or by the addition of preform technology such as through-thickness reinforcement.

REFERENCES

1. Draper Laboratory, "Cost Analysis Update for the Composite Armored Vehicle", Final Report prepared for United Defense, L.P., CSDL-R-2718, December 29, 1995

28th International SAMPE Technical Conference
November 4-7, 1996

NET RTM PREFORMING PROCESS FOR COST EFFECTIVE MANUFACTURING OF MILITARY GROUND VEHICLE COMPOSITE STRUCTURES

Robert L. Faiz
Charles Stark Draper Laboratory Inc.
Cambridge, MA 02139

ABSTRACT

While the strength, stiffness, and signature advantages of composite materials versus metals are well documented, their application to many military systems has been inhibited by the extreme cost-performance paradigm that has evolved in the industry, the poles of which are high performance and price insensitive aerospace components and low performance and price sensitive automotive components. Among the key requirements for military automotive composite parts are reasonably high fiber volumes ($\geq$ 40%) using a continuous reinforcement oriented in the principal stress directions. This attribute must be achieved at minimum cost in production volumes at or less than the 10,000 level. A process that potentially can satisfy the needs of military automotive composite production is Resin Transfer Molding (RTM). Cost studies in the aerospace and automotive industries -- and verified by Draper -- have indicated that RTM holds great promise for reducing many of the significant cost factors in composites manufacture especially if the high cost of preform preparation can be reduced.

This paper documents the successful definition and demonstration of the enabling technologies used for the development of cost effective continuous reinforcement, oriented fiber, RTM preforms satisfying all performance requirements. These technologies were developed by Draper supported by the U. S. Army Tank-Automotive Command (TACOM) in Warren, MI. Technologies address improvements in the application of innovative preform manufacturing techniques and equipment. Preliminary investigations, process laboratory experiments, and scale-up 3-D part demonstrations are documented.

The preform development efforts successfully demonstrated the enabling technologies for the application of cost effective advanced preforming techniques for a 3-D part. This demonstration part represents all the key features of typical military ground vehicle composite structures.

In order to verify that the net preforming process technology is production ready -- the sub-scale based activities must be scaled up to the fabrication of full size components. The fabrication of full scale components will provide data *verifying* the quality, performance, and cost levels that can be anticipated for any initial production start-ups.

KEY WORDS: Preforms, Diaphragm Forming, Composite Structures

1. INTRODUCTION

1.1 Cost Effective Performs For Military Ground Vehicle Composites Structures

Significant performance improvements per unit weight can be realized when using composite structures in military ground vehicles. Additionally, improvements in other areas can be realized as well (ballistic protection, signature reduction (radar cross-section, infrared and acoustic). cost reduction (through component integration), and corrosion control.

The development of cost effective, net shape, 3-dimensional preforms is one of the key technologies required to support the definition of manufacturing processes for high performance composite structures at an affordable cost. While the use of organic matrix composite materials has recently received much scrutiny for tank/AFV turret and hull armor, the focus of this paper is soft-skinned vehicular applications where component thicknesses (~1/10-1/2 inch) are amenable to a well-understood RTM process [1].

1.1.1 Barriers to adaptation of composites in military ground vehicles The principal barrier to composite adaptations in military ground vehicles is that these applications lie midway between the historical poles of the composites industry: high performance, price insensitive aerospace components on the one hand, versus low performance, price sensitive automotive and industrial products. Military vehicle applications require relatively high performance characteristics at low cost. An additional factor affecting cost, is low production rates of approximately 10,000 units per year. As a consequence of these requirements, the application of composites to military vehicles will require innovative processing technologies, combining in some measure the quality of aerospace technologies with the cost/quantity efficiency of commercial automotive processes.

The aerospace composites market demands the highest strength and stiffness performance (lowest weight) -- cost is secondary -- at low production rates. As a result of this constraint, the aerospace composite manufacturing processes used in material cutting, ply lay-up, and molding require large amounts of hand labor, a situation inimical to low cost/high rate process development [2].

By contrast, the automotive/industrial market has as its goal modest weight reduction using low performance/lowest cost resin systems and reinforcements [3]. Urethane and vinyl ester resins are typically used, since their rapid cure cycle supports the required high production rates. The structural performance of the finished composite is low when compared to the aerospace composite material database. All process development activities address component fabrication flow time measured in minutes, with production quantities in excess of 100,000 units. These quantities justify significant investment in process cost reductions for such low tech components: body panels hood and deck lids, interior trim panels.

1.1.2 Resin transfer molding using net shape preforms -- A promising composite fabrication approach Among the key requirements for military automotive composite parts will be reasonably high fiber volumes ($\geq$ 40%) using a continuous reinforcement oriented in the principal stress directions, integrated with a resin system combining good structural performance and acceptable fire/smoke/toxicity characteristics. These attributes must be achieved at minimum cost in production volumes at or less than the 10,000 level. A molding process which has the potential for providing many of these benefits is Resin Transfer Molding (RTM), illustrated on Figure 1 [4]. The optimum RTM process must be enhanced by a preforming process that provides net shaping and repeatability.

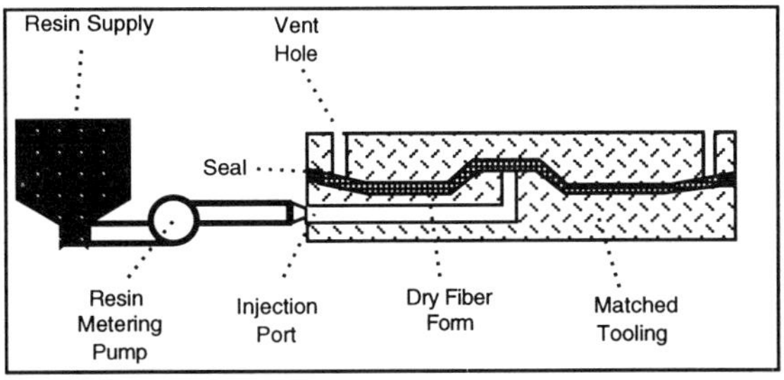

Figure 1. Schematic Of The Resin Transfer Molding Process

Ply laminate process control (a key to low cost and repeatability) is achieved during preforming. The Double Diaphragm Forming process, developed at Draper Laboratory, produces net shape forming. This development of this process is based on an in-depth analysis of fabric shaping and component geometric/structural design features [5, 6].

Cost analyses performed as part of the TACOM effort have identified that the introduction of net preforms in the RTM process can result in a component cost reduction up to 40% (reference Figure 2). These projected cost savings have been verified by sub-element fabrication experiments using technologies readily transferred to larger composite structures using a liquid molding process.

TACOM 3-D Part	
300/year	Unit Costs
labor	$2,300
capital	$100
tooling	$5
material	$100
total cost	$2,505
weight of part	4.5 kg
dollars per kg	$555

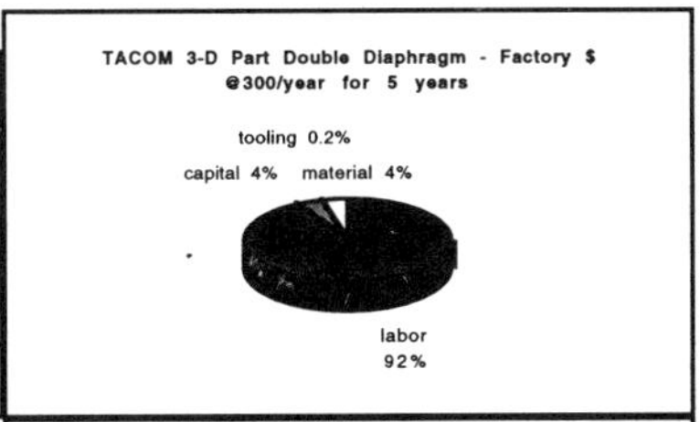

TACOM 3-D Part	
300/year	Unit Costs
labor	$3,900
capital	$0
tooling	$0
material	$300
total cost	$4,200
wieght of part	4.5 kg
dollars per kg	$930

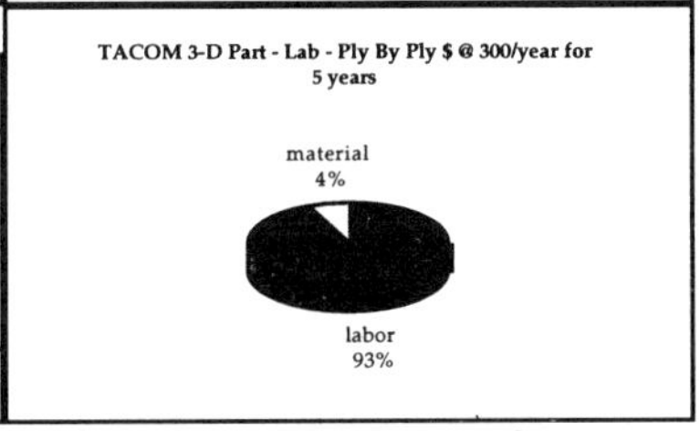

**Figure 2. Part Manufacturing Cost Summary
Net Preforming Vs Ply-By-Ply Preforming**

2. LABORATORY EXPLORATORY EFFORTS

Key to the successful development of cost effective, net shape preforming processes is the understanding of the forming characteristics and limitations dictated by the material form, design and equipment used. Draper Laboratory has been assessing these issues over the past several years, supporting the characterization of the Double Diaphragm Former -- net preform -- process.

2.1 Material Forming Characterization An understanding of the key drivers that affect preform processing performance, quality, and cost are assessed and empirically verified using simple laboratory experiments. The relationship of component geometry (laminate thickness, forming radii, severity of curvature, and sharp features) and reinforcement material architecture (weave, tow size, number of layers, and orientation) are established. Fabric architecture (weave) warpage models -- defining the predicted behavior of the material when formed -- are correlated with empirical data. A typical model is shown in Figure 3.

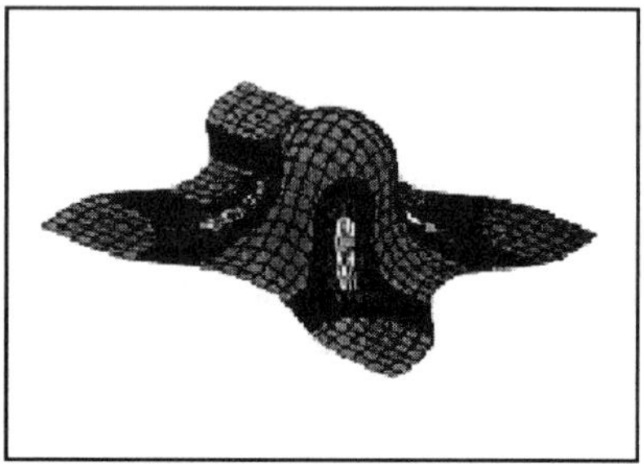

Figure 3. Model Predicting Fabric Warpage When Formed Over Sphere

2.2 Preliminary Guides For Design Trade-off Efforts The characterization data is used to establish engineering design rules to be used to assess the impact of component geometry, material choice, and processing techniques on laminate structure, performance, quality, and cost. These design rules and cost models focus on the relationship of part geometry and fabric architecture limitations when forming flat, 2-dimensional material over 3-dimensional shaped tooling surfaces assist the design process. Fiber locking, wrinkling onset, bridging, fiber tension, fiber displacement, and tow loss when forming into 3-D geometries are some of the key parameters investigated. The impact of forming technique and equipment as well as number of plies, orientation and ply trimming were also assessed. Figure 4 shows typical design guides addressing preform quality and ease of forming. Also shown is the output of a typical cost model establishing the key drivers of the component molding unit cost.

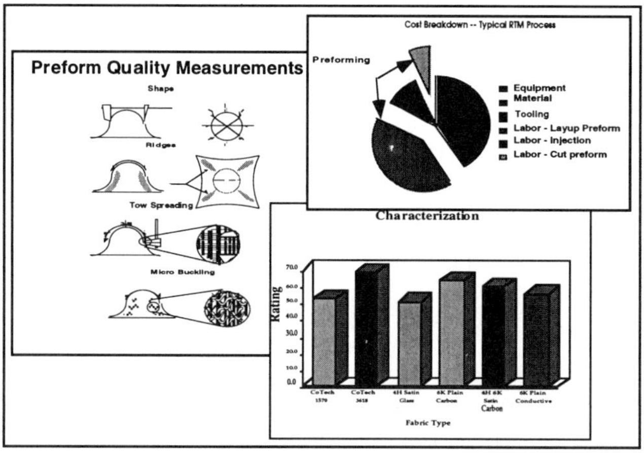

Figure 4. Design Guides & Cost Models Supporting The Development Of A Net Preforming Process

Both matched die and diaphragm forming processes were evaluated noting their impact on the formability of several candidate material forms. Ultimately, the formability and process data supported the selection of the double diaphragm forming process as the most promising process for the fabrication of the optimum net preform. This process offers the best solution for the fabrication of thick section fiberglass fabric preforms at acceptable cost for all types of structures and production scenarios associated with military ground vehicles. Sandwich panel preforms, preforms with inserts (as shown in Figure 5), have been successfully formed. Some of the key attributes of this process is the controlling of the fabric architecture (tension, deformation, compaction, edge control and temperature) during the complete vacuum forming process. Fabrication of the sub scale, 3-D component preform demonstrated the suitability of this preforming process.

Figure 5. Fabrication Of Net Shaped Preforms Using A Double Diaphragm Former.

3. DEVELOPMENT TASKS – NET SHAPED PREFORMING

Additional net shaped preforming activities, focusing on actual military ground vehicle component geometry, was performed by Draper Laboratory for the U.S. Army Tank-Automotive Command, Warren, MI 48397-5000, under Contract No. DAAE07-93-C-R084 "Cost Effective Composite Manufacturing Process For Military Ground Vehicles." This effort was aimed at bridging the cost-performance gap in the application of composites to military ground vehicles.

3.1 Review and Analysis of the HUMVEE Hood/Fender Design As part of the TACOM activity, Draper performed a design-for-manufacture assessment of an alternate method of fabricating the hood for the Army High Mobility Multi-purpose Wheeled Vehicle (HMMWV or also known as the "HUMVEE"). A Resin Transfer Molded (RTM) continuous fiber reinforced component was envisioned as a more durable replacement for the existing compression molded fiberglass reinforced hood (increased strength, stiffness at a reduced weight and procurement cost). Key to the determination of an optimum design, material and manufacturing process selection is an in-depth cost-benefit analysis supported by adequate laboratory verification testing.

3.1.1 Overview of the existing HMMWV hood/fender design The present design is basically equivalent to the original metal hood/fender fabricated from composite materials Figure 6. The outer skin is one piece with a uniform thickness of approximately 6.4 mm. A major portion of its stiffness and strength is derived from the step transition between the hood and fender top region. Additional stiffness and strength is provided by the vertical sides of the hood, the vertical forward sections, and the local pockets for mounting lights and other equipment. Structural stiffeners have been added in the flat hood areas adjacent to the top grill opening and across the top, rear area supported by the vehicle bulkhead.

Figure 6. HMMWV Hood

The material used for the HMMWV hood/fender is a fiberglass reinforced polyester sheet molding compound (SMC) structural skin with short, random orientation, fiberglass strand reinforcement. The same material is used for the hat section stiffeners. The compression molding process is used to form the SMC into the required contour using heated matched metal dies in a high capacity press. Post molding operations include trimming the hood skin and stiffeners to shape and bonding of several composite and metal stiffeners and reinforcements to selected areas of the skin. Drilling and machining for installation interfaces and painting are the final processing steps.

3.1.2 Overview of an improved RTM HMMWV composite hood/fender The proposed fabrication process for an improved HMMWV hood/fender is the Resin Transfer Molding (RTM) process. Resin Transfer Molding process technology has been developing rapidly over the past several years. The result has been the successful demonstration of this process in fabricating many high performance and cost effective composite structures. Many of the parts that have been successfully demonstrated to date are integrally stiffened shell structures similar to the HMMWV hood/fender. The proposed RTM hood/fender is interchangeable with the present component. The basic skin thickness is reduced because the RTM design takes advantage of the high performance matrix and fiber reinforcements used in the RTM process. The reinforcement is a woven fiberglass fabric inner and outer skin that provides increased strength and damage tolerance compared to SMC. The skin is stiffened with the addition of a long fiber random fiberglass mat core between the skins. The inner skin will also incorporate many of the stiffener details that are presently secondarily attached.

3.1.3 Comparison of Double Diaphragm Forming and hand lay-up The traditional manual RTM preforming process is not cost effective. The manual trimming and handling of the dry fiber reinforcement used in the RTM process is labor intensive. The dry material, which is not stable, makes cutting, trimming and placing the material into the mold cavity difficult. Similar preforming operations for the SMC process are less labor intensive as the SMC material is dimensionally stable. Reducing the cost of preforming is the focus of the development of cost effective, net shape preforms.

Cost estimates for the preform fabrication process were defined using existing cost models at Draper Lab. The results of the analysis supported the predicted 40% cost reduction identified earlier in Figure 2. In addition, the fiber architecture of the preforms produced using double diaphragm forming was more repeatable than that produced by hand lay-up. The repeatability of the double diaphragm forming process results in a higher quality, more consistent preform -- due to more uniform fiber volumes, better compactness and more consistent fiber paths. The repeatable compaction will also lead to further cost reductions associated with consistent mold closure and resin injection cycles. Mold closure time will be reduced due to the improved fit of the preform in the mold cavity and the fact that the double diaphragm formed preform is pre-compacted and wrinkle-free. Resin injection quality will be improved through the removal of variations in the preform's fiber volume which can lead to dry spots (variations can create high fiber volume regions which can be difficult to fill if not accounted for in selection of locations of ports and vents.)

3.2 Advanced Preforming Task Based on the successful preform characterization laboratory experiments, a commercial HMMWV (HUMMER®) hood/fender was purchased and used for the initial preform experiments. A mold was cast from the left front corner of the hood/fender. This corner has been selected because of the compound curvature of the front corner and the additional details that must be formed for the existing turn signal details. The mold is approximately 43 cm wide by 43 cm long by 30 cm deep. This mold enabled experiments forming fabrics and random mats using double diaphragm, single diaphragm, and match mold forming techniques.

Draper conducted single and double diaphragm forming experiments using the Lab's Double Diaphragm Former (see Figure 5 for a schematic) to form woven fiberglass fabric to the HMMWV hood/fender corner section. The goal of these experiments was to demonstrate that an advanced preforming method/technology can be used to cost effectively fabricate complex preforms using the HMMWV hood/fender corner section as a representative part. For the advanced preforming method to be successful, the preforming method must be able to form good quality preforms of the part at a cost lower than that of manual lay-up. A good quality preform is one that does not contain any darts (cuts in the fabric).

Before running the forming experiments in the Double Diaphragm Former, verification was needed that the HMMWV hood/fender corner section could be laid up without wrinkling the fabric. The current shape of the recessed region for the turn signal lights resulted in wrinkling when forming 8 harness fabric. To enable formation of the hood corner without wrinkling or cutting darts in the material (both of which lower part performance), filling in the turn signal light cavity to make it flush with the basic hood profile is required.

3.2.1 Matched tool forming experiments Several experiments were performed attempting to fabricate acceptable preforms with the matched tools for the hood/fender. All experiments were unable to successfully form the complex compound curvature of the corner section of the HMMWV hood/fender without creating severe wrinkles in the fabric (better described as folds in the fabric). Since the matched die approaches were unsuccessful at forming any single ply preforms, further attempts to form multi-ply stacks were abandoned.

3.2.2 Single diaphragm forming experiments Experiments were conducted to evaluate single diaphragm forming the hood/fender corner section. A review of Figure 7 will assist in the discussions of the key features of the diaphragm forming process. Tests were conducted in the Double Diaphragm Former by removing the bottom diaphragm and placing the fabric directly on the mold and forming the fabric over the mold using only the top diaphragm (i.e. single diaphragm forming). Draper has not been able to successfully form the hood/fender corner using single diaphragm forming due to wrinkles forming in the fabric during the forming operation. Since the fabric is not sandwiched between the upper and lower diaphragms, as it is with double diaphragm forming, the fabric is not tensioned during forming and is not constrained from out-of-plane buckling and thus it wrinkles when it is formed. Wrinkles have been deemed to be unacceptable due to the decreased part performance and the difficulties in closing the mold that accompany the wrinkled material.

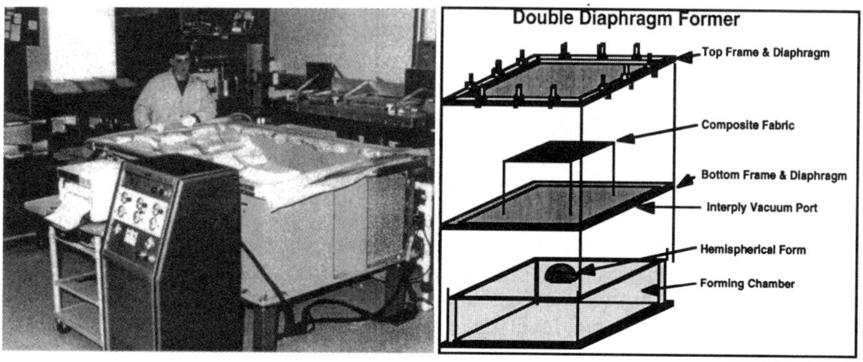

Figure 7. Draper Laboratory 4 X 6 Double Diaphragm Former Installation (Showing Chamber, Controller) & Schematic

3.2.3 Double diaphragm forming experiments Using Double Diaphragm Forming, Draper successfully formed ply of 8 harness fiberglass fabric (in both 0.55kg/m and 0.3kg/m weights) to the hood/fender without wrinkling the material as shown in Figure 8.

Figure 8. Successful Trial Forming of 8 Harness Fabric

The tensioning applied to the preform during double diaphragm forming, due to both rubber diaphragms stretching over the mold, assists the fabric in deforming to the mold without wrinkling. Stretching of the rubber diaphragm and fabric sandwich during forming also acts to constrain all deformation of the fabric to be in-plane deformation (trellising and slipping), generally inhibiting the fabric from wrinkling.

A successful hood corner preform was double diaphragm formed using a two-ply plain weave lay-up consisting of a 0°/90° and a ±45° ply. This lay-up is shown in Figure 9 (other unsuccessful preforming attempts are shown also). Table 1 presents the results for the single and double diaphragm forming tests in a tabular format. It is interesting to note that all of the materials tested with single diaphragm forming wrinkled on single ply forming tests, where as all of the double diaphragm forming single ply tests produced good single ply preforms. It should also be noted that the only multi-ply preforms to be formed without wrinkling was made from plain weave fabric. It is interesting that the plain weave fabric performed better than the other materials tested for this particular shape. It is believed that with some process modifications, the 8 harness satin materials should be capable of producing good preforms for this shape as well.

Table 1. Results Of Advanced Preforming Tests For Single And Double Diaphragm Forming

Material Tested	Single Diaphragm Forming (1 ply)	Double Diaphragm Forming	
		Single Ply	Multi-ply (2)
Plain Weave Glass, 0.3kg.m	Wrinkled	Good	Good
Plain Weave Graphite, 6K tows	Wrinkled	Good	Not Tested Yet
8 Harness Satin Glass, 0.3kg/m	Wrinkled	Good	Wrinkled
8 Harness Satin Glass, 0.55kg/m	Wrinkled	Good	Wrinkled
8 Harness Satin Graphite, 3K tows	Wrinkled	Good	Wrinkled
Biaxial CoTech NCF® Glass, 0.93kg/m	Wrinkled	Good	Wrinkled
5 Harness Satin Graphite, 6K tows	Not enough Material	Good	Slight Wrinkling

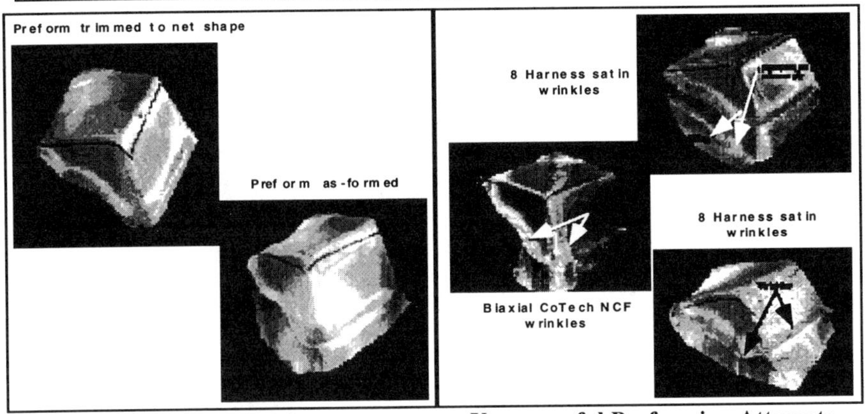

Successful Preforming Attempts **Unsuccessful Preforming Attempts**

Figure 9. Fabric Formability Development - Hood Preform

4. SUB-SCALE PREFORMING -- 3D PART GEOMETRY

To verify the Double Diaphragm Forming process, a sub-scale of a typical military ground vehicle upper hull structure was chosen as a demonstrator for the 3-D preforming process. The limitations in the capacity of the existing diaphragm former required the use of sub-scale overall length and width geometries to 51 cm maximum. However, the key feature (corner radii, thickness, and planar intersections) typical of military vehicle are not scaled. To perform these tests, developmental quality foam and plastic forming molds were fabricated for each of the candidates.

4.1 3-D Part Forming Experiments Initial double diaphragm forming preforming experiments with a 3-D forming tool have successfully formed the shape using a 0.25 kg plain weave in a $\pm 45^{\circ}$, 0/90 configuration. Although successful, a heavier weight fabric is desirable to reduce the number of plies in the lay-up.

4.1.1 Process operations for fabricating the 3-D inner preform Simultaneously formed two ply (0/90 and ±45° plies) with the 13 mm corner radii, resulted in wrinkling the plies at the edge of the highly deformed corner region of the plies (aft vertical corner of 3D shape). Increasing the radii of the sub-scale upper hull to 25.4 mm (from 20 mm) eliminated the problem. The increased radius was deemed as an acceptable design change for a full scale article. This change has reduced the severe deformation experienced in the aft corner regions and has significantly improved the ability to form multi-ply stacks (up to required 13 mm thickness).

A readily available material (0.8 kg/m) plain weave E and S-2 glass) formed very well. Figure 10 shows a 13 mm thick compacted preform formed from 20 plies of 0.8 kg/m material.

The deformed fiber architecture in the aft vertical corners of this preform result in increased fiber volume.. For a part with a nominal fiber volumes of 50% in the flat sections of the part, the aft corner regions have a predicted increased fiber volumes on the order of 67%. To account for the change in fiber volume the male mold contour is modified to provide increased mold gap in the aft vertical corner region to account for this increase in fiber volume.

Tests using the double diaphragm former and a fabricated, plastic forming mold have defined the limiting, two layer thickness that can be successfully formed without corner wrinkling for this particular configuration. Therefore the 13 mm thick inner preform is achieved by sequentially forming ten two-ply preforms over preceding layers. Figure 10 shows the resultant inner preform. The next step was the definition of the fabrication process for the outer preform with sharp corner radii.

Figure 10. Initial Inner Rtm Preform For The 3d Component

4.1.2 Process operations for fabricating the 3-D outer preform & assembly The technical challenge faced during the development of the complete preform was the definition and integration of the individual elements incorporating the continuous fiber inside radius corners and the sharp corners of the outside contour. Draper expanded the successful development of the inner preform to incorporate both of these critical geometric features. The resultant preform assembly is near net - easing the placement of the preform in the subsequent molding (conventional of vacuum, assisted RTM) cell. The preform assembly development effort is based on the forming of a separate, two ply outer skin in the outer mold cavity. The use of only two plys allows the easy formation of the material into the sharp outside corners of the mold upper cavity. Braided filler material is then placed against the inside surface of the outer skin in the sharp corner regions along the side and back of the mold cavity. A braided construction was chosen as it offers the optimum solution to providing the required fiber volume while providing a degree of conformability when the preform inner shell is assembled. The linear weight of the filler (2.0 to 2.4 g/cm) provides the desired fiber volume (45%) for the preform. Figure 11 defines a typical section cut of the preform assembly showing the relationship between the outer shell, inner preform and braided fillers. Of interest is the configuration of the braided fillers. The central braid is locked into position by the "tails" of the adjacent braided structures.

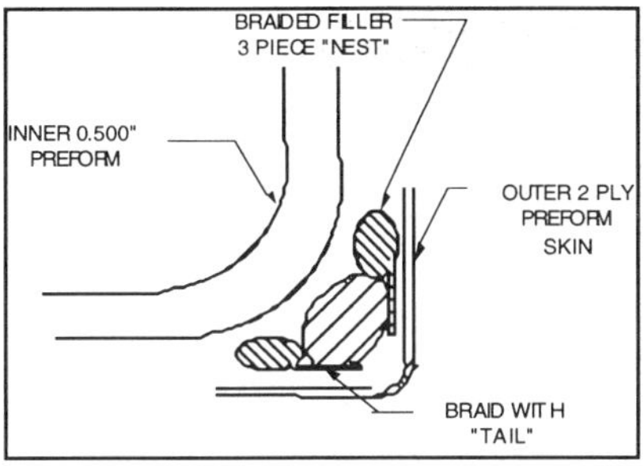

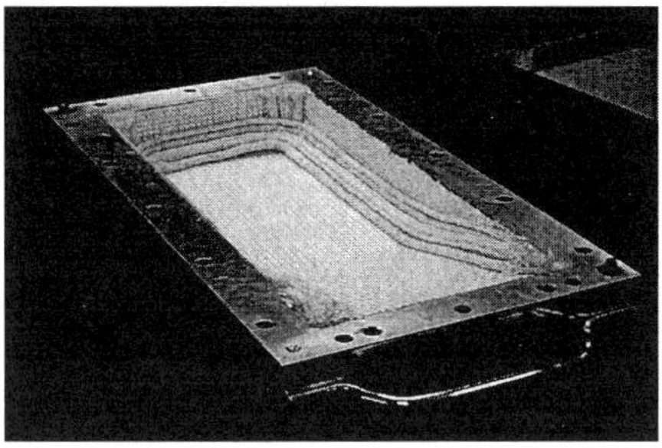

Figure 11. 3-D Preform Corner Filler Braid Concept.

4.1.3 Subsequent 3-D part preforming In order to verify that the preforming process was robust , additional preform assemblies were fabricated using a documented manufacturing procedure. This procedure was defined during the initial inner, outer, and assembly operations decried earlier. A total of four preform assemblies were fabricated -- all exhibiting good geometry, fiber architecture, and fiber volume repeatability.

The preform fabrication process involves a combination of;

- Double Diaphragm Forming (DDF) of the inner preform over a non-metallic male forming tool resulting in a net thickness and shape preform).
- Stamping (combined with a steel rule die) and forming the two-ply outer skin using a conveyor oven to heat the flat charge and a matched die platen press.
- The Double Diaphragm Forming process can fabricate a net preform with a thickness equivalent to 45%-55% fiber volume.
- Assembly of the outer skin, braided fillers and inner preform in a matched die prior to resin infusion.

The results of the preform development effort is the definition of a preform fabrication process that can be scaled up to full size components.

4.1.4. Fiber volume & preform thickness Structural fiber reinforced structures should have a fiber <u>*volume*</u> ratio of approximately 50% in order to demonstrate an acceptable strength to weight. A key to choosing the double diaphragm forming process is the demonstrated capability of fabrication of a thick section, net preform with an average thickness equal to a fiber volume of 50%.

The finished preforms exhibited fiber volumes ranging from 47% to 55% for the twenty ply lay-up used. These values are verified by weighing the net preform before and after molding and converting the fiber weight fractions (60% to 65%) to volume fractions based on the relative density ratios between fiberglass and the epoxy resin used. The 13 mm thickness (20 plys at 0.64 mm per ply) was used as an inspection criteria for the finished preform and is the basic dimension used for machining the mold cavity.

Figure 12 depicts a typical preform assembly - highlighting the net shape and outer sharp corner features.

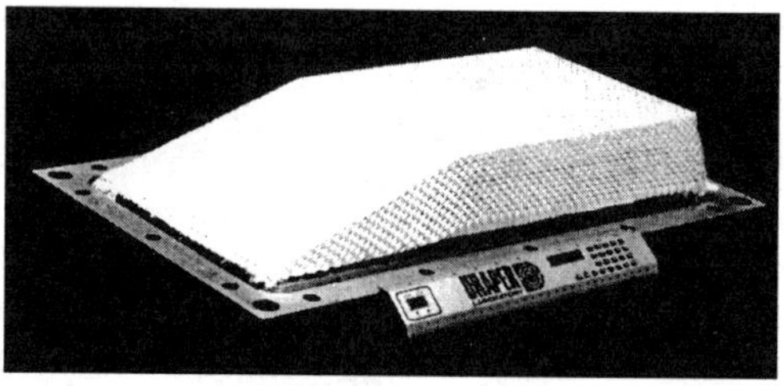

Figure 12. Final 3-D RTM preform assembly - ready for molding

5. CONCLUSIONS

In summary, improved quality, and lower cost are expected from the development of net shape preforms. The utilization of automated 2-D and 3-D preform shaping methods will bring significant cost reductions as compared to hand lay-up, RTM preforming approaches. These mechanized preforming techniques will further facilitate the production of reinforcement architecture having repeatable fiber volume and thickness throughout the part, thereby substantially improving part quality.

The key enabling technology supporting the development of a robust Resin Transfer Molding process has been demonstrated. Improvements in part quality, and repeatablilty while reducing product development cycles and cost can be anticipated at low risk. The development of a repeatable, net preforming process allows repeatable, predictable fiber placement in a form that is easily installed in complex, matched die cavity molds -- crucial to the introduction of process automation.

The net preform manufacturing process demonstrated on sub-scale components can be easily scaled up to a cost effective approach for the full-size composite components envisioned for future composite military ground vehicles (CAV, Scout, Mortar and Main Battle Tank).

6. REFERENCES

1. M. Foley, Techno-Economic Analysis of Composite Manufacturing Processes, *SAMPE Quarterly*, January 1991.
2. M. Foley and E. Bernardon, Cost Effective Automation for Manufacturing Composite Structures, Proceedings, SME Composites in Manufacturing 9 Conference, 1991.
3. S. Arnold and C. Johnson, Liquid Molding Process Design with the Help of Computer-Based Cost Modeling, Proceedings, Advanced Composite Materials: New Developments and Applications Conference, 1991.
4. J. Busch and D. Ragone, Composites in the U.S. Automotive Industry, Proceedings, ATAMAT 89.
5. S. Luby and E. Bernardon, Design of Fabric Preforms for Double Diaphragm Forming, Proceedings Fibertex 91 Conference, NASA Conference Publications, 1991.
6. M. Foley and E. Bernardon, Development of an Improved, Flexible Resin Transfer Molding (FRTM®) Process Proceedings, SAMPE Technical Conference, 1991.

28th International SAMPE Technical Conference
November 4-7, 1996

VERIFICATION OF A THREE-DIMENSIONAL RTM/RFI FLOW MODEL

Alfred C. Loos, John C. Fingerson and John D. MacRae
Department of Engineering Science and Mechanics
Virginia Polytechnic Institute and State University
Blacksburg, VA 24061-0219

ABSTRACT

A model of the RTM/RFI flow process is under development for use in tool design and process cycle optimization of complex shaped textile preforms. The model, based on the finite element/control volume technique, was developed to predict infiltration of resin into an anisotropic fiber preform. Experiments were performed to verify the accuracy of the flow model and the preform permeability measurements. A single blade-stiffened preform was infiltrated in an instrumented mold. The mold inlet pressure and the flow front positions were measured during infiltration by miniature flush mount pressure transducers mounted in the fixture. Results of the experiments indicate that the total preform infiltration time predicted by the model agrees well with the measured value. However, the model predicted mold inlet pressure is considerably lower than the pressure measured at the mold inlet during the filling experiment. The difference between the predicted and measured mold inlet pressures may be attributed to the actual fiber volume fraction in the skin directly beneath the stiffener being higher that the value input into the simulation model.

KEY WORDS: Resin Transfer Molding (RTM), Flow Modelling, Manufacturing/Fabrication/ Processing

1. INTRODUCTION

Fabrication of composite materials by processes such as Resin Transfer Molding (RTM) and Resin Film Infusion (RFI) involve the infiltration of fiberglass or carbon fiber preforms with resins such as epoxies, vinylesters, and polyesters. Recently, the development of automated textile technologies have allowed for three-dimensional preforms to be manufactured from both woven and stitched materials. The ability to produce a near-net-shaped preform reduces the mold layup time and can increase the product quality by reducing the variability in fiber orientation. Through-the-thickness reinforcements may be added to increase the damage tolerance of the finished part by strengthening the part in the transverse direction. Modeling of the resin infiltration process can be a critical step in the optimization of both tooling design and the processing cycle. One of the major considerations in tooling design is the location of injection ports. The characteristics of the resin flow are defined by the following material properties - the preform permeability tensor, preform compaction behavior, and the viscosity of the resin. Past research has been completed to verify flow characteristics in two dimensions (1-3). However, with the advent of more complicated three-dimensional woven and stitched preforms, the ability to verify the three-dimensional material properties is essential.

A three-dimensional model based on the finite element/control volume technique has been developed and can be use to predict flow front positions and pressure distributions during resin infiltration of a complex shaped textile preform (4-7). In this paper, the results of a program to verify the predictions of a three-dimensional RTM/RFI flow model will be presented. The preform selected for the verification experiments is a single blade-stiffened preform. Verification experiments included measurement of the permeability tensor values for each component of the preform (required as input to the model), and the flow front position and resin inlet pressure during actual resin infiltration of the preform mounted in an instrumented tool.

A three-dimensional finite element mesh of the preform was created utilizing PATRAN which allows for the permeability values to be entered for the individual components of the blade-stiffened preform, the application of boundary conditions and the choice of injection port locations. Visualization of results was completed using PATRAN as a post-processor.

2. Three-Dimensional Flow Model

2.1 Flow Model The three-dimensional resin flow model was developed to determine the position of the resin flow front and pressure distribution inside the preform. The following assumptions were made in the development of the resin flow model: 1) the textile preform represents a porous medium; 2) the preform is heterogeneous and anisotropic; 3) the resin is incompressible, and capillary and inertia effects are neglected (low Reynolds number flow); 4) the resin is considered to be Newtonian in behavior at the shear rates present within the preform; and 5) there is no leakage of resin out of the tooling assembly except at the specified vent locations in the tool.

For flow through a porous medium, the momentum equation can be replaced by Darcy's Law. Darcy's Law for an anisotropic medium can be written as

$$\vec{q} = -\frac{\vec{\vec{S}}}{\mu}\nabla P \qquad (1)$$

where $\vec{q}$ is the superficial velocity vector, $\vec{\vec{S}}$ is the permeability tensor, μ is the viscosity, and P is the resin pressure.

The continuity equation for incompressible flow is written as:

$$\nabla \bullet \vec{q} = 0 \qquad (2)$$

Combining Darcy's Law and the continuity equation, one obtains the form of the equation for resin flow within the fibrous preform.

$$\nabla \bullet \left(-\frac{\vec{S}}{\mu} \nabla P\right) = 0 \tag{3}$$

It should be noted that the permeability tensor represents the anisotropic material properties and can be formed by transforming the local material coordinates into the global coordinate system. Solution of equation (3) gives the pressure distribution P(x,y,z) within the region of the textile preform where the resin has infiltrated. Once the pressure distribution has been obtained, Darcy's Law is utilized to calculate the resin velocity at any point within the preform.

The interstitial resin velocity is calculated at the flow front and utilized for updating the flow front position. The interstitial resin velocities can be obtained from the following equation

$$\vec{q} = \vec{v}\phi \tag{4}$$

where $\vec{q}$ is the superficial velocity vector, $\vec{v}$ interstitial resin velocity vector, and ϕ is the porosity of the preform material.

2.1.1 Boundary Conditions The boundary conditions in the local coordinate system consist of the following:
At the mold boundaries,

$$\frac{\partial P}{\partial \eta} = 0 \tag{5}$$

and at the resin flow front,

$$P = P_{initial} \tag{6}$$

where $P_{initial}$ is the initial pressure within the preform.

For the RFI process, along the resin film- preform interface,

$$P = P_{comp} \tag{7}$$

where P_{comp} is the applied compaction pressure.

For the RTM process, at the mold inlets (gates), the boundary conditions specified are either,

$$P = P_o \tag{8}$$

for constant pressure mold filling, or,

$$\vec{v} = \vec{v}_o \tag{9}$$

for constant flow rate mold filling.

2.2 Finite Element/Control Volume Method For the modeling of the RTM/RFI mold filling/ resin infiltration processes, a finite element/control volume model was developed. The advantages of the finite element/control volume technique as compared to other numerical techniques are: ability to handle complex geometries; conservation of mass upheld; flow front tracking; numerical stability; and ease of handling multiple vents and gates.

The finite element/control volume technique requires that the mold geometry be divided into elements. At each node, a control volume is constructed by subdividing the element into smaller volumes and summing the contribution of adjacent volumes at each node. Element subdivision is carried out by breaking the element volume into smaller volumes connected at the centroid of each element. The finite element method solves for the pressure gradients. The pressure gradients are entered into Darcy's Law, to solve for the velocity field and the flow rate into the control volume. With the flow rate information, the fill factor for the control volume can be updated. The time step for the next iteration is determined by calculating the estimated time step to fill each control volume and using the smallest value. Each of the nodal quantities was marked as filled (fill factor = 1.0) or unfilled (fill factor =0.0) or somewhere in-between. This allows for tracking the flow front movement during the RFI processing without the need for expensive remeshing.

Eight node isoparametric hexahedron elements are used in the current version of the simulation model. The finite element/control volume formulation is described in further detail in references (4-7).

3. Experimental

3.1 Blade-stiffened preform Blade-stiffened preforms have been fabricated at NASA-Langley Research Center. The preform consists of a 54 ply multiaxial warp knit preform and a braided/woven stiffener stitched to it. Details of the fabrication are described in reference (8). The approximate outer dimensions of the preform are shown in Figure 1.

The resin flow model requires numerical values of the permeability tensor for each material type used in the construction of the preform. A complete characterization was performed to measure the transverse and in-plane permeabilities as a function of fiber volume fraction.

A system has been designed for accurate automated real-time measurement of fabric permeability. The system is comprised of a labVIEW virtual instrument created to compute permeability from pressure transducer, LVDT, and load cell measurements. Details of the permeability measurement system and numerical values for the permeabilities of the component materials that make up the preform are given in reference (8).

3.2 Blade-Stiffened Preform Flow Fixture An instrumented mold that would permit measurement of the resin flow patterns that develop during impregnation of the blade-stiffened preform was fabricated (see Figure 2). The mold was designed with the following objectives:

- Ability to change volume fractions of the components of the preform.
- Fluid injection that would result in a three-dimensional flow pattern.
- Insensitivity of flow front development to pressure transducers implemented to measure flow front position.

- Dimensional accuracy and stability, especially when preform compaction load is present.
- Creation of boundary conditions which can be accurately replicated in Finite Element model.

The mold consists of three plates - a lower mold plate and a left and right mold plate. The interior faces of the mold conform to the dimensions of the preform and are ground for dimensional accuracy. Ground spacers were implemented to allow for changes in fiber volume fraction. Steel was chosen as a structural material instead of materials such as plexiglass to meet the requirement of dimensional stability under preform compaction loads. The inlet of the mold was located at the center of the lower mold plate to allow for a three-dimensional flow pattern to develop.

The flow front position was measured by as many as six Entran EPX series pressure transducers. The sensor locations are shown Figure 1. The tip of each transducer is slightly offset from the interior mold face so that the transducer reading is not affected by the compaction load of the preform. The diameter of the transducer tip is 3.5 mm, which minimizes the effect of the transducer on the flow front.

The edges of the preform with the exception of the upper edge of the stiffener were potted with a two-part polyurethane (Hexcel Uralite 3138 Part A & B) to prevent flow through the edge of the preform

3.3 Test Procedures A total of three flow front tests were completed. The blade-stiffened preform was impregnated with corn oil at a constant inlet flow rate of 3 cc/min. The flow rate was determined by initial test results to be approximately the highest flow rate possible without compression of the preform at the inlet due to high inlet pressures. Corn oil was used as a test fluid and had a viscosity of 0.0581 Pa-s at room temperature. In order to maintain a fiber volume fraction of 55.0% across the entire preform, the spacers between the lower mold plate and the upper mold plates were 0.899 cm thick and the spacers between the upper mold plates were 1.107 cm thick. All edges were sealed with polyurethane, except for the top edge of the blade, which was left open to allow air to escape from the mold. The four corners of the base of the preform were vented by placing a small diameter wire under the polyurethane seal. The wires were removed when the fluid reached the corners.

The inlet pressure was monitored with an Omega pressure transducer, 0- 689 kPa. As many as six Entran transducers with pressure ranges of 0-345, 0-689, and 0-1,723 kPa were used to monitor the flow front position. Due to the low pressures in the saturated preform during mold filling, the transducers were only used to detect the time at which the flow front passes over the pressure tap. Also, since there were more pressure taps (see Figure 1) than pressure transducers, the transducers were initially placed at pressure taps that the flow front would reach first. After the flow front passed these pressure taps, the pump was stopped and the transducers were moved to the next series of pressure taps, replugging the initial pressure taps. This was repeated until the flow front had been measured at all pressure taps. Data was sampled every two seconds during the duration of the tests.

4. RESULTS

4.1 Finite Element Analysis PATRAN (PDA Engineering) was utilized to construct the geometry, discretize the geometry into elements, and apply boundary conditions and material properties. A PATRAN neutral file is output which contains all of the data required as input into the flow simulation model.

The finite element mesh (see Figure 3) generated was a quarter model of the entire preform and contained 2060 elements. The boundary conditions specified are zero flow rate normal to all surfaces except the inlet, and constant flow rate at the inlet. The initial pressure specified at all nodes was zero except for the nodes on the surface of the inlet where the pressure is not specified due to the constant flow rate condition.

The model was run on IBM RS-6000 (SP2) workstations at NASA-Langley Research Center. The computing time for a 2060 element model was 7443 CPU seconds.

PATRAN was utilized to input PATRAN formatted result files which contain nodal results of fill times, pressure values and flow front velocities at discrete time intervals. These results were post-processed to generate three-dimensional contour plots.

4.2 Test and Model Comparisions The FEM flow model can be used to predict and visualize the resin infiltration patterns and pressure distribution in the preform at selected times during mold filling. Shown in Figures 4 and 5 are the infiltration patterns at various times and the pressure distribution at 3242 s into the mold filling process. The model can be readily used to observe what effect changes in the boundary conditions (i.e., locations of the inlet ports and vents, inlet resin flow rate or injection pressure) and material properties (i.e., permeabilities, resin viscosity) will have on the resin fill patterns, pressure distributions and total infiltration time.

A comparison between the measured and model predicted inlet pressure for the blade stiffened preform is shown in Figure 6. The rapid rise in the inlet pressure observed at the beginning of mold filling was not predicted by the model. Also, the rate of the pressure increase after the initial rise was measured to be higher than the model predicted. Overall, the model underestimated the inlet pressure by approximately 50%.

The wet-out times (times for the flow front to reach the pressure taps located in the mold) for each of the pressure transducer location are shown in Figure 7. The flow front reached pressure tap locations 2/3 and 1/4, located on the bottom of the skin beneath the stiffener flange, much slower than predicted by the model. Similar results were noted for pressure tap locations 5-8 located on the top of the skin away from the stiffener flange region. However, the percent differences between the predicted and measured wet-out times were less than observed for the stiffener flange region. The measured and predicted wet-out times at the top of the stiffener (pressure taps 9-12) agreed reasonably well, with a maximum difference of less than 10%.

Some possible reasons for the mismatch in results are as follows:

1. Incorrect permeability data input to the model.
2. Assumption that the fiber volume fraction of the skin directly under the stiffener is maintained at 55%. This region is not supported on both faces by a mold surface, and the preform compaction may be affected by the mold bolting procedure.
3. Non-homogeneity of the region near the bottom of the stiffener and above the skin. The stiffener and stiffener flange are manufactured from the same textile preform as a single component. The stiffener flanges are formed by dividing the lower section of the preform, bending the two halves 90 ° and stitching the preform onto the skin. Continuous fiber rovings are placed in the gap under the stiffener before it is stitched to the skin. This is not accounted for in the model.

In order to determine why the predicted and measured wet-out times and mold inlet pressures did not agree, a parametric study is currently being performed. Preliminary results of this study indicate that the by reducing the permeability in the stiffener flange region the average percentage difference in wet-out times was reduced to 12%. Also, the predicted mold inlet pressure agrees well with the measure value.

The results of the parametric study indicate the need for better characterization of the materials that make up the stiffener flange region of the preform. As discussed earlier, this region is made up of three different materials: multiaxial warp knit skin; braided/woven stiffener that is divided and stitched onto the skin; and the unidirectional continuous fiber roving that is used to fill the gap that is formed when the stiffener is divided and folded over to attach onto the skin. Also, the procedure used to install the preform into the mold needs to be carefully controlled to ensure that the stringer flange region is properly compacted to the specified fiber volume fraction.

5. CONCLUDING REMARKS

An experimental program to verify the predictions of a three-dimensional RTM/RFI flow model has been performed. A single blade-stiffener preform was infiltrated in an instrumented mold. The mold inlet pressure and the flow front positions were measured during infiltration by miniature flush mount pressure transducers mounted in the fixture. The initial mold inlet pressure comparisions showed that the model predicted values were about 50% lower the measured values. The average error between the calculated and measured wet-out times was around 24%.

The preliminary results of a parametric study performed to determine why the model predictions did not agree with the data showed that the permeabilities in the stiffener flange region are too high which can be attributed to both the complexity of preform and the procedures used to install the preform in the mold. Reducing the permeabilities in this region resulted in a significant reduction in the average error between calculated and measured wet-out times. In addition, the calculated mold inlet pressures agreed well with measured values. These results show the importance of proper and accurate characterization of the preform.

6. ACKNOWLEDGMENTS

This research was supported by the NASA-Virginia Tech Composites Program, grant NAG 1-343. Mr. H. Benson Dexter was the project monitor.

7. REFERENCES

1. A. C. Loos, et.al., Proceedings of the Third NASA Advanced Composites Technology Conference, NASA Conference Publication 3178, Vol. 1, Part 1, NASA, 1992, pp. 361-379.

2. A. C. Loos, et. al., Proceedings of the Fourth NASA Advanced Composites Technology Conference, NASA Conference Publication 3229, Part 2, NASA, 1993, pp. 683-703.

3. A. C. Loos, et. al., in A. Poursartip and K. Street, eds., Proceedings of the Tenth International Conference on Composite Materials. Vol. III: Processing and Manufacturing, Woodhead Publishing Limited, 1995, pp. III-245 - III-252.

4. A. C. Loos and J. D. MacRae, Composite Science and Technology, 56 (3) 273 (1996).

5 J. D. MacRae, "Development and Verification of a Resin Film Infusion/Resin Transfer Molding Simulation Model for Fabrication of Advanced Textile Composites," M. S. Thesis, Department of Engineering Science and Mechanics, Virginia Tech, Blacksburg, VA, December 1994.

6 A. C. Loos, et.al., Proceedings of the Fifth/DoD NASA Advanced Composites Technology Conference, NASA Conference Publication 3294, Vol. 1, Part 2, NASA, 1995, pp. 811-833.

7. A. C. Loos, et.al., Proceedings of the 37th AIAA/ASME/ASCE/AHS/ASC Structures, Structural Dynamics and Materials Conference, Paper No. AIAA-96-1533-CP, Part 3, AIAA, Reston VA, 1996, pp. 1828-1837.

8. J. C. Fingerson, "Verification of a Three-Dimensional Resin Transfer Molding Process Simulation Model," M. S. Thesis, Department of Engineering Science and Mechanics, Virginia Tech, Blacksburg, VA, September 1995.

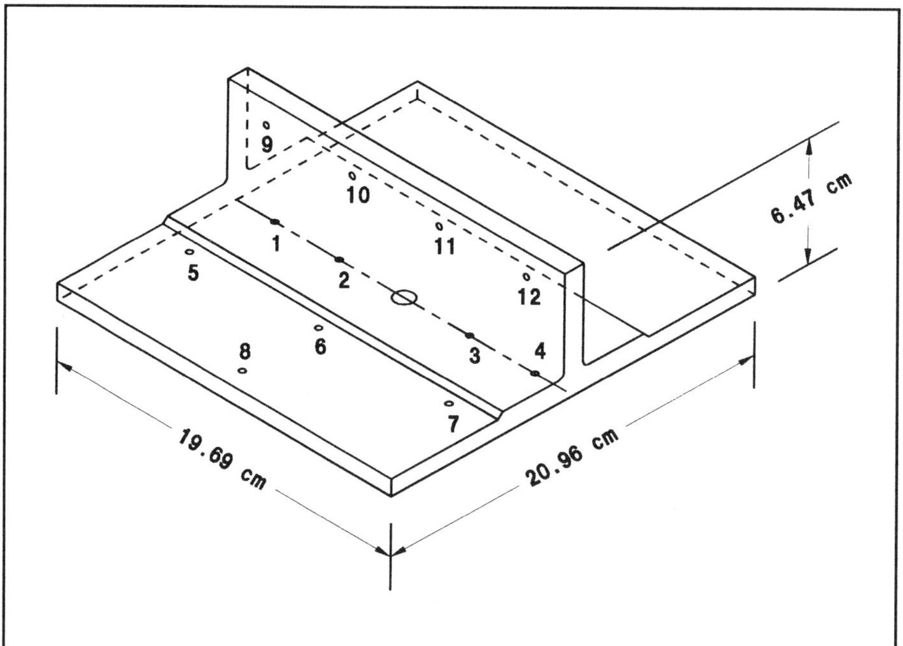

Figure 1. Blade-stiffened preform - approximate dimensions and pressure transducer locations.

Figure 2. Instrumented mold.

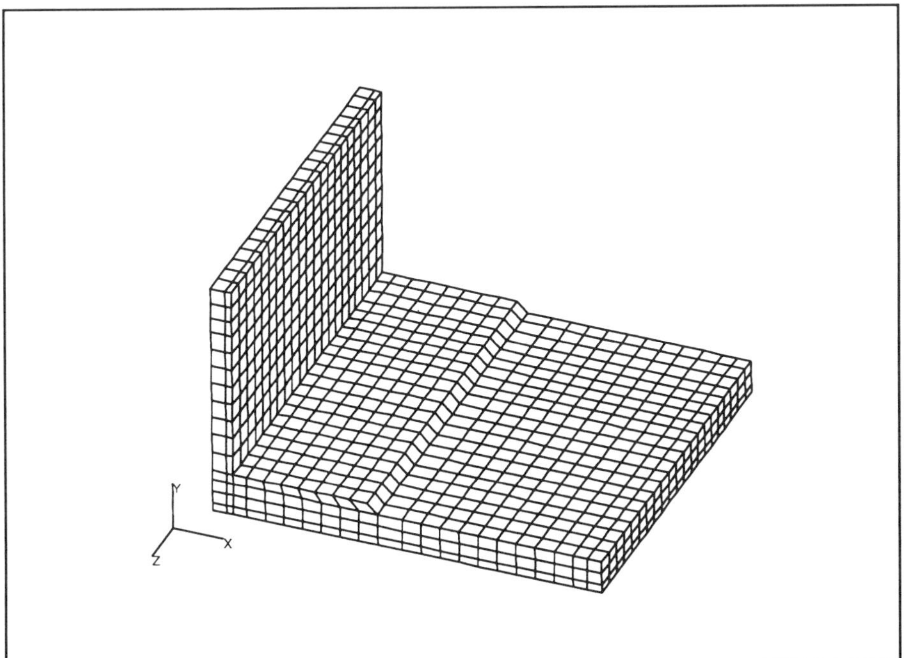

Figure 3. Finite element mesh of the blade-stiffened preform.

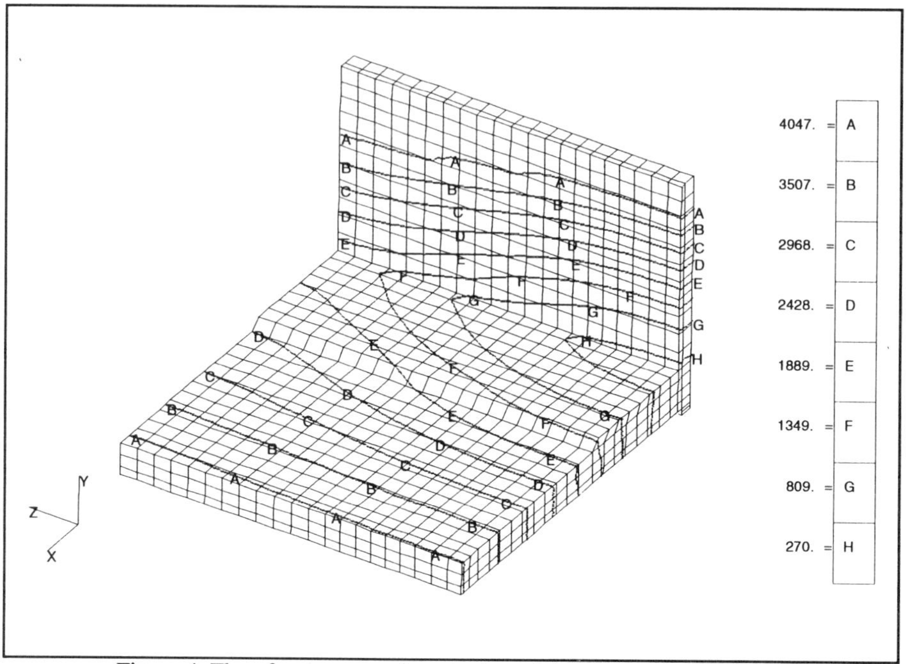

Figure 4. Flow front progression at selected time intervals (seconds).

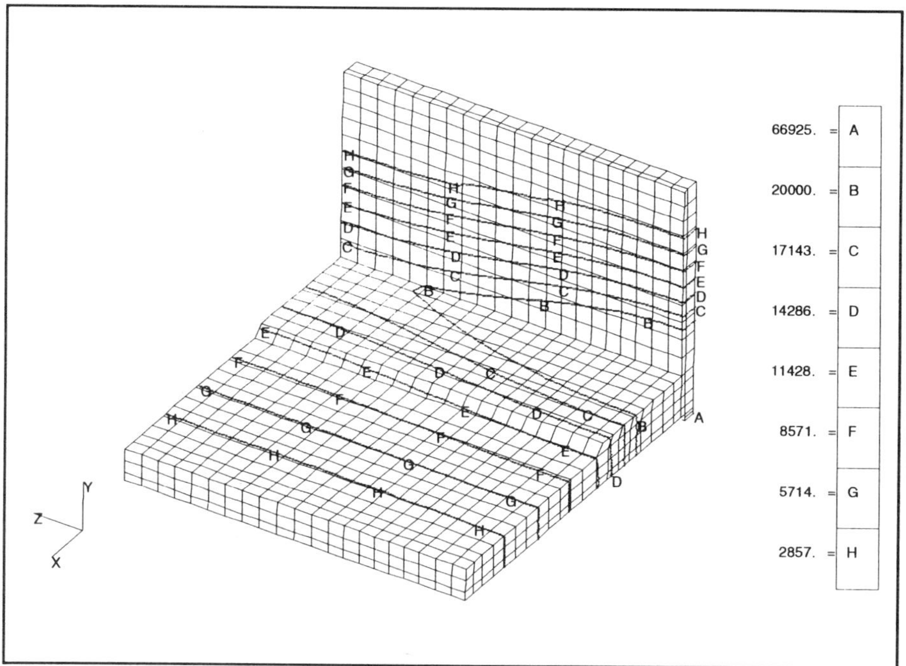

Figure 5. Pressure distribution at 3242 seconds. Pressure in Pa.

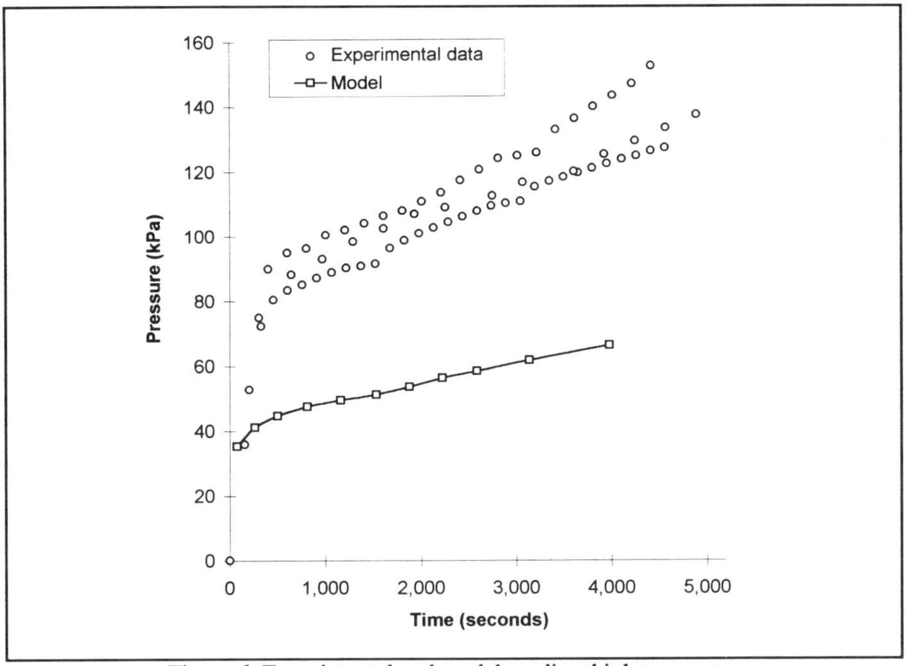

Figure 6. Experimental and model predicted inlet pressure.

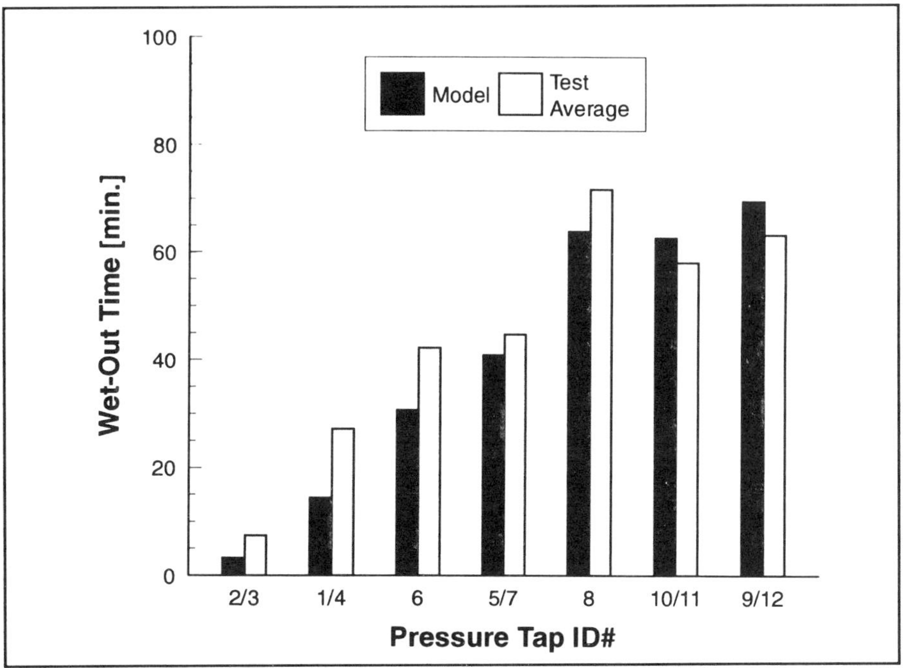

Figure 7. Experimental and model predicted wet-out times at different locations in the mold. The pressure tap locations are shown in Figure 1.

28th International SAMPE Technical Conference
November 4-7, 1996

INNOVATIVE TEXTILE REINFORCED COMPOSITE MATERIALS FOR AIRCRAFT STRUCTURES

H. Benson Dexter
NASA Langley Research Center
Hampton, Virginia 23681-0001

ABSTRACT

During the past 15 years NASA has taken the lead role in exploiting the benefits of textile reinforced composites for aircraft applications. Interim research results indicate that textile reinforced composites can save considerable cost and weight in wing and fuselage structural elements. Continued development and verification of this technology in full-scale wing and fuselage structures could lead to significant applications in next generation commercial transports. This paper discusses advanced material forms that have been developed, innovative machine concepts, and key technology advancements required for future application of textile reinforced composites in commercial transport aircraft. Multiaxial warp knitting, triaxial braiding, and through-the-thickness stitching are the three textile processes that have surfaced as the most promising for further development.

Textile reinforced composite structural elements that have been developed in the NASA Advanced Composites Technology(ACT) Program are discussed. Included are braided fuselage frames and window-belt reinforcements, woven/stitched lower fuselage side panels, stitched multiaxial warp knit wing skins, and braided wing stiffeners. Analytical models to predict resin flow and cure in textile preforms are also discussed. Selected test results for braided and knitted/stitched materials are discussed.

KEY WORDS: Textile Reinforced Composites, Braiding, Knitting, Stitching, Resin Film Infusion, Process Modeling

1. INTRODUCTION

Innovative textile material forms have been investigated by NASA for application to future composite airframe structures for more than a decade. NASA and the U.S. Navy sponsored a working group meeting on 3-D Composite Materials in 1985 to identify critical research issues that would further advance 3-D composites technology (1). During the next six years, NASA, several universities with textile engineering departments, textile machine producers, manufacturers of textile fabrics and preforms, and aerospace companies combined forces to further exploit the benefits of textile processes for fabrication of advanced composite materials. Annual conferences called FIBER-TEX were held to create a forum where technological advancements in the various textile disciplines could be presented (2-7). As specialty textile manufacturers learned how to fabricate aircraft-quality preforms with carbon fibers, the aircraft manufacturers began to show increased interest in textile processes. Compared to 1980's state-of-the-art composite prepreg fabrication methods, textile processes offered the potential for major cost reductions and performance gains. Some of the potential benefits are as follows: reduced material and assembly labor costs through automated fabrication of multilayer multidirectional preforms; reduced machining and material scrap through use of near net shape preforms; elimination of cold storage requirements and limits on shelf life; reduced tooling costs compared to autoclave processes; and improved damage tolerance and out-of-plane strength as a result of through-the-thickness reinforcements.

Working closely with the aircraft manufacturers, universities, and specialty textile fabricators, NASA began to develop a preliminary design database on available material forms and to focus research on key technology issues. The NASA Advanced Composites Technology(ACT) Program that started in 1989 provided the resources to move textile reinforced composites from a curiosity state to aircraft hardware development and demonstration. The NASA in-house and university research focused on development of a database on damage tolerance and mechanical properties of new material forms, development of analytical models to predict elastic properties and strength of textile reinforced composite materials, and development of analytical models to predict resin infiltration and flow in textile preforms. The aircraft manufacturers focused on development of new tooling concepts, fabrication of aircraft-quality textile reinforced composite structures, and structural testing to assess performance gains compared to laminated tape composites.

This paper summarizes the development of advanced material forms, textile machine advancements, fabrication of aircraft structural components, and lessons learned working with textile materials, machines and fabrication processes.

2. TEXTILE MATERIAL FORMS

Textile material forms that have shown the highest potential for application to airframe structures include low crimp uniweave, 2-D and 3-D integrally woven fabric shapes, multiaxial knitted fabrics, 2-D and 3-D braided preforms, and stitched combinations of the above materials. Figure 1 shows a schematic of some of the most promising materials evaluated. Figure 2 indicates some of the advantages and limitations of each of the textile material forms of interest. Low crimp unidirectional weaving (98-percent warp fibers and 2-percent fill fibers) is a high through-put process that gives mechanical properties that are close to those of unidirectional tape composites. However, uniweave has poor fabric stability, and is difficult to handle for off-axis plies. Although large carbon fiber tows can be used to produce uniweave, ply-by-ply lay-up for thick structures is labor intensive.

More than 20 years of experience has been accumulated in weaving 2-D carbon fabrics such as plain and satin weaves. The process is highly automated but tailorability is limited for off-axis properties. In addition, since there are no through-the-thickness fibers, damage tolerance and out-of-plane properties are only slightly better than those achievable with unidirectional prepreg material. Three-dimensional woven fabrics can be produced on conventional weaving equipment but the process is much slower than for 2-D fabrics. A limited number of 3-D woven shapes can be produced but drapability and tailorability for off-axis properties are poor.

Fabrication of two-dimensional braided preforms is highly automated and a good balance in off-axis properties can be achieved with 2-D triaxial configurations(0±Ø). The 2-D braided preforms are well suited for complex shapes and they have good drapability. Two limitations with 2-D braiding are size limitation due to lack of machine availability and low out-of-plane properties. Three-dimensional braiding was developed to provide improved damage tolerance and a good balance in in-plane and out-of-plane properties. However, current 3-D braiding processes are slow and preform size is limited by machine availability. Preliminary cost analyses indicate that 3-D braids cannot compete with 2-D braids on a cost basis.

Multiaxial warp knitting is a highly tailorable automated process that produces multilayer fabrics with a good balance in in-plane properties. Up to seven layers of carbon fabric can be produced with existing equipment. Since the knitting process is continuous, the fabric is well suited for large area coverage. The carbon fiber tow size and stiffness can be tailored in the 0-, 90-, and ±Ø-degree directions to meet design requirements.

Stitching is a highly automated process that can be used to improve damage tolerance of the textile materials discussed above. When used as an assembly aid, stitching can provide out-of-plane strength that is significantly higher than that achievable with adhesive bonding. Two limitations with stitching include a small reduction in in-plane mechanical properties and poor accessibility to complex curved shapes.

3. APPLICATION POTENTIAL OF TEXTILE PREFORMS

3.1 Fuselage Structures During Phases A and B of the NASA ACT program, several NASA contractors conducted trade studies to assess the potential of innovative material forms and manufacturing concepts. Textile preforms and resin transfer concepts were considered along with other material forms and processes. The structural elements shown in figure 3 were selected to determine the applicability of specific textile material forms and fabrication processes.

3.1.1 ***Skin/Stiffened Panels*** Northrop Grumman was selected to conduct design and fabrication studies for skin/stiffened fuselage side panels(8). Some of the design considerations included damage tolerance, buckling/postbuckling, and combined cyclic loads. Cross-stiffened or intersecting frames and stiffeners pose manufacturing problems for conventional prepreg tape materials. Primary load carrying fibers cannot pass through each other, hence only one plane can be continuous. However, 3-D weaving technology has progressed to the point where intersecting structures can be woven with continuous fibers in both directions. The textile process selected by Northrop Grumman included a combination of weaving and stitching. The lower fuselage side panel preform shown in figure 4 has four circumferential frames and four longitudinal stiffeners. The intersections of the woven stiffeners and the frames have continuous fibers through the intersection to provide structural continuity. Since the weaving process selected can only weave in the 0-and 90-degree directions, additional ±45-degree material had to be stitched onto the base fabric to provide shear load capability. The flanges of the woven frames and stiffeners were stitched to the woven skin material with Kevlar 29, 1600 denier thread. A resin film infusion(RFI) process was selected for fabricating composite panels(8).

3.1.2 ***Circumferential Frames*** The Boeing Company (9) and Lockheed-Martin(10) were selected to conduct design and fabrication studies on fuselage barrels, including internal frame structures. Some of the design considerations for the circumferential frames included combined loads, high interlaminar stresses, frame/stringer/skin attachments, and frame splice concepts. Because of their ability to produce curved structures with integral flanges, triaxial braiding and through-the-thickness braiding concepts were selected for preform manufacturing studies. The flanges were produced by braiding in pockets or bifurcations that could be folded out prior to resin application as shown in figure 5. Matched-metal tooling concepts and resin transfer molding(RTM) were selected to produce composite frames for structural evaluation. Several curved fuselage frames that were produced by the 2-D and 3-D braiding process are shown in figure 6. Due to cost considerations, the 2-D braiding process was selected for further manufacturing trials.

3.1.3 ***Keel Beam Frames*** The Boeing Company(11) and Lockheed-Martin conducted design and manufacturing studies on curved frames for the fuselage keel structure shown in figure 3. Some of the design considerations were impact damage tolerance, through penetration damage containment and durability of frame splices. The 2-D triaxial braiding and RTM fabrication process discussed above was used to fabricate curved frames for structural testing. Several frames that are bonded to a curved honeycomb sandwich keel panel are shown in figure 7.

3.1.4 ***Window Belt*** Lockheed Martin conducted design and manufacturing trade studies on fuselage window belt structural configurations. Some of the design considerations for the window belt include out-of-plane/interlaminar stresses, stability under combined loads, and damage tolerance/pressure containment. Because of complex stress states, design concepts with through-the-thickness reinforcements were selected for manufacturing trials(12). Both powder coated tow and resin transfer molding fabrication concepts were evaluated. Several braided and woven window belt frames are shown in figure 8.

3.2 Wing Structures During Phases A and B of the NASA ACT Program, McDonnell Douglas Aerospace was selected to investigate the potential application of textile reinforced composites to commercial transport wing structures. The structural elements shown in figure 9 were selected to determine the applicability of specific textile material forms and fabrication processes.

3.2.1 ***Upper and Lower Cover Panels*** Since most of the wing weight is in the cover panels they are a logical choice for pursuit of cost and weight savings. Some of the global design considerations for wing panels include strength, stiffness, and impact and discrete source damage tolerance. Other considerations include joints, access doors, bearing and open-hole strengths, out-of-plane loads, high load introduction points such as landing gear attachments, and lightning strike protection. Due to poor impact damage tolerance and the cost of ply-by-ply lay-up of prepreg tape, this material form has not proven to be a cost effective fabrication concept for primary composite aircraft structures. Since compression after impact strength has been a key design limitation with previous laminated tape composite structures, McDonnell Douglas and NASA investigated other material forms.

3.2.1.1 *Through-The-Thickness Stitching* As a result of tests conducted at NASA(13) during the years 1987-1990, it was determined that through-the-thickness stitching of dry textile preforms is an effective way to achieve improved damage tolerance. Early stitching studies included the effects of stitch type, stitching thread type, stitch pitch, and stitch spacing. Primary research focus was on improved damage tolerance compared to laminated tape composites, however the costs associated with stitching and composite consolidation methods were also of prime importance. Results of compression after impact (CAI) tests conducted on stitched uniweave fabrics, unstitched prepreg tape, and toughened matrix composites are presented in figure 10. The results indicate that for an unstitched brittle composite system, AS4/3501-6, the compression strength drops from 600 MPa to approximately 140 MPa due to a 68 J impact. The results for the toughened matrix composite system indicates a significant improvement in CAI strength compared to the unstitched brittle material, but the toughened materials are 2 to 3 times more expensive. In contrast, the stitched composite demonstrates a significant increase in CAI strength compared to the unstitched AS4/3501-6 composite(see the upper curve in figure 10). The results of these and other tests led McDonnell Douglas to select through-the-thickness stitching of dry textile preforms as the baseline for their advanced composite wing fabrication studies.

3.2.1.2 *Multiaxial Warp Knitting* To eliminate the labor intensity involved with conventional composite fabrication concepts, McDonnell Douglas and NASA sought alternative material forms for large area coverage. Initial attempts at using uniweave broadgoods proved to be unsuccessful due to fabric stability problems and labor costs associated with laying up thick wing skins. Although multiaxial warp knitting processes had been developed previously, this material form had quality control problems and was used primarily by fiberglass boat builders. However, as the market developed for multidirectional carbon broadgoods during the late 1980's and early 1990's, machine manufacturers and specialty textile companies modified knitting machines to produce aerospace-quality carbon fabrics. As a result of tests conducted by NASA(14-15), McDonnell Douglas became interested in multilayer multiaxial warp knitting as a process to produce cost effective wing structures. Knitted fabric stacks that had an areal weight equivalent to nine plies of uniweave fabric were chosen for development of a preliminary design data base. Some typical tension and compression properties for a multiaxial warp knit carbon fabric composite are shown in figure 11. These test results indicated that the knitted fabric produced by Saertex could meet the McDonnell Douglas wing design requirements.

3.2.1.3 *Wing Stub Box Fabrication* A 2.4m by 3.7m wing cover demonstration article that was fabricated by McDonnell Douglas using the stitched/RFI process(16) is shown in figure 12. The completed wing stub box with upper and lower covers is shown in figure 13. Successful design, fabrication and test of the stitched/RFI stub box and associated design development test articles, led McDonnell Douglas to selection of knitting, stitching, and RFI processes for development of a semi-span wing and a full-scale wing under Phase C of the NASA ACT Program.

3.2.2 ***Stiffeners and Spar Caps*** Blade stiffeners and integral spar caps were chosen by McDonnell Douglas as stiffening elements for the upper and lower wing cover panels. The upper skin stiffeners and spar caps are fabricated with Saertex knitted fabric with a material lay-up similar to that used in the basic skin. However, the curvatures are more severe on the lower skin where there is a marked dihedral contour along the span combined with airfoil camber across the wing chord. Saertex knitted stringers could not be forced to conform to the lower wing skin surface without significant fiber wrinkling. Braiding trials conducted by Fiber Innovations indicated that braided stringer preforms could be fabricated to closely match the lower surface contour. The braided stringers are fabricated by stacking tubular fabrics and partially stitching them together to form the upstanding blade. The unstitched portion of the tubular fabrics are folded left and right to form the stiffener flanges. The stiffener flanges are then stitched to the wing skin material to form an integral structure that can be infused with resin in one autoclave operation. Mechanical property tests were conducted on braided stiffener preforms to determine if the braided configuration could meet stiffener design requirements(17). Test results for Saertex warp knit and braided stiffener materials are summarized in figure 14. Based on these results, it was determined that braided stiffeners could meet the design requirements for the curved lower wing cover panels.

3.2.3 ***Intercostal Clips*** The assembly concept selected by McDonnell Douglas for the wing structure includes intercostal clips that attach rib structure to the wing cover panels. The intercostal clips must be designed to transfer compressive crushing and tensile fuel pressure forces. The intercostal clips can be fabricated with the Saertex knitted fabric or braided preforms. Flanges on the skin side of the clips are folded and stitched to the wing skin and flanges on each end of the clips are attached to blade stiffener webs(see figure 9). Due to limited access and the inability of existing stitching equipment to stitch horizontally, alternate concepts for attaching the intercostal flanges to the stiffener webs are under development.

3.2.4 ***Main Landing Gear Attachment*** The main landing gear attachment test article shown in figure 9 was designed, fabricated, and tested to verify that high point loads could be distributed in the stitched/RFI wing cover panels. The critical design load occurs from a "braked roll" condition in which the aircraft is braked to a stop just short of takeoff speed. Upper and lower bolted titanium doublers were used to transfer load into the upper and lower wing skin cover panels. Extra plies of carbon fabric were stitched in the wing skins to reduce the bolt bearing stresses in the regions under the titanium doublers. The test article successfully carried design ultimate load which demonstrated that large concentrated loads from the landing gear can be distributed into the stitched/RFI composite wing box.

4. ADVANCEMENTS IN MULTIAXIAL WARP KNITTING TECHNOLOGY

4.1 Advanced Machine Concepts Multiaxial warp knitting is a highly automated process for producing multilayer broadgoods. Compared to woven broadgoods, the knitted fabrics have less fiber crimp since the individual tows are not interlaced. Another benefit of multiaxial warp knitting is the ability to tailor the fiber architecture. In general, the off-axis ply orientations can range between 30 and 60 degrees and different fiber types and tow sizes can be used in the different directions to control fiber areal weight, stiffness, and strength. Also, the type of knit, the size of the knitting thread, and the volume fraction of the knitting thread can be varied to control fabric drapability.

The evolution of multiaxial knitting technology is shown in figure 15. Fabrics produced in the 1970's were made on malimo type machines. These fabrics had non-parallel yarns and they were used mostly by the boating industry for fiberglass boats. In the early 1980's Liba, a German-owned company, developed a new machine concept to produce fabrics that could be used by the aerospace industry for fabrication of composite structures. The early Liba machines were limited to 5-ply stacks but they could produce fabrics up to 2.5m wide. One drawback to the Liba machine was the inability to control fiber tension, which resulted in poor fiber alignment. In the early 1990's Saertex and Liba teamed up to upgrade tension control mechanisms. A schematic of the Saertex/Liba machine is shown in figure 16. The Saertex machine can produce 5-ply carbon fabrics in one pass through the machine. A two-step process is used by Saertex to produce a 7-ply fabric for the McDonnell Douglas ACT wing contract. Splicing concepts have been developed to produce fabrics up to 2.5m wide.

Some key technology advancements are required if multiaxial warp knitting is to play a major role in fabrication of next generation composite aircraft structures. To reduce cost, machines must be developed to produce high quality fabrics that are seven or more plies thick and over 2.5 meters wide with no splices. The advanced machines must have more flexibility in ply stacking sequence to allow optimization of the fiber architecture. If multiaxial warp knit fabrics are to see expanded applications to complex shapes, knitting processes must be developed for improved drapability. Also, stringent in-process inspection methods and process controls must be developed to ensure that aircraft-quality knitted fabrics can be cost competitive with other material forms.

5. ADVANCEMENTS IN STITCHING TECHNOLOGY

5.1 Advanced Machine Concepts Stitching studies in the 1970's and early 1980's focused on prepreg materials rather than dry fabric preforms. These experiments were mostly unsuccessful due to a significant loss in material properties that resulted from fiber breakage during the stitching operation. As specialty textile manufacturers learned how to fabricate aircraft-quality preforms with carbon fibers and as resin transfer molding methods matured, the aircraft manufacturers began to show renewed interest in stitching as a way to achieve damage tolerant composite structures. Stitching of dry fabrics proved to be more successful since lateral movement of the fibers allowed the needle to pass through the fabric with minimal fiber damage.

The evolution of stitching technology for application to aircraft structures is depicted in figure 17. Early stitching studies by McDonnell Douglas and NASA used a limited capacity extended arm single needle machine as shown in figure 18. This machine was slow and was limited to small preforms but good quality test samples were produced. It was evident that expanded stitching machine capability was needed and NASA and McDonnell Douglas contracted Pathe to develop a new machine. This first generation machine is a single needle computer controlled machine based on quilting technology. The machine shown in figure 18 is capable of stitching a 2.4m by 4.6m by 38.1mm preform at 200 stitches per minute. This machine has been used to produce manufacturing demonstration hardware and test articles during phases A and B of the NASA ACT Program. In anticipation of stitching requirements for full-scale wing structures to be fabricated during phase C of the ACT Program, Ingersoll and Pathe were contracted to develop a multi-head computer controlled gantry type stitching machine. The machine concept shown in figure 18 is in the final stages of development. The machine will have four stitching heads that will be upgraded with multiple needles per head. This second generation machine will be able to stitch a preform 3.0m by 15.2m by 38.1mm at 800 stitches per minute.

As stitching technology moves toward the next century, more stringent requirements will have to be met. In general, current stitching machines are limited to a vertical stitching plane. More complex structures will require off-axis stitching in limited access areas. To meet these needs, robotic multi-needle end effectors will be required.

6. PROCESSING SCIENCE

6.1 Analytical Modeling To eliminate trial and error processes, analytical models are required to predict resin flow into textile preforms. The models must be verified through precise experiments if they are to be used with confidence. Three-dimensional models are required to adequately capture response of complex preforms such as wing cover panels that contain stitched/knitted fabric skins and stitched/braided stiffeners. The objectives of the analytical model are to predict resin flow front position, resin viscosity, and degree of cure as a function of time. In addition, the model can be used to develop cure cycles for specific preform architectures and structural configurations. The various elements of the NASA processing science program for textile reinforced composites are shown in figure 19. An extensive experimental data base is required to support the analytical model development.

Textile preforms are deformable and anisotropic porous materials. Hence, the permeability depends both on the direction of flow and the degree of deformation of the preform. Thus, the compaction characteristics and the permeabilities in the principal material directions must be measured for each textile preform. In addition, cure kinetics of the resin must be characterized for input to the analytical model. Verification of the model is being performed with structural elements that have dielectric sensors and pressure transducers installed in the tooling to track resin flow front, resin viscosity, and degree of cure.

A 3-D RFI process simulation model is being developed under NASA grant to Virginia Polytechnic Institute(18). The RFI simulation model is divided into three main submodels. These include resin flow, heat transfer, and thermochemical submodels. A schematic of a 3-D finite element model for infusion of a stitched blade stiffener, including the tooling, is shown in figure 20. The preform, aluminum tooling, , graphite caul sheet, and resin are included in the model.

6.2 Preform Characteristics Due to the complexity of the textile architectures currently being considered in the NASA ACT Program, analytical models are not available to predict compaction and permeability response of the preforms. Experimental methods have been developed to measure fiber volume fraction as a function of compaction pressure and permeability as a function of fiber volume fraction. For stitched preforms, compaction and permeability are a function of the stitch density and stitch tension. Figure 21 shows compaction and permeability data for an unstitched Saertex multiaxial warp knit fabric. Compaction data is obtained by placing a sample of the fabric between two plates and applying compaction load through-the-thickness using a load frame. Defection is directly related to fiber volume fraction by the preform areal weight. Permeability is measured by placing the fabric sample in a fixture having a sealed chamber of variable cavity height. The surfaces touching the faces of the fabric are porous, and the cavity height is controlled by the load frame. Fluid is pumped through the fabric under various levels of compaction, and a record of fluid pressure drop and flow rate is made. Permeability is calculated using Darcy's Law. The data shown in figure 21 indicates that about 175 kPa of pressure is required to obtain a 60-percent fiber volume fraction and the permeability of the fabric is reduced by about 50-percent as the fiber volume fraction increases from 55-to 65-percent.

7. SUMMARY OBSERVATIONS AND LESSONS LEARNED

7.1 Preforms and Machine Concepts Early attempts to develop complex equipment to fabricate near net shape multidirectional multilayer fabrics was unsuccessful. This result was primarily caused by the fact that various textile technologies were being stretched beyond their technical boundaries. Limited success was achieved with triaxial weaving but the process was limited to thin flat fabrics. Limited success was also achieved with a manual multidirectional weaving apparatus but the process was labor intensive and funding for automation was not available. Two- and three-dimensional braiding showed a lot of potential but available machine capacity limited the architecture and the size of the preforms that could be achieved. Folding or postforming of triaxial braided preforms offered the most flexibility in achieving small cross-section complex shapes. Multiaxial warp knitting proved to be the best process for large area multiaxial multilayer broadgoods but structural shapes had to be achieved through postforming and stitching. Through-the-thickness stitching proved to be the best textile process to achieve improved damage tolerance. Compared to glass fibers, all textile processes investigated had to be slowed considerably when carbon fibers were used. Lessons learned to-date indicate that no one machine can produce the desired fiber architectures for all complex shape aircraft structural preforms. In addition, more stringent in-process controls and inspection techniques are needed to minimize scrap and to reduce costs.

7.2 Tooling Concepts and Process Modeling Analytical models are required to eliminate costly trial and error processes that are frequently used in tool design and development of processing cycles. Resin viscosity and cure kinetics must be characterized to consistently achieve high quality composite parts. Since compaction and permeability behavior are different for each fiber architecture and preform configuration, empirical relationships must be developed for input to analytical models. Tooling design flexibility is required to accommodate variations in preform quality and preform bulk. Tool-mounted sensors such as dielectric and pressure transducers can serve as an aid in tracking resin flow front position, resin viscosity and degree of cure.

8. CONCLUDING REMARKS

The development of textile preforms in concert with resin transfer tooling concepts has progressed to the point where it is now feasible to fabricate high quality damage tolerant aircraft structures with these material forms and processes. Recent developments in improved tension control of tows in the multiaxial warp knitting process have made this the material of choice for large area multidirectional broadgoods. Innovative tooling concepts have been developed to produce curved braided preforms for application to fuselage frames and wing stiffeners for commercial transport aircraft. NASA and McDonnell Douglas have demonstrated that through-the-thickness stitching of dry textile preforms can provide a 100-percent increase in damage tolerance of composite wing structures compared to laminated tape construction techniques. Recent investments in stitching machine technology are expected to pay-off in terms of improved quality, higher speed, and lower cost.

Analytical process models that are under development will minimize trial and error methods typically used in fabrication of composite structures. Fabrication processes that rely on science-based characterization of resin and textile preform properties will lead to more cost effective fabrication of large composite wing and fuselage structures. Lessons learned on limitations of textile machines will lead to selection of the most efficient preform fabrication methods in the future.

9.0 REFERENCES

1. H.B. Dexter, E.T. Camponeschi and L. Peebles, ed., Proceedings of Working Group Meeting on 3-D Composite Materials, NASA CP 2420, 1986.
2. J.D. Buckley, ed., Proceedings of FIBER-TEX 1987, NASA CP 3001, 1988.
3. J.D. Buckley, ed., Proceedings of FIBER-TEX 1988, NASA CP 3038, 1989.
4. J.D. Buckley, ed., Proceedings of FIBER-TEX 1989, NASA CP 3082, 1990.

5. J.D. Buckley, ed., Proceedings of FIBER-TEX 1990, NASA CP 3128, 1991.
6. J.D. Buckley, ed., Proceedings of FIBER-TEX 1991, NASA CP 3176, 1992.
7. J.D. Buckley, ed., Proceedings of FIBER-TEX 1992, NASA CP 3211, 1993.
8. J.R. Ewen and J.A. Suarez, Proceedings of the Third NASA ACT Conference, NASA CP 3178, Vol 1, Part 1, 1992, pp. 287-308.
9. L.B. Ilcewicz, et. al., Proceedings of the First NASA ACT Conference, NASA CP 3104, Part 1, 1991, pp. 127-155.
10. A.C. Jackson, R.E.Barrie and R.L. Chu, Proceedings of the Third NASA ACT Conference, NASA CP 3178, Vol 1, Part 1, 1992, pp. 79-96.
11. K. Wilden, et. al., Proceedings of the Fifth NASA/DoD ACT Conference, NASA CP 3294, Vol 1., Part 1, 1995, pp. 237-255.
12. A.C. Jackson, et. al., "Advanced Composite Structural Concepts and Materials Technologies for Primary Aircraft Structures," NASA CR 4726, July 1996.
13. R.J. Palmer, M.B. Dow and D.L. Smith, Proceedings of the First NASA ACT Conference, NASA CP 3104, Part 2, 1991, pp. 621-646.
14. H.B. Dexter, G.H. Hasko and R.J. Cano, Proceedings of the First NASA ACT Conference, NASA CP 3104, Part 2, 1991, pp. 589-619.
15. H.B. Dexter and G. H. Hasko, Proceedings of the Third NASA ACT Conference, NASA CP 3178, Vol 1, Part 1, 1992, pp. 231-261.
16. A. Markus, P. Thrash and B. Grossheim, Proceedings of the Fourth NASA/DoD ACT Conference, NASA CP 3229, Vol 1, Part 1, 1993, pp. 503-523.
17. K.W. Furrow, "Material Property Evaluation of Braided and Braided/Woven Wing Skin Blade Stiffeners," NASA CR 198303, April 1996.
18. A.C. Loos, et. al., Proceedings of the Fifth NASA/DoD ACT Conference, NASA CP 3294 Vol 1, Part 2, 1995, pp. 811-833.

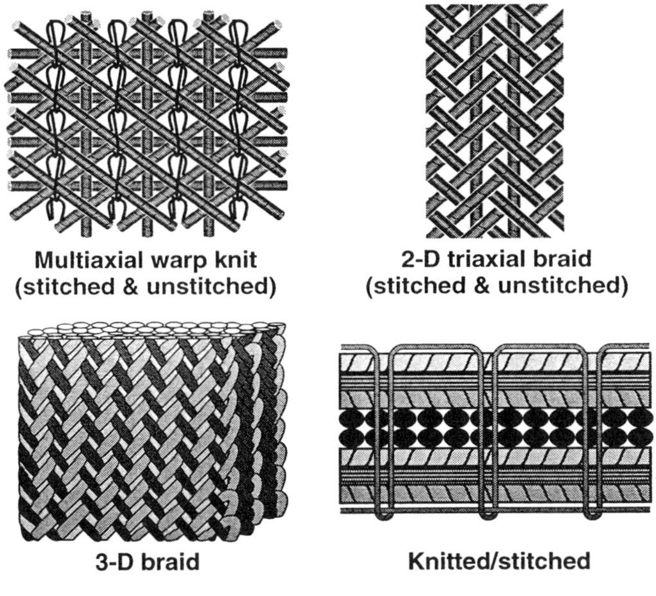

Figure 1. Textile material forms evaluated.

Textile Process	Advantages	Limitations
Low Crimp Uniweave	High in-plane properties Good tailorability Highly automated preform fabrication process	Low transverse and out-of-plane properties Poor fabric stability Labor intensive ply lay-up
2-D Woven Fabric	Good in-plane properties Good drapability Highly automated preform fabrication process Integrally woven shapes possible Suited for large area coverage Extensive data base	Limited tailorability for off-axis properties Low out-of-plane properties
3-D Woven Fabric	Moderate in-plane and out-of-plane properties Automated preform fabrication process Limited woven shapes possible	Limited tailorability for off-axis properties Poor drapability
2-D Braided Preform	Good balance in off-axis properties Automated preform fabrication process Well suited for complex curved shapes Good drapability	Size limitation due to machine availability Low out-of-plane properties
3-D Braided Preform	Good balance in in-plane and out-of-plane properties Well suited for complex shapes	Slow preform fabrication process Size limitation due to machine availability
Multiaxial Warp Knit	Good tailorability for balanced in-plane properties Highly automated preform fabrication process Multi-layer high throughput material suited for large area coverage	Low out-of-plane properties
Stitching	Good in-plane properties Highly automated process provides excellent damage tolerance and out-of-plane strength Excellent assembly aid	Small reduction in in-plane properties Poor accessibility to complex curved shapes

Figure 2. Application potential of textile reinforced composite materials for aircraft structures.

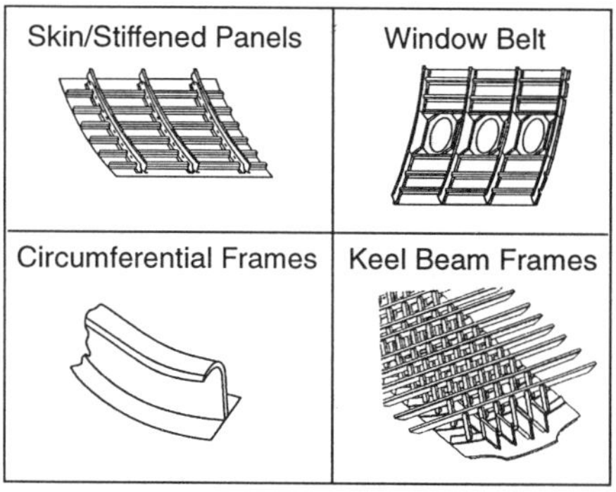

Figure 3. Application of textile reinforced composites in fuselage structures.

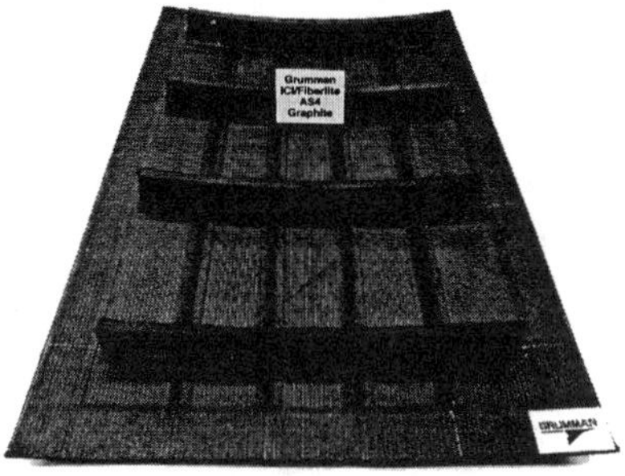

Figure 4. Woven/stitched lower fuselage side panel preform.

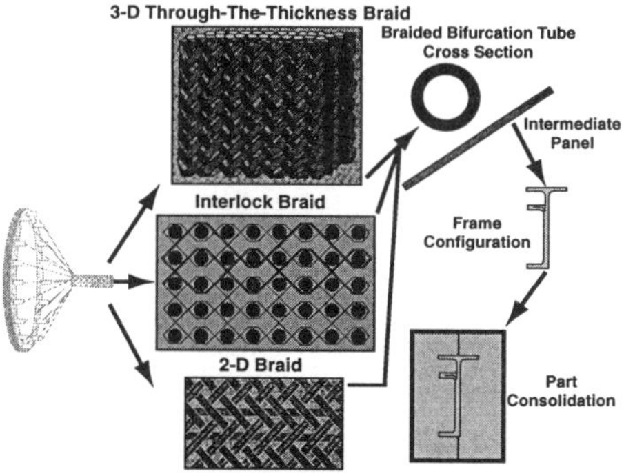

Figure 5. Fabrication process for braided fuselage frames.

Figure 6. 2-D and 3-D braided fuselage frames.

Figure 7. Curved textile frames for fuselage keel structure.

Figure 8. Braided and woven fuselage window frames.

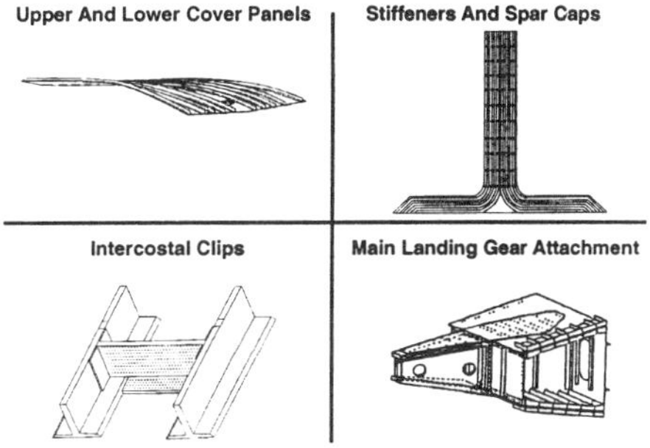

Figure 9. Application of textile reinforced composites in wing structures.

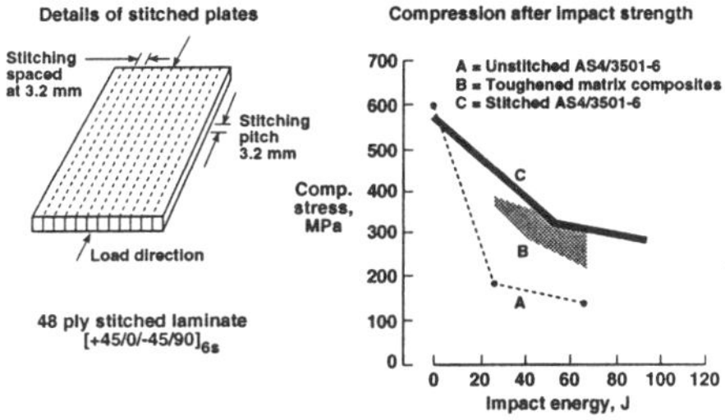

Figure 10. Effects of stitching on damage tolerance of composite materials.

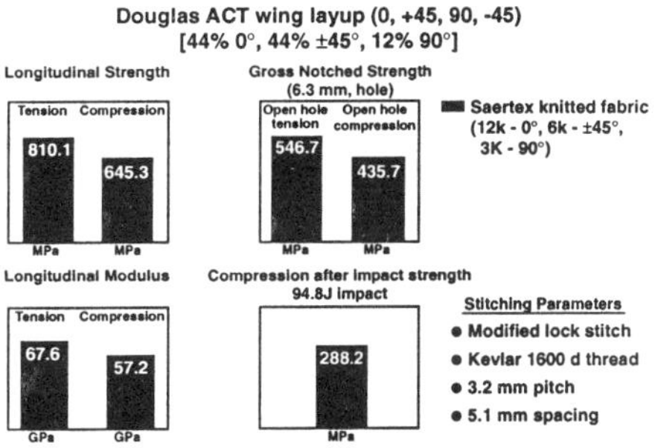

Figure 11. Performance of AS4/3501-6 Saertex knitted/stitched fabric composites.

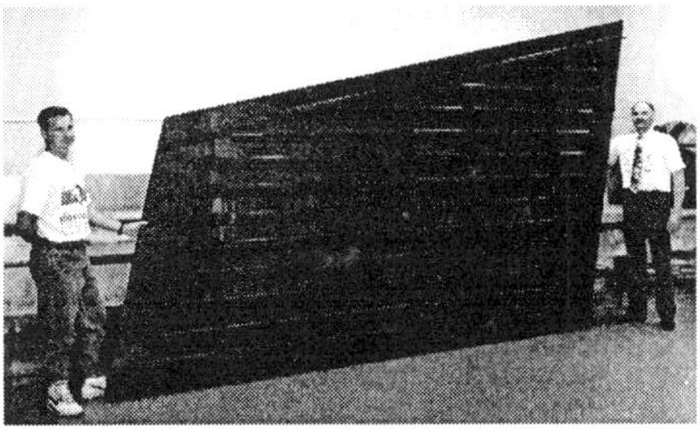

Figure 12. Stitched/resin film infused wing stub box cover panel.

Figure 13. Stitched/resin film infused wing stub box test article.

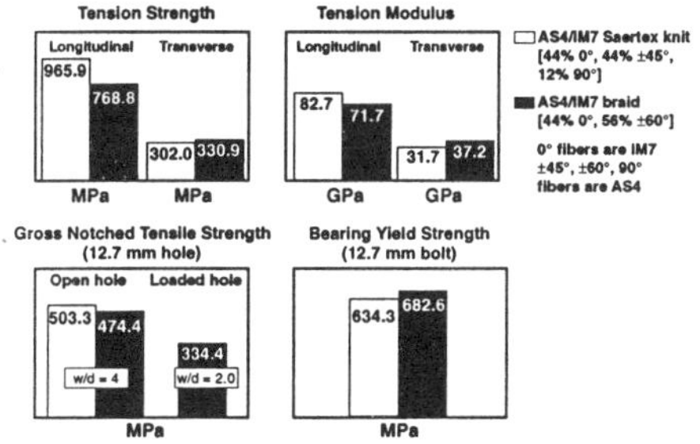

Figure 14. Performance of AS4/IM7 Saertex knitted and Fiber Innovations braided fabrics.

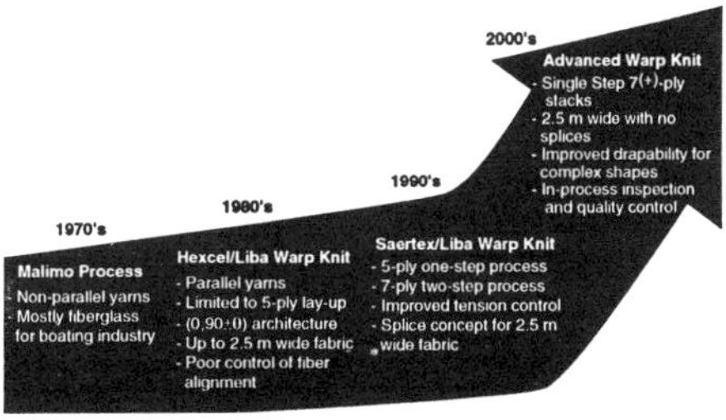

Figure 15. Evolution of multiaxial warp knitting technology.

Figure 16. Schematic of Saertex/Liba multiaxial warp knitting machine.

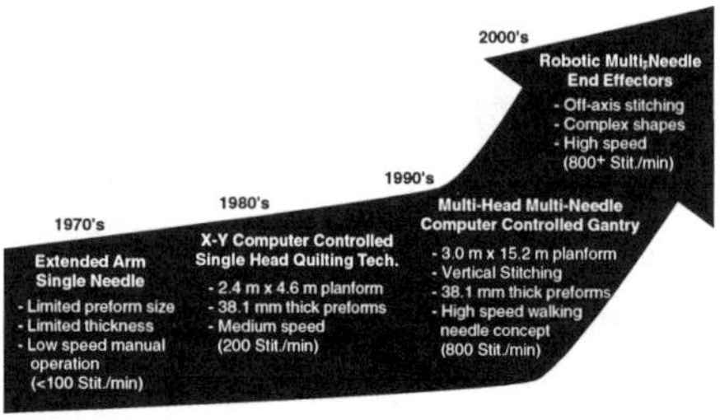

Figure 17. Evolution of stitching technology.

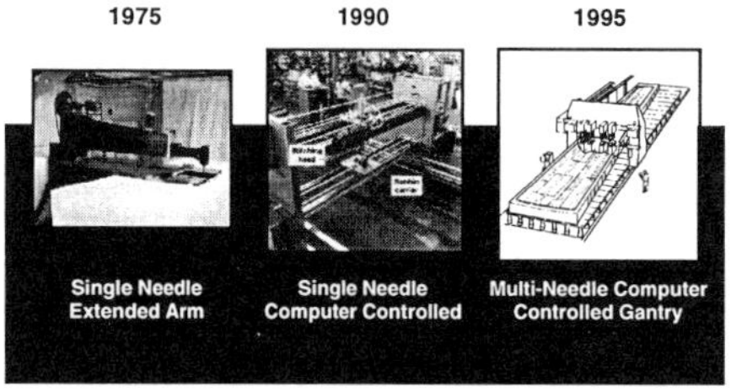

Figure 18. Stitching machine technology development.

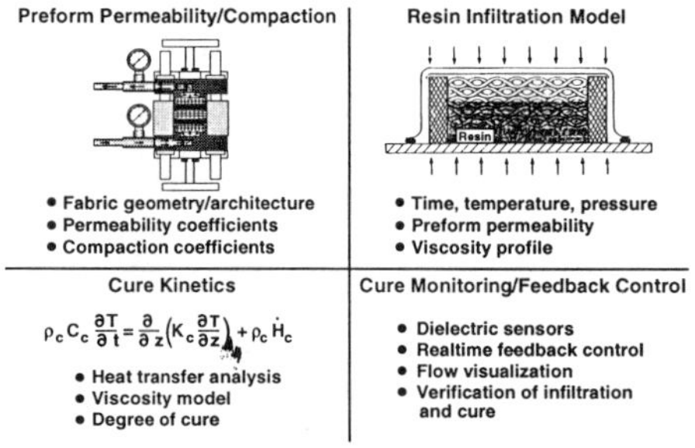

Figure 19. Processing science of textile reinforced composites.

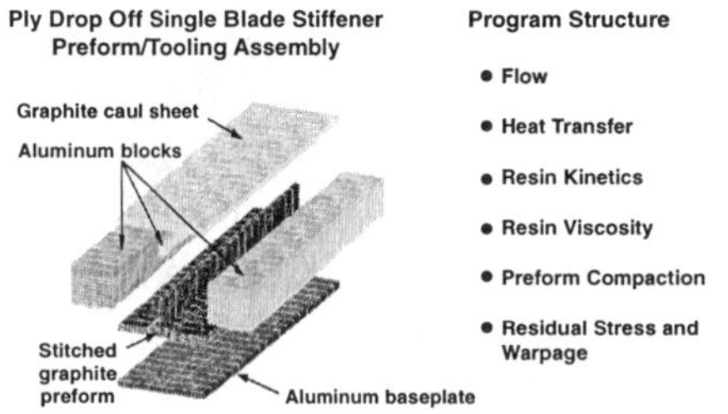

Figure 20. 3-D resin film infusion model.

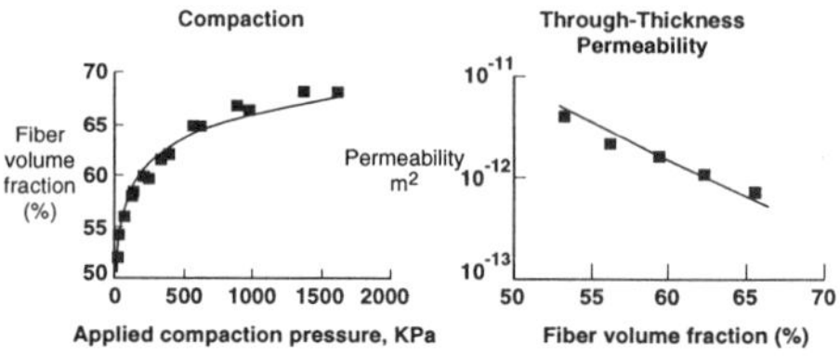

Figure 21. Compaction and permeability behavior of Saertex multiaxial warp knit fabric.

28th International SAMPE Technical Conference
November 4-7, 1996

RESIN COATED TOW MATERIALS FOR STRUCTURAL TEXTILE PREFORMS

Anthony C. Jackson
Lockheed Martin Aeronautical Systems
Marietta, GA 30101

ABSTRACT

A major factor in producing low cost composite structures is the size of structure that can be fabricated in one piece. Resin transfer molding and resin film infusion require high flow resin systems with remain low viscosity for periods of up to several hours. This limits the number of available resin systems. Handling of large dry preforms presents difficulties in maintaining shape and fiber orientation or requires considerable stitching. An alternative approach is to coat the fibers with resin, either in the form of powder or a solution. The fibers are gathered into tows which can be woven or braided into preforms. The resulting preforms contain all the resin required for production of the final parts. This resin is stable at room temperature for indefinite periods of time. The preforms maintain their shapes better than the dry preforms and can be assembled together to make large cocured structures. The problem of bulk factor, which can be two or three times that of a dry preform, can be minimized in the textile fabrication process or by the use of innovative tooling for compaction and cure.

KEY WORDS: Advanced Composites, Textiles, Powder Coated Tows, Weaving, Braiding.

1. INTRODUCTION

Recent advances in resin transfer molding(RTM) have created interest in the use of textile preforms for primary aircraft structures. Most resin systems do not possess the flow characteristics required for RTM so alternative methods for placing the resins are required. The use of powder coated tows extends the capabilities of textile preforms to many more resin systems, including some of the hard to process high temperature systems.

This work was performed as part of the NASA Advanced Composite Technology program under the direction of Dr. John G. Davis and Mr. William T. Freeman Jr.

2. POWDER COATING

2.1 Coating Processes There are several methods available for powder pregging. These include wet and dry methods . Both were investigated in this program.

The wet methods included slurry and solvent pregging both by Cytec Engineered Materials Inc and foam by Clemson. Water or an organic solvent was used as a powder carrier during the towpregging process. The slurry process required a suspension of powder particles in water with the aid of wetting and dispersing agents. The solvent process involved dissolution of the powder in a solvent which was fully removed after coating. These processes generally provide a smooth surface.

The dry methods included electrostatic coating developed by Georgia Tech and Custom Composites Inc, acoustical fluidizing by Michigan State University and mechanical fluidizing and curtain feed both by NASA LaRC. The tows are pneumatically or acoustically separated. They are passed through a curtain of powder, a fluidized bed, or a bubbling bed of powder, or a high voltage charged cloud of powder. The coated fibers are then fused in an oven and brought back together in tows.

The wet processes have the advantage of providing a very uniform coating with good adhesion to the fibers. The dry processes require no drying cycle and can achieve higher line speeds. However the particle size requires much tighter control, as does the humidity. Dielectric constant and polarity play an important role in the electrostatic method.

After an evaluation of the production readiness of the various processes, two were selected for further development, the Cytec slurry process and the Georgia Tech/Custom Composites electrostatic process. A schematic of a powder coating process is shown in Figure 1.

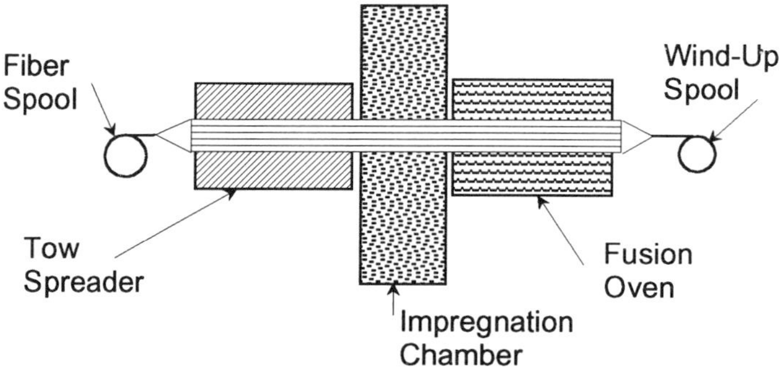

Figure 1. Schematic of Powder Coating Process

2.2. Process Evaluation The electrostatic process worked reasonably well on a small scale. However upon scale-up several problems were encountered. The resin system selected was the 3M PS-501 system which was the powder equivalent of the PR-500 resin transfer molding epoxy resin. The PS-501 went through various stages of development, becoming AMD0036 and finally PT-500. The powder had to be micronized and the particle size had to be controlled. This proved to be difficult. The powder tended to agglomorate on the fibers and to collect on the chamber walls. This slowed the process to a point were it was no longer cost effective. The tows produced tended to be extremely rough and because the individual fibers were coated and then reassembled into tows there was a very high bulk factor, in excess of 4 to 1. This material was weavable but could not be braided.

The Cytec processes were more forgiving. The slurry process produced a less bulky tow which was also smoother than the electrostatic coated tows. In this process 10 to 50 micron particles were suspended in water with a small amount of binder, to produce a slurry. The resulting towpreg had a crusty rough surface because the tows were not fully spread for coating. This material was weavable but also could not be braided.

The solvent coating process was then developed by Cytec. This process was selected for the program's main effort. The solvent is entirely removed in the drying cycle, thus playing no role in the storage or cure process. The smooth surface allows easy weaving and this material can be braided. This process permits high line speed thus minimizing the cost of coating. The material is storable at room temperature indefinitely and gives a lower bulk factor to the preforms than the dry processes. All work in the ACT program was performed using AS4/PT-500 and its earlier versions, PT-501 and AMD0036.

3. TEXTILE PROCESSES

Powder coated tows can potentially be used in most textile processes. Braiding was the most difficult process to develop due to the abrasiveness of the early tows. The friction caused not only powder loss but prevented the tows from collapsing properly on the mandrels. Fiber Innovations did considerable development work with Cytec and NASA LaRC assistance and succeeded in producing acceptable braids using a solution coated tow. Time constraints prevented this being pursued on the ACT program.

Weaving, both conventional 3-D by Fabric Development and the Techniweave 3-D processes, were accomplished with relative ease, even with the early bulky abrasive tows. Smoother, less bulky tows later improved the weaving quality and reduced the bulk factors.

Warp knitting was attempted but had to be abandoned due to equipment problems. The modifications required to the knitting machine were beyond the scope of this program. However, the problems were again related to friction and also to the more brittle nature of the coated tow. The standard guides were badly worn in a very short time. This could be solved by fabricating them with titanium oxide. The carriers, bobbin holders and brakes also needed redesign. Carriers and bobbin holders need to be capable of handling large bobbins. The brakes caused sharp tension loads when reversing direction which tended to break tows. Dust control was also a problem with the towpregs available at that time.

4. WOVEN POWDER COATED TOW PANELS

4.1 Panel Design A woven window belt concept was designed, consisting of a one-piece 3-D woven doubler. It incorporated integral longitudinal stiffeners above and below the window opening. The vendor selected to develop this concept was Techniweave Incorporated, of Rochester, New Hampshire. Taking advantage of the capabilities of the Techniweave 3-D weaving process permitted inclusion of integrally woven elliptical window frames. To minimize tool cost and complexity all the stiffening elements were blade sections. Another unique feature of this process was the ability to weave in the window openings, thus reducing material waste and costs. While this process could provide additional stiffening elements, such as, a circumferential frame attachment flange, the increased cost could not be justified for this program.

One of the unique capabilities of this 3-D weaving process was that the skin can include woven $\pm45°$ material. The integral stiffener elements as woven, however, were 0°/90° construction only. The 90° fibers were used to attach the stiffener to the preform skin through a series of loops as the stiffeners were woven. Therefore, $\pm45°$ fabric overwarps were added to the preform as angles on either side of the stiffeners to provide shear transfer capability and stability.

4.2 Process Development The initial process development was performed on coated tow woven 8-harness satin fabrics woven by Fabric Development, from AS4-6K carbon towpreg with 10 ends per inch in both warp and fill directions. The tows were coated initially by NASA LaRC and by Georgia Institute of Technology. Later batches were processed by BASF and Cytec. The material showed no advancement when stored at room temperature for six months. No advancement occurred during subsequent weaving and braiding operations.

The fabrics were laid up and cured using a 1.27 mm thick caul plate. Several breather plies were included to help remove trapped air. The resulting panel showed porosity in the center. The caul plate had bowed and pinched the edges which trapped moisture. A 12.7 mm thick caul plate eliminated this problem. All test panels were fabricated with this setup. The panels were cured in an autoclave at 450°K and 0.69 MPa. A heat-up rate of 2°-5°K per minute was used. The finished panels were ultrasonically inspected using a 12.7mm diameter focused transducer, at a frequency of 5 megahertz. Mechanical and physical property tests were run. The resin contents tended to be higher than the 55% desired and varied more than desired. This problem was eliminated with the subsequent change to the solvent coating process.

The next step was to have Techniweave fabricate some integrally stiffened Tee sections for use in debulking and consolidation trials. An aluminum tool was designed and fabricated which was intended to debulk the stiffener during the heat up and then to debulk the skin when pressure was applied and to consolidate the part. The tool containing section of the Tee preform is shown in Figure 2. Figure 3 shows a debulked preform on the left and an undebulked preform on the right. These Tee sections had no ±3% material added to the stiffener and as can be seen had no corner radii. The sliding of the tool on the skin was unreliable at best and it was obvious that when scaled up this approach would not work. The blocks tended to tip and dig into the skin thus locking in position. The resulting stiffener was tapered and the skin indented.

Figure 2. Development Tool

Figure 3. Stiffener Tee Before and After Debulking

4.3 Shear Panel Tool Design and Fabrication The shear panel was a flat panel with two stiffeners and a window frame. Figure 4 shows the preform. It was the first true preform fabricated by Techniweave and was approximately one meter by one meter in size.

Figure 4. Shear Panel Preform

The method for consolidating the powder window-belt preform was driven by the high bulk created by the combining of the 3-D weaving process and the powder resin. The original plan was for the tooling to slide as described above. Early process development work showed that this did not occur. Thus, a debulk cycle was added. The tooling had to be hard to avoid 'cupping' of the panel. A 6.35 mm thick aluminum plate was used for the constant thickness portion of both the shear and the later pressure test panel tools.

The stiffeners were debulked by preheating two aluminum bars to 394°K and clamping the stiffener between them. The bars had guide pins on either end to keep them parallel. The resulting stiffeners were near net thickness. The key was the debulking of the window frame. First, the powder overwrap plies were debulked and trimmed to the mandrel surface that defined the final inner surface. The assembly was then placed in position around the window frame. The first trial at debulking the preform used heat shrink tape around the window rame upstanding flange. This reduced the bulk factor from about three to approximately 1.5. A second debulk operation was performed using blocks preheated to 394°K and clamped to the center mandrel. This reduced the average bulk factor by about 1.2. The actual thickness varied along each of the four blocks due to the difficulty in clamping the curved blocks. The cure tooling was expected to bring the assembly to the final thickness. The overwrap plies were then put in place completing the debulking process.

The main cure tooling was then placed in position. A 0.127 mm thick shim was placed under each of the tool component intersections, to bridge any gaps. Considerable tool movement was expected during the cure, opening up these gaps. Figure 5 shows the tooling components being assembled. This assembly was placed into the autoclave and cured using the cure cycle shown in Figure 6. The final part is shown in Figure 7.

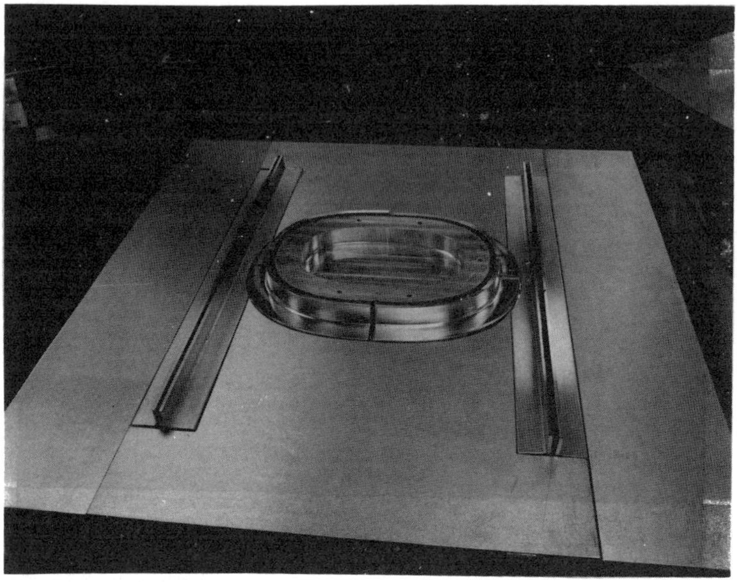

Figure 5. Tooling Components

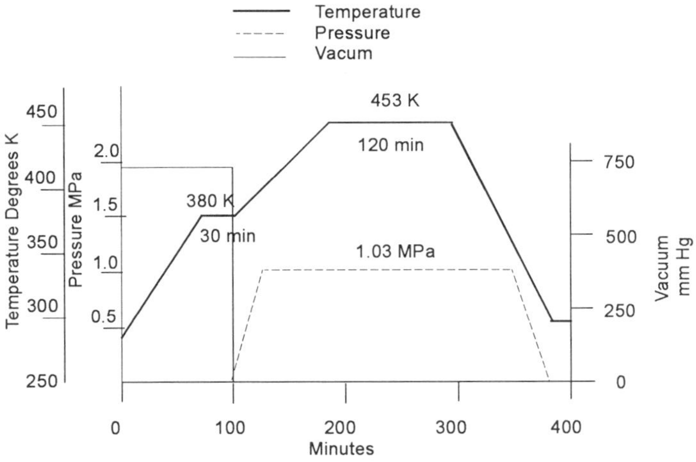

Figure 6. Cure Cycle

4.3.1 Evaluation When the part was first removed from the autoclave and the bag removed, resin ridges were obvious. They had formed where the tooling components joined. The 0.127 mm steel shim stock used was not rigid enough to prevent this from happening.

The panel was then ultrasonically inspected. C-scan and X-ray evaluation revealed some porosity around the top and bottom of the window frame. This was due to fiber pinching constraining the tooling from moving to the proper location during cure. Overall, the panel quality was good and it was accepted for testing.

A flat 1.05 m by 1.05 m window belt picture frame shear panel was prepared for test. Glass/epoxy doublers were bonded around the periphery of the panel for the picture frame fixture attachment. The test results satisfactorily demonstrated the structural integrity of the design.

Figure 7. Consolidated Shear Panel

4.4 Pressure Panel The lessons learned on the fabrication of the shear panel were applied to the pressure panel. Although the preforms were similar, the pressure panel was more complicated because of the curvature. The preform is shown in Figure 8.

4.4.1 Processessing Debulking of the two stiffeners was performed in the same manner as with the shear panel. Two straight aluminum blocks with guide pins at the ends were preheated to 394°K and clamped onto the stiffeners. After cooling the blocks were removed. The resulting thickness was within .051 mm of the predicted fully debulked preform.

Debulking of the pressure panel window frame area was complicated by the curvature. Four small wedge shaped pieces were fabricated to go between the four curved blocks. The shape allowed for easier tool movement. Flat areas were machined on these blocks as well as the mandrel to provide better control of the debulking. Figure 9 shows these blocks and other tooling details. The debulking was performed with the same preheating and clamping as the shear panel, but without using heat shrink tape. The window frame was well debulked, within 5% of the final thickness.

The curved portion of the tool was 6.35 mm aluminum plate rolled to the radius. A shim, near to the thickness of the overwrap plies, was cut to match the panel, butting up to the overwrap plies, thus eliminating a step in the tooling. The angle pieces used for the stiffeners extended beyond the end of the overwrap plies. This eliminated the gaps which had caused the resin ridges on the shear panel. Figure 10 shows the bagged and assembled tooling. The panel was bagged and cured using the same 1.034 MPa cure cycle as the flat shear panel.

During the cure cycle, at about 436°K, vacuum was lost and the bag blew up like a balloon. The cure was completed as the resin was too advanced to abort and rebag. Subsequent ultrasonic C-scan inspection revealed a significant amount of delamination and the panel was scrapped. The completed panel is shown in Figure 11.

Figure 8. Curved Panel Preform

Figure 9. Compaction Blocks

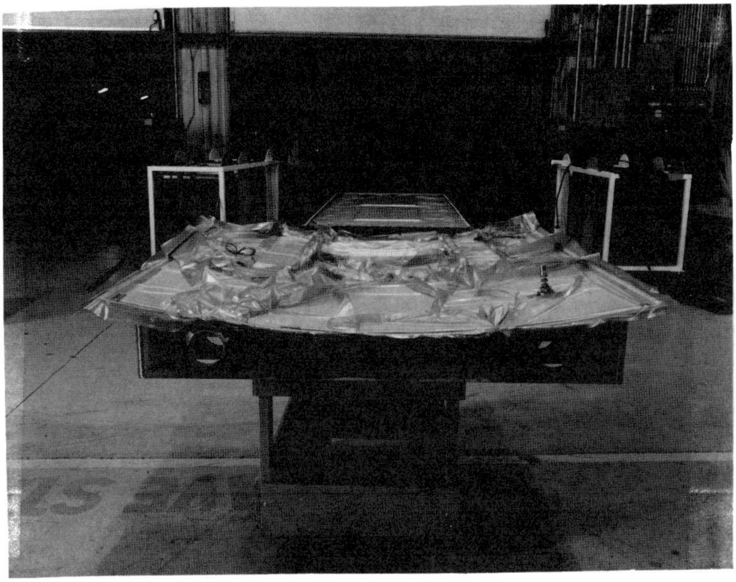

Figure 10. Curved Panel Bagged Assembled Tooling

Figure 11 Completed Curved Panel

4.5 Additional Preforms A second flat shear panel preform was fabricated by Techniweave and was processed by LMAS. Thicker shims were used at the tooling intersections. This resulted in minimum resin ridges. The outer compaction blocks around the window hung up during the cure and caused porosity in some areas of the window flange due to lack of pressure. The cured panel was visually excellent.

A second curved pressure panel preform was received and inspected. The inner flange around the window cut-out had some inconsistencies in the weave. The single glass stitching fiber did not complete the circumference on the preform. The area where the fiber has slipped off the structural graphite fibers raised concerns as to the resulting fiber orientation.

The overall qualities of this panel, a lack of funds and time constraints lead to the decision not to process this panel.

The weaving process at this time was entirely manual. Not only was it slow but it was impossible to maintain consistency over the many shifts involved. The stitching, being manual, could not provide the tension necessary for good debulking of the preform. Techniweave has since invested in a level of automation. This automation includes a system for laying down the tows in the xy plane and a stitching system. The stitching system has been shown to provide significant debulking. The system was not able to be implemented using the tools developed for this program. The pins were designed for manual layup and did not have sufficient strength to take the forces induced by the tows at turn around. Indexing requirements and gap tolerance requirements were tighter for the automated system. However, a flat panel was produced by Techniweave with this automated setup. It demonstrated the improvements in uniformity, overall quality and significantly lower bulk.

5. BRAIDING

The early towpregs were too rough to braid. The friction of the tows prevented proper collapse onto the mandrel and caused loss of powder. Various methods were tried for improving the sliding of tows.

Water was sprayed on the yarns at the cross-over to minimize friction. An open braid still resulted which caused voids in the resin pockets on cure, even when using pressures of 1.378 MPa.

Surface glazing by remelting the powder on the towpreg surface reduced friction but enhanced the resin adhesive properties, causing tows to stick together. Five percent zinc stearate powder was used as a dry lubricant. This worked extremely well but resulted in high void contents and low mechanical properties.

Surface application of less than one percent polyacrylic acid gel worked well on a 32 carrier braider but did not work when scaled up to a 72 carrier machine.

The smoother tows now being produced are proving much more adaptable to braiding. Preforms have now been successfully braided with some tows and in some configurations. While the potential has now been demonstrated additional work is needed before braiding of powder towpreg is generally accepted.

6. SUMMARY

Powder coated tow technology has been advanced significantly in the last few years. It has been demonstrated for an epoxy resin with woven preforms to be capable of producing excellent structural components. The bulk factor of the early tows was four to one, which presented some difficulties in processing. A void free panel was fabricated and tested successfully. The main problem was the bulk factor and designing tooling which would provide the necessary movement to debulk and provide the cure/consolidation pressure. Subsequent improvements in the coating process have reduced the bulk factor and the tow smoothness has been drastically improved. The new towpregs and automated weaving processes will provide lower cost, good quality structural components. The potential for braiding has been demonstrated.

28th International SAMPE Technical Conference
November 4-7, 1996

COINS
COMPOSITE INFORMATION SYSTEM MULTIMEDIA DATABASE

W. T. Freeman
NASA Langley Research Center

Shahid Siddiqi and Ralph Edlow
Analytical Services & Materials
Hampton VA

ABSTRACT

COINS provides a set of modern, easily accessible electronic databases for preliminary/conceptual design integrated product teams. COINS databases emphasize the application of high performance composites to airframe structures. Extensive data including aircraft prices and technical details, material properties for state of the art and developmental materials, multimedia examples of machines and fabrication methods, and collections of lessons learned from 2 decades of composite applications are being included. COINS provides the convenience of Internet or modem access from any computer with Microsoft Office.

1. INTRODUCTION

The COmposites INformation System (COINS) is a Windows NT server that hosts several databases of interest to scientists and engineers working with composite materials. The database was initiated at the beginning of the NASA Advanced Composites Technology (ACT) Program to provide a location for storage and transfer of electronic reports, test data and a summary of program accomplishments. This paper describes the software and hardware requirements to access COINS. Current content of the database is reviewed with descriptions of the text, image and multimedia information that can be queried and downloaded. A Data Abstraction Form (DAF) for unifying the reporting of composite hardware design, manufacturing, testing and cost data is also presented.

2. WORKSHOPS

Several workshops with airframe company cost estimators and engineers were held at the beginning of the ACT program (1989-1991) to determine what NASA should do to unify the way cost would be reported and predicted. Participants at these workshops defined issues and made the following recommendations (Ref. 1):

1. NASA should reinstitute and automate the collection of composite part fabrication costs in a format similar to the DoD/NASA Structural Composites Fabrication Guide (Ref. 2) with added emphasis on assembly cost. Standard forms for cost collection and reporting should be developed.

2. Establish a producibility guide to assist integrated product development teams in making decisions on a design concept. COINS will have an area devoted to describing the latest machines, fabrication methods and performance of potentially low cost processes. This area should serve as a modern educational tool for designers and individuals new to the latest in composites manufacturing.

3. Establish an "information at your fingertips" materials database for rapid access of preliminary or conceptual design material properties to be used in early sizing and concept development.

4. Software should be developed for IBM compatible computers since the companies represented unanimously agreed that this standard was used for research and development in their organizations. Commercially available, user friendly, software was recommended to reduce cost, maximize availability and increase the likelihood of future upgrade support. Some participants noted previous experience with specialized databases that have been lost when small specialty software companies went out of business or ceased to upgrade a product.

5. NASA should ensure that the database is kept current with long term support.

3. THE INTERNET

The profound impact of the Internet was not anticipated at the beginning of the ACT program. The advent of the world wide web with free and easy to use web browsers, servers, web publishing tools and secure Internet services will make COINS data easier to access and use. The rapid evolution of software for Internet support over the past year has been impossible to keep up with for supporting multiple hardware platforms with browsers and server support software. Microsoft's Internet Explorer and Netscape browsers are rapidly evolving into powerful graphical interfaces for access to COINS information. The COINS development team has baselined Internet Explorer since it can be downloaded free for all versions of Windows and Macintosh. The team assumes that helper application capabilities and interface uniformity of all browsers will eventually stabilize to allow all desktops to connect to the COINS information.

4. HARDWARE AND SOFTWARE

The COINS development team originally selected a high end database on a UNIX machine to host the COINS database. Microsoft (MS) Windows NT software with an Intel Pentium server was introduced in 1994 at a time when the originally selected UNIX database was now longer supported and the cost to update the software exceeded the cost to purchase an NT/Intel dual Pentium server system with several software packages.

MS Office 7.0 was selected as a low cost desktop package for collection, reporting and distribution of COINS data since it has integrated support for both IBM PC compatibles and Apple Macintosh desktops that represent the majority of ACT program participants computer hardware.

Additionally, free viewers are available for Word, Excel and PowerPoint (for users that don't own the full programs) to view and print documents created by the full programs. User's with these viewers, and the latest version of Internet Explorer, can view documents over the internet with the complete formatting and graphics of the Office application providing a unified low cost solution to access all COINS resources. This eliminates the need and time required to convert text and graphics into HTML (HyperText Markup Language) format.

The advent of MS Windows 95 with built-in clients for network and Intranet/Internet access provides a low cost solution for accessing report and data publishing that will expand the technology transfer options of ACT program research significantly.

Microsoft's Excel spreadsheet was selected to display summary data of interest to ACT program design teams to obtain the following benefits:

- To present data with a sophisticated but easy to use standard visual format.
- Excel spreadsheet data can be plotted and analyzed with sophisticated built in tools.
- User manuals and extensive how-to books have been written for Excel minimizing program documentation requirements.
- Excel comes with a Query program that can be used to search and compare data.
- The spreadsheet format has become a familiar and common work environment for the engineers, finance and management personnel that are anticipated to be interested in COINS.
- The software is commercially available, popular, relatively inexpensive, and well supported with frequent upgrades for more powerful capabilities.

5. EXISTING DATA AND REPORTS

The information that follows describes the knowledgebase that has been developed as of July 1996. Additional information and reports are being added as electronic versions become available. Information will be provided at the conference to describe Internet software security protocols and collect user identification for conference participants to access the COINS server.

1. Transport and Composite Aircraft

An Excel workbook (a portion of which is shown in Table 1 for Boeing Aircraft) has been developed for most of world's large commercial jet aircraft with special emphasis on carbon composite applications, aircraft performance and selling price. Price and direct operating cost information is provided to allow order of magnitude economic evaluation of various aircraft sizes, number of engines, etc. The workbook has spreadsheet tabs for Boeing, Douglas,

Airbus, the Commonwealth of Independent States and other information of interest to composites specialists. Spreadsheets have several note fields detailing information sources and relevant information for designers. Notes, video, photos and color are displayed on a computer display view of these workbooks. The purpose of these spreadsheets is to provide a quick look and order of magnitude for relative value analysis. All data has been obtained from public documents and verification of data for accuracy has been attempted. Aircraft changes or improvements that occur over time and specific model data may not be reflected in the data.

Transport Aircraft Data	MD-91/94	MD-90	MD-11	MD-81/82/83	MD-88
Airline Service Date	1996	1985	1990	1977	1977
Service Date Price $ Mil			$78.4	$23.5	$18.9
1991 Market Value $Mil			$107.5	$24.0	$27.0
1994 Market Value			$84.0	$15.9	$24.1
1999 Projected Value			$76.0	$14.8	$21.8
Number Built/Ordered by 6/94		49	129	1,141	
Engine Thrust	15,000	25,000	58,000	21,700	20,850
Number of Engines	2	2	3	2	2
Manufacturer/Model	GE36-C22	IAE V2525-D5	PW4460	JT8D-219	JT8D-217C
Stage3 TO Noise EPNdb		86.3	102.0	97.0	94.4
Max Seats	165	172	397	155	155
Typical Seats	114	16/136	34 / 287		
GTOW lbs	133,000	156,000	602,500	160,000	149,500
OEW lbs	64,000	88,171	277,000	78,420	78,420
Service Date Price/OEW - $/lb			$283	$300	$241
Max Land Wt lbs		142,000	430,000	128,000	128,000
Max Zero Fuel Wt lbs		130,000	400,000	118,000	118,000
Fuel Wt lbs		39,128	259,000	46,699	39,128
Fuel Capacity gals		5,983	39,602	6,929	5,805
Max Pass Load lbs			158,000	39,579	39,579
Max Load lbs	119		180,000	39,579	39,579
Max Cargo lbs			118,000		
Cargo Volume ft^3	727	1,300	6,231	1,253	1,253
Cruise Speed knots	448	436	511	499	499
Cruise Ceiling		41,000	45,000	37,000	37,000
Range nm Max Load	2,470	2,170	5,010	2,540	2,540
Max Range nm			8,500		
TO Length ft		7,000	10,100	7,250	7,250
Wing Span ft	93.4	107.8	169.5	107.8	107.8
Wing Area ft^2		1209.0	3674.0	1270.0	1270.0
Avg Chord ft		11.2	21.7	11.8	11.8
Aspect Ratio		9.6	7.8	9.2	9.2

TABLE 1 TRANSPORT SPREADSHEET EXAMPLE

2. Materials

An Excel workbook (a portion of which is shown in Table 2) of materials properties for Hercules (now Hexcel) fiber, fabric, resin, prepreg tape and prepreg fabric was developed for rapid property comparison. Relative properties for a wide variety of composite material forms can be quickly evaluated for comparison with alternative materials. Several commercial sources offer B-basis allowable data for specific material systems but the purpose of these spreadsheets is to provide screening and relational "quick look" information from materials supplier provided data.

		GRAPHITE FIBER PRODUCTS SUMMARY TYPICAL PROPERTIES									
		Aerospace Grade Fibers									
		American Units					S.I. Units				
		Tow Tensile			WPUL		Tow Tensile				
Fiber	Filament	Strength	Modulus	Strain	(lb/in)	Density	Strength	Modulus	Strain	MPUL	Density
Type	Count	(ksi)	(Msl)	(%)	E-6	(lb/in^3)	(MPa)	(GPa)	(%)	(g/m)	(g/cc)
AS4	3				11.8					0.21	
	6	570	32.8	1.6	24	0.064	3930	226	-1.6	0.428	1.79
	12				48					0.858	
HMS4	12	350	49	0.7	50	0.065	2410	338	0.7	0.893	1.8
HMU	1				4					0.071	
	3	440	52	0.8	12.4	0.0654	3030	3~9	0.8	0.221	1.81
	6				25.5					0.455	
	12				51					0.911	
IM6	12	760	40.5	1.7	25	0.0636	5240	279	1.7	0.446	1.76
IM7	6	770	39.5	1.8	12.5	0.0645	5310	272	1.8	0.223	1.79
	12				25					0.446	
IM8	12	780	44	1.6	25	0.0647	5380	303	1.6	0.446	1.79
UHMS	3				4.7					0.084	
	6	540	64	0.8	9.4	0.0675	3720	440	0.8	0.168	1.87
	12				18.8					0.336	

TABLE 2 MATERIALS PROPERTIES SPREADSHEET

3. Lessons Learned Reports

Several lessons learned and recommended reference documents of historical and educational value are being added to the COINS web site. "Composites Chronicles: A Study of Lessons Learned in the Development, Production, and Service of Composite Structures" (Ref. 4) and "Lessons Learned for Composite Aircraft Structures Qualification" (Ref. 5) are notable examples. Anyone with valuable composite design, fabrication, cost or important test data documents that they believe should be archived for availability to US companies should send them to the authors. Electronic data compatible with Microsoft Office is preferred, but valuable documents can be scanned if required. All documents provided must have the authors permission to publish them on the web without copyright ownership issues or other legal restrictions. COINS access will be restricted to US companies and all data will be encrypted for authenticated users.

4. Producibility Guide

The maturity of machine tools and automation in the aluminum aerospace industry is astonishing. Some of the largest machine tools in the world are used to reduce the fabrication cost of large commercial airframes. For composites to compete with aluminum primary structure, equivalent levels of machine tool hardware and software sophistication must be envisioned, prototyped and scaled for production readiness. Factories are systems of coupled process steps that result in the consumption of several types of resources. Summation of the recurring resource consumption and amortization of the capitalized nonrecurring expense determines the economic viability of an advanced technology when compared to a high volume, highly capitalized, highly automated existing technology.

5. Data Abstraction Form

Appendix A includes the Data Abstraction Form (DAF) that was developed from suggestions provided by several commercial and military company representatives. The form is a significant extension of the original Air Force Fabrication Guide DAF with the addition of detailed information on assembly, cost and test data if applicable. The recommended purpose of the DAF is to provide a standard form to guide collection and unify reporting of the most important information to acquire during the fabrication and testing of large, expensive, prototype hardware. The DAF is available as a self installing MS Visual Basic electronic form that can eliminate the paper format if desired.

6. CONCLUDING REMARKS

COINS provides the first Internet presence for extensive advanced composite materials and structures information. A variety of reports, pictures, and videos have been collected to describe the current state of the art for composite applications primarily on cost driven commercial aircraft. Special emphasis is given to manufacturing issues, composite machines and processes, and the maturity of automation in the aluminum industry.

Several challenges exist for COINS to be a valuable Internet source for designers. The continuous addition of useful information and updating of materials and cost data requires persistent support and funding in a era of downsizing, government reinvention, and extreme program plan changes. COINS has the potential to become a national resource knowledgebase for educational and technology transfer by the composites industry using the "information at your fingertips" potential of the Internet.

7. REFERENCES

1) W. T. Freeman, S. Siddiqi, L.F. Vosteen, A Unified Approach for Composite Cost Reporting and Prediction in the ACT Program, First NASA Advanced Composites Technology Conference, Seattle, WA, Oct. 29-Nov. 1, 1990, NASA CP-3104.

2) L.E. Meade, DoD/NASA Structural Composites Fabrication Guide-3rd Edition, Air Force Wright Aeronautical Laboratories, 1982

3) S. Siddiqi, L.F. Vosteen, R. Edlow, T. Kwa, COINS: A Composites Information Database System, Proceedings of the Ninth DOD/NASA/FAA Conference On Fibrous Composites In Structural Design, Lake Tahoe NV, Nov.4-7, 1991, DOT/FAA/CT-92-25 Vol. II

4) L.F. Vosteen, R.N. Hadcock, Composites Chronicles: A Study of Lessons Learned in the Development, Production, and Service of Composite Structures, NASA CR 4620, 1994

5) R. Whitehead, Lessons Learned for Composite Aircraft Structures Qualification, First NASA Advanced Composites Technology Conference, Seattle, WA, Oct. 29-Nov. 1, 1990, NASA CP-3104

8. APPENDIX A
DATA ABSTRACTION FORM

COINS

Questionnaire

INSIDE

About this Questionnaire

This questionnaire is prepared for the COINS participants who do not have access to the computerized version of the COINS software.
The COINS ADAF software for personal computer with windows is available from AS&M.

Please follow the following steps while filling out this questionnaire.

- Familiarize yourself with the definitions used in this questionnaire.
- Fill in the questionnaire.
- Follow the flow chart while filling out the questionnaire.
- Duplicate sections of the questionnaire if there is more than one part, or the questionnaire is used for a subassembly or major assembly.
- Staple the questionnaire together when finished.
- Fold the questionnaire in half, affix the proper postage to return the questionnaire to AS&M.
- If you have any questions about this questionnaire, please call Ralph Edlow or Shahid Siddiqi at AS&M (804) 865-7093 or fax (804) 865-7309.

Thank you for your participation.

Return this form to AS&M, Inc.
Mail or Fax to 804/865-7309
107 Research Drive
Hampton, VA23666

A 8/95

Please Answer All Questions.
Use COINS Definition Guide for explaination of terms. Follow the flowchart as a quide for completing a questionnaire.

1 Form Data

Submission Date ___/___/______ Fabrication Date ___/___/______
Contact Name ______________________________
Company ______________________________
Division ______________________________
Telephone No. (____)____________ Fax No. (____)____________
Part or Assembly Name ______________________________
Aircraft Component ______________________________
Program ______________________________
Aircraft Model No. ______________________________

Please duplicate (photo-copy) the questionnaire for additional forms and duplicate the dashed oval sections as denoted by the flow chart for completing questionnaires for subassemblies and major assemblies.

FOLD HERE

Return Address

____
____
____

POST OFFICE
WILL NOT DELIVER
WITHOUT PROPER
POSTAGE

A S & M, Inc
107 Research Drive
HAMPTON VA 23666

2 Part Construction Information

Part Name:__________

Shape __________

Construction Method __________

Substructure __________ Molding Method __________

Ply Information __________

Max Plies. __________ Min Plies __________

Raw Material Purchased __________

Trimmed Materials __________ Test Material __________

Scrap Weight __________ Composite Weight __________

Completed Part Weight __________ Ply Deposition __________

Ply Cutting Methods __________

Ply Cutting Mode __________ Compaction Methods __________

Bag Materials __________ Seal __________

Bleeder __________ Curing Consolidation __________

Maximum Cure Temperature __________ Maximum Cure Pressure __________

Total Cure Time __________ Maximum Heating Rate __________

Post Cure Temperatures __________ Post Cure Time __________

Part Trim __________ Cutting Tools Materials __________

Joining __________ Bonding Surface Preparation __________

Adhesive Trade Name __________ Fastener Trade Name __________

Alloy Name __________ Surface Treatments __________

Type of Paint __________

Erosion Protection __________ Lightning Protection __________

3 Assembly / Part Information

The following information may apply to either a Part or Assembly. If an Assembly is specified, the following will be requested for each Part making up the Assembly.

Quantity __________ Structural Type __________

Max. Length __________ Max. Width __________ Max.Thickness __________

Wetted Area __________ Weight __________

Maximum Service Temperature __________ Humidity __________

4 Assembly Process Information (Use only while filling out Subassembly or Major Assembly form)

Assembly Name:__________

Assembly Level __________ Bonding __________

Curing __________ Fasteners Used __________

Number of Fasteners __________ Trimming __________

Holes / Penetrations __________ Erosion Protection __________

Lightning Protection __________

5 Fabrication Tooling

Tool Fabrication Technique __________ Tool Function __________

Tool Description __________ Shape __________

Surface Support __________ Tooling Type __________

Surface Material __________ Surface Preparation __________

Caul Material __________ Caul Supplier __________

Mandrel __________ Life Expectancy __________

Maintenance __________

6 Assembly Tooling Information (Use only while filling out Subassembly or Major Assembly form)

Method of Assembly __________ Tool Maximum Temperature __________

Tool Maximum Pressure __________ Material(s) __________

Life Expectancy __________ Maintenance __________

Next Assembly __________

Material Information

Because of the large number of combinations of material configurations, this section is divided into several parts.

Reinforcement: Input of data on reinforcement fiber.

Matrix: Input of data on composite matrix materials.

Core: Input of data on core materials. Information required only for sandwich structures.
Architecture: Input of data on 1-D, 2-D, and 3-D structure of the fiber reinforcement.

7 Reinforcement

Reinforcement ______________________
Sizing ______________________
Designation or Types ______________________
Volume Fraction ______________________

8 Matrix Information

Matrix Type ______________________
Designation ______________________

9 Core Information

Core ______________________
Density ______________________
Designation ______________________

10 Uni-directional Ply Architecture

1-D Form ______________________
Spec. Document ______________________
Thickness ______________________
Areal Weight ______________________
Company Specification ______________________
Filament Count ______________________
Linear Density ______________________

11 2-D Product Architecture

Textile Process ______________________
Weave Pattern ______________________
Width, Length & Thickness ______________________
Warp Count ______________________
Warp Tow ______________________
Braid Angle / Braid Count ______________________
Braiding ______________________
Spec. Document ______________________
Style Specification ______________________
Areal Weight ______________________
Fill Count ______________________
Fill Tow ______________________
Number of Plies ______________________
Stitch Material / Stitch Pitch ______________________

12 3-D Product Architecture

Textile Process ______________________
Style Designation ______________________
Thickness Reinforcement Designation ______________________ Volumetric Weight ______________________

13 Quality Control

Tooling Periodic Inspection ______________________
Material Initial Inspection ______________________
Tooling Inspection Frequency ______________________
Periodic Reinspections ______________________

14 Cost Information

Cost Basis ______________________
Purchase Price of Elements ______________________
Quantity Purchased ______________________
Total Planned ______________________
Current Status ______________________
Raw Material Price ______________________
Production Rate ______________________
Total Produced ______________________

15 Assembly Cost Information (Use only while filling out Sub-assembly or Major Assembly form)

Cost Basis ______________________
Component Price per Assembly ______________________
Quantity Pounds ______________________
Production End Date ______________________
Total Planned ______________________
Current Status ______________________
Raw Material Price ______________________
Production Start Date ______________________
Production Rate ______________________
Total Produced ______________________

16 Direct Labor

Labor Operation	Labor Category	Hours

17 Direct Labor Assembly

Type Labor	Labor Category	Hours

18 Non-Recurring Cost Information

Cost For	Fabrication Labors (hours)	Design Labor (hours)	Inspection Time (hours)

19 Learning Curve Information

Projected ______ Actual ______

20 Experience Information

This will be a valuable lesson learned, information given, information specified for the production process.

Repair Documentation ______

Problems Re-work Box ______

Other References Box ______

Recommendations Box ______

21 Design Constraint

Loading Type ______
Design Conditions ______
Fuel Containment ______
Failure Loads ______
Load Factor ______
Max. Load ______
Operating Temperature ______
Loading Magnitude ______
Critical Flight Conditions ______
Safety Factors ______
Critical Design Conditions ______
Fatigue Criteria ______
Design/Analysis Codes Used ______
Test Performed ______
Failure Condition ______
Lessons Learned ______

22 Design Constraint Issues

Customer ______
Materials ______
Aircraft Services ______
Ground Crew Constraints ______
Policy ______
Aircraft Operation ______
Flight Crew ______
Repairs ______
Supportability ______

FLOW CHART FOR FILLING OUT THE FORM

Suggested sequence for completing the form for a part::

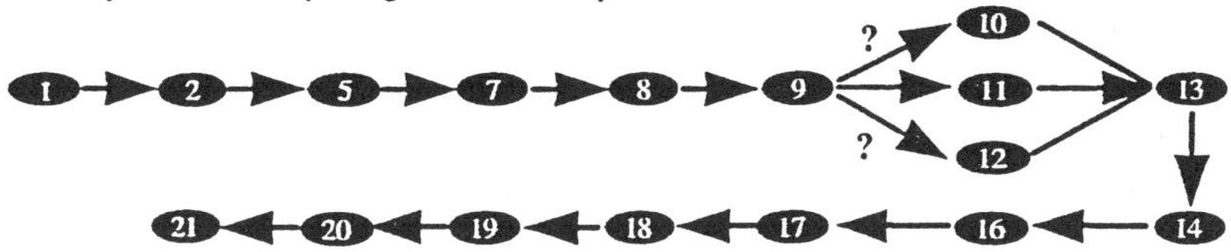

Suggested sequence for completing the form for a subassembly. Duplicate dashed ovals as needed:

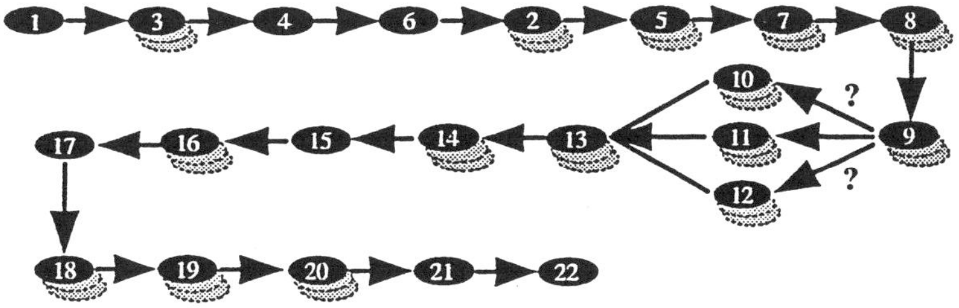

Suggested sequence for completing the form for a Major Assembly. Duplicate dashed ovals as needed:

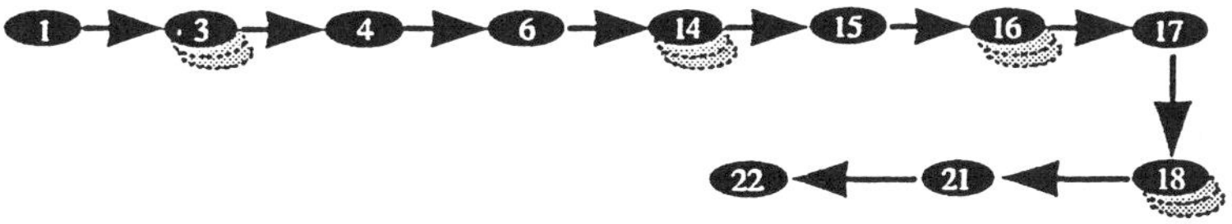

28th International SAMPE Technical Conference
November 4-7, 1996

SOL-GEL SURFACE TREATMENTS FOR ADHESIVE BONDING OF TITANIUM AND ALUMINUM STRUCTURES

K. Y. Blohowiak; J. H. Osborne; K. A. Krienke; and D. F. Sekits,
Boeing Defense & Space Group, Seattle, WA 98124-2499

ABSTRACT

A surface treatment process for titanium and aluminum alloys using sol-gel chemistry has been developed that produces strong, durable adhesive bonds without the toxic chemical usage and rinse water requirements of traditional anodizing or etching processes. In this process, an acid catalyzed sol composed of zirconium alkoxide and a silane coupling agent in water was applied to etched or grit blasted substrates by immersion or spraying. In laboratory testing, bond performance was comparable to the standard anodize controls. Parameters affecting performance include composition, Si/Zr ratio, concentration, solution age, catalysts, surface pretreatment, application method, curing, and primer used.

KEY WORDS: Sol-Gel, Adhesive Bonding, Surface Preparation, Titanium, Aluminum

1. INTRODUCTION

The conventional processes for preparing titanium and aluminum surfaces for bonding include application of electrolytic and chemical films. Electrolytic treatments such as anodizing, use chromic acid or an alkaline metal hydroxide for titanium, and chromic or phosphoric acid anodize for aluminum. Anodize systems produce a porous microrough surface into which an adhesive primer, a very dilute solution of adhesive, can penetrate. Adherence results from mechanical interlocking between the surface treatment and the adhesive primer. Chemical films include the phosphate-fluoride conversion coating for titanium substrates and the FPL and P2 etch processes for aluminum.

From an environmental viewpoint, current processes have the disadvantage of using strong acids or bases and toxic materials (such as chromates) in immersion tank processes. All require significant amounts of water to rinse excess process solution from parts. The rinse water and spent process solutions must be treated prior to discharge or reuse, generating hazardous waste materials. The goal of the research reported here is to develop a process that will produce structural adhesive bonds with equivalent strength and environmental durability to the existing processes without generating significant hazardous wastes or using environmentally undesirable materials. We are specifically targeting the surface treatment process as diagrammed in the process flow chart shown in Figure 1.

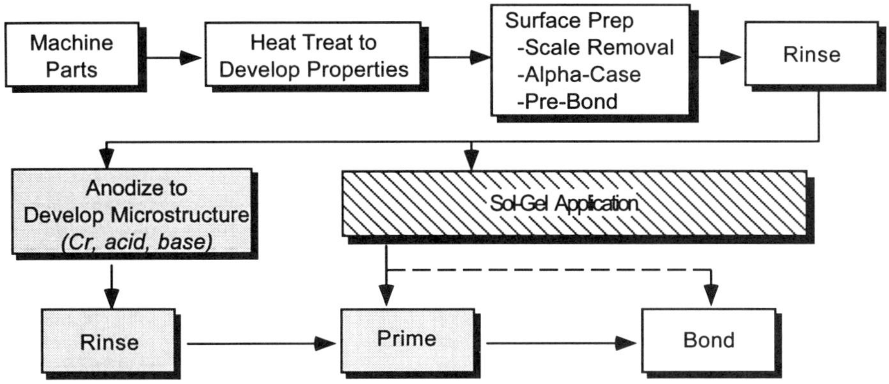

Figure 1. Process flow chart for structural adhesive bonding of metal parts

Many of the potential applications for sol-gel surface preparations involve the repair and retrofitting of older aircraft, in addition to the manufacture of new aircraft. This calls for a robust process that can be applied at one of the DoD depot maintenance facilities. The ideal process will not require the use of immersion tanks and will be field applicable. A water-based system alleviates many of the flammability, safety, toxicity, and environmental concerns associated with conventional anodic acid processes.

2.0 TECHNICAL APPROACH

In this study we use sol-gel chemistry to develop binder coatings that produce a gradient from the metallic surface through a hybrid inorganic/organic mix to the crosslinked adhesive. Sol-gel technology has recently been applied to bonding of metallic substrates.[1] Our approach applies a waterbased mixed metal alkoxide containing zirconium isopropoxide and a silane coupling agent. The coating is applied by either immersion, spraying, or drenching. No rinsing is required.

We make use of substrate metal surface chemistry, relative reactivity rates of sol-gel precursors, and coupling agent technology. By appropriately manipulating the sol-gel reaction conditions, we form a gradient coating from the metal surface through a zirconium oxide interfacial layer to a mixed silicon-zirconium region, with the organic coupling group oriented at the adhesive interface. The organic coupling group is chosen to be compatible with the adhesive system. The ideal system has covalent bonding throughout the gradient coating. Reducing dependence on Lewis acid/base and hydrogen bonding interactions for adhesive bond strength increases bond durability in a hot/wet environment.

2.1 Sol-Gel Chemistry The term "sol-gel," a contraction of solution-gelation, refers to a series of reactions where a soluble metal species (typically a metal alkoxide or metal salt) hydrolyzes to form a metal hydroxide.[2] Condensation reactions with the metal substrate couple the sol-gel with the surface producing a durable film. The sol-gel precursor species can also contain organic ligands tailored to a specific application. The use of typical coupling agents[3] creates a gradient film appropriate for adhesive bonding applications.

A clean and chemically active surface is required to bond the sol-gel coating to the metal surface. Bare, pure aluminum and titanium surfaces immediately oxidize in air or dry oxygen to form a barrier metal oxide film. In moist air, the oxide film on aluminum hydrolyzes to form the hydroxyl groups required for sol-gel condensation reactions on the surface. For titanium, oxide films are very tenacious and passive. Unlike aluminum, titanium surface oxides do not readily hydrolyze to form the active metal hydroxides. Sol-gel condensation reactions on titanium surfaces may be assisted by the presence of adsorbed water on the titanium oxide surface.

2.2 Test methods Screening studies were conducted on test panels of titanium 6Al-4V (Ti 6-4) and bare aluminum 2024 and aluminum 7075 alloys initially prepared by degreasing with an aqueous detergent. The panels were then either grit blasted with #180 grit alumina followed by a final aqueous detergent cleaning to minimize the presence of loosely adhered grit, or acid etched. The sol-gel system tested consisted of a dilute aqueous mixture of glycidoxytrimethoxysilane (GTMS) and tetra-n-propoxyzirconium (TPOZ) with an acetic acid catalyst. The panels were dip-coated with a 10 minute immersion time, held under ambient conditions for 15 to 30 minutes, and dried in a 110°C (230°F) oven for 15-30 minutes. A process flow chart for the optimized surface preparation technique using this sol-gel system is shown in Figure 2.

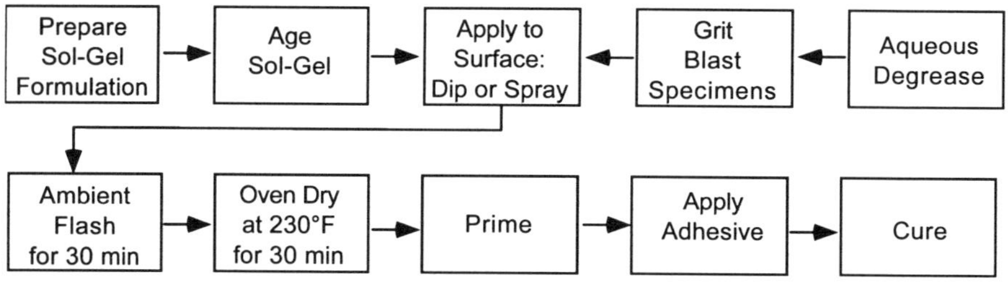

Figure 2. Process flow chart for sol-gel surface preparations for adhesive bonding on metal substrates

The dried panels were primed with either BMS 5-89 chromated adhesive primer (Cytec BR 127) or the nonchromated, waterborne version, Cytec XBR 6757, and bonded in an autoclave using BMS 5-101 Type II Grade 10 (Dexter-Hysol EA 9628) 121°C (250°F) cure adhesive. Samples were prepared for testing at Boeing and the University of Dayton Research Institute (UDRI).

Screening level wedge crack extension testing was conducted at Boeing using the ASTM D 3762, Wedge Test. Progress of the crack along the bondline was measured after the initial driving of the wedge, and after exposure to 60°C (140°F) and greater than 95% relative humidity for one hour, 24 hours, one week, and longer. Samples were monitored in the humidity chamber for over 2000 hours total exposure time.

Lap shear data was collected at UDRI on an Instron Series IX Automated Materials Testing System 6.04 per ASTM D 1002. Finger panels were primed and lap bonded with the same adhesive system as the wedge test samples. Measurements were taken at -54°C (-65°F), room temperature, and 74°C (165°F).

Floating-roller peel testing was conducted at UDRI on the aluminum specimens per ASTM D 3167. Data is reported as an average of five individual specimens and reported for testing at -54°C (-65°F) and room temperature.

Chromic acid anodized titanium control specimens were prepared at Boeing and primed with Cytec BR 127. Phosphoric acid anodize aluminum control specimens were prepared both at Boeing and at UDRI and were also primed with Cytec BR 127 and bonded with Hysol EA 9628.

2.3 Scale-up and Spray Application There are fundamental differences in film formation using an immersion versus a spray-on process. With the part immersed in the solution, reactants can convect or diffuse to the metal surface and reaction products can diffuse from the surface. The most thermodynamically stable coating will develop. In the case of the spray-on application only a thin film of the reacting solution contacts the part surface. Depletion of reactants can and likely does occur as the solution flows down the surface. Consequently, reaction products build up and may influence the chemistry that occurs. In addition, reaction products remain on the surface when the film dries. In all, the film developed in the spray-on application is dominated by kinetically accessible products. Advantages of a spray-on technique include coating thickness control and uniformity.

We also examined a drench method (flood coat) which combines elements of both the dip and spray processes. In this application technique, the surface of the part is wetted with a continuous stream of solution for a given period of time. The solution surface is wet with the solution for longer than in the spray process, but not as long as the dip process. One of the advantages of this technique is that it does not require the precise skills of an expert sprayer. It also uses significantly less solution than an immersion process. The coating thicknesses are controlled by the coating formulation itself and length of time that the surface is wet.

3.0 RESULTS AND DISCUSSION

3.1 Sol-Gel Processing The sol-gel system reported here meets many of the requirements for a factory and field applicable bond system. In most cases, substrate preparation is accomplished using aqueous cleaners and/or grit blasting. The sol-gel formulation itself is water-based and uses low hazard and low toxicity materials. Application of the sol-gel to the substrate is accomplished by drench and spray coating as well as immersion.

The water-based system has performance equivalent to our control samples. We chose a glycidoxy-functionalized silane because of its stability in solution and compatibility with various epoxy adhesive systems. Glycidoxytrimethoxysilane is acid-base neutral so its presence in the sol mixture does not increase the relative hydrolysis and condensation rates of the alkoxides, enabling greater pot-life stability. The coating solutions are relatively easy to prepare and apply with reproducible results.

3.2 Titanium Test Results Figure 4 shows wedge crack extension data plotted versus time. Each data point represents an average value for ten test specimens. The candidate sol-gel formulation compared very favorably with the chromic acid anodize standards even after 2000 hours hot/wet exposure. The crack growth rate levels off at approximately 2.5 mm (0.1") crack extension. Passing is generally considered to be less than 6.4 mm (0.25") extension after one hour humidity exposure.

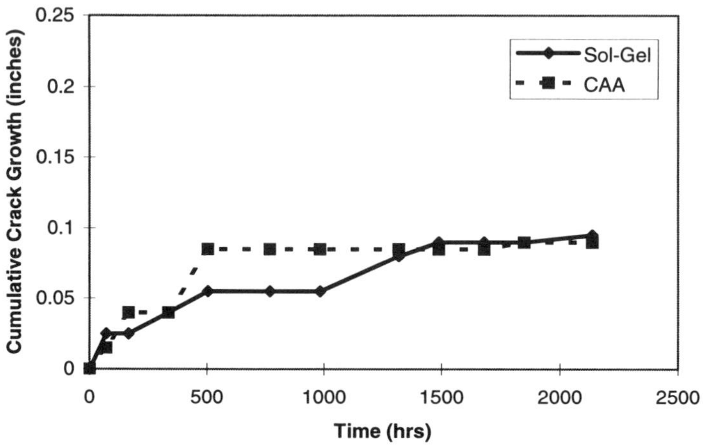

Figure 4. Boeing titanium 6Al-4V wedge crack extension results

Lap shear test results for the titanium substrates are shown in Figure 5. Results are an average of five finger test specimens per data point. The data for the CAA control and the sol-gel surface preparations with both primers was comparable. Analysis of the specimens showed the failure mode to be predominantly cohesive within the adhesive layer in all of the specimens at all of the temperatures. Specimens bonded with the water-based nonchromated primer performed as well as those primed with the solvent-based chromated primer.

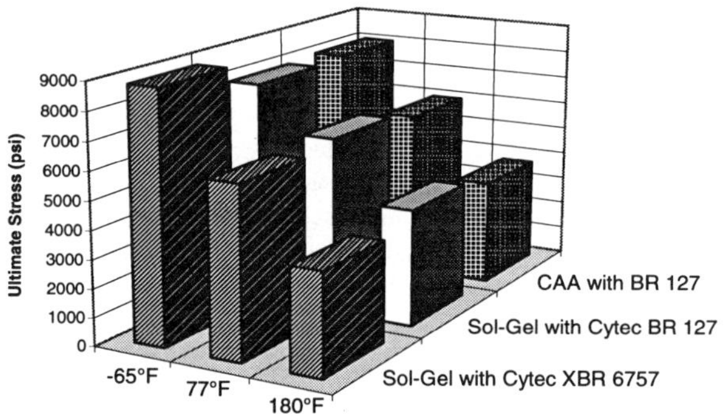

Figure 5. Titanium 6Al-4V lap shear test results

3.3 Aluminum Test Results The sol-gel used on the aluminum substrates was essentially the same as that developed for the titanium system. Figure 7 shows the cumulative crack extension data as a function of time for selected sol-gel treated specimens. In general, we saw some performance differences between the different alloys, with 7075 typically giving shorter crack extensions than 2024. The hot/wet durability of the sol-gel treated specimens was comparable with the PAA controls for the first 1000 hours. All of the panels were well within the passing criteria for BAC 5555, Phosphoric Acid Anodizing 7.6 mm (0.3") extension after one hour humidity exposure). Additional durability testing is required to assess the long term performance of the system.

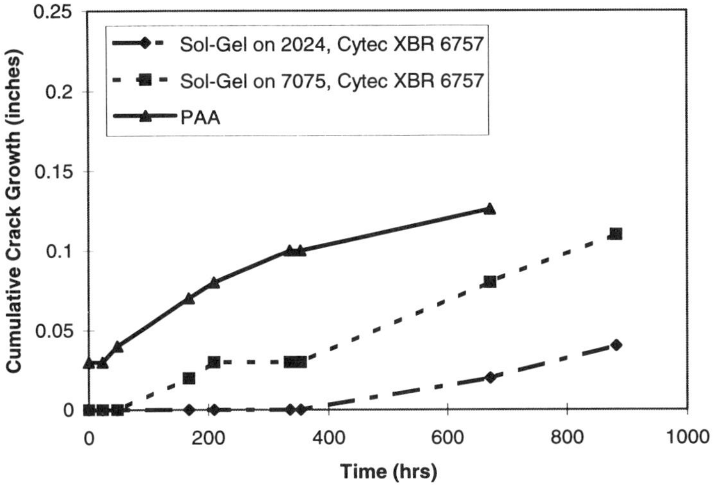

Figure 7. Wedge crack data for aluminum substrates

Lap shear data was collected on both 7075 and 2024 aluminum alloys. The results for the 7075 alloy and the 2024 alloy are shown in Figure 8. The room temperature lap shear values all surpass the minimum acceptance value of 4375 psi (BAC 5555). These preliminary results show that panels primed with the waterbased nonchromated primer, Cytec XBR 6757, performed as well as those primed with Cytec BR 127. As with the wedge crack specimens, the 7075 alloy tended to perform better than the 2024 alloy.

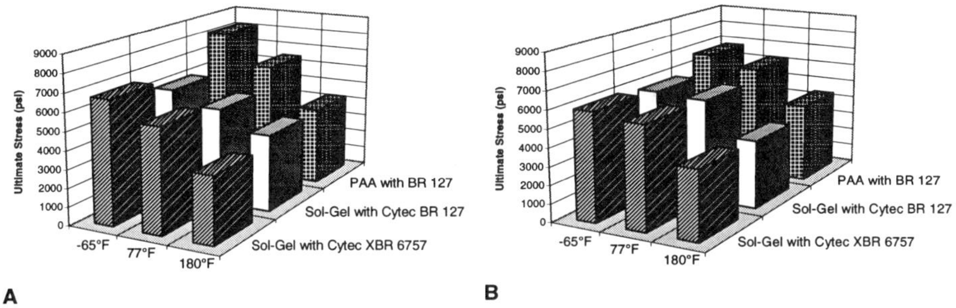

Figure 8. A) Aluminum 7075-T6 and B) aluminum 2024-T3 lap shear test data.

Figure 9 shows floating-roller peel test data with results for the aluminum alloys compared with a phosphoric acid anodize control. Data is reported as an average of five individual specimens. The sol-gel specimens compared favorably with the bond strengths exhibited by the phosphoric acid anodize in peel testing.

3.4 Spraying Studies Studies are currently being conducted on the effects of spraying and drench coating the current sol-gel formulations and comparing them with the immersion coating performance values. Preliminary data on aluminum substrates suggests little difference in performance between drench coating, mist spraying, and immersion application methods.

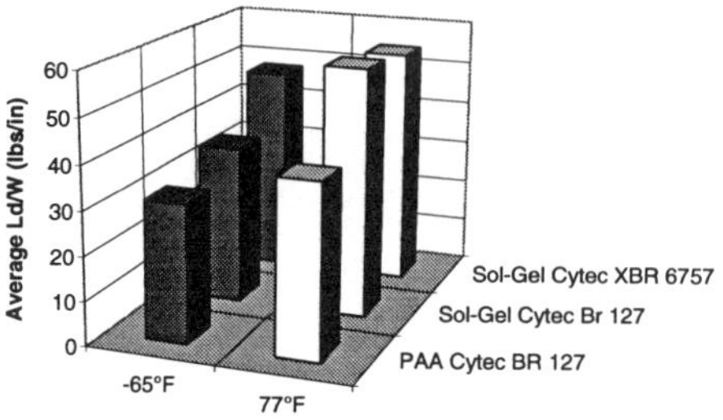

A

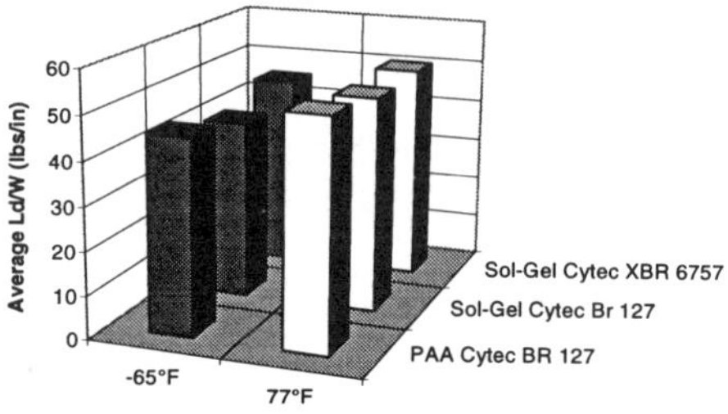

B

Figure 9. A) Aluminum 7075-T6 and B) aluminum 2024-T3 peel test data.

ACKNOWLEDGMENTS

We gratefully acknowledge the continued funding and support from Wright Laboratory Materials Directorate at Wright-Patterson AFB and the Strategic Environmental Research and Development Program (SERDP). This work was administered through WL Contract #F33615-93-C-5322; LAI Subcontract #94-5812-66-4; and SAIC Subcontract #4500038283. Technical oversight was provided by Jim Mazza of the Wright Laboratory Materials Directorate, WL/MLSE.

REFERENCES

[1] (a) S.R. Holmes-Farley, L.C. Yanyo, U.S. Patent 5,175,027, **1992**; (b) S.R. Holmes-Farley, L.C. Yanyo, *Mat Res Soc Symp Proc* **1990** 180, 439; (c) T. Sugama, C. Taylor, "Pyrolysis-Induced Polymetallosiloxane Coatings for Aluminum Substrates," preprint, 1993; (d) W. Dorogy, Jr., Foster-Miller, Inc., Personal Communication; (e) N.T. Castellucci, U.S. Patent 5,206,285, **1993**.

[2] C.J. Brinker and G. Scherer, "Sol-Gel Chemistry," Wiley **1991**.

[3] E.P. Pluddeman, "Silane Coupling Agents," Plenum, New York, **1991**.

28th International SAMPE Technical Conference
November 4-7, 1996

AN ENVIRONMENTALLY BENIGN ALUMINUM ALLOY SURFACE PRETREATMENT PROCESS FOR ADHESIVE BONDING

H. Zheng, C.J. Chu, C.J. Yang and M. Danron
CHEMAT TECHNOLOGY, INC.
19365 Business Center Drive, #9
Northridge, CA 91324

ABSTRACT

The wide application of aluminum alloys as structure materials requires the surface pretreatment for bonding and corrosion resistance. Traditional surface pretreatment of aluminum alloys involves usage of hazardous chemicals such as hexvalent chromium and large amount of water. In this research, an environmentally benign aluminum alloys surface pretreatment process has been developed. The new process produces high quality of the pretreated surface which has strong adhesive bonding and good corrosion resistance. In addition to the benefit for the environment, the new process is simple, low-cost and energy-saving.

KEY WORDS: Aluminum Alloys, Adhesive Bondings, Surface Pretreatment, Phosphoric Acid Anodizing(PAA), Water-Based

1. INTRODUCTION

Aluminum alloys have been widely used in all industries due to their lightness and high strength as well as corrosion resistance. One of the largest markets for aluminum alloys is in aerospace industry, and a wide application in automobile industry is expected in a few years. In these applications, aluminum alloys have been adhesively bonded. The surface of aluminum alloys has to be pretreated for good corrosion-resistance and adhesion.

An adhesive-bonded joint (Fig.1) between aluminum alloy parts can be considered as a chain consisting of a number of individual links, which include: (1) aluminum alloys (adhered); (2) adhesives; (3) primer; and (4) oxide layer (aluminum oxide) on an aluminum alloy substrate. In order to enhance the bonding strength and corrosion-resistance, the surface of aluminum alloys are pretreated. State-of-the-art processes for pretreatment of aluminum alloys for adhesive bonding or for corrosion protection involve chemically grown hydrated oxide (anodizing) and the use of chromate corrosion inhibitors.

The present practical processes include chromic acid anodizing (CAA), phosphoric acid anodizing (PAA), sulfuric acid anodizing (SAA), and oxalic acid anodizing processes. However, these processes suffer from the use of large amounts of water and power and the toxicity of chromates. Alternatively, it is possible to pretreat the surface of aluminum alloys using ceramic inorganic coatings where the necessary thickness, strength, adhesion and morphology can function as an adhesive bonding on corrosion protection layers. The sol-gel process is well known for its ability to produce tailorable oxide microstructure (e.g. pore size and pore volume). In addition, the sol-gel process has the added advantage of being environmentally benign, as well as being capable to coat large area on a mass production scale at low cost without the use of toxic chemicals and large amounts of water and power.

In this work, a process to deposit active organic modified alumina coatings on aluminum alloys as a surface pretreatment was developed. Aluminum oxide based coatings were deposited onto the surface of aluminum alloys via the aqueous-based sol-gel process. Both dip-coating and spray-coating processes have been used to deposit the coatings. By bonding with primers and adhesives, the jointed aluminum alloy coupons possess high shear strength (>40 MPa) with the same values as those samples prepared by commercial PAA processes. Sol-gel derived coatings have also showed good thermodynamic and hydrolytic stability as well as corrosion resistance.

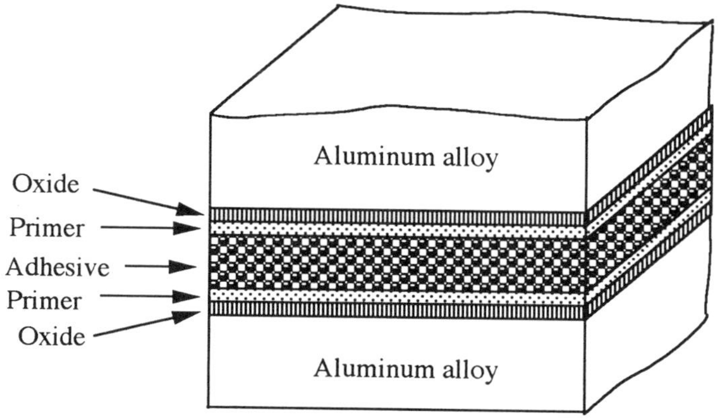

Fig. 1 Composite structure of typical aluminum-to-aluminum bonded joint

2. EXPERIMENTAL

2.1 Environmentally Benign Coating Solution The AL9201 coating solution is a water based, stable and thermally active colloidal solution made up of aluminum silicate. The shelf life of the solution is over one year. The pH value of the solution is about 10.5 and the viscosity is 2 centipoise. The following is a summary of the properties of AL9201 coating solution.

Table 1. The properties of AL9201 Coating Solution

Ingredient	Alumina, Organosilane
Physical State	Aqueous Colloidal Solution
Appearance	Milky White
Shelf Life	Over One Year
Specific Density	1.0
Oxide Content	5 wt%
Viscosity	2.0 centipoise
pH	10.5

2.2 Deposition of the Coatings and Bonding of the Aluminum Alloy Coupons The aluminum alloys 2024-T3 were cleaned by a standard cleaning procedure shown in Fig. 2. The cleaned aluminum alloys were coated in AL9201 solution by either dip-coating or spray-coating methods. Then, the coating was dried and cured for one hour. This was followed by applying BR127 (or BR6757) primer which was cured at 120°C for 30 minutes, the coated samples were bonded with FM73 film adhesives. The bonded coupon was cured at 120°C for one hour under 0.276 MPa(40 psi). The resultant coupon was subjected to various tests.

3. RESULTS

3.1 Simple Deposition Techniques The AL9201 coating solution can be applied to aluminum alloys surfaces by simple deposition techniques: dip-coating and spray-coating. Fig. 3 shows the wedge crack tests on the samples with either dip-coating or spray-coating. The initial crack lengths are about 2.90 cm and 2.95 cm for dip-coating and spray-coating, respectively, and all samples have cohesive failure. The crack growth is less than 0.25 inch, which is the limit of the crack growth length. The bonding strength of these two types of samples are around 41 MPa. It can be seen that both simple techniques produce strong bonding and good corrosion resistance. Fig. 4 is a photograph of a regular spray gun used here for deposition of the coating.

3.2 Thin and Homogeneous Coating The coatings resulting from either dip-coating or spray-coating are thin with thickness between 0.2 - 0.8 μm (Fig. 5). The coating does not have macropore, which is completely different from the morphology of alumina layers resulting from anodization. The Auger spectroscopy analysis of the coatings indicate the homogeneity of the composition in the whole coating.

3.3 High Mechanical Strength Fig. 6 shows the dependence of bonding strength of the bonded aluminum alloy coupons on withdrawing speed (i.e., coating thickness). All samples have cohesive failure and lap shear strength above 38.5 MPa. It is clear that the strength slightly decreases with the increase in the withdrawing speed. When the withdrawing speed is below 10 cm/min, the bonding strength is almost constant at 40.7 MPa.

3.4 Room Temperature Pretreatment The curing conditions: time and temperature have an effect in bonding strength, as shown in Fig. 7. At no curing, the lap shear strength is still strong (above 38.5 MPa) and the failure mode is cohesive. With increasing curing time, the strength increases and then decreases again. However, the lap shear strength is

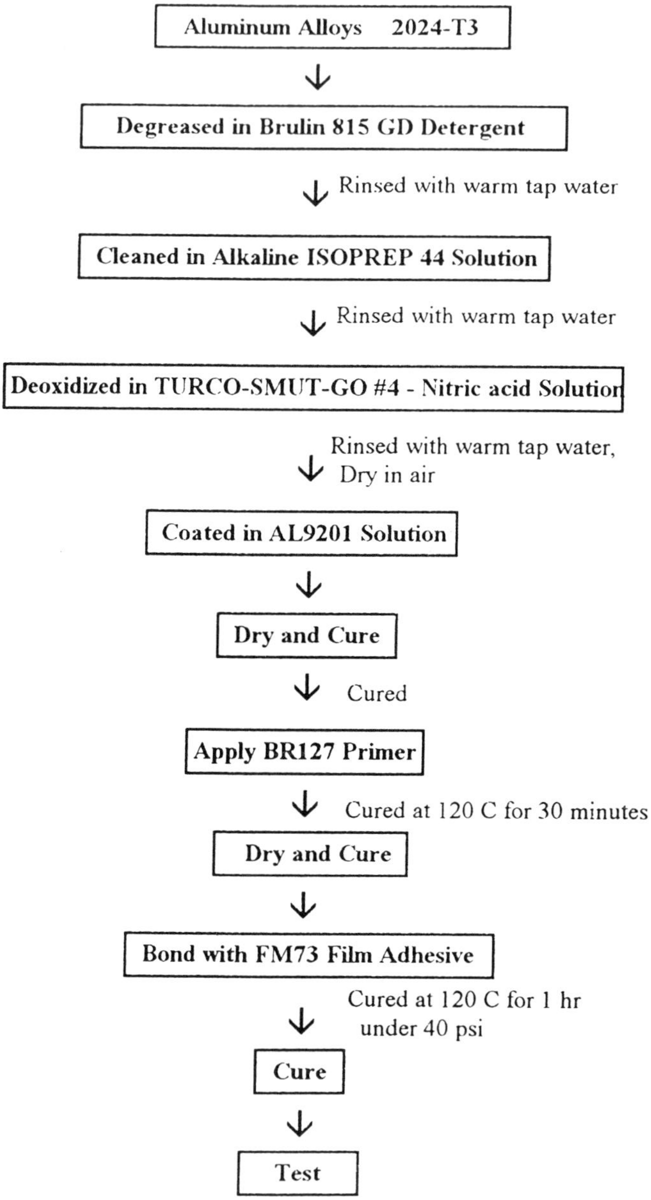

Fig. 2 The Surface Pretreatment and Bonding Procedure for Aluminum Alloys

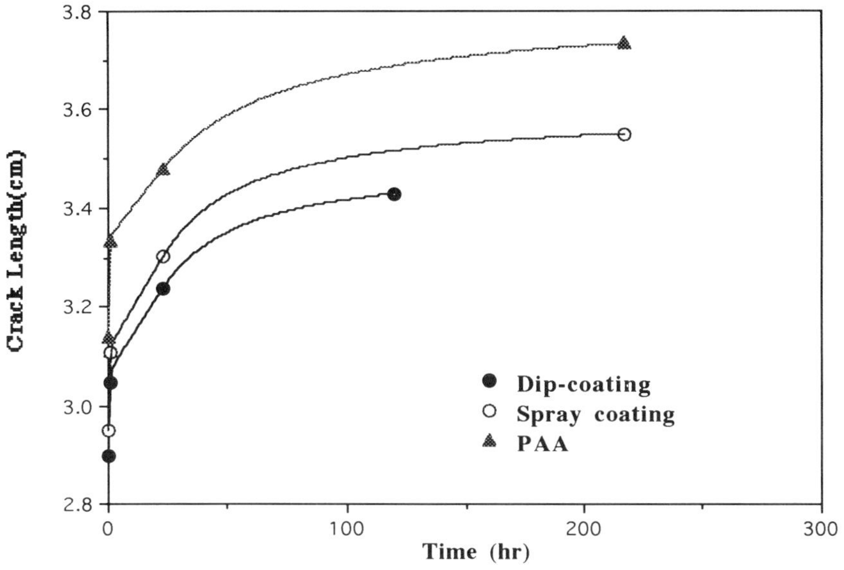

Fig. 3 The wedge crack tests on the samples with either dip-coating or spray- coating

Fig. 4 The portable spray gun for depositing the coating

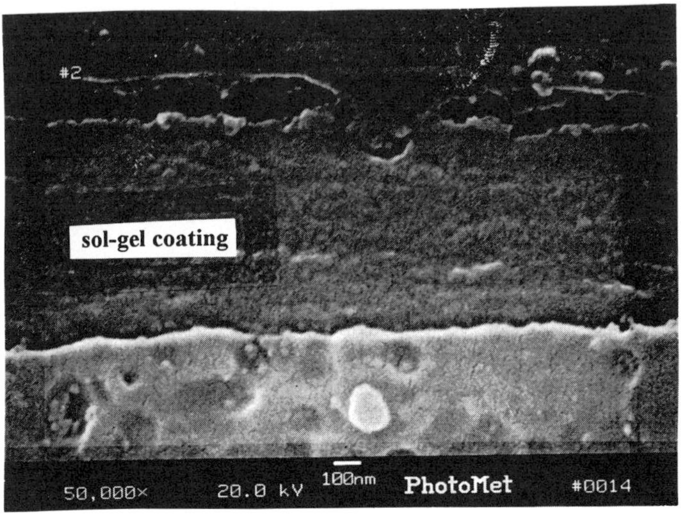

Fig. 5 Cross Section of the Coating

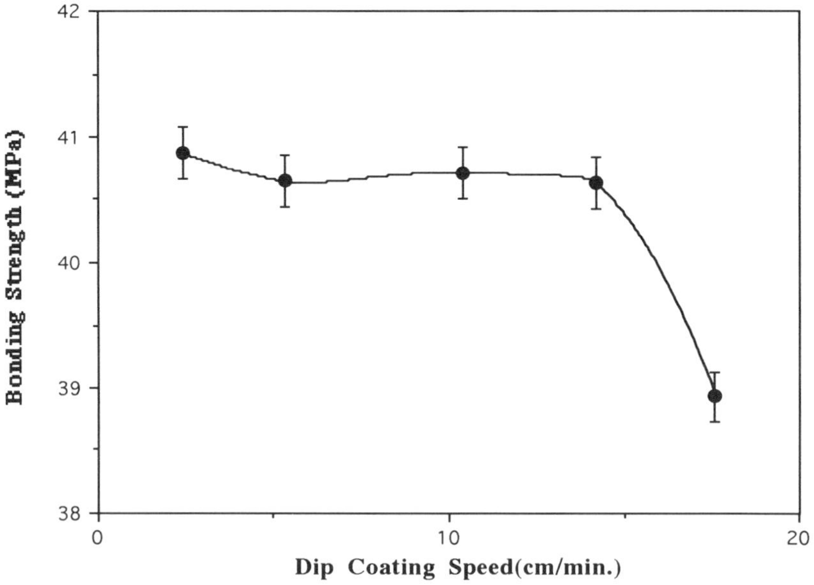

Fig. 6 The Bonding Strength vs. Withdrawing Speed

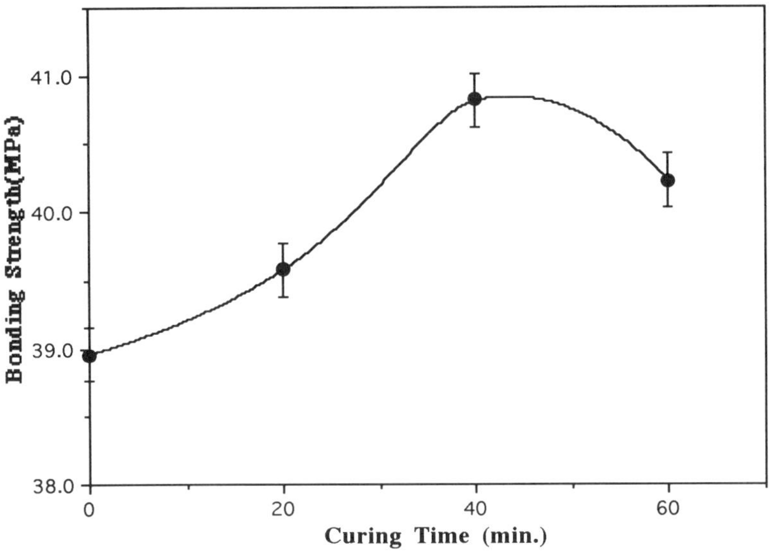

(A)

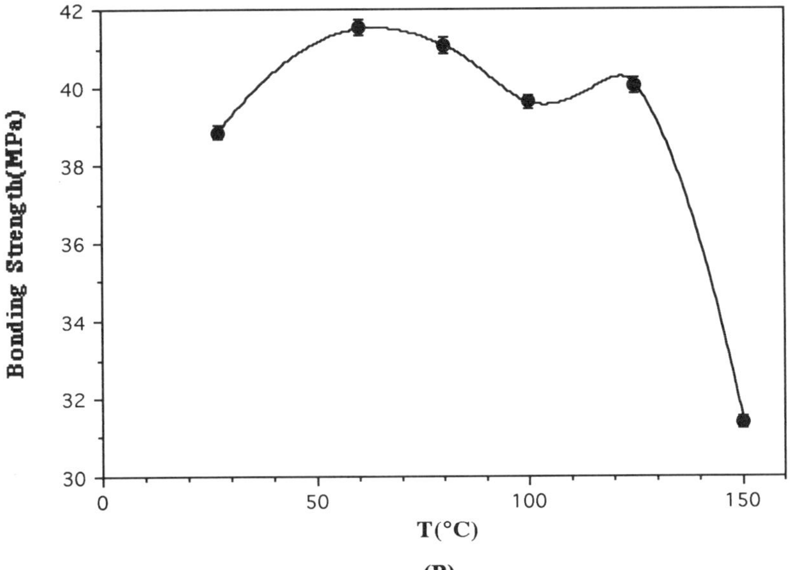

(B)

Fig. 7 The effect of curing conditions: time (A) and temperature (B) on the bonding strength of the joint

over 38.5 MPa within one hour of curing. When curing at room temperature, the mechanical strength is about 38 Mpa. The mechanical strength is the same or higher when curing temperature is below 125°C, and drops to below 31 MPa (but still cohesive failure) when curing temperature is 150°C.

3.5 Good Chemical Durability of the Joints Wedge crack test was performed in a humidity chamber with 95% humidity. Fig. 8 shows the crack growth of the bonded aluminum alloy coupons with different withdrawing speed. It can be seen that all crack growth is below 0.635 cm(0.25 inch.) after 24 hours at 60°C in 95% humidity. Long term environmental durability was also performed as in Fig. 9. The samples with active organic modified alumina coating showed smaller initial crack length then the PAA control samples with similar crack growth up to 35 days. The failure mode is generally cohesive (Fig. 10) even up to 35 days exposure, which means this active alumina coating can provide excellent environmental durability..

4. SUMMARY

It has been shown that active organic modified aluminum coatings can be deposited on aluminum alloys via the aqueous-based sol-gel solution AL9201. Both dip-coating and spray-coating processes have been used to deposit the coatings. By bonding with primers and adhesives, the jointed aluminum alloy coupons possess high shear strength (> 40 MPa) with the same values competitive to those prepared by commercial PAA processes. Sol-gel derived coatings have also showed good thermodynamic and hydrolytic stability as well as corrosion resistance. But most importantly, the sol-gel derived aluminum surface pretreatment process does not require large amounts of water and electrical power, and it is environmentally benign.

ACKNOWLEDGEMENT

The authors are grateful to the Air Force Wright Laboratory (AFMC), for support of this work under contract number F33615-94-C-5606. We would also like to thank James J. Mazza for technical assistance and helpful discussions.

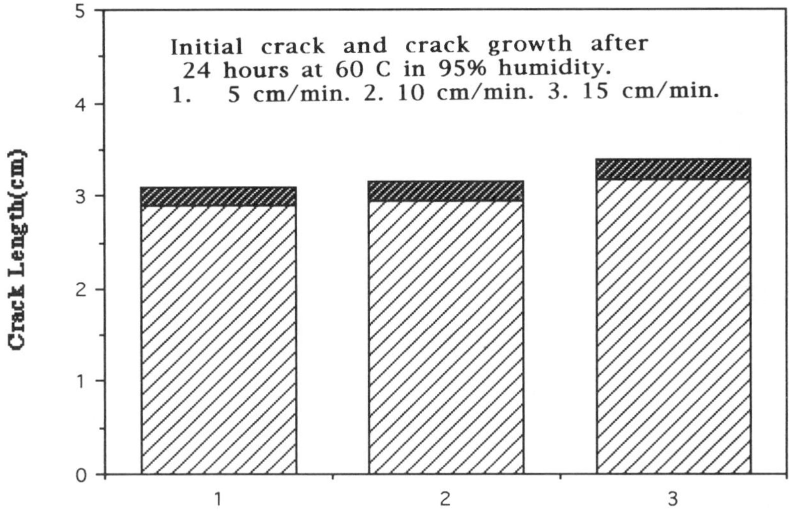

Fig. 8 The initial crack length and crack growth of the bonded aluminum coupons with different withdrawing speed

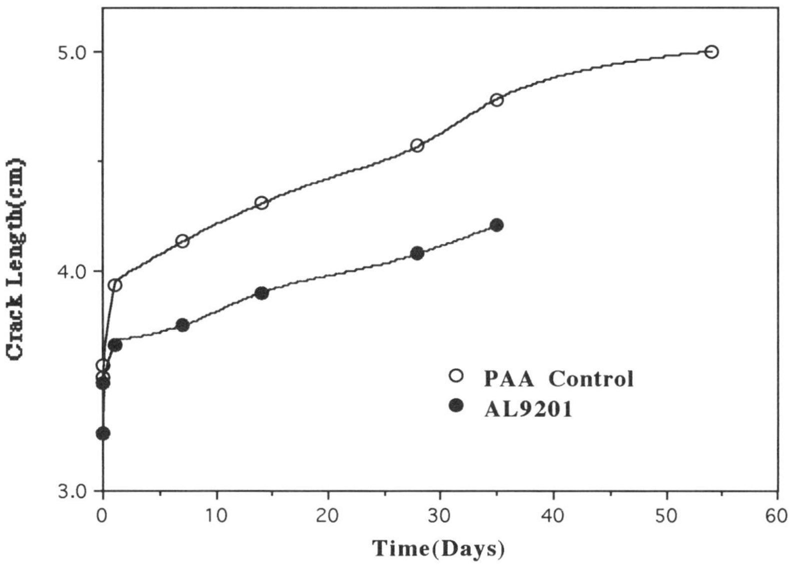

Fig. 9 Long term wedge crack test at 60°C in 95% humidity of bonded aluminum alloys with active alumina coating and PAA control samples.

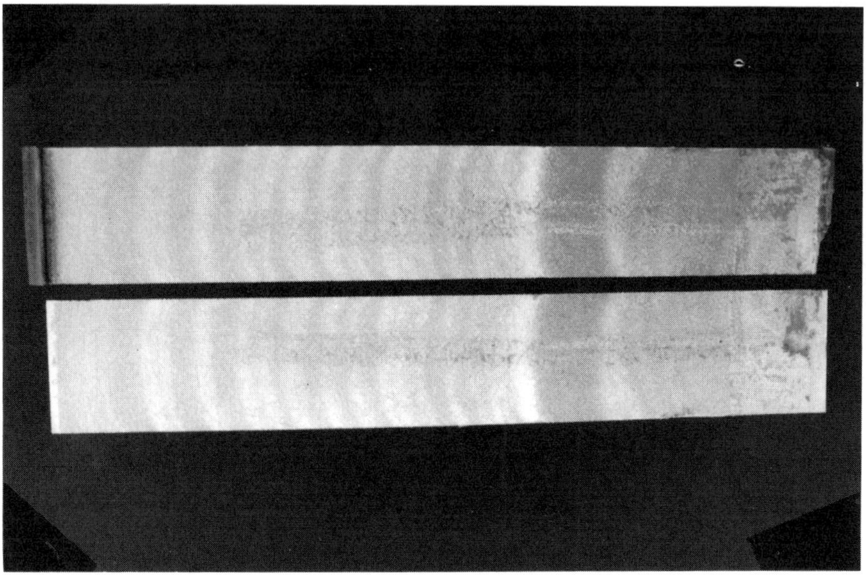

Fig. 10 A photo of fracture interface of the bonded aluminum alloy coupon showing the cohesive failure mode even after 35 days wedge crack test at 60°C in 95% humidity.

28th International SAMPE Technical Conference
November 4-7, 1996

A NON-CHEMICAL SURFACE TREATMENT FOR ADHESIVE BONDING OF AIRCRAFT ALUMINUM ALLOYS

Gerhardus H. Koch, and Gary L. Todd
CC Technologies Laboratories, Inc.
Dublin, Ohio

and

Arnold H. Deutchman
BeamAlloy Corporation
Dublin, Ohio

ABSTRACT

An environmentally friendly, non-chemical surface treatment has been developed to replace phosphoric acid anodizing (PAA) as pretreatment for adhesive bonding of aircraft aluminum alloys. This alternative surface treatment is based on ion beam enhanced deposition (IBED) of aluminum oxide onto the aluminum alloy surface. Results of experiments will be presented, demonstrating that the IBED surface treatment yields high bond strength and durability, comparable to those achieved with the traditional processes.

KEY WORDS: Adhesive Bonding, Anodizing, Ion Beam Enhanced Deposition

1. INTRODUCTION

Adhesively bonded structures are being used extensively as structural components for both military and civilian aircraft. It is essential to achieve a good bonding surface which will be the basis for high strength adhesive bonds with good durability in aggressive operating environments. Surface treatments based on wet chemical processes have been shown to create excellent surfaces for adhesive bonding and coating application, and are widely used throughout the industry. The most common surface treatment as pretreatment for adhesive bonding in the United States, is the phosphoric acid anodizing (PAA) process, forming a γ-Al_2O_3 film, which consists of a thin dense barrier film and a porous top film with a thickness of about 4 X 10^{-7} m (4,000 Å). The morphology of the micropores of the PAA film lends itself to excellent mechanical bonding (1). However, primers need to be applied prior to adhesive bonding to increase durability to an acceptable level.

Environmental Protection Agency (EPA) and Occupational Safety and Health Administration (OSHA) regulations have imposed increasingly strict limitations on the use of wet chemical surface preparation processes. Particularly, surface treatment processes such as pickling and anodizing, which rely on wet chemistry techniques and are large water users, are subject to regulation. These processes eventually contribute to the waste water, which must be treated. Of specific concern to the environment are acids and those chemicals which contain toxic heavy metals such as chromium. Moreover, many of the primers which are presently in use are solvent based, and are also subject to increasingly strict EPA and OSHA regulations.

In order to find replacements for the wet chemical surface treatment techniques, the Materials Directorate at Wright Laboratories (WL/MLSE) initiated a research program to develop alternative non-chemical techniques that do not produce waste, and are not detrimental to health and environment. An important requirement for such a technique is that it produces surfaces which result in mechanical strength and durability equal to or better than those produced by the wet chemical methods. In order to achieve the objective of this program, a non-chemical, physical technique based on ion assisted coating was applied.

2. EXPERIMENTAL APPROACH

In order to assess the mechanical strength and durability of adhesive bonds with the alternative surface treatment, peel and wedge testing were performed. In addition to these mechanical tests, various surface analytical techniques were used to aid in the understanding of the bonding mechanism. The surface treatments which were evaluated, include ion beam enhanced deposition (IBED), grit blasting, and phosphoric acid anodizing (PAA). The latter treatment served as bench mark for the non-chemical treatment.

2.1 Grit Blasting Grit blasting of the aluminum alloy 2024-T3 test panels was conducted with 50 µm Al_2O_3 grit at a gas pressure of 0.35 - 0.40 MPa (50-60 psi). The carrier gas was nitrogen (N_2). The nozzle was hand held at an angle of approximately 45 degrees. Prior to and after grit blasting, the surfaces were solvent cleaned.

2.2 Ion Beam Enhanced Deposition The oxide films formed on the surfaces of aluminum alloy 2024-T3 must satisfy three criteria in order to promote strong and durable adhesive bonding. These criteria are:

1) The oxide films must adhere well to the aluminum alloy substrate,
2) The oxide must be water and corrosion-resistant, and
3) The outermost layer of the surface film must have a morphology which promotes adhesive bonding.

Ion beam enhanced deposition is a process that combines a conventional thin film deposition technique such as vacuum evaporation or ion beam sputtering with the bombardment of the growing film by a secondary, high energy flux (2). A secondary ion flux, usually of an inert species such as Argon gas, is incorporated into the deposition process for two reasons. The secondary ion beam is used to mix the initial layers of the deposited film into the substrate surface to improve adhesion of the film to the substrate. The secondary ion beam is also used to control the morphology of the film as it grows from the substrate surface.

Ion beam enhanced deposition is carried out in a high vacuum, at pressures of 1.333×10^{-4} Pa (1×10^{-6} Torr) or below. With the proper choice of deposition parameters, the temperature of the parts being processed can be held below 50°C (120°F). A general diagram of the IBED process as implemented to deposit coatings is shown in Figure 1. The surface to be coated is first illuminated with a flux of high energy inert gas ions to remove surface oxides and contaminants.

This high energy flux is maintained, and once the surface is cleaned, a flux of filming atoms is directed simultaneously at the surface to be coated. The high energy inert gas ions are used to mix the initial few atom layers of the specific species into the surface being coated. This forms a mechanically alloyed bond layer in the surface that promotes excellent adhesion of the coating to the substrate.

Once the interfacial layer is formed properly, the film or coating is allowed to grow, see Figure 2. The high energy inert gas ion flux is used to control the morphology of the coating. This allows control over the grain structure of the film, as well as the film density and residual stresses. Conversely, the energy of the secondary beam can be reduced, and even be cut off to allow natural growth of the surface film.

Thus, the IBED process is used to deposit controlled-structure oxide films onto aluminum alloy surfaces. Using the IBED process, the oxide is initially mixed into the aluminum alloy surface for good oxide-substrate adhesion, and then grown in a manner such that the morphology of the oxide is highly controllable. By controlling the grain structure of the oxide film, bondability and corrosion resistance can be optimized.

2.3 Adhesive Bonding Immediately following the different surface treatments, the surfaces were primed with the epoxy polyamide primer BR 127 (Manufactured by Cyanimid), and were bonded with the epoxy polyamide adhesive FM 73. Per manufacturer's instructions, the adhesive bonding was done at a pressure of 0.276 MPa (40 psi) and at a temperature of 120 °C (250°F). Adhesive bonding was accomplished over a period of 75 minutes.

2.4 Peel Testing The floating roller peel test (ASTM D 3167) (3) was used to determine the relative peel resistance of adhesive bonds between a rigid and a flexible adherend. Due to the nature of the specimen configuration and loading mode, the highest stresses are exerted on the interface between the flexible adherend and the adhesive, which makes this test well suited for the evaluation of the quality of surface treatments. For the present tests, the flexible adherend was 0.635 mm (0.025 inch) thick 2024-T3 sheet with the various surface treatments, and the rigid sheet was kept the same for all tests, namely, 1.6 mm (0.063 inch) thick 2024-T3 sheet with a HNO_3-HF pickling treatment.

2.5 Wedge Testing A simple test method to determine the durability of bond lines stressed under Mode I condition, is the Boeing Wedge Test, which is also an ASTM standard test (ASTM D 3762) (4). The self stressing is induced by the insertion of a wedge, which creates the initial crack length. When the crack propagates as a result of exposure to the corrosive environment, the effective load or stress at the crack tip decreases. The decrease in effective load provides a self arrest capability for the wedge test, which enables the establishment of a threshold level. The environment to which the wedge test coupons were exposed, was a 60°C (140 °F) 98 %RH water vapor environment.

2.6 Microscopic Analyses In order to gain a better understanding of the mechanism of adhesive bonding to IBED treated surfaces, the surfaces were analyzed with high resolution scanning electron microscopy (SEM), and transmission electron microscopy (TEM).

3. RESULTS

The results of this study are divided into three separate sections: 1) mechanical strength, 2) durability, and 3) electron microscopy.

3.1 Mechanical Strength The peel strengths of IBED treated coupons were compared with that of anodized (PAA) coupons. Figure 3 shows that the typical peel strength of PAA specimens is in the range of 1.2×10^7 - 1.6×10^7 N/m (70-90 lbs/inch). These values are well above the minimum peel strength of 1×10^7 N/m (60 lbs/inch) specified by the manufacturer of the FM 73 adhesive.

When an Al_2O_3 oxide film thickness of approximately 4×10^{-7} m (4,000 Å) was deposited onto the aluminum alloy surface with IBED, the resultant peel strength varied significantly, and was generally low, see Figure 4. However, when the IBED film was allowed to continue to grow naturally, without the presence of the Argon augmenting beam, where the IBED barrier film thickness was 1×10^{-7} m (1,000 Å) and the grown top layer was about 3×10^{-7} m (3,000 Å), the peel strength could reach a value as high as that of the anodized specimens.

3.2 Durability In addition to the mechanical strength of adhesive bonds, the durability of the bonds in aggressive environments is of great concern. Water-containing environments are particularly detrimental to bond lines, and may cause unzipping of the bond at relatively low loads. Wedge panels with different surface treatments were exposed to a 60°C (140°F) 98% RH water vapor environment. The initial durability tests were conducted on IBED treated specimens using the process, which yielded high mechanical strength, namely, a 1×10^{-7} m (1,000 Å) thick amorphous or micro-crystalline barrier layer, with a 3×10^{-7} m (3,000 Å) thick crystalline top coat. The results of the wedge tests indicated that the durability of the bond line between this surface oxide and the primer/adhesive was extremely poor. In fact, Figure 5 shows that the crack jumps ahead a significant distance as soon as the loaded wedge specimen is exposed to the humid environment. A photograph of the fracture surfaces of one of these specimens suggests that fracture occurred along the oxide-adhesive interface, virtually unzipping the bond, see Figure 6 By comparison, the crack growth in PAA wedge specimens was minimal, and no or little disbonding was observed between the adhesive/primer and the substrate.

In order to improve the durability of the bond, the aluminum alloy surface was roughened prior to the IBED treatment by grit blasting with fine grit Al_2O_3. Figure 7, which shows the wedge test results for the grit blasted surfaces, demonstrates a marked improvement over the smooth IBED surface. The figure shows the crack growth of wedge specimens with grit blasted surfaces and with or without primer. Clearly, the presence of the chromate containing primer has again a significantly beneficial effect on the durability of the bond line.

Although marked improvement in durability of adhesive bonds could be achieved by promoting mechanical bonding, the objective of obtaining the consistency in durability of PAA treated aluminum alloy adhesive bonds was not achieved with the above described surface treatments. Since aluminum oxide is known to be very hygroscopic, it was therefore suspected that water would readily adsorb from the environment onto the surface. If not removed, a thin film of water could be trapped between the oxide film and the primer, and result in a weak interface. In order to achieve consistently good durability of the bond line, the surface water film needs to be removed either by solvent wiping or boiling off.

The reproducibility of the water removal process was evaluated with the post-IBED oven treatment. Figure 8 demonstrates that the IBED treatment over a grit blasted surface followed by 30 minutes in 110 °C (230°F) air, results in a strong and durable bond line. Moreover, specimens which were exposed to boiling water for 5 minutes, and were subsequently oven dried at 110 °C (230°F), demonstrated good durability as well.

3.3 Electron Microscopy In order to gain better insight into the mechanism of adhesion, surface analyses using various techniques were conducted. High resolution, high magnification scanning electron microscopy demonstrated the extreme smoothness of an IBED Al_2O_3 surface, see Figure 9. Electron diffraction of the surface oxide deposited by IBED confirmed that the structure was α-Al_2O_3. Electron diffraction patterns of the IBED + grown oxide, showed a mixture of amorphous or micro-crystalline and crystalline aluminum oxide structure, confirming an initial assumption that the IBED deposited barrier film was amorphous α-Al_2O_3, and the top film, which was grown naturally, was crystalline α-Al_2O_3.

4. DISCUSSION AND CONCLUSIONS

The results of this work have demonstrated that IBED can be used as a non-chemical alternative to PAA as pretreatment for adhesive bonding of aluminum alloy 2024-T3. The surface was designed such that adhesive bonding was achieved with a combination of mechanical and physical bonding. It was first shown that a smooth α-Al_2O_3, as obtained with the IBED process on as-received aluminum alloy surface, had very poor durability, despite high initial mechanical strength. The bond strength of these surface could be entirely attributed to physical or van der Waals forces.

Although the physical bonding is considered to be the weakest type of bonding, it is important in the development of adhesion. The van der Waals forces influence the degree of physical entanglement and surface wetness. Gledhill and Kinloch (5,6) developed an equation that relates the thermodynamic work of adhesion to the surface free energy:

$$W_A = \gamma_a + \gamma_b - \gamma_{ab}, \quad (1)$$

where W_A is the thermodynamic work of adhesion required to separate unit areas of two phases forming an interface, γ_a and γ_b are the surface free energies of the two phases, and γ_{ab} is the interfacial free energy. This relationship was used to explain the poor durability characteristics of the Van der Waals type bond. A positive value of W_A indicates that the interface is thermodynamically stable, whereas a negative value indicates an instable interface, where the primer or adhesive spontaneously separate from the substrate. Thus, the value of W_A can predict the stability of the interface in a wet environment. In the presence of water, the work of adhesion (W_A) can become negative. The change from positive to negative work of adhesion creates the driving force for displacement of the adhesive by water. This mode of failure by spontaneous dissociation would only occur if the bond relied solely on dispersive forces as is the case for microscopically smooth surfaces.

Another way that water can degrade the bond strength, is through hydration of the metal oxide. The typical natural or anodized amorphous γ-Al_2O_3 is readily converted to a hydrated oxide or hydroxide upon exposure to water. The composition of the hydrated oxides will range from α-$Al_2O_3 \cdot H_2O$ (boehmite) and β-$Al_2O_3 \cdot 2H_2O$ (pseudoboehmite). Analyses of failed adhesive bonds have indicated that the hydroxide is usually attached to the adhesive suggesting that the adhesion between the metal and the hydroxide is very weak (7). Brockman and coworkers (8,9) used ultramicrotomic techniques to determine the exact failure path. They observed essentially three modes of deterioration. The first is a mainly reversible weakening effect in the primer or adhesive layer near the metal oxide surface, the second is a slow transformation of the oxide by hydration and a diffusion of hydrated oxide into the polymer, and the third is the fast transformation of the oxide due to corrosion.

The durability of an adhesive joint can be improved by preventing water to enter the bond line or by making the bond line more resistant to hydration. This can be accomplished by increasing the barrier to water diffusion into the bond line, by providing an arduous path for water diffusion, by inhibiting hydration of the surface oxide, or by applying primers which contain corrosion inhibitors or coupling agents.

An obvious approach to increase the barrier to water diffusion is to select an adhesive which has low permeability and diffusivity, or to incorporate inert fillers into the adhesive, which can impede water diffusion. This approach is not often used, since by modifying polymers to decrease permeability and diffusivity, other important properties such as wettability, mechanical strength or toughness may be decreased. Other means of inhibiting water diffusion into the bond line is to apply a bead of sealant on the outer edge of the bond line.

Often, primers are applied to the adherend surface prior to adhesive bonding to improve the durability of the bond. The most common and most effective primers are chromate containing primers. The chromate in the primer is primarily present as Cr^{6+} (hexavalent chromium), which is initially hydrophilic, but becomes after 24 hours or heating to 50 °C (120 F) hydrophobic. It has been a general consensus that Cr^{6+} is the active species in chromate containing coatings, where corrosion inhibition is accomplished by reduction of Cr^{6+} to Cr^{3+}.

Based on the results of the experimental work, a process based on IBED was developed to prepare aluminum alloy surfaces for structural adhesive bonding. The adhesive bonds based on this surface treatment were shown to have mechanical strength and durability equal to or better than that based on the wet chemical PAA process. A general description of the surface treatment process, which is described in more detail elsewhere (10), is as follows:

1) Clean surface with degreasing medium,
2) Grit blast with 50 µm Al_2O_3 grit,
3) Clean again,
4) Sputter clean the surface,
5) Apply a 1×10^{-7} m (1,000 Å) thick α-Al_2O_3 barrier film by using the IBED process,
6) Allow a 3×10^{-7} m (3,000 Å) thick top layer of α-Al_2O_3 to grow,
7) Following oxide growth, backfill the vacuum chamber with dry N_2 gas,
8) Immediately prime or,
9) Oven dry at 120 °C and then prime.

Although the above described process was shown to provide surfaces for strong and durable adhesive bonds, further work is required to determine the engineering as well as economic parameters for scale up.

5. ACKNOWLEDGEMENT

This work was performed under U.S. Air Force Contract No. F33615-93-C-5323, which was administered under the technical direction of Mr. Theodore J. Reinhart and Mr. James Mazza, WL/MLSE. The effort was partially funded by the Strategic Environmental Research and Development Program (SERDP).

6. REFERENCES

1. J. D. Venables, et. al., "Oxide Morphologies on Aluminum Prepared for Adhesive Bonding Applications," Surface Science, Vol 3:1, 1979, p. 88.

2. J. A. Thorton, Coating Deposition by Sputtering Deposition Technologies for Films and Coating, R. F. Brunshah ed., Noyes Publication, New Jersey, 1988.

3. "Standard Test Method for Floating Roller Peel Resistance of Adhesives," ASTM D3167-76 (1986).

4. "Standard Test Method for Adhesive Bonded Surface Durability of Aluminum (Wedge Test)," ASTM D3762-79 (1988).

5. R. A. Gledhill and A. J. Kinloch, "Environmental Failure of Structural Adhesive Joints," Adhesion, 1974, Vol. 6, p. 315.

6. A. J. Kinloch, "Interfacial Fracture Mechanical Aspects of Adhesion Bonded Joints - Review," J. Adhesion, Vol. 10, 1979, p. 193.

7. J. D. Venables, "Review - Adhesion and Durablility of Metal Polymer Bonds," J. Material Science, Vol. 14, 1984, p. 243.

8. W. Brockmann, "Durability of Adhesion Between Metals and Polymers," Adhesion, 1989, Vol. 29, p. 53.

9. O. D. Hennemann and W. Brockmann, "Weak Boundary Zones in Metal Bonds," 14th National SAMPE Technical Conference, p. 302 (1982).

10. G. H. Koch and A. Deutchman, "Non-Chemical Surface Treatment for Aluminum Alloys," Contract No. F33615-93-C-5323, WL/MLSE, 1996.

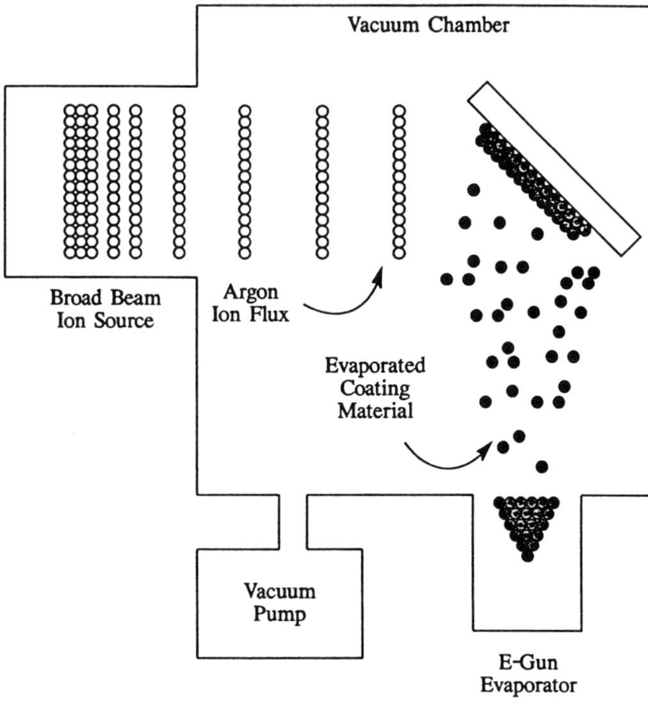

Figure 1. Schematic Drawing Of Ion Beam Enhanced Deposition.

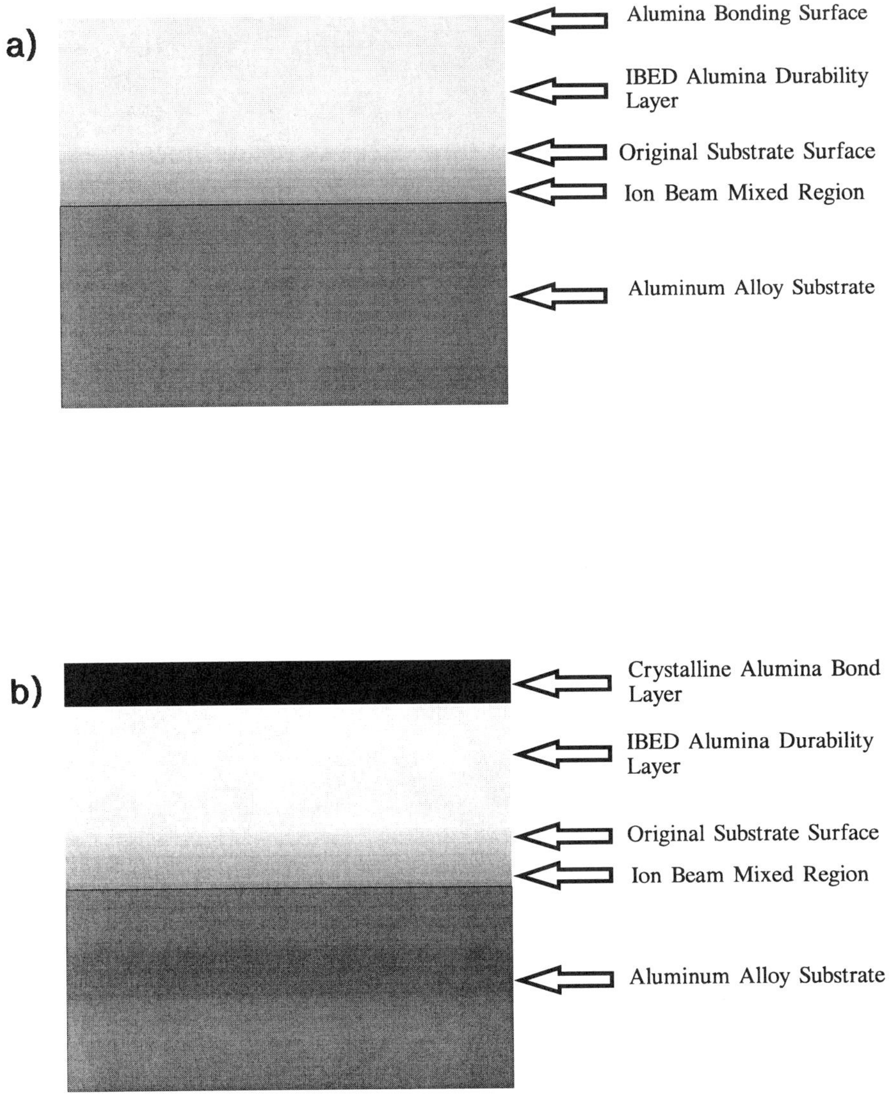

Figure 2. Schematic Diagram Showing Crystalline Amorphous (a) and (b) Bonding Surface, Resulting From Outer Oxide Growth With And Without The Influence Of The Augmenting Ion Beam, Respectively.

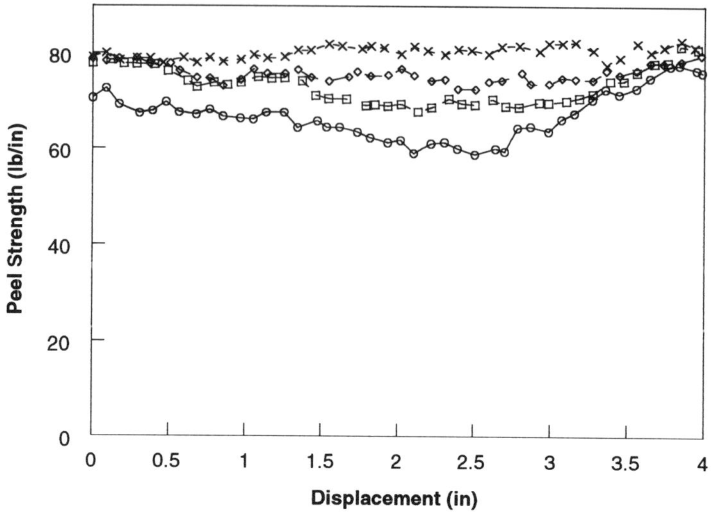

Figure 3. Peel Strength Diagram of Aluminum Alloy 2024-T3 With PAA Treated Surface.

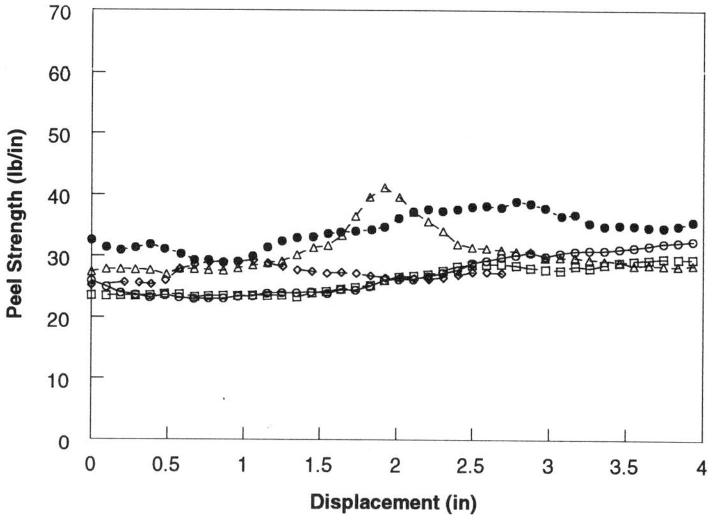

Figure 4. Peel Strength Diagram Of Aluminum Alloy 2024-T3 With A 5×10^{-7} m (5,000 Å) IBED Film Directly Deposited On Acetone - Methanoi Cleaned Surfaces.

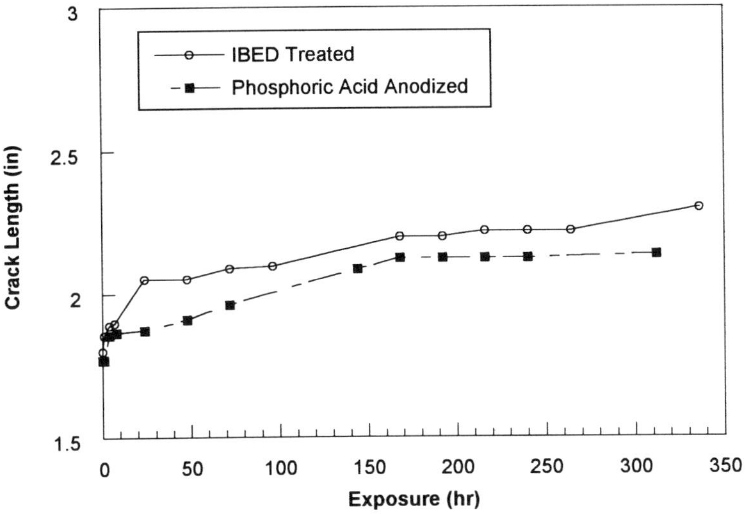

Figure 5. Crack Growth Data On IBED And PAA Treated Wedge Specimens Exposed To 60°C (140°F) 98% RH Water Vapor. Both Surfaces Were Primed Prior To Adhesive Bonding.

a)

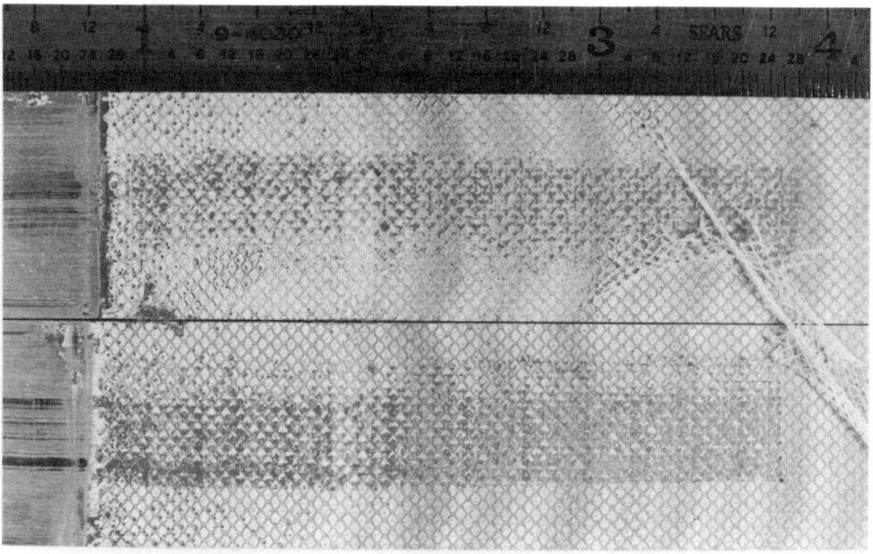

b)

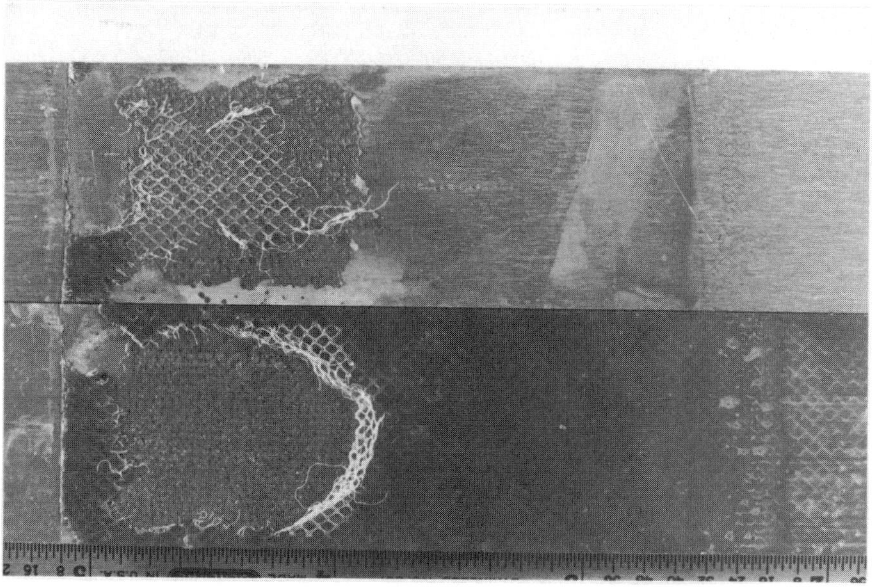

Figure 6. Photographs Of Fracture Surface Of PAA (a) and IBED (b) Treated Specimens (See Figure 5).

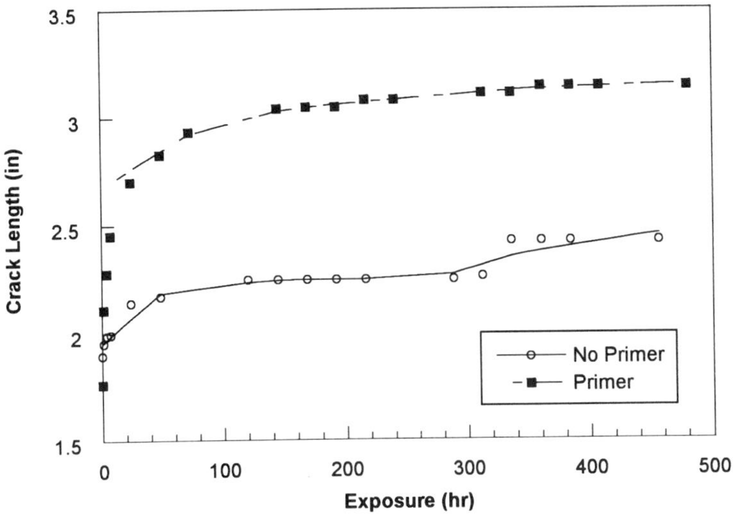

Figure 7. Crack Growth Data On Grit Blasted And IBED Treat Wedge Specimens, With And Without Primer. The Specimens Were Exposed To 60°C (140°F), 98% RH Water Vapor.

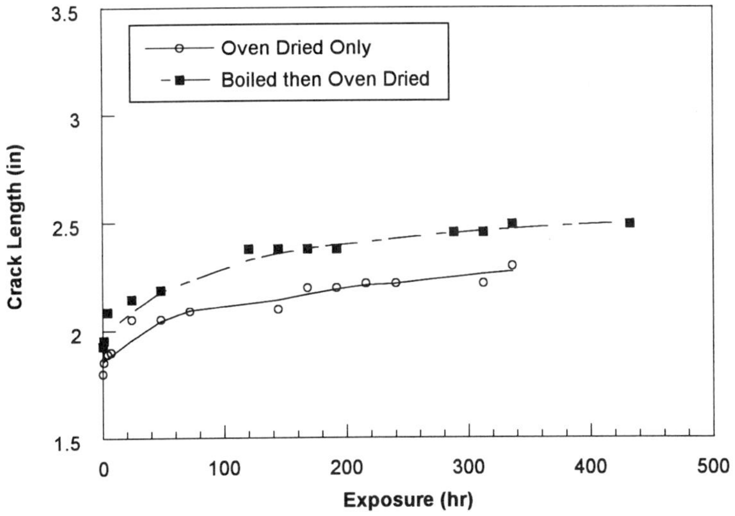

Figure 8. Crack Growth Data On Grit Blasted And IBED Treated Wedge Specimens Exposed To 60°C (140°F),98% RH Water Vapor. The Treated Surface Were Oven Dried At 120°C (250°F) And Exposed To Boiling Water And Then Oven Dried, Respectively, Before Priming.

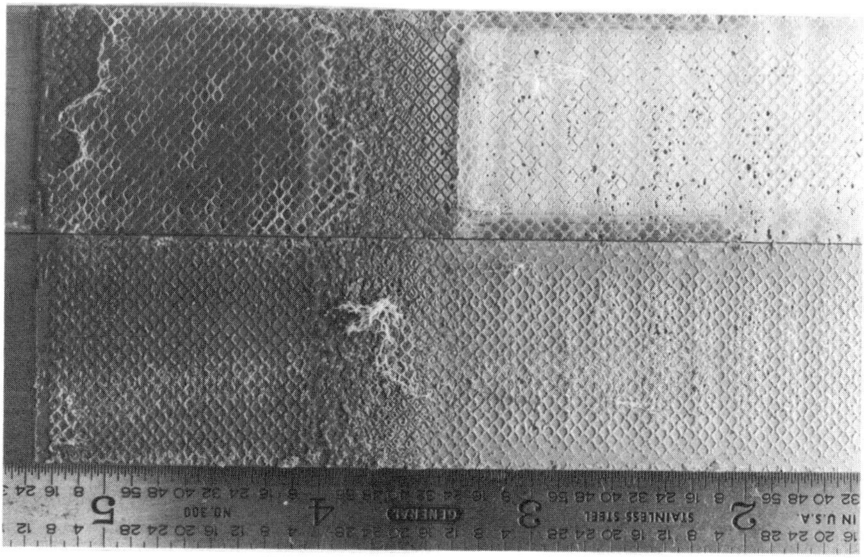

Figure 9. Photograph Of Grit Blasted, IBED Treated, And Oven Dried Wedge Specimens (See Figure 8).

Figure 10. Scanning Electron Micrograph Of IBED Treated Surface.

28th International SAMPE Technical Conference
November 4-7, 1996

NON-CHROMATIC/NON-CARCINOGENIC ETCHING OF TITANIUM ALLOY

Joseph A. Brescia, William S. De Piero, Robert B. Bonk
US Army Armament Research, Development and Engineering Center
Picatinny Arsenal, New Jersey

ABSTRACT

The objective of this effort was to investigate the feasibility of utilizing non-chromatic/non-carcinogenic surface preparation processes on commercially pure (CP) titanium substrates for structural adhesive bonding. Phase One of this effort consisted of a market survey, literature search, OSHA/EPA impact studies, and waste treatment cost analyses of potential non-chromatic alternatives to the CAA and Pasa-Jell processes. Based on Phase One results, the following processes were selected for laboratory evaluation: P2 etch anodization, sodium hydroxide (NaOH) anodization, and sodium hydroxide/hydrogen peroxide ($NaOH/H_2O_2$) anodization. In Phase Two of the study, CP titanium substrates treated with the three low toxicity ("lotox") processes were compared to the baseline CAA and Pasa-Jell 107 processes. Evaluations performed on a "fast track" iterative basis consisted of tensile lap shear strength tests at 24 °C (75 °F) and at 93 °C (200 °F, after soaking in water for 24 hours at 200 °F), and wedge crack extension tests. The data generated indicated that the NaOH anodization and the $NaOH/H_2O_2$ anodization processes exceeded the performance of the Pasa-Jell 107 process and were statistically equivalent to the CAA process. Results for several of the P2 anodization iterations appeared promising. However, due to sample size and data reproducibility, additional work is needed to fully determine optimum processing parameters for the P2 anodizing method.

KEY WORDS: Titanium, Surface Preparation

1.0 EXECUTIVE SUMMARY

1.1 *Introduction* The Polymers and Adhesives Branch (PAB) of the US Army Armament Research, Development and Engineering Center (ARDEC) and Concurrent Technologies Corporation (CTC) are investigating the feasibility of using a non-chrome acid etch, known as the P2 etch, to surface treat commercially pure (CP) titanium substrates for adhesive bonding. Phase One of this task consisted of a market survey, literature search, and technical evaluation of numerous low toxicity ("lotox") processes. The top three lotox alternatives were then selected for "fast track" testing studies in Phase Two of the project. This report discusses the results of Phase Two which compared the alternative processes to the baseline processes of chromic acid anodizing (CAA) and Pasa-Jell 107.

1.2 *Summary* Phase One of the study (reference 1) revealed the many regulatory advantages to the chromate-free systems, such as environmental, health, and safety benefits. It also analyzed the economic benefits, primarily from lower hazardous waste disposal costs, in switching to a P2 process. Tasks under Phase Two of the study included investigating the chemical and rheological effects of the lotox processes on titanium surfaces and the processing of titanium test coupons for subsequent bonding and testing on an iterative basis in an attempt to identify optimum lotox processing parameters. The benchmarking of the alternative processes to the baseline CAA and Pasa-Jell treatments consisted of ASTM D1002 lap shear strength tests at 24°C (75 °F) and 93 °C (200 °F) and wedge tests (ASTM D3762). These tests were utilized to demonstrate the feasibility to substitute any of the three alternatives for the baseline processes.

1.3 *Conclusions* The alternative processes have similar performance characteristics to the baseline processes while significantly reducing the effect on the environment. Chromium-free etching processes have properties similar to the Pasa-Jell and CAA processes. Elimination of chromium compounds from the bonding process will not significantly alter bond strengths.

2.0 PHASE ONE SUMMARY

The first phase of this project included a literature and market search to identify viable alternatives being explored in the military, industrial, and academic sectors. These alternative methods included anodizing, chemical etching, and other tank processes that are readily applicable to current bonding processes. The alternative techniques identified are currently available to DOD facilities manufacturing structurally-bonded titanium components.

2.1 *Regulatory Considerations* The regulations applying to surface treatments prior to titanium bonding originate from the Clean Water Act (CWA), the Resource

Conservation and Recovery Act (RCRA), the Clean Air Act (CAA), the Environmental Protection Agency (EPA) 33/50 program, the Emergency Planning and Community Right-to-Know Act (EPCRA), and the Occupational Safety and Health Administration (OSHA) Permissible Exposure Limit (PEL) guidelines for air contaminants. Chromium is not the only compound in the bonding process deemed hazardous. The alternative and control processes utilize other compounds considered hazardous such as hydrogen peroxide, nitric acid, and sulfuric acid. However, the alternative methods investigated are less of a concern than the two baseline methods. This is due to the lack of chromium in the alternative processes. The wastewater effluent limits of several compounds under the CWA are shown in Table 1 for reference.

Table 1. Clean Water Act Effluent Limits

Discharge	Direct Daily Maximum (mg/L)	Direct Monthly Average (mg/L)	Indirect Daily Maximum (mg/L)	Indirect Monthly Average (mg/L)
Chromium	**2.77**	**1.71**	**2.77**	**1.71**
Lead	0.69	0.43	0.69	0.43
Cyanide	1.20	0.65	1.20	0.65
Oils and Grease	52.00	26.00	52.00	26.00

Of the five treatment processes studied, there is a clear advantage for the alternative processes when regulatory compliance is considered. All three alternative methods have a favorable regulatory designation when compared with the chromatic processes. This can be best summarized by Table 2, showing the regulatory conditions for the five processes. The columns in the table reflect major areas of environmental, health, and safety laws and regulations including CWA, RCRA, the CAA's Hazardous Air Pollutant (HAP) regulations and Volatile Organic Compound (VOC) regulations, EPCRA, EPA 33/50, and OSHA PELs.

Table 2. Analysis of Regulatory Conditions

Process	CWA	RCRA	HAPs	VOCs	EPCRA	33/50	OSHA
Baseline							
CAA	3	3	3	1	2	3	3
Pasa-Jell 107	3	3	3	1	2	3	3
Alternative							
P2 Etch	4	4	1	1	2	1	2
NaOH	4	4	1	1	1	1	2
$NaOH/H_2O_2$	4	4	1	1	2	1	2

1 = favorable regulatory condition
2 = unfavorable regulatory condition
3 = most unfavorable regulatory condition
4 = regulated, but not as stringently as other substances

The EPA has targeted chromium, along with MEK and 1,1,1 trichloroethane, in their 33/50 Program which is designed to have industry voluntarily reduce the emissions of these compounds by 50 percent. Along with the 33/50 Program targeting chromate compounds, the reportable quantity from a spill is a mere 4.54 kilograms (10 pounds), while most of the other regulated compounds require a spill of near 454 kilograms (1,000 pounds) to be reportable. With this regulatory trend expected to continue in the future, it was this project's goal to find a suitable non-chromium containing replacement for current titanium pretreatment needs.

2.2 *Cost Effectiveness of Alternatives* The surveys and preliminary studies served to narrow the 21 alternatives for study down to 3 viable techniques for fast track testing. This group consisted of a P2 etch, which was developed by the ARDEC, Sodium Hydroxide Anodize, and a Sodium Hydroxide Hydrogen Peroxide Anodize. The mass balance analysis for estimating waste potential of these three processes and the baseline processes is summarized in Table 3. It is evident from the table that the cost differential between the baseline and alternative systems is not a key factor in choosing a pretreatment process. Acceptable physical properties determine the feasibility of switching processes to an alternative.

Table 3. Process Waste Analysis

Process	Chromium (kg/year)	Hazardous Wastewater (kg/year)	Hazardous Wastewater (m^3/year)	Hazardous Wastewater ($/year)	Hazardous Wastewater ($/$m^2$)
Baseline					
CAA	834	39,270	39.27	22,823	41.77
Pasa-Jell 107	427	19,371	19.37	11,258	54.36
Alternative					
P2 Etch	0	19,062	19.06	11,079	37.24
NaOH	0	19,062	19.06	11,079	37.24
$NaOH/H_2O_2$	0	19,283	18.83	10,944	37.03

3.0 PHASE TWO ANALYSIS

This section describes the processing of the panels and the physical testing accomplished under this task. The procedures for preparing the sample plates and how they were tested are discussed.

3.1 *Sample Preparation* Processing of the titanium panels was accomplished at CTC using laboratory-scale equipment. Each of the alternative and control processes were applied to CP grade titanium lap shear samples. All samples were first cleaned by wiping with acetone. The pretreatment steps were similar for all panels processed. The following describes the process sequence used in the preparation of the lap shear samples:

1. All sample plates were first degreased at 49 °C (120 °F) for ten minutes in an alkaline electrocleaner at a concentration of 60 to 120 grams per liter.

2. After rinse-off, the panels entered a pickling bath with target concentrations of 17 percent by weight HNO_3 and 3 percent by weight HF.

3. After rinsing, the panels were treated using either the baseline processes or the alternatives. The CAA panels were processed at an off-site facility, while the Pasa-Jell panels were processed by brush application in the CTC laboratory. The target concentrations for the alternative processes were:

   - NaOH (5.0 molar NaOH)
   - $NaOH/H_2O_2$ (0.5 molar NaOH and 0.2 molar H_2O_2)
   - P2 (20% vol. H_2SO_4 and 150 g/L $Fe_2(SO_4)_3$)

   A maximum of five test iterations was allowed for each process, with voltage and anodization time acting as the process variables. Table 4 summarizes these voltage and anodization time variations.

Table 4. Plating Conditions for Each Iteration

Process	Iteration 1	Iteration 2	Iteration 3	Iterations 4 and 5
Baseline				
CAA	10 V, 20 min.			
Pasa-Jell 107	10 V, 20 min.			
Alternative				
P2 Etch	10 V, 10 min.	15 V, 15 min.	10 V, 20 min.	14 V, 15 min.
NaOH	10 V, 20 min.	15 V, 20 min.		
$NaOH/H_2O_2$	10 V, 20 min.	15 V, 20 min.		

4. The panels were then rinsed, force-air dried, and baked at 60 °C (140 °F) for 30 minutes.

5. When the pretreatment process was complete, all of the panels were sprayed with EC-3960 adhesive primer within four hours, at an off-site facility. The panels were then air-dried at 24 °C (75 °F) for 30 minutes and baked at 121 °C (250 °F) for 60 minutes.

6. The panels were then packaged and shipped to ARDEC for fabrication and testing. All lap shear and wedge specimens were bonded with AF-163-2K film adhesive. The specimens were cured on a template fixture placed between the platens of a Dake press for 1 hour at 121 °C (250 °F) under 0.28 MPa (40 psi) pressure.

3.2 *Testing Performed on Panels* The two test methods utilized in this study were the Adhesive-Bonded Surface Durability "Wedge" Test (ASTM D3762) and the Strength Properties of Adhesives in Shear by Tension Loading Test (ASTM D1002). These methods were used to compare the bonding strength of the panels processed by the alternative and baseline techniques.

3.2.1 *Wedge Test* This method simulates in a qualitative manner the forces and effects on an adhesive bond joint at metal-adhesive/primer interface. It has proven to be highly reliable in determining and predicting the environmental durability of adherend surface preparations. In this test, a wedge is forced into the bondline of a flat-bonded specimen to create a tensile stress. This results in a crack forming and expanding into the bondline. The original length of the crack and the additional crack lengths at 4, 8, and 24 hours are recorded. The alternative processes were rated as being acceptable if the results were similar to the baseline processes.

3.2.2 *Shear Strength Test* This method covers the determination of the comparative shear strengths of adhesives for bonding metals when tested on a standard specimen and under specified conditions of preparation and test. In this test, the bonded material has shear stress applied along its axis until bond failure occurs. This specific test was performed on test panels at both 24 °C (75 °F) and 93 °C (200 °F) for all of the processes. The alternative processes were rated as being acceptable if the results were similar to the baseline processes.

3.3 *Test Results* The test results from the five iterations are summarized in Table 5.

Table 5. Summary of Test Data

Process	Original Crack Length (mm)	Crack Extension At 4 Hours (mm)	Crack Extension At 8 Hours (mm)	Crack Extension At 24 Hours (mm)	Total Crack Length After 168 Hours (mm)	Lap Shear Strength Results at 24 °C (MPa)	Lap Shear Strength Results at 93 °C (MPa)
Baseline							
CAA	16.9	9.1	2.2	0.9	32.4	41.7	10.9
Pasa-Jell 107	28.1	12.9	4.1	1.6	58.9	37.2	11.4
Alternative							
P2 Etch (It1)	30.2	22.6	2.1	1.1	61.3	39.1	9.9
P2 Etch (It2)	21.2	13.9	2.5	1.8	55.6	37.3	8.2
P2 Etch (It3)	30.7	25.8	2.6	3.4	72.0	34.0	8.1
P2 Etch (It4)	16.9	5.7	2.1	1.9	29.7	39.6	6.7
P2 Etch (It5)	33.5	20.0	6.0	3.4	66.5	38.4	7.8
NaOH (It1)	16.7	14.1	2.9	1.3	40.5	39.3	11.9
NaOH (It2)	18.5	9.9	2.5	2.0	38.7	39.6	9.0
NaOH/H_2O_2 (It1)	17.9	14.5	2.8	1.3	41.1	41.1	11.1
NaOH/H_2O_2 (It2)	16.0	3.5	2.1	1.4	25.8	40.3	10.4

The wedge tests, with crack extensions measured in millimeters (64ths in), show nearly identical results for the original crack length on the P2 and Pasa-Jell systems. However, both the CAA and NaOH systems significantly outperform the P2 and Pasa-Jell. While there are significant differences in the original and 4 hour crack lengths, values are similar for the 8 and 24 hour recordings. The shear strength tests reveal only slightly lower strength values for the non-chromatic processes when compared with the baseline processes. For the first iteration, there was only a five percent loss of strength on average, with significant overlap when the standard deviations are considered for all three non-chromatic processes at both 24 °C (75 °F) and 93 °C (200 °F).

Tables 6 and 7 show the results of the physical tests graphically. Both contain the baseline processes and the best performing iteration (in parentheses) of each of the three alternatives. This serves to illustrate the similar properties of the alternatives to the baseline processes. The axis labels identify which process is shown and the iteration number.

Table 6 is a chronological graph of the cumulative wedge crack length over a period of 24 hours. It shows that the alternatives demonstrate somewhat superior results; but due to a relatively small sample size, a definite "best" result cannot be determined.

Table 6. Wedge Test Results

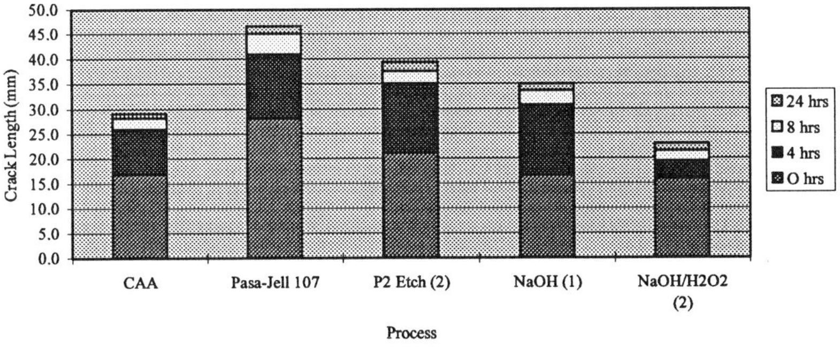

Table 7 is a graphical representation of shear strengths recorded at 24 °C (75 °F) and at 93 °C (200 °F). It shows that the alternative processes compare favorably to the baseline processes. The 24 °C (75 °F) tests show shear strengths in the 38 MPa (5500 psi) to 41 MPa (6000 psi) range and the 93 °C (200 °F) tests yield results in the 8 MPa (1200 psi) to 11 MPa (1600 psi) range. The acceptable values for alternative process testing at both temperatures are those numerically superior or equal to the values tabulated for the baseline processes. The three alternative processes have properties similar enough to the chromium-containing processes that none of them can positively be ruled out as a potential replacement.

Table 7. Lap Shear Test Results

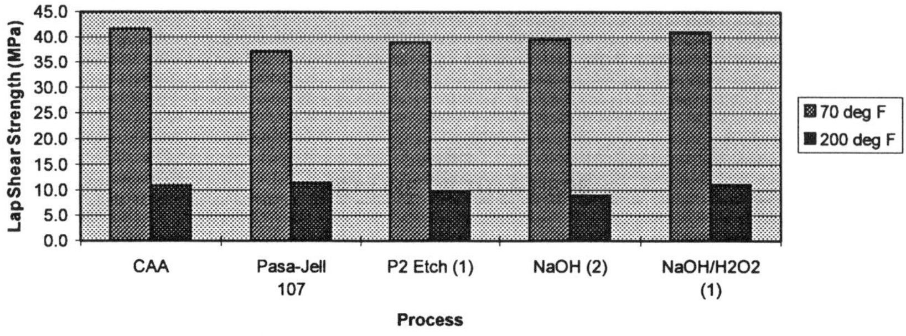

4.0 CONCLUSIONS

The three lotox processes provided performance equal to or better than Pasa-Jell 107 in tensile lap shear and wedge crack tests with the exception of the P2 etch tensile lap shear results at 93 °C (200 °F, 4 of 5 iterations gave lower values) and NaOH tensile lap shear results at 93 °C (200 °F, 1 of 5 iterations gave lower values).

The NaOH and NaOH/H_2O_2 processes provided performance equal to or better than CAA in tensile lap shear tests at 24 °C (75 °F) and 93 °C (200 °F). However, in the wedge crack tests, results for NaOH and NaOH/H_2O_2 were statistically lower than the results for CAA with the exception of NaOH/H_2O_2 in iteration 2.

The P2 process provided tensile lap shear performance values equal to or better than CAA in 2 of 5 iterations at 24 °C (75 °F) and only 1 of 5 iterations at 93 °C (200 °F). In the wedge crack tests, P2 provided statistically equivalent performance to that of CAA in only 1 of 5 iterations.

Hazardous waste treatment costs were found to be lower for the alternative processes when compared with CAA and Pasa-Jell. Also, the existing EPA and OSHA regulations that apply to surface treatments prior to titanium bonding favor the use of the alternative processes over CAA and Pasa-Jell.

5.0 RECOMMENDATIONS

The data generated in this study indicates that the NaOH and $NaOH/H_2O_2$ processes exceeded the performance of the Pasa-Jell 107 process and have the potential to match the performance of the CAA process, with lower hazardous wastewater treatment costs. The P2 process results were not as encouraging. The P2 processing variables/parameters require closer examination to determine if results comparable to Pasa-Jell 107 and CAA can be achieved on a more consistent basis.

6.0 ACKNOWLEDGMENT

The ARDEC PAB would like to acknowledge the technical contributions provided by CTC associated with this program and its continuing efforts to develop non-chromated surface preparation processes for the adhesive bonding of titanium.

7.0 REFERENCES

1. Concurrent Technologies Corporation, "Feasibility Study of Utilizing the P2 Etch or Modification in Identifying Viable Alternatives for Preparing Titanium for Adhesive Bonding," January 1995.

2. Concurrent Technologies Corporation, "Nonchromatic/Noncarcinogenic Etching for Bonded Materials," January 1996.

3. W.J. Russell and E.A. Garnis, "Chromate-Free Method of Preparing Aluminum Surfaces for Adhesive Bonding: An Etchant Composition of Low Toxicity," May 1978.

4. R.F. Wegman and S.A. Slota, "Evaluation of the Adhesive Bonding Processes Used in Helicopter Manufacture: Part 6, Evaluation of Improved Titanium Surface Preparations by Bond Durability Studies," March 1973.

5. J.A. Brescia and D.A. Martinelli, "Project #A131 Nonchromate/Non-carcinogenic Etching of Metals and Plastics - Interim Report," October 1994.

6. A.J. Kinloch, Durability of Structural Adhesives, Applied Science Publishers, Essex, England, 1983, pp. 269, 271-273.

7. MMM-A-132A, "Adhesives, Heat Resistant, Airframe Structural, Metal to Metal."

8. ASTM D3762, "Standard Test Method for Adhesive-Bonded Surface Durability of Aluminum (Wedge Test)."

9. ASTM D1002, "Standard Test Method for Strength Properties of Adhesives in Shear by Tension Loading (Metal-to-Metal)."

28th International SAMPE Technical Conference
November 4-7, 1996

STORAGE STABLE EPOXY RESIN ADHESIVES BASED ON ENCAPSULATED CURATIVES

D. K. Hoffman and D. V. Dellar
The Dow Chemical Company
Midland, MI 48674
and
M. L. Deviney and H. W. Schlameus
Southwest Research Institute
San Antonio, TX 78228-0510

ABSTRACT

An advanced one-part structural adhesive that (1) is stable under long-term (1-3 years), non-refrigerated storage conditions, (2) provides high strength service performance and (3) can be cured using conventional schedules, i.e., one hour at 121 - 177°C (250 - 350°F) would be highly desirable as a repair adhesive. The approach for this research program involved encapsulating a key epoxy resin curative in a protective shell. After screening and carefully selecting a curative, a microencapsulation process was developed and optimized. Using this approach, the curative is protected and remains inert in a one-part epoxy resin structural adhesive until it is thermally released during the cure cycle. The encapsulated curative and the total adhesive system have been characterized. Both paste and film adhesives have been formulated and evaluated for stability, reactivity and adhesive performance. Extensive comparisons have been made between formulations containing encapsulated curative and non-encapsulated curative. The results obtained with the adhesives containing microencapsulated curative compare favorably. Stabilities for the systems containing the encapsulated curatives are greatly increased for storage temperatures ranging from ambient to 49° (120°F). At the same time, cure rates are only slightly reduced. Adhesive performance approaches that for the formulations containing the non-encapsulated curative. This paper describes the performance of the paste and film adhesives.

KEY WORDS: Encapsulation, Epoxy Resins, Adhesives

1. INTRODUCTION

An advanced, one-part epoxy resin structural adhesive with long-term storage stability and high-strength service performance is of great interest for military and commercial aerospace applications. Areas for use include both initial fabrication of metallic and polymeric composite structures, and their rapid repair in the field. One-part formulations are preferred over two-part types that require careful mixing at the use site. Bonding problems can result with incorrect stoichiometry or air entrapment. However, one-part epoxy adhesives suffer from short shelf-life, even under costly refrigerated storage conditions.

Conceptually, microencapsulation[1,2] of selected reactive curing agents to separate them from the epoxy resin components is an especially attractive approach for developing storage stable one-part systems.

2. EXPERIMENTAL

2.1 MATERIALS

2.1.1 Curatives

The following curatives were evaluated for encapsulation; however, most of the manuscript discussion is about phenyldimethylurea.

1. Phenyldimethylurea (C.A.S. No. 101-42-8, solid, m.p. 129-135°C, water solubility 3850 ppm @ 25°C, Omicure® 94, Omicron Chemicals),
2. 2-Methylimidazole (C.A.S. No. 693-98-1, solid, m.p. 142-143°C, water solubility 100 mg/ml @25°C, Imicure® AMI-2, Pacific Anchor Chemical Corporation),
3. Benzyldimethylamine (N,N-dimethylbenzylamine, C.A.S. No. 103-83-3, liquid, m.p. -75°C, water solubility hot, Amicure® BDMA, Pacific Anchor Chemical Corporation),
4. Dicyandiamide (C.A.S. No. 461-58-5, m.p. ~ 98°C, water solubility 4.13 g/100 g at 25°C, soluble in ethanol and acetone, insoluble in ethyl ether and benzene, Amicure CG 1200 and CG 1400, Pacific Anchor Chemical Corporation),
5. Diaminodiphenyl sulfone (4-aminophenyl sulfone, C.A.S. No. 80-08-0, solid, m.p. 175-177°C, insoluble in water, soluble in ethyl alcohol, Aldrich Chemical Company, catalog no. A7,480-7), and
6. Para(aminocyclohexyl) methane (C.A.S. No. 1761-71-3, Amicure PACM, Pacific Anchor Chemical Corporation).

2.1.2 Epoxy Resins

The following epoxy resins were formulated into adhesives:

1. D.E.R.® 331 liquid epoxy resin (C.A.S. No. 025085-99-80), a diglycidyl ether of bisphenol A, a resin that is commercially available from The Dow Chemical Company,
2. D.E.R.® 755 liquid epoxy resin (50% C.A.S. No. 025036-25-3 and 50% C.A.S. No. 026142-30-3), a low viscosity liquid epoxy resin consisting of a blend of D.E.R. 337 liquid epoxy resin and D.E.R. 732 liquid epoxy resin, available from The Dow Chemical Company,
3. D.E.R.® 661 solid epoxy resin (C.A.S. No. 025036-25-3), a higher molecular weight diglycidyl ether of bisphenol A resin available from The Dow Chemical Company, and
4. XU-71790.04L experimental rubber modified epoxy resin (C.A.S. No. 121676-49-1), an acrylic rubber modified liquid epoxy resin available from The Dow Chemical Company.

® D.E.R. is a trademark of The Dow Chemical Company.

2.1.3 Additional Materials for Preparing Adhesives

Besides epoxy resins and curatives, the following ingredients were used for formulating paste and film adhesives: A-200 aluminum powder, Reynolds Metals; Cab-O-Sil M-5 fumed silica, Cabot Corporation; Amicure CG-1200 and CG-1400 dicyandiamide, Pacific Anchor Division of Air Products and Chemicals, and Byk R-605 polycarboxylic acid-amide rheology control agent, Byk Chemie.

2.2 ADHESIVE FORMULATIONS

2.2.1 Model Adhesive Formulations

Two- and three-component model formulations were made to evaluate the reactivity, stability and adhesive performance of curatives in epoxy resins. In the case of soluble liquid curatives, the mixture was stirred by hand until a clear solution formed. Solid curatives were blended into epoxy resin using a KitchenAid™ model K5SS heavy duty mixer using an anchor type stirrer. Formulations were mixed at a high shear rate for 15 minutes.

For two-component systems, mixtures of curative and epoxy resin were made by blending 1, 2, 5 and 10 parts curative into 100 parts epoxy resin ("phr", parts per hundred parts resin).

A 3-component model formulation is defined in Table 1. It is an "accelerated" dicyandiamide formulation in which dicyandiamide serves as the main curing agent. The accelerator, perhaps an imidazole or urea, speeds up the reaction rate of the dicyandiamide.

Table 1
Accelerated Dicyandiamide Model Paste Adhesive

Ingredient	*Amount, parts*
Epoxy Resin	100
Dicyandiamide	5
Accelerator	1

2.2.2 Fully Formulated Paste Adhesives

Table 2 shows an example of a recipe for a fully formulated paste adhesive. Besides epoxy resin and curatives, the adhesive contains "tougheners", "flexibilizers", fillers and rheology control agents.

The ingredients are formulated by the following procedure:

The D.E.R. 331 liquid epoxy resin (LER) and D.E.R. 755 LER are added to a KitchenAid mixer and blended for 15 minutes. Then, XU-71790.04L is added. This is followed by the aluminum powder, fumed silica, Byk R-605, dicyandiamide and phenyldimethylurea (PDMU) or encapsulated PDMU, which are added sequentially. Mixing is continued for 15 minutes after the addition of each ingredient.

The adhesive is transferred to a Ross planetary mixer and de-aired under vacuum.

Table 2
Fully Formulated Epoxy Resin Paste Adhesive

Ingredient	*Amount, parts*
D.E.R. 331 liquid epoxy resin	225
D.E.R. 755 liquid epoxy resin	225
XU-71790.04L acrylic rubber modified epoxy resin	450
Reynolds Metals A-200 aluminum powder	310
Cab-O-Sil M-5 fumed silica	50
Byk R-605 polycarboxylic acid-amide	15
Amicure CG-1400 dicyandiamide	50
"Active" Omicure 94 phenyldimethylurea	20

2.2.3 "Hot Melt" Adhesive Formulations

Table 3 illustrates an example of a "hot melt" adhesive formulation. For the best mixing results, ingredients for the hot melt adhesives were blended at 65°C using a Ross mixer and degassed before preparing film adhesive samples.

Table 3
Hot Melt Adhesive Formulation

Ingredient	*Amount, parts*
D.E.R. 661 solid epoxy resin	210
XU-71790.04L acrylic rubber modified epoxy resin	180
D.E.R. 755 liquid epoxy resin	60
Amicure CG-1200 dicyandiamide	18
Cab-O-Sil M-5 fumed silica	18
10 mil glass beads	4.7
"Active" Omicure 94 phenyldimethylurea	0.6

Using a 50:50 ratio of water and ethylene glycol for the heating jacket, a Ross planetary mixer was preheated to 65°C. D.E.R. 661 solid epoxy resin (210 g) was melted in an oven at a temperature of 90°C, and XU-71790.04L (180 g) was also heated to make it thinner for pouring. D.E.R. 755 LER (60 g), Amicure CG-1200 dicyandiamide (18 g), Cab-O-Sil M-5 fumed silica (25 g), 10 mil glass beads (24.6 g) and "active" accelerator were measured and stored in glass containers. When the D.E.R. 661 solid epoxy resin had been heated to a liquid state, all the ingredients were added to the heated (65°C) Ross mixer. The components were mixed at 50 (low rate) for 10 minutes, medium (75) for 15 minutes, and at high (100) for 20 minutes. The 10 mil glass beads (~ 1 weight percentage) were added to the mixture and mixed for 10 minutes at a medium speed of 75. Degassing the mixture was started as soon as the fumed silica and dicyandiamide were mixed and no longer free-flying. After adding the PDMU and glass beads, degassing was resumed for the remainder of the mixing.

2.2.4 Film Adhesives

Each film adhesive was prepared using a Mathis Coating Device (Model SV94392). The device was set to maintain a coating thickness of 0.03 cm (0.012"), including the release paper. The temperature of the coating blade was set at 55°C. To reduce viscosity, the hot-melt adhesives were preheated for 15 minutes in a 54°C oven.

The hot-melts were applied to the coating blade, then drawn across a piece of ~ 30 cm x 30 cm (12" x 12") release paper. The release paper was 0.005 cm (0.002") thick. Scrim was applied to the coated side of the paper after the adhesive had cooled to near room temperature. A second sheet of release paper was then applied to the other side of the film assembly. The film was then placed on a large-surface hot plate (61 cm x 91 cm, 2' x 3') and rolled with a steel cylinder to smooth any surface imperfections. This process also embedded the adhesive into the scrim.

Three types of scrim were investigated; NC-4, a polyester (Apex Mills Corp.), NC-48, a polyester/nylon blend (Apex Mills Corp.) and Scrim 104, a thin glass mat with a 504 finish from BGF Industries.

The thickness of each film was measured using a Dyer thickness caliper.

2.3 METAL COUPONS

The following metal substrates were used:

1. Polished cold-rolled steel (PCRS) was purchased from the Q-Panel Company. Tensile lap shear coupons measured 2.54 cm x 10.16 cm x 0.16 cm (1" x 4" x 0.063") of stock # RS-14; T-Peel coupons measured 15.24 cm x 30.48 cm x 0.08 cm (6" x 12" x 0.032") of stock # S-612 cut to 2.54 cm x 30.48 cm x 0.08 cm (1" x 12" x 0.032").
2. Unpolished cold-rolled steel (CRS), APR 17921, from Advanced Coating Technologies (ACT), Inc., 0.08 cm (0.032") thick.
3. Hot-dipped galvanized (HDG) steel, APR 10254, from Advanced Coating Technologies (ACT), Inc., 0.08 cm (0.032") thick.
4. Aluminum coupons, type 2024 T3 (bare), phosphoric acid anodized and primed with BR-127 chromate-based primer (American Cyanamid), were prepared by the Naval Air Warfare Center, Aircraft Division and shipped to Dow. The tensile lap shear test specimens were 2.54 cm x 10.16 cm x 0.16 cm (1" x 4" x 0.063") thick. The floating roller peel specimens were cut at 2.54 cm widths from 15.24 cm x 27.94 cm x 0.063 cm (6" x 11" x 0.025") substrate bonded to 15.24 cm x 25.4 cm x 0.22 cm (6" x 10" x 0.085") aluminum panels.

2.4 ADHESIVE TEST PROCEDURES

2.4.1 Tensile Lap Shear Strength

The tensile lap shear tests were performed in accordance with ASTM D 1002 and measured on a Universal Testing Machine (United Testing Systems, Inc.), Model TM-10 with a 22.25K Newton (5K lb) load cell. The crosshead rate was 0.254 cm/min (0.1 in/min) for the 0.16 cm (0.063") thick specimens and 1.27 cm/min (0.5 in/min) for the 0.08 cm (0.032") thick specimens.

2.4.2 T-Peel

T-Peel tests were performed in accordance with ASTM D 1876-72 (Reapproved 1983). The testing was conducted on an Instron Testing Machine with a crosshead speed of 25.4 cm/min (10 in/min).

2.4.3 Floating Roller Peel

Floating roller peel tests were performed in accordance with ASTM D 3167-76 (Reapproved 1983). The crosshead speed was 25.4 cm/min (10 in/min).

2.5 EVALUATION OF STABILITY AND REACTIVITY

Stabilities of model and fully formulated adhesives were determined by storing the formulations at ambient (nominally 22°C) and elevated temperatures (43°C and 49°C). Liquid model formulations and paste adhesives were monitored by following changes in viscosity as measured at 22°C using a Brookfield viscometer. For film adhesives, changes in the heat of cure were measured by Differential Scanning Calorimetry (DSC).

3. DISCUSSION

The objective of the project is to develop an advanced one-part structural repair adhesive that (1) is stable under long-term (1-3 years), non-refrigerated storage conditions, (2) provides high strength service performance, and (3) can be cured using conventional schedules. The concept chosen to meet the goal was an epoxy resin adhesive with an encapsulated curative to separate the reactive ingredients until thermally "triggered". Specifically, the requirements are:

1. One to three years ambient storage stability, without refrigeration including several excursions to 43 - 49°C (110 - 120°F).

2. Adhesive performance consisting of tensile lap shear strengths greater than 27.6 N/mm^2 (4,000 psi) on 0.16 cm (0.063") thick metal and bond line thicknesses of 0.013 cm (0.005") and peel strength greater than 3500 N/m (20 pli).

3. A cure rate of 1 hour at 121 - 177°C (250 - 350°F).

3.1 SELECTION OF CURATIVE

A variety of curatives were screened for their reactivity and stability. These materials represent not only individual species, but, different families of curatives. For example, some materials are catalytic, such as benzyldimethylamine (BDMA) and 2-methylimidazole (2-MI). Others are reacted at equivalent stoichiometry, such as para(aminocyclohexyl) methane (PACM) and diaminodiphenyl sulfone (DDS). Dicyandiamide (DICY) is usually used at about 40 - 60% stoichiometry based on epoxide groups and the ureas are generally used as accelerators for dicyandiamide-cured systems.

3.1.1 Stability Studies

A summary of the order of stability of the curatives in D.E.R.® 331 epoxy resin at ambient temperature is as follows:

DICY >> DDS > PDMU >> PACM > 2-MI > BDMA

Figure 1 shows viscosities of blends consisting of 10 parts 2-MI, PDMU or DICY in 100 parts D.E.R. 331 liquid epoxy resin as a function of time after blending. The blends were stored at ambient temperature.

Figure 1.
Stability of Curatives in D.E.R. 331 LER
(Stored at Ambient Temperature)

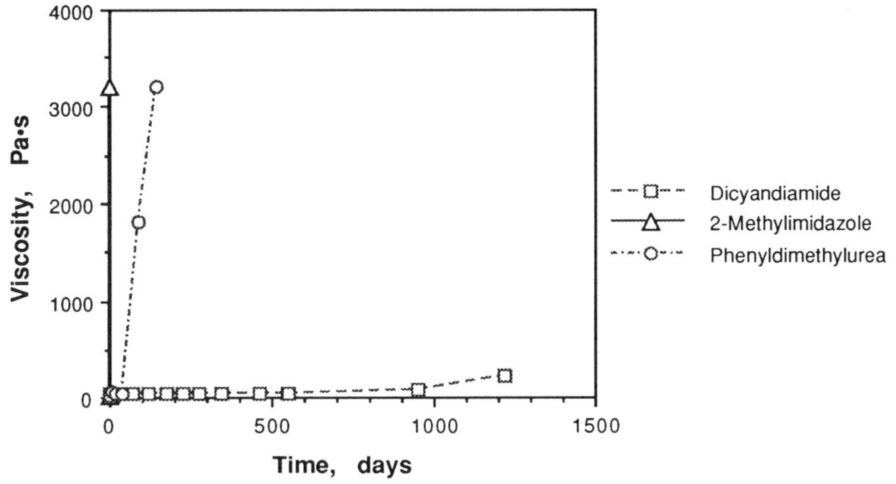

The results for the illustrated samples exemplify extremes in stability. DICY shows virtually no change in viscosity after 2 years when maintained at ambient temperature. At the other extreme, the sample of 2-MI had vitrified within one day, representing one of the most reactive materials. By comparison, PDMU had intermediate reactivity, showing virtually no increase in viscosity for 3 weeks but vitrifying after approximately 4 months.

Following the studies at ambient temperature, elevated stability studies at 43°C (110°F) were initiated for PDMU and DICY. As expected, increases in viscosity were accelerated. For phenyldimethylurea, vitrification occurred within two days versus the 4 months. By comparison, the vitrification of DICY took about one year.

Based on these results, no emphasis was placed on encapsulating DICY for this study.

3.1.2 Reactivity Studies

Reactivity information can be derived from differential scanning calorimetry (DSC) scans, which are illustrated in Figure 2. This shows DSC first scans of 2-component model adhesives containing 2-MI, PDMU or DICY in D.E.R. 331 liquid epoxy resin. Information that can be derived from these scans includes the temperature for the start of the reaction, the peak exotherm temperature and the total heat of reaction. A second scan can be used to determine the glass transition temperature (Tg) and if there is still residual heat of reaction. A reasonable indication of how well the adhesive cures can be approximated by measuring the total exotherm energy. The area under the peak is expressed in Joules per gram (J/g). For a 2-component system containing D.E.R. 331 epoxy resin and 10 phr PDMU, the energy for what appears to be complete cure was found to be approximately 450 J/g.

Figure 2.
DSC Scans of Curatives Blended in D.E.R. 331 Liquid Epoxy Resin

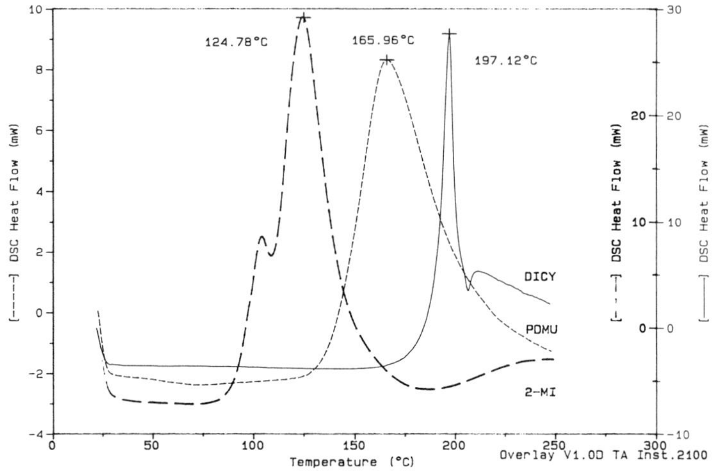

In addition to DSC scans, model formulations of the curatives were evaluated as adhesives and cured for 60 minutes at 121 - 171°C (250 - 350°F). A comparison of these results shows that a good correlation can be made between the necessary cure schedules and the DSC scans.

A summary of the reactivity data indicates that DICY is slow and cannot meet the cure requirements of one hour at 121°C (250°F) by itself. While 2-MI is fast enough, the poor stability makes meeting the storage requirements a difficult challenge for encapsulation. PDMU was chosen as the best compromise between reactivity and stability. However, its model formulations had poor adhesive performance. Therefore, a reasonable compromise was to use PDMU as an accelerator for dicyandiamide-based formulations. Fully formulated adhesives based on urea accelerated, dicyandiamide-cured epoxy resins had excellent adhesive performance and could be cured at temperatures ranging from 107 - 177°C (225 - 350°F).

3.2 CHARACTERIZATION OF ENCAPSULATED CURATIVE

A variety of encapsulation processes have been evaluated. An optimum has been selected for this study.[3] Samples of encapsulated PDMU were characterized for particle size, loading level, storage stability in liquid epoxy resin, and reactivity.

Particle sizes for the encapsulated curatives ranged between 10 and 1000 µm. Loading levels were between 10 and 45 % by weight. The results reported in this manuscript are for particles ranging from 30 - 200 µm and a loading level of about 25% by weight.

The stability of model formulations containing encapsulated PDMU is greatly improved. For accelerated DICY formulations, vitrification times ranged between that for the formulations containing non-encapsulated PDMU and formulations containing only dicyandiamide. As the encapsulation process was optimized, stability was improved.

Reactivity is only slightly reduced for formulations containing the encapsulated PDMU. For example, peak exotherm temperatures determined by DSC were reduced by 1 to 2°C.

Similar results were obtained for the fully formulated adhesives and will be discussed in more detail.

3.3 PASTE ADHESIVES

A fully formulated paste adhesive is described in Table 2. Similar formulations were prepared using the non-encapsulated and encapsulated PDMU. The ingredients were identical except for the presence of the encapsulant ("shell" material) in the formulation containing the encapsulated PDMU. Equivalent amounts of PDMU were compared.

3.3.1 Storage Stability

Figure 3 shows the elevated (43°C) storage stability of the fully formulated paste adhesives.

Figure 3.
Stability of Fully Formulated Adhesives
Stored at 43°C

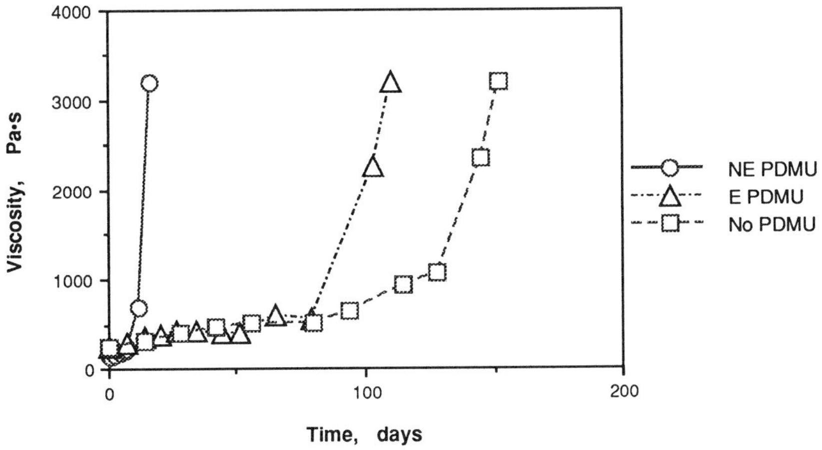

The three curves represent an optimized adhesive formulation containing the non-encapsulated accelerator (NE PDMU), no accelerator (No PDMU), and a formulation using the encapsulated PDMU (E PDMU).

The limit for stability is represented by the formulation containing no accelerator. Since the formulation contains non-encapsulated DICY as the curative, better stability than this must be obtained by either encapsulating DICY or using another approach. While the DICY formulation has acceptable stability, its cure rate is too slow.

The formulation accelerated with the non-encapsulated PDMU has an acceptable cure rate (which will be demonstrated in more detail later in this paper), but vitrifies after about a week at 43°C. For acceptable performance, a goal of greater than 5 weeks' stability was established.

By comparison, the formulation containing the encapsulated PDMU has greatly improved stability, that is better than 10 times that of the formulation with the non-encapsulated PDMU. Its stability is approaching the formulation without accelerator, which is the limiting case. At 43°C, vitrification occurred after 3 months for the adhesive containing the encapsulated PDMU. Ambient shelf lives greater than 1.5 years have been demonstrated for fully formulated paste adhesives.

3.3.2 Reactivity

Reactivity of the fully formulated adhesives was first evaluated by DSC.

The heat of reaction for the cured fully formulated adhesive containing the non-encapsulated PDMU was approximately 263 J/g and the peak exotherm temperature 161°C. Results for formulations containing encapsulated PDMU were similar. Heats of reaction for several samples ranged from 250 - 260 J/g, which is slightly lower than for the original and can be explained by the presence of a small amount of shell material that is additional in the storage stable formulations. The peak exotherm temperatures ranged from 163 to 165°C, which is slightly higher than the comparison formulation, but within experimental error. However, the temperatures were always slightly higher than that for the comparison formulation, indicating that the cure is slightly retarded.

3.3.3 Adhesive Performance

Table 4 shows a comparison of adhesive performance for fully formulated paste adhesives with encapsulated PDMU versus non-encapsulated PDMU. A number of cure schedules are shown.

Overall, the results for the two formulations compare favorably. For both formulations, cure can be obtained in a temperature range of 107 - 177°C (225 - 350°F). The adhesive performance is dependent upon the details of the cure schedule. Slightly lower tensile lap shear and peel strengths are obtained for the formulation containing the encapsulated PDMU. The reasons for this have not been fully investigated. However, two factors which affect this are the slightly retarded cure rate and the particle size of the encapsulated curative. Better adhesive performance has been correlated to smaller particle sizes.

3.4 FILM ADHESIVES

Preliminary studies on film adhesives have produced results similar to those for paste adhesives.

Table 4.
Adhesive Performance of a Fully Formulated Paste Adhesive

Cure Schedule*	Tensile Lap Shear, N/mm^2 on CRS	Tensile Lap Shear, N/mm^2 on PCRS	T-Peel, N/m on CRS	T-Peel, N/m on PCRS
Non-Encapsulated PDMU				
1. 177°C, C	16.6	33.7	6090	4100
2. 177°C, R	17.4	34.2	5580	5110
3. 149°C, R	16.9	34.4	6110	5020
4. 121°C, R	14.5	31.6	5020	3990
5. 107°C, R	11.5	23.9	3670	4410
6. 93°C, R	5.1	6.8	0.0	0.0
Encapsulated PDMU				
1. 177°C, C	16.5	28.6	4930	3430
2. 177°C, R	16.8	29.5	4530	3100
3. 149°C, R	16.2	28.1	4510	2960
4. 121°C, R	14.8	24.5	4360	2780
5. 107°C, R	12.6	24.8	3460	2380
6. 93°C, R	0.0	0.0	0.0	0.0

* "C" represents a constant temperature oven, preset at the indicated temperature. "R" represents a 1.67°C (3°F) per minute temperature ramp from ambient to the final oven temperature. Samples were cured for 30 minutes at temperatures of 149°C and above. Below 149°C, samples were cured for 1 hour.

The "hot melt" adhesive formulation described in Table 3 was processed into a film adhesive having a thickness of 0.025 cm (0.010"). The product was characterized for storage stability, reactivity and adhesive performance.

3.4.1 Storage Stability and Reactivity

Unlike paste adhesives, the storage stability of film adhesives is not easily determined by viscosity. However, changes in viscosity affect both tack and drape of a film adhesive. Our research indicates that a good correlation is obtained by measuring the residual heat of reaction of the aged hot melt adhesive by DSC and comparing it to the original. Figure 4 shows the storage stability of formulations containing the non-encapsulated accelerator (NE PDMU), no accelerator (No PDMU) and the same formulation using the encapsulated PDMU (E PDMU) stored at 49°C (120°F). For these particular formulations, poor tack and drape occur when the heat of reaction falls below ~ 170 J/g. This occurs for the formulation containing the non-encapsulated PDMU after about a week. By comparison, the time is extended to about 50 days for the formulation containing the encapsulated PDMU.

Similar to the paste adhesives, the DSC first scans for the hot melt adhesives can be evaluated for the initiation of the reaction, the peak exotherm temperature and heat of cure. The heat of reaction for the hot melt adhesives described in Table 3 is approximately 200 J/g. The peak exotherm temperature for hot melt adhesive containing the non-encapsulated PDMU was 161°C versus 164°C for the sample containing the encapsulated PDMU. The peak exotherm temperature for the non-accelerated sample (No PDMU) is ~ 200°C.

In agreement with the results for the paste adhesives, the storage stability for the film adhesives is greatly increased by encapsulating the PDMU. With the non-encapsulated PDMU, tack and drape are lost in about a week when stored at 49°C (120°F). With encapsulated PDMU, the stability of the sample was greater than about 5 weeks. At the same time, the reactivity is only slightly diminished.

Figure 4.
Storage Stability of Hot Melt Adhesives
Stored at 49°C

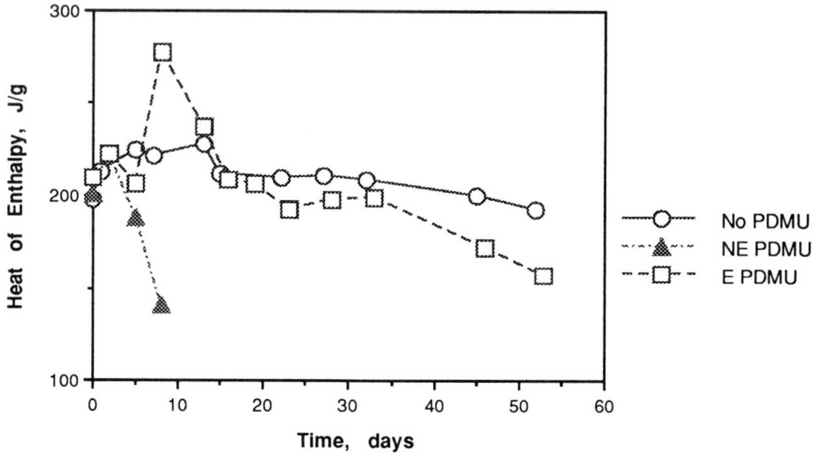

3.4.2 Adhesive Performance

So far, only a limited amount of adhesive data has been obtained for the film adhesives. Table 5 shows the adhesive performance of the film adhesives using the Apex Mills NC-48 polyester/nylon blend knit fabric scrim. The samples were cured at 177°C (350°F) for 30 minutes. A gradient heating schedule was used, i.e., 1.67°C (3°F) per minute from ambient temperature to 177°C (350°F). At lower temperatures such as 121°C (250°F), no cure is obtained for the sample containing no PDMU. The bond line thicknesses are 0.025 cm (0.010").

Table 5
Adhesive Properties of Film Adhesives on Aluminum

Sample Description	*Tensile Lap Shear, N/mm²*	*Floating Roller Peel, N/m*
Encapsulated PDMU	25.4	7450
Non-Encapsulated PDMU	24.1	3600
No PDMU	23.7	4090

The results show that all the samples have good adhesive performance under these cure conditions. The reason for the exceptional floating roller peel for the sample containing the encapsulated PDMU versus the others is unknown. However, the failure mode for this sample was excellent with extensive tearing of the scrim.

4. SUMMARY AND CONCLUSIONS

A series of curatives was evaluated for stability, reactivity and adhesive performance in epoxy resin formulations. Based upon the criteria for a Naval Air Warfare Center, Aircraft Division contract, phenyldimethylurea was selected as the curative for encapsulation. It is used as an accelerator for one-part, dicyandiamide-cured epoxy resin paste and film adhesives.

A process to encapsulate phenyldimethylurea has been successfully developed. One-part formulations containing the encapsulated accelerator have greatly improved non-refrigerated shelf lives while maintaining essentially the same cure rates. Adhesive performance approaches that of samples accelerated with non-encapsulated PDMU.

Shelf lives have increased from less than one week at 43 - 49°C (110 - 120°F) to greater than 5 weeks. Ambient shelf lives greater than 1.5 years have been demonstrated for paste adhesives. The stability approaches that of formulations containing no accelerator.

The adhesives can be cured at temperatures ranging from 107 - 177°C (225 - 350°F) over 30 - 60 minute schedules. This is the same cure schedule as for the formulations containing the non-encapsulated accelerator.

Under optimum conditions, tensile lap shear strengths greater than 27.6 N/mm^2 (4000 psi) and peel strengths greater than 3500 N/m (20 pli) were obtained.

6. REFERENCES

1. Curt Thies, "Microencapsulation" in The Encyclopedia of Polymer Science and Engineering, Volume 9, John Wiley and Sons, New York, © 1987, pp. 724-745.
2. Thomas R. Tice, Thomas O. Dappart, Robert J. Flores and David W. Mason, "Microencapsulation of Thermosetting Resins for use in Hot-Melt Adhesives", Final Report Naval Air Development Center Contract No. N62269-85-C-0251, Report No. NADC-90083-60, August 1, 1989.
3. Patents applied for.

7. ACKNOWLEDGMENTS

This work was conducted under Navy Contract N62269-91-0250. The authors thank the Naval Air Warfare Center, Aircraft Division for financial support and Georgette Gaskin and Stan Brown for their guidance in the project.

The authors thank Jack Trevino and Mike Seals for their dedication in the Southwest Research Institute labs. The authors also recognize the contributions of Mark Stachiowiak, Christi Bush, Erin Gatza and Michelle Gittins for their dependable laboratory work at The Dow Chemical Company.

8. BIOGRAPHIES

Dwight Hoffman is an Associate Scientist in the Adhesives Group of the Central and New Businesses R&D - Designed Products and Devices Laboratory of The Dow Chemical Company. He has a Ph.D. in Polymer Science from The University of Akron where he worked on ABA block copolymers. Since joining Dow in 1973, he has worked with a variety of thermoset polymer systems including polyurethanes, epoxies and vinyl ester resins. He has several patents and publications in such areas as copolymer polyols for flexible polyurethane foams, water-based coating systems and toughened epoxy resins. His general interests are in the area of phase separated polymers. In addition to being a member of SAMPE, he is a member of the Adhesion Society and the American Chemical Society.

David Dellar is a Senior Research Technologist in the Central and New Businesses R&D - Designed Products and Devices Laboratory of The Dow Chemical Company. He received a B.S. in Applied Biology with a minor in Industrial Chemical Technology from Ferris State University. Since joining Dow in 1980, he has worked in the areas of small molecule and polymer synthesis. Other areas of interest include polymer characterization by thermomechanical and microscopic techniques, structure/property relationships, and general adhesion science. He is a member of the American Chemical Society.

Dr. Marvin L. Deviney is an Institute Scientist at Southwest Research Institute. His scientific career includes twenty years of industrial managerial responsibilities involving R&D programs, government contracts, and joint multiclient programs with companies in the chemical, composites, oil, and aerospace areas. He has authored or co-authored more than 65 publications and patents in the fields of advanced polymeric composites, rubber technology, catalyst characterization, electrochemistry, fabric softeners, and radiotracer applications. He has received two "Best Paper" awards from the ACS Rubber Division, and has co-edited two books in the ACS Symposium Series on "Catalyst Characterization Science: Surface and Solid State Chemistry" and "Petroleum-Derived Carbons".

Wade Schlameus is a Staff Scientist in the Microencapsulation and Process Development Section of the Division of Chemistry and Chemical Engineering at Southwest Research Institute in San Antonio, Texas. He has a B.S. in Chemistry from Southwest Texas State University. Since joining Southwest Research Institute in 1960, he has worked in the field of microencapsulation. His work includes formulation development for controlled release of a wide range of products including industrial chemicals, pharmaceuticals, agricultural chemicals, biologicals, and food additives. He has eight patents in the encapsulation area and is a member of the Controlled Release Society.

28th International SAMPE Technical Conference
November 4-7, 1996

AN EVALUATION OF BONDING PROCESSES FOR GLASS FILLED PHENOLIC TO STEEL FOR PRODUCTION APPLICATIONS

Don Rorabaugh, Dean Martinelli, and William DePiero
U.S. Army
Armament Research Development and Engineering Center (ARDEC)
Picatinny Arsenal, Dover NJ, 07806-5000

ABSTRACT

Glass filled phenolic has been used extensively as a heat insulator and erosion inhibitor on a series of Rocket Assisted Projectile (RAP) artillery rounds. To achieve extended range, designers are increasing the potency of the rocket propulsion systems; this produces a more erosive environment due to the hot rocket gasses. Insulating materials become increasingly more important to protect electrical components, nozzle orifices, explosive components, and fuzes from hot gun and rocket gasses. Historically, phenolic insulators have been fastened with a variety of adhesive bonding materials/processes which were labor intensive and were not compatible with assembly line procedures. The operations tend to be time consuming, operator dependent and costly. This study evaluated the bond between a glass filled phenolic insulating material and a high carbon steel, using several different adhesives and processes over a wide temperature range. The study ranked best candidate systems not only on performance but on their ability to be incorporated into production. Production requirements taken into account were surface preparation, mixing, application, tooling/fixturing, pot life, and handling strength.

KEY WORDS: Adhesive Bonding, Artillery Rounds, Phenolics

1. INTRODUCTION

Rocket Assisted Projectiles (RAP's) utilize a rocket motor, located in the artillery projectile body, to provide additional propulsion to the round in order to achieve extended range. In many designs, glass filled phenolic is used to insulate and protect the mechanical, electrical, and energetic components of these projectiles from the heat of the burning rockets. Figure 1 illustrates a typical RAP artillery projectile design, in which the phenolic heat shield is adhesively bonded to the back of the explosively filled steel warhead, which is seated just ahead of the rocket propellant. The bond must be capable of withstanding set back forces of

22,000 g, as well as provide a seal against the hot propellant gasses. Recent experience with one RAP round design suggested that failure of the adhesive bond seriously reduced the reliability of the projectile's performance.

2. TECHNICAL OBJECTIVE

In response to the problem, the Polymers and Adhesives Branch (PAB) of the US Army Armament Research, Development and Engineering Center (ARDEC) initiated this study to evaluate and optimize the materials/processes used to bond the phenolic heat shields to the backs of the explosively filled warheads of the RAP rounds. The study was divided into two main parts: The first dealt with evaluating the surface preparation and adhesive mixing - application techniques used by the manufacturer at the production facility. The aim was to optimize a procedure that would be suitable for reworking a series of rounds that had less than adequate ballistic reliability. The second part of the study dealt with identifying, obtaining, testing and evaluating commercially available adhesives with the aim of recommending alternate adhesives for future improved RAP rounds.

3. EXPERIMENTAL

3.1 Materials The study dealt with four main components: the base of the warhead where the insulator was seated, the heat insulator disc, the ASTM tensile bond strength (T-Hat) test specimens, and the adhesives used for bonding. The warheads were made from MIL-S-50783, High Fragmentation (HF1) steel, and the ASTM T-Hat samples were made from ASTM A108, AISI 1020 cold finished bar stock steel. The insulators were made with 0.381 cm (1/8 in) glass reinforced phenolic. The adhesives materials listed below were evaluated in this study; the listing gives the name, type of adhesive, and curing cycle. The Armstrong epoxy was evaluated because it had been selected and specified for a RAP round prior to the initiation of this study. The other adhesives were selected with the aim of finding alternate materials with improved performance as compared to the Armstrong epoxy.

- Armstrong C-7/W (two part premeasured epoxy SEM kit) (RT/7 days)
- Ciba-Geigy Araldite 2014 (two part premeasured epoxy SEM kit) (RT/7 days)
- Dexter Hysol EA 907 (two part epoxy) (RT/3 days)
- Dexter Hysol EA 9396 (two part epoxy) (RT/5 days)
- Dow Corning Silastic J RTV (two part silicone and SS-4155 primer) (RT/24 h)
- GE 60 RTV (two part silicone and SS-4004 primer) (RT/7 days)
- GE 630 RTV (two part silicone and SS-4155 primer) (RT/7 days)
- Hernon 730 (modified methacrylate with activator) (RT/24 h)
- 3M Scotchweld DP 460 (two part premeasured epoxy SEM kit)) (RT/7 days)
- Permabond E32 (two part premeasured epoxy SEM kit) (RT/24 h)
- Permabond 665 (two part toughened premeasured acrylic SEM kit) (RT/12 h)

3.2 Procedures

3.2.1 Surface preparation of ASTM D897 Test Samples The surface preparation selected for the steel and phenolic materials, solvent wipe + abrasion + solvent wipe, is shown below:

Steel:
- Clean: Test samples were solvent-wiped with lint-free wipes saturated with methyl-ethyl-ketone (MEK), until no visible residue remained on the lint-free wipes.
- Abrade: Test samples were grit blasted with 180 grit aluminum oxide.
- Clean: Test samples were solvent-wiped with lint-free wipes saturated with methyl-ethyl-ketone (MEK), until no visible residue remained on the lint-free wipes.
- Inspection: A visual examination was conducted to insure that the surface was free of visible contaminants.

Insulator:
- Clean: Test samples were solvent-wiped with lint-free wipes saturated with isopropyl alcohol (IPA), until no visible residue remained on the lint-free wipes. The entire phenolic was cleaned.
- Abrade: Test samples were grit blasted with 180 grit aluminum oxide.
- Clean: Test samples were solvent-wiped with lint-free wipes saturated with isopropyl alcohol (IPA), until no visible residue remained on the lint-free wipes. The entire phenolic was cleaned.
- Inspection: A visual examination was conducted to insure that the surface was free of visible contaminants.

3.2.2 Preparation of Adhesives Armstrong C-7/W epoxy, prepared with a SEM kit supplied by the RAP fabrication contractor, was used to obtain base line tensile bond strength data. (The SEM kit is a two part, side by side, dual cartridge dispensing system that forces premeasured quantities of each component through a static 6 in mixing nozzle prior to bonding.) For comparison evaluations, the SEM nozzle material was further hand mixed; in addition, C-7 and W components were taken from non-SEM kit samples supplied by Armstrong and hand mixed at two different mixing ratios. Both of the curing temperatures and times that were recommended by the adhesive manufacturer, 74 °C (165 °F)/2 hr and room temperature (RT) for 7 days, were used in the study.

The remaining adhesives were all mixed in accordance with their manufacturer's instructions. The Ciba-Geigy Araldite 2014, 3M Scotchweld DP 460, and Permabond E32 epoxies had SEM kits with static mixing nozzles, as did the Permabond 665 acrylic adhesive. The Dexter Hysol materials (EA 907 and EA 9396) were two part epoxies that required hand measuring and mixing. The Hernon 730 required an activator to be applied to the bond surfaces prior to applying the methacrylate. The three room temperature vulcanizing (RTV) silicone adhesives (Dow Corning Silastic J, GE 60, and GE 630) required a vacuum degassing after hand mixing, and also required that a primer be applied to the bond surfaces and dried prior to applying the silicone material. All of these adhesives were tested with phenolic insulators bonded between their T-Hat samples, and all were RT cured. To insure that all adhesives were fully cured, samples were held for two weeks prior to testing.

3.2.3 Testing Adhesive tensile bond strengths, determined in accordance with ASTM Standard Test Method D897 (1), were used to evaluate and compare the adhesives. Determinations were made at 24 ° C and 63 ° C (75 ° F and 145 ° F) using an Instron tensile testing machine with a cross head speed of 0.123 cm/min (0.050 in/ min). Six bonded T-Hat specimens were evaluated for each test condition. The Armstrong SEM nozzle + 7 day RT cure material, plus all of the other adhesives evaluated in this study, had phenolic bonded between the T-Hats to simulate the warhead and to measure the tensile bond strength of the adhesive between phenolic and steel surfaces.

4. DISCUSSION

4.1 Review of Surface Preparation Procedures After the proper adhesive is selected, (based on technical data, experience, etc. and the understanding that selections must be tested for each specific use), the most important thing needed for effective bonding is the preparation of the surfaces to be bonded. For effective bonding, the adhesive must be able to wet the surface of the material. Without adequate surface preparation, contaminants such as grease, oil, dirt, dust, moisture, oxidation, etc. (which may not even be noticeable) can cause even a "good" adhesive to fail. Mechanical abrasion techniques are similar for both metals and plastics in that the objective is to clean and physically remove material from the surface to provide a clean, roughened surface for bonding.

Usually with poor surface preparation, the adhesive fails by adhesive (interfacial) failure, where the adhesive separates from the substrate. Ideally, the adhesive bond will be so good, that the material being bonded fails before the adhesive. The next best type of failure is cohesive failure of the adhesive, where the adhesive fails within itself and the bonds to the bonded materials (adherends) remain intact. In this last case, one can design the adhesive to have the needed strength properties for a particular application.

4.1.1 Original Surface Preparation Procedures A series of possible problems were discovered with the original preparation procedures used by the RAP fabricator: 1.) the steel and phenolic components probably were contaminated (oil, grease, mold release, etc.) during the manufacturing operations used to produce the warheads and insulators, and these parts were not solvent cleaned before or after abrading. 2.) the no. 36 aluminum oxide grit used to blast the warhead base was too large, and may have driven contaminants into the surface of the steel. 3.) Unfiltered shop air, which usually contains moisture and oil vapors, was used to "clean" the grit blasted warhead bases and could have further recontaminated the grit blasted surfaces. 4.) The surface of the grit blasted phenolic appeared to be too smooth for good adhesive bonding purposes.

4.1.1.1 Phenolic Insulator A light abrasion/grit blast, using aluminum oxide, has been shown to be an effective surface preparation for composites such as the glass filled phenolic insulator material (2). The superiority of abrasion/grit blasting over surface cleaning alone was documented by a number of aerospace companies (and reconfirmed by tests conducted as part of this study); after abrasion/grit blasting and subsequent surface cleaning, adhesive bond strengths are optimal and scatter is minimal (2-3). As shown in Figure 2, the clean unabraded phenolic has a smooth flat surface, while the grit blasted sample presents a rough surface (with significantly more surface area) for adhesive bonding. It should be noted, that too much roughening may weaken the phenolic and/or allow air to become entrapped under the adhesive, forming voids in the bond surface.

4.1.2 Modified Surface Preparation Procedures The optimized procedure of clean - abrade - clean used to process the T-Hat and phenolic test samples also was recommended for the plant remanufacturing surface preparation procedures. However, the unavailability of grit blasting equipment, for use on TNT loaded warheads, led to the development of this modified procedure. The warhead base was roughened with a drill mounted 2 in sanding disc (coarse no. 36 grit - 3M Co.), and the phenolic insulator was roughened in a cross hatch pattern with Scotchbrite very fine, general purpose hand pads (no. 47074 - 3M Co.). Additional ARDEC testing also showed that MEK was an acceptable alternate solvent cleaner (for IPA) for the phenolic material. Thus avoiding the need to use two different solvents on the rework line.

4.2 Phenolic Insulator Materials Glass fiber reinforced phenolic (molded plastic GPI -100) as listed in MIL-M-14/2 (4). was specified for the RAP round that utilized the Armstrong C-7/W epoxy adhesive, and was used as the centerpiece for the SEM nozzle + 7 day RT cure (with phenolic) ASTM T-Hat samples. Two other phenolic materials, which had been specified for a developmental RAP round, were used for the additional adhesives tested in this study. Both materials (MX 2625 and MX 2641) consisted of silica fibers impregnated with a MIL-R-9299 (5) phenolic resin. The MX 2625 phenolic, as listed on the manufacturer's data sheet, has strength properties higher than the MX 2641 phenolic. Both materials also were stronger than the GPI - 100 phenolic. The CIba-Geigy Araldite 2014 epoxy and the two silicone GE RTV 60 and GE RTV 630 adhesives used the MX 2625 as the centerpiece for the ASTM T-Hat test samples. The remaining adhesives, had the MX 2641 phenolic for their ASTM T-Hat centerpieces.

4.3 Evaluation of Mixing Techniques for Armstrong Adhesive The results of the ASTM D897 tensile bond strength testing are shown in Tables 1 and 2. Table 1 lists the bond strength values for samples tested at room temperature and at 63 $^{\circ}$ C (145 $^{\circ}$ F); it also shows the percentage loss in strength for the elevated temperature tests. Table 2 lists the types of failure found.

4.3.1 Room Temperature Testing Armstrong recommended either a 7 day RT cure or an elevated cure at 74 $^{\circ}$ C (165 $^{\circ}$ F)/2 h. The first two tests (A60 and A62) were conducted to determine if the 38:62 weight ratio mix of C-7:W used by the RAP round fabricator was an acceptable substitute for the specified ratio of 40:60. Both sets of test samples used hand mixed adhesive which was cured at 74 $^{\circ}$ C (165 $^{\circ}$ F)/2 h prior to testing at RT. New materials obtained from Armstrong and hand mixed to each of the two ratios were used for this evaluation. A slight problem occurred with the A60 samples. The test machine was set on the high gear ratio, and unknown to the test engineer, this setting had a safety shut off of 46.88 MPa. Thus, the recording chart did not record some of the higher break strengths for the A60 set, and the actual strength can only be listed as higher than 46.42 MPa. Subsequent tests were made on the low gear setting. Both the A60 and the A62 samples had bond strengths about four times higher than the minimum RT requirement of 11.72 MPa. Thus, the A62 mix ratio was equal to the A60 mix ratio.

The next two samples (SEM nozzle and SEM nozzle + hand mix) cured at 74 ° C/2 h, also had tensile bond strength properties about equal to four times the minimum RT requirement of 11.72 MPa. This indicates that the RAP round fabricator's SEM kit mixture, either by itself from the 6-in static mixing nozzle or with additional hand mixing, was equal to the A60 and A62 hand mixed materials. The SEM nozzle without additional hand mixing was about 6.89 MPa higher than its A62 counterpart, but the SEM nozzle mix had a standard deviation value about 8.27 MPa higher that the hand mixed sample. This large deviation may be an indication that, although the 6-in nozzle mixed the components sufficiently to obtain a good strength level, additional hand mixing may yield a more consistent uniform blend of epoxy and curing agent.

The SEM nozzle + 14 h RT cure sample set was fabricated to simulate the fabricator's procedure of bonding the insulator to the warhead, and allowing it to cure overnight before being packed for shipment. The 14 h cure gave a bond strength of 5.95 MPa; about half of the required 11.72 MPa. However, for an overnight cure, this seems to be more than adequate. The epoxy has hardened and should have sufficient strength to hold the phenolic insulator in place during normal handling procedures. A second follow-up set of samples was cured at 74 ° C/2 h after the 14 h RT cure, although Armstrong recommends that curing at 74 ° C should take place immediately after the epoxy is mixed and the substrates are bonded together. This test was to measure the bond strength that would be obtained if the elevated temperature cure were delayed. The 2 h elevated temperature cure produced a bond strength of 23.27 MPa, about four times greater that the 14 h cure value, and about half the value of the first four sets of samples cured at 74 ° C immediately after bonding. Thus, there is a significant loss in bond strength (50%) if the elevated temperature cure is delayed (14 h), but the strength is still equal to that obtained for samples cured 7 days at room temperature. The 14 h window may be useful for some manufacturing operations where it is inconvenient or costly to cure the bonded warhead and phenolic insulator components for 7 days at room temperature, and where they cannot be heat cured immediately after being joined together.

The 7 day RT cure SEM nozzle mix samples yielded tensile bond strength properties (24.77 MPa) about half that obtained for the comparable elevated temperature cure (50.79 MPa), and this is not in agreement with the Armstrong Company data which lists tensile bond strength properties of 18.82 MPa for the RT cure and 19.99 MPa for the elevated temperature cure. The reason for this large difference in strength properties is not known at this time. However, epoxies typically have better cross-linking and higher strength properties for elevated temperature cures as compared to RT curing cycles. The 7 day RT cured samples with the phenolic insulator centerpieces, simulating the warhead and insulator bond, had bond strength properties (15.93 MPa) lower than that exhibited by comparably cured samples without the phenolic centerpiece (24.77 MPa). However, this value was still higher than the minimum requirement of 11.72 MPa. In addition, all of the insulator samples maintained their adhesive integrity and failure occurred by phenolic delamination (the bonded item failed). It also should be noted that the phenolic samples used for this test were pie shaped wedges cut from larger insulator discs, and all of the failures initiated from one of the two cut edge surfaces. It is believed that with the actual insulators, which have smooth molded outer edges, failure would not be initiated by a cut surface and tensile bond strength should be greater than the measured 15.93 MPa value.

4.3.2 Elevated Temperature Testing All samples were conditioned at the test temperature of 63 ° C (145 ° F)/15 min prior to testing. The Armstrong samples showed a significant drop in bond strength as compared to similar samples tested at RT; the average loss was 80%. None of the samples equaled the minimum RT requirement of 11.72 MPa. It should be noted that there is no elevated temperature bond strength requirement for the adhesive, although the RAP round is required to function after being conditioned at 63 °C for 2 h. Being an epoxy, this loss was not unexpected, especially since the Armstrong C7/W property data sheet shows similar losses for tensile shear properties measured at elevated temperature versus RT.

4.3.3 Failure Mode Cohesive failures (85% to 95%) were observed for all of the samples cured at 74 ° C (165 ° F)/2 h and tested at RT (A60, A62, SEM nozzle, SEM nozzle + hand mix). When the same four conditions were tested at elevated temperature, the epoxy only had about a 50% cohesive failure. This was in agreement with the lower bond strength values, both indicated a breakdown in the epoxy's bonding performance at elevated temperature. Similar results were obtained for the SEM nozzle samples cured 14 h at RT and those given an additional 74 ° C/2 h thermal curing treatment. The SEM nozzle + 7 days RT cure samples had cohesive failures (60% to 85%) at RT, and 50% at the elevated temperature. Similar samples with phenolic bonded between the T-Hats had the phenolic fail at RT, and cohesive failure (30 % to 50%) at elevated temperature. The cohesive/phenolic failures indicate that the optimum surface preparation technique yielded a good bonding surface.

4.3.4 Epoxy Appearance The various sample mixes were applied to the surfaces of clean metal sheets, so that the cured material could be visually examined. Both the A60 and A62 hand mixed samples had about 10% to 15% of their surface areas covered with a coalesced collection of small bubbles which were introduced into the material during hand mixing. The SEM nozzle sample had about six tiny bubbles visible in the entire mixture. The SEM nozzle + hand mix had about as many bubbles as the A60 and A62, but these were mixed throughout the sample. All four of these samples, cured at 74 ° C/2 h, formed a clear slightly yellowish hard solid material. A sample of the SEM nozzle + hand mix was given a vacuum degas treatment (30 in mercury/5 min) in an effort to remove the entrapped air from the epoxy mixture. The cured material was cloudy and had many more bubbles than the comparable sample which hadn't been degassed. It was theorized, that air is dissolved in the mixture and the vacuum causes more bubbles to form in the epoxy.

The SEM nozzle mixtures aged 7 days at RT appeared cloudy with a lot of bubbles, as compared to the 74 ° C/2 h cured samples. The 7 day samples were similar to the vacuum degassed sample. One might assume that the 7 day RT cure would allow some extra time for the epoxy to lose bubbles. However, with a 90 min pot life, this was not the case. It may be that the elevated temperature cure allows the epoxy to become more fluid early in the curing cycle and thus allows the bubbles to outgas before the epoxy hardens.

4.4 Ballistic Testing Sixty regular TNT filled RAP rounds, fabricated using the modified surface preparation procedures and Armstrong C-7/W epoxy adhesive (nozzle mix + 14 h RT cure + 74 °C/2 h cure) were ballistically tested at an elevated temperature of 63 ° C (145 ° F). The first fifty rounds were fired in a "rocket on" condition, and there were no failures. The remaining ten rounds were fired with the rocket motor off so that the rounds could be recovered and examined; all of the insulators remained undamaged and were securely bonded to the warhead bases.

4.5 Adhesives for Future RAP designs

4.5.1 Room Temperature Testing As shown in Table 1, the methacrylate and all of the epoxy materials with the improved phenolic materials had bond strength values higher than the Armstrong epoxy with the GPI - 100 phenolic center piece. Improvements ranged from 87 % to 185%. The acrylic adhesive (Permabond 665) had a bond strength (10.09 MPa) lower than the Armstrong material (15.93 MPa), and also lower than the minimum requirement of 11.72 MPa. The three silicone adhesives also had low bond strength values. However, these values are considered tentative, because some test samples for each of the three materials (Dow Silastic J, GE RTV 60, and GE RTV 630) had small uncured patches of adhesive on their bond surfaces, and were scheduled for retesting.

4.5.2 Elevated Temperature Testing All samples were conditioned at the elevated test temperature of 63 ° C (145 ° F) for a minimum of 15 minutes prior to testing. As was seen with the Armstrong adhesive, the remaining adhesives also showed a significant drop in tensile bond strength, with an average loss of 76%. All but one of the other adhesives had tensile bond strength values lower that the room temperature requirement of 11.72 MPa. If the elevated temperature bond strength of the Armstrong + phenolic samples (4.32 MPa) is taken as a base line comparison value, since rounds test fired with this material had functioned correctly, then several of the adhesives (all tested with the improved phenolic materials) show tensile bond strength values ranging from notably better to about equal performance. The Ciba-Geigy Araldite 2014, with the high strength MX 2625 phenolic, had the highest bond strength at 21.49 MPa, This was followed by 3M Scotchweld DP 460 (7.76 MPa), Dexter Hysol EA 907 (7.61 MPa), Permabond E32 (5.74 MPa), and Hernon 730 (4.16 MPa); all of which had the MX 2641 phenolic..

4.5.3 Failure Mode The three silicone adhesives, with the partially uncured surfaces are scheduled for retesting. Three of the epoxies (Dexter Hysol EA 907, 3M Scotchweld DP 460, and Permabond E32) had MX 2641 phenolic failures at RT (45% to 100%), indicating that the adhesives were stronger that the material being bonded. The 3M and Permabond materials also had phenolic failure at elevated temperature (10 to 100%), while the Dexter adhesive had cohesive failure (15% to 45 %). The other epoxy adhesive (Ciba-Geigy Araldite 2014 with MX 2625) and the methacrylate (Hernon 730 with MX 2641), also with relatively high bond strength values, also had mixed cohesive/phenolic failures at both room and elevated temperature.The acrylic material (Permabond 665) had 100% cohesive at RT, and 100% adhesive failure at elevated temperature indicating a breakdown in performance at the higher temperature.

4.5.4 Epoxy Appearance This portion of the test plan has not been completed.

5. CONCLUSIONS

Surface cleaning, before and after surface abrasion provided acceptable metal and phenolic surfaces for adhesive bonding of the ASTM D897 test samples and for bonding the RAP round warhead base and glass filled phenolic insulator.

Ballistic testing indicated that the Armstrong C-7/W epoxy, properly mixed and properly applied, was acceptable for bonding the GPI - 100 phenolic insulator to the RAP warhead evaluated in this study.

Other adhesives evaluated in this study may offer improved performance as compared to the Armstrong material for bonding the phenolic insulator to the RAP warhead.

6. RECOMMENDATIONS

Additional low temperature testing is needed to measure the overall performance of the adhesives being evaluated in this study.

Metal filled epoxy materials may offer improved performance, especially at the high temperature firing range, and should be evaluated. Impact resistance is expected to be the limiting factor for the incorporation of these types of adhesives into future RAP designs.

7. REFERENCES

1. ASTM Specification D897, Standard Test Method for Tensile Properties of Adhesive Bonds, ASTM Committee on Standards, 1916 Race Street, Philadelphia, PA.

2 L. J. Hart-Smith et. al., Surface Preparation of Composites for Adhesive-Bonded Repair, Eng. Mat Handbook, Vol. 3: "Adhesives and Sealants", ASM International, 1990.

3. D. K. McNamara, H. M. Hand, and C. Arah, Surface Preparation Technologies for Improved Adhesive Bonding to Graphite/Epoxy Composites, " Proceedings from the 6th International Symposium on Structural Adhesive Bonding, pp 119-130, May 4-7, 1992.

4. Military Specification MIL-M-14/2, Molding Compounds, Phenolic, Thermosetting, Containing Mineral/Glass Fillers, Nov 90.

5. Military Specification MIL-R-9299, Resin, Phenolic, Laminating, Oct 91.

TABLE 1. ASTM D897 TENSILE BOND STRENGTH TEST RESULTS

Sample	Ambient (MPa)	63 ° C (MPa)	% Loss
A60 hand mix	> 46.42	10.76	77
A62 hand mix	50.79	8.99	82
SEM nozzle mix	52.40	8.50	84
SEM nozzle + hand mix	45.56	6.56	86
SEM nozzle + 14 h RT cure	5.95	1.21	80
SEM nozzle + 14 h RT cure + 74 ° C/2 h	23.27	------	--
SEM nozzle + 7 day RT cure	24.77	4.61	81
SEM nozzle + 7 day RT cure*	15.93	4.32	73
Ciba Geigy Araldite 2014*	45.42	21.49	53
Dexter Hysol EA 907*	34.47	7.61	91
Dow Corning RTV Silastic J*	2.90	0.86	70
GE RTV 60*	2.93	0.54	81
GE RTV 630*	4.47	1.65	63
Hernon 730*	29.82	4.16	86
3M Scotchweld DP 460*	37.55	7.76	79
Permabond E32*	33.73	5.74	83
Permabond 665*	10.09	1.85	82

* With phenolic sample between T-Hats

TABLE 2. ASTM D897 TENSILE BOND STRENGTH FAILURE MODE

Sample	Ambient	63 ° C
A60 hand mix	85 - 95 % Co	45 - 55 % Co
A62 hand mix	85 - 95 % Co	45 - 55 % Co
SEM nozzle mix	85 - 95 % Co	45 - 55 % Co
SEM nozzle + hand mix	85 - 95 % Co	45 - 55 % Co
SEM nozzle + 14 h RT cure	85 - 95 % Co	45 - 55 % Co
SEM nozzle + 14 h RT cure + 74 ° C/ 2 h	85 - 95 % Co	---------------
SEM nozzle + 7 day RT cure	60 - 85 % Co	45 - 55 % Co
SEM nozzle + 7 day RT cure*	100 % PS	30 - 50 % Co
Ciba Geigy Araldite 2014*	100 % Co/100% PS	70 - 100 % Co
Dexter Hysol EA 907*	45 - 100 % PS	15 - 45 % Co
Dow Corning RTV Silastic J*	95 % Co	55 - 90 % Co
GE RTV 60*	60 - 90 % Co	100 % Ad
GE RTV 630*	100 % Ad	0 - 80 % Co
Hernon 730*	95 - 100 % Co	75 - 100 % Co
3M Scotchweld DP 460*	100 % PS	50 - 100 % PS
Permabond E32*	45 - 100 % PS	10 - 75 % PS
Permabond 665	100 % Co	100 % Ad

* = with phenolic sample between T-Hats

Failure Modes

Co = Cohesive Failure
AdS = Adhesive failure with the steel
AdP = Adhesive failure with the phenolic
PS = Phenolic split, adhesive did not fail

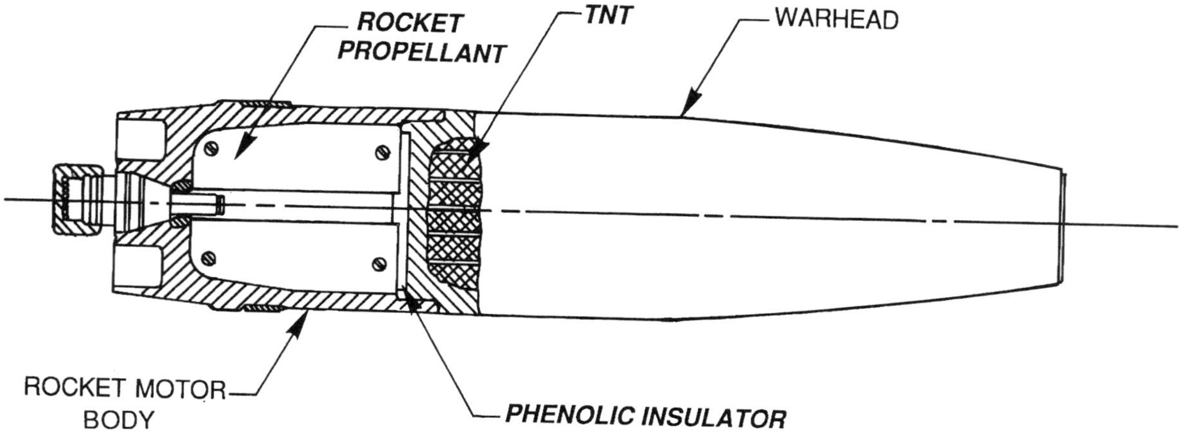

Figure 1. Cross-Sectional View of the Rocket Assisted Projectile

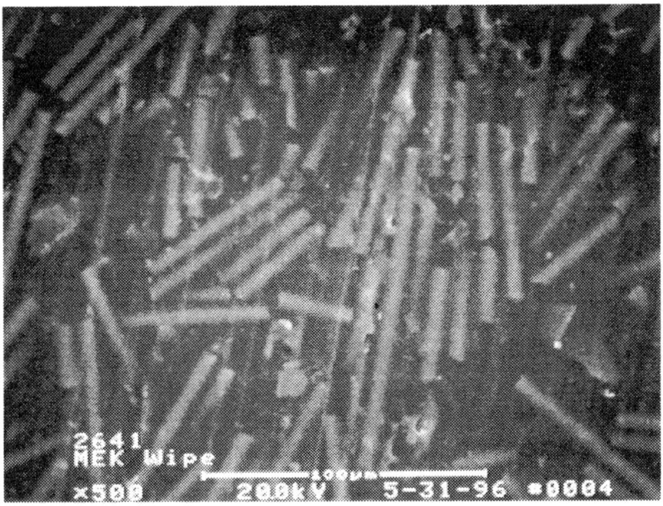

a. Solvent Wiped As Manufactured Surface

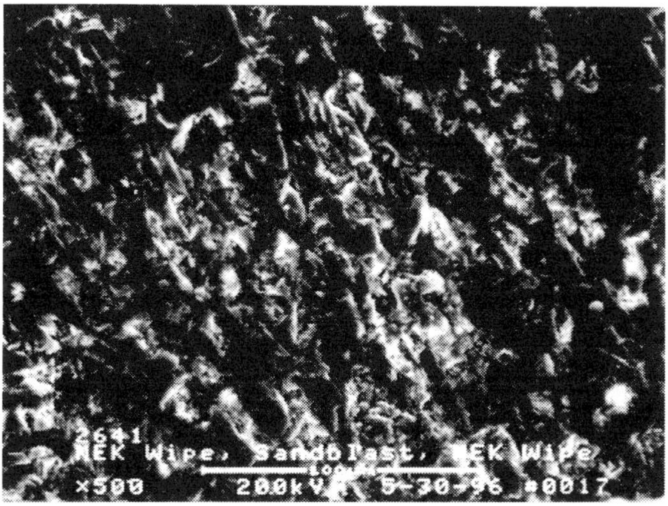

b. Grit Blasted Surface

Figure 2. Surface Views of Glass Reinforced Phenolic Insulators (500X)

28th International SAMPE Technical Conference
November 4-7, 1996

NON-OZONE DEPLETING ALTERNATIVE SEALANTS FOR SMALL CALIBER AMMUNITION

Dean Martinelli, Don Rorabaugh
U.S. Army
Armament Research Development and Engineering Center (ARDEC)
Picatinny Arsenal, Dover NJ, 07806-5000

ABSTRACT

In order to comply with the regulations set forth in the Montreal Protocol and the Clean Air Act, a challenge was presented to identify alternate materials, which are ozone depleting chemical (ODC) free, to seal the casemouth on small caliber ammunition. The existing casemouth sealant material is an asphalt resin with a 1,1,1 trichloroethane (TCA) solvent. A series of ODC free materials were evaluated, ranging from 100% solids to a variety of water-borne materials.

Top ranking alternates were categorized based on their ability to meet environmental regulations, OSHA safety requirements, compatibility with explosives and existing application equipment without compromising performance requirements. Initial fast track testing was used to down select best candidates. Fast track testing included viscosity, vacuum leak, bullet pull, and profile bulge. Chemical analysis and ballistic qualification testing were performed on the best candidates. Chemical analysis techniques used were FTIR and Mass Spectroscopy. Ballistic qualification testing was conducted under a variety of environmental conditions over a wide temperature range. Based on the results obtained, the most promising candidates were found to be water-borne asphalt resin materials.

KEY WORDS: Ozone Depleting Chemical (ODC), Volatile Organic Compound (VOC), 1,1,1 trichloroethane (TCA), Water-borne, Asphalt, Bullet Pull, Vacuum Leak, Profile Bulge, Viscosity, Flammability

1. INTRODUCTION

The Montreal Protocol and the 1990 Clean Air Act have prohibited the production of all class I ozone depleting chemicals (ODC's) as of 31 December 1995. The current material used to seal the projectile/cartridge case interface for all small caliber ammunition is a TCA- borne asphalt material with fillers. TCA has been designated as a class I ODC. In response, the Army initiated a study to identify/develop and evaluate alternate materials to seal the casemouth interface for all small caliber ammunition which encompasses 5.56mm, 7.62mm and .50 caliber, ball and trace. This study was funded through a Strategic Environmental Research Development Program (SERDP), which is a cooperative effort among DOD, DOE and EPA. To continue production of small caliber ammunition, as they await completion of this study, ammo factories have purchased about a one year supply of TCA-borne material.

Figure 1 illustrates the location of the casemouth sealant on a typical small caliber round. The main function of the casemouth sealant is to prevent moisture intrusion into the propellant which could potentially cause misfires or hangfires. The current application method specifies applying sealant material to the inside surface of the cartridge cases. The sealant dries as the cases sit in a random orientation inside a storage container for 4 to 72 hours prior to propellant loading and bullet insertion.

2. EXPERIMENTAL

2.1 Technical Objective The objective of this study was to either identify commercially available alternate sealants or request/contract potential vendors to develop a material that meets requirements listed in 2.2 below:

2.2 Sealant Requirements The acceptance requirements used to screen candidate materials were as follows:

2.2.1 Physical Properties
2.2.1.1 Flammability Materials must have a flashpoint greater than 100 ° F. Pass/fail criteria was used for evaluation. See Table 1 for rating system.
2.2.2.2 Viscosity Current material specifies a viscosity of 40 to 55 cps. See table 1 for rating system.
2.2.1.3 VOC Free In order to avoid further environmental issues such as permits and waste disposal, the desired material must be VOC free.
2.2.1.4 Curing The material must be a one part system which requires no mixing and cures (dries) at varying room temperature and humidity conditions.

2.2.2 Performance Requirements
2.2.2.1 Vacuum Leak The fabricated round shall not release more than one bubble of air when submerged under 1 in. of water and subjected to 7.5 psi for 30 sec. See table 1 for rating system.
2.2.2.2 Bullet Pull The force to extract the bullet from the cartridge case; shall not be less than 45 lbs. for 5.56mm, not less than 60 lbs. for 7.62mm, not less than 200 lbs. for .50 caliber. See table 1 for rating system.
2.2.2.3 Profile Bulge Rounds are tested for bulged necks with a calibrated ring gage. This is a pass/fail test. See table 1 for rating system.
2.2.2.4 Universal Sealant The desired material must exhibit successful results on all 5.56mm, 7.62mm and .50 caliber ammunition (ball/trace)
2.2.2.5 Reliability/Durability The desired material must maintain seal integrity throughout the life cycle of the ammunition similar to existing small caliber ammunition requirements.

2.2.3 Compatibility

2.2.3.1 Compatible With Existing Application Equipment The desired material must be capable of being applied with existing application equipment and existing material handling systems with little/no modifications and minimal capital equipment costs.

2.2.3.2 Propellant Compatibility The desired material must be compatible with the existing double base (nitrocellulose/nitroglycerine) propellant as tested IAW with MIL-STD-286B. Reference 1.

2.2.3.3 Weapons Systems The material cannot cause any solid particle build up which would cause weapon fouling.

2.3 Technical Approach Figure 2 shows the methodology used to technically screen potential candidate materials.

2.3.1 Literature Search A letter was sent to potential vendors listing desired properties for an alternate material requesting recommendations for potential candidates. Literature received from the vendors was reviewed and materials were down selected based upon being ODC free, VOC compliant, viscosity , flammability and potential to be applied with existing application equipment.

2.3.2 Fast Track Testing Samples of the most promising materials based upon the literature study were obtained for fast track testing which included vacuum leak, bullet pull and profile bulge. Initial fast track testing was conducted on 7.62mm ball ammunition. Down selected materials were subsequently evaluated on 7.62mm trace, 5.56mm ball/trace, and .50 caliber ball/trace using the same fast track test criteria.

2.3.3 Propellant Compatibility/Versatility Most promising candidates from fast track testing were subjected to vacuum stability testing to determine their relative compatibility with ammunition propellants.

2.3.4 Large Scale Fabrication for Qualification Test A best candidate sealant was selected based on screening procedures described above. Large quantities of ammunition were produced using the best candidate material, with some minor adjustments to existing application equipment. These rounds were used for qualification testing. Qualification tests evaluated modified ammunition for ballistic performance, weapon compatibility, environmental effects (temperature/humidity), and long term reliability/durability.

2.3.5 Specification/Purchase Description A material specification with acceptance requirements was prepared based on the best candidate material.

2.3.6 Modifications to Existing Application Equipment Modifications to existing application equipment will be made before incorporating alternate sealant into production.

2.4 Sealant Application/Round Fabrication Candidate materials were fabricated for fast track evaluations at both ARDEC and Lake City Army Ammunition Plant (LCAAP). All rounds which were evaluated at ARDEC and at LCAAP, except where specified, used hand application of sealant. Sealant was applied to the inside mouth of the cartridge case and allowed to dry for at least 24 hours with cartridges remaining in an upright position. Cartridges were then loaded, bullets inserted and brass cartridges were crimped. Cartridges at ARDEC were not loaded or crimped.

Sealant application and fabrication was performed on production application equipment with the most promising alternate materials. Current application equipment applies sealant to cartridge cases in the inverted position and moves along on a dry track. On the current plate loaded lines dry tracks allow for 34 seconds of drying before falling into a movable container where cartridges sit in a random position. Sealant is applied at a rate of 300 rounds per minute. It is extremely important that alternate materials become dry to the touch within 34 seconds, otherwise the sealant will splatter and flow or pool to one side of the cartridge. Slight modifications to the existing apparatus could be made to accommodate materials that would need a slightly longer drying time. Extended drying times could be extended 1 to 1.5 minutes.

After cartridge cases are coated, filled movable containers are moved to a waiting area before loading. The cartridges can sit for 4 to 72 hours in an environmentally uncontrolled area. The cartridges are loaded with propellant, bullets are inserted and the cartridges are mechanically crimped.

3. DISCUSSION

3.1 Background Search Anaerobic materials were investigated on a laboratory scale and proven to be a successful alternate, ref. 2. Anaerobics are a family of acrylics which polymerize in the absence of air and in the presence of active metal ions, ref. 3. They are desirable from an environmental standpoint since they are 100% solids and solvent free. Although anaerobics demonstrated the capability of providing an adequate seal on a laboratory scale, they require specific application equipment and a radical change in material handling procedures. Therefore implementation of anaerobics into production would require large expenditures, ref 4. In addition, anaerobics exhibited extremely high bullet pull values. Although the requirements for bullet pull do not specify a high limit, cartridges split and bullets remained bonded to cartridges during ballistic testing. Government and Contractor staffs at the production facility agreed not to pursue anaerobics and therefore shifted their focus to solvent systems containing low amounts of VOC's or preferably water-borne systems.

3.2 Vendor Search Initially over 200 potential vendors were contacted in writing, specifying a desired list of requirements. Approximately 100 vendors contributed literature on their recommendations. Half of the recommendations were eliminated after reviewing the manufacturer's supplied data. Many of these materials were either two part systems, had extremely high viscosity, contained high amounts of VOC's, had low flash points or were classified as hazardous. The remaining 50 or so were selected for initial fast track testing.

3.3 Initial Fast Track Testing Initial fast track testing was conducted on several water-borne/solvent-borne systems with either acrylic, synthetic polymer blends or elastomeric resins. Testing was conducted using 7.62mm ball rounds since this caliber was considered to be a "middle of the road" round. See table 1 for rating system and table 2 for results. Not all materials that failed the environmental requirements (either they contained low VOC's or were considered border line hazardous) were excluded from the test matrix. These materials were selected to remain in the original test matrix as back-ups in case a totally environmentally benign material could not be found for this application, however no testing was performed. Examples of these materials were Spartan Corp., SA96523, SA06832, Gumbs Corp., BS 8221, Red Devil , BASF H 1597, DEVAN 704.2, 3M 800, 3M -900 Duct SE. Dymax, UV 652 was left on the list and did pass all environmental requirements, but was eliminated since special explosion proof UV equipment would be needed in production.

The next set of materials either passed all physical requirements, had some very borderline VOC contents or were just slightly below the desired flammability range, but were still tested. In most cases these materials were of the water-borne synthetic polymer blends or water-borne elastomeric materials (see table 2 for results). Many of these materials caused cartridge cases to excessively bulge upon bullet fabrication thus failing the profile requirement. In general most of the materials from this family also exhibited unsuccessful vacuum leak results. Therefore these materials were not considered successful since they either failed the vacuum leak requirement or the profile gage requirement, if not both. Some of these materials were also evaluated for effect on bullet pull and in most cases were either rated as good or moderate; but, were not of significance since they failed to show any ability to seal and/or failed profile gage. Examples of these materials were Hernon 32222 and 29991, Loctite 17044 and Loctite L5-560, Pacer WB114-A and WB114-C, Ameritek 1004, BF Goodrich 5002 and 1610B, Bostik 8510, Bebetex, 8156, 8151, 7780, Dolphin 7030, HB Fuller 4009, Basic BR 3207, Ruscoe RB707, Carlisle CS 1158A and CS 1158B. Loctite Pipe Seal is an acrylic material which showed some promise but its viscosity far exceeded acceptable limits and therefore was not evaluated further. G.J. Nikolas 940725 and 941222 showed some promise as far as viscosity, vacuum leak, bullet pull and profile but they both contained VOC's. The G.J Nikolas materials were made of sugar resins. One material, Manus A700 showed some promise by passing the vacuum leak requirement and profile gage requirement, but appeared to contain some low level VOC's. This material was a water-borne (with some other VOC's) asphalt material, similar to the current standard production casemouth sealant without the TCA (table 2). This lead the program down a new path to search for water-borne asphalt materials. Vendors were contacted once again and although it appeared that most vendors did not have a commercially available water-borne asphalt to suit our needs several were willing to formulate one.

3.4 Further Fast Track Testing 3M 42 NF plus NV a water-borne asphalt material, showed acceptable results but was too viscous and resulted in a high number of profile gage failures. Monsey Products Inc. submitted several water-borne asphalt formulations; 10927-19, -32, -39, -55, -91, -95. All of these material were tested and all passed the requirements for 7.62mm ball ammunition seemingly well. The most promising being 10927-39 which was thinned with water and tested at two different viscosities. Beacon Chemical also supplied a series water-borne asphalt formulations 950728-14,-15, 950729-10, 950731-5. Like the Monsey Products formulations, all of the Beacon materials passed the requirements for 7.62mm ball ammunition. However, when we started to evaluate these materials using drying techniques which simulated production procedures, it was found that the water- borne materials, as expected, did not dry as quickly as the current material. Beacon Chemical 950728-14 was tested with cartridges dried mouthup and mouthdown, see results in table 2.

Hernon Manufacturing Inc. supplied the widest variety of water-borne asphalts in an attempt not only to meet the requirements for 7.62mm ball ammunition but a material which would be versatile enough for all three calibers and more closely compatible with the existing production equipment. First attempts were made with Hernon 32784, which was a highly filled material. This material met most requirements but drying and viscosity issues were still of concern. A series of Hernon materials (32942, 32943, 32944, 32945, 3462, 34142, 34143) which were modified water-borne asphalts were evaluated. Again results were acceptable but these materials, even though they dried to the touch in a reasonable period of time, remained tacky and when cartridges were loaded with propellant, grains adhered to the coated cartridges. Hernon subsequently submitted another modified water-borne asphalt, 34201. This low viscosity material passed all requirements, was not as tacky, (so that propellant grains did not stick to cartridges upon loading) and appeared to be compatible with the existing application process at the ammunition plant.

3.5 Fast Track Testing With Other Calibers Fast Track testing continued on the 7.62mm trace, 5.56mm ball/trace and .50 caliber ball/trace with materials which were found to be most promising on the 7.62mm ball round. Most of the materials were from Hernon Manufacturing Inc, Beacon Chemical, and Monsey Products. Some other Loctite Corp. and 3M formulations were also evaluated (data on other calibers not included). However Hernon 34201 still remained the best material across all three calibers as indicated with an "excellent" rating under caliber compatibility in table 2. The other candidates which were given a "good" rating under caliber compatibility were Beacon Chemical, 950728-14 and Monsey Products Inc, 10927-39.

3.6 Chemical Analysis All of the materials tested are proprietary formulations, therefore to obtain more knowledge of the best candidate alternates, FTIR and Mass Spectroscopy was conducted. Table 3 shows the results from the analysis. All of the alternates materials are water-borne and consist of asphalt resins with modifiers and fillers. For instance, Hernon 34201 is a modified asphalt with a styrene/acrylate copolymer. Similarly Beacon 950728-14 is modified with a methacrylate polymer.

3.7 Compatibility with Energetics The current asphalt resin used as a casemouth sealant has had no reported compatibility problems over the past 40 years. It was surprising then that it failed vacuum reactivity tests as did the proposed replacements. In light of these results, it was decided to rely upon subsequent qualification tests to surface any potential compatibility problems.

3.8 Lot Acceptance Testing A full lot acceptance evaluation was conducted on rounds modified with the best alternate, material Hernon 34201. Lot acceptance tests for ballistic performance which included velocity, pressure, action time as well as waterproofness, profile and bullet pull. Results from this evaluation are not included in this paper but all did pass the requirements.

3.9 Qualification Tests Large quantities of modified 7.62mm, 5.56mm and .50 caliber ammunition (ball/trace) were fabricated with existing production equipment with some modifications at LCAAP. All rounds were sealed, for this test, with the best alternate candidate, Hernon 34201. The purpose of this test is to determine whether modified ammunition will meet all performance, safety, human factors and logistic requirements contained in the applicable specifications for this ammunition. The testing will be performed by Aberdeen Test Center (ATC), Aberdeen Maryland and will consist of bullet pull, velocity, pressure, action time, dispersion, high-temperature storage, low temperature storage, high-temperature functioning, low-temperature functioning, temperature-humidity, chemical compatibility, thermal shock, reliability and fouling, waterproofness. Human factors, Safety and Health, and Logistics Supportability will also be assessed. Reference 5 contains a detailed description of the test plan. Results of this test were not available at time the manuscript was written.

3.10 Production Problems While fabricating large quantities of modified ammunition for the qualification tests several, high rate application problems were identified which must be addressed before full production can begin. Some of the problems identified were insufficient drying, settling of solids, and particle build up on applicators.

3.10.1 Drying Cartridges remain in the inverted position for 34 seconds before falling into containers which are later moved to the loading wings. Water- borne materials have a much lower vapor pressure than TCA-borne sealants and therefore do not dry as fast. Undried cartridges which fall into movable containers cause sealant to splatter, and contaminate the outside walls of neighboring cartridges, also this allows sealant to run or pool to one side of the cartridge since cartridges become randomly oriented once they fall into movable containers. High rates of leakers are detected on such rounds since sealant pools to one side leaving open paths on the other. Contractors at LCAAP and the Army has initiated a plan to alleviate this problem but will not be discussed in this article.

3.10.2 Applicator Build-Up Current application systems utilize a piston system which moves down into a reservoir of sealant then upwards into inverted cartridges and squeezes sealant which is deposited onto the cartridge. Application rates are as high as 300 cartridges per minute and the piston stroke is about 2 inches. Heat build up on the stainless steel piston (applicator) occurs as the piston rubs on a nylon bushing. It appears as the heat builds the water reaches its boiling point and solids start to build up near the bushing. Initial attempts allowed for 2 to 2.5 hours of production before build up started to inhibit application of sealant and the line had to be stopped to clean off the build up. Normal production runs are about 8 to 9 hours before clean up. Studies have been planned to address this problem at a later date.

3.10.3 Solids Settling Sealant in the reservoir showed signs of solids settling out of solution. The reason for this is unclear; however there are plans to investigate the problem at a later date.

4.0 Conclusions

The best alternate casemouth sealant evaluated in this study was Hernon Manufacturing Inc. 34201, now termed Hernon Bullet Proof. Other acceptable candidates were Beacon Chemical 950728-14 and Monsey Products Inc, 10927-39. All of these materials are water-borne modified asphalt materials.

5.0 Summary

The results from this study have provided several benefits to the Tri Services besides eliminating over 10,000 lbs of ODC's annually. LCAAP would have been faced with a potential shut down of ammunition manufacture due to the unavailability of a suitable alternate casemouth sealant. It is estimated that LCAAP will save approximately $0.5M annually by eliminating costs for disposal of wastes and permits for usage since the alternate materials do not contain any VOC's. Also since the alternate materials are not considered hazardous there is no need for pollution control equipment, also resulting in a significant cost savings. Although some modifications to the existing production line will be necessary in order to scale up to full production, capital investments will not be near that needed to apply anaerobic materials. Estimates to incorporate anaerobics into the existing operation were as high as $0.5M per unit.

6.0 References

1. Military Standard MIL-286B Method 403.1.2, Propellants Solid: Sampling Examination and Testing, Vacuum Stability Tests.
2. D. Martinelli and R. Bonk, Initial Evaluation of Environmental Casemouth Sealants for 7.62mm Ammunition, U.S. Army Technical Report, ARAED-TR-88035, U.S. Army Armament Research Development and Engineering Center 1988.
3. A. Landrock, Adhesive Technology Handbook, Noyes Publications, Park Ridge, NJ, 1985.
4. S. Parker, Alternate Casemouth Waterproofing Compound for 5.56mm and 7.62mm Ammunition, Lake City Army Ammunition Plant, 1994.
5. A. Rose and G. Steier, Detailed Test Plan for the Technical Feasibility Test (IFT) of the Alternate Casemouth Sealants for Cartridge, 5.56mm, 7.62mm and Caliber .50, U.S. Army Aberdeen Test Center, Aberdeen Proving Ground, 1996.

7.0 Biography

Dean Martinelli is a Materials Engineer with the Polymer and Adhesives Branch at the U.S. Army Armament Research Development and Engineering Center and has over 12 years experience in adhesives technology. Mr. Martinelli has conducted numerous studies on adhesive bonding/sealing for a variety of materials which includes polyolefins, engineering thermoplastics, elastomers, composites and metals. He has extensive application experience on armament munitons , weapons, armor and army vehicles which includes small caliber ammunition, mines, mortars, tank ammo, large caliber artillery, tanks and howitzers. He has been involved with many Red Team Root Cause Failure Analysis as related to armaments and is involved with several concurrent engineering teams involving future armament systems. Most of his recent work has centered around the development of environmentally benign alternative cleaners, degreasers, adhesives, sealants and processes. He has authored/co-authored over 40 technical reports for DOD and has authored/presented several articles at various symposia.

Donald T. Rorabaugh is a Materials Engineer with the Polymer and Adhesives Branch at the U.S. Army Armament Research Development and Engineering Center and has over 28 years experience in materials technology. Mr. Rorabaugh has conducted a wide range of materials studies leading to two patent awards, a patent disclosure and numerous Army/ DOD awards for improved materials for armament/military armor/anti-armor, shelters, mines, mortars, large/small caliber ammunition, vehicles, and aircraft. He has written over 100 reports and publications along with a chapter in the "Gun Propulsion Technology" Book, and has received over 100 suggestion awards. He also is a working member of the ASTM E06.53 Committee on Tactical Shelters. Current efforts are aimed at developing new/improved environmentally friendly adhesives/sealants/coatings and related application procedures for military applications.

8.0 Acknowledgment

The authors would like to thank Ms. Sheryl Parker, engineer, Olin Corp., LCAAP, Mr. John Middleton, engineer, Close Combat Armaments Center, ARDEC and Mr. James Tira, Environmentally Conscious Manufacturing Specialist, DOE/ Kansas City Plant, for their assistance in this study.

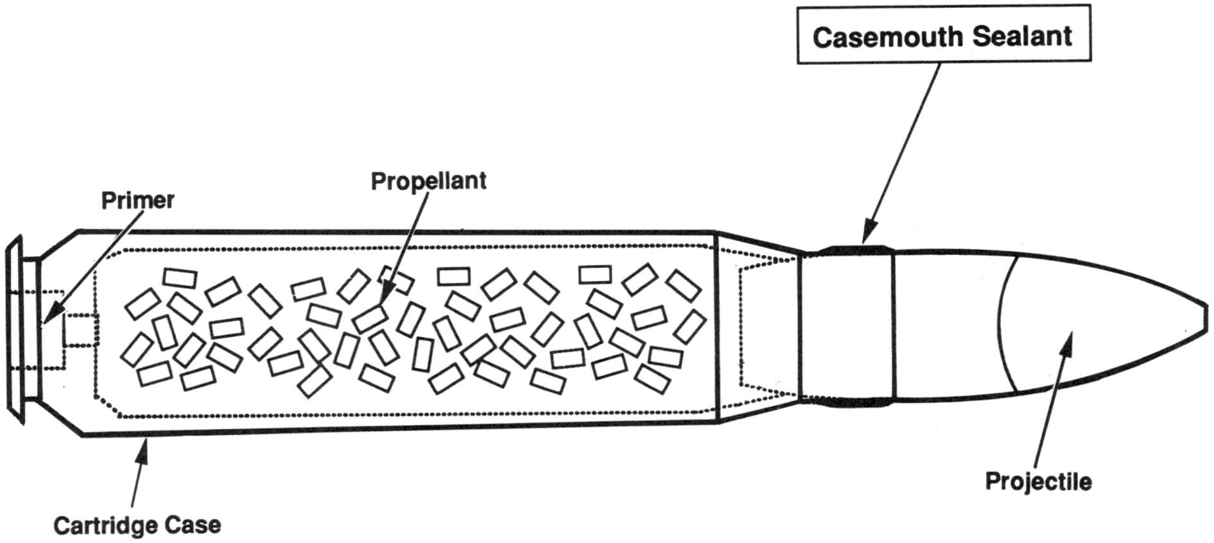

Figure 1 - Schematic of a Small Caliber Bullet

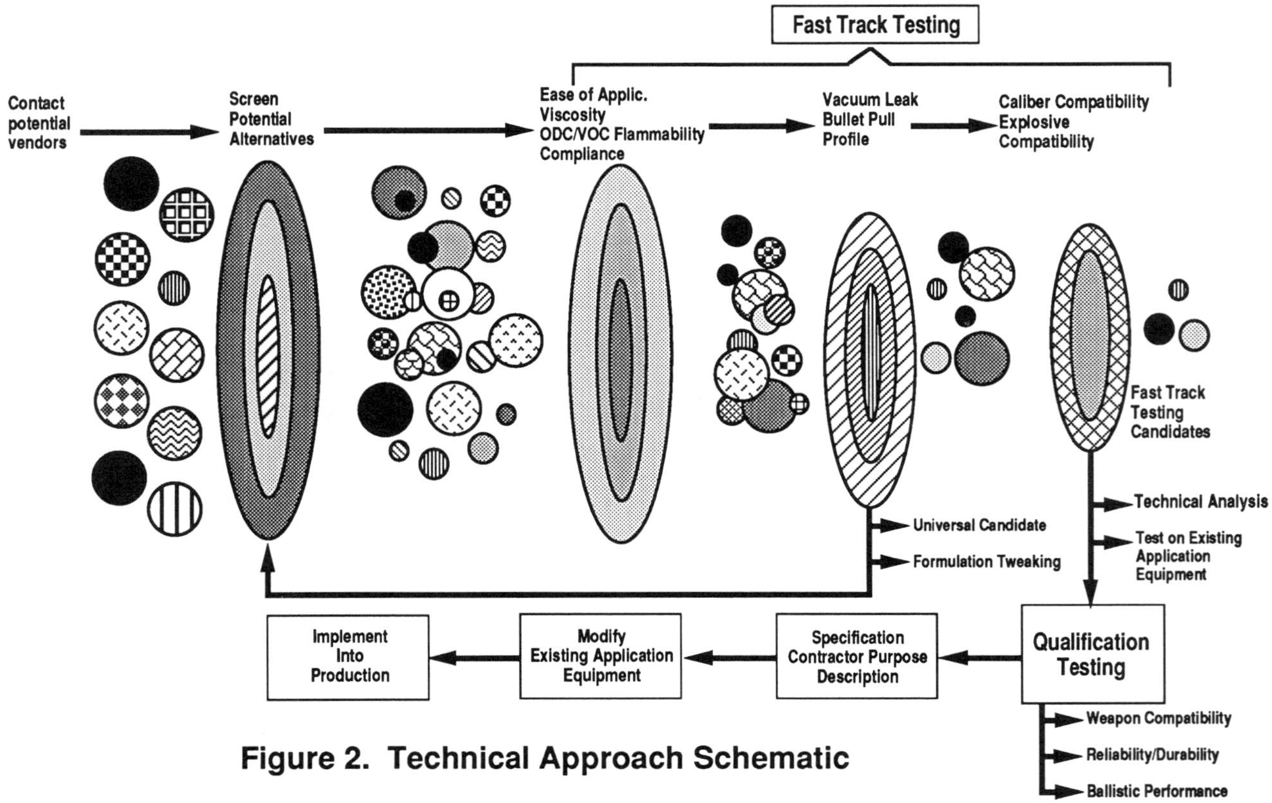

Figure 2. Technical Approach Schematic

CATEGORY	DESCRIPTION	RATING
TEST LOCATION	ARDEC (AR)	
	LAKE CITY AAP (LC)	
ODC'S	CONTAINS NO ODC'S	PASS (P)
	CONTAINS ANY ODC'S	FAIL (F)
VOC'S	CONTAINS NO VOC'S	PASS (P)
	CONTAINS ANY VOC'S	FAIL (F)
FLAMMABILITY	FLASHPOINT GREATER THAN 100 F	PASS (P)
	FLASHPOINT LESS THAN 100 F	FAIL (F)
VISCOSITY	LESS THAN 600 CP	EXCELLENT (E)
	BETWEEN 600-2000 CP	GOOD (G)
	BETWEEN 2000-4000 CP	MARGINAL (M)
	GREATER THAN 4000 CP	POOR (P)
VACUUM LEAK	LESS THAN 6% FAILURES	EXCELLENT (E)
(5.56MM/7.62MM)	BETWEEN 6-10% FAILURES	GOOD (G)
	BETWEEN 10-15% FAILURES	MARGINAL (M)
	GREATER THAN 15 % FAILURES	POOR (P)
(.50 CALIBER)	LESS THAN 2% FAILURES	EXCELLENT (E)
	BETWEEN 2-6% FAILURES	GOOD (G)
	BETWEEN 6-10% FAILURES	MARGINAL (M)
	GREATER THAN 10% FAILURES	POOR (P)
BULLET PULL	AVG.>45 LBS. AND AVG.-3 ST.DV. >45	EXCELLENT (E)
(5.56MM)	AVG.>45 LBS. AND AVG.- ST.DV. >45	GOOD (G)
	AVG.>45 LBS. AND AVG.- ST.DV.<45	MARGINAL (M)
	AVG.<45 LBS.	POOR (P)
(7.62MM)	AVG.>60 LBS. AND AVG.-3 ST.DV. >60	EXCELLENT (E)
	AVG.>60 LBS. AND AVG.- ST.DV. >60	GOOD (G)
	AVG.>60 LBS. AND AVG.- ST.DV.<60	MARGINAL (M)
	AVG.<60 LBS.	POOR (P)
(.50 CALIBER)	AVG.>200 LBS. AND AVG.-3 ST.DV. >200	EXCELLENT (E)
	AVG.>200 LBS. AND AVG.- ST.DV. >200	GOOD (G)
	AVG.>200 LBS. AND AVG.- ST.DV.<200	MARGINAL (M)
	AVG.<200 LBS.	POOR (P)
PROFILE	LESS THAN 6% FAILURES	EXCELLENT (E)
(5.56MM/7.62MM)	BETWEEN 6-10% FAILURES	GOOD (G)
	BETWEEN 10-15% FAILURES	MARGINAL (M)
	GREATER THAN 15 % FAILURES	POOR (P)
(.50 CALIBER)	LESS THAN 2% FAILURES	EXCELLENT (E)
	BETWEEN 2-6% FAILURES	GOOD (G)
	BETWEEN 6-10% FAILURES	MARGINAL (M)
	GREATER THAN 10% FAILURES	POOR (P)
EXPLOSIVE	PER MIL-STD-286B, METHODS 403.1.2	
COMPATIBILITY	0-3ML OF GAS	COMPATIBLE (C)
	3-5ML OF GAS	MARGINAL (M)
	5+ML OF GAS	INCOMPATIBLE (IC)
CALIBER	PASSING RESULTS ON THREE CALIBERS	EXCELLENT (E)
COMPATIBILITY	PASSING RESULTS ON TWO CALIBERS	GOOD (G)
	PASSING RESULTS ON ONE CALIBER	MARGINAL (M)
	PASSING RESULTS ON ZERO CALIBERS	N/A

Table 1: Fast Track Testing Rating System

CALIBER: 7.62MM
TYPE: M80 BALL

SEALANT	TESTED AT	ODC FREE	VOC FREE	FLAMM COMP	VISC	VAC LEAK	BULL PULL	PROF BULGE	EXPLO COMPAT	CALIBER COMPAT	NON VOLATILES	VOLATILES
CURRENT MATERIAL	AR/LC	F	F	F	G	E	E	E	IC	E	ASPHALT, ROSIN	1,1,1 THRICHLOROETHANE
HERNON 32222	AR	P	P	P	G	PR	M	M		N/A	ACRYLIC POLYMERS	WATER
HERNON 29991	AR	P	P	P	G	PR	M	M		N/A	ACRYLIC POLYMERS	WATER
LOCTITE 17044	AR/LC	P	P	P	E	M	PR	PR		N/A	NATURAL LATEX	WATER
LOCTITE L5-560	LC	P	P	P	E	E	G	E		M	ACRYLIC COPOLYMERS	WATER
LOCT PIPE SEAL	AR	P	F	P	PR	E		E		N/A	POLYGLYCOL, DIMETHACRYLATES	
G.J. NIKOLAS 940725	LC	P	F	P	E	E	E	E	IC	N/A	SUGAR RESINS	ETHYL ACETATE
G.J. NIKOLAS 941222	LC	P	F	P	E	E	E	E		N/A		ACETATES, WATER
3M-800	AR	P	F	F		PR	PR			N/A	ACRYLONITRILE/BUTADIENE POLY	MEK, MIBK
3M-900 DUCT SE	AR	P	F	F						N/A	SYNTHETIC RUBBER	KAOLIN,METHYLPENTANE
PACER WB114-A	AR	P	P	P	E	M				N/A	POLYCHLOROBUTADIENE	WATER
PACER WB114-C	AR	P	P	P	E	M		M		N/A	ACRYLIC RESIN	WATER
AMERITEK 1004	AR	P	P	P	G	M	M	PR		N/A		WATER
BF GOODR 5002	AR	P	F	F		PR	G	PR		N/A		MEK,TOLUENE,ALCOHOLS
BF GOODR 1610B	AR	P	F	F		G	PR			N/A		WATER, OTHER VOC'S
BOSTIK 8510	AR	P	P	P	G	M		PR		N/A		WATER
BOSTIK BEBETEX	AR	P	F	P	G	E		PR		N/A		WATER, XYLOL
BOSTIK 8156	AR	P	P	P	G	PR		PR		N/A	NEOPRENE	WATER
BOSTIK 8151	AR	P	P	P	G	M		PR		N/A	NEOPRENE	WATER
BOSTIK 7780	AR	P	P	P		PR				N/A	POLYURETHANE	ACETATE,MEK,ACETONE
DOLPHIN 7030	AR	P	F	F	G	PR	G	PR		N/A	THERMOPLASTIC RUBBER	NAPTHA, TOLUENE
HB FULLER 4009	AR	P	P	P	G	PR	G	PR		N/A	POLYURETHANE	WATER
BASIC BR 3207	AR	P	P	P	G	G	M	PR		N/A	ACETATE POLYMER	WATER
RUSCOE RB707	AR	P	P	P		G	G	PR		N/A	POLYVINYL ACETATE	WATER
CARLISLE												
CS1158A	AR	P	F	F		E		G		N/A	ETHYLENE PROPYLENE RUBBER	NAPTHA, TOLUENE
CS1158B	AR	P	F	F		M		G		N/A	POLYPROPYLENE,POLYBUTENE	TOLUENE
SPARTAN												
SA96523	AR	P	P	P						N/A	POLYMER LATEX	WATER
SA06832	AR	P	P	P						N/A	POLYMER EMULSION	WATER
GUMBS BS 8221	AR	P	F	F						N/A		
RED DEVIL	AR	P	F	P						N/A	BUTYL RUBBER	MINERAL SPIRITS
BASF H 1597	AR	P	F	P						N/A	STYRENE/ACRYLATE COPOLYMER	WATER
DEVAN 704.0	AR	P	F	P		PR				N/A	ACRYLIC POLYMER	WATER
DEVAN 704.2	AR	P	F	P		PR				N/A	ACRYLIC POLYMER	WATER
DYMAX -UV 652	AR	P	P	P						N/A	POLURETHANE,METHACRYLATE	NONE

BOLD INDICATES MACHINE APPLIED

Table 2: Fast Track Test Results

CALIBER: 7.62MM
TYPE: M80 BALL

SEALANT	TESTED AT	ODC FREE	VOC FREE	FLAMM COMP	VISC	VAC LEAK	BULL PULL	PROF BULGE	EXPLO COMPAT	CALIBER COMPAT	NON VOLATILES	VOLATILES
MANUS A700	AR	P	F	P		E		G		N/A	PETROLEUM ASPHALT	WATER, LOW LEVEL VOC'S
HERNON 32784											ASPHALT WITH FILLERS	WATER
MOUTHUP	LC	P	P	P	E	E	G	E	M	G		
MOUTHDOWN	LC	P	P	P	E	E	E	PR	M	G		
HERNON 34142	LC	P	P	P	E	E	M	E		N/A	ASPHALT WITH FILLERS	WATER
HERNON 34143	LC	P	P	P	E	E	E	M		N/A	ASPHALT WITH FILLERS	WATER
HERNON34201											ASPHALT, STYRENE/ACRYLIC	WATER
40CP	LC	P	P	P	E	E	G	E	IC	E		
40 CP MOUTHUP	LC	P	P	P	G	E	G	E	IC	E		
40 CP MOUTHDN	LC	P	P	P	G	E	G	E	IC	E		
HERNON 32942	AR/LC	P	P	P	G	E	E	E		M	ASPHALT WITH MODIFIERS	WATER
HERNON 32943	AR/LC	P	P	P	G	E	E	E		M	ASPHALT WITH MODIFIERS	WATER
HERNON 32944	AR/LC	P	P	P	G	E	E	E		M	ASPHALT WITH MODIFIERS	WATER
HERNON 32945	AR/LC	P	P	P	G	E	E	E		M	ASPHALT WITH MODIFIERS	WATER
HERNON 3462	AR/LC	P	P	P	G	E	E	E		M	ASPHALT WITH MODIFIERS	WATER
BEACON CHEM												
950728-14, 560 CP	AR/LC	P	P	P	E	E	E	E	IC	G	ASPHALT WITH MODIFIERS	WATER
950728-15, 560 CP	AR/LC	P	P	P	E	E	E	E		M	ASPHALT WITH MODIFIERS	WATER
950729-10	AR	P	P	P		E		G		N/A	ASPHALT WITH MODIFIERS	WATER
950731-5	AR	P	P	P		E		G		N/A	ASPHALT WITH MODIFIERS	WATER
950728-14, 253 CP											ASPHALT WITH MODIFIERS	WATER
MOUTH UP	LC	P	P	P	E	E	G	E		G		
MOUTHDOWN	LC	P	P	P	E	E	G	PR		G		
MONSEY												
10927-19	AR	P	P	P	PR	E		G		N/A	ASPHALT WITH FILLERS	WATER
10927-32	AR	P	P	P	PR	M		G		N/A	ASPHALT, BUTADIENE/STYRENE	WATER
10927-39, 1440 CP	AR/LC	P	P	P	G	G	E	E	C	G	ASPHALT WITH FILLERS	WATER
10927-39, 480 CP	LC	P	P	P	E	G	G	E	C	G	ASPHALT WITH FILLERS	WATER
10927-55, 190 CP	LC	P	P	P	E	G	M	E		G	ASPHALT WITH FILLERS	WATER
10927-91	AR	P	P	P	G	E		G		N/A	WAX	WATER
10927-95	AR	P	P	P	G	E		G		N/A	ASPHALT WITH FILLERS	WATER
3M 42 NF PLUS NV	LC	P	P	P	G	M	G	PR		N/A	ASPHALT,STYRENE/BUTADIENE	WATER

BOLD INDICATES MACHINE APPLIED

Table 2: Fast Track Test Results (continued)

MATERIAL	BASE RESIN	SOLVENTS	VISCOSITY (CPS)	% SOLIDS	MODIFIERS	FILLERS
CURRENT MATERIAL	ASPHALT MIXTURE	1,1,1 TRICHLOROETHANE AND OTHER VOC'S	40-55	40	-	ROSIN COMPONENTS
HERNON 32784	ASPHALT MIXTURE	WATER	-	49	BUTYL RUBBER	OXIDES OF TI & FE
HERNON 34201 (BULLET PROOF)	ASPHALT MIXTURE	WATER	15	54	STYRENE/ACRYLATE COPOLYMER	-
BEACON 950728	ASPHALT MIXTURE	WATER	99	45	METHACRYLATE POLYMER	-
MONSEY PRODUCTS 10927-39	ASPHALT MIXTURE	WATER	86	34	-	CLAY FILLER

Table 3: Results of Chemical Analysis

28th International SAMPE Technical Conference
November 4-7, 1996

MODULAR COMPOSITE SPACECRAFT STRUCTURES

Harry Dursch and Paul Nedervelt
Boeing Defense & Space Group
Seattle, WA 98124
206-234-2683

ABSTRACT

As part of an ongoing effort to decrease costs and increase acceptance of composite spacecraft structure, Boeing Defense & Space Group has developed a design and manufacturing process resulting in a low cost and high performance modular composite spacecraft structure. The design consists of a one piece multi-sided co-cured unibody with a mechanically attached upper deck and adhesively bonded lower deck. All structure is honeycomb sandwich using high modulus graphite fiber/121°C (250°F) cure cyanate ester prepreg and perforated aluminum core. Access to the interior of the unibody is achieved by removing the mechanically fastened top deck and through multiple access holes machined in the sides of the unibody. The co-cured manufacturing process used to fabricate the unibody results in significant cost savings compared to more traditional manufacturing technologies. This paper discusses 1) trade studies involved with the material selections, 2) process development required to achieve a low void content co-cured structure, 3) development of insert allowables for sandwich structures with thin facesheets, and 4) results from fabrication of full-sized prototype and flight hardware.

KEY WORDS: Composites, Composite Structures, Space Applications

INTRODUCTION

The All Composite Experimental Spacecraft Structure (ACESS) is a BMDO funded program to demonstrate the beneficial use of advanced composites in small spacecraft structures. The program goals were to 1) develop and verify low cost manufacturing approaches using advanced composites, and 2) reduce overall structure weight 40% as compared to an aluminum structure. The ACESS assembly is the module for all experiments and their associated hardware being flown on the Space Technology Research Vehicle (STRV-2) mission. STRV-2 is scheduled for launch in 1998 and will be a one-year mission in a 70-deg inclination, elliptical orbit (460 km x 1800 km). STRV-2 will carry several experiments including a mid-wave infra-red (MWIR) telescope and the associated vibration isolation system mounted to the interior surface of the bottom deck, space exposure and laser communication experiments mounted to the exterior surface of the top deck, and meteoroid and space debris experiments mounted to the exterior surfaces of the 12-sided unibody. In addition, ACESS adds a low-cost and lightweight composite structure to the other technologies being demonstrated on STRV-2. Figure 1 shows an exploded view of the ACESS assembly and the locations of the experiment hardware.

ACESS has been designed to accommodate launch loads for both the Taurus and Pegasus XL launch vehicles. The structure is primarily stiffness-designed (due to MWIR telescope jitter requirements) and all structures are thin composite facesheets/aluminum core sandwich panels. The design has been verified with detailed stress, dynamic and thermal analysis, along with the material and component testing discussed in this paper. At program completion, two identical assemblies will be delivered - one assembly being an engineering development model (EDM) and the second assembly being the structure launched on STRV-2. Both assemblies will have undergone centrifuge acceptance testing prior to delivery.

One primary design objective was low cost manufacturing of composites. The goal was to reduce costs by building a robust design that is modular and adaptable. A bottoms-up costing for small hexagonal composite configuration that was baselined early in the program indicated that it could be built for 960 fabrication and assembly hours. Similar pricing for an equivalent conventional aluminum design showed that it would take 2865 hours fabrication hours or 3 times as long to produce. Improved design and manufacturing techniques were used to reduce the aluminum design cost and still estimated it would require 2150 hours to fabricate and assemble the entire structure. In addition, the aluminum structure would weigh over twice as much as the 13.6 kg (30 lb) ACESS assembly.

ACESS DESIGN

The ACESS design consists of three subassemblies: 1) the one piece 12-sided unibody, 2) removable top deck, and 3) permanently attached bottom deck. All three sub-assemblies are sandwich construction using graphite fiber/cyanate ester facesheets bonded to perforated aluminum core. Specific ACESS design goals were as follows:

- Low cost
  - Over 50% reduction in total manufacturing hours compared to an equivalent aluminum structure.
  - Low part count
- Low weight
  - Total assembled weight < 13.6 kg (30 lb)
- Ultra high stiffness
  - Bottom deck > 80 Hz
- Passive thermal control
  - Bottom deck in plane conductance > 0.25 W/C
  - Bottom deck to unibody joint conductance > 1.0 W/C (per side)
- Factor of safety ≥ 2.0 with positive margins

Unibody - The 76 cm (30 in) diameter by 42 cm (16.6 in) tall unibody is a co-cured one piece structure laid up on a simple low cost aluminum tool. The sandwich construction consists of 0.6 mm (0.024 in) thick six ply facesheets with a 12.7 mm (0.5 in) thick perforated aluminum core. The facesheets are unidirectional prepreg tape consisting of Amoco's P100S graphite fiber and Bryte's EX 1515 (121°C cure cyanate ester resin). The structure is laid up on the male aluminum mandrel, bagged and then autoclave cured for 3 hrs at 121°C. This co-cured approach to fabricating the unibody takes only 120 labor-hours and results in a one-piece structure with continuity of mechanical and thermal properties around the 12 corners.

Decks - Both decks are sandwich construction using thin facesheets made from Mitsubishi's K13C2U graphite fiber and Bryte's EX 1515 cyanate ester prepreg tape. The core is 25.4 mm (1.0 in) thick lightweight perforated aluminum. The top deck facesheets use the same 6 ply layup as the unibody. Due to increased thermal and stiffness requirements, the bottom deck facesheets are eight plies and 0.82 mm (0.032 in) thick.

ACESS Assembly - Figure 2 shows the unassembled ACESS structure and the locations of the joints and inserts required to 1) join the decks to the unibody, 2) mount experiment hardware to the decks and unibody, and 3) attach the ACESS structure to the launch vehicle. The bottom deck is bonded to the unibody using 1.2 mm (0.048 in) thick K13C/EX 1515 angles (with 25.4 mm flanges) and Hysol's 9394 adhesive. The angles are bonded on both sides of the unibody and provide both a thermal and structure joint. The removable top deck is mechanically fastened to the unibody to permit access to the interior of the unibody and the bottom deck. Boeing designed thermal thru-inserts are potted into the top deck, threaded unibody edge inserts are bonded into the unibody, and machine screws used to bolt the assemblies together. The thru-inserts are designed to ensure contact between the thermally conductive facesheets and the unibody edge inserts, resulting in a joint possessing excellent thermal conductivity. The tested thermal conductance of a joint assembly using these inserts was three times the conductance of a joint using standard thru-inserts. These inserts (thru or threaded blind) will also be used to attach all ACESS experiment hardware.

Composite launch interface inserts are used to reinforce the twelve launch vehicle attachments points located on the bottom deck. These solid inserts, offering significant weight savings compared to aluminum or titanium, are machined from a 25.4 mm thick resin transfer molded (RTM) braided preform panel. The braided preform is comprised of 24 plies of 2D triaxial braid using T800 and M46J graphite fibers and an architecture designed to achieve in-plane stiffness equal to aluminum. To make the cured panel, the plies are placed in the mold and a 177°C (350°F) cure cyanate ester formulated for RTM is injected and cured.

MATERIAL SELECTION

The most critical material requirements were 1) graphite fibers having both a very high tensile modulus and thermal conductivity, enabling the design to meet the bottom deck first natural frequency of greater than 80 Hz and overall goal of passive thermal design, 2) a resin system able to meet the < 2.0% cured laminate void content after co-curing and also be low outgassing while in the space environment, and 3) cured laminate properties possessing both good strength to survive the launch loads and good dimensional stability while in orbit. Figure 3 shows the various composite materials used on ACESS along with the attributes specific to each composite part.

Amoco's P100S graphite fiber was initially selected for the ACESS structure. The fiber possesses a tensile modulus of 73 MPa (105 Msi) and a thermal conductivity of 310 W/mK. However, as the program matured it was determined that to meet MWIR vibration requirements, increased stiffness of the bottom deck was required. A design trade-off analysis between stiffness and weight was performed. The analysis showed that increasing the tensile modulus of the graphite fiber was the only approach that had no associated weight gain. Mitsubishi's K13C2U fibers offered the required increase, enabling the greater than 80 Hz goal to be achieved. To insure commonality, the K13C2U fiber was also used on the top deck. In addition, the K13C2U offers a higher thermal conductivity resulting in increased thermal efficiency in transferring interior heat loads to the unibody exterior surfaces. P100S was kept as the fiber of choice for the unibody because the unibody process specification was based on P100S prepreg handling characteristics.

Bryte's 121°C cure cyanate ester resin system was selected for all prepreg used on ACESS. Compared to epoxy systems, cyanate esters offer increased resistance to microcracking and lower moisture absorption, resulting in improved dimensional stability and lower outgassing rates. Several 177°C and 121°C cure cyanate ester resins were characterized in both the neat resin form and as prepregs. 121°C cure systems were desired due to the reduced post-cure residual stress associated with a 121°C vs 177°C cure resin systems. Resin flow characteristics were key to achieving a high quality, low void content unibody. Much lower than typical autoclave pressures were used to make co-cured structures due to the tendency of the brittle high modulus graphite fibers to break against the aluminum cell walls during high autoclave pressures. In addition, during a co-cure process only the cell walls are applying the autoclave pressure to the facesheets (in contrast to a flat plate tool surface used to consolidate laminates). A low viscosity resin was needed to achieve the cured unibody facesheet void content requirement. However, too low a viscosity would result in the resin flowing into the cell walls during cure. Bryte's EX 1515 has a viscosity developed for co-curing. This resin system provided the lowest facesheet void content (therefore the highest sandwich mechanical and thermal properties) during unibody process development.

CO-CURED UNIBODY PROCESS DEVELOPMENT

Two unibody prototypes were fabricated, using Boeing IR&D funds, to develop and verify a co-cure process (figure 4). The first prototype structure was 6-sided and fabricated using IM7/8552 prepreg. The second prototype structure was 8-sided and was fabricated using P100S/EX 1515. These two structures verified that the co-cured process produced a high quality part using a unique low cost manufacturing approach. Image analysis of the interior cutouts from both prototypes showed void contents less than the maximum allowable of 2%. A Boeing proprietary process specification was written describing tool preparation, core forming/splicing, ply application process, NDI acceptance criteria, and QA inspection requirements. This process spec was then used to make the deliverable hardware.

INSERTS/JOINING DESIGN/TESTING

To develop the necessary lamina, laminate, and sandwich design data, an extensive mechanical and thermal test program was performed for both the P100S/EX 1515 and K13C/EX 1515 material systems. Testing was performed at room temperature, -120°C and +100°C. B-basis allowables were developed and used in the analysis. Lamina and laminate testing included:

- Tensile strength and modulus
- Compression strength and modulus
- Isopesicu shear
- Thermal conductivity
- Coefficient of thermal expansion (CTE)

Testing of sandwich specimens determined flatwise tension, edgewise compression, and flexure strengths along with thru-thickness thermal conductivity. These specimens were representative of the both the co-cured P100S unibody and the K13C decks. Similar thermal and mechanical properties were achieved between unibody and deck specimens demonstrating that the co-cure process used in the ACESS program resulted in high quality structures. Insert and joint testing was also performed to develop the necessary design data.

Fabrication of both the EDM and flight hardware is underway. As of June 1, 1996 all hardware has been successfully fabricated and trimmed. Inserts are currently being installed. Centrifuge testing is planned for the end of July followed by delivery to JPL.

CONCLUSIONS

The ACESS program has demonstrated the variety of benefits offered by an all composite spacecraft structure. The use of materials and processes developed and characterized during both the ACESS program and related Boeing Defense & Space Group IR&D activities has enabled several ambitious goals to be achieved. The low weight ACESS assembly includes a bottom deck with a predicted stuffiness of > 80 Hz, extensive use of adhesive bonding for attaching the decks to the unibody, passive thermal control, low manufacturing costs, and has a factor of safety of ≥ 2 for all design requirements.

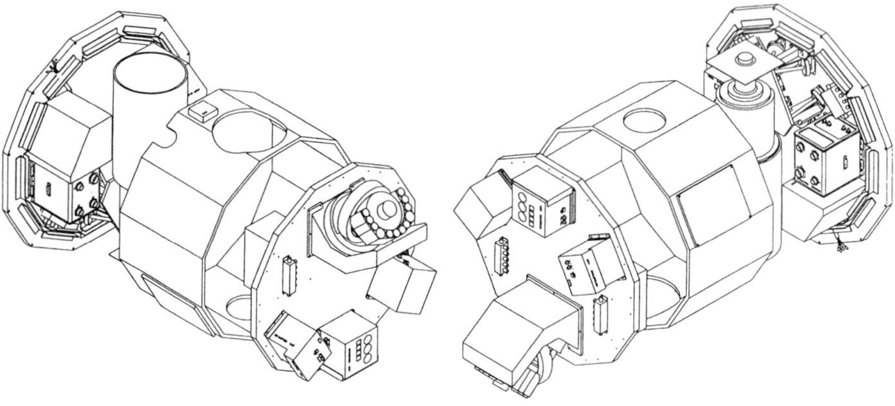

Figure 1. Exploded View of the All Composite Experimental Spacecraft Structure.

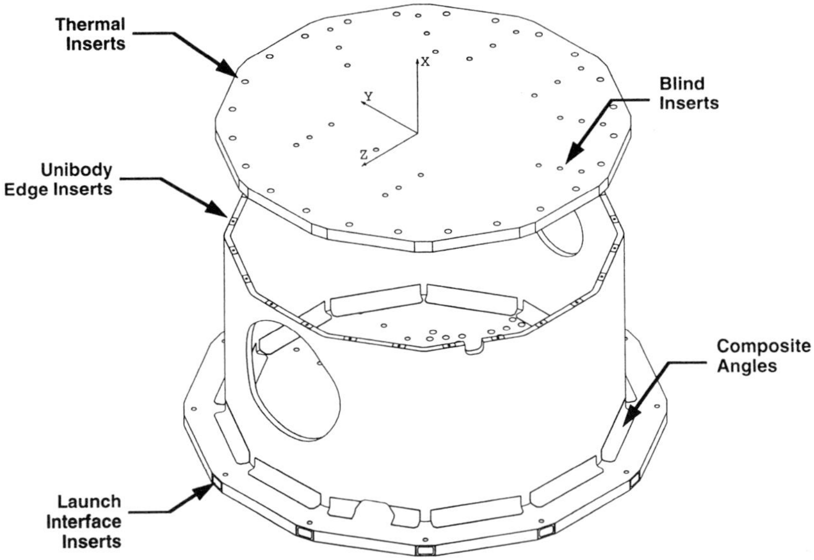

Figure 2. Unassembled ACESS Structure.

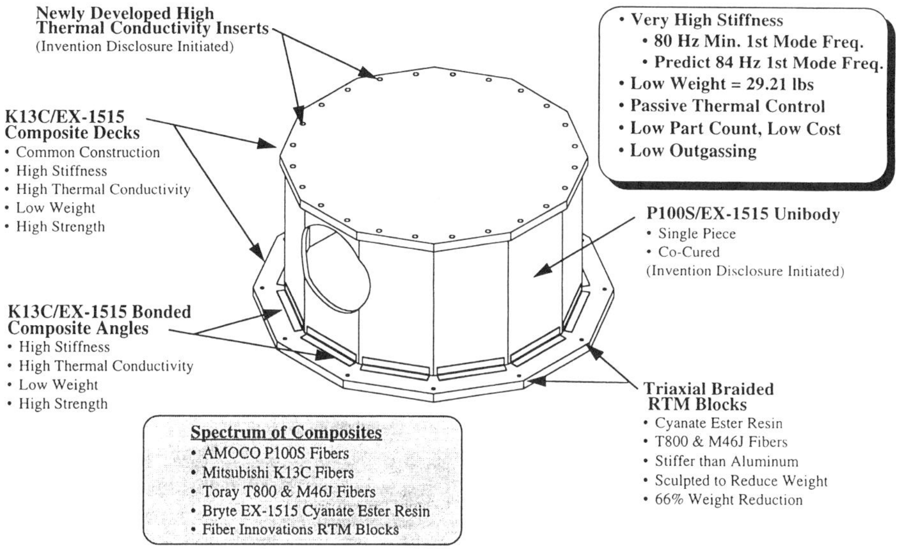

Figure 3. ACESS Material Selections

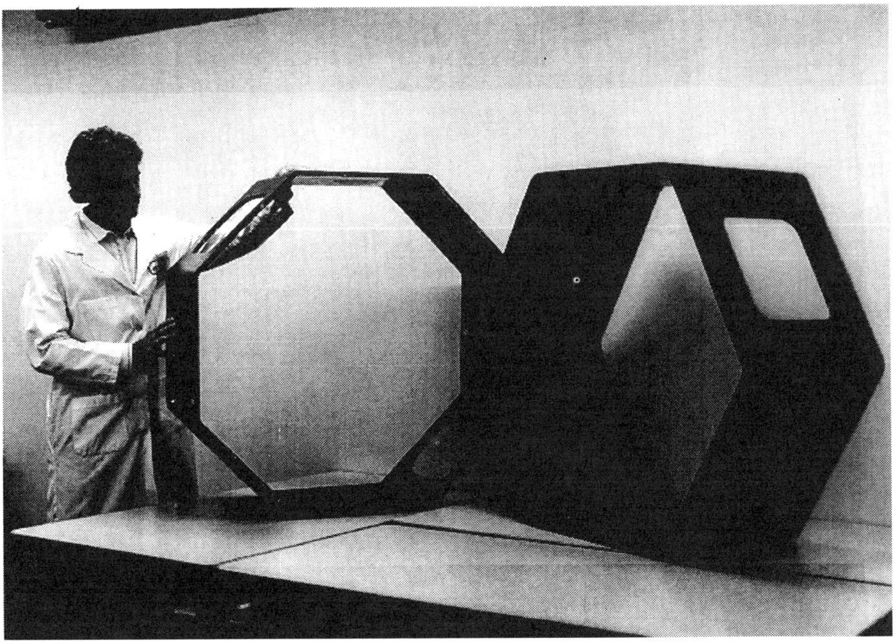

Figure 4. Unibody Prototypes.

28th International SAMPE Technical Conference
November 4-7, 1996

COMPOSITE STRUCTURES ON THE DC-XA REUSABLE LAUNCH VEHICLE

Michael J. Robinson
McDonnell Douglas Aerospace
Huntington Beach, California 92647

ABSTRACT

In August 1993, under the sponsorship of the Ballistic Missile Defense Organization (BMDO), McDonnell Douglas Aerospace (MDA) flew the first fully reusable rocket, the DC-X. It was a suborbital vehicle built primarily to demonstrate vertical take-off and vertical landing and the associated rapid, efficient, low cost operations achievable with this concept. Although significant portions of the DC-X were constructed using advanced composites, the vehicle was not intended to demonstrate the widespread use of these lightweight materials throughout the structure, a strictly necessary feature of an orbital reusable rocket. The aeroshell, base heat shield and control system pressure vessels were the only composite parts. Subsequently, the DC-X was incorporated into NASA's Reusable Launch Vehicle (RLV) program and has been rebuilt in a configuration called DC-XA, that flew for the first time on May 18, 1996. DC-XA is the first step in the development of an RLV that does incorporate lightweight structures throughout. DC-XA features (in addition to the original composite parts) an all-composite unlined liquid hydrogen tank, a composite intertank, an aluminum-lithium (Al-Li) liquid oxygen tank, and a composite liquid hydrogen feed line and valve. The development of these components is presented.

KEY WORDS: reusable launch vehicles (RLV), composites, liquid hydrogen tanks, intertanks, feed lines, valves

1. INTRODUCTION

The use of composite structures on launch vehicles has lagged behind their use on airplanes. One reason is that the best time to insert new technology (materials technology or any other kind) on an aerospace vehicle is during development, and launch vehicles are developed far less frequently than airplanes. But the time has come for new launch vehicles. The nation's current fleet of major expendable rockets, Titan, Delta and Atlas, have served well and will continue to place payloads in orbit for several years to come. But, although they have been incrementally upgraded over the years, they are all essentially very old designs originally developed as strategic weapon platforms. They are built with 40-year-old structures technology and they were designed with a forty-year-old attitude about cost. The expense of building, supporting and then discarding these vehicles after one use is getting harder for government launch customers to bear and is driving commercial customers elsewhere.

The Space Shuttle was an attempt to address high launch costs by building a partially reusable vehicle. Full reusability was not possible due, in part, to the limitations of 1970s structures technology. The necessarily expendable structures, the high margins and complexities required for manned systems, very extensive ground support operations, lower-than-planned launch rates, higher-than-planned refurbishment costs, and other factors combined to make the shuttle even more expensive than the old expendables.

Rapid, efficient launch operations and a *fully* reusable vehicle are the keys to reducing the cost of access to space. Two ways to reduce the operations costs are to adopt a minimalist aircraft-like approach to system support and to land the rocket vertically. The demonstration of these two features was the goal of the DC-X program when it was initiated in 1991, although more than cost concerns were involved. BMDO wanted to develop an inexpensive launch system capable of rapid deployment and recall of satellites to support US forces in major local conflicts such as the Persian Gulf war. Current launch systems need a minimum of several months to place a satellite in orbit, even assuming a suitable vehicle is readily available. MDA built and tested DC-X (Figure 1) to demonstrate the vertical takeoff/vertical landing and the rapid, efficient ground operations possible with this approach. DC-X flew successfully eight times from August 1993 to July 1995 using a flight crew of only three people and a maintenance crew of eight.

The other aspect of low cost space access, the fully reusable vehicle, presents a great technical challenge. For a launch vehicle to get to orbit in one piece and deliver a substantial payload, it must be about 90 percent propellant by weight at liftoff. Assuming about 1 percent for the payload, only about 9 percent of the total gross liftoff weight remains for the entire mass of all the vehicle structures and subsystems. This is theoretically impossible for a vehicle made with conventional aluminum structures. Partial use of lightweight materials makes it possible but the vehicle would be unrealistically large. In a workable RLV, widespread use of lightweight materials is absolutely necessary.

In early 1994, NASA initiated the RLV program and incorporated DC-X in it. As a part of the Phase I effort, NASA solicited industry to propose technology development projects and offered DC-X as a demonstration platform. Under a cooperative agreement with NASA, MDA designed and built a composite liquid hydrogen tank, a composite intertank, and an aluminum-lithium liquid oxygen tank to replace more conventional aluminum equivalents in the original DC-X design. An auxiliary propulsion system (APS) incorporating several new components, including a composite liquid hydrogen feed line and valve, has also been developed. All of these components were assembled into a new vehicle called DC-XA. It has the same aeroshell, base heat shield, engines and other subsystem components as DC-X. The new composite components and the aluminum-lithium tank are discussed below.

05765REU6

Figure 1. The DC-X Began Test Flight at the White Sands Missile Range in 1993

2. LIQUID HYDROGEN PROPELLANT TANK

The most prominent new composite structure on DC-XA, in both size and technology advancement, is the unlined liquid hydrogen tank. It is the largest single structure on the core vehicle, as it will be with subsequent RLVs. Thus the use of composite material in it provides a major weight benefit. To develop the tank, MDA drew upon several years of experience in the evaluation of composite material properties at cryogenic temperatures (including hydrogen permeability data), an improved version of the insulation developed for the Saturn S-IVB stage, and previous experience in the fabrication and testing of small pressure vessels and the National Aerospace Plane Task D demonstration tank.

The objective was not merely to build a structure that satisfied the requirements of DC-XA, but to demonstrate important technical features for a full scale RLV with the additional benefits of integrating the tank into an actual vehicle and subjecting it to real flight environments. As a result, the tank was designed to carry scaled RLV tank loads and has large margins for DC-XA loads. A full scale RLV hydrogen tank will be as much as 40 feet in diameter. For DC-XA, the hydrogen tank is a cylinder, 8 feet in diameter, 16 feet long with shallow domes and bonded skirts for attachment to the thrust structure and intertank.

No single scaling law works for all design variables in the tank, so decisions had to be made on the best approach to the most pressing technical questions. Structural integrity and hydrogen impermeability throughout tank working life at realistic stress levels are the most important issues. Stress is affected by temperature, wall thickness, internal pressure and diameter. If stress was the only concern, satisfactory scaling could be achieved by reducing the wall thickness in proportion to the diameter, provided minimum gauge limits are not exceeded, or by increasing the internal design pressure in inverse proportion to the diameter, provided the higher pressure does not greatly complicate the design of the closures. But permeability is affected by wall thickness, internal pressure and stress. Ideally, each of these parameters in the full-scale tank should be matched in the subscale demonstrator. But, for a given diameter, these parameters together are not independent so at least one must be scaled in relation to the diameter reduction. MDA decided to retain full scale RLV tank membrane thickness and scale up the internal design pressure to 135 psi. This resulted in a tank that was robust enough to more assuredly handle the DC-XA manufacturing and flight environments and serves as a conservative permeability demonstrator when fully pressurized.

The tank shell, skirts and joints were designed to full scale RLV tensile line loads. The design for axial compression is more complicated. Two issues exist: the strength of the skirt-tank joint and the buckling stability of the skirts and tank wall. MDA decided the strength of the joint was the more important of the two, so it was designed to full scale compression line loads. But this makes verification of buckling strength difficult. The full scale RLV tank will have a stiffened skin due to buckling criticality. Since MDA decided to use a full thickness skin on the smaller diameter, the tank is too stable to demonstrate buckling unless the stiffener design or arrangement was modified. MDA felt it more important to verify the joint strength as this feature represented something new and unusual, whereas the design of stiffened cylinders for buckling is well understood at MDA and elsewhere. With the addition of a few more plies to the shell, stiffeners were, in fact, unnecessary for DC-XA loads. MDA chose this option due to program schedule constraints.

The composite material used throughout the tank is IM7/8552 toughened epoxy from Hercules. The dome laminates consist of a quasi-isotropic layup using a specific five-harness satin cloth weave selected because its cured ply thickness is almost exactly twice that of unidirectional tape. This simplified the interleaving of the two material types as the layup transitioned from the cloth used in the domes to the tape used in the cylinder. The cylinder laminate was designed so that the strength in the hoop direction would be twice that of the axial direction to match the characteristic stress field in cylindrical pressure vessels.

The important design features of the tank are illustrated in Figure 2. Of particular concern were the circumferential Y joints where the skirts are attached to the shell. MDA wanted to avoid a bolted joint here so that there would be no holes in the composite that could leak hydrogen. Very strong bonded Y joints have been successfully developed on composite solid rocket motor cases but the challenge in this case was the cold temperature. In other bonded Y

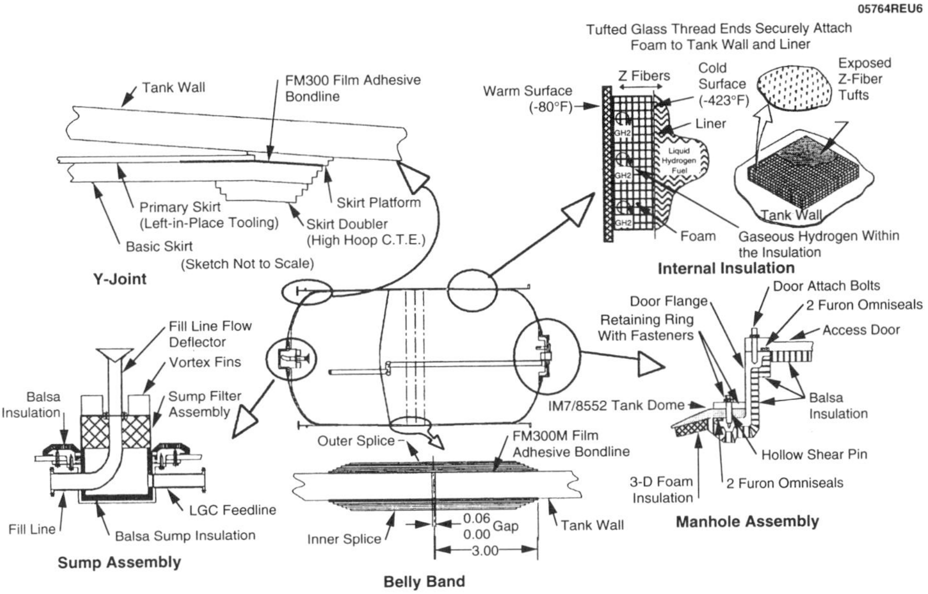

Figure 2. Key Features of the DC-XA Composite Liquid Hydrogen Tank

joints, rubber or some other "soft" material is used in the joint to relieve the shear peaking that develops in the bond line at the intersection. At the very cold temperatures this joint reaches, however, the conventional softening materials do not remain soft and their ability to reduce the shear peak is degraded. MDA went through a long phase of finite element modeling, coupon testing, and subcomponent testing to develop the Y joint design and verify its strength. In the end, strengths greater than 8500 pounds per inch were achieved in subcomponent tests—more than sufficient for the 1680 pound per inch requirement for DC-XA. Line loads for a full scale RLV, however, are expected to be on the order of 10,000 pounds per inch and will require improved designs. More advanced alternatives are being developed and tested in a separate program task.

Detailed analysis was also required for the sump and forward closure and a circumferential joint. The sump and forward closure were made of titanium for reduced weight and CTE compatibility with the composite. The primary concern with the closures was the integrity of the seals. The same redundant reinforced Teflon seal arrangement used on DC-X was retained with careful attention given to assuring smooth sealing surfaces of the titanium and composite.

The tank was fabricated in two halves (forward and aft) by hand layup and then the pieces connected by a circumferential double lap adhesive splice joint. Layup of one of the shells is shown in progress in Figure 3. Full scale tanks will be too large for their sections to be bonded in existing autoclaves, so MDA demonstrated adhesive bonding of the two sections outside of an autoclave using special tooling and localized application of the heat and pressure needed to effect the joint. In addition to a holding fixture for the tank, the tooling consisted of rubber bladders enclosed inside C channels placed around the perimeter of the tank (inside and out) to apply the pressure, and electric strip heaters located underneath the bladders to apply the heat. The setup for the bonding operation is shown in Figure 4. This technique had been demonstrated previously by MDA on a smaller diameter.

Composite skirts were cocured to the tank in a secondary layup process. A thin, precured shell was bonded to the tank, used as the layup surface for the bulk of the material, and left in place as part of the skirt. Aluminum interface rings were bolted to each skirt to provide the attachment to the thrust structure at the base and the intertank above.

The tank insulation consists of fiber-reinforced foam tiles bonded to the entire inside surface with a glass-reinforced semi-permeable urethane liner on the hydrogen side of the foam. MDA

05758REU6

05759REU6

Figure 3. Layup of the IM7/8522 Graphite/Epoxy Liquid Hydrogen Tank

Figure 4. Setup for Curing the Adhesive Double Lap Splice Joint in the Composite Tank Using Localized Heat and Pressure

developed this type of insulation for the Saturn SIVB stage in the Apollo program and used similar techniques to produce the new foam. The new foam, however, has a lower density than the Saturn material (2 pounds per cubic foot compared to 5) and is made with a water-based blowing agent instead of the chloroflourocarbon (CFC) blowing agents permitted in the Saturn era.

Neither the new composite tank nor the aluminum one it replaced were optimized for weight in the DC-X/XA configuration. The composite because of the scaling from the RLV design and the aluminum because of the tight schedule and conservatism of the original DC-X program. This makes a weight comparison dubious. Nevertheless, given that both were conservatively designed (albeit for different reasons), the fact that the composite tank weighs 67 percent of the aluminum tank weight (1908 pounds vs. 2851) does make a statement about the advantage of the composite material.

The completed tank, with bonded composite brackets for vehicle subsystems attachment, is shown in Figure 5. After assembly, the tank was shipped to NASA-MSFC for testing. After successful leak checks, it was subjected to 110% DC-XA body loads at 140% of DC-XA expected operating pressure at ambient temperature using nitrogen gas as pressurant. It was then filled and drained four times with liquid nitrogen and taken to the same load levels during each cycle. After another successful leak check, it was filled with liquid hydrogen twice and pressurized five times to DC-XA operating pressures. This test simulated flight temperature and pressure conditions and enabled characterization of the cryogenic insulation. The tank performed as expected in all tests and successfully passed a final leak check. It was then returned to MDA in Huntington Beach for DC-XA assembly.

3. COMPOSITE INTERTANK

The intertank (located between the hydrogen and oxygen tanks) on DC-X was made of aluminum box beam frame construction. The composite replacement is significantly different. Full scale RLV intertanks will be lightweight stiffened shell structures, so the DC-XA intertank was made of composite sandwich construction using T650/5250-4 graphite/BMI face sheets and 1-inch-thick aluminum honeycomb core. It is a 2-degree conical section, 36 inches tall, and was fabricated by MDA's St. Louis division. Like the hydrogen tank, the intent on the intertank was to design to scaled RLV loads. The maximum DC-XA compression load, however, was higher than the estimated RLV load, so the intertank was designed to DC-XA compression and RLV tension loads.

In another similarity to the hydrogen tank, the intertank was complicated by the concurrent goals to demonstrate a lightweight subscale RLV design and accommodate the DC-XA vehicle requirements and interfaces. In the case of the intertank, there were many vehicle subsystem hardware attachments carried over from the DC-X configuration, primarily the storage bottles and plumbing for the helium used to pressurize the propellant tanks. Accommodating this hardware inevitably resulted in design compromises. For example, the triangular panel shown in the foreground of the intertank in Figure 6, and a matching one on the other side, fit easily between the aluminum frame members of the DC-X intertank but could not be moved inboard or outboard to be attached to the wall of the new shell structure because of envelope constraints. Openings had to be designed into the composite sandwich shell so the panels, which provide the attachment for the helium regulating systems, could be recessed into the shell. Another opening (visible in the left background in Figure 6) was required to provide access to the interior after the vehicle is assembled and had to be big enough so that the bottles could be removed through it.

In addition to the helium pressurization system, the intertank interior must accommodate the bottom dome of the oxygen tank, the top dome of the hydrogen tank and the associated plumbing of each. From the sump at the base of the oxygen tank, four feed lines (one for each engine) come out, pass through the intertank wall and go down along the outside of the hydrogen tank. Two of the oxygen feed line openings are visible in Figure 6. Vent lines from the cover on the top of the hydrogen tank also pass through the wall of the intertank.

After fabrication in St. Louis, the intertank shell was shipped to NASA-MSFC and proof tested to 150 percent of expected DC-XA compression loads. It was then sent back to Huntington Beach for installation of the helium system and vehicle integration. Despite the weight compromises necessary to accommodate all the DC-XA vehicle interfaces, the new intertank was still 44 percent lighter than the aluminum frame structure on DC-X.

In order to further advance the state of the art for full scale RLV intertank structure, MDA's cooperative agreement with NASA includes the development and testing of additional subscale intertanks more representative of full scale RLV designs than was possible with the DC-XA

05757REU6

Figure 5. Completed Composite Liquid Hydrogen Tank Ready for DC-XA Integration

05756REU6

Figure 6. T650-42/5250-4 Graphite/BMI Intertank with Heliuim Pressurization Bottles and Lines Installed

constraints. A skin-stringer design has been developed and is being fabricated by MDA in St. Louis for ground testing at NASA. In cooperation with composites specialists from Russia's Central Research Institute for Special Machine Building, a filament-wound composite shell of integral rib/skin construction is also being developed for structural testing at NASA.

4. ALUMINUM-LITHIUM LIQUID OXYGEN TANK

Although not fabricated from composites, a lighter weight replacement for the DC-X aluminum liquid oxygen tank was developed. The DC-XA oxygen tank afforded MDA and NASA a great opportunity to evaluate the Russian 1460 Al-Li alloy, working with experts from the Russian Academy of Sciences, VIAM and RSC-Energia. Like domestic Al-Li alloys, 1460 has higher specific strength and stiffness than conventional aluminum. It also has a history of use on Russian (formerly Soviet) aircraft.

The DC-XA oxygen tank is basically a duplicate of the DC-X tank design using the different alloy. Wall thicknesses were reduced, however, to exploit the higher specific properties of the Al-Li. As a result, the new tank weighs 830 pounds less than the original. All internal slosh and vortex baffles and the interface hardware are the same design as the DC-X tank. Most of the conic section of the shell and skirts were fabricated from sheet that was roll-formed and seam welded. The Y-joint rings were machined from standard aluminum alloy ring forgings. The domes were formed from single pieces using a three-step process of press forming and explosive forming. Chem milling was used to achieve the final dimensions.

Although the long distance, time difference, language difference, engineering and testing philosophy differences and lack of previous experience developing hardware with the Russians combined to make this a challenging project, the tank was completed and delivered in time to support DC-XA assembly. It passed an ambient temperature proof test in Moscow at 110 percent of DC-XA loads and a liquid nitrogen temperature test at 110 percent loads in the US prior to DC-XA integration. The tank is shown being fitted to the intertank in Figure 7.

05755REU6

Figure 7. Russian 1460 Aluminum-Lithium Liquid Oxyen Tank During Fit-Up with the Intertank

5. COMPOSITE LIQUID HYDROGEN FEED LINE AND VALVE

The need to minimize weight on a fully reusable rocket is so great that the weight savings achievable by developing lightweight components in the secondary structures and subsystems is significant. Thus the development of composite liquid hydrogen feed lines and valves. The line and valve designed and built for DC-XA are for a 2-inch-diameter line. In a full-scale RLV, if long sections are made of the much larger lines required, thousands of pounds can be saved. While constituting only a small fraction of the total vehicle weight, this is still a worthwhile contribution in an RLV.

05752REU6

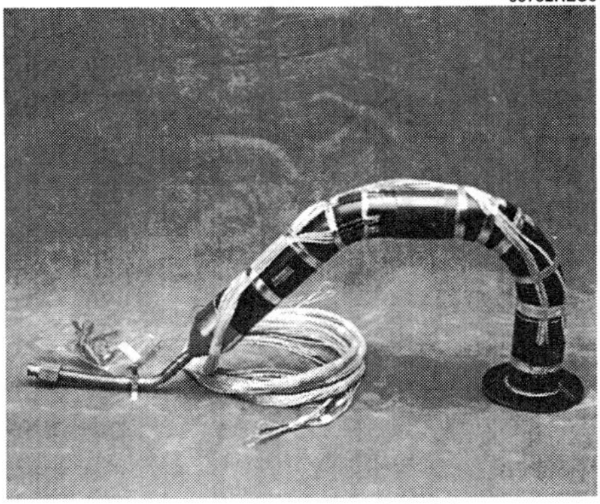

Figure 8. IM7/8552 Graphite/Epoxy Composite Liquid Hydrogen Feed Line

05754REU6

Figure 9. Molded Composite 2-in. Dia Liquid Hydrogen Ball Valve and Commercial Actuator

The DC-XA composite line and valve are illustrated in Figures 8 and 9. The valve is connected to an outlet on the sump of the liquid hydrogen tank, the composite line is connected to the other side of the valve, and the line runs down to a hydrogen liquid-to-gas converter in a new APS developed for DC-XA. The vehicle's reaction control system (RCS) uses four small gaseous hydrogen/oxygen thrusters for roll control. One of the goals of the APS development is to demonstrate a simplified system that draws liquid hydrogen directly from the main propellant tank and converts it to gas rather than having a separate gaseous hydrogen supply system for the RCS.

The composite line was designed by MDA and built by NASA-MSFC. It is 2 inches in diameter, about 2 feet long with one 90-degree elbow and one 45-degree elbow. The pressure in the line and the stresses that develop in the composite are very low, so the design was driven by minimum gauge limits. It was fabricated with the same IM7/8552 graphite/epoxy cloth used to fabricate the domes of the hydrogen tank. Ideally, lines like this would be fabricated in one piece, but because the lines used in an RLV would likely be made of several pieces, this one was too. The key is to build leak-tight joints between the sections. MDA and NASA-MSFC developed techniques to fabricate good adhesive splice joints in the line using paste epoxy adhesives. Several development articles were fabricated and tested for hydrogen leakage before the flight article and qualification test article were built.

The valve was designed, assembled and leak tested by Ketema Aerospace using requirements provided by MDA and molded composite components made by American Automated Engineering. Ketema modified a conventional 2-in. ball valve design. The molded composite parts are the valve body, outlet flange, ball, and stem. The molding material consisted of the same IM7/8552 cloth prepreg used in the hydrogen tank and feed line chopped into small squares to facilitate the molding. After initial molding efforts, a few design and fabrication method modifications were necessary to achieve the required dimensions within tolerance. In the final configuration, the assembled valve performed as well or better than comparable metal valves in leak tests.

After the line and valve were tested individually, the two components were assembled with the valve actuator and downstream titanium connecting hardware and subjected to rigorous qualification testing at Wyle Laboratories (Figure 10). The assembled qualification test unit was vibrated in each axis for 78 minutes above full flight vibration levels while liquid hydrogen was flowing through it. Leak checks were done throughout the test series using a helium mass spectrometer. Performance was comparable or superior to other valves and valve/line connections. The flight unit was proof tested in the same apparatus at 3 minutes per axis and then installed in DC-XA (Figure 11).

6. DC-XA ASSEMBLY AND FLIGHTS

The original DC-X aeroshell, shown in Figure 12, was used for DC-XA. This is a composite sandwich structure built by Scaled Composites and consists of four spliced sections made of

05753REU6

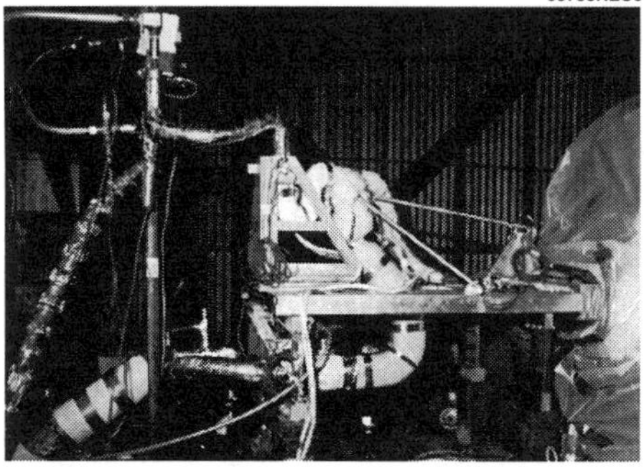

Figure 10. Insulated Composite Feed Line and Valve in Cold Flow Vibration Test at Wyle Laboratories

05763REU6

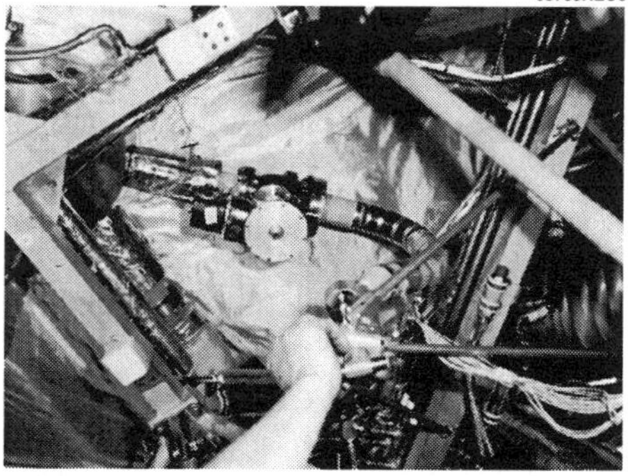

Figure 11. Composite Line and Valve Installed in DC-XA

graphite/epoxy face sheets and polyvinyl foam core. The original base heat shield, shown in Figure 13, was also retained. This too is a sandwich structure made with graphite/epoxy face sheets and syntactic foam core and fabricated in four quadrants and a central access panel. It was built by MDA in Huntington Beach. From the outside, DC-X and DC-XA appear identical, but Figure 14 illustrates the substantial differences in the core vehicle. The DC-XA roll out ceremony occurred on March 15, 1996 (see Figure 15). The assembled vehicle was then trucked from Huntington Beach to the Delta Clipper Site in the north central area of the White Sands Missile Range near Alamogordo, New Mexico.

DC-XA flew for the first time on May 18, 1996 at White Sands. It reached a height of 800 feet, moved laterally 350 feet and landed on target in a 62-second flight. The landing area is a 100 ft by 100 ft concrete pad. Prior to this flight, a 40 ft by 40 ft by 6-foot-deep pit was dug in the center of the pad and a steel grate placed over it. This new feature was intended to relieve some of the intense heat input to the base of the vehicle from the four RL-10 engines when the landing occurs on a flat concrete surface. Due to an actuator bracket failure, one of the four control flaps near the base of the vehicle opened unexpectedly shortly before landing. Hot exhaust from the engines, reflected upward from the bottom of the pit, ignited the flap, flap

05761REU6

Figure 12. DC-X/DC-XA Composite Aft Aeroshell Fabricated by Scaled Composites

05762REU6

Figure 13. DC-X/XA Composite Base Heat Shield

05751REU6

Figure 14. DC-X and DC-XA Core Vehicles After Assembly Showing New Composite Structures on DC-XA

05760REU6

Figure 15. Final Installation of Aeroshell on DC-XA

box and adjacent aeroshell area. But the core vehicle was unaffected. The flap and parts of the flap box and aeroshell had to be replaced but, in a demonstration of the robustness of vehicle and crew, this was completed in time for the second flight on the original scheduled date of June 7. Although the heating of the base heat shield was lower for this landing, peak heating of the lower aeroshell area was higher. Also, the vehicle slid about 11 feet across the grate after touching down, possibly due to the reflected engine exhaust acting on the open flap. For these and other reasons, MDA and NASA decided to abandon the pit for future flights.

During Flight 2 on June 7, DC-XA reached an altitude of 1950 feet and flew laterally 550 feet before landing (see Figure 16). This flight lasted 64 seconds. In another demonstration of robustness, the landing occurred on the white gypsum sand of the range. This had been done successfully before on an emergency landing during one of the DC-X flights. Flight 2 was flawless and the vehicle was in better condition afterward than on any previous flight. Substantial holes were dug in the sand by the engines, however, each of them uncomfortably close to a landing gear. MDA and NASA decided to land on a corner of the concrete pad on the next flight.

To demonstrate rapid turnaround of the vehicle, an attempt was made to fly again on the same day. The vehicle was moved to the flight stand and was ready to go when an approaching storm and the lateness of the day eliminated the opportunity. But Flight 3 did occur the next morning, June 8, just 26 hours after Flight 2. In another flawless flight lasting 140 seconds, the vehicle reached an altitude of 10,300 feet and landed safely on the corner of the pad. Additional flights are pending as of this writing.

7. THE NEXT STEP: X-33

The goal of the RLV program is to reduce the cost of access to space. DC-XA is the first of two intermediate steps in the development of a vehicle and operations philosophy that will achieve that goal. Phase II of the program, now underway, consists of three elements: (1) Design, construction and flight of a larger suborbital demonstrator called X-33, (2) demonstration projects developing additional technology that the X-33 alone can not provide, and (3) design and analysis of a full scale RLV based on the technology development and on system requirements generated by analysis of the projected launch market.

The X-33 vehicle is now being developed by Lockheed-Martin. Intended to fly for the first time in March 1999, it will reach much higher altitudes and fly at speeds greater than Mach 15. At this speed, critical engine and thermal insulation technology can be validated in addition to the structures technology that will be demonstrated building on the heritage of DC-XA.

As stated above, composites have been used very little in launch vehicle primary structure, especially in comparison to recently-developed military aircraft. In X-33 composites technology and the need for it have met in a launch vehicle whose primary structure will leapfrog to a very high percentage of composite material. In addition to being an important project in the nation's space program, X-33 will be an even greater showcase for composite materials than DC-XA.

05750REU6

Figure 16. DC-XA Flight 2, June 7, 1996

28th International SAMPE Technical Conference
November 4-7, 1996

Hybrid Tooling for Advanced Grid Stiffened (AGS) Structures

Troy E. Meink
Steven Huybrechts, Ph.D.
USAF Phillips Laboratory
Structures and Controls Division
Kirtland AFB, NM 87117

ABSTRACT

Lattices of rigidly connected ribs, known as *Advanced Grid Stiffened* (AGS) structures, have many advantages over traditional construction methods, which use panels, sandwich cores and/or expensive frameworks. A newly developed *Hybrid Tooling* process has been developed and researched in order to produce high quality AGS structures at low cost. This Hybrid Tooling concept involves a combination of hard and soft tooling to control part compaction, dimensional stability, and allow for automation. The tooling is comprised of a stiff, dimensionally stable *Base Tool*, overlaid with a high coefficient of thermal expansion *Expansion Tool* providing part compaction. Aspects of this tooling, specifically as they relate to construction of AGS structures, have been investigated. Flat, cylindrical and conical sections have been fabricated in order to optimize tooling parameters. An analytical thermal model has been developed to predict part consolidation. This novel tooling concept is presented along with experimental data, analysis of accompanying parameters, behavior and capability.

KEY WORDS: Tooling, Manufacturing/Fabrication/Processing, Thermal Analysis

1. INTRODUCTION

1.1 Advanced Grid Stiffened Structures Although it is widely accepted that composite materials will be used extensively in the next generation of launch vehicles, the specific structural type has yet to be determined. One possible structural type is the Advanced Grid Stiffened (AGS) structure, a skin-stringer configuration evolved from early isogrid stiffening concepts. AGS structures have recently gained popularity as a possible solution to many of the problems associated with traditional composite construction methods involving lamination of plates and sandwich type structures. AGS structures are characterized by a lattice of rigid, interconnected ribs, which proves to be an inherently strong and resilient arrangement for composite materials [2 & 3] and lacks the material mismatch associated with laminated structures. AGS structures possess inherent resistance to impact damage, delamination and crack propagation while showing a high potential for automation during fabrication.

All AGS structures suffer from the same manufacturing difficulty: for an AGS structure to have all fibers continuous through a rib crossing point, there must be twice as much material in each crossing point than in each rib, making out-of-plane compaction difficult or impossible. For most manufacturing methods, this difficulty leads to a buildup at the nodal points, which is undesirable for many reasons. Known manufacturing methods for AGS structures are listed below.

- Node Compaction: Brute force compaction through out-of-plane pressure using hard tooling.
- Fiber Cutting: Allowing only 50% of the fibers to pass through each node.
- Low Rib Fiber Volume: Having resin-rich ribs with 50% the fiber volume of the nodes.
- Lateral Compaction: Consolidating ribs from the sides, effectively causing the extra material at the node to lead to double node width rather than double node height.
- Nodal Spreading: Spreading of fibers as they enter the node allowing out-of-plane pressure to lead to double node width rather than double node height.

Of these methods, Nodal Spreading and Lateral Compaction are the only methods to experience high quality results; of these two, Lateral Compaction is the most cost effective. Lateral Compaction is typically performed with a silicon rubber tool that expands during the cure cycle, laterally compacting the AGS ribs. Unfortunately, this approach suffers from several drawbacks, including poor part geometry control, poor part stability, inability to manufacture large parts, inability to manufacture complex shapes and labor intensive processing.

1.2 Hybrid Tooling The concept of Hybrid Tooling solves many of the problems associated with Lateral Compaction while providing a number of additional advantages. Hybrid Tooling offers an AGS structure manufacturing solution that is low cost, is highly automated, provides for good part compaction and provides for good geometry control. Using Hybrid Tooling and standard filament winding technology, AGS tooling can be created, machined and wound in a series of seamless, automated steps. Also, the resulting tooling is, in most cases, reusable. An example of Hybrid Tooling is shown in Figure 1.

Figure 1: Example of Hybrid Tooling. *The Base Tool, in this case, is made of particle board and has grooves cut into it corresponding to the AGS rib locations. The silicon rubber Expansion Tool is laid into these grooves and graphite fiber prepreg is wound into the Expansion Tool grooves.*

The ideal Lateral Compaction tool combines high CTE, good predictability of thermal behavior, high tool stability, light weight, machinability and good part release. The Hybrid Tooling approach recognizes the inability of any existing tooling material to provide all these attributes. Therefore, Hybrid Tooling uses two different tooling materials, one as a Base Tool and one as an Expansion Tool, combining the advantages and mitigating the disadvantages of each material. An explanation of the two tooling components is given below.

- Base Tooling: The general geometry of the AGS part is determined by the base tool which must provide thermal/structural stability, part accuracy and skin release. The material selected for the Base Tool must be thermally stable (low CTE and thermally predictable), machinable and lightweight. In the majority of cases, the Base Tool defines the shape of the AGS structure and has machined grooves which hold the Expansion Tooling. Some materials that may be selected for the Base Tool are: particle board, tooling foam, tooling epoxy, aluminum and graphite.
- Expansion Tooling: The Expansion Tool is primarily responsible for lateral compaction of the AGS ribs. The material selected for the Expansion Tool must have a high CTE and provide good part release. The Expansion Tool is typically a fairly thin layer of silicon rubber resting in grooves in the base tool that correspond to the locations of the AGS ribs. To date, only silicon rubber (AirTech AirCast 3700) has been successfully used for Expansion Tooling.

Typically, the Base Tool is machined, including grooves for the Expansion Tool, which is separately cast. The Expansion Tool is laid into the grooves and prepreg tow is wound into this tool. The skin is wound and the entire part is then bagged and autoclaved. The machining of the base tool, the winding of the ribs and the winding of the skin can be accomplished in a fully automated manner. Both tooling sections are typically reusable.

2. THEORY

2.1 General Concepts An example schematic of Hybrid Tooling is shown in Figure 2.

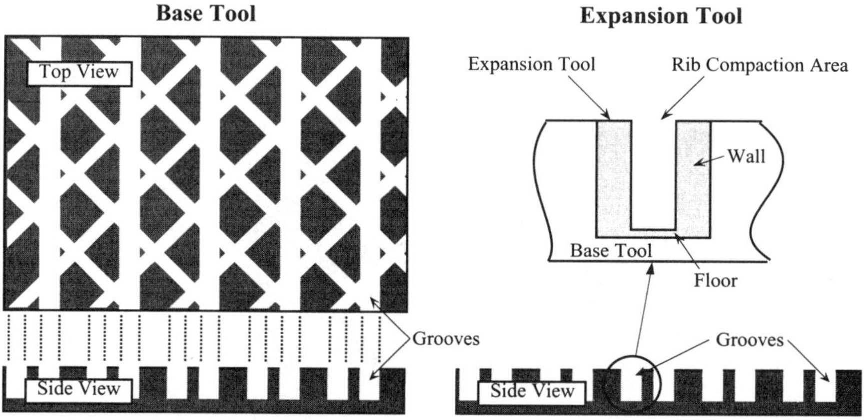

Figure 2: Schematic of Base and Expansion Tooling. *Hybrid Tooling consists of a Base and an Expansion Tool. The Base Tool is composed of a hard, thermally stable material with grooves cut into it. The Expansion Tool is composed of a high CTE material and is 'laid into' the Base Tool. Prepreg tow is laid or wound into the groove in the Expansion Tool to create the AGS structure's ribs. A skin is then laid up or wound over the top of the Base Tool.*

2.2 Tooling Material Selection Base Tooling materials have been evaluated as part of the Hybrid Tooling development process. The following list is a catalog of each potential base tooling material and the determined strengths and weaknesses.

- Particle Board: Particle Board is the cheapest available Base Tool material. Disadvantages include some warping during cure (limits but doesn't preclude reusability of the tool) and slower machining requirements.
- Tooling Foam: Tooling Foam provides for a relatively lightweight Base Tool that is highly machinable. Disadvantages include cracking and compressive failures during the cure cycle leading to non-reusable tooling.
- Tooling Epoxy: Tooling Epoxy is highly machinable and very thermally stable, providing for a high quality, reusable tool. Disadvantages include moderately high cost, high weight and low glass transition temperature. To date, Tooling Epoxy has been the most successful Base Tool Material discovered.
- Aluminum: Aluminum is highly predictable and fully reusable. Disadvantages include high cost, high weight and very slow (relative) machining requirements.
- Graphite: Graphite is the most stable tooling material available leading to very high tolerance parts. The main disadvantage of Graphite Tooling is the high associated cost.

2.3 Rib Sizing, Compaction and Imprinting In order to achieve high quality parts using Hybrid Tooling, careful attention must be paid to sizing of the Expansion Tool (wall thickness, groove width, groove depth) and the Base Tool (groove depth). When performing this sizing, minimization of imprinting must be considered. Imprinting is an undesired side effect of Hybrid Tooling where the Expansion Tool causes impressions on the outside of the AGS structure skin. Additionally, the tooling must be sized so that the desired rib dimensions are achieved and so that the Expansion Tool provides enough lateral pressure so as to fully compact the ribs. In order to predict and control these issues, a theory is developed below.

2.4 Plane-Strain Hybrid Tooling Thermal Model A good intuition into the behavior of the Expansion Tool can be gained from a closed form, plane strain solution. In this simple solution, the effect of the Expansion Tool Floor is not considered. The simplified situation is shown in Figure 3.

Neglecting the effect of the Expansion Tool Floor, there will be no shear deformation in the Expansion Tool and the constitutive equations will be [1, p.22]

$$\begin{bmatrix} \varepsilon_x \\ \varepsilon_y \\ \varepsilon_z \end{bmatrix} = \frac{1}{E}\begin{bmatrix} 1 & -\upsilon & -\upsilon \\ -\upsilon & 1 & -\upsilon \\ -\upsilon & -\upsilon & 1 \end{bmatrix}\begin{bmatrix} \sigma_x \\ \sigma_y \\ \sigma_z \end{bmatrix} + \alpha\Delta T\begin{bmatrix} 1 \\ 1 \\ 1 \end{bmatrix} \qquad (1)$$

where ε_i are the total strains in the i-direction, σ_i are the stresses in the i-direction, E, ν, α are the Young's Modulus, Poisson's Ratio, Coefficient of Thermal Expansion of the Expansion Tool and ΔT is the change in temperature of the system. Solving for the plane strain condition yields

$$\varepsilon_x = \frac{1+\upsilon}{E}\left((1-\upsilon)\sigma_x - \upsilon\sigma_y + E\alpha\Delta T\right) \qquad (2a)$$

$$\varepsilon_y = \frac{1+\upsilon}{E}\left(-\upsilon\sigma_x + (1-\upsilon)\sigma_y + E\alpha\Delta T\right) \qquad (2b)$$

where all strain in the z-direction has been set to zero.

Assuming that during the cure cycle the resin in the rib is in liquid form and that the Expansion Tool groove is sufficiently full of material to eliminate any membrane effect in the skin, the pressure on the top (y-direction) of the Expansion Tool and the uncured resin are both equal to the pressure being applied to the whole system. Due to the hydrostatic nature of liquids, the pressure applied laterally (x-direction) to the Expansion Tool is, therefore, also equal to the applied pressure. Therefore,

$$\sigma_x = \sigma_y = -\sigma$$
$$\varepsilon_x = \varepsilon_y = \varepsilon \qquad 3$$

where σ is the applied pressure and ε is the hydrostatic strain in the Expansion Tool. The resulting constitutive equation and its inverse are

$$\varepsilon = \frac{1+\upsilon}{E}\left((2\upsilon - 1)\sigma + E\alpha\Delta T\right)$$
$$\sigma = \frac{E}{2\upsilon - 1}\left(\frac{\varepsilon}{1+\upsilon} - \alpha\Delta T\right) \qquad 4$$

where the hydrostatic strain in the Expansion Tool can be written as

$$\varepsilon = \frac{\Delta h}{H} = \frac{\Delta w}{W} \qquad 5$$

where all symbols in Equation 5 are defined in Figure 3.

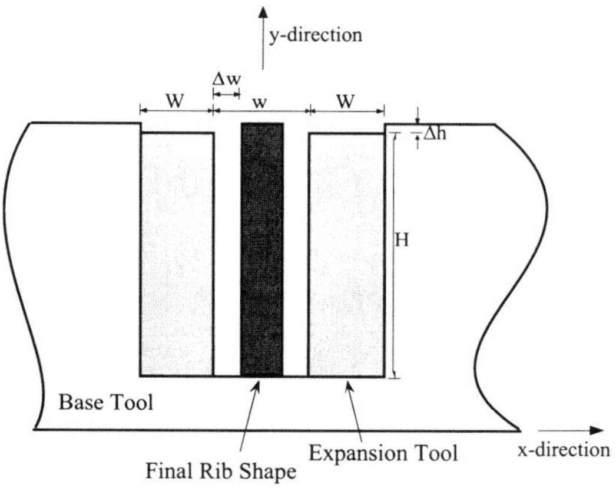

Figure 3: Thermal Model Diagram. *The Expansion Tool expands Δw and Δh to compact the rib as shown. This thermal model neglects the effect of the Expansion Tool Floor leading to a rectangular rib cross-section.*

2.5 Determining Expansion Tool Geometry Using the model developed in Section 2.4, it is possible to determine the Expansion Tool geometry by assuming that the hydrostatic strain is determined by material curing requirements (applied pressure (σ) and cure temperature (ΔT)) and that the groove width (w) is determined by the desired rib geometry. An additional constraint, that the groove width must be equal to or wider than the prepreg tow width, is necessary to facilitate winding. The desired rib width and rib height can be related to the Expansion Tool geometry parameters through

$$R_h = H + \Delta h \qquad 6$$

$$R_w = w - 2\Delta w \qquad 7$$

where R_h and R_w are the desired rib height and width respectively. Typically, for maximum flexibility, the groove width, w, is set to its minimum possible value, the width of the prepreg tow. Considering w to be known, therefore, Equations 5, 6 and 7 can be solved for the remaining Expansion Tool geometry parameters yielding

$$H = \frac{R_h}{1+\varepsilon} \quad 8$$

$$\Delta h = H\varepsilon \quad 9$$

$$W = \frac{w - R_w}{2\varepsilon} \quad 10$$

$$\Delta w = W\varepsilon \quad 11$$

where ε is a known parameter calculated from Equation 4.

2.6 Determining Tow Requirements Finally, it is necessary to calculate the number of tows required to obtain the desired rib geometry. The number of pre-preg tows required is

$$T_n = \frac{R_w R_h}{T_A} \quad 12$$

where T_A is the area of a single tow. Alternatively, substituting Equations 6, 7, 9 and 11 yields

$$T_n = \frac{H(w - 2W\varepsilon)(1+\varepsilon)}{T_A} \quad 13$$

which must be satisfied in order to validate the assumption that there is no membrane effect in the skin, the requirement for the simplification in Equation 3.

3. EXPERIMENTAL VERIFICATION

3.1 Theory Comparison to Experimental Results. The model in Section 2 relates Expansion Tool geometry to required cure cycle parameters and rib geometry. To validate the model, several Expansion Tool configurations where studied at several applied pressures. In each case, the required number of tows was calculated using Equation 13. Verification of the model was determined by comparing the experimentally achieved rib dimensions with those predicted by Equations 6 and 7.

The experimental setup is shown in Figure 4. No skin was used so that the achieved rib height could be accurately determined. For each Expansion Tool configuration tested, the predicted number of tows, T_n, was laid up into the tool. The applied pressure was varied from 207 to 620 KPa between different runs. The material used was IM7/977-2 and the cure temperature, driven by the manufacture's recommended material cure cycle, was 177 °C for all cases.

Figure 4: Thermal Model Test Setup. *Five Expansion Tool configurations were tested. In each case, the number of tows required was calculated and the resulting rib width and height compared to theoretical predictions.*

All results of the experiments were in good agreement with the model. The results obtained at 620 KPa applied pressure are shown in Figure 5. The actual experimental rib heights and widths are shown compared to those predicted by the theory. Very good agreement is shown in all cases. The largest error was in the prediction of the rib width in Groove 5. The predicted rib width in this case differed from the experimental rib width by 6% of the experimental rib width.

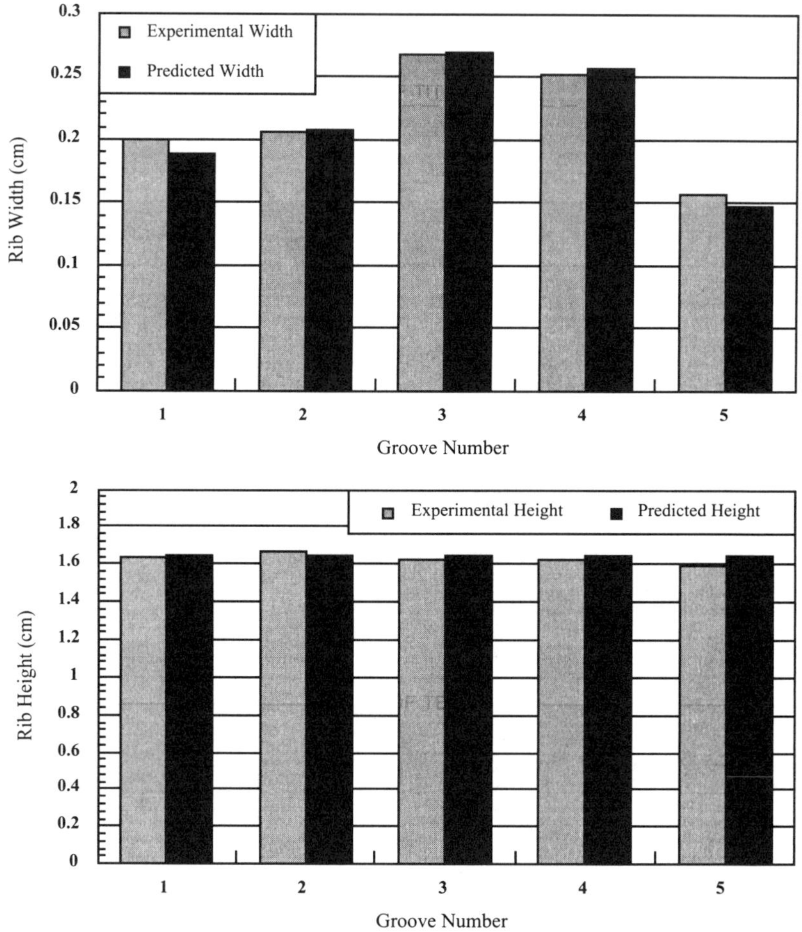

Figure 5: Results of Thermal Model Verification. *The results of the thermal model verification are shown for an applied pressure of 620 KPa. The predicted and experimental rib widths and heights are compared.*

4. DISCUSSION

4.1 Hybrid Tooling Thermal Envelope The results of Section 2.4 have implications on the *process envelope* within which the Hybrid Tooling concept can be applied. Composite curing techniques imply that the pressure applied to the tooling during cure must be some value greater than zero. Because of the hydrostatic nature of the strain in the Expansion Tool, this condition implies an upper limit on the amount that the Expansion Tool can compress the rib. In reality, compaction of the AGS structure skin mandates a minimum applied pressure that will be greater than zero so the possible groove closure of the Expansion Tool will be even further limited. Substituting Equation 5 into Equation 4, the maximum possible groove closure can be calculated as

$$\Delta w_{max} = W(1+\upsilon)\alpha\Delta T \qquad 6$$

where the applied pressure has been set to zero. Similarly, the maximum applied pressure is limited by the fact that the AGS rib requires that the Expansion Tool provide some compaction. This value can be calculated using Equation 4 and is

$$\sigma_{max} = \frac{E\alpha\Delta T}{1-2\upsilon} \qquad 7$$

where the strain in the Expansion Tool has been set to zero. These two limiting cases define a process envelope which is shown in Figure 6. For a given Hybrid Tooling case, values for the applied pressure and required compaction must be selected such that they fall inside this envelope.

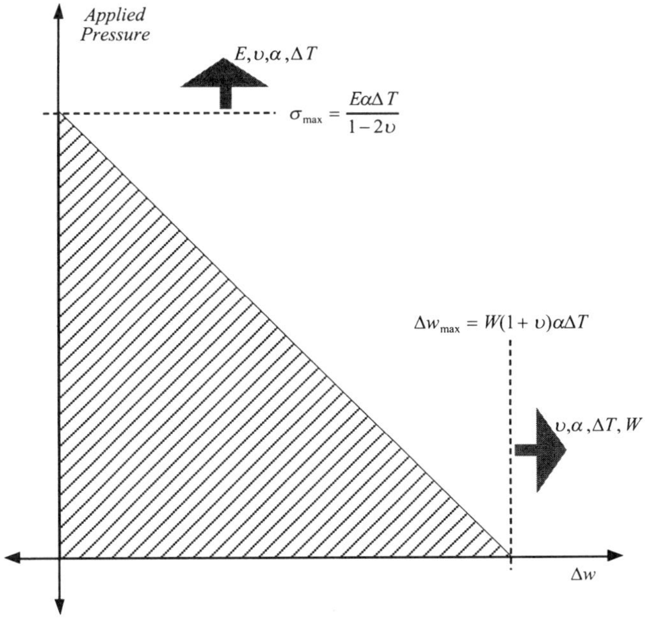

Figure 6: Hybrid Tooling Process Envelope. *The constraints of positive applied pressure and rib compaction imply that Hybrid Tooling must be used within a process envelope. The envelope can be expanded by increasing any of the parameters displayed next to the shaded arrows.*

4.2 Application of Hybrid Tooling to Launch Vehicle Structures In addition to several flat and cylindrical AGS parts, a section of a launch vehicle shroud has been fabricated. The Base Tool was machined on a filament winder and the Expansion Tool, cast separately, was laid into machined grooves. The filament winder was then used to wind ribs and skin for the structure. With exception of the laying of the Expansion Tool, this entire process proved to be fully automated and the resulting tooling proved to be reusable. Pictures of the constructed tooling, along with the resulting part, are shown in Figure 7.

Using this process, a full-scale launch vehicle part, such as a moderately sized shroud, can be constructed in 35-40 hours. This figure includes an autoclave cycle and fabrication of all associated tooling. Additional parts can then be wound, bagged and cured in 20-25 hours each, a large savings over traditional fabrication methods.

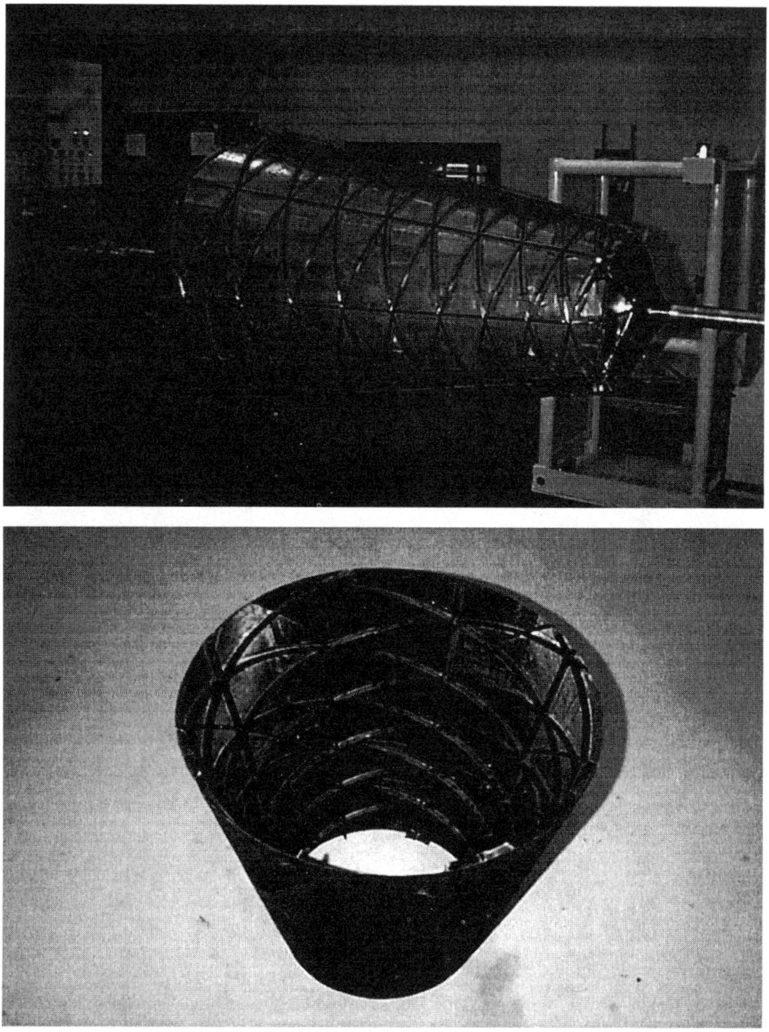

Figure 7: Hybrid Tooling and Resulting Part. *Low cost Hybrid Tooling was constructed in order to fabricate a launch vehicle shroud section.*

5. CONCLUSIONS

A newly developed *Hybrid Tooling* process has been developed and researched in order to produce high quality AGS structures at low cost. This Hybrid Tooling concept involves two distinct tooling sections; the *Base Tool* is primarily responsible for providing good part stability and geometry control while the *Expansion Tool* provides lateral compaction to the AGS structure ribs. Several materials have been studied for use as Base Tooling. Also, aspects of Hybrid Tooling have been investigated through experimental lab work and theoretical thermal models. Using these models, the required tooling geometry can be related to the desired rib geometry and the necessary processing conditions (temperature and pressure). This work shows that the available parameters governing the tooling geometry and process conditions must fall within a process envelope in order to achieve AGS compaction in both ribs and skin. Successful demonstration of the concept has been achieved through the fabrication of a composite AGS launch vehicle shroud section.

6. REFERENCES

1. Cook, D. Malkus and M. Plesha, Concepts and Applications of Finite Element Analysis, Third Edition, John Wiley & Sons, New York, 1989.

2. Huybrechts, Analysis and Behavior of Grid Structures, Stanford University, Stanford, CA, 1995.

3. Chen, Analysis and Optimum Design of Composite Grid Structures, Stanford University, CA, 1995.

28th International SAMPE Technical Conference
November 4-7, 1996

THE RAVEN: AN EDUCATIONAL PROJECT TO BUILD A WORLD RECORD SETTING HUMAN POWERED AIRCRAFT USING ADVANCED MATERIALS

S. H. Dillman
Western Washington University
Department of Engineering Technology
Bellingham, WA 98225

H. A. Costantino and P. R. Illian
The Raven Project
The Museum of Flight
Seattle, WA 98108

ABSTRACT

A human powered aircraft named the Raven has been designed. The Raven incorporates advanced materials to produce rigid wing and fuselage skins and a cantilever wing construction which should allow it to set four world records in human powered flight. Design specifications call for an aircraft with a wingspan of 35 m, a weight of only 34 kg, a cruising velocity of 32 km/hr, and a range of 160 km. The aircraft is currently under construction by over 350 volunteer students and industry professionals. Tooling for most of the structural parts of the aircraft is complete. Manufacture of prototype fuselage components required development of an unusual double vacuum bagging process. This process has been used to produce prototype parts that, while over design weight, suggest that the design specifications are achievable.

KEY WORDS: Aircraft, Education, Composite Structures

1. INTRODUCTION

1.1 Human Powered Flight

Mankind has been fascinated with flight since prehistory. Although the earliest attempts at flight used the pilot as the power source, these attempts were invariably failures. There were many factors that led to these failures including both limitations in the technology and the physiological limitations of human beings. Human beings are capable of producing little power per unit body weight. Consequently, for human powered flight to be possible an aircraft which is both extremely efficient aerodynamically and extremely light in weight is necessary. The myth of Icarus notwithstanding, only recently did aircraft and materials technology evolve to the point where human-powered flight over substantial distances became possible.

The first successful experiments in heavier-than-air human powered airplane were inspired by prizes offered by Robert Peugeot in 1912, only 9 years after the Wright brothers flew(1). By 1921 the record for distance had only reached 12 m. Since that time the record for distance in human powered flight has been reset several times. The current record stands at 119 km and was set by the Daedalus, an aircraft built at MIT, in 1988.

The Daedalus shared certain design features with most other successful human powered aircraft. Key among these were the use of a wing design that consisted of a spar and rib structure with stretched polyester film wing skins, and wire bracing running from the wing midspan to the bottom of the fuselage which allowed the use of a lighter weight more flexible wing without excessive deflection. While these design features allow the construction of a craft which is exceptionally light in weight, they carry with them a penalty. A stretched polyester wing skin does not hold a true airfoil shape, but scallops between the ribs, reducing the efficiency of the airfoil. Also, stretched polyester places a stress on the wing spar which tends to deform the airfoil. In addition, since stretched polyester is not rigid, it vibrates during flight and tends to wrinkle at corners, resulting in increased aerodynamic drag. Drag is further increased by the presence of the wire bracing. Furthermore, attachment of the wire bracing at the base of the fuselage necessitates a relatively tall fuselage, again resulting in increased drag.

1.2 The Raven Project

The Raven Project began in late 1993 with the educational goal of providing undergraduate students the opportunity to work closely with industrial professionals on a multidisciplinary project of larger scope than typically encountered in the undergraduate curriculum. Benefits to the students include: the opportunity to apply skills learned in school to a challenging problem; the opportunity to work closely with industrial professionals who can provide insight into the problem solving techniques used in industrial environments; experience working as part of a team on a long term goal; and experience working on a project in which several teams, each with different skills and in different locations, work on different aspects of the project. This last feature is perhaps the most important; it is unusual for students to work on projects in university courses in which they must interact with students of other disciplines working on other aspects of the same project. In industry, however, this is the norm. For example, engineers must work with sales and marketing, designers with engineers, research scientists with manufacturing, etc. Thus, a critical feature of the experience is the development of the ability of teams composed of individuals with greatly differing backgrounds to work together with a common goal.

The technical goal of the project is to build a human powered aircraft which will break four world records in human powered flight. These records will be distance and duration aloft for general flight, and distance and duration aloft for female flight. The goals for each of these are 160 km, 5 hours, 80 km, and 3 hours, respectively. In keeping with the tradition of naming human powered aircraft after birds, the craft will be named the 'Raven', a bird that figures prominently in the mythology of Native American cultures of the Pacific Northwest. The name is made more appropriate by the fact that the plane will be constructed predominantly from carbon fiber composites giving it the characteristic black coloration of the raven.

Nine colleges and universities in the Puget Sound area are involved in the project. A list of schools and their areas of responsibility is provided in Table 1. In addition over 100 industrial volunteers are providing technical assistance to the students. While giving an brief description of the key design features of the Raven, this paper will focus on the those areas for which Western Washington University has responsibility: the fuselage, wings, and tail boom.

TABLE 1: Schools Participating in the Raven Project

School	Responsibility
Art Institute of Seattle	Educational displays, logo
Bellevue Community College	Multi-media displays
Lake Washington Technical College	AutoCAD, Machining
North Seattle Community College	AutoCAD, Machining
Renton Technical College	Simulator, Wing tool, Machining
Seattle Pacific University	Human-performance
Seattle University	Drive train
University of Washington	Wind tunnel, Simulator, Transporter
Western Washington University	Tail boom, Fuselage, Wings

2. EXPERIMENTAL

2.1 The Design of the Raven

The critical factors in human powered aircraft design are weight and glide ratio. Glide ratio is simply the ratio of lift to drag. Under the same power, which is limited by the physical capability of a human pilot, a reduction of either weight or drag will result in an aircraft capable of flying faster, and therefore further before the pilot becomes too exhausted to continue. In order to achieve the reduction of drag while minimizing weight, the Raven incorporates many of the design features suggested by Schöberl in his discussion of the optimization of human powered aircraft (2). A schematic of the Raven in comparison to the Daedalus is provided in Figure 1. The weight of the Daedalus, at 31.8 kg, would be extremely difficult to improve upon significantly. However, the drag can be reduced by using a cantilever wing design, thereby eliminating the wire bracing of the wings and reducing the size of the fuselage, and by replacing the polyester wing skins with rigid wing skins. This is the design approach taken in the Raven. Unfortunately, use of a cantilever configuration requires a substantially stiffer wing. For a cantilever wing design to be sufficiently low in weight, materials of extremely high specific stiffness and specific strength must be used.

The are several other features of the Raven design that are unique. Because of the reduction of the size of the fuselage enabled by the use of the cantilever wing, the pilot of the Raven is seated in a recumbent position. Also, a landing gear strut is required to raise the aircraft high enough for the propeller to clear the ground. The landing gear and strut are designed to be retractable in order to reduce the drag of the aircraft after takeoff. In addition, the control system incorporates an autopilot feature in order to allow the pilot to concentrate on providing enough power to keep the plane in the air.

Table 2 presents some of the major design specifications for the Raven (3). The weight specifications are further broken down in Table 3. Note that these weight requirements are extremely challenging but, as shown in Table 3, the majority of parts constructed as prototypes have been produced within the specified weights.

TABLE 2: RAVEN Design Specifications			
PROJECT GOALS		DIMENSIONS	
Distance	160 km	Wingspan	35 m
Duration	5 hours	Height	3.0 m
		Length	9.1 m
PERFORMANCE REQUIREMENTS		Wing Area	32.5 m^2
Cruise Velocity	9.1 m/s	Empty Weight	34 kg
Power required	190 W	Take-off Weight	111 kg
Total Drag	20.5 N		

TABLE 3: Weight Specifications in kg. Numbers in parentheses represent the weight of prototype parts where they exist.			
Wing	20 kg	Control System	0.8 kg (1.4)
Spar	9.1	Computer	0.2
Skin	10.9	Servos	0.2
		Wiring	0.2
Fuselage	5.0 kg	Battery	0.2
Shell	1.8 (2.4)		
Structure	2.3 (1.3)	Tail Assembly	6.4 kg
Landing Gear	0.9 (0.3)	Boom	3.6
		Horizontal	1.4 (0.9)
Drive Train	1.8 kg	Vertical	1.4
Propeller	0.7 (0.3)		
Gears	0.2	Pilot	64 kg
Pedals	0.9		
		Water	6.4 kg

2.1.1 Fuselage Design

As illustrated in Figure 2, the fuselage consists of an internal load bearing structure surrounded by a rigid shell which provides the aerodynamic surface of the fuselage. The structure is essentially a C-channel configuration and will be constructed from carbon fiber reinforced epoxy composite. Figure 3 shows details of the fuselage shell construction. It consists of a sandwich structure with a 2.5 mm thick core of polystyrene foam having a density of 24 kg/m^3 and facings of carbon fiber mat with an areal weight of 7 g/m^2 impregnated with epoxy. Facing thickness are approximately 0.08 mm.

2.1.2 Wing Design

The airfoils used in the Raven are essentially the same as those used in the Daedalus (4,5). The wing will contain a spar which consists of a sandwich structure having carbon/epoxy caps and an aramid honeycomb core. Rib design and spacing will vary based on the load experienced in a particular section of wing. At the wing root ribs will be polystyrene foam core carbon mat sandwich structures, transitioning into unreinforced foam on the outer 20% of the wing. Rib spacing will vary from 15.2 cm at the root to 30.5 cm at the wing tip. The wing skins will be of the same construction as the fuselage shell. Total wingspan will be 35 m.

Construction of the Raven is currently in progress. As of this writing the tooling for the fuselage is complete and several prototype parts have been manufactured. The wing tooling has not yet been completed.

2.2 Fuselage Manufacture

Unlike the majority of human powered aircraft, which utilize a flexible skin stretched over an internal structure, the Raven's rigid skin design requires molding of the fuselage skin. Thus, the first step was construction of a fuselage master model. The model was constructed from fiberboard templates set 10 cm apart. The space between the templates was filled with polyurethane foam and sanded flush with the templates. A layer of fiberglass reinforced polyester was applied. After cure the model was sanded and filled with polyester filler and a filling primer repeatedly until an acceptable surface was obtained.

Four fuselage shell molds were constructed: the upper half of the fuselage, left and right lower quarters, and the nose bezel. The model was waxed and a polyvinyl alcohol based release film was spray applied. Each mold was prepared from fiberglass cloth impregnated with a room temperature curing epoxy resin. After removal from the model, the molds were filled, sanded, primed and painted.

Because of the fragile nature of the carbon/epoxy/polystyrene sandwich structure, the process of manufacturing the fuselage shell is extremely challenging. Perhaps the most challenging aspect is removal of the part from the mold. Several mold release agents were evaluated, but because of the large surface area and low strength of the sandwich construction, even a slight adhesion to the mold results in irreparable damage to the part during removal. Consequently, an unusual approach was developed. A high elongation vacuum bagging film, referred to as the release bag, is drawn over the mold prior to lay-up of the part. The part is then laid up on top of the release bag. Upon completion of cure the vacuum holding the release bag to the tool is removed and the part can be lifted from the tool using the release bag. The release bag is then peeled from the finished part.

Because the polystyrene foam is rigid and brittle and the mold contours complex, it is difficult to get it to lay into the mold without cracking. Thus, templates were prepared which produce individual pieces of foam which can conform to the mold. Early experiments with simple butt joints between the foam sections were failures as serious flaws invariably formed at the joints. Consequently each foam section is cut 1.25 cm oversize on all edges which intersect another section. The edges are then tapered using a hot wire cutter and a scarf joint results in the finished part. The polystyrene is perforated to allow resin to pass through it and to improved bonding to the carbon.

The first step in the construction of a fuselage shell part is preparation of the release bag. The high elongation bagging film is placed over the mold. The film is held in place by double sided tape under slight tension to prevent wrinkling when the film is drawn into the mold. Using vacuum bag sealing tape, the film is then sealed to a second film underneath the mold to form a bag which completely encloses the mold. A vacuum is drawn on this bag and the release bag is drawn into the mold. If any wrinkles are formed they are gently worked to the edge of the bag providing a smooth molding surface.

The carbon fiber mat is wet out with epoxy prior to laying it into the mold. Because weight is critical, excess resin must be removed from the mat prior to placing it into the mold. However, the fragile nature of this very light weight mat makes resin removal without fiber damage difficult. A technique that works well is to place the impregnated mat between two sheets of polyester film and squeegee the resin out at the edges. The polyester prevents tearing of the mat. The impregnated carbon is then laid into the mold. The polystyrene foam is then placed over the carbon and held in place temporarily by rice bags. A perforated release ply and a bleeder are placed over the part, followed by a vacuum bag. The vacuum bag must enclose the entire tool or it will result in wrinkling or lifting of the release bag. A cross-sectional view of the lay-up construction is provided in Figure 4.

Following cure the outer vacuum bag, release ply, and bleeder are removed. The upper skin is then laid-up, the release, bleeder and vacuum bag re-applied, and the upper skin cured. After cure the part is removed from the mold and the release bag peeled off.

Two prototype fuselage lower quarter panels and a fuselage upper half were produced using this processes. In the case of the fuselage upper half a thermoformed acrylic window of.approximately 0.4 mm thickness was epoxied in place prior to laying in the second layer of carbon.

Each of the two lower fuselage quarter panels weighed 0.5 kg. The prototype upper fuselage half had a weight of 1.4 kg. Thus, total weight of the prototype fuselage skins is 2.4 kg.

2.3 Method of Wing Manufacture

Because of the size of the wings, manufacture of a model followed by lay-up of a fiberglass mold is not practical. Therefore, the wing tools will be constructed by mounting templates at a spacing of 15.2 cm on a frame. 0.4 mm thick aluminum sheet will be forced to conform to the templates and will serve as the mold surface.

Lay-up of the wing skins will proceed in a way much like the fuselage skins. The concept of utilizing high elongation vacuum bagging film will also be utilized in the wing molds to allow for easy part removal.

2.4 Tail Boom

The tail of the Raven is supported by a 6.7 m long tail boom which tapers from a 10 x 7.6 cm elliptical cross section at the fuselage to a 5.1 cm diameter circular cross-section transitioning to a 7.6 x 7.6 cm square actuator housing at the tail. This tail boom will be made by roll wrapping carbon/epoxy pre-preg onto a 2-piece fiberglass mandrel. The part will then be shrink taped and oven cured. The mandrel is currently under construction.

3. DISCUSSION

Although there is considerably more work to be done before the Raven will be ready to fly, several important hurdles have been overcome. A process for successful manufacture of the delicate wing and fuselage skins has been developed. Furthermore, while the weight targets have not yet been met, the prototype fuselage skins are sufficiently close to the target weights to lead to confidence that these weights are attainable. This confidence is further enhanced by the successful construction of test specimens and tail components utilizing the same processes which meet the targets.

The prototype fuselage shell at 2.4 kg is approximately 33 % overweight. While the excess weight may be compensated by other parts which come in underweight, if the fuselage cannot be produced on target it is unlikely that the wings will be, as essentially the same process is used to manufacture both. Fortunately, inspection of the fuselage shell reveals that the weight reduction should be achievable. Lay-up test panels have shown that areal weights of 200 g/m^2 are easily attainable. The total surface area of the fuselage is approximately 7.4 m^2. Thus, neglecting the window, it should be possible to produce a fuselage shell having a weight of 1.5 kg. While the window, which has an areal weight of approximately 700 g/m^2 and occupies approximately 0.5 m^2 contributes approximately 0.2 kg of excess weight, the majority of the excess weight appears to be from excess epoxy resin in the sandwich facings. More careful control of the amount of epoxy used in the impregnation of the carbon should allow the production of parts with weights that are within specification. In addition, experience with lay-up of tail skin panels as well as the lower quarter fuselage panels shows that as experience is gained with production of a given part weight tends to decrease. Finally, if necessary the area of the window can also be reduced.

Surface inspection of the prototype fuselage parts revealed some minor defects. While the majority of the surface was adequately smooth, there were surface defects at some of the foam joints. It should be possible to eliminate these defects by more careful preparation of the foam. In addition, the upper fuselage shell exhibited an area of apparently dry fibers and delamination between the foam and skin. This was determined to be the consequence of bridging by the vacuum bag. Future lay-ups will require a pressure intensifier at this location.

Finally, it is important to remember that the primary objective of this project is an educational one. In that respect, it has already been a success. At Western Washington University alone, over 40 students have been involved in the project. All of these students have had the opportunity to interact with industrial professionals who are volunteering as advisors. In addition, several of these students have had the opportunity to serve in leadership roles and to interact with members of other teams on the project. As the project continues through design debugging, final parts manufacturing, flight testing, and finally the world record attempts, the students will continue to have new opportunities to practice their skills and learn new ones.

4. CONCLUSIONS

A human-powered aircraft, The Raven, has been designed which should be capable of a sustained cruising speed of 32 km/hr under a power of 190 W. The design, currently under construction, takes advantage of advanced composite materials to produce a craft which will have rigid wing and fuselage skins and a cantilever wing, while maintaining a weight within 10% of that of the Daedalus, the current world record holder for distance. The lack of wire wing bracing and the rigid skins are expected to produce less drag resulting in a significantly higher cruising velocity.

A process for manufacture of the delicate wing and fuselage skins has been developed and construction is currently in progress. Construction of prototype parts indicates that the weight specifications will be difficult to meet, but indications are that they can be met. However, testing to determine if the parts, most critically the wings, can withstand anticipated in-flight loads awaits completion of test parts.

5. REFERENCES

1. C. Roper, in A. V. Abbott and D.G. Wilson, ed., Human-Powered Vehicles, Human Kinetics, Champaign, IL, 1995, pp 217-245.

2. E. Schöberl, Human Power, 6, (1) (1987).

3. P. Illian, proceedings of the International Symposium for Human Powered Flight, 1994.

4. M. Drela, AIAA J. of Aircraft, 25 (8) 1988.

5. J. Langford, proceedings of AIAA Aircraft Design, Systems, and Operations Conference, 1989.

6. ACKNOWLEDGMENTS

The authors would like to thank all of the students and industrial advisors who have contributed to the Raven project. In addition we wish to thank all of the many companies which have donated materials, time, and money to this project. We would like to specifically acknowledge the contributions of the following companies for use at Western Washington University in the construction of the Raven fuselage and wings: American Institute of Aeronautics and Astronautics, C. R. Laurence, COMCO Plastics, DuPont, Fiberlay, General Plastics Manufacturing, Heath-Techna Aerospace, Innovative Composites Engineering, The Molen Company, VELO Graphics, Veratec Division - International Paper, West Coast Paper.

Figure 1: Top and Side View Comparisons of the Raven to the Daedalus (Raven is the upper drawing in each view)

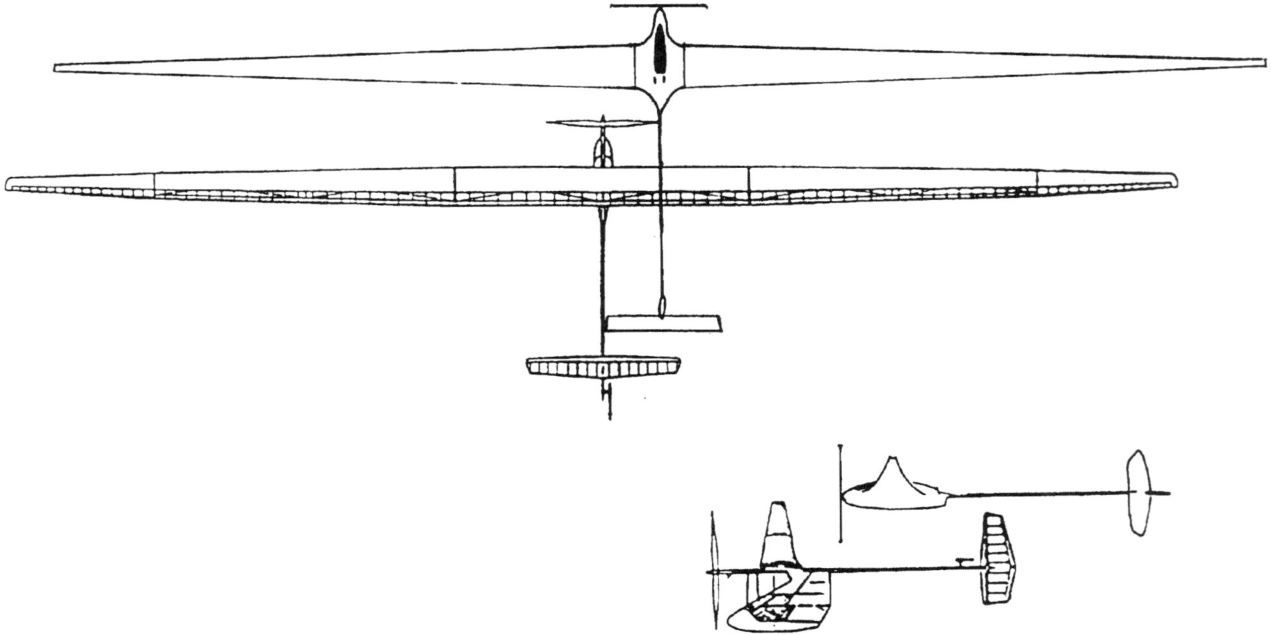

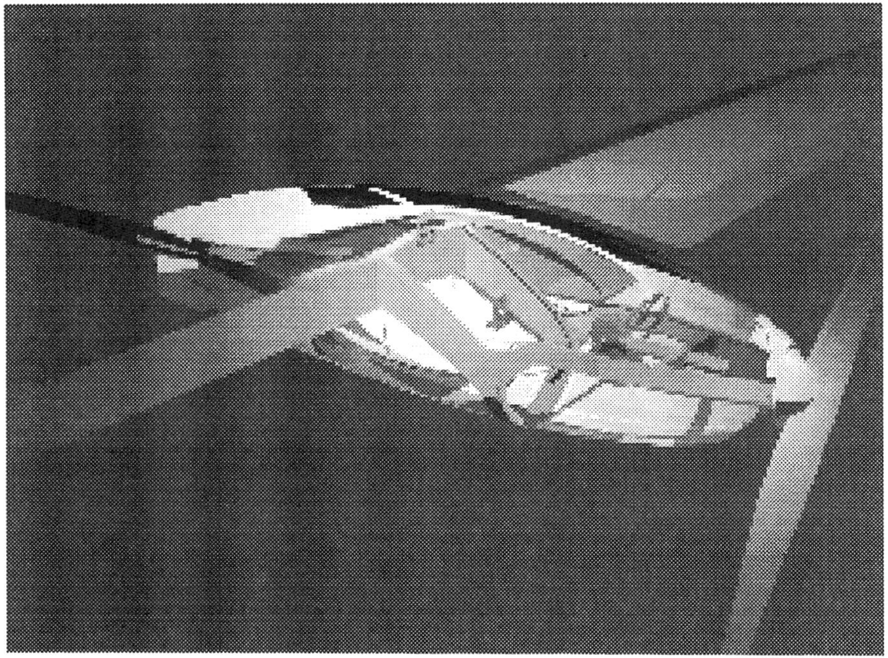

FIGURE 2: Cut-away view of the Raven fuselage showing the outer shell/internal structure configuration.

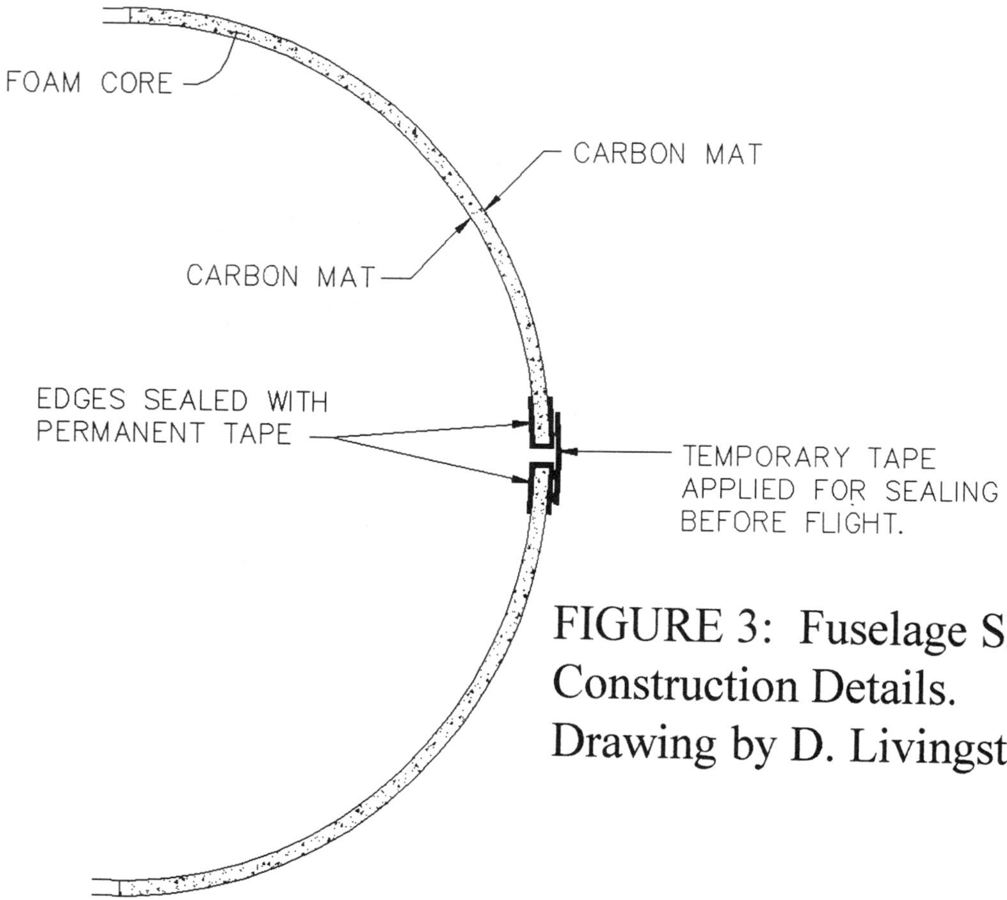

FIGURE 3: Fuselage Shell Construction Details. Drawing by D. Livingston.

Figure 4: Schematic Cross-section of Vacuum Bagged Fuselage Shell Component (Lay-up of exterior skin and core)

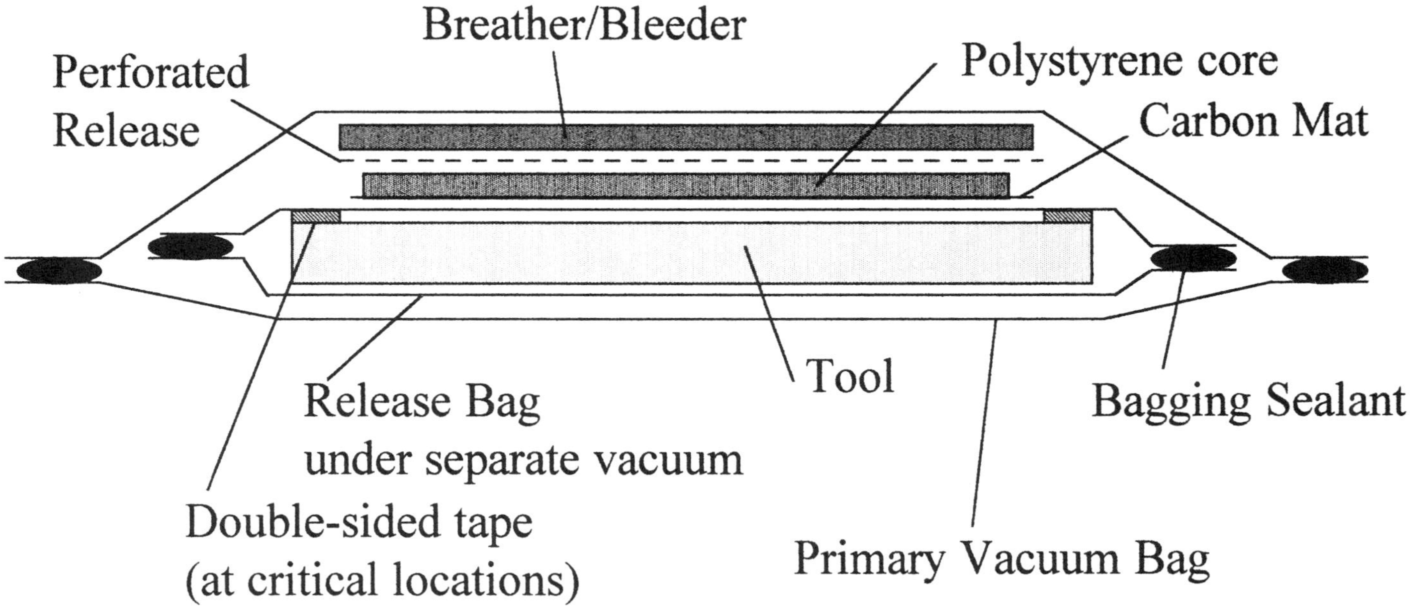

28th International SAMPE Technical Conference
November 4-7, 1996

THERMAL CONDUCTIVITY OF COMPOSITE MATERIALS

Christoph M. Pistor, M. Atıf Yardımcı, Kimberly J. Jasiak, Selçuk I. Güçeri
Mechanical Engineering Department
University of Illinois at Chicago
Chicago Illinois US, 60607-7022

ABSTRACT

A numerical model has been developed to predict the thermal conductivity of different types of composite materials. Effective intrinsic thermal properties, such as specific heat capacity, of particle-binder systems follow the rule of mixture, as do other intrinsic properties. Transport properties such as effective thermal conductivity or mass diffusivity, on the other hand, are strong functions of the moments of distributions, hence simplistic models do not predict their behavior. Random pattern generators are utilized to generate different morphological material patterns. The patterned mesh is utilized by a finite element model to solve steady-state heat conduction problem. Effective thermal conductivities are predicted for different volume percent loading of plastic, ceramic and metal particles. To validate the model several experiments have been carried out. Laser Flash technology has been used to determine the thermal conductivity of a silicon nitride/thermoplastic binder composite. The same material has been analyzed using Modulated DSC Scans of two samples. From Modulated DSC the thermal conductivity can be derived by comparing the apparent heat capacity of a thick and a thin test specimen. The comparison of the numerical results with those obtained by the various experiments shows the good performance of the model.

KEY WORDS: Thermal Analysis, Thermoplastic Resins/Polymers, Ceramic Materials/ Composites, Filament Winding

1. INTRODUCTION

Thermoplastic composite materials have a variety of advantages over traditional thermosets. Recycling issues, infinite shelf life, and especially the ability to go through several melting/ solidification cycles, made them interesting for new applications and processing techniques (1). A thermoplastic composite is assumed to consist of a thermoplastic matrix and some form of filler (fibers, particles, whiskers etc.). For these composites the final properties of the part depend largely on the thermal history of the thermoplastic matrix. It is therefore necessary to adjust the processing parameters such that the temperature as well as heating and cooling rates lie in a certain processing window (particular for every thermoplastic matrix) throughout the whole chain of processing steps. To predict such a window for tape winding of thermoplastic pre-preg tapes, numerical models have been developed in earlier studies (2,3,4,5,6). Literature values of the thermal conductivities longitudinal to and transverse to the fiber direction have been used for these simulations.

The laser assisted tape winding process has been recently adapted to process ceramic green parts (7), using a particular binder particle system of unknown thermal properties. The complete manufacturing process consists of three steps. After the green part has been processed it is placed in an oven and the thermoplastic binder is burned off. This production cycle is often referred to as Binder-Burn-Off (BBO). The part that remains consists of ceramic particles only and is now called a brown part. This part is again placed in an oven and heated to the sintering temperature of the particular ceramic material used. In the present investigations Silicon Nitride particles were used. The BBO and sintering procedure is described in (7).

The processing of the green part by the Laser Assisted Tape Winding (LATW) technique involves melting and solidification of the thermoplastic binder. In order to develop means that predict the optimal processing parameters, the knowledge of the thermal conductivity of the composite is of paramount importance.

Two experimental methods have been used to determine the conductivity of this composite system. First, Laser Flash technique (8) was used to determine the thermal diffusivity, which is widely used for fast measurement of materials with medium to low thermal diffusivities. This measurement was performed at the Physics Department University of Delaware directed by Prof. Onn. Next, Modulated Differential Scanning Calometry (MDSC) was used, which is a relatively new method that is based on comparison of measurement of two samples on a standard MDSC equipment (9). These tests have been performed since MDSC equipment from TA Instruments (DSC 2910, Thermal Analyst 2000, LNCA) was available and the feasibility of this method for the particular material was of interest. In comparison to previous attempts to use DSC equipment for thermal conductivity measurements (10), the MDSC technique has the advantage that no additional parts or physical changes of the DSC cell are necessary.

2. EXPERIMENTAL

Three types of Polymer based materials have been examined experimentally. First, the particle binder system has been examined by Laser Flash method. This material consists of 51 % volume fraction of silicon nitride particles of an average size of 0.5 μm and 49 % RU9 thermoplastic binder. The substances contained in the system are summarized in the following list.

Silicon Nitride
Yttrium Oxide
Aluminum Oxide
Ethylene Vinyl Acetate
Ethylene Ethyl Acrylate
Boron Nitride
Heavy Mineral Oil
Methoxypolyethylene Glycol

Next, the neat material, without particles and sintering agents, has been examined using MDSC. To calibrate the MDSC equipment, polystyrene samples (supplied by TA Instruments) have been used.

The Laser Flash method is described in the following. A light pulse (from the pulsed laser) is shined on one side of the material on which a layer of heater material is deposited. The energy of the light pulse can be absorbed and heats the heater material. This heat will then spread into the material while it is being measured. If simplified to a one-dimensional heat flow problem, a differential equation of heat propagation with these boundary conditions can be established and a solution can be found through series expansions. By observing the other surface of the measured material, and recording the time change (transient) of the back surface temperature, the diffusivity of the material can be determined. The laser used is a Nd-glass pulse laser at 1.06 μm wavelength. The laser unit has an output energy of 100 J, and a pulse has approximately 1 J of energy.

MDSC is a technique to characterize the thermal properties of materials and is patented by TA Instruments. In addition to linear heating of the sample (as in regular DSC), it superimposes a sinusoidal temperature oscillation. The thermal response of the sample is deconvoluted by a fast fourier transform and three components of heat flow can be separated: the reversing heat flow which is the portion of the signal that is out-of-phase with the temperature modulation; the non-reversing heat flow, which is the portion of the heat flow that is in-phase with the temperature modulation; and the total heat flow, which is the same response obtained from regular DSC. The reversing component becomes of primary interest for the conductivity determination, since it is related to the heat capacity of the material. This method of obtaining thermal conductivity from MDSC measurements has been recently reported (9). Two samples are needed to determine the thermal conductivity by MDSC, where their geometry (a right circular cylinder) is important for an accurate measurement. The sample faces are assumed to be flat and parallel. The thin sample is used to measure the apparent heat capacity. The dimensions are shown in Fig. 2.

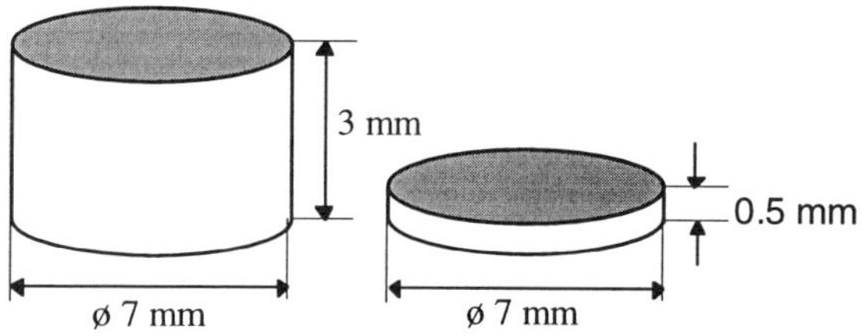

Figure 1: Samples for Thermal Conductivity Measurements

To obtain accurate results, several calibration steps must be taken. First, normal calibration of the MDSC is performed using indium metal and sapphire standard reference materials. Next, thermal conductivity calibration is achieved by using a reference material of known and low thermal conductivity, such as polystyrene.

Thermal conductivity is defined by the Fourier law of heat conduction (1). The assumption that the thermal conductivity is a positive number and the minus sign show the relation of this equation to the second law of thermodynamics, which states that the heat flow is always directed towards the lower temperature.

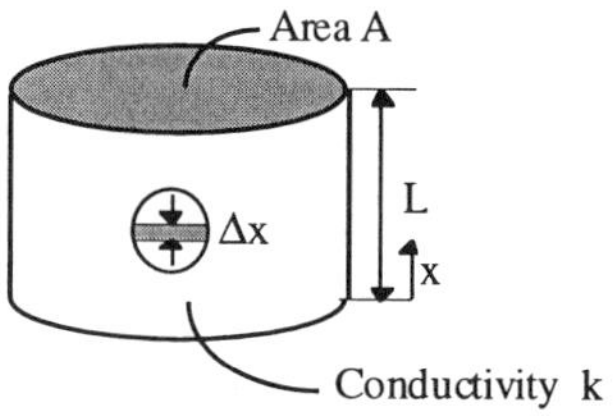

Figure 2: Thermal Conductivity

$$\frac{dQ}{dt} = -kA\frac{dT}{dx} \quad (1)$$

where:

Q: Heat [J]
t: Time [sec]
k: Thermal conductivity [W/(K*m)]
L: Length of specimen [m]
A: Cross sectional area [m^2]

According to Marcus et al. Eq. (1) can be expanded using the modulated heat flow generated by the MDSC, to yield:

$$\left(\frac{dQ}{dt}\right)^2 = 2(Z\theta_0 kA)^2 \frac{\left[1 - 2e^{2ZL}\cos(2ZL) + e^{4ZL}\right]}{\left[1 + 2e^{2ZL}\cos(2ZL) + e^{4ZL}\right]} \quad (2)$$

where:

dQ/dt: Heat flow amplitude [J/s]
Z^2: $(\omega\rho C_p)/(2k)$
C_p: Sample specific heat [J/K*g]
θ_0: Temperature modulation amplitude [K]
ω: $2\pi/P$ angular frequency [2π s^{-1}]
P: Period

For a right circular cylinder the thermal conductivity can then be expressed as:

$$k = \frac{8LC^2}{C_p M d^2 P} \quad (3)$$

where:

C: $(dQ/dt)/(\omega\theta_0)$ apparent heat capacity [J/K]
M: Mass of the sample [g]
L: Length of the sample [mm]
d: Diameter [mm]

The heat loss through the sides of the samples is accounted for by the introduction of a calibration constant. This constant can be found from the measurement of a material with known thermal conductivity:

$$D = \sqrt{k_0 k_r} - k_r \quad (4)$$

where:

D: Calibration constant
k_0: Observed reference material thermal conductivity
k_r: True reference material thermal conductivity

Using this constant the thermal conductivity can be determined as:

$$k = \frac{k_{0u} - 2D + \sqrt{k_{0u}^2 - 4k_{0u}}}{2} \quad (5)$$

where:

k_{0u}: Observed conductivity of the unknown specimen.

3. COMPUTATIONAL MODEL

Two motivations for thermal conductivity predictions are: relative cost of experimental methods and optimization of microstructure for required thermal conductivity. An additional push results from the observation of the theoretical similarities between heat diffusion and other diffusion processes, notably mass transfer. Hence, tools which have been developed in this study are easily applicable to other diffusion related design problems, such as membrane property prediction and design. Thermal conductivity prediction for heterogeneous systems has attracted considerable attention from diverse engineering disciplines ranging from composites thermal design to geothermal engineering (12, 13). The dependence of effective properties on micro-structure makes the property prediction problem a stiff one, due to the large degree of variation in physically observed and synthetically produced micro-structures. Micro-structure characterization, classification, and idealization problems must be solved before attacking the associated heat transfer problem. (14, 15).

The ability of computational techniques to generate random, quasi-random or ordered patterns, together with their ability to solve associated conduction heat transfer on heterogeneous finite element meshes, enables the researcher to predict the effective thermal conductivity (scalar or tensorial) for the given microstructure. In the present study, particle filled thermoplastic matrix material systems have been studied. Such systems are being used in Rapid Prototyping/Injection Molding of ceramics and metals and in Laser Assisted Tape Winding ceramic green bodies. In both cases, the lower thermal conductivity material, i.e. binder, is the continuous phase; whereas the filler material consists of higher thermal conductivity material. A rectangular model geometry has been chosen for the computational experiments. Standard random number generators and other geometrical design tools have been utilized to construct the required micro-structure. A finite element mesh has been produced based on the patterns that have been generated. A sample mesh is shown in Figure 3.

Figure 3: Sample Mesh Composed of Different Materials

The steady-state heat conduction in such a system may be written as:

$$\nabla \cdot \left(\underline{\underline{\mathbf{k}}} : \vec{\nabla}(\mathbf{T}) \right) = \mathbf{0} \tag{6}$$

Applying this partial differential equation to the heterogeneous medium produces infinite property gradients at phase interfaces due to piecewise linear property distribution. On the other hand, the finite element discretization of the above equation is based on the weak integral formulation of the above equation, allowing piecewise continuous property prediction. The computational experiment boundary conditions involve a prescribed temperature (Dirichlet) condition, and a prescribed heat flux condition along the opposite sides of the rectangle. Heat transfer occurs through these two surfaces. Adiabatic conditions are specified along the remaining boundaries. The boundary conditions may be summarized as:

$$T = T_0 \quad @\, x = 0, \qquad k\frac{\partial T}{\partial x} = q_0 \quad @\, x = L$$
$$\frac{\partial T}{\partial y} = 0 \quad @\, y = 0, H \qquad (7)$$

The model requires prior knowledge of the thermal conductivities of the particles and the binder. Finite element discretization was utilized to convert the PDE into a system of linear equations, which in turn was solved utilizing an iterative sparse Gauss-Seidel solver.

After the temperature field has been obtained, height averaged temperature, T_a, at the x=L boundary is calculated. Since for homogeneous media the temperature distribution would be linear, the effective thermal conductivity is calculated utilizing the following relation:

$$k_{eff} = \frac{q_0 \cdot L}{T_a - T_0} \qquad (8)$$

4. RESULTS

The value for the thermal diffusivity of the RU9 composite determined by Laser Flash method is 0.0049 cm^2/sec. The thermal conductivity can be found using the definition of the thermal diffusivity:

$$\alpha = \frac{k}{\rho C_p} \qquad (9)$$

where:

ρ: Density [kg/m^3]

The density of RU9 composite material was determined by the buoyancy method to be 2324 kg/m^3. Eq. (9) yields for the thermal conductivity 1.1 W/(K*m). It's specific heat was taken from the MDSC measurements. Table 1 shows the values obtained from the MDSC measurement of the polystyrene samples. These values were used to determine the calibration constant D. Table 2 and Table 3 show the values for the neat and composite RU9 samples respectively. The difference between the value of the thermal conductivity for Polystyrene and reported literature values (9,10,11) is 0.01 W/(K*m). All displayed values for thermal conductivity represent average values from 10 measurements. The variations in thermal conductivity correspond to 3.3 %, 5.7 % and 4.8 % experimental error for the measurement of polystyrene, RU9 neat and RU9 composite, respectively.

POLYSTYRENE	LENGTH [mm]: 3.56 DIAMETER [mm]: 6.76		MASS [g]: 0.131 PERIOD [1/sec]: 80	
SPECIFIC HEAT [J/K*kg]	HEAT CAPACITY [mJ/K]	THERMAL COND. OBSERVED [W/(K*m)]	CALIBRATIO N CONSTANT	THERMAL COND. w/CALIBRATION [W/(K*m)]
1134±187	61.1±4.7	0.17±0.1	0.02	0.15±0.00495

Table 1: Cumulated Results of MDSC on Polystyrene Samples

RU9 neat	LENGTH [mm]: 3.05 DIAMETER [mm]: 6.38		MASS [g]: 0.0905 PERIOD [1/sec]: 80	
SPECIFIC HEAT [J/K*kg]	HEAT CAPACITY [mJ/K]	THERMAL COND. OBSERVED [W/(K*m)]	CALIBRATIO N CONSTANT	THERMAL COND. w/CALIBRATION [W/(K*m)]
2480±257	57.43±2.97	0.11±0.0145	0.02	0.0716±0.0041

Table 2: Cumulated Results of MDSC on RU9 Neat Samples

RU9 composite	LENGTH [mm]: 4.06 DIAMETER [mm]: 7.06		MASS [g]: 0.2426 PERIOD [1/sec]: 80	
SPECIFIC HEAT [J/K*kg]	HEAT CAPACITY [mJ/K]	THERMAL COND. OBSERVED [W/(K*m)]	CALIBRATIO N CONSTANT	THERMAL COND. w/CALIBRATION [W/(K*m)]
964±43	170.09±0.2	1.157±0.052	0.02	1.122±0.054

Table 3: Cumulated Results of MDSC on RU9 Composite Samples

The RU9 composite material includes Allied Signal GS-44 silicon nitride particles. The thermal conductivity of silicon nitride was found from literature as k = 20 W/(K*m). The computational model with three different pattern generators was utilized to predict the thermal conductivity of RU9 composite at 56% volume loading. The results are shown in Table 4.

Conductivity from Laser Flash	Conductivity from MDSC	Conductivity from Model
1.1	1.122±0.054	1.2±0.1

Table 4: Comparison of Results Obtained by Experimentation and Computation

The computational model also has been utilized to study the dependence of effective thermal conductivity on particle volume fraction for a binder particle system where $k_b = 0.2$ W/(K*m) and $k_p = 10$ W/(K*m). Results of three sets of runs utilizing different random pattern generators are shown in Figure 4. The same figure also includes predictions obtained from harmonic averaging. The predicted signature has an inflection point, which at the same time denotes the percolation limit. After this point effective thermal conductivity signature approaches the limiting value for the configuration in which a block of particle material is sandwiched between thin layers of binder materials. High particle loading is required for green body production and subsequent powder processing (binder burnout and sintering) steps. Hence the present model will be useful as a selection tool in design of experiments for the development of new material systems.

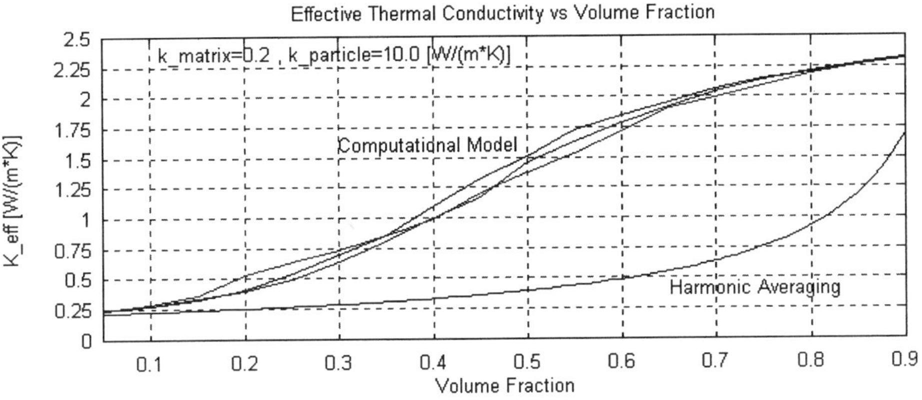

Figure 4: Dependence of Effective Thermal Conductivity on Particle Loading

To show the sensitivity of the effective thermal conductivity to the thermal conductivity of continuous phase, another set of computational experiments was performed utilizing binder conductivities k = 0.2 W/(K*m) and k = 0.4 W/(K*m). The result is shown in Figure 5 and compared to the previous signature predicted.

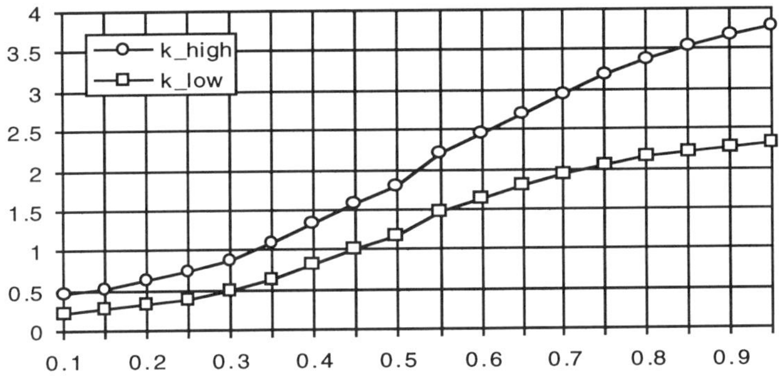

Figure 5: Sensitivity of Effective Thermal Conductivity on Binder Material

The effect of changing the particle material while keeping the binder thermal conductivity and particle volume percent constant, is shown in Figure 6. Three different materials are shown; PZT with k_p = 1.0 W/(K*m); silicon nitride with k_p = 20.0 W/(K*m); and tungsten-carbide cobalt particles with k_p = 100.0 W/(K*m).

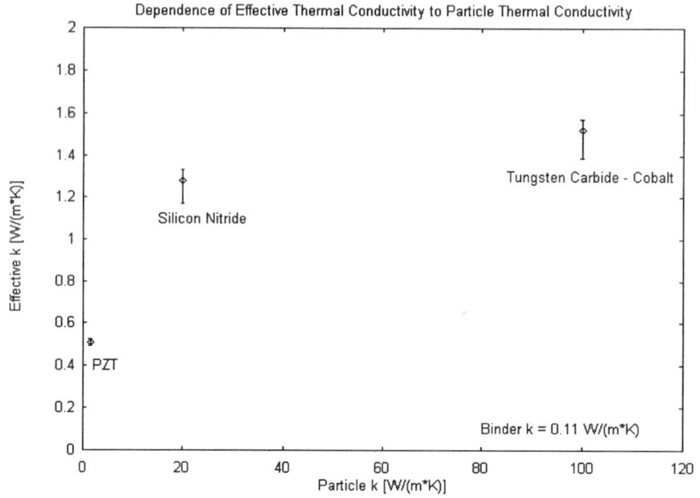

Figure 6: Sensitivity of Effective Thermal Conductivity on Particle Material

As can be seen, the effective thermal conductivity for a particular volume loading becomes less sensitive to particle thermal conductivity as this property is increased.

5. CONCLUSIONS AND ONGOING WORK

The determination of thermal conductivity of a composite system with thermoplastic matrix has been investigated using Laser Flash and Modulated DSC technology. MDSC values were 0.022 W/(K*m) higher than those obtained from the laser flash method. MDSC of neat polystyrene overpredicted its thermal conductivity by 0.01 W/(K*m) compared to literature values. These results show that Modulated DSC can be used to determine thermal conductivity of composites. The advantages of MDSC are that thermal conductivity can be measured in an effective and timely manner and that the existing DSC cell does not have to be changed physically.

Additionally, a computational model has been developed to determine the thermal conductivity of composite systems. Of particular interest were particle binder systems that are characterized by a lower thermally conductive matrix and higher conductive filler. The predictions of the model for the experimentally examined material match the experimental data.

Current research efforts include the generalization of the current model to three-dimensions using boundary elements/finite elements. On the application side, the efforts will be concentrated on constructing membrane models.

6. ACKNOWLEDGMENTS

The authors acknowledge the help of Professor Onn and his collaborators at the Physics Department of the University of Delaware and the support of this research by the National Science Foundation.

7. REFERENCES

1. Chang I. Y. and Lees, J. K., "Recent Development in Thermoplastic Composites: A Review of Matrix Systems and Processing Methods", Journal of Thermoplastic Composite Materials, Vol 1, pp. 277-295, 1988.

2. Beyeler E., Melting and Solidification in Processing and Applications to Thermoplastic Composites Manufacturing, Ph.D. Thesis, Department of Mechanical Engineering University of Delaware Newark DE, 1987.

3. Nejhad M. N., Cope R. D. and Güçeri S. I., "Heat Transfer Analysis for In-Situ Thermoplastic Filament Winding", Journal of Heat Transfer, Vol 113, pp.304-313, 1991.

4. Nejhad M.N., Three-Dimensional Thermal and Residual Stress Analysis of In- Situ Thermoplastic Composite Filament Winding, PhD Thesis, Department of Mechanical Engineering, University of Delaware, Newark, Delaware, 1992.

5. Agarwal V., Effect of Molecular Diffusion on the Bonding of Thermoplastic Composites, PhD Thesis, Department of Mechanical Engineering, University of Delaware, Newark, Delaware, 1990.

6. Irwin R.G. and Güçeri S.I., "Parametric Study of Laser-Assisted Processing of Thermoplastic Composites", Proceedings of the Winter Annual Meeting of ASME ,1994.

7. Pistor C., Yardımcı M. A., Güçeri S.I., Bandyopaddhay A., "Laser Assisted Tape Winding of Ceramic Green Parts", American Ceramic Society 1996 Cocoa Beach Meeting, Ceramics Science and Engineering Proceedings, Cocoa Beach,1996.

8. Agari Y., Ueda A. and Nagai, S., "Measurement of Thermal Diffusivity and Specific Heat Capacity of Polymers by Laser Flash Method", Journal of Polymer Science: Part B: Polymer Physics, Vol. 33, pp. 33-42, 1995.

9. Marcus S.M. and Blaine R.L., "Thermal Conductivity of Polymers, Glasses and Ceramics by Modulated DSC", Thermocimica Acta 243, pp. 231-239, 1994.

10. Chiu J. and Fair P. G., "Determination of Thermal Conductivity by Differential Scanning Calometry", Thermochimica Acta 27, pp. 267-273, 1979.

11. Duswalt A. A., 48th Calorimeter Conference, Durham NC, 1993.

12. Dasgupta A., Agarwal A.K. and Bhandarkar S.M., "Three Dimensional Modeling of Woven Fabric Composites for Effective Thermo-Mechanical and Thermal Properties", Composites Science and Technology, Vol. 56, No. 3, pp. 209-223, 1996.

13. Luo M., Wood J.R., and Cathles L.M., "Prediction of Thermal Conductivity in Reservoir Rocks Using Fabric Theory", Journal of Applied Geophysics, Vol. 32, No.4, pp. 321-334, 1994.

14. Zhang M.Q., Xu J.R., Zeng H.M., Huo Q., Zhang Z.Y., Yun F.C. and Friedrich K., "Fractal Approach to The Critical Filler Volume Fraction of an Electrically Conductive Polymer Composite", Journal of Materials Science, Vol. 30, pp. 4226-4232, 1995.

15. Choy, C., Kwok, K., Leung, W., Lau, F., "Thermal Conductivity of Poly(ether ether ketone) and its Short-Fiber Composites", Journal of Polymer Science: Part B: Polymer Physics, Vol 32, No 8, pp. 1389-1397, 1994.

28th International SAMPE Technical Conference
November 4-7, 1996

THERMAL STRESS IN RAPIDLY HEATED GRAPHITE/EPOXY COMPOSITES

Thomas Stephens and Douglas G. Blanchard
Naval Air Warfare Center Weapons Division
China Lake, CA 93555-6100

ABSTRACT

This paper describes testing and modeling to demonstrate the predictability of temperature and stress profiles in rapidly heated composite parts. Unidirectional filament wound, graphite/ epoxy composite coupons were subjected to rapid heating from one side, while surface temperatures and deflections were measured. From the surface temperatures and thermal properties, the internal temperature profiles were calculated. An analytical model was used to predict the stresses and strains in the coupon and the curvature of the coupon during the transient heating and subsequent cooling.

KEY WORDS: Heat Transfer, Modeling, Dimensional Stability

1. INTRODUCTION

Organic matrix composites can offer significant advantages in performance and cost over traditional materials such as metals. After launch, air-launched, supersonic tactical missile airframes can experience heating rates around 100°C/s and temperatures ranging from 300 to 800°C. Thermal and mechanical property data are needed under these conditions in order to perform structural and thermal analysis and design of composite airframes. Since current test methods are not adequate to provide these data for realistic environmental conditions, new methods are required. To design tests for measuring properties under rapid heating conditions, it is necessary to analyze temperature and stress profiles within test specimens. Thermal and stress analysis models should be validated by comparison with experiments to give confidence in model predictions.

In previous studies of thermal stresses in composites, models have been developed to predict thermal stresses in orthotropic materials with temperature-dependent properties (1, 2). Comparison of models developed to predict heat transfer and elastic response of rapidly heated composite structures with measured temperature profiles and mechanical response have shown the importance of time-dependence of properties for cases where polymeric matrix materials can undergo chemical reactions (3, 4, 5). Even in the absence of chemical reactions, viscoelasticity can produce time dependence in mechanical behavior when above or close to the glass transition temperature (6). In the present study, in order to isolate the time dependence due to the finite rate of heat transfer from time dependence due to chemical reaction or viscoelasticity, the response of composite coupons was measured under rapid heating to temperatures below the glass transition temperature. At such temperatures, no chemical reactions or viscoelasticity are expected, and thermal stresses should be predictable from a linear elastic analysis if temperature profiles can be accurately calculated.

The objective of this work was to investigate heat transfer and thermal stresses in graphite/ epoxy composite coupons under rapid heating conditions similar to the environmental conditions of air-launched missiles in free flight. Unidirectional composite coupons were subjected to rapid heating from one side, while surface temperatures and thermal deflections were measured. From the surface temperatures and thermal properties, the internal temperature profiles were calculated. An analytical model was used to predict the stresses and strains in the coupon and the curvature of the coupon during the transient heating and subsequent cooling. The predictions are based on thermal and mechanical property data measured as a function of temperature in isothermal tests. While in several cases, the predicted curvature agreed closely with the measured curvature, in several cases, there was a significant discrepancy (up to a factor of two) between measurements and predictions. The cause of the discrepancy is still under investigation and may be due to variability of properties from specimen to specimen, or even variability within the specimen. This work points out the necessity to adequately characterize materials and to validate computational methods used for predicting interior temperature profiles, heat fluxes and thermal stresses and strains in rapidly heated composite parts. These methods will be useful for designing further tests for thermal and property evaluation on rapidly heated composite materials for tactical missile airframes.

2. MATERIALS

Tests were performed on two specimens, both cut from filament wound panels of Hercules IM6 graphite fiber in Shell Epon 9405 toughened epoxy resin cured with Shell Epon 9470 aromatic amine hardener. This fiber and this resin are typical of materials used in filament wound missile airframes. In order to investigate the effects of voids on composite properties, two sets of panels were fabricated under different conditions in order to prepare composite panels with two different void contents. To achieve low voids (approximately 4 to 8% by volume), greater consolidation pressure was used during cure; lower pressure resulted in slightly higher voids (approximately 6 to 12%). It was noted that void content, as measured by acid digestion (ASTM D 3171) and judged by eye, varied from point to point in a given panel and was more variable in panels cured at low pressure.

Thermal conductivity, as measured by the University of Dayton Research Institute, is plotted in Figure 1. The specific heat, measured by differential scanning calorimetry, is plotted in Figure 2. The coefficient of thermal expansion (CTE), measured by thermomechanical analysis, was found to be nearly constant from room temperature up to 100°C. No statistically significant difference was seen between the CTE of samples from "high voids" panels and those from "low voids" panels. The uncertainties indicated in Table 1 are standard deviations; two to four replicate measurements were made for each sample. Tensile tests were performed on 2.54 cm wide specimens with fibers oriented perpendicular to the tension direction in order to obtain the tensile modulus transverse to the fiber direction, E_{22}. The transverse tensile modulus is plotted in Figure 3 for specimens cut from "high voids" and "low voids" panels. Although the modulus of specimens cut from the "low-voids" panels were on the average higher than those cut from the "high-voids" panel, post-test measurement of the void and fiber contents of the mechanical test specimens showed little correlation between void content and mechanical properties.

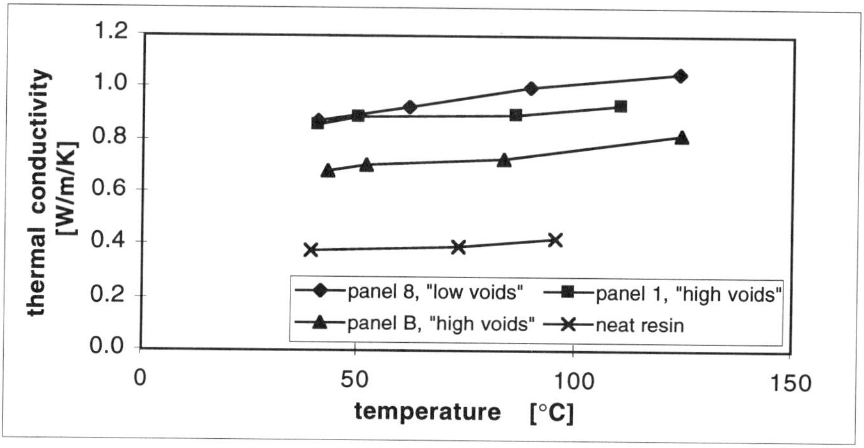

Figure 1. Thermal conductivity measured on three different panels of IM6/Epon graphite/epoxy composite and on neat Epon epoxy resin.

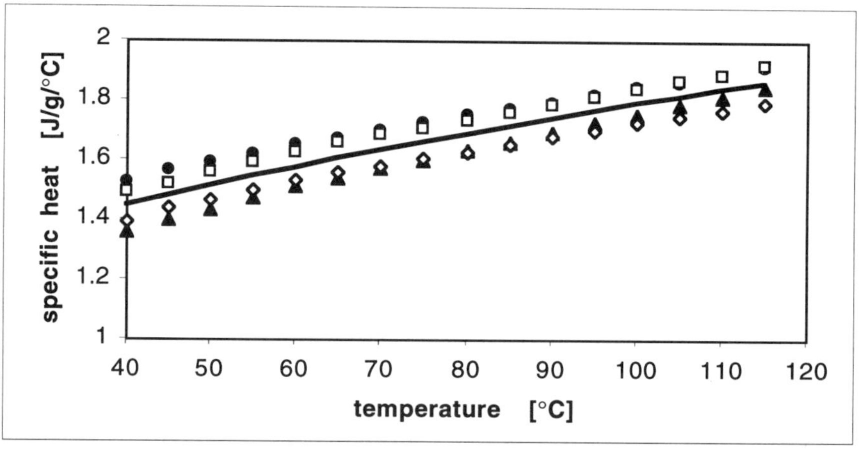

Figure 2. Specific heat measured on four different samples from the same, low-voids panel of IM6/Epon graphite/epoxy composite. The symbols show the experimental results, and the solid line is a second order polynomial fit.

Table 1. Coefficient of thermal expansion of Epon 9405/9470 epoxy and unidirectional Hercules IM6/Epon 9405/9470 epoxy composite.

Sample	Direction	CTE [K^{-1}]
neat resin	-	$63 \pm 1 \times 10^{-6}$
composite, "low" voids	fiber	$1 \pm 6 \times 10^{-6}$
composite, "low" voids	transverse	$40 \pm 2 \times 10^{-6}$
composite, "low" voids	thickness	$42 \pm 4 \times 10^{-6}$
composite, "high" voids	fiber	$4 \pm 3 \times 10^{-6}$
composite, "high" voids	transverse	$41 \pm 3 \times 10^{-6}$
composite, "high" voids	thickness	$41 \pm 5 \times 10^{-6}$

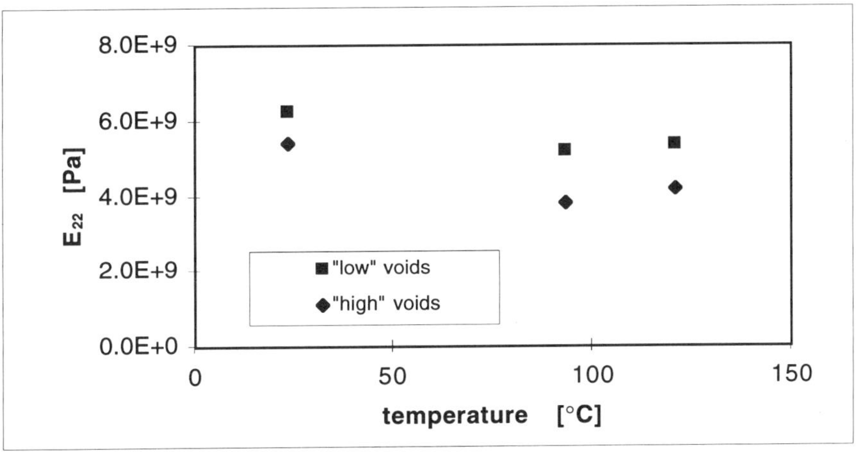

Figure 3. Transverse tensile modulus measured on specimens from low-voids and high-voids panels of IM6/Epon graphite/epoxy composite

The panels used in this project were filament wound with all the graphite fibers in the same direction, in order to study single ply behavior without the complications of interlaminar stresses that occur in multi-ply specimens. However, in the absence of crossplies, there was more freedom of movement of the tows during winding, and the uncured composite was more difficult to consolidate than a more conventional multi-ply part. The resulting properties of the panels studied may not be representative of multi-ply filament wound parts, and the unidirectional panels appeared to have more spatial variation in void content and degree of consolidation than is typical for filament winding.

3. EXPERIMENTAL

The apparatus used for rapid heating measurements is shown in Figures 4 and 5. A 2600 W infrared lamp was used to heat one side of a composite specimen. A shutter, opened and shut by a timer under computer control, allowed for reproducible exposure times. The lamp provided uniform heating over an area 12 to 14 cm long by 6 to 9 cm wide. Uniformity of heating was confirmed by using an infrared imaging camera to measure temperature contours on the rear side of a thin aluminum sheet, painted flat black, while the front side was heated by the lamp. Before each exposure, the lamp filament was allowed to pre-heat for a set time to give more repeatable heating conditions.

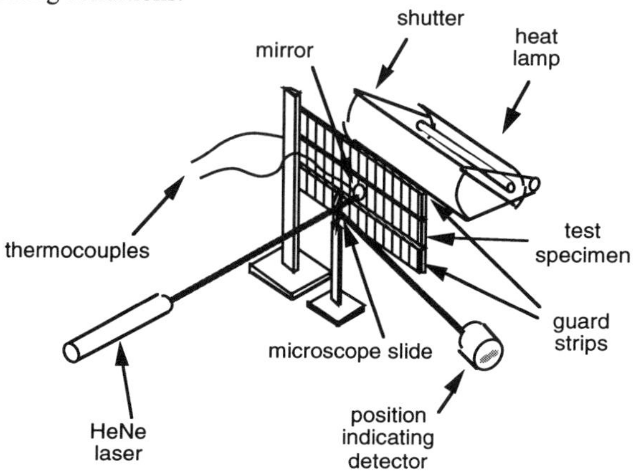

Figure 4. Rapid heating apparatus

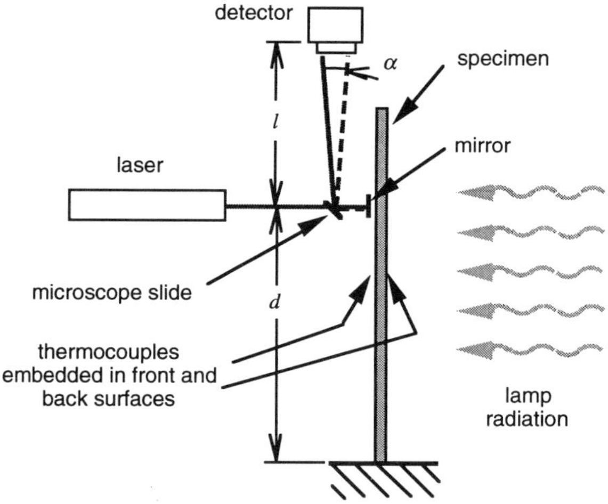

Figure 5. The rapid heating apparatus shown from above. The angular deflection of the specimen is measured at position d, then the mirror is relocated to another position and the measurement repeated. The angular deflection of the specimen is half of the angular deflection of the beam, α.

The composite specimens were 2.54 cm wide and 25 cm long, and were cut from unidirectionally reinforced, filament wound panels so that the fibers ran transverse to the long direction. Two identical specimens were held on either side of the test specimen and were heated in the same way as the test specimen. These additional specimens had the same temperature profile through the thickness as the test specimen and they served as guard plates to eliminate heat flow in the fiber direction. Foil thermocouples were embedded in the front and rear surfaces, and the signals from the thermocouples were recorded by using a digital storage oscilloscope. A small mirror was attached to the test specimen, parallel to the surface of the specimen at a distance d_1 from the clamped end, and a laser beam was directed onto the mirror, normal to the specimen. The beam then reflected off a stationary microscope slide to a position indicating detector that gave a signal from which the rotation of the mirror could be calculated. The rotation of the mirror, which was the same as the angular deflection at that point on the specimen, was measured for different heat exposures. Then the mirror was moved to a different point on the specimen at a distance d_2 from the clamped end, and the measurement was repeated for the same heat exposures. The angular deflection of the beam, α, was calculated from the change in the position of the beam indicated by the detector and the lever arm, l. The angular deflection of the mirror attached to the specimen was $\alpha/2$. The radius of curvature of the specimen between these two points was calculated from the difference in the angular deflections measured at the two points.

$$\frac{1}{R} = \frac{\alpha_1 - \alpha_2}{2(d_2 - d_1)} \qquad [1]$$

Surface temperatures measured in a typical run (run 74) are plotted in Figure 6. Front surface heating rates ranged from 5 to 40°C/s. The angular deflection which was measured at a point 72 mm from the clamped end in this run and at a second point 133 mm from the clamped end in run 75 (same heat exposure as in run 74) are plotted in Figure 7. The angular deflection was measured at one location during each run. The mirror was moved to another location on the specimen, and the heat exposure was repeated. Repeatability of heat exposures was good, with surface temperature histories replicated to within ±2°C, as long as the initial temperature (room temperature) was the same.

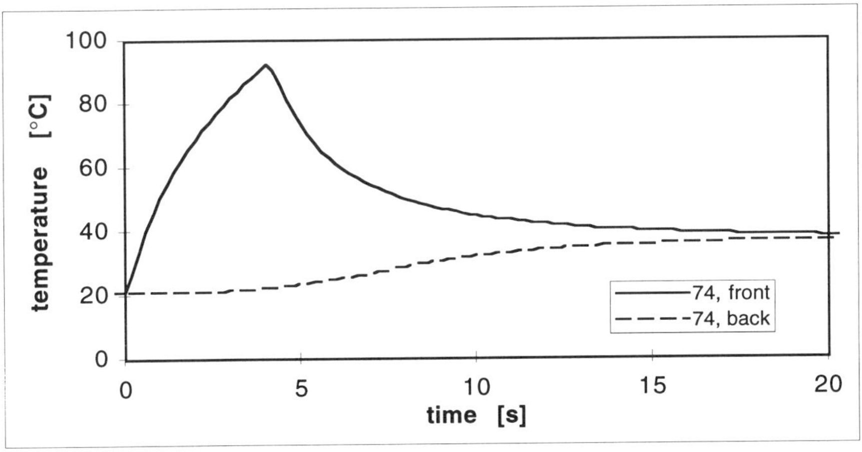

Figure 6. Measured surface temperatures during a rapid heating experiment (specimen 47, run 74).

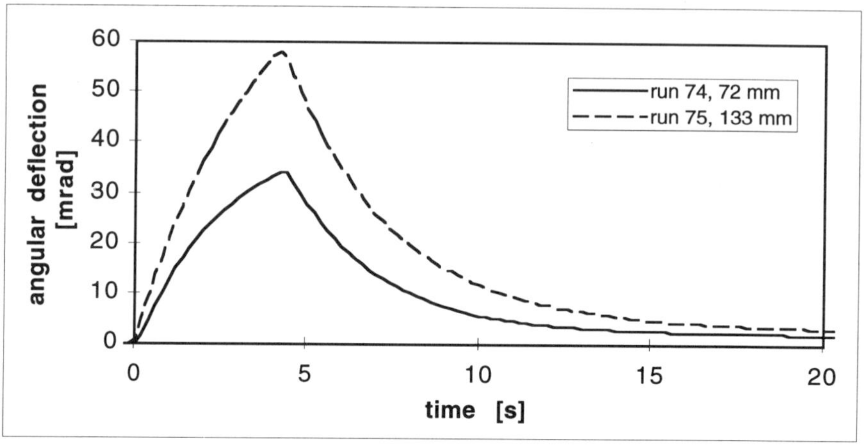

Figure 7. The angular deflection measured at two locations on the specimen heated as shown in Figure 6 (run 74, and replicate run 75). Distances from the clamped end to the measurement location are indicated.

Rapid heating measurements were performed on two specimens: specimen 36, cut from a "high voids" panel, and specimen 47, cut from a "low voids" panel. Specimen 36 was 6.48 mm thick, and specimen 47 was 3.76 mm thick. The surface temperatures for four different heat exposures of specimen 36 are shown in Figure 8. Curvatures measured in these experiments are plotted in Figure 9.

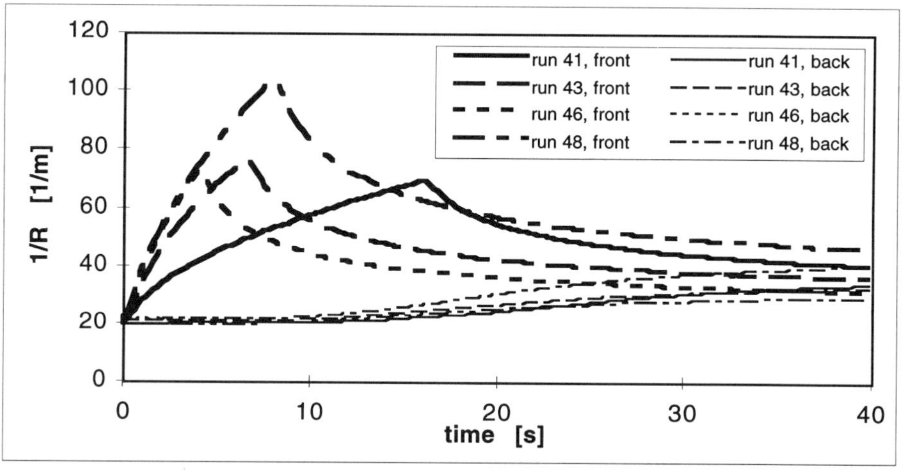

Figure 8. The measured front and back surface temperatures for four different heat exposures of specimen 36.

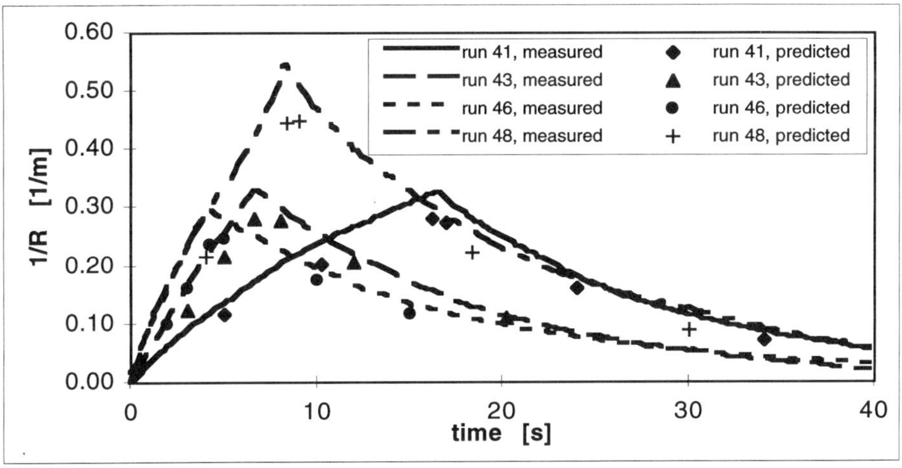

Figure 9. The measured curvature of "high voids" specimen 36, under the four different heat exposures shown in Figure 8. Curves are measured curvatures, symbols are predicted from calculated temperature profiles, as described below.

Figure 10 shows the surface temperatures measured during three different heat exposures of specimen 47. Curvatures measured in these experiments are plotted in Figure 11.

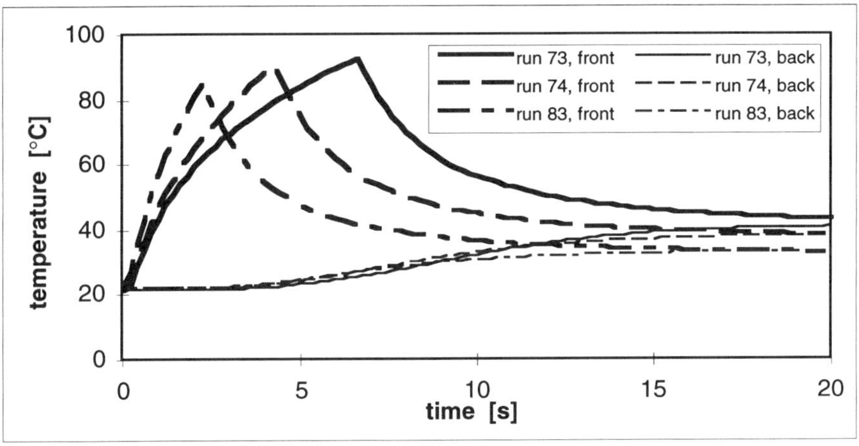

Figure 10. The measured front and back surface temperatures for three different heat exposures of specimen 47.

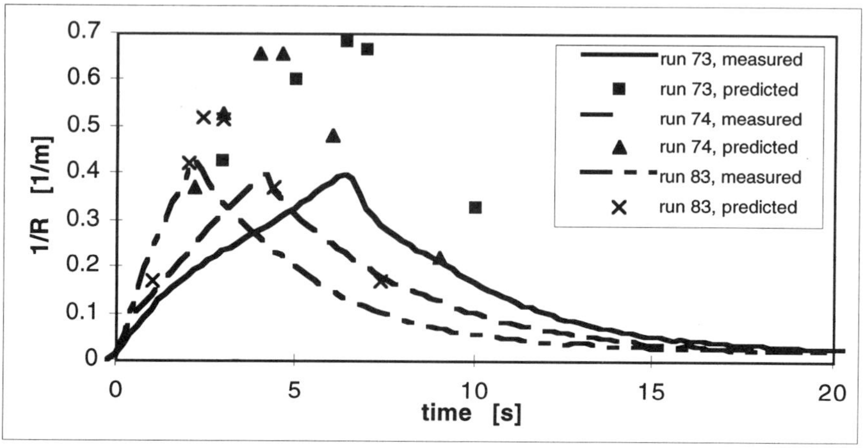

Figure 11. The measured curvature of "low voids" specimen 47, undergoing the three different exposures shown in Figure 10. Curves are measured curvatures, symbols are predicted from calculated temperature profiles, as described below.

4. TEMPERATURE AND STRESS CALCULATIONS

From the measured surface temperatures as a function of time and the specific heat, thermal conductivity and density, the temperature profile within the rapidly heated specimens was calculated. For the portion of the specimen where the front surface was heated uniformly, no heat transfer in the plane of the specimen was assumed, and the temperature profile through the thickness was calculated by using the finite difference code SINDA. The temperature profile satisfies the heat equation

$$\rho C_p \frac{\partial T}{\partial t} = \frac{\partial}{\partial x}\left(k \frac{\partial T}{\partial x}\right) \qquad [2]$$

where: ρ = density,
C_p = specific heat,
T = temperature,
t = time,
x = distance through the thickness, and
k = thermal conductivity.

Assuming an initially uniform temperature, the measured temperatures at the front and rear surfaces supply the boundary conditions needed to close equation 2. Temperature profiles calculated for the run shown in Figure 7 (run 74) are plotted in Figure 12.

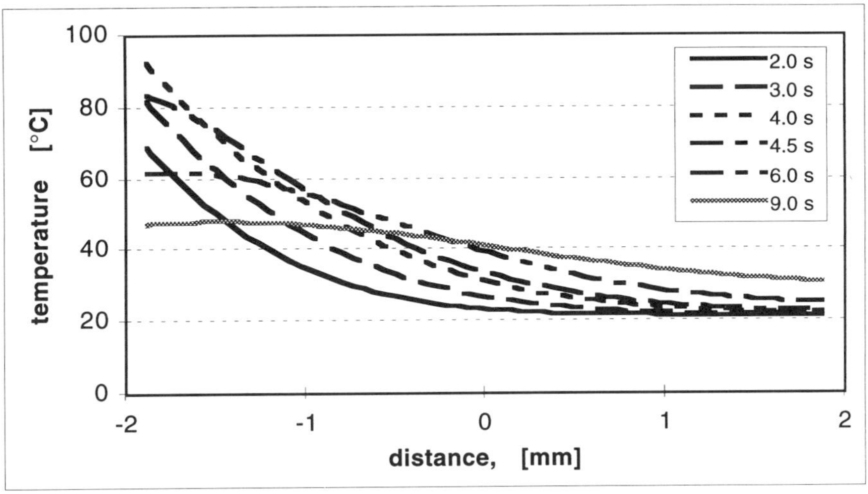

Figure 12. Temperature profiles through the thickness within specimen 47, during run 74, calculated at the times indicated. The surface temperatures as a function of time are shown in Figure 7.

From the calculated temperature profiles, the coefficient of thermal expansion, and the tensile modulus, the tensile stress in the specimen was calculated. Consider the stress in the specimen as shown in Figure 13.

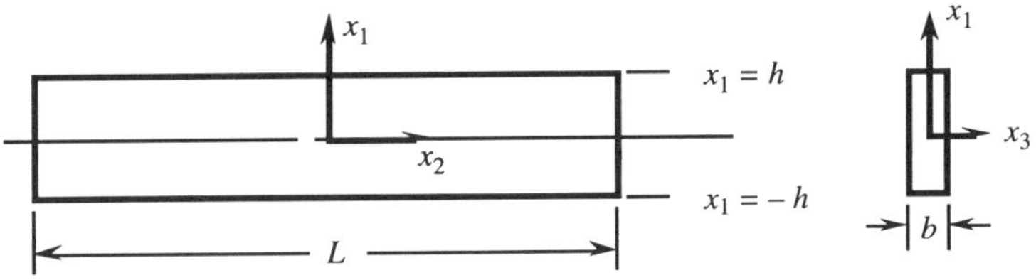

Figure 13. Coordinate system for analyzing thermal stress in a rapidly heated specimen. The surface $x_3 = -b/2$ is heated uniformly.

Since the specimen was freely supported and heated uniformly on the front surface, there was no net load or moment, and tensile stress in the direction perpendicular to the plane of the specimen (direction 3) was assumed to be zero, as were shear stresses. Since the coefficient of thermal expansion of graphite/epoxy composites is so small in the fiber direction, stresses and strains in the fiber direction (direction 1) were neglected. Assuming the only non-zero stress component of the stress to be the tensile stress in the 2-direction, this stress is governed by the conservation of linear momentum, which when combined with the compatibility equation may be written:

$$\frac{d^2}{dx_3{}^2}\left(\frac{\sigma_2}{E_{22}} + \alpha_2 T\right) = 0 \qquad [3]$$

where equation 3 may be integrated

$$\sigma_2 = -E_{22}\left(\alpha_2 T - C_1 x_3 + C_2\right) \qquad [4]$$

where: a_2 = CTE in the 2-direction, and the constants C_1 and C_2 are given by:

$$C_1 = \frac{\int_{-h}^{h} E_{22} x_3 dx_3 \int_{-h}^{h} \alpha_2 E_{22} T dx_3 - \int_{-h}^{h} E_{22} dx_3 \int_{-h}^{h} \alpha_2 E_{22} T x_3 dx_3}{\left(\int_{-h}^{h} E_{22} x_3 dx_3\right)^2 - \int_{-h}^{h} E_{22} x_3^2 dx_3 \int_{-h}^{h} E_{22} dx_3} \qquad [5]$$

$$C_2 = \frac{\int_{-h}^{h} E_{22} x_3 dx_3 \int_{-h}^{h} \alpha_2 E_{22} T x_3 dx_3 - \int_{-h}^{h} E_{22} x_3^2 dx_3 \int_{-h}^{h} \alpha_2 E_2 T dx_3}{\left(\int_{-h}^{h} E_{22} x_3 dx_3\right)^2 - \int_{-h}^{h} E_{22} x_3^2 dx_3 \int_{-h}^{h} E_{22} dx_3} \qquad [6]$$

where the modulus, E_{22}, and CTE, a_2, may depend on temperature.

From geometrical considerations, the curvature of the specimen may be related to the constants C_1 and C_2:

$$\frac{1}{R} = -\frac{C_1}{1 + C_2} \qquad [7]$$

From equation 7, the curvature can be calculated without explicitly calculating the stress profile. However, once the constants C_1 and C_2 are calculated, it is easy to calculate the stress profile from equation 4. Equations 5, 6, and 7 reduce to the proper forms for a beam with temperature independent properties (ref. 7). Curvatures were predicted from the temperature profiles calculated at various times for each of the rapid heating runs shown in Figures 9 and 11, and these predictions are plotted with the measured curvatures. The predictions for specimen 36 were consistently 5 to 15% lower than measured curvatures, while predictions for specimen 47 were up to 100% higher than measured values. This discrepancy is being examined, and calculations of the C_1 and C_2 are being checked. However, it appears at this time that the discrepancy is due to spatial property variations in the graphite composite specimens. Specimens used for property measurements were not the same specimens used in the rapid heating test, and differences in the properties used to calculate the curvature may have lead to disagreement with test results. This may demonstrate the importance of accurate material property data in modeling composite thermomechanical behavior.

5. CONCLUSIONS

Unidirectional filament wound, graphite/epoxy composite coupons were subjected to rapid heating from one side, while surface temperatures and deflections were measured. From the surface temperatures and thermal properties, the internal temperature profiles were calculated. An analytical model was used to predict the stresses and strains in the coupon and the curvature of the coupon during the transient heating and subsequent cooling. While in several cases, the predicted curvature agreed closely with the measured curvature, in several cases, there was a significant discrepancy (up to a factor of two) between measurements and predictions. The discrepancy may be due to variability of properties from specimen to specimen, or even within the specimen. This work points out the necessity to adequately characterize materials and to validate computational methods used for predicting interior temperature profiles, heat fluxes and thermal stresses and strains in rapidly heated composite parts. Validated methods are needed for designing further tests for thermal and property evaluation on rapidly heated composite materials for tactical missile airframes.

6. ACKNOWLEDGMENTS

This work was funded by the Office of Naval Research Airborne Weapons Materials Technology Project. The authors acknowledge helpful discussions with Mr. Theodore Hicks and valuable assistance with experimental measurements from Dr. Rena Yee, Dr. Ilzoo Lee, and Mr. Frank Hudson.

7. REFERENCES

1. W. Nowacki, Thermoelasticty, Pergamon Press, Reading, MA, 1962, pp. 589-595.
2. N. J. and H. T Hahn, in Composite Materials: Testing and Design (Fourth Conference), ASTM STP 617, American Society for Testing and Materials, Philadelphia, 1976, pp. 317-329.
3. I. Auerbach, D. A. Benson, S. G. Beard, and G. F. Wright, Jr, J. Thermophysics, 3, 395 (1989).
4. McManus, H. L. N., and G. S. Springer, J. Composite Materials, 26, 230 (1992).
5. R. M. Sullivan, J. Composite Materials, 27, 408 (1993).
6. R. M. Tuttle, A. Pasricha, and A. F. Emery, J. Composite Materials, 29, 2025 (1995).
7. B. A. Boley and J. H. Weiner, Theory of Thermal Stresses, John Wiley and Sons, New York, 1960, pp. 279-282.

8. BIOGRAPHIES

Tom Stephens has been a chemical engineer in the Research and Technology Division at the Naval Air Warfare Center Weapons Division at China Lake, California since 1988. He obtained a Ph.D. in Chemical Engineering from the University of Massachusetts in 1988, and a B.S. in Chemical Engineering from The University of Texas at Austin in 1980. His research interests include the flow and processing behavior of suspensions, and the mechanical and thermal properties of polymeric materials, including propellants, explosives, composites, adhesives, and other materials.

Doug Blanchard has worked for the past 9 years at the Naval Air Warfare Center Weapons Division at China Lake, California in the Aeromechanics and Thermal Analysis Section. He obtained his Master's Degree in Mechanical Engineering in 1994 from Texas A&M University and his Bachelor's Degree in Aerospace Engineering from the University of Arizona in 1986. He is a Senior Member of the American Institute of Aeronautics and Astronautics and actively participates in the Thermophysics Technical Committee. His recent work has been focused in the areas of aerothermodynamics and contact conductance.

28th International SAMPE Technical Conference
November 4-7, 1996

EFFECTS OF FIBER ORIENTATIONS ON HIGH STRAIN RATE BEHAVIOR OF GLASS/EPOXY COMPOSITES

A. Haque, U. K. Vaidya, H. Mahfuz and S. Jeelani
Tuskegee University Center for Advanced Materials
Tuskegee University
Tuskegee, Alabama 36088

ABSTRACT

The Split Hopkinson Pressure Bar (SHPB) technique have been used to characterize glass/epoxy composites under high strain rate compressive loading. Four types of fiber architecture namely, unidirectional, angle-ply, cross-ply and quasi-isotropic laminates are investigated. A recovery system have been incorporated into to the classical Hopkinson bar technique, to subject the specimen to a single predetermined stress pulse, and the damage modes at various stages of loading are identified. The quasi-static compression strength , strain to failure and elastic modulii are compared with the dynamic test results to determine the shift in the failure mechanism. The compressive stress pulse is applied through the thickness, as well as in the fiber direction. The failure modes at different levels of loading are determined using Optical and Scanning Electron Microscopes. Correlation between the failure modes and high strain rate effects on the various configurations of composite laminates has been established.

KEY WORDS : Strain Rate, Glass/Epoxy, Fiber Orientations

1. INTRODUCTION

Fiber reinforced polymer matrix composites are considered to be the new emerging materials for aircrafts, submarines and armor vehicles' structures due to their high specific strength and modulus. Such applications of composites require a thorough understanding of materials response under high velocity impact which eventually develops loading conditions of short duration (on the order of 100 μs) and produces stress(strain) wave pulse with strain rates up to a few hundred (m/m) per second [1-2]. It is observed that the failure strength of the materials increases with the increase of applied loading rate or strain rate [3]. Such phenomenon is known as strain rate sensitivity that is well developed for metals [4] on the basis of dynamic deformation process. The constitutive equations that are developed for metals are mostly empirical and require equation parameters from the experiments. The rate effects of metals with FCC, BCC, HCP crystalline structures usually occur in the strain rate range 50 to 300 sec^{-1} . The high strain rate effects of non-metallic composite materials are still at the stage of infancy due to inadequate materials data and lack of constitutive equations. The interaction of many parameters in composites related to material constituents, processing, fiber architecture, lay-up sequence and orientations made the formulation of constitutive equations more difficult for dynamic deformation process.

The issues related to strain rate behaviors of composites and many factors of various dynamic test methods are described by Sierakowski [1]. It is observed that Split Hopkinson Pressure Bar (SHPB) tests are widely used for applying load in compression, tension, flexural and shear mode at high strain rate upto a range of 10^4 sec^{-1}. Data obtained from these tests include information on strain rate sensitivity, stress pulse shaping, constituent properties, dynamic ultimate stress, failure or fracture mechanisms, constitutive equation modeling, pulse attenuation, and damage initiation. The effects of test temperature and strain rates on unidirectional glass and graphite reinforced polymer composites are studied by SHPB method [5-7]. Their work indicated increased failure strength at higher strain rate within a range of 0 to 2000 sec^{-1}. The loading rate effect on composites reinforced with various glass fiber architectures and manufactured through resin transfer molded (RTM) process is investigated by Karbhari[8]. The scaling effects of laminate thickness on dynamic stress-strain response of quasi-isotropic graphite/epoxy composites are investigated by Pintado[9]. Their results indicate that damage occurs for incident energy levels higher than a critical value that depends on the scale factor and the stiffness varies with incident velocity in a non-monotonic way. The SHPB method is applied to study the dynamic response of angle ply glass/epoxy composites under tensile loading by Stabb and Gilat [10]. It is shown that the maximum normal stress experienced by glass/epoxy laminates is higher for dynamic than for quasi-static loading conditions.

This paper investigates the influence various fiber orientations and lay-up sequence on the dynamic response of glass/epoxy composites. Laminate configurations such as unidirectional, cross-ply, angle ply and quasi-isotropic are considered. The stress-strain response at various strain rate in all three principle directions are investigated and the relevant failure modes are also studied by optical and scanning electron microscopic examinations.

2. MATERIALS AND TEST METHOD

Four panels of Scotchply 1003 glass/epoxy composite laminates were prepared by autoclave method following the curing procedure recommended by the prepreg supplier (3M). The laminate stacking sequence and fiber orientations were maintained at $[0]_{80}$, $[+45/-45]_{20S}$, $[0/90]_{20S}$, $[0/+45/-45/90]_{10S}$ configurations. The average panel thickness was 19 mm (0.75 in). The quality of the manufactured panels were checked by ultrasonic C-scans and microscopic examinations. The specimens of size 7.62 mm (0.3 in) cube were precisely machined from all the panels for high strain rate tests by Split Hopkinson Bar (SHPB). The opposite faces of the cube were maintained perfectly parallel with a high degree of accuracy (+/- 0.00254 mm /0.0001 in) which is a very critical factor in high strain rate test by SHPB.

A schematic of the modified version of the classical SHPB set-up for compression test is shown in Figure 1. The classical SHPB system mainly consists of a gas gun, an incident bar, a transmission bar and a trapping device. Pressurized nitrogen gas is released from the gas gun which operates a striker bar at predetermined velocity. Impact of the striker bar to the incident bar initiates a stress pulse which travels through the sample sandwiched between the incident and transmission bar. As a result of the interaction with the specimen, part of the stress pulse is reflected back into the incident bar as a tensile pulse, and rest of the pulse is transmitted as a compression pulse into the transmission bar. Two strain gages are mounted

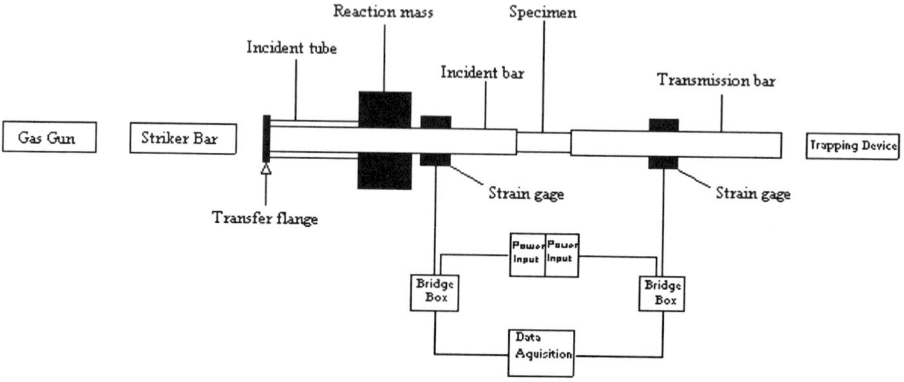

Fig 1. Schematic of the Classical Split Hopkinson Bar Test Set up with Reaction Mass

on the incident and transmission bar in order to record the incident pulse, reflected pulse and transmitted pulse. Formulation of one dimensional elastic waves in the bar provides basic equations to determine the stress, strain and strain rate in the specimen [11]. It has been established that the amplitude of the transmitted pulse is a measure of the stress in the specimen and the amplitude of the reflected pulse is proportional to the strain rate of the specimen. Integrating the strain rate with respect to time one can obtain the strain in the specimen.

In the present study an improved version of SHPB apparatus (Fig. 1) have been used as recommended by Nemat-Nasser et al. [11,12] in order to apply a predetermined level of impact loading, so that the microstructural damage evolutions in the materials are properly identified. Such technique is also shown to be effective in restricting the repeated compressive loading in the specimen which is usually developed in the classical Split Hopkinson Bar test method. This have been accomplished by introducing a reaction mass and a incident tube in the incident bar which has a transfer flange at its loading end. The incident tube rests between the reaction mass and transfer flange. A tensile pulse is developed in the transfer flange of incident bar after the compressive pulse travels toward the sample. This tensile pulse eventually isolates the subsequent repeated pulses to the specimen from the initial characteristic pulse.

3. RESULTS AND DISCUSSION

A typical response of the pulse that is received from the strain gage mounted on the incident bar for the case of unidirectional glass/epoxy composites is shown in Figure 2a. A compressive stress pulse was applied to a 7.62 mm (0.3 in) cubical specimen across the thickness. Initially the signal was recorded by a data acquisition software (Gagescope) . The data file stored by Gagescope software was converted into ASCII files and read by another signal processing software (Viewpoint) for further analysis of the data. The signal response in the form of strain rate versus time was obtained by modifying the data set with a predetermined strain rate scale factor. The strain rate scale factor was determined from the length of the specimen, longitudinal bar wave velocity and strain bridge voltage. The reflected pulse from the specimen (strain rate vs. time) was integrated to obtain compressive

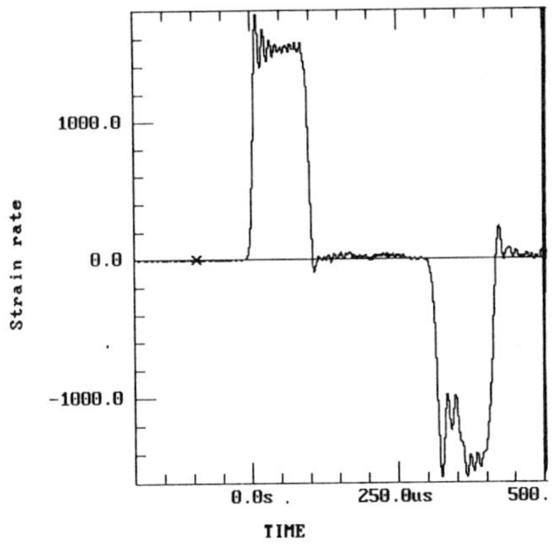

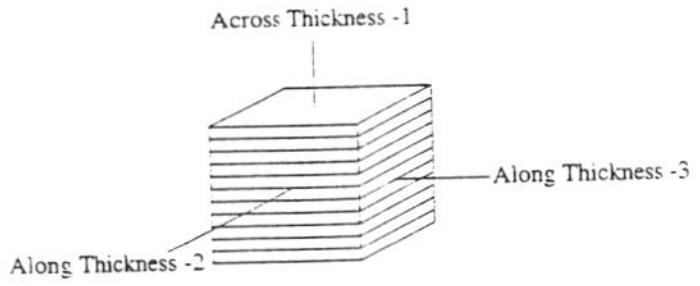

Fig 2a. A Typical Pulse received from the Incident Bar for Glass/Epoxy Composites

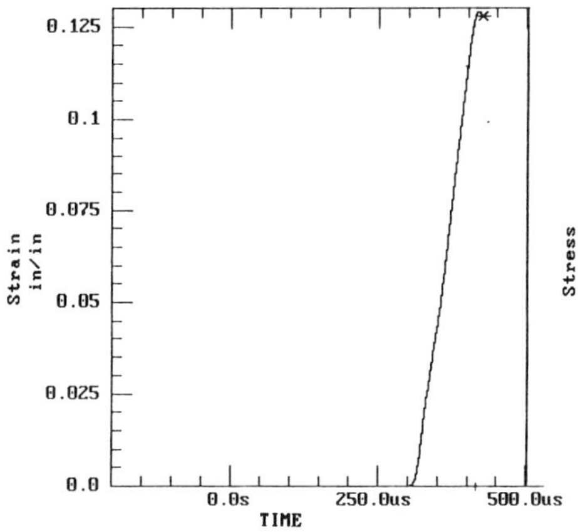

Fig 2b. A Typical Strain vs. Time plot for Glass/Epoxy

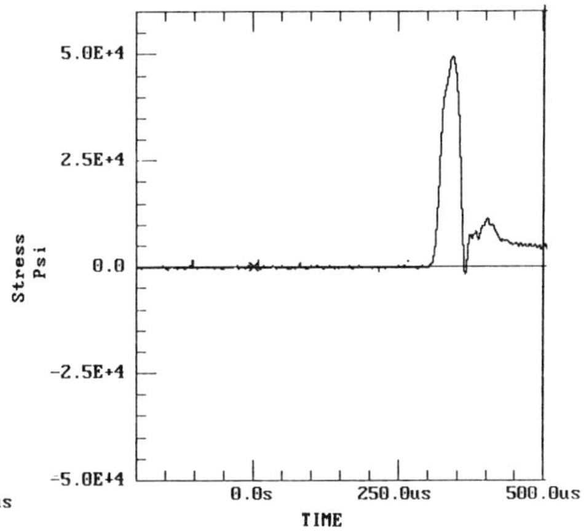

Fig. 2c A Typical Stress vs. Time plot for Glass/Epoxy

strain in the specimen as a function of time (Fig. 2b). The compressive stress in the specimen as a function of time is shown in Fig. 2c. This signal is received from the strain gage mounted on the transmission bar for the same glass/epoxy unidirectional composites loaded across the thickness. It is to be noted in Figs. 2 a-c that the time sequence of the strain rate, strain and stress pulse from the specimen should be same in order to verify the relevance of these signals.

The stress vs. strain plot for unidirectional glass/epoxy composites is shown in Fig. 3a and the micrograph of the fractured surface of the same specimen is shown in Fig. 3b. This stress vs. strain curve is basically generated from the plots of Figs. 2b-c and the corresponding strain rate is shown to be 1086 sec^{-1} . It is to be noted that this strain rate of loading is a materials response and a predetermined set value may not be accurate for all the specimens of the same materials. The stress vs. strain plot in Fig 3a was initially linear up to 165.36 MPa (24 ksi) and then gradually become nonlinear up to the point of ultimate failure at

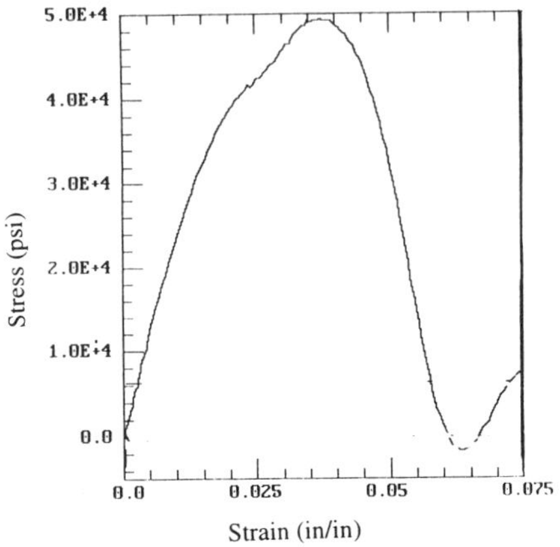

Fig. 3a Stress vs **Strain** plot for Unidirectional Glass/Epoxy (Strain rate : 1086 sec $^{-1}$, Loading : Across thickness-3

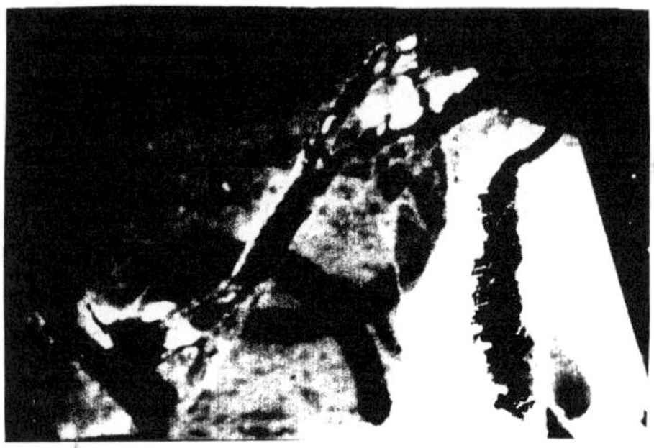

Fig. 3b. Micrograph of Fractured Surface for Unidirectional Glass/Epoxy (Strain rate : 1086 sec $^{-1}$, Loading : Across thickness-3)

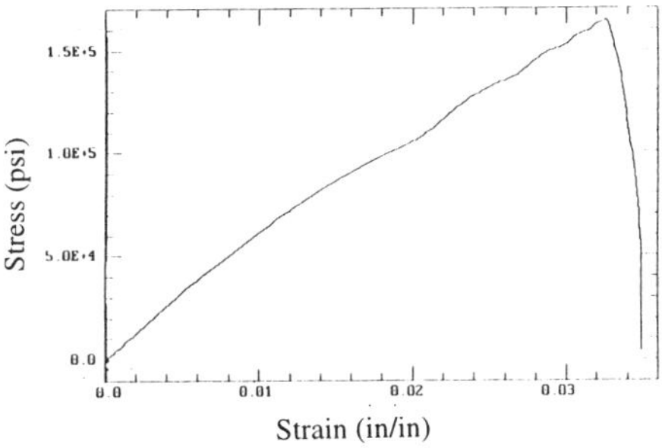

Fig. 4a Stress vs **Strain** plot for Unidirectional Glass/Epoxy (Strain rate : 1200 sec $^{-1}$, Loading : Along thickness -1)

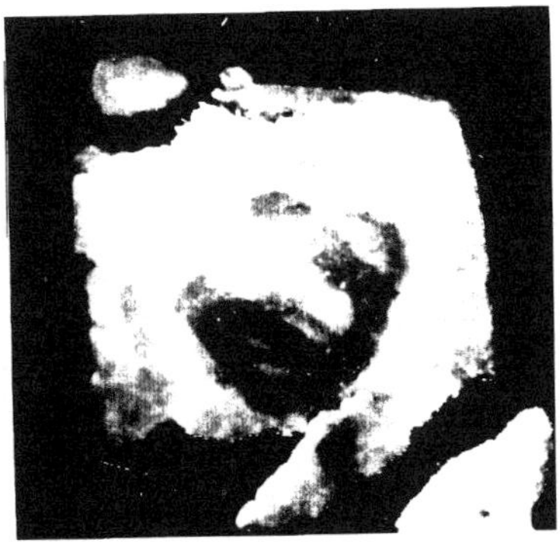

Fig. 4b. Micrograph of Fractured Surface for Unidirectional Glass/Epoxy (Strain rate : 1200 sec $^{-1}$, Loading : Along thickness-1)

323.83 MPa (47 ksi). The modulus of elasticity is shown to be 13.78 GPa (2 msi). The nonlinearity in stress vs. strain plot mostly arises due to matrix cracking and debonding which eventually resulted the final failure. The crack initially propagated at an angle 45^0 across the thickness and some cracks were also observed near the edges prior to final failure. Layer separations or delamination was not the predominant failure mode in this case. Figures 4a-b show stress vs. strain plot and the fractured surface of the unidirectional glass/epoxy composites loaded in the fiber direction (longitudinal). The load was applied on the same plane as viewed in the micrograph. The nature of the stress vs. strain plot was almost linear upto a stress level of 413.4 MPa (60 ksi) and then the curve showed nonlinearity prior to final failure at a stress level of 1143 MPa(166 ksi). The quasi-static compression strength for unidirectional composites is observed to be 880 MPa (128 ksi) as reported by 3M technical data manual of Scotch ply 1003. The modulus of elasticity and the strain rate were 43 GPa (6.25 msi) and 1200 s^{-1} respectively. It is shown in Fig. 4b that the failure mode mostly consists of debonding, delamination and fiber failure. The cracks were observed near the top and bottom corner edges of the loading plane. Figures 5a-b show the stress vs. strain plots of unidirectional glass/epoxy composites under transverse loading along the thickness. The ultimate failure strength and the modulus of elasticity is shown to be 323.83 MPa (47 ksi) and 13.78 GPa (2 msi). The nature of the curve is almost similar to Fig 3a in terms of failure strength and modulus of elasticity . This trend appears to be reasonable because in both the cases failure mode is matrix dominant.

Figure 6a shows stress vs. strain plot of cross-ply glass/epoxy composites under the compressive loading across the thickness. The micrograph of the same specimen is shown in Figure 6b. It is noticeable that the nature of the failure mode is totally different in comparison to unidirectional composites (Fig 3a). In unidirectional composite the cracks appear to propagate across the thickness in the loading direction with significant amount of matrix cracking and debonding whereas in angle ply composite no such cracks were observed in the loading direction across the thickness. The cracks in the form of debonding and delamination were observed in the direction perpendicular to the loading axis. Moreover, the extent of cracks are also shown to be much less in comparison to unidirectional composites. The ultimate failure strength (840.58 MPa, 122 ksi) and modulus

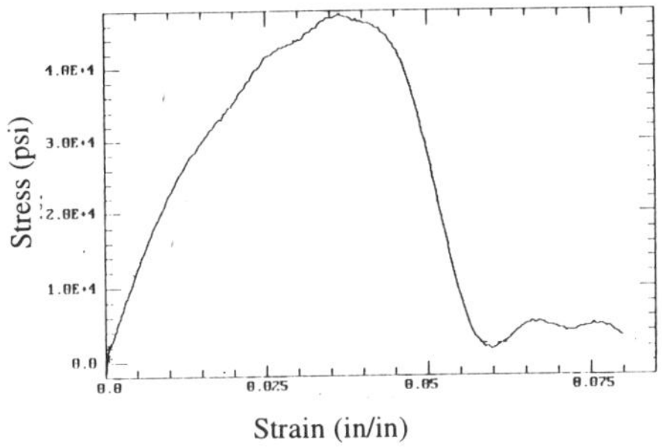

Fig. 5a Stress vs. Strain plot for Unidirectional Glass/Epoxy (Strain rate : 1000 sec^{-1} , Loading : Along thickness-2)

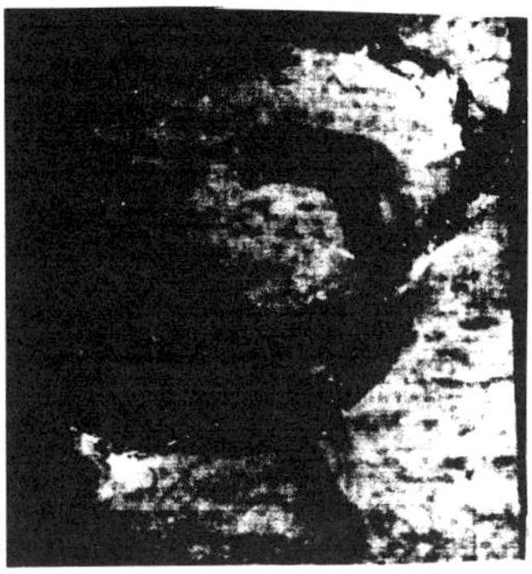

Fig. 5b. Micrograph of Fractured Surface for Unidirectional Glass/Epoxy (Strain rate : 1000 sec^{-1} , Loading : Along thickness-2)

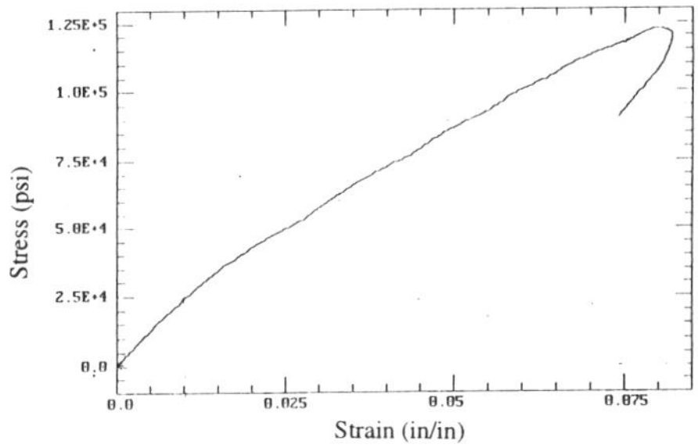

Fig. 6a Stress vs. Strain plot for Cross ply Glass/Epoxy (Strain rate : 1153 sec^{-1} , Loading : Across thickness-3)

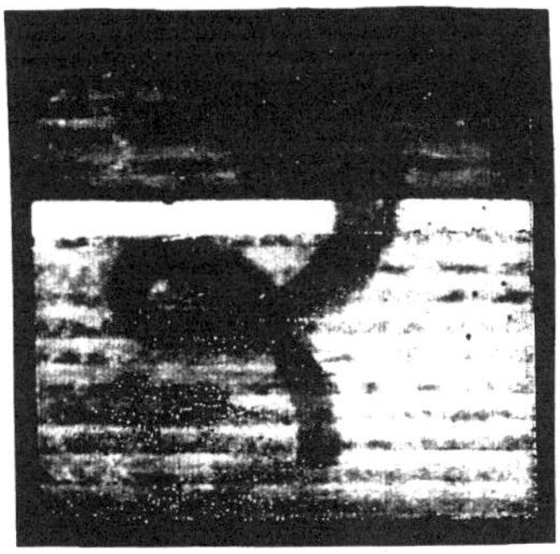

Fig. 6b. Micrograph of Fractured Surface for Cross-ply Glass/Epoxy (Strain rate : 1153 sec^{-1} , Loading : Across thickness-3)

of elasticity (16.94 GPa, 2.46 msi) were also shown comparatively higher than the unidirectional composites (323.83 MPa, 13.78 GPa). The strain rate of the same specimen as determined from strain vs. time plot was 1153 sec^{-1} . Figures 7 a-b show stress vs. strain plot and the micrograph of the fractured surface of cross ply glass/epoxy composites under compressive loading along the thickness. The fractured specimen is shown to be totally shattered with debonding, delamination and fiber breakage. The failure strength and modulus are shown to be 689 MPa (100 ksi) and 27.56 GPa (4 msi) at a strain rate of 588 sec^{-1}. The quasi-static compression strength for cross-ply glass/epoxy composites is observed to be 690 MPa according to ASTM D-3410. It is observed that the failure strength in this

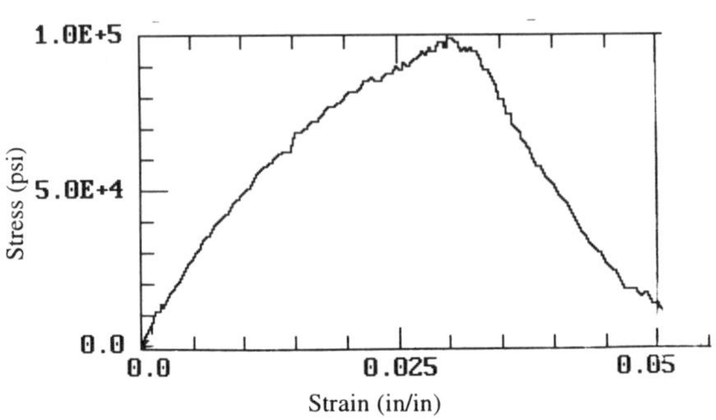

Fig. 7a Stress vs Strain plot for Cross ply Glass/Epoxy (Strain rate : 588 sec $^{-1}$, Loading : Along thickness-2)

Fig. 7b. Micrograph of Fractured Surface for Cross ply Glass/Epoxy (Strain rate : 588 sec $^{-1}$, Loading : Along thickness-2)

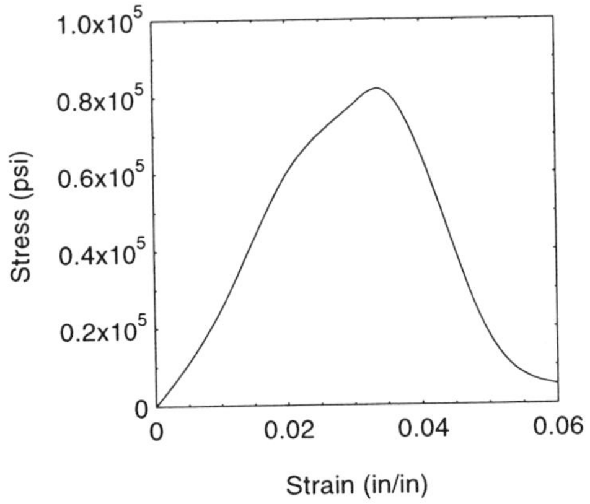

Fig. 8a Stress vs. Strain plot for Angle ply Glass/Epoxy (Strain rate : 545 sec $^{-1}$, Loading : Across thickness-3)

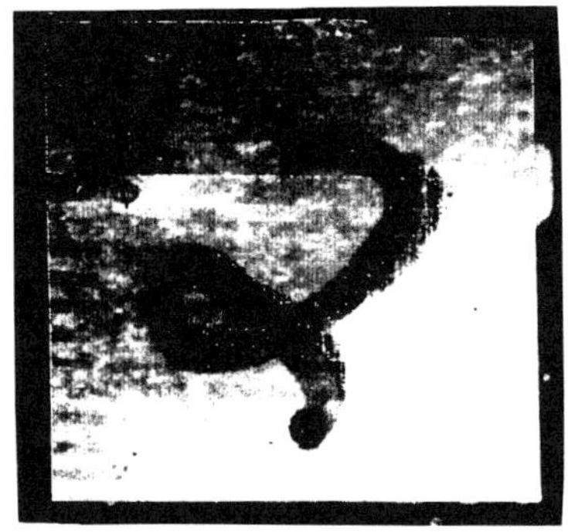

Fig. 8b. Micrograph of Fractured Surface for Angle ply Glass/Epoxy (Strain rate : 545 sec $^{-1}$. Loading : Across thickness-3)

case was lower than the strength of unidirectional composites loaded in the fiber direction (Fig. 4a) but it was higher than the strength loaded in the transverse direction (Fig. 5a). Such observations are reasonable in a sense that in cross-ply composites both matrix and fiber played dominant role in the failure process whereas in unidirectional composites the failure mode was either matrix or fiber dominant depending upon the direction of applied load.

The stress vs. strain plot of angle ply (+45/-45) glass/epoxy composites under compressive loading across the thickness is shown in Fig 8a. The failure strength and modulus are shown to be 578.76 MPa (84 ksi) and 20.25 GPa (2.94 msi) at a strain rate of 1276 s^{-1} . The fractured surface of the same specimen as shown in Fig. 8b indicates that cracks propagated in a direction perpendicular to the loading axis and no cracks were observed across the thickness. Such trend was also observed in cross ply composites and it was in contrast to the case of unidirectional composites. Figure 9a shows stress vs. strain plot of angle ply composites under compressive loading along the thickness. The failure strength and modulus of elasticity is shown to be 303.16 MPa (44 ksi) and 20.25 GPa (2.94 msi) at a strain rate of 909 sec^{-1} . These magnitudes appear to be significantly lower in comparison to unidirectional and cross-ply composites. The fractured surface of the same specimen is shown in Fig. 9b which indicates multiple delaminations and matrix cracking across the thickness and the crack propagated all through the specimen width. It is evident that the failure mode is mostly matrix dominant and it is more severe in the case of angle ply composites in comparison to unidirectional and angle-ply composites.

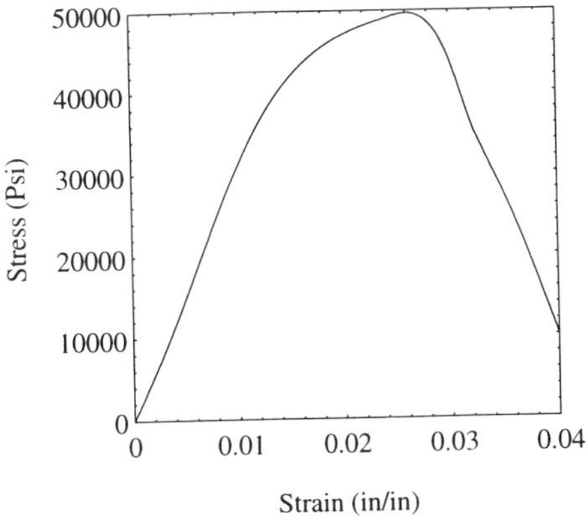

Fig. 9a Stress vs Strain plot for Angle ply Glass/Epoxy (Strain rate : 909 sec $^{-1}$. Loading : Along thickness -2)

Fig. 9b. Micrograph of Fractured Surface for Angle ply Glass/Epoxy Strain rate : 909 sec $^{-1}$, Loading : Along thickness-2)

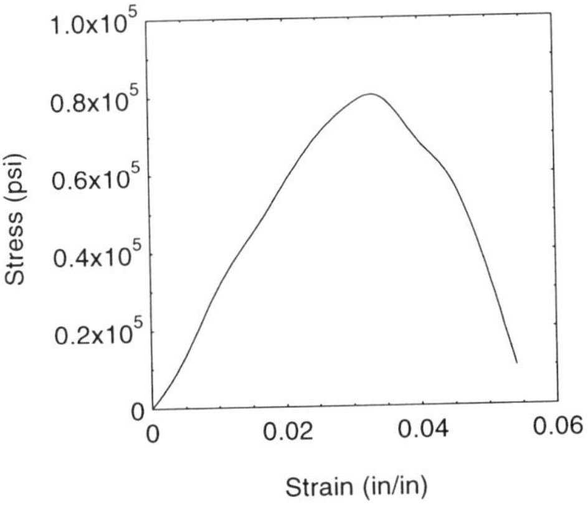

Fig. 10a Stress vs Strain plot for Quasi-isotropic Glass/Epoxy (Strain rate : 545 sec $^{-1}$, Loading : Across thickness-3)

Fig. 10b.. Micrograph of Fractured Surface for Quasi-isotropic Glass/Epoxy (Strain rate : 545 sec $^{-1}$, Loading : Across thickness-3)

Figure 10a shows stress vs. strain plot for quasi-isotropic glass/epoxy composites under compressive loading across the thickness. The failure strength and modulus are shown to be 564.98 MPa (82 ksi) and 17.43 GPa (2.53 msi) at a strain rate of 545 sec^{-1}. This modulus of elasticity was comparatively higher than the angle-ply composites but significant difference was not observed between the failure strength of angle ply and quasi-isotropic composites. The fractured surface of the same specimen is shown in Fig. 10b and the extent of cracks and failures are observed to be minimum in comparison to unidirectional, cross ply and angle ply composites. Some delamination and matrix cracking were observed across the thickness. Figure 11a. shows stress vs. strain plot for quasi-isotropic glass/epoxy composite under the compressive loading along the thickness. The failure strength and modulus are shown to be 578.76 MPa (82 ksi) and 23.70 GPa (3.44 msi) at a strain rate of 625 sec^{-1} . The quasi-static compression strength for quasi-isotropic glass/epoxy composites is shown to be 520 MPa. It is observed that the failure strength and modulus for quasi-isotropic laminates are comparatively higher than the angle ply composites but lower than the cross ply composites.

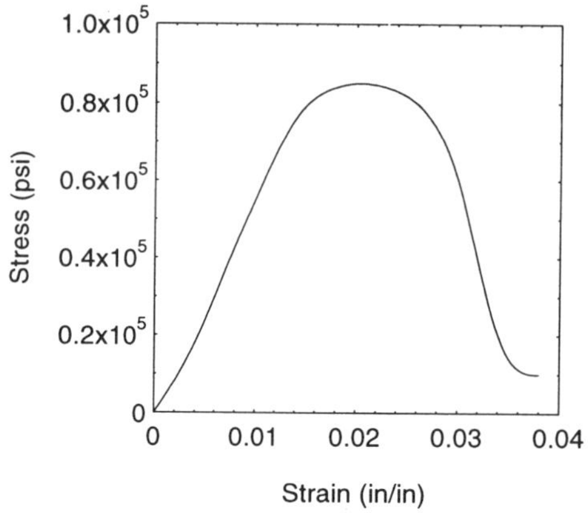

Fig. 11a. Stress vs. Strain plot for Quasi-isotropic Glass/Epoxy (Strain rate : 625 sec $^{-1}$, Loading : Along thickness-2)

Fig. 11b. Micrograph of Fractured Surface for Quasi-isotropic Glass/Epoxy (Strain rate : 625 sec $^{-1}$, Loading : Along thickness-2)

Figure 11b. shows the micrograph of the fractured surface of the same specimen. It is observed that the specimen is completely shattered with multiple delamination and complete separations between various layers.

4. CONCLUSIONS

The following conclusions can be drawn from the results of this investigation on the effects of fiber orientations on the high strain rate behavior of glass/epoxy composites.

1. Unidirectional composites of glass/epoxy showed comparatively 29 % higher compression strength in Split Hopkinson Bar Test method at a strain rate of 1086 sec $^{-1}$ as compared to quasistatic compression test (ASTM D-3410) results. It was observed that cracks initiated and propagated at an angle 45^0 across the thickness when the compressive load was applied across the thickness. Matrix cracking and debonding are shown to be the major failure modes in this case rather than delamination and fiber breakage.

2. Cross ply glass/epoxy composites showed maximum compression strength in comparison to unidirectional, angle ply and quasiisotropic laminates when the compressive loading was applied across the thickness. The cracks in the form of debonding and delamination were observed in a direction perpendicular to the loading axis. Similar failure nature was also observed in angle ply and quasiisotropic laminates in contrast to unidirectional composites. Compression strength data in both high strain rate (1153 sec $^{-1}$) test and quasi-static test (ASTM D-3410) is shown to be same when the compressive load was applied along the thickness.

3. Compression strength in both angle-ply and quasi-isotropic laminates showed similar results when the load was applied through the thickness. Quasi-isotropic laminates showed significantly higher compression strength in comparison to angle-ply laminates when the load was applied along the thickness. The compression strength of quasi-isotropic laminates was almost 11 % higher at high strain rate test (SHBT, 625 sec $^{-1}$) in comparison to quasi-static test (ASTM D-3410).

5. ACKNOWLEDGEMENT

The support for this research provided by the Army Research Office (ARO) under the grant DAAHO4-95-1-0369 is gratefully acknowledged.

6. REFERENCES

1. R. L. Sierakowski , AD-Vol. 48, High Strain Rate Effects on Polymer, Metal and Ceramic Composites and Other Advanced Materials, ASME , 1-5, (1995)
2. I. M. Daniel, W. G. Hamilton, R. H. LaBedz, Testing and Design (Sixth Conference), ASTM STP 787, I. M. Daniel. Ed. American Society for Testing and Materials, 393-413, (1982)
3. J. Harding, in Materials at High Strain Rates, T. Z. Blazynski, Ed. Elsevier Applied Science, London and New York, 133, (1987)
4. C. Y. Chiem, W. S. Lee, Materials Science and Engineering, 186, 43-50, (1994)
5. N. Takeda and L. Wan, ASME , High Strain Rate Effects on Polymer, Metal and Ceramic Matrix Composites and Other Advanced Materials, AD-Vol. 48 (1995)
6. J. F. Newill and J. R. Vinson, Proceedings of the Ninth International Conference on Composite Materials (ICCM9) Madrid, 12-16, July, 269-277, (1993)
7. M. A. El-Habak, Journal of Composites Technology & Research, JcTRER, Vol 15, No.4, pp.311-317, (1993)
8. V. M. Karbhari, Journal of Composite Materials, Vol.29, No.6, (1995)
9. P. Pintado, Proceedings of the Ninth International Conference on Composite Materials (ICCM9) Madrid, 12-16, July , 438-444, (1993)
10. G. H. Stabb and A. Gilat, Journal of Composite Materials, Vol. 29, No.10,1308-1320, (1995)
11. S. N. Nasser, J. B. Issacs, and J. E. Starrett , Proc. R. Soc. London, A. 435, pp. 371-391, (1991)
12. S. N. Nasser, JSME International Journal, Series I, Vol 34, No.2 1991, pp. 111-122, (1991)

28th International SAMPE Technical Conference
November 4-7, 1996

Superplastic Forming of Titanium Alloys Using PAM-STAMP™: Numerical Simulation and Industrial Validation

M. Jamjian[1], P.N. Comley[2], B.W. Varnell[2], A. Heath[3] and M. Pérennou[4]
[1]Engineering Systems International, North America,
5230 Carroll Canyon Road, Suite 224, San Diego, CA 92121.
[2]Murdock Inc.,
15800 South Avalon Blvd., Compton, CA 90220.
[3]Engineering Systems International GmbH,
Frankfurter Str 13-15, D-65760 Eschborn, Germany.
[4]Engineering Systems International, Middle-West,
570 Kirts Blvd., Suite 231, Troy, MI 48084.

ABSTRACT

The superplastic forming of Titanium alloys is simulated using PAM-STAMP™, an explicit non-linear finite element code, developed by Engineering Systems International. This code is used and validated for cold metal stamping and fiber-reinforced thermoplastic forming. Two algorithms are implemented in PAM-STAMP™ to allow the superplastic forming of Aluminum and Titanium alloys. The first algorithm prescribes the maximum allowable strain rate and regulates the pressure applied to the blank to achieve this strain rate. This yields a minimum forming time and the corresponding pressure history profile. Also, the blank thinning is obtained as a result of the numerical simulation. An alternative algorithm sets the gas flow rate history used to apply the pressure on the material resulting in part thickness and plastic strain rate distributions. The finite element shell material model includes membrane, bending and out-of-plane shear effects leading to a better characterization of the material behavior under forming conditions. The algorithms are implemented in PAM-STAMP™ and used to simulate the superplastic forming of a cone and two industrial parts. The experimental tests were performed by Murdock Inc. at their manufacturing facilities. Agreement between the numerical and the experimental results is shown.

KEY WORDS: Superplastic Forming, Numerical Simulation, SPF.

1. INTRODUCTION

The forming of complex thin sheet metal shapes such as automotive body parts is usually performed using metal stamping techniques in which an initially flat sheet, or blank, is pressed by a punch into a die cavity having the required shape. For the stamping of simple forms, experience is usually sufficient to design the required tooling so that a satisfactory part may be formed without tearing, wrinkling, or excessive elastic springback. In the case of more complex shapes, the stamping process may become very involved possibly requiring multiple stamping operations and the use of draw beads to force plastic stretching and to avoid wrinkling, or local lubrication to allow easier sliding of the blank between the tools and to reduce excessive local thinning. In these cases, previous knowledge is of limited use and invariably an expensive and time consuming trial and error design, manufacture and tryout processes are unavoidable.

Within the past ten years, numerical simulation techniques based on the finite element method have advanced tremendously in the prototyping of the stamping process allowing the design of the forming process to be computer assisted. The tryout process is now performed on the computer leading to substantial savings in tooling costs and the time to production. The numerical simulation process leads to a better understanding of the progressive forming and consequently, the opportunity to produce better quality parts. Figures 1 to 4 show an example of the state of the art for metal stamping simulation and compares the simulated final result to the test of an automotive part, a Renault Twingo dashpot cup designed and formed by Renault. Contours of thickness distribution and plastic strain and stress distribution as well as the forming limit diagram give vital information to the tool designer before any final decisions on the tooling and the forming process are made.

An extension of the metal stamping simulation techniques to the superplastic forming of aluminum and titanium alloys is presented in this paper and validated thorough industrial examples performed at Murdock Inc. and other stamping facilities. Through careful control of the pressure and the strain rate in the forming, complex shaped structural parts can be produced in one process thus leading to substantial savings in weight and avoiding the recourse to several stamping operations.

The additional complexities for the superplastic forming include the material model, the control of the strain rate or pressure loading to minimize forming time and thickness gradients in the final part. In the following sections, a brief description of the superplastic process is given, then material modeling and control processes in PAM-STAMP™ (1) are detailed. Finally industrial validation of the proposed modeling technique is provided through some examples and comparison between numerical simulations and industrial testing.

2. SUPERPLASTICITY REVIEW

Under appropriate conditions, certain metallic alloys deform to very large plastic strains of up to 1000% without fracturing. This phenomenon is known as superplasticity. Figure 5 illustrates the superplastic stretch of a titanium alloy bar under uni-axial loading.

Various alloys are known to exhibit superplastic behavior. The industrially important alloys include titanium, aluminum and iron as principal constituents. Necessary conditions for superplasticity are:

- a very fine grain size material (order of 10 μm),
- a high temperature (greater than half the melting temperature),
- a carefully controlled, low strain rate ($\dot{\varepsilon}$ =0.0001 to 0.01 s^{-1}).

Superplastic forming can be used to manufacture complicated components that could not be handled by conventional sheet forming methods. However, tools and sheet must be heated and maintained at a uniform temperature and the low strain rates lead to long process times in the order of hours, making it unsuitable for mass production. The process till now has been largely restricted to aerospace applications where production rates are low and the weight saving offered by the manufacture of complicated parts in a single piece is very important. Note that superplastic forming can be combined with diffusion bonding to manufacture hollow structures with internal stiffening members.

A number of techniques are applied to form components industrially. In a common method called blow/vacuum forming, a sheet of superplastic material is clamped in a rigid die and formed into the desired shape by a pressure differential applied by an inert gas (Figure 6). Forming pressures are of the order of 100 kPa for vacuum forming and 600 to 4000 kPa for blow forming. This process usually leads to a rather non-uniform thickness distribution with the greatest thinning occurring in regions where stresses are higher such as the center of the sheet. This can be controlled by pressing moveable tools into the sheet at appropriate stages of forming, thereby improving the stress distribution (Figure 7), or by starting with a thickness-profiled sheet (Figure 8).

Superplastic flow is a creep phenomenon which for uni-axial loading is approximately described by the equation:

$$\sigma = K\dot{\varepsilon}^m \varepsilon^n ,$$

where σ is the stress, ε the strain, and $\dot{\varepsilon}$ the strain rate. *K*, *m* and *n* are material parameters. *n* is the strain hardening exponent, taken equal to 0 in the following examples. This is a good approximation since *n* is small in reality. The main influencing factor is the exponent *m* which measures the strain rate sensitivity of the material. The extent of the thinning in the tensile specimen is strongly related to *m*. A large value of *m* is desirable to obtain more uniform thickness distribution after forming. In practice, the parameter *m* varies with the strain rate, typically exhibiting a maximum (Figure 9). For best forming conditions, it is important that the maximum strain rate in the sheet be regulated to stay in the optimum range where the *m* value is high. The book by Pearce (2) and the paper by Hamilton and Paton (3) are good references on superplasticity.

3. FINITE ELEMENT APPROACH

PAM-STAMP™ is an explicit non-linear finite element program, industrially validated for stamping processes. The explicit finite element method has been shown to be particularly robust and computationally efficient for the solution of large 3-dimensional, contact intensive, and highly non-linear problems. Also adaptive meshing techniques are implemented in PAM-STAMP™ allowing the use of a suitable coarse mesh at the start of the simulation and adaptively refining according to specified criteria in areas where a finer mesh is necessary for convergence to the correct solution. Also a rezone option, used to add and/or remove tools and change forming conditions without loss of previous material history variables is available to simulate multiple stamping operations. Friction, gas flow, draw beads as well as other forming conditions can be simulated leading to a better numerical prototyping of the superplastic forming process.

3.1 Material Modeling The superplastic sheet is modeled as a non-linear visco-elastic Maxwell material with the following constitutive equations:

$$s_{ij} = 2\mu\dot{e}_{ij}^{p}$$

$$s_{ij} = \sigma_{ij} - \frac{1}{3}\sigma_{kk}\delta_{ij}$$

$$e_{ij} = \varepsilon_{ij} - \frac{1}{3}\varepsilon_{kk}\delta_{ij}$$

$$\sigma_{ij} = \lambda\varepsilon_{kk}^{e}\delta_{ij} + 2G\varepsilon_{ij}^{e}$$

$$\varepsilon_{ij} = \varepsilon_{ij}^{e} + \varepsilon_{ij}^{p}$$

$$\varepsilon_{kk}^{p} = 0$$

$$\mu = \frac{1}{3}K\dot{\bar{e}}^{p^{m-1}}$$

where σ_{ij} is the Cauchy stress, ε_{ij} the total strain. The total strain can be divided into an elastic part ε_{ij}^{e} and a plastic part ε_{ij}^{p}. For superplastic forming, the elastic strain is negligible when compared to the plastic strain and can be neglected if no springback evaluation is needed. K and m are the superplastic material constants already described in the previous section. G is the shear modulus and λ the lame constant. $\bar{e}^{p}$ is the equivalent plastic strain. The above equations lead to the previous flow relation:

$$\bar{\sigma} = K\dot{\bar{e}}^{p^{m}},$$

where $\bar{\sigma}$ is the equivalent Von-Mises stress. This material law is incorporated into the formulations of three and four nodes shell elements. Three shell formulations are provided: a C^0 under-integrated Belytschko-Liu and Belytschko-Wong (4 and 5) shell elements with stiffness and viscous hourglass control options, and a Hughes-Tezduyar (6) shell element with four gauss points integration in the shell plane.

3.2 Solution Scheme In order to minimize the thinning and the forming time, a specific value of the strain rate to be achieved in the blank is desired. A constant pressure is seen to develop a variable strain rate, thus a variable forming pressure is necessary to maintain a constant strain rate. The pressure history is dependent on the strain history, consequently an automatic control process was incorporated in PAM-STAMP™ to regulate the pressure resulting in achieving a specified strain rate once it is reached in a part of the blank. Once the designer specifies a strain rate, both the pressure history and the total forming time are obtained as a result of the numerical simulation. This loading option is referred to as automatic pressure regulation to obtain maximum strain rate at prescribed value.

Another loading option, which can be used to retroactively check a pressure history, involves the prescription of the pressure-time history or the prescribed flow rate history of the gas in the fluid cell. The output of the simulation is a check on the formability of the sheet to the desired final shape.
In both loading options, stress, strain, strain rate and thickness contours are obtained from the numerical prototyping. Also, the tearing of the sheet can be detected if a corresponding criterion is specified.

4. INDUSTRIAL VALIDATION

The prescribed methodology is applied to several industrial examples and the results are compared to the tests conducted at Murdock Inc. and at Rohr.

4.1 Example 1 Four Ti-6Al-4V cones were superplastically formed at Murdock using boron nitride as a lubricant, at an imposed strain rate of 5e-4 s^{-1} according to a pressure history profile obtained from PAM-STAMP™. This pressure profile was obtained from a numerical simulation where a strain rate control option was used to minimize forming time and thinning gradients in the blank. The original thickness of the blank is 1.2954 mm. Figure 10 shows the initial mesh of the sheet and die as well as the pressure history. Due to symmetry, only a quarter of the blank was meshed with the appropriate symmetry boundary conditions. The deformed shape at 59 minutes is given in Figure 12 where thickness contours are also visible. The bulge height versus time histories from both numerical simulation and testing are also plotted in Figure 11. Finally the thickness values at different points of the cone are reported in Table 1 for both tests and finite element results. Both results are in good agreement with deviations of 2% to 10% between predicted and experimental values.

Table 1: Thickness Measurements [mm]

Position	*Test 1*	*Test 2*	*Test 3*	*Test 4*	*Num.*
A	.5588	.5334	.5588	.4826	.5700
B	.6096	.5842	.6096	.5842	.5880
C	.8128	.7366	.7366	.7874	.7400
D	.9652	.9144	.9144	.9652	.9300
E	1.0922	1.0414	1.0160	1.0160	1.0400
F	1.2954	1.2954	1.2954	1.2954	1.2954

4.2 Example 2 The forming of a headlight shell from a McDonnell Douglas aircraft using two blank variants: a uniform blank of constant thickness and a thickness-profiled blank is discussed in the next sections. Similarly, the comparison between experimental and simulation results is shown. A strain value of 2e-4 s^{-1} was used for the constant thickness blank resulting in pressure-time history while a pressure versus time curve applied by Murdock Inc. to form the part from the blank with varying thickness was specified in the numerical simulation. Half of the headlight shell was used in the finite element simulation due to symmetry conditions.

4.2.1 Constant Thickness Blank In a first approach, the superplastic forming of a uniform blank is simulated using PAM-STAMP™ with a prescribed strain rate control method to obtain pressure-time history and total forming time. The same part was formed at Murdock using the pressure-time history resulting from the finite element run (Figure 13). Figure 14 shows the final thickness profile of the headlight shell while Figure 15 shows the thickness profile both from the test and the numerical prototyping. Excellent agreement for both the forming time and the thickness profile is deduced from the obtained results. This industrial example verifies the possibility to predict the superplastic forming process using PAM-STAMP™ to minimize both forming time and thinning in the blank.

4.2.2 Thickness-Profiled Blank In order to minimize thickness gradients in the part and subsequently weight, the blank was profiled to obtain a more uniform thickness profile in the final shape. The blank profile is shown in Figure 16. This part was formed using a pressure time history given in Figure 16. The same blank was simulated in PAM-STAMP™ using a pressure control method. Figure 17 displays the progressive forming of the part with superimposed thickness profile. Again good agreement between numerical and test results is obtained as can be concluded from Figure 18 where numerical and test results are compared.

4.3 Example 3 As a part of the validation process, the superplastic forming of a MD90 aircraft vapor barrier designed by Rohr was simulated as a demonstration part using PAM-STAMP™. A maximum desired strain rate of 5e-4 s^{-1} was specified as well as a maximum pressure not to be exceeded. Figure 19 depicts the blank and the die at the initial forming stage. The final thickness contour is shown in Figure 20. The numerical results compared with the measurements supplied by Rohr are plotted in the graph of Figure 21 and show excellent agreement. Also the forming time of 53 minutes obtained from the numerical simulation agree with the real forming time of 50 minutes. The pressure time history of the real forming process and the predicted history are similar. The complexity of the part demonstrates the possibility to detect problems like thinning, tearing and incomplete part forming in the early design stages thus leading to better die design and less die tryouts. This leads to savings in cost and time to production.

5. CONCLUSIONS

The possibility to numerically simulate the superplastic forming process using PAM-STAMP™ has been demonstrated through several industrial examples. This offers the advantage to cut production time, design better tools and to save on the trial and error process when designing new dies. Additionally, the forming time as well as the pressure profile to be applied to guarantee a desired strain rate can be better monitored. Thickness profiles can also be obtained from the finite element simulation giving the die designer a tool to limit the thinning problems by small changes in the die that can be incorporated in the prototype and rerun numerically till a final satisfactory overall thickness is generated. The advantages of the numerical prototyping can be summarized as: cutting lead time and costs, assisting the die designer and a better understanding of the forming process.

6. ACKNOWLEDGMENTS

We would like to acknowledge the help provided by Murdock Inc., also, McDonnell Douglas and Rohr for the permissions to show the fabricated parts and the forming results.

7. REFERENCES

1. PAM-STAMP™, Users and Notes Manual, Engineering Systems International, 20 Rue Saarinen, Silic 270, 94578 Rungis-Cedex, France (1996).
2. C.H. Hamilton, and N.E. Paton, " *Superplasticity and Superplastic Forming*", Proceedings of the International Conference of the Minerals, Metals and Materials Society, pp. 706 (1988).
3. R. Pearce, Sheet Metal Forming, The Adam Hilger Series on New Manufacturing Processes and Materials (1991).
4. T. Belytschko, J.I. Liu and C. Tsay, " *Explicit Algorithms for the nonlinear Dynamics of Shells*", Computer Methods for Applied Mechanics and Engineering, 42, pp.225-251 (1984).
5. T.Belytschko, B.L. Wong, and H. Chiang, " *Advances in One Point Quadrature Shell Elements*", Computer Methods for Applied Mechanics and Engineering, 115, pp.277-286 (1992).
6. T.J.R. Hughes and T.E. Tezduyar, " *Finite elements based upon Mindlin Plate Theory with Particular Reference to the Four-Node Bilinear Isoparametric Element*", Journal of Applied Mechanics, 48, pp.587-591 (1981).

Figure 1: Final mesh of a Renault Twingo dashpot cup

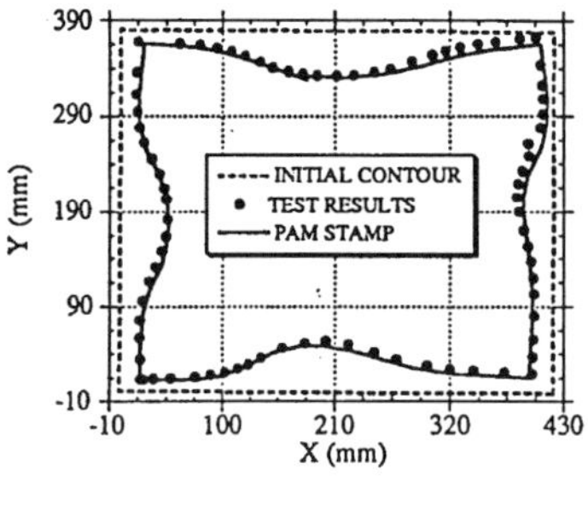

Figure 2: Draw-in comparison: experiment versus numerical

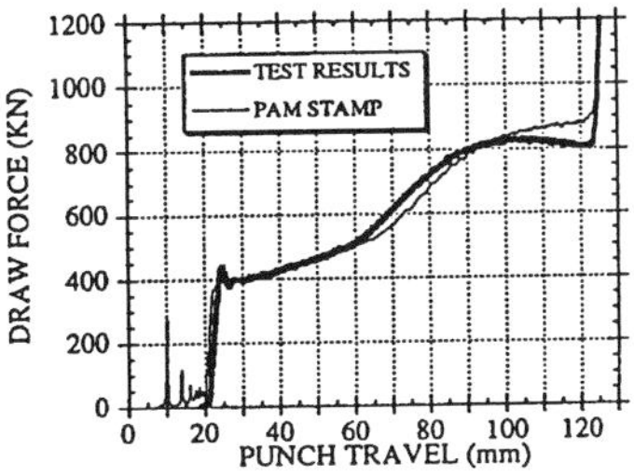

Figure 3: Stamping force

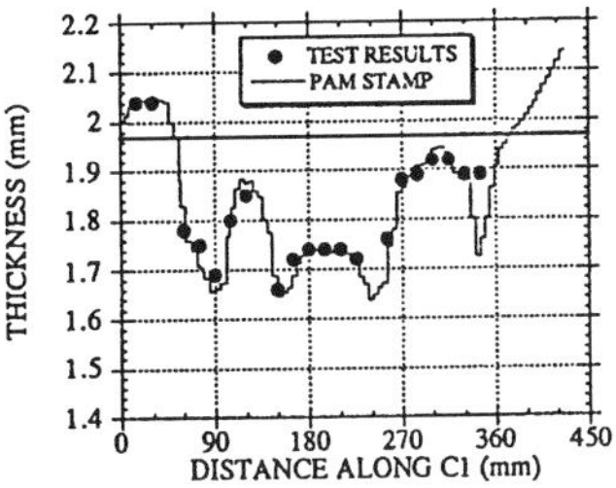

Figure 4: Thickness profile

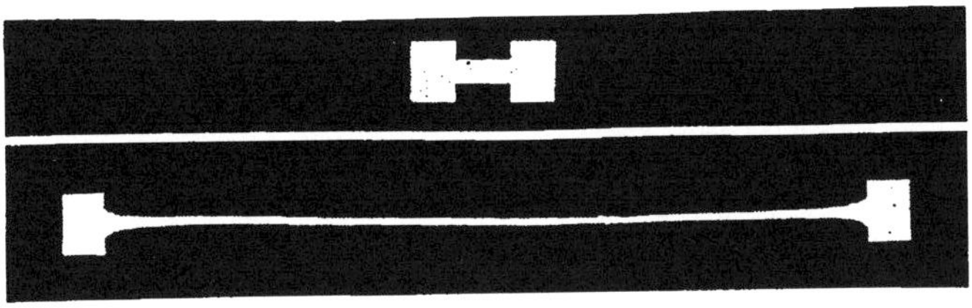

Figure 5: Stretch of Titanium alloy under uni-axial loading

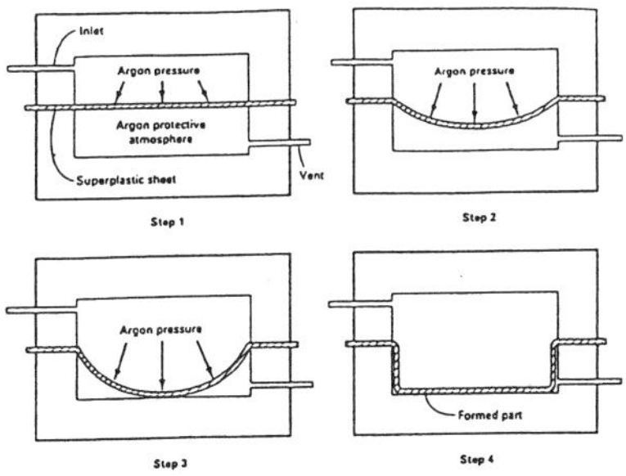

Figure 6: Vacuum/Blow superplastic forming

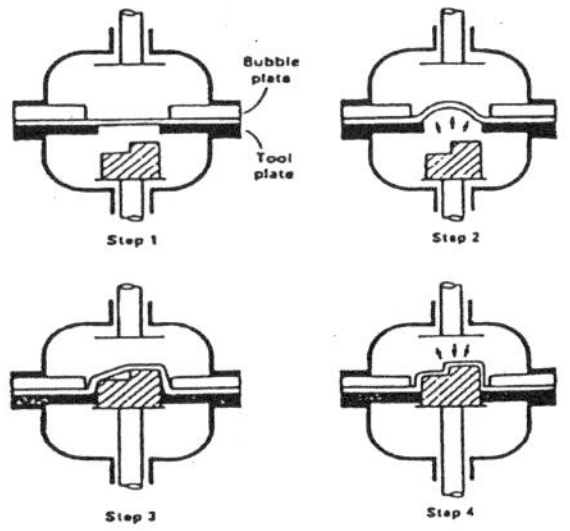

Figure 7: Superplastic forming using moveable tools

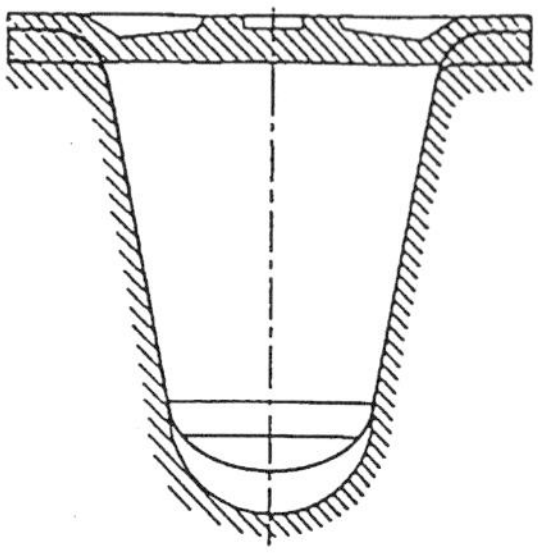

Figure 8: Thickness-profiled sheet

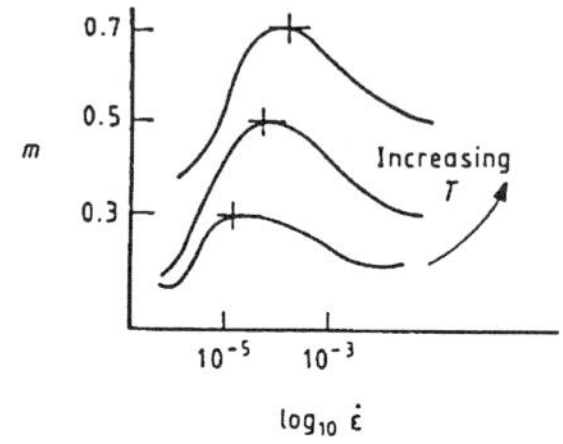

Figure 9: Variation of strain-rate sensitivity coefficient *m* with strain rate and temperature

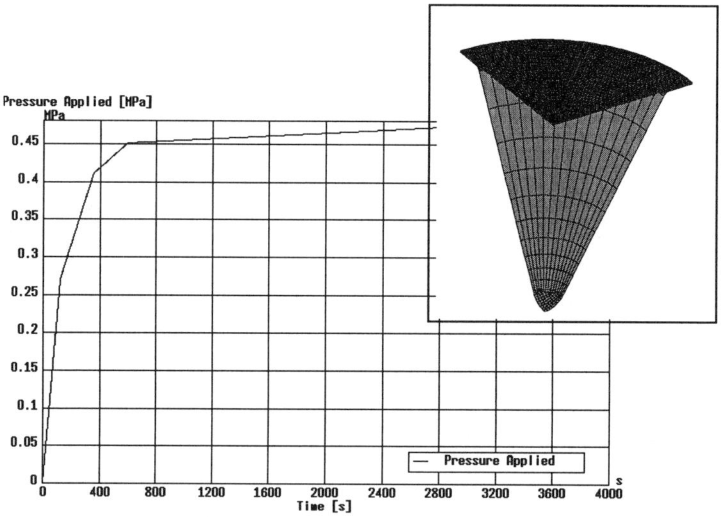

Figure 10: Pressure time history for superplastic forming of cone

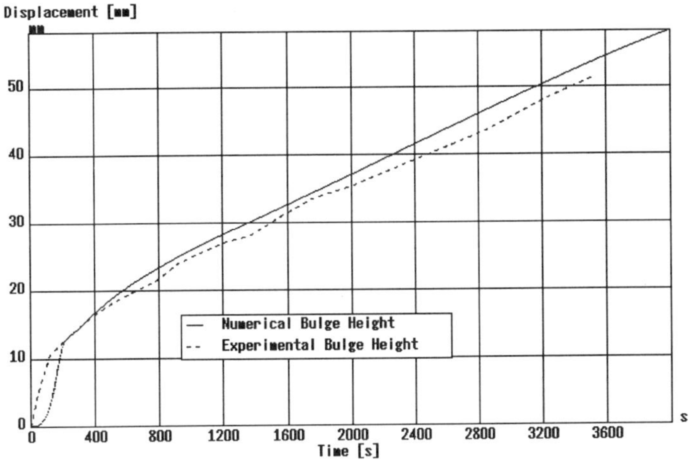

Figure 11: Bulge height versus time history for cone

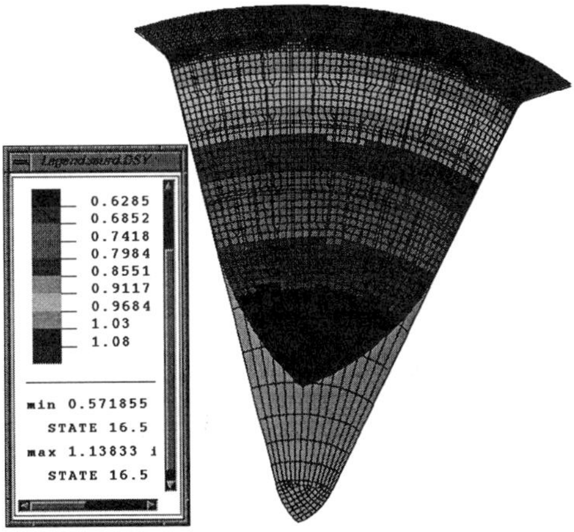

Figure 12: Thickness profile of cone after 59 minutes of forming

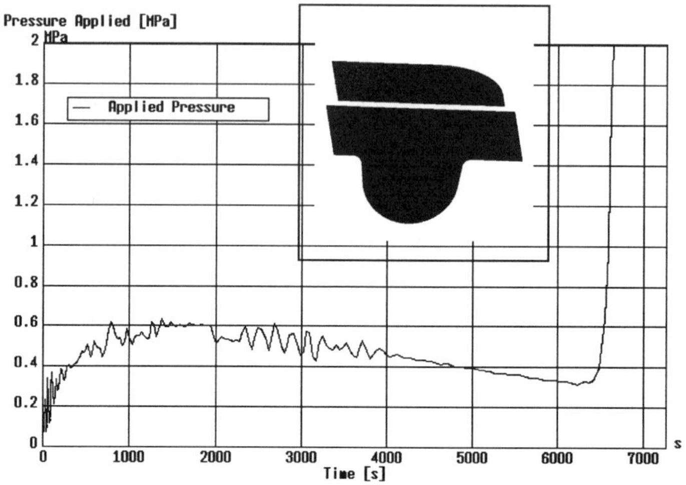

Figure 13: Pressure time history for constant thickness headlight shell

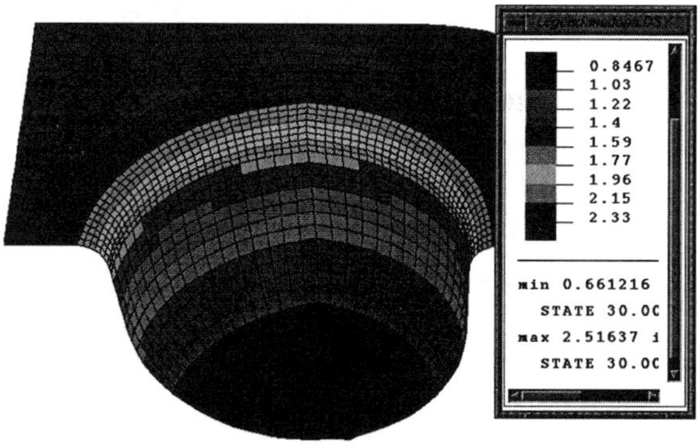

Figure 14: Thickness profile of constant thickness blank headlight shell at final forming

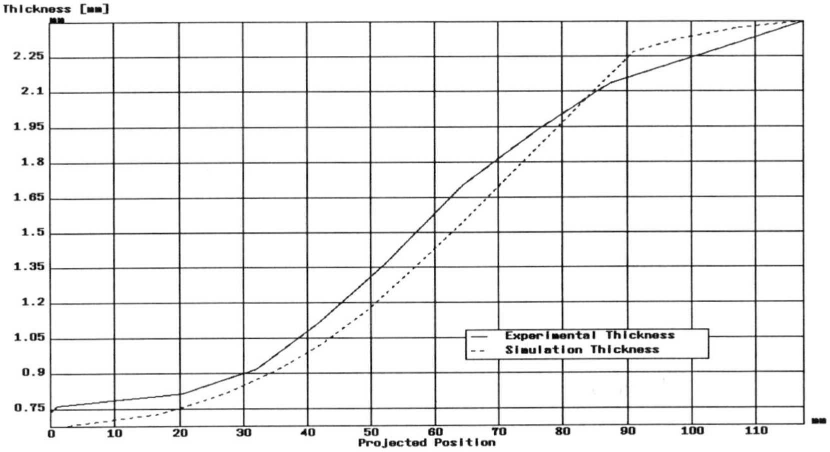

Figure 15: Experimental versus numerical simulation thickness of constant thickness sheet headlight shell

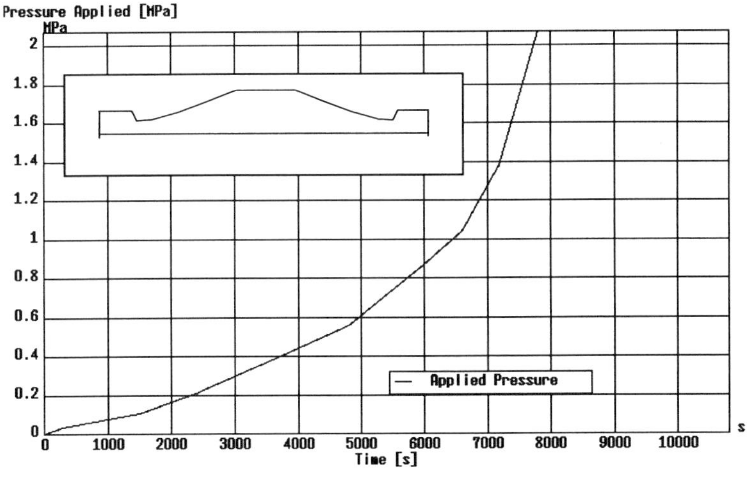

Figure 16: Pressure versus time history specified by Murdock Inc. for the forming of the variable thickness sheet headlight shell

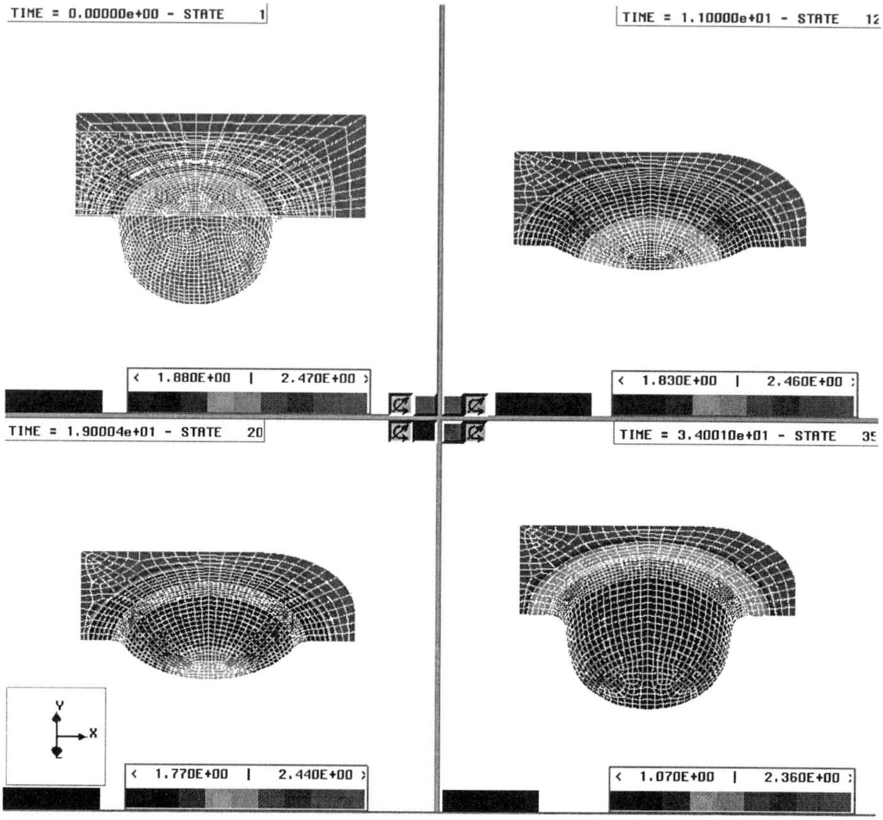

Figure 17: Progressive forming of variable thickness blank headlight shell

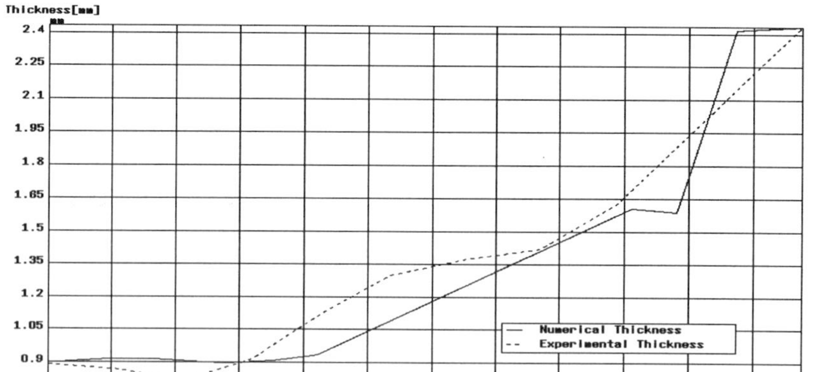

Figure 18: Experimental versus numerical simulation thickness of variable thickness blank headlight shell formed using the pressure profile specified by Murdock Inc.

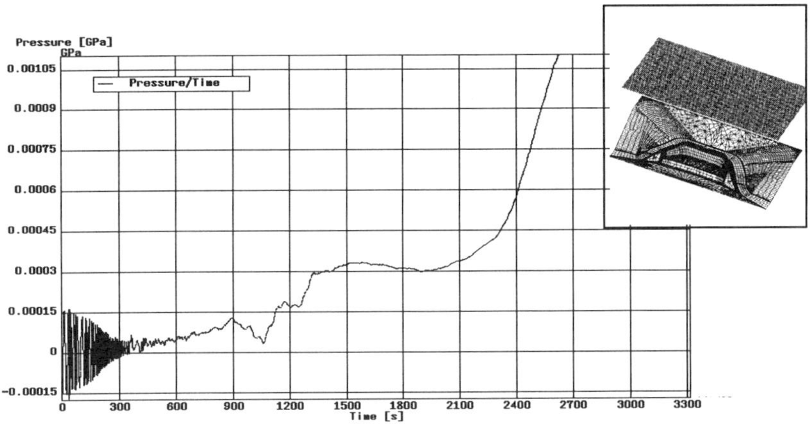

Figure 19: Pressure time history for the MD90 vapor barrier superplastic forming

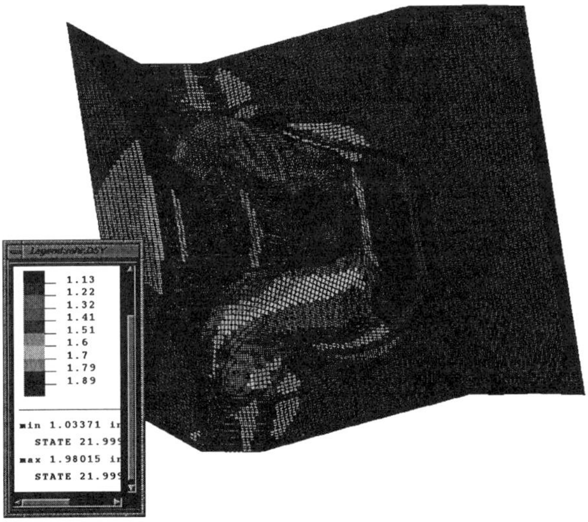

Figure 20: Final thickness profile for the MD90 vapor barrier

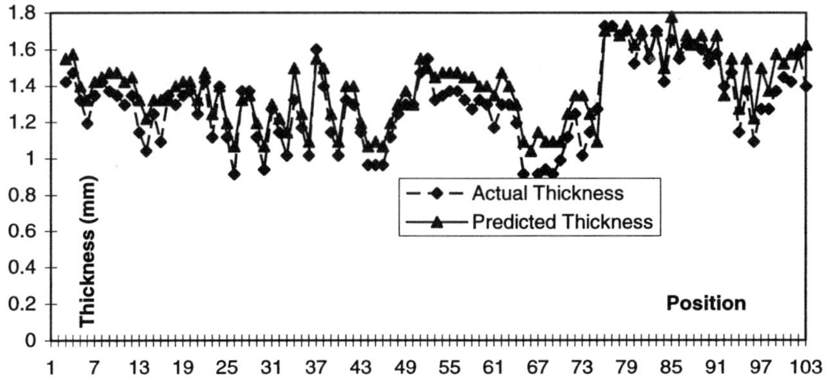

Figure 21: Comparison between experimental and numerical results for the superplastic forming of the MD90 vapor barrier

28th International SAMPE Technical Conference
November 4-7, 1996

SUPERPLASTIC PROCESS OPTIMIZATION OF INCONEL alloy 718SPF

by

Gaylord D. Smith and H. Lee Flower
Inco Alloys International, Inc.
3200 Riverside Drive
Huntington, WV 25705

and

R. B. Leholm
Rohr, Inc.
P.O. Box 878
Chula Vista, CA 91912

ABSTRACT

While superplastic forming of aerospace components has become a solidly entrenched manufacturing in recent years, its use has been restricted largely to the fabrication of aluminum and titanium components. It has only been in the last few years that a commercially practical nickel-base alloy has been available for use by aerospace designers and manufacturing engineers.

This paper seeks to establish that INCONEL® alloy 718SPF™ is ideally compatible with the superplastic forming process and current equipment. Processing parameters, including the employment of background pressure, are presented based on both laboratory and commercial practice, which seek to define an optimized temperature, preferred strain rate range and anticipated flow stresses for commercial exploitation of the process utilizing INCONEL alloy 718SPF. These data have been used to develop a nomograph correlating press tonnage requirements to part geometry. Parameters for achieving zero cavitation based on the use of background pressure and total strain limits are also presented.

KEY WORDS: Superplastic Forming, Superalloy, Processing, Mechanical Properties, Presses.

1. INTRODUCTION

Superplastic forming of titanium and aluminum alloys has been successfully used for years to fabricate complex, cost effective, integral structures. However, the maximum use of temperature is limited to 540°C for α/β titanium alloys, 650°C for near -α titanium alloys and an 80°C for conventional aluminum alloys. For applications above 650°C, nickel-based alloys such as INCONEL alloy 625 and INCONEL alloy 718 are generally used. INCONEL alloy 625, which is readily cold-formable, is not a superplastic forming alloy, and INCONEL alloy 718 is limited as a superplastic forming alloy because of its relatively large grain size (ASTM grain size #4-6) [(1-3)].

Recently, Inco Alloys International (Inco) developed a new nickel-based superplastic forming alloy, INCONEL alloy 718SPF, that combines the desirable high-temperature properties of the INCONEL alloys with the ease of superplastic forming. INCONEL alloy 718SPF is essentially a fine-gained (ASTM grain size #10 or finer) INCONEL alloy 718 where its fine-grain microstructure improves its superplasticity characteristics and increases its fatigue strength as compared to conventionally produced INCONEL alloy 718[(3)]. INCONEL alloy 718SPF is intended for use in various applications requiring part performance up to 820°C.

In order to more completely understand processing of this alloy, a research program to evaluate the cavitation characteristics of INCONEL alloy 718SPF, as a function of background pressure and true strain was initiated using the facilities at Rohr's Chula Vista facility. The details and results of this research program are reported in this paper.

2. MATERIALS

Three heats of INCONEL alloy 718SPF were selected for this study. The composition of each heat is presented in Table 1. Tensile properties and grain size are given in Table 2. These heats were produced by vacuum melting followed by electroslag remelting. The material was then hot worked using conventional practice. However, the cold working procedure was altered to assure the production of an ultrafine grain size in sheet.

Table 1. Composition of INCONEL alloy 718SPF Heats Studied (Wt %).

HEAT NUMBER	C	Mn	S	Fe	Ni	Cr	Al	Ti	Mo	Nb
HT3463EK	0.02	0.08	0.001	18.04	53.52	18.51	0.47	1.04	3.04	5.01
HT4262EK	0.03	0.08	0.001	17.97	53.47	18.47	0.52	1.04	2.88	5.11
HT4749EK	0.04	0.14	0.002	17.66	53.32	18.31	0.65	0.95	2.96	5.05

Table 2. Mechanical Properties of Mill Annealed INCONEL alloy 718SPF.

	Room Temperature Tensile Properties					
	Heat HT3463EK 0.51 mm gauge		Heat HT4262EK 1.22 mm gauge		Heat HT4749EK 0.76 mm	
Mill Annealed*						
0.2% Y.S., MPa	852		815		711	
U.T.S., MPa	1,121		1,114		1,114	
Elongation, %	29.0		33.0		33.0	
Hardness, Rc	25		32		32	
Grain Size ASTM #	13		12		12	
Aged**						
0.2% Y.S., MPa	1,461		1,323		1,271	
U.T.S., MPa	1,586		1,519		1,473	
Elongation, %	14.0		19.0		19	
Hardness, Rc	44		46		46	
Aged**	649°C Tensile Properties					
0.2% Y.S., MPa	1,116		1,105		1,014	
U.T.S., MPa	1,201		1,244		1,149	
Elongation, %	22.0		21.0		23	
	Life h	Elong. %	Life h	Elong.%	Life h	Elong. %
Stress, 793 MPa	12.2	9.8	-	-	-	-
Stress, 724 MPa	-	-	57.7	12.5	38.1	17.7

*Anneal: 927°C
**Aging Condition: 954°C 1h/AC plus 719°C/8h/FC at 56°C/h to 621°C plus 621°C /8h/AC.

Heat HT3463EK was evaluated for superplastic forming characteristics by Y. Ma and T. G. Langdon at the University of Southern California, Los Angeles, CA [(4)]. Heat HT4262EK was superplastically formed on conventional equipment by P. Comley of Murdock, Inc., Compton, CA, at 954°C using an argon gas pressure of 2.1 MPa (300 psi) and a time of approximately 4h. The superplastically formed part was evaluated for microstructure (grain size and area of cavitation) at various locations in the component. The third heat (HT4749EK) was used to make the test specimens superplastically formed by Rohr in their program to evaluate the effect of background pressure on cavitation.

3. DEFINING PROCESS PARAMETERS

3.1 ***Superplastic Forming Characteristics of INCONEL alloy 718SPF.*** 954°C was chosen as the aim temperature for characterization of the SPF parameters of INCONEL alloy

718SPF.[2] The grain size stability of INCONEL alloy 718SPF during the time for typical SPF of a conventional component should be excellent at 954°C as exemplified by the grain size data shown in Figure 1.

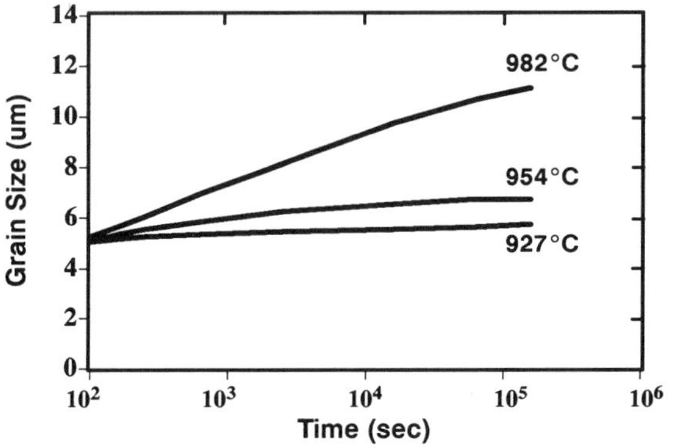

Figure 1. Plot of Grain Growth vs. Time at 927°C, 954°C and 982°C for INCONEL alloy 718SPF.

To conduct the superplastic forming process parameter evaluation, ten specimens were machined from 1.22 mm sheet of INCONEL alloy 718SPF and tested at 954°C to failure at various strain rates from 1.33 x 10^{-2} s^{-1} to 1.33 x 10^{-5} s^{-1}. The test results are presented graphically in Figure 2. A plot of engineering strain (total elongation) versus the initial strain rate is presented in Figure 3. Engineering strain increased from 150% at an initial strain rate of 1.3 x 10^{-2} s^{-1} to 760% at 3.3 x 10^{-5} s^{-1}. For the typical strain rates of 1 x10^{-3} to 1 x10^{-4} s^{-1}, the available engineering strains are approximately 280% at the faster strain rate to 480% at the slower strain rate.

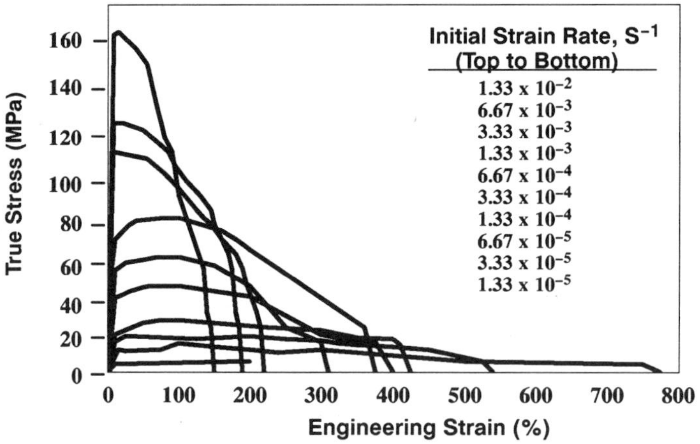

Figure 2. Plot of True Stress vs. Engineering Strain for 10 Strain Rates at 954°C for INCONEL alloy 718SPF, Heat HT3463EK.

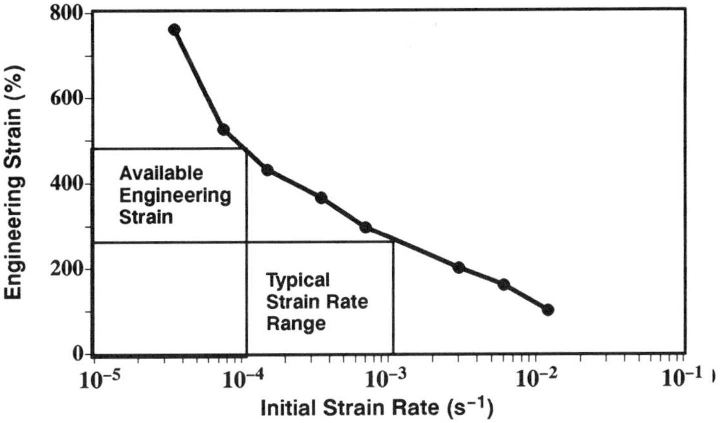

Figure 3. Engineering Strain vs. Strain Rate at 954°C for INCONEL alloy 718SPF, Heat HT3463EK. For Typical Strain Rates, The Available Engineering Strain is High-Lighted.

A plot of the true strain rate at maximum stress versus the true stress at maximum stress is presented in Figure 4. This plot defines the maximum stresses that must be generated during superplastic forming in order to deform INCONEL alloy 718SPF at the conventional strain rates of 1×10^{-3} to 1×10^{-4} s^{-1}. The necessary stress range is from 27.6 to 82.7 MPa. Also shown in Figure 4 is the changing value of n at different strain rates as determined by Equation 1:

$$\dot{\epsilon} = \kappa \sigma^n \quad (1)$$

where $\dot{\epsilon}$ = true strain rate, σ = true stress, κ = constant

The value of n is the slope of log $\dot{\epsilon}$ vs. log σ and is typically between 1 and 2.5 in the region of most practical interest.[3]

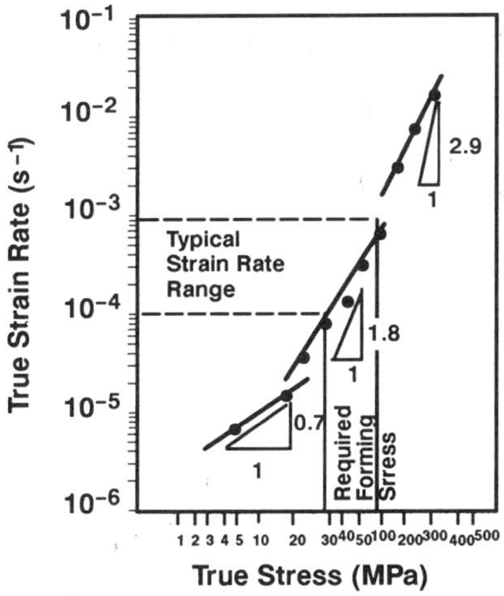

Figure 4. Plot of True Strain Rate vs. True Stress (Both at Maximum Stress) for INCONEL alloy 718SPF, Heat HT3463EK. For Typical Strain Rates, The Required Forming Stress Range is Highlighted.

The measure of strain rate sensitivity, m, is calculated from the data of Figure 3 and plotted in Figure 5. The m value of INCONEL alloy 718SPF becomes 0.4 at an initial strain rate of 1.3 x 10^{-4} s^{-1} and increases in value with decreasing strain rate.

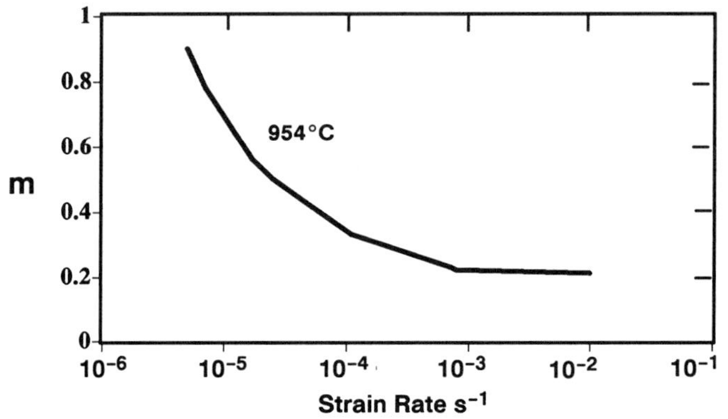

Figure 5. Plot of m Values, a Measure of Strain Rate Sensitivity vs. Strain Rate at 954°C for INCONEL alloy 718SPF.

3.2 ***Cavitation Under Uniaxial Deformation.*** The propensity of INCONEL alloy 718SPF to form pores during superplastic forming (cavitation) was measured (using an image analysis software package) by metallographic examination of the tensile specimens tested to fracture at various initial strain rates at 954°C as exemplified in Table 3. This alloy is subject to cavitation during processing with the degree of cavitation increasing with decreasing initial strain rate. At an initial strain rate of 1 x 10^{-2} s^{-1}, the area of cavitation, as measured near the fracture tip, is less than 0.01% for a tensile elongation of 150%. However, the degree of cavitation increases at a strain rate of 1 x 10^{-3} s^{-1} to 2.6% (369% elongation); becoming 10.3% at a strain rate of 1 x 10^{-5} s^{-1} (538% elongation). Table 3 shows that as the strain rate is decreased, the average cavity size and roundness coefficient increases. The average roundness coefficient is an indication of the degree of diffusion associated with cavity growth. A roundness coefficient of one indicates a perfect circular shape while a smaller value of the roundness coefficient relates more to cavity elongation. The increase in roundness coefficient at the strain rate of 1 x10^{-3} s^{-1} or less suggests that cavity growth by diffusion is becoming increasingly significant for decreasing strain rates (longer times at temperature).

Table 3. The Effect of Strain Rate at 954°C on The Size and Area of Cavitation of INCONEL alloy 718SPF. The Average Roundness Coefficient of The Cavities is Also Presented.

Strain Rate, s^{-1}	1.3 x 10^{-2}	1.3 x 10^{-3}	6.7 x 10^{-5}
Local Strain[(a)], %	410.0	590.0	550.0
Cavitation Area[(b)], %	<0.01	2.60	10.30
Avg. Cavity Area, um^2	5.20	145.0	580.0
Avg. Roundness Coefficient[(c)]	0.47	0.63	0.63

[(a)]Specimen tested to fracture. [(b)]Based on a measured area of 0.83 mm^2 near the fracture tip. [(c)]Defined as (4π area)/$(perimeter)^2$ where a value of 1.0 represents a perfect circle and values less than 1.0 represent deviation from a circle.

Because superplastically formed components normally are produced at strain rates of 1 x 10^{-3} s^{-1} to 1 x 10^{-4} s^{-1} with actual total elongations of less than 200%, four INCONEL alloy 718SPF specimens were evaluated at 954°C for strains of 86% and 194% at an initial strain rate of 1.3 x 10^{-3} s^{-1} and for strains of 73% and 132% at an initial strain rate of 1.3 x 10^{-4} s^{-1}. The area of cavitation was measured metallographically using an area of 1.07 mm2 near the gauge length. The results are presented in Table 4 and are combined with the results of Table 3 in Figure 6. The data show that increasing elongation and decreasing strain rate, increase the area of cavitation. However, at total elongations of less than 200% at either a strain rate of 1.3 x 10^{-3} or 1.3 x 10^{-4} s^{-1}, the area of cavitation is less than 0.5% and can be held below 0.1% for elongations of less than 100% at a strain rate of 1.3 x 10^{-3} s^{-1}. Typical cavity size, as shown in Table 4, is again larger for the slower strain rate and greater total elongation.

Table 4. The Effect of Strain Rate and Total Strain at 954°C on The Size and Area of Cavitation of INCONEL alloy 718SPF.

Initial Strain Rate, s^{-1}	1.3 x 10^{-3}	1.3 x 10^{-4}	1.3 x 10^{-3}	1.3 x 10^{-4}
Local Strain, %	86.0	73.0	194.0	132.0
Cavitation Area, %	0.04	0.18	0.45	0.34
Avg. Cavity Size, um2	2.40	8.90	10.50	11.20

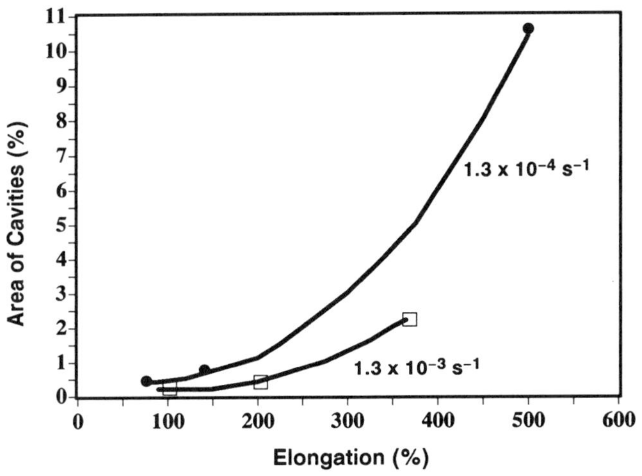

Figure 6. Plot of Percent Area of Cavitation vs. Total Elongation for Two Typcal Strain Rates at 954°C for INCONEL alloy 718SPF.

Cavitation tends to nucleate and grow around inclusions present in the alloy. Analysis of the number of cavities of a given size as a function of total strain shows a significant increase in the number of cavities within the smallest size range examined (0 to 2 μm2) as the elongation increases, supporting the concept that cavities are continuously being nucleated throughout the test. Thus, the extent and distribution of inclusions becomes an important issue in superplastic forming of INCONEL alloy 718SPF.

Analysis of the number of cavities of a given size as a function of total strain shows a significant increase in the number of cavities within the smallest size range examined (0 to 2 μm^2) as the elongation increases, supporting the concept that cavities are continuously being nucleated throughout the test. Thus, the extent and distribution of inclusions becomes an important issue in superplastic forming of INCONEL alloy 718SPF.

Since the application of superimposed background pressure has been successfully utilized to reduce cavitation in aluminum alloys, it seemed worthwhile to evaluate the effect of background pressure on reducing cavitation and cavitational damage on superplastically processed INCONEL alloy 718SPF. This interest resulted in the initiation of the project at Rohr to evaluate the effect of background pressure on the area of cavitation under biaxial flow conditions.

4. BACKGROUND PRESSURE SUPPRESSION OF CAVITATION

To conduct these tests, the apparatus depicted in Figure 7 was employed. Small disks 76 mm were machined from 0.76 mm ASTM grain size number 12 sheet. A conical die with a cone angle of 57 degrees was used as the die cavity. Argon gas pressurized the die cavity at levels dependent upon the strain rate desired. Fixed superimposed background pressures of 1.38 MPa, 2.76 MPa and 4.14 MPa were employed. The forming temperature was 954°C at a strain rate of $1 \times 10^{-4}\ s^{-1}$ and $3 \times 10^{-5}\ s^{-1}$. The clamping force exerted on the specimen by the upper and lower fixtures of the test apparatus was approximately 3400 kg. The tool sealing design used for the biaxial superplastic forming testing is shown schematically in Figure 8. This sealing concept successfully withstood the high pressure of the test apparatus without any argon leaks and without tearing the specimen at the seal. At the conclusion of each run, the apex area of each cone (the area of maximum biaxial strain) was used to determine maximum strain and the extent of cavitation. The percentage cavitation was determined using specific gravity measurements. The results are plotted in Figures 9 and 10 for the strain rates of $1 \times 10^{-4}\ s^{-1}$ and 3×10^{-5}, respectively. It is worthy of note that at the two higher background pressures, 2.76 MPa and 4.14 MPa, the rate of cavitation was less at the slower strain rate. This is opposite to the effect of strain rate on the area of cavitation at zero background pressure as depicted in Figure 6. Increasing superimposed background pressure at a given strain rate increasingly retards the initiation of cavitation, as is the case with aluminum alloys, and for any given total strain beyond "onset" significantly reduces cavitational formation. Figure 11 shows the predicted press tonnage requirements as a function of part cross section and planned background pressure generated from the data of the project from which this paper is prepared. The figure is a useful graphical tool for designers and manufacturing engineers for determining the producibility of a proposed part using INCONEL alloy 718SPF and a given background pressure.

5. CONCLUSIONS

The aim of this paper is to show the applicability of INCONEL alloy 718SPF to current superplastic forming practice and equipment; to describe the processing parameters and to define the microstructure of superplastically formed material. Based on this effort, a number of conclusions can be drawn:

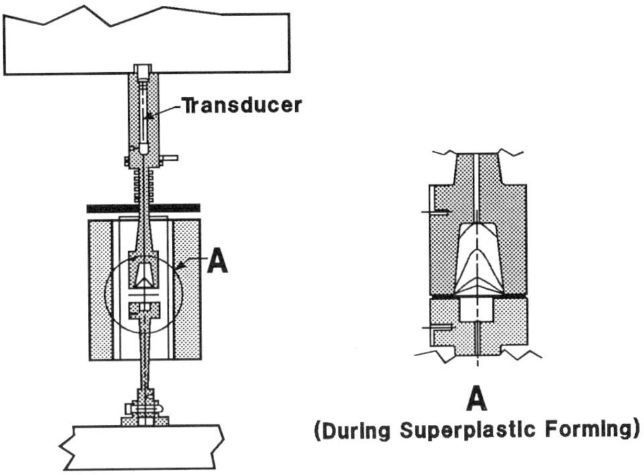

Figure 7. Schematic Drawing of The Biaxial Superplastic Forming Test Apparatus Used for This Study.

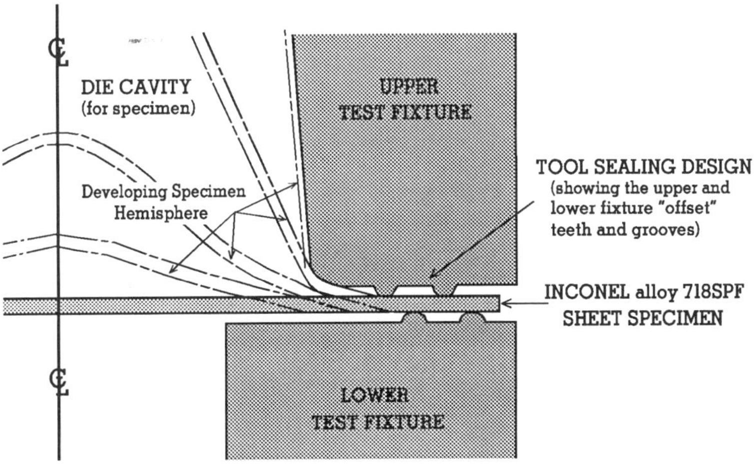

Figure 8. Sealing Design Used for The Biaxial Superplastic Forming Testing.

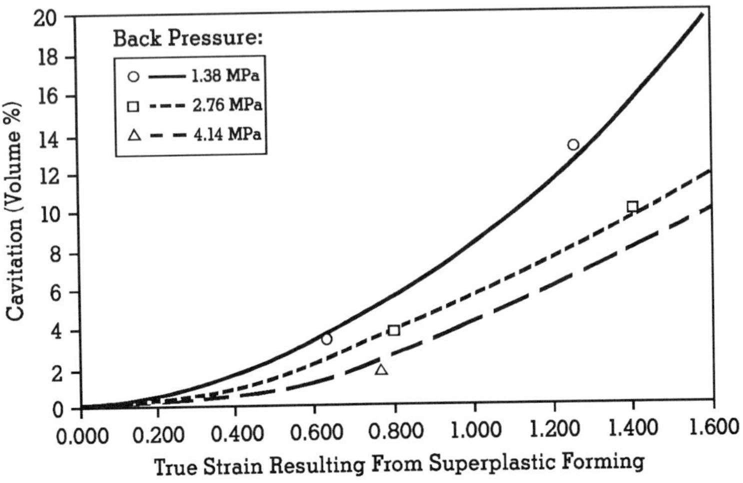

Figure 9. Cavitation Volume Percent in INCONELalloy 718SPF Sheet as a Function of True Strain and Background Pressure for a Strain Rate of 1 x 10^{-4} s^{-1}.

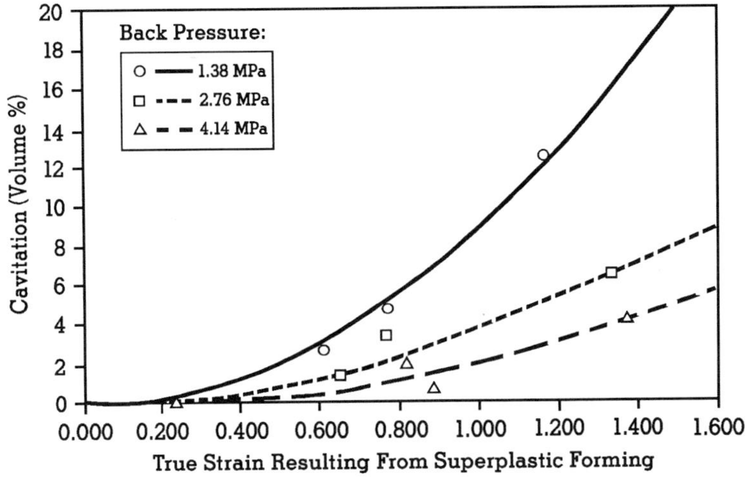

Figure 10. Cavitation Volume Percent in INCONEL alloy 718SPF Sheet as a Function of True Strain and Background Pressure for a Strain Rate of 3 X 10^{-5} s^{-1}.

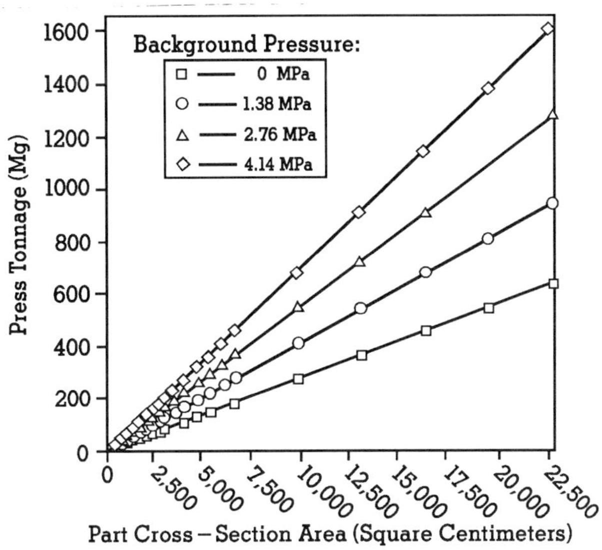

Figure 11. Press Tonnage Requirements for 0.76 mm INCONEL alloy 718SPF Sheet.

1. INCONEL alloy 718SPF sheet can be produced with an ASTM grain size of #10 or smaller and, subsequently, superplastically formed at 954°C using an argon gas pressure of 2.1 MPa.

2. For INCONEL alloy 718SPF, the elongation to failure increases markedly with decreasing strain rate, exceeding 350% at an initial strain rate of $1.3 \times 10^{-3}\ s^{-1}$, 450% at $1.3 \times 10^{-4}\ s^{-1}$ and 750% at $3.3 \times 10^{-5}\ s^{-1}$.

3. For INCONEL alloy 718SPF, the maximum true stress decreases with decreasing strain rate becoming less than 71.7 MPa at an initial strain rate of $1.3 \times 10^{-3}\ s^{-1}$ and less than 27.6 MPa at $1.3 \times 10^{-4}\ s^{-1}$.

4. The development of cavities at tensile failure is insignificant at a strain rate of $1 \times 10^{-2}\ s^{-1}$ (less than 0.01% by area) but becomes increasingly important as the strain rate decreases.

5. The application of superimposed background pressure during superplastic forming reduces or eliminates intergranular cavitation encountered at atmospheric background pressure.

6. The tool sealing concept (used as part of the biaxial superplastic forming test apparatus) was effective in containing high gas pressures without any gas leak and without causing the specimens to tear at the seal.

7. Press and background pressure superplastic forming guidelines (in graph format) were developed in order to assist designers and manufacturing engineers in determining:

   (a) whether or not a given INCONEL alloy 718SPF part can be superplastically formed with background pressure using an existing press;

(b) levels of background pressure that can be achieved for a given part using existing press;

(c) the largest size of an INCONEL alloy 718SPF part that can be superplastically formed with a given background pressure using an existing press;

(d) the press tonnage requirement for superplastic forming of an INCONEL alloy 718SPF part with a given background pressure.

6. REFERENCES

1. A. K. Ghosh and C. H. Hamilton, "Superplastic Forming of INCONEL alloy 718," from Report #ITR-79-033, Rockwell Science Center, October 1979.

2. M. W. Mahoney, "Superplastic Properties of INCONEL alloy 718," in Proceedings of the International Symposium on the Metallurgy and Applications of INCONEL alloy 718, ASM publishers, Pittsburgh, PA, June, 1989.

3. G. D. Smith and D. H. Yates, "Superplastic Forming of INCONEL alloy 718SPF", in Superalloys 718, 625, 706 and Various Derivatives, E. A. Loria, ed., The Minerals, Metals & Materials Society, New York, 1994, 355-364.

4. Y. Ma and T. G. Langdon, Private Communication, "Observations of Cavitation in Deformed INCONEL alloy 718", University of Southern California, February 1992.

5. R. C. Gifkins, "Mechanisms of Superplasticity." in Proceedings of Symposium on Superplastic Forming of Structural Alloys, June 1982, San Diego, CA, published by AIME.

6. Bampton, C. C. and Raj, R., "Influence of Hydrostatic Pressure and Multiaxial Straining on Cavitation in a Superplastic Aluminum Alloy," Acta Metallurgica 30, 1982.

28th International SAMPE Technical Conference
November 4-7, 1996

RAPID SUPERPLASTIC FORMING OF AL ALLOY 5083

Stephen J. Hales *
Analytical Services and Materials, Inc.
107 Research Drive, Hampton, VA 23666

ABSTRACT

Methods for decreasing the cycle time for superplastic forming of the commercially available aluminum alloy SKY 5083 have been explored. Using optimum forming conditions, a 1920 second (32 minute) cycle time is typical for forming a beaded web stiffener panel requiring a maximum true strain of $\approx$ 1.2. As a consequence of the current effort, the time for complete panel formation has been reduced by an order of magnitude without sacrificing part quality. Rapid differential pressure forming was achieved by regulating the pressure in the die cavity (below the sheet) and holding the pressure above the sheet constant. Thickness gradients and cavitation were controlled by using a combination of Al tooling, die lubrication and variable back pressure. In contrast to the traditional approach, the pressure-time profile employed was not designed to approximate a constant strain rate during biaxial deformation. Using non-optimum forming conditions, a linear pressurization ramp permitted nominal strain rates approaching 10^{-2} s^{-1} to be achieved. The implications of this simplified approach for the cost effective application of SPF technology are discussed.

KEY WORDS: Superplastic Forming (SPF), Aluminum (Al) Alloys

1. INTRODUCTION

Superplasticity describes the exceptional ductility and low flow stresses exhibited by a select family of materials [1]. Superplastic forming (SPF) technology has proved competitive with other fabrication processes for applications involving low to moderate volume production [2]. The high uniform deformation and low forming loads inherent to the process have permitted the production of components with deep recesses by gas pressure forming. In order to be economical, the low deformation rates (10^{-4}-10^{-3} s^{-1}) associated with the process have been offset by manufacturing complex shapes and/or multiple components in a single forming operation. More recently, there has been considerable interest in the technology for higher volume applications, such as in the automotive industry [3,4]. As a consequence, the technology is in direct competition with other commercial hot working practices with deformation rates from 10^{-1} s^{-1} to 10^{1} s^{-1} [5]. The capability of large forming strains is still a big advantage of SPF, but there is impetus to reduce cycle times for the technology to become more cost effective.

*** Research conducted at Langley Research Center under NASA Contract No. NAS1-19708.**

Al alloys tend to behave superplastically only in a narrow range of strain rates requiring careful process control to maintain the desired constant strain rate during SPF [1]. The shape of the pressure-time profile tends to reflect the geometry of the die and the deformation dependent characteristics of the material [6]. Traditionally, uniaxial testing has provided the correlation between temperature, strain rate, flow stress and strain hardening behavior, but the relationship of these data to biaxial conditions is complex. Superplastic forming operations are characterized by irregularly shaped die surfaces which create both variable stress states and strain rates in the deforming sheet. The situation is further complicated by frictional effects between the deforming sheet and the die surfaces. Consequently, the use of uniaxial data in combination with 2-D and 3-D models to formulate gas pressurization cycles for biaxial forming has met with limited success [6-9].

The relationship between strain rate and superplastic response for Al alloys produced by a variety of processing techniques is illustrated in Figure 1 [5]. The chart reveals that ingot metallurgy materials exhibit an upper strain rate capability of the order of 10^{-2} s^{-1}. Faster strain rates are possible using materials which require powder metallurgy or more exotic processing methods to produce highly refined microstructures. However, raw material costs will have a detrimental impact on the economic feasibility of SPF technology. Among the conventionally processed alloys, dynamically recrystallizing materials are formable at higher strain rates and exhibit higher ductility than statically recrystallized materials, such as alloy 7475 (Al-Zn-Mg). However, the commercially available Supral (Al-Cu) and 8090 (Al-Li) alloys tend to be more expensive as a result of specialized processing requirements [10].

There are a number of commercially available 5xxx (Al-Mg) alloys which are relatively inexpensive, the most common being alloy 5083 [11-14]. The alloy is non-heat treatable and is classified as a medium strength (σ_u= 300 MPa, σ_y= 145 MPa, El.= 20 %), weldable material with excellent corrosion resistance [3,4]. Common applications of superplastically formed 5083 include architectural cladding, ground transportation and interior fittings on aircraft. The maximum strain rate capability of fully recrystallized superplastic materials is considered to be primarily dependent on grain size [5]. Efforts aimed at expanding the superplastic envelope to higher strain rates have focussed on alloy modification of 5083 [15]. These have included lowering the impurity content and/or adding supplementary grain refiners (Mn, Zr). Both of these approaches result in increased raw material expenses due to the need for non-standard casting practices and rolling schedules [16].

In this investigation, an incremental, rather than a breakthrough, advance in technology was sought for the commercially available superplastic Al alloy SKY 5083. The raw material was purchased from Murdock, Inc. in the form of 2.45 m x 1.25 m x 2.0 mm sheet. The composition of the alloy, as provided by the vendor, is listed in Table 1. The material is fully recrystallized and possesses a uniform grain size of $\approx$ 10μm. Superplastic forming of SKY 5083 is normally conducted in the temperature range of 500-550°C at an optimum true strain rate of $\approx$ 2 x 10^{-4} s^{-1}, with a corresponding flow stress of 1.0-3.0 MPa [11-14]. The premise for this study is that the material can be formed with sufficient rapidity to be cost effective in applications involving higher volume production. The approach focussed on innovative biaxial forming procedures to reduce cycle times via forming near the upper strain rate limit. The factors explored include the use of Al tooling, linear pressurization ramps, back pressure and die lubrication.

Table 1. Composition of SKY 5083 (wt.pct.)

Element	Mg	Mn	Cr	Fe	Si	Cu	Ti	Al
Content	4.48	0.65	0.11	0.07	0.05	0.02	0.01	Bal

2. UNIAXIAL BEHAVIOR

The results of parametric uniaxial studies designed to assess the superplastic formability of the SKY 5083 material are presented in Figure 2. The effects of strain rate and temperature on the elongation-to-failure, the flow stress (at $\epsilon = 0.1$) and the strain hardening behavior of the material are illustrated. Compared to many superplastic materials, where maximum elongations are of the order of 1000% (Figure 1), the material is only moderately superplastic. The data also reveal a general trend of decreasing ductility with increasing strain rate. However, at strain rates greater than $\approx 2 \times 10^{-3}\ s^{-1}$, the elongation-to-failure levels out at $\approx$ 250 %. The superplastic ductility of the material appears relatively unaffected by temperature at the strain rates evaluated. Figure 2 also shows that the flow stress increases with increasing strain rate and decreasing temperature. The slope of the flow stress curve, which is a measure of formability, is highest at the lowest strain rates. At 535°C and $1 \times 10^{-4}\ s^{-1}$ the material exhibited the highest ductility (377 %) and lowest flow stress (1.2 MPa). Of significance is that the flow stress is affected much more by a decrease in temperature than is the elongation-to-failure.

The strain hardening behavior as a function of temperature and strain rate is presented in Figure 3. Figure 3(a) shows that increasing the temperature from 450°C to 535°C results in little change in the strain-to-failure, but causes more than a twofold decrease in flow stress. The slope of the stress-strain curve, which equates to the strain hardening exponent of the material, is unaffected by increasing temperature except at 535°C, where it is slightly lower. Figure 3(b) shows the effect of increasing the strain rate at a forming temperature of 535°C. Although the general trend comprises decreasing strain-to-failure with increasing strain rate, more important is the effect on flow stress. The data show that the flow stress increases by an order of magnitude between strain rates of $10^{-4}\ s^{-1}$ and $10^{-1}\ s^{-1}$. This becomes a major factor in efforts to reduce cycle time because the maximum pressure capability of the NASA-LaRC SPF facility is 3.45 MPa (500 psi) [17,18]. Above a certain strain rate, the high flow stress will render gas pressure forming impractical.

3. BIAXIAL FORMING

The SPF facility at NASA-LaRC, which has been described previously, is designed for a maximum operating temperature of 535°C (1000°F) [17,18]. Derivation of the pressurization cycles has been based on membrane theory where the forming pressure, P, may be expressed as [1]:

$$\mathbf{P = C \cdot \sigma_f \cdot (t_i / r_i)} \qquad \{1\}$$

In equation {1}, **C** is a geometry dependent constant, which is governed by the instantaneous shape of the deforming sheet (C= 2 for a sphere; C= 1 for a cylinder). $\boldsymbol{\sigma_f}$ is the strain dependent flow stress of the material and $\mathbf{t_i}$ and $\mathbf{r_i}$ are the instantaneous thickness and radius of curvature of the deforming sheet, respectively. The effect of deformation-induced changes to the material flow stress and, thus, the instantaneous pressure requirements can be accounted for through the relationship [1]:

$$\sigma_f = \mathbf{K} \cdot \dot{\epsilon}^m \cdot \epsilon^n \qquad \{2\}$$

In equation {2}, K is a constant, $\dot{\epsilon}$ is the strain rate, m is the strain rate sensitivity factor, ϵ is the strain and n is the strain hardening exponent. Although the change in m with strain is difficult to account for, the strain hardening effect has been incorporated through regression curve analysis of the uniaxial data.

The SPF die used to produce beaded web stiffener panels at NASA-LaRC is shown in Figure 4(a). The die is fabricated from Al alloy 6061 and is 61 cm long, 18 cm wide and 3.3 cm deep. The die material was selected based on the compatibility of its coefficient of thermal expansion with the Al superplastic sheet materials. Release of superplastically formed parts in the web area between the beads proved to be problematic with other die materials [17]. Previous activities on aerospace Al alloys have concentrated on biaxial forming at a temperature and constant strain rate which result in optimum superplastic response [18]. A schematic of the die cross-section is shown in Figure 4(b). Geometrical consideration of the die reveals that the draft angle causes the deforming sheet to make sidewall contact first. During this initial stage of forming, the radius of curvature decreases slowly, but faster than the decrease in thickness ($\Delta r > \Delta t$), and the ramp in forming pressure required is moderate. Subsequently, the sheet deforms down the sidewalls until it makes contact with the bottom of the die. During this stage, the radius of curvature and thickness of the sheet decrease slowly in unison ($\Delta r \approx \Delta t$) and the rate of pressurization is slow. After bottom contact, the radius of curvature decreases rapidly as the deforming sheet is forced into the corners of the die while the rate at which the thickness decreases remains the same ($\Delta r >> \Delta t$). Consequently, the pressure ramps rapidly to maintain a constant strain rate during this latter stage of forming.

In order to promote longevity of the Al tooling, a forming temperature of 525°C was chosen for the biaxial forming studies. As shown in Figure 3(a), the difference in behavior between 535°C, the temperature at which most of the uniaxial data was collected, and the forming temperature selected is small. Figure 5 shows theoretical pressurization cycles for superplastic forming of SKY 5083 at 525-535°C at a series of strain rates. The overall shape of the profiles reflects the geometrical effect of the die on biaxial forming at a constant strain rate. The model assumes sticking friction between the sheet and the die and that the membrane is secured around the perimeter of the die [17]. The profiles represent an adaptation of existing pressurization cycles developed for forming aerospace Al alloys on the NASA-LaRC equipment using the same die configuration [18]. The starting sheet thickness, flow stress and strain hardening data for SKY 5083 were compared with the characteristics of previously formed materials.

The baseline pressure-time profile for superplastic forming of SKY 5083 at a temperature of 525°C with a constant true strain rate of $5 \times 10^{-4}\ s^{-1}$ is shown in Figure 6. The pressurization cycle was designed for differential pressure forming taking full advantage of the hydrostatic pressure capabilities of the apparatus. In contrast to more orthodox procedures, the pressurization cycle was inverted. Instead of increasing pressure above the sheet, both sides were pressurized to 3.45 MPa and then the pressure below the sheet was regulated. The pressure differential and lower pressure constitute the forming pressure and the back pressure, respectively. As the lower pressure decreased, the forming pressure increased and, although decreasing, the maximum attainable back pressure was applied to the deforming sheet.

For a target strain rate of $5 \times 10^{-4}\ s^{-1}$ and using boron nitride as a lubricant, interrupted forming cycles established that sidewall contact occurs at $\epsilon = 0.1$ after 200 s, bottom contact occurs at $\epsilon = 0.3$ after 600 s, and corner contact occurs after 1900 s. These transition points in the cycles are identified as SWC, BC and CC, respectively. As shown in Figure 5, the pressure required tends toward infinity during the latter stages of forming as the radius of curvature of the deforming sheet becomes very small [1]. Empirical derivation of this latter section of the pressure-time profile using interrupted forming cycles has revealed that the panel is 95 % formed by the time that the pressure differential reaches a maximum. Based on experience, a specified dwell time at the ultimate forming pressure allows the sheet to conform with the die, i.e. effect complete panel formation [17].

The series of pressure-time profiles used in the forming trials aimed at cycle time reduction are shown in Figure 7. Again, the upper pressure is held constant and the lower pressure is regulated. In conceiving the experimental profiles, a simple approach was adopted which essentially ignored the complex effects of die geometry and deformation-induced changes in material behavior. The pressure-time profiles were designed for the sole purpose of reducing the cycle time with no consideration of the actual forming rate. In contrast to the shape of the baseline profile, which was formulated to maintain a constant strain rate, a linear ramp was employed when decreasing the lower pressure. As a result, the pressure differential across the sheet, the forming pressure, increases at a constant rate. Consequently, programming of the gas management system and control of the pressurization cycle are greatly simplified.

4. PART QUALITY

For structural applications, rapid differential pressure forming cannot be implemented if there is a degradation in part quality [2]. Therefore, an important aspect of the work involved assessment of the quality of the beaded web stiffener panels. It was recognized that, as the cycle time decreased, the deviation from the strain rate for optimum formability increased. Consequently, the use of non-optimum forming conditions was expected to have a detrimental impact. The criteria established for evaluation of the fast forming trials included: (i), no premature rupture at die entry radii; (ii), the degree of panel formation; (iii), no sheet perforation at points of maximum forming strain; (iv), minimal localized thinning; and (v), acceptable cavitation.

Premature rupture of the biaxially deforming sheet tends to be dominated by tool design rather than strain rate [6]. This can be circumvented by allowing more generous radii in areas of the die where there is large movement of the sheet over the die. This becomes particularly important when lubrication is used because the sheet-die friction is reduced [1]. As shown in Figure 4(b), the die entry radius at the cap/web intersection is approximately 3.2 mm and only 2.0 mm along the periphery of the die. As a consequence of this difference in dimensions, the sheet would be expected to rupture on the perimeter during fast forming trials. A complete forming cycle was accomplished without premature rupture of the sheet for all of the forming trials. As in the baseline profile, boron nitride was used as a sheet-die lubricant. An example of one of the beaded web stiffener panels fabricated from SKY 5083 in 195 s is shown in Figure 8.

Table 2. Results of Upper Limit Forming Evaluation

Run #	Time	r_{min}	t_{min}	r_{min}/t_{min}	ϵ_{max}	$\dot{\epsilon}_{nom}$
	sec	mm	mm			sec^{-1}
412C*	1920	0.66	0.66	1.00	1.11	5.8×10^{-4}
413A	1620	0.53	0.61	0.87	1.19	7.4×10^{-4}
413F	360	0.71	0.66	1.08	1.11	3.1×10^{-3}
414D	195	1.30	0.71	1.83	1.04	5.3×10^{-3}
414G	120	1.60	0.71	2.25	1.04	8.7×10^{-3}
414F	75	2.77	0.79	3.51	0.94	1.3×10^{-2}

* Baseline Profile: "Constant true strain rate" pressurization cycle

In order to assess the degree of panel formation for each forming trial, the minimum radius, r_{min}, and minimum thickness, t_{min}, were measured at the points of maximum forming strain, ϵ_{max}. In the panel, these locations corresponded to the web-flange intersection where the beaded web stiffener insert met the end wall. The results are compared with the baseline profile (412C) and summarized in Table 2. The t_{min} values were used to calculate ϵ_{max}, based on a starting thickness of 2.0 mm, which in turn was divided by the forming time to yield the nominal strain rate, $\dot{\epsilon}_{nom}$. Adopting $r_{min}/\ t_{min} \leq 2.00$ as the criterion for complete panel formation, Table 2 shows that Run # 414D, with a nominal strain rate of $5.3 \times 10^{-3}\ s^{-1}$, emerges as the best candidate. Therefore this profile was employed in auxiliary forming trials in order to assess the effect of lubrication and back pressure on fast forming. The results, which are summarized in Table 3, show that both lubrication and back pressure are required if perforation of the sheet at the point of maximum forming strain is to be avoided.

Table 3. Effect of Back Pressure and Lubrication on Formability

Forming Cycle	Variable Back Pressure	Lubrication Used		
		None	Boron Nitride	Graphite
Constant $\dot{\epsilon}$ (5.8×10^{-4})	Yes	228B*	412C	--
Non-Constant $\dot{\epsilon}$ (5.3×10^{-3})	Yes	504B*	414D	504D
	No	504A*	504E*	504C*

* Complete panel formation, but sheet perforated at point of maximum forming strain

As indicated in Figure 4(a), the elements of the 'top hat' structural stiffener comprise a cap with beaded webs and two flanges. The effects of strain rate, lubrication and back pressure on the thickness profile of the component are presented in Figures 9 and 10. Figure 9 shows that increasing the strain rate causes an increase in thickness in the cap and at the web/flange intersection and a decrease in thickness at the cap/web intersection. The thickness of the web and flange areas is little affected, but there is evidence of 'eyebrow' formation (indicated) adjacent to the cap/web intersection at the highest strain rate. The data confirm the increased tendency of the material to thin locally in regions of high stress concentration (i.e. at the die entry radius) at the faster strain rates. Figure 10 shows that the use of lubrication alone causes a decrease in the thickness of the cap, but little change elsewhere in the panel. Back pressure alone causes a decrease in thickness in the web and at the web/flange intersection, but the thicknesses of the cap and cap/web intersection are unaffected. The thickness profile of the flange area is different and the flange possesses a non-uniform thickness gradient rather than a smoothly tapered cross-section.

The use of lubrication and back pressure in combination causes a decrease in cap thickness equivalent to using lubrication alone. There is a noticeable decrease in thickness at the cap/web intersection, but the web thickness is unaffected. At the web/flange intersection, there is a decrease in thickness equivalent to using back pressure only. The thickness profile in the flange is similar to that obtained following the use of back pressure alone, but the effect is less pronounced. The effects of using lubrication are consistent with other observations where the sheet-die friction has been reduced allowing deformation to continue in the contact areas. However, the effects of using back pressure alone, or with lubrication, on the thickness profile remain uncertain. The effect appears to be greatest during the early stages of forming when the level of back pressure is at its highest.

It is generally accepted that the level of superimposed back pressure required to completely suppress cavitation corresponds to 0.5-0.75 σ_f of the material [19]. This is not practical for forming equipment with a 3.45 MPa maximum pressure capacity when the flow stress for SKY 5083 at 525°C and 5 x 10^{-4} s^{-1} is of the order of 7.0 MPa. The level of cavitation was assessed via quantitative image analysis and the results are summarized in Table 4. Five locations were selected corresponding to different levels of forming strain and biaxial stress state. The data reveal that the areal fraction of cavitation was ≤ 0.55 %, in all but the regions corresponding to ϵ_{max} (web-flange intersection). The cavitation behavior of the material appears to be relatively insensitive to $\dot{\epsilon}_{nom}$ when both lubrication and variable back pressure are employed.

Table 4. Areal Fraction Cavitation (%) vs. Forming Rate

Panel#	Forming Time* (sec)	Cap_{CL}	Cap/Web	Web_{CL}	$Flange_{CL}$	Web/Flange
		$\epsilon \approx 0.1$	$\epsilon \approx 0.3$	$\epsilon \approx 0.5$	$\epsilon \approx 0.5$	$\epsilon \approx 1.2$
413A	1620	0.10	0.45	0.55	0.48	0.40
413F	360	0.06	0.10	0.30	0.36	1.12
414G	120	0.08	0.32	0.14	0.12	3.38
414F	75	0.00	0.18	0.13	0.28	1.16

* "Non-constant strain rate" pressurization cycles employing boron nitride lubrication

It was recognized that the volume fraction of cavitation is strongly dependent on strain and, as such, cavity growth will not be suppressed at higher strains where the level of back pressure is approaching zero. However, the data suggest that suppression of cavity nucleation during the earlier stages of forming, when the level of back pressure was higher, may play a role. Furthermore, the levels of cavitation reported are acceptable for most purposes and appear relatively unaffected by the use of fast forming.

5. SUMMARY

It has been demonstrated that the cycle time for superplastic forming of a beaded web stiffener panel from SKY 5083 can be reduced by an order of magnitude. A pressure-time profile designed to simulate a constant strain rate of 5 x 10^{-4} s^{-1} was replaced by a linear pressurization ramp which resulted in complete panel formation at a nominal strain rate of 5 x 10^{-3} s^{-1}. Biaxial forming operations involved pressurizing both sides of the sheet to 3.45 MPa and then reducing the pressure below the sheet while holding the pressure above the sheet constant. The differential pressure across the sheet constituted the forming pressure and the gas pressure below the sheet constituted the back pressure.

Rapid superplastic forming, using pressure-time profiles which did not approximate a constant strain rate, was implemented without sacrificing part quality. The use of lubrication and variable back pressure played a role in maintaining thickness gradients and the volume of cavitation within acceptable margins. The use of Al tooling facilitated release of formed panels due to the matching thermal expansion coefficients of the sheet and die. The economic implications for application of SPF technology for larger volume production revolve around the potential to use commercially available material, shorter cycle times and low cost Al tooling. In addition, the use of linear pressurization ramps may permit the use of much simpler and less expensive gas management systems.

6. ACKNOWLEDGEMENTS

This work represents a team effort involving the author, Eric Hoffman, Rob Edahl, Ron Penner and Tom Bales. Assistance from Dorinda Humphries and Joel Alexa with manuscript preparation is also acknowledged. The various contributions are duly recognized and greatly appreciated.

7. REFERENCES

1. C.H. Hamilton, in **Superplasticity**, NATO/AGARD; LS-168, SPS Ltd., Loughton, England, 1989, pp. 2.1-2.23.
2. D. Stephen, in **Superplasticity**, NATO/AGARD; LS-168, SPS Ltd., Loughton, England, 1989, pp. 7.1-7.37.
3. R. Verma, et al., J. Mater. Engng. Perf., 4 (5), 543-550 (1995).
4. R.L. Hecht and K. Kannan in A.K. Ghosh and T.R. Bieler, eds., **Superplasticity and Superplastic Forming 1995**, TMS, Warrendale, PA, 1995, pp. 259-266.
5. K. Higashi in N. Ridley, ed., **Superplasticity: 60 Years After Pearson**, Institute of Materials, London, 1995, pp. 93-102.
6. A.K. Ghosh and C.H. Hamilton, in **Process Modeling - Fundamentals and Applications to Metals**, ASM, Metals Park, OH, 1978, pp. 303-331.
7. K.N. Shah and J.M. Story, J. Mater. Shaping Technol., 7 (3), 137-144 (1989).
8. H.S. Yang and A.K. Mukherjee, Int. J. Mech. Sci., 34 (4), 283-297 (1992).
9. N. Chandra and D. Kannan, J. Mater. Engng. Perf., 1 (6), 801-812 (1992).
10. R. Grimes, in **Superplasticity**, NATO/AGARD; LS-168, SPS Ltd., Loughton, England, 1989, pp. 8.1-8.16.
11. H. Imamura and N. Ridley in S. Hori, M. Tokizane and N. Furushiro, eds., **Superplasticity** in Advanced Materials, Japan Society for Research on Superplasticity, Tokyo, 1991, pp. 453-458.
12. H. Iwasaki, et al. in S. Hori, M. Tokizane and N. Furushiro, eds., **Superplasticity in Advanced Materials**, Japan Society for Research on Superplasticity, Tokyo, 1991, pp. 447-452.
13. J.S. Vetrano, et al., Scripta Metall., 30 (5), 565-570 (1994).
14. M. Matsuo in N. Ridley, ed., **Superplasticity: 60 Years After Pearson**, Institute of Materials, London, 1995, pp. 277-283.
15. C.A. Lavender, et al. in T.G. Langdon, ed., **Superplasticity in Advanced Materials ICSAM-94**, Materials Science Forum, Vols. 170-172 (1994), pp. 279-286.
16. H.S. Yang, M. Shaarbaf and K.R. Brown in A.K. Ghosh and T.R. Bieler, eds., **Superplasticity and Superplastic Forming 1995**, TMS, Warrendale, PA, 1995, pp. 17-24.
17. S.J. Hales, et al. in T.R. McNelley and H.C. Heikkenen, eds., **Superplasticity in Aerospace II**, TMS, Warrendale, PA, 1990, pp. 167-185.
18. S.J. Hales and J.A. Wagner, in **Technology 2000**, NASA Conference Publication 3109, vol. 2, 1991, pp. 200-209.
19. N. Ridley, in **Superplasticity**, NATO/AGARD; LS-168, SPS Ltd., Loughton, England, 1989, pp. 4.1-4.14.

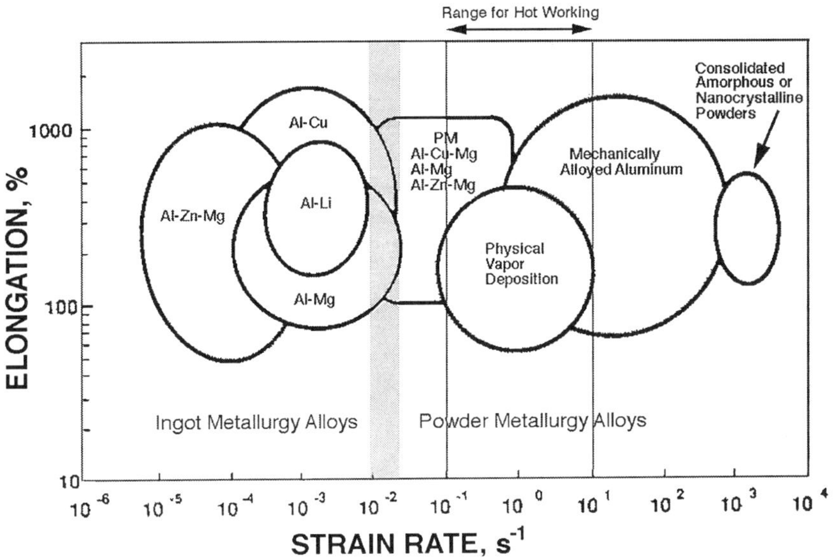

FIGURE 1. **The relationship between strain rate and superplastic response for Al alloys produced by a variety of processing techniques (After Higashi [5]).**

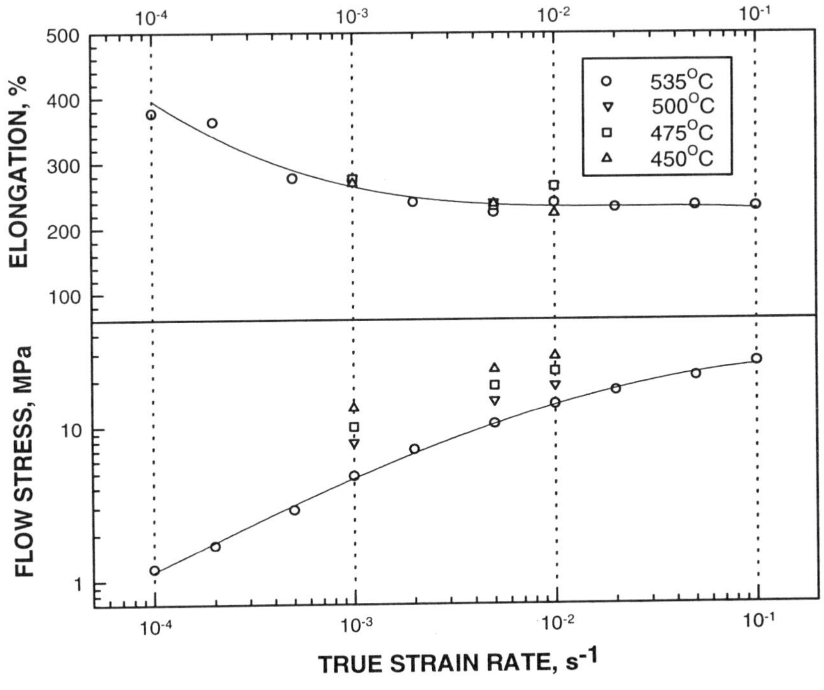

FIGURE 2. **Results of parametric studies to determine the effect of strain rate and temperature on the uniaxial behavior of SKY 5083.**

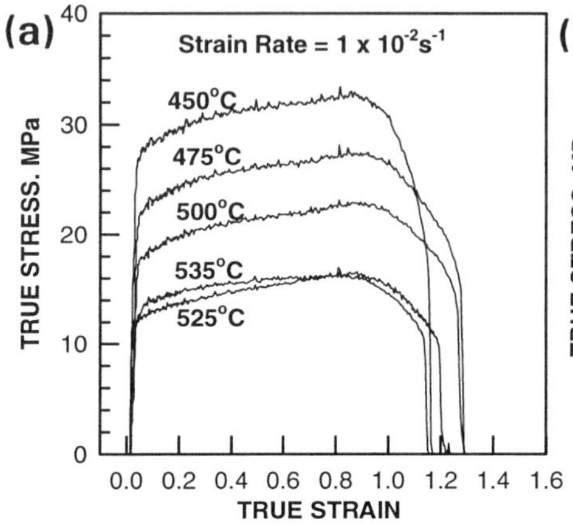

FIGURE 3. Stress-strain curves showing the effect of, (a), temperature at $1 \times 10^{-2}\ s^{-1}$ and, (b), strain rate at 535°C on uniaxial behavior.

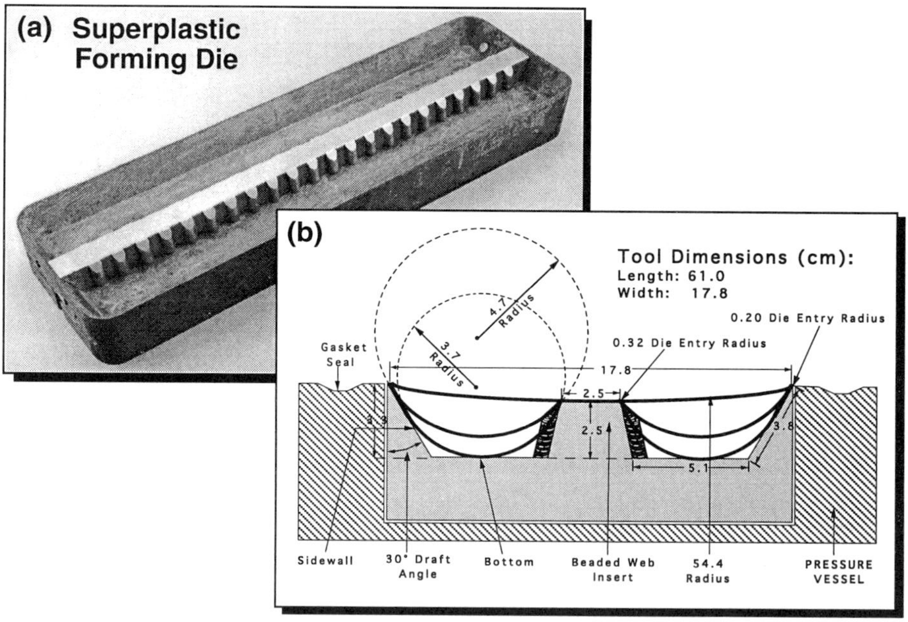

FIGURE 4. Aluminum tooling used in the NASA-LaRC SPF facility; (a), a photograph of the die and, (b), a schematic of the die cross-section.

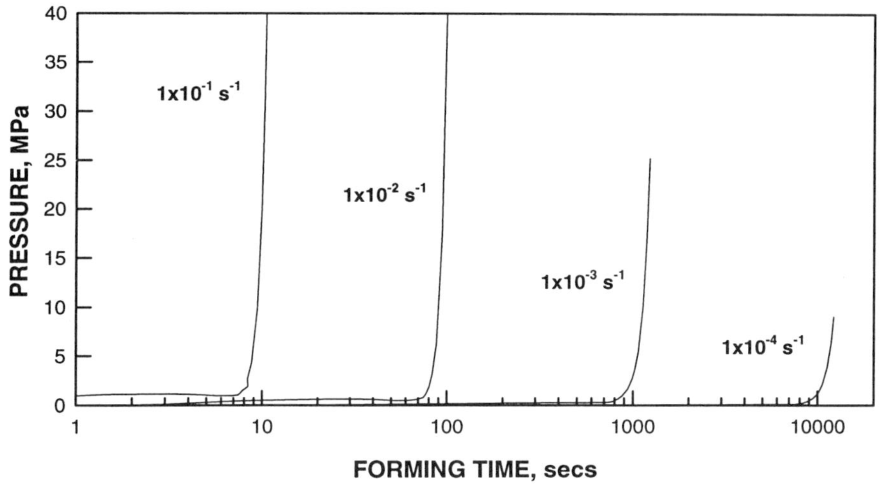

FIGURE 5. A series of theoretical pressure-time profiles for forming a beaded web stiffener panel at 535°C and different constant strain rates.

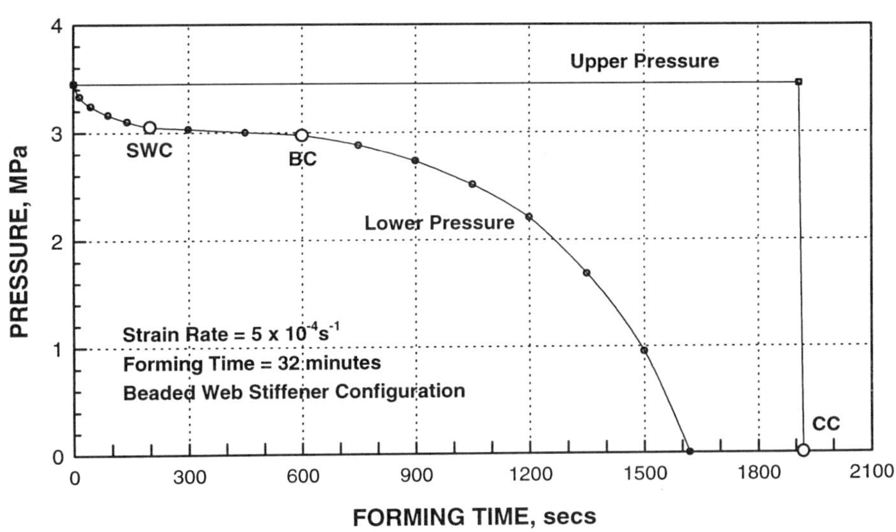

FIGURE 6. The baseline pressure-time profile for superplastic forming of a beaded web panel at 525°C and a constant strain rate close to that for optimum formability.

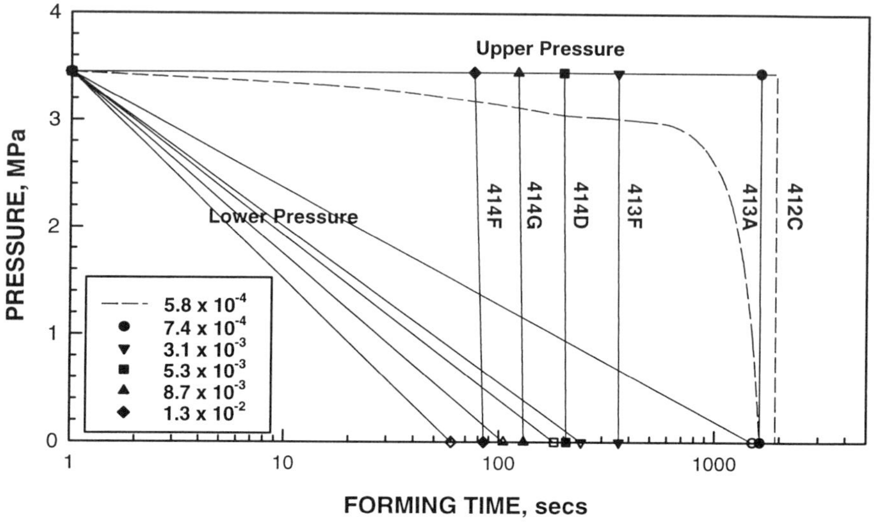

FIGURE 7. The series of pressure-time profiles developed for the biaxial forming trials aimed at reducing cycle time.

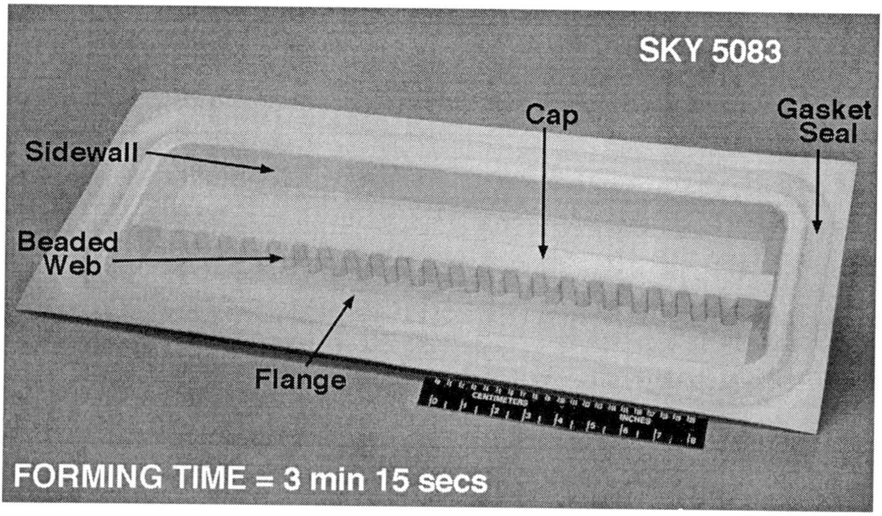

FIGURE 8. Photograph of a SKY 5083 beaded web stiffener panel superplastically formed in 195 seconds.

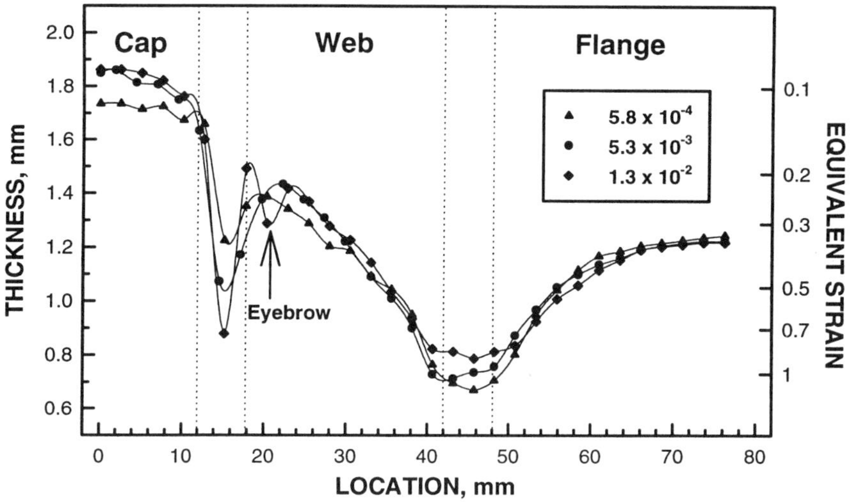

FIGURE 9. The effects of strain rate on the thickness profile of beaded web stiffener panels. Forming was conducted at 525°C using lubrication and variable back pressure.

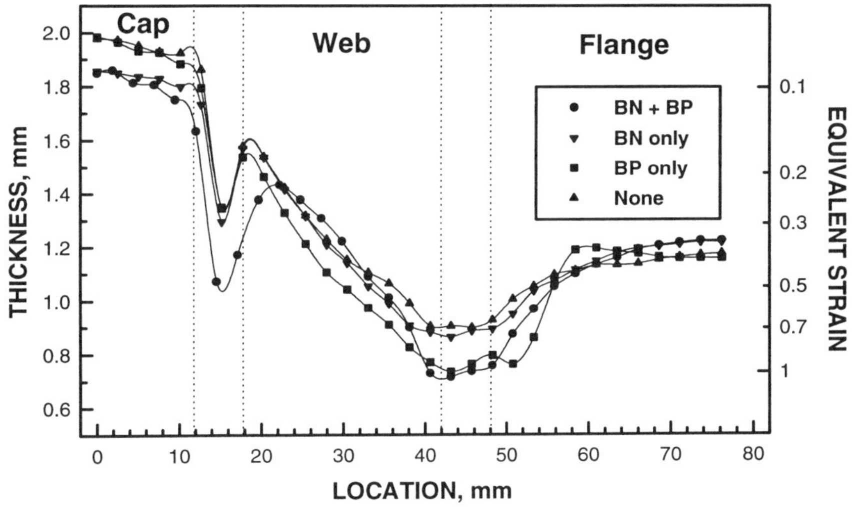

FIGURE 10. The effects of lubrication and variable back pressure on the thickness profile of beaded web stiffener panels. Forming was conducted at 525°C using a strain rate of $5.3 \times 10^{-3} s^{-1}$.

28th International SAMPE Technical Conference
November 4-7, 1996

OPTIMIZATION OF PRESSURE SCHEDULE IN SUPERPLASTIC FORMING

M. G. Zelin, D.K. Ebersole, R.S. Gampala, and P.K. Chaudhury
Concurrent Technologies Corporation, Johnstown, PA 15904

ABSTRACT

Superplastic gas forming is being used increasingly in the aerospace industry. Utilization of all the advantages of superplastic forming requires optimization of the pressure schedule in order to maintain the strain rate in various regions of the deformed component close to the optimum strain rate. In this work, first the existing FEM approaches to predict optimized pressure history were reviewed. Pressure schedules predicted by commercially available general purpose finite element codes, ABAQUS and MARC, were then compared with theoretical predictions and experimental results. Based on this analysis, a new approach is proposed combining the advantages of the methods used in ABAQUS and MARC to calculate the optimum pressure. In this new approach, the pressure is adjusted not only to control the average strain rate in the component, but also to control the maximum strain rate. Thus, the proposed method ensures that forming takes place under optimum superplastic conditions without excessive strains that lead to fracture.

KEY WORDS: Superplasticity, Finite Element Modeling, Pressure Optimization

1. INTRODUCTION

Because of the cost advantages of fabrication unitized, complex parts, superplastic forming (SPF) is being used increasingly in the aerospace industry. During superplastic forming, the pressure must be controlled carefully to maintain optimum strain rate and temperature in the component, and reach the desired thickness and microstructure. Therefore, optimizing the applied gas pressure schedule is critical to successful blow forming of complex shaped components.

The finite element method has proven to be a successful approach to simulating superplastic forming processes. Several finite element codes have implemented superplastic material models and these codes are capable of predicting the deformed shape and strain distribution in the formed components [1 - 3]. However, very few commercial general purpose finite element codes, including MARC and ABAQUS, or codes specifically developed for superplastic forming [4 - 11], are capable of generating an optimum forming pressure schedule.

Pressure schedule optimization is complicated since a number of variables control the SPF process. Consequently, a rigorous approach to forming pressure optimization has not yet been developed. In this paper, a method to calculate the optimum pressure schedule during superplastic blow forming is developed. This new procedure has also been implemented into the SPF subroutines of the MARC finite element code.

2. PRESSURE OPTIMIZATION IN FEM CODES

The two most popular approaches for predicting superplastic forming pressure schedules are: (i) calculating pressure as a solution variable in the equilibrium equations, and (ii) adjusting pressure continuously to maintain the strain rate in the component within the optimum superplastic regime. Both approaches use the optimum superplastic strain rate as the criterion to determine pressure. The target strain rate can either be the maximum strain rate among all elements, or an average of strain rates in the elements.

2.1 Incorporation of Pressure as a Solution Variable Bonet et al. [8, 9] incorporated pressure as a solution variable in the discretized equilibrium equations by augmenting them with an equation constraining the strain rate function to equal a target strain rate value. The strain rate in the constraint equation was either maximum strain rate [8] or an average strain rate for all elements [9]. The maximum strain rate approach has been found to be expedient; however, this approach may cause convergence problems if the element with the maximum strain rate undergoes rapid changes during the iterative solution.

2.2 Adaptive Pressure Change In the second approach, a small pressure, P_0, is initially applied to the sheet. The resulting strain rates in the sheet are calculated by using the standard FEM procedure. Then, the pressure is updated in the next fine increment using the following relation.

$$P = kP_0 \qquad (1)$$

where k is a constant that changes in each increment based on how far the applied pressure is from the optimum pressure. The value of k is determined differently in MARC and ABAQUS.

In MARC, an overall stress is calculated based on the average strain rate experienced by all elements using the relation:

$$\sigma = A\dot{\varepsilon}^m \qquad (2)$$

where A and m are experimentally determined material constrants. This stress value is then compared with a target stress, σ_t, corresponding to the optimum strain rate, $\dot{\varepsilon}_t$, from Eq. (2) to obtain the k value from:

$$k = \frac{\sigma_t}{\sigma} \qquad (3)$$

This k value is used to update the next pressure in Eq. (1).

In ABAQUS, the constant k is treated as a step function [10] and is determined by the ratio of maximum strain rate $\dot{\varepsilon}_{max}$ to the target strain rate $\dot{\varepsilon}_t$.

3. PRESSURE SCHEDULES PREDICTED BY ABAQUS AND MARC

In order to compare the pressure schedules calculated by the ABAQUS and MARC codes, two test cases were considered. Each of the tests was modeled in ABAQUS and MARC with identical geometry and material properties. The first case is the expansion of an infinitely long tube under an internal pressure and the second case is the forming of a rectangular pan. These two cases have theoretical solutions (see Appendix) for comparison of optimum pressure schedules.

3.1 Expansion of an Infinitely Long Tube The finite element mesh consisted of membrane elements. The material properties used were that of a superplastic aluminum alloy, AA 7475. The modeled strain rate distribution in a thin wall tube was fairly uniform. As a result, the pressure calculations of both ABAQUS (based on maximum strain rate) and MARC (based on average strain rate) gave very good predictions. Figure 1 shows the pressure

schedules predicted by ABAQUS and MARC and the pressure schedule obtained from theory (see Appendix). The slight difference between the pressure schedule predicted by MARC and theoretical calculations may be attributed to the presence of strain rate gradient through the wall thickness. While the strain rate gradient across the wall thickness is neglected by theoretical calculations, MARC takes this gradient into account.
The pressure schedule predicted by MARC is smooth, while that predicted by ABAQUS is not. The discontinuities in the pressure vs. time curve predicted by ABAQUS are caused by the parameter k (see Eq. (1)) in the pressure calculation algorithm implemented in ABAQUS. Obviously, this pressure adjustment algorithm used by ABAQUS is not adequate when the strain rate distribution is non-uniform.

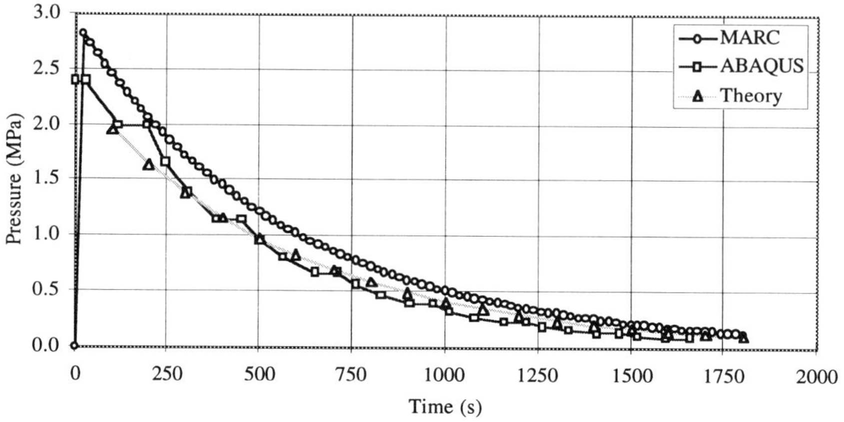

Figure 1. Optimized pressure schedules for expansion of infinitely long tube predicted by MARC, ABAQUS, and theory.

3.2 Forming of a Rectangular Pan An analytical solution for forming a rectangular pan is usually obtained assuming a plane strain state in the deformed sheet (see Appendix). The effect of the entry radius of the die is also neglected, and a solution for this simplified problem is obtained for the first stage of the forming process, which is characterized by free forming of the sheet (before the sheet comes into contact with the die walls). During this stage, the plane strain simulations were conducted in ABAQUS and MARC to be consistent with theoretical predictions. Figure 2 illustrates the geometry of the model and strain rate distribution in the deformed sheet. Initially, the strain rate is found to be highest in the elements bending around the entry radius. However, the location of the highest strain rate element changes with deformation. Figure 3 compares the pressure schedules predicted by ABAQUS, MARC, and theory. The general trends of these curves at the free forming stage are essentially the same. However, the forming schedule predicted by ABAQUS is not smooth. Additionally, in the plane strain case, ABAQUS predicts a much shorter forming time than that obtained theoretically.

A 3-D finite element model [12] showed that the forming time predicted by ABAQUS is much longer than that observed experimentally. This difference in prediction of the forming time by ABAQUS in the plane strain state case and the 3-D case is a result of controlling the pressure based on the maximum strain rate in the component. The forming time predicted by MARC is in good agreement with that found from the experiments [12]. When a majority of the elements in the component deform at a strain rate close to the target strain rate, the strain rate averaging technique yields very good results.

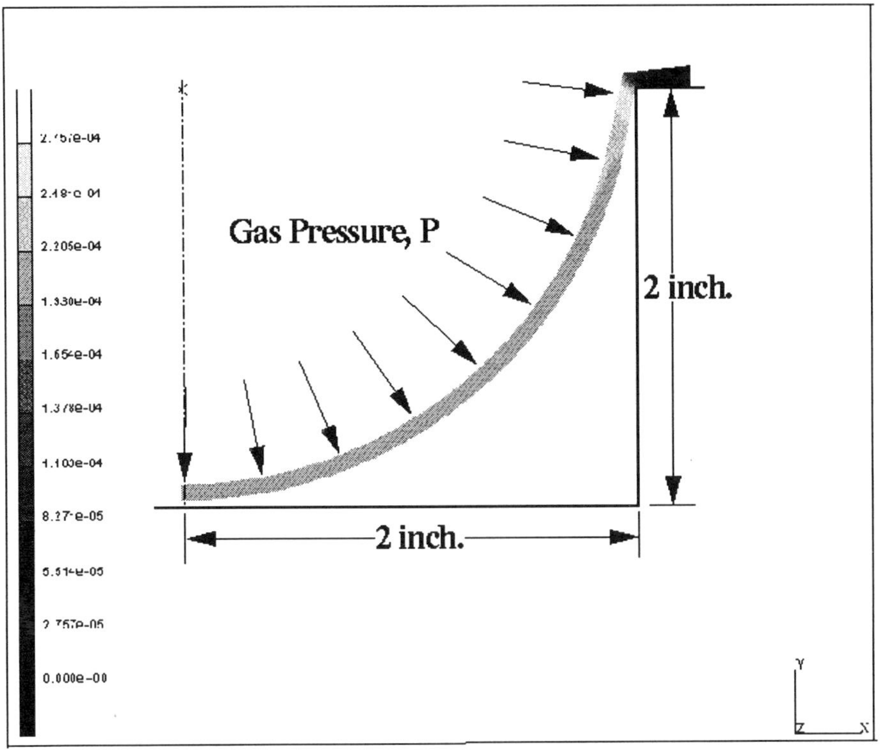

Figure 2. Strain rate distribution in the deformed sheet at the free forming stage predicted by MARC.

Thus, the pressure predictions of MARC are found to be more reliable than those of ABAQUS for the particular examples studied in this work. However, there are certain limitations to the pressure calculation algorithm used in MARC. For example, MARC analysis gives a poor prediction of pressure for forming a component that has a non-homogeneous strain rate distribution. An example is the bi-modal distribution found in forming a double well pan, which is discussed in the following section.

4. PROPOSED PRESSURE OPTIMIZATION PROCEDURE

Figure 4 shows the frequency of strain rates during forming of a double well pan, as predicted by MARC. This component consisted of a narrow well and a wide well of the same depth. The narrow well was two inches deep and four inches wide (Figure 2) and the wide well was two inches deep and eight inches wide. Even though the average strain rate of all the elements is close to the optimum strain rate $\left(2x10^{-4}s^{-1}\right)$, a significant number of elements have strain rates exceeding the optimum strain rate.

Along with the double well pan, forming of a narrow and a wide pan of the same dimensions as the narrow and wide wells were simulated. Figure 5 illustrates the differences in the predicted pressure history for the narrow, wide, and double well pans. The pressure schedule for the double well pan might appear unusual due to the pressure "bump" (see Figure 5). This sudden rise in the pressure prior to the deformed sheet coming into contact with the die may cause excessive strains in the sheet leading to fracture.

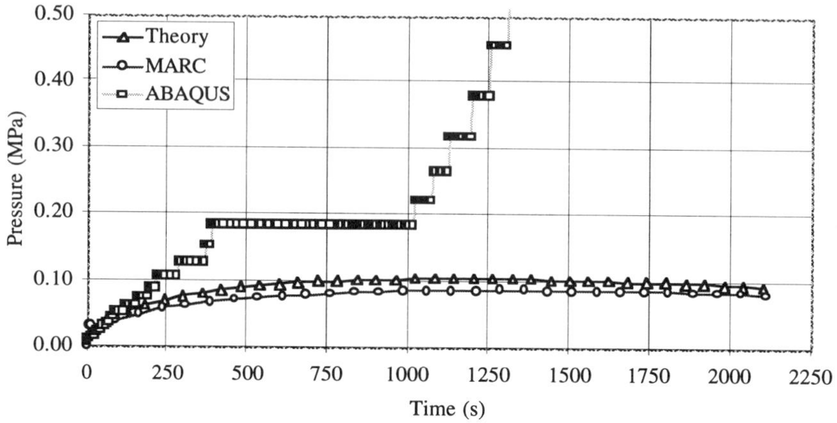

Figure 3. Optimized pressure schedules for gas forming of a rectangular pan predicted by MARC, ABAQUS, and theory.

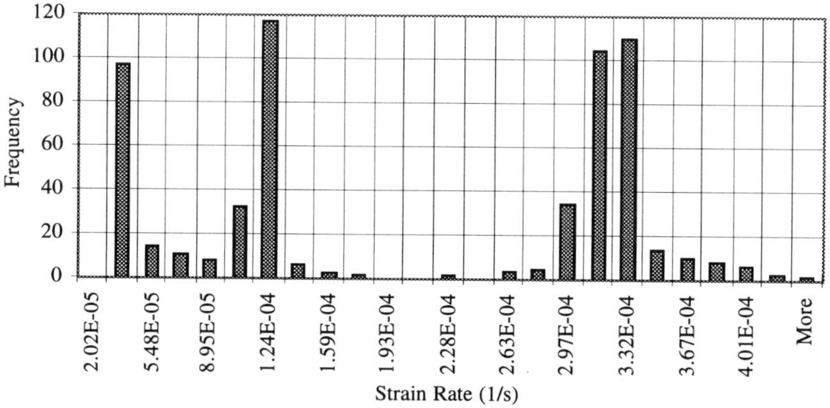

Figure 4. Strain rate frequency distribution predicted by MARC at the free forming stage of a double well pan.

In order to overcome this problem, a combiation of the MARC and ABAQUS approaches is proposed. The pressure schedule calculation in MARC is modified to ensure that the strain rate does not exceed a critical strain rate value, $\dot{\varepsilon}_c$, at which the ductility of the material sharply decreases. This critical strain rate might be defined as:

$$\dot{\varepsilon}_c = n\dot{\varepsilon}_t \tag{4}$$

where $\dot{\varepsilon}_t$ is the target strain rate and n is a constant greater than 1, determined by the material properties.

Pressure schedules predicted by the standard MARC procedure and the modified approach are shown in Figure 6. The pressure schedules are very similar, except that the modified approach predicts a slightly longer forming time. The difference between these two curves increases with increasing difference in the width of the wells. However, the modified approach ensures that the material does not fail by the strain rate exceeding the optimum strain rate.

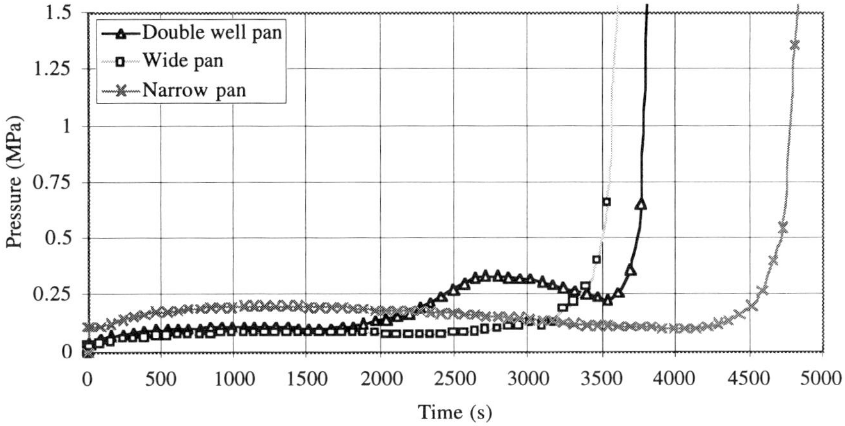

Figure 5. Optimized pressure schedules for forming a wide pan, narrow pan, and their combination (double well pan) predicted by MARC.

5. SUMMARY

Use of all advantages of superplastic forming requires optimization of pressure schedule in order to maintain the strain rate in various regions of the deformed component close to the optimum strain rate.

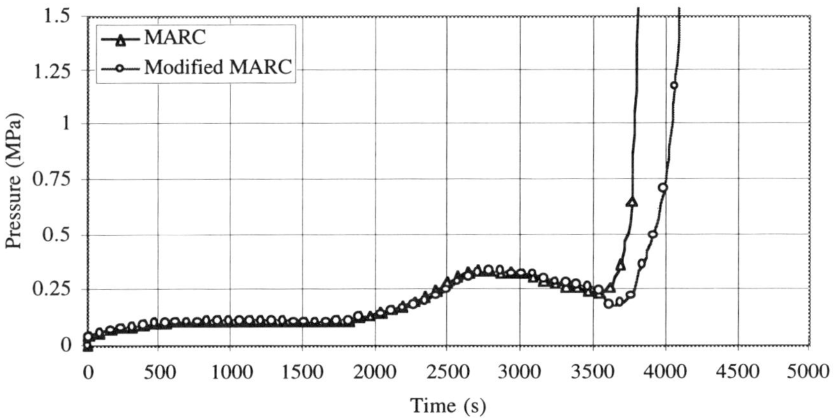

Figure 6. Optimized pressure schedules for forming a double well pan predicted by MARC using the standard algorithm and a modified algorithm.

Existing FEM approaches to predict optimized pressure schedule have been reviewed. Pressure schedules predicted by commercial general purpose finite element codes ABAQUS and MARC have been compared with theoretical predictions and experimental results. The pressure calculation algorithm of ABAQUS is based on maximum strain rate in the component and predicts a very long forming time. It also predicts stepped dependence of pressure on time when the strain rate distribution is significantly non-uniform. The pressure predictions in MARC are based on the average strain rate and are in good agreement with theoretical and experiment results. However, the pressure calculation algorithm of MARC can be unrealistic when the strain rates in the deforming sheet deviate significantly from the optimum superplastic strain rate.

A modified approach to calculate forming pressure has been proposed. In this approach, the pressure is controlled similar to that in MARC, using average strain rate in the component. The proposed approach also makes use of the ABAQUS pressure calculation algorithm, with the pressure being controlled such that the maximum strain rate in the component never exceeds a critical strain rate. This ensures successful forming of the component under conditions close to the optimum superplastic conditions without causing failure in the component.

6. REFERENCES

1. S.S. Gavriushin and O.C. Zienkievicz, Int. J. Num. Meth. Eng., 23, 1179 (1986).
2. W.S. Zhang, R.D. Wood and O.C. Zienkievicz, In: Aluminum Technology '86, The Institute of Metals, London, Book 4, Session C, 1986, pp. 111.1-111.6.
3. N. Chandrasekaran, R.E. Goforth and W.E. Haisler, American Society for Metals, Metals/Materials Technology Series, Materials Week, Paper 8511-004, 1985, pp. 1-8.
4. J.H. Argiris and J. St. Dolstinis, Comp. Methods Appl. Mech. Eng., 30, 83 (1984).
5. J. St Dolstinis, J. Phesique IV, 3, pp. 1187-1197 (1993).
6. M. Billet and J.L. Chenot, Proc. of Sixth World Conference on Titanium, France, 1988, pp. 1203-1208.
7. J.-H. Cheng, Jour. Chinese Soc. Mech. Eng., 13 (4), 289 (1992).
8. J. Bonet, A.H. Wargadipura and R.D. Wood, Comm. Appl. Numer. Methods, 5, 121 (1989).
9. J. Bonet, R.D. Wood and R. Collins, Int. J. Mech. Sci., 36 (4), 297 (1994).
10. MARC Analysis Research, Co., MARC User's Manual, 1994.
11. Hibbitt, Karlsson & Sorensen, Inc., ABAQUS User's Manual, 1994.
12. "Experimental Verification of SPF Simulations" NCEMT Report, 1996.

APPENDIX

Stresses, σ, in a thin shell subjected to the internal pressure, P, are defined as:

$$\sigma = PR/s \tag{A1}$$

where R and s are shell radius and thickness, respectively.

It can be assumed that both expansion of an infinitely long thin tube and sheet forming into a rectangular pan are under plane strain state. In the case of plane strain state:

$$\varepsilon_\theta = -\varepsilon_s \tag{A2}$$

where ε_θ is hoop strain and ε_s is thickness strain.
In both cases of tube expansion and forming a rectangular pan, thickness strain is defined as:

$$\varepsilon_s = -\ln s/s_0 \tag{A3}$$

where s_0 is the initial material thickness and the equivalent strain, ε, is defined as:

$$\varepsilon = \left(2\sqrt{3}\right)\varepsilon_s \tag{A4}$$

For the case of infinitely long tube expansion, hoop strain is defined as:

$$\varepsilon_\theta = \ln R/R_0 \tag{A5}$$

It follows from Eqs. A3-A5:

$$\ln R/R_0 = -\ln s/s_0 = \left(\sqrt{3}/2\right)\dot{\varepsilon}t \tag{A6}$$

where $\dot{\varepsilon}$ is the optimum strain rate and t is time.

Combining Eq. A1 through Eq. A8 and taking into account Eq. 2, one can obtain the following dependence of the optimized pressure on forming time:

$$P = A s_0 \dot{\varepsilon}^m \exp\left(-\sqrt{3}\,\dot{\varepsilon} t\right) / R_0 \tag{A7}$$

For the case of forming a rectangular pan, hoop strain is defined as:

$$\varepsilon_\theta = \ln\left(R\theta / a\right) \tag{A8}$$

where θ is the angle shown in Fig. 1A and a is one half of the pan's width.

It follows from the geometry of the deformed sheet that:

$$R = a / \sin\theta \tag{A9}$$

and the hoop strain could be defined as:

$$\varepsilon_h = \ln\left(\theta / \sin\theta\right) \tag{A10}$$

Taken into account Eqs. A4 and A10, the following expression could be obtained:

$$\theta / \sin\theta = \exp\left(\sqrt{3} / 2\dot{\varepsilon} t\right) \tag{A11}$$

Combining Eqs. A1, A4, and A11, the following expression for P could be obtained:

$$P = 2/\sqrt{3} A s_0 \theta \dot{\varepsilon}^m \exp\left(-\sqrt{3}\,\dot{\varepsilon} t\right) / a \tag{A12}$$

where time dependence of θ can be obtained from Eq. 6.

An approximate solution of Eq. A11 can be found if sin θ is expanded in series:

$$\sin\theta = \theta - \theta^3 / 6 \tag{A13}$$

Then,

$$\theta = \left(6\left(1 - \exp\left(-\sqrt{3}\,\dot{\varepsilon} t\right)\right)\right)^{1/2} \tag{A14}$$

Substituting Eqs. A14 into A12, the following equation for the stage of free forming of a rectangular pan can be obtained:

$$P = 2/\sqrt{3} A \dot{\varepsilon}^m s_0 \left(6\left(1 - \exp\left(-\sqrt{3}\,\dot{\varepsilon} t\right)\right)\right)^{1/2} \exp\left(-\sqrt{3}\dot{\varepsilon} t\right) / a \tag{A15}$$

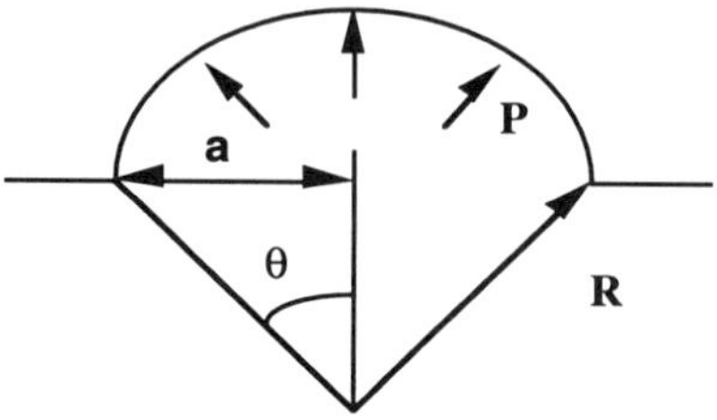

Figure A1. Schematic illustration of the forming of the rectangular pan at the stage of free forming.

ACKNOWLEDGEMENT

This work was conducted by the National Center for Excellence in Metalworking Technology, operated by Concurrent Technologies Corporation under contract No. N00140-92-C-BC49 to the U.S. Navy as part of the U.S. Navy Manufacturing Technology Program.

28th International SAMPE Technical Conference
November 4-7, 1996

DEVELOPMENTS IN LOW-COST CERAMIC TOOLING FOR AEROSPACE MATERIALS

John A. Griggs
Ceradyne Thermo Materials

ABSTRACT

The evolution and specific employment methodology of castable ceramic tooling materials as a direct replacement, in appropriate situations, for high strength steel tooling is reviewed. The thirty-year history of fused silica based, castable ceramic tooling systems used in the forming of such aerospace specific materials as titanium, aluminum and carbon based composite materials is briefly discussed. A detailed review of parameters which might suggest ceramic tooling as an economical alternative to high cost metal tooling is presented. The details of design, fabrication and employment of a typical ceramic tool are discussed.

KEY WORDS: Tooling, Ceramic Materials, Fused Silica

1. INTRODUCTION

The high temperature forming of titanium, and more recently aluminum and carbon based composite materials has resulted in the development of several technologies now in common use throughout the aerospace industry. Superplastic forming of titanium and aluminum and the forming of new carbon based composite materials are common throughout the industry today. Each of these technologies requires the used of tooling materials capable of withstanding repeated excursions to relatively high temperatures. The most common materials used in these applications are usually high alloy steels. These materials are expensive, require long lead times and extensive machining.

The use of fused silica based ceramic castable tools is, in many cases, an economical alternative to commonly utilized metal tools. Ceramic tools are inexpensive to produce and, in most cases, can be fabricated and placed in production within a fraction of the time required by the procurement and processing of high alloy steel tools.

2. HISTORY

The first large scale use of ceramic tooling was the self-heated and cooled, ceramic brazing fixtures utilized by Rockwell International to produce stainless steel honeycomb panels and other structural members for the B-70 bomber program in the early 1960's. Ceramic brazing fixtures as large as 7 m X 10 m were required as furnace heated metal tooling was impractical.

A ceramic tooling research project funded by the U.S. Air Force was awarded to the Lockheed-Georgia Co. In the late 1950's. The final report entitled "Non-Metallic Tooling for High Temperature Applications" was issued in 1961. The report confirmed the most promising ceramic material to be a hydraulically bonded castable utilizing fused silica as the primary constituent.

Another early application for fused silica based, ceramic tooling was the Boeing Company's development of the supersonic transport aircraft in the early 1960's. This was to be a one of a kind aircraft requiring extensive use of shaped titanium. High alloy steel tooling required for one or two, unique shapes would have been impracticable.

Initially, fused silica ceramic materials were employed in two basic forms: fabricated modular tools employing a very insulating foamed version of fused silica with a durable, high density working face and a castable material comprised of closely sized fused silica grains with a calcium aluminate binder. These materials were used extensively as direct replacements for alloy steel tools in the forming of titanium.

The advent of new carbon based composite materials resulted in curing temperatures significantly higher than those previously used. The standard plastic, fiberglass or other materials with low temperature capability were no longer sufficient to permit post curing at temperatures sometimes exceeding 370° C. The fused silica based ceramic castables were quite capable of these temperatures as well and could satisfy surface finish and accuracy requirements. In addition to these capabilities, new requirements imposed were a sealed surface to allow vacuum bagging, and since a great deal of the forming operations are accomplished in autoclaves, tools needed to be low mass. These problems were addressed with high temperature sealing materials and re-designing ceramic tools with the aim of greatly reducing the mass of ceramic.

Continued evolution of the castable eventually led to the current hybrid castable ceramic which employs a combination of fused silica, silicon carbide and calcium aluminate hydraulic cement. These castable formulations continue to be refined to enhance casting properties, improve tool strengths and prolong tool life.

3. FLEXIBILITY

3.1 Internally Heated Tools: The fabrication of ceramic tools allows the user to employ several options not normally available with steel tools. One of the most important benefits of modular or castable ceramic material is the capability to employ integrated, conformal heating systems eliminating the requirement for expensive heated presses. Perhaps the most common, and economical method of fabricating integrally heated ceramic tools is found in two common methods of titanium forming: heated, matched dies and vacuum creep forming. This capability is also applicable to composite forming operations requiring vacuum bagging or matched die systems.

3.2 Very Large Tools: Castable fused silica ceramic tools as large as 4.1 m x 4.1 m are in current use. One of the most important properties of fused silica is its extremely low coefficient of thermal expansion, 0.6 X 10^{-6} CM/CM/°C. Since fused silica remains essentially dimensionally stable throughout temperature ranges from ambient to forming temperatures, repeated excursions to forming temperatures do not adversely affect the tool. Another benefit is that tool geometry can be calculated and maintained with high accuracy. These tools are fabricated employing standard techniques used to fabricate tools only a fraction of their size. Most large tools are designed with integral, conformal heating elements eliminating the requirement for outsized, heated presses.

4. PROPERTIES OF FUSED SILICA CERAMIC CASTABLE

Perhaps the most beneficial property of fused silica is the aforementioned low coefficient of thermal expansion. Fused silica is an excellent insulator with a "K" factor of 5.0 BTU/hr./Ft.^{2}/In./°F at titanium forming temperatures. This property is an advantage in integrally heated, stand-alone or matched die tools allowing cooler external surfaces. Appropriate location of heating elements in proximity to the working surface of the tool allow efficient heat transfer to that surface allowing reasonable heating of the tool face.

The material maintains acceptable strength throughout operational temperature ranges. In the case of rebonded fused silica castings used as internally heated press platens, the material is designed to withstand crushing loads as high as 703 KG/CM^2. At temperatures of up to 1,100° C. Castable forming tools also demonstrate relatively high resistance to crushing loads of 350 - 500 KG/CM^2.

Tooling grade fused silica based castables are designed to allow extremely accurate replication of pattern geometry and surface finish. Most applications call for matte finishes which, when combined with standard parting agents like boron nitride, produce extremely high quality forming surfaces.

5. TOOL FABRICATION

5.1 Equipment and Mold Requirements: Tooling grade fused silica based castables are, by definition, refractory concretes. The methods used to fabricate high quality tools are quite similar to those employed in the concrete industry. A standard paddle type, or other high shear mixer is required. A vibrating table is necessary to provide sufficient agitation of the material during material placement. Drying and curing facilities are also desirable.

The primary objective is to provide an excellent tool surface with sufficient strength to withstand the physical requirements of the forming process. The mold should be designed such that the critical surface is as near to the horizontal plane as practicable. This will allow entrapped air to rise away from the surface as the castable material is placed in the mold.

The mold must be impervious to water. Radii, chamfers or other blending methods should be used in place of angles or corners. Castable mass reduction can be accomplished by the incorporation of the cores in areas where castable mass is to be reduced.

If conformal passages for heating elements are to be employed, appropriate templates, element guides and spacers must be accommodated in the basic mold design. Standard waxes and water resistant petroleum parting agents are used to part the castable ceramic from the mold.

Most tools will be exposed to some form of compressive loading. It is necessary that the back side of the tool be as flat as practicable to avoid bending loading which could cause premature failure of the tool. This feature can be accommodated by screeding the recently placed wet material, or in some instances, the post cured tool may be ground flat and re-dried prior to forming operations.

5.2 Casting and Placing the Castable: To maximize strength and minimize shrinkage, ceramic castable should be mixed using the least amount of water required to hydrate the cement and mobilize the mix. Properly proportioned and mixed castable is quite viscous and does not flow without the aid of vibratory casting techniques. Vibration is continued only briefly after final material placement. Since the castable is a chemically setting material, the casting should be covered with plastic or other material to inhibit the escape of water for the initial 24 hours.

5.3 Drying and Curing: Subsequent to removal from the mold, the tool should be inspected for imperfections on the working surface. Patching, designed specifically to match the castable materials are employed to resolve "Bubble" voids which may have been caused by incomplete vibration. Inserts or mass reduction materials should be removed immediately upon extraction of the tool from the mold.

The ceramic tool should be exposed to ambient drying conditions for 24 to 48 hours depending upon mass. The primary object is to remove the remaining mechanical water from the tool so slightly elevated temperatures combined with circulated air should be employed. Post curing steps which eliminate the water of hydration dictate that the tool should be taken to the eventual operational temperature at a moderate heating rate and stabilized for approximately 2 to 4 hours prior to cooling at any convenient rate.

If the tool is to be employed in the forming of metal, a seasoning coat of high purity boron nitride should be manually applied to the working surface of the ceramic tool. This serves to fill the exposed surface pores, which are no larger than a few microns, with refractory lubricant in preparation for the routine application of standard boron nitride slurries normally employed as hot die lubricants during the forming process.

If the tool is to be employed to form composites, a high temperature sealing coat is applied during the initial curing stages. Final coats of sealant are applied with spray painting techniques to achieve the highest quality tool surface finish. These sealants are then co-cured with the ceramic castable. Standard parting techniques normally employed with the specific composite to be formed are used on the sealed surface.

5.4 Ceramic Tool Practices: Shape geometries with relatively shallow elevations and generous radii are best suited to ceramic tools. Geometries with severe elevations or very sharp radii do not lend themselves to ceramic tooling if long production runs are anticipated. These guidelines are imprecise because virtually all shapes are different in scale and geometry. Empirical guidelines are developed in early prototype operations. The experiences gained in the initial stages usually dictate the limits of the ceramic tool design.

Ceramic tools, by their nature, are fragile in comparison to metal tools. Precautions must be taken to avoid undo physical abuse of the tools. Experience has indicated that hot die lubricants such as boron nitride should be used conscientiously and, in many cases more often than would normally be practiced with metal tools.

Superplastic operations are also limited in that ceramic tools with large, concave cavities are subjected to tensile stress. Ceramic materials do not exhibit high tensile strengths. Additionally, superplastic operations require a seal at the periphery of the tool against which the titanium or aluminum is clamped during forming operations. In most cases the ceramic is accommodated by clamping the forming sheet to a steel tool box containing the ceramic tool. Recent research and development is directed at designing sealing systems which allow the ceramic tool to be used on a stand-alone basis.

PROPERTIES
OF A TYPICAL
FUSED SILICA BASED TOOLING CASTABLE

Max. Service Temp. °C	1,100
Bulk Density Gms./cc	2.12
Absorption %	10.5
Linear Shrinkage	
Out of Mold, %	0.00
After 260°C, %	0.15
After 1,093°C, %	0.03
Compressive Strength	
After 260°C, Pascal X 10^6	48.3
After 1,093°C, Pascal X 10^6	51.7
Modulus of Rupture	
260°C, Pascal X 10^6	10.3
Coefficient of Thermal Expansion	
260°C, cm/cm/°Cx10^{-6}	1.6
1,093°C, cm/cm/°Cx10^{-6}	1.8
Thermal Conductivity	
at 870°C, Watt/(Meter x °C)	0.8

28th International SAMPE Technical Conference
November 4-7, 1996

FABRICATION OF LASER BEAM WELDED SUPERPLASTICALLY FORMED MULTI-SHEET STRUCTURE USING ADVANCED TITANIUM ALLOYS

Jeffrey D. Will and Matthew G. Kistner
The Boeing Company
P.O. Box 3999
Seattle, Washington 98124

ABSTRACT

Two methods for fabricating titanium sandwich structures using advanced titanium alloys (Titanium-6Aluminum-2Tin-2Zirconium-2Molybdenum-2Chromium, Titanium SP-700 and Timetal Beta 21S.) will be presented and discussed. The methods use laser beam welding (LBW) in conjunction with superplastic forming (SPF) to produce integrally stiffened panels. This paper includes a discussion of (a) LBW development, (b) heat treatment results, (c) forming results and (d) preliminary mechanical test data. This work demonstrated the ability to use various combinations of advanced titanium alloys to fabricate multi-sheet SPF structure.

KEY WORDS: Superplastic Forming, Laser Welding, Sandwich Construction

1. INTRODUCTION

Future high speed aircraft will require low-cost, light-weight sandwich structure for wing skin and fuselage applications. One method of producing complex titanium sandwich structure is multi-sheet superplastic forming (SPF). The ability to produce a variety of multi-sheet SPF structure using Titanium-6Aluminum-4Vanadium (Ti-6-4) has been established (1-4). However, to increase performance, these structures should incorporate improved alloys such as Titanium-6Aluminum-2Tin-2Zirconium-2Molybdenum-2Chromium (Ti-6-2-2-2), SP-700 and Timetal Beta 21S. These alloys possess much better mechanical properties than Ti-6-4 (5-9). Thus, it was the objective of this study to develop and demonstrate the application of these advanced titanium alloys to the multi-sheet SPF process.

Two multi-sheet SPF processes were selected for development; three-sheet laser beam welding/SPF (LBW/SPF) and four-sheet LBW/SPF/diffusion bonding (LW/SPF/DB). These processes have unique advantages as described in section 2.

2. BACKGROUND

Multi-sheet SPF processes typically involve welding or diffusion bonding of titanium sheets in an appropriate pattern. The "pack" of sheets is then superplastically expanded at elevated temperature into a variety of configurations. These structures have the biggest benefit when producing complex geometries for high temperature applications (greater than 300° C). Multi-sheet SPF is typically much lower in cost and weight than machined or built-up structures. Cost savings of 50% and weight savings of 30% have been achieved consistently (10).

2.1 Three-sheet LBW/SPF The three-sheet process is a low-cost method for fabricating multi-sheet structure. LBW/SPF is accomplished by first partial penetration laser welding through two sheets of a multiple-sheet stack of material in an appropriate pattern as shown in figure 1. The weld penetration depth is maintained with a very high degree of accuracy (+/-0.05 mm). The "pack" is then superplastically expanded using inert gas in a heated die. The resulting structure can assume very complex tool geometries.

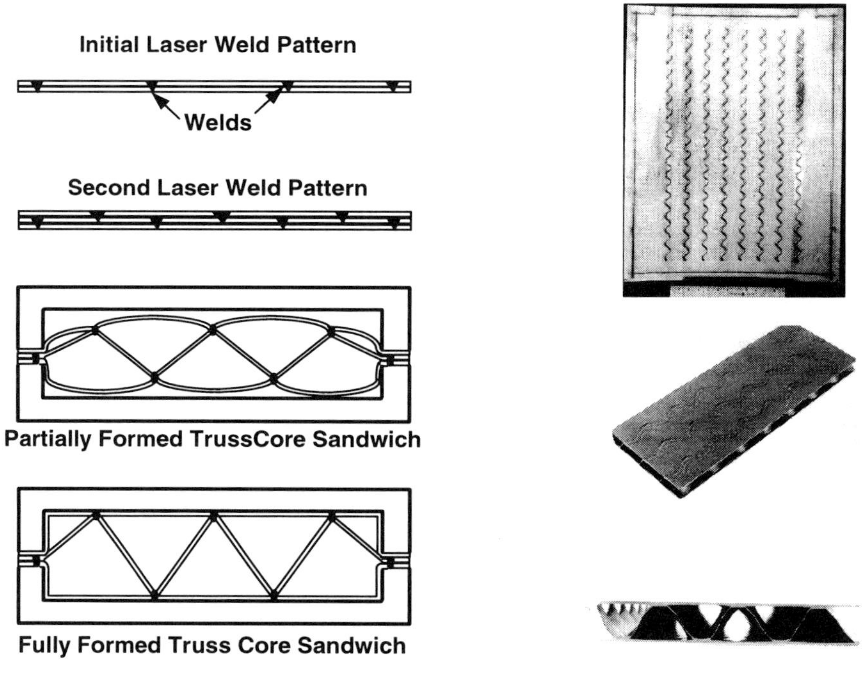

Figure 1- Three-Sheet LBW/SPF

The LBW/SPF process is particularly advantageous because DB is not required. This enables the use of materials that are difficult to DB, such as high strength Beta 21S. Additionally, the manufacturing difficulties, long processing time and inspection concerns associated with DB are eliminated. LBW/SPF processes are simple compared with alternative multi-sheet processes. The process requires only one pressure zone and very little superplastic elongation (typically 70%).

2.2 Four-sheet LBW/SPF/DB In the four-sheet process; first two "core-sheets" are welded in a pattern. These sheets are then placed in an envelope of two external "face-sheets". Two pressure zones are then used to expand the structure at a controlled strain rate as shown in figure 2.

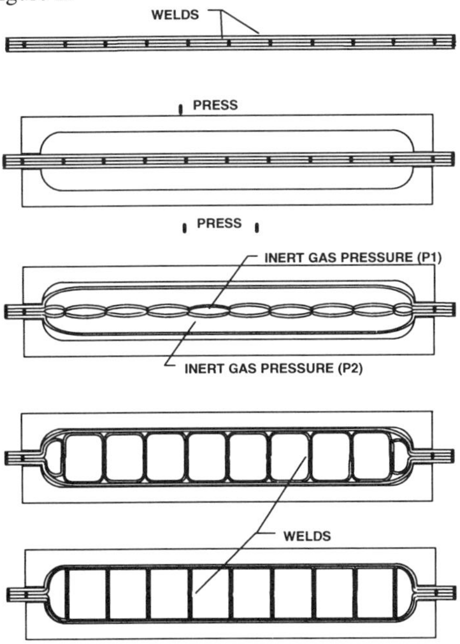

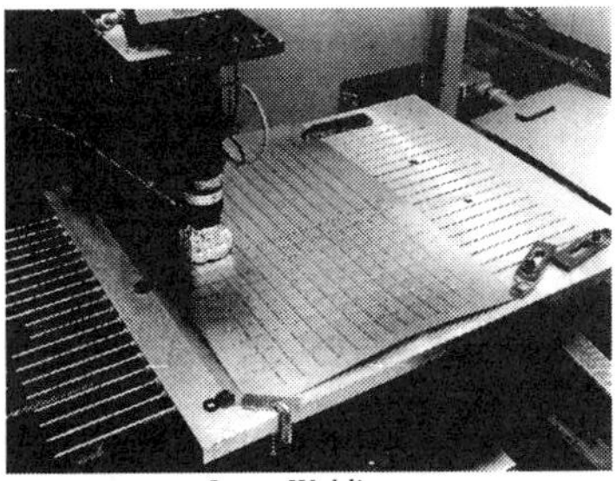

Laser Welding

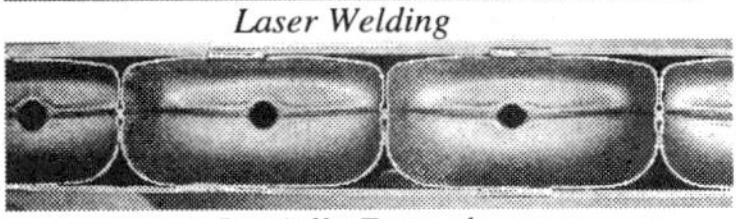

Partially Formed

Fully Formed

Figure 2- Four-Sheet LBW/SPF/DB

Techniques have been developed for applying LBW to four-sheet structure. LBW offers several advantages over conventional resistance welding in this application including the following:

1) LBW is easily computer numerically controlled (CNC) and is much more accurate and repeatable than the conventional manual feed resistance welding process.
2) LBW can produce very complex weld geometries, resulting in greater design freedom (e.g., sine wave rib patterns).
3) LBW can produce narrower welds than resistance seam welding. This results in much less localized thinning of the core webs as shown in figure 3.

(a) (b)

Figure 3- (a) Resistance Welded Core Structure, (b) Laser Beam Welded Core Structure

3. EXPERIMENTAL

Experiments were conducted to develop the processes necessary to fabricate advanced titanium alloy test panels. The fabrication of more conventional Ti-6-4 panels was also planned as a baseline for comparison purposes. The design of the advanced titanium alloy panels (figure 4) utilized the higher strength alloys (Beta 21S and Ti-6-2-2-2-2) in the face-sheet and used the high superplastic elongation alloy (SP-700) in the core-sheet.

3-Sheet LBW/SPF Test Panels

Panel Design	Face-sheets	Core-Sheets
All Ti-6-4	1.6 mm Ti-6-4	0.78 mm Ti 6-4
Ti-6-2-2-2-2/ SP-700 hybrid	1.6 mm Ti-6-2-2-2-2	0.78 mm SP-700
Beta 21S/ SP-700 hybrid	1.5 mm Beta 21S	0.78 mm SP-700

4-Sheet LBW/SPF/DB Test Panels

Panel Design	Face-sheets	Core-Sheets
All Ti-6-4	1.6 mm Ti-6-4	0.78 mm Ti 6-4
Ti-6-2-2-2-2/ SP-700 hybrid	1.6 mm Ti-6-2-2-2	0.78 mm SP-700

* Beta 21S was not used in the four-sheet structures, based on the reported difficulty in diffusion bonding the alloy.

Figure 4- Multi-sheet panel designs

Fabrication of these hybrid panels first required the development of LBW parameters and suitable heat treatment procedures. Forming trails were then conducted and specimens subjected to mechanical testing.

3.1 Laser Beam Welding Development LBW development was completed on Ti-6-4 (as a baseline), Ti-6-22-22, SP-700, and Timetal Beta 21S. The first three alloys were cleaned prior to welding using a standard Nitric-Hydrofluoric acid pickle. The Beta 21S was cleaned by hand wiping with a reagent grade solvent. Welding was accomplished using a 1500 watt carbon dioxide laser. A 6.35 cm focal length focusing optic was used for all of the welding experiments and test panels. The top and underside surfaces of the weld pool were shielded during welding by a cover of grade-5 helium and argon gas respectively.

Bead-on-plate welding development was conducted to provide tensile specimens for mechanical testing. These tests were used to evaluate the effect of laser processing on each titanium alloys. Partial penetration LBW development was then conducted for application to three-sheet LBW/SPF. Full penetration seam welding development was done for application to four-sheet LBW/SPF/DB.

3.1.1 Full Penetration Bead-on-Plate Welding Development Four-inch by eight inch strips were used for bead-on-plate welding experiments. The materials were clamped on an aluminum weld tool that provided backside shielding gas through a 6 mm wide slot machined along the axis of the weld. Several weldments were produced with various laser power and travel speed settings. Parameters were selected by examination of metallographic cross sections and visual examination of the top bead and under bead. The parameters chosen for the bead-on-plate weld tensile specimens are listed in figure 5.

Material	Laser Power	Travel Speed
1.6 mm Ti-6-4	850 watts	1.4 m/min.
1.6 mm Ti 6-22-22	850 watts	1.5 m/min.
1.5 mm Beta 21S	800 watts	1.5 m/min.
0.078 mm SP-700	500 watts	2.25 m/min.

Figure 5- Bead-on-Plate Welding Parameters

3.1.2 Partial Penetration Welding Development Four inch by eight inch strips of titanium were used for the partial penetration welding experiments. The sheets were clamped on an aluminum welding tool. Several laser power and weld travel speed combinations were examined for each of the different titanium alloy combinations.

Cross-sections of all the test welds were examined under an optical microscope. Selection of parameters for panel fabrication was made based on depth of penetration and weld width at the interface of the welded sheets. Figures 6a-c show the cross section of the selected welds for each of the alloy combinations. Figure 7 lists the selected parameters.

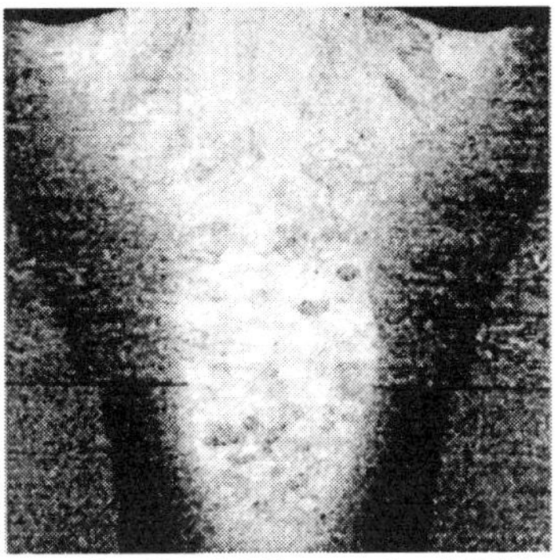

Figure 6a- Partial Penetration LBW Photomicrograph; 1.6 mm Ti-6-4 to 0.78 mm Ti-6-4.

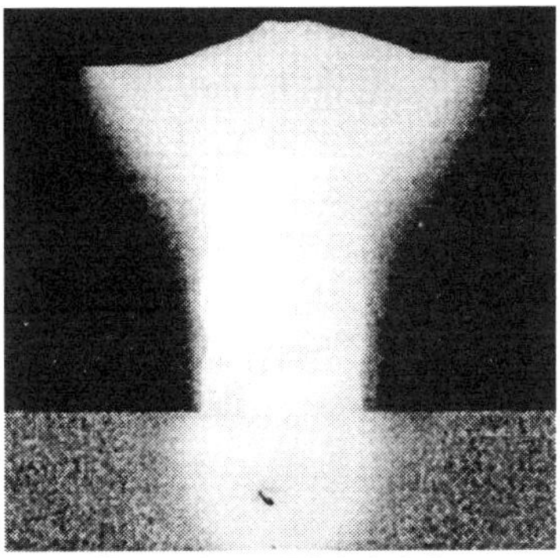

Figure 6b- 1.6 mm Ti-2-2-2-2 to 0.78 mm SP-700.

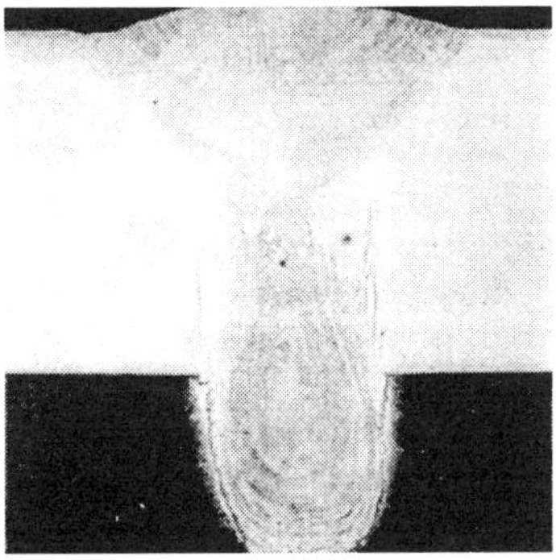

Figure 6c- 1.5 mm Beta 21S to 0.78 mm SP-700

Top Sheet	Bottom Sheet	Laser Power	Travel Speed
1.6 mm Ti-6-4	0.78 mm Ti-6-4	1100 Watts	2.0 m/min.
1.6 mm Ti-6-2-2-2-2	0.78 mm SP-700	1100 Watts	2.125 m/min.
1.5 mm Beta 21S	0.78 mm SP-700	1000 Watts	2.0 m/min.

Figure 7- Partial Penetration Welding Parameters

These weldments demonstrated a typical triangular shaped fusion zone with a slight amount of undercut at the top surface.

3.1.3 Full Penetration Seam Welding Development Full penetration seam welding development was conducted using the same fixtures and welding procedures used for the bead-on plate experiments. Cross-sections of all the test welds were examined under an optical microscope. Figures 8a-b show the cross section of the selected welds for the two the alloys. Figure 9 lists the parameters.

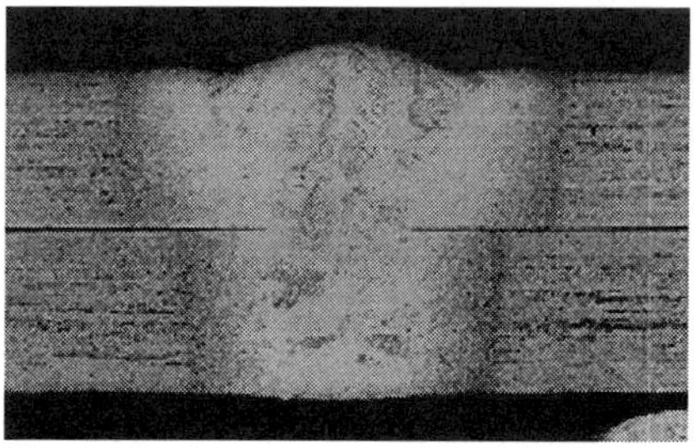

Figure 8a- Full Penetration Laser Seam Weld Photomicrograph,
0.78 mm Ti-6-4 to 0.78 mm Ti-6-4

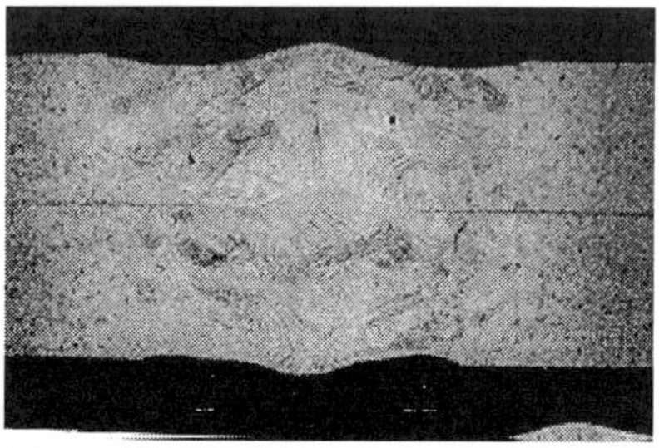

Figure 8b- Full Penetration Laser Seam Weld Photomicrograph,
0.78 mm SP-700 to 0.78 mm SP-700

Material	**Laser Power**	**Travel Speed**
0.78 mm SP-700	700 watts	1.75 m/min.
0.78 mm Ti-6-4	700 watts	1.75 m/min.

Figure 9- Laser Seam Welding Parameters

3.2 Heat Treatment Screening Tests The use of two-alloy hybrid structures requires some compromise of the optimum heat treatment for each alloy. Although the performance of the skin alloy is typically the most important; post-SPF heat treatments must consider the effects on both alloys. These heat treatments will limited by the thermal history associated with SPF (i.e., the forming time and temperature of the core alloy and multi-sheet SPF cooling rates). Ideally, post-SPF heat treatments will be limited to low temperature aging. This will minimize cost and eliminate the risk of possible distortion during high temperature solution treatment.

3.2.1 Heat Treatment Screening Matrix The effect of SPF thermal processing and several post-SPF heat treatments on room temperature tensile strength was evaluated to determine a suitable heat treatment for the hybrid multi-sheet panels. Figure 10 shows the tensile results for the heat treatments evaluated.

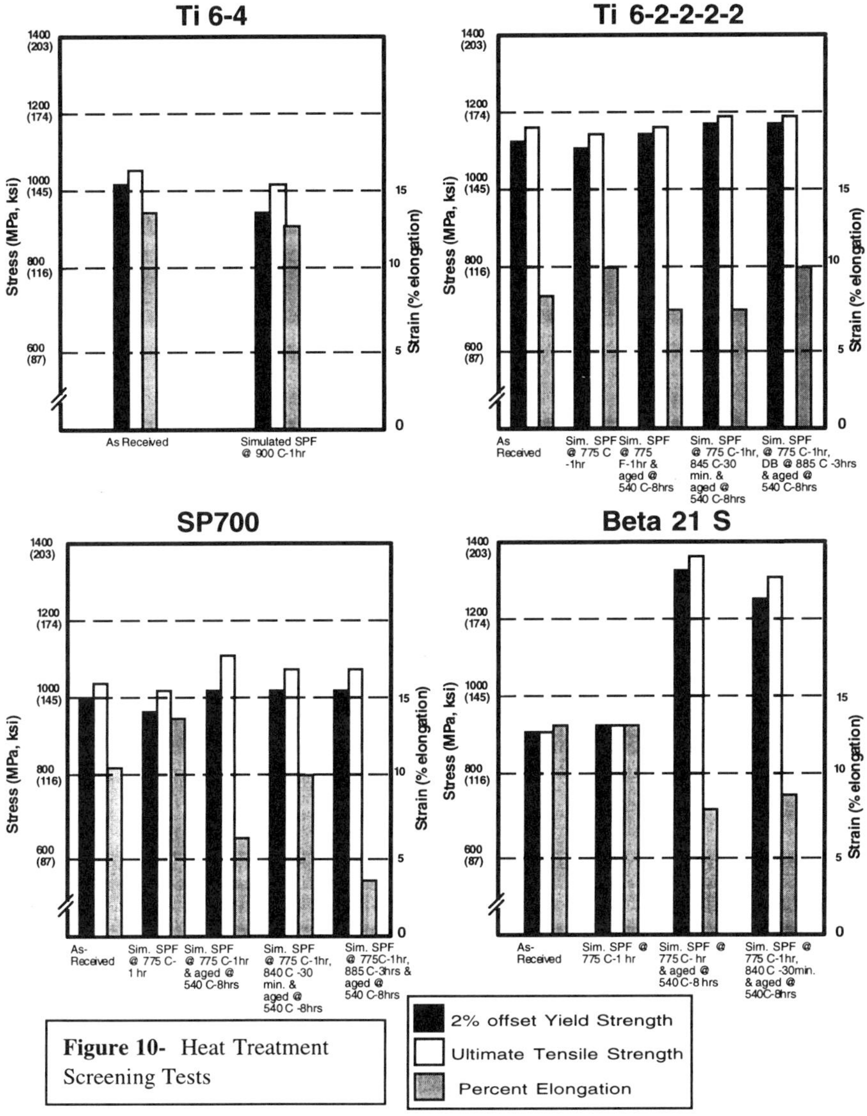

Figure 10- Heat Treatment Screening Tests

3.2.2 Microhardness Analysis A microhardness survey was conducted to determine the relative strength of the weld metal (WM), heat affected zone (HAZ) and base metal (BM) of the four titanium alloys.

The microhardness tests were performed with a Vickers diamond indenter at the mid-thickness of each material. The hardness survey was conducted from the center of the WM through the HAZ and into unaffected BM. The indenter measurements were converted to Rockwell C hardness numbers.

Three separate surveys were conducted in the four alloys after each of the following heat treat conditions: 1) as welded condition, 2) as SPF'd condition and 3) as SPF'd and aged condition. Figure 11 shows the hardness plots of the each of the alloys evaluated.

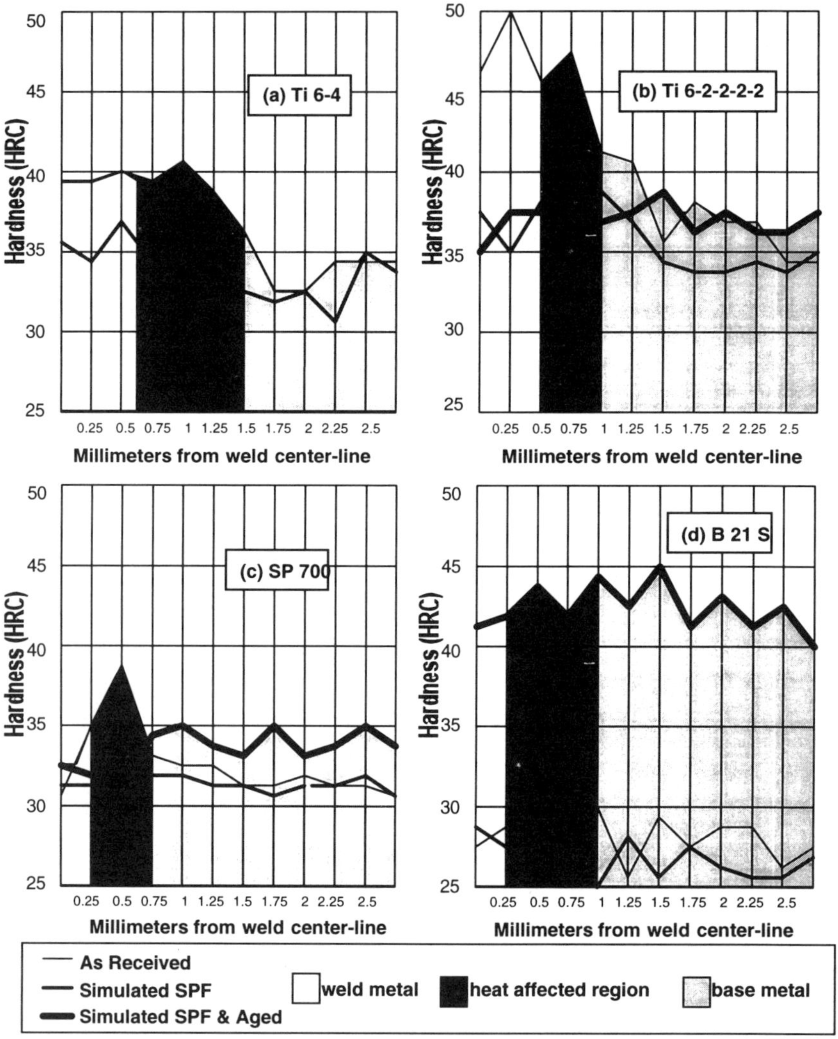

Figure 11- Hardness Survey: (a) Ti-6-4, (b) Ti-6-22-22, (c) SP-700, (d) Beta 21

3.3 Forming Trials Two panels for each of the five panel configurations described previously in figure 4 were fabricated (total of 10). The panels were formed using a 60 cm x 90 cm x 2.5 cm die.

3.3.1 Three-Sheet LBW/SPF A relatively simple 3-sheet design was selected as shown in figure 12. The design used 1.57 mm face-sheets and 0.78 mm core sheets. The total elongation requirement for this design is approximately 70 %.

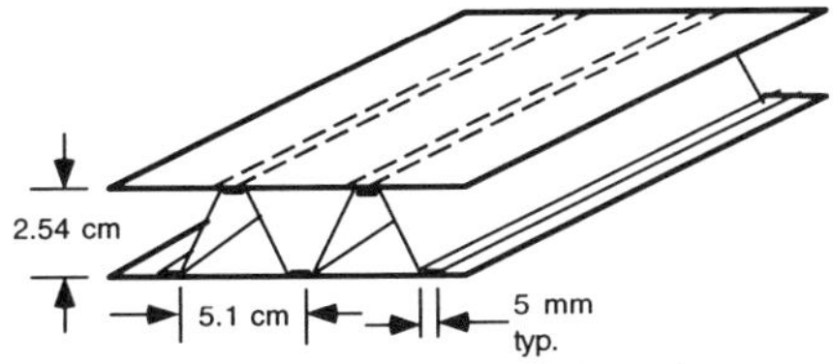

Figure 12- 3-Sheet Panel Design

3.3.1.1 Three-Sheet LBW Core Pack Fabrication LBW was done according to the parameters shown in figure 7. One titanium gas fitting was then attached and the part coated with boron nitride prior to forming.

3.3.1.2 Three-Sheet Superplastic Forming. The panels were loaded into a hot open die. The Ti-6-4/Ti-6-4 panels were formed at 900 °C. The Ti-6-22-22/SP-700 and the Beta 21S/SP-700 panels were formed at 775 °C which is the reported forming temperature for maximum elongation in SP 700 (11).

3.3.2 Four-sheet LBW/SPF/DB A typical 1.5" x 1.5" cell pattern was selected (figure 13). This design requires a core elongation of approximately 250 %.

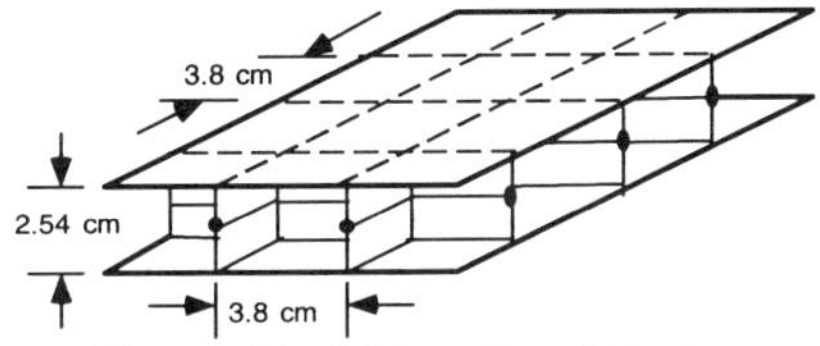

Figure 13- 4-Sheet Panel Design

3.3.2.1 Four-sheet Pack Fabrication Laser welding was done using the parameters previously developed (figure 9). Gas fittings were attached to the core pack and sealed using tungsten inert gas welding. The face-sheets were then added and gas fittings for the skin were attached.

3.3.2.2 Four-sheet Superplastic Forming/Diffusion Bonding The four-sheet processing parameters were established using a finite element modeling technique. For the Ti-6-4 panels, a flow stress of 8.27 MPa. and a strain rate of 3×10^{-4} mm/mm/sec was used. The Ti-6-4 panels were loaded in a hot open die and formed at 900 °C. Final DB was accomplished at 0.21 MPa , 915 °C for 3 hours and the parts were unloaded at 915 °C.

The Ti-6-2-2-2-2/SP-700 panels were formed at 775 °C. According to available data; SP-700 at 775 °C has a similar flow stress to that of Ti-6-4 at 900 °C (11). Also, SP-700 is reported to have a high strain rate sensitivity (m value) over a wide range of strain rates and temperatures (6). Therefore, an identical pressure schedule was used for the SP-700 panels as used for the Ti-6-4 panels. After core forming, the temperature of the Ti-6-2-2-2-2/SP-700 panels was programmed to increase while in the press to 845 °C during the DB portion of the cycle. This was done to improve DB and to increase the strength of Ti-6-2-2-2-2.

The 4-sheet Ti-6-4 panels achieved complete forming. However, although SP-700 has been reported to have much better superplastic elongation than Ti-6-4 the SP-700 thinned excessively in the highly strained areas of the core (figure 14). This ultimately resulted in a rupture of the core (figure 15) and incomplete forming.

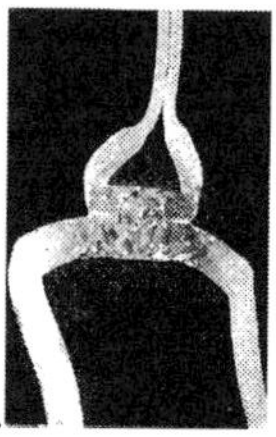

Figure 14- Excessive thinnout exhibited in SP-700 core.

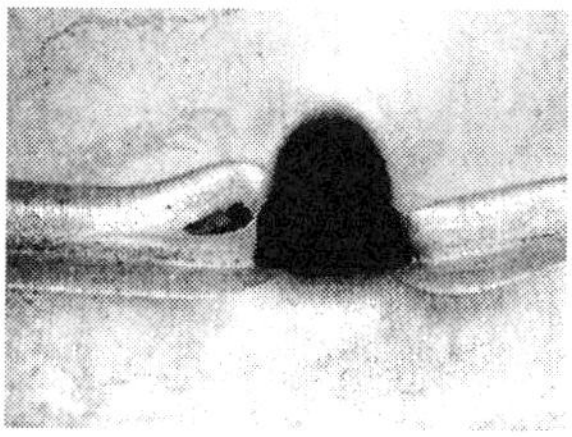

Figure 15- Rupture of SP-700 core at gas communication hole.

3.3.3 Post Processing The panels were sandblasted, chemically milled to remove alpha case and heat treated in a vacuum furnace. A temperature of 1000°/8 hours was chosen as the post-SPF heat treatment for all of the panels based on the heat treatment study. The completed multi-sheet SPF panels are shown in figure 16.

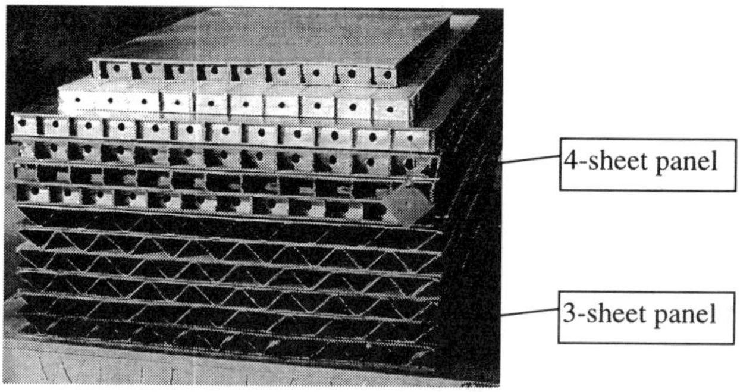

Figure 16- 3-Sheet LBW/SPF and 4-sheet LBW/SPF/DB panels

3.4 Subelement Mechanical Testing Four subelement test specimens of each panel configuration were subjected to static edgewise compression loading at room temperature (RT) and 177 °C (350 °F) using ASTM C634. Edgewise compression is accepted as a good test for evaluating sandwich structures. The test measures the ability of the core to stabilize the face-sheets during end loading.

Figure 17 shows the configuration of the edgewise compression specimens.
The test matrix and results are shown in figure 18.

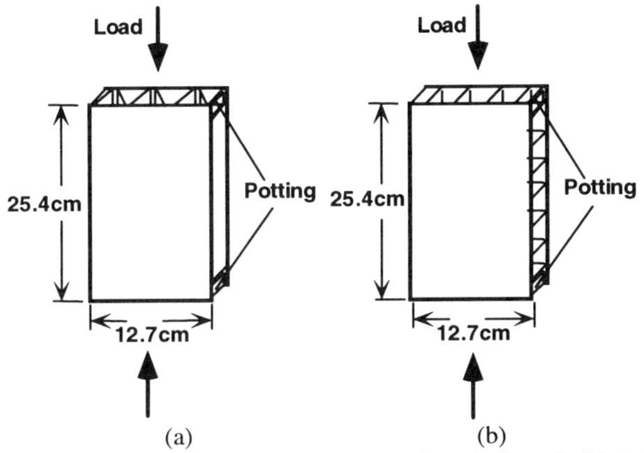

Figure 17 Mechanical test configuration of (a) 3-sheet & (b) 4-sheet

Specimen Type	Test Temp. °C (°F)	Ultimate Strength MPa (ksi)
3-Sheet LW/SPF		
Ti 6-4 Skin	RT	702 (102)
Ti 6-4 Core	RT	709 (103)
	176.6 (350)	578 (84)
	176.6 (350)	564 (82)
Ti 6-2-2-2-2 Skin	RT	778 (113)
SP-700 Core	RT	965 (140)
	176.6 (350)	737 (107)
	176.6 (350)	751 (109)
Beta 21S Skin	RT	971 (141)
SP-700 Core	RT	895 (130)
	176.6 (350)	641 (93)
	176.6 (350)	655 (95)
4-sheet LW/SPF		
Ti 6-4 Skin	RT	861 (125)
Ti 6-4 Core	RT	854 (124)
	176.6 (350)	744 (108)
	176.6 (350)	813 (118)
Ti 6-2-2-2-2 Skin	RT	1,226 (178)
SP-700 Core	RT	1,254 (182)
	176.6 (350)	571 (83)*
	176.6 (350)	578 (84)*

Ultimate Strength = Ultimate Load/Nominal Face-sheet Thickness(Actual width)
* Delamination failure (extracted from a panel that did not achieve full forming)

Figure 18- Edgewise Compression Test Data

The failure mode of the 4-sheet specimens was crippling within multiple cells. Crippling of the 3-sheet specimens led to face sheet/core sheet delamination and face sheet tearing. The results of the testing show that significantly higher strengths are achieved in the advanced alloy panels compared to the Ti-6-4.

5. CONCLUSIONS

Several important developments were achieved as a result of this study. Laser welding of various combinations of advanced titanium alloys was successfully developed. Additionally, the ability to use various combinations of these alloys to fabricate multi-sheet SPF structures was demonstrated. The demonstration of Beta 21S as face-sheets for three-sheet LBW/SPF structure may be especially significant. Other multi-sheet processes require diffusion bonding and cannot successfully use this high strength material. A summary of significant conclusions follows:

1. Ti-6-4, Ti-6-22-22, SP-700 and Beta 21S can be laser welded in several alloy combinations.

2. Three-Sheet LBW/SPF and four-sheet LBW/SPF/DB panels can be fabricated using several combinations of advanced titanium alloys.

3. A hybrid structure of SP-700 and Beta 21S can result in very high strengths in the Beta 21S (1375 MPa (199.4 ksi) UTS, 7.5% elongation) with usable properties in the SP-700 (1141 MPa (165.5 ksi), 10.8% elongation). Optimization of the aging process would likely result in improved properties for both alloys.

4. Hybrid Ti-6-22-22/SP-700 multi-sheet structures required compromising the heat treatments for each alloy. This lowered the strength compared to the individual optimum heat treatments. The preliminary heat treatment chosen for structure resulted in a room temperature ultimate tensile strength of 1189 MPa (172.5 ksi) (with 12.3% elongation) in the Ti-6-22-22 and 1141 MPa (165.5 ksi) (with 10.8% elongation) in the SP-700. This represents a modest improvement over the data collected for the baseline Ti-6-4 as-SPF'd properties (1017 MPa (147.5 ksi) UTS, 12.5% elongation).

5. Edgewise compression data indicates improved performance of up to 32% over Ti-6-4 structures.

6. The SP-700 exhibited excessive thinnout compared to Ti-6-4 in the 4-sheet structures. This ultimately resulted in premature core failures of highly strained areas during forming of the 4-sheet structure.

6. REFERENCES

1. Evaluation of SPF/DB of Ti-6Al-4V for Supersonic Cruise Aircraft Research, Douglas Aircraft Co., NASA contract NAS 1-15527, December 1985.

2. G. H. Arvin, L. Israeli, J. H. Stolpestad, G. Stacher, Evaluation of SPF/DB of Ti 6Al-4V Expanded Sandwich Structures, North American Aircraft Rockwell, NASA contract NAS1-15788, 1981

3. Application of SPF/DB Titanium Fabrication, Douglas Aircraft Co./McDonnell Douglas Corp., Final report AFWAL-TR-82-4059, 1982

4. R. Boyer, G. Welsch, E. W. Collings, Materials Properties Handbook Titanium Alloys, ASM International, 1994.

5. C. Ouchi et al, "Development of Beta-rich Alpha-Beta Titanium Alloy: SP-700," NKK Technical Review 65 pp. 61-67 1992.

6. A. Ogawa, H. Kukai, K. Minakawa, and C. Ouchi. "SP-700 Titanium Alloy Data Sheets," Beta Titanium Alloys in the 1990s, The Minerals, Metals & Materials Society, Edited by D. Eylon R.R. Boyer and D.A. Koss, 1993.

7. K Takahashi et al., "Microstructures and Mechanical Properties of High Strength Beta-rich Alpha-Beta Titanium Alloy; SP-700", NKK Corporation.

8. Boyer and A. Caddey, "The Properties of Ti-6Al-2Sn-2Zr-2Mo-2Cr Sheet", Titanium 1992, Science and Technology, The Minerals, Metals & Materials Society, Edited by F. Froes, 1993.

9. P. Bania, "Beta-21S: "A High Temperature Metastable Beta Titanium Alloy", Presented at the 1990 TDA International Conference, Orlando FL.

10. J.R. Williamson, "Diffusion Bonding in Superplastic Forming/Diffusion Bonding", Welding Technologies for the Aerospace Industry, American Welding Society, 1981. p.82.

11. Ref 5 p. 66.

7. ACKNOWLEDGMENTS

The authors thank John Lee and William Quist of Boeing Commercial Airplane Group and Dick Royster of NASA-Langley for sponsorship of this work. The authors would also like to gratefully acknowledge the help of several contributors: Fred Buldhaupt, John Packard, Jeffry Brewer, Brice Woods and Kin Li.

28th International SAMPE Technical Conference
November 4-7, 1996

ADVANCES IN LIGHTWEIGHT METALLIC MATERIALS FOR TRANSPORTATION

F.H. (Sam) Froes
Institute for Materials and Advanced Processes (IMAP)
University of Idaho, Moscow, ID, 83844-3026, USA

ABSTRACT

Further significant enhancements in the performance of lightweight metallic materials for transportation will come not from conventional ingot metallurgy (IM) approaches but rather from synthesis by "far from equilibrium" processes which allow development of novel crystal structures and microstructures. Three far from equilibrium synthesis techniques will be reviewed, with emphasis on the amount of departure from equilibrium: mechanical alloying (MA), plasma processing (PP), and physical vapor deposition (PVD). Scientific developments will be related to technological alloys with enhanced behavior, which could see future use in transportation including aerospace, automobile and rolling stock (trains) systems.

KEY WORDS: Transportation, Lightweight Metallic Materials, Synthesis

1. SYSTEM REQUIREMENTS/ADVANCED MATERIALS

Improved performance in aerospace, airframe, engine and missile systems requires in turn an enhancement in the behavior of the materials of construction (1-4). And the same is true for terrestrial applications such as automobiles and rolling stock (1-4). In recent years, improvements in materials characteristics have come mainly from improved synthesis or processing methods rather than from the more traditional approach of improved understanding of microstructure-property relationships. The goal is to move to an enhanced trend band as shown in Fig. 1, where parameters such as strength, ductility, and temperature capability should be maximized, but at the same time other characteristics such as "forgiveness" (ductility, fracture toughness, fatigue crack growth rate, etc.) and cost cannot be ignored.

Weight reduction also translates to material property enhancement with the value of weight savings varying considerably from one industry to another (Table I). By far the greatest weight reduction is achieved by a density decrease, which gives a direct weight reduction. Hence the importance of the low density structural alloys aluminum, magnesium, and titanium, and the potential for the intermetallic titanium aluminides (Ti_xAl, where x=1 or 3)*. This intermetallic class of materials represents a generally increased temperature capability compared to conventional metals, but below that of ceramics (Fig. 2).

* Throughout the text, terminal compositions will be given in wt.%, the intermetallics in at.%.

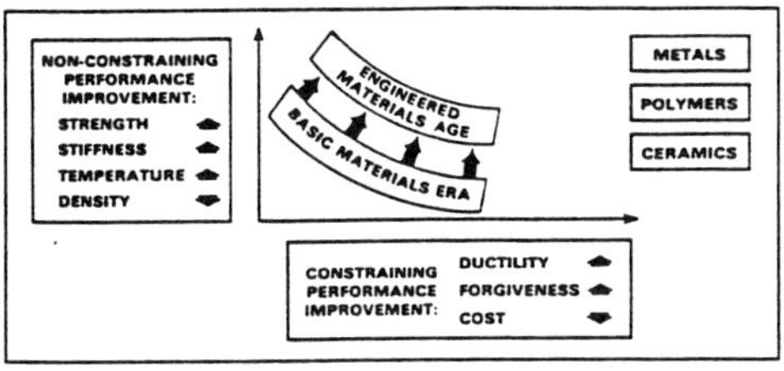

Figure 1 Trend Bands Exhibited by Basic Materials and Enhanced Trend Band Characteristic of Engineered Materials.

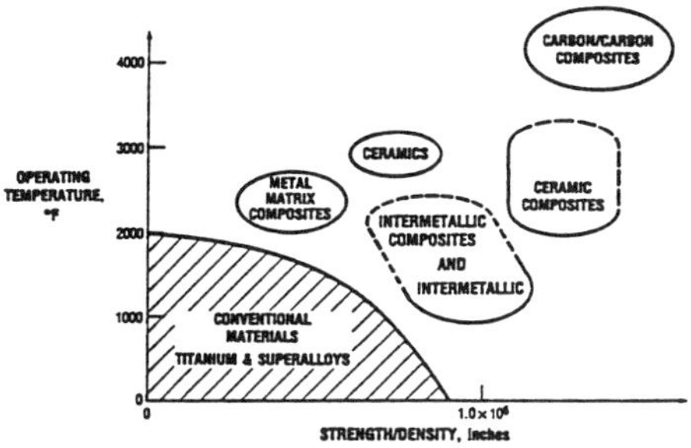

Figure 2 Temperature Capability of Various Material Classes.

Table I — Dollar Value of Weight Savings*

Industry	$
Automobile	3
Commercial Aerospace	300
Military Aerospace	3,000
Space	30,000

*One Pound Saved, Courtesy Boeing Company

1.1 Aerospace The increased performance requirement for aerospace systems applies both to airframes and engines. For example fighter aircraft must operate over broader combat envelopes with supersonic persistence, at greater turn rates, higher load factors and maximum Mach numbers, Fig. 3 (1-4). For this application, the major part that new and improved materials play in structural weight reduction is shown in Fig. 4 (1-4). The materials factor greatly exceeds the contribution made by other advanced technologies such as structural concepts, automated design and active load control, and flutter suppression.

1.2 Automobile New materials and processes are critical to achieve advances in automobile technology; with the major gains to be made in the power train - engine and transmission components. Lighter weight materials translate into enhanced responsiveness, space optimization and secondary advantages such as reduced friction for improved emission and fuel economy; a 10% reduction in vehicle weight translates to a 5.5% improvement in fuel economy (5,6). The major reciprocating and rotating mass components affecting fuel economy and emissions are shown in Fig. 5 (5).

1.3 Rolling Stock Traditional rolling stock does not emphasize light weight because of traction requirements (7). However, recent developments in the high speed arena (nominally > 100 mph [160 kph]) have shown the advantage of reduced weight equipment. The new Tokaido Shinkansen (bullet train) series 300, introduced onto the Tokyo-Osaka line in 1992, is shown with the earlier series 100 vehicle in Fig. 6. The newer model weighs 25% less than the older version, a result of extensive use of extruded aluminum. This has allowed an increase in speed from 140 mph (220 kph) to 170 mph (275 kph) and a significant savings in travel time between Tokyo and Osaka (7).

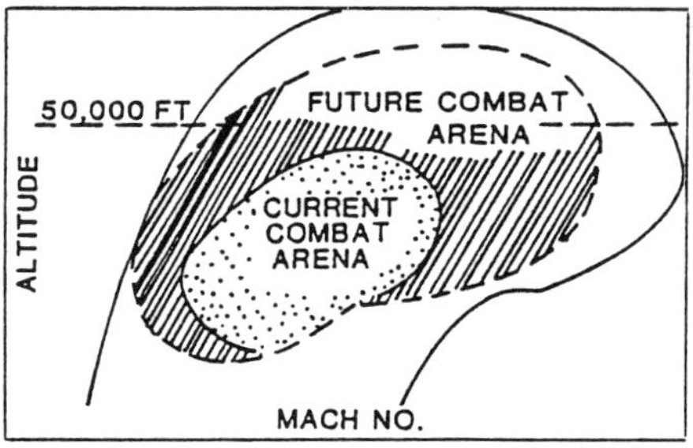

Figure 3 Combat Envelope for Future Fighter Aircraft.

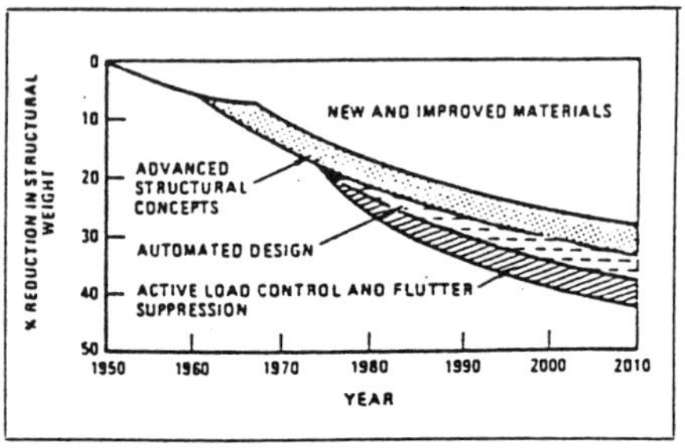

Figure 4 The Major Part that New and Improved Materials Play in Structural Weight Reduction.

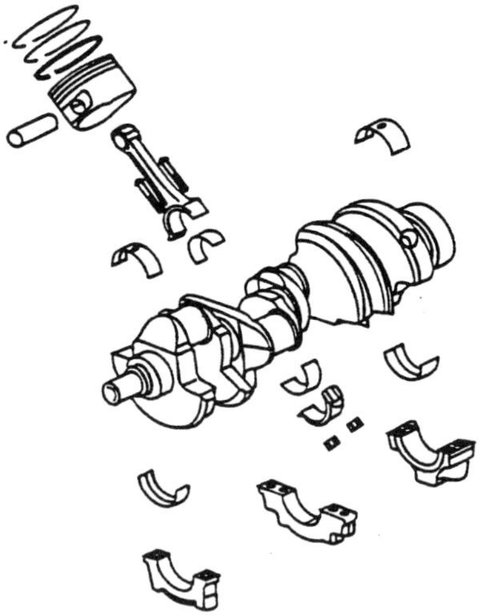

Figure 5 Major Reciprocating and Rotating Mass Components Affecting Fuel Economy and Emission in an Automobile Engine.

Figure 6 Japan Railways Tokaido Series 300 (left), the Newer Vehicle, and 100 (right) Shinkansen. Note the Low Profile of the Newer Vehicle.

2. SYNTHESIS/PROCESSING

In this paper three "far from equilibrium" (8,9) processes will be addressed: mechanical alloying (MA), plasma processing (PP), and physical vapor deposition (PVD). These various far from equilibrium synthesis techniques allow novel constitutional and microstructural effects, and hence potentially enhanced physical and mechanical properties. The departure from equilibrium achieved in these processes can be estimated (8,9) and is shown in Table II. Mechanical alloying and condensation from the vapor, especially the latter, can produce large departures from equilibrium, beyond those achievable with solid state processes such as quenching, leading to various combinations of attributes (Table III).

Table II — Departure From Equilibrium Achieved in Various Processes

Process	Maximum Departure from Equilibrium (kJ/N_a*)
Solid State Quench	16
Mechanical Alloying	30
Condensation from Vapor	160

*N_a is Avogadro's number of atoms.

Table III — Attributes of Far From Equilibrium Processes

1.	Production of a Fine Dispersion of Second Phase Particles
2.	Extension of Solubility Limits
3.	Refinement of the Matrix Microstructure down to the Nanometer Range
4.	Synthesis of Novel Crystalline Phases
5.	Development of Amorphous (Glassy) Phases
6.	Possibility of Alloying of Difficult to Alloy Elements
7.	Inducement of Chemical Reactions at Low Temperatures
8.	Scaleable Process

2.1 Mechanical Alloying Mechanical alloying leads to the production of ultrafine microstructures, and alloys not possible by conventional ingot metallurgy (IM) (1-4,10,11). In this process heavy working of powder particles results in intimate alloying by repeated welding and fracturing, Fig. 7.

2.1.1 Aluminum High temperature alloys with Ti additions, low density alloys with Mg additions and extra low density alloys with Li additions have been developed through MA (1-4,10,11). The MA Al-Li alloys are characterized by excellent general corrosion resistance that is 100 times better than that exhibited by the IM-7075-T73 alloy.

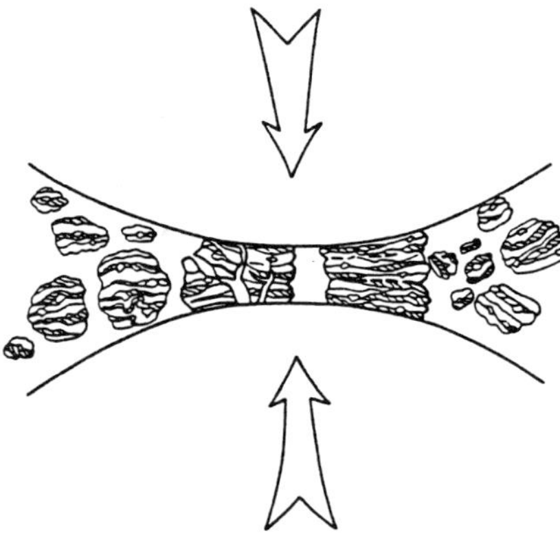

Figure 7 Schematic of Mechanical Alloying Process.

2.1.2 Magnesium MA has been employed in the development of "supercorroding" alloys for submarine applications, such as a heat source in diver suits and as a hydrogen gas generator. By tailoring composition, reaction rates can be precisely controlled in the device (1-4,10,11).

2.1.3 Titanium Application of MA to titanium alloys has resulted in supersaturated solid solutions, metastable crystalline and glassy phases as well as nanometer-sized grain structures (Fig. 8) (1-4,10,11). Recent work on the MA of titanium based alloys has shown the potential to produce virtually contamination free material (1-4,10,11), a crucial advance allowing sizable quantities of titanium based materials to be produced by MA.

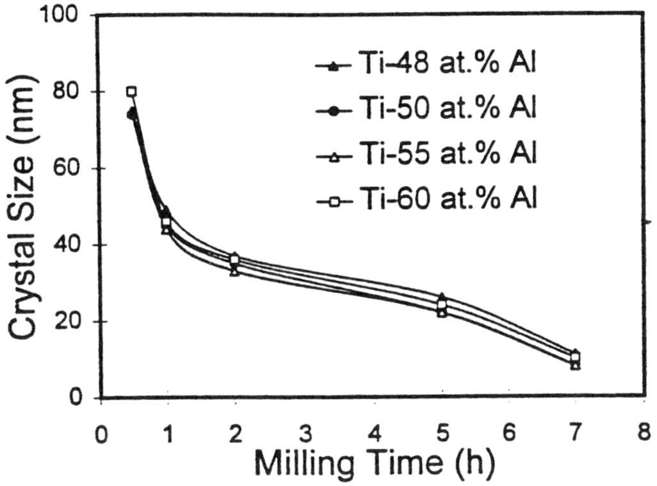

Figure 8 Reduction in Grain Size with MA Time Determined from X-ray Diffraction Line Broadening, for a Number of TiAl-type Alloys.

Examples of nanostructured grains in a mechanically alloyed and hot isostatically pressed TiAl alloy are shown in Fig. 9 (1-4,12-16). Not only can the nanostructured grains be produced in MA'd material but these grains do not show excessive growth during the compaction cycle; 25-40 nm grains can be retained with full density.

In comparison to rapidly solidified (RS) alloys, MA alloys exhibit finer grain structures, absence of dispersoid-free zones near grain boundaries (Fig. 10), and dispersoids that do not coarsen significantly up to 1100°C (1-4,10,11). Considerable work still needs to be done to optimize the processing parameters to produce these MA alloys consistently.

Figure 9 Nanostructured grains in MA and HIP'd Ti-47.5Al-3Cr (at.%).

2.2 Plasma Processing In plasma processing, solid feed-stock is converted to the vapor phase at elevated temperatures due to the highly concentrated enthalpy. Chemical reactions occur and a solid product nucleates and grows at a lower temperature (17,18). A number of studies have led to the formation of nanometer sized metallic powders in the transition metals, Ni and Fe, and in Al powders Fig. 11 (17,18).

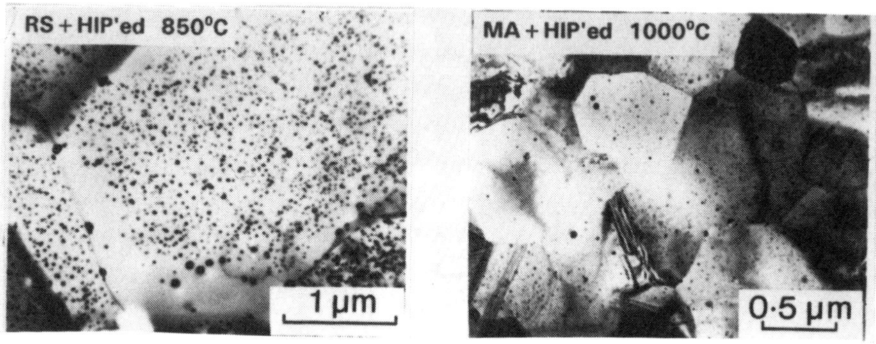

Figure 10 Comparison of Microstructure of Ti_3Al + 2Er Powder Compacts, (left) RS and (right) MA, HIP'd at 850°C and 1000°C, Respectively.

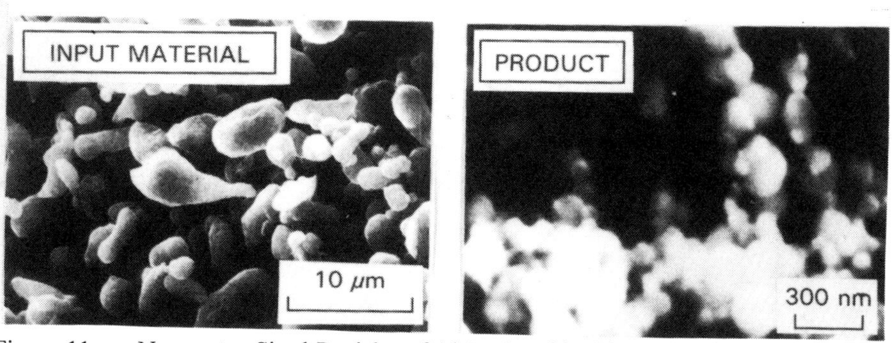

Figure 11 Nanometer Sized Particles of Al Produced by Plasma Processing: (left) Input Powder, (right) Product.

2.3 Vapor Deposition Physical vapor deposition (PVD) offers a route for the production of new alloys and microstructures, with advantages such as extremely fine grain sizes, extended solid solubilities, and freedom from segregation. In general, it is found that elements with low equilibrium solubilities (e.g. less that 1 at.%) can show large extensions of solid solubility by vapor deposition, whereas elements with larger equilibrium solubilities (and higher lattice mobilities) tend to show smaller extensions. Recent developments in high rate electron beam evaporation have shown that a wide range of new alloys, impossible to produce by conventional IM, can be processed by bulk PVD (1-4,10,11). Deposition rates up to 20 mm/hour have been achieved using evaporation from alloy baths, or by using multiple baths and vapor mixing.

2.3.1 Aluminum An aluminum alloy (designated RAE 72), containing the normally insoluble elements Fe and Cr, has been produced in slab form by high rate electron beam evaporation (Fig. 12) (10,11). The as-deposited alloy retained the Fe and Cr in a supersaturated solid solution; and ageing at 350°C was accompanied by the precipitation of sub-micron particles producing exceptionally high strength and stiffness. Table IV compares the properties of peak aged RAE 72 with other conventional aluminum alloys, and a titanium alloy.

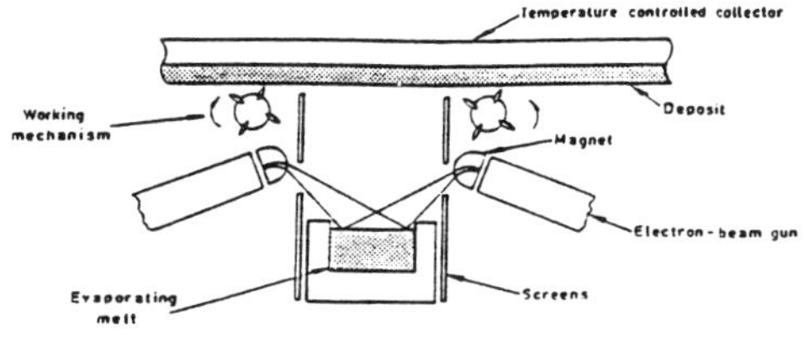

Figure 12 Schematic of the Vapor Depostion Process for Aluminum Alloys.

Table IV — Mechanical Properties of Vapor Quenched Al-Cr-Fe Alloy RAE 72 and Other Common Aerospace Alloys

Alloy	Density ($g.cm^{-3}$)	Ultimate Tensile Strength (MPa)	Elongation (%)	Young's Modulus (GPa)
RAE 72 (Al-7.5Cr-1.2Fe)	2.9	723	7.5	89
2014T6	2.8	410	8	74
7075	2.8	570	10	72
Ti-6Al-4V	4.4	1000	8	110

The vapor quenched alloy was found to be surprisingly stable at elevated temperature and retained a large portion of its room temperature strength up to 300°C. Fig. 13 compares the specific tensile strength of RAE 72 with conventional aluminum alloy 7075-T6 and titanium alloy Ti-6Al-4V.

Vapor deposition has also been used to produce nanostructure dimension layered materials with exceptional strengths (1-4,10,11). In this process different layer materials are evaporated from separate sources and collected on a heated rotating substrate. (Fig. 14)

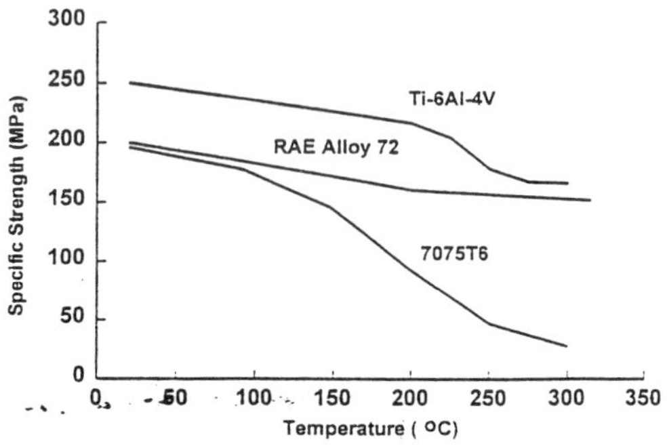

Figure 13 Effect of Temperature on Specific Strength for Vapor Quenched Al-Fe-Cr Alloy RAE72, Commercial Aluminum Alloy 7075-T6 and Titanium Alloy Ti-6Al-4V.

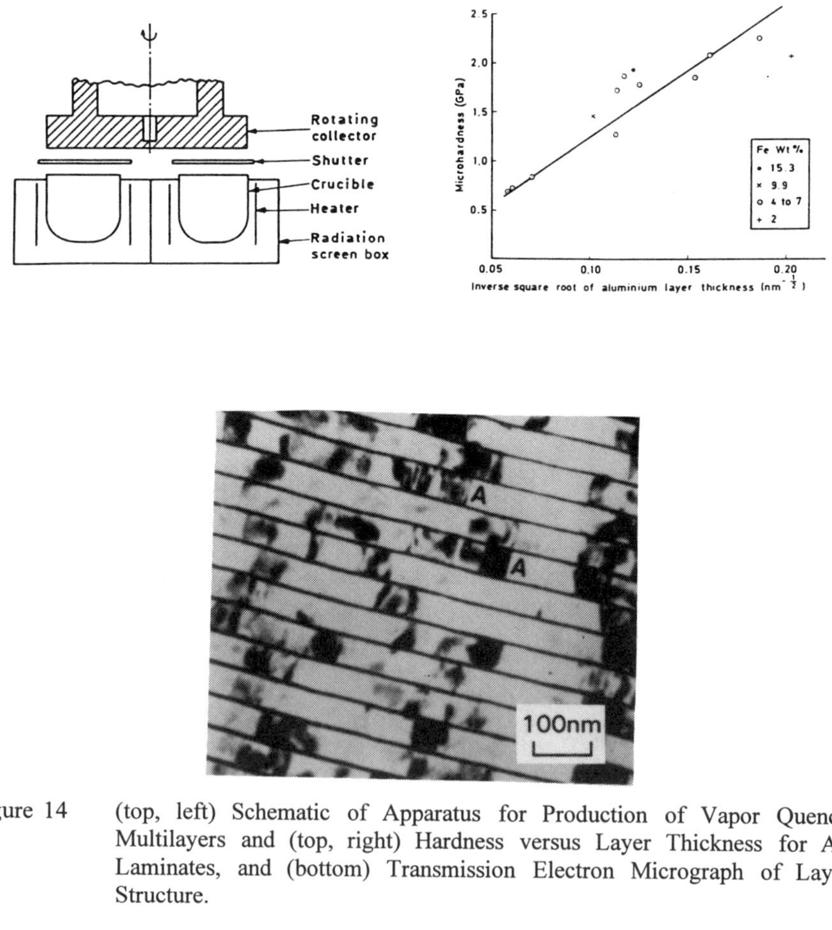

Figure 14 (top, left) Schematic of Apparatus for Production of Vapor Quenched Multilayers and (top, right) Hardness versus Layer Thickness for Al/Fe Laminates, and (bottom) Transmission Electron Micrograph of Layered Structure.

2.3.2 Magnesium Many elements that might be useful alloy additions to magnesium cannot be added by RS due to lack of solubility in liquid magnesium. High rate evaporation and vapor quenching avoids this problem and a range of new alloys has been produced using this alternative approach (1-4,10,11). The arrangement for producing these new magnesium alloys is illustrated in Fig. 15. Binary magnesium alloys containing Ti and Mn were found to exhibit 30-100 times lower corrosion rates in sodium chloride solution than pure magnesium (Table V).

Table V — Corrosion Rates for Vapor Quenched Magnesium-Titanium Alloys Compared with Various Commercial Magnesium Alloys

Alloy	Mg-22Ti	Mg-7.4Ti	ZK50A	WE43	AZ91E	EA55RS
Corrosion rate (mm/year)	0.03	0.005	3.25	0.42	0.24	0.14

2.3.3 Titanium Separate source evaporation and simultaneous vapor quenching has been used to produce low density Ti-Mg and Ti-Ca alloy deposits (1-4,10,11). Alloying titanium with these alkali metals is not possible by conventional IM due to their low melting points and very high vapor pressures. Deposits were produced with up to 28 weight % Mg retained in solid solution and magnesium was found to reduce the density of titanium by about 1% for each 1% addition. For the more dilute alloys (less than 10 wt% Mg) heat treatment in air or in vacuum up to 700°C was accompanied by a very substantial increase in hardness. Fig. 16 illustrates the effect of annealing temperature on microhardness for a series of samples heat treated for 30 min. in vacuum, and shows an almost three fold increase in hardness for a Ti-4Mg alloy annealed at 600°C. The hardening was found to be associated with the precipitation of fine magnesium particles in the 2-20 nm range with good thermal stability.

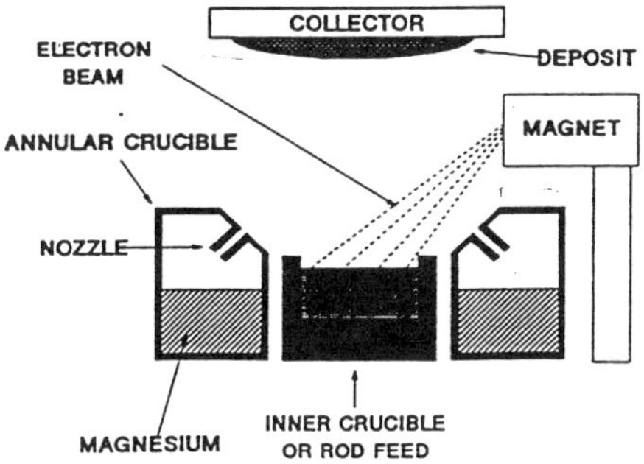

Figure 15 Schematic of Vapor Quenched Magnesium Alloy Process.

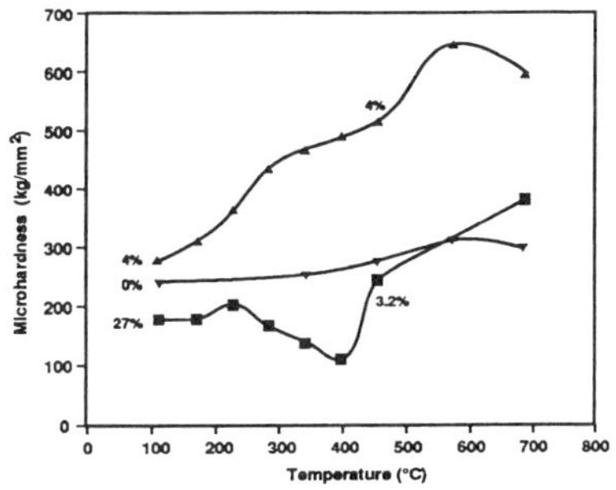

Figure 16 Effect of Annealing Temperature on the Hardness of Vapor Deposited Ti-Mg Alloys with 0, 4, and 28 wt.% Mg.

3. CONCLUSIONS

The science and applications of the "far from equilibrium" synthesis techniques, mechanical alloying, plasma processing, and physical vapor deposition, have been reviewed as applied to alloys based on the light metals. Enhanced mechanical properties compared with conventional processing methods can be obtained.

4. ACKNOWLEDGMENTS

The author would like to acknowledge useful discussions with Drs. P.R. Taylor, C. Suryanarayana and C.M. Ward-Close. In addition, the help of Mrs. Susan Goetz and Miss Jonica Johnson in manuscript typing and formatting is greatly appreciated.

5. REFERENCES

1. F. H. Froes and P.D. Desai, "Recent Developments in Powder Metallurgy Processing Techniques," MIAC, Purdue University, IN, Report #5, October 1994.
2. F.H. Froes et al., Light Materials for Transportation Systems, ed. N.J. Kim, Postech, Pohang, Korea, (1993) 27.
3. F.H. Froes, Material Science and Engineering, A184 (1994) 119.
4. F.H. Froes et al, Symposium on Powder Metallurgy in Aerospace, Defense and Demanding Applications (ADDA), edited by F.H. Froes, pub. MPIF, Princeton, NJ, 1995, 1.
5. J.E. Allison and G.S. Cole, JOM 45, No. 1 (1993) 19.
6. W.F. Jandeska, Jr., P/M in Aerospace, Defense and Demanding Applications - 1993, ed. F.H. Froes, pub. MPIF, Princeton, NJ, 45.
7. F.H. Froes, Light Metal Age, 52, Nos. 3 and 4, April 1994, 12.
8. F.H. Froes, C. Suryanarayana, K. Russell, and C.G. Li, Proceedings of Third International Conference on High Temperature Intermetallics, eds. D.P. Pope, S.H. Whang and C.T. Liu, ASM Int., Materials Park, OH, 2 (1995) 612.
9. F.H. Froes et al., Novel Techniques in Synthesis and Processing of Advanced Materials, eds. J. Singh and S.M. Copley, TMS, Warrendale, PA, 1995, 1.
10. F.H. Froes and C.M. Ward-Close, J. Ad. Mats., 25, No. 1, (1993) 20.
11. C.M. Ward-Close and F.H. Froes, JOM, 46, No. 1, (1994) 28.
12. Guo-Hao Chen, C. Suryanarayana and F.H. Froes, Seventh World Conference on Titanium, eds. F. H. Froes and I. Caplan, TMS, Warrendale, PA (1993) 903.
13. F. H. Froes, C. Suryanarayana, G. H. Chen, A. Frefer and D. K. Mukhopadhyay, P/M in Aerospace, Defense and Demanding Applications — 1993, ed. by F. H. Froes, MPIF, Princeton, NJ, (1993) 1.
14. C. Suryanarayana and F.H. Froes, Nanostructured Mater., 3 (1993), 147.
15. F.H. Froes et al., "Compaction of Nanograined Gamma TiAl by Hot Isostatic Pressing," PM^{2}TEC 1995, eds. M.A. Phillips and J. Porter, MPIF, Princeton, NJ, 1995. To be published.
16. J. Hebeisen et al., as ref. #4, 363.

17. P.R. Taylor and S.A. Pirzada, Advanced Performance Materials, 1 (1994) 37.

18. P.R. Taylor et al., Plasma Synthesis and Processing of Materials, ed. K. Upadhya, TMS, Warrendale, PA, (1992) 215.

28th International SAMPE Technical Conference
November 4-7, 1996

USE OF ALUMINUM-BERYLLIUM COMPOSITES ON THE ORBCOMM^{SM*} SATELLITE

Tom Dragone - Orbital Sciences Corporation, Dulles VA
Tom Parsonage - Brush Wellman, Inc., Orlando, FL
Rob Hardesty - Peregrine Falcon Corporation, CA

ABSTRACT

Newly developed metal matrix composites of aluminum and beryllium have been used in the primary structure of Orbital Sciences Corporation's ORBCOMM communication satellites. Hot isostatically pressed and extruded Al-Be alloys used in these spacecraft have a two-phase microstructure that combines the stiffness and light-weight of beryllium with the ductility and fabrication ease of aluminum and the synergistic strengthening effect of the dual phases. The ORBCOMM spacecraft has undergone a rigorous qualification testing program that included modal identification (to verify structural stiffness), ultimate static loading, vibration testing to levels typical of a Pegasus launch, separation shock loading to over 600 g's at greater than 1000 Hz, and thermal cycling between -35°C and +40°C. The aluminum-beryllium composite structure survived these environments with no cracking, no detrimental permanent deformation, and no degradation in material properties. Furthermore, the first two ORBCOMM satellites have been launched into low earth orbit by OSC's Pegasus launch vehicle, and have been operating for over a year. This paper will further describe the spacecraft structure, design drivers that led to the selection of this material, its fabrication process, results of qualification testing, and a cost comparison to other materials.

KEY WORDS: Metal Matrix Composites, Space Applications, Structures

1. INTRODUCTION

Looking for better and less costly methods of sending satellites into orbit is keeping materials scientists and metallurgists on a never-ending quest: finding metals and alloys that offer a combination of high strength, light weight, rigidity, and the ability to be fabricated into shapes and sizes to suit the application. Conventional metals, alloys, and some composites often don't have the properties to meet these demanding requirements. AlBeMet® alloys, a relatively new series of composites of aluminum and beryllium, combine the high modulus and low density of beryllium with the strength , ductility, and fabrication characteristics of aluminum. The composites promise to fulfill many of the needs of satellites and other spacecraft structures.

* ORBCOMM is a trademark, service mark, and trade name of ORBCOMM Global, L.P.

AlBeMet® composites produced by Brush Wellman, Inc. are being used for the major structural components of a satellite system being produced by Orbital Sciences Corporation (OSC). The initial system is expected to consist of a constellation of 28 small satellites in low earth orbit that will provide seamless, two-way, real-time messaging between any two points on the globe. In addition, with sophisticated global positioning system navigation technology, users will be able to accurately determine their position or that of their remotely located assets. Stranded boaters or hikers can send highly accurate position information to rescuers. Mobile equipment owners can monitor locations of all vehicles in a fleet of trucks, ships, or railroad cars with the ORBCOMM satellite system. Farmers can effectively manage large irrigation systems by monitoring temperature, water pressure and other parameters important to the irrigation program.

The first two ORBCOMM satellites were launched in April, 1995 on a Pegasus small launch vehicle. The satellites became fully operational a few months later, and limited service has already begun. Full service, with complete global coverage is scheduled for early in 1997 with three more Pegasus launches, each consisting of eight ORBCOMM spacecraft stacked on top of one another.

2. SPACECRAFT CONFIGURATION

With the exception of the first two spacecraft, each of the 28 spacecraft in the ORBCOMM constellation is planned to be identical and to consist of three major components identified in Figure 1 in the deployed configuration:

- a flat 1.04 m (41 in) diameter by 152.5 mm (6 in) high housing that contains the spacecraft's computers, attitude control system, propulsion system, and communications payload,
- two deployable, articulated solar panels for providing power to the spacecraft, and
- a segmented, deployable boom supporting the subscriber, UHF, and gateway antennas for communication with the ground.

For launch, the solar panels and the antenna boom are folded into the spacecraft structure, allowing multiple spacecraft to be stacked on top of each other. Non-explosive separable bolts attach the spacecraft to each other at three points on the interface.

The primary spacecraft ring consists of three arched 25 mm (1 in) thick AlBeMet® sandwich panels fastened together by three AlBeMet® separation brackets. The brackets incorporate 12.5 mm (0.5 in) thick upper and lower flanges to hold the separation bolts and the shear bearings to provide a stiff load path between spacecraft structures when they are stacked together for launch. Four dip-brazed vertical gusset panels connect the upper and lower flanges to provide a rigid structure that also allows access to the separation bolts during spacecraft integration. The outer two vertical gussets are machined into C-channels to accommodate the arched ring segments that are both bonded and riveted into the channels. Brazed AlBeMet® brackets connect payload shelf to the primary spacecraft ring. Since the payload mounting shelf is not in the primary stiffness path of the stacked configuration, it can be fabricated from standard aluminum. Bonded AlBeMet® pieces are also used for the rectangular tube used as the support structure for the antenna boom segments.

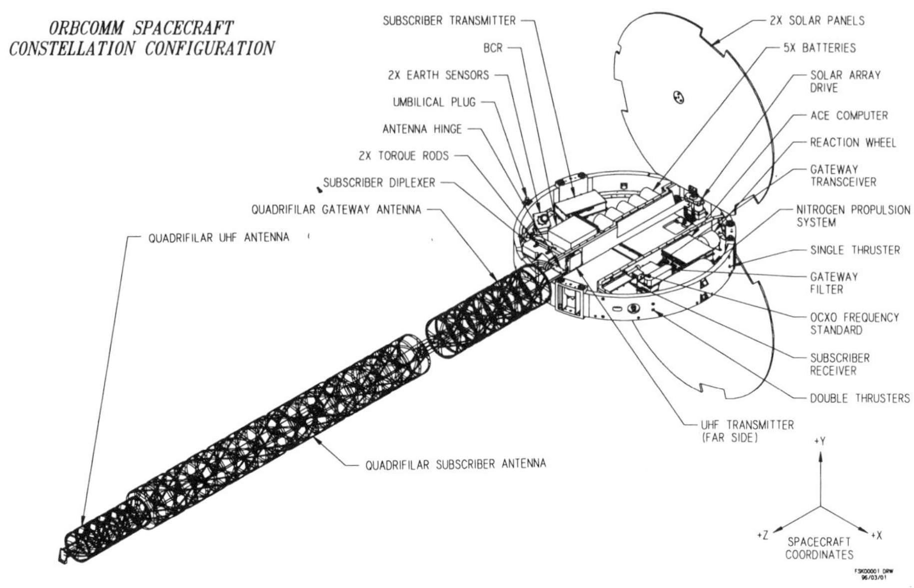

Figure 1 – ORBCOMM satellite in the deployed configuration, showing spacecraft structure, solar panels, and antenna.

3. DISCUSSION

3.1 Design Requirements The design of the ORBCOMM spacecraft and the choice of its materials was ultimately driven by the requirement to survive all launch vehicle imposed loads (1) with the minimum weight possible. From this system-level requirement, material requirements for stiffness, strength, and ductility were derived.

3.1.1 Stiffness Requirements To reduce the magnitude of the dynamic transient loads accompanying the launch and to avoid launch vehicle guidance problems, the combined stack of eight ORBCOMM satellites is required to have a bending frequency of no less than 20 Hz. In addition, under dynamic loading, the maximum deflection at the top of the stack of eight satellites can be no greater than 6.3 mm (0.25 in) to avoid violating the Pegasus fairing envelope.

Both of these requirements result in the need for high stiffness, as measured either by the natural vibration frequency (proportional to $[EI/M]^{1/2}$) or by deflection (proportional to 1/EI). Since the geometric stiffness (i.e. effective inertia, I, for the stacked configuration) is limited by the available payload fairing diameter and is further compromised by the added compliance of joints between spacecraft, a high intrinsic stiffness (i.e. modulus, E) is required. Initial stiffness sizing performed by OSC indicated that a modulus value higher than that of aluminum – greater than 69 GPa (10 Msi) – would be required.

3.1.2 Strength Requirements In addition to needing maximum rigidity in the satellite stack, the high loads resulting from the Pegasus launch transient mean that a high strength material is needed for the spacecraft construction as well. Initial sizing analyses conducted by

OSC indicated that yield values at least as high as those of 6061 aluminum – 273 MPa (40 ksi) or higher – would be required. In particular, high loads were expected for the flange pieces which support both the preload forces from the separation bolts and the dynamic bending loads associated with launch.

3.1.3 Ductility Requirements Although non-explosive separation bolts are used to connect the spacecraft together in the launch configuration, shock loads at the interface are high because a significant amount of stored strain energy in the preloaded bolts is released within a few milliseconds when separation is initiated. Therefore, brittleness was a concern in the design and material selection for the spacecraft. The dynamic nature of the shock loading could not be easily correlated to crack growth rates or material strength, so no requirement for fracture toughness could be derived from the loading. Instead, a substantial ductility – as evidenced by elongation to failure greater than 5% – was deemed indicative of sufficient resistance to shock loading and was desired for the materials selected for the spacecraft.

3.2 Materials Selection Initially, aluminum was considered for the primary spacecraft ring and bracket structure, but aluminum options could not meet the stiffness requirements within the allowable mass budget. Magnesium provided only a marginally stiffer structure for the same weight. A more costly beryllium structure could easily have met the stiffness requirements (with over six times the specific stiffness of aluminum), but beryllium presented considerable problems for the shock and dynamic loading because of its brittleness.

A unidirectional graphite / epoxy structure could be fabricated with high modulus fibers and result in a specific stiffness of over five times that of aluminum, but such a structure would also result in highly anisotropic properties. The ORBCOMM ring structure requires both hoop and shear stiffness to resist bending of the satellite stack; the specific stiffness of a composite layup tailored to these requirements would be reduced to less than twice that of aluminum. Furthermore, the ORBCOMM separation brackets require thick sections and nearly isotropic properties to resist the separation bolt preload forces. Consequently, there was significant concern about the ability to fabricate a graphite / epoxy composite into this configuration easily.

AlBeMet® composites ultimately provided the unique combination of stiffness, strength, and high ductility along with light weight that met all the ORBCOMM structure requirements. With a modulus in the range of 145-180 GPa (21-26 Msi) and a density in the range of 2.1 - 2.26 g/cc (0.076-0.082 lbm/in^3) AlBeMet® has a specific stiffness that is 2.5 to 3.5 that of aluminum. More importantly, the stiffness is isotropic, and ideally suited to the requirements of the ORBCOMM primary structure. AlBeMet®'s two-phase microstructure, shown in Figure 2, consists of elongated discrete or semi-continuous beryllium particles within a continuous aluminum matrix. Due to the mutual insolubility of the two phases in each other, the microstructure is quite stable, and little noticeable change can be observed after heat treatment (2).

The material can be characterized as an engineered composite material because of the potential synergistic combination of beryllium's high modulus and load bearing capability with aluminum's toughness and ductility. For example, strength of some AlBeMet® composite materials is higher than that of either of the parent metals because of the nature of the load sharing between the two phases and the impedance to dislocation motion and plastic flow presented by the phase boundaries. Strength and elongation properties of AlBeMet® composites are compared with those properties of 6061 aluminum and pure beryllium in Figure 3.

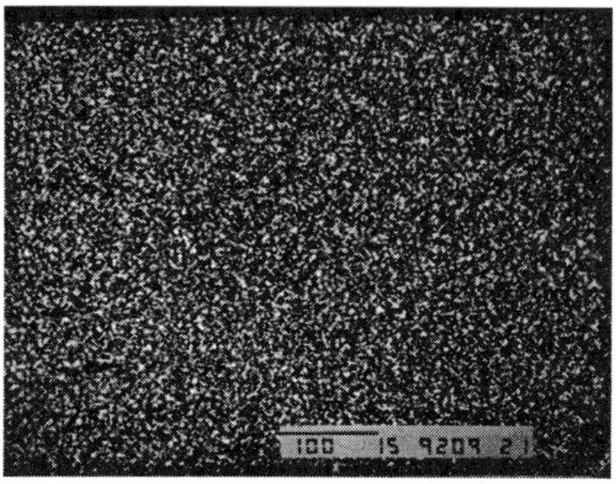

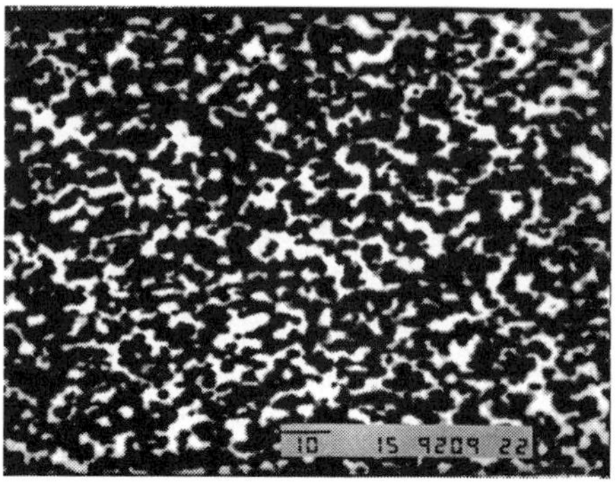

Figure 2 – AlBeMet® microstructure showing semi-continuous beryllium phase (dark) within continuous aluminum phase (light). Bars indicate 100μm (left) and 10 μm (right).

As in most composite systems, AlBeMet®'s atomic-based properties like density, modulus and coefficient of thermal expansion follow the Rule of Mixtures averaging, with properties varying monotonically with the proportion of constituent phases. Modulus and coefficient of thermal expansion (CTE) values for AlBeMet® are shown in Figure 4.

In addition to the stiffness and strength advantages that AlBeMet® has over aluminum, it also provides improved thermal properties as well. In particular, AlBeMet®'s thermal expansion coefficient is up to 40% lower than that of aluminum, resulting in lower thermally induced stresses and deflections. This property was of particular importance in the selection of a material for the ORBCOMM antenna boom. In this structure, the use of AlBeMet® significantly decreased the thermally induced deformation of the boom, and provided greater antenna accuracy.

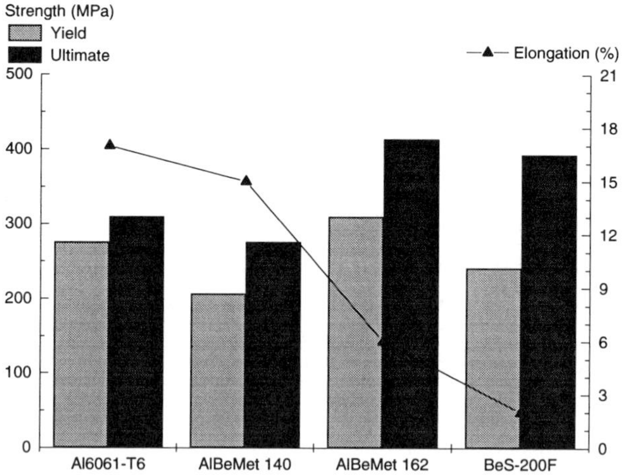

Figure 3 – Comparisons of typical strength and elongation to failure for 6061 aluminum, AlBeMet® alloys and pure beryllium.

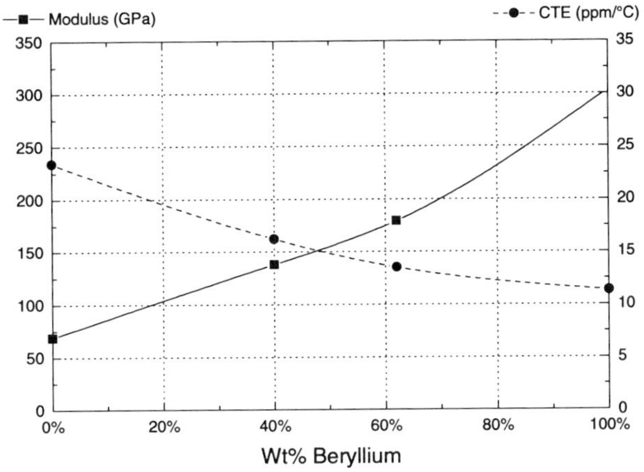

Figure 4 – Modulus and coefficient of thermal expansion (CTE) as a function of beryllium content for aluminum-beryllium alloys.

AlBeMet® grades 150 and 162 (corresponding to 50 wt% and 62 wt% Be respectively) were chosen for the ORBCOMM structures because they are nearly as stiff as steel (E = 145-180 GPa or 21-26 Msi), lighter than aluminum ($\rho \cong 2.19$ g/cc or 0.079 lbm/in^3) and less costly than pure beryllium. The AlBeMet® 162 grade is more completely characterized (3,4), and was specified for the critical components in the ORBCOMM spacecraft – the separation brackets and the payload deck brackets. These structures demanded the higher strength that the 162 grade could provide. Mechanical properties of the hot isostatically pressed (HIP'd), extruded, and cross-rolled sheet of AlBeMet® 162 are listed in Table 1. In the formed and heat treated condition, nearly isotropic strength was attainable with sufficient ductility (greater than 6%) to indicate that brittleness would not be a major concern. The less-demanding structures – face sheets for the honeycomb panels of the spacecraft ring and the antenna boom – are fabricated from the 150 grade. The honeycomb itself is 5052 aluminum.

Alloy Condition	**Heat Treatment**	**Yield Strength** *MPa (ksi)*	**Ultimate Strength** *MPa (ksi)*	**Elongation** %
HIP'd	None	242 (35.1)	306 (44.4)	2.56
	593°C/24hr	433 (32.8)	287 (41.6)	2.71
Extruded – L	None	562 (81.5)	609 (88.4)	1.44
	593°C/24hr	328 (47.6)	439 (63.7)	9.40
–T	None	558 (81.0)	607 (88.1)	0.24
	593°C/24hr	314 (45.6)	345 (50.0)	6.12
Rolled – L	None	399 (57.8)	463 (67.2)	5.09
	593°C/24hr	314 (45.5)	413 (59.9)	7.75
– T	None	374 (54.3)	453 (65.7)	2.62
	593°C/24hr	319 (46.3)	406 (58.9)	6.39

L = Longitudinal to work direction T = Transverse to work direction

Table 1 – Typical Properties of AlBeMet® 162 in the HIP'd, extruded and rolled condition.

3.2 Fabrication Fabricating AlBeMet® materials is quite similar to fabricating aluminum. The principal fabrication difference is the need for a facility that can handle beryllium-containing materials to remove the fine, airborne beryllium particles that could pose a health risk to some people. Forming is also similar to aluminum production in that the same tooling and temperature ranges can usually be used for forming at an elevated temperature – typically over 200°C (400°F). Forming rate is slightly slower for the AlBeMet® materials if severe deformation or bending is required. Gentle curves such as those of the ORBCOMM honeycomb ring panels were formed at the same rate as that used for aluminum panels.

3.2.1 Coating Like aluminum, the AlBeMet® materials form a protective oxide film on freshly machined surfaces, except the film on AlBeMet® is much more tenacious. For this reason, immediate coating after machining or grinding parts is critical. Coating the AlBeMet® surface resists corrosion and also provides a continuous bonding surface. Precleaning and proper coating application are critical for the primer – usually a phenolic-based epoxy compound – that is subsequently cured at 120°C (250°F).

Once the machined surfaces are stabilized by coating, coated parts may be stored for months, if necessary. After storage, a simple alcohol wipe removes any dust or fingerprints prior to further assembly operations. A quality fabricator can therefore furnish the product in a condition that the assembler/customer does not have to interface with the AlBeMet® materials in his subsequent assembly or mounting operations.

3.2.2 Joining The AlBeMet® materials can be joined by both electron beam welding and dip brazing methods. For the ORBCOMM spacecraft structure, brazing was chosen because of consistency and quality of the joints. Expert fabricators of AlBeMet® materials, such as Peregrine Falcon Corp., have developed proprietary programs involving higher temperatures and longer times then those used for aluminum to obtain high-quality brazed joints.

Design of joints for these alloys is quite different from those used for aluminum. Aluminum usually fails in a ductile manner, so bending occurs before failure, which usually occurs in the joint. With AlBeMet® materials, the metal is stiffer, so the joint is designed so the parent metal breaks before the joint fails. In this fail-safe design, the joints are not the weak members in the design and will therefore take the stress build-up without failing.

3.3 Testing Exhaustive testing of the ORBCOMM spacecraft structures was conducted by simulating all flight loads for all critical components including the separation event. These tests included modal identification testing, axial and lateral static loading conditions, anticipated axial and lateral vibration, shock loads, and thermal cycling loads.

Modal testing identified all the significant mode shapes and natural frequencies of the ORBCOMM spacecraft, and verified the in-situ stiffness of the AlBeMet® material. Modal properties agreed with predicted values and indicated that with AlBeMet®, the stacked spacecraft would have a natural frequency of 21Hz, meeting the 20Hz minimum requirement. In addition, the modal testing verified deflection analysis that indicated that the satellite stack would not violate the Pegasus fairing dynamic envelope.

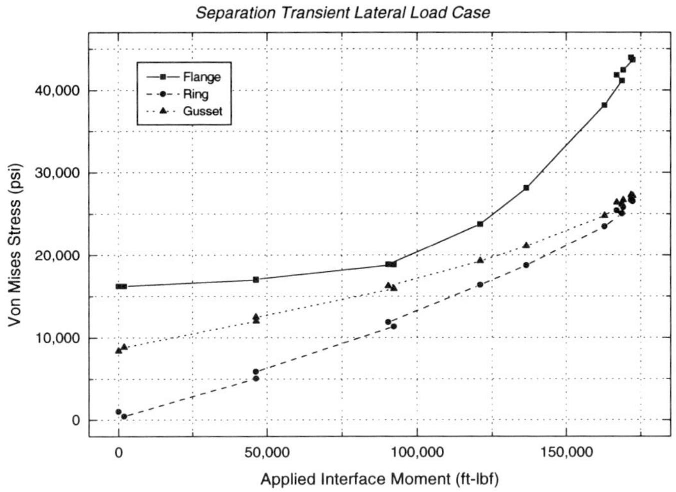

Figure 5 – Measured Von Mises stresses in the ORBCOMM separation bracket flange, cylindrical ring, and separation bracket gusset during simulated launch static loads testing.

Static load simulation of the launch environments was achieved using hydraulic actuators within a static loads test stand. Some of the recorded stresses for the separation transient lateral load case are shown in Figure 5. A peak stress of 300 MPa (43.9 ksi) was recorded in the lower flange under ultimate loading, resulting in a margin of safety of 0.45, based on UTS for the extruded bracket in the longitudinal direction (5). Slightly lower stresses were recorded in the ring panels and vertical gussets. Despite the nonlinear stress increase due to complex load transfer through the brackets, upon removal of the loads, stresses returned to their initial values indicating that no yielding or permanent deformation had occurred.

Random vibration testing was conducted over the range from 20 Hz to 2000 Hz in both the spacecraft axial and lateral directions. Typical results for an accelerometer placed on the payload shelf are shown in Figure 6. Resonance from the spacecraft's payload shelf diaphragm mode is clearly seen at 70 Hz, but at higher frequencies, the measured vibration levels are lower than the input levels. This phenomena was observed in most of the vibration responses and may indicate that the AlBeMet® composite material is providing a measure of internal damping to reduce the vibration (6).

Shock loading of the spacecraft was accomplished by simultaneously releasing the three separation bolts that connect two spacecraft and recording the resulting accelerations at the separation bracket flanges and at other locations on the spacecraft. While the release mechanism is non-explosive, shock levels are still high because of the stored strain energy in the bolts and in the brackets due to the preloading of the separation bolts. Typical accelerometer measurements at the separation bracket flange are shown in Figure 7. A peak acceleration of 600 g's was recorded at a frequency of about 1000 Hz. Despite this very high acceleration, this short duration shock pulse did not appear to affect the AlBeMet® material in any way. Even after repeated separation tests, no cracks were observed in the separation bracket flanges or vertical gussets originating with the shock loading. The strength and load bearing capacity of the AlBeMet composite did not seem to be compromised either, since no yielding or permanent deformation was observed during subsequent bolt preloading or further static loads testing.

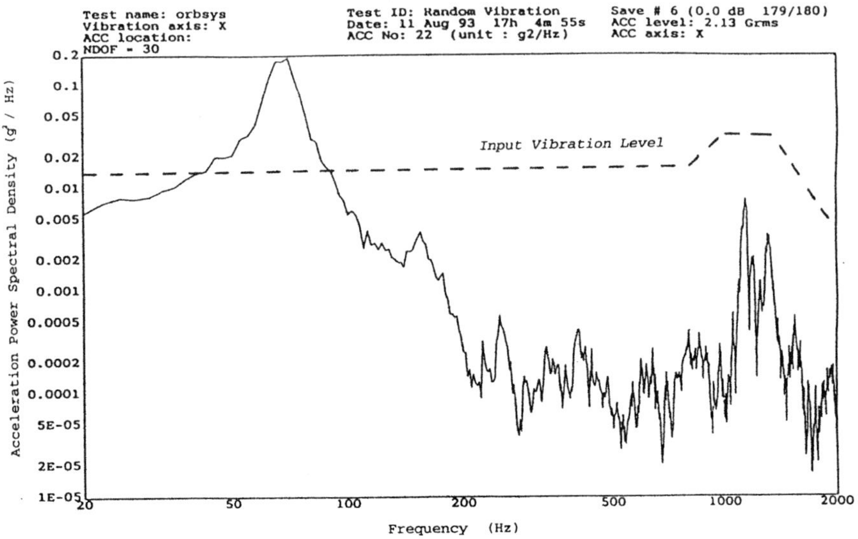

Figure 6 – Axial vibration response power spectral density at the ORBCOMM spacecraft payload shelf center under axial excitation.

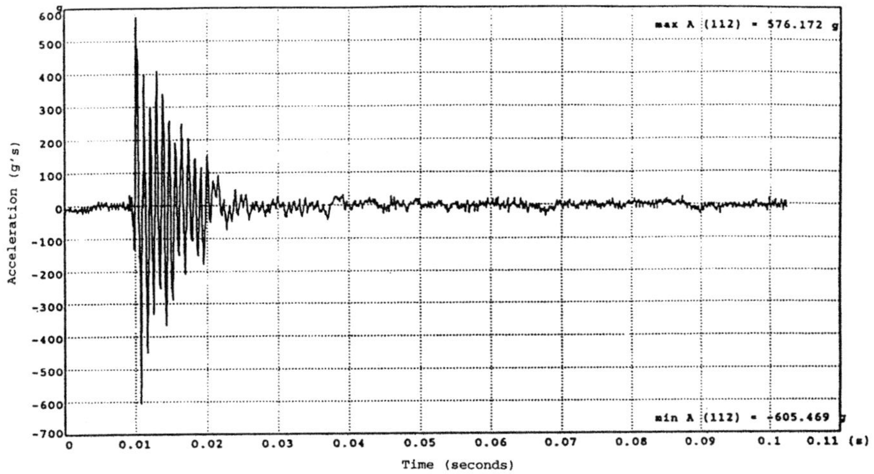

Figure 7 – Accelerometer response at the ORBCOMM spacecraft separation bracket flange during separation shock loading.

Thermal cycle qualification testing was conducted for twenty cycles over the range of -35°C (-31°F) to +40°C (104°F) with dwell times of over 1 hour at each end of the cycle. No detrimental deformation or material behavior was observed during this testing.

3.4 Cost Comparison Based on Brush-Wellman metrics for the fabricated material, AlBeMet® alloys provide about 75% of the properties of beryllium at about one half the price. A more specific comparison can be made for the finished structure, including machining and post-fabrication costs based on OSC data. Besides the ORBCOMM spacecraft, OSC designed, built, tested, and launched another spacecraft (called MicroLab) using the same general structural arrangement (ring structure, separation brackets, deployable solar panels, and deployable antenna boom). Although the payload was different (a optical sensor for a scientific experiment), MicroLab was also launched on Pegasus and was subjected to the same launch environments as the ORBCOMM spacecraft. However, separate constraints on the schedule and the payload led to the selection of magnesium for its primary structure. The two spacecraft primary structures are compared in Table 2. For a given weight of finished structure, the AlBeMet® structure is three times stiffer than the magnesium structure (per pound of material) at only slightly over 25% higher unit cost. In the demanding satellite market, this performance is well worth the slight additional cost.

Spacecraft	Unit	ORBCOMM	MicroLab
Material		AlBeMet®	Magnesium
*Cost**	$	$23,000	$30,000
Weight	kg (lbm)	4.54 (10.0)	7.62 (16.8)
Unit Cost	$/kg ($/lbm)	5000 (2300)	4000 (1800)
Modulus	GPa (Msi)	158 (23)	43.4 (6.3)
Density	g/cc (lbm/in^3)	2.1 (0.079)	1.7 (0.065)
Specific Stiffness	10^6 m (10^6 in)	7.39 (291)	2.46 (97)

* Finished primary structure cost including all machining and post-fabrication costs.

Table 2 – Comparison of cost and properties for ORBCOMM and MicroLab structures

4. CONCLUSIONS

The ORBCOMM qualification test program demonstrated that AlBeMet® materials can withstand the critical launch load environments and conditions for a structure of this type. In particular:

- AlBeMet®'s modulus allowed the stacked ORBCOMM spacecraft to meet the Pegasus' 20 Hz natural frequency requirement.
- AlBeMet®'s strength was more than sufficient to provide ample structural margin under the worst-case predicted static loading from Pegasus.
- AlBeMet® showed no degradation after extended vibration loads typical of a Pegasus launch.
- AlBeMet®'s ductility was sufficient to prevent cracking under the shock loading conditions of spacecraft separation.
- AlBeMet® showed no degradation after extended thermal cycling typical of the space environment.

In addition AlBeMet® proved to be cost-effective for this application. The ORBCOMM spacecraft's use of AlBeMet® materials may pave the way for more widespread use of the material in other space-related structures.

5. REFERENCES

1. Pegasus Launch System: Payload User's Guide, Release 3.00, Orbital Sciences Corporation, Dulles, VA, 1993.
2. F.C. Grensing and D. Hashiguchi, "Properties of Wrought Aluminum-Beryllium Alloys", Advances in Powder and Particulate Materials, MPIF, Vol.4, 1993.
3. F.C. Grensing and D. Hashiguchi, "Mechanical and Thermal Properties of Aluminum-Beryllium Alloy AM162", Advances in Powder Metallurgy and Particulate Materials, MPIF, 1995.
4. D. Hashiguchi and F.C. Grensing, "Ternary Aluminum-Beryllium Alloys", Advances in Powder Metallurgy and Particulate Materials, MPIF, 1995.
5. "ORBCOMM Qualification Static Loads Test Report", OSC Technical Document, TD-1711, June, 1994.
6. " ORBCOMM Development Random Vibration Test Report", OSC Technical Document, TD-1786, Jan. 1995.

28th International SAMPE Technical Conference
November 4-7, 1996

RESIN TRANSFER MOLDING OF POLYIMIDE(PMR15)/FABRIC COMPOSITES

Yoshihiro Soeda,Tsuyoshi Matsumoto
Research Institute of Advanced Material Gas-Generator
2-6, Kohinata 4-Chome, Bunkyo-ku, Tokyo 112, Japan

ABSTRACT

Thermosetting polyimide(PMR15)/fabric composites are important structural components in many high temperature applications. Most of those composites are fabricated from "prepregs". Because of polymer's high viscosity and difficulties of extract of methanol and condensed water from monomers or prepolymers. Generally, it is impossible to use resin transfer molding(RTM) for manufacturing fiber or fabric reinforced PMR15 composite. If RTM method would be usable for making PMR15/fabric composites, it will be very useful for manufacturing cost reduction and near netshape making. So, we tried new resin transfer molding method for the PMR15 composites. Finally, mechanical properties of the composites manufactured by the RTM method are almost as same as the properties of the composites manufactured by the prepreg lay-up method.

KEY WORDS:Resin Transfer Molding(RTM), Polyimide, Porous Filter

1.INTRODUCTION

Nowadays, high performance composites for airplanes and aerospace vehicles, are become popular. Those material are made by traditional method such as an autoclave method, prepreg method or filament winding method. But those methods need much hours and highly skilled personel. And also, they need very expensive manufacturing equipment and man powers. Recently, Resin Transfer Molding(RTM method; from here for shortness 'RTM') method becomes very popular, because of low cost, easiness of making, suitable for small lot production.

On the other hand, many kinds of metal have been used such structural parts. Also, organic material have been used some particular parts. Especially polyimide is a very useful material because of high heat resistance and modulus. So, the polyimide resin was selected for such parts' application. Many kinds of polyimide have been developed. The In-situ Polymerization Momoneric Reactants (from here for shortness 'PMR15') has very good mechanical property and easiness for manufacturing. In this study, PMR15 was chosen from many kind of polyimides. And, if this RTM method is useful for making composites, tremendous cost down and near net shape manufacturing should be achieved(refer Fig.-1).

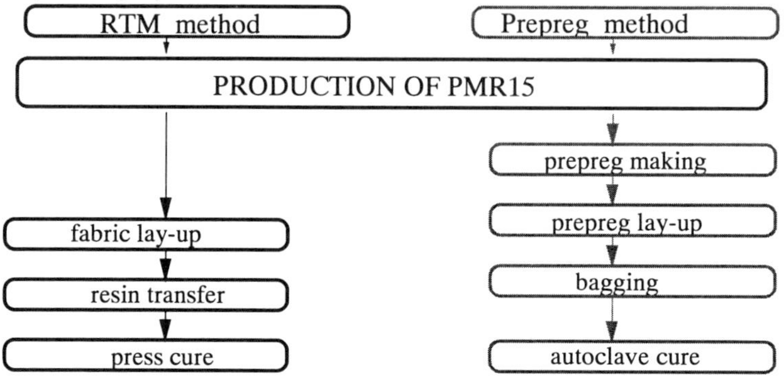

Fig.-1 The production comparison of RTM and Prepreg method

2.MATERIAL CONSIDERATIONS

2.1 Fabric Fabrics specification in this study is listed in Table-1. In this study, CL6649C(TORAY code No.) was used.

2.2 Resin The polymer used in this study was the high-temperature-resistant addition type of polyimide resin system (designated PMR for polymerization of monomer reactants) developed by Lewis Research Center)[1)] . This resin uses three monomer reactants(Fig.-2) :Nadic anhydride Monomethyl ester(NE), 4,4'-Methylene dianiline(MDA), and 3,3',4,4'-Benzophenone Tetracarboxylic Anhydride Dimethylester(BTDE). Usually, an alcohol solution of the monomers is used for reinforced fibers.

Table-1 Specification of the fabric

Name	TORAY TORAYCA fabric CL6649C	
Structure	8 harness satin weave	
Warp	TORAY TORAYCA T800HB-6K-40B	
Fill	TORAY TORAYCA T800HB-6K-40B	
Yarn	Tensile Strength(MPa)	5490
	Modulus(GPa)	294
	Elongation(%)	1.9
	Density(g/cm^3)	1.81
	Yield(g/1000m)	223
Fabric	Count of Warp(Ends/25mm)	17.5
	Count of Fill(Ends/25mm)	17.5
	Areal weight(g/m^2)	321
	Thickness(mm)	0.3

Let us mention the PMR15 polymerization reactions. First, the alcohol as a solvent is evaporates. In this process no chemical reaction occurs. Then the PMR15 varnish is heated to the onset temperature (160 ℃). The condensation reactions that the varnish

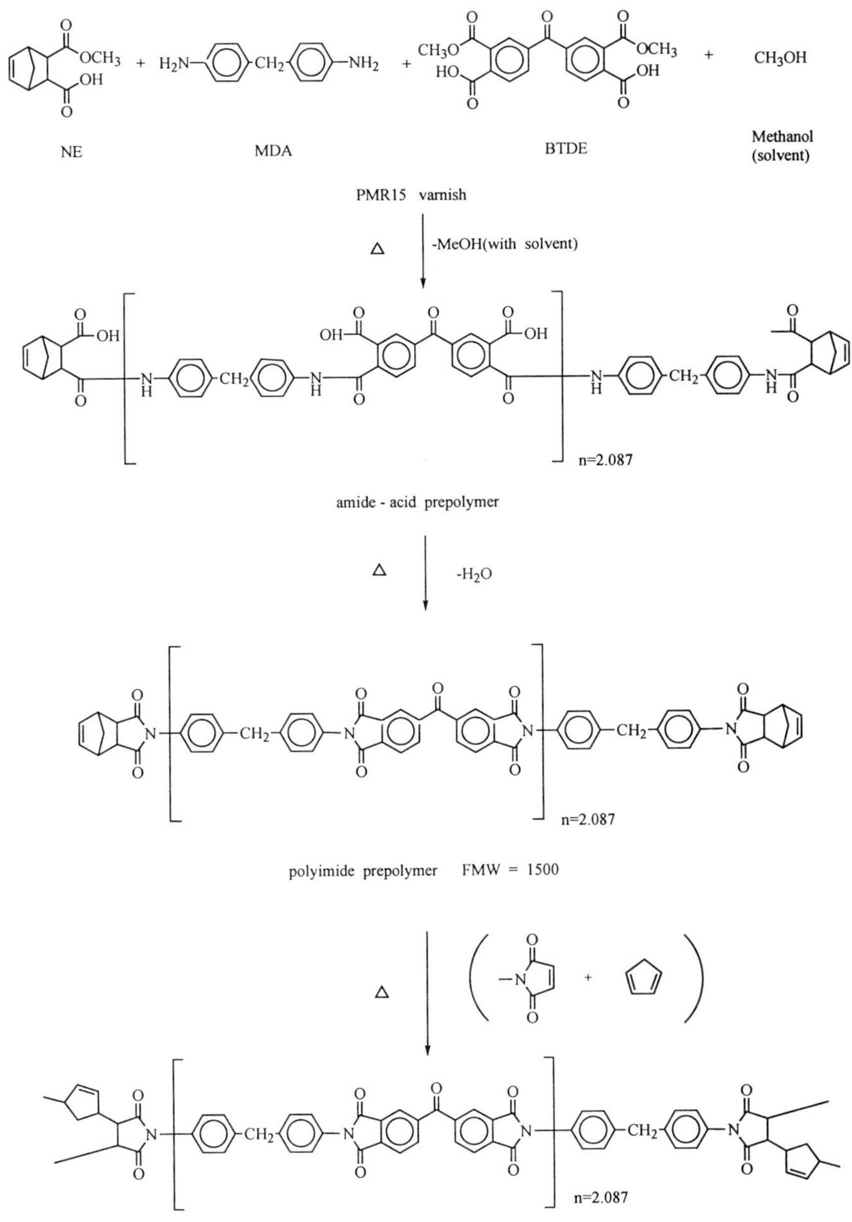

PMR15 : in-situ Polymerization of Monomeric Reactants(FMW=1500)
NE : Nadicanhydride Monomethylester
MDA : 4,4' -Methylene Dianiline
B T D E : 3,3',4,4' -Benzophenone Tetracarboxylic Anhydride Dimethylester

Fig-2 Reaction sequence for PMR15-polyimide resin

changes to amide-acid prepolymer occurs. Then the reaction that amide-acid prepolymer changes to polyimide prepolymer follows. In this stage, the average molecular weight becomes to 1500. Finally, imide oligomer starts crosslinking.

Fig.-3 shows polymer viscosity profiles of each steps. The abscissa axis shows measuring temperature. The vertical axis means measuring viscosity. Generally speaking, the upper limit viscosity of RTM manufacturing method is 10Pa·s. [2] So, it is said that the polyimide prepolymer is not suitable for the RTM method. The other materials can be used for the RTM method. The amid-acid prepolymer's viscosity goes down with temperature, but that viscosity is below 1Pa·s over 100 ℃ . Though viscosity is very low at that high temperature range, the reaction is proceeding in this stage continuously. Therefore, it is very difficult to control resin's viscosity. In consequence above, PMR15 varnish was selected to this study.

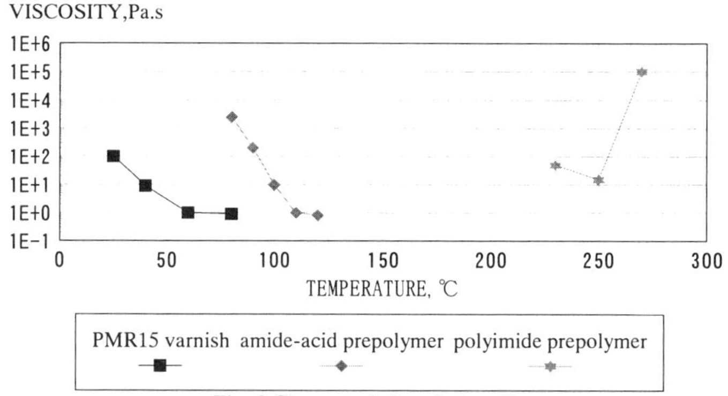

Fig.-3 Characteristics of viscosity

2.3 Filter Generally, because of high viscosity and creation of condensed water during polymerization reaction, it was said that the PMR15 was not suitable for RTM

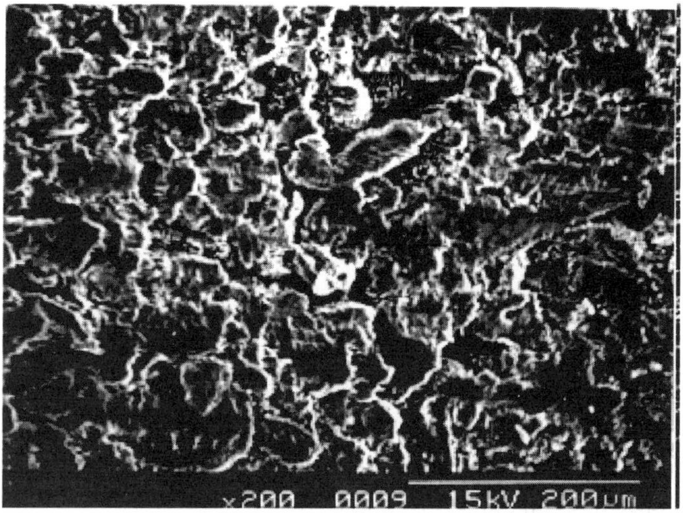

Fig.-4 The cross section photograph(SEM) of porous filter 200 times of magunification

manufacturing. [3] Recently, RTM simulation study for PMR15/fiber composites has been reported.[4] The aim of using porous filter is separation of solvents and vaporizing water from resin. The key point of selection was a difference of molecule size between vaporized water and polyimide prepolymer. There on, in case of this resin system, the curing temperature is very high, the temperature range is from 290 ℃ to 316 ℃ . The filter needs heat resistance too. Finally, release property between polymer and mold is also very important. In consequence of those discussion, polytetrafluoroethylene is the most suitable material for filter material.

Next, the design of porous filters is described here in detail. A polytetrafluoroethylene powder are sintered. The diameter of powder particles used are from 5 to 10 micrometers. Three types of porous filters are made with difference of density. The porous filter(type A) cross section surface SEM (scanning electron microscope) photograph is shown in Fig-4. The photograph of Fig.-4 is 200 times of magnification. According to this photograph, the polytetrafluoroethylene powders are harden into a mass. The diameters of this lump were from 10 to 40 micrometers. In order to check the performance of the filters, the following testes were carried out.

Table-2 Chracteristics of porous filter

filter type	specific gravity	void volume (%)	final vacuume value (hPa)	permeablity H2O	permeablity CH3OH
type A	1.43	33.2	600	GOOD	GOOD
type B	1.57	26.6	500	GOOD	GOOD
type C	1.65	22.9	340	GOOD	GOOD

Porous filter's permeability against vaporized H_2O, methanol and amide-acid prepolymers were evaluated. Characteristics of those three filters were listed in Table-2. And also, test results of air permeability were listed in Table-2. These tests were conducted in accordance with the following method.

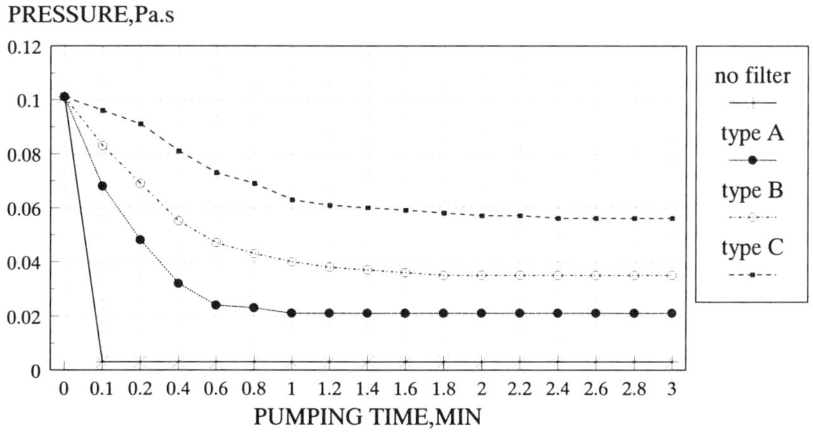

Fig.-5 Permeability of the air

The porous filter is located inside a vessel(volume is 106.6cm ^{3}), and air is pumped up by a vacuum pump. After 3 minutes pumping, pressure was measured by vacuum gauge. Fig.-5 shows all types of filters can pass through air.

Also, methanol and vaporized water were examined. These gases penetrated all kinds of porous filters.

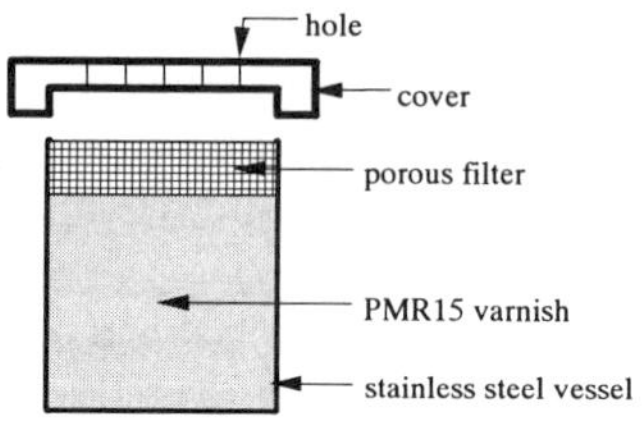

Fig.-6 The vessel for evaluation

Finally, the filter's permeability of amide-acid prepolymers was evaluated. The estimation method was written below. The vessel made of stainless steel was made(refer Fig.-6). First, PMR15 varnish were putted into that vessel. Second, each porous filter was set to the each vessel(cutting same diameter to the vessel). Finally, in order to evaluate gases permeability, a cap with hole was covered. Then put those vessels were set up in the oven. And they were kept for about one hour in it. The oven was set up at 160 ℃. After finishing this treatment, the remainder resins in each vessels were compared.

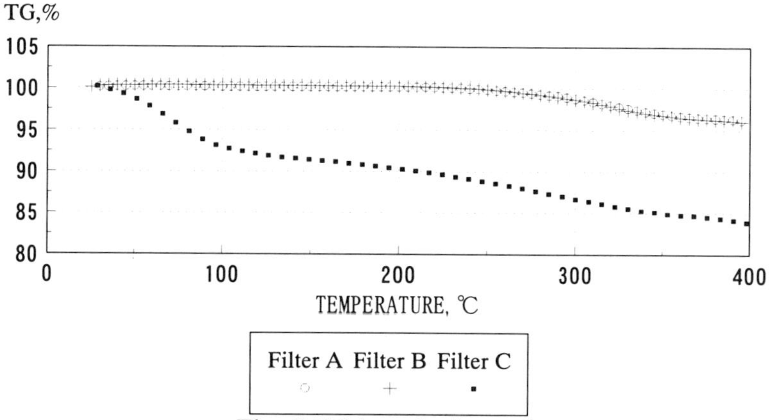

Fig.-7 TGA analysis result

From the observation of those remainders in the vessels, some difference were detected. The remainder of filter A and B were very similar at the point of the color. But the remainder of filter C was a little different from them. The difference of the color was a slightly white and brown and also not transparent.

So, it was very important to define the difference of those remainders more clearly. Thermal analysis(TGA: thermal gravimetric analysis) was done. This results were described in Fig-7. The remainder of the filter A and the filter B had same characteristics of TGA analysis. Both remainder's loss in quantity were almost the same, and the weight loss was about 4 percent. And the loss in quantity began at 250 ℃. On the other hand, the remainder's loss of filter C in quantity was very different from the others, and it was about 16 percent which was about four times of the others. And the loss in quantity began at 50 ℃.

3.EXPERIMENTAL

3.1 Manufacture of carbon fabric/PMR15 composites In order to estimate the effect of this new RTM method with porous filter, the composites were manufactured. After making those composites, the characteristics were evaluated.

3.1.1 Resin potting The temperature dependence of viscosity was already indicated in Fig.-2. The viscosity of PMR15 varnish was about 1Pa.s. Usually, resin below 10Pa.s is useful for RTM method. [2)]
The outline of the RTM mold system in this study is described in Fig-7. In this study, the PMR15 varnish was potted into the mold. The PMR15 varnish, carbon fiber fabric and the mold were heated at 70 ℃ . The fabrication process is as follows.

1st step: PMR15 resin was potted into the mold.
2nd step: 4ply carbon fabrics were set
3rd step: 1st step and 2nd step were repeated several times
4th step: Porous filter was set
5th step: Upper mold was set
6th step: Vacuum system was set up
7th step: Heated up to 160 ℃ and pumped up

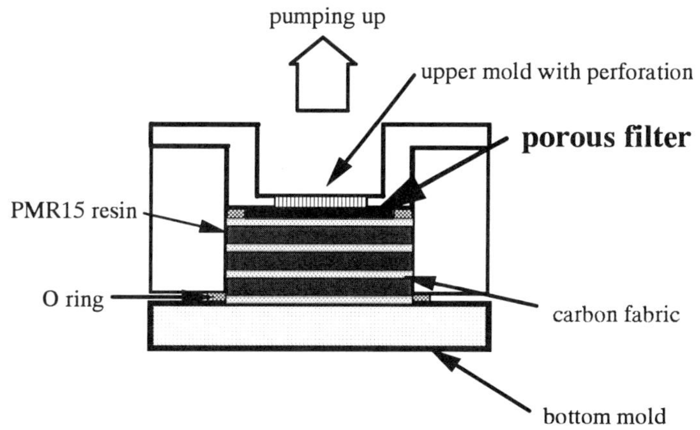

Fig.-8 Approximate illustration of RTM mold

3.1.2 Cure After the temperature reached 160 ℃ , it was kept one hour. In this process, methanol was vaporized, then was exhausted by vacuum pump. After finishing this process, vacuum system was removed and the mold was heated to 200 ℃ , it was kept 2.0hours. The condensed water was exhausted through the filter by its vapor pressure. Cure was performed by the following conditions. Temperature: 316 ℃ , Time: 3.5hours, pressure: 1.4MPa. After this sequence, post cure was performed by the following conditions. Temperature: 316 ℃ , Time: 30hours.

A IMIDIZE REACTION

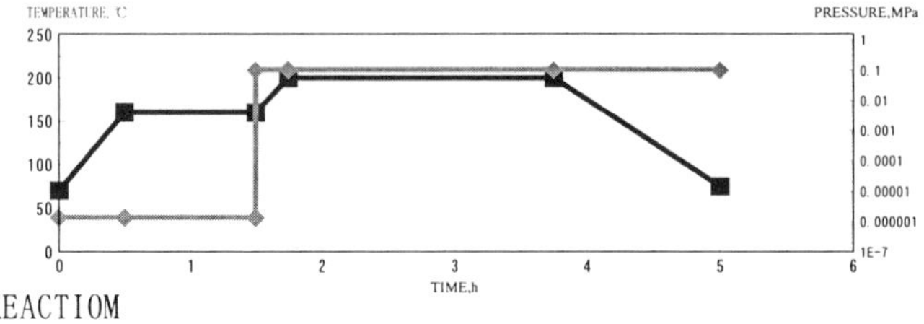

B CURING REACTIOM

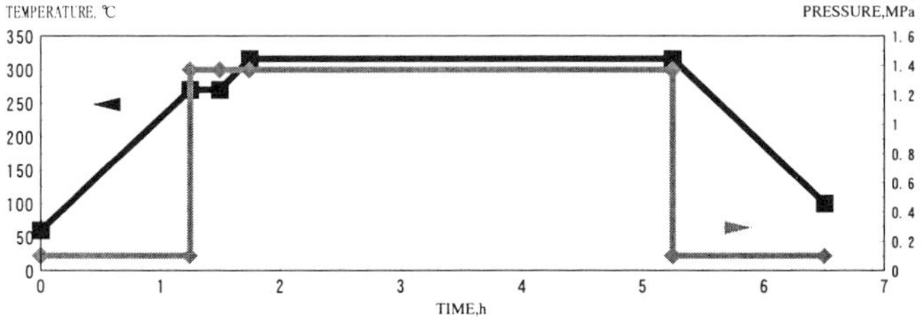

Fig.-9 the condition for manufacturing composite

3.2 Evaluation of test specimen The composite made by RTM method using porous filter was evaluated by both surface and in section surface. The photograph of carbon fabric/PMR15(8ply) and carbon fabric/PMR15(32ply) composites are shown in Fig-10.

a) surface view of 8ply composite

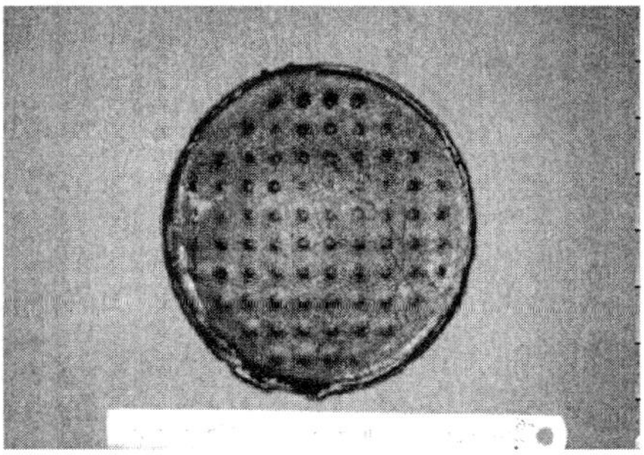

b) surface view of 32ply composite

Fig.-10 The surface view of the composites

Quality of these composites were good. The 32ply thick composite could be fabricated by this RTM method. But, its surface looks like a gooseskin. It appears only the side which is contacts with porous filter. Because, the shape of filter changes by the pressure.

3.3 Density, fiber volume content(Vf), void volume content(Vv)

The density was measured by liquid displacement method, then calculated from equation(1).

$$\rho = \frac{W \times \rho_W}{(W - W_W)} \quad \cdot\cdot\cdot\cdot\cdot (1)$$

here ρ_w is a density of distilled water, is equal to 0.9975g/cm ^{3} . W is a weight in air. W_w is a weight in methanol.

The V_f value was calculated by using density and fiber weight.

$$Vf=\frac{(Wf \times \rho)}{(\rho f \times W)} \times 100 \cdots\cdots (2)$$

in here ρ_f means the density of fiber, ρ_f is 1.81g/cm^3 . W_f is fiber weight.

Void volume content is calculated from equation(4) by using resin volume content(Vr) and fiber volume content(V_f).

$$Vr=\frac{(W-Wf) \times \rho}{(\rho r \times W)} \times 100 \cdots\cdots (3)$$

$$Vv=100-Vf-Vr \cdots\cdots\cdots\cdots (4)$$

in here ρ_r is the density of resin , ρ_r is 1.323g/cm^3 .

The results are summarized in Table-3. According to the result, Vr was 41.96%, Vv was 2.91%.

Table-3 Measurement result of composite

	weight in air (g)	weight in MeOH (g)	density ρ (g/cm3)	fiber weight Wf (g)	Vf	Vr	Vv
avg	1.8531	0.9030	1.5531	1.1910	55.140	41.96	2.91

3.4 Flexural strength and Modulus

According to the ASTM D790, flexural strength(Fb) and modulus was measured at room temperature. The composite of 16ply was tested by Test Method Ⅰ (A three-point loading system). On the other hand , the composite of 8ply was tested by method Ⅱ (A four-point loading system). Each flexural strength(Fb) and modulus(Eb) were calculated from the following equation.

Ⅰ Test Method Ⅰ

$$Fb=\frac{3PL}{2bd^2} \qquad Eb=\frac{L^3m}{4bd^3}$$

Ⅱ Test Method Ⅱ

$$Fb=\frac{3PL}{4bd^2} \qquad Eb=\frac{0.17L^3m}{bd^3}$$

here, b means width of beam tested, d means depth of beam tested, P means load at a given point on the load-deflection curve, L means support span, and m means slope of the tangent to the initial straight-line portion of the load-deflection curve.

Test result of the carbon fabric composite(16ply) and the carbon fabric composite(8ply) was summarized in Table-4. The property of the composite by using prepreg is also shown in Table-4. Fiber areal weight of composite made by RTM is the same as that of composite made by using prepreg.

Table-4 The mechanical properties of the RTM composites

test specimen	b (mm)	d (mm)	L (mm)	P (N)	Fb (MPa)	m (N/mm)	Eb (GPa)
16ply	12.978	5.321	85.15	1757	610.4	716.4	56.53
8ply	24.997	2.617	84.00	2304	848.3	296.1	66.68
preprg	13.003	4.537	72.60	2301	935.9	876.1	69.03

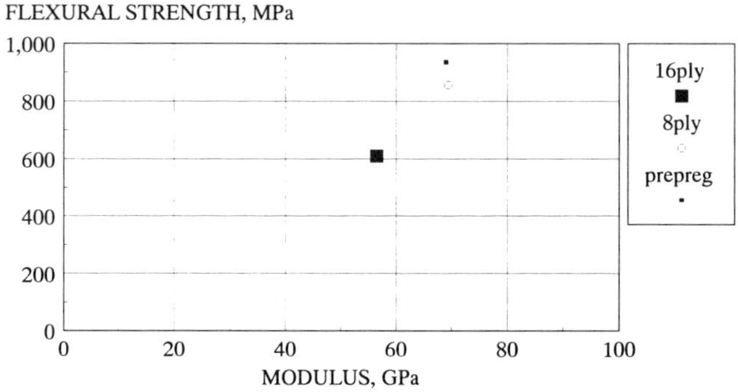

Fig.-11 Relation of flexural modulus and flexural strength

Furthermore, Fig-11 shows a relation of the modulus and strength. The composite made by RTM has same modulus as the composite made by prepreg. But, it's strength was a little lower than that of prepreg's. Its difference derives from respective ply thickness.

The ply thickness of composite made by prepreg was 0.2835mm. The ply thickness of 16ply composite was 0.3326mm. The ply thickness of 8ply composite was 0.3183mm. Recalculation of flexural strength using ply thickness of the composite made by prepreg was done. These results are shown in Fig.-12.

After compensation using this data, the flexural strength for 16ply was 716.1MPa, and also the flexural strength of 8ply was 958.1MPa.

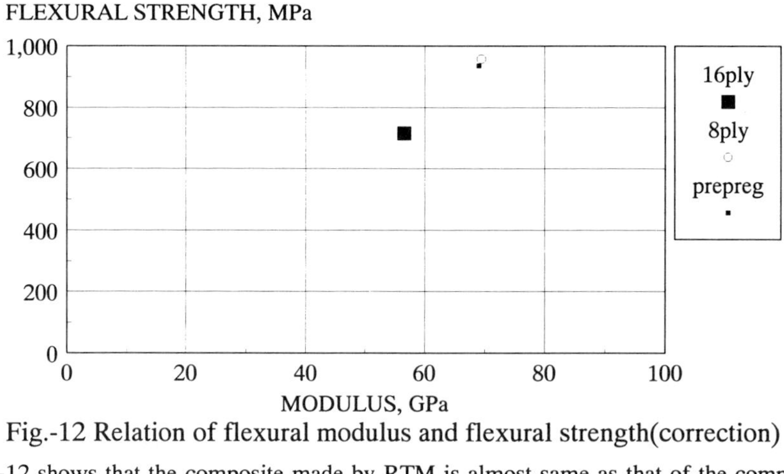

Fig.-12 Relation of flexural modulus and flexural strength(correction)

Fig.-12 shows that the composite made by RTM is almost same as that of the composite made by prepreg.

3.4.1 Glass transition temperature The measurement of glass transition temperature(Tg) was measured by TMA(thermal mechanical analysis). Tg were shown in Fig.-13.
Tg was between 340 ℃ and 350 ℃ . The glass transition temperature of the composite made by this method was a little higher than usual manufacturing method(prepreg method).

From this point of view, new RTM method using the porous filter is very good procedure.

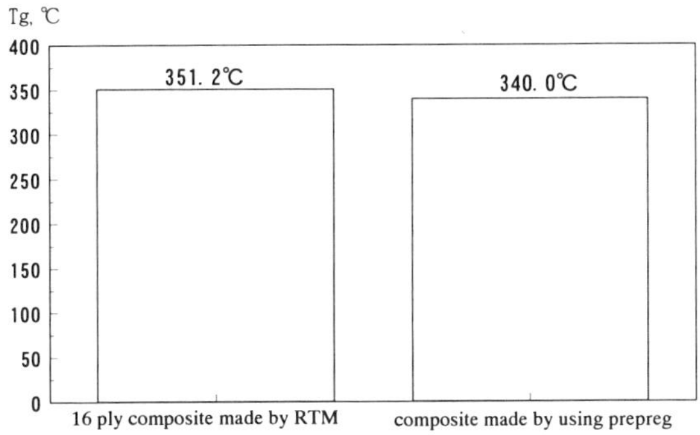

Fig.13 Glass transition temperature

4.CONCLUSIONS

According to this study, the new RTM method using the porous polytetrafluoroethylene filter is very useful and convenient for fabricating the PMR15 composite.

Ⅰ A porous filter is very useful for separating monomer, air, vaporizing water and prepolymer.

Ⅱ A porous filter should be selected considering molecular size of inter mediate products and filter's permeability.

Ⅲ The composite made by this RTM method has good mechanical properties and the property is almost same as the composite made by 'prepreg' method.

Ⅳ The composite made by this RTM method has higher Tg than that of the composite made by 'prepreg' method.

In this study, I appreciate Mr.Masato Tadokoro belongs to THE YOKOHAMA RUBBER CO. LTD..

5.REFERENCES

1. Serafini, Tito T.;Delvigs,Peter and Lightsey Geroge R., J.Apply.Polym.Sci., 16, (4), 905(1972).

2) E.B. Stark, W.V. Breitigam,R.D. Farris and D.G. Davis, SAMPE International Symposium, 35, 423(1990)

3) Doug Wilson, High Performance Polymers, 3, (2), 73(1991)

4) A. W. Chan and S. T. Hwang, Polymer Composites, 14, (6), 524(1993)

28th International SAMPE Technical Conference
November 4-7, 1996

THE EFFECT OF TIME AND TEMPERATURE ON THE FLEXURAL STRENGTH OF A SILICA PARTICLE FILLED EPOXY RESIN

Michael K. McMurray and Shigeo Amagi
The First Department of Materials Research, Insulating Materials Group,
Hitachi Research Laboratory, Hitachi, Ltd.,
Hitachi-shi, Japan 319-12

ABSTRACT

Composite materials that use an epoxy as a matrix resins have superior mechanical properties but these materials exhibit time and temperature behavior when used for long periods and under high temperatures. This time and temperature behavior is not yet fully understood. The purpose of this paper is to further describe this time and temperature behavior, increasing the reliability of these types of composite materials. The time and temperature dependence of flexural strength was examined by flexural static and fatigue testing. Static tests were carried out at various constant deflection rates and temperatures below the glass transition temperature of the epoxy resin. It was shown that the time-temperature superposition principle is applicable for the flexural static strength. Comparison of the neat resin and the composite results show that the neat resin is a major influence on the time and temperature behavior of the static strength behavior of the composite material. Fatigue tests were carried out at several stress ratios and various temperatures below the glass transition temperature of the matrix resin. It was shown that the time-temperature superposition principle is applicable for the initial flexural fatigue strength by using the time-temperature behavior of the matrix resin. As temperature and stress ratio increased the fatigue strength became more creep damage dependent.

KEY WORDS: silica particle, time-temperature superposition principle, flexural strength

1. INTRODUCTION

1.1 Purpose Epoxy resin with a filler, such as silica or alumina powder, because of its excellent electrical properties, small thermal expansion coefficient, strength and modulus, is being used in many electric products. For example, this composite is being used in molded transformers, the propulsion coils for the magnetic levitating vehicle and LSI encapsulation. In these applications the material undergoes complex loading, for example fatigue and creep loading. Propulsion coils and transformer applications require the composite to prevent the deflection of the windings of the coils due to magnetic forces. The material must also sustain its properties as the temperature increases during operation. It is commonly known that the mechanical behavior of epoxy resin, and composites made with these resins show viscoelastic behavior. This viscoelastic behavior is a major concern when designing structures for adverse environmental conditions and long periods, such as the applications mentioned above. This report is the study of the influence of time and temperature on the failure strength behavior of this material subjected to complex loading.

1.2 Previous research Many papers have been written on the effect of filler in polymer composites on their mechanical properties[1-5]. A major portion of the papers like Brown, discusses the effect of filler on the modulus and toughness of the composite. They have found that depending on the type of filler the modulus and toughness will increase with an increase in filler but this usually leads to a decrease in static and impact strength.

Static strength of a similar epoxy matrix resin with silica particles was tested by Nishimura et al.[6]. They found that this composite's static strength decreased with temperature but did not examine the deflection rate effect. They also found that at low temperatures the initiation of fracture began in the silica particles. A major cause of failure at high temperature was the interface between the particle and the resin.

Fatigue testing of filled thermoset composites has mainly been concerned with the effect of the amount and shape of the filler used in composites [1,6-7]. Many papers report the influence on the amount of particulate filler and its effects on the strength of their composite. Brown concluded that addition of glass bead filler decreased the flexural strength of the composite when conducting strain controlled fatigue tests. Adhesion between the filler and matrix is believed to significantly influence the fatigue properties because of the stress concentrations around the fillers [1]. Nishimura et al. found that fatigue strength of a silica filled epoxy resin was dependent on the type of stress input wave and mean stress [6]. They also determined that the silica filled epoxy resin's fatigue strength was independent of the frequency range tested (1-10Hz). They concluded that most of the fracture results could be attributed to the brittle fracture of the particles.

Research on the fatigue stress ratio influence on the strength of epoxy matrix FRP (Fiber Reinforced Plastics) has been performed previously [8-10]. Results from Miyano et al. and others showed that FRP fatigue strength dependence on stress ratio changes as temperature increases to the glass transition temperature of the matrix resin. Fatigue strength was measured with several ratios of mean stress to stress amplitude ranging from near zero to 1. In comparison with the other stress ratios the smallest fatigue strength at room temperature occurred when stress ratio was near zero for a given time. When the temperature neared the glass transition temperature, the stress ratio of near zero had the highest strength for a given time. Satin woven FRP composites showed a smaller difference of strength due to stress ratio when compared to unidirectional FRP. Therefore particle filled epoxy matrix composites are expected to also show this change in dependence due to temperature.

2. EXPERIMENTAL PROCEDURE

2.1 Specimen Preparation Specimens were an epoxy resin with crystalline silica particle filler. The epoxy resin and crystalline silica particles were mixed and poured into molds made of steel plates coated with a mold release of silicon oil, to produce specimens approximately 5mm thick. Air bubbles were removed by placing the molds in a vacuum. Curing temperature and time were 130°C for 3 hours then 150°C for 5 hours. These molds were then allowed to cool to 25°C at a rate of 30°C/hour. Specimens for dynamic viscoelastic testing were cut to dimensions of approximately 5×1×32mm (width, thickness, length). The specimens for static tests were then cut and sanded to dimensions of approximately 13×5×110mm (width, thickness, length). Specimens for fatigue tests were cut and polished to dimensions of approximately 13×5×110mm. These specimens were after cured at a constant temperature of 170°C for 5 hours with a cooled down at a rate of 30°C/hour. The silica content was 60% by volume with a maximum dimension of 80μm and an average size of 7μm, as given by the supplier.

2.2 Testing Conditions Data for the tension dynamic viscoelastic tests for storage modulus versus time curves were taken at frequencies of 1, 10, 100, 1000Hz at constant temperatures between 35 and 250°C, with a DVE Rhospectoler, DVE-V4 (Rheology Co. Ltd.). Static tests were performed in a 3-point bending jig at constant temperatures of 25, 40, 60, 80, and 100°C and deflection rates of 0.01, 0.1 and 100mm/min. with an Autograph IS-20T (Shimadzu Co.) with an attached constant temperature chamber. The span of the static tests was 80mm. Five specimens were tested for each static testing condition, a combination of temperature and deflection rate.

Data for the flexural fatigue maximum stress versus number of cycles to failure curves were taken at a frequency of 10Hz at constant temperatures of room temperature(24°C), 40, 60, 80, and 100°C. Preliminary testing showed that at 10Hz and 100°C, the surface temperature of the specimen only increased 1-2°C, therefore we assume there was negligible effect from heating do to testing frequency. These tests were conducted with a computer controlled servopulser EHF-FD05-4LA (Shimadzu Co.) with an attached constant temperature chamber. Fatigue tests were performed under stress control and a sine input wave. At least three specimens were tested for each stress level in 3-point bending. For comparison purposes fatigue test conditions are similar to the static tests done previously. Static testing span was 80mm but due to limitations of the fatigue testing machine's stroke, the span for fatigue test was changed to 60mm.

3. RESULTS AND DISCUSSION

3.1 Dynamic viscoelastic test results The logarithm of storage modulus E' versus time at several temperatures for the neat resin is shown in the left side of Figure 1. Time t is

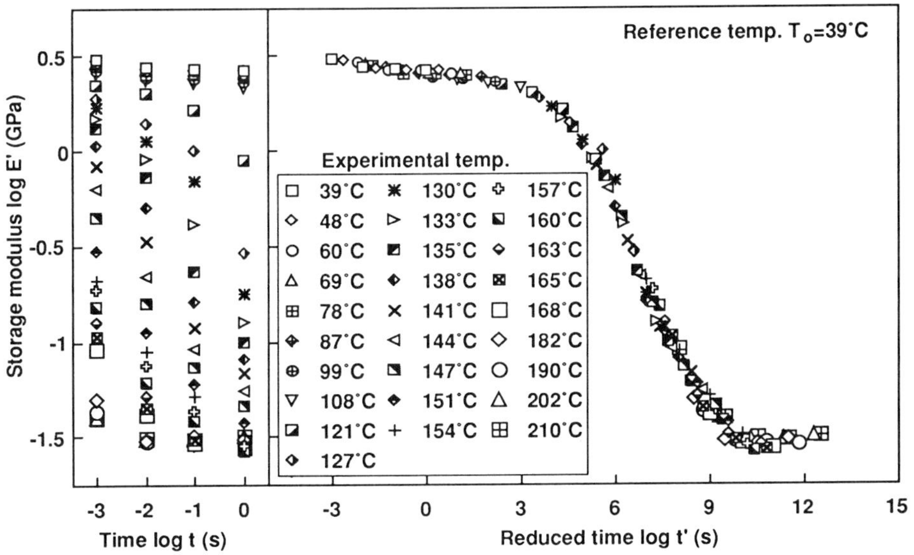

Figure 1. Master curve of storage modulus E' versus reduced time t' for the matrix resin after cured at 170°C for 5 hours.

equal to the inverse of frequency 1/f(1/Hz). Storage modulus E' for this neat resin specimen decreases considerably at 121°C, near T_g (T_g=119°C determined by TMA tests). The master curves of storage modulus E' versus reduced time t' at a reference temperature of T_o=39°C is shown on the right hand side of Figure 1. Construction of the master curve was accomplished by shifting the E' data points from left-hand side. Storage modulus E' at the various temperatures were shifted along the log scale of time until the data points overlapped, creating a smooth curve. The horizontal time-temperature shift factor $a_{T_o}(T)$ is defined as:

$$a_{T_o} = \frac{t}{t'} \tag{1}$$

where t is the experimental time data, and t' is the reduced time. Since a smooth curve is produced when shifting the E' data with respect to time, the time-temperature superposition principle is applicable.

3.2 Static Strength Typical load P deflection δ diagrams of flexural static tests for a constant deflection rate and 5 temperatures is shown in Figure 2. The figure shows that as the temperature increases failure load P decreases and deflection δ increases. It is also noticed that as the temperature increases the ratio P/δ decreases. Data taken at other deflection rates showed the same basic results but as deflection rate increased failure load increased. Even at temperatures near room temperature, a change is seen in static strength.

Figure 3 is the master curve of flexural static strength σ_{max} versus reduced time t' at a reference temperature of T_o=40°C. The left-hand side is the average values of 5 data points of maximum stress σ_{max} and log time to failure t. Static strength is dependent on time to failure t and temperature T, with the dependence becoming stronger as the temperature increases. This effect is a result of the temperature nearing the T_g of the matrix resin. The master curve of flexural

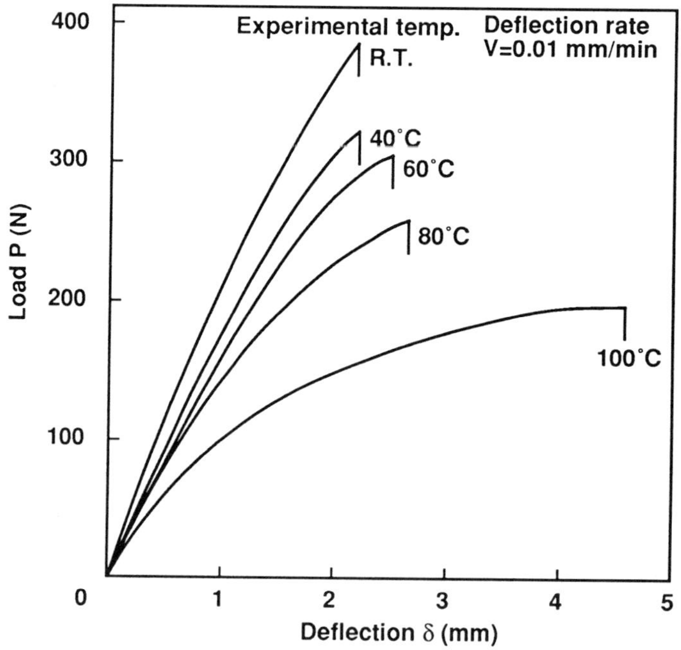

Figure 2. Load-deflection diagram for various temperatures and a deflection rate of V=0.01 mm/min.

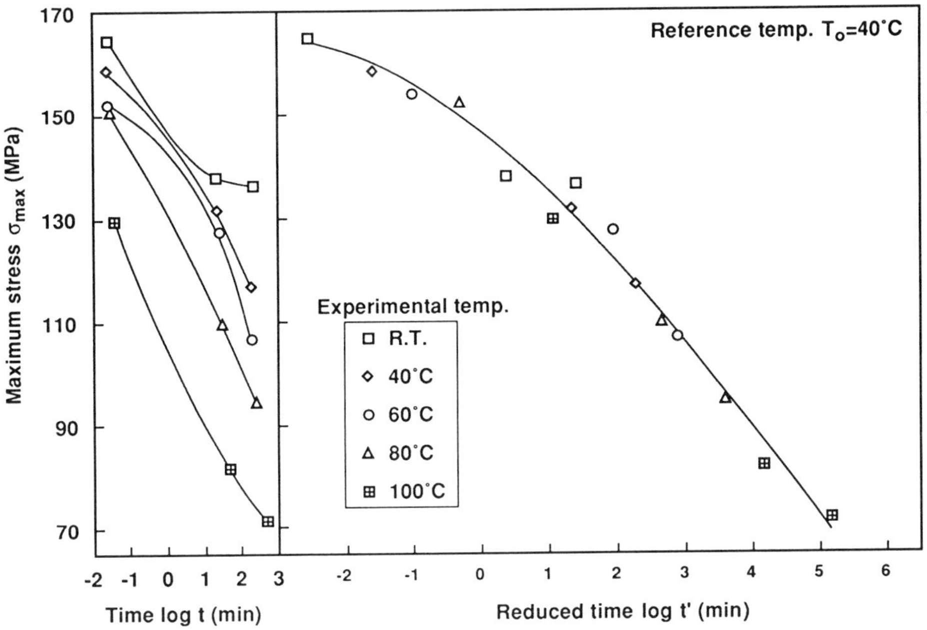

Figure 3. Master curve of flexural static strength for T_0=40°C.

static strength σ_{max} versus reduced time t' is seen in the right-hand side of the figure. Since σ_{max} at various temperature can be superimposed so that a smooth curve is created, the time-temperature superposition principle is applicable.

A comparison of the time-temperature shift factors a_{T_0} for static strength, storage modulus of the neat resin is seen in Figure 4. The time-temperature shift factors a_{T_0} change drastically with temperature near the T_g of the matrix resin. Shift factors a_{T_0} are in good qualitative agreement with the solid lines, which represent Arrhenuis' equation. Therefore a_{T_0} can be represented by using 2 different activation energies ΔH and the Arrhenuis' equation (2):

$$\log\ a_{T_0}(T) = \frac{\Delta H}{2.303R}\left[\frac{1}{T} - \frac{1}{T_0}\right] \tag{2}$$

where ΔH is the activation energy (kJ mol^{-1}), R= gas constant (8.314×10^{-3} kJ K^{-1} mol^{-1}), T is the testing temperature (K) and T_0 is the reference temperature (K)[11-12]. By using the time-temperature superposition principle the long term static strength behavior of this composite can be determined using short term and high temperature tests.

3.3 Explanation of Stress Ratio Fatigue tests were conducted at 3 stress ratios, R=0.05, 0.5 and 0.8. Stress ratio R is the ratio of minimum stress to maximum stress:

$$R = \frac{\sigma_{min}}{\sigma_{max}} \tag{3}$$

where σ_{min} is the minimum stress and σ_{max} is the maximum stress during fatigue testing. Minimum stress σ_{min} is defined as the smallest stress value during one fatigue cycle (1/ frequency). Maximum stress σ_{max} is defined as the largest stress value during one fatigue cycle.

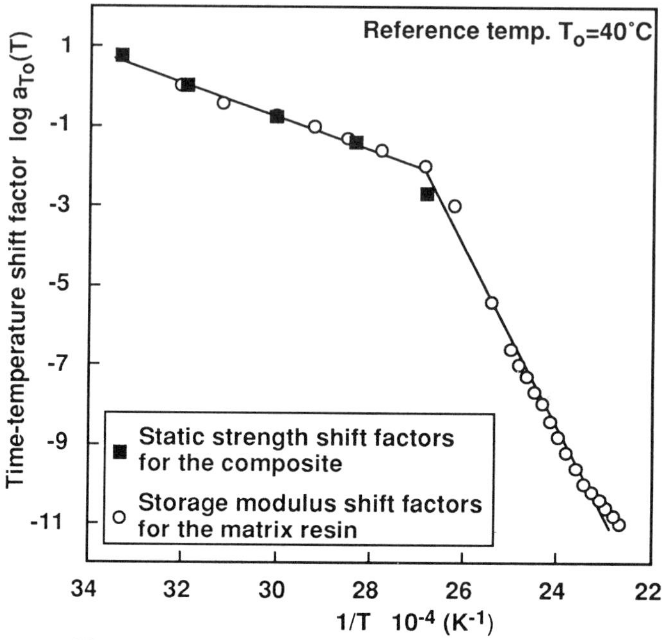

Figure 4. Time-temperature shift factors for flexural static strength and storage modulus for the matrix resin.

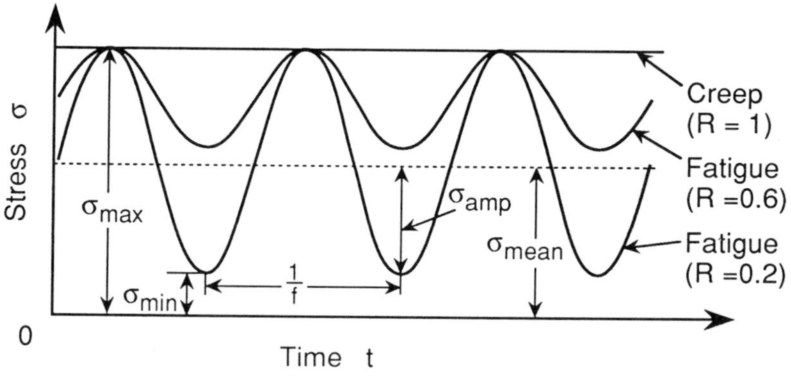

Figure 5. Explanation of stress ratio R.

Mean stress σ_{mean} is the average value between σ_{min} and σ_{max}, these definitions are explained graphicly in Figure 5. As R increases from 0 to 1 for a constant stress, the stress amplitude goes to 0 and mean stress becomes equal to maximum stress. When R is small, this means that the stress amplitude is quite large in comparison with a large R which has a small stress amplitude. This means that as R increases to 1 the stress amplitude becomes smaller but the mean stress becomes larger. Therefore damage due to creep loading should increase for large R values. This is shown in Figure 5. Stress ratio of R=0.05 is assumed to be equal to R=0 since it is difficult to test at a smaller stress ratio with a 3 point bending jig. This is due to the loading point losing contact with the specimen creating an impact type condition. To obtain an range of values, specimens were also tested at stress ratios of R=0.5 and R=0.8.

3.4 Fatigue Deflection and Modulus Curves Figure 6 is the maximum deflection versus time curves for 3 temperatures and various stress ratios. The values written inside the figure represent the maximum flexural testing stress for each curve. At T=24°C there was little increase in deflection seen before complete failure for all stress ratios. As temperature increased to T=60°C a small increase in deflection was seen for stress ratio of R=0.5 and R=0.8, but there was no increase in deflection before failure for R=0.05. When the temperature was increased to T=100°C, the deflection increased before failure for all stress ratios. Stress ratio of 0.8 had the largest rate of increase in deflection. This shows that creep influence is larger for larger stress ratios and at higher temperatures as described in 3.3.

3.3 Maximum fatigue stress versus time to failure curves Equation 4 was used to convert number of cycles to failure to time to failure so that static strength results could be compared with fatigue strength results:

$$t = \frac{N}{f} \tag{4}$$

where t is time (sec.), N is number of cycles (cycles) and f is testing frequency (Hz). This allows the comparison of fatigue with other properties usually measured with time, as in static strength [10].

Flexural fatigue strength results are plotted in Figures 7 and 8. Flexural fatigue strength is assumed to be equal to flexural static strength with a time to failure equal to 1/2 cycle. The points at a log time to failure of -3 and a stress ratio R=0.05 for each temperature is flexural static results. Since static tests could not be conducted at log time to failure equal to -3, stress values were arrived at by using the time-temperature superposition principle and the master curve of static strength. Using the time-temperature shift factors from the modulus data of the matrix resin, the strength at log time -3 was calculated.

As temperature increases, the difference among the 3 stress ratios decreases, showing that as temperature increases, stress ratio has less of an influence on fatigue strength. The slopes of the curves for R=0.05 and 0.8 are almost constant for all temperatures, except the slope for R=0.5 which changes drastically from 60°C. This is believed to because of a change in fracture mechanism. From the data shown, at low temperatures the composite fails with little deformation but as the temperature increases the composite becomes more ductile and creep damage occurs, which corresponds with the results from Figure 5.

The points at log time to failure -3, static strength, corresponds well with the curves that represent fatigue strength at R=0.05 for each temperature. This means, initial fatigue damage can be thought of a step-wise accumulation of damage equal to static strength at the same temperature with a time to failure equal to the time of 1/2 fatigue cycle. Therefore a type of damage accumulation prediction method may be appropriate for fatigue life. Since the time-temperature superposition principle, using the time-temperature shift factors from the matrix resin, held for the static strength which corresponds well with the curves representing the fatigue strength, then it can also be said that the time-temperature superposition principle also holds for the initial fatigue strength.

A single stress ratio for all temperatures is plotted in Figure 9. This shows clearly that the slope of the curves of all stress ratios changes little until 60°C. The amount that the strength decreases because of temperature becomes larger as stress ratio increases, this shows that creep damage increases with temperature and is an influence on fatigue strength. At 60°C the slope of the curves for R=0.5 changes drastically, but at 80 and 100°C the slope is similar to conditions below 60°C. The behavior of stress ratio R=0.8 is similar to that of R=0.5 but is not as extreme.

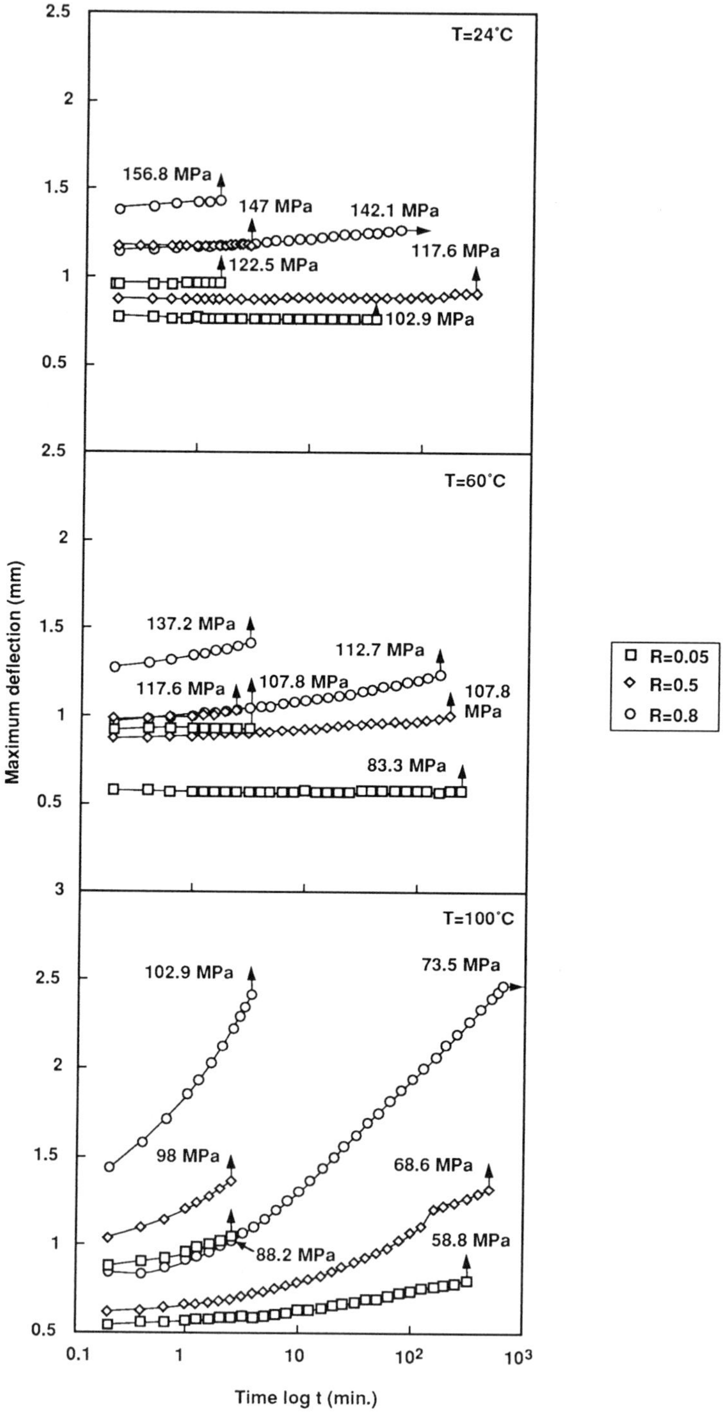

Figure 6. Deflection versus time for various stress ratios at various temperatures.

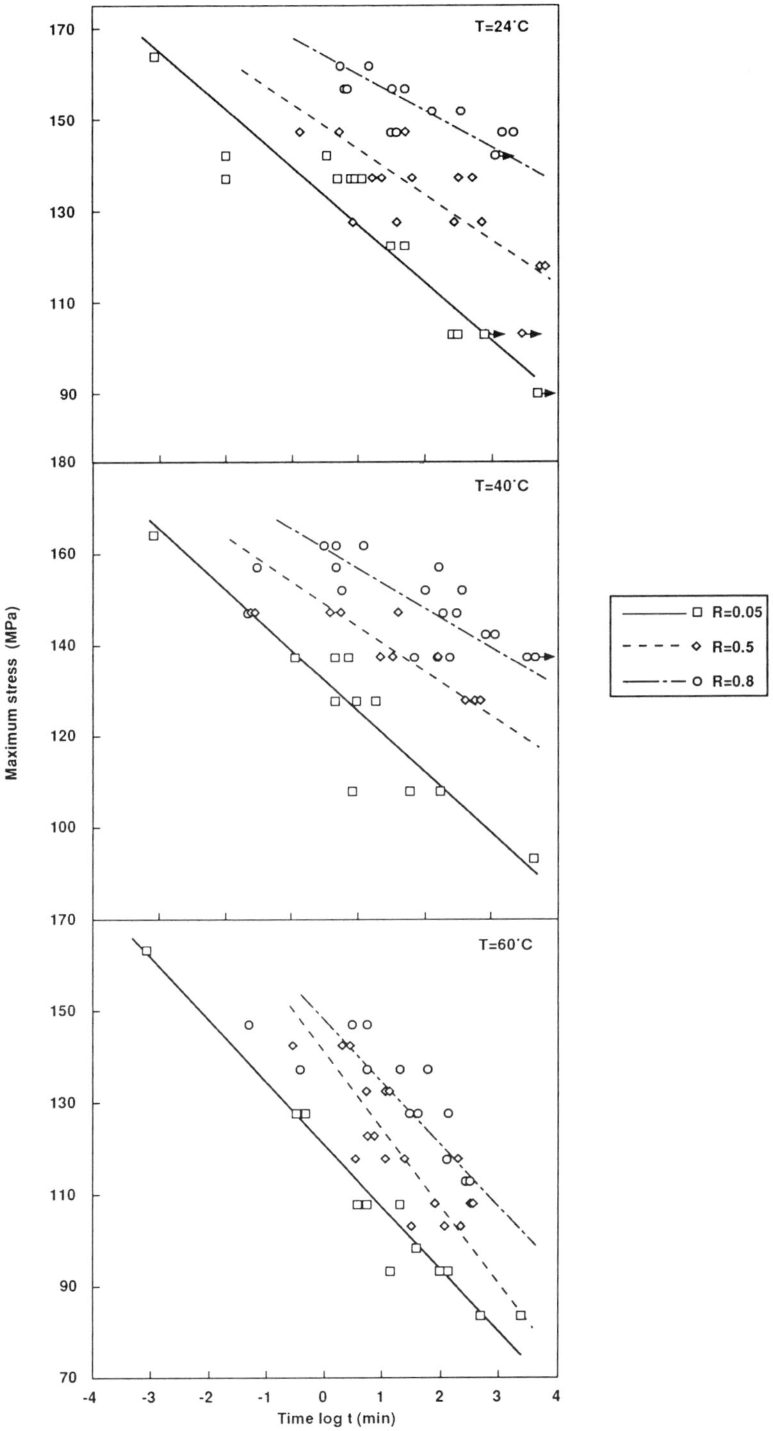

Figure 7. Maximum stress versus time to failure at various temperatures and stress ratios R.

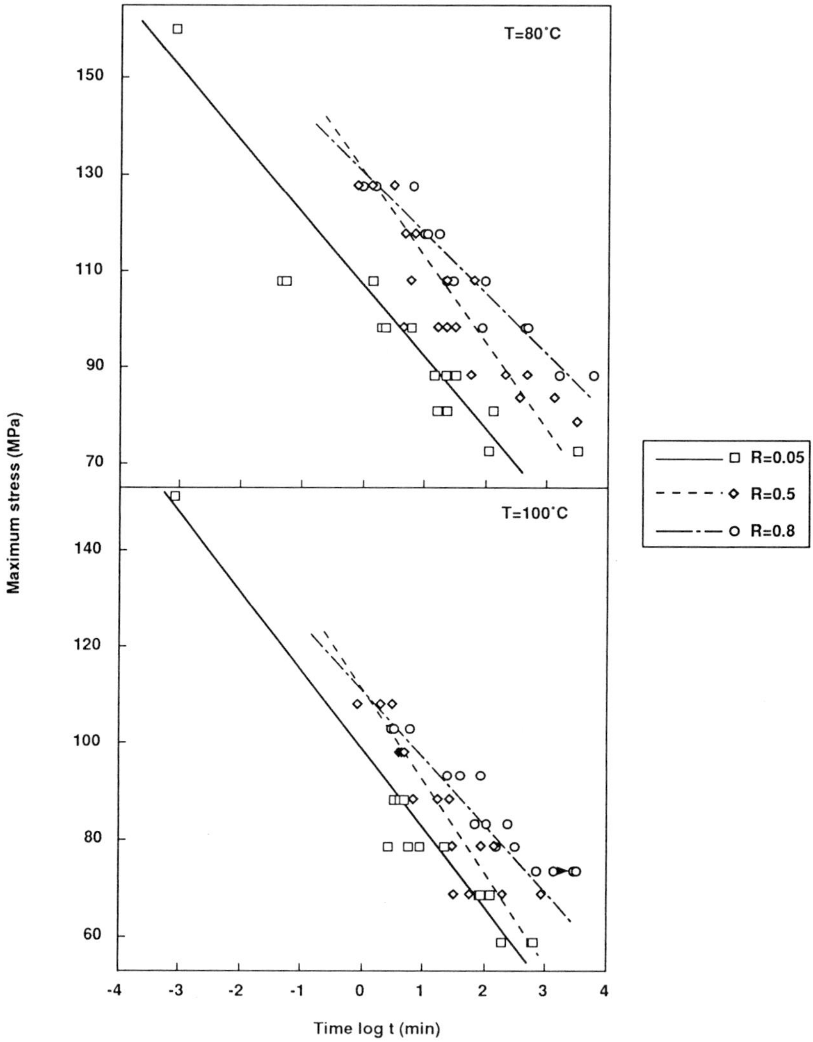

Figure 8. Maximum stress versus time to failure at various temperatures and stress ratios R.

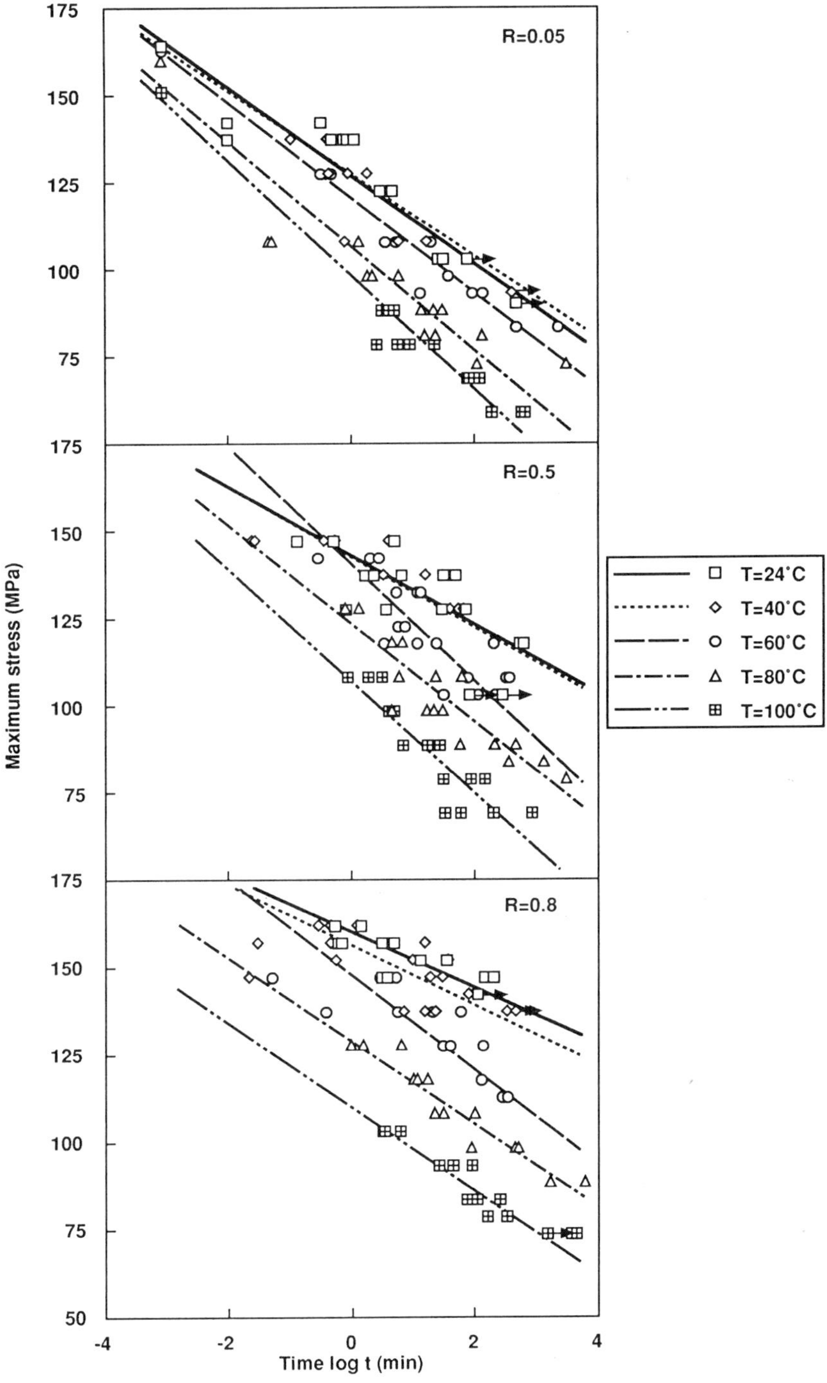

Figure 9. Maximum stress versus log time for stress ratio R at various temperatures.

4. CONCLUSIONS

The following conclusions were obtained from static and fatigue tests conducted at several deflection rates, temperatures and stress ratios on a epoxy resin with crystalline silica particle filler:

(1) The storage modulus is time and temperature dependent, becoming more viscoelastic near the glass transition temperature of the matrix resin. The time-temperature superposition principle for the matrix resin held for the storage modulus behavior of the composite material.

(2) The static strength of the composite is time and temperature dependent. It was shown that the time-temperature superposition principle held for the static strength results by using the same time-temperature shift factors as the storage modulus of the matrix resin.

(3) The largest deformation before failure occurred with increasing temperature and stress ratio, showing the increase of creep damage.

(4) As temperature and stress ratio increased the fatigue strength also becomes more creep damage dependent.

(5) The fatigue strength is time and temperature dependent, becoming more viscoelastic near the glass transition temperature of the matrix resin. The time-temperature superposition principle using the matrix resin's shift factors held for the initial fatigue strength behavior of the composite material.

5. REFERENCES

1. S. Brown, The British Polymer Journal, 14, 1 (1982).
2. M. Miwa, A. Takeno, K. Hara,et.al., Composites, 26, 371 (1995).
3. A. Moloney, H. Kausch, T. Kaiser, H. Beer, Journal of Material Science, 22, 381 (1987).
4. A. Evans, S. Williams, P. Beaumont, Journal of Material Science, 20, 3668 (1985).
5. H. Beer, T. Kaiser, Fillers, 16.1 (1986).
6. A. Nishimura, A. Yaguchi, S. Kawai, Zairyou, 38, (434), 92 (1989) (Japanese).
7. A. Yaguchi, A. Nishimura, Zairyou, 42, (472), 40 (1993) (Japanese).
8. Y. Miyano, M. McMurray, J. Enyama, et.al., Journal of Composite Materials, 28, (13), 1250 (1994).
9. M. McMurray, M. Nakada, Y. Miyano, Journal of Composite Materials, 29, (14), 1808 (1995).
10. M. McMurray, M. Nakada, Y. Miyano, Proc. 40th International SAMPE Symposium and Exhibition, 1316 (1995).
11. F. Schwarzl, A. Staverman, Journal of Applied Physics, 23, (8), 838 (1952).
12. M. Williams, R. Landel, J. Ferry, Journal of the American Chemical Society, 77, 3701 (1955).

28th International SAMPE Technical Conference
November 4-7, 1996

COMPATIBILITY OF CERAMICS WITH LIQUID SODIUM

Yoshiaki TACHI Yasushi HIRAKAWA
Eiichi YOSHIDA Shigeki KANO
Power Reactor and Nuclear Fuel Development Corporation (PNC)
Oarai Engineering Center
4002 Narita, Oarai-machi, Ibakari-ken 311-13, Japan

ABSTRACT

The compatibility of ceramics with liquid sodium being used as coolant for fast breeder reactors (FBR) has been investigated for the creation of advanced ceramics which could be used under FBR environment. Various sintered ceramics were tested for 3.6Ms and 14.4Ms in elevated temperature liquid sodium of 823K and 923K. The sodium corrosion behavior and mechanism of several ceramics were clarified according to the experimental results and thermodynamic consideration. Alumina and Si ceramics such as Si_3N_4, Sialon and SiC, which were typical engineering ceramics, were corroded mainly at grain boundaries by sodium. It was thought that grain boundary corrosion were caused by sodium penetration to material and chemical reactions between penetrated sodium and additives and/or impurities included in grain boundaries.

It is very important for good compatibility of sintered ceramics with sodium to control the micro-structures of grain boundaries and remove impurities included in there. Based on these countermeasures, some improved test pieces were trially manufactured for above-mentioned ceramics. Thereafter, the effects of these improvements were evaluated by sodium corrosion test. As a result, some kinds of ceramics showed improved corrosion resistance, but others were needed to be more improved by additional methods.

KEY WORDS: Engineering Ceramics, Sodium Corrosion, Modified Ceramics

1. INTRODUCTION

Most main components constructing fast breeder reactors (FBR), such as pump, piping, valve and nuclear reactor vessel etc., have been made of stainless steel, which had good compatibility with FBR environment. However, it is thought that the innovative development of more superior structural materials to stainless steel will enable to make the rapid progress in performance, safety, reliability and economy of FBR.

Among several advanced materials having been developed, ceramics materials attract to notice by their many excellent properties. Compared with stainless steel, ceramics have higher melting point, smaller heat expansion rate, lower density, higher strength at elevated temperature, more excellent heat-resisting and tribological property, etc.. For the investigation of the application of ceramics to some parts of FBR's components, the Oarai Engineering Center of PNC has conducted the research and development on advanced ceramics having good compatibility with FBR environment.

As the first step of the research, the compatibility of typical engineering ceramics such as Al2O3, Si3N4, Sialon and SiC, which were on sale, with liquid sodium was evaluated through sodium corrosion tests. As a result, it was found as some papers reported[1-5] that almost all the engineering ceramics were seriously corroded at grain boundaries by sodium, and that the structure and constituents of grain boundaries influenced largely on sodium corrosion. The outline of sodium corrosion behavior and mechanism of ceramics was clarified[4,5] by mutual comparison with results of test pieces with different purities and structures, and single crystal, CVD and sintered ceramics.

In order to improve sodium corrosion resistance to same corrosion rate to that of stainless steel and prevent grain boundary corrosion, for applying ceramics to some parts of FBR components, such as tribological parts, thermal liner, etc., some modified ceramics were trially manufactured by using different additives, removing impurities and crystallizing grain boundaries, etc.. This paper describes the compatibility between modified ceramics and liquid sodium through evaluation of modification effects on ceramics.

2. Design of Modified Ceramics

Generally, grain boundaries of ceramics are amorphous phase consisting of some additives and impurities. It is already known in previous report[4,5] that sodium corrosion of ceramics result from the penetration of liquid sodium to material inside through amorphous grain boundaries and the reaction of sodium with constituents of ceramics. This behavior of sodium leads ceramics to degradation of their grain boundaries by constituent dissolution into sodium.

The structure and condition of grain boundaries influences largely sodium corrosion of ceramics, therefore, the modification of grain boundaries, such as crystallizing amorphous grain boundaries and removing impurities from them, is thought to be more effective method to prevent sodium corrosion.

For Al2O3, Si3N4, Sialon and SiC, which were typical engineering ceramics, some modified ceramics were trially manufactured. Their modification methods are described as follows.

In Al2O3, it was clarified[4,5] that sodium corrosion depended largely on its purity, that was, the lower the purity of Al2O3 was, the more serious its sodium corrosion was. Impurities included in Al2O3 were mainly Si, Mg and Ca. Silicon content increased as the purity of Al2O3 was low, as shown in Figure. 1. Because it is well known that sodium reacts with Si or Si compounds like SiO_2, it is thought that low purity Al_2O_3 having much Si impurity is attacked seriously by sodium. In order to prevent Al_2O_3 from sodium corrosion, therefore, high purity Al_2O_3 of 99.99% (AL-1) was trially manufactured by prevention of contamination with impurities. It was sintered by using very pure and minute Al_2O_3 powders and no additive at lower temperature than usual.

In Si_3N_4, same as the case of Al_2O_3, sodium corrosion was caused by amorphous grain boundaries and Si compounds included in there[4,5]. Two kinds of modified Si_3N_4 were trially manufactured. One was grain boundary crystallized Si_3N_4 (SN-1) for the prevention of sodium penetration to inside. Amorphous grain boundaries were crystallized by heat treatment after hot-press sintering. The other was grain boundaries crystallized Si_3N_4 (SN-2) using AlN as additive to change grain boundary constituent. By using AlN additive instead of some oxides, it was expected that Si-O compounds at grain boundaries would be changed to Si-N-Al-O compounds, then the interaction between sodium and Si compounds would be prevented.

In Sialon, to improve the resistance against sodium corrosion, three kinds of test pieces were trially manufactured. One was Sialon (SL-1) , whose grain boundaries had small volume same to about 1/10 of that of non-modified. Sialon is usually made from Si_3N_4, AlN and SiO_2 powder, etc. During pressureless sintering, these powders react each other, and compose Sialon ceramics. Therefore, it is expected that grain boundaries of Sialon enable their volume to reduce by using Sialon powders as one of raw materials. The second was Sialon (SL-2) using AlN additive, same as the case of Si_3N_4, to change Si-O compounds to Si-N-Al-O compounds at grain boundaries. The third was Sialon (SL-3) using raw powders of Sialon made by sol-gel method. Sol-gel method enables to make high purity Sialon powders.

In SiC, three kinds of test pieces were trially manufactured. To improve the resistance against sodium corrosion, some different additives, such as Al_2O_3, B_4C and C, were used. Sintering atmosphere was controlled to remove Si-O compounds. One was SiC (SC-1) using Al_2O_3 additive. The second was SiC (SC-2) using B_4C and C as additives. This was manufactured by pressureless sintering in deoxidization atmosphere and, thereafter, HIP. The third was hot-pressed SiC (SC-3) using Al_2O_3 additive. Hot-press was applied in vacuum to remove Si-O compounds.

All modified test pieces are showed in Table 1.

3. Experimental

Corrosion tests of modified ceramics were carried out in liquid sodium at elevated temperatures of 823K and 923K for 3.6Ms and 14.4Ms by using loop type of sodium experimental apparatus. The temperature of 823K is approximately same to the maximum design temperature of structural materials contacted with liquid sodium except for core materials such as cladding tube, wrapper tube, etc.. The temperature of 923K is more severe condition to accelerate corrosion test. Oxygen level in liquid sodium was kept about 1 ppm by continuous operation of cold trap at 393K. The test pot of apparatus is showed in Figure 2.

After corrosion test, all ceramics test pieces was washed by ethyl-alcohol and sodium was eliminated. Tested ceramics were investigated on their sodium corrosion behaviors by some micro-analysis methods. Weight changes before and after corrosion test were measured on all tested ceramics. Appearances and micro-structures of some ceramics mainly tested at 923K were observed by using SEM and TEM, and the contents of sodium and constituent elements included in tested ceramics were analyzed by EPMA or GDS[6] etc..

4. Experimental results

(1) Al_2O_3

Figure 3 shows the weight change of modified Al_2O_3 after sodium corrosion test with those of non-modified ones. After the sodium corrosion test at 823K and 923K for 3.6Ms, modified Al_2O_3 gained in weight of ave. 0.007mg and 0.003mg per $1cm^2$, respectively. These quantities were significantly smaller than those of non-modified Al_2O_3 and same to that of single crystal.

In modified Al_2O_3 tested at 923K for 3.6Ms, the observations by SEM and TEM showed no traces of grain boundaries attack. The analysis by EPMA showed no penetration of sodium inside material as shown in Figure 4. Further, in modified Al_2O_3 tested at 923K for 14.4Ms, the analysis of sodium and constituent elements by GDS showed quite small penetration, less than 1 μm depth, of sodium at short distance of nearly surface as shown in Figure 5.

(2) Si_3N_4

Figure 6 shows weight changes of some modified Si_3N_4 with non-modified and CVD Si_3N_4. Compared with non-modified Si_3N_4, weight change of all modified Si_3N_4 were smaller, about 0.4 - 0.6mg per 1 mm^2, and the difference of weight changes between test pieces were smaller too. The surface micro-structures of Si_3N_4 (SN-2) before and after test at 923K for 3.6Ms is showed in Figure 7. Compared with test piece before test, the surface of tested Si_3N_4 was corroded by sodium. However, in all modified Si_3N_4 tested at 923K for 3.6Ms, the trace of serious grain boundaries corrosion like that of non-modified was not found from cross-sectional observation by SEM. Analysis of TEM-EDS and EPMA indicated no existence of sodium inside modified Si_3N_4. Furthermore, GDS profiles showed quite small penetration, ca. 0.1 μm depth, of sodium on nearly surface of modified Si_3N_4 tested at 923K for 3.6Ms and 14.4Ms. Figure 8 showed GDS profiles of SN-2 tested at 923K for 14.4Ms.

(3) Sialon

Figure 9 shows weight change of some modified Sialon with non-modified. Compared with non-modified Sialon, weight changes of modified Sialon were smaller like those of Si_3N_4. These quantities of weight changes were nearly same each other. The surface micro-structures of tested modified Sialon were same as those of Si_3N_4. No trace of grain boundaries corrosion was found by SEM observation in modified Sialon tested 923K for 3.6Ms. Further, the observation by TEM indicated no existence of sodium at grain boundaries. However, GDS profiles of modified Sialon (SL-1) tested at 923K for 14.4Ms showed penetration, ca. 20μm depth, of sodium and dissolution of constituent elements, as shown in Figure 10.

(4) SiC

Weight changes of modified, non-modified, CVD and single crystalline SiC are shown in Figure 11. Weight changes of all modified SiC were large same as those of non-modified ones. The observation by SEM showed traces caused by serious sodium attack to grain boundaries. Micro-structure of SiC after corrosion test at 923K for 3.6Ms is shown in Figure 12. The result of EPMA and GDS analysis showed sodium penetration to modified SiC inside tested at 923K for 3.6Ms and 14.4Ms. Figure 13 shows GDS profiles of SiC tested at 923K for 14.4Ms.

5. Discussion

It is well known from some experiments and previous papers[1-5] that sodium corrosion of Al_2O_3 and Si ceramics such as Si_3N_4, Sialon and SiC were caused by sodium penetration through grain boundaries and the interaction between sodium and Si-O compounds in grain boundaries. It is considered that Si-O compounds are formed in Al_2O_3 by Si as impurity and oxygen in matrix, and in Si ceramics, by Si in matrix and oxygen in additives. From the consideration on formation free energy and phase diagram of Si-O-Na system as shown in Figure 14 and Figure 15, it can be predicted that Si-O compounds form complex oxides like Na_2SiO_3 through the reaction with sodium[1-3]. These complex oxides dissolve into sodium, then constituents of grain boundaries leach out in sodium. This proceeds sodium corrosion at grain boundaries.

In modified Al_2O_3, sodium corrosion test showed that the purification was very effective method to prevent sodium corrosion. This indicated that sodium corrosion of Al_2O_3 was caused by Si impurity and that Al_2O_3 grain was stable against sodium as known from Figure 14.

In Si_3N_4 and Sialon, the observation of micro-structure showed that their corrosion states were similar each other. Crystallization or volume reduction of grain boundaries were seemed to be slightly effective from the comparison of weight change with non-modified ceramics. Though weight change quantities of Si_3N_4 were larger than those of stainless steel and high purity Al_2O_3, GDS analysis showed that sodium was prevented to penetrate to Si_3N_4 inside. In Sialon tested at 923K for 14.4Ms, it was clarified by GDS analysis that sodium penetrated to Sialon and, Si and N as constituents of matrix leached out from the surface. Therefore, it is considered that Si-N-Al-O system is no good for sodium, because it was showed by corrosion test results of CVD Si_3N_4 that Si-N system had good compatibility with sodium. This trend coincided with no effect of Si-N-Al-O grain boundary of Si_3N_4 (SN-2) using AlN as additive. It is thought enough that sodium corrosion of Sialon was caused mainly by chemical interaction between sodium and constituents in material.

In SiC, from results of SEM observation and GDS analysis, as shown in Figure 12 and

Figure 13, grain boundary of SiC was corroded seriously regardless of the difference of additives, AlN, Al_2O_3 and B_4C, and the control of sintering atmosphere. As single crystalline and CVD SiC showed good compatibility with sodium, most of SiC were attacked at grain boundary and SiC grains came off the surface into sodium by degradation of grain boundaries. Therefore, it is thought that weight change of SiC after test were significantly large.

6. Conclusions

Through the manufacturing modified ceramics and their sodium corrosion tests, many acknowledges were clarified about the compatibility between ceramics and liquid sodium.

1) Purification of Al_2O_3 was an effective method for improvement of sodium corrosion resistance. High purity Al_2O_3 has excellent compatibility with liquid sodium.

2) In the case of Si_3N_4 and Sialon, crystallization or volume reduction of amorphous grain boundaries prevented that sodium penetrated to material inside and grain boundaries attack was generated. Resultantly, their resistance against sodium corrosion was improved. However, it was clarified that compatibility of Si-N-Al-O system with sodium was not good, because Si and N, in Si-N-Al-O system, leached out into sodium by their interaction with sodium.

3) Sodium corrosion of SiC largely depended on sodium penetration to material inside. Grain boundaries were degraded by sodium penetration, and then SiC grains came off into sodium. Sodium penetration was not prevented by any additive and removal of Si-O compounds by the control of sintering atmosphere.

Modified high purity Al_2O_3 is thought to be practical use in FBR environment. Though the sodium corrosion resistance of Si ceramics such as Si_3N_4 and Sialon was improved by trial manufacturing, comparing with those of conventional materials, furthermore, additional countermeasures are needed in manufacturing process for practical use. In SiC, the essential search is inevitable for effective methods to prevent sodium penetration.

References

1) H.G.BARKER, T.K.LEUNG, ENERGY AND CERAMICS, Elseriuer Scientific Co. 1980, pp.870-78.

2) J.JUNG, A.RECK, et.al., J.Nucl.Mater.,119, 339(1983).

3) G.DE.WITH, P.J.VURGT, et.al., J. Mater.Sci., 20, 1215(1985).

4) E.YOSHIDA, S.KANO, et.al., The International Symposium on Material Chemistry in Nuclear environment, 901(1992).

5) S.KANO, E.YOSHIDA, et.al., The second Seminar on Liquid Metal Systems, 85 (1995).

6) K.SUZUKI, Zairyo-to-Kankyo, 42, 384(1993).

Table 1 Modified ceramics

	material	modification points
AL-1	Al_2O_3	high purity
SN-1	Si_3N_4	grain boundary crystallization
SN-2	Si_3N_4	grain boundary crystallization & AlN additive
SL-1	Sialon	reduction grain boundary volume
SL-2	Sialon	grain boundary crystallization & AlN additive
SL-3	Sialon	sol-gel Sialon powder
SC-1	SiC	Al_2O_3 additive
SC-2	SiC	B_4C, C additives & removal Si-O compounds
SC-3	SiC	Al_2O_3 additive & removal Si-O compounds

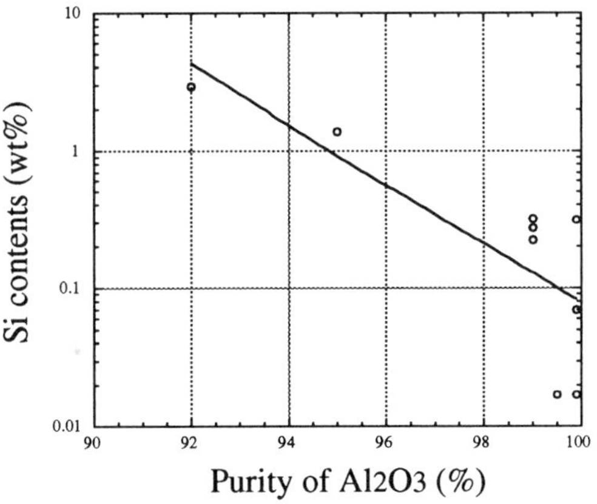

Figure 1 Si contents of Al_2O_3

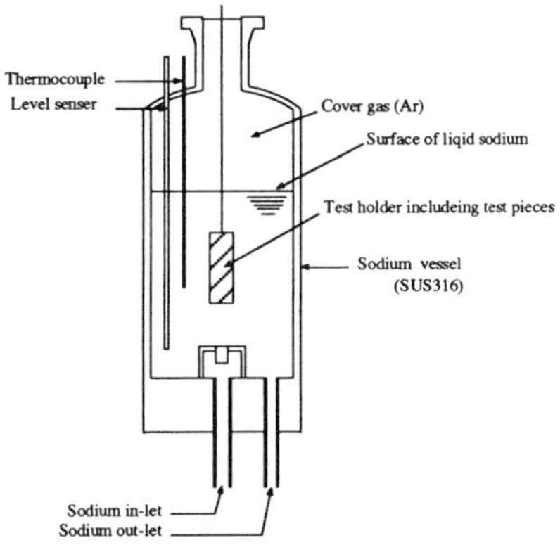

Figure 2 Test pot of apparatus

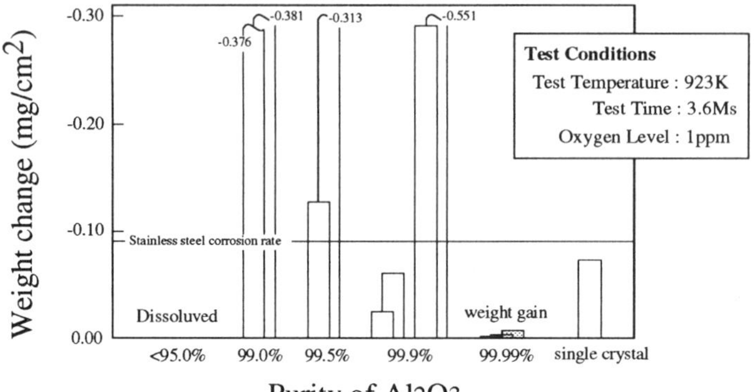

Figure 3 Weight change of Al2O3

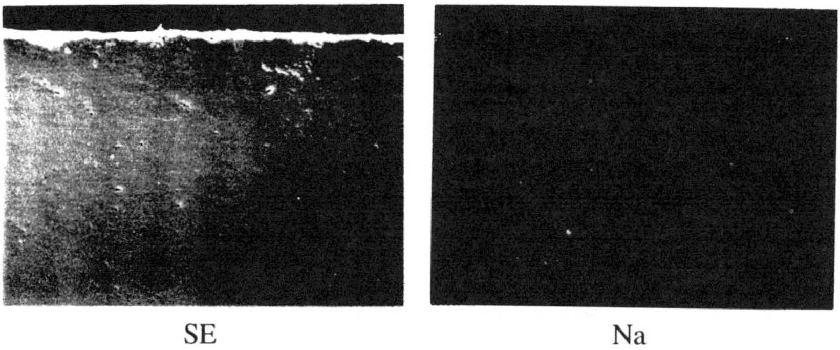

Figure 4 EPMA result of Al2O3

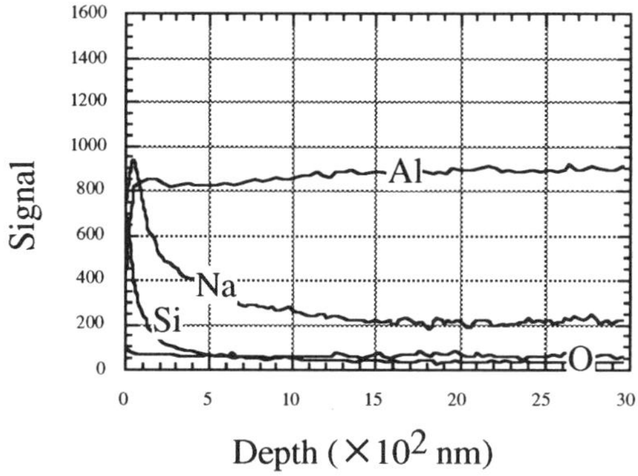

Figure 5 GDS profiles of Al2O3

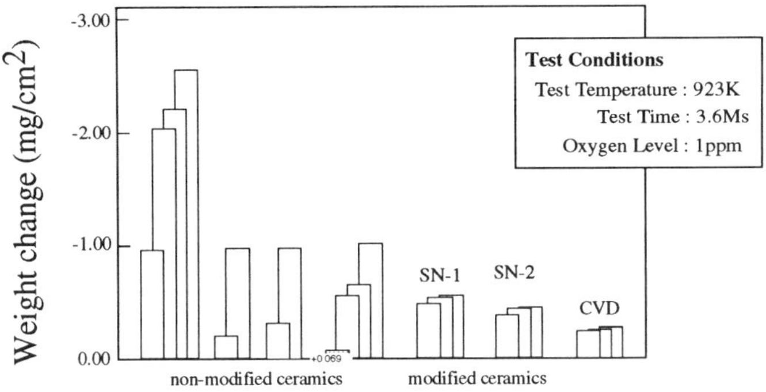

Figure 6 Weight change of Si3N4

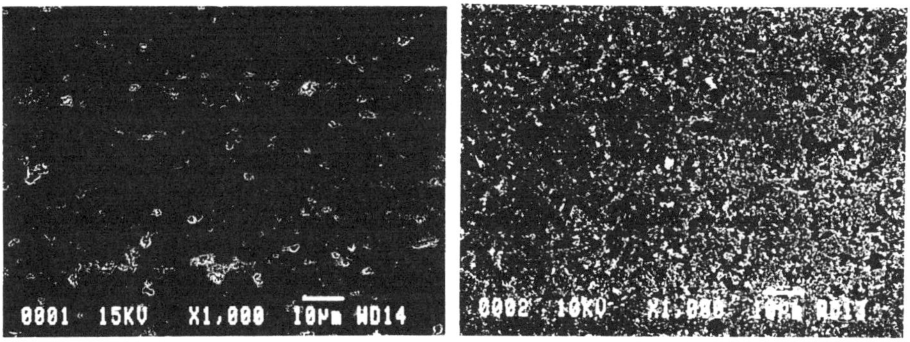

Before Test After Test (923K, 3.6Ms)

Figure 7 Microstructure of Si3N4 (Surface)

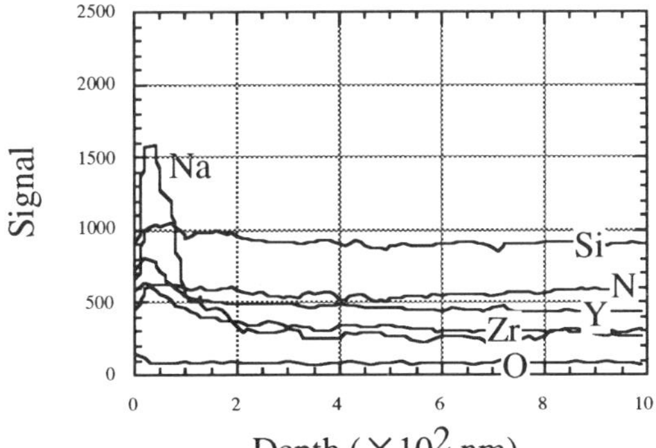

Figure 8 GDS profiles of Si3N4

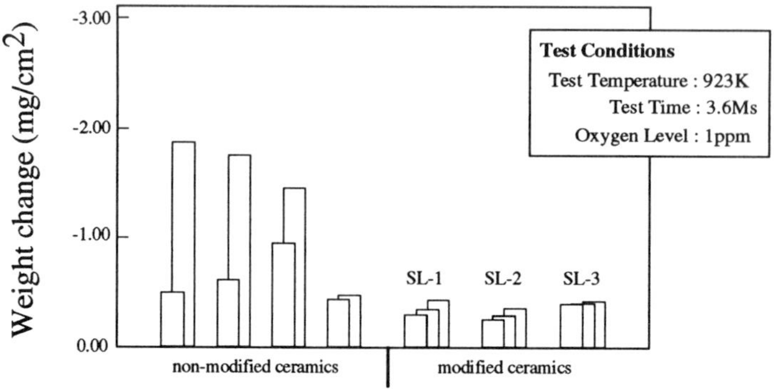

Figure 9 Weight change of Sialon

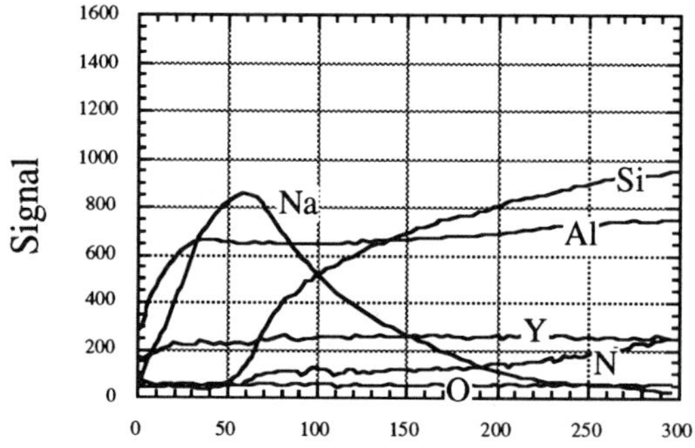

Depth (×10^2 nm)

Figure 10 GDS profiles of Sialon

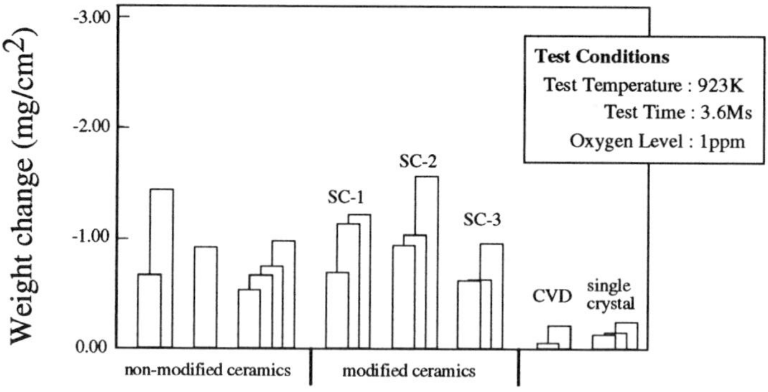

Figure 11 Weight change of SiC

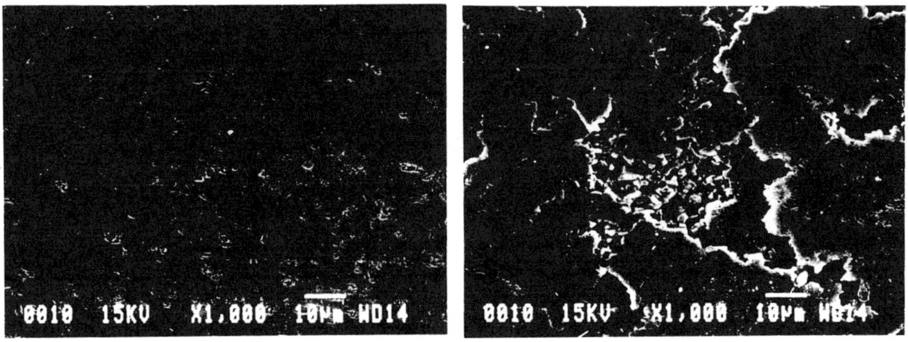

Before Test After Test (923K , 3.6Ms)

Figure 12 Microstructure of SiC (surface)

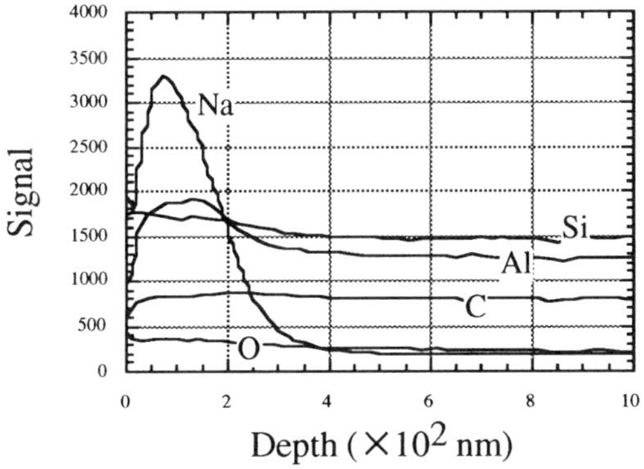

Figure 13 GDS profiles of SiC

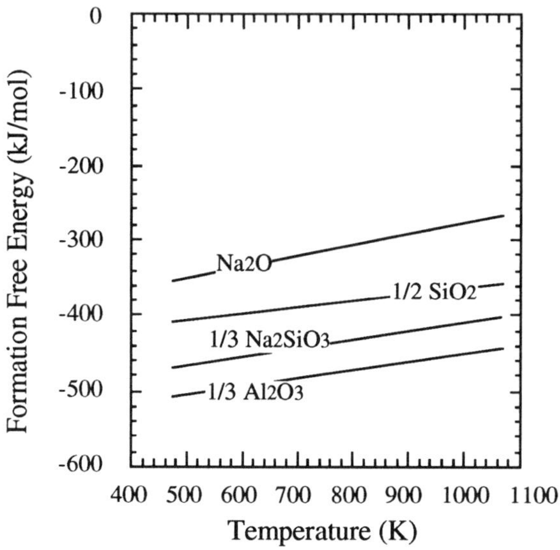

Figure 14 Formation Free Energy of Oxides

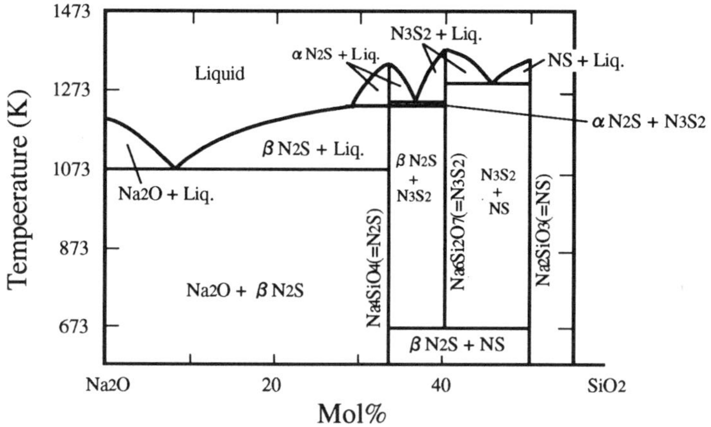

Figure 15 Phase Diagram of Na2O-SiO2 syatem

Biography

Yoshiaki TACHI belongs to the Frontier Technology Section, the advanced Technology Division, Oarai Engineering Center, PNC. He is research engineer who studies the sodium environmental effect engineering ceramics, constructs of materials database on advanced materials including ceramics and develops high performance transparent neutron shielding materials since 1991.

Yasushi HIRAKAWA belongs to the Structure Safety Engineering Section, the Safety engineering Division, Oarai Engineering Center, PNC. He is research engineering who has studied the strength properties and the sodium environmental effect on structural materials and core materials since 1982. Currently, he is in charge of in-sodium test and analysis of ceramics.

Eiichi YOSHIDA belongs to the Structure Safety Engineering Section, the Safety engineering Division, Oarai Engineering Center, PNC. He is assistant senior research engineering who has studied the strength properties and the sodium environmental effect on structural materials and core materials since 1970. Currently, he is in charge of in-sodium test and analysis of ceramics.

Shigeki KANO belongs to Frontier Technology Section, the Advanced Technology Division, Oarai Engineering Center, PNC. He is senior research engineer who has studies the sodium environmental effect on the structural materials and core materials for fast reactor since 1969. Currently, he is in charge of programming and managing of the research on creation of advanced materials such as ceramics, functionally graded materials and super heat-resisting materials for fast breeder reactor's application.

28th International SAMPE Technical Conference
November 4-7, 1996

MECHANICAL PROPERTIES OF INCONEL 600 BRAZED JOINTS WITH Pd-Ni-Co ALLOYS

Hiroshi Izui and Yoshifumi Suezawa
Nihon University, College of Science and Technology,
Department of Aerospace Engineering
Chiba, Japan 274

ABSTRACT

Brazing filler metals of the Pd-Ni-Co system were considered for use in brazing for high temperature service applications. Brazed joint specimens fabricated from nickel-based alloy Inconel 600 using boron-free brazing filler metals, 40Pd-50Ni-10Co and 30Pd-50Ni-10Co-10Cu, boron-containing ones, 40Pd-47Ni-10Co-3B and 40Pd-49Ni-10Co-1B. Palladium-containing brazing filler metals were selected because of their oxidation resistance, ductility and relatively high melting temperatures. The mechanical properties and microstructures of these brazed joints were investigated. The brazed joint specimens were tested in air at room temperature and 473K, 673K, 873K, and 1073K. Mechanical tests included tension and creep rupture tests. The joints brazed with the boron-free brazing filler metals, 40Pd-50Ni-10Co and 30Pd-50 Ni-10Co-10Cu had high tensile strengths of 580 MPa and 560 MPa receptively at room temperature. The joints brazed with 40Pd-50Ni-10Co had average tensile strengths of 473 to 130 MPa from 473 to 1073K. The nature of the reactions and microstructures in the brazed joints will be discussed.

KEY WORDS:Brazing, Palladium-containing Brazing Filler Metals, Nickel-based Alloy, Tensile Strength, High Temperature Strength, Creep Rupture Strength

1. INTRODUCTION

Brazing filler metals for high temperature brazing are nickel- based, gold-based ,and palladium-containing ones selected, because of their high strength at elevated temperatures, and excellent corrosion and oxidation resistance properties. Compared with nickel-based brazing filler metals, the gold-based and palladium-containing noble brazing filler metals exhibit lower hardness, better ductility, less tendency towards intergranular penetration, and no intermetallics, but are more expensive. Reports of joints brazed with palladium-containing have been relatively few compared with ones on joints brazed with nickel- and gold-based brazing filler metals (Refs.1-6). The mechanical properties of SUS316L brazed joints with Pd-Ag-Cu, Pd-Cu-Co, and Pd-Ni-Co brazing filler metals have been described in earlier papers(Refs.7-12)

Four types of Pd-Ni-Co brazing filler metals used in this research were boron-free ones, 40Pd-50Ni-10Co and 30Pd-50Ni-10Co-10Cu, boron-containing ones 40Pd-49Ni-10Co-1B and 40Pd-47Ni-10Co-3B. This study was undertaken to investigate the properties and microstructure of nickel-based alloy Inconel 600 joints brazed with these brazing filler metals. Brazed joint properties were evaluated at room temperature and at elevated temperatures using tensile tests and creep rupture tests. The brazing filler metal suitable for high temperature brazing will be deiced in this study.

2. EXPERIMENTAL PROCEDURE

2.1 Materials The base metal used was commercially available nickel-based alloy Inconel 600 of nominal composition Ni-Cr-Fe. The chemical compositions and mechanical properties of the base metal are shown in Table 1 and Table 2 respectively.

The brazing filler metals used in this research were the Pd-Ni-Co system alloys, the boron-free brazing filler metals, 40Pd-50Ni-10Co and 30Pd-50Ni-10Co-10Cu, the boron-containing ones, 40Pd-47Ni-10Co-3B and 40Pd-49Ni-10Co-1B. The compositions of these brazing filler metals were decided using ternary phase diagram as shown in Fig. 1(Ref.13). As the melting temperatures of the boron-free brazing filler metals are very high, boron was used as the melting temperature depressant. The compositions of brazing filler metals are shown in Table 3. The tensile strengths of 40Pd-50Ni-10Co and 30Pd-50Ni-10Co-10Cu at room temperature are 616 MPa and 603 MPa respectively.

Table 1 Chemical compositions of Inconel 600 (wt%)

	C	Mn	Fe	S	Si	Cu	Ni	Cr
Inconel 600	0.04	0.23	8.14	0.001	0.34	0.22	Bal.	16.22

Table 2 Mechanical properties of Inconel 600

Yield strength	319 MPa
Tensile strength	701 MPa
Elongation	42 %
Hardness	86 RB

2.2 Brazing Procedure Figure 2 shows the schematic of a brazed joint. In preparation for braze processing, the surfaces to be brazed were polished in direction z with #400 emery paper, and the base metals and braze alloys were degreased in acetone with an ultrasonic-washer. For braze processing, two base metals were butted lengthwise as shown in Fig.2 and set in a brazing jig after the brazing filler metal was inserted between them. Brazing was conducted using a resistance-heated vacuum furnace at degree of vacuum from 2.6 to 6.7 mPa. The furnace temperature was held at that level for 5 min to permit the base and braze alloys and furnace temperatures to equilibrate. Table 4 shows the brazing conditions for each braze alloy.

2.3 Metallurgical Analysis After braze processing, excess braze alloy around the brazed area was removed. Brazed joint microstructures were characterized using an optical microscope and scanning electron microscopy. The content distributions of brazed joints were investigated with an electron microprobe analyzer(EPMA) .

2.4 Mechanical Testing The tensile tests of brazed joints were conducted in air at room temperature, 473K, 673K, 873K and 1073K using an Instron material test machine. For the elevated-temperature tests, the specimens and grips were heated using an electrically heated furnace. Creep rupture tests of brazed joints were performed in air at 673K, 873K, and 1073K,

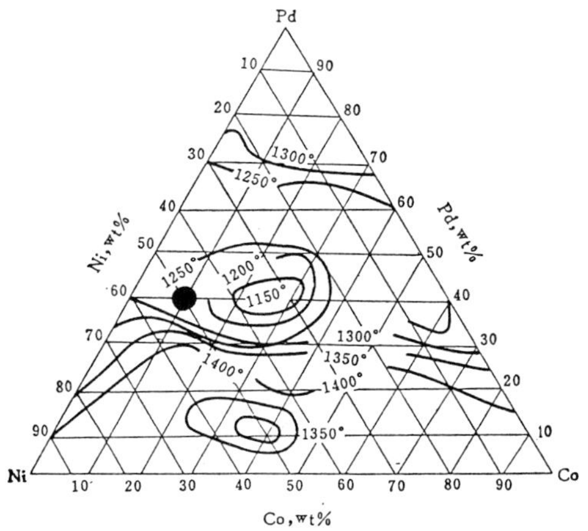

Fig.1 The composition of the brazed alloys used plotted on the liquidus surface of the ternary Pd-Ni-Co phase diagram (wt%)

Table 3 Chemical compositions of brazing filler metals (wt%)

	Pd	Ni	Co	Cu	B
40Pd-47Ni-10Co-3B	40.0	47.0	10.0	----	3.0
40Pd-49Ni-10Co-1B	40.0	49.0	10.0	----	1.0
40Pd-50Ni-10Co	40.0	50.1	9.9	----	----
30Pd-50Ni-10Co-10Cu	30.0	50.0	10.0	10.0	----

and the times to failure were recorded. As shown in Fig.3, specimens for tensile and creep rupture tests machine-ground to rods of 4.5mm in diameter and 10 mm in length at brazed parts.

3. RESULTS AND DISCUSSION

3.1 Microstructures of Brazed Joints Figures 4-7 show representative microstructures, electron microprobe traces of each of brazed joints. These traces showed palladium, nickel, cobalt, and copper x-ray line traces across the brazing layer. As shown in Fig.4A and 5A, these

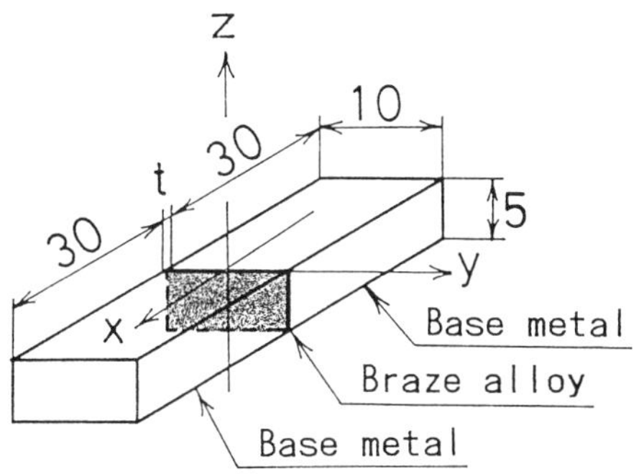

Fig.2 Schematic of joint geometry

Table 4 Brazing conditions

Filler metal	Brazing temp.	Brazing time	Pressure
40Pd-47Ni-10Co-3B	1298 (K)	5 (min)	2.6×10^{-3} ~ 6.7×10^{-3} (Pa)
40Pd-49Ni-10Co-1B	1373 (K)		
40Pd-50Ni-10Co	1563 (K)		
30Pd-50Ni-10Co-10Cu	1563 (K)		

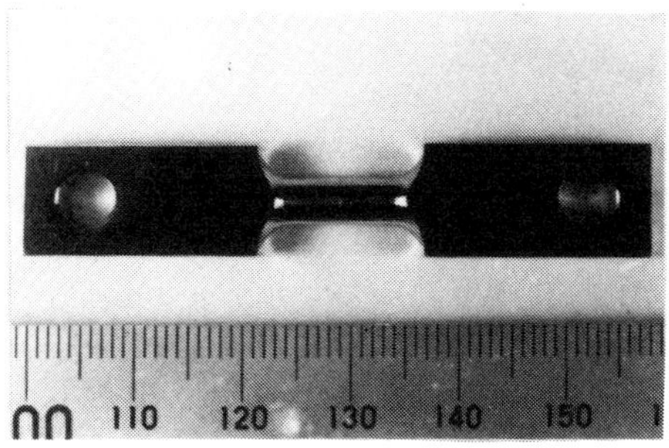

Fig.3 Tensile tests specimen

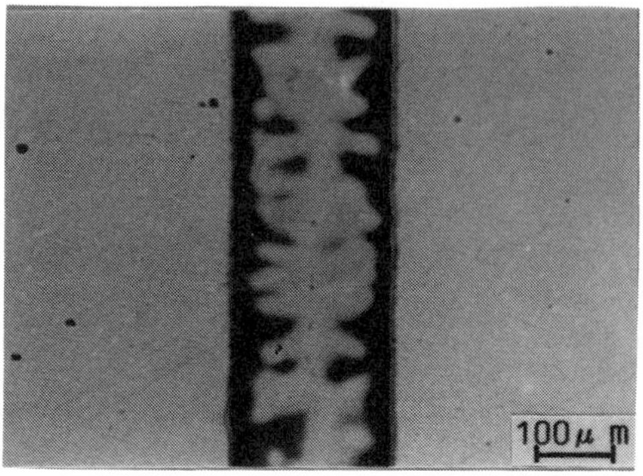

(A) Microstructure

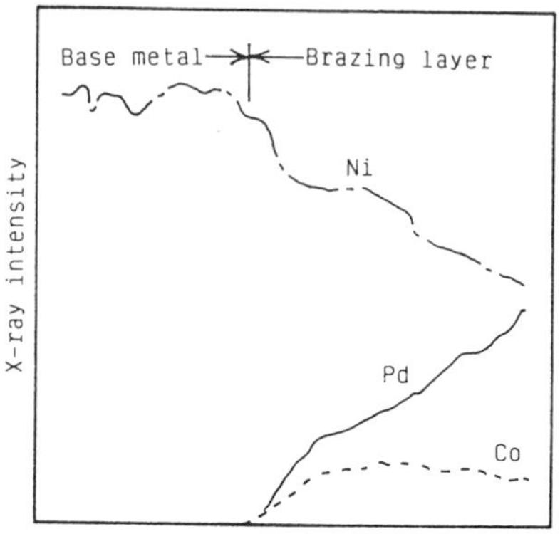

(B) EPMA analysis

Fig.4 Microstructure and EPMA analysis of a brazed joint with 40Pd-50Ni-10Co

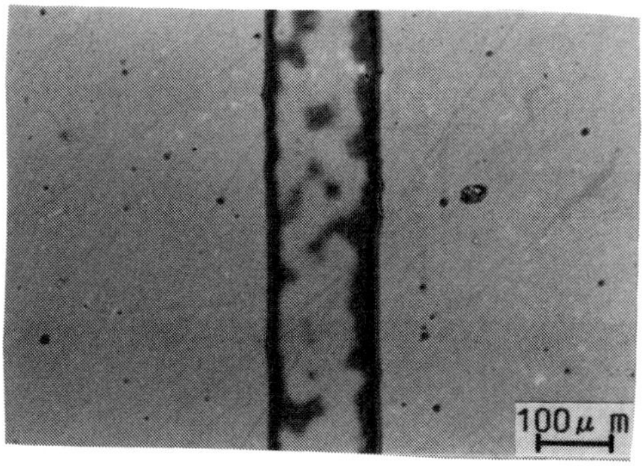

(A) Microstructure

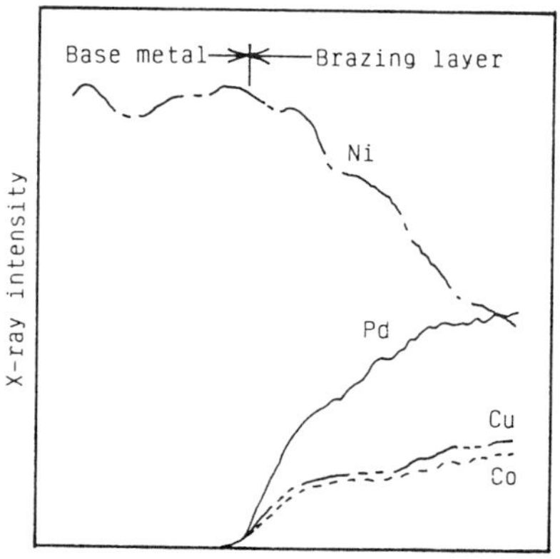

(B) EPMA analysis

Fig.5 Microstructure and EPMA analysis of a brazed joint with 30Pd-50Ni-10Co-10Cu

(A) Microstructure

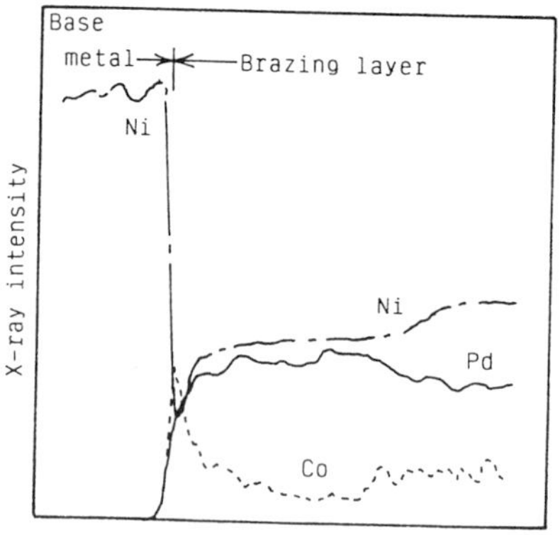

(B) EPMA analysis

Fig.6 Microstructure and EPMA analysis of a brazed joint with 40Pd-49Ni-10Co-1B joints brazed with the boron-free brazing filler metals exhibited a darker phase at the interface and a light and microscopically homogeneous structure at the center of the brazing layer. These brazed joints had similar microstructures. Electron probe microanalysis results of Fig.4B and 5B showed that palladium and cobalt contents increased at the center of the brazing layers. In the darker phase at the interface, as nickel diffused through brazing layer, its contents increased compared to the nickel content of brazing filler metals.

Figures 6 and 7 show the microstructures and EPMA analysis of brazed joints with boron-containing brazing filler metals. As shown in 6A and 7A, the brazing filler metals diffused into the base metals near the interfaces. The series of white particles in the brazing layer were a boride. EPMA analysis of a brazed joint with 40Pd-49Ni-10Co-1B is shown in Fig.6B. The nickel and palladium were uniformly distributed in the brazing layer. The Co-rich layer was formed near the interface. In a brazed joint with 40Pd-47Ni-10Co-3B, The nickel, palladium, and cobalt were uniformly distributed---Fig.7B.

3.2 Tensile Strength of Brazed Joints at Room Temperature Figures 8A, 8B, 8C and 8D show the tensile strengths of brazed joints with each of the brazing filler metals at room temperature. The dependence of tensile properties on joint clearance was established by tensile strength/joint clearance graphs. Joints brazed with the boron-free brazing filler metals, 40Pd-50Ni-10Co and 30Pd-50Ni-10Co-10Cu, had higher tensile strengths than those of joints brazed with the boron-containing brazing filler metals, 40Pd-47Pd-10Co-3B and 40Pd-49Pd-10Co-1B. Brazed joints with 40Pd-50Ni-10Co and 30Pd-50Ni-10Co-10Cu show nearly the same constant strength values, 580 MPa and 560 MPa receptively. Tensile test specimens brazed with these boron-free brazing filler metals fractured at the base metal. Therefore the strengths at brazed joints had higher than those at base metals. Therefore, the tensile strengths of joints brazed with 40Pd-50Ni-10Co and 30Pd-50Ni-10Co-10Cu did not depend on the joint clearance. Fig.9 shows the tensile test specimen brazed with 40Pd-50Ni-10Co that fractured at the base metal. With the boron-containing brazing filler metals, tensile test specimens fractured at the brazing layer. As the boron-containing brazing filler metals were brittle because of the presence of intermetallics, the brazed joints with these brazing filler metals show lower strength compared to ones with the boron-free brazing filler metals.

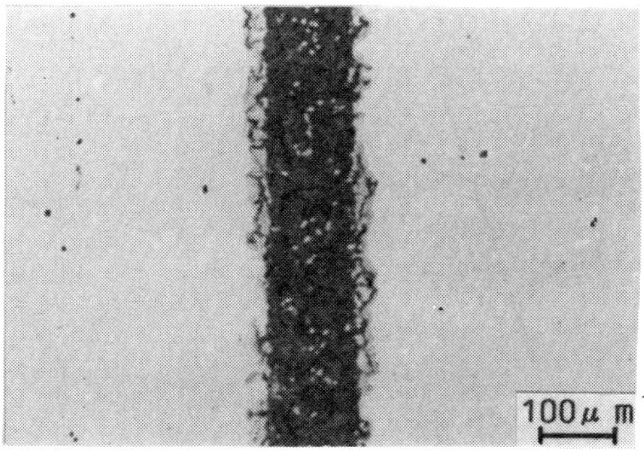

(A) Microstructure

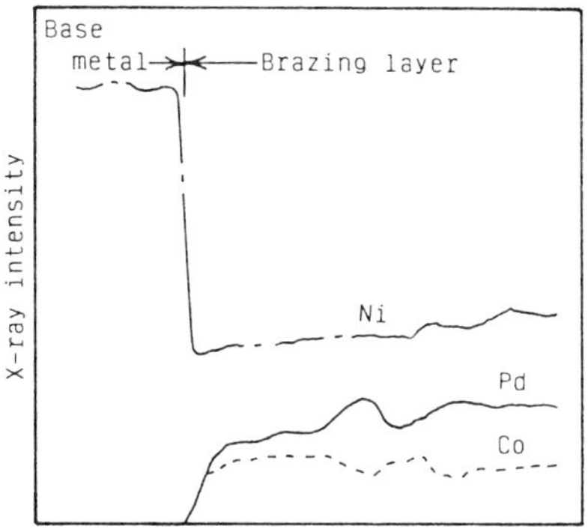

(B) EPMA analysis

Fig.7 Microstructure and EPMA analysis of a brazed joint with 40Pd-47Ni-10Co-3B

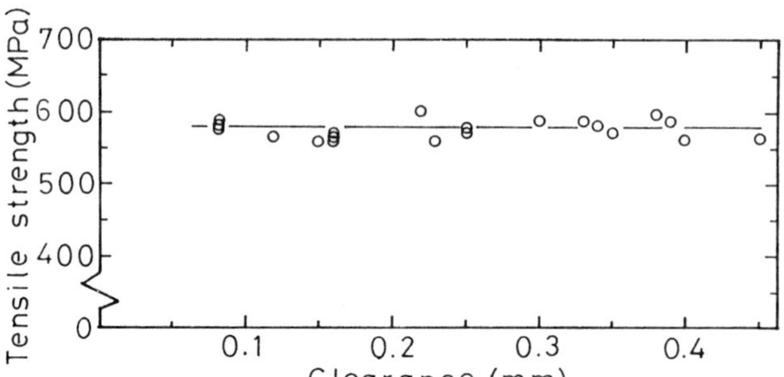

(A) 40Pd-50Ni-10Co

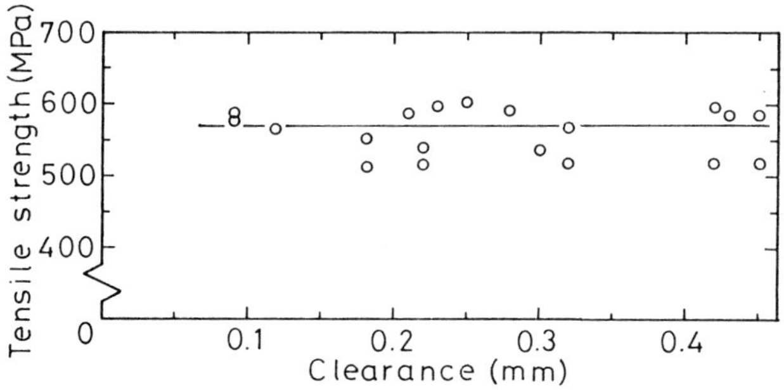

(B) 30Pd-50Ni-10Co-10Cu

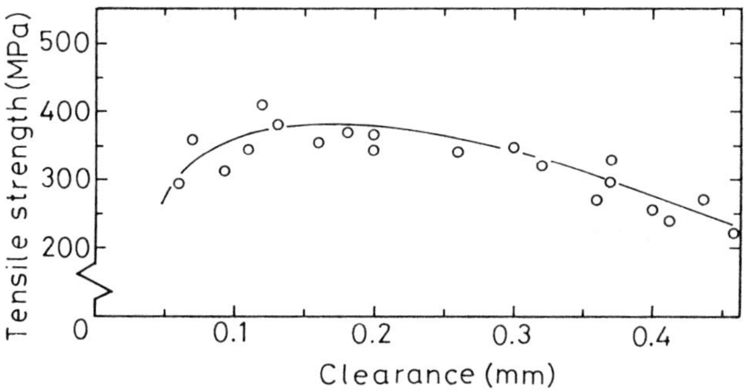

(C) 40Pd-49Ni-10Co-1B

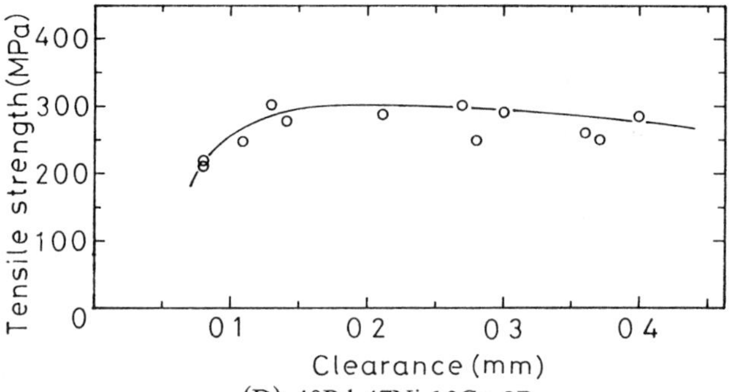

(D) 40Pd-47Ni-10Co-3B

Fig.8 Relation between tensile strength and joint clearance

3.3 Tensile Strengths of Joints at Elevated Temperatures The results of the tensile strengths on joints at elevated temperatures are shown in Fig.10. As the test temperatures increased, the strengths of brazed joints decreased. All of the tensile specimens fractured at the brazing layers. Brazed joints with 40Pd-50Ni-10Co showed the highest strength at elevated temperatures except for at 473K. The average tensile strengths were 407 MPa at 673K, 207 MPa at 873K, and 130 MPa at 1073K, respectively. At 473K, the joint brazed with 30Pd-50Ni-10Co-10Cu had the highest tensile strength of 509 MPa.

3.4 Creep Rupture Strengths of Joints Figure 11 depicts the creep rupture strengths of brazed joints with 40Pd-50Ni-10Co and 30Pd-50Ni-10Co-10Cu. This figure indicated the relationship between applied stress and time to failure. These tests continued until the time to failure reached over 1000 hours. The arrows represent the time to failure of 1000 hours or over. Joints brazed with 40Pd-50Ni-10Co had higher creep rupture strength than joints brazed with 30Pd-50Ni-10Co-10Cu. The joints brazed with 40Pd-50Ni-10Co had the creep rupture strengths with a time to failure of 1000 hours or over at applied stresses of 325 MPa at 673K, 200MPa at 873K, and 50 MPa at 1073K respectively.

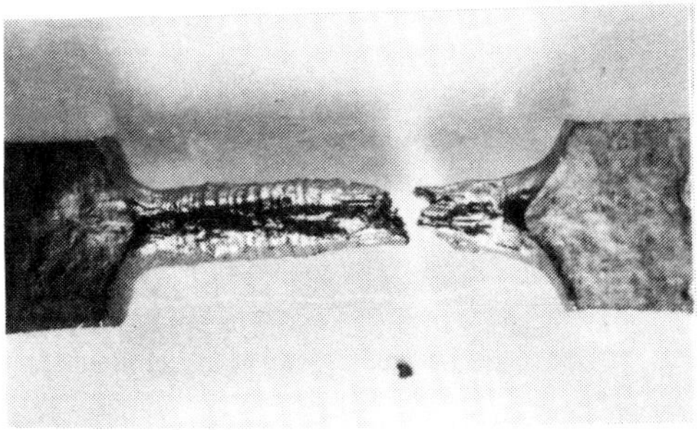

Fig.9 Tensile test specimen fractured at base metal

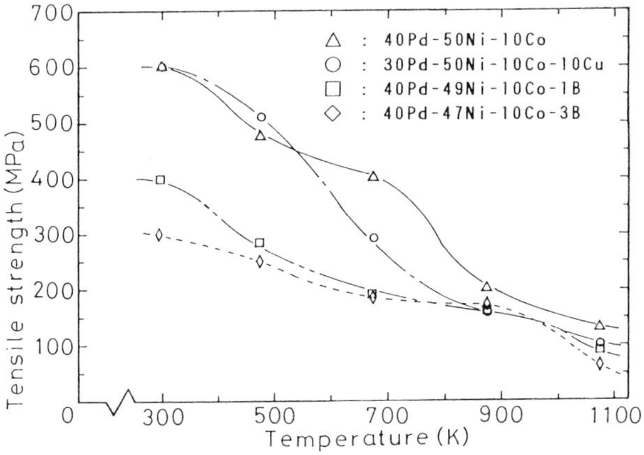

Fig.10 Tensile strengths of brazed joints at various temperatures

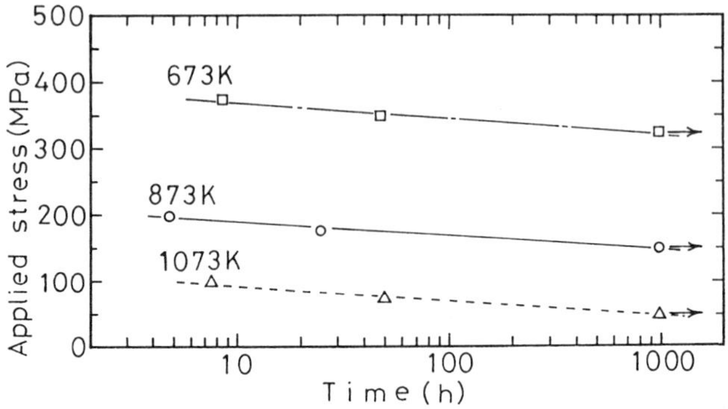

(A) 40Pd-50Ni-10Co

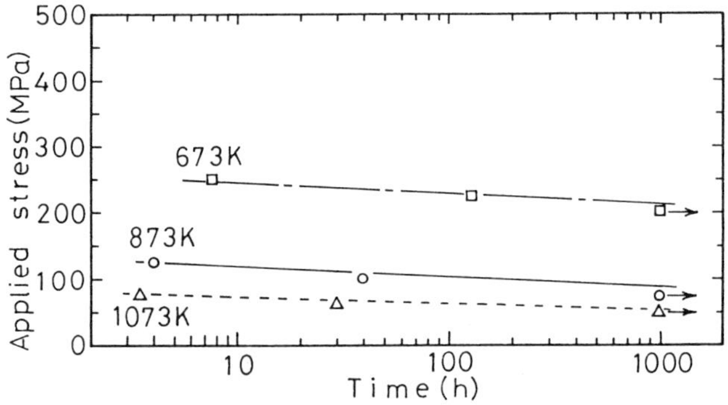

(B) 30Pd-50Ni-10Co-10Cu

Fig.11 Creep rupture strengths of brazed joints

4.CONCLUSIONS

Four types of brazing filler metals, boron-free ones, 40Pd-50Ni-10Co and 30Pd-50Ni-10Co-10Co, boron-containing ones, 40Pd-49Ni-10Co-1B and 40Pd-47Ni-10Co-3B, were developed to satisfy the requirements of high-temperature performance for Inconel 600 brazed joints.

Joints brazed with the 40Pd-50Ni-10Co brazing filler metal had the highest tensile strength of the tested joints, and the mean tensile strength was 580 MPa. All of the tensile test specimens fractured in the brazing layer.

The average 30Pd-50Ni-10Co-10Cu joint strength was 509 MPa at 473K. Brazed joints with 40Pd-50Ni-10Co showed the highest tensile strength of 407 MPa at 673K, 207 MPa at 873K, and 130 MPa at 1073K.

Joints brazed with 40Pd-50Ni-10Co had good creep rupture strengths. These joints showed the times to failure of 1000 hours or over at applied stresses of 325 MPa at 673K, 200 MPa at 873K, and 50 MPa at 1073K.

REFERENCE

1. E. Lugscheider and O. Knotek, Welding Journal, 55(10), 314s(1976)
2. I. Kawasatsu, T. Osawa, and H. Saito, Journal of the Japan Institute of Metal, 43(11), 1001(1979)
3. E. Lugsheider and K.D. Parts, Welding Journal, 62(6), 160s(1983)
4. E. Lugsheider and H. Karppitz, Welding Journal, 65(10), 261s(1986)
5. W.F. Gale and E.R. Wallach, Welding Journal, 69(2), 107(1990)
6. J.H. Selverian, D.A. O'Neil, and S.Kang, ASME Journal of Engineering for Gas Turbines and Power, 116(6), 622(1994)
7. T.L. D'silva, Welding Journal, 58(10), 283s(1979)
8. J.H. Selverian, and S. Kang, Welding Journal, 71(1), 25s(1992)
9. H. Izui and Y. Suezawa, Transaction of Japan Welding Society, 24(2), 29(1994)
10. H. Izui and Y. Suezawa, Transaction of Japan Welding Society, 24(2), 35(1994)
11. H. Izui and Y. Suezawa, Theoretical and Applied Fracture Mechanica, 23(1), 67(1995)
12. H.Izui and Y.Suezawa, PVP, Fitness-For-Service and Decisions for Petroleum and Chemical Equipment, 315, 467(1995)
13. C. He, G.Ma, W.Wang, and H.Zhao, Phase Diagrams of Precious Metal Alloys, " The Metallurgical Industry Press, 1988

28th International SAMPE Technical Conference
November 4-7, 1996

FINITE ELEMENT ANALYSIS OF AUTOCLAVE PROCESSING OF LARGE COMPOSITE STRUCTURES USING A SUBSTRUCTURING TECHNIQUE

A. Johnston[1], P. Hubert[1], K. Nelson[2], and A. Poursartip[1]
1 - Composites Group, Department of Metals and Materials Engineering,
The University of British Columbia, Vancouver, B.C., CANADA
2 - Boeing Defense & Space Group, Seattle, WA, USA

ABSTRACT

The full potential of composite materials in commercial aerospace applications can only be realized if the relatively high material costs are offset by exploiting their inherent suitability for direct manufacture into large monolithic structures, thereby significantly reducing production costs. However, the manufacture of such structures presents a number of difficult challenges, of which one of the most important is the prediction and control of process-induced deformation. Modelling the processing of such structures using the plate and shell elements typically employed in finite element analysis is problematic due to the very local action of most important processing phenomena. This paper discusses the application of a comprehensive 2-D finite element processing code, COMPRO, to the prediction of process-induced deformation in stiffened substructures. Predictions are compared to experimental measurements and a technique for using the substructure analysis results in a full-scale F.E. model of the complete structure is presented.

KEY WORDS: Composite Structures, Process Modelling, Warpage

1. INTRODUCTION

Empirical tooling and process cycle optimization techniques are unsuitable for large structures, particularly in commercial applications where low cost and rapid development times are essential. To minimize the cost and uncertainty of this phase of the design process, there is a need to implement a science-based approach to composites manufacturing.

Over the past two decades, a number of increasingly capable computational models for the processing of polymeric composite materials have been developed [e.g. 1-8]. Some of these models have focused largely on a single phenomenon such as heat transfer and resin reaction kinetics [e.g. 4,6,7], resin flow [e.g. 2], or stress development [e.g. 8,9] while others have been more broad in scope examining a number of important processing phenomena and their interactions [e.g. 1,3,5]. Currently only models by Bogetti and Gillespie [5] and White and Hahn [9] incorporate analyses of stress development *throughout* processing. While potentially very useful for research, neither model can be used for components of appreciable complexity.

1.1 Size Scales of Process Modelling

The scales of interest in processing and process modelling range from the level of fibre and matrix to large-scale structures (Fig. 1). The analysis of the entire structure requires large-scale or 'global discretization' models, employing plate and/or shell elements. However, with such models it is generally not possible to capture the effect of small-scale geometric variations (e.g. local resin flow or through-thickness effects) and mechanism interactions that are often important contributors to warpage. Between these extremes there is an intermediate 'local discretization' level at which it is still practical to use elements of small enough size to adequately examine the most important contributors to process-induced warpage. The challenge is to develop a practical method for transferring the predictions of such intermediate 'local discretization' to the 'global discretization' of the large structure of interest.

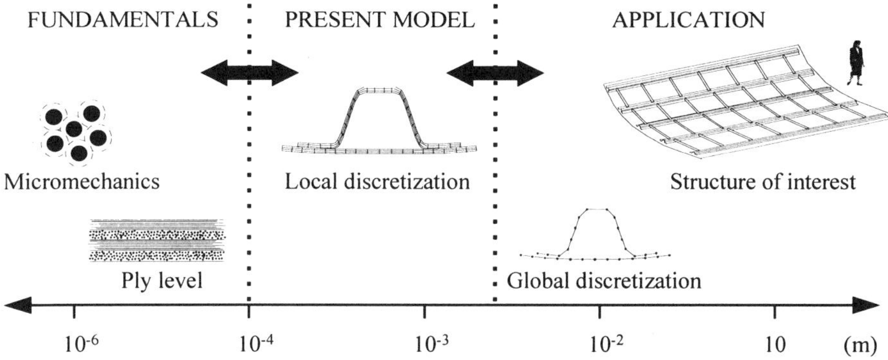

Figure 1 Range of scales of interest in modelling the processing of composite structures (from [10])

1.2 Process Modelling Software

In this work we use COMPRO, a comprehensive two-dimensional finite element process modelling software developed by the Composites Group at The University of British Columbia specifically to analyze industrial autoclave processing of composite structures of intermediate size and complexity. COMPRO incorporates analyses of a number of important processing mechanisms including component internal temperature, resin degree of cure, resin flow and the development of residual stress and deformation. This commercially available software can examine complex 2-D structures with multiple composite and non-composite materials, including the effects of process tooling and autoclave characteristics. A more detailed description is given in [10].

A case study is presented to demonstrate the application of COMPRO to process-induced warpage of a large-scale composite structure by employing a substructuring method.

2. CASE STUDY

2.1 Modelling of a Large-Scale Composite Structure

As part of the product development phase for a large composite structure such as that shown in Fig. 2, finite element analyses will normally be performed to study the in-service mechanical behaviour under a variety of loading conditions. The initial conditions for such analyses will normally assume the as-designed part geometry and a nominal initial stress state, perhaps including a hygro-thermal contribution. This approach may not be adequate since there may be significant warpage and/or residual stresses induced during processing. In addition, even when warpage is small, significant pre-loads may be required to force adjacent components to fit together.

In our proposed approach, the first step in predicting the post-processing state of a large structure is to start with the same FE mesh as used in later analyses, but with a shape based on process tooling rather than the as-designed shape. If the structure is too large to be discretized fully, then an intermediate 'local discretization' must be performed.

This can often be accomplished by taking advantage of the fact that many large, complex structures, such as the one in Fig. 2, are composed of a small number of simple repeating units or substructures. In these cases, it is feasible to use small-scale models to analyze each different type of substructure individually and to use their predictions as input to the larger-scale discretization.

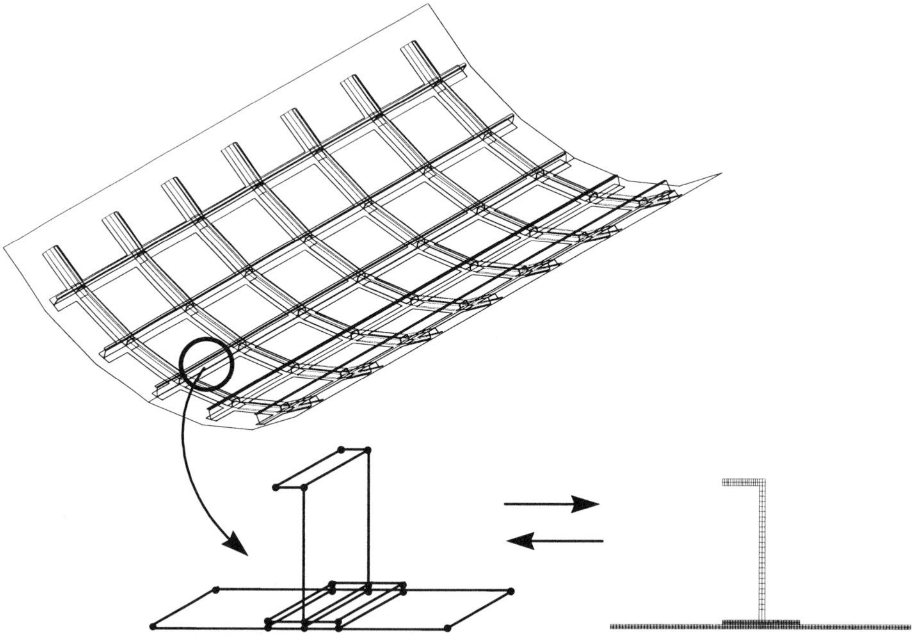

Figure 2 A large complex structure and a typical substructure showing 3-D shell element 'global' discretization and 2-D 'local' (COMPRO) discretization

2.2 Representative Substructures

Substructures must be similar both in geometry and in the boundary conditions which they see during processing. Thus a substructure in the interior of the part may have to be examined separately from one on the edge, even if they have identical geometry. For our case study, as shown in Fig. 2, a local-scale finite element description of each J-frame substructure is developed and the processing boundary conditions determined.

An analysis of two typical J-stiffened substructures was performed. These substructures are illustrated in Figs. 3 and 4. The face sheets for both parts were made of 12-plies of Hercules AS4/8552 grade 190 prepreg tape with $[45/90/-45/0/45/90]_S$ lay-up. A single pre-cured braided J-frame made by RTM was centered on the top face sheet and bonded to it using Cytec Metlbond 1515 structural adhesive, which is also used to bond the skins to the Hexcel HRP 0.1 kg/m^3 (8 lb/ft^3) honeycomb core. No special tooling was used to prevent the frames from shifting during cure.

The parts were assembled on a large aluminum tool with a constant radius of curvature of 3.10 m (122") in the y-direction and cured using the process cycle shown in the inset in Fig. 5. To examine part-to-part variability, three samples of both components were built. As a baseline, three samples of skins and honeycomb structures with no frames were also built.

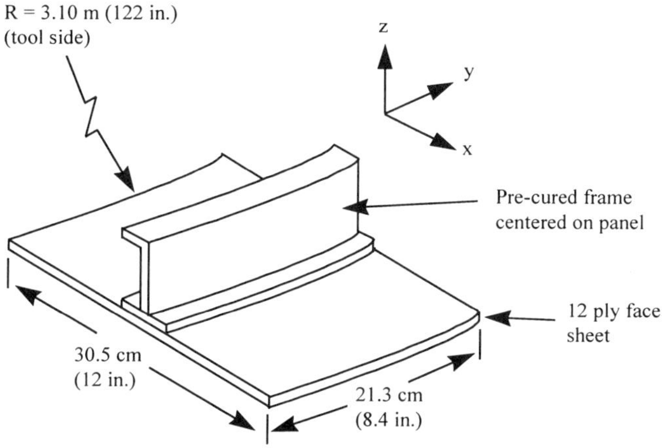

Figure 3 Skin and J-frame Substructure

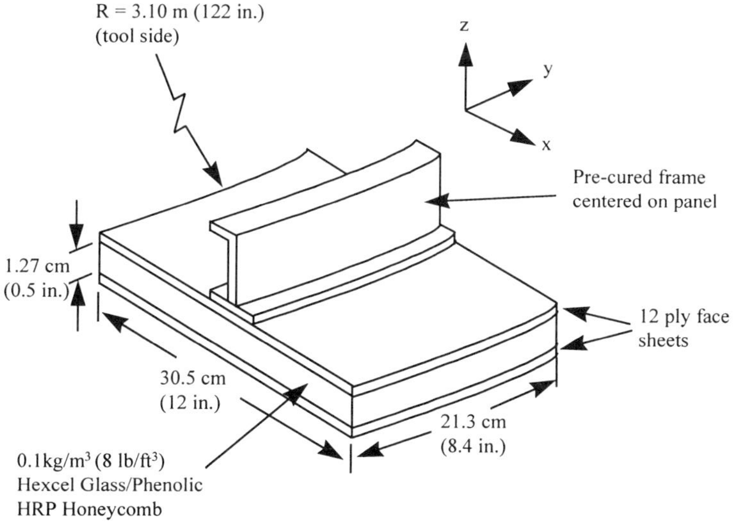

Figure 4 Honeycomb and J-frame Substructure

2.3 Finite Element Modelling

Finite element descriptions of all parts were generated for COMPRO using the PATRAN preprocessor. An example of the FE discretization employed is illustrated in Fig. 5. As shown in this figure, cross-sections of the representative parts were discretized using 4-noded isoparametric plane-strain elements. A simplified, rectangular representation of the lower section of the J-frame (the 'foot') was employed in which the adhesive 'noodle' was ignored. The adhesive layer between the foot and the skin was considered, but the honeycomb/skin adhesive was not modelled. The complex geometry of the tool was replaced with a thin rectangular section with 'effective' properties chosen to mimic the actual thermal and in-plane mechanical behaviour. A 'sliding' condition was employed for the lower surface of the tool to simulate the high bending rigidity of the tool actually used. Interaction between the tool and the lower surface of the part was modelled using an elastic 'shear layer'. Flow of the matrix resin was not calculated. As illustrated in Fig. 5, a convective heat transfer boundary condition was assumed at all external model surfaces, with a heat transfer coefficient calculated from tool heating rate measurements.

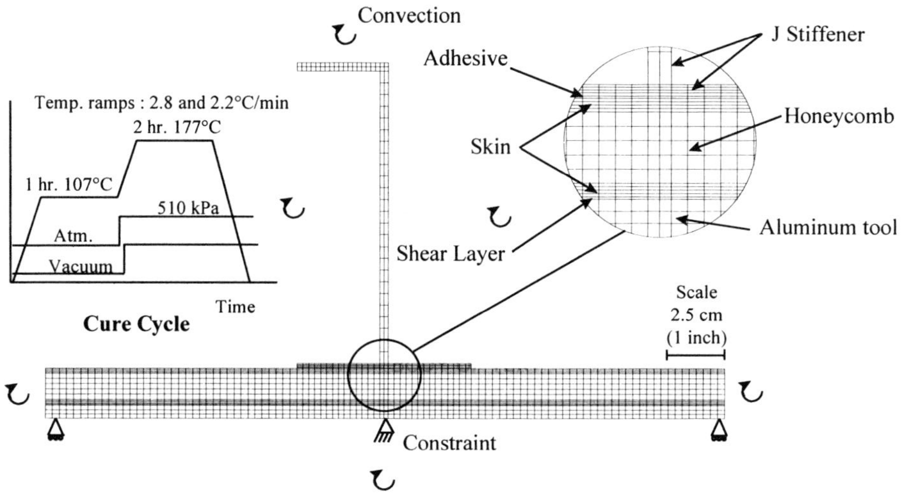

Figure 5 Finite Element Representation of Honeycomb and J-frame Substructure

3. RESULTS

3.1 Experimental Results

Measurements of the process-induced warpage of the representative parts were performed by comparing the components' tool-side surface profile after processing to the tool shapes (i.e. a cylinder with radius of 3.10 m). Approximately 100 points were used to map the surfaces of each component.

A typical displacement contour plot is shown in Fig. 6 for a honeycomb and J-frame substructure. The maximum deformation of this substructure was measured to be about 0.15 mm (0.006"), with both magnitude and warpage profile being very consistent for all three specimens. For comparison with the predictions, it is more convenient to plot the linear variation of the warpage with distance from the centre-line (J-frame) in the X-Z plane, as defined in Fig. 3.

The warpage of the unstiffened skin was 'outwards' (towards the tool) with a parabolic shape, as seen in Fig. 7. This shape is indicative of tool-part interaction. The warpage of the unstiffened honeycomb panel was much smaller, and barely within the limits of resolution of the measurement system, as seen in Fig. 8.

The shape of the warpage was found to be quite similar for both stiffened substructures, as seen in Figs. 9 and 10. In both cases, the tool-side surfaces warped 'inwards' (away from the tool) with warpage being approximately symmetric about the center line of the J-frame and varying approximately linearly with distance from this line. As expected, the skin and J-frame substructure exhibited a much higher level of warpage than the stiffer honeycomb and J-frame substructure with maximum deformation averaging about 0.9 mm (0.035") for the three skin specimens.

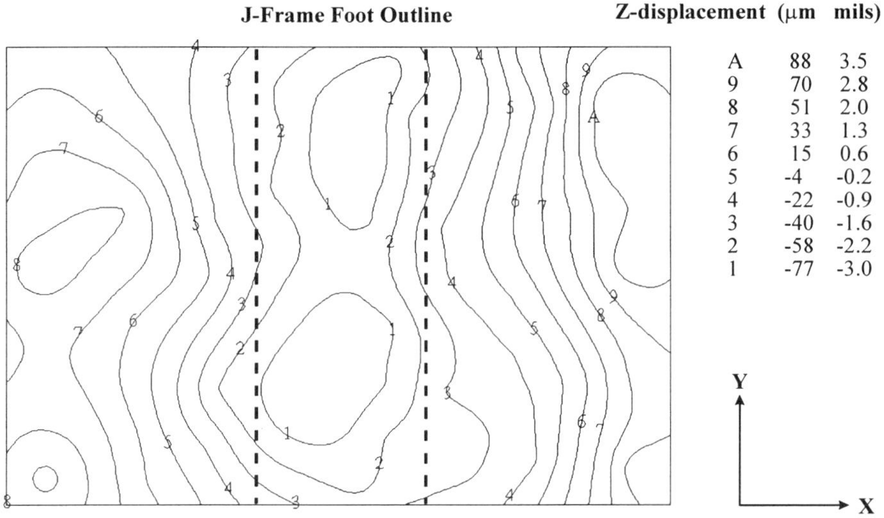

Figure 6 Typical Contour Plot of Warpage of Honeycomb and J-Frame Substructure

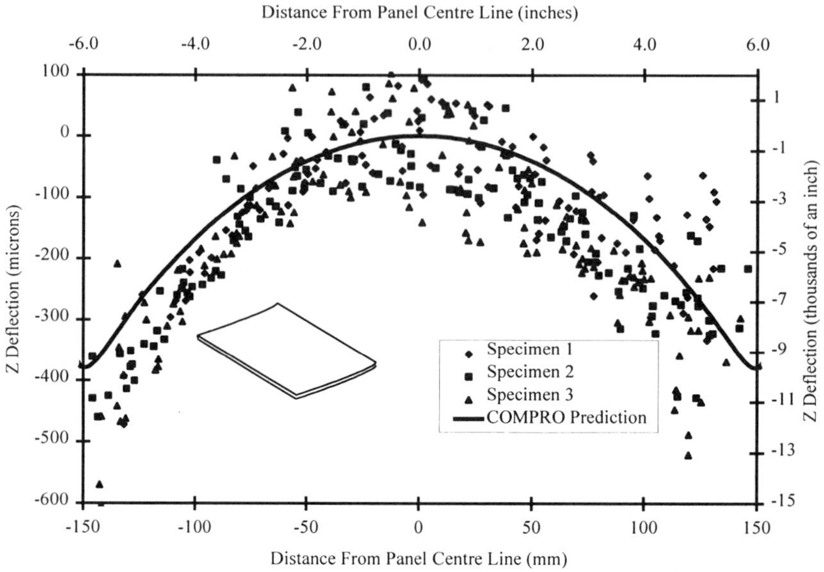

Figure 7 Warpage of Unstiffened Skin Substructure (3 specimens)

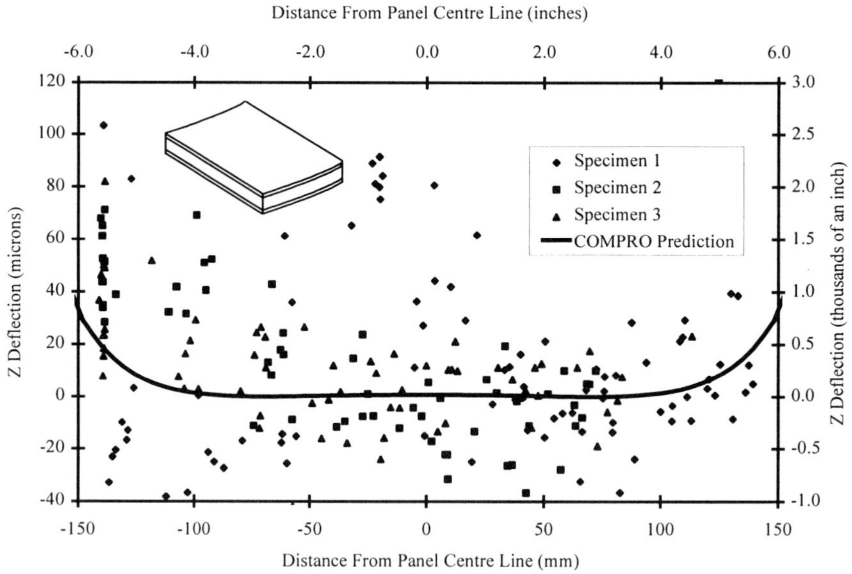

Figure 8 Warpage of unstiffened honeycomb substructure (3 specimens)

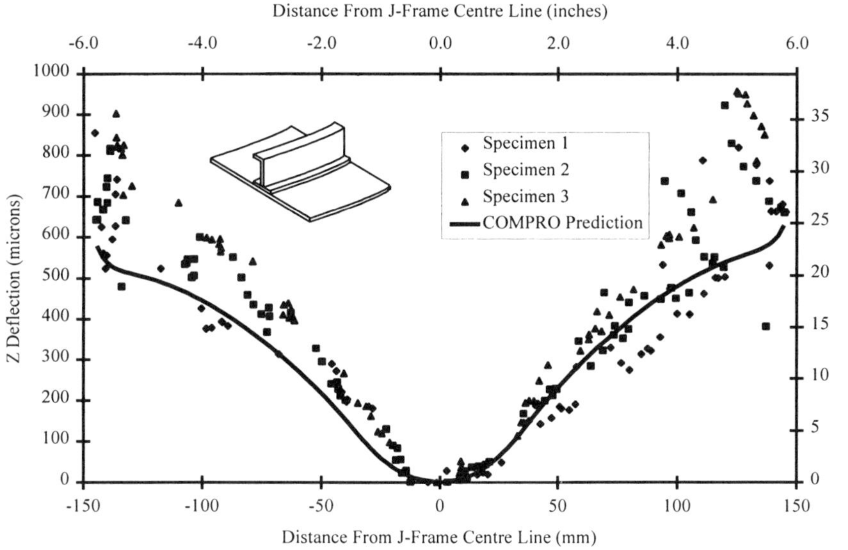

Figure 9 Warpage of skin and J-frame substructure (3 specimens)

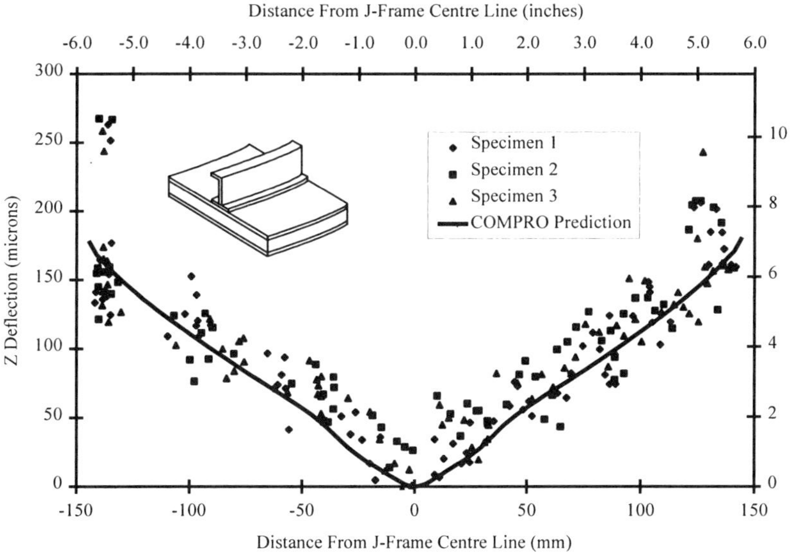

Figure 10 Warpage of honeycomb and J-frame substructure (3 specimens)

The difference in warpage between the unstiffened and stiffened panels indicates that there are two significant sources of process-induced deformation. One is a lower face/tool interaction and the other is an interaction between the J-frame and the underlying skin. The effect of the lower face/tool interaction is to warp the panel out towards the tool. The effect of the J-frame is to pull the panels inwards and away from the tool, and in the case of the stiffened skin, the effect is large enough to change the sign of the overall warpage.

3.2 Model Predictions

Tool/part interaction, which can be thought of as a frictional or adhesive interaction has not been modelled previously in the literature. Although a good understanding of this mechanism is presently lacking, COMPRO allows for the effect with a thin shear layer of adjustable modulus. In the present case study, good agreement with the experimental results for the unstiffened skin substructure is found when a shear layer modulus of 1 MPa (145 psi) is used. This agreement is shown in Fig. 7, and this value of shear layer modulus is used in the rest of the case study.

Using this value of shear modulus, the warpage of the unstiffened honeycomb substructure is predicted and presented in Fig. 8. The predictions are consistent with the measurements. We do not say that the agreement is good, since the measured values are very small and at the limit of the measurement system.

The other unknowns are the properties of the adhesive and the coefficient of thermal expansion of the foot. The adhesive layer is assumed to be a resin which cures later than the prepreg material, and is found not to affect significantly the shape or magnitude of the predictions. The coefficient of thermal expansion of the foot is adjusted until there is good agreement with the skin and J-frame experiments (Fig. 9). This value is then used to predict the behaviour of the honeycomb and J-frame substructure. The predictions for this part are presented in Fig. 10 and show very good agreement with the experimental results. Since the only two unknown properties of importance were adjusted and verified with the other substructures, the good agreement that is observed in Fig. 10 is an independent verification of the validity of the approach.

4. TRANSFERRING SUBSTRUCTURE PREDICTIONS TO LARGE-SCALE MODELS

The case study presented in the previous section shows that it is possible to predict the behaviour of representative substructures. Considering the range of scales shown in Fig. 1, we now consider the transfer of the predictions from the 'local discretization' domain to the 'global discretization' domain of a larger structure. The following technique is proposed:

- The element nodes of the undeformed global model must correspond to strategically chosen positions ('reference nodes') in the undeformed local model, as shown in Fig. 11a.
- The process-induced u, w displacements of the 'reference nodes' are determined from the local discretization model.
- Strains and curvatures (e.g. ε and κ) are determined such that the global element nodes see these u, w displacements (Fig. 11b).
- In the global model, the calculated strains and curvatures are specified as element 'properties' such as thermal expansion (or other) coefficients. These strains and curvatures can then be applied by increasing the component temperature, for example, by $\Delta T = 1$.

Given the local reference nodes, strains and curvatures in the global model can be determined for each element using only the local finite element mesh, predicted nodal displacements, and the global element shape functions.

In the present case study, the effect of the J-frame can be isolated from the effect of the tool/part interaction and the skin/honeycomb thickness effect, and thus the procedure described above would have reasonable generality.

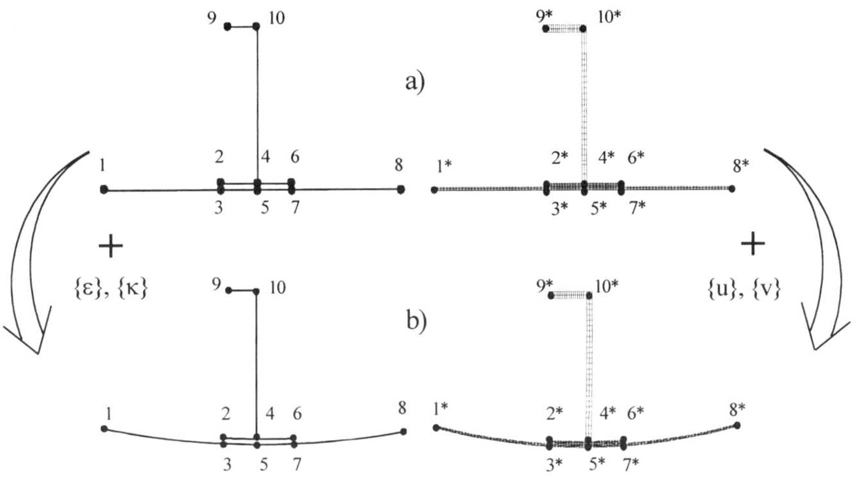

Figure 11 The behaviour of the global model elements (left hand side) is matched to give an equivalent response to the output of the local model (right hand side)

5. DISCUSSION AND CONCLUSIONS

Prediction of process-induced warpage of complex composite structures requires modelling of the component in its entirety, yet computational limitations prevent modelling of large structures on a fine enough scale to capture important local effects. However, by employing the substructuring technique presented here, this task can be accomplished by modelling the processing of a small number of intermediate-scale components, then using the results in a FE model of the complete component. The processing of such intermediate-scale structures can be modelled using COMPRO, a 2-D finite element process modelling software whose application has been demonstrated with a case study.

Process modelling of composite structures, especially the large, complex structures that are the focus of this paper, offers a number of advantages over trial-and-error techniques. The final part shape may be predicted prior to construction of tooling which increases greatly in cost as structures increase in size. Then, by employing sensitivity analyses, areas of process sensitivity can be pinpointed and eliminated, resulting in much more robust manufacturing processes. Such analyses can also be used to predict process variability without requiring that a large number of components be built. Finally, by knowing the level of process-induced stresses and warpage, one may develop improved models of structural loading response and improved component design.

A number of important challenges to full implementation of the techniques outlined in this paper and Ref [10] include the fact that a large number of material properties are required by such complex modelling, many of which are difficult and expensive to obtain. Furthermore, for some mechanisms such as tool/part interaction, an adequate physical understanding and description may still need to be developed. Finally while many substructures can be represented by 2-D models, some are likely to require 3-D models to be represented accurately.

6. ACKNOWLEDGMENTS

This paper summarizes work performed over a number of years under funding by the Science Council of British Columbia, the Natural Sciences and Engineering Research Council of Canada, The Boeing Company, The National Aeronautics and Space Administration and Integrated Technologies Inc. We would also like to gratefully acknowledge the significant interaction and support from our colleagues at The Boeing Company and The University of British Columbia. A special word of thanks to Mr. Robert Courdji for his help in preparing this paper.

7. REFERENCES

1. A.C. Loos and G.S. Springer, J. of Composite Materials, 17, (2), pp. 135-169 (1983).
2. R. Davé, J.L. Kardos and M.P. Dudukovic, Proceedings of the American Society for Composites, 1st Technical Conference, Technomic Publishing, pp. 137-153 (1986).
3. A.R. Mallow, F.R. Muncaster and F.C. Campbell, Proceedings of the American Society for Composites, Third Technical Conference, Technomic Publishing, pp. 171-186 (1988).
4. J. Mijovic and J. Wijaya, SAMPE J., 25, (2), 35-39 (1989).
5. T.A. Bogetti and J.W. Gillespie Jr., J. of Composite Materials, 26, (5), pp. 626-660 (1992).
6. T.A. Bogetti and J.W. Gillespie Jr., J. of Composite Materials, 25, (3), pp. 239-273 (1991).
7. J.M. Kenny, Proceedings of the Third Conference on Computer Aided Design in Composite Materials Technology, pp. 530-544 (1992).
8. S.R. White and H.T. Hahn, J. of Composite Materials, 26, (16), pp. 2402-2422 (1992).
9. H.T. Hahn and N.J. Pagano, J. of Composite Materials, 9, 91-108 (1975).
10. P. Hubert, A. Johnston, R. Vaziri, A. Poursartip, in A. Poursartip and K.N. Street, eds., Proceedings of The 10th International Conference on Composite Materials (ICCM-10), Woodhead Publishing, pp. 149-156, (1995).

28th International SAMPE Technical Conference
November 4-7, 1996

HIGH VOLUME PROCESSING OF LAMINATED CERAMIC SUBSTRATES FOR MICROCIRCUIT MODULES

Wei Koh
Express Packaging Systems
Palo Alto, California 94303

ABSTRACT

The design and fabrication of a microcircuit module consisting of laminated thin film ceramic substrates is described. These laminated modules contain cavities to form multichip modules. To achieve high density interconnect and low package profile, both the ceramic substrate and the laminating adhesive layers are kept very thin. Epoxy preforms, thermoplastic pastes, and low temperature firable glass pastes have been evaluated for lamination. The glass paste materials were found to be the most suitable for automated, high volume processing despite a higher cost. Using the glass-ceramic laminates, a new multilayer fabrication process for high rate production, called fusion-lamination, is developed.

KEY WORDS: Adhesives, Ceramic Composites, Glass Laminates

1. INTRODUCTION

Multichip modules (MCMs), particularly the high density, deposited thin film type (MCM-D), have been considered to be expensive and are typically limited to high end applications for military and space. Thus, low cost MCM manufacturing is receiving serious attentions for broadening their applications to workstations and automotive [1,2] MCM-Ls, on the other hand, have been perceived as a lower cost process with their use of organic laminates and high rate processing [3]. This paper describes a new approach to fabricate modules comprised of laminated thin film ceramic substrates using high volume, automated process. The resulting microcircuit modules maintain the high density interconnection and performance advantages of the MCM-Ds but cost less to fabricate--approaching that of the MCM-Ls. Modules laminated by this new method may further contain two-sided cavities and can be used as stackable building blocks for 3-D memory packages and infrared focal plane arrays. The original concepts and fabrication processes described here are based on the work completed at Grumman Aerospace Corporation for its z-axis mosaic focal plane modules[4].

2. DESIGN DESCRIPTION

2.1 Module Dimensions and Layout The design layout and dimensions can be quite flexible, depending on the chip size and number in a cavity. Routing of the signal and power lines affect the number of layers required although a 4-metal layer is typical. Vertical interconnect is accomplished by laser drilled fine vias that are sputter coated with metals. A representative cross-sectional five-layer structure is shown in Fig. 1, where the power and ground planes are located within the two base layers and the signal lines are on the upper tiers. A wire bond chip is shown inside the cavity, however, other types of die attach such as tape automated bonding and flip chip can also be used to further lower the package profile. The chips may have to be thinned down to less than 0.25mm to fit inside the cavity.

Due to the thin and fragile nature of the substrates, the practical limit of a layer panel is about 6x8 cm. Figure 2 illustrates a representative design where three modules are assembled from a single panel. After dicing and trim, each module's final dimension becomes 2.5x7.5 cm, each containing cavity for eight wire bond ICs. It is also possible to laminate a module consisting of two-sided cavities as shown in Fig. 3. A high reliability glob top encapsulant is applied to seal the cavities after the module and chip assembly is completed.

2.2 Thin Film Substrates High purity 99.6% alumina substrates are used. A typical metal thin films consists of sputtered base metals, TiW and gold. The thin film microcircuits are fabricated by photodelineation to obtain fine lines and spaces as small as 25 μm. Table 1 summarizes the characteristic features of these layers. Each substrate is first lapped to meet the thickness requirement of <0.2 mm. Such thin ceramics are quite weak and fragile; they must be placed in trays and handled with great care to reduce breakage and chipping.

Table 1. Thin Film Layer Properties

Property	Typical values
Substrate material	99.6% Al_2O_3
Panel size	6x8 cm
Thickness	0.2 mm
Coefficient of thermal expansion	7 ppm/°C
Thermal conductivity	35 W/mK
Surface finish, lapped (CLA)	0.12 μm
Thin film metals	TiW/Au
Typical line width/spacing	30/25 μm
Via interconnect diameter	30 μm

2.3 Lamination Adhesives Three types of materials—epoxy films, thermoplastic pastes, and low temperature firable sealing glass pastes were evaluated for high rate processing. Table 2 lists the key properties of a representative material from each type.

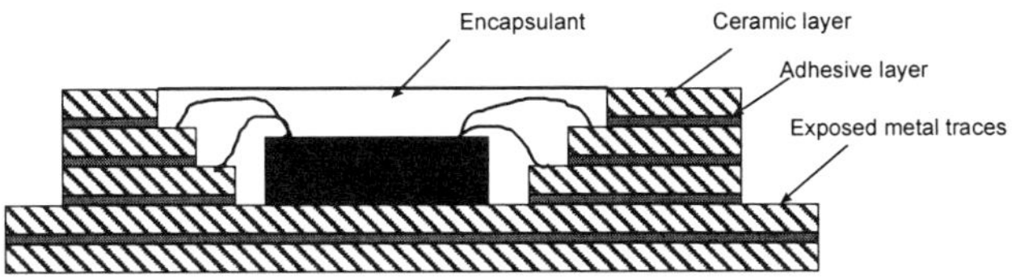

Fig. 1 Cross-section of Module with Wire Bond Chip in Cavity

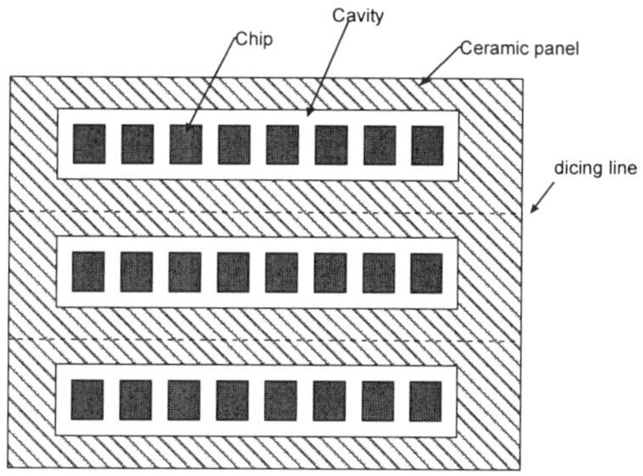

Fig. 2 Laminated Panel Containing 3 Modules Each Having 1 Cavity and 8 Chips

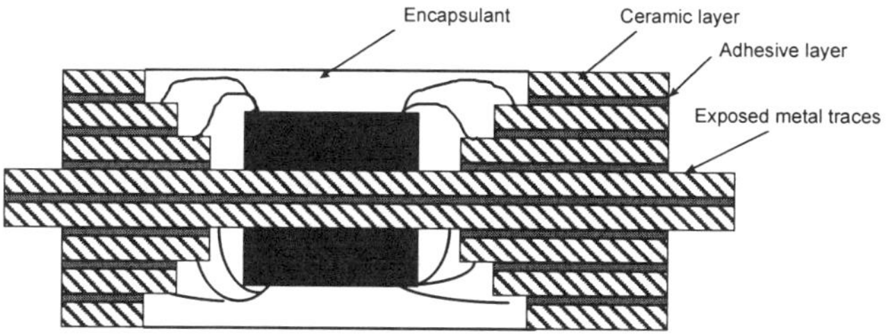

Fig. 3 Cross-section of Two-sided Cavity Module

Table 2. Adhesive Material Properties

Property	Epoxy	Thermoplastic	Glass
Glass transition temperature (°C)	90	150	550
CTE (10^{-6} mm/mmK)	60	50	10
Moisture absorption (wt%)*	2	1	0.2
Dielectric constant @ 1 KHz	3.8	3.3	6.9
Thermal conductivity (W/mK)	0.2	0.3	1.2
Pre-lamination thickness (μm)	25	35	25
Post-lamination thickness (μm)	20	30	20

*Post lamination 85°C/85%RH for 168 h

2.3.1 CTE match One critical consideration in selecting these laminating materials is their close match in the coefficient of thermal expansion (CTE) to that of the ceramic substrate over a wide temperature range. Otherwise, thermal stresses due to the CTE mismatch could lead to module warpage and even delamination. The two organic adhesives obviously do not have closely matched CTE to the alumina, however, their low glass transition temperatures and compliant nature serve to relieve the stresses arisen from the CTE mismatch. For the glass material, on the other hand, thermal mechanical analysis (TMA) measurement indicated that it exhibited a very close CTE match to the alumina in the temperature range between 0 and 450°C, as shown in Figure 4.

2.3.2 Thermal cycling Samples laminated from all three materials have been subjected to temperature cycling from -55 to 150°C. The average ramp rate was 5°C/min and soak time was 10 min. No delamination was found for any of the three lamination materials after 100 cycles.

2.3.3 Moisture absorption Moisture absorption was measured in a 85/85 humidity chamber for the lamination materials after each was coated to a ceramic layer and fully cured. Both the epoxy and thermoplastic materials have moisture gain considered undesirable for high reliability applications. The glass laminate also showed a small amount of weight gain, due mostly to tiny voids that may have trapped some moisture.

3. FABRICATION PROCESSING

The main processing steps in the module fabrication consists of adhesive layer application, lamination, trim, and testing and inspection. For the three adhesives the applicable techniques are also different, as compared in Figure 5. The epoxy preforms must be applied first to each bare ceramic layer by vacuum and cured to form a B-staged coating. The vacuum cure is necessary to avoid forming trapped bubbles. After the B-stage cure, any excess materials around the layer edges must be removed by trimming. The trimmed layers are then stacked and laminated in a lamination press. The thermoplastic paste, on the other hand, can be screen printed to a bare layer. The paste materials, however, are not as easy to print as the glass pastes to be discussed next. Frequent equipment cleaning was also necessary. Due to the

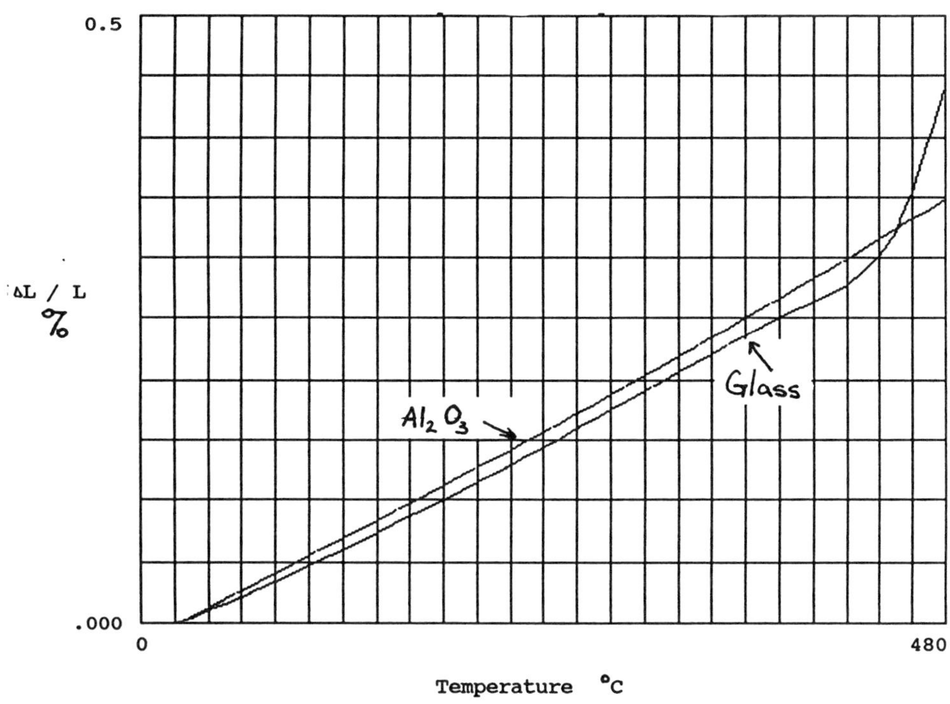

Fig. 4 Thermal Expansion Coefficient Comparison Between Alumina and Glass Laminate

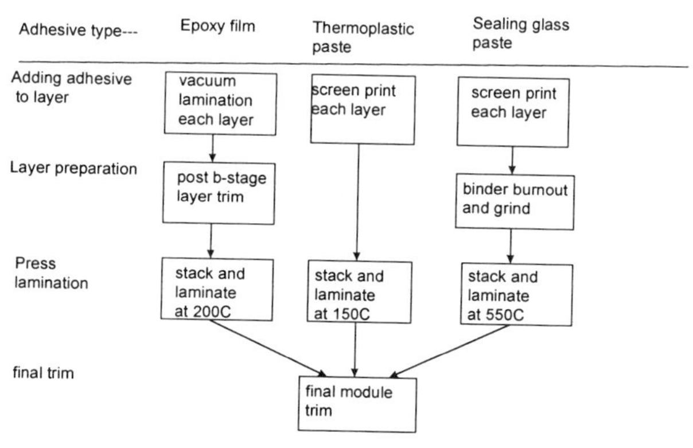

Fig. 5 Lamination Process comparison

thermoplastic nature of this, the screen printed layers can be stacked and laminated directly without the need for the burn-out and grinding step that are necessary for the glass material.

Although the glass pastes are easily screen printed, they require a burnout process to remove the organic binder and to form a glazed coating that can be ground in the subsequent step to arrive at a predetermined thickness. During lamination, they further require a much higher temperature and pressure to achieve a complete layer-to-layer fusion lamination. Figure 6 shows a photomicrograph of a section of the glass/ceramic composite after fusing-lamination. As can be seen in the figure, the dimensions of each bond gap is closely controlled. The thicker bondline at the bottom is between the two base layers and the thinner one at the top is between the upper tier layers as shown earlier in Fig. 1.

Table 3 compares the lamination conditions used for the three different materials. The least costly processes are for the thermoplastic laminate, whereas the glass pastes require the most elaborate equipment and processing parameters.

Table 3 Lamination Conditions

Lamination conditions	Epoxy	Thermoplastic	Glass
Pressure (MPa)	4	2	10
Temperature (°C)	200	160	550
Total time (h)	2	1.5	3

4. Conclusions

Three different adhesive materials were compared for use in high volume lamination of thin ceramic substrates for high rate fabrication of microcircuit modules. While the thermoplastic pastes required the most simple processes, their processing yield in screen printing and final module dimension stability were relatively poor. Low temperature firable sealing glass pastes required more elaborate processing equipment and steps that incur higher cost than the other two. However, the resulting module products have the best mechanical strength, rigidity, and dimensional stability. Further, they retained many high performance attributes such as no warpage, no outgassing, and low moisture absorption. Their long term reliability was excellent due to a close CTE match between the glass and the alumina in temperatures as high as 450°C. The compact size and high density interconnect rendered these modules for high performance space and military applications. An automated fusion-lamination process of fabricating ceramic thin film modules was developed. When produced in high quantity, the per unit cost should be lowered substantially.

5. ACKNOWLEDGMENT

The author would like to thank Grumman Aerospace Corporation (now part of Northrop-Grumman) where the initial fusion-lamination concept and processes were developed.

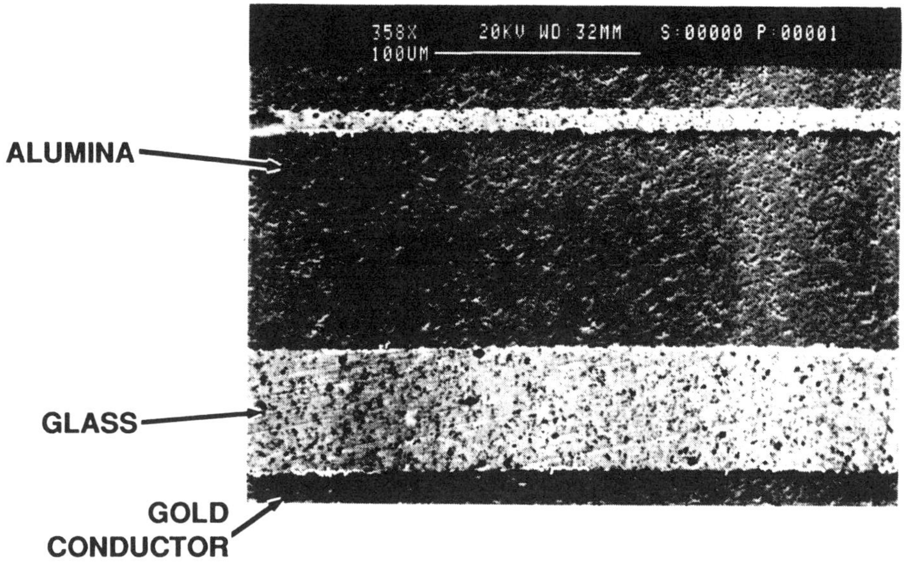

Fig. 6 SEM of Laminate Cross-section

6. REFERENCES

1. E. J. Vardaman, "Multichip Module Outlook: A Perspective on Low-Cost Solutions," Proceedings of 1995 International Conference on Multichip Modules, Denver, CO, April 19, 1995, pp. 262-267.
2. L. Matthew et al., "Production Update: Report on Volume MCM Implementation," Proceeding of 1995 International Conference on Multichip Modules, Denver, CO April 19, 1995, pp. 229-234.
3. L. Lemke, "Recent Developments in the Production and Application of Advanced Laminate Based MCMs and Chip Carriers," Proceedings of 1995 International Conference on Multichip Modules, Denver, CO, April 19, 1995, pp. 235-240.
4. W. Koh and B. Hornback, "Fused Ceramic Multichip Modules," Du Pont Symposium on High Density Interconnect Technology, Wilmington, DE, 1991.

Biography

Dr. Wei Koh is Director of Packaging Technology at Express Packaging Systems, where he is responsible in the engineering design and development of advanced single chip and multichip packages. Before that, he was a program manager at Dexter Electronic Materials for developing plastic package encapsulants. He also worked in various engineering and research capacities at Grumman Aerospace Corporation and Lockheed. Dr. Koh received his Ph.D. in Chemical Engineering from Cornell University and B.S. in Chemical Engineering from University of Washington. He has over 60 technical publications and 15 US patents.

28th International SAMPE Technical Conference
November 4-7, 1996

REAL-TIME CONTROL OF ADVANCED SOLDERABILITY PRESERVATIVES

Michael Frederickson
Electronics Manufacturing Productivity Facility
Indianapolis, Indiana

Brett Piekarski
Army Research Laboratory
Adelphi, Maryland

1. ABSTRACT

Organic Solderability Preservatives (OSPs) are being applied over bare copper as an approach to improve the co-planarity of circuit boards used in surface mount assemblies. In addition, the coatings offer a more environmentally friendly approach to circuit board solderability preservation in lieu of tin-lead hot air leveled coatings. Current methods of electroplating and hot-air leveling do not provide the co-planarity needed to manufacture high density assemblies and are typically more costly than applying OSP coatings over bare copper. Current problems associated with the OSP technology is its inability to assess the solderability of the bare copper and the integrity of the coating itself. The OSP coatings present an even more critical concern due to their reduced shelve life and their inability to survive mishandling in manufacturing. Real time, nondestructive methods of rapidly assessing the integrity of these coatings are currently not available to the electronics industry.

Surface spectroscopy measurement techniques have the potential of measuring the structure and characteristics of the organic coatings and surface oxides that develop with time and temperature. These measurement techniques are rapid, non-contact, and are relatively inexpensive to make as compared to existing methods of measuring solderability. Finally, this technology provides critical surface information that is needed for troubleshooting solderability problems.

KEY WORDS: Organic Solderability Preservative (OSP), Spectroscopy

2. INTRODUCTION

Surface spectroscopy measurements are concerned with the interaction of radiant energy with the surface of materials. Materials, based on their surface structure and constitutive nature, will absorb radiant energy at different wave lengths. Spectroscopy measurements impart radiant energy on the surface of materials and measure the reflective or absorption character across the spectrum. Fundamentally, the absorption character at different wavelengths provides insight as to the atomic or molecular energy levels which relates to molecular geometries, chemical bonding, and the interaction of molecules in solution. There are essentially 4 regions within the electromagnetic spectrum that are of interest for most inorganic and organic materials: Infrared, Near Infrared, Visible, and Ultraviolet. Each of these regions are sensitive to different constitutive properties within a material.

The infrared spectrum is defined as electromagnetic radiation in the range from 1 μm to 300 μm and induces vibrational and rotational energy level transitions in materials. The vibrations that atoms undergo within a molecule are on the same order of magnitude as the frequencies found in the infrared (IR) spectrum. And in particular, any molecular vibration that causes a change of dipole moment will absorb, by resonance, infrared radiation at that frequency and give rise to an absorption peak in the IR spectrum. Almost all organic compounds meet this requirement and absorb infrared radiation.

Table I

Group	Frequency Range (cm^{-1})
C=C stretching	1575-1700
Aromatic C=C stretching	1450 - 1540 singlet
Aromatic C=C stretching	1550 - 1650
Benzene Substitutions	600 - 1000
N-H stretching	3100 - 3700
N-H rocking	700 - 900
C=N	1580 - 1700
C-H stretching	2700 - 3300
C-H bending	1300 - 1475
C-H rocking	600 - 1000
C-H out-of-plane def.	725 - 775
C-O stretching	900 - 1300
C=O stretching	1640 - 1850
O-H	3100 - 3700
O-H bending	1200 - 1450

Several types of vibrations can occur between atoms and each causes a distinct fundamental vibration frequency for that group of atoms. The vibration types that can occur are classified as either stretching (back and forth) or bending. Bending vibrations can further be classified as deformation, wagging, rocking, and twisting vibrations. The IR spectrum of any given molecule is typically made up of absorptions from these small groups of atoms. These groups tend to have the same vibrational frequencies regardless of the molecule in which they occur and therefore the combination in which they appear can be used to characterize and identify many compounds. Table I gives a brief list of some relevant group frequency ranges for bonds typically found within organic compounds. [1,2]

Electromagnetic radiation in the 0.8 μm to 1 μm range is defined as the near infrared region. This region tends to be made up of harmonics from the infrared region. Bands in this region tend to be quite sharp and can be of great value in evaluating materials with functional groups that contain hydrogen atoms. Based on the hydrocarbon nature of the OSP used in this study and the absorption peaks of Cu_2O in the near infrared region, this will prove to be a critical spectral region of interest.[6]

Absorption or reflection in the visible region is caused by the excitation or energy state transfer of the outer electrons of molecules or atoms. Visible light spectroscopy is sensitive to changes in metal surfaces due to oxidation or chemical reactions with organics. Typically, changes that are detected in the visible spectrum are those changes that either effect the hue or color of the surface or that effect the overall bulk reflectance of the surface. There are two measurement techniques that can be used when evaluating the effect of materials in the visible spectrum: (1) unreferenced visible light reflectometry (color analysis) and (2) differential reflectometry (referenced to the unoxidized or uncoated material).

The ultraviolet spectrum is sensitive to both inorganic and organic materials. Ultraviolet spectroscopy has been useful in evaluating organic materials such as greases, coatings, or contaminants. Organic compounds that absorb ultraviolet radiation will have at least one double bond (e.g. C=O or C=C). Again, the OSPs of interest will generally respond well in the ultraviolet spectrum due to the double bond structure inherent within them. There is some limited sensitivity to inorganic materials in that they generally will absorb ultraviolet radiation simply due to the numerous valence electrons available for electron transitions.

3. EXPERIMENTAL

The focus of the data presented here is with Organic Solderability Preservatives (OSP) coating materials. There are two predominate OSP compounds that are used through out the electronics industry to preserve the solderability of copper: (1) benzotriazole and (2) substituted benzimidizole. The benzotriazole material is a monomolecular layer that passivates the copper surface and is typically unsuitable for manufacturing processes with multiple heat cycles. The substituted benzimidizole materials, however, range in thickness from 2000 to 5000 angstroms and are generally suitable for multiple heat cycles. The substituted benzimidizole coating materials were used in collecting visible and infrared data.

The experimental approach was established to determine the effect of time and temperature on OSP coated copper substrates. In addition to differing time and temperature excursions, there were 3 different states of OSP coatings that were investigated: (1) Not Coated (e.g. Bare Copper), (2) Degraded Coating, and (3) Fully Coated. The degraded coating was accomplished by simply turning off the pumps in the OSP coating stage, whereby significantly reducing the amount of coating on the substrate surface and weakening its protective nature. This matrix of substrates and environmental excursions were investigated using both IR and visible light spectroscopy in an attempt to understand changes in both the organic coating and the metal surfaces.

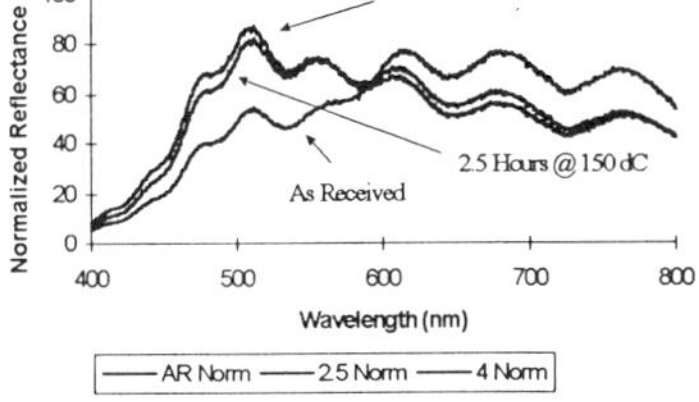

Figure 1. Visible Spectrum of Copper Surface from Test Samples Pulled from a Board Shop Manufacturing Line. Note that the Boards were Measured As Received With No Further Preparation or Etching.

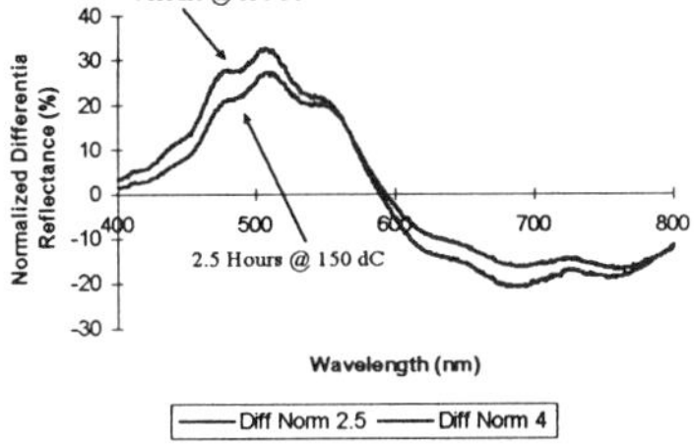

Figure 2. Differential Reflectance of Copper Oxide within the Visible Spectrum When Exposed to 2.5 and 4 Hours at 150 dC.. Note the Radical Change in Differential Reflectance as Bare Copper Oxidizes at These Accelerated Temperatures.

4. DISCUSSION

Visible Light Spectroscopy. It is theorized that the visible spectrum will be sensitive to the changes in the electronic state of the surface of the copper under an organic coating. The assumptions are that the organic coating will be somewhat translucent to the visible light and that the surface of the copper or the copper-organic interface will be the primary absorber of the radiant energy. In order to establish a baseline as to the behavior of copper in the experiment conducted, an unprotected copper surface was exposed to the same time and temperature excursions. Figure 1 provides the reflectance behavior of unprotected copper. From figure 1 it can be seen that the natural color change that occurs when copper oxidizes can be readily measured and quantified using the visible spectrum. The value of this information is that if the copper surface begins to oxidize under a protective coating, it can be potentially quantified using the visible spectrum. It is understood, based on the benzimidizole structure (Cu(1)BIA), the oxide of interest will be the Cu_2O species [7]. The visible spectrum has both an electronic reflective character and a molecular absorption character within the visible spectrum.

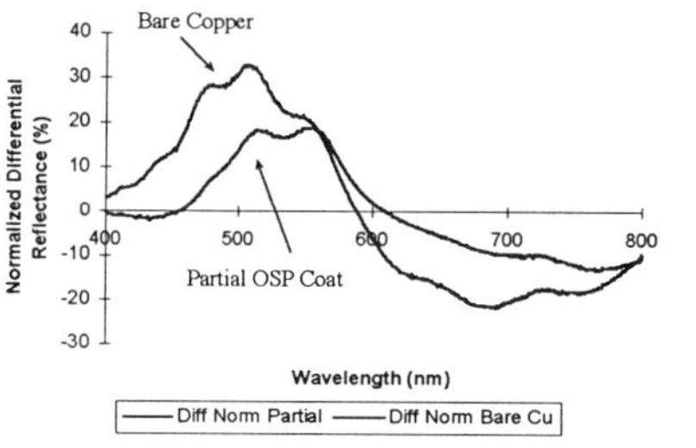

Figure 3. Changes in the Differential Reflectance Between Bare Copper and a Partial OSP Coating at 2.5 Hours @ 150 dC in Air.

In order to measure the effective change on the surface of the material when exposed to time and temperature, differential techniques can be employed. Differential reflectometry measurement techniques were suggested by Hummel, Shanley, and Verink in the 1980's as a technique to evaluate copper surfaces in both predominate oxidation states Cu_2O and CuO [5]. Hummel suggested that the differential reflectance at a specific wavelength between the oxidized and unoxidized copper be evaluated by dividing the absolute difference in their reflectance by their average. (i.e $|R_{sample} - R_{standard}| / ((R_{sample} + R_{standard})/2)$).

In the case of evaluating copper oxide surfaces under semi-transparent OSP coatings where there is both an increase and decrease in the reflective character relative to the unoxidized copper and the chemical interactions between the copper and the OSPs, it was found that the data became more meaningful when evaluated using a normalized difference. The normalized difference between the standard copper reference and the sample of interest is calculated using the following formula:

$$\text{Normalized } \Delta R = \frac{R_{sample}}{\int_{\lambda_{400}}^{\lambda_{800}} R_{sample}(\lambda)d\lambda} - \frac{R_{standard}}{\int_{\lambda_{400}}^{\lambda_{800}} R_{standard}(\lambda)d\lambda}$$

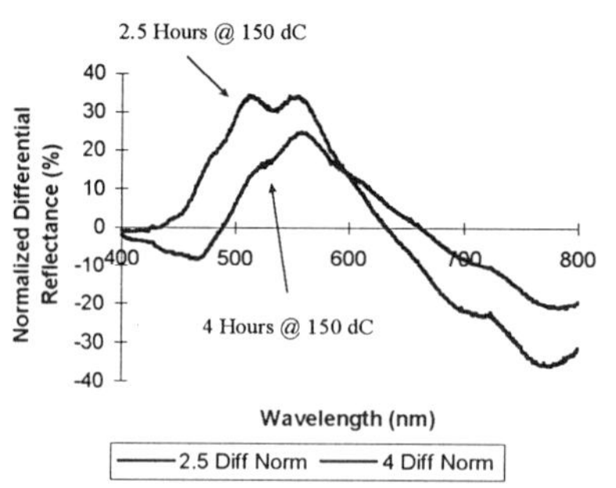

Figure 4. Differential Reflectance Changes for the Partial Coat OSP at Different Time Exposures @ 150 dC in Air.

Figure 2 illustrates the normalized differential reflectance across the visible spectrum for bare copper. From this figure, it shows that the color change exhibited in figure 1 can be seen as shifts in the reflectance (either increasing or decreasing) relative to unoxidized copper across the spectrum. The increase in reflection at wavelengths less than 600 nm are attributed to electronic transitional energies while the decrease in reflection above 600 nm can be attributed to molecular energy absorption. From figure 2, it can be seen that the discriminating reflective wavelengths for the copper oxidation process are at approximately 480, 500, and 560 nm. From Hummel's work, it was found that the critical wavelengths based on electronic transitional energies were 455, 500, and 560 nm for Cu_2O and 500 nm for CuO. [5] The data in figure 2 is in agreement with Hummel's work with a prominate peak at or near 500 nm and a shoulder peak at 560 nm. In addition to the increase in reflection (i.e. reflectance peaks), there are absorption peaks (i.e. a significant decrease in reflection) at approximately 680 and 760 nm and a shoulder peak at 650 nm. These molecular absorption peaks for Cu_2O are in agreement with Prevot's work where near infrared absorption peaks were found for Cu_2O at 650, 760, 880, 950, 1100, and 1280 nm, with the strongest peak at 1280 nm. [6]

The next item of interest is the change in the electronic state of the OSP coated copper surface when exposed to excessive temperatures. Figure 3 provides a comparison between the bare copper and a partially coated copper substrate. From this data, it is seen that there is a loss of the 480 nm peak in the presence of the OSP coating and that the 500 and 560 nm peaks are present but of lesser magnitude under the same time and temperature conditions. In addition, the molecular absorption peaks also possess less discrimination at 650 and 680 nm and magnitude at 750 nm. It is clear from this data that the Cu_2O peaks are the dominate characterization peaks for tracking changes in the OSP coatings under accelerated temperature excursions.

Figure 4 provides the behavior of the partially coated OSP over time and temperature. From figure 4, it appears as though the discriminating electronic reflective peaks are at or near 500 and 560 nm while the molecular absorption peak ir near 760 nm. There was a shift in the magnitude of the peaks at 500, 560, and 760 nm over time. In addition, there was a fairly substantial shift in both the magnitude and prominence at 500 nm.

Infrared Spectroscopy. The infrared region was investigated to determine potential changes in both the organic material and the surface of the copper. Figure 5 shows a grazing angle (75°) FTIR spectrum for the organic protective coating over a bare copper substrate. Several peaks can be identified and potential bond groups (e.g., C-H, N-H, C=N, and C=C) assigned based on the ranges listed in Table I. The large peaks at approximately 2350 cm^{-1} are due to CO_2 within the system. Since copper does not absorb in the infrared, these bond groups can all be attributed to the organic coating. Based on figure 5, it is clear that IR techniques can be used as a

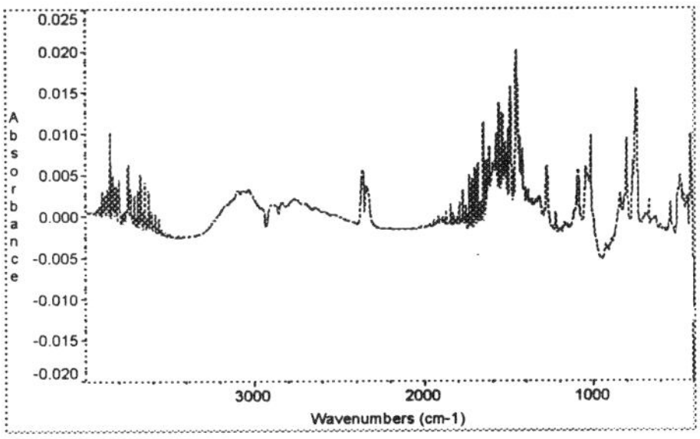

Figure 5. IR Spectrum of the OSP Coated Material. Several Peaks Can be Identified in a Number of Bond Groups.

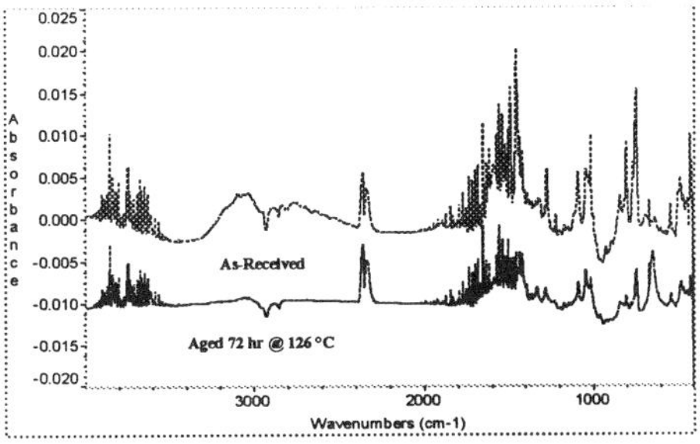

Figure 6. IR Spectrum of the OSP Coated Copper Before and After Aging (72 Hours @ 126 dC). Note the Degradation in the Organic Coating and the Potential Reduction of Material Due to the Reduction of the Peak Height. (Zero Reference for Aged Sample is at -0.010)

process control tool for detecting the presence of organic coatings and particular bond groups that may have an impact on the effectiveness of the coating.

The height of the peaks, or intensity of the absorption, is a function of the absorptivity, the path length in the sample, and the concentration of the sample. Since the thickness of the organic coatings tend to be on the order of 2000 to 5000 angstroms, the absorptions for protective organic coatings tend to be very small.

Figure 6 shows the FTIR spectrum of a sample treated with an organic coating and a similar sample that was aged 72 hours at 126 °C. The curves are offset for clarity. The data shows that as the coating degraded during the aging process, a reduction of the amount of material occurred as indicated by a reduction of peak heights for several of the peaks. One peak of interest, however, actually increases in height. This peak occurs at approximately 650 cm^{-1} and can be attributable to the growth of copper oxide on the surface and is not associated with the organic coating [3,4]. Figure 6 shows that IR techniques can be used to show a degradation in the organic coating and also shows the formation of copper oxides which indicates the coating is no longer protecting the surface from oxidation. Further, figure 6 shows that the IR technique could potentially be used for monitoring the coating thickness.

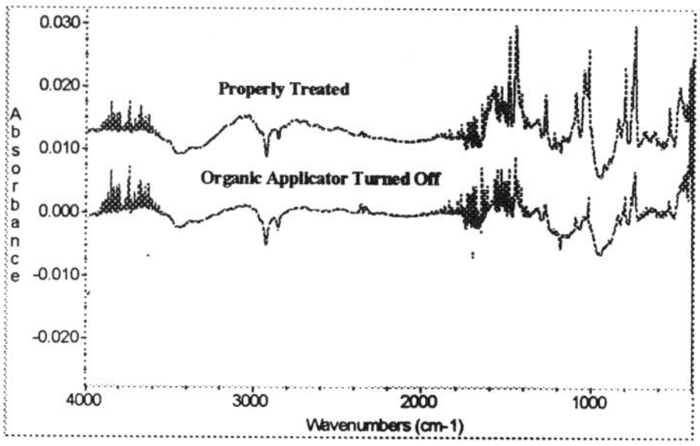

Figure 7. IR Spectrum of the OSP and Partial OSP Coated Copper Substrate in the "As Received" State. Note the Ability of the IR Spectrum to Detect the Amount of OSP Coating. (Zero Reference for Properly Treated Sample is at 0.013)

Figure 7 shows the offset FTIR curves for a properly organically coated sample and the partial coated OSP sample (i.e. the sample that was run through the application process with the applicator turned off). The curve for the partial OSP coating still indicates the presence of an organic coating suggesting that even though the applicator was off, there was enough organic material within the system to partially treat the sample. This is a good example of how the IR technique could be used as a process control tool to give information on how the process is working since the organic coatings are typically transparent and their presence is difficult to ascertain.

5. CONSORTIA APPROACH

In an effort to bring about an instrument that can provide "in situ" measurements of advanced organic coatings, a consortia is being developed. Areas that are to be investigated include the Raman Scattering, UV, Visible, Near Infrared, and the Infrared spectrum. The consortia is composed of coating suppliers, end users, research entities, and a commercializer. The objective is to establish a natural food chain ranging from suppliers to end users along with a development group to bring about a functional and affordable instrument. The goal of the project is to have a prototype ready for Beta Site implementation by the first part of 1997 and to have a commercial instrument by the end of 1997.

6. CONCLUSIONS

It was clear that between the visible and infrared spectrum, the organic and copper oxidation character can be evaluated. With both the visible and the IR spectrum, the prominate peaks detected were representative of what one would expect from copper oxides and benzimidizole materials. Within the visible spectrum, there was an electronic-based reflective behavior along with a molecular absorption behavior. In the infrared region, where molecular behavior is dominate, it was clear that both the copper oxide structures and the organic medium could be evaluated. From both of these measurement techniques, it was clear that process controls could be put into place to control the state and quantity of OSP coated copper materials.

Future work will include correlating the changes in the Cu_2O structure and its associated spectral response and solderability. Once this correlation is established, then control limits can be put in place to allow real-time process control of these coatings in current manufacturing processes. Surface spectroscopy measurement techniques provide a lot of promise in monitoring the integrity of the OSP materials and establishing controls in manufacturing where these materials are used.

7. REFERENCES

1. S. Mohan, N. Sundaraganesan, "FTIR and Raman studies on benzimidazole," Spectochimica Acta, Vol 47A, No. 8 pp. 1111-1115, 1991.

2. David N. Kebdall, "Survey of Practical Information," Applied Infrared Spectroscopy, Ed. by David Kendall, 1966, Reinhold Publishing Corp., New York.

3. Fredrick P. Eng, Hatsuo Ishida, "New Corrosion Inhibitors for Copper - Polyvinylimidazoles," Proceedings of the ACS Division of Polymeric Materials: Science and Engineering, Volume 53, Fall Meeting 1985.

4. G.W. Poling, "Infrared Reflection Studies of the Oxidation of Copper and Iron," J. Electrochem. Soc.:Solid State Science, Vol 116, No. 7, July 1969.

5. C. Shanley, R. Hummel, E. Verink, "Differential Reflectometry - A New Optical Techniques to Study Corrosion Phenomenon", Corrosion Science, Vol. 20, pp. 467-480, Pergamon Press Ltd., 1980.

6. B. Prevot, C. Carabatos, M. Sieskind, "Near Infrared Optical and Photoelectric Properties of Cu_2O", Phys. Stat. Sol., Vol. 10, pp 455. 1972.

7. V. Sastri, R. Packwood, "Surface Analysis in Corrosion Inhibition Mechanisms", Werkstoffe und Korrosion, Vol 38, pp 77-82, 1987.

28th International SAMPE Technical Conference
November 4-7, 1996

DETERMINATION OF CONDITIONS TO AVOID THE FORMATION OF WHITE RESIDUES DURING MANUFACTURING OF PRINTED CIRCUIT BOARDS

J. D. Metz, M. A. Palmer
Rensselaer Polytechnic Institute
Troy, New York 12180

ABSTRACT

It is well known that white residues may be formed during the solder process. These residues, which can be conductive, corrosive, or benign, can grow if the soldered board is exposed to elevated temperature or humidity. Using designed experiments, we have examined how the formation of these residues is effected by process variables such as preheat temperature, preheat time, solder time, and time before cleaning. These parameters were chosen to simulate conditions that might be found in a manufacturing environment. The results of these experiments, which show that the most significant factor is the interaction between preheat time and preheat temperature, will serve as the basis for future designed experiments investigating the effect of the type of flux and permanent solder mask that is used. The combined results of these designed experiments will serve as the basis for a study in a manufacturing environment.

KEY WORDS: Cleaning, Residue Elimination

1. INTRODUCTION

1.1 ***Background*** Soldering is one of the most versatile and reliable methods of joining electronic components. In order to create such reliable bonds, the surfaces to be joined must first be cleaned. Fluxes are used for this purpose, as they remove the layer of oxide and tarnish present on the surface of the components to be joined, allowing the molten solder to wet the interconnect and the pad on the printed wire assembly. After the electrical components are placed on the printed wire board (PWB), flux is applied, the printed wire assembly (PWA) is then heated to the flux activation temperature. At this temperature, the flux breaks up the oxides at the metallic joints. Solder is then applied to the PWB and during soldering most of the flux is removed and replaced by solder, which bonds to the now clean surfaces to be joined. After soldering there is a flux residue which may require removal. These flux residues can be classified into three classes: benign, corrosive, and conductive. The first category can be left on the board, and additional expense will not be incurred in their removal. The corrosive class attacks the metal leads and lines on the printed wire board. In addition to destroying the metal leads, the corrosion particles can cause a reduction in the distance between leads, increasing the possibility of short circuiting the board. The conductive residues left are often in the form of polar ions typically left by RMA fluxes.(1) These ions become conductive in solution or when exposed to humid environments, however there is little actual "attack" on the metal traces. Due to their destructive nature, the last two types of residues must be removed. Many different methods of cleaning these residues are available, and which specific process is determined by the manufacturer.

Until 1992 the "solvents of choice" were Freon TMS and 111 trichloroethane. In 1992 the Montreal Protocol and subsequent Amsterdam regulations imposed a ban on ozone depleting chemicals for use in manufacturing operations in developed nations by January 1, 1996. The elimination of chlorofluorocarbons (CFC) has banned the industry's "solvent-of-choice"(2) for cleaning soldered PWA's. This has opened up the field of research to find alternative methods. Efforts have been focused on finding alternative cleaners for CFCs, alternative fluxes that leave less residue, or combinations of the two. The alternative cleaners are simply replacements for a system that had been widely used and effective. The driving force behind attempts to find a "drop-in replacement" is a desire to minimize capital investment costs for replacement machinery. These investigations have identified several possible methods for removing flux residue, such as alternative solvents, aqueous, or semiaqueous cleaning. In addition to the alternative cleaners developed, No-Clean fluxes have been introduced as an alternative to the entire cleaning process. These fluxes usually have low solids content and are designed to leave no harmful residues after soldering, precluding the need for a cleaning step. Alternatively, one could modify the manufacturing process to eliminate or reduce the likelihood of residue formation.

As a by-product of the elimination of chlorofluorocarbons for electronics defluxing, the incidence of "white residue" formation (typically seen after machine soldering) has increased. As white residues are difficult to clean, the prevention of residue formation is critical to printed wire assembly fabrication and compliance to government regulation.

1.2 *Planned Research* This study attempts to define the process parameters that are indeed influencing residue formation, and how to control them to eliminate the residues.

The three main objectives of this study are:

1. Determine the factors which contribute to residue formation, focusing on the solder process itself along with varying flux chemistries (rosin, resin, "no clean" and water soluble) and permanent solder mask types (screenable epoxies, dry film and liquid photoimagible.)
2. Determine whether the residues, specifically white residue, left after cleaning PWB's are detrimental to the performance of the PWB.
3. Once the factors are identified, adjust the process parameters to eliminate the formation of the residues.

The primary concern about the residues is their effect on electrical performance. Surface insulation resistance (SIR) testing is a common method employed to determine these effects. SIR testing is usually a test where the boards in question are subjected for one week to elevated temperature at high humidity, with a voltage bias. This is called a temperature, humidity, bias (THB) environment, and the resistance between test leads is measured frequently to monitor the changes over time. Test boards with Class II classification as specified by IPC codes §2.6.3.3 are subjected to SIR testing. The results show whether the residues are detrimental.

This investigation consists of experiments designed to determine, with a minimum amount of tests, what are the significant factors that affect the formation of residues on printed wire assemblies. Many possible parameters that could affect residue formation such as type of solder mask, type of flux, amount of preheat, solder time, board thickness, type of cleaner, and length of cleaning. From the long list of possible parameters, four thermal history conditions were chosen: preheat time (t_1) , preheat temperature (T_1), soldering time (t_2), and the time between soldering the PWA and cleaning (t_3). The effects of these parameters are analyzed by a 2^4 factorial experiment with two replicates, using a single flux (water white rosin RMA) chosen for its high solids content on bare copper test boards. Levels are taken to represent common settings in a wave solder process at both high and low values.

2. PROCEDURE

2.1 ***Test Pattern*** A picture of the test pattern is shown in Figure 1. This is a standard honeycomb pattern. The electrical resistance between the "forks" in the pattern are measured in the SIR Test. If the residue does not form, then current will not flow and the resistance will approach infinity. If a corrosive or a conductive residue forms, then current will flow, and the resistance will be finite.

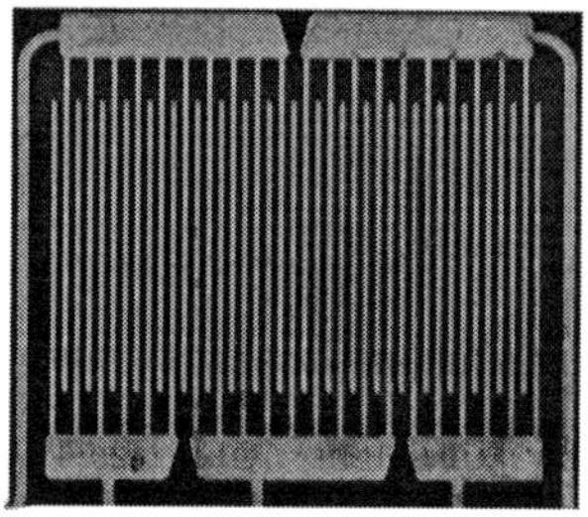

Figure 1: IPC-A-24 Test Patterns used in this Study

2.2 ***Preparation*** Each board is subjected to the same process, although the process parameters are varied. The PWBs were pre-cleaned (to remove handling residues) by wiping with the alcohol/DI water mix. Each board is then dipped vertically into the flux until the SIR patterns on both sides of the board are completely, evenly, coated. This has been determined to give the best distribution of flux compared to other methods such as spraying or floating. The boards are removed in the same manner as they are dipped, allowing excess flux to drip off the board briefly. The board is then transferred to a preheating oven for the specified time (t_1) and temperature (T_1). A similar board with an attached thermocouple is placed beside the test board to measure the maximum temperature experienced by the test board. The two boards hang vertically from a rack in the oven, allowing excess flux to drain from the test board. The temperature measurement board is not fluxed. The final temperature is recorded for later comparison with the target preheat temperature.

2.3 ***Soldering*** When the preheat time has been reached, careful note of the board temperature is taken. The test board is immediately removed from the preheat oven to the solder pot. The solder is kept at a constant 280°C (selected as representative of the high end of the normal wave soldering temperature range).

2.4 ***Cleaning*** All boards, regardless of treatment, are cleaned in the same manner. The cleaner used for the experiments is an aqueous cleaner designed to remove rosin and water soluble fluxes. All manufacturer directions are followed in preparing the cleaner for use. The cleaner is diluted 1:20 with deionized water solution for use. According to directions, the cleaner is heated to 70°C (160 °F) prior to use. To avoid complications, the deionized rinse water is also heated to 70°C (160°F).

2.5 ***Testing*** After cleaning, the test boards are immediately placed in a card edge connector and the resistance readings are taken for the zero hour measurement. They are then placed in the environmental chamber four at a time to begin the SIR test at 85°C and 85% relative humidity. The SIR Test Circuit is shown in Figure 2. A 100V bias was applied to the circuit and the current was measured, allowing the resistance to be calculated. Measurements are taken after 24, 96, and 168 hours. At the end of the week long test, the boards are reexamined to identify residue growth.

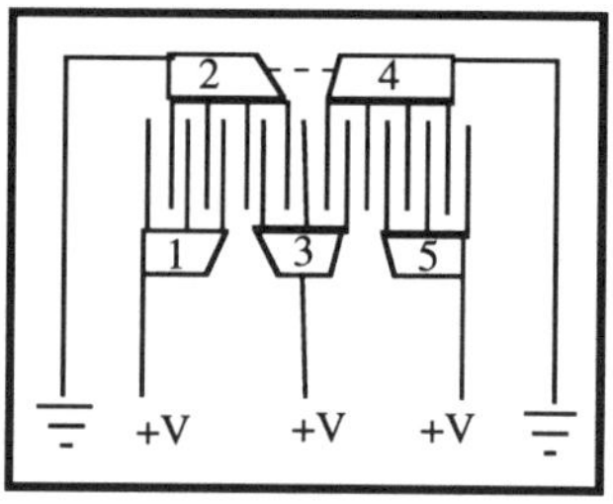

Figure 2: Schematic of SIR Circuit

3. RESULTS

3.1 ***Factors Affecting SIR Measurements*** This designed experiment identifies which factors effect residue formation. An RMA flux with 30% solids content was used throughout the experiment to simulate a difficult cleaning condition. Table 1 lists the four parameters varied, and their low and high values.

Table 1 - Process Effects / Levels Tested

Preheat Temperature:	100°C	130°C
Preheat Time:	130 sec	210 sec
Solder Time:	3 sec	7 sec
Wait Time:	5 min.	60 min.

The data used during the SIR tests was used to check for statistically significant parameters. Table 2 summarizes the analysis of variance.

Table 2 - Summary of ANOVA Results

Factor	Significance at			
	0 hr.	24 hr.	96 hr.	168 hr
Preheat Temp	99%	79%	99%	71%
Preheat Time	89%	89%	60%	96%
Solder Time	55%	48%	79%	79%
Wait Time	89%	97%	40%	81%
Preheat Time - Temperature Interaction	93%	82%	99%	98%

Note that for statistical purposes a factor should be at least 95% significant. It is noteworthy that although the significance of preheat time and preheat temperature vary, the significance of their interaction is always high. In fact the most significant factor is that of the preheat time - preheat temperature interaction. The wait time was not very significant, nor was the solder time. It was also observed that the best conditions occurred with a low pre-heat temperature and a high preheat time, and the worst conditions occurred with a high a high preheat time. A summary of our findings for one run is shown in Table 3.

Table 3 - Data From Replicate #1 of study

Temperature °C	time s		time min.	Leakage Current (mA) at time (hr) in Chamber			
Preheat	Preheat	Solder	Wait	0	24	96	168
100	135	3	5	0.020	0.058	0.072	0.052
			60	0	0	0	0
		7	5	0	0	0	0
			60	0.076	0.137	0.144	0.131
	210	3	5	0.079	0.164	0.176	0.162
			60	0.061	0.330	0.161	0.102
		7	5	0	0	0	0
			60	0	0.240	0.106	0.103
130	135	3	5	0	0.140	0	0
			60	0.071	0.060	0	0
		7	5	0.070	0.228	0	0
			60	0.073	0.306	0.221	0.220
	210	3	5	0.073	0.167	0.140	0.134
			60	0	0	0	0
		7	5	0	0	0	0
			60	0	0	0	0

Note that a high current reading indicates that the resistance is low. Two observations can be made from the data presented in the table. First the current reaches a maximum at an intermediate time, typically 24 hours. Second if the initial current is high, indicating an amount of residue has formed, then the current decreases with time.

3.2 *Correlation With White Residue Formation* A correlation between white residue formation and the maximum current passed shows four distinct regimes. To assess the amount of white residue formed an arbitrary, but well established scale of 1-5 was used, 1 represents no white residue visible at a 10X magnification, and 5 represents that white residue is visible to the naked eye.

We categorized the residue as inconclusive (low residue number, low current), benign (high residue number, low current), corrosive (low residue number, high current), and conductive (high residue number, high current). This was done as the species responsible for a conductive residue are likely to be different from those of a corrosive residue. Before the boards were placed in the humidity chamber they were scanned and the amount of residue was recorded. As the boards were removed from the furnace the amount of residue was also recorded, this residue rank was used for the last three readings. Table 4 delineates those boards which are conductive or corrosive.

Table 4 - Initial vs. Final Residue Effect Classification

	Treatment	0 hr	24 hr	96 hr	168 hr
		Boards that start out Inconclusive			
1	abcd	Inc	Inc	Corr	Inc
1	acd	Inc	Ben	Ben	Ben
2	ad	Inc	Inc	Inc	Inc
2	abcd	Inc	Inc	Inc	Inc
		Boards that start out Benign			
1	.(1)*	Ben	Cond	Cond	Cond
1	ac*	Ben	Ben	Ben	Ben
1	ab*	Ben	Cond	Cond	Cond
1	abc*	Ben	Cond	Cond	Cond
1	ad	Ben	Inc	Inc	Inc
2	ac	Ben	Corr	Corr	Corr
2	bc	Ben	Corr	Corr	Inc
2	acd	Ben	Inc	Corr	Inc
2	c	Ben	Inc	Corr	Inc
		Boards that start out Corrosive			
1	bcd	Corr	Corr	Corr	Corr
2	bd	Corr	Ben	Ben	Ben
		Boards that start out Conductive			
1	c*	Cond	Cond	Cond	Ben
1	d	Cond	Cond	Cond	Cond
2	a	Cond	Ben	Ben	Ben
2	abc	Cond	Inc	Inc	Inc
2	d	Cond	Corr	Corr	Corr

In Table 4 the letters a, b, c, and d refer to the "high" value conditions for the preheat temperature, preheat time, solder time and wait time respectively. Those boards listed with an asterisk (*) use the residue ranking after testing for the zero hour classification as initial rankings are not available. Note that the residue characteristics do not appear to change, that is from conductive to corrosive or vice versa. Typically a residue either becomes benign with time, or if the initial residue is benign become either conductive or corrosive with time. This data must be interpreted in the context of Table 3, which shows how the amount of current changes with exposure time.

4. DISCUSSION

The purpose of this work was to identify the conditions under which white residue formation is likely, to detect the growth of white residue "in-situ", and to determine the effect of residue formation on the performance of the board. The work completed in this paper represents the first of a series of designed experiments to identify those manufacturing parameters which are significant in effecting the formation of white residue. Two aspects of this experiment will have to be refined for further studies. First the wiring of the IPC-24 Board will need to be changed. Note that in this experiment the two leads specified by the IPC codes §2.6.3.3 to be negative shared a common ground, shown in Figure 2 as a dashed line. This means there is only one negative terminal, which makes the connection of lead 3 larger than either connection 1 or 5. In this analysis readings from the third lead were not considered. In future studies it is suggested that one only test between leads 1-2 and 5-4. This way two circuits of identical size can be examined, and they will not share a common ground.

In assessing the results, it was noted that the values of measured resistance were too low, on the order of $10^6\Omega$ to $10^8\Omega$. This is due in part to the limited precision of the measuring instrument, a Hewlett Packard 3478A multimeter. In order to examine the effects of small amounts of white residue a more precise SIR tester which has a resolution of up to $10^{12}\Omega$ was obtained to support future work. If kinetic issues are to be addressed or studied in the future this refinement is necessary so that the growth of a residue can be better monitored.

Our method of identifying the type of residue was somewhat arbitrary. Small amounts of residue were defined as having a rank of less than 3. Appreciable current was defined as 0.05 mA. However, the data presented in Table 4 show that two distinct types of residue are forming conductive and corrosive. This interpretation can be used to give greater insight to what is happening.

In almost all cases where an appreciable amount of residue (benign or conductive) is created through the solder process, there are reactions which occur in the hot humid environment. As no residues were observed as being benign throughout the process, this means they either became conductive, or the amount of residue decreased causing it to be identified as corrosive or inconclusive. In any event a chemical reaction is occurring between the initial residue and the humid environment.

In most cases where the preheat temperature was high, large amounts of residue formed. The notable exceptions being when the solder time was long, or the wait time was long. This could mean that the residue which is formed by the preheat conditions is burned off by the longer exposure to solder. Additionally the reaction with air could be similar to that which occurs in the chamber at a smaller scale. The longer the board is exposed to the solder temperature, the longer it will take to cool. This means that if the solder time and the wait time are long, the residue could react with the moisture in air, at an elevated temperature. Note that this study did not examine the amount of residue formed at each stage of the process. These results indicate that the residue which is noted to be a problem could be a product of a residue which forms during the solder process which then reacts with water.

These results are supported by the observation that the most significant factors are the pre-heat temperature, and the interaction of preheat temperature and preheat time. It is likely that the initial residue is a product of the fluxing reaction, which is very complex. There may also be a reaction with the solder. In any case it is likely that the kinetics of this reaction can be described by the Arrhenius equation [1], which describes the progress of a thermally activated process,

$$x(t) = x(0) + te^{-Q/RT} \qquad [1]$$

where 'x' represents the amount of product formed, 'Q' is the activation energy, 'R' is the ideal gas constant, 'T' the absolute temperature, and 't' the time. In reactions where the kinetics are described by the above reaction, temperature is the dominant factor.

5. CONCLUSIONS

It appears that destructive residues form in a series of steps. The first of these is a precursor which forms during the preheat stage, where the flux is activated and the flux reaction occurs. This is supported by the evidence that the two most significant factors are the preheat temperature and the interaction between preheat temperature and preheat time. Additionally it has been shown that frequently when large amounts of residue are formed these residues react with a humid environment and form a conductive or corrosive residue.

6. REFERENCES

1. R. Schreitmueller and R. Burke, Richard, "The Proof is in the Cleaning!" SMT August 1994

2. W.G. Kenyon, "No-Clean Designs, HFCs & In-Line Emulsion Cleaning," Surface Mount Cleaning, SMT November 1995

7. ACKNOWLEDGMENTS

The authors would like to acknowledge the efforts of Lawrence Felton at the start of this project and the support of the Office of Naval Research under Grant N00163-96-C-0005._ Additionally, special thanks are offered to James Maguire and Sarah Keough of the Boeing Company for their support and comments on this project.

28th International SAMPE Technical Conference
November 4-7, 1996

SpecRite
A PROGRAM MANAGER'S WORKSTATION (PMWS) TOOL FOR ASSISTING IN THE DEVELOPMENT OF PERFORMANCE SPECIFICATIONS

Brian Willoughby
BMP Program, Computer Sciences Corporation
College Park, Maryland 20740

ABSTRACT

SpecRite is a performance specification generator developed by the Navy's Best Manufacturing Practices (BMP) Program. The latest electronic tool in the BMP's Program Manager's WorkStation, SpecRite assists users in creating functional specifications. The resulting specifications are fully compatible with commercial applications while also compliant with the MIL-STD-961D format. This paper describes the engineering focus as well as the open architecture used to develop SpecRite. The resultant tool is easy to use and tailor to specific applications. A menu based process to assist the user is described. Additionally, as further improvements are introduced to SpecRite, this knowledge-based tool will effectively support those developing specifications while promoting best practices in technology, design, test, and production.

KEY WORDS: SpecRite, Performance Specifications, Engineering Processes

SpecRite HERITAGE

The Best Manufacturing Practices (BMP) Program was established by the Navy in 1985 with the mission to collect, analyze and disseminate information on proven techniques used by the government and industry throughout the design, development, and production process. The BMP program has four main products: Facility surveys (and resultant benchmarking database entries), development of engineering guidance documents, the annual BMP workshop, and the Program Manager's WorkStation (PMWS).

The PMWS is an electronic consultant, consisting of a series of expert systems that provide:

KNOWLEDGE - Through KnowHow expert systems handbooks; and SpecRite performance specification generator.
INSIGHT - Through TRIMS, Technical Risk Identification & Mitigation System.
EXPERIENCE - Through the Best Practices database.

UNIQUE CAPABILITIES

The PMWS is uniquely different then existing program management "tools" and is meant to fill a long standing gap left by them. Other tools (Microsoft Project, Defense System Management College's PMSS, AFAM, etc.) center mainly on cost, schedule, and documentation relationships (typically Contract Data Requirements List or Work Breakdown Structure oriented) and attempt to manage a development based on their status. There are many good programs of this type on the market today and the PMWS is not attempting to duplicate their functions. While these tools have there place, and "master schedules and documentation" are critical for long lead planning and coordination of the interactions between organizations, we still have many development programs "failing" despite the use of these types of tools.

PMWS - FOCUSED ON ENGINEERING PROCESSES

PMWS tools are focused on the engineering process itself, and hence are very process oriented. If all engineering processes are understood, proper, and under control, the results will be as good as the state of the art will allow.

Cost and schedule are downstream indicators of technical problems, this is why PMWS tools manage technical and process risks, so that engineering problems are surfaced and given visibility at the earliest possible point. For example, if the risk management portion of the PMWS (TRIMS) indicates that your program has / is incurring a risk due to a Design Reference Mission Profile not being conducted; the user can go directly to the knowledge base and get FULL details on why a Design Reference Mission Profile is needed and how to develop one, including example text for a performance specification. This type of problem would typically be indicated by the PMWS early in the design process, yet cost and schedule indicators of this problem typically will not surface until the operational test and evaluation process, when the system will perform properly because the specification and resultant design environments do not reflect current life cycle operational environments.

SpecRite is the newest PMWS tool and is an out growth of our KnowHow Expert system handbook technology and Modular Handbook Language (MHL). Currently SpecRite is a DOS program like all PMWS software (which runs well under Windows including cut and paste, etc.). However, all PMWS software is currently undergoing an upgrade (to improve it's capabilities) and conversion to Windows programs.

WHAT IS A PERFORMANCE SPECIFICATION?

In it's simplest terms it is a specification that defines WHAT YOU WHAT and how it is expected to perform (and in some cases how it will be tested) instead of telling the contractor HOW TO BUILD IT.

SpecRite PHILOSOPHY- EXPERT ASSISTANCE

SpecRite's goal is to help a user, or more often group of users, to develop quality performance specifications compliant with the MIL-STD-961D format. While 961D is a military specification, it is functional in nature, and is completely compatible with commercial applications.

SpecRite is an important advancement to two communities of users:

1. Users developing specifications for the procurement / development of systems and equipment.
2. Organizations with expertise in designing, maintaining, testing, and producing specific equipment and systems.

SpecRite has two main parts, the knowledge base / logic system, and the specification development manager.

As shown in Figure 1, the user follows the special menu structure to locate and accept their performance values for a variety of criteria (environment, temperature, shipping, reliability, etc.). The menu structure has unique features which alert a user when they are about to go down a path

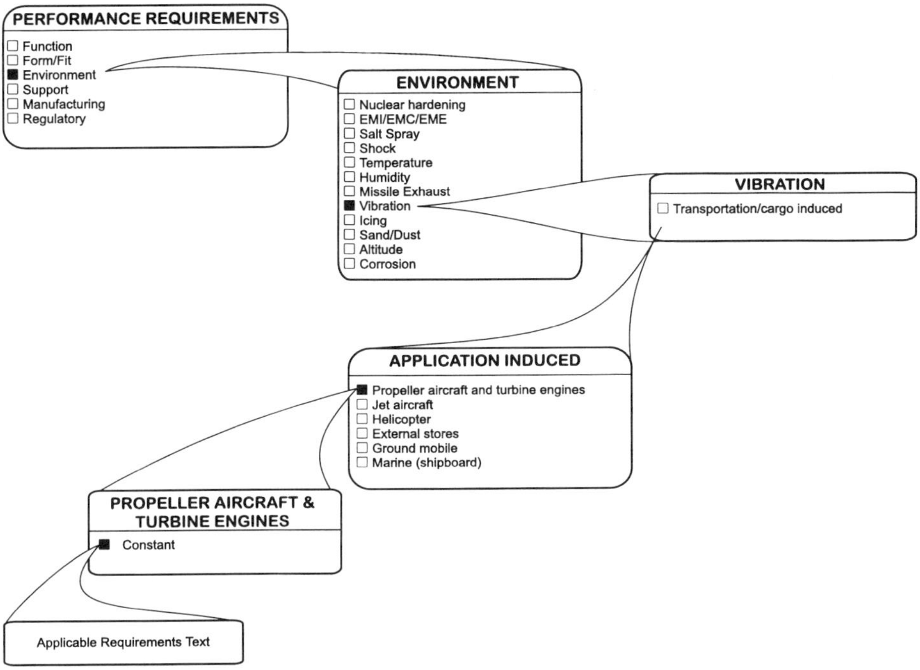

Figure 1. User follows menus to define performance values

that will result in an attempt to add to the specification a value that is mutually exclusive to a value already chosen. For example you may have chosen a temperature range previously of 20 to 50 degrees; and you are now about to define operating temperatures as low as -30 degrees; the program will remove the previous values since they cannot co-exist.

Another important aspect of the SpecRite methodology is ownership. While we will never have performance values for every type of hardware in existence (the combinations are almost infinite), we can achieve a very substantial and valuable knowledge-base with an open architecture / ownership approach.

Any organization with unique expertise can "own" a section of the logic tree in SpecRite. Figure 2 shows an example of how a user might step though the logic system to the performance values that are applicable. In this case, environmental limits for computers in a sheltered environment Naval Surface Warfare Center, Crane would have developed (and maintains) this section of the logic systems and performance data. It is impossible for one organization to have expertise in all areas of systems performance; but by allowing ownership a SpecRite user has the expertise of potentially hundreds of organizations at his disposal. This methodology also allows SpecRite to be able to tailor a larger portion of a users specification automatically.

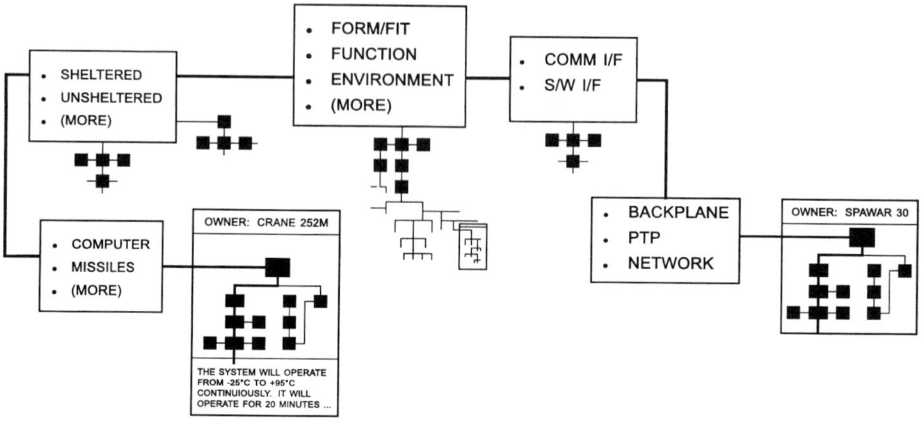

Figure 2. The expertize in SpecRite can come from many organizations

SpecRite is not all knowing; and in the areas where automatic tailoring is not available SpecRite provides the user with directions on how to develop that section.

While it takes very little time to step through the logic system and tailor your specification, in the real world their are many people involved in the development of a performance specification (each doing sections based on their expertise). To handle this situation SpecRite has a development manager menu.

The management menu, shown in Figure 3, among other things, allows the user to manage and review the specification development process. A manager can see what sections of the specification have been developed, if they are draft or final (alerts are given if you are about to overwrite existing information), and if they have been modified (an editor is also available for editing the selected text to better suit specific program needs). If a section has been reviewed (or edited) the last reviewed date is displayed. The Sections titles shown in the management menu are generic in nature (not 961D specific) at this stage.

KnowHow v2.30e BMP AUTOMATED SPEC GENERATION LIBRARY CSC

Review/Edit Spec

Section	Reviewed	Mod	Status
Form	01/16/95	X	FINAL
Fit	01/17/95	X	draft
Environment	01/17/95		draft
Temperature [+]	n/a		n/a
Operating Temperature	n/a		n/a
Non-Operating Temperature	n/a		n/a
Vibration	...		draft
PCB Vibration	...		...
Other Vibration	...		draft
Shock	...		draft
Mechanical Shock	...		draft
.	...		...
.	...		...
.	...		...

F1-Help F5-Print F10-Exit

Figure 3. Sample management screen

Once a user is satisfied with the development process an output file can be produced. SpecRite will then produce a WordPerfect 5.1 file based on the information selected in a MIL-STD-961D format. This includes on the fly section numbering and the removal of any blank or unused section titles.

It is at this point that the specification is put into the 961D formate, this de-coupling of the selection process from the document formatting process is a key to reduced maintenance costs. If the 961D format changes we simply need to update the format relationships to the generic categories; we do not have to re-map the entire knowledge base to the new section titles.

WHAT DOES ALL THIS MEAN TO ME

So what can SpecRite do for you and why should you use it?

If you are in the position of defining and developing systems, SpecRite will provide you with directions on how to define your specification in a performance nature; and it will actually develop many of the sections of the specification for you as you simply answer a series of questions. If there is very little or nothing in SpecRite's knowledge base that is suitable for your type of product, contact the BMP Center Of Excellence (COE) and see how you can customize the program for your future use so you will not have to keep "reinventing the wheel".

If you are an expert in an area of technology, design, test, or production you can take ownership (or simply provide the BMP COE with a one time data input) of a section of the SpecRite logic system, increasing the data available to all users, and making SpecRite "smarter" so it can automatically help others in defining their systems requirements in a new area. This will also give you great visibility as an expert in your field.

Want to know more? Please do not hesitate to contact me at the BMP COE 301-403-8100 or BRIANW@BMPCOE.ORG

BIOGRAPHY

Mr. Willoughby has developed many state of the art systems, and algorithms to insure that weapons systems exhibit reliability and producibility by design. He is known world-wide for his risk management experties speaking at DSMC, Naval Post Graduate School, U.K. MOD, and Canadian DND. He also teaches TQM and risk management courses for DoD. He developed the architecture, algorithms, and leads the team that built and maintains the Navy's Program Manager's WorkStation expert systems and BMPnet. Several of his designs have earned U.S. Patents.

28th International SAMPE Technical Conference
November 4-7, 1996

REDUCED LEAD USE IN AIRCRAFT ELECTRICAL SEALANTS

John A. Emerson
Sandia National Laboratories
Albuquerque, NM 87185-0958
Stanley Brown
Naval Air Warfare Center
Patuxent River, MD 20650-5304
Alan Fletcher
Wright-Patterson Air Force Base
Dayton, OH 45433-7717

ABSTRACT

Electrical grade polysulfide is used extensively in high performance applications for sealing, bonding and coating. However, this material has a number of environmental drawbacks which make it unacceptable for future use from both processing and disposal. Electrical grade polysulfide contains lead oxide, a hazardous material which will potentially undergo very strict regulations (Lead Exposure Reduction under the Clean Air Act of 1991) in the near future. Also, it has cure rate which is sensitive to humidity. The excess material from processing must be disposed of as a hazardous waste and the cure rate conditions are uncontrollable because vendors are unable to obtain consistent lead oxide. An interagency government task force was formed in late 1992 to develop a lead-free drop-in material, i.e., an environmentally safe material meeting as many of the desirable material and processing characteristics as possible. We report on the interagency progress and discuss two property measurement techniques that allow rapid screening of potential drop-in replacements.

KET WORDS: Sealants, Electrical Insulator, Polysulfides, Encapsulation/Potting, Elastomers

1. INTRODUCTION

Sealants are elastomeric polymers that have adhesive like properties along with the ability to seal and fill gaps. Their ability to expand and contract over wide temperature ranges and still maintain an effective seal limits the choice of polymers. Furthermore, in applications where conditions such as excellent electrical and good fuel resistance are required, the selection is limited to polysulfides or polyphosphazenes. It is this reason why polysulfides are found universally in most commercial products and also because of their availability and low-cost of the starting precursors. This type of polysulfide is used because it has numerous desirable processing and material property characteristics: low mix viscosity, low cure temperature (often room temperature), long working life, stress relief for encapsulates, good adhesion to many materials, excellent elastomeric properties at low temperatures (-55°C), and the best electrical properties within all other classes of polysulfide materials. A specification, MIL-S-8516F, states the sealant requirements, application parameters, type of cure conditions, and testing procedures. The only class of elastomers that meet the specification is lead peroxide cured polysulfides. Lead is a hazardous material which will potentially undergo very strict regulations (Lead Exposure Reduction under the Clean Air Act of 1991) in the near future. Also, it has cure rate which is sensitive to humidity. The excess material from processing must be disposed of as a hazardous waste and the cure rate conditions are uncontrollable because vendors are unable to obtain consistent lead oxide. In the past finding a direct drop-in replacement for the 8516 material (as it is commonly called) has not been successful because of the difficulty in meeting the electrical requirements as maintaining the fuel resistance requirement. We have been addressing the issue by working directly with the sealant manufactures to develop environmental friendly drop-in sealant.

This work describes how we evaluated new replacement sealants as they were being formulated by the vendors. A rapid screening procedure examines what we believe to be the critical properties–electrical resistance, easy of use, and curing profile. Quick evaluation of potential formulations allowed the material formulator to map a larger material/property space than their normal practice. We discuss two of these property measurement techniques.

2. RESULTS AND DISCUSSION

2.1. Curing Profile Measurements

The determining the curing profile of a material that has a long cure schedule is difficult. Normal techniques such as DSC, which measures the extent of reaction, cannot be applied for materials of materials as polysulfide sealants. For example, DSC depends upon on developing a thermogram. The amount of heat generated by these material systems is near non existent during the cumulative cure time. Monitoring the glass transition, Tg, increase is a possibility. The Tg of these sealants are usually well below -55°C. It would be a difficult technique because the starting Tg would be substantially below the ultimate Tg or it may not exist until the cure is fairly advanced. Another possible approach is to follow the rise in modulus from the liquid state by measuring the viscosity. Then when the sample has solidified, switching to a mechanical spectroscopy (torque) technique which would track the modulus to the final cured elastomer. We believe there is an easier and more sensitive technique by following ion mobility as macroscopic property that characterizes the extent of cure by measuring ionic conductivity from dielectric data. The ionic mobility in a polymer has been related to viscosity before gelation and rigidity after gelation (1). The application of this dielectric analysis technique for following the dynamic cure by of a silicone elastomer was done by Wong (2). We applied a similar procedure in following the cure of a 8516 material and its possible replacement.

Dielectric measurement during temperature ramps and isothermal of the 8516 and 8516 replacement materials were collected using a Micromet System III Microdielectrometer. The dielectric loss factor consists of energy losses due to the conduction of ions plus losses due to dipole orientation:

$$\varepsilon''(\omega) = \frac{\sigma}{\omega\varepsilon_0} + \varepsilon''_0 \quad (1)$$

where σ is the ionic conductivity, ω is the frequency, ε_0 is the permittivity of free space that equals 8.85 x 10^{-14} farads/cm, and ε'_d is the contribution to ε'' arising from losses due to dipole orientation. In the frequency range where ε'' is proportional to $1/\omega$, the losses associated with dipole orientation (3), and thus the ionic conductivity can be expressed as:

$$\sigma = \varepsilon'' \omega \varepsilon_0 \quad (2)$$

Ionic conductivity measurements are done by placing the uncured sample on interdigitated electrodes fabricated on a silicon chip. A temperature sensor using a microthermal diode is included on the microdielectric sensor. Local temperature measurement is obtained along with the material's dielectric readings. The silicon sensor is packaged in a Kapton® cable for connection to external signal processing (4). The sensor/sample is mounted in Delta Design 9039 oven that is controlled by a LabVIEW® computer program via an IEEE-448 interface.

By sweeping frequency, ω, from 0.01 Hz to 10KHz during the cure of material A, a purchased 8516 material, we observed different sensitivities in the ionic conductivity data, σ. As seen Figure 1, the 0.1Hz and 1 Hz data show the greatest sensitivity over the cure range. Other frequencies were not plotted as they showed only limited ionic conductivity data over the cure as shown with 10 Hz in this plot. All subsequent dielectric measurements were at 1 Hz.

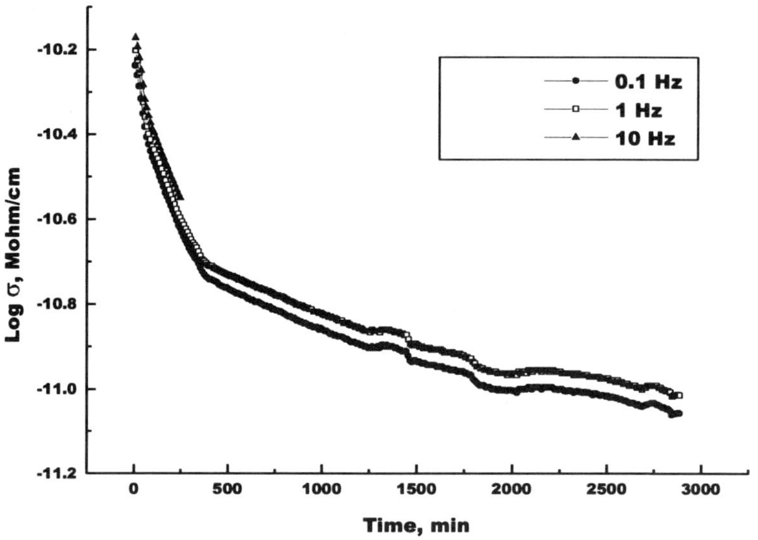

Figure 1. Ion Conductivity for 8516 Material at 25°C

In Figure 2, the ion conductivity data of a possible replacement material, B, is plotted. When compared to Figure 1, the two materials behave quite differently. Material B levels off after 500 min indicating the cure has ceased. However, after 1000 min the conductivity starts to drop until 2000 min. Because the cure temperature is at 70°C, we expect a faster curing profile than material A which was cured at 25°C. Adjustments were made in the curative to resin ratio and subsequent dielectric analysis allowed the proper stoichiometric ratios to be established.

Rapid information about the curing profile allows mapping of larger material formulation regions - an important factor for introducing new materials into application.

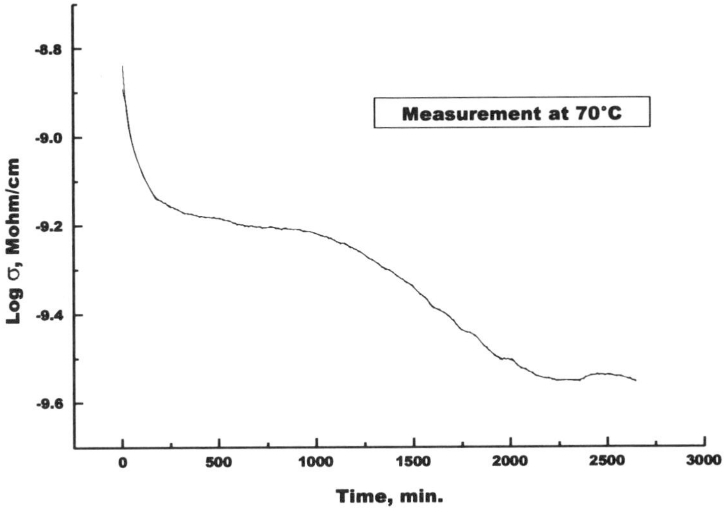

Figure 2. Ion Conductivity Plot for Possible Replacement Material

2.2. Electrical Measurements

One of the critical properties for the drop-in replacement material is the electric performance. The electrical dielectric constant and dissipation factor were determined by the ASTM D-150-94 method (5). Some modifications in electrode geometries were made to accommodate the MIL-S-8516F specification. Instead of using lead or tin foil mounted onto brass plates for the electrodes which is specified in MIL-S-8516F, we used a 5 cm diameter guarded brass electrode, with a 10 cm guard-ring. The brass counter electrode was 10 cm in diameter. The volume resistivity was determined by ASTM D-257-93 method (5). A snug fit assured excellent electrical contact. Frequency steps of 1 KHz, 10 KHz, 100 KHz, and 1 MHz were applied to electrodes for determining electrical dielectric constant and dissipation factor. For surface and volume resistivity, 500 V was applied. The current was read by a Keithy 617 Electrometer.

Table 1 summarizes the electrical data collected from the described procedure. When the two materials are compared to the specification, as in some cases the purchased 8516 does not meet the spec. Often the vendor has difficulty in making acceptable material because of difficulties in procuring the proper lead peroxide curative. As the example shows, material B may be good as the material A. However, it also is not passing the specification. Again, suggestions from this data to the formulator of how to reformulated particularly for the higher temperatures was constructive.

	Specification Requirement	Material A (8516)	Material B
Dielectric Constant, Max			
1 KHz at 25°C	9.5	7.78	9.26
10 KHz at 25°C	9.5	7.76	9.14
100 KHz at 25°C	9.5	7.73	8.92
1 MHz at 25°C	9.5	7.77	8.58
1 KHz at 85°C	9.0	7.13	7.87
10 KHz at 85°C	9.0	7.00	7.83
100 KHz at 85°C	9.0	6.97	7.79
1 MHz at 85°C	9.0	7.03	7.88
Dissipation Factor, Max			
1 KHz at 25°C	0.01	0.0031	0.0088
10 KHz at 25°C	0.01	0.0023	0.0132
100 KHz at 25°C	0.01	0.0049	0.0273
1 MHz at 25°C	0.03	0.0245	0.0667
1 KHz at 85°C	0.03	0.2263	0.1539
10 KHz at 85°C	0.03	0.0265	0.0187
100 KHz at 85°C	0.03	0.0046	0.0064
1 MHz at 85°C	0.03	0.0042	0.0124
Volume Resistivity, Min, Ohm-cm			
500V at 25°C	1×10^{10}	3.91×10^{11}	1.89×10^{11}
500V at 85°C	2×10^{10}	2.79×10^{9}	3.63×10^{9}
Surface Resistivity, Min, Ohm			
500V at 25°C	1×10^{11}	9.48×10^{12}	1.70×10^{13}
500V at 85°C	2×10^{11}	7.08×10^{10}	4.63×10^{11}

Table 1. Comparison of Electrical Data for a 8516 and Possible Replacement Material

3. SUMMARY

In finding a drop-in replacement for MIL-S-8516F electrical grade sealant, two techniques are being used to rapidly screen prospective materials. Curing profiles for the elastomer were determined by dielectric measurements. The ionic conductivity calculated from the ionic mobility is very sensitive as compared to other established techniques in obtaining curing conditions. The second screen test are electrical measurements at 25°C and 85°C. In previous attempts to find a replacement material, it was often the higher temperature requirement that could not be meet. This problem still has not been solved. However, the gap is being greatly reduced between the specification and prospective materials.

4. ACKNOWLEDGMENTS

This work was performed at Sandia National Laboratories supported by the U.S Department of Energy under Contract DE-AC0-94AL85000. The Strategic Environmental Research and Development Program (SERDP), contract PP-429, also funded this effort. The authors thank Paul Beeson and Carl Robinson for collecting the experimental data. Samples were supplied by Courtlaulds Aerospace and Morton International.

5. REFERNCES

1. S.D. Senturia and N.F. Sheppard, Adv. Polym. Sci., 80, 1 (1986).
2. C.P. Wong in J.H. Lupinski and R.S. Moore, eds., Polymeric Materials for Electronics Packaging and Interconnection, ACS Symposium Series 407, American Chemical Society, Washington, DC, 1989, 220.
3. S.A. Bidstrup and S.D. Senturia, Polym. Eng. Sci., 29, 290 (1989).
4. D.R. Day, "Dielectric Properties of Polymeric Materials," Micromet Instruments, Newton Centre, MA, 1988.
5. American Society of Testing and Materials, Philadelphia, PA.

28th International SAMPE Technical Conference
November 4-7, 1996

HYGROTHERMAL STUDIES ON FLUORINATED POLYIMIDES - A PHYSICAL CHARACTERIZATION

Brian P. Rice
University of Dayton Research Institute
300 College Park
Dayton, OH 45469-0168

ABSTRACT

Recently, concerns have been raised regarding the hydrolytic stability of fluorinated polyimides such as AFR700B and Avimid® N. The service environment of these materials can be broadly summarized as consisting of four stages. First the material is exposed to the ambient environment when not in service, second the material can be subjected to rapid heating rates of 20-50°C/min when operation is initiated, third the material is subjected to prolonged exposure at temperatures exceeding 300°C during operation, and fourth the material is returned back to ambient conditions.

Moisture trapped within polyimide composite parts during rapid heating can lead to plasticization, hydrolysis and blistering. These phenomena have prompted several initiatives to develop an understanding of the mechanisms responsible for and consequences of hygrothermal degradation. This paper summarizes a number of efforts undertaken to address these concerns. An accelerated hydrolytic test has been developed to explore the range of conditions under which hydrolysis will occur and the subsequent effect on resin properties. AFR700B was selected as the model compound for this study.

KEY WORDS: High-Temperature Polyimides, Hygrothermal Performance

1. INTRODUCTION

In advanced vehicle designs high-temperature organic matrix composites, HTOMCs, are typically selected because of either their dielectric performance or high specific mechanical properties. In particular, it is desirable to utilize these materials in and around jet engines to increase the thrust-to-weight ratios. The service environment of these materials can be broadly summarized as consisting of four stages. First the material is exposed to the ambient environment when not in service, second the material can be subjected to rapid heating rates of 20-50°C/min when operation is initiated, third the material is subjected to prolonged exposure at high temperatures ranging from 230-370°C, and fourth the material is returned back to ambient conditions.

Traditionally, the selection of candidate HTOMCs has been based on property retention after oxidative exposure, such as 100 hours at 371°C in air. Generally it was assumed that measurement of hot-wet properties was not important because at the high temperatures of operation, the composite parts dry very quickly. In the case of relatively thin parts measuring less than 2 mm, this assumption is generally correct; however recent studies conducted by Cornelia (1) on composites exceeding 3 mm in thickness have shown that moisture retained in the composite while heating can cause significant property reductions. Further studies conducted by Price, et al. (2) also showed that during rapid heating moisture trapped in thick polyimide matrix composites can lead to plasticization, hydrolysis and blistering.

A research program has been initiated at the U.S. Air Force Materials Directorate to develop a fundamental understanding of the mechanisms leading to hydrolytic degradation in HTOMCs. Because hydrolytic degradation of HTOMCs can be a function of many variables (resin formulation, fiber, fiber sizing, weave, part geometry, heating profile, ...), it was determined that studies would be conducted on neat resins under well-controlled conditions. In particular, the hydrolytic cycling conducted by Cornelia and Price (transferring composite plates between a humidity chamber and oven) is resource-intensive. Because the chemical degradation occurs primarily during rapid heating when the composite moisture content and temperature are high, an accelerated test was developed which satisfies these conditions.

2. EXPERIMENTAL

2.1 Materials The hygrothermal cycling tests were conducted on composite laminates fabricated from AFR700B polyimide resin (3) (Figure 1) reinforced with a uniweave (95% warp/5% fill) fabric constructed of G30-500 fiber with an HTS sizing. The composite laminates were warp aligned with 16 plies of prepreg, and the final cured thickness was 3.7 mm. Accelerated hygrothermal studies were conducted on AFR700B neat resin plates which were fabricated from preimidized powder which was consolidated between two Kapton® release plies in a press. The final plate thickness was 1.3 mm.

2.2 Hygrothermal Cycling The composite laminates were hygrothermally cycled by storing them in a humidity cabinet at 85% RH/29.5°C, followed by exposure to a daily oven cycle shown in Figure 2. The peak temperature of the oven cycle was approximately 240°C. During the weekends the laminates were stored continuously in the humidity cabinet. The composites were weighed before and after each oven cycle to monitor the moisture content throughout the cycling. During hygrothermal cycling the sample begins to lose moisture as it is being heated, and a moisture gradient develops where the highest moisture content exists near the laminate midplane. Figure 3 illustrates the composite moisture content throughout the process. The upper values were recorded prior to oven exposure, while the lower values were measured after exposure. After several weeks of cycling the composite moisture content tends to cycle between 0.8% and 0.2%. Of course these values are a function of polymer chemistry, moisture diffusivity, sample geometry and hygrothermal cycling conditions.

2.3 Accelerated Hydrolytic Conditioning A controlled characterization method for evaluating the susceptibility of various high-temperature resin systems to hygrothermal degradation is required. Because hydrolytic degradation is a function of the material moisture content, it is important to limit moisture diffusion and drying, while the sample is maintained at an elevated temperature. The method is intended to simulate the worst-case condition during operation, which is maximum moisture content at elevated temperatures. During accelerated hygrothermal conditioning neat resin samples are first saturated in a humidity cabinet (85%RH/29.5°C) and then placed in sealed stainless-steel hydraulic tubing which has been filled with a sufficient quantity of distilled water to ensure that the sample will be exposed to saturated steam at the desired test temperature. The samples are placed on top of a packing to ensure that they are not immersed in liquid. The saturated steam surrounding the specimen will prevent drying and thus preserve a uniform moisture content. The sample weight is measured "dry" prior to exposure, "wet" after exposure in the humidity cabinet and after exposure to steam, and finally after redrying in a vacuum oven at 110°C. Ideally a mass balance could then be used to determine the degree of hydrolysis as will be discussed later.

2.4 Glass Transition Temperature Measurement The glass transition temperature, Tg, of each specimen was measured under dynamic torsion by a Rheometrics RDS-II. Samples were vacuum dried to constant weight at 110°C prior to testing. The strain rate was 100 rad/sec, the strain was 0.1% and the heating rate was 10°C/minute. In most cases two samples were measured per condition. The Tg was defined as the loss modulus, G", peak.

2.5 Composite Interlaminar Shear Strength Measurement The composite interlaminar shear strength was determined from a four-point bend test with quarter-point loading and a span-to-depth ratio of 16 to 1. Testing was conducted at 24°C and 332°C on specimens cut from a larger laminate which was hygrothermally cycled and then vacuum dried at 25°C.

2.6 Neat Resin Tensile Properties Measurement Tensile properties of AFR700B neat resin were measured on 15x1.27 cm, straight-sided specimens using an Instron 5582 test machine. Testing was conducted on hydrolytically-aged specimens which were vacuum dried at 110°C. A clam shell oven was used to heat the samples for testing at 250°C and 300°C. A special extender fixture was developed to permit strain measurement of the sample with a conventional clip-on extensometer placed outside the clam shell oven. Multizone temperature control of the oven was used to ensure that the test temperature was highest at the 2.5 cm gage section. The Instron cross-head displacement was 1.27 mm/minute.

3. RESULTS AND DISCUSSION

3.1 Composite Hygrothermal Cycling Hygrothermal cycling of high-temperature polyimide composites can lead to a substantial reduction in Tg and mechanical properties as reported by Cornelia (1) and Price et al. (2). Figure 4 summarizes the results following

96 days of hygrothermal cycling of G30-500/AFR700B composites. The initial Tg of 452°C was reduced by 85°C to 367°C, however the shear strengths measured at 24°C and 332°C showed little change. The failure mode during the test was a combination of compression under the top load pins and shear. Following the first 96 days of hygrothermal cycling, the samples were aged in the humidity cabinet for two weeks and then were subjected to an oven cycle with a maximum temperature of 332°C. This led to severe blistering of the test laminates at the midplane. Inspection of the laminate surface also revealed severe microcracking resulting from high internal steam pressure. Further testing under these conditions was thus terminated. It is important to note that these results occur due to the extended residence time of moisture in thick parts. A four-ply, 0.93 mm thick, composite which went through the same cycle process showed no visible degradation.

3.2 Accelerated Hydrolytic Conditioning of AFR700B The study of hygrothermal degradation of composites as presented above tends to be material and profile specific because of the numerous variables and failure mechanisms involved. For this reason further studies were conducted on AFR700B neat resin using the accelerated hydrolytic conditioning method discussed in section 2.3.

AFR700B neat resin samples were aged in saturated steam at 200°C, 150°C, and 100°C. The reduction in dry Tg of the conditioned specimens is shown in Figure 5. After 16 hours of exposure at 200°C, 150°C, and 100°C, the Tg decline was 110°C, 40°C and 4°C, respectively. Perhaps the most surprising result was that a reduction in Tg of 16°C occurred after 72 hours of aging at 100°C. This indicates that attempts to dry thick parts at a "low" temperature prior to use can still lead to property reduction, although at a much slower rate than typical operating temperatures.

It is also noteworthy that only eight hours of exposure at 200°C (Figure 5) were required to achieve the same Tg reduction which accumulated after 96 days of hygrothermal cycling (Figure 4). Although during the hygrothermal cycling the maximum temperature was 240°C, the moisture content of the AFR700B resin in the laminates decreased throughout the oven cycle from a high of 4.5% down to 0.5%. During the accelerated procedure the moisture content of the resin increased slightly from 4.2% (after aging in the humidity cabinet) to 4.8% (after exposure to saturated steam). As previously stated the goal of the accelerated hydrolytic conditioning is to simulate the worse-case condition when moisture content and temperature are at their highest. There is little reason to believe that the accelerated test results in unrealistic chemical degradation; however, prolonged exposure at 200°C did result in unexpected damage. Figure 6 shows the "redried" resin weight change after the exposure to saturated steam. During the first eight hours of exposure at 200°C, there was a weight increase followed by a steep weight loss. This resulted from the extraction of low molecular weight components from the hydrolytically-degraded resin which were readily observed through a discoloration of the distilled water used in the pressure vessel. Samples aged at 150°C and 100°C showed only an increase in dry weight after aging, and no discoloration of the distilled water was observed. Following this observation it was decided that long-term exposures at 200°C were too severe, and a standard exposure temperature of 150°C was selected.

It is believed that the dry weight gain after aging results from chemically-bound water in the resin. NMR studies conducted by Curliss (4) indicated that hydrolysis of imide linkages is occurring. Theoretically, an increase in weight of 0.4% would result in an average of one hydrolyzed imide linkage per chain out of a possible 17 (AFR700B MW=4400, H_2O MW=18). Forty-eight hours of aging in saturated steam at 150°C resulted in a dry weight increase of 0.58% and a decline in Tg of 63°C. The fact that a relatively small percentage of possible sites are readily hydrolyzed indicates that specific sites may be more vulnerable. It is also interesting to note that TGA-mass spectroscopy data taken during the postcure of as-cured AFR700B indicate that water does evolve. This data suggests that imide linkages formed during the postcure may be highly strained and thus more vulnerable to hydrolysis.

The reduction in Tg is reversible when the resin is heated to its typical postcure temperature of 400°C. Figure 7 illustrates the increase in storage modulus, G', of a torsion specimen which was aged for 48 hours at 150°C and then held isothermally at 400°C for six hours. The Tg of the postcured specimen nearly returned to its original value of 432°C. A similar study was conducted with an isothermal hold at 300°C to determine if it is likely that an AFR700B composite part would recover its Tg in service. After a prolonged hold of 66 hours at 300°C in nitrogen, the sample Tg increased only slightly from 365°C to 370°C. Although the chemical hydrolytic degradation is reversible, it is unlikely to occur within an operational environment.

3.3 AFR700B Tensile Properties Composite shear strength data presented in Figure 4 show only a slight reduction in strength occurs even after a Tg decline of 84°C. Because of the mixed failure mode discussed in Section 3.1, the test may not have been sensitive to hydrolytic degradation of the resin. Therefore, it was decided to test neat resin tensile properties after accelerated hygrothermal conditioning. The AFR700B neat resin samples had a postcured Tg of 432°C which was reduced to 384°C and 368°C after aging at 150°C in saturated steam for 24 and 48 hours, respectively (Figure 8). The tensile properties shown in Figures 9 and 10 were evaluated at 250°C and 300°C. The ultimate stress at 250°C was reduced 38% after 48 hours of exposure, while a 49% reduction occurred at a test temperature of 300°C. The secant modulus at 250°C remained unchanged after 48 hours of exposure, while a 43% reduction occurred at a test temperature of 300°C. This is a surprising reduction in properties considering that the specimen Tg was 368°C. Figures 11 and 12 show representative stress-strain curves before and after aging. The sample yielding observed in the stress-strain curve resulted from necking in the gage section.

4. CONCLUSIONS

Hygrothermal cycling of thick HTOMCs can lead to both physical and chemical degradation. Such cycling, generally required for component testing resource intensive. In addition the hygrothermal degradation is not only a function of polymer chemistry but sample geometry and moisture diffusion as well. An accelerated hydrolytic conditioning method was developed which allows for evaluation of the effects of chemical degradation on neat resin samples. For a similar reduction in Tg, the reduction in neat resin tensile properties was considerably more severe than that of the composite interlaminar shear strength.

5. REFERENCES

1) Dick Cornelia, "Hygrothermal Performance of Polyimides," 39th International SAMPE Symposium Proceedings, pp. 917-929, April 1994.

2) W.A. Price, B.P. Rice, A.S. Crasto, K.A. Thorp, "Hygrothermal Aging of Imide Composites," Proceedings of the High Temple Workshop XV, pp. S1-S24, Jan. 1995.

3) B. P. Rice & K. Johnson, "AFR700B - An Overview," Proceedings of the HITEMP Review, pp. 10-1 - 10-9, October 1993.

4) D.B Curliss, "Environmental Effects on High Performance Polymers and Composites," Proceedings of the High Temple Workshop XVI, pp. H1-H16, January 1996.

6. BIOGRAPHY

Brian Rice is an Associate Research Engineer at the University of Dayton Research Institute. He received his B.S. degree in chemical engineering from The Ohio State University in 1986 and received a M.S. degree in materials engineering from the University of Dayton in 1990. His research interests include the processing and characterization of advanced composite materials.

7. ACKNOWLEDGMENT

This work was supported under U.S. Air Force Contract No. F33615-95-D-5029. The author wishes to thank James Lute for fabricating and conditioning neat resin specimens, Kenneth Lindsay for conducting Rheometrics torsion experiments, and Charles Fowler for conducting tensile tests.

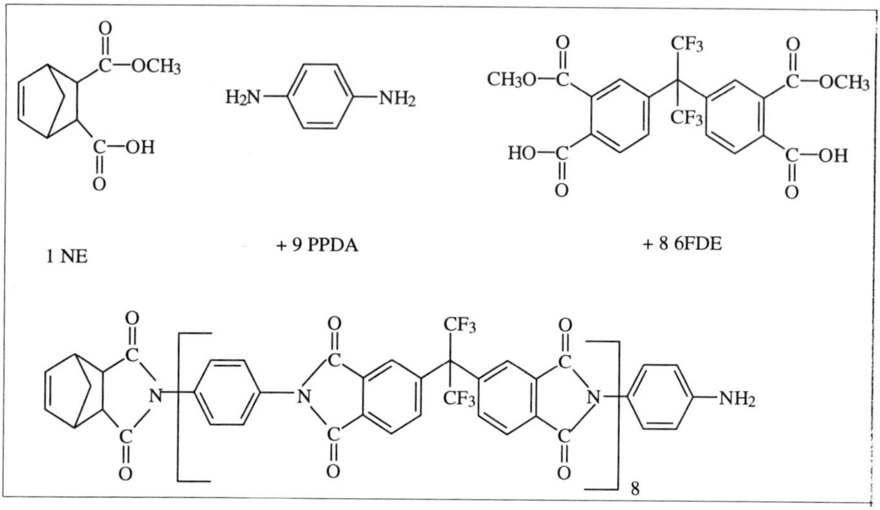

Figure 1. Formulation for AFR700B Resin.

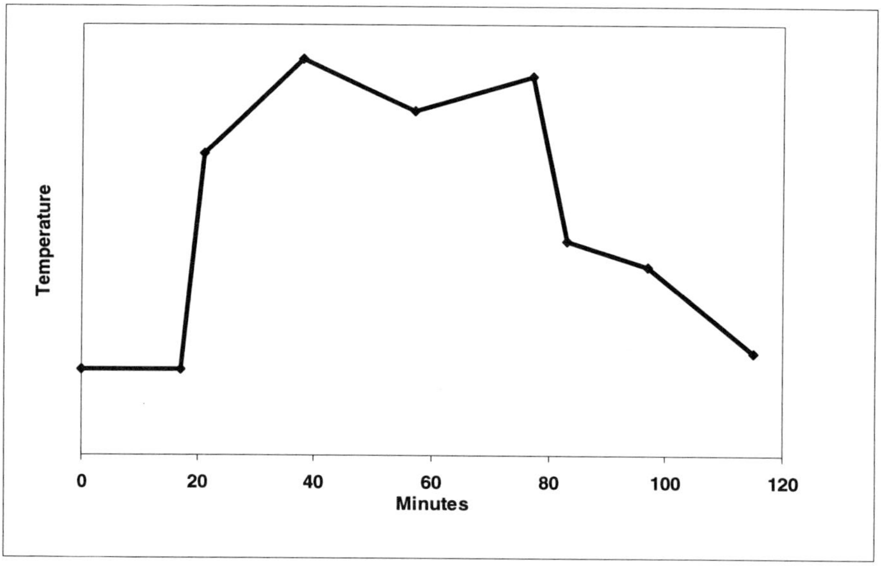

Figure 2. Sample thermal history for HTOMC parts used in cycling studies.

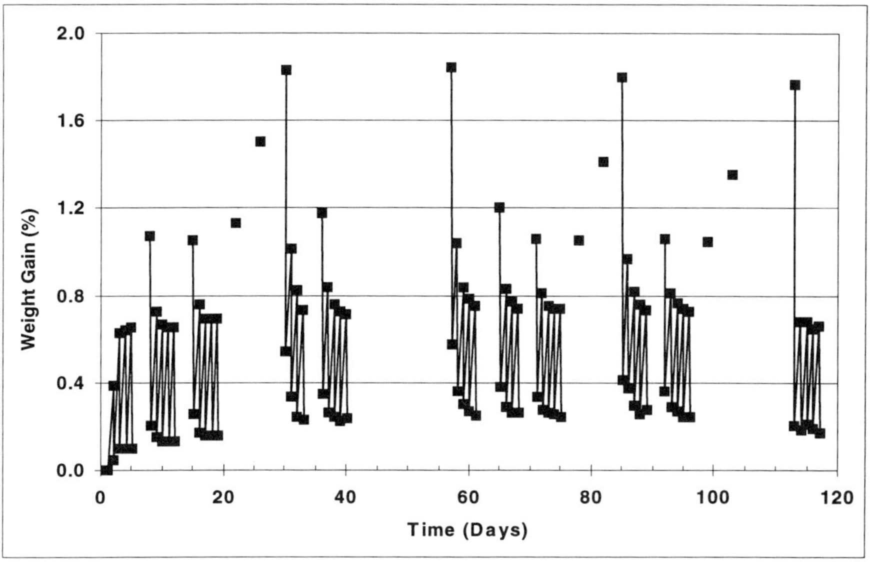

Figure 3. Moisture content of G30-500/AFR700B composite during hygrothermal cycling.

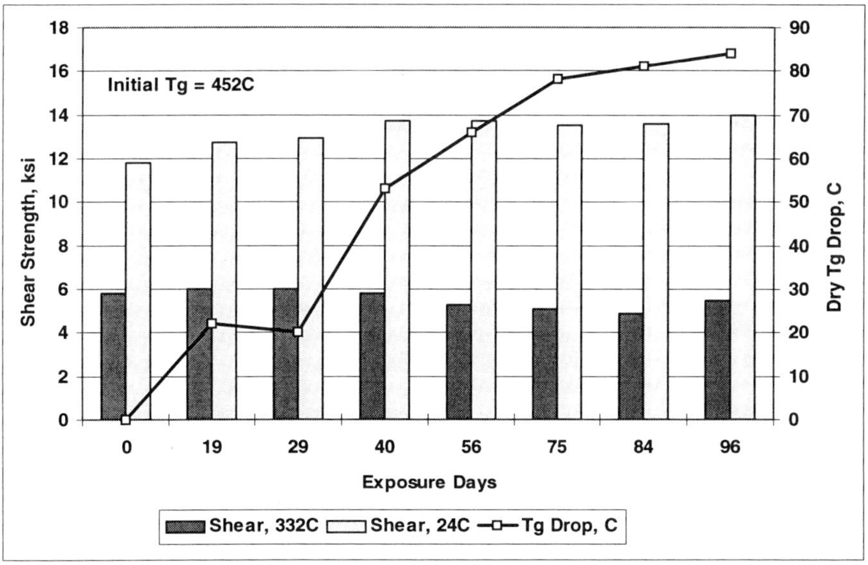

Figure 4. Composite shear strength and dry Tg drop during hygrothermal cycling.

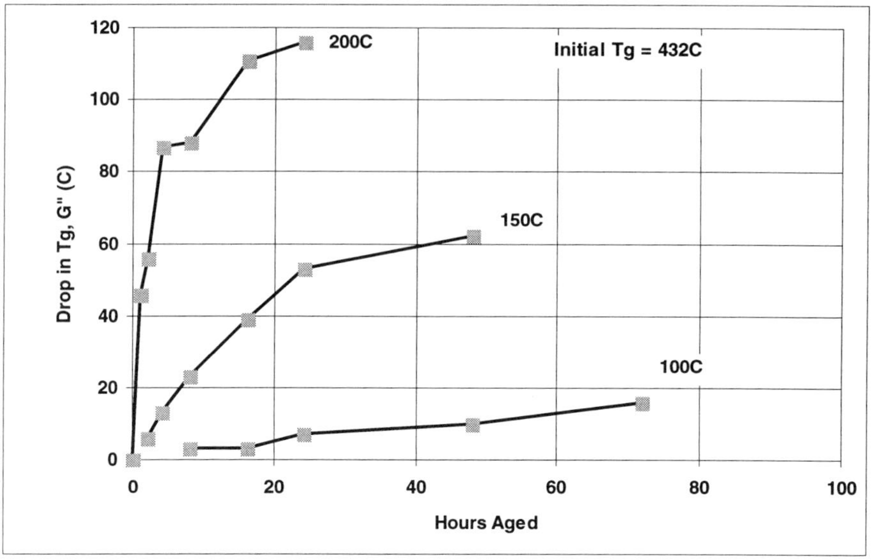

Figure 5. Drop in dry Tg of AFR700B resin after exposure to saturated steam.

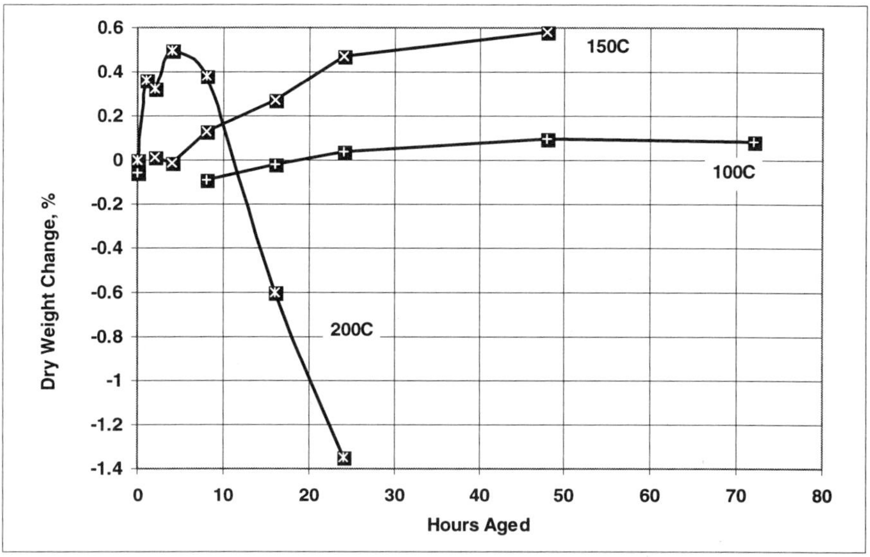

Figure 6. Weight change of vacuum-dried AFR700B samples after steam exposure.

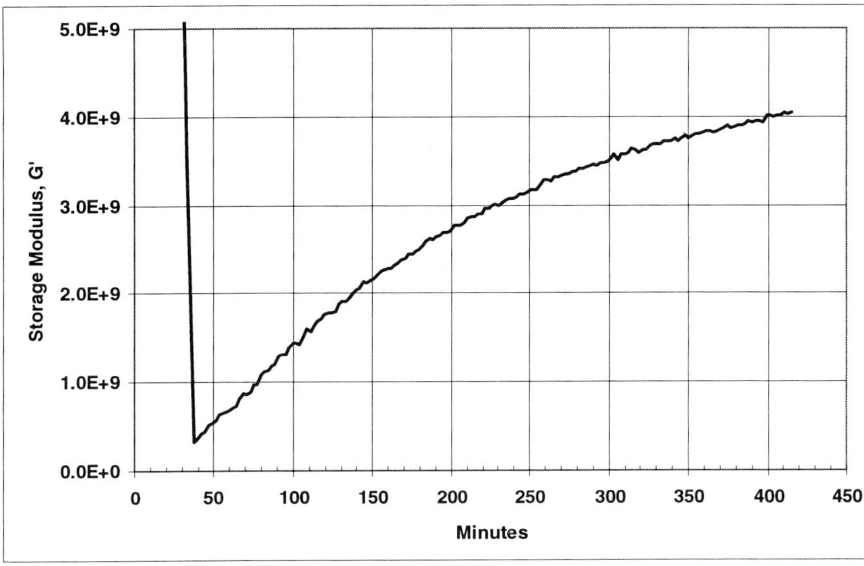

Figure 7. Storage modulus increase during an isothermal hold at 400°C of vacuum-dried AFR700B after 48 hours of steam exposure at 150°C.

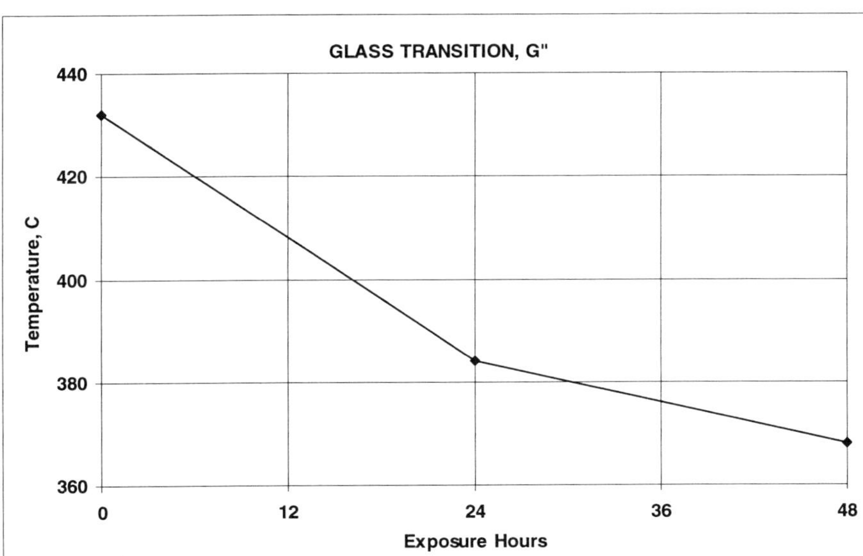

Figure 8. Tg of AFR700B tension samples after exposure to saturated steam at 150°C.

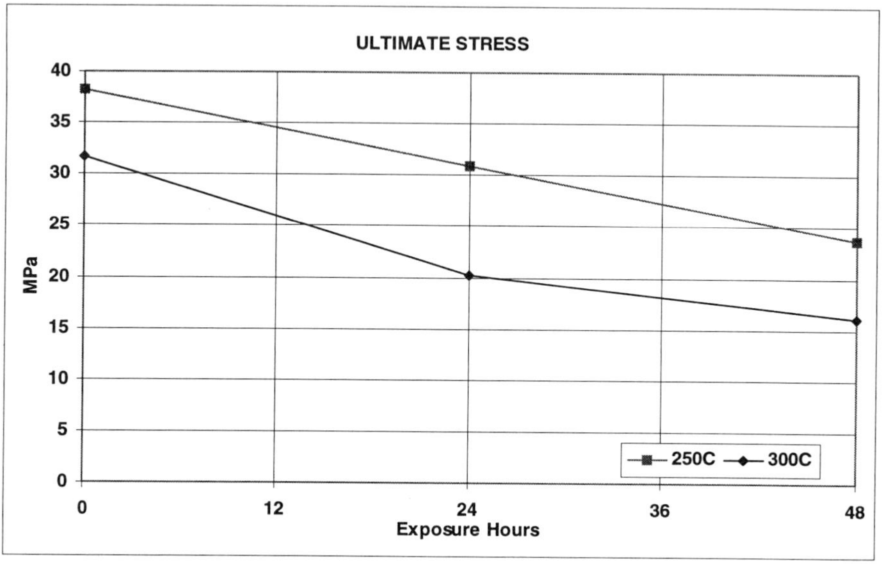

Figure 9. Ultimate stress of AFR700B tension samples after exposure to saturated steam at 150°C.

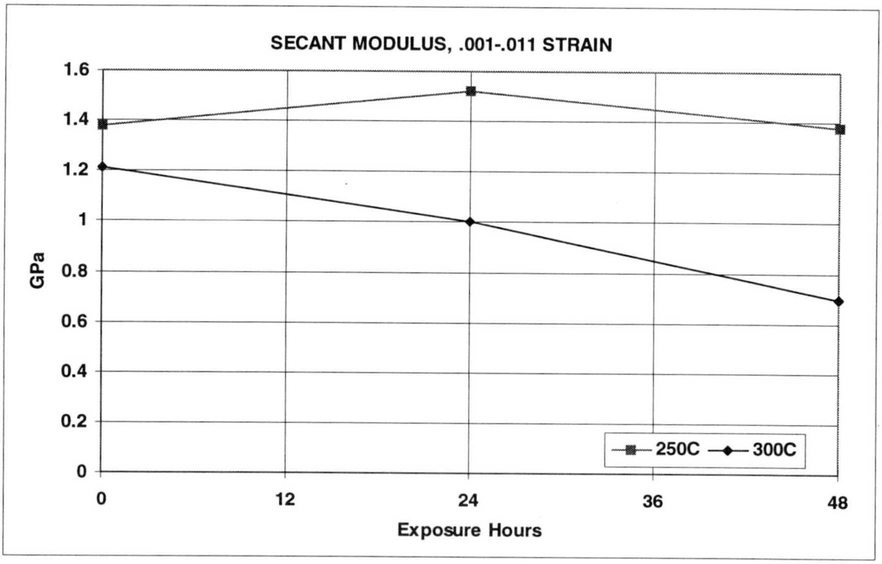

Figure 10. Secant modulus of AFR700B tension samples after exposure to saturated steam at 150°C.

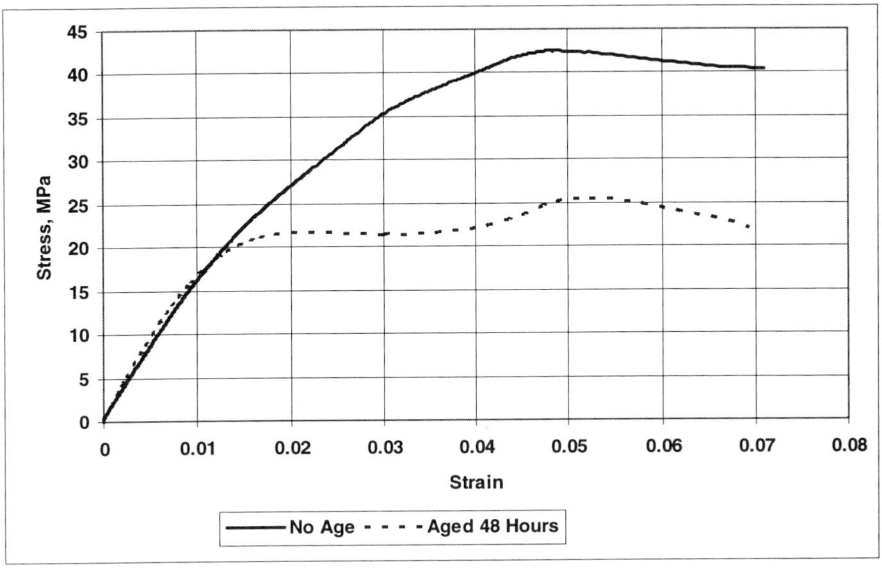

Figure 11. Stress strain curves, at 250°C, of AFR700B tension samples after exposure to saturated steam at 150°C.

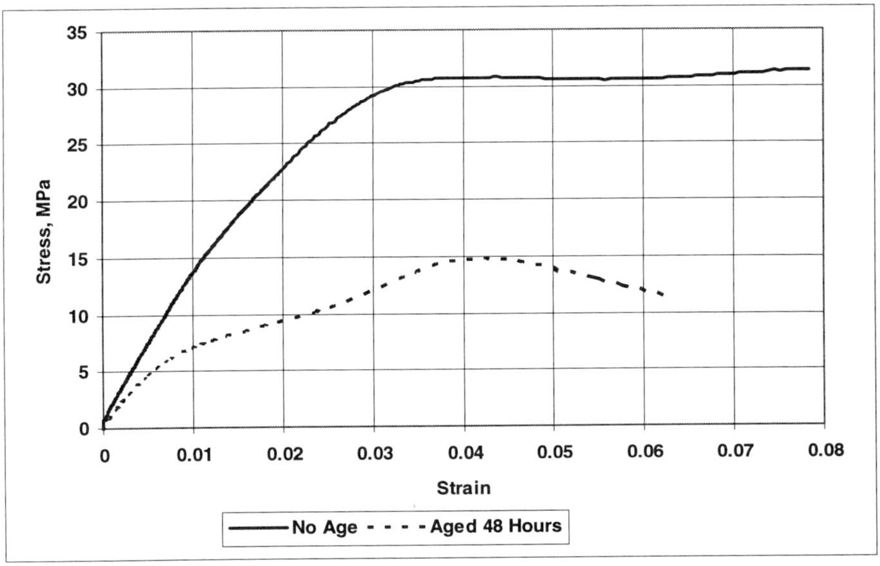

Figure 12. Stress strain curves, at 300°C, of AFR700B tension samples after exposure to saturated steam at 150°C.

28th International SAMPE Technical Conference
November 4-7, 1996

HYGROTHERMAL STUDIES ON FLUORINATED POLYIMIDES - MODEL COMPOUND STUDIES OF CURE AND DEGRADATION

David B. Curliss
U.S. Air Force Wright Laboratory
Materials Directorate
Wright-Patterson Air Force Base, Ohio 45433

ABSTRACT

We have used solid state nuclear magnetic resonance (NMR) spectroscopic techniques to study the network formation reactions and degradation reactions in service environments for a number of high performance imide-containing polymers. ^{15}N spectroscopy has proven to be most useful in the studies of imide polymers since the reactions of interest generally involve changes in the ^{15}N NMR spectrum. The techniques allows us to clearly identify and quantify reactions of the polymer systems. For example, we have investigated the curing, post-curing, and hydrolytic degradation of the polyimide resin AFR700B, a normally intractable material by other techniques, using ^{15}N NMR spectroscopy and can clearly observe reactive species in both the cure reactions and in the hydrolysis of a fully cured and post-cured resin. To further understand hydrolytic susceptibility exhibited by this class of polymers a variety of model compounds simulating the reaction products in the addition polyimide were prepared. NMR spectroscopy was performed on these model compounds after being subjected to simulated service environments in order to gain insight into the mechanisms and kinetics of the hydrolytically reactive sites in these classes of polymers.

KEY WORDS: Solid State Nuclear Magnetic Resonance (NMR), Polyimide, Degradation

1. INTRODUCTION

Addition polyimides offer high temperature capability with relatively benign thermoset processing. Because of their commercial importance they have been the subject of a great many research efforts aimed at understanding their processing chemistry and their service performance. In this effort I have employed solid state nuclear magnetic resonance (NMR) spectroscopy to investigate the cure and degradation of a model high performance addition polyimide. Studies of the ^{15}N nucleus have proven to be most useful to elucidate the reactions of interest in these systems.

2. EXPERIMENTAL

A number of commercially important addition polyimide systems are based on the 4,4'-Hexafluoroisopropylidene bisphthalic compound (6F). AFR700B is one such 6F-based formulation whose formulation is illustrated in Figure 1. Many of the known 6F-based addition polyimides exhibit similar processing characteristics and degradation mechanisms as AFR700B and thus we expect it to be generally representative of the family of resins with similar chemistry and have used it extensively as the basis for our model studies.

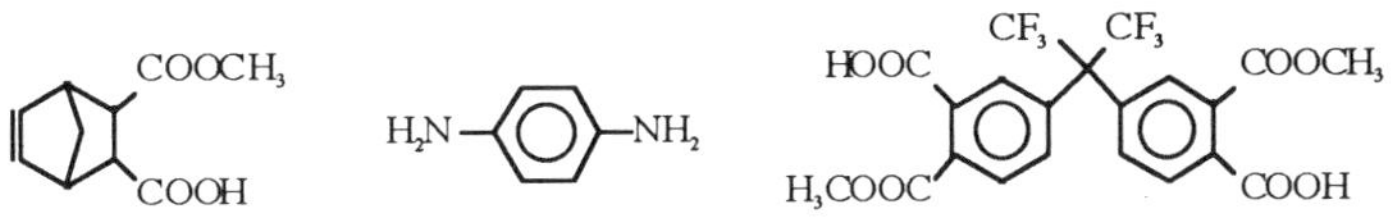

CH3OH

1 NE | 9 p-PDA | 8 6-FDE | Solvent

Figure 1. Monomers of AFR700B Polyimide Formulation.

The resin samples were always prepared using standard procedures and highly purified monomers and solvents. The ^{15}N enriched compounds were prepared in-house using either 100% $^{15}N_1$ 1,4-diaminobenzene (p-PDA) or 100% ^{15}N aniline, both obtained from Cambridge Isotope Laboratories. All compounds synthesized were chromatographically purified and characterized using NMR.

The solid state NMR experiments were performed on a General Electric Omega 400 system using a Doty 5mm high speed CP-MAS probe. All CP-MAS experiments were done under Hartmann-Hahn matched conditions with CP contact times of 6ms used for imide nitrogens and 2ms for amine or amide nitrogens. Spinning rates were approximately 4000 Hz, which was found to eliminate virtually all sidebands.

3. RESULTS AND DISCUSSION

All nitrogen-containing groups in the AFR700B resin arise from reactions of the diamine, making the preparation of the ^{15}N enriched resin straightforward, we only need replace unlabelled p-PDA with ^{15}N enriched p-PDA to completely label all nitrogen-containing product groups. In Figure 2 is the ^{15}N solid state NMR spectrum for the AFR700B formulation cured at 225°C for 2 hours and 20 hours. The peaks are readily identified from their relative chemical shifts: 1) the peak at 24 ppm is due to the primary amine; 2) the peak at 100 ppm is due to the amide-acid groups; 3) the peak at 140 ppm is due to the backbone imide groups; and 4) the peak at 160 ppm is most likely due to the norbornene endcap group. The assignment of the norbornene endcap group is due to its aliphatic character causing the upfield shift relative to the aromatic backbone imide or maleimide. The only discernible difference between the 2 hour and 20 hour imidization is the qualitatively relative smaller peak area for the amide acid and amine regions. This is expected as these two groups will be consumed as imidization proceeds. It must be remembered that in the fully imidized oligomeric system we expect an equal number of primary amine nitrogens and norbornene endgroup nitrogens. and that the CP experiment exaggerates the intensity of those nitrogens with more efficient polarization kinetics. We can thus perform an ^{15}N SP-MAS proton decoupled experiment on the 20 hour imidization specimen to better understand the quantitative relative peak areas. The critical issue in performing quantitative SP spectra is the T_1 of the ^{15}N nuclei of interest. We performed a T1 study of the 20 hour imidization specimen and found that the 140 ppm ^{15}N resonance had the longest T_1 of 100 seconds. We then obtained the ^{15}N SP-MAS spectrum shown in Figure 3.

Although S/N is low we can readily resolve the backbone imide peak at 140 ppm, the primary amine peak at 24 ppm and the norbornene endgroup peak at 160 ppm. The theoretical ratios of amine endgroup to norbornene endgroup to backbone imide in the fully imidized system would be 1:1:18 based on the initial stoichiometry. The free amine and norbornene nitrogen resonances should thus represent approximately 5% of the total nitrogen in the ideally imidized system. The S/N was too low to accurately fit the areas of the amine and norbornene endcap, but we can visually estimate the peaks as 5% to 10% of the backbone peak area. Of greater significance, however, is no evidence of an amide acid peak. We know that some amide acid remains due to the resonance in the ^{15}N CP-MAS spectrum, however, it must be of significantly lower than 5% concentration. Upon post-cure of the fully imidized AFR700B at approximately 360°C for 24 hours under N_2 both the free amine and the norbornene endgroup resonance disappear. The ^{15}N CP-MAS spectrum for the cured and post-cured AFR700B resin is shown in Figure 4.

We would expect the primary amine to have reacted to form some of a Michael addition product and yield a secondary amine peak. Proposed reaction mechanisms always consider that Michael addition products result from the endgroup reactions and so either the amines have become so broadened as to be lost in the baseline noise or they are undergoing other reactions and are possibly masked by the backbone imide peak. Likewise, the norbornene endgroup is most likely involved in a variety of reactions that distribute its resonance to the point of obscurity.

While quantitative experiments of the hydrolytic degradation have not been possible due to insufficient sensitivity, we have shown qualitatively via CP-MAS analysis that chemical degradation of the polyimide formulation does occur. Figure 5 gives the ^{15}N spectrum for the cured and post-cured AFR700B specimen after exposure to a water saturated nitrogen atmosphere at 200°C for 2 hours We see the appearance of the expected amide acid and primary amine hydrolysis products and the emergence of a new resonance at 121 ppm. Interestingly, it is well established that the rate of hydrolysis of an imide to amide-acid is far greater than the hydrolysis of the amide-acid to the free amine and di-acid. However, the relatively great primary amine resonance intensity Indicates that not all imides are hydrolyzable at the same rate. The resonance at 121 ppm is clearly a degradation product. Its chemical shift indicates that it could be an amide-acid, but different from the one occurring at 100 ppm. One possible interpretation is that the two forms of amide-acid groups arise from backbone and endgroup imides respectively.

To explore both the cure and degradation mechanisms and kinetics further and to help elucidate the distinction between endgroup and backbone reactions, ^{15}N enriched model compounds have been prepared specifically to highlight the chemistry that we are observing with the fully enriched AFR700B formulation. Three of the model compounds and their ^{15}N solid state NMR spectra are shown in Figure 6. The reactions of N-Phenylnadimide and N-Phenylmaleimide with the free amine of the addition polyimide oligomer will help reveal the specific cure and post-cure mechanisms and environmental durability of the endgroup reaction products. The phenyl-terminated 6F compound will be used to investigate the inherent hydrolytic stability of the backbone imide groups characteristic of these polyimides.

It is inconclusive at this time what the specific mechanisms of cure, post-cure, and hydrolytic degradation are in the 6F based family of addition curing systems. The primary difficulty is that the key reactions that lead to outstanding thermal properties of the system through cure and post-cure involve the terminal groups. These groups are present in very low concentration relative to the "backbone" of these systems and tend to be obscured. Our model compound approach and use of ^{15}N solid state NMR techniques will be the most fruitful approach to date in understanding these systems.

BIOGRAPHY

David B. Curliss received a B.S. with honors in Chemical Engineering from the University of Notre Dame in 1985 and a M.S. in Chemical Engineering from Purdue University in 1987. Following his M.S. David served four years in the U.S. Air Force at Wright Laboratory, Wright-Patterson AFB where he became acquainted with structural composite materials. He returned to Purdue Univeristy and received a Ph.D. in Chemical Engineering in 1995 and is now employed in the Nonmetallic Structural Materials Branch at the U.S. Air Force Wright Laboratory, Wright-Patterson AFB where he works in the areas of environmental durability, viscoelasticity, and advanced processing.

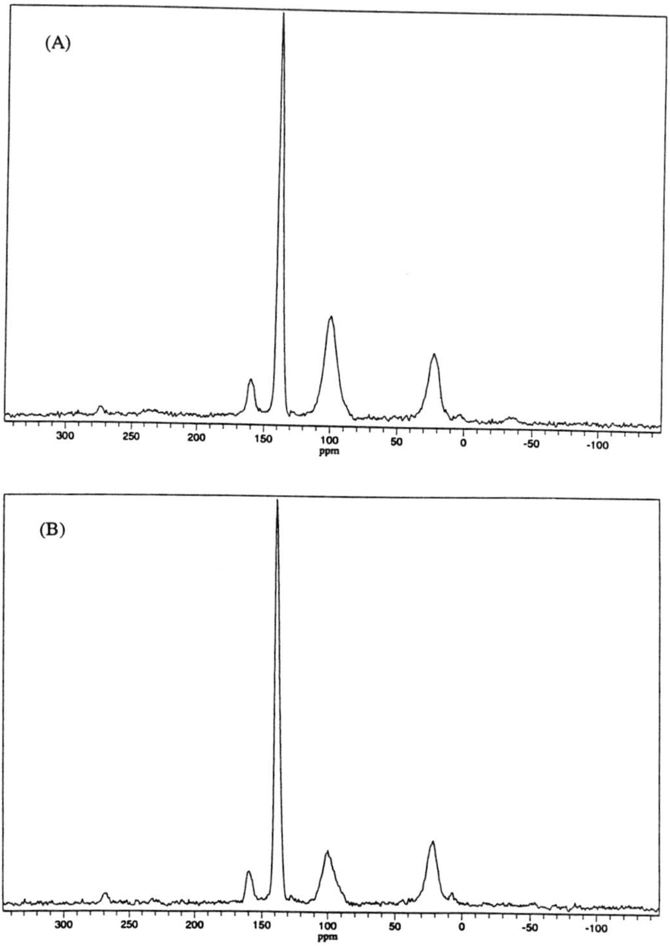

Figure 2.

A) ^{15}N CP-MAS spectrum for AFR700B after cure for 2 hours at 225°C

B) ^{15}N CP-MAS spectrum for AFR700B after cure for 20 hours at 225°C

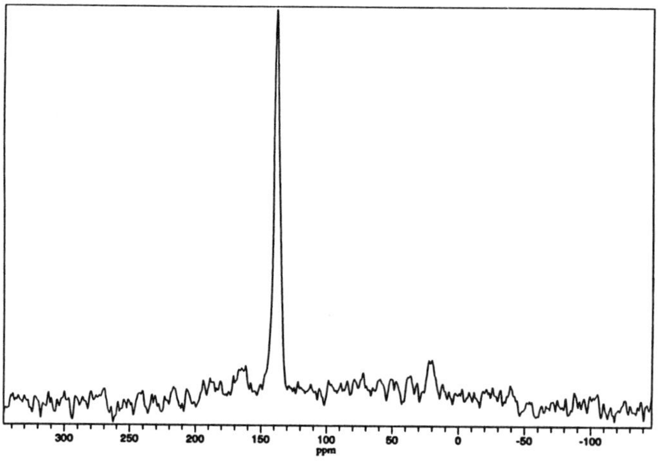

Figure 3. ^{15}N SP-MAS spectrum for "as-cured" AFR700B.

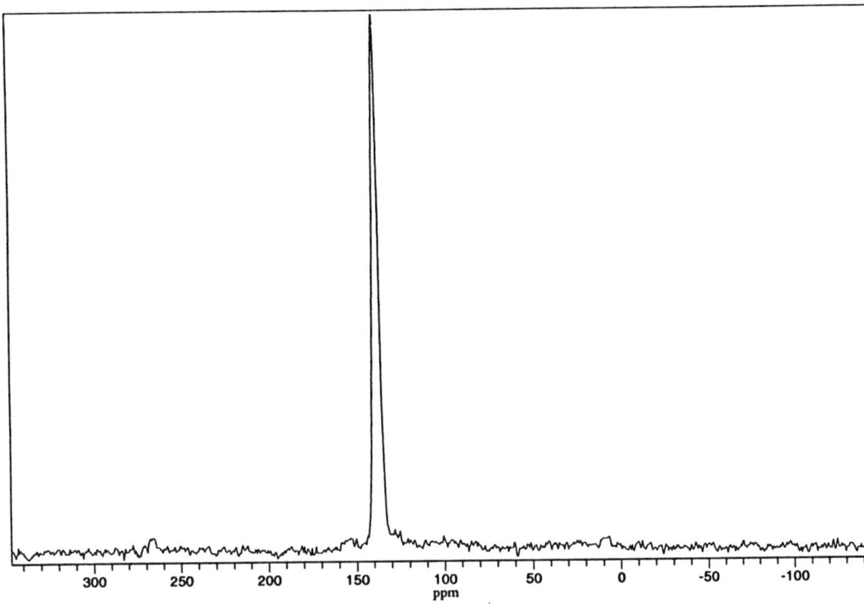

Figure 4. ^{15}N CP-MAS spectrum of AFR700B after cure and postcure.

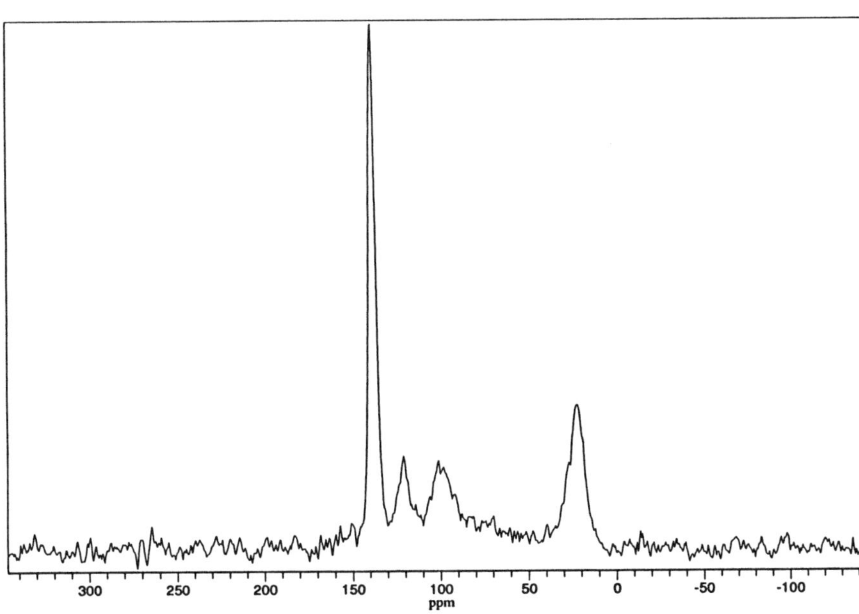

Figure 5. ^{15}N CP-MAS spectrum of AFR700B cured and postcured after saturated moisture exposure at 200°C.

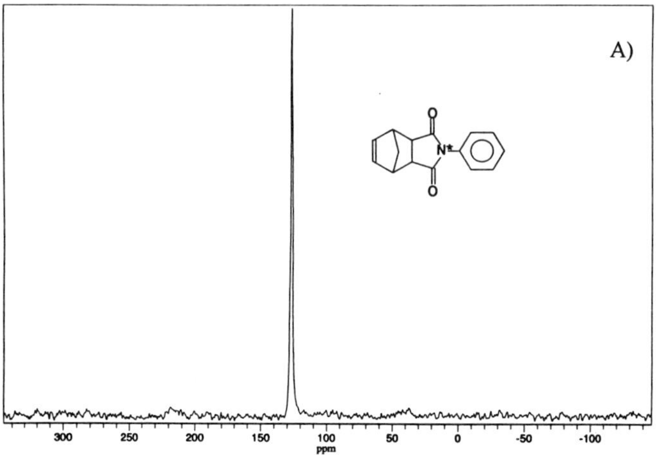

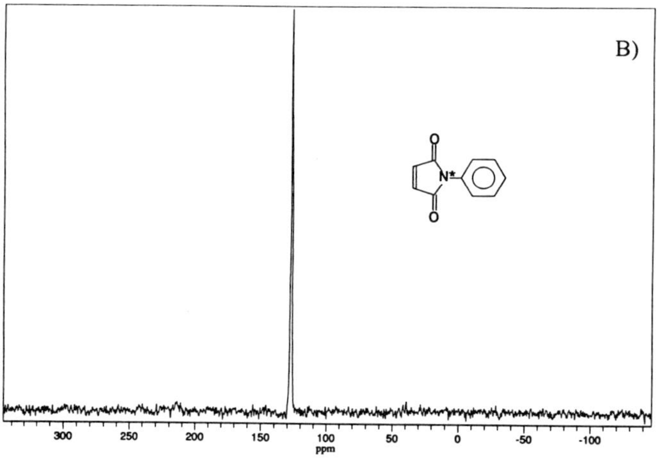

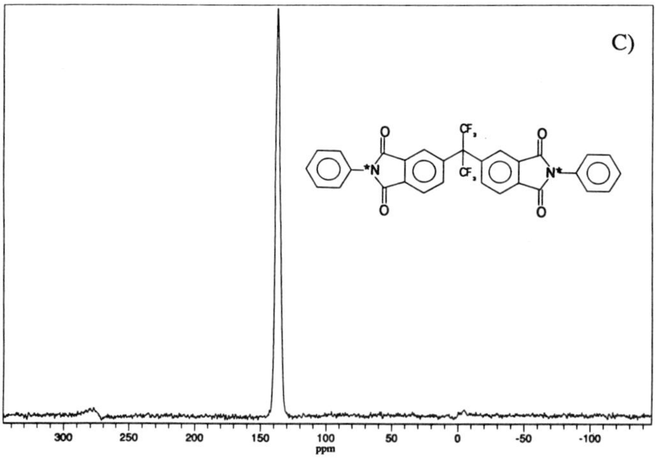

Figure 6.

A) ^{15}N CP-MAS spectrum of N-Phenylnadimide

B) ^{15}N CP-MAS spectrum of N-Phenylmaleimide

C) ^{15}N CP-MAS spectrum of Phenyl-terminated "6F backbone"

28th International SAMPE Technical Conference
November 4-7, 1996

THE EFFECT OF ISOTHERMAL AGING ON THE RELAXATION SPECTRA OF AFR700B

Katie E. G. Thorp, Ajit K. Roy and Allan S. Crasto
University of Dayton Research Institute
Dayton, Ohio 45469-0168

ABSTRACT

A variety of environmental factors, acting individually or synergistically, can have a profound influence on the performance of advanced polymers and composites. In particular, exposure to a combination of moisture and elevated temperature can significantly alter properties; the extent of chemical and physical changes is a function of exposure time and the temperature and humidity of the environment. Changes in the relaxation time spectrum can be related to chemical or physical alterations which occur in the polymer. Chemical degradation can result in a broadening of the relaxation time spectrum due to a widening of the molecular weight distribution resulting from the generation of lower-molecular weight species. Physical aging may be reflected in a shift in the relaxation time spectrum to longer times. This paper discusses the effect of chemical and physical aging on the relaxation time spectra of AFR700B, a high-temperature polyimide resin. Neat resin samples were isothermally aged for 500 hours at 250°C in dry air, humid air and dry nitrogen. Changes in the dynamic mechanical properties of the resin after these different exposures are discussed.

KEY WORDS: Aging, Polyimide, Relaxation Spectra

1. INTRODUCTION

The current trend in the aerospace industry is toward lighter-weight materials with improved high-temperature properties. As a result, fiber-reinforced polymeric composites are being applied in structures which are exposed to elevated service temperatures, in excess of 300°C. These high-temperature exposures often occur after, or in conjunction with, exposure to humid environments. The combined effects of these environments can lead to an alteration of the physical and chemical structure of the resins as well as the fibers and fiber-resin interfaces. These changes can, in turn, result in a decline in the mechanical properties of the composite material. Because of their relatively high glass transition temperatures and high thermooxidative stability, polyimide resins have been utilized in a variety of high-temperature structural applications. They have been shown to withstand extended exposures to elevated temperatures without a significant decrease in mechanical stability [1, 2]. The high-temperature stability of these resins has been attributed to their high aromatic content and efficient crosslinking [3]. Recently, however, several of these resins have been shown to undergo degradation when exposed to hygrothermal conditions [4, 5]. Under these conditions the glass transition temperature (T_g) decreases with time. This decrease has been attributed to physical [4] and chemical [6, 7] changes occurring in the resin.

Physical changes in the molecular structure of a resin can occur during aging at a temperature below its glass transition temperature. Subsequent to processing at elevated temperatures, polymer molecules are frozen into nonequilibrium conformations as they are cooled below their glass transition temperature. The lack of mobility of these long-chain molecules at these lower temperatures results in the polymer remaining in a nonequilibrium state. Entropy promotes the continuous rearrangement of the molecules towards lower-energy conformations. The molecular motions associated with this decrease in entropy are termed physical aging and can lead to an alteration of the mechanical and physical properties of the material. During aging the rigidity of the resin increases with time. This change is small but measurable. The change in shear modulus of the polymer with aging can be used as a measure of the effect of physical aging on the material. The changes in modulus are particularly localized in the vicinity of the aging temperature, T_a, and decrease in magnitude with decreasing T_a [8]. The changes associated with physical aging are reversible if the material is exposed to a temperature above T_a or T_g.

Many polymers undergo chemical changes when exposed to humid environments, particularly at elevated temperatures. Polyimides have been shown to undergo hydrolysis under such conditions [6, 9]. Nuclear magnetic resonance studies have shown that polyimides differ in their susceptibility to hydrolytic degradation. Variations in backbone structure and steric hindrance can greatly affect the hygrothermal stability of these polyimide resins [6]. The exact nature of the chemical changes which these materials undergo in service environments is an area which requires further study. In addition the effects of these chemical changes on the mechanical and physical properties of the material are still unclear. In the present study the effects of chemical and physical aging on the thermomechanical and relaxation spectra of a high-temperature polyimide, AFR700B, are investigated.

2. EXPERIMENTAL

2.1 Materials and Sample Preparation Studies were conducted on a neat polyimide resin, AFR700B, derived from a mixture of nadic ester, paraphenylenediamine (PPDA), and dimethyl 4,4'-(hexafluoroisopropylidene)-bis(phthalate) (6FDA), and obtained from Daychem, Inc. as a 70 weight percent solution in methanol. Neat resin plates of AFR700B, 2.3 mm thick, were processed in a hot press to a T_g of approximately 360°C in the as-cured state. Some of these plates were further postcured in a nitrogen atmosphere to a T_g of approximately 430°C. These resin plates were then cut into rectangular specimens, measuring 4.5 cm x 0.6 cm, which were dried to a constant weight in a vacuum oven at 50°C.

2.2 Environmental Exposure Specimens were placed in three glass cloth-lined steel bags, with a gas inlet port at the base of each bag. These bags were placed in an oven maintained at 250°C, and continuous streams of dry air, humid air and dry nitrogen gas, respectively, were allowed to flow through them. The bags were initially purged with the gas, and the flow was then adjusted to maintain a slight positive pressure in each bag. The sources of the inlet gas streams were (1) compressed air filtered through a desiccant, (2) compressed air percolated through a heated water bath, and (3) liquid nitrogen, and the gas flow through each port was maintained at approximately 62 cc/minute throughout the experiment. After 500 hours at 250°C the specimens were removed, weighed and dried to constant weight in a vacuum oven at 50°C. All specimens were stored under these conditions prior to testing.

2.3 Property Evaluation Specimens were tested in a Rheometrics dynamic mechanical analyzer operating in the torsion mode. Temperature scans were conducted to determine the T_g of specimens using operating conditions of 100 radians/second and a heating rate of 2°C/minute. All the glass transition temperatures were derived from the maximum in the loss modulus (G") peak, and the values reported are the average of two specimens. Frequency sweeps were conducted over the range of 0.1 to 100 radians/second, between 50°C and 490°C, at temperature intervals of 20°C. Time-temperature superposition was used to generate a master curve for each specimen. The relaxation spectra were generated from master curves of the storage moduli and the method of Tschoegl, as described by Ferry [10].

3. DATA ANALYSIS

A first approximation of the relaxation spectrum was first generated using

$$H(\tau) = dG'/d \ln \omega + (1/2)\, d^2G' /d(\ln \omega)^2 \tag{1}$$

where $1/\omega = \sqrt{2}\tau$. When the slope of this curve is negative, the alternate equation

$$H(\tau) = dG'/d \ln \omega - (1/2)\, d^2G' /d(\ln \omega)^2 \tag{2}$$

where $1/\omega = \tau/\sqrt{2}$, was used. Calculation of the terms $dG'/d \ln \omega$ and $d^2G'/d(\ln \omega)^2$ in the above equations requires smoothing of the experimental data prior to numeric differentiation. In some practical problems interpolation of a single polynomial is not a suitable approximation of a function or data set; to obtain a good approximation by an *n-th* degree polynomial, it may be necessary to use a large value of n. Unfortunately, polynomials of a large degree often have highly oscillatory behavior which is not desirable in approximating functions or data that are reasonably smooth. To overcome the oscillatory behavior of polynomials and still provide a smooth approximation, the use of cubic splines has proven quite effective. In this technique, instead of approximating the whole data range with a single spline, several basic cubic spline functions, defined in segments of the data range, are used. The approximation procedure used here is outlined below.

Consider an interval $a \le x \le b$ over which a function f(x) is to be approximated with spline functions. A set of spline functions defined in subdivisions $(a=x^{(1)}<x^{(2)}<x^{(3)}<...<x^{(m)}=b)$ of the interval yield a better approximation than the use of one spline function for the whole interval. The spline functions $X_{n,i}(x)$ used were developed by Iarve [11] using the following recursive procedure:

$$X_{n+1,i}(x) = \frac{I_{n,i-1}(x)}{I_{n,i-1}(x_m)} - \frac{I_{n,i}(x)}{I_{n,i}(x_m)}, \quad i = 2, \ldots, n+m; \tag{3}$$

$$X_{n+1,1}(x) = 1 - \frac{I_{n,1}(x)}{I_{n,1}(x_m)}, \quad X_{n+1,n+1+m}(x) = \frac{I_{n,n+m}(x)}{I_{n,n+m}(x_m)}, \tag{4}$$

where

$$I_{n,i}(x) = \int_a^x X_{n,i}(\xi)\, d\xi$$

$$X_{0,i}(x) = 1,\ x_i \le x \le x_{i-1} \ \ and \ \ X_{0,i}(x) = 0,\ x_{i-1} > x,\ x > x_i\,,\ i = 1, \ldots, m.$$

The subscript n used in the notation of the spline function, $X_{n,i}(x)$, indicates the order of the spline. For cubic spline functions, n is equal to 3. Using these cubic splines as the basis, the approximated function for $f(x)$ is defined as

$$F(x) = \sum_{1}^{m+3} C_i\, X_{3,i} \tag{5}$$

where C_i are arbitary constant coefficients. For m subdivisions in the interval $(a=x^{(1)}<x^{(2)}<x^{(3)}<...<x^{(m)}=b)$, the number of cubic spline functions defined in the interval is $m+3$. If the function f(x) is known at p distinct points $(x_1, x_2, \ldots, x_p)$, then f(x) is best approximated by minimizing the error in the least square sense

$$\varepsilon = \sum_{1}^{p} [f(x_i) - F(x_i)]^2 \tag{6}$$

The coefficients C_i in equation (5) are determined by minimizing the error ε by setting $\frac{d\varepsilon}{dC_i} = 0$. This leads to a set of systems of equations which can be solved for C_i [12]:

$$A^T A C = A^T f \tag{7}$$

where

$$A = \begin{bmatrix} X_{3,1}(x_1) & X_{3,2}(x_1) & \cdots & X_{3,m+3}(x_1) \\ X_{3,1}(x_2) & X_{3,2}(x_2) & \cdots & X_{3,m+3}(x_2) \\ \cdots & \cdots & \cdots & \cdots \\ X_{3,1}(x_p) & X_{3,2}(x_p) & \cdots & X_{3,m+3}(x_p) \end{bmatrix}$$

$$C = [\; C_1 \quad C_2 \quad \ldots \quad C_p \;]^T$$

$$f = [\; f(x_1) \;\; f(x_2) \quad \ldots \;\; f(x_p) \;]^T$$

and A^T is the transposed matrix of A. Equation (7) is an (m+3) x (m+3) linear system of equations, and the coefficients C_i are determined by solving numerically, using the LU decomposition technique in this case [13]. The determination of C_i thus yields the best-fitted curve in the least square sense, as given by equation (5). This spline variational technique provided excellent agreement between the experimental and approximated data, an example of which is shown in Figure 1. This allowed smoothing of the experimental data and subsequent numeric differentiation.

4. RESULTS AND DISCUSSION

4.1 Exposure Conditions The elevated temperature at which the aging was conducted prohibited direct measurement of the relative humidity in each environment. Room temperature measurements of humidity were conducted to estimate the amount of moisture present in each gas stream. The "dry" air had a relative humidity of 19% at 23°C, the humid air had a relative humidity of 100% at 25°C, and the nitrogen had a relative humidity of 19% at 22°C. Each of these conditions is equivalent to a relative humidity at 250°C of less than 1%. However, the humidified air had five times more moisture than either of the other gas streams which translates to a significantly higher amount of water vapor flowing over the specimens at 250°C. The rationale for the selection of these three aging environments is as follows. Material property changes in dry nitrogen should reflect the influence of physical aging; comparison of the results from specimens aged in dry air and nitrogen should allow an estimate of the added influence of thermal oxidation; and comparison of results from specimens aged in dry and humid air should provide insight on the influence of moisture-induced degradation (such as hydrolysis) on material properties.

4.2 Effect of Aging on Glass Transition Temperature The initial T_g of the as-cured specimens, prior to aging, was approximately 360°C. After aging at 250°C the T_g of all these specimens increased significantly. The specimens aged in humid air displayed a smaller increase than did the specimens aged in dry air or nitrogen (Figure 2). The initial T_g of the postcured specimens was approximately 430°C. After 500 hours at 250°C the T_g of these specimens exposed to dry air and nitrogen remained relatively unchanged. However, the specimens aged in humid air showed a decline in T_g of approximately 13°C (Figure 2).

The increase in the T_g of as-cured specimens suggests additional crosslinking during aging, even though the aging temperature of 250°C is considerably lower than the cure temperature of 380°C. The smaller increase in humid air, compared to dry air or nitrogen, could be due to an inhibition of the crosslinking by the moisture present, or more likely due to competition between crosslinking and concurrent chemical degradation in the hygrothermal environment. This conclusion appears to be corroborated by trends observed in the thermomechanical spectra (section 4.3). None of the postcured specimens experienced an increase in T_g as a result of aging at 250°C. While the T_g of specimens aged in nitrogen and dry air remained relatively unchanged, the T_g of specimens aged in humid air declined by approximately 13°C after 500 hours at 250°C. This decline is significant and suggests a breakdown in the polymer molecular structure by the moisture present. The exact nature of this chemical breakdown was not addressed in this study. However, the fact that it occurred in the presence of moisture while specimens aged similarly in dry air experienced no change (as reflected by the measured Tg) suggests imide hydrolysis.

4.3 Effect of Aging on Thermomechanical Spectra The thermomechanical spectra used to determine the glass transition temperature revealed additional information about the effect of aging on the molecular structure of the resin. For the as-cured specimens, the shape of the storage modulus (G') curve in the thermomechanical temperature sweep changed after aging; the aged specimens displayed appreciably larger dynamic shear moduli than the baseline specimen. Due to the susceptibility of these resins to additional crosslinking above the T_g, a true rubbery modulus cannot be directly obtained from this data. However, from the plots in Figure 3, the trend above the Tg is clearly apparent; the baseline specimen has the lowest modulus in this region followed by the specimen aged in humid air, while the curves for specimens aged in dry air and nitrogen overlap at higher moduli values. This trend is similar to that observed in Tg measurements (Figure 2) and reflects additional crosslinking during aging. The baseline postcured specimen, as expected, has a modulus substantially higher than any of the (unaged or aged) as-cured specimens due to its higher degree of crosslinking.

The storage moduli generated from temperature sweeps of unaged and aged postcured specimens are shown in Figure 4. In the vicinity of the aging temperature (250°C), there is an expected increase in modulus for all aging environments, attributed to physical aging. This effect disappears as the temperature scan exceeds the aging temperature. Above the T_g, specimens aged in dry air display a significant increase in dynamic shear modulus compared to baseline specimens, although there is no increase in T_g (see Figure 2); for specimens aged in nitrogen, however, the increase in modulus in this region is barely perceptible. These results suggest that thermal oxidative crosslinking during aging in dry air may be responsible for the increase in modulus in this region. Aging in humid air produces a drop in modulus in this region, which points to the greater influence of moisture-induced chemical breakdown compared to oxidative crosslinking. Regardless of the aging environment or the initial cure state, the shape of the α-peak in the G" and Tan δ curves was altered after aging. In each case a shoulder appeared on this curve in the region below the T_g (Figure 5) suggesting the occurrence of additional molecular motions in this temperature range. A change in T_g with aging resulted in a lateral shift of this shoulder along with the α-peak but with no change in shape or intensity. Similar shoulders have been previously attributed to stresses frozen into polymers as a result of processing or quenching [14, 15] These changes in the shape of the thermomechanical spectra prompted an investigation of the relaxation spectra and the effects of physical aging.

4.4 Effect of Aging Environment on Relaxation Spectra Relaxation spectra, $H(\tau)$, provide additional insight into the physical and chemical alterations occurring in polymers as a result of environmental exposure. Figure 6 shows the relaxation spectra of unaged and aged postcured specimens. The initial peak corresponds to ß-transitions in the resin occurring deep in the glassy state. Physical aging is not expected to alter the ß-transition, since the amount of molecular motion possible at those temperatures and/or frequencies is generally considered too small to be measurable [16]. As expected, therefore, the ß-transition in Figure 6 showed no changes for all the aging conditions on the short-time edge and at the peak. The long-time edge of the ß-transition for the aged specimens showed a decrease in $H(\tau)$ compared to the the baseline specimen. A corresponding increase in the $H(\tau)$ over the baseline value was observed at the short-time edge of the α-transition for all the aged specimens. The area enclosed between the curves for baseline and aged specimens in the short-time region of the α-transition is similar to the area between the same curves in the long-time region of the ß-transition. These results suggest that physical aging causes a shift in the relaxation times toward longer times, independent of the aging environment. The equivalence of these shifts suggests that the rates and mechanisms of physical aging are the same in all the aging environments. The specimen which aged in humid air showed an additional slight increase in $H(\tau)$ on the short-time edge of the α-transition, over the baseline specimen and those aged in nitrogen and dry air. This suggests the generation of additional relaxation processes which were not present in any of the other specimens.

5. SUMMARY AND CONCLUSIONS

Neat as-cured and postcured AFR700B polyimide resin was aged for 500 hours at 250°C in environments of dry air, humidified air and dry nitrogen. Thermomechanical and relaxation spectra suggest that physical aging occurred in all the aged specimens in a similar manner, regardless of the aging environment. As-cured specimens displayed evidence of additional crosslinking during aging, through increases in glass transition temperature and storage modulus in the rubbery region. The magnitude of these changes was reduced in the specimens aged in humid air, probably due to the competing influence of moisture-induced chemical breakdown of the molecular structure. The decline in the glass transition temperature of postcured specimens, and the appearance of additional features in the relaxation spectra of specimens aged in humid air, support the suggestion of molecular breakdown under prolonged hygrothermal exposure. These changes were not observed in specimens aged in dry air or nitrogen. The use of a cubic spline data approximation technique proved critical for accurate data analysis.

6. ACKNOWLEDGMENTS

The authors would like to acknowledge Bill Price and Ken Lindsay of the University of Dayton Research Institute for assistance with sample exposure and testing. This work was sponsored by the U.S. Air Force under Contract No. F33615-95-D-5029.

7. REFERENCES

1. D. Scola and J. Vontell, "High Temperature Polyimides, Chemistry and Properties," Polymer Composites, 9 (6), 443-452 (1988).

2. S. Stoessel, et al., "High-Performance Composites from Polyimide Matrix Resins that Utilize a Crosslinking Reaction of Biphenylene End-Caps with Acetylene Units in the Main Chain," Journal of Applied Polymer Science, 36 (8), 1847-1864 (1988).

3. J. Sutter, et al., "N-CYCAP Polyimide Matrix Composites in High Temperature Environments," Polymer Preprints, 33 (1), 366-367 (1992).

4. D. Cornelia, "Hygrothermal Performance of Polyimides," 39th International SAMPE Symposium, 917-929 (1994).

5. K. E. G. Thorp and A. S. Crasto, "Hygrothermal Stability of Polyimide Resins and Composites," Proceedings of the American Society for Composites, 10th Technical Conference, 601-612 (1995).

6. P. Murphy, et al., "Nitrogen-15 Solid State NMR Studies of the Cure and Degradation of Polyimide Films under Temperature and Humidity Stresses," Macromolecules, 27, 279-286 (1994).

7. D. Curliss, "Environmental Effects on High Performance Polymers and Composites," High Temple Proceedings, H (1996).

8. R. A. Venditti and J. K. Gillham, "Physical Aging Deep in the Glassy State of a Fully-Cured Polyimide," Journal of Applied Polymer Science, 45, 1501-1516 (1992).

9. C. A. Pryde, "Polyimide Hydrolysis; Measurement by Fourier Transform-IR Spectroscopy," Polymeric Materials for Electronics Packaging and Interconnection, 57-66 (1989).

10. J. D. Ferry, Viscoelastic Properties of Polymers, Third Edition, John Wiley & Sons, Inc., 1980, pp. 83-84.

11. E. V. Iarve, "Spline Variational Three Dimensional Stress Analysis of Laminated Composite Plates With Open Holes," Intl. J. Solids Structures, 33, (14), 2095-2118 (1996).

12. L. W. Johnson and R. D. Riess, Numerical Analysis, 2nd Ed., Addison-Wesley Publishing Co., 1982, Section 2.5, pp. 71.

13. W. H. Press, et. al., Numerical Recipes, Cambridge University Press, New York, 1989, Section 2.3, pp. 31.

14. E. R. Harrell, Jr., and R. P. Chartoff, "Effects of Thermal and Mechanical History on the Viscoelastic Properties of Rigid Poly(vinyl Chloride)," J. Macromol. Sci. - Phys., B14, (2), 277-305 (1977).

15. C. Bauwens-Crowet and J-C. Bauwens, "The Relationship Between the Effect of Thermal Pre-Treatment and the Viscoelastic Behaviour of Polycarbonate in the Glassy State," Journal of Materials Science, 14, 1817-1826 (1979).

16. L. C. E. Struik, Physical Aging in Amorphous Polymers and Other Materials, Elsevier Scientific Publishing Co., Amsterdam, 1978, pp. 22-27.

7. BIOGRAPHIES

Katie E. G. Thorp received her B.S. in Ceramic Engineering from the University of Washington in 1989 and her M.S. in Materials Science and Engineering from the University of Washington in 1991. Since that time she has worked as an Associate Research Engineer for the University of Dayton Research Institute on an on-site contract in the Structural Materials Branch of the Materials Directorate at the U.S. Air Force's Wright Laboratory. Her primary research activities currently involve environmental effects on polymer-reinforced composites.

Ajit K. Roy is a Research Engineer with the University of Dayton Research Institute. He received his Ph.D. in Mechanics from the University of Minnesota in 1985. His current research interests include mechanics of fabric-based composites, carbon-carbon composites, and bonded composite joints.

Allan Crasto received his Ph.D. in Materials Science & Engineering from Washington State University in 1986. He spent a year conducting postdoctoral research on interphase modification in carbon-fiber composites and another year as a Visiting Scientist at the U.S. Air Force Materials Laboratory where he worked on the processing and characterization of molecular composites. He joined the University of Dayton Research Institute in 1988 and is currently a Senior Materials Scientist. His research interests include the investigation of structure/property relationships in polymer composites, measurement of residual stresses, development of composites for space, and the application of composites to infrastructure.

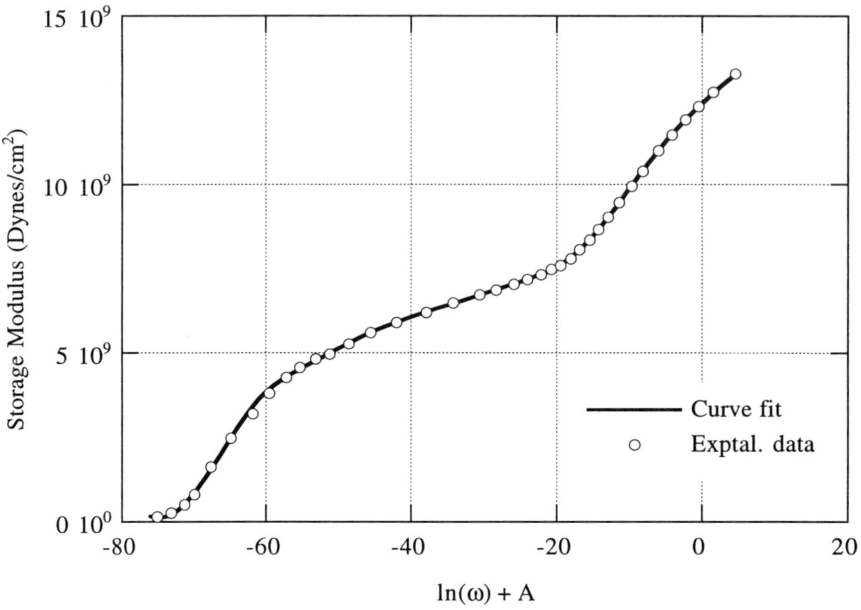

Figure 1: Cubic spline curve fit of raw data

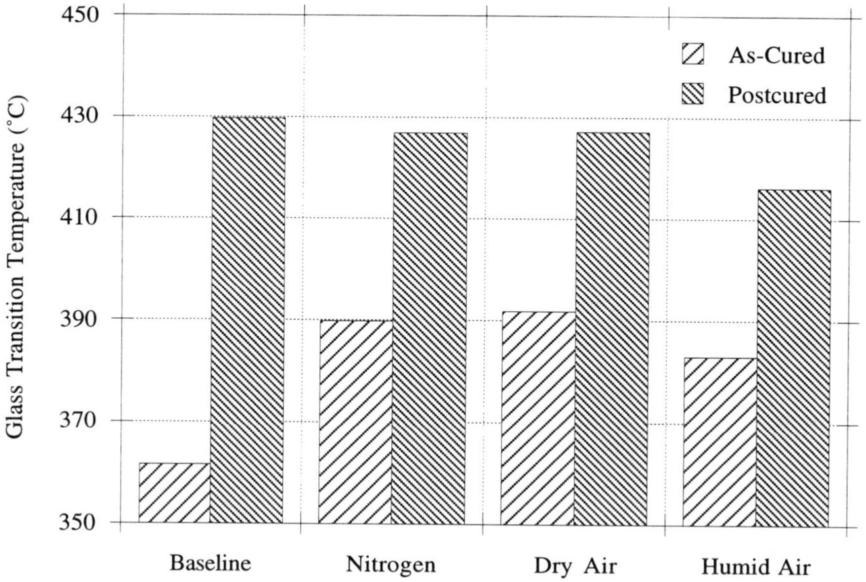

Figure 2: Glass transition temperature as a function of aging environment

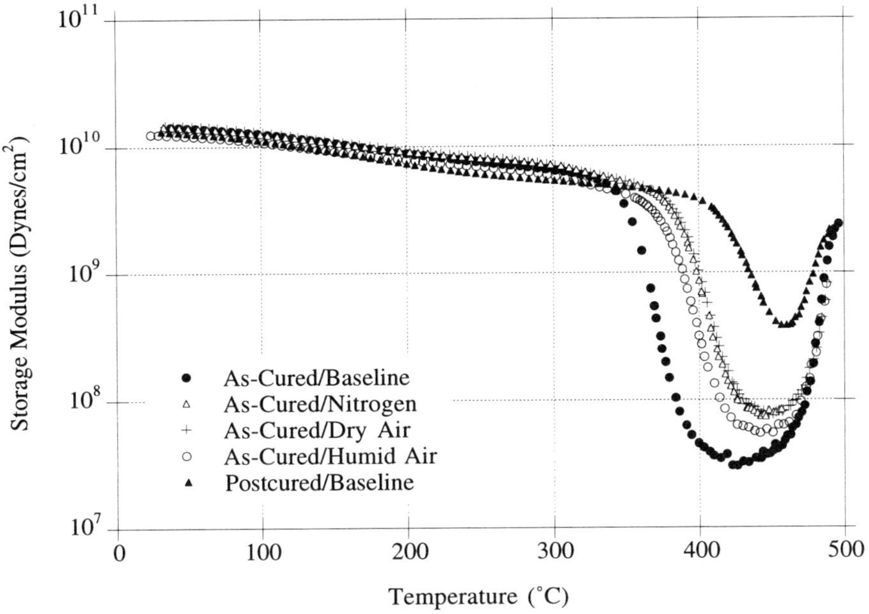

Figure 3: Storage modulus as a function of temperature for the as-cured specimens and the postcured baseline specimen

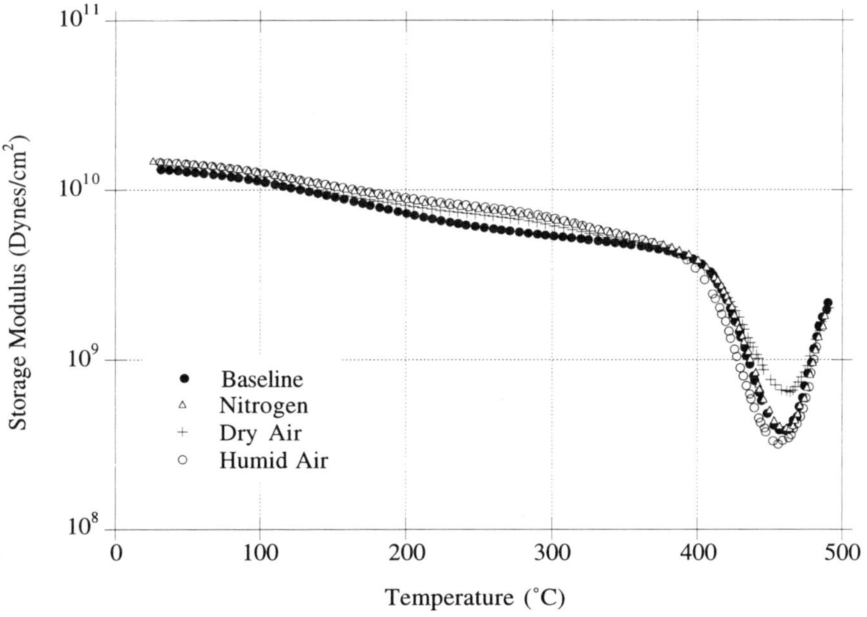

Figure 4: Storage modulus as a function of temperature for the postcured specimens

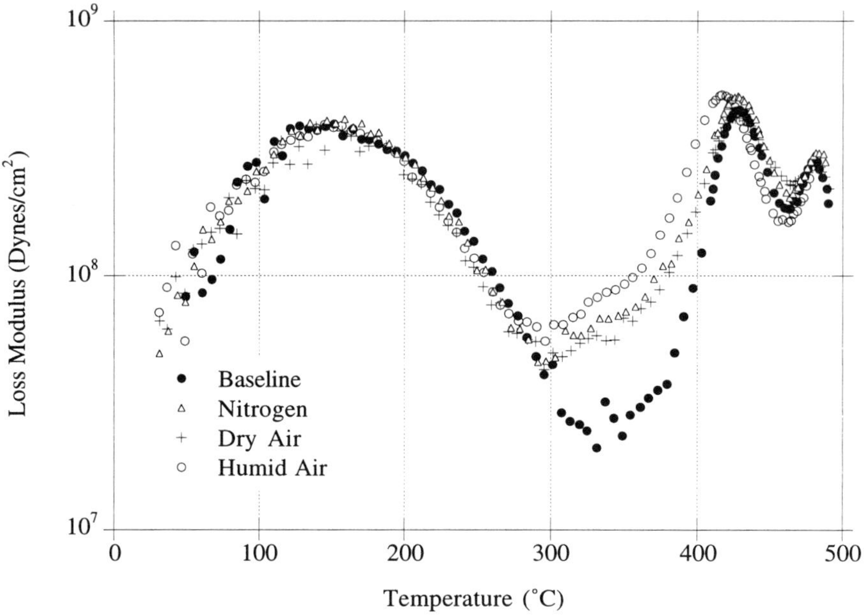

Figure 5: Loss modulus as a function of temperature for the postcured specimens

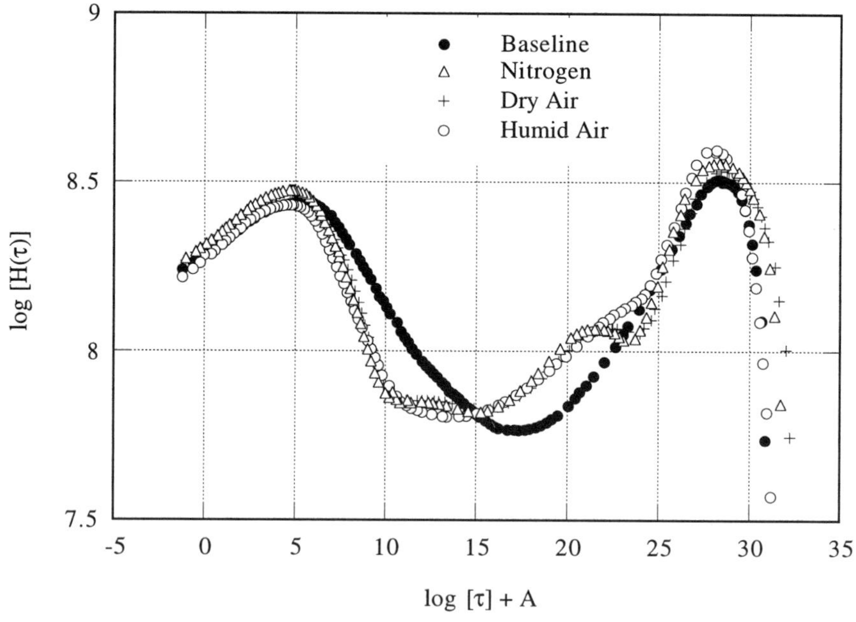

Figure 6: Relaxation spectra for the postcured specimens

28th International SAMPE Technical Conference
November 4-7, 1996

PROCESS DISCOVERY: AUTOMATED PROCESS DEVELOPMENT FOR AUTOCLAVE CURING

Frances Abrams, PhD
WL/MLBC
Wright-Patterson AFB OH

ABSTRACT

In order to expedite the development process for Self-directed, State-based Qualitative Control (SSQC) of polymer curing, an algorithm for data-based, real-time discovery called Process Experiment Theory Evaluator (PETE) was developed. PETE learned new relationships between process controls and process behavior, both during and between process runs and attempted to optimize process cycles based on simple rules for evaluation of process outcomes. In addition, PETE recorded the learned information in a form that could be used by the process engineer user to enhance understanding of the process and further improve the system performance.

KEY WORDS: composites processing, intelligent processing, artificial intelligence, qualitative reasoning, discovery

1. INTRODUCTION

Aerospace production is rapidly evolving into a high technology, low-volume industry where flexibility and adaptability are at a premium. Non-recurring development costs are becoming a significant part of the life cycle cost of systems. At the same time the introduction of advanced composites, co-curing, and bonding means that larger, more complex parts are being proposed to replace assemblies of many smaller parts. While there is an obvious advantage to reducing assembly costs, there is an increased risk in making these larger, more complex structures, since the failure to properly process any one part represents a greater loss of investment.

The autoclave curing of advanced composites is a batch process. Characteristically, batch processes are (a) not in equilibrium; (b) complicated by multiple inputs and outputs; (c) dynamic; (d) nonlinear; and (e) require additional control functions such as safety interlocks, batch management, and scheduling (1). The autoclave cure also involves sequences of irreversible changes to the material. Further, significant time lags between control input and process response, create a control problem with delayed feedback (2).

The non-equilibrium, dynamic nature of batch processes is usually modeled as a sequence of states. Consequently, control of the autoclave cure has typically consisted of a pre-planned cure cycle consisting of a sequence of temperature and pressure setpoints. Prediction of material response to these pre-planned cure cycles is complicated by the non-linear relationships between multiple process parameters and complex material behavior, as well as variations in material, processing vessel, tooling, part geometry and autoclave loading. These non-linearly related inputs and outputs can be modeled with multiple interrelated partial differential equations. However, inversion of these equations is difficult, as is direct, quantitative measurement of some critical parameters. Strict control of incoming material properties and other possible perturbations is used to minimize variations in material response. Even so, there may be variation in quality and post-process inspection must be used to detect unacceptable products.

Self-directed State-based Qualitative Control (SSQC) is an alternative to preplanned cure cycles. SSQC uses information from multiple sensors to detect patterns and changes in patterns that identify intermediate material states and dynamic trends during the process. Qualitative reasoning is then used to adjust the process inputs to affect the process dynamics and drive the process towards a predefined goal state. SSQC has been shown to increase the efficiency and efficacy of control in materials processing as compared to traditional control methods (3-8). Production times are often reduced. Rejection rates and post-process inspection requirements may also be reduced. In addition, these controllers can adapt to variability in starting materials and can handle some unexpected interruptions or deviations in the process. SSQC can also provide the engineer with information on the process that relates directly to quality concerns.

However, SSQC requires sensors to detect material changes and enough knowledge of the process to identify states and trends and respond appropriately. Sensors cost money and add to labor costs for layup, and to operate and maintain the measurement equipment. Further, when rates, second derivatives and spatial gradients are calculated, even a few sensors create a combinatorial explosion in possible states. Therefore it is desirable to minimize the number of sensors, placing them in the most critical positions, and choosing the sensors which provide the most critical and least ambiguous information.

The knowledge necessary for sensor placement, for interpretation, and for the control strategies necessary for SSQC has been developed by a modified trial-and-error method (4 and 9). While the knowledge about control actions and their effects on material behavior is often available, there are interactions (nonlinearities) which complicate these effects. Consequently, traditional wisdom on control strategies may not be available, or, if available, correct.

The purpose of this work was to develop an SSQC system which could adapt to unexpected process behavior and retain the rules and states used. This system, in addition to being more flexible and self-improving, could be used by the process engineer to develop knowledge bases for use in production.

2. APPROACH

The approach chosen to develop a self-improving SSQC is built on a combination of qualitative reasoning and discovery. A qualitative model of the process, or envisionment, provides an initial process control strategy and a means to assess progress towards the goal(s) of the process. During each time cycle, the discovery system takes data, parses the data into symbolic form, compares the current state and trends (CURRENT.AFFAIRS) to the envisionment, determines progress, and makes a new set of control decisions. If the current conditions do not match any of the expected next states, an expectation failure is generated and the controller adds the new state and/or transition rule to the map while making control adjustments appropriate to bring the process back on track. Discovery of unexpected states and transitions change the behavior of the controller during the run, but only within safe limits. At the end of the process, a new envisionment exists. The engineer can use this envisionment as a new starting point, either as it is, or after alteration.

2.1 Discovery Systems The artificial intelligence (AI) community has developed a variety of reasoning algorithms to "discover" trends and correlations in data and recognize exceptions to these trends and correlations. These programs have recognized relationships between data (10, 11), recombined knowledge to create new ideas (12), and formed theories (13). Some can adaptively change their behavior and knowledge by "learning" from experience (14). It seems logical to extend these "discovery" systems to further automate the development and enhance the performance of SSQC systems.

An advantage of discovery systems is that they not only recognize new patterns but also "name" and record the patterns and can record a trace of the reasoning leading to the discovery. When evaluating the usefulness of new sensors, or combinations of sensors, and process control strategies, it is helpful to have a record of discovered states and control reasoning. Neural networks, which are also excellent for pattern recognition, do not name or explain the patterns automatically. Discovery systems can also be used to recognize causes and effects which are separated in time, such as the transfer of heat with a large thermal lag, a common problem in batch processing.

No discovery system, however, had been used to learn during interactive control of a dynamic process. In-process control requires the ability to act on incomplete knowledge. Many of the discovery algorithms also consume considerable computational time, making them unlikely candidates for real-time control. One commercial discovery system, The Scholar's Companion, (TSC) was found which had some of the necessary attributes for process control, but many of the algorithms in PETE were developed during this project.

2.2 Assumptions of the Model There were actually two models used in this project. One was a simple finite difference model of the autoclave cure of a thermosetting composite. This model was used as a simulation of the autoclave. Details of the simulator can be found in Springer and Loos (4). The other model was the qualitative model, the envisionment, which was the map for the controller. These two models had to correspond well enough so that the parsed data would make sense when the controller attempted to match process data to the envisionment.

The physical situation being modeled by both is sketched in Figure 1. The temperature of the autoclave, AUTO.TEMP, is controlled by, and identical to, the SETPOINT. The laminate temperatures are measured at the top surface of the laminate, TOP.TEMP and in the middle of the laminate, MID.TEMP. The laminate slab is heated on both sides and surrounded by the insulating bag and tool. A calculated dielectric loss factor at the location of the MID.TEMP (DIELECTRIC) is simulated as well.

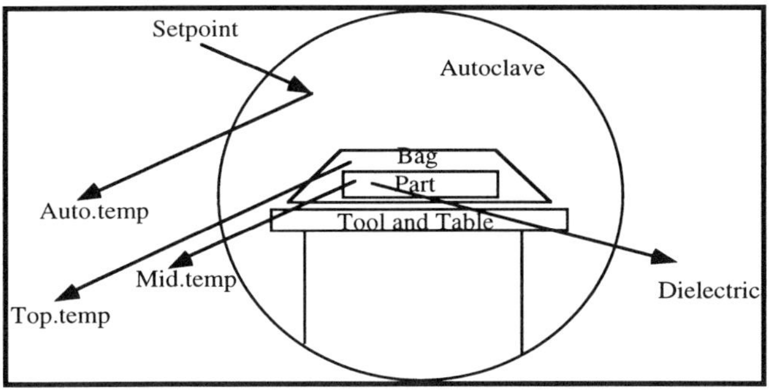

Figure 1. Sensors and control for the simulated autoclave cure. All process states and relationships are derived from the temperature and dielectric sensors, and the setpoint is used to change the AUTO.TEMP.

The equation for unsteady state heat transfer in a slab, including a term for heat generation by the exothermic reaction of the thermosetting polymer, symbolized as the product of a constant heat of reaction (H_R) and the reaction rate (the change in degree of cure with time, $d\alpha / dt$), is shown in Equation 1. This equation describes the temperature change at MID.TEMP.

$$\rho C_p \frac{\partial T}{\partial t} = -\frac{\partial}{\partial x}(-k\frac{\partial T}{\partial x}) + H_R \frac{d\alpha}{dt} \qquad [1]$$

At the surface of the laminate (TOP.TEMP) the heat flux is modeled by Equation 2.

$$q = -k\frac{\partial T}{\partial x} \qquad [2]$$

The qualitative inequalities for the same situation are greatly simplified by the assumption that thermal diffusivity (k), heat capacity (C_p) and density (ρ) are relatively constant, and by the assumption that the laminate and layup are symmetric so that the temperature at the bottom of the laminate is the same as that at the top. The corresponding qualitative equations are shown in Equations 3 and 4, where [x] is the qualitative value of x. Qualitative values are either zero, positive or negative (15).

$$[\Delta\text{MID.TEMP}] = [\text{TOP.TEMP}] - [\text{MID.TEMP}] + [\text{REACTION HEAT}] \quad [3]$$

$$[\Delta\text{TOP-TEMP}] = [\text{Q}] = [\text{AUTO.TEMP}] - [\text{TOP.TEMP}] + [\text{REACTION HEAT}] \quad [4]$$

In the envisionment used for this work, the REACTION HEAT term was purposely omitted to determine the ability of the system to "discover" an unexpected behavior. The differences TOP.TEMP - MID.TEMP and AUTO.TEMP - TOP.TEMP, since they are drivers for heat transfer, are treated as variables and named TOP-MID and AUTO-TOP, respectively, reducing Equations 3 and 4 to Equations 5 and 6.

$$[\Delta\text{MID.TEMP}] = [\text{TOP-MID}] \quad [5]$$

$$[\Delta\text{TOP-TEMP}] = [\text{Q}] = [\text{AUTO–TOP}] \quad [6]$$

Another assumption made in both models was that AUTO.TEMP followed SETPOINT exactly. In a real process, there would be a time lag between SETPOINT and AUTO.TEMP but it would be significantly smaller than the other temperature gradients in the system in a working autoclave.

2.3 The Envisionment. The envisionment is a means of representing the possible qualitative states of the process. These states of the process, called episodes, are "snapshots" which, when sequenced represent the dynamic process much as a motion picture is made up of a series of photographs. A typical episode is shown in Figure 2. The first list of statements describe the relationships (greater.than, less.than and about.equal) between the measured variables and a set of landmark temperatures, ROOM.TEMP, FLOW.TEMP, CURE.TEMP, and MAX.TEMP. There are also about.equal statements to indicate when the differences between temperatures are small, i.e., MID.TEMP about.equal TOP.TEMP. The states list contains information about directions of rate and differences. At the bottom is a control statement which describes the control action which causes this episode to succeed the previous episode and a list of the PREV.EPISODE(s) and of the possible NEXT.EPISODE(s).

```
EP_2
   RELATES      ( ( MID.TEMP about.equal TOP.TEMP )
                ( SETPOINT less.than FLOW.TEMP )
                ( SETPOINT greater.than ROOM.TEMP )
                ( AUTO.TEMP greater.than ROOM.TEMP )
                ( AUTO.TEMP less.than FLOW.TEMP )
                ( AUTO.TEMP about.equal ROOM.TEMP )
                ( SETPOINT about.equal ROOM.TEMP )
                ( TOP.TEMP about.equal ROOM.TEMP )
                ( MID.TEMP about.equal ROOM.TEMP ) )
   STATES       ( ( SETPOINT increasing ) ( AUTO.TEMP increasing )
                ( TOP.TEMP steady ) ( MID.TEMP steady )
                ( AUTO-TOP positive ) ( AUTO-TOP increasing )
                ( TOP-MID zero ) ( TOP-MID steady )
                ( DIELECTRIC steady ) )
   CONTROL      ( SETPOINT up )
   PREV.EPISODE       EXPERIMENT.E1
   NEXT.EPISODE       EP_6 EP_7 EP_8
```

Figure 2. EP_2, an example of an episode from an envisionment.
This syntax is a simplification of the actual code.

Transitions from one episode to another occur when one of the statements in an episode change. Considering EP_2 in Figure 2, there are at least three possible next episodes. If the setpoint is changed to decreasing or steady the state of SETPOINT and AUTO.TEMP will change to decreasing or steady. If the SETPOINT continues going up, it (and AUTO.TEMP) will exceed the range delineated as about.equal to ROOM.TEMP and that statement will be changed to false. Each new episode creates a new branch. Some episodes may have many branches in theory. In actual runs, the number of branches is usually more limited. Eventually the process terminates, either with successful cure of the composite (goal.episode) or an unsatisfactory state (bad.episode). A list of sequential episodes from the starting conditions to a goal episode is a path for the controller to use as a plan.

In the simple envisionment of Figure 3, for instance, one path is to go from EXPERIMENT.E1 (the starting episode) to EP_2, to EP_5, ending at the goal.episode EP_12. As it happens, this is also the shortest path through this particular envisionment. The envisionment in Figure 3 is not a complete list of possible transitions by any means. It is a seed envisionment containing some of the more important episodes and transitions. It was used as a basis for the data-driven development of a more complete and accurate envisionment by PETE. In the TSC browser there are two rectangles for EP_12,. The second one is shown in dotted lines to indicate that it is a duplicate. There is no difference between the two episodes but if every episode were shown only once, the number of line crossings could make things very confusing in some of the more complex envisionments.

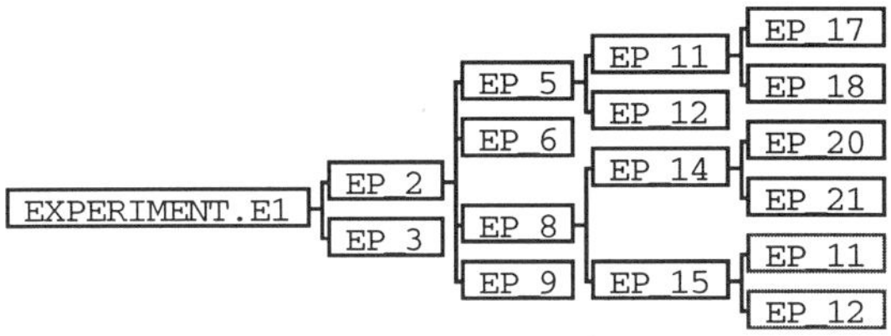

Figure 3. TEE as it is seen in the TSC format.
An episode that can follow two different episodes, such as EP_12, is shown as occurring twice, even though there is really only one EP_12.

2.4 How PETE Works At the beginning of the run, PETE selects the shortest path to one of the goal episodes. It then attempts to follow that path by executing the control instructions required to get to the next desired episode, determining if the expected episode has been reached and deciding on the next step. The path chosen at the outset of a process run may not be the one on which the controller remains throughout the process. Just as there are crossroads in a travel map where more than one path converge, so there are crossroads in envisionments. For each cycle of the run PETE: 1) acquires and encodes the data from the simulator; 2) finds or creates an episode that explains the current combined conditions of the sensors (current affairs); and 3) decides upon and executes a control action.

Current and past data are used to parse the data coming from the process (simulated in this case) into a frame called CURRENT.AFFAIRS. An example of a CURRENT.AFFAIRS frame is shown in Figure 4. The data from the previous cycle is stored in a similar frame called PAST.AFFAIRS. In this manner, TSC keeps track of at least two cycles of data at all times. Both CURRENT.AFFAIRS and PAST.AFFAIRS look like episodes, except for the extra slots shown in bold type in Figure 4.

The **current.data** and **previous.data** slots are lists of the values for the variables named in **data.definition**. **Current.path** is the path that the controller is currently attempting to follow. The **control** slot holds the most recent decision of the control functions. In Figure 4, the entry in the control slot is "up," indicating that the setpoint was increased in the previous cycle. **Found.episode** is the episode that PETE has found to explain the data. A complete list of these episodes is kept in **my.founds**, starting with the most recent episode found and ending with the first. **Index.episode** keeps track of the last episode that was on a desirable path. In this case, it is the same as the found.episode because the process is on the current.path. Following is a brief explanation of how PETE finds the entries to these slots and uses them in control.

```
CURRENT.AFFAIRS
   CURRENT.DATA        ( 90.00 80.00 80.00 2865 10.00 0.00 90 1 )
   PREVIOUS.DATA       ( 80.00 80.00 80.00 2865 0.00 0.00 80 0 )
   DATA.DEFINITION     ( AUTO.TEMP TOP.TEMP MID.TEMP DIELECTRIC
                         AUTO-TOP TOP-MID SETPOINT TIME )
   CURRENT.PATH        ( EXPERIMENT.E1 EP_2 EP_5 EP_12 )
   CONTROL             UP
   FOUND.EPISODE       EP_2
   MY.FOUNDS           EP_2   EXPERIMENT.E1
   INDEX.EPISODE       EP_2
   RELATES           ( ( AUTO.TEMP about.equal ROOM.TEMP )
                       ( SETPOINT about.equal ROOM.TEMP )
                       ( TOP.TEMP about.equal ROOM.TEMP )
                       ( MID.TEMP about.equal ROOM.TEMP )
                       ( SETPOINT greater.than ROOM.TEMP )
                       ( AUTO.TEMP greater.than ROOM.TEMP )
                       ( TOP.TEMP greater.than ROOM.TEMP ) )
   STATES            ( ( AUTO.TEMP increasing ) ( SETPOINT increasing )
                       ( TOP.TEMP increasing ) ( MID.TEMP steady )
                       ( AUTO-TOP positive ) ( TOP-MID positive )
                       ( AUTO-TOP increasing ) ( TOP-MID increasing )
                       ( DIELECTRIC steady ) )
```

Figure 4. The CURRENT.AFFAIRS frame.

PETE has several options in trying to match the statements of CURRENT.AFFAIRS to the envisionment. First, the match could be to some episode that is already connected to the past episode in the envisionment, an expected episode. Second, the new match could be to an episode that is in the envisionment but not connected directly to the past episode found. Third, there may be no match between the data and the envisionment. To shorten the time required for explanation of the current state, PETE searches the envisionment in this order, taking the smallest set of most likely episodes first and expanding the search only when necessary. If no match is found in the envisionment, a new episode is conjectured and added to the envisionment. If an episode is found to follow one which it was not previously known to follow, a new link is conjectured. This process of combined search and explanation is diagrammed in Figure 5.

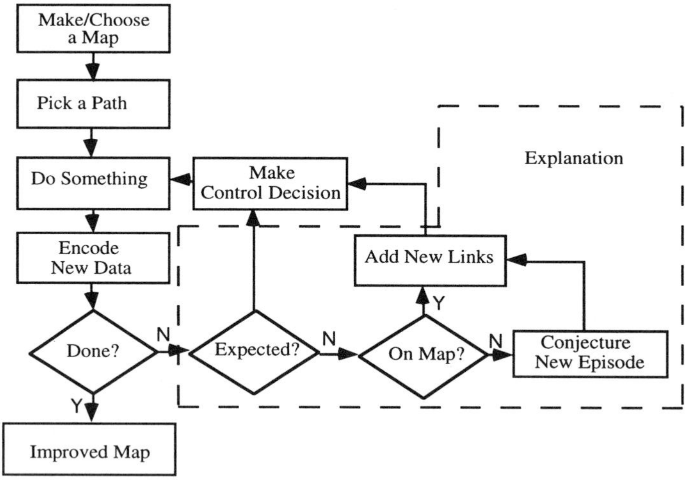

Figure 5. The search and explain behavior of PETE.
The episode found, or conjectured, to match CURRENT.AFFAIRS, will be the new found.episode.

The most important thing to remember about the control behavior of PETE is that control decisions are based on what the previous episode was and what kind of path the episode is on rather than on the control information that is in the found.episode. In addition, the operating limits of the equipment are included in the control functions, especially when the current.affairs do not match a known episode.

In general, episodes on desirable paths trigger control actions found in the episode following them in the path; episodes which are not on desirable paths cause a reversal of the control behavior; and conjectured episodes maintain the prior control strategy. Because of this, a conjectured episode may cause different control behavior at different times during a run. Repeated episodes continue the same control strategy until it is proven ineffective, at which time they may cause a change in strategy. In the interest of safety however, there are limits on the setpoint and experimentation is approached with caution. There are stricter limits on the setpoint for conjectured episodes and, if a sequence of conjectured episodes occurs, there is a backup cure strategy based on the relationship between the setpoint and the setpoint limits.

The variety of responses possible allows PETE to do some experimentation, especially when the envisionment and the process do not match. This experimentation, although it seldom produces an optimum cure cycle on the first run, quickly expands the envisionment so that subsequent trials have a more complete envisionment and respond differently.

3. RESULTS

PETE learns in two different ways. During the run it automatically adds conjectured episodes and links to the envisionment. These new additions are recognized when they are encountered again and may alter the behavior of the envisionment, but they are not treated the same as the episodes which are part of the original envisionment. Between runs, however, these conjectured links and episodes may be permanently added to the envisionment. When this happens, a totally new envisionment has been created. This envisionment can then be used on subsequent runs.

The runs which will be discussed below were on a simulated 64-ply graphite epoxy laminate. The laminate was considered acceptable if the temperatures stayed below 204°C (400°F) or MAX.TEMP. Time cycles were 30 simulated seconds. The first run with the small envisionment of Figure 3 is shown in Figure 6. There are some large oscillations in this run which are typical of what happens when PETE is "lost" and conjecturing mostly new episodes. The middle of the laminate exceeds the 204°C (400°F) mark briefly and the cure takes about 220 minutes.

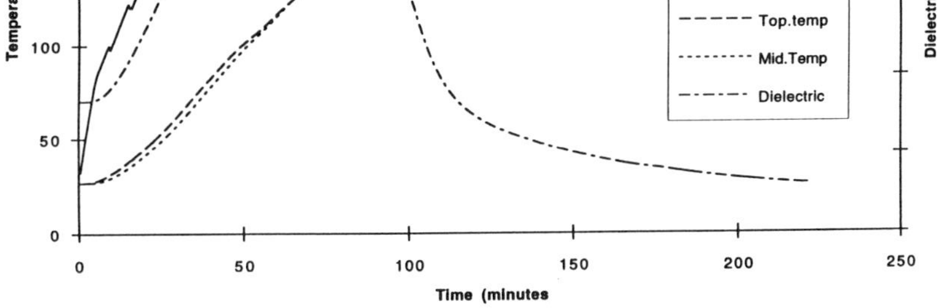

Figure 6. First Run on TEE.

Fairly large "seeking" oscillations around the cure temperature are a symptom of the immaturity of the envisionment.

3.1 Real-Time Learning: During the run, two types of learning take place: new episodes are conjectured; and new links are formed between episodes. In the envisionment of Fig. 3, many possible episodes were purposely left out to see if PETE would "discover" them. These included any episode which might give an indication of the heat of reaction. Conjectured episode CON_69, first encountered at 65 minutes into the cure run, is an episode in which the MID.TEMP is increasing even though [TOP–MID] is negative. In order for Equation 5 to match the situation in CON_70 an additional positive term, represented as the unknown qualitative heat flux [Q2], must be added to the left side. This additional heat source is shown in Equation 7. Equation 7 is actually equivalent to Equation 3, although the present system would not recognize the connection between this unknown Q2 and the REACTION.HEAT of Equation 3.

$$[\Delta\text{MID.TEMP}] = [\text{TOP–MID}].+ [\text{Q2}] \qquad [7]$$

PETE is not clever enough to "understand" the meaning of this discrepancy between the envisionment and the process and conjectures the new episode. However, the user could recognize the conflict and PETE recognizes that it is not present in any episode in the known envisionment. Even earlier, at 37 minutes into the cure run an episode is conjectured (CON_43) in which TOP-MID is decreasing, or ΔTOP-MID is negative. CON_43 is a more ambiguous sign of an exotherm. A sudden decrease in the rate of TOP.TEMP has the same effect as a sudden increase in the rate of MID.TEMP, but, given the immediate past history, in which no such decrease in TOP.TEMP occurred, there is indication of exotherm in this episode. Eventually, starting with the conjectured episode CON_135, there is a series of conjectured episodes in which the MID.TEMP exceeds the MAX.TEMP (204°C). Since the autoclave, the only known heat source in the original envisionment, never gets that hot, there is clearly another heat source at work here, more evidence for [Q2].

Another "discovery" that PETE made was that the temperature gradient between the autoclave temperature and the top temperature was much larger than the gradient within the part. This could be used to infer the significant difference between the thermal transfer coefficients at the boundary and within the laminate.

About 100 episodes were added to TEE in the first run. The new envisionment is shown in Figure 7. The shaded episodes were in the original envisionment. All others were "learned" during the run. Episodes enclosed in dotted lines are duplicate episodes. The complexity of the envisionment makes it difficult to capture all the relationships between episodes.

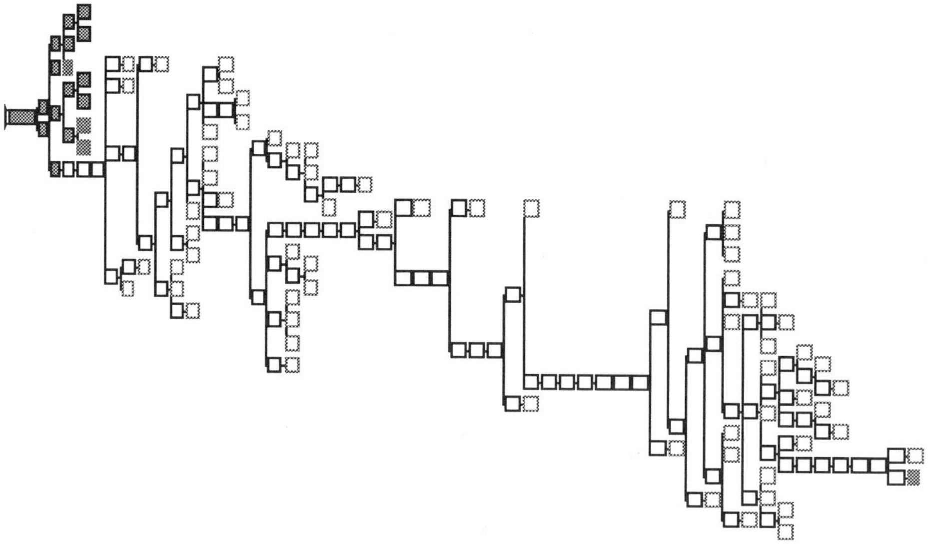

Figure 7. TEE.A, the envisionment after TEE-1.
Only the shaded episodes were in the original envisionment.

Subsequent runs added fewer episodes. Although all episodes conjectured by PETE were representative of actual process events and so essentially correct, there are some limitations to the current approach of deriving the episodes from the data. One of these is that the episodes conjecture are very specific. Another is that, because only two cycles of information are represented, delayed responses to control inputs are not captured very well.

The specificity of episodes may result in a series of conjectured episodes with minor variations. An example is the following sequence from the latter part of TEE-1: (.... CON_223 CON_59 CON_38 CON_223 CON_59 CON_223 CON_38 CON_223...). The only difference between CON_223, CON_59 and CON_38 is whether TOP–MID is increasing, decreasing or steady. What is happening is that TOP-MID is nearly zero but is oscillating slightly. One more general episode could cover all three cases.

In summary, the episodes discovered by PETE include episodes which reflect the exotherm and the thermal gradients, although the explanation is not very deep. The number of episodes added decreases sharply with subsequent runs, approaching a number of episodes that is considerably smaller than would be necessary to describe all possibilities. The episodes created by TSC are highly specific and can be used to improve the encoders used to derive information from the data.

Discovery of conjectured episodes is only half of the learning process. PETE must also decide what to do about these episodes. Control is determined by the links which are established between the conjectured episodes and the rest of the envisionment. Each time a conjectured episode is encountered during a run, the control response that PETE takes is an experiment. The results of those experiments become the next episodes for the conjectured episode. The control response depends on the previous episode and that episode determines what the SETPOINT is doing. This makes control more context sensitive for conjectured episodes than for regular episodes. Experimentation is a significant factor in all discovery, human or machine. New paths, the key to PETE's ability to learn from run to run, are formed when the combined results of all the discoveries within the run are converted to named episodes. The results of these experiments are then incorporated into a new envisionment after the run, creating new paths which may guide the control.

3.2 Run to Run Learning. At the completion of a run, the new envisionment is saved to the log. Before reusing, the conjectured episodes are renamed. PETE always uses the same names for conjectured episodes (CON_xx) so, if the names are not changed, the episodes will be written over. This new envisionment can be used as is, or as altered by the user. In either case, the behavior of the system may change radically.

The cure resulting from the new envisionment shown in Fig. 7 is shown in Figure 8. The SETPOINT went up much faster in the second run than it did in the first because the map now includes the episodes for the heat up part of the cycle. In the first run, most of these episodes were conjectured which caused PETE to slow the rate of increase and to occasionally decrease the setpoint when an episode was repeated too many times in a row.

The sharp peaks seen in Figure 8 between 10 and 30 minutes were from one time excursions to above the cure temperature when new episodes were conjectured or when bad episodes were encountered, but there were no long period oscillations like those seen in Figure 6 where conjectured episodes made up most of the run.

The most noticeable difference between the Figures 6 and 8 is the large decrease in setpoint which starts about 50 minutes in Figure 8. This is initiated when PETE finds FIRST_43, an episode in which the two laminate temperatures, TOP.LAM and MID.LAM are greater than flow temp and increasing; and the difference between them, TOP-MID, is decreasing. This is an episode very similar to one of the early states used to recognize the exotherm with QPA. It is also an episode which is on an undesirable path and so PETE turns the heat down instead of increasing the setpoint. A series of conjectured episodes after that keeps the temperature on the decrease.

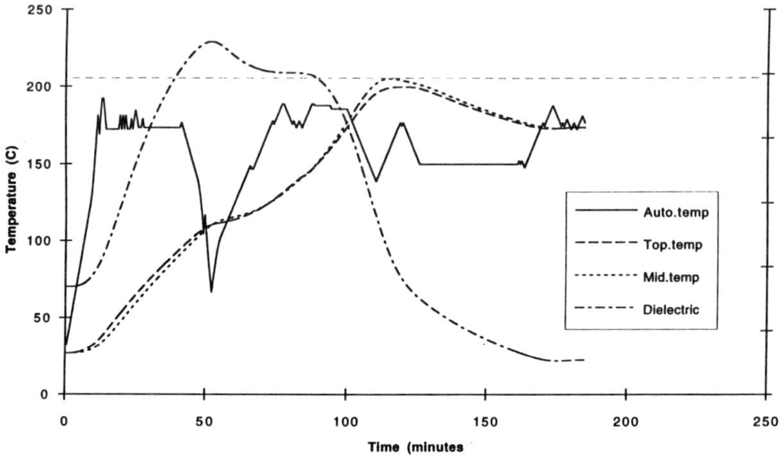

Figure 8. First Run of TEE.A (TEE.A-1).
The cure time was faster than TEE-1 and part temperatures did not exceed the maximum laminate temperature.

3.3 Problems with PETE. Unfortunately, not all of the paths found by PETE are improvements. This is because of the conservative way PETE treats some conjectured episodes. While it is desirable to avoid harming the material while controlling the process, it is also necessary to move forwards towards completion. The run in Figure 9 resulted from the use of another starting envisionment . It is important to note the time scale on Figure 9. The holds early in the run, although they reduced the exotherm, made for a very long cure cycle.

Currently, the path is "optimized" by choosing the path with the fewest number of episodes. This does not count repeats of a given episode, however. In QPE.A, one of the early episodes in the path caused a temperature hold. Until conditions changed enough for a new episode to be conjectured, the setpoint remained at that temperature, with brief excursions whenever an episode was repeated too long. Consequently, although the path was short, the cycle was very long.

Interestingly, the cure cycle in Figure 9 resembles some commercial cycles for thick laminates. It is not unusual, when a manufacturer is having problems with production of composite parts, to add another hold at low temperatures or to decrease the heating and/or cooling rates of the process. Yet it has been demonstrated that this is often exactly the wrong sort of adjustment, in terms of both exotherm control and processing time. This run provided the most dramatic demonstration that the shorter path (number of episodes) did not necessarily lead to a shorter run. The engineer, however, has full access to the envisionment evolved from a run, however, and can easily alter it by changing episodes or eliminating inappropriate links. After some minor changes to links, QPE could be made to behave as shown in Figure 10. This user modified envisionment demonstrates the power of PETE as an engineering aid for process development.

4. SUMMARY AND CONCLUSIONS

PETE was demonstrated on a simulator and shown to learn both during and between runs while interactively controlling the process. During the run PETE actively experimented with the process, learned new links between episodes, and added new conjectured episodes. These new episodes included the exothermic behavior of the material missing in the original envisionments.

At the end of each run, a new envisionment was created, incorporating the episodes and links formed in the run. The number of episodes added was large for the first run but tapered off quickly with succeeding runs. The information in the new envisionments is available to the user and the derivation of each episode as well as the path of the process can be explained using the data in the log. The algorithm for cycle "optimization," however, needs some work. PETE improves its performance based on the shorter path, but that criteria has shortcomings.

The user can access any of the episodes or links learned by PETE and change them if desired. This allows the engineer to use PETE as an active observer during the process and analyze the results when convenient.

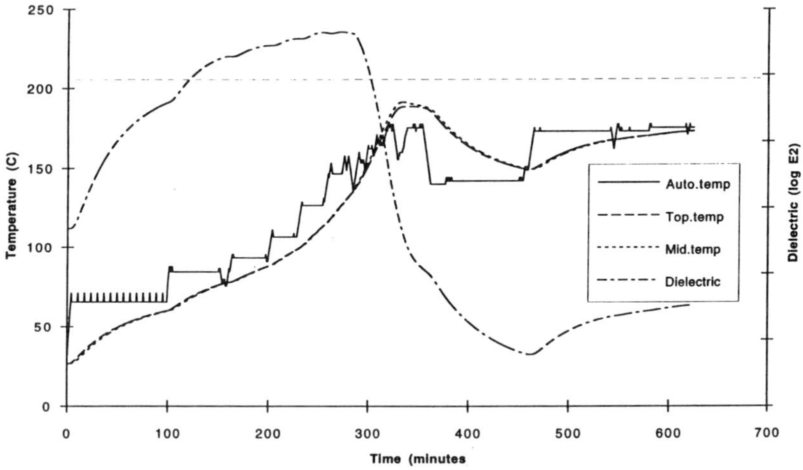

Figure 9. QPE.A-1, first run with QPE.A.
This run was extended by a path that had a hold episode early in the cycle.

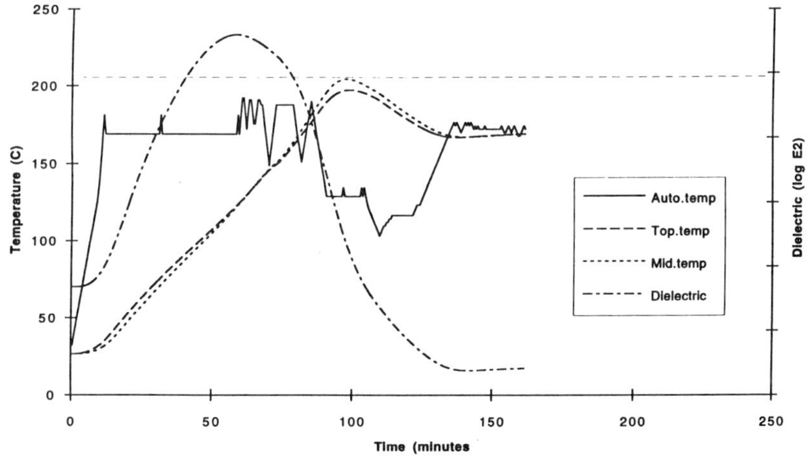

Figure 10. QPE-MOD-1 on a 64-ply laminate.
This envisionment used for the run in Figure 9 was altered by the engineer for much improved results.

5. FUTURE ACTIVITIES

The first use for PETE will probably be as an observer to detect patterns and sequences of patterns in new sensors and sensor combinations. This may be done on real processes when there is little danger of harm to the laminate from PETE's experimentation. It is also proposed to operate on some improved simulators.

The criteria by which PETE evaluates performance should be more closely tied to laminate quality rather than merely on shortening the path, or even the cure time. Eventually PETE should be able to evaluate performance on the basis of multiple criteria.

Additional knowledge can be incorporated into PETE. Arranged in levels, some knowledge could be accessed by PETE that might aid it in making a more intelligent choice of control options when new episodes were discovered. Initially, this knowledge will probably take the form of rules, although it could also be in the form of a larger, more complete envisionment.

Statistics on the number of hits for links and episodes can easily be added to the envisionment, making it possible for PETE to "forget" those links and episodes that are never found and increasing the expectation of links and episodes more often encountered.

6. REFERENCES

1. A. Ghosh, "Why Batch Process Control is not Chemical Engineering," Presentation at AIChE Annual Meeting, Los Angeles CA, (1991).

2. Y. Pao, S.M. Phillips, and D.J. Sobajic, International Journal of Control, 56, (2), 263 (1992).

3. S.R. LeClair and F.L. Abrams, Intl. J. of Computer Integrated Manufacturing, 2 (4) 205 (1989).

4. A.C. Loos and G.S. Springer, J. of Composite Materials, 17 (2) 135 (1983).

5. J.D. Hajicek, The Automotive Challenge and Plastics Response, Society of Plastics Engineers, 120, 1987.

6. R. Warnock and S.R. LeClair, "Qualitative Process Automation for Autoclave Curing of Composites," WL-TR-92-4085, May 1992.

7. R.Y. Kim, F.L. Abrams, and M. Knight, Proceedings of the American Society for Composites, Technomic Pub Co, Lancaster PA, 711 (1988).

8. B.K. Fussell, QPA Control of the End Milling Process, Report on Air Force Office of Scientific Research Contract F49620-87-R-0004, 1988.

9. C.W. Lee and F.L. Abrams, Proceedings of the Fifth Japan-US Conference on Composite Materials, 19907.

10. P. Langley, H.A. Smith, G.L. Bradshaw, and J.M. Zytkow, Scientific Discovery, MIT Press, Boston MA, 1987.

11. J.M. Zytkow, J. Zhu, and A. Hussam, Proceedings of the 8th Conference of the AAAI, MIT Press, Cambridge MA, pp 889-894, 1990.

12. P. Thagard, Computational Philosophy of Science, MIT Press, Cambridge MA, 1988.

13. D.B. Lenat and R.V. Guha, Building Large Knowledge-Based System-Representation and Inference in the Cyc Project, Addison-Wesley Publishing Company, Inc, 1989.

14. B.G. Buchanan and E.A. Feigenbaum, Artificial Intelligence, 11 (5) 24 (1978).

15. J. de Kleer and D.G. Bobrow, Proceedings of AAAI-84, MIT Press, Cambridge MA, pp 86-91 (1984).

7. ACKNOWLEDGMENTS

This work would not have been possible without the assistance of Jack Park of Thinkalong Software who helped me learn to use and program TSC and Dr. C. William Lee of the University of Dayton who provided the simulator used and served an eight-year sentence as my advisor. Dr. Robert Trelease of the University of California and Dr. Steven LeClair of Wright Laboratory also provided considerable encouragement and support.

28th International SAMPE Technical Conference
November 4-7, 1996

NEW DEVELOPMENTS IN THE CURE MONITORING OF THERMOSETTING RESINS AND COMPOSITES UTILIZING ULTRASONIC MEASUREMENTS

David D. Shepard and Kim R. Smith
Micromet Instruments, Inc.
7 Wells Avenue
Newton Centre, MA 02159

ABSTRACT

Measurements of the ultrasonic sound speed of thermosetting resins and composites can be used as an in-process cure monitoring technique. Ultrasonic measurements have an advantage over other in-process techniques in that ultrasonic sensors do not make contact with the part (thus leaving no imbedded sensor or witness mark) and can make true bulk measurements of the part. A new commercially available ultrasonic cure monitoring system has been developed which easily enables ultrasonic measurements to be made in compression molding, resin transfer molding, and autoclave processes. Advancements in ultrasonic sensor technology enable the sensor to maintain good coupling to the part during thermal cycling to 260°C. Data is presented showing the change in ultrasonic sound speed during the compression molding of a graphite-epoxy prepreg. The data shows a good inverse relationship to the ionic conductivity data collected via dielectric cure monitoring.

KEY WORDS: Cure Monitoring, Ultrasonic, Thermoset Resins/Composites

1. INTRODUCTION

Measurements of the ultrasonic sound speed of thermosetting resins and composites can be correlated to the material's viscosity and cure state. The material under test will change state as it undergoes cure. This change of state can be monitored by measuring the velocity of acoustic waves in the material. Previously, use of ultrasonic measurements as a cure monitoring technique had been hindered by poor reliability of ultrasonic transducers under typical processing conditions, especially thermal cycling at elevated temperatures. A new in-mold, reusable ultrasonic sensor has been developed. This robust sensor is capable of withstanding hundreds of thermal cycles up to 260°C. Ultrasonic sound speed measurements may now be routinely made during the processing of thermosets using this new sensor technology. An electronics instrumentation and software package has also been developed for the specific application of making a complete ultrasonic cure monitoring system.

The new ultrasonic measurement technology was developed under the National Center for Manufacturing Sciences (NCMS) Adaptive Process Control of Compression Molded Composites Project. The project team members consisted of United Technologies, Pratt & Whitney Aircraft; Erie Press Systems; and Allen Bradley, a Rockwell Automation business. Upon successful development and demonstration of the ultrasonic cure monitoring system in a compression molding process by the project team, Micromet Instruments, Inc. became a subcontractor for the project to transfer the technology into a commercially viable sensor and instrumentation system. Micromet Instruments has acquired a license for the ultrasonic sensor system from NCMS and is currently marketing the product it has developed, the UCMS-200 Ultrasonic Cure Monitoring System. Pratt & Whitney and Sikorsky Aircraft are currently evaluating the system for use in controlling the molding of composite parts.

The ultrasonic sensor is located within the mold and does not contact the part. Once the sensor is installed, hundreds of runs can be made with no incremental sensor cost and no lay up of sensors in each part. The ultrasonic technique is ideal for applications where there can be no sensor imbedded in the part or no witness mark on the outside of the part. The system makes a through-transmission measurement which measures the bulk properties of the part, even on thick parts of over 5 cm (2 inches) in thickness. The reusable, non-intrusive nature of the sensor, along with the ability to measure bulk material properties of thick parts, provide unique features over other in-process cure monitoring techniques such as dielectric or fiber optic.

2. ULTRASONIC MEASUREMENT TECHNIQUE

The ultrasonic measurement system uses piezoelectric elements to generate an acoustic wave. High voltage pulses excite the piezoelectric element in the ultrasonic transducer, causing this element to oscillate at 5 MHz. This creates the acoustic wave that propagates at a material specific velocity. Whenever the wave reaches a boundary some of the wave is transmitted through the boundary into the second material, and some of the wave is reflected back. The transmitted and reflected waves are then collected by the ultrasonic transducers, which also act as ultrasonic sensors. When an acoustic wave reaches the piezoelectric element, the element produces a voltage which is the sensor signal. Measurements of sound velocity by through-transmission require two in-line ultrasonic transducers/sensors. One is used to generate the acoustic pulse while the opposing transducer/sensor records the arrival time of the acoustic wave. The excitation and detection function is alternated between two transducers. The electronics allow for multiple signal averaging thereby creating clean signals. The distance between the opposing sensors is monitored by the system in order to provide accurate distance measurements. The ultrasonic sound speed is then calculated from the time and distance information.

Sensor and sample geometry are critical to this measurement technique. The wave will change direction at a boundary if the wave is not perpendicular to the boundary. Therefore the molds must be carefully machined to insure the required perpendicularity. Holes for the sensors should be 2.54 cm in diameter and bored to within approximately 0.6 cm of the mold inner surface. The bottom surface of the hole must be very smooth to insure good coupling between the piezoelectric element of the sensor and the mold. Silicone grease is used to eliminate any air between these mating surfaces, resulting in good coupling.

3. ULTRASONIC MEASUREMENT SYSTEM

The UCMS-200 is a complete stand-alone system consisting of sensors and cabling, a rack-mountable electronics package, and a Windows based LabVIEW software package. A computer is integrated into the electronics package complete with monitor, keyboard, mouse, and disk drives. Figure 1 shows a schematic of the ultrasonic measurement system installed in a compression molding process. An ultrasonic sensor set consists of two sensors located on opposite sides of the mold. For autoclave applications, an ultrasonic sensor is mounted in the tool and the second sensor is mounted in a cull plate which is placed on the sample, beneath the vacuum bag.

The system can measure the ultrasonic sound speed at two different locations using two sets of ultrasonic sensors. Part thickness is monitored using up to four capacitive non-contact displacement probes or linear voltage displacement transducers (LVDT). Accurate measurement of the part thickness is important in accurately calculating the sound speed, especially in processes where the part thickness changes during the cure cycle. Temperature can be measured from two thermocouples.

The LabVIEW based software package enables easy operation of the system and allows the display screen to be customized by the user. Algorithms can be easily written by the user to perform special functions and closed loop control of the process.

4. EXPERIMENTAL RESULTS

The change in sound speed has been correlated with the change in the viscosity of the prepreg (1). Heating the prepreg decreases the material's viscosity and the acoustic wave velocity is decreased. As the material cures, the molecular structure is constrained and supports a higher acoustic wave velocity. The change in ultrasonic sound speed of an epoxy-graphite fiber prepreg during a compression molding process is plotted in Figure 2. The initial decrease in sound speed shows the decrease in viscosity as the prepreg increases in temperature. A broad minimum is seen as the temperature cycle enters a

121°C hold period. As the temperature is increased to 177 °C, the sound speed increases as curing causes the viscosity of the prepreg to increase. The rate of increase in sound speed then slows as the rate of cure slows and the reaction nears completion.

Dielectric cure monitoring is a widely used, in-process cure monitoring technique involving the measurement of changes in the electrical conductivity of the material. Dielectric measurements have been shown to correlate with the change in viscosity and cure state of epoxy and other thermosetting resins and composites (2,3,4). Dielectric cure monitoring data thus provide a good comparator to the validity of the ultrasonic sound speed in the cure monitoring. Figure 3 compares the ionic conductivity as measured by a Micromet Instruments Eumetric System II Microdielectrometer to the ultrasonic sound speed of an epoxy-graphite fiber prepreg (1). The ultrasonic sound speed exhibits a good inverse relationship to the log conductivity data. This inverse relationship is expected since ionic motion is reduced with higher cross-linking of the material, whereas the velocity of the acoustic wave is increased.

5. FUTURE WORK

Work is currently under way to redesign the ultrasonic sensor so that it may be more easily installed in the mold by the user. In the current sensor design, the sensor must be assembled by individually installing the components into the sensor cavity in the mold. The sensor must then be tuned by tightening the sensor until a good pulse/echo response is obtained. The re-designed sensor will be of a "Plug-In" design where the sensor components will be pre-installed within the sensor housing and tuned. The sensor will then be screwed into the sensor cavity and tightened to a preset level using a torque wrench.

Future applications work will concentrate on resin systems where dielectric cure monitoring does not provide useful data, such as phenolic resins and various rubber materials. Also to be studied will be the sensitivity of the ultrasonic sound speed to the end of cure.

6. CONCLUSIONS

A new commercially available ultrasonic cure monitoring system has been developed which enables ultrasonic measurements to be easily made during the processing of thermosets in presses, autoclaves, and RTM processes. New ultrasonic sensors maintain good coupling to the mold during thermal cycling to 260°C. The non-intrusive ultrasonic sensors do not contact the part and thus leave no mark on the part or sensor embedded in the part. The ultrasonic measurements provide a bulk measurement which correlates with the change in material viscosity and correlates well with dielectric cure monitoring data. The ultrasonic cure monitoring system can be used for production closed-loop control of thermoset composite processing.

7. REFERENCES

1) W. Veronesi, 41st Sagamore Conference, (1994.)

2) J. Gotro, NATAS Conference, 19, 523 (1990)

3) D.R. Day and D.D. Shepard, NATAS Conference, 16, 52 (1987)

4) D.R. Day, D.D. Shepard and K.J. Craven, International SAMPE Technical Conference, 22, 724 (1990).

8. BIOGRAPHIES

Dave Shepard was graduated from The Pennsylvania State University with a B.S. in chemistry in 1982. He was employed as an Analytical Chemist at Reichhold Chemicals, Inc. until 1986 when he joined Micromet Instruments, Inc. as an Applications Engineer. He is currently the General Manager of Micromet Instruments. He has written several papers and articles on the dielectric analysis of polymers

Kim Smith has many years of experience in materials characterization. He obtained a B.S. in Physics from the University of Massachusetts-Amherst in 1987 and is pursuing a M.S. in Physics at the University of Massachusetts-Lowell. Currently he is a Product Development Engineer at Micromet Instruments, Inc. Previous positions he has held have been Engineer in the Sensor Systems Group at Geo-Centers, Inc.; Electron Microscopist at Industrial and Environmental Analysis (IEA), Inc. and Materials Characterization Engineer at ThermoTrex Corporation.

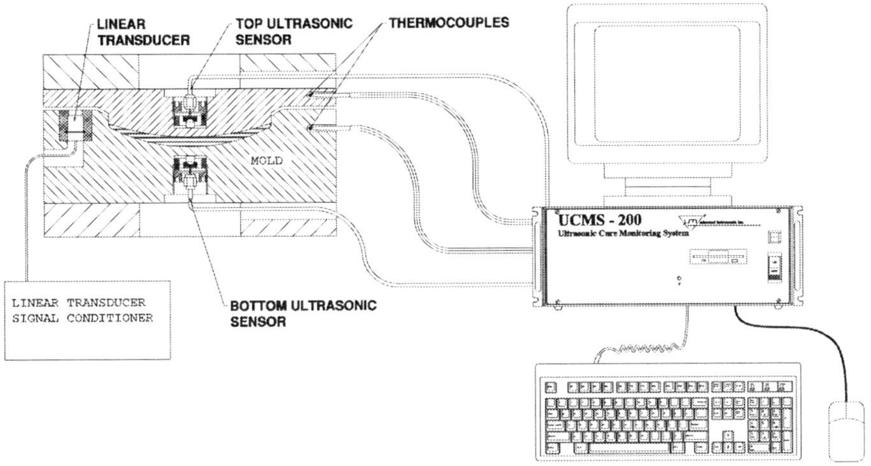

FIGURE 1: A schematic representation of the UCMS-200 installed in a compression molding process.

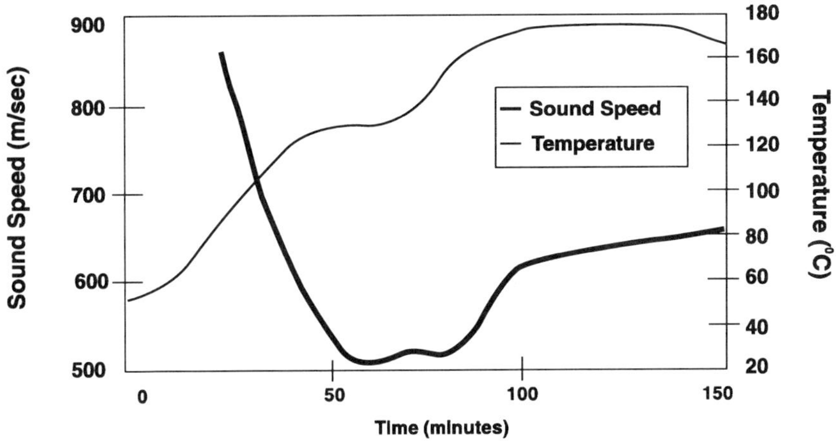

FIGURE 2: The change in ultrasonic sound speed and temperature of an epoxy-graphite fiber prepreg during a compression molding process.

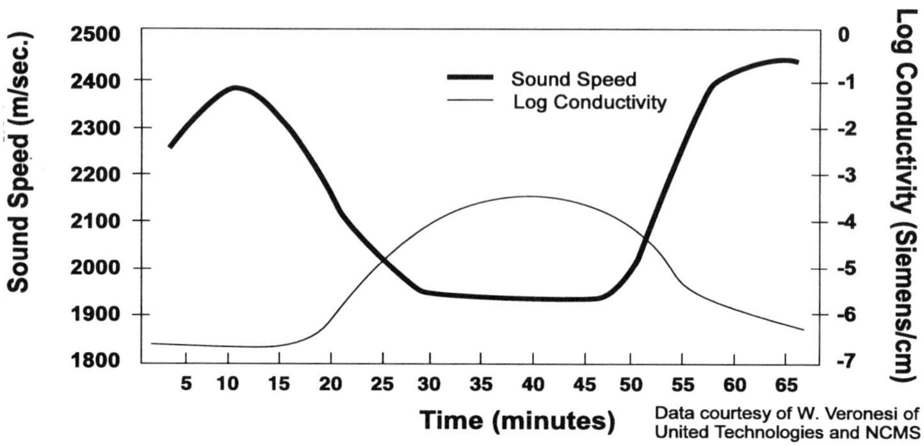

FIGURE 3: A comparison of the ultrasonic sound speed with the ionic conductivity during the compression molding of an epoxy-graphite fiber prepreg.

28th International SAMPE Technical Conference
November 4-7, 1996

MODELING OF EPOXY CURE REACTION RATE BY NEURAL NETWORK

C. William Lee
and
Brian P. Rice

University of Dayton Research Institute
300 College Park Avenue
Dayton, OH

ABSTRACT

For model-based temperature control of many composite materials, it is necessary to have the reaction kinetics model for the exothermic cure reaction. Most analytical models require elaborate data analysis to determine the many parameters contained in the model. A neural network model offers simplicity in construction. In a back-propagation neural network, the analysis is automated. With four sets of DSC data obtained at four different heating rates, a network containing two hidden layers with 5 and 4 neurons, respectively, was successfully trained to predict reaction rates of an epoxy resin undergoing cure. The results are verified against reaction rate data obtained from DSC following a standard cure cycle. The neural network model performed considerably better than the Springer-Loos analytical model in predicting reaction rate and degree of cure.

KEY WORDS: Degree of Cure, Epoxy/Graphite Composites, Neural Network, Reaction Rate

INTRODUCTION

Although the reaction rate of a graphite/epoxy (Hercules AS4/3501-6) composite can be determined on-line through the use of a thermopile sensor [1], it is desirable to be able to predict the reaction rate of the resin. For the case of thick composite parts where heat transfer is limited, detection of acceleration in reaction rate [2] can be too late for a feedback controller to prevent a large temperature overshoot in the part. Being able to predict the reaction rate would allow for feed-forward control of temperature. In addition, from the reaction rate history, one can obtain the degree of cure of the resin which influences the resin viscosity and hence the optimal time for pressure application to consolidate the composite prepreg [3]. This paper presents a back-propagation neural network reaction rate model for 3501-6 epoxy resin. The results predicted by the neural network model are compared to experimentally-measured reaction rate and degree of cure. The performance of the network is also compared to an analytical model developed by Loos and Springer [4].

NEURAL NETWORK MODEL

Only a very brief description of back-propagation neural network (BNN) is given in this paper. Detailed explanations, applications, and references can be found in a recent text by Baughman and Liu [5]. Su et al. [6] developed a neural network model for the same epoxy resin as that used in this study. Their model uses time, temperature, and the Damköhler number as inputs to predict the degree of cure with all data based on the Loos-Springer model.

We will use the network structure employed for this study to briefly outline the workings of a BNN. As shown in Figure 1, the network consists of four layers of neurons or nodes (shown as circles). The first layer is the input layer (layer 0) which contains two neurons. There are two hidden layers (layers 1 and 2) which contain five and four neurons, respectively. The last layer is the output layer (layer 3) which contains a single neuron. In general, there can be an arbitrary number of hidden layers (although usually not more than two) with any number of neurons; however, the simplest structure that produces satisfactory results should be used. Using a large number of neurons is analogous to having a very large number of parameters for a curve fit.

In this study we want the network to predict a reaction rate (output from the network) given the temperature and the degree of cure (inputs to the network) of an epoxy resin. The inputs are rescaled, or mapped, to the range [0, 1]. These scaled values are the activities of the input neurons ($a_{0,j}$, where the first index is the layer number and the second, the neuron number). For each layer except the output layer, there is an additional bias neuron whose activity is always 1 ($a_{i,0} = 1$, for $i = 0$ to $L–2$, where L is the total number of layers). The neurons in the subsequent hidden layer are connected to the input neurons through links as shown in Figure 1. Each link has a weight ($w_{i,j,k}$). Each neuron in the hidden layer receives and sums the weighted activities of all neurons in the previous layer according to

$$s_{i,j} = \sum_{j=1}^{N_i} \sum_{k=0}^{N_{i-1}} \left(a_{i-1,k} \cdot w_{i-1,k,j} \right), \tag{1}$$

where the upper limit of summation, N_i, is the total number of neurons, excluding the bias neuron, in layer i. The activity of this neuron is obtained by transforming this sum by a transfer (or activation) function

$$a_{i,j} = \frac{1}{1+e^{-s_{i,j}}} \tag{2}$$

such that the activity is in the range of [0, 1]. The transfer function in Equation 2 is used for this study although there are other transfer functions. These calculations are performed for all neurons ($j = 0$ to N_i) and propagated through all hidden layers to the output layer ($i = 1$ to $L–1$). Finally, the activities of the output neurons are mapped from a range of [0.1, 0.9] to the corresponding ranges of the output data; in this case, the reaction rate. The choice of [0.1, 0.9], rather than [0, 1], for mapping yields better results [7].

In neural network modeling the objective is to find the link weights that will predict the correct output given the input data. This is done by "training" the net with known input and output data. The known inputs and the corresponding known outputs form a training pattern. A set of initial weights are assigned, and the inputs of a training pattern are propagated through the network. The network outputs are compared to the desired outputs and errors are calculated. Then the errors are propagated backwards (hence the name back-propagation neural network) to adjust the weights so as to minimize the errors. This part of the computation is done by a single subroutine call in the neural network development tool, NeuroWindows™ [7], used for this study. The details of the optimization procedure can be found in Baughman and Liu [5]. A collection of training patterns, called a training set, is presented to the network repeatedly during the training phase until an acceptable level of average error in the predicted values is reached. At this point the link weights are the model parameters. One can test the recall capability of the BNN model by comparing the predicted and the desired outputs of the training data; however, the more important test is to use input data not included in the training set to see whether the network can generalize and predict correctly.

TRAINING THE NEURAL NETWORK

In order to train the BNN reaction rate model, experimental data of reaction rate at various temperatures and degrees of cure are required. The reaction rates were measured by differential scanning calorimetry (DSC) at four temperature ramp rates of 0.2, 1, 2, and 5°C/min. These heat rate data were normalized by the total heat of reaction to yield the reaction rates. Integrating this normalized reaction rate over time gives the degree of cure. By so doing we have the inputs (the temperature and the degree of cure) and the desired output (reaction rate) for the BNN model. Although the DSC data were recorded at fixed time intervals, the data used for training the neural network were chosen at about 1°C increments. The training set consists of a total of 654 patterns – about 160 patterns from each DSC scan. The data in the training set are shown in Figure 2. Before the training patterns were presented to the network, the order of the patterns in the training set was randomized so that they were not in any particular time or temperature sequence. The BNN was trained for 10,000 iterations, and the resulting link weights are shown in Table I.

VERIFICATION OF THE NEURAL NETWORK MODEL

The link weights in Table I and Equations 1 and 2 comprise the BNN model for predicting the reaction rate of 3501-6 epoxy resin at a given temperature and degree of cure. The input temperature is mapped from a range of [50, 250] to a range of [0, 1] to yield $a_{0,1}$. The input degree of cure, which is already in the range of [0, 1], is $a_{0,2}$. Equations 1 and 2 are then used to propagate these two activities through the hidden layers. The activity of the last layer, $a_{3,1}$, is then mapped from a range of [0.1, 0.9] to [0, 0.001] to obtain the predicted reaction rate.

A comparison of the BNN model predicted reaction rate and the actual values contained in the training set is shown in Figure 3. It can be seen that at high reaction rates, the BNN model underpredicts. For most processing conditions, however, the reaction rate does not reach such high levels. Figure 4 shows that the relative errors can be very large in the regions of very low and very high degree of cure. However, in these ranges the reaction rates are so low that the errors in the predicted reaction rates do not contribute much to the errors in predicted degree of cure. This is important since the reaction rate prediction itself is dependent on the predicted degree of cure. Nearly all of the relative errors in predicted reaction rates in the range of 0.2 to 0.8 degree of cure are below 10%.

Testing the recall capability of the BNN model is not sufficient, however. It is important to test whether the network model can predict reaction rates of conditions different from those in the training set. We then tested the model's capability in predicting the reaction rate and degree of cure of 3501-6 epoxy resin when it was subjected to a standard cure cycle. The actual reaction rates were measured by DSC. The neat resin sample was first cured in DSC following a standard cure cycle as shown in Figure 5 (heat at 2.8°C/min to 116°C; hold for 60 minutes; heat at 2.8°C/min to 177°C; hold for 90 minutes, which is shorter than common practice; cool down). The sample was then reheated at 4°C/min to 300°C to complete the cure in order to determine the total heat of reaction. Using this total heat of reaction, data analysis followed that for preparation of the training set.

The BNN model used the actual temperature recorded by DSC (at 9-second intervals) as the first input. The second input, the degree of cure, was assumed to be zero initially. The predicted rate is multiplied by the time step size and summed to the previous degree of cure to yield the predicted degree of cure at the end of the current time step, which becomes the input for the next time step. This simple numerical integration scheme is sufficient for the small time step size used. Figure 5 shows a comparison of the BNN model predicted reaction rate and degree of cure and the DSC measured results. The BNN model predicted reaction rate and degree of cure are in very good agreement with the measured results. As can be seen in Figure 6, for the most part of the cure, the errors in predicted degree of cure are less than 0.03, and the maximum error is less than 0.04.

COMPARISON OF NEURAL NETWORK MODEL AND ANALYTICAL MODEL

Also presented in Figures 5 and 6 are comparisons of the performance of the BNN model and an analytical model by Loos and Springer [4]. The analytical model also uses the same temperature cycle as its input. It predicts a reaction rate and integrates the rate to obtain the degree of cure. A drawback of the analytical model is that it has a discontinuity in reaction rate because of the change in rate expression. It can be seen in Figure 5 that the BNN model predicts the reaction rates more accurately. As a consequence the BNN model predicted degree of cure is also more accurate. As can be seen in Figure 6, the errors in degree of cure predicted by the analytical model are much greater than those predicted by the BNN model – a factor of four or greater for degrees of cure below 0.8. Perhaps more important is the ability of the BNN model in predicting an undercure situation, as it was in this test case, while the analytical model failed.

It should be pointed out, however, that the BNN model has far more parameters than does the analytical model, and so it is expected to give a better fit of the data. In the BNN model there is a total of 44 link weights, while the analytical model has only eight parameters (six for the three Arrhenius equations and two having to do with gel point). On the other hand, having fewer parameters does not necessarily make it easier to determine these parameters. There are in fact a total of 20 parameters in the general analytical model, so that considerable analysis is required to determine the analytical model parameters. The BNN approach, while limited to interpolation and providing no mechanistic insight, allows very rapid development of an empirical reaction rate model based on four sets of DSC data that cover the range of processing conditions. Since the processing conditions are usually quite narrow, the limitation of interpolation does not present a serious constraint.

SUMMARY AND CONCLUSION

The results from this study show that a simple back-propagation neural network can be trained with four sets of DSC data to model the reaction rate of an epoxy resin with a high degree of accuracy. Although the BNN model has more model parameters than does an analytical model, it requires much less effort to obtain these model parameters. It is a simple method for quickly obtaining a reaction rate model for simulation or model-based control purposes. It should be kept in mind, however, that the BNN is limited to interpolation within the range of the training set.

REFERENCES

1. Lee, C. W., and B. P. Rice, "On-Line Determination of Composite Laminate Thermal Diffusivity and Heat Release Rate," SAMPE Quarterly, 24 (3), 30, (1993).
2. Abrams, F., P. Garrett, T. Lagnese, S. LeClair, C. W. Lee, J. Park, "Qualitative Process Automation for Autoclave Curing of Composites," AFWAL-TR-87-4083, Nov. 1987.
3. Buczek, M. B., and C. W. Lee, "Self-Directed Process Control System for Epoxy Matrix Composites," Proceedings of the 40th International SAMPE Symposium, May 8-11, 1995, Anaheim, CA.
4. A. C. Loos and G. S. Springer, "Curing of Epoxy Matrix Composites," J. Composite Materials, 17 (2) 135-169 (1983).
5. Baughman, D. R. and Y. A. Liu, Neural Networks in Bioprocessing and Chemical Engineering, Academic Press, 1995.
6. Su, H. B., L. T. Fan, and J. R. Schlup, "On-Line Determination of the Degree of Cure of Epoxy/Graphite Composites with a Neural Network," Proceedings of ISPE '95, Intelligent Systems in Process Engineering, July 9-14, 1995, Snowmass Village, CO.
7. NeuroWindows™, Neural network dynamic link library, Users' Manual, Ward Systems Group, Inc. 1993

ACKNOWLEDGMENTS

This work was funded under U.S. Air Force Contract No. F33615-95-D-5029. The authors wish to thank Mr. William A. Price for the DSC data.

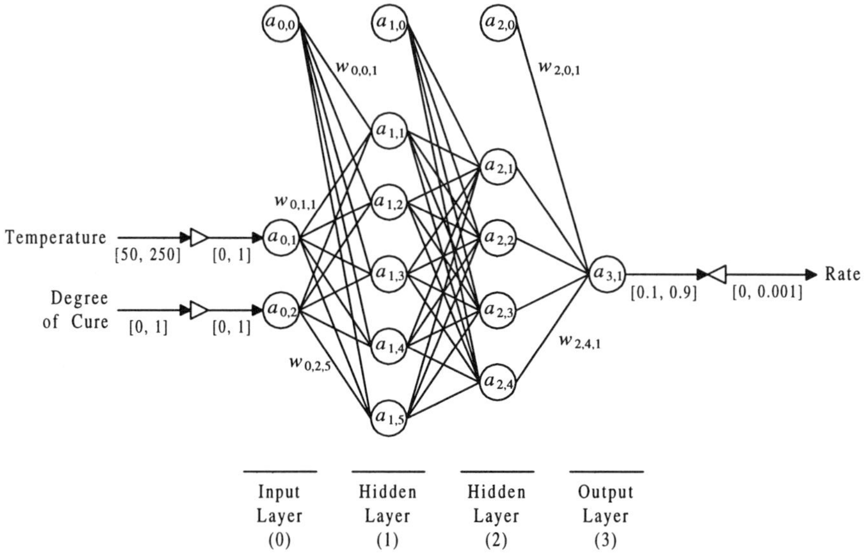

Figure 1. The structure of the backpropagation neural network reaction rate model.

Table I. Link weights ($w_{i,j,k}$) of the BNN reaction rate model for 3501-6 epoxy resin.

		$k = 1$	2	3	4	5
$i = 0$	$j = 0$	–28.5113	–5.2629	–7.2896	5.1274	–4.6926
	1	6.2032	0.8277	10.7357	9.6543	11.9382
	2	20.6651	21.9899	0.1172	–14.9512	–4.0963

		$k = 1$	2	3	4
$i = 1$	$j = 0$	–5.5719	–4.1387	–4.3776	0.8924
	1	–3.8342	14.0272	0.8261	0.6289
	2	–7.2250	3.0481	4.1117	7.3470
	3	4.9659	0.7964	4.7585	2.5187
	4	4.3182	–2.1499	–1.8305	–0.6892
	5	5.3127	–1.6449	2.9860	–4.1794

		$k = 1$
$i = 2$	$j = 0$	–5.5859
	1	5.8232
	2	–7.1419
	3	1.3189
	4	3.6830

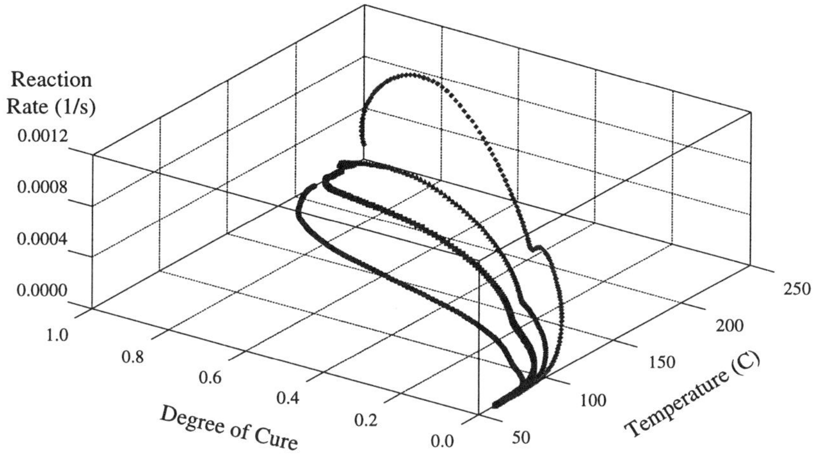

(a)

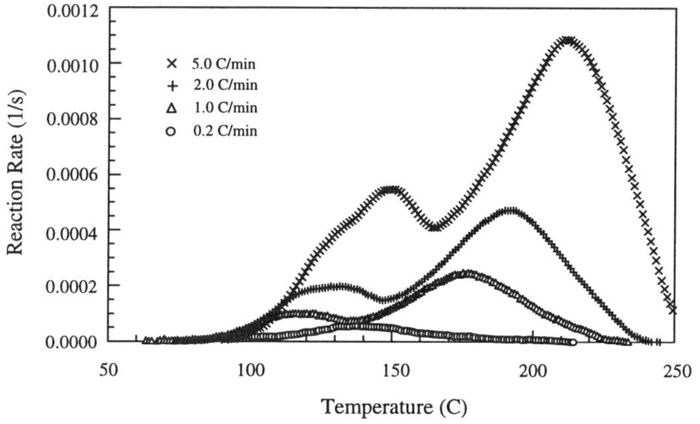

(b)

Figure 2. (a) The training set from DSC scans at 0.2, 1, 2, and 5°C/min. (b) Projection of the training set on the reaction rate–temperature plane.

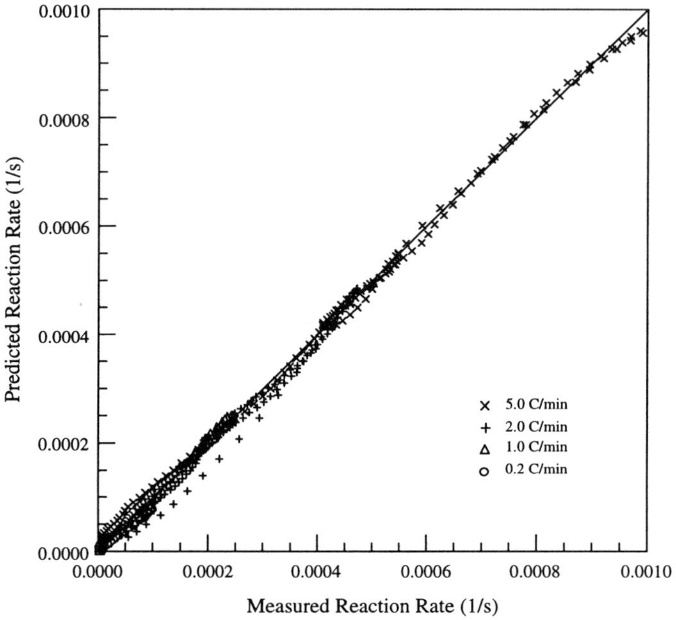

Figure 3. Test of the recall capability of the BNN model.

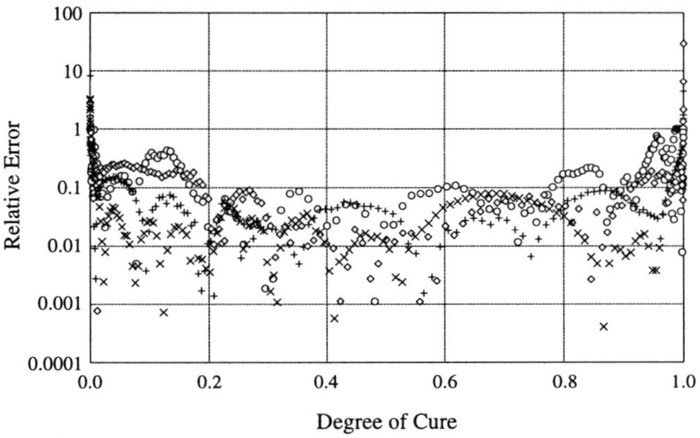

Figure 4. Relative errors of BNN predicted reaction rates in recall test.

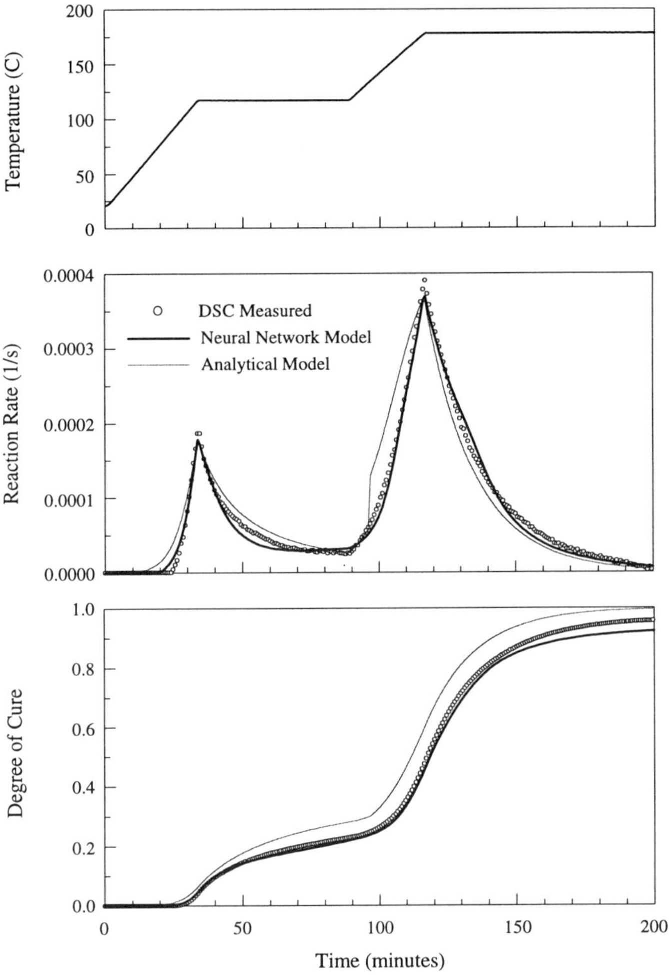

Figure 5. Verification of BNN model and comparison to an analytical model.

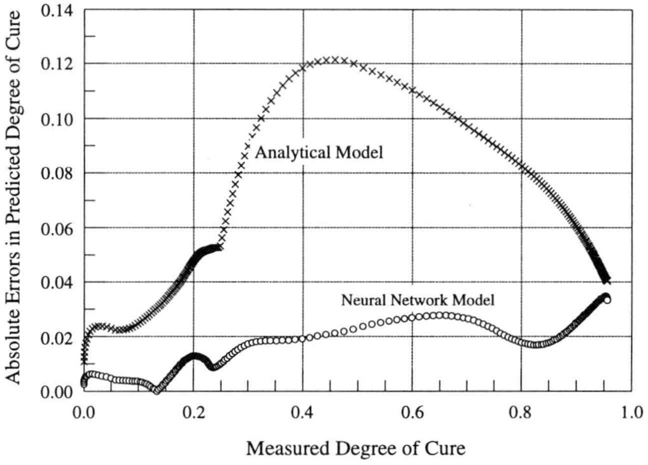

Figure 6. Absolute errors in predicted degrees of cure.

28th International SAMPE Technical Conference
November 4-7, 1996

ENVIRONMENTAL DURABILITY OF A COMPOSITE-TO-COMPOSITE ADHESIVE BOND IN INFRASTRUCTURE APPLICATIONS

Allan S. Crasto and Ran Y. Kim
University of Dayton Research Institute
Dayton, OH 45469-0168

ABSTRACT

Fiber-reinforced composites have proven to be effective in the rehabilitation of concrete bridge beams as an externally-bonded tensile reinforcement. Practical requirements for bonding precured composite plates to large concrete beams in the field necessitate some form of composite joint to limit the size of the plates to be transported to the site. In earlier work to demonstrate the feasibility of this approach to strengthening concrete beams in bridge decks [1], the composite selected was a graphite/epoxy, AS4C/1919, bonded to the concrete with Hysol EA9460, a two-part ambient-cure epoxy adhesive. Composite joints were introduced at periodic intervals, both in the lab studies and in the field, to simulate techniques necessitated by the actual application. From tests conducted to date it appears that this joint is more susceptible to failure under mechanical loading than the composite itself or the composite-to-concrete adhesive bond. In the present study, therefore, the durability of such a composite-to-composite joint was evaluated under simulated outdoor environmental exposure using a double-lap shear test coupon. Specimens were subjected to various combinations of moisture, thermal cycling and fatigue, and the residual lap shear strengths determined at various temperatures. The results are compared with those of unaged specimens.

KEY WORDS: Adhesives, Environmental durability, Infrastructure

1. INTRODUCTION

An accepted method of repair of weak or damaged concrete structural members is flexural strengthening or stiffening with bonded steel plates. The advantages of this method over others, such as external post-tensioning or additional supports, are lower cost, ease of application and maintenance, elimination of special anchorages, and the ability to strengthen the structure while it remains in use. The bonded plate serves as a second layer of reinforcement that can increase flexural strength by as much as 40% [2] and provide crack control for the concrete member. Epoxy-bonded steel plates have been used extensively in the field to strengthen buildings in Switzerland and bridges in England, the former USSR, Poland and Japan. An important drawback of this structural system, however, is the potential for corrosion at the epoxy/steel interface with consequent reduction in the adhesive bond strength. This is evident from studies conducted on unreinforced concrete beams with steel plates bonded to one face [3], in which outdoor exposure caused the steel plate to corrode significantly, resulting in a loss in bond strength at the steel/epoxy interface. In another study, residual strength tests of reinforced concrete beams conducted at the Swiss Federal Laboratories for Materials Testing and Research after 15 years of exposure to weathering confirmed corrosion at the steel/adhesive interface.

Fiber-reinforced composites offer distinct advantages over steel as an externally-bonded plate reinforcement for concrete structural members. Carbon fiber-reinforced plastics (CFRP) in particular have higher specific stiffnesses and strengths, demonstrate excellent durability in a saline environment, are resistant to corrosion by acids, alkalis and salts over a wide range of temperatures, and can be tailored to the desired mechanical properties. Lab tests in Switzerland and Germany have also shown that replacement of steel plates with CFRP in these applications can reduce the total cost of a reinforcing project by about 20% [4]. Although the CFRP itself is much costlier than steel, the lighter weight and better corrosion resistance can result in significant reductions in fabrication and long-term costs. Research in this area originated in the mid-eighties in Switzerland [5] and has recently mushroomed, for example [6-8], and the results indeed confirm the promise of this avenue of repair.

In any structural repair there are two issues which need to be addressed: the mechanical performance of the repaired structure and the durability of the repair over the life of the structure. The mechanical performance can be readily assessed using the appropriate mechanics analysis and experimental validation; evaluating the durability, however, is more complex. In its application the structure may be subjected to different combinations of temperature and humidity on which are superimposed various monotonic and fatigue loads, exposure to chemicals, etc. To accurately assess the influence of the application environment on repair durability, periodic tests after real-time exposures are required. For structures with long projected lifetimes, such an approach is impractical and leads to long cycle times for the introduction of new materials and processes for repair. Consequently, accelerated aging tests must be conducted. In the compromise between reduced test time and appropriate acceleration of environmental effects, skepticism creeps in on the value of the data generated and its true reflection of long-term, more realistic aging. This is especially true in the civil engineering industry where liability is a major concern. The solution lies in initiating exposure to the actual application environment and comparing the results obtained at periodic exposure intervals with accelerated test data to build confidence in the latter.

In a recent field demonstration project [1], composite plates were bonded to the lower face of concrete beams in a bridge deck with materials and processes identified from laboratory tests on appropriately scaled-down (8.5-ft) reinforced concrete beams [9]. Precured composite plates were utilized and bonded to the 28-ft concrete bridge beams with an ambient-cure epoxy adhesive. Composite joints are both practical and necessary in field implementation of this scheme, and doubler joints were introduced at four locations along the beam's length. In tests conducted to date, the composite plate reinforcement has been successfully designed to allow the concrete to fail initially in compression under flexural loading; subsequent failure occurs in the joint and not through tensile fracture of the composite plate or failure at the concrete/adhesive interface [1]. The aim of the project was to determine the durability of this reinforcement scheme in the actual application environment by removing these (28-ft) composite-reinforced beams after suitable intervals of field exposure, measuring their flexural properties, and comparing the results with unaged control beams (with and without bonded composite plates). The time intervals selected for removal of the first two beams were one and five years. Although the results from such real-time exposure to the actual application environment are preferable, the concurrent development and selection of materials and processes require an acceleration of the outdoor exposure.

In the present study the durability of the weakest link in the field rehabilitation scheme [1], viz. the composite joint, is evaluated after accelerated exposure to outdoor environmental conditions.

2. EXPERIMENTAL

2.1 Materials

The materials selected for this study were identical to those utilized in an earlier field demonstration project [1] on composite rehabilitation of concrete bridge beams. The adherends for this adhesive study were sectioned from $[0]_5$ composite plates (nominal thickness 0.9 mm) made from a commercial 121°C-cure graphite/epoxy unitape prepreg (AS4C/1919 from Hercules, Inc.). The composites were cured between plies of porous

Teflon-coated release nylon fabric which imparted its texture to the composite plate surface. Prior to bonding, the composite surface was cleaned with water and acetone. The adhesive used in this study was a two-part epoxy adhesive (EA9460 from Dexter Hysol, Inc.) which cures under ambient conditions.

2.2 Specimen Fabrication

The specimen selected for this study was a double-lap shear test coupon. The two-part adhesive was mixed in accordance with the manufacturer's instructions, and composite plates were bonded at ambient temperature and under a pressure of 8 psi (applied with dead weights) for a minimum of 48 hours. This resulted in a uniform adhesive layer with a nominal thickness of 0.15 mm. These bonded plates, with a lap length of 25 mm, were then sectioned into test coupons, each 25 mm wide. A sketch of the test coupon is shown in Figure 1. Strips of Kapton release tape were utilized during bonding to prevent the composite/adhesive bondline from extending out of the lap area.

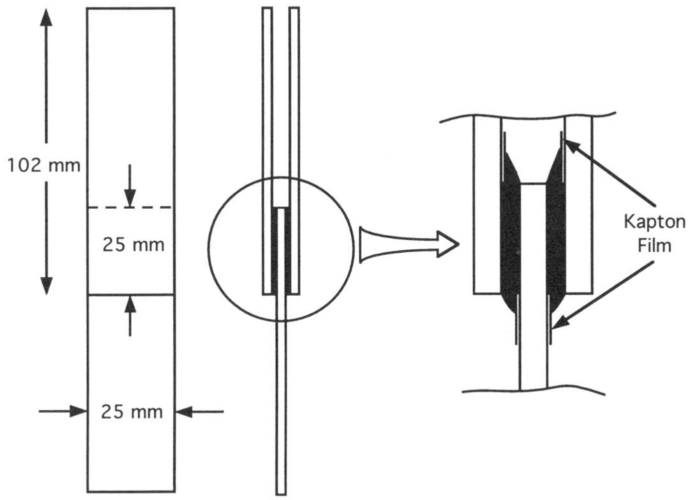

FIGURE 1. Sketch of a double-lap shear test specimen

2.3 Specimen Conditioning

Prior to testing, the lap shear coupons were subjected to various conditioning treatments to simulate some of the environmental conditions which could possibly be encountered in the intended application, albeit under an accelerated schedule. One batch of specimens was retained as a control under ambient lab conditions. A second batch was immersed in water at ambient temperature until the specimens were saturated with moisture as determined from periodic measurements of weight gain. A third batch of specimens, immersed in a container of water, was placed in a chamber which was thermally cycled for up to 1000 cycles between -23°C and 27°C. The typical specimen and chamber temperature profiles over one thermal cycle are shown in Figure 2. As is evident from this figure, the entire volume of water in the specimen container was allowed to freeze and then thaw in each cycle. Specimens were periodically withdrawn during the thermal cycling for mechanical testing.

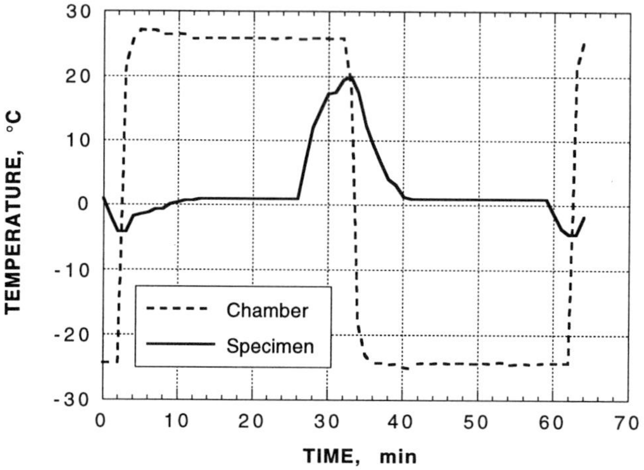

FIGURE 2. Plot of chamber and specimen temperature during a typical freeze-thaw cycle

2.4 Specimen Testing

Lap shear specimens from the three batches conditioned as described above were tested at temperatures ranging from -23°C to 66°C under monotonic loading at a rate of 10 lb/second. Some specimens were also subjected to fatigue loading (R=0.1, 10 Hz) at loads of 1800 and 2300 lbs, corresponding to 40 and 50 percent, respectively, of the ultimate strength of control specimens. Table 1 describes the details of the test matrix. The glass transition temperature of cured adhesive was determined in a Rheometrics RDS-II dynamic mechanical analyzer in the torsion mode at 100 radians per second with a scan rate of 10°C/min. The coefficients of thermal expansion (CTEs) of the composite plate, neat adhesive and concrete were determined over the temperature range of -40°C to 66°C; 12.7-mm strain gages were employed to measure the composite and adhesive thermal strains, while 50.8-mm gages were used for the concrete specimen.

TABLE 1. Test Matrix for Mechanical Characterization of Lap shear Specimens

Test Conditions / Specimen conditioning	Static Tension (Temperature, °C)					Fatigue (23°C, R=0.1) (Stress, psi)	
	-23	4.4	23.9	51.7	65.6	900	1150
Ambient (control)	4*	3	5	3	5	7	9
Moisture-saturated	3	5	5	5			
Thermally-cycled			17				4

* No. of specimens tested

3. RESULTS AND DISCUSSION

The use of composites in infrastructure repair will be facilitated by decreasing the influence of processing variables on the mechanical performance of the bonded structure, thereby increasing the consistency of performance of the same material system from one field application to another. These variables include surface preparation of the adherends, mixing and application of the adhesive, and cure conditions (temperature, pressure and humidity). For example, an *ideal* composite surface preparation could involve a mere peeling away of a porous release ply against which the composite plate is cured and application of adhesive directly to the textured surface. Additional surface preparation, such as mechanical abrasion, has the disadvantages of increasing the cost, introducing or spreading contaminants unless care is taken, and standardizing the operation to obtain reproducible adhesive bond strengths. The composite adherends used in this study were fabricated against a peel-ply with the surface texture shown in Figure 3A. The composite surface in this figure displays the imprint of the fabric texture of the nylon peel-ply. Additional mechanical abrasion with 240-grit sandpaper resulted in the surface texture shown in Figure 3B. Tests with both these adherend surface preparations showed a marginal increase in average double-lap shear strength (about 7%) with sanding, although the scatter in the data was significant. Wingfield [10] also reports a marginal (< 5%) increase in single-lap shear strength with mechanical abrasion compared to a peel-ply in a study on carbon/epoxy composite adherends (with mechanical, energetic and chemical surface treatments) bonded with an ambient-cure epoxy paste. Matz [11], on the other hand, showed a 5% increase in lap shear strength with a peel-ply over mechanical abrasion of the adherends for specimens with CFRP adherends and an ambient-cure epoxy paste adhesive.

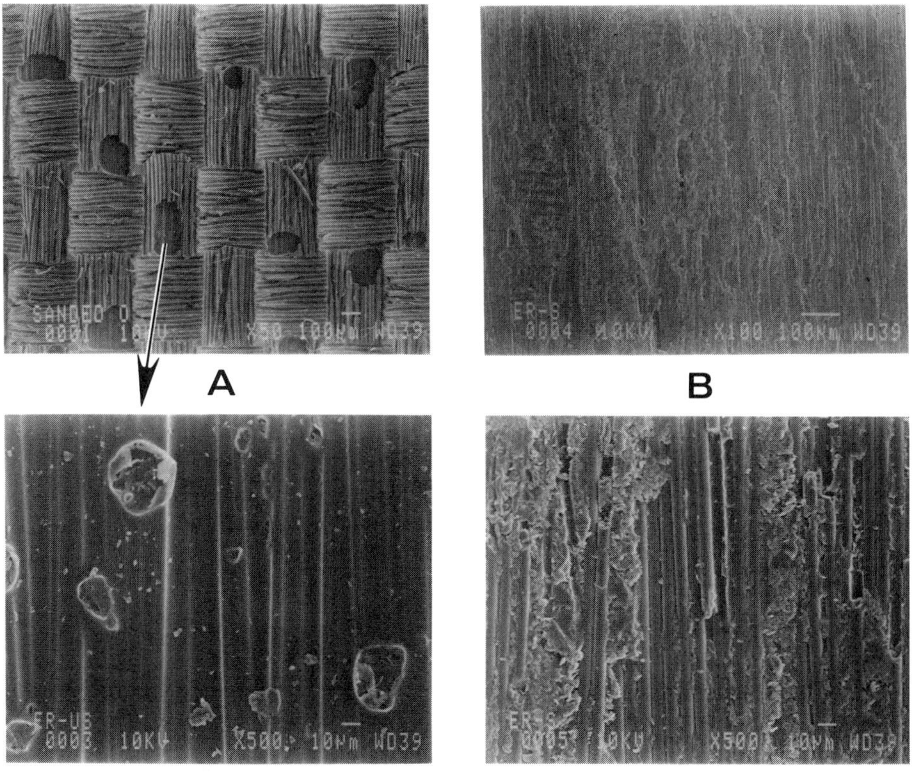

FIGURE 3. Scanning electron micrographs of a composite adherend surface (A) with a peel-ply texture and (B) the same surface after mechanical abrasion

An *ideal* adhesive should have a long pot life, be able to cure under a wide range of outdoor ambient temperatures (to eliminate the inconvenience of providing thermal energy for elevated-temperature cure at the repair site) and maintain its required performance under the extreme temperatures of the specific outdoor environment. The two-part epoxy adhesive selected for the field demonstration project [1] and for this study is a non-sag, low-viscosity (1500-2500 poise) paste with an adequate working life of 40 minutes. The pressure applied during adhesive cure in fabrication of the lap shear test specimens was similar to that applied with a vacuum bag in the field, resulting in a quality bondline with negligible voids. Lap shear tests of the control batch of specimens under ambient conditions resulted in an average lap shear strength of 2240 psi. In general, each lap failed through a combination of shear fracture of the adhesive layer and (predominantly) interfacial failure between the fractured adhesive sections and the two adherends. The outer adherends from both laps of one test specimen are shown in Figure 4A. The adhesive layer near the composite free edges appears rough, suggesting some cohesive failure, while the adhesive layer away from these free edges is characteristic of clean interfacial failure. An SEM examination of both the failed adhesive and corresponding adherend (Figure 5) confirmed these observations. Residual traces of adhesive can be seen on the adherend surface near a free edge in Figure 5A. Away from the free edge, in Figure 5B (and going from left to right), an intact layer of adhesive (bearing the clean textured imprint of the composite surface from which it debonded) can be seen, as well as its shear fracture and the clean mating adherend below. The data from the various lap shear tests conducted are summarized in Table 2.

TABLE 2. Summary of Lap Shear Strength Data

Conditioning	Test Temperature (°C)				
	-23.3	4.4	23	51.7	65.6
None	1175 (70)	1335 (115)	2235 (270)	1620 (35)	995 (40)
Moisture saturated	1305 (110)	1960 (495)	2745 (75)	1355 (60)	
200 thermal cycles			2305 (315)		
400 thermal cycles			2390 (185)		
600 thermal cycles			2510 (195)		
800 thermal cycles			2600		
1000 thermal cycles			2550 (100)		

3.1 Thermal Mismatch in Infrastructural Repair

Key to the successful repair of concrete structures with composites is allowance for the large mismatch in thermal expansion between adhesive, composite and concrete in the material selection and design. Whereas thermal expansion of the adhesive and concrete are isotropic, the CTE of a unidirectional composite is highly anisotropic. The CTE of composite plate reinforcement can therefore be tailored with appropriate ply orientation. From earlier studies [2], a unidirectional composite was found to provide adequate reinforcement with a minimum amount of material with off-axis plies being superfluous. Since this plate does provide the greatest mismatch in the structural constituent CTEs, however, the stresses resulting from variations in the outdoor temperature (between -23°C and 66°C) were analyzed. Typical thermal strains for concrete and the repair materials evaluated in this study, over this temperature range, are plotted in Figure 6. The low stiffness and high CTE of the adhesive (relative to concrete and [0] composite adherends in structural repair) will result in compressive stresses in this layer and at the corresponding interfaces as the temperature is raised above the (ambient) cure temperature and tensile stresses as the temperature is lowered

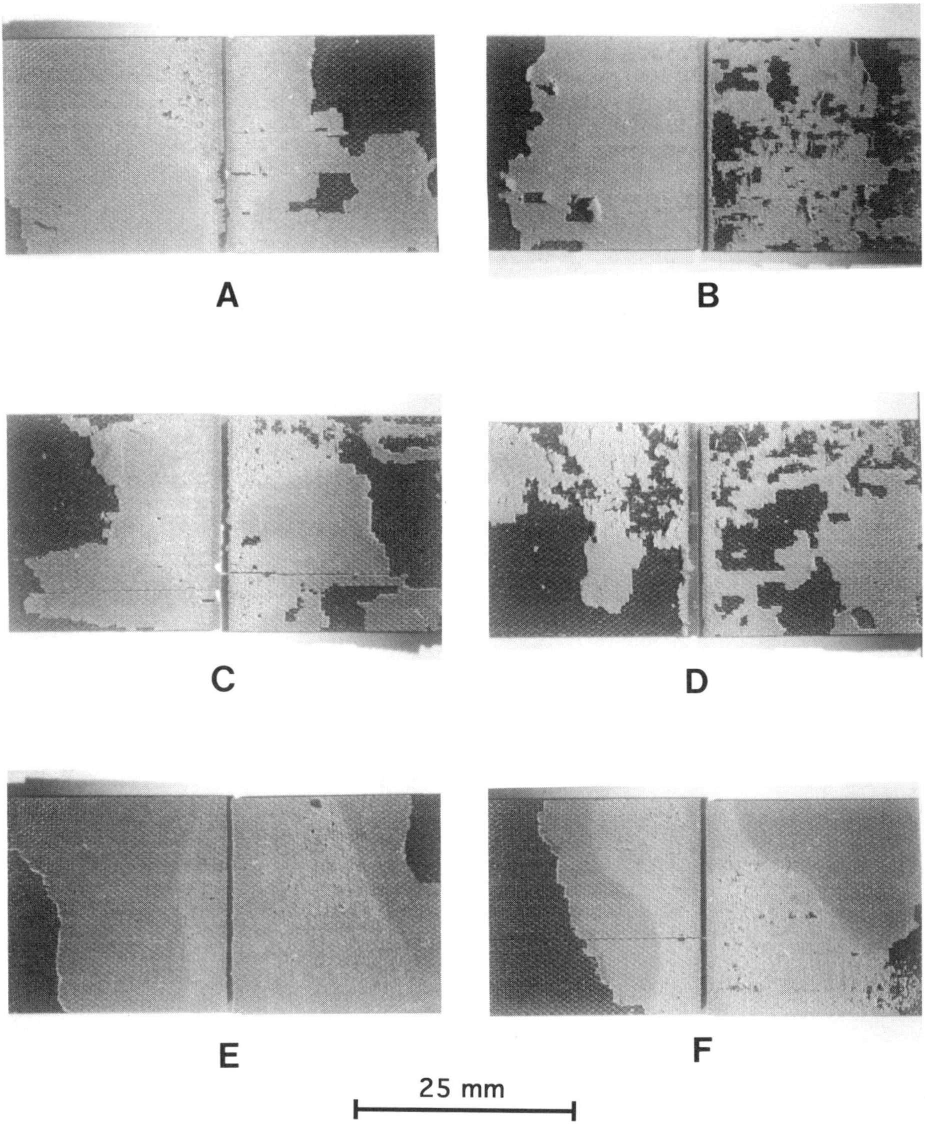

FIGURE 4. Photographs of outer composite adherend pairs from double-lap shear specimens showing the failed lap regions. (Adherends are aligned horizontally in the figure with the free edges butted up against each other). Test conditions: (A) control, 23°C; (B) dry, 66°C; (C) wet, 23°C; (D) wet, 52°C; (E) dry, fatigued to failure at 900 psi; (F) fatigued to failure at 1150 psi after 1000 freeze-thaw cycles.

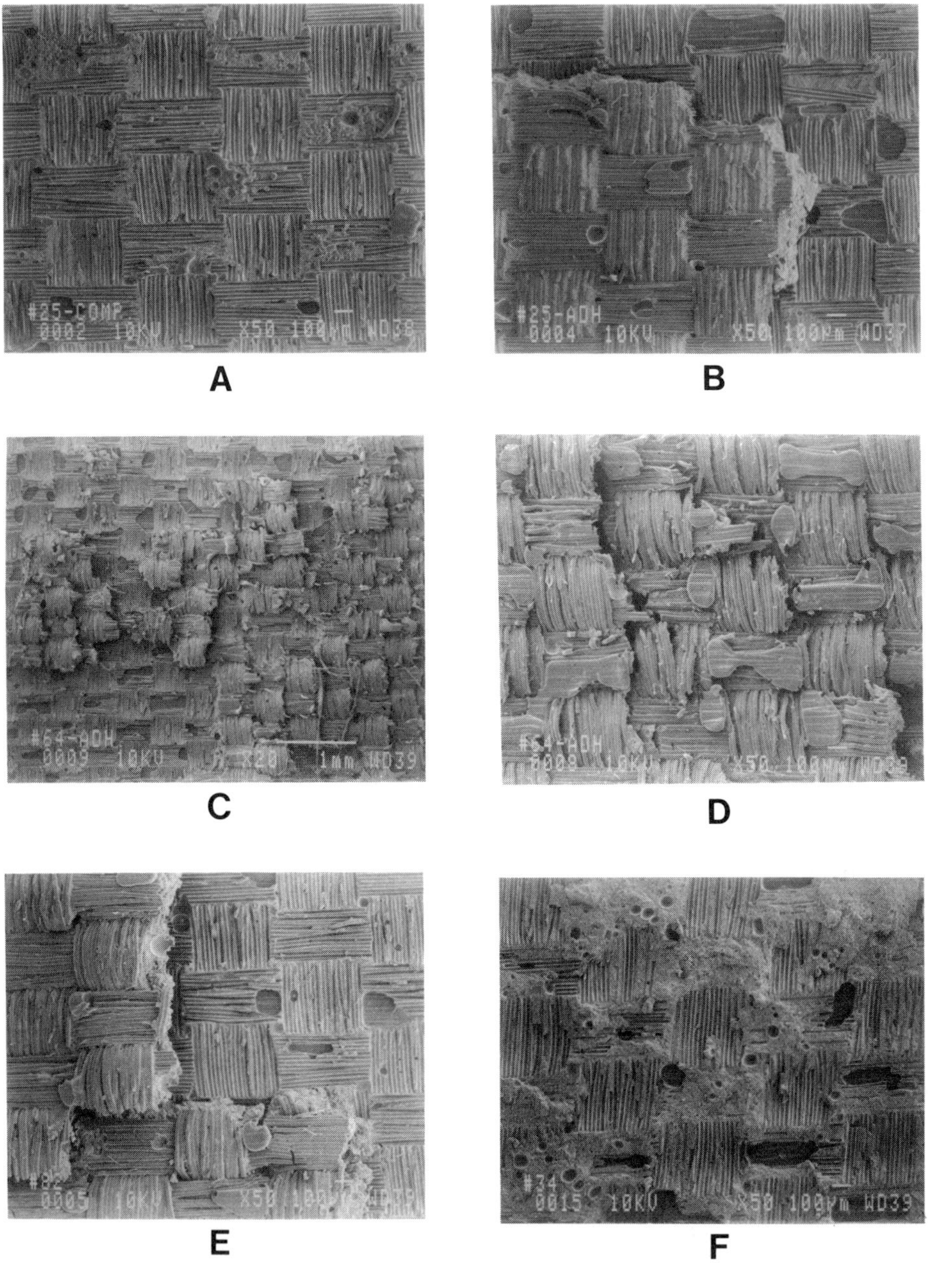

FIGURE 5. Scanning electron micrographs of failed lap shear test specimens. Test conditions: (A&B) control, 23°C; (C&D) dry, 66°C; (E) wet, 52°C; (F) dry, fatigued to failure at 1150 psi.

below ambient. These thermal stresses in turn influence the magnitude of the ultimate stresses under mechanical loading, for example in the lap shear tests conducted in this study.

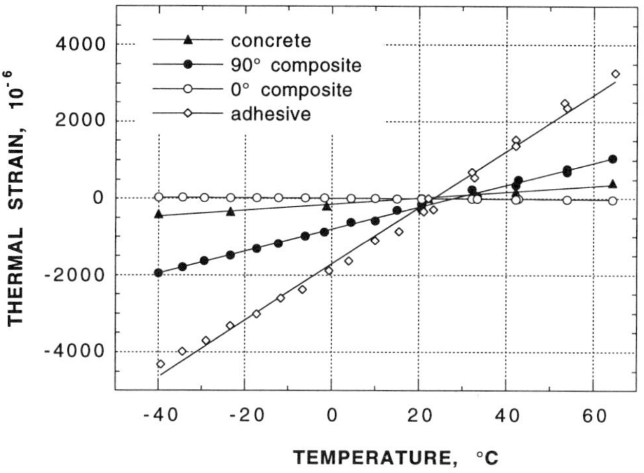

FIGURE 6. Comparative thermal strains of concrete and repair materials

3.2 Effect of Temperature on Lap Shear Strength

Lap shear strengths of unaged (control) specimens were determined over the possible range of outdoor temperatures for this repair application (-23°C to 66°C), and the data are shown graphically in Figure 7. The shear strength appears to peak in the vicinity of 23°C, decreasing by almost 50 percent as the temperature is increased or decreased to the extremes of the test

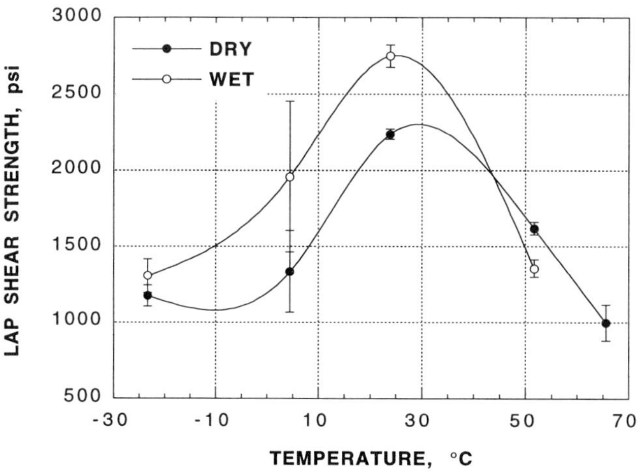

FIGURE 7. Lap shear strength of CFRP specimens as a function of temperature

range. Similar trends were observed by other researchers for CFRP composite/epoxy adhesive specimens in double-lap shear tests over the temperature range of -55°C to 132°C [12] and in single-lap tests over the range of 23°C to 90°C [13]. At temperatures below 23°C the failure appears to be predominantly interfacial with the adhesive layer of one lap remaining intact on the outer adherend. The lower strengths with decreasing temperature are in part due to increased tensile stresses in the adhesive due to the mismatch in CTE between the adhesive and $[0]_5$ composite adherend. At temperatures above 23°C the adhesive fractured in a more ductile manner and at random in the lap in combination with interfacial failure. The outer adherends from a typical failure at 66°C are shown in Figure 4B. Scanning electron micrographs of failed surfaces (Figures 5C & 5D) show evidence of adhesive ductility and tearing. The reduced adhesive strength at this temperature (which is 5°C above its Tg of 61°C) is primarily responsible for the decline in lap shear strength.

3.3 Effect of Moisture on Lap Shear Strength

The lap shear strength of specimens saturated with moisture at 23°C was measured over the temperature range of -23°C to 66°C, and the results are shown in Figure 7. The general trend is similar to that observed with the dry specimens with the strength peaking around 23°C. At higher temperatures the resin fractured in a ductile manner in combination with interfacial failure (see Figures 4 and 5). Plasticization of the adhesive by the absorbed moisture increases its ductility at these temperatures contributing to the lower shear strengths. As with the dry specimens, failure at temperatures below 23°C appeared to be predominantly interfacial with little or no shear fracture of the adhesive layer. In comparing the test results of the wet and dry specimens, however, lap shear strengths of wet specimens below 23°C appear to be significantly higher than those of dry specimens. This may be explained by plasticization of the adhesive by the absorbed moisture, which enhances its toughness at room temperature and below when the adhesive is in a glassy state and which reduces residual curing stresses. At temperatures above 23°C the wet shear strength drops off faster than the dry shear strength, since the plasticizing effect of the absorbed moisture reduces the adhesive Tg to near or even below the test temperature. Similar trends were observed by Ahn, et al. [13] in lap shear tests of CFRP/epoxy specimens, before and after hot/wet conditioning, over the temperature range of 23°C to 90°C.

3.4 Effect of Thermal Cycling on Lap Shear Strength

Differences in CTE between the concrete, adhesive layer and composite plate reinforcement induce stresses in these materials and at the corresponding interfaces when a repaired structure is exposed to temperatures above or below the ambient (or cure) temperature. Cycling between outdoor temperature extremes, in the presence of moisture, puts additional stress on the structure due to the expansion of absorbed moisture as it freezes. This thermal cycling is akin to the imposition of mechanical fatigue loading which may degrade the adhesive bond performance in service. To simulate the influence of the outdoor thermal environment on composite joint strength, albeit in a more aggressive manner, specimens were subjected to a number of accelerated freeze-thaw cycles between -5°C and 20°C as illustrated in Figure 2. Specimens were withdrawn periodically from this environment and tested at 23°C. Failure in all cases was predominantly interfacial with very little adhesive residue on the mating adherend. Figure 8 shows the lap shear strength as a function of the extent of thermal cycling. Contrary to expectations, there is a steady increase in lap shear strength with thermal cycling until it appears to plateau after about 800 thermal cycles at a value approximately 15 percent higher than the control sample. One possible explanation for this trend is an increase in adhesive strength over time in the thermal cycling chamber. From the manufacturer's data, however, the adhesive achieves 90 percent of its ultimate strength after 24 hours and full strength after three days with a room-temperature cure, while thermal cycling and testing were conducted over several weeks. A more likely explanation is the increase in adhesive toughness with increasing moisture pickup during thermal cycling through plasticization of the resin in its glassy state at room temperature. In section 3.3 it was seen that the lap shear strengths of moisture-saturated specimens at 23°C are 23 percent higher than corresponding (dry) control specimens. In contrast to the results discussed above, Chajes, et al. [14] obtained a decrease in flexural performance of concrete beams, with CFRP bonded to the face in tension, after freeze-thaw cycling, and attributed it to deterioration of the concrete-adhesive-composite bond.

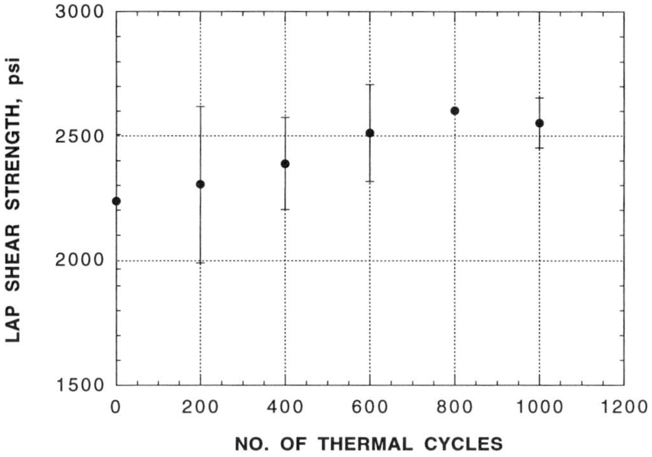

FIGURE 8. Lap shear strength of CFRP specimens as a function of thermal cycling

3.5 Fatigue Performance of Lap shear Specimen

Composite joints in plate reinforcement of concrete bridge beams will be subjected to fatigue loading as live loads pass over the bridge, and their fatigue life is therefore an important consideration in material selection and design. The fatigue life of lap shear specimens was determined at fatigue stresses of 40 percent and 50 percent of the lap shear strength of control specimens, and the data are shown in Table 3. The cycles to failure at each stress level are low in comparison to the expected life of the structural repair of a bridge beam. However, the

TABLE 3. Fatigue Life of Lap shear Test Specimens

Conditioning	Cycles to Failure at 900 psi	Weibull Parameters	Cycles to Failure at 1150 psi	Weibull Parameters
None	35938 61547 69269 70039 90884 91332 408340	$\alpha = 1.2$ $\beta = 127{,}014$	961 4287 4603 4837 5108 5200 20037 24150 28546	$\alpha = 1.1$ $\beta = 11{,}308$
1000 thermal cycles			4402 5357 8842 34067	$\alpha = 1.2$ $\beta = 14{,}077$

fatigue stresses are substantially higher than a joint might experience in this particular application and were selected primarily to determine the influence of environmental exposure. Moisture-saturated specimens tested to failure at the higher stress level displayed no appreciable difference in fatigue life compared to the corresponding control specimens. Failure of all fatigued specimens appeared to have initiated at the free edge, in the adhesive near the composite surface, across the entire specimen width. This failure appeared to propagate inward with increasing fatigue cycles until the specimen failed catastrophically at the interface. The demarcation between these failure modes can be seen (and is identified) in Figure 4E. Figure 5F shows a magnified view of a representative adherend surface near the composite free edge with a substantial amount of residual fractured adhesive.

4. SUMMARY AND CONCLUSIONS

The use of precured composite plates as externally-bonded reinforcements for concrete bridge beams necessitates the incorporation of composite joints in the design. In this study the environmental durability of such joints was investigated with the double-lap shear test using $[0]_5$ CFRP adherends and an ambient-cure epoxy adhesive similar to the materials employed in an earlier field rehabilitation scheme. The individual and combined effects of temperature, moisture, thermal cycling and tensile fatigue on the lap shear strength were determined. Both wet and dry shear strengths peak around room temperature and decrease to approximately 40 to 50 percent of this value at the extremes of the test temperature range (-23°C to 66°C). At ambient and lower temperatures the wet shear strength is higher than the corresponding dry shear strength, and this is attributed to the increase in adhesive toughness due to plasticization by moisture. At higher temperatures, however, the wet shear strength decreases more rapidly than the dry shear strength. Freeze-thaw cycling in a moist environment between -5°C and 20°C results in an steady increase in lap shear strength over time, and increased toughness due to increased moisture absorption is thought to be responsible for this trend. From the limited fatigue studies conducted, it appears that moisture saturation has negligible influence on the fatigue life of these specimens. Analysis of the stresses developed in the structural constituents (as a result of their CTE mismatch) at the extreme temperatures of the application environment indicate that they are not large enough to adversely affect the strength of a [0] composite-to-concrete bond. From the results, therefore, it appears that the simulated application environment causes significant degradation in the properties of this composite joint. Experimental and analytical studies are necessary to determine if the shear stresses in the composite joint of a rehabilitated beam exceed the minimum shear strengths obtained in this study. The outcome of this study may determine if adhesives with elevated temperature cures are required to raise the Tg and thereby decrease the reduction in lap shear strength at elevated temperatures. Failure in low-temperature lap shear tests was predominantly interfacial which suggests the need to improve adherend surface treatment for increased shear strength.

5. ACKNOWLEDGMENTS

The authors gratefully acknowledge the assistance of the following individuals from the University of Dayton Research Institute: Ron Esterline for specimen fabrication and testing; John Camping for specimen thermal cycling; and Ron Cornwell and Bill Ragland for optical and scanning electron microscopy, respectively. This work was sponsored by the U.S. Air Force under Contract No. F33615-95-D-5029.

6. REFERENCES

1. A. S. Crasto, R. Y. Kim and J. P. Mistretta, "Rehabilitation of Concrete Bridge Beams with Externally-Bonded Composite Plates. Part II," Int. SAMPE Symp., 41 (1996), pp. 1269-1279.

2. R. N. Swamy and R. Jones, "Plate Bonding Technology - The Painless Technique of Structural Rehabilitation," Proc. ACI Int. Conf., SP-128, Vol. II, American Concrete Institute, Detroit (1991).

3. M. D. MacDonald and A. J. Calder, "Bonded Steel Plating for Strengthening Concrete Structures," Int. J. of Adhes. and Adhes., 2 (2) (1982), pp. 119-127.

4. A. A. Mufti, et al., "Introduction and Overview," in Advanced Composite Materials with Application to Bridges, A. A. Mufti, et al., eds., Canadian Society for Civil Engineering, Montreal, Canada (1991), p. 2.

5. H. P. Kaiser, "Strengthening of Reinforced Concrete with Epoxy-Bonded Carbon-Fiber Plastics," Doctoral Thesis, Diss. ETH No. 8918, ETH Zurich, Switzerland (1989).

6. U. Meier, et al., "Strengthening of Structures with CFRP Laminates: Research and Applications in Switzerland," in Advanced Composite Materials in Bridges and Structures, K. W. Neale and P. Labossiere, eds., Canadian Society for Civil Engineering, Montreal, Canada (1991), p. 243.

7. H. Saadatmanesh and M. R. Ehsani, "Fiber Composite Plates can Strengthen Beams," Concrete International (Mar. 1990), pp. 65-71.

8. A. Sharif, et al., "Strengthening of Initially Loaded Reinforced Concrete Beams Using FRP Plates," ACI Structural Journal (Mar.-Apr. 1994), pp. 160-168.

9. A. S. Crasto, R.Y. Kim, C. Fowler and J. P. Mistretta, "Rehabilitation of Concrete Bridge Beams with Externally-Bonded Composite Plates. Part I," Proc. First Int. Conf. on Composites for Infrastructure (ICCI-I), Phoenix, AZ (Jan. 1996).

10. J. R. Wingfield, "Treatment of Composite Surfaces for Adhesive Bonding," Int. J. of Adhes. and Adhes., 13 (3) (1993), pp. 151-156.

11. C. W. Matz, "Adhesion on Carbon Fiber-Reinforced Plastics," J. Adhes., 22 (1987), pp. 61-65.

12. J. A. Bishopp, L. Davies and J. J. Haslam, "Chemical and Mechanical Characterization of Adhesive Matrices," Int. J. of Adhes. and Adhes., 13 (2) (1993), pp. 111-119.

13. K. J. Ahn, Y. S. Eom, Y. T. Shim and J. C. Seferis, "Characterization of Structural Adhesives for Load-Bearing Composite Structural Applications," Composite Structures, 27 (1994), pp. 57-64.

14. M. J. Chajes, et al., "Durability of Composite Material Reinforcement," in Infrastructure: New Materials and Methods of Repair, K. D. Basham, ed. (ASCE, New York, NY 1994), pp. 598-605.

7. BIOGRAPHIES

Allan Crasto received his Ph.D. in Materials Science & Engineering from Washington State University in 1986. He spent a year conducting postdoctoral research on interphase modification in carbon-fiber composites and another year as a Visiting Scientist at the U.S. Air Force Materials Laboratory where he worked on the processing and characterization of molecular composites. He joined the University of Dayton Research Institute in 1988 and is currently a Senior Materials Scientist. His research interests include the investigation of structure/property relationships in polymer composites, measurement of residual stresses, development of composites for space, and the application of composites to infrastructure.

Ran Kim was born in Korea and graduated from Seoul University with a B.Sc. and M.Sc. He obtained his Ph.D. from Penn State University in 1971 and joined the staff of the University of Dayton Research Institute in 1974. His research interests include fatigue damage and life prediction, strength and progressive failure in laminates, environmental effects on mechanical performance including curing residual stresses, and the mechanical behavior of composite materials.

28th International SAMPE Technical Conference
November 4-7, 1996

Induction Welding of Thermoset Composite Adherends Using Thermoplastic Interlayers and Susceptors

Alan R. Wedgewood
E. I. DuPont de Nemours & Co.
Wilmington, DE 19898

Patrice E. Hardy
Boeing Defense & Space Group
PO Box 3707 M/S 4X-59
Seattle, WA 98124

ABSTRACT

An approach for induction welding thermoset composite laminates using a combination of thermoplastic interlayers and susceptors has been demonstrated by joining "T" sections to flat laminates. In this approach, a thermoplastic interlayer is first co-cured to the thermoset laminate surfaces to be joined. A thermoplastic impregnated metallic susceptor is placed between the laminate surfaces to be joined. Power is supplied to a work coil which generates an alternating magnetic field. Current induced by the magnetic field heats the weld line which then fuses the resin rich layers together. Joint quality, as assessed by NDI, photomicrographs and pull-off strength determinations, have shown that high quality joints can be readily produced. Typical failures observed for the pull-off tests were not in the weld line. The failures either initiated as peel at the edge of the weld line that traversed into the skin plies or initiated as a crack in the radius region of the "T" section that traversed across the cap. These observations show that for this joint design the thermoset material is the weakest component.

KEY WORDS: Induction Welding, Thermoplastic Resins/Polymers, Thermoset Resins/Polymers

1. INTRODUCTION

One of the obstacles to greater commercial and military adoption of polymer matrix composites is the cost of detail part fabrication and assembly. Industry studies have shown that assembly costs alone compose 19 to 42% of the final aircraft composite structures cost. Current composite aircraft structures are primarily assembled using fasteners. A cost comparison performed by Boeing Defense and Space Group showed labor savings greater than 61% could be achieved by welding a composite wing structure when compared with the current bolted wing structure assembly methods. These labor savings should reduce the assembly cost for future aircraft which are forecasted to use large amounts of composites to meet range and payload mission requirements.

2. ELEMENTS OF INDUCTION WELDING

In the controversy between bonding and welding versus autoclave consolidation, welding can be compared to autoclave consolidation which is considered stronger than bonding. Bonding is the joining of two solid interfaces through the use of an adhesive that forms a physio-chemical attachment to the adherend surfaces and results in distinct separation between bonded surfaces. However, welding requires a close approach of surfaces to be joined where interpenetration of polymer chains occurs across an interface resulting in disappearance of the surfaces and development of the ability to transfer loads equivalent to the bulk polymer. Autoclave consolidation consists of through the thickness compaction, again resulting in disappearance of ply interfaces due to macroscopic resin mixing.

3. EXPERIMENTAL

3.1 Induction Welding Set-up The induction welding set-up consists of: 1) skin panel-susceptor-"T" spar details, 2) induction work coil and power supply components, and 3) X-Y-Z robot gantry. Each detail part must have a resin rich interlayer which provides a mating surface in the area to be welded. A resin encapsulated copper susceptor is placed between the spar and skin as seen in Figure 1.

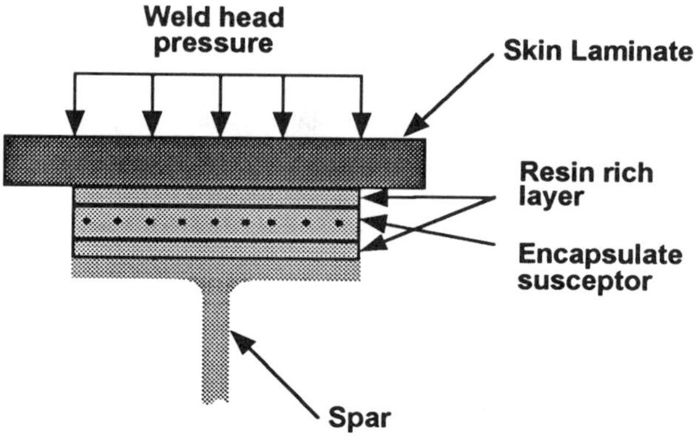

Figure 1 Induction Welding Set-up

A power amplifier delivers an alternating current to the induction work coil generating an oscillating magnetic field. This field is focused by the coil to induce eddy currents in the laminate and susceptor as shown in Figure 2.

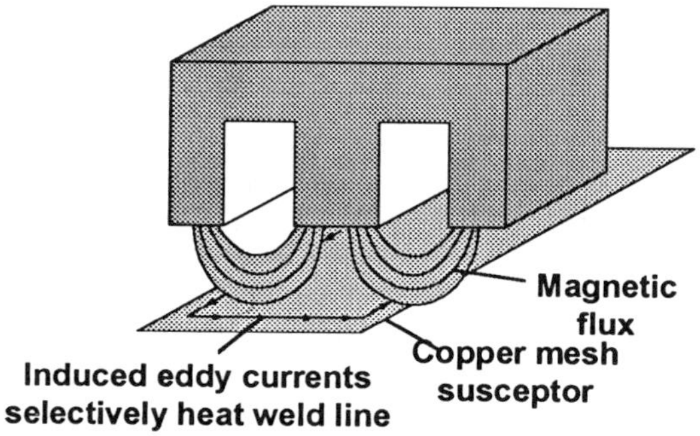

Figure 2 Eddy Currents Induced in Susceptor by Magnetic Field

Resistive heating, which occurs as a result of the current flow in the susceptor, softens and fuses together the encapsulated susceptor and the resin rich interlayers on the detail parts. The time required to develop the full strength capability of the weld is governed by intimate contact and healing processes (1). Pressure is applied to the weld line to achieve intimate contact and compaction of the weld line. Therefore, the spars require support to eliminate material deformation and induced stresses. The coil itself is self-shielding and minimizes stray magnetic fields in nearby metal structure. A copper susceptor is used to preferentially heat the weld line to the temperature required for fusion. Without a susceptor, the weld line remains cold relative to the temperature at the part surface closest to the coil which can be high enough to degrade the thermoset composite adherends.

3.2 Lap Shear Testing The lap shear coupons tested here were prepared by placing PS or PES adhesive film between 3501-6 unidirectional tape lay-ups (i.e. two [0]20) and curing with a standard 3501-6 autoclave cycle. A single lap joint was made by machining notches into this laminated assembly, following the test protocol described in ASTM Designation D3165-91. The machined notch quality is one factor that can cause the failure to occur away from the relatively tough TP bond line. Overnotching often results in failure initiation at the notch tip. For strong interfaces, such as those achieved with PS and PES, this often results in delamination of the brittle 3501-6 epoxy. Undernotching, results in a tensile test configuration that results in failure of the unnotched region of the epoxy laminate. Even when the lap-shear coupon is notched correctly, the failure tended to migrate from the bond line into the brittle epoxy laminate. For these soluble TP resins, the toughness gradient through the thickness in the bond line region appears to be the main reason for migration of the failure from the center of the bond line towards the less tough semi-interpenetrating network or even into the brittle epoxy laminate.

3.3 Interlaminar Fracture Toughness The double cantilever beam (DCB) coupons tested here were prepared by placing the PS or PES adhesive film between 3501-6 and 8552 lay-ups (i.e. two [0]20). For the 3501-6 epoxy both unidirectional tape and plain weave fabric lay-ups were prepared, whereas only plain weave fabric lay-ups were prepared for the 8552 toughened epoxy. All of these lay-ups were cured using standard autoclave cycles. A 25.4 mm (1.0 inch) wide piece of 0.051 mm (0.002 inches) thick Kapton film was inserted as a crack starter. The test coupons were 152 mm (6.0 inches) long. Following the test protocol described in ASTM Designation: D5528-94a, each of these coupons typically gave 8-10 toughness measurements. For all of the data reported a minimum of 3 and as many as 6 DCB coupons were tested.

3.4 Pull-off Strength Pull-off test specimens were cut out of each welded "T" section, as shown in Figure 3.

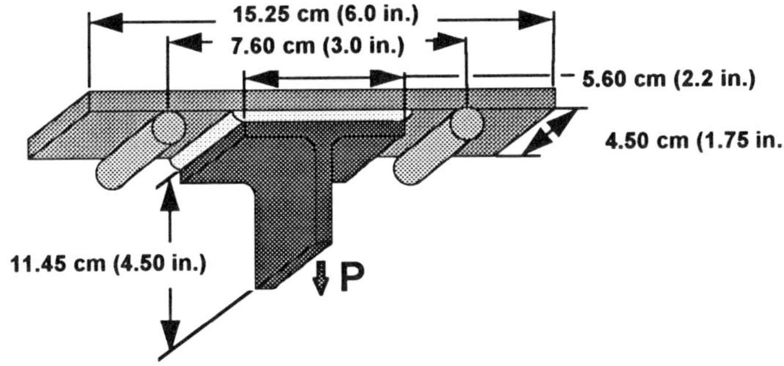

Figure 3 Pull-off Test Configuration

These specimens were simply supported and centered on the "T" section while loading the "T" section perpendicular to the skin panel at room temperature.

4. DISCUSSION

4.1 Interlayer Material Selection and Development The approach used for welding thermoset (TS) laminates together involves the attachment of a resin rich thermoplastic (TP) interlayer to both TS adherend mating surfaces. A number of factors must be considered when selecting the TP interlayer, including its compatibility with the TS adherend, its mechanical and environmental performance, and its ability to welded without degrading the TS adherend. Compatibility refers to the ability to attach the TP adhesive interlayer to the TS laminate in a way that achieves sufficiently high bond strength. A governing guideline in this study was that the TP adhesive interlayer must be attached in a manner that does not alter the standard TS cure cycle. In this study, where Hercules 3501-6 and Hercules 8552 epoxies were welded, this guideline limits the process time (straight ramp at 2.8°C/min (5°F/min)., typically with no intermediate holds), maximum temperature 177°C (350°F) and maximum pressure 0.62 MPa (90 psi) that can be used to attach the TP interlayer.

A number of approaches for attaching a TP adhesive interlayer to the TS mating surfaces are being investigated. Here, results are reported for the Soluble TP Resin approach. This approach relies on penetration of the epoxy monomer reactants into the TP adhesive layer to form a semi-interpenetrating network that bonds the TS laminate to the TP adhesive interlayer (2,3). Two soluble TP interlayer candidates were identified and investigated (Polysulfone (PS), Tg = 190°C (374°F), Union Carbide - Udel 1700) and (Polyethersulfone (PES), Tg = 225°C (437°F), Amoco - Radel A). A potential short-coming of this approach is that these soluble TP candidates are expected to have inherently poor resistance to aerospace solvents. Despite his potential short-coming, this approach was still pursued for a number of reasons, including: 1) potential interlayer candidates were commercially available, 2) these candidates are expected to meet the elevated temperature (82°C (180°F) at 95% RH) performance requirements, and 3) the relatively high melt processing temperatures of these candidates require us to weld at the high end of the expected epoxy processing window. Welding at these higher temperatures will enable us to study the epoxy adherend degradation limits.

4.2 Susceptor Material Selection and Development The encapsulated susceptor package was prepared by melt impregnating a copper mesh screen with the same TP adhesive as used for the interlayer. Though using the same TP adhesive for both the interlayer and susceptor is conceptionally simpler, concepts that use combinations of different TP adhesives are presently being investigated.

4.3 Interlayer Mechanical Performance Interlayer attachment was assessed using both lap shear strength and interlaminar fracture toughness measurements, which confirmed the excellent attachment achieved between the soluble PS and PES interlayers and both the 3501-6 and 8552 epoxies. Unfortunately, as discussed in section 3, a number of test considerations can make it difficult to interpret the results of these mechanical tests. Despite these testing concerns, with careful analysis these tests can be indicative of the quality of the TP/TS interface

4.3.1 Lap Shear Strength Lap shear strengths measured for PS and PES are reported in Table 1. The room temperature (RT) lap-shear strengths appear to be slightly higher than the epoxy control consistent with the tendency for the failure to migrate towards the brittle epoxy. The Hot/Dry and Hot/Wet test results indicate that the PS and PES retain their properties as expected for these high Tg polymers relative to the 82°C (180°F) epoxy requirements.

Table 1. Lap-Shear Strength Measured for PS and PES

Interlayer	Test Condition	Stress (MPa)	Stress (ksi)
3501-6 Control	RT	7.2 +/- 0.5	1.05 +/- 0.08
PES	RT	8.2 +/- 1.3	1.19 +/- 0.19
PES	82°C (180°F)Dry	9.0 +/- 1.0	1.30 +/- 0.14
PES	82°C (180°F) Wet*	8.4 +/- 1.0	1.22 +/- 0.15
PS	RT	8.2 +/- 0.4	1.19 +/- 0.06
PS	82°C (180°F)Dry	9.1 +/- 1.1	1.33 +/- 0.16
PS	82°C (180°F) Wet*	8.1 +/- 0.6	1.18 +/- 0.08

* Wet - 14 day exposure at 95% RH

In general, lap shear strengths have proven to be less than desirable for differentiating interface strength improvements and often lead to inconsistent and unclear results.

4.3.2 Interlaminar Fracture Toughness In the case of the 3501-6 tape laminates, the crack tends to migrate away from the tough TP interlayer and into the region of the semi-interpenetrating network. The symmetrical loading in this test geometry helps to keep the crack from migrating entirely into the 3501-6 region. Thus an increase in the fracture toughness, when compared to the epoxy control, was observed (see Table 2.). Examination of the fracture surface indicated that diffusion of the epoxy monomers into the TP resins had occurred to the extent that they were totally dissolved.

Since the surface plies on the T-spar and skin laminates welded together in this study are plain weave fabric, DCB testing was done both 3501-6 and 8552 plain weave fabric laminates with PES interlayers (see Table 2.).

For the 3501-6 laminates, the toughness values measured for the plain weave fabric coupons were significantly higher that those measured for the tape coupons. A comparison of the failures observed for these tape and fabric laminates gives some insight into these test results. As discussed above, in the case of the unidirectional tape lay-ups, the toughness gradient, which has its maximum near the center of the bond line results in the crack migrating from the center of the bond line towards the less tough semi-interpenetrating network or even into the epoxy laminate. In contrast, for the fabric coupons the crack propagated between the two fabric plies adjacent to the center of the bond line. These failures were often accompanied by a delamination between adjacent fabric plies on one side of the crack. These observations suggest that the mechanical integrity of the fabric layers helps to contain the crack in the region where the PES resin concentration and thus the toughness is expected to be the highest.

For the 3501-6 and 8552 fabric controls, the 8552 toughened epoxy yielded a higher toughness value as expected. A similar comparison between the 3501-6 and 8552 coupons made with the PES interlayer also yielded higher toughness values for the 8552 toughened epoxy. For both adherends, the fabric coupons made with the PES interlayer were twice tough as the respective control.

Table 2. Mode I Fracture Toughness for 3501-6 and 8552 Adherends Measured For Unidirectional Tape and Plain Weave Fabric Laminates

Adherend	Interlayer	Lay-up	G1c (KJ/m^2)	G1c (in-lb/in^2)
3501-6	Control	Tape	0.44	2.53
" "	PS	" "	0.59	3.36
" "	PES	" "	0.77	4.40
3501-6	Control	PW Fabric	0.59	3.35
" "	PES	" "	1.22	6.94
8552	Control	PW Fabric	0.82	4.67
" "	PES	" "	1.63	9.29

4.4 Weld Process Development In order to evaluate the weld strength of the TP interlayer materials, extensive testing was first performed at the coupon level welds, before a final evaluation of a continuous 127 cm (50 in.) weld of a skin to "T" spar assembly was done. Process parameters for induction welding of epoxy adherends, including, but not limited to, coil current, skate velocity and pressure, were developed. The welded assembly was inspected using non-destructive ultrasonic techniques to assess weld line quality.

4.4.1 Detail Part Fabrication The skin panel and spar adherends fabricated for induction welding trials consisted of a 42 ply symmetric lay-up flat skin panel and a 16 ply symmetric lay-up "I" beam spar made from "Hercules" 3501-6/AS4 fabric and 3501-6/IM6 tape for initial trials, and Hercules 8552/IM7 3K-70-PW fabric and 8552/IM7 tape for the adherend material redirection trials. Various TP interlayer materials were co-bonded onto the mating surfaces of each detail during a 177° C (350°F) autoclave cure. After ultrasonic inspection, the skin panel was cut into 50.8 cm x 15.2 cm x 0.635 cm (20 in. x 6 in. x 0.25 in.)specimens. The "I" beam spar was cut into two "T" spars and sectioned into 45.7 cm x 5.6 cm x 0.345 cm (18 in. x 2.2 in. x 0.136 in.) (cap dimension) specimens. After oven drying, all weld surfaces were lightly sanded and cleaned with isopropyl alcohol prior to welding.

4.4.2 Weld Trials - Hercules 3501-6 Epoxy Pull-off strengths measured for the Hercules 3501-6 epoxy adherend details welded using PS and PES are listed in Figure 4 as a function of peak weld temperature.

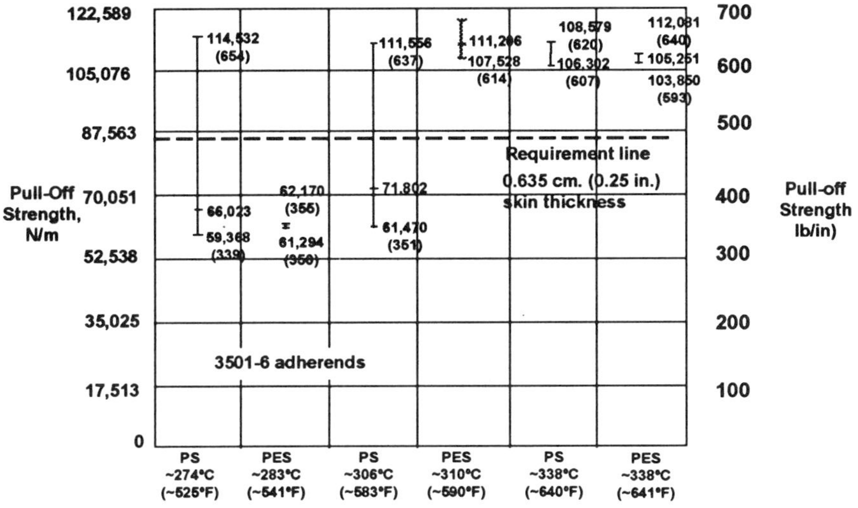

Figure 4 Pull-off Strength Results for PES and PS at Various Weld line Temperatures

An examination of the test specimens showed two distinct modes of failure: 1) Cap failure that initiates in the radius filler and results in delamination of the spar cap and 2) Skin failure that initiates at the wildlife as interfacial peel resulting in delamination of the skin panel. Figure 5 schematically shows the cap and skin failure modes. The lower temperature welds (i.e. PS at 274°C (525°F) and 306°C (583°F) and PES at 283°C (541°F)) all displayed skin failure. In comparison, all but one of the higher temperature welds (i.e. PS at 338°C (640°F) and PES at 310°C (590°F) and 338°C (641°F)) displayed cap failure. The failure mechanism for higher temperature welds precludes exercising the weld and thus the full capability of the weld can not be determined.

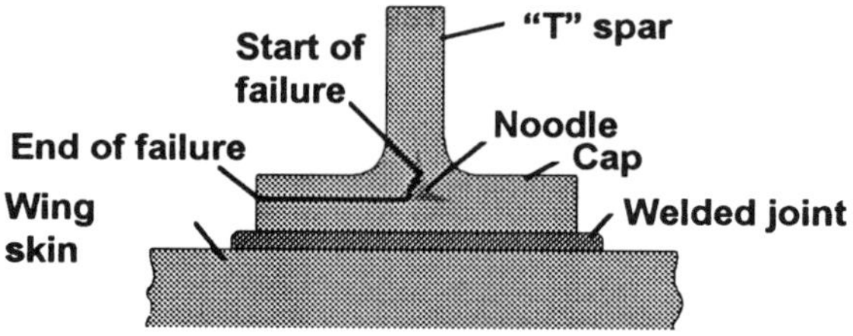

Cap Failure initiating in radius filler

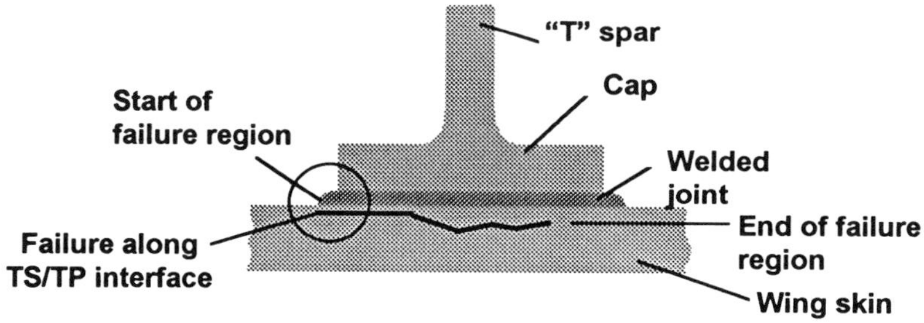

Skin Failure initiating as peel at the weldline edge

Figure 5 Two Modes of Failure in Pull-off Specimens

An investigation was conducted to determine why the failures were occurring in the 3501-6 epoxy adherends rather than the weld line. Overheating of the epoxy adherend was eliminated for various reasons. First, thermal analysis predicted that the radius filler would not exceed 204°C (400°F) during the weld process which is below the temperature at which significant degradation would occur. This analysis was confirmed with actual thermocouple readings. Also, microscopic infrared measurements of the welded adherend, near and far away from the heated weld line, detected no significant differences, indicating no degradation or chemical changes occurred during welding. In addition to thermal analysis and IR measurements, the epoxy details were adhesively bonded and tested to compare failure modes. One set-up used the original radius filler and one set-up used a modified radius filler in an attempt to strengthen the filler and reduce chances of failure in this region. The modified radius filler consisted of the original rolled tape filler wrapped with 0.0127 cm (0.005 in.) thick BMS 8-245 adhesive. The original radius filler test yielded an average pull-off strength of 103,500 N/m (591 lb/in) with failure initiating in the radius filler and traversing across the cap. The modified radius filler test yielded an average value of 136,424 N/m (779 lb/in) with the failure initiating as peel at the bond line and traversing into the skin. In a final attempt to fail the weld line, a thinner skin (32 ply) was welded to a "T" spar with the thought that thinner skin welds fail at lower pull-off strengths. However, the radius did not fail but peel did initiate at the weld line with the crack extending into the skin laminate. In all cases, the specimens failed as a result of the brittle nature of the 3501-6 epoxy material system.

To demonstrate the capability to continuously induction weld, PES was selected for fabricating a 127 cm (50 in.) long "T"-spar (with modified radius filler)/skin assembly. The assembly was welded at a peak weld line temperature of 310° C (590°F). NDI showed that the entire joint was welded. The pull-off tests yielded an average strength value of 126,617 +/-7,005 N/m (723 +/-40 lb/in) with the failure initiating as peel at the edge of the weld line then traversing into the skin laminate.

4.4.3 Weld Trials - Hercules 8552 Toughened Epoxy In an attempt to exercise the weld line, two different material systems were evaluated: 1) Hercules 8552, a toughened epoxy system and 2) Cytec 5250-4, a BMI system. The two material systems were screened by adhesively bonding a "T" spar to a skin using the same lay-up as for 3501-6 (including the modified radius filler). The 8552 toughened epoxy system yielded the highest average pull-off strength: 174,251 N/m (995 lb/in) compared to 5250-4 which gave a 113,307 N/m (647 lb/in).

Based on this evaluation, 8552 was selected for additional welding trials. In addition to this change, the radius size was changed from 0.9525 cm (0.375 in.) to 0.4763 cm (0.1875 in.) in an attempt to facilitate processing of the spars. The objective was to reduce the area required to be filled with the radius filler, and thus, reduce the weakest zone of the spar. Since PES performed the best with 3501-6, PES was selected as the interlayer and susceptor material. Weld trials were conducted on the coupon level at low, medium and high peak weld line temperatures (277° C (530°F), 293° C (560°F) and 310° C (590°F)). Figure 6 shows that 8552 welded at both the low and the high weld line temperatures greatly out-performed 3501-6, with the best pull-off strengths between 192,114 N/m (1097 lb/in) and 270,921 N/m (1547 lb/in).

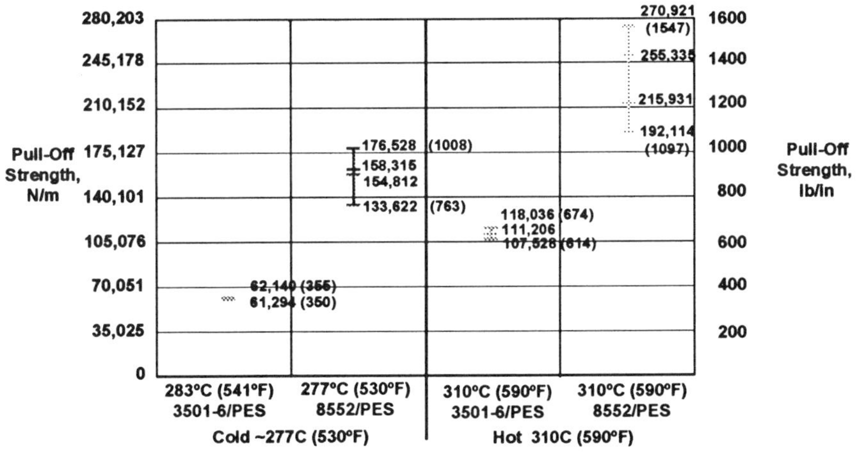

Figure 6 Comparison of Pull-Off Strength for 8552 vs. 3501-6

After coupon testing, a 127 cm (50 in.) continuous demonstration weld was performed at the high peak weld line temperature (310° C (590°F)). The pull-off test results ranged from a minimum of 262,690 N/m (1500 lb/in) to a maximum of 367,766 N/m (2100 lb/in). Again, two types of failure occurred. The failure initiates in the radius filler and traverses across the cap or begins as peel at the edge of the weld line and traverses into the skin. In either case, the material system is still not allowing the weld line capability to be fully determined

5. CONCLUSION AND SUMMARY

Soluble thermoplastic interlayers and susceptors can be used to induction weld thermoset composite adherends. NDI, photomicrographs and pull-off strength determinations showed that high quality welded joints were readily produced. When evaluated by a pull-off strength testing, the samples did not fail in the weld line indicating that the thermoset adherend was the weakest component in this joint configuration. The observed failures initiated as peel at the edge of the weld line or as a crack at the radius filler in the "T" spar. Since the failures observed were in the adherend material, the full capability of the weld could not be determined. Still, the strength values obtained were sufficiently high and clearly show that induction welding has great potential for use in future aircraft structure.

The 8552 toughened epoxy "T" spar/skin assembly welded with PES a peak weld line temperature of 310° C (590°F) gave the highest pull-off strengths resulting in two failure modes, both of which are material dependent. Since the failure modes observed involve phenomena such as peeling, crack initiation, crack propagation and delamination, it is a reasonable result that the 8552 toughened epoxy yielded significantly higher pull-off strengths than the brittle 3501-6 epoxy. Though, as discussed in detail, a number of test considerations can make it very difficult to interpret the DCB test results, the Mode 1 interlaminar fracture toughness values measured were indicative of the measured pull-off strengths. For example, a comparison between the 3501-6 and 8552 coupons made with the PES interlayer also gave significantly higher toughness values for the 8552 toughened epoxy adherend.

6. REFERENCES

1. C. A. Butler, R. Pitchumani, J. W. Gillespie, Jr., and A. R. Wedgewood, Proceedings of the 10th Annual ASM/ESD Advanced Composites Conference and Exposition, Dearborn MI, November 1994, pp. 595

2. R. C. Don, S. H. McKnight, and J. W. Gillespie, Jr., Proceedings of the 10th Annual ASM/ESD Advanced Composites Conference and Exposition, Dearborn, Michigan, November 1994, pp. 605

3. S. H. McKnight, R. C. Don, M. Scott. A. Braem, and J. W. Gillespie, Jr., Proceedings of the ANTEC 1994, Boston MA, May 1994

7. ACKNOWLEDGMENT

This work was supported (i.e. 50% cost share) by ARPA (Order No.:FY1457-95-020000) and administered under the technical direction of Ms. Tia Benson-Tolle, AFWL/MLBC, Structural Materials Branch, Wright Patterson Air Force Base.

28th International SAMPE Technical Conference
November 4-7, 1996

FEASIBILITY STUDY OF ELECTRON BEAM CURED COMPOSITES

Tuan Q. Cao, Anthony Falcone, James R. Beymer,
Dennis A. Russell, Charles E. Vicks and Erica D. Smith-Hicks
Boeing Defense & Space Group, Seattle, Washington

Abstract

In the past five years, many development projects have demonstrated that the electron beam (EB) curing process can potentially be used to fabricate graphite composites for structural aerospace applications. Economic analysis of the process revealed that the cost of the electron beam curing is potentially very attractive for manufacturing low-cost structural composites. When EB curing is coupled with automated layup equipment such as filament winding or resin transfer molding (RTM), the manufacturing cost can be further reduced. A Boeing funded EB composites project has tested both free radical and cationic EB curable composite matrices. Mechanical properties of some matrices are equivalent to those of 177°C (350°F) thermally cured epoxy systems. This paper discusses the results of the Boeing technical development effort that demonstrates the feasibility of the EB curing process to manufacture affordable structural composites.

1. Introduction

In the past five years, industry has investigated the electron beam (EB) process for manufacturing affordable structural composites. Advances in the design of accelerators have reduced the size and cost of EB equipment. Thus EB process has become more attractive for manufacturing structural composites. Many investigations have indicated that the EB process is technically feasible for manufacturing structural composites[1,2]. Matrices containing acrylate and vinyl ether functional groups have been used successfully in ultraviolet (UV)

Key Words: Adhesives, Composites, Electron Beam Processing, Affordable Composites.

curing processes such as protective coatings. These matrices are cured by a free radical mechanism, and they can also be cured by EB process at a much faster rate. Unlike UV photocatalytic curing process, an electron beam can penetrate deep into materials for curing. A 10 Mev electron beam can penetrate 2.54cm (1 inch) thick fiberglass or carbon fibers reinforced composites for curing polymeric matrices. The EB Cooperative Research and Development (CRADA), sponsored by the Department of Energy Defense programs and ten industrial partners, has been developing EB cured composite technology for manufacturing affordable composites [3]. Its focus is to develop cationic EB curable epoxy systems for manufacturing structural graphite composites. These new cationic epoxy systems are more efficient EB curable matrices that may reduce the processing cycle time of composites. When coupled with automated manufacturing equipment such as filament winding and resin transfer molding, as suggested by Walton and Crivello [4], the manufacturing cost can be further reduced.

Advantages of the EB curing process include low-temperature curing, controllable energy delivery, fast process, and short cycle time. The low-temperature curing capability of EB would allow low-cost tooling such as wood or aluminum to be used, and the end products would have less internal stresses than those from a thermal cured process. Since energy delivery of the electron beam is quite controllable, the part can be cured slowly and uniformly to control warpage.

In the past four years, Boeing Defense & Space Group in Seattle, Washington, has investigated the feasibility of the EB curing process for manufacturing affordable composites for subsonic and high-performance aircraft. The purpose of the investigation was to determine the feasibility of both free radical and cationic cured systems for manufacturing structural composites. This paper describes the investigation and preliminary test results.

2. Experimental

2.1 Test Materials Table 1 shows the candidate matrices of the investigation. These materials were formulated by and purchased from Applied Poleramic Inc (API). FW3 and FW4 are two free radical cured systems and Boecat-C is a cationic cured system. Boecat-C is a proprietary formulation of high aromatic epoxy catalyzed with (4-Octyloxyphenyl)-phenyliodonium hexafluoroantimonate (OPPI), which is a General Electric aryl fluoroantimonate, product 479-2092C.

Table 1. Candidate Matrices

Matrix ID	Resin	Neat Resin	Prepreg AS4-5SH
FW3	Epoxy-Acrylate	X	X
FW4	Toughened Epoxy-Acrylate/ bismaleimide	--------	X
Boecat-C	proprietary formulation	X	X

2.2 Specimen Fabrication Boecat-C neat resins and small samples of its prepregs were screened in ^{60}Co gamma cell. A series of samples were irradiated with different dosages to determine the optimum dosage for curing matrices and laminates. Dosages of 50-200kGy were used. Two samples from the Boecat-C laminate specimens exposed to 100 and 150 kGy were postcured at 177°C (350°F) for one hour to compare to the as-irradiated specimens. Thermal analyses were performed on these cured samples to determine the degree of cure and their Tg. Differential Scanning Calorimetry (DSC) and Dynamic Mechanical Analysis (DMA) were used.

Unidirectional laminates measuring 25.4cm X 10.16cm (10" X 4") and 16 ply thick of three materials were laid up. The layups were placed between two plies of porous armalon and then two plies of 1781 glass fabric breather. The whole assembly of each layups was laid on an aluminum caul plate and vacuum bagged using Kapton bagging materials and high-temperature sealant. The layups were preconsolidated in a vacuum oven at 71°C (160°F) for 30 minutes, and were then stored at room temperature with vacuum applied to the bag until EB treatment. A dosage of 100-150 kGy was deposited on these specimens while vacuum pressure was applied on them by the 10MeV Linac described in section 2.3 below. TTU inspections were performed on cured laminates, and they were machined into open-hole compression coupons for mechanical tests. Cross-section microscopy and thermal analyses were performed on trim pieces from each specimen.

2.3 EB Equipment and Electron Treatment Specimens were exposed to electron beam and gamma ray radiation at the Boeing Radiation Effects Laboratory. The ^{60}Co gamma cell produced an approximately 1 MeV gamma ray at a dose rate of 18-20 Gy/sec. Screening specimens were placed in this chamber for curing and were exposed to gamma ray radiation in the chamber at different durations to achieve the required dosage deposit.

The source of electrons for the curing was a 10 MeV Linear Accelerator (Linac). The S-band Linac electron beam was directed horizontally in air at the layups mounted vertically on an "x-y"-translation table as shown in figure 1. Since the EB was fixed, the panel was moved in both horizontal and vertical directions in front of the electron beam in order to irradiate the entire laminate with uniform dosage. This method is similar to the more traditional technique of electrically scanning the electron beam in one direction and mechanically moving the laups to be irradiated in the other.

Electron beam dosimetry was performed by first centering the "x-y"-translation table on the electron beam and then measuring the beam profile. Figure 2 is a typical profile made by scanning a dosimetry diode across the beam and displaying the spatial results as equal dose lines. Having achieved a symmetrical profile, a set of thermo luminescent dosimeters (TLD) are placed over the area to be scanned, and the x-y translating table is set up under computer control to move the samples in an x-y pattern through the beam as it pulses. Typical accelerator and scanner parameters used are listed in table 2. The dose delivered to the samples varied from 820 krads (8.2 kGy) to 900 krads (9.0 kGy) per scan.

Table 2. Typical Accelerator and Scanner Parameters

Beam Energy	10 MeV
Electron Pulse Height	230 milliamperes
Electron Pulse Width	5 microseconds
Electron Pulse Rate	15 pulses/min.
Scanner Speed (x & y)	2 in./min
Scanner Acceleration (x&y)	50 rev/sec^2
Scanner Vertical Step	0.2 in./trace
X travel	12 in.
Y travel	7 in.
Dose per scan	0.8-0.9 Mrad

2.4 Thermal Analysis A Perkin-Elmer Series 7 Thermal Analysis System was used for the DSC and DMA analyses to determine glass transition temperatures (Tg) of the EB cured specimens. The DSC analysis was scanned at 10°C/min from 50°C to 350°C. For those specimens exhibiting residual heat of reaction, a second scan was performed after quenching to 50°C to determine thermal postcure Tg.

Mechanical properties were measured using the Dynamic Mechanical Analyzer (DMA) of the Perkin-Elmer Series 7 Thermal Analysis System. Three point-bending mode was used. Samples of 2mm wide, 1mm thick and greater than 15mm long were mounted on a 15mm wide gold-plated stainless steel bending platform. A typical strain of 0.25% was applied to the sample at a frequency of 1hertz using a 10mm knife probe tip, and with a tension control setting of 120 percent. Temperature was ramped from 20°C to 250°C at a rate of 3°C per minute.

2.5 TTU Inspection Waterjet through transmission ultrasonic (TTU) inspection of the cured specimens was performed at 5Mhz.

2.6 Cross-Section Microscopy Representative samples from cured specimens were submitted for cross-section microscopy. Samples were mounted in clear epoxy and then polished to obtain micrographs. Micrographs of 5X, 20X, and 50X magnifications were obtained from each sample.

2.7 Open-hole Compression Specimens and Testing Two open-hole compression specimens were machined from FW3 and FW4 laminates to dimensions of 3.175cm (1.25 in.) X 22.86cm (9 in.). A hole of 0.635cm (0.25in.) diameter was drilled at the center of the specimens. Open-hole compression testing was performed on an MTS mechanical test instrument using a cross-head speed of 0.127 cm/min (0.05 in/min).

3. Results and Discussions

3.1 FW3 and FW4 Matrix Systems Table 3 summarizes the thermo-mechanical test results of screening EB cured specimens of AS4/FW3 and AS4/FW4. They were exposed to different dosages of EB irradiation to optimize the EB exposure of these free radical matrices. DSC analysis of the EB cured FW3 matrix did not show any useful data, and Tg could not be extracted from DSC thermograms. Only DMA could reveal the Tg of the EB cured FW3 samples. DSC thermograms of FW4 matrix (figure 3) did show a Tg around 49-54°C and an exothermic peak at 200°C. This exothermic peak may be due to the crosslinking of the bismaleimide (BMI) component in the FW4 matrix. The DMA's Tg is rather low, but this may be due to the amount of the thermoplastic toughener added in the FW4 matrix. Since no thermal postcure was performed on EB cured FW4 smaples for the DMA or second heat scan by DSC, the ultimate Tg of FW4 is unknown.

Table 3. Thermal Analysis Data of Screening AS4/FW3 and AS4/FW4 Specimens

		DSC		DMA	
Matrix	Dosage kGy	Residual Heat (J/g)	Tg (°C) (First Heat)	Tg (°C) (G')	Tg(°C) tan δ
FW3	75	-	-	<100	125, 175
	100	-	-	115	160, 210
	150	-	-	118	150, 200
FW4	75	-19	49	45	105
	100	-18	52	58	115
	150	-14	54	48	125

Six laminates measuring 25.4cmX10.16cm (10"x4") by 16 plies thick were fabricated and cured by EB at a dosage of 100 kGy. Three of those were AS4/FW3 and three were AS4/FW4. An average temperature rise during the irradiation was 27°C, but this only reflected the equilibrium between the heat generated by the exotherm of the crosslinking and the heat loss to the aluminum caul plate underneath the laminate being cured. TTU inspections were performed on these specimens. The average attenuation of the FW3 laminate was 36 ±3dB and of FW4 was 45±3dB. Cross-section micrographs of FW3 and FW4 laminates indicated that the porosity of all panels was more than 10%, and FW4 panels (figure 4) were consolidated better than FW3 panels(figure 5). Open-hole compression coupons were fabricated from these six laminates and tested at 82°C (180°F) wet. The average open-hole compression strengths were 93 MPa (13.5 ksi) for FW3 coupons and 120 MPa (17.4 ksi) for FW4 coupons. Improved consolidation of FW4 panels might explain their higher tested open-hole compression strengths than those of FW3.

Another two laminates measuring 25.4cmX10.16cm (10"X4") by 16 plies thick were fabricated for open-hole compression test at room temperature. A 16-ply laminate of FW3 irradiated with 107 kGy, and a 16-ply laminate of FW4 irradiated with 130kGy were made. TTU inspections of these laminates had shown that the attenuation of the FW3 panel was detected at 34±4 dB and of the FW4 panel was 29±3dB (figure 6). These results are lower than those of previous laminates indicating improved laminate quality. Photomicrographs of these two laminates (figure 7 and figure 8) also showed an estimated porosity of 5-10% by visual inspections. The laminates appeared very dry, showing no resin flow as compared to those of fresh prepregs. The micrographs showed high resin content in the laminate due to the lack of adequate consolidation pressure during EB curing. Open-hole compression coupons were machined from these laminates and tested. The average open-hole compression strength at room temperature of FW3 coupons was 137.9 MPa (20 ksi) and of FW4 was 175.1 MPa (25.4 ksi). These test results were considered very good for fabric laminates of 5-10%

porosity. A typical open-hole compression strength at room temperature of thermal cured system such as BMS 8-256 fabric of IM7977-3 fiber is 117.2 MPa (17 ksi). Based on these data, FW3 and FW4 matrices are potential candidates for the EB curable structural composites.

3.2 Boecat-C Matrix System Table 4 summarizes thermal-mechanical test results of the screening EB cured Boecat-C neat resin samples. Five samples of neat resin were exposed to gamma ray radiation at dosages of 60, 80, 100, 150 and 200 kGy.

Figure 9 showed a typical DSC thermogram of the EB cured Boecat-C neat resin. All EB cured neat resin samples showed two exothermic peaks, one at around 120°C and another at 200°C. This indicated that Boecat-C contained two reactive epoxy components having crosslinking kickoff temperatures at these two exothermic temperatures.

Table 4. Thermal-Mechanical Analysis of Neat Resin Boecat-C

	DSC		DMA	
Dosage kGy	Residual Heat (J/g)	Tg (°C) second heat	Tg (°C) (G')	Tg(°C) tan δ
60	-238	158	153	202
80	-176	162	147	200
100	-125	159	145	190
150	-137	164	149	**
200	-109	154	147	180

**missing data

Table 5 is the summary of the thermal-mechanical test data of the screening AS4/Boecat-C specimens cured by gamma ray radiation. DSC thermograms of all EB cured Boecat-C laminates (figure 10) showed a residual heat of reaction around 17-19 J/g for all dosages, and two peak exothermic temperatures at 125°C and 200°C. Theoretically, these laminates should have 35% by weight resin contents, and therefore, their residual heat of reactions should be around 44 - 48 J/g. The measured residual heat of reaction of the composite laminates was less than half of calculated values. This indicated that the resin in graphite composites can be cured at less dosage than neat resin, and the uncured epoxy functionality that has a total of 17-19 J/g heat of reaction may not be cured by EB regardless the levels of irradiation.

Table 5. Thermal-Mechanical Analysis of AS4/Boecat-C

	DSC		DMA	
Dosage (kGy)	Residual Heat (J/g)	Tg (°C) second heat	Tg (°C) G'	Tg(°C) tan δ
100	-16	175	170	200
100 kGy and 1hr 177°C postcure	-	175	173	200
150	-17	179	167	189
150 kGy and 1hr 177°C postcure	-	180	159	200
200	-19	184	159	189

Boecat-C marix had a purple-red color after being cured by EB, but this color turned to the usual amber color of cured epoxy after a thermal postcure at 177°C. The DSC data of table 5 showed that the Tg's did not increase after thermal postcure, although it removed the residual heat of reaction. DMA data also shows a similar effect. A dosage of 100kGy and a short thermal postcure appeared to be sufficient for the Boecat-C to reach the Tg for structural applications.

Three 16-ply laminates of AS4/Boecat-C fabric were fabricated and cured with 150kGy using 10MeV EB. An average temperature rise during the irradiation was 45°C, but again, this only reflected the equilibrium between the exotherm due to the polymerization and heat loss to the aluminum caul plate underneath the samples. The attenuation of the TTU inspections of these laminates was detected at an average of 48±5 dB, and the TTU images indicated lower porosity than that of FW3 and FW4 panels. Boecat-C of the EB cured panels appeared to flow very well. On-going tests are being done on these panels.

4. Conclusion

Preliminary data of this investigation has indicated that both FW3 and FW4, which are free radical cured systems, produced excellent open-hole compression strengths, exceeding that of BMS 8-256 which is a thermal cured system. Both systems could potentially be used for structural composites. Further testings are needed to fully characterize the properties of these free radical systems for structural applications.

Preliminary screening works of the Boecat-C indicated that this cationic cured system can develop Tg high enough for use in application areas where 177°C thermally cured epoxy composites can be used. Boecat-C may need a thermal postcure to improve its mechanical properties.

5. Acknowledgments

The author would like to extend thanks to Rich Moulton from API who provided many consultations, to Jennie Chen and Gary Duschl who provided thermal analyses of the samples, and to John Packard and Jeff Brewer who provided photomicrographs of the cured laminates.

6. References

1. Saunders, C.B., V.J. Lopata, W. Kremers, M. Tateishi and A. Singh; SAMPE International Symposium, 37, 944(1992)

2. Saunders, C.B., V.J. Lopata, W. Kremers, T.E. McDougall, M. Tateishi and A. Singh; SAMPE International Symposium, 38, 1681(1993)

3. C. J. Janke, G F. Dorsey, S.J. Havens, V.J. Lopata; SAMPE International Symposium, 41, 196(1996)

4. T. C. Walton, J V. Crivello; SAMPE International Symposium, 44, 1266(1995)

7. Figures

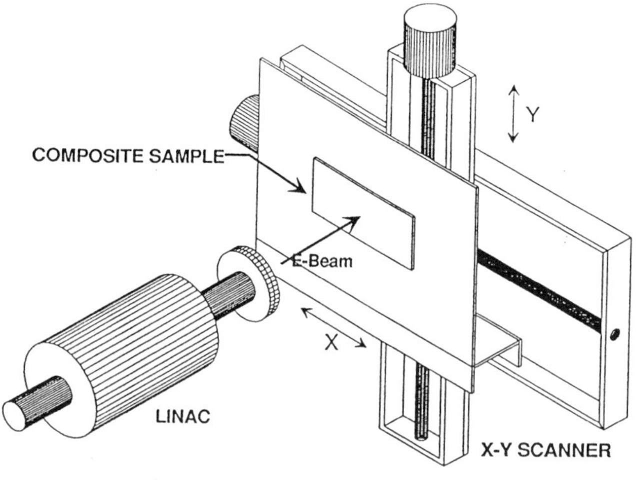

Figure 1. Equipment Configuration for EB Curing

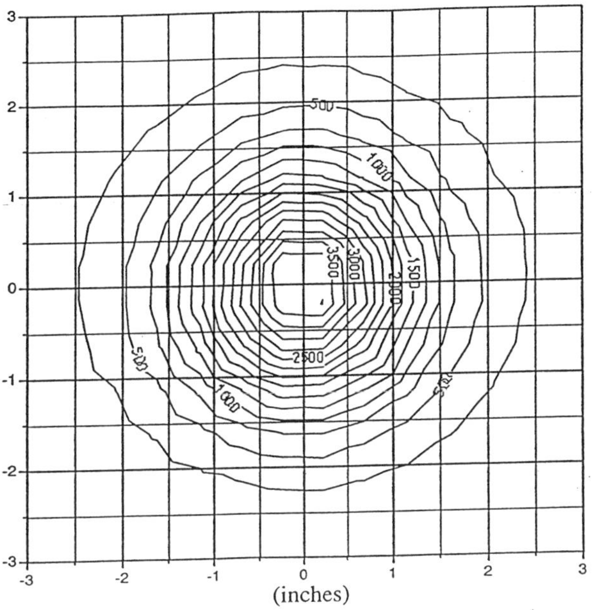

Figure 2. Typical E-Beam Profile

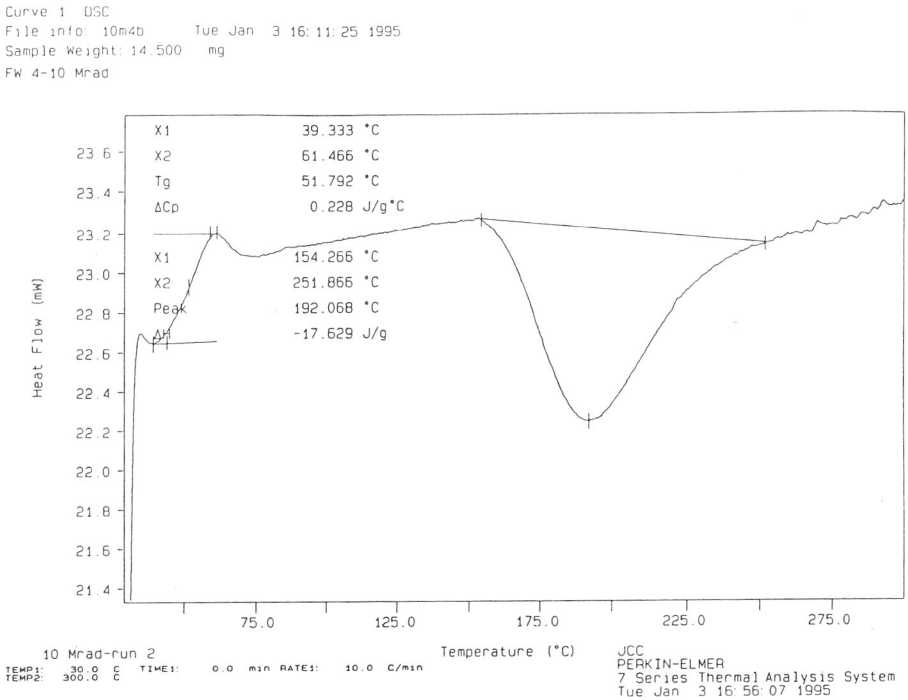

Figure 3. Typical FW4 DSC Thermogram

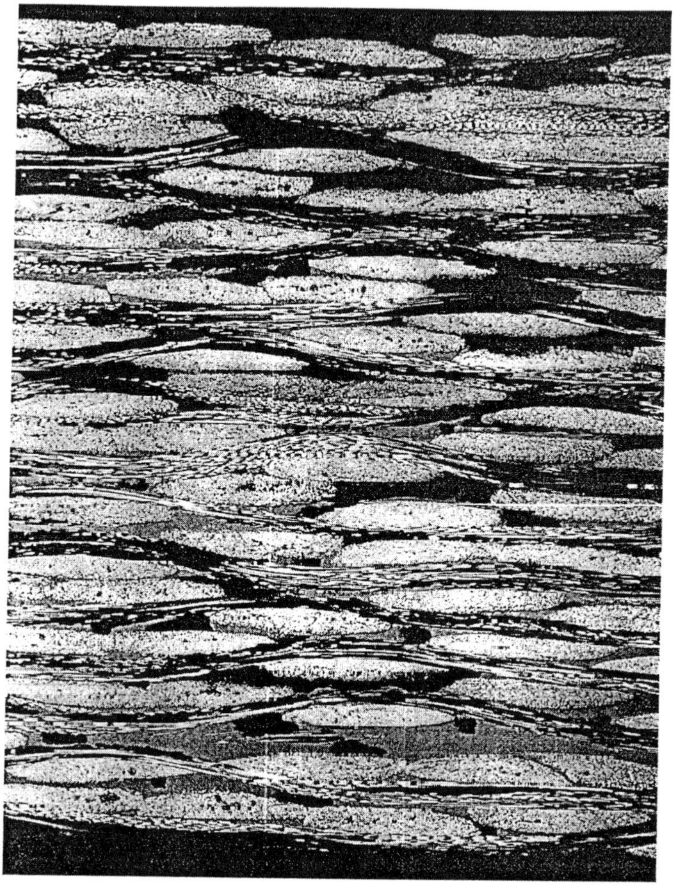

Figure 4. First FW4 Laminate Cross-Section Micrograph

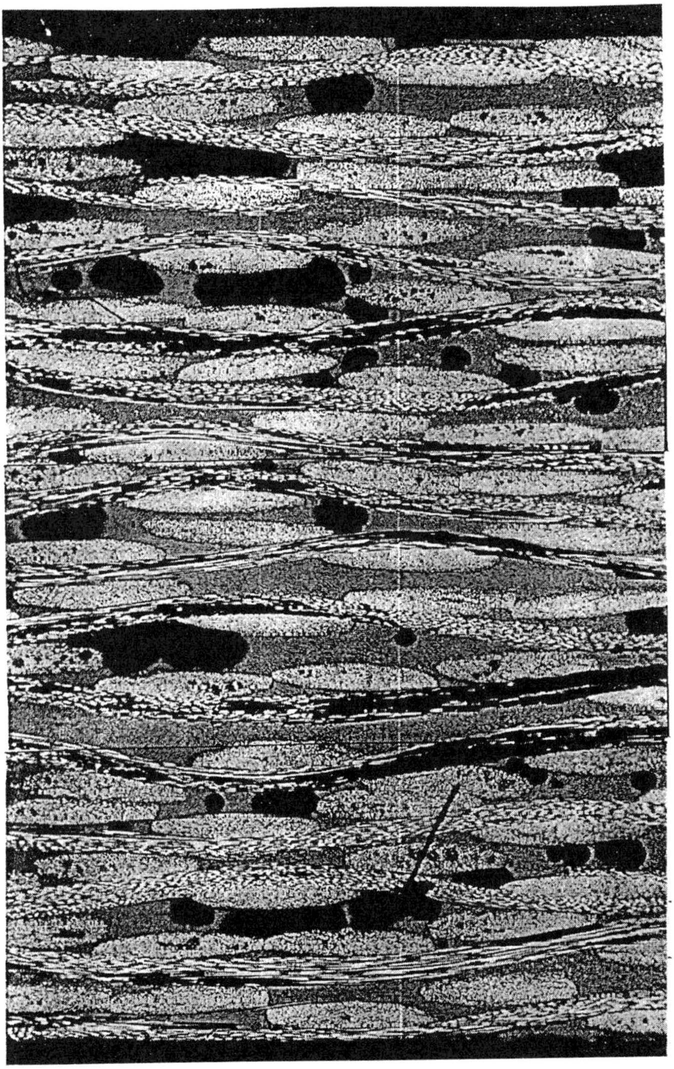

Figure 5. First FW3 Laminate Cross-Section Micrograph

Figure 7. Second FW3 Laminate Cross-Section Micrograph

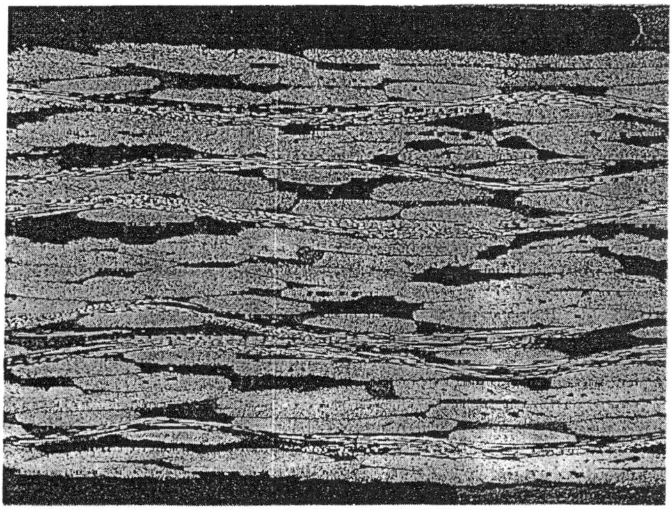

Figure 8. Second FW4 Laminate Cross-Section Micrograph

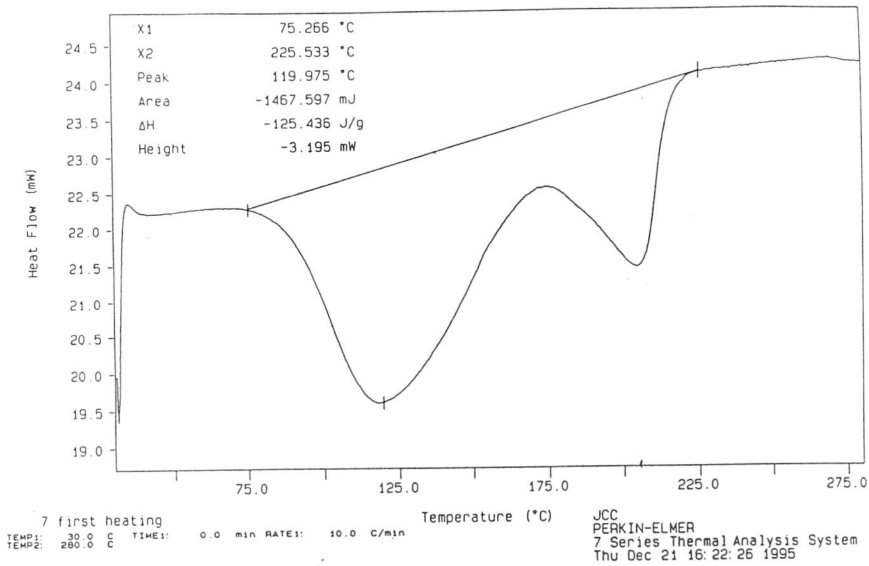

Figure 9. DSC Thermogram of Boecat-C Neat Resin Exposed to 100kGy Radiation

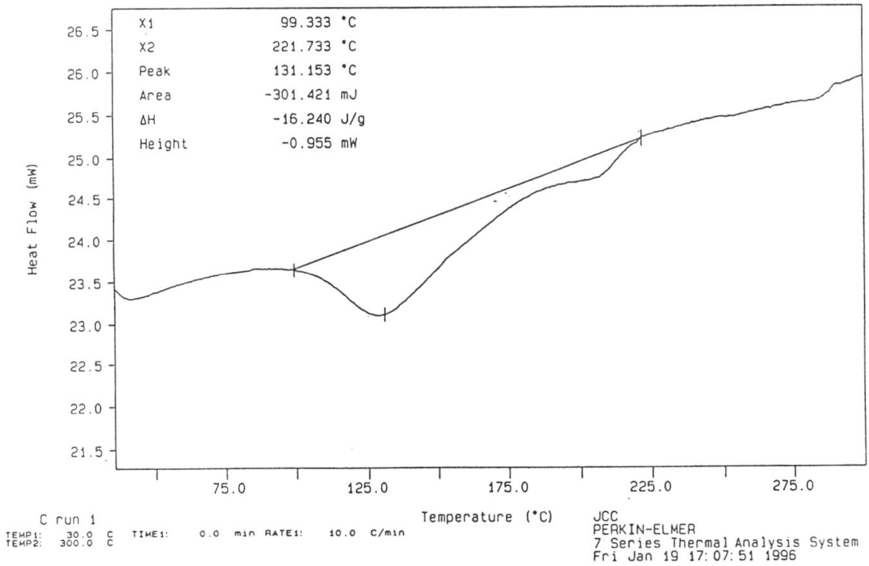

Figure 10. DSC Thermogram of AS4/Boecat-C Exposed to 100kGy Radiation

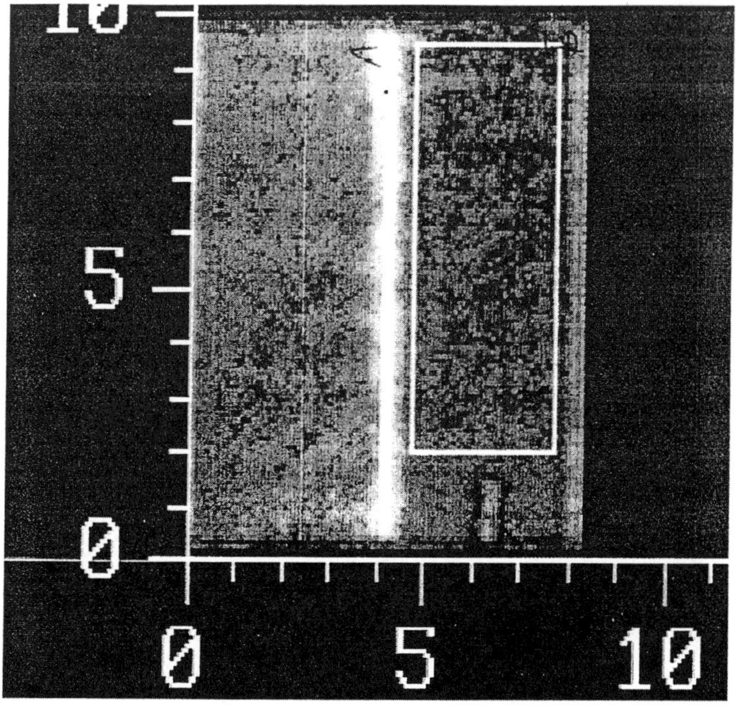

Figure 6. TTU Images of EB Cured FW3 Laminate (left) and FW4 Laminates (right)

28th International SAMPE Technical Conference
November 4-7, 1996

TOUGHENED EPOXY RESINS CURED BY ELECTRON BEAM RADIATION

Christopher J. Janke
Oak Ridge Centers for Manufacturing Technology
Oak Ridge National Laboratory*
Oak Ridge, Tennessee 37831-8048

George F. Dorsey
Lockheed Martin Energy Systems, Inc.**
Oak Ridge, Tennessee 37831-8097

Stephen J. Havens
Oak Ridge Institute for Science and Education,
Oak Ridge, Tennessee 37831-0117

Vincent J. Lopata
Atomic Energy of Canada Ltd. - Whiteshell Laboratories,
Pinawa, Manitoba, Canada ROE 1LO

ABSTRACT

Preliminary research has determined that conventional epoxy resins can be cured by exposure to Electron Beam (EB) radiation as provided by a high energy/power electron beam accelerator to provide materials with high glass transition temperatures and mechanical properties comparable to thermally cured epoxies. A cationic photoinitiator at a concentration of 1-3 parts per hundred of the epoxy resin is required for this process. However, these materials tend to be brittle. In order to increase the fracture toughness (K_{IC}) of the EB cured materials, a variety of toughening agents were incorporated into the epoxy resin/photoinitiator blend in concentrations ranging from 5 - 30 percent. High molecular weight thermoplastics such as Udel P-1700, Radel A-200, and Ultem 1000 and hydroxy-terminated polysulfone oligomers nearly doubled K_{IC} with no reduction in glass transition temperature. Viscosities of the thermoplastic/epoxy resin/photoinitiator blends were extremely high. Reactive flexibilizers and rubbers also provided tougher materials, but sometimes with reduction in glass transition temperature. For example, a blend of 10

*Managed by Lockheed Martin Energy Research Corporation for the U.S. Department of Energy under contract DE-AC05-96OR22464.
**Managed by Lockheed Martin Energy Systems Incorporated for the U.S. Department of Energy under contract DE-AC05-84OR21400.

percent Shell Kraton FG 1901 X (a maleic anhydride terminated SEBS copolymer) with Dow Tactix 123 (bisphenol A epoxy) containing 3 phr (4-octyloxyphenyl) phenyliodonium hexafluoroantimonate was irradiated at a total dosage of 150 kGy.Glass transition temperature (tan delta) of the cured material was 171 °C and the K_{IC} was 0.92 $MPa \cdot m^{1/2}$.

KEY WORDS: Electron Beam Curing; Epoxy Resins; Toughened Epoxy Resins

1. INTRODUCTION

Previously the authors described a process by which conventional epoxy resins can be cured by electron beam (EB) radiation at selectable temperatures to provide materials with high glass transition temperatures (essentially the same as with thermal curing) (1). A cationic photoinitiator at a concentration of 1-3 parts per hundred of the epoxy resin is required. This process was developed as part of a Cooperative Research and Development Agreement (CRADA) sponsored by the Department of Energy Defense Programs and ten industrial partners. One of the primary objectives of this project is the development of high performance EB curable matrix resins for use in DOE Defense Programs and United States aircraft and aerospace industries. Although epoxy resins cured by this process have flexural strengths and moduli comparable to thermally cured epoxy resins, they tend to be somewhat brittle. Resins intended for use in aircraft, aerospace, and many other applications must have sufficient toughness and impact resistance to endure many years of harsh service. What is needed is a means by which EB cured cationic epoxies can be toughened and still retain the good thermal and mechanical properties of the original composition. Additionally, the toughening agent must not interfere with the cationic curing mechanism and must be miscible with or form a stable suspension with the epoxy resin-photoinitiator composition.

A variety of toughening agents and flexibilizers have been used with thermally cured epoxy resins, which typically use an amine coreactant or hardener for crosslinking. These tougheners include: rubbers or elastomers, either as discrete particles or reacted with epoxy resin (2-7), epoxy functionalized flexibilizers (8), engineering thermoplastics (9-12), and amine or hydroxy-terminated thermoplastic oligomers (13-16). There are no references in the literature to the use of these or any other materials to toughen an EB curable cationic epoxy composition (patents are pending). Our approach has been the evaluation of a wide variety of toughening agents in a baseline EB curable cationic epoxy composition, with the expectation that the results obtained would be applicable to higher performance epoxy compositions. EB irradiation of a mixture of Dow Tactix 123 (a high purity, low viscosity diglycidyl ether of bisphenol A) containing three parts per hundred (4-octyloxyphenyl) phenyliodonium hexafluoroantimonate (OPPI) resulted in a cured material with a glass transition temperature of 182 °C (1). The initial low viscosity of this composition (~ 6000 cP at 25 °C) made it very amenable to the addition of toughening agents. The addition of high molecular weight thermoplastics in particular significantly increased the viscosity of the epoxy resin composition. Compositions of the epoxy resin, cationic photoinitiator, and toughener were cured using a high energy EB accelerator. Thermal and mechanical properties, including fracture toughness (K_{IC}), were determined.

2. EXPERIMENTAL

2.1 Materials

2.1.1 Epoxy Resins Commercial epoxy resins were obtained from the manufacturer. These are described in Table 1.

2.1.2 Photoinitiator (4-Octyloxyphenyl) phenyliodonium hexafluoroantimonate (OPPI) was purchased from General Electric Silicones Company as Aryl Fluoroantimonate, Product 479-2092 C.

2.1.3 Toughening Agents Samples of commercial materials were obtained from the various manufacturers (Table 1). Hydroxy-terminated polysulfone oligomers were either generously provided by NASA Langley Research Center or prepared as described in the literature (17).

2.2 Epoxy-Toughening Agent Blending Many of the toughening agents were easily stirred into the epoxy resin at approximately 80°C along with the photoinitiator to form a homogeneous mixture. Udel P-1700 and the hydroxy-terminated polysulfone oligomers were slowly dissolved in the epoxy at 140-150°C with vigorous stirring. The viscous solutions were allowed to cool to approximately 80°C before the photoinitiator was added. Other high molecular weight thermoplastics such as Radel A and Ultem 1000 were dissolved in methylene chloride, the solution mixed with the epoxy resin, and the methylene chloride removed under vacuum at elevated temperature. Solid Kraton rubber was either similarly blended into the epoxy using toluene or melt blended into the epoxy at 190-200°C for 30 minutes using a high shear mixer. Again, the photoinitiator was not added until the temperature cooled to approximately 80°C. Nylon particles such as Orgasol 2001 and polyimide particles such as IMITEC X-902 were insoluble and dispersed to form a suspension. Rubber particles such as Paraloid EXL-2330 were blended into the epoxy as a dispersion in acetone, with subsequent removal of the acetone under vacuum at elevated temperatures.

2.3 EB Accelerator All electron beam irradiation was performed at the Whiteshell Laboratories of AECL using the I-10/1 Electron Linear Accelerator (Energy, 10MeV; Power, 1kW).

2.4 Fracture Toughness The previously degassed toughened epoxy-photoinitiator mixture was poured hot into an aluminum mold (specimen thickness, 6.25 mm) which was treated with release agents. Thickness of the side plates was 1.6 mm. Specimens were placed in a vacuum oven maintained at approximately 80°C to remove any remaining bubbles, allowed to cool, then irradiated in multiple passes of 50 kGy per pass for a total of 150 kGy. Cured panels were cut into 12.5 x 90 mm test specimens and fracture toughness (K_{IC}) determined according to ASTM E 399 using 6-10 specimens per test condition.

2.5 Dynamic Mechanical Analysis Dynamic Mechanical Analysis (DMA) was carried out using a Rheometrics Solids Analyzer RSA. II. The sample was placed in a dual cantilever tool, with an initial stress and strain of zero. The strain used for measuring the rheological properties was set at 0.1 percent. The temperature was raised at 2.5°C per step with a dwell time of 12 seconds. The frequency of the applied strain was 6.28 radians per second. Sample size was normally 3 mm x 7 mm x 55 mm. Measurements were obtained using the software provided by Rheometrics for the instrument.

2.6 Scanning Electron Microscopy (SEM) Micrographs of the fracture surfaces were taken using a JEOL JSM-35 CF Scanning Electron Microscope operated at 25 kV.

3. RESULTS AND DISCUSSION

Untoughened EB cured bisphenol A-based epoxy resin was quite brittle. The addition of any toughening agent resulted in some increase in fracture toughness (K_{IC}). Addition of high molecular weight, high glass transition temperature (T_g) thermoplastics to the uncured resin composition resulted in increased toughness with little reduction and some increased T_g's in the cured resin (Table 2). Only small decreases in moduli were observed. In several cases (Radel and Ultem) increasing toughener concentration was ineffective. Concentrations of Udel up to 20 percent were increasingly effective. However, the viscosity of the toughened epoxy resin composition increased exponentially with Udel concentration (e.g. 7000 and 50,000 cP for 10 and 20 percent solutions, respectively, of Udel in Tactix 123 at 60°C). Udel was more useful since it could be dissolved into the epoxy without the use of solvents. Insoluble nylon and polyimide particles (Orgasol and IMITEC) did not substantially increase resin viscosity, but did increase toughness without decreasing T_g or modulus.

In general the hydroxy-terminated polysulfone oligomers were more effective as their concentration and molecular weight increased (Table 3). For the same degree of toughness, viscosities of the hydroxy-terminated polysulfone compositions were comparable to high molecular weight thermoplastic compositions. For example, the viscosity of the 30 percent solution of 6000 g/mole polysulfone oligomer in Tactix 123 was approximately 29,000 cP at 60°C. For the most part the addition of reactive flexibilizers only increased toughness by 30-40 percent (Table 4), often with decreased T_g's. However the low viscosity flexibilizers kept the viscosities of the resin compositions low and suitable for resin transfer molding (RTM) or filament winding applications. Tactix 123 containing 10 percent Vikoflex 7170 had a viscosity of 3600 cP at 24°C. The effect of rubbers and elastomers was varied (Table 5). All aided in increasing toughness, but many significantly reduced T_g. Of particular interest is the 10 percent mixture of Kraton FG1901X with Tactix 123. Fracture toughness of the EB cured composition was more than double that of untoughened Tactix 123 (control) and T_g and initial service temperature were slightly higher than for the control. No low temperature transitions were observed in the DMA curves. However, when the Kraton concentration was increased to 20 percent, a slight decrease in fracture toughness and the appearance of a transition at 86°C in the DMA curve occurred. Viscosity of the 10 percent Kraton FG1901X Tactix 123 mixture was 1100 cP at 60°C. Kraton FG1901X is a styrene-ethylene-butylene-styrene (SEBS) block copolymer which is terminated with maleic anhydride. This rubber was specially designed for the toughening of epoxy resins.

Weight gain after 48 hours in boiling water was also determined for a number of the specimens. Values were typically low (1-1.5 percent). Incorporation of toughening agents such as Udel P-1700 and Radel A-200 resulted in less water absorption than obtained with untoughened epoxy resin.

For comparison purposes, the T_g (tan delta) of the thermally cured 977-3 toughened resin system is reported as 190°C and the fracture toughness is 0.90 MPa·m$^{1/2}$. Weight gain when exposed to 65°C and 85% relative humidity to equilibrium is approximately 1.1%.

Fracture surface of the untoughened epoxy specimens were smooth and glassy when viewed under an optical microscope (60 X). This morphology is also seen in the SEM of this surface (Figure 1). The fracture surface of the thermoplastic (Udel P-1700) toughened specimens also looked smooth, but there is some indication of "tough crack initiation." The SEM of these fracture surfaces is also much rougher (Figure 2). The fracture surface of a specimen toughened with core shell rubber particles (Rohm and Haas Paraloid EXL-2330) exhibits crack front growth which suggests a ductile-brittle advance of the fracture (run-arrest). A very rough surface with evidence of "pull out" is seen in the SEM (Figure 3). The specimen toughened with Kraton rubber exhibited an extremely rugged fracture surface morphology with evidence of large voids created by "pull out" of large pieces of material. The fracture appears to be a ductile failure. The SEM of this surface also exhibits large irregular morphology (Figure 4). This probably accounts for the high fracture toughness of this composition.

4. CONCLUSIONS

Thermoplastics, hydroxy-terminated polysulfone oligomers, reactive flexibilizers, rubbers, and elastomers all increased the fracture toughness of an EB cured epoxy resin composition. The use of thermoplastics such as Udel P-1700 and rubbers such as Kraton 1901X provided significant increases in fracture toughness without adverse reduction of glass transition temperature and modulus.

5. REFERENCES

1. C.J. Janke, G.F. Dorsey, S.J. Havens, and V.J. Lopata, SAMPE International Symposium, 41, 196 (1996).

2. F.J. McGarry and A.M. Willner, Amer. Chem. Soc. Div. Org. Coating Plast. Chem., Pap., 28,(1), 512 (1968).

3. A.C. Meeks, Polymer, 15, 675 (1974).

4. A.J. Kinloch, S.J. Shaw, D.A. Tod, and D.L. Hunston, Polymer, 24, 1341 (1983).

5. C.K. Riew and J. . Gillham, Ed., Advances in Chemistry Series, Vol. 208: Rubber Modified Thermoset Resins, American Chemical Society, Washington, DC, 1984.

6. C. K. Riew and A.J. Kinloch, Advances in Chemistry Series, Vol. 233: Toughened Plastics I, Science and Engineering, American Chemical Society, Washington, DC, 1993.

7. L. Leonard, Advanced Composites, 6 (4), 35 (1991).

8. T.F. Mika and R.S. Bauer in Epoxy Resins - Chemistry and Technology, Second Ed., C. May, Ed., Marcel Dekker, 1988, p. 506.

9. G.R. Almen, P. Mackenzie, V. Malhotra, R.K. Maskell, P.T. McGrail, and M.S. Sefton, SAMPE International Technical Conference, 20, 46 (1988).

10. R.S. Bauer, SAMPE International Symposium 34, 899 (1989).

11. C.B. Bucknall and A.H. Gilbert, Polymer, 30, 213 (1989).

12. R.S. Bauer, H.D. Stenzenberger, and W. Römer, SAMPE International Symposium, 35, 395 (1990).

13. J.L. Hendrick, I. Yilgor, J.C. Hedrick, G.L. Wilkes, and J.E. McGrath, SAMPE International Symposium, 30, 947 (1985).

14. J.A. Cecere, J. L. Hendrick, and J.E. McGrath, SAMPE International Symposium, 31, 580 (1986).

15. R.S. Bauer, SAMPE International Technical Conference, 18, 510 (1986).

16. J.C. Hedrick, J.L. Hedrick, J.A. Cecere, S.C. Liptak, and J.E. McGrath, Polym. Mater. Sci. Eng., 63, 190 (1990).

17. J.E. McGrath, T.C. Ward, E. Shehori, and A.J. Wnuk, Polym. Eng. Sci., 17, 647 (1977).

18. 977-3 Epoxy Resin, “A Data Sheet from ICI Fiberite”, March 1, 1995 (Rev.E).

Table 1. Description of Materials

Epoxy Resins	Description
Tactix 123	Dow Tactix 123: High purity, low viscosity diglycidyl ether of bisphenol A; Epoxy equiv. wt., 172-176; Viscosity at 25°C, 4400-5600 cP.
Tactix 556	Dow Tactix 556: Glycidyl ether of the condensation product of dicyclopentadiene and phenol; Epoxy equiv. wt., 215-235; Viscosity at 85°C, 1000-1500 cP.
Tactix 742	Dow Tactix 742: Triglycidyl ether of tris (hydroxyphenyl)methane; Epoxy equiv. wt., 150-170; Viscosity at 79°C, 2860 cP.
DER 661	Dow D.E.R. 661: Solid diglycidyl ether of bisphenol A; Epoxy equiv. wt., 500-560.
ERL 2258	Union Carbide ERL 2258 (Discontinued): 50:50 Blend of bis (2, 3-epoxy cyclopentyl) ether isomers (ERL 4205) and diglycidyl ether of bisphenol A.
Toughening Agents	**Description**
Udel P-1700 NT	Amoco Udel P-1700 NT: Polysulfone; Glass Transition Temperature (Tg) 190°C.
Radel A-200 NT	Amoco Radel A-200 NT: Polyethersulfone; Tg, 220°C.
Ultem 1000 - 1000	General Electric Ultem 1000 - 1000: Polyetherimide; Tg, 220°C.
Hydroxy - Terminated Polysulfone Oligomers	Polysulfone oligomers based on the Udel P-1700 backbone were synthesized by the reaction of a calculated excess of bisphenol A with bis(4-chlorophenyl)sulfone in the presence of potassium carbonate. Tg's for calculated molecular weights of 1500, 2650, 5360, 6000, and 8890 g/mole were 125, 146, 170, 176, and 180°C, respectively.
Orgasol 2001	Elf Atochem Orgasol 2001 UD NAT 2: Nylon 12 in ~ 5 μ particles; mp, 175°C.
Vikoflex 7010	Elf Atochem ATO Vikoflex 7010: Methyl epoxy soyate; Viscosity at 25°C, 80 cP.
Vikoflex 7170	Elf Atochem ATO Vikoflex 7170: Epoxidized soybean oil; Viscosity at 25°C, 380 cP.
Vikoflex 7190	Elf Atochem ATO Vikoflex 7190: Epoxidized linseed oil; Viscosity at 25°C, 660 cP.
Vikolox 14	Elf Atochem ATO Vikolox 14: 1,2-Tetradecane oxide.
DER 736	Dow D.E.R. 736: Polyglycoldiepoxide; Epoxy equiv. wt., 175-205; Viscosity at 25°C, 30-60 cP.
Heloxy 32	Shell Heloxy Modifier 32: Polypropylene glycol glycidyl ether; Epoxy equiv. wt., 305-335; Viscosity at 25°C, 45-90 cP.

Table 1. Description of Materials Continued

Heloxy 68	Shell Heloxy Modifier 68: Neopentyl glycol diglycidyl ether; Epoxy equiv. wt., 130-140, Viscosity at 25°C, 13-18 cP.
Heloxy 71	Shell Heloxy Modifier 71: Dimer acid diglycidyl ester; Epoxy equiv. wt., 390-490; Viscosity at 25°C, 400-900 cP.
Heloxy 505	Shell Heloxy Modifier 505: Castor oil polyglycidyl ether; Epoxy equiv. wt., 550-650; Viscosity at 25°C, 300-500 cP.
Poly bd 600	Elf Atochem ATO Poly bd 600 VM 312051: Internally epoxidized 1,3-butadiene homopolymer; Viscosity at 25°C, 68 cP.
Tone 301	Union Carbide Tone 301: ε-Caprolactone triol; Hydroxyl equiv. wt., 98-103; Viscosity at 25°C, 2250 cP.
PY 322	Ciba-Geigy PY 322: Modified bisphenol A glycidyl ether.
Dow XU 71790.04L	Dow XU 71790.04L: Bisphenol A glycidyl ether reacted with acrylic rubber.
EXL -2330	Rohm and Haas Paraloid EXL-2330: Methacrylate-butadiene core shell rubber
Epon 58134	Shell Epon 58134: Bisphenol A glycidyl ether blended with castor oil glycidyl ether and a polyurethane resin.
Kraton FG 1901X	Shell Kraton FG 1901X: Styrene-ethylene-butylene-styrene block copolymer terminated with maleic anhydride.

Table 2. Toughening of Radiation Cured Epoxy Resins with Engineering Thermoplastics[1,2]

Toughening Agent	Toughening Agent Concentration (%)	Epoxy Resin	Glass Transition Temperature (°C)		Initial Service Temperature[3] (°C)	Weight Gain (48h H_2O Boil) (%)	Fracture Toughness, K_{IC} ($MPa \cdot m^{1/2}$)
			Tan Delta	G"			
Control	None	Tactix 123	163	145	120	1.41	0.411 (25°C)
Udel P-1700 NT	10	Tactix 123	168	156	132	1.25	0.680 (25°C)
Udel P-1700 NT	20	Tactix 123	158	144	128	0.95	0.746 (25°C)
Radel A-200 NT	10	Tactix 123	177	162	141	1.38	0.549 (25°C)
Radel A-200 NT	20	Tactix 123	168	143	129	1.37	0.494 (25°C)
Ultem 1000-1000	10	Tactix 123	175	157	136	1.37	0.696 (25°C)
Ultem 1000-1000	20	Tactix 123	189	174	155	1.69	0.691 (25°C)
Orgasol 2001	20	Tactix 123	163	138	79	2.49	0.667 (25°C)
IMITEC X-902	10	Tactix 123	161	144	124	1.63	0.774 (25°C)
IMITEC X-902	20	Tactix 123	161	146	128	1.81	0.481 (25°C)
Udel P-1700 NT	20	ERL 2258	164		87		0.755 (25°C)
Udel P-1700 NT	11	Tactix 556/ ERL 2258 (44.5:44.5)	157		75		0.805 (25°C)
Udel P-1700 NT	10	Tactix 742/ ERL 2258 (45:45)	106		65		0.643 (25°C)

1. All formulations contained OPPI at a concentration of 3 phr.
2. All specimens were cured using a total dosage of 150 kGy.
3. Initial service temperature is defined as the temperature at which the modulus is at 50% of its value at 25°C.

Table 3. Toughening of Radiation Cured Epoxy Resins with Hydroxy-Terminated Polysulfone Oligomers[1,2]

Hydroxy-Terminated Polysulfone Oligomer (Mn, g/mole)	Toughening Agent Concentration (%)	Epoxy Resin	Glass Transition Temperature (°C)		Initial Service Temperature[3] (°C)	Weight Gain (48h H_20 Boil) (%)	Fracture Toughness, K_{IC} (MPa·m^{½})
			Tan Delta	G"			
Control	None	Tactix 123	163	145	120	1.41	0.411 (25°C)
1500	5	Tactix 123	173	160	137	1.06	0.508 (25°C)
1500	20	Tactix 123	155	143	132	1.21	0.536 (25°C)
1500	30	Tactix 123	160	145	120	2.56	0.557 (25°C)
2650	20	Tactix 123	149	134	122	1.57	0.700 (25°C)
5360	20	Tactix 123	150	134	123	2.38	0.731 (25°C)
6000	5	Tactix 123	171	156	130	1.15	0.475 (25°C)
6000	20	Tactix 123	159	150	135	1.26	0.593 (25°C)
6000	30	Tactix 123	161	149	134	1.64	0.705 (25°C)
8890	20	Tactix 123	154	136	122	1.77	0.738 (25°C)

1. All formulations contained OPPI at a concentration of 3 phr.
2. All specimens were cured using a total dosage of 150 kGy.
3. Initial service temperature is defined as the temperature at which the modulus is at 50% of its value at 25°C.

Table 4. Toughening of Radiation Cured Epoxy Resins with Reactive Flexibilizers[1,2]

Toughening Agent	Toughening Agent Concentration (%)	Epoxy Resin	Glass Transition Temperature (°C)		Initial Service Temperature[3] (°C)	Weight Gain (48h $H_2$0 Boil) (%)	Fracture Toughness, K_{IC} ($MPa \cdot m^{1/2}$)
			Tan Delta	G"			
Control	None	Tactix 123	163	145	120	1.41	0.411 (25°C)
DER 736	10	Tactix 123					0.569 (25°C)
Heloxy 32	10	Tactix 123					0.487 (25°C)
Heloxy 68	10	Tactix 123	158	135			0.534 (25°C)
Heloxy 71	10	Tactix 123					0.529 (25°C)
Heloxy 505	10	Tactix 123					0.480 (25°C)
POLY bd 600	10	Tactix 123	166	151	123	1.28	0.594 (25°C)
Vikoflex 7010	10	Tactix 123	139	121	101		0.592 (25°C)
Vikoflex 7170	10	Tactix 123	151	133	117	2.02	0.598 (25°C)
Vikoflex 7170	20	Tactix 123	134		113		0.559 (25°)
Vikoflex 7190	10	Tactix 123	172		137		0.479 (25°C)
Vikolox 14	10	Tactix 123	130		112	1.12	0.491 (25°C)
Tone 301	10	Tactix 123					0.810 (25°C)
Vikoflex 7170	10	Tactix 556	175		150		0.555 (25°C)

1. All formulations contained OPPI at a concentration of 3 phr.
2. All specimens were cured using a total dosage of 150 kGy.
3. Initial service temperature is defined as the temperature at which the modulus is at 50% of its value at 25°C.

Table 5. Toughening of Radiation Cured Epoxy Resins with Rubbers and Elastomers[1,2]

Toughening Agent	Toughening Agent Concentration (%)	Epoxy Resin	Glass Transition Temperature (°C)		Initial Service Temperature[3] (°C)	Weight Gain (48h H_2O Boil) (%)	Fracture Toughness, K_{IC} ($MPa \cdot m^{1/2}$)
			Tan Delta	G"			
Control	None	Tactix 123	163	145	120	1.41	0.411 (25°C)
PY 322	20	Tactix 123	132	119	107	1.06	0.630 (25°C)
Dow XU 71790.04L	20	Tactix 123	166	151	121	2.00	0.565 (25°C)
Dow XU 71790.04L	30	Tactix 123/ Tactix 556 (28:42)	176	161	137	2.22	0.411 (25°C)
EXL-2330	10	Tactix 123	132	132	109	2.59	0.585 (25°C)
EXL-2330	15	Tactix 123	129	111	92	2.47	0.600 (25°C)
EXL-2330	20	Tactix 123	122	104	81	2.19	0.572 (25°C)
Epon 58134	10	Tactix 123	183		158		0.618 (25°C)
Kraton FG 1901X	10	Tactix 123	171		131		0.919 (25°C)
Kraton FG 1901X	20	Tactix 123	86/176		111		0.841 (25°C)

1. All formulations contained OPPI at a concentration of 3 phr.
2. All specimens were cured using a total dosage of 150 kGy.
3. Initial service temperature is defined as the temperature at which the modulus is at 50% of its value at 25°C.

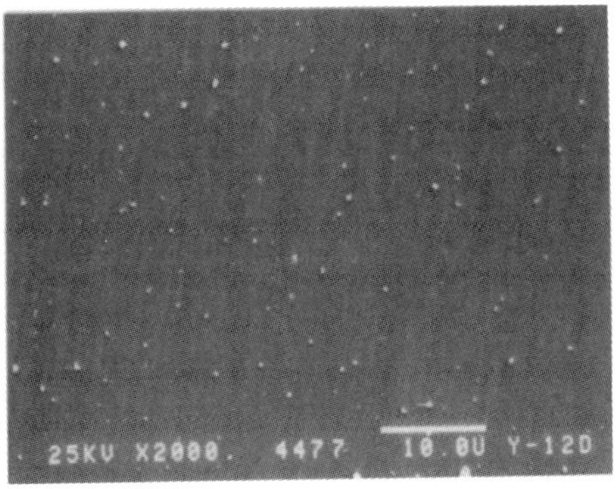

Figure 1

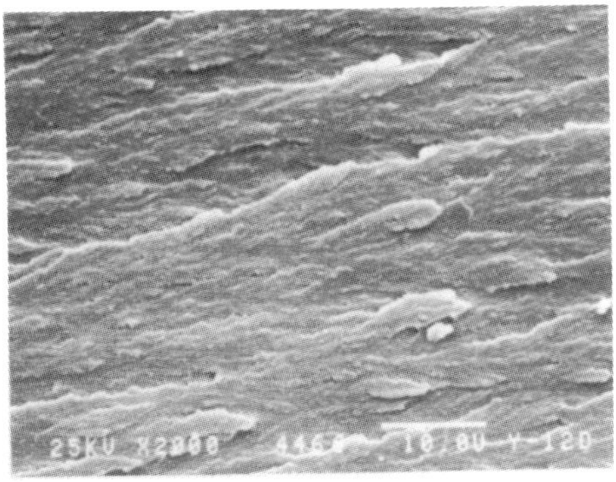

Figure 2

Figure 3

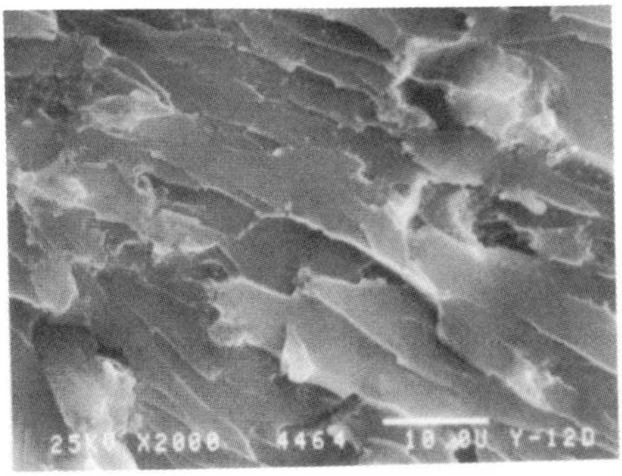

Figure 4

Figures 1-4. Scanning Electron Photomicrographs of : 1. Untoughened Tactix 123. 2. Tactix 123 toughened with 10% Udel P-1700. 3. Tactix 123 toughened with 10% Paraloid EXL-2330. 4. Tactix 123 toughened with 10% Kraton FG-1901X, respectively.

28th International SAMPE Technical Conference
November 4-7, 1996

SELECTION OF MATERIALS TO BE UTILIZED IN FABRICATION TOOLING FOR ELECTRON-BEAM CURING OF COMPOSITES

Robert E. Norris
Oak Ridge Centers for Manufacturing Technology
Oak Ridge National Laboratory*
Post Office Box 2009
Oak Ridge , Tennessee 37831-8048

Mark D. Schulz
Lockheed Martin Tactical Aircraft Systems
Post Office Box 748
Fort Worth, Texas 76101

ABSTRACT

A Cooperative Research and Development Agreement (CRADA) sponsored by the Department of Energy and 10 industrial partners has been established to develop high performance electron beam (E-beam) curable polymer matrix composites. Since E-beam curing is a nonthermal curing method, the use of lower cost, low temperature tools made of foam, wood, or other materials becomes possible. As part of the CRADA, E-beam irradiation/stability studies on low-cost EB tooling have been performed. Approximately 50 potential tooling materials representing a broad cross-section of both commonly used and specialty materials were tested. Coupons from each material were exposed to the equivalent of 5 E-beam curing cycles of 75 kGy each. At the conclusion of each 5-cycle exposure set, the materials were evaluated for dimensional and hardness changes. The testing continued for each material until obvious breakdown from radiation exposure or handling. The goal is ultimately to extend testing to approximately 150 simulated cure cycles (11,250 kGy) which represents a fairly large aircraft/aerospace structure production run.

KEY WORDS: Tooling; Electron Beam Curing; Radiation Stability

---

* The Oak Ridge National Laboratory is managed by Lockheed Martin Energy Research Corporation for the U.S. Department Of Energy under contract DE-AC05-96OR22464.

1. INTRODUCTION

Electron-beam (E-beam) curing is a non-thermal method that uses high-energy electrons and/or X-rays as ionizing radiation to initiate polymerization and crosslinking reactions at controlled dose rates in polymeric materials such as those used as matrices in composites. Radiation curing has been employed in industry for a number years - typically for curing thin films or coatings. Some polymers - particularly acrylates - naturally cross link via EB energy, while others such as most high-performance epoxies, require modification to initiate curing. It has been shown that epoxies modified by the addition of photoinitiators can achieve thermomechanical properties comparable to (and in some cases better than) those typical of thermal curing (1). With this technological breakthrough created a Cooperative Research and Development Agreement (CRADA) sponsored by the Department of Energy and 10 industrial sponsors, high performance composites (as defined by thermomechanical properties) can now be fabricated via the E-beam process.

E-beam curing has several advantages over conventional thermal curing methods including: improved part quality/performance; reduced environmental, safety, and health concerns; improved material handling; ability to combine various materials and functions in a single operation; ability to utilize lower cost tooling; capability to produce unique parts that cannot be fabricated any other way; and reduced cure times. All of these factors combine to significantly reduce overall manufacturing costs.

Fabrication tooling is one area that frequently drives the costs of composite products, especially in prototypes or production lots of a couple of hundred or less (2). Traditional tooling materials utilized in fabricating composites for high performance applications have been graphite, steel, aluminum, and, more recently, Invar. These materials are relatively expensive to form and machine and the resulting tools can be fairly heavy. These materials are utilized primarily for their durability at the elevated temperatures normally required to cure these parts.

2. TOOLING MATERIALS SELECTION FACTORS

Materials selection for tool design is very much interrelated with the design of the part to be fabricated and is inseparable from design of the tool itself. There are many factors involved in both materials selection and tool design including (but not limited to):

- Fabrication Process Itself
- Dimensional Stability/Achievable Tolerances
- Mold Fabrication Process (Number Of Steps)
- Maintenance, Transportability, Handling
- Repair
- Alterability
- Venting/Removal Of Trapped Air And Volatiles
- Need For Resin Flow
- Desired/Required Thermal Expansion
- Wear/Durability/Toughness
- Part Release
- Weight
- Thermal Mass
- Thermal Conductivity
- Required Part Curing Pressure
- Temperature Level And Distribution

Cooling	Surface Finish
Undercuts, Hardware, Inserts	Ease Of Mold Duplication

As indicated earlier, the E-beam curing process is not thermally driven. This offers a major advantage in allowing a "choice" of curing temperatures from well below 0°C to as high as thermally cured materials. This advantage certainly expands the number of materials that may be chosen for utilization in tooling, as operating temperature requirements limit and frequently exclude a number of good, relatively inexpensive structural materials. E-beam curing does, however, introduce some factors that are not normally a consideration in the conventional process. Although E-beam curing does not require achievement of specific temperatures for completion, the radiation is a form of energy that is attenuated or absorbed to varying degrees as it passes through materials. Therefore, there is a temperature rise in both tool and part due to this absorption. (Temperature rise due to exothermic cure reaction is a function of the material being cured, and must be considered separately.) Another factor is that for closed mold processes such as RTM or injection molding, the ability of the radiation to penetrate through the tooling material to effectively cure the part must be considered. Finally, because radiation can alter the atomic structure of materials, the stability of a potential material candidate due to radiation exposure must be considered.

3. RADIATION INDUCED HEATING AND RADIATION PENETRATION

Maximum temperature rise due to radiation exposure can be estimated from a simple analytical expression based on conservation of energy. It is:

$$\Delta T = \frac{D}{Cp} \tag{1}$$

where
ΔT = temperature rise in C°
D = exposure dosage in kGy, and
Cp = material specific heat in J/(g x C°).

This relationship assumes that all absorbed radiation is evenly and instantaneously converted into heat and corresponding temperature rise. In reality, the dosage is absorbed over a finite time period in which some heat is transferred away by the usual mechanisms, primarily conduction and convection. Therefore, the estimate should be considered conservative in actually predicting a worst-case scenario.

Penetration capability is not quite as simple. In selecting materials for tooling, this is an issue only when a portion of the tooling is between the source (accelerator) and the part to be cured. Computer codes have been developed by Sandia National Laboratory and others utilizing monte carlo simulations to predict the energy absorption distribution within a target material subjected to radiation of various types. These codes can handle various materials, cross sections, and other geometries. The Integrated Tiger Series (ITS) codes (3) were used to generate generalized curves (4) showing the distribution of energy from beams having energies of 2, 5, and 10 MeV through the thickness of a material having a density of 1.6 g/cm^3 are shown in Figure 1. Note that the absorbed dose initially increases as the

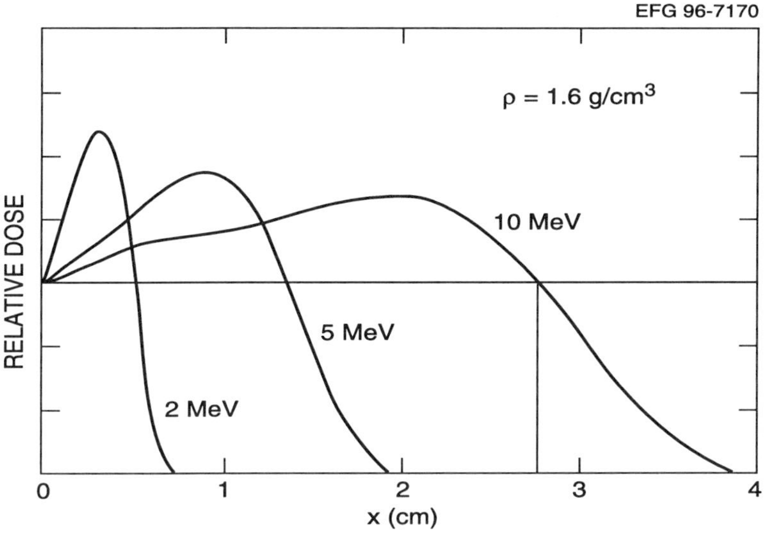

Figure 1. Relative dose deposited in a target by 2, 5, and 10 MeV electron beams.

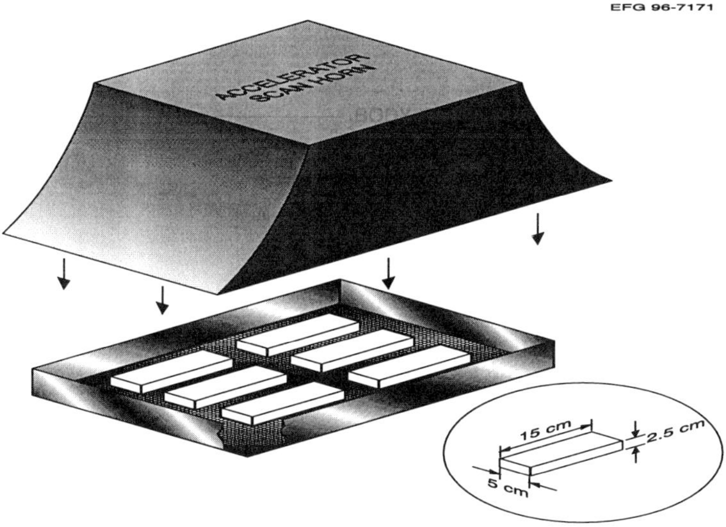

Figure 2. Radiation stability testing of coupons under the accelerator beam.

electrons in the beam slow down and are better absorbed. The dose then decreases as the rate of absorption becomes equal to that at the surface and then exponentially tapers off. If the "useful deposition range" is defined as the thickness at which the dose absorption equals that at the target surface, this relationship can be estimated from the previously referenced generalized curves by the equation:

$$\mathbf{W_a = \frac{E}{2.2}} \tag{2}$$

where W_a = material areal weight in g / cm^2, and
E = beam energy in MeV.

This relationship allows one to estimate the total useful deposition range or "uniform" penetration dosage from the combination of the areal weights of the material being cured and the tooling. For example, the useful deposition range for a 10 MeV accelerator is an effective material areal weight of ~4.5 g/cm^2. If the material being cured is graphite fiber composite having a density of 1.6 g/cm^3 and is 2 cm thick, the material has a W_a of 3.2 g/cm^2. This leaves an effective W_a for the tooling of 1.3 g/cm^2. If the tooling were also graphite fiber composite of 1.6 g/cm^3, the allowable tooling thickness to have uniform penetration dosage would then be 1.3/1.6 or ~ 0.8 cm. Note that by double sided curing (turning the part over and curing from the backside), the effective thickness would be essentially doubled. As will be shown later, most work indicates that epoxies and several other polymers are stable to many times the dosage expected in normal curing. The important parameter is to achieve the minimum dosage required for curing. This would provide capability for small increases in effective thickness above the useful deposition range to be compensated with additional exposure in curing. Also, converting the electron beam to X-rays can increase the effective thickness about an order of magnitude with a resultant increase in cure time of about the same magnitude.

Examples of how various materials commonly used in the composites industry compare in terms of temperature rise and penetration capability with respect to radiation exposure are shown in Table 1.

Worth noting in Table 1 is that steel is the least attractive candidate material from both the penetration capability and temperature rise criteria. Aluminum is much better than steel, but in most cases is not as desirable as the polymers and polymer composites.

4. RADIATION STABILITY OF CANDIDATE MATERIALS

The third factor in tooling material selection is radiation stability. An informal survey was conducted to identify potential materials, materials systems, or concepts that take advantage of unique capabilities of E-Beam curing and evaluate these options. As much information as possible was obtained from material vendors on various products that were thought to have some potential for use as a tooling material. A database of this information is being compiled and includes information such as CTE's, hardness, cost, moisture absorption, etc. The goal is to organize the data such that the database can be used as a

Material	Density	Penetration Capability (E=10 MeV)*	Specific Heat	Heat Rise (D=35 kGy)**
	g / cm^3	cm	J / (g x C°)	C°
Aluminum	2.74	1.63	0.97	36
Steel	7.68	0.58	0.46	76
Epoxy	1.24	3.59	1.84	19
PVC Foam	0.24	18.56	0.97	36
Polyurethane Foam	0.77	5.79	1.67	21
Phenolic	1.55	2.87	1.30	27
Carbon fiber/epoxy	1.59	2.80	1.13	31
Glass fiber/epoxy	2.17	2.05	0.63	56

* Assumes equation E = 2.2 x Wa, where Wa is in g/cm^2
** Assumes adiabatic process ΔT = Dose / Specific Heat

Table 1. Comparison of estimated temperature rise and penetration thickness for various types of material.

users' guide to evaluate material options when designing tools for E-beam processing. While radiation stability of metals is not infinite, metals are assumed to have a useful lifetime of $>10^7$ kGy (5). This is many orders of magnitude greater than the comparable lifetime of polymers and other organics. Based on the assumption that radiation stability is not an issue for metals and that cost, temperature rise, penetration capability, and other factors make alternate materials more attractive in deriving the full benefits of E-beam processing, metals were not considered in the evaluation of radiation stability of potential tooling materials.

A series of tests were conducted to screen the material candidates for radiation stability. The ultimate goal of testing was exposure of 11,250 kGy, representing 150 cure cycles. (Early tests had indicated that ~75 kGy would completely cure most composites.) This level was chosen as representing transition from prototyping or low-level production into what would be considered high production for aircraft usage. As mentioned previously, at this range, the cost of tooling becomes much easier to amortize and is not as significant in the overall cost of production.

Sample materials were obtained from as many vendors as practical and cut into 2.5-cm x 5-cm x 15-cm coupons. Six hardness measurements were taken with a Shore-D or Shore-A Durometer on the top (5 x 15) surface. Six dimensional measurements were taken to the closest 0.01-mm with a digital caliper in the center of the surfaces defining the length (15-cm) and width (5-cm). The coupons were placed in trays with the 5 x 15 surface facing up as shown in Figure 2. The trays were then processed through an accelerator beam at a rate calculated to yield dosages equivalent to approximately 30-40 kGy per pass, which is representative of typical industry processing of polymeric materials. (As shown previously in Table 1, an instantaneous 35 kGy dose would limit heating of most polymeric materials to ~35C°.) Because it was desired that these tests be used to screen the radiation effects alone and not unnecessarily complicate the results with the effects of heating, the materials were allowed to cool back to ~40°C before re-exposing. After verifying the coupons had sufficiently cooled between each pass, the exposure was repeated until the coupons had received 375 kGy since the last measurements. (375 kGy simulates the exposure that the tooling material might receive in curing 5 parts.) Both dimensional and hardness measurements were taken at each 375 kGy increment and the sequence repeated until the ultimate goal was reached or the material obviously degraded beyond productive use. This condition was typically identified by extreme changes in dimensions or hardness or even disintegration of the sample itself.

Table 2 shows the materials that were selected for screening in this project. Most of the materials are polymeric-based or depend on polymers as matrices or binders. Although initial experience with wood (especially pine) had not been positive, a few other woods were included to quantify capabilities. Also included were plasters and bonded ceramics. Most of the materials are marketed for tooling purposes by vendors; however, a few products are not classified specifically as tooling materials. Examples are the foams, which have general sandwich construction applications, and the Aquapreg, which could be classified as a novelty material in search of an application.

5. DISCUSSION OF RESULTS

Table 3 shows the status of the testing as of June 1996. For comparison purposes, failure criteria was selected as a hardness change of >10% or a dimensional change of >0.25-mm. Even after these limits were surpassed, the materials were continued in the testing if they visually maintained structural integrity. Although none of the materials have yet completed the entire goal of 11,250 kGy, about 2/3 of the materials have <u>not</u> demonstrated significant change in dimensions or hardness at well over 5,000 kGy.

As classes of materials, some of the PVC foams dropped out in the early stages of testing. However, the stability depended on the exact configuration and formulation as the H250 significantly outlasted the H030, H080, and HT050. Similar, but wider, discrepancies were found with the polyurethane data. In literature references (6,7), polyurethane is characterized as among the most stable polymers. In terms of hardness, it was found to be stable in this testing also. This is not necessarily the case for dimensional stability. After a few thousand kGy, a large portion of the polyurethanes became dimensionally unstable and demonstrated pronounced warpage. However, some of the polyurethanes have remained stable above 5,000 kGy in terms of both dimensional and hardness data. Apparent data descrepancies such as those with the polyurethanes should be studied in more detail, but are presumed to be caused by variations in the exact chemistry of the material and by the diversity of filler materials employed. Typically, this information is considered proprietary to the material manufacturer and may be difficult to obtain.

It should be noted that the objective of this study was not to downselect a single, best

Material ID	Manufacturer or Supplier	Material Description
Aquapreg	Crest Products Corp.	Fiberglass reinforced isocyanate (cures with water)
BryteCor™ EX-1541	Bryte Technologies, Inc.	Cyanate ester syntactic foam
CareMold 800	Composites Horizons, Inc.	Proprietary castable and wash out composite
Castable 220	J.L. Peters Co.	Single component, castable fused silica
Cercon T-11	Cercon, Inc.	Chemically bonded ceramic composite
CibaTool (XB-5143)	CIBA-GEIGY Corporation	Acrylated stereolithography resin
DP-1051	Tool Chemical Co., Inc.	Aluminum filled polyurethane
FASTool	Integrated Composites	Castable epoxy/hollow glass beads for structure back
FP-1050	Tool Chemical Co., Inc.	Urethane
FR-3706	General Plastics Manufacturing Co.	Rigid polyurethane closed cell foam
FR-3710	General Plastics Manufacturing Co.	Rigid polyurethane closed cell foam
FR-3715	General Plastics Manufacturing Co.	Rigid polyurethane closed cell foam
FR-3718	General Plastics Manufacturing Co.	Rigid polyurethane closed cell foam
FR-3720	General Plastics Manufacturing Co.	Rigid polyurethane closed cell foam
FR3730	General Plastics Manufacturing Co.	Rigid polyurethane closed cell foam
Freeman Wax	Freeman Mfg. & Supply	Machinable wax
H030	Divinycell International, Inc.	PVC, closed cell foam
H080	Divinycell International, Inc.	PVC, closed cell foam
H250	Divinycell International, Inc.	PVC, closed cell foam
Hexcelite5003	Hexcel Corporation	Epoxy syntactic foam tooling board
Himeko BL-3	Fujipoly Inc.	Epoxy
HT050	Divinycell International, Inc.	PVC, closed cell foam
Jelutong	McCausey Lumber Co.	Indonesian white wood, medium hardness
Mahogany	McCausey Lumber Co.	Relatively hard wood, readily available, easy to work
MB2000	BCC Products, Inc.	Polyurethane
Modelmaster A160	Airtech International, Inc.	Epoxy tooling board
Modulan	McCausey Lumber Co.	Polyurethane foam tooling board
MP-1053	Tool Chemical Co., Inc.	Urethane
MP-42-400	Tool Chemical Co., Inc.	Bis A Epoxy
PFP (1124 surface coat)	(Internally Produced)	Plastic-faced-plaster
PFP (Epolite 1350)	(Internally Produced)	Plastic-faced-plaster
Pine Perfect Plank		Soft wood sold without knots
Polyboard 140	Fiber-Resin Corp.	Filled polyurethane
PP-1052	Tool Chemical Co., Inc.	Filled urethane
REN 350	CIBA-GEIGY Corporation	Polyurethane foam tooling board
REN 350 w/epoxy coat	CIBA-GEIGY Corporation	Polyurethane foam tooling board
REN 450 w/epoxy coat	CIBA-GEIGY Corporation	Polyurethane foam tooling board
REN 450 w/epoxy surface	CIBA-GEIGY Corporation	Polyurethane foam tooling board
REN 460	CIBA-GEIGY Corporation	Polyurethane foam tooling board
REN 550	CIBA-GEIGY Corporation	Polyurethane foam tooling board
REN 550 w/epoxy surface	CIBA-GEIGY Corporation	Polyurethane foam tooling board
Richlite	The Henry Company	Phenolic laminate sheets
T-11	Cercon, Inc.	Chemically bonded ceramic composite
TB 620	Advanced Composites Group	Epoxy based syntactic foam
TB 650	Advanced Composites Group	Epoxy based syntactic foam
UCB	UCB Chemicals	Acrylated stereolithography resin

Table 2. Materials selected for radiation stability screening.

Material	Exposure to Date	Stability at Dose		Test Status ***	Comments/Observations
	kGy	Hardness *	Dimension **		
Aquapreg	4,125	LC	LC	IP	
Bryte Synpak	5,250	LC	LC	IP	
Castable 220	3,000	LC	LC	IP	Testing to resume
CerCon CBC (T-11)	2,250	LC	LC	IP	1 broke in handling after 1500 kGy
DP-1051	6,000	LC	LC	IP	
FASTool	4,125	NA	LC	IP	
FP-1050	5,625	LC	750	TT	All samples bent or warped
FR-3710	6,000	1,125	3,375	IP	Hardness increasing, small dimensional changes
FR-3715	5,250	LC	LC	IP	
FR-3720	5,250	LC	LC	IP	
FR3730	5,625	LC	LC	IP	
Freeman Wax	6,000	LC	375	IP	Small surface defects on 1 sample @ 4875
H030	4,125	375	750	TT	Samples disintegrating
H080	3,375	0	750	TT	Samples disintegrating
H250	5,250	4,500	4,125	TT	
Hexcelite5003	7,125	LC	LC	IP	
Himeko BL-3	5,625	LC	LC	IP	
HT050	375	0	0	TT	Samples disintegrating
Jelutong	5,250	750	3,375	TT	Samples destroyed at final dosage
Mahogany	4,875	3,750	4,125	TT	Length more stable than width, 1 cracked @3375
MB2000	4,875	LC	4,500	TT	Surface bubble on #4 @ 4125
Modelmaster A160	5,250	LC	LC	IP	
Modulan	5,625	LC	3,000	TT	Bowing @ 4875
MP-1053	5,250	LC	LC	IP	
MP-42-400	5,625	LC	4,500	IP	
PFP (1124 surface coat)	5,250	4,825	4,125	TT	
PFP (Epolite 1350)	5,625	LC	LC	IP	Sample 4 broke @ 2625
Polyboard 140	4,875	LC	1,125	TT	
PP-1052	5,625	LC	LC	IP	
REN 350	5,250	LC	LC	IP	
REN 450 w/epoxy surface	4,875	LC	2,250	TT	
REN 460	4,875	LC	1,500	TT	
REN 550	5,625	LC	LC	IP	
Richlite	4,500	3,750	3,750	TT	Some delamination, 3 samples removed @ 4125
T-11	6,000	LC	LC	IP	Sample 1 broke @ 2625
TB 620	5,250	LC	LC	IP	
TB 650	5,250	LC	LC	IP	
UCB	5,250	LC	LC	IP	Appearance of fine cracks @ 3750

* Dose at which hardness reduction exceeds 10%, LC = Little Change

** Dose at which dimensional changes exceed 0.010-in, LC = Little Change

*** IP = In Progress

TT = Tests Terminated

Table 3. Radiation stability test status as of June 1996.

material for E-beam tooling, but to identify significant factors different from thermal curing, establish a screening process for radiation stability, and generate data for material candidates. Based on this data, most of the materials offer potential for utilization in tooling for at least some applications. At this point, the epoxies, the cyanate ester, and the ceramics appear to be the most stable. However, it should be noted that one of the disadvantages of the ceramics is their brittleness. Several coupons from different batches were damaged and lost to testing due primarily to bumps in shipping and handling between exposure sequences.

6. CONCLUSIONS

As pointed out earlier in this text, tooling design and associated material selection are dependent on a number of factors. Most of the materials selection methodology is similar to that which is employed in conventional thermal curing with the exception of the environmental conditions encountered. Although radiation stability and durability may or may not be suitable for large production runs, significant savings can be achieved in reducing the costs of prototype tooling. E-beam curing appears to open new opportunities to materials that previously could not be utilized because of temperature restrictions.

Although testing has not been completed, several conclusions can be drawn from this work:

1. A number of materials can be used in E-beam curing of composites. Most polymers appear capable of stability in radiation through at least ten's of prototypes if they can meet other design requirements. They offer advantages in terms of weight, penetration capability, and frequently radiation heating. This study did not address factors other than those that are by definition different from factors encountered in conventional thermal curing or combined effects.

2. Existing literature on radiation stability is a good place to start in terms of screening materials for tooling applications. However, because most testing documented in the literature is for other unrelated purposes and the formulations and configurations are likely different, representative material samples should be tested before committing to large expenses. This study provides some comparative data that allows the user a head-start in the selection process. The testing is continuing and more data will be released as it becomes available.

3. For radiation stability as applied to tooling, hardness is a good tool for screening purposes, but should not be used alone. Materials can be dimensionally unstable without appreciable changes in hardness. Likewise, materials can demonstrate appreciable changes in hardness (which may indicate loss of mechanical strength) without dimensional changes.

7. ACKNOWLEDGEMENTS

The authors would like to gratefully acknowledge the efforts of a number of people at Fort Worth and Oak Ridge who helped take measurements and compile the large amount of data. Special thanks go to our CRADA partners Atomic Energy of Canada Limited (AECL) and Nicolet Electron Services (NES) who provided the large amounts of beam time under the CRADA to make this study possible.

8. REFERENCES

1. C.J. Janke, et. al., "Electron Beam Curing of Epoxy Resins by Cationic Polymerization", 41st International SAMPE Symposium/Exhibition, March 25-28, 1996.
2. "Composite Makers Teach Old Processes New Tricks", Aerospace America, June 1994, pp. 34-42.
3. ITS 3.0, RSIC Computer Code Collection, Radiation Shielding Information, Oak Ridge National Laboratory, CCC-467 (1994).
4. R.J. Colchin,"Electron and X-Ray Transport in Materials", Internal ORNL correspondence, February 7, 1996.
5. L.W. Ricketts, "Fundamentals of Nuclear Hardening of Electronic Equipment", Radiation Effects on Electronic Materials and Passive Components, Wiley-Interscience, 1972, p. 128.
6. M.H. Van de Voorde and C. Restat, "Selection Guide to Organic Materials for Nuclear Engineering", CERN 72-7, May 17, 1972.
7. L.K. English, "How High-Energy Radiation Affects Polymers", Materials Engineering, May 1986, pp. 41-45.

28th International SAMPE Technical Conference
November 4-7, 1996

ELECTRON CURING OF EPOXY RESINS: INITIATOR AND CONCENTRATION EFFECTS ON CURING DOSE AND RHEOLOGICAL PROPERTIES

Vincent J. Lopata and Minda Chung
Atomic Energy of Canada Limited, Whiteshell Laboratories
Pinawa MB R0E 1L0
Canada

Christopher J. Janke
Oak Ridge Centers for Manufacturing Technology
Lockheed Martin Energy Research Systems Corporation
Oak Ridge National Laboratory
Oak Ridge TN 37831-8048
U.S.A.

Stephen J. Havens
Oak Ridge Institute for Science and Education
Oak Ridge TN 37831-0117
U.S.A.

ABSTRACT

A Cooperative Research and Development Agreement (CRADA) using Electron Beam (EB) Polymer Matrix Composite (PMC) curing technology has, as one of its major objectives, the development of high-performance matrix resins. The program has been very successful in expanding the number of available electron-curable matrix resins, with a wide variety of properties. This major development in radiation-curable resins was brought about by the use of various cationic initiators. This report describes the effects of different initiators and initiator concentrations on the curing dose and rheological properties of selected resins such as EPON 862 and Tactix 123.

KEY WORDS: Advanced Composites, Cationic Initiators, Electron Curing

1. INTRODUCTION

As part of the Cooperative Research and Development Agreement (CRADA) on advancing Electron Beam (EB) Polymer Matrix Composite (PMC) curing technology, one of the major objectives was the development of high-performance electron-curable matrix resins (1). Electron curing of polymer matrix composites is a nonthermal curing method that uses either high-energy electrons or electron-generated X-rays to initiate polymerization and crosslinking. Electron-curing of PMCs offers several advantages such as reduced curing times, fewer volatiles, reduced thermal stresses, curing at various temperatures, and lower production costs (2,3,4). Cost estimates have consistently shown that 20% to 65% savings in production costs can be achieved depending on the part being manufactured (5). Acrylate and methacrylate monomers and oligomers have been used extensively as electron-curable coatings and adhesives (6). Recently, these resins have been studied for use as matrices for the manufacturing of PMCs (7,8). However, these resin systems produce low-performance properties, unsuitable for aircraft and aerospace applications. The limitations include low glass transition temperatures (T_gs), low fracture toughness, and high shrinkage when cured.

Curing epoxy resins, which are widely used for high-performance PMCs using electron-curing, would be extremely useful. Ultraviolet curing of thin films of cycloaliphatic epoxies is well known (7,8). Mechanical and thermal properties of these materials are low, and the curing of the epoxy is incomplete. Recently, Crivello reported a series of highly reactive cycloaliphatic silicon epoxies which are electron-curable in the presence of certain cationic photoinitiators (diaryliodinium and triarylsulfonium salts) (9-13). These resins gave satisfactory mechanical properties and high T_gs, while curing at low doses (20 to 60 kGy).

Our group studied a wide range of cationic initiators and commercially available epoxy resins using both electron treatment and gamma irradiation. Our initial work used gamma calorimetry to screen a wide variety of epoxy resins and cationic initiator mixtures (14,15). This study screened over 700 initiator/resin combinations as possible candidates for more extensive testing as PMCs. Further work, using electron-curing was performed on select mixtures of cationic initiators and epoxy resins. This work has been filed for patent. In this paper, we will show how the curing dose and rheological properties for Tactix 123 and EPON 862 vary with the use of different cationic initiators, as well as how the cationic initiator concentration affects the rheological properties and curing dose for a given epoxy resin system.

2. EXPERIMENTAL

2.1 Materials

2.1.1 Epoxy Resins Samples of commercial epoxy resins were obtained from various manufacturers. These are described below:

> Tactix 123: High purity, low viscosity diglycidyl ether of Bisphenol A; Epoxy Equivalent weight: 172-176; Viscosity: 4400-5600 cps (25°C); Supplier: Dow Chemical.
>
> Epon 862: Diglycidyl ether of Bisphenol F; Epoxy Equivalent weight: 166-177; Viscosity: 3000-4500 cps (25°C); Supplier: Shell Chemicals.

2.1.2 Cationic Initiators Samples of commercial cationic initiators were obtained from various manufacturers. Others were synthesized. These are described below:

UVI 6990: 50% solution of triaryl-sulfonium hexafluorophosphate in propylene carbonate. Supplier: Union Carbide Chemicals and Plastics Co, UCAR Coatings Resins.

UVI 6974: 50% solution of triaryl-sulfonium hexafluoroantimonate in propylene carbonate. Supplier: Union Carbide Chemicals and Plastics Co, UCAR Coatings Resins.

CD1012: Diaryliodonium Hexafluoroantimonate. Supplier: Sartomer.

OPPI: Aryl Fluoroantimonate, Product 479-2092 C; (4-Octyloxyphenyl)phenyliodonium hexa-fluoroantimonate; mp: 55-60°C. Supplier: General Electric Silicones

DW1: Diphenyl iodonium tetra kis[3,5-bis(trifluoromethyl)phenyl]borate. Synthesized by Dave Wheeler, Sandia National Laboratories;

2.2 Gamma Calorimetry: This method of evaluating the reactivity and dose required for curing various resin-cationic initiators has been discussed previously (14,15). A sample with an imbedded thermocouple was placed in an insulated jacket and exposed to gamma irradiation. The reference temperature was monitored using a sample of cured resin. The temperature profile of each resin was monitored during curing using a data acquisition system.

2.3 Gammacell: The gammacell used for the gamma calorimetry experiments was an AECL Model 220 Gammacell. The dose rate was approximately 100 Gy min^{-1}.

2.4 Electron Accelerator: All electron treatments were carried out using the I-10/1 Electron Accelerator (10-MeV; 1 kW) at AECL's Whiteshell Laboratories. The instantaneous dose rate was 1.5 MGy s^{-1}. All electron-cured samples were treated with a dose of 150 kGy.

2.6 Dynamic Mechanical Analysis (DMA) Dynamic mechanical analysis was performed using a Rheometric Solids Analyzer RSA II. The resin samples were trimmed to fit a dual cantilever tool, with an initial stress and strain of zero. The applied strain used for measuring the rheological properties was set at 0.1%. The temperature was raised at 2.5°C per step with a dwell time of 15 seconds. The frequency of the applied strain was 6.28 radians per second. Measurements were obtained using the software provided by Rheometrics.

3. RESULTS AND DISCUSSION

Most of the available cationic initiators are diaryliodonium or triarylsulfonium salts of weak bases. In previous work (14) it was shown that the nature of the anion influences the effectiveness of the initiator in catalyzing electron curing. The anion effect is inversely related to the nucleophilicity of the anion:

$$SbF_6^- > AsF_6^- > PF_6^- > BF_4^-$$

The less nucleophilic anion is less attracted to the Bronstead acid (proton) generated by the radiation degradation of the diaryliodonium or triarylsulfonium groups. This results in producing very strong or "super" acids that initiate cationic epoxy curing. The aryl groups also have a profound influence upon the reactivity.

3.1 Gamma Calorimetry: The gamma calorimetry method was developed to screen large numbers of electron-curable resins very quickly (14,15). The polymerization and crosslinking reactions of resins are similar for electron, X-ray, or gamma curing. Polymerization and crosslinking reactions are exothermic processes. Therefore, monitoring the temperature rise of a resin/initiator solution during gamma irradiation could provide relevant information. Resin systems that did not show any temperature rise were eliminated from further screening. The lack of temperature rise suggests that these resin systems would not cure by irradiation. By examining the gamma calorimetry temperature curve, we were able to obtain the induction dose, the dose/cure profile, and the estimated curing dose.

Figure 1 shows a typical temperature profile obtained from gamma calorimetry. As well as the information listed above, the profile can give some further indication of the curing process. The profile shows that resin curing occurs over a range of doses. Because the dose delivered to the composite part can be controlled, this profile shows what doses are required to partially cure resins or B-stage curing. B-Staging allows a composite part to be partially cured but still have sufficient stiffness to be handled. Once the final alterations have been made to the part, it can be electron treated to full cure.

We have established in previous work that the induction dose for a resin is the same, whether gamma radiolysis or electron curing is used. This dose is dependent upon the resin and initiator combination used. The induction dose for epoxy resins with cationic initiators can range from 0.5 to 10 kGy. Once the induction dose has been reached crosslinking/polymerization takes place. This is indicated by the initial temperature rise in the resin being studied. This temperature rise can be a slow process occurring over tens of kGy or very rapid (1 to 2 kGy). The shape of the temperature rise can indicate whether there are single or multiple reactions occurring during curing processes. Multiple reaction curves are usually seen with multi-component resin systems. When multiple reactions are occurring, the shape of the curve will deviate from a normal Gaussian distribution shape. The dose associated with the maximum temperature rise indicates the point at which 50% of the curing has been complete. As a rule of thumb, we have found multiplying this dose by 10 will give the approximate curing dose for that resin system. The drop in temperature after the maximum temperature rise is due primarily to the heat loss from the apparatus. Heat generated from polymerization and/or crosslinking is still occuring; however, there are insufficient reactions occurring to overcome the heat losses from the apparatus.

3.2 Different Initiator Effects In this study we used five different cationic initiators to demonstrate the effects on curing dose with Tactix 123 and EPON 862 epoxy resins. Information on the cationic initiators has been given above (Section 2.1.2). For the comparison between the various initiators a concentration of 3 phr (parts per hundred resin) was used. The same solutions were used for the gamma calorimetry and electron curing. Figure 2 shows the gamma calorimetry temperature curves for EPON 862 with the five cationic initiators studied. The figure shows that each initiator produces a different temperature profile for the same epoxy resin. Similar effects occurred for all the epoxy resins studied. Table 1 shows the gamma calorimetry results for various initiators with EPON 862 and Tactix 123. From the table it can be seen that the initiator plays a significant role in the dose required to cure the resin.

Table 1. Effect of Various Cationic Initiators on the Curing Dose and Temperature Rise of EPON 862 and Tactix 123 Epoxy Resins

Resin	Initiator	Dose at Peak Temperature (kGy)	Calculated Curing Dose (kGy)	Temperature Rise (°C)
EPON 862	UVI 6990	2.0	20	10
	UVI 6974	2.0	20	12
	CD1012	11.9	119	38
	OPPI	7.6	76	65
	DW1	4.0	40	118
Tactix 123	UVI 6974	2.2	22	8
	CD1012	7.4	74	42
	OPPI	5.2	52	69
	DW1	3.2	32	80

The efficiency of the initiator to cure a resin system is demonstrated by the temperature rise; the higher the temperature, the more efficient the initiator for curing. We found that the temperature rise is not only affected by the initiator used, but also by the viscosity and type of epoxy used. The higher the viscosity, the lower the temperature rise. The initiator also plays a role in the dose required to cure. Even though the triarylsulfonium initiators appear to have lower curing doses, their rheological properties are poorer than the arylfluoroantimonate initiators (See later). Among the fluoroantimonate initiators, the radiation curing efficiency is CD1012 < OPPI < DW1. The initiator effects are similar for both EPON 862 and Tactix 123. The iodonium initiators are reported to be more efficient than the sulfonium initiators (16). Our work is consistent with that finding. The iodonium initiators are probably more efficient because they lead to much higher temperatures during curing than the sulfonium initiators. Within an initiator family, antimonate initiators are more efficient than phosphate initiators. The UVI 6974 is a antimonate initiator, whereas the UVI 6990 is a phosphate initiator. We found within a particular family of initiators, the dose required to cure drops with the efficiency of the initiator. However, we cannot explain why the less efficient sulfonium initiators gave lower curing doses than the iodonium initiators.

The samples used to determine the rheological properties were irradiated to a dose of 150 kGy using the I-10/1 electron accelerator at AECL's Whiteshell Laboratories. Table 2 summarizes the rheological properties for various initiators with EPON 862 and Tactix 123. The rheological properties for a particular epoxy resin are affected by the efficiency of the initiator used. The more efficient initiator yields higher glass transition temperatures and service temperatures. For example, EPON 862, with 3 phr of initiator UVI 6974, gave a service temperature of 104°C, whereas with OPPI the service temperature was 134°C. It was found that the rheological properties for Tactix 123 and EPON 862 remained the same when OPPI, CD1012, and DW1 were used (see Table 2). However, these same initiators produce different temperature rises and different curing doses. When choosing one of these initiators, one must consider the cost and availability of the particular initiator.

Table 2. Effect of Different Initiators and Their Concentration on Electron-Curing of Tactix 123 and EPON 862 Epoxy Resins

Resin	Initiator	Conc. (phr)	Service Temp. (°C)	Glass Transition Temperature, (°C)		Dose at Max. Temp. Rise (kGy)	Temp. Rise (°C)
				Tan(δ)	E"		
EPON 862	UVI6974	0.5	75	102	79		
		1	89	124	92		
		2	95	134	107		
		3	104	149	102	2.0	12
		4	117	144	124		
		5	100	131	107		
	CD1012	3	143	168	156	12	38
	OPPI	0.25				7.3	20
		0.5	74	104	76	6.6	29
		1	146	167	156	6.2	39
		2	147	167	154	6.0	55
		3	134	164	141	7.1	64
		4	133	161	134	8.1	66
		5	121	162	129		
Tactix 123	UVI6974	0.5	59	82	59		
		1	77	102	77		
		2	89	121	92		
		3	94	134	92	2.2	42
		4	87	134	92		
		5	82	129	79		
	CD1012	3	165	195	180	7.4	42
	OPPI	0.25				5.4	18
		0.5	92	124	92	5.1	26
		1	142	207	142	4.4	48
		2	164	194	164	4.8	58
		3	156	182	156	6.9	69
		4	156	179	156	8.8	65
		5	146	174	146		
	DW1	1	139	179	152	3.3	59
		2	151	187	160	2.9	75
		3				2.7	78
		4				2.6	80
		5				2.7	75

3.3 Initiator Concentration Effects Gamma calorimetry shows that an optimum concentration produces the minimum dose required to cure a particular epoxy resin. For most epoxy resins this optimum concentration is between 2 and 3 phr, regardless of the initiator used. The initiator concentration also affects the rheological properties of a particular resin. Table 2 summarizes the rheological properties for various concentrations of the initiators with EPON 862 and Tactix 123. We found that each initiator has an optimum concentration, which results in a minimum dose required to cure and, at the same time maximize the rheological properties such as the glass transition and service temperatures. In the case of Tactix 123, the optimum concentration for UVI 6974 was found to be between 3 and 4 phr (active component). When OPPI was used as the initiator with Tactix 123, the optimum concentration was approximately 2 phr. The glass transition and service temperatures for UVI 6974 at 3 phr were found to be 134 and 95°C, respectively, whereas OPPI at 2 phr gave glass transition and service temperatures of 194 and 164°C, respectively.

5. CONCLUSIONS

We have found that within the same family of cationic initiators, the more efficient the initiator the less the dose required to cure the resin; the more efficient the initiator, the higher the glass transition and service temperatures obtained. When iodonium cationic initiators from the same family were used, it was found that the glass transition and service temperatures were the same; however, the dose required to cure the resin decreased with initiator efficiency. Each initiator has an optimum concentration for curing dose and maximizing rheological properties. This concentration was found to be between 1 and 3 phr for all the cationic initiators studied.

6. REFERENCES

1. "Electron Beam Curing of Polymer Matrix Composites" CRADA Annual Report Y/AMT-258, April 1994 to March 1995.

2. A. Singh, C.B. Saunders, V.J. Lopata, W. Kremers and M. Chung; 1995, "An Emerging Alternative to Thermal Curing: Electron Curing of Fibre-Reinforced Composites", Proceedings of the International Conference on Composite Materials and Energy - ENERCOMP 95, 389-394.

3. A. Singh, C.B. Saunders, V.J. Lopata, W. Kremers, T.E. McDougall, M. Tateishi, and M. Chung; 1996, "An Emerging Technology to Cure Fiber-Reinforced Composites: Electron Curing", Advanced Performance Materials, 3, 57-73.

4. Ajit Singh, Chris B. Saunders, Vince J. Lopata, Walter Kremers, Tom E. McDougall, Miyoko Tateishi, and Minda Chung, 1996, "Electron Processing of Carbon-Fibre-Reinforced Advanced Composites: A Status Report", Irradiation of Polymers: Fundamentals and Technological Applications, edited Roger L. Clough and Shalaby W. Shalaby, American Chemical Society, ACS Symposium Series 620, 197-205.

5. Don Sidwell, Lockheed Martin Advanced Development Corporation, personal communication.

6. J.G. Woods, 1992. "Radiation-Curable Adhesives", Radiation Curing, Science and Technology, edit: S.P. Pappas, Plenum Press, New York.

7. C.B. Saunders, A. Singh, V.J. Lopata, W. Kremers, and T.E. McDougall; 1994, "Electron Curing of Advanced Composites", Final Report, Defence Research Establishment Pacific, Department of National Defence, FMO Victoria, BC Canada.

8. C.B. Saunders, A. Singh, V.J. Lopata, W. Kremers, and T.E. McDougall; 1991, "Electron Curing of Advanced Composites", Final Report, Defence Research Establishment Pacific, Department of National Defence, FMO Victoria BC Canada.

9. J.V. Crivello, 1984; Adv. In Polym. Sci., 61, 1.

10. Cycloaliphatic Epoxides, Cationic UV Cure, Union Carbide pamphlet.

11. J.V. Crivello, M. Fan, and D. Bi, 1992; J. Appl. Polym. Sci., 44, 9.

12. J.V. Crivello, U.S. Patent 5,260,349, Nov. 9 1993 (Polyset Corp.).

13. T.C. Walton and J.V. Crivello, 1994; "Innovative Composite Fabrication Using Electron-Beam Rapidly Cured, Micrometeoroid and Atomic Oxygen Resistant EFS Polymers", 39th SAMPE, Anaheim, CA.

14. C.J. Janke, G.F. Dorsey, S.J. Havens, and V.J. Lopata; 1996. "Electron Beam Curing of Epoxy Resins by Cationic Polymerization", 41th SAMPE, Anaheim, CA.

15. A. Singh, C.B. Saunders, J.W. Barnard, V.J. Lopata, W. Kremers, T.E. McDougall, M. Chung, and M. Tateishi, 1996; "Electron Processing of Fibre-Reinforced Advanced Composites, Radiat. Phys. Chem., in press.

16. J.V. Crivello, J.H.W.Lam, J.E. Moore, and S.H. Schroeter, 1978, J. Radiat. Curing, 5, 2.

Figure 1. Typical Gamma Calorimetry Curve

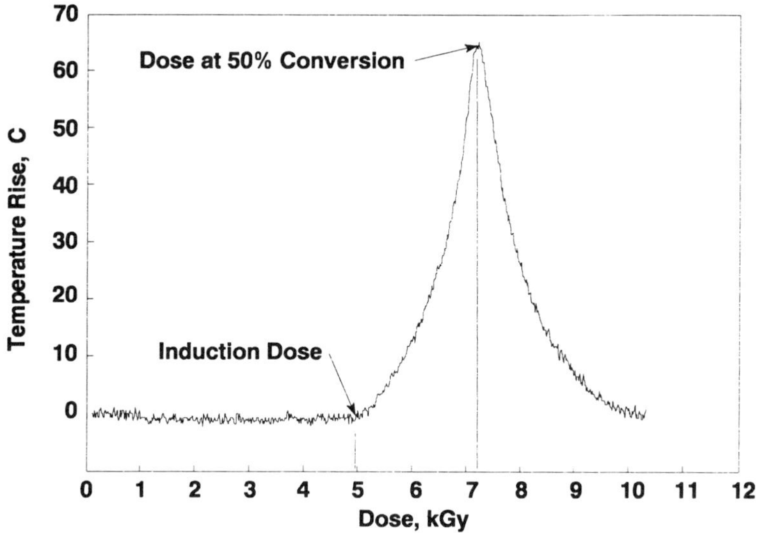

Figure 2. Effect of Various Cationic Initiators on the Curing of EPON 862

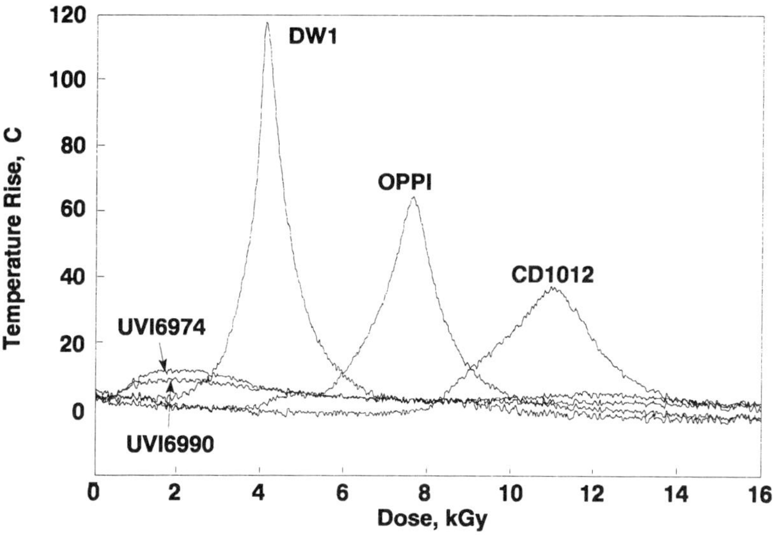

Figure 3. OPPI Concentration Effect on the Curing Dose of EPON 862

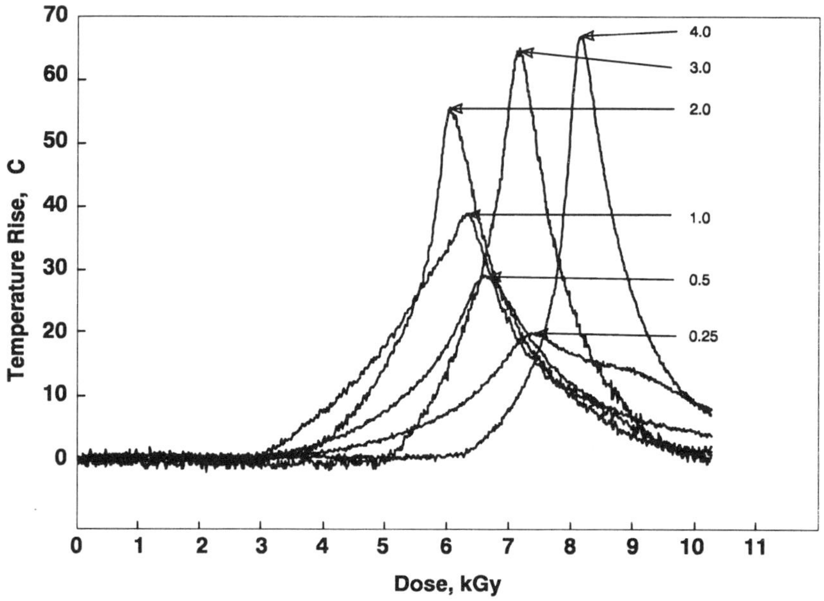

28th International SAMPE Technical Conference
November 4-7, 1996

COMPOSITE CURING WITH HIGH ENERGY ELECTRON BEAMS: NOVEL MATERIALS AND PROCESSES

Daniel L. Goodman and Catherine A. Byrne
Science Research Laboratory, Somerville MA 02143
Giuseppe R. Palmese and Satyendra Sarmah
University of Delaware Center for Composite Materials, Newark, DE 19716

ABSTRACT

Curing of fiber-reinforced polymer composites using high energy electron beams (HEEB) offers significant advantages for fabricating a variety of aerospace and ground vehicle components. The HEEB curing process greatly reduces the time required to cross-link the polymer matrix compared to conventional heat curing. The process allows curing at room temperature for high throughput and reduced residual stresses for good part fit-up.

Science Research Laboratory (SRL) has developed materials, processes and electron accelerator technology for high energy electron beam curing of advanced composites. One fabrication process under development has significant processing and material property advantages. In this process, the composite is a handleable solid at B-stage during curing. A novel resin system appropriate for this process is described. The resin system exhibits very low cure shrinkage, reducing residual stresses, and good thermal and mechanical properties.

KEY WORDS: Composites, Electron Beam, Accelerator, Polymer Matrix, Curing.

1. INTRODUCTION

Radiation curing using electron beams has the potential to cure composites and coatings quickly, without releasing environmentally hazardous solvents. Curing is at low temperature, reducing the residual mechanical stresses which can accompany conventional thermal curing.

Although there are a number of processing methods which can be used to cure composites with radiation, they can be divided into two general categories: Curing while the resin matrix is in a liquid state or in a solid state.

To cure a composite while the matrix is in a liquid state, the assembly must be enclosed

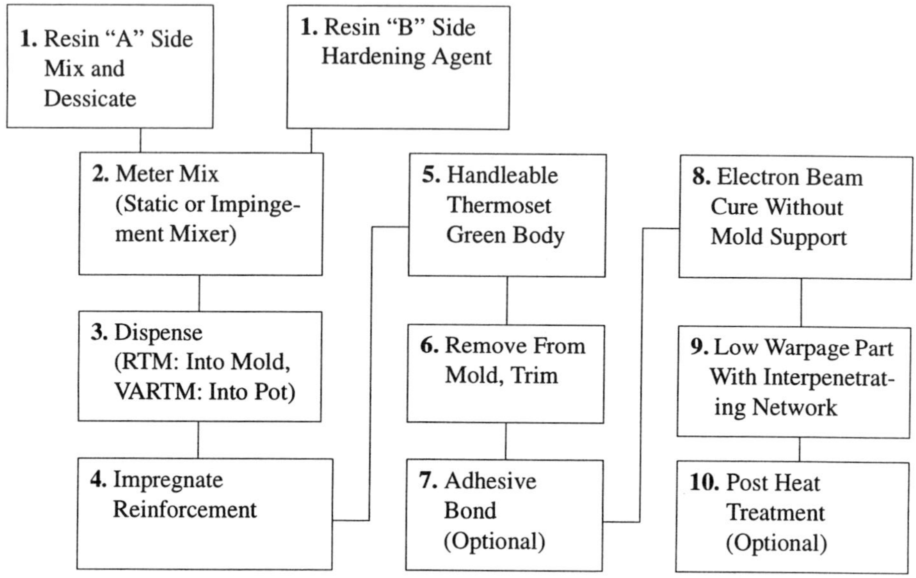

Figure 1: A flow diagram for fabrication of composites curable with electron beam. We have designated this process as "Tool-free" curing because the composite needs no supporting at EB curing step **8**.

in a mold, which determines the composite's final shape after it is crosslinked to a solid. Electron beams penetrate into the mold to crosslink the liquid resin matrix.

Alternatively, one may produce a handleable solid whose properties are improved by subsequent radiation crosslinking. This self-supporting solid is referred to as a "green body" at the process "B-stage." For the purpose of this patent, curing while the resin matrix is in a solid state will be designated as "free-standing" or "tool-free" curing, because no mold is required during the radiation curing process.

There are a number of processing and cost advantages to free-standing curing. Reference 2 describes how free-standing curing can be used to assemble a hollow section assembly. The cost advantages of using free-standing curing instead of conventional steel stamping at moderate production quantities are described in Ref. 3.

A generic problem in fabricating large structural composite parts is the difficulties of part fit-up. In applications involving the assembly of multiple parts, such as the case in automotive or aircraft applications, good part fit-up is necessary to allow final assembly, as well as for aerodynamic and aesthetic reasons. Unlike ductile metals, whose final shape can be modified by mechanical means at the final stages of assembly, rigid polymer matrix composites must achieve part fit-up by maintaining accurate fabrication tolerances.

In conventional composite processing (such as autoclave curing of aircraft composites), part fit-up problems are primarily caused by residual thermal stresses due to thermal expansion mismatch between fiber reinforcement and resin matrix upon cooling.

A major advantage of low-temperature curing processes such as electron beam curing is the reduction in residual thermal stresses. However radiation curing has an analogous problem: resin shrinkage during cure. In composites, resin shrinkage leads to internal stresses which can cause warpage of the composite part and poor part fit-up. It also leads to higher process engineering costs, since molds must be oversized or odd-shaped to account for shrinkage and warpage. In coatings, resin shrinkage can contribute to poor adhesion, especially to some metal substrates.

Conventional resin systems used in radiation curing shrink significantly when cured. Volumetric shrinkage values for common acrylic monomers are given in Ref. 4 and range from 11-30%. Linear shrinkage of acrylated oligomers are listed in Ref. 5 and correspond to volumetric shrinkages of 7-25%. A typical resin formulation, incorporating such oligomers and monomers, will have a volumetric shrinkage of 9-15%.

One possible approach to reducing shrinkage of radiation-curable resins is the use of cationic polymerization [6] via cationic initiators possessing highly non-nucleophilic anions (PF_6^-, AsF_6^- or SbF_6^-). These are an alternative to the more common radiation-curable resin systems of the free-radical type, which possess carbon-to-carbon double bonds that cross-link upon exposure to radiation.

Radiation-induced cationic polymerization has shown lower shrinkage than free-radical systems (typically ~6% shrinkage). Cationic catalysts are, however, easily poisoned by nucleophilic resin components, contaminants or fiber sizings, making radiation curing via cationic mechanism difficult. This is especially true for free-standing curing, where the presence of weakly nucleophilic urethane groups or of highly nucleophilic resin "B' components such as amines are incompatible with the use of cationic catalysts.

There is thus needed a method for formulating radiation-curable resins with low shrinkage for use in electron beam composite curing and coating applications, especially in the case where a free-standing radiation cure is desired.

We have developed materials which addresses this need, using an interpenetrating network (IPN) to reduce the polymer cure shrinkage and warpage of parts to ensure good part fit-up. An IPN is formed from a combination of two polymers in which at least one of the polymers has been synthesized and/or crosslinked in the presence of the other [7]. It has been shown that the properties of IPNs can differ significantly from those of the constituents and that often they cannot be predicted by simple mixing rules. For example, the glass transition temperature of polyester-acrylate semi-interpenetrating networks can be higher than that of either constituent [8].

2. PROCESS AND MATERIAL DESCRIPTION

A flow chart for a novel process, which we have designated as Tool-free or Free-standing composite curing, is shown in FIG. 1. The reason for this designation can be seen at step **5**, because the composite is already a self-supporting handleable solid which does not necessarily require supporting tooling or a mold when it is cured by high energy electron beams at step **8**. The 10 steps of the Tool-free process, including two optional steps are listed in FIG. 1.

At process step **1**, the components of resin side "A" are first uniformly mixed and desiccated to remove residual moisture. The resin side "A" consists of a mixture of oligomer, reactive diluent monomer and optionally one or more of the following: low molecular weight multifunctional crosslinking compounds, catalysts, pigments, surfactants and mold release agents.

We have obtained good neat resin and composite properties using an oligomer composition containing the chemical structure of a typical aromatic epoxy backbone, hydroxyl groups for reaction at step **2** and carbon-to-carbon double bonds for crosslinking via free-radical polymerization at step **8**. Such a material can be synthesized by the reaction of multifunctional phenolic compounds with molecules containing both epoxy groups and double bonds in the presence of base or by the addition of (meth)acrylic acid to a multifunctional epoxy. Typical chemical diagrams are shown in FIG. 2, and the numbered stages correspond to the steps of FIG. 1.

The reactive diluent monomer is a low viscosity ($<$ 500 cps) liquid containing carbon-to-carbon double bonds for reaction at step **8** but not containing hydroxyl groups. Appropriate monomers are available from Sartomer Company, Inc and UCB Chemicals Corp.

The desired weight fraction of the oligomer and monomer components of resin side "A" depends on the choice of composite fabrication method **3**. In general, the fraction of monomer will be the minimum which provides a sufficiently low resin viscosity for impregnation **4**. This is because a high oligomer fraction generally maximizes the physical and thermal properties of the fabricated composite. Typical values for composite fabrication are 50-90% oligomer, 10-45% monomer and 0-5% other components. Typical values for coatings are 40-70% oligomer, 20-50% monomer and up to 40% pigments and other compounds.

Several optional components can be added to the resin side "A" to increase throughput or for other processing advantages. Low molecular weight multifunctional crosslinking aids containing hydroxyl groups at a 0-5% weight fraction can be added to the resin side "A" to increase the composite hardness at B-stage step **5**. Metal catalysts at concentrations from 0.005-0.1% can be added to decrease the gelation time between steps **2** and **5**. Mold release agents can be added to facilitate removal of the composite from the mold at step **6** without sticking.

The resin side "B" hardening agent is a reactive species capable of gelling the polyol side "A" to a green body at B-stage. We usually use one of the Dow Chemical PAPI 2000 series mixture of di- and multifunctional isocyanates.

We have also experimented with other epoxy-based systems which are capable of forming an electron beam-curable interpenetrating network. For example, we have tried mixtures of epoxy resins (such as the Shell Company Epon 828 bisphenol A epoxy) mixed with EB-curable reactive diluent monomers. We reacted the epoxy to B-stage with several commercial epoxy curing agents such as amine, anhydride and thiols. The properties of these systems are not yet as good as those made using an isocyanate cure to B-stage.

In production, mixing of the resin sides "A" and "B" at step **2** in an appropriate ratio to produce gelation would be done using high pressure meter-mix equipment with a static or impingement mix head. For laboratory experiments, we have done the mixing either by hand or using meter-mix equipment with a replaceable static mix head. Mixing by hand is sufficient for vacuum assisted resin transfer molding (VARTM) at step **3**, but the

1.

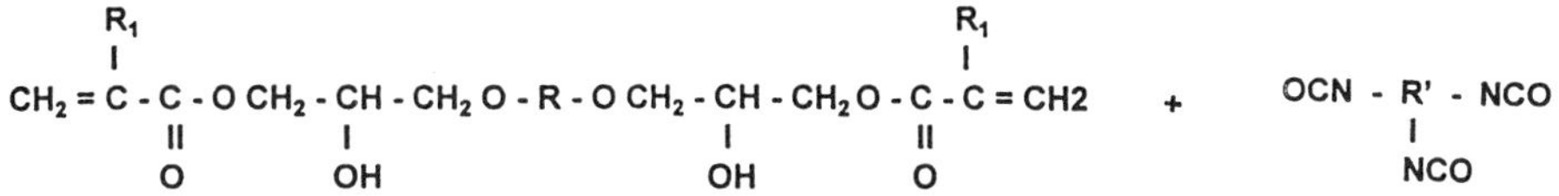

5.

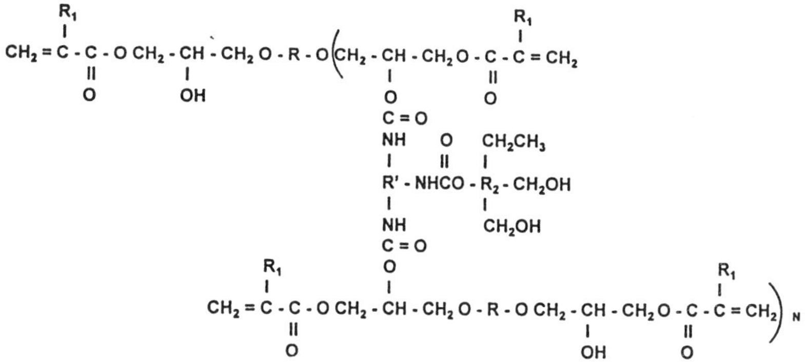

9.

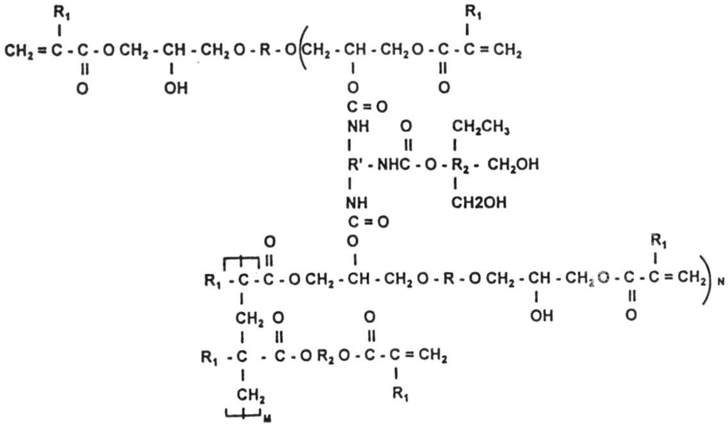

Figure 2: Typical chemical diagrams for a B-stageable resin system which undergoes free-radical polymerization by electron beam irradiation. The numbered stages correspond to the steps of FIG. 1.

higher pressures ($\sim$ 100 psi) needed for resin transfer molding (RTM) requires a resin injection molding machine with static mixer.

Following mixing, the resin is dispensed **6** in order to impregnate reinforcement material. Using RTM and VARTM, we have made composites with glass reinforcement for vehicle applications and carbon fiber reinforcement for aerospace applications.

Gellation of the resin encapsulates the reinforcement material, creating a thermoset green body at B-stage. The resin composition is chosen such that this green body will have a glass transition temperature above room temperature, (typically $\sim$50°C). This is done so that the material will be a handleable solid, and will maintain its shape when removed from the mold at step **6** and trimmed to net-shape. The gelation network can be seen in FIG. 2.

If multiple parts are to be separately molded, and then joined, as might be the case for a complex structure such as an aircraft wing, adhesive bonding using a chemically compatible system can be done at step **7**. Co-curing and co-bonding processes using electron beam curing allow the fabrication of large integrated structure, which is a key cost reduction strategy for vehicles and aircraft. Electron curable adhesives are available from a number of manufacturers including the Loctite Company of Newington CT. Curing dose requirements, shrinkages and adhesive strength for many of the Loctite products cured with electron beam are listed in Ref. 9.

It is preferable to bond at step **10**, rather than after step **13** for reasons of bond strength and part fit-up. Applying the adhesive and simultaneously curing the adhesive and the composite should produce a strong joint due to the creation of carbon bonding across the adhesive/composite interface. Such a bond might be expected to be stronger than one between two previously cured composites because of the ability of the polymer, and particularly the polymer chain ends in the B-staged resin, to move during curing, allowing a greater number of bonds per unit area to be formed. The movement of the polymer occurs because the glass transition temperature of the B-staged system is chosen to be lower than the temperature rise of the adhesive-composite interface during cure. The adhesive temperature rise during cure is due to both direct electron beam heating and to the exothermic heat of reaction. A proper choice of adhesive chemical composition will allow the adhesive to heat and react at a lower electron beam dose than the composite to be joined.

The green body at B-stage at step **5** is flexible and has a lower modulus and glass transition temperature than at step **9**. Components are thus more easily joined at this stage. If necessary, some localized heating can be used to facilitate part fit up. Using bond jigs, several parts may be joined together at this step. An activator capable of initiating the free radical cure of the adhesive may be sprayed onto the composite. The bond jigs may be removed after 2-5 minutes and the assembly is available for curing by electron beam.

After the (optional) adhesive bonding step, the part is cured by electron beam at step **8**. Typically, the part is brought to the electron beam accelerator using a material handling system such as a conveyor belt or cart-on-track. Since the part is a handleable solid, no mold enclosure is required, although some tooling may be required to support the part during transport to the electron accelerator. This tooling can be fabricated from low cost

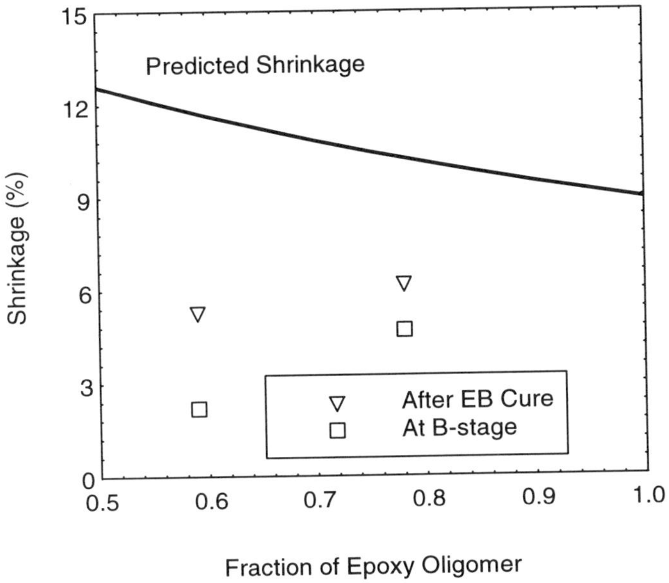

Figure 3: Measured resin shrinkage values are less than calculated values (curve), leading to reduces cure stresses, less part warpage and better part fit-up. Error bars for the measured points are between 0.1-0.5% shrinkage

materials such as wood or plaster, since its temperature remains quite low (typically less than 70°C) during curing.

Electron accelerators with the necessary parameters for high energy electron beam composite curing are available from a number of manufacturers including Science Research Laboratory of Somerville, MA, Atomic Energy of Canada Limited of Ontario, Canada and Radiation Dynamics, Inc. of Edgewood NY. A summary of the types of accelerators and their operating principles are given in Ref. 10.

Electron beam irradiation causes the formation of a crosslinked carbon bond network which interpenetrates the existing urethane network produced by the reaction of the isocyanate with the hydroxy-containing compounds at step **2**. These two networks form an interpenetrating network which we believe reduces the shrinkage of these composite to the values needed for good part fit-up. Our theory is that the urethane network restricts the shrinkage of the carbon-carbon double-bond-containing compounds, producing localized stresses on the molecular length-scale.

The final (optional) step in fabrication of a composite using this process is a post heat treatment. By heating the composite part to an elevated temperature, the final use temperature can be increased. During this step, a part would typically placed in an oven for approximately one hour at a temperature in the range 120-180°C. This process will typically increase the final use temperature (before softening) by 40°C. The heat treatment process promotes polymer diffusion, reducing the configuration free energy and utilizing residual free radicals to fully crosslink any unconverted carbon-carbon double bonds.

3. EXPERIMENTAL RESULTS

3.1 Neat Resin Results

Resin shrinkage leads to composite warpage and poor part fit-up. The shrinkage of neat resin samples was measured by displacement in water (Archimedes Method) in accordance with ASTM Standard D792-96. The results are shown in FIG. 3.

Two resin compositions, with differing ratio of epoxy-derived oligomer to reactive diluent monomer were prepared. The resin with a lower fraction of epoxy-based oligomer has lower viscosity, and has been used successfully for composite manufacture using the vacuum assisted resin transfer molding (VARTM) process. The material with a higher fraction of epoxy-based oligomer and higher viscosity has been used for composite fabrication using the high pressure resin transfer molding (RTM) process.

The shrinkage at B-stage (step **5**) and after electron beam cure (step **8**) is shown in FIG 3. (As used here, percentage shrinkage is given by $100 * (\rho_s - \rho_l)/\rho_l$, where ρ_l and ρ_s are the liquid and solid densities.) Also shown is the expected shrinkage of these resins using published and measured shrinkage data for the resin components, combined using a law of mixtures approach. The predicted shrinkage curve in FIG. 3 was calculated using all components, their densities, functionalities and known radiation-cure shrinkages. The equation for the predicted change in resin volume upon shrinkage in FIG. 3 is:

$$\frac{\Delta V}{V} = \frac{(\Sigma S_i M_i/\rho_i)(\Sigma M_j \rho_j)}{(\Sigma M_j)^2}$$

where:

S_i, M_i and ρ_i are the fractional shrinkage, equivalent weight and density of the ith component, and the sum over j includes all resin components, but the sum over i includes only those which shrink upon electron beam irradiation.

The lower viscosity VARTM resin with a 60% epoxy oligomer fraction produces a more flexible network at B-stage. The data in FIG. 3 shows that at B-stage, this system has shrunk by less than half that of the higher viscosity RTM resin with 78% epoxy oligomer fraction. After electron beam crosslinking, the two systems have similar shrinkage relative to the liquid resin. The data in FIG. 3 shows that the actual shrinkage of resins produced using these resins is significantly less than would be expected based on a rule of mixtures. This reduced shrinkage shows up in composite properties, as described below.

3.2 Composite Results

Composite test panels to measure warpage were fabricated using this technique. Reinforcement material was woven S-2 glass roving at a fiber volume fraction of approximately 50%. To maximize the composite warpage, half the plys were laid up at a 45° angle. Warpage of the composite was measured by edge displacement. Warpage of the panels made using this technique was less than half that of identical panels made using a conventional resin matrix (Dow Derakane Vinyl Ester 11C-50, Dow Chemical Co, Midland MI.)

Composite test panels to measure mechanical and thermal properties were made using this technique. Reinforcement materials included woven S-2 glass roving and graphite fiber cloth, at fiber volume fraction of approximately 50%. Glass transition temperature

Sample Type	Test Type	Value	Notes
Neat resin	Viscosity	300-1500 cP	By varying epoxy/reactive diluent ratio.
Neat resin	Modulus	300 ksi	Typical curing dose 20 MR.
Neat resin, without heat postcure	Tg from DMA loss modulus peak	110-170°C	Low temperature during cure (50°C) Tg depends on composition and dose.
Neat Resin, with free-standing heat postcure	Tg from DMA loss modulus peak	156-212°C	Hold at 130°C for 1 hour or 200°C for 5 min.
Composite (S2 glass)	Tensile Strength	100 ksi	ASTM D3039, 50% fiber volume fraction.
Composite (S2 glass)	Tensile Modulus	3.6-4.2 Msi	0.13 in. thick laminate, 933 fiber sizing.

Table I: Resin and composite properties. Resin system is the EB-curable B-stageable system described in the text and shown in FIG. 2.

(a measure of final use temperature) was in the range of 110-170°C depending upon exact composition and radiation dose, increasing to 156-212°C after a post heat cure. The values given correspond to the peak in the loss modulus as measured using dynamic mechanical analysis.

Tensile tests made on the glass fiber-reinforced composite panels showed higher values (100 ksi) than tests on similar panels with conventional vinyl ester (Dow Derakane resin) matrix. Composite properties for glass-reinforced laminates are listed in Table 1.

4. CONCLUSIONS

We have developed materials for reducing warpage and improving part fit-up in structural composites fabricated using electron beam-curable resins.

The process shown in FIG. 1 produces a rigid, fiber reinforced, radiation-curable composite at B-stage using a resin injection process in which a two-component thermosetting resin is used to impregnate the fiber reinforcement. After reaction to a handleable state, the green body is removed from the mold and subjected to high energy electron beam radiation to complete the resin cure and achieve full thermal and mechanical properties.

A novel two-component reactive resin system has been developed to allow use of this "Tool-free" process. The resin consists of an "A" side containing a polyol oligomer and a "B" side containing polyaromatic isocyanate which are mixed in an appropriate ratio using either a static or impingement mixer prior to injection. The polyol oligomer contains the carbon backbone of an aromatic epoxy as well as hydroxyl groups which react with the isocyanate to create a gelled urethane thermoset as well as carbon-to-carbon double bonds, which cross-link in the presence of the electron beam. Resin side "A" also contains low-viscosity reactive diluents with unsaturated carbon bonds and additional crosslinking agents to tailor the final composite properties, and a metal catalyst to determine the gelation rate to B-stage. The combination of the two reactions (the initial urethane formation and the radiation-induced cross-linking) produces an interpenetrating network (IPN) which restricts shrinkage, and thus reduces warpage of the structural composite part.

References

[1] Work supported by the Advanced Research Projects Agency under Contract MDA972-95-C-0031.

[2] U.S. Patent 5,173,142 (Dec. 22, 1992) C.R. Billiu (to Wellman Machinery Co.).

[3] D.L. Goodman, D.L. Birx, G.R. Palmese and A. Chen, SAMPE International Symposium, 41, 207 (1996).

[4] Volume Shrinkage During Polymerization of Acrylic Monomers, Rohm and Haas Company, Washington Square, PA, 1964.

[5] UV/EB Curable Oligomers and Shrinkage Behavior, UCB Chemical Company, Radcure Business Unit, Smyrna, GA, 1996.

[6] S.P. Pappas, Radiation Curing, Science and Technology, Plenum Press, New York, 1992, pp. 5, 378.

[7] L.H. Sperling, Interpenetrating Polymer Networks and Related Materials, Plenum Press, New York, 1971, pp. 1-8

[8] J.J Fitzgerald and C.J.T. Landry, Journal of Applied Polymer Science, 40, 1727 (1990).

[9] C.B. Saunders, V.J. Lopata, W.Kremers, M. Chung and J.W. Barnard, SAMPE International Symposium, 40, (1995).

[10] M.R. Cleland in A. Singh and J. Silverman, ed., Radiation Processing of Polymers, Oxford University Press, New York, 1992, pp. 23-46.

28th International SAMPE Technical Conference
November 4-7, 1996

MECHANICAL PROPERTIES OF ELECTRON BEAM CURED FIBERGLASS LAMINATES

Barbara G. Devney
DAMILIC Corporation
14670 Southlawn Lane
Rockville, Maryland 20850

Anthony J. Vizzini and Walter J. Chappas
University of Maryland
College Park, Maryland 20742

ABSTRACT

For the past several years, electron beam curing has been evaluated in the authors' laboratories for the high speed manufacture of a vehicle such as the Army's Composite Armored Vehicle (CAV). This paper describes tension, compression, and shear test results for fiberglass laminates prepared with five newly developed electron beam curable resin formulations. The resin systems include cationically cured epoxies and free-radical cured phenolic acrylate and aromatic urethane acrylate systems. Inspection of the failed test coupons revealed two distinct failure modes, with the aromatic urethane acrylate resin/matrix system exhibiting the more complicated failure mechanism. The urethane acrylate formulation also exhibited the highest tensile strength and highest shear strength.

KEY WORDS: Electron Beam, Fiberglass Reinforced Plastic (FRP), Advanced Composites, Cationic Resins, Epoxies, Acrylates, Phenolics, Radiation Cure

1. INTRODUCTION

The commercialization of energy efficient transport vehicles could be accelerated if the manufacturing costs for fiber reinforced composites were significantly lowered. Electron beam curing has the potential to reduce manufacturing costs through lower tooling costs, increased production speeds, and lowered electric utility costs (when compared to autoclave cures). Moreover, the low temperature electron beam cure avoids the buildup of thermal stresses in large heterogeneous structures, which may facilitate design concepts that integrate primary through tertiary structures. [1-5]

DAMILIC Corporation, in a Phase I Small Business Innovative Research (SBIR) grant sponsored by the Army's Tank and Automotive Command (TACOM), previously demonstrated the feasibility of using gamma (γ) radiation to cure composite parts by fabricating a 3.8 cm thick armor plate and a 2.6 cm thick angle bracket composed of alternating layers of glass and graphite (woven roving) with an outer copper mesh simulating an EMI shield.[1] This paper describes the mechanical performance of five new electron beam curable resin/matrix systems. From these results, the top performing resin, an aromatic urethane acrylate mixture, was incorporated into the production of a prototype part for the Army's Composite Armored Vehicle (CAV) in support of TACOM's Advanced Technology Demonstrator (ATD) program.

2. MATERIALS SELECTION

2.1 Overview Earlier work in this program evaluated glass laminates prepared with an acrylated diglycidal ether of bisphenol A (DEGBA) modified with multi functional acrylates (monomers) to reduce the viscosity and lower the cure dose. This study utilized an acrylated phenolic and a hard aromatic urethane acrylate oligomer in two new resin formulations. This program also evaluated three new electron beam curable epoxy formulations, developed by Applied Poleramics.

The resins in this study utilize a range of chemistries that can be divided into two categories, those that cure by a free-radical addition polymerization and those that cure by a cationic initiated homopolymerization. Formulations that cure via free radical polymerization typically exhibit better thermal stability, a longer shelf-life, and little postcure. These formulations do not require the addition of any catalysts of sensitizers to be electron beam curable. A disadvantage of free-radical curing resins is their reactions with oxygen that degrades the polymer network and the composite's mechanical performance. In order to attain the best performance from a free-radical formulation, vacuum bagging is required.

The cationic resin formulations included in this study, CAT-10, BD-2 and BD-3, were formulated with conventional epoxy resins, commonly used in filament winding and resin transfer molding applications. CAT-10 contains a multi-functional epoxy and BD-3 contains a mixture of multi-functional and cycloaliphatic epoxies. The BD-2 resin is a mixture of a cationically curable multi-functional epoxy and an acrylated resin that cures by a free-radical mechanism. By combining resins that polymerize independently, an interpenetrating polymer network with improved toughness can result. All the cationic formulations contain an organometallic catalyst that, when exposed to high energy electrons (or UV light), initiate an acid catalyzed homo-polymerization of the epoxy group. The polymerization reactions that follow the electron beam initiated cation formation is a thermally activated process. With time, further polymerization may occur and with it dimensional changes. The net shrinkage expected for these systems is typically lower than vinyl esters and acrylates that cure by a free-radical polymerization.

2.2 Preliminary Screening Process As a preliminary screening, several resins were irradiated at doses ranging from 20-100 kGy. A prerequisite for inclusion in this program was that the cure dose be less than 50 kGy. The elastic modulus and the glass transition temperature as measured by dynamic mechanical analysis (DMA) for some of the resins utilized in our work are shown in Table 2. Based on the preliminary screening process, the novolak acrylate and aromatic urethane acrylate oligomers were chosen for further development and testing. Processing requirements for the manufacture of the CAV hull section necessitated the addition of reactive diluents to lower the viscosity while maintaining an economical (<50 kGy, 1Gy =1J/kg) cure dose.

Table 2 Glass Transition Temperature and Elastic Modulus of Selected Resins

Resin	T_g °C	Modulus (MPa)
60% epoxy acrylate[a]	178 ± 5	1282 ± 20%
60% novolak acrylate[b]	180 ± 5	1655 ± 20%
60% aromatic urethane[c]	N.A.	1027 ± 10%[§]
multi functional epoxy (CAT-10)	167 ± 5	N.A.

[a] Sartomer Product CN120E60, diluted with 40% ethoxylated trimethylol propane triacrylate.
[b] Sartomer Product CN112C60, diluted with 40% ethoxylated trimethylol propane triacrylate.
[b] Sartomer Product CN970E60, diluted with 40% trimethylol propane triacrylate.
[§] Manufacturer's data for UV cured resin.

Table 3 CN970 Resin Mixture

Component	Weight %	Product Name[‡]
60% aromatic urethane acrylate[†]	65	CN970E60
ethoxylated trimethylolpropane triacrylate (EOTMPTA)	28	SR454
dipentaerithyritol pentacrylate (DPEA)	7	SR399

[‡] Sartomer Product
[†] diluted with 40% ethoxylated trimethylol propane triacrylate

Table 4 CN112 Resin Mixture

Component	Weight %	Product Name[‡]
60% novolak acrylate[†]	57	CN112C60
2-phenoxy acrylate	34	SR339
dipentaerythritol pentacrylate (DPEA)	9	SR399

[‡] Sartomer Product
[†] diluted with 40% trimethylol propane triacrylate

2.3 Resin Formulations The acrylate formulations evaluated in this study are shown in Tables 3 and 4. The monomers trimethylol propane triacrylate (viscosity, 0.106 Pa•s) and 2-phenoxy acrylate (viscosity, 0.0012 Pa•s) were added to lower the viscosity of the oligomers. Dipentaerythritol pentacrylate was added to reduce the cure dose. The resulting formulations have viscosities between 2 and 5 Pa•s @ 25° C. The CN970 resin mixture has an elastic modulus of 794 ± 50 MPa and a glass transition temperature of 110° ± 15 C as measured by dynamic mechanic analysis.

3. EXPERIMENTAL PROGRAM

3.1 Manufacture of Test Panels For each of the formulations, two 4-ply (0°, 90°, 0°, 90°) test panels measuring 35.6 cm x 45.7 cm, were manufactured. The fabrication protocol included steps to degas the resin and carefully weigh the amount of resin applied per fabric ply. A five sided mold was used to prevent motion of the resin at the edge of the laminate. The laminate was vacuum bagged and consolidated at 100 kPa for at least four hours prior to cure.

The fabric used was style 6781, an 8 harness satin weave constructed from continuous filament, single-strand high-strength S2 glass yarn, 360 μm in diameter, 150 m/kg. The S2 glass fabric had a silane finish applied after a chemical scour. This finish is a proprietary blend of organic functional silanes having dual functionality (manufacturer JPS, no. 9827). This finish is recommended for use with epoxies, polyimides, and bismaleimides.

Curing took place within 48 hours after vacuum bag consolidation. Vacuum was maintained until transport to Irradiation Industries Inc., a 2.7 MeV electron beam contract irradiation facility located one half hour from the University of Maryland Composites Research Laboratory, where the lay up took place. The uncured composite test panels arrived at the electron beam facility with approximately 200 mm Hg vacuum. The irradiations were fractioned, with each sample receiving a dose of 20 kGy per pass beneath the beam until a total surface dose of 100 kGy was received. Typically the accelerator was operated with a 4.0 - 5.0 mA beam current and was scanned over a 1.2 meter width. In one case, the irradiation of BD-3, a 20.5 mA beam current was used. High beam current implies a higher dose rate that can affect the polymerization kinetics.

To insure uniformity, each plate was rotated after every 20 kGy pass and placed on a different cart every alternate pass. Several radiochromic dosimeters, manufactured by Far West Technologies, were placed on each plate to confirm the total dose. These dosimeters are 1 cm square nylon films (approximately 50 microns thick) that contain a radiation sensitive dye. Upon irradiation, the dye changes from colorless to blue. The degree of color change, measured through simple spectrophotometric techniques, is proportional to the dose. The dosimeters for this work were calibrated at both 510 nm and 601 nm before use, and were handled according to a protocol that is fully traceable to the National Institute of Standards and Technology.

After the initial cure to 100 kGy, some of the plates felt mildly tacky and had some residual odor. The tackiness and odor was most notable for the CN970 and CN112 acrylate mixtures. Upon further irradiation, the tackiness and odor were eliminated. The net dose applied to each plate is as follows: BD-3, 200 kGy; BD-2, CN112 mix, CN970 mix, 300 kGy.

3.2 Specimen Preparation Machining of the composite panels was performed using a water cooled diamond grit cutting wheel. For the compression and tensile test samples, epoxy tabs were then applied. Tabs were prepared from 0.32 cm thick plates of epoxy glass reinforced plastic. Tabs were bonded to the test panels with FM-123 film adhesive at 115°C and 275 kPa. After tab bonding, the tensile and compression samples were cut to their final dimension of 3.8 cm x 35.6 cm and 1.3 cm x 14.0 cm, respectively. The shear specimens were 1.3 cm x 3.8 cm with two central V-notches. Prior to attachment of the strain gages, the thickness and width of each test coupon was measured in several locations. Strain gages were bonded to both the front and back of each test coupon. For both the tension and compression specimens a single longitudinal strain gage was bonded to one side of the test specimen, and a 0/90 rosette strain gauge was bonded to the other side, enabling simultaneous measurement of the modulus and the major Poisson's ratio. For the Iosipescu shear specimens, a single notched shear gauge was used. The thicknesses and widths of the specimens are provided in Table 5.

Table 5 Average Thicknesses/Width of Composite Specimens

Resin System †	Tension Thickness/Width (cm) (C.V.)	Compression Thickness/Width (cm) (C.V.)	Shear Thickness/Width (cm) (C.V.)
CN970 mix	0.094 (3.80%)	0.092 (1.19%)	0.096 (1.29%)
CN112 mix	0.109 (3.80%)	0.112 (1.20%)	0.112 (1.18%)
CAT-10	0.112 (3.80%)	0.122 (1.21%)	0.124 (1.12%)
BD-2	0.124 (3.80%)	0.130 (1.20%)	0.127 (1.24%)
BD-3	0.104 (3.80%)	0.102 (1.20%)	0.102 (1.25%)
RT epoxy ††	0.147 (1.28%)	0.140 (1.28%)	0.142 (1.25%)

† S2 glass fabric style 6781; 4 plies @ 0.023 cm/ply [0.092 cm total thickness]
†† Epon 828, Versamid 125/V-25 (1:1)

All mechanical tests were performed on a 55 kip universal testing machine equipped with hydraulic grips and interfaced to a computer that recorded the applied load, stroke, and strain at 0.25 second intervals. In all tests the specimens were strained at a constant stroke rate of 0.00254 cm/second.

4. RESULTS/DISCUSSION

Three standardized tests were performed on each radiation curable resin/matrix system as a means of comparing the different resin systems and to selecting materials for the CAV ATD program. In addition, the mechanical properties of a conventional, room temperature cured, glass laminate was measured to provide a baseline reference.

The tests included:

1. Tensile Properties of Fiber-Resin Composites (ASTM D 3039),
2. Compressive Properties of Unidirectional and Crossply Fiber-Resin Composites (ASTM D 3410), and
3. Shear Properties of Composite Materials by the V-Notched Beam Method (ASTM D 5379).

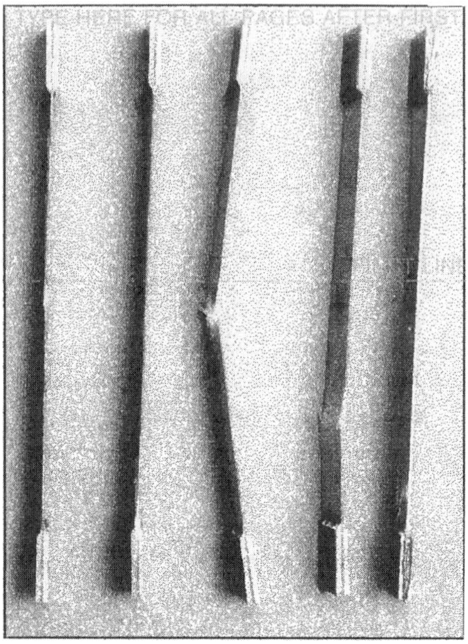

Figure 1. BD-3 Tension Specimen

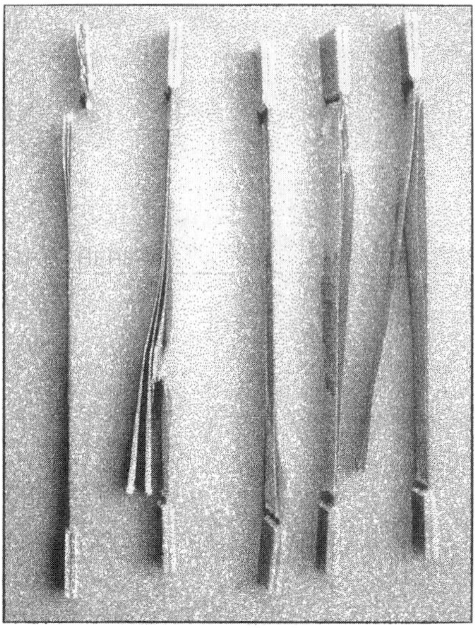

Figure 2. BD-2 Tension Specimen

4.1 Tensile Testing Uniaxial tensile tests on 3.8 cm x 35.6 cm x 0.11 cm coupons were performed under the conditions specified in the preceding section. The average tensile strengths of the electron beam cured test coupons ranged from 230 MPa to 383 MPa. The room temperature cured epoxy performed somewhat better with a failure strength of 415 MPa. The performance of the electron beam cured systems can be ranked from high to low as follows: CN970 mix> CN112 mix> BD2> CAT10> BD3 (See Table 6).

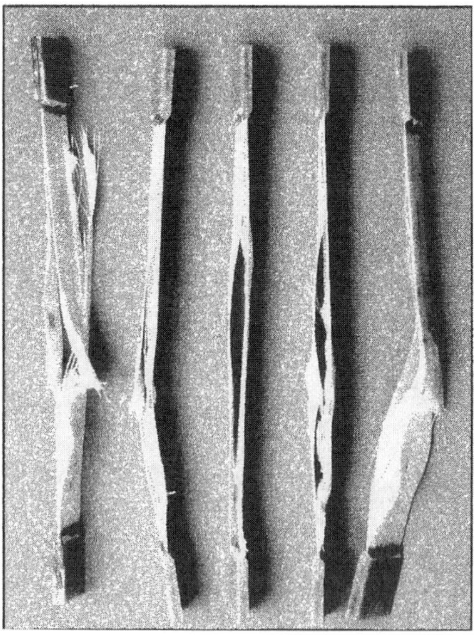

Figure 3. CN970 Tension Specimen

Inspection of the failed test coupons revealed two distinct failure patterns. The simplest failure pattern observed was a localized tear perpendicular to the applied load. Both CAT10 and BD3 exhibited this type of failure (See Figure 1). The specimens made from resins containing acrylate functional groups CN970, CN112, and BD-2 exhibited a more complicated damage progression. For instance, the BD-2 specimen failed when an initial tear oriented perpendicular to the applied load created sufficient shear force to cause widespread delamination (See Figure 2).

Delamination also occurred in the CN970 (acrylated aromatic urethane) and CN112 (acrylated phenolic) specimens. In these systems a tear propagated diagonally across the fabric plies and lead to broken fibers throughout and along delaminations (Figure 3).

Table 6 Tensile Properties of Ebeam Cured Laminates

Resin	Strength (MPa) (C.V.)	Modulus (GPa) (C.V.)		Poisson Ratio (C.V.)
CN970 mix	383 (9%)	24.9 (2%)	18.8 (2%)	0.113 (13 %)
CN112 mix	342 (9%)	24.4 (3%)	18.4 (3%)	0.101 (29 %)
CAT-10	259 (3 %)	25.0 (5%)	18.8 (3%)	0.085 (33 %)
BD-2	314 (10%)	24.7 (12%)	18.6 (4 %)	0.074 (18 %)
BD-3	230 (8%)	23.6 (9%)	16.6 (12%)	0.106 (15 %)
RT epoxy†	415 (4%)	21.8 (4%)	17.0 (12%)	0.089 (21 %)

† Epon 828, Versamid 125/V-25 (1:1)

The stress strain curves for the five systems tested all showed initially elastic behavior with similar slopes, indicating fairly comparable tensile moduli. Knees in the stress strain curves can be seen at around 500μstrain. The values measured for the tensile modulus (below 500μstrain) were approximately 24 GPa. These results are consistent with the fact that the tensile behavior is largely fiber dominated.

4.2 Compression Tests The compressive properties were measured, utilizing an IITRI fixture according to ASTM D-3410 on coupons measuring (nominally) 14 cm x 13 cm x 0.11 cm. The stress strain curves for all the systems indicated bending without buckling prior to failure. The compressive strengths showed very little variation from system to system. The five electron beam cured resin/matrix systems performed similarly and failed at approximately 183 MPa. The results for compressive modulus were approximately for 24 GPa for all five systems tested (Table 7).

The close proximity of the strain gauges and the smallness of the area to which the compressive forces were applied made inspection of the failure mode difficult. A top performer could not be

Table 7 Compressive Properties of Ebeam Cured Laminates

Resin	Compressive Strength (MPa) (C.V.)	Compressive Modulus (GPa) (C.V.)	Poisson Ratio (C.V.)
CN970 mix	179 (7%)	24.6 (6%)	0.109 (21%)
CN112 mix	179 (7 %)	23.5 (3%)	0.111 (26%)
CAT-10	188 (5 %)	23.7 (3 %)	0.101 (13%)
BD-2	182 (13%)	24.3 (5 %)	0.088 (16%)
BD-3	191 (6%)	23.5 (4 %)	0.102 (37%)
RT epoxy†	236 (12 %)	22.7 (8 %)	0.092 (27%)

† Epon 828, Versamid 125/V-25 (1:1)

distinguished from these tests due to the closeness in results. The superior performance exhibited by the room temperature cured epoxy laminate may have resulted because these specimens contained more resin, measuring 33% thicker than the average electron beam cured specimens.

4.3 Shear Tests Shear properties correlate strongly with the load transfer capability of the fiber/resin matrix and is therefore a sensitive test of a composite performance. The shear properties were measured on V-notched (7.6 cm x 1.9 cm x 0.11 cm) specimens mounted in an Iosipescu test fixture. Careful control of the notch dimensions ensured equivalent shear stress concentrations at the notch roots. The measured shear strengths exceeded 40 MPa for CN970, CN112, CAT10, and BD3 resin systems (Table 8). Further analysis of the stress strain curves yielded moduli values that varied significantly from one system to another. The room temperature cured epoxy performed 25% better than the best performing electron beam cured samples under shear. The shear strengths for the systems tested can be ranked according to performance as follows: RT epoxy >CN970 mix> CN112 mix> BD2> BD3>CAT10. It was also observed, that improved shear modulus strongly correlated with improved tensile strength (Figure 4).

Table 8 Shear Properties of Ebeam Cured Laminates

Resin	Strength (MPa) (C.V.)	Modulus (GPa) (C.V.)
CN970 mix	53.2 (24 %)	4.72 (4 %)
CN112 mix	46.0 (7 %)	3.60 (17 %)
CAT-10	45.6 (16 %)	0.83 (15 %)
BD-2	37.7 (6 %)	2.90 (17 %)
BD-3	47.6 (8 %)	2.66 (10 %)
RT epoxy†	66.9 (7 %)	1.72 (13 %)

† Epon 828, Versamid 125/V-25 (1:1)

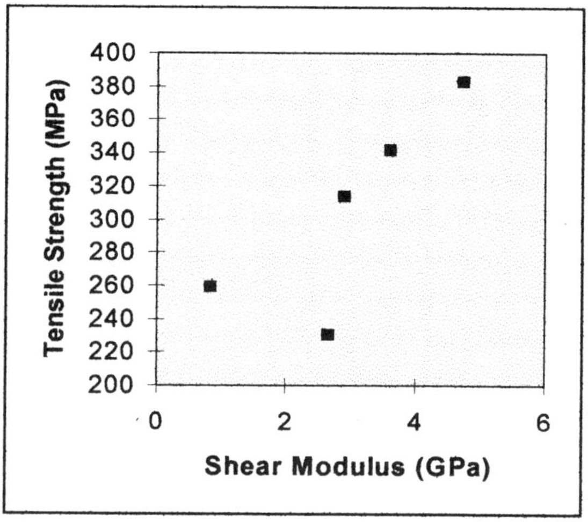

Figure 4. Tension v. Shear for the 5 formulations

5. SUMMARY

The relative performance of five newly developed electron beam curable resins under standard tensile, compression, and shear loads were evaluated in order to select an electron curable system for incorporation into the CAV prototype. Consistently superior composite performance, in both tension and shear strength, was exhibited by an aromatic urethane acrylate resin system (CN970 mix). However, the tensile strength of the CN970 composite laminate was only 90% of the epoxy reference material, and its shear properties were only 80% of the room temperature cured conventional composite. This resin was chosen for further testing and incorporation into the CAV ATD program.

Acknowledgments: The authors are grateful to Rich Moulton of Applied Poleramics, Inc. for his support in the resin development.

6. REFERENCES

1. W. J. Chappas, N. Grossman and B. Pourdeyhimi, 10th Annual ASM/ESD Advanced Composites Conference, Dearborn, Michigan, USA, (1994).

2. Chris B. Saunders, et al., "Recent Developments in the Electron Beam Curing of Fiber-Reinforced Composites," 37th SAMPE International Symposium, 944 (1992).

3. Chris B. Sanders, et al., "Electron and X-Ray Curing of Thick Composite Structures," 39th SAMPE Conference, Anaheim, California, (1994).

4. A. Singh, et al., "An Emerging Alternative to Thermal Curing: Electron Curing of Fibre-Reinforced Composites," Proceedings of the International Conference on Composite Materials and Energy - Enercomp 95, (1995).

5. Daniel L. Goodman and Daniel L. Birx, "Composite Curing with High Energy Electron Beams," 41st International SAMPE Symposium, Anaheim, California, (1996).

6. P. A. Lagace and A. J. Vizzini, "The Sandwich Column as a Compressive Characterization Specimen for Thin Laminates," Composites Materials Testing and Design: Eighth Symposium, ASTM STP 972, American Society for Testing and Materials, Philadelphia, Pennsylvania, 143-160 (1988).

BIOGRAPHY

Dr. Barbara Gans Devney is currently a staff scientist and program manager at ***DAMILIC*** Corporation. She obtained her Ph.D. in Chemistry from the University of Chicago where she designed and constructed a state of the art molecular beam surface scattering apparatus. Her thesis work examined the subtleties in the relationship between surface forces, surface structure, and surface vibrational properties. She held a National Research Council Postdoctoral Fellow at the Naval Research Laboratory where she designed and built a frictional force microscope for the study of surface adhesion. Her current position has expanded her physical chemistry background to the field of radiation chemistry.

Dr. Anthony J. Vizzini is an Associate Professor in the Department of Aerospace Engineering at the University of Maryland and the Director of the Composites Research Laboratory. He has been the principal investigator on such sponsored research projects as delamination prevention through structural tailoring techniques; short duration, high temperature effects of composites; and crushing mechanisms of composites. He is, or has been, an active researcher in other programs such as the Center of Rotocraft Education and Research funded by the Army Research Office and the Office of Naval Research, University Research Initiatives Program on thick-section composites. He is the author of over 60 articles on such topics as delamination prevention, crashworthiness, structural integrity of fire-exposed composites, expanding monomers, and advanced manufacturing techniques.

Dr. Walter J. Chappas is currently Director of the University of Maryland's Radiation Facilities and a faculty member of the Department of Materials and Nuclear Engineering. His research interests include the radiation modification of polymers and elastomers, radiation damage to materials, the manufacturing of fiber reinforced composites, and UV/EB coatings technology. He is an active consultant to the United Nations International Atomic Energy Agency and has served as a technical advisor to the governments of Mexico, Sri Lanka, and Ecuador. He has published in the fields of radiation dosimetry/process quality control, industrial radiation manufacturing, and radiation chemistry and engineering.

28th International SAMPE Technical Conference
November 4-7, 1996

PORTABLE LINEAR ELECTRON ACCELERATORS FOR INDUSTRIAL APPLICATIONS

A.V. Mishin, R.G.Schonberg, Schonberg Research Corporation, Santa Clara, California

ABSTRACT

This paper discusses microwave portable linear accelerators designed and manufactured at SRC. Along with the standard systems, we are developing a number of new products and broadening the range of parameters. Light weight linac heads deliver electron beam in an energy range from 1 to 13 MeV and peak current from 0.1 to 1 A, depending on a power source. Average beam power varies from 0.5 to 10 kW. Magnetrons are the preferred power sources due to low cost and reliability. It is possible to mount any of the portable linac heads described in the article on a robotic or C- arm and use it for radiation processing purposes. Several new products are under consideration which will make it possible to build comparatively inexpensive electron beam sources with average beam power from 10 to 500 kW and energy from 1 to 10 MeV.

1. X-BAND ACCELERATORS

1.1 Minac Portable Linear Accelerator Systems. MINAC systems were initially built for radiographic applications. They are available in three standard models: MINAC-1.5, MINAC-4 and MINAC-6 [1]. Each system can be delivered either as a source of X-rays or electron beam source.

Each MINAC is a fully self-contained device with the exception of external electrical power. The system is comprised of several subassemblies, including a control console, modulator unit, RF head, X-ray head, water circulating unit, remote interlock unit and power transformer unit, which are interconnected by a set of power and signal cable, hoses and flexible RF waveguide.

The RF Head contains a mechanically tunable magnetron, a pulse transformer, magnetron pulse modulator, electron gun filament isolation transformer, RF circulators, and other RF components. The pulse transformer steps the pulse voltage up to 36 kV. A voltage divider network is used to provide 18 kV to pulse the electron gun injector. An isolation transformer supplies power to the electron gun filament. Beam current can be regulated by varying electron gun grid voltage.

The X-ray head which contains the accelerator unit is separate from the RF head. It is interconnected by means of flexible RF waveguide, power cables and cooling system hoses. Following interconnection, the flexible RF waveguide must be pressurized with sulfur hexafluoride (SF_6) gas to 30 PSI before use.

X-RAY ENERGY - OUTPUT - PENETRATION

Energy	Radiation Output (max. intensity)	Leakage	Maximum Penetration (2R exposures, ≈1 hour)
1.5 MeV	3 R/min. @ 1 m.	1%	3 inches steel (Fe)
4.0 MeV	100 R/min. @ 1 m.	1%	11 inches steel (Fe)
6.0 MeV	300 R/min. @ 1 m.	1%	14 inches steel (Fe)

ELECTRON BEAM CURRENT AND POWER CHARACTERISTICS

Energy Level (MeV)	1.5	4.0	6.0
X-band RF Frequency (MHz)	9303	9303	9303
Accelerator Peak Current (mA)	30.0	50.0	50.0
Accelerator Avg. Current (μA)	30.0	50.0	50.0
E-Beam Peak Power (kW)	45.0	200.0	300.0
E-Beam Avg. Power (W)	45.0	200.0	300.0
Magnetron Peak RF Power (MW)	0.2	1.5	1.5
Magnetron Avg. Power (kW)	0.2	1.5	1.5
Pulse Voltage (kV)	23.0	35.0	35.0
P.R.F. (pulses/second, nom.)	250.0	250.0	250.0
Pulse Width (microseconds)	4.0	4.0	4.0
Duty Cycle	0.001	0.001	0.001
Injection Voltage (kV)	17.0	17.0	17.0
Gun Fil Voltage (Volts)	6.0	6.0	6.0
Gun Fil Current (Amperes)	1.6	1.6	1.6
Voltage Requirements (VAC)	115/220 1 phase	208/220 3 phase	208/220 3 phase
Power Requirements kVA	11	11	11

BEAM GEOMETRY

System	Spot Size	Emittance	Phase Angle
1.5 MeV	1.7mm	3.7 π mm milliradians	15 degrees
4.0 MeV	1.7 mm	3.0 π mm milliradians	15 degrees
6.0 MeV	1.7 mm	2.0 π mm milliradians	15 degrees

The same control, power supply, RF generator, heat exchanger, dosimeters and interconnect cables are used both for MINAC 4 and MINAC 6 X-ray heads.

Fig.1 4 MeV and 6 MeV linac (left) and RF head (right)

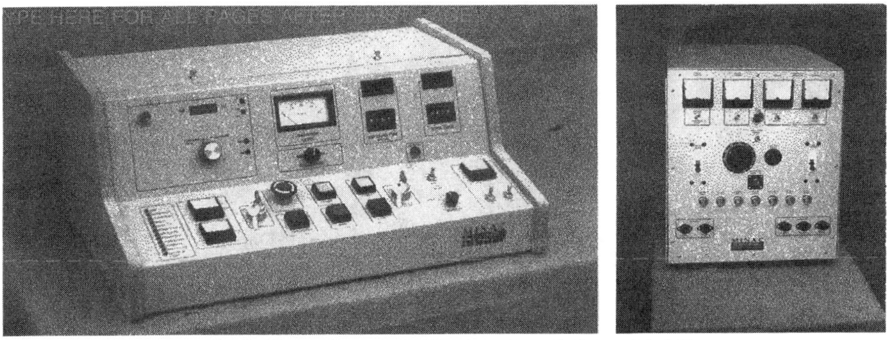

Fig.2. Control console (left) and modulator (right)

X-ray head contains the accelerator guide, gridded electron gun, beam collimator, magnetic shield, ion chamber and a protective mechanical support. Electrons are injected into the accelerator guide from the gridded electron gun and accelerated to relativistic speed.

1.2 New Developments. Efficiency of X-band portable linacs for various applications has recently been improved. Advanced design and some improvements to the existing MINAC provided encouraging experimental results using 1.5 MW magnetron [2].

1.2.1 MINAC-5. We tested the linac using only 0.4 m active RF length. The conventional 6 MeV guide with detuned last eight cavities was used for this purpose.

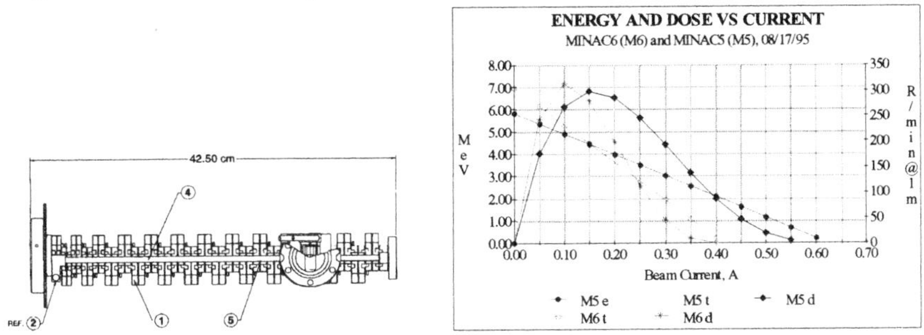

Fig.3 View of shortened 5 MeV 100 mA section and characteristics of MINAC-5 and MINAC-6. Dose rate is calculated at 0.0006 duty factor.

1.2.2. MINAC-8. Another interesting result was obtained during testing of a prototype of X-band linac designed for intraoperative therapy [3]. We were able to accelerate 100 mA current to energy of approximately 4 MeV using only 50 % of the power, available from the magnetron. Section utilizes principle of short Alvarez cavities coupled through coupling cavities and microwave structure prototype which first was described and tested in [4, 5, 6] with on-axis cavities and then used in [3] with side coupling cavities. It allowed boosting of bunching efficiency at lower particle velocity in a range 0.1 c to 0.5 c. Additional presentation, describing the approach was made at LINAC96 [7].

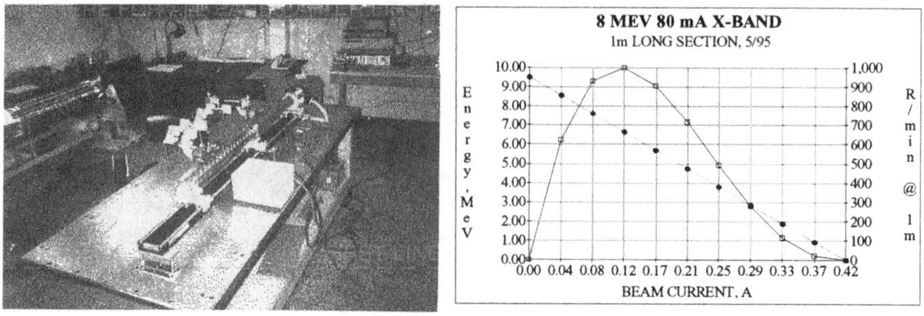

Fig.5 Two-section linac prototype capable of smooth energy regulation from 4 to 9 MeV. Measured beam efficiency of this section exceeded 40 %. Projected load line and dose rate at 1 m are presented on the right graph. Based on the results obtained for the first section tested at 100 mA, we have proposed a linac with expected characteristics, shown in Fig.8.

MINAC-5 and MINAC-8 characteristics are summarized in the following Table.

PARAMETER	UNITS	MINAC5	MINAC8
Maximum Energy	MeV	6	9*
Maximum Current	mA	110	100
Nominal Energy	MeV	5	7.6
Nominal Current	mA	100	80
Linac RF Length	m	0.4	1
Max.DoseRate@1m	R/min	600	1000
AverageBeamPower	W	520	700
Approx.HeadWeight	kg	20	30

* - Smooth regulation from 4 to 9 MeV

Presented data are based on both design and experimental numbers. Some values are predicted using the experimental results. Both linac heads do not use any external focusing.

1.2.3 Multiple Beam Coupled Cavity Structure. Operating at higher frequency permits a reduction in outline dimensions and weight of the final package. However, in higher frequency ranges one faces a problem of beam current restriction among a number of other complications. One of the solutions is a multiple beam system [8, 9]. MBS can be formed by combination of parallel structures formed by TM_{020} cavities, coupled to each other using electrical or magnetic coupling by common TM_{010} "side" cavities, which in this case are positioned on the central axis of the formed structure.

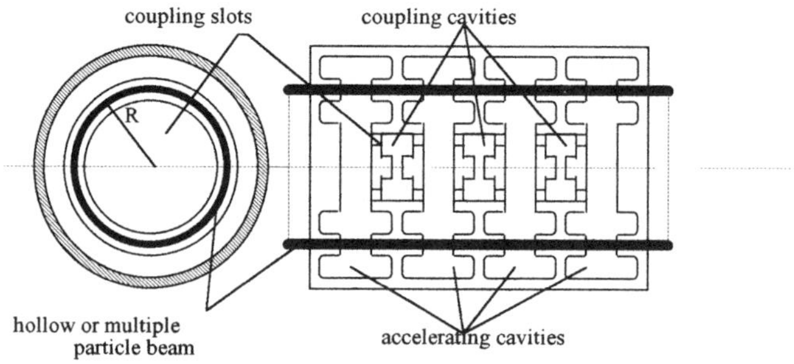

Fig. 6. Magnetically Coupled MBS.

Example of Four Beam Structure (FBS) formed by combination of four beam centerlines with common coupling cavities shown in Fig.7. Section of linear accelerator is divided into four segments as described above using an electrical circuit concept. Accelerating cavities in a "ring" should be coupled together.

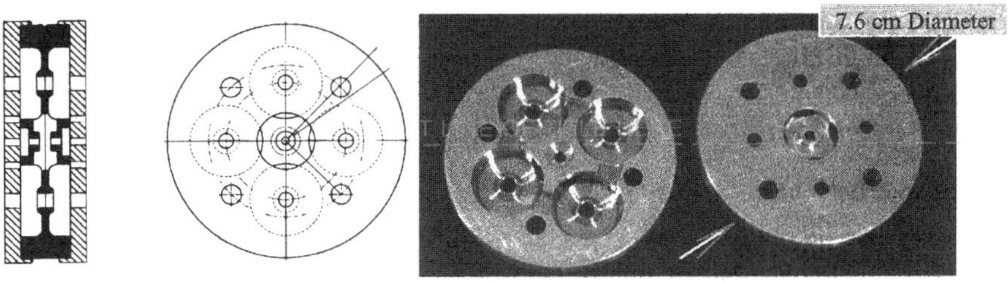

Fig.7 Four Beam Structure (FBS) and Manufactured X-band Cavity

The proposed concept and design could be called an "inside-out" version of a conventional side coupled structure. Maximum beam current is expected to increase proportionally to a number of beam apertures compared to a single aperture conventional cavity.

Presently, the plan we plan to built ultralight electron beam head prototype which will deliver 1.2 MeV electron beam at 1 A current in four beams in 12 cm long X-band structure [5]. The goal is to achieve 1 kW average power stored in four beams at 0.001 duty factor and 70 percent beam efficiency.

The concept has an extensive list of potential applications, discussed in [8,9]. One of them is a multipass linac with bending magnets allowing beam to be directed from one aperture to another. Obviously, the bending magnets would have magnetic fields of different amplitudes. Another possible application is for klystrons or traveling wave tubes. In particular, one could combine a power amplifier and a linac into a single RF "envelope" and reduce the size and weight of the overall package. This approach is certainly applicable for wavelengths differing from those at X-band.

2. S-BAND ACCELERATORS

2.1 Minatron Portable Linear Accelerator Systems. Using an S-band power source and a corresponding accelerator structure permits accelerating higher currents compared to the beam current achieved for X-band linacs, as presented in a Table below[1]. However, the accelerating gradients reduce proportionally to the square root of operating frequency and the dimensions grow linearly with wavelength. Fig. 8 shows the difference between S and X band standing wave cavities.

MINATRON SYSTEM X-RAY AND ELECTRON BEAM CHARACTERISTICS

X-RAY ENERGY/OUTPUT/PENETRATION

Energy	Radiation Output (max. intensity)	Leakage	Maximum Penetration 2R exposures, »1 hour, in steel @ 3 meters	Spot	HVL
2.5 MeV	250 R/min. @ 1 m.	0.1%	8.5 inches steel (Fe)	2 mm	0.91"
6.0 MeV	1000 R/min. @ 1 m.	0.1%	14.0 inches steel (Fe)	2 mm	1.17"
9.0 MeV	3000 R/min. @ 1 m.	0.1%	16.0 inches steel (Fe)	2 mm	1.29"

ELECTRON BEAM CURRENT AND POWER CHARACTERISTICS

	250 R/min.	1000 R/min.	3000 R/min.
Energy Level (MeV)	2.5	6.0	9.0
S-band RF Frequency (MHz)	2998	2998	2998
Accelerator Peak Current (mA)	350	130	100
Accelerator Avg. Current (mA)	300	130	100
E-Beam Peak Power (kW)	875	780	900
E-Beam Avg. Power (W)	750	780	900
Magnetron Peak RF Power (MW)	2.5	2.5	2.5
Magnetron Avg. Power (kW)	2.0	2.5	2.5
Pulse Voltage (kV)	40	40	40
P.R.F. (pulses/second, nom.)	250	250	250
Pulse Width (μsec)	3.5	3.5	3.5
Duty Cycle	0.0009	0.001	0.001
Injection Voltage (kV)	17	17	17
Gun Fil. Voltage (V)	6	6	6
Gun Fil. Current (A)	1.6	1.6	1.6
Voltage Requirements (VAC)	208/220 3 phase	208/220 3 phase	208/220 3 phase
Power Requirements (kVA)	11	11	11

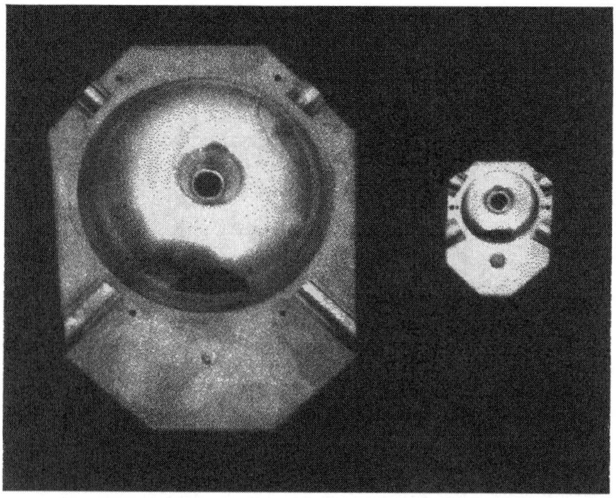

Fig.8 S and X band cavities for standing wave side coupled structure.

2.2 New Developments. Recently, we have completed work on a portable traveling wave linear accelerator designed for 2 MeV at 1 A of accelerated current. A klystron was used with 3 MW peak RF power output. This linac was later incorporated into MINILAC™ electron beam system with an average beam power of 10 kW[8]. Studies were performed to determine whether a traveling wave or standing wave structure would produce the highest RF to beam efficiency. After careful study it was decided that both structures were capable of nearly the same efficiency. The customer chose traveling wave because of prior favorable experience with traveling wave structures. The system is comprised of a low voltage triode electron gun, a prebuncher and the accelerator structure with a tapered β section integral with a β=1 set of cavities. The accelerator design uses 2π/3 cavity geometry with constant potential gradient over all 13 cavities. The electron gun is a traditionally used at SRC triode structure operating at voltages between 10 and 20 kV. Low injection voltage is one of the advantages of the system. The gun is comparatively inexpensive and replacement could be made in few hours. The total length of the accelerator centerline is 82 cm measured from the e-gun housing to the output flange. The 37 cm long traveling wave structure has a buncher section with a tapered β starting at β=0.5 and increasing to β = 1.0 at the fifth cavity. The resultant electron beam capture exceeds 50% of high energy electrons out the exit window. The peak RF power at the input RF coupler of the accelerator is 2.7 MW.

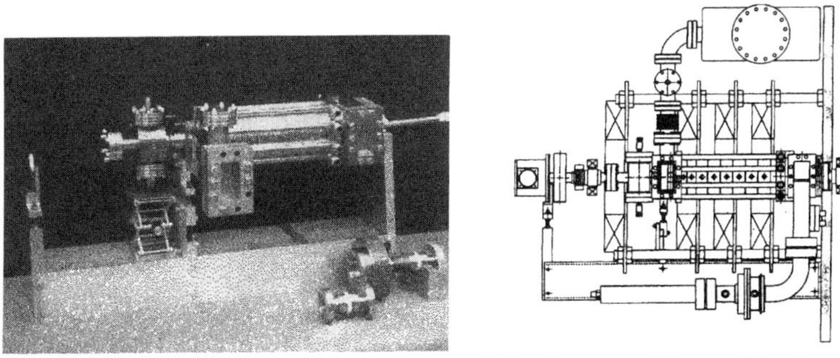

Fig.9 MINILAC™ beam centerline designed for the possibility of both vertical and horizontal operation.

The following is a table of beam design characteristics.

Beam Energy	2.0 MeV
Beam Current Peak	1.0 Amps
Beam Current Average	5 mA
Energy Spread	± 10%
Spot Size	10 mm at the end of accelerator
Frequency	2998 MHz
RF Peak Power	2.7 MW at accelerator
RF Average	15 kW at accelerator
Pulse Rep. Frequency	500 PPS
RF Pulse Length	10 μsec.
VSWR	1.05, not more than 1.15 within 5 MHz

The system operates at very nearly the design level with a 71% RF to beam conversion. The operation was commenced within a year and final testing completed within a one month period. All aspects of the system from the modulator, water cooling, controls, monitor point, scanner, vacuum, etc. function in a suitable manner with few changes needed to obtain stable, full performance ratings.

3. OTHER SYSTEMS

3.1 High energy and high power systems. We developed a variety of linacs for a broad range of applications including systems for medical applications, synchrotron light sources, SLED - cavities and some L-band accelerator sections for sterilization, electron free lasers, etc. Example of L-band section is described below[1].

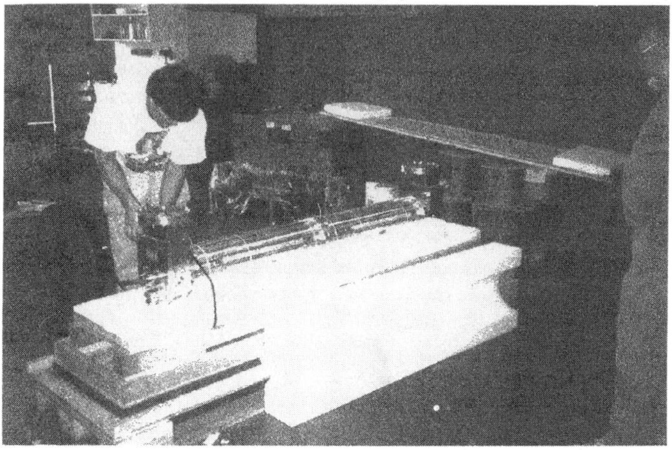

Fig. 10 L-band section is being prepared for shipment

Mode	$2\pi/3$ constant gradient, 24 cavities
Frequency	1300 MHz at operating temperature
Couplers	Iris, compensated for asymmetry
RF Windows	Chip type, removable, 2 each
Power Rating RF	60 kW average 25 MW peak, with 7 μsec RF pulse
Output Beam	With 20 MW RF peak at 12 MeV Ipeak=1 A
	With 20 MW RF peak at 16 MeV Ipeak= 0.78 A
VSWR	Less than 1.21 through windows into matched load; Max. 1.5:1 for f 0.5 MHz band width. VSWR matched for 1.0 amp peak beam current
Magnetic Field	1 kGs (Helmholz coils)
Design Properties	Shunt impedance: 41 MOhm/m
	Initial attenuation coefficient: 0.14 nepers/m
	Length: 1.845 m (24 cavities)
	Nominal operating frequency 1300 MHz (23.06 cm wavelength)
	Q 19,400
Performance Properties (steady state)	Beam loading factor: -11 MeV/A
	Filling time: 1.73 μsec
	Energy sensitivity: 0.82 %/C
	1.78 1/ MHz
	For RF input 20 MW
	No load energy: 26 MeV
	Max. efficiency current 1.15 A
	Max. field strength: 170 kV/cm

We are also considering a development of comparatively inexpensive accelerator models for very high average power in a range from 10 to 200 kW at energy from 1 to 10 MeV.

3.2 Medical linacs. Two medical systems Neurotron and Mobetron were developed by "spin-off" companies Accuray and Intraop, correspondingly.

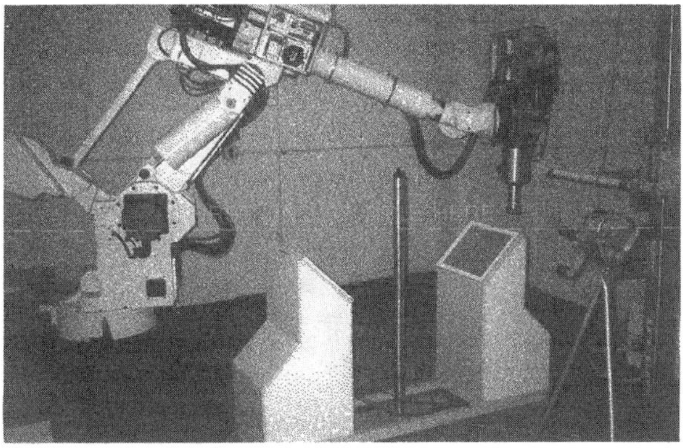

Fig. 11 Neurotron system utilizes MINAC-6 incorporated into irradiator which, in turn, is mounted on a robotic arm.

4. CONCLUSIONS

We have successfully used our 4 MeV linac for radiation processing of composites for armored vehicles.Polymer curing tests have been performed using a MINAC-4 electron beam unit. This highly portable linac delivers between 160 and 200 Watts of electron beam. The electron beam is passed through an aluminum scattering foil and applied to the composite materials up to 0.65 inch thick. Irradiation has been done using two side beam application. Doses between 25 and 200 kGy have been used to obtain cures of various resins.

The approach of having an irradiator mounted on a robotic arm or C-arm used for medical linacs can be applied for radiation processing of composites. X-band linacs or the described above lightweight S-band linacs are appropriate electron beam sources to be used for electron beam curing of the systems with complex shape and developed surface.

5. REFERENCES

1. Advertising brochures, SRC.
2. H.Deruyter, A.V. Mishin, T. Roumbanis, R.Schonberg, D.Skowbo, Schonberg Research Corporation, Latest Results on Portable X-Band Linac Development, to be published in the Proceedings of the 5th European Particle Accelerator Conference, IEEE, 1996.
3. A.V. Mishin, R.G. Schonberg, H.Deruyter, T.Roumbanis, Intraoperative X-band Accelerator Microwave Structure Design, Proceedings of the 4th European Particle Accelerator Conference,London, 1994.
4. A.V. Mishin, V.P.Komarov Accelerating Structure for Charged Particle Accelerator, Patent of Russia#1577678, issued 3/8/90, filed 12/1/88.
5. A.V. Mishin, Portable X-band Linear Accelerator, Ph.D. Thesis, Moscow, MEPhI, 1992.
6. A.V. Mishin, Accelerator Structure for Low-Energy Electron Beam,Proceedings of the 1993 Particle Accelerator Conference, Washington D.C., p.p. 971-973, 1993.
7. A.V. Mishin, Design and Application of Microwave Structure for Low Velocity Particles, to be published in Proceedings of the LINAC96.
8. A.V.Mishin, Multiple Beam Coupled Cavity Concept and Structure, Proc. of EPAC96.
9. H.Deruyter, A.V.Mishin, T.Roumbanis, R.Schonberg, R.Miller, J.Potter, Properties and Possible Applications of Multiple Beam Coupled Cavity Structure*, Proceedings of EPAC96.
10. H.Deruyter, A. Mishin, T. Roumbanis, R. Schonberg, S. Farrell, R. Smith, R.Miller, High Power 2 MeV Linear Accelerator Design Characterristics, Proceedings of EPAC96.

28th International SAMPE Technical Conference
November 4-7, 1996

AFFORDABLE COMPOSITES FOR PROPULSION

PROGRESS REPORT ON THE FIRST 24 MONTHS

Richard J. Holman
Charles B. Webster
Pratt & Whitney, Division of UTC
West Palm Beach, FL 33410-9600

ABSTRACT

The progress of the first 24 months of activity on the Affordable Composites for Propulsion (ACP) program funded through DARPA will be reviewed. Affordable polymer matrix composites (PMC) provide critical enabling technology for the next generation aerospace propulsion systems. The ACP program is a strategic pull through program for the U.S. composite industries. In order to accelerate U.S. advanced engine market opportunities and restore jobs, defense conversion funding is crucial. The ACP Team currently produces 60% of America's commercial and DoD aerospace composite needs. ACP is a critical dual use initiative working to:

- Attack the high costs of commercial/defense composite
- Bolster the production volume of DoD suppliers
- Apply Government investments to U.S. economic growth

For the past 24 months the ACP Team, shown in Figure 1-1, has focused on demonstrating

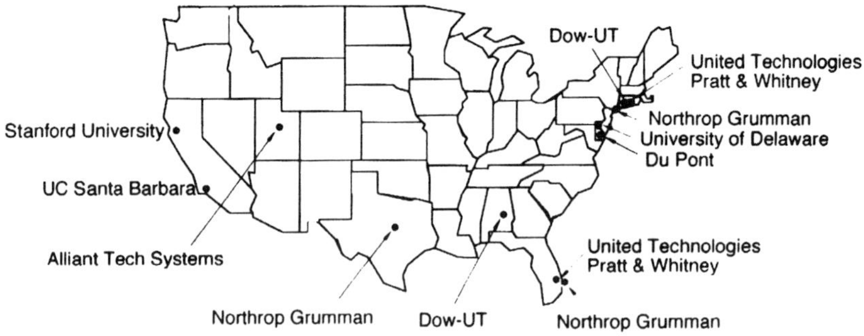

Figure 1-1

affordable manufacturing processes for composite structures leading to weight reductions in military and commercial aerospace engines.

KEY WORDS: Composites Structures, Composite Components, Resin Transfer Molding, Auto Tow Placement

1. INTRODUCTION

Commitment to affordable composites, the ACP Team is comprised of industry leaders whose state-of-the-art capabilities have spanned over 20 years of Government investment. The Team is committed to composites and is prepared to accomplish the objectives of the Technology Reinvestment Project (TRP) by tackling three overriding national priorities:

1) Jobs restoration

2) Dual-use industrial manufacturing base stability

3)Restoration of traditional aerospace trade balance.

The ACP program addresses the technology development objectives of three broad areas of TRP Statutory programs: Defense Dual-Use Critical Partnerships, Defense Advanced Manufacturing Technology Partnerships, and Commercial/ Military Integration Partnerships. Prior to the defense downturn, the advanced composite industrial base was poised to satisfy an annual DoD composites need of 2.2 million lbs. This planned industrial output has declined to about 15% of this value, a loss of over 10,000 jobs. ACP is a critical TRP program for conversion of the DoD composites supplier base to commercial opportunities. Success will lead to lower acquisition costs for future DoD requirements.

2. PROGRAM UPDATE

Integration of the ACP Team resources and technology are key to achieving affordability. The ACP program is taking action to obtain affordability by focusing on three elements:

- Improving the state-of-the-art in manufacturing process reliability
- Using the labor savings attributes of new automated low cost processes
- Implementing concurrent product development involving the entire ACP team

This approach over the past 24 months has laid the ground work for an integrated propulsion and composites industry defense conversion plan. Our technical plan is unique compared to other comparable efforts due to our integrated approach to technology sharing among industry leaders and a near-term production focus.

ACP has defined low-cost, producible designs and manufacturing approaches. Trade studies conducted within the first six months began the refining process of production cost estimates and served as the basis for manufacturing processes selection. A specific ADP engine size and installation, reflecting the first potential production opportunity, has been selected for hardware demonstrations.

The ACP Team has produced ADP parts to develop manufacturing cost data and demonstrate manufacturing feasibility as well as repeatability, Figure 2-1. The ACP program focused on hardware demonstrations of resin transfer molding (RTM), automated tow placement (ATP), and isogrid winding to validate their cost reduction potential.

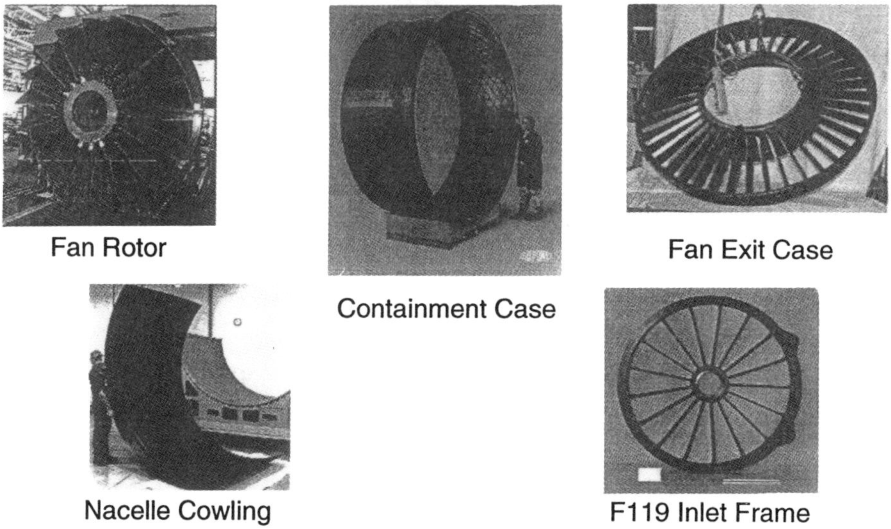

Figure 2-1

Critical blade-out rig tests, using full scale hardware, are to be conducted to verify blade containment and structural integrity of the assemblies. Manufacturing data will be used to validate production cost forecasts and will lead to a Production Transition Plan.

Special emphasis has been placed on developing reliable processes and manufacturing methods which build consistent, acceptable parts with high yield. An effects-of-defects program was implemented parallel to the manufacturing development program and has provide statistical data to establish process windows and accept/reject standards. A focus on building both cost and rule based design data bases and processes will support future production decisions and challenge traditional designs and requirements. Full-scale component fabrication and test will validate the developed approach.

To ensure efficient exchange of information, the ACP Team established an electronic communication and database transmission system between members. In addition, the Team collocated engineering and manufacturing personnel at P&W during the affordability trade studies and preliminary design program phases. During critical phases of component manufacturing this process supplemented by P&W engineering and manufacturing personnel collocating at Teammate's facilities.

The ACP design and process selection was conducted by Integrated Product Development Teams (IPDT) consisting of all manufacturing, engineering, and quality assurance disciplines, Cost implications were fully considered in each design and process decision. Component configurations have been tailored to the characteristics for the process, Figure 2-2.

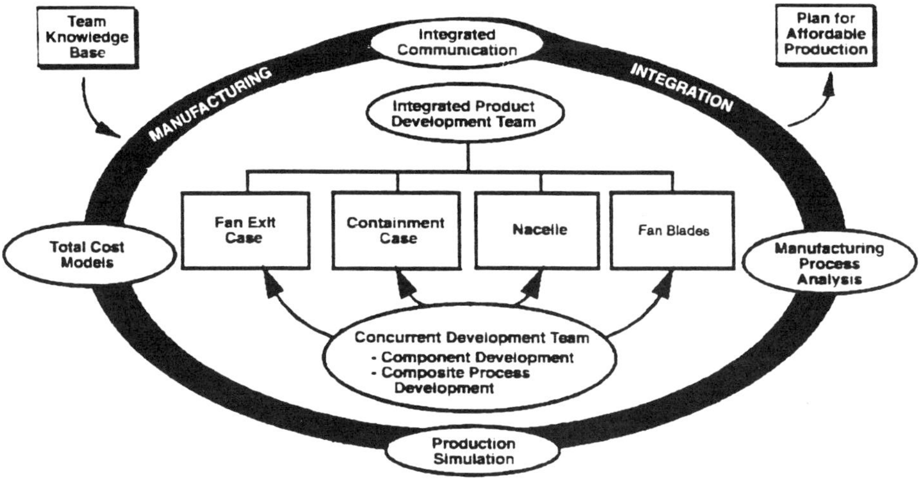

Figure 2-2

As an important part of the IPDT efforts, cost and weight metrics were initiated early in the program. Baseline values were established for each component. Design and process trades have compared cost reduction concepts to the cost baseline and weight savings to the baseline, Figure 2-3.

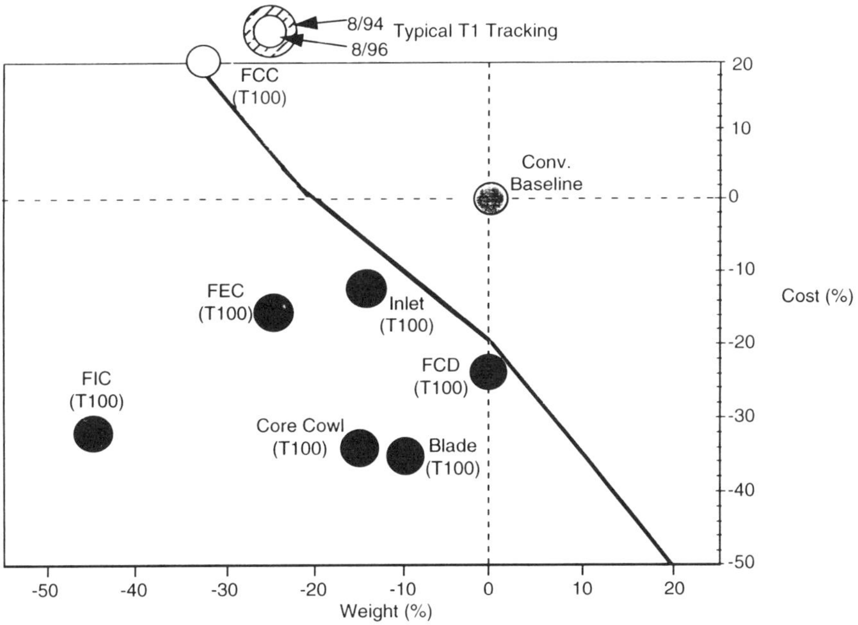

Figure 2-3

A Gated Approach is being used throughout the program to measure both technical and financial progress, Figure 2-4

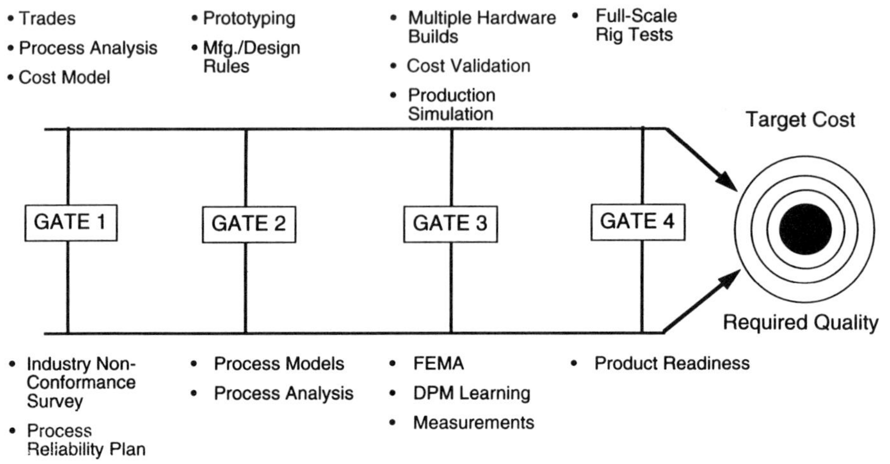

Figure 2-4

3. Summary

The ACP program's first 24 months have yielded significant progress and production insertion opportunities, including:

- Commercial engine fan module
- Commercial nacelle components
- C-17/F117 nacelle components
- F22/F119/JSF fan module

The ACP Program is currently funded through fiscal year 1996. FY 1997 and FY 1998 depend the future of the "Dual-Use Military Application" program.

28th International SAMPE Technical Conference
November 4-7, 1996

FABRICATION CHALLENGES AND ACCOMPLISHMENTS FOR RTM ENGINE COMPONENTS

Lawrence A. Gintert, Steven R. Hayse,
and Susan St.Germain
Dow-United Technologies Composite Products, Inc.
Wallingford, Connecticut 06492

ABSTRACT

Aggressive cost and weight goals for large, complex structural engine components are met using the Resin Transfer Molding (RTM) process. Full scale static testing of the 2.7 m (9 ft) diameter ARPA Affordable Composites for Propulsion (ACP) Fan Exit Case (FEC) comprised of assembled RTM components, has successfully passed multiple ultimate loading conditions demonstrating a robust structural design. Critical processing evaluations and detailed cost modeling efforts have provided an in-depth understanding of processing capabilities and design features which drive cost. This discussion presents the results of the first eighteen (18) months of the FEC team efforts, including design, processing and cost modeling overviews as well as detail examples of process and cost improvements. This effort was conducted as part of the Pratt & Whitney ACP program (ARPA Contract No. MDA972-94-3-0029).

KEY WORDS: Aircraft Engine, Composite Structures, Resin Transfer Molding

1. INTRODUCTION

The challenge of reduced cost and lower weight has been posed for a composite FEC for the large diameter jet engine family. The metrics employed to evaluate whether this challenge has been met or not include substantiation of design and analyses through structural testing, and critical evaluation of fabrication processes and materials for reliability and cost. The ACP FEC team has successfully demonstrated through rigorous analyses, design trades, coupon and subelement testing, full scale testing, and detailed scrutiny of process capability and cost analyses that agressive goals are achievable using current materials and manufacturing approaches. Production-type tooling and certified production processes have been utilized for RTM component fabrication. Multiple iterations of designs, analyses, fabrication and testing have been completed showing progressive improvements in product performance and cost reduction. This paper documents the design evolution of the composite FEC as well as the approach to process reliability and cost modeling.

2. DESIGN CONFIGURATION

2.1 The Jet Engine Environment The conditions that exist within a jet engine can be very demanding on a component. The RTM composite FEC is no exception. The component must withstand environmental effects including operating temperatures ranging from -54°C (-65°F) to 121°C (250°F), corrosive fluids, and vibration. It must contain a high level of damage tolerance since it is exposed to foreign object damage from rocks, hail, erosion, tool drops, and bird strike. It must be durable enough to withstand these effects and function for 50,000 engine take off and landing cycles. This establishes the need to design for and define various repair limits and procedures since occasional excursions outside the design envelope tend to occur.

The FEC component also has certain functional requirements that it must satisfy. The FEC serves a dual role in controlling the airflow through the engine bypass duct as well as supporting the fan containment case (FCC) and other engine components that include tubes, wiring harnesses, electrical boxes, and other peripheral equipment. The FEC RTM 4-pack subcomponents define the flowpath duct walls and the flow directing struts. To properly control airflow the FEC struts require a very closely controlled airfoil tolerance. To properly support the FCC the 4-pack provides a very stiff load path through a complex arrangement in the strut to box joints. FCC concentricity with the fan requires precision molding and machining of the shroud box interfaces. The FEC must also provide for safe engine shutdown in the event of a fan blade loss, further demanding a sound joint design.

The RTM process is uniquely capable of providing the configuration necessary in a composite assembly to achieve all the demanding requirements for this system while doing so in a cost effective and low weight manor. Complex structural arrangements are possible that allow excellent fiber control and optimized tailoring providing the most weight effective load carrying design. Areas of the part that are subject to damage in service can be readily repaired. Repair methods are currently in development and are being verified by repair trials.

2.2 The Cost/Weight Trade The primary incentive for introducing the RTM composite FEC into the turbofan engine is cost, though weight is always a factor. This program looks to achieve an aggressive 30% cost reduction and a 30% weight reduction over the conventional design. The market for jet engines has shifted from peak performance to lower cost. Good engines sell when their cost is low and their weight is reasonable. It is therefore imperative to understand all the costs associated with the product. Life cycle cost is a key factor. This cost takes into account not only the acquisition cost that the airline customer must pay for the engine, but also the maintenance cost. It also has a component of operating cost associated with weight since extra weight results in lost revenue and increased fuel consumption for the airline. Cost driver relationships have been developed for each manufacturing step that breakdown the fabrication cost into many contributing subcategories. Direct labor may only be a small component of the overall cost of one particular fabrication step. Machine cell overhead, material utilization, direct support, quality, rework, and scrap are significant others. It has been shown that the manufacturing solution that produces the lowest direct labor may not be the lowest cost approach for that manufacturing step. Similarly, the design configuration that produces the lowest weight component usually does not produce the lowest cost one.

Maintenance cost is reduced by designing into the structure the most durable features possible. A replaceable metal leading edge and repairable flowpath surface erosion protection layer are provided to extend the life of the component. Structural through-the-thickness stitching is provided in the structural strut to improve its impact toughness thereby shifting the onset of damage beyond the .55 kg (2.5 lb) birdstrike range and potentially eliminating the need to repair the strut in routine service.

2.3 Fan Exit Case Design The FEC is the main structural static component in a gas turbine engine. This unique FEC design combines the functions of a traditional aerodynamic fan exit guide vane case and the structural FEC rear into a single, structural ring/strut/ring assembly. Multiple 4-pack subcomponents greatly reduce part count and provide inherent stiffness and optimum load carrying capability. The demonstrated Pratt & Whitney ADP fan exit case has 40 structural struts with a maximum thickness of 1.5 cm (.62 in) and a chord length of 23 cm (9 in). The outer ring is 2.7 m (9 ft) in diameter.

The main component of the FEC is the 4-pack (**Figure 1**). It is fabricated with the RTM process using IM-7 and AS-4 graphite and PR-500 resin. The OD Ring provides hoopwise continuity. It is made with the Auto-Tow Place (ATP) process using IM-7 graphite and 8552 resin prepreg tow.

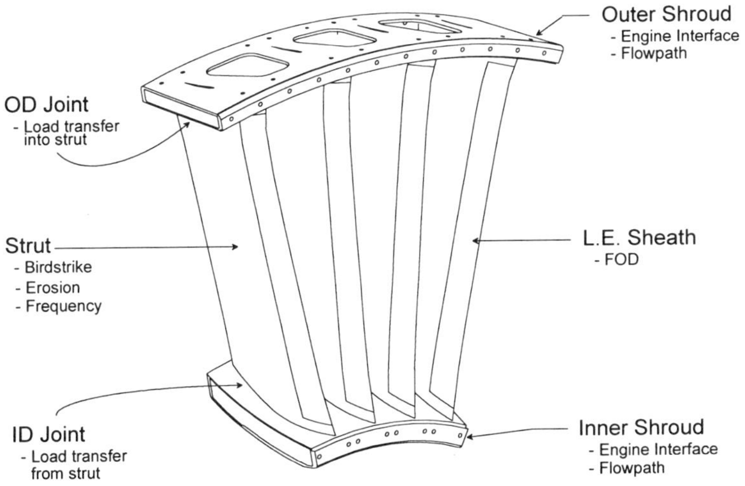

Figure 1 - ADP 4-Pack Subcomponent Design Features

3. STRUCTURAL TEST - DESIGN AND PROCESS VALIDATION

3.1 Subelement Test Plan The composite FEC was designed using a preexisting materials database supplemented with additional materials characterization tests more specific to the operation of these engine components. Flange joint subelements were built and tested to help understand the capabilities of the design, materials, and process in similar loading conditions to actual operation. Room temperature dry and environmentally conditioned specimens were tested in a full range of tests including static loading, low cycle fatigue (LCF) and high cycle fatigue (HCF) including the capability of repaired specimens and defect specimens. The design was also supplemented with strut subelement testing which included bird strike impact, hail impact and blunt body impact to understand the effectiveness of the through-the-thickness stitching and the size and location of the hollow

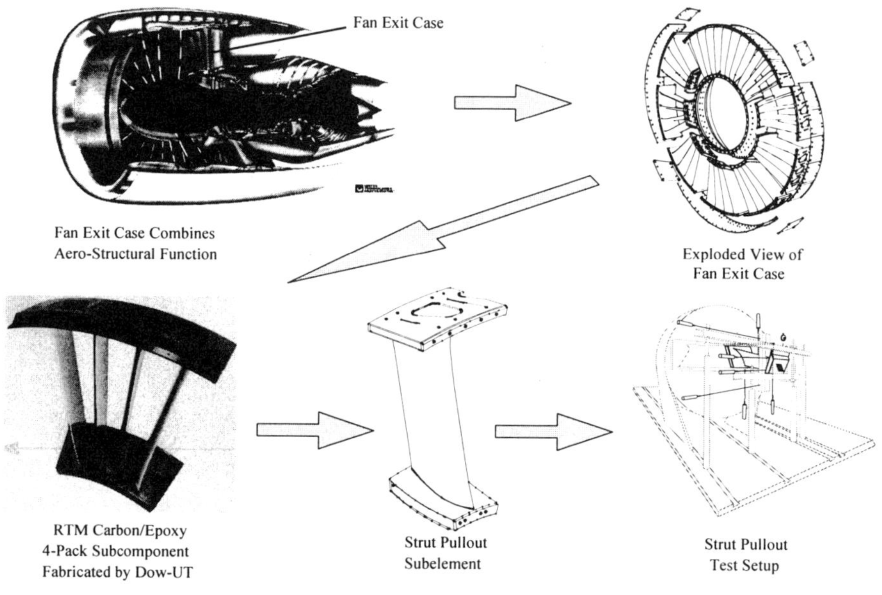

Figure 2 - Fan Exit Case Strut Pullout Test Program

cavity in the strut. Hard body impact and an extensive erosion test program also paralleled design efforts to influence design improvements.

A major portion of the test plan validation approach involved fabricating and testing strut pullout subelements (**Figure 2**). Full-size 4-pack subcomponents were molded and machined to produce two pullout subelements per molding. These subcomponents were also tested in a full range of tests. Tests included pure axial load, pure tangential load [1], combined axial/tangential loading in proper ratios as seen in normal operation, and combined load LCF testing. Tests were performed at room temperature and at elevated temperature (121°C / 250°F). Some specimens were subjected to bird strike impact and blunt body impact prior to testing. The purpose of this level of testing was two-fold: to understand capabilities of the design and to understand the capabilities of the analysis techniques by studying the correlation between the data and the analytical predictions.

3.2 Full Scale FEC Static Test The early and ongoing materials characterization and subelement testing provided valuable design improvement input. Design changes and configuration changes were incorporated into the next set of fabricated 4-pack components and a full scale FEC, including both the ID Ring and the OD Ring, was fabricated and assembled. The test program continued with a full scale static test on this case (**Figure 3**). The FEC was subjected to pure torsion and pure trunnion bending to obtain valuable data for correlation with the analysis techniques, as with the subelements above. Maximum combined operating loads were also applied to the structure. After successfully demonstrating its operational capabilities, the FEC was subjected to its most severe structural test, the fan blade-loss event. The FEC is required to withstand the fan blade impact on the wall of the Fan Containment Case (FCC) and fan rub torque that result when a

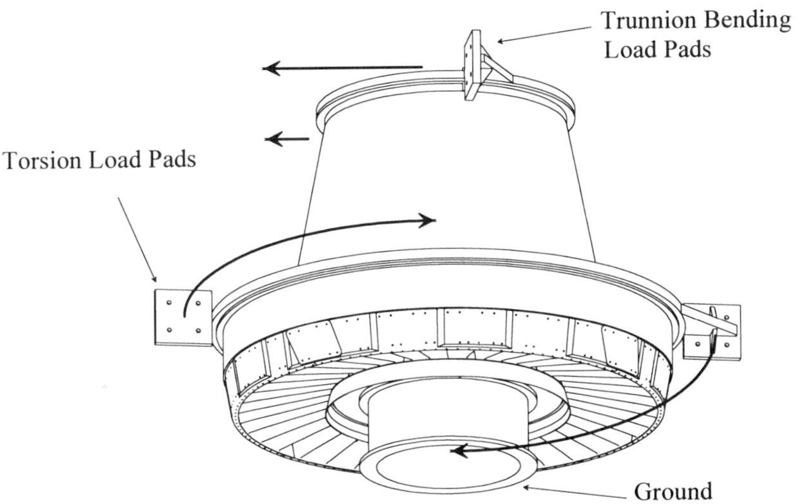

Figuire 3 - Fan Exit Case Static Test

full fan blade breaks off at its root. Damage will ensue from this event, but the FEC which supports the FCC and inlet must retain sufficient strength to allow safe engine shutdown and support the fan system to achieve a safe landing.

The first step in demonstrating the fan blade-loss capability is a static test using the maximum blade-loss loads that occur during the first few milliseconds of the event. Two conditions were tested. Maximum torque combined load and maximum trunnion bending moment combined load tests were executed. The next step, after incorporating the lessons learned and making design improvements, is to proceed with actual dynamic testing in the Horizontal Containment Rig (HCR). The program is currently working toward this test in the future.

3.3 Test Results and Analysis The test program has shown that the design approach taken and the materials and process employed have produced a viable FEC system for use in these turbofan engines. It has been a valuable addition to the design process providing concurrent feedback along the way.

Material characterization tests verify the capabilities of the selected resin and fiber system under extreme hygrothermal environmental cycling. Strut subelement ballistic testing identified some early design flaws, providing the necessary insight on strut design to reach successfully the 2.5 pound birdstrike capability. It also identified necessary changes to the metal leading edge design for hard body (gravel) ingestion and hail impact.

Strut pullout subelements consistently perform over 125% of ultimate at 121°C (250°F) up to 175% of ultimate at room temperature. They demonstrate excellent post-failure and post-damage strength and residual stiffness, attributable to the multiple load paths present in the complex co-molded strut/box joints. They also identified that strut trailing edge compression was the weak spot in the design. The most impressive perhaps of the pullout tests was the specimen that, after sustaining 50,000 flight load cycles, the design life of the part, it sustained 175% of the ultimate combined load in pullout.

Full scale static testing further confirmed that the trailing edge would benefit from local improvements near the inner shroud joint. It also identified high loads in the OD Ring most likely resulting from the large strains after the strut trailing edge began to pull out. However, the FEC satisfied all the requirements of the test by sustaining 100% maximum trunnion bending moment combined blade-loss loading for 60 seconds and then by sustaining 100% maximum torque combined blade-loss loading with only minor damage

Full scale static testing is continuing on this same FEC with two replacement 4-packs and one replacement OD-Ring segment. This test will be run at 121°C (250°F) elevated temperature.

3.4 Design Improvements The test program greatly influenced many design changes. Hard body impacts, hail, and erosion testing drove leading edge changes.

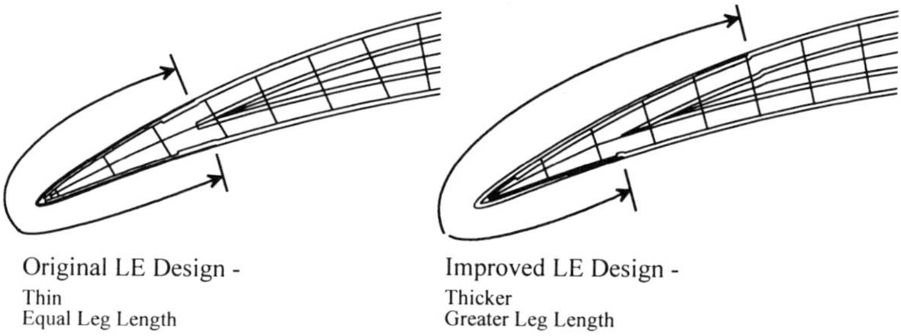

Figure 4

Birdstrike testing drove changes for a more robust design for this size airfoil at this aspect ratio. Cost trades show that the more durable solid strut costs less over the component lifetime. Process reliability studies show improved quality in the solid strut design.

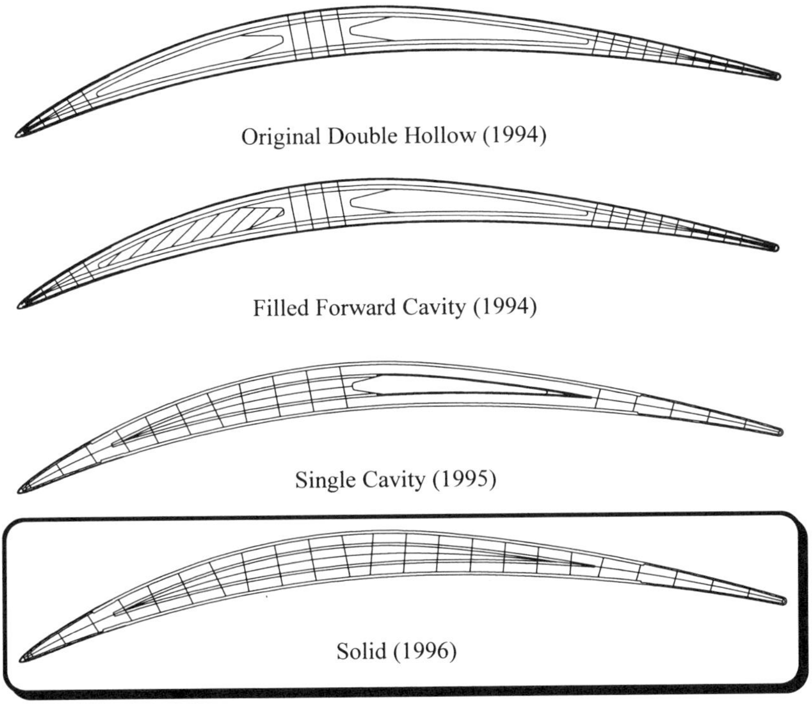

Figure 5 - Strut Design Evolution

Flange joint testing and configuration improvements drove FEC to FCC joint changes.

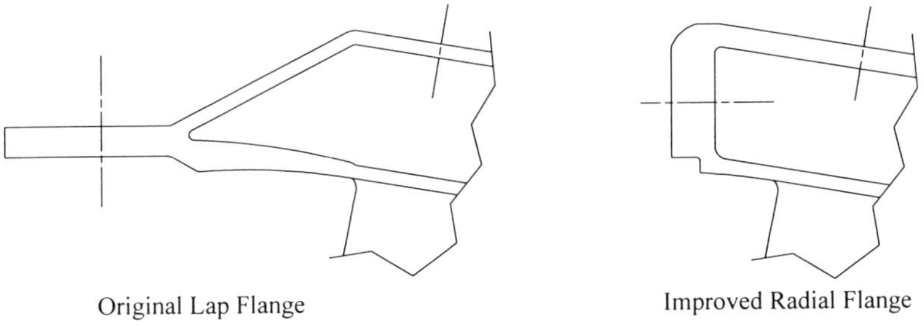

Figure 6 - Fan Containment Case Flange Mount

Structural pullout testing defined improvements to internal co-molded structural arrangement.

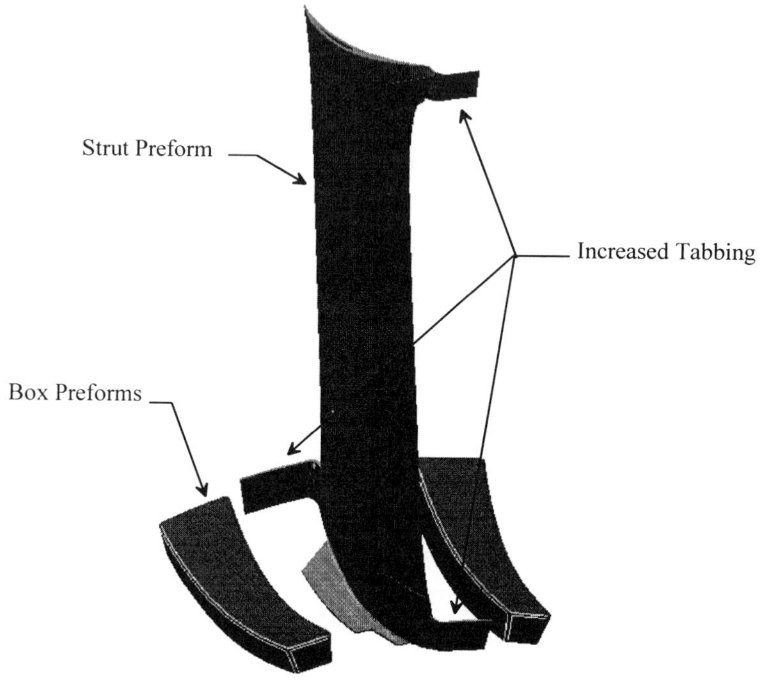

Figure 7 - Strut-to-Shroud Joint Design: Increased co-molded structure

4. PROCESS RELIABILITY

4.1 Process Reliability Plan The steps to develop a reliable process for the 4-strut-pack are outlined in the flow chart in **Figure 8**. Key Part Characteristics or (KPC's) are identified from the part design in relation to flight safety, durability / reliability, interchangeability, and cost. A process is then developed which will produce a part which meets the defined KPC's. An evaluation of the process cost and capability is performed to define the areas which currently do not meet the part requirements. The next step is to perform a cause and effect analysis, where the top level is the KPC's and the lowest level defines the Critical Process Parameters (CPP's). Process improvements to improve process control and reduce process costs are identified and implemented. Actual parts are then fabricated to evaluate the process improvements incorporated. This developmental process is repeated several times to evaluate the process improvements incorporated through each iteration. During the course of the ACP program three iterations of this process are taking place as shown in the flow chart in 1994, 1995, and 1997. This plan will guide us through the developmental process to obtain a production ready status after the last iteration of the 4-strut-pack fabrication process.

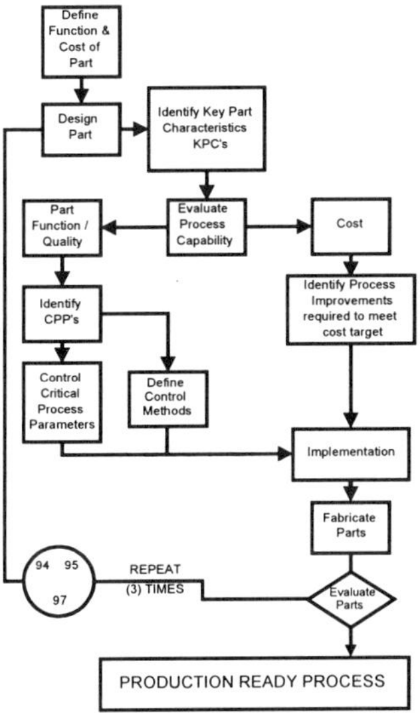

Figure 8 - Process Reliability Plan Flow

4.1.1 Definition of Key Part Characteristics The Key Part Characteristics are defined with respect to function of the part, environmental conditions, aerodynamic efficiency, durability and damage tolerance, and cost and weight. Detail design analysis is performed to define the construction of the part to meet these requirements. Specific requirements are typically listed on the part drawing.

4.1.2 Cause and Effect Analysis The cause and effect analysis consists of up to four levels. The top level is the Key Part Characteristics (KPC's). If one of these requirements is not met it will become a cause for a possible failure of the part. A listing of all the possible causes for failure to meet a KPC was developed and makes up the second level of the cause and effect analysis. The example in **Figure 9** shows "Safe Engine Shut Down after blade loss" as one of the requirements for the FEC. If safe engine shut down after blade loss did not occur it would be categorized as an effect due to a "structural failure during or after a fan blade loss". A cause and effect analysis was then performed on the second level. This example shows the "structural failure during or after a fan blade loss" then becomes an effect due to a "strut ID joint failure". The "strut ID joint failure" makes up the third level of the cause and effect analysis. This particular example needs to be taken one more step to identify all the possible causes for a strut ID joint failure. This forth level will begin to identify the causes in the manufacturing process which will have an effect on the overall part performance. This example shows that one of the causes for a strut ID joint failure is a misplaced core preform (a core preform is a specialty woven preform where two cores preforms are used in the strut preform). The cause and effect analysis is performed on all possible failure modes of the part.

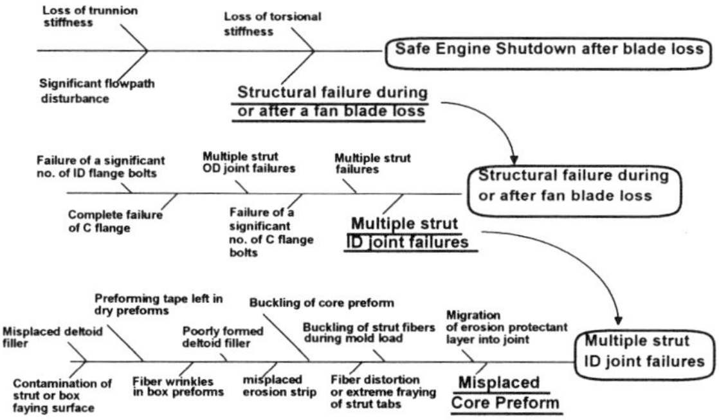

Figure 9 - Cause and Effect Analysis

4.1.3 Identification of Process Controls The next step of the process development is to identify the process controls required to prevent a failure from occurring. The causes identified in the lowest level of the cause and effect analysis are the process parameters which require process controls. A table was generated which listed each of the possible process causes for failure. **Table I** contains one example from the complete table, the example shown is for a misplaced core preform. The table is broken into nine basic columns.

The first column describes the failure mode and the second column describes the effect of the failure mode. The third column gives the cost if the defect were to occur. This example shows that the entire part would need to be scrapped if the core was misplaced.

The next column represents whether or not the defect is inspectable. There are two areas of inspection for the occurrence of a defect, in process and final inspection. The fifth column gives the acceptance criteria for the defect. It is extremely important to define the acceptance criteria to be able to define the degree of process controls required. The sixth column lists the process controls required to control the process to minimize the occurrence of the defect.

The seventh column addresses the effect the incorporation of each process control will have on the cost target. The first column under cost looks at the type of cost the incorporation of a particular control will have, i.e.. nonrecurring or recurring. The next columns under cost break out the cost into different categories which are direct labor, raw material cost, tooling cost, and capitol equipment cost. A "+" indicates that there will be an increase in the final cost target, a "-" indicates there will be a decrease in the final cost target, and a "0" indicates the final cost target will not be effected by the incorporation of the process controls.

The eighth column is a measure of confidence of the process to produce a part without the particular defect. "DPO" stands for defects per opportunity. Nonconformance data from previous RTM production programs is used to define DPO where ever possible. In this example two different DPO numbers are given. The first is for the process which has been used up to this point to manufacture 4-packs. This process uses preforming technology which is 3-4 years old. New preforming and tooling technology developed over the past year will be implemented into the tooling and manufacturing plan for the next set of n-packs to be fabricated. This technology has been used over the last year where ply placement defects have been eliminated. The next column under the confidence level gives the amount of opportunities for the defect to occur in the 4-pack. The third column gives a calculated sigma value. This value is calculated using the Six Sigma method. The goal is to have a sigma value of six.

The last column rates the priority. A "High" priority is given if the defect cost is 100% scrap, if there is no in-process or final inspection methods, or if there is a low Sigma value.

4.1.4 Implementation Plan A plan was generated for the implementation of the process controls once the controls were identified, reference **Table II**. The implementation plan identifies what action has to take place to incorporate specific process controls and who will be implementing the identified actions. The implementation plan also identifies the success of a process control. A sampling plan and sample size is identified for the next set of n-packs to be fabricated. A new Sigma value will be calculated from the data collected from the sampling plan identified. This Sigma value will be used to define the yield rate for a production process. This yield rate in turn will be used to update the final part cost estimate.

Table I. Identification of Process Controls

Failure Mode	Effect	Defect Cost	Inspectable		Acceptance Criteria	Process Control	Process Improvement Cost					Confidence Level			Priority
			in process	final			R vs N[1]	Cost R[2]	M[2]	T[2]	E[2]	DPO[3]	M[4]	Sigma[5]	
Strut ID Joint Failure Modes															
misplaced core causing low fiber volume in LE or TE	reduced joint strength	100% - scrap	Yes	Yes	Ply placement tolerance +/- 0.090"	1. pick & place strut preform 2. incorporate ply pinning	- R - N	0 0	0 +	0 +	+ 0	'95 Process: lay-up 1% load 4% (4% from Vane data)	8 8	3.46	High
												'96 Process: lay-up .001% load .001% (0% from 1995 FEGV data which utilizes ply pinning)	8 8	5.76	

1. Recurring cost (R) versus nonrecurring cost (N)
2. Cost: "0" = No additional cost to production part; "+" = Additional cost to production estimate; "-" = Reduction in cost to production estimate
3. DPO: Defects per opportunity (this number is obtained from actual data where noted)
4. M: No. of opportunities on the 4-pack
5. Sigma: calculated value (Six Sigma Method)

Table II. Process Control Implementation Plan

Failure Mode	Acceptance Criteria	Process Control Improvements	Action	Responsibility	Compl Date	Actual Effect on Cost	Sigma calc for improved process				Priority
							Insp Method	Sample Size	Est. Sigma[1]	New Sigma[1]	
Strut ID Joint Failure Modes											
misplaced core causing low fiber volume in LE or TE	Ply placement tolerance +/- 0.090"	1. pick & place strut preform 2. incorporate ply pinning	1. Design flat patterns to be P&P / incorp P&P into mfg paperwork 2. Incorp pins into flat pat & tool design	1. Hayse / St.Germain 2. Hayse / Simkulak	1. 12/1/96 / 1/15/97 2. 12/1/96	TBD	Visual / FV	196 / 20	5.76	TBD	High

1. Sigma: calculated value (Six Sigma Method)

5. COST MODELING

5.1 Cost Model The fabrication process is modeled in detail. Manufacturing steps are classified into various categories. For RTM manufacturing, these categories are weaving, tackifying, cut & kit, eutectic mandrel fabrication, preforming, preform assembly, molding, demolding, mandrel melt-out, and tool prep. In addition, machining, NDI, inspection, and assembly apply to all the components of the FEC assembly.

The basic cost model is comprised of two key parts. The first is the detailed fabrication cost model that establishes the cost driver relationships effecting direct manufacturing labor and material. The second is the fabrication simulation model. This model identifies points in the process that limit the productivity of the factory. It helps define rate tooling requirements, equipment requirements, and manpower requirements. Used together, these tools optimize the design, process, and factory for producing the lowest cost components for a given production rate. They help steer the design and process definition to take advantage of cost saving opportunities.

To best represent the actual manufacturing processes, the original cost model was refined with input from experienced manufacturing personnel. This improved cost model was then verified to assure that it produced the expected results from known input changes. The task of validating the accuracy of the fabrication cost model is currently under way. The simulation model is in detailed development. Improvements and updates continue in parallel with the design and fabrication of the latest FEC design headed for HCR testing.

5.2 Cost Target The ACP Program has set a cost target for the 100th production FEC. The cost target is generated from an estimate of the cost for each individual process step required to fabricate the entire FEC. That is, each process step for the fabrication of the 4-pack has an estimated cost associated with it. The estimate includes material cost, waste material cost, non recurring rework, nonrecurring scrap, equipment overhead, nonrecurring tooling, nonrecurring engineering hours, recurring support labor, recurring direct labor. These estimates were obtained by taking actual cost data from the fabrication of the development 4-packs in 1995 and 1996 and extrapolating what the cost per process step would be in production. A comparison between the developmental 4-pack cost (T1) and the production cost estimate (T100) is shown in **Figure 10**. The cost estimate is broken down into individual process steps and each step is broken into different cost categories. Specific process improvements, equipment upgrades, tooling modifications, and learning curves are used to derive the T100 cost estimate. The cost estimate is reviewed and updated during each iteration of the development process (as shown in **Figure 8**).

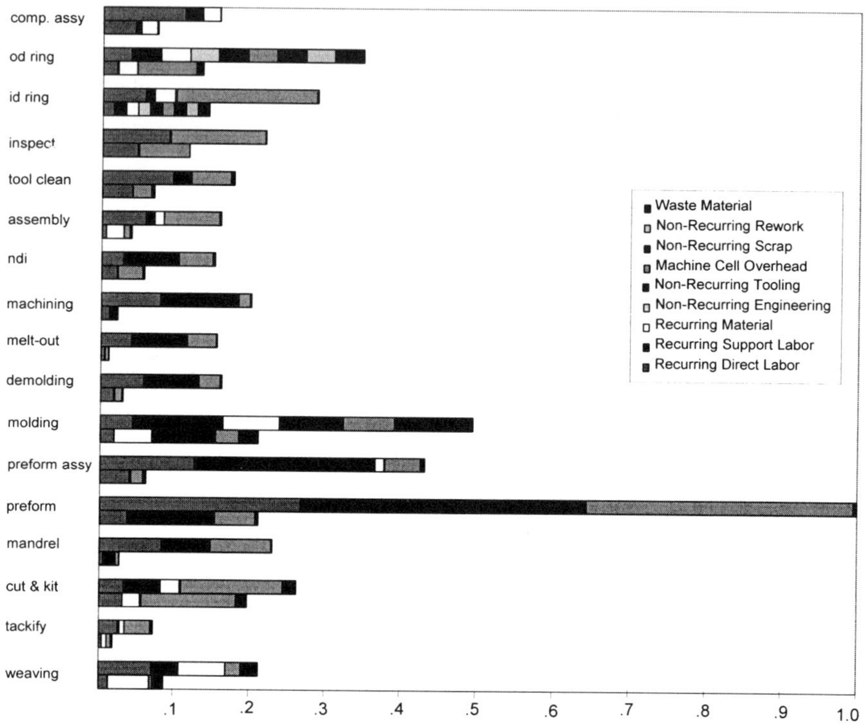

Figure 10 - T1 versus T100 Cost Breakdown

5.3 Implementation Plan The implementation plan defines the process improvements required to meet the T100 cost target, reference **Table III**. The process improvements are prioritized according to pay off. The time in which the process improvement will be implemented has been identified. Cost tracking is ongoing to verify the estimate of cost benefit for specific process improvements as these process improvements are implemented.

Table III. Cost Improvement Implementation Plan

Process	Category	Method	Implementation 95	97	Production
Preforming	Equipment	Reduced number of heated debulks	X	X	X
		Fully loaded ovens			X
		Larger Size Ovens			X
Preforming	Recurring Labor	Reduced debulks for 3D boxes	X	X	X
		Multiple 3D ply preforming			X
		Developed preforming concept		X	X
		Improved preform tools		X	X
		Higher build rate improves cell efficiency			X
		Automated stitching process			X
		Improved stitching fixtures	X	X	X
		Improved preform trimming		X	X
		Technician learning	X	X	X
Inner Ring	Equipment	Improved AGFM cutting time with multiple ply cutting			X
		Autoclave cost burdened over fewer preforming hours			X
Preform Assy	Recurring Labor	Improved tooling concept			X
		Improved mandrel location & positioning		X	X
		Technician learning	X	X	X
Mandrel Mfg	Recurring Labor	Batch manufacturing with higher build rate			X
Mandrel Mfg	Equipment	Multiple cavity mold versus single cavity			X

6. SUMMARY

Within the first two (2) years of this effort, full scale hardware has evolved to meet the demands of the ACP program, demonstrating reduced cost and lower weight compared to a conventional FEC design configuration. Significant testing has been conducted to substantiate analyses and design features, producing a database upon which future designs may evolve. Critical evaluation of processing capabilities has been conducted to identify areas for further improvement in manufacturing and tooling details for consideration in future and ongoing applications of RTM. Plans have been established for the continuation of this effort to support the transition of this affordable technology into future large jet engine structural applications.

7. REFERENCES

1. G. Griesheim, et.al., SAMPE International Symposium, 27 , 436 (1995).

28th International SAMPE Technical Conference
November 4-7, 1996

BUILDING BLOCK TEST PROGRAM AND RESULTS FOR AN RTM PRIMARY STRUCTURAL COMPONENT

Gary E. Griesheim, R. Bruce Berkley, D. S. Murphy
Pratt & Whitney
East Hartford, Connecticut 06108

and

Richard T. Reavely
Dow-United Technologies
Wallingford, Connecticut 06492

ABSTRACT

Polymer matrix composite materials are being investigated for use in primary structural components in advanced gas turbine engines to increase performance and save weight. Pratt & Whitney's Advanced Ducted Propulsor (ADP) Fan Exit Case (FEC) lends itself to composite application. Therefore a technology program has been established to develop a composite design that consists of concentric structural rings connected by a multitude of aerodynamic/structural airfoil struts.

The technologies necessary to design, analyze, fabricate, and test integral strut and concentric structural ring composite structures must be developed and validated before these structures can be introduced into advanced gas turbine engines. As part of this process, a full scale composite fan exit case was designed for ADP fan blade-loss ultimate loads. The fan exit case is built up from ten 4 - strut detail components, each of which was fabricated by Dow-United Technologies (Dow-UT) using the resin transfer molding (RTM) process. Subcomponent test programs were used to verify the structural integrity of the design. A "building block" testing approach was developed which began with coupon testing and culminated in a simulated static fan blade-loss test on a full scale FEC assembly. Results showed the RTM ring-strut-ring concept to be a structurally viable design. This effort was conducted as part of the Pratt & Whitney Affordable Composites for Propulsion (ACP) program (ARPA Contract No. MDA972-94-3-0029).

KEY WORDS: Aircraft Engine, Composite Structures, Resin Transfer Molding (RTM)

1. INTRODUCTION

The FEC is a key static structural component in a gas turbine engine. The unique Pratt & Whitney (P&W) ADP FEC design combines the functions of a traditional aerodynamic fan exit guide vane case and structural FEC frame into a single, structural ring/strut/ring assembly, as shown in Figure 1. This design reduces part count, eliminates manufacturing joints, and is well suited for composite material construction. The FEC supports the fan containment case (FCC) and inlet through an outer ring joint and must control fan blade tip clearance by maintaining case/rotor concentricity and minimizing deflections under service loads. The FEC must also continue to support the FCC and inlet during a fan blade-loss event.

Polymer matrix composite materials are being investigated for use in advanced gas turbine engine primary structural components to increase performance and reduce weight. A technology program has been established to develop a composite material FEC. The FEC is designed to be assembled from ten detail components, each one of which consists of four struts and the inner and outer flow paths between the struts (4-pack). The 4-packs are mounted between inner diameter (ID) ring and outer diameter (OD) rings, as shown in Figure 1. The rings provide hoopwise rigidity and strength to the assembly while providing the ability to build the assembly as a separate component from the engine. Methodologies necessary to design, analyze, fabricate and test integral strut and concentric structural ring composite structures must be developed and validated before these components can be introduced into advanced gas turbine engines. As part of this process, full scale RTM carbon/epoxy 4-pack FEC components were designed to withstand ADP fan blade-loss ultimate loads and were fabricated by Dow-UT as shown in Figure 1. [1]

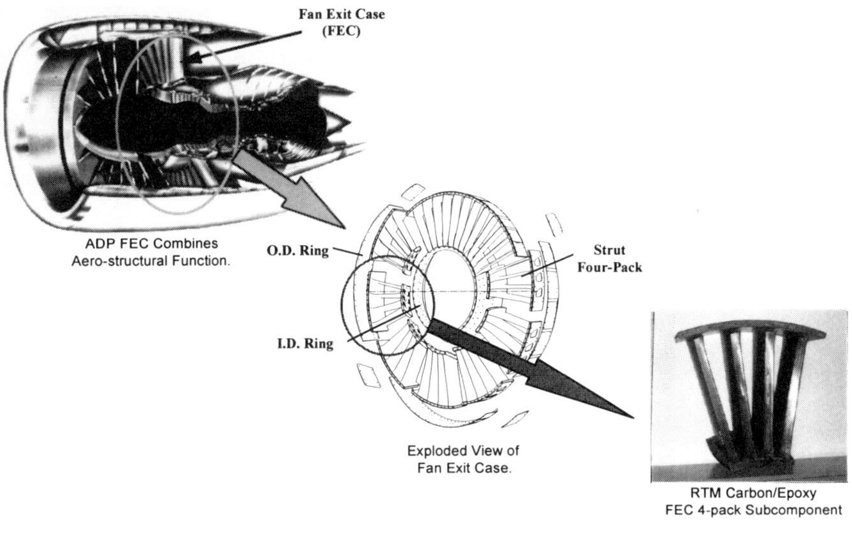

Figure 1. Composite Fan Exit Case Design

This paper summarizes the "building block" test program approach being conducted to develop and calibrate both a composite FEC design system and a manufacturing process to meet requirements of the P&W ADP. The objective of this "building block" test program is to determine environmental effects on failure modes, interaction effects of failure modes, scale-up effects and component critical stress effects.

2. CRITICAL LOAD CONDITION

The design-limiting load condition for the FEC occurs during a fan blade-loss event. Because of the high centrifugal forces imposed by the fan angular velocity, a liberated blade would strike the FCC at a high tangential and radial velocity, imposing an impact load which results in trunnion bending, shown as Load 1 in Figure 2. This impact load also produces large local deflections as a shock wave travels through the structure. In addition to the case impact load, fan blade-loss causes a severe rotating imbalance which introduces large bending moments into the FEC as it supports the overhung masses of the FCC and inlet during rotor deceleration. Abnormally high deflections caused by the large rotating imbalance also results in the fan blade tips rubbing on the FCC. This imparts a torque (about engine centerline) to the FCC which is then reacted by the FEC.

Critical FEC regions under fan blade-loss loads were identified through finite element analysis and a complete FEC component failure modes and effects analysis (FMEA). These critical regions are summarized and identified in Figure 2.

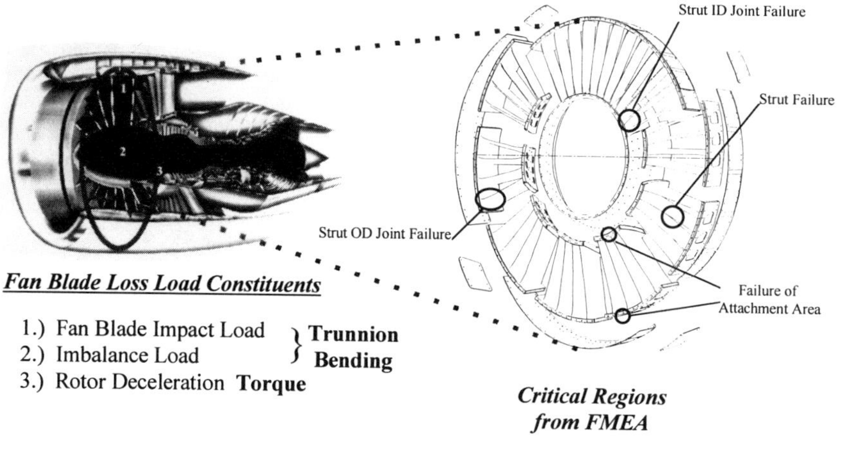

Figure 2. Fan Blade-Loss Is Critical Load Condition.

3. FABRICATION

The 4-pack was fabricated using an AS4®, IM7®/PR500® carbon/epoxy construction and the RTM manufacturing process. This approach used carbon fabric preforms that were assembled and injected with resin in a closed cavity mold.

The preforms were net shape details, assembled from laminated fabric plies held together by an epoxy powder "tackifier." Pressure and heat were used to press the preforms to net shape over forming blocks. Use of the tackifier resulted in dimensionally accurate and stable preforms that could be assembled into mold details requiring no further movement of fibers in the mold during subsequent assembly or resin injection.

Several preforming techniques were used in making the 4-pack. The RTM process allowed the use of specially woven fiber architectures that were designed to the specific requirements of the FEC component. The struts were fabricated by combining four of these sub-preforms with broadgoods which were then consolidated under heat and pressure to form a stable preform assembly. Through-thickness stitching with Kevlar® thread was used through the strut preform laminate for improved toughness to enable the strut to withstand large bird, hail, and other foreign object impacts in service. Other preforms in the part were made using tackified broadgoods laid up and formed over preform tools, and then compacted by vacuum under a diaphragm. Oven and infra-red (IR) heating were used during the preform compaction operation.

The basic construction of the 4-pack combined four strut preform assemblies with five outer and five inner shroud box preforms. The box preforms bridged the gap between the struts and carried loads from the struts to the mounting face of the assembly. Also included were three outer and two inner shroud preforms that span the entire arc of the 4-pack. These provided strut-to-strut load sharing capability to the structure. A total of 35 fabric preforms was used to construct the 4-pack component shown in Figure 1.

The RTM manufacturing approach used in the 4-pack provided the composite designer unique design flexibility through component-specific specialty woven fiber architectures. This specialty woven fiber architecture presented a challenge when developing a traditional "building block" test program to verify component structural integrity.

4. BUILDING BLOCK APPROACH

The "building block" testing approach addresses the complex nature of composite materials, specifically, their sensitivity to out-of-plane stresses and multiple failure modes. Applying such an approach to composite aerospace structures has been well established, is used currently for Federal Aviation Authority (FAA) component certification, and has been successful in certifying primarily prepreg / autoclave manufactured aerospace structure.[2 through 5] Here, the approach presented is adapted for an RTM composite FEC structure.

The "building block" approach began with complete structural analysis of the composite component to identify critical regions for test evaluation. Failure modes were then determined for these areas. The "worst" environmental conditions were based on the predicted service use. A test pyramid was then developed to verify critical regions and predicted failure modes and to understand effects of "worst" case environmental conditions as shown in Figure 3. This design verification pyramid begins at the lowest geometric complexity coupon level to verify fundamental material behavior and its repeatability and environmental effects, and culminates at the full scale most complex design component level, as described below.

4.1 Coupon Level Coupon level testing typically involves five batch qualification testing at critical environmental conditions for the material system and manufacturing process used in the component. Five batch qualification testing establishes the mechanical property repeatability of the material system. Critical conditions for the component are typically at the maximum operating temperature both "wet" and "dry" and at the minimum operating temperature. Design allowables for tension, compression, in-plane shear, and flexure are generally developed at this level.

4.2 Element Level Element level testing focuses on the component's specific laminates (or preforms, for RTM). This level also includes damage tolerance and design specific attachment testing. Design allowables are developed at the element level typically with one batch qualification at critical operating temperature and condition. Laminate or textile architecture analysis codes are calibrated with results from this level.

4.3 Subelement Level Subelement level testing is component specific. This level includes design features representative of the actual part and typically machined from actual full-scale RTM manufactured components. Actual joint attachment hardware is tested to critical loading and "worst" case environmental conditions. Damage tolerance, repairs, effects of manufacturing defects and mission fatigue requirements at critical regions are also demonstrated at this level.

4.4 Subcomponent Level Subcomponent level testing simulates actual engine hardware boundary conditions and verifies analytical models and previously predicted failure modes. This level also evaluates component durability to in-service use and associated foreign object damage (FOD) from birds, hail, gravel, and erosion. Spectrum fatigue testing is also used at this level to verify structural adequacy and evaluate repairs and inspection limits and intervals.

4.5 Full Scale Level Full-scale level testing involves actual engine attachment hardware for test boundary conditions and verifies full-scale finite element analysis and predicted failure modes. This level also validates component structural integrity through mission "normal operation' spectrum fatigue loading with expected in-service damage and repairs at critical regions followed by a full-scale static strength test at ultimate loading conditions. In addition, the full-scale level includes a full engine dynamic test of an ultimate fan blade-loss loading conditions to verify component structural integrity.

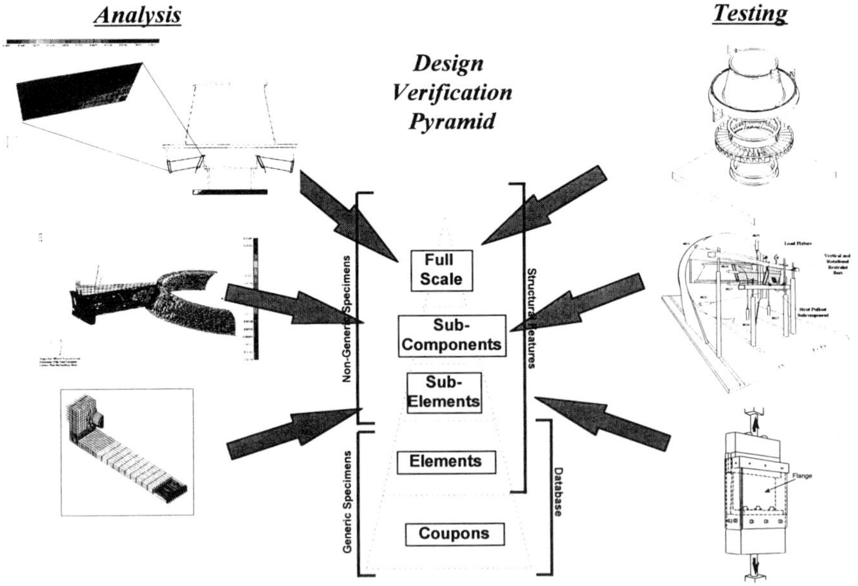

Figure 3. Resin Transfer Molding Building Block Approach.

5. BUILDING BLOCK RESULTS

Figure 3. provides examples of FEC analysis and tests at each level of the design verification pyramid.

5.1 Coupon Level Coupon testing for the FEC included traditional composite in-plane (tension, compression, shear, and flexure) as well as out-of-plane (interlaminar tension and shear) testing. Results revealed the RTM manufacturing method provides similar mechanical properties to prepreg / autoclave carbon / epoxy material systems. Environmental knockdowns observed were also typical for carbon / epoxy materials.

5.2 Element Level Element level testing completed for the FEC included bolt bearing tests, compression after impact, and laminate testing. Results of these tests verified similar critical temperature condition mechanical property knockdowns also observed at the coupon testing level and validated the laminate analysis and textile architecture analysis programs used for design allowable predictions.

5.3 Subelement Level Detailed joint attachment testing was the focus of the subelement test program. Flange subelements were selected to represent the attachment detail between the FEC and the mating FCC and were of the exact flange geometry and attachment hardware except for case curvature. Figure 3. illustrates the subelement test set-up and associated three-dimensional (3-D) finite element analysis. Results again revealed similar environmental effects, as observed with both coupon and element testing and verified joint margin of safety and detailed joint and contact finite element analysis methods. Several manufacturing effects of defects subelements were also tested, revealing less than a 10 percent reduction in joint strength.

5.4 Subcomponent Level Testing a single strut at a time was chosen over an entire four pack to simplify load paths for post test analysis at this level. The strut subcomponent was fixtured in a simulated engine case ID fixture. Results verified the failure mode and location previously predicted by finite element analysis. All struts exhibited a gradual failure mode indicating multiple redundant load paths. Predicted strut stiffness was also verified through test and strut frequency checks.

5.5 Full Scale Level A full scale FEC assembly was tested in a steel test fixture constructed to apply static equivalent blade-loss loads. The fixture was configured to ensure the two components of the load (trunnion bending and centerline torque, shown in Figure 2.) could be introduced into the case at four points. The locations were chosen to approximate the proper shear-to-moment ratio. A computer controlled set of hydraulic actuators was used to apply the different load cases. The FEC was grounded at the ID through a steel cylinder, and load was introduced into the case through the containment case flange as shown in Figure 4.

The FEC was originally designed to the requirements of a 40,000-lb thrust ADP propulsion system configured with an Integrated Outer Cowl (IOC), in which the inlet was attached directly to the pylon and only the containment case was attached to the FEC. This concept was later replaced with a more conventional turbofan/nacelle design, with the inlet attached to the engine through the containment case and, ultimately, the FEC. Blade-loss loads corresponding to both configurations were applied to the case during the full-scale testing. Initially, the case was cycled to low load levels corresponding to maximum take-off conditions. Deflection was carefully monitored during this cycling to check for any evidence of yielding or settling, which would lead to problems maintaining the proper fan blade tip clearance in service. No

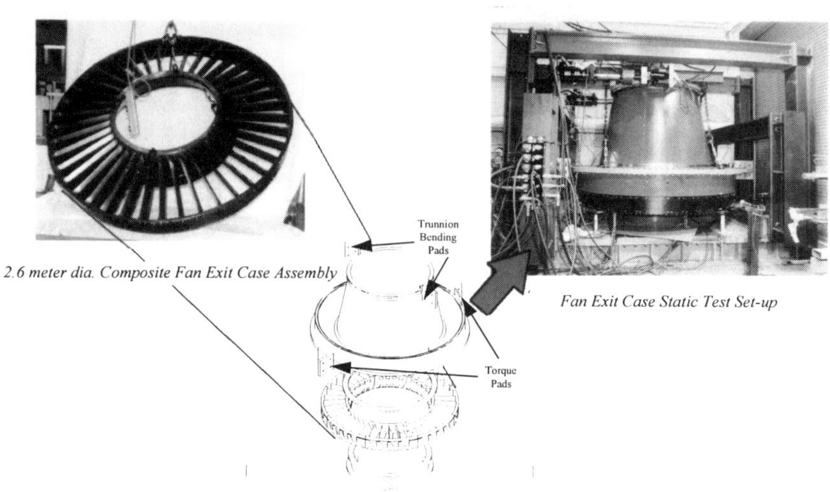

Figure 4. Full-Scale FEC Assembly Test Set-up.

permanent displacement was observed. The blade-loss loads corresponding to the IOC configuration were then applied and held for 60 seconds. No observable damage and only a very small residual deformation resulted. A residual strength test was performed to simulate a 6 g maneuver with the engine shutdown (returning to base). Subsequently loads corresponding to the current engine configuration were applied representative of a 46,000-lb thrust turbofan engine with a conventional nacelle. After the blade-loss loading was completed, the same residual load case was performed to the level of a 6 g maneuver. The FEC sustained all the above loadings without significant damage.

6. CONCLUSIONS

This composite FEC "building block" test program can be summarized as follows:

Results:

1.) Mechanical property environmental knockdowns for the RTM manufactured coupons were similar to prepreg/autoclave manufactured carbon/epoxy coupons.
2.) Mechanical property environmental knockdowns were consistently verified at all levels of the test pyramid.
3.) Analytical methods used for RTM manufactured structures were successful in predicting component structural behavior.
4.) Manufacturing-induced defects tested did not reduce component ultimate strength by more than 10 percent.
5.) No apparent loss in component stiffness or strength was observed during fatigue testing.
6.) All FEC component testing demonstrated a slow progressive failure mode with a multiplicity of load paths.

Conclusions:

1.) All tested RTM components exceeded design ultimate load requirements.
2.) Resin Transfer Molding is a viable process to produce a composite FEC.
3.) The "building block" test approach is a technically feasible method to evaluate the structural integrity of RTM composite components.

7. BIOGRAPHIES

Gary E. Griesheim is a Materials Technologist in the Polymeric Materials Group with Pratt & Whitney in East Hartford, CT. His present responsibilities include composite materials technology development for commercial aircraft engines. He previously worked as a structural research engineer with advanced composite structures in the Center of Hypersonic Excellence at McDonnell Aircraft Company. He began his career in the aerospace industry after receiving his B.S. and M.S. in Engineering Mechanics, with a specialization in composite materials, and an M.B.A. from Southern Illinois University at Carbondale and Edwardsville.

R. Bruce Berkley is a structural test engineer working in the Applied Mechanics Group of Pratt & Whitney in E. Hartford, CT. His present duties include static testing, low cycle fatigue testing and rotating component testing. He joined the aerospace industry after receiving his B.S. degree in both Chemical Engineering and Mechanical Engineering from Arizona State University in Temple, AZ. and a M.S. degree in Mechanical Engineering from the Hartford Graduate Center in Hartford, CT.

David S. Murphy is a materials testing engineer working in the Applied Mechanics Group of Pratt & Whitney in E. Hartford, CT. His present duties include overseeing specimen testing in the areas of low cycle fatigue, fracture mechanics, and composites(PMC, CMC). He joined the aerospace industry after receiving his B.S. degree in both Materials Science and Mechanical Engineering from the University of Connecticut, and a M.S. degree in Metallurgy from the Hartford Graduate Center in Hartford, CT.

Richard T. Reavely joined Sikorsky Aircraft in 1972 where he worked in the Rotor Systems Design & Development Group and later in the New Composite Business Group. While in the Rotor Systems group he participated in the design and development of several production and experimental composite main and tail rotor blades. While in the New Composite Business Group, which would later become Dow-UT Composite Products, he participated in the design and development of a variety of products, including several turbine engine components, missile components, airframe components and a dome for a motion base simulator. He has a B.S. degree in Aerospace Engineering from Syracuse University and a M.S. Degree in Mechanical Engineering from the University of Bridgeport.

8. REFERENCES

1. C. Cimini, G. Griesheim, D. Murphy, R. Kersey, "Design and Testing of Composite Radial Flanges,"ASC Technical Conference Proceedings, (1996).

2. R. Whitehead, R. Deo, "A Building Block Approach to Design Verification Testing of Primary Composite Structure", AIAA - 83-0947, (1983).

3. K. Kedward, P. Beaumont, “The Treatment of Fatigue and Damage Accumulation in Composite Design”, International Journal of Fatigue, 0142-1123/92, (1992).

4. D. Wilkins, “A Preliminary Damage Tolerance Methodology for Composite Structures”, Procedural Workshop on Failure Analysis and Mechanisms of Failure of Fibrous Composite Structures, NASA-CP-2278, (1982).

5. K. Sanger, “Certification Testing Methodology for Composite Structure”, Report No. NADC-86132-60, (1986).

28th International SAMPE Technical Conference
November 4-7, 1996

DESIGN AND MANUFACTURING CONCEPTS FOR DAMAGE TOLERANT NACELLE STRUCTURE

Stephen W. Rice, Northrop Grumman , Dallas, TX;
Carroll Grant, Alliant Techsystems, Magna, UT

ABSTRACT

Composite structure on commercial aircraft is subject to impact damage from numerous sources as the aircraft are handled daily. This paper will evaluate a composite design that will provide hardware with increased damage resistance. In addition, the manufacturing process required to economically produce hardware to this design will be explored.

KEY WORDS: Damage Tolerant Composites, Tow Placement, Cocured Composites

1. INTRODUCTION

Data collected by one major airline indicates that millions of dollars are lost each year due to impact damage to nacelle structure. The cost incurred is due to down time of the equipment as well as the actual cost of repair and / or replacement of the nacelle hardware. With this liability, it is natural to expect that they would prefer a more durable material system for this vulnerable area of the aircraft.

The baseline for our study was the relatively thin-skinned honeycomb structure that represents most of the composite hardware used in this application. Although not part of our comparison, we have also included some data on metallic skins to provide an indication of how that material system would respond to impact.

2. DESIGN

The key element of the hardware designed to resist damage due to impact is a hybrid composite skin of 75% AS4 Carbon and 25% S-2 Glass. A description of the entire assembly is shown in section 3.0. The skin is produced by laying tows of composite materials via a fiber placement machine in specific patterns. Basically, three tows of carbon flanked by one tow of glass. For our test, each tow is 0.157 inches wide by 0.0084 inches thick.

2.1 Background Before we get into the specifics of the design, let us propose a heuristic look as to why the "Hybrid" skin offers increased damage tolerance.

Carbon fiber is the composite material of choice for primary aircraft structure. Unfortunately for many aerospace applications, a laminate designed with carbon must grow in thickness to gain the necessary damage tolerance. Increased weight: not good! Some of this added weight was avoided on the Affordable Composites for Propulsion (ACP) program by selecting a hybrid (carbon and glass) laminate for the skins. Glass in the hybrid is surprising since it is often treated as a secondary material. A non technical explanation follows for the use of glass in hybrids and its corresponding success for impact resistance.

The hybrid laminate's improvement in impact resistance comes at a cost. The cost is a loss in static capabilities like strength and stiffness. As a definition of terms, a material's "impact resistance" can be assessed by the damage area resulting from a given impact event; "strength" of a material is its ability to carry load; and "stiffness" of a material is its ability to resist deflection under load. The table below contains a comparison of static capabilities of carbon, glass, and the hybrid materials.

STATIC CAPABILITY					
PROPERTY	CARBON (AS4)c	GLASS (S2)g	HYBRID (3c TO 1g)	CHANGE (HYBRID)	PERCENT REDUCTIO N
STIFFNESS (MSI)	**18.62**	**6.82**	**15.56**	**3.06**	**16.4 %**
STRENGTH (KSI)	**290**	**273**	**246**	**44**	**15.2 %**
STRAIN (%)	**1.56**	**4.00**	**1.58**	-	-

Carbon's superior stiffness and strength is obvious. What is interesting, and subtle, is glass having twice the strain (meaning deflection) capability of carbon but one-third the stiffness. The hybrid material's stiffness may be understood as a blending of both, carbon and glass, and so is a reduction in static capabilities from the carbon material. Also note that the carbon material with its lower strain capability limits the hybrid material to a strain more typical of carbon; around 1.5%. For reasons explained later, the high strain and low stiffness capabilities of glass are key to understanding the hybrid's impact resistance.

The success of the hybrid is that it is a thinner (lower weight) assembly relative to the thick (heavier) carbon facesheets while providing the same impact resistance. The all carbon laminate designed for impact requirements (impact resistance) is thick and therefore has excess strength and stiffness capability. It is the presence of excess static capability that allows glass to be substituted for carbon. Moreover, for the ACP door design, substituting glass adds impact resistance quicker than the laminate loses strength and stiffness. Hybrid skins are thinner than the durable skin in carbon, but thicker than the skin designed for static loads in carbon. The end result is a hybrid design that meets the static *and* the impact resistance requirements. Even though the tow-preg hybrid laminate is at least 16 percent less stiff and 15 percent weaker than the pure carbon design of the same configuration, the ability to resist damage more than doubles.

Along with impact resistance, another important consideration for including glass is its lower purchase price. Much like the clothing industry; the suit made of "all wool" gained impact resistance and cost less when it became "75% wool and 25% man made fiber".

The traditional approach for improving the impact resistance of composites include:

- Toughen resin (sophisticated control of delaminations)
- Increase fiber strain capability (squeezing blood out of turnip)
- Thickened laminate (brute force)
- Improved reparability (apologetic)
- Ignore it as a failure (head in the sand)
- Lower the impact requirement (denial)

These approaches have met with various degrees of success.

Tow placement adds another technique to the composite community's bag-of-tricks for impact resistance. Tow placement allows differing fibers to be mixed in the same ply. Prior to now, materials were purchased as either a fabric or tape consisting of one type of fiber and resin. The pre-preg supplier is reluctant to supply material forms with a mix of fibers. Now, mixing is done right at the machine during detail fabrication. (This manufacturing technique will be explored in more detail later in the paper.)

Our program did not try to look at the damage tolerance issues from a microscopic standpoint; however, we tested to evaluate the macroscopic benefits that could be realized from the hybridization concept.

2.2 Testing In an attempt to improve the damage resistance of composite nacelle structures, without increasing weight or cost, a frame stiffened solid laminate composite skin design approach was selected. To evaluate the damage resistance of the solid laminate skin design, a limited test program was conducted using a baseline, thin-skin honeycomb sandwich construction, as a comparison. A minimum threshold of 0.05 inch dent depth at 10 ft-lb. impact energy was established as a design requirement. Testing was conducted at the Northrop Grumman Commercial Aircraft Division in Dallas, Texas.

Testing focused on skin design and explored the relationships between carbon fiber type (AS4 and IM7), matrix toughness (3502 and 8552), textile form (tape, fabric and tow), and their affect on the resistance to impact damage.

Material system selection was based on test results using a 0.50 inch diameter impactor. Sensitivity testing was conducted to evaluate the effect on damage by varying impactor diameter, panel thickness, and panel support.

Test panels were flat and measured 20 x 20 inches. All tape and fabric panels were fabricated by Northrop Grumman Commercial Aircraft Division in Dallas, Texas. Tow placed panels were produced by Alliant Techsystems in Magna, Utah with their fiber placement machine. Panel thickness for the different material/matrix/form combinations were representative of fan cowl doors which were designed to satisfy a no buckling at 115% limit load requirement. Each panel had external plies of 0.0038 inches thick fiberglass cloth which serve to decrease the damage to laminates when being drilled and act as a galvanic isolator.

Baseline thin skin honeycomb panels were fabricated by Pratt & Whitney and consisted of three plies of AS4 fabric over one inch, 1/8-inch cell, 4 PCF Nomex honeycomb core.

Testing was performed with a Dynatap 8250 test machine . Impact energy levels for this test ranged from 1 to 60 ft-lb. Impactor weight ranged from 13 to 28 pounds.

Test panels were clamped along two edges to the test support fixture. A 20-inch span was chosen to represent a mid-bay between frame locations on the fan cowl door.

Damage size was measured in visible depth of impact using a dial indicator. Ultrasonic measurements were conducted utilizing a portable hand-held pulse-echo instrument to determine delamination area.

2.3 Dent Depth Test Results The 0.5-inch diameter impactor results for all material and structural configuration systems are presented in Figure 1. Data is plotted as dent depth versus impact energy. As can be seen, all solid laminate material systems, satisfy the 0.05-inch dent at 10 ft-lb. requirement. It further shows how superior all the solid laminate configurations are compared to the thin-skinned honeycomb baseline.

2.4 Test Conclusions From the results of this limited impact resistance test program, the following conclusions can be drawn:

1) The high modulus fiber (IM7) provided increased impact resistance over the medium modulus fiber (AS4).
2) The toughened matrix (8552) exhibited increased impact resistance over the untoughened matrix (3502).
3) Tape provided slightly more impact resistance than tow for the mid-bay type locations (20-inch span).
4) Fabric offers significant improvement in impact resistance over both the AS4 and IM7 tape and tow.
5) Syntactic core, sandwiched into the laminate, resulted in reduced impact resistance.
6) The AS4/S-2 hybrid (tow) significantly improved impact resistance over the all-carbon fabric, the all carbon tape, and the all-carbon tow.

Figure 1. DAMAGE ASSESSMENT OF "THIN" PANELS
0.5 inch Diameter Impact Head

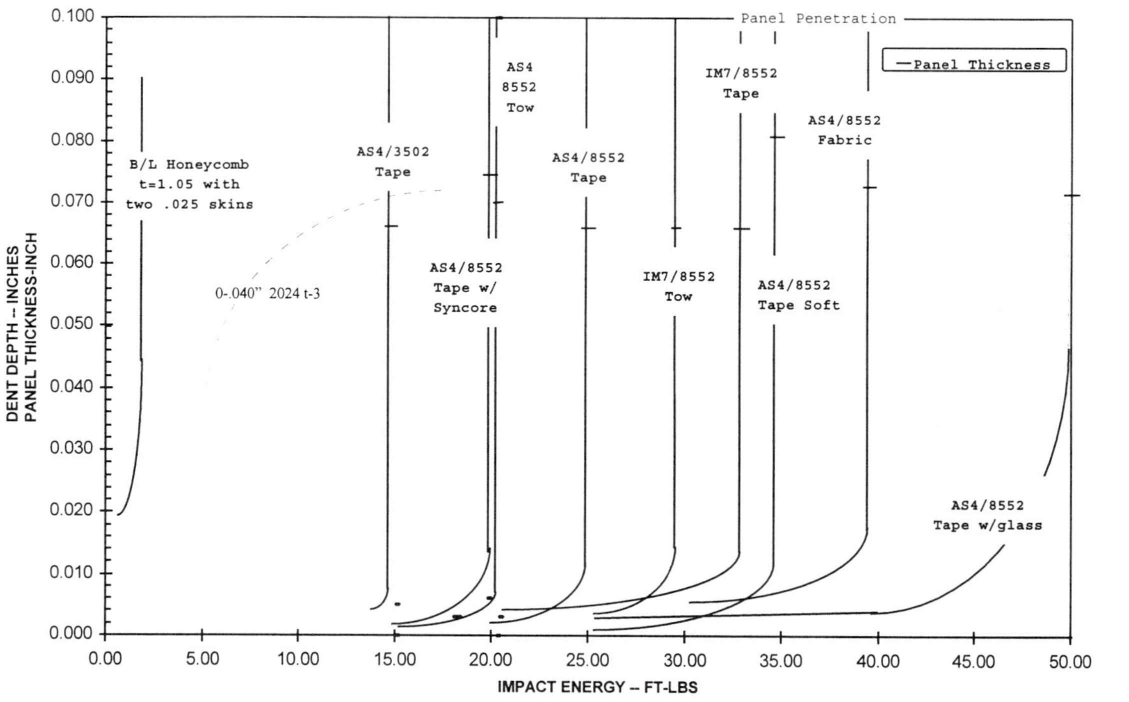

3.0 HARDWARE DESCRIPTION

The hardware designed under the ACP program is depicted in Figure 2. The composite part of the door is a skin with three "hat" section stiffeners in the circumferential or hoop direction. The "skin" portion of the laminate is composed of a hybrid laminate of 75% AS4 carbon tow and 25% S2 glass tow. Each ply has a repeating pattern of 3 tows of carbon and 1 tow of glass. This design incorporates tow material that is 0.157 inches wide and 0.0084 inches thick. For our purposes, the tow material equates to very narrow bands of unidirectional tape. When these bands are positioned side by side, they resemble full plies of tape with a repeating band of glass fiber representing 25% of the surface area, (See Figure. 3).

Figure 2 ACP Fan Cowl Demonstration Door Design

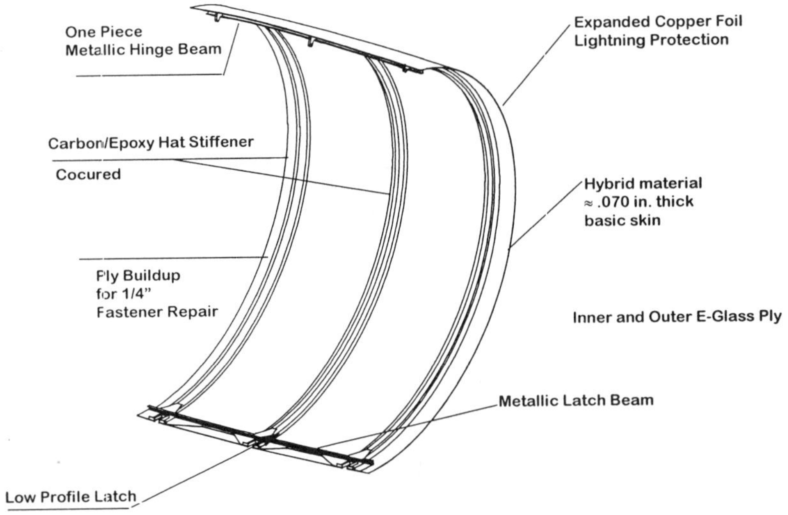

Figure 3. Carbon / Glass Hybrid

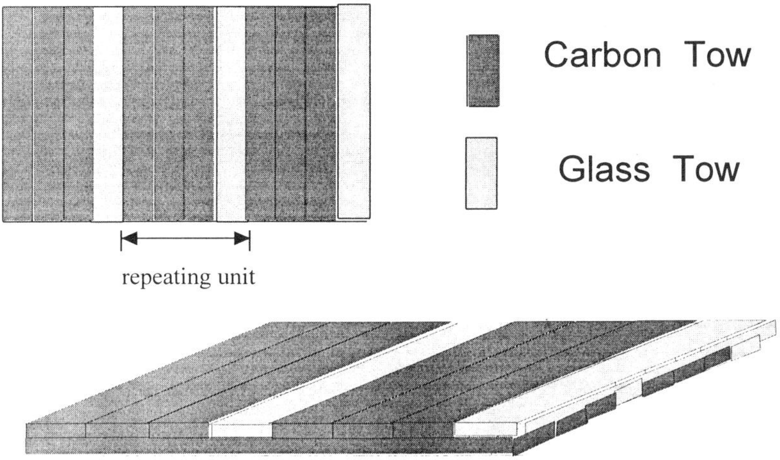

The hat stiffener is made of AS4 fabric laid at + and - 45 degrees with unidirectional tape reinforcing only the cap area. This combination offers the stiffness required while significantly improving the producibility over the same hat made of 0 / 90 fabric and / or full plies of tape (See Figure. 4)

Figure 4. Hat stiffener

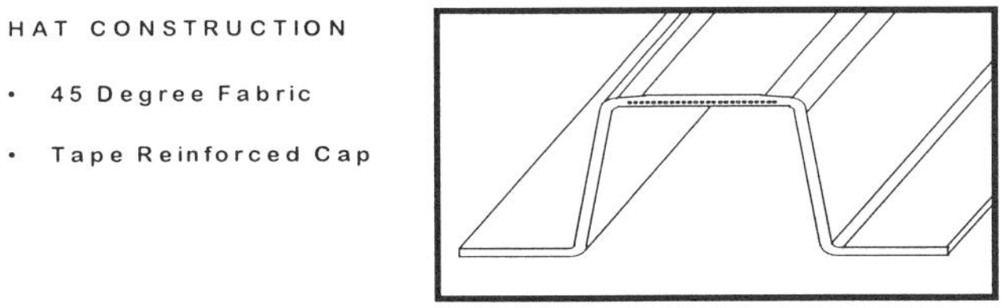

4.0 MANUFACTURING CONCEPTS

There are two unique areas of manufacturing that will be addressed here. The first deals with the lay-up of the hybrid skin via automated tow placement. The other deals with cocuring of the skin and hat stiffener assembly.

4.1 Tow Placement Automated tow placement utilizes a numerically controlled machine tool to manipulate a delivery head over the surface of a tool. The tool can be either an intermediate mandrel or an actual cure mold. The delivery head designs vary from machine to machine depending on the requirements. For our development activity, we used machines at Alliant Tech Systems in Salt Lake City for our test panel work, and the Ingersol machine at McDonnell Douglas in St Louis for our full scale hardware. The various heads are capable of delivering as few as one tow and up to as many as 32 tows on the St. Louis machine. Tow widths can also vary. The three common widths are 0.125, 0.157, and 0.182 inches. The Alliant machines have heads that can utilize any of the common widths while the 32-tow head on McDonnell Douglas's Ingersol machine is set up for the 0.182 inch tow. Figure 5 is a conceptual sketch of the delivery head system.

Figure 5. Tow Placement Delivery Head System

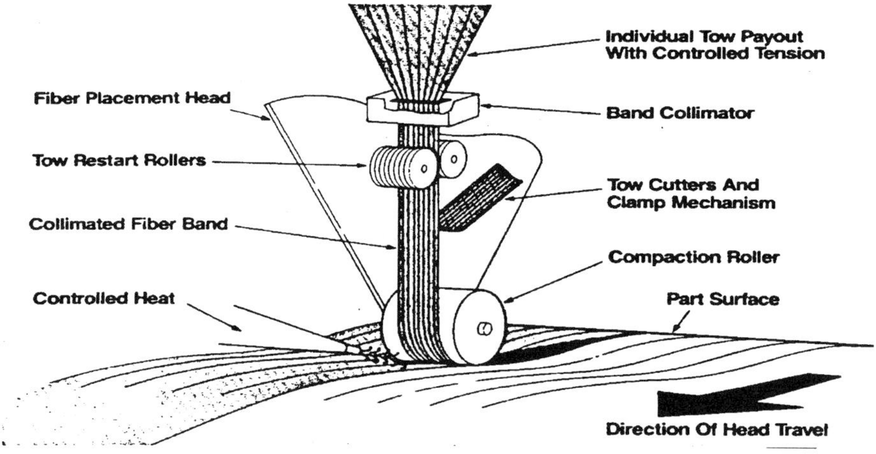

The unique ability of the tow placement equipment to assemble narrow widths of composite prepreg Tow directly onto the surface of a tool so that it resembles wider bands of prepreg Tape offers several advantages. First, the individual narrow bands of tow allow the head to apply the material to a compound contour tool surface without having to use the "Natural Path" laydown technique required with Tape laying equipment. Each tow can be dispensed through the head at its own rate so that the band can actually be "steered" over any surface in almost any direction. Secondly, the fact that each tow width of prepreg is supplied from its own spool allows the "hybridization" to be accomplished simply by replacing spools of carbon with spools of the different material, in our case S2 glass. A third feature of this approach is that the N/C control of this type equipment will accurately and repeatably locate each ply and its corresponding ply terminating areas so that the thickness of the laminate will accommodate a "pseudo" matched die tooling concept that will be discussed in section 4.2.

For our manufacturing demonstration, we utilized a male mandrel machined to match the inside skin surface of the nacelle hardware (fan cowl door). To maximize the utilization of the tow placement machine, the tool was designed to accommodate four (4) parts at one time (two left.-hand and two right-hand parts). The part size is approximately 170 degree arc of a 10-foot diameter nacelle and 68 inches in length; consequently, the mandrel was fairly large but well within the limits of McDonnell Douglas's machine.

4.2 Cocure Cocure operations quite often produce undesirable side effects when applied to complex configurations. Figure 6 is an illustration of some of the side effects associated with the hat stiffener concept.

Figure 6. Cocure Problems

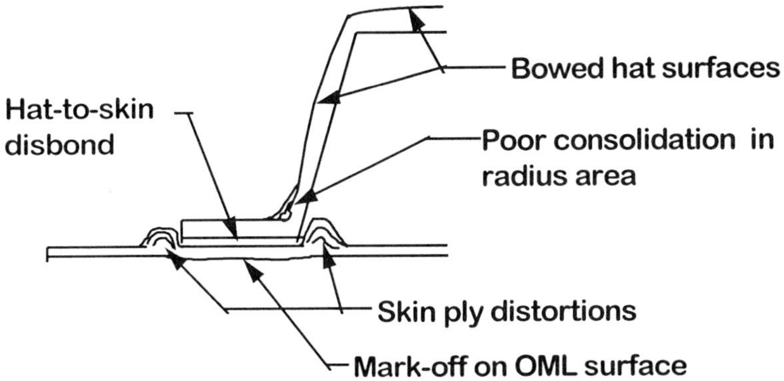

Specific tooling concepts must be employed to overcome some of these phenomena. Tooling was designed to provide uniform consolidation pressure during cure. A bladder type elastomeric mandrel was designed to fit inside or the hat stiffener to provide uniform pressure to the hat stiffener. The mandrel was enclosed in a bag and plumbing was installed to vent the inside of the bladder to autoclave pressure. As the autoclave is pressurized, the same pressure is allowed to fill the bladder mandrel as is being applied to the skin and the outside of the stiffener. This provides the uniform pressure required for consolidation of the laminate. To control the stiffener shape and location, a semi-rigid carbon/epoxy "caul" was developed to duplicate the exact Inner Mold Line (IML) surface of the part including ply drop off areas, stiffener flanges, and stiffeners. The caul supports the stiffener "shape" during cure, yet allows uniform pressure to be applied over the entire part surface.(See Figure. 7)

Figure 7. Cocure System Pressure Profile

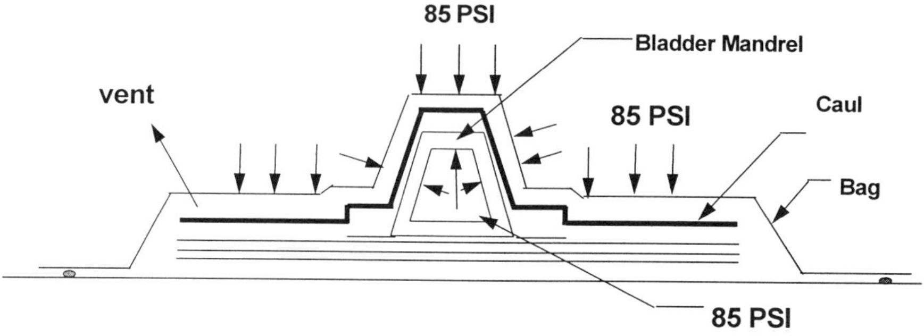

Using this approach makes it imperative that the materials under this caul are of the correct thickness and in the correct location. This concept simulates the function of a "matched die" tooling approach and any mislocated thickness will keep the die from closing properly.

The fabrication of the cocured assembly begins with the uncured tow placed skin being located into the cure mold. The hat stiffeners are laid up by hand in their own lay up tools. The bladder mandrels are located in the hat sections and then transferred as a unit to the IML caul. For ease of installation, the caul is supported in an inverted position until all of the stiffeners have been positioned. The caul along with the stiffeners are then accurately positioned on to the skin. Coordination tool pins in the mold maintain alignment through holes in the skin, and the caul. This ensures accuracy and repeatability. The lay up is then bagged and cured.

5.0 CONCLUSIONS

1. A hybrid laminate fabricated with carbon and fiberglass tows is an effective way to increase the impact resistance over all carbon laminates.

2. Tow placement is a unique manufacturing technique that along with being an efficient lay up process, facilitates combining tow widths of unlike materials into a single laminate to form a hybrid. For the ACP fan cowl demonstration doors, the hybrid of choice was 75% AS4 carbon / 25% S2 fiberglass.

3. The cocuring process with the use of semi- rigid cauls is an attractive approach for manufacturing large, contoured, hat stiffened composite assemblies.

28th International SAMPE Technical Conference
November 4-7, 1996

AUTOMATED TOW PLACEMENT OF FAN COWL DOOR SKINS

Roger Neddo
Alliant Techsystems

Chris Kogstrom
Pratt & Whitney

August 1996

INTRODUCTION

This paper discusses the automated tow placement (ATP) of the fan cowl door skins for the Affordable Composites for Propulsion (ACP) Program. The objective and background of this program, the roles of the company teams, the progress to date at McDonnell Douglas Aerospace (MDA) where the skins are being tow placed, and the part design will be addressed. Discussion of the process description includes the tooling, process flow, off-line programming and fiber path selection, and prepreg tow. Learning curve data associated with the fan cowl door skins will be presented, along with discussion of the challenges, process improvements, and lessons learned.

BACKGROUND AND OBJECTIVE

The objective of this program is to fabricate fan cowl door skins to assess ATP as an affordable, capable process for the production of cowl-type structures. Equivalent weight to a honeycomb baseline design and greatly increased damage resistance and tolerance were key design drivers for the fan cowl doors. Affordability and reliability were key process selection drivers for the door skin fabrication. The ATP of a glass-carbon hybrid skin meets the design and process objectives.

Several companies teamed to design and fabricate the doors, collect data, evaluate results, and project production costs. Pratt and Whitney oversaw the overall ACP Program and provided the initial baseline configuration for the cowl door. Alliant Techsystems designed and built the ATP mandrel and cure mold, submitted part design changes to utilize the advantages of tow placement, and fabricated numerous test panels at Alliant Techsystems' facilities for the fan cowl door evaluation. Alliant Techsystems contracted and guided the fan cowl door skin fabrication at MDA in St. Louis. MDA-East fabricated the fan cowl door skins using the ATP process and shipped them in an uncured condition to Northrop Grumman Commercial Air Division. Northrop Grumman is responsible for the design, assembly, and cost model of the doors. They fabricated both the hat stiffener co-cure tooling and caul, and co-cured three hat stiffeners to each skin.

Tow placement of all the demonstration skins was completed in the third quarter of 1996. Twelve fully assembled left-side demonstration doors are planned. The door dimensions are roughly 6 ft wide by 15 ft in circumference, with a mildly complex-contoured surface. Twelve left demonstration skins and five right skins were fiber placed. These include three 2-up configurations and one 4-up (Figure 1) configuration. As the skin fabrication continued, process and efficiency improvements and refinements were implemented each time.

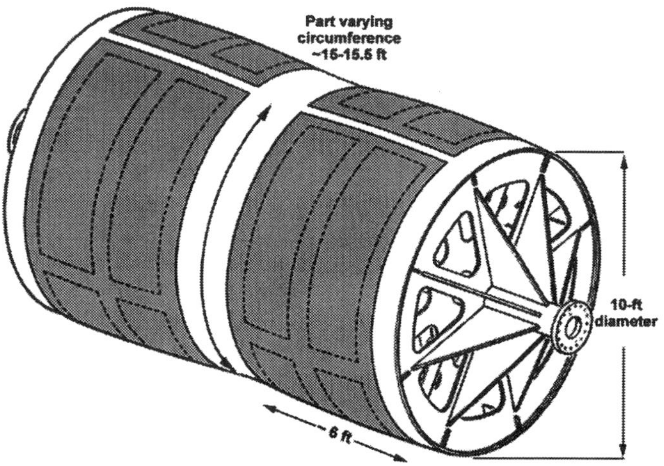

Figure 1. Four demonstration skin configuration on tow-placement mandrel

PART DESIGN

The ACP fan cowl demonstration doors were designed to utilize the capabilities of tow placement. With the tow placement process, S2 glass and AS4/8552-1 prepreg tow are employed in the same ply in a 1:3 ratio to provide the damage tolerance and damage resistance that the design required. Twelve 0.0072-in.-thick plies, including three 90° padup plies, were tow placed for each door. The number of 90° plies (90° is in the hoop direction) were maximized because the machine provides the most efficient laydown in that orientation. All three of the padup plies were grouped together with the largest padup ply (which has the smallest window area) being collated first (Figure 2), thereby allowing all of the padup end-of-plies (EOPs) to be inspected at the same time. The prior ply reference also helps the machine operator detect EOP errors during the tow placement process.

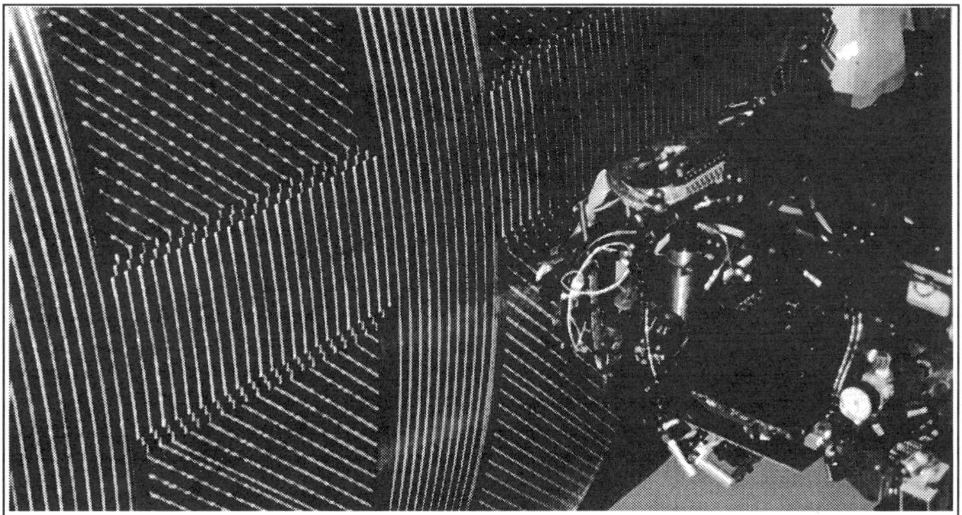

Figure 2. Three sequential 90° padup plies tow placed.

A 0.5-in. spacing of the cutout EOPs with a ± 0.2-in. tolerance was a design feature that provided flexibility. This spacing and tolerance enabled any of the common prepreg tow widths of 0.128, 0.157, or 0.182 in. to be used. The generous tolerance allows acceleration of the tow cut-and-add process and the EOP inspections. Further efficiency gains with cut and add on the fly are expected. Designing the internal EOPs perpendicularly to the laydown direction also reduced tow cut-and-add time because fewer stops were needed for separate cuts and adds.

PROCESS FLOW

The general process flow steps are outlined below. These process steps are ready to implement in production:

- Prepare mandrel—apply release film
- Install mandrel into machine
- Cut and apply inner E-glass fabric ply
- Prepare delivery head and load prepreg tow into creel
- Tow place 12 plies, in-process inspect as needed
- Cut and apply outer E-glass fabric and copper mesh
- Machine debulk
- Roughly trim edges and drill skin-locating holes
- Remove skin from mandrel and transfer

The steps of applying release film and E-glass fabric to the mandrel can be accomplished in the mandrel cart before the installation of the mandrel into the machine, or these steps can be accomplished after the mandrel is installed into the machine.

Frekote was originally applied to the mandrel, but, after encountering difficulty with skin removal, the mandrel preparation step was changed to taping release film to the mandrel surface. Using the release film proved to be a more efficient and trouble-free process. In Figure 3, the release film is being applied onto the mandrel while it is in the cart, but, for recurring part operations, it is preferred to accomplish this step in the fiber placement machine. Using an overhead crane, the 12,000-lb aluminum 4-up production mandrel is installed into the fiber placement machine (Figure 4).

Thirty-two spools of AS4/8852-1 and S2/8552-1 prepreg tow are loaded into the refrigerated creel of the machine and threaded through the delivery head (Figure 5). These 32 tows—each 0.182 in. wide—are delivered to the mandrel as a 5.8-in.-wide band. Tow placement provides the ability to apply both types of material at the same time, which gives the part its desired damage tolerance and damage resistance. The Hercules 8552-1 toughened epoxy resin provides more than 30 days of working life.

After initializing the machine to the mandrel surface, 12 hybrid plies of varying orientations—including three padup plies—are tow placed. As needed, in-process inspections for gapping, tow twisting, angle orientation, and EOP accuracy are accomplished. The tow placement of 45° plies over previously placed 0° plies for four fan cowl door skins is shown in Figure 6.

E-glass/8552 fabric is manually collated over the completed tow-placed plies (Figure 7). Lightning strike protection is then manually applied.

To accomplish a machine compaction operation after the lightning strike protection is applied, the delivery roller applies 250 to 300 lb of force using the machine control program for the spiral 90° plies (Figure 8). The first demonstration skin made had two vacuum compactions. Later, only one vacuum compaction was accomplished—just after the lightning strike protection material was applied. Finally, by using the machine debulk, no vacuum compactions are needed.

The skin operations of roughly trimming the edges and drilling skin locating holes come next (Figure 9). The cut grooves in the mandrel guide the manual trim of the uncured skins.

A drilling fixture locates the holes in the uncured skin. Corresponding holes are designed in the cure mold and caul. Alliant Techsystems has used this method extensively for locating a skin in its cure mold and MDA and Northrop Grumman CAD have found the method to be quite satisfactory in fabricating the fan cowl doors.

The uncured skin removes from the mandrel with the release film underneath (Figure 10). This release film method proved to be more reliable and efficient than release coating the mandrel, which was tried on the initial units. For demonstration skin No. 12, the release film used for demonstration skin No. 11 remained on the mandrel and was used for subsequent tow placement. The number of times the film will remain serviceable before replacing it is currently not determined, but five to ten cycles are considered reasonable with proper care.

After removing the skins from the mandrel, they were transferred to a shipping box for frozen shipment.

Figure 3. Prepare mandrel. Apply release film.

Figure 4. Install mandrel into machine.

Figure 5. Prepare delivery head and load tow in creel.

Figure 6. Tow place 12 plies; in-process inspect as needed.

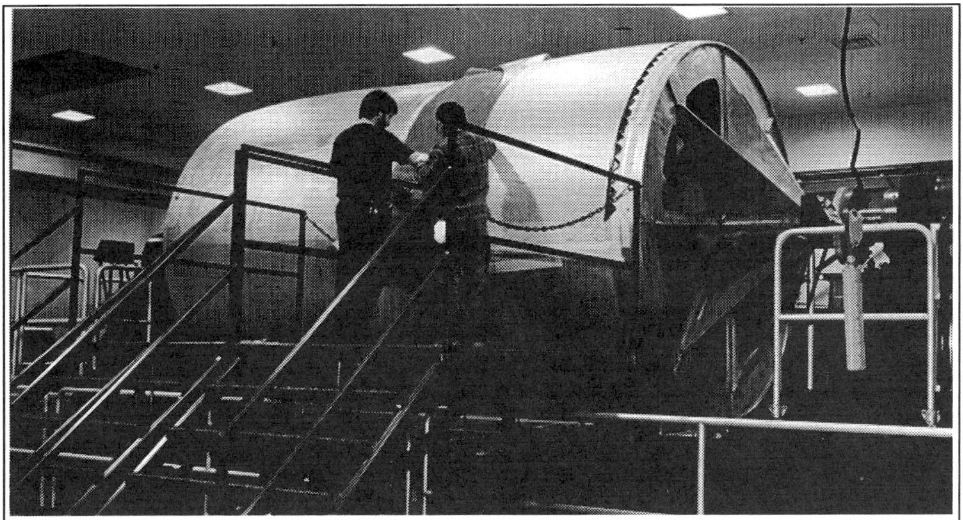

Figure 7. Apply outer E-glass fabric.

Figure 8. Machine debulk.

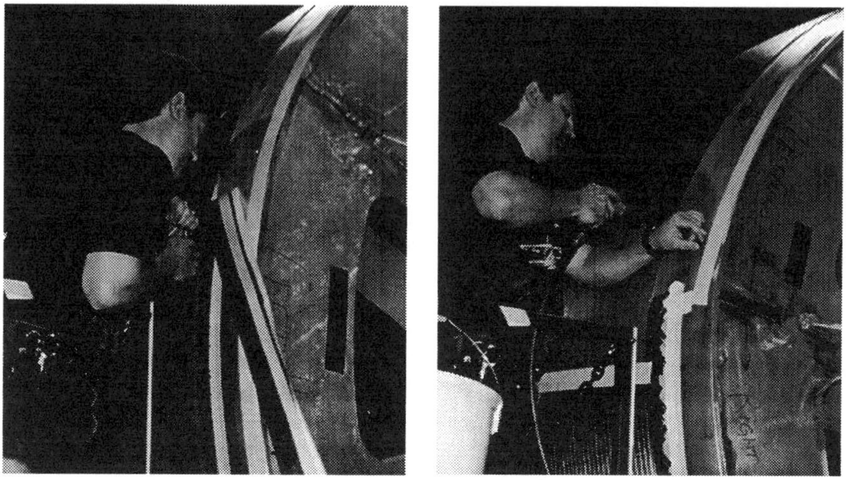

Figure 9. Roughly trim edges and drill skin-locating holes.

Figure 10. Move skin from mandrel and transfer.

Process refinements were made from skin to skin to improve reliability, efficiency, or part quality. The three examples shown in Table 1 are representative of many implemented improvements that occurred along the way. Some refinements (e.g., learning to apply the lightning strike protection without darting) were minor in and of themselves. Others (e.g., eliminating vacuum compactions, improving the mandrel release method) were more significant. Combined together, significant overall improvements were realized.

TABLE 1. PROCESS CHANGES BETWEEN SKIN NOS. 1 AND 9 SHOW PROGRESS

Skin No. 1	Skin Nos. 8 and 9
Mandrel coated with Frekote	Release film applied to mandrel
Darted lightning strike protection	Applied lightning strike protection without using darts
Two vacuum compactions	Machine compact only—no vacuum

FIBER PATHS AND MACHINE PROGRAMMING

In addition to the general process refinements, numerous refinements to the off-line program for the fan cowl doors were accomplished to increase fabrication efficiency or increase part quality. For example, on the first skin, controlled angle programming was employed on all 12 plies. Programming changes were implemented between skins to culminate in the refined programming approaches for each ply. In addition, like-ply staggers and the amount of tow gapping/overlapping of converging tows were adjusted to achieve the desired results. For the 90° padup plies in Figure 11, the 1.1 tow stagger prevents the glass tows from stacking on top of themselves.

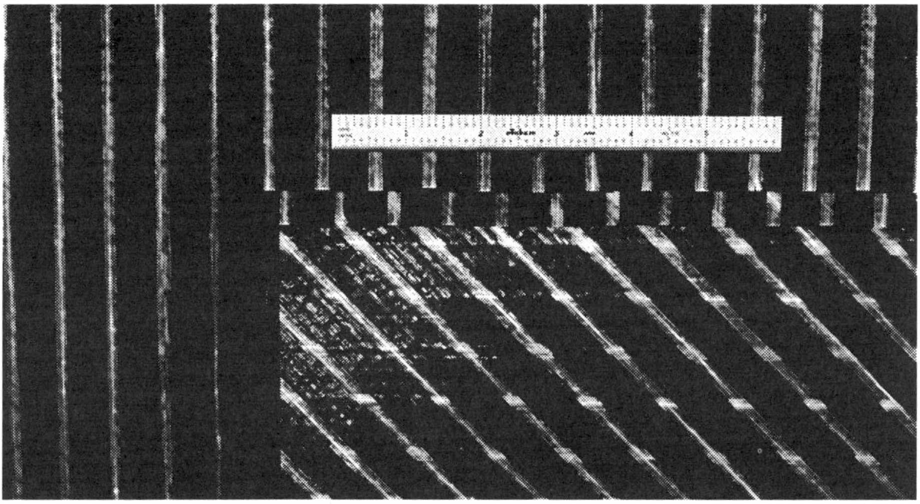

Figure 11. Padup ply stagger of 1.1 tows prevents glass-on-glass stacking.

Some of the machine programming changes are shown in Table 2. Please note that the changes did not occur in one step just before demonstration skin Nos. 8 and 9, but changes were made along the way between skin fabrications. From an efficiency perspective, one of the major programming and machine motion improvements was graduating from 1-up unidirectional laydown to bi-directional 4-up laydown. Bi-directional motion is the type of motion when using a paint brush. In unidirectional motion, the head lifts from the mandrel after laying its course, returns "empty," and then lays down another course in the same direction.

As depicted in Figure 12, and listed in the right column of Table 2, numerous fiber path approaches and orientations were used for the 4-up fan cowl door tow placement. Spiral paths were used for the full 90° plies in the 2-up and 4-up configurations because they provided the most efficient laydown by not having tow cut-and-add or head retractions inside the lateral boundaries. A true 90° guide path was used for the partial (padup) 90° plies, which provides minimal tow drops and adds at the window-side boundaries and places the internal EOPs perpendicular to the laydown direction. Controlled angle programming was employed for the 0° and 45° plies because it maintains precise ply orientation by adding/dropping tows as the circumference changes and it is more efficient than band offset for 2-up and 4-up configurations. In the latch and hinge ends of only the 0° plies, the programming method was switched to band offset to eliminate band-to-band gapping that would occur with converging tows if the controlled angle programming method was used.

TABLE 2. MACHINE PROGRAMMING CHANGES BETWEEN SKINS SHOW PROGRESS

Skin No. 1 (1-Up)	Skin Nos. 8 and 9 (4-Up)
All plies controlled angle	• Band offset partial 90° plies and spiral full 90° plies • Controlled angle 45° plies • Controlled angle 0° plies <u>except</u> band offset for hinge and latch ends
Full-gap/no-overlap control of converging tows	Half-gap/half-overlap control of converging tows
Unidirectional 1-up laydown of all plies	Four-up laydown with bi-directional 0° and 45°s, spiral 90°s, and unidirectional partial 90°s across adjacent two parts
Same orientation plies staggered 1.1 tows	90° plies staggered 1.1 tows, 45° and 0° plies staggered roughly half a band

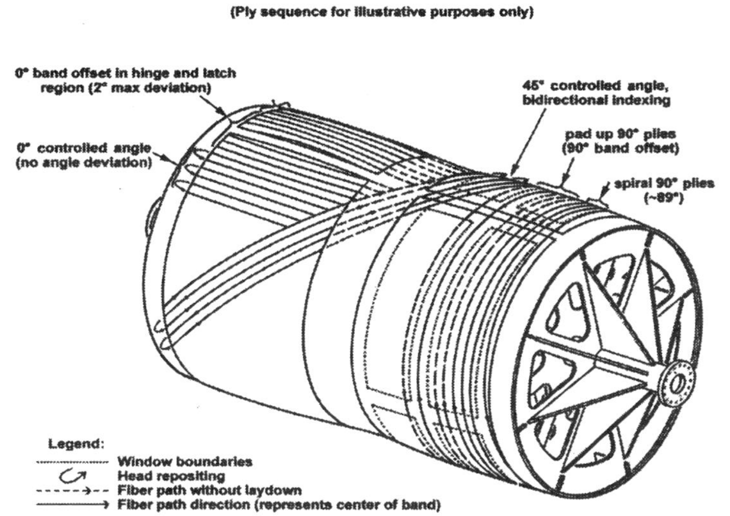

Figure 12. Numerous fiber path approaches used for FCD tow placements

MATERIAL DELIVERY

In MDA and Alliant Techsystems experience, the supplied prepreg tow was among the best material forms used in ATP machines, behaving better than prepreg tow and slit tape formerly used on most other programs. However, the tow placement experience in this program has further identified some needed consistency improvements for the material and delivery system improvements for the ATP machine.

Tow twisting was not uncommon during ATP of the fan cowl door skins. Tow twists are a function of the material type, tow width-to-thickness ratio, machine delivery system, ply orientation, and laydown speed. The S2/8552-1 tows exhibited fewer twists than the AS4 did. The 90° plies exhibited fewer tow twists than occurred on either the 45° or 0° plies, which is related to the redirection of tows in the delivery system of the machine. In July 1996, tow placement of demonstration skin Nos. 10 through 12 demonstrated fewer twists than any other skin previously fabricated. This improvement was attributed to the changed redirect and tow-tensioning system. Because sufficient data to prove that tow twists were not degrading the designed properties did not exist, a conservative approach was initially taken of not allowing any tow twists. With this requirement, tow twists were the limiting factor for laydown speed and were the major cause of rework. Effect-of-defect testing is now being accomplished to define the relationship between the frequency of tow twists and the degradation of mechanical properties.

Infrequent AS4 tow difficulties were encountered with broken tows, strippers (i.e., filaments of tow adhering to and staying with the underlying tow during despooling), and the inability to find the end of a tow on a spool. In general, however, fewer delivery system difficulties (e.g., tows jamming in the head) were experienced with the AS4 than the S2 tow.

Although the majority of the AS4 and S2 prepreg tow spools had good width control, improved consistency is desired. Out-of-width areas result in delivery head difficulties and gapping between the laid tows. Improved width control of latter prepreg tow lots contributed to the fewer gaps observed after demonstration skin No. 4.

Larger spool sizes between 6 to 8.5 lb were tried with satisfactory results at both Alliant Techsystems and MDA. Larger spools can reduce time lost because of spool change-out when much material is being used. For the four-up tow placement of the fan cowl door skins, AS4 spools of eight pounds or more were placed without requiring spool changeout.

Excellent material utilization was realized in the ATP of the fan cowl door skins (Table 3). Excluding planned scrap realized after trimming the cured parts, the planned and unplanned scrap factors are about 5 % or better. Material utilization is a key advantage of ATP.

TABLE 3. ATP DEMONSTRATES EXCELLENT MATERIAL UTILIZATION

	Total Material Used (lb)[a]	**Scrap (lb)**		**Scrap Factor (%)**
		Planned	**Unplanned**	
Skin No. 2	67.13	3.44	0.16	**5.4**
Skin No. 3	65.33	3.11	0.36	**5.3**
Skin No. 4 (2-up)	67.08	2.18	0.68	**4.3**
Skin No. 5	65.51	3.52	0.53	**6.2**
Skin No. 6	65.51	2.81	0.42	**4.9**
Skin No. 7	65.51	1.72	0.64	**3.6**
Skin No. 8 (4-up)	65.25	1.36	0.26	**2.5**
Skin No. 9 (4-up)	65.25	1.36	0.26	**2.5**
a. Discarded nearly empty spools and trade-out spools not included.				

Figure 13 is a MDA chart showing the excellent improvement curve realized through demonstration skin No. 9 in tow placing the fan cowl door skins. Machine downtime includes rethreading tows as necessary but does not include the periodic head cleaning needed after running 65 to 70 pounds of material. In process inspection and tow rework times are included in the charted area for quality. Significant improvements can still be achieved in reducing downtime and in-process inspection time. Likewise, one can see the overall downward trend in tow placement anomalies (Figure 14). It is important to understand that these anomalies or occurrences are not part defects. In addition, please note that the gap inspection requirement was relaxed starting with demonstration skin No. 5, and the use of the mylar templates was reduced for the latter skins (Figures 13 and 14). The favorable trends for fabrication efficiency and observed anomalies continued during the fabrication of demonstration skin Nos. 10, 11, and 12.

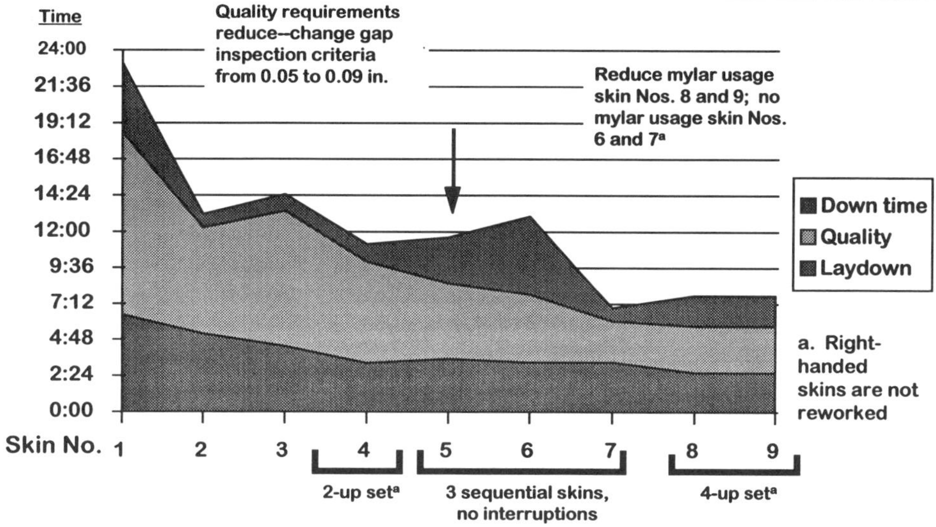

Figure 13. Steep efficiency learning curve realized during initial fabrication of skins.

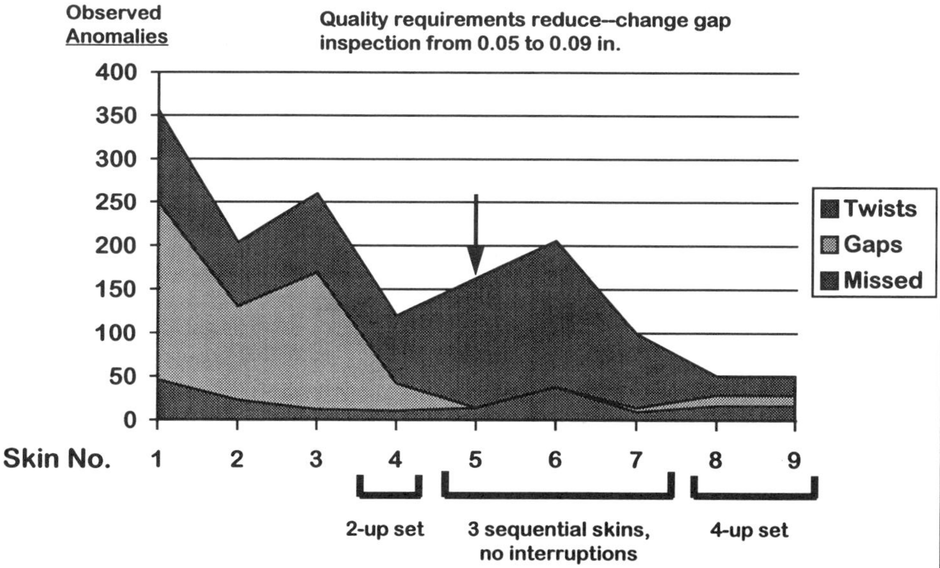

Figure 14. Observed anomalies decrease significantly.

PROCESS IMPROVEMENTS AND LESSONS LEARNED

Besides the process improvements previously addressed, numerous machine programming and machine control improvements were implemented during the course of this program which improved the process efficiency. The head retract distance during repositioning was reduced and program-controlled accelerated motion for head indexing was implemented, thereby saving seconds on each course. Increased cut/add zone feed rates were employed. The time taken to extend and compact the delivery roller on each pass was reduced—again saving seconds on each course. Further shortening of time to extend and compact the roller is still possible. Reducing the desired tow cut and add precision to use more of the drawing tolerance reduced the tow cut and add times on the pad-up plies. The use of the tow tensioning system with a modified redirect apparatus in the delivery system is a key contributing factor for the vastly reduced tow twists observed on demonstration skin Nos. 10, 11, and 12.

Other ATP machine and programming improvements that will increase manufacturing efficiency while maintaining part quality can still be developed and implemented. Implementing tow cut and add on the fly and reducing the time to compact the pay-out roller will knock seconds off each course. Reducing the time required for head cleaning and machine down time will provide significant productivity improvements as will reducing items causing in-process rework. Although development is needed, automating the layup of the glass fabric plies and the copper mesh used for lightening strike protection could provide manufacturing economies. Providing machine control features to let the machine stop and restart during the spiral ply laydown will substantially lessen rework time for those particular plies.

The 4-up tow placement mandrel worked well except an efficiency improvement could be made for a future mandrel design. This mandrel was designed with a forward-to-aft-dead space between parts to accommodate any machine, delivery head size, or type of delivery roller. This flexibility was needed at the time; however this space could be reduced by optimizing the delivery roller compliance of the specific machine used with the mandrel concavity. The forward-to-aft dead space would then be similar to the circumferential dead space, resulting in less wasted machine motion and the elimination of tow cutting and adding between parts.

In-process inspection is another area where significant productivity improvement can be achieved. Improved prepreg tow width control would facilitate the inspection for gaps in addition to reducing rework time. Decreasing the occurrence of tow twists and other anomalies requiring rework will further improve productivity. Furthermore, completing the effect-of-defect testing to quantify the allowable frequency and severity of gaps, laps, and tow twists will likely allow the number of these anomalies to increase; we are currently being overly conservative. In addition, our intended future use of a laser projection system to verify EOP and ply angle accuracy will provide accuracy and efficiency improvements over the mylar templates.

Another separate inspection topic is the incorporation of a laser video in-process inspection system that measures gaps, overlaps, tow twists, and other occurrences. Pratt and Whitney, Alliant Techsystems, and MDA were pleased with the initial feasibility demonstrations of this inspection system employed on the Alliant Techsystems and MDA tow placement machines. Further development is needed to provide a production-worthy system for ATP.

SUMMARY

From experience with the fabrication of these fan cowl doors, the following key elements are recommended in production component processes:

- Apply a release film on the ATP mandrel instead of coating it with a liquid release agent
- Employ 2-up or 4-up skin configuration for high production tow placement
- Use bi-directional laydown for long course lengths
- Replace vacuum compactions with machine debulk using the delivery roller
- Use machine work cell for manual operations if it is available; use the cart for these operations if the machine is not available

In conclusion, the automated tow process is ready to fabricate affordable cowl door structures. Tow placement processes AS4 fiber and S2 glass 8552-1 epoxy prepreg tow concurrently to meet design requirements. The tow placement process has demonstrated that it is capable of high quality production with low material scrap. The 4-up skin scenarios provide efficiencies over the 1-up and 2-up scenarios for this fan cowl door configuration. Further process improvement, especially with regard to reducing down time, can increase productivity while maintaining quality.

ACKNOWLEDGMENTS

The authors gratefully acknowledge the cooperative efforts of several MDA-East personnel who contributed to the fabrication of the fan cowl doors. In addition to the hands-on and engineering efforts to successfully tow place numerous demonstration skins, the collected data and fabrication reports provided by James Thomas were particularly useful.

28th International SAMPE Technical Conference
November 4-7, 1996

THE EFFECT OF HYGROTHERMAL CYCLING ON THE MICROCRACKING BEHAVIOR OF FABRIC LAMINATES

Andreas W. Obst, Mark R. VanLandingham, Rushad F. Eduljee and John W. Gillespie, Jr.
Center for Composite Materials and
the Materials Science Program
University of Delaware
Newark, DE 19716

Gary E. Griesheim and Kenneth F. Tosi
United Technologies Pratt & Whitney
East Hartford, CT 06108

ABSTRACT

Microcracking is usually the first form of damage to manifest itself in composites. Microcracks can propagate other forms of damage under hygrothermal and mechanical cycling. They also form pathways for the accelerated ingress of moisture and chemicals into the materials, resulting in a possible reduction in the residual material properties and hence, a reduced lifetime. This study characterizes the evolution of microcracks under static and fatigue loading in AS4/PR500 five-harness satin weave laminates and the effect of microcracks on the mechanical performance. The laminates were tested in the as-manufactured condition as well as after 16 weeks of hygrothermal cycling. The effect of the presence of microcracks on the mechanical performance was evaluated through extensive mechanical testing at room temperature as well as at elevated temperatures. The presence of microcracks reduces the measured bearing strength as well as the interlaminar shear strength of the composite. The effect of microcracks on the diffusion of moisture into the laminates was also characterized, showing a significant increase in the apparent diffusion coefficient in microcracked specimens.

KEY WORDS: Microcracking, Hygrothermal Cycling, Environmental Durability

1. INTRODUCTION

PR500 epoxy resin reinforced with a carbon fiber (AS4) five-harness satin weave (5HS) fabric is a potential composite material system being investigated for use on the fan exit case (FEC) of the Advanced Ducted Prop (ADP) engine being developed by the United Technologies Pratt & Whitney team through the DARPA-funded Affordable Composites for Propulsion (ACP) program. This work characterizes the microcracking behavior of the PR500/AS4 5HS material under static and fatigue loading, and the effect of the microcracks on the mechanical performance, and on the diffusion of moisture in the material.

Microcracking is usually the first form of damage to manifest itself when composite laminates are subjected to either static or fatigue loads. Microcracks can have a potentially significant impact on the design and processing of the composite structure. A design philosophy that does not tolerate the presence of microcracks may be overly conservative, resulting in a weight-inefficient structure. If the presence of microcracks is to be accepted, then the structural design should be based on allowables determined in the presence of microcracks. Previous work on bismaleimide/carbon fiber laminated systems showed that the bearing strength decreased by 12%, while the open-hole compression strength decreased by 13% in the presence of microcracks [1]. During fatigue cycling, these microcracks can lead to the formation of delaminations.

Microcracks may also initiate and grow in the absence of any mechanical loads, due to environmental effects. Burcham et al [2] has characterized the microcracking in 5260/IM7 laminates due to thermal aging. The absorption of moisture and solvents leads to reductions in the matrix- and interphase-dominated properties and the formation of microcracks [3-7]. Microcracks can form pathways for the accelerated diffusion of moisture, oxygen, solvents and other agents which could lead to an acceleration in the degradation of the material. This could lead to the excitation of other failure modes such as freeze/thaw delamination, skin-core debonding or adhesive failures, resulting in a premature structural failure of the component.

The detrimental effects of hygrothermal cycling on polyimides has been characterized by Cornelia [8]. Characterization of microcracking behavior and the impact of microcracks on mechanical and transport properties is an important step in the evaluation of relatively new composite material systems targeted for use in hostile environments characteristic of structural aircraft engine components.

2. EXPERIMENTAL CHARACTERIZATION

The test program characterized the initiation of microcracking and the increase in crack density during fatigue and static loading. The tests were conducted on the as-manufactured material as well as on material which had undergone sixteen weeks of hygrothermal cycling. The hygrothermal cycle, which is representative of a typical flight cycle, is shown in Figure 1. The microcracking characterization was conducted at room temperature as well as at 121°C (250°F). Bearing strength and interlaminar shear strength were measured to evaluate the impact of microcracking on the mechanical performance of the composites.

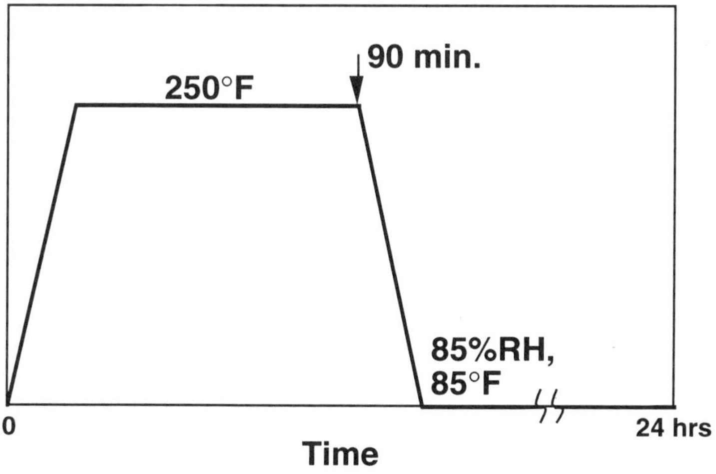

FIGURE 1: The Hygrothermal Cycle Representative of a Typical Flight Cycle.

The test panels were provided by Pratt & Whitney and manufactured using the Resin Transfer Molding (RTM) process. The panels were made from eight plies of AS4 five-harness satin weave fabric, which were laid-up in a mold and infused with PR500 epoxy resin. The panels had a nominal thickness of 3 mm.

2.1 Microcrack Development Under Static Loading The static testing characterized the initiation strain level for microcracking as well as the evolution of microcracking with a monotonically increasing load. The static test specimens had a 0°/90° lay-up with nominal dimensions of 203 mm (L) x 25.4 mm (W) x 3 mm (T) (8in x 1in x 0.12in), and had been end-tabbed. One edge of the specimen was polished to enable the observation of the microcracks under a microscope. The microcracks were counted in each ply of the composite over a 76 mm (3in) gage length, and the crack density was computed as the number of cracks per ply per centimeter of specimen. Each specimen was incrementally loaded to various strain levels, and the strain level at which microcracks first appeared was recorded as the crack initiation strain. Upon further loading, the crack density increased until specimen failure.

The crack density evolution with increasing strain for the as-manufactured specimens and for the hygrothermally cycled specimens is shown for the room-temperature tests in Figure 2, and for the elevated temperature tests in Figure 3. The static test data is also summarized in Table 1. Three specimens were tested for each test condition.

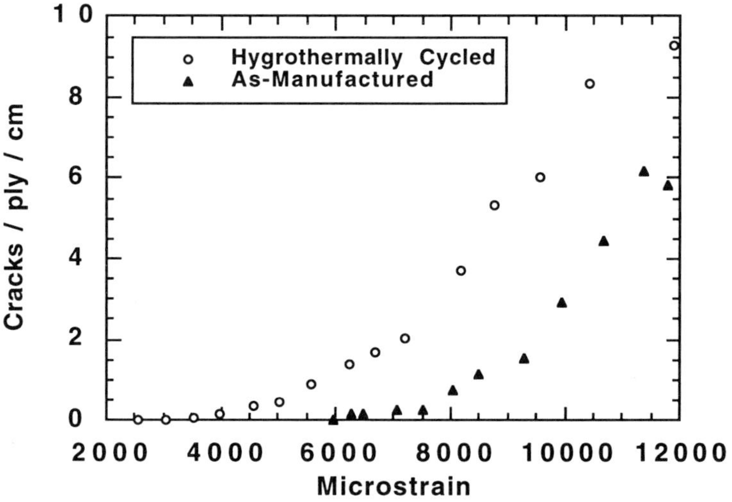

FIGURE 2: Microcrack Development at Room Temperature in AS4/PR500 5HS Material in the As-Manufactured Condition, and after Sixteen Weeks of Hygrothermal Cycling.

TABLE 1. Summary of Static Test Data

Material/Test Condition	Modulus (GPa)	Microcrack Initiation Strain ($\mu\varepsilon$)	Microcrack Density near Failure (cracks/ply/cm)
As-mfg., RT	60.1 ± 0.9	6,000	6.1
Hygro. cycled, RT	61.6 ± 1.5	3,000	9.3
As-mfg., 93°C	60.3 ± 1.0	6,900	1.9
Hygro. cycled, 121°C	60.7 ± 0.8	4,500	2.1

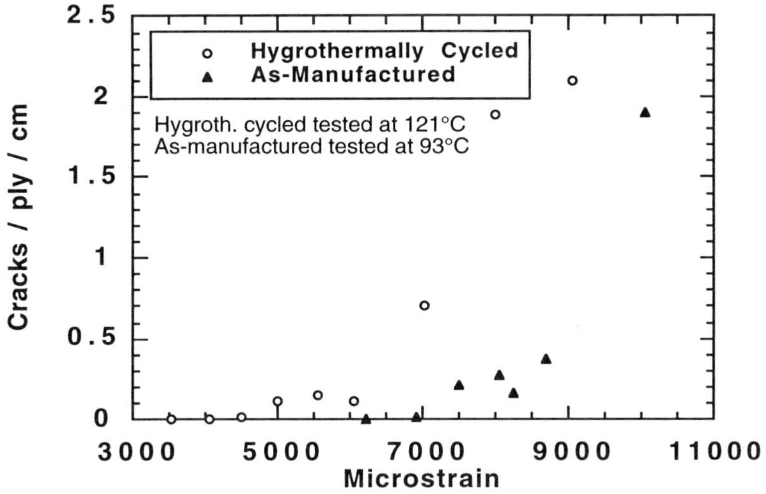

FIGURE 3: Microcrack Development in AS4/PR500 5HS Material in the As-Manufactured Condition Tested at 93°C (200°F), and after Sixteen Weeks of Hygrothermal Cycling Tested at 121°C (250°F).

The stress-strain curves were found to be linear up to failure, indicating a negligible effect of microcracking on the composite modulus. As can be seen from the data in Table 1, the sixteen weeks of hygrothermal cycling produces a significant change in the crack initiation strain both at room temperature as well as at elevated temperatures. A modest increase in the microcrack density just prior to failure is also observed in the hygrothermally cycled specimens. The crack initiation strain was found to increase at the elevated temperature. This may be due to the fact that at the higher temperature, the residual stress in the material is lower and hence can sustain a higher mechanical load before cracking. A similar behavior has also been observed in carbon fiber-reinforced bismaleimide and polyimide cross-ply laminates [9]. The hygrothermal cycling, however, has a negligible effect on the modulus, since it is a fiber-dominated property in the laminates tested.

2.2 Microcrack Development under Fatigue Loading Fatigue tests evaluating the microcracking behavior of the 0°/90° laminates under dynamic load conditions were conducted at room temperature as well as at 121°C (250°F). Based on the modulus measurements from the static tests, the load level corresponding to the desired strain amplitude was determined, and the test was run under load control with an R-ratio of 0.01 and a frequency of 7 Hz. No heating of the sample was observed at this frequency. The specimens tested at room temperature were cycled to 500,000 cycles, while those tested at 121°C were cycled to 100,000 cycles (100,000 cycles represents two times the fatigue life of the component). Microcracks were counted after 10, 100, 1,000, 10,000, 50,000, 100,000 and 500,000 cycles. The plotted data represents an average of three specimens.

Figure 4 compares the microcrack density evolution with the number of fatigue cycles at room temperature at a strain level of 6,000με and 4,000με, in the as-manufactured condition. 6,000με corresponds to the crack initiation strain under static loading, while a value of 4,000με represents the low-cycle potential operating strain, as provided by the ACP team. At 6,000με, the microcrack density increases rapidly with the number of fatigue cycles. After 500,000 cycles, the crack density is greater than 10 cracks/ply/cm., significantly higher than the 6 cracks/ply/cm. prior to failure under static loading. At 4,000με, no microcracking is seen up to 10,000 cycles, after which the crack density increases slowly with the number of fatigue cycles. After 500,000 cycles the crack density was observed to be less than 2.4 cracks/ply/cm.

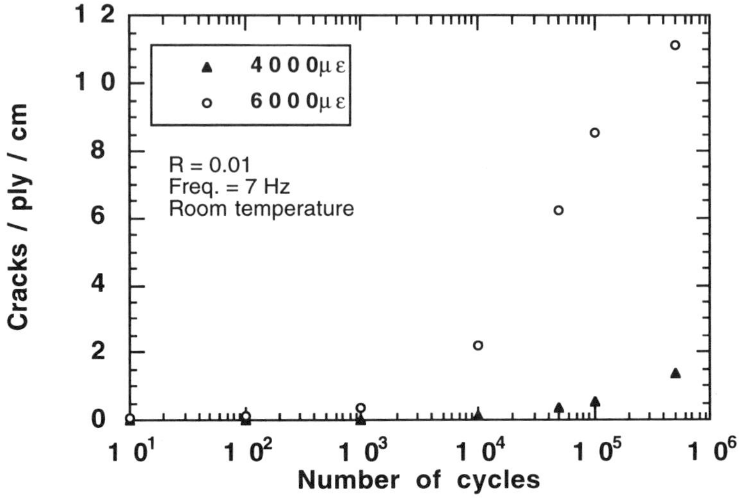

FIGURE 4: Microcrack Development in AS4/PR500 5HS Material During Fatigue at Room Temperature at Strain Levels of 4,000με and 6,000με.

Figure 5 compares the microcrack density evolution at 4,000με at room temperature and at 121°C, for both the as-manufactured and hygrothermally cycled materials. This strain level is higher than the static initiation strain of the hygrothermally cycled material at room temperature and close to the static initiation strain of the hygrothermally cycled material at 121°C. Hence, during fatigue cycling, the crack density of the hygrothermally cycled material increases rapidly. For both the as-manufactured and hygrothermally cycled materials, the crack density is seen to develop faster at 121°C than at room temperature.

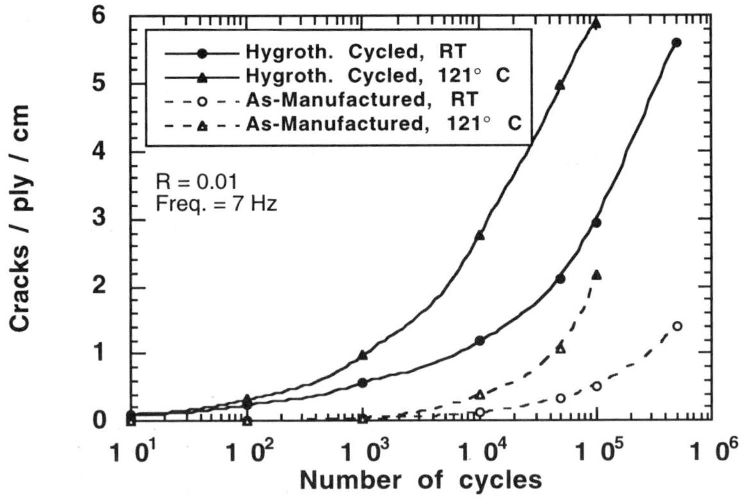

FIGURE 5: Microcrack Density Evolution at 4,000με at Room Temperature and at 121°C, for the As-Manufactured and Hygrothermally Cycled Materials.

2.3 Effect of Microcracking on Mechanical Performance The effect of microcracking on mechanical performance was evaluated by measuring the bearing strength and the interlaminar shear (ILS) strength. Test coupons were loaded statically to a point close to static failure to introduce microcracks. Bearing and ILS test specimens were then cut from these microcracked coupons. Some of the microcracked coupons were also subjected to the sixteen weeks of hygrothermal cycling (following the introduction of the microcracks), and then tested for bearing strength and ILS strength. Tests on the as-manufactured material provided the baseline properties. The ILS tests were conducted at room temperature and 121°C (250°F), while the bearing tests were done at room temperature only.

2.3.1 Bearing Strength The bearing tests were conducted on specimens with a quasi-isotropic layup. The specimens were 50.8 mm (2in) wide and 22.9 cm (9in) long. The specimens were loaded in tension through a 6.35 mm (0.25in) diameter pin placed at a fixed distance 'e' from one end of the specimen. The first drop in the load-displacement curve was defined as the bearing strength. Since it was observed that the as-manufactured specimens failed by shear-out for the recommended end-distance to pin diameter ratio, e/d = 2.5, this e/d ratio was increased to 6 to ensure a bearing failure. The results of the bearing test are shown in Figure 6. Although there is some scatter in the data, the bearing strength decreases by 6–8% from the as-manufactured baseline.

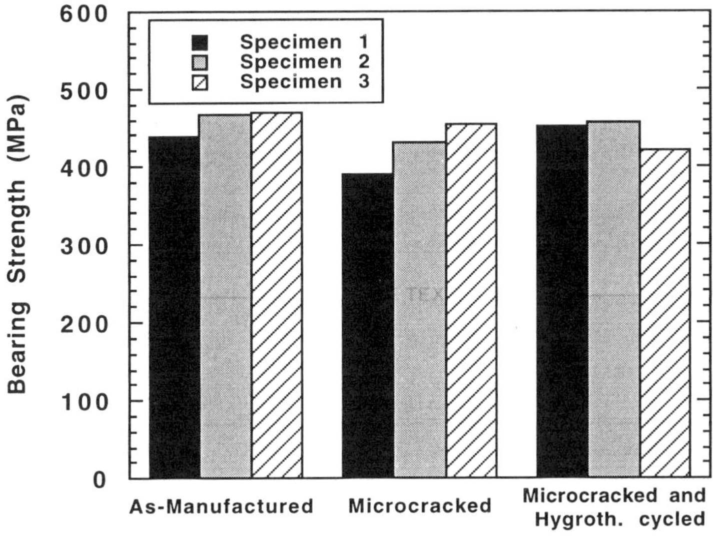

FIGURE 6: Bearing Strengths for Material in the As-manufactured, Microcracked, and Microcracked Followed by Hygrothermal Cycling Conditions.

2.3.2 Interlaminar Shear Strength The ILS strength was measured by the short-beam four-point bend test. The tests were conducted at room temperature as well as at 121°C. Due to the unavailability of suitable fixtures, this ILS testing was conducted at the DuPont Pencader facilities. The measured ILS strengths are shown in Table 2. At room temperature, the presence of microcracks has a negligible effect on the ILS strength. However, hygrothermally cycling a microcracked specimen results in a 25% decrease in the ILS strength compared to the as-manufactured baseline. On the other hand, hygrothermally cycling the uncracked material results in only a marginal decrease in the ILS strength for room-temperature data. This underscores the additional degradation due to a combination of hygrothermal cycling and microcracking. At 121°C, there is a 10% reduction in the ILS strength compared to the as-manufactured baseline at 121°C. After hygrothermally cycling the microcracked material, the ILS strength shows a 15% reduction compared to the as-manufactured baseline at 121°C. Comparison of this strength reduction with the marginal decrease in ILS strength after hygrothermally cycling the uncracked material emphasizes the detrimental interaction between microcracks and hygrothermal cycling.

TABLE 2. ILS Strengths (MPa) at Room Temperature and 121°C (250°F).

Material	Room Temperature	121°C
As-manufactured	65.29 ± 1.59	50.06 ± 0.62
Microcracked	64.67 ± 2.21	44.95 ± 0.55
Hygrothermally cycled [10]	63.40	53.78
Microcracked, then hygro. cycled	48.81 ± 0.76	42.33 ± 0.55

2.4 Effect of Microcracking on Moisture Diffusion Besides their impact on the mechanical performance, microcracks can also form pathways for the accelerated diffusion of moisture, which could lead to the excitation of alternate failure mechanisms. The effect of microcracking on diffusion was evaluated by exposing samples to a 70% relative humidity environment at 40°C and tracking the percentage change in the weight of the specimens. As-manufactured, statically cracked (having a crack density of 6 cracks/ply/cm) and fatigue cracked (having a crack density of 10 cracks/ply/cm) specimens were exposed. The average % change in weight of three specimens per condition is shown in Figure 7. The microcracked specimens exhibit an accelerated diffusion of moisture compared to the as-manufactured, uncracked specimens. The cracked specimens also exhibit a higher moisture saturation level. This could be partly due to macroscopic water droplets trapped within the microcracks. Based on the initial slope of the moisture uptake curve and the saturation level, the apparent diffusion coefficient was computed for the three cases and is shown in Table 3. As can be seen from this table, the diffusion coefficients increase rapidly with increasing crack densities.

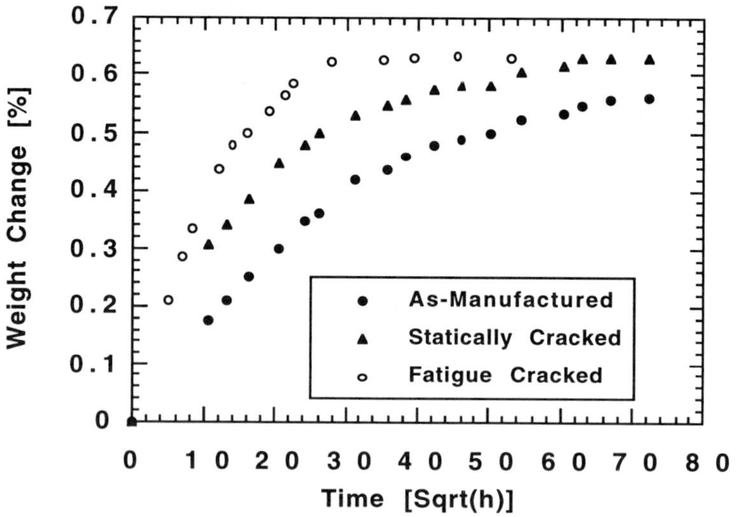

FIGURE 7: Moisture Diffusion Curves for As-manufactured, Statically Cracked (6 cracks/ply/cm) and Fatigue Cracked (10 cracks/ply/cm) Specimens in 40°C, 75% Relative Humidity Environment.

TABLE 3. Effect of Microcrack Density on the Moisture Saturation Levels and on the Diffusion Coefficients.

	As-manufactured	Statically cracked	Fatigue Cracked
Moisture saturation level	0.57%	0.65%	0.63%
Diffusion Coefficient (cm^2/s)	3.88×10^{-9}	8.76×10^{-9}	20.05×10^{-9}

3. CONCLUSIONS

This effort characterized the microcracking behavior of the PR500/AS4 5HS material under static and fatigue loading as well as the effects that the microcracks had on the mechanical performance and on the diffusion of moisture in the material. The microcrack initiation strain under static loading was found to be significantly reduced after sixteen weeks of hygrothermal cycling. The elastic modulus, however, was unaffected by the presence of microcracks or by the hygrothermal cycling. Increasing the test temperature increased the initiation strain.

Upon fatigue cycling at a strain amplitude equal to the crack initiation strain at room temperature, the crack density increased rapidly, reaching 10 cracks/ply/cm after 500,000 cycles (compared to 6 cracks/ply/cm obtained prior to failure during static loading). At the low-cycle fatigue potential operating strain of 4000με, the hygrothermally cycled material exhibits a faster increase in the crack density at room temperature and at 121°C, compared to the as-manufactured material.

The impact of microcracks on the mechanical properties was evaluated by measuring the ILS strength and the bearing strength. A microcrack density of 6 cracks/ply/cm resulted in a decrease of 6–8% in the bearing strength compared to the as-manufactured baseline. The synergistic effect between microcracking and hygrothermal cycling is evident from the measured ILS strengths. Hygrothermal cycling of an uncracked specimen has a negligible effect on the ILS strengths, both at room temperature and at 121°C. However, hygrothermal cycling of a microcracked specimen reduces the ILS strength by 25% at room temperature (from the as-manufactured baseline at room temperature) and by 15% at 121°C (from the as-manufactured baseline at 121°C).

The strong impact of microcracking on the diffusion behavior is demonstrated by the measurement of the moisture uptake on the as-manufactured (uncracked) material and on material having various microcrack densities. The microcracked specimens absorbed more moisture and exhibited diffusion coefficients over five times higher than the uncracked, as-manufactured material.

Thus, the criticality of microcracks on the durability and life of the material is two-fold. The structural issues relate to the reduction in the mechanical properties due to the presence of microcracks. A design philosophy that accepts the presence of microcracks will require assessment of the impact of microcracks on the design allowables. The environmental issues revolve around the increased susceptibility of the material to environmental attack in the presence of microcracks. Accelerated diffusion of moisture, oxygen and other solvents through the pathways formed by the microcracks could excite other failure mechanisms, resulting in a premature structural failure of the component.

4. ACKNOWLEDGMENTS

The authors would like thank R. H. Cornelia, DuPont Chemical Company, for conducting the ILS strength tests and for many useful discussions in interpreting the test data.

5. REFERENCES

1. J. W. Gillespie, Jr. and R. F. Eduljee, "Long-Term Behavior of High-Performance Composites for HSCT Applications," presented at the Boeing Commercial Airplane Company, Seattle, WA, September 1993.

2. L. J. Burcham, R. F. Eduljee and J. W. Gillespie, Jr., Polymer Composites, 16, 507 (1995).

3. C. E. Browning, G. E. Husman and J. M. Whitney, Composite Materials: Testing and Design, ASTM STP 617, American Society for Testing and Materials, Philadelphia,1977, pp.418-496.

4. S. F. Jackson and Y. Weitsman, International Journal of Solids and Structures, 23, 131 (1987).

5. A. C. Loos and G. S. Springer, Journal of Composite Materials, 13, 131 (1979).

6. G. S. Springer, Composite Materials: Testing and Design, ASTM STP 674, American Society for Testing and Materials, Philadelphia,1979, pp.291-312.

7. J. M. Whitney and C. E. Browning in J. R. Vinson, ed., Advanced Composite Materials – Environmental Effects, ASTM STP 658, American Society for Testing and Materials, Philadelphia,1978, pp.43-61.

8. R. H. Cornelia, SAMPE International Symposium, 39, 917 (1994).

9. X. Huang, R. F. Eduljee and J. W. Gillespie, Jr., "Effect of Temperature on the Transverse Cracking Behavior of Cross-Ply Composite Laminates," Composites Engineering, accepted for publication (1996).

10. R. H. Cornelia, "Long Term Hygrothermal Durability of Composites," to be published as a deliverable for ACP and NASA programs (1996).

28th International SAMPE Technical Conference
November 4-7, 1996

Alternative Ribbon Cross-Sections for Automated Fiber Placement

R.W. Grenoble and J.M. Marchello
Old Dominion University

H.L. Belvin and T.W. Towell
NASA Langley Research Center

ABSTRACT

Widespread use of automated fiber placement (AFP) with in-situ consolidation in a manufacturing environment will require high material placement rates and accurate, repeatable material positioning. As part of ongoing composite fabrication research at NASA Langley Research Center, ribbons of alternative cross-sectional architecture are being investigated as a means to increase the potential productivity of AFP while improving ribbon alignment during placement. Use of shaped ribbon will close placement laps and gaps.

Current AFP practice utilizes unidirectional ribbons or slit tape typically 6.35 mm (0.250 in) wide by 0.13 mm (0.005 in) thick with a rectilinear cross-sectional shape. In the present study the NASA Langley Ribbonizer hardware and ribbon processing conditions were modified to permit manufacture of trapezoidal cross-section ribbon.

Experimental quantities of 0.66 mm. (0.026 in) thick by 6.19 mm. (0.244 in) wide composite ribbon with a trapezoidal cross-sectional shape were fabricated. The ribbon was found to have low dimensional variation and void content. Several panels were fabricated from the ribbon using the NASA Langley Robotic Fiber Placement Facility. The short beam shear testing of a panel placed at the most favorable processing condition indicated that interlaminar strength was comparable to that obtainable with conventional ribbon. Placement of thick-section ribbon was shown to be sensitive to overheating of the ribbon with post-compaction deconsolidation resulting. Void expansion was controlled with reductions in compaction roller temperature. A method of ribbon preheating, together with nip point heating, will be required to fully exploit the nesting advantages of shaped ribbon.

KEYWORDS: Automation, Composite, Customized Ribbon

The use of trademarks or names of manufacturers in this report is for accurate reporting and does not constitute an endorsement, either expressed or implied, of such products or manufacturers by NASA.

1. INTRODUCTION

Fiber placement is envisioned as a high accuracy, labor saving method of composite fabrication. Automated placement of plies with complex geometry can be carried out with a minimum of human intervention through the use of numerical controlled machine platforms. The cost of a fiber placement machine is high and high productivity rates are therefore necessary to justify its expense. Coffenberry et al. showed that high placement rates are possible when thick and/or wide tapes are placed [1].

A potential advantage of AFP over hand layup is high accuracy and repeatability of the machine motion. However the overall accuracy of the AFP process is hindered by the lack of precise ribbon position control during placement, particularly at the start of placement. On the NASA tow placement robot [2] a band of five 6.35 mm (0.250 in) wide ribbons is held in alignment by guide slots as they are directed over the feed rollers, through a cutter blade, and toward the nip point. Between the guide slots and nip point, however, little ribbon guiding takes place. As a consequence, considerable misalignment of ribbons can occur. Further complicating ribbon alignment are variations in individual ribbon widths.

Lack of full positional control causes laps and gaps among the ribbons and between adjacent ribbon bands. This in turn lowers laminate consolidation quality. Ribbon lapping causes local variations in the height of the part surface. Incomplete consolidation results in areas around lapped ribbons as the raised surface of the overlap bears a disproportionate share of the compaction load. Gaps between adjacent ribbons must be filled during placement of subsequent plies, a difficult task especially when fiber angle changes between plies.

Sandusky and Marchello have suggested the use of contact forces between inclined edges of adjacent ribbons to encourage nesting of the ribbons and so prevent lap/gap problems [3-5]. Ribbon shaped for nesting can be produced either by quenching a molten tow into the desired shape or by slitting broad tape with inclined cutters in a secondary process. This work is an attempt to manufacture experimental quantities of ribbon, specially shaped for AFP, using a profiled open roller die and to evaluate its effect on placement.

2. RIBBON FABRICATION

2.1 Constituent Materials

PIXA high temperature thermoplastic polyimide (Mitsui Toatsu), powder-coated onto unsized (12k) IM-7 carbon fiber (Hercules), was obtained from CYTEC as the ribbon precursor. Each spool of towpreg contained two co-wound ends, that is, two 12k ends wound together on the same spool. Four spools, eight ends total, were used to form the ribbon.

2.2 Ribbon Shape

Generation of aligning forces among ribbons requires that the ribbon edges be oblique to the substrate on which the ribbons are being laid. For this work a trapezoidal ribbon cross-sectional shape was chosen with a 45° inside angle on the large base. The trapezoidal shape required only minor modifications to the NASA LaRC ribbonizer line and NASA Langley's fiber placement robot for its manufacture and placement.

Conventional ribbon for AFP is approximately 6.35 mm (0.205 in) wide and approximately 0.127 mm (0.005 in) thick. Fabrication of such a ribbon requires two powder-coated 12k-IM7 tows. The nesting of inclined surfaces on adjacent ribbons requires the horizontally projected width of the inclined surface to be of approximately the same magnitude as the required lap/gap closing. For this work, the ribbon centerlines were assumed to vary by ± 0.254 mm (0.010 in). Since the ribbon is formed by discrete 12,000 filament tow bundles, the ribbon cross-sectional area will be an integer multiple of that of the towpreg precursor. Based on the above information, it was determined that eight ends of 12k powder-coated IM7 towpreg, 96k in all, would form a ribbon of the required cross-sectional geometry. The resulting ribbon shape is shown in Figure 1.

2.3 Ribbonizer Equipment and Hardware Modifications

The NASA LaRC ribbonizer line [6-8] consists of a supply creel, two electromagnetic tow tensioners, a three-zone tube furnace, a set of air cooled, modular, tongue-in-groove nip rollers, and pulling/takeup system. A single tensioner was adapted to tension all eight ends entering the system. Based on previous experience manufacturing conventional ribbon of similar materials, the oven temperature profile was set at 520/490/480°C for the exit, middle and entrance zones respectively. Oven temperature profile was held constant through all but the last trial run.

Required modifications of the current ribbonization process were determined with trial runs. Initial trials entailed fabrication of thick rectangular ribbon with the existing equipment used in conventional ribbon manufacture. Ribbon was fabricated with 84k total fiber content, short of the 96k needed for the trapezoidal ribbon. Trial run experience showed that a method of decreasing the tow band width was needed upstream of the nip roller. Narrowing of the tow band was accomplished by installing a set of vertical bars inside the oven approximately two inches behind the impregnation bars, Figure 2. The bars form a vertical S-wrap which causes the tows on the edges of the tow band to fold over onto the center, forming a thicker, narrower band. The bars were made of fire polished quartz secured in parallel by a steel pinch clamp. The clamp was attached to the impregnation bar mounting fixture by a single cap screw to permit adjustment of the tow band width.

A second important hardware modification was installation of a trapezoidal profile nip roller center ring in place of the one used to form conventional ribbon. Nip roller center rings were fabricated to form a trapezoidal ribbon 0.635 mm (0.025 in) thick by 6.35 cm (0.250 in) wide and were installed in place of the conventional center rings.

2.4 Ribbonization Trials

2.4.1 *Trapezoidal Ribbon Trials*

Several 96k ribbon trial runs were made using oven settings and line speed identical to those used to make 84k rectangular ribbon. Samples were taken from along the length of each trial ribbon and potted in epoxy. Optical microscopy was used to determine ribbon void content.

Processing thick ribbon was found to be much more difficult than conventional ribbon manufacture due to severe heat transfer requirements in the nip rollers. Conventional ribbon is effectively quenched in the nip rollers at ambient temperature. However, the trapezoidal ribbon, due to its fourfold increase in bulk, had a far lower surface area to volume ratio and required a much greater heat flux to quench. During ribbonizing trials line speed and roller temperature were varied to optimize fiber wetting and ribbon quenching.

Line speeds of 4.57, 7.63 and 12.2 m/min (15, 20 and 25 ft/min) were run. Initial trials began with room temperature, air-cooled rollers. Post-nip deconsolidation of the resulting ribbons was obvious in micrographs, Figure 3.

Ribbon was manufactured with chilled rollers in an attempt to reduce deconsolidation. Dry bottled nitrogen was cooled through a copper coil in a slurry of dry ice in acetone and plumbed into the nip roller air lines. Improvements were seen in cross-sectional shape and void content was lower. However, post-nip void expansion was still apparent, especially in the ribbon center where agglomeration of neighboring voids can be seen, Figure 4.

Cooling of the rollers with dry nitrogen chilled in a liquid nitrogen bath yielded ribbon with greatly improved cross-sectional shape with the exception of incomplete flow into one corner of the roller die. Significant reductions in void content were realized due to reduced void expansion. As a final process parameter adjustment the oven temperature profile was lowered to 480/490/480°C in an attempt to reduce the post-nip ribbon temperature. Figure 5 shows ribbons made with the lowered oven temperature profile.

2.4.2 *Large Batch Processing*

The ribbonizer line was subsequently reconfigured to process sufficient quantities of trapezoidal ribbon to permit placement studies. The oven temperature profile was set at 480/490/480°C and allowed to stabilize. Roller cooling was configured as in the last trial with the addition of post-nip cooling to assure complete ribbon quenching. A copper tube was mounted behind the nip rollers such that a stream of cold nitrogen gas impinged on the ribbon just as it left the nip. Approximately 61 m (200 ft) of trapezoidal ribbon was made with approximately equal lengths made at line speeds of 7.63 m/min (20 ft/min) and 12.2 m/min (25 ft/min).

2.5 Ribbon Characterization

2.5.1 *Dimensional Variation*

The ribbon made in the large batch was examined for dimensional variations. Six measurements of ribbon width and thickness were taken at five approximately equally spaced locations along the ribbon length for a total of 30 samples. Mean width was 6.188 mm (0.2436 in) with a standard deviation of 0.054 mm (0.0021 in). The slight narrowness of the ribbon was caused by incomplete filling of the sharp corners of the roller die. Mean thickness was found to be 0.658 mm (0.0259 in) and its standard deviation was 0.002 mm (0.0009 in).

2.5.2 *Microstructure and Void Content*

Two specimens were taken from each of the aforementioned five locations. Void content was determined by acid digestion. The mean void content was determined to be 1.6% with a standard deviation of 2.9%. Of the ten samples taken, the two from ribbon made at 7.63 m/min (20 ft/min) showed void contents of 7.7% and 6.2% by volume. The void content of all other samples together averaged 0.22% with a standard deviation of 0.4%. The first two samples were taken from ribbon formed early in the process when thermal conditions had not yet equilibrated. The reduction in void content of the latter samples can be attributed to stabilization of the process.

Several cross-sections of ribbons formed at both 7.63 and 12.2 m/min (20 and 25 ft/min) were potted and polished for microscopy. The cross-sections are shown in Figure 6 with the upper two ribbons formed at 7.63 m/min (20 ft/min) and the lower one at 12.2 m/min (25 ft/min). Resin-fiber distribution for all ribbons appears uniform with the exception of some resin rich areas between adjacent tows. Resin rich tow boundaries are expected as the resin distribution of the constituent towpreg is higher near the tow outer surface.

2.5.3 *Curvature*

Despooling of the ribbon for testing revealed that considerable lengths of the ribbon exhibited visibly discernible curvature. The in-plane and out-of-plane curvatures of the ribbon were measured at five sample points along the ribbon length. In-plane (flatwise) radius of curvature varied between 213 cm and 762 cm (84 and 300 in) with an average of 597 cm (231 in). Out-of-plane radius of curvature varied from 86.4 cm to 223 cm (34 to 88 in), averaging 120 cm. (47.2 in).

Two probable causes of ribbon curvature were identified. Given the severe heat transfer requirement experienced during processing it is suspected that the center of the ribbon was still above T_g after exiting the nip rollers. Laboratory space limitations required the nip rollers to be less than 10 cm (4 in) away from a pair of vertical takeup alignment rollers. This close proximity caused the ribbon to twist through a 90° angle shortly after exiting the nip. Relaxation of residual stresses and cooling of the ribbon in this twisted position may have contributed to the observed curvature. Differential tow tension can also cause curvature by creating an asymmetric state of residual stress in the ribbon. The use of a single tensioner permitted some tows to pass over one another during processing which isolated the lower tow from the tensioning roller and caused the other to carry a greater tensile load. Additionally, differential tension in the co-wound tows could be observed as the towpreg unwound from the spool. All of the aforementioned causes of ribbon curvature will need to be addressed if consistently straight ribbon is to be made.

2.6 Summary of Ribbonization Efforts

Several trial runs were performed and processing conditions were identified to manufacture low void content trapezoidal cross-section ribbon from PIXA/IM7 towpreg. Approximately 61 m (200 ft) of continuous ribbon was made. Low void content and uniform fiber distribution were obtained for the ribbon fabricated under steady-state processing conditions. Ribbon width was slightly but consistently narrower than expected due to the difficulty of squeezing the cooling material into the sharp corner of the roller die that forms the ribbon edge. The greater bulk and low aspect ratio of trapezoidal ribbon dramatically increased roller cooling requirements compared to conventional ribbon. The ribbon exhibited varying amounts of curvature along its length which is primarily attributable to too short distances between the nip and takeup rollers and differential tension among the constituent tows.

3. TRAPEZOIDAL RIBBON PLACEMENT

The goal of the placement phase of this work was to evaluate the effectiveness of the cross-sectional shape in encouraging ribbon alignment and closing laps and gaps. The trapezoidal ribbon made in the ribbon manufacture phase was used in the automated placement study.

3.1 Ideal Placement Process

Exploitation of ribbon contact forces requires that the ribbons maintain their overall shape and elasticity during placement. The ideal ribbon heating method will therefore heat the ribbon surface to a temperature far above the matrix T_g to permit establishment of high interfacial bond strengths in very short times (~0.1 sec) while keeping the bulk of the ribbon near T_g. Since the ribbons are poor transverse thermal conductors, establishment of a high thermal gradient at the ribbon surface just prior to consolidation should be achievable with an appropriately designed heating method.

3.2 Placement Machine and Process Constraints

Primary process parameters for NASA Langley Fiber Placement Facility are placement speed, compaction force, hot gas torch (HGT) temperature, heated compaction roller temperature and heated tool temperature. The HGT's direct a stream of hot gas into the gap formed by the ribbon band and the panel [2]. It was shown in reference 2 that convective heating rate, as measured near the surface of the ribbon facing away from the gas stream, decreases as the ribbons approach the panel surface due to stagnation of the gas flow in the nip point. Appreciable conductive heating of the ribbon bulk due to heated roller contact was demonstrated as well. Convective ribbon heating takes place over several centimeters of ribbon length. Additionally, heat transfer from the roller preferentially heats the ribbon top and center. The current heating method consequently results in unavoidable bulk heating of the ribbon well above the resin T_g. This is not the preferred thermal condition for placement of this material form. The non-ideal nature of the heat source complicated the placement.

The fiber placement robot is configured for use with conventional 6.35 mm (0.250 in) wide ribbon. As a result, several additional equipment difficulties were encountered in placement of the trapezoidal ribbon. Testing showed that the cutter was incapable of shearing the thick trapezoidal ribbon. In the present study the need to cut the ribbon at the end of a ply was avoided by placing ribbons shorter than expected by the machine's trajectory file such that actuation of the cutter blade occurred just after passage of the end of the ribbon.

The greater thickness of the trapezoidal ribbon necessitated installation of shims to increase the depth of the ribbon guide grooves and offset the position of the ribbons in their guides to reduce ribbon bending around the feed rollers. Ribbon curvature previously mentioned caused additional drag between the ribbons and their guide chutes.

3.3 Ribbon Alignment

At the start of ply placement the ribbons are driven down toward the nip by the ribbon feed motor. Since the ribbons are cantilevered from the guide chute to the nip, laps and gaps are most evident at the start of placement. After establishment of the initial bond of a band of ribbons to the part, ribbon alignment is controlled by inter-ribbon contact, ribbon tension (less than 4.5 N (1 lbf)) and by contact with the ribbon guides. In this work short ribbon sections were used. As the placement of each ribbon section progressed, a decreasing length of ribbon was engaged in the ribbon guides which quickly reduced the guiding effect of the ribbon guides and the ribbon tension caused by ribbon/guide contact. Ribbon guides and ribbon tension therefore had little influence on ribbon position as placement of a ply progressed. Ribbon alignment in the final 20 cm (8 in) of a ply was controlled primarily by ribbon lateral stiffness and ribbon contact forces.

3.4 Panel Description

The panels were formed of either two, three or five unidirectional plies. Each ply was offset from the one below by one half ribbon width in order to stagger the ribbon edges through the panel thickness. All panels were formed of plies five ribbons wide. Adjacent ribbons within each ply were oriented in an alternating up/down arrangement to allow transverse ribbon nesting to eliminate intraband laps and gaps. Each ply consisted of ribbons from each of the sample locations along the length of the ribbon made in the final large batch.

3.5 Panel Placement

The trapezoidal ribbons were robotically placed onto a heated flat tool in the NASA Langley Robotic Fiber Placement Facility. The HGT temperature and tool temperature was held constant during placement of all panels. As a limited amount of material was available, process optimization was restricted primarily to varying the heated roller temperature. The first ply of all panels was placed onto a starter ply of APC-2 (Fiberite) prepreg with a fiber angle oriented at 45° to the placement direction. The starter ply was coated with a thin layer of mold release to permit its removal from the completed panel.

Minimum roller temperature was determined by placing single plies onto a well consolidated substrate and then trying to peel the ribbons from the substrate by gripping the ribbon ends with a pair of pliers and pulling upward at a 45° angle in the fiber direction. Ribbons deemed successfully placed fail cohesively. It was found that at roller temperatures below 320°C, regardless of placement speed and hot gas torch temperature, ribbons could not be bonded to a composite substrate with sufficient interfacial strength to prevent their being peeled.

Compaction force and placement speed were altered in two panels in attempts to maintain ribbon cross-sectional shape during placement. Compaction loads of 334 N (75 lbf) and 890 N (200 lbf) were applied with a 4.45 cm (1.75 in) diameter heated roller. Roller temperatures of 320°C, 360°C, and 400°C were subsequently selected for the placement study. Panel void content was determined by optical image analysis of panel cross-sections. A summary of the processing conditions and the resulting panel void content is given in Table 1.

Table 1: Panel Processing Conditions[a] and Resulting Void Content

Panel Number	Heated Roller Temperature °C (°F)	Placement Speed cm/sec (in/sec)	Compaction Force N (lbf)	Void Content[b] (area percent)
1	400 (752)	2.54 (1)	334 (75)	4.4
2	400 (752)	2.54 (1)	890 (200)	1.4
3	360 (680)	2.54 (1)	890 (200)	6.0
4	320 (608)	2.54 (1)	890 (200)	4.4
5	320 (608)	1.69 (0.67)	890 (200)	7.2
6[c]	400 (752)	2.54 (1)	890 (200)	0.8

[a] HGT temperature: 900°C, tool temperature: 120°C
[b] Determined by optical image analysis
[c] Two roller passes made on top ply.

Panel 1 was fabricated from three plies at 2.54 cm/sec (1 in/sec) with a roller temperature of 400°C (752°F) and 334 N (75 lbf) of compaction force. Figure 7 shows a typical cross-section. Void content was approximately 4.4% by area. Voids in the ribbons appear mainly between adjacent tows and are probably due to expansion of pre-existing ribbon voids. Interply voids most likely resulted from insufficient compaction pressure to fully establish intimate interfacial contact. Voids are concentrated in the top ply, especially near the upper surface. The large void size in the top ply and the unevenness of the panel upper surface suggest that considerable void expansion occurred in the unrestrained upper ply after roller passage. Post-compaction void expansion suggests that ribbon bulk temperatures, as mentioned previously, were well above T_g after consolidation. Adhesion of broken fibers and resin to the compaction roller was observed during placement, necessitating occasional cleaning of the compaction roller. Roller sticking contributed to roughening of the panel surface and complicated void removal and establishment of intimate contact during placement of subsequent plies.

Panel 2 was fabricated of five plies with a 400°C (752°F) roller at 2.54 cm/sec (1 in/sec) with the compaction force increased to 890 N (200 lbf) in an attempt to remove interfacial voids. Greatly reduced void content was apparent throughout the laminate with the exception of the upper ply, Figure 8. Areal void content averaged 1.4 %. The upper surface of the laminate was roughened by roller sticking and void expansion. Interfacial voids were nearly eliminated with the increased pressure. Since subsequent laminates were to be placed at equal or lower roller temperature, all remaining panels were processed with 890 N (200 lbf) of compaction load.

The third panel was formed of two plies placed at 2.54 cm/sec (1 in/sec) with a roller temperature of 360°C (752°F) and 890 N (200 lbf) of compaction force. The smooth surface of the top ply indicates that sticking of the fibers to the compaction roller was significantly reduced by the lower roller temperature, Figure 9. Note that voids are again concentrated in the upper ply, indicating that post-compaction void expansion occurred. Void content increased to 6.0 area percent. Note that the discernability of the ply interfaces and the increase in void content suggests a reduction of squeeze-flow under the nip roller due to the lower roller temperature. Decreased flow is thought to have impeded removal of existing ribbon voids resulting in higher void content. Further reductions in roller temperature were deemed necessary to control post-compaction deconsolidation.

The roller temperature was reduced further by 40°C to 320°C in Panel 4. Placement speed was 2.54 cm/sec (1 in/sec) with 334 N (200 lbf) of compaction force. As seen in Figure 10, ply interfaces are more sharply defined than those in the previous panels. The lower roller temperature improved retention of ribbon cross-sectional shape during placement. In fact, several intact individual ribbons could be seen in the panel cross-sections. Void content averaged 4.4 percent by area. The reduced void content and uniform void distribution through the panel thickness suggest that the lower roller temperature greatly reduced post-consolidation void expansion. Delaminations near the panel edge, however, suggest that interfacial bond strength was low.

Placement speed was reduced to 1.69 cm/sec (0.66 in/sec) for Panel 5 in an attempt to increase interfacial bond strength. Roller temperature and compaction force were unchanged. Three plies were placed. Void content increased to an average 7.2 area percent. The apparent increase in void content, especially in the surface ply, Figure 11, indicates that post-compaction deconsolidation was again a problem due to overheating of the ribbon. Note that the interface between two nested ribbons is visible in the top ply near the center of the photograph.

A five ply panel was subsequently placed at the best apparent processing conditions permitted by the current heating method. Panel 6 was placed with a 400° C (752°F) roller temperature, 2.54 cm/sec (1 in/sec) placement speed and 890 N (200 lbf) compaction force. Deconsolidation of the top ply was overcome by reheating and recompacting the panel surface by making a 'dry run' over the panel surface without placing additional ribbons. The top ply was thus reconsolidated without disturbance of the material below. Average void content near the panel end was found to be 0.8 area percent. A cross-section of the panel is shown in Figure 12. No ply or ribbon interfaces were visible. Resin-fiber distribution was generally uniform. The voids present were evenly distributed throughout the cross-section and probably resulted from expansion of pre-existing ribbon voids.

3.6 Mechanical Testing

Panel 6 was tested in interlaminar shear in accordance with ASTM Short Beam Shear method D-2344. Six specimens were cut from the panel between approximately 15.2 cm (6 in) and 30.4 cm (12 in) from the panel end. Short beam shear (SBS) strength averaged 39.2 MPa (5.68 ksi) with a maximum of 54.3 MPa (7.87 ksi). Standard deviation was 10.6 MPa (1.53 ksi).

Visual examination of panel remnants revealed the presence of significant fiber waviness in the first ply. Apparently the starter ply wrinkled during initial ply placement. Examination of the tested specimens showed that all were tested with the wavy side in compression. Furthermore, micrographs of the tested specimens indicate that at least partial failure occurred in the interface between the first and second plies in all of the samples. This would indicate that waviness induced by the wrinkled starter ply reduced the apparent strength of the SBS test specimens. SBS strengths of panels made by automated fiber placement [9] have shown SBS strengths of 62.7 MPa (9.1 ksi) when placed from conventional thin ribbon of similar material in a highly optimized process at 50.8 mm/sec (2 in/sec). Though the SBS strengths measured in this study are lower than those in reference 9, it is expected that elimination of starter ply wrinkling would result in significantly improved SBS values.

4. CONCLUSIONS

The NASA Langley ribbonizer line hardware was changed to permit the successful manufacture of thick-sectioned trapezoidal ribbon. As part of the changes trials were carried out to identify specific processing challenges in non-conventional ribbon manufacture. The low aspect ratio of the trapezoidal ribbon imposed severe cooling requirements on the ribbonizer roller dies. Approximately 61 m (200 ft) of continuous trapezoidal ribbon was made. The ribbon was shown to have low void content but possessed dimensional variability. Ribbon curvature was traced to the hardware configuration and towpreg spooling method.

The trapezoidal ribbon was placed using the NASA Langley Fiber Placement Robot to form unidirectional panels. Panel fabrication required one fifth the number of passes needed for conventional ribbon. Processing conditions to form low void content panels were identified. Ribbon cutter problems necessitated use of short ribbon segments which resulted in a reduction in machine related ribbon guiding. Nesting of inclined ribbon edges closed the resultant ribbon gaps.

Upper ply deconsolidation experienced during placement was caused by the use of a high compaction roller temperature. Lowering the compaction roller temperature was shown to reduce roller sticking and control post-compaction deconsolidation. However, poor interply bonding resulted from insufficient heating of the nip region.

A panel placed under the best processing conditions found exhibited low void content and uniform resin-fiber distribution. Short beam shear testing and examination of specimen microstructure indicated that trapezoidal ribbons could be placed to form panels with interlaminar shear strength comparable to that of panels formed from AFP of conventional ribbon. However, attainment of high interfacial bond strength and low void content will require a supplemental nip point heating method.

5. FUTURE WORK

The present work provides a basis for further optimization of shaped ribbon manufacture and placement. Future ribbon manufacturing efforts in the near term will focus on use of lower oven temperatures with slower line speeds which should reduce entrainment of air and lower ribbon void content. An additional benefit of slower speeds is an increase in dwell time of the ribbons in the nip to improve filling of the roller die and reduce the severity of roller cooling requirements. Use of nip roller center rings made from a high thermal conductivity material should further improve ribbon quenching and again permit operation at higher line speeds. Ribbon curvature will be eliminated by using individual tensioners on single tow spools of towpreg and by increasing the distance between the nip rollers and the takeup winder guide rollers.

Several equipment modifications to the placement head will be made improve its suitability for shaped ribbon placement. A ribbon cutter will be designed to permit shearing of thick ribbons. Guide plates will also be designed with groove depths to reduce ribbon-guide friction during placement. A supplemental infrared radiant heat source will be designed to produce a high thermal gradient near the ribbon bottom surface just prior to consolidation. A subsequent reduction in roller temperature should reduce adhesion between the panel surface and the compaction roller and eliminate post-compaction deconsolidation. Lower compaction roller and gas torch temperatures together with nip heating provided by an IR lamp would then permit placement of an elastic ribbon with a melted exterior which would nest effectively in a building block fashion with its neighbors and yield high interfacial bond strength.

Additional tailoring of ribbons to the AFP process will take the form of surface treatments to encourage the establishment of intimate interfacial contact with lower compaction forces and at higher placement speeds. Experiments will be carried out in smoothing of ribbon surfaces by mechanical post-manufacturing processes and by coating ribbons with a thin adhesive layer. The adhesive layer would be formed of the same polymer as the ribbon matrix but with a slightly lower T_g and a much lower melt viscosity. Given adequate ribbon and substrate smoothness and thin surface layer of a compatible low viscosity resin, ribbon placement may be possible at temperatures near the ribbon resin T_g. High speed, cost effective fabrication of void-free high strength in-situ consolidated composites with lower cost, higher speed equipment may then be possible.

6. REFERENCES

1. B.S. Coffenberry, D.E. Hauber, and M. Cirino, Low-Cost Alternative: In-situ Consolidated Thermoplastic Composite Structures. International SAMPE Symposium and Exhibition, Science of Advanced Materials and Process Engineering Series, 38, 1640-1650, (1993).

2. T.W. Towell, N.J. Johnston, R.W. Grenoble, J.M. Marchello and W.R. Cox, Thermoplastic Fiber Placement Machine for Materials and Processing Evaluations. International SAMPE Symposium and Exhibition, Science of Advanced Materials and Process Engineering Series, 41, 1701-1707, (1996).

3. D.A. Sandusky, J.M. Marchello, R. M. Baucom and N.J. Johnston, Customized ATP Towpreg, 24th International SAMPE Technical Conference. International SAMPE Technical Conference Series, 24, 591-605. (1992)

4. D.A. Sandusky, Alternative Cross-Section Towpregs for Robotic Layups, NASA LaRC Technical Briefs, October, 1993.

5. D.A. Sandusky, J.M. Marchello, and R. M. Baucom, Non-rectangular Towpreg Architectures and Required Manufacturing Procedures. US. Patent 5,405,683; April 11, 1995.

6. D.A. Sandusky, Apparatus for Providing a Uniform Consolidated Unidirectional Continuous Fiber Reinforced Polymeric Material and Method Relating Thereto. U.S. Patent 5,395,477; March 7, 1995.

7. D.A. Sandusky, J.M. Marchello, N.J. Johnston and R.M. Baucom. Ribbonizing Powder-Impregnated Towpreg. International SAMPE Symposium and Exhibition, Science of Advanced Materials and Process Engineering Series, 39, 2604-2611 (1994).

8. S.J. Claus, R.W. Grenoble, D.A. Sandusky, J.M. Marchello, and N.J. Johnston, Composite Ribbon Fabrication with Reactive Polymers, International SAMPE Technical Conference Series, 27, 854-866. (1995).

9. R. Evans, T. Tsuchiyama, D. Lund and D.L. Grande, Structures and Material Technology for Aircraft Composite Structures, Task 5, (January 20, 1993-March 31, 1995) , prepared for NASA under contract NAS1-193494.

Figure 1: Ribbon Cross-sectional Shape

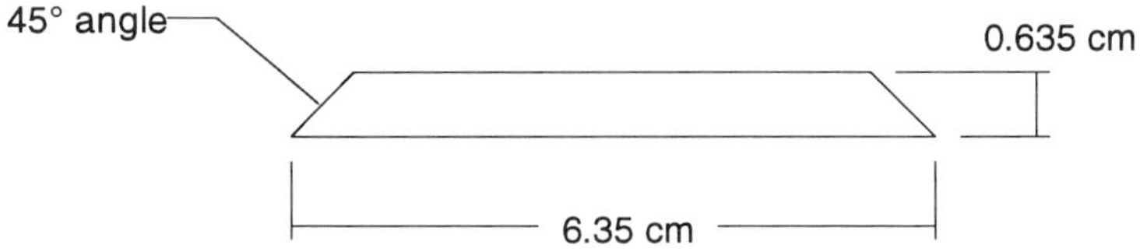

Figure 2: Vertical Bar Assembly, Top View

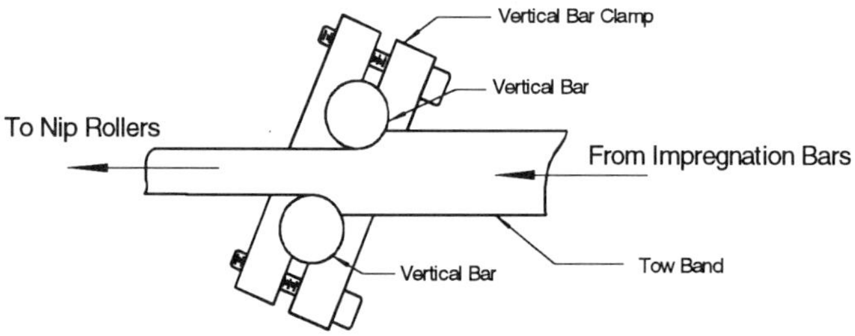

Figure 3: Ribbon Made with Air-Cooled Rollers

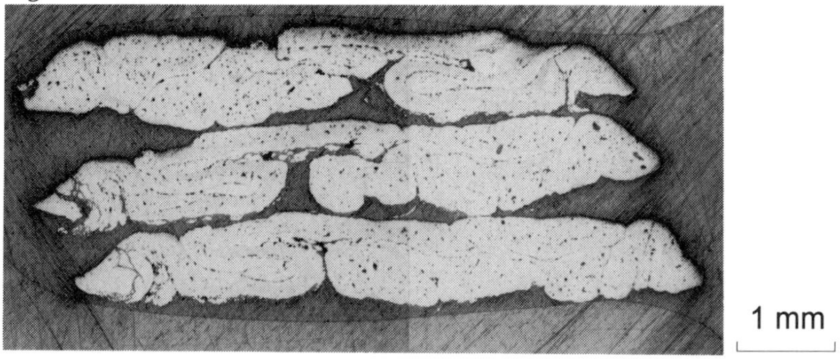

Figure 4: Ribbon Made with Acetone Bath Cooled Rollers

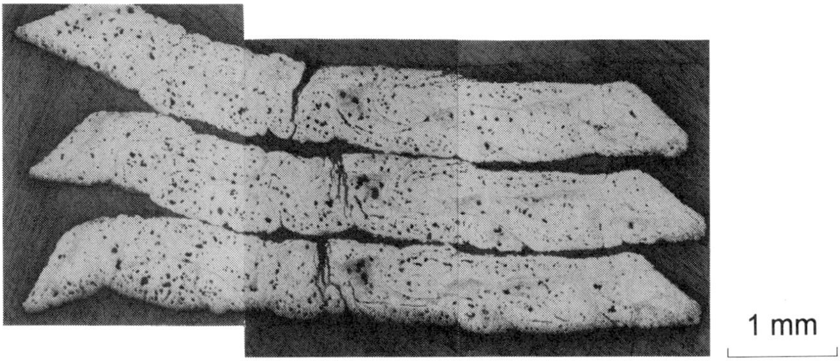

Figure 5: Ribbon made with LN_2 Bath Cooled Rollers

Figure 6: Large Batch Ribbon

Figure 7: Panel 1; 2.54 cm/sec, 400°C Roller and 334N Compaction Force

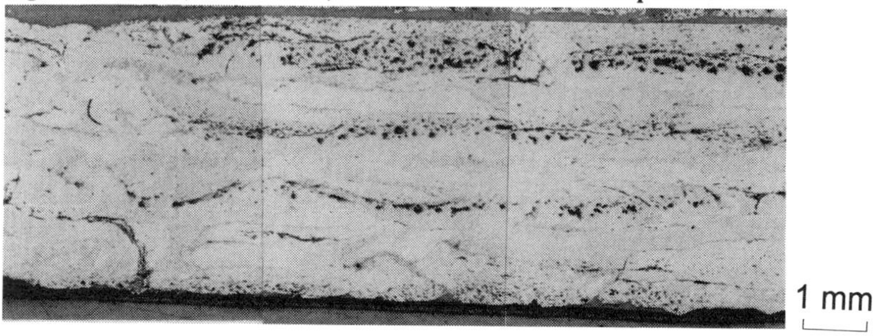

Figure 8: Panel 2; 2.54 cm/sec, 400°C Roller and 890N Compaction Force

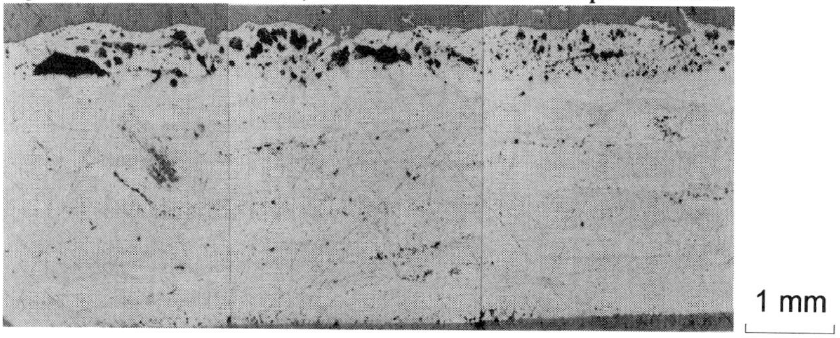

Figure 9: Panel 3; 2.54 cm/sec, 360°C Roller and 890N Compaction Force

Figure 10: Panel 4; 2.54 cm/sec, 320°C Roller and 890N Compaction Force

Figure 11: Panel 5; 1.27 cm/sec, 320°C Roller and 890N Compaction Force

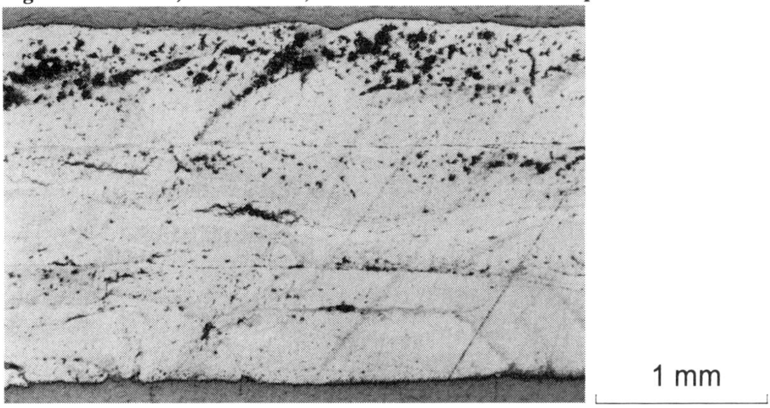

Figure 12: Panel 6; 2.54 cm/sec, 400°C Roller and 890N Compaction Force

28th International SAMPE Technical Conference
November 4-7, 1996

FABRICATION OF NONCIRCULAR SHAPES WITH LASER ASSISTED TAPE WINDING: UNSTEADY PROCESS ANALYSIS

M. Atıf Yardımcı, Christoph M. Pistor, Selçuk I. Güçeri
Mechanical Engineering Department
University of Illinois at Chicago
Chicago Illinois US, 60607-7022

ABSTRACT

Laser Assisted Tape Winding of thermoplastic composites has been shown to be feasible for circular or elliptic ring/helix production. The employment of localized heat sources for production, together with unique melting/solidification behavior of thermoplastic materials; enables the manufacturer to build thick cross section parts by lowering thermally induced residual stresses. It is also possible to wind on non-geodesic paths on the mandrel surface. Applicability of this process would be greatly enhanced if arbitrary cross-sectioned parts could be produced. The winding of such general shapes requires the production of special mandrels, and also extensive off-line process planning due to non-constant winding speeds and laser power requirements. A set of computational tools has been developed to generate machine control commands for noncircular cross-section parts. Previous quasi-steady-state winding models neglected the effects of cyclic nonuniform heating of the part, the evolution of temperature field due to growing thermal mass, and most importantly the effect of varying winding speed during the process. An unsteady thermal model has been developed to address these issues and to analyze the thermal behavior of the part during production. The developed set of computational tools is capable of predicting the unsteady temperature response, interlaminar bonding quality across the part, and required laser power history for constant nip-point temperature operation.

KEY WORDS: Filament Winding, Process Planning, Thermal Analysis

1. INTRODUCTION

Thermoplastic matrix composites offer a competitive advantage over thermoset matrix composites in manufacturability due to their response to thermal loads. A thermoplastic material can go through multiple melting/solidification cycles without losing relevant engineering properties, whereas the cross-linking of polymer chains in thermosets involves an exothermic and practically irreversible process. The thermally reversible processability of thermoplastic materials enables them to be thermoformed into intermediary solid products, such as sheets/bars/tapes with or without reinforced particles or fibers, and to be processed further into the final product by thermal processing. Tape winding and filament winding are affordable processing techniques for fast production of round composite parts. These processes traditionally included a thermoset resin bath where fibers were pulled through. Due to low viscosities and wetting characteristics of thermosets the resin filled most of the interfiber voids. The produced structure was cured further in ovens to pyrolyze the thermoset component. Thermal stress build-up, due to exothermal reaction of the resin during curing cycle , made the production of thick cross-section parts impossible. Thermoplastics were not considered as an alternative because their high viscosities caused poor wetting behavior of the fibers and voids to occur. The elimination of the wet step in tape winding, with the introduction of preprocessed reinforced thermoplastic tapes, made the production of thick parts possible (1, 2, 3).

The most important process parameters in laser assisted tape winding are winding speed and laser power. These parameters, together with tape material properties, determine nip-point temperature, cooling rate, and bonding quality. The process window on the parameter winding speed / laser power parameter plane is determined with two bounding curves; the first one representing inadequate interlayer bonding and the second one representing polymer degradation (2). The consolidation region in the processing window is rather narrow. Process modeling efforts have been driven by the need to minimize process tuning costs for different material systems and complex geometries. Beyeler (4) restricted his analysis to the production of fixed size rings. A two-dimensional steady-state thermal model was utilized to study the heat transport phenomena during tape winding. Computational process windows have been predicted for Ryton and APC-2 materials. Nejhad (5) developed a three-dimensional steady-state Eulerian model which is capable of also predicting the temperature distribution inside the mandrel. The most important contribution of this model was to include angular laydown configurations. Agarwal (6) developed the bonding quality prediction capability for tape winding process applying nonisothermal diffusion bonding theories to the quasi-steady state process models at hand. Unsteady process models have been also developed for thermoset filament winding of circular geometries by Lee and Springer, and Sun et.al (7, 8). Thermosets and thermoplastics differ in thermal, physical, and chemical routes to bond generation, i.e. heat generating and irreversible vs. heat absorbing and reversible. Therefore, a novel modeling approach was needed to analyze the unsteady thermal behavior of product during laser assisted tape winding (LATW). Methods originally developed for the thermal process modeling of Rapid Prototyping/Solid Freeform Fabrication techniques (9) have been applied to LATW of ceramic green bodies with circular geometry (10). The process modeling approach taken in this study is based on that work.

The current research is aimed first at the development of process planning tools. These geometrical design and machine control generation tools are required for the production of noncircular convex geometries with tape winding. Section 2 presents these tools, and their applications to an elliptical mandrel geometry. The capabilities of the planning tools are demonstrated by predicting the mandrel centerpoint trajectories, corresponding linear stage translations, winding speed distribution, and laser power requirements. Also, an unsteady heat transfer model has been developed to address the issues regarding dynamic thermal response during the tape winding process. Section 3 outlines the development of the thermal model, the particular grid topology, and also the bonding quality prediction model built on top of the heat transfer simulation. The models are applied to an elliptical geometry. Dynamic temperature evolution, bonding quality distribution across the part, as well as predicted laser power requirements have been determined. The summary of computational tools and their capabilities is reported in Section 4 together with some remarks on the ongoing work in this area.

2. OFF-LINE PROCESS PLANNING FOR NONCIRCULAR SHAPES

The current set-up that is present in the Manufacturing Processes Laboratory of the University of Illinois at Chicago (Figure 1), employs two motion stages. A rotary stage which carries the mandrel is mounted onto a linear stage. The translational stage is itself mounted on linear bearings in the normal direction, and is constrained by an air cylinder. The air cylinder acts as a constant force device, ensuring the contact of mandrel to the consolidation head. The prepreg tape is fed to the consolidation point from a spool. Portions of fresh tape and previously deposited material are melted with the heat energy coming from the laser beam at the nip-point. The molten material cools down very rapidly, bonding the fresh tape to the substrate. The compaction roller ensures appropriate consolidation and reduces the occurrence of voids.

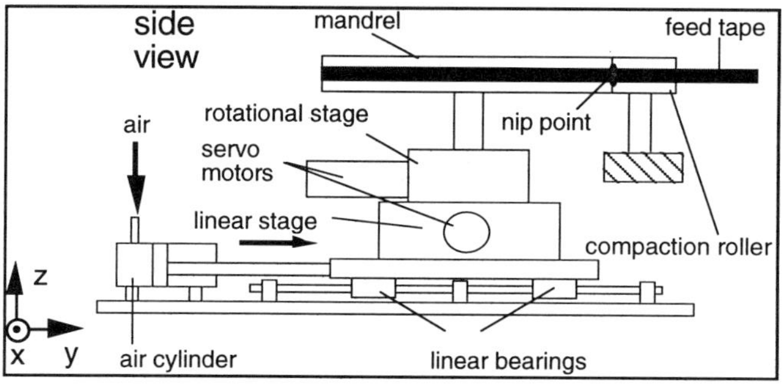

Figure 1: Laser Assisted Tape Winding Setup at UIC-MPL

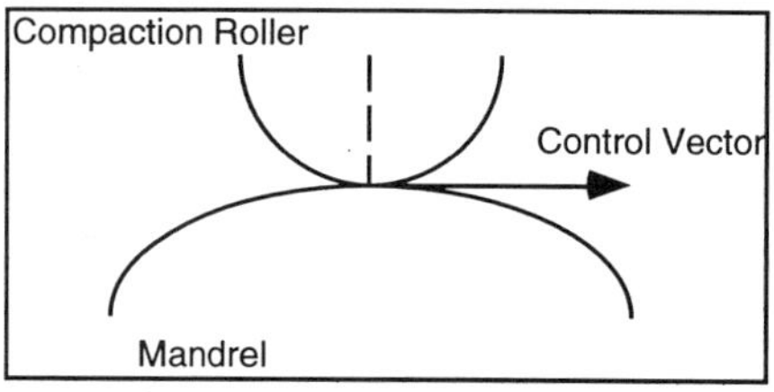

Figure 2: Horizontal Alignment for Roller/ Mandrel Pair

The laser beam is manipulated and transferred to the nip-point by a series of lenses and mirrors. This delivery process requires a certain opening angle dependent on beam width and mandrel and compaction roller diameters. One way of ensuring a minimum opening angle for convex noncircular geometries is to force the instantaneous tangent vector to be aligned with the tangent vector of compaction roller (control vector) at the nip-point (Figure 2).

More conservative bounds on the opening angle may be set by rotating the control vector away from the compaction roller tangent towards the mandrel side, and enforcing the alignment rule of the mandrel tangent vector to this modified control vector (Figure 3). It should be emphasized that the nip-point stays at the same location, while the control vector is rotated towards the mandrel. Hence, for the given roller/mandrel geometries, there will be a bound on the maximum angle the control vector can make with the horizontal.

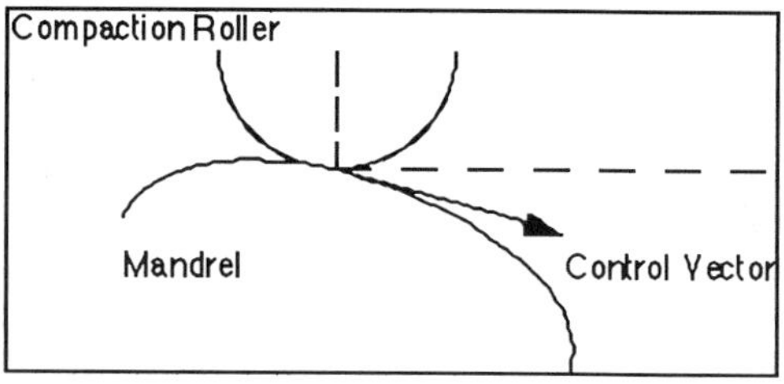

Figure 3: Roller/Mandrel Configuration for Modified Control Vector

The mandrel geometry is specified with cubic splines, e.g. for elliptic or circular shapes four points with proper tension values suffice to describe the geometry. Intermediate points need to be generated for each spline segment for subsequent calculations. A chord length parameterization is utilized for the determination of cubic splines for each segment, to ensure nearly equi-spaced point placement (11, 12). After the generation of a sufficient number of intermediate points the spline representation is dropped, and the global geometry is represented by the collection of points.

The wound tape is basically a spiral which steps outwards at the end of each layer. For the rest of the analysis it will be assumed that the rotational stage has constant angular velocity throughout the production. Hence, the linear stage which carries the mandrel needs to be controlled to achieve the necessary motion during winding in accordance with the minimum opening angle constraint; therefore, the trajectory of mandrel centerpoint during production needs to be determined for motion control command generation. This trajectory can be calculated by the determination of the planar location of the mandrel centerpoint for each point on the mandrel boundary .

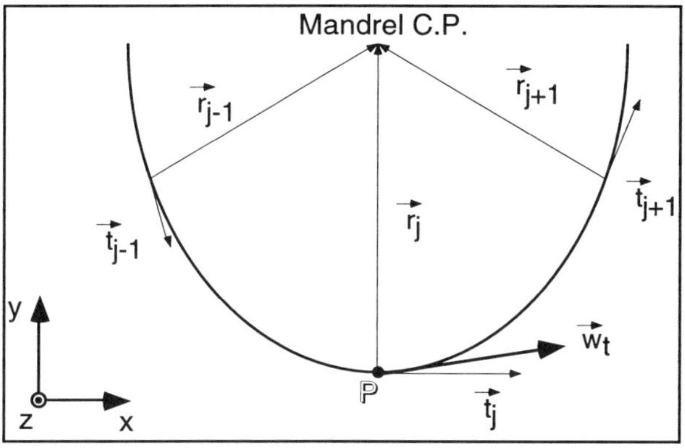

Figure 4: Vectors for Mandrel Centerpoint Trajectory Calculation

First, the geometric center of the mandrel is calculated, then relative position vectors for the desired boundary points are constructed together with the corresponding curve tangent vectors. The spatial location of the nip-point, point P in Figure 4, is fixed, as is the control vector w_t . Individual boundary points are made to coincide with point P, together with the constraint that the tangent vectors should be aligned with the control vector. Since the relative position of the mandrel geometric center with respect to boundary points are known, the required centerpoint trajectory can be easily predicted utilizing the following vector relationships.

$$\vec{\mathbf{n}}_j = \vec{\mathbf{t}}_j \times \vec{\mathbf{k}}\,, \quad \vec{\mathbf{w}}_n = \vec{\mathbf{w}}_t \times \vec{\mathbf{k}}$$

$$_{w_t,w_n}\left(\vec{\mathbf{r}}_j\right) = \left(\vec{\mathbf{r}}_j \cdot \vec{\mathbf{t}}_j\right) \vec{\mathbf{w}}_t + \left(\vec{\mathbf{r}}_j \cdot \vec{\mathbf{n}}_j\right) \vec{\mathbf{w}}_n \qquad (1)$$

$$\vec{\mathbf{r}}_{cj} = {}_{x,y}\left(\vec{\mathbf{r}}_j\right) = \left(\vec{\mathbf{r}}_j \cdot \vec{\mathbf{t}}_j\right)\left((\vec{\mathbf{w}}_t + \vec{\mathbf{w}}_n)\cdot \vec{\mathbf{i}}\right)\vec{\mathbf{i}} + \left(\vec{\mathbf{r}}_j \cdot \vec{\mathbf{n}}_j\right)\left((\vec{\mathbf{w}}_t + \vec{\mathbf{w}}_n)\cdot \vec{\mathbf{j}}\right)\vec{\mathbf{j}}$$

The substrate thickening during the winding process can be accounted for by calculating the mandrel centerpoint trajectories for dynamically modified effective mandrel shapes. The winding operation is semi-discrete, i.e. one can identify individual layers except at the interlayer transitions. Thus, the total winding trajectory may be found as the sum of these discrete shapes.

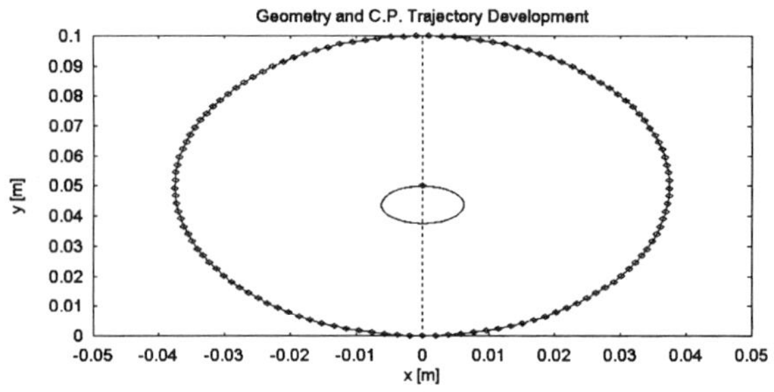

Figure 5: Elliptic Mandrel

Figure 5 depicts an elliptical mandrel having major radius of 50 mm and minor radius of 37.5 mm. Also the predicted mandrel centerpoint trajectory is shown. The trajectory is periodic since the mandrel shape is symmetric and closed.

For the more realistic case of Figure 6, a constant thickness tape of 0.15 mm (e.g. APC-2 tape) is being wound onto the mandrel of Figure 5. The outer figure shows the semi-discrete spiral generated by winding and the inner figure shows three selected trajectories for the given configuration.

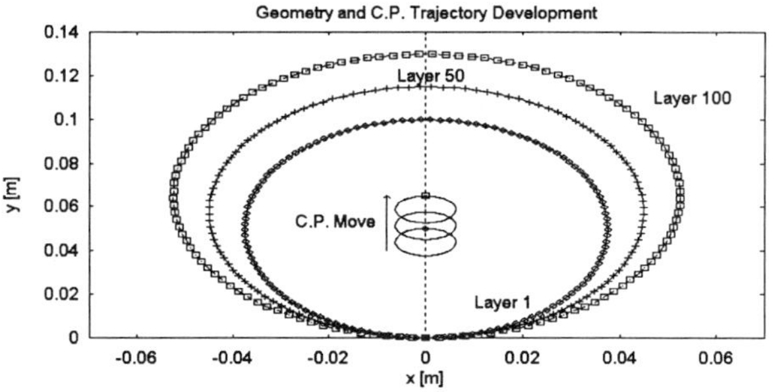

Figure 6: Semi-Discrete Spiral and Centerpoint Trajectory

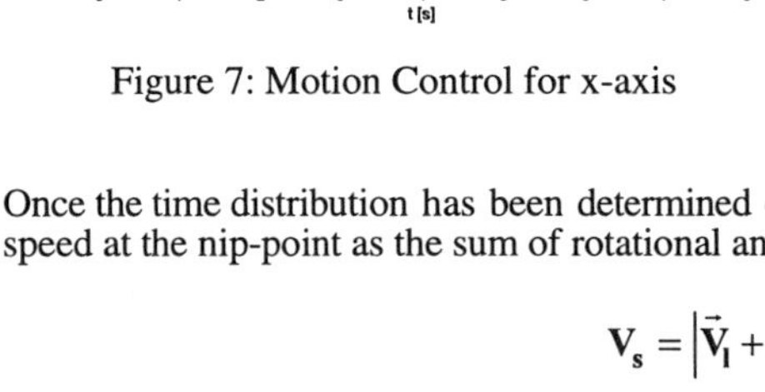

Figure 7: Motion Control for x-axis

One can assume a time distribution, based on the constant rotational stage angular speed, for the points shown in the mandrel centerpoint graphs of Figures 5 and 6. Hence, motion commands can be generated for the linear stage controller. The sample command set for the constant geometry mandrel is shown in Figure 7.

Once the time distribution has been determined one can easily determine the relative winding speed at the nip-point as the sum of rotational and linear components as

$$\mathbf{V_s} = \left|\vec{\mathbf{V}}_{\mathbf{l}} + \omega\, \vec{\mathbf{k}} \times \vec{\mathbf{r}}_{\mathbf{j}}\right| \tag{2}$$

where V_l denotes the contribution due to mandrel centerpoint movement, and the second term denotes the contribution from rotation of the contact point around the mandrel centerpoint.

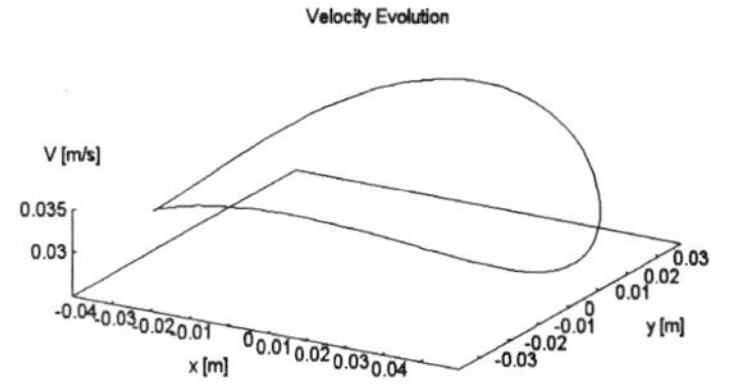

Figure 8: Winding Speed for Thin Ring

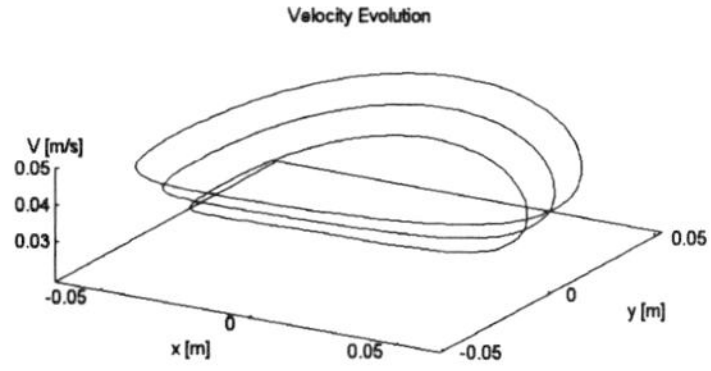

Figure 9: Winding Speed for Thick Ring

Figures 8 and 9 show the velocity calculations for the thin and thick cross sectional rings depicted in Figures 5 and 6. Again, constancy of rotational stage angular speed was utilized for time distribution calculations. The increase in winding speed at major axis ends and for outer layers is evident.

Previous quasi-steady state models produced operation windows for different materials, including winding speed versus required laser power curves. Figures 10 and 11 show the result of applying these data to the velocity distributions obtained in Figures 8 and 9. Since winding speed/power relationship was nearly linear, the calculations based on the above mentioned models produced only a scaling in the z-axis of the figures. It should be noted however that such quasi steady-state models exclude the dynamic thermal behavior of the system, by decoupling every material point on the mandrel from each other rather unrealistically.

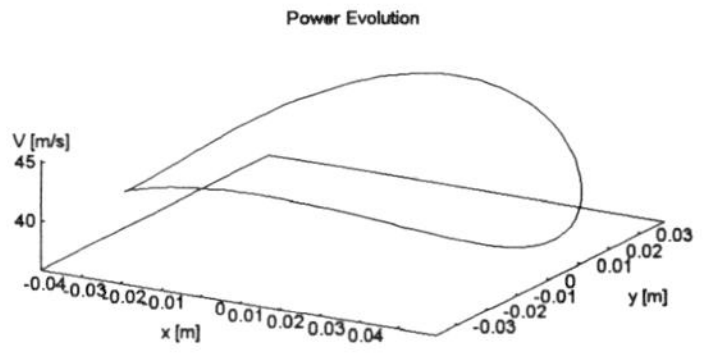

Figure 10: Laser Power for Thin Ring

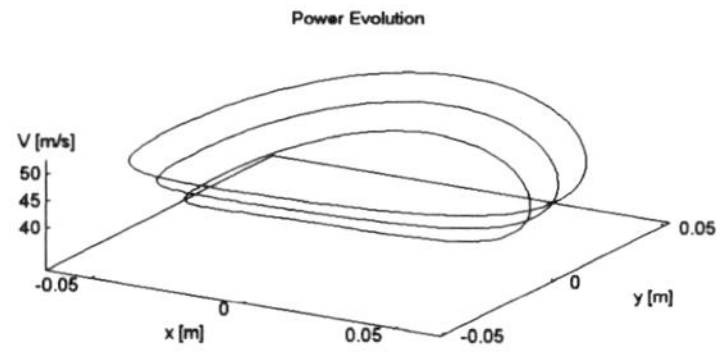

Figure 11: Laser Power for Thick Ring

3. UNSTEADY HEAT TRANSFER MODEL

Previous heat transfer models of Beyeler and Nejhad for tape winding involved the calculation of quasi-steady state temperature field in the part (and in the mandrel for Nejhad's model). These models assumed that the dynamic thermal behavior of the part did not matter for the particular geometries that were of interest. That is, either the heat transfer phenomena remained localized due to rapid quenching or a steady-state thermal wave traveling angularly at winding speed set in the part. It is one of the theses of this paper that dynamic thermal behavior of the part is important for noncircular shapes, small parts, and even for production environments where a low thermal conductivity mandrel is utilized. The dynamic behavior expands the previously defined operation windows significantly by allowing high temperature winding or winding under oscillating mean temperature conditions.

The winding tape is characterized as a semi-discrete spiral. The energy equation is integrated across the tape cross-section and is cast into one-dimensional form in the following way:

$$\rho \frac{\partial q_{ent}}{\partial t} = k \frac{\partial^2 T}{\partial s^2} + S_n - S_c$$
$$S_n = \sum_n h\, A_n(t)\,[T - T_n] \qquad (3)$$
$$S_c = h\, A_c(t)\,[T - T_\infty]$$

where the left hand side denotes the change in internal energy, the first term in the RHS denotes the diffusive redistribution of temperature, the second term denotes convective cooling to the ambient, and the third term describes the conductive interaction with the material radially above or below the point of interest.

The boundary conditions may be written as

$$\text{i.} \quad T = T_{n.p.} \quad @\, \vec{r} = \vec{r}_0(t),$$
$$\text{ii.} \quad -k \frac{\partial T}{\partial s} = hA_{c0}[T - T_\infty] \quad @\ \text{tip} \qquad (4)$$

The most important boundary condition is i, namely the constant temperature at the moving nip-point. Since the governing equation has been cast in Lagrangian form, this moving boundary condition was necessary to describe the effect of winding. A sub-grid movement model has been employed to capture the thermal behavior of deposited material to a traveling Dirichlet type boundary condition. Also, it is assumed that the other end of the spiral is open to the ambient, and the innermost layer has thermal contact to the mandrel as well as the ambient. The mandrel was idealized as a big isothermal mass with an effective heat transfer coefficient based on the mandrel material conductivity.

The spiral is effectively divided into curvilinear one-dimensional grids representing separate layers. The ends of each grid are connected to the beginning of the next layer to ensure that the spiral is geometrically connected. New sets of equations are added as the number of deposited layers grows, and also the number of points inside a given layer increases. An evolving grid had to be implemented to track the movement of the deposition head.

The above set of equations and boundary conditions was discretized spatially using finite volume method, and temporally using explicit time integration. The thermal simulator predicts grid growth as well as temperature field evolution. Once the temperature history is known, one can calculate other field variables based on the thermal history of the part. The kinetic crystallization model (2) may be utilized to predict the degree of crystallinity inside the wound part. One can also utilize an incremental thermoelastic solver to calculate the total amount of residual stresses and strains in the body. A very important and difficult to measure quantity is the degree of interlayer bonding. A bonding metric, temperature-weighted residence time over a critical temperature, had been developed (9) as:

$$\varphi = \int_0^t \langle T - T_b \rangle \, d\tau \qquad (5)$$

The dynamic power consumption is calculated by constructing an open control point around the nip-point and by summing up the power requirements for preheating of tape to nip-point temperature. Other possible calculations are: the required total laser power to bring the tape to nip-point temperature; convective losses to air; and conductive losses to previously wound material.

Figure 12 shows the results of a simulation for a material with c_p=1200 J/(kg*K), ρ=1600 kg/m^3, k_l=2.0 W/(m*K),k_t=0.6 W/(m*K). Process planning tools together with the elliptical geometry of Section 2 were utilized for the simulations. Nip-point temperature was set to 100 °C, the convection heat transfer coefficient was taken as 5 W/(m^2*K) for natural convection, ambient temperature was 20 °C, and critical bonding temperature was also selected to be the same. A steel mandrel maintained at 20 °C, with an effective heat transfer coefficient of as 1000 W/(m^2*K) was utilized.

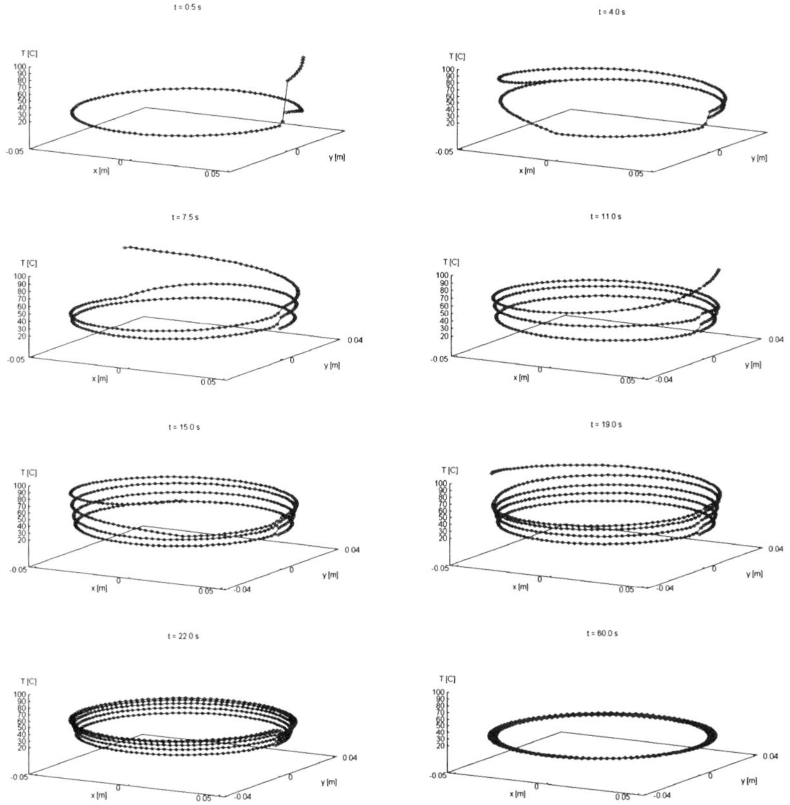

Figure 12: Evolution of Temperature Field

The nip-point travels along a semi-discrete spiral outwards from the center. The thermal mass of the part keeps increasing during the operation, also the winding speed is a function of azimuthal and radial coordinates. As the thermal mass increases, the heat sink effect from the mandrel is shielded by the wound layers. The increased thermal mass begins to act as a heat sink itself after a threshold has been passed. The nonconstant winding speed, together with the constant nip-point temperature, influence instantaneous temperature distributions across the part. The most important observation to be made however, is the fact that the dynamic average part temperature is above the ambient. The corresponding bonding metric distribution is shown in Figure 13.

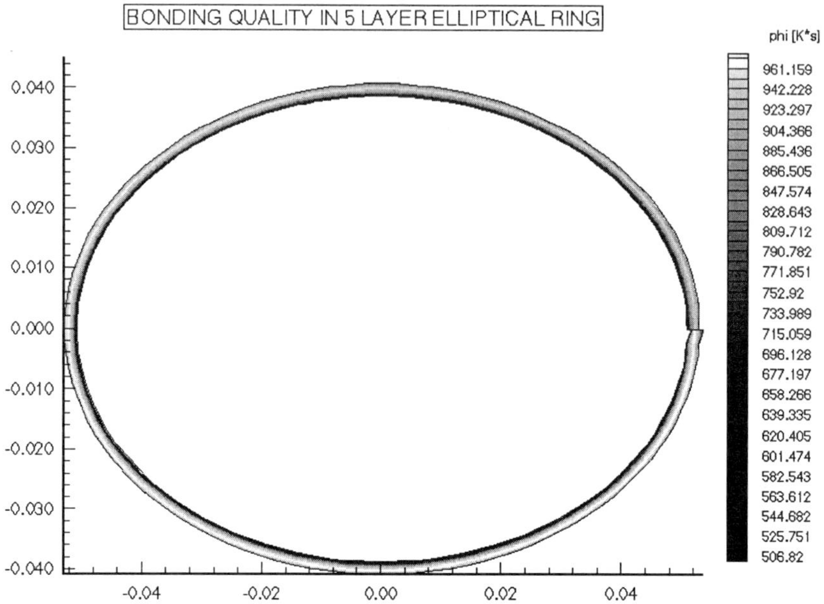

Figure 13: Bonding Quality Distribution

In Figure 13 light areas correspond to good bonding, and dark areas to poor bonding. The innermost layer interface, i.e. between layers 1 and 2, is poorly bonded due to the rapid quenching of mandrel and low thermal mass of wound material. Furthermore, there is an evident azimuthal structure to the bonding quality; it increases along the winding direction.

The dynamic power loss prediction is shown in Figure 14.

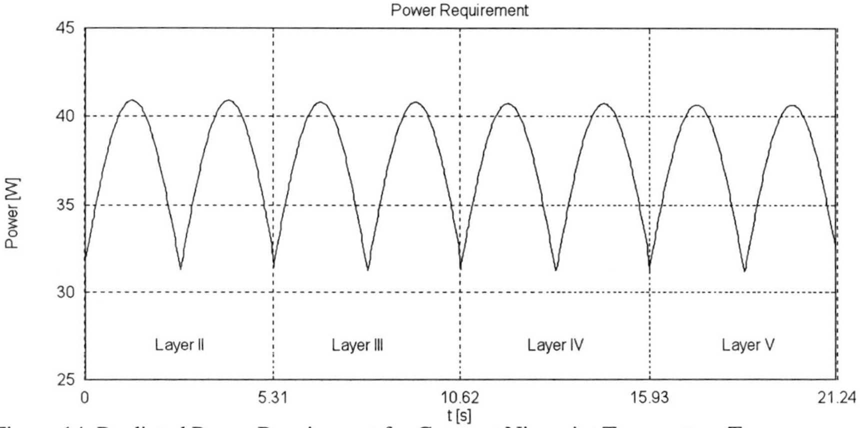

Figure 14: Predicted Power Requirement for Constant Nip-point Temperature Tape Winding

It is interesting to note that the required power changes at twice the frequency of layer deployment. This is due to the axial symmetry of the elliptical mandrel, and the corresponding winding speed distribution. Although it is not evident from Figure 14, it has been observed that the maxima of power requirement cycles decay as the number of layers grow. This decay is due to the increase in average part temperature, which reduces conductive losses to the previously wound material.

The average and minimum part temperatures vary during the part production. An exemplary temperature history from two virtual thermocouples is shown in Figure 15.

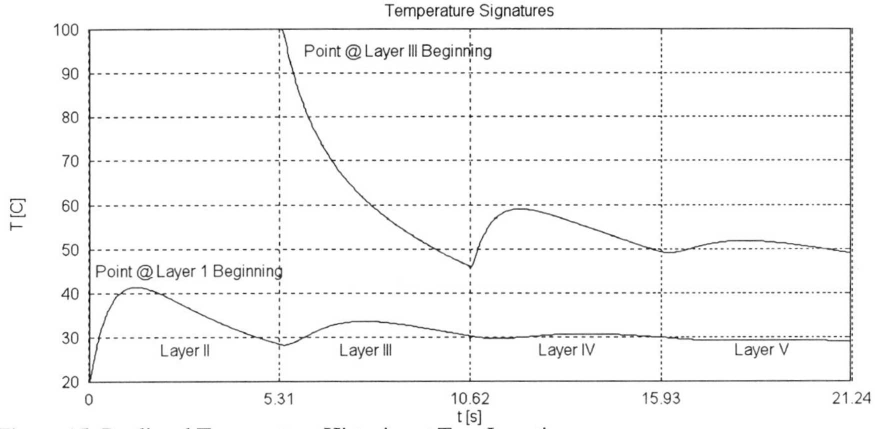

Figure 15: Predicted Temperature Histories at Two Locations

These temperature signatures in Figure 15 demonstrate the cooling behavior of two material points in the part. The first layer is always wound with the laser turned off. The cyclic heating of the material in the first layer slows down the cooling process and keeps the local temperature around 10 °C higher than the ambient temperature. Even more interesting, the material point in the third layer settles at a higher temperature, creating a considerable temperature gradient through a layer in the part.

4. SUMMARY AND ONGOING WORK

Geometrical process planning tools have been developed to automate laser assisted tape winding of thermoplastic matrix composites in convex, noncircular cross-section geometries. These tools are able to generate motion control commands and diagnostics for the given part geometry. Also an unsteady heat transfer and diffusion bonding model has been developed to assess the importance of dynamic thermal behavior in tape winding. The process planning tools have been utilized to calculate winding speed evolution, which in turn was utilized in the unsteady thermal model. Bonding quality distributions as well as laser power requirements for constant nip-point temperature winding have been developed.

Ongoing research includes the development of a radiative laser power deployment model based on the theory developed by Agarwal (6), and Irwin (13). Since the material time histories are easily predicted, the present model will be coupled with crystallinity prediction models on a growing mesh, as well as growing domain thermoelastic models for process induced stress/strain calculations.

5. ACKNOWLEDGMENTS

The authors acknowledge the support of this research by the National Science Foundation, DMI program.

6. REFERENCES

1. Beyeler E.P., Melting and Solidification in Processing and Applications to Thermoplastic Matrix Composites Manufacturing, PhD Thesis, Department of Mechanical Engineering, University of Delaware, Newark, Delaware, 1987.

2. Ghasemi Nejhad M.N., Three-Dimensional Thermal and Residual Stress Analysis of In-Situ Thermoplastic Composite Filament Winding, PhD Thesis, Department of Mechanical Engineering, University of Delaware, Newark, Delaware, 1992.

3. Hümmler J., Lee S.K., Steiner K.V., "Recent Advances in Thermoplastic Robotic Filament Winding", Proceedings of 36th International SAMPE Symposium, April 15-18 1991, San Diego, California

4. Beyeler E.P. and Guceri S.I., "Thermal Analysis of Laser-Assisted Thermoplastic Matrix Composite Tape Consolidation", ASME Journal of Heat Transfer, Vol. 110, pp. 424-430, 1988.

5. Ghasemi Nejhad M.N., Cope R.D. and Güçeri S.I., "Thermal Analysis of In-Situ Thermoplastic-Matrix Composite Filament Winding", ASME Journal of Heat Transfer, Vol. 113, pp. 304-313, 1991.

6. Agarwal V., The Role of Molecular Mobility in the Consolidation and Bonding of Thermoplastic Composite Materials, PhD thesis, Materials Science Program, University of Delaware, Newark, Delaware.

7. Lee S.Y. and Springer G.S., "Filament Winding Cylinders: I. Process Model", Journal of Composite Materials, Vol. 24, pp. 1270-1298, 1990.

8. Sun L., Mantell S.C., Banerjee A. and Cohen D., "Thermoset Filament Winding Process Model and Parametric Study", Proceedings of The ASME Materials Division, MD-Vol. 69-1, pp. 11-23, ASME-IMECE 1995, San Francisco California, US.

9. Yardımcı M.A., Güçeri S.I., Danforth S.C., Agarwala M., Safari A., "Numerical Modeling of Fused Deposition Processing", Proceedings of The ASME Materials Division, MD-Vol. 69-2, pp. 1225-1235, ASME-IMECE 1995, San Francisco California, US.

10. Pistor C., Yardımcı M.A., Güçeri S.I., Bandyopadhay A., "Laser Assisted Tape Winding of Ceramic Green Parts", American Ceramic Society 1996 Cocoa Beach Meeting January 1996, Ceramics Science and Engineering Proceedings, Cocoa Beach.

11. Rogers D.F., Adams J.A., Mathematical Elements for Computer Graphics, pp. 255-271, McGraw Hill, 1990.

12. Farin G., Curves and Surfaces for Computer Aided Geometric Design - A Practical Guide, pp. 125-138, Academic Press, 1990.

13. Irwin R.G. Jr. and Güçeri S.I., "Laser Assisted Processing of Thermoplastic Composites", Proceeding of First International Conference on Transport Phenomena in Processing, March 22-26 1992, Honolulu, Hawaii.

28th International SAMPE Technical Conference
November 4-7, 1996

A Simple Filament Winding Pattern Generation Algorithm

Y. D. Liang
State Key Laboratory of CAD/CAM
Dept. of Mathematics, Zhejiang University, P.R.C.
Gene Luo
Geodesk Engineering Systems
724 Sunrise Dr., Fremont, CA 94539

Abstract

The production of high performance composite product requires an accurate placement of fiber path controlled by computer programming. In the design of fiber path, the algebraic patterns deal with the distribution of fiber around the mandrels and play a key role in the computer aided design system of filament winding. Unfortunately, the commonly used definitions of filament wining patterns are incomplete and ambiguous ([1], [2], [3], [4]), and the algorithms for generating all feasible patterns have not been well developed. This paper presents a systematic analysis of filament winding patterns. Based upon which, a simple pattern generation algorithm is proposed. This pattern generation algorithm employs two recursive formulas to greatly simplify the procedure and reduces the computation by half with another half to be obtained as their conjugates.

KEYWORDS: Composites, Filament Winding Patterns, Conjugate Patterns, Euclidean Algorithm, Pattern Generation Algorithm

1. Introduction

Filament winding is being increasingly used as a method of fabricating high performance pressure vessels and other composite structures. The production of such composites requires an accurate placement of fiber path. Mathematically, the pattern of fiber path can be classified into two categories: geometric patterns and algebraic pattern [1]. The geometric patterns are mainly involved with certain types of differential equations which specify the geometric characteristics of fiber paths (for example, the geodesics),while the algebraic patterns deal with the distribution of fiber path around the mandrel, and can be regarded as the initial or boundary conditions of the differential equations specifying the corresponding geometric pattern. Therefore the studies of the related mathematical models and the algorithms are more and more important to filament winding techniques. Unfortunately, as mentioned in [1] the most commonly used definitions of the filament winding patterns by Lessie[2] or by Weck[3] are incomplete and ambiguous. In this paper, a precise definition and systematic analysis of the algebraic patterns are given, and a simple, fast pattern generation algorithm is developed which costs the least computation and least memory space. Based on the pattern generation algorithm, all feasible algebraic patterns can then be easily obtained. Thus, the pattern generation algorithm will serve as an important tool in the computer aided filament winding design system. Some graphical representations of the algebraic pattern of the filament winding are shown to illustrate the algorithm[1]. Since in this paper, only algebraic patterns are concerned (not geometric patterns), for convenience, the term "patterns" is usually used instead of algebraic patterns without special mention.

2. The Definition and Analysis of Filament Winding Patterns.

The filament winding patterns or more exactly the algebraic patterns of filament winding are widely used in the design and analysis of the filament winding techniques, the concept and the definition of the algebraic patterns still have not been well studied. First of all, a precise definition of the algebraic patterns is given in this section.

Filament winding is a process or operation in which resin impregnated fibers are systematically wound over a rotating mandrel. The shape of the mandrels can be of great variety. In the studies of the algebraic patterns, it is convenient to assume that the mandrel can be represented as a surface of revolution and the fibers are uniformly distributed on the cross sectional circles of the mandrel. Without loss of generality, it is also assumed that the radius of a certain circle of the cross sections is 1. Let Π be a complex plane and C be a unit circle with origin as the center representing a cross sectional circle of a mandrel. Let $P = \{W_t \mid t = 0,1,2,\cdots\}$ be a sequence of points on C representing the center of each fiber band on C, to be placed in the time sequence and let ψ $(0 < \psi < 2\pi)$ be a given angle by which each fiber is shifted over another, then the sequence P can be expressed in the complex plane Π as follows.

$$P = \{ W_t = e^{it\psi} \mid t = 0, 1, 2, \cdots \} \tag{2.1}$$

Where $e^{it\psi}$ is the Euler expression of the corresponding complex points in sequence P in Π (complex plane is a plane constituted by all complex points representing complex variables) (see Fig.1), i is the imaginary unit and t is the index of the point W_t in the time sequence.

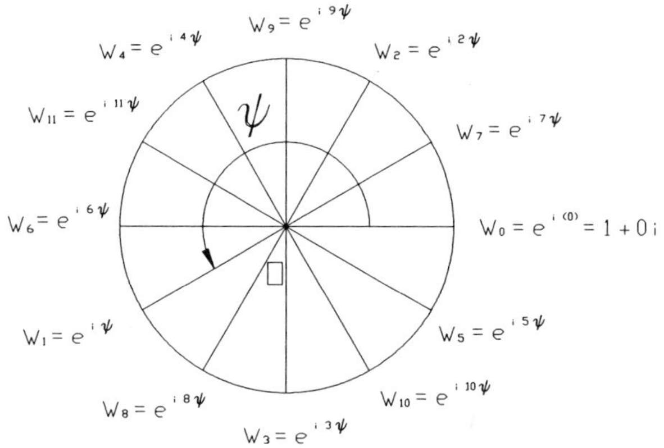

Fig.1 A uniformly distributed sequence $\{W_t \mid t \geq 0\}$ in Π, where $e^{it\psi}$ is the Euler expression of W_t

the Euler expression can be written in the following complex form.

$$e^{it\psi} = \cos \psi + i \sin \psi \tag{2.2}$$

Where i is the imaginary unit and ψ is the corresponding argument. The number of the points in the sequence P can be of infinitely many. For better control of a design system, it is naturally to require that there are only a finite number of points in P. It is easy to show that the requirement of the number of points in P being finite is that there is a unique pair of numbers M and K (0 < K < M) such that

$$M\psi = 2\pi K \tag{2.3}$$

with M being the minimal one satisfying (2.3). It is easy to show that P contains exactly M points, and (2.1) can be expressed as

$$P = \{ W_t = e^{it\psi} \mid t = 0, 1, 2, \cdots, M-1\} \tag{2.4}$$

with $W_M = e^{iM\psi} = e^{i0} = W_0$. Since P is uniformly distributed, the circle C is split into M arcs with equal central angle $2\pi/M$. Denote the central angle by $\theta = 2\pi/M$, the set of the points in P can then be expressed and denoted by U as follows,

$$\begin{aligned} U &= \{ u_k = e^{ik\theta} \mid k = 0, 1, 2, \cdots, M-1\} \\ &= \{ e^{iM\theta} = e^{i0}, e^{i\theta}, e^{i2\theta}, \cdots, e^{i(M-1)\theta} \} \end{aligned} \tag{2.5}$$

The set of P equals the set U, but the points in P and U are in different orders. The two closest points to W_0 are $e^{+i\theta}$ and $e^{-i\theta}$, thus, there is an integer m > 0 such that the m-th points in P reaches $e^{+i\theta}$ or $e^{-i\theta}$, that is

$$e^{im\psi} = e^{\pm i\theta} \tag{2.6}$$

notice that

$$e^{i\psi} = e^{i\frac{2\pi K}{M}} = e^{iK\theta} \tag{2.7}$$

$$e^{imK\theta} = (e^{i\psi})^m = e^{mi\psi} = e^{\pm i\theta} \tag{2.8}$$

Comparing both sides of (2.8), there is an integer $n \geq 0$, such that

$$mK\theta = 2\pi n \pm \theta \tag{2.9}$$

Substituting $2\pi = M\theta$, we have,

$$mK - Mn = \pm 1 \tag{2.10}$$

Suppose that m and n can be positive or negative, then the sign in (2.10) can be absorbed and (2.10) is written as

$$mK - Mn = 1 \tag{2.11}$$

This shows that for a given uniform sequence P, M and K are uniquely determined, and there is a solution (m, n) to (2.11). Of course, the solutions (m, n) to (2.11) can be of greatly many. It is easy to show that the solution (m,n) with minimal |m| is unique and the condition of minimization of |m| is equivalent to

$$-\frac{M}{2} < m < \frac{M}{2} \tag{2.12}$$

Thus, for a given uniform sequence P, there is a unique (M, K, m, n) satisfying (2.11), (2.12) and

$$1 < K < M \text{ or } K = M + 1 \tag{2.13}$$

Therefore, for a given uniform sequence P in (2.1), there exist four uniquely determined numbers M, K, m and n satisfying the equation (2.11) with (2.12) and (2.13). On the contrary, for given four numbers M, K, m, n satisfying (2.11) with (2.12) and (2.13), there is one and only one uniform sequence P defined by (2.1) with $\psi = 2\pi K / M$. The equation(2.11) with (2.12) and (2.13) is called the fundamental equation of the filament winding, and M is called the spilt number, K is called the skip number, m and n are called the pattern numbers.

Definition 1. An algebraic pattern of the filament winding with split number M, skip number K and the pattern numbers m and n is a sequence of points on a unit circle corresponding to the time sequence t and is denoted by P or P(M, K, m, n) as follow

$$P = \{ W_t = e^{it\frac{2\pi K}{M}} \quad | \, t = 0, 1, 2, \cdots, M-1\} \tag{2.14}$$

Where M and K satisfy the fundamental equation with the split number M, skip number K and the pattern numbers m and n. The corresponding four numbers M, K, m and n are called the parameters of P.

As $m = \pm 1$ we have two trivial cases corresponding to P(M, M+1, 1,1) and $\overline{P}$(M, M-1, -1, -1) as shown in Fig. 2.

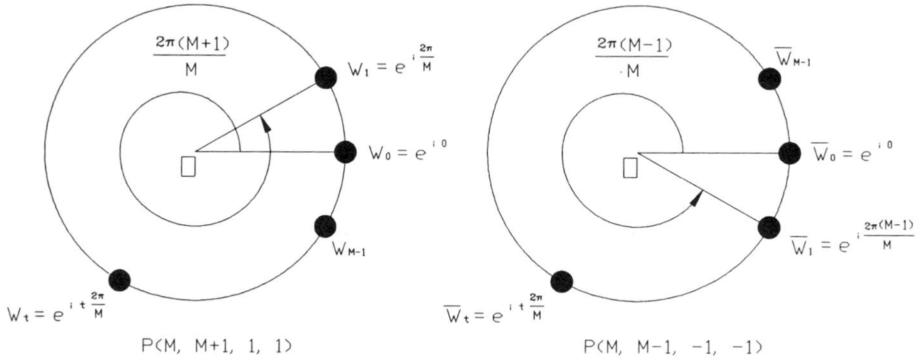

Fig. 2 Two trivial algebraic patterns with $m = \pm 1$.

For example, the following sequence is an algebraic pattern of filament winding.

$$P = \{ W_t = e^{it\frac{2\pi 7}{12}} \quad | \, t = 0, 1, 2, \cdots, 11\} \tag{2.15}$$

with the parameters M = 12, K = 7, m = -5 and n = -3 and is denoted by P(12, 7, -5, -3) as shown in Fig.3a. It is easy to examine that these parameters satisfy the fundamental equation.

Obviously, the following statement is true

$$P \Leftrightarrow (M, K) \Leftrightarrow (M, K, m, n) \Leftrightarrow (M, m) \tag{2.16}$$

where the notation $\Leftrightarrow$ means that the correspondence between the right hand side and the left hand side of it is one-to-one. It concludes from this statement that the algebraic patterns can be defined by the four parameters M, K, m, n satisfying the fundament equation of the filament winding. Thus, for convenience, the notation P(M, K, m, n) is used to represent the corresponding algebraic pattern with split number M, skip number K and the pattern numbers m and n.

Definition 2. An algebraic pattern is said to be a positive pattern if its first pattern number $m > 0$, and is said to be a negative pattern otherwise.

The geometric interpretation of m is as follow: The |m|-th point in P is the first one to reach the point next to $W_0 = 1$, which is rotated over an angle θ from W_0 anti-clockwise if $m > 0$ and is rotated clockwise if $m < 0$. For example, the pattern P (12, 7 -5, -3) is negative, W_5 is the first one in P to reach the point next to W_0 and is rotated over $\pi/6$ from W_0 clockwise as shown in

Fig.3a. the pattern P(12, 5, 5, 2) is positive, W_5 is the first point in P to reach the point next to W_0 and is rotated from W_0 counter-clockwise as shown in Fig. 3b.

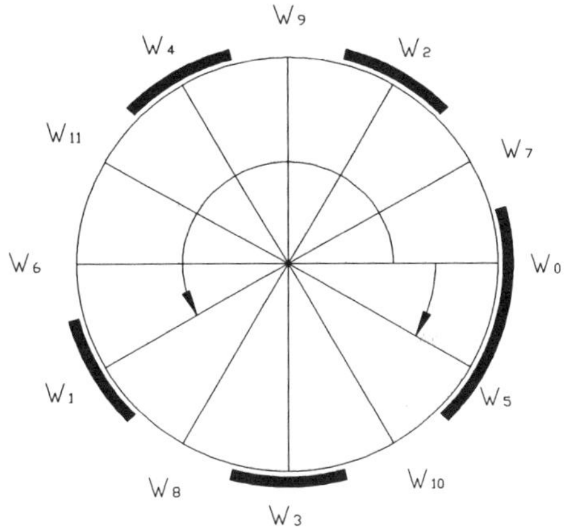

Fig. 3a A negative pattern P(12, 7, -5, -3) with m = -5 < 0 and band width shown in black.

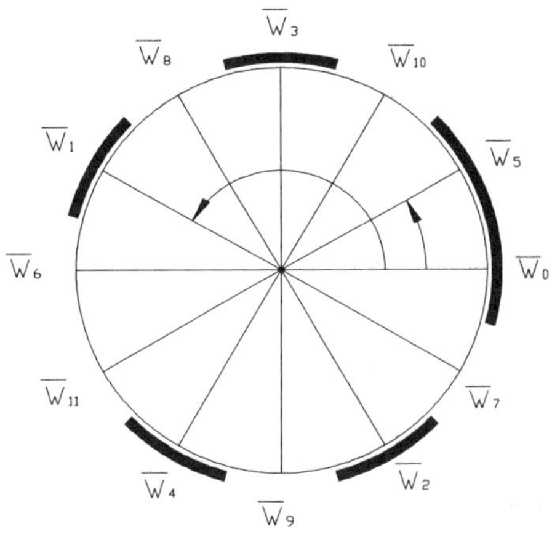

Fig. 3b A positive pattern $\overline{P}$(12, 5, 5, 2) with m = 5 < 0 and band width shown in black.

Definition 3. The algebraic patterns with identical |m| and opposite signs and are called the conjugate patterns. A pair of conjugate patterns are usually denoted by P and $\overline{P}$.

It is easy to show that P(M, M-K, -m, -m+n) is the conjugate pattern to P(M, K, m, n) ,and of course, P(M, K, m, n) is the conjugate pattern to $\overline{P}$(M, M-K, -m, -m+n). For example, P(12, 7 ,-5, -3) and P(12, 5, 5, 2) are conjugate patterns. Especially, two trivial patterns P(M, M+1, 1, 1) and $\overline{P}$(M, M-1, -1, -1) are conjugate patterns.

The significance of the conjugate patterns is in that it reduces half of the computation in finding all feasible patterns. In fact, the condition (2.12) can be replaced by a stronger condition as follows,

$$0 < m < M/2 \tag{2.17}$$

With another half of the feasible patterns to be obtained as their conjugates. For example, the pattern P(12, 7, -5 ,-3) can be directly obtained from its conjugate pattern P(12, 5, 5, 2) by replacing (5) with 12 – 5 =7 as K replacing (-5) with (+5) as m and replacing (-3) with 5 – 3 = 2 as n. Since the interval of the first pattern number m (2.17) is cut into half of the original one, the computation cost is reduced by half.

3. The Pattern Generation Algorithm

As mentioned above, an algebraic pattern is uniquely determined by the corresponding parameters (M, K, m, n). These four parameters are not independent and they satisfy the fundamental equation of the filament winding. Suppose that the split number M is given. Since M and K satisfy the fundamental equation, according to the theory of modern algebra, M and K are relatively prime meaning M and K have no common divisor except 1. Thus, once M is given, K is no longer free, and M and K are relatively prime. A well known Euclidean Algorithm in modern algebra can be used to check whether two numbers are relatively prime, and find all numbers that are less than M and are prime to M. Let a = M, and b = K. Dividing a by b, we have

$$a = bq + r \qquad 0 \le r \le b \qquad (3.1)$$

where q is called the quotient and r is called the remainder in the division of a by b. Generally, we have,

$$\begin{aligned} a &= bq_0 + r_1 && 0 \le r_1 < b \\ b &= r_1q_1 + r_2 && 0 \le r_2 < r_1 \\ r_1 &= r_2q_2 + r_3 && 0 \le r_3 < r_2 \\ &\vdots && \vdots \\ r_k &= r_{k+1}q_{k+1} + r_{k+2} && 0 \le r_{k+2} < r_{k+1} \end{aligned} \qquad (3.2)$$

Since the integer r_1, r_2, ... , r_{k+2} are decreasing and are all nonnegative, there is a smallest integer n such that $r_{n+1} = 0$

$$r_{n-1} = r_nq_n + r_{n+1}\,, \qquad r_n \neq 0 \text{ and } r_{n+1} = 0 \qquad (3.3)$$

It is proved that r_n is the common divisor of M and K. Thus, the condition of M and K being relatively prime is equivalent to $r_n = 1$. Suppose that M and K are relatively prime, the Euclidean Algorithm can also be used to find the pattern numbers m and n. In fact, from(3.2) we have

$$\begin{aligned} r_1 &= a - bq_0 \\ r_2 &= b - r_1q_1 \\ r_3 &= r_1 - r_2q_2 \\ &\vdots \\ r_{k+2} &= r_k - r_{k+1}q_{k+1} \\ r_n = 1 &= r_{n-2} - r_{n-1}q_{n-1} \end{aligned} \qquad (3.4)$$

Substituting the remainders from the preceding equations successively until a and b are present in the equation.

$$1 = r_{n-2} - r_{n-1}q_{n-1} = -\,q_{n-1}\,r_{n-3} + (1 + q_{n-1}q_{n-2})\,r_{n-2} = \cdots = mK - Mn \qquad (3.5)$$

It is easy to show that m obtained from (3.5) satisfies (2.12). Thus, m and n are uniquely determined and are the pattern numbers of the corresponding pattern P(M, K, m, n). For example, let M = 27, K = 19, first of all, check if M =27 and K=19 are relatively prime by performing the arithmetic for Euclidean Algorithm as follows,

$$\begin{aligned} 27 &= 1 \cdot 19 + 8 && (q_0 = 1,\ r_1 = 8) \\ 19 &= 2 \cdot 8 + 3 && (q_1 = 2,\ r_2 = 3) \\ 8 &= 2 \cdot 3 + 2 && (q_2 = 2,\ r_3 = 2) \\ 3 &= 1 \cdot 2 + 1 && (q_3 = 1,\ r_4 = 1) \end{aligned} \qquad (3.6)$$

Thus, the last non-zero remainder is $r_n = r_4 = 1$, then 27 and 19 are relatively prime. Expressing all remainders as follows

$$\begin{aligned} 8 &= 27 - 9 \\ 3 &= 19 - 2 \cdot 8 \\ 2 &= 8 - 2 \cdot 3 \\ 1 &= 3 - 1 \cdot 2 \end{aligned} \qquad (3.7)$$

Substituting the remainders from preceding equations successively, we obtain,

$$\begin{aligned} 1 &= 3 - 1 \cdot 2 = 3 - 1 \cdot (8 - 2 \cdot 3) = 3 \cdot 3 - 1 \cdot 8 = 3\,(19 - 2 \cdot 8) - 1 \cdot 8 \\ &= 3 \cdot 19 - 7 \cdot 8 = 3 \cdot 19 - 7 \cdot (27 - 1 \cdot 19) = 10 \cdot 19 - 7 \cdot 27 \end{aligned} \qquad (3.8)$$

Comparing with the fundamental equation with M = 27 and K = 19 as follows,

$$m \cdot 19 - n \cdot 27 = 1 \tag{3.9}$$

hence

$$m = 10, \quad n = 7 \tag{3.10}$$

notice that (2.12) is satisfied . Thus P(27, 19, 10, 7) is a filament winding pattern. Based on the above procedure, a much simpler and faster algorithm for generating the filament winding patterns can be developed. Let Q be an array for all the quotient q_i obtained in the process of the Euclidean Algorithm.

$$Q = \{ q_i \mid i = 0, 1, 2, \cdots, n-1\} \tag{3.11}$$

notice that

$$1 = r_{n-2} - r_{n-1}q_{n-1} \tag{3.12}$$

denote Q(n) = 1, Q(n+1) = 0 and rewrite (3.12) as

$$1 = Q(n)r_{n-2} + Q(n-1)r_{n-1} \tag{3.13}$$

it requires that $Q(n-1) = -q_{n-1}$, then the (n-1)-th variable in the array Q is now assigned with a new value $-q_{n-1}$ (it used to be q_{n-1}). Generally, suppose we have

$$1 = Q(k+2)r_k + Q(k+1)r_{k+1} \qquad (-1 \le k \le n-2) \tag{3.14}$$

where $r_{-1} = a = M$, and $r_0 = b = K$. Substituting

$$r_{k+1} = r_{k-1} - r_k q_k \tag{3.15}$$

into (3.14), we have

$$1 = Q(k+1)r_{k-1} + [Q(k+2) - Q(k+1)Q(k)]r_k \tag{3.16}$$

comparing with

$$1 = Q(k+1)r_{k-1} + Q(k)r_k \tag{3.17}$$

a recursive formula for the coefficients in the linear combination (3.5) is obtained as follows,

$$Q(k) = Q(k+2) - Q(k+1)Q(k) \quad (k = n-1, n-2, \cdots, 1, 0) \tag{3.18}$$

Since

$$1 = mr_0 - nr_{-1} = Q(1)r_{-1} + Q(0)r_0 \tag{3.19}$$

then,

$$m = Q(0), \quad n = -Q(1) \tag{3.20}$$

The algorithm is briefly described as follows

<u>The Filament Winding Pattern Generation Algorithm:</u>

For given split number M and skip number K (< M), perform the arithmetic for the Euclidean Algorithm to obtain q_i, r_n and n, If $r_n \neq 1$ then no solution, otherwise, form the array Q.

$$Q = \{ q_0, q_1, \cdots, q_{n-1}, 1, 0\} \tag{3.21}$$

Evaluate Q(k) by recursive formula (3.18) and (3.20) to obtain the pattern numbers m and n. Thus, a filament winding pattern can be easily determined by the pattern generation algorithm. The algorithm costs least computation and memory space that it has only 2n multiplications / divisions and n subtractions and employs only one array.

For example, let M = 27 and K = 19. Perform the arithmetic (3.2) for q_i. Since $r_n = r_4 = 1$, there exists a filament winding pattern with M = 27 , K = 19. Set the array Q containing n+2 = 6 numbers as follows,

$$Q = \{1, 2, 2, 1, 1, 0\} \quad (3.22)$$

using the recursive formula (3.18), the new values of the array Q are obtained as

$$Q = \{10, -7, 3, -1, -1, 0\} \quad (3.23)$$

which gives

$$m = Q(0) = 10, \qquad n = -Q(1) = 7 \quad (3.24)$$

Thus, P(27, 19, 10, 7) is the filament winding pattern with M = 27 and K = 19. Of course, for a given split number M, not any number can be taken as the skip number K. M and K must be relatively prime. For example, let M = 27 and K = 6, from the above algorithm, r_n =3 that means 3 is the greatest common divisor of 27 and 6, then no non-overlapping pattern exists with M = 27 and K = 6, and the answer is " No Solution".

The pattern number m has its geometric interpretation and is important to the design of filament winding procedure as shown in [1] that the density of the interweaving degree of a filament winding pattern is uniquely determined by the split number M and the pattern number m.

4. The Design of Filament Winding Patterns

To provide more flexibility in a design system, it is required to find all feasible patterns for a given split number M. Recall that the skip number K is in the interval of [2, M-1] or K = M+1, and P(M, M+1, 1, 1) and $\overline{P}$(M, M-1, -1,-1) are two trivial patterns. Thus, it suffices to assume that K is in the interval [2, M-2]. Notice that, the conjugate transformation is as follows,

$$P(M, K, m, n) \Rightarrow \overline{P}(M, M-K, -m, -m+n) \quad (4.1)$$

then it suffices to consider K in the following intervals

$$K \in [2, \frac{M-2}{2}] \qquad \text{if M is even} \quad (4.2)$$

$$K \in [2, \frac{M-1}{2}] \qquad \text{if M is odd} \quad (4.3)$$

Thus, it reduces half of the computation cost with another half of the feasible patterns to be obtained by the conjugates as shown in (4.1). The algorithm is summarized as follows.

<u>The Algorithm for Generating Feasible Filament Winding Patterns</u>

Let M be a given split number and K be a number in (4.2) or (4.3) depending on whether M is even or odd. Set

$$r_0 = M \quad \text{and} \quad r_1 = K \quad (4.4)$$

find q_i and assign them in an array Q by the following recursive formula,

$$r_k = r_{k+1}q_{k+1} + r_{k+2} \qquad (k = 0, 1, 2, \cdots) \quad (4.5)$$

Let r_{n+1} be the last nonzero number in the above loop. Return if $r_{n+1} \neq 1$ (no solution!), otherwise, set the array Q as

$$Q = \{ q_1, q_2, \cdots, q_n, 1, 0\} \quad (4.6)$$

reassign array Q by the following back recursive formula

$$Q(k) = Q(k+2) - Q(k+1)Q(k) \qquad (k = n, n-1, \cdots, 1) \quad (4.7)$$

it follows

$$m = Q(1), \quad n = -Q(2) \tag{4.8}$$

the pattern numbers $\overline{m}$ and $\overline{n}$ of the conjugates are obtained as

$$\overline{m} = -Q(1), \quad \overline{n} = -Q(1) - Q(2) \tag{4.9}$$

as K runs through the interval in (4.2) or (4.3) depending on whether M is even or odd, all feasible filament winding patterns are obtained.

This program employs two recursive formulas to greatly simplify the procedure and reduces the computation by half with another half of the feasible patterns obtained as their conjugates as shown in (4.1) The computation for q_i costs only n·divisions although in formula (4.3) there is an addition. The computation for Q(k) costs n multiplications and n subtractions. Therefore, for each pattern generation the algorithm costs 2n multiplications /divisions and n subtractions where n is the repetition number in the recursive loop for q_i.

Take M = 27 for example, all feasible filament winding patterns with $2 \leq K \leq (M-1)/2$ and K = M + 1 are shown in Table 1.

K	28 or (1)	2	4	5	7	8	10	11	13
m	1	-13	7	11	4	-10	-8	15	-2
n	1	-1	1	2	1	-3	-3	2	-1

Table 1. The feasible filament winding patterns with M = 27, $2 \leq K \leq 13$ or K = 28 (or 1)

The conjugate patterns are shown in Table 2.

K	14	16	17	19	20	22	23	25	26
m	2	-5	8	10	-4	-11	-7	13	-1
n	1	-3	5	7	-3	-9	-6	12	-1

Table 2. The conjugate patterns of above with M = 27 and $14 \leq K \leq 26$

All feasible patterns with M = 27 are in table 1 and table 2. It is clear that altogether there are 18 different feasible patterns with M = 27, and K = 3, 6, 9, 12, 15, 18, 21 and 24 are not included as the skip numbers, since they have common divisor 3 with M = 27. The graphical representation of a pair of conjugate patterns P(27, 19, 10, 7) and $\overline{P}$(27, 8, -10, -3) are shown in Fig. 4, Fig. 5, Fig. 6 and Fig. 7 [1].

It is a strong desire to have as many feasible patterns as possible so that it provides more flexibility in the computer aided filament winding design system. Let w be the width of the fiber band and α be the winding angle between fiber band and the central axis, and let θ_0 be the central angle corresponding to the band width w on circle C as shown in fig. 8, then

$$\theta_0 = \frac{w}{(r \cdot \cos\alpha)} \tag{4.10}$$

The central angle θ corresponding to the uniform partition of circle C by the split number M is that

$$\theta = \frac{2\pi}{M} \tag{4.11}$$

The coverage of a filament winding pattern P(M, K, m, n) is defined and denoted by C(P) as follows,

$$C(P) = \frac{M\theta_0}{2\pi} \tag{4.12}$$

from (4.10) and (4.11), we have

$$C(P) = \frac{M\theta_0}{M\theta} = \frac{\theta_0}{\theta} = \frac{w}{(r \cdot \theta \cdot \cos\alpha)} \tag{4.13}$$

if $\theta_0 = \theta$ then $C(P) = 1$ that means the filament winding pattern P uniformly cover the entire cross-sectional circle C. Generally, it is required that the filament winding patterns fully cover the entire mandrel, the condition for which is

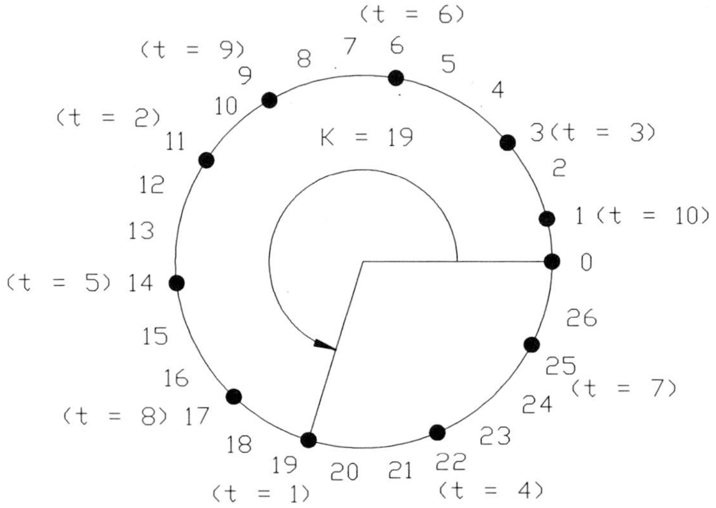

Fig. 4 One dimensional graphical representation of P(27, 19, 10, 7), where t is the index.

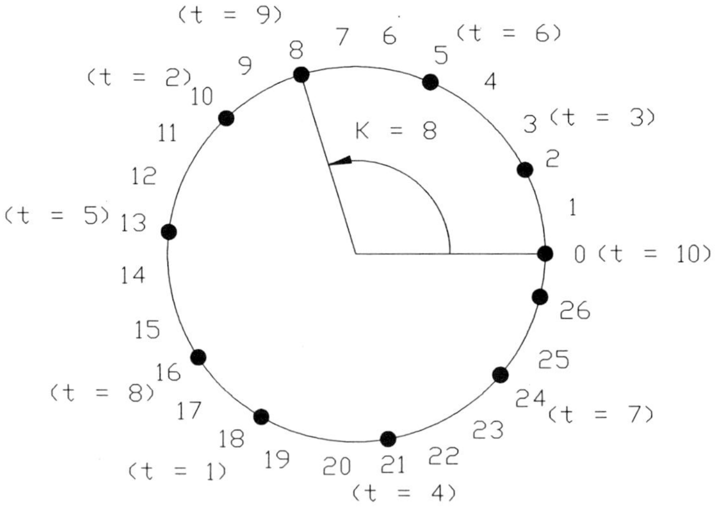

Fig. 5 One dimensional graphical representation of $\overline{P}$(27, 8, -10, -3), where t is the index.

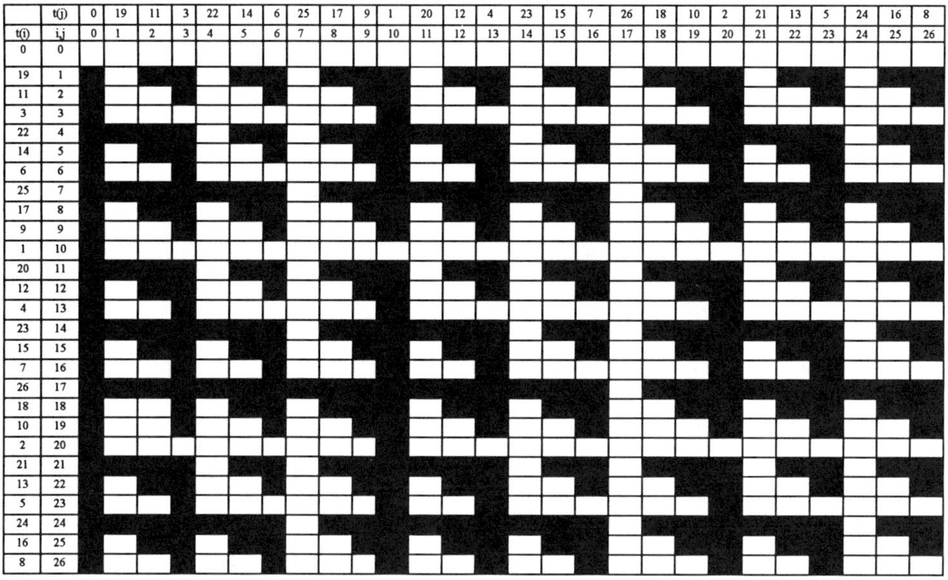

Fig.6 Two dimensional graphical representation of P(27, 19, 10, 7), where i , j are Location indexes and t(i) and t(j) are Power indexes [1].

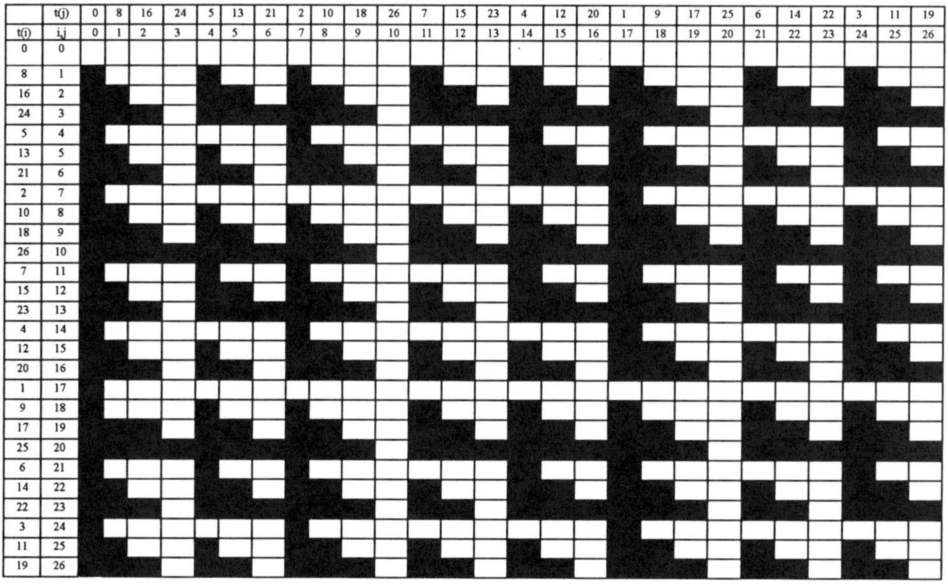

Fig.7 Two dimensional graphical representation of $\overline{P}(27, 8, -10, -3)$, where i , j are Location indexes and t(i) and t(j) are Power indexes [1].

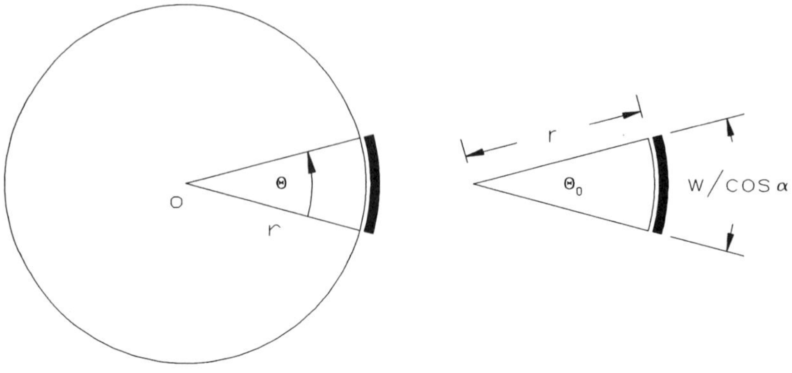

Fig. 8 The central angles θ and θ_0.

$$C(P) = \frac{\theta_0}{\theta} \geq 1 \tag{4.14}$$

or

$$2\pi = M\theta \leq M\theta_0 \tag{4.15}$$

The condition (4.15) is equivalent to

$$M \geq \tilde{M} = \begin{cases} [\frac{2\pi}{\theta_0}] & \text{if } \frac{2\pi}{\theta_0} \text{ is an integer} \\ [\frac{2\pi}{\theta_0}] + 1 & \text{otherwise} \end{cases} \tag{4.16}$$

where $[\frac{2\pi}{\theta}]$ is the greatest integer that is less than or equal to $\frac{2\pi}{\theta}$.

Thus, (4.16) is the condition of a filament winding to be of full coverage. It is obvious that $\tilde{M}$ is the minimal number that can be taken as the split number for full coverage. In fact, if $M = \tilde{M}$, we have

$$\theta = \frac{2\pi}{\tilde{M}} \tag{4.17}$$

then

$$C(P) = \frac{\tilde{M}\theta_0}{2\pi} \geq 1 \tag{4.18}$$

and (4.14) is not valid if M is less than $\tilde{M}$. To provide more flexibility, one can take $M = \tilde{M} + i$, $(i \geq 1)$, such that the coverage

$$C(P) = \frac{(\tilde{M} + i)\theta_0}{2\pi} \tag{4.19}$$

is acceptable to the users and consistent with all other requirements.

The authors would like to express their deep thanks to Chinese National Science Foundation for the partly support of this work.

References

[1] Y. D. Liang, Gene Luo, Yong Chen, Kuo Xian Qiu, " Graphical representation and Calculation of Interweaving Degrees of Filament Winding Patterns", Proceedings of the 41st International SAMPE Symposium, March 1996.

[2] Lossie M., " Production oriented design of filament wound composite", Ph.D. dissertation 90D5, faculty of applied science, Division PMA, K. U. Leuven, 1990.

[3] Weck M., Zender M., Herberg F., " Automatisierte fertigung von faserverbundbauteilen im wiekelverfahven", - VDI-Z, 131/5, 1989.

[4] Curtis P. T., Bishop S. M., " An assessment of the potential of woven carbon fiber reinforced plastics for high performance application", Composites, Vol. 15 No. 4, Oct. 1984

28th International SAMPE Technical Conference
November 4-7, 1996

DYNAMIC SYSTEM CHARACTERISTIC OF FIBER PLACEMENT TOOLING FOR AIRCRAFT FUSELAGE SHELL.

Ly D. Nguyen
Kevin M. Retz
Premier I Fuselage Design Built Team
Raytheon Aircraft Company
Wichita, Kansas 67201

ABSTRACT

One of the newest and most promising automated composite lay up processes is the fiber placement system (FPS) by Cincinnati Milacron. A CNC driven FPS can lay prepreg tow down on any large male mandrel that contains complex contoured geometry. However, manufacturing a large complex part often requires complicated tooling. Tooling size, weight, functional requirements such as low deflection, high rotational frequency, and high torsional stiffness to accommodate the manufacturing process must be taken into consideration when designing FPS tools. These requirements usually approach the limit of conventional material properties and design techniques due to cost, functional requirements and structural integrity. A methodology has been developed to understand the sources of critical variables that play a vital role in the design and development of fiber placement tooling. This paper describes the mathematical modeling technique used to analyze the dynamic system characteristics of a large mandrel in motion, critical design variables, and analysis. It also demonstrates the important role of the ratio of total tool torsional stiffness to mass movement of inertia and the rotational stability technique used to mitigate tool deflection and unbalancing.

1. INTRODUCTION

Producing modern composite parts using fiber placement has emerged into a sophisticated manufacturing process. This technology is being applied in the aerospace industry where large advanced composite parts are being fabricated with precision controlled aerodynamic curves and surfaces. Fabricating such complex parts often requires a complicated mandrel design for the machine to lay material onto. The tool concepts can be solid, inflatable, break-away and retractable or pogo, all of these tool types are used in industry today. However, as parts are becoming larger and more complex in shape, the selection of which concept to use becomes more difficult due to stringent process requirements. Advanced analysis for understanding part geometry, machine requirements, tool structures sizing and functional requirements will lead to a less complex mandrel design. The importance of understanding the tooling structural sizing process is illustrated in this aircraft fuselage study. It must be pointed out that, finding a perfect tool concept while minimizing weight, deflection and maximizing torsional stiffness is often very difficult and sometimes impractical due to cost, material properties and process requirements.

2. OBJECTIVES

The objective of this work is to design a mandrel that will fabricate a composite forward fuselage shell. The fuselage shell includes the nose, cockpit and cabin section. Figure 1 depicts the general geometry of a forward aircraft fuselage shell.

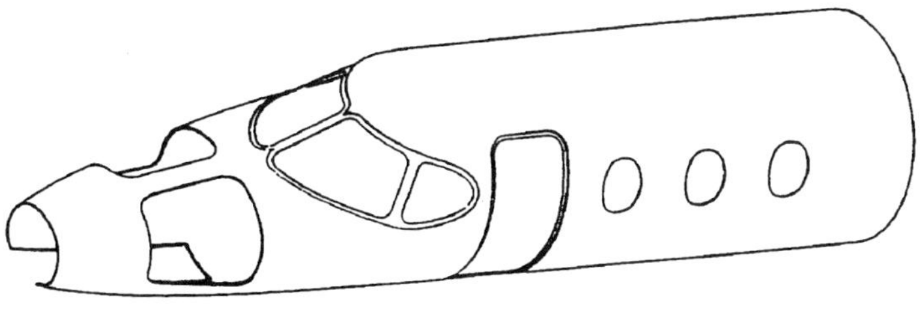

Figure 1 - Forward Fuselage Shell

The composite shell has laminate prepreg fiber materials with multiple fiber orientations such as 0, 90,+/-45 degree. The fiber band is approximately 3 inches wide and is mechanically compacted onto the mandrel surface with a variable load from 100 to 300 lbf.

3. NOMENCLATURE

L	= Length of the mandrel (in)	y_1	= Local deflection due to L_p (in)
CG	= Center of gravity (in)	a	= Width of segment (in)
CLR	= Center line/axis of rotation	b_1	= Width of roller (in)
d	= Distance from CG to CLR (in)	a_1	= Length of roller (in)
y	= Longitudinal deflection (in)	t	= Skin thickness (in)
E	= Modulus of Elasticity (lb/in^^{2})	P_1	= Point of contact at head stock
G	= Shear Modulus (lb/in^^{2})	P_2	= Point of contact at tail stock
θ	= Total angle of twist (degree)	W	= Total weight of the tool (lb)
J_i	= Polar moment of inertia (in^^{4})	λ	= CLR off set angle (degree)
θ_l	= Local angle of twist (degree)	q	= Shear flow (lb/in)
R	= Tool outer radius (in)	ds	= Differential skin surface
r	= Tool inner radius (in)	A	= Total skin surface area (in^^{2})
ψ	= Linear slope approximation	m	= Total mass of the tool (lb)
n,k	= No. of components in tool	β	= Rate of twist
R(x)	= Rate of change in tool radius at the cockpit section		
T_m	= Torque produce by the machine (lb-ft)		
$d\lambda$	= The rate of change in CLR off set angle (degree)		
a_t	= Tangential acceleration/deceleration (in/sec^2)		
F	= Load produced during acceleration and deceleration (lb)		
α	= Ratio of flat plate formula approximation		
b	= Length of segment or distance required between headers (in)		
τ	= Local shear due to tortional stress (lb/in^^{2})		
η	= Ratio of mass of inertia to torsional stiffness		

I_i = Mass movement of inertia (lb.Sec$^{\wedge 2}$.in)
M_{off} = Moment produce by off set load (ft-lb)
L_p = Compaction load from machine compaction roller (lbf)
T = Total torque acting on the tool (ft-lbf)
δ = Machine radial acceleration (Rad/sec^2)
a_t = Tangental acceleration (In/sec$^{\wedge 2}$)
Γ = Differential area enclosed by the medial line (in$^{\wedge 2}$)

4. GENERAL TOOL GEOMETRY

The description of the mandrel structure is devided into two cagegories, the external and internal. The external structure includes the skin, torque box and the shaft as shown in figure 2. Figure 3 shows the internal structure with reinforcement such as longerons, headdres, guset and truses. It also illustrates the basic free body diagram. The design utilized CATIA®[1].

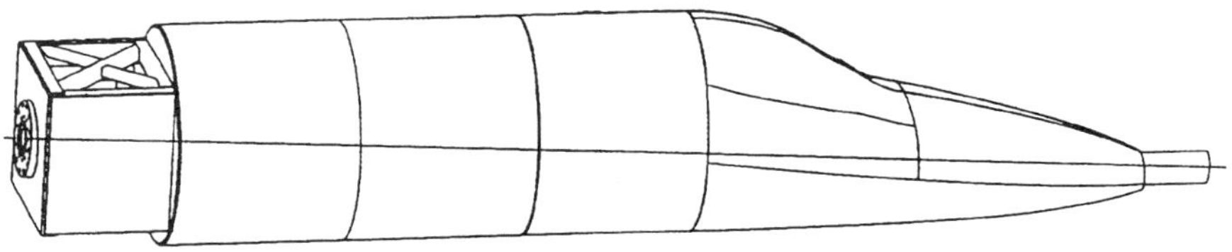

Figure 2- Forward Fuselage External Structures

5. ANALYSIS MODELING

The tool structural integrity and key characteristic are primarily depended on the combined weight of the tool and part, surface skin stiffness, deflection due to off set load, sag of the tool due to its own weight and the rotational natural frequency of the tool.

5.1 Tool Surface Stiffness The tool skin must have sufficient stiffness to carry its own weight as well as the compaction load (L_p). If the tool surface deflects in bending due to L_p, the machine will place fiber in the wrong position and orientation. The laid tows will also buckle after the skin springs back into it's original position once L_p is removed. If the tows buckle they will not carry the desired design load. Figure 4 demonstrates the header span and sizing technique for supporting the skin.

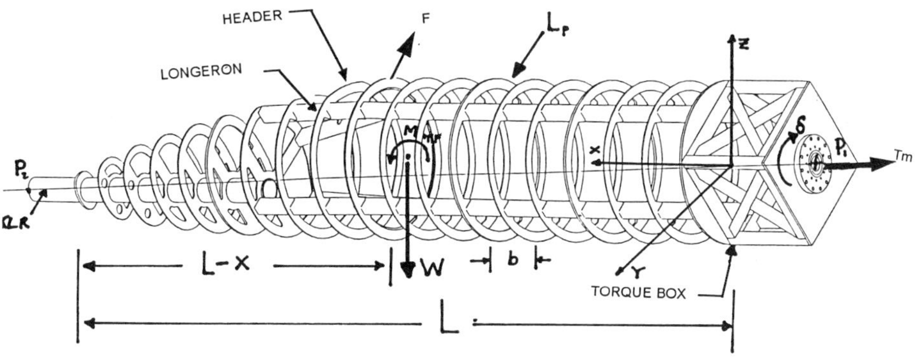

Figure 3- Forward Fuselage Mandrel Internal Structures

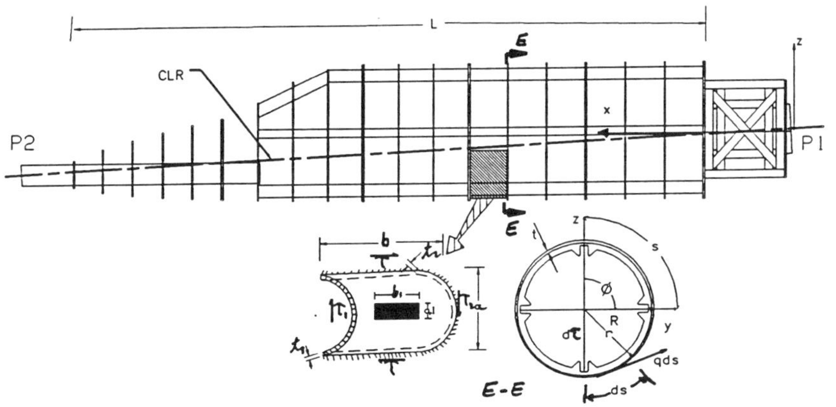

Figure 4 - Differential Element of Skin Between Support

The conservative assumptions used are: (1) the edges of the skin are simply supported, (2) the load L_p is uniform over the central area between a_1 and b_1, (3) the radius of the skin is gradual enough to be flat. By applying the basic stress formula [1,3], interpolating α , and using a required deflection of 0.001 along with using iterations of various t and L_p we can determine the best header span (b) and the skin thickness (t) from equation (2).

$$y_l = \frac{\alpha L_p b^{\wedge 2}}{E t^{\wedge 3}} \quad (1)$$

and solve for header span

$$b = \frac{\sqrt{y_l E t^{\wedge 3}}}{\alpha L_p} \quad (2)$$

The b distance can be used to determine the number of headers required over the span of the tool. Having the skin thickness t value, we can further determine the torsional stiffness of the skin by integration of the basic mass moment of inertia over the entire length of the skin. By approximating the mandrel (figure 4) as a thin-walled hollow tube with a cross section as shown the shear stress and local skin angle of twist can be derived based on the membrane analogy. The shear stress is

$$\tau_1 t_1 dL = \tau_2 t_2 dL \quad (3)$$

from equilibrium forces and the increment of torque, we have:

$$dT = r(qds) \quad (4)$$

From (3) and (4), we see that, the τt is constant around the mandrel. The torque produced by moment arm r that has q independent of s yields the total torque as:

$$T = \int r(qds) \quad (5)$$

By performing the integration process around the medial line of the cross sectional area yields:

$$\Gamma = r\, ds/2 \quad (6)$$

so the integral of rds is twice the area Γ enclosed by the medial line. Note that the area Γ is not the cross section A of the tube of material since this is a thin-walled closed tube, $\Gamma >> A$.
The rate of local twist during acceleration and deceleration θ_l is the work done by the strain energy from zero to a final value of T. Simultaneously, the angle of twist increases from zero to a final value $\theta = \beta L$, The work done by T is $T\theta_l/2$. Strain energy per unit volume is $\tau^2/2G$. The differential skin element can be taken as Ltds, which yields the energy balance as

$$T\,\theta_l/2 = \int (\tau^2/2G) Ltds \quad (7)$$

$$\theta_l = (T/G)\int (\tau^2 Ltds \quad (8)$$

Which is known as Bredt' Formula. By integration, it will yield the local angle of twist for sizing the skin parameter.

To calculate the total tool and skin mass moment of inertia, we use the method of approximation as described in figure 5.

$$J = \int [R^{\wedge 2} dA] + 2\pi R^{\wedge 2}\, t \quad (9)$$

The slope of the cockpit section is approximated as a cone by using R(x) = ψx+b and substitute into equation (9) to get

$$J = 2\pi \int^{L}_{L-x} \left[\left(\psi^3 x b^2 + 2\psi^2 x^2 b^2 + \psi x b^3 \right) dx \right] + 2\pi R^{\wedge 2} t \qquad (10)$$

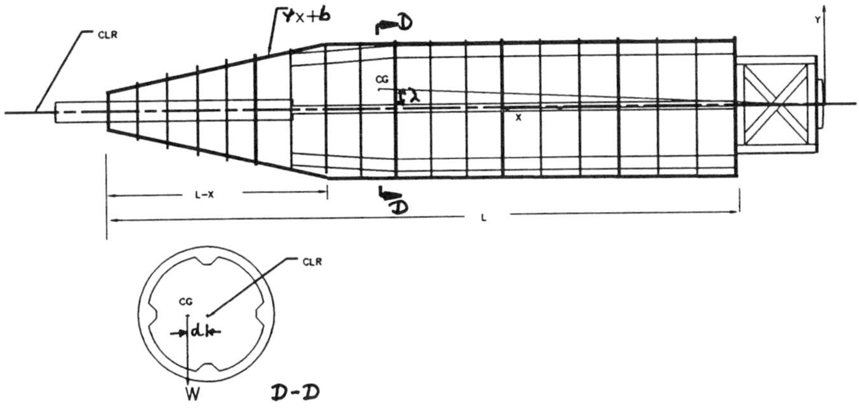

Figure 5 - Geometric Approximation

5.2 Tool Unbalancing - A large mandrel can weight several tons and is often not symmetrical about the CLR. The entire tool skin and torque box can be twisted due to torque generated by an unbalanced load, in couple with the inertia during acceleration and deceleration of the tool. This fluctuation and deflection can cause the machine to lay fiber tow in the incorrect position, possibly buckling the fiber, and create difficulty for locating and controlling other reinforcement details such as cabin window, cabin door, cockpit window frames and other attachments. Another constraint on the off set load is that it can not be greater than the break load of the head stock motor on the machine. Therefore, this unbalancing variable must be minimized. Several techniques such as using lower density material at the outer skin or add additional weight as a counter balance have been applied on smaller applications. Using lower density composite material often results in internal tool structure assembly difficulty and additional tooling required for fabricating the tool components. Additional mass for balancing will yield additional inertia, longitudinal and torsional deflection. Selecting the CLR is critical in minimizing the unbalancing problem in selecting the CLR. During the process of selecting the CLR, one should consider the overall geometry of the part, the tool and the working envelope of the machine. The CLR should go through the CG of the tool whenever possible, a small amount of off set can be exempted to reduce the movement of the fiber placement machines arm. Figure 5 shows how the off set load relates to the off set angle (λ) and causes the moment on the tool.

$$M_{off} = W x \tan\lambda \qquad (11)$$

To illustrate the important of the relationship between the off set angle λ to the weigth acting at the C.G of the tool, the moment is equal to zero as λ approach zero.

$$\lim_{\gamma \to 0} M_{off} = 0 \tag{12}$$

This not only minimizes the off set load, but it also reduces the isolation and inertia of the mandrel.

5.3 Mounting Initial Condition - The head stock P_1 and tail stock P_2 of the fiber placement machine are important factors that influence the deflection and torsional stiffness of the mandrel. One should design a tool head stock with the largest circumference or diameter possible to carry the torsional load produced by the machine during acceleration and deceleration. The method of mounting will give the initial condition for calculation of the total tool deflection. In a typical condition P_1 is fix supported and P_2 is simply supported. The simply supported end will compensate for any small misalignment in the tool structures as well as reduce the torsional load of the tapered end. The standard deflections and slopes of the uniform beam formula can be applied to calculate the deflection due to shear, moment and distributed loading. The total deflection should not be greater than 0.050 inch [3,4].

5.4 Angle of Twist - The total angle of twist is depend on the acceleration and deceleration of the tool which can be calculated from the following formulas. The tangential acceleration of the tool is a variable of the tool radius and the machines radial acceleration, with the basic body force in motion we have:

$$a_t = R\delta \tag{13}$$

$$F = m\, a_t \tag{14}$$

Combining the tangential force to the differential moment arm in the radius (from the center of the mandrel) and the offset load coupled with the CLR offset angle will yield the total torque on the mandrel as:

$$T = F\, dR + Wx\tan\gamma \tag{15}$$

The total angle of twist is the total torque divided by the total torsional stiffness:

$$\theta = TL/GJ \tag{16}$$

5.5 Ratio of Mass of Inertia to Torsional Stiffness - The mass of inertia and the torsional stiffness are very important variables that determine the natural frequency of the tool. While the tool is in motion, the rotational acceleration and deceleration will generate a torsional load that causes rotational deflection and oscillation. The rotational deflection if large enough will cause variation in fiber position. The oscillation of the tool is very difficult to calculate precisely because of the overall tool behavior in motion. However, the best way to increase the natural frequency is to increase the torsional stiffness, by designing an internal structure to carry loads directly into the skin. In other words, designing a tool that utilizes the outer skin as primary structure with internal reinforcement. The process for calculating the constant η of this ratio is:

$$\eta = \frac{\sum_{i=1}^{n} I_i}{G \sum_{j=1}^{k} (J_j/L_j)} > 0.001 \qquad (17)$$

The η value should be equal to or slightly greater than 0.001 since it is preferred that the mandrel have as small of a mass of inertia as possible to avoid handling problems as well as excessive longitudinal and torsional deflection while maintaining maximum torsional stiffness.

6. CONCLUSIONS

The fiber placement process shows great promise in its ability to make large complex composite parts. This ability requires accurate tooling to accommodate the manufacturing requirements. As this process becomes more applicable in aerospace, a FEA code should be developed to analyze a large tool with the above mentioned variables. This paper has illustrated the methodology used in preliminary sizing of the forward fuselage mandrel and it's design requirements.

7.0 ACKNOWLEDGMENTS

It is always a pleasure for us to express our appreciation for the support and technical advise from Jerry Lunde, Jerry Acker, Robert Johnson, Mike Edmonson, Surrinder Aggarwal, Feaq Skaikh, Steve Young, Ric Abbott, Tom Fagan, Don Harris and the staff at Cincinnati Milacron. The authors also would like to thank the Raytheon Aircraft Company for providing the opportunity to publish this paper.

8.0 REFERENCES

[1] Timosshenko, S.P. and Gere, J.M, Theory of Elasticity Stability, McGraw-Hill, New York, 1961, pp.225-250.

[2] Nguyen, L.D. and Jones, S.P, Methodology to Assess Assembly Variation in Aircraft Fuselage Structures, MED-Vol, 2-2, Manufacturing Science and Engineering 1995, pp 1157-1161.

[3] Cook, R.D. and Young, W.C, Advanced Mechanics of Materials, Macmillan, New York 1989, pp 193-198.

[4] Nguyen, L.D, Compressed Air Refrigeration Technology For Aerospace Coating Applications, 94-WA/HT-8, International Mechanical Engineering Congress & Exhibition, Chicago,Il, (1994).

[5] Raymond J. Roark, Young, W.C, Formulas for Stress and Strain, 5th Ed. McGraw-Hill Book Company., New York, 1975

[1] .CATIA is registered trade mark of the Dasualt Company

28th International SAMPE Technical Conference
November 4-7, 1996

COMPOSITE TOOLING VIA LAMINATED OBJECT MANUFACTURING, A RAPID AND AFFORDABLE METHOD

Richard P. Chartoff
Brian Priore
Donald A. Klosterman
Rapid Prototype Development Laboratory and
Center for Basic and Applied Polymer Research
University of Dayton
Dayton, OH 45469-0130
and
Sung S. Pak
Helisys, Inc.
Torrance, CA 90505

ABSTRACT

An innovative, rapid prototyping (RP) technology is being adapted for the automated manufacture of composite tools and molds. The technology is a direct extension of the Laminated Object Manufacturing (LOM) process. LOM is an established technique that is used to create solid prototypes by sequentially cutting and laminating layers of adhesive paper. With this process a full scale, geometrically complex prototype can be created directly from a CAD image in less than a day. This report provides a description of how LOM is being applied for the production of functional composite laminates, such as composite tools and molds. Several material systems have been examined, including monolithic ceramics (SiC), ceramic matrix composites (SiC/SiC), and polymer matrix composites (glass/epoxy). Realistic tools and molds are being created from these materials using the LOM process with little modification. However, post processing (ceramic densification, polymer post cure) is necessary to obtain articles with good mechanical properties. Technical solutions are being developed for maintaining geometrical accuracy during the post processing cycles. The mechanical properties and performance of the LOM-produced parts are reported herein. Overall, this technology holds great promise for lowering the cost of tooling due to the ability to create geometrically complex objects rapidly using a highly automated processing sequence.

KEY WORDS: Rapid Prototyping, Composites, Tooling, Molds, Ceramic Materials/Composites, Advanced Manufacturing

1. INTRODUCTION

Originally rapid prototyping technology was used to build physical models solely for product visualization and subjective design evaluation. However, as the technology has evolved the emphasis has shifted to producing functional parts. An area of particular emphasis has been production of tools to facilitate limited run, secondary part manufacturing for industrial products in the development stage. Recently industry has demonstrated that models obtained through rapid prototyping can be used to produce investment casting shells for metal molding [1,2]. Additionally, attempts have been made for the direct manufacture of molds for limited run plastic injection molding [3,4,5], resin transfer molding [6], spin casting [7], and die casting [8]. This work is now being extended into directly forming polymer composite and ceramic parts that are suitable for tooling applications in forming high temperature composites.

Laminated Object Manufacturing (LOM) is an established rapid prototyping technique that is being innovatively applied to the direct manufacture of structural composites. With this technology three dimensional articles can be produced directly from CAD files in less than a day. A key to this evolution has been the development of suitable tapes or prepregs which can be processed via LOM into useable objects such as composite tools. Specifically, two glass fiber/epoxy compositions and several monolithic and composite ceramic compositions have been demonstrated to date. In this paper, a description will be given of the LOM process and the parts that have been built with these prepreg and tape materials. The demonstrated parts are of suitable quality for many tooling applications.

2. EXPERIMENTAL DEVELOPMENT AND DISCUSSION

2.1 Laminated Object Manufacturing

The LOM process (see **Figure 1**) builds layered, three dimensional objects by stacking and fusing thin layers of paper that are trimmed to shape using a CO_2 laser. Essential to this process are software algorithms that slice a CAD file into cross-sectional layers and coordinate this information with system hardware. The building sequence begins with the lamination of a fresh layer of paper to a solid base. The paper is coated on the lower side with a thermally activated adhesive, and lamination is accomplished with a heated, stainless steel roller. After the lamination step, the laser cuts the part boundary contained within the cross-section, cutting only one layer deep. Next, the laser cuts the excess material in a crosshatch pattern (the excess material is not removed so as to provide support for subsequent layers). Finally, an overall rectangular outline is cut, freeing the cross-section from the paper roll. The platform moves down and the feed paper advances.

The sequence repeats when the platform returns to the paper level. The process continues, layer by layer, until all of the cross-sections have been deposited and cut. The product comes out of the machine as a rectangular block. The excess material surrounding the part, already sectioned into crosshatched columns, is removed manually at the end of the process. The final LOM parts exhibit a wood-like appearance, are dimensionally stable, and can be finished by sanding/sealing if necessary. The LOM 2030 can create parts with maximum dimensions of 51 cm x 76 cm x 51 cm high (20 in. x 30 in. x 20 in.), but in theory the technology is not size-limited.

2.2 Production of Polymer Composites

The feasibility of adapting the LOM process for the rapid prototyping and fabrication of polymer composite parts has been demonstrated. It was found that arbitrary, complex shaped parts can be made from a wide range of composite prepregs with an ease and versatility similar to the original LOM paper material. However, due to the transient temperature and pressure application associated with the LOM process, a secondary, post processing step is required for final part consolidation and cure advancement. Without the post cure, the functionality of LOM composite parts is severely reduced and the parts are not suitable for the targeted tooling applications. Currently, technical solutions are being developed for maintaining geometrical accuracy during the post processing cycles.

The performance of two composite prepreg systems has been evaluated in the LOM process. Both are epoxy/glass fiber systems. The first material was specifically designed for the LOM process to duplicate the handling characteristics of paper; it required no modifications to the current LOM machine or build sequence. This prepreg was supplied on nine inch wide rolls, similar to the LOM paper, with a layer of highly B-staged epoxy resin coated on the lower surface of a non-woven glass fiber mat. Because the prepreg was primarily designed to process as similar to the LOM paper as possible, the measured properties of composites made from this material were not good. Due to the one-sided distribution of highly B-staged resin, LOM processing resulted in a layered structure with alternating resin rich areas and layers of unimpregnated fibers (see **Figure 2**). Since the resin did not flow significantly during processing, consolidation was poor. Additionally, fully cured parts were dimensionally stable only at temperatures up to 110^{0}C, as indicated by laboratory mechanical analysis results in **Figure 3**, because the fully cured matrix resin had a particularly low glass transition temperature.

Despite these findings there have been several preliminary field reports noting the adequacy of this prepreg for producing prototype injection molds. At one beta test site, a user was able to inject up to a total of 40 shots using polypropylene, acetal, glass filled acetal, neoprene, and nylon. The prototype mold, however, had to be treated differently than traditional tool steel because of its lower thermal conductivity and use temperature. For example, the runner length had to be extended to prevent hot, molten plastic from contacting the mold prematurely. Additionally, a cycle time as long as 5 minutes was required due to long cooling times. In an attempt to increase the wear resistance of LOM molds, Helisys has developed a metal coating process involving hydroforming of thin copper sheets.

The second material system, a commercially available aerospace-grade prepreg, was much more promising for creating high performance, functional polymer composite parts using LOM. This prepreg consisted of unidirectional "E" glass fibers in an epoxy matrix and was supplied on twelve inch wide rolls. Some modifications to the LOM build sequence were required to process this type of material. Most notably, fully automated processing was not possible with the current LOM machine. The prepreg was a "high flow" system, since the resin was minimally B-staged and evenly distributed through the prepreg thickness. Thus a carrier film was needed on top of the layup to prevent the prepreg from sticking to the heated roller during the lamination step. As a part's cross-section was cut and the next layer applied, the layup cycle needed to be interrupted to remove the carrier film. However, the improved performance of the material compensated for the additional processing steps. The reduced B-staging of the resin, compared to that of the first prepreg, resulted in a better bond between layers (see **Figure 4**) which minimized delamination. Fully cured parts exhibited a use temperature up to 200^{0}C and mechanical properties comparable to more traditionally made epoxy/glass composites (see **Figure 3** and **Table 1**, respectively). However, this material system has not yet been tested in field applications such as injection molding and autoclave tooling.

The use of an aerospace-grade composite material in the LOM process combines the best aspects of traditional polymer composite fabrication and rapid prototyping. The enhanced mechanical properties associated with polymer composites are not sacrificed to achieve the complex, freeform shapes common to rapid prototyping. But in order to further develop this promising new fabrication method, additional improvements are needed in both the LOM process and prepreg formulation. These developments include methods for automating the layup procedure and fine-tuning the prepreg's degree of B-staging and fiber volume to facilitate processing without forfeiting material performance.

2.3 Production of Ceramic Articles

The rapid prototyping and manufacturing of monolithic and composite ceramics is being addressed by an industry/university collaborative team funded by the Defense Advanced Research Projects Agency and the Office of Naval Research. This technology, based on LOM, will provide a capability for producing high density, near-net-shape ceramic matrix composites. Among several others, tooling for polymer composites is a promising application. To date, monolithic SiC and AlN parts have been produced. The favorable thermal conductivity (TC) of AlN or SiC would provide a durable tool with the inherent TC advantage that a metal tool has to offer, but with less thermal expansion, and at a lower cost.

Appropriate ceramic preforms were developed using a standard tape casting process. These "tapes" contain a ceramic powder at approximately 60 vol% and a polymeric binder system. Ceramic tapes were made with thicknesses of 150-175 μm and 300-325 μm, widths of 20 cm, and lengths of 1 m. A significant effort was required to obtain the proper binder formulation for effective tape lamination. Three ceramic systems have been investigated: a coarse, infiltratable SiC composition with average diameter of 30 μm; a bimodal SiC composition with coarse and fine particles; and a sinterable AlN composition with average particle size near 2 μm.

Ceramic tapes were formed into three dimensional objects using a LOM 2030 machine at the University of Dayton. The placement of tapes was performed manually because the tapes were not long enough to form continuous rolls nor strong enough to survive the feed mechanism. Otherwise, the building of ceramic parts was conducted with the same degree of automation as with paper parts.

"Green" is a term that refers to the state of a ceramic article prior to densification. Green objects tend to be fragile and are prone to damage. Most LOM-produced green objects are soft and easily deformed due to the flexibility of the ceramic tapes. Therefore, a partial binder burnout cycle was used to stiffen the parts in order to facilitate green state handling. This cycle involved careful heating of a part to slowly volatilize the plasticizer while leaving the polymer intact. Prior to densification, the polymer binder was removed with a higher temperature burnout cycle developed using thermogravimetric analysis data.

SiC parts were densified by two separate processes, either by using silicon infiltration or reaction bonding (reactive metal infiltration). These methods of densification result in net shape parts with little or no dimensional change. The sintering of AlN parts has not yet been addressed. However, sintering shrinkage and warpage are expected to be primary issues.

Examples of parts made on the LOM system are given in **Figure 5** and in previous reports [9, 10]. The mechanical properties of the infiltrated and reaction bonded SiC parts are currently being measured and will be available at the SAMPE meeting.

Helisys, the manufacturer of the LOM 2030, has developed several hardware and software changes for improved ceramic handling. These modifications will be implemented on an updated LOM machine that will be made commercially available. To facilitate more efficient laser cutting of refractive materials, a pulsed CO_2 laser with a 700 W peak power will be used. To prevent oxidation of carbides and nitrides during laser cutting, a blower that supplies a thin layer of inert gas over the cutting surface has been implemented. Since ceramic prepregs are costly, the material feed mechanism has been redesigned to minimize waste while providing full automation.

2.4 Ceramic Matrix Composites

Work is also progressing toward the ultimate goal of producing ceramic matrix composites with the LOM system. Al_2O_3 and SiC compositions containing SiC whiskers or SiC continuous fibers are being considered. The initial challenge here is tape casting. Incorporating reasonable percentages of fibers uniformly distributed in tapes presents obvious challenges. Novel approaches to this are being developed in the current program and will be reported at the SAMPE meeting.

A second challenge in LOM processing of advanced fiber prepregs is cutting the fibers. CO_2 lasers will not do this efficiently. The laser cutting of fiber prepregs is being investigated with a 3W, pulsed copper vapor laser that delivers 2.7 mJ per pulse at a repetition rate of 8000 Hz. The laser can be focused to a spot size of around 50 μm and is quite effective in "cleanly" cutting SiC fibers, as illustrated in **Figure 6**. The cutting mechanism involves photoablative erosion rather than pyrolysis.

3. FUTURE COMPOSITE DEVELOPMENTS

The process development component of the program is focused on machine modifications. These include a flat tape feed mechanism, a rotating building stage for changing fiber orientation layer-by-layer, an in-situ curing/heat treating device, and a curved layer layup capability. The latter is a highly innovative, key concept for composite fabrication by the LOM process. In conventional composites, the laminate plies conform to the surface of a tool. To facilitate such lamination on the LOM, in place of the current planar layup geometry, a "curved surface" algorithm is being developed. This algorithm also will improve surface finish and allow the fiber reinforcement to remain continuous in the plane of curvature. The curved layer building concept is illustrated in **Figure 7**.

Also under development are various processing methods for improving the ease of part removal from the crosshatched support material.

4. CONCLUDING COMMENTS

Laminated Object Manufacturing is an established and accepted technique for rapid prototyping. Until recently, LOM has been used to produce arbitrarily shaped objects which were of limited usefulness due to material property limitations. However, several new prepregs and tapes have been developed that enable the fabrication of more robust, high temperature materials with the current LOM machine. One promising application of LOM is tooling and molds for high temperature composites.

A strength of LOM technology is its generic nature. One machine can be used to build paper prototypes, high temperature structural composites, and functional ceramic articles. It may be possible as well to create metallic objects starting from a metal powder tape. Another possibility with LOM technology is the production of functionally graded materials formed by alternating or sequential material layers. To enable this process, a multiple cartridge feed mechanism is envisioned, comparable to that of a paper copying machine. Rather than continuous material rolls, flat sheets could be used. This idea would also enable the LOM machine to accommodate thermoplastic prepregs. Compared to other rapid prototyping technologies, LOM is particularly suited for the manufacture of composite articles due to its inherent capability for tape and sheet handling and laminating.

Another inherent strength of rapid prototyping technology is speed; three dimensional objects can be created in a day. This advantage alone is capable of eliminating one of the main cost drivers of tooling--lead time. Because objects are created directly from CAD files, rapid prototyping also saves material cost and fabrication time by eliminating several manufacturing steps, i.e. there is no need to fabricate negatives. Therefore, LOM technology holds promise for both rapid *and* affordable tooling.

5. REFERENCES

1. Sarkis, B.E. and S. Kennerknecht, "Rapid Prototype Casting (RPC): The Fundamentals of Producing Functional Metal Parts from Rapid Prototype Models Using Investment Casting," Proceedings of the 5th International Conference of Rapid Prototyping, University of Dayton, June, 1994, pp. 291-300.

2. Jacobs, P.F., "Quickcast™ 1.1 & Rapid Tooling," Proceedings of the 6th International Conference of Rapid Prototyping, University of Dayton, June, 1995, pp. 237-257.

3. Badrinarayan, B. and J.W. Barlow, "Manufacture of Injection Molds Using SLS," Proceedings of Solid Freeform Fabrication Symposium, University of Texas, August, 1994, pp. 371-386.

4. Yeung, M.J. McKeen, "Rapid Manufacturing of Plastic Injection Molds for Prototype ABS Parts," Proceedings of the North American Stereolithography User Group Conference and Annual Meeting, Tampa, Florida, March, 1995, pp. 2-5.

5. Venus, A.D., and S.J. van de Crommert, "Rapid Product Development with SLS Injection Tooling," Proceedings of the International Conference on Rapid Product Development, Messe Stuttgart, Germany, June 10-11, 1996, pp. 199-208.

6. Ruud, N.A., "Rapid Tooling for Resin Transfer Molding, RTM," Proceedings of the 6th International Conference of Rapid Prototyping, University of Dayton, June, 1995, pp. 205-215.

7. Schaer, L., "Spin-Casting Fully Functional Metal and Plastic Parts from Stereolithography Models," Proceedings of the 6th International Conference of Rapid Prototyping, University of Dayton, June, 1995, pp. 217-236.

8. Warner, M.C., "Using Rapid Prototyping to Build Prototype Die Cast Tooling Results in Significant Savings in Tooling Costs and Reduces Time for Product Development," Proceedings of the 6th International Conference of Rapid Prototyping, University of Dayton, June, 1995, pp. 201-203.

9. Klosterman, D.A., R.P.Chartoff, and S.S. Pak, "Affordable, Rapid Composite Tooling via Laminated Object Manufacturing," Proceedings of the 41st International SAMPE Symposium and Exhibition, Anaheim, CA, March 24-28, 1996, pp. 220-229.

10. Klosterman, D.A., R.P. Chartoff, N.R. Osborne, G. Graves, and A. Lightman, "Structural Ceramic Components via Laminated Object Manufacturing,"Proceedings of the International Conference on Rapid Product Development, Messe Stuttgart, Germany, June 10-11, 1996, pp. 247-256.

6. ACKNOWLEDGMENTS

The work reported here was performed under a grant by DARPA/ONR, N00014-95-1-0059, with Dr. William Coblenz and Dr. Steven Fishman acting as program officers. A portion of the glass fiber/epoxy work was performed under U.S. Army Aviation and Troop Command Contract DAAJ02-95-C-0020, "Rapid Fabrication of Fully Functional Models by LOM," with Clay Ames acting as program officer.

7. BIOGRAPHIES

Richard P. Chartoff is Professor of Materials Engineering at the University of Dayton (UD) and a Senior Polymer Engineer in the University of Dayton Research Institute's (UDRI) Center for Basic and Applied Polymer Research. He teaches graduate courses in polymer science and engineering and supervises laboratory activities in the Polymer Center. Dr. Chartoff is also a co-founder of the UDRI's Rapid Prototype Development Laboratory, which is engaged in various research programs associated with materials processing in automated layered manufacturing systems. UD is one of the only Universities carrying out R&D programs in materials processing for rapid prototyping. The program at UD in this area is internationally recognized. Dr. Chartoff has numerous publications and patents and has served as a consultant for many American and European industrial organizations. He has been recognized as a research fellow by the Royal Norwegian Council for Scientific and Industrial Research and has been elected as a Fellow of the North American Thermal Analysis Society (NATAS). His activities in technical societies include service on the executive board of NATAS and the Dayton Section of the American Chemical Society as well as President of the Miami Valley Section of the Society of Plastics Engineers.

Brian Priore recently received an M.S. degree in Chemical Engineering from the University of Dayton where he conducted the first in-depth investigation of the processing science of polymer composites via LOM. He received a bachelor's degree in Chemical Engineering from the University of Dayton in 1995. Mr. Priore was a national finalist in the 1996 SAMPE student paper competition in Anaheim.

Donald A. Klosterman is an Associate Polymer Engineer for UDRI's Rapid Prototyping Development Laboratory and Center for Basic and Applied Polymer Research. Dr. Klosterman received a Ph.D. in Materials Engineering from the University of Dayton (1994) where he studied composite processing and sensor technology. He received a Master's degree in Chemical Engineering from the University of Dayton (1991) where he investigated the high temperature, photolytic destruction of hazardous compounds using simulated, concentrated solar radiation. Dr. Klosterman is currently the chairman of the SAMPE Midwest chapter.

Sung S. Pak, Manager of R&D at Helisys, Inc. holds a B.S. in Ceramic Engineering and an M.S. in Materials Science from the University of Washington. Mr. Pak has held research positions at General Atomics and Babcock & Wilcox, where he developed various thin film, fiber, superconductor, and high temperature composites. Ten publications and three U.S. Patents bear his name.

8. ILLUSTRATIONS

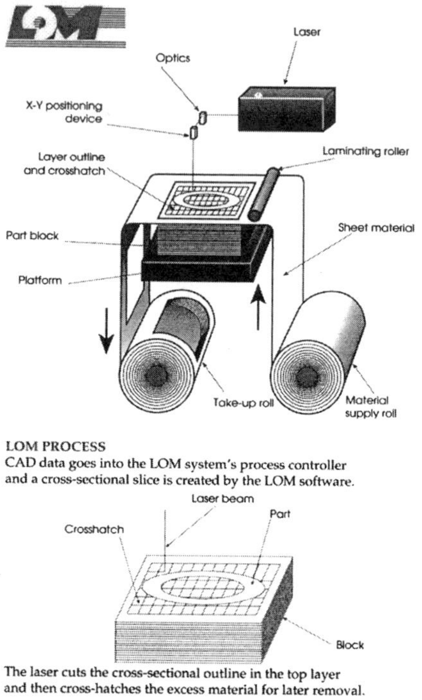

Figure 1 : schematic of the LOM process.

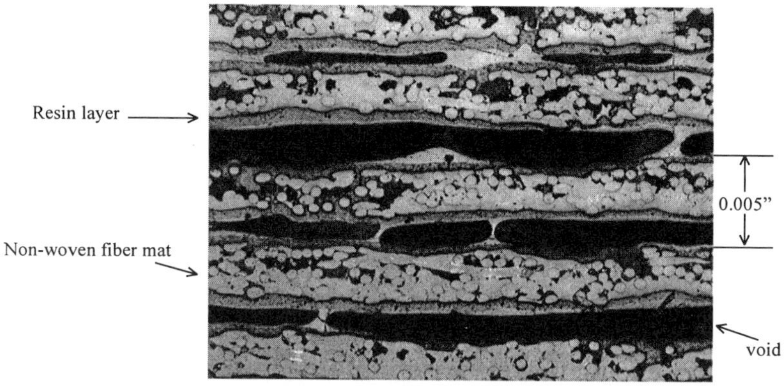

Figure 2 : photomicrograph of a composite panel made with the LOM-designed prepreg. Note the resin's inability to penetrate into the fiber mat and the large scale voids entrapped in the resin rich layers.

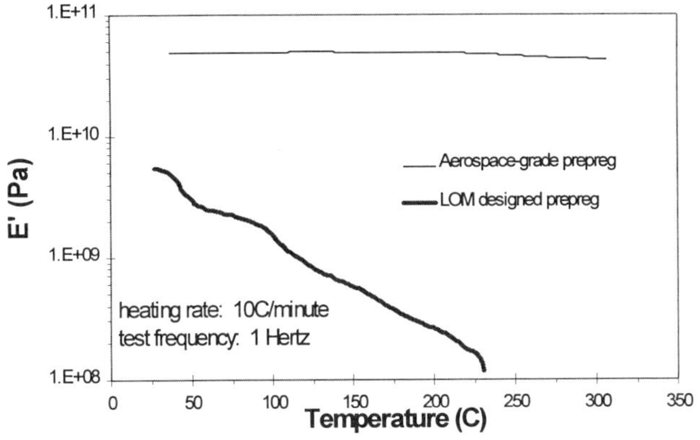

Figure 3 : Dynamic Mechanical Analysis, three point bend test method, of post cured LOM composites indicating useable temperatures for both materials.

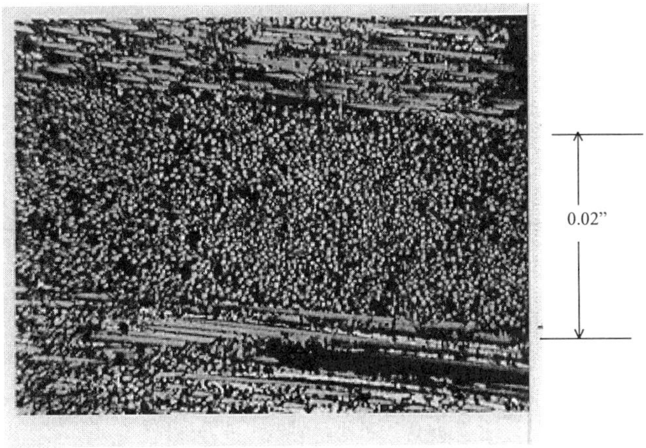

Figure 4 : photomicrograph of a cross section of a LOM processed, aerospace-grade composite part. Shown in the field of view are four layers: 90°/0°/0°/90° (from top to bottom).

Table 1 : Room temperature mechanical properties of LOM processed and post cured aerospace-grade composites.

Test Method	LOM Processed Ultimate Strength (MPa)	Supplier Data Ultimate Strength (MPa)
Tension (0° orientation) ASTM-D-3039	712.9 ± 37.2	1000
Compression (0° orientation) ASTM-D-695	895.6 ± 57.9 357.1 ± 80.0 (no post cure)	792
Flexure (0° orientation) ASTM-D-790	1189.3 ± 51.0	1310
Interlaminar Shear ($0^{o}/90^{o}$ orientation) ASTM D-2344	42.6 ± 3.9	55

Figure 5 : SiC and AlN (left to right) demonstration widgets made with LOM at the University of Dayton. These parts appear in the "green state", but effective, near-net-shape densification procedures have been developed for SiC.

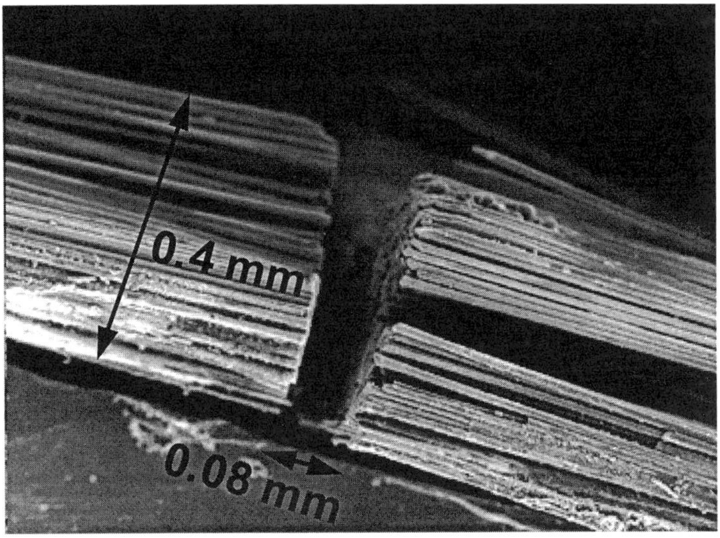

Figure 6 : Thick, SiC fiber prepreg cut with a copper vapor laser operating at 8000 pulses/sec and 2.7 mJ per pulse. Such high quality fiber cutting will be required to enable the LOM of polymer and ceramic matrix composites.

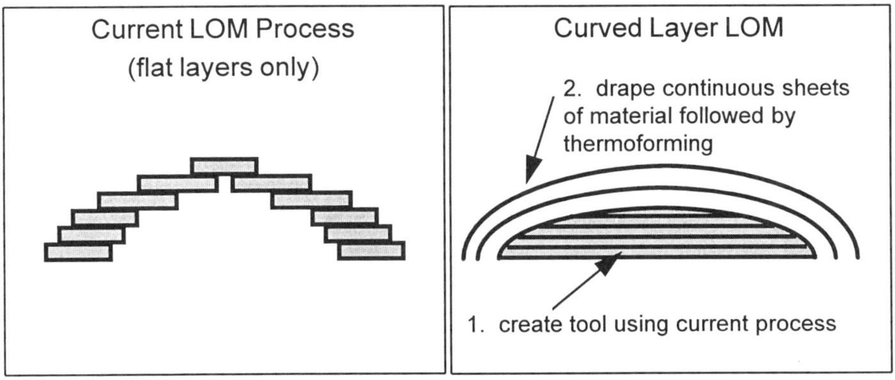

Figure 7 : Curved Layer LOM concept. Curved layer parts will be built by draping the composite preforms over a tool that has been premade by the existing flat layer process. The draped sheets will be rolled or thermoformed into place.

28th International SAMPE Technical Conference
November 4-7, 1996

BONDED, MULTI-SHELL COMPOSITE PRESSURE VESSEL DEVELOPMENT

D.W. Radford and E.D. Carlson
Composite Materials, Manufacture and Structures Laboratory
Mechanical Engineering
Colorado State University
Fort Collins, CO 80523-1374

ABSTRACT

Conventional cylindrical composite pressure vessels are designed with continuous fiber hoop wraps to counter the stresses developed in the shell. In certain applications it would be advantageous to be able to assemble a composite pressure vessel around internal components; however, this forces a discontinuity in the hoop reinforcing fibers or may lead to complex fabrication. This research investigates the design considerations involved in the development of a bonded, multi-shell composite U-tube pressure vessel, the determination of critical design parameters, and hydroburst testing of prototype pressure vessels. The basic concept was to bond together 4 half-shells with the seams of the inner and outer pair rotated 90 degrees. This approach results in a maximized lap area and therefore the strongest possible bond geometry. Preliminary testing indicated that composite failure occurred at regions of concentrated stress due to a minor design flaw related to mismatched shell compliance. Considering these preliminary test results, and finite element modeling, an approach was developed to overcome the localized failure. The final test vessels were hydroburst tested to pressures beyond 8.2 MPa without composite failure, closely matching predictions and indicating the feasibility of this approach to bonded, multi-shell composite pressure vessels.

KEY WORDS: Advanced Composites, Composite Pressure Vessels, Bonded Joints

1. INTRODUCTION

Pressure vessels are one of the oldest structural applications of fiber reinforced composite materials, and in recent years there has been a significant increase in the number and variety of composite pressure vessels. Due to the relatively simple state of stress in common cylindrical pressure vessels with hemispherical endcaps uniaxial fiber has been able to be efficiently utilized, primarily as circumferential wraps, to resist hoop stresses (1). This, along with fact that simple pressure vessels are based on strength limited design, has allowed effective composite pressure vessels to be produced from relatively low cost fiberglass reinforcement using high volume manufacturing techniques such as filament winding (2,3,4). These factors have resulted in the acceptance of composite materials in pressure vessels as cost competitive alternatives to metals. However, difficulties do arise in this approach to pressure vessel design when numerous penetrations have to be made, or when the geometry deviates significantly from a circular cylinder (5,6).

As the geometry becomes more complex, composite pressure vessel design becomes more complicated, with the resulting vessels often incorporating a thin metallic liner with integral fittings and a composite overwrap (5). Also, as the geometry deviates from a circular cylinder the approach begins to take on aspects of a stiffness limited design, leading to the use of higher modulus fibers, such as carbon and aramids. Carbon fiber reinforcement has also been recently favored for long term, high pressure containment because of the extremely low inherent creep compared to other reinforcements (2).

Unfortunately, a large number of desired pressure vessel geometries do not lend themselves to the use of composite materials. This is especially true for applications that require the vessel to be assembled around internal hardware. The approach in such cases has been to utilize metallic pressure vessels formed in segments and welded together, surrounding the necessary internal components. It is the development of a composite replacement pressure vessel for this type of application that is the topic of this discussion.

2. THE BONDED MULTI-SHELL PRESSURE VESSEL CONCEPT

2.1 Pressure Vessel Configuration In an effort to decrease the cost of a stainless steel secondary containment vessel for a commercial Coriolis mass flow controller, a bonded, multi-shell composite pressure vessel concept has been developed. Since the present stainless steel design uses two U-shaped half cylindrical shells welded together around internal components no obvious composites approach that incorporated continuous fiber hoop wraps was obvious. Thus, it became clear that to develop this complex geometry, cost effectively, and to be able to assemble it around internal hardware already in place and quality tested, some form of bonded unit would be necessary. To ensure the highest performance bonded joint it was determined that as large an overlap as possible should be incorporated.

The concept that resulted was to mold two matching sets of half shells, with the seams rotated 90°, to achieve the desired maximum joint overlap. Further, to meet cost and quality requirements it was determined that shells would be prefabricated so that only shell-to-shell bonding would be necessary during final assembly. Thus, in addition to generating the largest possible overlap area this design results in shells that can be conveniently manufactured by a number of high volume, low cost composite fabrication technologies.

For the U-tube geometry the requirement for rotated shell seams results in the need for two side shells, plus a top and bottom shell. These four shells will require three separate sets of tooling for production level fabrication. While the concept shown in figure 1 uses only four

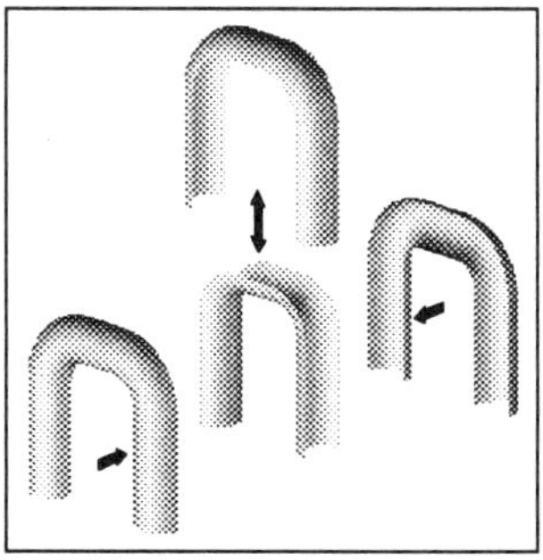

Figure 1 - Bonded multi-shell U-tube pressure vessel assembly concept.

shells, further improvements in ultimate performance should be possible with more shells with consecutively overlapping seams. However, the incorporation of more sets of shells was not desirable for this application due to increased cost to manufacture and assemble.

2.2 Design Approach Performance of the proposed bonded, multi-shell pressure vessel can be determined by considering the two primary structural areas of concern, the shell strength and the shell-to-shell bond strength. As a first estimation of the optimized pressure capability of this design, the hoop strength of the shells is equated to the shell-to-shell bond strength. The hoop and axial stresses generated in the pressure vessel walls can be calculated using the following equations:

$$\sigma_1 = \frac{pr}{t} \qquad [1]$$

$$\sigma_2 = \frac{pr}{2t} \qquad [2]$$

where,

σ_1 = hoop stress,
σ_2 = axial stress,
p = internal fluid pressure,
r = radius of vessel, and
t = thickness of vessel wall.

It is clear from these equations, for a cylindrical pressure vessel, that the hoop stress is the critical requirement for the shell. The predicted performance of the adhesive joint between the shells is based on the shear strength of the adhesive and the shell overlap area. For this design, the overlap area is the product of the vessel length and 25% of the circumference, which is:

$$A = (0.25\pi D)L = (0.5\pi r)L \qquad [3]$$

where,

A = shell overlap area
L = length of pressure vessel, and
D = diameter of pressure vessel.

Thus, to determine critical values of internal pressure, necessary shell wall strength, thickness or required adhesive performance, the load carrying ability per unit length of pressure vessel for these two contributions can be equated. The result is:

$$p = 0.5\pi\tau \qquad [4]$$

where,

τ = inherent shear strength of the adhesive.

Equation 4 indicates that the bonded shell design limits the internal pressure to approximately 1.5 times the shell-to-shell adhesive strength. To evaluate the effects of shell strength, thickness and dimension, equation 4 can be substituted into equation 1, resulting in:

$$\sigma_1 = \frac{(0.5\pi\tau)r}{t} \qquad [5]$$

By using equations 4 and 5, in sequence, the necessary adhesive strength can be determined and the corresponding shell properties can be calculated. In a more detailed design many other factors may need to be incorporated into the analysis. Other factors may include the effect of flaws, material non-linearities and geometry variations; however, these effects will primarily affect the shell strength and bondline performance and not the critical design relationship for the multi-shell configuration (7,8,9). To test the design approach, an experimental hydroburst program was initiated to evaluate the potential of this bonded, multi-shell pressure vessel concept.

3. EXPERIMENTAL APPROACH

For this initial experimental program, five multi-shell pressure vessels were produced and hydroburst tested. The general goal of these tests was to demonstrate the feasibility of a multi-shell composite pressure vessel that can be assembled around internal hardware in a fashion conceptually similar to the welding of a metallic enclosure. The specific goal of the test program was to achieve an internal pressure in excess of 6.9 MPa (1,000 psi). Each set of tests gave new insight into the difficulties inherent in this new composites design.

3.1 Specimen Geometry Straight cylindrical composite test sections were used, capped on each end by bonded-in stainless steel welded fabrications that included the necessary pressure fittings as shown in figure 2. The test specimen diameter (~14 cm) was determined by the current size of the commercial steel vessel under consideration. The 30 cm length of the straight test section was chosen out of convenience, since a sufficient length of large diameter aluminum tubing had to be procured and turned to precise dimensions.

Figure 2. Bonded pressure vessels and stainless steel endcaps.

3.1.1 Composite Shells The composite test shells were designed with a diameter roughly equivalent to the actual metal component under review. Two shell configurations were tested. The first configuration used inner and outer shells of uniform thickness, which resulted in a thickness "step" at each of the shell seams. To evaluate possible stress concentration at these seam "steps" a second configuration, using inner shells with tapered thickness, was also produced. Outer shells were formed with flanges at the seams as an aid to clamping the shells during final assembly. Figure 3 indicates the two different shell configurations evaluated. All test shells were roughly 30 cm in length.

3.1.2 Stainless Steel Endcaps The stainless steel endcaps incorporate a tubular bond area for mating to the inner composite shells. These bond areas were calculated to provide

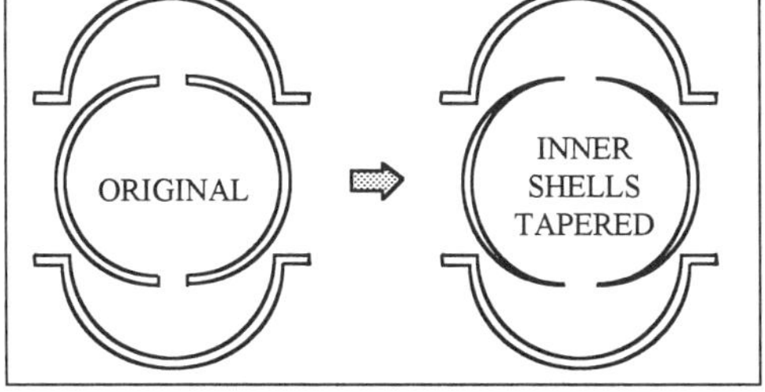

Figure 3. Composite shell configurations tested.

sufficient performance based on the shear strength of the planned endcap to shell adhesive. Two O-ring grooves were cut into each bond area to locate O-rings that insured correct bond thickness and consistent bond length. End fittings were provided on both ends to simplify the filling procedure. The outer diameter of the endcaps was roughly 13.6 cm, with a 6 mm wall thickness and a groove-to-groove bond length of 5.1 cm. The O-ring grooves were centered in the bond area and the edge of the insert tube was chamfered to reduce the possibility of composite component damage in assembly.

3.2 Composite Pressure Vessel Test Specimen Manufacture All test vessels made use of Shell Epon 828 epoxy resin and 3274 hardener as a matrix material. The use of epoxy was necessary to model the performance of a composite component that will ultimately be required to be rated for operation at temperatures in excess of 175°C. The reinforcement chosen for these tests was woven 7.5oz fiberglass plain weave fabric (E-glass) to minimize cost. The endcaps were bonded to the composite shells, throughout most of the testing, with PPG HC5099 which is a 177°C cure epoxy based adhesive specifically designed for high fatigue capabilities in steel bonded joints. The final test specimen used Plexus MA320 methylmethacrylate adhesive to investigate the effect of a bondline with much higher strain to failure capabilities than PPG HC5099 and simplified "in-production" application.

3.2.1 Composite Shell Fabrication A length of large diameter aluminum tubing was turned to precise diameter of the shell-to-shell joint and then cut to a half cylinder. From this section of tubing, a 38 cm long composite female mold was fabricated. Thus, the length of this mold limits the pressure vessel test length to approximately 32 cm. The aluminum tool and the composite mold act as a matched set of male and female tools.

The approach to the fabrication of the composite shells involved the use of four separate semi-cylindrical segments. Two of these shells were produced on the male tooling and two were produced on the matching female tooling. The result of this approach was a molded bondline at the interface between the inner and outer shells that was intended to enable a high quality shell-to-shell bond. Since these shells were produced in contact molds, the resulting inner and outer surfaces of the completed pressure vessel were not mold surfaces. For the outer shells this was of little importance, other than cosmetics; however, for the inner shells this directly impacted the ability to bond to the inserted metal endcaps. The molding approach also fixed the maximum allowable wall thickness for a given diameter endcap and bondline thickness. Test vessels were produced using the matching male and female tooling with shell walls of uniform thickness and with tapered inner shell thickness.

To produce the pressure vessel shells, fiberglass sheets of plain weave were cut 30 cm^2 and wet out with approximately 60% resin and 40% fiber volume ratio to obtain good sealing and leak resistance. The inner shells were constructed from 7 plies in the female mold, which resulted in an inner shell thickness of roughly 1.75 mm. The two innermost plies were oriented at (±45) degrees from the centerline and the other five plies were oriented at (0/90) degrees. The innermost (±45) degree plies were incorporated to more closely match the modulus of the endcap and adhesive for stress concentration reduction.

The basic outer shells were constructed on the male aluminum tooling with four (0/90) plies and one outermost (±45) ply for surface finish. In one of the five test specimens the outer shell was fabricated with an additional four (0/90) plies to evaluate increased shell strength. All composite components were vacuum bagged for the duration of their 24 hour room temperature cure. The shells were then postcured at 120°C for 1 hour.

3.2.2 Pressure Vessel Assembly Components were removed from the molds after a 24 hour room temperature cure, cut to a length of 30 cm and trimmed. The inside surface of the inner shells and the cut edges were then given a coating of resin for enhanced sealing. Each set of inner shells was then lightly sanded and acetone cleaned at the endcap bond surfaces. The bond surfaces of the endcaps were bead blasted and acetone cleaned, then fitted with O-rings. Both mating surfaces of the endcaps and composite were wet out with the endcap adhesive. The inner shells were applied to the endcaps and clamped against the O-rings. The bonds were cured for 30 minutes at 177°C. After cooling from the initial cure, any gaps along the seams between the inner shells were sealed with epoxy. The sealing resin was allowed to cure, then the outer surfaces of the inner shells and corresponding outer shells were lightly sanded and acetone wiped. The mating surfaces of the inner and outer shells were wet out with the matrix epoxy and clamped together using the flanges of the outer shells. Clamps were first applied to the center of the vessel and added outward to force air from the bondline between the inner and outer shells. The vessel was allowed to cure and the seams were sealed with additional resin. The complete vessel was then given a final 1 hour postcure at 120°C.

The final pressure vessel of this test group was assembled in a modified manner to evaluate the higher strain to failure methylmethacrylate endcap adhesive. The higher strain to failure of the MA320 was predicted to be helpful in response to any transverse tensile stresses being generated across the bondline. The shells were bonded together and postcured at 120°C over aluminum tooling prior to endcap insertion. Holes were drilled and tapped near the ends of the assembled composite tube to line up with the bond region of the endcaps. Endcaps, with O-rings in place, were then inserted into the composite pressure vessel. The adhesive was injected into each hole, which was then sequentially sealed with a nylon screw. Injection proceeded until each endcap bondline was complete. In addition to generating a specimen to check the desirability of a higher strain adhesive, this approach demonstrated a manufacturing technique that promises improved quality control in volume production.

3.3 Hydroburst Test Technique Pressure tests were performed using a hydroburst apparatus. Each test vessel was filled with tap water and purged of air. The vessel was then connected to a high pressure water pump fed from the domestic water line. Data output was transferred to a computer acquisition system and the pressure history was recorded. To minimize the impact of any pin-hole leaks in the pressure vessels the pressure was applied as rapidly as possible. Failed vessels were visually inspected for form of failure.

4. RESULTS AND DISCUSSION

Due to the size of the pressure vessel specimens and the difficulties involved in creating the composite shells by hand fabrication, the number of vessels tested was limited to the 5 specimens described previously, and shown in figure 5. In general, continued improvement in pressure was seen throughout the test set, with a final pressure of over 8.5 MPa (1,200 psi) achieved without composite shell failure. The following discussion covers the various findings that resulted from this bonded, multi-shell pressure vessel development program.

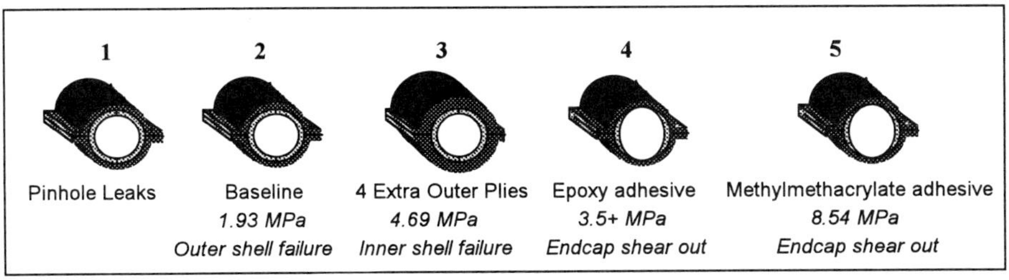

Figure 5. Hydroburst test program summary.

4.1 Shells of Uniform Thickness The first pair of specimens was planned to generate baseline pressures for future comparison, but due to insufficient consideration of shell porosity during fabrication of the first vessel, composite failure pressures could not be attained. Test procedures other than hydroburst have been suggested that would be effective even with pinhole defects, but were not applied in this research (10). Pressure vessels 2 and 3 both held pressure until composite failure occurred. Vessel 2 gave the baseline pressure data for a 12 ply composite, while vessel 3 was included to investigate the effect of additional outer shell plies.

4.1.1 Hydroburst Results In vessels 2 and 3 composite failure was achieved; however, at pressures substantially lower than preliminary estimates had suggested. For the baseline pressure vessel thickness of 12 plies a maximum pressure of approximately 1.93 MPa (280 psi) was achieved. Failure occurred as a 10 cm long break in the outer shell along the inner shell seam. Vessel 3, with an additional 4 outer shell plies, reached 4.69 MPa (680 psi) prior to failure. Failure of vessel 3 was similar to the baseline failure, with a shell fracture several centimeters long but in the inner shell, at the outer shell seam. Figure 6 illustrates the pressurization curves for these two vessels and shows the relative locations of shell failure. Since in each case failure was localized along a shell seam, it became clear that the thickness discontinuity was resulting in a large concentrated stress, and it was theorized that this was actually a flexural stress concentration. Follow-up computer modeling of the cross-section confirmed the belief that a flexural concentration was occurring and therefore indicated a need for design changes that would alleviate the problem. The design change that resulted was the incorporation of thickness tapered inner shells.

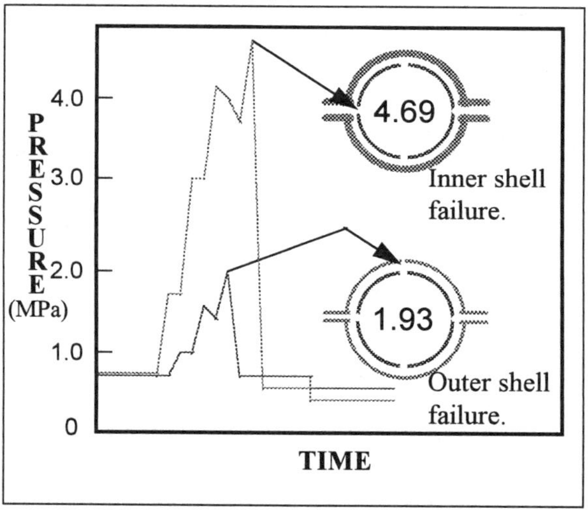

Figure 6. Pressurization charts and location of failure.

4.1.2 Computed Performance Using equations 1, 4 and 5 from Section 2 a predicted maximum pressure can be calculated for the test vessels. Further, these equations should give an indication of whether failure should occur in the composite shells or in the shell-to-shell adhesive for the test vessels. Adhesive strengths can vary dramatically; however, single lap shear tests using the Epon 828 as an adhesive consistently generate values greater than 14 MPa. Since the quality of the shell-to-shell bond is difficult to guarantee, a value of 7 MPa (1,000 psi) is used for initial calculation. Laminate tensile tests of material representative of the fiberglass shells resulted in a tensile strength exceeding 345 MPa (50,000 psi).

Computing the shell and bond failure pressures for vessel 2, using equations 1, 4 and 5, results in an internal pressure of 11 MPa (1,570 psi) for bondline failure and 9 MPa (1,300 psi) for outer shell fracture. The outer shell in this case used fewer plies and was expected to be the weakest shell. Thus, it is seen that outer shell fracture is the predicted mode of failure; however, this value is more than four times greater than the experimental burst pressure. Similarly for vessel 3 an internal pressure of 11.7 MPa (1,690 psi) is predicted for inner shell failure while the bondline performance should remain unchanged. Again, shell fracture is the predicted mode of failure and in this case the predicted value is more than twice the experimental burst pressure.

4.1.3 Significance of Stress Concentration Based on the predicted burst pressures and the location of fracture it seems clear that some form of concentrated stress is occurring at the thickness steps in the bonded pressure vessel shells. Since a uniform internal pressure exists, it was suggested that the thin areas of the shell were actually flexing rather than the vessel uniformly expanding in diameter. Preliminary finite element evaluation confirmed this flexural stress concentration. Thus, the use of tapered inner shells was expected to improve the hydroburst performance.

4.2 Tapered Shells The final two test vessels used tapered inner shells and uniform thickness outer shells. The ply configuration of the shells matched the 12 ply configuration of the baseline vessels, with 7 inner plies and 5 plies in the outer shell. Shells were produced to the same specifications for each pressure vessel with only the final assembly varying. Tapered inner shells were created by dropping plies during the layup procedure. The computed performance for these final two test vessels is the same as that of vessel 2 since the shell wall thicknesses are unchanged. Thus, the principal uses of vessels 4 and 5 were to investigate the stress concentration effect and the effect of changing the endcap adhesive on the burst pressure.

4.2.1 Hydroburst Results In vessels 4 and 5 composite failure was not achieved. In each case the failure was in the shell-to-endcap bond. Vessel 4, using the high temperature epoxy-based adhesive, failed in the endcap bond at approximately 3.5 MPa (500 psi) while vessel 5, using the methylmethacrylate adhesive, failed at 8.5 MPa (1,240 psi). This final result is still slightly below the inner shell burst pressure of 9 MPa calculated in the previous section. It is useful to discuss these two bonded joints in more detail since the only design difference between the two test vessels is the endcap adhesive used.

4.2.2 Effect of Shell Thickness Taper It is clear from the dramatic increase in pressure capability of vessel 5 over vessel 2 that the shell thickness taper reduces the concentrated stresses. Further, since the pressure attained in vessel 5 generates a shell hoop stress of approximately 325 MPa (47,200 psi), which closely approaches the predicted shell failure stress of 345 MPa, it seems that any stress concentration has been effectively negated. However, this approach does result in an elliptical inner surface which is still expected to result in some shell flexure under pressure. This tapered inner surface also increases the manufacturing complexity by forcing the end bond areas of the inner shells to be produced without taper to facilitate bonding.

4.2.3 Effect of Endcap Adhesive Lap shear tests of the MA320 and HC5099 adhesives joining stainless steel and fiberglass adherends indicated a much lower absolute shear strength for the MA320 methylmethacrylate than the epoxy based HC5099 (11.0 versus 17.9 MPa). However, the strain to failure measured was roughly 4 times greater for the methylmethacrylate and all failures occurred within the adhesive rather than at an adherend surface. The HC5099 specimens also indicated stainless steel adherend preparation difficulties as all failures were at this interface. With a hydroburst specimen endcap shear area of approximately 225 cm^2 and an end plate area of 125 cm^2, even the 11 MPa shear strength of the MA320 was predicted to enable 20 MPa of internal test pressure. This was considered high enough that even with a factor of safety, sufficiently high internal pressures could be achieved to prove the acceptability of the composite pressure vessel. However, since the experimental endcap failure pressures were significantly lower than predicted and the lower strength methylmethacrylate adhesive performed better than the higher performance epoxy, another explanation for the failures was suggested.

Transverse tensile stresses across the shell-to-endcap bondline are a concern based on differences in the hoop stiffness of the bonded composite pressure shells and the stainless steel endcap inserts. This is true partially because of the relatively low modulus of the fiberglass reinforcement, but also due to the stresses supported by the shell-to-shell adhesive bond. As the vessel is pressurized it is suggested that a "barreling" takes place as the unsupported central region expands more than the ends which are constrained by the thick stainless steel endcaps. If a significant transverse tensile strain is being generated it will exceed the abilities of the highly filled HC5099 adhesive much sooner than those of the methylmethacrylate adhesive. Thus, the use of the much more compliant MA320 is expected to help shift the mode of endcap bond failure from transverse tension back to the desired axial shear, therefore enabling a higher internal pressure prior to failure. Considering the difference in endcap bond failure for vessels 4 and 5 the use of the higher strain adhesive seems to have proven successful. Unfortunately, this methylmethacrylate adhesive is not appropriate for the final design due to poor elevated temperature performance; however, its use clearly indicates that the endcap-to-shell bond is a critical design issue that must be addressed.

4.3 Future Design and Evaluation The hydroburst results substantiate the advantage of the tapered design and prove that pressures required for the external containment vessel can be achieved using the multi-shell composite design approach. Future efforts are planned to investigate the use of tapered outer composite shells in addition to the already implemented tapered inner shells. At the same time it is planned to modify the form of manufacture to generate an elliptical bondline between the inner and outer shells so that the surfaces will be circular. Detailed finite element analysis of this multi-shell geometry is also planned, and will include effects of the U-tube geometry to ensure that no significant secondary effects exist.

The endcap region is also to be re-evaluated in future work. The present endcap design was specifically chosen for the straight-section test vessels, but was not intended as the only option for the actual commercial design. Substantially more design effort is required in this area of composite-to-metal attachment prior to commercial fabrication. The design effort must include modeling taking into account deformation of the pressure shell that could result in a transverse tensile stress across the bondline in addition to the expected shear stresses.

5. CONCLUSIONS

The concept of a bonded multi-shell composite pressure vessel that can be assembled around internal hardware has been successfully demonstrated. A maximum burst pressure of 8.54 MPa was achieved. However, even at this pressure the failure was in the test fixturing at the endcap-to-shell bond and was not a composite failure.

The shell failures of the baseline multi-shell pressure vessels indicated a significant stress concentration at the shell seams. This effect reduced the composite shell performance by more than fourfold from the predicted maximum pressure capability. The stress concentration was confirmed by testing of the modified, tapered inner shell geometry that resulted in pressures approaching those predicted and nullified failure in the composite. Simple modeling of the stresses in the shells and in the bondline indicates that internal pressures of roughly 1.5 times the shell-to-shell adhesive strength should be possible with this pressure vessel concept. Further, the relationships presented give an initial procedure for determining an optimal composite shell thickness. Precision of these predictions was not tested as the high pressure failures occurred in the shell-to-endcap bonds rather than in the composite shells or the shell-to-shell bonds.

The significant improvement in ultimate performance of the shell-to-endcap bond through the use of a lower strength, but higher strain, adhesive is a clear indication that the endcap bond is not failing in pure shear. This is surmised to be related to the relatively low hoop stiffness of this pressure vessel configuration, leading to transverse tensile stresses being generated across the bondline. Thus, future efforts must investigate alternate endcap designs.

6. REFERENCES

1. S. Adali, E.B. Summers and V.E. Verijenko, "Optimisation of Laminated Cylindrical Pressure Vessels Under Strength Criterion," Composite Structures, 25 (1), 305 (1993).
2. B.W. Tew, "Preliminary Design of Tubular Composite Structures Using Netting Theory and Composite Degradation Factors," Journal of Pressure Vessel Technology, 117 (4), 390 (1995).
3. C.S. Mao, M.F. Yang,, D.G. Hwang and H.C. Wang, "An Estimation of Strength for Composite Pressure Vessels," Composite Structures, 22 (3), 179 (1992).
4. D. Cohen and K.A. Lowe, "The Influence of Epoxy Matrix Properties on Delivered Fiber Strength in Filament Wound Composite Pressure Vessels," Journal of Reinforced Plastics and Composites, 10 (2), 112 (1991).
5. B. Mouhamath, "Stress Concentrations with Singularities in Filament-Wound Spherical Pressure Vessels," Composite Structures, 23, (2), 131 (1993).
6. H. Karandikar, R. Srinvasan and F. Mistree, "Compromise: An Effective Approach for the Design of Pressure Vessels Using Composite Materials," Computers and Structures, 33 (6), 1465 (1989).
7. L. Minnetyan, C.C. Chamis and P.L.N. Murthy, "Structural Durability of a Composite Pressure Vessel," Journal of Reinforced Plastics and Composites, 11, (11), 1251 (1992).
8. D. Cohen, "Application of Material Nonlinearity to a Composite Pressure Vessel Design," Journal of Spacecraft and Rockets, 28 (3), 339 (1991).
9. W.M. Banks, A.S. Tooth and P.M. Wilson, "Design of Composite Cylindrical Vessels for Fluid Loading," Journal of Pressure Vessel Technology, 111 (2), 102 (1989).
10. D. Cohen, Y.T. Toombes,.A.K. Johnson and M.F. Hansen, "Pressurized Ring Test for Composite Pressure Vessel Hoop Strength and Stiffness Evaluation," Journal of Composites Technology and Research, 17 (4), 331 (1995).

ACKNOWLEDGMENT

This research was supported by a Research Initiation grant from the Colorado Advanced Materials Institute in conjunction with Micro Motion, Inc. of Boulder, CO. Hydroburst testing was performed at Micro Motion.

BIOGRAPHIES

Donald W. Radford is an Associate Professor of Mechanical Engineering at Colorado State University and is the Director of the Composite Materials, Manufacture, and Structures Laboratory. He received his PhD. in Materials Engineering at Rensselaer Polytechnic

Institute, his MASc. in Metallurgical Engineering and BASc. in Mechanical Engineering at the University of British Columbia. Recent research programs have included approaches to measuring and predicting distortion during composites manufacture, dual resin bonded carbon fiber reinforced PEEK/PEI composites, impact fatigue of metal-composite joints, ultra-lightweight EMI shielding composites and morphological studies of composites.

Eric D. Carlson is presently pursuing a Master of Science in Mechanical Engineering at Colorado State University. He received a Bachelor of Science in Mechanical and Aerospace Engineering at Virginia Polytechnic Institute in 1994. His research involves design optimization of composite components.

28th International SAMPE Technical Conference
November 4-7, 1996

Quasi-Geodesics — A New Class of Simple and Non-Slip Trajectories on Revolutional Surfaces

Y. D. Liang, Z. Q. Zou and Z. F. Zhang
The State Key Laboratory of CAD/CG
Department of Applied Mathematics,
Zhejiang University, P.R.C.

Gene Luo
Geodesk Engineering Systems
724 Sunrise Dr., Fremont, CA 94539

Abstract

Precise placement of fibers is the main problem in computer aided filament winding design. To prevent slippage, the geodesics and semi-geodesics are commonly used as trajectories of fiber paths. Unfortunately, the differential equations of the semi-geodesics are more complicate and more difficult to solve in a simple form. This paper presents a new class of simple and non-slip trajectories on revolutional surfaces, called quasi-geodesics, the equation of which is obtained in a simple and analytical form and can be regarded as an extension of the well known "Clairaut equation".

KEYWORDS: Composites, Filament Winding, Geodesics, Semi-Geodesics, Quasi-Geodesics, Friction coefficient, Slippage Resistance.

1. Introduction

In the filament winding process, one of the most important elements is the determination of the fiber orientation and trajectories on the mandrel surfaces. It is well known that the greatest winding stability is obtained following the geodesic line on the mandrel surfaces. The geodesic is uniquely determined by an initial point and its winding angle. This severely limits the choice of fiber paths and is an obstacle to the production of optimized composite components ([1], [2], [3]). In these two decades, several non-geodesic filament winding techniques, including semi-geodesics and other methods based on mathematical model or experiment have been proposed([2], [3], [4], [5], [6]). Unfortunately, all the methods now available are related with a certain type of differential equations and is difficult to solve and control.

In this paper, a new class of non-slip trajectories on revolutional surfaces, called the quasi-geodesics is proposed, which is defined as the limit of the piecewise semi-geodesics on the conical segments obtained by the subdivision of a revolutional surface. It is proved that a curve defined by the piecewise semi-geodesics converges to a new stable trajectory with an appropriate restriction. The equation of the quasi-geodesics on revolutional surfaces is so simple that not only the fiber paths can be easily calculated but also can be easily controlled to meet various kinds of requirement in the filament winding design. Especially, the equations of the quasi-geodesics for cylinders, cones, spheres and ellipsoids are directly obtained as the special cases. Some examples of the filament winding design using quasi-geodesics are illustrated.

2. The Differential Equation of Semi-Geodesics on Revolutional Surfaces

Filament winding on surfaces of revolution is usually based on geodesic lines. For more flexibility, the semi-geodesic is proposed and defined as a curve with constant slippage resistance. Recently, the semi-geodesics are widely used in the filament winding design, since the semi-geodesic offers more freedom and makes it possible to wind divergently from the geodesic lines([1], [4]).

Let $\boldsymbol{S}$ be a given mandrel surface, $\boldsymbol{C}$ be a fiber path, $\boldsymbol{K_g}$ and $\boldsymbol{K_n}$ be the geodesic curvature and normal curvature at a point $\boldsymbol{P}$ on $\boldsymbol{C}$ as shown in Fig. 1.

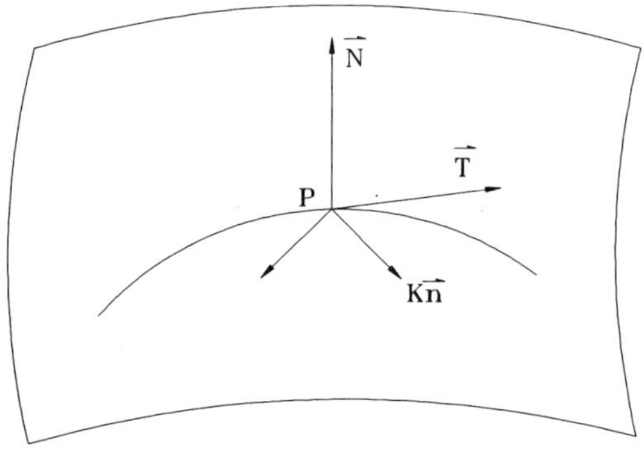

Fig. 1. $\boldsymbol{K_g}$ and $\boldsymbol{K_n}$ — projection of curvature vector $k\vec{n}$ on tangent plane and normal line.

In differential geometry, the slippage resistance λ expressed by

$$\lambda = \boldsymbol{K_g} / \boldsymbol{K_n} \tag{2.1}$$

In a coordinate reference system, the revolutional surface of a mandrel is described by the following equation and is shown in Fig. 2.

$$S(x,\theta) = \begin{pmatrix} x \\ r(x)\cos(\theta) \\ r(x)\sin(\theta) \end{pmatrix} \tag{2.2}$$

The differential equations of a curve $\boldsymbol{C}$ with constant slippage resistance λ can be derived as follows,

$$\frac{dx}{ds} = \frac{\cos(\alpha)}{\sqrt{1+(r')^2}} \tag{2.3}$$

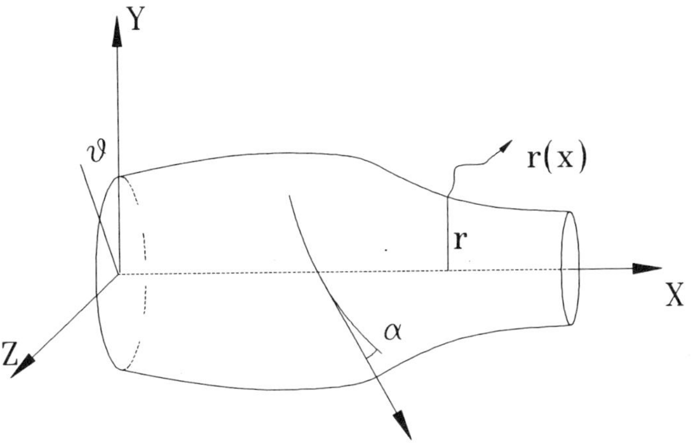

Fig.2. A revolutional surface and a trajectory with winding angle α.

$$\frac{d\theta}{ds} = \frac{\sin(\alpha)}{r} \tag{2.4}$$

$$\frac{d\alpha}{ds} = \frac{\lambda[1+(r')^2\sin^2(\alpha) - r'r''\cos^2(\alpha)] - [1+(r')^2]r'\sin(\alpha)}{r[1+(r')^2]^{\frac{3}{2}}} \tag{2.5}$$

where s is the arc length and α is the winding angle, r' and r'' are the first and second derivatives of r with respect to x. The solution curve C of (2.3), (2.4) and (2.5) is called the curve with slippage resistance λ or semi-geodesic line for short. The semi-geodesic is stable if

$$\lambda \leq \mu \tag{2.6}$$

where μ is the coefficient of friction. The condition (2.6) ensures that the slippage of the filament along the semi-geodesic is to be prevented. The semi-geodesics are in many cases the basis for the control of the filament winding machine.

3. The Quasi-geodesics

For a given point $\boldsymbol{P}$ on $\boldsymbol{S}$ and an initial winding angle α, the solution curve to the differential equations of semi-geodesics is uniquely determined and can be obtained by numerical method, but the calculation is complicate and is difficult to control to meet various technique requirement. In this section, a new class of simple and non-slip trajectories on revolutional surfaces is defined based on the semi-geodesics on the conical surfaces.

For a conical surface

$$r(x) = kx + b \tag{3.1}$$

where k and b are constant. We have $r' = k$ and $r'' = 0$ and (2.5) becomes

$$\frac{d\alpha}{ds} = \frac{\lambda \sin^2(\alpha) - k \sin(\alpha)}{r\sqrt{1+k^2}} \tag{3.2}$$

notice that

$$\frac{dx}{dr} = \frac{1}{k} \tag{3.3}$$

$$\frac{ds}{dx} = \frac{\sqrt{1+k^2}}{\cos(\alpha)} \tag{3.4}$$

then

$$\frac{ds}{dr} = \frac{\sqrt{1+k^2}}{k\cos(\alpha)} \tag{3.5}$$

and

$$\frac{dr}{d\alpha} = r\left(\frac{\lambda}{\lambda \sin(\alpha) - k} - \frac{1}{\sin(\alpha)}\right)\cos(\alpha) \tag{3.6}$$

then r is expressed as

$$r = \left(\frac{r_0 \sin(\alpha_0)}{\lambda \sin(\alpha_0) - k}\right)\left(\lambda - \frac{k}{\sin(\alpha)}\right) \tag{3.7}$$

where r_0 and α_0 are initial values. Thus the quasi-geodesics are calculated from (3.7) and

$$\frac{d\theta}{dx} = \frac{\sqrt{1+(r')^2}}{r}\tan(\alpha) \tag{3.8}$$

For a given surface of revolution $\boldsymbol{S}$, subdivide $\boldsymbol{S}$ into conical sections as shown in Fig. 3.

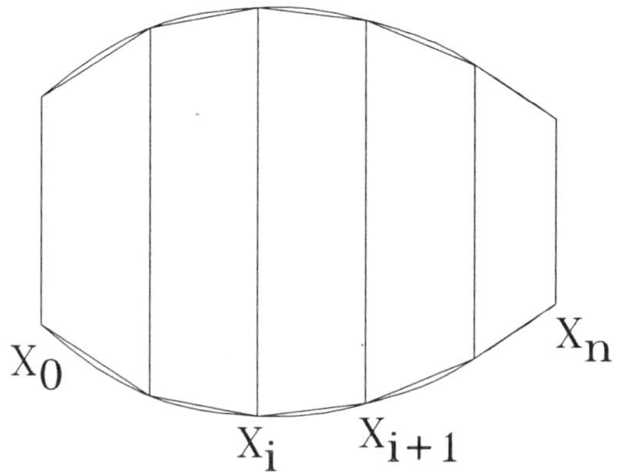

Fig. 3. Subdivision of a revolutional surface into conical sections

It is proved that a curve defined by the piecewise quasi-geodesics with a certain joint condition at nodes x_i converges to a new stable trajectory with an appropriate restriction and is expressed as follows,

$$r\sin(\alpha) = \frac{C}{[1-\lambda C f(r)]} \tag{3.9}$$

where C is a constant, λ is the coefficient of friction and $f(r)$ is a uniquely determined function of r. The limit curve defined by (3.9) is called the quasi-geodesic and (3.9) is called the equation of the quasi-geodesic. Equation (3.9) can be regarded as an extension of the well known "Clairaut equation". In fact, as $\lambda = 0$, the equation of quasi-geodesics becomes "Clairaut equation".

$$r\sin(\alpha) = C \tag{3.10}$$

Of course, the quasi-geodesic is an extension of the geodesic. The equation of the geodesics on revolutional surfaces is so simple that not only the fiber paths can be easily calculated but also can be controlled to make it much easier turn-around and bridging-free.

4. The Quasi-Geodesics on Cylinders, Spheres and Ellipsoids

As special cases, the calculation of quasi-geodesics on cylinders, spheres and ellipsoids can be much simplified as follows.

1. Cylinders

For cylinders, since

$$r' = r'' = 0,\ r = \text{constant} \tag{4.1}$$

from (2.3), (2.4) and (2.5), the quasi-geodesics are determined by

$$x = x_0 + \frac{r}{\lambda}\left(\frac{1}{\sin(\alpha_0)} - \frac{1}{\sin(\alpha)}\right) \tag{4.2}$$

$$\theta = \theta_0 + \frac{1}{\lambda}\left(\ln\left|\tan(\frac{\alpha_0}{2})\right| - \ln\left|\tan(\frac{\alpha}{2})\right|\right) \tag{4.3}$$

where (x_0, θ_0) is the initial point, α_0 is the initial winding angle and λ is the slippage resistance. Obviously, the quasi-geodesic is coincident to the semi-geodesic on cylinders.

2. Spheres

The equation of a sphere with radius R is expressed as

$$S(\theta,\ \varphi) = \begin{pmatrix} R\cos\varphi \\ R\sin\varphi\sin\theta \\ R\sin\varphi\cos\theta \end{pmatrix} \tag{4.4}$$

then we have

$$\frac{d\varphi}{ds} = \frac{1}{R}\cos\alpha \tag{4.5}$$

$$\frac{d\theta}{ds} = \frac{\sin\varphi}{r} \tag{4.6}$$

$$\frac{d\theta}{d\varphi} = \frac{\tan\alpha}{\sin\varphi} \tag{4.7}$$

and

$$r(t) = R\sin\varphi \tag{4.8}$$

$$t = R\cos\varphi \tag{4.9}$$

$$dt = -R\sin\varphi d\varphi \tag{4.10}$$

then the quasi-geodesic lines on spheres are determined as follows,

$$\frac{1}{\sin\alpha} = \left(\frac{R\sin\varphi}{r_0\sin\alpha_0} + \frac{\lambda\sin\varphi}{2}\ln\left|\frac{1-\cos(\varphi)}{1+\cos(\varphi)}\right|\right) \tag{4.11}$$

$$\theta = \theta_0 + \int_{\arccos\varphi_0}^{\arccos\varphi} \frac{dt}{(1-t^2)\sqrt{(1-t^2)(\frac{R}{r_0\sin\alpha_0} + \frac{\lambda}{2}\ln\frac{1-t}{1+t})^2 - 1}} \tag{4.12}$$

3. Ellipsoids

The equation of an ellipsoid with long and short axes a and b is expressed as

$$S(\theta\,,\ \varphi) = \begin{pmatrix} a\cos\varphi \\ b\sin\varphi\sin\theta \\ b\sin\varphi\cos\theta \end{pmatrix} \tag{4.13}$$

then we have

$$\frac{d\varphi}{ds} = \frac{\cos\alpha}{\sqrt{a^2\sin^2\varphi + b^2\cos^2\varphi}} \tag{4.14}$$

$$\frac{d\theta}{ds} = \frac{\sin\alpha}{\sqrt{a^2 + b^2}} \tag{4.15}$$

$$\frac{d\theta}{d\varphi} = \frac{\sqrt{a^2\sin^2\varphi + b^2\cos^2\varphi}}{\arctan\alpha} \tag{4.16}$$

and

$$r(t) = b\sin\varphi \tag{4.17}$$

$$t = a\cos\varphi \tag{4.18}$$

$$dt = -a \sin\varphi d\varphi. \tag{4.19}$$

Thus, the Quasi-geodesics on ellipsoids are as follows,

$$\frac{1}{\sin\alpha} = b\sin\varphi\left(\frac{1}{r_0 \sin\alpha_0} + \frac{\lambda a}{b^2}\int_{\frac{\varphi}{2}}^{\varphi}\frac{dt}{\sin t}\right) \tag{4.20}$$

$$\theta = C + \int_{\varphi_0}^{\varphi} \frac{\sqrt{a^2 \sin^2 t + b^2 \cos^2 t}}{(1-t^2)\sqrt{b^2 \sin^2 t \,(\frac{1}{r_0 \sin\alpha_0} + \frac{\lambda a}{b^2} \ln \frac{du}{\sin u})^2 - 1}} dt \tag{4.21}$$

Obviously, on the conical surfaces the equation (3.8) becomes (3.7) and the quasi-geodesic turns out to be the semi-geodesic.

5. Filament Winding Patterns Using Quasi-Geodesics

The trajectories of quasi-geodesics previously described can be applied to axi-symmetric mandrels. In [7], [8], a filament winding pattern is classified into two categories: the algebraic patterns and the geometric pattern. These two kinds of patterns have to be consistent and incorporated in a filament winding design system. A well designed algorithm consistent with the algebraic patterns has been developed to calculate the fiber path using quasi-geodesics. Three examples of the filament winding patterns using quasi-geodesics are illustrated and shown in the following figures.

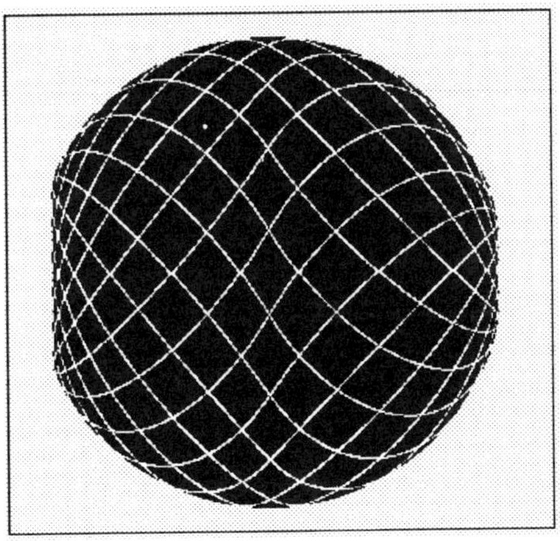

Fig. 4. A filament winding pattern on a sphere like mandrel

Our future work will be the extension of quasi-geodesics to the mathematical model for non-axi-symmetric filament winding, that will allow an increased flexibility of manufacturing process.

We would express our deep thank to Chinese Science foundation for the partly support of this work.

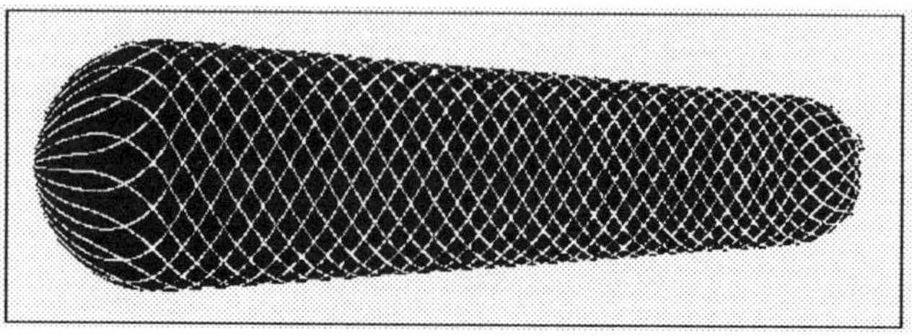

Fig. 5. A filament winding with $\alpha = 60°$ and $\mu = 0.2$.

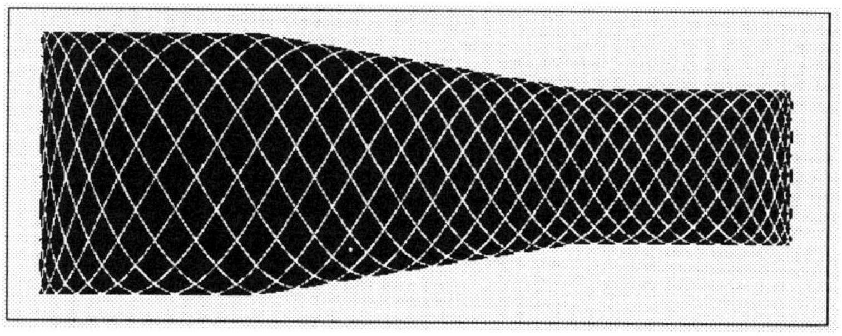

Fig. 6. A filament winding with $\alpha = 60°$ and $\mu = 0.2$.

Reference

[1] Menges, G., Wodicka, R. and Barking, H. L., Non-geodesic winding on a surface of revolution, 33rd Ann. Tech. Conf. of the Reinforced Plastics/Composites Inst. S. P. I. 1978.

[2] G. M. Wells, K. F. McAnnlty, "Computer Aided Filament Winding Using Non-Geodesic Trajectories", ICCM-VI, pp. 1-161-1.173, Proceedings of the 6th International Conference on Composite Materials London, UK, 1987.

[3] G. Di Vita, M. Grimaldi, M. Marchetti, and P. Moroni, "The Filament Winding Manufacturing Technique: studies of the Determination of the Friction Coefficient and on the Optimization of the Feed-eye Motion", pp. 972-979, 22nd International SAMPE Technical Conference, November 6-8, 1990.

[4] X. L. Lin, D. H. Liu, " Non-Geodesic Winding Equation on a General surface of Revolution", Proceedings of the 6th International Conference on Composite Materials, London, 1987.

[5] M. Marchetti, D. Cutolo, G. Di Vita, " Design of Domes by Use of the Filament Winding Technique", ECCM-III, PP. 401-408, Proceedings of the 3rd European Conference on Composite Materials, Bordeaux, France, 1989.

[6] F. Bullock, S. Kowalski, R. Young, "Automated Prepreg Tow Placement for Composite structures", 6th International SAMPE Symposium, April 2-5, 1990.

[7] Y. D. Liang, Gene Luo, Y. Chen, Kou-xìan Qiu, " Graphical Representation and calculation of Interweaving Degrees of Filament Winding Patterns", 41st International SAMPE Symposium March 24-28, 1996.

[8] Y. D. Liang, Gene Luo, " A Simple Filament Winding Pattern Generation Algorithm", Submitted to SAMPE this year.

28th International SAMPE Technical Conference
November 4-7, 1996

EMITTANCE OF THERMAL CONTROL MATERIALS BETWEEN 100 AND 400 K

Donald A. Jaworske
NASA Lewis Research Center
Cleveland, Ohio 44135

ABSTRACT

The hemispherical total emittances of several common thermal control tapes were evaluated calorimetrically over a wide temperature range. The calorimetric technique used here to evaluate thermal control materials allows a thermally isolated sample to cool solely through radiant heat transfer to a liquid nitrogen cooled cold wall. The mechanism of cooling is similar to that found in space, providing a functional evaluation of hemispherical total emittance. The five samples that were evaluated included several first and second surface mirrored Kapton tapes, one carbon-filled Kapton tape, and one black film. These tapes were adhesively bonded to an aluminum substrate, which provided the sensible heat for the calorimetric calculation. Temperature-time data were collected as heat flowed from the aluminum, through the adhesive layer, through the thermal control tape, and ultimately through the surface to the surroundings. Hemispherical total emittances were calculated over the temperature range of 100 to 400 K.

KEY WORDS: Thermal Management, Optical Testing

1. INTRODUCTION

Materials used for thermal control play an important role in the thermal management of spacecraft because they provide the only radiative pathway for heat rejection. The optical property of interest for heat rejection in optically thin materials is hemispherical total emittance, ε_h. Many techniques have been developed to evaluate ε_h and many materials have been evaluated, mostly at or near room temperature. Many surfaces of spacecraft in low Earth orbit often vary widely in temperature. Hence, evaluating the emittance of materials over a wide temperature range is important.

Much work has been done characterizing the emittance and emissivity of metals (1) and nonmetals (2). Of particular interest is the emittance of thermal control tapes specifically manufactured for use on spacecraft (3). Such tapes will experience a wide range of temperature during use. Hence, evaluating the emittance of these thermal control tapes in a functional manner over a wide temperature range is of interest.

There are many techniques available for calculating the hemispherical total emittance of surfaces and coatings. Both optical techniques and calorimetric techniques provide emittance data over a wide temperature range. For the thermal control tapes studied here, a calorimetric technique was chosen. There are many different calorimetric techniques available, including a variety of steady state (4) and transient (5) techniques. The technique chosen to evaluate the thermal control tapes of interest in this paper was a transient technique (6-7) that utilizes either a collimated beam from a quartz halogen lamp or a nicrome wire electric heater for heating the sample and a liquid nitrogen cooled cold wall for cooling the sample. This technique provides a functional evaluation of emittance over a wide temperature range because the thermal control coating of interest is actually heated and cooled to the desired temperatures.

2. MATERIALS AND METHODS

2.1 Sample Description The five thermal control tape samples evaluated for emittance consisted of: a first surface aluminum mirror on 0.0254 mm Kapton, a second surface aluminum mirror on 0.0127 mm Kapton, a second surface aluminum mirror on 0.0254 mm Kapton, a carbon-filled Kapton tape which appeared black, and a proprietary black film. Each sample tape was cut to fit the surface of a 2.54 cm diameter aluminum disk, 1.27 cm in height, such that all surfaces of the disk were covered. Acrylic adhesive on the back side of each tape bonded the tape to the 6061-T6 aluminum disk. All of the tape samples were kindly provided by the Harris Corporation, Melbourne, FL.

2.2 Calorimetric Characterization A transient calorimetric technique was chosen to measure the emittance of these thermal control tapes (6-7). This technique measures the emittance based on heat flow from the interior of a substrate disk through the surface of the thermal control tape and to a surrounding cold wall, and provides hemispherical total emittance values as a function of temperature. Normally, the sample is illuminated with a collimated light beam from a quartz halogen light source. However, in the case of the three mirrored tape samples, a small bucket arrangement was needed because of the low absorption coefficient of these samples. The bucket was made from nichrome wire and aluminum foil and easily heated the sample to 400 K. By using a remotely activated fusible link, the bucket could be dropped from around the sample to allow the sample to see the surrounding cold wall and begin the cooling process. The collimated light source was used to heat the black tape samples.

Evaluation of the temperature-time data was performed according to equation [1]. The

$$\epsilon_h = \frac{-mC_{p,Al}\frac{dT}{dt} - mC_{p,Coating}\frac{dT}{dt} + \sigma\alpha A_s T_o^4 + \sigma\alpha\epsilon\rho_o \frac{A_s^2}{A_o} T_s^4 - q_{losses}}{\sigma A_s T_s^4} \qquad [1]$$

key:

ϵ_h	= hemispherical total emittance	A_o	= surface area of the wall
m	= mass	A_s	= surface area of the sample
C_p	= specific heat	T_o	= temperature of the wall
dT/dt	= change in temperature with time	T_s	= temperature of the sample
σ	= Stefan-Boltzmann Constant	ϵ	= emittance (estimated)
α	= absorptivity of sample	ρ_o	= reflectivity of the wall
		q_{losses}	= heat losses by conduction

first term in the numerator represents the energy leaving the aluminum, the second term represents the energy leaving the coating, the third term represents the energy absorbed by the sample from the cold walls, the fourth term represents the energy radiated by the sample, reflected off the cold walls and reabsorbed by the sample, and the fifth term represents other losses due to conduction. In solving equation [1] for each temperature-time increment, values of unity were assigned to α, ε, and ρ_o. Unity as the maximum allowable value for each of these parameters would tend to overestimate ε_h; however, the contribution of these terms to the total energy balance is small at room temperature, with only the third term becoming important at low temperatures. The q_{losses} term is difficult to evaluate and was considered negligibly small at all temperatures. The term in the denominator accounts for energy that could leave the sample if it were a blackbody emitter. This approach is more rigorous than has been used previously (7).

The four measured parameters used to calculate emittance are: mass, specific heat, surface area, and the change in temperature with time. Mass was measured on an analytical balance, surface area was calculated from dimensions measured with calipers, and temperature was measured to the nearest 0.1°C with a Keithley model 740 programmable thermometer. The specific heat of aluminum was found from Touloukian (8). Care was taken to use temperature dependent specific heat data for the aluminum. The specific heat of Kapton at room temperature was used to account for the energy contribution from the thermal control tape. By using a large aluminum substrate (compared to the tape) the mass of the aluminum substrate accounts for over 98.5% of the mass of the entire sample. The hemispherical total emittance was calculated for each temperature increment between 100 and 400 K.

2.3 Thermocouple Installation The aluminum disks were suspended in the vacuum chamber by 0.127 mm diameter copper and constantan thermocouple wires. The type T thermocouple wires were attached to the tape-covered aluminum samples by two 4-40 set screws imbedded in the aluminum. The high thermal conductivity of the aluminum substrate insures that the surface of the tape is near uniform in temperature, providing there are no adhesion problems between the thermal control tape and the aluminum.

3. RESULTS AND DISCUSSION

The calorimetric technique provides a functional evaluation of emittance in that the sample is allowed to cool to the surroundings in a way similar to that found on spacecraft. Heat flows from the interior of the aluminum disk, through the adhesive and Kapton, and out of the surface to the surroundings. This process continues for all temperatures studied. The efficiency of this process depends on the emittance or emissivity of the surface and the emittance or emissivity of the surface depends on the composition and morphology of the surface.

Figure 1 summarizes the emittance as a function of temperature for the three mirrored samples. As expected, the mirror with no Kapton overlayer has the lowest emittance, while the mirror with the 0.0127 mm layer of Kapton has an intermediate emittance and the mirror with the 0.0254 mm layer of Kapton has the highest emittance. Each sample has a different dependence on temperature. The mirror with no Kapton overlayer has little to no dependence on temperature, while the mirrors with 0.0127 and 0.0254 mm of Kapton have a similar, larger, dependence on temperature. The slight upward trend in emittance below 230 K for the first surface mirror may be an artifact of the simplifying assumptions made in the emittance calculations, because heat absorbed by the sample from the walls becomes an important contributor to the energy balance at low temperatures.

Figure 2 summarizes the emittance as a function of temperature for the two black thermal control tapes. Both tapes have similar emittances as a function of temperature and are higher at all temperatures than the mirrored Kapton samples.

Given the large variety in emittance exhibited in the thermal control tape samples summarized here, one quickly reaches the conclusion that, by appropriate material selection, emittance can be tailored for almost any application and temperature range. Future evaluaton of emittance with the calorimetric technique will include the characterization of silver coated teflon and aluminum coated teflon.

4. CONCLUSIONS

The emittance of several thermal control tapes was evaluated utilizing a calorimetric technique, where the samples were allowed to cool solely by radiant heat transfer to a surrounding liquid nitrogen cooled cold wall. This mode of cooling allows heat to flow from the interior of an aluminum disk, through the materials and interfaces contained in the tape, and out of the surface to the surroundings, in a way similar to that encountered by spacecraft. An energy balance based on energy contributions from the aluminum substrate, the thermal control tape, and the surroundings was used to calculate emittance at temperature increments between 100 and 400 K. The thermal control tape composed of a first surface aluminum mirror had the lowest emittance while a carbon-filled Kapton tape had the highest. Second surface mirrors on Kapton had intermediate values, suggesting that emittance may be tailored over a wide range by proper materials selection.

5. REFERENCES

1. Y. S. Touloukian and D. P. DeWitt, Thermophysical Properties of Matter, Vol. 7, IFI/Plenum, New York, 1970.
2. Y. S. Touloukian and D. P. DeWitt, Thermophysical Properties of Matter, Vol. 8, IFI/Plenum, New York, 1972.
3. J. H. Henninger, "Solar absorptance and thermal emittance of some common spacecraft thermal control coatings," NASA RP 1121, 1984.
4. F. DiFilippo, M. J. Mirtich, B. Banks, C. Stidham and M. Kussmaul, "Total hemispherical emittance measured at high temperatures by the calorimetric method," NASA Tech. Memo. 102322, 1989.
5. W. B. Fussell, J. J. Triolo and J. Henninger, "A dynamic thermal vacuum technique for measuring the solar absorptance and thermal emittance of spacecraft coatings," NASA Tech. Note D-1716, 1963.
6. D. A. Jaworske, Thin Solid Films, 236 (1993) 146-152.
7. D. A. Jaworske, Thin Solid Films, 253 (1994) 233-237.
8. Y. S. Touloukian and E. H. Buvco, Thermophysical Properties of Matter, Vol. 4, IFI/Plenum, New York, 1970.

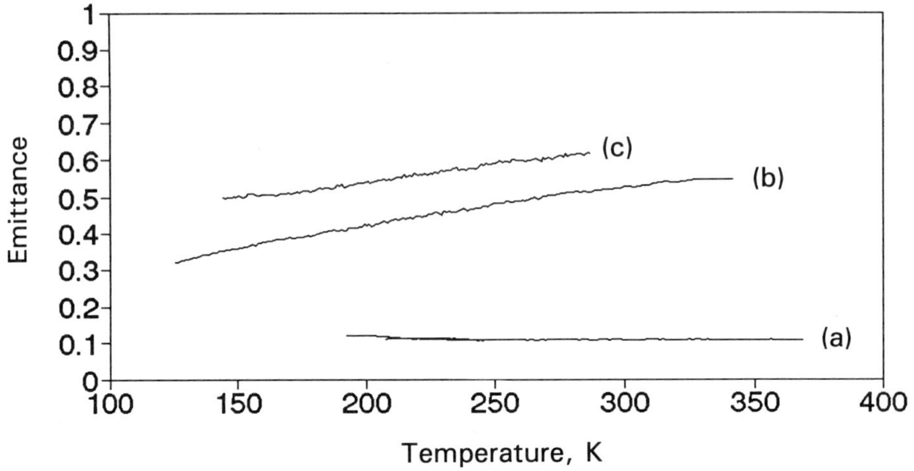

Figure 1.--Emittance versus temperature for (a) 1st surface aluminum on 0.0254 mm Kapton, (b) 2nd surface aluminum on 0.0127 mm Kapton, and (c) 2nd surface aluminum on 0.0254 mm Kapton.

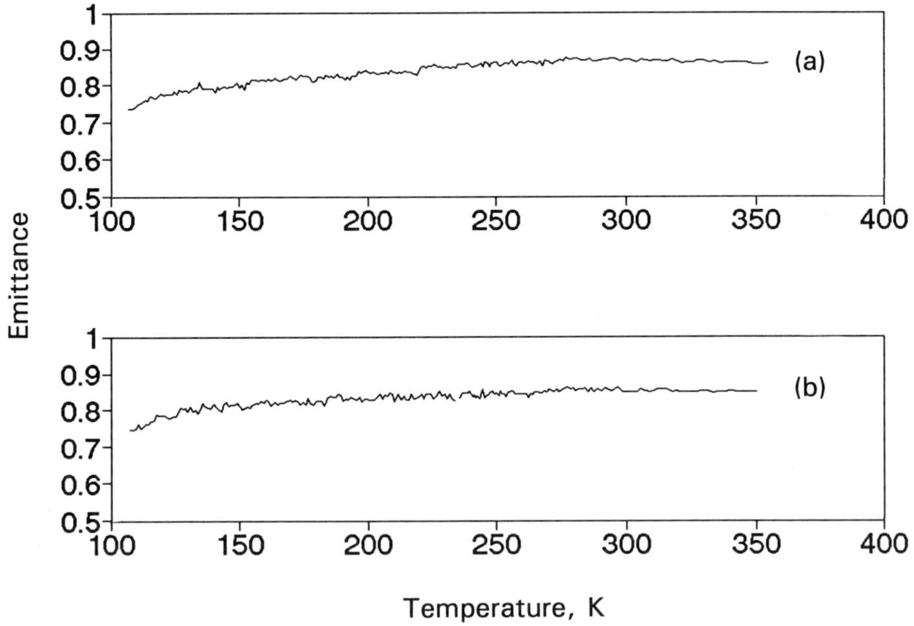

Figure 2.--Emittance versus temperature for (a) carbon-filled Kapton thermal control tape and (b) black film.

28th International SAMPE Technical Conference
November 4-7, 1996

INVESTIGATIONS OF SPACE ENVIRONMENT EFFECTS ON ELECTRICALLY CONDUCTIVE THERMAL CONTROL COATINGS AT MSFC

Miria M. Finckenor, Rachel R. Kamenetzky, Jason A. Vaughn
EH12/Physical Science and Environmental Effects
Marshall Space Flight Center, AL 35812

Richard Mell, John Sims, Randy Thompson, Paul LeMaster
AZ Technology
Huntsville, AL 35805

ABSTRACT

Electrically conductive thermal control coatings must maintain their electrical and optical properties in the space environment to ensure mission success. This paper details the investigations of electrically conductive coatings developed by AZ Technology for MSFC under contract NAS8-40557. Also included are the results of the RM400 coating development at MSFC for the Tethered Satellite System. Results from laboratory exposures of RM400 to atomic oxygen, ultraviolet radiation, thermal cycling, and other flight qualification data for both TSS-1 and TSS-1R missions are presented. Also included are post-flight analyses from the TSS-1 flight in August 1992 during the STS-46 mission. Emphasis of the program is on maintaining electrical conductivity, especially in vacuum, and the optical properties of solar absorptance and infrared emittance. Data from simulated space environment exposures at MSFC is presented for International Space Station candidate coatings. Also presented are results of pre-flight testing of a white and a black conductive coating selected for flight on the Passive Optical Sample Assembly-I experiment.

KEY WORDS: Space environment, thermal control, conductive coatings

1. INTRODUCTION

As a spacecraft moves through the low Earth orbit (LEO) polar attitudes or the geosynchronous orbit (GEO) attitudes, the high energy particulate radiation from the solar wind causes chargebuild-up on the surface of insulating materials but does not charge conducting materials. The potential difference could lead to electrical discharges which have caused many documented failures[1]. Many of these failures could have been avoided if the insulating material was replaced by a conducting material. Spacecraft surfaces usually cannot be metallic simply for the elimination of these charging problems because metals tend not to have the required optical properties needed to maintain temperatures or may react unfavorably with the space environment. Therefore thermal control coatings, such as paints and anodizes, are used. For many years, the only available thermal control coatings had surface resistivities ranging from 10^{11} to 10^{13} Ω/□, which will not prevent charging in the LEO polar environment or GEO. Electrically conductive or dissipative thermal control coatings are needed.

In non-polar LEO, the charging problems do not exist because the spacecraft is immersed in the ionosphere. However, several spacecraft require thermal control coatings to collect charge from the ionosphere for special purposes, such as generating power in tether applications, controlling charge collected by solar arrays, or for specific science missions. One example of this special case is the Tethered Satellite System (TSS) which required a highly conductive thermal control coating to collect electrons from the LEO ionosphere to generate power. Another potential application of conductive thermal control coatings would be the International Space Station (ISS) radiators for control of the charging caused by the high powered solar arrays. A suitable conductive thermal control coating did not exist at the time of design, so ISS has addressed their charging problems by using a costly plasma contactor for active charge control.

At MSFC, the Physical Science and Environmental Effects Branch studies charging due to plasma interactions and qualifies candidate materials for use on spacecraft. MSFC currently has a SBIR contract (NAS8-40557) with AZ Technologies to develop innovative thermal control coatings to meet future spacecraft needs.

2. TEST FACILITIES AND ANALYTICAL CAPABILITIES

The LEO Space Plasma Simulator (Fig. 1) located at MSFC is used to measure conductivity of materials in vacuum. This system has proven invaluable in evaluating new conductive thermal control materials such as RM400, which was used on the external surface of the Tethered Satellite System (TSS) satellite for thermal control while collecting 1 A of current. Also, the system has been used to predict the anodized aluminum arcing problem caused by the negative grounding scheme initially baselined by ISSA. The LEO Space Plasma Simulator is a 4 ft. dia. x 6 ft long vacuum chamber capable of a base pressure in the high 10^{-8} Torr. An argon plasma is produced in the chamber using a hollow cathode plasma source[2]. This type of source is used in the chamber because it produces a fairly uniform low density ($10^{6}/cm^{3}$), low temperature (1 eV) plasma which is a close simulation of the LEO environment. Argon is used as the carrier instead of oxygen because it is easier and safer to use. Because the chamber is used to study mechanisms for charge collection, argon ions are a good first order approximation. The chamber also contains both langmuir and emissive probes to measure the plasma properties in the chamber and any electric field variations.

A variety of instruments are used for pre- and post-test characterization. Solar absorptance is measured for the wavelength range of 250 nm to 2500 nm using either an AZ Technology Laboratory Portable Spectroreflectometer or a Beckman DK-2A Spectroreflectometer. Measurements have also been made using a Gier Dunkle MS251 portable reflectometer. Infraredreflectance measurements are made per ASTM E408-71 method A using either an AZ Technology TEMP 2000 Infrared Reflectometer or a Gier Dunkle DB100 Infrared Reflectometer.

Bulk resistivity is measured by two different methods. When candidate conductive thermal control coatings are provided only on aluminum substrates, resistivity is measured using the setup shown in figure 2. Electrical connection is made by using a nylon bolt for a good connection to the substrate aluminum while not producing a short to the painted surface. The connection to the painted surface is made using a 2.5" diameter aluminum disk, and a 2 kg weight is applied to the top of the disk to produce equal pressure on the paint. Both the electrical leads to the aluminum substrate and to the painted surface are connected to a Keithley 617 digital electrometer to directly measure bulk resistance. The Keithley 617 electrometer uses a constant current method to measure the resistance of the sample. By employing this method the electrometer is capable of measuring resistances as low as 0.1 Ω and as high as 200 GΩ. The bulk resistance is recorded, and this procedure is repeated at various points on the sample to produce the average bulk resistance. The bulk resistivity of the coating is calculated using equation (1).

$$\rho = \frac{R \pi d^2}{4 \tau} \qquad \textbf{(1)}$$

Where, R is the average resistance measured from the witness panels, d is the diameter of the disk (5.08 cm), and τ is the thickness of the coating which varied on each witness plate. These measurements were highly dependent on the pressure applied to the painted surface and the smoothness of the coating. Also, this bulk resistivity measurement includes the coating/aluminum interface.

Where possible, glass substrates are coated with candidate thermal control coatings, and surface resistance is measured using a Magne-Tron Instruments model M-700 four point probe. The M700 uses a in-line four point probe and automatic computation of geometric factors for a surface resistivity measurement range of 10^{-3} to 10^{10} Ω/□. Calibration of the instrument is checked prior to sample measurements by measuring the surface resistivity of carbon-impregnated Kapton, also known as black Kapton. The sample of black Kapton is certified to be 500 ± 10% Ω/□. Then, the sample surface resistivity is measured and multiplied by the coating thickness for bulk resistivity. Coating thickness is measured using profilometer or eddy current methods.

Surface resistivity measurements in vacuum are made on a 6" x 6" glass substrate coated with the candidate thermal control coating. Four concentric copper rings are vapor deposited on the paint surface. Leads are soldered to the copper rings, taking care to insure good connection both between the leads on the copper, the copper and the paint. A 10 to 1000 volt potential is applied to the two inner rings, and the two outer rings are grounded. The outermost ring is used as a guard to prevent stray electric fields. The potential difference between the inner and outer rings induces charge flow between the rings through the sample coating. The current produced by a single bias voltage is measured as a function of time in vacuum. The surface resistivity ρ_s of thepaint is calculated using equation 2:

$$\rho_s = \frac{2 \pi V}{I \ln (r_1/r_2)} \qquad \textbf{(2)}$$

Where, V is the applied voltage, I is the measured current, r_1 is the radius of the outer concentric copper ring, r_2 is the radius of the inner concentric copper ring. Bulk resistivity can be calculated if the paint thickness is known.

3. RM400 Thermal Control Coating

3.1 Pre-Flight Qualification The thermal control coating with the greatest amount of testing performed at MSFC is RM400. RM400 is a doped tin oxide pigment in an epoxy binder that was developed in a short time frame for the Tethered Satellite in 1992. The Tethered Satellite System (fig. 3) flew on STS-46 from July 31 - August 8, 1992. Despite problems with the tether reel mechanism, the TSS generated 40 volts during its first mission. This power generation proved the utility of the RM400 conductive coating. The paint was tested again in 1994, since the reflight mission, TSS-1R, was planned for nearly twice the duration of the first flight, TSS-1. TSS-1R flew on STS-75 in February 1996 and resulted in the loss of the satellite when the tether was unreeled to 19.7 km (12.2 mi). Before satellite separation occurred, power of -3500 volts at about 1 amp had been generated by its rapid motion through the Earth's magnetic field lines, again proving the utility of the coating. Details of the TSS hardware and mission objectives may be found in reference 3.

The components of RM400 are used in other space qualified paints, so no difficulties were encountered during the qualification tests. As part of the qualification of RM400 for flight, paint samples were exposed to atomic oxygen, ultraviolet radiation, thermal cycling, and hard vacuum. After these exposures, analyses were performed to ensure that optical, electrical, and adhesive properties were maintained within the flight requirements. Data for the TSS-1 and TSS-1R qualification tests can be found in Tables 1 and 2, respectively. Qualification also included tests for flammability, humidity, and volatile condensable materials.

A sample of RM400 paint on an aluminum substrate was subjected to flammability testing per NHB 8060.1C, Test 1, upward configuration. A 2.0 mil thick RM400 coating was applied to a 0.063" thick aluminum substrate. The burn length of the sample was 0 inches in a burn environment of 20.9% oxygen. The coating was discolored at the ignition point but did not propagate flame. This test was repeated in 1994 with a 3.0 mil coating on an aluminum substrate with a burn length of 0.125" in air. RM400 was given an "A" rating for both tests.

Humidity testing was judged non-critical and was not performed in 1992 due to the tight timeline required for qualifying RM400. The original TSS coating, NS43C, is a water-based paint which may be more susceptible to the effects of humidity. RM400, which is epoxy-based, is cured for 24 hours at 100 °C and was presumed unsusceptible to humidity effects. A 1994 sample of RM400 paint on an aluminum substrate was subjected to 90% ± 5% relative humidity for 48 hours while being held at a temperature of 40 °C. Solar absorptance and thermal emittance were measured after the exposure and found within specified parameters.

Small samples of RM400 removed from the glass substrates were subjected to outgassing tests per JSC-SP-R-0022A, "Vacuum Stability Requirements of Polymeric Material for Spacecraft Application", at 125 °C under 10^{-6} torr vacuum for 24 hours. Collector plates for volatile condensable material (VCM) content determination were held at 25 °C. Test results for 1992 included a total mass loss (TML) of 0.31% and a volatile condensable material (VCM) of 0.10%. Test results for 1994 included a TML of 0.20% and a VCM content of 0.05%. Both are considered acceptable per JSC-SP-R-0022A. Visible films of contaminant were noted on collector plates after both tests, therefore, this material should not be used in the vicinity of contamination sensitive surfaces, i.e. mirror optics or windows, without further testing.

3.2 STS-46 Flight Exposure of RM400 After flight on STS-46, two battery doors (part # 3150D-101-401, NC002 and NC003) were returned to MSFC for evaluation of the RM400 coating. Various tests were performed to ensure the paint's critical properties would resist space environment degradation during the reflight of TSS. These tests evaluated adhesion, solar absorptance, infrared emittance, bulk resistivity, and changes in thickness. Witness samples from the RM400 application procedure were used as control samples.

The battery doors were examined under a portable stereo microscope for signs of pitting, cracking, delamination, or other failure of the paint. Examination under 50X magnification revealed no obvious defects. The surface of the paint appeared to be uniform with the exception of a few minor scratches. Surface morphology was described as rough and diffuse with no apparent pitting or cracking.

Solar absorptance was measured on the control sample and directly on the battery doors using the LPSR. Solar absorptance of the control samples averaged 0.530, while solar absorptance measured on the NC002 and NC003 battery doors averaged 0.545 and 0.541, respectively. These numbers vary from the pre-flight measurements because: 1.) a different instrument was used, 2.) more stray light interference due to the curvature of the test article, and 3.) possible aging of the paint.

Infrared emittance measurements were also made on the samples using a Gier-Dunkle DB 100 per ASTM E408-71 method A. The infrared reflectometer was calibrated prior to each measurement using gold and black standards. Then, data readings were taken on the RM400 control sample and the battery doors. Again, some stray light effects were seen during emittance measurements due to curvature of the battery door. Infrared emittance of the control samples was measured to be 0.89. Emittance measurements on the NC002 and NC003 battery door averaged 0.93 and 0.88, respectively.

The bulk resistivity was measured using the Keithley electrometer set-up, except the 2.5" diameter aluminum disk was machined to closely match the curvature of the battery doors. The average bulk resistivity for the NC002 battery door was 7.97×10^6 ohms-cm, with a range of 4.11×10^6 to 1.90×10^7 ohms-cm. The average bulk resistivity for the NC003 battery door was 3.10×10^7 ohms-cm, with a range of 1.92×10^7 to 5.08×10^7 ohms-cm. Control samples from the 1992 batch averaged 2.26×10^6 ohms-cm.

Coating thickness measurements were made using an eddy current probe. The instrument was calibrated using a bare aluminum plate, then a thickness standard. After each set of measurements, the probe was checked again for calibration. Ten measurements were made on each surface tested. The flight panel was acceptable if 80% of the measurements were between 1.5 and 5.0 mils. None of the flight hardware measurements failed to meet this requirement, with measurements ranging from 1.60 to 2.65 mils.

3.3 Other Tests of RM400 While the Tethered Satellite is deployed, the electrons from the ambient space plasma are accelerated and impact the coating with up to a few keV of energy. Samples of RM400 were exposed in the lab to electrons with energies ranging from 0.1 to 1 keV to study the effect on the coating. During those tests, the coating was found to luminesce, and prolonged exposure of RM400 to high energy electrons caused the coating to darken[4].

4. INNOVATIVE CONDUCTIVE THERMAL CONTROL COATINGS

Under the SBIR contract NAS8-40557, AZ Technology has been developing conductive thermal control coatings similar to RM400, but with greater environmental stability for long-term exposure in LEO. Both white and black conductive coatings have been developed under this contract.

4.1 White Conductive Thermal Control Coatings Whereas RM400 contained doped tin oxide, AZ Technologies' current efforts are in using doped zinc oxide. Zinc oxide is the pigment found in Z-93, a commonly used spacecraft thermal control coating with low solar absorptance and high infrared emittance. Experimentation with conductive binders has also been performed.

A test series of doped zinc oxides explored the relation between surface resistivity, solar absorptance, and the processing temperature of the pigment. Doped zinc oxide powder was produced then fired at temperatures varying from 700 to 1000 °C then pressed into powder sample holders. This permits measuring surface resistance without including the effects of a binder. With both gallium- and aluminum-doped zinc oxide, the surface resistivity decreases as the processing temperature increases, since the dopant more thoroughly migrates into the ZnO matrix[5]. Processing temperature appeared to have little or no effect on solar absorptance.

4.1.1 Gallium-Doped Zinc Oxide After promising resistivity measurements made in air, a gallium-doped zinc oxide coating was tested for resistivity in vacuum. The pigment was calcined at 700 °C for 4 hours, mixed with a potassium silicate binder, and sprayed onto a glass substrate for the test. The solar absorptance of this sample was 0.186. The initial surface resistivity was 1.05 x 10^7 Ω/□, but as the vacuum dropped to the 10^{-7} torr range, the resistivity increased. Resistivity after 30 hours was 5 x 10^{12} Ω/□. This increase was due to water desorption from the potassium silicate binder. After measurements were completed, the test chamber was backfilled with nitrogen. The resistivity remained high until ambient air was introduced into the chamber. Following the test, cracking of the paint was noted. The evaporation of the water from the potassium silicate binder causes mechanical stresses in the coating, forming microscopic cracks and interfering with the electrical path through the coating.

After potassium silicate was dismissed as an unsatisfactory binder for conductive paints, polydimethylsiloxane was tested. Dimethylsiloxanes are one of the more stable polymers used in spacecraft applications and, with proper distillation and curing procedures, are not significantly contaminating. Various pigment-to-binder ratios (PBRs) were tested to optimize physical, optical, and electrical properties. The gallium-doped zinc oxide coating formulated with a 6:1 PBR had good physical properties and a solar absorptance as low as 0.159.

4.1.2 Aluminum-Doped Zinc Oxide Aluminum-doped zinc oxide coatings showed the same promise as the gallium-doped samples but incurred the same problems with cracking of the potassium silicate binder in vacuum. The resistivity in vacuum sample had a solar absorptance of 0.178. The initial surface resistivity was 9 x 10^6 Ω/□, but as the chamber was evacuated, the resistivity increased to 5 x 10^{13} Ω/□. Again, polydimethylsiloxane was tested in a 6:1 PBR with aluminum-doped zinc oxide pigment. Solar absorptance of 0.155 was measured.

4.2 Black Conductive Thermal Control Coatings The requirements of this contract are for AZ Technologies to develop black electrically conductive coatings that resist the space environment and do not rely on graphite for conductivity. Desired optical properties of this coatings are solar absorptance and infrared emittance greater than or equal to 0.85.

Antimony and tin oxides were cured in phenylmethyl silicone to form the antimony tin doped silicon oxycarbide, an electrically conductive black glassy material. The solar absorptance and infrared emittance were 0.96 and 0.91, respectively, meeting the requirements. The doped silicon oxycarbide coating was tested for resistivity changes in vacuum with promising results. Surface resistivity was initially 8.3 x 10^6 Ω/□ and rose to 2.68 x 10^{9} Ω/□ after 1.5 days in vacuum before stabilizing. Even with the three orders of magnitude change in resistivity, it is better than the five to seven orders of magnitude seen with the potassium silicate binders used in the conductive white coating test series.

4.3 Space Environment Exposure of AZ Technologies Conductive Coatings While it is imperative that the conductive thermal control coatings maintain their electrical properties in vacuum, the other aspects of the space environment must be considered. Future laboratory testing of coatings developed by AZ Technologies will include atomic oxygen and ultraviolet radiation exposures to ensure that electrical and optical properties are maintained.

To complement the laboratory testing, one set each of white and black conductive coatings were selected to be flown as part of the Passive Optical Sample Assembly-I (POSA-I) experiment. The samples chosen are the gallium-doped zinc oxide and the antimony tin doped silicon oxycarbide coatings. POSA-I (Fig. 4), a suitcase-size collection of optics, thermal control coatings, and other candidate spacecraft materials, is part of the Mir Environmental Effects Payload. This experiment was launched on STS-76 in March 1996 and is attached to the docking module of the Russian space station MIR. One side of the experiment faces MIR; the other side, with a duplicate set of samples, faces space. The expected duration of this exposure is 17 months. Post-exposure characterizations of optical and electrical properties are planned.

5. CONCLUSIONS

RM400 is well-suited for short term low Earth orbit (LEO) exposure and was flight-proven on both TSS-1 and TSS-1R missions. Tests of adhesion, optical properties, electrical properties, and resistance to detrimental space environment effects were completed to prove that the paint was suitable for missions up to 14 days and ensure that its critical properties would not change outside of desired parameters. The tests performed were to determine adhesion, solar absorptance, thermal emittance, bulk resistivity, outgassing, flammability, and resistance to thermal cycling, atomic oxygen, ultraviolet radiation, vacuum, and humidity. RM400 paint passed all of these tests satisfactorily.

New conductive coatings are being developed at AZ Technologies. Doping of zinc oxide pigment with gallium and aluminum, along with other organo-metallics, has produced conductive white pigments. Work is still underway to produce a binder that will withstand vacuum without cracking. Antimony tin doped silicon oxycarbide holds promise as a conductive black thermal control coating. Selected conductive thermal control coatings are currently being exposed to the space environment as part of the POSA-I experiment.

6. REFERENCES

1. R.D. Leach and M.B. Alexander, eds., "Failures and Anomalies Attributed to Spacecraft Charging," NASA Reference Publication 1375, August 1995.
2. D.E. Siegfried, "A Phenomenological Model for Orificed Hollow Cathodes," NASA publication CR-168026, Dec. 1982.
3. C. Blake Powers, Charlotte Shea, and Tracy McMahan, The First Mission of the Tethered Satellite System, U.S. Government Printing Office 1992-324-999.
4. J.A. Vaughn, M. McCollum, and M. R. Carruth, Jr., "Laboratory Electron Exposure of TSS-1 Thermal Control Coating," NASA publication TM-108503, Dec. 1995.
5. R.J. Mell, Quarterly Technical Progress Report, Contract NAS8-40557, May 1996.

Table 1. RM400 Qualification Tests and Results for TSS-1

	Requirement	Control	AO + UV 3.6×10^{20} atoms/cm² 100 ESH	Thermal Cycle 30 cycles -95 to 74 °C in vacuum	Vacuum 45 hrs at 10^{-6} torr
Solar Absorptance	≤0.50	0.49	0.47	0.46	
Infrared Emittance	0.91±0.03	0.90	0.91	0.89	
Adhesion	Pass	Pass	Pass	Pass	N/A
Bulk Resistivity	<10^6 Ω-cm	580 Ω-cm**	8.3×10^5 Ω-cm†	9.1×10^5 Ω-cm†	890 Ω-cm**

† As calculated using equation 1 from surface resistivity measured by Keithley electrometer.
** As calculated using equation 2 from surface resistivity measured on coated glass substrate with concentric copper rings.

Table 2. RM400 Qualification Tests and Results for TSS-1R

	Requirement	Control	AO‡ 8.7×10^{21} atoms/cm²	UV 205 ESH	Thermal Cycle 80 cycles -95 to 74 °C in vacuum	Vacuum 48 hrs at 10^{-6} torr
Solar Absorptance	≤0.55	0.47	0.47	0.52		
Infrared Emittance	0.91±0.03	0.90	0.91	0.89		
Adhesion	Pass	Pass	Pass	Pass	Pass	N/A
Bulk Resistivity	<10^6 Ω-cm	695 Ω-cm**	910 Ω-cm§			340 Ω-cm**

** As calculated using equation 2 from surface resistivity measured on coated glass substrate with concentric copper rings.
‡Exposed in the MSFC Atomic Oxygen Drift Tube System, with <0.1 eV energy atomic oxygen. A safety factor of 10 was added to ensure that desired properties remained stable even with this dose.
§ As calculated from surface resistivity measured on coated glass substrate using Magnetron four-point probe.

Figure 1 LEO Plasma Simulator at MSFC

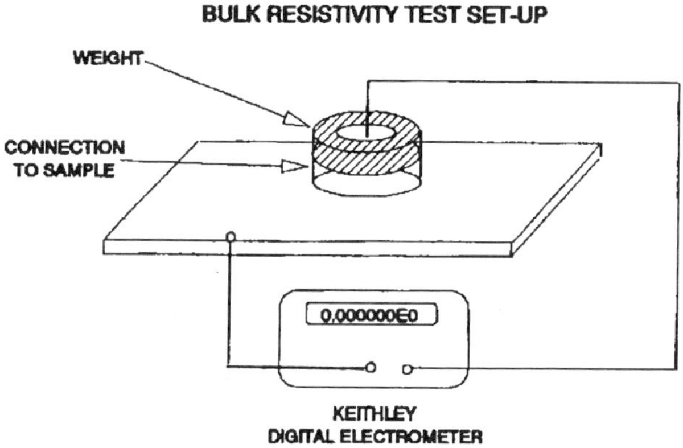

Figure 2 Bulk resistivity set-up for coated aluminum and TSS hardware

Figure 3 Deployment of Tethered Satellite painted with RM400

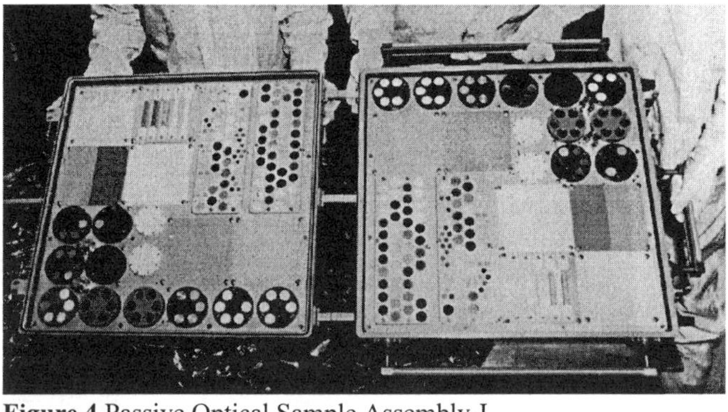

Figure 4 Passive Optical Sample Assembly-I

28th International SAMPE Technical Conference
November 4-7, 1996

PASSIVE OPTICAL SAMPLE ASSEMBLY II EXPERIMENT INTEGRATION

Gail L. Bohnhoff-Hlavacek and H. Gary Pippin
Boeing Defense & Space Group
P.O. Box 3999 M/S 8H-01
Seattle, WA 98126
(206)773-6892

ABSTRACT

The MIR Environmental Effects Payload experiments are a set of joint U.S./Russian experiments to assess the performance and degradation of materials placed on the exterior of the MIR Space Station. The effects of contamination, solar exposure, meteroid and orbital debris impacts, atomic oxygen, and charged particle radiation will be evaluated. A summary of sample production, hardware design and manufacture, cleaning and assembly procedures, and documentation of the POSA II experiment will be presented. The information processing aspects of the experiment will also be emphasized. This includes the development and structure of the data base, pre-flight data acquired and plans for on-orbit photo surveys, deintegration, and post-flight measurements.

KEY WORDS: MIR Environmental Exposure Package (MEEP), Passive Optical Sample Assembly (POSA), Space Environmental Effects

1. INTRODUCTION

The Passive Optical Sample Assembly (POSA) II experiment is part of the MIR Environmental Exposure Package (MEEP) suite of experiments being flown on the MIR Space Station between March, 1996 and September, 1997. The experiments were chosen to be part of an International Space Station (ISS) risk mitigation study. The performance of baseline and candidate materials for the ISS is being assessed in this experiment.

The POSA II experiment is a companion experiment to the POSA I experiment is directed by the NASA Marshall Space Flight Center (MSFC). The POSA II experiment team includes Boeing, McDonnell-Douglas, Rocketdyne, NASA Langley Research Center (LaRC), NASA Lewis Research Center (LeRC), MSFC, Johnson Space Center (JSC), RSC Energia, Illinois Institute of Technology Research Institute (IITRI), and AZ Technology. In addition to these organizations, Sheldahl, Parker-Hanefin, and Castrol Corporations each provided materials for evaluation. The POSA II experiment was integrated at Boeing, shipped to NASA LaRC, where it was integrated into the Passive Experiment Carrier (PEC), put through a launch simulation vibration test, re-inspected, and then closed and shipped to KSC for the March 22, 1996 launch of STS-076.

2. EXPERIMENT DESIGN

The POSA II experiment was proposed and agreed to as an additional experiment, by the ISS Phase I Integrated Products Team sponsoring the MEEP payload, because the original proposed materials list for the POSA I flight experiment was too large to be accommodated completely on one experiment. It was also desired that POSA I would include no silicone containing materials. POSA II provided for duplicates of some high priority specimens from POSA I, multilayer insulation blankets (MLI), a number of silicone coatings, a selection of "pre-contaminated" surfaces, materials under accelerated exposure conditions, as well as selected ISS second-source materials and a few experimental coatings.

The performance/degradation of the materials will be correlated with their space environmental exposure. Kapton films and optical solar reflectors were flown as atomic oxygen and contamination monitors. NASA MSFC provided a solar coulombmeter for monitoring solar UV. A few plastic nuclear track detectors were also flown. Upon return, surfaces will be examined for micrometeroid and debris impact damage. Computer models for prediction of atomic oxygen and equivalent sun hours of solar exposure will also be used to augment the environment monitors.

The POSA I and II experiments are each essentially two baseplates with samples, mounted on the inner walls of the PEC, an approximately 2-foot square by 6-inch high "suitcase". The PEC is sealed by a fluorosilicone elastomer o-ring. The experiments were deployed by astronaut extra-vehicular activity (EVA). When opened, the PEC halves folded back upon themselves and the two sets of samples are exposed 180° apart. One surface of the POSA II experiment is nominally 45° from ram.

To organize the data being obtained from this experiment, a common structure for a data base was developed and is being used by all members of the experiment team. Figure 1 shows a template of the database, which was developed using Filemaker Pro[R] software from the Claris Corporation. The sample nomenclature system incorporates the name of the specimen, individual companies internal identification codes, whether the sample is a flight specimen or a ground control specimen, and the exact location of the flight specimens by side, tray or plate, tray location, and company. All of the preflight data is included in the data base presently, and post-flight data will be input once analysis has begun. The same system is being used for POSA I and it is expected that a combined POSA database will be distributed after the flight. Many of the samples have reflectance data. The reflectance data is configured to be loaded into an Excel spreadsheet and plotted. Weights, outgassing measurements and descriptions of essential features of sample preparation are also included in this database.

3. MATERIAL CONSIDERATIONS

Individual samples were provided by, or manufactured at several different locations. While care was taken in the handling of each specimen, the specimens were not manufactured or produced in "clean room" conditions. The exception to this is the optical specimens with the thin anti-reflective coatings, which were produced by vacuum deposition, and maintained under clean room conditions prior to launch. Many of the flight specimens and ground control samples were exposed to laboratory environments during pre-flight measurements. Specimens were handled using nylon clean room gloves and tweezers which had been cleaned and isopropyl alcohol (IPA) wiped. Specimens were stored in cleaned Fluoroware containers. The large majority of specimens on POSA II are 1-inch diameter disks. Four sample trays, each capable of holding forty-six 1-inch diameter specimens, were provided by NASA from the Energetic Oxygen Interaction with Materials (EOIM) III experiment. Four more trays were made by Boeing from the JSC drawings for the EOIM trays. Four smaller trays, each capable of holding thirty 1-inch diameter specimens, were also made following the JSC drawings as closely as possible. A 6" x 6" housing was made to hold nine surfaces, each covered by a selective wavelength UV filter. Each surface had a specific compound deposited prior to flight. This housing was also maintained under clean room conditions prior to integration. A small frame was also made to hold the McDonnell-Douglas 3" x 3" specimens.

Half of the specimen trays were chromic acid anodized and half the trays were boric-sulfuric acid anodized. The anodized trays, backplates and tray covers were detergent washed, rinsed twice with distilled water, and solvent-rinsed first with high pressure liquid chromatography (HPLC) acetone, then with ethanol, and finally with a mixture of ethanol and tri-chloroethane (1:3), and dried in a laminar flow clean bench. The tray pieces were then double-bagged and brought to another clean bench for sample integration.

The POSA II baseplates, each a square approximately 23.75" on a side, were machined at NASA Langley Research Center, shipped to Boeing, and cleaned and bagged in the same manner as the tray pieces. The solar diodes and coulombmeters were assembled at NASA Marshall Space Flight Center and shipped to Boeing for integration.

4. INTEGRATION STEPS

Individual 1-inch diameter samples were taken from their Fluoroware containers and loaded into their respective trays in a class 100 clean bench. The NASA LeRC specimens were marked on the back side to document their orientation. Each tray containing LeRC specimens was photographed from the back, prior to bolting the tray backplate on to the tray, to document their orientation in the tray. In addition, a written description of the position of each sample was recorded as the sample was being loaded into its location on a given tray. After the samples were loaded into the trays, and the trays and backplates bolted together, each tray was photographed again from the front, still within the clean bench. Close-ups covering about half of a tray and photos of five to six samples at a time were also taken, as shown in Figure 2. A complete set of photos will again be taken just prior to disassembly after the flight.

The integration of the specimen trays onto the PEC baseplates was carried out in a class 100 clean tent. The integration sequence for each POSA II baseplate was observed and each step signed off by NASA Quality Assurance (QA). The following paragraphs described the steps taken to integrate the samples onto the trays.

The solar diode was first mounted to the underside of the baseplate. The solar coulombmeter was then mounted to the top side of the baseplate. The remaining hardware was then mounted from the center of the tray outward to avoid moving hardware over previously mounted hardware.

Twelve compound parabolic focusing concentrators for accelerated atomic oxygen exposure were mounted near the coulombmeter. Then trays E2, E2S, E1, E1S, E3, and E4 were attached in that order. After each tray was bolted to the PEC baseplate, two photographs were taken, one with the specimens exposed, and then one with the tray cover in place. After all the trays were mounted on a baseplate, the 6" x 6" piece of MLI was bolted directly to the baseplate. The contamination package with its chambers covered by selected UV filters was then bolted to the baseplate. Finally, the McDonnell-Douglas 3" x 3" specimens were bolted onto the baseplate. Because the expanded Gortex material was much thicker than the other three thin film specimens, those specimens were shimmed by a 0.125" thick piece of aluminum. The second baseplate was integrated in exactly the same sequence as the first. After all the experiment hardware was loaded on each baseplate, the baseplates were double-bagged, wrapped in polypropylene padding, and placed into wooden containers. The sealed containers were then shipped to NASA Langley Research Center for the next phase of the integration.

The baseplates with loaded sample holders were received at NASA Langley Research Center and unpacked in a class 100 clean room, with personnel donned in full clean room attire, and placed into the PEC using sixteen standoffs for each half of the PEC as shown in Figure 3. All fasteners were torqued, the experiment covers were removed, and the PEC was closed and latched as shown in Figure 4. The PEC was carried to a vibration table, where it was mounted to a side-wall carrier and vibration tested. A small tent was constructed around the PEC to protect it from contamination during the vibration test.

The PEC was removed from the vibration table, rebagged, and walked to the cleanroom, where the PEC was unbagged, opened and inspected. The PEC passed the inspection, and then the PEC was closed, and the pip pins holding the PEC closed were secured with tie wraps. The integrated experiment was double-bagged, placed in transportation containers, and driven to Kennedy Space Center with the other hardware for integration onto the Orbiter. Once the hardware was unpacked at KSC, the PEC was placed in the side wall carrier of the Orbiter payload bay and latched, then covered with a protective film until one day before the launch. The day before the launch, the PEC was uncovered and the tie wraps were undone.

On March 22, 1996, the Space Shuttle Atlantis carried the POSA experiments to orbit as part of Mission STS-76, as shown in Figure 5, and then docked with the MIR Space Station as shown in Figure 6. During an extravehicular activity (EVA), astronauts Linda Godwin and Michael Clifford removed the POSA experiments from the Space Shuttle payload bay, and transferred them to the MIR Space Station (Figure 7), where the PECs were placed on hand rails near the scuff plates on the MIR docking module and secured. Once secured, the PEC was then opened to expose the samples to the space environment. A view of the POSA II experiment mounted on the docking module shortly after deployment is shown in Figure 8.

5. CONCLUSION

The samples will remain on the MIR for about 18 months, until an EVA on Mission STS-80 when an astronaut will retrieve POSA II and stow it on the Shuttle for return to JSC for deintegration and subsequent analysis by scientists at Boeing, McDonnell-Douglas, Rocketdyne and NASA. During each shuttle/MIR rendezvous mission which will occur during the 18 months while the POSA is on-board MIR, overall views and close-up photographs of the samples will be taken. Within 3-6 months after retrieval, initial "first-look" observations are expected to be reported. As data is collected during the detailed post-flight analysis, results will be recorded on the POSA II database and available for review.

6. ACKNOWLEDGMENTS

The authors would like to express their appreciation to Mr. Greg Stover, MEEP Payload Manager, and NASA QA Inspector Richard Cooper for their assistance in the integration process of the POSA II experiment. The authors would also like to acknowledge the support of the National Aeronautics and Space Administration who funded portions of this experiment integration process under contract number NAS 151-0000.

POSA Sample Serial Number: Aluminum- FL-1ALM-(posa2-E1S-26-A)-LeRC
Sample Type: Aluminum Reflector
Internal Sample Number: 1ALM

MIR ENVIRONMENTAL EFFECTS PAYLOAD
Passive Optical Sample Assembly (POSA)
SAMPLE DATABASE

Electron beam deposited SiO2 (1800 Angstroms)/Al (1500 Angstroms)/Polyimide/Al Substrate

Fabricated by LeRC

Weighing done on Sartorius balance, tare before each sample
Elapse time for entire ALM series: 66 min. 68.4 F 57% R H
96 hrs in Vacuum

DB-100: Standard black = .089
Standard gold = .976
Calibrated with black & gold standard prior to each sample
Sensitivity = 2 Course gain = 5

Sample Weight: 0.71730 g, 0.71731 g, 0.71732 g

Sample Dimensions: 1 inch diameter, 0.020 inch thick

Company ID (Boeing = B, MacDac= MD, Rocketdyne = RD): LeRC

Sample Carrier: posa2

Sample Tray: ES1

Tray Position Number: 26

Drawing Numbers:

Optical Measurements

Spectral Reflectance Curve (yes/no): Yes

Total Reflectance (Specular + Diffuse): T = 0.855 D = 0.016 S = 0.839

Solar Absorptance: lambda-9 = 0.145 LPSR 200 = 0.149

Thermal Emittance: Reflectivity = 0.976 Emittance = 0.024

Absorptance to Emittance Ratio (a/e):

BRDF/BTDF Curve (yes/no): No

Spectral Transmittance Curve (yes/no): No

Transmittance: No

Chemical Characterization

Auger Spectroscopy Scan (yes/no): No

Mass Spectrometry Scan (yes/no): No

Total Mass Loss (TML):

Collected Volatile Condensable Materials (CVCM):

Fourier Transform Infrared (FTIR) Scan (Yes/no): No

Electron Spectroscopy for Chemical Analysis (ESCA) Scan: No

Imaging Techniques

Microscopy, Photodocumentation, Surface Analysis Results:

Photograph Numbers:

Mechanical Properties

Tensile Strength:

Other Tests Performed:

Observations from Preflight Examination:

FIGURE 1. DATABASE TEMPLATE FOR PRE-FLIGHT, FLIGHT, AND POST-FLIGHT DATA ON POSA EXPERIMENT SAMPLES

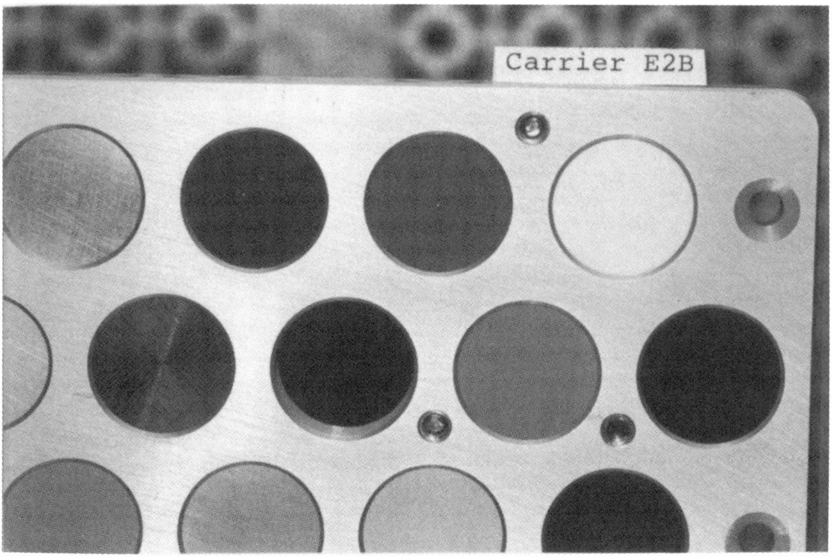

FIGURE 2. CLOSE-UP VIEW OF A BOEING SAMPLE TRAY

FIGURE 3. INTEGRATION OF POSA II SAMPLES ONTO THE PEC
(NASA PHOTOGRAPH)

FIGURE 4. PHOTOGRAPH OF A CLOSED PASSIVE EXPERIMENT CARRIER (NASA PHOTOGRAPH)

FIGURE 5. LOCATION OF SIDE WALL CARRIERS WITH SECURED PECs IN THE SPACE SHUTTLE ATLANTIS PAYLOAD BAY (NASA PHOTOGRAPH)

FIGURE 6. MIR SPACE STATION JUST PRIOR TO DOCKING WITH ATLANTIS SPACE SHUTTLE (NASA PHOTOGRAPH)

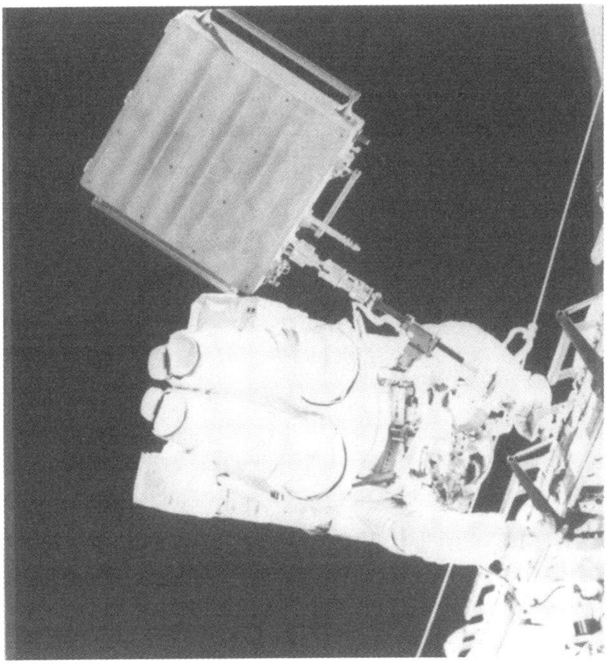

FIGURE 7. TRANSFERRING THE POSA II EXPERIMENT FROM SPACE SHUTTLE ATLANTIS TO THE MIR SPACE STATION BY ASTRONAUT GODWIN DURING AN EVA (NASA PHOTOGRAPH)

FIGURE 8. POSA II EXPERIMENT MOUNTED ON THE MIR DOCKING MODULE SHORTLY AFTER DEPLOYMENT (NASA PHOTOGRAPH)

28th International SAMPE Technical Conference
November 4-7, 1996

STATUS OF THE PASSIVE OPTICAL SAMPLE ASSEMBLY FLIGHT EXPERIMENTS

H.G. Pippin and G.L. Bohnhoff-Hlavacek
Boeing Defense & Space Group
P.O. Box 3999
Seattle, Wa 98124-2499
(206)-773-2846

ABSTRACT

The Passive Optical Sample Assembly Experiments (POSA I & II) are a collection of approximately 900 passive material specimens comprising part of a set of experiments designed primarily for space station risk mitigation. The POSA experiments are to be launched on Space Shuttle flight STS-076 and mounted on the docking module attached to the MIR space station. The POSA experiments will remain on-orbit for about thirteen months. Return is scheduled for May, 1997. Two meteoroid & debris experiments complete the Materials Environmental Experiment Payload (MEEP). This project is a collaboration between NASA LaRC, MSFC, JSC, Russia's space agency, and Rocketdyne, McDonnell-Douglas, and Boeing. NASA LeRC, IITRI, AZTEK, Sheldahl, and several other organizations have also provided materials and/or test specimens.

A description of materials flown, exposure conditions, techniques used to vary exposure times, and the mission profile will be presented. Sample selection will be covered with respect to contamination issues, significance to the International Space Station, and value to other future flights. Results from periodic on-orbit monitoring will be reported as available.

1. INTRODUCTION

Under the leadership of an International Space Station (ISS) Integrated Products Team (IPT) a series of "risk mitigation" experiments are being carried out on the Russian space station (MIR). The Materials Exposure Experiment Package (MEEP) includes five experiments, the Passive Optical Sample Assembly (POSA) I and II experiments, the Polished Plate Micrometeoroid Debris (PPMD), the Orbital Debris Collector (ODC), and four Passive Experiment Carriers (PEC), each of which contains one of the above mentioned experiments.

The POSA I and II experiments are contamination risk mitigation experiments designed to evaluate the effects of contamination and natural environmental factors on selected materials. The performance of specific material types used on Space Station is of primary interest. The results are also of general interest to spacecraft design. Many of the materials selected have flown on other flights as test specimens and/or in engineering service. The range of conditions on the MIR orbit represents an extension to the previous exposure range of many of these materials. The exposure period of about eighteen months is the fourth longest returned passive materials exposure experiment by NASA. The Long Duration Exposure Facility experiments endured 69 months of low Earth orbit (LEO) exposure, the Solar Maximum satellite flew for about 52 months prior to hardware being returned, and the Hubble Space Telescope flew for over three years prior to hardware return. In terms of the number of specimens and variety of materials, only the LDEF experiments were more comprehensive. Russian experimenters have also conducted long term passive materials exposures on the external surface of MIR.

The POSA I and II experiments are a co-operative effort involving four NASA centers, Langley, Marshall, Johnson, and Lewis, and several companies, Boeing, MacDonnell-Douglas, Rocketdyne, AZ Technology, Sheldahl, and IITRI. Several other companies have also supplied materials and test specimens.

2. FLIGHT CONDITIONS

Materials from the POSA experiments will be examined for changes in optical and surface properties due to the flight exposure. The POSA experiments were carried to MIR via Space Shuttle mission STS-076. Space Shuttle Atlantis was launched on march 22, 1996 and rendevoued with MIR on march 23, 1996. On march 27, 1996, the POSA I and II experiments were removed from their carriers in the Space Shuttle payload bay and mounted on opposite sides of the Space Shuttle-MIR docking module. The two orbital debris experiments were also deployed at this time. The locations of the POSA experiments are near the scuff plates at the end of the docking module closest to the Krystall science module. POSA I was on the nominal wake side and POSA II was on the nominal ram side. One surface of POSA II was oriented 45° to ram with a significant view factor to MIR and the other side views the docking module and space. POSA I has one side facing the MIR and the other side essentially viewing space with a small view factor to the docking module.

3. PHOTOGRAPHIC DOCUMENTATION

An extensive photographic record is being made of the POSA I and II experiments on-orbit. During deployment overall views of the POSA with respect to MIR and to the docking module will be documented. Close-ups of each side of each POSA experiment in the PEC's will be obtained to verify the condition of the hardware at the start of mission and to provide a reference for changes. During each shuttle/MIR rendezvous mission which will occur while the POSA experiments are on-board (3 missions), overall views and close-ups will be photographed in as much detail as practical. These flights are scheduled for august and December, 1996, and may, 1997. During the recovery activity, scheduled for September, 1997, the photographic survey will again be repeated. The recovery process will be captured on video as was the deployment.

4. SPECIMENS

The material specimen types include passive specimens exposed directly to the environment for the entire 18 months, and a variety of specimens whose exposure or preparation is designed to create some form of "enhanced" or indirect exposure. Passive specimens include anodizes and alodynes on aluminum, white paints, elastomeric materials, lubricants, thermal control films and blankets, and contamination witness plates. Environment monitors include a set of CR-39 polycarbonate particle track detectors, a solar coulombmeter designed and built by NASA MSFC, and Kapton® for atomic oxygen fluence determination.

Four sets of three accelerated atomic oxygen exposure focusing devices are being flown on each side of the POSA II experiment. The compound parabolic concentrators are designed to provide 4x, 9x, and 16x acceleration factors. The acceptance angle is an issue for these devices because the "ram" side is actually 45° to ram. However, the orientation of MIR may change at certain times and in practice both the "leading" and "trailing" sides of POSA may see atomic oxygen. Materials being subjected to the accelerated exposure are Kapton®, FEP Teflon®, Z93-P white thermal control paint, and KEL-F800®. A set of optical solar reflectors, mounted under selected UV filters, are being exposed to the outgassed product from selected compounds to examine the wavelength dependence of fixing contaminants on a surface. Both MacDonnell-Douglas and Boeing have provided pre-contaminated specimens on certain materials. The MacDonnell-Douglas specimens are primarily silicone deposits of differing thickness with some pre-flight UV exposure to enhance the on-orbit exposure. The Boeing specimens were chosen to represent typical contamination deposits, outgassed materials from a Kevlar®-epoxy rocket motor casing, a silicone grease, apiezon L grease, and a VacSeal resin.

The sample holders are either trays used on the Energetic Oxygen Interaction with Materials III (EOIM-III) Space Shuttle experiment of 1992, or built essentially to the EOIM-III drawings. POSA I has a number of specimens mounted into carousels which were designed to fit into a UV spectrometer. The sample trays are either chromic acid anodized or boric-sulfuric acid anodized.

Selected materials were covered with thin Kapton® (0.3 mil, 0.5 mil, 0.6 mil) which will essentially erode away during the flight to expose fresh surfaces. This will provide some time variation in the exposure of a few material types. The Kapton® thicknesses were chosen based on an expected recession rate of 3.0×10^{-24} cm^3/atom. The atomic oxygen should completely remove the Kapton® covering specimens on the "ram" side after 2 to 6 months exposure. Kapton® covered specimens on the "wake" side are likely to remain covered during the entire mission, although depending on orientation of MIR for the remainder of the mission, the possibility for complete removal does exist. Preliminary estimates using a Boeing atomic oxygen fluence prediction computer program are $\sim 2.3 \times 10^{+21}$ atoms/cm^2 ram fluence over the mission duration. This should provide about $1.6 \times 10^{+21}$ atoms/cm^2 at the nominal 45° from ram orientation.

A large variety of anodized and painted specimens are being flown by the major ISS contractors. The paints include the old and new versions of the ISS baseline white thermal control paint, Z93 and Z93P, and old and new versions of IITRI paints with ceramic and silicone binders, YB-71 and YB-71P, and S13/GLO and S13/GLO-P. Selected additional white thermal control paints were flown, including Z93 applied by different organizations, and paints with 0.5% or 2.0% sodium salicylate by weight. The sodium salicylate was added in an attempt to mitigate damage due to the UV portion of the solar spectrum.

Rocketdyne has provided many colored anodizes which are candidates to be used as exterior markings. Coatings produced using three different alodyne processes are being flown as well as chromic, sulfuric, and boric-sulfuric acid anodize specimens. Some of these test specimens are being flown on either POSA I or both POSA trays.

Elastomeric materials being flown include viton 835, silicone rubber 383-70, and a low outgassing fluorosilicone. Selected viton and silicone rubber specimens are coated with thin films of Braycote 601 or Braycote 803 space-grade perfluoroether based lubricant greases. A selection of FEP, TFE, and Gortex® fluorocarbon specimens are being flown.

Pre-flight optical properties measurements have been made on many flight and ground control specimens. Results of these measurements, together with sample weights, processing details and identification codes are stored in an electronic data base. Reflectance data is stored in a digitized format for easy loading into an EXCEL spreadsheet and subsequent plotting.

5. SUMMARY

The deployment, on march 27, 1996, was successful and all hardware appeared to be in place when the POSA II PEC was opened. Figure 1 shows a photograph of the POSA II specimens mounted in the Passive Experiment Carrier. Figure 2 shows a view of the POSA II experiment mounted on the docking module, shortly after deployment. A "quick-look" summary of findings is expected to be released within 3-6 months after retrieval. Post-flight analysis of this hardware will not be complete until at least 1998.

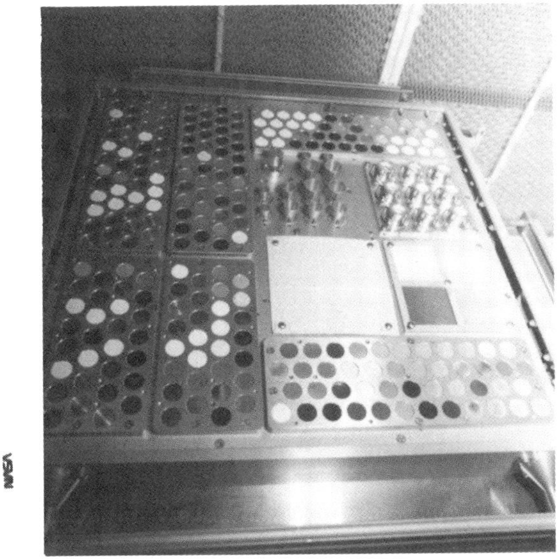

Figure 1. Photograph of the POSA II specimens mounted in the Passive Experiment Carrier.

Figure 2. POSA II experiment mounted on the docking module (NASA Photograph).

28th International SAMPE Technical Conference
November 4-7, 1996

INTEGRATING EMI SHIELDING INTO COMPOSITE STRUCTURE

D. A. Olivero and D. W. Radford
Composite Materials, Manufacture and Structures Laboratory
Mechanical Engineering
Colorado State University
Fort Collins, CO 80523-1374

ABSTRACT

Current trends in aerospace applications indicate an increasing demand for improved economy and payload capability. Most aerospace designs therefore emphasize the reduction of mass in all components, including ENI shielding enclosures. Previous work indicates that high volume fraction silver coated ceramic microballoon filled resins provide excellent EMI shielding at extremely low density, while overcoming many of the problems associated with metallic coatings. Unfortunately, poor mechanical properties make these materials unsuitable for most applications. The incorporation of these fillers into the matrix of a fiberglass/epoxy composite is investigated to address the poor mechanical properties of the unreinforced shield material. The volume fraction required to establish continuous particle networks is driven down by the segregation of filler particles to discreet regions external to the fiber reinforcement. Mechanical properties are seen to be largely unaffected by this approach, while shielding effectiveness in excess of 50 dB is obtained at a filler volume fraction of only 19%. Tests indicate that this filler configuration is mechanically superior to a layer of filler material at the mid-plane region. The use of conductive short fiber is also investigated as a way to introduce tailorability for specific EMI design requirements, although no significant changes in mechanical or electrical performance were observed relative to conductive microballoon fillers. A composite materials approach is therefore presented which simultaneously offers excellent mechanical and electrical properties at a fraction of the mass of conventional metallic shielding materials and with none of the durability issues associated with metallic coatings.

KEY WORDS: EMI Shielding, Electrically Conductive Materials, Microballoons

1. INTRODUCTION

Electromagnetic interference (EMI) shielding of composite materials is a subject receiving increasing attention as applications for composites in the aerospace and telecommunications industries are continuously discovered. In many applications (most notably structural) where a reduction in mass is desired, the high specific strength and stiffness of fiber reinforced polymer composites are particularly attractive. This is especially true in modern aerospace design where improvements in economy and payload capability are high priorities. However, most polymer matrix composite materials lack the electrical conductivity necessary to provide an adequate shield against unwanted electromagnetic radiation. Of these materials, carbon fiber reinforced plastics (CFRP) have perhaps the best shielding potential, but even this electrically superior reinforcement suffers from poor conductivity compared to metals. Further, reinforcing fibers have highly anisotropic shielding effectiveness (SE) resulting from the polarizing effect of fiber orientation (1).

A well-established method of creating a conductive composite is to incorporate an electrically conductive particulate filler material into a plastic. Common fillers include metal-coated short fibers, flakes and microballoons, as well as metallic and carbon powders. The advantages of including filler materials into a plastic versus applying a metallic surface coating are that the resulting shield is relatively impervious to micro-impact damage, wear, surface defects and environmental degradation. Additionally, if a low-mass filler such as microballoons is used, the density of the composite can be reduced.

Previous work has employed metal-coated microballoons in an epoxy matrix to create very high performance EMI shielding with densities as low as 0.5 g/cm^3 at high conductive filler volume fraction, V_{cf} (2). Such high values of V_{cf} are necessary to establish electrical conductivity because of the need for continuous networks of particles in the conductor/ insulator system, allowing the electrical current to "percolate" through the composite. At high V_{cf}, however, mechanical properties of the resulting composite are found to degrade to the point where the material is no longer a suitable alternative for structural applications (3). Therefore to make included conductive fillers a viable technique for structural shielding applications, a method of incorporating both conductive filler and fiber reinforcement into a composite has been investigated. Such a shielding approach has the potential to achieve excellent shielding, electrical surface contact and mechanical strength, as well as a reduction of final part cost, weight and complexity.

2. PERCOLATION PHENOMENA AND CONDUCTIVITY

Many important mechanical and physical properties, such as strength and conductivity, of any filled plastic material are a strong function of V_{cf}. Typically, these properties exhibit a sharp transition behavior at a certain volume fraction. It has been well demonstrated that this transition, for randomly distributed filler material, is governed at the most fundamental level by percolation phenomena (4-6). The percolation threshold (or "critical volume fraction"), ρ_c, is defined as the V_{cf} where the random location of filler material provides a continuous interconnected network across the composite. This allows externally applied phenomena, such as heat flux or an electric current, to percolate through a system. In this way the bulk properties of the material depend on percolation. The percolation transition is illustrated schematically for spherical fillers in figures 1(a) and (b). In figure 1(a), clustering is a local phenomenon that does not affect bulk material properties, while in figure 1(b) the gray coloring indicates the percolating, or "infinite", cluster. Since this percolation phenomenon is partly geometric in nature, fillers with different shapes or aspect ratios can have very different percolation thresholds even in otherwise identical systems (7).

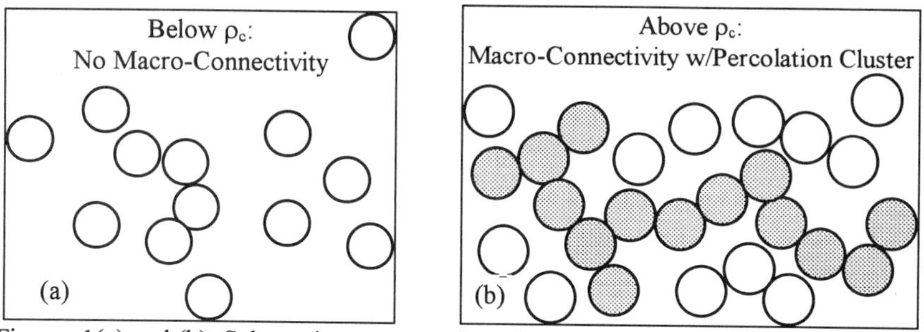

Figures 1(a) and (b): Schematic representation of 2-D networking above and below ρ_c

While it is generally accepted that all particulate composite systems will display some sort of percolation behavior, ρ_c can vary greatly in the same filler/matrix system under different processing conditions (5). That is, establishment of macroscopic electrical conductivity is a function of not only filler conductivity σ_f, volume fraction (V_{cf}) and shape or aspect ratio (A_r), but also of the relative particle arrangement (8):

$$\sigma = f(\sigma_f, V_{cf}, A_r, \text{arrangement}) \qquad [1]$$

One intuitively expects that the establishment of highly connected particle structures, as opposed to purely random distributions, can potentially reduce ρ_c. For example, it has been demonstrated that by introducing a small amount of carbon black (CB) into a blend of nonmiscible polymers, polyethylene and polystyrene, a selective localization of the CB takes place at the phase boundaries. This localization was shown to establish connected networks at 0.1% volume CB (9). As another example, continuous networks of short nickel-coated carbon fibers were created at 0.5 wt.% by aligning the fibers in a strong magnetic field during cure (10). Such methods of favoring spatial inhomogeneities are particularly attractive given the reduced quantity of filler required to achieve conductivity, and therefore the reduced effect on final cost and composite properties.

Potentially, the inclusion of conductive fillers into the matrix of a fiber reinforced composite laminate can yield a similar effect, favoring spatial inhomogeneities and creating highly connected particle structures at lower V_{cf} than in a purely random dispersion. More specifically, the warp/weft crossover points of a woven fabric can be thought of as creating a three dimensional network throughout the laminate of "negative space". Since the filler constituent size is much too large to penetrate between fibers in a bundle, the fillers should be retained in these matrix rich spaces, creating a highly connected network at a relatively low volume fraction. Such an approach might not seriously complicate fabrication or increase weight and cost. It could also eliminate the need for secondary structures, i.e. EMI shielding enclosures, providing primary structural integrity is not compromised by included fillers.

Another potential benefit of the use of composite shielding materials is the ability to optimize certain variables for specific EMI design requirements. This tailorability of final properties can include such variables as filler volume fraction, aspect ratio, and filler material (2,3). Beyond the effect these variables have on composite bulk conductivity as mentioned above, different materials and configurations can potentially be employed in situations where different emission frequencies require specific shielding mechanisms.

Thus, the purpose of this study is to investigate the EMI shielding potential of conductive fillers incorporated into the matrix of a woven fiber-reinforced composite laminate, and to determine the resulting effects on both mechanical properties and filler percolation. To gain further insight into the effects of some tailorability variables, both conductive microballoon and short fiber fillers at different volume fractions were investigated.

3. EXPERIMENTAL

3.1 Constituent Materials The ceramic Metalite SF-20 microballoons used in this study range in size from 10 to 100 μm diameter with an average particle diameter of roughly 74 μm. These microballoons are coated with a 1 to 2 μm layer of silver. The specific gravity of these fillers, obtained experimentally, was estimated at 0.54. Ag Clad 32 grade silver coated glass fibers were also investigated, having a range of aspect ratios from 2 to 25 (average of 10), and specific gravity of 2.55. These fillers were chosen specifically for their silver coating, which provides excellent conductivity and oxidation resistance. Particle contact resistance is thus much lower for silver than for other metals such as aluminum. In addition, the relatively high cost of silver is offset by the extremely small amount needed for particle coating.

The composite panels were a 10 oz. fiberglass plain weave fabric laid up with Epon 813/TETA resin/hardener system mixed 100:14, cured under vacuum at room temperature for 24 hours and postcured for two hours at 110°C. Glass was chosen as the fiber reinforcement for its negligible conductivity, and thus any measurable shielding effectiveness would result from the presence of connecting filler material.

3.2 Conductive Composite Laminates To minimize cost and weight penalties as well as any reduction of mechanical properties, the smallest quantity of filler that yields bulk conductivity in the composite is the focus of interest. The fiberglass fabric was laid up as four plies of (±45) relative to the axis of tensile loading. This ply orientation was chosen to load the matrix rather than the reinforcement during tensile testing. Filler materials were thoroughly blended with the uncured matrix material prior to lay-up. For each filler material, four panels were created to explore different filler volume fractions, 5%, 10%, 20% and 30%. Note that these volume fractions are defined as the filler fraction of the matrix material prior to lay-up and not for the final composite volume fraction. An estimate of the final composite volume fraction of the conductive filler can be obtained with a conversion factor of all matrix constituents over the total composite volume:

$$V\%_{f,comp} = V\%_{f,matrix}\left[\frac{\sum V_{constituents,matrix}}{\sum V_{constituents,total}}\right] \quad [2]$$

The conversion factor in eq. [2] gives the final composite volume fraction of conductive filler as approximately 3%, 6%, 12% and 19% for the four matrix volume fractions mentioned above. Composite volume fractions will be used in all further discussions. Note that the composite volume fractions are all well below the threshold for conductivity ($\rho_c \approx 0.25$ to 0.30) established for random dispersions of microballoons in epoxy (2). A panel containing no conductive fillers was also prepared to serve as a baseline for mechanical properties. Lastly, a panel was fabricated containing the equivalent of 19% composite volume microballoons, concentrated in the mid-plane interlaminar region. This panel was created to investigate the effect of interlaminar versus intralaminar filler distribution on shielding and mechanical properties.

3.2.1 Evaluation of Properties Along Reinforcement Direction Of primary interest, with any fiber reinforced composite material, are the properties along the direction of reinforcement. To investigate the effects of incorporating conductive fillers into a laminate on its true tensile performance, several test specimens (0/90 ply orientation) were prepared from both mid-plane concentrated and uniformly distributed 19% microballoon panels. As before, specimens were also prepared using unfilled resin to serve as a baseline.

3.3 Tensile Testing Specimens and Technique Each panel was cut into several strips 1.91 cm wide by 17.8 cm long, such that the fiber orientation is preserved relative to the specimen axis. The specimens were then machined to conform to ASTM type #1 dimensions from standard D-638-89, a dog-bone shape with a 1.27 cm wide, 6.35 cm long gauge length. The cut edges of the gauge lengths were then progressively sanded down to minimize machining-induced surface defects. Thickness measurements were averaged across the length of the gauge section for computation of stress.

A universal tensile tester fitted with wedge lock grips was used, with the crosshead rate set to 0.25 cm/min. An extensometer with a 2.54 cm gage length was used to measure strain. All data were collected via computer data acquisition.

3.4 Evaluation of Conductivity and Shielding Effectiveness Shielding effectiveness test samples were produced from the same panels used to create the tensile test coupons. Each sample was carefully machined to completely fill an X-band waveguide, 22.86 mm wide by 10.16 mm high. These samples were then inserted into a X-band calibrated Hewlett Packard 8015A network analyzer capable of measuring signal transmission. All data were collected at a frequency of 10.0 Ghz. The magnitude and phase of the transmission coefficient, S21, was recorded, thus yielding the relative strength of the transmitted electric field. The shielding effectiveness (SE) is then defined as:

$$SE(dB) = 20\log\left[\frac{E_i}{E_t}\right] \qquad [3]$$

where E_i and E_t are the incident and transmitted electric field magnitudes, respectively. This equation is equivalent to the standard definition of SE involving the ratio of incident to transmitted signal power (11).

4. RESULTS AND DISCUSSION

4.1 Tensile Test Results Tensile testing did not indicate any major changes in mechanical performance with V_{cf} for the (±45) laminates investigated. Further, results for all V_{cf} in the (±45) configuration indicated at least 2.5% strain to failure. To reduce the impact of processing parameters as contributing variables, mechanical properties are compared whenever possible to a baseline laminate containing no conductive filler materials.

4.1.1 Failure Load, Strength and Volume Fraction While the amount of reinforcing fiber was held constant at four plies in all test panels, the inclusion of conductive fillers into the laminate, even at the low levels of V_{cf} (below 20%) investigated, was found to increase the laminate thickness by as much as 40%. This increase in thickness with V_{cf} manifests itself in a change in strength vs. V_{cf} relative to failure load vs. V_{cf}. The plots of load and strength vs. V_{cf} for the (±45) laminates with uniformly distributed fillers are found in figures 2 and 3:

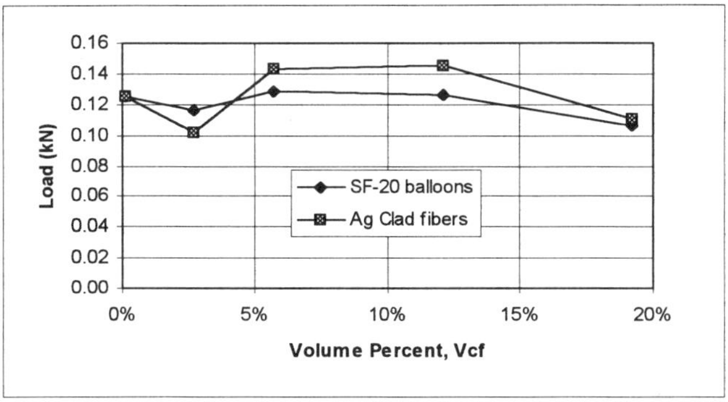

Figure 2: Average Failure Load vs. V_{cf} for Uniformly Distributed Fillers

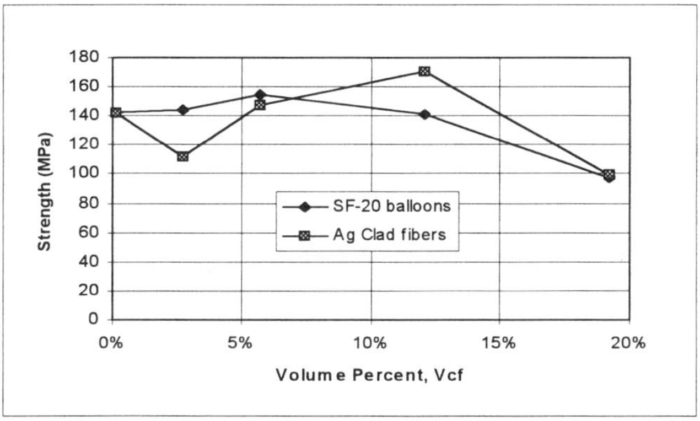

Figure 3: Average Failure Strength vs. V_{cf} for Uniformly Distributed Fillers

Note in figure 2 that for V_{cf} below 19% there is no obvious trend in failure load with V_{cf}, especially for the microballoon-filled laminates. Fluctuations in the average failure loads of these samples are attributed to processing anomalies rather than the presence of filler specifically. While fiber-filled samples have greater variability, the two filler constituents have very similar performance. This result is consistent with the fact that the critical aspect ratio should be much greater than the aspect ratio of the Ag Clad fibers. At V_{cf} = 19%, however, both filler constituents show a drop in load. This drop is magnified in the strength plot of figure 3, indicating a relatively sharp increase in the laminate thickness from 12% to 19%. Since previous work has indicated that silver-filled epoxy composites suffer from poor wetting and porosity (3), it is possible that a build-up of filler material between plies at high V_{cf} is responsible for the decrease in strength, due to a reduction of interlaminar shear strength.

4.1.2 Effect of Filler Distribution on Strength Table 1 compares the performance of the three (±45) test panels created with 19% V_{cf}; the uniformly distributed Ag Clad fibers, uniformly distributed microballoons, and mid-plane layer of microballoons. While all three sample groups show property degradation relative to the unfilled panel, the layer sample has the lowest load to failure. More significantly, the computed failure strength is quite low. This is caused by the substantially larger thickness of the mid-plane layer laminate compared to its uniformly distributed counterpart. Since both laminates are compositionally identical, this thickness difference suggests that the so-called uniformly distributed filler materials in reality concentrate in "negative space" in the woven fabric, that is, at the matrix rich bundle crossovers.

TABLE 1: Mechanical Performance of 19% V_{cf} (±45) Laminates

Sample Description	Average Failure Load (kN)	Average Strength (MPa)
Unfilled Matrix	0.1249	142.7
Mid-plane, 19% V_{cf} SF-20	0.1060	92.7
Uniform Dist., 19% V_{cf} SF-20	0.1067	97.8
Uniform Dist., 19% V_{cf} Ag Clad	0.1106	99.1

4.1.3 Effect of Filler on Strength of (0/90) Laminates Three sets of specimens were tensile tested in the (0/90) ply orientation; 19% V_{cf} microballoons uniformly distributed, in a single layer at mid-plane, and also with unfilled resin. Table 2 shows the results of the tensile tests. As with the (±45) ply orientation, it appears that locating a balloon layer at mid-plane is a weaker configuration. This result is probably caused by poor interlaminar shear strength at the mid-plane and/or a translaminar stiffness mismatch that is avoided in the case of uniform distribution. The tensile strength of the composite in the (0/90) orientation seems slightly improved by the presence of microballoons uniformly distributed throughout the laminate.

TABLE 2: Average Failure Loads of Filled Composites in (0/90) Orientation

Filler Configuration	Average Load (kN)
Unfilled Matrix	0.175
19% Balloons, Mid-plane Layer	0.160
19% Balloons, Uniform Dist.	0.180

4.2 Laminate Microstructure Selective localization of particulate fillers in the test panels is of great interest, given that understanding the relative arrangement of conducting particles in a conductor/insulator system is a critical part of predicting the behavior of percolating properties such as electrical conductivity. With respect to the laminates under investigation, the small change in laminate thickness at low V_{cf} suggests that filler material accumulates in bundle crossover regions in the weave, normally occupied by matrix. Obviously, as V_{cf} is increased a point is reached where these regions can no longer contain all filler material, and particles begin to occupy interlaminar regions. This would have the effect of connecting the bundle crossover regions, and is conceptually equivalent to a percolation threshold.

4.2.1 Micrographic Verification Figures 5(a) and (b) are micrographs of microballoon-filled laminate cross sections at V_{cf} = 6% and 19%, respectively. These micrographs employ polarized light so that the fiberglass bundles appear as dark regions. Plies are oriented horizontally in these images. Figures 6(a) and 6(b) are fiber-filled laminate cross sections at identical V_{cf}. Note that even at 6% there is significant clustering of filler material at gaps in the weave, and that these gaps are not quite connected by small interlaminar clusters of material. At 19%, however, the amount of material in the weave gaps does not appear to increase, but rather the quantity of interlaminar material, which now clearly forms a network on a macro-scale.

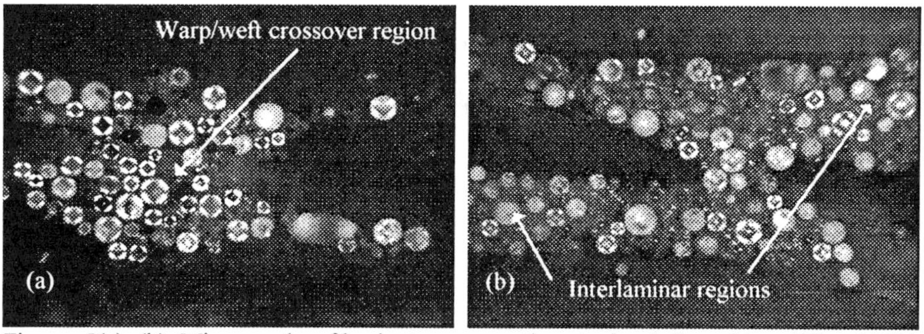

Figures 5(a), (b): Micrographs of laminate cross-section at microballoon V_{cf} = 6% and 19%

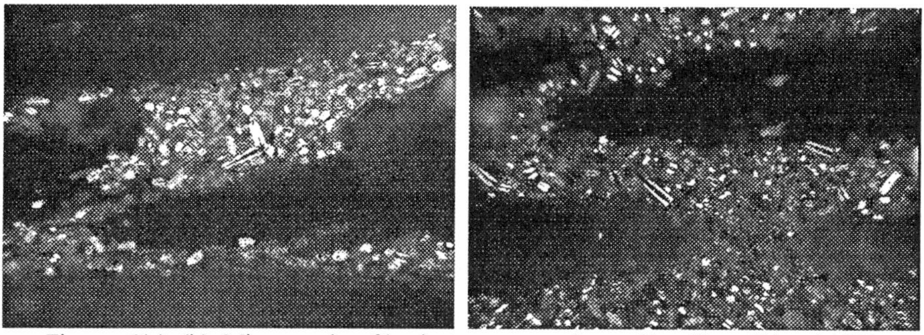

Figures 6(a), (b): Micrographs of laminate cross-section at fiber V_{cf} = 6% and 19%

4.2.2 Mid-plane Layer Distribution While the laminates with uniform distributions of filler material seem to exhibit a percolation threshold as discussed above, percolation within a packed layer such as the mid-plane balloon panel is assured, assuming that the layer is at least one or two particle diameters in thickness (6). Figure 7 is a micrograph from the mid-plane microballoon layer panel with an equivalent V_{cf} of 19%. Plies in this image are oriented vertically, and the dark regions on the right and left hand sides of the balloon layer are adjacent fiberglass plies. The degree of particle connectivity in this configuration is of course optimal, since the vacuum bag cure acts to pack the filler particles together and the number of contacts is relatively high. However, figure 7 also shows that the local volume fraction of microballoons at the mid-plane is much greater than 19%, and the local tensile properties of the laminate at mid-plane are probably degraded substantially due to the diminished resin content (3). This observation is consistent with the slight decrease in strength of the mid-. plane configuration compared to uniform distribution.

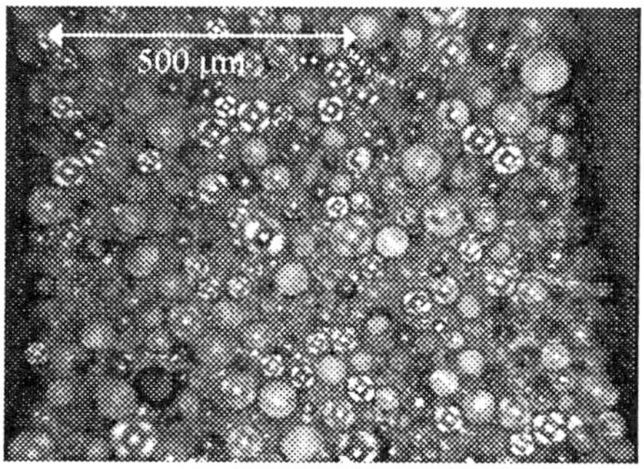

Figure 7: Micrograph of microballoon fillers concentrated at mid-plane, V_{cf} = 19%

4.2.3 Theoretical versus Measured Density Figure 8 shows the theoretical and measured values of laminate density for all uniformly distributed filler laminates. Some degree of entrained air was observed in every laminate, although it appears most significant at V_{cf} values of 3% and 6%. At high V_{cf}, measured values of density are much closer to prediction. Figure 8 also indicates that in weight-critical applications, microballoon fillers may be preferable to fibers, especially recalling figures 2 and 3 and the lack of significant strength difference between the two filler materials. Indeed, at 19% V_{cf} the reduced density of the microballoon-filled laminate offsets the strength reduction and any change from the specific strength of the unfilled laminate.

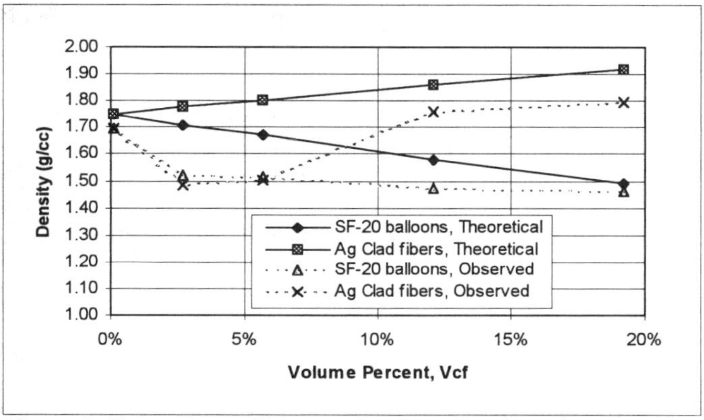

Figure 8: Comparison of Measured and Predicted Laminate Density

4.3 Shielding Effectiveness Results Since both the epoxy resin and glass fiber reinforcement are essentially EMI transparent materials, negligible shielding effectiveness is observed in the unfilled case. As V_{cf} increases, shielding effectiveness will remain constant or slightly increases until conductive filler content reaches ρ_c, at which point a sharp increase in effectiveness occurs.

4.3.1 Electrical Continuity Given the low interparticle contact resistance of silver, the existence of a percolating pathway is easily verified with an electrical continuity check. Such a continuity check with a hand held multimeter revealed that all sample configurations (uniformly distributed microballoons, fibers and the mid-plane balloon layer) were electrically conductive at $V_{cf} = 19\%$. No percolating pathways were detected at values of V_{cf} lower than 19% for any filler configuration.

4.3.2 X-Band Attenuation Shielding effectiveness values, as defined in eq. [2], are shown in figure 8 for the various laminates. The data is in clear agreement with the establishment of electrical continuity observed at $V_{cf} = 19\%$, with a sharp transition evident between 12% and 19%. The latter value is therefore a good qualitative estimate of the percolation threshold of the composite system. The fact that this value of ρ_c is substantially lower that that observed for the random system (2) confirms the relationship between percolation and filler arrangement. It is interesting to note, however, that the shielding effectiveness of the mid-plane microballoon layer is not higher than the uniform distribution, as was anticipated. This fact suggests that the quality of interparticle contact in uniform distributions is as good as the mid-plane layer case.

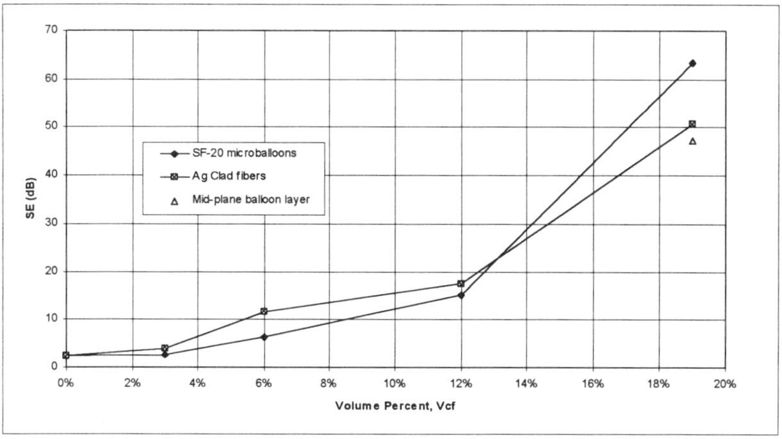

Figure 8: Shielding Effectiveness (eq. [3]) of All Laminate Configurations at 10.0 GHz

4.3.3 Effects of Filler Shape on Shielding Effectiveness Since the aspect ratio of the Ag Clad fibers is so low, it is assumed that no bulk orientation of the fibers takes place and that SE anisotropy can be neglected. For such short fibers one might expect that the shielding performance, specifically the percolation transition, would look similar to microballoons. Figure 8 clearly shows this, although below the percolation threshold the fiber-filled samples have consistently higher shielding effectiveness, possibly due to interactions between fibers and waves aligned on the fiber axis.

5. CONCLUSION

Prior to this study, adequate shielding effectiveness was obtained in microballoon-filled epoxy composites only by raising the filler volume fraction to a point where strength and strain to failure became severely limiting factors to application. In addition, the high cost of silver-coated microballoons made these composites less attractive. This study has attempted to address both of these issues, as well as investigate other parameters such as filler type and arrangement. The three key benefits of this study are as follows:

First, it has been demonstrated that the percolation threshold of a conductor/insulator system is a function not only of the constituents but also of their relative arrangement. Compared to a purely random distribution of the constituents, ρ_c has been significantly reduced by favoring spatial inhomogeneities in the filler distribution. This has been accomplished by using the fiber reinforcements themselves to segregate the filler materials to discreet regions. Mechanically and electrically, this configuration seems to be slightly superior to a separate layer configuration of conductive filler located at the mid-plane of the laminate.

Second, at V_{cf} as low as 19%, shielding effectiveness in filled laminates becomes high. Percolation appears to occur by the establishment of interlaminar networks which connect regions of concentrated filler at bundle crossover regions in the weave. In the uniformly distributed filler case, shielding effectiveness comes at no strength expense in the (0/90) ply orientation and only a slight expense in the (±45) orientation. For microballoon fillers, the slight decrease in shear strength is offset by a decrease in density, maintaining the high specific performance of the composite. The structural integrity of this material therefore greatly improves upon the fragility of the balloon-filled epoxies from previous work, at a fraction of the weight of conventional metallic shielding.

Lastly, a method has been demonstrated to make an essentially EMI transparent material (i.e., glass fiber/epoxy laminate) extremely conductive which avoids standard metallic coating approaches and the associated problems of wear and degradation. Instead, the filled laminate is expected to have a long service life without need for inspection.

Differences in the mechanical and electrical properties of the two filler types, balloons and flakes, appear to be negligible. Micrographs of the two constituents taken at the same V_{cf} look very similar. Future work with tailorability variables should investigate less similar filler types and material as well. For the weight critical applications this work addresses, microballoons seem the more attractive filler material.

The incorporation of conductive fillers directly into the matrix material of a structural composite has thus shown promise to address questions raised in previous work of mechanical performance and cost, while still maintaining significant weight savings over metallic shields. Future work in this area should employ very different constituent materials to further investigate tailorability, as well as the potential for use of this approach with preimpregnated materials.

6. REFERENCES

1. M.-S. Lin and C. Chen, IEEE Transactions on Electromagnetic Compatibility, 35 (1), 21 (1993).
2. D. Radford and B. Cheng, SAMPE Quarterly, 24 (4), 54 (1993).
3. D. Radford, Journal of Advanced Materials, 26 (1), 45 (1994).
4. G. Ruschau, S. Yoshikawa, and R. Newnham, Journal of Applied Physics, 72 (3), 953 (1992).
5. F. Lux, Journal of Materials Science, 28 , 285 (1993).
6. D. Stauffer and A. Aharony, Introduction to Percolation Theory, Taylor and Francis, London, 1992, ch. 2.
7. Jana et al., IEEE Transactions on Electromagnetic Compatibility, 34 (4), 487 (1992).
8. M. Zhang et al., Journal of Materials Science, 30, 4226 (1995).
9. F. Gubbels et. al., Macromolecules, 28 ,1559 (1995).
10. Y-S. Ho and P. Schoen, Journal of Materials Research, 11 (2), 469 (1996).
11. N. Colaneri and L. Shacklette, IEEE Transactions on Instrumentation and Measurement, 41 (2), 291 (1992).

ACKNOWLEDGMENT

The authors extend their gratitude to Dr. Reza Zoughi, Carl Bois and the Applied Microwave Nondestructive Testing Laboratory of the Department of Electrical Engineering at Colorado State University for their assistance with Shielding Effectiveness measurements. Appreciation is also extended to the PQ Corporation for the conductive filler materials and to Shell for the Epon resin.

BIOGRAPHIES

David A. Olivero is presently pursuing a Master of Science in Mechanical Engineering at Colorado State University. He received a Bachelor of Science in Mechanical Engineering at Pennsylvania State University in 1994. His research involves development and modeling of ultralightweight composite EMI shielding. Other recent research has focused on the monitoring of multiple impact damage in composite plates using millimeter wave nondestructive techniques.

Donald W. Radford is an Associate Professor of Mechanical Engineering at Colorado State University and is the Director of the Composite Materials, Manufacture, and Structures Laboratory. He received his Ph.D. in Materials Engineering at Rensselaer Polytechnic Institute, his MASc. in Metallurgical Engineering and BASc. in Mechanical Engineering at the University of British Columbia. Other recent research programs have included approaches to dual resin bonded carbon fiber reinforced PEEK/PEI composites, impact fatigue of metal-composite joints, distortion measurement and modeling of composites during manufacture and morphological studies of composites.

28th International SAMPE Technical Conference
November 4-7, 1996

METALLIZATION OF POLYCYANATES

William E. Converse and Frank P. Lary
Composite Optics, Incorporated
San Diego, California 92121

ABSTRACT

Polycyanates that are cross linked with a triazine ring are very stable and particularly resistant to oxidation. Therefore, this material is difficult to activate in order to accept an adherent metal coating, especially when applying the metal from plating solutions.

Since it is desirable to metallize composite materials for space hardware (e.g., mirrors and waveguides), this has been a technology that Composite Optics, Incorporated (COI) has seen a need to develop for future requirements for these components.

The theoretical approach and the method for resolving this issue will be discussed. Moreover, the quality and adhesion of the plated metal will be demonstrated with Scanning Electron Microscope (SEM) photographs, photomicrographs as well as surface finish and Flat Wise Tensile (FWT) data.

KEY WORDS: Composite Components, Surface Preparation, Plating

1. INTRODUCTION

Metallized composites are more in demand for space hardware applications since mirrors, reflectors, feedhorns and waveguides made from these materials can significantly reduce the payload mass.

Polycyanates using graphite fiber reinforcement has become a composite of choice for many space hardware applications because of its resistance to oxidation (higher service temperature), improved thermal stability and lower affinity for moisture.

Because of the resistance to oxidation, the hydrophobic nature of polycyanate composites makes it difficult to coat with metal, especially when the metal is applied using a plating process.

Plating metal onto composites is most desirable when a thickness of the metal is required because it is the most economical method. Therefore, COI became more interested in this technology in the last ten years since its customers have required some form of metallized composite for mirrors, reflectors, feedhorns and waveguides.

Initially bonding thin metal to surfaces, using vacuum deposition or abrading the surface for plating the metal were typically used to attain a metal surface. As components became more complex, these methods were no longer adequate for all applications.

With complex composite components such as feedhorns and waveguides, it became necessary to develop a method to prepare the surface without mechanically abrading it. Consequently, a process was developed for chemically etching the surface which allows the material to accept the metal as a result of a plating process.

This resulted in good adherent plated metal with a smoother surface that exceeded previous quality standards. Improved adhesion was important to allow thicker metal and/or electroplated silver for better conductivity of microwaves. In addition, the better surface finish minimized the attenuation of a microwave signal as it travels through the plated composite RF component.

Examples of several types and configurations of components that have been plated can be seen in Figures 1, 2, 3, 4a & 4b.

Figure 1. Silver plated graphite composite waveguides for the ASAR™ program.
(Matra Marconi Space)

Figure 2. Silver plated graphite composite feedhorns and connectors for INDOSTAR™ (CTA, Incorporated®)

Figure 3. Copper plated graphite composite feedhorn. (A2100™ program-Lockheed Martin)

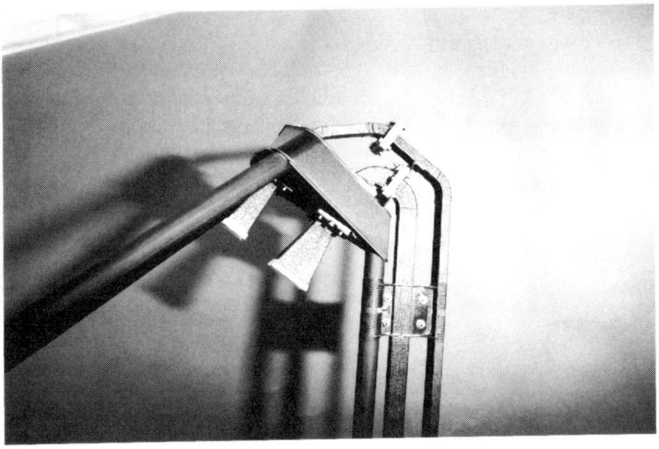

Figure 4a. Silver plated composite waveguides and feedhorns in a structure for SeaWinds™
(Jet Propulsion Laboratory)

Figure 4b. SeaWinds™ antenna assembled with silver plated composite waveguides and feedhorns
(Jet Propulsion Laboratory)

2. MATERIAL CONSIDERATIONS

Most of the initial work was accomplished using YLA's RS-3/T300 material. However, testing has shown 954-3 resin responds equally as well. The polycyanate structure (Figure 5) is designed to inhibit oxidation as evidenced by the stable triazine component in the polymer. Because of this strong cross linking structure, it is very difficult to render the surface hydrophilic.

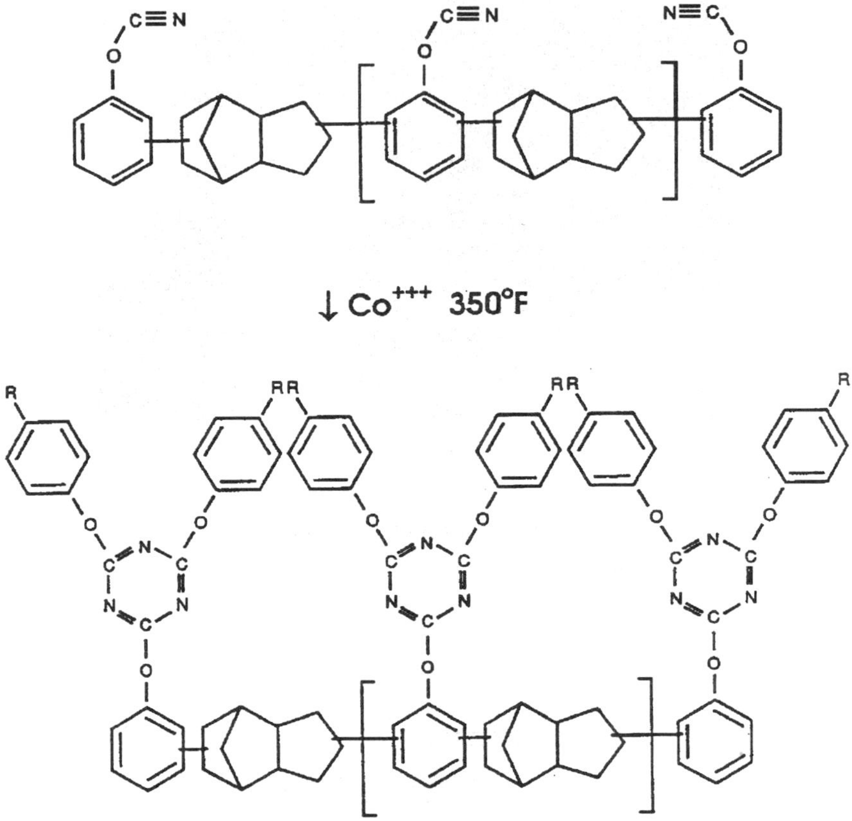

Figure 5. Polycyanate Chemistry
(Courtesy of YLA Incorporated)

Several attempts to oxidize the surface were not successful, including using a plasma media. With technical support from Dow Chemical, a viable approach was discovered to activate the surface for plating.

Consequently, from these basic ideas, a process was developed that chemically removed small amounts of the material and left the surface in an active condition. This process allows the surface of the polycyanate to interact with the activation and plating processes.

Although most of the chemicals used in the process are commercially available, the details of the sequence of the process, the times and temperatures remain proprietary.

3. EXPERIMENTAL

Several methods were investigated to convert the polycyanate to a hydrophilic condition at the surface. Techniques, such as oxidation, plasma etching and chemical etching with acids and caustic solutions were not effective in obtaining good adhesion with the plated metal. However, these experiments revealed a high resistance to oxidation of the polycyanates, which is an excellent attribute of the material.

Since the plasma etched surface was hydrophilic but did not accept the plating as an adherent coating, it became apparent that just making the surface hydrophilic did not guarantee good plating adhesion.

All the initial experiments, were conducted using 2" X 2" coupons. A control of the original base line process, which was a grit blasted surface, was processed concurrently with the various experimental processes.

From these experiments a chemical etch process was developed which removed 30-80 micro inches of polycyanate material and exhibited acceptable adhesion. (Table 1)

Table-1. Adhesion Tests Of Plated Copper On A Composite

ENVIRONMENTAL CONDITIONS	BASELINE PROCESS	MODIFIED B.L. PROC	INT. ETCH PROCESS	MOD ETCH PROCESS	IMPROVED PROCESS
1 Hr bake @ 150ºF	5B	5B	4B	5B	5B
Vacuum bake	5B	5B	4B	5B	5B
Thermal Cycle	5B	5B	4B	5B	5B
Thermal Shock	4B	5B	3B	5B	5B
Humidity Test	3B	4B	2B	4B	5B

PROCESSES DEFINED: (Copper plating on a RS-3/T300 composite substrate)

Baseline process: Grit blasted surface with standard catalyst and accelerator.
Modified B. L. process: Grit blasted surface with standard catalyst and accelerator with Alloy X.
Initial Etch process: Chemically etched with standard catalyst and accelerator.
Modified Etch process: Chemically etched with standard catalyst and accelerator, with Alloy X.
Improved process: Chemically etched with standard catalyst and accelerator with improved release process.

ENVIRONMENTAL CONDITIONS DEFINED:

Vacuum Bake: Vacuum-less than 10 µ; Temperature-212°F; Time-8+ hours
Thermal Cycle: Ten cycles (-175° to 175°F)
Thermal Shock: Five cycles (-320° to 300°F)
Humidity Test: 100 hours @ 100% humidity and 150°F

CLASSIFICATION OF ADHESION TEST: (ASTM D 3359-90)

Classification	5B	4B	3B	2B	1B	0B
Metal removed	none	1%	2%	10%	40%	>65%

Once the desired adhesion was attained, actual composite waveguides with various configurations were plated both with copper and silver. These were then cut apart to determine the uniformity of the plating in the waveguide using a Coating Measurements Instruments' (CMI) Beta backscatter instrument. (Figure 6).

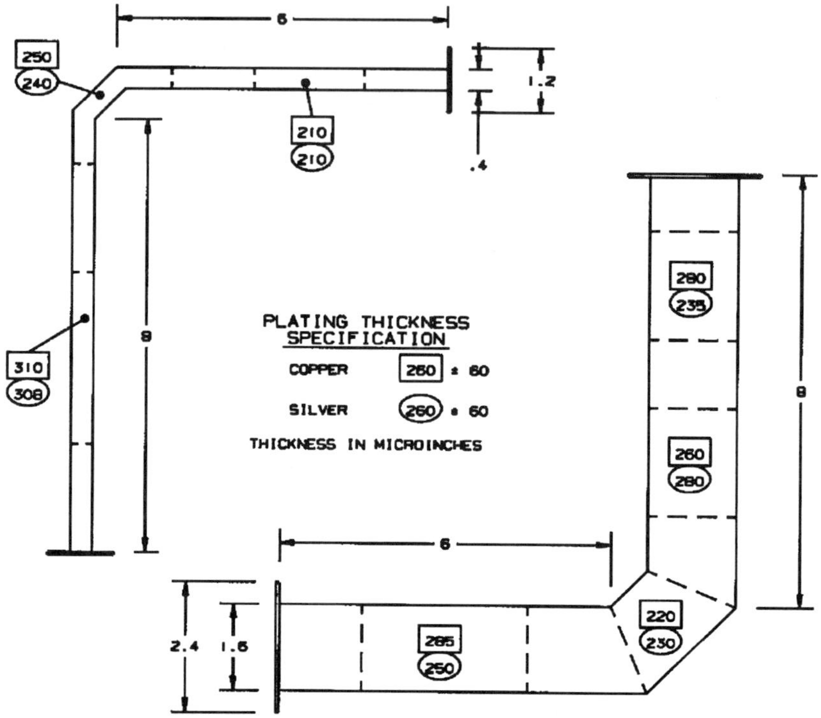

Figure 6. Profile of Plated Metal Thickness on a WR159 Waveguide

When the waveguides were tested for their ability to conduct the microwave, it was determined that the surface finish played an important part in the attenuation of the signal. At this time it became apparent that the etching process was more valuable for many applications, since degradation of surface finish was minimal. (Table 2)

Table-2. Affects Of Surface Roughness On Microwave Attenuation

DESCRIPTION 1 Meter straight waveguide (WR-75)	INSERTION LOSS dB No plating	 With copper	 With silver	COMMENTS
Theoretical value	--	0.12	0.10	100% metal
Grit blasted surface	--	0.24	--	64-120 RMS (μ in.)
Etched surface	6. 0	0.18	0.12	20-32 RMS (μ in.)

NOTES: 1. These values are averaged from limited data to give a trend of the affects of surface roughness for a microwave signal at this frequency
2. Measurement uncertainty estimate is 0.05 dB

The result of these tests was a much better conductivity of the microwave. However, more work is required in this area because the substrate itself has a major affect on surface roughness which in turn depends on the selection of graphite fiber and the manner the material was released from the tool. The significance of this data is that the better surface finish that is achieved by etching, also improves microwave conductivity.

Adhesion of the plating was initially evaluated with a scribe and tape test according to ASTM D 3359-90. When this test was performed, it became apparent that the razor cut was damaging the fibers. Therefore, it was determined to be too severe for this type of substrate. Consequently, a Flat Wise Tensile (FWT) test was developed which is much more quantitative and better represents the adhesion on a composite substrate. The relative values of the two methods are given below in Table 3.

Table-3. Relative Values Of Adhesion Tests

Classification	Flat Wise Tensile (PSI)
5B	1500 and above
4B	1000 to 1500
3B	500 to 1000
2B	300 to 500
1B	Below 300

To fully characterize this process required over 400 coupons and 50 waveguides. The time to do this, plus the analysis and evaluation of data, required a two man year effort.

4. DISCUSSION

This etch process has made it possible to plate 0.2 to 0.5 mils (0.0002 - 0.0005 in.) of copper on composite space hardware as well as 0.2 to 0.4 mils of silver when needed. The adhesion of the copper to the substrate was evaluated using Flat Wise Tensile (FWT). This gave an adhesive value of 1000 to 2000 PSI depending on the type of substrate and the surface preparation used. (Table 4)

Table-4. Flat Wise Tensile Measurements With Various Conditions

Surface Preparation	Flat Wise Tensile (PSI)
Grit Blasted	1000 to 1800
Etched	1200 to 1800
Grit Blasted & Etched	1200 to 2000
Intermediate Polymer/Etched	1200 to 2400

The FWT values have a broad range because it is dependent on the composite material. For example, a fabric such as T300 with a RS-3 polycyanate and a resin rich surface would give the higher values. A higher modulus graphite fiber such as M55J with the same polymer would give a lower FWT because the integrity of the composite is not as great and the failure is interlaminar. Therefore, the best adhesion is with a composite with good cohesive strength and a resin rich surface.

The following pictures show SEM's and cross sections of various surface treatments and the resulting plated layer of copper. These will be explained sequentially.

Figure 7 shows an SEM of a typical grit blasted surface with a fabric composite (T300/RS-3). The fiber damage is not as extensive (Figure 7) as the damage with a similar process for the high modulus fiber in the M55J/RS-3 composite (Figure 8) because of the brittle nature of the M55J/RS-3 fiber.

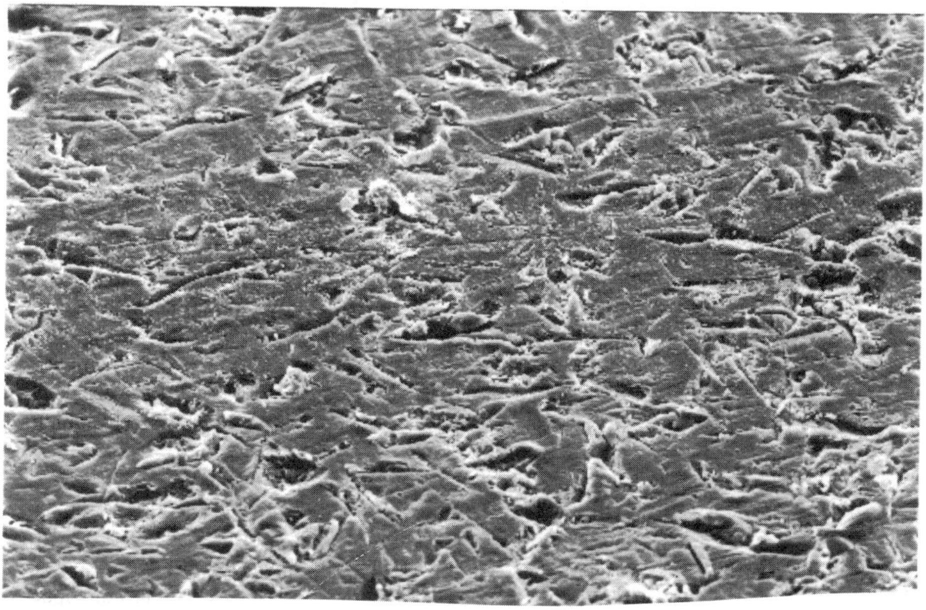

Scale: ___ 20 μ m

Figure 7. SEM (250X) of a grit blasted surface of a graphite composite (T300/RS-3)

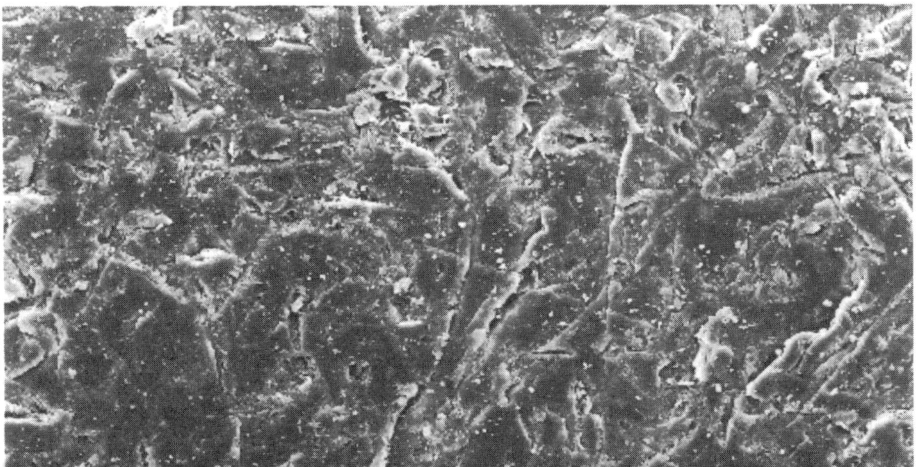

Scale: ___ 20 μ m

Figure 8. SEM (250X) of a grit blasted surface of a graphite composite (M55J/RS-3)

The SEM's in Figures 9 and 10 both show a surface that has no damaged fibers since these surfaces were only etched. However, the fibers will be exposed if the surface is resin poor which is the case in Figure 10.

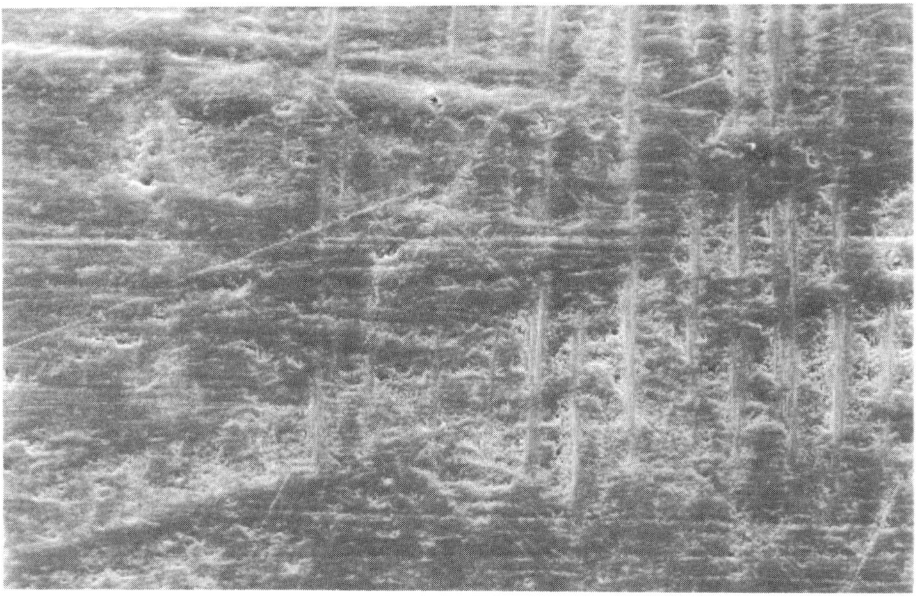

Scale _____ 10 μ m

Figure 9. SEM (1000X) of an etched surface (T300/RS-3 composite)

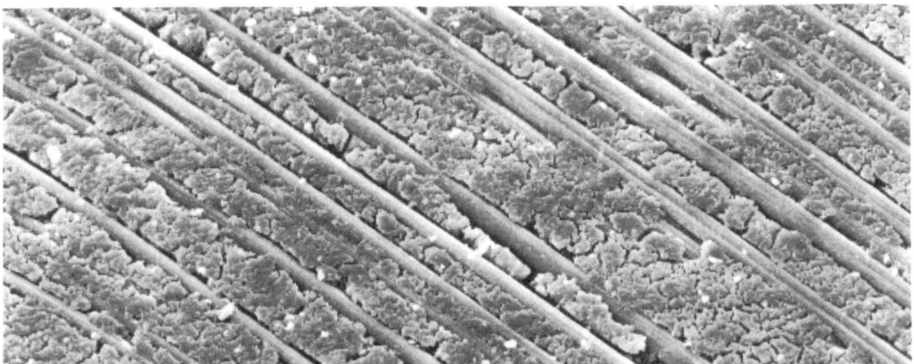

Scale _____ 10 μ m

Figure 10. SEM (1000X) of an etched surface (M55J/RS-3 composite)

Photomicrographs of cross sections illustrated in Figures 11 and 12 are grit blasted and etched surfaces, after they have been plated. These demonstrate the advantage of the etch process. A much smoother surface can be maintained without the grit blast.

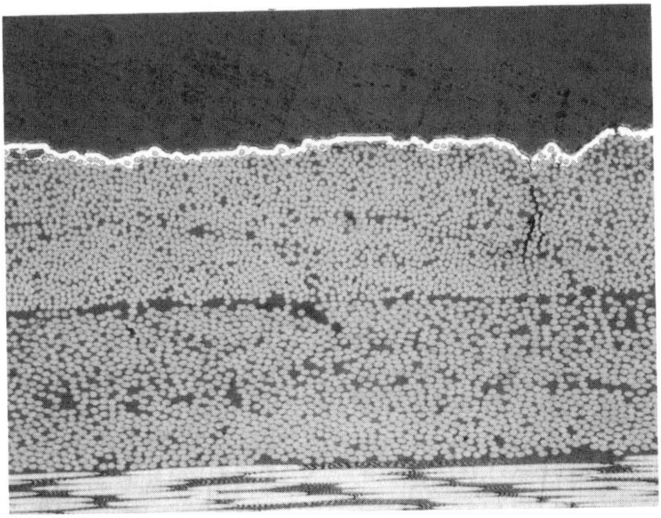

Scale: ___ 25 μ m

Figure 11. Microphotograph (200X) showing cross section of a grit blasted surface after plating. (M55J/RS-3 composite)

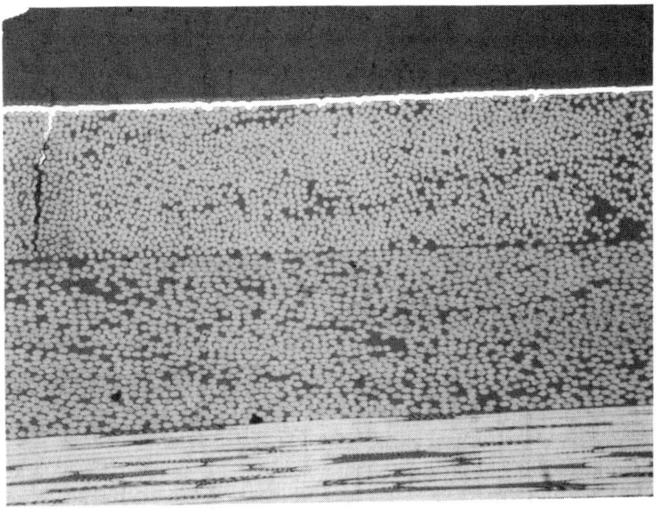

Scale: ___ 25 μ m

Figure 12. Microphotograph (200X) showing cross section of an etched surface after plating. (M55J/RS-3 composite)

Figure 13 shows the cross section of a resin rich surface which is the most desirable when a polished surface is required for waveguides and especially for mirrors.

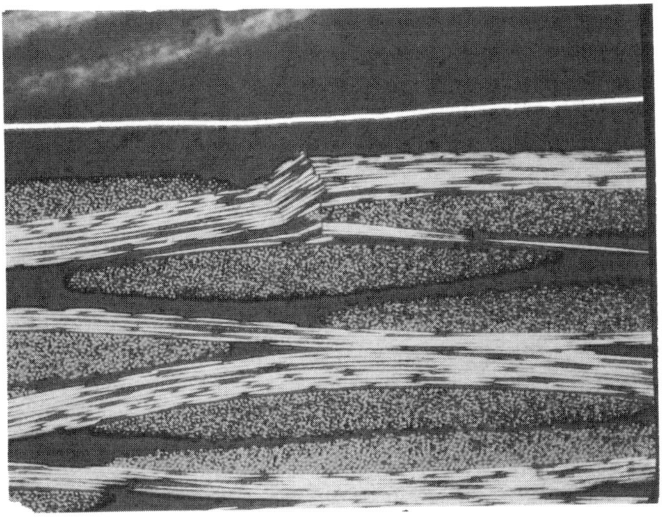

Scale: __ 25 μ m

Figure 13. Microphotograph (200X) showing cross section of an etched resin rich surface after plating. (T300/RS-3 composite)

Figure 14 illustrates the ability for the copper to fill in the voids around the fibers, which also increases the bond strength of the plating, to the substrate. However, if there are excessive voids in the substrate as with the M55J/RS-3 composite, a premature failure will occur in the laminate.

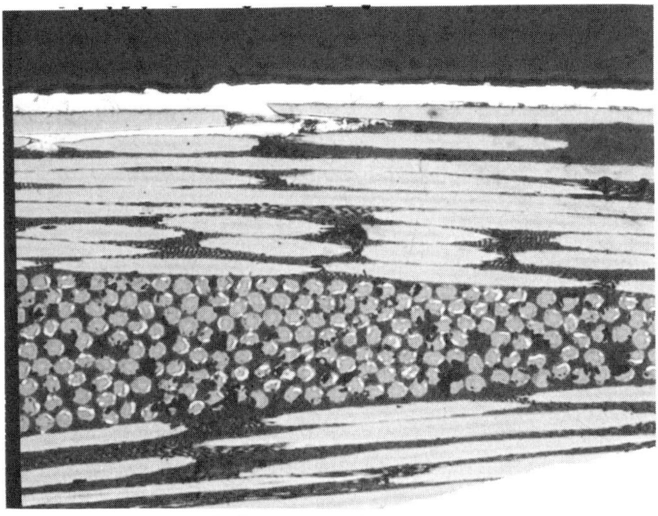

Scale _____ 20 μ m

Figure 14. Microphotograph (500X) showing cross section of an etched resin poor surface after plating. (T300/RS-3 composite)

5. CONCLUSION

COI has been successful in developing this etch process to a useful application with space hardware. This has resulted in fabricated and plated component parts that have qualified for space flight. Consequently, some of this hardware is presently flying.

This has opened the door to convert a large number of metallized composite space hardware from the metal counterpart. The weight reduction will have a significant impact on payloads in the future.

6. SUMMARY

The key to the success of metallizing composite hardware for space is understanding the surface chemistry of the polycyanate resin. This has been achieved by developing a process which activates the cured polycyanate resin so that the surface becomes hydrophilic and reactive with the chemicals in the plating process.

Another factor that enters into the formula for success, however, is the experience of effectively plating on non-metallic surfaces. Additionally, one must have extensive process control of the complete operation. This is absolutely necessary because of the complexity of the procedure. There are many variables in this process and since there is a sensitivity in some critical steps, it is necessary to have sufficient control of these parameters.

7. REFERENCES

1. Krumweide, G.C., Derby, E.A. and Chamberlin, D.N., "The Performance of the Effective Moisture Barriers for Graphite/Epoxy Instrument Structure," Proc. Of SAMPE, Atlantic City, NJ, (1989).

2. Brand, R.A. and Marks, J.E., "Composite Structures for Optical Mirror Applications", SPIE, Vol. 1303, Advances in Optical Structure Systems (1990).

8. ACKNOWLEDGEMENTS

We would like to thank the technicians in the plating shop at COI for their support and their insight into plating techniques. Specifically, to Carl Houston, who conducted most of the initial experiments and during that time helped resolve many plating issues.

Also, we appreciate the input given by Karen Spinar and John Marks on the previous effort and data for surface preparations on polycyanates. Moreover, Susan Robitaille (YLA, Incorporated) was very helpful in explaining the polymer chemistry of polycyanates and assisting us in obtaining the information necessary to successfully develop the etch process.

9. BIOGRAPHIES

William E. Converse received his BS degree in Chemistry from San Diego State University. He has 40 years of a diverse experience in Materials and Processes in both the Aerospace and Electronics industries which included surface chemistry for plating, vacuum deposition and chemical vapor deposition (CVD) processes.

Frank Lary has 25 years experience in the electroplating, metal finishing industry. His experience includes Electroless deposition and Electroplating in the Aerospace, Plastic-Injection Mold, Electronics, and Electroforming industries. He is a member of the Electrochemical Society and American Electroplaters and Finishers Society.

28th International SAMPE Technical Conference
November 4-7, 1996

DURABILITY OF INTERCALATED GRAPHITE EPOXY COMPOSITES IN LOW EARTH ORBIT

James R. Gaier*, Michelle L. Davidson**, and Rhonda K. Shively***
*NASA Lewis Research Center, Cleveland, OH 44135
**The Ohio State University, Columbus, OH 43201
***Manchester College, North Manchester, IN 46962

ABSTRACT

The electrical conductivity of graphite epoxy composites can be substantially increased by intercalating (inserting guest atoms or molecules between the graphene planes) the graphite fibers before composite formation. The resulting high strength, low density, electrically conducting composites have been proposed for EMI shielding in spacecraft. Questions have been raised, however, about their durability in the space environment, especially with respect to outgassing of the intercalates, which are corrosive species such as bromine. To answer those concerns, six samples of bromine intercalated graphite epoxy composites were included in the third Evaluation of Oxygen Interaction with Materials (EOIM-3) experiment flown on the Space Shuttle Discovery (STS-46). Changes in electrical conductivity, optical reflectance, surface texture, and mass loss for SiO_2 protected and unprotected samples were measured after being exposed to the LEO environment for 42 hr. SiO_2 protected samples showed no degradation, verifying conventional protection strategies are applicable to bromine intercalated composites. The unprotected samples showed that bromine intercalation does not alter the degradation of graphite-epoxy composites. No bromine was detected to have been released by the fibers, allaying fears that outgassing could be disruptive to the sensitive electronics the EMI shield is meant to protect.

1. INTRODUCTION

One of the important trends in the manufacture of aerospace structures has been the replacement of conventional metallic components with polymer matrix composites. When high strength and stiffness are required, carbon fiber polymer matrix composites have often been the material of choice. Mass savings in excess of 20 percent are common, resulting in significant performance improvements.

There has been resistance, however, to substituting composites in structural components which also have an electrical role, such as electromagnetic interference (EMI) shielding enclosures and ground planes, because the electrical resistivity of carbon fiber polymer composites is typically three orders of magnitude higher than that of metals. High performance pitch-based graphite fibers, such as Thornel P-100 (Amoco) have lower resistivity, and so reduce the margin to two orders of magnitude. The process of intercalation, the insertion of guest atoms or molecules between the graphene planes within the fiber, can reduce the resistivity by another factor of five (1).

Studies have shown that applications such as EMI shielding may indeed be feasible when intercalated pitch-based fiber composites are used. Such composites could reduce the mass of satellite power systems by as much as 17 percent and provide similar mass savings within the communications systems. The combined mass of the power and commnications systems typically account for 40 percent of the spacecraft mass. Reducing the mass of these systems would provide the opportunity to significantly increase payload capacities (2).

One of the reservations about using intercalated graphite fiber composites has been the possibility of chemical instability of the fibers. Since the most stable intercalates are strong oxidizing agents (such as bromine) the concern is justified. Bromine vapor is corrosive enough that even at low concentrations it could wreak havoc on an electrical system. Although studies have shown that bromine intercalated pitch-based fibers are stable under ambient conditions, high humidity, high vacuum, and temperatures from 150 to 300 °C (depending on the fiber host) for several years (3), intercalated graphite fiber composites had not been directly exposed to the launch and space environments.

The low Earth orbit (LEO) environment experienced by spacecraft is particularly hostile for organic compounds, such as intercalated graphite composites. There is rapid thermal cycling as the spacecraft flies in and out of the Earth's shadow. Short wavelength ultraviolet light which is blocked from the Earth's surface by the atmosphere (vacuum ultraviolet) is not blocked, and facilitates many chemical reactions of organic compounds. Perhaps most significantly at altitudes between 300 to 500 km, where Space Shuttles operate, the most common chemical specie is atomic oxygen (4). This atomic oxygen is the outer remnant of the Earth's atmosphere, and as such follows the 24 hr rotational motion of the Earth. Shuttle however, orbits about the Earth about once every 90 min (8 km/s), resulting in significant relative motion between the spacecraft and the atomic oxygen. This leads to an average relative kinetic energy of 4 to 5 eV for the atomic oxygen. This highly energetic, reactive oxygen quickly attacks any compound that is susceptible to reaction with oxygen.

The consequence of this environment is that any surface which is susceptible to atomic oxygen attack must be protected. The strategy which has proved most successful is to apply thin films of materials which are already highly oxidized. Sputter deposited SiO_2 and Al_2O_3 films which are a few tenths of a micrometer thick have proved effective for protecting smooth surfaces.

The effects of the LEO environment on materials have been subject to specific investigation by NASA for more than a decade. An important component has been the Evaluation of Oxygen Interaction with Materials (EOIM) experiments flown on the space shuttle (5). The first (EOIM-1) was flown in November, 1982 on the STS-5 mission. Forty-five material samples were exposed for 44 hr to ram atomic oxygen at an altitude of 300 km. The resulting atomic oxygen fluence was determined to be 1×10^{20} atoms/cm^2. EOIM-2 was flown aboard STS-8 in August 1983. Over 300 samples were flown for 42 hours at 250 km, resulting in an atomic oxygen fluence of 3.5×10^{20} atoms/cm^2. The most recent of the Evaluation of Oxygen Interaction with Materials experiment (EOIM-3) was launched on the Space Shuttle Discovery (STS-46) on July 31, 1992. On August 6, at 5:30 AM EST the experiment was initiated where the materials were exposed to atomic oxygen in the ram direction for 42 hr and an altitude of 230 km. The resulting fluence was 2.6×10^{20} atoms/cm^2 (6). EOIM-3 carried many active and passive materials experiments including more than 300 samples. Six were designed to evaluate the durability of intercalated graphite composites to LEO. Additionally, control samples were subjected to simulated LEO conditions in an attempt to determine how closely ground based testing simulates the true LEO environment for these materials.

2. EXPERIMENTAL

Three grades of Amoco pitch-based graphite fibers, Thornel P-100, P-75, and P-55 were selected for this study because of their availability and the large data base available on the properties of their intercalation compounds. Bromine was selected as the intercalate because it forms reproducible and very stable intercalation compounds with all three types of fibers. Mutual Industries (Philadelphia, PA) wove P-55 and P-75 fibers in a five harness satin weave with 11 × 11, 6000 filament tows per inch. Fabric Development, Inc. (Quakertown, PA) wove P-100 fabric in a 10 × 10, 2000 filament tows per inch. A plain weave was used to minimize fiber breakage during weaving. Four ply 0 to 90° laminar composites were fabricated by Rohr Industries (Chula Vista, CA) using Hysol-Grafil EAG101-1 film epoxy. The details of the intercalation and composite fabrication and characterization have been reported elsewhere (7).

Silicon dioxide (SiO_2) coatings were applied to some of the samples by ion beam sputtering using an Ion Tech dual ion source with argon ions and an SiO_2 target. Circular disks (2.50 cm) were cut from the composite sheets using electron discharge machining (EDM). These fit into the sample holder in a hexagonal array as shown in Figure 1. These were flown in tray 6 as samples 6-40 to 6-45.

Properties of the six flight samples are summarized in Table 1. In addition to the flight samples there was a complete set of back-up samples, which traveled with the flight samples, but were not mounted. There were also two sets of control samples which were prepared at the same time, but which stayed at the lab during the flight.

Sample 6-40 was a pristine P-100 composite with no coating. To our knowledge P-100 composites had not been flown before, but it was expected to behave similarly to other carbon fiber epoxy composites that have been flown in the past. This sample was to act as a baseline to which the intercalated fiber samples could be compared.

Sample 6-41 was a bromine intercalated P-100 composite sample. Its comparison with sample 6-40 would show the effects of intercalation on the durability of an unprotected sample in LEO. It would also show whether subjecting an intercalated composite sample to the LEO environment would deintercalate the sample or change its physical or chemical properties.

Sample 6-42 was a bromine intercalated P-100 composite sample that was protected with 0.10 μm of SiO_2. Its comparison with sample 6-41 would indicate whether intercalated graphite composites could be protected using the same techniques used for other carbon fiber polymer composites.

Samples 6-43, 6-44, and 6-45 were similar except that the composites were made with intercalated P-100, P-75, and P-55 fibers respectively. One half of each of these samples was coated with SiO_2, so that the effect of the coating could be observed directly. In addition, an expanded copper foil screen was placed over each of these samples. This was done to provide a "bromine trap." If bromine was released from the composites the bromine would be expected to react with the copper and could be detected in the post-flight analysis. A photo of one of these samples is shown in Figure 2.

Optical images with a Kodak MP-4 camera were obtained before and after the flight. Masses of the samples (both pre-flight and post-flight) were measured using a Mettler H315 analytical balance. Reflectance over the 250 to 2500 Å wavelength range was determined using a Perkin-Elmer Lambda-9 UV-Vis spectrophotometer with an integrating sphere. Electrical conductivity was determined using a Leheighton Electronics 1010A eddy current conductivity device which was factory modified to operate at 55.55 kHz, and calibrated with NIST standards. Analysis using a scanning electron microscope (SEM) was carried out using a Cambridge 200 which allowed the sample to be imaged either in transmission or backscatter modes. This instrument was also fitted with energy dispersive spectroscopy (EDS) which enabled elemental analysis at the resolution of the microscope. As much of the SEM analysis as possible was done on the samples as received, but the high resolution work required about 20 nm of gold to be sputtered onto the surfaces.

Approximate values for the erosion yields (mass loss per impinging oxygen atom) were also determined using the SEM. Dust particles adhered to the surface of the composite protect small portions of

TABLE 1.—SAMPLES FLOWN ON STS-46

Sample	Fiber type	Intercalate	Coating	Br_2 trap
6-40	P-100	none	none	none
6-41	P-100	Br_2	none	none
6-42	P-100	Br_2	SiO_2	none
6-43	P-100	Br_2	none/SiO_2	Cu
6-44	P-75	Br_2	none/SiO_2	Cu
6-45	P-55	Br_2	none/SiO_2	Cu

the composite from erosion, and the depth of the material eroded is directly measured from the micrographs. In addition to the dust particles, the interfaces where the sample was protected by the sample holder, the SiO_2 coatings, or the bromine traps, could also be utilized. The technique is fraught with assumptions, but given enough data points, consistent values for the erosion depths do emerge.

Control samples were subjected to ground based atomic oxygen simulation in a Plasma Prep 2 radio frequency plasma asher. These samples received an oxygen fluence equal to that received by the flight samples as determined by Kapton witness coupons (6).

3. RESULTS AND DISCUSSION

Images of the samples before and after the flight test are shown in Figure 3. The outer ring on each of the samples is where the surface was protected by contact with the sample holder, and appears to be very similar to the pre-flight condition. It is immediately apparent that the unprotected portions of the samples were changed by the LEO environment. Those areas have a velvet texture and appear to be darker than the unexposed areas. This same phenomenon was reported by Slemp, et al. for graphite epoxy composites flown on STS-8 in EOIM-2 (8). There are no obvious differences either between the pristine and intercalated samples (6-40 and 6-41), or among the different grades of pitch-based fiber (samples 6-43, 6-44, and 6-45). Those portions of the sample which were under the SiO_2 coating are visually indistinguishable from those areas protected by the sample holder (note especially the fully protected sample 6-42). The cross-hatch pattern visible in samples 6-43, 6-44, and 6-45 corresponds to the copper screen which was placed over each of those samples as a bromine trap.

These qualitative observations were quantified using the reflectance measurements. Figure 4 shows the total hemispherical reflectance from 250 to 2500 nm before and after LEO exposure for the pristine P-100 composite (6-40), the bromine intercalated composite (6-41) and the protected bromine intercalated composite (6-42). A change in total reflectance factor, ΔR was calculated point by point for each of these samples (Table 2) using:

$$\Delta R = \frac{R_{after} - R_{before}}{R_{before}}$$

There was a wavelength dependence in the reflectance change for the pristine samples with the low frequency ΔR being about -0.4 and in the high frequency about -1.0. The ΔR of the intercalated fiber composite showed no frequency dependence except for the the 250 to 500 nm region. The ΔR of the SiO_2 protected composite showed a slight increase in the reflectance which increases from negligible (between 900 and 1000 nm) to 0.04 at 2500 nm. The important technological conclusion however, is that all three composites maintained low reflectance values (below 0.2) over the entire measured spectrum. These results are consistent with other unprotected and protected carbon fiber-polymer composites which have been exposed to the LEO environment (9).

The mass of each of the samples was measured pre-flight and post-flight and compared. Although there were indications from the appearance and the optical reflectance data that there were mass losses, Table 3 shows that they were below our detection limit. Composite samples readily absorb water from the atmosphere, and the uncertainty of the mass due to absorbed water was larger than that lost due to space exposure. The mass losses were estimated from the erosion

TABLE 2.—CHANGE IN TOTAL HEMISPHERICAL REFLECTANCE ON EXPOSURE TO LEO

Sample	ΔR
6-40	−0.64 ± 0.03
6-41	−0.51 ± 0.09
6-42	0.17 ± 0.12

TABLE 3.—MASS OF SAMPLES FLOWN OF STS-46

Sample	$m_{initial}$	m_{final}	$m_{final}/m_{initial}$
6-40	0.5842	0.5850	1.001
6-41	0.7405	0.7401	0.9995
6-42	0.7452	0.7482	1.004
6-43	0.7497	0.7515	1.002
6-44	0.9050	0.9058	1.001
6-45	0.8886	0.8928	1.005

TABLE 4.—RESISTIVITY OF SAMPLES FLOWN OF STS-46

Sample	$\rho_{initial}$ μΩ-cm	ρ_{final} μΩ-cm	$\rho_{final}/\rho_{initial}$
6-40	1420	1695	1.194
6-41	352	350	0.994
6-42	362	369	1.019
6-43	362	363	1.003
6-44	910	945	1.038
6-45	1520	1529	1.006

apparent in the SEM analysis (discussed below), but the uncertainty of knowing the original surface morphology on a microscale introduces large errors into any calculation. That the mass losses were very small is the only statement that can be made with confidence.

The resistivity of the samples before and after the flight is shown in Table 4. There were no significant changes, with the possible exception of sample 6-40. This is the pristine P-100 composite, and no plausable explanation for its increase has been reached. These are critical measurements because the principal motivation for using intercalated graphite fiber composites is their low resistivity. The fact that the intercalated fiber composites are unchanged is also a good indication that large amounts of intercalate were not lost, since a small amount of intercalate loss increases the resistivity of the fibers dramatically.

The pristine P-100 composite sample, 6-40, when imaged with the SEM showed evidence of being etched by the atomic oxygen on orbit (Figure 5). The ram direction is apparent by the direction of the cone-shaped structures. This type of structure has been seen before in graphite-epoxy composites subjected to LEO conditions (10).

When the pristine flight sample, 6-40, (Figure 5) is compared to the intercalated sample, 6-42, (Figure 6), it is apparent that bromine intercalation has no major affect of the degradation of the composite in LEO. The sizes and shapes of the etch artifacts are qualitatively the same. Further, when the unprotected sides of samples 6-43, 6-44, and 6-45, are compared (Figure 7), there are no major differences among the degradation of composites made from the three grades of graphite fibers. To a first approximation, neither fiber type nor intercalation affect the degradation of graphite-epoxy composites in LEO.

When the SiO_2 protected intercalated sample, 6-42, (Figure 8) is compared to the unprotected intercalated sample the effectiveness of the coating can be judged. The sample looked essentially the same as the unexposed samples. Further, comparing the protected sides of 6-43, 6-44, and 6-45, (Figure 9) it can be seen that changing fiber types does not alter the coating performance.

Erosion yields for the samples are shown in Table 5. Sample 6-42 and the sides of samples 6-43, 6-44, and 6-45 which were protected by the SiO_2 coating showed no noticeable erosion.

TABLE 5.—EROSION YIELDS FOR SAMPLES FLOWN ON STS-46

	# of matrix photos	AVG erosion yield matrix (cm^3/atom)	# of fiber photos	AVG erosion yield fiber (cm^3/atom)
6-40 Pristine P-100	6	$3.5 \times 10^{-24} \pm 0.6$	11	$1.0 \times 10^{-24} \pm 0.4$
6-41 P-100+Br	7	$4.1 \times 10^{-24} \pm 0.9$	11	$0.8 \times 10^{-24} \pm 0.4$
6-42 P-100+Br+SiO_2	0	none	0	none
6-43 P-100+Br[1]	---	-----------	---	-----------
6-44 P-75+Br[2]	7	$3.9 \times 10^{-24} \pm 0.5$	11	$0.7 \times 10^{-24} \pm 0.2$
6-45 P-55+Br[2]	7	$2.8 \times 10^{-24} \pm 0.5$	11	$0.9 \times 10^{-24} \pm 0.4$

[1]Erosion expected to be same as 6-41.
[2]Erosion yield for unprotected side, protected side showed on erosion.

The epoxy matrix showed an average of about 3.5×10^{-24} cm^3 of matrix lost for every atomic oxygen impact. The graphite fibers were somewhat more durable, with an average erosion yield of about one-fourth (0.85×10^{-24} cm^3/atom) that of the matrix. As a result, the matrix tends to erode away and leave the fibers on the surface connected to the composite only by the matrix that is shadowed by the fiber. Both the fiber and the matrix tend erode in such a way as to form cone-shaped structures, but the appearance of matrix structures is more complex, with holes and curled points. Intercalation did not appreciably effect the erosion yield of the fiber. Though it is interesting to note that the highest erosion yield was for the pristine fiber, within the error the yields, were the same. The erosion yield calculated under much more controlled conditions for pyrolytic graphite was 0.65×10^{-24} cm^3/atom, and EOIM-1 found the erosion yield of pyrolytic graphite to be 1.2×10^{-24} cm^3/atom (6).

Given the erosion yields above, the mass loss (Δm) for an unprotected sample can be calculted from

$$\Delta m = (fYFAd)_{fiber} + (fYFAd)_{matrix}$$

where f is the volume fraction (0.48 and 0.52 for fiber and matrix respectively), Y the erosion yield, F the fluence of atomic oxygen, A the sample area exposed (4.9 cm^2) and d is the mass density (2.293 and 1.265 for fiber and matrix respectively). This equation assumes that the eroded layer composition is the same as the bulk, which is probably a poor assumption. The surface is known to be resin rich, with at least a thin (1 μm) layer of matrix resin covering all of the surface. Given the uncertainties in the rest of the factors, this effect was considered to be below the error. The total mass lost from an unprotected sample was calculated to be about 1.2 mg. This is within the error introduced by the fluctuation of the water content within the composites. The mass loss gives an average etch depth of 4.8 μm.

Samples exposed to the plasma asher induced simulated LEO environment show similar amounts of erosion, but the morphology of the degradation is different (Figure10). Whereas the flight samples show a cone-shaped degradative morphology, in the asher a more rounded, pillow-shaped morphology appears. This has been seen in samples before and has been attributed to the flight geometry providing oxygen from an essentially unidirectional source (ram direction through the atmosphere), where as the asher is omni-directional (9).

One of the most important aspects of this series of experiments was to determine whether bromine would outgas from the intercalated fibers and corrode electrical components. Bromine is not found in detectable quantities on the surface of the fibers, but in a gradient that is at a maximum in the center of the fibers (11). However, even within the 40 hr period of LEO exposure, some individual fibers were eroded nearly half-way through, so some bromine must have been lost to the environment around the spacecraft. No detectable amount of bromine was found on any of the three copper traps using EDS. A small concentration of oxygen was the only impurity found on the traps.

EDS and backscattering were used to determine if bromine leached from the fibers into the epoxy. This scenario was thought unlikely since there is no bromine on the fiber surface, but since the erosion yield of the epoxy was four times that of the fibers, it has the potential to be a potent mechanism for releasing bromine during erosion. The EDS and backscattering results are shown in figure 11, and demonstrate conclusively that the bromine is confined to the fiber.

This raises the question of the fate of bromine that was once present in the fibers which were eroded. Since it is expected that free bromine would have reacted with the copper traps, the bromine must be released in a chemically inactive form, perhaps as brominated carbon oxides (Br_2CO, etc.), though there are no data to support this speculation.

4. CONCLUSIONS

Bromine intercalated graphite fiber epoxy composites were fabricated and flown on the Space Shuttle Discovery as part of the EOIM-3 experiment on STS-46. They were subjected to 42 hr of ram atomic oxygen which corresponded to a fluence of about 2.6×10^{20} atoms/cm^2. Unprotected samples degraded with a matrix erosion yield of about 3.5×10^{-24} cm^3/atom, and a fiber erosion yield of about 0.85×10^{-24} cm^3/atom. The bromine intercalated erosion was indistinguishable from the pristine composite erosion. There was also no erosion yield differences among the various grades of intercalated graphite fibers. The mass loss for all of the samples was below the detectable limit using a conventional analytical balance, but was calculated from the erosion yield to be about 1.2 mg. Total reflectance of these samples decreased by a factor of 50 to 65 percent. The electrical conductivity of the samples was unchanged. Bromine was found to remain within the fibers with no detectable amounts leaching into the epoxy or out onto copper bromine traps. SiO_2 coatings were found to be effective in preventing the erosion of the composites without adversely affecting conductivity or reflectance. Control samples exposed to the same fluence in a plasma asher were also found to exhibit no detectable changes in mass loss and conductivity, but the surface morphology was quite different than that built up during low Earth orbit.

This experiment allays fears that intercalated graphite fiber composites might outgas large quantities of corrosive materials which would destroy sensitive electronics in the vicinity of the composites. It also shows that there is no particular oxidation resistance conferred on composites by intercalation. Rather, it reinforces the perception that the only significant difference between pristine and intercalated graphite fiber composites is the latter's enhanced electrical conductivity.

5. ACKNOWLEDGMENTS

The authors would like to express our gratitude to M. Mirtich of NASA Lewis Research Center for coordinating the EOIM-3 samples and K. deGroh of NASA Lewis Research Center for assistance with the MP-4 photographs and advice with SEM images. Expanded foil copper screens were kindly provided by N. Brink of Rohr Industries. MLD was supported by NASA Lewis Research Center through a fellowship with the Ohio Aerospace Institute.

6. REFERENCES

1. J.R. Gaier, Extended Abstracts of the 17th Biennial Conference on Carbon (American Carbon Society, St. Marys, PA, 1985) 80.
2. J.R. Gaier, IEEE Transactions on Electromagnetic Compatibility (1992) 351.
3. J.R. Gaier, M.L. Slabe, and N. Shaffer, Carbon 26(3) (1988) 381.
4. U.S. Standard Atmosphere, 1976 (NOAA, NASA, USAF, Washington D.C., 1976)
5. R.H. Liang, Proceedings of the EOIM-3 BMDO Experiment Workshop, Jet Propulsion Laboratory, Pasadena, CA (1993).
6. B.A. Banks, Proceedings of the EOIM-3 BMDO Experiment Workshop, Jet Propulsion Laboratory, Pasadena, CA (1993).
7. J.R. Gaier, P.D. Hambourger, and M.E. Slabe, Carbon 29(3) (1991) 313.
8. W.S. Slemp, B. Santos-Mason, G.F. Sykes, Jr., and W.G. Witte, Jr., AIAA Paper 85–0415 Presented at 23rd Aerospace Sciences Meeting (1985).
9. B.A. Banks, S.K. Rutledge, J.D. Hunt, E. Drobotij, M.R. Cales, and G. Contrell, NASA TM–106769 (1995).
10. B.A. Banks, S.K. Rutledge, P.E. Paulsen, and T.J. Steuber, NASA TM–101971, (1989).
11. X.W. Qian, S.A. Solin, and J.R. Gaier, Physical Review B 35(5) (1987) 2436.

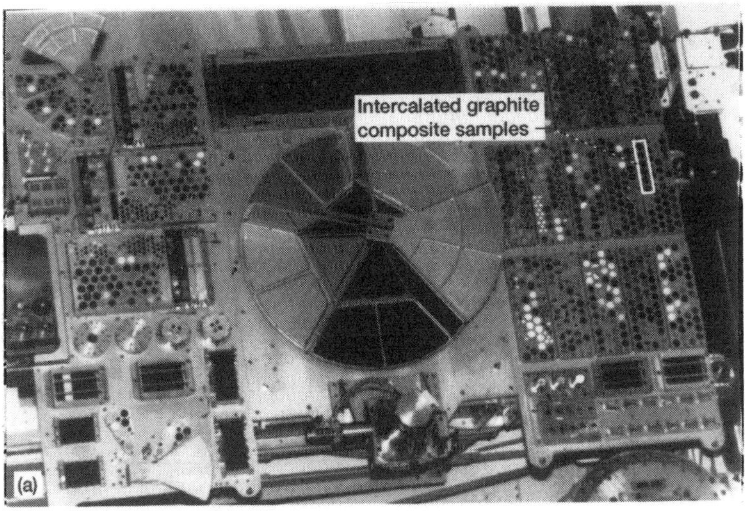

Figure 1.—EOIM III sample trays with intercalated graphite composite samples (a) indicated; (b) in shuttle bay.

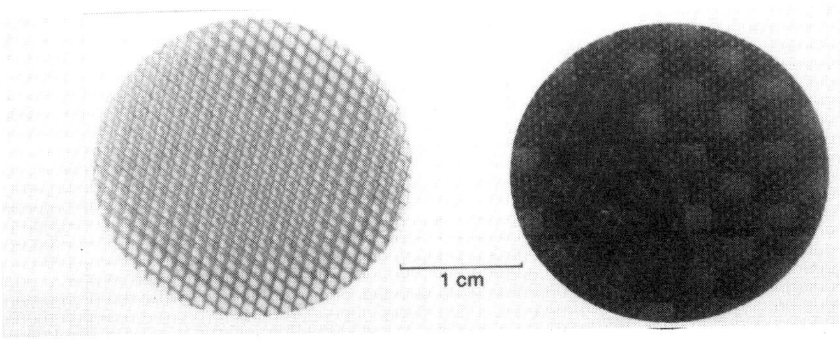

Figure 2.—Flight sample 6-43 showing the configuration of the protective film and the bromine trap.

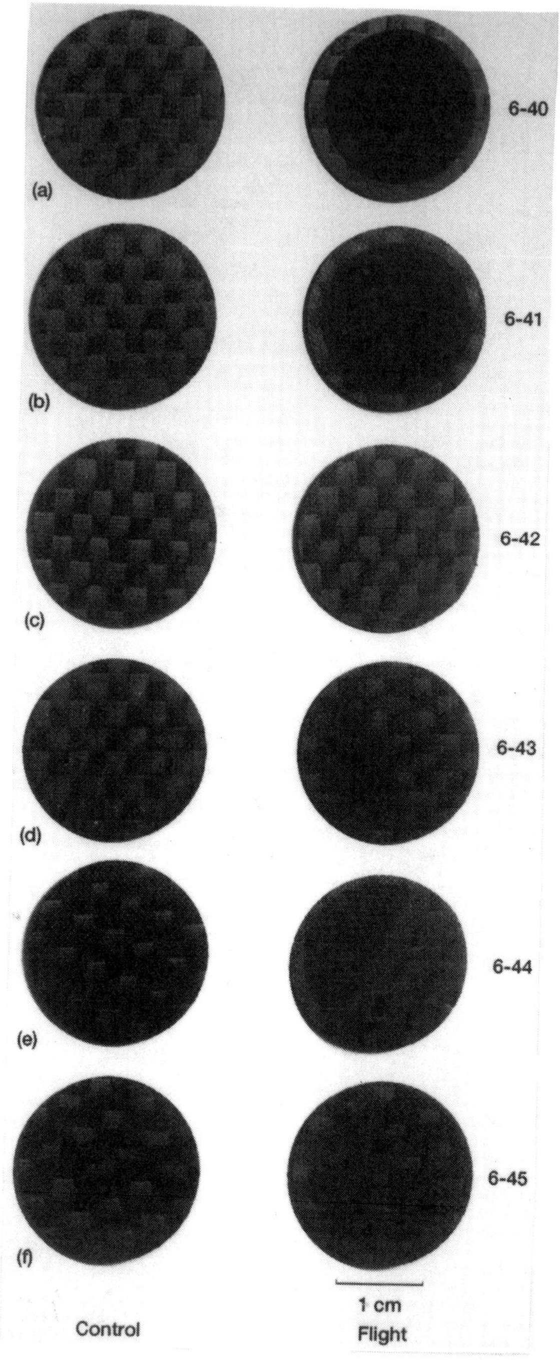

Figure 3.—Optical images of the six flight samples after exposure to LEO and their control samples. (a) Pristine P-100. (b) P-100 + Br. (c) P-100 + Br (coated w/SiO_2). (d) P-100 + Br (1/2 coated w/SiO_2). (e) P-75 + Br(1/2 coated w/SiO_2). (f) P-55 + Br (1/2 coated w/SiO_2).

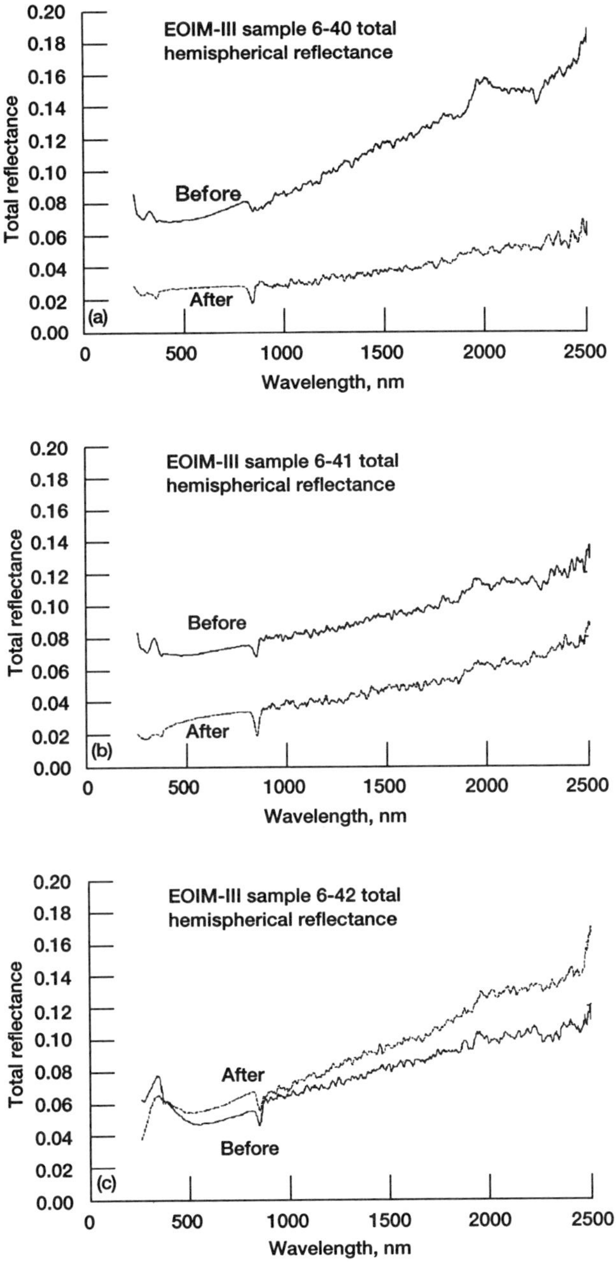

Figure 4.—Reflectance of; (a) pristine P-100 composites, (b) bromine intercalated P-100 composites, (c) SiO_2 protected bromine intercalated P-100 composites; before and after LEO exposure.

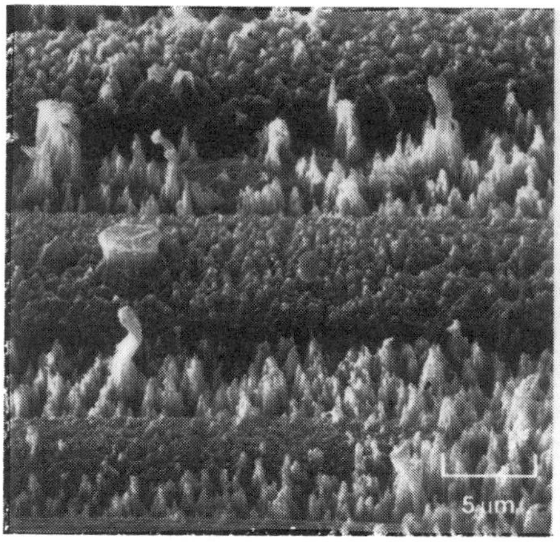

Figure 5.—Photomicrograph of pristine P-100 composite after LEO exposure.

Figure 6.—Photomicrograph of bromine intercalated P-100 composite after LEO exposure.

Figure 7.—Photomicrograph of unprotected bromine intercalated (a) P-100; (b) P-75; (c) P-55; composite after LEO exposure.

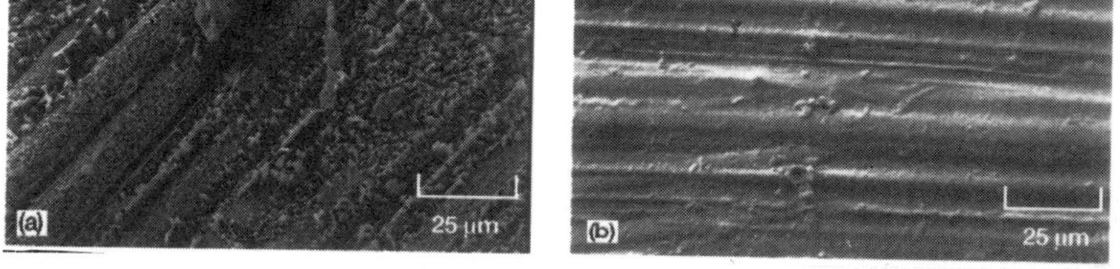

Figure 8.—Photomicrograph of (a) P-100 + Br; (b) P-100 + Br (coated w/SiO_2) composite after LEO exposure.

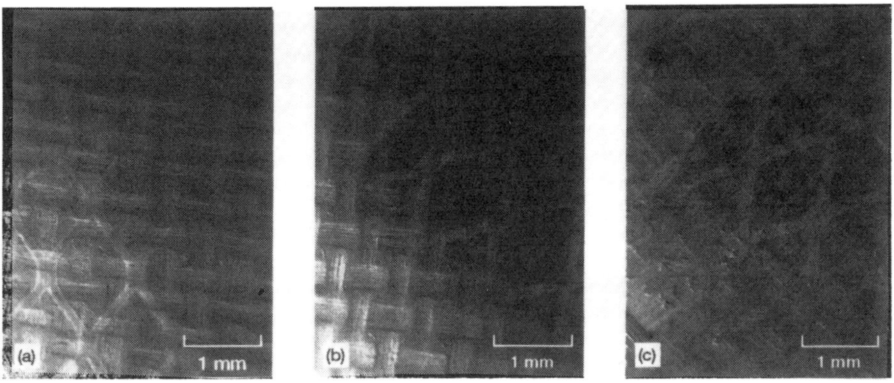

Figure 9.—Photomicrograph of SiO_2 protected bromine intercalated (a) P-100; (b) P-75; (c) P-55; composite after LEO exposure.

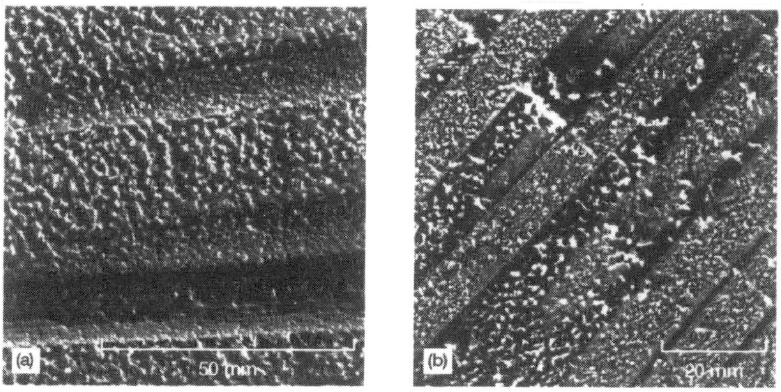

Figure 10.—Photomicrograph of bromine intercalated P-100 composite (a) after LEO exposure; (b) after plasma asher exposure.

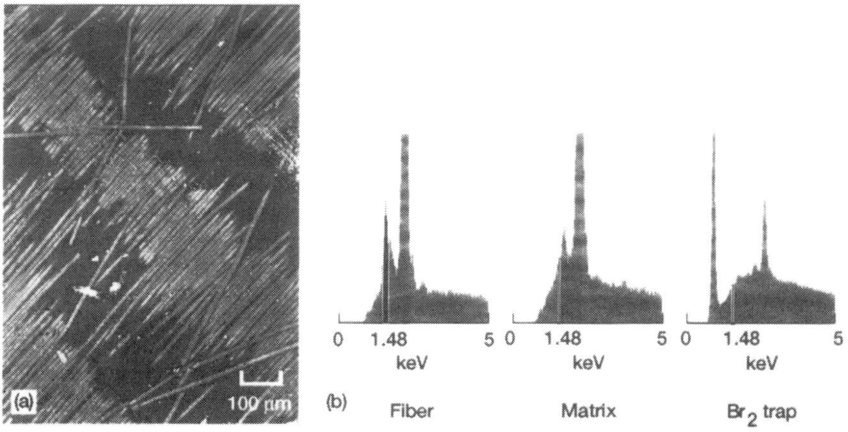

Figure 11.—Confinement of bromine to the fibers, (a) backscattering; and (b) energy dispersive x-ray spectroscopy.

28th International SAMPE Technical Conference
November 4-7, 1996

EFFECTS OF SIMULATED ORBITAL ATOMIC OXYGEN ON GERMANIUM-COATED KAPTON FILMS

Michael Hanichak
Lockheed Martin Missiles and Space
East Windsor, NJ

Miria Finckenor
EH12/Physical Science and Environmental Effects
Marshall Space Flight Center, AL

ABSTRACT

Germanium-coated black Kapton and germanium-coated Kapton film samples were exposed to ground simulated atomic oxygen environments at NASA Marshall Space Flight Center. One group of specimens was exposed to a total fluence of 3.5×10^{22} atoms/cm² along with 740 equivalent sun hours of vacuum ultraviolet radiation using the Atomic Oxygen Drift Tube System (AODTS). The second group of samples received an initial exposure ranging from 5.7 to 7.9×10^{20} atoms/cm^2 using the Atomic Oxygen Beam Facility (AOBF) prior to exposure in the AODTS with the other group of samples. A comparison of the mass, solar absorptance, infrared emittance, and surface resistance before and after exposure indicates no significant changes due to reaction with atomic oxygen. ESCA measurements indicated germanium oxide thickness of about 20 Å for the samples exposed only in the AODTS and about 40 Å for samples receiving exposure in both AOBF and AODTS. ESCA measurements of control samples which were not exposed yielded oxide thickness of about 10 Å. Oxide formation in laboratory simulation compared to flight sample oxide formation is discussed. Future investigations of germanium-coated Kapton, including long-term exposure on the Passive Optical Sample Assembly-I, are discussed.

KEY WORDS: Atomic oxygen, polyimides, space environment.

1. INTRODUCTION

Germanium-coated black Kapton is currently used as an RF-transparent antenna sunshield membrane (U.S. Pat. No. 5,373,306). Depending on the thickness of germanium coating, this configuration provides favorable thermal radiative properties (α_s/ε as low as 0.48) while maintaining high RF transparency in the UHF (30 to 300 MHz) and Ku (26 to 40 GHz) ranges. The conductive nature of the germanium coating ($< 10^9$ $\Omega/\square$) also serves to dissipate electrical charge buildup which might otherwise result in electrostatic discharge (ESD). In addition to antenna membrane applications, germanium-coated black Kapton can also be used for the backside surface of low earth orbit (LEO) solar arrays which are often coated with white paint. Although germanium-coated black Kapton has been fully qualified for use in geosynchronous orbit, its use in LEO applications requires testing to determine its ability to withstand exposure to atomic oxygen.

In low Earth orbit, atomic oxygen has been shown to damage materials used on orbiting spacecraft through erosion or oxidation. Spacecraft traveling in LEO, defined as that region of space between 200 and 1000 km (124 - 621 miles), have orbital velocities of 8 km/s (18,000 mph). At this velocity, atomic oxygen impacts the leading edge of the spacecraft with an energy of approximately 5-7 eV. Material durability data is generally reliable when this energy level is achieved. However, because of the large number of candidate materials for spacecraft, plasma chambers are still used for screening materials even though the energy level may be an order of magnitude lower or more. Synergism between atomic oxygen and ultraviolet radiation must also be considered when qualifying materials for use in the LEO environment. In addition to laboratory simulations, designers and engineers also use the wealth of data provided by retrieved long term space environment exposures (LDEF, Solar Max) and short-term Space Shuttle flight experiments (EOIM-I, II, III).

In order to assess the atomic oxygen stability of the germanium coating and its ability to protect the underlying Kapton, the following sample configurations were tested:

- 600 Å Ge on 1-mil non-conductive black Kapton (Type 100CB)
- 2000 Å Ge on 2-mil natural Kapton (Type HN)
- 4000 Å Ge on 2-mil natural Kapton (Type HN)

The 600 Å germanium coating was selected as the baseline coating for RF-transparent antenna sunshield membranes because it provides the best combination of thermal radiative properties and surface conductance. Since it was not known prior to atomic oxygen testing what thickness of germanium would be affected by exposure to atomic oxygen, it was decided to test thicker coatings as well. Unfortunately, the only materials available at that time in the desired thickness range were coatings on natural Kapton. Since the purpose of this effort was to determine the atomic oxygen resistance of germanium coatings, it was agreed that coatings on natural Kapton would be acceptable. All germanium depositions were performed by Courtaulds Performance Films, Canoga Park, CA.

The target atomic oxygen fluence for the test was 1×10^{21} atoms/cm^2. This fluence is representative of LEO spacecraft having similar mission duration (6 years) and altitude (705 km) as that of the Earth Observing System (EOS) spacecraft. Samples were evaluated for changes in mass, surface resistance, solar absorptance, and infrared emittance.

2. TEST DESCRIPTION

The Atomic Oxygen Beam Facility (Figure 1) located at MSFC produces a 5 eV neutral atomic oxygen beam by placing a metal plate in contact with magnetically confined atomic oxygen plasma. Because of the facility geometry, the atomic oxygen plasma is magnetically confined such that a 1 cm (0.39 in.) dia. plasma column is produced on centerline of the test chamber. The plasma column interacts with an electrically biased metallic plate. The bias applied to the plate accelerates ions from the plasma to the plate. During the acceleration process, the ions gain energy equal to the difference in the plasma potential and the neutralizer plate bias. Once the ions hit the plate, they collect an electron from the metal lattice and become neutral. Following collision with the neutralizer plate, the atoms are reflected towards the test specimen at a fraction of their pre-collision energy. The facility is operated in a duty cycle (generally 5%) to prevent overheating of the RF electrode. Details of how the atomic oxygen beam is produced at a similar facility developed for MSFC at the Princeton Plasma Physics Laboratory (PPPL) may be found in Reference 1.

The atomic oxygen flux produced by the AOBF system ranges from 5×10^{15} atoms/cm^2·sec to 1×10^{16} atoms/cm^2·sec. During production of the atomic oxygen plasma, the system produces electromagnetic radiation. This radiation is produced primarily during the dissociation and ionization process. Attempts to identify and quantify the radiation using a photodiode with appropriate bandpass filters indicated that the primary radiation line was 130 nm, the atomic oxygen resonant peak in the VUV region. The VUV intensity was determined to be nearly 200 times the sun's intensity averaged over the duty cycle. In order to eliminate possible magnetic interactions, appropriate shielding was placed around the diode during these measurements.

One sample of each configuration of germanium-coated Kapton was exposed in the AOBF to a fluence ranging from 5.7 to 7.9×10^{20} atoms/cm^2. Mass loss, solar absorptance, and infrared

Figure 1 MSFC Atomic Oxygen Beam Facility (AOBF)

emittance measurements were made at 6.2 x 10^{19} atoms/cm^2 and 2.5 x 10^{20} atoms/cm^2 fluences.

Unlike the AOBF facility which is capable of producing neutral 5 eV atomic oxygen, the Atomic Oxygen Drift Tube System (AODTS) (Figure 2) produces a thermal atomic oxygen plasma. Generated by a 14.7 MHz RF field, the AODTS plasma contains monatomic neutral and ionic oxygen, molecular neutral and ionic oxygen, excited state atoms and electrons. The energy of the oxygen atoms is generally less than 0.1 eV, and the flux produced is approximately 5 x 10^{16} atoms/cm^2·sec.

The AODTS facility is designed such that samples are exposed outside the RF field, eliminating sample exposure to any plasma charged particles and unwanted sample heating. Vacuum UV was added through a window port for synergism study; near-UV radiation can also be added.

The germanium-coated Kapton samples exposed previously in the AOBF were also exposed in the AODTS system, along with additional samples of each coating configuration, to a total fluence of 3.5 x 10^{22} atoms/cm^2 along with 740 equivalent sun hours of vacuum ultraviolet radiation.

Figure 2 Atomic Oxygen Drift Tube System (AODTS)

3. TEST RESULTS

Raw data from the surface resistance, solar absorptance, and emittance measurements may be found in Table 1. Visual observations showed no cracking, pitting, or other sign of damage on any exposed sample.

Mass was measured at MSFC using a Mettler AT261 balance. Because Kapton is hygroscopic, the mass of the sample may vary with humidity. To cancel out the effects of humidity, samples were placed individually in a vacuum chamber and evacuated to 10^{-3} torr to remove the water. Mass was measured at one-minute intervals after removal from the vacuum chamber, and the zero-time "dry" mass was calculated using regression analysis. No significant change in mass was noted for either AOBF or AODTS samples.

Surface resistance was measured at Lockheed Martin using a General Radio 1863 Megohmeter. Only slight changes were noted; the material is still within specifications for use as an antenna sunshield membrane.

Solar absorptance (α_s) was calculated from reflectance measurements (ρ_s) made with calibrated standards in accordance with ASTM E903 at a near-normal (8 degrees) angle of incidence. At Lockheed Martin, a Perkin Elmer Lambda 9 UV/VIS/NIR spectrophotometer with Labsphere DRTA-9A Integrating Sphere Attachment was used over a wavelength range of 225 nm to 2220 nm. The most significant change was the 4000 Å germanium on Kapton HN exposed in the AOBF and AODTS, with solar absorptance increasing from 0.461 to 0.479. Solar absorptance measurements were repeated at MSFC using an AZ Technology Laboratory Portable Spectroreflectometer over a wavelength range of 250 to 2500 nm. All changes in solar absorptance as measured by MSFC were less than 0.01.

Lockheed Martin calculated normal emittance (ε_n) from spectral reflectance measurements (ρ_{IR}) made at one-micron intervals over a wavelength range of 2.5 to 24.5 microns. A Gier Dunkle PR310 Paraboloid Reflectometer with Perkin Elmer 210 Grating Monochromator Attachment was employed using a calibrated gold standard. Since the natural Kapton samples were not opaque in the infrared region, transmittance measurements (τ_{IR}) were made using a Perkin Elmer 580B Infrared Spectrophotometer. Normal emittance was then calculated by subtracting reflectance and transmittance from unity for each wavelength interval, and integrating using blackbody weighting factors. MSFC measured infrared emittance using an AZ Technology TEMP 2000 Portable Infrared Reflectometer, using calibrated gold and black standards. Infrared transmittance measurements were taken by a Nicolet Magna-IR 750 Spectrometer. The transmittance for each wavelength interval was calculated using the same integrating blackbody weighting factors as Lockheed Martin, then integrated for the entire wavelength range as measured by the TEMP infrared reflectometer. Normal emittance was then calculated by subtracting reflectance and transmittance from unity.

Following the atomic oxygen exposure, ESCA analysis was performed on the samples to determine the presence and thickness of germanium oxide. Control samples were also measured. Standard ESCA analysis revealed the presence of oxidized germanium, as expected, plus a thin carbon overlayer. Small concentrations of fluorine, nitrogen, and silicon were present on the surfaces of some samples. The binding energy of the GeO_2 and metallic germanium was determined to be 32.7-33.1 eV and 28-29 eV, respectively.

Angular resolved ESCA analysis was also performed. For the samples exposed in the AOBF, additional peaks of tantalum oxide (Ta_2O_3 and Ta_2O_5) were detected at 26.3 - 28.9 eV. The atomic concentration of tantalum ranges from 1.23% for the 600 Å germanium sample to 1.76% for the 2000 Å germanium sample. It appears that a thin layer of tantalum oxide is on top of the oxidized germanium. The electrically biased neutralizer plate used in the AOBF during these tests is tantalum. The tantalum oxide and germanium oxide layers are thick enough that metallic germanium was not detected during normal ESCA analysis.

The thicknesses of the oxide layers were determined by angular resolved ESCA for the control and AODTS samples and by sputter depth profiling for the AOBF samples. The angular resolved ESCA procedure consisted of recording peak data at six different angles then fitting the data to a non-linear least squares procedure. From this fit, the ratio of the oxide thickness to the electron attenuation length was determined. Attenuation length was derived from the Tanuna, Powell, and Penn equation for the electron inelastic mean free path, and the oxide thickness was calculated (Ref. 2, 3). The results are in Table 2. Control sample oxide thicknesses were generally 10 Å, AODTS sample oxide thicknesses were 20 Å, and AOBF sample oxide thicknesses were 40 Å.

4. DISCUSSION

Both exposures represented accelerated conditions, especially the second exposure in the AODTS. For materials such as Kapton which erode upon exposure to atomic oxygen, testing can be accelerated to a fairly large extent with confidence. Alternatively, materials such as germanium (as well as silver and other materials) do not erode, but instead form an oxide, for which the rate of reaction is controlled by diffusion. In this case, the rate at which testing can be accelerated may be limited since oxygen atoms will not remain adsorbed on a surface which is already saturated with oxygen atoms.

It is noteworthy to compare the results of this test with previous investigations with germanium. A bulk germanium sample was flown by Peters et al. (Ref. 5) on STS-8 where it was exposed to a fluence of 3.5 x 10^{20} atoms/cm^2. The germanium specimen was slightly discolored following exposure, and ellipsometric measurements indicated a 159-nm thick film had formed with a refractive index n = 1.53. (Germanium has a refractive index of about 4.0.)

Sputtered thin films of germanium on one-inch-diameter fused silica optical flats were exposed to a total fluence of 9 x 10^{21} atoms/cm^2 on the Long Duration Exposure Facility (LDEF). The thickness of the films were 22 and 67 nm. One half of each specimen was masked during flight to serve as a control. Following exposure (Ref. 6), the exposed portion of the germanium films possessed much higher transmittance than the unexposed portions. Stylus profilometry and x-ray diffraction measurements indicated the entire 22-nm film had been oxidized to GeO_2. Although x-ray diffraction was not performed on the thicker film specimen, transmittance measurements and stylus profilometry measurements indicated that about 50 nm of the 67-nm original germanium thickness was converted to GeO_2.

Germanium/Kapton specimens were also flown on the heated trays and passive tray of the JPL EOIM-3 experiments on the STS-46 Shuttle flight (Ref. 7) where they were exposed to a total fluence of 2 x 10^{20} atoms/cm^2. Germanium thickness was 1500 Å while the Kapton thickness was 2 mils. Thermal property results indicated insignificant changes in both solar absorptance and emittance. ESCA measurements showed GeO_x thickness ranging from 20 to 60 Å.

The thickness of oxide formed in the current test is similar to that reported for the STS-46 specimens although the total atomic oxygen fluence was only 2 x 10^{20} atoms/cm^2 for STS-46. Alternatively, the thickness of oxide formed on the thicker germanium coating sample on LDEF is considerably higher. It is not clear as to why the disparity is so great. One explanation is that the reaction of germanium is diffusion-limited, and that the exposure times of the accelerated tests (days, weeks) were not sufficient in relation to the LDEF exposure time (years) to permit diffusion to take place. Similar differences between accelerated and long-term testing (i.e. LDEF) have been shown for FEP Teflon.

Comparison of the current test results to that of the STS-8 bulk germanium sample is even more perplexing. Peters et al. (Ref. 5) report the formation of a film that was 159 nm thick with an index much lower than that of germanium. It is not clear whether this film was shown to be entirely germanium oxide; the investigator believes that silicon oxide, possibly due to contamination, may be present as well.

5. CONCLUSIONS

In summary, a comparison of the solar absorptance (α_s), normal emittance (ε_N), and surface resistance in the table for before and after exposure indicates no significant reaction with atomic oxygen. ESCA analysis shows formation of a thin film of GeO_2 due to exposure. Slight increases in emittance (measured at both Lockheed Martin and MSFC) as well as small increases in surface resistance are consistent with this oxidation.

Direct comparisons of current test results with prior flight data is complicated by variations in sample fabrication (sputtered films vs. bulk specimens, etc.) as well as measurement techniques. For comparison of accelerated laboratory and flight conditions, 1" dia. samples of the 600 Å germanium on black Kapton and the 4000 Å germanium on Kapton HN were submitted as guest samples on the POSA-I experiment. These Kapton samples are from the same batch as the AOBF- and AODTS-exposed samples.

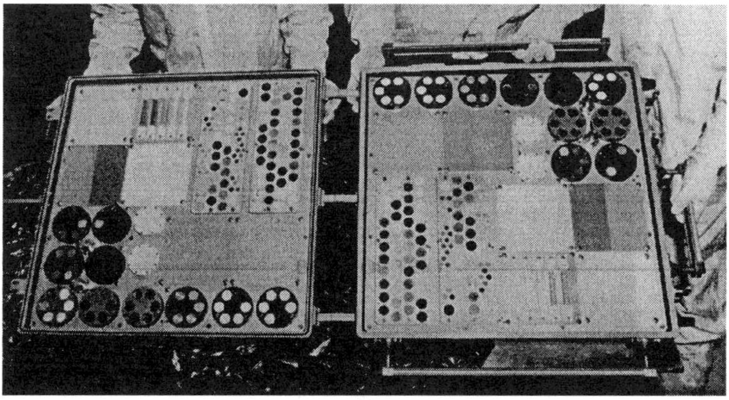

Figure 3 Passive Optical Sample Assembly

POSA-I (Passive Optical Sample Assembly), a suitcase-size collection of optics, thermal control coatings, and other candidate spacecraft materials (Fig. 3), is part of the Mir Environmental Effects Payload. It was attached to the docking module of the MIR space station in March 1996 during the STS-76 mission. One side of the experiment faces MIR; the other side, with a duplicate set of samples, faces space. POSA-I will remain in space for approximately 17 months, orbiting at 51.6° inclination and an altitude of 300 to 400 km. The minimum expected atomic oxygen fluence is 2 x 10^{20} atoms/cm^2; however, the fluence could be as high as 5.6 x 10^{21} atoms/cm^2, depending on MIR's altitude and attitude during POSA-I's exposure. This flight exposure should be comparable to the AOBF study, and similar post-exposure characterizations of optical properties, surface resistance, and ESCA are planned.

6. ACKNOWLEDGMENTS

The authors gratefully acknowledge the contributions of the following: Dr. Carl Miller, Raytheon Company, Research Division, Lexington, MA, for ESCA analysis and interpretation performed under contract NAS8-39924; Jason Vaughn and Rachel Kamenetzky, MSFC, for assistance with the atomic oxygen exposures; Dr. Palmer Peters, also of MSFC, for sharing his previous experiences with germanium exposed to atomic oxygen; and Joseph Antinovitch of Lockheed Martin for sample characterization.

7. REFERENCES

1. R.R. Kamenetzky, J.A. Vaughn, M.M. Finckenor, and R.C. Linton, "Evaluation of Thermal Control Coatings and Polymeric Materials Exposed to Ground Simulated Atomic Oxygen and Vacuum Ultraviolet Radiation", NASA Technical Paper 3595 (1995).
2. Dr. Carl Miller, Raytheon memo, Feb. 16, 1996.
3. Julia E. Fulghum, "Determination of Overlayer Thickness by Angle-resolved XPS: a Comparison of Algorithms," Surface and Interface Analysis, 20, 161-173 (1993).
4. Dr. Carl Miller, Raytheon memo, June 18, 1996.
5. P.N. Peters, J.C. Gregory, and J. T. Swann, "Effects on Optical Systems from Interactions with Oxygen Atoms in Low Earth Orbits," Applied Optics, 25, 1290 (1986).
6. John C.Gregory, et al., "Surface Processing of Semiconductor Materials with Fast Atomic Oxygen," AIP Conference Proceedings, NASA Conference on Centers for Commercialization and Development of Space, Albuquerque, NM (1995).
7. Edward M. Silverman, "Space Environmental Effects on Spacecraft: LEO Materials Selection Guide," NASA Contractor Report 4661, Part 1, p. 8-16 (1995).

8. ABBREVIATIONS

AO	Atomic oxygen
AOBF	Atomic Oxygen Beam Facility
AODTS	Atomic Oxygen Drift Tube System
EOIM	Evaluation of Oxygen Interaction with Materials
EOS	Earth Observing System
ESCA	Electron spectroscopy for chemical analysis
ESD	Electrostatic discharge
ESH	Equivalent sun hours
LEO	Low Earth orbit
LDEF	Long Duration Exposure Facility
LPSR	Laboratory Portable Spectroreflectometer
MSFC	Marshall Space Flight Center
NUV	Near ultraviolet
POSA	Passive Optical Sample Assembly
PPPL	Princeton Plasma Physics Laboratory
RF	Radio frequency
STS	Space Transportation System
UHF	Ultra-high frequency
VUV	Vacuum ultraviolet

Table 1. Atomic Oxygen Exposure Results

	Before Atomic Oxygen Exposure							After Atomic Oxygen Exposure						
	ρ_s	τ_s	α_s	ρ_{IR}	τ_{IR}	ε_N	Surface Resistance (Ω/□)	ρ_s	τ_s	α_s	ρ_{IR}	τ_{IR}	ε_N	Surface Resistance (Ω/□)
600 Å Ge (AOBF + AODTS)	0.551	0.000	0.449	0.115	0.000	0.884	1.5E+08	0.543	0.000	0.457	0.099	0.000	0.900	1.8E+08
MSFC			0.445	0.157		0.843				0.449	0.148		0.852	
600 Å Ge (AODTS only)	0.551	0.000	0.449	0.115	0.000	0.884	1.5E+08	0.548	0.000	0.452	0.099	0.000	0.900	1.5E+08
MSFC			0.443	0.156		0.844				0.439	0.149		0.851	
2000 Å Ge (AOBF + AODTS)	0.460	0.115	0.425	0.270	A	A	2.0E+07	0.439	0.119	0.442	0.265	0.283	0.452	3.5E+07
MSFC			0.453			0.421				0.454			0.417	
2000 Å Ge (AODTS only)	0.460	0.115	0.425	0.270	A	A	2.0E+07	0.467	0.119	0.414	0.266	0.283	0.451	3.0E+07
MSFC			0.451			0.421				0.443			0.413	
4000 Å Ge (AOBF + AODTS)	0.467	0.072	0.461	0.494	A	A	5.5E+06	0.445	0.076	0.479	0.485	0.175	0.340	7.0E+06
MSFC			0.470			0.300				0.476			0.309	
4000 Å Ge (AODTS only)	0.467	0.072	0.461	0.494	A	A	5.5E+06	0.459	0.076	0.465	0.488	0.175	0.338	8.0E+06
MSFC			0.472			0.300				0.471			0.308	

Notes:

A) An error was made while running the 100% baseline for the IR transmittance measurements (i.e. the needle was "pegged" over a significant wavelength interval). Therefore, the transmittance (τ_{IR}) and subsequent ε_N calculations were in error. However, a transmittance scan performed after AO exposure using the same settings as before AO exposure (i.e. "pegged" pen for the 100% baseline) indicated no change in τ_{IR}. Since IR reflectance (ρ_{IR}) changes were not significant, it can be safely assumed that ε_N did not change significantly. The ε_N values shown in the table above for the "after" condition were measured with a corrected 100% baseline.

B) All measurements made at Lockheed Martin except for those rows labeled "MSFC".

C) Samples labeled "600 Å Ge" were on 1-mil black Kapton (Type 100CB); other samples were on 2-mil natural Kapton (Type HN).

Sample	Atomic Concentration			Angular Resolved/Depth Profiling ESCA Results		
	Ge 2p3 %	O 1s %	C 1s %	Oxide Thickness/ Attenuation Length	Oxide Thickness for Tetragonal GeO_2 (Å)	Oxide Thickness for Hexagonal GeO_2 (Å)
Control 600 Å Ge	37.08	44.32	16.88	0.262	5.5	6.2
Control 2000 Å Ge	30.54	45.99	19.29	0.351	7.4	8.4
Control 4000 Å Ge	33.67	43.23	22.76	0.301	6.3	7.2
AODTS 600 Å Ge	25.86	53.00	20.33	0.783	16.4	18.6
AODTS 2000 Å Ge	17.94	48.66	30.73	0.898	18.9	21.4
AODTS 4000 Å Ge	20.96	55.99	21.51	0.900	18.9	21.4
AOBF 600 Å Ge	6.68	53.47	37.31		29 (Depth profile)	29 (Depth profile)
AOBF 2000 Å Ge	17.20	56.87	17.87		41 (Depth profile)	41 (Depth profile)
AOBF 4000 Å Ge	17.22	71.79	7.52		38 (Depth profile)	38 (Depth profile)

Table 2. ESCA Results

Data from Refs. 2 and 4.

28th International SAMPE Technical Conference
November 4-7, 1996

SURFACE ANALYSIS OF MOLECULAR CONTAMINATION ON SELECTED LONG DURATION EXPOSURE FACILITY TRAY SURFACES

H.G. Pippin and V.A. Loebs
Boeing Defense & Space Group
P.O. Box 3999
Seattle, Wa 98124-2499
(206)-773-2846

ABSTRACT

A detailed surface analysis using ESCA, sputter depth profiling, and imaging techniques has been carried out for selected areas of trays A4 (trailing edge), C6 (side), and E10 (leading edge) from the Long Duration Exposure Facility (LDEF). Surfaces near vent openings in the thermal control blankets on each of these trays were examined for silicon content.

The trays selected were part of the Ultra-heavy Cosmic Ray Experiment. The thermal control blankets for this experiment were fastened to the experiment using a Velcro material, which was bonded to the undersides of the blankets and to a lattice structure in the tray using DC6-1104 silicone based adhesive.

This adhesive and the silicone gaskets used for individual tray covers pre- and post-flight are the sources of silicon contamination on the tray walls. The experiment geometry, environmental exposures as a function of position, materials properties, and flight profile are known. The results of the surface analysis and significance for understanding spacecraft contamination will be presented. Results will be correlated with on-orbit conditions.

1. INTRODUCTION

This paper describes results to date from an examination of selected experiment tray surfaces from the Long Duration Exposure Facility (LDEF) for <u>silicone based</u> molecular contamination. This work is being carried out under NASA Contract NAS8-40581, "Comparison of Spacecraft Contamination Models with Well-Defined Flight Experiment."

The results of these analyses will be used to compare with the predictions of two different spacecraft molecular contamination models. The detailed geometry of the hardware is known. Essential properties of the anodized aluminum, velcro™, silverized teflon™, tray cover silicone gaskets, and DC6-1104™ silicone adhesive used to bond the velcro fasteners to the tray frame and thermal control blanket have been determined.

Selection of areas for analysis from trays A4, C6, and E10 has been completed, and surface analysis at many locations on the selected tray walls is completed.

2. FLIGHT

The Long Duration Exposure Facility flew in a fixed orientation with respect to the direction of motion for 69 months, at 28.5 degrees, in low Earth orbit. The LDEF was a 12-sided structure, with trays along a given row labeled A-F, with tray A at the Earth end, and rows numbered from 1-12. Row 9 was closest to the velocity vector. The fixed orientation over time allows well defined environmental exposure conditions. Trays from the Ultra Heavy Cosmic Ray experiment (UHCRE) at LDEF locations A4(trailing edge, UV exposure), E10(leading edge, UV and atomic oxygen exposures) and C6(side, UV and small amounts of atomic oxygen) were selected for examination. The tray geometry's are well-defined and virtually identical, the silicone sources are well characterized and defined, and the anodized aluminum surface contains only aluminum oxide as a background.

3. GEOMETRY OF THE EXPERIMENT

Each analysis concerns essentially three surface regions, referred to as the tray lip, which is the surface around the edge of the trays, the interior tray walls, and the curved region between the lip and walls. Along the sides of the tray the angle between lip and wall is about 75 degrees. At the Earth and space ends of the tray, the angle between lip and wall is 90 degrees. The UHCRE trays were covered by a silverized teflon™ blanket. The blanket was held in place by pairs of velcro™ strips, one strip bonded to the underside of the blanket and one strip bonded to an aluminum lattice framework on the top of three steel canisters, which contain the actual experiment. Figure 1 is a drawing showing the location of the velcro™ fastener pairs and associated silicone adhesive relative to the keyhole shaped vents around the edge of the silverized teflon™ thermal control blankets which covered each tray of the UHCRE. At the center vent locations along each of the long sides, and at the end locations (along the shorter sides of each tray) the velcro™ fasteners are positioned so as to allow line-of-sight outgassing normal to the tray wall. At the other vent locations on the long side of the trays, the line of sight from silicone adhesive source through the vent slot to the tray wall is not normal to the tray wall. However the discolored areas on C6 and E10 walls are essentially directly across from the vent slots indicating that the silicone molecules took multiple bounces before exiting at the vent and condensing on the tray wall.

Figure 2 is from a European Space Agency drawing showing details of the experiment assembly. In particular, detail A shows how the silverized teflon™ blanket contacts the tray side. This detail shows that the tray wall below the edge of the Ag/FEP blanket is exposed to outgassing products from silicone adhesive patches holding the velcro™ strips, but not to atomic oxygen or solar UV, except at vents cut into the blanket. Below certain vent locations on trays E10 and C6, a plume of faint discoloration can be seen extending down the tray wall several centimeters below the end of the Ag/FEP blanket. Discolored areas appear only where contaminated surfaces have been exposed to significant amounts of both atomic oxygen and solar UV.

The numbering scheme for the vent locations around a tray is shown in figure 3. An x,y coordinate system is defined for each tray vent location so that the distribution of silicon with respect to the vent and rivets is known.

4. CONTAMINATION SOURCES

Silicone based contamination films are of primary interest. Organic based contamination is examined only as it may mask or alter the apparent amount of silicon present. In particular, the Z306 black thermal control paint, used on the UHCRE stainless steel cyclinders within the tray, is polyurethane based. This material outgasses carbon based products, but contains no silicones.

The potential silicon contamination sources for the surfaces being examined under this contract include pre-flight outgassing from the tray cover gaskets, outgassing from the Space Shuttle during deployment and retrieval, outgassing from other LDEF experiments, outgassing from the DC6-1104™ adhesive used on the UHCRE, and post-flight outgassing from new tray cover gaskets. The tray side walls near vent locations were chosen for examination because this gives the best opportunity to minimize all sources of contamination except the DC6-1104™.

The geometry around the vent locations at the corners of the trays allows for a large potential contribution from the tray cover gaskets. These specific locations are therefore being excluded from the current study. Previous examination of LDEF stainless steel bolt heads(ref. 2) indicates some silicon based material is outgassed from the Space Shuttle. However, even near the top of tray interior walls the view factor to the Space Shuttle was very small. Thermal control blankets physically blocked any potential Space Shuttle sources from over 90% of the tray wall. Previous studies on LDEF hardware do not indicate tray-to-tray molecular cross-contamination was significant. Significant contamination sources are the DC6-1104™ adhesive and the silicone gaskets for the tray covers. The outgassed gasket material may extend down onto the curved portion of the tray lip.

5. OUTGASSING

The distributed nature of the adhesive locations within the individual trays suggests the outgassed product is likely deposited around the entire tray interior. The 12 inch depth of the trays, the metal lattice supporting the thermal control blanket, and the large stainless steel cylinders served to partially block exposure to the lower portions of the walls. However, visual plumes, created by the interaction of the environment with contamination on the tray walls, extend to near the bottom of the tray walls, and also suggest that the contamination layer is well distributed.

The gasket material used for the tray covers was not space grade material and a silicon based film was left where this material contacted the tray lips. This material also had line-of-sight to the tray interior and exterior walls at the corner of each tray and left an outgassed layer which was darkened by the on-orbit environmental exposure, particularly on the exterior surfaces where the exposure was most severe.

Average values of outgassing properties of the DC6-1104™ adhesive and velcro used to hold silverized teflon blankets to tray frames of LDEF experiment AO138 are reported in NASA CR 4646, "Evaluation of Adhesive Materials Used on the Long Duration Exposure Facility" (ref. 3).

6. EXPOSURES

The row 4 atomic oxygen(AO) exposure is essentially zero and the solar UV exposure is 10500 Equivalent Sun Hours(ESH). For row 6 the exposure is 4.94E+19 atoms/cm^2 AO and 6400 ESH solar UV radiation. The LDEF row 10 baseline exposure is 8.43E+21atoms/cm^2 AO and 10700 ESH solar UV radiation. The trays are subject to mild thermal cycling each orbit. The specific exposure levels are a very strong function of location on each surface. For a given tray, the side wall primary surfaces are about 75 degrees, and the tray end walls about 90 degrees, to the plane of the tray surface. Detailed model calculations have been made for sections of each tray where the surface analysis is being carried out. The rivets, vent slots, and underlying framework visible through the vent slots are included in the geometry modeled for the exact exposure predictions.

7. TRAY A4

Figure 4 is a NASA post-flight photograph of tray A4. Position 9 is at the top left of the photo and is marked by the small arrow. No visible discoloration was seen on the interior walls near any of the vent areas on this tray.

Figure 5 shows the surface silicon percent around tray area A4-9. The vent position was roughly centered between the two rivets. One group of measurements was conducted on the surface directly across from the large circular portion of the vent. Background measurements were taken at surface locations 4-5 inches below, and well to the right of the vent at this location.

To make ESCA measurements on the interior walls of the tray, small pieces about 1-1.5 cm by 2 cm were cut from the roughly 25 cm by 30 cm sections. The small pieces were cut by hand with a jewelers saw to minimize the amount of material destroyed in the process. An area approximately 1mm wide is destroyed by the cutting.

Each ESCA analysis area is a 0.6 mm diameter circle centered on a location defined by its (x,y) coordinate pair. The silicon percent generally increases toward the top of the tray wall. The measurements around the rivet on the right show some shadowing due to this structure. The 25+ percent silicon measurements at the top are from locations on the tray lip surface (~90 degrees from the tray wall). Two pairs of measurements were also made about 60 and 30 degrees from the tray wall. Background silicon levels measured at locations on the lower tray wall several inches from the vent are about 3-3.5%.

Repeated cycles of sputtering to remove material followed by ESCA measurements build up a profile of the composition of the material as a function of depth. The silicon containing films on tray A4 appear to be quite thin, surface measurements show 10-15% aluminum and the silicon % drops rapidly within 5-10 nm of the surface at most locations not directly aligned with the vent. Locations near the A4-9 vent which have been depth profiled show thicker silicon containing films relative to areas further from the vent. Light sputter locations are immediately adjacent to areas which have previously been sputtered. These "adjacent" areas were exposed to the edge of the sputtering beam and some surface material is removed prior to any ESCA measurements. The effect is essentially to "clean" the surfaces by partially removing the layer of carbon based contamination found on any surface which has been exposed to ambient atmosphere.

The post-flight exposure to air allows deposition of carbon based material. This layer is extremely thin, as seen by examining the sputter depth profiles. For the A4 locations it is difficult to separate the contributions from the adhesive and the tray gasket because there was no conversion of silicone to silicate material due to atomic oxygen. Comparisons of tray lips and curved surface areas at, and away from, vents indicate that the gasket background probably doesn't extend far onto the lip. Measurements to confirm this are planned.

8. TRAY C6

Areas near vents on the tray C6 side closest to row 5, but with some direct exposure to atomic oxygen, show discoloration due to the contaminant layer. Surface analysis measurements have been carried out in the C6-2 vent area of the tray. The discolored region at C6-2 extends into the curved portion of the tray lip and is well centered between the side wall rivets.

Figure 6 shows the surface silicon percent for locations on C6-2 measured using ESCA.

For locations 1, 2, and 3 on tray C6 the potential exists to partially separate contributions from the tray gaskets and from the DC6-1104™ adhesive. Silicone from the gasket deposited pre-flight should have oxidized to silicate. Silicone deposited from post-flight gasket use must be identified by background measurements away from the vents. High resolution ESCA measurements can distinguish between silicone and silicate based material. The assumption of two peaks, approximately 1.3 eV apart, will be used to separate silica from silicone peaks.

Measurements on the tray lip at locations away from the vent should show silicate material from the pre-flight gasket exposure and silicone material from the post-flight gasket contact. On both E10 and C6 the remainder of the curved regions and tray lips do not show the same discoloration observed near the vents. This indicates much thinner silicon based deposition in these areas. It is therefore likely that the gasket contribution is relatively small in the areas near vents. At certain vent locations, the discoloration pattern is so distinct as to show the shape of the vent.

There are several assumptions which must be considered in this analysis. IF the outgassing occurred exclusively during the mission, then enough time was available to oxidize all the outgassed silicones from the adhesive to silicate material. If the outgassing was gradual then some silicone may remain from outgassing near the end of mission. Measurements of the silicon based film thickness on the tray lip and then around the curved portion onto the tray wall at a location some distance from any of the vents should also provide an indication of relative contributions. Similar considerations should hold true for all vent areas on tray E10.

9. TRAY E10

The deposition pattern from vent location E10-7 extended over a copper grounding strap and along the top edge of the wall. Figure 7 shows a post-flight close up of this area. A gap in the discoloration is seen in the area from where the copper strap has been removed. It is clear that the blanket material around the vent is skewed so as to provide a better line of sight to the right of the vent (toward the discolored area) than to the left, where little discoloration is seen.

Surface analysis measurements have been carried out in the E10-3 vent area of the tray. The central portion of the discolored area shows a very clear image of the shape of the vent hole in the blanket. The interior tray wall shows a distnct plume directly below the vent hole. Measurements are planned to examine the plume area and compare the surface deposition with adjacent, but not discolored, locations.

Figure 8 shows the surface silicon percent obtained by ESCA measurements for locations around the E10-3 vent.

10. SUMMARY AND CONCLUSIONS

Linear fits to the data for both A4-9 and C6-2 at locations on the tray wall between the two rivets(no shielding) closest to the vent opening show strong correlations with distance from the top of the wall (y-direction), but not with distance along the wall(x-direction). These preliminary findings are based on locations near the vent openings. The silicon sources were in a plane with the top of the tray side wall. Experiment hardware within the tray blocks the line of sight of some source areas from the lower portions of the tray walls

For the A4-9 locations which were sputter depth profiled, the higher on the tray wall, the "thicker" the silicon containing film. However, all the silicon containing films are extremely thin. By 30 nm depth the levels of silicon are quite low, and give the appearance of filling the voids in the anodize layer. Areas with over 30% silicon show no aluminum peaks in the ESCA spctra. As the silicon % drops, the % aluminum detected increases. This is as expected for very thin or patchy contaminant films.

No silicon source induced discoloration appears on the interior tray walls of A4 around the vent areas. The most discoloration observed on C6 and E10 tray walls are on the sides which receive the most direct atomic oxygen. The tray lip on the side of C6 which is closest to row 7 curves away from the ram oxygen. No discoloration is seen on this side of the tray. The contaminant film on the examined area of A4 is extremely thin.

The selected tray surfaces have proved to be valuable sources of information in our attempts to understand spacecraft contamination. The role of the environment in determining the final state of the deposited films is significant. Finally, the data being obtained under this contract should provide a well-documented, common basis for comparison of a variety of computer models used to predict spacecraft contamination.

The aouthors would like to thank the NASA Space Environments & Effects Program for the funding required to carry out the efforts described in this paper. This work was carried out under NASA contract NAS8-40581.

References

1. NASA CP 3134, "LDEF-69 Months in Space First Post-retrieval Symposium," Part 1, June, 1991.

2. NASA CR 4662, "Effects of Space Exposure on Metals Flown on the LDEF," June, 1995.

3. NASA CR 4646, "Evaluation of Adhesive Materials Used on the Long Duration Exposure Facility," March, 1995.

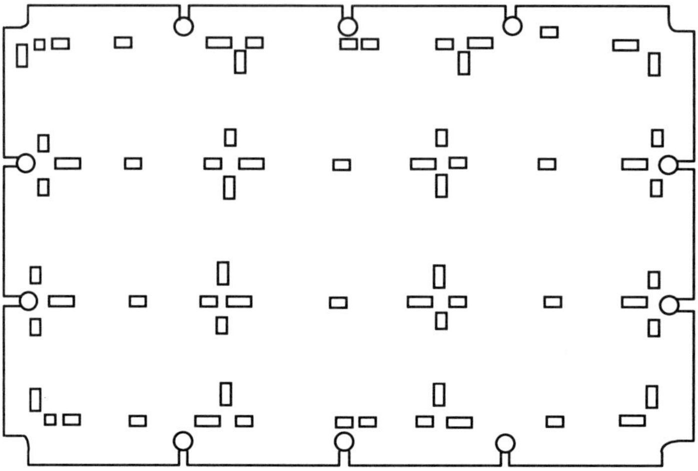

Figure 1. Drawing of LDEF thermal blanket assembly showing Velcro™ attachment locations and vent locations.

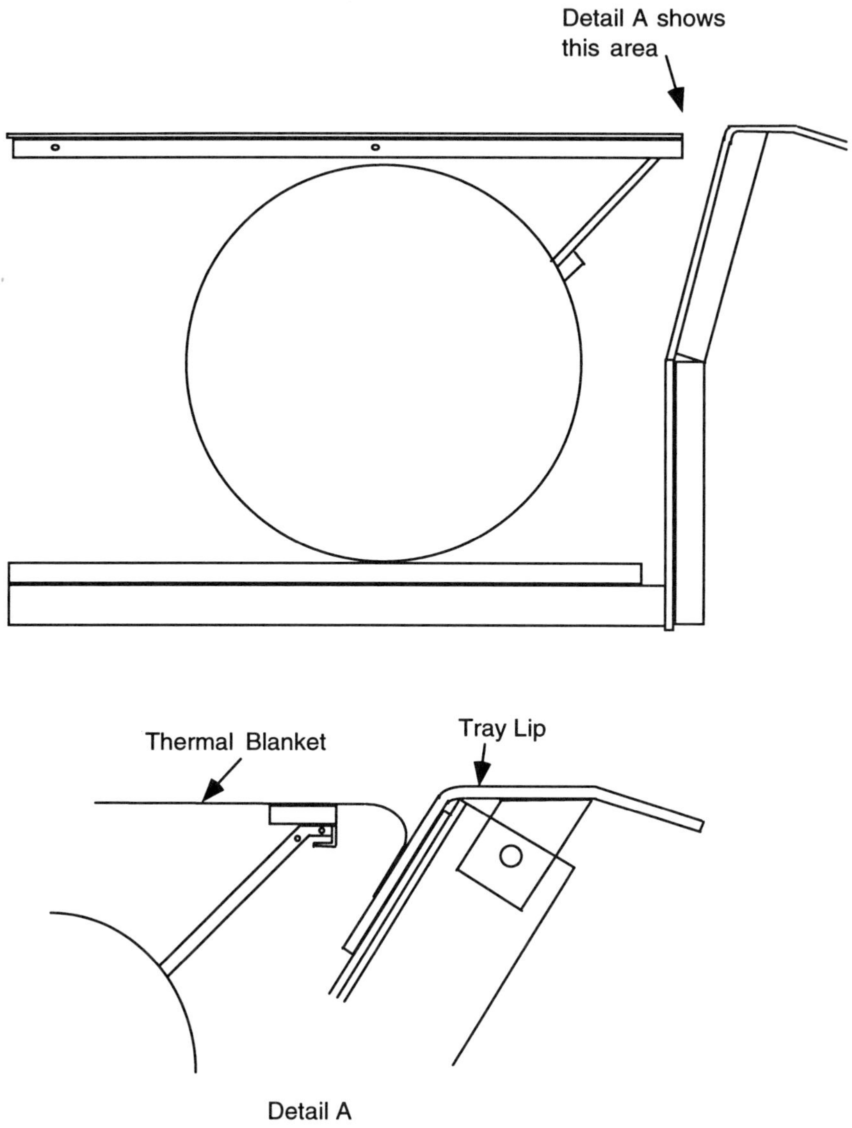

Figure 2. Cross-section drawing of details of the UHCRE trays showing orientation of blanket with respect to tray walls.

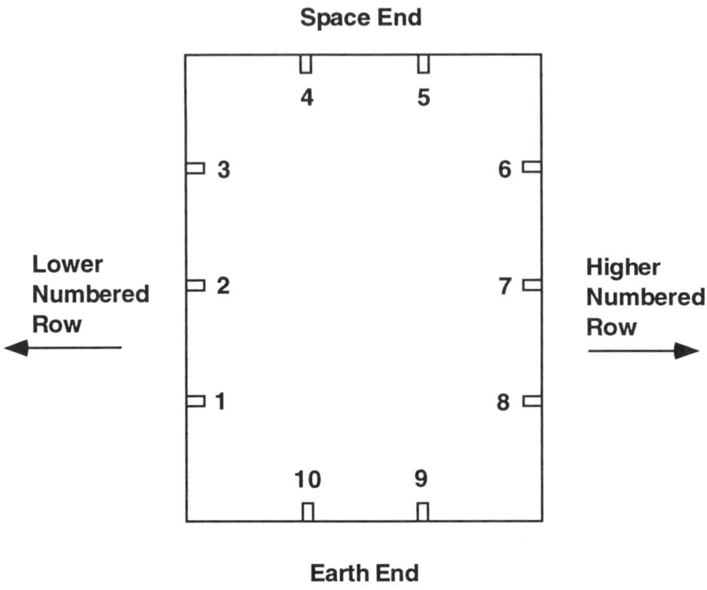

Figure 3. Vent location numbering system for LDEF trays.

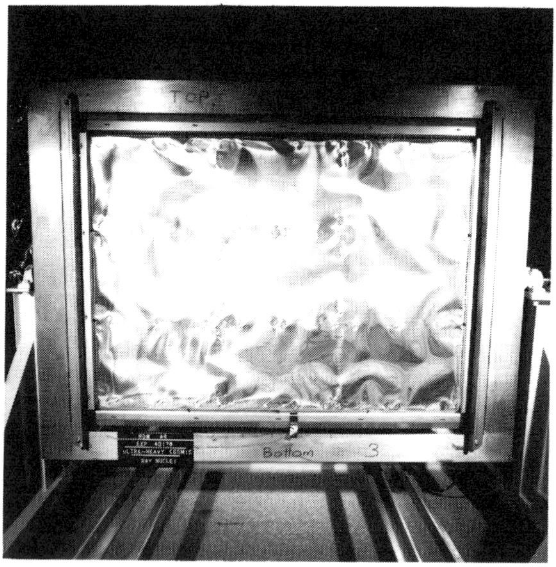

Figure 4. NASA post-flight photograph of tray A4.

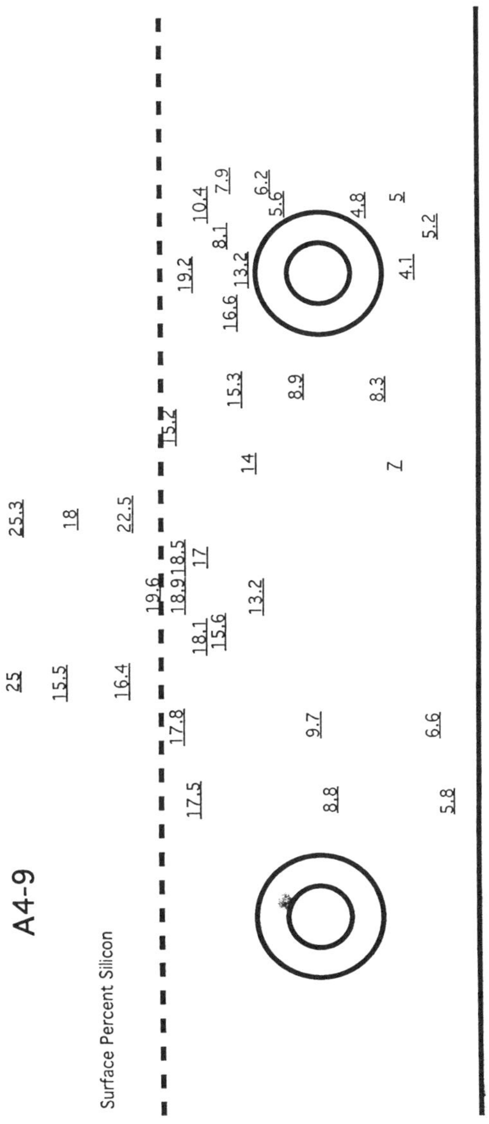

Figure 5. Map of surface silicon distribution from ESCA measurements on A4-9.

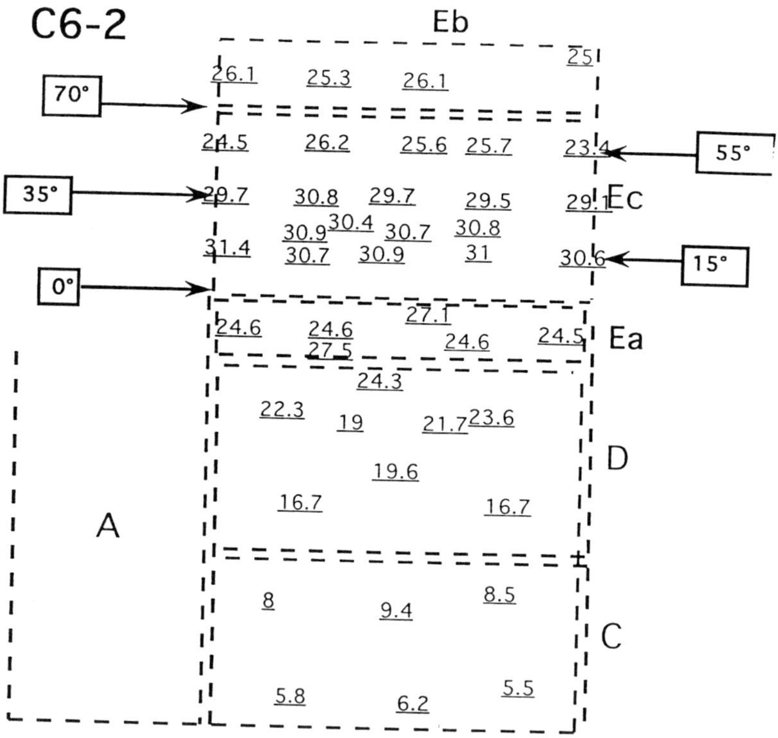

Figure 6. Map of surface silicon distribution from ESCA measurements on C6-2.

Figure 7. NASA post-flight photograph showing close-up of blanket vent at location E10-7.

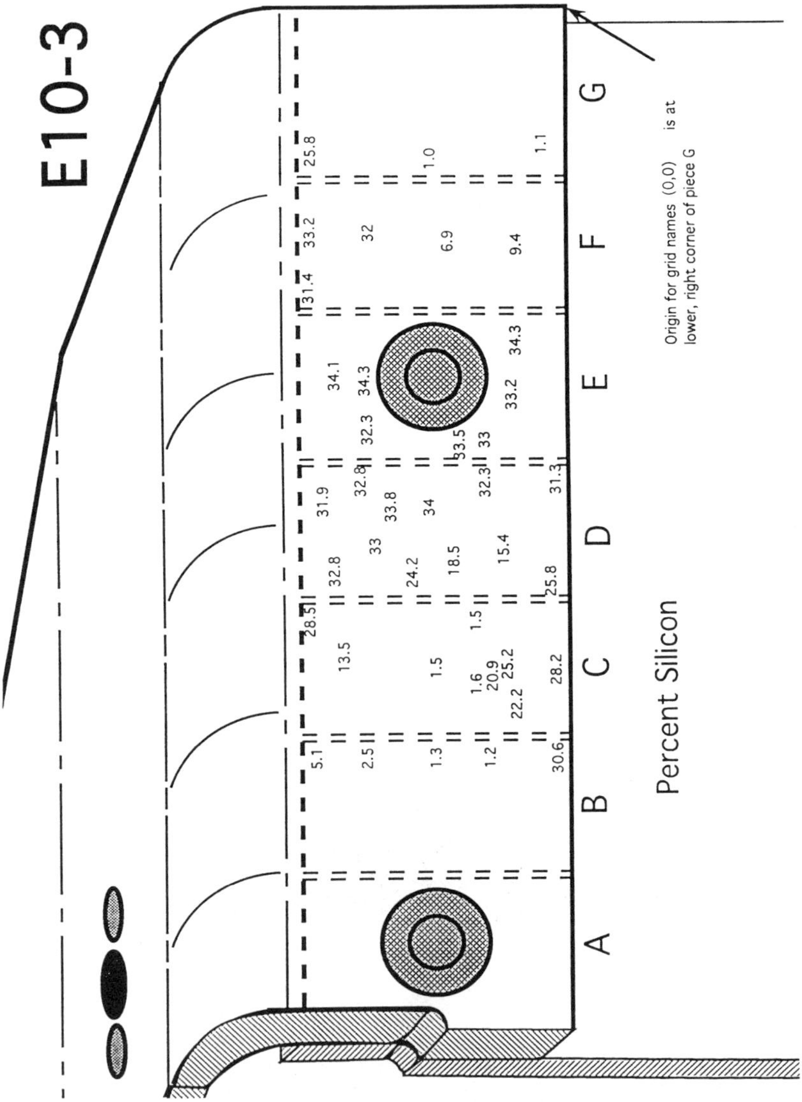

Figure 8. Map of surface silicon distribution at ESCA measurement locations on E10-3.

28th International SAMPE Technical Conference
November 4-7, 1996

NONDESTRUCTIVE INSPECTION OF ORGANIC FILMS ON SANDBLASTED METALS USING DIFFUSE REFLECTANCE INFRARED SPECTROSCOPY

G. Louis Powell[a], Richard L. Cox[b], Tye. E. Barber[c], and John. T. Neu[-]
[a]Oak Ridge Y-12 Plant[e], Oak Ridge, TN 37831-8096
[b]Oak Ridge National Laboratory, Oak Ridge, TN 37831
[c]Sam Houston State University, Huntsville, TX 77341-2117
[d]Surface Optics Corporation, San Diego, CA 92131

ABSTRACT

Diffuse reflectance infrared spectroscopy is a very useful tool for the determination of surface contamination and characterization of films in manufacturing applications. Spectral data from the surfaces of a host of practical materials may be obtained with sufficient insensitivity to characterize relatively thick films, such as paint, and the potential exists to detect very thin films, such as trace oil contamination on metals. The SOC 400 Surface Inspection Machine / InfraRed (SIMIR) has been developed as a nondestructive inspection tool to exploit this potential in practical situations. This SIIR is a complete and ruggidized Fourier transform infrared spectrometer with a very efficient and robust barrel ellipse diffuse reflectance optical collection system and operating software system. The SIMIR weighs less than 8 Kg, occupies less than 14 L volume, and may be manipulated into any orientation during operation. The surface to be inspected is placed at the focal point of the SIM/IR by manipulating the SIMIR or the surface. The SIMIR may or may not contact the surface being inspected. For flat or convex items, there are no size limits to items being inspected. For concave surfaces, the SIMIR geometry limits the surface to those having a radius of curvature greater than 0.2 m. For highly reflective metal surfaces, the SIMIR has a noise level approaching 1×10^{-4} absorbance units, which is sufficient for detecting nanometer thick organic film residues on metals. The use of this nondestructive inspection tool is demonstrated by the spatial mapping of organic stains on sand blasted metals in which organic stains such as silicone oils, mineral oils, and triglycerides are identified both qualitatively and quantitatively over the surface of the metal specimen.

KEY WORDS: Inspection, Surface Analysis, Cleaning, Thin Films, Infrared Spectroscopy

---

[e]Managed for the U. S. Department of Energy by Lockheed Martin Energy Systems, Inc. Under Contract No. DE-AC05-84OR21400.

1. INTRODUCTION

Alternative industrial cleaning processes are increasingly being implemented due to the environmental banning of halogenated cleaning solvents. The effectiveness of the new (or old) cleaning processes often requires that the level of cleaning be verified prior to the continuation of manufacturing processes such as painting, bonding, or surface plating. Many of these tests are time consuming, do not provide definitive results, and will not quantify levels of cleanliness to a high level of accuracy.

Diffuse reflectance mid-infrared spectroscopy is a very useful tool for the inspection of surfaces in manufacturing applications (1), particularly those related to the cleaning of metals. The nominal sensitivity of Fourier transform infrared spectroscopy (FTIRS) is optimum for analyzing organic films in the micrometer thickness range, and the better laboratory instruments have signal-to-noise >10,000 yielding detection limits approaching monolayer quantities of organic compounds. Furthermore, this spectral data is explicitly chemical in nature with similar sensitivities for most chemical moieties. Thus, FTIRS has both qualitative and quantitative analytical applications. In collaboration with Harrick Scientific Corporation, remote-sensing, barrel-ellipse diffuse- reflectance accessories have been developed for FTIRS and field demonstrated to be practical means of surface inspection for organic contamination on metals (2,3), lithium hydroxide on lithium hydride (4), and heat damage to graphite-epoxy laminates (5). The method is also applicable to identifying organic solids, and for characterizing textiles, paper, wood, coal, and adhesives (1,6). This collaboration ultimately resulted in laboratory instruments with sensitivities better than 0.0001 Absorbance units (a.u.) for light reflected from highly-reflective, sandblasted metals such as gold and aluminum, and a prototype portable Surface Inspector (based on the barrel ellipse diffuse reflectance collection optics and a MIDAC Illuminator spectrometer) having an order of magnitude less sensitivity.

These systems were used to demonstrate the feasibility of diffuse reflectance mid-infrared spectroscopy for surface inspection and the need for user-friendly, field hardened, and portable inspection machines with sensitivities matching the laboratory instruments. They were also used to demonstrate such fundamental concepts as surface inspection in the field and sample positioning technology for surface mapping. Recent collaboration between the Lockheed Martin Energy Systems, Inc., NASA/MSFC, the Army, and Surface Optics Corporation has extended these developments to the extent that a practical, user-friendly, highly-sensitive infrared surface inspection machine is in service at sites of these collaborators, and is available in the market place. The Surface Optics Corporation SOC 400 Surface Inspection Machine / InfraRed (SIMIR) (6) was designed to meet a NASA requirement that the sandblasted inner surface of solid rocket motor casings be free of both hydrocarbon grease and silicone oils at levels approaching 10 nm thick films (~1 mg ft^{-2} or ~10 mg m^{-2}).

This paper describes the application of hand-held (or remotely positioned) Fourier transform infrared spectrometer that uses diffuse reflectance optics to inspect sandblasted metal surfaces for contamination by common oils. The performance is demonstrated in terms of the signal-to-noise of the instrument (i.e., the sensitivity of the SIMIR in spectrometric units) when used to inspect highly reflective sandblasted surfaces. The use of the technology to map surfaces for oil contamination, with qualitative and quantitative detection of the oil is demonstrated using the Harrick/MIDAC Surface Inspector.

2. EXPERIMENTAL

2.1 Mid-Infrared Surface Inspection The key to the performance of a mid-infrared surface inspection machine is the integration of an intense source, very efficient collection of diffusely scattered light, and an infrared light detector having very high signal-to-noise assembled in a durable package that is easy to operate. Diffuse reflectance is the preferred mode for a general purpose surface inspection since most surfaces do not have the mirror-like qualities necessary for high specular reflectance yields. In the case of sandblasted metals, specular reflection is the dominant process, but it occurs from micromirrors randomly oriented on the surface of the metal. During this reflection, the light passes twice though residual films on the metal. The barrel ellipse optical configuration solves the problem of light collection efficiency by collecting ~70% of the diffusely scattered light from a typical FTIR spectrometer collimated beam focused on a perfect diffuse scatterer from a direction approximately normal to the surface. Coupling the barrel ellipse with a high-performance laboratory spectrometer, such as a BIO-RAD FTS-60, with a very intense (water-cooled) source allows the achievement of signal-to-noise levels better than 10,000. The achievement of a particular signal-to-noise level for a given collection time interval (typically one minute) is the best method for defining performance in a FTIR spectrometer. Indeed, the performance of the spectrometer, perceived by the operator as the rate at which a spectrum of a particular quality can be obtained, goes as the square of the signal-to-noise since multiple spectra are usually averaged to yield a final spectrum. The Harrick/MIDAC Surface Inspector (the coupling of a Harrick barrel ellipse to a MIDAC Illuminator) was developed to demonstrate the principles of portability and flexibility in surface inspection applications. It formed an excellent test bed for developing surface mapping techniques and demonstrating field applications, but its size and weight were too great to be easily fielded, and its signal-to-noise (<1000) limited its operational speed. In order to meet the NASA/MSFC requirement for certifying sandblasted metal surfaces clean of organic contamination at levels approaching 10 nm thick films (~1 mg ft^{-2}), a specification was issued based on the operational performance of the Harrick/MIDAC Inspector, but with the size and weight each reduced by a factor of 2 and with an order of magnitude increase in signal-to-noise. In response to this specification, Surface Optics Corporation developed the SOC 400 Surface Inspection Machine / InfraRed (SIMIR).

2.2 The SOC 400 Surface Inspection Machine / InfraRed (SIMIR) The SOC 400 Surface Inspection Machine / InfraRed (SIMIR) (6) shown in Figure 1 weighs less than 8 Kg, occupies less than 14 L, and may be operated at any orientation. The focal point of the barrel ellipse diffuse reflectance optics is located in the plane of an 12-mm D opening in the center of the face plate of the instrument. The barrel ellipse optical configuration (1) is located behind the black circular (85-mm D) face-plate shown in Figure 1. The SIMIR uses the MIDAC FTIR spectrometer, but the components have been reconfigured to reduce the size and weight of the system, the optical throughput has been increased, the barrel ellipse is half-scale, and the detector performance has been improved. The system is supported by either 12 VDC or 120 VAC and a computer (a 90 MHz Pentium and a 486 laptop were used interchangeably) using MIDAC GRAMS/386 software for the operating system. The SIMIR operates at <40 W, slightly less than the power requirements for the laptop computer. This "point-and-shoot" type of instrument allows rapid in-field analyses to be made with sensitivity comparable to laboratory FTIR spectrometers. Dedicated scanning and collection routines, and the robust characteristics of the barrel ellipse focal point with respect to sample positioning substantially enhance the utility of the SIMIR. When the capability for having the spectrometer software instruct positioning devices in a noncontact mode is utilized (3,6), dramatic increases in productivity and utility are achieved.

Figure 1. The SOC 400 Surface Inspection Machine/InfraRed (SIMIR).

The SOC 400 operates in a similar fashion to most laboratory FTIR spectrometers, a characteristic that makes its operation and data evaluation very familiar to most experimental chemists. The principal difference with respect to laboratory instruments is that aligning accessories and waiting for the spectrometer to purge no longer exist. The SIMIR and the specimen surface are brought into appropriate proximity, made easy by the contact face plate, and the spectrum is collected while observing the spectrum develop in nearly real time (1.3 s phase lag). Normally, the specimen may be located within a millimeter of the faceplate without loss in calibration or sensitivity. The operating system may also be customized using MIDAC/Grams386/Array Basic or MIDAC Visual Basic software for dedicated inspection operations. This is done to make the inspection operation user friendly, hardened against accidental loss of data and changes in operational parameters, to introduce dedicated post processing of data, and to instruct external devices by serial communications to reposition the specimen and execute other operations.

2.3 Mid-Infrared Surface Measurements An infrared spectrum is a measure of light absorption in one substrate relative to another, usually called the reference or background spectrum. With respect to certifying the cleanliness of metals, the inspector must know that the reference spectrum substrate is clean within the detection limits of the method. Contamination must not be transferred from one substrate to another, an event that is easily avoided by operating in a noncontact mode. In the process of cleaning metals, this is the technique intensive process of keeping a clean part clean. With respect to observing stains on metals, a fortuitous property of sandblasted metals (the archtypical surface for adhesive bonding to metals) is that applying a nonvolatile oil or powder to sandblasted metal substrates, followed by vigorous scrubbing of the substrate with a simple tissue to remove as much residue as possible, leaves an optimum residue (typically 1 μm) for producing a mid-infrared spectrum. Figure 2 compares SOC 400 SIMIR measurements of typical (mineral, vegetable, and silicone) oil stains on sandblasted gold, prepared as described above, to similar

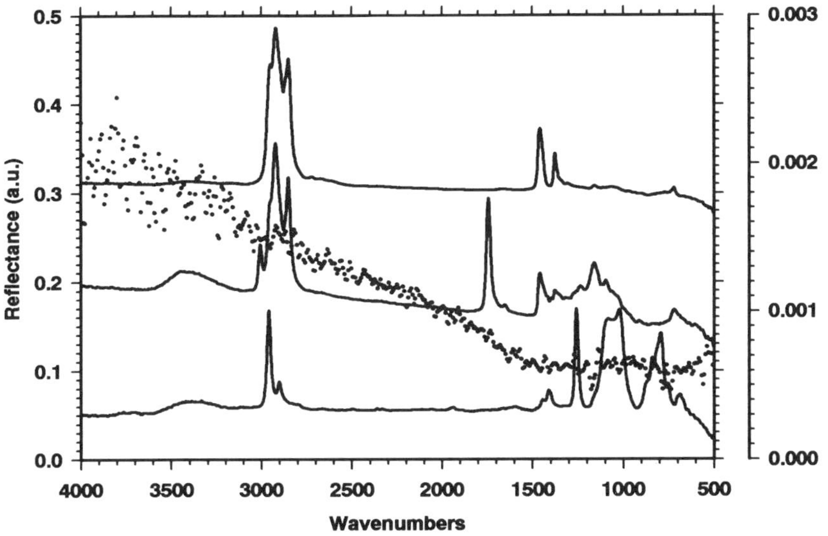

Figure 2. An infrared spectra obtained from stained sandblasted gold specimens using 16 cm^{-1} resolution, 2 times zero filling, and 48 coadded scans (1 min.) referenced to the same specimen background spectrum collected for 256 coadded scans.

Dotted curve: clean gold - right ordinate.
Upper curve: light mineral oil-ordinate offset +0.27 a.u.
Middle curve: safflower (vegetable cooking) oil-ordinate offset +0.14 a.u.
Lower curve: silicone oil (Dow-Corning DC200-5cpi).

measurements on clean gold. These measurements demonstrate the unique patterns for each type oil that allows for the qualitative distinction between types of oil. The amplitude of the spectral bands in Figure 2 are related to the film thickness with an appropriate calibration. A comparison of the stained specimens to the clean specimen indicates that spectral quality adequate for qualitative analysis exists at an 100X decrease in band amplitude and that the presence or absence of organic contamination can be detected at >500X decrease in band amplitude.

2.4 Substrate Materials Sandblasted gold is the optimum substrate material for metal cleanliness studies because gold is easily cleaned. Gold can be cleaned with very harsh chemical methods. Freshly sandblasted gold that was ultrasonically cleaned in Micro detergent, rinsed under a stream of distilled water, and dried in a stream of high-purity nitrogen was used in following experiments. Gold is one of the most reflective of metals and displays no features related to oxide films. Thus, sandblasted gold returns to the infrared detector a true representation of the spectral distribution of the incident beam. Sandblasted aluminum is also a good substrate, but the aluminum oxide film can significantly contribute to the spectral features. Highly anodized aluminum strongly absorbs over much of the mid infrared spectral range. Sandblasted stainless steel is also a good substrate in that it is very durable and cleanable. It does contribute some oxide film features to the spectra. Again, the emphasis is on this form of surface inspection as a comparative technique suggesting that the background spectrum for surface cleanliness approximate as closely as possible the clean substrate of the surface being analyzed.

2.5 Surface Mapping Surface mapping is a powerful tool for approaching the difficult task of establishing calibration factors for relating the amplitudes of the spectral bands (Figure 2) to the quantity of film that contaminates the substrate. Mapping is also important in inspection operations for determining the distribution of a stain over a surface. For spectral mapping operations, the MIDAC GRAMS/386 operating system would collect spectra at fixed time intervals. Immediately upon collecting a spectrum, the software would calculate the instructions to reposition the substrate according to a predetermined grid, and send that instruction by serial communication to a VELMEX NF90 controller that directed two VELMEX Unislides used as X-Y translators. The substrate was a flat, sandblasted specimen of gold, aluminum, or stainless steel approximately 100-mm by 100-mm located ~0.5 mm beneath the contact surface of the barrel ellipse sampling port mounted in a downward looking orientation. The X-Y translator moved the specimen horizontally in a noncontact mode. At the end of the experiment, all of the spectra are packaged as a single data file along with a second file containing the X-Y grid definition. An Array Basic program operates on the spectral data using the X-Y grid definition to determine the band amplitudes and to array these amplitudes along with the appropriate grid dimensions in a format that can be directly converted to a surface map using a spread sheet.

2.6 Quantitative Stain Application and Measurement Within 10%, a microliter syringe can deposit 0.85 μg of oil on a substrate. This drop rapidly spreads from its initial ~1 mm size to a stain that may be tens of millimeters in diameter after a period of days to weeks. Once the stain has stabilized at typically a 30-mm to 50-mm size, spectra are obtained at each point on an X-Y grid that includes the stain and clean regions outside the stain. The amplitude of a spectral band is measured relative to its baseline and used to produce a three dimensional image of the stain on the substrate. Dividing the numerical integral of these spectral amplitudes over the substrate surface into the weight of oil in the initial drop yields the calibration factor. This calibration factor, when multiplied by a spectral band amplitude measurement, will convert the spectral amplitude (usually in absorbance units, a.u.) to weight per unit area for that particular location. The reciprocal of this calibration is the sensitivity of the method.

3. RESULTS

3.1 Safflower Oil on Sandblasted Gold A number of experiments have been carried out to use the spreading of a microliter drop of mineral oil, safflower oil, and silicone oil on sandblasted gold, aluminum, and stainless steel substrates in an attempt to determine the relationship between calibration factors, a particular oil, and the substrate. The spread of a safflower oil drop on sandblasted gold will be described in detail to demonstrate the application of surface mapping based on reduced spectral data to the establishment of calibration factors. The stain was visible to the eye, spread in a circular geometry rapidly at first and more slowly later, and kept a distinct visible boundary at the edge of the stain. The aliphatic hydrogen stretching band at 2925 cm^{-1} was taken as the band to monitor since it represents a very general indication of organic contamination. The spectra were linearly baseline corrected between the average value over 3050 cm^{-1} and 3075 cm^{-1} and the average value over 2700 cm^{-1} and 2725 cm^{-1}, and the band amplitude was taken as the "T" average over a 25 cm^{-1} range centered near 2825 cm^{-1}. Figure 3 shows this baseline corrected portion

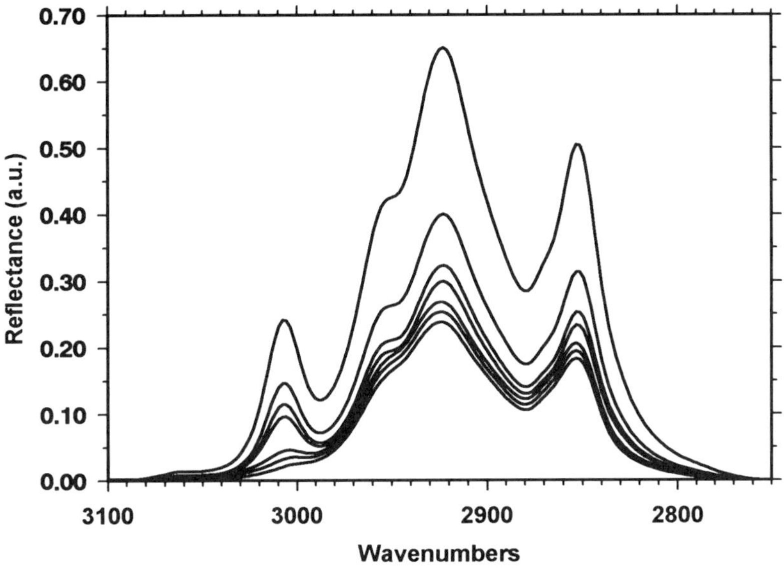

Figure 3. Baseline corrected aliphatic hydrocarbon bands from spectra monitored over time at the point were the microliter safflower oil drop was applied. Elapsed time top-to-bottom: 15 ks, 45 ks, 88 ks, 135 ks, 410 ks, 551 ks, 1450 ks.

of the safflower stain as a function of time after the drop was applied. The aliphatic hydrocarbon bands (those below 3000 cm^{-1}) decreased over the first day maintaining their relative amplitudes and decreased only slightly after the first day. The alkene band at 3011 cm^{-1}, that indicates the degree of unsaturation of the safflower oil, continued to decay and completely disappeared after 2 weeks. The disappearance of this band was the result of oxidative polymerization that resulted in the stain being very difficult to remove.

3.2 Contour Maps and Calibration Factors The spectra in Figure 3 were taken from a set of spectral maps obtained by collecting a spectrum every 50 s at 2 mm intervals along a line through the center of the safflower oil drop with the drop located near the middle of the line. Figure 4 shows a contour map over the stain diameter and over time constructed from the amplitude of the aliphatic hydrocarbon band as described above. During the first 10 ks, the aliphatic hydrocarbon band appeared to saturate near 1 a.u., after which this band decayed monotonically as the stain spread. After 138 ks, the stain had stabilized at a fixed diameter with circular symmetry. The edge of the stain, as defined by the contour map, approaches the spatial resolution of the focal point of the barrel ellipse. Once the spread of the stain had stabilized, a full X-Y map of the stain was obtained over a 50 by 50 spectrum grid with 1 mm

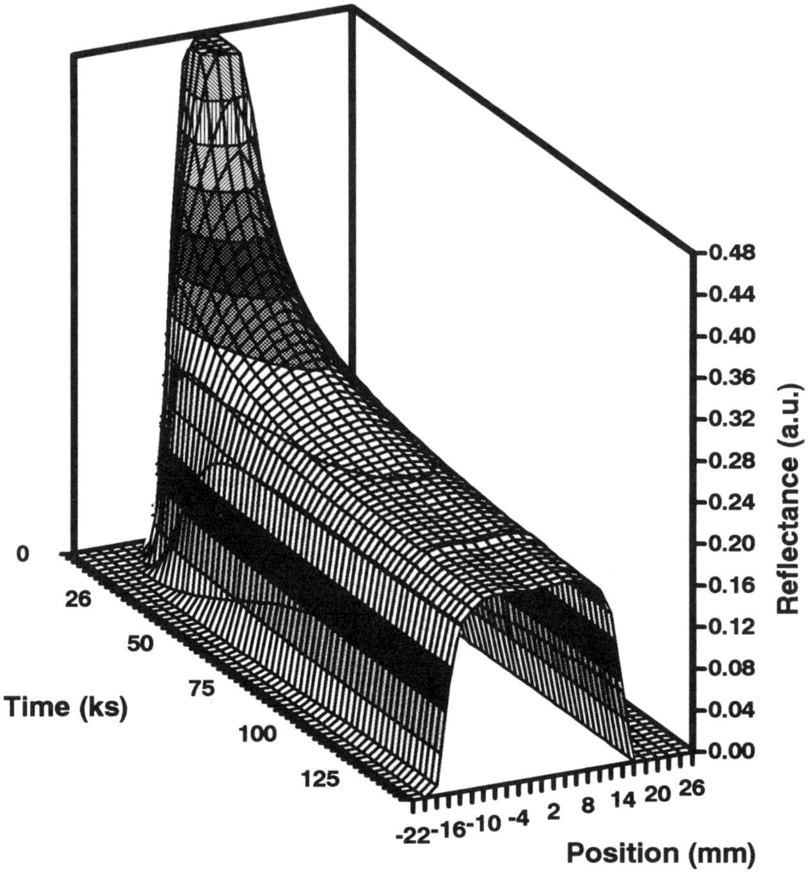

Figure 4. A contour map over time and the diameter of the spreading safflower oil stain based on the baseline corrected aliphatic hydrocarbon bands.

intervals between grids. This contour map, shown in Figure 5, demonstrates the circular symmetry and the uniform thickness of the stain. Numerical integration of this stain yielded a calibration factor of 6.9 g m^{-2} $a.u.^{-1}$ or 7.5 μm $a.u.^{-1}$ indicating this stain to be 1.3 μm thick (1.2 mg m^{-2}). The contour map can now be redone using the calibration factor to convert band amplitude to stain thickness. Recognizing the circular symmetry of the spreading oil drop, each contour in Figure 4 was integrated under the assumption of circular symmetry to yield a calibration factor as a function of time or band amplitude. These results are given in Figure 6. Note that over the band amplitude range of 0.18 to 0.5, the calibration factor shows some systematic variation, but the full range of this variation is less than 5% of the calibration factor. This indicates that the band amplitude is a nearly linear indicator of stain thickness. On sandblasted aluminum and stainless steel, safflower oil stains spread in a manner very similar to that observed here for gold. Calibration factors for safflower oil on these substrates was a factor of 2 larger than that observed for gold. Whether this is due to surface reflectivity or surface texture of the steel and aluminum substrates is not known. Light mineral oil on

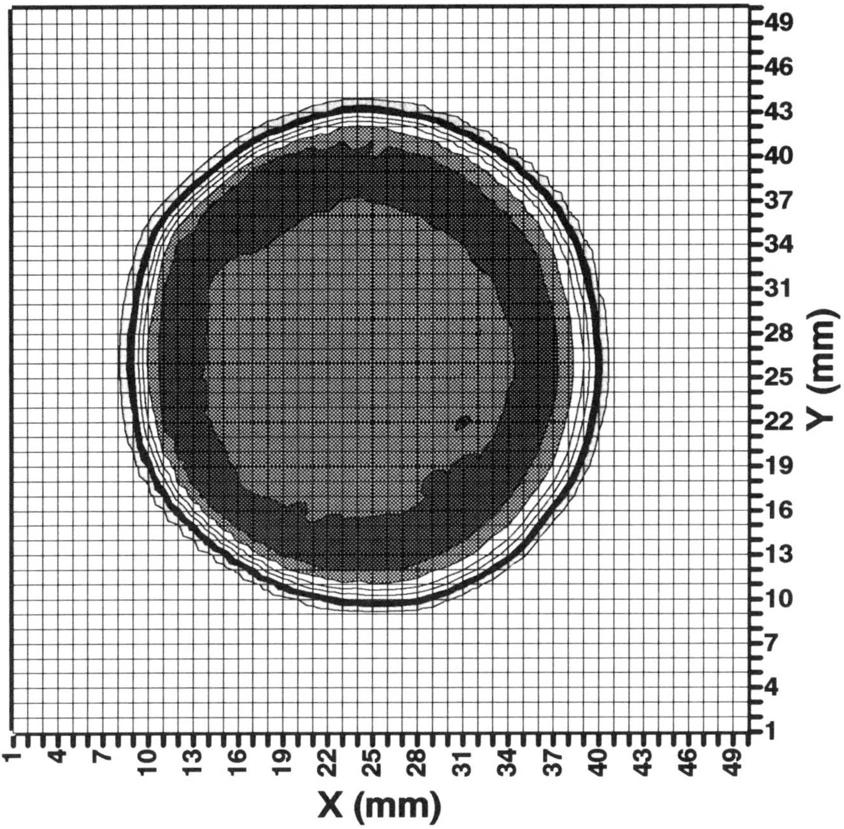

Figure 5. Contour map over a safflower oil stain based on the baseline corrected aliphatic hydrocarbon bands. Contour range: 0.00 to 0.18 a.u., intervals: 0.02 a.u.

gold spread in a pattern that was less uniform across the stain and spread to a larger diameter than the safflower oil. This could be due to fractionation by molecular weight of the components in the oil and to the absence of oxidative polymerization,respectively, that occurred in the safflower oil. The calibration factor for mineral oil on sandblasted gold was found to be 5.1g m^{-2} $a.u.^{-1}$ indicating that the substrate (as in the case of steel and aluminum) has a greater influence on the calibration than does the particular aliphatic organic compound.

3.3 Limits of Detection Having demonstrated that calibration factors can be determined for organic stains on sandblasted metals, the ultimate use of this technology is to determine the lower limits for which stains can be detected. For this, the contour map for safflower oil on sandblasted gold (Figure 5) was modified such that the contour lines indicated the limits with which the band amplitude could be measured as determined by the variation of the aliphatic hydrocarbon band amplitude outside the region of the substrate containing the stain. This is demonstrated in Figure 7 where the limit of the stain is clearly defined by sharp contours. In Figure 7, the region of "clean" substrate near the origin (X = 1, Y = 1) has a consistently lower value than the region near the corner of the substrate opposite the origin (X = 50, Y =50). This could be due to hydrocarbon contamination or to some imperfection with the

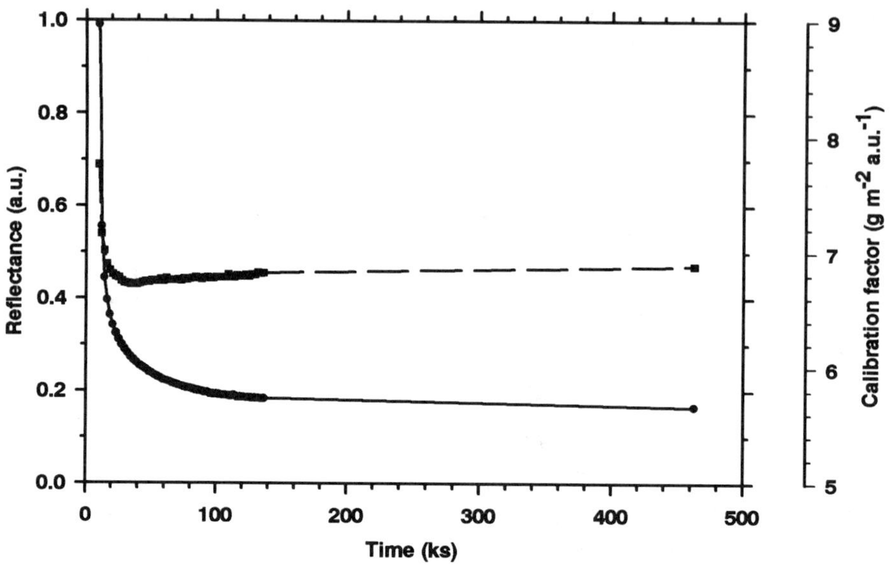

Figure 6. The calibration factor and the band amplitude for a safflower oil stain based on the baseline corrected aliphatic hydrocarbon bands used to construct Figures 4 and 5. Calibration factor - squares and broken line. Band amplitude - circles and solid line.

baseline correction/band amplitude measurement. To investigate this further, 100 spectra representing a 10-mm by 10-mm area from the origin corner and from the corner opposite the origin were averaged to yield a single spectrum for each corner. The spectrum for the low corner (near the origin) was subtracted from that for the high corner (far from the origin) to yield an average difference spectrum of the high corner relative to the low. If the difference across the map is due to real spectral features, then the averaging operation followed by subtraction should reduce the spectral noise by an order of magnitude. Figure 8 shows the resulting spectrum in the baseline corrected form similar to Figure 3. The variation is in the "clean" region of the substrate is due to the presence of a hydrocarbon film in the region outside the safflower stain that has a band amplitude of 0.0007 a.u. (5 nm thick using the calibration factor for 1 μm thick safflower oil). This data was taken with the Harrick/MIDAC Surface Inspector. Based on Figure 8 and the calibration factor for 1 μm thick safflower oil, this instrument has a detection limit of approximately 1 nm for hydrocarbon stains using areal averaging of 100 spectra. Had the measurement been done with the SOC 400 Surface Inspection Machine/Infrared using the same collection times, each spectrum would resemble that in Figure 8, and another order of magnitude in detection limit could be achieved by averaging.

4. CONCLUSIONS

The use of diffuse reflectance, mid-infrared spectroscopy to inspect sandblasted metals for stains has been demonstrated with detection limits approaching nanometer film thickness' for aliphatic hydrocarbons. Surface inspection machines have been developed for exploiting this technology in the field. The use of this technology with surface mapping techniques introduces a wide range of options for qualitatively and quantitatively mapping and monitoring of organic stains and residues.

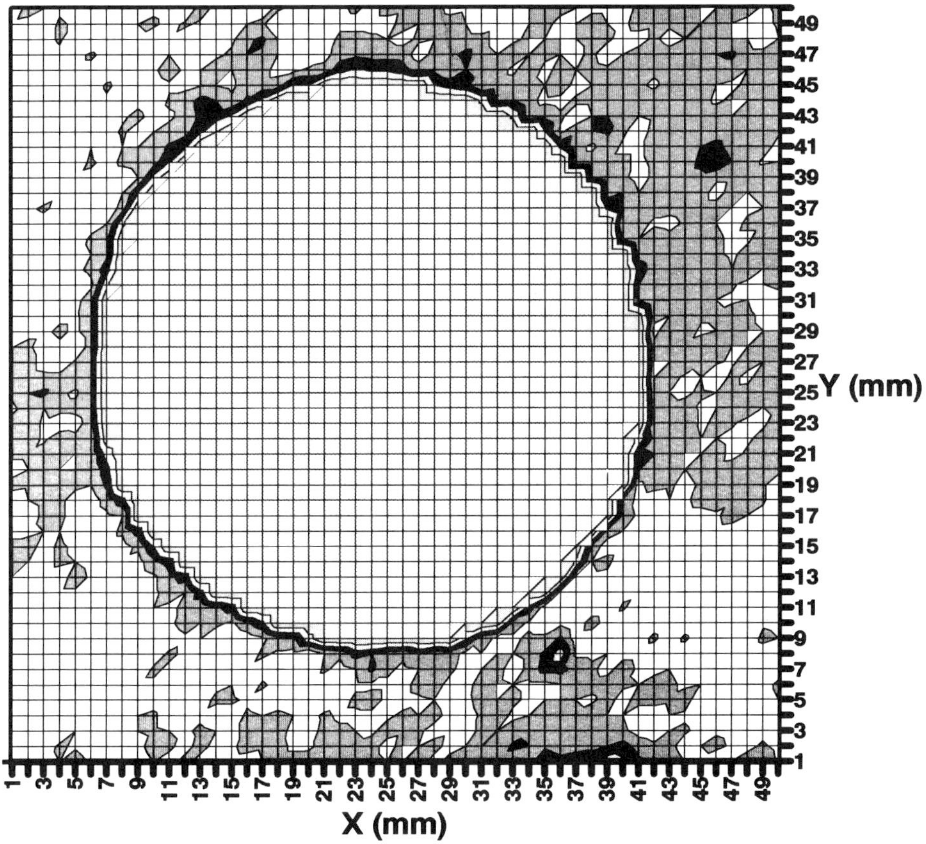

Figure 7. Contour map over a safflower oil stain based on the baseline corrected aliphatic hydrocarbon bands. The contour limits are set to show the outer range of the stain and the contour interval set to detect 2 standard deviations in the ability to measure the band amplitude. Contour range - -0.0005 to 0.0020 a.u. , intervals - 0.0005 a.u.

5. ACKNOWLEDGMENTS

The authors gratefully acknowledge B. H. Nerren, National Aeronautics and Space Administration, Marshall Space Flight Center, Alabama, and E. R. Engberg, U. S. Army Environmental Center, Aberdeen Proving Grounds, Maryland for their efforts to implement this inspection technology.

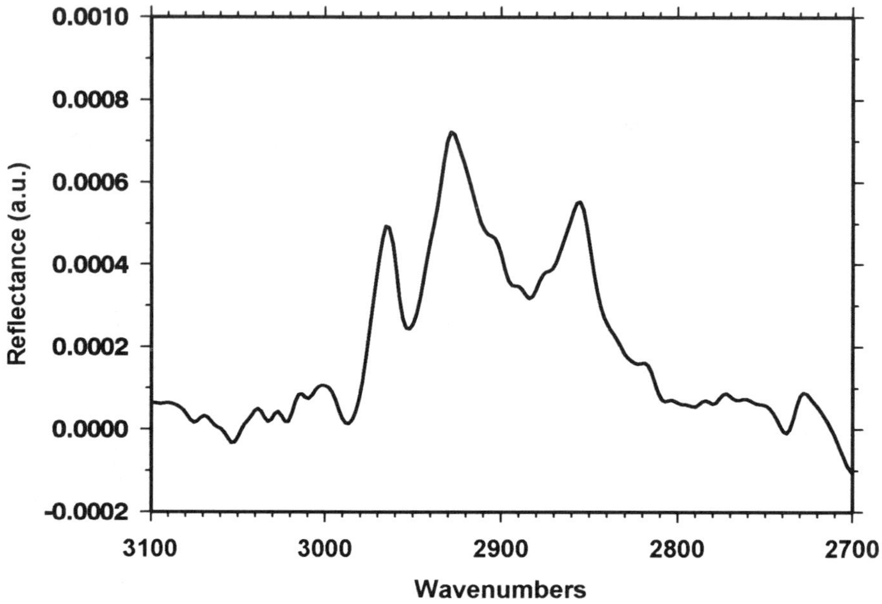

Figure 9. Baseline corrected aliphatic hydrocarbon bands from a spectrum that is the difference of the average of 100 spectra representing the clean corner of a substrate and the average of a similar number of spectra representing a more contaminated corner of of the substrate.

6. REFERENCES

1. G. L. Powell, M. Milosevic, J. Lucania, and N. J. Harrick, Appl. Spectrosc., 46, 111-125 (1992).
2. G. L. Powell, N. R. Smyrl, D. M. Williams, H. M. Meyers, III, T. E. Barber, M. Marerro-Rivera, in A. F. Whitaker, ed., Aerospace Environmental Technology Conference, NASA Conference Publication 3298, MSFC, Alabama, 563 (1995).
3. G. L. Powell, T. E. Barber, M. Marrero-Rivera, D. M. Williams, N. R. Smyrl, and J. T. Neu, Micrichimica Acta, in press.
4. G. L. Powell, N. J. Harrick, ed., Reflections, Harrick Scientific, Inc., Ossining, NY (1990).
5. G. L. Powell, N. R. Smyrl, C. J. Janke, E. A. Wachter, W. G. Fisher, J. Lucania, M. Milosevic, and G. Auth, in The Proceedings of the Conference on Characterization and NDE of Heat Damage in Graphite Epoxy Composites, NTIAC, Austin, Texas, 97 (1993).
6. G. L. Powell, T. E. Barber, and J. T. Neu, in J. C. Stover, ed., Proceedings of SPIE's 40th Annual Meeting and International Symposium on Optical Science, Engineering, and Instrumentation, 2541, Bellingham, Washington, 142 (1995) .

28th International SAMPE Technical Conference
November 4-7, 1996

DRYING METHOD FOR COMPOSITE HONEYCOMB STRUCTURES

Bob Geyer
Boeing Commercial Airplane Group
Seattle, WA 98124

ABSTRACT

It is widely recognized that some airplane honeycomb structures absorb and retain moisture. Often these structures are graphite/epoxy or fiberglass/epoxy skins over non-metallic honeycomb core.

In 1993, the Boeing Commercial Airplane Group issued a Service Bulletin explaining a method for drying B757 spoilers. Due to some reports of water remaining in spoilers, Boeing undertook a study of alternative drying methods in 1995. This presentation discusses the statistically significant results of that study which lead to an improved method of drying water-contaminated spoilers. The improved method involves drilling a pattern of holes through one skin, poking into the honeycomb through the holes, then heating the part under vacuum. Details of this procedure, the repair process for the drilled skin, and the factors that are most critical in drying honeycomb structures are given.

KEY WORDS: Moisture Removal, Honeycomb/Sandwich Core, Composites

1. INTRODUCTION

It was a warm and stormy night. As flight #202 lifts off from Bangkok International Airport, rain water is driven through small channels to the interior of a composite honeycomb part. At cruise conditions, the water freezes to solid ice, trapping it in the structure. Eventually, low pressure equalizes throughout the part. As the airplane descends, the increasing pressure in the atmosphere slowly but steadily pushes more moisture into the structure through any available openings in the 757 spoiler. The plane lands and prepares for another flight. The cycle repeats itself over and over, eventually accumulating copious amounts of water inside the spoiler. Sometime later that year during C-check, a mechanic removes the spoiler from the airplane to repair the delaminated skin. As he handles the spoiler, water literally runs out of the part, puddling on the floor of the hangar.

Though fictional, (that is, undocumented) similar versions of the above scenario have been told and retold over the past several years. Many airline operators have experienced moisture ingression into composite honeycomb structure. 757 Spoilers are one high profile example.

The problem of trapped water in composite structures is two fold. Initially, the major concern is strength degradation of epoxy bonded structures. Secondarily, water increases the chance of skin-to-core delamination from pressure increases during follow-on repairs or expanding water during freezing. Both of these problems contribute to longer down times and higher repair costs.

In the case of the 757 spoiler, water is most likely entering the structure through holes created by improper sealing around the metal actuator hinge fitting at the center of the forward edge of the spoiler (See Figure 1 for a diagram of a spoiler and the typical location of water.) As the above scenario depicts, it is thought to accumulate over long periods of time through the pumping action of flight cycles.

The Boeing Company issued a Service Bulletin in 1993 to address removing water from 757 spoilers. This Service Bulletin explains two water removal methods, each with its own advantages and disadvantages. One method is removing the entire bagside skin, then heating the exposed core in an oven to quickly evaporate the water. This is the quickest but more expensive method since it requires autoclaves, tooling and original prepreg raw materials. The second method involves drilling a gridwork of .4 cm (5/32 inch) diameter holes through the bagside skin, then puncturing six times into each hole at about 45 degrees down towards the opposite skin, (which creates multiple vent paths in the honeycomb) creating a vacuum, then heating the part for about 24 hours.

Some operators using the drill and puncture method reported excessive dry times (50-100 hours). In response to these operator concerns, the Boeing Company funded a research program to develop an improved moisture removal process: one that would take less flow time to dry the part, and remove all significant water from the honeycomb structure in about 24 hours.

The purpose of this paper is to describe the improvements of the moisture removal process for the drill and puncture method. This process is specific to the Service Bulletin for the 757 spoiler, though it can be applied to many honeycomb sandwich parts.

2. IMPROVED METHOD

The improved drying method is explained below, but also shown visually in the Boeing produced videotape #16850, titled *Water Inspection and Removal*. This video highlights the improvements to the Service Bulletin in a simple and straight forward training format.

2.1 Detect and map wet areas. The first step in removing water from a composite part is to identify the water. Frozen water in the core of a spoiler will create frost immediately after an airplane lands. This is an indication that water is present, but should not be the method used to map out the pockets of water. Several non-destructive techniques are available and discussed in the Service Bulletin and appropriate Structural Repair Manual (SRM). The least expensive, and one that was recently added to the 757 Non-Destructuive Test (NDT) manual, (Chapter 51-00-02, part 9) is liquid crystal sheets. These thermally sensitive plastic films make use of the high heat capacity of water to indicate the location of wet core or pockets of water in composite parts. The sheets have been shown to be very accurate and relatively inexpensive.

After the water is detected, its location in the spoiler should be mapped out by marking directly onto a cover film or onto the bagside of the spoiler.

2.2 Mark and Drill a gridwork of holes. Mark a gridwork of dots spaced at 2.5 cm (1 inch) intervals extending just beyond the wet areas that were mapped out previously. Most operators have found that a plastic template outlining this one inch pattern is a useful guide for this task. Using a .4 cm (5/32 inch) diameter maximum bit, drill through the skin at each of the marked dots. See Figure 2. Drill straight down through the bagside skin about .6-1.3 cm (.25 to .5 inch) using a drill stop. If a septum is present, drill far enough down to go through it. Be careful not to drill into or damage the aero skin.

2.3 Puncture six times into each hole in a specific pattern. The next step is to puncture down through each hole six times at about a 45 degree angle with an .15-.22 cm (.060-.090 inch) diameter sharpened ice pick style tool. A plan view of the puncture direction and a typical tool are shown in Figures 3 and 4. Note the rotated puncture pattern. Every adjacent hole going in a horizontal line will have the alternate pattern of puncturing.

In some areas, the foaming adhesive or previous repairs may interfere with the puncturing. The foaming adhesive forms a wall that is difficult to puncture through. To continue the puncture pattern, use a one-sixteenth inch diameter drill as the puncture tool to drill through these barriers.

2.4 Create a vacuum over the wet area. Using standard composite bagging technology, build a vacuum bag covering the drilled and punctured area. Do not envelope bag the entire part as this may contribute to part warpage. The recommended bagging method consists of two plies of polyester mat breather covered by nylon bagging film.

Testing on the polyester breather has shown higher rates of air flow when compared to other breathers.

2.5 Flip the part over. Pull vacuum for at least ten minutes with the part bag side up. Then slowly rotate the part to position the bag side down (flip the part over), and wait another ten minutes with full vacuum. This step helps to remove a significant portion of water. As the water runs to the other end of the core cell and passes by the punctured holes, it gets drawn out by the vacuum.

2.6 Heat to 93°C (200 F) for about 24 hours. Still under full vacuum, place the part in an air circulating oven, and ramp the temperature to 93°C. The part may be checked frequently with the liquid crystal sheet. Since water will hold its heat much longer than the surrounding honeycomb core, a wet area will remain hot longer than a dry area. Place the liquid crystal sheets on the aerodynamic side (the other side has a bag covering the wet area) of the part and observe any color changes that indicate moisture. Return the part to the oven if more time is needed to dry.

2.7 Repair (3 ply wet lay-up) over drilled area. After the water is removed, the drilled and punctured area needs to be repaired per the Service Bulletin. Typically, the drill holes are filled with a thickened resin, then three wet lay-up plies are cured over the holes. A final coat of resin is applied over the repaired area to seal that surface prior to painting. The Service Bulletin also explains where to apply extrudable sealant around the fitting and bolt holes of a spoiler to prevent water from entering the part in the future.

3. EXPERIMENTAL DISCUSSION

3.1 Design of Experiment. The Boeing company used a team approach to improve the drying process. Representatives of several organizations met to brainstorm improved drying methods. From this list of methods, the team ran a Designed Experiment (DOE) to investigate drying factors.

The DOE required over 130 consistent test panels. We fabricated and dried 30x30x2.5 cm (12x12x1 inch) panels with 3 ply graphite epoxy skins and 3 pound density Nomex core. One skin of the test panel was autoclave cured onto the core. The other side was exposed core. This gave us the ability to pour in 300 grams of water into a 15 cm (6 inch) diameter circle in the middle of each panel. The second skin which had been precured separately was then bonded on over the exposed core with a structural repair epoxy.

After the DOE testing, the following factors were identified as the most critical. These factors and a brief conclusion of the statistical analysis are included below.

3.2 Drying temperature. As expected, the higher drying temperatures were statistically shown to dry the test panels faster (77°C -104°C). Though on first thought this is inherently obvious, the data was useful in demonstrating a lack of spikes in the drying rate at any one temperature. For the Improved Method, the temperature of 93°C was chosen since it gave a good compromise between dry rate and the safety of remaining below the boiling point of water.

3.3 Puncture Pattern. This was the most significant parameter. A puncture pattern similar to the one described in the Service Bulletin was used as a baseline. See Figure 5. This puncture pattern exposes 60% of all the cells to the vacuum used during drying. A new pattern was designed to increase this amount to shorten the drying time. The new method punctures 73% of the cells. This higher percentage of exposed cells is statistically significant in decreasing the drying time. The new pattern is shown in Figure 4.

3.4 Honeycomb Core Density. Three and 9 pound Nomex core (.048 and .144 Mg/m3) and 4 and 12 pound (.064 and .192 Mg/m3) fiberglass core were evaluated for their effect on drying rate. The only significant decrease in rate (approximately 10 % less water removed in 24 hours) was seen with the 12 pound core. This is most likely due to the thicker cell wall and heavier resin coating of the high density core which slowed diffusion through the cell wall.

3.5 Heat Blanket. In theory, the method of heat application should not affect the drying rate. An oven should dry a spoiler in the same time as two heat blankets, one on each side. In reality, we had a very difficult time controlling the rate of heat-up and the consistency of heat during the drying process when using heat blankets. Heat sinks are unavoidable in a wet spoiler. The pockets of water, the metal fittings, the varying densities in core and foaming adhesive, the change in water mass as one area dries out and the overall tapering thickness of the part all combine to make consistent control of a heat blanket nearly impossible.

Statistically, the DOE identified that heat from a conductive source such as a heat blanket dried a test panel as quickly as heat from a convective source such as a panel placed in an air circulating oven. However, in all the tests on actual wet spoilers, both we and subcontractors had difficulty maintaining a consistent heat.

3.6 Thickness. As the volume of a honeycomb cell increases due to height, the vapor pressure generated inside the cell by heating a constant amount of water will necessarily decrease. This decrease in pressure, combined with the increased distance the vapor must travel to be removed, will slow down the drying rate. Approximately 10% less water was removed from a 5 cm (2 inch) thick test panel than from a 2.5 cm (1 inch) thick test panel over a 24 hour drying period.

3.7 Vacuum versus flow. A container of water was experimentally shown to dry twice as fast under full shop vacuum than it did when under reduced vacuum. The initial theory was that a leak path in the part (which will reduce the vacuum pressure) would draw in warm, dry air which would pick up moisture as it was pulled through the structure and contribute greatly to overall speed of drying. However, test results indicate that reducing the vapor pressure to a minimum by maintaining full vacuum over the drilled holes is more important to quick drying.

3.8 Septum. The presence of an internal ply of film adhesive separating two layers of core does not significantly slow down the drying process as long as the puncturing pattern is continued through the septum.

3.9 Vendor Evaluations. These statistical methods and information were combined into an improved drying method. After laboratory testing, the method was evaluated on a wet spoiler that Boeing purchased from an airline operator. The lessons learned from this effort included heat blanket control, foaming adhesive barriers, and detection and removal of water hidden under the metal actuator fitting.

Secondary evaluations were performed during visits to two companies that do repair work on 757 spoilers using the Service Bulletin drill and puncture method. These real world evaluations refined and validated the laboratory work which led to the recommended Improved Method.

4. SUMMARY

An Improved Drying Method for honeycomb structures has been developed using statistical methods and verified in spoiler repair shops.

The technique involves mapping and then drilling into the wet areas of a skin to create a gridwork of one inch spaced holes. Each hole is punctured six times in a specific pattern which is designed to maximize the quantity of core cells that get punctured. Puncturing the cells creates pathways for the moisture to be removed. Full vacuum is then created over the holes and the wet area is heated to 93°C. Carefully following these techniques has been shown to dry wet spoilers in less than 24 hours of heating time.

A videotape has also been made which demonstrates this method and is available from Boeing. This drying method will be incorporated into Service Bulletin 757-57-0047, revision 3, expected to be released in late 1996.

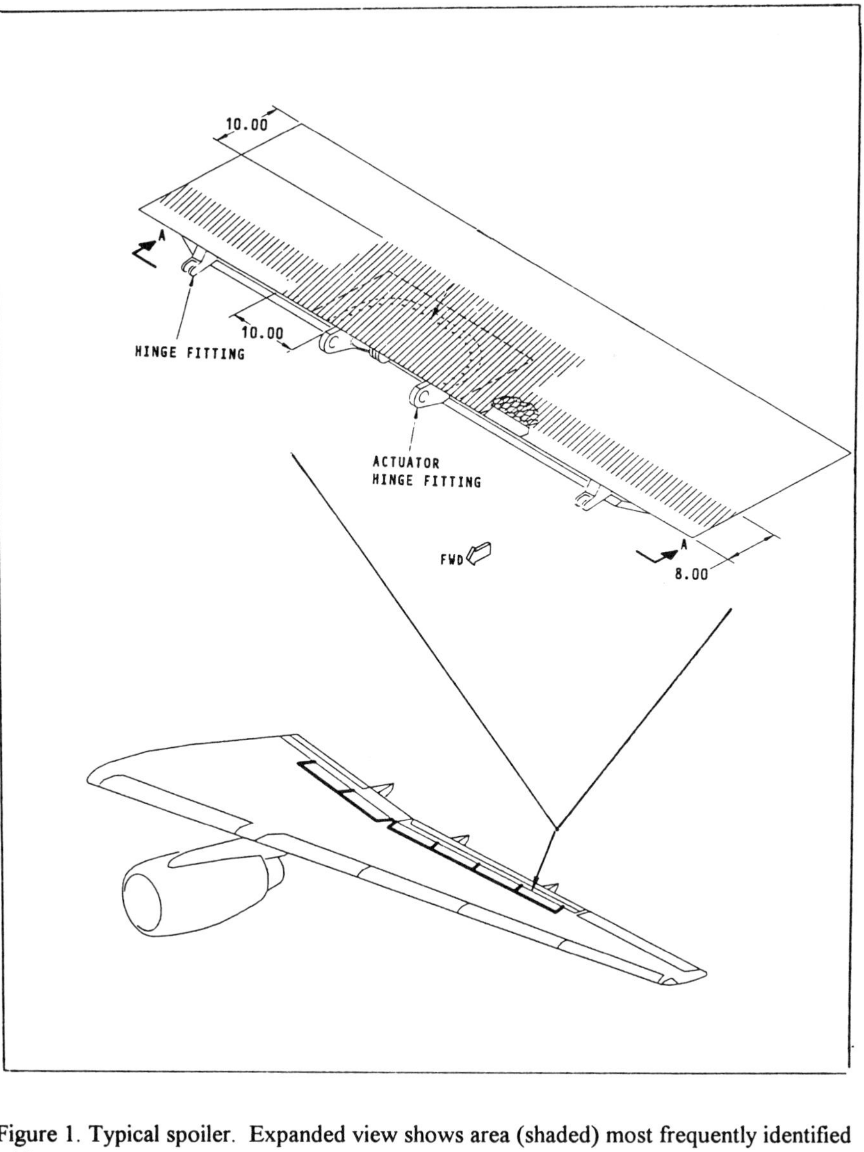

Figure 1. Typical spoiler. Expanded view shows area (shaded) most frequently identified with water.

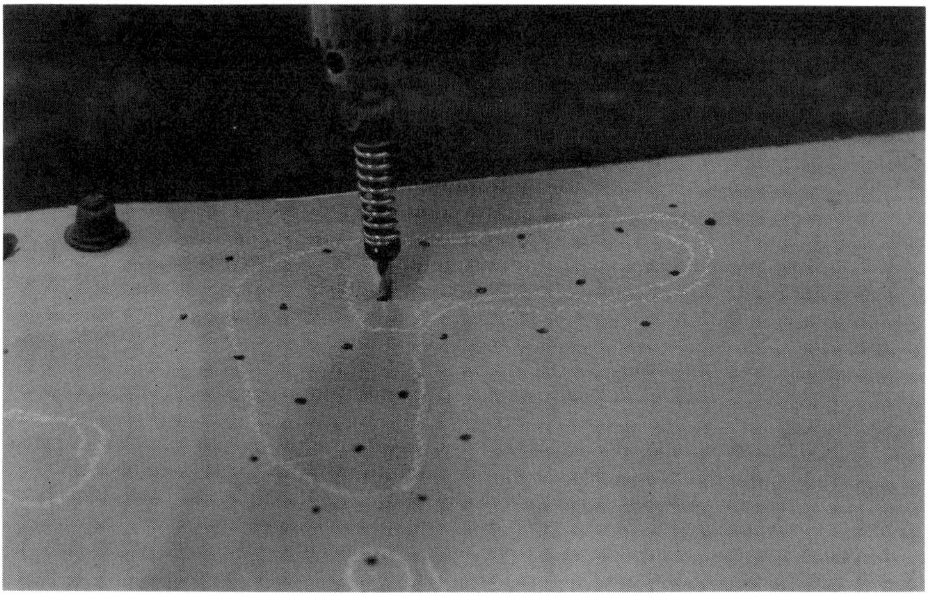

Figure 2. Drilling through bagside skin at marked dots. The wet area is mapped out in white. Hole pattern continues just beyond the wet area.

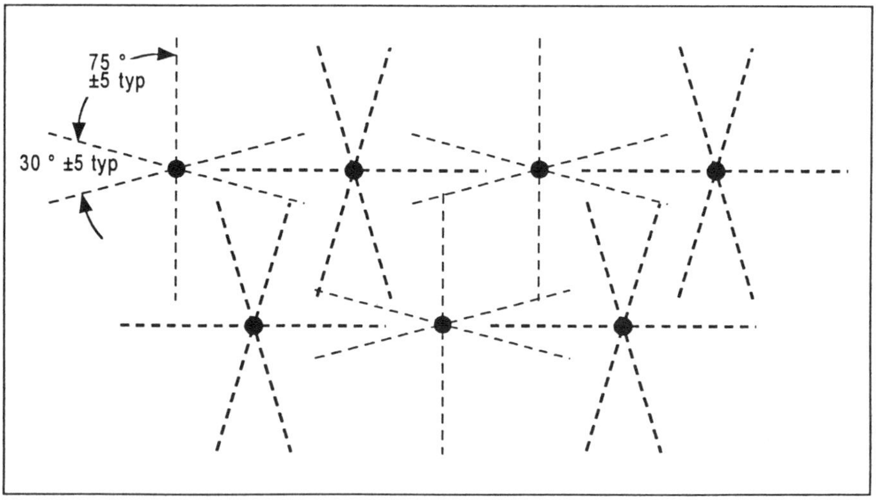

Figure 3. Plan view of improved puncture pattern that exposes 73% of cells. Dotted lines represent where the core has been punctured, solid dots represent drilled holes. Note every adjacent hole has puncture pattern rotated 90 degrees.

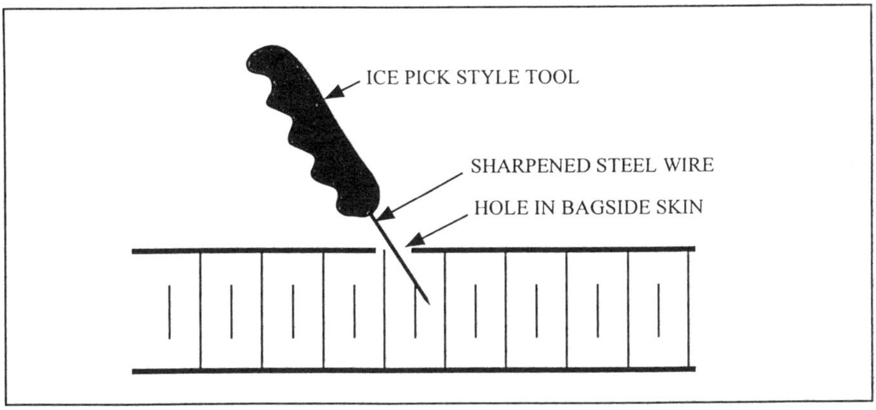

Figure 4. Cross section view of honeycomb being punctured.

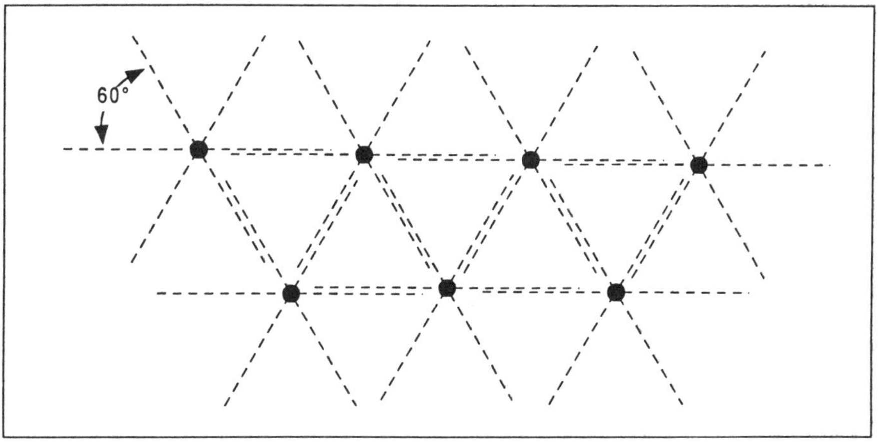

Figure 5. Plan view of original puncture pattern that exposes 60% of cells. Dotted lines represent where core has been punctured, solid lines represent drilled holes.

BIOGRAPHY

Bob Geyer has worked in various manufacturing research groups at the Boeing Company for 12 years, focusing on producibility of composite structures. Composite repair has been an interest for the past several years.

Prior to joining the Boeing Company, he worked for 3M Company in St Paul, Minnesota, where he worked to develop more durable retro-reflective pavement markings and highway signs.

A native of the Seattle area, his personal interests include biking, gardening and playing with his three children.

28th International SAMPE Technical Conference
November 4-7, 1996

TDA: A MAJOR STEP FORWARD IN COMPOSITE REPAIR HEATING BLANKETS

David W. Kelsey
WichiTech Industries, Inc.

ABSTRACT

In the repair of composite components and in metal-metal bonding, non-uniform heating of the repair area is a significant problem. Use of conventional silicone/electric blankets may produce significant temperature variations. This paper reports on a new blanket that significantly reduces temperature variations. The new blanket employs a Thermal Diffusion Apparatus (TDA), which is technology patented by The Boeing Company. The new blanket, when applied to a test panel with a demonstrated temperature range of 39°C (70°F) and a standard deviation of 11°C (20°F) when heated to 177°C with a conventional blanket, brought the range down to +/- 12°C (22°F) with a standard deviation of 3°C (5.4°F). The temperature leveling process is passive, requiring no special equipment. The blanket can be used in conjunction with any controller. The TDA components are environmentally friendly and non-toxic. This blanket, which will be available commercially from WichiTech Industries, Inc., offers a simple and economical means to dramatically improve repair consistency.

KEYWORDS: Composite Repair, Electric Heat Blanket, Bonding

1. INTRODUCTION

Many composite repairs require the application of external heat to cure the resin. Typical sources of heat are electric blankets attached to a computer control unit which regulates the ramp rate, soak temperature and cool down rate. The temperature on the surface of the repair area may be non-uniform. Temperature variations result from the thermal characteristics of the object being heated, non-uniformity in the insulating effect of the materials used in the lay-up, uneven heating from the heat source, or a combination of these factors. These heat variations may result in a non-serviceable repair. If the resin does not reach specified cure temperatures, it may not fully cure. Conversely, an overheated area may damage the resin, or in extreme cases, cause a fire hazard. Detection of hot or cold spots depends on the fortuitous placement of the TC's used to track the temperatures during the cure process. Hot and cold spots are a deterrent to uniformity in the quality of repairs.

Boeing first explored the effect of using a Thermal Diffusing Apparatus (TDA) to passively redistribute the heat energy to provide a more even temperature. Boeing has patented the use of this method. This paper reports the results of efforts to demonstrate and optimize the TDA effect.

2. MATERIALS AND METHODS

In order to demonstrate temperature variations, a test panel of fiberglass and polyester resin was constructed. On the back of the flat panel were attached an iron heat sink, a sheet metal vertical fin to cause a steady radiant heat loss, and a trapped air chamber filled with a single layer of polyester bleeder cloth to create an area with reduced heat loss. The heat sink and radiating fin were covered with an insulating layer of micro-balloons and resin to adjust the surface temperature variations to simulate field conditions.

Heat control and vacuum were provided by a WichiTech HB-2 composite repair system. This system was programmed to raise the temperature in two stages. The first stage increased temperature at a ramp rate of 8.3°C (15°F) per minute to 121°C (250°F) degrees, where the panel was allowed to soak for 5 minutes. The heat was further increased at 2.78°C (5°F) per minute to 177° (350°F) degrees, followed by a thirty minute soak stage. Temperature was monitored at fixed locations on the panel with eight thermocouples. The HB-2 was programmed to use all eight thermocouples to control the heat, keeping the blanket at a temperature at which the hottest and coldest TC's bracketed the target temperature. The HB-2 provided vacuum through an internal electric vacuum pump which was regulated to 74 Kpa (22 in Hg). These restrictive conditions provided results which could be compared between runs.

The Temperature Diffusion Apparatus (TDA) consists of a containment bag filled with a heat transfer medium sandwiched between two identical conventional 5 watt / 6.45 cm. sq. (five watt/sq. in.) silicone-electric blankets wired in parallel. In these tests the diffusion medium selected was a environmentally friendly oil. The oil was contained in a high temperature plastic bag. The design included a valve that permitted oil to be added or removed between tests. Comparison tests were conducted with various weights of oil to demonstrate the effect of the diffusion medium and to determine the relationship of quantity on effectiveness.

The test lay-up simulated a field repair. The thermocouples were attached to the surface of the panel with tape. The test area was then covered with a layer of bleeder cloth. The electric blankets and TDA were placed over the bleeder cloth. A breather layer covered the top electric blanket. Finally, a vacuum bag was sealed over the lay-up. Between test runs the panel was permitted to cool to less than 48° (120°F). To adjust oil quantity to test parameters, the vacuum bag was opened and the top bleeder cloth and blanket were carefully peeled back to permit the TDA to be removed, adjusted, and replaced. The lower blanket, bleeder cloth, and thermocouples were not disturbed. This procedure permitted extremely uniform test conditions, as indicated by close correspondence between tests with identical TDA parameters.

Readings of the eight thermocouples were taken at intervals during the ramp and soak phases. The HB-2 provided a real time reading of the mid point between the hottest and coldest observed temperature. This temperature was used to determine the interval between readings. Measurements were made at 66°C (150°F), 93°C (200°F), at the conclusion of a five minute soak at 121°C (250°F), 149°C (300°F), and at the beginning of the 177°C (350°F) soak period, and 10, 20, and 30 minutes into the soak period.

In practice small differences in the actual average temperature from the nominal temperature interval resulted from fluctuations in control temperatures and sampling time during the ramp stage. In order to permit consolidation of results, the temperatures at each reading were normalized with the assumption that the observed individual temperatures tended to rise together but at slightly different rates. The midpoint between the hottest and coldest TC's was calculated and the difference between the midpoint and the nominal temperature was used as a correction factor. Finally, the correction factor was applied to each individual thermocouple reading. For example, if at the nominal 66°C reading, the midpoint of the hottest and coldest TC was 68°C degrees, 2 degrees was subtracted from each individual TC reading. Temperature divergence that might occur over a few degrees was negligible. This method permitted readings from different runs to be combined and included as part of the population of potential readings at each interval. The impact of the TDA blanket was so dramatic that sampling errors were negligible.

3. RESULTS

Baseline tests were conducted to demonstrate the thermal characteristics of the panel, and the materials used in the lay-up. The initial tests were made with a single conventional blanket. The results from a single heat cycle are shown in figure 1.

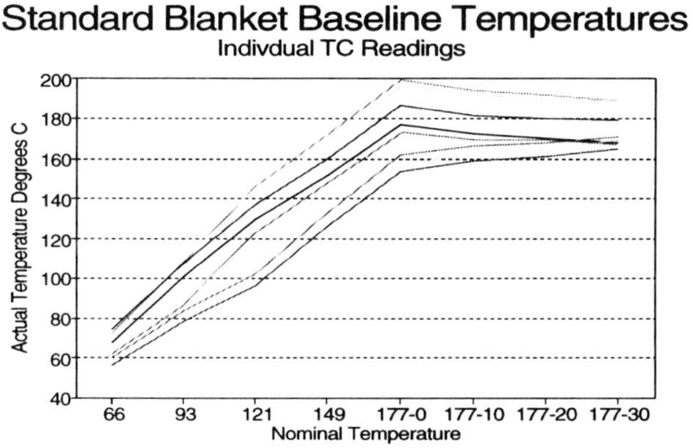

Figure 1

Figure 2 shows the observed TC readings of a single heat cycle made with a TDA blanket containing 1200 g of oil.

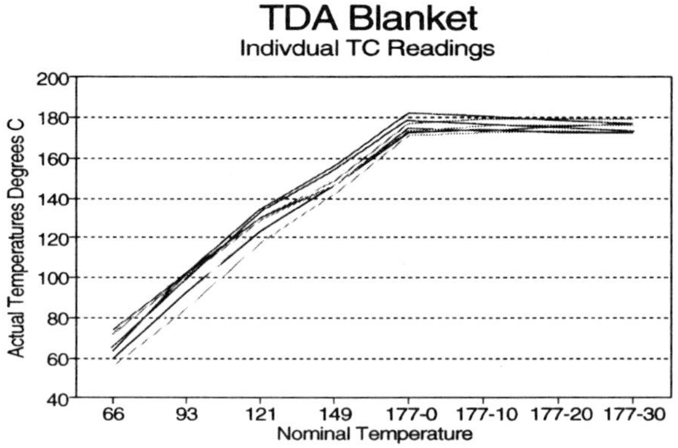

Figure 2

The impact of the TDA is dramatic. Figure 3 is a chart of the maximum and minimum observed temperatures for a standard blanket and a TDA blanket containing 1200g of oil. The chart shows two heat cycles for each blanket.

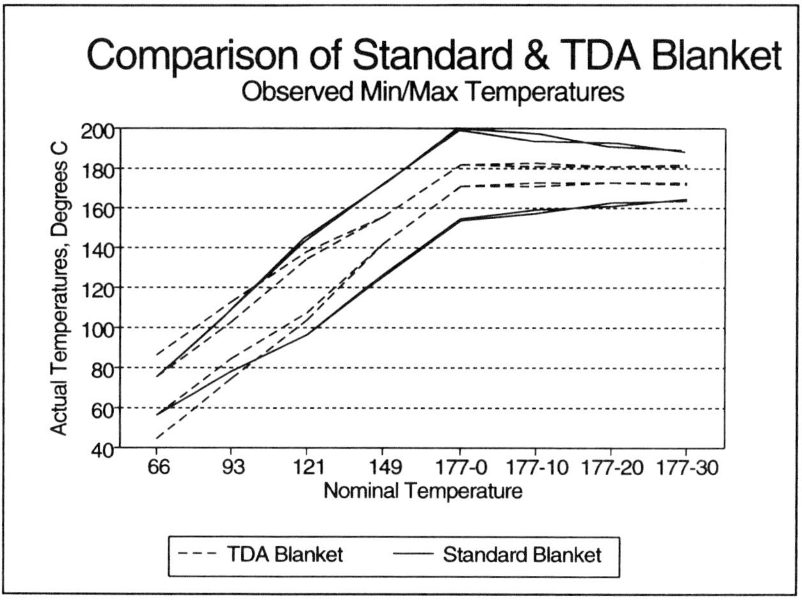

Figure 3.

Since two blankets were used in the TDA device, it was necessary to identify the effect of the two blankets, without the effect of the TDA device. Tests were run with a single blanket, two heat blankets with both blankets providing heat, two blankets with just the top blanket providing heat, and two blankets with just the bottom blanket providing heat. All tests were repeated once and the results of duplicate tests combined. The results presented below in figure 4 indicate very little impact attributable to the heating blankets alone.

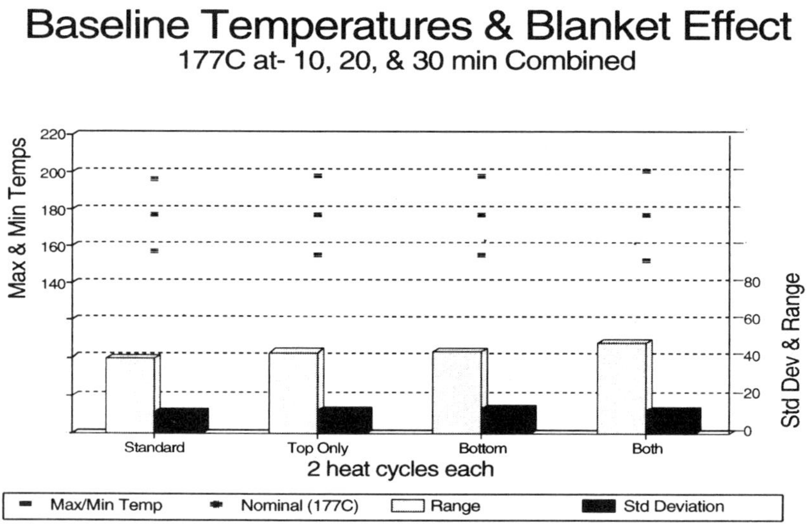

Figure 4

The remaining tests were all made with the TDA device between the two electric blankets. Tests were run with 400, 600, 800, 1000, 1100, 1200, 1300, 1400, and 1500 grams of oil in the bag. In each instance, two tests were run to verify the results. Figure 5 demonstrates the effect of quantity on effectiveness. All normalized TC readings observed during the soak phase at 10, 20 and 30 minutes from both tests were combined. The absolute high and low are plotted against the left Y-axis and the range and standard deviation are plotted against the right Y-axis.

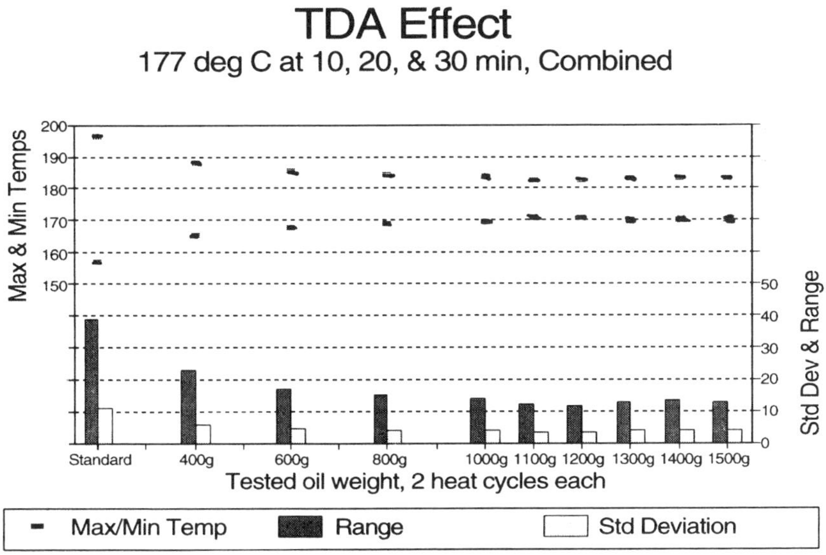

Figure 5.

4. CONCLUSIONS

The effect of the TDA is dramatic. It was possible to reduce a temperature range in excess of 39°C (70°F) to 12°C (22°F). The effect is noted at all observed intervals during the ramp and soak stages. The process is entirely passive and requires no special equipment. The material used in the TDA is environmentally friendly and non-toxic. The blanket is similar to conventional blankets in both appearance and use. This blanket, which will be available commercially from WichiTech Industries, Inc., offers a simple and economical means to dramatically improve repair consistency.

BIOGRAPHY

David Kelsey resides in Warrenton, Virginia. He received his undergraduate degree in biology from the University of Virginia in 1966. He went on to do graduate study in developmental botany at the University of Alabama and the University of Virginia. Mr. Kelsey is a retired pilot from the District of Columbia Air National Guard where is served for 22 years flying F-100, F-105, and F-4 fighter aircraft. He is currently the Marketing Director for WichiTech Industries, Columbia, MD USA.

28th International SAMPE Technical Conference
November 4-7, 1996

A PORTABLE AUTOCLAVE FOR FIELD LEVEL REPAIR

MSgt B.C. Cramer
Systems Support Division, Materials Directorate
Wright Laboratories, Wright-Patterson AFB, OH 45433
and
W.M. Purcell
University of Dayton Research Institute
Dayton, OH 45469

ABSTRACT

Autoclave curing of reinforced composites is universally accepted to yield the highest quality, most uniform laminate at the most attractive cost. While autoclaves are widely used in factory operations the initial cost, operating complexities, and recurring cost have prevented their use in field repair operations. Field repair operations have relied upon bolted or bonded precured patches, or vacuum bag cure of prepreg or wet layup patches. The quality limitations commensurate with vacuum bag curing of laminates or adhesives have relegated field level repair to relatively simple applications like cosmetic or lightly loaded repairs. This paper describes the design and development of an autoclave specifically intended to be compatible with field repair operations. The autoclave is simple, inexpensive, and adapted to the facilities already available at the field level maintenance organization. This paper describes the portable autoclave and illustrates how it can be integrated into the field repair arena. It also discusses the field maintenance operations needed to make maximum use of the portable autoclave capability.

KEY WORDS: Field Repair, Autoclave Curing, Autoclave

1. BACKGROUND

UDRI was tasked to design and prototype a self-contained mobile repair station for the Air Force. The effort was funded by Warner Robins ALC through the Materials Directorate at Wright-Patterson AFB. Both Air Force organizations contributed technically to the design effort. The repair cart contained a diesel generator, a compressor, a freezer, and storage space for all of the materials, tools, and facilities necessary to make advanced composites repairs to metal or composite structure. The cart also has surface preparation materials and facilities for PAA and silane techniques. Since the cart was intended to repair corrosion grind out damage, materials and facilities were included to prepare plaster splash master tools, lay-up/cure tools, and curing facilities. Hot bonders with electric resistance heaters were included on the cart for vacuum bag curing operations. The cart was also fitted with two portable autoclaves. All of the facilities and tools supplied with the cart were designed to be used at a remote site without logistics support other than fuel. The remainder of this paper will discuss the design and use of these two portable autoclaves.

Autoclave curing of reinforced composites is universally accepted to yield the highest quality, most uniform laminates. Autoclave curing is the standard by which all alternate curing techniques are judged. Repair designers and planners have resisted employing autoclaves at field and intermediate repair facilities for some obvious and some not no obvious reasons. Autoclaves are usually large, heavy, expensive devices which require regular maintenance. Autoclave technicians require special skills and training to safely operate autoclaves. Autoclaves have high initial cost and high operational costs. Finally autoclaves with their requisite pumps and controls are too large to meet the mobility requirements imposed upon most military organizations. As a result field and intermediate maintenance operations have been required to rely upon wet-layup, pre-cured patches, or prepreg layup repairs. Each of these techniques have shortcomings which inhibit field repair operations. Wet layup repairs are made using two-part paste resins and adhesives and dry fabric and cured in a vacuum bag (Figure 1). They result in laminates of high porosity and thereby less quality than autoclaved patches. High porosity patches also inhibit post-repair inspection using ultrasonic techniques. Pre-cured patches are autoclave quality laminates and result in high quality repairs, but this technique results in serious supply and resupply problems. Each distinct pre-cured patch requires its distinct part number. Any deviation in number of plies, ply orientation, stacking sequence, size, shape, ply drop-off, or curvature requires a new and distinct part number. It is obvious how the multiplicity of repair patch variables could result in an inordinate number of part numbers and result in an unmanageable supply problem. Preimpregnated materials used to fabricate repair patches have many of the same problems encountered with wet lay-up materials (Figure 2). Prepregs cured under vacuum will exhibit unacceptable levels of porosity. It is also difficult to control final resin content and thereby per ply thickness. Prepreg materials also usually cure at higher temperatures than wet lay-up materials and require low temperature storage prior to curing.

In light of the many advantages of autoclave curing (Figure 3) over other techniques it is appropriate to develop a device which could provide the advantages of autoclave curing while circumventing the disadvantages of autoclaves for field maintenance operations.

2. DESIGN CRITERIA AND DETAILS

The UDRI-designed portable autoclave was designed to be easy to use, safe, low cost (acquisition and operation), and to take advantage of the facilities already available at a field repair site. For example, only air-powered tools are used in aircraft structural repair. The same source of compressed air used to power the hand tools was used to pressurize the portable autoclave. Additionally the portable autoclave was designed to be heated using an electric resistance heater identical to the heaters used in bonded or laminated repair operations. The autoclave temperature control is provided by the same hot bonder used to control the conventional composite repairs. Vacuum is required to make field level bonded repairs. This vacuum is supplied by the hot bonder or by an external pump or ejector. This vacuum source is also available to the portable autoclave. The autoclave vessel is, therefore, the only part of the system not already available. The repair technicians are already available at the field repair facility. This design concept greatly reduces the initial cost of the portable autoclave in that many of the expensive components of an autoclave are already familiar with the existing components which reduces training to a minimum and eliminates the need for additional training updates. Additionally maintenance of the autoclave is held to an absolute minimum.

The pressure vessel can be sized to accommodate the specific repair activity anticipated for the aircraft involved. The portable autoclave is intended to fabricate repair patches at the field level so the vessel size can be held to a minimum. The autoclave design pursued at UDRI resulted in two pressure vessels. The first was 57 x 57 cm (22.5 x 22.5 in) to accommodate 50 x 50 cm (20 x 20 in) patches. The second was 25 x 102 cm (10 x 40 in) to accommodate 22 x 95 cm (9 x 38 in) patches. Both vessels were 7.5 cm (3 in) deep to accommodate flat or mildly contoured patches and the requisite tooling. The vessel (Figure 4) consists of two similar clamshells separated by a silicone diaphragm. The two halves are held together by a series of bolts. The silicone diaphragm acts as the seal between the halves and separates the high pressure and the negative pressure chambers of the vessel.

The lower half of the clamshell has three penetrations; one for vacuum, one for thermocouples, and one for electrical power to the heater blanket. The upper clamshell has a single penetration for high pressure gas. Both halves of the clamshell are reinforced to limit deflection of the clamshell when pressure is applied. The high pressure side of the chamber is regulated by a dual pressure regulator with a maximum low side of 34.5 MPa (50 psi) and a pop-off valve set to open at 44 MPa (65 psi). The clave is further protected by using non-standard fittings to reduce the possibility of a hook-up to the wrong regulator or pressure source. Additionally when the clamping bolts are properly torqued, the seal tends to leak at modest overpressures. The small internal volume results in relatively low stored energy in the event of a leak.

The two portable autoclaves installed on the repair cart are:

- Inside dimensions: 57 x 57 x 7.5 cm (22.5 x 22.5 x 3 in)
  25 x 102 x 7.5 cm (10 x 40 x 3 in)
- Operating pressure: -0.103 to +34.5 MPa (-15 to 50 psi)
- Operating temperature: 177°C (350°F)

This autoclave gives the field repair technician the capability to fabricate precision high quality composite patches in the field using the identical cure process employed in the factory. The exact aircraft contour can be reproduced by employing any one of several tool molding techniques. This tool can be used to autoclave cure the exact patch for each repair application.

The portable autoclave, while intended for field repair, has found considerable use in the laboratory. The system is relatively small, easy to use, and inexpensive to operate. The portable autoclave has been shown to be quite useful in curing small flat test specimens when the number of specimens in a set does not warrant using the larger clave. The portable autoclave is also useful in conducting processing studies. The double bag curing technique can be performed quite easily in this autoclave with no modifications. Double bag debulking of uncured materials is also possible using this autoclave.

Operation of the portable autoclave is simple. Figure 5 shows the autoclave vessel ready to be opened. Figure 6 shows the autoclave opened with the interior components set aside. The insulated lower platen is exposed. Figure 7 shows the autoclave with the lower platen covered with a separator film with the patch in place and thermocouples located. Figure 8 shows the resistance heater in place over the patch and unincorporated materials. Figure 9 shows the silicone bladder in place. Figure 10 shows the portable autoclave bolted closed with both high pressure and vacuum lines attached. Figure 11 shows the portable autoclave in operation with a Briskheat ACR 6000 hot bonder.

3. SUMMARY

UDRI has designed and prototyped a portable autoclave intended for field repair operations. The autoclave is compatible with on-aircraft repair techniques. The design takes advantage of the compressed air, computer controlled hot bonders, and vacuum sources currently available at Air Force repair stations. The autoclave is easy to use, safe, and requires little maintenance. The acquisition and operating costs are also quite modest compared to conventional stationary autoclaves. The portable autoclave has been used in simulated remote field repair operations in the Composites Supportability Laboratory at Wright-Patterson AFB and in the Plastics, Adhesives and Composites Laboratory at UDRI.

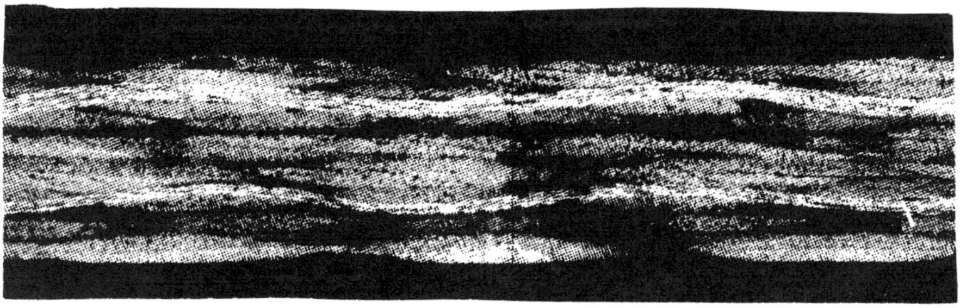

Figure 1. Wet Lay-Up Vacuum Bag Cured Laminate.

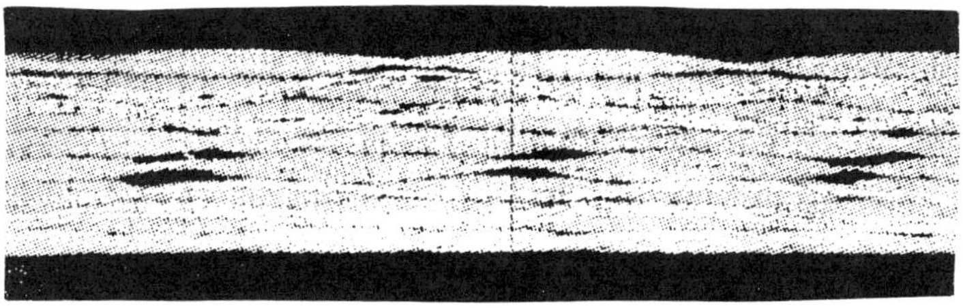

Figure 2. Pre-Impregnated Material Vacuum Bag Cured Laminate.

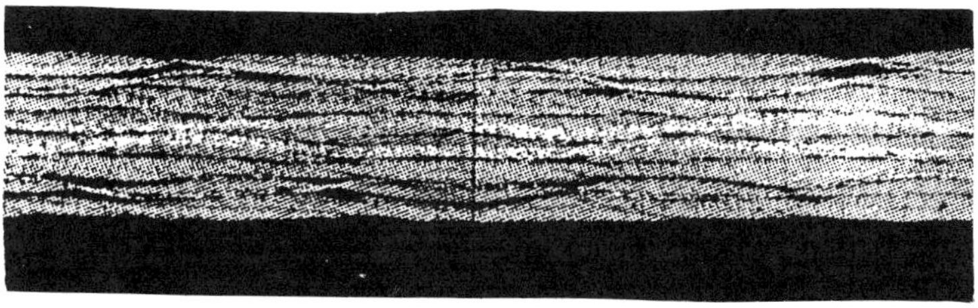

Figure 3. Pre-Impregnated Material Portable Autoclave Cured Material.

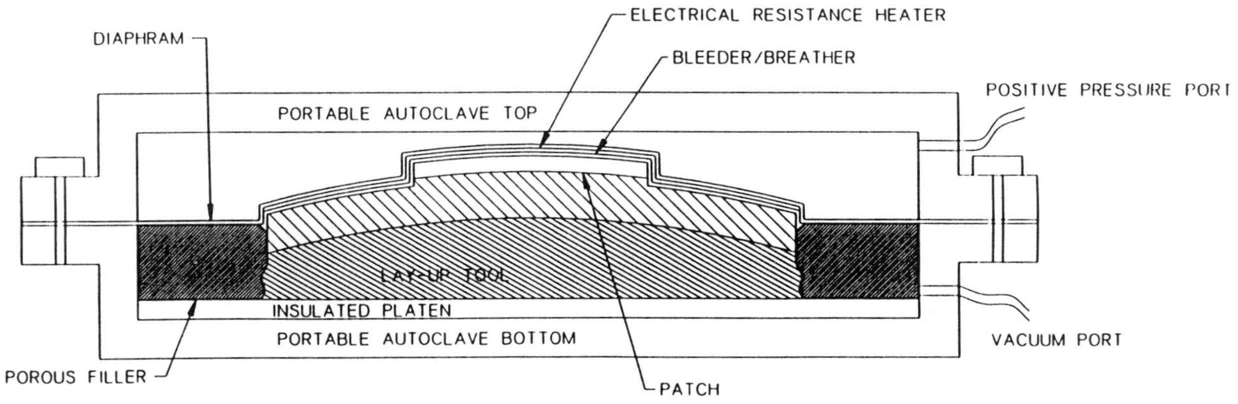

Figure 4. Portable Autoclave.

Figure 5. Portable Autoclave Ready To Open.

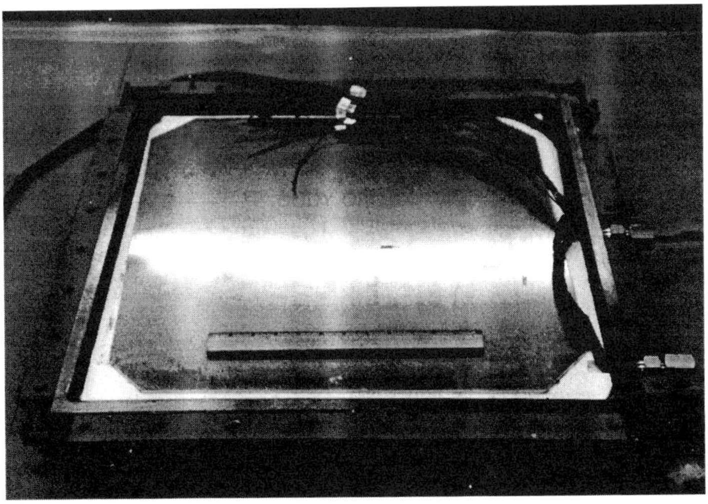

Figure 6. Autoclave Open Showing Insulated Platen.

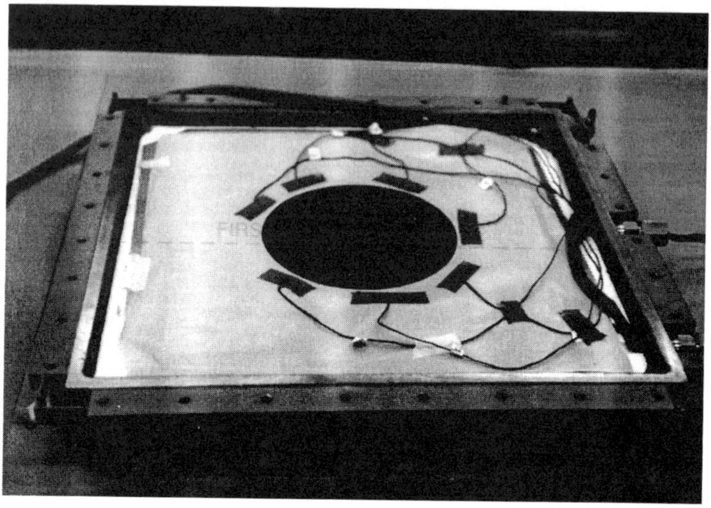

Figure 7. Portable Autoclave With Patch and Thermocouples Installed.

Figure 8. Portable Autoclave With Heater Blanket Installed.

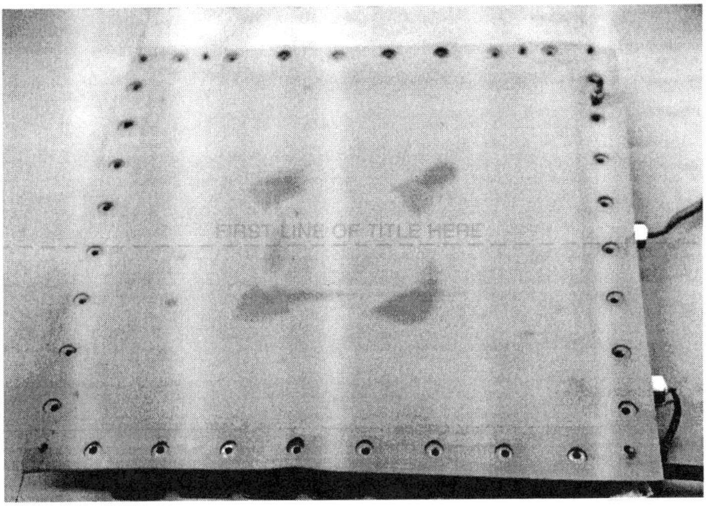

Figure 9. Portable Autoclave With Silicone Bladder in Place.

Figure 10. Portable Autoclave Closed With Lines Installed.

Figure 11. Portable Autoclave in Operation.

28th International SAMPE Technical Conference
November 4-7, 1996

Boric/Sulfuric On-Equipment Anodizing

James R. Bopp
Boeing Defense & Space Division
Seattle, Washington 98124

ABSTRACT

The Air Force's KC-135(R) CFM-56 engine leading edge cowling is being corroded by use in a high salt environment.. Boeing's task was to take a "Tank Quality" hard surface anodizing process and adapt it for on-equipment use by field level personnel. The Boric/Sulfuric anodizing process (1,3) was chosen as the best answer for this application. Application methods, durability, and scale-up were major challenges. The capability of the Boric/Sulfuric anodizing solution to provide sufficient coating thickness on 2219 alloy aluminum was investigated due to the alloy's high copper content, 5.8-6.8%. The coating thickness and weight were crucial to ensure sufficient corrosion protection. Testing on a small section of the 2219 alloy was accomplished to prove the proper coating thickness and weight were achievable and meet a minimum coating weight of 21.5 mg/dm^2. The next portion of the effort was to scale-up the process from .093 m^2 to .58m^2, the area of one segment of the cowling. The coating system provided complete wetout of the large area thus achieving an anodized surface of the proper weight and thickness.

KEY WORDS: Anodizing, Corrosion, Corrosion Resistant

1. INTRODUCTION

Corrosion causes weapon system's reduced availability and high maintenance costs. This paper will address one of the Air Force's corrosion problems, the KC-135(R) aircraft CFM-56 engine inlet cowling and the developed solution. The leading edge of the cowling is experiencing increasing amounts of corrosion caused by a high salt environment and trapped moisture. Cost and downtime were major drivers in locating a process to treat the corrosion while the parts are still on the aircraft. Boeing was requested to investigate possible solutions for this problem (2).

Several different anodizing processes were investigated to locate the most efficient process to treat the corrosion while the part was still installed. Tank quality anodizing processes were analyzed for ease of application outside of the tank. The best process identified was a Boeing patented Boric/Sulfuric hard surface anodizing solution. Air Force requirements stated that the process must be usable in an on-equipment mode, so a vacuum bag configuration was used for application.

2. REQUIREMENTS

2.1 On Aircraft Requirements The requirement for on-equipment application was levied due to the vast number of fasteners which would need to be removed to repair the cowling in an off-equipment mode. This along with the amount of time required to repair the corrosion made on-equipment application the only choice for the repair process for minor corrosion damage. Also the non-availability of the aircraft during any off-equipment repairs was totally unacceptable. Current time-to-repair at a depot facility is between three to five weeks plus 200-300 man-hours.

2.2 Hazardous Waste Reduction The overall reduction of waste is a key driver for any process used in the Air Force today. This includes any consumable materials used in a repair process. The solutions used in a process must render themselves to simple neutralization for disposal. Present corrosion treatments generate a varying amount of waste which must be reduced or eliminated.

3. DAMAGE & REPAIR HISTORY

3.1 Unique Problems A need to develop sufficient anodic coating on the 2219 alloy and provide an on-equipment process for use over large areas are the most difficult problems. The level of copper in this alloy makes it a prime choice for many applications except for use in a marine environment. The salt air causes corrosion problems in a very short time frame. The addition of an Air Force requirement that the engine's inlets be covered to preclude Forigen Object Damage during downtime aggravates the problem. The constant movement of the cover over the surface grinds in the salt which breaks down the surface protection much quicker which causes poultice corrosion to form. This progresses into more damaging crevice corrosion if left untreated.

3.2 Air Force Technical Order Repair The corrosion technical order requires the technician to grind out the surface pits and treat the area using a brush Alodine solution. This process leaves a noticeable discoloration on the cowling and has limited effectiveness. If the corrosion is deeper than approximately .011 mm or has fastener involvement, then the section of cowling must be removed and a new item installed. The majority of the corrosion is <.0058 mm in depth and confined to the nose section of the inlet cowling. The minimal effectiveness of the Alodine solution causes this process to be repeated several times before the limit of the damage forces the removal of the cowling section.

4. SEARCH FOR RESOLUTION

4.1 Application Several different techniques were investigated for applicability to this unique problem. The treatment had be simple and still provide a hard and durable surface to be acceptable. An application system which allowed on-equipment application was the next step in the search. The use of a vacuum bag to contain the solution was the best application method for large repair areas. The Phos-Acid Anodizing Containment System (4), PACS, was the best answer due to its ability to control the flow of solution, the vacuum bag pressures and provide the electrical power required to accomplish the anodizing process. The only concern associated with the PACS unit was the corrosive nature of the solution on the internal components. Several test were conducted to ensure the PACS system would provide a usable life using the boric/sulfuric acid anodizing solution. Both ATACS, the manufacture, and Boeing performed several tests on the components in the PACS system to ensure the long term exposure to the acid would not have a harmful effect.

4.1.1 PACS System The ATACS 0810 PACS system is currently used to surface treat aluminum for adhesive bonding. This process was patented by Boeing (3)to improve the adhesion for metal bonded repairs and components. The PACS system is designed to automate the anodizing process on small areas. PACS controls the amount of vacuum applied, the voltage and the flow of any solution used in the process. PACS controls the amount of fluid passing through a vacuum bag by metering to ensure sufficient flow in the vacuum bagged area. There is also a power supply inside the unit which provides sufficient power for the anodizing process. These two function of the unit provide the capability to accomplish the corrosion rework in an on-equipment mode.

4.1.2 Boric/Sulfuric Anodizing The process is similar to any other anodizing process with one exception, the boric acid tends to retard the attack of the sulfuric acid to allow a uniform coating to be achieved. Total wetting of the area is crucial for formation of the proper coating thickness. Isolation of the cathode must be complete to preclude any shorting to the surface which will reduce any possibility of smutting Figure 1 depicts the dilution of the acid components to provide the proper solution concentration.

4.2 Process Testing Initial testing required numerous steps of the process to be proven for use in the vacuum bag environment. The following steps were tried and the changes made to ensure success in an on-equipment mode:

- Surface activation methods.
- Time for total wetting of the surface.
- Level of power required.
- Solution flow rates.
- Amount of deionized water required for the rinse cycle.
- Time required for the rinse cycle.
- Simple inspection methods.

The Boeing process specification BAC5632 (1) process parameters were used as a starting point for all testing. Testing on a small scale was accomplished first to prove the capability. A piece of 2024 alloy aluminum, 9.3 cm^2 in size, was used as the initial screening material. The aluminum was weighed before any tests were run. The first test on the aluminum was accomplished using the tank process parameters. 10 minutes was added to the initial time for total wetting of the surface to ensure complete coverage of the repair area. This procedure resulted in a reduced level of coating weight, reversing this trend required a higher power setting. Power was increased from 3-4 volts to 5-6 volts. The time and flow rates were also manipulated until the optimum coating thickness was achieved. Figure 3 depicts the testing for optimun thickness of the coating using different power settings. Different breathers were investigated for their capability to isolate the cathode and their porosity for proper wetting of the surface.

Application of these parameters were then applied to the same size specimen of 2219 alloy aluminum. Each step of the process was investigated for cause and effect to understand how they would work on the 2219 material. Each test showed that the developed parameters could provide sufficient coating thickness on this unique material.

The next step was to scale up the process to provide coverage over one section of the four section inlet cowling. The inlet section has a total area of about .58 m^2 not including the fasteners. A representative section of the 2219 alloy aluminum was cleaned and the process performed. There was an increase of 10 minutes in wet out time due to the larger area to be anodized. The results of the process were within the acceptable parameters for coating thickness.

Inspection was the last real issue. Using the differential in weight technique used for the testing cycle could not be employed in an on-equipment mode. Several techniques were investigated and two methods provided desirable results. The first was the use of high intensity light and a polarizing filter. The light is applied at a 45 degree angle to the surface and looked at using the polarizing filter. The surface quality can be seen through the filter.

The next inspection technique used was Eddy current. This technique yielded the most consistent results for understanding the coating thickness and coverage. Eddy Current inspection of the surface provided the best results and since the Air Force already uses this type of equipment in the field it was the best choice.

5. RESULTS

Final process steps were tested on the actual inlet cowling section installed on a K(C)-135(R) and the results were as predicted from the lab testing. Figure 2 shows the steps required to accomplish the anodizing process including the times and power requirements. The Air Force has incorporated the process into their technical order system for use on this aircraft. On-equipment processing of the damage provides the field level personnel an ability to reduce the total repair time to a single eight hour shift.

As a final note, the reduction of hazardous waste was achieved by using the weakest acid solution which could provide sufficient coating thickness and the use of a rinse agent which diluted the acid to below required limits for simple disposal.

6. REFERENCES

1. Boeing specification BAC5632, Boric Acid - Sulfuric Acid Anodizing, Mar 15, 1993

2. Contract: U.S. Air Force, EA 93-135A1-49, Inlet Cowl/TAI Exhaust Rework and Repair.

3. U.S. Pat. 4,504,325 (12 Mar 1985) Yukimori Moji (to the Boeing Co.), 4,894,127 (16 Jan 1990) Chun-Ming Wong (to the Boeing Co.)

4. U.S. Pat. 4,988,414 (29 Jan 1991) E.A. Westerman (to the Boeing Co.)

EQUIPMENT REQUIRED	PURPOSE	ALTERNATIVE
ATACS 0810 PACS Unit	Provide a central tool for flow, amperage, voltage, and fluid switching control	All the individual pieces of equipment needed to accomplish the same functions
Acid - BORIC/SULFURIC (PER BAC5632 Specification) Sulfuric Acid - 4.5% by weight Boric Acid - 0.8% by weight Mixed with 3.7 liters of deionized water to make 3.8 liters of solution	Act as the anodizing fluid	None
Deionized Water	Seal for the anodized surface	Chromic Acid solution
Vacuum Bag material and equipment (2 mils thick)	Containment of the solution	Any compatible nylon bagging material
Breather cloth (Airweave N-4)	Isolate the cathode and assist in the wetting process	Suitable open weave material
Cathode (Stainless Steel 30-32 mesh screen)	Serve as electrical component for current flow	
Surface thermometer (0°C-40°C)	Test surface temperature for specification requirements	Thermocouple
Teflon tape (Permacel - P442-03163 CL)	Mask anodizing area and isolate cathode from surface	Equivalent tape
Bagging tape	Seal the vacuum bag and tubing to repair surface	Equivalent material
Aluminum tape	Secure anode to surface	Equivalent metallic tape

Figure 1. Equipment Listing

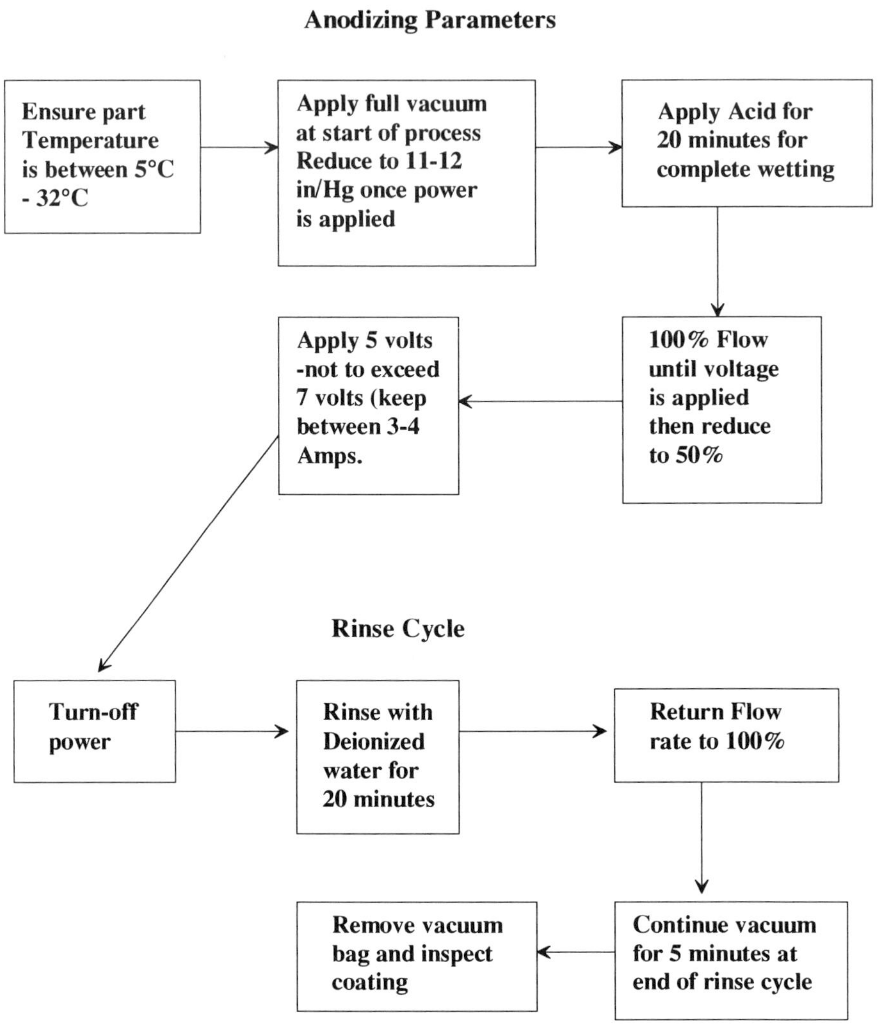

Figure 2. Process Parameters and Flow

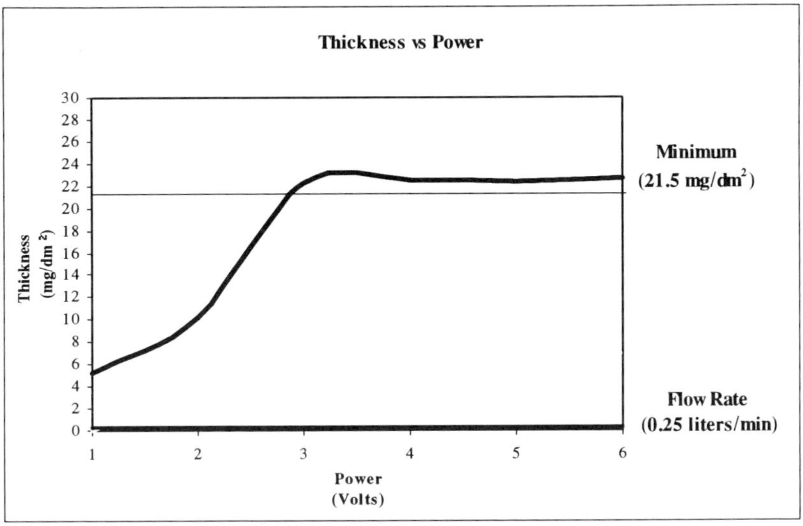

Figure 3. Testing Results

BIOGRAPHY

James (Bob) Bopp has been involved in various levels of repair, rework and inspection of both composite and metalic structures since coming to Boeing in 1990. His recent work includes simple and repeatable field level inspection methods for composites and as a maintaninability specialist for the Aft Fuselage Product Team on the F-22 aircraft. His past experience as a structural repair technician for 21 years in the United States Air Force and 9 years in composite repair have provided insight into the real life workings of aircraft maintenance. While in the Air Force he earned an AA degree from the Community College of the Air Force in Metal Technology.

28th International SAMPE Technical Conference
November 4-7, 1996

ISSUES IN COMPOSITE COMPATIBILITY FOR REPAIR

J. E. Shafizadeh and J. C. Seferis*
Polymeric Composites Laboratory
Department of Chemical Engineering
University of Washington
Seattle, Washington 98195

and

C. Blohm
Lufthansa Technik AG
Hamburg, Germany

ABSTRACT

With increasing calls for material standardization in the airline industry, the topic of composite repair compatibility is becoming of great importance. In this paper, issues regarding the effects of water ingression, hot bonding and cocuring repair techniques, tack, and flatwise tensile strength are discussed for various prepreg/adhesive combinations and honeycomb composite sandwich structures with the intent of outlining some of the critical factors necessary in developing a standardized repair methodology. Issues of material compatibility were not predictable from the rule of mixtures alone and prepreg/adhesive interactions played significant roles in the tack and glass transition temperature of the final material form. The repair technique employed was also found to affect the glass transition temperature. For adhesive/prepreg laminates, the effect of the repair technique on glass transition temperature was material dependent. In prepreg/adhesive systems, the rate of moisture ingression was initially controlled by the properties of the laminate, but eventually dominated by the properties of the adhesive. Finally, factors of compatibility in honeycomb sandwich structures were found to be independent of the material properties of the core up to temperatures of 93°C (200°F).

KEY WORDS: Repair, Composite Compatibility, Standardization Methodology

1. INTRODUCTION

Composite honeycomb sandwich structures are light-weight structural materials which have found extensive use in the aerospace and sporting goods industry[1]. Honeycomb composite structures are the materials of choice in these fields because of their low density and high shear and bending moduli. As the use of composites continues to grow, issues of repairability are becoming increasingly important. Composite compatibility, or how different composite structures interact with each other, both in manufacturing and repair, has been the focus of countless standardization teams formed by the airline industry.

* Author to whom correspondences should be addressed

In the airline industry, there are three methods currently available to repair damaged composite structures: externally bolted patches, bonded patches, and flush aerodynamic plug repairs[2]. Flush aerodynamic repair, as shown in Figure 1, is the most widely used technique in repair situations.

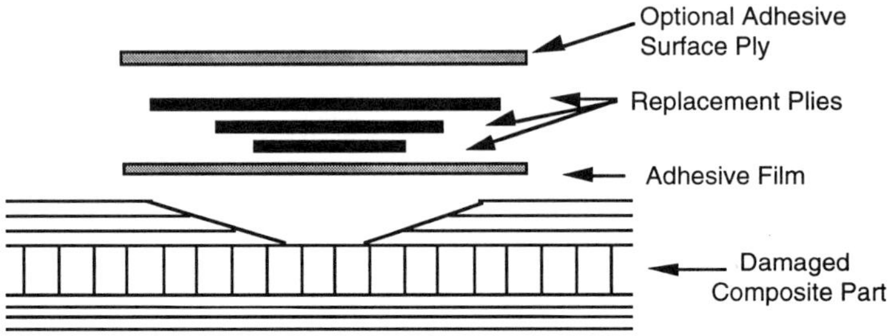

Figure 1. A Typical Schematic of a Flush Aerodynamic Plug Repair

The main drawback with this method is that each composite part must be repaired using the original material specified by the manufacturer. This forces large multi-vendor airlines to stock as many as 65 different repair materials to cover the different types of repairs regularly seen[3]. In most cases, these composite materials must be bought in bulk due to vendor packaging and difficulties in scheduling airline demand, even though only small quantities are needed for most repairs. Most of these repair materials also have a limited shelf life of six to twelve months. The net effect of large material procurement followed by limited shelf life results in a great deal of waste and a call for material standardization throughout the industry. The call for repair material standardization was made to allow materials other than those of the original manufacturer to be used in repair situations thus minimizing the volume and number of materials needed for composite repairs. Such an effort is currently under way through the Commercial Aircraft Composite Repair Committee (CACRC) under ATA/IATA/SAE sponsorship.

Although there is a need for standardization, a lack of fundamental understanding concerning the compatibility of various composite repair materials has made it difficult to develop a rational standardization methodology. With a better understanding of composite compatibility, a framework of the critical chemical, physical, and mechanical parameters affecting material standardization can be established. The focus of this paper is to investigate some of the factors which influence composite compatibility including tack, glass transition temperature, flatwise tensile strength and environmental conditioning. It should be kept in mind although these four factors are crucial in developing a standardized repair methodology, they are only a few of the many factors which influence composite compatibility.

2. EXPERIMENTAL

A variety of repair materials were selected for this analysis. The materials investigated include three types of carbon fiber prepreg (Prepreg A, B, and C). Prepreg A was a high viscosity, rubber modified epoxy impregnated into a carbon fiber 3k70 plain weave fabric. Prepreg B was an interlayered toughened, unidirectional, low viscosity epoxy prepreg. Prepreg C used the same resin system as Prepreg B, but was impregnated into a carbon fiber 3k70 plain weave fabric. Three types of adhesive (Adhesive X, Y, and Z) were used. These adhesives were modified epoxies on a supported film. Finally two types of aramid fiber honeycomb core (Core J and K) were used (both three pound per cubic foot, 1/8" cell size).

Cocured repair structures were prepared by laying-up a laminate consisting of 4 plies of prepreg and 1 ply of adhesive, as shown in Figure 2. Samples were cured using a standard autoclave cycle of 0.586 MPa (85 psi) with a two hour hold at 177°C (350°F). Bonded repair structures were constructed in a two step cycle. A two ply laminate was first prepared in the same manner as the cocured structures using the standard autoclave cycle described above. The top surface was sanded and cleaned as specified in typical repair procedures[4,5]. An adhesive ply was laid-up on the laminate followed by two uncured prepreg plies, as shown in Figure 2. This construction was then cured for two hours at a compaction pressure of 0.083 MPa (12 psi) and 177°C using an ATACS 8024-2 hot bonder.

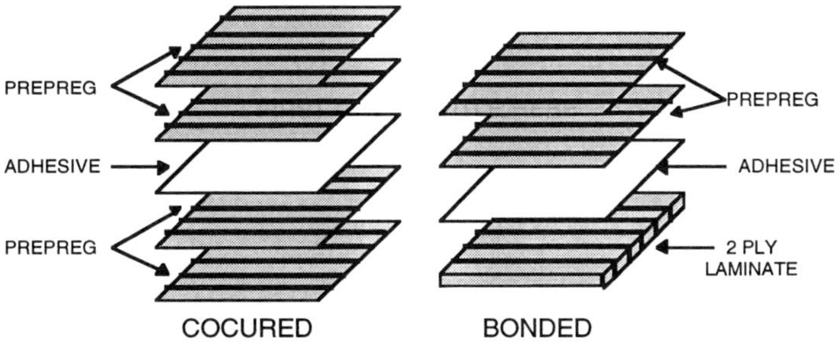

Figure 2. Fabrication Technique of Cocured and Bonded Repair Samples

Some of the cocured, hot bonded and individual repair samples were immersed in a 23°C (73°F) distilled water bath for 46 days to examine the effects of water absorption on the glass transition temperature. The samples were periodically removed, blotted dry, and weighed to measure their weight gain as a function of time.

Mixtures of adhesive/prepreg neat resin samples were also prepared. The adhesive and epoxy resins were extracted using acetone. The acetone was subsequently evaporated from the mixtures in an air circulating oven at 50°C (122°F). The adhesive resin samples were cured into plaques using the cocure autoclave cycle described above.

In these series of experiments, glass transition temperatures were measured using dynamic mechanical analysis. A TA Instruments DMA 983 interfaced to a TA 2000 Thermal Analysis System was used with a heating rate of 5°C (9°F) per minute, at a frequency of 1 Hz, and an amplitude of 0.2 mm.

Tack tests performed on adhesive/prepreg samples were based upon the methodology developed by Seferis and coworkers[6,7]. Prepreg/adhesive samples were laid up in an alternating fashion to give the final construction consisting of 3 plies of prepreg and 2 plies of adhesive as shown in Figure 3. The tack test procedure consisted of three steps. First, the adhesive/prepreg samples were compressed to 0.103 MPa (60 psi) at a rate of 0.254 mm/min (0.01 inches/min). The samples were then held in compression for 30 seconds and finally pulled apart in tension at 0.324 mm/min (0.0134 inches/min) until the samples were 1.27 mm (0.05 inches) apart. Tests were performed on an Instron 4505 mechanical testing machine interfaced with an IBM AT computer.

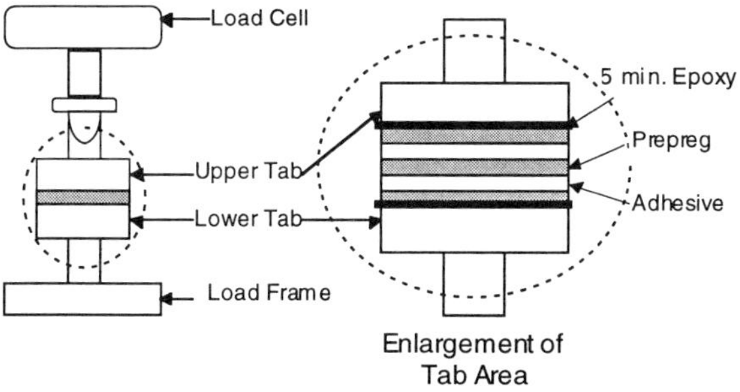

Figure 3. Tack Test Configuration and Sample Setup

Flatwise tensile tests, in accordance with ASTM C-297, were performed on a series of prepreg/ adhesive/honeycomb sandwich structures. Test coupons were fabricated using three plies of prepreg followed by a ply of adhesive on each honeycomb surface. The honeycomb sandwich structure was cured in an autoclave using a standard honeycomb composite manufacturing cycle of 0.310 MPa (45 psi) with a two hour hold at 177°C (350°F). Flatwise tensile testing was performed on 1 inch square composite coupons in an Instron testing machine 4505 interfaced with an IBM AT computer using a cross-head speed of 0.127 mm/min (0.005 inches/min). The test coupons were bonded to carbon steel tabs with BFG MA2500 epoxy and cured at 150°F for 1 hour. Six coupons from each prepreg/adhesive/core combination were tested at temperatures of 24°C (75°F), 48°C (113°F), 71°C (160°F), and 93°C (200°F).

3. RESULTS AND DISCUSSION

In order to begin developing a methodology for composite repair and compatibility, it is necessary to fully understand how different composite materials interact in a repair environment. It is not sufficient to simply add the contributions of the individual components. If composite properties and issues on compatibility were additive and not synergistic qualities, the combined properties of the repair materials should be predictable through the rule of mixtures, which is a weighted average of physical properties of the pure materials. Figure 4[8] demonstrates that mixtures of resins from Prepreg A and Adhesive Y have higher glass transitions temperatures than predicted by the rule of mixtures. This indicates the adhesive/prepreg combinations enhance each other when mixed and constitute compatible systems. The degree of difference between the individual composite components and the actual adhesive/prepreg mixtures also implies these two systems have a sizable degree of interaction.

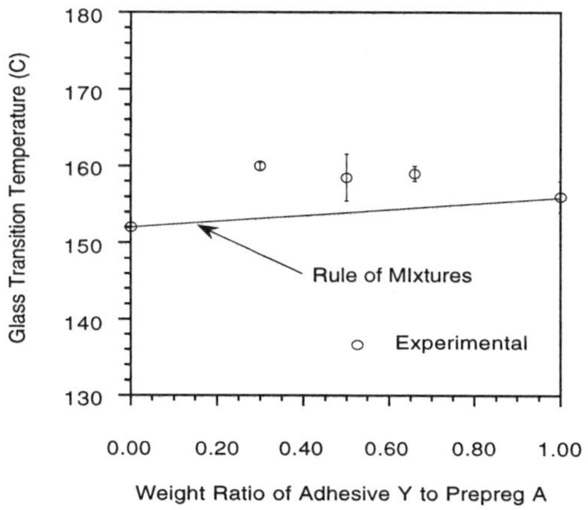

Figure 4. Glass Transition Temperatures for Cured Mixtures of Adhesive Y and Prepreg A Resin

But prepreg/adhesive resin systems are not always synergistic. In Figure 5[9], a mixture of Adhesive Z, an adhesive system not qualified in composite repair, and Prepreg A shows a lower glass transition than predicted from the rule of mixtures, indicating two incompatible systems.

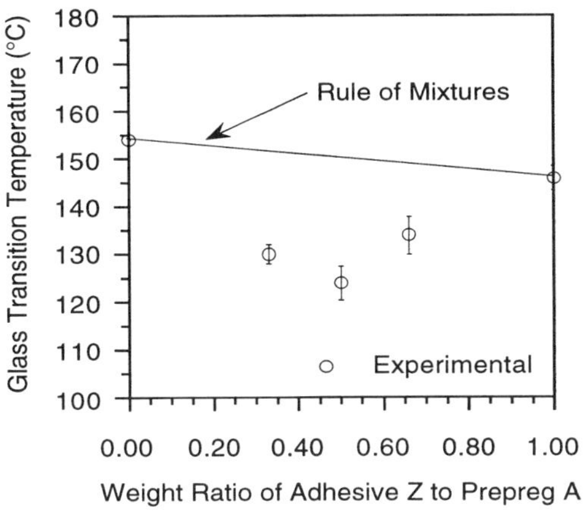

Figure 5. Glass Transition Temperatures for Cured Mixtures of Adhesive Z and Prepreg A Resin

This incompatibility might be partially explained by the nature of Adhesive Z versus Adhesive Y. Adhesive Z has a higher rubber content than Adhesive Y. The higher rubber content may lower the crosslink density of the material and is most likely the underlying factor for the observed differences in compatibility between the Adhesive Y and Adhesive Z systems. Another possible explanation for the large drop in glass transition temperature is that the different prepreg/adhesive systems could be poisoning each other, thus limiting the extent of reaction and lowering the glass transition temperature.

In developing a standardized repair methodology, it is not only important to understand that materials interact differently as a system than as individual components, as demonstrated above, but it is also significant to understand how different repair techniques affect the properties of these structures.

The interaction of the adhesive and the resin systems not only affects the properties of the final material properties of the composite, but also influences the material properties of the individual composite components. In Tables 1 and 2[2], a direct comparison is made of the cocuring and hot bonding repair techniques for different adhesives and prepregs, and the effect of these techniques on the different composite substructures.

Table 1. Glass Transition Temperatures for Different Cocured and Hot Bonded Repair Materials

Sample	**Materials**		
	Prepreg C	Adhesive Y	Adhesive X
Individual	215°C	177°C	180°C
Cocured	212°C	163°C	167°C
Hot Bonded	212°C	157°C	172°C

Prepreg C showed little change in glass transition temperature after being cocured or hot bonded with the two different adhesives indicating that the prepreg material was not greatly affected by the presence of the adhesive.

The effect on the adhesives was quite different. Adhesive Y showed significant drops of 14°C and 20°C in glass transition temperature after being cocured and hot bonded, respectively. The glass transition point of Adhesive X dropped 13°C and 8°C after being cocured and hot bonded, respectively. The effects of hot bonding and cocuring affected the two adhesives in different manners. Adhesive Y dropped to a lower glass transition temperature than Adhesive

X after cocuring and hot bonding, and was more seriously affected by hot bonding rather than cocuring. Adhesive X possessed better compatibility with Prepreg C then Adhesive Y after simulated repair. Adhesive X showed a greater degree of degradation with cocuring, while the Adhesive Y favored a hot bonding repair technique. The general inference which can be drawn from this is that the degree of adhesive/prepreg interaction is not only dependent on the interactions at the interface, but also on the processing conditions at that interface. In developing a repair methodology the issue of repair technique plays a significant role in the physical and chemical properties of the final composite material.

Although repair technique and adhesive/prepreg synergy have been considered as necessary variables in developing a standardized repair methodology, it is also necessary to consider the effects of water in order to fully understand the significant factors which affect repair material properties.

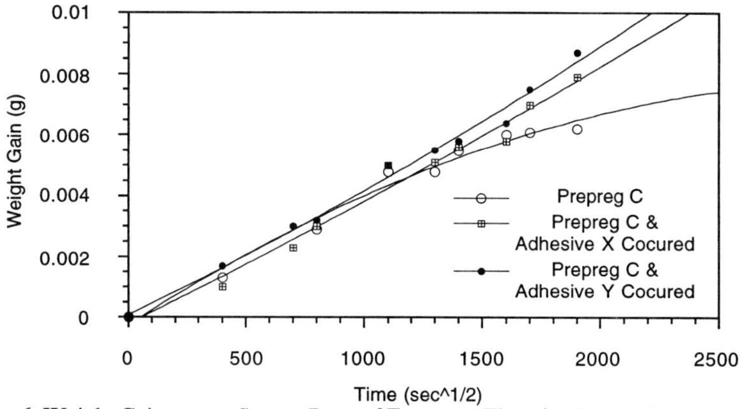

Figure 6. Weight Gain versus Square Root of Exposure Time for Cocured Repair Samples

In Figure 6[2], the water adsorption of a four ply laminate of Prepreg C (prepared in the same manner as the cocured samples) was compared with Prepreg C samples cocured with Adhesive X and Adhesive Y. From time 0 to time 1300 square root seconds, the moisture uptake of the cocured samples was the same as the moisture uptake of the Prepreg C laminate, indicating the main mechanism of moisture ingression is through the skin or laminate. Once water reached the adhesive, or saturated the laminate, the amount of water in the cocured samples increased dramatically. Similar trends were seen for the hot bonded samples. This indicates that moisture ingression into composite materials is two staged. In the first stage, the characteristics of the laminate structure dominate the water penetration behavior. Once water reaches the adhesive, absorption by the adhesive begins to dominate the process. Attempts to model the moisture ingression into a carbon fiber epoxy structure must first take into account the individual effects of the graphite skin, and secondly the interactions of the laminate with the adhesive at the bond line.

Although the glass transition temperature is a critical factor in developing a methodology for composite repair, the tack is an equally important factor. During composite lay-up, the tack between the adhesive and prepreg is the most critical material characteristic. Tack assesses the ability of the materials to be repositioned during lay up and wet each other once contacted. From the tack tests performed, the toughness factor and the elongation factor were measured for different prepreg/adhesive combinations as shown in Figures 7-10[2]. The toughness factor, as measured by the tack test, is a type of material wettability measurement which is equal to the energy required to separate the samples from the zero stress point to the point of maximum load. The elongation factor is a type of repositionability measurement which is mathematically defined as the displacement from the zero stress point to the point of maximum stress.

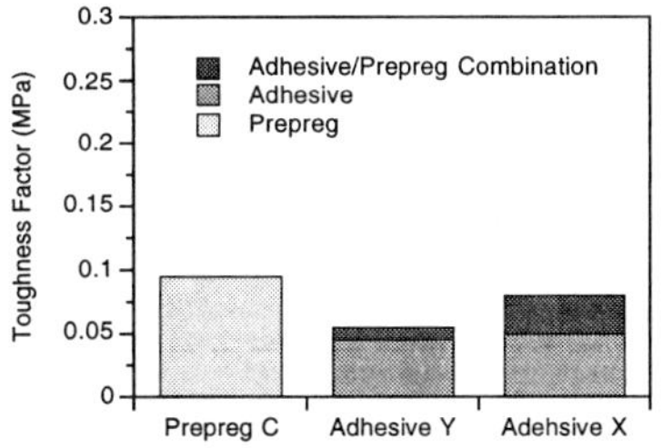

Figure 7. Toughness Factors for Prepreg C Combined with Adhesive Y and Adhesive X

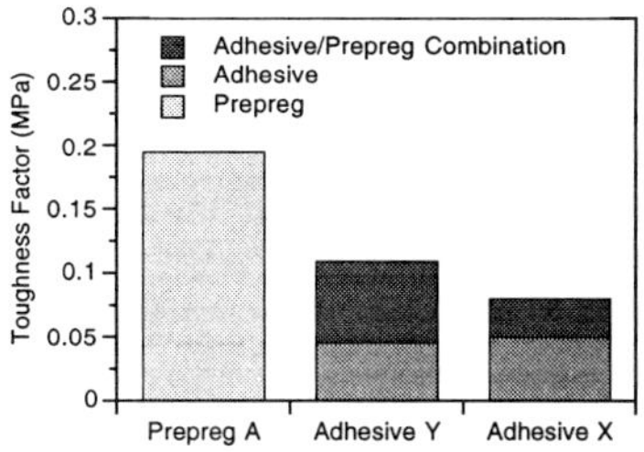

Figure 8. Toughness Factors for Prepreg A Combined with Adhesive Y and Adhesive X

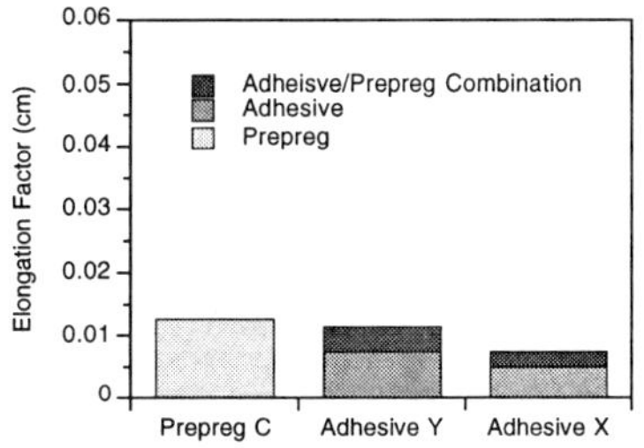

Figure 9. Elongation Factors for Prepreg C Combined with Adhesive Y and Adhesive X

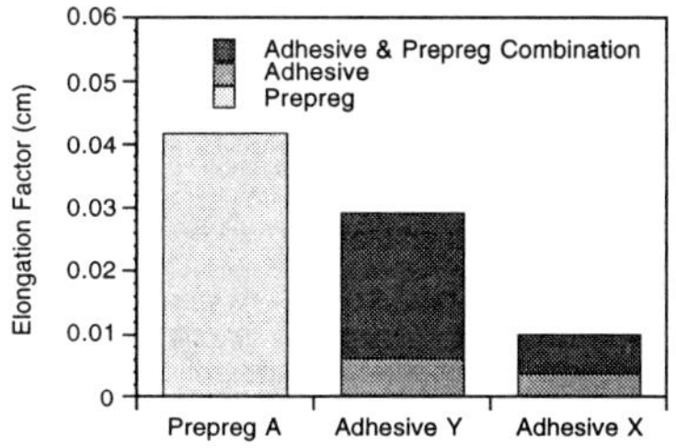

Figure 10. Elongation Factors for Prepreg A Combined with Adhesive Y and Adhesive X

The Prepreg C system, when rated by qualitative shop standards, had less tack then the Prepreg A system. This is reflected in the higher toughness and elongation factors of the Prepreg A material with respect to Prepreg C. With lower toughness and elongation factors, the materials are easier to reposition during a lay-up process, which could also affect the number and size of gas pathways in the material subsequently influencing the porosity and degree of honeycomb core crush[9].

Prepreg C/adhesive combinations had noticeably different toughness factors despite the fact that the toughness factors of both adhesives were roughly equal. The lower toughness of the combined samples indicates a decrease in the wetting of the surfaces when combined. The Prepreg A/Adhesive X combination showed similar toughness characteristics to those of the Prepreg C/adhesive systems. Prepreg A and Adhesive Y together had a higher toughness than that of the adhesive alone, suggesting the combination had a greater degree of adhesion than the adhesive alone.

The adhesive systems displayed differing degrees of elongation, while the resultant elongation factor of prepreg/adhesives combinations were about equal for both adhesives. For the Prepreg C and Prepreg A systems, the elongation factors of both the adhesive/prepreg system and the adhesive alone differed from one adhesive to another implying the elongation factor for both systems is dependent upon the interaction of the adhesive and prepreg and cannot be attributed to one material alone.

Finally, the issue of honeycomb core compatibility needs to be addressed. Flatwise tensile tests were performed on different combinations of Prepreg A and B, Adhesive X and Y, and Cores J and K.

Table 2. Flat wise Tensile Strength (MPa) of Various Prepreg/ Adhesive/Honeycomb Sandwich Structures

Core	J				K			
Prepreg	A		B		A		B	
Adhesive	X	Y	X	Y	X	Y	X	Y
24°C	2.08 ±0.08	2.23 ±0.16	2.19 ±0.14	2.26 ±0.11	3.49 ±0.19	3.61 ±0.17	3.34 ±0.23	3.39 ±0.22
48°C	2.14 ±0.20	2.21 ±0.08	2.05 ±0.15	2.27 ±0.08	3.20 ±0.29	3.46 ±0.22	3.20 ±0.22	3.74 ±0.60
71°C	1.83 ±0.07	1.81 ±0.14	2.19 ±0.05	2.17 ±0.10	3.53 ±0.27	3.22 ±0.17	3.11 ±0.36	3.11 ±0.33
93°C	1.79 ±0.08	2.12 ±0.12	2.03 ±0.11	2.19 ±0.11	3.50 ±0.25	3.32 ±0.29	3.60 ±0.33	3.11 ±0.18

The data presented in Table 2 shows no significant changes in the flatwise tensile strengths of the honeycomb composite structures due to changes in the adhesive or prepreg. All tensile failures occurred in the center of the core, away from the adhesive and laminate interfaces. The average tensile strength of Core J and Core K were found to be 2.10 ± 0.11 MPa and 3.37 ±0.26 MPa, respectively, in the temperature range of 24°C (75°F) to 93°C (200°F). For all temperatures at which tests were performed, the flatwise tensile strength of honeycomb composites seemed to be independent of temperature, prepreg and adhesive. The general conclusion which can be drawn is that core compatibility is not a significant material factor in composite compatibility or in developing a methodology for repair standardization, but may directly influence the properties and performance of the final material form.

4. CONCLUSIONS

Several conclusions can be drawn about the important factors necessary to create a standardized composite repair methodology. Issues of compatibility and composite repair are more than issues of composite characterization. How different adhesives and prepregs interact cannot be predicted based solely on the characteristics of the individual components. Prepregs and adhesives interact to form complex systems which, when compatible, result in higher glass transition temperatures than predicted from the rule of mixtures indicating significant chemical and physical interaction between the prepreg and adhesive. Different repair techniques have different effects on the glass transition temperature. Hot bonded and cocured repair techniques were found to have little influence on the properties of the laminate, but greatly affected the properties of the adhesive. The effect of the repair technique on the glass transition temperature of the adhesive was found to be material dependent. Studies of water absorption in prepreg/adhesive combinations indicated that moisture uptake is a two stage process. The first stage is dominated by moisture adsorption through the laminate, while in the second stage water uptake is dominated by the adhesive. Tack testing was also performed on prepreg/adhesive combinations. The adhesive/prepreg samples showed differing degrees of elongation and toughness which are rough measures of material repositionability and wetting. Finally the compatibility of honeycomb composite structures were explored through flatwise tensile testing. The honeycomb core was found to have little influence on material compatibility, although it may directly influence the properties and performance of the final material form. In conclusion, issues regarding tack, resin/adhesive mixing properties, repair technique, and water adsorption all play key and critical roles in developing a methodology for standardizing repair.

5. ACKNOWLEDGMENTS

The authors would like to express their appreciation to members of the Polymeric Composites Laboratory, Louise Peterson, Brian Hayes, Cary Martin, and Jeremy Klug for helpful discussions. Special acknowledgment is extended to the members of the United Airlines team under the University of Washington's Team Certificate Program as well as to the Alexander von Humboldt Foundation for support of Professor Seferis' collaborative activities with Lufthansa.

6. REFERENCES

[1] L. J. Gibson, Materials Science and Engineering, 110, 1, (1989)

[2] J. A. Fenbert, Characterization of Dissimilar Matrix Composite Materials, Master's Thesis in Engineering, University of Washington, Seattle, Washington, (1992)

[3] V. P. McConnell, Advanced Composites, 60, May/June, (1989)

[4] K. A. Armstrong, Conference on Bonding and Repair of Composites, July, (1989)

[5] Guidance Material for the Design, Maintenance, Inspection and Repair of Thermosetting Epoxy Matrix Composite Aircraft Structures, International Air Transport Association, (1991)

[6] K. J. Ahn, J. C. Seferis, T. Pelton and M. Wilhelm, Polymer Composites, 13, 197, (1992)

[7] T. M. Pelton, M. Wilhelm, K. J. Ahn, and J. C. Seferis, Proceedings SAMPE Technical Conference, 196, (1991)

[8] S. Zeng, Layered Advanced Thermosetting Polymer Systems, Ph.D. Dissertation in Chemical Engineering, University of Washington, Seattle, Washington, (1994)

[9] S. Zeng, J. C. Seferis, and C. L. Pederson, Journal of Advanced Materials, 25, 9, (1993)

7. BIOGRAPHY

Jahan Shafizadeh received his Bachelor of Science in Chemical Engineering at the University of California, Davis in 1994. In the Fall of 1995, he joined the Polymeric Composites Laboratory in the Department of Chemical Engineering at the University of Washington under the guidance of Dr. J. C. Seferis.

James C. Seferis is the Boeing/Steiner Professor of Polymeric Composites at the University of Washington and Director of the Polymeric Composites Laboratory in the Department of Chemical Engineering. Professor Seferis has pioneered several important concepts for the design, manufacturing and performance of composite materials in the aircraft environment. Having published extensively in the open literature, he has recently begun implementation of his developed methodologies in collaboration with the aircraft and chemical industries in a teaming environment created under the graduate Team Certificate Program of the University of Washington.

Carlos Blohm is a distinguished aircraft engineer with over twenty years of experience in composite utilization and maintenance at Lufthansa. He has lectured extensively on the state of composites in the aircraft environment and has mentored several professionals in aircraft maintenance from the U. S. and Europe, both academically and with "hands-on" exposure. He is the current vice chairman of the Commercial Aircraft Composite Repair Committee sponsored jointly by the American Transport Association (ATA), the International Airline Transport Association (IATA), and the Society of Automotive Engineers (SAE). He also serves as a mentor to teams in the Graduate Team Certificate Program at the University of Washington.

28th International SAMPE Technical Conference
November 4-7, 1996

AN OVERVIEW OF THE DEVELOPMENT OF THE ARPA/BIR COMPOSITE ASSAULT BRIDGE

John B. Kosmatka
Structural Systems Research
Department of Applied Mechanics & Engineering Science
University of California
San Diego, CA 93093-0085

Fred Policelli
Engineering Consultant
Salt Lake City, Utah 84117

Brian Hornbeck
U.S. Army Research, Development, and Engineering Center
Warren, MI 48397

Karl Bernetich
Hardcore Du Pont Composites
New Castle, DE 19720

ABSTRACT

The U.S. Army is interested in developing new light-weight short-span mobile assault bridging systems that will greatly improve their tactical mobility (rapid deployment/ retrievals, carry multiple bridges per launcher, minimum profile depth). One concept currently being investigated as part of the ARPA sponsored Bridge Infrastructural Renewal (BIR) program involves the development of an all-composite assault bridge (CAB). In this paper, an overview of the CAB development program including; Army requirements, bridge design, analysis, low-cost fabrication techniques using the SCRIMP process, and experimental testing program. A detailed weight and cost analysis of the current composite bridge reveals that it will weigh 75% of a similar existing metallic (Aluminum) bridge and also cost 20% less than the same metallic bridge.

KEY WORDS: Military, Composite Structures, Design/Analysis/Testing

1. INTRODUCTION

The U.S. Army is interested in developing new light-weight short-span mobile assault bridging systems. Most of the current bridging systems developed in the 1960's are not capable of carrying many of today's heavier tracked vehicles (MLC 70 vs older MLC 60), are not serviceable or upgradable [1], or are so heavy that they limit the force's mobility. The need for rapidly deployable light-weight bridging remains as important today as ever in order for armored vehicles and support vehicles to accomplish their mission. It has been shown by NATO obstacle plans of Western Europe[2] that through a combination of simple natural features and a coordinated plan of man-made obstacles that the mobility of an entire armored formation can be restricted, channeled, or even blocked. For the defender (home team), the creation of a coordinated obstacle plan is central to forcing the enemy to a prearranged target spot, while at the same time providing security from flanking maneuvers. For example, it was established that in the absence of bridges, 80% of water obstacles cannot be crossed or would prove difficult to cross even with modern water mobile vehicles and/or armored tracked vehicles. In most cases, it was not the depth of the water or the width of the gap, but rather the condition of the bank that stopped the vehicle. In fact, over 92% of the natural gaps in Europe and over 51% of the natural

natural gaps in Southeast Asia are less than 40 feet wide. In addition, almost all man-made obstacles of "tank-traps" are less than 30 feet wide. Thus, in these areas short-span bridging is just as important, if not more important than long-span bridging.

A detailed review study was performed to assess the current state of military bridging, worldwide. Single and multi-piece bridging systems of 15 different countries were reviewed. These countries include: Brazil, China, Czechoslovakia, France, East Germany, West Germany, Italy, Japan, Netherlands, Poland, Soviet Union, Sweden, Switzerland, United Kingdom, and the United States. Based upon this study it was concluded that: 1.) there are only eight different types of single-span bridges on the market, where the bridge length varies from 40 to 60 feet and the weight varies from 6,000 to 20,000 lbs, 2.) there are 14 different multi-span bridges, 3.) there are **no** composite bridges or even composite bridge components in commercial existence, 4.) the launching systems vary from country to country, where the preferred approach involves placing the bridge in a horizontal (low-profile) operation as opposed to a sweeping arc-like vertical motion (high-profile), 5.) the United Kingdom based Thompson Defense Projects, a subsidiary of Rolls-Royce Group manufacturers the best single-span bridge with a length of 44 feet, weight of 12,500 lbs, height of 28 inch, a capacity of 140,000 lbs (fully loaded M1-A1 tank), and 6.) the United States does not have a manufacturer for Army bridges, instead it has teaming agreements where the U.S. Government buys bridging from U.S. companies that act as suppliers of foreign bridging systems.

The above information was used to define the critical "metrics" for our proposed Composite Assault Bridge (CAB). Our bridge is being designed as a low-cost and low weight approach, that must compete with the existing best single-span (Thompson Defense Projects) bridge in a number of categories including; weight (9000 lbs), and capacity (150,000 lbs for a fully outfitted M1-A1 vehicle). In order to take full advantage of the high stiffness/weight ratios and strength/weight ratios of the composites, it was decided that the overall height of the bridge would be limited to a maximum of 24 inch so that a typical M1-Al launcher could carry two bridges at a time and still be within the 4 meter maximum Western Europe height limit (See Fig. 1). Thus, doubling the range, so that the launcher would have the capability of placing a bridge over the first gap and then placing the second bridge over a second gap before needing to be restocked (See Fig. 2). Alternatively, both bridges could be partially stacked to cross a complicated longer single gap.

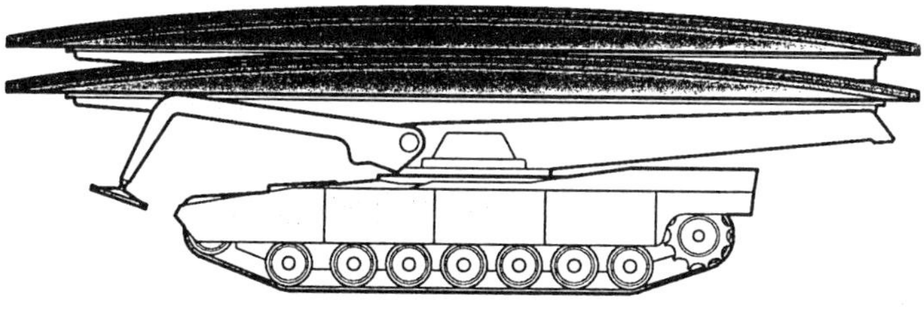

Fig. 1. Side-view of launcher carrying two Composite Assault Bridges (CAB)

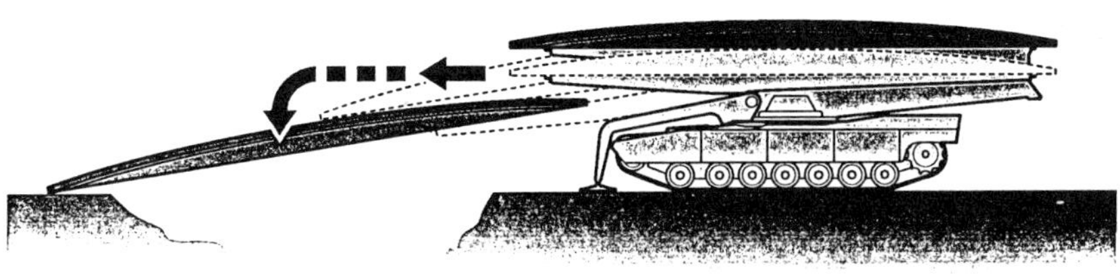

Fig 2. Launcher placing first CAB over a gap.

The overall form of the bridge is dictated by the need for low-weight, transportability, and quick placement from the launcher without direct human contact. The bridge has a very simple form that is characterized by two parallel treadways (planks) that are connected by separator bars. These separator bars are detachable so that the bridge can be stacked as treadways in five foot widths for storage or transportation. Each treadway has a flat, but open, bottom surface for crossing gaps from zero to maximum length. The upper surface of metallic bridges is typically defined as either two flat faceted surfaces (i.e., up-ramp and down-ramp) or three flat faceted surfaces (up-ramp, flat, down-ramp), where the corners of the facets experience large structural loads. For the CAB, the upper (roadway) surface was designed as a constant radius camber form to maximize the bending stiffness without producing any highly loaded corners or mechanical joints (See Fig. 3).

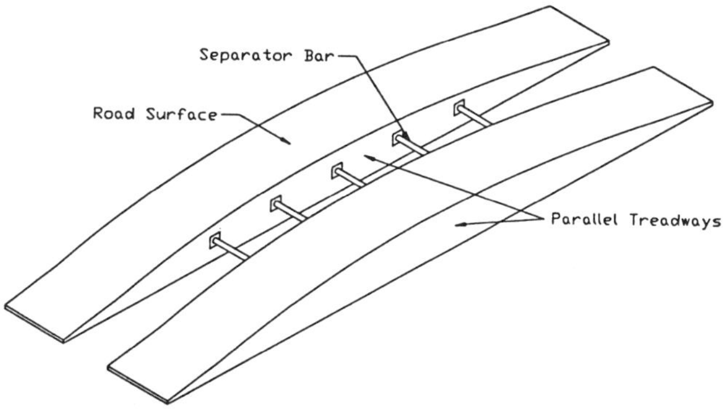

Fig. 3 Schematic of Composite Assault Bridge

It is obvious that using composite materials will improve the performance of the bridge (load carrying capability per pound of bridge weight) and the mobility of the launcher (light-weight bridging), but one must be careful to do it in a cost-effective manner and also make sure that the design is optimized for composites, not metallics. For example, two studies in the 1980's attempted to use composites as light-weight replacement components for upgrading existing metallic bridges. These studies, replaceable roadway deck panels and tension chord overlays, exposed the weaknesses of composites (instead of their strengths) by producing costly failures as a result of bolted joint failure, abrasion, and unplanned compression loadings. For structural components of this size; filament winding, pultrusion, and autoclave cures using prepreg materials are generally too expensive. Instead, the SCRIMP process was used. This process is a patented semi-automated extension of the vacuum assisted resin transfer molding (V/RTM) process. In this process, the structure is layed-up using dry fabric, vacuum bagged, and then infused with an resin. Because of the need for higher strength properties at a high operating temperature (120 degree F), an oven cured (150 degree F) epoxy resin whose properties improved with a post-cure was selected.

The objective of this paper is to present an overview of the ARPA/BIR CAB development program including; requirements, design, analysis, fabrication, and experimental testing program.

2. REQUIREMENTS

The critical loads envelope for the Composite Assault Bridge was defined using the Trilateral Design and Test Code for Military Bridging [3]. This code was developed based upon a Memorandum of Understanding for joint studies on military bridging which was signed by the United States, the Federal Republic of Germany, and the United Kingdom of Great Britain in 1972. The code is updated periodically, where the current edition was published in January,

1996. The loads envelope, which was developed for our single-span 15 meter bridge included: (1) vehicle weights including both military load class 70 tracked vehicles (M1-A1 Abrams 70-ton tank) and class 100 (100-ton wheeled vehicles), (2) maximum number of crossings (5000 crossings with an MLC-70 vehicle), (3) bridge geometry (maximum slope, treadway width, spacing), (4) bank conditions (uneven slopes (1:10 uphill/downhill, 1:20 twisting), irregularities, bearing pressures), (5) material properties for metallics, composites, and adhesives (safety factors (1.8 based upon first-ply Tsai-Wu failure criteria) thermal/environmental effects), and (6) other loading (side slope, eccentricity, launching, wind, thermal, mud (16 lb/ft2), braking, impact).

3. DESIGN ISSUES

The design of the CAB evolved to a two parallel treadway design, where each treadway is characterized by a superstructure, deck, wear surface, Aluminum end-caps for impact resistant end, and Aluminum launch rails (See Fig. 4 for an overall view and Fig. 5 for an exploded view of half of the bridge-length). The dimensions of each treadway are 46 feet long, by 5 feet wide, by a maximum of 2 feet deep at the mid-length, where the upper surface is characterized by an arc.

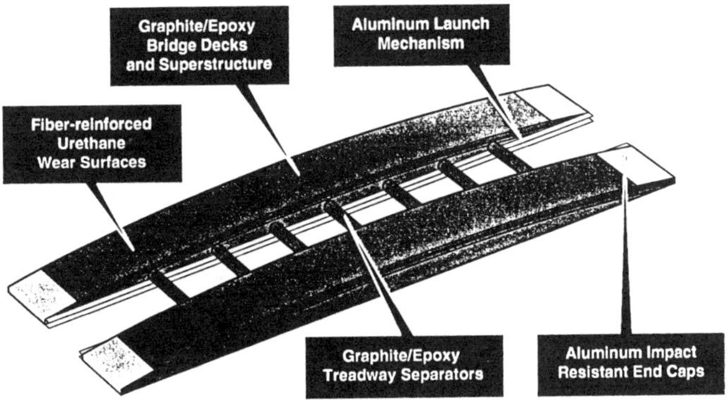

Fig. 4 Overall View of Composite Assault Bridge

The superstructure of a treadway is an open bottom W-shaped design having approximately 1.17 inch thick stitched unidirectional graphite on the lowest-most surface for tension stiffness, stitched [±45/90] graphite on the vertical and angled side-walls. The unidirectional graphite is thru-thickness stitched tape quilt packs in order to make it easy to handle and more importantly to provide a resin feed path in the SCRIMP process. Internal transverse [±45/90] graphite/epoxy (2 inch) balsa sandwich bulkheads are placed in 30 inch centers for shear and buckling stability. Hull side-wall [0/90] graphite/epoxy "hat"stiffeners (3 inch deep stringers) are secondarily bonded to the vertical and angled side-walls to greatly increase the potential buckling load to a value five times greater than the expected maximum load. The W-shape in the bottom section has a constant side-wall angle along the length which means that the apex of the (W) is much wider at the toe of the bridge than at the deep mid-length section. At the toe of the bridge, solid composite termination blocks are used to provide adequate strength from potential crushing.

The graphite used in the bridge is either Hercules AS4-D in the following forms: 0 degree stitched unidirectional, or 5-harness satin fabric, or AMOCO T-300 [±45/90] stitched fabric, or AMOCO T-650 [±45] stitched fabric. The resin for the SCRIMP cure is a Epon 862 Epicure, whereas the resin for secondarily bounding the bulkheads to the superstructure is a Ciba TDT 177-151, which is an experimental low viscosity epoxy system for V/RTM usage that cures at room temperature and has excellent properties. Lastly, the adhesive for bonding the side-wall stiffeners and the bulkhead caps is an Araldite AV 8531/hardener HV 8531.

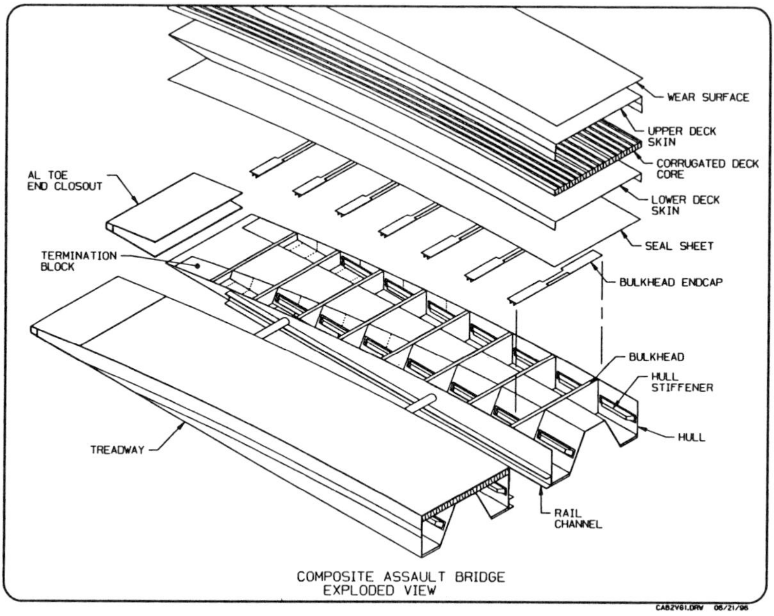

Fig. 5 Exploded view of one-half bridge length

Above the superstructure is an upper deck that is designed to: (1) provide the overall treadway compression-side bending stiffness, and (2) withstand local puncture (point) loads associated with roadway debris, carried in the tank treads, that gets through the wear surface. The entire deck is a symmetric graphite/epoxy lay-up that is defined as: two plies of a 5-harness satin AS4-D fabric then five alternating plies of unidirectional AS4-D and 5-harness satin AS4-D, then two plies of a 5-harness satin AS4-D, then a three inch thick core section, and then the aforementioned nine-ply laminate is repeated. The three-inch deep 46 foot long core section is a foam filled pre-cured (SCRIMP) corrugation composed of four plies of 5-harness satin AS4-D. The one-piece corrugation is composed of graphite fabric [0/90] which provides much needed axial stiffness to the compression bridge side, so that the overall bridge bending stiffness increased. In addition, cross-fibers in the fabric provides transverse stiffness for crush resistance. Originally, pultrusions were investigated for the corrugation sections, but were abandoned because of the difficulty in getting good transverse fiber placement for deck core crush resistance.

The wear surface could not easily be studied using analysis or even simple experimentation due to the severe nature of the tank loads. Alternatively, we surveyed the field and developed a list of 14 different potential composite, polymer, and/or metallic (reference) materials. See Section 6.2 for materials list and current results. The materials were tested at the U.S. Army Aberdeen Test Center, Aberdeen, Maryland using actual M1-A1 tank crossings. Currently, the wear materials have been exposed to crossings during the ice, snow, and freeze thaw conditions for 1200 crossings, and this summer they are experiencing the heat and rain for 2000 crossings.

The Aluminum end-caps (end close-outs) are secondarily bonded to the completed treadway to provide impact and abrasion resistant for the upper entry surface and the lower bearing surface, where jagged rocks could puncture the lower structure. The launch rails are not being addressed during this first phase of the program.

4. STRUCTURAL ANALYSIS

A detailed linear static and free vibration finite element analysis of a treadway was performed using NASTRAN, where the global model was composed of 22,000 elements and had 28 different property sets (see Fig. 6 where the model is shown without the deck for clarity). Two different vehicle loads were considered (MLC-70, MLC-100), at three different positions on the bridge (entrance ramp, mid-span centered on bridge, and mid-span with tank treads off-centered). In addition, three different boundary conditions were considered: level, (1:10) uphill, and (1:20) twisting. In addition to these defined load cases, the magnitudes of the loads were increased by the following three factors: (1.2) impact factor, (1.26) eccentricity factor (off-center tank C.G. due to even loading or rotated turret), and (1.8) safety factor. For example, the load distribution of the 7-axle M1-A1 tank (75-ton) and the tank tread pad definition are given in Fig. 7. Under these worst case loadings, the bridge maximum center span deflection is 3.5 inch, and the calculated safety margin (above the safety factor) is 1.5 in the upper roadway deck, 1.3 in the side-walls, and 4.5 in the lower-most tension members.

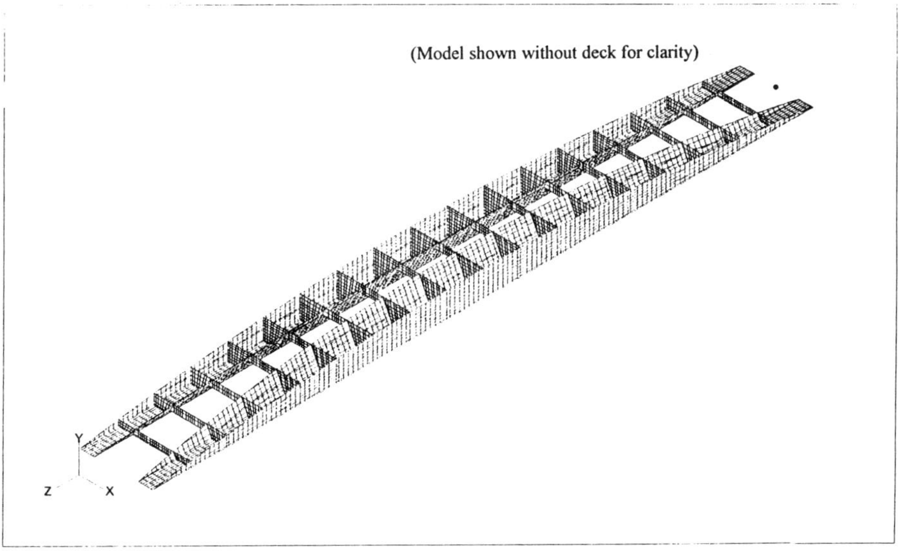

Fig. 6: Finite element model of treadway (decked removed for clarity).

Fig. 7: Load distribution of an M1 A1 tank

In addition to the large models, a series of small local models were constructed in order to size and locate the bulkheads, to design the "buckling resistant" side-wall stiffeners (3 inch deep stringers), and define the corrugation of the deck.

Based upon the finite element results and the positive safety margins throughout, the weight of the bridge can be calculated to be: 9000 lbs, where the detailed breakdown is 1355 lbs for the weight of each treadway superstructure, 424 lbs for the bulkheads in each treadway, 2088 lbs for the upper deck of each treadway, 300 lbs for the wear surface, and 660 pounds for the treadway separator bars and launch rails. Thus, the current design is less than 73% of the weight of the current competing metallic bridge.

5. FABRICATION

The fabrication steps for the treadway are defined in this section, where the treadway components are shown Figs. 4 & 5. First a large 50 foot steel female tool was fabricated for laying up the lower super-structure section (See Fig. 8). In addition, a matching 60 foot oven was fabricated for curing the treadway and all subcomponents at 150 degrees F. Second, the tool was prepared and the lower laminate was laid-up. This part was bagged, SCRIMPed, cured, and then removed from the tool. Third, grooved slots were cut transversely into the lower section and the premade sandwiched bulkheads were secondarily bonded into place. Four, the sidewall stiffeners were bonded to the vertical and angled sidewalls. As part of this step, 57 rosette strain gages were mounted to all of the significant stress regions inside the treadway. Fifth, a cured sealing skin (2-plies of 20 oz 5-harness satin graphite fabric) was secondarily bonded to the lower hull assembly. This sealing skin becomes the tool surface for fabricating and attaching (SCRIMPing) the upper deck. The treadway deck consists of the following 3 components. (1) a 9-ply lower deck laminate composed of 3-plies of unidirectional graphite and 6-plies of 20 oz 5-harness satin graphite fabric, where the two lower plies are the actual sealing skin, (2) a 3 inch foam-filled corrugated core, and (3) a 9-ply upper deck laminate composed of 3-plies of unidirectional graphite and 6-plies of 20 oz 5-harness satin graphite fabric. All of the deck laminates tie into the treadway sidewalls with 5 inch of bond-length. Lastly, manufactured 6061-T6 Aluminum end-caps (close-outs) are bonded to the treadway ends to provide impact and wear resistance. The end-ramps are custom made for the treadway using the actual manufactured dimensions to insure accurate bond-line thicknesses.

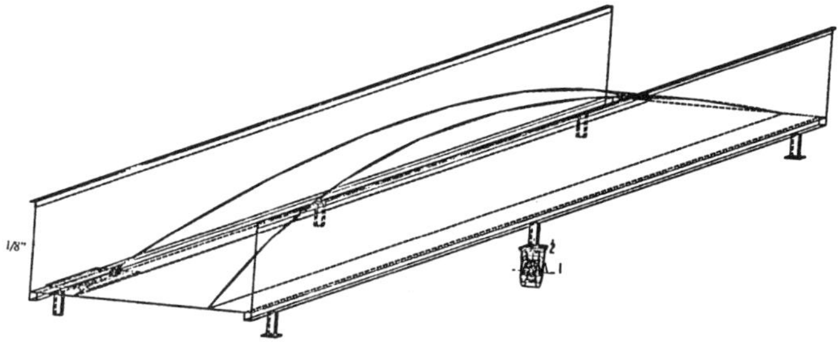

Fig. 8 Fifty-foot steel tool for lower structure

6. EXPERIMENTAL STUDIES

Three sets of experiments were performed as part of this program, where the fourth set (full-scale treadway testing) is planned for fall, 1996.

6.1 ***Material Testing*** A material fabrication study was performed at Hardcore-Dupont to assess the quality and performance of different resin and fiber material types. The goal of this task was to arrive at the best resin and fiber combination for "SCRIMPed" bridge components. The resins that are being investigated include: Dow 411-350 (vinyl ester), Dow 8084 (rubber toughened vinyl ester), Dow DER 332 Resin/DCH 99 Hardener (epoxy), Ciba TDT 177-151 (epoxy), and the Shell Epon 862 Epicure (epoxy). The different fiber types include: AS4D (Hercules) Unidirectional (24 oz/yd) and 5-Harness Satin (24 oz/yd), T300 (AMOCO) Unidirectional (24 oz/yd) and 5-Harness Satin (24 oz/yd), and A370 (Hercules) 5-Harness Satin (10.94 oz/yd). Twenty-five different laminate pairs were fabricated using the SCRIMP process and most of the panels were tested. The laminate stiffness and strength properties are given in Table 1, where it is observed that the cheapest vinyl ester resin (DOW 8084) provides good laminate properties in tension, but are unacceptably low in compression and shear. This occurs because the graphite fibers are not correctly sized for the vinyl ester resins so that during the cure the vinyl esters undergo significant shrinking so that the bond to the graphite fibers is limited. Next, the room temperature epoxies (Ciba TDT 177-151, Dow DER 332 Resin/DCH 99) provide similar tension properties as the vinyl ester, as expected, but the compression properties are significantly better. These resins were given serious consideration but it was felt that their "pot-life" was not long enough at room temperature to fully wet-out the 46 foot treadway. Next, 150 degree F cure epoxy (Shell Epon 862 Epicure) provided the best tension, compression and shear properties and were only slightly lower than autoclave cured laminates. As a comparison, the final column is of a 5 harness satin laminate composed of E-glass/vinyl ester. It is interesting to observe that while the stiffness properties of the glass/vinyl ester laminate are lower than the graphite/vinyl ester laminates, the strengths of the glass/vinyl ester laminate is higher in both tension and compression. This occurs because the glass fibers are correctly sized for vinyl ester resins.

Table 1: Composite Material Test Results

	AS4-D (5H-satin) /Dow 8084	T300 (5H-satin) /TDT 177-151	AS4-D (5H-satin) /TDT 177-151	AS4-D (5H-satin) /DER 383	AS4-C (5H-satin) /Epon 862	CM 5005 /Dow 411-350
E_x (Msi)	8.28	7.31	8.02	8.66	9.59	3.92
E_y (Msi)	7.70	7.45	8.32	8.28	8.72	3.72
G_{xy} (Msi)		0.49	0.46		0.759	0.61
ν_{xy}	0.110	0.0636	0.1037		0.0485	0.071
X_t (ksi)	82.10	70.46	88.07	79.45	120.1	85.8
X_c (ksi)	28.00	49.88	50.52	57.3	94.08	46.90
Y_t (ksi)	61.48	81.99	85.51	80.73	117.5	72.91
Y_c (ksi)				48.5		45.90
S (ksi)		11.55	11.45		17.20	11.14
V_f (Msi)		52.70	53.96		56.23	54.24
Void (%)		1.28	1.66		0.40	0.84

In addition to the material coupon tests, a series of SCRIMP "adhesion" studies were performed to assess the quality of the bond between embedded metallic (or precured composite) subcomponents and the SCRIMPed epoxy resin. These studies were performed with five different surface preparations including; as finished, bead-blasted, hand-sanded, acid-wipe, and sanding/acid-wipe. The results for the study revealed that the hand sanding with the acid wipe of the laminate and metallics produced the best shear bond of nearly 3000 psi, whereas the other procedures produced shear bonds on the order of 1500-2000 psi.

6.2 *Wear Surface Testing* A broad survey of potential bridge deck wear surface materials was also performed. These materials fall into two general types of categories: metallics and non-metallics, where metallics (Aluminum, Titanium, ARALL, GLARE) provide well understood behavior, low cost, and structural considerations, but with the concerns of corrosion and thermal mismatch with the deck composite materials. Non-metallics offer a low-cost solution, but have concerns that include UV stability, abrasion and puncture resistance, high temperature problems, limited life, and flammability. These materials include thermoplastic resins (AZDEL) which offer structural stiffness and strength with reparability/weldability, but suffer from a low coefficient of friction, urethane sheets/coatings (Diamondback, Rhino linings) which offer resistance to friction abrasion but are heavy with no structural advantage, and rubber sheets (Santoprene) which provide toughness and wear resistance, but also are heavy with no structural advantage. Alternatively, the composite bridge deck could be constructed of a thermoplastic (PPS, nylon) composite which may offer the wear and impact resistance needed to withstand the abrasive tank-tread loads without the need for a separate wear surface, but at a much higher cost (currently prohibitive).

Based upon these results the following 14 material combinations (Table 2) were wear tested using an 1200 actual M1-A1 tank crossings during the winter months to experience the freeze/thaw conditions at U.S. Army Aberdeen Test Center, Aberdeen, Maryland. Currently, the wear materials are experiencing the 2000 crossings in the summer heat and rain.

Table 2: Candidate Wear Surface Materials

Biaxial [0/90] E-Glass/Vinyl Ester Quad[0/±45] E-Glass /Vinyl Ester	Biaxial [0/90] E-Glass/Vinyl Ester (felt skin) Quad[0/±45] E-Glass /Vinyl Ester (felt skin)
Biaxial [0/90] Graphite/Epoxy Quad[0/±45] Graphite/Epoxy	Biaxial [0/90] Graphite/Epoxy (felt skin) Quad[0/±45] Graphite/Epoxy (felt skin)
Urethane Sheet Rubber (Santoprene)	Spray-On Polyurethane AZDEL (Thermoplastic/glass)
Titanium (Ti-6Al-4V)	Aluminum (7072-T6-T651)

The graphite is a Hercules AW-370 11 oz 5-harness satin. The glass is a E-glass CM-5005 biaxial and QM6408 quad. The vinyl ester is a rubber toughened product; DOW 8084. The epoxy is a Ciba TDT-177-151. Approximate initial thickness of all laminates is 0.25-0.28 inch. The first four on the right and left side of the table represent potential upper deck materials, where the right side has an added felt layer to act as a sacrificial ply. The two on the right and left side of the table represent potential added nonmetallic wear surfaces. The final material on the right and left side of the table represent the metallics. These could be considered as possible add-on wear surfaces or as experimental controls for comparison.

After 1200 crossings, only one material has completely failed; Rubber Santoprene. Two other materials; Quad glass and Quad graphite with the sacrificial felt skin were showing signs of delamination into the felt skin. Whereas the composites without the felt skin had no appreciable signs of damage. Lastly of the add-on nonmetallic wear surfaces, the Spray-on Polyurethane is holding up very well.

6.3 *Sub-Scale Components* Static tests have been performed on three sub-scale treadway like structures. These structures were 10 feet long by 1 foot wide with a mid-span depth of 10 inches (See Fig. 9). These tests were to characterize the structural behavior of a treadway-like structure and correlate the results to a finite element analysis (See Fig. 10 for three-point bend test set-up). The first two beams were manufactured by Hardcore- Dupont and the third by Seemann Composites. The first Hardcore Dupont fabricated enclosed beam, which weighs 119 lbs, was made using a Hercules supplied graphite fabric, AMOCO supplied pultruded "ribbon" graphite rod for the upper and lower bending stiffness components, and a room temperature cured epoxy (Ciba). The test results correlated well with the finite element predictions. The measured beam stiffness was 46.9 kips/in which is slightly larger than the predicted stiffness of 43 kips/in. The measured failure load was 75,180 lbs which was less than the predicted failure load of 98,000 lbs. The failure was characterized by a premature side-wall failure, which could be associated with a manufacturing defect. The second Hardcore-Dupont beam was considered a *low-cost* equivalent beam. It was constructed of E-glass/vinyl ester with AMOCO supplied embedded pultruded "ribbon" graphite rod. Thus the beam was designed to carry the tension and compression loads in the pultruded graphite rods and the shear loads in the E-glass sidewalls. The beam weighed 191 lbs, a stiffness of 37.6 kips/inch, and a failure load of 74, 340 lbs. This failure load is nearly identical to the previous graphite beam, but this time the failure was characterized by shear bond failure between the pultruded rod and the vinyl ester resin. From a simple hand calculation, the shear stress at failure is nearly 1300 psi, which is reasonable since the surface prep was only a light hand sand. The finite element correlation was very good with a predicted stiffness of 32 kips/in and a predicted failure load of 83,000 lbs at the pultruded rod bond. Finally a Seeman Composites fabricated beam composed of Hercules supplied graphite fabric, AMOCO supplied embedded pultruded "ribbon" graphite rod, and cured using an elevated temperature epoxy (Shell Epon 862). The measured stiffness was 50.22 kips/inch, which is significantly higher than the predicted stiffness of 43 kips/inch. It was the identical design as the aforementioned Hardcore-Dupont carbon beam and was built to assess different manufacturing variables. This beam weighed 132 lbs due to the use of a higher density internal foam core.

Table 3: Enclosed 10 foot Beams - Experimental Results

Type	Manufacturer	weight (lb)	Stiffness (Kip/in)		Strength (Kips)		
			predicted	measured	predicted	measured	description
carbon/ epoxy	Hardcore-Dupont	119	43.0	46.9	98.0	75.0	side-wall collapse
carbon/ epoxy	Seemann Composite	132	43.0	50.2	-	-	-
glass/ epoxy	Hardcore-Dupont	191	32.1	37.6	83.0	74.7	shear failure @ ribbon rod bond $\tau = 1300$ psi

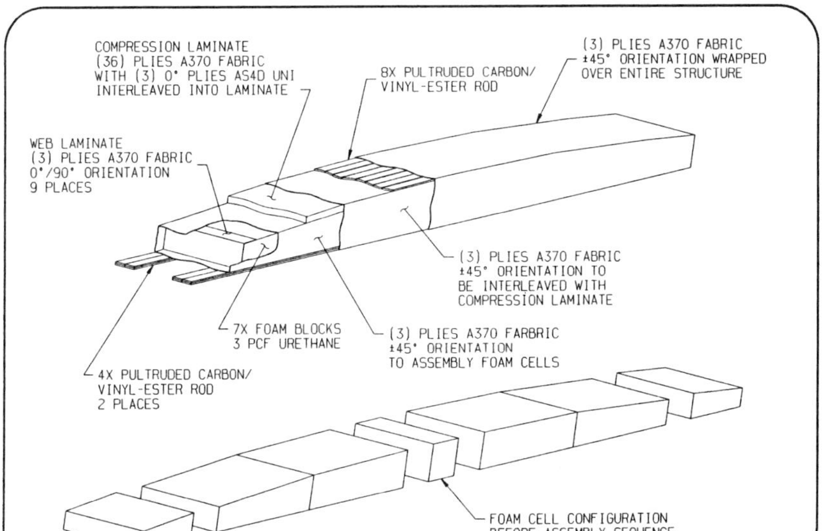

Fig. 9 Exploded view of ten foot beam specimen

6.4 *Full-Scale Treadway*

Two full-scale treadways are currently being manufactured at Hardcore-Dupont to be completed during August 1996. Both treadways will initially sent to UCSD for structural characterization. One will remain at UCSD for static failure (strength) testing. The other one gets delivered to Army for dynamic testing in the fall of 1996. The static characterization and strength test will use the Army's M1-A1 whiffle tree load distribution system to simulate the actual tank tread loading on the treadway (Fig. 11). Six different experimental load cases are planned for validating the treadway analytical models and determining the treadway ultimate load (Table 4).

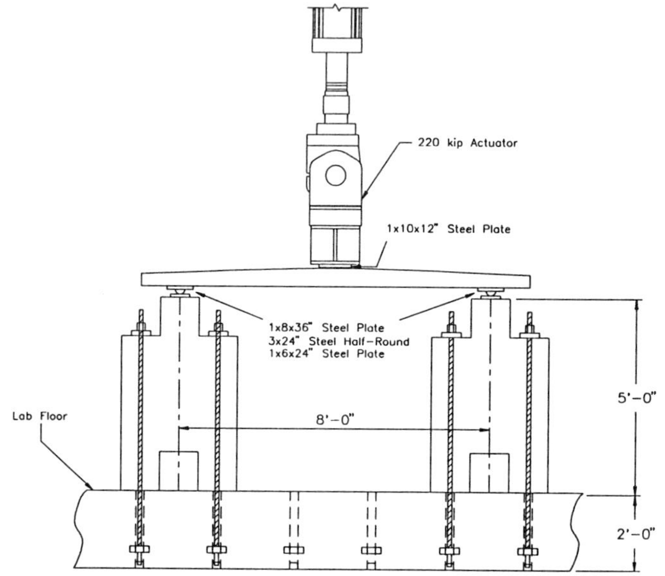

Fig. 10 UCSD set-up for three point bend test of ten foot beam

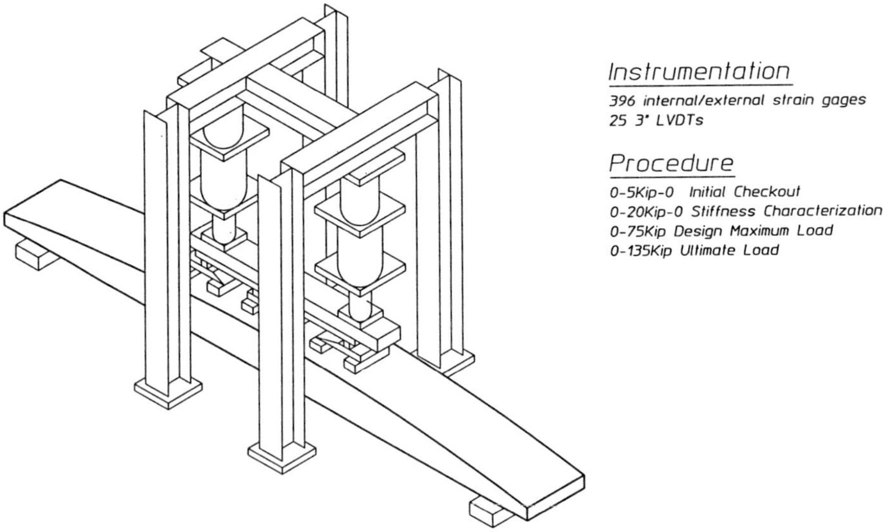

Fig. 11 Full-scale treadway test set-up showing Army whiffle tree assembly (UCSD Powell Structures Laboratory).

Table 4: Test Plan for the Full-Scale Treadway

Load (Kips)	Location		Description
5	mid-span	center	initial check-out
20	mid-span	center	stiffness characterization
20	mid-span	edge	stiffness characterization
20	entrance ramp	center	stiffness characterization
75	mid-span	center	Design Load
135	mid-span	center	Ultimate Load

8. ACKNOWLEDGMENTS

The authors gratefully acknowledge the support of this research from the ARPA TRP contract MDA972-94-3-0030 and the technical guidance by the monitors: Steve Wax and Jon DeVault. In addition, the authors wish to acknowledge the efforts of Ralph Nuismer at Alliant Corp and Chris Spragg at AMOCO for supplying the graphite materials, Tim Grady at Composite-Optics Inc. and Professor J. Gillespie at the University of Delaware for performing the material testing, and Trans-Science Corp for many of the engineering drawings. Lastly, the authors wish to acknowledge David Dillon, Cindy George, Frank Ho, and David Wolfe, of UCSD for performing the finite element analyses and the experimental tests, and Chris Thompson and Mike Grunza of Hardcore-Dupont for analyzing and fabricating the hardware.

9. REFERENCES

1.) Bustin, I., "Military Bridging Programmes for Armored Vehicles," Military Technology, vol. XVII, Issue 2, pp. 50-61, February, 1993.

2.) Bischmann, P; "River Crossing in the Central Region," NATO's Sixteen Nations, vol. 3, pp. 76, 1985.

3.) "Trilateral Design and Test Code for Military Bridging and Gap-Crossing Equipment," Trilateral Design and Analysis Group of the United States, Federal Republic of Germay, and the United Kingdom of Great Britain, January, 1996.

28th International SAMPE Technical Conference
November 4-7, 1996

SCRIMP FABRICATION OF BRIDGE DECKS

K. Bernetich, D. Briggs, S. Viskocil
Hardcore DuPont Composites L.L.C.
New Castle Delaware 19720

ABSTRACT

During the last year and a half, Hardcore DuPont Composites has produced several glass fiber reinforced plastic (GFRP) bridge decks using the Seeman Composite Resin Infusion Molding Process (SCRIMP). These decks have been produced and tested through funding from the DARPA/BIR and NIST/ATP programs. This paper presents the relative structural performances of the deck configurations relative to their manufacturing labor and material costs in order to assess which configurations are the most economically viable in the bridge infrastructure marketplace. The conclusion is that the most cost efficient design is a function of the support configuration and deck size. Triangle and trapezoid web sandwich decks are most efficient in 1-D beam like support configurations, the wrapped cube deck was most efficient in a 2-D low aspect ratio plate support configuration, and the balsa deck was most cost effective of all in the particular size tested, but could not be economically scaled up in depth for superstructure applications where the others could.

KEY WORDS: SCRIMP, Bridge-Deck, Sandwich-Core

1. SUMMARY

This paper documents the selection of one of the following four configurations for a glass fiber reinforced plastic (GFRP) bridge deck to be used in the Delaware River and Bay Authority (DRBA) Magazine Ditch Bridge.

1) Triangle Core Deck
2) Trapezoid Core Deck
3) Cube Core Deck
4) Balsa Core Deck

All four of these deck configurations were built by Hardcore DuPont Composites L.L.C. and tested under the DARPA/BIR and NIST/ATP programs over the last year and a half. Dr.

Vistasp Karbhari of the University of California, San Diego, and Dr. Jack Gillespie of the University of Delaware directed the testing at their respective institutions.

The stiffness and bending (not shear) strengths of the decks are governed by the face skins, not the core, since the span to depth ratios for real applications will be greater than 8:1. Since all candidates were essentially sandwich structures, and therefore could have the same face skins, the core shear strength and overall manufacturing cost of the deck core became the primary differentiating aspect of the designs.

In order for the GFRP bridge decks to be widely applied, they must be cost competitive with traditional materials. Therefore, the key selection criteria employed in this study was core shear strength per overall manufactured cost

After applying normalizing functions for test configurations and design weights, it was found that the triangle and trapezoid cores had the highest specific strength (strength per $) for a 1-D support configuration (i.e. an end supported beam), and a cube core was best for a 2-D support configuration (i.e. a low aspect ratio plate supported on all four edges). The balsa core was by far the least expensive and performed adequately, though it was not cost competitive when scaled up to the size to be used in Magazine Ditch.

2. TESTED DECK DESCRIPTION

All decks used the off-the-shelf stitch-bonded glass fabrics whose fiber aerial weight architecture, strengths and moduli are shown in Table I.

Table 1 - GFRP Fabric Architecture and Properties

	Direction	Fiber Aerial Weight g/m2 (oz/yd2)	Design Tensile & Shear Strength MPa (Ksi)	Tensile & Shear Modulus GPa (Msi)
UM 2403	Warp	813 (24)	410 (59)	39 (5.7)
	Fill	0	24 (4)	15 (2.3)
	Shear	0	45 (7)	6 (0.9)
CM 5005	Warp	848 (25)	232 (34)	26 (3.7)
	Fill	848 (25)	192 (28)	28 (4.1)
	Shear	0	26 (4)	5 (0.7)
TVM 3408	Warp	542 (16)	271 (39)	27 (3.9)
	Fill	0	67 (10)	15 (2.2)
	Shear	610 (18)	75 (11)	8 (1.2)
QM 6408	Warp	813 (24)	164 (24)	25 (3.6)
	Fill	542 (16)	156 (23)	22 (3.2)
	Shear	813 (24)	75 (11)	7 (1.0)

Figure 1 shows the general construction of the four deck types. Table 2 lists the ply lay-up of all decks tested, grouped by configuration.

Figure 1 Deck Core Construction

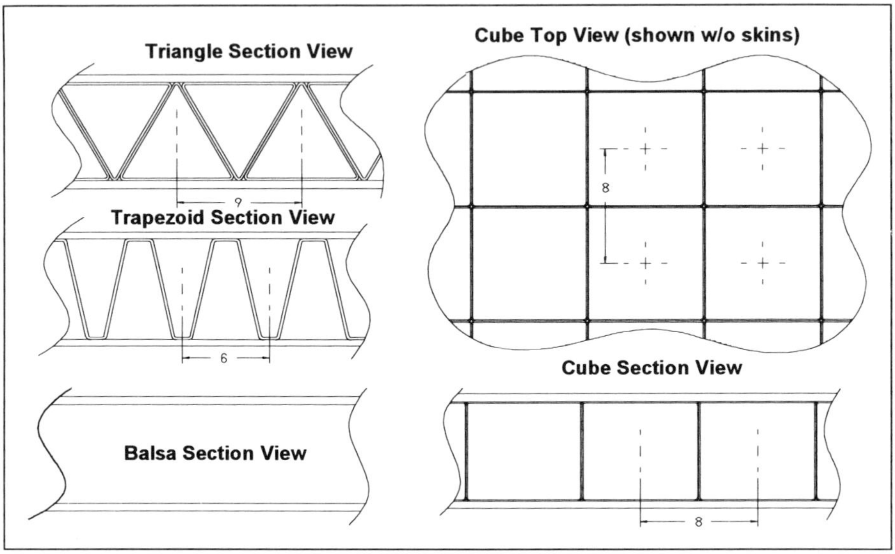

Table 2 Skin and Core GFRP Lay-up

Deck	Face Skins	Webs	Face Skin thick incl. adj webs mm (in)	Web Aerial wt (excl faces) Kg (lb)	Total Aerial weight Kg (lb)
Tri GT1B transverse	7 ply CM5005 2 ply QM6408	2 adjacent 4 ply TVM3408 wraps	17.7 (0.697)	3.08 (6.8)	9.46 (20.9)
Tri GTS1 transverse	7 ply CM5005 2 ply QM6408	2 adjacent 4 ply TVM3408 wraps	17.7 (0.697)	3.08 (6.8)	9.46 (20.9)
Tri GT1A transverse	7 ply CM5005 2 ply QM6408	2 adjacent 4 ply TVM3408 wraps	17.7 (0.697)	3.08 (6.8)	9.46 (20.9)
Triangle Long	7 ply CM5005 2 ply QM6408	2 adjacent 4 ply TVM3408 wraps	17.7 (0.697)	3.08 (6.8)	9.46 (20.9)
Trap 1 transverse	6 ply UM2403 4 Ply QM6408	5 ply TVM3408	11.4 (0.450)	2.81 (6.2)	6.93 (15.3)
Trap 2 transverse	6 ply UM2403 4 Ply QM6408	5 ply TVM3408	11.4 (0.450)	2.81 (6.2)	6.93 (15.3)
Trap 1 Long	6 ply UM2403 4 Ply QM6408	5 ply TVM3408	11.4 (0.450)	2.81 (6.2)	6.93 (15.3)
Box 1	7 ply CM5005 2 ply QM6408	4 plyTVM3408, 2 Ply CM5005	20.5 (0.807)	4.63 (10.2)	12.00 (26.5)
Box A	2 Ply CM5005 2 Ply QM6408	2 adjacent 2 ply TVM3408 wraps	8.6 (0.338)	1.41 (3.1)	4.50 (9.9)
Box B	2 Ply CM5005 2 Ply QM6408	2 adjacent 2 ply TVM3408 wraps	8.6 (0.338)	1.41 (3.1)	4.50 (9.9)
Box 441	2 Ply CM5005 2 Ply QM6408	2 adjacent 2 ply TVM3408 wraps	8.6 (0.338)	1.41 (3.1)	4.50 (9.9)
Box 10% Styrene	2 Ply CM5005 2 Ply QM6408	2 adjacent 2 ply TVM3408 wraps	8.6 (0.338)	1.41 (3.1)	4.50 (9.9)
GB2 Balsa	7 ply CM5005 2 ply QM6408	Balsa	17.7 (0.697)	0.00 (0.0)	6.37 (14.1)

The Triangle Decks are composed of isosceles triangles (2 equal length sides) with the height of the foam being about 16.5 cm (6.5"). The triangles were individually wrapped with four layers of TVM 3408. The face skins consisted of seven layers of CM5005, two layers of QM6408 on the outside and the four wrapped core layers which were oriented against the faces.

The Trapezoid Decks were made using a simpler manufacturing technique than the Triangle decks involving laying interstitial plies only between the trapezoid shaped foam pieces. This represents a manufacturing improvement, because wrapping individual cells is time consuming and labor intensive. The webs contained 5 layers of TVM 3408 , and the face skins had 4 layers of QM6408 and 6 layers of UM2403. The bond area between the webs and the face skin is only 3.81 cm (1.5 in) on the top surface, and 2.54 cm (1.0 in) on the bottom surface.

The Box Decks were made by hand wrapping layers of TVM3408, (and CM5005 on Box 1 only), onto 20.32 cm (8 in) square cubes. Box 1 Deck had 6 layers of fabric on each face from 9 individual pieces. Decks A, B, 441, and 10% Styrene, had 2 layers of TVM3408 on each face from 3 individual pieces. The cubes were then assembled by hand and face skins put on. The face skins on Box 1 consisted of 7 layers of CM5005, and 2 layers of QM6408. The face skins on the rest of the box decks consisted of 2 layers of CM5005, and 2 Layers of QM6408.

The Balsa Deck was made of 250 Kg/m^3 (15 lb/ft^3) balsa core with face skins on each side. The face skins consisted of seven layers of CM5005 and two layers of QM6408 on the outside.

3. TEST PERFORMANCE

This section will present the actual performance of the decks in test, and Section 4 will discuss the significance of this data. Table 3 shows the failure loads of each of the tested decks. The normalization function equivalenced all decks to an 8' span, 24" wide simply supported configuration. Then the specific failure load was derived by taking the normalized load and dividing by the aerial weight shown in Table 2.

Table 3 Summary of Deck Test Failure Loads

Deck	Span Tested m (in)	Actual Failure Load t (Kips)	Normalized Failure Load t (Kips)	Normalized Specific Failure Load t/Kg/m2 (Kips/lb/ft2)
Tri GT1B X'verse	1.1 (45)	24 (52)	24 (52)	37 (8)
Tri GTS1 X'verse	1.1 (44)	34 (74)	34 (74)	53 (11)
Tri GT1A X'verse	1.1 (44)	20 (44)	20 (44)	32 (6)
Tri Longitudinal	2.4/.9 (96/36)	72 (158)	91 (201)	144 (30)
Trap 1 X'verse	1.5 (60)	10 (23)	10 (23)	18 (4)
Trap 2 X'verse	1.5 (60)	11 (24)	11 (24)	19 (4)
Trap 1 Longitude	2.4/.9 (96/36)	98 (215)	124 (274)	216 (44)
Box 1	1.0 (38)	49 (108)	71 (157)	75 (15)
Box A	1.6 (63)	50 (110)	50 (110)	173 (35)
Box B	1.6 (63)	44 (98)	44 (98)	154 (32)
Box 441	1.6 (63)	53 (117)	53 (117)	184 (38)
Box 10% Styrene	1.6 (63)	51 (112)	51 (112)	176 (36)
Balsa GB2	1.2 (46)	41 (90)	41 (90)	N/A

All decks failed in core shear except the box decks A, B, 441 and 10%. These four decks failed in face buckling because their skins were thinner than the rest of the decks (see Table 2). It may be conservatively assumed that they were on the verge of failing in core shear since Box 1, with thicker face skins failed in shear at about the same load.

Because all decks failed in shear (or nearly so), normalization of the compound span data to a simple span only requires the normalization of the shear force, not the bending moments. Figure 2 shows the compound bending test setup which was only used for two triangle and one trapezoid decks. All other decks were tested in the simple span configuration shown in Figure 3. Equations 1 through 4 derive the maximum shear in the compound span setup. Ratioing the compound span maximum shear of .63P by the simple span maximum shear of .5P gives a normalization factor of 1.26.

Figure 2 Compound Span Test

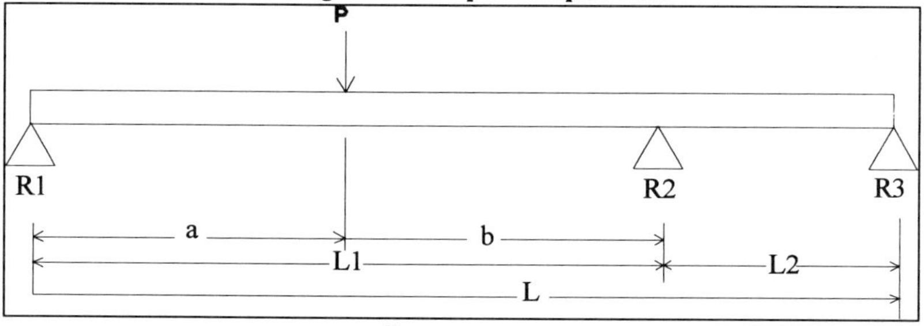

$$(1)\ R_1 = P - R_2 + R_3 \quad (2)\ R_2 = \frac{Pa}{2L_1^2 L_2}\left[L_1^2 + 2L_1L_2 - a^2\right] \quad (3)\ R_3 = \frac{Pab(L_1 + a)}{2L_1L_2L}$$

Figure 3 Simple Span Test

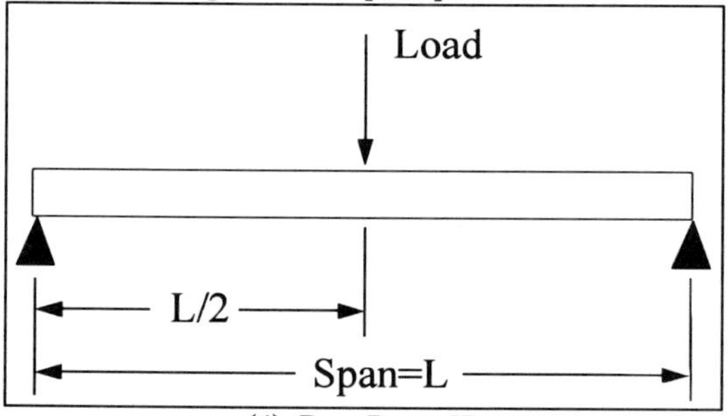

$$(4)\ R_1 = R_2 = .5P$$

All test decks were approximately 24" wide except Box 1 which was 16.5" wide. The width normalization factor for that deck is 24/16.5 = 1.455.

Each of Figures 4 through 8 are a representative load deflection curve from each of the four types of decks. For the triangle and trapezoid decks, data from the simple support configuration was chosen for presentation because it is more comparable to the other deck types which were only tested in simple support configurations.

Figure 4 Triangle Deck GT1B

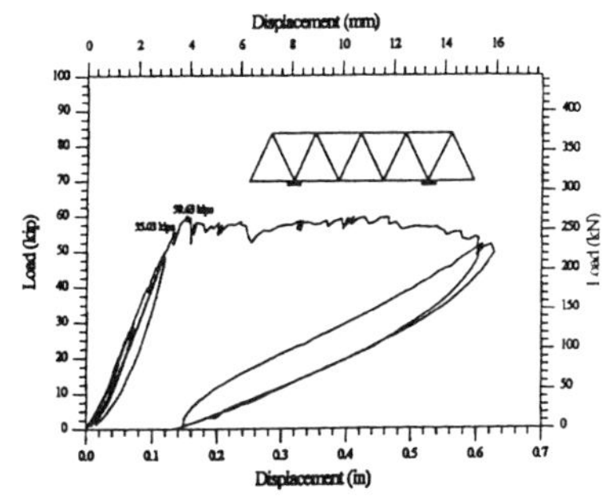

Load deflection curve for GT1B

Figure 5 Trapezoid Deck

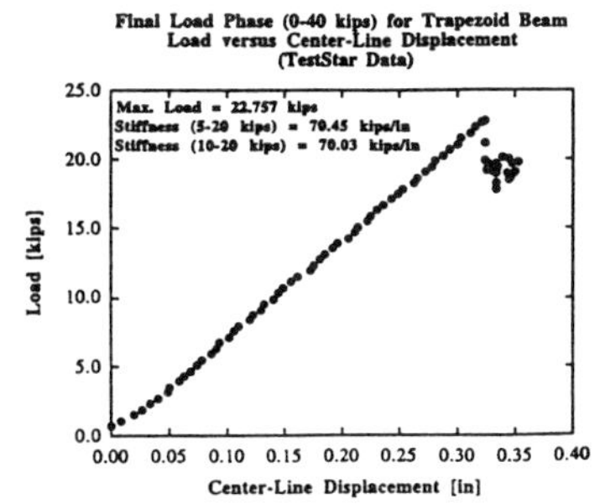

Load deflection curve for Trapezoidal deck.

Figure 6 Box Deck A

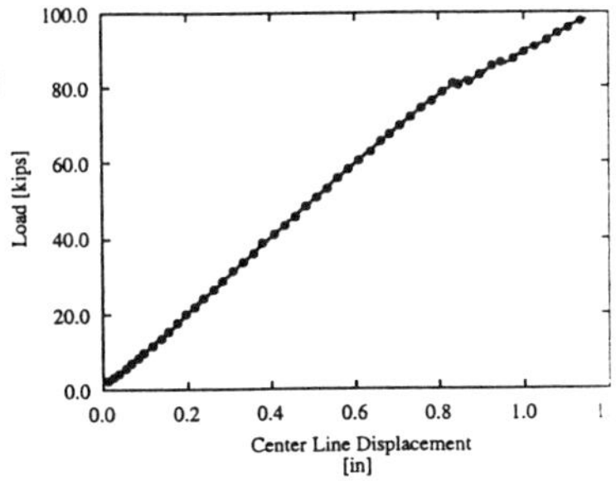

Figure 7 Balsa Deck GB2

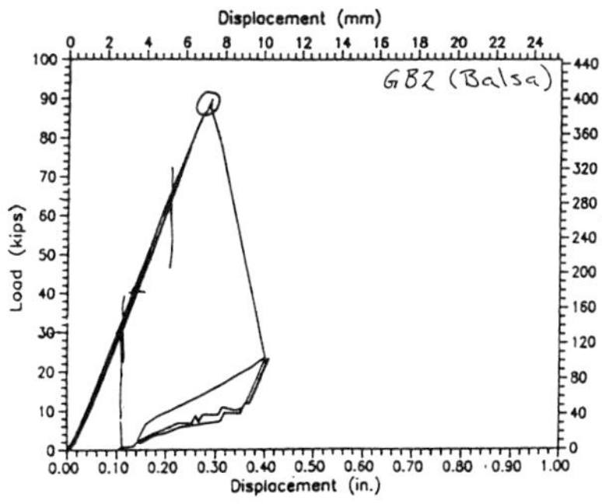

4. PERFORMANCE DISCUSSION

On a raw performance basis, the triangle and trapezoid decks in the longitudinal direction had the highest load capacity by a margin of > 50% over all decks other than Box 1. However, the transverse triangles, and especially the transverse trapezoids, had the lowest load capacity of all decks, and that also by a wide margin.

The tests were all beam type tests, that is they supported the decks only on the short sides of a plate which had a high aspect ratio. In the cases where the triangle and trapezoid webs were running along the beam axis (longitudinally), there was no truss action of the inclined webs, and the web to face interface was in pure shear. A calculation of the shear stress in the longitudinal connection between the webs and the skins shows that the shear stress in the joint was about 24.1 MPa (3.5 Ksi), which is near the short beam shear values which have been obtained in coupon testing

In the cases where the triangle and trapezoid inclined webs were running across the beam axis (transversely), the deck failed at a lower shear stress value. Analysis of the joint at failure, shows that the interlaminar tensile stresses are key to both the failures. In the transverse direction the deck behaves like a truss, where the members are in either tension or compression. This produces significant tensile stresses at the bond line which causes the failure. The failure occurred at an interlaminar tensile stress of less than 13.8 MPa (2 Ksi). This is even lower than the neat resin tensile strength of about 27.6 MPa (4 Ksi). From a fatigue standpoint, interlaminar tension is undesirable since cracks grow in tension.

Obviously, the triangle and trapezoid decks fit in applications where the load path is primarily in one direction, such as a deck on girder bridge, and the triangles or trapezoids could be oriented in their favorable direction.

The box decks were constructed such that no favored axis existed. The resultant strengths were intermediate between the strong and weak axes of the triangle and trapezoid decks. The lightweight box decks (all but Box 1) had the highest specific strength (strength per core aerial weight) of all decks except the trapezoid in the longitudinal direction.

The Balsa Deck was failed in core shear. A calculation of the shear stress in the deck at failure shows a stress level of 1.52 MPa (221 psi). This is well below the average ultimate shear stress claimed by the manufacturer for 250 Kg/m^3 (15 lb/ft^3) balsa of 4.94 MPa (717 psi). The reason for this discrepancy is the large standard deviations in the density and hence the properties which are proportional to the density. The lower bound on the balsa density is 96 Kg/m^3 (6 lb/ft^3) with a claimed average shear strength of 1.85 MPa (268 psi) which is close to the test failure shear stress. Based on this and other tests, Balsa properties should always be assumed to be those of the lowest density blocks, regardless of the density used.

5. MANUFACTURING COMPARISON

Table 4 shows a breakdown of the cost of manufacturing associated with each deck. The Balsa Deck is the least expensive to produce, and is a viable alternative for thin (23 cm, 9 in) decks. The factor of safety needs to be increased, and this could be done by adding internal fiberglass webs at the joints in the wood. This would slightly increase the cost, but increase the shear strength significantly.

The Light Box decks were the second least expensive deck to produce. The deck has good properties in each direction. Making of the deck currently requires an excessive number of man hours. The Box 1 deck currently requires even more time and material and is not economical to produce, costing over four times as much as the Balsa deck.

The Trapezoid and Triangle decks have similar manufacturing costs, differing by only twelve percent. This is surprising since the Triangle deck required wrapping of the cells, and the Trapezoid webs were simply placed between the foam pieces. Both deck types had high longitudinal strengths, and low transverse strength.

Table 4 Summary of Labor Costs for Deck Manufacture

Deck (hours per task)					
Task	Triangle	Trapazoid	Box 1	Box (Light)	Balsa
Prep	1	1	1	1	1
Cut Fabric	3	2	8	6	2
Wrap	2	0	14	6	0
Assemble	3	3	4	4	2
Shot Set Up	3	3	3	3	3
Total Man Hours	12	9	30	20	8
Labor Cost @ 50 $/Hr	600	450	1500	1000	400
Material Costs Core @ 4.4 $/Kgf	2122	1935	3184	968	600
Total Mfg Costs	2722	2385	4684	1968	1000

6. RECOMMENDATIONS

From the table it can easily be seen that the balsa core is the least expensive too produce of the deck designs. Additionally Balsa cored decks offer good strength properties in both directions. Balsa decks don't scale well to thicker designs when compared to the other decks. The cost of the wood used in the deck goes up linearly with the thickness of the deck. The shear strength required from a deck does not change with span, it is dependent only on the design load. This means that the minimum volume of glass required in the core is independent of the deck thickness, assuming that web buckling can be prevented by adding core to the webs. This limits balsa decks to thin deck applications. Higher strength can be obtained by adding fiberglass webs if the application calls for it.

The wrapping of individual boxes required to make the Box Decks, makes their manufacturing costs high. The Box 1 deck's high manufacturing, and material costs combined to make it the most expensive deck design to build. The lighter version had higher assembly costs, but its lower weight makes it the most efficient design with acceptable strength in each direction.

The Trapezoid Deck is easy to manufacture and has sufficient strength in the longitudinal direction, but has low strength in the transverse direction. The bond line for each of the trapezoids has to be lengthened, and/or the web angle changed to reduce the tensile stress at the interface to a sufficiently low level that failure does not occur until a much higher load has been achieved.

The Triangle Deck's performance in the longitudinal direction is comparable to the Trapezoid Deck's, but the Triangle design does much better in the transverse direction. This can be attributed to the continuous bond lines formed by wrapping the triangles. Failure of the triangles at the resin rich nodes may be overcome with changing the shape so the fabric preform does not have to conform to such a large angle.

It is Hardcore/Dupont's contention that the wrapped trapezoid deck design will offer the best performance to price ratio of all the decks for applications where strength in only on direction is required.. Thicker fabrics will reduce the time required to wrap the individual trapezoids. Resizing the trapezoids and wrapping will increase the bond line and reduce the stress levels at the web to skin interface. The Trapezoid shape will allow the fabric to conform better to the foam around the corners, improving the fiber orientation and reducing the amount of excess resin in the node area where the triangles typically fail. Finally knocked down allowable for tensile stresses at the bond line, both longitudinal and transverse, should be used in resizing the trapezoid.

For applications where the deck is required to carry high loads in both directions a box type deck will offer the best performance. The light weight box decks had the highest efficiency in two directions, and sufficient strength in both directions. The design can economically move to thicker sections such as Magazine Ditch and reducing the cost of wrapping the cubes with automation will make the decks a more attractive alternative to traditional building approaches.

28th International SAMPE Technical Conference
November 4-7, 1996

REHABILITATION OF STEEL BRIDGE GIRDERS THROUGH APPLICATION OF COMPOSITE MATERIALS

John W. Gillespie, Jr., Dennis R. Mertz,
William M. Edberg, Nouredine Ammar
University of Delaware
Newark, Delaware 19716

Kazuhiko Kasai, Ian C. Hodgson
Lehigh University
Bethlehem, Pennsylvania 18015

ABSTRACT

A complete system approach to the rehabilitation of degraded steel bridge girders through the adhesive bonding of composite materials is presented and demonstrated considering a particular bridge girder removed from service in Pennsylvania as a case study. Issues of design, durability, field implementation, and affordability are addressed. The flexural characteristics of steel bridge girders are shown to be improved through testing of various reinforcement schemes on scale steel sections and testing of a selected scheme on the corroded bridge girder taken out of service. The stiffness of a degraded member is restored to its initial state and the load capacity of the member is increased significantly beyond the initial state. A methodology for selection of the most suitable adhesive based on the long-term durability for any particular rehabilitation scheme is developed and the results of the methodology are presented. The necessary surface treatments for the steel and the composite material are addressed as well as techniques for performing an accelerated cure through field heating methods. A case study was performed to examine the economic advantages of rehabilitation of the steel girders with graphite pultrusions vs. replacement of the girders.

KEY WORDS: Rehabilitation, Infrastructure, Composites

1. INTRODUCTION

The national bridge inventory is in need of a comprehensive system approach to extend the life of deteriorated bridges at a minimum inconvenience to the public. The use of composite materials to rehabilitate corroded steel girders can satisfy this need. This project has focused on repairing girder-type members, which have been determined to have less than the desired flexural strength or stiffness; typically this would be due to corrosion of the flange section. Due to the extreme light weight of composite materials and to adhesive or fusion bonding methods, installation of composite rehabilitation can be faster, resulting in lower labor costs, than rehabilitation with traditional materials. More important than the direct costs is the reduction in delay of bridge traffic offered by using composite materials. Due to their flexible nature, composite retrofits can be tailored to various field conditions such as extensively corroded members or riveted members.

2. APPROACH

A comprehensive system approach to the problem of deteriorated steel bridge girders has been developed. The affordability of performing a rehabilitation using composite materials must be compared to that of a rehabilitation or replacement using steel for specific bridge. When the rehabilitation using composites is the most affordable option a composite rehabilitation scheme with a geometry suitable to the field conditions is selected and designed to restore the required stiffness and load capacity to the degraded members. The appropriate adhesive and surface treatments are chosen to provide long-term durability and to conform to the conditions in the field. The materials for the rehabilitation are procured and finally installed in the field.

2.1 Design This step requires assessment of the condition of the deteriorated girders in the field in order to provide input for the rehabilitation design. A comparison of the degradation of corroded girders as gauged by a bridge inspection performed by a DelDOT bridge inspector versus laboratory testing of the members removed from the field, Mertz and Gillespie [1], indicates the current field evaluation practices provide adequate input to the design. Based upon the condition of the members to be rehabilitated the deteriorated stiffness and strength of the sections are determined through modeling by reductions in the section areas based on the losses reported from the field evaluation. In cases of localized damage it may only be necessary to design the composite materials to be fully developed and effective at a particular section. In these cases the design can be based on restoring the moment of inertia of a particular location. Corrosion damage is usually distributed along the length of a girder and the composite materials should be designed with the intent of restoring the overall stiffness of the member as opposed to the properties at a particular section.

With the help of PennDOT, arrangements were made to procure steel girders from a highway bridge that was demolished in August, 1995. Constructed circa 1940 in Valley View, Pennsylvania in western Schuylkill county, the bridge spanned approximately 9754 mm over Rausch Creek. The deterioration of the girders had progressed to such a point that the bridge had to be temporarily shored at midspan with wooden braces prior to demolition. Corrosion of the steel sections was severe enough on the flanges to warrant the demolition of the bridge. The case study girder was 610 mm deep with a 229 mm flange width. The original thicknesses were assumed based on historical records of rolled beams. The web thickness was 11 mm, the flanges tapered from 23 mm at the intersection with the web to 14 mm at the flange tips. The original moment of inertia about the strong axis (I_x) was 935 x 10^6 mm^4.

A 6096 mm long section of the girder was tested in three point bending. The girder was determined to have 87% of its original stiffness. The plastic moment capacity was determined to be 88% of original.

Corrosion of steel bridge girders in the field is concentrated on the bottom flange. Under the governing bending actions this section experiences force effects primarily in the axial direction, for this reason the most efficient rehabilitation with composite materials will utilize unidirectional members oriented with the fibers in the axial direction to maximize the axial stiffness and strength of the component. An exception is cases where the geometry of the flange is such that either resin infusion or wet lay-up with a formable fabric is necessary.

The composite materials used in the rehabilitation of the case study were unidirectional pultruded T-300 carbon fiber/vinyl ester matrix strips donated by Amoco. The strips were 6.4 mm thick and 38.1 mm wide. The modulus of elasticity in the fiber direction (E_1) was 116 x 10^6 kPa. Strain to failure in the composites was approximately 8000 me. Six strips placed on the interior and exterior of the tension flange were chosen to restore the original section moment of inertia.

Affordable composite materials of interest have lower moduli of elasticity in the primary direction than steel but most have higher failure stresses in this direction. For this reason the replacement of lost steel with composite materials will typically be controlled by matching the stiffness of the composite materials to that of the lost steel. This results in a repaired member with comparable stiffness to the original and greater load capacity.

2.2 Adhesive and Pretreatment Selection An adhesive for a particular rehabilitation scheme must perform three functions. First, the adhesive must have adequate bond strength so that the composite material can be optimally utilized, this requires the failure mode of the system to be governed by the ultimate stress capacity of the composite and not by a premature bond failure. Second, the system must be sufficiently durable in the design environment for bridges to match the life expectancy of the structure (typically 75 years). Environments tested for bridge applications were a 65 C hot water bath, a chloride solution to model deicing salts, a commercial deicing solution without chlorides, and a freeze-thaw atmosphere (-20 C to 60 C). An accelerated test method using the wedge test where crack tips are continuously exposed to the environment was used (ASTM D-3762). Joints that debonded in less than 6 months in any of the environments tested were rejected. Third, the adhesive must be easy to utilize under field conditions. This consideration includes the glass transition temperature, T_g, after saturation in hot water (should be greater than 60 C for infrastructure applications), the pot-life of the two-component epoxies after mixing, and also the cure cycle for adhesive cure. Adhesives like PEEK with high processing temperatures require further process development to achieve required temperatures in the field. While T_g of the adhesive in various environments is a necessary criteria, the pot-life and the cure cycle requirements are application specific, and hence low pot-life values are admissible for small-scale operation but single component epoxies with infinite pot lives are a must for large-scale structures.

Surface preparation is the key to a strong and durable adhesive bond. Since the rehabilitation takes place on-site, the surface treatment must be environmentally friendly, easily accomplished in field conditions and durable enough to last for several days. Silane coupling agents are designed specifically to enhance bonding between the surface oxides of metals and polymers. They were found to be necessary for long term durability.

The results of an extensive evaluation of various thermoset and thermoplastic adhesives is given in [2]. For the case study girder the two-part epoxy Ciba-Geigy AV8113 was chosen as the optimal adhesive. A pretreatment of paint removal by sandblasting, a solvent wipe and application of silane 6040 for use with epoxy adhesives was selected. A room temperature cure is available with this epoxy and was selected. The cure time is 16 hours for the room temperature cure, by applying increased temperature (100 C) the cure time can be reduced to 20 minutes.

2.3 Large Scale Field Demonstration The inside and outside of the tension flange of the case study girder were sandblasted to remove the oxide layer. This revealed an extremely pitted and nonuniform surface. The girder was then transported to the ATLSS laboratoory at Lehigh University for a simulation of a field rehabilitation. The girder was wiped with a solvent prior to bonding. The composite strips were sandblasted with a portable unit on-site and cleaned with a solvent. The two part epoxy was mixed and applied to the surface of the girder and to the surface of the carbon strips. The 6096 mm assemblies of composite strips were lifted into place by hand and fastened to the steel girder by means of C-clamps and wooded tabs as shown in Figure 1. The composites strips were successfully clamped to the beams within the working time of the epoxy. A team of six graduate students performed the rehabilitation in six working hours. The feasibility of using heating methods to increase the curing rate of the adhesive was demonstrated by using both heating blankets and induction heaters to elevate temperature in the member at discrete locations. The elevated temperatures were monitored by using infra-red equipment.

Figure 1. Large Scale Rehabilitation

2.4 Economic Analysis The first decision to make in determining if there is an economic advantage in the use of composite materials over other traditional materials (e.g. steel). When comparing rehabilitation versus replacement the costs associated the structure as a whole can offset the high material costs of composite materials. When considering a rehabilitation using composite materials versus a rehabilitation using steel there are two considerations that favor composites. First the costs associated with rehabilitation are often more associated with the time the project requires, such as labor costs and the costs to divert traffic, than with material costs. Due to the light weight of composite materials it is expected that they could be installed in less time than equivalently stiff steel reinforcement would require. The second factor that favors composite materials is that by matching the stiffness of a steel component a composite member will have a greater strength, this results in greater load carrying capacity for the deteriorated member.

3. VALIDATION

3.1 Scale Beam Tests The base member used in evaluating the reinforcement schemes was a W8x10 beam of A709 grade 36 steel. The length of the beam was 1524 mm (60 in). This member was chosen as a lower bound on the type of members that can be found in use on bridges and since its dimensions are proportionally similar to those of steel beams typically used as large primary girders. The reinforcement schemes developed are designed to improve the flexural characteristics of steel girder shapes. A schematic of the four basic reinforcement schemes geometries is shown in Figure 2. The first reinforcement scheme (Figure 2a) fabricated was a unidirectional carbon-fiber (IM-7)/epoxy (8551-7) composite plate bonded directly to the tension flange of a beam using a two-part epoxy adhesive (AV8113). The fabricated laminate had a nominal fiber volume fraction of 62 percent. Two beams of this type were fabricated. The second reinforcement scheme (Figure 2b) used the same composite plate but took advantage of an aluminum honeycomb structure to space the composite plate farther away from the steel section; this forms a composite material system referred to as sandwich construction, where two materials are separated by a core that has practically no influence on the flexural behavior. In these two cases the thickness of the composite was 4.6 mm (0.18 in), which is comparable to the flange thickness of the base member of 5.1 mm (0.2 in). For cases where a variable surface might be present, such as due to riveted construction, where a rigid composite member could not be employed a different strategy (Figure 2c) was developed. A composite fabric was wrapped from a uniform section on the web of the beam down over the flange and a foam core spacer and up onto the web on the other side. The fabric was a ±45 degree E-Glass. Three beams of this type were fabricated with differences in the bonding method. The process differed in that a thin layer of epoxy was applied to the steel for the second and third beam and the first layer of fabric was placed over that epoxy before wrapping the subsequent layers. The wrapped sections were then infused with a vinyl ester resin (8084) and vacuum bagged for curing. The fourth scheme (Figure 2d) utilized an E-Glass pultruded channel adhesively bonded and mechanically connected with self-tapping screws. The fifth scheme is geometrically similar to the first. Three discrete adhesively bonded DFl pultruded, Amoco T-300 carbon fiber/vinyl ester matrix strips were attached to the tension flange. The strips were 6.4 mm (0.25 in) thick and 38.1 mm (1.5 in) wide.

Figure 2. Rehabilitation Geometries

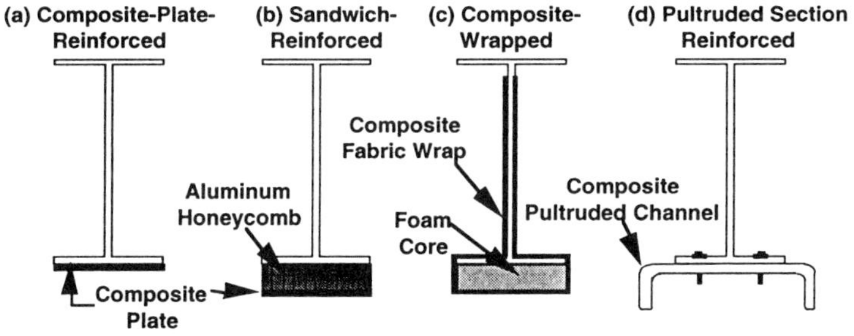

Specimens of each rehabilitation scheme were tested to determine increases in stiffness and increases in the first yield and ultimate strength, results are given in [1]. These varied rehabilitation schemes allow a greater range in the types of damage that can be repaired using composite materials. Typical corrosion losses in bridges are in the range of 20% to 30% decreases in stiffness. Each of the rehabilitation geometries provided increases in this range.

The various rehabilitation schemes were modeled by beam bending using transformed areas as well as by finite element models. The results of the finite element models matched the experimental tests well. The simple model was accurate for all the schemes except the fabric reinforced beam which displayed nonlinear elastic behavior.

A primary concern is the potential for failure of the system due to a debonding at the termination of the composite materials due to the stress concentration at that location. This failure occurs due to concentrations of shear and peeling stresses acting at the termination of the composite material. This mode is not dominant in cases where shear stresses are small relative to bending stresses and curvatures are small at the composite termination, girders that are sufficiently long compared to the section depth where the composite reinforcement is carried close to the supports are not subject to this failure mode. Tapering of the composite rehabilitation can help to relieve the stress concentration and prevent this failure.

Based on the results of the scale beam tests and the desire to typically keep the girder depth as constant as possible for underpass clearance the first rehabilitation geometry was selected to be applied to full scale members.

3.2 Case Study Girder The case study girder was tested to failure at the ATLSS laboratory at Lehigh University. Midspan deflections as well as strains along the cross section at three locations were recorded. Material tests on coupons cut from the girder indicated the flanges had yield stresses of 272 MPa and ultimate stresses of 439 MPa, the web had a yield stress of 307 MPa and an ultimate stress of 443 MPa. The beam was tested in three point bending as shown in Figure 3. More details on the testing are found in [3].

Figure 3. Test Setup of Rehabilitated Girder

The girder was restrained to prevent lateral torsional buckling. Shortly after the onset of compression flange yielding the compression flange underwent local buckling. This resulted in a gradual decrease in load carrying capacity with increased deflection instead of the gradual increase that would be expected if the local buckling had not occurred. The presence of a concrete deck slab in the field would contribute toward preventing the local compression flange buckling. The load deflection curve for the case study girder is given in Figure 4. The girder was unloaded at a deflection of approximately 125 mm in order to reset the midspan deflection measurement.

Figure 4. Test Results of Rehabilitated Girder

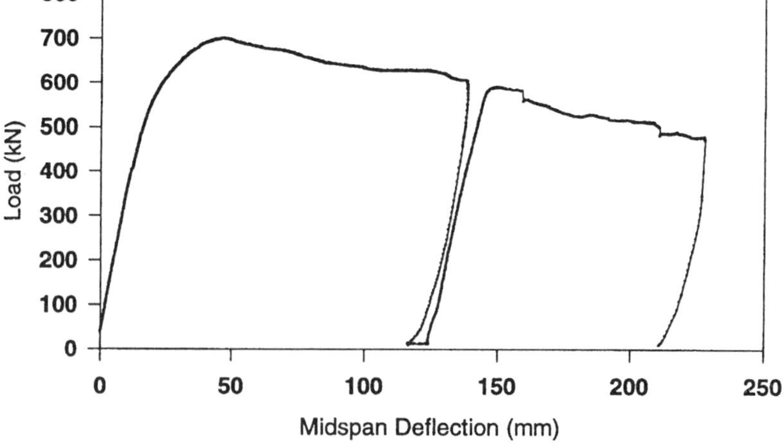

The data acquired in the elastic region showed that the desired goal of restoring the stiffness of the member to its original state was achieved. The stiffness of the girder after rehabilitation was 32.0 kN/mm, 97% of the original girder stiffness. Despite the occurence of local buckling in the compression flange the ultimate load capacity of the rehabilitated section reached 1119.0 kN-m, 113% of the original section capacity. If the local buckling had not occured the ultimate load capacity would have been even greater (133% of original) as can be seen in Figure 5. The girder underwent inelastic deformations of over 220 mm before the composite strips began to fail. The large deformations achieved demonstrate that the ductility of the girder was not adversely affected by the addition of the composite materials.

Figure 5. FEM Analysis of Rehabilitated Girder

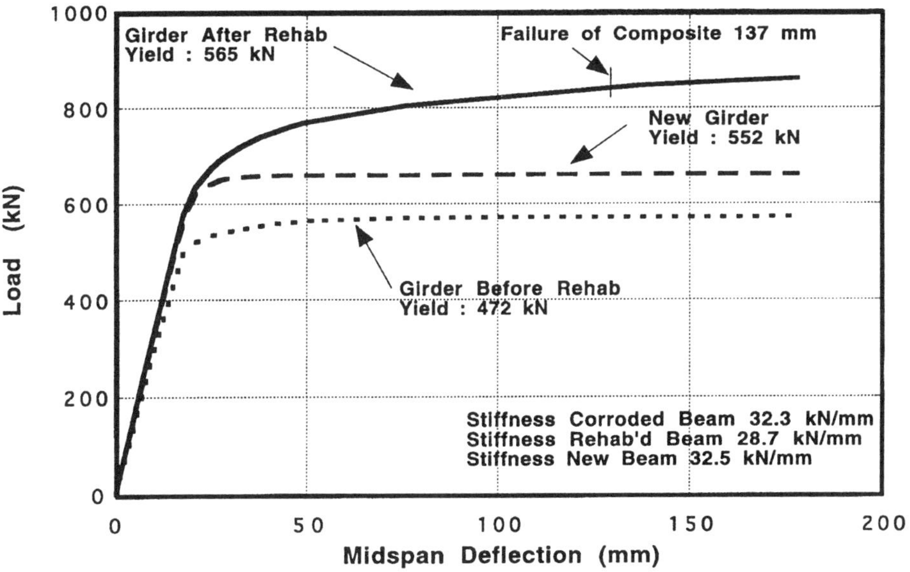

3.3 Economic Case Study The ideal economic comparison for the rehabilitated case study girder would be a comparison of the costs to rehabilitate the bridge the girder was taken from in Pennsylvania versus the cost that the rehabilitation would have required. The costs for the replacement of that particular bridge are not yet available and therefore the costs of a replacement project in Delaware have been compared to the case of rehabilitating the girders with the costs and required time for the rehabilitation of the case study girder as a basis. Bridge 509 in Delaware suffered extensive damage and 180.1 meters of girders were replaced.

Originally the girder dimensions were:
Web: 1600 mm x 9.5 mm
Upper Flange: 381 mm x 15.9 mm
Lower Flange: 381 mm x 34.9 mm

Figure 6 compares the cost of replacement to the cost of rehabilitation at an assumed 25% section loss. The replacement costs on Bridge 509 represent the awarded repair bid. They were obtained from the Delaware Department of Transportation (DelDOT).

Traditional approaches to girder replacement require removal of the deck system to access the supporting girders. The proposed rehabilitation method allows the deck to remain in place providing a large cost savings.

The rehabilitation costs and labor timing were based on the study of the rehabilitation of the PennDOT girders at the University of Delaware and Lehigh University. Except for the graphite pultrusion, most of the relevant costs were converted into costs per square meter. Then these costs, as well as the timing, were extrapolated to the rehabilitation of Bridge 509. The PennDOT girders have 2.79 m^2 of "bondable" area of their tension flange, while the repairable area on the tension flanges of Bridge 509 is 135.6 m^2 . Cost List :

Graphite pultrusion (DFI actual price)	$ 33.75 per kg
Adhesive (CIBA-GEIGY actual price)	$ 44.13 per m^2
Steel preparation (200 man.hours @ $25/hr)	$ 36.81 per m^2
Graphite preparation (200 man.hours @ $25/hr)	$ 36.81 per m^2
Set-up and installation(7008 man.hours @ $25/hr)	$ 1297.59 per m^2

It was estimated that the job could be done by a crew of 12 on 8 hour work day. The entire job would then be completed in 7008 hours ie.73 days. A support system cost was included in case the bridge needed to be shored. A very conservative estimate is that the bridge would be closed for 73 days if necessary. It is worth noting that in this case, the curing time can be reduced with heating devices. It must be noted that costs to society were not included in Fig. 6. Although they are hidden, they must nonetheless be taken into account. Some estimate that these costs can reach as much as 25 % of the entire budget, since these costs are time based the affordability of the composite rehabilitation would be improved by their incorporation since the time to complete repairs is reduced by an order of magnitude.

Figure 6. Economic Case Study

	PRICE OF TOTAL REPLACEMENT		**COST OF REHABILITATION WITH GRAPHITE PULTRUSION** 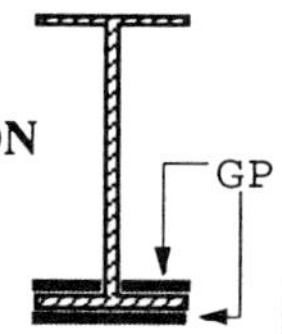	
	Items	Cost	Items	Cost
Materials Costs	Concrete Deck	$ 530,350	Graphite	$ 168,300
			Adhesive	$ 6,000
	Steel Structures	$ 356,410		
	Support System	$ 110,610	Support System	$ 50,000
	Total	**$ 997,370**	**Total**	**$ 188,000**
Labor Costs	Actual Costs	$ 614,500	Steel Prep.	$ 5,000
	Steel Salvage	$ 62,679	Graphite Prep.	$ 5,000
			Set-up/Install.	$ 175,200
	Total	**$ 677,179**	**Total**	**$ 185,200**
Traffic Costs	160 days@$1704/day Flaggers, Signs ...	**$ 272,605**	73 days@$1704/day Flaggers, Signs ...	**$ 124,400**
Other Costs	Initial Expense	$ 122,900	Initial Expense	$ 25,400
	Engineering	$ 9,930	Engineering	$ 9,930
	Total	**$ 132,830**	**Total**	**$ 35,330**
TOTAL COSTS		**$ 2,079,984**		**$ 569,230**
Cost Per Meter		**$ 11,555**		**$ 3,162**

This case study demonstrates that the rehabilitation strategy which can restore stiffness and load capacity is a viable economic solution. The cost of composite rehabilitation is 28% of the cost using traditional materials.

4. CONCLUSIONS

The developed methodology was successfully applied to a bridge girder removed from service due to corrosion. The structural performance of bonded composite graphite pultrusions to the tension flange of corroded steel girders was successful in renewing stiffness and load capacity in large scale, corroded girders. The PennDOT girder used in this study could have been successfully repaired with this method and left in service at a significant cost savings over replacement of the entire bridge superstructure.

5. REFERENCES

1. D.R. Mertz and J.W. Gillespie, Jr., "Rehabilitation of Steel Bridge Girders Through Application of Advanced Composite Materials," NAS NCHRP Ideas Program Final Report, July 1996.
2. J.W. Gillespie, Jr., G. Rajagopalan, and K. Immordino, ASC 11th Technical Conference, October 1996.
3. J.W. Gillespie, Jr., D.R. Mertz, W.M. Edberg, J.R. Demitz, N. Ammar, K. Kasai, and I.C. Hodgson, ASC 11th Technical Conference, October 1996.

6. ACKNOWLEDGEMENTS

The authors gratefully acknowledge the support of the ARPA Bridge Infrastructure Renewal and the National Academy of Sciences IDEA programs. The support of Amoco through the donation of the pultruded carbon composites and the donation of the bridge girders by PennDOT as well as the bridge inspection information provided by DelDOT are all greatly appreciated.

28th International SAMPE Technical Conference
November 4-7, 1996

STRUCTURAL BEHAVIOR OF CONCRETE FILLED CARBON FIBER COMPOSITE TUBULAR COLUMNS

F. Seible[1], A. Davol[2], R. Burgueño[2], R.J. Nuismer[3], and M.G. Abdallah[4]

ABSTRACT

This paper investigates the use of premanufactured filament wound carbon tubes as reinforcement for concrete columns. Two initial design concepts are considered, a ductile design with steel anchorage bars to allow for large inelastic rotations at the column base and a strength design with the carbon shell continued into the concrete footing to give an essentially linear elastic response. The structural behavior of the two design concepts is investigated experimentally under simulated seismic loads and the performance of the test units is compared to that of a conventionally reinforced concrete column. The ductile design response was almost exactly the same as the conventionally reinforced reference column. The strength design suffered an early failure due to high stress concentrations at the footing interface. Analytical expressions for predicting the stress and strain state in the shell and concrete are presented for compression loading and compared to the test results from the strength design column under axial load only. Bending behavior is discussed qualitatively with an analytical approach described for the case when the shell and core act together (no slip). The ability to predict the strains in the section with existing analytical techniques is investigated. Issues pertaining to the case of slip between the shell and core are briefly presented.

KEYWORDS: Carbon Fibers, Concrete, Columns

1. INTRODUCTION

The use of premanufactured filament wound carbon shell tubes to replace the conventional steel reinforcement in concrete columns is currently under investigation at the University of California, San Diego (UCSD) (Fig.1). This system offers the advantage of acting as the formwork for the concrete as well as taking over the structural actions normally performed by the steel reinforcing cage. Columns in seismic regions must be designed to ensure both the serviceability and survival limit states are satisfied. Under seismic loading conditions extremely large forces can be generated if columns are designed to respond elastically. A capacity based design approach is often used which selects areas of inelastic action and ensures that the rest of the structure remains elastic throughout the loading. This design approach may require large inelastic deformations for the system to survive an earthquake. The inelastic action for columns is in the form of plastic hinges that allow large rotations at specified locations. Brittle shear failure modes are then precluded by ensuring that the shear

[1] Professor of Structural Engineering, [2] Graduate Research Assistant
Division of Structural Engineering, University of California, San Diego, La Jolla, California 92093-0085
[3] Program Manager, Alliant Techsystems Inc., Composite Structures, Magna, Utah 84044-0098
[4] Senior Research Scientist, Hercules Inc., Composite Products Division, Salt Lake City, Utah 84118-0748

strength of the column is greater than the shear corresponding to the maximum moment capacity that can be developed at the plastic hinge locations. In a conventional column steel reinforcement provides the flexural capability, increases the shear capacity, and provides confining pressure for the enclosed concrete. This confining pressure becomes an important issue as the strength and deformation capacity of concrete have been shown to increase greatly under the presence of confining pressure. All of these functions can be performed by a carbon shell.

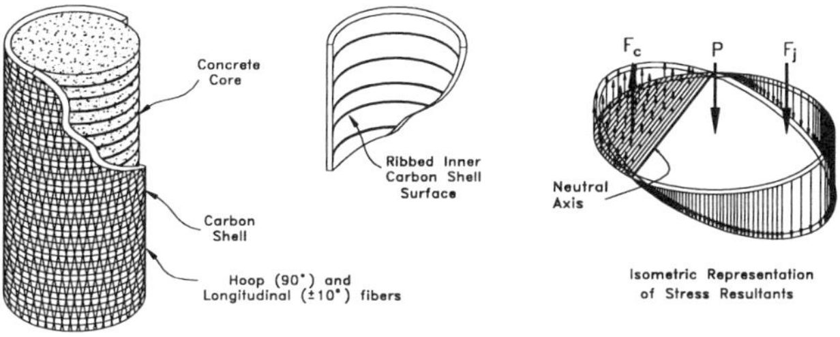

Fig.1 Schematic of Concrete Filled Carbon Shell System

The development of a concrete filled carbon shell system for columns demands that the lateral response of the new system be comparable to or better than that of conventionally reinforced concrete columns. The preliminary research focus considered the design and testing of two design concepts for seismic response, namely one based on ductility and the other one on strength. A conventionally reinforced "as-built" concrete column lateral load test was used for direct comparison. The general geometry of the three pilot test units is shown in Fig.2. The ductile design concept, illustrated in Fig.2b, relies on the inelastic deformation capacity provided by steel starter bars that extend from the footing into the column for the development of a conventional ductile anchorage detail. In the strength design concept the carbon shell jacket continues into the footing without the use of steel bars for anchorage, as shown in Fig.2c.

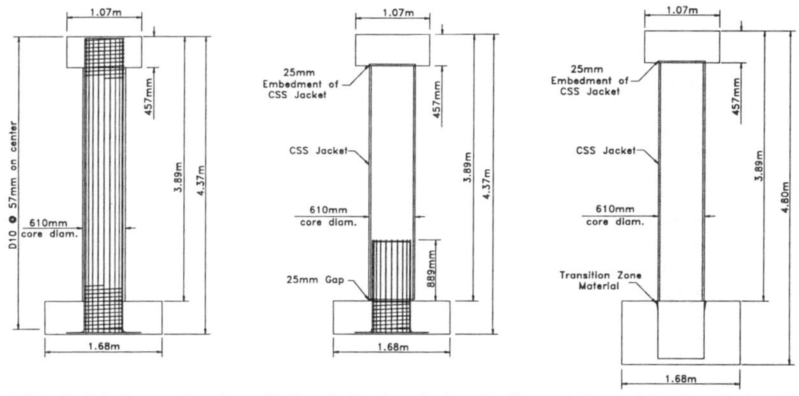

a) "As-Built": Conventional steel reinforced column

b) Ductile Design: Carbon Shell System with steel starter bars

c) Strength Design: Carbon Shell System with built in connection

Fig.2 Test Units

2. STRUCTURAL BEHAVIOR

2.1 DESIGN MODELS For design purposes several assumptions were made regarding the interaction between the longitudinal flexural stress, the circumferential hoop tension stress resulting from confinement of the concrete, and the shear stress carried by the shell. This allowed a simple but rational approach to determine analytical models to characterize the response of the system, which would allow rapid experimental investigation and provide the basis for further more rigorous analytical modeling. Design and assessment models developed at UCSD for retrofit measures constitute the basis for the design modeling [1]. Most of these models are considered applicable to the carbon shell system, while simple modifications are proposed where appropriate. The design equations and assumptions can be found in references [2] and [3].

2.2 ANALYTICAL METHODS

2.2.1 Variable Definitions

ε_l - axial strain in concrete
ε_r - radial strain in concrete
ε_L - longitudinal strain in shell
ε_H - hoop strain in shell
ε_{cu} - ultimate strain for unconfined concrete
σ_l - axial stress in concrete
σ_r - radial stress in concrete
σ_L - longitudinal stress in shell
σ_H - hoop stress in shell
E_c - Young's modulus of concrete
E_L - Young's modulus of shell in longitudinal direction
E_H - Young's modulus of shell in hoop direction
ν_o - initial Poisson's ratio of concrete
ν_c - Poisson's ratio of concrete (function of axial strain)
ν_{HL} - Poisson's ratio for shell (longitudinal strain due to loading in the hoop direction)
ν_{LH} - Poisson's ratio for shell (hoop strain due to loading in the longitudinal direction)
R - radius to midplane of shell
t - shell thickness

2.2.2 Shell Properties For analysis purposes equivalent properties for the laminate are used to describe the system behavior (see section 3.1.2). Once a suitable layup is defined the lamina properties are used to derive equivalent laminate properties through the use of classical lamination theory. A biaxial stress state is assumed to exist in the shell. Thus the longitudinal and hoop strains can be described by the following

$$\varepsilon_L = \frac{\sigma_L}{E_L} - \nu_{HL}\frac{\sigma_H}{E_H} \qquad (1)$$

$$\varepsilon_H = \frac{\sigma_H}{E_H} - \nu_{LH}\frac{\sigma_L}{E_L} \qquad (2)$$

2.2.3 Concrete Properties The concrete model employed neglects the concrete tension carrying capability and considers the compression response to be dependent on the confining pressure. It has long been known that the strength and displacement capacity of concrete can be substantially improved by the presence of confining pressure. Many models have been proposed that calculate the confining pressure in a reinforced concrete column as a function of the transverse steel reinforcement [4,5,6]. Some models, because of the ductile nature of the steel reinforcement, calculate the confining pressure based on the yield stress of the steel. These models are referred to as active confinement models because the pressure is assumed constant throughout the loading history. The linear elastic behavior of the carbon shell does not allow for such a derivation. The pressure developed is proportional to the expansion of the concrete core and is therefore regarded as passive confinement. For this analysis an active confinement model is used incrementally to account for the passive nature of the confining action due to the carbon shell [5]. The longitudinal and radial strains in the concrete are given respectively as

$$\varepsilon_1 = \frac{1}{E_c}(\sigma_1 - 2\nu_c\sigma_r) \tag{3}$$

$$\varepsilon_r = \frac{1}{E_c}((1-\nu_c)\sigma_r - \nu_c\sigma_1) \tag{4}$$

The Poisson's ratio of the concrete is assumed to vary with the axial strain as [5,7,8]

$$\nu_c = \nu_0\left[1.0 + 1.3763\frac{\varepsilon_1}{\varepsilon_{cu}} - 5.36\left(\frac{\varepsilon_1}{\varepsilon_{cu}}\right)^2 + 8.586\left(\frac{\varepsilon_1}{\varepsilon_{cu}}\right)^3\right] \tag{5}$$

with a limiting value of 0.5.

2.2.4 System Behavior The carbon shell works to provide confining pressure by limiting the expansion of the concrete core. Carbon fiber wraps have been successfully used to retrofit existing bridge columns that were not designed with sufficient transverse reinforcement to confine the concrete core [1]. In the case of a retrofit measure for circular columns the fibers are usually all aligned in the circumferential direction to give the maximum confining pressure without increasing the bending stiffness of the column. For this configuration the radial expansion of the shell due to axial deformations is very small because the Poisson's ratio for loading transverse to the fiber direction is very small. For the system here being considered we no longer have only circumferential fibers and thus the expansion of the shell due to longitudinal deformation can become considerable. If the Poisson's ratio of the shell is greater than that of the concrete (.18-.3) then no confining pressure is generated as the shell will expand faster than the concrete core. This effect is mitigated by the fact that the Poisson's ratio of the concrete is assumed to increases with axial deformation up to a limiting value of .5 (Equation 5).

Initially the system is assumed to work homogeneously with no slip between the shell and the concrete core. This is referred to as full composite action. At some level the shell will slip past the concrete and the following assumptions must be revised to account for this effect.

By assuming full composite action between the concrete core and the composite shell we have:

$$\varepsilon_L = \varepsilon_1$$
$$\varepsilon_H = \varepsilon_r \tag{6}$$
$$\sigma_H = -\sigma_r\frac{R}{t}$$

Utilizing equations 1-5 the following relations can be derived,

$$\sigma_r = -\frac{(\nu_{LH}\varepsilon_1 + \varepsilon_r)E_H t}{(1-\nu_{HL}\nu_{LH})R} \tag{7}$$

$$\varepsilon_r = \varepsilon_1 \left[\frac{\nu_c E_c R(1-\nu_{HL}\nu_{LH}) + \nu_{LH} E_H t(1-\nu_c - 2\nu_c^{\,2})}{-E_H t(1-\nu_c - 2\nu_c^{\,2}) - E_c R(1-\nu_{HL}\nu_{LH})} \right] \qquad (8)$$

Equations 7 and 8 can be used incrementally to establish a relation between ε_r and ε_1. Equation 6 can then be used to calculate the confining pressure on the concrete corresponding to the given axial strain. Once the confining pressure is known, an active confinement model can be used to calculate the axial stress in the concrete. Equations 5 and 3 can then be used to calculate the longitudinal stress in the shell (Equation 9).

$$\sigma_L = E_L \left(\varepsilon_L - \sigma_r \frac{\nu_{HL}}{E_H} \frac{R}{t} \right) \qquad (9)$$

Existing models that do not take the hoop expansion of the shell due to axial deformations into account work well for shells with predominantly circumferential fibers but will over predict the confining pressure and thus over predict the strength of the concrete core for shells with a higher Poisson's ratio for loading in the longitudinal direction [9,10]. This can be observed by comparing the confining pressure, Fig.3a, and the resulting stress strain curves, Fig. 3b, for the current model with and without the axial Poisson's effect. Fig.3 shows results for the strength design test unit utilizing the active confinement model proposed by Mander et.al. [4] incrementally.

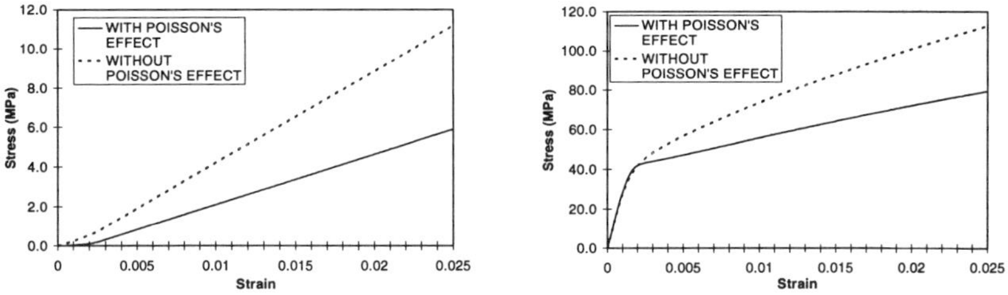

a) Confining Pressure vs Axial Strain　　　b) Concrete Stress vs Strain

Fig.3 Comparison of model with and without Poisson's effect of shell taken into account for strength design test unit

3. EXPERIMENTAL PROGRAM

3.1 TEST UNITS AND TEST SETUP

3.1.1 Unit Configuration The experimental large scale validation of the system to characterize both strength and inelastic deformation capacity potential is of primary importance towards the development of the carbon shell system concept for columns in seismic zones [3]. The selected pilot test unit configuration consisted of a 0.40 scale model of a typical circular bridge column. A column aspect ratio susceptible to a flexure dominated response was chosen, and an axial load level representing what is considered to be a practical upper bound expected in single column bridge bents was provided. The typical geometry of the test units, as shown in Fig.2 consisted of a circular cantilever column with an effective height of 3.66 m, and a concrete core diameter of 0.61 m.

The "as-built" column (see Fig.2a) contained 20 #7 G60 bars of continuous reinforcement, corresponding to a longitudinal steel ratio of 2.66% with a 25 mm cover to the main bars.

Transverse reinforcement was provided by #3 G60 spirals with a pitch of 57 mm, or a 0.82% transverse volumetric reinforcement ratio.

As mentioned previously, the Ductile Design Concept (see Fig.2b) used steel starter bars for anchorage into the footing. With the purpose of maintaining the same moment capacity of the "as-built" column at the critical section, the starter bar details also consisted of 20 #7 G60 steel bars but without any transverse steel reinforcement, which was in this case substituted by the carbon shell. The required starter bar length was assumed to be forty times the longitudinal starter bar diameter, thus obtaining a starter bar length of 889 mm. An anchorage region, or transition zone, in the carbon shell was identified to be 914 mm from the bottom of the shell to accommodate the starter bar length. The inside surface of the carbon shell along this transition region was manufactured with internal ribs to aid in the force transfer mechanism to the starter bars by multiple compression strut development. A 25 mm gap was provided between the bottom of the shell and the top of the footing to avoid crushing of the carbon shell at the compression toe of the column under large deformations and allow plastic hinge rotations without added strength and stiffness from the carbon shell.

The Strength Design Concept (see Fig.2c) continued the carbon shell into the footing for anchorage avoiding the use of steel starter bars. The carbon shell tube had a constant thickness of 12 mm along the member length. The reaction moment at the footing connection depends only on the compression force couple developed inside the footing. This situation gives rise to high joint shear stresses in the footing at ultimate response. Careful consideration was given to this issue and without any guidelines for joint shear strength assessment for this new system, it was decided to minimize the shear induced principal tensile stresses in the joint to the allowable levels in conventionally reinforced concrete systems. The footing depth was increased to a height of 914 mm and postensioned. The configuration of this design concept also brought concerns regarding normal stress concentrations on the carbon shell at the column-footing interface. Finite element analysis indicated the need to provide a transition zone inside the joint to alleviate the stress concentrations generated on the compression side of the column. The geometry and material properties for this region were determined by performing parametric studies. A compliant transition zone distributed vertically along the top 152 mm of the footing connection, and tapering down linearly from a 13 mm thickness at the top was provided. The selected material for this interface region was a structural adhesive with a compression elastic modulus of approximately half that of concrete.

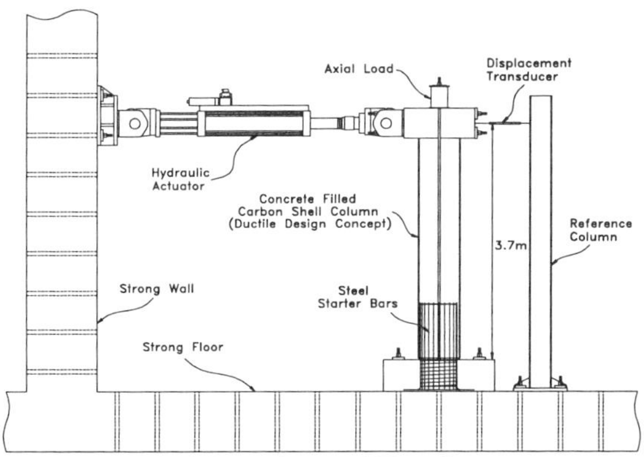

Fig.4 Experimental Test Setup

3.1.2 Carbon Shell Configuration and Properties The carbon shell tubes were filament wound with Hercules AS4D-GP(12k) carbon fibers and a HBRF-55A epoxy resin system utilizing a wet winding operation. Details of the manufacturing process are presented in references [3] and [11]. The layup for the ductile design column consists of two distinct areas. The main section with $[90/\pm10/90/\pm10_2/90/\pm10/90]$ and an additional $[90_{20}]$ in the plastic hinge region extending 914 mm up from the footing interface to provide an appropriate confinement level to the anchorage bars [3]. For the strength design column the layup is

[$90/\pm10_2/90/\pm10_2/90/\pm10_2/90/\pm10_2/90/\pm10_2/90$]. The 90° plies are in the hoop direction. The thickness of the 90° plies is .254 mm and the thickness of the ±10° plies is 1.016 mm (for each pair). The ply properties in the material coordinate system are E_{11}=130.3 Gpa, E_{22}=8.205 Gpa, G_{12}=4.344 Gpa, ν_{12}=0.445. The equivalent properties for the manufactured shells tested at UCSD are shown in Table 1.

Table 1 Equivalent Mechanical Properties of Manufactured Carbon Shells

	E_L (GPa)	E_T (GPa)	G_{LT} (GPa)	ν_{LT}
Ductile Design[1]	100	32.92	7.05	0.19
Ductile Design[2]	54.6	82.2	5.69	0.061
Strength Design[3]	100	24.2	7.52	0.23

[1]General region, [2]Plastic hinge region

[3]Values from manufacturer not calculated from above properties

3.1.3 Test Setup The test set-up, shown in Fig.4, was designed to subject the columns to a constant axial load and cyclic lateral displacements. The unidirectional seismic attack was simulated by fully reversed cyclic loading to the top of the column load stub by a single hydraulic actuator.

3.2 OBSERVED BEHAVIOR The ductile design concept displayed a stable histeretic response up to a displacement ductility level of 8 (maximum displacement divided by displacement at first yield), corresponding to a top displacement of 315 mm, or a drift ratio (DL/L) of 8.6% (see Fig.5 and Fig.7a). The maximum sustainable lateral load capacity was 390 kN. Failure was caused by concentration of inelastic flexural actions in the gap between the carbon shell jacket and the top of the footing, leading to the fracture of the longitudinal starter bars. As judged from the excellent response of the system, the force transfer mechanism between the carbon shell and the anchorage bars with the aid of the internal ribs provided in the shell was successful. A profile of the longitudinal strain profiles in both the shell and the reinforcing bars on the compression side can be seen in Fig.7c. The strain transfer from the concrete core can be visualized by the incremental difference in gage readings in both the shell and the reinforcing bar at the same level. Only limited curves are presented here since many longitudinal gages in the carbon shell were lost due to microcracks in the resin at the outer hoop layer. Post test evaluation of the concrete core by removal of the carbon shell jacket over the lower 1.2 m of the column showed that the concrete core, with the exception of the plastic hinge zone was essentially intact. The concrete core exhibited a minimum number of 'hair-line' cracks, and very few equally thin vertical splitting cracks. The ribs in the concrete core were in excellent condition, with only signs of friction with the carbon shell in the bottom part due to the debonding of the shell and the greater demand for rotational deformation in this region.

The response of the strength design concept was essentially linear elastic, and failure was observed at a load of 512 kN with a corresponding top displacement of 77.5 mm, or a 2.12% drift ratio, during the first cycle in the push direction (see Fig.6 and Fig.7b). The system exhibited a slight non-linear response believed to be caused by debonding of the concrete core from the shell, and slipping out of the carbon shell from the footing on the tension side, as it is observed in Fig.7b. The system failed in a brittle fashion at the footing interface in the compression side of the column at an unexpectedly low horizontal load level. Longitudinal strain profiles along the compression side are shown in Fig.7d. Similar profiles were observed on the tension side of the column with slightly higher strains at higher load levels. This is an indication of debonding between the carbon shell and the concrete core, and the lack of moment contribution from concrete tensile stresses. The failure mode is attributed to a combined stress state of in-plane compressive stresses at the toe of the column and high interlaminar shear stresses due to boundary effects. This stress state is generated at the compression toe of the column due to a combination of a Poisson's effect at the connection singularity and high demands for localized transverse shear deformations at the critical section. The transition zone material proved to be effective as far as avoiding a crushing failure of the carbon shell, but it possessed a higher modulus than that required to minimize both the in-plane compressive and interlaminar shear stresses.

Fig.5 Ductile Design Test Unit at μ_Δ=8

Fig.6 Strength Design Test Unit at Failure Load of 512 kN

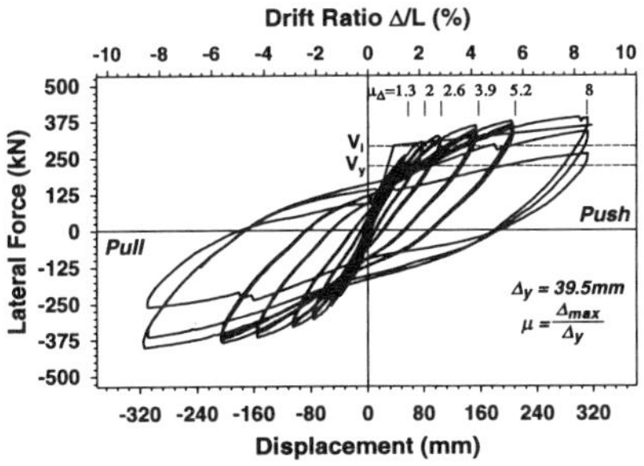

a) Ductile Design: Force-Displacement Hysteresis

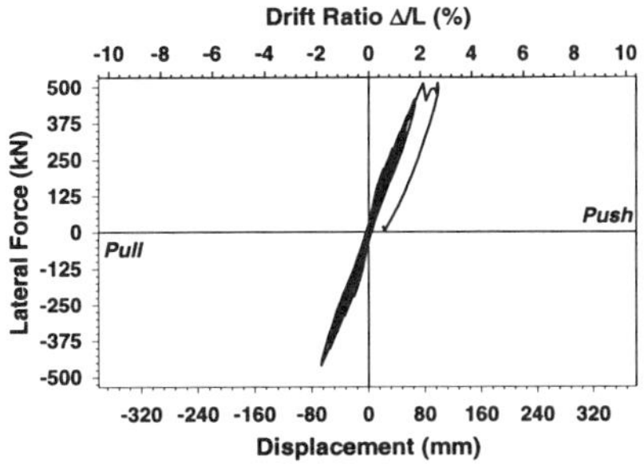

b) Strength Design: Force-Displacement Hysteresis

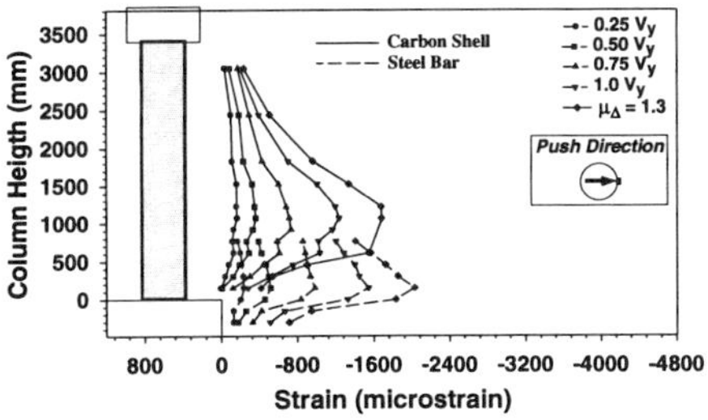

c) Ductile Design: Compression Strain Profiles

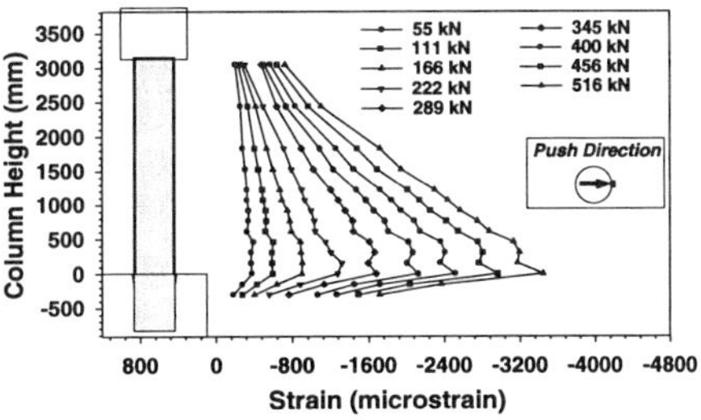

d) Strength Design: Compression Strain Profiles

Fig.7 Test Results

The comparative force-displacement envelopes shown in Fig.8 indicate how the behavior of the carbon shell ductile design concept is almost an exact duplicate of the "as-built" column response. The histeretic behavior of the ductile carbon shell column showed slightly thinner loops than those obtained in the testing of the "as built" column. This can be attributed to debonding of the cross section, and reduced regions of inelastic deformation for the case of the carbon shell column. The strength design concept displayed essentially a linear elastic response. The failure of the carbon shell jacket below the predicted capacity represents a main concern with the overall performance of this system. As can be observed from the force-displacement envelopes in Fig.8, the system failed at approximately 42% of the required load needed to obtain equal energy absorption to the ductile design. The observed failure behavior is directly associated with the low interlaminar shear strength of the two-dimensional laminated shell. However, the connection detail can be improved by providing a more flexible transition zone at the footing top and by additional hoop fibers in the critical transition region. Further details on the design, construction, instrumentation, and data analysis aspects of the experimental research can be found in reference [12].

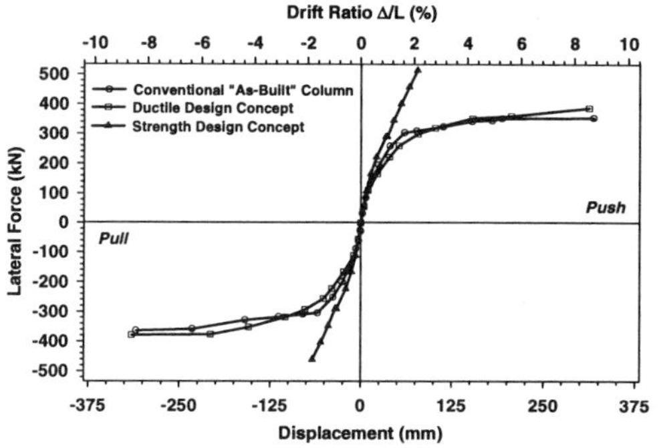

Fig.8 Force-Displacement Envelope Comparison

3.3 ANALYTICAL MODEL CORRELATION

3.3.1 Compression The analytical methods presented in Section 2 have been used to calculate the axial and hoop strains in the shell for the strength design test unit during the application of the axial load (1780 kN). The theoretical strains and experimental strains are plotted against the axial load in Fig.9. The experimental strains shown are the average of four gages circumferentially distributed at a section 762 mm up from the footing interface (Fig.10). This level was chosen to avoid boundary effects at the footing interface which are not accounted for in the theoretical procedure. The longitudinal and hoop strains are shown in Fig.9a and Fig.9b respectively. For the hoop strains two theoretical curves are shown. The curves are developed assuming a concrete Poisson's ratio of 0.2 and 0.25. These values straddle the calculated Poisson's ratio of the shell (see Table 1) which is 0.23. The lower curve represents the case of no confining pressure being present because the shell is expanding faster than the concrete core. For this case the shell expands in the hoop direction with the Poisson's ratio of the shell. The upper curve shows the increased hoop strain due to the confining of the concrete core. This occurs because with the higher concrete Poisson's ratio the core if unrestrained would expand faster than the shell. Good agreement is observed between the experimental and theoretical axial strains which shows that at this loading level the assumption of composite action is justified. The experimental average hoop strain of the four gages starts out negative at low load levels which is not expected, however the theoretical curves do follow the increasing hoop stress trend.

3.3.2 Bending An analytical moment-curvature response of the concrete filled carbon shell section was evaluated based on the developed design models [3]. The procedure assumes a linear strain field distribution across the section, equivalent orthotropic shell properties, and an active confined concrete model. Furthermore, the flexural capacity evaluation of the cross

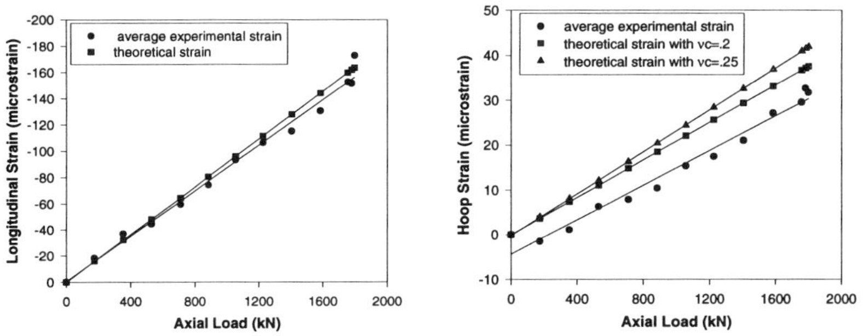

a) Longitudinal Strains

b) Hoop Strains

Fig. 9 Comparison of Theoretical and Experimental Strains in the Shell

section neglects the contribution from tensile stresses in the concrete core. With the objective of evaluating the response of the concrete filled carbon shell section, small internal bars were located in the concrete core only for instrumentation purposes. The internal instrumentation cage consisted of 4 #2 smooth steel bars extending 1.2m above the footing level as shown in Fig.10. The bars were instrumented with longitudinal gages matching the same height as longitudinal gages provided on the carbon shell to determine section strain profiles. The enhancement to flexural strength due to these steel bars is negligible. As above, the section located 762mm from the footing interface was selected for comparison. The strain gage locations at this section are indicated in Fig.10. An experimental moment-curvature trace was taken to be that given by the measurements of the extreme longitudinal strain gages on the carbon shell. The theoretical and experimental moment-curvature traces for the selected section are shown in Fig.11a. The theoretical curve falls below the experimental curve for the lower load levels, which is to be expected since the theoretical procedure does not take into account the moment contribution from the concrete tensile stresses. As previously discussed, the instrumentation on internal bars in the section allowed the measurement of section strain profiles and the monitoring of the slipping between the carbon shell and the concrete core. Measured strain section profiles at different load levels are compared with the predicted strain profiles obtained from the moment-curvature analysis in Fig.11b. The strains in the shell are predicted well even for the load levels where the shell has slid past the concrete core. This is true because at these high load levels, with the high stiffness of the carbon shell in the

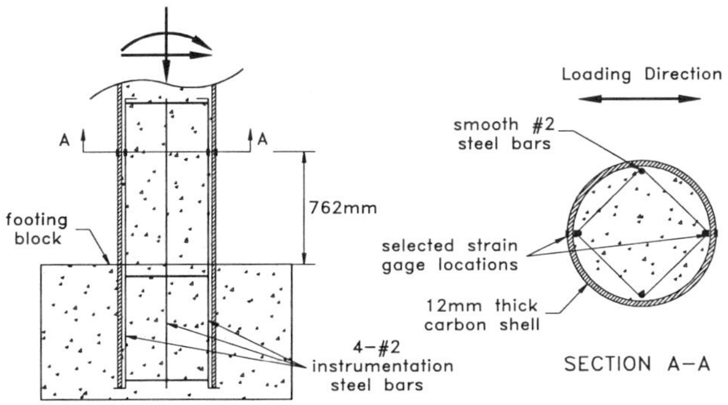

Fig.10 Strain Gage Locations For Comparisons

longitudinal direction for the strength design concept, the moment contribution from the concrete is small compared to the moment contribution from the shell. From these results it seems that the structural behavior of the system in bending is not very dependent on the concrete core. The core does however prevent the shell from buckling and in cases where the shell is not so stiff in the longitudinal direction the contribution from the concrete will become more important. If the concrete contribution is small the interactions described in Section 2.2 do not significantly influence the response of the system. These interactions are however very important for predicting the failure of the shell since the total stress state, longitudinal and hoop, must be known to determine a rational failure criteria. After the shell has slipped past the concrete new methods must be developed to account for the change in the structural response. The load transfer between the shell and the concrete after slippage will depend on the friction between the surfaces and on the normal force which generates the confining pressure on the concrete core. Work is currently in progress to quantify these effects so that the bending behavior may be fully characterized.

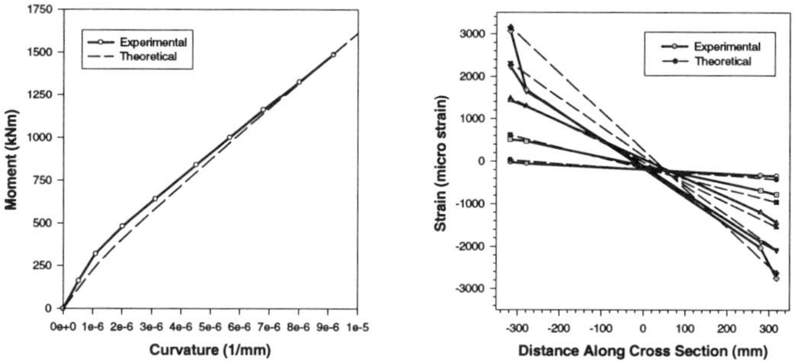

a) Moment-Curvature Comparison b) Strain Profile Comparison

Fig.11 Bending Behavior of Section

4. CONCLUSION

The concrete filled carbon shell concept represents an innovative approach to column design by utilizing premanufactured carbon tubes both as permanent formwork and reinforcement for the concrete core. The system combines the mechanical properties of advanced composite materials in directional tension, with the characteristics of concrete in compression, and steel in inelastic deformation. The preliminary research focused on the development of two design concepts which addressed both ductility and strength. The ductile design concept was successful in duplicating the response of the "as-built" column. The desired response of the strength design concept was limited by failure of the carbon shell due to its low interlaminar shear strength. It is believed that the response for the strength design concept can be improved by the strengthening of the shell at the critical section with additional shell hoop reinforcement and the provision of a more compliant transition region. The preliminary analytical tools developed were shown to adequately predict the section system response for compression and bending loading. Current work is concentrated on the development of a rational failure criteria for the shell taking into account the biaxial nature of the loading, the characterization of the interaction between the shell and the concrete core, and the development of other potential structural systems that may benefit from this technology.

5. REFERENCES

[1] M.J. N. Priestley, F. Seible and M. Calvi, Seismic Design and Retrofit of Bridges, John Wiley & Sons, New York, 1996.

[2] R. Burgueño, , J. McKinley, F. Seible and G.H. Hegemier, "Carbon Shell Systems for Concrete Piers Design of Pilot Test Specimens," Advanced Composites Technology Transfer Consortium Report No. ACTT-95/01, University of California, San Diego. La Jolla, California, January 1995.

[3] F. Seible, R. Burgueño, M. G. Abdallah, R. Nuismer, "Advanced Composite Carbon Shell Systems for Bridge Columns under Seismic Loads", Proceedings from the National Seismic Conference on Bridges and Highways, San Diego, California, December 1995.

[4] J.B. Mander, M. J. Priestley, and R. Park, "Theoretical Stress-Strain Model for Confined Concrete.", J. Struct. Engr. , Vol. 114, No. 8, August, 1988.

[5] P. Madas, and A.S. Elnashai, "A New Passive Confinement Model For the Analysis of Concrete Structures Subject to Cyclic and Transient Dynamic Loading.", Earthquake Engr. and Structural Dynamics, Vol. 21, 409-431, 1992.

[6] S. H. Ahmad, and S. P. Shah, "Stress-Strain Curves of Concrete Confined by Spiral Reinforcement.", ACI Journal, Nov. Dec. 1982, pp. 484-490.

[7] H. Kupfer, H. K. Hilsdorf and H. Rüsch, "Behavior of Concrete Under Biaxial Stresses", J. Am. Concrete Inst. 66, 656-666, 1969.

[8] A. Elwi and D. Murray "A 3D Hypoelastic Concrete Constitutive Relationship" J. Engr. Mech. Div., Vol 105, No. 4 , August 1979

[9] C. P. R. Hoppel, T. A. Bugetti, John W. Gillespie Jr., I. Howie, V. M. Karbhari, "Analysis of a Concrete Cylinder with a Composite Hoop Wrap" , Proceedings of the ASCE Materials Engineering Conference, San Diego, CA 1994, pp. 191-198

[10] A. Mirmiran, M. Shahawy, "A Novel FRP-Concrete Composite Construction for the Infrastructure." , Proceedings, Structures Congress XIII, ASCE, Boston MA, pp. 603-606. 1995.

[11] M. G. Abdallah, "Task 5: Shell Systems for Renewal and New Construction - Shell Systems Manufacturing Report," ARPA/BIR Program Report, Hercules Incorporated, Composite Products Division, Bacchus Works, Magna, Utah. May 1995.

[12] R. Burgueño, F. Seible, and G. H. Hegemier, "Concrete Filled Carbon Shell Bridge Piers under Simulated Seismic Loads - Experimental Studies," Advanced Composites Technology Transfer Consortium Report No. ACTT-95/12, University of California, San Diego. La Jolla, California, September 1995.

28th International SAMPE Technical Conference
November 4-7, 1996

HYBRID COMPOSITE I-GIRDER PEDESTRIAN BRIDGE

Issam E. Harik, Peter J. Szak, Bradley N. Robson, and Theodore H. Hopwood
University of Kentucky
Dan Witcher
Morrison Molded Fiber Glass
Bruce Brailsford
Great Lakes Composites Consortium

ABSTRACT

An 18.29 m (60 ft) pedestrian I-girder bridge is analyzed, designed, tested, and constructed. The main load carrying members for the bridge are two 60.96 cm (24 in) deep fiber glass reinforced plastic I-girders with embedded carbon fibers in the flanges. Laboratory testing quantified the increased stiffness of the hybrid carbon beams over those manufactured solely from fiberglass. Typical joint connections were also tested to investigate their performance. A vibrational analysis was accomplished to check the bridges response to pedestrian traffic. The bridge will be monitored to obtain both long term structural integrity and creep performance.

KEY WORDS: Composite Structures, Carbon Fibers, Pultrusion

1. INTRODUCTION

Although composites have been widely used in the aerospace industry as principal structural components, they have not gained wide acceptance in the civil engineering community. The objective of this study is to design a typical transportation structure using advanced composites in place of traditional civil engineering materials. An 18.29 m (60 ft) long by 1.83 m (6 ft) wide pedestrian I-girder bridge is designed, analyzed, and drawings prepared for its eventually construction in the Daniel Boone National Forest near Cave Run Lake in Bath County, Kentucky. The main load carrying members are produced by Morrison Molded Fiber Glass (MMFG) and are 60.96 cm (24 in) deep fiber glass reinforced plastic I-girders with embedded carbon fibers in the flanges. This bridge will be the first structure of its kind to utilize this type of hybrid composite I-beams as its main load carrying members. The bridge utilizes two hybrid I-beams, with the decking made of fiberglass grating, and is simply supported on concrete abutments. Due to the low modulus of elasticity of the beams, E = 39.3 GPa, ($5.7x10^6$ psi), control of deflection limits, not strength, governs the design under a uniform pedestrian loading. With the addition of fiberglass rods running from the I-beams to anchor points in the abutments, a simple and efficient way of achieving the deflection limit is found. Utilizing composites for a pedestrian bridge in this remote location allows it to be constructed with light equipment or to be prefabricated in pieces and then lifted into place. Remote site locations with limited access for both material and machinery delivery are ideally suited for this type of structure. The design, testing, and construction of this pedestrian bridge will show how off-the-shelf composite materials can be used in innovative ways, which could, in the future, be applied to the construction of vehicular bridges.

2. DESIGN PARAMETERS

The American Association of State Highway and Transportation Officials (AASHTO) requirements call for a pedestrian bridge to sustain a live loading of 415.0 kg/m^2 (85 psf). This loading, along with the bridge's length and width, governed the sizing of the girders. A two-girder bridge was chosen to facilitate construction and transportation, and to minimize cost. After consultation with Forest Service representatives, a deflection limit of L/180 was assigned, which allows for a total deflection of 10.16 cm (4.0 in) for the 18.29 m (60 ft) bridge.

The load carrying capability of FRP beams may be limited by considerations of strength, stability, or deflection. The strength capacity is characterized by an allowable working stress; the stability of the beam is characterized by its resistance to twisting or buckling laterally; and the deflection of the beam is usually limited by functional requirements. The allowable flexural stress of the I-beam is usually governed by local buckling of the outstanding flange. After extensive testing, MMFG determined the value of the ultimate flexural stress for laterally supported I shapes to be 227.5 MPa (33,000 psi). The allowable stress is based on a factor of safety of 2.5 and is 91.01 MPa (13,200 psi). For shear, the allowable stress is 10.34 MPa (1500 psi). Deflections for composite beams are a combination of both flexural and shear effects, where for short span beams, the shear deformation may contribute a significant part of total deflection.

By doing a simple example case, it may be easier to see how bending, shear, and deflection requirements may control the choosing of a FRP section for design considerations. The maximum allowable uniform load for the pedestrian bridge based on flexural, shear, and deflection using a 60.96 cm (24 in) MMFG FRP section (properties listed in Table 1) will be solved for. The allowed stress based on flexural limits is found to be 18.59 kPa (0.388 kips/ft). This is much larger than the distributed load stress of 12.21 kPa (0.255 kips/ft) that results from the 415.0 kg/m^2 (85 psi) pedestrian loading for the 2 beam composite bridge. Based on shear, the allowable maximum distributed load stress is 20.21 kPa (0.422 kips/ft), which is again greater than the applied load. It is seen that the controlling factor is the serviceability limit of deflection, L/180=10.16 cm (4 in). The allowable stress based on this deflection limit is 3.26 kPa (0.068 kips/ft), which is substantially lower than that being designed for, 12.21 kPa (0.255 kips/ft).

When the current modulus and loads are used for the deflection calculation, a total deflection of 42.44 cm (16.71 in), or L/43 is found. This is much higher than the required limit of 10.16 cm (4 in). Reduction of the deflection to the acceptable limit of 10.16 cm (4 in) will require the beam to be stiffened in some measure. For this particular case, a modulus of 72.39 GPa (10,500 ksi) would be required to meet the deflection limitations imposed. Stiffening of the beams was the impetus of the inclusion of high strength and stiff carbon fibers into the flanges of the I-beam.

3. MATERIAL PROPERTIES

The main load carrying members of the pedestrian bridge are 60.96 cm (24 in) hybrid FRP I-beams, whose geometric dimensions are the same as those of the purely FRP beam, shown in Table 1. These beams are notable due to the addition of carbon fibers in the flanges of the beam, which can be seen as the dark horizontal lines in Figure 1. One of the short comings of FRP beams are their low stiffness. Typical series 625 MMFG beams have a modulus of elasticity of 17.24 GPa ($2.5x10^6$ psi), as compared to steel, which has a modulus of 199.95 GPa ($29.0x10^6$ psi); consequently, serviceability, not strength, governs section selection. As a means of addressing this issue, MMFG has pioneered the integration of high strength and stiffer carbon fibers into their pultrusion process. Along with the typical mats and strands of fiberglass, Zoltek carbon fibers were included in a layered manner into the upper and lower flanges. Utilizing carbon fiber in the manner of including them into the flanges where they are most beneficial, did indeed help significantly in increasing the beams stiffness, and these results are presented in the next section.

Laboratory testing was conducted on the 60.96 cm (24 in) deep hybrid FRP I-beams in order to quantify their increased stiffness. All test were performed at the University of Kentucky in the structures lab. Three beams of lengths 7.62 m, 7.92 m, and 9.14 m (25, 26, and 30 ft), were used as test specimens, all 60.96 cm (24 in) deep, with carbon fibers in their flanges. The testing machine is a SATEC 110, which has a 489.3 kN (110 kip) compression/tension capacity, and a 33 cm (13 in) stroke. A 4-point bending test was conducted, with the load points at the third points, this is shown in Figure 2. Lateral bracing was provided, and web stiffeners were used at the supports. Loading was at a rate of 0.51 cm/min (0.2 in/min), and data was taken at 4.45 kN (1 kip) intervals. Data included loading, and deflections at the third points and center span. From this, both shear and bending moduli were found for the beams. Although ASTM standards recommends a span to depth ratio of at least 60 to reduce shear effects to a minimum, this was not possible due to a design restraint. The zero shear and constant moment area between the loading points allows for the separation in the calculation of the shear and bending modulus. An average deflection between both third points was first calculated, and the difference between this and the mid-point deflection was used to separate the shear and bending modulus. A modulus of elasticity of 39.3 GPa, ($5.7x10^6$ psi), and a shear modulus 4.83 GPa ($0.70x10^6$ psi) were calculated for these hybrid beams. These test results showed an increase of 128% in flexure, and an 65% increase in shear modulus. This first trial in the inclusion of carbon fibers is quite promising, and further tests into the incorporation of carbon fibers in the pultrusion manufacturing process are planned.

4. BRIDGE DESIGN

The bridge will be located at Cave Run Lake, in the Daniel Boone National Forest, located 120.7 km (75 miles) east of Lexington, Kentucky. The proposed site spans Clear Creek, which connects the Clear Creek parking and picnic area to the Sheltowee Trace Trail. The positioning of the bridge along this heavily used trail will provide it with high public visibility, aiding in the campaign for community acceptance of FRP.

A team of engineers from UK surveyed the site to determine the best location for the positioning of the bridge which would provide easy access from the parking area, over the creek, and to the main trail. The main factors governing the decision about placement of the bridge were a desire to blend the bridge into the natural surroundings while also limiting the impact of construction on the environment. The bridge's position in shown if Figure 3.

A distance of 18.29 m (60 ft) was deemed sufficient to allow construction space for the abutments and adequate clearance from the face of the abutments to the river. Typical pedestrian bridge width for the Forest Service is 1.83 m (6 ft) clear between a set of 0.914 m (3 ft) high hand rails. Figure 4 shows the bridge dimensions and general layout. In order to better blend into the natural environment, the bridge uses both timber hand rails and posts.

The large amount of runoff carried by Clear Creek during peak drainage times governed the height of the bridge above the creek bed. The high water level during peak flood stage was calculated and the abutments were designed to provide a clearance of 0.61 m (2 ft) from the bottom of the beams to the water level of the river.

Since the resulting 39.3 GPa (5.7×10^6 psi), flexural modulus still fell short of the 68.95 GPa (10×10^6 psi) required to achieve a deflection limitation of L/180=10.16 cm (4 in), the incorporation of fiberglass sucker rods with a 3.81 cm ($1^1/_2$") diameter were employed. These rods were used as suspension members, being anchored to the abutments and draped to a tie-down location beneath the beams (Figure 4). At this tie-down point, the rods are attached to a 20.32 cm (8 in) diameter pressure treated log, which support the I-girder. A second rod then attaches from the log support to below the center of the bridge, where it is threaded into a turn-buckle assemblage. The hinged connection of the rods at their attachment to the abutment and to the log support allows for vertical movement of the bridge in response to loadings. And the turnbuckle located at the center of the bridge provides a means for removing any deflection due to dead load or creep. All connections are made from 304 stainless steel, for ease of maintenance.

The basic bridge configuration of two rods on either side of the beam, anchored to an abutment wall, was modeled in a finite element design code. Parametric studies, which included different positions of the supporting cross member, resulted in the optimized configuration.

These results also provided the force being transmitted through the rods and into the abutments, which governed abutment sizing. The overturning force caused by the rods, resulted in increasing the heel and toe of the abutment, as well as the increase of stem thickness and steel content. Alternate methods of securing the rods using dead-man anchors were investigated, but owing to the position of the bridge in a forest environment, the proximity of large trees precluded this measure. And since abutments were required to raise the bridge to above flood stage levels, increasing its size was of minor inconvenience.

The cross bracing on the beams resulted from the requirements of adequate lateral bracing for the beam against lateral torsional buckling, and for limiting the lateral deflection due to wind loads. The spacing of the bracing was ultimately decided by the hand rail spacing requirements. In order to maintain a rustic type of structure, the forest service chose to use wooden hand rails, mounted to timber posts. The spacing of these posts are usually 1.52 m (5 ft), and this agreed closely with the required bracing of the beams for lateral-torsional buckling. Combing the attachment of the timber post to the bridge at the same location as that of the cross-bracing, allows for only a minimum number of holes which need to be drilled through the web. The wooden spacer between the timber post and the beam is also used as a web stiffener. 5.08 cm (2 in) angle sections were used for bracing members, and are attached at the beam using a Tee section. For bracing requirements due to wind loading, various bracing schemes were run in a finite element code to limit the lateral deflection to 2.54 cm (1 in) with a loading of 344.8 kPa (50 psi) applied to the bridges spanwise surface area. The resulting cross bracing consisted of 10.16 cm (4 in) angles.

Owing to the fact that a 18.29 m (60 ft) beam length could not be used, a beam splice was required. Splicing the beam between the cross bracing facilitated fabrication, and provided for the pultrusion of manageable lengths of beam. The splice plate thickness was kept to a minimum to avoid interference with the walkway grating, so instead of FRP plating, stainless steel was used. Design of the plate was similar to that for typical beam splices, but for the attachment of the plate to the beam. With FRP beams, bearing stress can control connection design. This was the case with the splice plate, and resulted in an increase in both bolt diameter and number of bolts in order to keep the bearing stress within allowable limits. Lab tests of the splice plate showed acceptable performance.

Bridge decking consists of 5.08 cm (2 in) deep series T-5000 MMFG DURADEK grating. The decking panels are 1.83 m (6 ft) wide by 1.52 m (5 ft) in length, and are attached by stainless steel hold downs which are screwed directly into the upper flange with self-tapping bolts. Although no composite action was assumed to result from the decking/beam interaction, this is one area of research that is continuing.

5. DYNAMIC ANALYSIS

The dynamic response of a pedestrian bridge to walking vibrations is always of major concern. Oscillatory vibrations of almost any magnitude will usually have an undue effect on a person's perception of a bridges structural stability. To insure that this bridge was in an acceptable range for acceleration limits, an analysis was conducted using a finite element code. The time history analysis of the bridge was conducted using a pedestrian loading function as given in [3]. and the fundamental natural frequency was determined to be 4.35 Hz. According to the annoyance criterion of International Standards Association (ISO, 1992), the peak acceleration should be less than or equal to 5% of gravity for a fundamental frequency in the range of 4.0-6.3 Hz, and for critical damping ratios larger than or equal to 4%. The oscillatory vibrations of the bridge are within the prescribed comfort limits.

6. CONCLUSION

A pedestrian bridge has been constructed using hybrid FRP beams and has shown exceptional performance. Its light weight, ease of constructability, and low maintenance requirements offers the engineer a new choice of building materials other than steel, wood, or concrete. Although composites have shortcomings, namely in the way of stiffness, using innovative design techniques and advanced hybrid FRP beams, the deficiencies can be overcome. Success of this type of structure shows how these materials may be used for the construction of vehicular bridges.

7. ACKNOWLEDGMENTS

This project was completed under the ARPA Technology Reinvestment Program with the work being managed by the Great Lakes Composites Consortium.

8. REFERENCES

1. EXTREN Design Manual, Morrison Molded Fiber Glass, Bristol, VA, 1993.

2. ASTM D 790-86, Standard Test Methods for Flexural Properties of Unreinforced and Reinforced Plastics and Electrical Insulating Materials, 1986.

3. D. E. Allen and T. M. Murray, Engineering Journal, Fourth Quarter, 117 (1993).

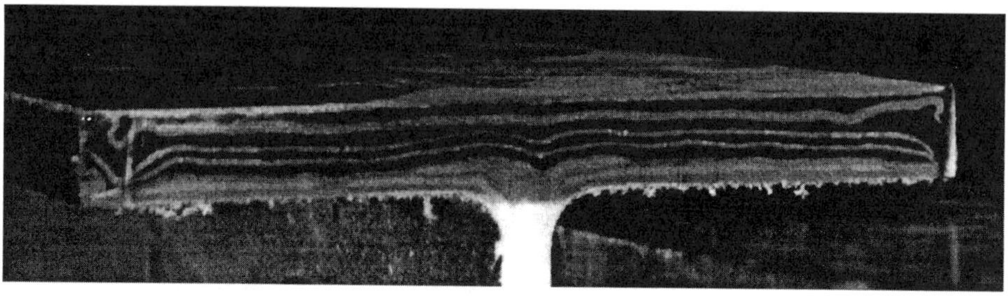

Figure 1. Cross-section of hybrid composite beam showing carbon fibers in the flanges.

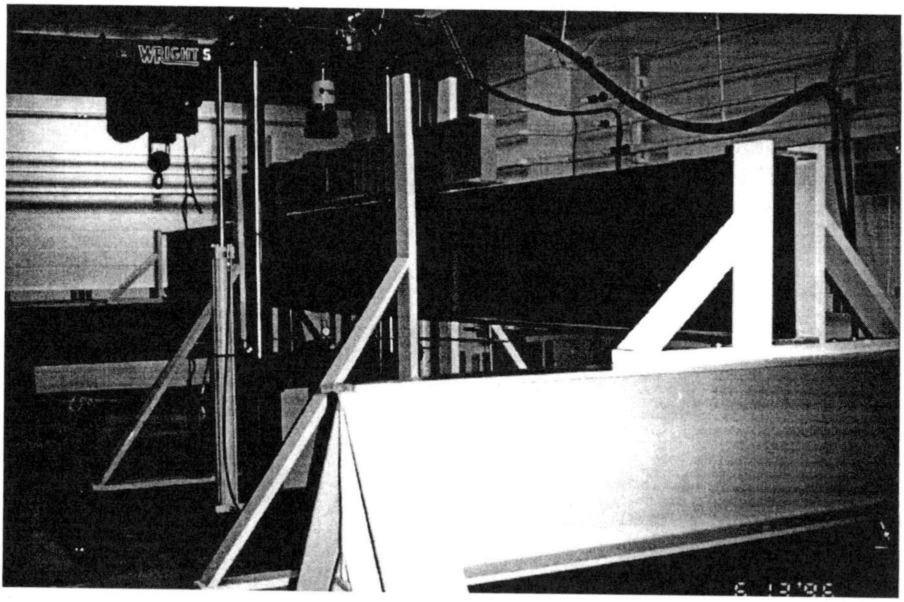

Figure 2. Laboratory set up for the testing of 24" hybrid fiber reinforced plastic beam.

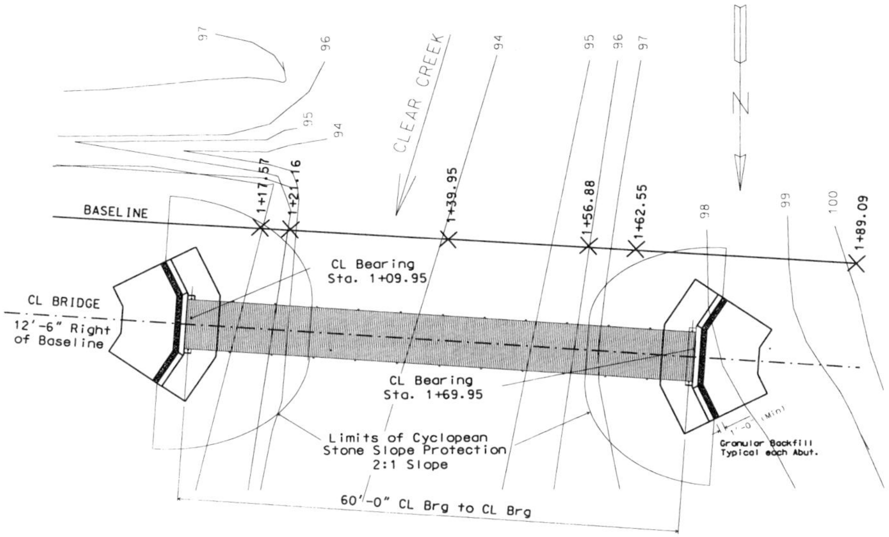

Figure 3. Plan view showing the composite bridge's placement over Clear Creek, Daniel Boone National Forest.

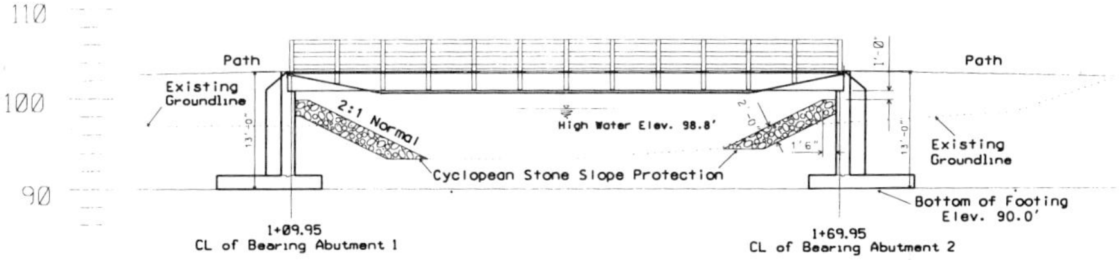

Figure 4. Elevation view of the pedestrian bridge and abutments.

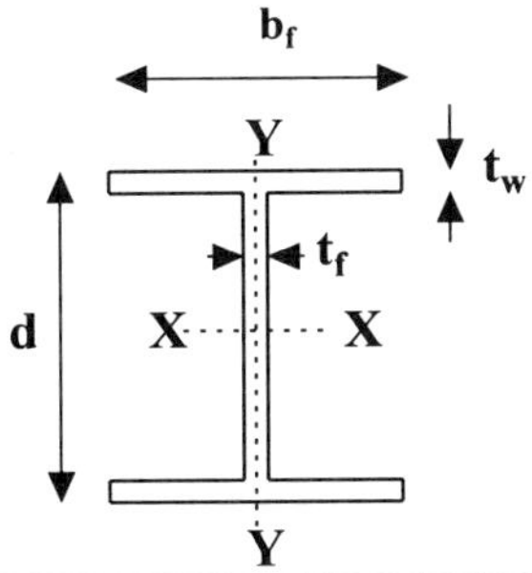

PHYSICAL PROPERTIES						SECTION PROPERTIES			
SIZE				AREA	NOM	AXIS X-X		AXIS Y-Y	
d	b_f	t_w	t_f		WT	I	S	I	S
cm	cm	cm	cm	cm^2	kg/cm	cm^4	cm^3	cm^4	cm^3
60.96	19.05	0.9525	1.91	128.39	222.6	79,227.1	2,599.3	2,199.0	230.9

Table 1. Geometric properties of the MMFG hybrid composite I-beam.

28th International SAMPE Technical Conference
November 4-7, 1996

Long-Term Durability of Composites in Secondary Infrastructure Applications

Kin Liao[1], Robert I. Altkorn[1], Scott M. Milkovich[1], Jose Gomez[2], Carl R. Schultheisz[3], L. Catherine Brinson[4], John M. Fildes[1], and Bruce Brailsford[5]

This presentation contains data generated under a contract with the U.S. Navy's Center of Excellence for Composites Manufacturing Technology, funded by DARPA through the Technology Reinvestment Program, and managed by GLCC, Inc., West Columbia, South Carolina.

Abstract: Pultruded glass-fiber-reinforced composite coupons and I-beams were subjected to four-point-bend environmental fatigue, freezing and thawing, and salt spraying in the laboratory to study their long-term durability for infrastructure applications. Specimens have survived to over five million cycles under four-point bend fatigue at 30% of the flexural strength in air without showing signs of degradation. However, preliminary results suggested that water is detrimental to the life of the material during long-term fatigue. Freezing and thawing in 2% (by weight) NaCl solution resulted in a decrease in dynamic modulus, flexural modulus, flexural strength, and toughness of the composite. Spraying of composite I-beams and hybrid I-beams with carbon fiber outer layers with salt solution for nearly 1,300 hours at 19°C does not seem to change their flexural moduli. Two composite structures are monitored in addition to laboratory testing. Field data collected from an all-composite pedestrian bridge for over 7 months does not suggest any change in its flexural modulus, and two composite walkway sections on a steel lift bridge have performed well in corrosion resistance since they were installed in November, 1995.

Key Words: glass-fiber-reinforced composites, composite bridge, composite walkway, durability, four-point-bend fatigue, water, salt spray, freezing/thawing, infrastructure.

Introduction

Glass-fiber-reinforced composites have seen limited use in the building and construction industry for decades [1-4]. Because of the need to repair and retrofit rapidly deteriorating infrastructure in recent years, the potential for using composites for a wider range of applications is now being realized [5-21]. The state of the "industry" has been described in several review articles and in the proceedings from recent technical conferences pertaining to the subject [5-11].

Fiber-reinforced composites offer excellent resistance to environmental agents and fatigue as well

[1]Basic Industrial Research Laboratory, Northwestern University, 1801 Maple Ave., Evanston, IL 60201-3135.
[2]Virginia Transportation Research Council, 530 Edgemont Road, Charlottesville, VA 22903.
[3]Polymers Division, National Institute of Standards and Technology, Gaithersburg, MD 20899.
[4]Mechanical Engineering Department, Northwestern University, Evanston, IL 60208-3111.
[5]GLCC Inc., 103 Trade Zone Dr., Suite 26C, West Columbia, SC 29170.

as the advantages of high stiffness-to-weight and strength-to-weight ratios when compared to conventional construction materials. However, one of the obstacles preventing the extensive use of composites has been a lack of long-term durability and performance data. Although there have been numerous studies in the areas of creep, stress corrosion, fatigue, environmental fatigue, moisture effects, and natural weathering of composite materials in the past three or four decades, most of these have been aimed at applications in the aerospace industry. At this time, the construction industry has focused predominantly on lower-cost glass reinforcement rather than the carbon fiber reinforcement used in aerospace applications. In addition, the expected service life is much longer in infrastructure applications. For instance, some bridges are designed to last for over 50 years. Hence the infrastructure community must be concerned with longer-term behavior as well as different materials and service environments than the aerospace industry. As a result, although data and experience gained from the past may serve as a general guideline, new studies and data pertaining to infrastructure applications are in great demand, especially for composites produced by low-cost, large-volume processing methods such as pultrusion.

The present work is concerned with obtaining long-term durability data for pultruded glass-fiber composites in the laboratory, performed in parallel with field studies to assess the long-term performance of composite structures in the outdoor environment. The goal of the field studies is to introduce composites into secondary structural applications so that confidence in using the material can be developed prior to applications in large scale and/or primary structures. We will present results of laboratory experiments studying environmental fatigue, freeze/thaw cycling, and salt spray cycling of pultruded composite coupons and I-beams. We will also discuss field data collected from an instrumented composite pedestrian bridge in Antioch, Illinois, and a project with the Chicago Department of Transportation (CDOT) in which two pultruded composite sidewalk sections were installed on the high-traffic LaSalle Street lift bridge in Chicago, Illinois.

Laboratory Studies of Long-Term Durability[#]

Four-point bend environmental fatigue, freezing/thawing, and salt spray tests are carried out in the laboratory to study the long-term environmental durability of glass-fiber-reinforced composites in an accelerated time scale. Results of salt spray tests for hybrid I-beams, which have had layers of carbon fiber bonded to the flanges of pultruded glass-reinforced I-beams in an effort to increase their stiffness, are also included.

Four-Point-Bend Fatigue Test

Four-point-bend fatigue tests are being carried out at the National Institute of Standards and Technology (NIST). The material under study is pultruded E-glass fiber-reinforced vinyl ester composite (Morrison Molded Fiber Glass (MMFG) EXTREN® 625). The as-received materials were 30.5 cm x 122 cm x 0.64 cm (12 in x 48 in x 0.25 in) plates. Bar specimens of dimensions 12.7 cm x 1.3 cm x 0.64 cm (5.0 in x 0.5 in x 0.25 in) were cut from the plates according to the ASTM standard (D790-92) for four-point bend tests. The structure of the composite consists of unidirectional fiber roving embedded in layers of continuous fiber strand mat. Specimens

[#]Certain commercial materials and equipment are identified in this paper in order to specify adequately the experimental procedure. In no case does such information imply endorsement by the National Institute of Standards and Technology, Northwestern University, Virginia Transportation Research Council, GLCC, or DARPA, nor does it imply necessarily that the items are the best available for the purpose.

were loaded sinusoidally at 10 Hz, with R=10 (where R is the ratio of the maximum to the minimum cyclic load) at room temperature under two different conditions, air (dry) and water (wet). For tests carried out in water, specimens were sealed in nylon bags filled with de-ionized water. Specimens were loaded at 85, 65, 45, and 30% of the mean flexural strength. Three specimens were tested at each load level. The fatigue tests were periodically interrupted to measure the flexural modulus of the specimen. Results of the four point bend fatigue tests are summarized in the stress-life (SN) curve shown in Fig. 1.

The mean quasi-static flexural strength of the material in the fiber direction is 314 MPa with a standard deviation of 23.8 MPa. Failure of the specimen begins with the development of matrix cracks transverse to the tensile stress on the tensile surface, followed by failure of the unidirectional fiber roving and propagation of delaminations between the fiber roving and the strand mat. Similar failure processes occur under fatigue loading. For specimens that failed within 10^5 cycles, it seems that the SN curves are not distinguishable for dry and wet specimens.

As has been extensively documented, stiffness change is a strong indicator of the state of the material [22-25]. Several stiffness degradation curves for both wet and dry specimens failed within 10^5 cycles are shown in Fig. 2. Degradation of flexural modulus during (short term) cyclic loading appears to be similar in pattern for various load levels (Fig.2). Here, the abscissa is normalized life. The curve is typified by a small drop in stiffness within the first 10% of life, followed by a relatively flat region, indicating a slower rate of modulus change, to about 85% of life, then succeeded by a much faster drop during the last stages of life. Such a damage-related pattern of stiffness degradation is typical of fiber-reinforced composites [22-25].

Three wet specimens failed at 7.96×10^4, 2.89×10^6, and 3.74×10^6 cycles under cyclic load of 30% flexural strength. All the dry specimens tested at the same load level have survived to over 5 million cycles with apparent damage. Several flexural modulus degradation curves for high cycle fatigue are shown in Fig. 3. For the dry specimens, the average modulus remains approximately the same. These preliminary results suggest that water did exert a detrimental effect on the fatigue life during long-term loading. At the time of this writing, additional long-term fatigue tests are being carried out, the results of which will be reported later.

Freeze/Thaw Cycling

A study has been carried out at the Virginia Transportation Research Council on the resistance of two pultruded composite systems subjected to freezing and thawing. One is E-glass fiber reinforced vinyl ester (MMFG EXTREN® 625), similar to that used in the fatigue test. The other is E-glass fiber-reinforced isophthalic polyester (MMFG EXTREN® 525). The specimens are 30.5 cm x 2.54 cm x 0.95 cm (12 in x 1.0 in x 3/8 in) coupons, cut from as-received flat sheets. One set of specimens from each system was sealed with an epoxy sealant on the cut edges to restrict moisture infiltration prior to testing while another set was not sealed. The coupons were subjected to cycles of freezing and thawing while submerged in a solution of 2% (by weight) NaCl and water, according to ASTM standard C-666, a method to test the resistance of concrete to rapid freezing and thawing. A complete cycle consisted of a ramp down from 4.4°C (40°F) to -17.8°C (0°F), at a rate of about 0.5°C/min, followed by a hold at -17.8°C for about 100 min, a ramp to 4.4°C at 0.5°C/min, and another hold at that temperature. It takes about three hours to complete one such freeze/thaw cycle.

The coupons were tested to determine the dynamic modulus, using ASTM C-215, Standard Test Method for Fundamental Transverse, Longitudinal, and Torsional Frequencies of Concrete

Specimens, since there is no similar method for testing composites. In this test, the specimens are placed on elastic supports. A small accelerometer is then placed on the specimen. This accelerometer (PCB Piezotronics, Inc.) is attached to a Fast-Fourier Transform analyzer, which transforms the response of the accelerometer from the time domain to the frequency domain. The specimen is then impacted with a small force. The results yield the fundamental frequency for the specimen. The dynamic modulus is then back-calculated using an equation derived from classical vibration analysis for the fundamental frequency of a simply supported beam. The coupons were also tested to failure under three-point-bend flexural test, according to ASTM D-790. During such tests, flexural modulus and toughness (area under the load-deflection curve) were also recorded in addition to the flexural strength.

The results of the freeze/thaw tests are summarized in Figs. 4a-d. Values are all normalized to the initial value. Each data point represents the average value for 9 coupons for the edge-sealed specimens and the average for 5 coupons for the unsealed. The unsealed vinyl ester matrix coupons lost 32% of their initial flexural strength at 300 cycles, while the sealed coupons lost about 22%. Both sealed and unsealed coupons lost an average of 3% of their initial dynamic modulus values. Both also lost between 22% and 32% of their initial toughness, again indicating an increase in internal damage with freeze/thaw cycles. The moisture content at the termination of cycling was determined to be 0.14% (by weight) for the unsealed coupons and 0.07% (by weight) for the sealed.

The unsealed isophthalic polyester matrix coupons lost 19% of their initial flexural strength at the termination of cycling, and the sealed specimens lost 24%. The unsealed coupons lost about 3-4% of their initial dynamic modulus whereas the sealed coupons lost between 2-3%. Losses in toughness were observed to be 20% for the unsealed coupons and 27% for the sealed. The moisture contents at the end of cycling were 0.15% (by weight) and 0.14% (by weight) for unsealed and sealed coupons, respectively.

Salt Spray Test

One of the critical issues in applying fiber-reinforced composites in transportation infrastructure (such as bridges) in colder climates is the resistance to long-term exposure under salt spray for melting snow. A study is underway at Northwestern University's Basic Industrial Research Laboratory (BIRL) using pultruded I-beams made from E-glass reinforced vinyl ester. These as-received I-beams are 91 cm (36 in) long, 3.8 cm (1.5 in) wide in the flange, 7.6 cm (3.0 in) high in the web, and 0.64 cm (0.25 in) thick. One such I-beam (labeled as B1), is tested in as-received condition. For two of the I-beams, thin layers of unidirectional carbon-fiber-reinforced epoxy were bonded to the top and bottom flange after two layers of glass/vinyl ester of the same thickness were removed so that there was no change in the cross sectional area and moment of inertia. Two 0.13 cm (0.05 in) thick layers of carbon/epoxy were used for one of the hybrid beams (labeled as B2), and two 0.25 cm (0.10 in) thick layers were used for the other (labeled as B3). The carbon/epoxy composite layers were bonded to the I-beams using Hysol 934-NA adhesive (Dexter Corporation), cured at room temperature under pressure. The main purpose of using the carbon/epoxy layer is to increase the stiffness of the I-beam while at the same time enhance its performance under environmental exposure, because carbon-fiber-reinforced composites are more resistant to moisture effects than glass-fiber-reinforced composites [26-28]. Each salt spray cycle, performed in a Q-FOG environmental chamber (Q-Panel, Cleveland, OH), consisted of 90-minute fogging time by a salt solution (0.05g NaCl / liter of water and 4g $(NH_4)_2$ SO_4 / liter of water), followed by a 60-minute drying time, at a constant temperature of 19°C. The I-beams were taken out periodically and their flexural modulus measured under three-point-bend loading.

Results of the salt spray test are shown in Fig. 5, where the normalized effective flexural modulus is plotted against salt spray time. The initial effective flexural moduli are 20.1, 30.9, and 37.2 GPa for I-beams B1, B2 and B3, respectively. The hybrid I-beams B2 and B3 offered a 54 and 85% increase in flexural stiffness over the glass/vinyl ester I-beam (B1). It is seen that there is essentially no change in flexural modulus after nearly 1,300 hours of salt spray cycling.

Structural Monitoring / Response

Antioch bridge

The Antioch bridge is a 13.7-meter (45-foot) long, 3-meter (10-foot) wide structure designed by E.T. Techtonics, Philadelphia, PA. A photograph and a drawing of the bridge are shown in Figs. 6-7. The load rating of the bridge is five tons. The bridge is subjected to traffic by pedestrians and golf carts. The primary support elements are four girders constructed of back-to-back sections of pultruded E-glass/vinyl ester C-channel (20.3 cm (8 in) web, 6.4 cm (2.5 in) flange, 0.95 cm (3/8 in) thick) spaced by square cross-section E-glass/vinyl ester tubing. Each of the half-girders consists of two 3.2-meter and a 7.3-meter long C-channels, connected together by splice plates. Below the girders are eight transverse beams constructed of the same C-channel and tubing. The deck is made of wooden planks and is supported directly on the girders.

The bridge was instrumented with eight strain gages (Model CEA-00-250UW-350, Measurements Group, Inc, Raleigh, NC), five of which are located at the center span on one of the interior girders. Of these, one is located on the underside of the top flange, two on the web (above and below the mid line), one on the top of the bottom flange, and one on the bottom of the bottom flange. In addition one strain gage was mounted on the bottom side of the lower flange near an abutment and one was mounted on each side of a splice plate on an exterior girder. Tilt meters (Lucas AccuStar analog electronic clinometer, Schaevitz Products, Pennsauken, NJ) were placed adjacent to the strain gages straddling the splice plate. The strain gages and tilt meters were monitored by a Somat Model 2100 field computer (Somat Corp., Champaign, IL). The bridge has been visited monthly since November 1995 to make strain and deflection measurements. Data is acquired from strain gages and tilt-meters as a vehicle (automobile) of known weight is driven across the bridge.

The normalized stiffness of the girder on which the strain gages were installed is shown in Fig.8, for a period of 200 days. The strain gage data were corrected according to the temperature at the day of visit. The data shown were calculated using the strain gage on the bottom of the bottom flange at the mid-span. As can be seen, the stiffness of the girder remains approximately the same at the end of 200 days.

LaSalle Street lift bridge

The LaSalle Street bridge over the Chicago river was originally constructed in 1928 and underwent renovation most recently in 1995. As part of the renovation, the existing pedestrian walkways were removed, and both the walkways and the wooden stringers supporting the walkways were replaced. The new walkways are 3.66 meter (12 foot) wide and made up of approximately 120 1.22 meter (4 foot) long by 3.66 meter (12 foot) wide sections supported on three steel stringers (two W14x30 I-beams and one C12x25 C-channel). The walkway is cantilevered outside the main bridge trusses, which also serve to protect it from vehicular traffic. All but two of walkway sections are made of steel. The remaining two are composite, consisting of T-5000-2" Duradek® E-glass-fiber reinforced vinyl ester composite grating covered with 1.3 cm (½ in) E-glass-fiber reinforced EXTREN® vinyl ester composite sheet (both from MMFG, AFC Division Chatfield, MN). The T-5000-2"Duradek® grating is made up of 51 mm (2 in) tall pultruded composite I-bars spaced on 51 mm (2 in) centers connected by pultruded composite cross-rods on 305 mm (12 in) centers. Each composite section was bolted to the stringers using 12 pairs of stainless-steel hold-downs (each pair "sandwiching" an I bar) and 9.5 mm (3/8 in) screws. EXTREN® shims were used under the composite sections to match the height of the steel sections. The two composite sections, as well as the steel sections, were installed by engineers and ironworkers with CDOT in November 1995 and have since been subjected to heavy pedestrian and light maintenance vehicle traffic, winter conditions, and occasional raising and lowering of the bridge (Fig. 9).

The composite sections appear to be in excellent condition at this time while some of the steel sections (13 out of 118) have already shown noticeable corrosion on their surfaces since installation. The composite walkway sections offer several advantages over steel. The most obvious of these, and the primary motivation for their use in this application, is corrosion resistance and related reduction in maintenance costs. Beyond this, however, the composites sections are much lighter than their steel counterparts (approximately 180 kg vs 500 kg) and offer the possibilities of reducing both counterweight size and lifting stresses in lift bridges and installation without the use of cranes. Further, the use of bolting as opposed to welding could further reduce installation costs.

Conclusions

It is shown that at relatively short term (around 10^5 cycles), pultruded glass/vinyl ester behaves similarly under four-point-bend fatigue in air and in water. At longer term (beyond 10^6 cycles), however, preliminary results indicate that water has a significant detrimental effect on the life of the coupon. Salt spray tests of I-beam mad of glass/vinyl ester and hybrid I-beams with carbon/epoxy outer layers to nearly 1,300 hours do not seem to cause any degradation in flexural modulus. Results of freeze/thaw cycling for glass/vinyl ester and glass/isophthalic polyester over 300 cycles all show marked degradation in flexural modulus, flexural strength, and toughness.

We have monitored the performance of a composite pedestrian bridge in Antioch, IL. and two composite pedestrian walkway sections on the LaSalle Street bridge over the Chicago river in Chicago, IL. Both installations appear to be performing well, with no signs of deterioration. However, severe laboratory tests indicate that weathering can adversely affect the durability of composite structures. Therefore, more long-term testing will be performed and monitoring of the installations will continue.

Acknowledgments

The authors thank Mr. Jeff Mainzer (Northwestern University) for his work on the Antioch and LaSalle Street bridges as well as on the salt spray tests. We also thank Mr. Richard Cole (Morrison Molded Fiberglass Corp., AFC Div., Chatfield, MN), Mr. Daniel Witcher (Morrison Molded Fiberglass, Bristol, VA), and Mr. Glenn Barefoot (Morrison Molded Fiberglass, Bristol, VA), for their expertise and the donation of pedestrian walkway sections and pultruded composite plates. In addition, we thank Mr. Christopher Holt (Chicago Department of Transportation) for his efforts on the installation of the LaSalle Street Bridge pedestrian walkway sections.

References

1. Chambers, R. E., "Structural Fiber Glass-Reinforced Plastics for Building Applications," *Plastics in Buildings*, I. Skeist Ed., Reinhold Publishing Co., New York, 1965, pp. 72-118.
2. Hollaway, L., *Glass Reinforced Plastics in Construction*, John Wiley & Sons, New York, 1978.
3. Makowski, Z. S., "Symbiosis of Architecture and Engineering in the Development of Structure Uses of Plastics," in *Plastics in Material and Structural Engineering*, Elsevier Scientific Publishing Company, New York, 1982, pp. 59- 72.
4. Starr, T. F., "Structural Applications for Pultruded Profiles," *Composite Structures*, 2, I. H. Marshall Ed., Applied Science Publishers, New York, 1983, pp. 192-211.
5. Barbero, E. and GangaRao, H. V. S., "Structural Applications of Composites in Infrastructure, Part I," *SAMPE Journal*, Vol. 27, No. 6, November/December 1991, pp. 9-16.
6. Barbero, E. and GangaRao, H. V. S., "Structural Applications of Composites in infrastructure, Part II ," *SAMPE Journal*, Vol. 28, No. 1, January/February 1992, pp. 9-16.
7. GangaRao, H. V. S. and Barbero, E., "Construction, Structural Applications," in *International Encyclopedia of Composites*, Vol. 6, VCH, New York, 1991, pp. 173 - 187.
8. *Fiber Composites in Infrastructure, Proceedings of the First International Conference on Composites in Infrastructure (ICCI'96)*, H. Saadatmanesh and M. R. Ehsani Eds., Tucson, Arizona, January 15-17, 1996.
9. *Advanced Composite Materials in Civil Engineering Structures*, S. L. Iyer and R. Sen Eds., American Society of Civil Engineers, New York, 1991.
10. *Infrastructure: New Materials and Methods of Repair, Proceedings of the Third Materials Engineering Conference*, K. D. Basham Ed., San Diego, California, November 13-16, 1994.
11. *Proceedings of Composites Institute's 51st Annual Conference*, The Society of the Plastics Industry, Inc., Cincinnati, OH, February 5-7, 1996.
12. Salama, M. M., "Advanced Composites for the Offshore Industry: Applications and Challenges," *Revue De L'Institut Francais Du Petrole*, Vol. 50, No. 1, January - February, 1995, pp. 19 -26.
13. Karbhari, V. M., "Composites and Infrastructure: A Perspective Influence by a Panel Study in Japan," *SAMPE Journal*, Vol. 29, No. 3, May/June, 1993, pp. 74-75.
14. Tarricone, P. "Composite Sketch," *Civil Engineering*, May 1995, pp. 52- 55.
15. *Composite News: Infrastructure*, Issue No. 29, 17 July, 1995.
16. Sprecher, N., "Composites and Buildings: New Applications," *Composites, Plastiques Reinforcses, Fibres De Verre Textile*, Vol. 32, No. 3, 1992, pp. 250-263.
17. Green, A., Bisarnsin, T., and Love, E. A., "Pultruded Reinforced Plastics for Civil Engineering Structural Applications," *Proceedings of the 47th Annual Conference, Composites Institute*, The Society of the Plastics Industry, Inc., February, 1992, section 15-B.
18. *High-Performance Construction Materials and Systems: An Essential Program for America and Its Infrastructure*, Executive Report 93-5011.E, The Civil Engineering

Research Foundations, April, 1993.

19. Loud, S., "Three Steps Toward a Composites Revolution in Construction," *SAMPE Journal* Vol. 32, No. 1, January/February, 1996, pp. 30- 35.

20. Green, A. "Glass-Fiber-Reinforced Composites in Building Construction," *Transportation Research Record 1118*, Transportation Research Board, Washington, D. C., 1987, pp. 73-76.

21. Tsuji, Y., Kanda, M., and Tamura, T., "Applications of FRP Materials to Prestressed Concrete Bridges and Other Structures in Japan," *PCI Journal*, July-August 1993, pp. 50-58.

22. Dharan, C. K. H., "Fatigue Failure Mechanisms in a Unidirectionally Reinforced Composite Material," *Fatigue of Composite Materials, ASTM STP 569*, American Society for Testing and Materials, 1975, pp. 171-188.

23. Reifsnider, K. L., Schulte, K., and Duke, J. C., "Long-Term Fatigue Behavior of Composite Materials," *Long-Term Behavior of Composites, ASTM STP 813*, T. K. O'Brien, Ed., American Society for Testing and Materials, Philadelphia, 1983, pp. 136-159.

24. Jamison, R. D., Schulte, K., Reifsnider, K. L., and Stinchcomb, W. W., "Characterization and Analysis of Damage Mechanisms in Tension-Tension Fatigue of Graphite/Epoxy Laminates," *Effects of Defects in Composite Materials, ASTM STP 836*, American Society for Testing and Materials, 1984, pp. 21-55.

25. Echtermeyer, A. T., Engh, B., and Buene, L., "Lifetime and Young's Modulus Changes of Glass/Phenolic and Glass/Polyester Composites Under Fatigue," *Composites*, Vol. 26, No. 1, 1995.

26. French, M. A. and Prichard, G., "Strength Retention of Glass/Carbon Hybrid Laminates in Aqueous Media," *Durability of Polymer Based Composite Systems for Structural Applications*, A. H. Cardon and G. Verchery, Eds., Elsevier Applied Science, 1993, pp. 345-354.

27. Ishai, O., Garg, A., and Nelson, H. G., " Hygrothermal Effects on the Mechanical Behavior of Graphite Fibre-Reinforced Epoxy Laminates Beyond Initial Failure," *Composites*, Vol. 17, No. 1, January 1986, pp. 23-32.

28. Weitsman, Y. J., "Effects of Fluids on Polymeric Composites - A Review," *Office of Naval Research Contract Report, ONR N00014-90-J-1556*, July, 1995.

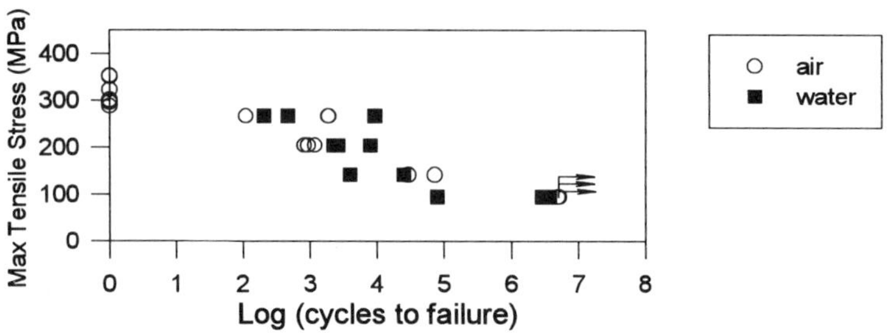

Fig. 1 - SN curve of four-point bend fatigue of glass/vinyl ester.

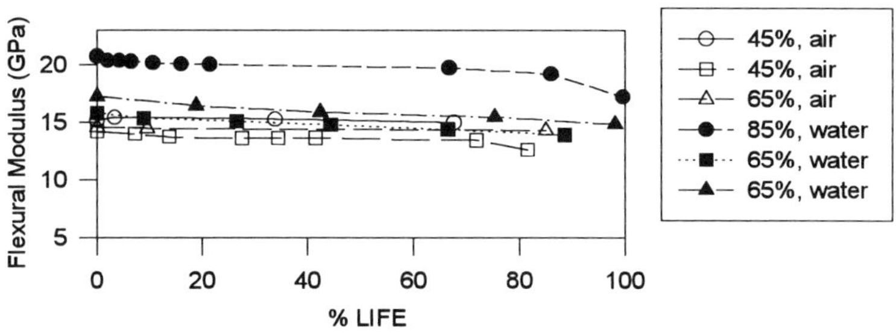

Fig. 2 - Stiffness degradation curves for specimens loaded between 45-85% flexural strength.

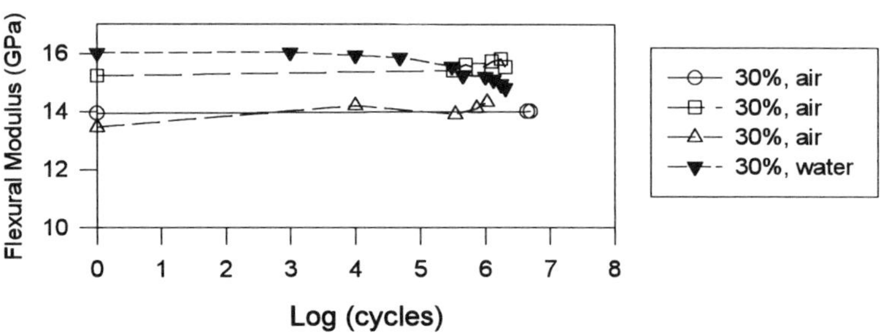

Fig. 3 - Stiffness degradation curves for specimens loaded at 30% flexural strength.

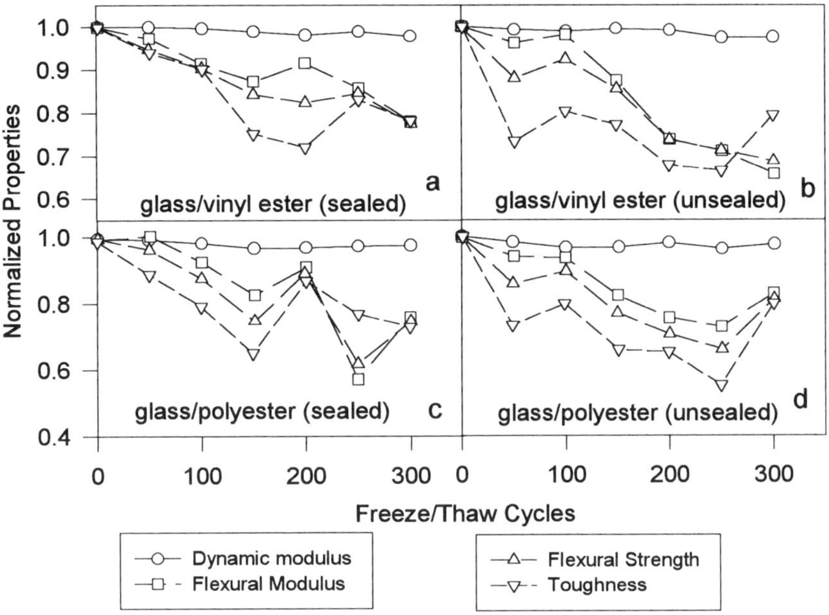

Fig. 4 - Results of freeze/thaw cycling of glass/vinyl ester and glass/isophthalic polyester composites.

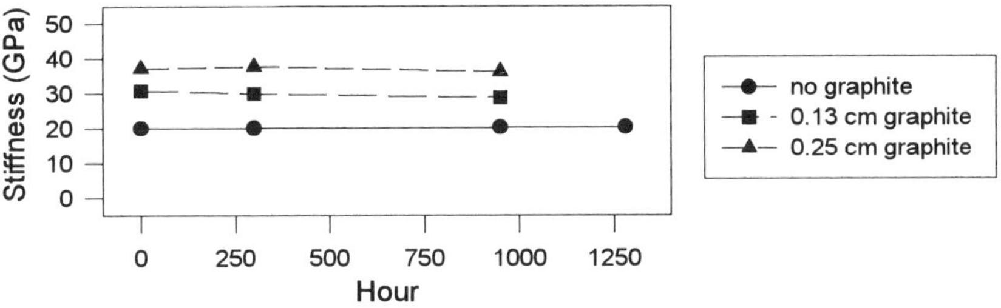

Fig. 5 - Effect of salt spray on flexual modulus of composite I-beams.

Fig. 6 - Composite bridge at Antioch, Illinois.

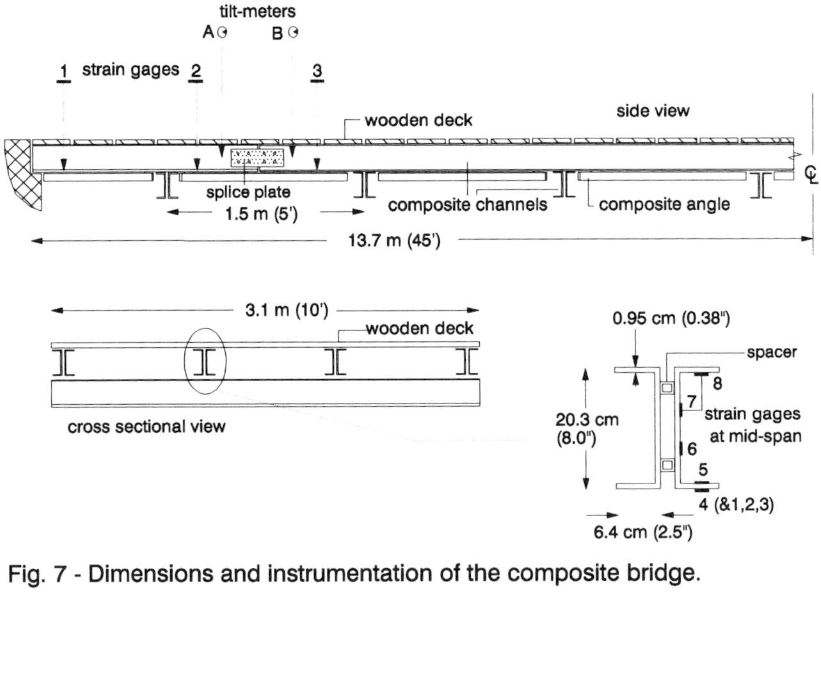

Fig. 7 - Dimensions and instrumentation of the composite bridge.

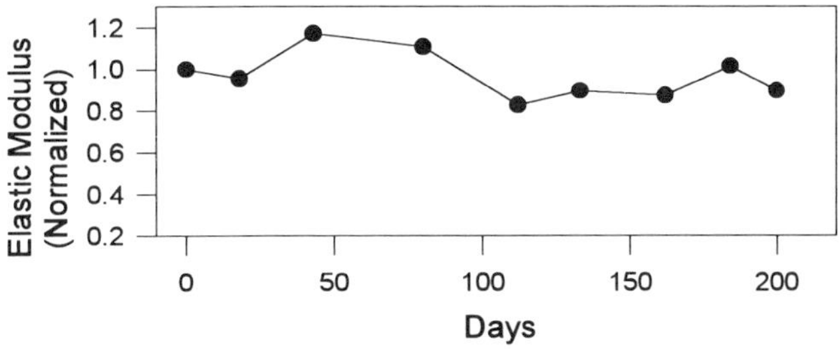

Fig. 8 - Change in elastic modulus of the girder of the composite bridge.

Fig. 9 - Installation of the composite walkway sections in the LaSalle Street Bridge.

28th International SAMPE Technical Conference
November 4-7, 1996

B-STAGING OF TOUGHENED EPOXY COMPOSITES

Fred E. Arnold and Steve Thoman
Naval Air Warfare Center
Aircraft Division, Polymers and Composites Branch
Patuxent River, MD 20670-5304

ABSTRACT

B-staging as an alternative processing scheme for toughened epoxy composites is presently being explored. The objective is to eliminate the effect of room temperature advancement as well as reduce interlaminar stresses during processing, and promote the proper conditions for adhesion of new material to the composite. Modulated differential scanning calorimetry (MDSC) and dynamic mechanical analysis (DMA) were utilized to characterize cure behavior and vitrification temperatures. Subambient Tgs were measured on specimens exposed to a temperature of 23°C for approximately 90 days. Significant advancement starts to occur after eighteen days and levels off at a glass transition temperature of 44°C. Specimens staged to glass transition temperatures above 50°C show negligible advancement at room temperature. Hercules 8552 neat resin gels at a conversion of 0.43 measured by differential scanning calorimetry and dynamic mechanical analysis. A time temperature transformation diagram [1,2] was generated for conversions of less than 0.50. Two assumptions for calculating the isothermal cure diagrams were, the reaction prior to vitrification is kinetically controlled and there is a one to one relationship between the glass transition temperature and degree of conversion. An activation energy of 69.9 kJ/mol was calculated for the reaction.

KEY WORDS: modulated differential scanning calorimetry (MDSC), Hercules 8552 resin, activation energy, b-staging

1. INTRODUCTION

Hercules 8552 sulfone toughened epoxy is presently being applied in the manufacturing of aircraft components. B-staging refers to a process in which different components are cured to a pre-determined degree of conversion and later united and co-cured to form the final structure. The objective of this process is to reduce interlaminar stresses during processing, eliminate the effect of room temperature advancement of the neat resin and promote proper conditions for adhesion of new material onto the lay-up. Modulated differential scanning calorimetry (MDSC) and dynamic mechanical analysis (DMA) were utilized to characterize cure behavior and glass transition temperatures. Subambient glass transition temperatures were measured on specimens exposed to a temperature of 23°C for approximately 90 days. The resin has been modified with a reactive thermoplastic toughener improving fracture toughness. Dry and wet glass transition temperatures range from 197-205°C (388-401°F) and 150-156°C (302-313°F) respectively. A recommended cure temperature of 177°C (350°F) guaranties chemical degrees of conversion (into the fully cured thermoset) in excess of 0.90. Lower curing temperatures may be used but a post cure cycle should be added inorder to maximize mechanical strength and solvent resistance.

2. EXPERIMENTAL

A TA 2920 differential scanning calorimeter equipped with a modulated DSC cell was used to characterize both the cure behavior and glass transition temperatures of staged and advanced 8552 epoxy resin and pre-preg. DSC runs were recorded using a heating rate of 3°C/min with a modulation frequency of 0.5°C. Sample sizes varied between 3-5 mg for the neat resin and 25-35 mg for pre-preg. A Rheometric RDA II analyzer was used to determine gelation times and correlated with DSC results. Parallel plates were used with a diameter of 25 mm and an average gap distance of 0.5 mm.

3. RESULTS AND DISCUSSION

3.1 Rheological Behavior Determining processing variables starts with establishing the rheological behavior and its relationship to the chemical rate of cure. 8552 neat resin was found to gel at a chemical degree of conversion of 0.43 determined from dynamic mechanical analysis (DMA) and differential scanning calorimetry. Typically amine hardened epoxy systems gel at conversions in between 0.40 and 0.60. The reaction is autocatalytic and controlled by the amount of amine hardener [3]. The measured degree of conversion at gelation corresponds to literature values supplied to NAWCAD by Hercules.

3.2 Out-Life Determination It is imperative that the physical and chemical state of a system be identified before processing variables are determined. During manufacturing 8552 pre-preg tape is often left exposed to ambient room temperature for extended periods of time. This practice may pose potential problems with processing due to chemical advancement at room temperature. Modulated differential scanning calorimetry (MDSC) was utilized to characterize the effect of room temperature advancement on the glass transition temperature. Equations 1 and 1a describe the temperature dependence of the sample measured by standard DSC and modulated DSC.

$$T(t) = T_0 + \beta t \tag{1}$$

$$T(t) = T_0 + \beta t + A_t (\sin \omega t) \tag{1a}$$

The time dependent temperature is determined by β the linear heating rate, T_0 the initial temperature and ω the modulation frequency. The heat flow through the sample is described in equations 2 and 2a,

$$dQ/dt = C_p\beta + f(t,T) \tag{2}$$

$$dQ/dt = C_p(\beta + A_t\omega * \cos \omega t) + f(t,T) + A_k (\sin \omega t) \tag{2a}$$

where $(\beta + A_t\omega * \cos \omega t)$ is the measured heating rate, f(t,T) is the kinetic response without the temperature modulation and A_k is the amplitude of the kinetic response to temperature modulation [4]. The reversing component from the modulated mode eliminates effects from crystallization and any non-reversible process including heat from residual unreacted polymer.

Figure 1 illustrates the effectiveness of using the modulated heating rate verses the normal linear rate. The glass transition temperature can easily be extracted from the reversing component using the modulated mode. The influences of residual heat release due to incomplete cure can readily be eliminated. Out-life measurements, degree of conversion and the glass transition temperature of pre-staged composites were determined using MDSC. Figure 2 illustrates room temperature advancement of the resin measured from an increase of the subambient glass transition temperature over a 90 day time period. After 18 days significant advancement starts to occur and after 40 days the material's glass transition temperature has risen above room temperature. One notes that after 18 days approximately 5% of the system has reacted and after 40 days close to 20% has reacted. From Figure 3 we established the relationship between the glass transition temperature and degree of chemical conversion. Gelation for 8552 occurs at a chemical degree of conversion of 0.43 which corresponds to a glass transition temperature of 86.3°C (187.3°F). The gel point represents a physical and chemical state in which no further flow is possible and essentially one observes a rigid glass that can not be processed. The processing window for this system is reduced as the neat resin or pre-preg advances at room temperature, so it is vital that the pre-preg tape is stored under an environmentally controlled climate. If proper storage of the pre-preg is not practical a careful record of out time should be recorded so processing variables can be adjusted.

3.3 B-Staging 8552 Composites B-staging is a process in which the system is pre-reacted to a pre-determined degree of chemical conversion and later combined with staged composites and co-cured to form the final part. Components staged to conversions above 50°C can be machined at room temperature and left on the shop floor for prolonged periods of time eliminating the concern of room temperature advancement. The objective of b-staging is to eliminate the effects of room temperature advancement of the resin, reduce interlaminar stresses during processing and to promote the proper conditions for adhesion of "green" resin and other staged components. Figure 4 shows a plot of chemical degree of conversion as a function of isothermal time. The recommended cure temperature of 177°C (350°F) obtains conversions of over 0.90 after approximately 60 minutes. The lowest experimental isothermal run was at 110°C (230°F) and approximately 50% of the material reacted. Corresponding to the degree of chemical conversion we also looked at the glass transition temperature as a function of isothermal time (Figure 5). For the isothermal temperatures investigated the glass transition temperature levels off and for all practical purposes becomes independent of time at the isothermal temperature being investigated.

It was decided that for b-stage processing the 8552 composites would be staged to a glass transition temperature of 54°C (130°F) which corresponds to a chemical degree of conversion of 0.29. Composites staged to this conversion should possess enough mechanical strength at room temperature to avoid warping the material but exhibit sufficient flow at elevated temperatures to ensure the formation of an interphase with new material or staged components. A time temperature transformation diagram was generated to estimate gelation times at lower temperatures. There are two primary assumptions that are necessary for the generation of an isothermal time temperature transformation diagram (TTT). The reaction prior to vitrification is kinetically controlled and there is a one to one relationship between Tg and chemical degree of conversion. Our time temperature transformation treatment was only concerned with low degrees of conversion since the objective of this effort was to establish processing variables for under cured composites. Degree of conversion as a function of time is described by equation 3,

$$dX/dt = k(T) \times f(X) \qquad (3)$$

where k(T) is the reaction rate constant, X is the fractional conversion and t is the cure time. Upon separation of variables followed by integration one obtains equation 4.

$$\ln\left(\int dX/f(X)\right) = \ln(k(T)) + \ln(t) \qquad (4)$$

The left side of equation 4 is only a function of conversion I. e. a function of Tg. Simplifying equation 4 we obtain equation 5.

$$F(Tg) = \ln(k(T)) + \ln(t) \quad (5)$$

According to equation 5 and illustrated in equation 6, Tg will vary at time t_1 for cure temperature T_1 and with t_2 for cure temperature T_2.

$$F(Tg) = \ln(k(T_1)) + \ln(t_1) = \ln(k(T_2)) + \ln(t_2) \quad (6)$$

Rearranging the time and temperature variables a relationship arises in which we can execute a time temperature shift along the time axis. The resulting shift factors A(T) can be plotted verse reciprocal temperature. The slope yields an activation energy for the process as expressed in equation 7.

$$A(T) = -E/RT + E/RT_{ref} \quad (7)$$

We took 149°C (310°F) as our reference temperature. The experimental procedure consists of generating several isothermal diagrams of Tg verses time as expressed in Figure 5. Secondly, these curves are shifted along the time axis towards the reference temperature. The shift factors are then plotted against reciprocal temperature generating an activation energy. The activation energy in conjuncture with the times needed at the reference temperature to reach a specified Tg are used to generate the times needed to reach the same Tg but at a different isothermal temperature. Equation 8 lists the equation used to generate the theoretical isothermal cure diagrams.

$$\ln(t_2Tg) = -19.923 + (8407.51)/T_2 + \ln(t_1Tg) \quad (8)$$

Equation 8 includes the calculated activation energy of 69.9 kJ/mol and the times required to reach a particular Tg at the reference temperature of 149°C. Numerical values for [$\ln(t_1Tg)$] are listed in Table 1.

Figure 6 shows the generated theoretical cure times for 8552 neat resin. The theoretical time to reach a Tg of 86.3°C is 10.4 minutes. This corresponds to a conversion of 0.43 or rather the gelation point. The experimental time for gelation as measured from dynamic mechanical analysis is 12.0 minutes. The discrepancy between the values may be the result of not subtracting the ramp time before isothermal cure. Figures 7 and 8 illustrate theoretical predications and experimental results for isothermal cure temperatures of 135°C (275°F) and 163°C (325°F). Results were analyzed for degrees of conversion less than 50%. For isothermal temperatures of 135°C and 163°C, as molecular weight of the network increases a variance in the measured and theoretical Tgs appears. The variance is small but it is important to keep this fact in mind when selecting cure temperatures and dwell times.

4. CONCLUSIONS

Modulated differential scanning calorimetry enables investigations of thermal transitions at a better sensitivity compared with the standard method. The reversing component from the modulated mode eliminates any affect of residual heat release at or near the measured glass transition temperature. From out-life measurements it is essential to put into place strict quality control procedures inorder to ascertain the physical and chemical state of the system prior to processing. A b-staged condition of 54°C was selected inorder to ensure sufficient mechanical strength at room temperature. A time temperature transformation diagram was produced and found to approximate cure times at various isothermal temperatures. B-staged components can now be prepared for adhesive compatibility with staged and "green" material.

5. REFERENCES

1. Gilham, J. K., Enns, J. B., "Trends in Polymer Science", Vol. 2, No. 12, 1994.

2. Gilham, J. K., American Chemical Society Proceeedings PMSE Vol 74, New Orleans Louisiana, Spring 1996.

3. Turi, Edith, Thermal Characterization of Polymeric Materials, Academic Press Inc., San Diego, California, 1981.

4. Reading, R., "Trends in Polymer Science", Vol. 1, No. 8, 1993.

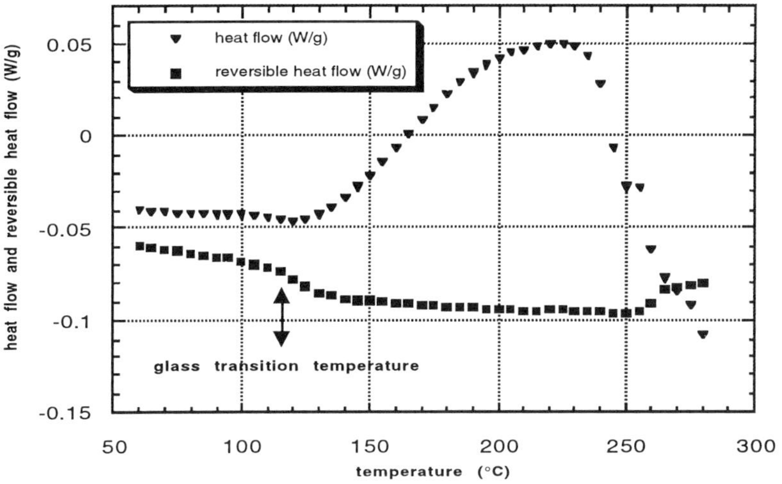

Figure 1. Comparison between modulated and standard heat flow (DSC and MDSC).

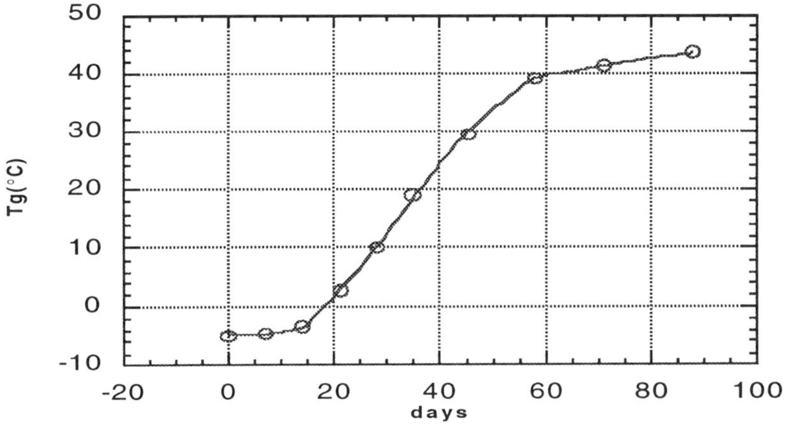

Figure 2. Room temperature advancement of 8552 neat resin.

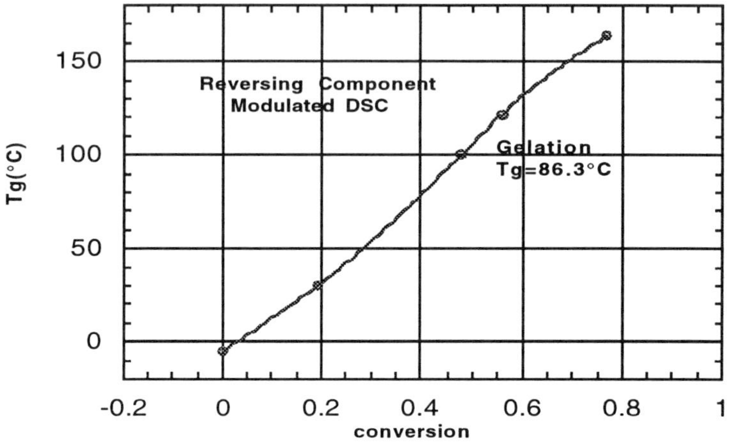

Figure 3. Glass transition temperature (8552 neat resin) as a function of chemical degree of conversion.

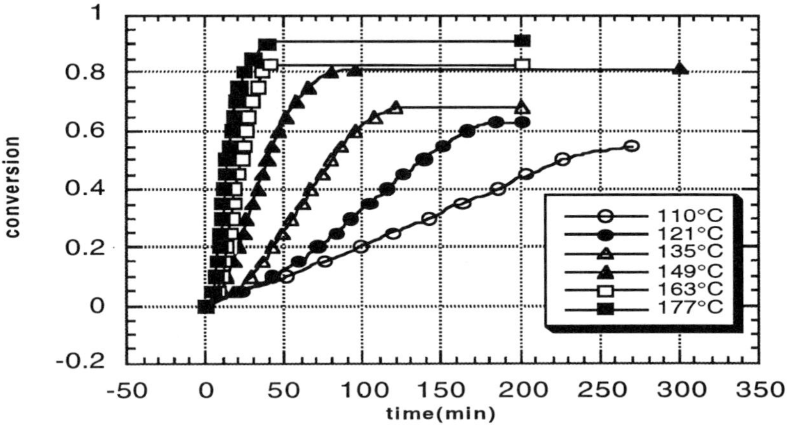

Figure 4. Chemical degree of conversion as a function of isothermal cure (8552 neat resin).

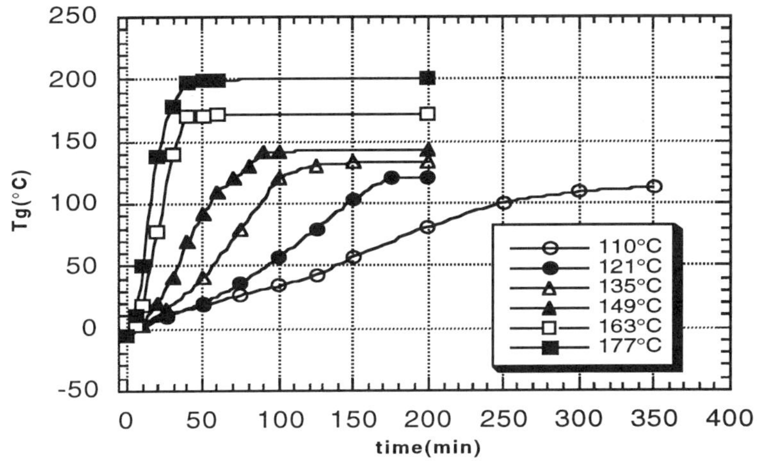

Figure 5. Glass transition temperature as a function of isothermal cure (8552 neat resin).

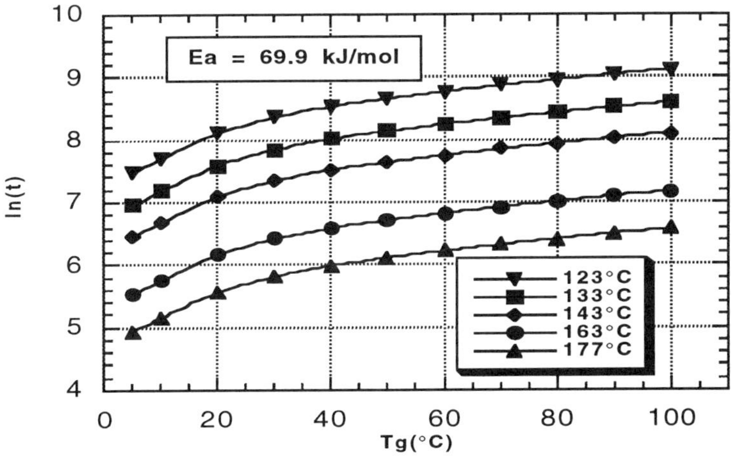

Figure 6. Time temperature transformation diagram for 8552 neat resin.

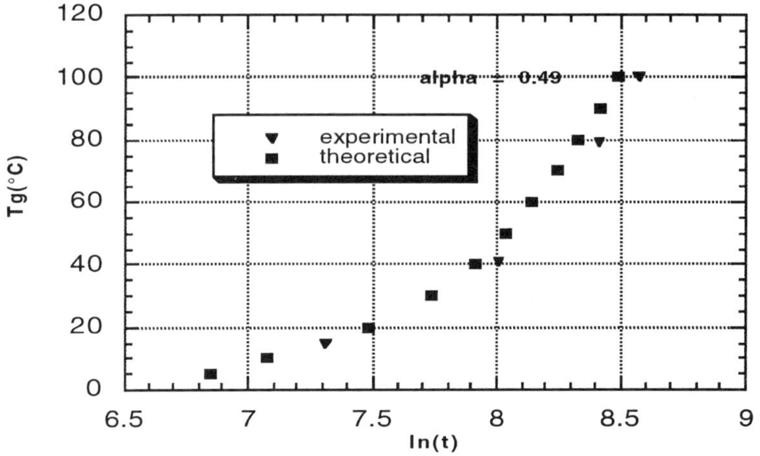

Figure 7. Theoretical prediction of cure time for 8552 isothermally cured at 135°C (275°F).

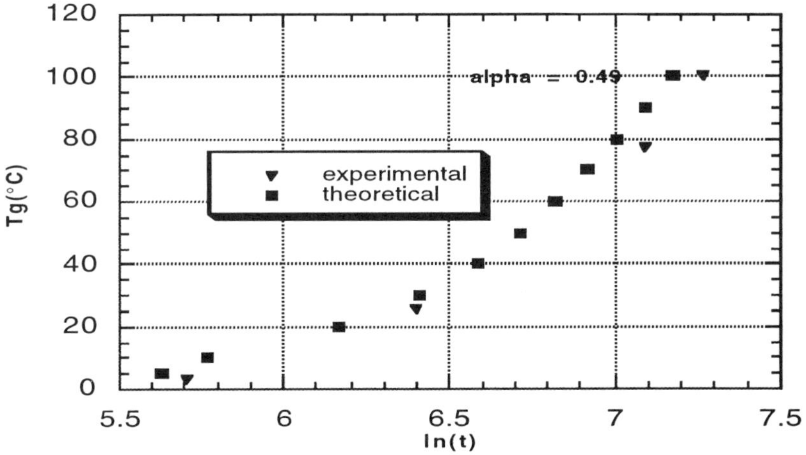

Figure 8. Theoretical prediction of cure time for 8552 isothermally cured at 163°C (325°F).

Table 1. Times to reach a specified Tg at the chosen reference temperature.

Tg(°C)	t_1Tg (sec)	$\ln(t_1Tg)$
5	480	6.1737
10	600	6.3969
20	900	6.8023
30	1158	7.0544
40	1380	7.2298
50	1560	7.3524
60	1740	7.4616
70	1920	7.5600
80	2100	7.6496
90	2292	7.7371
100	2466	7.8104

28th International SAMPE Technical Conference
November 4-7, 1996

A SYSTEMS APPROACH FOR IMPROVEMENT OF THE MANUFACTURING PROCESS FOR COMPOSITE HONEYCOMB SANDWICH PANELS

Thomas H. Brayden, Northrop Grumman Commercial Aircraft Division, Dallas, Texas
Donald C. Darrow, Boeing Commercial Airplane Group, Wichita Division

ABSTRACT

Problems arising from the manufacture of composite parts can cause impacts on cost, weight, and schedule. These problems include core movement, surface pitting, resin exotherms, inhibited resin flow, and internal porosity. Correction of these defects by improved processing requires an intimate and detailed knowledge of the physics and chemistry connected with the materials and structures involved. Chemistry and chemistry related phenomena include chemical composition of the resin, rheology and moisture sensitivity. Important physical characteristics include lateral compressive strength of the core, pressure buildup inside the core with temperature, flow and filleting, and the temperature at which minimum viscosity occurs. An improved process uses a regulated vacuum, at a temperature corresponding to minimum viscosity, in a devolatilization dwell. After the devolatilization dwell, high pressure is applied to reduce the size of gas bubbles and to assure a good face-to-core adhesion. The end results are sandwich panels with smoother faces, stronger face-to-core fillet bonds and no core movement.

All of these considerations can be combined into a multi-level chart that summarizes all of the different aspects involved in and affecting the manufacture of composite honeycomb core panels.

KEYWORDS: Honeycomb/Sandwich Core, Improved

INTRODUCTION

In any field of scientific endeavor, there is a consistent and persistent search for unifying principles. These principles tend to be simple straight-forward concepts that order and explain large bodies of facts and observations. Once these are enunciated, they usually become obvious and are then widely accepted and used. The explanation for the behavior of the daily weather becomes obvious once one has seen a satellite weather map showing highs, lows, fronts and the movement of the jet streams. Although it is not as grand and global as an explanation of weather, such a model can be formulated for the process of manufacturing composite honeycomb sandwich panels.

The model is built up using conceptual components that relate to the details of the process. An awareness of the necessary details comes from an understanding of the product and its manufacturing sequence. This should extend down to the manner in which the most basic laws of chemistry and physics are involved. Operating in an empirical Edisonian manner without a clear understanding of the cause and effect relationships can result in a loss of product and less than an optimal degree of control of the process.

The process for the manufacture of composite honeycomb sandwich panels often results in less than perfect parts. Common defects include; core collapse, porosity, dimpling, frosting, poor filleting, weak skin to core bonds and local exotherms. In order to gain an understanding of the reasons behind these and other processing problems, one must simultaneously command a comprehensive overview as well as an in-depth detailed understanding.

OVERVIEW

Figure 1 gives both the high level overview and the associated underlying ideas. In this form, the model divides into three different aspects, the end goal and the sub-goals, the essential elements dictating the selection of the sub-goals, and the improvements arising from the synergistic interaction between the essential elements. The consequences of the things happening in each section of the lower panel have effects that are shown in the middle panel. Relationships and connections are shown by the alphabetic characters at the heads and the ends of the flow chart arrows.

A common device in work planning is to assume the final goal and then work backwards to more fundamental sub-goals at lower levels. In the case of composite honeycomb sandwich panels, the ultimate goal is an improved honeycomb sandwich panel that does not possess any of the enumerated defects. Presuming the same materials and methods of construction, the only variation possible is to change the cure cycle in order to attain a better product. Therefore, the immediate sub-goal is an improved cure cycle. The components contributing to an improved cure cycle are a devolatilization dwell [1], a balance of forces [2] and an increase of pressure after the devolatilization dwell [3]. The word improved is meant to convey the fact that whatever changes are made, from the standard cure cycle, will result in an improved honeycomb sandwich panel. A common cycle [2] for 177°C(350°F) curing graphite/epoxy honeycomb core panels, calls for the application of 3.10×10^5Pa (45psi) autoclave pressure at the start, venting the bag, a 1.4-2.4°C/min (3-5°F/min) rate of rise to 121°C(250°F), a hold at 121°C(250°F) for 60 minutes, a 1.4-2.4°C/min (3-5°F/min) rate of rise to 177°C(350°F), a hold at 177°C(350°F) for a minimum of 2 hours. See Figure 2 for additional details.

A modified cure cycle [2] for 177°C(350°F) curing graphite/epoxy honeycomb core panels is shown in Figure 3. From an examination of Figures 2 and 3, it is seen that the process parameters are vacuum, time, heat-up rate, temperature and pressure.

The difference between these two cure cycles lies in how these parameters are manipulated to produce the desired end result. Variation of these parameters can either be done in a cookbook manner or on an informed insightful basis.

This insight requires an operational model that is descriptive of the process. Figure 1 is just that required model. The upper panel is an out-growth of the actions described in the lower panel. The lower panel also feeds the middle panel which in turn has a connection with the upper portion. The overall model is best understood by focusing on the components of the improved cure cycle. Previous work [2] indicates that an improved cure cycle has major subcomponents consisting of a devolatilization dwell, a balance of forces and an increase of pressure after the devolatilization dwell. Each of these subgoals is in itself a major consideration.

Additional information to keep in mind about a composite honeycomb core panel is that it is a system being acted on or modified by another system. The panel system is a container [2] of fixed volume composed of the prepreg and its resin system, and the core and its mechanical properties. The system acting on this composite honeycomb core panel is the autoclave which serves as a source of elevated temperature and pressure.

The panel is intentionally described as a system to emphasize the fact that the different components are interrelated and must be considered as a whole. Selecting an individual component and ignoring the others leads to non-optimal processing. For example, increasing the autoclave pressure to condense gas bubbles, while ignoring the potential for core collapse, can result in core movement.

DEVOLATILIZATION DWELL

Many prepreg specifications require that the volatiles present in the as-received material be less than two percent [4,5]. These volatiles may consist of residual solvents from the manufacturing process, entrapped air and adsorbed moisture. At the processing pressures and temperatures, these volatiles are present as gases. At the relatively low pressures existing under a vented bag, the large increase in molar volume on going from a liquid to a gas will convert small volumes of liquid into large volumes of gas. If this gas remains in contact with the curing resin, it can become incorporated into the final product causing and creating undesired effects.

Therefore, it is reasonable to seek methods for the removal of these volatiles at an early stage in the process before they can create undesirable results. A traditional method [1,2] for the removal of volatiles is to apply a partial vacuum to the layup during the period of time that the volatiles are being released. To maximize the removal of these volatiles, it is best that the vacuum be applied during the time that the resin is at its minimum viscosity. However, there are other sources of gases such as the core and the fiber reinforcement in the prepreg material. These gases may be released at varying rates as compared to those originating from the resin. With this in mind, the length of time [1] over which this partial vacuum is applied becomes an important consideration. The objective is to maintain the resin at its minimum viscosity while allowing enough time for the removal of the majority of all volatiles. The time span between minimum viscosity and beginning of gelation is a function of what the temperature is and how rapidly the temperature is changing. A rapid rate of rise past the temperature of minimum viscosity up to the gel point results in a relatively short time span at minimum viscosity. Therefore, the desire is to find a temperature at which the rein's viscosity remains low over a significant length of time.

Figures 2 and 3 show 121°C(250°F) as the temperature for the devolatilization dwell. However, this temperature is appropriate for the particular material involved. The correct temperature for the devolatilization hold needs to be determined from rheology curves for that specific material. For several different representative materials it has been observed that appropriate devolatilization dwell temperatures can range between 88±5.6°C and 121±5.6°C.

A devolatilization dwell then consists of holding the lay-up at a constant temperature over a span of time during which a vacuum is applied to facilitate the removal of volatiles originating from various portions of the lay-up. At the end of the specified time period, the vacuum can be released and the heat up towards final cure can be continued. There is an alternative technology [6], for the removal of volatiles, called mold side breathing in which the vacuum may be applied, to a region between the outermost ply and the tool surface, throughout the cure cycle.

The benefits of eliminating these volatiles are several fold. Looking at the left-hand section of the bottom panel of Figure 1, it is seen that the removal of volatiles and entrapped air at minimum viscosity has multiple effects. It results in reduced porosity and an improved core to skin fillet bond [2].

Reduced porosity results from the absence of the gas bubbles in the resin at the time of cure. Likewise, the elimination of bubbles in the material forming the core to skin fillet bond results in a strengthening of this bond.

Another equally important volatile is adsorbed moisture. Adsorbed moisture can change viscosity/time profiles [7], surface morphology [1,2] and the tendency to exotherm [8]. Removal of this volatile promotes better flow and wetting of the resin, reduced porosity and an improved core to skin fillet bond.

BALANCE OF FORCES

A necessary condition for an object to remain stationary is that the vector sum of the forces and moments acting on that object are zero. A composite honeycomb core panel, is acted on by a combination of external and internal forces. If the external forces acting on the panel are greater than the opposing internal forces, there must be movement. This is most often seen as a compression of the core in its lateral dimension - i.e. core collapse.

The lateral component of the force due to the autoclave pressure acts to compress the core in its lateral dimension. An obvious strategy then is to diminish this vector force so that it will not overpower the resisting forces. The resisting forces are relatively small at the beginning of the time/temperature/pressure profile. In order to compensate for this, it is best to maintain the autoclave pressure at a low value. Just enough positive pressure is maintained to assure that the details of the panel assembly are firmly seated in place.

As the temperature increases, the gas pressure in the interior of the panel increases. The internal forces are better able to resist the external compressive force. The force corresponding to the internal pressure enables the panel to resist dimpling and core movement [2]. At this point, it then becomes possible to increase the autoclave pressure to the final processing pressure.

An ability to keep the core stationary has the benefit of eliminating core movement and reducing the distortion of edgebands and surface plies [2]. As a consequence, the panels internal condition is improved. This in turn leads to an improved mechanical strength [2]. See item C in the middle section of the bottom panel and the results of item C in the middle panel of Figure 1.

The autoclave pressure that is normal to the beveled face of the core can be decomposed into a component parallel to the cell walls and a component perpendicular to the cell.

The autoclave pressure that is normal to the beveled face of the core can be decomposed into a component parallel to the cell walls and a component perpendicular to the cell. The magnitude of this lateral force being dependent upon the autoclave pressure and the ramp angle of the core. Materials of construction, the core ramp angle, its lateral compressive strength, autoclave pressure and the build-up of pressure inside the core can vary for different panel systems. The inequality shown under the balance of forces section takes into account these diverse factors.

INCREASE OF PRESSURE AFTER DEVOLATILIZATION DWELL

At the end of the devolatilization dwell, the vacuum is released, the heat up to the final cure temperature is begun and the full processing pressure is applied. The devolatilization dwell should remove the majority of gas bubbles from the resin. Upon release of the vacuum, the diameter of any remaining bubbles will be reduced in size due to the increase in pressure. Increasing the autoclave pressure results in a greater force being exerted on the fluid resin. The increased pressure can either drive the remaining bubbles into solution or dramatically reduce their size prior to gelation. As consequence, it is expected that the size of the porosity will be greatly reduced. See item D of Figure 1.

CONCLUSION

Each of the major components, the devolatilization dwell, the balance of forces and the increase or pressure after the devolatilization dwell, has a beneficial effect by itself. However, it is noted that these effects can be multiple and overlapping. This creates a synergistic interaction between these essential elements. This synergy is best presented and summarized as shown in Figure 1. A complete overview of the entire subject is shown. Organizing and presenting the subject in this manner provides a unified comprehensive view. This overview illustrates the interplay among the various aspects and how they combine to affect the desired end product. This chart then presents the subject in such a way that one can see the individual elements and their ultimate impact on the end product.

REFERENCES

1. S. E. Susan, 25th SAMPE Symposium, 251, (1980).

2. T. H. Brayden and D. C. Darrow, 34th SAMPE Symposium, 861, (1989).

3. Y. Yamaguchi and S. Saki, 9th International Conference on Composite Materials, 616, (1993).

4. Vought Aircraft Material Specification 207-7-501J.

5. Vought Aircraft Material Specification 207-7-515D .

6. T. H. Brayden, D. C. Darrow and P. A. Poropatic, J. of Advanced Materials, **27**(1), 41(1995).

7. R. Hinrichs and J. Thuen 25th SAMPE Symposium, 126, (1980).

8. T. H. Brayden, D. C. Darrow and J. A. Ayres 31st SAMPE Symposium, 374, (1986).

BIOGRAPHY

Thomas H. Brayden is a Senior Quality Assurance Engineer in Northrop Grumman's Materials Test Technology unit. Current responsibilities include directing work in the analysis of resin systems and the use of a sensor array to monitor in-autoclave cure of composites. Academic background includes a Ph.D in physical chemistry from LSU.

Donald C. Darrow is presently working as an advanced technology development analyst in MR&D at the Boeing Commercial Airplane Group's Wichita Division. He is carrying out research and introducing new composite manufacturing technologies at Boeing. His academic background includes a Bachelor's Degree in Chemistry from the University of Hawaii.

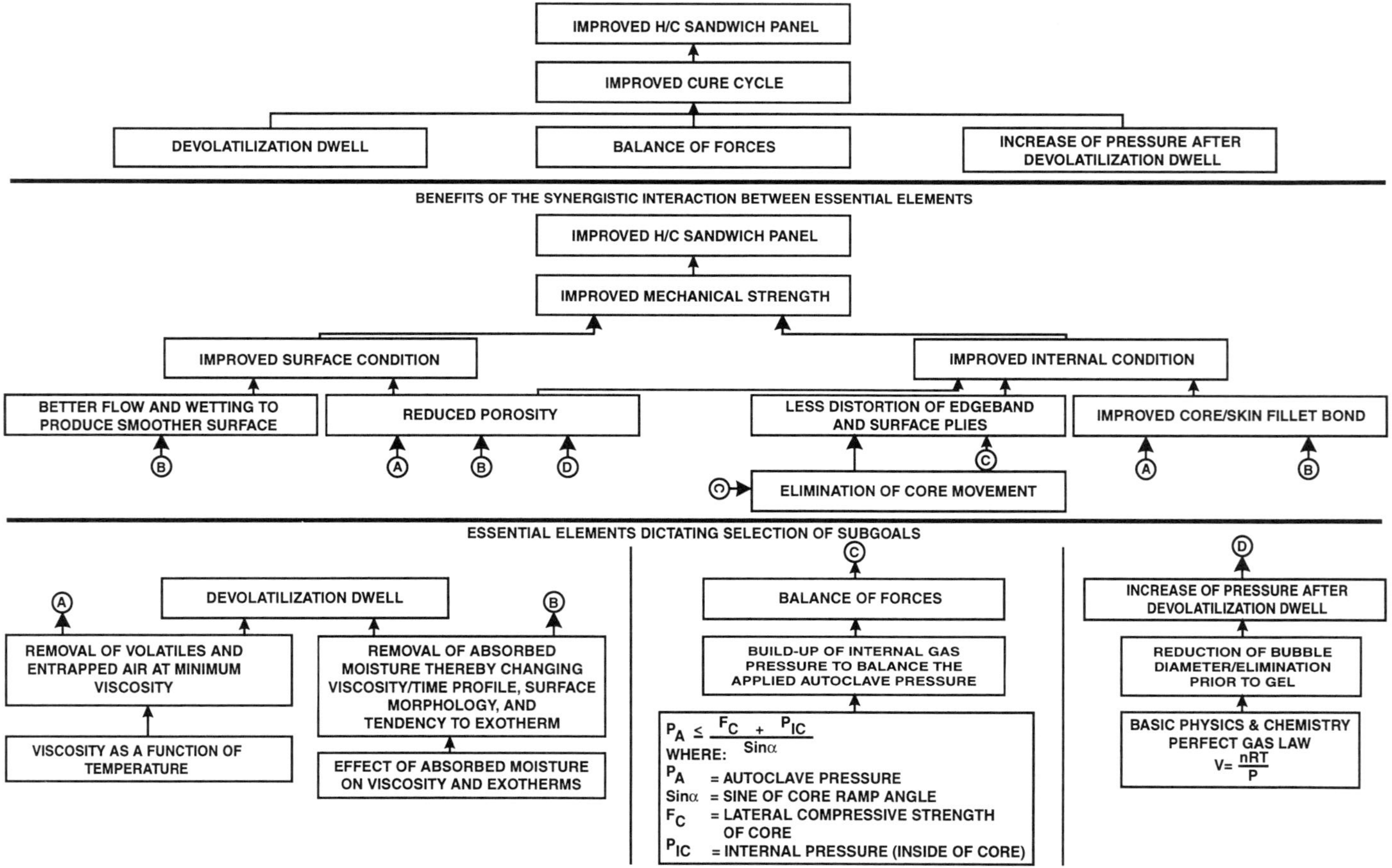

Figure 1. Overview of Manufacturing Process Showing Details and Synergistic Intractions

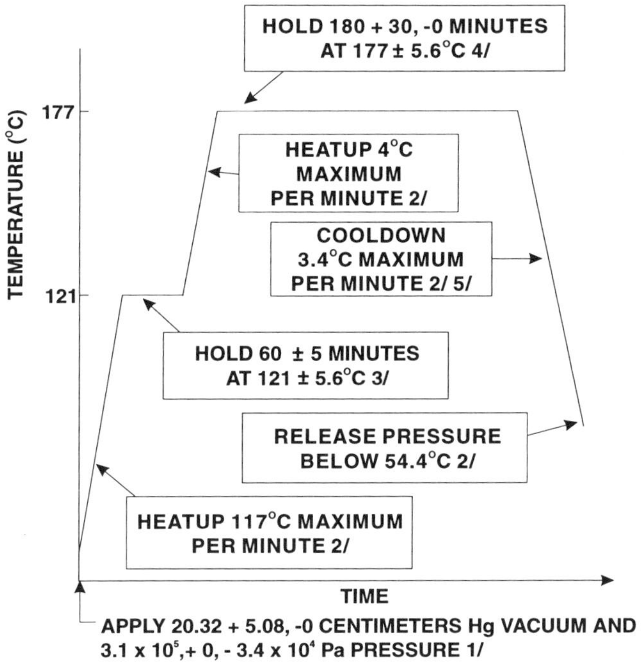

NOTES:

1/ VENT VACUUM BAG TO ATMOSPHERE AT A RATE OF 5.08 CENTIMETERS Hg EVERY 2 MINUTES WHEN PRESSURE REACHES 6.9×10^4 PA. CONTINUE INCREASING DURING VENTING.

2/ HEATUP AND COOLDOWN RATES FOR ALL PART THERMOCOUPLES SHALL BE IN THE SPECIFIED RANGE AND SHALL BE BASED ON THE LAGGING PART THERMOCOUPLE.

3/ THE 121°C HOLD PERIOD SHALL BEGIN WHEN THE LAGGING PART THERMOCOUPLE IS AT 115.6°C.

4/ THE 177°C HOLD PERIOD SHALL BEGIN WHEN THE LAGGING PART THERMOCOUPLE IS AT 171°C. NATURAL DEPRESSURIZATION IS PERMISSIBLE AT THE END OF THE HOLD.

5/ BACK PRESSURE REQUIREMENTS ARE NOT APPLICABLE DURING COOLDOWN SEGMENT.

Figure 2. Common Cure Cycle for 177°C Curing Graphite/Epoxy Composite Honeycomb Core Panel.

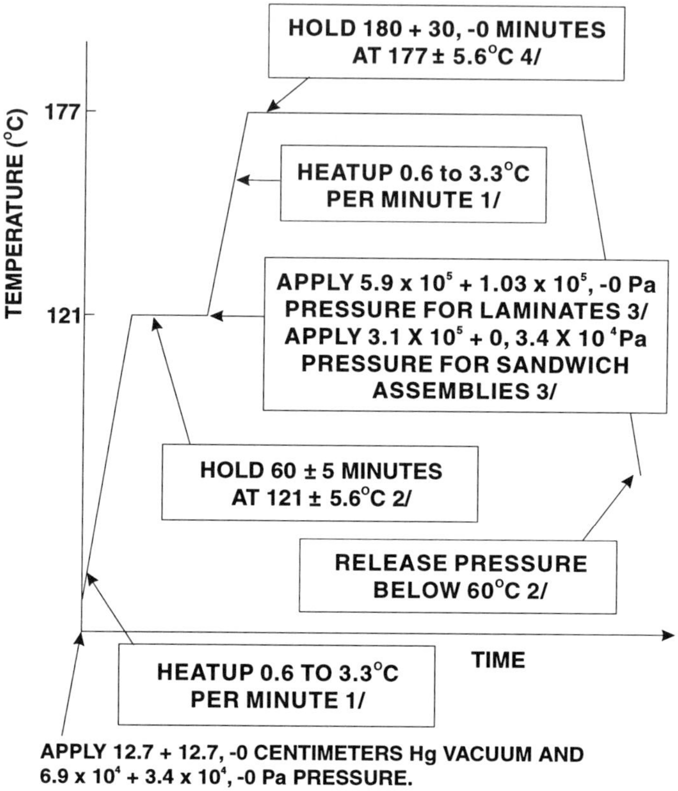

NOTES: 1/ HEATUP RATES FOR ANY PART THERMOCOUPLE SHALL BE IN THE SPECIFIED RANGE.

2/ BASED ON LEADING PART THERMOCOUPLE.

3/ VENT VACUUM BAG TO ATMOSPHERE AT A RATE OF 5.08 CENTIMETERS Hg EVERY 2 MINUTES WHEN PRESSURE REACHES 1.38×10^5 Pa.

4/ BASED ON LAGGING PART THERMOCOUPLE.

Figure 3. Modified Cure Cycle for 177°C Curing Graphite/Epoxy Composite Honeycomb Core Panel

28th International SAMPE Technical Conference
November 4-7, 1996

FABRICATION OF COMPOSITE TAPE FROM THERMOPLASTIC POWDER-IMPREGNATED TOWS*

Harry L. Belvin and Roberto J. Cano

Mail Stop 226
NASA Langley Research Center
Hampton, VA 23681

Ray W. Grenoble and Joseph M. Marchello

College of Engineering
Old Dominion University
Norfolk, VA 23509

Steven J. Claus

Steve Claus Code 4.3.4.3 MS 3 Bldg. 2188
Naval Air Warfare Center Aircraft Div.
22541 Millstone Road
Patuxent River, MD 20670-5304

ABSTRACT

Recent automated manufacture of thermoplastic matrix composites has focused on the use of multi-tow heated head placement machines employing narrow consolidated ribbon. Use of narrow ribbon allows small turning radii and less edge scrap when cutting and adding individual tows. However, in applications where only moderate curvatures exist, such as aircraft wings and fuselages, narrow ribbon placement is inefficient. Placement of wide tape is one way to achieve the high lay-down quantities per hour needed for commercial applications. Benefits also include fewer gaps and laps between sequentially placed material and correspondingly better quality.

As part of ongoing composite materials forms research at NASA Langley Research Center, a previously developed process for making narrow thermoplastic ribbon was modified and extended to make a 7.6 cm (3 in)-wide tape. Improvements were made in tow heating, design of the holder for the stationary bars, and design of the contoured nip-rollers. A processing window for producing high quality well-consolidated tape was established. Tape samples were hand laid and compression molded into laminates that exhibited good mechanical properties.

KEYWORDS: Tape Fabrication, Thermoplastic, Automated Placement

---

1.0 INTRODUCTION

As part of an effort to develop high performance composite technology for the aircraft industry, NASA Langley Research Center (LaRC) has investigated ways to make intermediate composite material forms and develop the processes required for the manufacture of aircraft structure. These composite research and development activities include powder impregnation of carbon fiber tow bundles, dry (solventless) polymer matrix composite (PMC) ribbon/tape fabrication from powder-coated towpreg, and automated tow placement (ATP) of ribbon/tape to form elements and built-up structure. The research at LaRC is coordinated with several large airframe manufacturers in a united effort to develop lightweight materials for next generation aircraft.

During the past decade several methods have been developed for the powder impregnation of carbon fiber tow bundles. The method in operation at NASA Langley is the powder curtain process (Fig. 1) [1]. These powder coated tows have been successfully processed into fully consolidated 6.1 mm (0.250 in)-wide, 127 mm (0.005 in)-thick ribbon for use in heated head automated tow placement [2,3,4]. This narrow ribbon allows small turning radii and less edge scrap during the automated placement operation. In applications where only moderate curvatures exist, such as aircraft wings and large fuselages, the use of narrow ribbon during the placement process is inefficient. Use of wider PMC tape provides potentially greater lay-down quantities per hour, requires greatly simplified placement head hardware and has the benefit of reducing gaps and laps between sequentially placed composite material.

The object of this study is to develop fabrication procedures for the manufacture of a wide dry PMC tape. A previously developed process at LaRC [2,3,4] for making narrow ribbon provided the basis for the development of a method for making a wider dry PMC tape. During the course of the work, a new tape line was built and a process window for the production of high quality, fully consolidated, 7.6 cm (3 in)-wide dry composite tape was established.

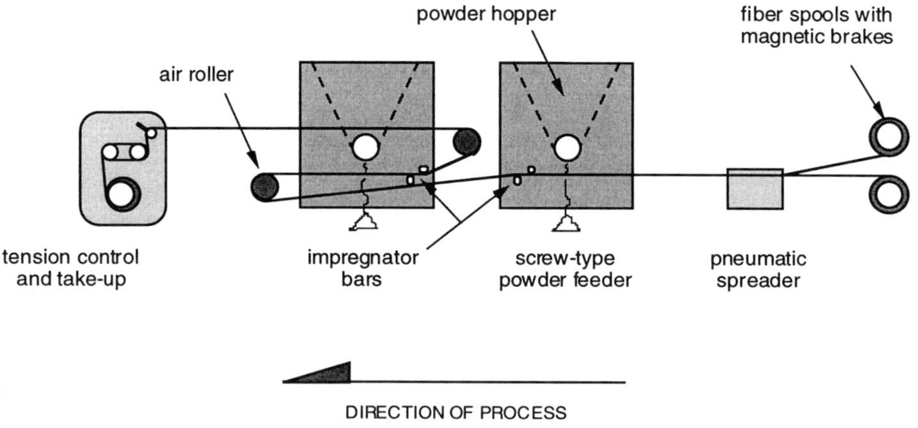

Figure 1 : The NASA LaRC powder curtain process for powder impregnating carbon fibers

2.0 EXPERIMENTAL

2.1 Ribbonizing Process The general procedure is described below and shown schematically in Figure 2. The individually tensioned powder impregnated tows are drawn through a tube furnace where the polymer is heated above its melt temperature (T_m). Near the exit end of the furnace, the tows are drawn over and under a set of stationary ceramic bars in an S configuration which facilitates the flow of polymer on both sides of the tow. At this point, air is squeezed from the tows as the molten polymer wets the fiber bundle. As it exits the furnace, the heated tow is relatively consolidated and impregnated but unconstrained, and a forming and cooling stage is required to shape it to a desired configuration. This is accomplished with a set of tongue and grove air-cooled nip rollers placed at the exit of the furnace. After exiting the nip rollers, the cooled ribbon is below its glass transition temperature (T_g) and is wound onto spools.

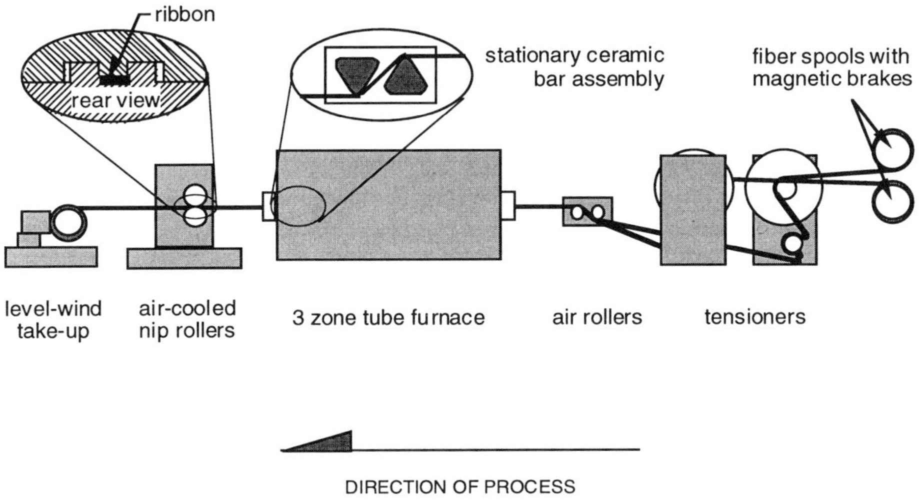

Figure 2 : The ribbonizing process for converting powder-coated towpreg into fully consolidated 6.1 mm (0.250 in)-wide ribbon

2.2 Material Properties** The process for making narrow ribbon and wide tape is largely dependent upon the physical properties of the polymer powder, the type of fiber, and tow size. Values of T_g, T_m and melt viscosity over a range of temperatures are essential data to help develop process parameters. In this study, Hercules unsized 12k IM7 carbon fibers were used. The polymer powder utilized was Mitsui Toatsu PIXA-M. The fiber was powder-coated at Cytec Engineered Materials Inc., Anaheim, California via a proprietary method.

Differential scanning calorimetry (DSC) data for PIXA-M is shown in Figures 3 and 4. Figure 3 shows the initial scan; Figure 4 is the second scan of the PIXA-M powder. The former shows a double endotherm or coalescence of two peaks, the shoulder being the T_g and the major peak possibly a crystalline melt. The latter peak disappears after the material is heated above 300°C during the first DSC scan, leaving only the T_g

**The use of trademarks or names of manufacturers in this report for accurate reporting does not constitute an endorsement, either expressed or implied, of such products or manufacturers by NASA.

endotherm at 235°C (Fig. 4). A sample of the tape was processed through the same DSC cycle and verified that it contained only amorphous material. The melt flow characteristics of the PIXA-M polymer are shown in Figure 5. Minimum viscosity occurs in the 350° - 400°C range which was used to establish the operating parameters for dry tape fabrication.

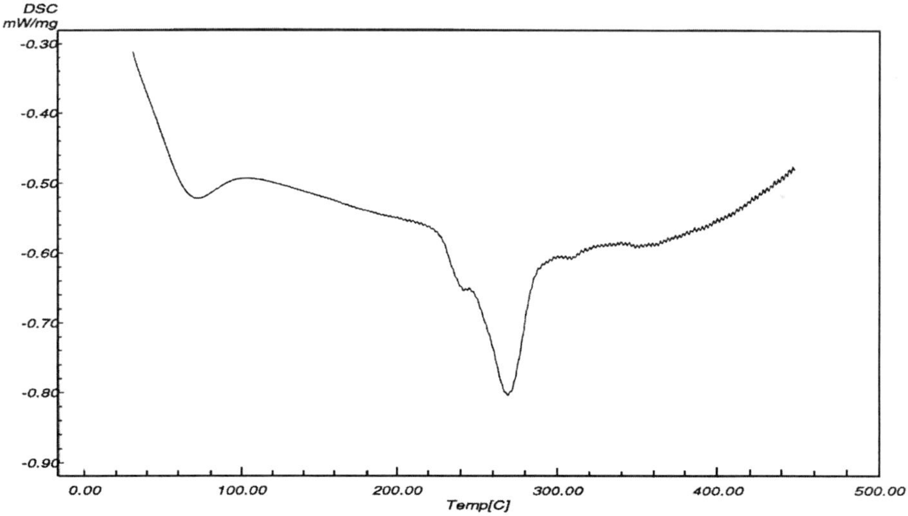

Figure 3 : DSC of PIXA-M powder, initial scan

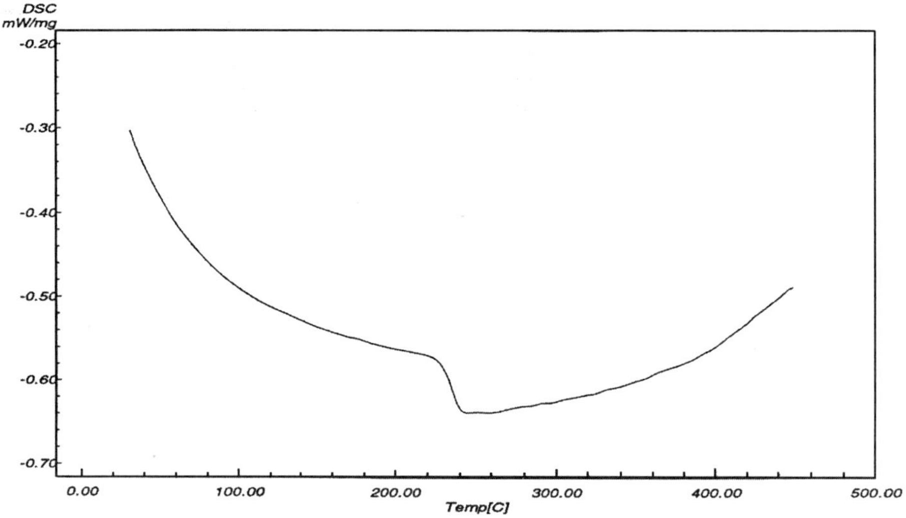

Figure 4 : DSC of PIXA-M, second scan

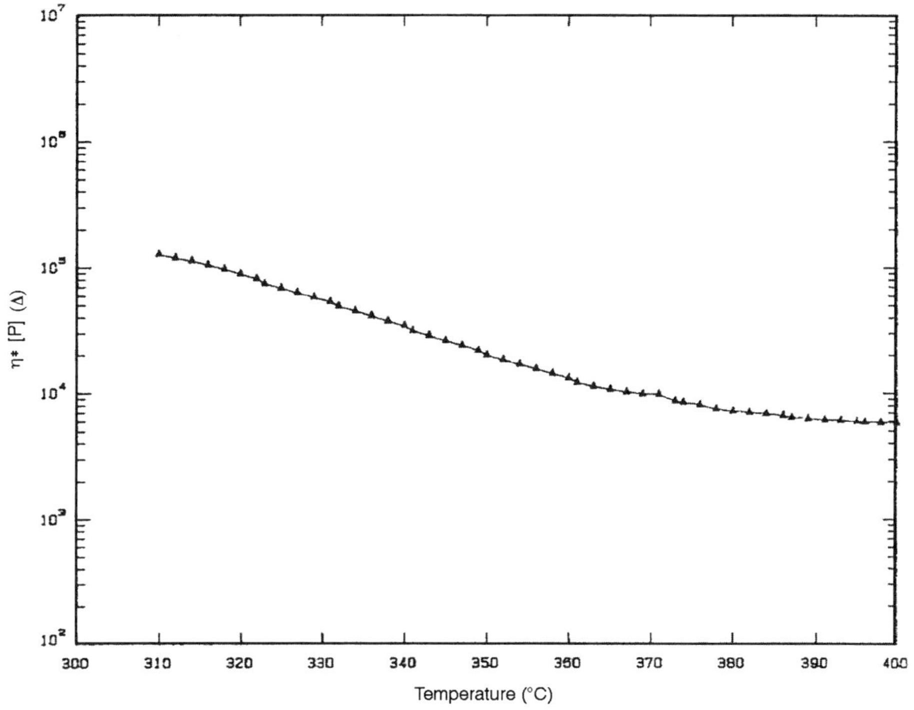

Figure 5 : Melt viscosity vs. Temperature of PIXA-M

2.3 Equipment Modifications To meet the objective of fabricating 7.6 cm (3 in)-wide tape, a new processing line was assembled (Fig. 6). The line was modeled after the existing ribbonizing line shown in Figure 2, but with major modifications to the nip rollers, bar assembly, and the drive system. An additional ceramic bar was added for wrap angle adjustment. Wrap angle determines the amount of time the powder-coated towpreg makes contact with the ceramic bars [2,3,4]. The bottom of the two outside bars were fixed at a predetermined height (centerline of the tube furnace). The center bar height was set to give the desired wrap angle. The air-cooled nip rollers were widened to 7.6 cm (3 in) to accommodate the larger number of powder-coated tows. Rubber-coated pull rollers were designed and fabricated at LaRC to provide uniform tow tension across the tape width. A 15.2 cm (6 in) diameter furnace was used to accommodate the wider tape. Meticulous equipment alignment was necessary to achieve proper tow spreading, tape shape and split-free tape product during processing and wind-up.

2.4 Tape Fabrication To conserve material, initial studies were conducted on the ribbonizer (Fig 2). To optimize process parameters, the ribbonizer was operated over a range of bar temperatures and line speeds. Bar temperatures ranged from 350° to 420°C (662° to 788°F) using 10°C (18°F) increments. Line speeds ranged from 4.57 m/min (15 ft/min) to 9.14 m/min (30 ft/min) using 1.52 m/min (5 ft/min) increments. The parameters found to fabricate good quality ribbon were then used as the starting point for processing material through the newly constructed 7.62 cm (3 in) line.

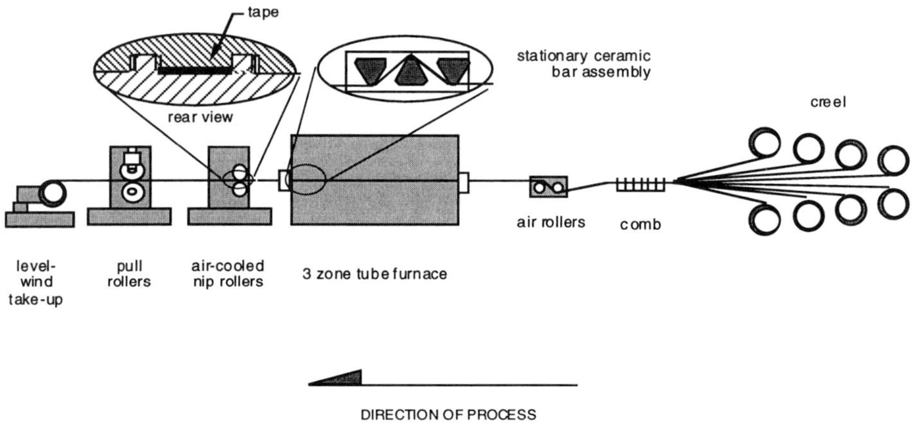

Figure 6 : The tape process for converting powder-coated towpreg into fully consolidated 7.6 cm (3 in)-wide tape

Initial tape runs consisted of 24 ends of IM7 12k powder-coated tow. Line speed was maintained at 9.14 m/min (30 ft/min). Two additional tows were added to fill gaps and develop the full width of the tape. Line speed was reduced to 7.62 m/min (25 ft/min) to provide additional heating of the tows and ensure good polymer flow, tow consolidation and shaping. The stationary bar temperature was maintained at 380°C (716°F).

3.0 RESULTS and DISCUSSION

3.1 Tape Characteristics. Approximately 5.45 kg (12 lbs) of 7.62 cm (3.00 in)-wide tape were produced during this study. In general the tape seemed well consolidated and broke cleanly. In the initial runs with twenty-four (24) tows, tape width varied from 7.37 cm (2.90 in) to 7.62 cm (3.00 in). When two (2) additional tows were added, the nip rollers were filled and a tape was produced that ranged in width from 7.52 cm (2.96 in) to 7.62 cm (3.00 in). Thickness across the width ranged from 0.114 mm (0.0045 in.) to 0.228 mm (0.009 in.). Photomicrographs showed the excessive thickness was confined to one edge and resulted from the over-lapping of two tows. This suggests that a tape of twenty-five (25) tows would have completely filled the rollers and maintained a more uniformed thickness across the width.

3.2 Tape Morphology The SEM of the fracture surface (Fig. 7) show the fibers to be well wet-out. The photomicrograph of the tape cross section (Fig. 8) shows a void fraction of less than one percent. These results demonstrate that the tape was of good quality.

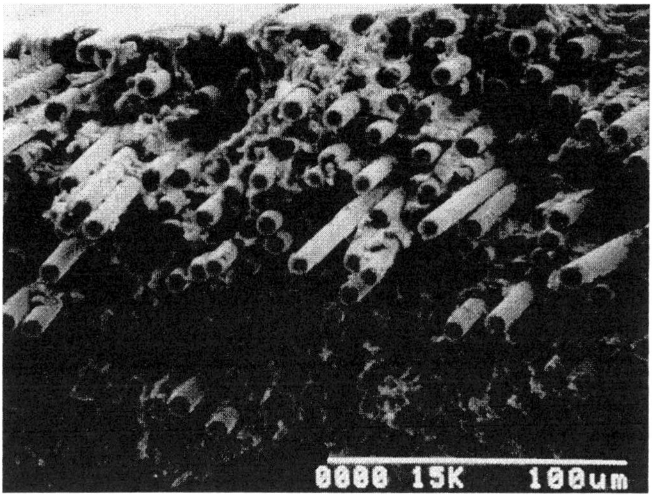

Figure 7 : SEM of fracture surface of PIXA-M 7.6 cm (3 in)-wide tape

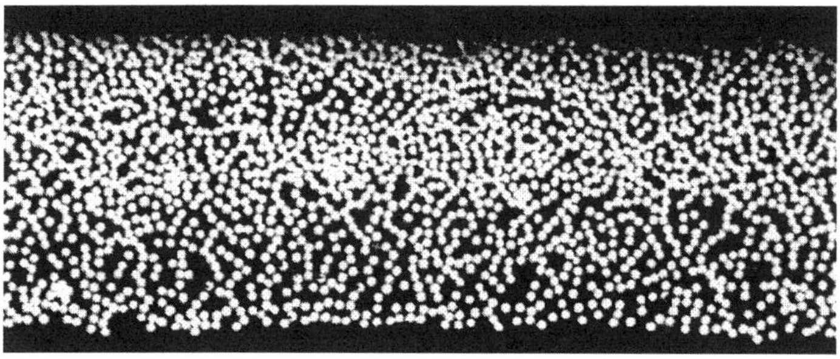

Figure 8 : Photomicrograph of cross-section of PIXA-M 7.6 cm (3 in)-wide tape

3.3 Mechanical Properties. Pieces of tape were hand laid into steel molds and compression molded into unidirectional coupons. Specimens were cut for short beam shear strength and 0° and 90° flexural measurements. The short beam shear panel was fabricated with 20 plys and the 0° and 90° flexural panels were fabricated with 12 plys. The panels were fabricated using a cure cycle which ramped from room temperature to 382°C (720°F) held for an hour then cooled to room temperature. 200 psi was applied at 288°C (550°F) during heat up. Pressure was maintained throughout the hold and cool down. The mechanical properties for PIXA-M/IM7 7.6 cm (3 in)-wide tape are shown in Table 1. Material properties are generally considered good if the flexure moduli are between

110-131 GPa and short beam shear strengths are above 97 MPa (at room temperature)[6]. Although mechanical testing for this material is limited, determined test values are considered acceptable by the above criteria.

Table 1 : Nominal mechanical properties for PIXA-M 7.6 cm (3 in)-wide tape

Test Temp.	Mechanical Test	Avg. Stress MPa (ksi)	Standard Deviation MPa (ksi)	Avg. Modulus GPa (ksi)	Standard Deviation GPa (ksi)
RT	Short Beam Shear	97.08 (14.08)	2.14 (0.31)	——	——
RT	0° Flex	1368.08 (198.42)	55.05 (7.98)	122.96 (17834.32)	4.33 (628.11)
RT	90° Flex	115.32 (16.73)	28.37 (4.12)	4.19 (607.12)	.229 (33.18)
177°C	Short Beam Shear	46.26 (6.71)	.28 (0.04)	——	——
177°C	0° Flex	1005.70 (145.86)	59.36 (8.61)	123.36 (17893.21)	6.44 (933.38)
177°C	90° Flex	46.85 (6.79)	21.90 (3.18)	1.73 (250.96)	1.35 (196.00)

4.0 CONCLUSIONS

A method was successfully developed for fabricating thermoplastic carbon fiber reinforced tape from powder-coated tow bundles by significantly modifying a previously developed ribbonizing process. Processing parameters were identified that will be essential in any scale-up process for the fabrication of 7.6 cm (3 in)-wide composite tape. 5.45 kg (12 lbs) of good quality PIXA-M of 7.6 cm (3 in)-wide composite tape was fabricated at LaRC. This bench-scale pilot study demonstrates the feasibly of commercial fabrication of high quality thermoplastic tape for use in heated head automated tow placement equipment for the fabrication of composite structure.

7.0 REFERENCES

1. R. M. Baucom And J.M. Marchello, J. of Adv. Materials, 25 (4), 31 (1994)
2. D. A. Sandusky, "Apparatus for Consolidating A Pre-impregnated, Filament-Reinforced Polymeric Prepreg Material", U.S. Patent Number 5,395,477, March 7, 1995.
3. D. A. Sandusky, J. M. Marchello and N. J. Johnston, SAMPE International Symposium, 40, 180 (1995).
4. S. J. Claus, R. W. Grenoble D. A. Sandusky, J. M. Marchello, and N. J. Johnston, SAMPE Technical Conference, 27, 854 (1995)
5. R. W. Grenoble, H. L. Belvin and S. J. Claus, "Modular Nip Roller Design and Systematic Fabrication Procedures for Composite Ribbon Manufacture", NASA Patent application LAR 15430-1-CU, October 5, 1995
6. Leif A. Carlsson and R. B. Pipes, ed., Composite Materials Series, Thermoplastic Composite Materials, Vol. 7, Elsevier Science Publishing Company Inc., New York, 1991, pp. 47

28th International SAMPE Technical Conference
November 4-7, 1996

PRODUCTION OF HIGH INTERNAL PHASE EMULSION POLYMERIZED FOAMS FOR COMPOSITE PROCESSES

Mark A. Hoisington and Joseph R. Duke
Los Alamos National Laboratory
Polymers and Coatings Group
Los Alamos, New Mexico 87545

ABSTRACT

Heat resistant, polymeric foams have potential applications in numerous composite structures including core material for sandwich panels and reinforcing filler of honeycomb cores. This work focused on the development of high internal phase emulsion (HIPE) polymerized foams with thermal and mechanical performance designed for use in composite processes. In contrast to typical blown foam systems, HIPE foams were investigated in order to take advantage of their unique combination of processing and property benefits. Thermal properties of typical styrene/divinylbenzene HIPE foams were dramatically improved by copolymerizing styrene with N-cyclohexylmaleimide and bismaleimide monomers in a HIPE process. This process produced polymeric, structural foams with glass transition temperatures ranging from 130 - 220°C (265 - 425°F) and compression strengths of 550 - 700 kPa (80 - 100 psi) at foam densities under 80 mg/cm^3 (5.0 lb/ft^3).

KEY WORDS: Foams, Sandwich Core

1. INTRODUCTION

Polymeric foams have potential applications in numerous composite structures including uses as core material for sandwich panel construction and as a reinforcing filler in the cells of honeycomb core. These foam systems must possess thermal and mechanical properties capable of withstanding composite curing conditions that typically consist of 177°C (350°F) cure temperatures and consolidation pressures up to 700 kPa (100 psi). Commercially available foam systems generally considered for these composite processes include: Rohacell (polymethacrylimide), Divinylcell (polyvinylchloride), Airsan (polyetherimide), and others.

In contrast to the commercially available blown foam systems, high internal phase emulsion (HIPE) polymerized foams can provide a unique combination of processing and property advantages. HIPE foams can be processed over a wide range of densities (30-350 mg/cm^3), the emulsion produced during processing allows for near-net-shape casting capabilities, and the final foam possesses isotropic properties and morphology. In addition, the emulsion produced during processing allows for easy and homogeneous incorporation of secondary phases such as powders, particles, or fibers for modification of mechanical, electrical, or other specialized properties.

HIPE foam processing consists of three steps (1-8). First, a monomer based oil phase is mixed with a deionized water phase to produce a water-in-oil emulsion. Internal water phase concentrations of at least 76% of the emulsion by volume are used during processing. Generally, water phase concentrations exceed 90% of the emulsion by volume in order to produce low density foam materials. After the emulsion has been produced, the oil phase monomers are free radically polymerized around the water phase droplets to create an open cell, polymeric microstructure. The final foam is obtained by removal of the water phase. Although the final foam contains on open cell morphology, the 10 μm diameter cells and 1-3 μm diameter interconnecting pores are much smaller than typical open cell foam systems.

Recently, heat resistant HIPE foam materials have been produced by copolymerizing styrene with maleimide based monomers (9-11). Although styrene/maleimide copolymers have been widely used to produce heat resistant thermoplastic materials, styrene/maleimide copolymerization presents difficulties in a HIPE processes (12-14). Maleimide based monomers present solubility difficulties in both the oil and water phases. HIPE foams are polymerized at low temperatures, and obtaining a stable HIPE is more difficult with the use of maleimide monomers. This work focused on the development of heat resistant HIPE foam systems capable of meeting composite processing requirements by copolymerizing styrene with N-cyclohexylmaleimide and a bismaleimide based monomer.

2. EXPERIMENTAL

2.1 Materials N-cyclohexylmaleimide was obtained from the Ken Seika Corporation. Bis(3-ethyl-5-methyl-4-maleimide-phenyl)methane was obtained from the Ihara Chemical Industry Company, with the help of Ken Seika Corporation, under the product name BMI-70. Stabilized styrene was used as received from VWR Scientific, and a 55% mixture of DVB isomers was used as received from Polysciences Incorporated. Potassium persulfate was obtained from Aldrich Chemical Corporation and used as the free radical initiator. Sorbitan Monooleate (SMO) was used as the emulsifying agent and was obtained from the Henkel Corporation as Emsorb 2500. Monomer chemical structures are shown in Figure 1.

2.2 Processing The foams were processed by first dissolving the cyclohexylmaleimide and BMI-70 into the styrene monomer. SMO was then added to the monomer mixture. In all the foams processed, a total oil weight of 7.5 g was prepared and placed in a 100 ml Qorpak jar. The oil phase was then mechanically stirred at approximately 600-800 rpm while 92.5 g of distilled water containing 1.5 g/l potassium persulfate was added dropwise. After all the water had been added, the emulsion was further stirred for 5 minutes at a speed of approximately 900 rpm and for 5 minutes at a speed of 1000 rpm. Emulsions were capped and put in a convection oven at 60°C to cure overnight. After polymerization, the foams were removed from the Qorpak jars and returned to the convection oven to dry at 60°C. Drying times varied from several hours to 48 hours depending on the morphology of the cells. The dried foams were then postcured at 250°C under vacuum overnight.

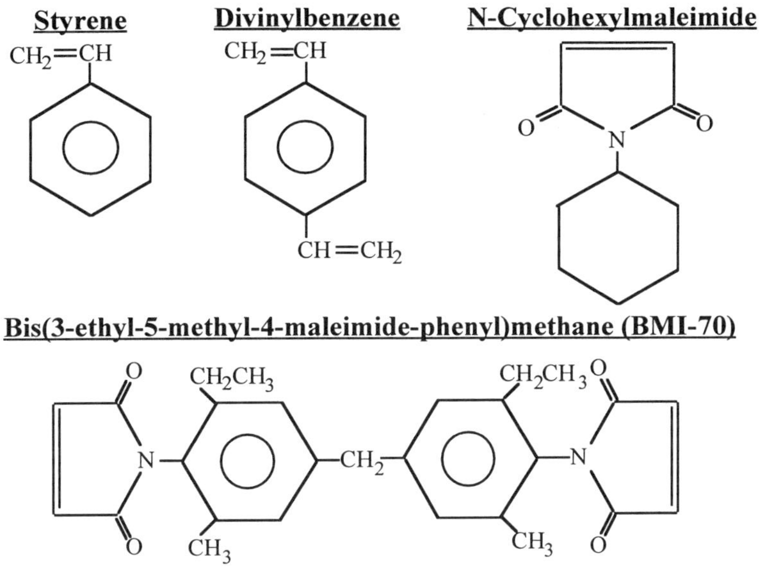

Figure 1. Oil Phase Monomer Chemistry

2.3 Analysis Dynamic mechanical analysis (DMA) was performed on a Rheometrics RDA II mechanical spectrometer. The foam samples glass transition temperature (Tg) was defined as the peak in the tan delta curve. A clamping fixture capable of gripping onto the foam samples was machined to fit into the normal Rheometrics adapters. Testing was done at a fixed frequency of 1 Hz, and the temperature was stepped at 5°C increments from 30°C to a maximum temperature of 350°C in a nitrogen environment.

Foam compression strengths were evaluated at 25°C on an Instron Model 4483 testing frame. Only one sample of each composition was tested to quickly screen mechanical performance. Samples were cylindrical in shape, approximately 25 mm (1.0 in) high and 45 mm (1.77 in) in diameter, and tested at rate of 0.25 mm/min (0.1 in/min).

Foam microstructures were investigated by Scanning Electron Microscopy (SEM).

3. RESULTS AND DISCUSSION

Two series of foam samples were processed and compared to a standard HIPE foam containing 90% styrene/10% DVB. The first series examined styrene/cyclohexylmaleimide copolymer crosslinked with 10% DVB, and the second series examined styrene/cyclohexylmaleimide copolymer crosslinked with 20% BMI-70. All compositions produced a foam after processing. Sample identifications and compositions are presented in Table 1.

Table 1. Sample Identifications and Compositions of Cyclohexylmaleimide Modified Foams

	% by Weight of Each Component			
Sample	**Styrene**	**DVB**	**Cyclohexylmaleimide**	**BMI-70**
CYMI-1	**90**	**10**	**0**	**0**
CYMI-2	**80**	**10**	**10**	**0**
CYMI-3	**70**	**10**	**20**	**0**
CYMI-4	**60**	**10**	**30**	**0**
CYMI-5	**50**	**10**	**40**	**0**
CYMI-6	**40**	**10**	**50**	**0**
CYMI-7	**80**	**0**	**0**	**20**
CYMI-8	**70**	**0**	**10**	**20**
CYMI-9	**60**	**0**	**20**	**20**
CYMI-10	**55**	**0**	**25**	**20**
CYMI-11	**50**	**0**	**30**	**20**

3.1 Microstructure Analysis SEM photomicrographs of all DVB crosslinked foams are shown in Figure 2. Sample CYMI-1 represents the morphology of a baseline styrene/DVB foam system. A very open, interconnected cell morphology typical of HIPE foam systems is easily observed. The remaining SEM's demonstrate the effect of adding increasing concentration of cyclohexylmaleimide to the HIPE process. All cyclohexylmaleimide modified samples (Samples CYMI-2 to CYMI-6) have similar microstructures; however, they differ slightly from the base styrene/DVB system. Samples CYMI-2 to CYMI-6 appear to have slightly smaller cell sizes and larger interconnecting pores that result in a strut configured cell with very little wall. The baseline styrene/DVB system cells have more of a cell wall on each strut creating smaller interconnecting pores.

Figure 3 presents the SEM photomicrographs of the BMI-70 crosslinked foam systems. As in the DVB crosslinked foams, only small changes in morphology are observed. In comparison to the baseline styrene/DVB system (Sample CYMI-1), the cell sizes are slightly smaller, but the interconnecting pores appear to be smaller and more numerous than in all the DVB crosslinked foams. The smaller interconnecting pores lead to a cell morphology with more of a wall and less of a strut configuration.

Overall, the addition of cyclohexylmaleimide and BMI-70 to the HIPE polymerization process had only minor disruptions to the cellular microstructure. The foam morphologies were consistent with typical HIPE microstructures and appeared to be independent of cyclohexylmaleimide concentration.

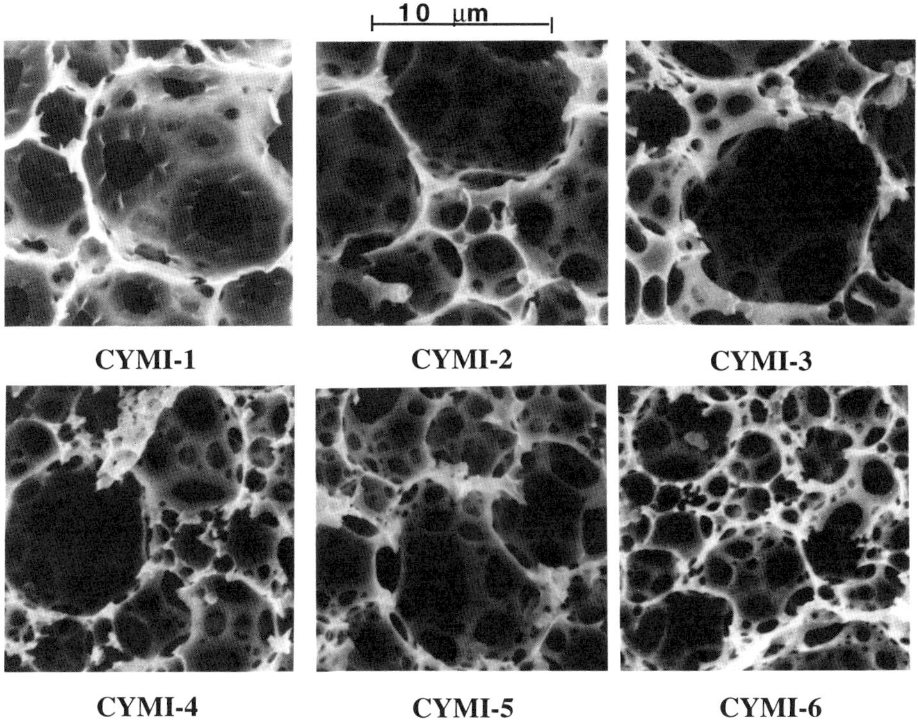

Figure 2. SEM Photomicrographs of DVB Crosslinked Foams.

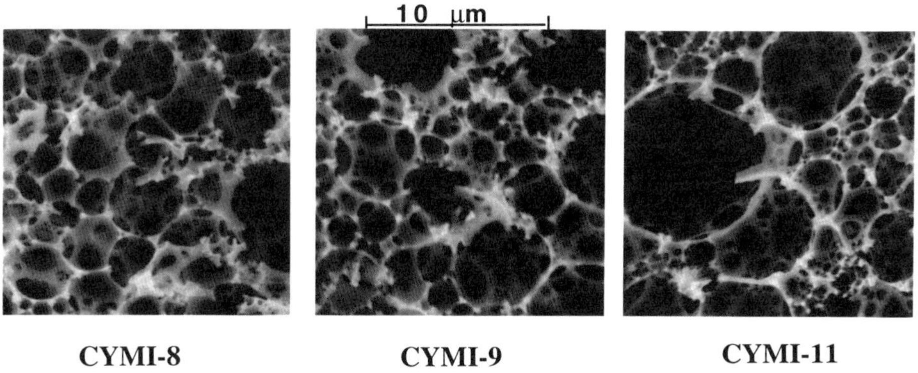

Figure 3. SEM Photomicrographs of BMI-70 Crosslinked Foams.

3.2 Dynamic Mechanical Analysis The influence of cyclohexylmaleimide and BMI-70 on the foams thermal mechanical properties was evaluated by Dynamic Mechanical Analysis (DMA). Figure 4 presents the DMA results for the DVB crosslinked foam systems in terms of shear storage modulus (G') and tan delta as a function of temperature. The increase in Tg of the foam systems is easily observed with the addition of increasing concentrations of cyclohexylmaleimide. The baseline styrene/DVB system (Sample CYMI-1) produces a peak in tan delta at 129°C. With every 10% increase in cyclohexylmaleimide concentration, a corresponding increase of 10-20°C in the tan delta peak is obtained for a maximum of 199°C with the addition of 50% cyclohexylmaleimide.

The DMA data also suggests that the cyclohexylmaleimide is reacting uniformly with the styrene throughout the oil phase to create a random copolymer. Only one transition region is observed in the G' and tan delta curves which indicates that no phase separation is occurring during polymerization.

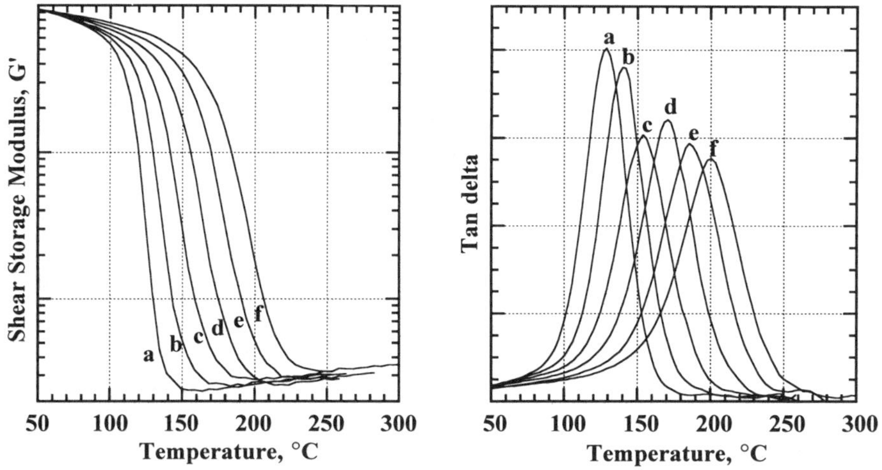

Figure 4. Dynamic Mechanical Analysis of DVB Crosslinked Foams: a) Sample CYMI-1, 129°C b) Sample CYMI-2, 144°C c) Sample CYMI-3, 154°C d) Sample CYMI-4, 160°C e) Sample CYMI-5, 184°C f) Sample CYMI-6, 199°C

Different DMA results are observed when BMI-70 is used during processing. Figure 5 presents the G' and tan delta results for all BMI-70 crosslinked systems. When no cyclohexylmaleimide is used during processing (Sample CYMI-7), an extremely broad transition region is obtained between 100°C and 350°C. This broad transition region is due to the lower Tg of the large styrene blocks between BMI-70 crosslinks. However, the broad Tg transition and the ability of the foam to survive a postcure at 250°C under vacuum demonstrates that the BMI-70 was able to free radically polymerize with styrene to create a crosslinked polymer system. As increasing concentrations of cyclohexylmaleimide are incorporated into the HIPE process, a narrowing of the glass transition region is observed (Samples CYMI-8 to CYMI-11). All cyclohexylmaleimide modified samples exhibit approximately the same peak in tan delta (218°C), but the width of the tan delta peak becomes narrower with increasing cyclohexylmaleimide concentration. This narrowing is due to the increased Tg of the styrene/cyclohexylmaleimide copolymer between BMI-70 crosslinks.

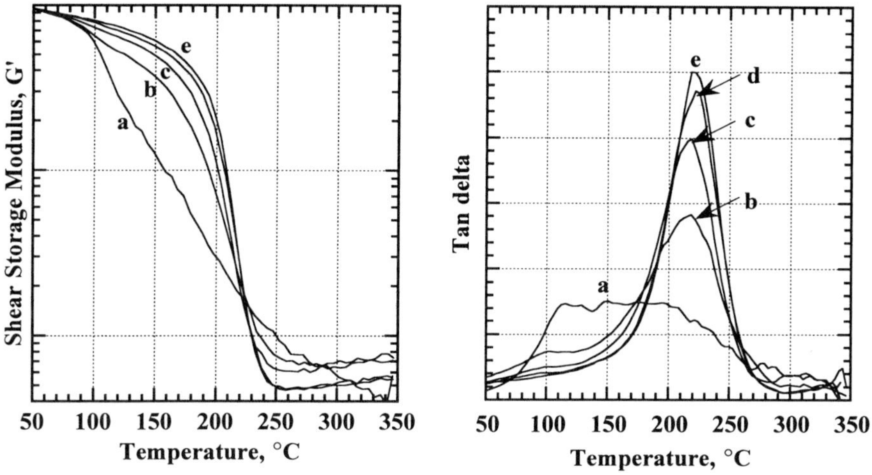

Figure 5. Dynamic Mechanical Analysis of BMI-70 Crosslinked Foams: a) Sample CYMI-7, b) Sample CYMI-8, c) Sample CYMI-9, d) Sample CYMI-10, e) Sample CYMI-11. All tan delta peaks, except CYMI-7, at approximately 218°C.

These DMA results demonstrate that cyclohexylmaleimide and BMI-70 are able to free radically copolymerize with styrene in a HIPE polymerization process. This copolymerization of styrene with maleimide based monomers produces dramatic increases in the Tg of the final polymer system that can be tailored to meet individual composite processing thermal requirements by controlling maleimide modifier concentration.

3.3 Mechanical Property Analysis Room temperature compression strengths of all foam samples were measured to evaluate mechanical performance. Table 2 presents the compression strength results along with foam densities and specific compression strengths, which were calculated by dividing the compression strength by the density. Sample CYMI-1, the baseline styrene/DVB foam, produced the best performance with a compression strength of 779 kPa (113 psi) and a specific compression strength of 10.6 kPa cm^3/mg (24.7 psi ft^3/lb). As cyclohexylmaleimide is added to the DVB crosslinked foams, a slight decrease in the measured compression strength is observed with Sample CYMI-6 (50% cyclohexylmaleimide) producing a compression strength of 589 kPa (85.4 psi). A larger decrease is observed in the specific compression strengths due to increasing foam densities with increasing cyclohexylmaleimide concentrations. The increasing densities were produced by increased volume shrinkages occurring during postcure.

The BMI-70 crosslinked foam samples exhibited more consistent compression strength and density values with cyclohexylmaleimide concentration. Samples CYMI 8-11 produced compression strengths ranging from 585-635 kPa (85-92 psi). However, the BMI-70 crosslinked foam without cyclohexylmaleimide addition (Sample CYMI-7) produced poor compression strength performance of only 311 kPa (45.1 psi).

Table 2. Compression Strengths of Cyclohexylmaleimide Modified Foams

Sample	Density, mg/cm^3 (lb/ft^3)	Compression Strength, kPa (psi)	Specific Compression Strength, kPa cm^3/mg (psi ft^3/lb)
CYMI-1	73.4 (4.58)	779 (113)	10.6 (24.7)
CYMI-2	74.8 (4.67)	675 (97.9)	9.02 (21.0)
CYMI-3	78.4 (4.89)	672 (97.5)	8.57 (19.9)
CYMI-4	79.0 (4.92)	654 (94.9)	8.28 (19.3)
CYMI-5	82.1 (5.12)	620 (89.9)	7.55 (17.6)
CYMI-6	96.0 (5.98)	589 (85.4)	6.14 (14.3)
CYMI-7	75.6 (4.72)	311 (45.1)	4.11 (9.56)
CYMI-8	68.1 (4.25)	462 (67.0)	6.78 (15.8)
CYMI-9	72.8 (4.54)	635 (92.2)	8.72 (20.3)
CYMI-10	72.5 (4.52)	624 (90.5)	8.61 (20.0)
CYMI-11	73.5 (4.58)	585 (84.8)	7.96 (18.5)

In this study, the cyclohexylmaleimide modified foam samples examined were only processed at one density. In order to evaluate potential mechanical performance capabilities, Figure 6 presents the cyclohexylmaleimide modified foam compression strengths in comparison to styrene/10% DVB foam systems before and after postcure over a wide density range. Both postcured and no postcure systems produced compression strengths proportional to the foam densities raised to a power. No postcure systems were proportional to the density raised to the 1.5 power, and postcured systems were proportional to the density raised to the 1.6 power. By postcuring the styrene/DVB foams, an increase in compression strength is observed at all densities due to the removal of the emulsifying agent (SMO) during postcure. The SMO, which acts as a plasticizer, volatilizes and is collected in a cold trap during postcure at 250°C under vacuum. In comparison to the styrene/DVB systems, the cyclohexylmaleimide modified foams appear to produce slightly lower compression strengths than the postcured styrene/DVB foams but significantly higher compression strengths than the styrene/DVB foams without postcure. As an approximation, the cyclohexylmaleimide modified foams should produce compression strengths similar to the postcured styrene/DVB foams as a function of density.

Therefore, with proper processing and microstructure development, maleimide monomer modified foams can provide adequate mechanical performance for composite processes. Foam densities can be varied to optimize the trade-off in mechanical performance and weight savings.

4. CONCLUSIONS

HIPE foams have been produced with thermal and mechanical properties capable of meeting many composite processing requirements. Cyclohexylmaleimide and BMI-70 monomers were able to free radically polymerize with styrene in a HIPE polymerization process to produce structural, polymeric foams with Tg's of 130 - 220°C (265 - 425°F). Only minor disruptions in typical HIPE foam microstructures were observed, and compression strengths of 550 - 700 kPa (80 - 100 psi) at densities under 80 mg/cm^3 (5.0 lb/ft^3) were obtained. Combination of the improved thermal performance with HIPE processing advantages may provide significant benefits in comparison to blown foam systems.

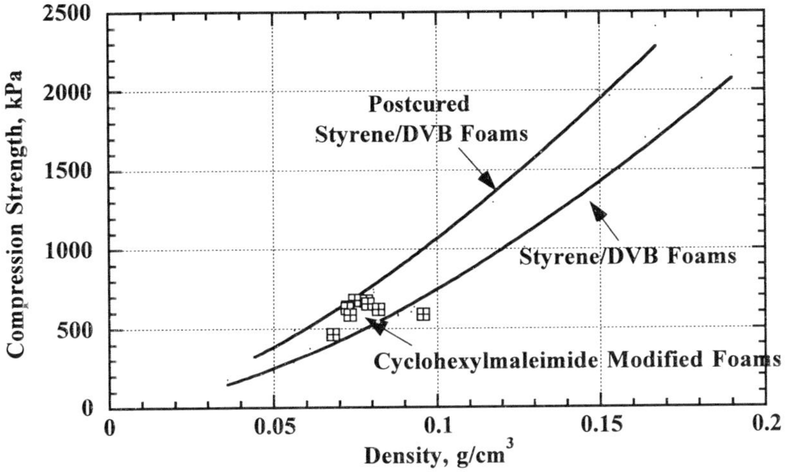

Figure 6. Compression Strengths of HIPE Foams as a Function of Density

5. ACKNOWLEDGMENTS

The authors are grateful to Vivian Gurule for her efforts in taking the SEM pictures for this project. This work was funded under contract W-7405-ENG-36 for the Department of Energy as part of the Laboratory Directed Research and Development program at Los Alamos National Laboratory.

6. REFERENCES

1. D. Barby, A. Haq, European Patent 0,060,138 (1982)
2. J. M. Williams, Langmuir, 4, 44 (1988)
3. J. M. Williams, D. A. Wrobleski, Langmuir, 4, 656 (1988)
4. J. M. Williams, D. A. Wrobleski, Journal of Materials Science Letters, 24, 4062 (1989)
5. A. M. Nyitray, J. M. Williams, Journal of Cellular Plastics, 25, 217 (1989)
6. J. M. Williams, A. J. Gray, M. H. Wilkerson, Langmuir, 6, 437 (1990)

7. J. M. Williams, M. H. Wilkerson, Polymer, 31, 2162 (1990)

8. J. M. Williams, Langmuir, 7, 1370 (1991)

9. M. A. Hoisington, J. R. Duke, P. G. Apen, Polymer (submitted December 1995)

10. M. A. Hoisington, J. R. Duke, P. G. Apen, Proceedings American Chemical Society PMSE Division, 74 (1996)

11. M. A. Hoisington, J. R. Duke, P. G. Apen, Proceedings Materials Research Society (Spring 1996)

12. L. E. Coleman Jr., J. A. Conrady, Journal Polymer Science, 38, 241 (1959)

13. G. V. Paesschen, D. Timmerman, Makromolekulare Chemie, 78, 112 (1964)

14. R. C. P. Cubbon, Polymer, 6, 419 (1965)

28th International SAMPE Technical Conference
November 4-7, 1996

PREPREG SYSTEM MODELING METHODOLOGY

Brian S. Hayes, and James C. Seferis*
Polymeric Composites Laboratory
University of Washington
Seattle,WA 98195

ABSTRACT

A methodology has been developed to model prepreg systems used in the commercial airplane industry. This methodology consists of a two step process appropriately termed simulation engineering and re-engineering. Simulation engineering is made up of two sections consisting of reverse engineering and model engineering. Reverse engineering involves complete analysis and characterization of the commercial prepreg system from the neat resin through fabricated parts. Accordingly, the objective is to fundamentally understand the prepreg system through processing, structure, and property information, not to determine the exact ingredients of the formulation. The actual development of the model prepreg system is performed through model engineering which consists of developing a model resin and manufacturing model prepreg as well as composite parts. Once the model prepreg system has been developed, the knowledge acquired from the entire process can be used to re-engineer a customer driven prepreg system.

KEY WORDS: Epoxy Resins, Prepreg Characterization, Composites

1. INTRODUCTION

The usefulness of using model prepreg systems to provide a fundamental understanding of commercial prepreg systems has become more apparent to prepreg consumers and composite manufacturers. Because commercial epoxy-based prepreg systems used in the airplane industry are mostly proprietary, issues or problems concerning prepreg handling and composite part quality are not fully understood. This is true even for commercial systems that have been in use for many years. Some of the phenomena that are of present interest are variations in prepreg tack, self-adhesive characteristics, internal void formation, and honeycomb core crush. To understand commercial prepreg systems from neat resin through end-use, model prepreg systems have been developed so that proprietary issues are not a concern.

Seferis and co-workers have previously developed model prepreg systems to investigate commercial prepreg material (1-7). The usefulness of model prepreg systems is that they allow the resin components, fiber type/weave, and prepreg processing parameters to be changed to better understand their effects on prepreg and final composite characteristics. In addition, it is also possible to use model systems to investigate the possibility of changing the method of prepreg processing. In any case, the prepreg user can use the non-proprietary information

* Author to whom correspondence should be addressed

generated from model systems to discuss relative issues with the prepreg manufacturer who might not have the time, money, or equipment for the investigation.

An integrated methodology has been established for developing model prepreg systems which can in turn be used to create a customer driven prepreg system. The modeling methodology was developed based on traditional chemists analytical tools combined with bulk prepreg analytical methods using an engineering approach. This methodology was based on the non-traditional viewpoint of designing for prepreg characteristics first, followed by cured mechanical properties. This is a logical progression of design when considering that prepreg is the precursor to composite parts and only through optimized prepreg characteristics will composite manufacturing and final mechanical properties be optimized, therefore adding value to the customer and prepreg manufacturer. Prepreg systems, however, still need to posses the necessary final cured mechanical properties but it should be realized that this only one part of many in the design of high performance prepreg systems.

The modeling methodology is divided into two sections termed simulation engineering and re-engineering. Simulation engineering also contains two sub-sections termed reverse engineering and model engineering. Reverse engineering involves the complete analysis and characterization of the commercial prepreg system from neat resin through final parts. Reverse engineering of commercial prepreg systems are performed to fundamentally understand the prepreg system from a processing, structure, and property viewpoint, not to determine the exact ingredients of the formulation. Therefore, from the processing, structure, property information acquired from the commercial prepreg system, the development of a model resin system followed by manufacturing model prepreg and model composite parts can be performed through model engineering. After the model prepreg system has been developed, it can be used to aid in re-engineering an existing commercial prepreg system to improve customer satisfaction. The process from the existing commercial prepreg to a re-engineered customer driven prepreg system can be observed in Figure 1.

2. BACKGROUND

The development of model epoxy-based resin formulations require an understanding of the epoxy resins, curing agents, and additives. An introduction to these different components are discussed and methods of stoichiometry determination are described.

2.1 Resin Component Epoxy resins are used for a wide variety of applications because they are easy to process and can be tailored to provide a wide range of properties. Molecular weight, functionality, and backbone structure all greatly influence the properties and processing capabilities of the epoxy resin. As the drive to use epoxies for increasing applications, different epoxy resins have been developed to provide specific properties and characteristics such as flame and weather degradation resistance. Since resin systems may require various properties which one epoxy can not provide, combinations of several epoxies are often used. Once the combination of epoxy resins are selected, the epoxide equivalent weight (EEW) needs to be determined so that the appropriate amount of curing agent can be used. The epoxy molecular weight divided by its functionality (number of epoxide groups per molecule) is the epoxide equivalent weight. The EEW of a combination of epoxy resins can be theoretically determined by using the manufacturers suggested EEW with the following equation:

$$\frac{(W+X+\ldots)}{EEW_{combined}}=\frac{W}{EEW_W}+\frac{X}{EEW_X}+\ldots \qquad [1]$$

where

$$W + X + \ldots = 100 \quad [2]$$

and

W, X, etc. are weight percents of the epoxies

The combinations of epoxies can be titrated for a more accurate EEW determination (8).

2.2 Curing Agent Component The transformation of an uncured epoxy resin to a crosslinked three-dimensional network is developed by using a curing agent. There are many curing agents used for epoxy formulations which promote cross-linking through different mechanisms. Curing agents work as either a catalyst and promote homopolymerization of the epoxy or co-react forming a heteropolymer (9). The type and quantity of curing agent depends on the desired processing, pot-life and final part properties. To achieve the desired characteristics, more than one curing agent may be required for a certain formulation. In any case, the co-reactant curing agent equivalent weights (CEW) can be determined theoretically, similar to the epoxy equivalent weights. To determine the theoretical ratio of equivalents of the curing agent to the equivalents of epoxide the following equation can be used:

$$\left(\frac{\frac{\text{wt\% Curing Agent}}{\text{CEW}}}{\frac{\text{wt\% Epoxy}}{\text{EEW}}}\right) = \frac{\text{Curing Agent Equivalents}}{\text{Epoxy Equivalents}} \quad [3]$$

Ideally, the stoichiometric ratio of curing agent equivalents to epoxy equivalents should be equal to one if all epoxide groups are to be reacted. A stoichiometric ratio of 1:1 is seldom used however, because the reaction becomes diffusion limited as the cure proceeds. As a result, some of the curing agent may remain unreacted. Therefore, lower stoichiometric quantities of curing agents are typically used because of problems associated with unreacted curing agents. Also, in some epoxy systems, specific reactions involving a greater quantity of secondary reactions are desired so the curing agent is used in substantially lower quantities (10, 11).

2.3 Reactive and Non-Reactive Additives The additives used in formulations can be numerous. An additive is defined as anything that is not a part of the base epoxy resin(s) or the base curing agent(s). There are a variety of additives used in developing commercial resin formulations that produce large differences in resin characteristics even though their weight percentage is small. These additives may be included to increase the toughness, modify the viscosity, promote better wetting and adhesion, increase flame and chemical resistance, etc. These additives are generally categorized into reactive and non-reactive additives.

Reactive additives can be reacted during formulating, after prepreg processing, or during the final curing process. Examples of these additives are epoxy diluents, carboxyl modified elastomers, and reactive intermediates. Sometimes these components are pre-reacted during the formulating stage to reduce problems of curing agent selectivity and reactivity (12). These reactive additives, however, affect the stoichiometry of the total resin system. If the reactive components are pre-reacted with the base epoxy resins, the EEW is increased and needs to be re-calculated to adjust the stoichiometric ratio of the curing agent used in the resin system. This can be performed using the following equation:

$$\frac{(W+X+Y+Z\ldots)}{EEW_{combined}} = \frac{W}{EEW_W} + \frac{X}{EEW_X} - \frac{Y}{Eq.wt._{comp.1}} - \frac{Z}{Eq.wt._{comp.2}} - \ldots \quad [4]$$

where

$$W + X + Y + Z + ... = 100 \quad [5]$$

and

W, X, are weight percents of the epoxies
Y, Z, etc. are the weight percents of the reactive components
$Eq.wt._{comp}$ is the weight per functionality of the reactive component

The resulting EEW can be verified through titration due to incomplete or excess reactions. In addition, titration makes it possible to determine how much of the reactive additive was reacted during the pre-reaction stage (13). This is necessary for determining the amount of catalyst needed and the time required to fully react. If the reactive additive is not pre-reacted, an adjustment in the stoichiometry needs to be performed as in equation [4], but the reactive additive contributions are now added to the base epoxy resin. As was mentioned previously, this can lead to problems in the curing agent reactivity and selectivity depending on the reactive group and structure (12).

Non-reactive additives do not significantly covalent bond to the epoxy network and therefore no EEW corrections are required. Non-reactive additives include various thermoplastics, elastomers, fillers, flame retardants, deareating agents, etc. In general, these components do not change the stoichiometry of the resin system, however, they may change the kinetics of the reactions which should be taken into account when determining the final part processing conditions.

3. MODEL METHODOLOGY

3.1 Reverse Engineering A complete understanding of commercial prepreg systems from neat resin through final parts is accomplished through reverse engineering. In this methodology, reverse engineering involves both literature searches and rigorous experimental work. A procedure of this process is shown in Figure 2.

Once the commercial system is selected to be modeled, an initial understanding of the system can be found through many resources. A source that is common to all materials in commercial use is a Material Safety Data Sheet (MSDS). Some of the components used (usually very general), as well as an amount range of the components, are given due to their potential hazardous nature. If the material is used in the airplane industry, a Material Specification Document can be obtained which provides information on the resin content of the prepreg, maximum allowable volatile content, fiber type, mechanical properties,etc. More information on the commercial prepreg system can potentially be found with a thorough literature search to see if any papers on the material have been published. These sources, however, usually do not offer information about how the prepreg was processed, its structure, or uncured properties. However, they do provide general information on the resin components which are necessary for model resin development. Once sources such as these are investigated, it is necessary to perform experiments on the material to get processing, structure and property information to define the requirements of the model resin system. The experimental techniques used to analyze and characterize the commercial prepregs are shown in Figure 2, and described in the following sections.

3.1.1 Commercial Resin Analysis and Characterization The initial experiments are typically performed on the resin of the commercial prepreg. The commercial resin can be obtained by squeezing several prepreg plies, wrapped in a screen, at high pressures and slightly elevated temperatures for a short period of time to prevent reaction. This method is preferred over solvent extraction because it is not always possible to remove all of the solvent from the resin. Another possible method to obtain resin is by the flash material (excess resin) on the backing paper of the prepreg. In all cases, since the resin is obtained from the prepreg, it is very likely that the resin was B-staged during or after prepreg processing. B-staging is the advancement of the cure of the resin in the prepreg. This may be done to reduce resin flow during final part processing or to improve handling characteristics.

The extracted resin from the commercial prepreg is characterized by thermal analytical techniques including differential scanning calorimetry (DSC), simultaneous differential thermal analysis-thermogravimetric analysis (TGA/DTA), hi-resolution thermogravimetric analysis (TGA), dielectric analysis (DEA), and rheometry. To completely analyze and characterize a commercial prepreg system with these techniques, many dynamic and isothermal experiments need to be performed, along with simulated cure cycles. Some of the information generated from these techniques include the glass transition temperature (T_g), heat of reaction, reaction onset-peak-end, percent cured, gelation, viscosity, weight loss, storage and loss moduli, etc. Although these thermal techniques do not specify the exact components used in the commercial resin, the information can be used to identify the components necessary for development of a model resin system.

The prepreg modeling methodology was not developed to determine the exact ingredients of the commercial formulation, but to determine information necessary for developing a characteristically similar model prepreg system based on processing, structure, and property information. However, if there are uncertainties in developing a representative model system, more information on the components used in the commercial formulation can be determined from Fourier Transform Infra Red Spectrometry (FTIR) and/or Gas Chromatography/Mass Spectrometry. In the past, methodologies used for quality control have been established to determine the exact types and quantities of the components of prepreg systems (11, 14, 15).

If the manufacturer of the prepreg is willing, a sample of the resin before impregnation will provide additional useful information. By performing various thermal analytical experiments on this resin, valuable processing information can be gained by determining the amount of B-staging and the initial processing viscosity. All of this information aids in the development of a model prepreg system. After the resin has been characterized, it is necessary to investigate the prepreg characteristics. The analysis and characterization methods are shown in Figure 2.

3.1.2 Commercial Prepreg Analysis and Characterization In order to investigate the effects of the fibers on the resin system, it is necessary to characterize the prepreg as a bulk material. Processing and property information obtained from the resin extracted from the prepreg may not correspond to the resin information while it is in the prepreg. This is due to the differences in thermal conductivity, degradation, and fiber/resin interactions caused by the heterogeneous and anisotropic structure of the prepreg. Also, if some components of the resin system can not be extracted from the resin, or become oriented during extraction, prepreg analysis may be the only means to obtain information on the resin characteristics. The techniques described previously for the resin analysis are used to identify the different phenomena occurring in the resin as compared to the prepreg. The results from these techniques need to be combined with bulk physical prepreg analysis to further characterize a commercial prepreg system.

There are several bulk prepreg properties which are not specified by the prepreg consumer in the Material Specification Document (MSD) that must be investigated to fully characterize the prepreg system. The MSD usually defines the required resin content, fiber type, and areal weight. However, the resin content of the prepreg may vary significantly batch to batch, as well as across the width of the prepreg. Therefore, experiments are performed to determine resin content by cutting two inch squares across the width of the roll, extracting the resin with solvent, and weighing the resulting dried fibers. Other bulk prepreg characteristics that are not specified in the MSD are the tack, and degree of impregnation of the prepreg. The tack and drape of a prepreg determine the handling characteristics. Tack is loosely defined as the ability of prepreg plies to stick together while drape can be defined as the prepregs ability to conform to different contours. Prepreg tack is determined quantitatively through a compression-tension experiment developed by Seferis et al. (16-20). Seferis and co-workers have determined that tack is a process descriptor that provides information on the processing conditions in relation to prepreg structure and properties (5, 21).

The degree of impregnation of a prepreg can significantly effect the prepreg handling characteristics as well as final part quality. For unidirectional prepreg material, the permeation test gives quantitative values of how well the prepreg is impregnated (16, 22-24). These quantitative values can be verified by qualitative results generated from polishing prepreg samples and taking optical photomicrographs of the prepreg cross-section and surface, therefore providing essential prepreg structure information developed through processing.

3.1.3 Commercial Prepreg Consolidation Analysis and Characterization In the airplane industry, the Material Specification Document (MSD) provides a wealth of information on cured composite properties. This is because prepreg systems used in the airplane industry are almost exclusively qualified from their cured composite properties. Although cured composite properties are specified in the MSD, laminates and/or honeycomb structures are manufactured with the commercial prepreg and subjected to mechanical tests. The sample preparation and experiments are shown in Figure 2. For commercial prepreg systems used in the airplane industry, the cure cycle and lay-up procedure can be found in a Processing Specification Document. The commercial prepreg is cured in the form of a laminate or honeycomb structure in either an oven, heated press, or autoclave using the specified standard cure cycle. Samples are cut from these parts and prepared for analysis. Some of the mechanical tests which are performed on the cured composites include G_{IC} and G_{IIC} fracture tests, short beam shear, and tensile as well as compressive tests. These tests provide information on the resin and composite damage resistance, strength, and stiffness. Other useful experiments which provide information on cured composites are dynamic mechanical analysis (DMA) and thermal mechanical analysis (TMA). These thermal analytical methods provide information on the cured prepreg glass transition temperature (T_g), loss and storage moduli, and the coefficient of thermal expansion. Samples are also subjected to various environments (hot-wet) and the effects are investigated through the previously mentioned techniques.

Lastly, it is essential to understand the composite morphology. The use of optical microscopy can identify whether the material has voids or if various phases are present in the cured part. Thereby, yielding insight into the prepreg processing method as well as possibly identifying additives used in the prepreg system. If optical microscopy does not provide this information, the use of other types of microscopy or etching may be required.

3.2 Model Engineering After completing the reverse engineering process, the next stage involves modeling the commercial prepreg system which is accomplished through model engineering. Model engineering consists of developing a model prepreg system from the knowledge acquired from reverse engineering. Model engineering is performed in three steps which include model resin formulation, model prepreg processing, and model laminate as well as model honeycomb processing. Throughout these developments, many iterations may be required to produce characteristics similar to the commercial system. Flow charts are provided and discussed to help clarify the entire process. These are shown in Figures 3, 4, and 5 for model resin formulating, model prepreg development, and model laminate and honeycomb structure development, respectively.

3.2.1 Model Resin Formulating Model engineering begins with developing a model resin system. The model formulation process is shown in Figure 3. Background on epoxy resin formulating was previously provided for clarification of model resin development. The ease of resin formulation, including the procedure and the techniques, depends greatly on the mixing equipment available. Using the information acquired from reverse engineering, the base epoxy resin(s) are selected. In most cases, a combination of epoxies are used to obtain the desired prepreg as well as cured composite properties. After the epoxies have been selected and combined, a sample should be tested dynamically using DSC, from -50°C, to determine the glass transition temperature (T_g). This can be compared with the commercial resins T_g, but the value will change when other combinations of epoxies or components are added. It is also necessary to determine the epoxy equivalent weight (EEW), through titration for the combination of epoxies. This information is necessary to determine the curing agent quantities. All the experiments mentioned above should be performed before developing a model formulation, and each time the combinations of components are changed, as shown in Figure 3.

Adding curing agents to epoxies is sometimes a very difficult task. Although some curing agents dissolve readily in epoxy resins, many do not. The manner in which these curing agents are added is usually by making a paste of the curing agent with another epoxy resin or diluent. Some types of curing agents can be melted and then added to the resin but this usually reduces the latency of the resulting epoxy resin. The proper method of adding the curing agent to the epoxy resin may be provided by the manufacturer, however, special formulating techniques may need to be determined experimentally. Depending on the epoxy resins and the available mixing equipment, elevated resin temperatures may be required to lower the viscosity enough so that adequate mixing of the curing agent and epoxy resin can take place. Therefore, experiments should be performed using DSC and rheometry to determine the curing agent reactivity and gelation time at certain formulating temperatures. If the method of formulating does not allow for adequate mixing of the curing agent and components due to viscosity limitations, the viscosity can be lowered with the addition of solvents. If this is the case, possible changes in the curing agent reactivity and solvent recovery should be investigated.

The appropriate curing agents for the model resin can usually be determined through the initial resources combined with the experimental evaluation of the commercial resin. This is usually not to difficult since there are relatively few curing agents used for commercial airplane epoxy prepreg resin systems. The use of the exact curing agent(s) as in the commercial resin is not necessary, because model systems are developed to be characteristically, not chemically similar to the commercial system. However, the cure characteristics need to be similar for both the model and commercial resin system. This is accomplished by comparing DSC thermograms of trial model resin formulations which contain various quantities and ratios of curing agents, with the commercial resin, using the same dynamic heating rates. In most cases, 10°C/min heating rate DSC experiments are used to screen the initial trial model resin formulations. This provides a quick indication of the similarities between the trial model resins and the commercial resin. A comparison the reaction onset temperature, peak exotherm temperature, heat of reaction, and ending reaction temperature should show similar results. Many trial model resin formulations may be necessary, utilizing different amounts of the curing agents, before similar cure characteristics are observed. This process is shown in Figure 3. In general, the quantities of the curing agent(s) selected for use in the final model system may be far from the stoichiometric ratio. Although this is generally true, the first quantities of the curing agents incorporated should be close to a 1:1 stoichiometric ratio, since this is typically the upper limit. If the quantities of the curing agents exceeds the stoichiometric equivalence or their are other physical or chemical uncertainties, the base epoxy resin combination and/or curing agent(s) may need to be changed obtain similar cure kinetics and characteristics. Typically, the curing agents and base epoxies are first combined without any other additives so that an approximate ratio can be determined before proceeding further with the modeling. This allows for less formulation time, especially if pre-reactions are carried out during formulating, and less material consumption. The greatest benefit is then to observe how the addition of various components affect the cure characteristics.

As shown in Figure 3, once a trial model resin is developed that shows similar dynamic DSC cure characteristics as the commercial resin, the next step is to determine if it has similar isothermal cure characteristics. This is important because in many instances, dynamic DSC thermograms of the trial model systems are very similar to the commercial system, but isothermal DSC experiments show different behavior. It is also necessary that cure cycle experiments are performed with DSC, and a similar residual cure exotherm and degree of cure are exhibited within reason. This is usually determined by performing the normal cure cycle on the model resin in a DSC, cooling to room temperature, and performing a dynamic experiment at 5°C/min through degradation. If there is a significant residual cure, an exotherm will exist below the decomposition temperature and can be compared with the commercial systems response. If the isothermal or residual cure DSC thermograms are not similar, the types or quantities of the curing agents, additives, or even the base epoxy resin(s) may need to be changed. As a result, the process starts over from the beginning by comparing the commercial and trial model resin dynamic DSC thermograms and so on, as shown in Figure 3. Consequently, several iterations of trial model resin formulations may be required to get the dynamic, isothermal, and residual cure DSC thermograms to be similar to the commercial system.

The additives that have been selected for the model formulation can now be added to see their effect on the cure kinetics. This can be performed earlier if the additives are known to alter the kinetics significantly. The method most commonly used to add the additives, especially for hot-melt applications, involves addition of the additives before the curing agent(s) are incorporated into the resin. This is absolutely necessary if elevated temperatures are required to incorporate the additives or pre-reactions are to be carried out.

In many cases, pre-reactions of an additive with a portion of the base epoxies are used in formulations to eliminate the problems of selectivity and reactivity of the curing agents in the final cure reactions (12). This is done through the use of reaction specific catalysts. Pre-reactions involving carboxyl and hydroxyl functionalities are used extensively in formulations to develop epoxy-adducts. The time and temperature necessary to perform the pre-reaction may be supplied by the manufacturer, but should be experimentally determined for each epoxy combination. This is typically performed through titration experiments, in which the extent of reaction can be determined as a function of time at a certain temperature. It is also necessary to determine the resulting EEW after the pre-reactions have been completed to determine curing agent/epoxy-adduct stoichiometry. After the incorporation of the additives, the quantity of the curing agent(s) or epoxies may need to be adjusted to get similar DSC thermograms for the dynamic, isothermal, cure cycle, and residual cure experiments as compared to the commercial system.

Some additives may require long periods at elevated temperatures to dissolve in the base epoxy resins. In some cases, additives may need to be dissolved or pre-reacted in a solvent. Problems, however, arise when trying to get solvents out of a high viscosity resins. Experimental formulations can be developed to determine which method is the most feasible for the system and the method of impregnation.

Once the cure kinetics of the model and commercial resin systems are found to be similar from dynamic and isothermal thermal analysis experiments, it is necessary to determine the volatile content and weight loss of the model resin. This is an intermediate stage that must be performed especially if no more solvents are to be used in the formulation. If it is determined that the volatile content is too large, a different formulation scheme may be necessary. It should also be noted that some curing agents do not work as well, or at all, when a solvent is included in the formulation. The volatile content and weight loss are examined using Hi-Res® TGA or TGA/DTA as the reaction proceeds and compared with the commercial resin.

Finally, the completion of the model resin system relies on similar rheological characteristics. As shown in Figure 3, this is the last step in developing a model resin. Some of the necessary information include similar viscosity trends, minimum viscosity, and gelation using both dynamic and isothermal experiments. If the results of the commercial and model system differ significantly, the addition of more or different additives may be necessary along with modifying the amount of the epoxy resins and/or curing agents. Thus, the process begins again starting with DSC and continues until all criteria are met. By using this methodology, a representative model resin system can be efficiently developed with relatively few iterations.

3.2.2 Model Prepreg Processing Developing a model prepreg system involves combining the model resin system with either unidirectional or woven fibers. The process of developing model prepreg is shown in Figure 4. The development of the model prepreg can be performed with either hot-melt impregnation or simulated solution impregnation. Hot-melt impregnation, used in this investigation, involves producing resin film(s) using doctor blade(s). This resin film is further heated in an impregnation zone to lower the viscosity. The resin is then combined with fibers under roll pressure where up to six rollers can be used to provide the impregnation pressure. The prepreg is then cooled on a chill plate to prevent resin flow, B-staging, and relaxation of the fibers. Finally, the upper release paper is separated from the prepreg and collected on a take-up roll. Simulated solution prepregging involves the same process as the hot-melt prepregging, except the model resin is diluted with solvent. The solvent used is usually in the amount of 20 wt.%, which is enough to lower the impregnation viscosity while still providing filming capabilities. This type of impregnation is especially useful when solvent effects are to be investigated.

The fiber type and areal weight of the commercial prepreg are defined in the Material Specification Document. In addition, the resin content is defined in this document and can also be determined experimentally. Although the resin content, fiber type and areal weight are specified, the final prepreg characteristics are developed from the prepreg processing parameters. The most important processing parameters include line speed (web speed), fabric/tow tension, and impregnation pressure and temperature (25, 26). Furthermore, the number of rollers and pressure on each roller needs to be determined. A design of experiments (DOE) is usually performed to understand how the prepreg processing parameters affect the final prepreg characteristics. The characteristics investigated are the prepreg tack, permeation, degree of impregnation, resin content/fiber content, surface characteristics, and volatile content. These prepreg characteristics are investigated for each of the processing conditions and compared with the commercial prepreg characteristics. If the characteristics are not similar to the commercial prepreg, another DOE may be performed with different processing parameters until a model prepreg is developed with similar characteristics. Through the prepreg experiments, the effect of different prepreg processing conditions as they relate to prepreg characteristics can be investigated.

3.2.3 Model Laminate and Honeycomb Processing The development and characterization of model laminate and honeycomb structures is the last phase of the model system investigation. This process is shown in Figure 5. Depending on the structure being developed, the number of plies, orientation, core material, lay-up technique, cure cycle, compaction pressure, etc., need to be the same as used in the commercial system experiments, the only difference being the use of the model prepreg. Once the laminates and/or honeycomb structures have been developed, thermal-mechanical experiments, mechanical tests, and microscopy needs to be performed. These experiments should be performed in exactly the same manner as those on the commercial laminates and honeycomb structures. The results can then be compared with the commercial prepreg system. Since the model resin was primarily developed to simulate commercial neat resin and prepreg processing characteristics, the results of these experiments may be found to be somewhat different.

3.3 Re-Engineering After the development of the model prepreg system, the knowledge acquired through the model development can be further used to re-engineer a customer driven prepreg system. This is only possible if the prepreg customer is willing to provide their input. The customers desires or requirements of prepreg systems can be investigated through the use of model prepreg systems. Furthermore, sometimes the desires or problems associated with a commercial prepreg system are only identified by crossing the traditional prepreg specification boundaries.

4. CONCLUSIONS

An integrated methodology has been developed for modeling commercial epoxy-based prepreg systems. The modeling methodology was developed based on traditional chemists analytical tools combined with bulk prepreg analytical methods using an engineering approach. This methodology was based on the non-traditional viewpoint of designing for prepreg characteristics first, followed by cured mechanical properties. The design of prepreg still needs to include the required final cured mechanical properties but it should be realized that this is only one part in the design of high performance composites.

The prepreg modeling methodology consists of simulation engineering and re-engineering in which simulation engineering is composed of sub-sections of reverse engineering and model engineering. Reverse engineering involves the complete analysis and characterization of the commercial prepreg system from the neat resin through fabricated parts. This, however, is performed to understand the prepreg through process, structure, property information, not to determine the exact prepreg ingredients. The actual development of the model prepreg system is performed through model engineering which consists of developing a model resin and manufacturing model prepreg as well as composite parts. Throughout the model system development, many iterations may be required to produce a representative model system. The number of iterations, however, can be minimized with an increased understanding of the commercial prepreg system and previous formulating, prepregging, and modeling experience. This entire process and the techniques utilized are described with flow charts to provide clarification. Once the model system has been developed, the knowledge acquired from the entire process can be used to re-engineer a customer driven prepreg system. This can be performed in a cooperative environment between the prepreg manufacturer and the prepreg customer. With the input of the customer, the problems or desires concerning the present commercial prepreg system can be defined. Collectively, the prepreg modeling methodology developed in this study can be used to develop model prepreg systems based from processing, structure, property information of a commercial prepreg system.

5. REFERENCES

1. W. J. Lee, J. C. Seferis, and D. C. Bonner, SAMPE Qtr., 17 (2), 58 (1986).
2. M. A. Hoisington, and J. C. Seferis, SAMPE Qtr., 24 (2), 10 (1993).
3. S. Zeng, M. Hoisington, and J. C. Seferis, Polym. Eng. Sci., 14 (6), 458 (1992).
4. M. A. Hoisington, Process/Property Interrelations of Layered Structured Composites, University of Washington, Doctorate Thesis (1992).
5. J. W. Putnam, B. S. Hayes, and J. C. Seferis, J. of Adv. Mat., (in Review) (1995).
6. B. S. Hayes, and J. C. Seferis, North American Thermal Analysis Society, 23, 354 (1994)
7. B. S. Hayes, J. C. Seferis, and J. S. Chen, Polym. Comp., (in Review) (1995).
8. Determination of Epoxide Content, Shell Analytical Method HC-427E-91.
9. H. S. Goodman in H. S. Goodman, ed., Handbook of Thermoset Plastics, Noyes Publications, Park Ridge, N.J., 1986, 132.
10. N. S. Schneider, J. F. Sprouse, G. L. Hagnauer, and J. K. Gillham, Polym. Eng.and Sci. 19 (4), 304 (1979).
11. C. A. May, Epoxy Resins Chemistry and Technology, Marcel Dekker, Inc., New York, 1988.
12. C. K. Riew, E. H. Rowe, and A. R. Siebert in D. R. Deanin, and A. M. Crugnola, ed., Toughness and Brittleness of Plastics, American Chemical Society, Washington, DC, 1976, 326.
13. B. L. Lee, C. M. Lizak, C. K. Riew, and R. J. Moulton, 12th National SAMPE Technical Conference, October 7-9, 1116 (1980)
14. J. F. Carpenter, and T. T. Bartels, SAMPE Qtr., 7 (2), 1 (1976).
15. C. A. May, Proceedings of the 24th National SAMPE Technical Conference, 24, 390 (1979)
16. K. J. Ahn, J. C. Seferis, J. O. Price, and A. J. Berg, SAMPE J., 27 (6), 19 (1991).
17. K. J. Ahn, L. Peterson, J. C. Seferis, D. Nowacki, and H. G. Zachmann, J. Appl. Polym. Sci., 45 (3), 399 (1992).
18. K. J. Ahn, J. C. Seferis, T. Pelton, and M. Wilhelm, SAMPE Qtr., 23 (2), 54 (1992).
19. K. J. Ahn, J. C. Seferis, T. Pelton, and M. Whilhelm, Polym. Comp., 13 (3), 197 (1992).
20. J. C. Seferis, and J. Meissonnier, SAMPE Qtr., 20 (3), 55 (1989).
21. J. W. Putnam, J. C. Seferis, T. Pelton, and M. Wilhelm, Sci. and Eng. of Comp. Mat., (in press) (1995).
22. K. J. Ahn, J. C. Seferis, J. O. Price, and A. J. Berg, Proceedings: 34th International SAMPE Symposium, 35, 2260 (1990)
23. K. J. Ahn, J. C. Seferis, and J. C. Berg, Polym. Comp., 12 (3), 146 (1991).
24. J. W. Putnam, and J. C. Seferis, Journal of Advanced Materials, 26 (3), 35 (1995).
25. K. J. Ahn, and J. C. Seferis, Polym. Comp., 14 (4), 346 (1993).
26. K. J. Ahn, and J. C. Seferis, Polym. Eng. Sci., 33 (18), 1177 (1993).

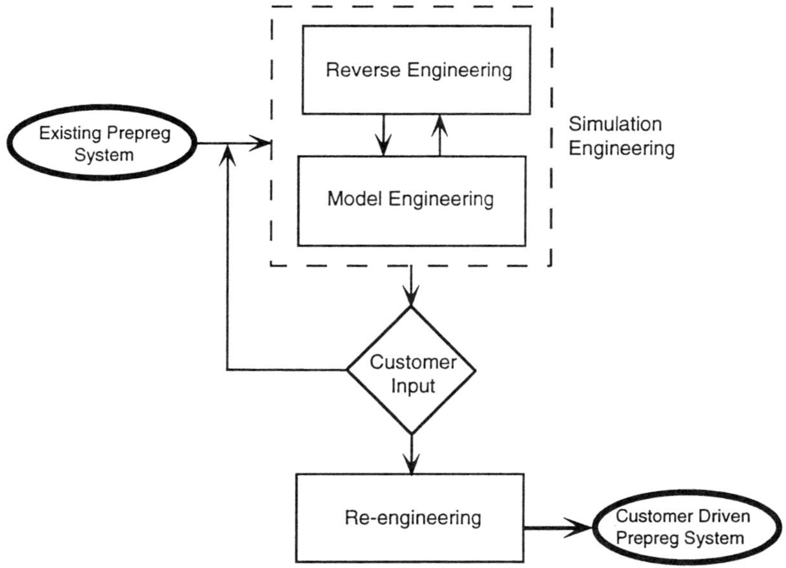

Figure 1 Schematic of the prepreg modeling methodology from existing commercial prepreg to a customer driven prepreg system.

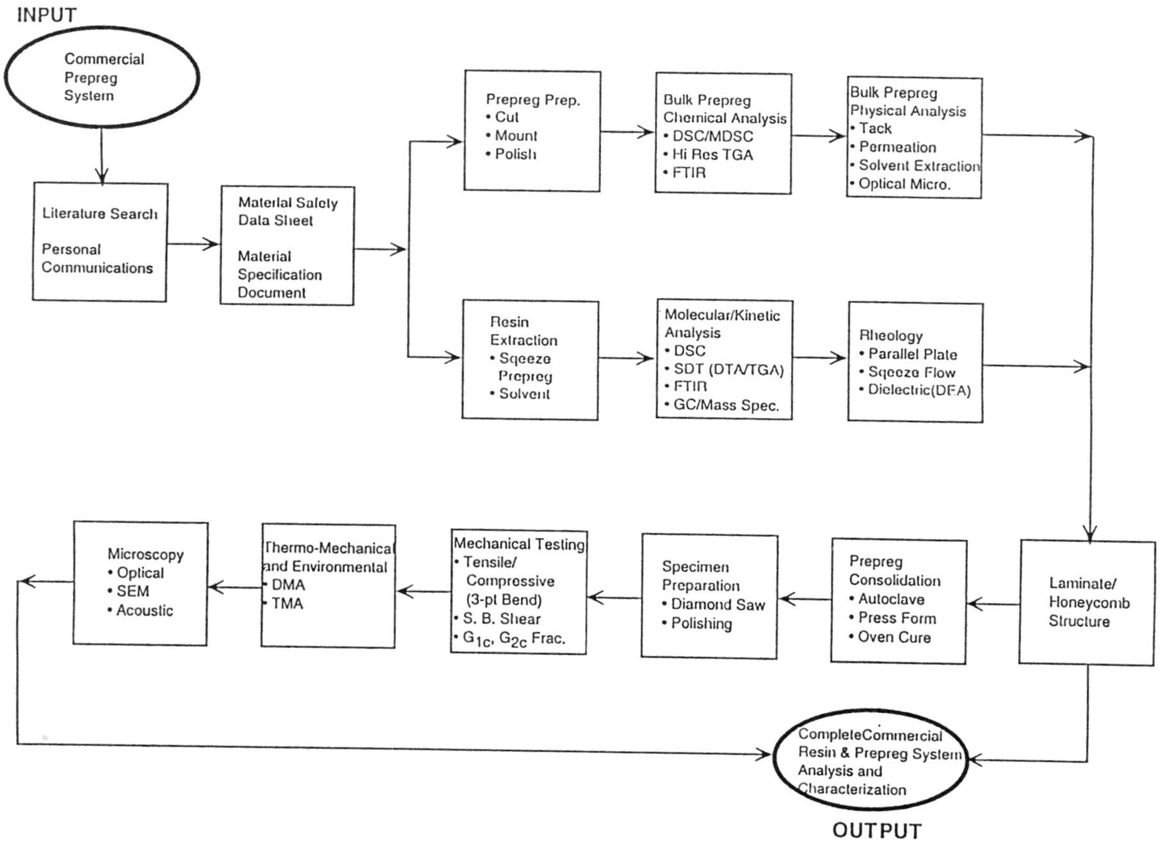

Figure 2 Flow chart of the analysis and characterization techniques used in reverse engineering commercial prepreg systems.

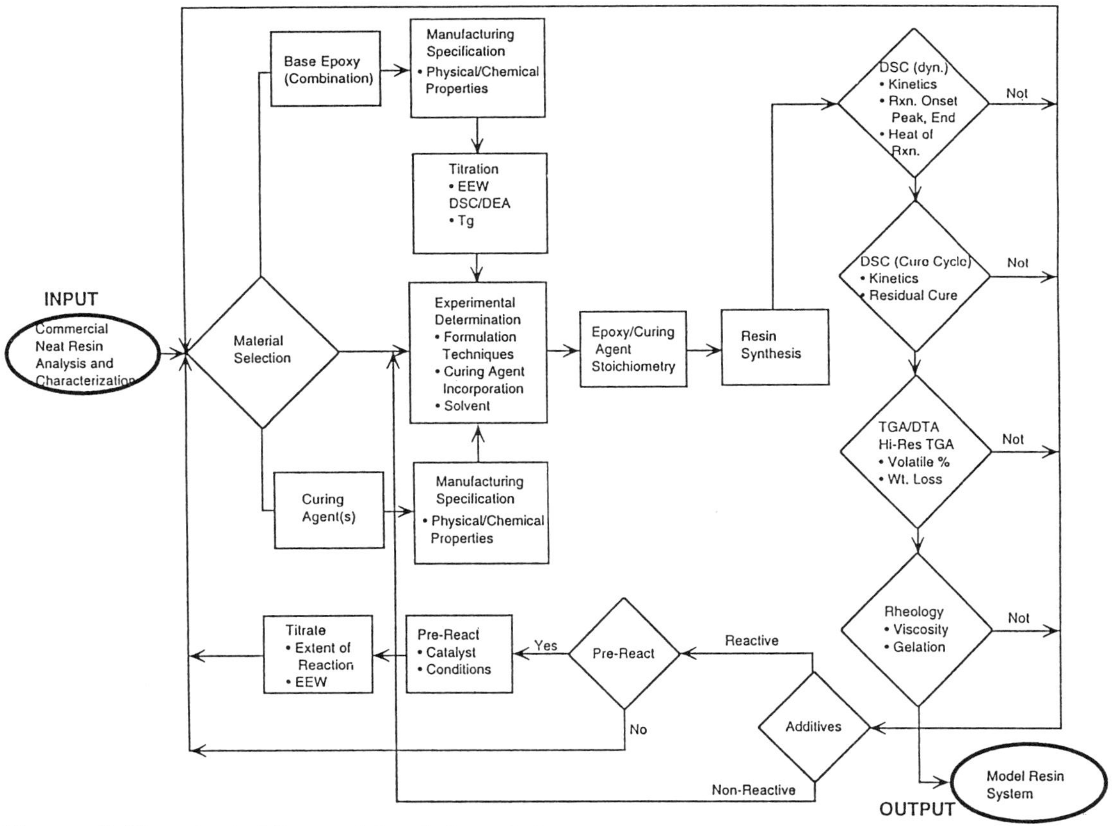

Figure 3 Flow chart of the model resin formulating procedure (step 1-model engineering).

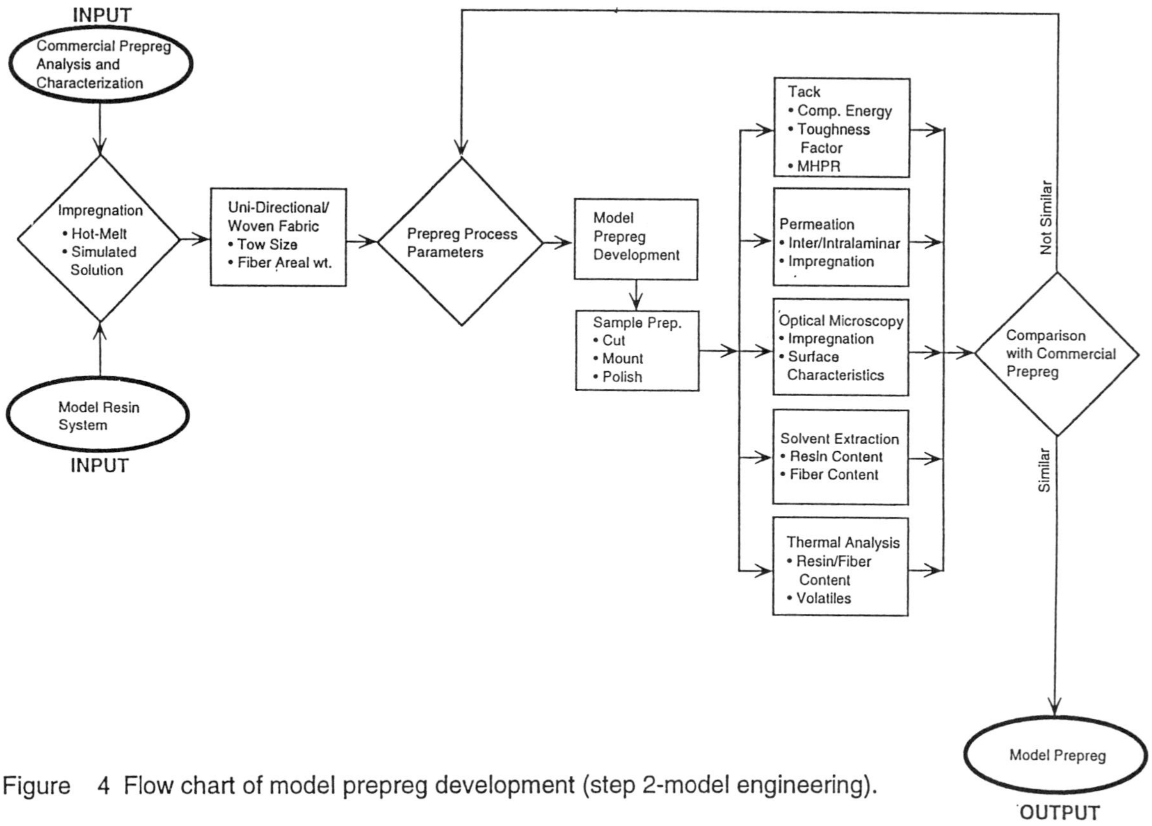

Figure 4 Flow chart of model prepreg development (step 2-model engineering).

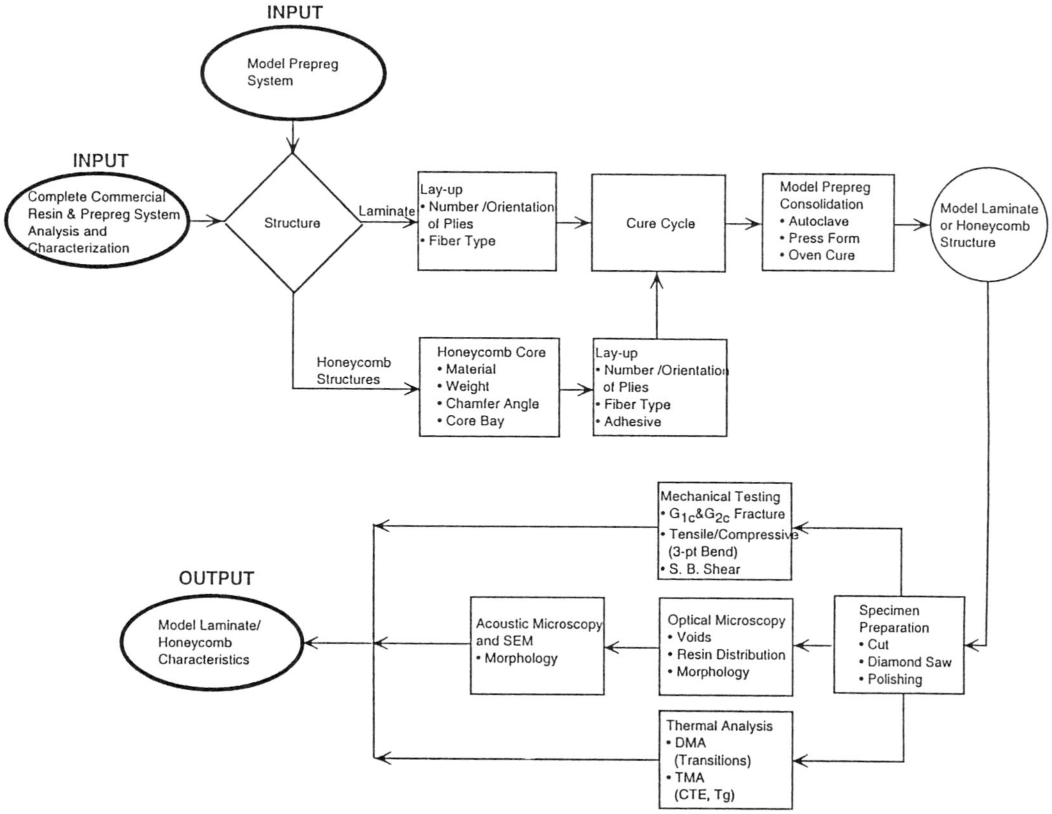

Figure 5 Flow chart of model laminate and model honeycomb structure development (step 3-model engineering)

28th International SAMPE Technical Conference
November 4-7, 1996

PREPREG MANUFACTURING PROCESS DESCRIPTORS

Frédéric U. Buehler and James C. Seferis*
Polymeric Composites Laboratory
University of Washington
Seattle, WA 98195

ABSTRACT

A model epoxy resin was used to impregnate unidirectional carbon fibers with a hot-melt process under different processing conditions. The impregnation step of the process was the main focus of the study. In this step, resin and fibers are combined together on an impregnation plate under an impregnation roll. The impregnation roll can be controlled by either constant pressure, or by setting a constant gap between the roll and the impregnation plate. Ten different prepregs were manufactured by varying either gap or pressure, keeping all other processing parameters constant. Prepregs were characterized in terms of resin content, grade, thickness, tack, permeability, and impregnation quality. Differences in prepreg characteristics between gap and pressure control were assessed. Both gap and pressure were found to have an influence on resin content and grade. Under pressure control, it was observed that variations in thickness were severe. Tack was also affected by impregnation gap and impregnation pressure. The specific compressive power correlated well with the power input necessary for impregnation. This correlation could allow an on-line measurement of the specific compressive power, which could be very useful in prepreg manufacturing for quality control.

KEY WORDS: prepreg manufacturing, tack, impregnation

1. INTRODUCTION

Although prepreg materials have been manufactured for nearly half a century, the relationships between processing parameters and performance characteristics are not fully understood. Two main processes are used to produce prepreg: hot melt impregnation and solution dip (1). Both methods consist of combining unidirectional fibers or woven fabric with either a thermoplastic or a thermoset resin to make a preimpregnated laminate material. Figure 1 shows a schematic of a hot-melt process. In such a process, a continuous film of resin is produced by either doctor blades or reverse roll coaters. The resin and the reinforcement fibers are then combined together under application of temperature and pressure. The final product is then cooled on a chill plate to prevent resin flow and to impede chemical reactions.

---

* Author to whom correspondence should be addressed

In a hot-melt prepreg process the quality of impregnation is influenced by four parameters, which include line speed, impregnation temperature, fiber tension, and pressure or gap on the impregnation roll (2, 3). Line speed affects the residence time of the resin and fibers in the different stages of the process. Impregnation temperature affects the resin viscosity, hence resin flow in the fibers. Fiber tension has an effect on the permeability of the fibers. Impregnation pressure provides the mechanical driving force for the resin to impregnate the fibers. The effects of these parameters and their influence on the prepreg quality has been investigated by Seferis and co-workers (2, 3).

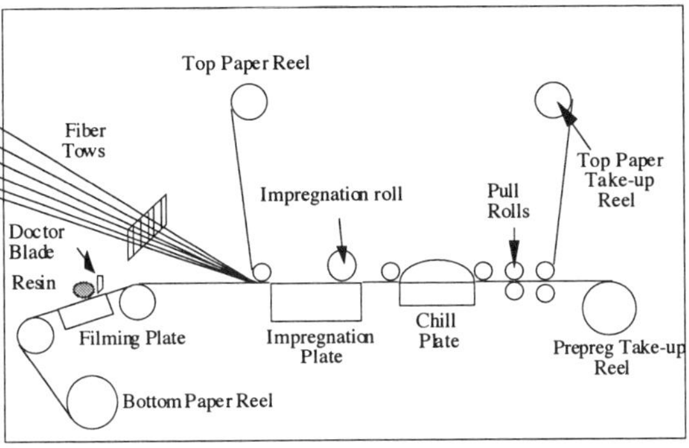

Figure 1. Schematic of a hot melt impregnation unidirectional prepregging process

As a continuation of this work, the focus of the present investigation was to identify the differences between gap control and pressure control as they relate to the prepreg quality. Gap controlled prepregging involves maintaining a constant gap between the impregnation plate and the impregnation roll, whereas the pressure control process involves a constant pressure on the impregnation roll. A Design of Experiment (DOE) was then performed using a commercial scale hot-melt prepregging machine. A total of ten different prepregs were produced in this study. Five different prepregs were produced by setting the gap at different values, and five were produced by varying impregnation pressure. All other parameters, line speed, impregnation temperature, fiber tension, were kept constant throughout the experiment.

The second objective was to examine the possibility of relating some of the prepreg characteristics to the electrical power involved in the prepregging process. Prepreg characteristics assessed were resin content, grade (fiber areal weight), thickness, impregnation quality, and tack. Specifically, tack was determined by the method developed by Seferis et al.(4).

2. BACKGROUND / THEORY

Among the several end-user attributes of prepreg, the one with the highest concern is probably tack. Tack is perhaps the most difficult characteristic to control and to measure. Tack is traditionally defined as the ability of two prepreg plies to adhere to one another. The measurement and quantification of tack is not an easy issue, as it depends on the testing method (5). Many tack measurement techniques have been proposed and developed (4, 6, 7). In industry, prepreg tack is measured qualitatively by hand. Therefore, Seferis and co-workers developed a test which gives a quantitative tack value well correlated to this qualitative rating (8). Tack is determined by a compression-tension test performed between two flat square tabs (Figure 2). In a standard cycle, five prepreg plies are compressed at constant displacement rate, held at constant load for a short time, and then pulled apart in tension at constant displacement rate. The stress-strain data obtained during such a cycle yields a compressive-tensile curve such as the one shown in Figure 3.

Previous studies have determined that the most representative tack values are the compressive energy and the toughness factor (9-11). Compressive energy is the total energy under the compressive part of the stress-strain curve. Toughness factor was defined as the amount of energy under the tensile portion of the stress-strain curve up to maximum load (Figure 3). The compressive energy is the result of a compressive step and a hold step. The toughness factor is the amount of energy required to separate the prepreg stack after it has been compressed.

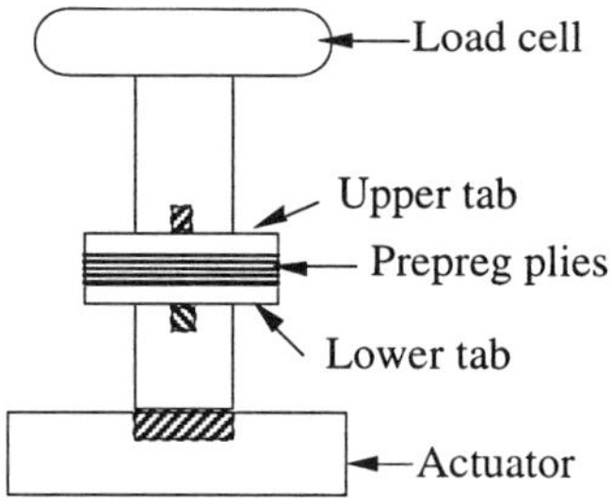

Figure 2. Schematic of the tack test apparatus.

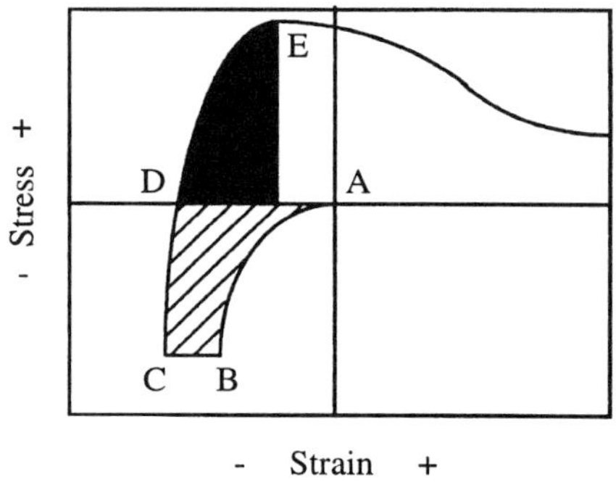

Figure 3. The tack test compression-tension curve. The bottom shaded surface area is compressive energy and the black-shaded surface area is the toughness factor.

Tack values are influenced by all of the prepreg processing parameters. Studies have shown that in general compressive energy decreases as the impregnation temperature and pressure increase. The effect of line speed was not found to be significant (12). This behavior can be explained by the impregnation level of the prepreg. Higher impregnation temperature decreases resin viscosity and allows greater impregnation of the resin into the fibers (12). This effect is more pronounced under high impregnation pressures since the pressure is the driving force for resin flow. A well impregnated prepreg has dry surface fibers which are more elastic and therefore require less compressive energy. The toughness factor on the other hand generally shows a maximum for medium impregnation temperatures and low or medium pressures (12). At medium temperatures and pressures the prepreg impregnation is fairly good, but there is still some resin left on the surface. This surface resin allows adequate wetting between the plies, which in turn increases the amount of energy required to pull the specimen apart (12). At high temperatures and pressures there is less resin left at the surface, which means less wetting between the plies, hence lower toughness factor. At low temperatures and pressures, the impregnation level is low and most of the resin is on the surface of the material, which leads to cohesive failure.

The present study focuses on the impregnation roll control. There are two different ways to control the impregnation roll: by using a constant pressure (pressure control) or by setting a constant gap between the roll and the impregnation plate (gap control). Whichever control is selected and the parameter used, it will dictate the prepreg quality. A schematic of the impregnation area is shown in Figure 4. Pressure control means that the impregnation roll can move up if the pressure build-up under the roll exceeds the pressure applied on the piston. In such a case, when an inhomogeneity (fiber threading, fiber tow overlap, resin heterogeneities, air bubbles, etc.) occurs, the prepreg will still be processed under the same pressure. Note however that the contact area of the impregnation zone is not well defined or known, and the pressure distribution under the roll is not known. In the case of gap control, the presence of an inhomogeneity will cause a pressure build-up under the roll, but the roll will not move. This means an increased tension on the fibers on the post-impregnation side of the roll since an extra amount of force will be required to enable the inhomogeneity to go through the gap.

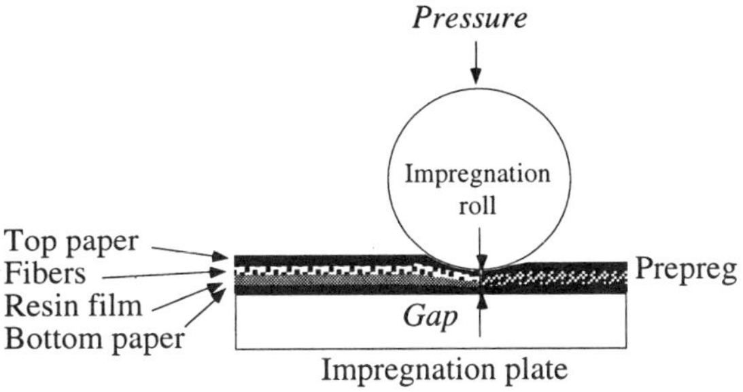

Figure 4. Schematic of the impregnation plate and impregnation roll.
Either gap control or pressure control can be applied to the impregnation roll.

As can be seen in Figure 4, paper, fibers, and resin have to be compacted in order to go under the roll, which will necessitate a certain amount of power. The power requirement for the impregnation step can be decomposed as being the sum of the power required to compress the materials so that they can go under the roll, and the power required to force the resin into the fiber bundles. This power will be provided by the tension applied on the paper and fibers. The paper, for obvious reasons will not bear a lot of the load. Therefore most of it will be carried by the fibers. The power input can be compared with compressive energy if the latter is expressed on the same scale, i.e. power units. Compressive power and toughness power can be defined and calculated from the compressive energy and toughness factor by dividing these latter values by the time taken to complete their respective cycles (strain rate), including the hold step. From Figure 3, the time to complete the compressive part of the tack test is the time to go from point A to point D following the A-B-C-D path. It yields the following formula for specific compressive power:

$$\text{Specific Compressive Power (W/m}^3\text{)} = \frac{\text{Compressive Energy (kPa)}}{\text{Time to go from A to D (s)}} \qquad [1]$$

Similar calculation can be applied to the toughness factor to yield equation [2] below:

$$\text{Specific Toughness Power (W/m}^3\text{)} = \frac{\text{Toughness factor (kPa)}}{\text{Time to go from D to E (s)}} \qquad [2]$$

The specific compressive power is expected to be correlated to power input since both of these figures represent the same kind of process: compacting fibers and resin. However the compressive power is measured with a load applied perpendicularly to the fibers. Impregnation power is on the other hand calculated with a load along the fibers.

3. EXPERIMENTAL

The model resin consisted of a combination of tetrafunctional epoxy resin, difunctional epoxy resin, thermoplastic and curing agent. The tetrafunctional epoxy used was tetraglycidyl diamino diphenyl methane (TGDDM), MY 9512 manufactured by Ciba Geigy. The other epoxy, diglycidyl ether of bisphenol A (DGEBA), was provided by Shell Chemical Co. (trade name Epon 828). The thermoplastic was polysulfone (PS) Udel P1800 manufactured by Amoco.

The curing agent used was diamino-diphenylsulfone (DDS), from Ciba Geigy (HT 976). The two epoxy resins were mixed in equal amount of 150 g each for 15 minutes at 140°C. Then 30 g of PS was slowly added over a 30 minute period under constant mixing. After one hour of mixing at this temperature, the epoxy solution was cooled down to 120°C. Upon which 120 g of the melted curing agent was added. The resin was quickly cooled down to -20°C and stored at this temperature.

Unidirectional prepreg was manufactured using a California Graphite® hot-melt prepreg machine described by Hoisington et al. (13). The carbon fibers used were Toray T800 12K sized fibers, impregnated with the above described resin. Forty tows were used, and the line speed was set at 46 cm/min (1.5 ft/min). The filming plate and impregnation plate temperatures were 66°C (150°F) and 88°C (190°F), respectively. Doctor blade gap was set to 0.254 mm (10 mils), which provides resin content of about 30%. Pressure on the impregnation roll was varied from 138 kPa to 276, 414, 552, and 621 kPa (20 psi to 40, 60, 80, and 90 psi, respectively). Then gap controlled prepreg was produced using gaps of 305, 330, 356, 406, and 457 μm (12, 13, 14, 16, and 18 mils, respectively). Prepregs were tested for resin content, fiber areal weight, compressive energy, toughness factor, thickness, and permeability.

Prepreg samples to be tested for resin content and grade were cut into 5.1 cm by 5.1 cm (2 x 2 inch) squares. Resin content and fiber areal weight were determined by weighing a coupon of prepreg, extracting the resin out of the prepreg in acetone, and weighing the dried fibers. Five samples were tested and averaged for every prepreg prepared.

Permeability was measured on prepreg sheets cut into 2.5 x 12.7 cm (1 x 5 inch) rectangles with the fibers parallel to the length of the sample. Intralaminar permeability was measured using the technique described by Putnam et al.(12). One end of the sample was exposed to atmospheric pressure while the other one was exposed to vacuum in a vacuum bag. Vacuum was applied for five minutes on a 84 cm^3 bag after which the vacuum line was closed. Permeation of air in the prepreg along the fiber direction caused the pressure to increase in the bag. This pressure was monitored versus time to calculate the permeability constant. Six to nine permeability measurements were assessed for each prepreg sample.

Quantitative tack measurements were carried out on all samples according to the method described and developed by Seferis et al. (8-10, 12). All tack tests were performed on a screw-type Instron 4505 tensile/compressive testing machine. To perform the tack test, five plies of 5.1 x 5.1 cm (2 x 2 inch) prepreg are bonded between two metal tabs with double back-tape, and subject to a compression to tension cycle. In a standard cycle, the prepreg is compressed to 27 kg (60 lb) at a displacement rate of 0.254 mm/min (0.01 inch/min), held at 27 kg (60 lb) compression for 30 seconds, and then pulled apart in tension. Stress/strain data are recorded during the test . Analysis of the data is then performed and five test samples are averaged together for each tack value. From an analysis of the stress/strain graph collected during the tack test, both compressive energy and toughness factors were obtained to quantitatively define the tack of the materials.

Sample thickness was evaluated using the same instrument set-up as for the tack test. Displacement was set to zero when the tabs of the Instron were in contact under a very small load. Then a 5.1 x 5.1 cm (2 x 2 inch) piece of prepreg was placed between the two tabs and a compressive load was applied. Prepreg thickness was measured when a small but measurable load of 270 g (0.6 lb) was applied on the sample. Five samples were tested and averaged to determine thickness values. The thickness measured was the thickness of the finished prepreg and did not include the top and bottom paper thicknesses, which were respectively 120 μm (45 mils) and 100 μm (40 mils).

Prepreg photo-micrographs were taken to determine the level of impregnation in the different prepreg samples. The samples were mounted between two pieces of thermoplastic with five-minute epoxy and were polished using a technique described by Ahn et al.(14).

The power input of the electric motor driving the pull roll was evaluated by monitoring the difference of potential at the poles of the motor and by monitoring the current flow to the motor.

4. RESULTS AND DISCUSSION

After the prepreg DOE was completed, the first type of characterization performed was resin content. This measurement provides information on how much resin is present in the prepreg. The effect of the impregnation gap versus the resin content was investigated. Resin content was found to decrease as the gap decreased. This trend was expected since a smaller gap will force a greater amount of resin to be squeezed out of the fiber bed. When the process was pressure controlled, the resin content dropped as the pressure increased. This is also explained by the fact that more resin was squeezed out of the fiber bed to the side of the paper as impregnation pressure was increased .

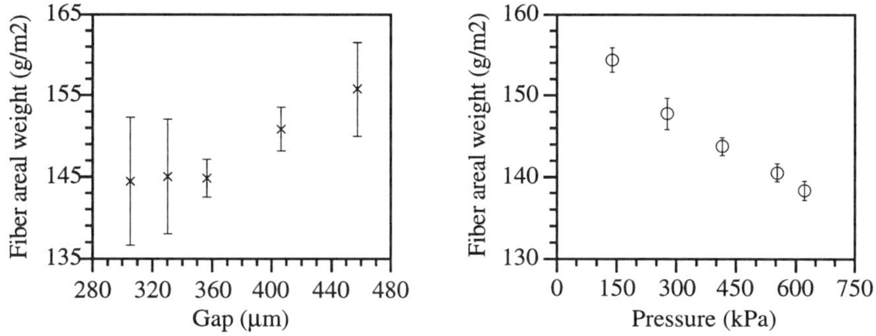

(a) (b)

Figure 5. Grade as a function of a) impregnation gap, b) impregnation pressure. Error bars represent standard deviations.

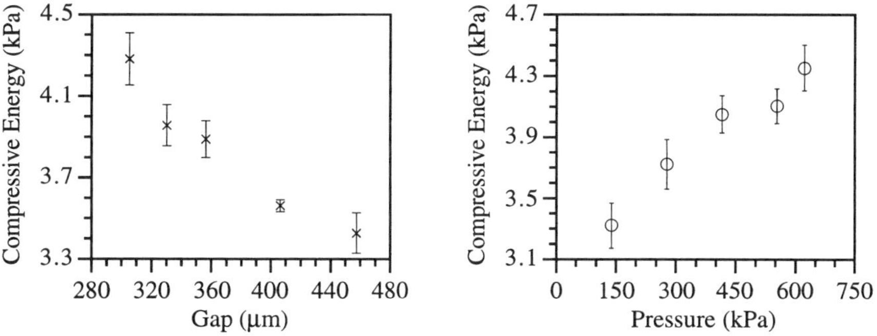

(a) (b)

Figure 6. Compressive energy as a function of a) impregnation gap, b) impregnation pressure. Error bars represent standard deviations.

Figure 5a shows the variation of the fiber areal weight along with the gap. The grade decreased as the gap became smaller and then leveled off. The grade decrease was caused by fiber spreading. A smaller gap did not allow as many fibers to go through along the same width, therefore spreading occurred and lowered the grade. However, after a certain point, fiber spreading stopped and instead more resin was squeezed out. Figure 5b shows the fiber grade - pressure relationship for a pressure controlled prepregging process. In this case the grade decreased with increasing pressure, which is readily explained by spreading. However in this case, the curve did not flatten out at any point. This might be due to the fact that the pressure controlled roll allowed better fiber spreading through the possibility of having the roll to move up and down, providing space for the fibers to rearrange rather than being pinched.

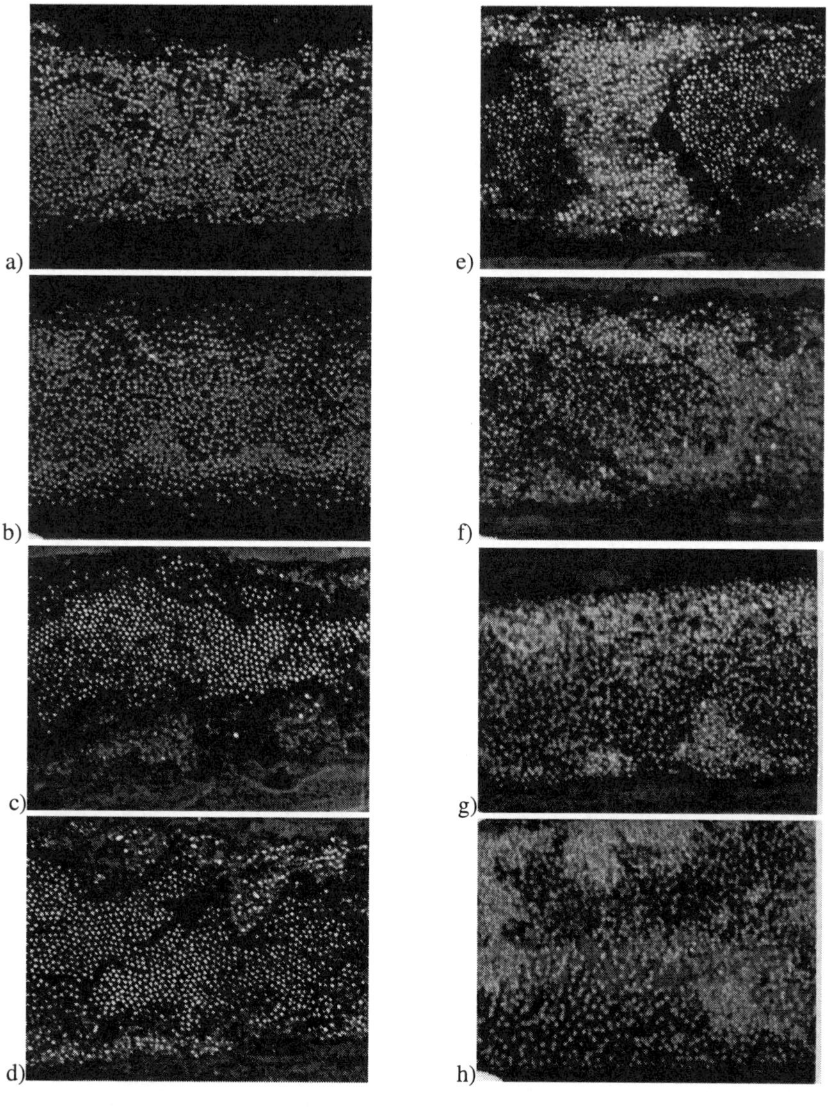

Figure 7. Photomicrograph of prepreg. a-d is 12, 14, 16, 18 mils gaps respectively. e-h are 20, 40, 80 and 90 psi pressures respectively. White area represent carbon fibers, light gray is the resin matrix and black areas represent voids. Scale: ├────┤ = 100 μm.

Prepreg thickness variation with gap and pressure was the next characteristic to be investigated. Prepreg thickness decreased at first with decreasing gap, but remained roughly constant for gaps between 280 and 360 μm. Thickness did not show a linear relationship to impregnation pressure. The thinnest prepreg was not obtained under the highest pressure. This behavior may be explained by the fact that prepreg sprang back after its travel under the impregnation roll.

The thickest prepreg was obtained under the highest pressure, leading to the suggestion that at high impregnation pressures more resin was squeezed out, resulting in dry fibers on the top and bottom of the prepreg. Dry fibers were more likely to spring back since they were not maintained by the resin matrix. Prepreg thickness was not found to be constant along the width of a prepreg tape, leading to fairly large standard deviations.

Figure 6 displays the compressive energy dependence on gap and pressure. Compressive energy was highest for the smallest gap and for highest pressure. Under low pressure or large gap, the resin did not penetrate very well into the fiber bundles, leaving voids. This can be seen on the photomicrographs on Figure 7 d) and e). Compression of such prepregs may not require a lot of energy because of the numerous voids. As gap was decreased (or pressure increased), the resin penetrated deeper into the fiber bundles (Figure 7 a-d and e-h) and the compressive energy increased.

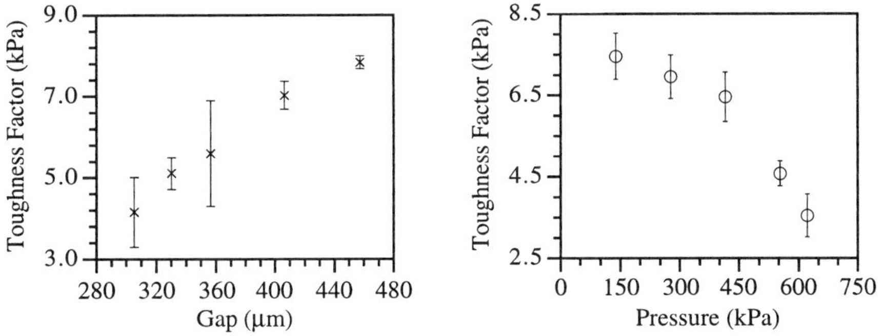

(a) (b)

Figure 8. Toughness factor versus a) impregnation gap, b) impregnation pressure . Error bars represent standard deviations.

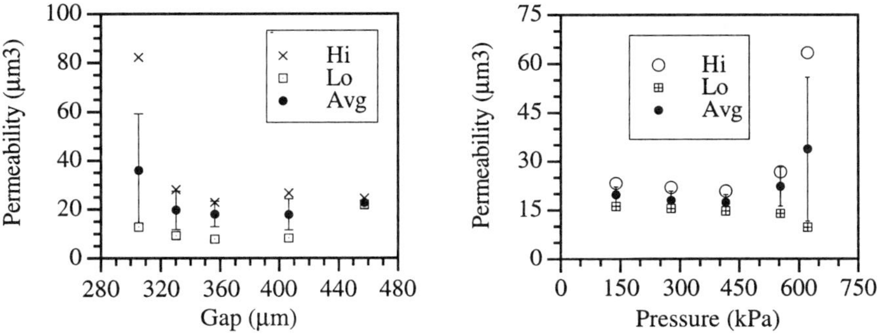

(a) (b)

Figure 9. Prepreg permeability versus a) impregnation gap, b) impregnation pressure. The highest value, lowest value and average value out of 6 to 9 samples are shown. Error bars are standard deviations.

The toughness factor, which represents the energy per volume to pull apart the five-ply sample after the compression and hold steps in the tack test, shows a trend opposite to the compressive energy. Toughness factor was found to increase linearly with gap and decrease linearly with pressure (Figure 8). This means that a sample with high compressive energy had a low toughness factor, and vice-versa. The explanation for this phenomena may be as follows: under high impregnation pressure or small impregnation gap, most of the resin is forced into the fiber bundles (Figure 7a and 7h), leaving the prepreg surface mostly dry. The dry surface fibers impaired wetting and yielded a low toughness factor, while the good impregnation in the center of the bundle yielded a high compressive energy. Under low pressure or large gap, the resin will not impregnate the fibers very well (Figure 7d and 7e) which leaves a significant amount of resin on the surface of the prepreg. This resin provided adequate wetting between the plies and more energy was required to separate the plies.

Permeability versus impregnation gap and impregnation pressure are shown in the Figure 9a and 9b respectively. It shows that prepreg produced under extreme conditions (i.e. small gap or high pressure) had a wide spectrum of permeability constants and that on average the permeability of such prepregs was higher than that of prepregs produced under more moderate conditions. It was suggested previously that prepreg produced under high pressure was thicker because of dry surfaces. These dry fibers tend to increase the air permeation, hence the higher permeability observed at extreme conditions.

Figure 10 is a plot of the compressive power per unit volume versus the motor power input. The compressive power per unit volume is calculated from the compressive energy divided by the time taken to complete the compressive portion of the tack test, from zero load up to 27 kg (60 lb) and back to zero load. This value correlated well to the pull roll motor power input. The power input is related to the force required to pull the fibers, paper and resin beyond the roll. Therefore, it should represent the energy of impregnation minus some friction and heat losses, which are thought to be constant or to vary linearly with the power input. The energy of impregnation has two contributions: the energy required to compress the materials so that they can go under the roll, and the energy required to force the resin into the fiber bundles. During a tack test a comparable quantity, compressive energy, is measured. Therefore, a relationship between the two is expected. This relationship should not depend on the type of prepregging process (gap or pressure control). This is observed in Figure 10, where gap and pressure control data points fall on the same line.

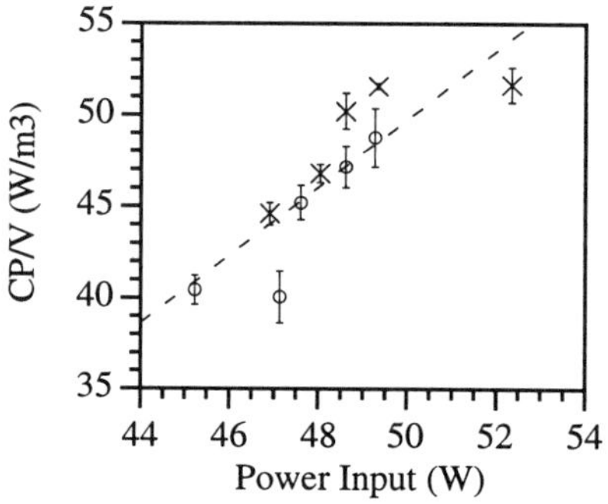

Figure 10. Compressive power per unit volume versus pull roll motor power input. Crosses are gap controlled data points and circles are pressure controlled data points. Dashed line is best least square fit and error bars are standard deviations.

Such a similarity between gap and pressure control is not observed for toughness power to power input relationship. Specific toughness power was calculated similarly to specific compressive power. The toughness factor obtained by the tack test was divided by the time required to complete the tensile part of the tack test up to maximum load. No correlation between specific toughness power and power input was found. Moreover, at equal power input, the toughness power was greater for the prepreg produced with the gap controlled prepregging process than for the prepreg produced with the pressure controlled process. The lack of correlation may be due to the fact that specific toughness power is measured on the tensile part of the tack test, and power input is related to the compression rather than tension.

5. CONCLUSION

An assessment of prepreg manufacturing descriptors was completed by hot-melt impregnating unidirectional carbon fibers with a model epoxy resin. A DOE was utilized to fundamentally understand the differences between two different methods of controlling the impregnation roll. Five impregnation gaps and five impregnation pressures were set to manufacture ten different prepregs. After the prepreg was processed, it was characterized in terms of resin content, grade, thickness, tack, permeability, and impregnation quality. Tack was measured quantitatively through a compression to tension test. Data obtained from this test was also compared with impregnation power requirements.

In general decreasing the impregnation gap had the same effect as increasing the impregnation pressure. Decreasing impregnation gap or increasing impregnation pressure were found to force more resin out of the fiber bed and to spread the fiber bed wider. The consequences were lower resin content and lower grade. Prepreg thickness was found to decrease along with the gap at large gaps. At smaller gaps thickness was roughly constant. In a pressure control process, prepreg thickness was smallest around 300 kPa (40 psi) and was largest under high pressure. It was expected that prepreg would be thinnest under the highest pressure or the smallest gap. It was not the case because fibers had a tendency to spring up after impregnation, thus changing the thickness. Under high pressure or small gap, more resin was squeezed out, leaving the top and bottom surface fibers dryer. Dry fibers sprang back more than impregnated ones because they did not have the resin matrix to maintain them in place. Permeability was increased by dry fibers since these provide an easier path for air permeation. Dry fibers also decreased the toughness factor because of poor wetting between prepreg plies, making it easier to separate them by pulling the plies apart. Compressive energy on the other hand decreased with gap and increased with pressure. This was explained by the improved level of impregnation at small gaps and high pressures. Having less voids inside the prepreg requires more energy to compress the prepreg. A characteristic very close to compressive energy, specific compressive power, was found to correlate well with the impregnation power input. This direct relationship was expected since both terms measure similar quantities: a quantity of energy per unit time required for compressing the material. Specific toughness power on the other hand did not correlate with power input. Specific toughness power is related to the amount of resin on the prepreg surface because of its relationship to toughness factor. It is therefore not related to any compressive feature. Relating impregnation power requirements to prepreg characteristics is important for developing an on-line quality control system.

Methods of analysis such as the one presented here can be used to determine prepreg quality. This can benefit the prepreg manufacturer who can then have an improved quality control. It is also beneficiary to the part manufacturer because constant quality reduces the number of problems such as delamination and wrinkling as well as shimming requirements in final part assembly.

6. ACKNOWLEDGMENTS

The authors express their appreciation to Brian Hayes, Louise Peterson, and Mark Wilhelm for their editorial help and especially to Brian Hayes for his help in the DOE. Special thanks also go to Dave Gary and Roger Pick of the Chemical Engineering Department for help in modifying the equipment. Financial support for this work was provided by the Boeing Commercial Airplane Group through project support to the Polymeric Composites Laboratory at the University of Washington.

7. REFERENCES

1. W. J. Lee, J. C. Seferis, and D. C. Bonner, SAMPE Qtr., 17 (2), 58 (1986).
2. M. A. Hoisington, and J. C. Seferis, SAMPE Qtr., 24 (2), 10 (1993).

3. K. J. Ahn, and J. C. Seferis, Polym. Eng. Sci., 33 (18), 1177 (1993).
4. J. C. Seferis, and J. Meissonnier, SAMPE Qtr., 20 (3), 55 (1989).
5. C. A. Dahlquist, Adhesives Age, 2 , 25 (1959).
6. R. F. Bull, C. N. Martin, and R. L. Vale, Adhesives Age, 11 (5), 20 (1968).
7. A. M. Gillanders, S. Kerr, and T. J. Martin, Int. J. Adhesion and Adhesives, 1 (1), 125 (1981).
8. J. W. Putnam, J. C. Seferis, T. Pelton, and M. Wilhelm, Sci. Eng. Comp. Mat., 4 (3), 143 (1995).
9. K. J. Ahn, J. C. Seferis, T. Pelton, and M. Wilhelm, SAMPE Qtr., 23 (2), 54 (1992).
10. K. J. Ahn, J. C. Seferis, T. Pelton, and M. Wilhelm, Polym. Comp., 13 (3), 197 (1992).
11. K. J. Ahn, L. Peterson, J. C. Seferis, D. Nowacki, and H. G. Zachmann, J. Appl. Polym. Sci., 45 (3), 399 (1992).
12. J. W. Putnam, B. S. Hayes, and J. C. Seferis, J. Adv. Mater., in press (1995).
13. M. A. Hoisington, and J. C. Seferis, 37th International SAMPE Symposium and Exhibition, 37, 264 (1992)
14. K. J. Ahn, Thermosetting Matrix Based Composite Impregnation Science and Engineering, University of Washington, Doctorate Thesis (1990).

8. BIOGRAPHY

Frédéric U. Buehler received his Diploma degree in Chemical Engineering from the Swiss Federal Institute of Technology in Lausanne, Switzerland, in 1995. The same year he joined the Polymeric Composites Laboratory at the University of Washington in Seattle under the supervision of Dr. Seferis for advanced graduate work. Fred's research focuses mainly on the prepregging process parameters and their relationship to the prepreg characteristics.

James C. Seferis is the Boeing/Steiner Professor of Polymeric Composites at the University of Washington and Director of the Polymeric Composites Laboratory in the Department of Chemical Engineering. Professor Seferis has pioneered several important concepts for the design, manufacturing, and performance of composite materials in the aircraft environment. Having published extensively in the open literature, he has recently begun implementation of his developed methodologies in collaboration with the aircraft and chemical industries in a teaming environment created under the graduate Team Certificate Program of the University of Washington.

28th International SAMPE Technical Conference
November 4-7, 1996

"PLB, A NEW HIGH PERFORMANCE PHENOLIC RESIN FOR FST COMPOSITES"

Axel Boettcher, Bakelite AG, Duisburg, Germany and
Louis A. Pilato, Bound Brook, New Jersey 08805, USA

ABSTRACT

A new high performance phenolic resin for FST fiber reinforced composites has been developed. Modification of the phenolic microstructure allows the introduction of an elastomeric phase into the phenolic continuous phase. Upon curing, the elastomeric modified phenolic composition leads to a material whose property profile approaches that of an epoxy resin.

The PLB phenolic resin exhibits similar flexural strength, interlaminar shear resistance, and peel strength as is shown by flame retarded epoxy or epoxy-phenolic co-cured resin systems.

Utility of this novel phenolic resin is being actively pursued in those composite applications where FST characteristics are mandated such as aircraft interiors, trains, buses and boats.

KEY WORDS: FST COMPOSITES, FR EPOXY, CO-CURE EPOXY/PHENOLIC,PHENOLICS, AIRCRAFT INTERIOR, MASS TRANSPORTATION, TRAINS, BUSES, NAVAL-VESSELS, HONEYCOMB,

1. INTRODUCTION

The use of fiber reinforced composite materials in many segments of mass transportation is dictated by the optimization of selected components or, in some cases, the structure itself. As these fiber reinforced composite materials are transformed into the desired component or structure, materials, type of fabrication, design and manufacturing processes must be considered. The aircraft and aerospace industries have been the leaders in developing innovative composite materials and techniques for light weight structural components. Other mass transportation industries have capitalized on these "high tech" aircraft/aerospace developments and have implemented some of these developments in those areas that are cost effective for these less demanding industries.

A. COMPOSITES FOR MASS TRANSPORTATION

High quality, high strength materials and related manufacturing processes that are used to obtain complex, light weight components invariably always lead to some increased cost. Thus the use of fiber reinforced composite systems in various modes of mass transportation requires an evaluation of the cost/performance benefits of these anisotropic materials as they apply to each industry. Fuel costs per mass unit or individual transported must be considered in relation to the lifetime of the components within the structure. The savings of reduced weight per kilogram in a passenger car, for instance, is only a few dollars. Other means of transportation (land or sea) with greater longevity and greater speeds (rail) may provide savings of $20/kg.; for commercial aircraft the financial benefit exceeds $500/kg.

The major distinguishing features of these composite materials are high strength to weight ratios and design flexibility. Fatigue behavior, crack resistance, corrosion resistance and, more importantly, combustibility of these materials are critical and relate to the end use application. It is noteworthy that "safety," an aspect that is particularly important in mass transportation and can not be compromised, is achieved, in some cases, by "over design" where some additional weight is allowed.

B. COMPOSITE COMPONENTS

Fiber reinforced composite materials are based on a high performance reinforcing fiber and a matrix resin(1). These fibers include glass, carbon fiber and aramid. A large variety of thermosetting and thermoplastic matrix resins can be considered. The excellent processability combined with excellent chemical and thermal resistance of epoxy and phenolic thermosetting resins are distinct advantages in comparison to thermoplastic resins for fiber reinforced composite materials. Epoxy resins continue to enjoy prominence as a preferred matrix resin in high performance composites because of ease of processing, flexural and toughness properties. An extensive data base of epoxy reinforced fiber composite systems are available and encompass many performance driven applications especially in military and aerospace areas. Phenolics, however, are too brittle for use in structural applications. Yet the attractive FST characteristics of phenolic resins (low flame, low smoke, low toxicity) continue to be the stimuli to develop new and improved phenolics with enhanced properties for use in fiber reinforced composite systems.

2. DISCUSSION

A. MATRIX RESINS

1. EPOXY RESINS

Difunctional and poly functional epoxy resins are cured by hardeners such as amines or anhydrides, with amines preferred. The curing of multifunctional epoxy resins results in materials with higher thermal and chemical resistance than the corresponding difunctionally cured epoxy resins. This enhanced behavior is attributable to the very high cross link density that occurs in the multifunctional system. Curing is conducted with aromatic amines such as diamino diphenyl methane or diamino diphenyl sulfone accelerated by dicyandiamide. Use of accelerators or catalysts in the curing reaction affects processability, composite properties, and long term strength/thermal resistance.

Epoxy resins with a good balance of strength/toughness combined with chemical/thermal resistance are attractive matrix resins in many high quality, demanding fiber composite materials in commercial aircraft, military aircraft, and aerospace market areas. Their use in mass transportation interior components is limited due to combustibility/smoke or toxic emissions that can occur. In the event of an accidental fire, mass transportation interior components are required to emit minimum smoke density and toxicity. These requirements are summarized in the international FST test procedures that have been adopted globally for commercial aircraft interiors (2). A comparison of smoke density of various polymeric materials identifies phenolic resin as an outstanding resin system (Table 1).

Various techniques have been evaluated to minimize ignition or suppress continued burning of epoxy resins. The use of tetrabromo Bis A (TBBA) and, in some instances, combined with antimony oxide synergist reduces cross link density/properties of the cured product due to addition of the difunctional bromo epoxy. Other flame retardants such as phosphorous compounds, aluminum hydroxide, boron compounds, nitrogen compounds, and elementary red phosphorous have been evaluated (3). Most of these additives (powders in some cases) embrittle the matrix resin and compromise epoxy toughness (a very desirable matrix resin characteristic) at the expense of flame retardance. Furthermore these resulting heterogeneous matrix resin systems exhibit reduced mechanical stability.

Thus throughout the aviation industry flame retardant epoxy resins are used for structural components and aircraft interior use. In the cargo or passenger area side walls and floor panels are fabricated with these epoxy resins. Presently flame retardant epoxy composites systems do not meet the demanding safety standards for aircraft interior areas. About 2000t/yr. of epoxy resins are consumed by the aircraft and aerospace industry in Western Europe. Ciba, Dow, Shell and Bakelite are the leading producers of these resins (4).

2. PHENOLIC RESINS

a. Current Phenolic Resin Technology

Phenolic resins have been known for over 130 years. The pioneering efforts of Baekeland (5) in the early 1900's identified the use of intermediate condensation products of phenol formaldehyde reactants for further incorporation into fiber composites. Baekeland's commercialization of phenolic resins spawned the "Plastics Industry" as it is known today.

Phenolic resins are prepared by the condensation of phenols and aldehydes, preferably phenol and formaldehyde. Phenolic resins are differentiated from each other on the basis of phenol or substituted phenols, the phenol/formaldehyde ratio, and the resulting chemical or physical modifications (5). Resole resins prepared under basic conditions with varying F/P ratios are heat reactive materials which undergo crosslinking by thermal or strong acid/base conditions. Novolaks are prepared under acidic conditions with less than a molar ratio of F/P. Novolaks are thermoplastic and require a source of F, preferably hexamethylene tetramine, to crosslink them into a cured system. Phenolic resins exhibit outstanding combustion behavior (Table 1). These features relate to phenolic resin non-flammability as well as low amounts of smoke and toxic gases being emitted under fire conditions. Additionally low flame spread and high ignition delay (difficult to burn) are further attributes. During combustion phenolics carbonize/char and arrest the formation of volatile components resulting in a "self-extinguishing" trait. These favorable FST characteristics are contrasted by the unfavorable brittleness of phenolic resins resulting in reduced utility of phenolics as matrix for fiber reinforced composite systems.

Either resoles or novolaks are brittle resin compositions and require some modifications as matrix resins in fiber reinforced composite materials. Known techniques include the use of substituted phenols (alkyl or bisphenol A), etherification of resole methylol, reaction with unsaturated natural oils like tung or linseed, reaction with inorganic acids, physical modification or curing with epoxies. Thus some reduction in brittleness occurs by modification of these phenolics into fiber reinforced composite structures for applications involving minimal mechanical stress. Some of the FST characteristics of the phenolics are compromised depending on the modifier. Phenolics combined with glass fiber are being used for a variety of air, land and marine transportation systems such as aircraft interior side wall panels, baggage racks, ceiling racks as well as interiors in trains, buses and marine ferries. Annual consumption in Western Europe for phenolic resins in the aircraft interior application area is currently estimated to be about 1000t/yr. Leading phenolic resin producers are Bakelite, BP, GP, Borden and Cray Valley.

b. Co-Cure Epoxy/Phenolic

In the late 1980's Bakelite developed a system that combined epoxy resin curing with phenolic resin condensation (6). The objective was to combine the toughness of epoxies with the FST characteristics of phenolics into a co-cured system suitable for fiber composite structures with high strength to weight ratio and mechanical/thermal stability. The patented technology involved a single step combining epoxy prepreg with phenolic prepreg to the cured laminate. A judicious matching of polymerization rates and resin viscosity curves of the two different curing processes led to the co-cured composition (6). Thus the toughness and the high peel strength of the epoxy segment combined with the favorable FST features of the phenolic component resulted in an attractive laminate facing for honeycomb panels. The novel co-cured system with high peel strength eliminated the use of a film adhesive. Since mid 1994 these co-cured panels have been installed as flooring panels in the new Airbus 340.

c. PLB Phenolic Resin

Modification of phenolic microstructure for fiber reinforced composites can occur during resin manufacture or by prepreg manufacturer. The controlled co-cure epoxy/phenolic conditions were extended by introducing other resin phases into the phenolic microstructure phase. These studies led to a unique phenolic/elastomeric composition known as PLB (internal Bakelite designation). PLB is a recent discovery. Since patents were recently filed in October 1995, no disclosure of PLB experimental conditions can be presented. The type and method of introduction of selected resin phases is important for enhanced improvement in mechanical properties of the resulting modified phenolic resin. Preliminary evaluation of the newly developed PLB resins has been conducted in selected aircraft interior applications. Laminate properties of PLB-14 such as flexural strength and interlaminar shear strength (Table 2, entries 1 and 2) compare quite favorably with similar properties of flame retardant epoxy resin and co-cured Epoxy/Phenolic system. Peel strength of facings on aluminum honeycomb also compare favorably (Table 2 entry 3). PLB-14 exhibited 405 Newtons peel strength versus 400-410 Newtons for the epoxy systems.

The PLB-14 phenolic resin exhibits mechanical strength, shear and peel strength comparable to the other epoxy containing systems. Yet there is no epoxy phase in PLB. Two different PLB phenolic resins (PLB-14 and PLB-15) with structurally different resin phases added to the phenolic phase were evaluated for flammability characteristics (Table 3). Resins, PLB-14 and PLB-15, are examples of different resins phases introduced into phenolic microstructure under similar reaction conditions. The flammability results demonstrate that the resin phase of PLB-15 is superior to PLB-14 but both compositions are below the values of typical phenolic resin as well as the low flammability standards for aircraft interior use. Thus the PLB phenolic resin system exhibits similar mechanical strength characteristics as epoxy resins combined with outstanding FST features. These combined high performance mechanical and FST characteristics are expected to result in new and expanded opportunities for these newly developed PLB resins. These include use in additional aircraft interior structures, buses, trains, and naval vessels.

Although phenolic resins represent a mature resin system approaching a century of commercial existence, the newly developed PLB phenolic resin is only one of many, new and exciting technical developments that are emerging from Bakelite laboratories.

III CONCLUSIONS

A newly patented system known as PLB phenolic resin is reported. Expanding Bakelite patented co-cure epoxy/phenolic technology, selected resin phases are introduced into the phenolic microstructure phase resulting in PLB phenolic resins with mechanical strength properties similar to epoxy resins and outstanding FST features. These combined high performance mechanical and FST characteristics of the PLB phenolic resins are expected to provide new and expanded opportunities for use in mass transportation industries where FST combustibility characteristics are mandated. These include aircraft interiors, buses, trains, and naval vessels.

IV REFERENCES

1. L.A. Pilato and M.J. Michno, "Advanced Composite Materials," Springer-Verlag, Berlin, 1994.

2. U. Sorathia, R. Lyon, R. Gann, and L. Gritzo, SAMPE J. 32 (3) 8, (1996)

3. J. Troitzsch, "International Plastics Flammability Handbook," Vol 2, Auflage, Hansa Munich, 1990.

4. Frost I. Sullivan, "The European Commercial and Industrial Composites Market," Market Study, 1993.

5. A. Knop and L.A. Pilato, "Phenolic Resins," Springer-Verlag, Berlin, 1985.

6. Bakelite Patent EP 0242512

Table 1

Comparison of Smoke Density of Several Plastics

Plastic	Smoke Density (Ds)[a]	
	Non-Flaming	Flaming
Phenolic resin	2	16
Epoxy resin	132 to 206	482 to 515
Vinyl ester resin	39	530
Polyvinyl chloride	144	384

a Optical Density @ 4 minutes

Table 2

Comparison of Mechanical Strength of PLB with Epoxy Containing Resins

Properties	Epoxy Resin, flame retardant	Epoxy/Phenol Resin Co-Curing System	PLB Phenolic Resin (PLB-14)
Flexural Strength(MPa)	540	562	551
ILSS(MPa)	48	45	43
Peel strength (N)	400	410	405

Table 3

Flammability Behavior of PLB Phenolic Resins

Phenolic Resin Systems	OSU[a]	NBS[b]	FAR[c]
Phenolic Control	61/30	2.0	102
PLB - 14	46.7/27.0	1.0	90
PLB - 15	27.3/19.3	1.0	77

a Heat Release KW.min/m^2; KW/m^2 MAXIMUM 65/65
b Smoke Density Flaming Mode at 4 min. MAXIMUM 200
c FAR 25.853(a) Burn Length (mm.) MAXIMUM 152

DUAL-USE PROJECT TEAMS LESSONS LEARNED

by
Sidney H. Goodman
Principal Engineer/Scientist
Hughes Aircraft Company
El Segundo, CA 90245

ABSTRACT

This paper highlights some of the successes and pitfalls encountered with dual-use project teams. We examine reasons why a diverse group of technologists came together to solve a common problem in this new, nontraditional manner. Then we examine how cultural rather than technical differences can sometimes complicate routine project management.

In the beginning, confusion about this new, untried approach caused contradictory, conflicting interpretations and expectations among the project participants. We discuss the results of these early proposal efforts and how the difficulties were successfully mastered.

A presumption on the part of most dual-use participants continues to be that defense industry technology must be adapted to commercial productization. We show that more often a commercial technology needs the right kinds of resources to be adapted to the new low-cost, high-performance defense industry manufacturing paradigm.

In this case, successful consortia were formed and achieved positive results. Outright competitors found ways to work together, share critical information, and protect patentable data. The role of a champion as a critical element in the process is described.

KEY WORDS: Dual-use Technology, Government/Industry Partnerships, R&D-General

INTRODUCTION

This conference's theme is research conducted by government/industry partnerships. Over the last five years, a new paradigm has evolved for government sponsorship of technology and research, that is, partnerships with industry. Traditional support for R&D programs is based on the U.S. Government directly funding individual organizations through grants and contracts. The basis for this support is the need to develop technologies of generic concern to the United States that no enterprise would find worthy (profitable) to address on its own. If the problem was one of basic scientific understanding, it was left almost exclusively to universities or national research institutes to establish expertise in the field that could solve the problem using government-provided resources.

BACKGROUND

Industry, however, has long maintained an R&D base, not just for its long-term economic self interest, but in some well-known instances, to probe the frontiers of science. Bell Laboratories and General Motors Research are excellent examples. But no matter the motive, the resource support of science and basic engineering still rose and fell as the economy of the country or company changed.

During the post-World War II ascendancy of the so-called "military-industrial" complex, a new factor entered the equation: high-risk technology. The objective of R&D became more clearly defined. It was not to improve man's knowledge of the world or to develop a new apparel fiber to replace some scarce natural material, but to develop and build primarily military devices using technology that was so new as to be uncertain of success. Breakthrough was now an expectation: a laser, an infrared guidance mechanism, an ultracompact electronic package, extraordinarily lightweight and strong structures. Since profit margins were less assured for industry and Nobel awards less likely for university researchers, the government created a new infrastructure, that is, take the risk and we will, to some extent, indemnify you against failure through the use of our money, facilities and equipment.

The social consciousness of the 1960s produced a broadening of this concept. Small and minority-owned companies were acknowledged as needing special assistance to participate in the new high-tech revolution. The small business program from the Department of Commerce and the small/minority business set-aside from the Department of Defense were introduced to ensure that all possible sources of new development and scientific prowess were included in the technology revolution.

Through the 1980s the infrastructure took on the structure shown in simplified form in Figure 1. Notice that commercial industry was seen as an outsider, driven by competitive product/profit incentives, only occasionally interfacing with research facilities where the payoff was unusual. For example, to this day, research facilities continue to hope their work will result in royalty income from a new Xerox, Polaroid, Apple, HP, etc. For the U.S. Government, however, the focus was on the military: aircraft, armaments and weaponry, electronics, etc. Commercial companies wanting in on this business had to learn "how to do business with the government," going to seminars or learning the hard way, from audits and paperwork blizzards. More often than not, particularly among smaller companies, the bureaucracy burden wasn't worth the trouble and so the government was cut off from a significant source of new technology. In many other cases, commercial companies created government/defense divisions and sectors to become part of the military/industrial complex with the necessary infrastructure to participate competitively. Parenthetically, the record of defense industry companies going "commercial" was dismal, despite the conventional wisdom that ultimately all this high-tech development would significantly benefit citizen consumers.

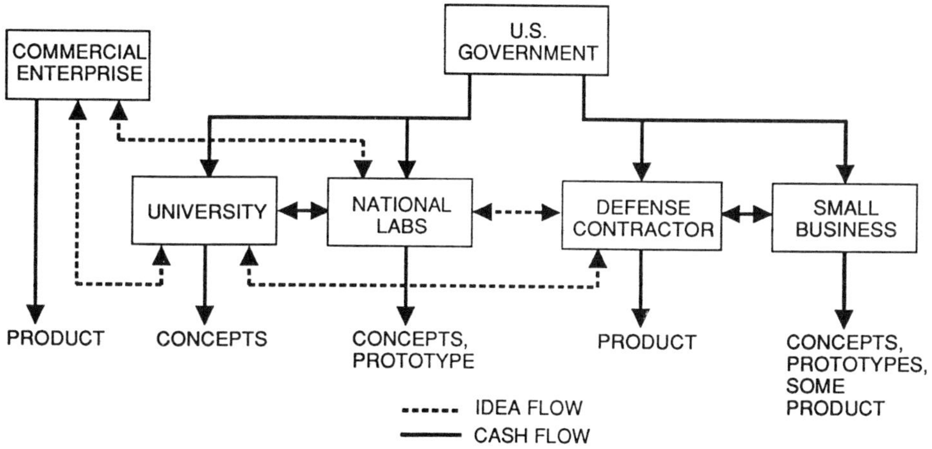

Figure 1. Traditional Flow of Technology

PARADIGM SHIFT

Then came the end of the Cold War, and the whole picture changed. As in earlier times, R&D suddenly was impacted by economic events. Now however, the change was unexpected, seemingly more abrupt, and in terms of hard available dollars, of much greater impact on a wider community of participants. Not only was funding reduced, but it was often eliminated. Facilities were to be closed, equipment duplication terminated, priorities reassessed. Business and the research community were being asked to adjust to five years of erosion of an environment it took over thirty years to construct.

Management and government began looking at new and alternative approaches to mitigate the impact on the operations of contractors and their subtier supply chain. Defense company managers were told to diversify and reduce dependence on military support. Government agencies were told to stretch the budget to get broader bang for the buck. Terms such as *defense conversion* and *dual-use technology* were coined to respond to these directives. Conventional wisdom continued to adhere to the concept that the "real technology" resided in the military/industrial complex R&D network. Although there were exceptions to this thinking, the pattern of programs that began to proliferate seemed to have as their base objectives mechanisms for transferring high-tech defense industry knowledge to those able to apply it to consumer/commercial/industrial applications.

An alphabet soup of new initiatives came out, each with its own spin and interest base: TRP from ARPA, ATP and MEP from NIST, SBIR from the Department of Commerce, CRADA with National Labs. Simultaneously, a series of projects focusing on improving the automobile were announced that incorporated many of the same infrastructure features, for example, cost sharing, cooperative consortia among competitors, specific time lines for implementation, etc. Although differing often in the detail, the new paradigm was becoming defined: let us pool our resources, expedite the informational interchange, and commit to a productization plan. Then natural market forces will take over and drive the success of the technology.

Suddenly, the separation of commercial industry and government/defense-based industry evaporated. In fact, as economic realities took hold it became clear that many of the fundamental processes and constraints imposed in defense work needed to be revisited. Everyday cost-driven production processes produced highly reliable, durable products. Specifications, test requirements, and design concepts could be "relaxed" and still be used to build high performance military goods. Commercial-off-the-shelf became something to strive for, not deride. And by the way, you wouldn't believe the neat technologies found hiding among those consumer-focused commercial enterprises.

The new paradigm has come full circle in defining the productization and commercialization of America's technology. Everybody, the lab researcher, the university, the entrepreneur, the for-profit business, the U.S. Government, contributes to the common development goal; in-kind resources, facilities, intellectual property, and the like. We share and allot responsibility in accelerating the process through its steps. We reach agreement on how to share the benefits of the outcome, through usage, license, or equity earnings.

I believe it is important that we understand what has happened over the past few years that makes the theme of this session — technology developments from government/industry partnerships — have value. My talk is not intended to be a critique of the paradigm. On the contrary, I view it as a refreshing and needed change in the way we in the defense establishment do business. I will not enumerate a series of war stories gleaned from others which tend to overplay the problems, oversights, and misunderstandings that occurred in the beginning, and thus lose the significance of the successes about which other speakers will tell us today.

In fact, my story is a personal one. It talks to the events I participated in during the last three years to achieve an objective we haven't yet reached, the successful transfer of a commercial technology into the military/defense mainstream and then back to the commercial arena. My company, Hughes Aircraft Company, has been singularly successful in diversifying its business base. Indeed, its success in winning TRP, ATP, and other programs has played a significant role in this diversification. Our corporate parent company, GM, is participating in so many different CRADAs, Future Car, Clean Car, and other programs I can't keep track of them all. So my talk will not reflect the official views of GM or Hughes, but the issues encountered by the teams I worked with that illustrate the complexity of establishing the new paradigm.

When GM first bought Hughes, the desire to exchange technologies and cross-fertilize development projects was intense. The processes needed to efficiently accomplish these transfers were either not in place or were being defined on the fly. Personnel from both organizations came from different corporate cultures. They had different objectives, used methods traditional and language peculiar to their own environment, and were saddled with preconceptions about how the other group operated.

My first exposure to the GM culture was during a four-month assignment to what was then called Central Foundry. My job was to manage a series of small projects that Hughes materials and process engineers were tasked to perform in support of some developmental manufacturing activities for the Foundry. I was also directed to become acquainted with the broader Central Foundry and GM environment with the objective of increasing the type and amount of engineering activity between our organizations.

- Routinely puts men in space and goes surfin' on Sunday
- Too expensive
- Bogged down in red tape and minutiae
- Inventiveness exemplary, but can't build paper clips

Figure 2. The Misconception of the Aerospace Engineer

Figure 2 summarizes, somewhat tongue in cheek, what the typical automotive engineer thought about California aerospace engineeers. There was a belief that, if only we knew where to look, most electronic wizardry able to solve nearly any sensing, measuring, or instrumenting problem ever, had already been developed. However, we never seemed to have just the right black box that could be taken directly to the manufacturing line, be installed and become operational for $129.99.

- Prehistorically low-tech at best, no-tech at worst
- Miserly attention to costs to fractions of a cent
- Unstructured/open loop program control

Figure 3. The Misconception of the Automotive Engineer

Figure 3 similarly describes the false stereotype of U.S. auto industry engineering. The reality, of course, is nothing at all like this. One could argue that the defense industry still is too expensive and that the development cycle takes forever, but if you watch the numbers, you can see the changes occurring. The automotive industry is now routinely delivering quality at best-cost based in fact on some rather sophisticated and vanguard technologies.

TECHNOLOGY REINVESTMENT PROGRAM

About three years ago, my colleagues at GM described a new approach for incorporating aluminum metal matrix composites into cars. The process consists of a synergistic combination of three technologies: a competitively priced ceramic reinforcement, a novel approach to preform the reinforcement, and a high-volume molding process that could produce exceptionally high quality, selectively reinforced parts. However, the time line for the successful unification of all three developments was askew, largely due to cost reduction pressures, competing technologies and projects, and available manpower. In fact, similar scenarios were occurring throughout the U.S. industrial base, often as a result of the issues described earlier. Because Hughes had the understanding of and infrastructure for capturing and performing DoD programs, we could create a union of Hughes, GM foundry, and other appropriate parties and avail ourselves of the resources being offered in the new government/industry partnership programs. Each participant could leverage its individual resources and share the workload, expediting the development process through productization and introduction into the national industrial base.

At that time, our M&P labs were looking at ways to commercialize a new environmentally benign extraction process that appeared compatible with the GM preform manufacturing process. With this commonality of interests we decided to pursue one of the first TRPs from ARPA. With many others we attended the briefings and seminars to understand the ground rules and award criteria. Consortia were favored. Participation with universities, small businesses, and particularly the National Laboratories, appeared helpful. Above all, the need for productization was emphasized. Cost-sharing was required. New allowances for maintenance of intellectual property were identified. Spin-off was stressed, spin-on was acceptable almost as an afterthought, and dual-use was essential.

Recall my earlier comments that conventional wisdom had it that the "real technology" lay with the defense contractor and that commercialization was the translation of this product into a consumer market. Significantly, the limiter on all proposals was the productization step. The program was not to provide the resources to commercialize. It was designed to take the development/technology through prototype, then the "market" would take over, presumably with the commercial partner providing the resources, will, and ultimate motive to reap the anticipated payoff.

Without belaboring the point, our team was one of 258 losers in a field of 259 applicants. Figure 4 lists some of the reasons we were unsuccessful.

- Team too big and unwieldy
- Didn't have a business plan
- Too expensive
- Weak case for spin-on
- Tried to do too much

Figure 4. Post Mortem Analysis of Our TRP Proposal

CURRENT PROJECTS

Towards the end of the TRP campaign, we began a smaller scale, parallel effort for the National Automotive Center of TARDEC (the U.S. Army Tank Automotive & Armament Command,

Research, Development, and Engineering Center). The objective of this effort was to demonstrate the application of this new technology to the vehicular needs of the military. The Army had identified many areas for improvement in its vehicle characteristics as a consequence of Desert Storm maneuvers. Foremost among these needs was the impact of weight on vehicle performance. Increased mileage, carrying capacity, armor, armament, etc. are linked to the reduction of total vehicle weight. It appeared that our new, low-cost approach to MMC parts might be a logical area for sponsorship.

The most significant lesson learned from winning our first TARDEC proposal was the need to validate the dual nature of the "dual-use" technology. The argument boiled down to this: GM needs this technology for its new powertrain insertions. Because the technology is new, the ceramic reinforcement supplier has a developed product that is lower cost than the state of the art, but not low enough to gain immediate entry into the marketplace, let alone the ultra-low cost required by the automotive companies. Some kind of market must be created that will allow the scale-up of the fiber manufacturing process to a consistent, intermediate-cost state. Much like the old chicken and egg conundrum, cost will come down with the increased purchases, but nobody will buy until the cost comes down. Therefore, the new technology will either sit idle, of no use to anyone, or will be sold offshore for exploitation in the EEC or Pacific Rim, as shown in
Figure 5.

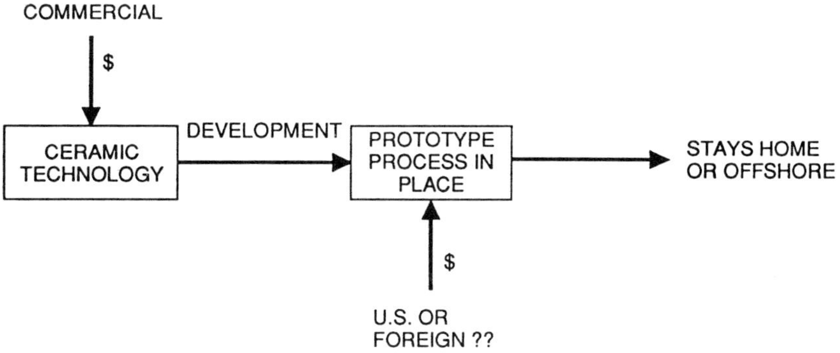

Figure 5. Many Resources Are Available to Get Technology to Prototype Stage

If, on the other hand, a parallel development effort can demonstrate that an equivalent MMC part for military powertrains can be made, then we can focus on the supply to the military as the market pull to this desirable intermediate state. The commercial technology is spun-on to the defense establishment where it becomes a self-sustaining economic entity.

Subsequently, as the automotive industry goes on line with its MMCs, the sheer size of its market pull gives the incentive for normal financial resources to be applied for final scale-up/minimization of cost. Indeed, the generic marketplace (recreation, medical, electronic, etc.) now provides the medium for growth and stabilization of the technology, as shown in Figure 6.

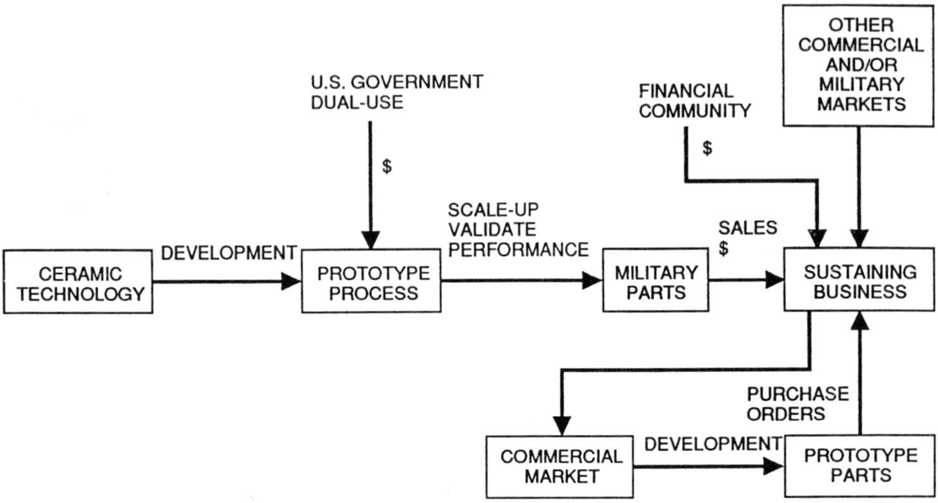

Figure 6. Government Resources Bridge the Gap Between Development and Full Productization – Both Military and Commercial Benefit; Technology Stays Home

This argument looks great on paper. As we learned the hard way, however, it was an uphill battle to convince people that our technology was "high" enough for the military. The Army agreed to support a small R&D task requiring us to demonstrate the quality of our MMC through a series of mechanical tests. We reported the results of those tests in earlier conferences. Suffice it to say, we showed equal to or better than state-of-the-art formulations.

LESSONS LEARNED

More to the point, the performance of this contract taught us a great deal about the benefits and pitfalls of a shared military/commercial program. Our commercial brethren knew nothing of CDRL requirements, accounting for labor and nonlabor expenditures, delays in getting approvals to scope changes — all business as usual for those of us who do it every day. It was hard to convince a financial executive of the biggest company in the world that we need a special timekeeping process to validate labor charges. It was even more difficult to convince government auditors (before the advent of the new acquisition reform) that a commercial process in place is effective and sufficient to provide us with the data we need. The fact that my company had a strong and capable infrastructure in place (contracts, finance, project management, etc.) enabled the whole team to minimize differences, mitigate confusion, and ultimately bring the contract to a successful conclusion, on time, within budget, meeting performance goals.

We learned another major lesson from this program. Seek out and enlist champions in each organization who will stay with the effort from preproposal to conclusion. These champions performed two invaluable services for us. First, they carried the corporate/technological memory with them so that we didn't reinvent wheels or travel down tangential paths. Second, they each had a personal stake in the success of the program, providing a unified force to confront the inevitable bureaucratic impediment or management offer of help. Indeed, the same core group was used to propose our follow-on with TACOM, an insertion program to demonstrate we can make a series of diverse military parts using our technology. Ultimately, this core group will act as the focal point for the transfer of this technology into the mainstream United States industrial base.

These programs also demonstrated that all the participating professionals, both automotive and aerospace, are equally and highly competent, can learn much from each other, and that the solution to the tasked problem is the primary motivator for success. The clearly defined dual-use nature of the project softened the distinction between spin-on/spin-off so that it became a non-issue.

Finally, each team member felt a mutual and vested interest in the outcome of the program. Each team member has something to gain by supporting each element of the program. In the long run, the gain will be the earnings coming from new, competitive products. But in the short run, it will be the satisfaction of promulgating something never done before with a new composite (excuse the pun) team of industry and government experts rarely assembled before.

BIOGRAPHY

Sidney H. Goodman has a B.S. and M.S. in Chemical Engineering. He is currently a principal engineer scientist in the Component & Materials Center of Hughes Aircraft Company. He has been with Hughes for 18 years leading developmental program and management teams for the triservices and NASA. For the last 5 years he has been a technical liaison with General Motors Powertrain Castings Operations. Mr. Goodman is a member of SAMPE and senior member of the Society of Plastics Engineers. He is a senior lecturer at the University of Southern California, has published 2 books, 12 papers, and has 1 patent.

Technical Conference Author Index

Technical Conference Author Index

Technical Conference Company Index

Technical Conference Company Index

Technical Conference Company Index

Technical Conference Company Index

Technical Conference Subject Index

Technical Conference Subject Index

Technical Conference Subject Index

Technical Conference Subject Index

Technical Conference Subject Index

Technical Conference Subject Index

Technical Conference Subject Index

Technical Conference Subject Index

Technical Conference Subject Index